**Translational Medicine
Molecular Pharmacology and
Drug Discovery**

*Edited by
Robert A. Meyers*

Translational Medicine

Molecular Pharmacology and Drug Discovery

Edited by
Robert A. Meyers

Volume 1

Verlag GmbH & Co. KGaA

The Editor

Dr. Robert A. Meyers
Editor-in-Chief
Ramtech Limited
122, Escalle Lane
Larkspur, CA 94939
USA

Cover

Cover picture based on an artistically abstracted version of figure 5 from chapter 10 "RNA Interference to Treat Virus Infections" by Karim Majzoub and Jean-Luc Imler.

All books published by **Wiley-VCH** are carefully produced. Nevertheless, authors, editors, and publisher do not warrant the information contained in these books, including this book, to be free of errors. Readers are advised to keep in mind that statements, data, illustrations, procedural details or other items may inadvertently be inaccurate.

Library of Congress Card No.: applied for

British Library Cataloguing-in-Publication Data
A catalogue record for this book is available from the British Library.

Bibliographic information published by the Deutsche Nationalbibliothek
The Deutsche Nationalbibliothek lists this publication in the Deutsche Nationalbibliografie; detailed bibliographic data are available on the Internet at <http://dnb.d-nb.de>.

© 2018 Wiley-VCH Verlag GmbH & Co. KGaA, Boschstr. 12, 69469 Weinheim, Germany

All rights reserved (including those of translation into other languages). No part of this book may be reproduced in any form – by photoprinting, microfilm, or any other means – nor transmitted or translated into a machine language without written permission from the publishers. Registered names, trademarks, etc. used in this book, even when not specifically marked as such, are not to be considered unprotected by law.

Print ISBN: 978-3-527-33659-3
ePDF ISBN: 978-3-527-68719-0
ePub ISBN: 978-3-527-68721-3
Mobi ISBN: 978-3-527-68722-0

Cover Design Adam Design, Weinheim, Germany

Typesetting SPi Global, Chennai, India

Printing and Binding C.O.S. Printers Pte Ltd, Singapore

Printed on acid-free paper

10 9 8 7 6 5 4 3 2 1

Contents

Preface		IX
Volume 1		
Part I	Biopharmaceuticals	1
1	Analogs and Antagonists of Male Sex hormones Robert W. Brueggemeier	3
2	Annexins Carl E. Creutz	77
3	Genetic Engineering of Antibody Molecules Cristian J. Payés, Tracy R. Daniels-Wells, Paulo C. Maffía, Manuel L. Penichet, Sherie L. Morrison, and Gustavo Helguera	85
4	Growing Mini-Organs from Stem Cells Hiroyuki Koike, Tamir Rashid, and Takanori Takebe	137
5	Hemoglobin Maurizio Brunori and Adriana Erica Miele	159
6	Immune Checkpoint Inhibitors Laura Mansi, Franck Pagès, and Olivier Adotévi	199
7	Molecular Mediators: Cytokines Jean-Marc Cavaillon	229
8	Neural Transplantation: Evidence from the Rodent Cerebellum Ketty Leto and Ferdinando Rossi	267
9	RNA Interference in Cancer Therapy Barbara Pasculli and George A. Calin	291

| 10 | RNA Interference to Treat Virus Infections | 345 |

Karim Majzoub and Jean-Luc Imler

| 11 | Stem Cell Therapy for Alzheimer's Disease | 383 |

Rahasson R. Ager and Frank M. LaFerla

| 12 | Immunotherapy with Autologous Cells | 411 |

Andrew D. Fesnak and Bruce L. Levine

| 13 | Targeted Therapy: Genomic Approaches | 439 |

Tim N. Beck, Linara Gabitova, and Ilya G. Serebriiskii

Volume 2

| Part II | Drug Discovery Methods and Approaches | 493 |

| 14 | Pharmaceutical Process Chemistry | 495 |

Michael T. Williams

| 15 | High-Performance Liquid Chromatography of Peptides and Proteins | 531 |

Reinhard I. Boysen

| 16 | Hit-to-Lead Medicinal Chemistry | 575 |

Simon E. Ward and Paul Beswick

| 17 | Mass Spectrometry-Based Methods of Proteome Analysis | 595 |

Mihir Jaiswal, Michael P. Washburn, and Boris L. Zybailov

| 18 | Natural Products Based Drug Discovery | 649 |

Shoaib Ahmad

| 19 | Neurological Biomarkers | 683 |

Henrik Zetterberg

| 20 | Pharmacokinetics of Peptides and Proteins | 697 |

Chetan Rathi and Bernd Meibohm

| 21 | Physical Pharmacy and Biopharmaceutics | 725 |

M. Sherry Ku

| 22 | Prions | 773 |

Vincent Béringue

| 23 | RNA Metabolism and Drug Design | 827 |

Eriks Rozners

| 24 | Structure-Aided Drug Design and NMR-based Screening | 871 |

Lee Quill, Michael Overduin, and Mark Jeeves

| 25 | Tuberculosis Drug Development | 899 |

Kingsley N. Ukwaja

Part III Nanomedicine — 941

26 Microfluidics in Nanomedicine — 943
YongTae Kim and Robert Langer

27 Nanoparticle Conjugates for Small Interfering RNA Delivery — 969
Timothy L. Sita and Alexander H. Stegh

28 Quantum Dots for Biomedical Delivery Applications — 995
Abolfazl Akbarzadeh, Sedigheh Fekri Aval, Roghayeh Sheervalilou, Leila Fekri, Nosratollah Zarghami, and Mozhdeh Mohammadian

Index — 1009

Preface

The approach we pursued in this compendium is to provide the latest insights into cutting edge methodology, approaches and results in the molecular and cellular basis of human disease as well as the discovery, evaluation, formulation and production of new drugs across the widest possible range of diseases. Two important factors are increasingly recognized in the field of translational medicine: i) despite considerable progress in the field, we still don't know enough about the mechanisms of many if not all of the critical diseases and conditions (Alzheimer's is an example), and ii) the new drug pipeline is in a shrinking mode with a concomitant decrease in R&D productivity. Our compendium aimed at capturing the *state-of-the-art* in the field and is designed to offer both answers and pathways to drug discovery and testing.

The *Biopharmaceuticals* section covers the latest developments and experimental approaches in the field from biologics extracted from living systems (*e.g.*, hormones, annexins, hemoglobin, cytokines), recombinant protein and stem cell therapies (*e.g.*, neural transplantation), as well as RNA interference in cancer therapy (nanodelivery of microRNAs), to growing organs from stem cells utilizing *in vitro* approaches to model human organogenesis producing self-organizing three-dimensional (3D) tissues so-called organoids or organ buds. This section also includes immunotherapy with autologous cells and targeted genomic approaches. Biologics approaches as applied to specific diseases such as cancer (immune checkpoint inhibitors and RNA interference), viral infections, atherosclerosis and malaria and Alzheimer's are covered in detail.

The *Drug Discovery* section provides the latest on advanced methodologies for drug discovery from cutting edge analytical techniques, *e.g.*, multidimensional HPLC and also Mass Spectrometry-Based Methods of Proteome Analysis), to drug discovery methodology, *e.g.* Hit-to-Lead and Structure-Aided Drug Design and NMR spectroscopy-based Screening as well as Neurological Biomarkers. A number of articles in this section are directed to drugs for specific diseases such as the prion family of diseases, cancer, tuberculosis, Parkinson's, Huntington's, schizophrenia, frontotemporal dementia, and Alzheimer's. *Natural Products Based Discovery* covers extraction processes from plants, animals, bacteria and associated "smart screening" methods, robotic separation with structural analysis, metabolic engineering, and synthetic biology. This section is completed with chapters on preformulation, biopharmaceutics, drug absorption, nanotechnology, pharmacokinetics and drug delivery systems design and performance including targeted drug delivery—application of physical chemistry principles to the area of pharmacy in the design of drug molecules and drug products.

The emerging field of nanomedicine is the medical application of nanotechnology for the treatment and prevention of major ailments. The *Nanomedicine* section concerns the preparation of nanomaterials based drug delivery systems. These are nanoparticles *"which have been formulated using a variety of materials that includes lipids, polymers, inorganic nanocrystals, carbon nanotubes, proteins, and DNA origami. The ultimate goal of nanomedicine is to achieve a robust, targeted delivery of complex assemblies that contain sufficient amounts of multiple therapeutic and diagnostic agents for highly localized drug release, but with no adverse side effects and a reliable detection of any site-specific therapeutic response"*—Kim and Langer (authors of our article on Microfluidics in Nanomedicine). The Microfluidics article provides an overview of highly compatible platforms to create new nanomedicine development pipelines that include the required methodologies. Importantly, microfluidics presents a number of useful capabilities to manipulate very small quantities of samples, and to detect substances with a high resolution for a wide range of applications. Then there are articles on two exciting cutting edge nanomedicine approaches *Nanoparticle Conjugates for Small Interfering RNA Delivery* and *Quantum Dots for Biomedical Delivery Applications*, where these therapeutic approaches are discussed for cancer, hereditary disorders, heart disease, inflammatory conditions, and viral infections. In addition, quantum dots probes accumulate at tumors, due both to an enhanced permeability and retention at tumor sites, and also by the binding of antibodies to cancer-specific cell-surface biomarkers.

Larkspur, California, October 2017

Robert A. Meyers
Editor-in Chief
RAMTECH LIMITED

Part I
Biopharmaceuticals

1
Analogs and Antagonists of Male Sex Hormones

Robert W. Brueggemeier

The Ohio State University, Division of Medicinal Chemistry and Pharmacognosy, College of Pharmacy, Columbus, Ohio 43210, USA

1	**Introduction** 6	
2	**Historical** 6	
3	**Endogenous Male Sex Hormones** 7	
3.1	Occurrence and Physiological Roles 7	
3.2	Biosynthesis 8	
3.3	Absorption and Distribution 12	
3.4	Metabolism 13	
3.4.1	Reductive Metabolism 14	
3.4.2	Oxidative Metabolism 17	
3.5	Mechanism of Action 19	
4	**Synthetic Androgens** 24	
4.1	Current Drugs on the Market 24	
4.2	Therapeutic Uses and Bioassays 25	
4.3	Structure–Activity Relationships for Steroidal Androgens 26	
4.3.1	Early Modifications 26	
4.3.2	Methylated Derivatives 26	
4.3.3	Ester Derivatives 27	
4.3.4	Halo Derivatives 27	
4.3.5	Other Androgen Derivatives 28	
4.3.6	Summary of Structure–Activity Relationships of Steroidal Androgens 28	
4.4	Nonsteroidal Androgens, Selective Androgen Receptor Modulators (SARMs) 30	
4.5	Absorption, Distribution, and Metabolism 31	
4.6	Toxicities 32	

Translational Medicine: Molecular Pharmacology and Drug Discovery
First Edition. Edited by Robert A. Meyers.
© 2018 Wiley-VCH Verlag GmbH & Co. KGaA. Published 2018 by Wiley-VCH Verlag GmbH & Co. KGaA.

5	**Anabolic Agents** 32	
5.1	Current Drugs on the Market 32	
5.2	Therapeutic Uses and Bioassays 33	
5.3	Structure–Activity Relationships for Anabolic Agents 34	
5.3.1	19-Nor Derivatives 34	
5.3.2	Dehydro Derivatives 35	
5.3.3	Alkylated Analogs 36	
5.3.4	Hydroxy and Mercapto Derivatives 38	
5.3.5	Oxa, Thia, and Aza Derivatives 39	
5.3.6	Deoxy and Heterocyclic-Fused Analogs 40	
5.3.7	Esters and Ethers 41	
5.3.8	Summary of Structure–Activity Relationships 42	
5.4	Absorption, Distribution, and Metabolism 43	
5.5	Toxicities 45	
5.6	Abuse of Anabolic Agents 45	
6	**Androgen Antagonists** 46	
6.1	Current Drugs on the Market 46	
6.2	Antiandrogens 47	
6.2.1	Therapeutic Uses 47	
6.2.2	Structure–Activity Relationships for Antiandrogens 47	
6.2.3	Absorption, Distribution, and Metabolism 51	
6.2.4	Toxicities 51	
6.3	Enzyme Inhibitors 51	
6.3.1	5α-Reductase Inhibitors 52	
6.3.2	17,20-Lyase Inhibitors 53	
6.3.3	C19 Steroids as Aromatase Inhibitors 56	
7	**Summary** 58	
	Acknowledgments 60	
	References 60	

Keywords

Androgens
Steroid hormones responsible for the primary and secondary sex characteristics of the male, including the development of the vas deferens, prostate, seminal vesicles, and penis.

Testosterone
The C_{19} steroid hormone that is the predominant circulating androgen in the bloodstream and is produced mainly by the testis in males.

Dihydrotestosterone
The C_{19} steroid hormone that is the 5α-reduced metabolite of testosterone. It is produced in certain androgen target tissues and is the most potent endogenous androgen.

Anabolics
Compounds that demonstrate a marked retention of nitrogen through an increase of protein synthesis and a decrease in protein catabolism in the body.

Antiandrogens
Agents that compete with endogenous androgens for the hormone-binding site on the androgen receptor and thus block androgen action.

Selective androgen receptor modulators
Agents that may act as an androgen antagonist or weak agonist in one tissue, but as a strong androgen agonist in another tissue type.

5α-Reductase inhibitors
Compounds that inhibit the conversion of testosterone to its more active metabolite, dihydrotestosterone.

The steroid testosterone is the major circulating sex hormone in males and is the prototype for the androgens, the anabolic agents, and androgen antagonists. Endogenous androgens are biosynthesized from cholesterol; the majority of the circulating androgens are produced in the testes under the stimulation of luteinizing hormone (LH). The reduction of testosterone to dihydrotestosterone is necessary for androgenic actions of testosterone in many androgen target tissues such as the prostate; the oxidation of testosterone by the enzyme aromatase produces estradiol. The androgenic actions of testosterone and dihydrotestosterone are due to their binding to the androgen receptor, followed by nuclear localization, dimerization of the receptor complex, and binding to specific DNA sequences. This binding of the homodimer to the androgen response element leads to gene expression, stimulation, or repression of new mRNA synthesis, and subsequent protein biosynthesis. The synthetic androgens and anabolics were prepared to impart oral activity to the androgen molecule, to separate the androgenic effects of testosterone from its anabolic effects, and to improve on its biological activities. Novel nonsteroidal androgens, termed selective androgen receptor modulators, were developed to impart agonist activity in selective tissues. Drug preparations are used for the treatment of various androgen-deficient diseases and for the therapy of diseases characterized by muscle wasting and protein catabolism. Androgen antagonists include antiandrogens, which block interactions of androgens with the androgen receptor, and inhibitors of androgen biosynthesis and metabolism. Such compounds have therapeutic potential in the treatment of acne, virilization in women, hyperplasia and neoplasia of the prostate, and baldness.

1
Introduction

Androgens are a class of steroids responsible for the primary and secondary sex characteristics of the male. In addition, these steroids possess potent anabolic or growth-promoting properties. The general chemical structure of androgens is based on the androstane C_{19} steroid, which consists of the fused four-ring steroid nucleus (17 carbons atoms, rings A–D) and the two axial methyl groups (carbons 18 and 19) at the A/B and C/D ring junctions. The hormone testosterone (**1**) is the predominant circulating androgen and is produced mainly by the testis in males. 5α-Dihydrotestosterone (**2**) is a 5α-reduced metabolite of testosterone produced in certain target tissues and is the most potent endogenous androgen. Other endogenous androgens are produced by the adrenal gland in both males and females.

(**1**)

(**2**)

These two steroids and other endogenous androgens influence not only the development and maturation of the male genitalia and sex glands, but also affect other tissues such as kidney, liver, and brain. In this chapter, the endogenous androgens, synthetic analogs, various anabolic agents, and the androgen antagonists employed in clinical practice or animal husbandry in the United States and elsewhere will be discussed. Modified androgens that have found use as biochemical or pharmacological tools also are included. More extensive presentations of the topic of androgens, anabolics, and androgen antagonists have appeared in several treatises published over the past four decades [1–11].

2
Historical

The role of the testes in the development and maintenance of the male sex characteristics, and the dramatic physiological effects of male castration, have been recognized since early times. Berthold [12] was the first to publish (in 1849) a report that gonadal transplantation prevented the effects of castration in roosters, suggesting that the testis produced internal secretions exhibiting androgenic effects. However, the elucidation of the molecules of testicular origin responsible for these actions took almost another century. The first report of the isolation of a substance with androgenic activity was made by Butenandt [13, 14], in 1931. The material, isolated in very small quantities from human male urine [15], was named androsterone (**3**) [16]. A second weakly androgenic steroid hormone was isolated from male urine in 1934; this substance was named dehydroepiandrosterone (**4**) because of its ready chemical transformation and structural similarity to androsterone [17]. A year later, Laqueur [18, 19] reported the isolation of the testicular androgenic hormone, testosterone (**1**), which was 10-fold more potent than androsterone in promoting capon comb growth. Shortly after this discovery, the first chemical synthesis of testosterone was reported by Butenandt and Hanisch [20] and confirmed by Ruzicka [21, 22].

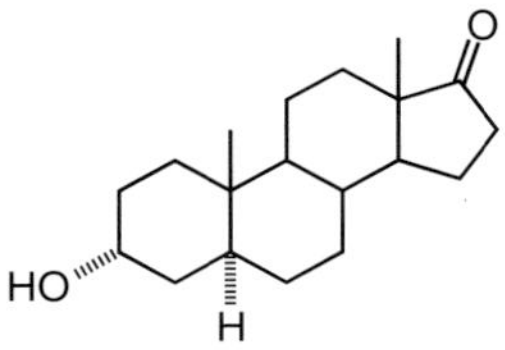

(3)

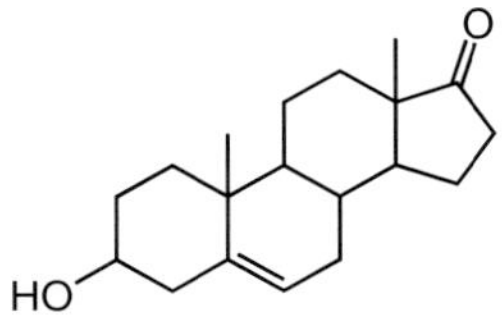

(4)

For many years, it was believed that testosterone was the active androgenic hormone in man. In 1968, however, research in two laboratories demonstrated that 5α-dihydrotestosterone (DHT, 2), also referred to as stanolone, was the active androgen in certain target tissues such as the prostate and seminal vesicles, and was formed from testosterone by a reductase present in these tissues [23, 24]. Shortly thereafter a soluble receptor protein was isolated and shown to have a greater specificity for DHT and related structures [25, 26]. In general, DHT is thought to be the active androgen in tissues that express 5α-reductase (e.g., the prostate), whereas testosterone appears to directly mediate these effects in muscle and bone.

The anabolic action of the androgens was first documented by Kochakian and Murlin in 1935 [27]. In their experiments, extracts of male urine caused a marked retention of nitrogen when injected into dogs fed a constant diet. Soon afterwards, testosterone propionate was observed to produce a similar nitrogen-sparing effect in humans [28]. Subsequent clinical studies demonstrated that testosterone was capable of causing a major acceleration of skeletal growth and a marked increase in muscle mass [29–31]. This action on muscle tissue has been referred to more specifically as the myotrophic effect.

The first androgenic-like steroid used for its anabolic properties in humans was testosterone. Unfortunately, its use for this purpose was limited by the inherent androgenicity and the need for parenteral administration. 17α-Methyltestosterone (5) was the first androgen discovered to possess oral activity, but it too failed to show any apparent separation of androgenic and anabolic activity. The promise of finding a useful, orally effective, anabolic agent free from androgenic side effects prompted numerous clinical and biological studies.

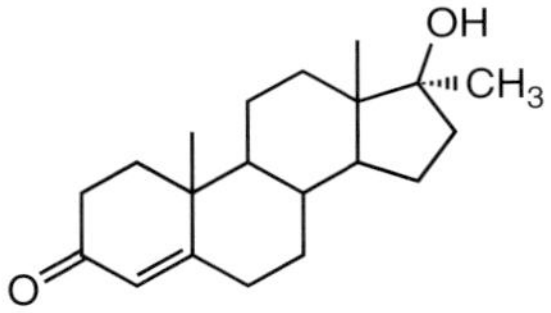

(5)

3
Endogenous Male Sex Hormones

3.1
Occurrence and Physiological Roles

The hormone testosterone affects many organs in the body, its most dramatic effects being observed on the primary and secondary sex characteristics of the male. These actions are first manifested in the developing male fetus, when the embryonic testis begins to secrete testosterone. Differentiation of the Wolffian ducts into the vas deferens, seminal vesicles, and epididymis occurs under this early androgen influence, as does the development of external genitalia and the prostate [32]. The reductive metabolism of testosterone to DHT is critical for virilization during this period of fetal development, as demonstrated dramatically in patients with a 5α-reductase deficiency [33].

At puberty, further development of the sex organs (prostate, penis, seminal vesicles, and vas deferens) is again evident and under the control of androgens. Additionally, the testes now begin to produce mature spermatozoa. Other effects of testosterone, particularly on the secondary sex characteristics, are observed; hair growth on the face, arms, legs, and chest is stimulated by this hormone during younger years. In later years, however, DHT is responsible for a thinning of the hair and recession of the hairline. At puberty, the larynx develops and a deepening of the voice occurs, the male's skin thickens, the sebaceous glands proliferate, and the fructose content in human semen increases. Testosterone influences sexual behavior, mood, and aggressiveness of the male at the time of puberty.

In addition to these androgenic properties, testosterone also exhibits anabolic (myotropic) characteristics. A general body growth is initiated, including increased muscle mass and protein synthesis, a loss of subcutaneous fat, and increased skeletal maturation and mineralization. This anabolic action is associated with a marked retention of nitrogen brought about by an increase in protein synthesis and a decrease in protein catabolism. The increase in nitrogen retention is manifested primarily by a decrease in urinary rather than fecal nitrogen excretion, and results in a more positive nitrogen balance. For example, the intramuscular administration of 25 mg testosterone propionate twice daily causes nitrogen retention to appear within 1–3 days, reaching a maximum in about 5–8 days. This reduced level of nitrogen excretion may be maintained for at least a month, and depends on the patient's nutritional status and diet [34].

Androgens influence skeletal maturation and mineralization, which is reflected in an increase in skeletal calcium and phosphorus [35]. In various forms of osteoporosis, androgens decrease urinary calcium loss and improve the calcium balance in patients; this effect is less noticeable in normal patients. Moreover, the various androgen analogs differ markedly in their effects on calcium and phosphorus balance in man [35]. Androgens and their 5β-metabolites (e.g., etiocholanolone) markedly stimulate erythropoiesis, presumably by increasing the production of erythropoietin and by enhancing the responsiveness of erythropoietic tissue to erythropoietin [36]. The effects of androgens on carbohydrate metabolism appear to be minor, and secondary to their primary protein anabolic property, but the effects on lipid metabolism seem unrelated to this anabolic property. Weakly androgenic metabolites such as androsterone have been found to lower serum cholesterol levels when administered parenterally.

3.2
Biosynthesis

The androgens are secreted not only by the testis in males, but also by the adrenal cortex in males and females, and the ovary in females. Testosterone is the principal circulating androgen and is formed by the Leydig cells of the testes. Other tissues, such as liver and human prostate, form testosterone from precursors, but this contribution to the circulating androgen pool is minimal. Since dehydroepiandrosterone and androstenedione are secreted by the adrenal cortex and ovary, they indirectly augment the circulating testosterone pool because they can be rapidly converted to testosterone by peripheral tissues. This local production of testosterone from circulating adrenal androgens can significantly contribute to local androgen concentrations in certain tissues, such as prostate.

Plasma testosterone levels for men usually range between 6 and 11 ng ml^{-1}, and are between fivefold and 100-fold the values in females [37]. The circulating level of DHT in normal adult men is about one-tenth the testosterone level [38]. Daily testosterone production rates have been estimated at 4–12 mg for young men and 0.5–2.9 mg for young women [39]. Although attempts have been made to estimate the secretion rates for testosterone, these studies have been hampered by the number of tissues capable of secreting androgens and the considerable interconversion of the steroids concerned [40, 41].

The synthesis of androgens in the Leydig cells of the testes is regulated by the gonadotropic hormone, luteinizing hormone (LH). The other pituitary gonadotropin, follicle-stimulating hormone (FSH), acts primarily on the germinal epithelium and is important for sperm development. Both of these pituitary gonadotropins are under the regulation of a decapeptide hormone produced by the hypothalamus. This hypothalamic hormone is luteinizing hormone-releasing hormone (LHRH), also referred to as gonadotropin-releasing hormone (GnRH). In adult males, a pulsatile secretion of LHRH, and subsequently of LH and FSH, occurs at a frequency of 8–14 pulses in 24 h [42]. The secretions of these hypothalamic and pituitary hormones are, in turn, regulated by circulating testosterone and estradiol levels in a negative feedback mechanism. Testosterone will decrease the frequency and amplitude of pulsatile LH secretion [43], whereas both testosterone and a gonadal peptide, inhibin, are both involved in suppressing the release of FSH [44].

The present understanding of steroidogenesis in the endocrine organs has advanced considerably during the past four decades, based largely on initial investigations with the adrenal cortex and subsequent studies also of the testis and ovary [45]. Figure 1 outlines the following sequence of events known to be involved with steroidogenesis in the Leydig cells. LH binds to its receptor located on the surface of the Leydig cell and, via a G protein-mediated process, activates adenylyl cyclase to result in an increase in intracellular concentrations of cyclic AMP (cAMP). cAMP activates a cAMP-dependent protein kinase, which subsequently phosphorylates and activates several enzymes involved in the steroidogenic pathway, including cholesterol esterase and cholesterol side-chain cleavage [46]. Cholesterol esters (present in the cell as a storage form) are converted to free cholesterol by cholesterol esterase, and free cholesterol is translocated to the mitochondria where a cytochrome P450 mixed-function oxidase system, termed cholesterol side-chain cleavage, converts cholesterol to pregnenolone. Several non-mitochondrial enzymatic transformations then convert pregnenolone to testosterone, which is secreted.

The conversion of cholesterol (**6**) to pregnenolone (**7**) has been termed the rate-limiting step in steroid hormone biosynthesis. The reaction requires NADPH and molecular oxygen, and is catalyzed by the cholesterol side-chain cleavage complex. The latter enzyme complex is comprised of three proteins: cytochrome P450$_{SCC}$ (also called cytochrome P450 11A1); adrenodoxin; and adrenodoxin reductase. Three moles of NADPH and

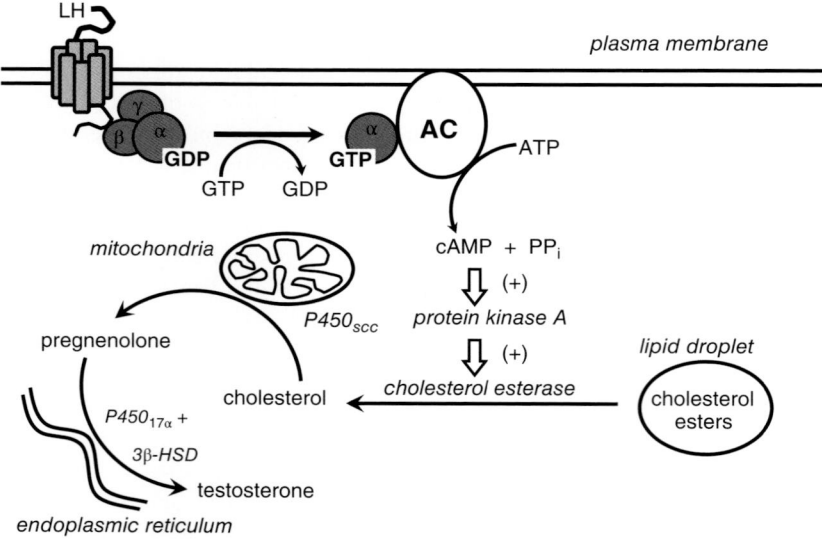

Fig. 1 Cellular events in steroidogenesis in the Leydig cell.

oxygen are required to convert one mole of cholesterol into pregnenolone (Fig. 2).

Tracer studies have shown that two major pathways known as the "4-ene" and "5-ene" pathways are involved in the conversion of pregnenolone to testosterone. Both of these pathways and their requisite enzymes are shown in Fig. 2. Earlier studies tended to favor the "4-ene" pathway, but more recent studies have disputed this view and suggest that the "5-ene" pathway is quantitatively more important in man. When Vihko and Ruokonen [47] analyzed the spermatic venous plasma for free and conjugated steroids, all intermediates of the "5-ene" pathway were identified but progesterone (**8**), an important intermediate of the "4-ene" pathway, was not found. In addition, sulfate conjugates were present in significant quantities, especially androst-5-ene-3β,17β-diol 3-monosulfate. These data strongly suggest that this intermediate and its unconjugated form constitute an important precursor of testosterone in man. This view, however, was not supported by a kinetic analysis of the metabolism of androst-5-ene-3β,17β-diol (**12**) in man [48]. Further evidence that the predominant pathway appears to be the "5-ene" pathway was provided by *in-vitro* studies in human testicular tissues [49].

Another important step is the conversion of the C-21 steroids to the C-19 androstene derivatives. Whereas, the enzymes for side-chain cleavage are localized in the mitochondria, those responsible for cleavage of the C_{17}-C_{20} bond (C_{17}-C_{20} lyase) reside in the endoplasmic reticulum of the cell. Early studies implicated 17α-hydroxypregnenolone (**9**) or 17α-hydroxyprogesterone (**10**) as obligatory intermediates in testosterone biosynthesis [50], and the C_{17}-C_{20} bond was subsequently cleaved by a second enzymatic process to produce the C-19 androstene molecule. This view of the involvement of two separate enzymes in the conversion of C-21 to C-19 steroids existed until purification of the proteins during the 1980s. The 17α-hydroxylase/17,20-lyase cytochrome

Analogs and Antagonists of Male Sex Hormones | 11

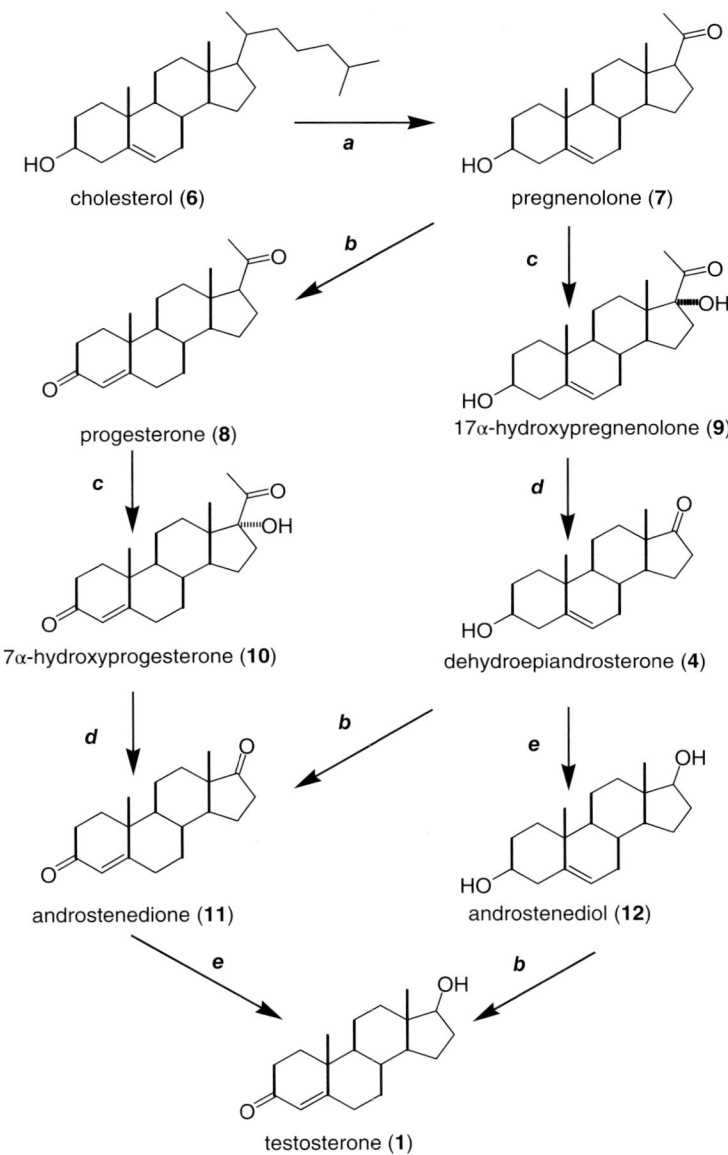

Fig. 2 Enzymatic conversion of cholesterol to testosterone. Enzymes are denoted as: (*a*) side chain cleavage; (*b*) 3β-hydroxysteroid dehydrogenase; (*c*) 17α-hydroxylase; (*d*) 17,20-lyase; (*e*) 17β-hydroxysteroid dehydrogenase.

P450 (abbreviated cytochrome P450 17 or cytochrome P450$_{17\alpha}$) was first isolated from neonatal pig testis microsomes by Nakajin and Hall [51]. Cytochrome P450$_{17\alpha}$ possessed both 17α-hydroxylase and 17,20-lyase activity when reconstituted with cytochrome P450 reductase and phospholipid. Identical full-length human cytochrome P450$_{17\alpha}$ complementary DNA (cDNA) sequences were independently isolated and reported in 1987 [52, 53]. Extensive reviews of the molecular biology, gene regulation, and enzyme deficiency syndromes have been published [46, 54].

Two additional enzymes are necessary for the formation of testosterone from dehydroepiandrosterone. The first is the 3β-hydroxysteroid dehydrogenase/$\Delta^{4,5}$-isomerase complex, which catalyzes the oxidation of the 3β-hydroxyl group to the 3-ketone and isomerization of the double bond from $C_5=C_6$ to $C_4=C_5$. Again, these processes were originally thought to involve two different enzymes, but purification of the enzymatic activity demonstrated that a single enzyme catalyzes both reactions [55]. The final enzyme in the pathway is the 17β-hydroxysteroid dehydrogenase, which catalyzes the reduction of the 17-ketone to the 17β-alcohol.

3.3
Absorption and Distribution

Although considerable research has been devoted to the biochemical mechanism of the action of natural hormones and the synthesis of modified androgens, little is known about the absorption of these substances. It is well recognized that a steroid hormone might have a high intrinsic activity but exerts little or no biological effect because its physico-chemical characteristics prevent it from reaching the site of action. This is particularly true in humans, where slow oral absorption or rapid inactivation may greatly reduce the efficacy of a drug. Even though steroids are commonly given by mouth, little is known of their intestinal absorption. One study in rats showed that androstenedione (**11**) was absorbed better than testosterone or 17α-methyltestosterone, and conversion of testosterone to its acetate enhanced absorption [56]. Results with other steroids have indicated that lipid solubility is an important factor for intestinal absorption, and this may explain the oral activity of certain ethers and esters of testosterone.

Once in the circulatory system, either by secretion from the testis or absorption of the administered drug, testosterone and other androgens will reversibly associate with certain plasma proteins, the unbound steroid being the biologically active form. The extent of this binding is dependent on the nature of the proteins and the structural features of the androgen.

The first protein to be studied was albumin, which exhibited a low association constant for testosterone and bound less-polar androgens such as androstenedione to a greater extent [57–59]. α-Acid glycoprotein (AAG) was shown to bind testosterone with a higher affinity than albumin [60, 61]. A third plasma protein to bind testosterone is corticosteroid-binding α-globulin (CBG) [62]. However, under normal physiological conditions these plasma proteins are not responsible for an extensive binding of androgens in plasma.

A specific protein termed sex hormone binding β-globulin (SHBG) or testosterone-estradiol binding globulin (TEBG) was found in plasma that bound testosterone with a very high affinity [63, 64]. The SHBG–sex hormone complex serves several functions, such as a transport or carrier system in the bloodstream, a storage site or reservoir for the hormones, and protection of the hormone against metabolic transformations [65]. SHBG has been purified and contains high-affinity, low-capacity binding sites for the sex hormones [66]; the protein has subsequently been cloned and crystallized [66]. Dissociation constants of approximately 1×10^{-9} M have been reported for the binding of testosterone and estradiol to SHBG, and are two orders of magnitude less than values reported for the binding of the hormone to the cytosolic receptor protein [67–69]. The plasma levels of SHBG are regulated by the thyroid hormones [70] and remain fairly constant throughout adult life in both males and females [71]. SHBG is not present in the plasma of all animals [65, 72]; for example, SHBG-like activity is notably absent in the rat, and testosterone may be bound in the rat plasma to CBG.

Numerous studies have been performed on the specificity of the binding of steroids to human SHBG [65, 71–77]. The presence of a 17β-hydroxyl group is essential for binding to SHBG. In addition to testosterone, DHT, 5α-androstane-3β,17β-diol (**20**), and 5α-androstane-3α,17β-diol (**21**) bind with high affinity, and these steroids compete for a common binding site. Binding to SHBG is decreased by 17α-substituents such as 17α-methyl and 17α-ethinyl moieties and by unsaturation at C-1 or C-6. Also, 19-nortestosterone derivatives have lower affinity. The steroid-binding site and the dimerization domain of SHBG, referred to as the amino-terminal laminin G-like domain, has been crystallized and demonstrated important hydrogen bonding of the C_3 and C_{17} moieties of steroidal ligands with Ser[42] and Asp[65] of SHBG [78].

Another extracellular carrier protein which exhibits a high affinity for testosterone, is found in seminiferous fluid and the epididymis and originates in the testis, is called androgen binding protein (ABP) [79–81]. This protein is produced by the Sertoli cells on stimulation by FSH [82, 83], and has very similar characteristics to those of plasma SHBG produced in the liver [82].

The absorption of androgens and other steroids from the blood by target cells was usually assumed to occur by a passive diffusion of the molecule through the cell membrane. However, studies conducted during the early 1970s, using tissue cultures or tissue slices, suggested entry mechanisms for the steroids. Estrogens [84, 85], glucocorticoids [86, 87] and androgens [88–91] exhibit a temperature-dependent uptake into intact target cells, suggesting a protein-mediated process. Among the androgens, DHT exhibited a greater uptake than testosterone in human prostate tissue slices [92], and it was found that estradiol or androstenedione interfered with this uptake mechanism [93, 94]. In addition, cyproterone competitively inhibited androstenedione, testosterone and DHT entry, whereas cyproterone acetate enhanced the uptake of these androgens [91]. Little is known regarding the exit of steroids from target cells; the only reported studies have investigated the active transport of glucocorticoids out of cells [92, 93].

3.4
Metabolism

For decades, the primary function of metabolism was thought to be an inactivation of testosterone, an increase in hydrophilicity, and a mechanism to facilitate excretion of the steroid into the urine. However, the identification of metabolites of testosterone formed in peripheral tissues, as well as the potent and sometimes different biological activities of these products, has emphasized the importance of

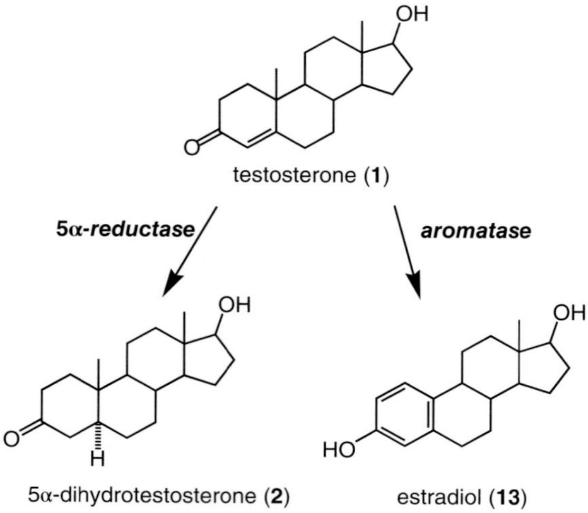

Fig. 3 Enzymatic conversion of testosterone to biologically active metabolites, 5α-dihydrotestosterone and estradiol.

metabolic transformations of androgens in endocrinology. Two active metabolites of testosterone have received considerable attention, namely the reductive metabolite 5α-dihydrotestosterone (**2**) and the oxidative metabolite estradiol (**13**) (Fig. 3).

3.4.1 Reductive Metabolism

The metabolism of testosterone in a variety of *in-vitro* and *in-vivo* systems has been reviewed [50, 94–96]. The principal pathways for the reductive metabolism of testosterone in man are shown in Fig. 4. Human liver produces a number of metabolites, including androstenedione (**11**), 3β-hydroxy-5α-androstan-17-one (**17**), 5α-androstane-3β,17β-diol (**20**), and 5α-andro-stane-3α,17β-diol (**21**) [97, 98]. In addition, cirrhotic liver was shown to produce more 17-keto-steroids than normal liver [99]. Human adrenal preparations, on the other hand, produced 11β-hydroxytestosterone as the major metabolite [100]. The intestinal metabolism of testosterone is similar to transformations in the liver [95], while the major metabolite in lung is androstenedione [101].

Studies on testosterone metabolism conducted since the late 1960s have centered on steroid transformations by prostatic tissues. Normal prostate, benign prostatic hypertrophy (BPH), and prostatic carcinoma all contain 3α-, 3β-, and 17β-hydroxysteroid dehydrogenases, and 5α- and 5β-reductases, capable of converting testosterone to various metabolites. Prostatic carcinoma metabolizes testosterone more slowly than does BPH or normal prostate [102]. On the other hand, recent studies have shown that adrenal androgens can be converted into testosterone and dihydrotestosterone in prostate cancer cells [103, 104]. K. D. Voigt *et al.* [105, 106] have performed extensive studies of *in-vivo* metabolic patterns of androgens in patients with BPH by injecting them (intravenously) with tritiated androgens 30 min before prostatectomy. Tissues from the prostate and surrounding skeletal muscle, as well as blood plasma, were then analyzed for

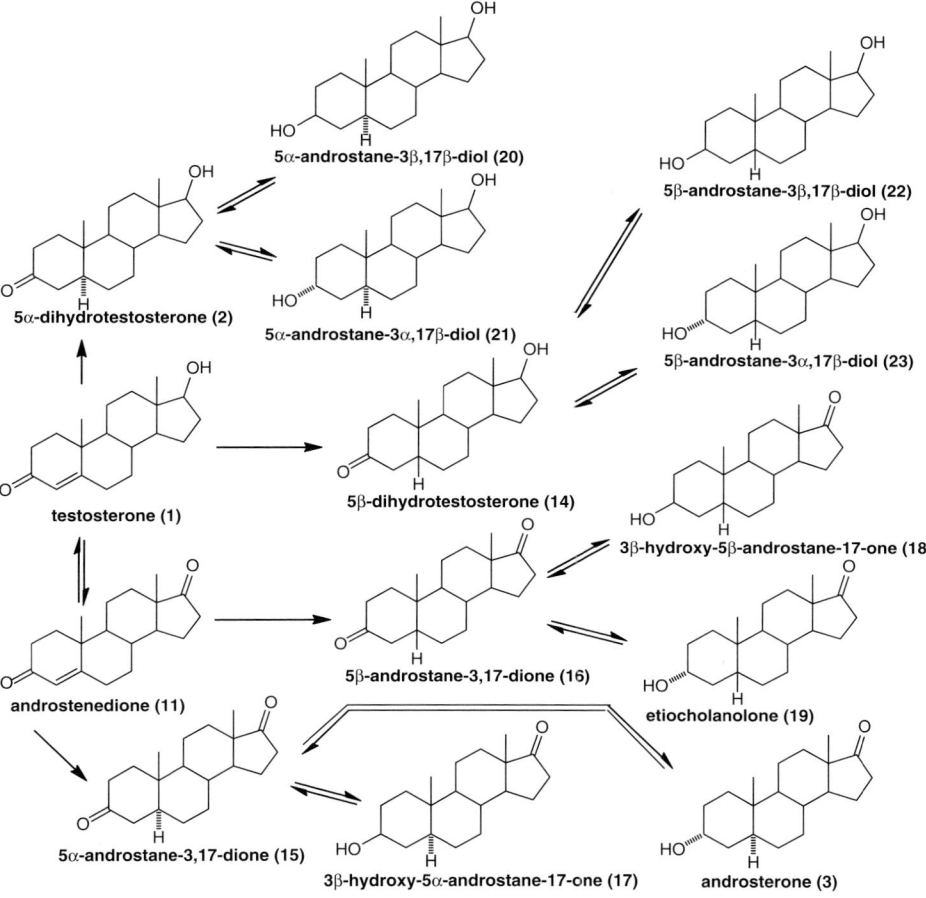

Fig. 4 Reductive metabolites of testosterone.

metabolites. The major metabolite of testosterone found in BPH tissues was DHT, with minor amounts of diols isolated. Skeletal muscle and plasma contained primarily unchanged testosterone.

Androsterone (**3**) and etiocholanolone (**19**), the major urinary metabolites, are excreted predominantly as glucuronides, and only about 10% as sulfates [37, 107]. These conjugates are capable of undergoing further metabolism. Testosterone glucuronide, for example, is metabolized differently from testosterone in man, giving rise mainly to 5β-metabolites [108]. Only a relatively small amount of the urinary 17-ketosteroids is derived from testosterone metabolism. In men, at least 67% and in women about 80% or more, of the urinary 17-ketosteroids are metabolites of adrenocortical steroids [39]. This explains why a significant increase in testosterone secretion associated with various androgenic syndromes does not usually lead to elevated levels of 17-ketosteroid excretion.

Although androsterone and etiocholanolone are the major excretory products, the exact sequence whereby these 17-ketosteroids arise is still not clear. Studies with radiolabeled androst-4-ene-3β,17β-diol and the epimeric 3α-diol in humans showed that oxidation to testosterone was necessary before reduction of the A-ring [109]. Moreover, 5β-androstane-3α,17β-diol (**23**) was the major initial liver metabolite in rats, but this decreased with time with a simultaneous increase of etiocholanolone [110]. This formation of saturated diols agrees with studies using human liver [97] and provided evidence that the initial step in testosterone metabolism is a reduction of the α,β-unsaturated ketone to a mixture of diols, followed by oxidation to the 17-ketosteroids.

Until 1968, it was generally thought that the excretory metabolites of testosterone were physiologically inert, but subsequent studies have shown that etiocholanolone has thermogenic effects when administered to man [111]. Hypocholesterolemic effects of parenterally administered androsterone have also been described [112].

The conversion of testosterone to DHT by 5α-reductase is of major importance in the mechanism of action of the hormone, as this enzyme has been found active in the endoplasmic reticulum [113, 114] and the nuclear membrane [23, 115–120] of androgen-sensitive cells. In addition, levels of 5α-reductase are under the control of testosterone and DHT [120]; 5α-reductase activity decreases after castration and can be restored to normal levels of activity with testosterone or DHT administration [121].

Early biochemical studies of 5α-reductase were performed using a microsomal fraction from rat ventral prostate. The irreversible enzymatic reaction catalyzed by 5α-reductase requires NADPH as a cofactor, which provides the hydrogen for carbon-5 [122]. The 5α-reductase from rat ventral prostate tissues exhibited a broad range of substrate specificity for various C_{19} and C_{21} steroids [99]; this broad specificity was also observed in inhibition studies [123]. However, more detailed studies of the enzyme were limited due to the extreme hydrophobic nature of the protein, its instability upon isolation, and its low concentrations in androgen-dependent tissues [96].

Investigations of the molecular biology of 5α-reductase resulted in the demonstration of two different genes and two different isozymes of the enzyme [124–126]. The first cDNA to be isolated and cloned that encoded 5α-reductase was designated Type 1, and the second Type 2. The gene encoding Type 1 is located on chromosome 5, while the gene encoding Type 2 is located on chromosome 2. The two human 5α-reductases have approximately 60% sequence homology. The two isozymes differ in their biochemical properties, tissue location, and function [126, 127]. For example, Type 1 5α-reductase exhibits an alkaline pH optimum (6–8.5) and has micromolar affinities for steroid substrates, whereas Type 2 5α-reductase has a sharp pH optimum at 4.7–5.5, a higher affinity (lower apparent K_m) for testosterone, and is more sensitive to inhibitors than the Type 2 isozyme. The latter isozyme is expressed primarily in androgen target tissues, the liver expresses both types, and Type 1 is expressed in various peripheral tissues. Type 2 5α-reductase appears to be essential for masculine development of the fetal urogenital tract and the external male phenotype, whereas the Type 1 isozyme is primarily a catabolic enzyme. In certain cases of human male pseudohermaphroditism, mutations in the Type 2 5α-reductase gene have been observed that resulted in significant decreases in DHT levels needed for virilization [128].

3.4.2 Oxidative Metabolism

Another metabolic transformation of androgens leading to hormonally active compounds involves their conversion to estrogens. Estrogens are biosynthesized in the ovaries and placenta and, to a lesser extent, in the testes, adrenals and certain regions of the brain. The enzyme complex that catalyzes this biosynthesis is referred to as aromatase, and the enzymatic activity was first identified by Ryan [129] in the microsomal fraction from human placental tissue. The mechanism of the aromatization reaction was first elucidated during the early 1960s and continues to be the subject of extensive studies. Aromatase is a cytochrome P450 enzyme complex [130] that requires 3 mol of NADPH and 3 mol of oxygen per mole of substrate [131]. Aromatization proceeds via three successive steps, the first two of which are hydroxylations. The observation by Meyer [132] that 19-hydroxyandrostenedione (**24**) was a more active precursor of estrone (**27**) than the substrate androstenedione led to its postulated role in estrogen biosynthesis. This report, as well as numerous subsequent studies, led to the currently accepted pathway for aromatization (as shown in Fig. 5).

The first two oxidations occur at the C_{19} position, producing the 19-alcohol (**24**) and then the 19-*gem*-diol (**25**), originally isolated as the 19-aldehyde (**26**) [133, 134]. The exact mechanism of the last oxidation remains to be fully determined. The final oxidation results in a stereospecific elimination of the 1β and 2β hydrogen atoms [135–137] and a concerted elimination of the oxidized C_{19} moiety as formic acid [134]. Hydroxylation at the 2β-position was suggested as an intermediate in this final oxidation, as this substance is spontaneously aromatized to estrone [138]. However, investigations using $^{18}O_2$ and isotopically labeled steroidal intermediates failed to show any incorporation of the 2β-hydroxyl group into formic acid under enzymatic or nonenzymatic conditions [139]; neither was it demonstrated that the oxygen atoms from the first and third oxidation steps were incorporated into formic acid [140–142]. These results led to the proposal that the last oxidation step

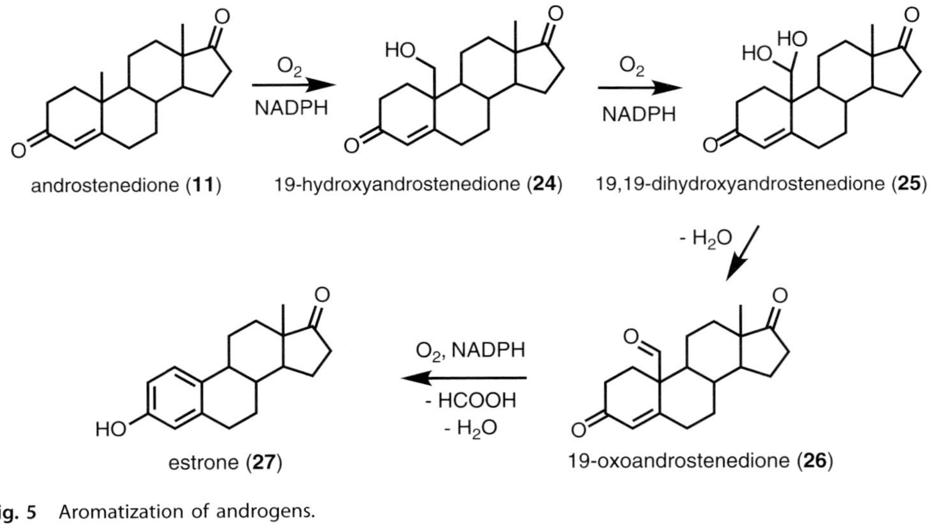

Fig. 5 Aromatization of androgens.

is a peroxidative attack at the C_{19} position [143–145]. However, recent computational chemistry studies have suggested the involvement of a cytochrome P450 oxene intermediate in the final catalytic step of aromatase, resulting in 1β-hydrogen atom abstraction and the release of formic acid [146].

The incubation of a large number of testosterone analogs with human placental tissue [147, 148] has provided some insight into the structural requirements for aromatization. Whereas, androstenedione was converted rapidly to estrone, the 1-dehydro and 19-nor analogs were metabolized slowly, and the 6-dehydro isomer and saturated 5α-androstane-3,17-dione remained unchanged. Hydroxyl and other substituents at 1α, 2β, and 11β interfered with aromatization, whereas similar substituents at 9α and 11α seemingly had no effect. Among the stereoisomers of testosterone, only the 8β, 9β, 10β-isomer was aromatized, in addition to compounds having the normal configuration (8β, 9α, 10β). Thus, the substrate specificity of aromatase appears to be limited to C_{19} steroids with the 4-en-3-one system. Inhibition studies with various steroids have provided additional insights into the structural requirements for the enzyme [149–151]; steroidal aromatase inhibitors are described later in Sect. 6.3.3.

Recent investigations of aromatase have focused on the biochemistry, molecular biology and regulation of the aromatase protein. Aromatase is a membrane-bound cytochrome P450 monooxygenase consisting of two proteins: aromatase cytochrome P450 (P450$_{arom}$); and NADPH-cytochrome P450 reductase. Cytochrome P450$_{arom}$ is a heme protein which binds the steroid substrate and molecular oxygen and catalyzes the oxidations. The reductase is a flavoprotein, is found ubiquitously in endoplasmic reticulum, and is responsible for transferring reducing equivalents from NADPH to cytochrome P450$_{arom}$. The purification of cytochrome P450$_{arom}$ proved to be very difficult because of its membrane-bound nature, instability, and low tissue concentration. The reconstitution of a highly purified cytochrome P450$_{arom}$ with NADPH-cytochrome P450 reductase and phospholipid resulted in a complete conversion of androstenedione to estrone, thus, demonstrating that one cytochrome P450 protein catalyzes all three oxidation steps [152]. The first report of the three-dimensional (3-D) crystal structure of human aromatase, published over two decades later, provided a molecular understanding of androgen substrate specificity and the unique enzyme reaction [153]. Knowledge of the molecular biology of aromatase has advanced greatly during the past two decades. A full-length cDNA complementary to messenger RNA (mRNA) encoding cytochrome P450$_{arom}$ was sequenced, and the open reading frame (ORF) encodes a protein of 503 amino acids [154]. When this cDNA sequence was inserted into COS1 monkey kidney cells, aromatase mRNA and aromatase enzymatic activity were detected in the transfected cells. The entire human cytochrome P450$_{arom}$ gene is greater than 70 kb in size [155, 156] and is located on chromosome 15 [157]. Clones have been utilized to examine the regulation of aromatase in ovarian, adipose, and breast tissues [158–161].

The metabolism of androgens by the mammalian brain has also been investigated under *in-vitro* conditions. In 1966, Sholiton *et al.* [162] were the first to report the metabolism of testosterone in rat brain, while later studies demonstrated the conversion of testosterone to DHT, androstenedione, 5α-androstane-3,17-dione, and 5α-androstane3β,17β-diol

[163–168]. The aromatization of androgens to estrogens was also found to occur in the hypothalamus and the pituitary gland [169–174]. The full significance of these metabolites on various neuroendocrine functions, such as the regulation of gonadotropin secretion and sexual behavior, is not yet fully understood [175, 176].

3.5
Mechanism of Action

It would indeed be impossible to explain all the varied biological actions of testosterone by one biochemical mechanism. Androgens, as well as the other steroid hormones adrenocorticoids, estrogens and progestins, exert potent physiological effects on sensitive tissues, yet are present in the body in only extremely low concentrations (e.g., 0.1–1.0 nM). The majority of investigations to elucidate the mechanisms of action of androgens have dealt with actions in androgen-dependent tissues and, in particular, the rat ventral prostate. The results of these studies have indicated that androgens act primarily to regulate gene expression and protein biosynthesis by the formation of a hormone–receptor complex, analogous to the mechanisms of action of estrogens and progestins. Extensive studies directed at elucidating the general mechanism of steroid hormone action have been performed for over three decades, and several reviews have emerged on this subject [177–190].

Jensen and Jacobson [191], using radiolabeled 17β-estradiol, were the first to show that a steroid was selectively retained by its target tissues. Investigations of a selective uptake of androgens by target cells performed during the early 1960s were complicated by a low specific activity of the radiolabeled hormones and the rapid metabolic transformations. Nonetheless, it was noted that target cells retained primarily unconjugated metabolites, whereas conjugated metabolites were present in nontarget cells such as blood and liver [192, 193]. With the availability of steroids of high specific activity, later studies demonstrated the selective uptake and retention of androgens by target tissues [23, 24, 115, 194, 195]. In addition, DHT was found to be the steroidal form selectively retained in the nucleus of the rat ventral prostate [23, 115]. This discovery led to the current concept that testosterone is converted by 5α-reductase to DHT, which is the active form of cellular androgen in androgen-dependent tissues such as the prostate. In general, DHT is thought to be the active androgen in tissues that express 5α-reductase (e.g., the prostate), whereas testosterone appears to directly mediate these effects in muscle and bone where 5α-reductase is absent.

The rat prostate has been the most widely examined tissue, and current hypotheses on the mode of action of androgens are based largely on these studies (see Fig. 6). The lipophilic steroid hormones are carried in the bloodstream, with the majority of the hormones reversibly bound to serum carrier proteins and a small amount of free steroids. The androgens circulating in the bloodstream are the sources of steroid hormone for androgen action in target tissues. Testosterone, synthesized and secreted by the testis, is the major androgen in the bloodstream and the primary source of androgen for target tissues in men. Dehydroepiandrosterone (DHEA) and androstenedione also circulate in the bloodstream and are secreted by the adrenal gland under the regulation of adrenocorticotrophic hormone (ACTH). DHEA and androstenedione supplement the androgen sources in normal adult men, but these

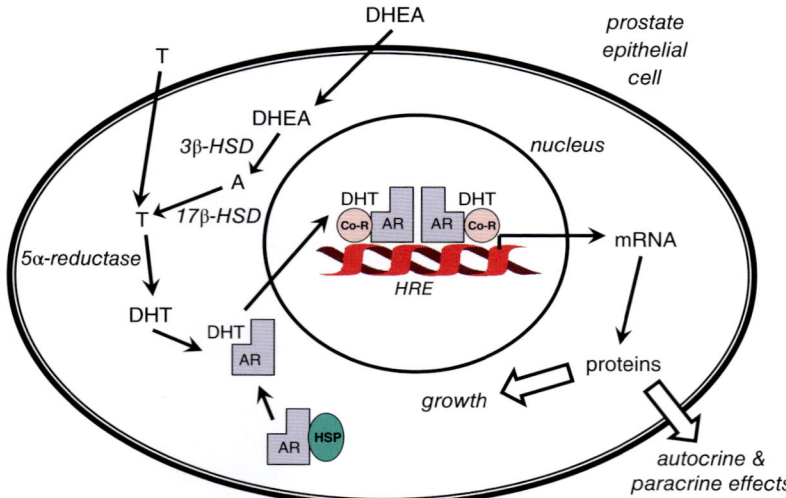

Fig. 6 Mechanism of action of 5α-dihydrotestosterone (DHT). T = testosterone; A = androstenedione; DHEA = dehydroepiandrosterone; AR = androgen receptor; HSPs = heat shock proteins; Co-R = coregulators/coactivators; HRE = hormone response element; 3β-HSD = 3β-hydroxysteroid dehydrogenase; and 17β-HSD = 17β-hydroxysteroid dehydrogenase.

steroids are the important circulating androgens in women. The free circulating androgens diffuse passively through the cell membrane and are converted to the active androgen 5α-DHT within the target tissues that express the enzyme.

The androgens act on target cells to regulate gene expression and protein biosynthesis via the formation of steroid–receptor complexes. Those cells sensitive to the particular steroid hormone (referred to as target cells) contain high-affinity steroid receptor proteins capable of interacting with the steroid [25, 196]. The binding of androgen with the receptor protein is a necessary step in the mechanism of action of the steroid in the prostate cell. The results of early studies suggested that the steroid receptor proteins were located in the cytosol of target cells [191] and, following formation of the steroid–receptor complex, the latter would be translocated into the nucleus of the cell. More recent investigations on androgen action have indicated that the unoccupied receptor proteins are present in the cytoplasm bound to various heat shock proteins (Hsps) and chaperones such as Hsp70 and Hsp90 to prevent degradation [197]. Binding of androgen with the androgen receptor results in a conformational change of the receptor complex, a disassociation of the Hsp proteins, and nuclear localization of the steroid–receptor complex.

In the nucleus, the steroid–androgen receptor complex is activated, resulting in the formation of a homodimer [186]. The homodimer then interacts with particular regions of the cellular DNA that are referred to as androgen-responsive elements (AREs), and also with various coactivator/coregulator proteins and other nuclear transcriptional factors [197].

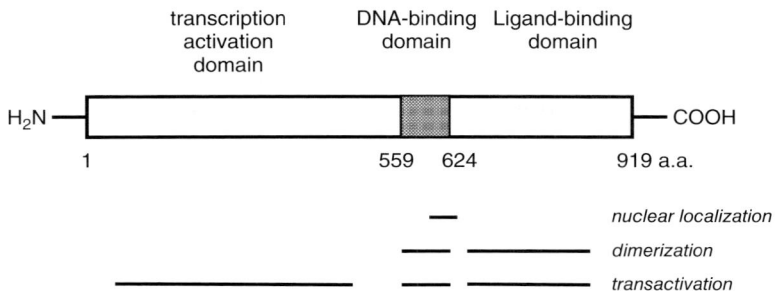

Fig. 7 Schematic diagram of the androgen receptor.

Binding of the nuclear steroid–receptor complex to DNA initiates a transcription of the DNA sequence to produce mRNA. Finally, the elevated levels of mRNA lead to an increased protein synthesis in the endoplasmic reticulum; the proteins synthesized include enzymes, receptors and/or secreted factors that subsequently result in the steroid hormonal response regulating cell function, growth, and differentiation.

Extensive structure–function studies on the androgen receptor (AR) have identified regions critical for hormone action. The AR is encoded by the *AR* gene located on the X chromosome, and the *AR* gene is comprised of eight exons. The human AR contains approximately 900–920 amino acids, and the exact length varies due to polymorphisms in the NH_2-terminal of the protein. The primary amino acid sequences of AR, as well as of the various steroid hormone receptors, were deduced from cloned cDNAs [186, 188]. The calculated molecular weight of AR is approximately 98 kDa, based on amino acid composition; however, the AR is a phosphoprotein and migrates higher at approximately 110 kDa in sodium dodecyl sulfate (SDS) gel electrophoresis. The steroid receptor proteins form part of a larger family of nuclear receptor proteins that also include receptors for vitamin D, thyroid hormones, and retinoids. The overall structural features of the AR have strong similarities to the other steroid hormone receptors (Fig. 7), with proteins containing regions that bind to the DNA and bind to the steroid hormone ligand [189, 198, 199]. A high degree of homology (sequence similarities) in the steroid receptors is found in the DNA-binding region that interacts with the hormone response elements (HREs). The DNA-binding region is rich in cysteine amino acids and chelate zinc ions, forming finger-like projections called zinc fingers that bind to the DNA. The hormone-binding domain (or ligand-binding domain; LBD) is located on the COOH-terminal of the protein. Structure–function studies of cloned receptor proteins have also identified regions of the molecules that are important for nuclear localization of the receptor, receptor dimerization, interactions with nuclear transcriptional factors, and the activation of gene transcription. Importantly, two regions of the AR protein are identified as transcriptional activation domains; the domain on the NH_2-terminal region may interact with both coactivators and corepressors, while the COOH-terminal domain initiates transcriptional activation only upon binding of an agonist such as 5α-DHT. The interactions necessary for formation of the steroid–receptor complexes and subsequent activation of gene transcription are complicated, involve multiple protein partners referred to as coactivators and corepressors, and leave many unanswered questions.

Although the tertiary structure of the entire AR has not been determined, the crystallographic structure of the LBD has been reported [200, 201]. The AR LBD consists of an α-helical sandwich, similar to the LBDs reported for other nuclear receptors, and contains only 11 helices (no Helix 2) and four short β-strands. Minor differences in the two reported crystallographic structures are likely due to limits of experimental resolution, differences in data interpretation, and the use of different ligands for crystallization. The endogenous ligand DHT (**2**) interacts with helices 3, 5, and 11, and the DHT-bound AR LBD has a single, continuous helix 12. Similar interactions are observed with metribolone (methyltrienolone, **55**); however, helix 12 is split into two shorter helical segments. Overall, the binding of steroidal ligands to amino acid residues of the AR LBD involves two hydrogen bonds with the 3-ketone function, and two hydrogen bonds with the 17β-hydroxyl group. Hydrophobic interactions of several amino acid residues with the steroid scaffold are also observed. Investigations on selective androgen receptor modulator (SARM binding to the LBD have provided further insights into the molecular interactions of nonsteroidal agents with AR [202].

Additional information on receptor structure–function has been obtained by analyzing AR mutations in patients with various forms of androgen resistance and abnormal male sexual development [189, 199, 203–205]. Two polymorphic regions have been identified in the NH_2-terminal region, encoding a polyglycine repeat and a polyglutamate tract. Certain polymorphic regions have recently been shown to significantly alter AR levels, stability, or transactivation [199]. These repeats are useful in the pedigree analysis of patients [189]. Mutations in the AR have been identified in patients with either partial or full androgen insensitivity syndrome (AIS), with the majority of mutations identified in exons 4 through 8 encoding the DNA-binding domain and the hormone-binding domain. Finally, studies with the human LNCaP prostate cancer cell line have provided interesting results regarding receptor protein structure and ligand specificity. The LNCaP cells exhibited an enhanced proliferation in the presence of androgens, but these cells unexpectedly proliferated in the presence of estrogens, progestins, cortisol, or the antiandrogen flutamide [206, 207]. Analysis of the cDNA for the LNCaP AR revealed that a single base mutation in the LBD was present, and this resulted in the increased affinity for progesterone and estradiol [208]. The crystallographic structures of the LBD with the T877A mutation confirm that the mutated AR LBD can accommodate larger structures at the C-17 position [200, 201].

The ultimate action of androgens on target tissues is the stimulation of cellular growth and differentiation through the regulation of protein synthesis, and numerous androgen-inducible proteins have been identified [199]. One prominent androgen-inducible protein is prostate-specific antigen (PSA), a serine protease that is expressed by secretory prostate epithelial cells and utilized as blood test in screening for possible prostate diseases such as prostate cancer. Three AREs have been identified in the promoter regions of the *PSA* gene [209–211]. Another androgen-regulated gene which has been examined extensively in rats is the gene encoding the protein probasin [212, 213], a 20-kDa secretory protein from the rat dorsolateral prostate that is structurally similar to serum globulins. Recently, a transmembrane serine protease called TMPRSS2 was identified in human prostate cells that may have a role in male reproduction and is overexpressed in poorly differentiated prostate cancer [214]. Other proteins induced by androgens include spermine-binding protein [215], keratinocyte growth factor (KGF or FGF-7) [216], androgen-induced growth factor (AIGF or FGF-8) [217, 218], nerve-growth factor [219], epidermal growth factor (EGF) [220], c-myc [221], protease D [222], β-glucuronidase [223], and $α_{2u}$-globulin [224, 225]. Studies of these proteins have suggested that androgens act by enhancing the transcription and/or translation of specific RNAs for the proteins. The AR also represses the gene expression of certain proteins such as glutathione S-transferase, TRPM-2 (which is involved in apoptosis), and cytokines such as interleukin (IL)-4, IL-5, and γ- interferon (IFN) [199, 226].

While most biochemical studies have been focused on the rat ventral prostate, some groups began to investigate the presence of cellular receptor proteins in other androgen-sensitive tissues. ARs have been reported in seminal vesicles [227, 228], sebaceous glands [229–231], testis [230, 232], epididymis [227, 233, 234], kidney [235], submandibular gland [236, 237], pituitary, and hypothalamus [238–244], bone marrow [245, 246], liver [247], and androgen-sensitive tumors [248, 249]. Although DHT is the active androgen in rat ventral prostate, it is not the only functioning form in other androgen-sensitive cells. In ventral prostate and seminal vesicles, DHT is readily formed but is metabolized only slowly and therefore can accumulate and bind to receptors.

A comparison of the binding kinetics for testosterone and DHT also showed that testosterone dissociates faster, implying an extended retention of DHT by the AR [250]. In other tissues, such as brain, kidney or skeletal muscle, DHT is not readily formed and is metabolized quickly compared to testosterone. Species variations have also been demonstrated, the most striking example being the finding that 5α-androstane-3α,17α-diol interacts specifically with cytosolic receptor protein from dog prostate [251] and may be the active androgen in this species [252]. Apparently, the need for a 17β-hydroxyl is not essential in all species.

Thus, current findings indicate that AR proteins vary in steroid specificity among different tissues from the same species, as well as among different species. Nevertheless, the basic molecular mechanism of action of the androgens in androgen-sensitive tissues is consistent with the results of the studies on rat ventral prostate.

The manner whereby the androgens exert their anabolic effects has not been studied so extensively. Indeed, the conversion of testosterone to DHT was shown to be insignificant in skeletal and levator ani muscles, which suggests that the androgen-mediated growth of muscle is due to testosterone itself [253, 254]. Classical steroid receptors for testosterone are found in the cytoplasm of the levator ani and quadriceps muscles of the rat [255, 256]. Unlike the AR in the prostate, DHT had a lower affinity than testosterone for the AR in muscle. Notably, ARs have also been identified in other muscle tissues, including cardiac muscle [257–262].

In addition to the genomic mechanisms, nongenomic pathways for androgen action through the AR have been reported in various tissues, including spermatogenesis

[263], oocytes [264], skeletal muscles [265], and prostate cancer cells [266]. Several characteristics of possible nongenomic pathways include a rapid timeframe for effects (varying from seconds to hours), the regulation of androgen-responsive genes that do not contain androgen response elements, and alterations of intracellular signaling pathways. The rapid activation of kinase signaling pathways, such as the activation of MAP kinase and ERK kinase pathways, and the modulation of intracellular calcium levels, are two examples of the nongenomic mechanisms of action of androgens and ARs.

4
Synthetic Androgens

4.1
Current Drugs on the Market

Currently available synthetic androgens used as therapeutics are listed in the following table.

Generic name (structure)	Trade name	U.S. manufacturer	Chemical class	Dose
Testosterone enanthate (**31**)	Delatestryl	Various suppliers	Androstane	Injection: 200 mg ml^{-1}
Testosterone cypionate (**33**)	Depotestosterone, Andronate	Various suppliers	Androstane	Injection: 100 mg ml^{-1} 200 mg ml^{-1}
Testosterone pellets (**1**)	Testopel	Bartor Pharmacal	Androstane	Pellets: 75 mg
Testosterone transdermal system (**1**)	Androderm, Testoderm	Various suppliers	Androstane	Transdermal: 12.2 mg 24.3 mg
Testosterone gel (**1**)	AndroGel 1%	AbbVie	Androstane	Gel: 1% testosterone
	Testim 1%	Auxilium	Androstane	Gel: 1% testosterone
Testosterone, buccal system (**1**)	Striant	Columbia	Androstane	Mucoadhesive: 30 mg
Methyltestosterone (**5**)	Methyltestosterone	Various suppliers	Androstane	Tablets: 10 mg 25 mg Tablets (buccal): 10 mg
	Methitest	Global	Androstane	Capsules: 10 mg
	Testred	Valeant	Androstane	Capsules: 10 mg
	Android	Valeant	Androstane	Capsules: 10 mg
	Virilon	Star	Androstane	Capsules: 10 mg
Fluoxymesterone (**36**)	Halotestin	Pfizer	Androstane	Tablets: 2 mg 5 mg 10 mg
	Fluoxymesterone	Various suppliers	Androstane	Tablets: 10 mg
Danazol (**77**)	Danazol	Various suppliers	Androstane	Capsules: 50 mg 100 mg 200 mg
Testolactone (**38**)	Teslac	Bristol-Myers Squibb	Androstane	Tablets: 50 mg

4.2
Therapeutic Uses and Bioassays

The primary uses of synthetic androgens are the treatment of disorders of testicular function and of cases with decreased testosterone production. Several types of clinical condition result from testicular dysfunction. Information on the biochemistry and mechanism of action of testosterone that has accumulated over the past 30 years has greatly aided in the elucidation of the underlying pathophysiology of these diseases. Two reviews describe in greater detail the mechanisms involved in disorders of testicular function and androgen resistance [267, 268].

Hypogonadism arises from the inability of the testis to secrete androgens, and can be caused by various conditions. These hypogonadal diseases can, in many cases, result in disturbances in sexual differentiation and function and/or sterility. Primary hypogonadism is the result of a basic disorder in the testes, while secondary hypogonadism results from the failure of pituitary and/or hypothalamic release of gonadotropins and thus a diminished stimulation of the testis. Usually, primary hypogonadism is not recognized in early childhood (with the exception of cryptorchidism) until the expected time of puberty. This testosterone deficiency is corrected by androgen treatment for several months, at which time the testes are evaluated for possible development. Long-term therapy is necessary if complete testicular failure is present. Patients with Klinefelter's syndrome, a disease in which a genetic male has an extra X chromosome, have low testosterone levels and can also be treated by androgen replacement.

Male pseudohermaphroditism incorporates disorders in which genetically normal men do not undergo normal male development:

- *Testicular feminization* is observed in patients who have normal male XY chromosomes but the male genitalia and accessory sex glands do not develop; rather, the patients have female external genitalia. These patients are unresponsive to androgens and have defective ARs [269–271].
- An alternative *male pseudohermaphroditism* results from a deficiency of the enzyme 5α-reductase [272, 273]. Since DHT is necessary for early differentiation and development, the patients again develop female genitalia; later, some masculinization can occur at the time of puberty due to elevated testosterone levels in the blood.
- *Reifenstein syndrome* is an incomplete pseudohermaphroditism. In these patients, the androgen levels are normal, 5α-reductase is present, and elevated LH levels are found. Partially deficient ARs are present in these patients [269, 271].

In most cases of male pseudohermaphroditism, androgen replacement has little or no effect, and thus steroid treatment is not recommended.

Deficiencies of circulating gonadotropins lead to secondary hypogonadism. This condition can be caused by disorders of the pituitary and/or hypothalamus, resulting in diminished secretions of neurohormones. The lack of stimulation of the seminiferous tubules and the Leydig cells due to the low levels of these neurohormones decreases androgen production. Drugs such as neuroleptic phenothiazines and the stimulant marijuana can also interfere with the release of gonadotropins. The use of androgens in secondary hypogonadism is symptomatic.

Synthetic androgens have also been used in women for the treatment of endometriosis, abnormal uterine bleeding, and menopausal symptoms, but their utility is severely limited by the virilizing side effects of these agents. Two weak androgens – calusterone and 1-dehydrotestolactone – have been used clinically in the treatment of mammary carcinoma in women. The mode of action of these drugs in the treatment of breast cancer is unknown, but it is not simply related to their androgenicity [274]. More recent evidence on the ability of these compounds to inhibit estrogen biosynthesis catalyzed by aromatase suggests that they effectively lower estrogen levels *in vivo* [150].

The various analytical methods used to establish the androgenic properties of steroidal substances have been reviewed by Dorfman [275]. Traditionally, androgens have been assayed using the capon comb growth method, and by using the seminal vesicles and prostate organs of rodents. Increases in the weight and/or growth of the capon comb have been used to denote androgenic activity following injection or topical application of a solution of the test compound in oil [276]. A number of minor modifications of this test have been described [277–279]. Increases in the weight of the seminal vesicles and ventral prostate of immature castrated male rats has provided another measure of androgenic potency [280–283]. In this case, the test compound is administered either intramuscularly or orally and the weight of the target organs is compared with those of control animals. *In-vitro* evaluations of the relative affinity of potential androgens for the AR have also become an important tool in assessing the biological activity of androgens [123, 284].

4.3 Structure–Activity Relationships for Steroidal Androgens

4.3.1 Early Modifications

Most of the early structure–activity relationship studies concerned minor modifications of testosterone and other naturally occurring androgens. Studies in animals [285] and humans [286] showed the 17β-hydroxyl function to be essential for androgenic and anabolic activity. In certain cases, esterification of the 17β-hydroxyl group not only enhanced but also prolonged the anabolic and androgenic properties [287].

(28)

Reduction of the A-ring functional groups has variable effects on activity. For example, the conversion of testosterone to DHT has little effect or may even increase potency in a variety of bioassay systems [288–290]. The 1-dehydro isomer of testosterone (**28**) and related compounds are potent androgenic and anabolic steroids [285]. On the other hand, changing the A/B *trans* stereochemistry of known androgens such as androsterone (**3**) and DHT to the A/B *cis*-etiocholanolone (**19**) and 5β-dihydrotestosterone (**14**), respectively, drastically reduces both the anabolic and androgenic properties [291–293]. These observations established the importance of the A/B *trans* ring juncture for activity.

4.3.2 Methylated Derivatives

The discovery that C-17α-methylation conferred oral activity on testosterone prompted the synthesis of additional C-17α-substituted analogs. Increasing the chain length beyond methyl invariably led to a decrease in activity [294]. However, as a result of these studies, 17α-methylandrost-5-ene-3β,17β-diol (methandriol, **29**) was widely evaluated in humans as an anabolic agent and showed no clinical advantage of methandriol over 17α-methyltestosterone (**5**) [295].

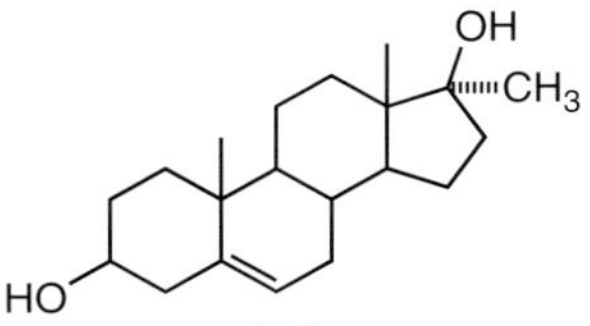

4.3.3 Ester Derivatives

As early as 1936 it was recognized that the esterification of testosterone at the 17β-hydroxy moiety markedly prolonged the activity of this androgen when it was administered parenterally [296]. This modification enhances the lipid solubility of the steroid and, after injection, permits a local depot effect. The acyl moiety is usually derived from a long-chain aliphatic or arylaliphatic acid such as propionic, heptanoic (enanthoic), decanoic, cyclopentylpropionic (cypionic), or β-phenylpropionic acid (**30–34**).

4.3.4 Halo Derivatives

In general, the preparation of halogenated testosterone derivatives has been therapeutically unrewarding. 4-Chloro-17β-hydroxyandrost-4-en-3-one (chlorotestosterone, **35**) and its derivatives are the only chlorinated androgens that have been used clinically, albeit sparingly [297]. The introduction of a 9α-fluoro and an 11β-hydroxy substituents (analogous to synthetic glucocorticoids) yields 9α-fluoro-11β, 17β-dihydroxy-17α-methylandrost-4-en-3-one (fluoxymesterone; Halotestin, **36**), which is an orally active androgen exhibiting an approximately fourfold greater oral activity than 17α-methyltestosterone. Early clinical studies with fluoxymesterone indicated an anabolic potency that was 11-fold that of the unhalogenated derivative [298–300], but nitrogen balance studies revealed an activity that was only threefold that of 17α-methyltestosterone [301]. Because of the lack of any substantial separation of anabolic and androgenic activity, halotestin is used primarily as an

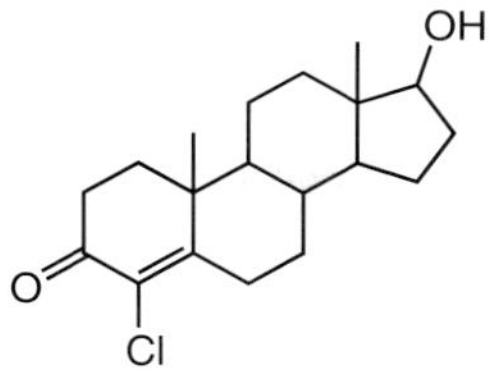

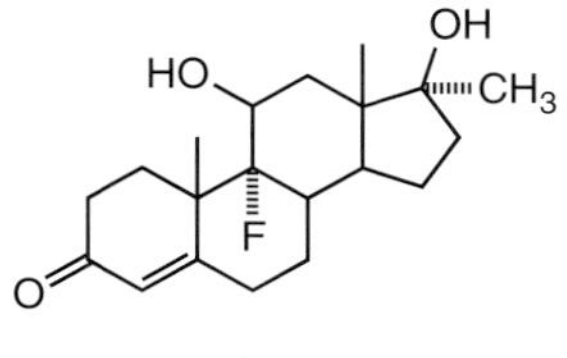

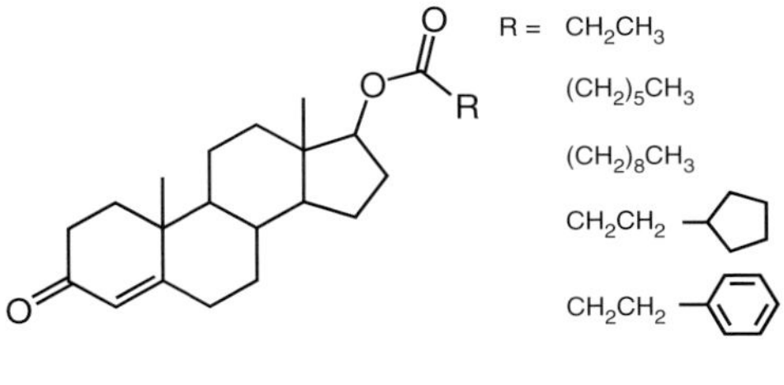

(30) propionate
(31) heptanoate (enanthate)
(32) decanoate
(33) cyclopentylpropionate (cypionate)
(34) β-phenylpropionate

orally effective androgen, particularly in the treatment of mammary carcinoma [302, 303].

4.3.5 Other Androgen Derivatives

Several synthetic steroids having weak androgenic activity have also been utilized in patients. 7β,17α-Dimethyltestosterone (calusterone, **37**) and 1-dehydrotestolactone (Testlac, **38**) are very weak androgenic agents that have been used in the treatment of advanced metastatic breast cancer [304–306].

(**37**)

(**38**)

4.3.6 Summary of Structure–Activity Relationships of Steroidal Androgens

As with other areas of medicinal chemistry, the desire to relate chemical structure to androgenic activity has attracted the attention of numerous investigators. Although it is often difficult to interrelate biological results from different laboratories, androgenicity data from the same laboratory afford useful information. In evaluating the data, care must be taken to note not only the animal model employed but also the mode of administration. For example, marked differences in androgenic activity can be found when compounds are evaluated in the chick comb assay (local application) as opposed to the rat ventral prostate assay (subcutaneous or oral). The chick comb assay measures "local androgenicity," and is believed to minimize such factors as absorption, tissue distribution, and metabolism, which complicate the interpretation of *in-vivo* data in terms of hormone–receptor interactions.

Furthermore, although the rat assays correlate well for various C_{19} steroids with what is eventually found in humans, few studies of comparative pharmacology have been performed. Indeed, DHT may not be the principal mediator of androgenicity in all species. For example, a cytosol receptor protein has been found in normal and hyperplastic canine prostate that is specific for 5α-androstane-3α,17α-diol [250].

Since the presence of the 17β-hydroxyl group was demonstrated at a very early stage to be an important feature for androgenic activity in rodents, most investigators interested in structure–activity relationships maintained this function and modified other parts of the testosterone molecule. Three observations can be made based on these studies: (i) the 1-dehydro isomer of testosterone is at least as active as testosterone; (ii) the 1- and 4-keto isomers of testosterone and DHT have variable activities; and (iii) the 2-keto isomers of testosterone and DHT consistently lack appreciable activity.

The first attempt to ascertain the minimal structural requirements for androgenicity was made by Segaloff and Gabbard [307]. Whereas, the oxygen function at position 3 could be removed from testosterone with little reduction in androgenic activity, removal of the hydroxyl group from position 17 sharply reduced the androgenicity. As a continuation of these studies, the hydrocarbon nucleus, 5α-androstane (**39**), was synthesized [307], and it too was found to possess androgenicity when applied topically or given intramuscularly in the chick comb assay (albeit at high doses). On the other hand, it subsequently emerged that the 19-nor analog, 5α-estrane (**40**), had less than 1% of the androgenic activity of testosterone propionate in castrated male rats [308].

Nonetheless, the studies of Segaloff and Gabbard set the stage for a more thorough analysis of 3-deoxy testosterone analogs by Syntex scientists [309, 310]. The relative androgenicity of the isomeric A-ring olefins of 3-deoxy testosterones was the order $\Delta^1 > \Delta^2 > \Delta^3 > \Delta^4$. The Δ^2-isomer displayed the greatest anabolic activity and the best anabolic-to-androgenic ratio.

On the basis that sulfur is bioisosteric with CH=CH, Wolff and coworkers [311] synthesized the thio, seleno, and tellurio androstanes, all of which displayed androgenic activity. When the heteroatom was oxygen, however, the compound (**41**) was essentially devoid of androgenicity [312]. The oxygen analog was said to be inactive because oxygen is isosteric with CH_2 rather than $CH_2=CH_2$. Thus, a minimum ring size was found to be required for activity. When the oxygen atom was introduced as part of a six-membered A-ring, an active androgen resulted [312].

As with the case of the double-bond isomers, the position of the oxygen atom was found to be important. The substitution of oxygen at C-2 gives rise to the most active compound, and the order of activity was $2 > 3 \gg 4$. As pointed out by Zanati and Wolff [312], these and earlier results are consistent with the concept that "…the activity-engendering group in ring A is wholly steric and that, in principle, isosteric groups of any type could be used to construct an androgenic molecule." Further support for this idea has been obtained from X-ray crystallographic structure determinations [313]. Zanati and Wolff [314] reported that even the full steroid nucleus is not essential for activity, with 7α-methyl 1,4-seco-2,3-bisnor-5α-androstan-17β-ol (**42**) having 50% of the anabolic activity of testosterone.

Studies by Segaloff and Gabbard [315] illustrated the marked enhancement of androgenicity achieved when a double bond was introduced at C-14. Both, 14-dehydrotestosterone and the corresponding 19-nor analog, were found to be potent androgens when applied topically. The introduction of a 7α-methyl substituent also resulted in active androgens [316]. The effects of either 7α-methyl or 14-dehydro modifications are more pronounced for 19-nortestosterone than for testosterone; the 14-dehydro modification had a greater effect on local androgenicity, whereas 7α-methylation had a more positive effect on systemic androgenicity. A marked synergism resulted when both the 14-dehydro and 7α-methyl modifications were present.

(**39**) R = CH$_3$
(**40**) R = H

(**41**) X = O

(**42**)

Tab. 1 Binding affinity of various androgens for rat ventral prostate receptor protein.

Steroid	K_B (M^{-1})
5α-DHT	6.9×10^8
5β-DHT	6.4×10^7
17β-Testosterone	4.2×10^8
17α-Testosterone	2.1×10^7
Androstenedione	1.3×10^7
5α-Androstanedione	3.5×10^7
19-Nortestosterone	8.6×10^8
14-Dehydrotestosterone	4.4×10^8
14-Dehydro-19-nortestosterone	5.9×10^8
7α-CH$_3$-14-Dehydro-19-nortestosterone	5.0×10^8

The characterization of a specific receptor protein in androgen target tissues has made it possible to directly analyze the receptor affinity of various testosterone analogs. Liao and coworkers [284] were the first to employ this parameter for comparison with systemic androgenicity. As would be expected, the receptor affinity data did not necessarily correlate with the systemic androgenicity although, in some cases, such as with 7α-methyl-19-nortestosterone, there was a good agreement. This was not the case, however, for 19-nortestosterone. Receptor binding analyses of androgens were also performed by other groups [123, 317], and their findings are summarized in Table 1. Whereas, the importance of the A/B *trans* ring fusion and 17β-hydroxyl prevailed, the data failed to demonstrate the potency previously noted for 7α-methyl-14-dehydro-19-nortestosterone. Moreover, 19-nortestosterone displayed a receptor affinity greater than that of DHT, yet its androgenicity was much less than that of DHT.

These differences in correlations between receptor assays and *in-vivo* data should not cloud the importance of the receptor studies. The receptor assays measure affinity for the receptor protein, and this property is shared by androgens as well as antiandrogens. Moreover, such assays cannot predict the disposition and metabolic fate of an androgen following administration. A summary of the structure–activity relationships for androgens is provided in Fig. 8.

4.4
Nonsteroidal Androgens, Selective Androgen Receptor Modulators (SARMs)

Until the mid-1990s, androgen agonistic activities were limited to steroidal molecules. Synthetic modifications of

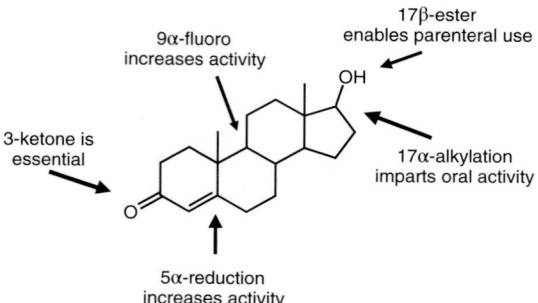

Fig. 8 Summary of structure–activity relationships for androgens.

nonsteroidal androgen antagonists (see Sect. 6.2.2.2) resulted in molecules that exhibited androgen agonist activities in various tissues. These molecules are now referred to as selective androgen receptor modulators, which describes compounds that act as antagonists or weak agonists in the prostate but exhibit agonist activities in other tissues such as muscle and bone (for reviews, see Refs [319–321]).

The pharmacophores associated with SARM activity include N-arylpropionamides, bicyclic hydantoins, and various quinolinones. Extensive research has been performed on the N-arylpropionamides, derived from structural modifications of the antiandrogen bicalutamide [320]. Initial studies identified the sulfide analogs as being effective *in vitro* when binding to the AR [321], but these analogs lacked significant *in-vivo* activity due to a rapid oxidative metabolism of the sulfide [322]. Replacement of the sulfide linkage with an ether bridge resulted in potent agonist activity both *in vitro* and *in vivo* for andarine (**43**) [322–324]. Currently, these N-arylpropionamide SARMs are under clinical investigations for the treatment of muscle wasting, osteoporosis, and other conditions associated with aging or androgen deficiency.

Another pharmacophore that has been extensively investigated are the bicyclic and tricyclic quinolinones with several analogs, including LGD-3303 (**44**), demonstrating oral bone anabolic activities [325, 326]. Several bicyclic hydantoin derivatives have shown potent muscle anabolic activities [327, 328]. Clinical investigations of SARMs are currently underway in patients with various androgen-dependent disorders to determine beneficial pharmacological androgenic activities without unwanted side effects.

(**43**)

(**44**)

4.5
Absorption, Distribution, and Metabolism

Numerous factors are involved in the absorption, distribution, and metabolism of the synthetic androgens, and the physico-chemical properties of these steroids greatly influence the pharmacokinetic parameters. The lipid solubility of a synthetic steroid is an important factor in its intestinal absorption. The acetate ester of testosterone demonstrated an enhanced absorption from the gastrointestinal tract over both testosterone and 17α-methyltestosterone (**58**). Injected solutions of testosterone in oil result in a rapid absorption of the hormone from the injection site; however, rapid metabolism greatly decreases the biological effects of the injected testosterone. The esters of testosterone are much more nonpolar and, when injected intramuscularly, are absorbed more slowly. As a result, commercial preparations of testosterone propionate are administered every few days. Increasing the size of the ester functionality enables testosterone esters such as the ethanate or cypionate to be given in a depot injection lasting two to four weeks.

Once absorbed, the steroids are transported in the circulation, primarily in a protein-bound complex. Testosterone and other androgens are reversibly associated with certain plasma proteins, and the unbound fraction can be absorbed into target cells to exert its action. The structure–binding relationships of the natural and synthetic androgens to SHBG have been extensively investigated [63–66, 71–77]. A 17α-hydroxyl group is essential for binding, while the presence of a 17α-substituent such as the 17α-methyl moiety decreases its affinity. The 5α-reduced androgens bind with the highest affinity. A much smaller quantity of the androgen is bound to other plasma proteins, principally albumin and transcortin, or CBG.

The metabolism of synthetic androgens is similar to that of testosterone, and has been extensively studied [284, 315–327]. Introduction of the 17α-methyl group greatly retards the metabolism, thus providing oral activity. Reduction of the 4-en-3-one system in synthetic androgens to give the various α- and β-isomers occurs *in vivo* [329, 330]. Finally, aromatization of the A-ring can also occur [146–148]. One analog that demonstrates an alternate metabolic pattern is 4-chlorotestosterone, which in humans gave rise to an allylic alcohol, 4-chloro-3α-hydroxyandrost-4-en-17-one [331]. A number of other halogenated testosterone derivatives subsequently were found to take this abnormal reduction path *in vitro* [332]. It was proposed that fluorine or chlorine substituents at the 2-, 4-, or 6-position in testosterone interfere with the usual α,β-unsaturated ketone resonance so that the C-3 carbonyl electronically resembles a saturated ketone.

4.6
Toxicities

The use of androgens in women and children can often result in virilizing or masculinizing side effects. In boys, an acceleration of the sexual maturation is seen, while in girls and women the growth of facial hair and a deepening of the voice can be observed [333, 334]. These effects are reversible when medication is stopped, but prolonged treatment can produce effects that are irreversible. An inhibition of gonadotropin secretion by the pituitary can also occur in patients receiving androgens.

Both, males and females experience salt and water retention resulting in edema. This edema can be treated by either maintaining a low-salt diet or by using diuretic agents. Liver problems are also encountered with some of the synthetic androgens, with clinical jaundice and cholestasis often developing after the use of 17α-alkylated products [335–338]. Various clinical laboratory tests for hepatic function, such as bilirubin concentrations, bromosulfophthalein (BSP) retention and glutamate transaminase and alkaline phosphatase activities, are affected by these androgen analogs.

5
Anabolic Agents

5.1
Current Drugs on the Market

Currently available anabolic agents used as therapeutics are listed in the following table.

Generic name (structure)	Trade name	U.S. manufacturer	Chemical class	Dose
Nandrolone decanoate (**45**)	Deca-durabolin	Organon	Estrane	Injection (In Oil): 200 mg ml^{-1}
Oxandrolone (**67**)	Oxandrin	Savient	Androstane	Tablets: 10 mg 2.5 mg
Oxymetholone (**65**)	Anadrol-50	Alaven	Androstane	Tablets: 50 mg

5.2 Therapeutic Uses and Bioassays

Many synthetic analogs of testosterone were prepared in order to separate the anabolic activity of the C_{19} steroids from their androgenic activity. Although the goal of a pure synthetic anabolic that retains no androgenic activity has not been accomplished, several preparations are now available commercially that have high anabolic:androgenic ratios. Extensive reviews on anabolic agents are available [4, 10].

The primary criterion for assessing the anabolic activity of a compound is the demonstration of a marked retention of nitrogen. This nitrogen-retaining effect is the result of an increased protein synthesis and a decreased protein catabolism in the body [339]. Thus, the urinary nitrogen excretion – particularly of urea – is greatly diminished. The castrated male rat serves as the most sensitive animal model for nitrogen retention, although other animals have been used [340–342]. Another bioassay for anabolic activity involves monitoring the increase in levator ani muscle mass in rats following the administration of an anabolic agent [281, 282]. This measure of the myotrophic effect correlates well with the nitrogen retention bioassay, and the two assays are usually performed together when determining anabolic activity.

Anabolic steroids also exert other effects on the body, with skeletal mineralization and bone maturation notably being enhanced by androgens and anabolic agents [35]. Such agents decrease calcium excretion by the kidney, and this results in an increased deposition of both calcium and phosphorus in the bones. Androgenic and anabolic agents also can influence erythropoiesis (red blood cell formation) via two mechanisms of action, namely an increased production of erythropoietin and an enhanced responsiveness of the tissue [36].

These various biological activities of anabolic agents have prompted their use in a range of treatment protocols, albeit with varying success. Previously, clinical trials have demonstrated the effectiveness of anabolic steroids in inducing muscle growth and development in some diseases [343]. Anabolic steroids are effective in the symptomatic treatment of various malnourished states due to their ability to increase protein synthesis and decrease protein catabolism. The treatment of diseases such as malabsorption, anorexia nervosa, emaciation and malnutrition as a result of psychoses includes dietary supplements, appetite stimulants and anabolics [344–349]. An improved postoperative recovery with the adjunctive use of anabolic agents has been demonstrated in numerous clinical studies [344, 350–354]. However, the usefulness of these agents in other diseases such as muscular dystrophies and atrophies, and in geriatric patients, has not been observed.

Anabolic steroids also have the ability to lower serum lipid levels *in vivo* [354–357]. The most widely studied agent in this respect is oxandrolone, which dramatically lowers serum triglycerides and, to a lesser extent, cholesterol levels at pharmacological doses [358–360]. The proposed mechanism of this hypolipidemic effect includes both an inhibition of triglyceride synthesis [361] and an increased clearance of the triglycerides [362]. The androgenic side effects of the anabolics and their lack of superiority over more efficacious hypolipidemic agents have curtailed their use in the treatment of these conditions, however.

Although now supplanted by the availability of recombinant erythropoietin, the stimulation of erythropoiesis by anabolics has resulted in the use of these agents for the treatment of various anemias [363–366]. Anemias arising from deficiencies of the bone marrow are particularly responsive to pharmacological doses of anabolic agents. The treatment of aplastic anemia with anabolics and corticosteroids has been proven effective [363–366], while secondary anemias resulting from inflammation, renal disease or neoplasia have also been shown responsive to anabolic steroid administration [36, 366–370]. Finally, synthetic anabolic agents have been prescribed for women with osteoporosis [35] and for children with delayed growth [361]. Although these applications have produced limited success, the virilizing side effects that occur have severely limited their usefulness, particularly in children.

The methods employed to determine the anabolic or myotrophic properties of steroids have been reviewed [371]. Generally, these are based on an increase in nitrogen retention and/or muscle mass in various laboratory animals. The castrated male rat is the most widely used [340], but dogs and ovariectomized monkeys have also been employed [341, 342]. Although it is generally agreed that variations in urinary nitrogen excretion relate to an increase or decrease in protein synthesis, nitrogen balance assays are not without their limitations [372]. This is partly because such studies fail to describe the shifts in organ protein and measure only the overall status of nitrogen retention in the animal [373].

The easily accessible levator ani muscle of the rat has provided a valuable index for measuring the myotrophic activity of steroidal hormones [280]. By comparing the weight of levator ani muscle, seminal vesicles and ventral prostate with those of controls, it is possible to obtain a ratio of anabolic to androgenic activity [280, 282]. There also appears to be some correlation between the levator ani response and urinary nitrogen retention [280]. A modification of this muscle assay utilizes the parabiotic rat [374, 375] and allows for the simultaneous measurement of pituitary gonadotrophic inhibition and myotrophic activity. The suitability of the levator ani assay has been questioned on the possibility that its growth is more a result of androgenic sensitivity than of any steroid-induced myotrophic effect [375–378]. Thus, this assay is usually performed in conjunction with nitrogen balance studies or an acceleration of body growth [379].

5.3
Structure–Activity Relationships for Anabolic Agents

5.3.1 19-Nor Derivatives
An important step towards developing an anabolic agent with minimal androgenicity was taken when Hershberger and associates

[282], and later others [380, 381], found 19-nortestosterone (17β-hydroxyestr-4-en-3-one, nandrolone, **45**) to be equally myotrophic as testosterone, but only about one-tenth as androgenic. This observation prompted the synthesis and evaluation of a variety of 19-norsteroids, including the 17α-methyl (normethandrone, **46**) [382] and the 17α-ethyl (norethandrolone, **47**) [383] homologs of 19-nortestosterone.

R = H   (**45**) nandrolone
R = CH$_3$ (**46**) normethandrolone
R = CH$_2$CH$_3$ (**47**) norethandrolone

Nandrolone, in the form of a variety of esters (such as decanoate and β-phenylpropionate), and norethandrolone have been widely used clinically. The latter agent (trade name Nilevar®) was the first to be marketed in the United States as an anabolic steroid, but its androgenic [384] and progestational [385] side effects eventually led to it being replaced by other agents.

These studies stimulated the synthesis of other norsteroids, but both 18-nortestosterone [386] and 18,19-bisnortestosterone [387] were found essentially to be devoid of androgenic and anabolic properties. A contraction of the B ring led to B-norsteroids which were also lacking in androgenicity but, unlike the foregoing, this modification at least resulted in compounds with antiandrogenic activity.

Among the number of homoandrostane derivatives (those having one or more additional methylene groups included in normal tetracyclic ring system) that have been synthesized, only B-homodihydrotestosterone [388, 389] and D-homodihydrotestosterone [390, 391] have shown appreciable androgenic activity. A D-bishomo analog (**48**) was reported to be weakly androgenic [392].

(**48**)

5.3.2 Dehydro Derivatives

The marked enhancement in biological activity afforded by the introduction of a double bond at C_1 of cortisone and hydrocortisone prompted similar transformations in the androgens. The acetate of 17β-hydroxyandrosta-1,4-dien-3-one (**49**) [393] was as myotrophic as testosterone propionate but much less androgenic. Furthermore, 17α-methyl-17β-hydroxyandrosta-1,4-dien-3-one (methandrostenolone, **50**) had equal to twofold the oral potency of 17α-methyltestosterone in rat nitrogen retention [394, 395] and levator ani muscle assays [396, 397]. In clinical studies, methandrostenolone produced a marked anabolic effect when given orally at doses of 1.25–10 mg per day, and was several-fold more potent than 17α-methyltestosterone [398].

In contrast to the 1-dehydro analogs, the introduction of an additional double bond at the 6-position (**51**) markedly decreased both androgenic and myotrophic activities in the rat [393, 399]. Moreover, removal of the C_{19}-methyl [400], inversion of the configuration at C_9 and C_{10} [401] and at C_8 and C_{10} [402], and reduction of the C_3-ketone failed to improve the biological properties [403].

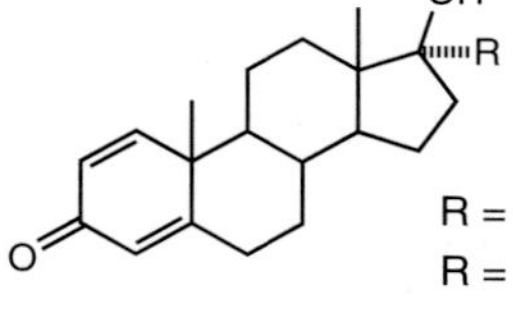

R = H (**49**)
R = CH₃ (**50**)

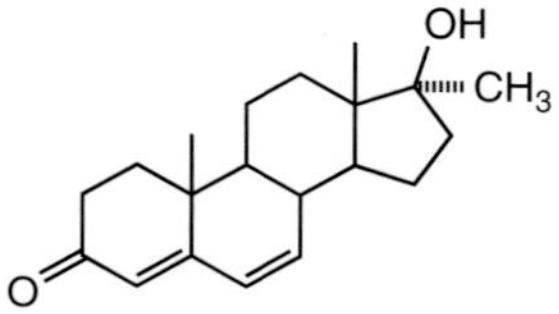

(**51**)

On the other hand, the introduction of unsaturation into the B, C, and D rings has given rise to compounds with significant androgenic or anabolic activities. Ethyldienolone (**52**), for example, displayed an anabolic:androgenic ratio of 5 and was slightly more active than methyltestosterone when both were given orally [404]. Segaloff and Gabbard [315] showed that the introduction of a 14–15 double bond (**53**) increased androgenicity when compared to testosterone applied locally in the chick comb assay. In contrast, a 25% decrease in androgenicity was identified in the rat ventral prostate following subcutaneous administration. Although conversion to the 19-nor analog (**54**) increased androgenicity, the anabolic activity was significantly enhanced [316].

Among a variety of triene analogs of testosterone that have been tested, only 17α-methyl-17β-hydroxyestra-4,9,11-trien-3-one (methyltrienolone, **55**) showed significant activity in rats. Surprisingly, this compound had 300-fold the anabolic potency and 60-fold the androgenic potency of 17α-methyltestosterone when administered orally to castrated male rats [405]. In this instance, however, the potent hormonal properties in rats did not correlate with later studies in humans [406–408]. In fact, a study in patients with advanced breast cancer showed methyltrienolone to have weak androgenicity and to produce severe hepatic dysfunction at very low doses [408].

5.3.3 Alkylated Analogs

An extensive effort has been directed towards assessing the physiological effect of replacing hydrogen with alkyl groups at most positions of the steroid molecule. Although methyl substitution at C_3, C_4, C_5, C_6, C_{11}, and C_{16} has generally led to compounds with low anabolic and androgenic activities, similar substitutions at C_1, C_2, C_7, and C_{18} have afforded derivatives of clinical significance.

1-Methyl-17β-hydroxy-5α-androst-l-en-3-one (methenolone, **56**) as the acetate (methenolone acetate) was about fivefold more myotrophic, but only one-tenth androgenic, as testosterone propionate in animals [409]. In addition, this compound or the free alcohol represented one of the few instances of a C_{17} nonalkylated steroid

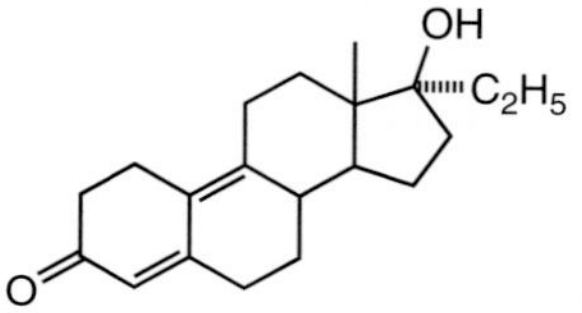

(**52**)

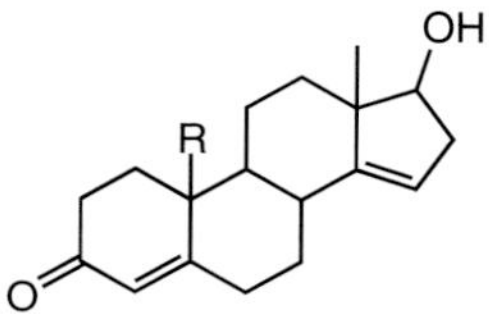

(**53**) R = CH₃
(**54**) R = H

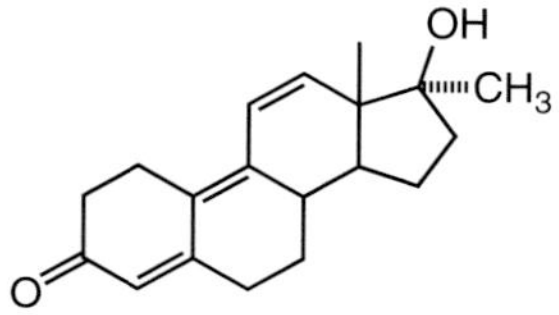

(**55**)

that possessed significant oral anabolic activity in animals [410] and in humans [411]. This effect may be related to the slow *in-vivo* oxidation of the 17β-hydroxyl group when compared with testosterone [412]. At a daily dose of 300 mg, methenolone acetate caused little virilization [413] or BSP retention [414]; by contrast, the dihydro analog, 1α-methyl-17β-hydroxy-5α-androstan-3-one (mesterolone, **57**), was found to possess significant oral androgenic activity in the cockscomb test [415] and also in clinical assays [416]. A comparison of the anabolic and androgenic activity of **56** with its A-ring congeners revealed that the double bond was necessary at C_1 for anabolic activity. For example, 1α-methyl-17β-hydroxyandrost-4-en-3-one had a much lower activity [417]. Furthermore, either reduction of the C_3 carbonyl group of **57** [418] or removal of the C_{19} methyl group [419, 420] greatly reduced both anabolic and androgenic activities in this series.

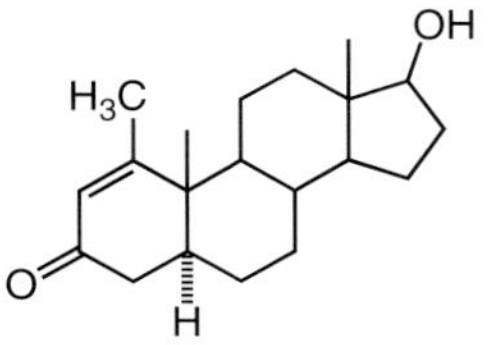

(56)

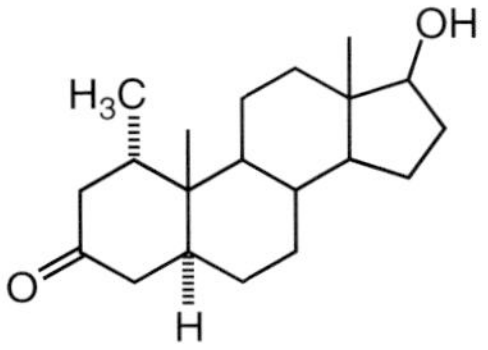

(57)

Among the C_2-alkylated testosterone analogs, 2α-methyl-5α-androstan-17β-ol-3-one (drostanolone, **58**) and its 17α-methylated homolog (**59**) have displayed anabolic activity both in animals [421] and in man [422]. In contrast, 2,2-dimethyl and 2-methylenetestosterone or their derivatives showed only low anabolic or androgenic activities in animals [421, 423, 424].

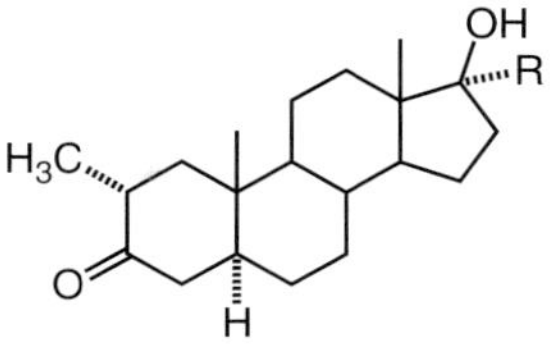

(58) R = H
(59) R = CH₃

7α,17α-Dimethyltestosterone (bolasterone, **60**) had 6.6-fold the oral anabolic potency of 17α-methyltestosterone in rats [425]. A similar activity was observed in humans at 1–2 mg per day, without many of the usual side effects [426]. Moreover, the corresponding 19-nor derivative was 41-fold as active as 17α-methyltestosterone as an oral myotrophic agent in the rat [427]. Segaloff and Gabbard [316] found 7α-methyl-14-dehydro-19-nortestosterone (**61**) to be approximately 1000-fold as active as testosterone in the chick comb assay, and about 100-fold as active as testosterone in the ventral prostate assay.

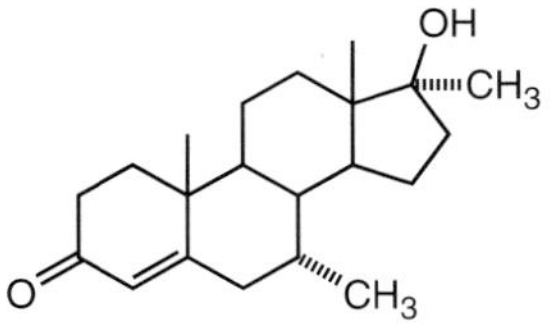

(60)

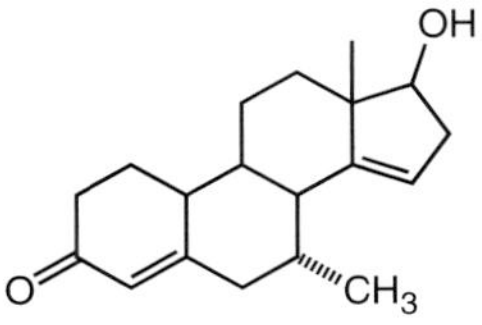

(61)

Certain totally synthetic 18-ethylgonane derivatives possessed pronounced anabolic activity. Similar to other 19-norsteroids,

13β,17α-diethyl-17β-hydroxygon-4-en-3-one (norbolethone, **62**) was found to be a potent anabolic agent in animals and in humans [428, 429]. Since it is prepared by total synthesis, the product was isolated and marketed as the racemic DL-mixture; notably, the hormonal activity resides in the D-enantiomer.

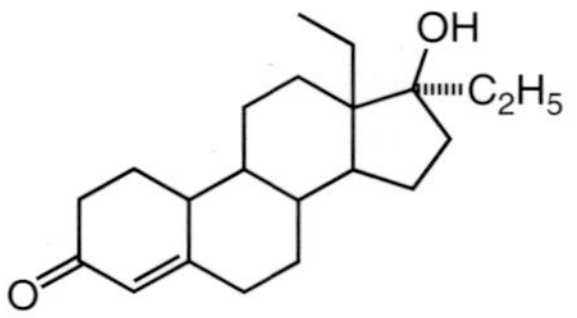

(**62**)

5.3.4 Hydroxy and Mercapto Derivatives

Testosterone has been hydroxylated at virtually every position on the steroid nucleus. For the most part, nearly all these substances possess no more than weak myotrophic and androgenic properties, though two striking exceptions to this are 4-hydroxytestosterone and 11β-hydroxytestosterone. 4-Hydroxy-17α-methyltestosterone (oxymesterone, **63**), for instance, had three- to fivefold the myotrophic activity but only half of the androgenic activity of 17α-methyltestosterone in rats [430]. In clinical studies, oxymesterone produced nitrogen retention in adults at a daily dose of 20–40 mg, and no adverse effect on liver function were observed [431, 432]. The introduction of an 11β-hydroxyl group in many instances resulted in a favorable effect on biological activity; for example, 11β-hydroxy-17α-methyltestosterone (**64**) was more anabolic in rats than was 17α-methyltestosterone [298], and 1.5-fold as myotrophic in humans [398].

To date, one of the most widely studied anabolic steroids has been 2-hydroxymethylene-17α-methyl-5α-androstan-17β-ol-3-one (oxymetholone, **65**). In animals, this compound was found to be threefold as anabolic but only half as androgenic as 17α-methyltestosterone [433, 434]. These results were confirmed in clinical studies [432–435].

The substitution of a mercapto for a hydroxyl group has generally resulted in decreased activity. However, the introduction of a thioacetyl group at C_1 and C_7 of 17α-methyltestosterone afforded 1α,7α-bis(acetylthio)-17α-methyl-17β-hydroxyandrost-4-en-3-one (thiomesterone, **66**), a compound with significant activity. Thiomesterone was 4.5-fold as myotrophic and 0.6-fold as androgenic as 17α-methyltestosterone in rats [436], and has been used clinically as an anabolic agent [437].

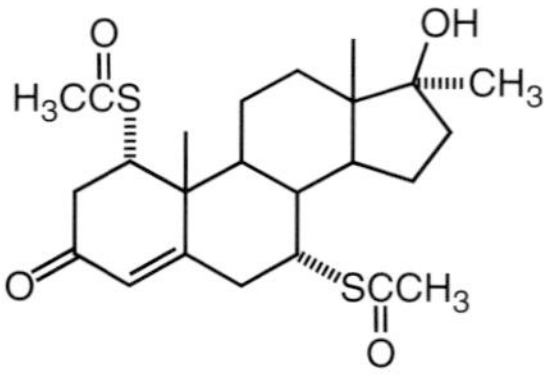

(**66**)

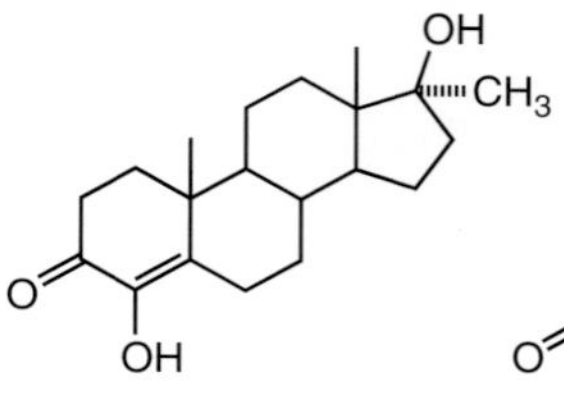

(**63**) 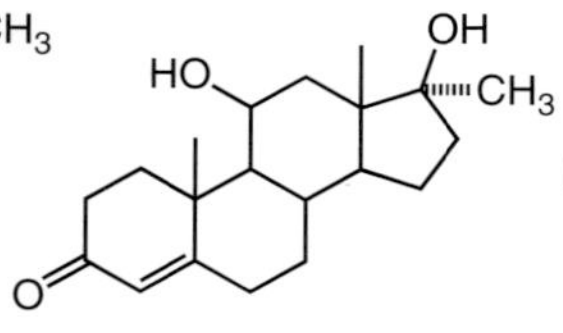   (**64**) 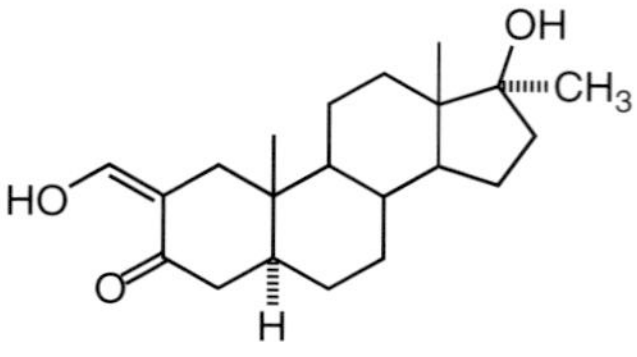   (**65**)

Moreover, numerous 7α-alkylthio androgens have exhibited anabolic–androgenic activity similar to that of testosterone propionate when administered subcutaneously [438, 439]. Even though no clinically useful androgen resulted, similar 7α-substitutions proved to be advantageous in the development of radioimmunoassays now employed in clinical laboratories [440]. In addition, certain 7α-arylthioandrost-4-ene-3,17-diones are effective inhibitors of estrogen biosynthesis (see Sect. 6.2.3).

5.3.5 Oxa, Thia, and Aza Derivatives

A number of androgen analogs in which an oxygen atom replaces one of the methylene groups in the steroid nucleus have been synthesized and evaluated biologically. Of these derivatives, 17β-hydroxy-17α-methyl-2-oxa-5α-androstan-3-one (oxandrolone, **67**) [441] was threefold as anabolic but only 0.24-fold as androgenic as 17α-methyltestosterone in the oral levator ani assay [442]. By contrast, only minimal responses were obtained following intramuscular administration. The 2-thia [443] and 2-aza [444] analogs were essentially devoid of activity by both routes. The 3-aza-A-homoandrostene derivative **68** displayed only 5% of the anabolic-to-androgenic activity of methyltestosterone [445].

The clinical anabolic potency of oxandrolone was considerably greater than that of 17α-methyltestosterone, and provided perceptible nitrogen-sparing at a dose as low as 0.6 mg per day [446]. Moreover, at dosages of 0.25–0.5 mg kg^{-1}, oxandrolone was effective as a growth-promoting agent, without producing androgenically induced bone maturation [447]. Because of this favorable separation of anabolic from androgenic effects, oxandrolone has been one of the most widely studied anabolic steroids. Its potential utility in various clinical hyperlipidemias was discussed in Sect. 5.2.

The significant hormonal activity noted for estra-4,9-dien-3-ones such as **52** (see Sect. 5.3.2) prompted the synthesis of the 2-oxa bioisosteres in this series. Despite the lack of a 17α-methyl group, **69** had 93-fold the oral anabolic activity of l7α-methyltestosterone, and was also 2.7-fold as androgenic. As might be expected, the corresponding 17α-methyl derivative, **70**, was the most active substance in this series, with myotrophic and androgenic effects that were 550-fold and 47-fold, respectively, that of 17α-methyltestosterone [448]. These two compounds differed dramatically in their progestational activity, however, with the activity of **69** being only one-tenth that of progesterone but the activity of **70** being 100-fold in the Clauberg assay [448]. The pronounced oral activity of **69** suggests that it is not a substrate for the 17β-alcohol dehydrogenase, but does represent an interesting finding.

(**67**)

(**68**)

(**69**) R = H
(**70**) R = CH$_3$

5.3.6 Deoxy and Heterocyclic-Fused Analogs

Early studies conducted by Kochakian [449] indicated that the 17β-hydroxyl group and the 3-keto group were essential for maximum androgenic activity. Based on these observations, the C_3 oxygen function was removed in the hope of reducing the androgenic potency while maintaining the anabolic activity [450]. Unfortunately, the results obtained failed to substantiate the rationale, and 17α-methyl-5α-androstan-17β-ol (**71**) was found to be a potent androgen in animals [451] and humans [452]. However, Wolff and Kasuya showed that this substance is extensively metabolized to the 3-keto derivative by rabbit liver homogenate [453]. Other deoxy analogs of testosterone have been synthesized and tested. For example, a 19-nor derivative, 17α-ethylestr-4-en-17β-ol (estrenol, **72**) had at least fourfold the anabolic activity but only one-fourth of the androgenic activity of 17α-methyltestosterone in animals [454], and was effective in humans at a daily dose of 3–5 mg [455–457]. 17α-Methyl-5α-androst-2-en-17β-ol (desoxymethyltestosterone, **73**) also offered a good separation of anabolic from androgenic activity [451, 458].

Since sulfur is considered to be isosteric with -CH=CH-, Wolff and Zanati [326] reasoned that 2-thia-A-nor-5α-androstane derivatives such as **74** should have androgenic activity. Indeed, this compound possessed high androgenic and anabolic activities, which served to verify that steric rather than electronic factors are important in connection with the structural requirements at C-2 and/or C-3 in androgens [459]. Interestingly, the selenium and tellurium isosteres in the same series were found to have good androgenic activity [460, 461]. Moreover, experiments with a ^{75}Se-labeled analog have shown **75** to selectively compete with DHT for binding to the AR in rat prostate [462].

The high biological activity noted for the 3-deoxy androstanes prompted numerous investigators to fuse various systems to the A-ring. The simplest such changes were 2,3-epoxy, 2,3-cyclopropano, and 2,3-epithioandrostanes. The 2,3α-cyclopropano-5α-androstan-17β-ol was as active as testosterone propionate as an anabolic agent [463]. While the epoxides had little or no biological activity, certain of the episulfides possessed pronounced anabolic and androgenic activities [464]. For example, 2,3α-epithio-17α-methyl-5α-androstan-17β-ol (**76**) was found to have approximately equal androgenic activity but 11-fold the anabolic activity of methyltestosterone after oral administration to rats. The 2,3β-episulfide, on the other hand, was much less active. 2,3α-Epithio-5α-androstan-17β-ol has been shown to have long-acting antiestrogenic activity, as well as some beneficial effects in the treatment of mammary carcinoma [465].

Other heterocyclic androstane derivatives have included the pyrazoles. Thus, 17β-hydroxy-17α-methylandrostano-(3,2-c)-pyrazole (stanozolol, **77**) was 10-fold as active as 17α-methyltestosterone in improving nitrogen retention in rats [466], but the myotrophic activity was only twice that of 17α-methyltestosterone [467]. Stanozolol, at a dose of 6 mg per day, produced an adequate anabolic response with no lasting adverse side effects [468, 469].

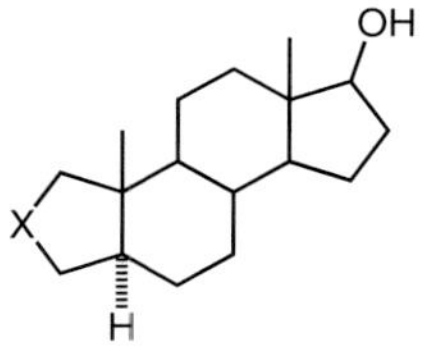

inhibiting gonadotropin release, a direct inhibition of Leydig cell androgen synthesis was also observed. Other studies have shown danazol to be effective in the treatment of endometriosis, benign fibrocystic mastitis and precocious puberty [473]. Several reports have appeared relating to its disposition and metabolic fate [473, 474].

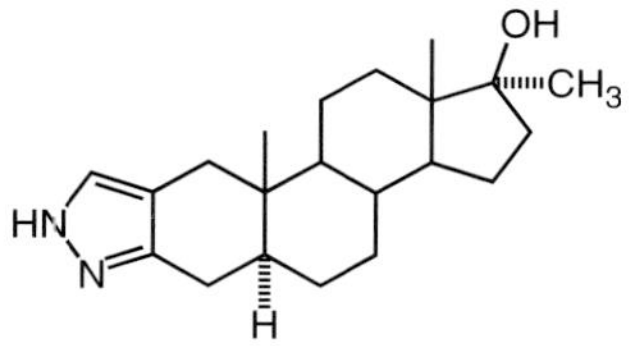

(74) X = S
(75) X = Se

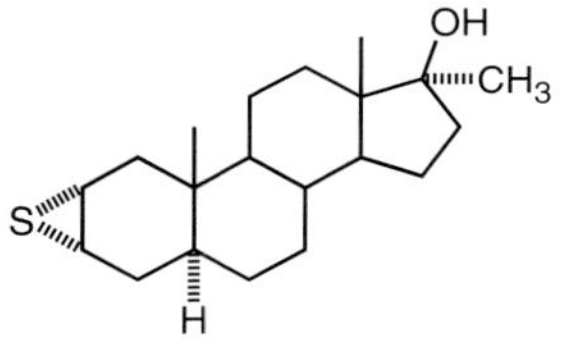

(76)

(77)

(78) R = H
(79) R = C≡CH

The high activity of the pyrazoles instigated the synthesis of other heterocyclic fused androstane derivatives including isoxazoles, thiazoles, pyridines, pyrimidines, pteridines, oxadiazoles, pyrroles, indoles, and triazoles. One of the most potent was 17α-methylandrostan-17β-ol-(2,3-d)-isoxazole (androisoxazol, **78**), which exhibited an oral anabolic-to-androgenic ratio of 40 [470]. The corresponding 17α-ethynyl analog (danazol, **79**) has been of most interest clinically, as this compound is known to impede androgenic activity and inhibit pituitary gonadotropin secretion [471]. Since danazol depresses blood levels of androgens and gonadotropins, it has been studied as an antifertility agent in males [472]. Indeed, daily doses of 200 or 600 mg caused a dose-related lowering of plasma levels of testosterone and androstenedione. Moreover, in addition to

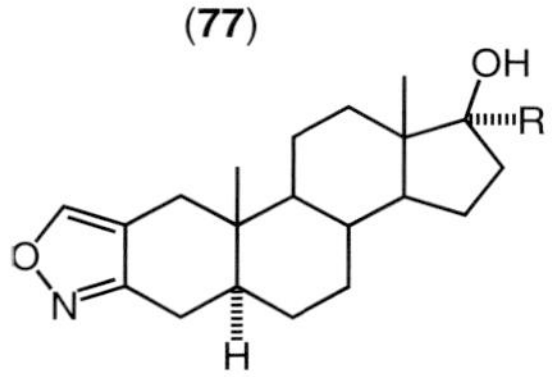

5.3.7 Esters and Ethers

Since the esterification of testosterone markedly prolongs its activity, it was unsurprising that this approach to increasing drug latency would be extended to the anabolic steroids. The acyl moiety is usually derived from a long-chain aliphatic or arylaliphatic acid such as heptanoic (enanthoic), decanoic, cyclopentylpropionic, and β-phenylpropionic. For example, no less than 12 esters of 19-nortestosterone (nandrolone) have been used clinically as long-acting anabolic agents [475, 476].

In the case of nandrolone, the duration of action and the anabolic-to-androgenic ratio were each increased with the chain length of the ester group [477, 478]. The decanoate and laurate esters, for instance, were active at six weeks after injection. Clinically, nandrolone decanoate appeared to be the most practical, as a weekly dose of 25–100 mg produced a marked nitrogen retention [479, 480].

Since the 17α-alkyl group has been implicated as the cause of hepatotoxic side effects of oral preparations, the effect of esterification on oral efficacy has attracted much attention. For example, the esterification of dihydrotestosterone with short-chain fatty acids resulted in oral anabolic and androgenic activities in rats [481]. Moreover, esters of methenolone possessed appreciable oral anabolic activity [482]. Unfortunately, follow-up studies with steroid esters in humans have not been reported.

The manner by which steroid esters evoke their enhanced activity and increased duration of action has puzzled investigators for many years. The classical concept has been that esterification delays the rate at which the steroid is absorbed from the site of injection, thus preventing its rapid destruction. However, other factors must also be involved as the potency and prolongation of action are known to vary markedly with the nature of the esterifying acid.

James and coworkers shed much light on this problem by studying the effect of various aliphatic esters of testosterone on rat prostate and seminal vesicles, and correlating androgenicity with lipophilicity and the rate of ester hydrolysis by liver esterases [483, 484]. Peak androgenic response was observed with the butyrate ester, which was also the most readily hydrolyzed, while the more lipophilic valerate ester was slightly less androgenic (in a quantitative sense) but its action was longer-lasting. James and colleagues concluded that the ease of hydrolysis controls the weight of the target organs, whereas lipophilicity was responsible for the duration of the androgenic effect. These results also explained the low androgenic activity that had been previously noted for hindered trimethylacetate (pivalate) esters, which would be expected to be resistant to in-vivo hydrolysis.

The effect of etherification on anabolic or androgenic activities has been studied less rigorously. Replacement of the 17β-OH with l7β-OCH_3 markedly reduced androgenic activity but had no great effect on the ability to counteract cortisone-induced adrenal atrophy in male rats [485]. A series of 17β-acetals [486, 487], alkyl ethers [488], and 3-enol ethers [489, 490], however, showed significant activity when given orally [491–493]. For example, the cyclohexyl enol ether of 17α-methyltestosterone, when given orally, was fivefold as myotrophic as 17α-methyltestosterone [493].

Other ethers such as the tetrahydropyranol [494, 495] and trimethylsilyl have oral anabolic and androgenic activities in animals. The trimethylsilyl ether of testosterone (silandrone) showed a protracted activity following injection [496], and after oral dosing had twice the anabolic and androgenic activities of 17α-methyltestosterone. Solo et al. [497] evaluated a variety of ethers that would not be expected to be readily cleaved in vivo, and found them to be almost devoid of anabolic and androgenic activities. These findings provided additional support for the need of a free 17β-hydroxyl for androgenic activity.

5.3.8 Summary of Structure–Activity Relationships

Synthetic modifications of C_{19} steroids have resulted in the enhancement of anabolic activity, even though a pure synthetic anabolic agent which retains no androgenic activity has not yet been accomplished. Structural changes in two regions of the testosterone molecule have resulted in the greatest enhancement of the anabolic/androgenic ratio. The first region is the C-17 position of the testosterone

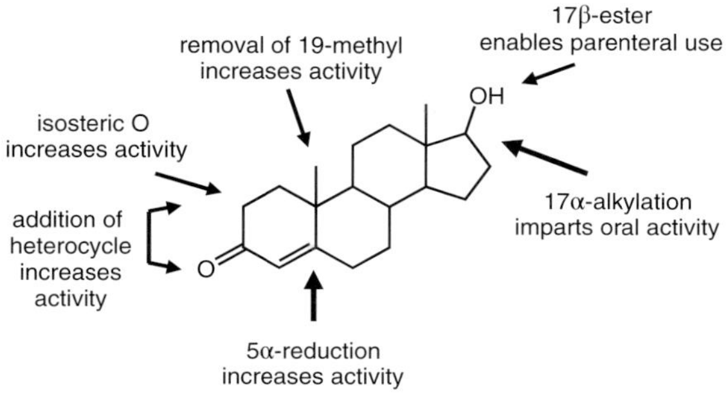

Fig. 9 Summary of structure–activity relationships for anabolic agents.

molecule. The introduction of a 17α-alkyl functionality, such as a 17α-methyl or a 17α-ethyl group, greatly increases the metabolic stability of the anabolic and decreases the *in-vivo* conversion of the 17β-alcohol to the 17-ketone by 17β-hydoxysteroid dehydrogenases. In addition, esterification of the 17β-alcohol enhances the lipid solubility of the steroids, thus providing injectable preparations for depot therapy.

The A ring of testosterone is the second region in which structural modifications can be made to increase anabolic activity. Removal of the C-19 methyl group results in the 19-nortestosterone analogs, which have slightly higher anabolic activities. A major impact on the structure–activity relationships of anabolic agents can be observed with modifications at the C-2 position. For example, a bioisosteric replacement of the carbon atom at position 2 with an oxygen provides a threefold increase in anabolic activity, as is seen with oxandrolone. Finally, the greatest effects were observed with the addition of heterocyclic rings fused at positions 2 and 3 of the A ring. The two heterocycles that have led to the greatest changes are the pyrazole and the isoxazole rings, as seen in stanozolol and androisoxazole, respectively. In these anabolics, the 3-ketone of testosterone is replaced by the bioisosteric 3-imine.

Stanozolol, which contains the pyrazole ring at C-2 and C-3, shows the greatest increases when compared to testosterone. The anabolic activities of nitrogen retention and myotrophic activity for several common anabolic agents are listed in Table 2, while a summary of their structure–activity relationships is shown in Fig. 9.

5.4
Absorption, Distribution, and Metabolism

The absorption, distribution, and metabolism of the various anabolic steroids is quite similar to the pharmacokinetic properties of the endogenous and synthetic androgens discussed earlier in the chapter [498]. Again, lipid solubility is critical for the absorption of these agents following oral or parenteral administration. The 17α-methyl group retards the metabolism of the compounds and provides orally active agents. Other anabolics such as methenolone are orally active without a 17α-substituent, indicating that these steroids are poor substrates for 17α-hydroxysteroid dehydrogenase [411, 499]. Reduction of the 4-en-3-one system in synthetic anabolics to give the various α- and β-isomers occurs *in vivo* [329]. The 3-deoxy agent 17α-methyl-5α-androstan-17α-ol was shown to be

Tab. 2 Comparison of anabolic activities.

Compound	Number	Trade names	Anabolic activity	
			Nitrogen retention	Myotrophic activity
Testosterone	1	Android-T, Malestrone, Oreton, Primotest, Virosterone	1.0	1.0
19-Nortestosterone Nandrolone	45	Nerobolil, Nortestonate	0.8	1.0
Normethandrone	46	Methalutin, Orgasteron	4.0	4.5
Norethandrolone	47	Nilevar, Solevar	3.9	4.0
Methandrostenolone Methandienone	50	Danabol, Dianabol, Nabolin, Nerobil	0.6	1.4
Drostanolone	58	Drolban, Masterone	—	1.3
Oxymetholone	65	Adroyd, Anadrol, Anadroyd, Anapolon, Anasterone, Nastenon, Protanabol, Synasteron	2.75	2.8
Oxandrolone	67	Anavar, Provita	3.0	3.0
Estrenol	72	Duraboral-O, Maxibolin, Orabolin, Orgaboral, Orgabolin	1.7	2.0
Stanozolol	77	Stanozol, Winstrol, Tevabolin	10.0	7.5
Androisoxazole	78	Androxan, Neo-ponden	1.5	1.7

extensively converted to the 3-keto derivative by liver homogenate preparations [453]. The metabolic fates of stanozolol and danazol have been reported [473, 474], with the major metabolites being heterocyclic ring-opened derivatives and their deaminated products. Finally, both the unchanged anabolics and their metabolites are primarily excreted in the urine as glucuronide or sulfate conjugates.

5.5 Toxicities

The major side effect of anabolic steroids is the residual androgenic activity of the molecules, with virilizing actions being undesirable in adult males as well as in females and children. Many anabolic steroids can also suppress the release of gonadotropins from the anterior pituitary, leading to lower levels of circulating hormones and potential problems of reproductive function. Headaches, acne and elevated blood pressure are common symptoms in individuals taking anabolics, while the salt and water retention induced by these agents can produce edema.

The most serious toxicities resulting from the use of anabolic steroids relate to subsequent liver damage, including jaundice and cholestasis, that can occur after the administration of the 17α-alkylated C_{19} steroids [335–338]. Individuals who have received anabolic agents over an extended period have also developed hepatic adenocarcinomas [500–502]. Such clinical reports serve to underscore the inherent risks associated with anabolic steroid use in amateur athletes, for no demonstrable benefits.

5.6 Abuse of Anabolic Agents

The myotrophic effects of testosterone and other anabolic steroids have led to the use and abuse of these agents by athletes [503–506]. Conflicting reports on the effectiveness of anabolics to increase strength and power in healthy males have resulted from early clinical trials. Several groups have found no significance differences between groups of male, college-age students receiving anabolics and weight-training, and other (control) students receiving placebo plus weight-training in double-blind studies [507–510]. Whilst other studies have reported some improvement in strength and power, they involved only small numbers of subjects or were single-blind in design [511–514]. More recently, a small randomized study on "supraphysiologic" doses of an anabolic (testosterone enanthate, 600 mg per week) in normal men demonstrated an enhancement of muscle size when such doses were combined with strength training [515]. Studies using higher doses of an anabolic or using multiple forms of anabolic agents simultaneously – a practice referred to as "anabolic stacking" – are lacking. Anabolic steroids also exhibit an anticatabolic effect; that is, a reversal of the catabolic effects of glucocorticoids released in response to stress. Such effects would enable individuals to recover more quickly following strenuous workouts.

Currently, an alarming percentage of professional and amateur athletes utilize anabolic steroids [516, 517], which are readily available "on the street." The abuse of these steroids for increasing strength and power is banned in intercollegiate and international sports, and very sensitive assays such as radio-immunoassay and gas chromatography/mass spectrometry (GC/MS) have been developed for measuring levels of anabolic agents in urine and blood [356]. More recently, "designer" anabolic steroids were introduced in attempts to evade anti-doping detection methods. Between 2002 and 2005, laboratories accredited by the World Anti-Doping Agency (WADA) developed sensitive GC/MS and liquid chromatography (LC)/MS/MS assay methods for designer steroids such as norbolethone **62**, desoxymethyltestosterone **73**, trenbolone **80**, and tetrahydrogestrinone **81** [517–520].

6
Androgen Antagonists

A majority of recent research efforts in the area of androgens has concentrated on the preparation and biological activities of androgen antagonists. An androgen antagonist is defined as … a substance which antagonizes the actions of testosterone in various androgen-sensitive target organs and, when administered with an androgen, blocks, or diminishes the effectiveness of the androgen at various androgen-sensitive tissues. Androgen antagonists may act to block the action of testosterone at several possible sites. First, androgen antagonists may block the conversion of testosterone to its more active metabolite, DHT. Second, competition for the high-affinity binding sites on the AR molecule may account for antiandrogenic effects. Finally, certain agents such as LHRH agonists can act in the pituitary to lower gonadotropin secretion via gonadotropic receptor downregulation, and thus diminish the production of testosterone by the testis. The substances described in the following section act through at least one of these mechanisms. Several reviews on androgen antagonists are available [521–527].

6.1
Current Drugs on the Market

Currently available androgen antagonists used as therapeutics are listed in the following table.

Generic name (structure)	Trade name	U.S. manufacturer	Chemical class	Dose
Antiandrogens				
Cyproterone acetate (**82**)	Androcur	Schering AG	Pregnane	
Flutamide (**95**)	Eulexin	Various suppliers	Nonsteroidal	Tablet: 125 mg
Nilutamide (**97**)	Nilandron	Aventis	Nonsteroidal	Tablets: 50, 150 mg
Bicalutamide (**101**)	Casodex	AstraZeneca	Nonsteroidal	Tablet: 50 mg
Enzalutamide (**106**)	Xtandi	Astellas	Nonsteroidal	Capsule: 40 mg
5α-Reductase inhibitors				
Finasteride (**108**)	Proscar	Merck	Androstane	Tablet: 5 mg
	Propecia	Merck	Androstane	Tablet: 1 mg
Dutasteride (**112**)	Avodart	GlaxoSmithKline	Androstane	Capsule: 0.5 mg
17,20-Lyase inhibitors				
Abiraterone (**132**)	Zytiga	Cadia	Androstane	Tablet: 250 mg
Aromatase inhibitors				
Exemestane (**140**)	Aromasin	Pfizer	Androstane	Tablet: 25 mg
Anastrozole (**141**)	Arimidex	AstraZeneca	Nonsteroidal	Tablet: 1 mg
Letrozole (**142**)	Femara	Novartis	Nonsteroidal	Tablet: 5 mg

6.2 Antiandrogens

6.2.1 Therapeutic Uses

Antiandrogens are agents that compete with endogenous androgens for the hormone-binding site on the AR. These agents have therapeutic potential in the treatment of acne, virilization in women, hyperplasia and neoplasia of the prostate, baldness and male contraception, and clinical studies have demonstrated their potential therapeutic benefits. The application of antiandrogens to the treatment of prostatic carcinoma and for the treatment of BPH has also been investigated. Antiandrogens are effective for the treatment of prostate cancer when combined with androgen ablation, such as surgical (orchiectomy) or medical (LHRH agonist) castration.

6.2.2 Structure–Activity Relationships for Antiandrogens

Steroidal Agents Several steroidal and nonsteroidal compounds with demonstrated antiandrogenic activity have been utilized clinically [523]. The first compounds to be used as antiandrogens were the estrogens and progestins [528], and steroidal estrogens and diethylstilbesterol are still used for the treatment of prostatic carcinoma [529–532], exerting their action via a suppression of the release of pituitary gonadotropins. Progestational compounds have also been utilized for antiandrogenic actions, but with limited success [533]. The inherent hormonal activities of these compounds, and the development of more selective antiandrogens, have limited the clinical applications of estrogens and progestins as antiandrogens.

One modified progestin that is a potent antiandrogen and has minimal progestational activity is the agent cyproterone acetate (**82**). This compound was originally prepared in the quest for orally active progestins, but was quickly recognized for its ability to suppress gonadotropin release [25, 534–541]. It was shown later that cyproterone acetate also bound with high affinity to the AR, and thus competed with DHT for the binding site [25, 166, 542, 543]. Cyproterone acetate has received the most clinical attention of steroidal agents in antiandrogen therapy [544–555], having provided satisfactory results in the treatment of acne, seborrhea, and hirsutism [544–550]. The therapeutic effectiveness of cyproterone acetate in treating prostatic carcinoma has been reported [551–555], and it was also shown to be a good alternative to estrogens for treating prostate cancer, when combined with androgen ablation [556, 557]. Unfortunately, this combination did not improve disease-free survival or overall survival when compared to castration alone.

Other pregnane compounds that exhibit antiandrogenic actions by binding to the AR include chlormadinone acetate (**83**), medroxyprogesterone acetate (**84**), medrogestone (**85**), A-norprogesterone (**86**), and gestonorone caproate (**87**) [558–562]. Medrogestone also exerts antiandrogenic effects by inhibiting 5α-reductase and thus preventing the formation of DHT [563, 564]. Gestonorone caproate interferes with the uptake process in target cells [562].

Several androstane derivatives have demonstrated antiandrogenic properties. In 1964, 17α-methyl-β-nortestosterone (**88**) was the first to be prepared and tested for antihormonal activity [565], but within the next decade several other androstane analogs were synthesized and found to possess antiandrogenic activity [566–572]; these included BOMT (**89**), R2956 (**90**), and oxendolone (TSAA-291; **91**). As expected, the mechanism of antiandrogenic action

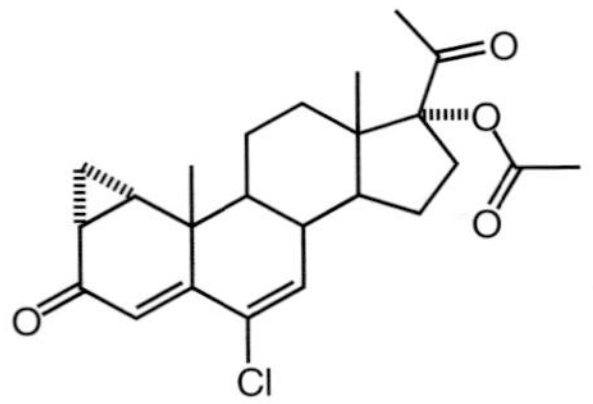

(82) 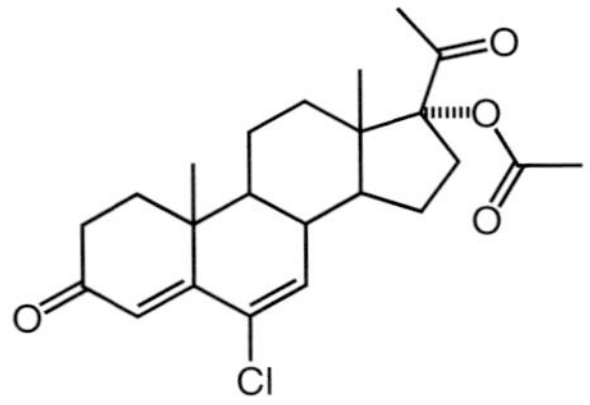 **(83)**

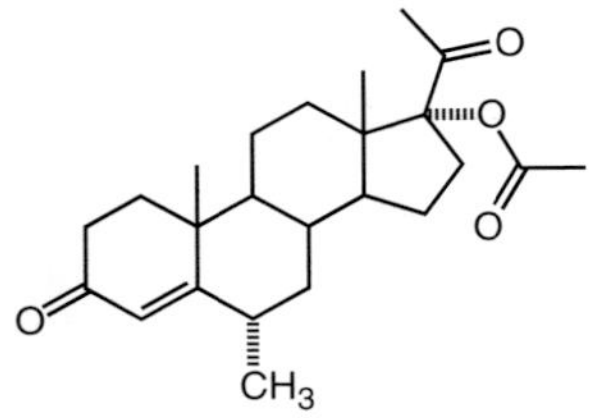

(84) 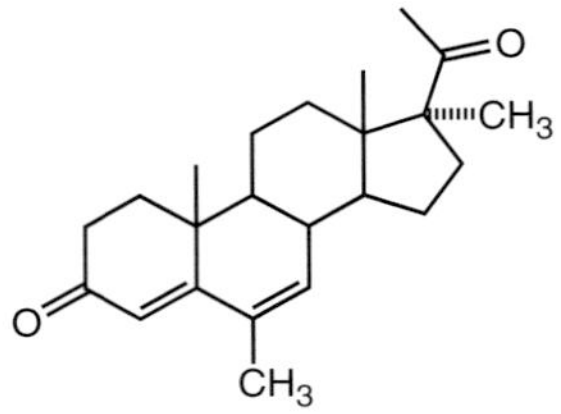 **(85)**

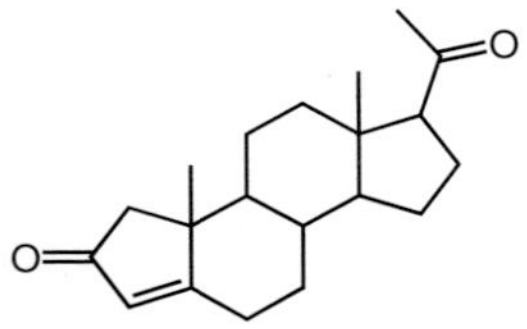

(86) 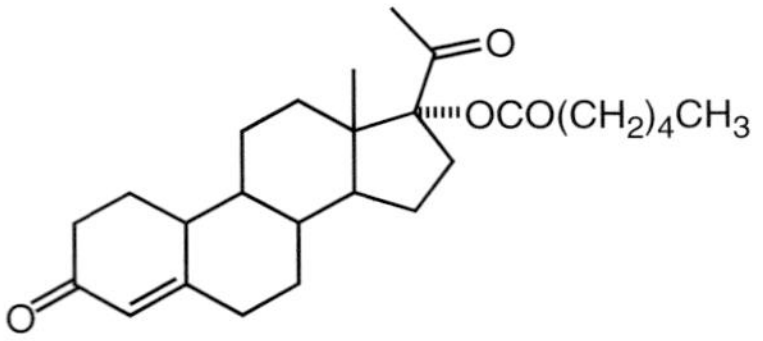 **(87)**

of these synthetic steroids is competition with androgens for the binding sites on the receptor molecule [543, 573–576]. Numerous A and B ring-modified steroids were examined for their antiandrogenic activities and the ability to bind to the AR [577, 578], thus confirming that the structural requirements of the AR binding site could accommodate some degree of flexibility in the A and/or B rings of antiandrogenic molecules. Heterocyclic-substituted A-ring antiandrogens such as zanoterone (WIN 49,596; **92**) further supported these conclusions on the structure–activity relationships of steroidal antiandrogens. Additional A-ring heterocycles identified as novel antiandrogens are the thiazole (**93**) and oxazole (**94**) [579, 580]. The optimal substitutions on the A-ring heterocyclic androstanes for *in-vivo* antiandrogenic activity are the methylsulfonyl group at the N-1′ position and a 17α-substituent (e.g., 17α-methyl or 17α-ethinyl).

Nonsteroidal Agents The absolute requirement of a steroidal compound for interaction with the AR was invalidated when the potent nonsteroidal antiandrogen flutamide (Eulexin, **95**) was introduced [581, 582]. Subsequent receptor studies [576, 583, 584] showed that this compound competed with DHT for the binding sites. The side chain of flutamide allows sufficient flexibility for the molecule to assume a structure similar to an androgen. In addition, a hydroxylated metabolite (**96**)

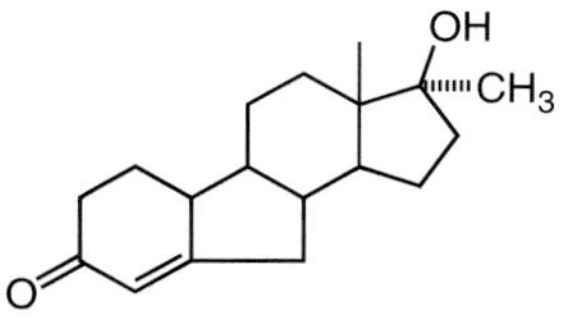

Analogs and Antagonists of Male Sex Hormones | 49

(88) 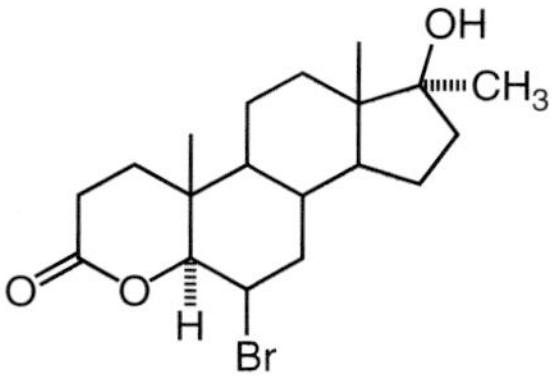 **(89)** 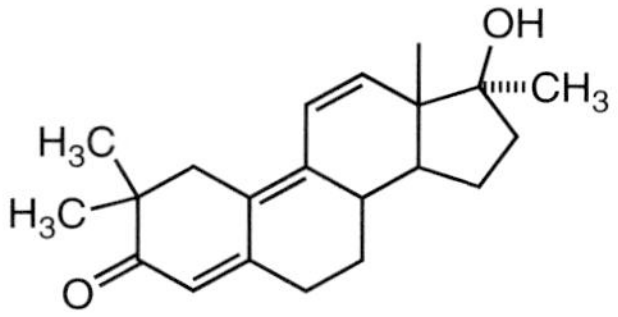 **(90)**

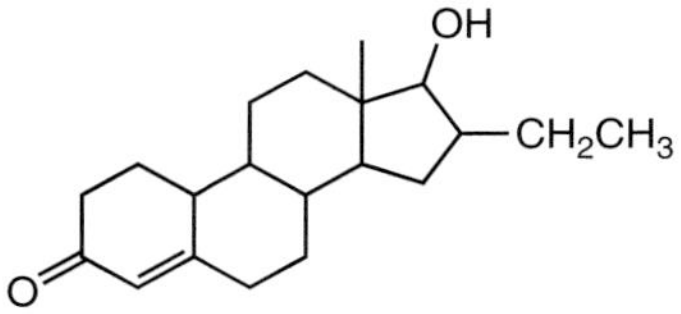

(91) 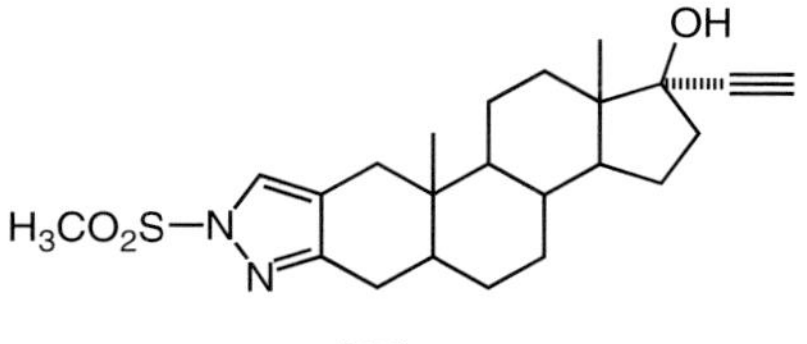 **(92)**

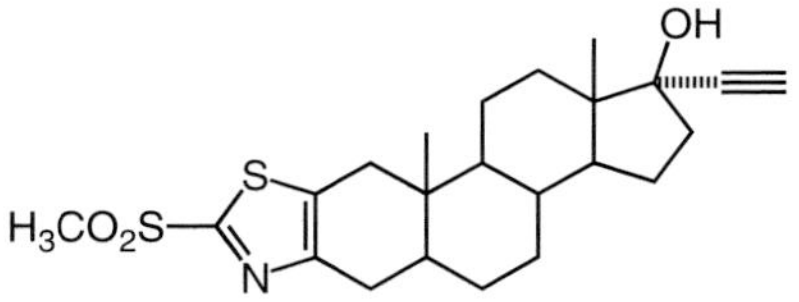

(93) 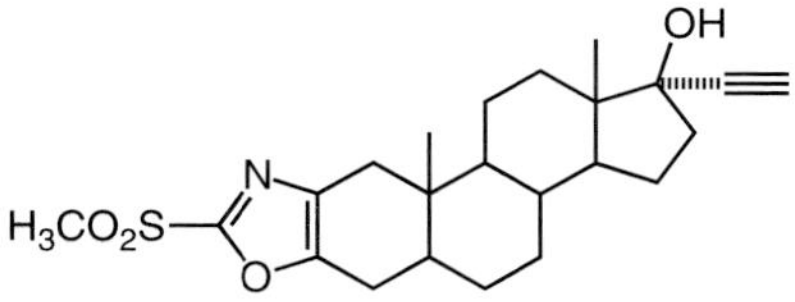 **(94)**

has been identified, is a more powerful antiandrogen *in vivo*, and has a higher affinity for the receptor than the parent compound [576, 585]. Important factors in the structure–activity relationships of flutamide and analogs are the presence of an electron-deficient aromatic ring and a powerful hydrogen bond donor group.

(95) X = H
(96) X = OH

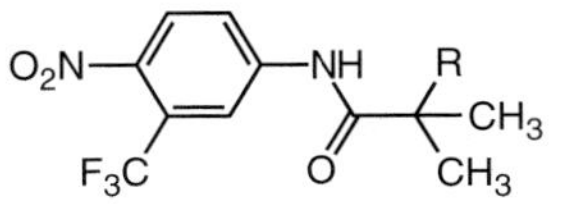

Flutamide has been evaluated extensively for the treatment of prostate cancer. Notably, large double-blind studies in prostate cancer patients, using a combination of flutamide with an LHRH agonist (as a medical castration), resulted in an increased number of favorable responses and increased overall survival when compared to an LHRH agonist or surgical castration [586, 587].

Nilutamide (Anandron, **97**) and related nilutamide analogs (**98–100**), and bicalutamide (Casodex, **101**) are other nonsteroidal antiandrogens with a similar electron-deficient aromatic ring, and have been shown to interact with the AR to varying degrees [528, 588]. Nilutamide and bicalutamide are pure antiandrogens, and are effective in suppressing testosterone-stimulated cell proliferation [589]. Both compounds have demonstrate an effective therapy against prostate cancer [590–594].

Other aryl-substituted nonsteroidal compounds have also been identified as antiandrogens. DIMP (**102**) is a phthalimide derivative that showed a weak affinity for the AR and poor *in-vivo* activity [576, 595]. A series of tetrafluorophthalimides (e.g., **103**) demonstrated moderate activity

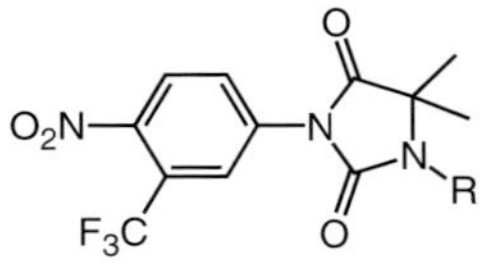

(97) R = H
(98) R = (CH₂)₄OH

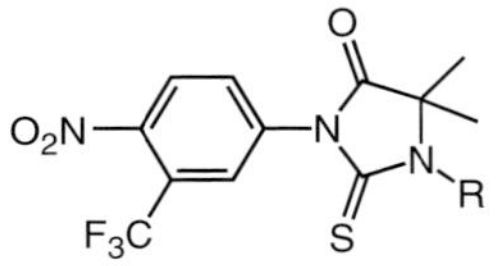

(99) R = CH₃
(100) R = (CH₂)₄OH

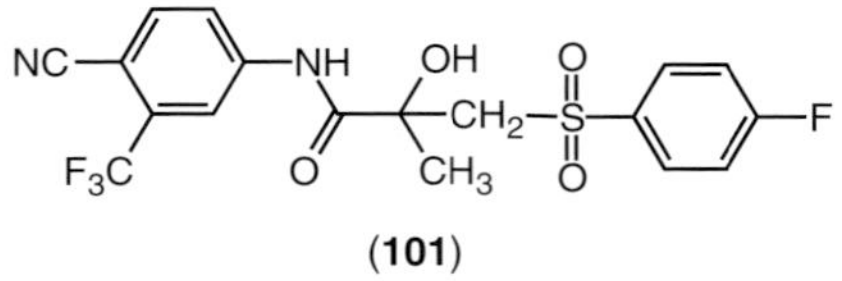

(101)

as antiandrogens in cell proliferation assays [596].

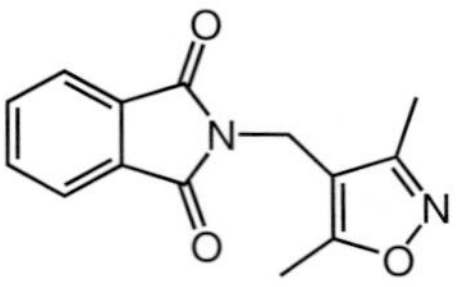

(102)

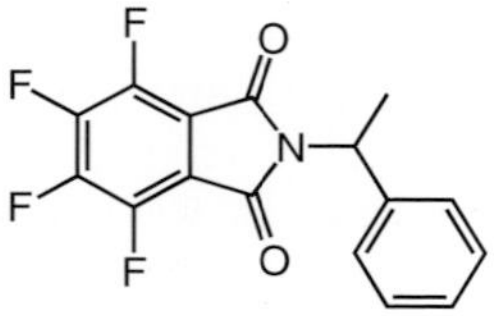

(103)

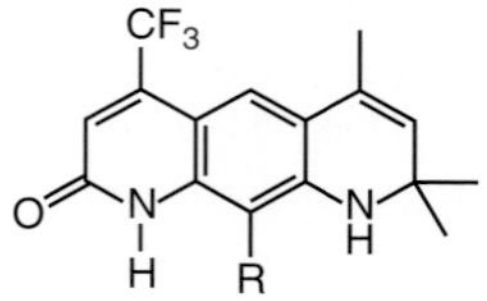

(104) R = H
(105) R = CH₃

(106)

(107)

A series of 1,2-dihydropyridono[5,6-g]quinolines was identified as novel nonsteroidal antiandrogens, employing a cell-based screening approach [597]. Several analogs (e.g., **104** and **105**) demonstrated excellent *in-vivo* activity, reducing rat ventral prostate weight without affecting serum levels of gonadotropins and testosterone.

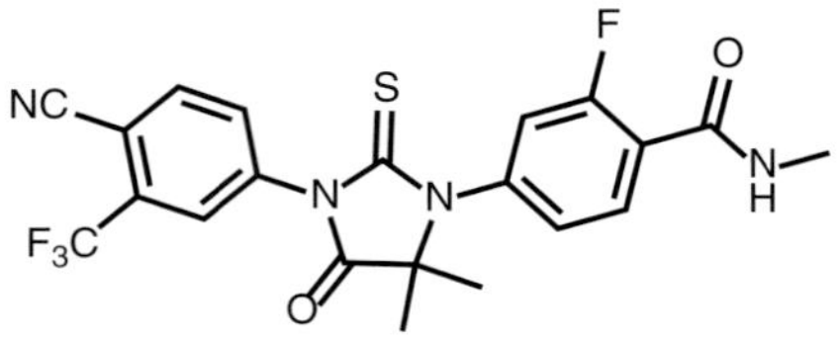

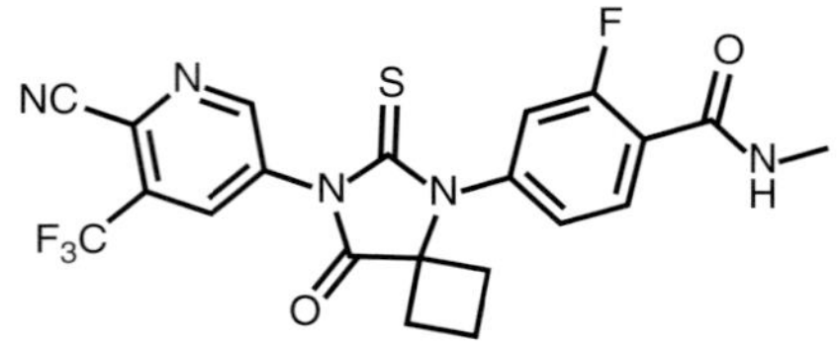

Several biarylthiohydantoins have demonstrated potent androgen antagonist activities in preclinical and clinical studies. Enzalutamide (formerly referred to as MDV3100; **106**) exhibited a high affinity for the AR, antagonized androgen-induced gene expression, and did not display any agonist activities in LNCaP prostate cancer cells *in vitro* and in xenograft mouse models [598, 599]. Among related analogs, enzalutamide exhibited excellent pharmacokinetic properties and was selected for clinical trials. In both Phase I and II trials, enzalutamide demonstrated antitumor activity in patients with castration-resistant prostate cancer [600], and was approved for use by the FDA in 2012. A related analog, ARN-509 (**107**), has shown greater *in-vivo* activity in xenograft models [601].

6.2.3 Absorption, Distribution, and Metabolism

The steroidal antiandrogens exhibit similar pharmacokinetic properties to the androgens and anabolic agents, with the lipophilicity of the compounds influencing absorption both orally and from injection sites [602–606]. Reduction of the 3-ketone and 4,5 double bond are common routes of metabolism [602]. An unusual metabolite of cyproterone acetate, 15α-hydroxycyproterone acetate, was isolated and identified in both animals and humans [607]. The nonsteroidal antiandrogen flutamide is rapidly absorbed and extensively metabolized *in vivo* [608, 609] and, as noted above, the hydroxy metabolite (**96**) of flutamide is a more potent antiandrogen [576, 585]. The major metabolite of bicalutamide is the glucuronide, which has comparable *in-vivo* activity [610]. Finally, the antiandrogens are primarily excreted as glucuronide and sulfate conjugates in the urine.

6.2.4 Toxicities

Adverse side effects of these agents have been identified from various clinical trials. In particular, testicular atrophy and decreased spermatogenesis have been observed during treatment with cyproterone acetate [611, 612]. Antiandrogens can also impair libido and result in impotence [613], while certain antiandrogens (e.g., cyproterone acetate and medrogesterone) also exhibit inherent progestational activity, suppress corticotropin release, and have some androgenic effects [614–616]. No hormonal activities were observed for the nonsteroidal antiandrogens, such as flutamide [609]. On the other hand, many nonsteroidal antiandrogens exhibit other endocrine side effects, such as elevated serum levels of gonadotropins and testosterone. Gynecomastia, nausea, diarrhea and liver toxicities have been observed in patients receiving nonsteroidal antiandrogens [617]. Resistance to antiandrogen therapy has also been observed in prostate cancer patients [618].

6.3 Enzyme Inhibitors

Enzymes involved in the biosynthesis and metabolism of testosterone represent attractive targets for drug design and drug development. To suppress of the synthesis of androgenic hormones and androgen precursors offers a viable therapeutic approach for the treatment of various androgen-mediated disease processes, and is an important endocrine treatment for prostate cancer. A potent inhibition of Type 2 5α-reductase in androgen target tissues, and the resultant decrease in DHT levels, will provide a selective interference with androgen action within those target tissues and no alterations of other effects produced by testosterone, other structurally related steroids, and other hormones such as corticoids and progesterone. The cytochrome $P450_{17\alpha}$ enzyme complex displays two enzymatic activities: 17α-hydroxylation to produce 17α-hydroxysteroids; and C_{17}-C_{20} bond cleavage (17,20-lyase activity) to produce androgens. In the male, this enzyme is found in both testicular and adrenal tissues, with these organs providing circulating androgens in the blood. An effective inhibition of this microsomal enzyme complex would eliminate both testicular and adrenal androgens and remove the growth stimulus to androgen-dependent prostate carcinoma. Synthetic androgen analogs that inhibit the oxidative metabolism of androgens to estrogens can serve as potential therapeutic agents for controlling estrogen-dependent diseases such as hormone-dependent breast cancer.

6.3.1 5α-Reductase Inhibitors

The most extensively studied class of 5α-reductase inhibitors is the 4-azasteroids [619], which includes the drug finasteride (Proscar, **108**). Finasteride, which is the first 5α-reductase inhibitor to be approved in the United States for the treatment of BPH, has an approximately 100-fold greater affinity for Type 2 5α-reductase than for the Type 1 enzyme, demonstrating an IC_{50} value of 4.2 nM for Type 2 5α-reductase [620]. In humans, finasteride decreases prostatic DHT levels by 70–90% and reduces prostate size [621], while testosterone tissue levels are increased. Clinical trials have demonstrated a sustained improvement in BPH disease and a reduction in PSA levels [622, 623]. Related analogs (**109–111**) have also demonstrated effectiveness both *in vitro* and *in vivo* [624–628]. These agents were originally designed to mimic the putative 3-enolate intermediate of testosterone and serve as transition-state inhibitors [625, 626]. Subsequently, finasteride was shown to produce time-dependent enzyme inactivation [627] and to function as a mechanism-based inactivator. The structure–activity relationships for the 4-azasteroids illustrate the stringent requirements for the inhibition of human Type 2 5α-reductase [628]. The 5α-reduced azasteroids are preferred, a 1,2-double bond can be tolerated, and the nitrogen can be substituted with only hydrogen or small lipophilic groups. Lipophilic amides or ketones are preferred as substituents at the C-17β position.

Several 6-azasteroids, such as dutasteride (**112**) and **113**, were prepared as extended mimics of the enolate transition state and have also demonstrated a potent inhibition of 5α-reductase [629]. Although the 6-azasteroids are more effective inhibitors of Type 2 5α-reductase, some analogs also exhibit a good inhibition of Type 2 5α-reductase. Alkylation of the nitrogen can be tolerated, but a 1,2-double bond

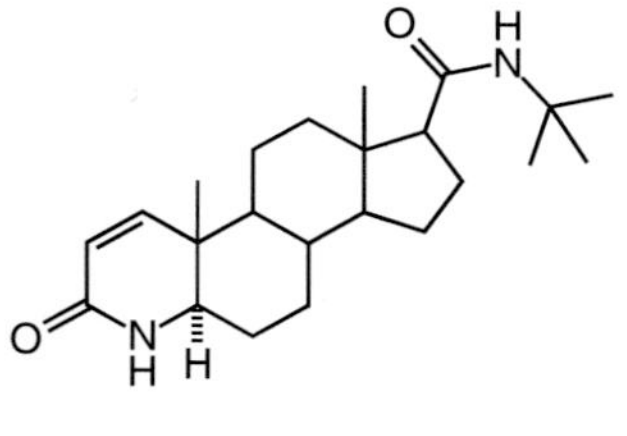

(**108**)

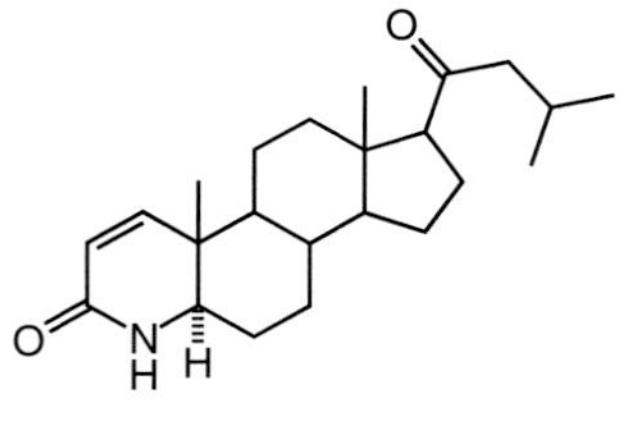

(**109**)

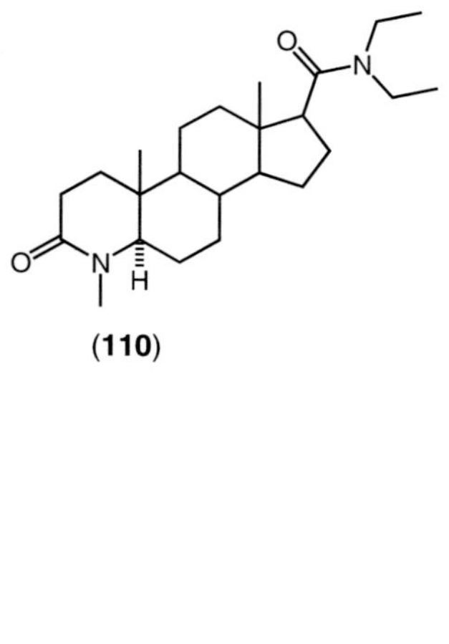

(**110**)

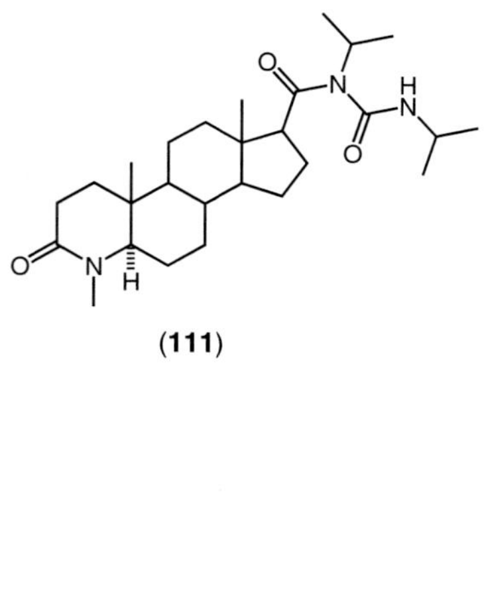

(**111**)

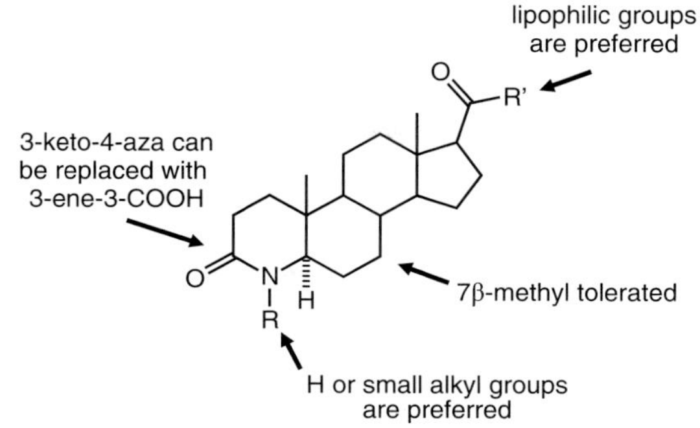

Analogs and Antagonists of Male Sex Hormones | 53

Fig. 10 Summary of structure–activity relationships for 5α-reductase inhibitors.

- lipophilic groups are preferred
- 3-keto-4-aza can be replaced with 3-ene-3-COOH
- 7β-methyl tolerated
- H or small alkyl groups are preferred

decreases the inhibitory activity in this series. The best inhibitors contain large lipophilic substituents at the C-17β position. A summary of the structure–activity relationships for steroidal 5α-reductase inhibitors is provided in Fig. 10.

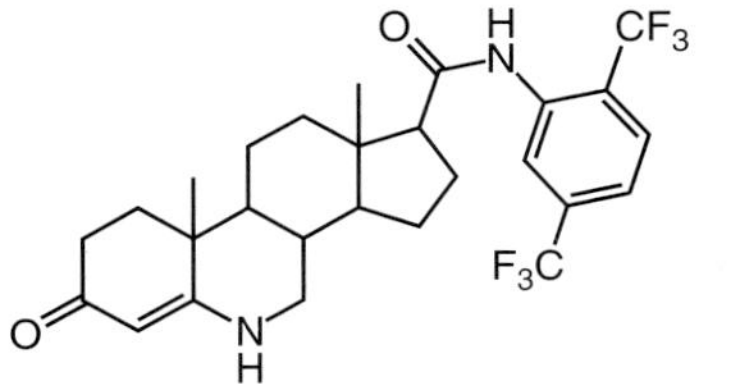

(112)

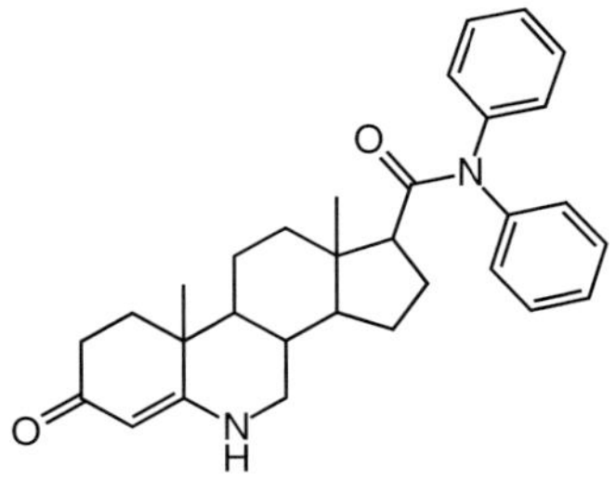

(113)

Androstadiene 3-carboxylic acids **114** and **115** were designed as transition-state inhibitors, and have demonstrated a potent uncompetitive inhibition of Type 2 5α-reductase [630, 631]. Epristeride (SK&F 105,657; **114**) has demonstrated the ability to lower serum DHT levels by 50% in clinical trials [632, 633]. Other analogs with acidic functionalities at the C-3 position include other androstene carboxylic acids (**116**, **117**) and estratriene carboxylic acids (**118**) [634]. The allenic secosteroid (**119**) has been demonstrated as a potent irreversible inhibitor of 5α-reductase, even though it was originally developed as an irreversible inhibitor of 3β-hydroxysteroid dehydrogenase/Δ4,5-isomerase [635–637]. Finally, selective and potent inhibitors of Type 1 5α-reductase were developed based on the 4-azacholestane MK-386 (**120**) [638].

Several nonsteroidal 5α-reductase inhibitors have been developed based on the azasteroid molecule, or from high-throughput screening methods. Examples of these nonsteroidal inhibitors include the benzoquinolinone (**121**), an aryl carboxylic acid (**122**), and FK143 (**123**) [639–641].

6.3.2 17,20-Lyase Inhibitors

Both, nonsteroidal and steroidal agents have been examined as inhibitors of 17α-hydroxylase/17,20-lyase. The nonsteroidal agents studied most extensively are aminoglutethimide (**124**) and ketoconazole (**125**), both *in vitro* and in clinical trials. Objective response rates for treatment of prostate cancer in relapsed patients were observed with high doses of aminoglutethimide [642]

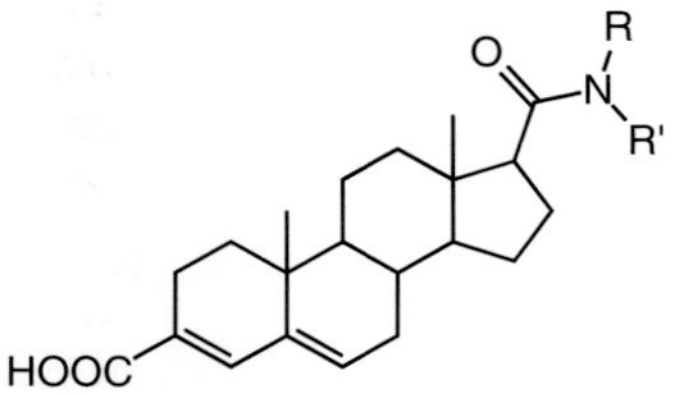

(114) R = H; R' = C(CH₃)₃
(115) R = R' = CH(CH₃)₂

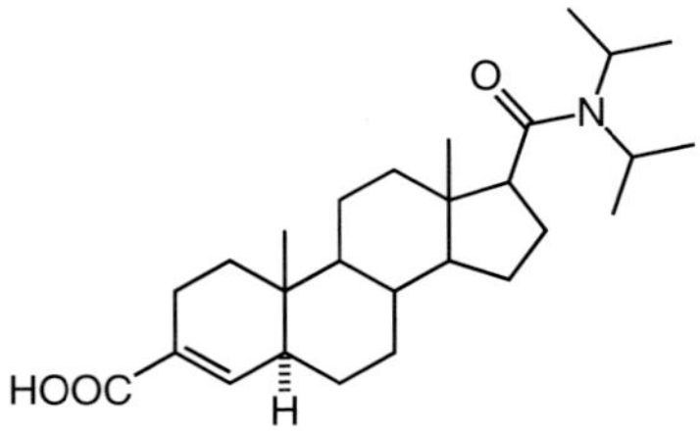

(116)

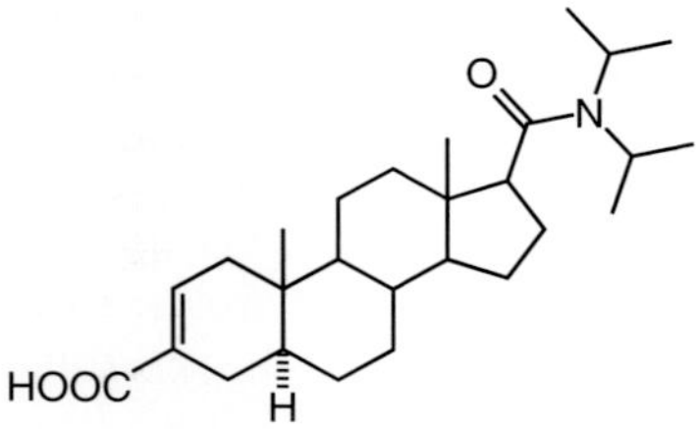

(117)

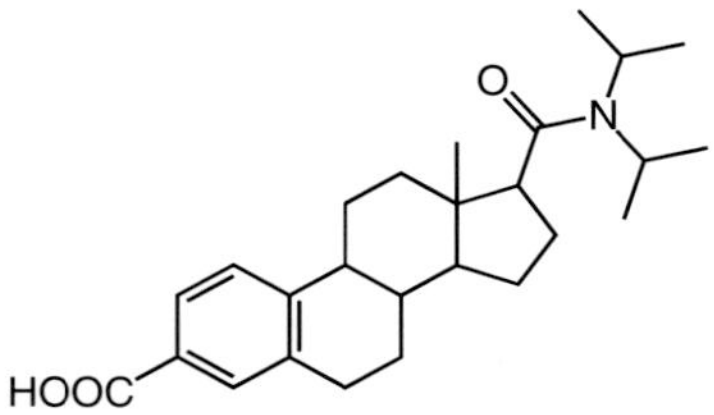

(118)

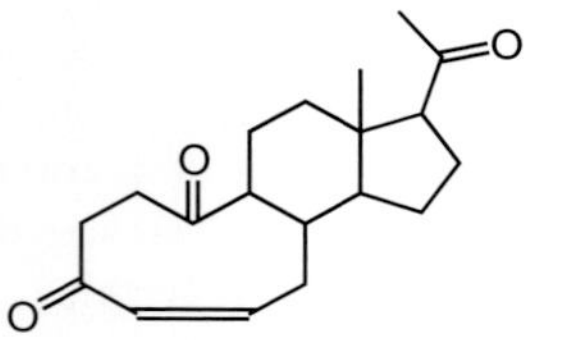

(119)

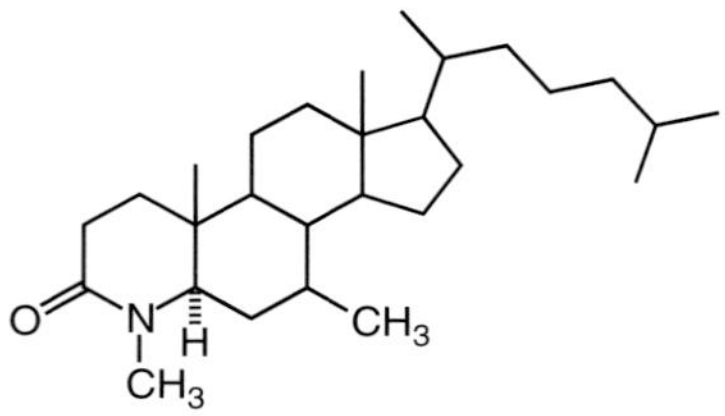

(120)

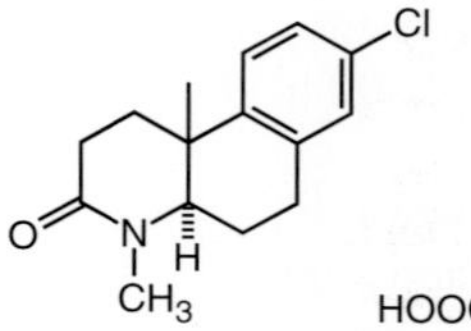

(121)

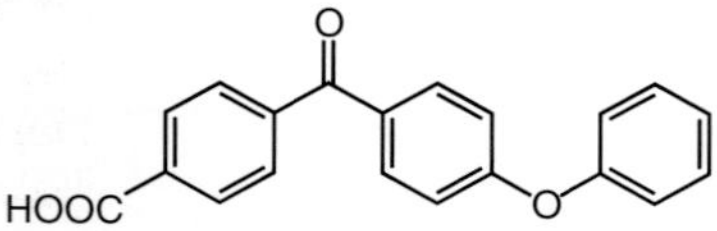

(122)

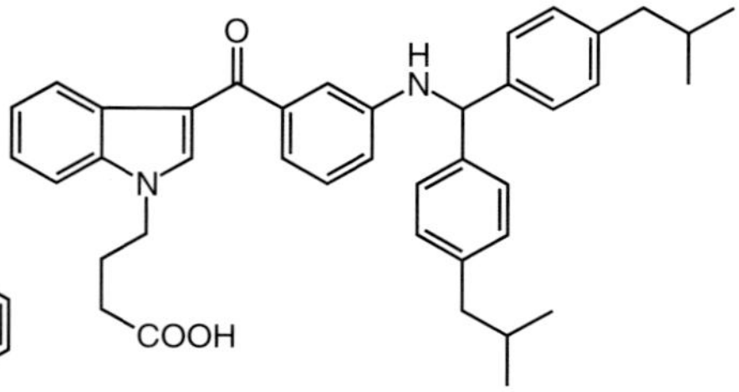

(123)

and high doses of ketoconazole [643], but both agents produced frequent adverse side effects. A third nonsteroidal agent that has received extensive preclinical evaluation is the benzimidazole analog, liarozole (**126**), which produced a reduction in plasma levels of testosterone and androstenedione *in vivo* [644]. Other nonsteroidal agents

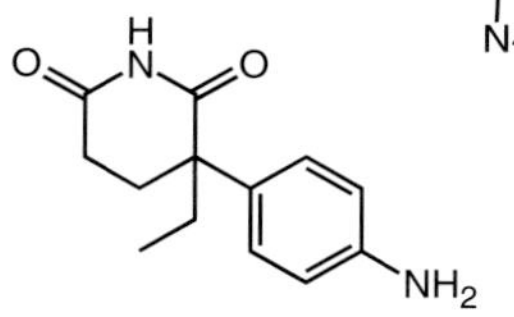

Analogs and Antagonists of Male Sex Hormones | 55

(124) 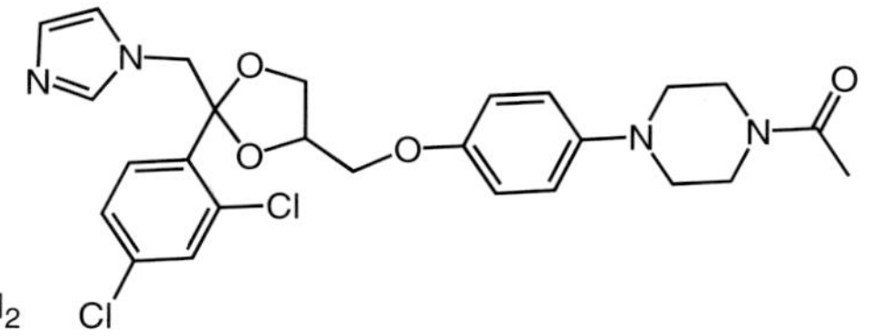 (125) 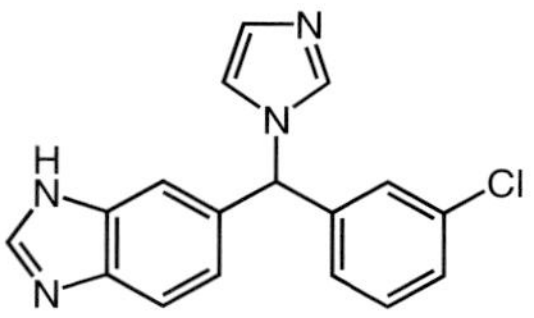 (126)

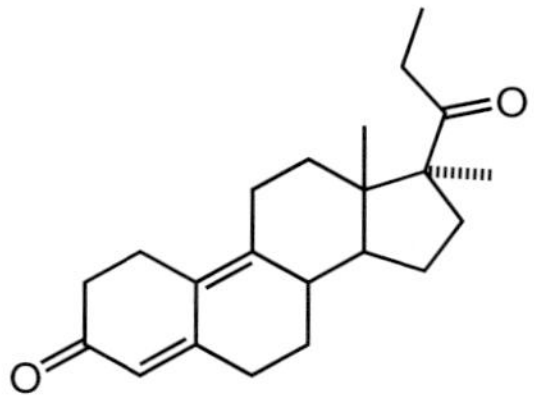

(127) 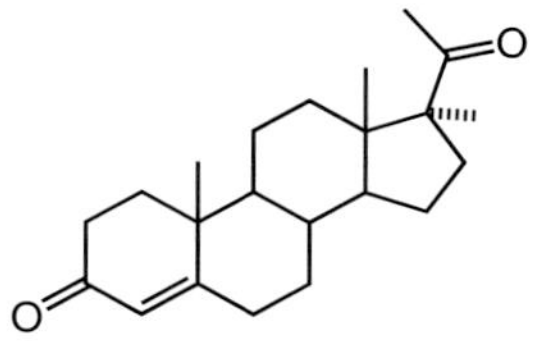 (128)

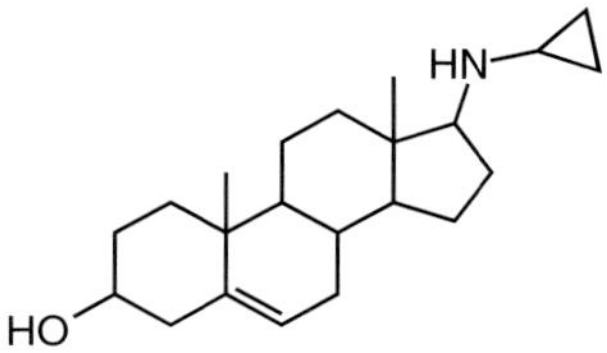

(129) 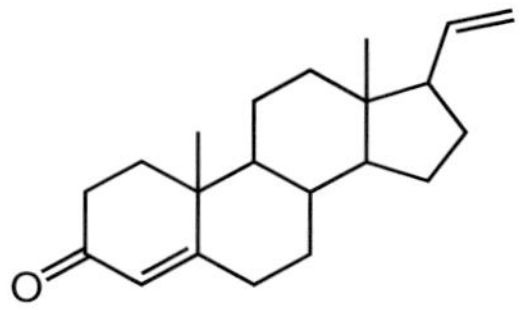 (130)

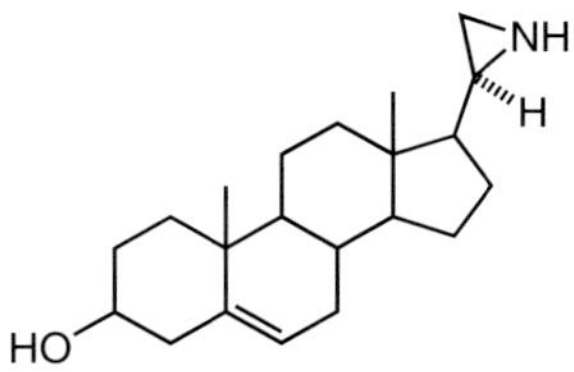

(131) 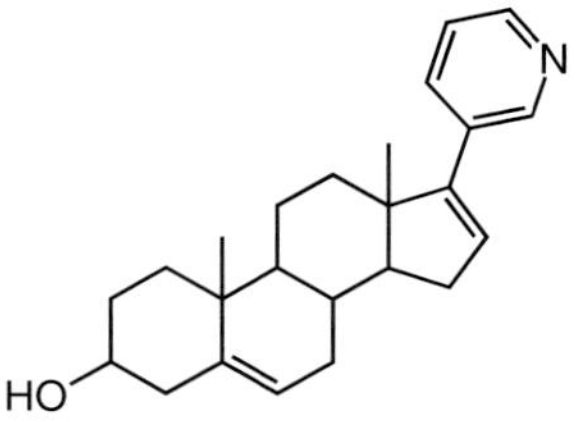 (132)

reported to exhibit 17α-hydroxylase/17,20-lyase inhibitory activity *in vitro* include other imidazole analogs [645], nicotine [646], bifluranol analogs [647], and pyridylacetic acid esters [648]. In general, high doses of nonsteroidal agents are needed to produce significant *in-vitro* or *in-vivo* activity. Another potential problem with these agents is the nonspecific inhibition of other cytochrome P450 enzymes involved in either steroidogenesis or liver metabolism.

A few studies of steroidal inhibitors of 17α-hydroxylase/17,20-lyase have been reported. An extensive analysis of the specificity of steroid binding to testicular microsomal cytochrome P450 identified several steroids exhibiting binding affinity [649]. One of these, promegestrone (**127**), has been utilized in the kinetic analysis of purified cytochrome P450$_{17\alpha}$ [650]. An affinity label inhibitor, 17-bromoacetoxyprogesterone (**128**), alkylates a unique cysteine residue on purified cytochrome P450$_{17\alpha}$ [651]. Potential mechanism-based inhibitors include 17β-(cyclopropylamino)-5-androsten-3β-ol (**129**; 652) and 17β-vinylprogesterone (**130**; 653). To date, all of these inhibitors exhibit apparent K_i-values in the micromolar (µM) range, while the apparent K_m for progesterone is 140 nM. The 17β-aziridinyl analog (**131**) and 17β-pyridyl derivative (abiraterone, **132**) also exhibited similar inhibitory activity [654, 655].

In a small clinical study, abiraterone (**132**) demonstrated clinical efficacy in suppressing testosterone levels in males with prostate cancer [656]. In patients with metastatic castration-resistant prostate cancer who had already received docetaxel, abiraterone plus low-dose prednisone improved overall survival [657], leading to FDA approval in 2011. Subsequently, a large clinical trial in men with metastatic castration-resistant prostate cancer without previous chemotherapy demonstrated that abiratone supplemented with prednisone resulted in progression-free survival and improved clinical outcomes [658].

6.3.3 C$_{19}$ Steroids as Aromatase Inhibitors

Aromatase is the enzyme complex that catalyzes the conversion of androgens into the estrogens. This enzymatic process is the rate-limiting step in estrogen biosynthesis, and converts C$_{19}$ steroids, such as testosterone and androstenedione, into the C$_{18}$ estrogens, estradiol and estrone, respectively. The inhibition of aromatase has been an attractive approach for examining the roles of estrogen biosynthesis in various physiological or pathological processes. Furthermore, effective aromatase inhibitors can serve as potential therapeutic agents for controlling estrogen-dependent diseases such as hormone-dependent breast cancer. Investigations on the development of aromatase inhibitors began during the 1970s and have expanded greatly over the past four decades, with summaries of research into steroidal and nonsteroidal aromatase inhibitors having been presented at several international conferences on aromatase [659–667] and several reviews also being published [668–675].

Steroidal inhibitors that have been developed to date build upon the basic androstenedione nucleus, and incorporate chemical substituents at varying positions on the steroid. These inhibitors bind to the aromatase cytochrome P450 enzyme in the same manner as the substrate androstenedione. Even though the steroidal aromatase inhibitors are C$_{19}$ steroids, these agents exhibit no significant androgenic activity. A limited number of effective inhibitors with substituents on the A ring have been reported. Several steroidal aromatase inhibitors contain modifications at the C-4 position, with 4-hydroxyandrostenedione (4-OHA; formestane; **133**) being the prototype agent. Initially, 4-OHA was thought to be a competitive inhibitor, but was later shown to produce enzyme-mediated inactivation [149, 676, 677]. In vivo, 4-OHA inhibits the reproductive process [678] and causes a regression of hormone-dependent mammary rat tumors [679, 680]. 4-OHA is also effective in the treatment of advanced breast cancer in postmenopausal women [681–683], and has been approved in the United Kingdom for the treatment of breast cancer. Thus, the spatial requirements of the A-ring for binding of the steroidal inhibitor to aromatase are rather restrictive, permitting only small structural modifications to be made.

(133)   (134)   (135)

An incorporation of the polar hydroxyl group at C-4 enhances inhibitory activity. 1-Methyl-1,4-androstadiene-3,17-dione (**134**) is a potent inhibitor of aromatase *in vitro* and *in vivo* [684], whereas bulky substituents at the 1α-position led to poor inhibitors [685]. At the C-3 position, replacement of the ketone with a methylene provided an effective inhibition [686].

More extensive structural modifications may be made on the B-ring of the steroid nucleus. Bulky substitutions at the C-7 position of the B-ring have provided several very potent aromatase inhibitors [685]. For example, 7α-(4′-amino)phenylthio-4-androstene-3,17-dione (7α-APTA; **135**) is a very effective competitive inhibitor, with an apparent K_i of 18 nM. This inhibitor has also been shown effective in inhibiting aromatase in cell cultures [687, 688] and in treating hormone-dependent rat mammary tumors [688, 689]. The evaluation of various substituted aromatic analogs of 7α-APTA provided no correlation between the electronic character of the substituents and the inhibitory activity [690]. Investigations of various 7-substituted 4,6-androstadiene-3,17-dione derivatives [691, 692] have suggested that only those derivatives which can project the 7-aryl substituent into the 7α pocket are effective inhibitors. Overall, the most effective B-ring-modified aromatase inhibitors are those with 7α-aryl derivatives, with several analogs having two- to tenfold greater affinity for the enzyme than the substrate. These results suggest that additional interactions occur between the phenyl ring at the 7α-position and amino acids at or near the enzymatic site of aromatase, resulting in an enhanced affinity.

Numerous modified androstenedione analogs have been developed as effective mechanism-based aromatase inhibitors. The first of these compounds to be designed, 10-propargyl-4-estrene-3,17-dione (PED; MDL 18,962; **136**), was synthesized and studied independently by three research groups [693–695]. MDL 18,962 has an electron-rich alkynyl function on the C-19 carbon atom, the site of aromatase-mediated oxidation of the substrate. Although the identity of the reactive intermediate formed is not known, an oxirene and a Michael acceptor have been suggested. This agent is an effective inhibitor *in vitro* and *in vivo* [695–700]. Other approaches to C-19-substituted, mechanism-based inhibitors containing latent chemical groups have provided a limited number of inhibitors, including the difluoromethyl analog (**137**) [701] and a thiol (**138**) [702, 703]. Another series of mechanism-based inhibitors has been developed from more detailed biochemical investigations of several inhibitors which originally were thought to be competitive inhibitors. These inhibitors can be grouped into general categories of 4-substituted

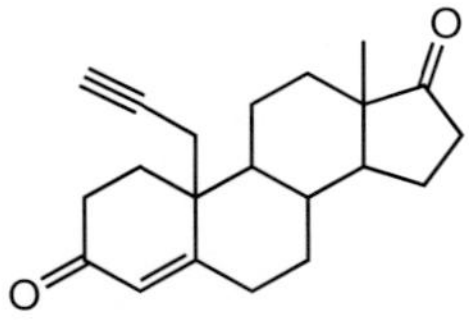

(136) 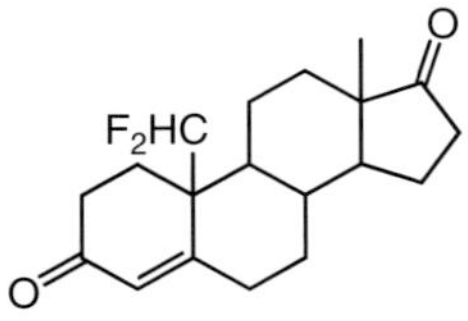 **(137)** 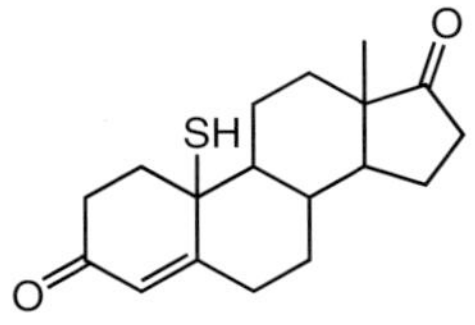 **(138)**

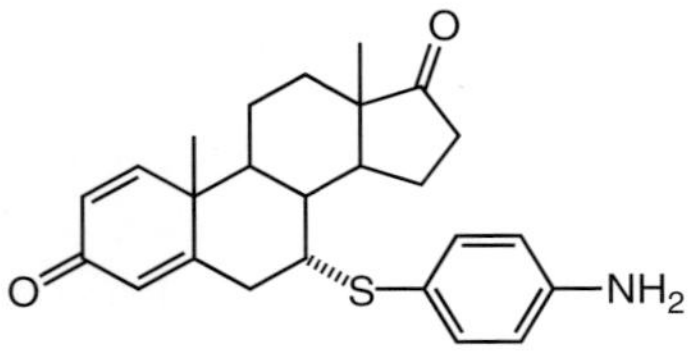

(139) 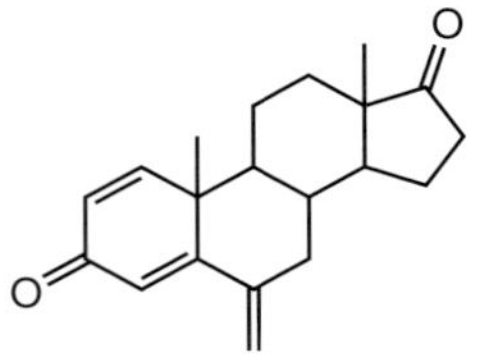 **(140)**

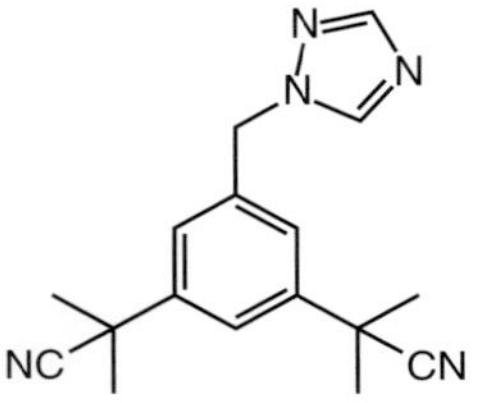

(141) 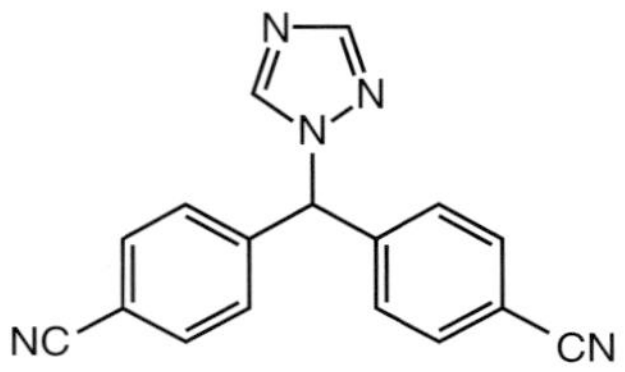 **(142)**

androst-4-ene-3,17-diones such as 4-hydroxyandrostenedione (**133**) [677], substituted androsta-1,4-diene-3,17-diones such as 7α-(4′-amino)-phenylthioandrosta-4,6-diene-3,17-dione (7α-APTADD; **139**) [704], and 6-methyleneandrost-4-ene-3,17-dione (exemestane; **140**) [705]. Exemestane (Aromasin®) was originally marketed as second-line therapy for the treatment of breast cancer patients in whom tamoxifen treatment had failed. Exemestane has been approved as a first-line therapy in women with advanced breast cancer, and as an adjuvant therapy in hormone-dependent breast cancer patients [706]. Two nonsteroidal aromatase inhibitors, anastrozole (Arimidex; **141**) and letrozole (Femara; **142**), were developed simultaneously and approved as first-line therapy in women with advanced breast cancer and as an adjuvant therapy in hormone-dependent breast cancer patients [706].

7
Summary

The steroid testosterone is the major circulating sex hormone in males, and serves as the prototype for the androgens, the anabolic agents, and androgen antagonists. Endogenous androgens are biosynthesized from cholesterol in various tissues of the body, with the majority of the circulating androgens produced in the testes under stimulation of the gonadotropin, LH. One critical aspect of testosterone and its biochemistry is that this steroid is converted in various cells to other active steroidal agents. For example, the reduction of testosterone to DHT is necessary for the androgenic actions of testosterone to be effected in some androgen target tissues such as the prostate, whereas testosterone itself appears to be the active androgen in muscle. On the other hand, the oxidation of testosterone by the enzyme aromatase to yield estradiol is crucial for certain actions in the central nervous system. Investigations of these enzymatic conversions of circulating testosterone continue to be a fruitful area of biochemical research on the roles of steroid hormones in the body. Additionally, the elucidation of the mechanism of action of androgens in various target tissues is receiving on-going attention. The actions of androgens are considered due to their binding to the androgen nuclear receptor, followed by dimerization of the receptor complex and binding to a specific DNA sequence. This binding of the homodimer to the androgen response element leads to gene expression, stimulation of the synthesis of new mRNA, and subsequent protein biosynthesis. Other actions of testosterone, particularly the anabolic actions, appear to be mediated through a similar nuclear receptor-mediated mechanism. Nongenomic pathways for androgen action through the androgen receptor have also been reported, and include a rapid activation of kinase signaling pathways, such as the activation of MAP kinase and ERK kinase pathways, and the modulation of intracellular calcium levels. Many of the intricate biochemical events that occur during the action of androgens in their target cells remain to be further clarified. Nevertheless, receptor studies of new agents represent an important biological tool in the evaluation of the compounds for later, in-depth, pharmacological testing.

Synthetic androgens and anabolic agents were first created to impart oral activity to the androgen molecule, to separate the androgenic effects of testosterone from its anabolic effects, and to improve upon its biological activities. Over the years, these research efforts have provided several effective drug preparations for the treatment of various androgen-deficient diseases, for the therapy of diseases characterized by muscle wasting and protein catabolism, for postoperative adjuvant therapy, and for the treatment of certain hormone-dependent cancers. Unfortunately, however, some of these synthetic anabolics have also been abused by athletes. During the past decade, nonsteroidal androgen agonists have been identified that are referred to as selective androgen receptor modulators. These compounds exhibit agonist activities in tissues such as muscle and bone, yet act as antagonists or weak agonists in the prostate gland. Finally, steroidal and nonsteroidal agents have both been developed as androgen antagonists, the two major categories of which are the antiandrogens (which block the interactions of androgens with the androgen receptor) and the inhibitors of androgen biosynthesis and metabolism. These compounds have therapeutic potential in the treatment of acne, virilization in women, hyperplasia and neoplasia of the prostate, baldness, and also for male contraception. A number of androstane derivatives are also currently being developed as inhibitors of aromatase for the treatment of hormone-dependent breast cancer. Thus, the numerous biological effects of the male sex hormones testosterone and dihydrotestosterone, and the varied chemical modifications of the androstane molecule, have resulted in the development of effective medicinal agents for the treatment of androgen-related diseases.

Acknowledgments

The present text is a modified and updated version of the chapter written by the present author, entitled "Male Sex Hormones, Analogs, and Antagonists," in *Burger's Medicinal Chemistry and Drug Discovery, Seventh Edition, Volume 5* (eds D. Abraham and D. Rotella), John Wiley & Sons, New York, pp. 153–217, 2010 [6].

References

1. J.A. Vida. *Androgens and Anabolic Agents. Chemistry and Pharmacology.* New York, Academic Press, 1969.
2. K.B. Eik-Nes, *The Androgens of the Testis*, Marcel Dekker, New York, 1970.
3. P.L. Munson, E. Diczfalusy, J. Glover, and R.E. Olsen (eds), *Vitamins and Hormones*, Vol. **33**, Academic Press, New York. 1975.
4. C.D. Kochakian, *Anabolic-Androgenic Steroids*, Springer-Verlag, New York, 1976.
5. L. Martini, M. Motta (eds), *Androgens and Antiandrogens*, Raven Press, New York, 1997.
6. R.W. Brueggemeier, Male Sex Hormones, Analogs, and Antagonists, in *Burger's Medicinal Chemistry & Drug Discovery*, 7th edn, Vol. **5**, D. Abraham, D. Rotella (eds), John Wiley & Sons, Inc., New York, pp. 153–217, 2010.
7. P. J. Snyder, Androgens, in *Goodman and Gilman's The Pharmacological Basis of Therapeutics*, 12th edn, L.L. Brunton, B.A. Chabner, B.C. Knollmann (eds), New York, McGraw-Hill Medical Publishing, 2011, pp. 1195–2017.
8. F.J. Zeelen, *Medicinal Chemistry of Steroids*, Elsevier, Amsterdam, 1990, p. 177.
9. G.D. Braunstein, Testes, in *Greenspan's Basic & Clinical Endocrinology*, 9th edn, D.G. Gardner, D. Shoback (eds), New York, McGraw-Hill Medical Publishing, 2011, pp. 395–422.
10. A. T. Kicman, *Br. J. Pharmacol.* **154**, 502 (2008).
11. S. Bhasin, H.L. Gabelnick, J.M. Spieler, R.S. Swerdloff, C. Wang, C. Kelly (eds), *Pharmacology, Biology, and Clinical Applications of Androgens*, Wiley-Liss, New York, 1996.
12. A.A. Berthold, *Arch. Anat. Physiol. Will. Med.*, **16**, 42 (1849).
13. A. Butenandt, *Angew. Chem.*, **44**, 905 (1931).
14. A. Butenandt, K. Tscherning, *Z. Physiol. Chem.*, **229**, 167 (1934).
15. T.F. Gallagher, F.C. Koch, *J. Biol. Chem.*, **84**, 495 (1929).
16. A. Butenandt, *Naturwissenschaftlicher*, **21**, 49 (1933).
17. A. Butenandt, H. Dannenberg, *Z. Physiol. Chem.*, **229**, 192 (1934).
18. K. David, E. Dingemanse, J. Freud, E. Laqueur, *Z. Physiol. Chem.*, **233**, 281 (1935).
19. K. David, *Acta Brevia Neerl. Physiol. Pharmacol. Microbiol.* **5**, 85, 108 (1935).
20. A. Butenandt, G. Hanisch, *Berichte*, **68**, 1859 (1935); A. Butenandt, G. Hanisch, *Z. Physiol. Chem.*, (1935), **237**, 89.
21. L. Ruzicka, *J. Am. Chem. Soc.*, **57**, 2011 (1935).
22. L. Ruzicka, A. Wettstein, H. Kagi, *Helv. Chim. Acta*, **18**, 1478, (1935).
23. N. Bruchovsky, J.D. Wilson, *J. Biol. Chem.*, **243**, 5953 (1968).
24. K.M. Anderson, S. Liao, *Nature (London)*, **219**, 277 (1968).
25. S. Fang, K.M. Anderson, S. Liao, *J. Biol. Chem.*, **244**, 6584 (1969).
26. W.I.P. Mainwaring, *J. Endocrinol.*, **45**, 531 (1969).
27. C.D. Kochakian, J.R. Murlin, *J. Nutr.*, **10**, 437 (1935).
28. A.T. Kenyon, I. Sandiford, A.H. Bryan, K. Knowlton, *et al.*, *Endocrinology*, **23**, 135 (1938).
29. R.K. Meyer, L.G. Hershberger, *Endocrinology*, **60**, 397 (1957).
30. S.L. Leonard, *Endocrinology*, **50**, 199 (1952).
31. J.M. Loring, J.M. Spencer, C.A. Villee, *Endocrinology*, **68**, 501 (1961).
32. D.B. Villee, *Human Endocrinology, A Developmental Approach*, W.B. Saunders, New York, 1975.
33. J.E. Griffin and J.D. Wilson, The Androgen Resistance Syndromes: 5α-Reductase Deficiency, Testicular Feminization, and Related Syndromes, in *The Metabolic Basis of Inherited Diseases*, 6th edn, C.R. Scriver, A.L. Beaudet, W.S. Sly, D. Valle (eds), McGraw-Hill, New York, 1989, p. 1919.
34. R. L. Landau, Metabolic effects of anabolic-steroids in man, in *Anabolic-Androgenic*

Steroids, C.D. Kochakian (ed.), Springer-Verlag, New York, 1967, p. 48.
35. H. Spencer, J.A. Friedland, I. Lewin, Effects of androgens on bone, calcium, and phosphorus metabolism, in *Anabolic-Androgenic Steroids*, C.D. Kochakian (ed.), Springer-Verlag, New York, 1976, p. 419.
36. C.W. Gurney, The hematologic effects of androgens, in *Anabolic-Androgenic Steroids*, C.D. Kochakian (ed.), Springer-Verlag, New York. 1976, p. 483.
37. F.T.G. Prunty, *Br. Med. J.*, **2**, 605 (1966).
38. A. Vermeulen, *Acta Endocrinol. (Copenhagen)*, **83**, 651 (1976).
39. M.B. Lipsett, S.G. Korenman, *J. Am. Med. Assoc.*, **190**, 757 (1964).
40. A. Chapdelaine, P.C. MacDonald, O. Gonzalez, E. Gurpide, R.L. et al., *J. Clin. Endocrinol. Metab.*, **25**, 1569 (1965).
41. J.F. Tait, R. Horton, *Steroids*, **4**, 365 (1964).
42. R.J. Santen, C.W. Bardin, *J. Clin. Invest.*, **52**, 2617 (1973).
43. A.M. Matsumoto, W.J. Bremmer, *J. Clin. Endocrinol. Metab.*, **58**, 609 (1984).
44. D.M. Robertson, R.I. McLachlan, H.G. Burger, Inhibin-Related Proteins in the Male, in *The Testis*, 2nd edn, H. Burger, D. de Kretser (eds)., Raven Press, New York, 1989, p. 231.
45. F.F.G. Rommerts, B.A. Cooke, H.J. Van der Mulen, *J. Steroid Biochem.*, **5**, 279 (1974).
46. W.L. Miller, *Endocr. Rev.*, **9**, 295 (1988).
47. R. Vihko, A. Ruokonen, *J. Steroid Biochem.*, **5**, 843 (1974).
48. C.E. Bird, L. Morrow, Y. Fukumoto, S. Marcellus, et al., *J. Clin. Endocrinol. Metab.*, **43**, 1317 (1976)
49. T. Yanaihara, P. Troen, *J. Clin. Endocrinol. Metab.*, **34**, 783 (1972).
50. R.I. Dorfman, F. Ungar, *Metabolism of Steroid Hormones*, Academic Press, New York, 1965.
51. S. Nakajin, P.F. Hall, *J. Biol. Chem.*, **256**, 3871 (1981).
52. B. Chung, J. Picado-Leonard, M. Haniu, M. Bienkowski, et al., *Proc. Natl Acad. Sci. USA*, **84**, 407 (1987).
53. K.D. Bradshaw, M.R. Waterman, R.T. Couch, E.R. Simpson, et al., *Mol. Endocrinol.*, **1**, 348 (1987).
54. T. Yanese, E.R. Simpson, M.R. Waterman, *Endocr. Rev.*, **12**, 91 (1991).
55. H. Ishii-Ohba, H. Inano, B.-I. Tamaoki, *J. Steroid Biochem.*, **27**, 775 (1987).
56. H.P. Schedl, J.A. Clifton, *Gastroenterology*, **41**, 491 (1961).
57. K.B. Eik-Nes, J. Schellmann, A.R. Lumry, L.T. Samuels, *J. Biol. Chem.*, **206**, 411 (1954).
58. B.H. Levedahl, H. Bernstein, *Arch. Biochem. Biophys.*, **52**, 353 (1954).
59. J. Schellmann, A.R. Lumry, L.T. Samuels, *J. Am. Chem. Soc.*, **76**, 2808 (1954).
60. J. Kerkay, U. Westphal, *Biochim. Biophys. Acta*, **170**, 324 (1968).
61. J. Kerkay, U. Westphal, *Arch. Biochem. Biophys.*, **129**, 480 (1969).
62. P.I. Corvol, A. Chrambach, D. Rodbard, C.W. Bardin, *J. Biol. Chem.*, **246**, 3435 (1971).
63. W.H. Pearlman, O. Crépy, *J. Biol. Chem.*, **242**, 182 (1967).
64. J.L. Guériguan, W.H. Pearlman, *Fed. Proc.*, **26**, 757 (1967).
65. B.E.P. Murphy, *Can. J. Biochem.*, **46**, 299 (1968).
66. G.V. Avvakumov, I. Grishkovskaya, Y.A. Muller, G.L. Hammond, *J. Biol. Chem.*, **277**, 45219 (2002).
67. M.C. Lebeau, C. Mercier-Bodard, J. Oldo, D. Bourguon, et al., *Ann. Endocrinol.*, **30**, 183 (1969).
68. W. Rosner, N.P. Christy, W.G. Kelley, *Biochemistry*, **8**, 3100 (1969).
69. W.H. Pearlman, I.F.F. Fong, K.J. Tou, *J. Biol. Chem.*, **244**, 1373 (1969).
70. F. Dray, I. Mowezawicz, M. J. Ledru, O. Crépy, et al., *Ann. Endocrinol.*, **30**, 223 (1969).
71. T. Kato, R. Horton, *J. Clin. Endocrinol. Metab.*, **28**, 1160 (1968).
72. P. DeMoor, O. Steeno, W. Heyns, H. Van Baelen, *Ann. Endocrinol.*, **30**, 233 (1969).
73. C. Mercier-Bodard, A. Alfsen, E.E. Baulieu, *Acta Endocrinol. (Copenhagen)*, **147**, 204 (1970).
74. R. Horton, T. Kato, R. Sherino, *Steroids*, **10**, 245 (1967).
75. A. Vermeulen, L. Verdonck, *Steroids*, **11**, 609 (1968).
76. C. Mercier-Bodard, E.E. Baulieu, *Ann. Endocrinol.*, **29**, 159 (1968).
77. B.E.P. Murphy, *Steroids*, **16**, 791 (1970).

78. I. Grishkovskaya, G.V. Avvakumov, G. L. Hammond, M. G. Catalano et al., *J. Biol. Chem.* **277**, 32086 (2002).
79. E.M. Ritzen, S.N. Nayfeh, F.S. French, M.C. Dobbins, *Endocrinology*, **89**, 143 (1971).
80. V. Hansson, O. Djoseland, *Acta Endocrinol. (Copenhagen)*, **71**, 614 (1972).
81. F.S. French, E.M. Ritzén, *J. Reprod. Fertil.*, **32**, 479 (1973).
82. V. Hansson, O. Trygotad, F.S. French, W.S. McLean et al., *Nature (London)*, **250**, 387 (1974).
83. R.G. Vernon, B. Kopec, I.B. Fritz, *Mol. Cell. Endocrinol.*, **1**, 167 (1974).
84. D. Williams, J. Gorski, *Biochem. Biophys. Res. Commun.*, **45**, 258 (1971).
85. E. Milgrom, M. Atger, E.E. Baulieu, *Biochim. Biophys. Acta*, **320**, 267 (1973).
86. R.W. Harrison, S. Fairfield, D.N. Orth, *Biochemistry*, **14**, 1304 (1975).
87. R.W. Harrison, S. Fairfield, D.N. Orth, *Biochem. Biophys. Res. Commun.*, **61**, 1262 (1974).
88. E.P. Gorgi, J.C. Stewart, J.K. Grant, R. Scott, *Biochem. J.*, **122**, 125 (1971).
89. E.P. Gorgi, J.C. Stewart, J.K. Grant, I.M. Shirley, *Biochem. J.*, **126**, 107 (1972).
90. E.P. Gorgi, J.K. Grant, J.C. Stewart, J. Reid, *J. Endocrinol.*, **55**, 421 (1972).
91. E.P. Gorgi, I.M. Shirley, J.K. Grant, J.C. Stewart, *Biochem. J.*, **132**, 465 (1973).
92. S.R. Gross, L. Aronow, W.B. Pratt, *Biochem. Biophys. Res. Commun.*, **32**, 66 (1968).
93. S.R. Gross, L. Aronow, W.B. Pratt, *J. Cell Biol.*, **44**, 103 (1970).
94. P. Ofner, *Vitamins and Hormones*, Vol. **26**, R.S. Harris, I.G. Wool, J.A. Lorraine (eds), Academic Press New York, 1968, p. 237.
95. K. Hartiala, *Physiol. Rev.*, **53**, 496 (1973).
96. J.D. Wilson, Metabolism of Testicular Androgens, *Handbook of Physiology: Section 7, Endocrinology*, Male Reproductive System Vol. **5**, R.O. Greep, E.B. Astwood (eds), American Physiology Society, Washington, 1974, p. 491.
97. M.I. Stylianou, E. Forchielli, NI. Tummillo, R.I. Dorfman, *J. Biol. Chem.*, **236**, 692 (1961).
98. D. Engelhardt, J. Eisenburg, P. Unterberger, H.J. Karl, *Klin. Wochenschr.*, **49**, 439 (1971).
99. B.P. Lisboa, I. Drosse, H. Breuer, *Z. Physiol. Chem.*, **342**, 123 (1965).
100. E. Chang, A. Mittelman, T.L. Dao, *J. Biol. Chem.*, **238**, 913 (1963).
101. K. Hartiala, W. Nienstedt, *Int. J. Biochem.*, **7**, 317 (1970).
102. A. Vermeulen, R. Rubens, L. Verdonck, *J. Clin. Endocrinol. Metab.*, **34**, 730 (1972).
103. M. Stanbrough, G.J. Bubley, K. Ross, T.R. Golub et al., *Cancer Res.* **66**, 2815 (2006).
104. F. Labrie, L. Cusan, J.L. Gomez, C. Martel et al., *J. Steroid Biochem. Mol. Biol.* **113**, 52 (2009).
105. H. Becker, J. Kaufmann, H. Klosterhalfen, K.D. Voigt, *Acta Endocrinol. (Copenhagen)*, **71**, 589 (1972).
106. H.J. Horst, M. Dennis, J, Kaufmann, K.D. Voigt, *Acta Endocrinol. (Copenhagen)*, **79**, 394 (1975).
107. A.E. Kellie, E.R. Smith, *Biochem. J.*, **66**, 490 (1957).
108. P. Robel, R. Emiliozzi, E. Baulieu, *J. Biol. Chem.*, **241**, 20 (1966).
109. N. Kundu, A.A. Sandberg, W.R. Slaunwhite, Jr, *Steroids*, **6**, 543 (1965).
110. T. El Attar, W. Dirscherl, K.O. Mosebach, *Acta Endocrinol. (Copenhagen)*, **45**, 527 (1964).
111. A. Kappas, R.H. Palmer, in *Methods in Hormone Research*, Vol. **4**, Part B, R.I. Dorfman (ed.), Academic Press, New York, 1965, p. 1.
112. L. Hellman, H.L. Bradlow, B. Zumoff, D.K. Fukushima et al., *J. Clin. Endocrinol.*, **19**, 936 (1959).
113. K. Nozu, B.I. Tamaski, *Biochim. Biophys. Acta*, **348**, 321 (1974).
114. K. Nozu, B.I. Tamaski, *Acta Endocrinol. (Copenhagen)*, **76**, 608 (1974).
115. N. Bruchovsky, J.D. Wilson, *J. Biol. Chem.*, **243**, 2012 (1968).
116. J. Shimazaki, N. Furuya, H. Yamanaka, K. Shida, *Endocrinol. Jpn.*, **16**, 163 (1969).
117. J. Shimazaki, I. Matsushita, N. Furuya, H. Yamanaka et al., *Endocrinol. Jpn.*, **16**, 453 (1969).
118. D.W. Frederiksen, J.D. Wilson, *J. Biol. Chem.*, **246**, 2584 (1971).
119. R.J. Moore, J.D. Wilson, *J. Biol. Chem.*, **247**, 958 (1972).
120. R.J. Moore, J.D. Wilson, *Endocrinology*, **93**, 581 (1973).
121. J.P. Karr, R.Y. Kirdani, G.P. Murphy, A.A. Sandberg, *Life Sci.*, **15**. 501 (1974).
122. D.C. Wilton, H.J. Ringold, *Third International Congress of Endocrinology*, Excerpta Medical Foundation, Amsterdam, 1968, p. 105.

123. R.W.S. Skiner, R.V. Pozderac, R.E. Counsell, P.A. Weinhold, *Steroids*, **25**, 189 (1975).
124. S. Andersson, R.W. Bishop, D.W. Russell, *J. Biol. Chem.*, **264**, 16249 (1989).
125. S. Anderson, D.M. Berman, E.P. Jenkins, D.W. Russell, *Nature*, **354**. 150 (1991).
126. D.W. Russell, D.M. Berman, J.T. Bryant, K.M. Cala et al., *Rec. Prog. Horm. Res.*, **49**, 275 (1994).
127. D.W. Russell, J.D. Wilson, *Ann. Rev. Biochem.*, **63**, 25 (1993).
128. J.D. Wilson, J.E. Griffin, D.W. Russell, *Endocr. Rev.*, **14**, 577 (1993).
129. K.J. Ryan, *J. Biol. Chem.*, **234**, 268 (1959).
130. E.A. Thompson, P.K. Siiteri, *J. Biol. Chem.*, **249**, 5373 (1974).
131. E.A. Thompson, P.K. Siiteri, *J. Biol. Chem.*, **249**, 5364 (1974).
132. A.S. Meyer, *Biochim. Biophys. Acta*, **17**, 441 (1955).
133. M. Akhtar, S.J.M. Skinner, *Biochem. J.*, **109**, 318 (1968).
134. S.J.M. Skinner, M. Akhtar, *Biochem. J.*, **114**, 75 (1969).
135. J.D. Townsley, H.J. Brodie, *Biochemistry*, **7**, 33 (1968).
136. H.J. Brodie, G. Possanza, J.D. Townsley, *Biochim. Biophys. Acta*, **152**, 770 (1968).
137. Y. Osawa, D.G. Spaeth, *Biochemistry*, **10**, 66 (1971).
138. J. Goto, J. Fishman, *Science*, **195**, 80 (1977).
139. E. Caspi, J. Wicha, T. Aninachalam, P. Nelson et al., *J. Am. Chem. Soc.*, **106**, 7282 (1984).
140. D. Arigoni, R. Battaglia, M. Akhtar, T. Smith, *J. Chem. Soc., Chem. Commun.*, **185**, (1975).
141. M. Akhtar, M.R. Calder, D.L. Corina, J.N. Wright, *J. Chem. Soc., Chem. Commun.*, **129** (1981).
142. M. Akhtar, M.R. Calder, D.L. Corina, J.N. Wright, *Biochem. J.*, **201**, 569 (1982).
143. P.A. Cole, C.H. Robinson, *J. Am. Chem. Soc.*, **110**, 1284 (1988).
144. M. Akhtar, V.C.O. Njar, J.N. Wright, *J. Steriod Biochem. Mol. Biol.*, **44**, 375 (1993).
145. S.S. Oh, C.H. Robinson, *J. Steriod Biochem. Mol. Biol.*, **44**, 389 (1993).
146. J.C. Hackett, R.W. Brueggemeier, C.M. Haddad, *J. Am. Chem. Soc.*, **127**, 5224–5237 (2005).
147. K.J. Ryan, *Recent Prog. Horm. Res.*, **21**, 367 (1965).
148. C. Gual, T. Morato, M. Hayano, M. Gut et al., *Endocrinology*, **71**, 920 (1962).
149. W.C. Schwarzel, W. G. Kruggel, H.J. Brodie, *Endocrinology*, **92**, 866 (1973).
150. P.K. Siiteri, E.A. Thompson, *J. Steroid Biochem.*, **6**, 317 (1975).
151. F.L. Bellino, S.S. H. Gilani, S.S. Eng, Y. Osawa et al., *Biochemistry*, **15**, 4730 (1976).
152. J.K. Kellis, L.E. Vickery, *J. Biol. Chem.*, **262**, 4413 (1987).
153. D. Ghosh, J, Griswold, M. Erman, W. Pangborn, *Nature*, **457**, 219–223 (2009).
154. C.J. Corbin, S. Grahan-Lorence, M. McPhaul, J.I. Mason et al., *Proc. Natl Acad. Sci. USA*, **85**, 8948 (1988).
155. M.S. Mahendroo, C.R. Mendelson, E.R. Simpson, *J. Biol. Chem.*, **268**, 19463 (1993).
156. E.R. Simpson, M.S. Mahendroo, G.D. Means, M.W. Kilgore et al., *Endocr. Rev.*, **15**, 342 (1994).
157. S. Chen, M.J. Beshna, R.S. Sparkes, S. Zollman et al., *DNA*, **7**, 27 (1988).
158. M. Steinkampf, C.R. Mendelson, E.R. Simpson, *Mol. Endocrinol.*, **1**, 465 (1987).
159. C.T. Evans, J.C. Merrill, C.J. Corbin, D. Saunders et al., *J. Biol. Chem.*, **269**, 6914 (1987).
160. E.R. Simpson, J.C. Merrill, A.J. Hollub, S. Grahan-Lorence et al., *Endocr. Rev.*, **10**, 136 (1989).
161. E.B. Bulun, E.R. Simpson, *Breast Cancer Res. Treat.*, **30**, 19–29 (1994).
162. L.S. Sholiton, R.T. Mornell, E.E. Werk, *Steroids*, **8**, 265 (1966).
163. R.B. Jaffe, *Steroids*, **14**, 483 (1969).
164. L.S. Sholiton, E.E. Werk. *Acta Endocrinol. (Copenhagen)*, **61**, 641 (1969).
165. L.S. Sholiton, I.L. Hall, E.E. Werk, *Acta Endocrinol. (Copenhagen)*, **63**, 512 (1970).
166. J.M. Stern, A.J. Eisenfeld, *Endocrinology*, **88**, 1117 (1971).
167. R. Massa, E. Stupnicka, Z. Kniewald, L. Martini, *J. Steroid Biochem.*, **3**, 385 (1972).
168. E.D. Lephart, S. Andersson, E.R. Simpson, *Endocrinology*, **127**, 1121 (1990).
169. F. Naftolin, K.J. Ryan, Z. Petro, *J. Clin. Endocrinol. Metab.*, **33**, 368 (1971).
170. F. Naftolin, K.J. Ryan, Z. Petro, *Endocrinology*, **90**, 295 (1972).
171. F. Flores, F. Naftolin, K.J. Ryan, *Neuroendocrinology*, **11**, 177 (1973).
172. F. Flores, F. Naftolin, K.J. Ryan, R.J. White, *Science*, **180**, 1074 (1973).

173. J.A. Canick, D.E. Vaccaro, K.J. Ryan, S.E. Leeman, *Endocrinology*, **100**, 250 (1977).
174. E.D. Lephart, E.R. Simpson, M.J. McPhaul, M.W. Kilgore et al., *Mol. Brain Res.*, **16**, 187 (1992).
175. G. Perez-Palacios, K. Larsson, C. Beyer, *J. Steroid Biochem.*, **6**, 999 (1975).
176. E.D. Lephart, *Mol. Cell. Neurosci.*, **4**, 473 (1993).
177. B.W. O'Malley, A.R. Means, *Receptors for Reproductive Hormones*, Plenum Press. New York, 1974.
178. R.J.B. King, W.I.P. Mainwaring, *Steroid-Cell Interactions*, University Park Press, Baltimore, 1974.
179. S. Liao, *Int. Rev. Cytol.*, **41**, 87 (1975).
180. H.G. Williams-Ashman, A.H. Reddi, Androgenic Regulation of Tissue Growth and Function in *Biochemical Actions of Hormones*, Vol. **2**, G. Litwack (ed.), Academic Press, New York, 1972, p. 257.
181. L. Chan, B.W. O'Malley, *N. Engl. J. Med.*, **294**, 1322, 1372, 1430 (1976).
182. J.-A. Gustaffson, J. Carlstedt-Duke, L. Poellinger, S. Okret et al., *Endocr. Rev.*, **8**, 185 (1987).
183. R.M. Evans, *Science*, **240**, 889–895 (1988).
184. G. Ringold (ed.), *Steroid Hormone Action*, Alan R. Liss, New York, 1988.
185. M. Beato, *Cell*, **56**, 335 (1989).
186. B. O'Malley, *Mol. Endocrinol.*, **4**, 363 (1990)
187. M.A. Carson-Jurica, W.T. Schrader, B. O'Malley, *Endocr. Rev.*, **11**, 201 (1990).
188. D.J. Mangelsdorf, C. Thummel, M. Beato, P. Herrlich et al., *Cell*, **83**, 835–839 (1995).
189. J.M. Kokontis, S. Liao, *Vitam. Horm.*, **55**, 219–307 (1999).
190. A.K. Roy, Y. Lavrovsky, C.S. Song, S. Chen et al., *Vitam. Horm.*, **55**, 309–352 (1999).
191. E.J. Jensen, H.I. Jacobson, *Rec. Progr. Hormone Res.*, **18**, 387 (1962).
192. W.H. Pearlman, M.R.I. Pearlman, *J. Biol. Chem.*, **236**, 1321 (1961).
193. B.W. Harding, L.T. Samuels, *Endocrinology*, **70**, 109 (1962).
194. K.J. Tveter, A. Attramadal, *Acta Endocrinol. (Copenhagen)*, **59**, 218 (1968).
195. W.I.P. Mainwaring, *J. Endocrinol.*, **44**, 323 (1969).
196. S. Fang, S. Liao, *J. Biol. Chem.*, **246**, 16 (1971).
197. H.V. Heemers, D.J. Tindall, *Endocr. Rev.*, **28**, 778–808 (2007).
198. Z.-X. Zhou, C.-L. Wong, M. Sar, E.M. Wilson, *Rec. Prog. Horm. Res.*, **49**, 249 (1994).
199. T. Brown, Androgen Receptor Structure, Function, Regulation, and Dysfunction, in S. Bhasin, H.L. Gabelnick, J.M. Spieler, R.S. Swerdloff, C. Wang, C. Kelly (eds), *Pharmacology, Biology, and Clinical Applications of Androgens*, Wiley-Liss, New York, 1996, p. 45.
200. P.M. Matias, P. Donner, R. Coelho, M. Thomaz, et al., *J. Biol. Chem.*, **275**, 26164–26171 (2000).
201. J.S. Sack, K.F. Kish, C. Wang, R.M. Attar et al., *Proc. Natl Acad. Sci. USA*, **98**, 4904–4909 (2001).
202. (a)C.E. Bohl, D.D Miller, J. Chen, C.E. Bell et al., *J. Biol. Chem.* **280**, 37747 (2005); (b)C.E Bohl, W. Gao, D.D. Miller, C.E. Bell et al., *Proc. Natl Acad. Sci. USA*, **102**, 6201 (2005)
203. M.J. McPhaul, M. Marcelli, W.D. Tilley, J.E. Griffin et al., *FASEB J.*, **5**, 2910 (1991).
204. A.O. Brinkmann, J. Trapman, *Cancer Surv.*, **14**, 95 (1992).
205. M.J. McPhaul, *J. Steroid Biochem. Mol. Biol.*, **69**, 315–322 (1999).
206. C. Sonnenschein, N. Olea, M.E. Pasanen, A.M. Soto, *Cancer Res.*, **49**, 3474 (1989).
207. G. Wilding, M. Chen, E.P. Gelmann, *Prostate*, **14**, 103 (1989).
208. S. Harris, M.A. Harris, Z. Rong, in *Molecular and Cellular Biology of Prostate Cancer*, J.P. Karr, D.S. Coffey, R.G. Smith, D.J. Tindall (eds), Plenum Press, New York, 1991, p.315.
209. P.H. Riegman, R.J. Vlietstra, J.A. van der Korput, A.O. Brinkmann et al., *Mol. Endocrinol.*, **5**, 1921–1930 (1991).
210. K.B. Cleutjens, C.C. van Eekelen, H.A. van der Korput, A.O. Brinkmann et al., *J. Biol. Chem.*, **271**, 6379–6388 (1996).
211. K.B. Cleutjens, H.A. van der Korput, C.C. van Eekelen, H.C. van Rooij et al., *Mol. Endocrinol.*, **11**, 148–161 (1997).
212. Y. Matuo, P.S. Adams, N. Nishi, H. Yasumitsu et al., *In Vitro Cell. Dev. Biol.*, **25**, 581–584 (1989).
213. A. M. Spence, P. C. Sheppard, J. R. Davie, Y. Matuo et al., *Proc. Natl Acad. Sci. USA*, **86**, 7843–7847 (1989).
214. Y.-W. Chen, M.-S. Lee, A. Lucht, F.-P. Chou et al., *Am. J. Pathol.*, **176**, 2986–2996 (2010).

215. T. Liang, G. Mezzetti, C. Chen, S. Liao, *Biochim. Biophys. Acta*, **542**, 430–441 (1978).
216. G. Yan, Y. Fukabori, S. Nikolaropoulos, F. Wang et al., *Mol. Endocrinol.*, **6**, 2123–2128 (1992).
217. A. Tanaka, K. Miyamoto, N. Minamino, M. Takeda et al., *Proc. Natl Acad. Sci. USA*, **89**, 8928–8932 (1992).
218. M. Koga, S. Kasayama, K. Matsumoto, B. Sato, *J. Steroid Biochem. Mol. Biol.*, **54**, 1–6 (1995).
219. I. Schenkein, M. Levy, E.D. Bueker, *Endocrinology*, **94**, 840 (1974).
220. P.L. Barthe, L.P. Bullock, I. Mowszowicz, *Endocrinology*, **95**, 1019 (1974).
221. D.A. Wolf, P. Schulz, F. Fittler, *Br. J. Cancer*, **64**, 47–53 (1991).
222. M.F. Lyon, I. Hendry, R.V. Short, *Endocrinology*, **58**, 357 (1973).
223. C.W. Bardin, L.P. Bullock, R.J. Sherins, *Rec. Progr. Hormone Res.*, **29**, 65 (1973).
224. M. Kumar, A.K. Roy, A.E. Axelrod, *Nature*, **223**, 399 (1969).
225. A.K. Roy, *Endocrinology*, **92**, 957 (1973).
226. C. Chang, T.-M. Lin, P. Hsiao, C. Su et al., Androgen-Responsive Genes, in S. Bhasin, H.L. Gabelnick, J.M. Spieler, R.S. Swerdloff et al. (eds), *Pharmacology, Biology, and Clinical Applications of Androgens*, Wiley-Liss, New York, 1996, p. 45.
227. K.J. Tveter, O. Unhjem, *Endocrinology*, **84**, 963 (1969).
228. J.M. Stern, A.J. Eisenfield, *Science*, **166**, 233 (1969).
229. K. Adachi, M. Kano, *Steroids*, **19**, 567 (1972).
230. W.I.P. Mainwaring, F.R. Mangan, *J. Endocrinol.*, **59**, 121 (1973).
231. S. Takayasu, K. Adachi, *Endocrinology*, **96**, 525 (1975).
232. V. Hansson, Reusch E, Trygstad O, Torgersen O, et al., *Steroids*, **23**, 823 (1974).
233. D.J. Tindall, F.S. French, S.N. Nayfeh, *Biochem. Biophys. Res. Commun.*, **49**, 1391 (1973).
234. J.A. Blaquier, R.S. Calandra, *Endocrinology*, **93**, 51 (1973).
235. E. M. Ritzén, S.N. Nayfeh, F.S. French, P.A. Aronin, *Endocrinology*, **91**, 116 (1972).
236. J.F. Dunn, J.L. Goldstein, J.D. Wilson, *J. Biol. Chem.*, **248**, 7819 (1973).
237. G. Verhoeven, J.D. Wilson, *Endocrinology*, **99**, 79 (1976).
238. P. Jouan, S. Samperez, M.L. Thieulant, L. Mercier, *J. Steroid Biochem.*, **2**, 223 (1971).
239. P. Jouan, S. Samperez, M.L. Thielant, *J. Steroid Biochem.*, **4**, 65 (1973).
240. M. Sar, W.E. Stumpf, *Endocrinology*, **92**, 251 (1973).
241. D.P. Cardinali, C.A. Nagle, J.M. Rosner, *Endocrinology*, **95**, 179 (1974).
242. J. Kato, *J. Steroid Biochem.*, **6**, 979 (1975).
243. O. Naess, V. Hansson, O. Djoseland, A. Attramadal, *Endocrinology*, **97**, 1355 (1975).
244. T.O. Fox. *Proc. Natl Acad. Sci. USA*, **72**, 4303 (1975).
245. L. Valladares, J. Mingell. *Steroids*, **25**, 13 (1975).
246. J. Mingell, L. Valladares, *J. Steroid Biochem.*, **5**, 649 (1974).
247. A.K. Roy, B.S. Milin, D.M. McMinn, *Biochim. Biophys. Acta*, **354**, 213 (1974).
248. N. Bruchovsky, J.W. Meakin, *Cancer Res.*, **33**, 1689 (1973).
249. N. Bruchovsky, D.J.A. Sutherland, J.W. Meakin, T. Minesita, *Biochim. Biophys. Acta*, **381**, 61 (1975).
250. E.M. Wilson, F.S. French, *J. Biol. Chem.*, **251**, 5620 (1976).
251. C.R. Evans, C.G. Pierrepoint. *J. Endocrinol.*, **64**, 539 (1975).
252. K.B. Eik-Nes, *Vitam. Horm.*, **33**, 193 (1975).
253. R.W. Glovna, J.D. Wilson, *J. Clin. Endocrinol. Metab.*, **29**, 970 (1969).
254. V. Hansson, K.J. Tveter, O. Unhjem, O. Djoseland, *J. Steroid Biochem.*, **3**, 427 (1972).
255. I. Jung, E.E. Baulieu, *Nat. New Biol.*, **237**, 24 (1972).
256. M.G. Michel, E.E. Baulieu, *C.R. Acad. Sci. Paris*, **279**, 421 (1974).
257. S.R. Max, S. Mufti , B.M. Carlson, *Biochem. J.*, **200**, 77 (1981).
258. M. Krieg, K.D. Voigt, *J. Steroid Biochem.*, **7**, 1005 (1976).
259. S.R. Max, *J. Steroid Biochem.*, **18**, 281 (1983).
260. H.C. McGill, V.C. Anselmo, J.M. Buchanan, P.J. Sheridian, *Science*, **207**, 775 (1980).
261. E. Dahbberg, *Biochim. Biophys. Acta*, **717**, 65 (1982).
262. R. Hickson, T. Galessi, T. Kurowski, D. Daniels et al., *J. Steroid Biochem.*, **19**, 1705 (1983).
263. W.H. Walker, *Steroids*, **74**, 602 (2009).

264. L.B. Lutz, M. Jamnongjit, W.H. Yang, D. Jahani et al., *Mol. Endocrinol.*, **17**, 1106 (2003).
265. M. Estrada, A. Espinosa, M. Muller, E. Jaimovich, *Endocrinology*, **144**, 3586 (2003).
266. E. Unni, S. Sun, B. Nan, M. J. McPhaul et al., *Cancer Res.*, **64**, 7156 (2004).
267. W.D. Odell, R.S. Swerdloff, *Clin. Endocrinol.*, **8**, 149, (1978).
268. J.E. Griffin, J.D. Wilson, *N. Engl. J. Med.*, **302**, 198, (1980).
269. J.E. Griffin, K. Punyashthiti, J.D. Wilson, *J. Clin. Invest.*, **57**, 1342, (1976).
270. M. Kaufman, C. Straisfeld, L. Pinsky, *J. Clin. Invest.*, **58**, 345, (1976).
271. J.E. Griffin, *J. Clin. Invest.*, **64**, 1624, (1979).
272. P.C. Walsh, J.D. Madden, M.J. Harrod, J.L. Goldstein et al., *N. Engl. J. Med.*, **291**, 944, (1974).
273. J. Imperato-McGinley, L. Guerrero, T. Gautier, R.E. Peterson, *Science*, **186**, 1213, (1974)
274. A. Segaloff, *Rec. Prog. Horm. Res.*, **22**, 351, (1966).
275. R.I. Dorfman, in *Methods in Hormone Research*, Vol. **2**, A. Dorfman (ed.), Academic Press, New York, 1962, p. 275.
276. T.F. Gallagher, F.C. Koch, *J. Pharmacol. Exp. Ther.*, **55**, 97 (1935).
277. A.W. Greenwood, J.S.S. Blyth, R.K. Callow, *Biochem. J.*, **29**, 1400 (1935).
278. C.W. Emmens, *Med. Res. Council. Spec. Rep. Ser.*, **234**, 1 (1939).
279. D.R. McCullagh, W.K. Cuyler, *J. Pharmacol. Exp. Ther.*, **66**, 379 (1939).
280. A. Segaloff, *Steroids*, **1**, 299 (1963).
281. E. Eisenberg, G.S. Gordan, *J. Pharmacol. Exp. Ther.*, **99**, 38 (1950).
282. F.J. Saunders, V.A. Drill, *Proc. Soc. Exp. Biol. Med.*, **94**, 646 (1957).
283. L.G. Hershberger, E.G. Shipley, R.K. Meyer, *Proc. Soc. Exp. Biol. Med.*, **83**, 175 (1953).
284. S. Liao, T. Liang, S. Fang, E. Casteneda et al., *J. Biol. Chem.*, **248**, 6154, (1973).
285. C.D. Kochakian. *Rec. Progr. Hormone Res.*, **1**, 177 (1948).
286. C. Huggins, E.V. Jensen, *J. Exp. Med.*, **100**, 241 (1954).
287. K. Junkmann, *Rec. Progr. Hormone Res.*, **13**, 389 (1957).
288. R.E. Counsell, P.D. Klimstra, F.B. Colton, *J. Org. Chem.*, **27**, 248 (1962).
289. J.D. Wilson, R.E. Gloyna, *Rec. Progr. Hormone Res.*, **26**, 309 (1970).
290. F.J. Zeller, *J. Reprod. Fertil.*, **25**, 125 (1971).
291. I.H. Harris, *J. Clin. Endocrinol. Metab.*, **21**, 1099 (1961).
292. C. Huggins, E.V. Jensen, A.S. Cleveland, *J. Exp. Med.*, **100**, 225 (1954).
293. R.B. Gabbard, A. Segaloff, *J. Org. Chem.*, **27**, 655 (1962).
294. V.A. Drill, B. Riegel, *Rec. Progr. Hormone Res.*, **14**, 29 (1958).
295. J.W. Partridge, L. Boling, L. DeWind, S. Margen et al., *J. Clin. Endocrinol. Metab.*, **13**, 189 (1953).
296. K. Miescher, E. Tschapp, A. Wettstein, *Biochem. J.*, **30**, 1977 (1976).
297. G. Sala, G. Baldratti, *Proc. Soc. Exp. Biol. Med.*, **95**, 22 (1957).
298. S.C. Lyster, G.H. Lund, R.O. Stafford, *Endocrinology*, **58**, 781 (1956).
299. R.M. Backle, *Br. Med. J.*, **1**, 1378 (1959).
300. T.H. McGavack, W. Seegers, *Am. J. Med. Sci.*, **235**, 125 (1958).
301. G.H. Marquardt, C.I. Fisher, P. Levy, R.M. Dowben, *J. Am. Med. Assoc.*, **175** 851 (1961).
302. B.J. Kennedy, *N. Engl. J. Med.*, **259**, 673 (1958).
303. H. Nowakowski, *Deut. Bed. Wochenschr.*, **90**, 2291 (1965).
304. G.S. Gordan, S. Wessler, L.V. Avioli. *J. Am. Med. Assoc.*, **219**, 483 (1972).
305. I.S. Goldenberg, N. Waters, R.S. Randin, F.J. Ansfield et al., *J. Am. Med. Assoc.*, **223**, 1267 (1973).
306. R. Rosso, G. Porcile, F. Brema, *Cancer Chemother. Rep.*, **59**, 890, (1975).
307. A. Segaloff, R. Bruce Gabbard, *Endocrinology*, **67**, 887 (1960).
308. R.E. Counsell. *J. Med. Chem.*, **9**, 263 (1966).
309. A. Bowers, A.D. Cross, J.A. Edwards, H. Carpio et al., *J. Med. Chem.*, **6**, 156 (1963).
310. F.A. Kincl, R.I. Dorfman, *Steroids*, **3**, 109 (1964).
311. M.E. Wolff, G. Zanati, *J. Med. Chem.*, **12**, 629 (1969).
312. G. Zanati, M.E. Wolff, *J. Med. Chem.*, **14**, 958 (1971).
313. W.L. Duax, M.G. Erman, J.F. Griffin, M.E. Wolff, *Cryst. Struct. Commun.*, **5**, 775 (1976).
314. G. Zanati, M.E. Wolff, *J. Med. Chem.*, **16**, 90 (1973).
315. A. Segaloff, R.B. Gabbard, *Steroids*, **1**, 77 (1963).

316. A. Segaloff, R.B. Gabbard, *Steroids*, **22**, 99 (1973).
317. S.A. Shain, R.W. Boesel, *J. Steroid Biochem.*, **6**, 43 (1975).
318. A. Negro-Vilar, *J. Clin. Endocrinol. Metab.*, **84**, 3459 (1999).
319. L. Zhi, E. Martinborough, *Annu. Rep. Med. Chem.*, **36**, 169 (2001).
320. M. L. Mohler, C.E. Bohl, A. Jones, C.C. Coss et al., *J. Med. Chem.*, **52**, 3597 (2009).
321. J.T. Dalton, A. Mukherjee, Z. Zhu, L. Kirkovsky et al., *Biochem. Biophys. Res. Commun.*, **244**, 1 (1998).
322. (a)D. Yin, H. Xu, Y. He, L.I. Kirkovsky et al., *J. Pharmacol. Exp. Ther.*, **304**, 1323 (2003);(b)D. Yin, W. Gao, J.D. Kearbey, H. Xu et al., *J. Pharmacol. Exp. Ther.*, **304**, 1334 (2003).
323. C.A. Marhefka, W. Gao, K. Chung, J. Kim et al., *J. Med. Chem.*, **47**, 993 (2004).
324. (a)W. Gao, P.J. Reiser, C.C. Coss, M.A. Phelps et al., *Endocrinology* **146**, 4887 (2005);(b)W. Gao, J.D. Kearbey, V.A. Nair, K. Chung et al., *Endocrinology* **145**, 5420 (2004)
325. E. Martinborough, Y. Shen, A. Oeveren, Y.O. Long et al., *J. Med. Chem.* **50**, 5049 (2007).
326. E.G. Vajda, F.J Lopez, P. Rix, R. Hill et al., *J. Pharmacol. Exp. Ther.* **328**, 663 (2009).
327. C. Sun, J.A. Robl, T.C. Wang, Y. Huang et al., *J. Med. Chem.* **49**, 7596 (2006).
328. X. Zhang, G.F. Allan, T. Sbriscia, O. Linton et al., *Bioorg. Med. Chem. Lett.* **16**, 5763 (2006).
329. L.L. Engel, J. Alexander, M. Wheeler, *J. Biol. Chem.*, **231**, 159 (1958).
330. A. Segaloff, B. Gabbard, B.T. Carriere, E.L. Rongone, *Steroids*, 5 (Suppl. I), 419 (1965).
331. E. Castegnaro, G. Sala, *Folia Endocrinol.*, **14**, 581 (1961).
332. H.J. Ringold, J. Graves, M. Hayano, H. Lawrence, Jr, *Biochem. Biophys. Res. Commun.*, **13**, 162 (1963).
333. H.A. Plantier, *N. Engl. J. Med.*, **270**, 141 (1964).
334. H. Hortling, K. Malmio, L. Husi-Brummer, *Acta Endocrinol. (Copenhagen)*, Suppl. 39, 132 (1962).
335. A.A. de Lorimier, G.S. Gordan, R.C. Lowe, J.V. Carbone, *Arch. Intern. Med.*, **116**, 289 (1965).
336. I.M. Arias, in *Influence of Growth Hormone, Anabolic Steroids, and Nutrition iii Health and Disease*, F. Gross (ed.), Springer-Verlag, Berlin, 1962, p. 434.
337. H.A. Kaupp, F.W. Preston, *J. Am. Med. Assoc.*, **180**, 411 (1962).
338. D. Westaby, S.J. Ogle, F.J. Paradinas, J.B. Randell et al., *Lancet*, **2**, 261, (1977).
339. R.L. Landau, The Metabolic Effects of Anabolic Steroids in Man, in *Androgenic-Anabolic Steroids*, C.D. Kochakian (ed.), Springer-Verlag, New York, 1976, p. 45.
340. R.O. Stafford, B.J. Bowman, K.J. Olson, *Proc. Soc. Exp. Biol. Med.*, **86**, 322 (1954).
341. E. Henderson, M. Weinberg, *J. Clin. Endocrinol.*, **11**, 641 (1951).
342. J.C. Stucki, A.D. Forbes, J.I. Northam, J.J. Clark, *Endocrinology*, **66**, 585 (1960).
343. A.D. Mooradian, J.E. Morley, S.G. Korenman, *Endocr. Rev.*, **8**, 1 (1987).
344. H. Kopera, Miscellaneous Uses of Anabolic Steroids, in *Anabolic-Androgenic Steroids*, C.D. Kochakian (ed.), Springer-Verlag, New York, 1976, p. 535.
345. C. Huseman, A. Johanson, *J. Pediatr.*, **87**, 946 (1975).
346. L. Tec, *Am. J. Psychiatr.*, **127**, 1702 (1971).
347. P.M. Sansoy, R.A. Naylor, L.M. Shields, *Geriatrics*, **26**, 139 (1971).
348. B.O. Morrison, *J. Mich. St. Med. Soc.*, **60**, 723 (1961).
349. A.L. Kolodny, *Med. Tms. (N.Y.)*, **91**, 9 (1963).
350. H. Buchner, *Wien. Med. Wschr.*, **111**, 576 (1961).
351. R.M. Konrad, U. Ammedick, W. Hupfauer, W. Ringler, *Chirung*, **38**, 168 (1967).
352. U. Ammedick, R.M. Konrad, E. Gotzen, *Med. Ernähr.*, **9**, 121 (1968).
353. D.E.F. Tweedle, C. Walton, I.D.A. Johnston, *Br. J. Surg.*, **59**, 300 (1972).
354. A.A. Renzi, J.J. Chart, *Proc. Soc. Exp. Biol. Med.*, **110**, 259 (1962).
355. R.P. Howard, R.H. Furman, *J. Clin. Endocrinol.*, **22**, 43 (1962).
356. J.F. Dingman, W.H. Jenkins, *Metabolism*, **11**, 273 (1962).
357. S. Weisenfeld, S. Akgun, S. Newhouse, *Diabetes*, **12**, 375 (1963).
358. C.J. Glueck, *Clin. Res.*, **17**, 475 (1971).
359. C.J. Glueck, *Metabolism*, **20**, 691 (1971).
360. A.E. Doyle, N.B. Pinkus, J. Green, *Med. J. Aust.*, **1**, 127 (1974).
361. B.A. Sachs, L. Wolfman, *Metabolism*, **17**, 400 (1968).

362. C.J. Glueck, S. Ford, P. Steiner, R. Fallat, *Metabolism*, **17**, 807 (1973).
363. N.T. Shahidi, *N. Engl. J. Med.*, **289**, 72 (1973).
364. A. Killander, K. Lundmark, S. Sjolin, *Acta Paediatr. Scand.*, **58**, 10 (1969).
365. R.F. Branda, T.W. Amsden, H.S. Jacob, *Clin. Res.*, **22**, 607A (1974).
366. D.W. Hughes, *Med. J. Aust.*, **2**, 361 (1973).
367. E.D. Hendler, J.A. Goffinet, S. Ross, R.E. Longnecker et al., *N. Engl. J. Med.*, **291**, 1046 (1974).
368. A. Blumbert, H. Keller, *Schweiz. Med. Wochenschr.*, **101**, 1887 (1971).
369. W. Fried, O. Jonasson, G. Lang, F. Schwartz, *Ann. Intern. Med.*, **79**, 823 (1973).
370. J. Keyssner, C. Hauswaldt, N. Uhl, W. Hunstein, *Schweiz. Med. Wochenschr.*, **104**, 1938 (1974).
371. F.A. Kincl, *Methods in Hormone Research*, Vol. **4**, R.I. Dorfman (ed.), Academic Press. New York, 1965, p. 21.
372. G.O. Potts, A. Arnold, A.L. Beyler, *Endocrinology*, **67**, 849 (1960).
373. M.E. Nimni, E. Geiger, *Endocrinology*, **61**, 753 (1957).
374. J.N. Goldman, J.A. Epstein, H.S. Kupperman, *Endocrinology*, **61**, 166 (1957).
375. F.A. Kincl, H.J. Ringold, R.I. Dorfman, *Acta Endocrinol.*, **36**, 83 (1961).
376. M.E. Nimni, E. Geiger, *Proc. Soc. Exp. Biol. Med.*, **94**, 606 (1957).
377. R.O. Scow, *Endocrinology*, **51**, 42 (1952).
378. J. Leibetseder, K. Steininger, *Arzneim.-Forsch.* **15**, 474 (1965).
379. R.A. Edgren, *Acta Endocrinol. (Copenhagen)*, **44** Suppl. 87, 3 (1963).
380. F.J. Saunders, V.A. Drill, *Endocrinology*, **58**, 567 (1956).
381. L.E. Barnes, R.O. Stafford, M.E. Guild, L.C. Thole et al., *Endocrinology*, **55**, 77 (1954).
382. C. Djerassi, L. Miramontes, G. Rosenkranz, F. Sondheimer, *J. Am. Chem. Soc.*, **76**, 4092 (1954).
383. F.B. Colton, L.N. Nysted, B. Reigel, A.L. Raymond, *J. Am. Chem. Soc.*, **79**, 1123 (1957).
384. E.B. Feldman, A.C. Carter, *J. Clin. Endocrinol. Metab.*, **20**, 842 (1960).
385. J. Ferrin, *Acta Endocrinol. (Copenhagen)*, **22**, 303 (1956).
386. K.V. Yorka, W.L. Truett, W.S. Johnson, *J. Org. Chem.*, **27**, 4580 (1962).

387. W.F. Johns, *J. Am. Chem. Soc.*, **80**, 6456 (1958).
388. H.J. Ringold, *J. Am. Chem. Soc.*, **82**, 961 (1960).
389. A. Zaffaroni, *Acta Endocrinol. (Copenhagen)*, **34** Suppl. 50, 139 (1960).
390. I.W. Goldberg, J. Sicé, H. Robert, Pl.A. Plattner, *Helv. Chim. Acta*, **30**, 1441 (1947).
391. H. Heusser, P.T. Herzig, A. Furst, Pl.A. Plattner. *Helv. Chim. Acta*, **33**, 1093 (1950).
392. G. Eadon, C. Djerassi, *J. Med. Chem.*, **89**, (1972).
393. G. Sala, G. Baldratti, R. Ronchi, V. Clini et al., *Sperimentale*, **106**, 490 (1956).
394. G.S. Gordan, *Arch. Intern. Med.*, **100**, 744 (1957).
395. A. Arnold, G.O. Potts, A.L. Beyler, *Endocrinology*, **72**, 408 (1963).
396. P.A. Desaulles, *Helv. Med. Acta*, **27**, 479 (1960).
397. R.I. Dorfman, F.A. Kincl, *Endocrinology*, **72**, 259 (1963).
398. G.W. Liddle, H.A. Burke, Jr, *Helv. Med. Acta*, **27**, 504 (1960).
399. A.L. Beyler, G.O. Potts, A. Arnold, *Endocrinology*, **68**, 987 (1961).
400. Colton, F.B. (1959) 6-Dehydro-17-alkyl-19-nortestosterones, US Patent 2,874,170.
401. A. Smit, P. Westerhof, *Rec. Trav. Chim. Pays-Bas*, **82**, 1107 (1963).
402. R. Van Moorselaar, S.J. Halkes, E. Havinga, *Rec. Trav. Chim. Pays-Bas*, **84**, 841 (1965).
403. J.S. Baran, *J. Med. Chem.*, **6**, 329 (1963).
404. R.A. Edgren, D.L. Peterson, R.C. Jones, C.L. Nagra et al., *Rec. Progr. Hormone Res.*, **22**, 305 (1966).
405. J. Tremolieres, E. Pequignot, *Presse Med.*, **73**, 2655 (1965).
406. H.L. Kruskemper, G. Noell, *Steroids*, **8**, 13 (1966).
407. H.L. Kruskemper, K.D. Moraner, G. Noell, *Arzneim.-Forsch.*, **17**, 449 (1967).
408. A. Halden, R.M. Watter, G.S. Gordan, *Cancer Chemother. Rep.*, **54**, 453 (1970).
409. G.K. Suchowsky, K. Junkmann, *Acta Endocrinol. (Copenhagen)*, **39**, 68 (1962).
410. B. Pelc, *Collect. Czech. Chem. Commun.*, **29**, 3089 (1964).
411. Weller, O *Endokrinologie*, **42**, 34 (1962).
412. H. Langecker. *Arzneim.-Forsch.*, **12**, 231 (1962)
413. O. Weller, *Endokrinologie*, **41**, 60 (1961).

414. H.L. Kruskemper, H. Breuer, Studies on the anabolic effect and metabolism of 1-methyl-1-androstene-17β-ol-3-one, in International Congress on Hormonal Steroids, Excerpta Medica International Congress Series, No. 51, p. 209 (1962).
415. F. Neumann, R. Wiechert, M. Kramer, G. Raspe, *Arzneim.-Forsch.*, **16** 455 (1966).
416. O. Weller, *Arzneim.-Forsch.*, **16** 465 (1966).
417. B. Pelc, J. Jodkova, *Collect. Czech. Chem. Commun.*, **30**, 3575 (1965).
418. B. Pelc, *Collect. Czech. Chem. Commun.*, **30**, 3408 (1965).
419. C. Djerassi, R. Riniker, B. Riniker, *J. Am. Chem. Soc.*, **78**, 6377 (1956).
420. A. Bowers, H.J. Ringold, E. Denot, *J. Am. Chem. Soc.*, **80**, 6115 (1958).
421. O. Abe, H. Herraneu, R.I. Dorfman, *Proc. Soc. Exp. Biol. Med.*, **111**, 706 (1962).
422. D. Berkowitz, *Clin. Res.*, **8**, 199 (1960).
423. R.E. Counsell, P.D. Klimstra. *J. Med. Chem.*, **6**, 736 (1963).
424. R.I. Dorfman, A.S. Dorfman, *Acta Endocrinol. (Copenhagen)*, **42**, 245 (1963).
425. A. Arnold, G.O. Potts, A.L. Beyler, *J. Endocrinol.*, **28**, 87 (1963).
426. D.R. Korst, C.Y. Bowers, J.H. Flokstra, F.G. McMahon, *Clin. Pharmacol. Ther.*, **4**, 734 (1963).
427. J.A. Campbell, S.C. Lyster, G.W. Duncan, J.C. Babcock, *Steroids*, **1**, 317 (1963).
428. R.A. Edgren, H. Smith, G.A. Hughes, *Steroids*, **2**, 731 (1963).
429. R.B. Greenblatt, E.C. Jungck, G.C. King, *Am. J. Med. Sci.*, **318**, 99 (1964).
430. G. Baldratti, G. Arcari, V. Clini, F. Tani et al., *Sperimentale*, **109**, 383 (1959).
431. G. Sala, A. Cesana, G. Fedriga, *Minerva Med.*, **51**, 1295 (1960).
432. A.A. Albanese, E.J. Lorenze, L.A. Orto, *N. Y. State J. Med.*, **63**, 80 (1963).
433. G. Sala, *Helv. Med. Acta*, **27**, 519 (1960).
434. R.M. Myerson, *Am. J. Med. Sci.*, **241**, 732 (1961).
435. W.W. Glas, E.H. Lansing, *J. Am. Geriatr. Soc.*, **10**, 509 (1962).
436. H.G. Kraft, H. Kieser, *Arzneim.-Forsch.*, **14**, 330 (1964).
437. H.L. Kruskemper, *Arzneim.-Forsch.*, **16**, 608 (1966).
438. R.E. Schaub, M.J. Weiss, *J. Org. Chem.*, **26**, 3915 (1961).
439. H. Kaneko, K. Nakamura, Y. Yamato, M. Kurakawa, *Chem. Pharm. Bull. (Tokyo)*, **17**, 11 (1969).
440. A. Weinstein, H.R. Lindner, A. Frielander, S. Bauminger, *Steroids*, **20**, 789 (1972).
441. R. Pappo, C.J. Jung, *Tetrahedron Lett.*, **9** 365 (1962).
442. H.D. Lennon, F.J. Saunders, *Steroids*, **4**, 689 (1964).
443. Sollman, P.B. (1967) 17-oxygenated-2-thia-5α-androstan-3-ones and the corresponding mercapto acids and esters thereof, US Patent 3,301,872, Chem. Abstr., 66, P95299d.
444. Mazur, R.H., Pappo, R. (1964) 17-Oxygenated 2-azaandrostan-3ones, Belg. Patent 631,372, Chem. Abstr., 61, P705.
445. A.P. Shroff, C.H. Harper, *J. Med. Chem.*, **12**, 190 (1969).
446. M. Fox, A.S. Minot, G. Liddle, *J. Clin. Endocrinol. Metab.*, **22**, 921 (1962).
447. C.G. Ray, J.F. Kirschvink, S.H. Waxman, V.C. Kelley, *Am. J. Dis. Child.*, **110**, 618 (1965).
448. E.F. Nutting, D.W. Calhoun, *Endocrinology*, **84**, 441 (1969).
449. C.D. Kochakian, *Am. J. Physiol.*, **145**, 549 (1946).
450. C.D. Kochakian, *Proc. Soc. Exp. Biol. Med.*, **80**, 386 (1952).
451. E.F. Nutting, P.D. Klimstra, R.E. Counsell, *Acta Endocrinol. (Copenhagen)*, **53**, 627, 635 (1966).
452. I.A. Anderson, *Acta Endocrinol. (Copenhagen)*, Suppl. 63, 54 (1962).
453. M.E. Wolff, Y. Kasuya, *J. Med. Chem.*, **15**, 87 (1972).
454. G.A. Overbeck, A. Delver, J. deVisser, *Acta Endocrinol. (Copenhagen)* Suppl. 63, 7 (1962).
455. J.L. Kalliomaki, A.M. Pirila, I. Ruikka, *Acta Endocrinol. (Copenhagen)*, Suppl. 63, 124 (1962).
456. H. Kopera, The therapeutic usefulness of some esternols, in International Congress on Hormonal Steroids, Excerpta Medica International Congress Series, No. 51, p. 204 (1962).
457. A. Walser, G. Schoenenberger, *Schweiz. Med. Wochenschr.*, **92**, 897 (1962).
458. J.A. Edwards, A. Bowers, *Chem. Ind. (London)*, 1962 (1961).

459. M.E. Wolff, G. Zanati, G. Shanmagasundarum, S. Gupte et al., *J. Med. Chem.*, **13**, 531 (1970).
460. M.E. Wolff, G. Zanati, *Experientia*, **26**, 1115 (1970).
461. G. Zanati, G. Gaare, M.E. Wolff. *J. Med. Chem.*, **17**, 561 (1974).
462. R.W.S. Skinner, R.V. Pozderac, R.E. Counsell, C.F. Hsu et al., *Steroids*, **25**, 189 (1977).
463. M.E. Wolff, W. Ho, R. Kwok, *J. Med. Chem.*, **7**, 577 (1964).
464. P.D. Klimstra, E.F. Nutting, R.E. Counsell, *J. Med. Chem.*, **9**, 693 (1966).
465. M. Fujimuri, *Cancer*, **31**, 789 (1973).
466. G.O. Potts, A. Arnold, A.L. Beyler, Comparative nitrogen retaining and androgenic activities of certain orally active steroids, in International Congress on Hormonal Steroids, Excerpta Medica International Congress Series, No. 51, p. 211 (1962).
467. G.O. Potts, A.L. Beyler, D.F. Burnham, *Proc. Soc. Exp. Biol. Med.*, **103**, 383 (1960).
468. P.C. Burnett, *J. Am. Geriatr. Soc.*, **11**, 979 (1963).
469. W.G. Mullin, F. diPillo, *N.Y. State J. Med.*, **63**, 2795 (1963).
470. A.J. Manson, F.W. Stonner, H.C. Neumann, R.G. Christiansen et al., *J. Med. Chem.*, **6**, 1 (1963).
471. G.O. Potts, A. Beyler, H.P. Schane, *Fertil. Steril.*, **25**, 367 (1974).
472. R.J. Sherrins, H.M. Gandy, T.W. Thorsland, C.A. Paulsen, *J. Clin. Endocrinol. Metab.*, **32**, 522 (1971).
473. D. Rosi, H.C. Neumann, R.G. Christiansen, H.P. Shane et al., *J. Med. Chem.*, **20**, 349 (1977).
474. C. Davison, W. Banks, A. Fritz. *Arch. Int. Pharmacodyn. Ther.*, **221**, 294 (1976).
475. G.A. Overbeck, *Anabole Steroide*, Springer-Verlag, Berlin, 1966.
476. H.L. Kruskemper, *Anabolic Steroids* (translated by C.H. Doering), Academic Press, New York, 1968.
477. G.A. Overbeck, J. Van Der Vies, and J. de Visser, in *Protein Metabolism*, F. Gross (ed.), Springer-Verlag, Berlin, 1962, p. 185.
478. J. de Visser, G.A. Overbeck, *Acta Endocrinol. (Copenhagen)*, **35**, 405 (1960).
479. G.A. Overbeck, J. de Visser, *Acta Endocrinol. (Copenhagen)*, **38**, 285 (1961).
480. H. Nowakowski, *Acta Endocrinol. (Copenhagen)* Suppl. 63, 37 (1962).
481. A. Alibrandi, G. Bruni, A. Ercoli, R. Gardi et al., *Endocrinology*, **66**, 13 (1960).
482. K. Junkmann, G. Suchowsky, *Arzneim.-Forsch.*, **12**, 214 (1962).
483. K.C. James, *Experientia* **28**, 479 (1972).
484. K.C. James, P.J. Nicholl, G.T. Richards, *Eur. J. Med. Chem.*, **10**, 55 (1975).
485. R. Gaunt, C.H. Tuthill, N. Antonchak, J.H. Leathem, *Endocrinology*, **52**, 407 (1953).
486. P. Borrevang, *Acta Chem. Scand.*, **16**, 883 (1962).
487. R. Huttenrauch, *Arch. Pharm.*, **297**, 124 (1964).
488. Colton, F.B., Ray, R.E. (1962) Alkyl ethers of 17-(hydrocarbon substituted)estr-4-ene-3,17-diols, US Patent 3,068,249.
489. A. Ercoli, R. Gardi, R. Vitali, *Chem. Ind. (London)*, **1962** 1284 (1962).
490. R. Vitali, R. Gardi, A. Ercoli, New derivatives with ether linkage at C17 in androstane and estrane series, in International Congress on Hormonal Steroids, Excerpta Medica International Congress Series, No. 51, p. 128 (1962).
491. R.I. Dorfman, A.S. Dorfman, M. Gut, *Acta Endocrinol. (Copenhagen)*, **40**, 565 (1962).
492. R. Vitali, R. Gardi, G. Falconi, A. Ercoli, *Steroids*, **8**, 527 (1966).
493. A. Ercoli, G. Bruni, G. Falconi, F. Galletti et al., *Acta Endocrinol. (Copenhagen)*, Suppl. 51, 857 (1960).
494. A.D. Cross, I.T. Harrison, P. Crabbe, F.A. Kincl et al., *Steroids*, **4**, 229 (1964).
495. A.D. Cross, I.T. Harrison, *Steroids*, **6**, 397 (1965).
496. F.J. Saunders, *Proc. Soc. Exp. Biol. Med.*, **123**, 303 (1966).
497. A.J. Solo, N. Bejba, P. Hebborn, M. May, *J. Med. Chem.*, **18**, 165 (1975).
498. C.D. Kochakian, N. Arimasa, The Metabolism *in vitro* of Anabolic-Androgenic Steroids by Mammalian Tissues, in *Anabolic-Androgenic Steroids*, C.D. Kochakian (ed.), Springer-Verlag, New York, 1976, p. 287.
499. B. Pelc, *Collect. Czech. Commun.*, **29**, 3089 (1964).
500. J.T. Henderson, J. Richmond, M.D. Sumerling, *Lancet*, **1**, 934 (1973).
501. F.L. Johnson, J.R. Feagler, K.G. Lerner, P.W. Majerus et al., *Lancet*, **2**, 1273 (1972).
502. K.G. Ishak, Hepatic Neoplasms Associated with Contraceptive and Anabolic Steroids, in *Carcinogenic Hormones*, C.H. Lingeman

503. A.J. Ryan, Athletics, in *Anabolic-Androgenic Steroids*, C.D. Kochakian (ed.), Springer-Verlag, New York, 1976, p. 516.
504. J.D. Wilson, *Endocr. Rev.*, **9**, 181 (1988).
505. G.C. Lin, L. Erinoff (eds), *Anabolic Steroid Abuse*, NIDA Research Monograph, Vol. **102**, p. 29 (1990).
506. S.E. Lukas, *Trends Pharmacol. Sci.*, **14**, 61 (1993).
507. S. Casner, R. Early, B.R. Carlson, *J. Sports Med. Phys. Fitness*, **11**, 98 (1971).
508. T.D. Fahey, C.H. Brown, *Med. Sci. Sports*, **5**, 272 (1973).
509. L.A. Golding, J.E. Freydinger, S.S. Fishel, *Phys. Sports-Med.*, **2**, 39 (1974).
510. S.B. Strömme, H.D. Meen, A. Aakvaag, *Med. Sci. Sports*, **6**, 203 (1974).
511. G. Ariel, W. Saville, *J. Appl. Physiol.*, **32**, 795 (1972).
512. G. Ariel, *J. Sports Med. Phys. Fitness*, **13**, 187 (1973).
513. L.C. Johnson, G. Fisher, L.J. Silvester, C.C. Hofheins, *Med. Sci. Sports*, **4**, 43 (1972).
514. M. Steinbach, *Sportarzt Sportmed.*, **11**, 485 (1968).
515. S. Bhasin, T.W. Storer, N. Berman, C. Callegari et al., *N. Engl. J. Med.*, **335**, 1–7 (1996).
516. N. Wade, *Science*, **176**, 1399 (1972).
517. L.D. Bowers, R.V. Clark, C.H.L. Shackleton, *Steroids*, **74**, 285 (2009).
518. D.H. Catlin, B.D. Ahrens, Y. Kucherova, *Rapid Commun. Mass Spectrom.*, **16**, 1273 (2002).
519. D.H. Catlin, M.H. Sekera, B.D. Ahrens, B. Starcevic et al., *Rapid Commun. Mass Spectrom.*, **18**, 1245 (2004).
520. M.H. Sekera, B.D. Ahrens, Y.C. Chang, B. Starcevic et al., *Rapid Commun. Mass Spectrom.*, **19**, 781 (2005).
521. H. Steinbeck, F. Neumann, Androgen Antagonists: Chemistry and Influence on Neural-Gonadal Function, in *Reproductive Endocrinology*, R. Vokaer, G. DeBock (eds), Pergamon Press, Oxford, 1975, p. 135.
522. L. Martini, M. Motta (eds), *Androgens and Antiandrogens*, Raven Press, New York, 1977.
523. J.P. Raynaud, The Mechanism of Action of Anti-Hormones, in *Advances in Pharmacology and Therapeutics*, Vol. **1**: Receptors, J. Jacob (ed.), Pergamon Press, Oxford, 1979, p. 259.
524. G.H. Rasmusson, J.H. Torrey, *Annu. Rep. Med. Chem.*, **29**, 225 (1994).
525. A.D. Abell, B.R. Henderson, *Curr. Med. Chem.*, **2**, 583–597 (1995).
526. M. Jarman, H. J. Smith, P. J. Nicholls, C. Simons, *Nat. Prod. Rep.*, **15**, 495–512 (1998).
527. S. M. Singh, S. Gauthier, F. Labrie, *Curr. Med. Chem.*, **7**, 211–247 (2000).
528. J.P. Raynaud, B. Azadian-Boulanger, C. Bonne, J. Perronnet et al., Present Trends in Antiandrogen Research, in *Androgens and Antiandrogens*, L. Martini, M. Motta (eds), Raven Press, New York, 1977, p. 281.
529. C. Huggins, C.V. Hodges, *Cancer Res.*, **1**, 293 (1941).
530. C. Huggins, R.E. Stevens, Jr, C.V. Hodges, *Arch. Surg.*, **43**, 209 (1941).
531. E.A.P. Sutherland-Bawlings, *Br. Med. J.*, **111**, 643 (1970).
532. E.C. Dodds, *Biochem. J.*, **39**, 1 (1945).
533. J. Geller, B. Fruchtman, C. Meyer, H. Newman, *J. Clin. Endocrinol. Metab.*, **27**, 556 (1967).
534. F. Neumann, Methods for Evaluating Antisexual Hormones, in *Methods in Drug Evaluation*, P. Mantegazza, F. Piccinini (eds), North-Holland, Amsterdam, 1966, p. 548.
535. K. Mietkiewski, L. Malendowicz, A. Lukaszyk, *Acta Endocrinol. (Copenhagen)*, **61**, 293 (1969).
536. R.O. Neri, *Adv. Sex Horm. Res.*, **2**, 233 (1976).
537. F. Neumann, *Horm. Metab. Res.*, **9**, 1 (1977).
538. U. Fixson, *Geburtsh. Frauenheilk*, **23**, 371 (1963).
539. K. Junkmann, F. Neumann, *Acta Endocrinol. (Copenhagen)*, Suppl. 90, 139 (1964).
540. F. Neumann, K.J. Gräf, S.H. Hasan, B. Schenck et al., Central Actions of Antiandrogens, in *Androgens and Antiandrogens*, L. Martini, M. Motta (eds), Raven Press, 1977, p. 163.
541. F. Neumann, R. von Berswodt-Wallace, W. Elger, H. Steinbeck et al., *Rec. Progr. Horm. Res.*, **26**, 337 (1970).
542. S. Fang, S. Liao, *Mol. Pharmacol.*, **5**, 420 (1969).
543. F.R. Mangan, W.I.P. Mainwaring, *Steroids*, **20**, 331 (1972).

544. J. Hammerstein, J. Meckies, I. Leo-Rossberg, L. Moltz et al., *J. Steroid Biochem.*, **6**, 827 (1975).
545. J.L. Burton, U. Laschet, S. Shuster, *Br. J. Dermatol.*, **89**, 487 (1973).
546. J. Hammerstein, B. Cupceancu, *Dtsch. Med. Wochenschr.*, **94**, 829 (1969).
547. A.A. Ismail, D.W. Davidson, A.R. Souka, E.W. Barnes, et al., *J. Clin. Endocrinol. Metab.*, **39**, 81 (1974).
548. E. Cittadini, P. Barreca, Use of Antiandrogens in Gynecology, in *Androgens and Antiandrogens*, L. Martini, M. Motta (eds), Raven Press, New York, 1977, p. 309.
549. V.B. Mahesh, Excessive Androgen Secretion and Use of Antiandrogens in Endocrine Therapy, in *Androgens and Antiandrogens*, L. Martini, M. Motta (eds), Raven Press, New York, 1977, p. 321.
550. F.J. Ebling, Antiandrogens in Dermatology, in *Androgens and Antiandrogens*, L. Martini, M. Motta (eds), Raven Press, New York, 1977, p. 341.
551. J. Geller, B. Fruchtman, H. Newman, T. Roberts et al., *Cancer Chemother. Rep.*, **51**, 441 (1967).
552. U. Bracci, F. DiSilverio, *Prog. Med.*, **29**, 779 (1973).
553. U. Bracci, *J. Urol. Nephrol.*, **79**, 405 (1973).
554. F. DiSilverio, V. Gagliardi, *Boll. Soc. Urol.*, **5**, 198 (1968).
555. U. Bracci, F. DiSilverio, Role of Cyproterone Acetate in Urology, in *Androgens and Antiandrogens*, L. Martini, M. Motta (eds), Raven Press, New York, 1977, p. 333.
556. C. Labrie, L. Cusan, M. Plante, S. Lapointe et al., *J. Steroid Biochem.*, **28**, 379–384 (1987).
557. F. Sciarra, V. Toscano, G. Concolino, F. DiSilverio, *Mol. Biol.*, **37**, 349–362 (1990).
558. C. Labrie, J. Simard, H.F. Zhao, G. Pelletier et al., *Mol. Cell. Endocrinol.*, **68**, 169–179 (1990).
559. T. Ojasoo, J. Delettre, J.P. Mornon, C. Turpin-VanDycke et al., *J. Steroid Biochem.*, **27**, 255–269 (1987).
560. N. Jagarinec, M.L. Givner, *Steroids*, **23**, 561 (1974).
561. L.J. Lerner, A. Bianchi, A. Borman, *Proc. Soc. Exp. Biol. Med.*, **103**, 172 (1960).
562. F. Orestano, J.E. Altwein, P. Knapstein, K. Bandhauer, *J. Steroid Biochem.*, **6**, 845 (1975).
563. W.I.P. Mainwaring, Modes of Action of Antiandrogens: A Survey, in *Androgens and Antiandrogens*, L. Martini, M. Motta (eds), Raven Press, New York, 1977, p. 151.
564. S.Y. Tan, *J. Clin. Endocrinol. Metab.*, **39**, 936 (1974).
565. H.L. Saunders, K. Holden, J.F. Kerwin, *Steroids*, **3**, 687 (1964).
566. A. Boris, M. Uskokovic, *Experientia*, **26**, 9 (1970).
567. A. Boris, L. DeMartino, T. Trmal, *Endocrinology*, **88**, 1086 (1971).
568. E.E. Baulieu, I. Jung, *Biochem. Biophys. Res. Commun.*, **38**, 599 (1970).
569. G.H. Rasmusson, A. Chen, G.F. Reynolds, D.J. Patanelli et al., *J. Med. Chem.*, **15**, 1165 (1972).
570. J.R. Brooks, F.D. Busch, D.J. Patanelli, S.L. Steelman, *Proc. Soc. Exp. Biol. Med.*, **143**, 647 (1973).
571. K. Hiraga, A. Tsunehiko, M. Takuichi, *Chem. Pharm. Bull. (Tokyo)*, **13**, 1294 (1965).
572. G. Goto, K. Yoshiska, K. Hiraga, M. Masouka et al., *Chem. Pharm. Bull. (Tokyo)*, **26**, 1718 (1978).
573. G. Azadian-Boulanger, C. Bonne, J. Sechi, J.P. Raynaud, *J. Pharmacol. (Paris)*, **5**, 509 (1974).
574. C. Bonne, J.P. Raynaud, *Mol. Cell Endocrinol.*, **2**, 59 (1974).
575. P. Corvol, A. Michaud, J. Menard, M. Freifeld et al., *Endocrinology*, **97**, 52 (1975).
576. A.E. Wakeling, B.J.A. Furr, A.T. Glen, L.R. Hughes, *J. Steroid Biochem.*, **15**, 355 (1981).
577. L. Starka, J. Sulcova, P.D. Broulik, J. Joska et al., *J. Steroid Biochem.*, **8**, 939 (1977).
578. L. Starka, R. Hanapl, M. Bicikova, V. Cerny et al., *J. Steroid Biochem.*, **13**, 455 (1980).
579. R.G. Christiansen, M.R. Bell, T.E. D'Ambra, J.P. Mallamo et al., *J. Med. Chem.*, **33**, 2094–2100 (1990).
580. J.P. Mallamo, G.M. Pilling, J.R. Wetzel, P.J. Kowalczyk et al., *J. Med. Chem.*, **35**, 1663–1670 (1992).
581. R. Neri, K. Florance, P. Koziol, S. van Cleave, *Endocrinology*, **91**, 427 (1972).
582. R.O. Neri, M. Monohan, *Invest. Urol.*, **10**, 123 (1972).
583. E.A. Peets, M.F. Henson, R. Neri, *Endocrinology*, **94**, 532 (1974).
584. S. Liao, D.K. Howell, T. Chuag, *Endocrinology*, **94**, 1205 (1974).

585. R. Neri, E.A. Perts, *J. Steroid Biochem.*, **6**, 815 (1975).
586. E.D. Crawford, M.A. Eisenberger, D.G. McLeod, J.T. Spaulding et al., *N. Engl. J. Med.*, **321**, 419–424 (1989).
587. L. Denis, G.P. Murphy, *Cancer*, **72**, 3888–3895 (1993).
588. G.H. Rasmusson, G.F. Reynolds, N.G. Steinberg, Walton, E., et al., *J. Med. Chem.*, **29**, 2298 (1986).
589. J. Simard, S.M. Singh, F. Labrie, *Urology*, **49**, 580–586 (1997).
590. L.M. Eri, K.J. Tveter, *J. Urol.*, **150**, 90 (1993).
591. U. Fuhrmann, C. Bengston, G. Repenthin, E. Schillinger, *J. Steroid Biochem. Mol. Biol.*, **42**, 787 (1992).
592. C.J. Tyrell, *Prostate*, **4**, 97 (1992).
593. R.A. Janknegt, C.C. Abbou, R. Bartoletti, L. Bernstein-Hahn, et al., *J. Urol.*, **149**, 77–82 (1993).
594. P. Iversen, K. Tveter, E. Varenhorst, *Scand. J. Urol. Nephrol.*, **30**, 93–98 (1996).
595. A. Boris, J.W. Scott, L. DeMartino, D.C. Cox, *Acta Endocrinol. (Copenhagen)*, **72**, 604 (1973).
596. H. Miyachi, A. Azuma, T. Kitamoto, K. Hayashi et al., *Bioorg. Med. Chem. Lett.*, **7**, 1483–1488 (1997).
597. L.G. Hamann, R.I. Higuchi, L. Zhi, J.P. Edwards et al., *J. Med. Chem.*, **41**, 623–639 (1998).
598. C. Tran, S. Ouk, N.J. Clegg, Y. Chen, et al. *Science*, **324**, 787–790 (2009).
599. M.E. Jung, S. Ouk, D. Yoo, C.L. Sawyers et al., *J. Med. Chem.*, **53**, 2779–2796 (2010).
600. H.I. Scher, T.M. Beer, C.S. Higano, A. Anand, et al. *Lancet*, **375**, 1437–1446 (2010).
601. N.J. Clegg, J. Wongvipat, J.D. Joseph, C. Tran, et al. *Cancer Res.*, **72**, 1494–1503 (2012).
602. S. Tanayama, K. Yoshida, T. Kondo, Y. Kanai, *Steroids*, **33**, 65 (1979).
603. U. Speck, H. Wendt, P.E. Schulze, D. Jentsch, *Contraception*, **14**, 151 (1976).
604. M. Hümpel, H. Wendt, P.E. Schulze, G. Dogs et al., *Contraception*, **15**, 579 (1977).
605. M. Hümpel, H. Dogs, H. Wendt, U. Speck, *Arzneim.-Forsch.*, **28**, 319 (1978).
606. M. Frölich, H.L. Vader, S.T. Walma, H.A.M. De Rooy, *J. Steroid Biochem.*, **13**, 1097 (1980).
607. A.S. Bhargava, A. Seeger, P. Günzel, *Steroids*, **30**, 407 (1977).
608. B. Katchen, S. Buxbaum, *J. Clin. Endocrinol. Metab.*, **41**, 373 (1975).
609. R.O. Neri, Studies on the Biology and Mechanism of Action of Nonsteroidal Antiandrogens, in *Androgens and Antiandrogens*, L. Martini, M. Motta (eds), Raven Press, New York, 1977, p. 179.
610. H. Tucker, J.W. Crook, G.J. Chesterson, *J. Med. Chem.*, **31**, 954–959 (1988).
611. F. Neumann, R. Von Berswordt-Wallrabe, *J. Endocrinol.*, **35**, 363 (1966).
612. J. Hammerstein Male Contraception, in *Androgens and Antiandrogens*, L. Martini, M. Motta (eds), Raven Press, New York, 1977, p. 327.
613. H.J. Horn, Role of Antiandrogens in Psychiatry, in *Androgens and Antiandrogens*, L. Martini, M. Motta (eds), Raven Press, New York, 1977, p. 351.
614. F. Neumann, K. Junkmann, *Endocrinology*, **73**, 33 (1963).
615. R.O. Neri, M.D. Monahan, J.G. Meyer, B.A. Afonso et al., *Eur. J. Pharmacol.*, **1**, 438 (1967).
616. L.J. Lerner, *Pharmacol. Ther. B*, **1**, 217 (1975).
617. L.A. Dawson, E. Chow, G. Morton, *Urology*, **49**, 283–284 (1997).
618. E.J. Small, P.R. Carroll, *Urology*, **43**, 408–410 (1994).
619. J.R. Brooks, G.S. Harris, G.H. Rasmusson, Steroidogenesis Pathway Inhibitors, in *Design of Enzyme Inhibitors as Drugs*, Vol. 2, M. Sandler, H.J. Smither (eds), Oxford University Press, Oxford, 1994, p. 495.
620. G. Harris, B. Azzolina, W. Baginsky, Cimis, G., et al., *Proc. Natl Acad. Sci. USA*, **89**, 10787 (1992).
621. J.D. McConnell, J.D. Wilson, F.W. George, Geller, J. et al., *J. Clin. Endocrinol. Metab.*, **74**, 505 (1992).
622. E. Stoner and Study Group, *Urology*, **43**, (1994).
623. H.A. Guess, J.F. Heyse, G.J. Gormley, *Prostate*, **22**, 31 (1993).
624. J. Schwartz, O. Laskin, S. Schneider, Meeter, C.A, et al., *Clin. Pharmacol. Ther.*, **53**, 231 (1993).
625. A.A. Geldof, M.F. Meulenbroek, I. Dijkstra, S. Bohiken, et al., *J. Cancer Res. Clin. Oncol.*, **118**, 50 (1992).

626. E. diSalle, D. Guidici, G. Briatico, Ornati, G., et al., *J. Steroid Biochem. Mol. Biol.*, **46**, 549 (1993).
627. H.G. Bull, *J. Am. Chem. Soc.*, **118**, 2359 (1996).
628. B. Kenny, S. Ballard, J. Blagg, D. Fox, *J. Med. Chem.*, **40**, 1293–1315 (1997).
629. (a) S.V. Frye, C.D. Haffner, P.R. Maloney, Mook RA, et al., *J. Med. Chem.*, **36**, 4313 (1993); (b) S.V. Frye, C.D. Haffner, P.R. Maloney, Mook RA, et al., (1995) *J. Med. Chem.*, **38**, 2621.
630. M.A. Levy, M. Brandt, J.R. Heys, Holt, D.A., et al., *Biochemistry*, **29**, 2815 (1990).
631. M.A. Levy, B.W. Metcalf, M. Brandt, J.M. Erb, et al., *Bioorg. Chem.*, **19**, 245 (1991).
632. P. Audet, H. Nurcombe, Y. Lamb, Jorkasky, D., et al., *Clin. Pharmacol. Ther.*, **53**, 231 (1993).
633. R.E. Johnsonbaugh, B.R. Cohen, E.M. McCormack, F.W. George, et al., *J. Urol.*, **149**, 432 (1993).
634. D.A. Holt, M.A. Levy, Erb, J.M., Heaslip, JI, et al., *J. Med. Chem.*, **33**, 943 (1990)*J. Med. Chem.*, (1990), **33**, 937.
635. G.M. Cooke, B. Robaire, *J. Steroid Biochem.*, **24**, 877 (1986).
636. B. Robaire, D.F. Covey, C.H. Robinson, L.L. Ewing, *J. Steroid Biochem.*, **8**, 307 (1977).
637. W. Voigt, A. Castro, D.F. Covey, C.H. Robinson, *Acta Endocrinol.*, **87**, 668 (1978).
638. R.K. Bakshi, G.F. Patel, G.H. Rasmusson, W.F. Baginsky et al., *Chemistry*, **37**, 3871–3874 (1994).
639. C.D. Jones, J.E. Audia, D.E. Lawhorn, L.A. McQuaid et al., *J. Med. Chem.*, **36**, 421–423 (1993).
640. D.A. Holt, D.S. Yamashita, A.L. Konialian-Beck, J.I. Luengo et al., *J. Med. Chem.*, **38**, 13–15 (1995).
641. H. Kojo, O. Nakayama, J. Hirosumi, N. Chida et al., *Mol. Pharmacol.*, **48**, 401–406 (1995).
642. J.R. Drago, R.J. Santen, A. Lipton, T.J. Worgul et al., *Cancer*, **53**, 1447 (1984).
643. G. Williams, D.J. Kerle, H. Ware, A. Doble et al., *Br. J. Urol.* **58**, 45 (1986).
644. J.P. Van Wauwe, P.A. Janssen, *J. Med. Chem.*, **32**, 2231 (1989).
645. M. Ayub, M.J. Levell, *J. Steroid Biochem.*, **32**, 515 (1989).
646. J. Yeh, R.L. Barbieri, A.J. Friedman, *J. Steroid Biochem.*, **33**, 627 (1989).
647. S.E. Barrie, M.G. Rowlands, A.B. Foster, M. Jarman, *J. Steroid Biochem.*, **33**, 1191 (1989).
648. R. McCague, M.G. Rowlands, S.E. Barrie, J. Houghton, *J. Med. Chem.*, **33**, 2452 (1990).
649. W.N. Kuhn-Velten, I. Meyer, W. Staib, *J. Steroid Biochem.*, **33**, 33 (1989).
650. Kuhn-Velten, W.N., Bunse, T., Forster, M.E.C. *J. Biol. Chem.* **266**: 6291, 1991.
651. M. Onoda, M. Haniu, K. Kanagibashi, F. Sweet et al., *Biochemistry*, **26**, 657 (1987).
652. M.R. Angelastro, M.E. Laughlin, G.L. Schatzman, P. Bey et al., *Biochem. Biophys. Res. Commun.*, **162**, 1571 (1989).
653. J. Stevens, J. Jaw, C.T. Peng, J. Halpert, *Biochemistry*, **30**, 3649 (1991).
654. V.C. Njar, M. Hector, R.W. Hartmann, *Bioorg. Med. Chem.*, **4**, 1447–1453 (1996).
655. G.A. Potter, S.E. Barrie, M. Jarman, M.G. Rowlands, *J. Med. Chem.*, **38**, 2463–2471 (1995).
656. A. O'Donnell, I. Judson, M. Dowsett, F. Raynaud, et al. *Br. J. Cancer*, **90**, 2317–2325 (2004).
657. J.S. de Bono, C.J. Logothetis, A. Molina, K. Fizazi, S. North, et al. *N. Engl. J. Med.*, **364**, 1995–2005 (2011).
658. C.J. Ryan, M.R. Smith, J.S. de Bono, A. Molina, et al. *N. Engl. J. Med.*, **368**, 138–148 (2013).
659. R.J. Santen, S. Santner, A. Lipton, *Cancer Res.*, **42** Suppl. 8, 3461s–3467s (1982).
660. R.J. Santen, (ed.) *Steroids*, **50**, 1–665 (1987).
661. A.M.H. Brodie, H.B. Brodie, G. Callard, C. Robinson et al., (eds), *J. Steroid Biochem. Mol. Biol.*, **44**, 321–696 (1993).
662. A. Bhatnagar, A. Brodie, R. Brueggemeier, S. Chen et al., (eds), *J. Steroid Biochem. Mol. Biol.*, **61**, 107–425 (1997).
663. E.R. Simpson, J.R. Pasqualini (eds), *J. Steroid Biochem. Mol. Biol.*, **79**, 1–314 (2001).
664. A. Brodie, M. Dowsett, N. Harada, P. Lonning et al. (eds), *J. Steroid Biochem. Mol. Biol.*, **86**, 217–507 (2003).
665. W.R. Miller, J.R. Pasqualini (eds), *J. Steroid Biochem. Mol. Biol.*, **95**, 1–127 (2005).
666. A.M.H. Brodie, J.R. Pasqualini, (eds), *J. Steroid Biochem. Mol. Biol.*, **106**, 1–186 (2007).
667. S. Chen, J. Adamski, (eds), *J. Steroid Biochem. Mol. Biol.*, **118**, 195–315 (2010).
668. J.O. Johnston, B.W. Metcalf Aromatase: A Target Enzyme in Breast Cancer, in *Novel Approaches to Cancer Chemotherapy*,

P. Sunkara (ed.), Academic Press, New York, 1984, pp. 307–328.
669. L. Banting, H.J. Smith, M. James, G. Jones et al., *J. Enzyme Inhib.*, **2**, 215–229 (1988).
670. D.F. Covey Aromatase Inhibitors: Specific Inhibitors of Oestrogen Biosynthesis, in *Sterol. Biosynthesis Inhibitors*, D. Berg, M. Plempel (eds), Ellis-Horwood, Chichester, 1988, pp. 534–571.
671. L. Banting, P.J. Nichols, M.A. Shaw, H.J. Smith, *Prog. Med. Chem.* **26**, 253–298 (1989).
672. R.J. Santen, A. Manni, H. Harvey, C. Redmond, *Endocr. Rev.*, **11**, 221–265 (1990).
673. R.T. Blickenstaff, *Antitumor Steroids*, Academic Press, San Diego, 1992, pp. 68–78.
674. R.W. Brueggemeier, J.C. Hackett, E.S. Diaz-Cruz. *Endocr. Rev.*, **26**, 331–345 (2005).
675. R.J. Santen, H. Brodie, E.R. Simpson, P.K. Siiteri et al., *Endocr. Rev.*, **30**, 343–375 (2009).
676. D.A. Marsh, H.J. Brodie, W. Garrett, C.-H. Tsai-Morris et al., *J. Med. Chem.*, **28**, 788–795 (1985).
677. A.M.H. Brodie, W. Garrett, J.R. Hendrickson, C.-H. Tsai-Morris et al., *Steroids*, **38**, 693–702 (1981).
678. A.M.H. Brodie, W.C. Schwarzel, A.A. Shaikh, H.J. Brodie, *Endocrinology*, **100**, 1684–1695 (1977).
679. L.Y. Wing, W. Garrett, A.M.H. Brodie, *Cancer Res.*, **45**, 2425–2428 (1985).
680. A.M.H. Brodie, W. Garrett, J.R. Hendrickson, C.-H. Tsai-Morris, *Cancer Res.*, **42**, Suppl. 8 3360s–3364s (1982).
681. R.C. Coombes, P. Goss, M. Dowsett, J.C. Gazet et al., *Lancet*, **2**, 1237–1239 (1984).
682. P.E. Goss, T.J. Powles, M. Dowsett, G. Hutchison et al., *Cancer Res.*, **46**, 4823–4826 (1986).
683. M. Dowsett, D. Cunningham, S. Nichols, A. Lal et al., *Cancer Res.*, **49**, 1306–1312 (1989).
684. D. Henderson, G. Norbisrath, U. Kerb, *J. Steroid Biochem.*, **24**, 303–306 (1986).
685. R.W. Brueggemeier, E.E. Floyd, R.E. Counsell, *J. Med. Chem.*, **21**, 1007–1011 (1978).
686. S. Miyairi, J. Fishman, *J. Biol. Chem.*, **261**, 6772–6777 (1986).
687. R.W. Brueggemeier, N.E. Katlic, *Cancer Res.*, **47**, 4548–4551 (1987).
688. R.W. Brueggemeier, P.-K. Li, C.E. Snider, M.V. Darby et al., *Steroids*, **50**, 163–178 (1987).
689. R.W. Brueggemeier, P.-K. Li, *Cancer Res.*, **48**, 6808–6810 (1988).
690. M.V. Darby, J.A. Lovett, R.W. Brueggemeier, M.P. Groziak et al., *J. Med. Chem.*, **28**, 803–807 (1985).
691. P.-K. Li, R.W. Brueggemeier, *J. Med. Chem.*, **33**, 101–105 (1990).
692. P.-K. Li, R.W. Brueggemeier, *J. Enzyme Inhib.*, **4**, 113–120 (1990).
693. B.W. Metcalf, C.L. Wright, J.P. Burkhart, J.O. Johnston, *J. Am. Chem. Soc.*, **103**, 3221–3222 (1981).
694. D.F. Covey, W.F. Hood, V.D. Parikh, *J. Biol. Chem.*, **256**, 1076–1079 (1981).
695. P.A. Marcotte, C.H. Robinson, *Steroids*, **39**, 325–344 (1982).
696. J.O. Johnston, C.L. Wright, B.W. Metcalf, *Endocrinology*, **115**, 776–785 (1984).
697. J.O. Johnston, C.L. Wright, B.W. Metcalf, *J. Steroid Biochem.*, **20**, 1221–1226 (1984).
698. C. Longcope, A.M. Femino, J.O. Johnston, *Endocrinology*, **122**, 2007–2011 (1988).
699. J.O. Johnston, *Steroids*, **50**, 106–120 (1987).
700. S.J. Ziminski, M.E. Brandt, D.F. Covey, C. Puett, *Steroids*, **50**, 135–146 (1987).
701. P.A. Marcotte, C.H. Robinson, *Biochemistry*, **21**, 2773–2778 (1982).
702. P.J. Bednarski, D.J. Porubek, S.D. Nelson, *J. Med. Chem.*, **28**, 775–779 (1985).
703. P.J. Bednarski, S.D. Nelson, *J. Med. Chem.*, **32**, 203–213 (1989).
704. C.E. Snider, R.W. Brueggemeier, *J. Biol. Chem.*, **262**, 8685–8689 (1987).
705. D. Giudici, G. Ornati, G. Briatico, F. Buzzetti et al., *J. Steroid Biochem.* **30**, 391–394 (1988).
706. E.P. Winer, C. Hudis, H.J. Burstein, A.C. Wolff, et al. *J. Clin. Oncol.*, **23**, 619–629 (2005).

2
Annexins

Carl E. Creutz
*University of Virginia, Department of Pharmacology, 1340 Jefferson Park Avenue,
Charlottesville, VA 22908, USA*

1 Diversity and Functions 78

2 Structure and Mechanism of Action 79

3 Regulation 82

4 Applications 83

References 83

Keywords

Exocytosis
The release of a secretory product from a cell when an intracellular vesicle containing the product fuses with the cell-surface membrane.

Membrane fusion
The merger of two biological membranes to form a single membrane; an example is when the membrane of a secretory vesicle fuses with the surface membrane of the cell.

Phospholipid
This common constituent of biological membranes is composed of two fatty acids esterified to a glycerol backbone, with a head group containing a phosphate ester.

Phosphorylation
The covalent attachment of a phosphate group, as in the enzyme-catalyzed phosphorylation of serine or tyrosine amino acid side chains in a protein.

Translational Medicine: Molecular Pharmacology and Drug Discovery
First Edition. Edited by Robert A. Meyers.
© 2018 Wiley-VCH Verlag GmbH & Co. KGaA. Published 2018 by Wiley-VCH Verlag GmbH & Co. KGaA.

The annexins are a family of structurally related, calcium-dependent, phospholipid-binding proteins that have been postulated to mediate calcium-dependent activities at membrane surfaces, including membrane fusion, lipid metabolism and reorganization, and ion permeation. The basic annexin structure consists of four homologous 70 amino-acid repeats and a unique N-terminal domain. These repeats do not contain sequences found in other intracellular calcium-binding proteins; therefore the annexins represent a novel class of calcium-binding proteins. The "core" domains of all annexins are comprised of the four 70-amino acid repeats, and are 40–60% identical in sequence. One annexin family member (annexin A6) has been formed as a result of gene duplication and comprises eight of the 70-amino acid repeats. Another type of duplication has occurred with annexin A2, in which two 36 kDa molecules (the "heavy" chains), each of which contains four repeats bind to a dimer of a 10 kDa protein (the "light" chains) to form a tetramer. In contrast to other lipid-binding proteins, such as protein kinase C or phospholipase A2, the annexins are unique in that most are bivalent; that is, they can attach to two membranes rather than just one membrane which, as a consequence, draws them together.

1
Diversity and Functions

Members of the annexin family of proteins, of which 12 mammalian variants have now been recognized were discovered independently in a number of contexts [1–5]. As a consequence, various names have been used for the different family members, such as synexins, chromobindins, lipocortins, calcimedins, calphobindins, and calpactins. Subsequently, however, a standard reference nomenclature has been adopted which utilizes the term annexin, followed by the letter A and a number which reflects the gene names; for example, annexin A1 is encoded by the human *ANX1* gene. A schematic of several representative human annexins is shown in Fig. 1.

Annexins have been isolated from a wide variety of animals, plants, as well as from the slime mold *Dictyostelium* and the fungus *Neurospora*, which attests to the universality of the functions performed by annexins. However, annexins are not present in ciliates, nor in yeasts, such as *Saccharomyces* sp. Because of the bivalent activity of the annexins, one of their postulated roles is in the promotion of membrane fusion in exocytosis [4]. Following the aggregation of secretory vesicle membranes by the action of an annexin, the vesicles undergo fusion if exposed to *cis*-unsaturated fatty acids [6, 7]. In a cell, such fusogenic fatty acids might be made available through the activation of lipases. Alternatively, the annexins might function in concert with other proteins, such as the SNARE proteins, that promote membrane fusion and remodeling after the membranes have been brought into close apposition by the annexin.

In addition to a possible role in exocytosis and membrane trafficking, some annexins have been suggested to be mediators of the anti-inflammatory effects of steroids, to serve as components of the submembranous cytoskeleton, as inhibitors of blood coagulation, as transducers of signals generated by tyrosine kinases at the cell membrane, as mediators of bone formation and remodeling, as mediators of cell–matrix interactions, voltage-dependent ion channels, as actin-bundling proteins, as regulators of calcium-release

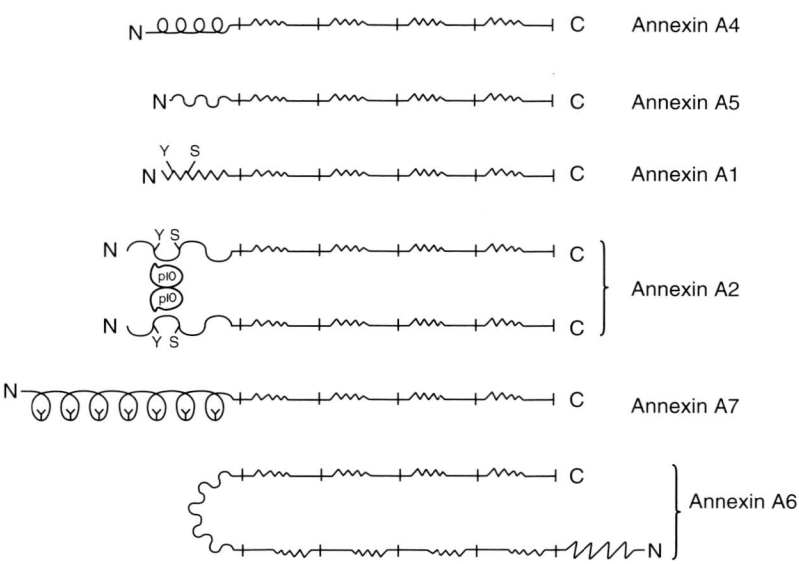

Fig. 1 Schematic illustration of the primary structures of six annexins. Each of the four (or eight) homologous domains contains the 17-amino acid "endonexin fold" sequence represented by a sawtooth line (sequence: KGhGTDExxLIpILApR: h, hydrophobic residue; p, polar residue; x, variable residue) [10]. The unique N-terminal structures are on the left (or right in the case of annexin A6); Y and S represent phosphorylation sites in the tails of annexins A1 and A2. The annexin A2 tetramer is drawn showing the association of the N termini of the heavy chains with the light chain (p10 or S100A10) dimer. The Ys inside the loops in the tail of annexin A7 represent a pro-beta helix [11].

channels in muscle and of DNA polymerase activity, as extracellular binding sites for plasminogen, and as agents for the repair of plasma membrane injuries [1, 2, 8, 9]. Some proposed sites of action of annexins in membrane organization and trafficking are summarized in Fig. 2.

This diversity of hypotheses for annexin functions attests to the ubiquitous and abundant nature of these proteins. In many cases, the data supporting a role for the annexins in these various processes are derived from *in-vitro* systems involving calcium and lipid membranes, and it has been difficult to relate these activities to true cellular functions. However, it seems likely that this family of proteins has radiated to perform a variety of cellular functions, as so many are coexpressed in single cells and their individual differences have been highly conserved during evolution. In that sense, the annexins may be similar to the other major class of calcium-binding proteins, the "EF-hand" family, which includes calmodulin and troponin C and is responsible for a variety of nonoverlapping cellular phenomena.

2
Structure and Mechanism of Action

The annexins, in general, have proven to be relatively easy to isolate in highly pure forms, and this advantage has led to several successful attempts to crystallize members of the family. The structure of annexin A5 revealed that each of the four

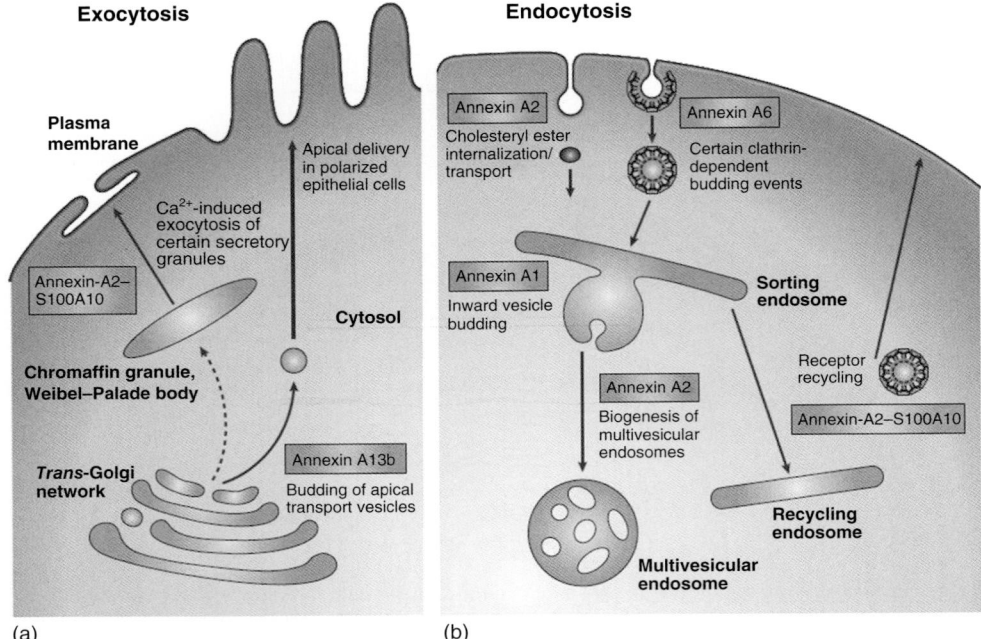

Fig. 2 Schematic representation of various membrane-trafficking steps, showing the involvement of annexins. (a) In the biosynthetic pathway, annexin A2 in complex with S100A10 has been shown to participate in the Ca^{2+}-evoked exocytosis of chromaffin granules and endothelial Weibel–Palade bodies. The complex probably functions at the level of the plasma membrane, possibly by linking the large secretory vesicles to the plasma membrane or by organizing plasma-membrane domains so that efficient fusion can take place. Annexin A13b is required for the budding of sphingolipid- and cholesterol-rich membrane domains at the *trans*-Golgi network, and therefore the delivery of such material to the apical plasma membrane in polarized epithelial cells; (b) In the endocytic pathway, annexin A6 has been proposed to be involved in clathrin-coated-pit budding events that depend on the activity of a cysteine protease that is required to modulate the spectrin membrane skeleton. Annexin A2, which can associate with caveolae, has been shown to form a lipid–protein complex with acylated caveolin and cholesteryl esters that seems to be involved in the internalization/transport of cholesteryl esters from caveolae to internal membranes. Annexin A2 is also found on early endosomes, where it is required, in complex with S100A10, to maintain the correct morphology of perinuclear recycling endosomes. Moreover, its depletion can interfere with the proper biogenesis of multivesicular endosomes from early endosomes. Annexin A1 also seems to function in multivesicular endosome biogenesis, more specifically, in the process of inward vesicle budding. Reproduced with permission from Ref. [1]; © 2005, Nature Publishing Group.

70-amino acid repeats forms a compact bundle of alpha-helices, and these four bundles are arranged in a plane (Fig. 3) [12]. Viewed from the side, the molecule displays a slightly convex face, which is thought to be the membrane-binding face, and a concave face, which is thought to face the cytoplasm. Each homologous domain forms one or two potential calcium-binding sites, which are complex in that they involve convergent loops from different parts of the domain. In addition, the positions of

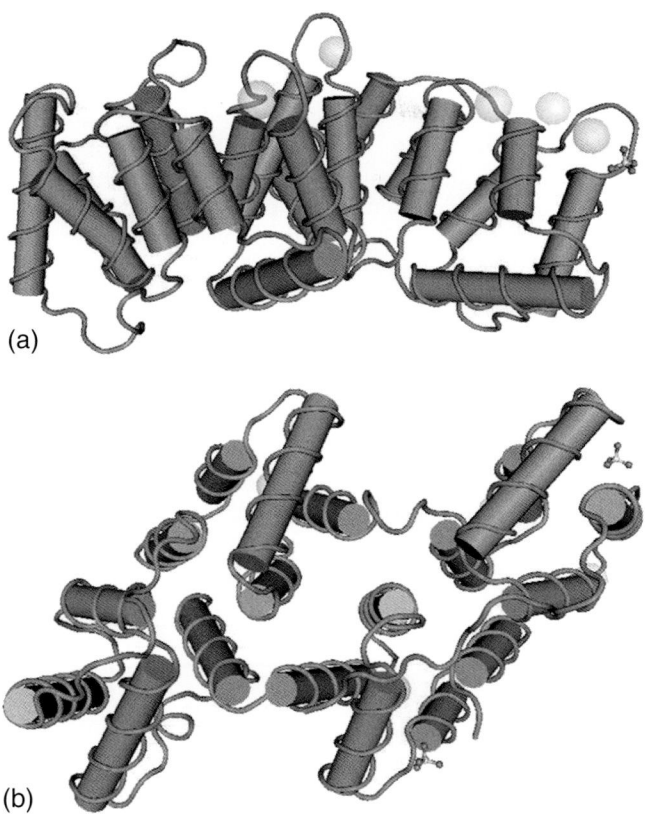

Fig. 3 The structure of annexin A5 as determined by X-ray diffraction (Molecular Modeling Database ID: 55379; Protein Data Bank ID: 1AVR) [12]. (a) "Side" view of the molecule, with the calcium-binding sites, and the membrane-binding face at the top. Calcium ions are represented by the spheres. The-high affinity sites are represented by the first, second, and fifth spheres from left to right; the third and fourth spheres indicate low-affinity ion-binding sites that were identified by lanthanide binding. The extended N terminus is at the bottom; (b) View of the "cytoplasmic" side of the molecule and the N terminus. The calcium-binding sites are on the opposite face. Note the open architecture and potential ion channel in the center of the molecule.

sulfate ions in the loops (derived from the crystallization medium) suggest that the phosphate of a lipid head group may also sit in the binding pocket and coordinate the calcium ion. This geometry would explain the interdependence of calcium and lipid binding by the annexins. As all of the lipid-binding sites of the annexin core domains face a single membrane, they may be able to integrate information about membrane lipid composition. Although, in general, all annexins bind with a higher affinity (i.e., at lower levels of calcium) to acidic phospholipids, some variation has been reported for different annexins with regards to the order of preference for different acidic lipids. In addition, the four homologous repeats may have different specificities, as the repeats are only 40–50% identical in sequence.

The crystal structure of annexin A5 has also been interpreted in terms of the ion channel-forming properties of the protein. The molecule has a hydrophilic pore perpendicular to the face of the membrane (Fig. 3). Because the molecule has no hydrophobic external surfaces, it is assumed it cannot actually enter the bilayer in this conformation. However, it has a calculated external electrical field similar to the field strength necessary to punch a hole through a membrane by electroporation [13]. Therefore, it has been suggested that the molecule sits on the surface of the membrane, which leads to a disordered state in the lipids immediately below it. In this disordered state the membrane conducts ions that must also pass through the hydrophilic pore in the molecule. In this way, the molecule provides the ion selectivity of the overall transmembrane channel.

An alternative mechanism for the ability of annexins to promote ion permeation has been proposed that involves a significant reorganization of the molecule at low pH to form a transmembrane channel [14]. However, a crystallographic model of this proposed structure has not yet been determined. It is also not known if the channel activity that annexins exhibit *in vitro* corresponds to a specific physiological channel in cells.

Since the calcium/lipid-binding sites all appear to be on one side of the annexin molecule, it is not clear how the annexins may cause membranes to aggregate. It seems likely that a self-association of the annexin molecules attached to different membranes may be required for membrane aggregation, and that during such self-association the concave faces of the molecules which face the cytoplasm may become interlocked. In the case of annexin A7 (synexin) acting *in vitro* on chromaffin granules (the secretory vesicles of the adrenal medulla), annexin binding to the membranes occurs at low levels of calcium ($<10\,\mu M$), though membrane aggregation and fusion depend on higher levels of calcium ($>100\,\mu M$) [15]. This calcium dependence of granule aggregation correlates exactly with the calcium dependence of annexin A7 self-association in the absence of membranes [16], and suggests that the mechanism of membrane aggregation might involve membrane-bound annexin A7 molecules undergoing self-association to bring the two membranes together. If this mechanism is correct, then the "bottleneck" in the system would be the high level of calcium needed to promote annexin self-association.

The annexin A2 (calpactin) tetramer promotes the aggregation and fatty acid-dependent fusion of chromaffin granules at the lowest level of calcium of any annexin (ca. $1\,\mu M$). This may occur because the tetramer represents a permanently self-associated annexin (see Fig. 1). The two heavy chains might be associated through the light chain dimer in such a way that each heavy chain can bind a different membrane so that they are pulled together. Thus, the processes of overall membrane aggregation and fusion are catalyzed by the low levels of calcium needed simply to promote membrane binding.

3
Regulation

The N-terminal domains of the annexins are the major sites of sequence divergence in the family (see Fig. 1). They may be the site of interaction with other proteins, as in the case of the annexin A2 heavy chain, which associates with p10 (S100A10), and annexin A11, which associates with the small calcium-binding protein, calcyclin. The N-terminal domains are also potential regulatory sites for the remainder of the molecule. For example, cleavage of the N-terminus of the isolated annexin A2 heavy chain is necessary for this protein to aggregate chromaffin granules at low levels of calcium [7]. Furthermore, the calcium sensitivity of membrane aggregation by such cleaved molecules is strongly affected by the exact site of cleavage. In addition, annexins A1 and A2 are phosphorylated in the N-terminal domain by tyrosine- or serine/threonine-specific kinases. Phosphorylation of the N-terminus of annexin A1 by protein kinase C blocks the ability of this protein to aggregate chromaffin granules, while slightly enhancing its ability to bind the granule membrane [17]. These biochemical observations on the importance of the annexin N-terminal structure can be interpreted in terms of the crystal structure of the annexins. As the N terminus is located on the "cytoplasmic" side of the membrane-bound annexin, it may be in a position to participate in the annexin–annexin self-association that is essential for membrane aggregation. This may explain why alterations in the tails of other annexins have a great effect on membrane aggregation, but not on membrane binding. Such alterations, resulting from either proteolysis or phosphorylation, may provide additional mechanisms for the cellular control of these calcium-dependent proteins.

4
Applications

Although the complete breadth and details of the biological roles of annexins have yet to be determined, a number of biotechnological and medical applications are currently under development. Most of these applications relate to the signature ability of annexins to bind to acidic phospholipids in the presence of calcium. Members of this protein family have potential utility as anticoagulants because they can prevent the association of clotting factors with acidic lipids on the surface of activated platelets and endothelial cells. Annexins may also function as immunosuppressive agents by shielding phospholipids from attack by phospholipase A_2, preventing the release of arachidonic acid, which is the precursor for a number of inflammatory lipid mediators.

Annexins are a staple of any research laboratory studying apoptosis (programmed cell death) *in vitro*, since one hallmark of the apoptotic pathway is an exposure of phosphatidylserine on the cell surface, which can be marked by the binding of a fluorescent annexin. Labeled annexins have also been tested in whole animals for the ability to localize occult thromboses or apoptotic cells in whole-body imaging techniques, while annexins fused to clot-dissolving enzymes have been assessed for their ability to direct these enzymes towards thromboses.

References

1. Gerke, V., Creutz, C.E., Moss, S.E. (2005) Annexins: linking Ca^{2+} signalling to membrane dynamics. *Nat. Rev. Mol. Cell Biol.* **6**, 449–461.
2. Creutz, C.E. (2003) Reflections on Twenty-Five Years of Annexin Research, in: Bandorowicz-Pikula, J. (Ed.) Annexins: Biological Importance and Annexin-Related Pathologies, Landes Bioscience, Georgetown, TX, pp. 1–20.
3. Creutz, C.E., Pazoles, C.J., Pollard, H.B. (1978) Identification and purification of an adrenal medullary protein (synexin) that causes calcium-dependent aggregation of isolated chromaffin granules. *J. Biol. Chem.*, **253**, 2858–2866.

4. Creutz, C.E. (1992) The annexins and exocytosis., *Science* **258**, 924–931.
5. Morgan, R.O., Martin-Almedina, S., Iglesias, J.M., Gonzalez-Florez, M.I., et al. (2004) Evolutionary perspective on annexin calcium-binding domains. *Biochim. Biophys. Acta*, **1742**, 133–140.
6. Creutz, C.E. (1981) *cis*-Unsaturated fatty acids induce the fusion of chromaffin granules aggregated by synexin. *J. Cell Biol.*, **91**, 247–256.
7. Drust, D.S., Creutz, C.E. (1988) Aggregation of chromaffin granules by calpactin at micromolar levels of calcium. *Nature*, **331**, 88–91.
8. Draeger, A., Monastyrskaya, K., Babiychuk, E.B. (2011) Plasma membrane repair and cellular damage control: the annexin survival kit. *Biochem. Pharmacol.*, **81**, 703–712.
9. Creutz, C.E., Hira, J.K., Gee, V.E., Eaton, J.M. (2012) Protection of the membrane permeability barrier by annexins. *Biochemistry*, **51**, 9966–9983.
10. Kretsinger, R.H., Creutz, C.E. (1986) Cell biology. Consensus in exocytosis. *Nature*, **320**, 573.
11. Creutz, C.E. (2009) Novel protein ligands of the annexin A7 N-terminal region suggest pro-beta helices engage one another with high specificity. *Gen. Physiol. Biophys.*, **28**, Focus issue, F7–F13.
12. Huber, R., Berendes, R., Burger, A., Schneider, M., et al. (1992) Crystal and molecular structure of human annexin V after refinement. Implications for structure, membrane binding and ion channel formation of the annexin family of proteins. *J. Mol. Biol.*, **223**, 683–704.
13. Huber, R., Berendes, R., Burger, A., Luecke, H., et al. (1992) Annexin V-crystal structure and its implications on function. *Behring Inst. Mitt.*, **91**, 107–125.
14. Langen, R., Isas, J.M., Hubbell, W.L., Haigler, H.T. (1998) A transmembrane form of annexin XII detected by site-directed spin labeling. *Proc. Natl Acad. Sci. USA*, **95**, 14060–14065.
15. Creutz, C.E., Sterner, D.C. (1983) Calcium dependence of the binding of synexin to isolated chromaffin granules. *Biochim. Biophys. Res. Commun.*, **114**, 355–364.
16. Creutz, C.E., Pazoles, C.J., Pollard, H.B. (1979) Self-association of synexin in the presence of calcium. Correlation with synexin-induced membrane fusion and examination of the structure of synexin aggregates. *J. Biol. Chem.*, **254**, 553–558.
17. Wang, W., Creutz, C.E. (1992) Regulation of the chromaffin granule aggregating activity of annexin I by phosphorylation. *Biochemistry*, **31**, 9934–9939.

3
Genetic Engineering of Antibody Molecules

Cristian J. Payés[1], Tracy R. Daniels-Wells[2], Paulo C. Maffía[3], Manuel L. Penichet[2,4,5,6], Sherie L. Morrison[4,5,6], and Gustavo Helguera[1]

[1]*Laboratory of Pharmaceutical Biotechnology, Institute of Experimental Biology and Medicine, Vuelta de Obligado 2490, Buenos Aires, Argentina*

[2]*University of California, Division of Surgical Oncology, Department of Surgery, David Geffen School of Medicine, 10833 Le Conte Ave, Los Angeles, CA 90095, USA*

[3]*Laboratory of Molecular Microbiology; Department of Science and Technology, National University of Quilmes, Roque Sáenz Peña 352, Bernal, Buenos Aires, Argentina*

[4]*University of California, The Molecular Biology Institute, 611 Charles E. Young Dr., Los Angeles, CA 90095, USA*

[5]*University of California, Jonsson Comprehensive Cancer Center, 700 Tiverton Ave., Los Angeles, CA 90095, USA*

[6]*University of California, Department of Microbiology, Immunology, and Molecular Genetics, David Geffen School of Medicine, 10833 Le Conte Ave, Los Angeles, CA 90095, USA*

1	**Antibody Structure and Engineering**	88
1.1	The Basic Structure of Antibodies	88
1.2	Classes and Subclasses of Antibodies	90
2	**Technological Milestones in Antibody Engineering**	93
2.1	Murine Monoclonal Antibodies	93
2.2	Chimeric Antibodies	96
2.3	Humanized Antibodies	96
2.4	Fully Human Antibodies	101
3	**Expression Systems**	105
3.1	Prokaryotic Expression Systems	105
3.2	Yeast Expression Systems	106
3.3	Insect Cell Expression Systems	106
3.4	Mammalian Cell Expression Systems	111
3.5	Antibody Expression in Transgenic Plants	112
3.6	Transgenic Mammals	112

Translational Medicine: Molecular Pharmacology and Drug Discovery
First Edition. Edited by Robert A. Meyers.
© 2018 Wiley-VCH Verlag GmbH & Co. KGaA. Published 2018 by Wiley-VCH Verlag GmbH & Co. KGaA.

4	**Variable Region Engineering** 113
4.1	Improving the Affinity of Recombinant Antibodies 113
4.2	Reduction of Immunogenicity 114
5	**Constant Region Engineering 115**
5.1	Fc Engineering of IgGs for ADCC and ADCP Functions 115
5.2	Fc Engineering of IgGs for CDC Function 117
5.3	Antibody Engineering for Altered Half-Life 117
5.4	Antibodies of the IgE Class 118
6	**Structurally Modified Antibodies 118**
6.1	Antibody Fragments: Fabs and scFvs Monovalent, Bivalent, and Multivalent 118
6.2	Bispecific Antibodies 120
6.3	Polymers of Monomeric Antibodies 120
6.4	Antibody Fusion Proteins 122
7	**Conclusions 126**
	Acknowledgments 127
	References 127

Keywords

Antibody/antigen
Antibodies are glycoproteins (also known as immunoglobulins; Igs), which are produced by plasma cells (terminally differentiated B lymphocytes) of the immune system in response to the presence of a foreign substance. Antigens are molecules present on cells, viruses, fungi, bacteria, and some non-living substances such as toxins, chemicals, and drugs which are recognized and specifically bound by antibodies. The humoral immune system recognizes certain antigens (also known as immunogens) as foreign and produces antibodies that can neutralize or destroy substances or organisms containing these antigens.

Antibody-dependent cell-mediated cytotoxicity (ADCC)
Cell-killing activity in which an antibody simultaneously binds the Fc receptor-bearing effector cells and antigen on the surface of target cells, leading to the destruction of targeted cells by the effector cells.

Antibody-dependent cell-mediated phagocytosis (ADCP)
Mechanism of cell killing in which an antibody simultaneously binds the Fc receptor-bearing effector cells, such as macrophages, and antigen on the surface of target cells leading to the destruction of the target cells via phagocytosis.

Bacteriophage
Bacteriophages or phages are viruses that infect bacteria. Although bacteriophage λ is a temperate phage, in most cases a lytic cycle ensues following infection of the bacteria and

a clear plaque forms on a lawn of bacteria. Infection with filamentous phages such as M13 is not lethal and the host bacteria are not lysed. Instead, their rate of growth slows and they form turbid plaques on the bacterial lawn.

Complement
A group of plasma proteins that participates in the lysis of foreign cells and pathogens (a process known as complement-dependent cytotoxicity; CDC). The complement system also interacts with pathogens to mark them for destruction by phagocytes and participates in the induction of an inflammatory response.

Constant region
Portion of the antibody molecule exhibiting little variation and determining the isotype (class or subclass) of the antibody.

Fab
Also known as the "fragment antigen binding," the Fab fragment is a monovalent antigen-binding fragment of an antibody that consists of one light chain and part of one heavy chain (the variable region and the first constant region domain). It can be obtained by the digestion of intact antibody with papain or by genetic engineering techniques.

Fc
Also known as "fragment crystallizable," the Fc region is the portion of the antibody constant region at the carboxy-terminus of both heavy chains. It can be obtained by papain digestion of an intact antibody. Two Fabs and one Fc fragment comprise a complete IgG antibody.

Fv/scFv
The Fv is a monovalent antigen-binding fragment of an antibody composed of the variable regions from the heavy and light chains. Single-chain Fv (scFv) fragments are composed of the variable domains of heavy and light chains (V_L and V_H) joined by a synthetic flexible linker peptide. Thus, the scFv provides a fully functional antigen-binding domain expressed as a single polypeptide.

Hybridoma
A hybrid cell derived by a fusion between a normal cell (usually a lymphocyte) and a malignant cell (usually a myeloma cell). Commonly, a normal B lymphocyte producing a specific antibody is fused to a myeloma cell that does not produce any antibody; this results in an ability of the B cell to proliferate indefinitely (immortalized). The resulting hybridoma produces large quantities of a monoclonal antibody of the same specificity as that produced by the normal B lymphocyte.

Isotype
Isotypes are the different forms of the constant regions of the immunoglobulin heavy and light chains. In humans, there are two isotypes of light chains (κ and λ). There are five human classes of heavy chain: μ (IgM), δ (IgD), γ (subclasses IgG1 IgG2 IgG3, and IgG4), α (subclasses IgA1 and IgA2), and ε. Both, classes and subclasses are isotypes as they have different constant region sequences.

Variable region
The amino-terminal portion of the heavy and light chain of an antibody molecule that is variable and is responsible for antigen binding, determining the specificity of the antibody.

■ Antibodies are proteins produced by the immune system to combat pathogens and have long been appreciated for their exquisite specificity. The development of the hybridoma technology made it possible to immortalize single B cells, resulting in the production of unlimited quantities of antibodies of a single, well-defined antigen-binding specificity, known as *monoclonal antibodies*. However, the initial hybridoma-derived monoclonal antibodies were murine and highly immunogenic in humans. Advances in genetic engineering and expression systems have been used to overcome problems of the immunogenicity of rodent-produced antibodies, and to improve the ability of the antibodies to trigger human immune effector activity. The development of chimeric, humanized, and totally human antibodies, as well as antibodies with novel structures and functional properties, has further expanded the potential use of monoclonal antibodies as targeted therapeutics. As a consequence, recombinant antibody-based therapies are now used to treat a variety of diverse conditions that include infectious diseases, inflammatory disorders, and cancer. Today, these therapies are one of the fastest growing classes of biopharmaceutical therapeutics. The different strategies for developing recombinant antibodies and their derivatives are summarized and compared in this review.

1
Antibody Structure and Engineering

1.1
The Basic Structure of Antibodies

Antibodies, also known as *immunoglobulins*, are unique proteins with multiple properties that make them a critical component of the immune system. Their most notable property is that they exhibit high affinity and specificity for a vast array of different molecules known as *antigens*. In addition, antibodies have the ability to interact with and activate the host immune effector system. The basic structure of an antibody consists of two identical light chain polypeptides and two identical heavy chain polypeptides, linked together by disulfide bonds (Fig. 1a). The heavy (H) and light (L) chains are encoded by separate genes (Fig. 1b) and are organized into discrete globular domains separated by short peptide segments (Fig. 1a).

The antibody molecule typically exhibits a "Y" shape, with two "fragment antigen bindings" (Fabs) forming the arms, linked via a hinge to the "fragment crystallizable" (Fc) domain that forms the stem. The domain at the amino-terminus of both heavy and light chains contains the antigen-binding site. These domains are characterized by their sequence variability (variable region or V), and are termed the V_H and V_L regions to indicate the heavy and light chains, respectively (Fig. 1a) [2, 3]. The specificity of an antibody is dependent on the variable region sequence and the total number of different variable regions available in humans is estimated to be at least 10^{11}. However, the actual number of antibody specificities in an individual

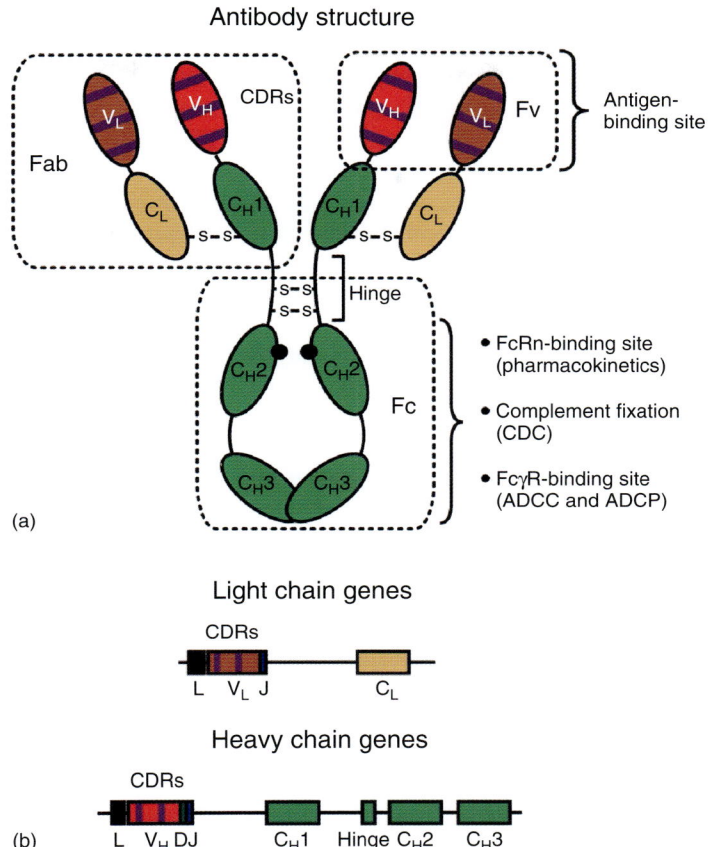

Fig. 1 IgG structure. (a) The general structure of the immunoglobulin G (IgG) molecule (the most abundant antibody in blood) and the active fragments that can be derived from it. The antibody protein is made of two light chains and two heavy chains with discrete domains: two domains constitute the light chain (V_L and C_L), while four domains make up the heavy chain (V_H, C_H1, C_H2, and C_H3). The variable regions are designated as the Fv region and include the complementarity-determining regions (CDRs) that are the critical amino acid residues for the affinity and specificity of the antibody-binding sites. The effector functions of the antibody are properties of the Fc fragment. The carbohydrate units (black circles) present at N297 within the C_H2 domains contribute to the functional properties of the antibody. The hinge region provides flexibility to the antibody molecule, facilitating antigen binding, and effector functions. The enzyme papain cleaves the antibody into two Fab fragments containing the antigen-binding sites, and an Fc fragment responsible for the effector functions; (b) Genes that encode the heavy and light chains. In the genes, each domain is encoded by a discrete exon (indicated by boxes) separated by intervening sequences (introns) indicated by the line. Both, heavy and light chains contain hydrophobic leader sequences (L) necessary for their secretion, a variable domain that contains three CDR sequences, which provides variability in recognition sites between the antibodies, and a joining segment (J). In the heavy chains there is also a diversity segment (D). Adapted from Ref. [1].

(known as the *antibody repertoire*) is limited to the total number of B cells and to the individual's history of exposure to antigens. The extraordinary variability of the antibody response is the result of processes that include V(D)J recombination and somatic hypermutation (SHM). V(D)J recombination is the process by which the V, D, and J gene segments are randomly rearranged, leading to the development of great variable region diversity [4]. Within the heavy and light chain variable regions are three hypervariable or complementarity-determining regions (CDRs) that encode the antigen-binding pocket of the antibody. In humans and rodents, CDR1 and CDR2 of both the heavy and light chains are formed from the sequences of germline variable (V) gene segments. In contrast, the CDR3 loop is produced by the joining of the V segment with the joining (J) segment of the light chain, or with the diversity (D) and joining (J) segments of the heavy chain. For both the heavy and light chain, diversity is further amplified by the process of SHM that occurs within the variable regions of heavy and light chains. These mutations occur mainly by single nucleotide substitutions, with occasional nucleotide additions and deletions. During the course of the immune response, the binding affinity of the antibodies for antigen increases, a process that is process is known as *affinity maturation* and is the result of the selection within the germinal center of B cells producing antibodies of increasingly higher affinity [5]. The rate of SHM in mice and humans is approximately 10^{-5} to 10^{-3} mutations per base pair per generation, which is about one million-fold higher than the spontaneous rate of mutation in most other genes [6].

The remainder of the heavy and light chains has a relatively constant structure. The constant region of the light chain consists of only one domain, termed the C_L region. In contrast, the constant region of the heavy chain (C_H) is further divided into discrete structural domains stabilized by intrachain disulfide bonds: C_H1, C_H2, and C_H3 for IgG (see Fig. 1). The domain structure of antibodies is well suited for genetic engineering because it allows the exchange of functional domains carrying antigen-binding activities (Fabs or Fvs) or effector functions (Fc). The hinge region, a segment of the heavy chain between the C_H1 and C_H2 domains in IgG, provides flexibility to the molecule. The Fc region is responsible for the pharmacokinetic properties of the antibody and for effector functions such as complement-dependent cytotoxicity (CDC), antibody-dependent cell-mediated cytotoxicity (ADCC), antibody-dependent cell-mediated phagocytosis (ADCP), and placental transmission. Antibodies are glycoproteins and the carbohydrates present in the C_H2 domain of IgG have been shown to be essential for its effector functions.

1.2
Classes and Subclasses of Antibodies

The constant region of the heavy chain determines the class or isotype of the antibody. In humans, there are five different heavy-chain classes (Fig. 2) that are designated by lower-case Greek letters: γ for IgG; δ for IgD; ε for IgE; α for IgA, and μ for IgM. Differences in the isotype of the heavy chain determine the glycosylation and the ability to engage in various effector functions such as CDC, ADCC, and ADCP, as well as placental transmission [7]. Human IgG antibodies can be subdivided into four subclasses: IgG1, IgG2, IgG4 (all three with a molecular weight of approximately 146 kDa), and IgG3 with a molecular weight of approximately 165 kDa. The higher molecular weight exhibited by human IgG3

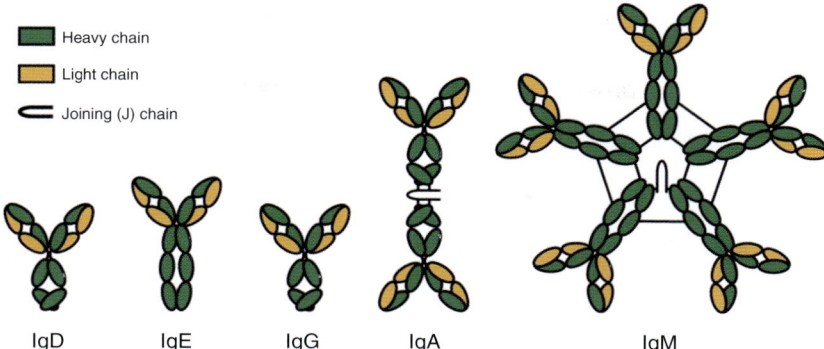

Fig. 2 The comparative structure of the human antibody classes. IgD (184 kDa), IgE (189 kDa), IgG (146 kDa for IgG1, 2, and 4, and 165 kDa for IgG3), IgA (dimer 300–360 kDa), and IgM (pentamer 970 kDa). Antibodies can be monomeric (IgG, IgE, and IgD) or polymeric (IgA: dimeric and IgM: pentameric), where the polymeric antibodies contain a joining (J) chain. Adapted from Ref. [10]. IgM can also be a hexamer lacking the J chain.

is due to the presence of an extended hinge region, which provides greater flexibility compared to the other subclasses. The properties of the human IgG subclasses are listed in Table 1. In humans, there is only one subclass of IgD, IgE, and IgM, but there are two subclasses of human IgA, namely IgA1 and IgA2. In contrast to the other classes, IgM and IgE heavy chains contain an extra C_H domain (C_H4) and lack the hinge region found in IgG, IgD, and IgA. The absence of a hinge region does not imply that the Fabs of IgM and IgE are rigid. Electron micrographs of IgM molecules bound to ligands have shown that the Fab arms can bend relative to the Fc domain [8]. In the case of IgE, the C_H2 domain pair replaces the hinge region of IgG and allows for a bent confirmation of the Fc region [9].

There are two different light-chain isotypes, which are designated by the lower-case Greek letters κ and λ. Light chains of both isotypes are found associated

Tab. 1 Functional properties of the human IgG subclasses.[a]

Characteristic or function	Subclass			
	IgG1	IgG2	IgG3	IgG4
Approximate molecular weight (kDa)	146	146	165	146
Hinge length (number of amino acids)	15	12	62	12
ADCC	+++	+/−−	++	+/−−
ADCP	+	+	+	+/−−
C1q binding	++	+/−	+++	−
Complement-mediated cell lysis	++	+/−	++	−
FcRn binding	+	+	+/−	+
Plasma half-life (days)	21	21	5–7.5	21
Approximate average plasma concentration (mg ml^{-1})	9	3	1	0.5

[a] Adapted from Refs [11] and [12].

with all the heavy chain classes. Differences in the isotype of the light chain do not appear to significantly influence either the structure or the effector functions of the antibody molecule.

Although all antibody molecules are constructed from the basic unit of two heavy and two light chains (H_2L_2), both IgA and IgM form polymers (Fig. 2). The IgM and IgA heavy-chain constant regions contain a "tailpiece" of 18 amino acids with a cysteine residue that is essential for polymerization. The J-chain is a 15 kDa polypeptide produced by B lymphocytes and plasma cells (antibody-secreting cells) that promotes polymerization by linking to the cysteine of the tailpiece. IgA forms a dimeric structure in which two H_2L_2 units are joined by a J-chain $[(H_2L_2)_2J]$, and IgM forms either a pentameric structure with five H_2L_2 units joined by a J-chain $[(H_2L_2)_5J]$ or a hexameric structure $[(H_2L_2)_6]$, which does not have the J-chain. Secretory IgA found at the mucosal surfaces contains an additional

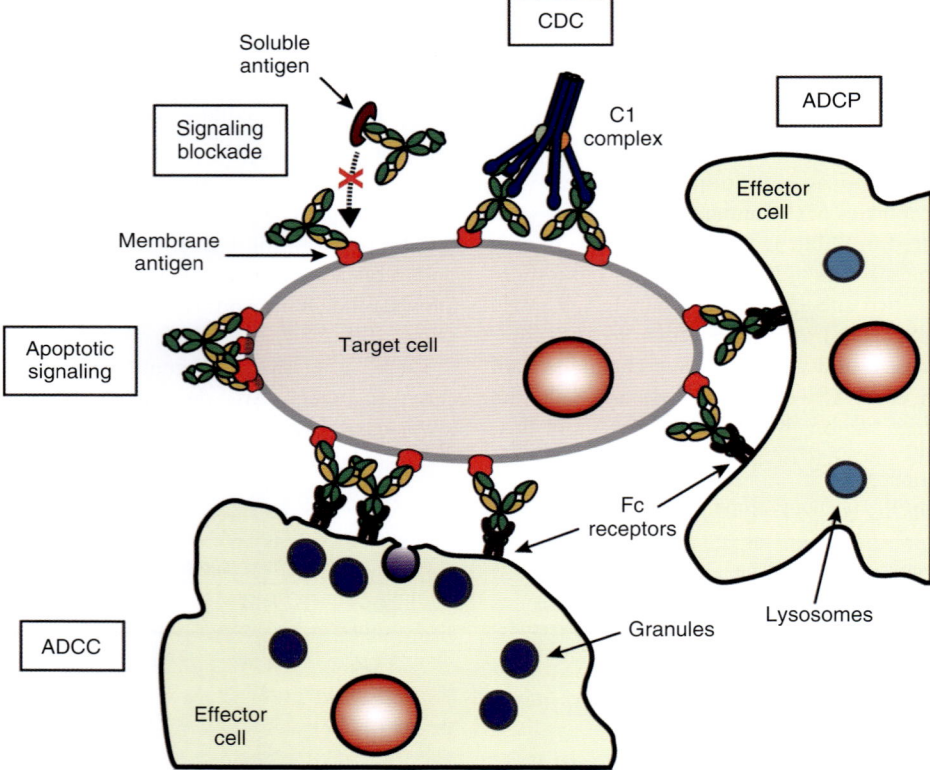

Fig. 3 Potential modes of action of antibodies. Antibodies can elicit their protective activity by blocking the interaction of soluble factors (ligands) to surface receptors by either targeting the free soluble factor or the receptor on the target cells, which may be a cancer cell. Targeting antigens on the surface of the cells may alter signaling and induce an antiproliferative and/or proapoptotic activity that can be favored by crosslinking. In addition, antibodies such as IgG1 can activate immune effector functions such as ADCC, ADCP, and CDC.

polypeptide, the secretory component (SC), a cleavage fragment of the polymeric immunoglobulin receptor used to transport it to the mucosal surface.

Antibodies exhibit different biological activities that make them effective therapeutics (Fig. 3) [11]. Because of their exquisite specificity, antibodies can prevent soluble mediators such as cytokines, toxins, or growth factors from reaching their receptors, blocking their interaction either by binding to the factor itself or to its receptor. Antibodies can also crosslink the cell-surface receptors, thereby activating or inhibiting responses. The Fc region of IgG binds to the neonatal Fc receptor (FcRn), also known as the "*salvage receptor*" or *Brambell receptor*. This receptor is expressed by endothelial cells and has an important role in extending the half-life of antibodies by binding and internalizing IgG and recycling it back into the blood, thereby protecting the IgG from catabolism [13]. In addition, depending on the antibody class, the Fc domain can elicit immune activities such as ADCC though the interaction between the Fc region of IgG and the Fcγ receptor (FcγR) expressed on the surface of cells of the immune system including natural killer (NK) cells, macrophages, and neutrophils [14]. FcγRs can also lead to the phagocytosis of antibody-coated cellular targets by phagocytic cells such as macrophages and dendritic cells, a phenomenon described as ADCP [15, 16]. NK and other effector cells, once activated, release factors that can directly kill targeted cells, as well as cytokines and chemokines that may inhibit cell proliferation, enhance the immunogenicity of target cells, and attract other immune effector cells. In addition to ADCC and ADCP, the Fc domain of certain antibodies, can activate the complement cascade leading to inflammation, complement-mediated opsonization of targets, and CDC [16]. Although many therapeutic monoclonal antibodies can induce complement-mediated tumor cell killing *in vitro* [17], its relevance in the clinical setting for cancer treatment remains to be fully established.

Because of their properties, antibodies – in particular, recombinant IgGs – are currently one of the fastest growing classes of biotherapeutics. In blood, IgG is the most abundant class of antibody and the most stable, with a half-life of approximately 21 days for all subclasses except IgG3, whose half-life is approximately 7 days. As shown in Table 1, IgG1 exhibits significantly superior ADCC and CDC activities compared to IgG2 and IgG4 [11, 12]. Therefore, since the antibody class determines the nature of the immune effector functions, IgG1 has been the class of choice for most therapeutic antibodies [18].

2
Technological Milestones in Antibody Engineering

During the past 40 years, continuous technological breakthroughs have improved the therapeutic properties of antibodies, starting from the original murine monoclonal antibody ultimately resulting in fully human immunoglobulins (Fig. 4).

2.1
Murine Monoclonal Antibodies

In the early twentieth century, Paul Ehrlich coined the term "*magic bullets*" to describe the ability of antibodies to specifically react against disease agents and eliminate them without affecting normal tissues [20]. In the natural immune response, an antigen elicits a polyclonal response composed of a variety of antibodies (i.e., they are the products of

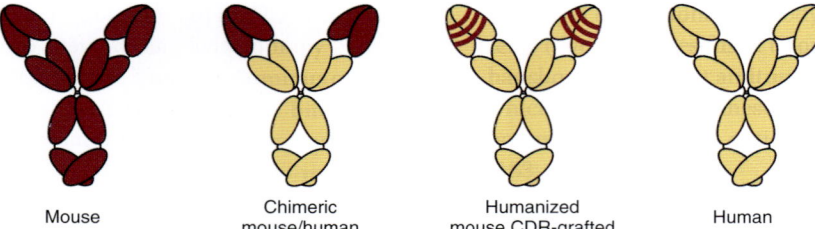

Fig. 4 Schematic representation of a murine monoclonal antibody, a mouse/human chimeric antibody, a "CDR-grafted" or humanized antibody, and a fully human monoclonal antibody. Chimeric antibodies have variable regions and binding specificities derived from a murine monoclonal antibody and human constant regions with their corresponding effector functions. Humanized antibodies are composed mostly of human sequences, except for the areas in contact with the antigen (CDRs), which are derived from mouse sequences. Adapted from Ref. [19].

many different antibody-producing cells) that includes antibodies with different variable regions as well as antibodies with the same variable regions with different constant regions. The humoral response is individualistic and is rarely the same in different people. This heterogeneity in the immune response, plus ethical and safety concerns, has made it difficult to effectively use human polyclonal antibodies for most therapeutic applications. However, a major breakthrough was made in 1975 when Georges Köhler and César Milstein developed a method for producing stable cell lines (named *"hybridomas"*) that secrete a homogeneous antibody with a known specificity (Fig. 5) [21]. This was achieved by fusing a normal B cell from the spleen of an immunized animal (initially a mouse or a rat) with a myeloma cell. The resulting hybridoma possesses the "immortality" of the malignant myeloma cell and the unique antibody characteristic of the normal B cell. Antibodies produced by hybridomas are monoclonal (i.e., they are the product of a single antibody-producing cell), and therefore have a single variable region associated with only one constant region. The "immortality" of the hybridoma ensures the continued availability of a well-characterized antibody. Once a hybridoma cell line is developed, it can be grown *in vitro* or *in vivo* for the large-scale production of monoclonal antibodies.

Due to their high affinity and exquisite specificity, murine monoclonal antibodies seemed to be the ideal *"magic bullets"* for diagnosis or therapy of multiple diseases, including cancer. However, murine monoclonal antibodies were rarely effective [22]. This lack of success can be attributed to several reasons, including the high immunogenicity of the foreign mouse proteins that results in the induction of a "human anti-mouse antibody" (HAMA) response, leading to the inactivation and elimination of the mouse antibody [23]. The immunogenicity of murine monoclonal antibodies also prevents their being administered multiple times, as is required to treat several diseases, including cancer. In addition, the human immune effector system interacts poorly with murine antibodies, and key immune effector functions mediated by the Fc region of the antibody, such as ADCC, are absent or diminished [24].

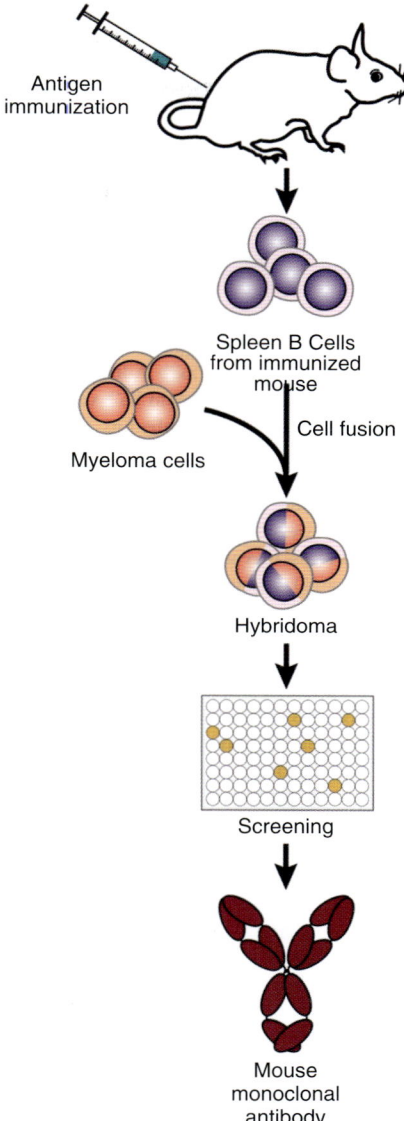

Fig. 5 Hybridoma production for the generation of mouse monoclonal antibodies. Mice immunized with an antigen of interest elicit an adaptive humoral immune response against the antigen. Splenic B cells from the immunized mice are harvested and fused with murine myeloma cells. The resulting antibody-secreting cells are screened to identify fused cells (hybridomas) specific for the antigen of choice. Single clones secreting the antibody with the desired properties are isolated using selection media that kill the unfused myeloma cells, while the unfused B cells die because they have a short lifespan. The new hybridomas are later subcloned to obtain a homogeneous cell line. Adapted from Ref. [3].

2.2 Chimeric Antibodies

To overcome the problems associated with the administration of murine monoclonal antibodies in humans, genetic engineering has been used to convert murine monoclonal antibodies to mouse/human chimeric antibodies (Fig. 4). In chimeric antibodies, the mouse heavy and light chain variable regions are joined to human constant regions [25]. Cloning of the DNA encoding the antibody variable regions has been greatly facilitated by polymerase chain reaction (PCR)-based procedures. A limited number of different hydrophobic leader sequences are found associated with the different variable regions, which makes it possible to design sets of oligonucleotide primers on the basis of either the framework or the leader regions coupled with downstream constant region primers that will amplify virtually all mouse variable regions from cDNAs generated by reverse-transcription polymerase chain reaction (RT-PCR) directly from hybridoma mRNA [26] (Fig. 6).

These products are cloned and sequenced, and then ligated into expression vectors containing the heavy and light chain constant regions for the construction of full-length chimeric antibodies, or for further genetic modifications (as described below). These constructs can be transfected into mammalian cells such as Chinese hamster ovary (CHO) cells or murine myeloma cell lines, such as P3X63Ag8.653, Sp2/0-Ag14, and NSO/1 that possess the machinery required to produce a functional antibody, including its correct assembly, glycosylation, and secretion [27]. Generally, chimeric antibodies retain their target specificity, while the human constant regions help to reduce the immunogenicity and HAMA response. In addition, the human constant regions are more effective in interacting with the human immune effector system, which results in improved ADCC, ADCP and CDC activities [28]. An example of a successful mouse/human chimeric IgG1 antibody which the United States Food and Drug Administration (FDA) and European Medicines Agency (EMA) have approved for clinical use is rituximab (Rituxan®/Mabthera®), which targets the CD20 antigen and is now widely used to treat non-Hodgkin lymphoma and autoimmune diseases. Other examples of chimeric antibodies are listed in Table 2.

2.3 Humanized Antibodies

Chimerization was a major advance in the engineering of therapeutic monoclonal antibodies, as it dramatically reduced immunogenicity and resulted in antibodies with improved effector functions. However,

Fig. 6 Recombinant technology for the generation of chimeric monoclonal antibodies from murine hybridomas. mRNA obtained from a mouse hybridoma cell line is reverse-transcribed and variable regions amplified (RT-PCR) using a set of primers flanking the V_L and V_H regions. The PCR products are cloned into intermediate cloning vectors, digested with restriction enzymes flanking the variable regions, and ligated into expression vectors containing human light and heavy constant region sequences. Here, the yellow rectangles are selectable markers for bacteria (amp – ampicillin) and the beige rectangles are selectable markers for eukaryotic cells (neo – neomycin/his – histidinol). The green segments are the eukaryotic expression promoters, the red arrows are the coding sequences for constant region κ and γ1, the violet arrow is the variable light (V_L), and the blue arrow is variable heavy (V_H). These vectors are transfected into mammalian cells, which are then screened for antibody production.

Genetic Engineering of Antibody Molecules

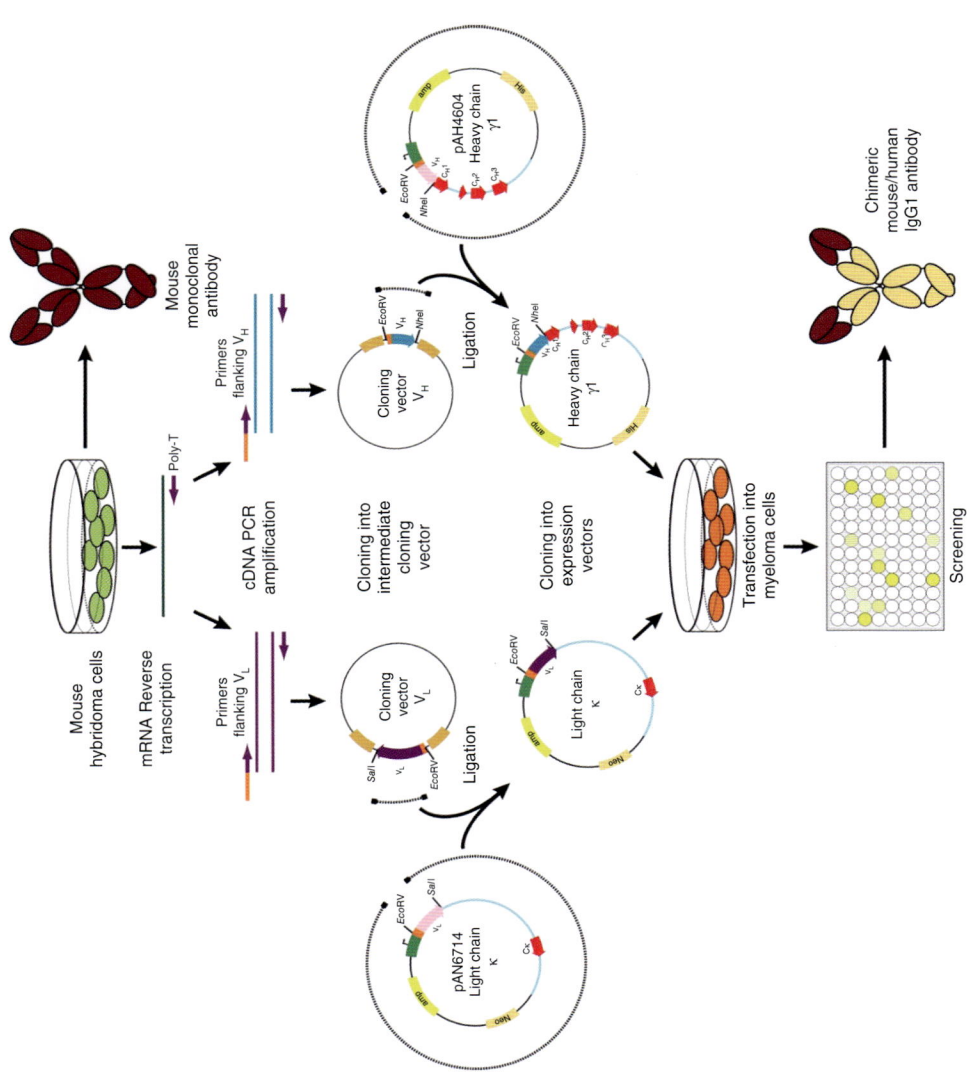

Tab. 2 Therapeutic monoclonal antibodies marketed in the European Union and United States.

International non-proprietary name (trade name)	Manufacturing cell line	Type	Target	Major indications	First EU (US) approval year
Abciximab (Reopro®)	Sp2/0	Chimeric IgG1 Fab	GPIIb/IIIa	To prevent the formation of blood clots	NA (1993)
Rituximab (MabThera®, Rituxan®)	CHO	Chimeric IgG1	CD20	Non-Hodgkin lymphoma, chronic lymphocytic leukemia, rheumatoid arthritis	1998 (1997)
Basiliximab (Simulect®)	Sp2/0	Chimeric IgG1	IL-2R	To prevent organ rejection after a kidney transplant	1998 (1998)
Palivizumab (Synagis®)	NS0	Humanized IgG1	RSV	To prevent serious lung disease caused by respiratory syncytial virus	1999 (1998)
Infliximab (Remicade®)	Sp2/0	Chimeric IgG1	TNF-α	RA, psoriatic arthritis, ulcerative colitis, Crohn's disease, ankylosing spondylitis, psoriasis	1999 (1998)
Trastuzumab (Herceptin®)	CHO	Humanized IgG1	HER2/*neu*	Breast and stomach cancers	2000 (1998)
Etanercept (Enbrel®)	CHO	Fc IgG1-(TNFR)2 fusion protein	TNF-α	Psoriatic arthritis	2000 (1998)
Alemtuzumab (MabCampath, Campath-1H®)	CHO	Humanized IgG1	CD52	B cell chronic lymphocytic leukemia	2001 (2001)
Adalimumab (Humira®)	CHO	Human IgG1	TNF-α	RA, JRA, psoriatic arthritis, ankylosing spondylitis, plaque psoriasis, and Crohn's disease, ulcerative colitis	2003 (2002)
Tositumomab-^{131}I (Bexxar®)	Hybridoma	Murine IgG2a	CD20	Non-Hodgkin lymphoma	NA (2003)
Cetuximab (Erbitux®)	Sp2/0	Chimeric IgG1	EGFR	Colon and rectum cancers, head, and neck cancers	2004 (2004)

Name	Cell line	Format	Target	Indication	Year
Ibritumomab tiuxetan (Zevalin®)	CHO	Murine IgG1	CD20	Non-Hodgkin lymphoma	2004 (2002)
Omalizumab (Xolair®)	CHO	Humanized IgG1	IgE	Asthma	2005 (2003)
Bevacizumab (Avastin®)	CHO DP-12	Humanized IgG1	VEGF	Metastatic carcinoma of the colon or rectum	2005 (2004)
Natalizumab (Tysabri®)	NSO	Humanized IgG4	α4-integrin	Crohn's disease (ulcerative colitis)	2006 (2004)
Ranibizumab (Lucentis®)	E. coli	Fab humanized IgG1	VEGF	Neovascular age-related macular degeneration, macular edema following Retinal vein occlusion, diabetic macular edema	2007 (2006)
Panitumumab (Vectibix®)	CHO	Human IgG2	EGFR	EGFR-expressing, metastatic colorectal cancer	2007 (2006)
Eculizumab (Soliris®)	NSO	Humanized IgG2/4	C5	Paroxysmal nocturnal hemoglobinuria, atypical hemolytic uremic syndrome	2007 (2007)
Abatacept (Orencia®)	CHO	Fc IgG1-(CTLA4)2 fusion protein	CD80/CD86	Rheumatoid arthritis	2007 (2005)
Certolizumab pegol (Cimzia®)	E. coli	Humanized IgG1 Fab, pegylated	TNF-α	Crohn's disease, rheumatoid arthritis, psoriatic arthritis	2009 (2008)
Golimumab (Simponi®)	Sp2/0	Human IgG1	TNF-α	Rheumatoid arthritis, psoriatic arthritis, ankylosing spondylitis	2009 (2009)
Canakinumab (Ilaris®)	SP2/0-Ag14	Human IgG1	IL-1β	TNFR-associated periodic syndrome	2009 (2009)
Catumaxomab (Removab®)	Hybrid hybridoma	Rat IgG2b/mouse IgG2a bispecific	EpCAM/CD3	Ascites ovarian cancer	2009 (NA)
Ustekinumab (Stelara®)	Sp2/0	Human IgG1	IL-12/23	Chronic plaque psoriasis	2009 (2009)
Tocilizumab (RoActemra®, Actemra®)	CHO	Humanized IgG1	IL-6R	Chronic idiopathic arthritis	2009 (2010)

(continued overleaf)

Tab. 2 (Continued.)

International non-proprietary name (trade name)	Manufacturing cell line	Type	Target	Major indications	First EU (US) approval year
Ofatumumab (Arzerra®)	NS0	Human IgG1	CD20	Chronic lymphocytic leukemia	2010 (2009)
Denosumab (Prolia®, Xgeva®)	CHO	Human IgG2	RANK-L	Prevent bone fractures in people with tumors that have spread to the bone	2010 (2010)
Belimumab (Benlysta®)	NS0	Human IgG1	BLyS	Active autoantibody-positive SLE	2011 (2011)
Ipilimumab (Yervoy®)	CHO	Human IgG1	CTLA-4	Advanced melanoma	2011 (2011)
Raxibacumab (ABthrax®)	NS0	Human IgG1	*B. anthracis* PA	Inhalational anthrax	NA (2012)
Brentuximab vedotin (Adcetris®)	CHO	Chimeric IgG1 conjugated to monomethyl auristatin E	CD30	Hodgkin lymphoma	2012 (2011)
Pertuzumab (Perjeta®)	CHO	Humanized IgG1	HER2/*neu*	HER2/*neu*-positive metastatic or locally recurrent unresectable breast cancer	2013 (2012)
Obinutuzumab (Gazyva®)	CHO	Humanized IgG1	CD20	Chronic lymphocytic leukemia	NA (2013)
Trastuzumab emtansine (Kadcyla®)	CHO	Humanized IgG1-DM1	HER2/neu	Metastatic breast cancer	2013 (2013)

Note: Information current as of 30 January 2014.

BLyS, B-lymphocyte stimulator; C5, component 5; CD, cluster of differentiation; CHO, Chinese hamster ovary; CTLA-4, cytotoxic T-lymphocyte antigen 4; DM1 or emtansine, derivative of maytansine; EGFR, epidermal growth factor receptor; EpCAM, epithelial cell adhesion molecule; Fab, antigen-binding fragment; GP, glycoprotein; IL, interleukin; IL-2R, interleukin-2 receptor; IL-6R, interleukin-6 receptor; JRA, juvenile idiopathic arthritis; NA, not approved; PA, protective antigen; RA, rheumatoid arthritis; RANK-L, receptor activator of NFκB ligand; RVO, retinal vein occlusion; SLE, systemic lupus erythematosus; TNF, tumor necrosis factor; TNFR, tumor necrosis factor receptor; and VEGF, vascular endothelial growth factor.

Cell names listed as in sources: European Medicines Agency public assessment reports, United States Food and Drug Administration (drugs@fda) and Ref. [18].

in some cases chimeric antibodies may elicit a significant human anti-chimeric antibody (HACA) response to the murine variable regions of the antibody, thereby decreasing the therapeutic efficacy of the chimeric antibody [29].

Each variable domain consists of a β-barrel with seven anti-parallel β-strands connected by loops, among which are the CDRs. One approach to further decrease immunogenicity is to make the variable region more human by transferring the CDRs of the murine monoclonal antibody to a scaffold of human origin, thereby creating "CDR-grafted" or "humanized" antibodies (Fig. 4). The process of CDR-grafting, developed in 1986, yields a recombinant antibody where only the CDRs are of murine origin [30]. However, frequently transferring only the CDRs from a murine antibody onto a completely human framework may not be sufficient because the resulting variable region has a diminished or no binding activity [31]. This issue can be addressed by transferring key murine framework residues along with the murine CDRs, with the compromise of a potential increase in immunogenicity. An example of a successful humanized IgG1 monoclonal antibody that has been approved for clinical use is trastuzumab (Herceptin®), which has demonstrated significant antitumor activity in patients affected with breast cancer overexpressing the tumor-associated antigen HER2/*neu* [32, 33]. Other examples of humanized antibodies are listed in Table 2.

2.4
Fully Human Antibodies

The next step in improving the properties of therapeutic antibodies was the generation of fully human monoclonal antibodies (Fig. 4). One approach to making completely human antibodies is to use phage display [34–37], where the heavy and light V-genes obtained from peripheral blood mononuclear cells (PBMCs) of a naïve or immunized donor are expressed as variable region fragments such as Fab or scFv (single-chain Fv) fused to proteins of filamentous bacteriophages (viruses that infect bacteria) (Fig. 7). Both the scFv and Fab fragments can be expressed on the surface of filamentous bacteriophages (f1, M13, and fd) as either single or multiple copies, depending on the phage protein used for fusion. Phage libraries can be generated using variable antibody gene repertoires from any species including humans, or even synthetic sequences [28]. The development of scFv libraries using a human donor is illustrated in Fig. 7. When functional antibody variable domains are displayed on the surface of filamentous phages, the resulting phages bind specifically to antigen, and rare phages can be isolated on the basis of their ability to bind antigen. Several rounds of enrichment consisting of binding to immobilized antigen, expanding the bound phage, and then further enriching by binding again to immobilized antigen, can yield specific phages [38]. The selected variable regions may have affinities similar to monoclonal antibodies and can be expressed as antibody fragments in the bacterium *Escherichia coli*, or they can be used to produce complete antibodies and expressed in mammalian hosts. In addition, specific variable regions can be further mutated and phages that express variable regions with increased affinity can be selected.

The decision of whether to produce scFv or Fab libraries depends partly on the intended use. The single-gene format of the scFv is an advantage for the construction of fusion proteins such as "immunotoxins" (antibodies or antibody fragments bound

Fig. 7 Generation of fully human monoclonal antibodies using recombinant DNA technology and phage display libraries. The sequences encoding the V_L and V_H are amplified from the mRNA of peripheral blood mononuclear cell-derived donor B cells using RT-PCR and used to construct a scFv library. The genetic sequence encoding each scFv is cloned into a phage display vector for expression on the surface of the bacteriophage. Bacteriophage from the resulting library of scFv clones are selected for their ability to bind antigen. Multiple rounds of binding, elution, and amplification result in a library that is highly enriched for phage displaying scFvs that recognize the target of interest. The scFvs from these phage are isolated and cloned into antibody expression vectors and transfected into myeloma or other cells appropriate for expression. Adapted from Ref. [10].

to a toxin), or for targeted gene therapy approaches where the scFv gene is fused to a viral envelope protein gene. The use of Fab fragments offers the advantage that heavy- and light-chain libraries can be produced independently and the genes encoding for the heavy and light chains randomly combined and tested for their ability to bind the desired antigen [34]. In theory, Fab libraries may be preferred where the final product will be an intact antibody since, in some instances, removal of the scFv linker might alter the antigen-binding properties. Another limitation of phage display is that it requires further subcloning into mammalian cells in order to obtain a fully functional antibody. Moreover, since the antibody fragments are generated in prokaryote systems, they lack the expression machinery present in the endoplasmic reticulum of mammalian cells, and scFvs can be susceptible to the dimerization of V_H and V_L segments or aggregate formation in bacteria [39]. However, scFvs have been successfully converted into complete antibodies. Phage display has proved to be very effective in generating a variety of antibodies that are difficult to obtain using the hybridoma technology. Phage display can be used to obtain antibodies that are entirely human in sequence, but which bind to the same part of the antigen (epitope) as existing mouse monoclonal antibodies [28]. Antibodies to targets previously not feasible for use in immunization approaches (e.g., self-antigens, ubiquitous compounds, or toxic compounds) have been isolated by the selection of phage that bind antigen *in vitro*. The first human antibody developed from phage display libraries and approved for clinical use is adalimumab (Humira®), a human IgG1 specific for tumor necrosis factor-α (TNF-α). Adalimumab is both FDA- and EMA-approved for the treatment of several autoimmune diseases, including rheumatoid arthritis, ankylosing spondylitis and Crohn's disease [40].

The production of human monoclonal antibodies using animals is a viable alternative to the phage display technology (Table 3). Today, transgenic mice with a "humanized" humoral immune response can create antigen-specific antibodies that are totally human. One of these transgenic mouse strains, the *XenoMouse*™, has been modified to carry portions of the human IgH and Igκ coding sequences [41] (Table 3). This strain contains 80% of the human variable region repertoire, the genes for Cμ, Cδ and Cγ1, Cγ2, or Cγ4, as well as the *cis* elements required for their function in a murine background with partially disrupted endogenous mouse heavy and light chains. The human transgenes are compatible with mouse enzymes responsible for SHM, and class switching from IgM to IgG, supporting the generation of a diverse, affinity-matured immune repertoire similar to that of adult humans. Importantly, the immune system of the *XenoMouse*™

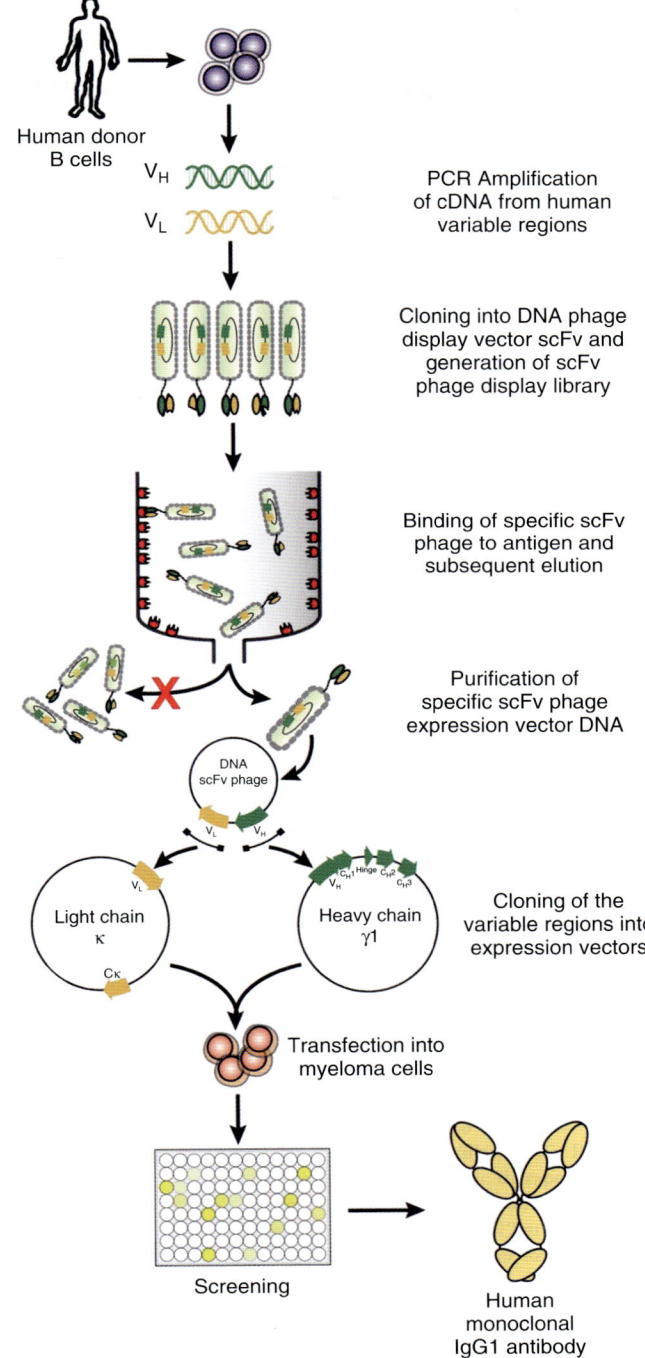

Tab. 3 Strategies for the generation of fully human antibodies.

Strategy	Description	Advantages	Disadvantages	Reference(s)
XenoMouse™	Mouse carrying YACs partially encoding the human IgH and κ loci that can produce human IgG1, IgG2, or IgG4 antibodies	Allows for the production of fully human antibodies with high affinities and does not produce murine heavy and κ light chains.	*in vivo* generation is time-consuming. Contains a limited repertoire of variable region genes and can only produce certain antibody classes.	[41, 42]
TC-Mouse™	Mouse carrying the entire unrearranged human heavy chain and κ loci	Similar to the *XenoMouse*™, but contains full repertoire of human heavy chain and κ light chain loci, which can produce a full polyclonal response.	*in vivo* generation is time-consuming. Hybridoma generation efficiency 10-fold lower than that of normal mice due to instability of κ chromosomal fragment.	[43]
KM-Mouse™	TC-Mouse™ carrying half the κ region stably integrated into its genome	A more stable and efficient hybridoma production compared with *TC-Mouse*™ and with polyclonal antibody yield similar to that of normal mice.	*in vivo* generation is time-consuming. Contains only half of the κ locus stably integrated into the genome.	[43, 44]
TC-Calf	Calf carrying a HAC encoding for the entire unrearranged human loci of both the heavy chain and λ light chain	A very high volume of human polyclonal antibodies produced from the entire human λ repertoire.	*in vivo* generation is time-consuming. The calf may also retain the bovine immunoglobulin loci.	[45–49]
Phage display	A library of PCR-amplified V heavy and light gene segments from PBMCs of a human donor cloned into and displayed on the surface of a phage	Rapid technique not subject to immune tolerance or specific for immunodominant epitopes. These can be retro-engineered to generate full-length human antibodies from libraries with greater than 10^{11} clone diversity.	scFvs can be susceptible to dimerization of scFvs V_H and V_L segments or aggregate formation. They may also require *in vitro* affinity maturation through the use of random mutagenesis.	[28, 34, 37]
Yeast cell display	Human scFv library genetically fused with α-agglutinin receptors on yeast cell wall	Rapid technique similar to phage display. The selection of antibodies of high affinity using flow cytometry allows for fast screening and production of scFvs.	Similar to phage display, requires the use of random mutagenesis to artificially mimic affinity maturation process. Has low transformation efficiency compared to phage display.	[50–52]

HAC, human artificial chromosome; PBMCs, peripheral blood mononuclear cells; scFvs, single-chain Fvs; YAC, yeast artificial chromosome.
Adapted from Ref. [3].

recognizes human antigens as foreign and, in conjunction with standard hybridoma procedures, this has resulted in human IgG monoclonal antibodies with a high affinity for human antigens (Fig. 8). In 2006, the FDA approved the anti-epidermal growth factor receptor (EGFR) IgG2 panitumumab for the treatment of colorectal cancer; this was the first fully human antibody to be generated with the *XenoMouse*™ technology.

Another transgenic mouse used for the production of human antibodies is the TransChromo mouse (*TC-Mouse*™) [43] (Table 3). The production of these mice was a multistep process that required the fusion of human fibroblasts with mouse embryonic stem (ES) cells. In the *TC-Mouse*™, the endogenous mouse immunoglobulin genes have been disrupted and therefore the mouse does not produce any murine antibodies. In addition, in contrast to the *XenoMouse*™, the *TC-Mouse*™ contains the entire human unrearranged heavy chain and the κ light chain loci. The *TC-Mouse*™ produces a very diverse repertoire of all classes of human antibodies, with a robust response to immunization with human antigens. The human antibody levels in these mice are similar to the murine antibody levels in normal mice, which makes it possible to construct hybridomas using the spleens of these mice. However, the stable hybridoma production efficiency is about 10-fold lower than that of normal mice, most likely due to the instability of the κ chromosomal fragment. A step forward in the improvement of this technology has been the generation of a transgenic mouse, the *KM-Mouse*™. This new strain contains only half of the κ region stably integrated into the mouse genome, and has a hybridoma production success rate that is similar to that of wild-type mice [43].

3
Expression Systems

A large variety of expression systems are currently in use for the production of genetically engineered antibodies and antibody fragments (Table 4). The variety of recombinant production approaches currently under development is rapidly growing in systems ranging from bacteria, yeast, plants, baculovirus and mammalian cells to transgenic organisms.

3.1
Prokaryotic Expression Systems

Bacteria are an important recombinant protein expression system because of their high yields (which are on the gram per liter scale), rapid growth, and ease of genetic manipulation. Gram-negative bacteria, such as *E. coli* have the ability to produce multi-ton yields of antibody annually, with the advantage of rapidly generating high levels of product at low-cost, using simple media (Table 4). However, proteins expressed in bacteria lack the glycosylation of mammalian cells, are frequently insoluble and/or inactive, and refolding may be required to obtain functional fragments [53, 54]. For this reason, bacteria tend to be favored for expression of small, non-glycosylated Fab and scFv fragments. When antibody fragments are expressed in the cytoplasm of *E. coli*, insoluble inclusion bodies, which require efficient refolding and renaturation processes associated with high production costs, often result. However, when preceded by a signal sequence, the functional product is secreted into the bacterial periplasm or culture supernatant without the need for refolding [55, 56]. In addition, Gram-positive bacteria such as *Bacillus brevis* have been successfully used for the production of antibody fragments;

Fig. 8 Generation of fully human monoclonal antibodies using the *XenoMouse*™ technology. Mouse embryonic stem (ES) cells are produced in which either the endogenous heavy or light chain locus has been disrupted by homologous recombination or in which yeast artificial chromosomes (YACs) encoding for the human IgH and Igκ loci have been introduced. These modified ES cells can give rise to either immune-compromised mice that are incapable of producing endogenous antibodies, or transgenic mice that can make both human and mouse immunoglobulins. Breeding the two mouse strains and backcrossing their progeny can generate a *XenoMouse*™ that produces only fully human antibodies. This mouse can be immunized with an antigen and generate fully human monoclonal antibodies using the hybridoma technology. Adapted from Ref. [3].

these have the advantage that, as they lack an outer membrane, they can secrete the recombinant protein directly into the medium [61] (Table 4). An example of successful expression in a bacterial system is the anti-vascular endothelial growth factor (VEGF) Fab ranibizumab, which has been approved for the treatment of age-related macular degeneration. In this case, the variable region of bevacizumab has been modified to improve affinity and, instead of production in mammalian CHO cells, it is expressed in *E. coli*, with significantly lower production costs [91].

3.2
Yeast Expression Systems

Eukaryotic cells naturally possess the machinery for correct protein folding, post-translational modifications, and secretory pathways required to produce proteins such as antibodies. Yeasts such as *Pichia pastoris* and *Sacchcaromyces cerevisiae* combine the properties of eukaryotic cells with the short generation time, ease of genetic engineering, and simple medium requirements of prokaryotes [62]. For example, *P. pastoris* can produce and secrete recombinant proteins, which simplifies the purification process, and can also grow at very high density, yielding more than 1 g/l of IgG [63]. Like other eukaryotes, yeasts attach oligosaccharides to proteins; however, the attached glycoforms differ from those present in humans and the resulting protein may be immunogenic, have altered functions, and a short half-life. To address this problem, glycoengineered *P. pastoris* strains have been generated that reproduce humanized glycosylation patterns via the deletion of yeast genes that encode for enzymes producing fungal-type or high-mannose glycans, and the insertion of enzymes for pathways that produce human-like glycosylation structures, except fucose [64]. Antibodies lacking fucose have been shown to exhibit an increased ADCC activity [92]. A version of the anti-HER2/*neu* antibody trastuzumab produced in glycoengineered yeast, which achieved a higher ADCC activity than the original trastuzumab produced in mammalian cells, has been described [65]. The yeast-produced antibody also exhibited pharmacokinetics and *in vivo* tumor growth inhibition comparable to that of the antibody produced in mammalian cells, which suggests that glycoengineered *P. pastoris* has potential utility for the production of therapeutic antibodies.

3.3
Insect Cell Expression Systems

The production of complex recombinant proteins in insect cells has great potential because insect cells perform signal-peptide cleavage, glycosylation, and efficient secretion (Table 4). Insect cells can be transfected with baculoviruses,

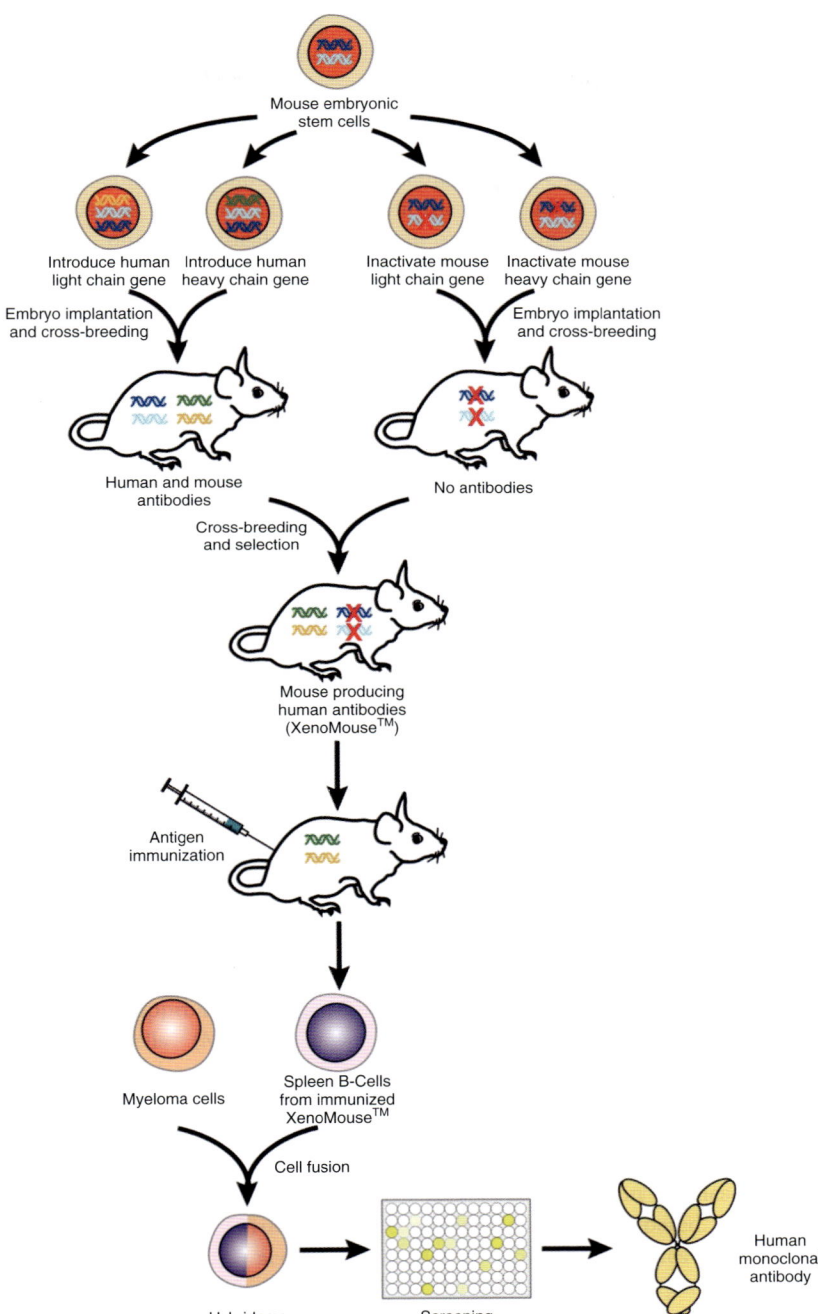

Tab. 4 Expression systems of recombinant antibodies.

Host class		Species	Antibody configuration	Advantages	Disadvantages	References
Prokaryote	Gram (−)	E. coli	IgG, Fab, VHH, F_V, V_L dAb, scFv, scFv-scFv	Low production cost. Possibility of product secretion in the culture supernatant	Does not make post-transcriptional modifications necessary for full antibody biological activity. Production of recombinant antibodies in the reducing cytoplasmic compartment results mostly in non-functional aggregates	[53–60]
	Gram (+)	Bacillus brevis	Fab	Directly secrete proteins into the medium due to the lack of an outer membrane. Some "generally regarded as safe" (GRAS) microorganisms can directly be used for oral application in humans	Does not make post-transcriptional modifications necessary for some biological activities in humans	[61]

Eukaryote	Yeast	*P. pastoris*	scFv, scFv-Fc, Fab, VHH-Fc, IgG1, sc(Fv)2, diabody, triabody	Makes post-transcriptional modifications necessary for biological activity. Have an advanced folding, post-translational, and secretion apparatus, which enhances the production of antibodies, including intact immunoglobulins, compared to bacteria. Do not produce bacterial endotoxins and have gained the GRAS status. All of these characteristics reduce the cost of production	The glycosylation is different from that of human cells. Yeasts tend to hyper glycosylate	[62–70]
		S. cerevisiae	IgG, Fab, VHH		heterologous proteins, which can influence activity of antibodies and is a potential source of immunogenicity or adverse reactions in human patients	[71, 72]
	Insect cells	*Spodoptera frugiperda*, SF-9	IgG, scFv	Makes post-transcriptional modifications required for biological activity. They can be efficiently transfected with insect-specific viruses from the family of Baculoviridae	Glycosylation is different from that of human cells	[73–77]
		Trichoplusia ni BTI-TN5B1-4	IgG			[78, 75]

(continued overleaf)

Tab. 4 (Continued.)

Host class	Species	Antibody configuration	Advantages	Disadvantages	References
Mammalian cells	CHO	IgG, scFv, VHH-Fc	Makes post-transcriptional modifications necessary for biological activity that are very similar to humans. Received regulatory approval for recombinant protein production	High production costs and difficult in handling. Small differences in the production process can Influence the final product, with implications in pharmacokinetics and effector functions of antibodies	[79–81]
	NSO/1	IgG			[82, 83]
	Per. C6	IgG, IgM			[84, 85]
Transgenic plants	*Nicotiana benthamiana*	scFv, IgG	Makes post-transcriptional modifications necessary for biological activity. Up-scaling of this production system can be achieved easily	Glycosylation is different from that of human cells. This procedure requires several months of transformation and special regeneration protocols	[86–88]
	Nicotiana tabacum	scFv			[89, 90]
Transgenic animals	Cattle	IgG	Makes post-transcriptional modifications necessary for biological activity. Have a competitive cost, a lower capital investment, the flexibility to modulate capacity, and the ability to assemble complex antibodies	Glycosylation is different from that of human cells. The generation of transgenic animals requires several months and requires a high initial cost	[45–47, 49]

dAbs, domain antibodies; Fv, Fragment variable; sc, single chain; V_H, variable heavy chains; VHH, camelid single domain antibodies; V_L, variable light chains.

an expression system that can be easily engineered to transiently produce recombinant proteins [73]. Full-length chimeric mouse/human recombinant monoclonal antibodies targeting the pre-S2 surface antigen of the hepatitis B virus (HBV) have been produced in baculovirus-infected insect cells [93]. This chimeric antibody, when expressed in insect cells, exhibited the same binding affinity as did the antibody expressed in murine myeloma cells. In addition, both antibodies showed the same C1q (the first protein in the complement cascade) binding capacity [93]. A human IgG1 antibody against factor rhesus (D) expressed in Sf9 insect cells also had the capacity to mediate effector functions, including ADCC [74]. However, the glycans attached to proteins expressed in insect cells lacked sialic acid and contained a higher mannose content compared to protein produced in mammalian cells, which may result in an increased clearance by the reticuloendothelial system and an increased risk of allergic reactions in humans [78].

In other studies, the recombinant human monoclonal antibody anti-gp41 transiently expressed in baculovirus insect cells was compared with the same antibody produced in CHO cells. *Trichoplusia ni "High Five"* insect cells express large amounts of antibody, comparable to the amount produced in some CHO transient expression systems [75]. Antibodies expressed in the insect cell line showed specific antigen binding. Also, the *"High Five"* produced antibodies that carried (as did the CHO-produced form) fucosylated N-glycans, including high levels of core α1,3-fucose. Although there is a great potential in this antibody production system, glycoengineering may be required to produce antibodies with optimal glycoforms for therapeutic use [75].

3.4
Mammalian Cell Expression Systems

Despite the high production costs and technical challenges, 60–70% of all recombinant proteins for therapeutic use, and 95% of the currently approved therapeutic antibodies, are produced in mammalian cell lines [94] (see Table 2). Complete functional antibodies have been most successfully expressed in mammalian cells, as these cells possess the systems required for correct antibody assembly, post-translational modification, glycosylation, and secretion (Table 4). Correct post-translational modifications are required to obtain the appropriate biologic properties and effector functions of full-length antibodies.

Examples of mammalian cells that have been used successfully to express correctly assembled and glycosylated antibodies and antibody fusion proteins are the mouse myeloma cell lines P3X63Ag8.653, Sp2/0-Ag14, and NSO/1 that lack the capacity to produce endogenous IgGs. Antibodies produced in cell lines such as CHO, HeLa, C6, and PC12 are also correctly assembled and glycosylated, with antibody structure and biologic functions that are virtually indistinguishable from those of the human immune system. Although mammalian cells grow slowly, require a longer time frame for production, and are more expensive than bacterial or yeast expression systems, they have been preferred when complete functional antibodies with correct glycosylation and disulfide bonds are required. These advantages result in a high product quality, which in turn reduces both effort and costs in any downstream processing steps. Current production following well-documented Good Manufacturing Practice (GMP)-compliant cell substrates and chemically defined media without animal protein components have eliminated the risks of contamination by pathogens or transmissible spongiform encephalopathy/bovine spongiform encephalopathy (TSE/BSE) agents [79]. IgG production at industrial levels often exceeds 12 g/l, and has even been reported to reach 27 g/l, obtained using the human embryonic retinal cell line Per.C6 [84]. The steady advance in mammalian cell culture technology has mainly resulted from the use of optimized production media, improved producer cell lines, and prolonged production times at high cell density.

3.5
Antibody Expression in Transgenic Plants

The use of plants to produce antibodies has several advantages that include low costs, fewer health risks from pathogen contamination, and the possibility to agriculturally scale-up manufacturing to produce (at least potentially) many tons of antibodies per year (Table 4). Plants have the machinery for complex N-glycosylation, the capacity to efficiently fold and assemble complex secretory antibodies, and the ability to target antibodies for storage in stable compartments such as seeds or tubers. However, differences exist in the glycosylation pattern between mammals and plants and these differences represent a significant barrier to producing therapeutic antibodies in plants [95]. In particular, plants attach 1,2-xylose and 1,3-fucose that can lead to immunogenicity of the therapeutic proteins and they also lack the capacity to add sialic acid. Consequently, the "humanization" of plants would require both the elimination of plant-specific sugars and the insertion of human glycosylation pathways [96]. Additionally, the use of transgenic plants will require issues regarding the environmental control or disposal of biomass with residual antibody to be addressed.

Most of the antibodies and antibody fragments expressed in plants, which are referred to as *"plantibodies"* have been produced in tobacco or corn. Recently, however, an *Agrobacterium*-mediated transient expression system has been used to produce a functional version of the anti-HER2/*neu* antibody, trastuzumab, in the tobacco plant *Nicotiana benthamiana*, using intron-optimized tobacco mosaic virus and potato virus X-based vectors [86]. Full-size antibodies, when extracted and purified from plant tissues, exhibited an equivalent binding to HER2/*neu* on the surface of SK-BR-3 human breast cancer cells compared to the original version of the antibody expressed in mammalian systems. The plant-produced antibodies also inhibited the proliferation of SK-BR-3 HER2/*neu*-expressing human cancer cells and, when injected intraperitoneally into mice, retarded the growth of xenografted tumors of SKOV3 human ovarian cancer cells overproducing HER2/*neu*. These results suggested that this approach might provide an attractive option for the large-scale production of therapeutic monoclonal antibodies, once the issues of glycosylation and immunogenicity have been fully resolved [86].

3.6
Transgenic Mammals

The use of passive immunotherapy with human polyclonal antibodies represents a therapeutic option for the treatment of infectious diseases, Rh incompatibility during pregnancy, some types of cancer, and medical emergencies such as snake and spider bites. In the past, polyclonal antibodies have only been available from human donors that cannot be repeatedly boosted with antigen. Efforts to overcome this problem have led to the development of large transchromosomic domestic animals (Tables 3 and 4). One of the most promising sources of human polyclonal antibodies is *transchromosomic calves* [45], in which antigen-specific human polyclonal antibodies can be produced in very large volumes without the participation of human subjects. A human artificial chromosome (HAC) containing the entire unrearranged loci of both the human heavy chain and λ light chain was created by fusing cells containing a HAC carrying a fragment of human chromosome 22 (λ locus) and a HAC with the fragment of human chromosome 14 heavy chain locus with cells that have both of the bovine immunoglobulin μ heavy chains (bIgHMs) and bovine immunoglobulin μ light chains (bIgMLs) homozygously inactivated [46]. However, these transchromosomic cattle still produce bovine IgG, and the production of human antibodies is low because B-lymphocyte development has been shown to be compromised by human–cattle species incompatibilities. This issue was addressed by engineering a new HAC containing both the hIgH, hIgK, and hIgL chromosome loci with the entire human immunoglobulin gene repertoire and the human VpreB (hVPREB1) and λ5 (hIgLL1) genomic loci from human chromosome 22 (hChr22). In addition, part of the C_H and transmembrane domains of the hIGHM gene were replaced by the corresponding bovine gene sequences. In this way, "bovinization" of the hIgM constant domain by replacing a part with the corresponding bovine IgM (bIgM) constant region sequence, led to the functionality of hIgM in supporting B-cell activation and proliferation being improved. Moreover, the new transchromosomic cattle produced physiological levels of antigen-specific human IgG upon immunization, which suggests that this strategy may be an effective alternative for the generation of human hyperimmune polyclonal serum for the treatment of human disease [47].

4
Variable Region Engineering

One of the essential properties of therapeutic antibodies is their binding affinity for the targeted antigen, which is determined by the variable regions. Given the critical role of this activity in antibody function, a major effort has focused on its optimization for increased therapeutic efficacy.

4.1
Improving the Affinity of Recombinant Antibodies

The mammalian immune system naturally produces antibodies with high-affinity antigen-binding receptors selected *in vivo* during affinity maturation. However, the generation of antibodies with extremely high (low picomolar) affinity *in vivo* can be difficult due to an "affinity ceiling" effect of the B-cell response that thermodynamically limits the on- and off-rates of the antibody–antigen binding interaction [97]. Nevertheless, *in vitro* affinity maturation can be used to isolate antibodies with up to femtomolar (fM) affinity [98]. Random mutagenesis is frequently used for *in vitro* affinity maturation, with random mutations being introduced into the variable regions of the light and heavy chains using an *E. coli* mutator bacterial strain [99], saturation mutagenesis [100], or error-prone PCR [101]. Targeted mutagenesis using alanine-scanning or site-directed mutagenesis can also be used to generate limited collections of specific variants of the parental antibody [102].

DNA shuffling [103], CDR shuffling [104] or light chain shuffling [105] can be used to generate variants of the parental antibody. Using *in-silico* strategies for affinity maturation, computational design has been used to achieve a more than 400-fold improvement in affinity compared to the parental antibody [106]. An iterative computational design method with emphasis on electrostatic interactions was used to successfully improve the affinity of recombinant monoclonal antibodies several fold compared to the parental antibody [107].

Antibodies against poorly immunogenic antigens and their fragments with improved affinity can also be derived using yeast cell display technology [108]. Human antibodies have been produced by yeast cell display using scFv with human germline variable regions genetically fused with the α-agglutinin adhesion receptor located on the cell wall of the yeast *S. cerevisiae*. Random mutagenesis can be used to simulate affinity maturation and to isolate antibodies with higher affinity. The cells are co-stained with fluorescently labeled antigen and an anti-epitope tag reagent. The advantage of this method is that, since the antibody expression is on the cell surface, it allows antibody selection by fluorescence-activated cell sorting (FACS). A potential drawback of this method is that libraries of only $10^6 - 10^7$ clones can be generated because of the low transformation efficiency of the yeast [50]. Efforts to diversify these libraries using combinatorial yeast mating have been successful, resulting in libraries of 5×10^9 clones. Screening a library with random mutations of a parental antibody yielded a high affinity variant with a 10,000-fold improved dissociation constant of 48 fM [51, 109].

4.2
Reduction of Immunogenicity

Because of its intrinsic characteristics, the variable region can be immunogenic. Additionally, short peptide sequences from antibodies bound to the major histocompatibility complex (MHC) class II on the surface of human antigen-presenting cells (helper T-cell epitopes) are often recognized by helper T cells. These T-cell epitopes are immunogenic and activate helper T cells, with the subsequent activation of B cells leading to a HAMA or HACA response. T-cell epitopes are frequently introduced in the process of engineering the variable regions to improve various properties. Several *in-silico* tools such as T-cell epitope databases can help in the identification of these motifs in the antibody sequence. For *De-Immunization*™, helper T-cell epitopes on an antibody are identified and removed by genetic alteration, yielding sequences with the minimum number of T-cell epitopes [110–112].

Most engineering techniques operate on the assumption that a greater global sequence identity of the humanized framework compared to the non-human counterpart will result in a decreased immunogenicity. The humanization strategy of grafting murine CDR in a human framework was a major advance [30]. A further improvement in the process of humanization consisted of the identification of specificity-determining residues (SDRs), which are critical for antigen binding, and the development of a monoclonal antibody variant that retained only the SDRs of murine origin [113]. However, global identity is only an estimation of the potential immunogenicity, and therefore, a method has been developed to identify the framework and CDR peptides or conformational motifs that may activate

T-helper cells. This method allows the determination of the human string content or "humanness" of murine variable regions [114]. The human string content is determined by using as a template a structurally equivalent region of the human variable sequences, thus comparing the proportion of human peptide strings within the variable regions of the murine antibody. The substitution of non-human regions with human regions can improve the human string content. It is expected that these antibodies will be more "immunologically human" compared to their CDR-grafted counterparts [114]. It is important to note that immunogenic murine variable regions show a low degree of "humanness," which means that they have low human content strings.

Even though the above-described *in-silico* tools are helpful, they tend to be overpredictive. On the other hand, the identification of T-cell epitopes can be performed using an *in vitro* CD4$^+$ helper T-cell assay. In fact, it was shown that modifying the CDR regions to reduce the number of T-cell epitopes identified when using this *in vitro* assay could serve as a valuable approach for generating antibodies with less immunogenicity [112].

5
Constant Region Engineering

Clinical experience, as well as research into effector mechanisms and the pharmacology of monoclonal antibodies, has identified limitations in their therapeutic action, due either to inadequate effector function or to a limited half-life. These insights suggest approaches for further improving antibody functions via modifications in the constant region, thereby creating a new generation of therapeutic recombinant antibodies [115].

5.1
Fc Engineering of IgGs for ADCC and ADCP Functions

The therapeutic action of the monoclonal antibody depends on the biology of the target, the role of the epitope it binds and the binding affinity, the effector function of the Fc domain and how this impacts the biological activity of the antibody, its immunogenicity and safety profile, biodistribution, pharmacokinetics, and pharmacodynamic profile.

The IgG subclass of choice for the development of a therapeutic antibody will be influenced by the effector functions that are required in order to achieve the intended therapeutic activity. The effector functions of the four human IgG subclasses are different and depend mainly on the binding affinities they have for the activating FcγR, FcγRI, FcγRIIa, and FcγRIII, the inhibitory receptor FcγRIIb, and the complement protein C1q [116]. Examples of Fc engineering for increased or decreased ADCC, ADCP, CDC, and half-lives, are listed in Table 5. An example of genetic engineering of the Fc region to enhance the effector functions is an IgG1 with the mutations S239D and I332E. This mutant IgG1 exhibits stronger binding to FcγR, and exhibits improved ADCC and ADCP activities compared to the wild-type antibody, with more effective cytotoxic activity against cancer cells [117].

Almost two-thirds of therapeutic antibodies currently in the market or in development are targeted at receptors or other cell-surface proteins. For many of these cell-surface targets the effector functions of ADCC, ADCP, and/or CDC would be detrimental and may pose a safety risk. A more effective antibody might result if the Fc function were to be reduced. Marketed products such as eculizumab

Tab. 5 Examples of Fc sequence changes for functional modifications.

Function	Effect	Subclass	Mutations	Reference(s)
ADCC	Increased	IgG1	S298A/E333A/K334A	[118]
			S239D/I332E	[117]
			S239D/A330L/I332E	[117]
			D280H, K290S	[119]
			F243L/R292P/Y300L	[120]
			F243L/R292P/Y300L/V305I/P396L	[120]
	Decreased	IgG1	G236A	[121]
			N297A	[122]
			L234A/L235A	[123]
			C220S/C226S/C229S/P238S	[124]
			C226S/C229S/E233P/L234V/L235A	[125]
			M252Y/S254T/T256E	[133]
			K326W	[131]
		IgG4	L235A/G237A/E318A	[126]
ADCP	Increased	IgG1	S239D/I332E	[117]
			S239D/A330L/I332E	[117]
			G236A	[121]
	Decreased	IgG1	C226S/C229S/E233P/L234V/L235A	[125]
CDC	Increased	IgG1	K326W	[128]
			K326W/E333S	[128]
		IgG2	E333S	[128]
	Decreased	IgG1	S239D/A330L/I332E	[117]
			C226S/C229S/E233P/L234V/L235A	[125]
			L234A/L235A	[130]
			D270A, K322A, P329A, P331G	[129]
Half-life	Increased	IgG1	M252Y/S254T/T256E	[130]
			T250Q/M428L	[131]
			N434A	[132, 133]
			T307A/E380A/N434A	[133]
			M428L/N434S	[134]
			L235A/G237A/E318A	[126]
		IgG3	R435H	[135]
	Decreased	IgG1	I253A	[133]
			P257I/N434H, P257I/Q311I or D376V/N434H	[136]

ADCC, antibody-dependent cell-mediated cytotoxicity; ADCP, antibody-dependent cell-mediated phagocytosis; CDC, complement-dependent cytotoxicity.

and the fusion protein abatacept address this issue. In the case of eculizumab, which is used to treat paroxysmal nocturnal hemoglobinuria, the Fc effector function has been largely reduced by the use of an IgG2–IgG4 fusion Fc that is known for its low effector function compared to human IgG1 (see Table 1) [137].

Another approach is to alter the glycosylation pattern of IgGs, since glycans are known to play an important role in the effector function of antibodies. The

N-linked oligosaccharide attached at N297 in the Fc region of human IgGs is important for both structure and function. The mutation N297A eliminates the glycosylation of IgG1 and renders the antibody unable to bind the FcγRs, and therefore, should not be able to elicit ADCC [122]. Otelixizumab is an aglycosylated anti-CD3 IgG1 antibody that has reached Phase II clinical development for the treatment of Type I diabetes [138].

5.2
Fc Engineering of IgGs for CDC Function

Cytotoxicity mediated by complement is one of the mechanisms that therapeutic antibodies can use to kill target cells (see Fig. 3) [139]. Improving this activity may render antibodies more effective as antitumor agents. It has been reported that the mutants K326A/E333S of human IgG1 can significantly increase the CDC activity of the anti-CD20 antibody rituximab [128]. Moreover, when those mutations were introduced into IgG2, which naturally lacks CDC activity, it gained C1q binding and CDC function [128]. These results suggest that antibody molecules containing point mutations at K326 and E333 may be useful for producing therapeutic antibodies that are more effective in complement activation.

In contrast, the generation of some recombinant monoclonal antibodies may require abrogation of the CDC function, especially as complement-mediated activity has been linked with injection site reactions [140]. An Fc mutant has been developed in which the S239D/A330L/I332E mutation eliminates C1q binding in the context of IgG1 [117]. A therapeutic FDA-approved antibody with a CDC-engineered Fc region is abatacept, a cytotoxic T-lymphocyte antigen 4 (CTLA-4)-Fc fusion product that binds CD80 and CD86 on T and B cells. In this recombinant protein the C1q- and FcγR-binding sequences were modified to eliminate CDC and ADCC activities [124].

5.3
Antibody Engineering for Altered Half-Life

Another important approach to improving antibody therapeutic performance is modification of the Fc region in order to prolong the half-life, with the aim of reducing the frequency of administration. Examples of these modifications are listed in Table 5. One of the most relevant predictors of a long half-life for IgG is its pH-dependent binding to FcRn (*Brambell receptor*). Some antibodies that preferentially bind FcRn at pH 6.0, but not pH 7.4, show an extended half-life in animal models. A good example is the "YTE" mutant, which contains three mutations (M252Y, S254T, T256E) and exhibits an eight- to ten-fold improvement in binding to the FcRn at pH 6.0. Interestingly, compared to the wild-type version of the same antibody, the YTE mutant exhibited a four-fold longer half-life in non-human primates [130]. The main therapeutic candidate containing this mutation is the anti-respiratory syncytial virus (RSV) antibody MEDI-557, a third-generation therapeutic antibody derived from motavizumab (second-generation), an affinity-improved version of palivizumab (first-generation) [141]. Pharmacokinetic studies of this Fc-modified antibody in healthy adults have shown it to have a two- to four-fold longer half-life, reaching up to 100 days compared to the parental antibody motavizumab, which has a serum half-life of about 24 days in children [142]. The pattern of pH-dependent binding of the Fc region of the antibody to FcRn may not always predict the half-life, however. For example, an anti-HBV IgG1 with the T250Q/M428L mutations exhibited a two-fold increase in half-life [131], while an anti-(TNF-α) IgG1 with the same mutations failed to show an increase in half-life. This suggests that the sequence of the variable chain may also contribute to the half-life in antibodies with this set of mutations [143].

5.4
Antibodies of the IgE Class

Although IgG is currently the class of choice for most antibodies used for clinical applications, IgE is emerging as an interesting therapeutic option for cancer therapy. There is a growing interest in the use of IgE in cancer treatment as part of the field of research termed "AllergoOncology," that aims to reveal the function of IgE-mediated immune responses against tumor cells and to develop novel IgE-based treatment options against malignancies [144]. A significant advantage of IgE compared to IgG is the exceptional affinity of IgE for FcεRI; this affinity is two to five orders of magnitude higher than that of IgG for the FcγRs (FcγRI-III). The affinity of IgE for FcεRII (CD23) in its trimeric form is as high as that of IgG for its high-affinity receptor FcγRI. Both, FcεRI and FcεRII can be found on the surface of relevant human immune effector cells; in fact, it has been reported that tumor-specific IgE can elicit effector functions such as ADCC and ADCP [145–147]. A further advantage is that IgE is present in the serum at low levels compared to IgG, providing less competition for Fc receptor occupancy. A fully human anti-HER2/*neu* IgE containing the variable regions of the scFv C6MH3-B1 specific for HER2/*neu*, which binds a different epitope than that of trastuzumab, has been developed and shown to elicit an *in vitro* and *in vivo* degranulation of effector cells [148]. In addition, this anti-HER2/*neu* IgE enhances antigen presentation in human dendritic cells *in vitro*, prolongs the survival of human FcεRIα transgenic mice bearing intraperitoneal murine mammary tumors expressing human HER2/*neu*, and was well tolerated in a preliminary study in non-human primates [148]. A chimeric IgE targeting the human prostate-specific antigen (PSA) has also been developed and, when complexed with PSA, significantly prolongs the survival of human FcεRIα transgenic animals challenged with PSA-expressing tumors in a prophylactic vaccination setting [149].

6
Structurally Modified Antibodies

The growing number of therapeutic targets and special requirements for the different therapeutic interventions has favored the emergence of a broad spectrum of antibody configurations, with properties beyond those of the natural antibodies, tailor-made for specific applications. The domain structure of antibodies allows the generation of alternative formats, where the different constant domains are removed or rearranged to alter their biological properties.

6.1
Antibody Fragments: Fabs and scFvs Monovalent, Bivalent, and Multivalent

A primary binding block of the antibody is the Fab fragment, which consists of a light chain covalently linked via a disulfide bond to a heavy chain segment that includes the V_H and C_H1 domains (Fig. 9) [150]. An example of a therapeutic monovalent antibody fragment is the anti-VEGF Fab ranibizumab, an inhibitor of angiogenesis that has been FDA- and EMA-approved

Genetic Engineering of Antibody Molecules | 119

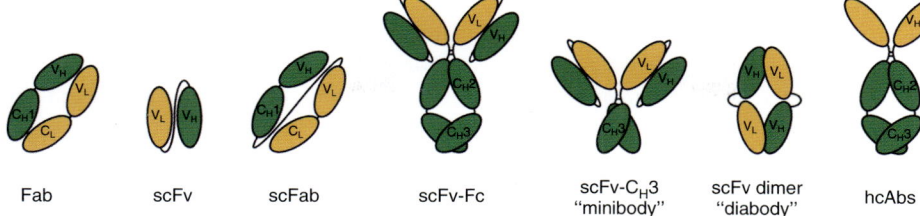

Fig. 9 Schematic representation of various genetically engineered antibody fragments. The relative size and domain relationships between engineered Fab, single-chain Fv (scFv), single-chain Fab, scFv-Fc, scFv-C_H3 or minibody, scFv dimer or diabody, and hcAbs fragments are shown. Adapted from Refs [10] and [152].

for the treatment of age-related macular degeneration [151].

As explained previously, scFv fragments consist of the variable domains of heavy and light chains (V_L and V_H) genetically fused by a flexible linker peptide. The molecular weight of scFvs is much lower than that of intact antibodies (Fig. 9) (25–27 kDa versus approximately 146 kDa for intact IgG1), but the scFv still may retain the specificity and the affinity of the Fab of the full-length antibody. However, fragments lacking the Fc region or containing truncated Fc regions have no immune effector functions and are incapable of binding to the FcRn salvage receptor with the consequent reduction in half-life (hours compared to 21 days for IgG1) [153].

scFv fragments display a combination of rapid, high-level tumor targeting associated with a rapid clearance from normal tissues and the circulation. Radiolabeled scFvs are important tools for the detection and treatment of cancer metastasis in both preclinical models and patients [150]. Unlabeled scFvs can also have therapeutic efficacy. An unlabeled scFv that is currently being evaluated in the clinic is DLX105, a humanized scFv specific for TNF-α, a major pro-inflammatory cytokine important for inducing and perpetuating peripheral hyperalgesia and cartilage degeneration in patients with osteoarthritis. DLX105, expressed in bacteria, has been used in two clinical trials for the treatment of severely painful osteoarthritis of the knee (NCT00819572) and mild to moderate *psoriasis vulgaris* (NCT01595997). Both trials showed that DLX105 met initial safety and efficacy endpoints.

One limitation of scFvs compared to full-length antibodies is their monovalent binding to antigen. Because of the presence of two antigen-binding sites, intact antibodies exhibit improved avidity compared to scFvs. To address this issue, dimers of scFv named "diabodies" have been produced by incorporating a carboxy-terminal cysteine residue to form a disulfide bridge, yielding (scFv)$_2$ fragments (Fig. 9) [154]. Another configuration with small size is the heavy chain antibodies (hcAbs) derived from camelids, which lack light chain and C_H1 domain, are bivalent, and bind its target through V_HH domains [152] (Fig. 9). The avidity of antibody fragments can be further increased in the diabody approach by decreasing the length of the interdomain linker peptide, which may result in the formation of triabodies and tetrabodies [155]. Alternately, scFv dimers have been constructed using a very short linker peptide to connect the antibody variable regions, resulting in the formation of "cross-paired" dimers, in which the V_L of one molecule associates with the V_H

of a second, while the V_L of the second molecule associates with the V_H of the first [156] (Fig. 9). These noncovalent dimers have a molecular weight similar to that of the antibody Fab fragments (55–60 kDa) but contain two antigen-binding sites and are capable of bivalent binding to antigen, exhibiting a significant improvement in *in vivo* targeting compared to standard scFv. Another bivalent configuration of scFv fragments includes their fusion to an immunoglobulin C_H3 domain, resulting in a self-assembling "*minibody*" [157] (Fig. 9). Moreover, by fusing scFvs to protein domains normally involved in protein association, such as helix bundles or leucine zippers, it is also possible to obtain larger bivalent or multivalent fragments [158].

6.2
Bispecific Antibodies

In contrast to natural antibodies, bispecific antibodies have two different variable regions, each one with the capacity to recognize one distinct antigen [159]. Originally, the generation of these molecules required the fusion of two different hybridomas yielding a "*hybrid hybridoma*" or "*quadroma*" [160]. However, the bispecific antibodies produced in this way are part of a mixed population of ten different potential pairings, making it difficult to isolate the desired molecule. One strategy to address this problem is the "knobs-into-holes" approach [159], where a mutation is made in one of the C_H3 domains to insert a large amino acid. This results in a protrusion or knob in one side, while in the other C_H3 domain a small residue is inserted in the same region, creating a hole that constitutes a novel site which favors heterodimeric Fc formation. However, light chain mispairing still occurs in these molecules [159]. An alternative approach for producing bispecific antibodies is to fuse a scFv of one specificity after the hinge (Hinge-scFv) or at the carboxy-terminus (C_H3-scFv) of an antibody with a different specificity [166].

Catumaxomab is the first bispecific monoclonal antibody approved in the European Union for therapeutic use. This hybrid hybridoma was produced via quadroma technology, and contains one mouse IgG2a heavy chain and one rat IgG2b heavy chain targeting the human epithelial cell adhesion molecule (EpCAM) present on the tumor cells and CD3 present on T cells [161]. The interplay of several immune effector mechanisms in the tumor microenvironment results in a multifaceted immune reaction, leading to the elimination of tumor cells. In preclinical studies several killing mechanisms were identified, including T-cell-mediated lysis, cytotoxicity by released cytokines, ADCP, and ADCC. As expected, a HAMA and human anti-rat antibody (HARA) response was observed in the majority of patients (>70%) after the last infusion [162]. It is important to note that a combination of the parental antibodies showed a much lower antitumor activity compared with the bispecific antibodies *in vitro*, as well as in a mouse model, in which long-lasting antitumor immunity was seen with the bispecific antibody [162]. Since bispecific antibodies have already shown promise in clinical trials and in the clinic, it is expected that they will continue to be used as effective therapeutic strategies.

6.3
Polymers of Monomeric Antibodies

Multivalent antibodies are associated with more efficient antigen binding which, together with multiplied effector functions, may increase the potential therapeutic

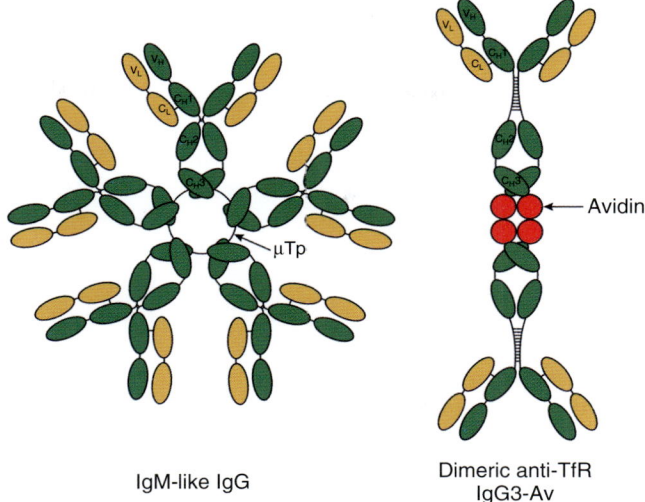

Fig. 10 Schematic representation of the structure of two different polymeric IgGs. IgM-like IgG molecules form a covalent assembly of several IgG monomers due to the presence of a cysteine residue in the μ tailpiece (μTp) of human IgM that is added to the carboxy-terminus of the heavy chain of human IgG through genetic engineering. A polymeric antibody can also be generated when a non-antibody partner such as avidin is fused to the carboxy-terminus of the C_H3 domain of an IgG, as in the case of the dimeric anti-TfR IgG3-Av. Since avidin exists naturally as a non-covalent tetramer, the IgG–avidin fusion protein has a dimeric structure. The molecule shown schematically is IgG3 with its long extended hinge region.

benefits. IgM is a naturally occurring polymeric antibody (see Fig. 2), but it does not bind FcγRs [8]. In contrast, IgG effectively triggers effector functions mediated through FcγRs. Therefore, for several applications it is desirable to have antibodies such as IgG in an IgM-like (polymeric) structure. Polymeric human IgG has been developed by fusing the 18-amino acid carboxy-terminal tailpiece from human μ chain to the carboxy-terminus of the γ constant region (Fig. 10) [164]. Using IgG1, IgG2, IgG3, and IgG4, IgM-like antibodies were generated with up to six Fc regions and 12 antigen-binding sites. These antibodies showed an increased avidity for the antigen and the FcγRs. Not surprisingly, the complement activity of normally active IgG1 and IgG3 and somewhat less-active IgG2 antibodies is dramatically enhanced upon polymerization. Unexpectedly, the polymeric IgG4 was able to direct the complement-mediated lysis of target cells almost as effectively as the other polymers, even though monomeric IgG4 lacks complement activity. This strategy has been used to develop IgM-like IgG3 anti-Rh antibodies as direct hemagglutinating agents [165].

An alternative strategy for creating polymers of IgG is to genetically fuse chicken avidin to the carboxy-terminus of each of the heavy chains of an IgG [166–168]. This approach is based on the fact that both streptavidin and avidin are tetramers of four non-covalently linked monomers. As each antibody–avidin protein contains two molecules of avidin (one genetically fused at the carboxy-terminus of each heavy chain), two independent antibody-fusion proteins bind to each other through their respective avidins forming a dimeric (tetravalent) structure (Fig. 10). One example of this approach is a mouse/human chimeric IgG3-avidin fusion protein specific for the human transferrin receptor 1 (TfR1, also known as *CD71*) [167–169]. Anti-hTfR IgG3-Av (also known as *ch128.1Av*) was designed to function as a universal vector to deliver different biotinylated compounds into cancer cells overexpressing TfR1 [167, 169–171]. Moreover, this fusion protein possesses enhanced antiproliferative/proapoptotic activity against malignant hematopoietic cells compared to the original monoclonal or the parental IgG3 without avidin [167, 168, 170]. Given its dimeric structure, it is possible that crosslinking of the surface TfR1 may be responsible, at least in part, for the cytotoxic activity. It is important to note that the presence of avidin fused to the C_H3 domain does not interfere with the ability of the fusion protein to bind FcγRs and C1q [170]. Moreover, ch128.1Av elicits significant anticancer activity in two multiple myeloma xenograft models, suggesting that ch128.1Av could be an effective therapy against multiple myeloma and other hematopoietic malignancies [170].

These findings demonstrate that it is possible to transform an antibody specific for a cell-surface receptor that exhibits minimal inhibitory activity into a novel drug with significant intrinsic cytotoxic activity against selected cells by fusing it with avidin. More importantly, since the antibody fusion protein is a delivery system, this property can also be exploited to deliver biotinylated therapeutics into cancer cells, rendering cells that were resistant to the antibody fusion protein alone sensitive to treatment [171, 172]. Further development of this technology may lead to effective therapeutics for the *in vivo* eradication of hematological malignancies and the *ex vivo* purging of cancer cells in autologous transplantation.

6.4
Antibody Fusion Proteins

Anti-hTfR IgG3-Av, as mentioned above, is an example of an antibody fusion protein. However, antibodies can be fused to other biological agents such as cytokines, which represents another promising avenue in the field of immunotherapy. The extraordinary specificity of antibodies, together with their functional properties, facilitates the targeting of therapeutic agents to improve their efficacy. These antibody fusion proteins have been prepared in a broad variety of configurations exploiting the variable regions of the antibody for its binding properties, harnessing the effector functions (including an increase in half-life of the cytokine due to its fusion with the antibody), or taking advantage of the sum of all these activities (Fig. 11).

Non-antibody partners such as cytokines can be genetically fused to the end of the C_H3 domain (C_H3-cytokine) (Fig. 11a, A1), combining the specificity and antibody-related effector functions to deliver the biological activity of the fused moiety to the intended target. Antibody fusion proteins with this configuration have been developed to target cytokines such as interleukin-2 (IL-2), granulocyte macrophage-colony stimulating factor (GM-CSF), and interferon-α (IFN-α) to tumor cells [7, 173–176]. This strategy can be used to concentrate the cytokine in the tumor microenvironment and to enhance the direct tumoricidal effect of the cytokine and/or the antibody and potentiate the host immune response against the tumor. At the same time, it extends the half-life of the cytokine while limiting the severe toxic side effects associated with a high dose of

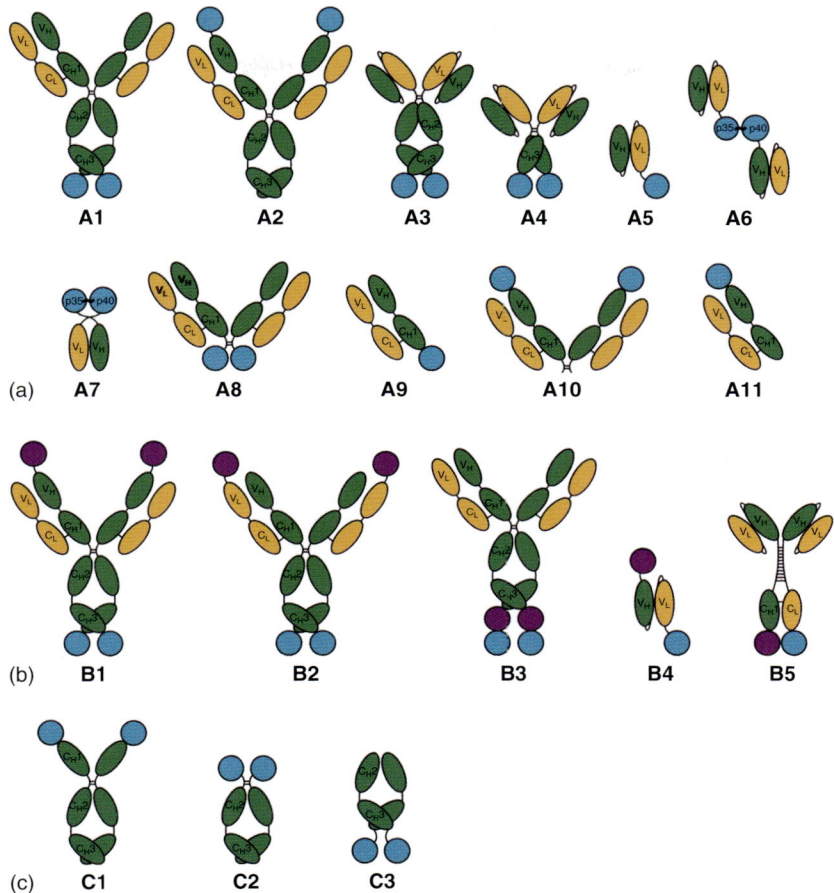

Fig. 11 Schematic representation of recombinant fusion proteins. Non-antibody proteins such as cytokines, ligands, or toxins can be fused to different fragments of the antibody to generate an antibody fusion protein. (a) A non-antibody protein (blue circle) can be genetically fused to the carboxy- (A1) or amino-terminus (A2) of the full-length antibody or to different antibody fragments (A3–A11) to generate an antibody–fusion protein; (b) Two different non-antibody proteins (blue and purple circles) can be fused in a single (B1–B2 and B4–B5) or in tandem format (B3) at the ends of the same antibody or antibody fragment. The domain structure of the antibodies facilities their use to develop bifunctional antibody fusion proteins that can consist of non-antibody protein (fused to the amino-terminus of the heavy (B1) or light chain (B2)) and the second non-antibody protein (fused to the carboxy-terminus of the heavy chain). The heterominibody (DCH) (B5), consists of one protein fused to the C_H1 domain, the second different protein fused to the C_L domain linked both to two scFvs by the IgG3 hinge region; (c) C1–C3 represent three fusion proteins that lack antibody variable regions with the non-antibody protein fused to the amino-terminus of the C_H1 domain (C1), immediately before the hinge (C2), or fused to the carboxy-terminus to the C_H3 domain (C3). Adapted from Refs [10] and [19].

Tab. 6 Examples of antibody–cytokine fusion proteins (AbFPs) in clinical development.

AbFP	Cytokine	Target antigen	Format	Indication	Intervention	Clinical trials identifier	Status	First received year
hu14.18-IL-2	IL-2	GD2	IgG1-IL-2	Melanoma	Alone	NCT00590824	Phase II	2007
				Neuroblastoma	Alone	NCT00082758	Phase II	2004
				Neuroblastoma	Sargramostim and isotretinoin	NCT01334515	Phase II	2011
Dl-Leu16-IL2	IL-2	CD20	IgG1-IL-2	B-cell non-Hodgkin lymphoma	After rituximab	NCT00720135	Phase I	2008
				B-cell non-Hodgkin lymphoma	Alone	NCT01874288	Phase I/II	2013
F16IL2	IL-2	Tenascin-C A1	scFv-IL-2	Advanced solid tumor/breast cancer	Doxorubicin	NCT01131364	Phase I/II	2010
				Merkel cell carcinoma	paclitaxel	NCT02054884	Phase II	2014
				Merkel cell carcinoma	paclitaxel	EUCTR2012-004018-33-DE	Phase II	2012
L19IL2	IL-2	Fibronectin ED-B	scFv-IL-2	Advanced solid tumors	Alone	NCT01058538	Phase I/II	2010
				Pancreatic cancer	Gemcitabine	NCT01198522	Phase I	2010
				Metastatic melanoma	Dacarbazine	NCT01055522	Phase II	2010
				Melanoma	Alone	NCT01253096	Phase II	2010
				Metastatic melanoma	Dacarbazine	EUCTR2012-004495-19-DE	Phase I/II	2012
				Melanoma	L19TNFα	EUCTR2012-001991-13-IT	Phase II	2012
				Melanoma	Alone	EUCTR2009-014799-23-DE; EUCTR2009-014799-23-AT; EUCTR2009-014799-23-IT	Phase II	2009
				Pancreatic cancer	Gemcitabine	EUCTR2007-001609-81-DE	Phase I/II	2007

Name	Cytokine	Target	Fusion format	Indication	Combination	ClinicalTrials.gov identifier	Phase	Year
EMD 521873 (NHS-IL2-LT)	IL-2	DNA	IgG-IL-2	B-cell non-Hodgkin lymphoma	Alone or with cyclophosphamide	NCT01032681	Phase I	2009
				Lung cancer and non-small-cell lung cancer	Radiotherapy	NCT00879866	Phase I	2009
AS1409 (HuBC1-IL12)	IL-12	Fibronectin ED-B	IgG-scIL-12	Metastatic renal cell carcinoma and metastatic melanoma	Alone	NCT00625768	Phase I	2008
L19TNFα	TNFα	Fibronectin ED-B	scFv-TNFα	Solid tumors/colorectal cancer	Alone	NCT01253837	Phase I/II	2010
				Melanoma	Melphalan	NCT01213732	Phase I	2010
VRC-ADJDNA004-IL2-VP	IL-2		Fc-IL-2	HIV Infections	VRC-HIVDNA009-00-VP	NCT00069030	Phase I	2003
F8-IL10 (Dekavil)	IL-10	Fibronectin ED-A	scFv-IL-10	RA	Methotrexate	Recruiting	Phase I	

Note: Information current as of 30 January 2014.
IL, interleukin; RA, rheumatoid arthritis; TNF, tumor necrosis factor.

Source: *ClinicalTrials.gov* and *Apps.who.int*.

cytokine administration [19, 177]. It is also possible that the non-antibody partner can be fused to the amino-terminus of the heavy chain without disrupting binding to the antigen. This may be necessary for proteins such as the co-stimulatory molecule B7.1 or IL-12 that require N-terminal processing or special folding to retain activity (Fig. 11a, A2). Fusion proteins containing the cytokine fused to the amino-terminus of the antibody have been shown to exhibit both the ability to bind antigen and the activity of the non-antibody partner [178, 179]. It should be noted that, although in Fig. 11 the non-antibody partner is fused to the heavy chain, it could also be fused to the light chain. The non-antibody partner can also be fused to the carboxy-terminus of antibody fragments containing the antigen-binding domain to retain antibody specificity (Fig. 11a, A3–7) [19]. The same approach can be used to create Fab-cytokine fusion proteins with the non-antibody partner fused immediately after the hinge (Fig. 11a, A8), to the C_H1 domain (Fig. 11a, A9), or to the V_H domain (Fig. 11a, A10–11). These configurations may be useful when the antibody-related effector functions are unnecessary or harmful. For many applications such as tumor targeting, the small size of the antibody fragments fused to the cytokine may be an advantage. Antibody–cytokine fusion proteins have shown significant antitumor activity in mice bearing tumors, with some candidates leading to clinical trials [180–182] (Table 6). As some combinations of cytokines have shown improved antitumor activities, more complex molecules have been constructed in which two cytokines are fused to one antibody (Fig. 11b, B1–5). The combination treatment in either the form of coadministration of two different antibody–cytokine fusion proteins, or by the use of bifunctional antibody fusion proteins (two cytokines fused to one antibody) has shown significant antitumor protection, which suggests that these strategies may be useful treatment options in humans [183–186].

The Fc region has also been fused via the amino-terminus to non-antibody partners to gain antibody-associated properties such as improved pharmacokinetics and effector functions. In this case, the non-antibody sequence usually exhibits an affinity for a target protein or receptor (Fig. 11c, C1–3). These constructs have been termed *immunoadhesins* because they present an adhesive molecule linked to the immunoglobulin Fc domain. An example is etanercept, a TNF-α receptor Fc fusion protein, which binds to TNF-α (a mediator of inflammation) and neutralizes its activity. This molecule has been approved by the FDA for the treatment of autoimmune diseases such as rheumatoid arthritis [187].

7
Conclusions

Rapid progress has been made in producing genetically engineered antibodies. The ability to express foreign DNA in a variety of host cells has made it possible to produce chimeric, humanized, and human antibodies, as well as antibodies with novel structures and functional properties in quantities sufficient for multiple applications, including therapy. Importantly, previous experience suggests that antibody-based therapies can be successfully developed for use in clinical situations in which no alternative effective therapy is available. However, continued progress in the development of antibody-based therapies will require extensive research to further define the mechanism of antibody action and to optimize the use of these novel proteins with unique functional properties.

Acknowledgments

These studies were supported in part by NIH/NCI R01CA136841, R01CA107023, K01CA138559, R21CA179680, R01CA16 2964, ANPCyT-FONARSEC PICT-PRH 2008–00315, CONICET-PIP no. 114-2011-01-00139, and UBACYT 200-2011-02-00027. C.P. is supported by a CONICET Fellowship, G.H. and P.C.M. are members of the National Council for Scientific and Technological Research (CONICET), Argentina.

References

1. Helguera, G., and Penichet, M.L. (2005) Antibody-cytokine fusion proteins for the therapy of cancer. *Methods Mol. Med.*, **109**, 347–374.
2. Alt, F.W., Blackwell, T.K., and Yancopoulos, G.D. (1987) Development of the primary antibody repertoire. *Science*, **238** (4830), 1079–1087.
3. Helguera, G., Daniels, T.R., Rodríguez, J.A., Penichet, M.L. (2010) Monoclonal antibodies, human, engineered, in: *Encyclopedia of Industrial Biotechnology: Bioprocess, Bioseparation, and Cell Technology*, 2010 edition John Wiley & Sons, Inc., Hoboken, NJ, pp. 3526–3542.
4. Dudley, D.D., Chaudhuri, J., Bassing, C.H., and Alt, F.W. (2005) Mechanism and control of V(D)J recombination versus class switch recombination: similarities and differences. *Adv. Immunol.*, **86**, 43–112.
5. Di Noia, J.M. and Neuberger, M.S. (2007) Molecular mechanisms of antibody somatic hypermutation. *Annu. Rev. Biochem.*, **76**, 1–22.
6. Rajewsky, K., Forster, I., and Cumano, A. (1987) Evolutionary and somatic selection of the antibody repertoire in the mouse. *Science*, **238** (4830), 1088–1094.
7. Arnold, J.N., Wormald, M.R., Sim, R.B., Rudd, P.M. *et al.* (2007) The impact of glycosylation on the biological function and structure of human immunoglobulins. *Annu. Rev. Immunol.*, **25**, 21–50.
8. Davis, A.C., Roux, K.H., and Shulman, M.J. (1988) On the structure of polymeric IgM. *Eur. J. Immunol.*, **18** (7), 1001–1008.
9. Wan, T., Beavil, R.L., Fabiane, S.M., Beavil, A.J. *et al.* (2002) The crystal structure of IgE Fc reveals an asymmetrically bent conformation. *Nat. Immunol.*, **3** (7), 681–686.
10. Penichet, M.L., Morrison, S.L. (2004) Genetic engineering of antibody molecules, in: Meyers, R.A. (Ed.) *Encyclopedia of Molecular Cell Biology and Molecular Medicine*, 2nd edn , Wiley-VCH Verlag GmbH & Co. KGaA, Weinheim, pp. 329–351.
11. Strohl, W.R. and Strohl, L.M. (2012) Therapeutic antibody classes, in *Therapeutic Antibody Engineering: Current and Future Advances Driving the Strongest Growth Area in the Pharmaceutical Industry*, 1st edn, Woodhead Publishing, Sawston, Cambridge, pp. 197–223.
12. Bruggemann, M., Williams, G.T., Bindon, C.I., Clark, M.R. *et al.* (1987) Comparison of the effector functions of human immunoglobulins using a matched set of chimeric antibodies. *J. Exp. Med.*, **166** (5), 1351–1361.
13. Ghetie, V. and Ward, E.S. (2000) Multiple roles for the major histocompatibility complex class I- related receptor FcRn. *Annu. Rev. Immunol.*, **18**, 739–766.
14. Ravetch, J.V. and Bolland, S. (2001) IgG Fc receptors. *Annu. Rev. Immunol.*, **19**, 275–290.
15. Horton, H.M., Bernett, M.J., Pong, E., Peipp, M., Karki, S., Chu, S.Y. *et al.* (2008) Potent in vitro and in vivo activity of an Fc-engineered anti-CD19 monoclonal antibody against lymphoma and leukemia. *Cancer Res.*, **68** (19), 8049–8057.
16. Nimmerjahn, F. and Ravetch, J.V. (2006) Fcgamma receptors: old friends and new family members. *Immunity*, **24** (1), 19–28.
17. Weiner, L.M. (2007) Building better magic bullets – improving unconjugated monoclonal antibody therapy for cancer. *Nat. Rev. Cancer*, **7** (9), 701–706.

18. Reichert, J.M. (2012) Marketed therapeutic antibodies compendium. *MAbs*, **4** (3), 413–415.
19. Ortiz-Sanchez, E., Helguera, G., Daniels, T.R., and Penichet, M.L. (2008) Antibody-cytokine fusion proteins: applications in cancer therapy. *Expert Opin. Biol. Ther.*, **8** (5), 609–632.
20. Winau, F., Westphal, O., and Winau, R. (2004) Paul Ehrlich – in search of the magic bullet. *Microbes Infect.*, **6** (8), 786–789.
21. Kohler, G. and Milstein, C. (1975) Continuous cultures of fused cells secreting antibody of predefined specificity. *Nature*, **256** (5517), 495–497.
22. Reichert, J.M., Rosensweig, C.J., Faden, L.B., and Dewitz, M.C. (2005) Monoclonal antibody successes in the clinic. *Nat. Biotechnol.*, **23** (9), 1073–1078.
23. Shawler, D.L., Bartholomew, R.M., Smith, L.M., and Dillman, R.O. (1985) Human immune response to multiple injections of murine monoclonal IgG. *J. Immunol.*, **135** (2), 1530–1535.
24. Ober, R.J., Radu, C.G., Ghetie, V., and Ward, E.S. (2001) Differences in promiscuity for antibody-FcRn interactions across species: implications for therapeutic antibodies. *Int. Immunol.*, **13** (12), 1551–1559.
25. Morrison, S.L., Johnson, M.J., Herzenberg, L.A., and Oi, V.T. (1984) Chimeric human antibody molecules: mouse antigen-binding domains with human constant region domains. *Proc. Natl Acad. Sci. USA*, **81** (21), 6851–6855.
26. Wang, Z., Raifu, M., Howard, M., Smith, L. *et al.* (2000) Universal PCR amplification of mouse immunoglobulin gene variable regions: the design of degenerate primers and an assessment of the effect of DNA polymerase 3′ to 5′ exonuclease activity. *J. Immunol. Methods*, **233** (1-2), 167–177.
27. Yoo, E.M., Chintalacharuvu, K.R., Penichet, M.L., and Morrison, S.L. (2002) Myeloma expression systems. *J. Immunol. Methods*, **261** (1-2), 1–20.
28. Maynard, J. and Georgiou, G. (2000) Antibody engineering. *Annu. Rev. Biomed. Eng.*, **2**, 339–376.
29. Kuus-Reichel, K., Grauer, L.S., Karavodin, L.M., Knott, C., Krusemeier, M., and Kay, N.E. (1994) Will immunogenicity limit the use, efficacy, and future development of therapeutic monoclonal antibodies? *Clin. Diagn. Lab. Immunol.*, **1** (4), 365–372.
30. Jones, P.T., Dear, P.H., Foote, J., Neuberger, M.S. *et al.* (1986) Replacing the complementarity-determining regions in a human antibody with those from a mouse. *Nature*, **321** (6069), 522–525.
31. Riechmann, L., Clark, M., Waldmann, H., and Winter, G. (1988) Reshaping human antibodies for therapy. *Nature*, **332** (6162), 323–327.
32. Vogel, C.L., Cobleigh, M.A., Tripathy, D., Gutheil, J.C. *et al.* (2002) Efficacy and safety of trastuzumab as a single agent in first-line treatment of HER2-overexpressing metastatic breast cancer. *J. Clin. Oncol.*, **20** (3), 719–726.
33. Baselga, J., Tripathy, D., Mendelsohn, J., Baughman, S. *et al.* (1996) Phase II study of weekly intravenous recombinant humanized anti-p185HER2 monoclonal antibody in patients with HER2/neu-overexpressing metastatic breast cancer. *J. Clin. Oncol.*, **14** (3), 737–744.
34. McCafferty, J., Griffiths, A.D., Winter, G., and Chiswell, D.J. (1990) Phage antibodies: filamentous phage displaying antibody variable domains. *Nature*, **348** (6301), 552–554.
35. Marks, J.D., Hoogenboom, H.R., Bonnert, T.P., McCafferty, J. *et al.* (1991) By-passing immunization: human antibodies from V-gene libraries displayed on phage. *J. Mol. Biol.*, **222** (3), 581–597.
36. Bradbury, A.R. and Marks, J.D. (2004) Antibodies from phage antibody libraries. *J. Immunol. Methods*, **290** (1-2), 29–49.
37. Carter, P.J. (2006) Potent antibody therapeutics by design. *Nat. Rev. Immunol.*, **6** (5), 343–357.
38. Hoogenboom, H.R., Griffiths, A.D., Johnson, K.S., Chiswell, D.J., Hudson, P., and Winter, G. (1991) Multi-subunit proteins on the surface of filamentous phage: methodologies for displaying antibody (Fab) heavy and light chains. *Nucleic Acids Res.*, **19** (15), 4133–4137.
39. Georgieva, Y. and Konthur, Z. (2011) Design and screening of M13 phage display cDNA libraries. *Molecules*, **16** (2), 1667–1681.
40. Lin, J., Ziring, D., Desai, S., Kim, S. *et al.* (2008) TNFalpha blockade in human

diseases: an overview of efficacy and safety. *Clin. Immunol.*, **126** (1), 13–30.
41. Green, L.L. (1999) Antibody engineering via genetic engineering of the mouse: XenoMouse strains are a vehicle for the facile generation of therapeutic human monoclonal antibodies. *J. Immunol. Methods*, **231** (1-2), 11–23.
42. Cohenuram, M. and Saif, M.W. (2007) Panitumumab the first fully human monoclonal antibody: from the bench to the clinic. *Anticancer Drugs*, **18** (1), 7–15.
43. Ishida, I., Tomizuka, K., Yoshida, H., Tahara, T. *et al.* (2002) Production of human monoclonal and polyclonal antibodies in TransChromo animals. *Cloning Stem Cells*, **4** (1), 91–102.
44. Newcombe, C. and Newcombe, A.R. (2007) Antibody production: polyclonal-derived biotherapeutics. *J. Chromatogr. B: Anal. Technol. Biomed. Life Sci.*, **848** (1), 2–7.
45. Kuroiwa, Y., Kasinathan, P., Choi, Y.J., Naeem, R. *et al.* (2002) Cloned transchromosomic calves producing human immunoglobulin. *Nat. Biotechnol.*, **20** (9), 889–894.
46. Kuroiwa, Y., Kasinathan, P., Sathiyaseelan, T., Jiao, J.A. *et al.* (2009) Antigen-specific human polyclonal antibodies from hyperimmunized cattle. *Nat. Biotechnol.*, **27** (2), 173–181.
47. Sano, A., Matsushita, H., Wu, H., Jiao, J.A. *et al.* (2013) Physiological level production of antigen-specific human immunoglobulin in cloned transchromosomic cattle. *PLoS One*, **8** (10), e78119.
48. Kacskovics, I., Kis, Z., Mayer, B., West, A.P. Jr., *et al.* (2006) FcRn mediates elongated serum half-life of human IgG in cattle. *Int. Immunol.*, **18** (4), 525–536.
49. Kuroiwa, Y., Kasinathan, P., Matsushita, H., Sathiyaselan, J. *et al.* (2004) Sequential targeting of the genes encoding immunoglobulin-mu and prion protein in cattle. *Nat. Genet.*, **36** (7), 775–780.
50. Feldhaus, M.J., Siegel, R.W., Opresko, L.K., Coleman, J.R. *et al.* (2003) Flow-cytometric isolation of human antibodies from a non-immune *Saccharomyces cerevisiae* surface display library. *Nat. Biotechnol.*, **21** (2), 163–170.
51. Blaise, L., Wehnert, A., Steukers, M.P., van den Beucken, T. *et al.* (2004) Construction and diversification of yeast cell surface displayed libraries by yeast mating: application to the affinity maturation of Fab antibody fragments. *Gene*, **342** (2), 211–218.
52. Hoogenboom, H.R. (2005) Selecting and screening recombinant antibody libraries. *Nat. Biotechnol.*, **23** (9), 1105–1116.
53. Nadkarni, A., Kelley, L.L., and Momany, C. (2007) Optimization of a mouse recombinant antibody fragment for efficient production from *Escherichia coli*. *Protein Expr. Purif.*, **52** (1), 219–229.
54. Simmons, L.C., Reilly, D., Klimowski, L., Raju, T.S. *et al.* (2002) Expression of full-length immunoglobulins in *Escherichia coli*: rapid and efficient production of aglycosylated antibodies. *J. Immunol. Methods*, **263** (1–2), 133–147.
55. Nesbeth, D.N., Perez-Pardo, M.A., Ali, S., Ward, J. *et al.* (2012) Growth and productivity impacts of periplasmic nuclease expression in an *Escherichia coli* Fab' fragment production strain. *Biotechnol. Bioeng.*, **109** (2), 517–527.
56. Ward, E.S. (1993) Antibody engineering using *Escherichia coli* as host. *Adv. Pharmacol.*, **24**, 1–20.
57. Friedrich, L., Stangl, S., Hahne, H., Kuster, B. *et al.* (2010) Bacterial production and functional characterization of the Fab fragment of the murine IgG1/lambda monoclonal antibody cmHsp70.1, a reagent for tumour diagnostics. *Protein Eng. Des. Sel.*, **23** (4), 161–168.
58. Veggiani, G. and de Marco, A. (2011) Improved quantitative and qualitative production of single-domain intrabodies mediated by the co-expression of Erv1p sulfhydryl oxidase. *Protein Expr. Purif.*, **79** (1), 111–114.
59. Cossins, A.J., Harrison, S., Popplewell, A.G., and Gore, M.G. (2007) Recombinant production of a VL single domain antibody in *Escherichia coli* and analysis of its interaction with peptostreptococcal protein L. *Protein Expr. Purif.*, **51** (2), 253–259.
60. Zhao, J.B., Wei, D.Z., and Tong, W.Y. (2007) Identification of *Escherichia coli* host cell for high plasmid stability and improved production of antihuman ovarian carcinoma × antihuman CD3 single-chain bispecific antibody. *Appl. Microbiol. Biotechnol.*, **76** (4), 795–800.

61. Inoue, Y., Ohta, T., Tada, H., Iwasa, S. et al. (1997) Efficient production of a functional mouse/human chimeric Fab' against human urokinase-type plasminogen activator by *Bacillus brevis*. *Appl. Microbiol. Biotechnol.*, **48** (4), 487–492.
62. Jeong, K.J., Jang, S.H., and Velmurugan, N. (2011) Recombinant antibodies: engineering and production in yeast and bacterial hosts. *Biotechnol. J.*, **6** (1), 16–27.
63. Potgieter, T.I., Kersey, S.D., Mallem, M.R., Nylen, A.C. et al. (2010) Antibody expression kinetics in glycoengineered *Pichia pastoris*. *Biotechnol. Bioeng.*, **106** (6), 918–927.
64. Potgieter, T.I., Cukan, M., Drummond, J.E., Houston-Cummings, N.R. et al. (2009) Production of monoclonal antibodies by glycoengineered *Pichia pastoris*. *J. Biotechnol.*, **139** (4), 318–325.
65. Zhang, N., Liu, L., Dumitru, C.D., Cummings, N.R. et al. (2011) Glycoengineered *Pichia* produced anti-HER2 is comparable to trastuzumab in preclinical study. *MAbs*, **3** (3), 289–298.
66. Wang, D.D., Su, M.M., Sun, Y., Huang, S.L. et al. (2012) Expression, purification and characterization of a human single-chain Fv antibody fragment fused with the Fc of an IgG1 targeting a rabies antigen in *Pichia pastoris*. *Protein Expr. Purif.*, **86** (1), 75–81.
67. Ji, X., Lu, W., Zhou, H., Han, D. et al. (2013) Covalently dimerized Camelidae antihuman TNFα single-domain antibodies expressed in yeast *Pichia pastoris* show superior neutralizing activity. *Appl. Microbiol. Biotechnol.*, **97** (19), 8547–8558.
68. Liu, J., Wei, D., Qian, F., Zhou, Y. et al. (2003) pPIC9-Fc: a vector system for the production of single-chain Fv-Fc fusions in *Pichia pastoris* as detection reagents in vitro. *J. Biochem.*, **134** (6), 911–917.
69. Jafari, R., Holm, P., Piercecchi, M., and Sundstrom, B.E. (2011) Construction of divalent anti-keratin 8 single-chain antibodies (sc(Fv)(2)), expression in *Pichia pastoris* and their reactivity with multicellular tumor spheroids. *J. Immunol. Methods*, **364** (1–2), 65–76.
70. Schoonooghe, S., Kaigorodov, V., Zawisza, M., Dumolyn, C. et al. (2009) Efficient production of human bivalent and trivalent anti-MUC1 Fab-scFv antibodies in *Pichia pastoris*. *BMC Biotechnol.*, **9**, 70.
71. Horwitz, A.H., Chang, C.P., Better, M., Hellstrom, K.E. et al. (1988) Secretion of functional antibody and Fab fragment from yeast cells. *Proc. Natl Acad. Sci. USA*, **85** (22), 8678–8682.
72. Frenken, L.G., van der Linden, R.H., Hermans, P.W., Bos, J.W. et al. (2000) Isolation of antigen specific llama VHH antibody fragments and their high level secretion by *Saccharomyces cerevisiae*. *J. Biotechnol.*, **78** (1), 11–21.
73. Jarvis, D.L. (2009) Baculovirus-insect cell expression systems. *Methods Enzymol.*, **463**, 191–222.
74. Edelman, L., Margaritte, C., Chaabihi, H., Monchatre, E. et al. (1997) Obtaining a functional recombinant anti-rhesus (D) antibody using the baculovirus-insect cell expression system. *Immunology*, **91** (1), 13–19.
75. Palmberger, D., Rendic, D., Tauber, P., Krammer, F. et al. (2011) Insect cells for antibody production: evaluation of an efficient alternative. *J. Biotechnol.*, **153** (3-4), 160–166.
76. Liang, M., Dubel, S., Li, D., Queitsch, I. et al. (2001) Baculovirus expression cassette vectors for rapid production of complete human IgG from phage display selected antibody fragments. *J. Immunol. Methods*, **247** (1-2), 119–130.
77. Kurasawa, J.H., Shestopal, S.A., Jha, N.K., Ovanesov, M.V. et al. (2013) Insect cell-based expression and characterization of a single-chain variable antibody fragment directed against blood coagulation factor VIII. *Protein Expr. Purif.*, **88** (2), 201–206.
78. Hsu, T.A., Takahashi, N., Tsukamoto, Y., Kato, K. et al. (1997) Differential N-glycan patterns of secreted and intracellular IgG produced in *Trichoplusia ni* cells. *J. Biol. Chem.*, **272** (14), 9062–9070.
79. Wurm, F.M. (2004) Production of recombinant protein therapeutics in cultivated mammalian cells. *Nat. Biotechnol.*, **22** (11), 1393–1398.
80. Mader, A., Prewein, B., Zboray, K., Casanova, E. et al. (2013) Exploration of BAC versus plasmid expression vectors in recombinant CHO cells. *Appl. Microbiol. Biotechnol.*, **97** (9), 4049–4054.
81. Agrawal, V., Slivac, I., Perret, S., Bisson, L. et al. (2012) Stable expression of chimeric

heavy chain antibodies in CHO cells. *Methods Mol. Biol.*, **911**, 287–303.

82. Spens, E. and Haggstrom, L. (2007) Defined protein and animal component-free NS0 fed-batch culture. *Biotechnol. Bioeng.*, **98** (6), 1183–1194.
83. Burky, J.E., Wesson, M.C., Young, A., Farnsworth, S. *et al.* (2007) Protein-free fed-batch culture of non-GS NS0 cell lines for production of recombinant antibodies. *Biotechnol. Bioeng.*, **96** (2), 281–293.
84. Jarvis, L.M. (2008) A technology bet. *Chem. Eng. News Arch.*, **86** (29), 30–31.
85. Tchoudakova, A., Hensel, F., Murillo, A., Eng, B. *et al.* (2009) High level expression of functional human IgMs in human PER.C6 cells. *MAbs*, **1** (2), 163–171.
86. Komarova, T.V., Kosorukov, V.S., Frolova, O.Y., Petrunia, I.V. *et al.* (2011) Plant-made trastuzumab (herceptin) inhibits HER2/Neu + cell proliferation and retards tumor growth. *PLoS One*, **6** (3), e17541.
87. Fecker, L.F., Kaufmann, A., Commandeur, U., Commandeur, J. *et al.* (1996) Expression of single-chain antibody fragments (scFv) specific for beet necrotic yellow vein virus coat protein or 25 kDa protein in *Escherichia coli* and *Nicotiana benthamiana*. *Plant Mol. Biol.*, **32** (5), 979–986.
88. Huang, Z., Phoolcharoen, W., Lai, H., Piensook, K. *et al.* (2010) High-level rapid production of full-size monoclonal antibodies in plants by a single-vector DNA replicon system. *Biotechnol. Bioeng.*, **106** (1), 9–17.
89. McCormick, A.A., Reddy, S., Reinl, S.J., Cameron, T.I. *et al.* (2008) Plant-produced idiotype vaccines for the treatment of non-Hodgkin's lymphoma: safety and immunogenicity in a phase I clinical study. *Proc. Natl Acad. Sci. USA*, **105** (29), 10131–10136.
90. Almquist, K.C., McLean, M.D., Niu, Y., Byrne, G. *et al.* (2006) Expression of an anti-botulinum toxin A neutralizing single-chain Fv recombinant antibody in transgenic tobacco. *Vaccine*, **24** (12), 2079–2086.
91. Reichert, J.M. (2008) Monoclonal antibodies as innovative therapeutics. *Curr. Pharm. Biotechnol.*, **9** (6), 423–430.
92. Shinkawa, T., Nakamura, K., Yamane, N., Shoji-Hosaka, E. *et al.* (2003) The absence of fucose but not the presence of galactose or bisecting N-acetylglucosamine of human IgG1 complex-type oligosaccharides shows the critical role of enhancing antibody-dependent cellular cytotoxicity. *J. Biol. Chem.*, **278** (5), 3466–3473.
93. Jin, B.R., Ryu, C.J., Kang, S.K., Han, M.H. *et al.* (1995) Characterization of a murine-human chimeric antibody with specificity for the pre-S2 surface antigen of hepatitis B virus expressed in baculovirus-infected insect cells. *Virus Res.*, **38** (2-3), 269–277.
94. Frenzel, A., Hust, M., and Schirrmann, T. (2013) Expression of recombinant antibodies. *Front Immunol.*, **4**, 217.
95. Strasser, R., Stadlmann, J., Schahs, M., Stiegler, G. *et al.* (2008) Generation of glyco-engineered *Nicotiana benthamiana* for the production of monoclonal antibodies with a homogeneous human-like N-glycan structure. *Plant Biotechnol. J.*, **6** (4), 392–402.
96. Ko, K., Brodzik, R., and Steplewski, Z. (2009) Production of antibodies in plants: approaches and perspectives. *Curr. Top. Microbiol. Immunol.*, **332**, 55–78.
97. Foote, J. and Eisen, H.N. (1995) Kinetic and affinity limits on antibodies produced during immune responses. *Proc. Natl Acad. Sci. USA*, **92** (5), 1254–1256.
98. Fukunishi, H., Shimada, J., and Shiraishi, K. (2012) Antigen-antibody interactions and structural flexibility of a femtomolar-affinity antibody. *Biochemistry*, **51** (12), 2597–2605.
99. Irving, R.A., Kortt, A.A., and Hudson, P.J. (1996) Affinity maturation of recombinant antibodies using *E. coli* mutator cells. *Immunotechnology*, **2** (2), 127–143.
100. Chowdhury, P.S. and Pastan, I. (1999) Improving antibody affinity by mimicking somatic hypermutation in vitro. *Nat. Biotechnol.*, **17** (6), 568–572.
101. Martineau, P. (2002) Error-prone polymerase chain reaction for modification of scFvs. *Methods Mol. Biol.*, **178**, 287–294.
102. Rajpal, A., Beyaz, N., Haber, L., Cappuccilli, G. *et al.* (2005) A general method for greatly improving the affinity of antibodies by using combinatorial libraries. *Proc. Natl Acad. Sci. USA*, **102** (24), 8466–8471.
103. Jermutus, L., Honegger, A., Schwesinger, F., Hanes, J. *et al.* (2001) Tailoring in vitro evolution for protein affinity or stability. *Proc. Natl Acad. Sci. USA*, **98** (1), 75–80.

104. Marks, J.D. (2004) Antibody affinity maturation by chain shuffling. *Methods Mol. Biol.*, **248**, 327–343.
105. Yoshinaga, K., Matsumoto, M., Torikai, M., Sugyo, K. *et al.* (2008) Ig L-chain shuffling for affinity maturation of phage library-derived human anti-human MCP-1 antibody blocking its chemotactic activity. *J. Biochem.*, **143** (5), 593–601.
106. Barderas, R., Desmet, J., Timmerman, P., Meloen, R. *et al.* (2008) Affinity maturation of antibodies assisted by in silico modeling. *Proc. Natl Acad. Sci. USA*, **105** (26), 9029–9034.
107. Lippow, S.M., Wittrup, K.D., and Tidor, B. (2007) Computational design of antibody-affinity improvement beyond in vivo maturation. *Nat. Biotechnol.*, **25** (10), 1171–1176.
108. Mondon, P., Dubreuil, O., Bouayadi, K., and Kharrat, H. (2008) Human antibody libraries: a race to engineer and explore a larger diversity. *Front. Biosci.*, **13**, 1117–1129.
109. Boder, E.T., Midelfort, K.S., and Wittrup, K.D. (2000) Directed evolution of antibody fragments with monovalent femtomolar antigen-binding affinity. *Proc. Natl Acad. Sci. USA*, **97** (20), 10701–10705.
110. Koren, E., De Groot, A.S., Jawa, V., Beck, K.D. *et al.* (2007) Clinical validation of the "in silico" prediction of immunogenicity of a human recombinant therapeutic protein. *Clin. Immunol.*, **124** (1), 26–32.
111. Van Walle, I., Gansemans, Y., Parren, P.W., Stas, P. *et al.* (2007) Immunogenicity screening in protein drug development. *Expert Opin. Biol. Ther.*, **7** (3), 405–418.
112. Harding, F.A., Stickler, M.M., Razo, J., and DuBridge, R.B. (2010) The immunogenicity of humanized and fully human antibodies: residual immunogenicity resides in the CDR regions. *MAbs*, **2** (3), 256–265.
113. Tamura, M., Milenic, D.E., Iwahashi, M., Padlan, E. *et al.* (2000) Structural correlates of an anticarcinoma antibody: identification of specificity-determining residues (SDRs) and development of a minimally immunogenic antibody variant by retention of SDRs only. *J. Immunol.*, **164** (3), 1432–1441.
114. Lazar, G.A., Desjarlais, J.R., Jacinto, J., Karki, S. *et al.* (2007) A molecular immunology approach to antibody humanization and functional optimization. *Mol. Immunol.*, **44** (8), 1986–1998.
115. Walsh, G. (2004) Second-generation biopharmaceuticals. *Eur. J. Pharm. Biopharm.*, **58** (2), 185–196.
116. Bruhns, P. (2012) Properties of mouse and human IgG receptors and their contribution to disease models. *Blood*, **119** (24), 5640–5649.
117. Lazar, G.A., Dang, W., Karki, S., Vafa, O. *et al.* (2006) Engineered antibody Fc variants with enhanced effector function. *Proc. Natl Acad. Sci. USA*, **103** (11), 4005–4010.
118. Shields, R.L., Namenuk, A.K., Hong, K., Meng, Y.G. *et al.* (2001) High resolution mapping of the binding site on human IgG1 for Fc gamma RI, Fc gamma RII, Fc gamma RIII, and FcRn and design of IgG1 variants with improved binding to the Fc gamma R. *J. Biol. Chem.*, **276** (9), 6591–6604.
119. Watkins, J.D., Allan, B. (2004) Fc region variants. Organization, W.I.P. WO 2004/074455 A2, filed February 20, 2004 and issued September 2, 2004.
120. Stavenhagen, J.B., Gorlatov, S., Tuaillon, N., Rankin, C.T. *et al.* (2007) Fc optimization of therapeutic antibodies enhances their ability to kill tumor cells in vitro and controls tumor expansion in vivo via low-affinity activating Fcgamma receptors. *Cancer Res.*, **67** (18), 8882–8890.
121. Richards, J.O., Karki, S., Lazar, G.A., Chen, H. *et al.* (2008) Optimization of antibody binding to FcgammaRIIa enhances macrophage phagocytosis of tumor cells. *Mol. Cancer Ther.*, **7** (8), 2517–2527.
122. Bolt, S., Routledge, E., Lloyd, I., Chatenoud, L. *et al.* (1993) The generation of a humanized, non-mitogenic CD3 monoclonal antibody which retains in vitro immunosuppressive properties. *Eur. J. Immunol.*, **23** (2), 403–411.
123. Alegre, M.L., Peterson, L.J., Xu, D., Sattar, H.A. *et al.* (1994) A non-activating "humanized" anti-CD3 monoclonal antibody retains immunosuppressive properties in vivo. *Transplantation*, **57** (11), 1537–1543.
124. Davis, P.M., Abraham, R., Xu, L., Nadler, S.G. *et al.* (2007) Abatacept binds to the Fc receptor CD64 but does not mediate complement-dependent cytotoxicity or antibody-dependent cellular cytotoxicity. *J. Rheumatol.*, **34** (11), 2204–2210.

125. McEarchern, J.A., Oflazoglu, E., Francisco, L., McDonagh, C.F. et al. (2007) Engineered anti-CD70 antibody with multiple effector functions exhibits in vitro and in vivo antitumor activities. *Blood*, **109** (3), 1185–1192.
126. Hutchins, J.T., Kull, F.C. Jr., Bynum, J., Knick, V.C. et al. (1995) Improved biodistribution, tumor targeting, and reduced immunogenicity in mice with a gamma 4 variant of Campath-1H. *Proc. Natl Acad. Sci. USA*, **92** (26), 11980–11984.
127. Xu, D., Alegre, M.L., Varga, S.S., Rothermel, A.L. et al. (2000) In vitro characterization of five humanized OKT3 effector function variant antibodies. *Cell. Immunol.*, **200** (1), 16–26.
128. Idusogie, E.E., Wong, P.Y., Presta, L.G., Gazzano-Santoro, H. et al. (2001) Engineered antibodies with increased activity to recruit complement. *J. Immunol.*, **166** (4), 2571–2575.
129. Idusogie, E.E., Presta, L.G., Gazzano-Santoro, H., Totpal, K. et al. (2000) Mapping of the C1q binding site on rituxan, a chimeric antibody with a human IgG1 Fc. *J. Immunol.*, **164** (8), 4178–4184.
130. Dall'Acqua, W.F., Kiener, P.A., and Wu, H. (2006) Properties of human IgG1s engineered for enhanced binding to the neonatal Fc receptor (FcRn). *J. Biol. Chem.*, **281** (33), 23514–23524.
131. Hinton, P.R., Xiong, J.M., Johlfs, M.G., Tang, M.T. et al. (2006) An engineered human IgG1 antibody with longer serum half-life. *J. Immunol.*, **176** (1), 346–356.
132. Yeung, Y.A., Leabman, M.K., Marvin, J.S., Qiu, J. et al. (2009) Engineering human IgG1 affinity to human neonatal Fc receptor: impact of affinity improvement on pharmacokinetics in primates. *J. Immunol.*, **182** (12), 7663–7671.
133. Petkova, S.B., Akilesh, S., Sproule, T.J., Christianson, G.J. et al. (2006) Enhanced half-life of genetically engineered human IgG1 antibodies in a humanized FcRn mouse model: potential application in humorally mediated autoimmune disease. *Int. Immunol.*, **18** (12), 1759–1769.
134. Zalevsky, J., Chamberlain, A.K., Horton, H.M., Karki, S. et al. (2010) Enhanced antibody half-life improves in vivo activity. *Nat. Biotechnol.*, **28** (2), 157–159.
135. Stapleton, N.M., Andersen, J.T., Stemerding, A.M., Bjarnarson, S.P. et al. (2011) Competition for FcRn-mediated transport gives rise to short half-life of human IgG3 and offers therapeutic potential. *Nat. Commun.*, **2**, 599.
136. Datta-Mannan, A., Witcher, D.R., Tang, Y., Watkins, J. et al. (2007) Humanized IgG1 variants with differential binding properties to the neonatal Fc receptor: relationship to pharmacokinetics in mice and primates. *Drug Metab. Dispos.*, **35** (1), 86–94.
137. Rother, R.P., Rollins, S.A., Mojcik, C.F., Brodsky, R.A. et al. (2007) Discovery and development of the complement inhibitor eculizumab for the treatment of paroxysmal nocturnal hemoglobinuria. *Nat. Biotechnol.*, **25** (11), 1256–1264.
138. Hale, G., Rebello, P., Al Bakir, I., Bolam, E., Wiczling, P., Jusko, W.J. et al. (2010) Pharmacokinetics and antibody responses to the CD3 antibody otelixizumab used in the treatment of type 1 diabetes. *J Clin Pharmacol*, **50** (11), 1238–1248.
139. Harjunpaa, A., Junnikkala, S., and Meri, S. (2000) Rituximab (anti-CD20) therapy of B-cell lymphomas: direct complement killing is superior to cellular effector mechanisms. *Scand. J. Immunol.*, **51** (6), 634–641.
140. van der Kolk, L.E., Grillo-Lopez, A.J., Baars, J.W., Hack, C.E. et al. (2001) Complement activation plays a key role in the side-effects of rituximab treatment. *Br. J. Haematol.*, **115** (4), 807–811.
141. Weisman, L.E. (2009) Respiratory syncytial virus (RSV) prevention and treatment: past, present, and future. *Cardiovasc. Hematol. Agents Med. Chem.*, **7** (3), 223–233.
142. Robbie, G.J., Criste, R., Dall'acqua, W.F., Jensen, K. et al. (2013) A novel investigational Fc-modified humanized monoclonal antibody, motavizumab-YTE, Has an extended half-life in healthy adults. *Antimicrob. Agents Chemother.*, **57** (12), 6147–6153.
143. Datta-Mannan, A., Witcher, D.R., Tang, Y., Watkins, J. et al. (2007) Monoclonal antibody clearance. Impact of modulating the interaction of IgG with the neonatal Fc receptor. *J. Biol. Chem.*, **282** (3), 1709–1717.
144. Penichet, M.L. and Jensen-Jarolim, E. (eds) (2010) *Cancer and IgE, Introducing the Concept of Allergooncology*, Springer, New York.

145. Karagiannis, S.N., Bracher, M.G., Beavil, R.L., Beavil, A.J. et al. (2008) Role of IgE receptors in IgE antibody-dependent cytotoxicity and phagocytosis of ovarian tumor cells by human monocytic cells. *Cancer Immunol. Immunother.*, **57** (2), 247–263.

146. Teo, P.Z., Utz, P.J., and Mollick, J.A. (2012) Using the allergic immune system to target cancer: activity of IgE antibodies specific for human CD20 and MUC1. *Cancer Immunol. Immunother.*, **61** (12), 2295–2309.

147. Karagiannis, P., Singer, J., Hunt, J., Gan, S.K. et al. (2009) Characterisation of an engineered trastuzumab IgE antibody and effector cell mechanisms targeting HER2/neu-positive tumour cells. *Cancer Immunol. Immunother.*, **58** (6), 915–930.

148. Daniels, T.R., Leuchter, R.K., Quintero, R., Helguera, G. et al. (2012) Targeting HER2/neu with a fully human IgE to harness the allergic reaction against cancer cells. *Cancer Immunol. Immunother.*, **61** (7), 991–1003.

149. Daniels-Wells, T.R., Helguera, G., Leuchter, R.K., Quintero, R. et al. (2013) A novel IgE antibody targeting the prostate-specific antigen as a potential prostate cancer therapy. *BMC Cancer*, **13**, 195.

150. Beckman, R.A., Weiner, L.M., and Davis, H.M. (2007) Antibody constructs in cancer therapy: protein engineering strategies to improve exposure in solid tumors. *Cancer*, **109** (2), 170–179.

151. Folk, J.C. and Stone, E.M. (2010) Ranibizumab therapy for neovascular age-related macular degeneration. *N. Engl. J. Med.*, **363** (17), 1648–1655.

152. Wesolowski, J., Alzogaray, V., Reyelt, J., Unger, M. et al. (2009) Single domain antibodies: promising experimental and therapeutic tools in infection and immunity. *Med. Microbiol. Immunol.*, **198** (3), 157–174.

153. Kim, S.J., Park, Y., and Hong, H.J. (2005) Antibody engineering for the development of therapeutic antibodies. *Mol. Cell*, **20** (1), 17–29.

154. Albrecht, H., Burke, P.A., Natarajan, A., Xiong, C.Y. et al. (2004) Production of soluble ScFvs with C-terminal-free thiol for site-specific conjugation or stable dimeric ScFvs on demand. *Bioconjugate Chem.*, **15** (1), 16–26.

155. Hudson, P.J. and Kortt, A.A. (1999) High avidity scFv multimers; diabodies and triabodies. *J. Immunol. Methods*, **231** (1-2), 177–189.

156. Perisic, O., Webb, P.A., Holliger, P., Winter, G. et al. (1994) Crystal structure of a diabody, a bivalent antibody fragment. *Structure*, **2** (12), 1217–1226.

157. Hu, S., Shively, L., Raubitschek, A., Sherman, M. et al. (1996) Minibody: a novel engineered anti-carcinoembryonic antigen antibody fragment (single-chain Fv-CH3) which exhibits rapid, high-level targeting of xenografts. *Cancer Res.*, **56** (13), 3055–3061.

158. Weisser, N.E. and Hall, J.C. (2009) Applications of single-chain variable fragment antibodies in therapeutics and diagnostics. *Biotechnol. Adv.*, **27** (4), 502–520.

159. Carter, P. (2001) Bispecific human IgG by design. *J. Immunol. Methods*, **248** (1-2), 7–15.

160. Milstein, C. and Cuello, A.C. (1983) Hybrid hybridomas and their use in immunohistochemistry. *Nature*, **305** (5934), 537–540.

161. Seimetz, D., Lindhofer, H., and Bokemeyer, C. (2010) Development and approval of the trifunctional antibody catumaxomab (anti-EpCAM x anti-CD3) as a targeted cancer immunotherapy. *Cancer Treat. Rev.*, **36** (6), 458–467.

162. Linke, R., Klein, A., and Seimetz, D. (2010) Catumaxomab: clinical development and future directions. *MAbs*, **2** (2), 129–136.

163. Coloma, M.J., and Morrison, S.L. (1997) Design and production of novel tetravalent bispecific antibodies. *Nat Biotechnol*, **15** (2), 159–163.

164. Smith, R.I. and Morrison, S.L. (1994) Recombinant polymeric IgG: an approach to engineering more potent antibodies. *Biotechnology (NY)*, **12** (7), 683–688.

165. Montano, R.F., Penichet, M.L., Blackall, D.P., Morrison, S.L. et al. (2009) Recombinant polymeric IgG anti-Rh: a novel strategy for development of direct agglutinating reagents. *J. Immunol. Methods*, **340** (1), 1–10.

166. Shin, S.U., Wu, D., Ramanathan, R., Pardridge, W.M. et al. (1997) Functional and pharmacokinetic properties of antibody-avidin fusion proteins. *J. Immunol.*, **158** (10), 4797–4804.

167. Ng, P.P., Helguera, G., Daniels, T.R., Lomas, S.Z. et al. (2006) Molecular events contributing to cell death in malignant human hematopoietic cells elicited by an IgG3-avidin fusion protein targeting the transferrin receptor. *Blood*, **108** (8), 2745–2754.
168. Ng, P.P., Dela Cruz, J.S., Sorour, D.N., Stinebaugh, J.M. et al. (2002) An anti-transferrin receptor-avidin fusion protein exhibits both strong proapoptotic activity and the ability to deliver various molecules into cancer cells. *Proc. Natl Acad. Sci. USA*, **99** (16), 10706–10711.
169. Ortiz-Sanchez, E., Daniels, T.R., Helguera, G., Martinez-Maza, O. et al. (2009) Enhanced cytotoxicity of an anti-transferrin receptor IgG3-avidin fusion protein in combination with gambogic acid against human malignant hematopoietic cells: functional relevance of iron, the receptor, and reactive oxygen species. *Leukemia*, **23** (1), 59–70.
170. Daniels, T.R., Ortiz-Sanchez, E., Luria-Perez, R., Quintero, R. et al. (2011) An antibody-based multifaceted approach targeting the human transferrin receptor for the treatment of B-cell malignancies. *J. Immunother.*, **34** (6), 500–508.
171. Daniels, T.R., Ng, P.P., Delgado, T., Lynch, M.R. et al. (2007) Conjugation of an anti transferrin receptor IgG3-avidin fusion protein with biotinylated saporin results in significant enhancement of its cytotoxicity against malignant hematopoietic cells. *Mol. Cancer Ther.*, **6** (11), 2995–3008.
172. Daniels-Wells, T.R., Helguera, G., Rodriguez, J.A., Leoh, L.S. et al. (2013) Insights into the mechanism of cell death induced by saporin delivered into cancer cells by an antibody fusion protein targeting the transferrin receptor 1. *Toxicol. In Vitro*, **27** (1), 220–231.
173. Harvill, E.T., Fleming, J.M., and Morrison, S.L. (1996) In vivo properties of an IgG3-IL-2 fusion protein. A general strategy for immune potentiation. *J. Immunol.*, **157** (7), 3165–3170.
174. Dela Cruz, J.S., Trinh, K.R., Morrison, S.L., and Penichet, M.L. (2000) Recombinant anti-human HER2/neu IgG3-(GM-CSF) fusion protein retains antigen specificity and cytokine function and demonstrates antitumor activity. *J. Immunol.*, **165** (9), 5112–5121.
175. Xuan, C., Steward, K.K., Timmerman, J.M., and Morrison, S.L. (2010) Targeted delivery of interferon-alpha via fusion to anti-CD20 results in potent antitumor activity against B-cell lymphoma. *Blood*, **115** (14), 2864–2871.
176. Huang, T.H., Chintalacharuvu, K.R., and Morrison, S.L. (2007) Targeting IFN-alpha to B cell lymphoma by a tumor-specific antibody elicits potent antitumor activities. *J. Immunol.*, **179** (10), 6881–6888.
177. Helguera, G., Morrison, S.L., and Penichet, M.L. (2002) Antibody-cytokine fusion proteins: harnessing the combined power of cytokines and antibodies for cancer therapy. *Clin. Immunol.*, **105** (3), 233–246.
178. Challita-Eid, P.M., Penichet, M.L., Shin, S.U., Poles, T. et al. (1998) A B7.1-antibody fusion protein retains antibody specificity and ability to activate via the T cell costimulatory pathway. *J. Immunol.*, **160** (7), 3419–3426.
179. Peng, L.S., Penichet, M.L., and Morrison, S.L. (1999) A single-chain IL-12 IgG3 antibody fusion protein retains antibody specificity and IL-12 bioactivity and demonstrates antitumor activity. *J. Immunol.*, **163** (1), 250–258.
180. Gillies, S.D., Lan, Y., Williams, S., Carr, F. et al. (2005) An anti-CD20-IL-2 immunocytokine is highly efficacious in a SCID mouse model of established human B lymphoma. *Blood*, **105** (10), 3972–3978.
181. Hank, J.A., Gan, J., Ryu, H., Ostendorf, A. et al. (2009) Immunogenicity of the hu14.18-IL2 immunocytokine molecule in adults with melanoma and children with neuroblastoma. *Clin. Cancer Res.*, **15** (18), 5923–5930.
182. Shusterman, S., London, W.B., Gillies, S.D., Hank, J.A. et al. (2010) Antitumor activity of hu14.18-IL2 in patients with relapsed/refractory neuroblastoma: a Children's Oncology Group (COG) phase II study. *J. Clin. Oncol.*, **28** (33), 4969–4975.
183. Halin, C., Gafner, V., Villani, M.E., Borsi, L. et al. (2003) Synergistic therapeutic effects of a tumor targeting antibody fragment, fused to interleukin 12 and to tumor necrosis factor alpha. *Cancer Res.*, **63** (12), 3202–3210.
184. Helguera, G., Rodriguez, J.A., and Penichet, M.L. (2006) Cytokines fused to antibodies and their combinations as therapeutic

agents against different peritoneal HER2/neu expressing tumors. *Mol. Cancer Ther.*, **5** (4), 1029–1040.

185. Helguera, G., Dela Cruz, J.S., Lowe, C., Ng, P.P. *et al.* (2006) Vaccination with novel combinations of anti-HER2/neu cytokines fusion proteins and soluble protein antigen elicits a protective immune response against HER2/neu expressing tumors. *Vaccine*, **24** (3), 304–316.

186. Helguera, G., Rodriguez, J.A., Daniels, T.R., and Penichet, M.L. (2007) Long-term immunity elicited by antibody-cytokine fusion proteins protects against sequential challenge with murine mammary and colon malignancies. *Cancer Immunol. Immunother.*, **56** (9), 1507–1512.

187. Moreland, L.W., Cohen, S.B., Baumgartner, S.W., Tindall, E.A. *et al.* (2001) Long-term safety and efficacy of etanercept in patients with rheumatoid arthritis. *J. Rheumatol.*, **28** (6), 1238–1244.

4
Growing Mini-Organs from Stem Cells

Hiroyuki Koike[1], Tamir Rashid[2], and Takanori Takebe[1,3,4]
[1]*University of Cincinnati, Division of Pediatric Gastroenterology, Hepatology, and Nutrition, Department of Pediatrics, Cincinnati Children's Hospital Medical Center, 3333 Burnet Avenue, Cincinnati, OH 45229-3039, USA*
[2]*King's College London, Centre for Stem Cells and Regenerative Medicine and Institute of Liver Studies, London SE5 9RS, UK*
[3]*Yokohama City University Graduate School of Medicine, Department of Regenerative Medicine Yokohama, Kanagawa 236-0004, Japan*
[4]*PRESTO, Japan Science and Technology Agency, 4-1-8 Honcho Kawaguchi-shi, Saitama 332-0012, Japan*

1	Introduction	138
1.1	Stem Cell Technologies	139
1.2	Paradigm Shift from "Cell Induction" to "Mini-Organ Generation"	140
2	**Spatiotemporal Control of Mini-Organ Structure and Differentiation**	**140**
3	**Organoid Technology**	**143**
3.1	Ectodermal Organoids	143
3.1.1	Cerebral Organoids	143
3.1.2	Retinal Organoids	144
3.1.3	Pituitary Organoids	144
3.1.4	Salivary Organoids	144
3.2	Endodermal Organoids	145
3.2.1	Intestinal Organoids (Somatic Stem Cells)	145
3.2.2	Intestinal Organoids (Pluripotent Stem Cells)	145
3.2.3	Gastric Organoids	146
3.2.4	Pancreatic Organoids	146
3.2.5	Thyroid Organoids	147
3.2.6	Biliary Organoids	147
3.3	Mesodermal Organoids	147
3.3.1	Renal Organoids	147
3.3.2	Cardiac Organoids	148

Translational Medicine: Molecular Pharmacology and Drug Discovery
First Edition. Edited by Robert A. Meyers.
© 2018 Wiley-VCH Verlag GmbH & Co. KGaA. Published 2018 by Wiley-VCH Verlag GmbH & Co. KGaA.

4	Missing Cues in Current Organoid Technology	148
5	Organ Bud Technology	149
5.1	Liver Buds	149
5.2	Kidney Buds	150
5.3	Mesenchymal Condensations	151
6	The Future of Mini-Organ Technologies	151
	Acknowledgments	153
	References	153

Keywords

Mini-organs
A tissue generated in a dish through a self-organization principle by activating both internal and extrinsic controls of collective cell behaviors.

Organoids
A tiny organized tissue structure generated from a stem cell aggregate, which is mostly reconstituting a uniform epithelial tissue. Most of these tissues lacked the supporting cell components such as blood vessels.

Organ buds
A more complex and highly organized tissue made from a heterogeneous mixture of stem cells by using specific four-dimensional culture such as "self-condensation" method, which is capable of evolving vascularized and functional tissue elements *in vivo*.

■ The discovery of human induced pluripotent stem cells has offered new opportunities to study human health and diseases, and ultimately therapy. Yet, their realistic application remained a challenge simply due to the failure in recapitulating natural differentiation process with currently available methods. In development, the process of directed differentiation requires spatiotemporal interactions among multiple supporting cells and their environments. Recently, novel three-dimensional (3D) stem cell culture methods for recapitulating the organogenesis process were reported and proved successful for self-organizing 3D mini-organs, so-called an organoid or organ bud. Here, the latest culture technologies to use stem cells for generating functional 3D mini-organ are reviewed and their potential for future applications is discussed.

1 Introduction

Stem cells possess the important capacity of being able to differentiate into various different cell types of the body and to replicate themselves for long periods while maintaining their undifferentiated state [1]. These multi-differentiation and self-renewal capabilities have resulted in considerable interest for using stem cell-derived cells in disease modeling, drug screening, and regenerative medicine. For all these applications, it is critical to establish robust techniques for inducing target cells by logically directing stem cell fate in a stepwise manner. In contrast to relatively simple tissues such as cornea and cartilage, the pharmaceutical and clinical application of complex organs using stem cells still faces major challenges because of the lack of efficient methods for inducing target cell differentiation. Towards this aim, *in vitro* approaches to model human organogenesis are an intense area of stem cell-related research; thereby self-organizing three-dimensional (3D) tissues so-called organoids or organ buds. Here, the emerging technologies for generating such mini-organs *in vitro* will be reviewed and their possible future applications discussed.

1.1
Stem Cell Technologies

The rare undifferentiated cell that exists among differentiated cells of an established tissue or organ, is called a "somatic stem cell" [2]. Somatic stem cells have multipotency; that is, they can generate a limited number of differentiated cell lineages *in vitro*. Somatic stem cell-based regenerative medicine approaches, using mesenchymal stem cells (MSCs), for example, have already been tested in several clinical trials for diseases such as osteoarthritis [3, 4]. However, because of their lineage-committed nature, somatic stem cells can only produce cells and tissues with simple composition and structure, making it much more difficult to accommodate this technology into larger, more complex organs. Furthermore, because somatic stem cell collection and subsequent expansion is only possible for very limited organs, such as intestinal stem cells, it is considered generally very difficult to procure sufficient quantity for their applications.

In contrast, pluripotent stem cells have the ability to give rise to all cells derived from the three germ layer lineages (mesoderm, endoderm, and ectoderm). Cells established in *ex vivo* culture originating from the early embryo's inner cell mass (which *in vivo* go on to become the fetus) are called "embryonic stem cells (ESCs)." These can be cultured for long time periods while maintaining a pluripotent state. Because the human ESC, established by Thomson *et al.* in 1998, can differentiate into all cells of the human body and be prepared in large quantities *in vitro*, its application toward various regenerative contexts has been eagerly anticipated [5]. However, establishing ESC lines holds not only some technical difficulties, but also ethical problems to consider in association with the donor egg cell collection. In addition, the problem of immune rejection following the allogeneic transplantation of ESC-derived cells increases the hurdle for regenerative medicine applications.

In 2007, Drs Takahashi K. and Yamanaka S. reported a cell reprogramming technique to create human induced pluripotent stem cells (iPSCs) that are biologically similar to ESCs [6]. iPSC technologies enabled the derivation of patients' own pluripotent stem cells; hence, they are expected to be a powerful tool for drug discovery and a new cell source for regenerative medicine. The imminent realization of these expectations is globally anticipated. However, decades of laboratory studies have so far failed to produce the target functional cell populations, either from ESCs or iPSCs. The inductive method of human

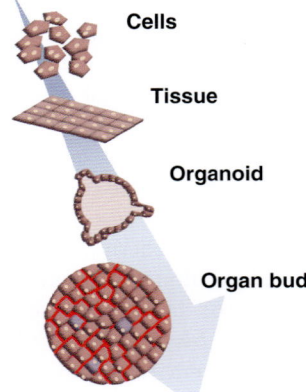

Fig. 1 Paradigm shift in stem cell differentiation. Studies of induction of stem cell differentiation have begun to aim at generating 2D cell or engineered tissue. Recently, the target has dynamically shifted to generating complex 3D mini-organs by recapitulating organogenesis, such as organoids and organ buds.

pluripotent stem cell is largely different and often more complicated than that of somatic stem cell-based technologies, because there are numerous amounts of key events to model organogenesis in culture to direct cell fates that occur during the long history (ca. 10 months) of human embryogenesis.

1.2
Paradigm Shift from "Cell Induction" to "Mini-Organ Generation"

Knowledge of the *in vivo* developmental processes was extensively translated into a two-dimensional (2D) differentiation strategy *in vitro*, especially at a molecular level based on an entirely reductionism approach. As a result, many types of tissue cell-specific differentiation protocols have been reported with the combinatory use of different inductive factors [7–9]. These 2D culture platforms have indeed enabled drastic advancements in controlling the fate specification of pluripotent stem cells, though many of the studies failed to provide evidence of the adult level functionality of generated cells, rather showing fetal phenotype in most cases. More importantly, even if it possible to obtain full functionality in culture, cell transplantation approaches such as hepatocyte transplantation have been proven to offer limited therapeutic efficacy, posing a question with conventional 2D differentiation protocols.

Alternatively, a paradigm shift away from conventional strategies has been emerging in various contexts by modeling the holistic nature of human organ development. An epoch-making study – that is, a mini-organ-based approach – is now becoming a promising field by designing the organoid or organ buds in stem cell culture (Fig. 1).

2
Spatiotemporal Control of Mini-Organ Structure and Differentiation

Mini-organs, which structurally recapitulate aspects of native tissues, but are created mainly from tissue-specific 3D stem cell culture systems have been reported for many types of cell and tissue [10] (Table 1). Cell culture is generally performed on 2D standard laboratory plates or dishes, but biological tissues *in vivo* are formed in 3D structure by experiencing time- and space-dependent complex phenomena.

Tab. 1 Current stem/progenitor cell-derived mini-organ technology.

Germ layer	Organ	Cell source	Species	Reference(s)
Ectoderm	Brain	ESC	Mouse	[11, 12]
			Human	[12, 13]
		iPSC	Human	[14, 15]
	Optic cup	Neonatal cell	Rat	[16]
		ESC	Human	[17–23]
			Mouse	[24, 25]
		iPSC	Human	[22, 23, 26, 27]
	Pituitary	ESC	Mouse	[28]
	Salivary gland	Embryonic cell	Mouse	[29, 30]
		Adult stem cell	Mouse	[31]
	Mammary gland	Progenitor cell line	Mouse	[32]
		Adult stem cell	Human	[33, 34]
			Mouse	[35]
	Inner ear	iPSC	Mouse	[36]
	Skin	Adipose-derived stem cell	Rat	[37]
	Hair follicle	Embryonic cell	Mouse	[38]
		Adult stem cell	Mouse	[39]
	Tooth	Embryonic cell	Mouse	[40]
	Lacrimal gland	Embryonic cell	Mouse	[41]
Endoderm	Small intestine	Adult stem cell	Mouse	[42–45]
			Human	[46, 47]
		Embryonic cell	Rat	[48]
		Embryonic progenitor cell	Mouse, human	[49]
		ESC	Human	[50]
		iPSC	Human	[49–51]
	Colon	Adult stem cell	Human	[52]
			Mouse	[53, 54]
	Stomach	Adult stem cell	Mouse	[55, 56]
			Human	[57]
		ESC and iPSC	Human	[58]
	Pancreas	Embryonic progenitor cell	Mouse	[59]
		Adult stem cell	Mouse	[60]
		ESC	Mouse	[61–63]
			Human	[64, 65]
		iPSC	Mouse	[66]
			Human	[67]
	Thyroid	ESC	Mouse	[68]
	Liver	Adult stem cell	Mouse	[69]
			Human	[70]
		iPSC	Human	[71]
	Lung	Embryonic cell	Mouse	[72]
		Adult stem cell	Human	[73]
			Mouse	[73, 74]
		iPSC	Human	[75]
	Prostate	Adult progenitor cell	Mouse	[76]

(*continued overleaf*)

Tab. 1 (Continued)

Germ layer	Organ	Cell source	Species	Reference(s)
Mesoderm	Kidney	Embryonic cell	Mouse	[77–84]
			Rat	[78, 85]
		Adult stem cell	Rat	[86]
		ESC	Mouse	[87, 88]
			Human	[89, 90]
		iPSC	Human	[87, 89, 90]
	Heart	Neonatal cell	Rat	[91]
		Adult progenitor cell	Mouse	[92]
		Adult stem cell	Mouse, human	[93]
			Dog	[94]
		ESC and iPSC	Human	[95]
	Skeletal muscle	Myoblast progenitor cell	Mouse	[96]
		ESC	Human	[97]
	Bone	MSC	Human	[98, 99]
		ESC	Human	[100]
		iPSC	Human	[101]

ESC, embryonic stem cell; iPSC, induced pluripotent stem cell; MSC mesenchymal stem cell.

Therefore, it is considered that 3D culture systems would be more similar to the natural biological environment; for example, cell polarity and cell-to-cell transmission between each cells are three-dimensionally conducted, while the nutrient and medium supply pathways are strictly controlled according to their specific spatial orientation between the inside and the outside of the developing organ.

In biological systems, multicellular self-organization is, in principle, composed of self-assembly, self-patterning, and self-morphogenesis, as elegantly summarized by Dr Sasai's previous review [102] (Fig. 2). Briefly, self-assembly involves the time-evolving control of cell positions relative to each other, such as in the formation of a layered pattern or vascular sprouting [71, 77]. Self-patterning is the spatiotemporal control of cell status, so that cells acquire heterogeneous properties in a region-specific manner from a homogeneous cell population. Self-driven morphogenesis is the spatiotemporal control of intrinsic tissue mechanics, shown as a representation of retinal epithelium deformation [17, 24].

Thus, modeling organogenesis in culture utilizes self-organizing activities involving a combination of self-assembly, self-patterning, and self-driven morphogenesis. In the developing tissue, morphogenesis is confined to the cell in a process of patterning events that are spatially and temporally coordinated with other developmental changes. In this manner, to gain enough functional organ from stem cell, it is necessary to reconstruct the cells through orchestrated communication of tissue-level 3D instructions with intricate patterns, rather than to differentiate into just 2D functional cells. During recent years, the generation of mini-organs in culture has become the new cornerstone, where there are numerous methods that enable self-organization by combining multicellular interactions, soluble chemical factors actuation, and extracellular matrix

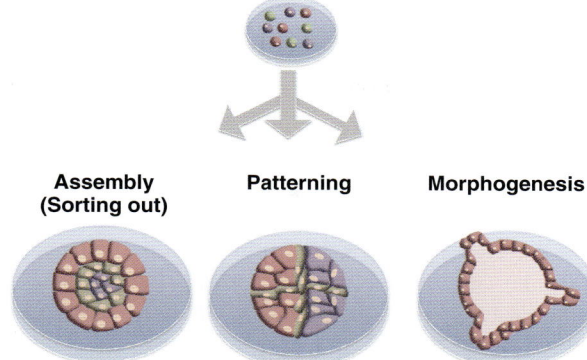

Fig. 2 Principles of self-organization phenomena. For simplicity, the principles of self-organization can be divided into three major phenomena. *Assembly* is an intrinsic cell-sorting process in a time course manner. *Patterning* is a region-specific cell fate specification in a tissue so as to evolve a heterogeneous cell population. *Morphogenesis* is a spatiotemporal control of tissue mechanics accompanied by dynamic morphological change.

support. Recent advances in mini-organ approaches will be valuable tools to manipulate 3D structure derived from stem cells in a spatiotemporally and highly controlled manner within the mini-organ.

3 Organoid Technology

Self-organization into complex architectures has been extensively observed in classical embryonic cell re-aggregation experiments in various organs of sea sponges [103], early amphibian embryonic tissues [104], and bipinnaria larvae of starfish [105]. From this classical observation, remarkable progress has been made in understanding self-organization using mammalian stem cell cultures that allow the *in vitro* recapitulation of organogenetic processes, which is well described in the nervous system [11–15, 17–27, 36] and gut epithelial tissues [42–47, 49–58]. Recent identification of conditions would permit the *in vitro* self-organization into a structurally ordered micro-tissue – referred to as an organoid – using 3D cultures of homogeneous stem cells.

3.1 Ectodermal Organoids

3.1.1 Cerebral Organoids

By mimicking the process of organogenesis in the early stages of cerebral development, Sasai and colleagues confirmed that ESC-derived embryoid bodies produced neural ectoderm, and self-organized the 3D brain tissue using various inhibitors of anti-neuralizing signals [wingless-int (WNT), Nodal, and bone morphogenetic protein (BMP)], Dkk1, LeftyA, and BMPRIA-Fc, respectively, and inhibitor of Rho-associated coiled-coil forming kinase (ROCK) in a floating culture system called SFEBq culture [12, 13]. In the formation of these 3D organoids, the phenomenon of self-organization caused advanced morphogenesis in the entire organoid that includes distinct zones, such as ventricular, cortical-plate, and Cajal–Retzius cell zones by spontaneous interactions between ESC-derived cerebral cortex precursors.

Recently, Lancaster et al. reported the generation of 3D self-organized brain tissue by using the human iPSC-derived neuroectodermal cells with a modification of the method described above by Eiraku and Sasai et al. [15]. In this way, human iPSC-derived organoids of 4 mm in size could be cultured for over six months in a bioreactor. As, in principle, the mouse models failed to model aspects of human pathology, these authors investigated whether the organoid could be used for human disease modeling. By using a microcephaly patient-derived cerebral organoid, they showed that a defection of the correct expansion of the neural progenitor cell caused the patient tissue phenotype. In time, this technology is expected to provide a human-specific disease analysis of the brain, thereby reducing the dependence on experiments in mice.

3.1.2 Retinal Organoids

Eiraku and Sasai et al. succeeded in recapitulating the process of optic cup formation from the optic vesicle, *in vitro*, by using the same approach as cerebral organoid generation [17, 24]. A remarkable aspect of their study was to define the controversial morphogenetic mechanisms of optic cup deformation by using their organoid culture.

Recently, they also succeeded in producing a complex 3D retinal tissue that contained the ciliary margin as a stem cell niche from human ESCs [21]. It was found that the human ciliary margin included the stem cell, and that the retina was derived by proliferation of this stem cell *in vitro*. This novel technology for producing a 3D retina in a robust manner is useful for clarifying the role of the human ciliary margin in retinal development. The 3D retina is expected to aid regenerative medicine efforts in diseases such as retinitis pigmentosa.

3.1.3 Pituitary Organoids

The pituitary gland is a small endocrine organ located at the base of the brain, and plays a major role as the control center for a variety of hormones, such as adrenocorticotropic hormones and pituitary hormones [106]. Whilst dysfunction of the pituitary gland can cause serious systemic diseases, it has been difficult to form pituitary tissue from stem cells because this structure is created through a very complex developmental process. Both, Suga and Sasai et al. reported the establishment of a methodology for recapitulating the developmental process of pituitary gland formation *in vitro* using the previously mentioned 3D self-organization technology employed for cerebral and retinal organoid formation [28]. They succeeded in generating Rathke's pouch, the pituitary primordium and artificial pituitary tissue with hormone-producing capacity. This research will be a useful approach for regenerative medicine therapies of endocrine tissues.

3.1.4 Salivary Organoids

Salivary gland stem cell therapy is an attractive field for treating xerostomia patients, but the limitation in availability of adult human tissue interrupts their application. Ogawa and Tsuji et al. have accordingly challenged murine salivary gland regeneration by organoid technologies [29]. They induced a salivary primordium by co-culturing gently dissociated fetal cells from the submandibular gland and sublingual gland, and eventually found that both primordia formed the secretory-specific branched structures after extended culture *in vitro*. As a result of acidity-induced stimulation after transplantation of the generated salivary glands into the mouse mouth, salivary secretion was significantly increased. Amylase was found in the secreted saliva, suggesting the functionality of the reconstituted salivary gland from transplanted organoid. These authors' approach will facilitate the use of salivary organoid transplantation for a variety of scientific and therapeutic purposes, including the treatment of severe type Sjögren's syndrome. The same group showed that the culture method is applicable also to other tissues, such as hair follicle [38, 39], tooth [40], and lacrimal gland [41], to form 3D mouse tissues.

3.2
Endodermal Organoids

3.2.1 Intestinal Organoids (Somatic Stem Cells)

Intestinal stem cells represent undoubtedly one of the most intense areas of stem cell research, and have been studied extensively for organoid systems. Somatic stem cells of the intestine are reproducibly shown to exist at the bottom of intestinal crypts, expressing Leucine-rich repeat-containing G protein-coupled receptor 5 (Lgr5) [107]. The establishment of an *in vitro* culture method for intestinal stem cells affords the possibility of numerous studies, such as in-depth analyses of proliferative processes regulating stem cells and the effects of chemicals. Indeed, Sato and Clevers *et al.* succeeded in culturing the human epithelial organoids from various gut systems by adding combinatory proteins, WNT3A, gastrin, nicotinamide, Alk4/5/7 inhibitor, and p38 inhibitor [53], following the establishment of a single intestinal stem cell-derived organoid culture system based on using R-spondin-1 (a ligand for Lgr5), epidermal growth factor (EGF), Noggin, and laminin-rich Matrigel [43]. These authors expanded the single colonic stem cells *in vitro* and found that the expanded cells could regenerate the colonic epithelium, and that the formed tissue could survive for over six months in recipient mice after transplantation.

One attractive application of these epithelial organoid technologies is to analyze the precise evolutional process of malignancy, which is otherwise inaccessible. As an initial demonstration, Matano and Sato *et al.* developed a technology to generate gene-modified colonic stem cells by inducing multiple mutations of tumor suppressor genes *APC*, *SMAD4*, and *TP53*, and of oncogenes *KRAS* and *PIK3CA*, into human colonic organoids via the CRISPR-Cas9 genome-editing system for an analysis of colonic carcinogenesis [52]. It was revealed that human gut epithelial organoids carrying gene mutations produced tumors after transplantation, but that the five mutations were insufficient to achieve the metastatic ability. Furthermore, Drost and Clevers *et al.* revealed that sequential mutations of *APC*, *TP53*, *KRAS*, and *SMAD4* can initiate the growth of stem cells in the absence of stem cell niche factors, WNT, Noggin, R-spondin, and mutant organs grow as tumors with features of invasive carcinoma [47]. These culture methods could be applied not only for normal intestinal epithelium but also for carcinogenesis of the intestine. This makes it applicable for various fields of research such as disease modeling and pharmacological testing.

3.2.2 Intestinal Organoids (Pluripotent Stem Cells)

The use of pluripotent stem cells will be important just because they have a possibility to reconstitute the other supportive cell components such as mesenchymal cells, which is currently lacking in the above-mentioned epithelial organoid technology.

Furthermore, it is also attractive in the context of disease modeling, especially for studying congenital developmental disorders. The recent identification and cultivation methods for human intestinal stem cells from postnatal tissues have opened a new possibility for the study of pluripotent stem cell (PSC)-derived intestinal organoids with some modification in *ex vivo* culture conditions [108]. Spence and Wells *et al.* identified conditions for spontaneous hindgut tubular tissue formation from iPSC-derived gut endodermal cells by adding growth factors WNT3A and fibroblast growth factor 4 (FGF4), and showed that further culture in Matrigel containing R-Spondin1, Noggin, and EGF initiated maturation into premature intestine like tissues with a crypt architecture and a layered mesenchymal structure [51]. Watson *et al.* revealed that the expanded intestinal organoid based on this technique became vascularized by host vessels and grew into further mature intestinal tissue, similar to the human biological intestine at six weeks after transplant into the kidneys of immunodeficient mice [50]. The generated tissue developed a typical complex intestine structure such as crypt and villus, and included multiple types of cell such as lamina propria, muscularis mucosa, and submucosa. Moreover, the tissue also indicated digestive and absorptive functions. Notably, it was shown that the engrafted human tissue had grown and adapted after surgical removal from part of the mouse intestine, which in turn suggested that this human intestinal tissue could respond to signals in mouse blood.

3.2.3 Gastric Organoids

Human gastric tissue is largely different from that of other species, and there is no appropriate way to study human gastric disease [109]. McCracken and Wells *et al.* succeeded in inducing a 3D foregut and gastric organoid from a human PSC-derived definitive endodermal cell *in vitro* by activating the FGF, WNT, BMP, retinoic acid, and EGF signaling pathways [58]. This 2–4 mm human gastric organoid included an antrum cell population, such as LGR5-expressing cells, mucous cells, and gastric endocrine cells. The gastric organoid recapitulated the inflammation that occurs in actual stomach tissue after exposure to *Helicobacter pylori* via the CagA pathway, which suggests that this organoid technique could be used as a very useful model of human gastric disease.

3.2.4 Pancreatic Organoids

Although drug therapy for diabetes is quite advanced, a fundamental cure for type 1 diabetes caused by insulin deficiency still requires pancreas or islet transplantation [110]. Despite the large numbers of patients waiting for a transplant, there remains a chronic shortage of donors.

In order to overcome this shortage, investigations into the generation of functional islets and pancreatic tissues with a 3D structure have been extensive, using PSCs [61–67]. Pagliuca *et al.* reported methods for an efficient generation of functional islet tissue by inducing the differentiation of sphere-like iPSCs in spinner culture vessels [67]. The authors were successful in identifying 11 factors, namely protein kinase C activator, BMP receptor inhibitor, keratinocyte growth factor, hedgehog signaling inhibitor, retinoic acid, Notch inhibitor, betacellulin, high glucose, heparin, Alk5 inhibitor, and triiodothyronine, for the efficient production of stem cell-derived β cells from over 150 candidates. It was suggested that these islets showed a greater insulin secretion *in vitro* compared to stem cell-derived islets produced by conventional differentiation methods.

A glucose-sensitive insulin secretion ability was also demonstrated *in vivo*. Such islets, when transplanted into diabetic mice, led to the generation of β-cells that could produce insulin for a few months and improve the symptoms caused by a hyperglycemic state. This technique was established using spinner culture vessels with mass production in mind, making the clinical possibility of a practical novel therapeutic for type 1 diabetes more tangible.

3.2.5 Thyroid Organoids

Antonica *et al.* reported a method for differentiation into functional thyroid follicles from mouse ESCs *in vitro* [68]. The differentiation to thyroid follicle cells was induced by the overexpression of transcription factors *NKX2-1* and *PAX8*, after which self-organization was initiated by thyroid-stimulating hormone. The 3D thyroid follicles obtained by this method showed characteristic thyroidal function *in vitro*, and rescued symptoms after transplantation into mice whose thyroid tissue had previously been removed. The results of this study expanded the knowledge of molecular mechanisms responsible for thyroid gland development, and also helped pave the way towards a replacement treatment for congenital hypothyroidism.

3.2.6 Biliary Organoids

Huch *et al.* established the methods for long-term culture of mouse [69] and human adult liver stem cells, which are obtainable even from a tiny biopsy sample [70]. It was found that cells derived from human liver could be stably proliferated about once every 60 h, using a 3D culture system with transforming growth factor-β (TGFβ) inhibition and cyclic adenosine monophosphate (cAMP) pathway induction [70]. The isolated stem cell from adult tissue was derived from the hepatic biliary ductal system, which has been considered as the origin of liver stem cells in other studies [111]. Cultured liver stem cell organoids were genetically stable and safe in that they showed less genomic mutation in clones from single stem cells after three months of culture. Expanded organoids could also be differentiated into hepatic tissue by treating with Notch inhibitor, FGF, and BMP signals. Differentiated cells derived from organoids could secrete albumin for over two months and also rescued liver failure in a mouse model. This organoid technology is expected to greatly improve hepatocyte- and cholangiocyte-related disease modeling.

3.3 Mesodermal Organoids

3.3.1 Renal Organoids

The kidneys filter blood through the glomeruli, and produce urine by reabsorbing necessary molecules via the renal tubules. They play an essential role in the maintenance of body fluid balance and the regulation of blood pressure [112]. Many diseases that cause disorders of the kidney cannot be treated, and in most cases kidney function never returns to normal. Consequently, a human kidney model derived from stem cells could realize a deeper scientific understanding of the kidney in health and disease.

The generation of renal organoids with a 3D structure has been successful using rodent adult and fetal stem/progenitor cells [78–86]. For human cells, Xia *et al.* created a ureteric bud with a 3D structure that could grow into collecting duct tubes [89]. In this case, a simple and efficient method was developed to generate the kidney organoid using both iPSC- and ESC-derived kidney precursor cells. The precursors received a proper chaperone, allowing them to grow into the kidneys and eventually to develop into a 3D kidney structure by co-culturing with mouse fetal cells. iPSCs from a polycystic kidney disease patient were also established and showed that kidney tissue could similarly be generated from patient-specific iPSCs. Based on previous medical experience, gene therapy and immunotherapy are not realistic treatments for the disease; therefore it is expected this kidney model will become a powerful tool for developing new pharmaceutical treatments.

Taguchi *et al.* generated a 3D kidney organoid accompanied with glomeruli and renal tubules from mouse ESCs and human iPSCs [87]. The glomeruli and tubules were generated from fetal kidney progenitor cells, in parallel with mouse study. It was found that the posterior precursor population could be the origin of the nephron progenitors in mouse, and stage-specific fundamental signals of activin A, BMP, WNT, retinoic acid, and FGFs were identified for the induction of progenitors. The same group succeeded in generating a nephron progenitor from mouse ESCs and human iPSCs by adding these factors sequentially. Further culture of the progenitors reconstituted the 3D renal structure with glomeruli and renal tubule *in vitro*.

Kidney development is preceded by interactions between metanephric mesenchyme and ureteric epithelium that were derived from the intermediate mesoderm. Takasato *et al.* reported the creation of a 3D renal organoid from human ESCs by tracing this process *in vitro* [90]. They introduced intermediate mesoderm from ESCs via the primitive streak stage, and demonstrated further differentiation into metanephric mesenchyme and ureteric epithelium. The re-aggregation culture of these cells generated the self-organized 3D complex tissue, including the proximal tubule. The results of these studies should help to reveal mechanisms in kidney development, and are expected to be useful for drug development by recapitulating renal disease *in vitro*.

3.3.2 Cardiac Organoids

Regenerative medicine strategies for heart diseases have been studied utilizing cell sheet technologies [113]. However, research in this field is still developing because, at present, there is no technology to build the myocardium in a manner that functionally integrates into a recipient muscle cell with coordinated mechanical contraction. Alternatively, the technologies for building 3D structure to form organoids by culturing three-dimensionally *in vitro* has attracted much interest [92–95].

Zhang *et al.* have focused the critical requirements that are mediated by the correct signaling pathways of FGF, BMP and WNT for efficient cardiac differentiation from human PSCs [95]. After four weeks of 3D culture, human PSC-derived cardiomyocytes were shown to be in a transcriptionally stable state. Although far from recapitulating a heart morphology which is functionally essential, the established 3D cardiac differentiation protocol of human PSCs is expected to be applied to cardiac disease modeling, as well as subsequent pharmaceutical applications such as drug toxicity testing.

4
Missing Cues in Current Organoid Technology

Much of the reported organoid technologies described above relies on their high levels of intrinsic self-organizing capacity. Furthermore, the described methods generally reconstitute only an epithelial or neural lineage starting from a homogeneous stem cell population, and have lacked supportive cell components such as blood vessels. In contrast, in order to obtain functionally biological tissue a critical requirement is to incorporate multiple cell lineages, including at least endothelial cells, into complex architectures. Indeed, the results of emerging studies have demonstrated the important roles of endothelial cells in organogenesis and regeneration, at least through angiocrine support, and beyond simply nutrients and oxygen delivery [114–116].

5
Organ Bud Technology

One clear limitation in previous organoid technology is the failure to artificially reconstitute different progenitor intercommunications in a 3-D and time-evolving manner. Therefore, to recapitulate the more complex tissue self-organization *in vitro*, there is a need to develop a specific culture by examining communications across different lineages or germ layers. Remarkably, a recently developed culture principle using a "self-condensation" method has enabled the study of aspects of early human organogenesis in culture.

5.1
Liver Buds

During liver development, liver bud formation is initiated by the orchestrated interaction among hepatic endoderm cell, endothelial, and mesenchymal progenitors (Fig. 3). These developmental cellular interactions have been recapitulated *in vitro*, and have been successful in generating self-organizing liver buds from human iPSCs [71]. Because the generated liver buds self-assembled the developing endothelial

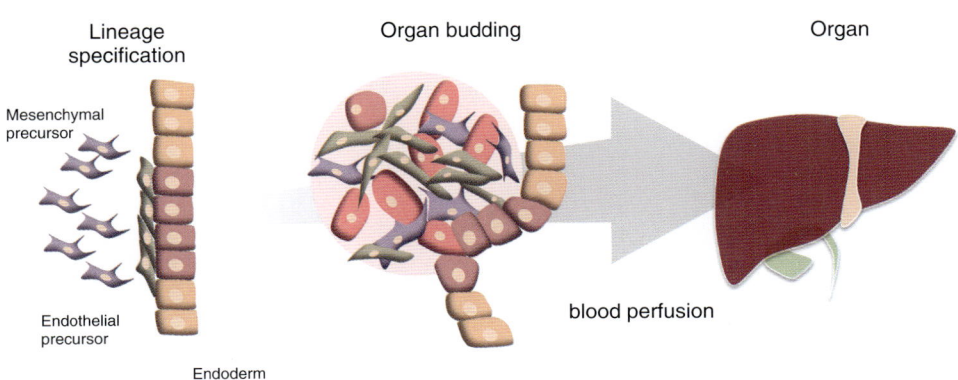

Fig. 3 Recapitulating a key organogenetic event in culture. Pioneering studies have implied that intercellular communication plays an important role in directing liver bud delamination from a primitive gut. In the mouse, liver organogenesis is initiated at embryonic day 8.5 via hepatic specification, which is driven by an interaction between the ventral portion of the foregut endoderm and the constituents of the adjacent immature endothelial cells and the surrounding septum transversum mesenchyme. Interestingly, *in vivo* liver budding process progresses prior to blood perfusion.

networks inside in culture, they are capable of generating a functional human vascular network within only 48 h of transplantation, and eventually developed functional liver tissue, similar to hepatogenesis (Fig. 4). In fact, the transplanted liver buds recapitulated aspects of expressional and structural features specific to adult-type liver tissues. Human-specific protein production and drug metabolism capacity in immunodeficient mouse was also demonstrated. Furthermore, human liver bud transplants significantly improved survival following induced liver failure in a murine model, paving new routes to future human clinical applications. The establishment of a scalable current Good Manufacturing Practices (cGMP) culture and transplantation system could eventually lead to organ bud transplantation therapy being realized in a clinical setting for currently intractable diseases such as congenital liver disorders.

5.2 Kidney Buds

Investigations have also been made as to whether the self-condensation culture principles can be adapted to reconstitute other organ systems of health and disease. To this aim, the first attempts were to clarify the essential factors to reproduce organ bud formation in culture. Combined with biomolecular and mathematical analysis, sequential image data analysis of organ bud formation revealed that the mesenchymal cell-dependent contraction of multiple lineage cells stimulated a dynamic and drastic process of self-condensation. Further analysis, coupled with modified environmental conditions, showed that the specific physical properties (stiffness) of culture substrate is essential for maximizing MSC-dependent self-condensation towards subsequent organ bud self-organization [77] (Fig. 4).

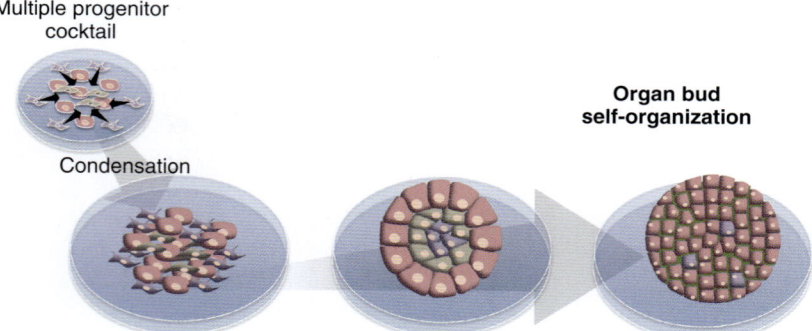

Fig. 4 Organ bud self-organization via mesenchymal cell-driven condensation. This liver bud self-organization was recently achieved by recapitulating the liver budding process as follows: human iPSC-derived hepatic endoderm cells were mixed with human endothelial and mesenchymal progenitor cells on a soft substrate *in vitro*; the mixed cells self-condensed to form a 3D mass, approximately 4–5 mm in diameter. Additional culture of these condensed masses eventually self-organized into a tissue that had an inner-branched endothelial structure, showing features specific to early liver bud in terms of structures and gene-expression properties. A self-condensation approach could be adapted to self-organize kidney bud, pancreatic bud, and mesenchymal condensation.

Considering that the MSC-driven contraction force on defined stiffness is central to the observed self-driven behavior, the extendibility of this principle to organ systems including kidney was then demonstrated. In fact, co-culturing mouse fetal kidney cells with human MSCs resulted in 3D organ bud self-organization with endothelial networks similar to liver bud culture [77]. The kidney bud thus generated *in vitro* was transplanted and reconstituted the functional human blood vessel networks, as well as self-organized renal tissue with urine-producing capability within the glomerular structures, including the presence of podocytes and slit membrane. It is greatly anticipated that future adaptation using iPSC-derived kidney precursors will allow disorders of kidney development to be examined in detail, and ultimately for therapy to be achieved.

5.3
Mesenchymal Condensations

The other unique example of such an organ bud-based approach is related to cartilage engineering. Because adult cartilage tissue lacks blood vessels and a neural system, cartilage is considered to be a field in which the promise of regenerative medicine compared to more complex organ targets such as liver and kidney could soon be realized [117]. Although, conventional methods for cartilage differentiation from stem cells are performed by adding various combinations of growth factors TGFs or FGFs, their efficiency is not sufficiently high to regenerate the large cartilage tissue by a cell-based approach [118]. Alternatively, while scaffolding materials have been extensively studied they still hold many scientific and clinical limitations, including post-transplant inflammation.

In order to improve differentiation efficiency into cartilage, attention has been focused on the advanced culture method to precisely recapitulate the process of normal cartilage development [119]. The findings of the present authors and others' studies have greatly suggested that, during early chondrogenesis, cartilage progenitors interact with nascent endothelial cells by providing some unknown mutual instructive signals towards maturation. Subsequent blood perfusion stimulated their additional proliferation and differentiation by a temporary existence of blood vessels at the early stage of organogenesis. To recapitulate this phenomenon *in vitro*, the "self-condensation" culture was adapted to realize spatiotemporal interactions between chondrogenic and endothelial progenitors. 3D cartilage organ buds, defined as mesenchymal condensation, were successfully generated in a similar way. Surprisingly, after transplantation into immunodeficient mice, the *in vitro*-generated mesenchymal condensation recapitulated the transient vascularization process, which was observed in normal chondrogenesis. Functional vascularization and subsequent regression eventually resulted in efficient human cartilage generation to a greater extent than conventional cartilage engineering approaches. Thus, these reports strongly suggest that even a simple organ generation from stem cells requires a sophisticated time-evolving control of multiple cell collectives during the structural organization, let alone more complex organs, highlighting the importance of mini-organ-based approaches.

6
The Future of Mini-Organ Technologies

Thus far, emphasis has been placed on the need to focus on reconstituting miniature

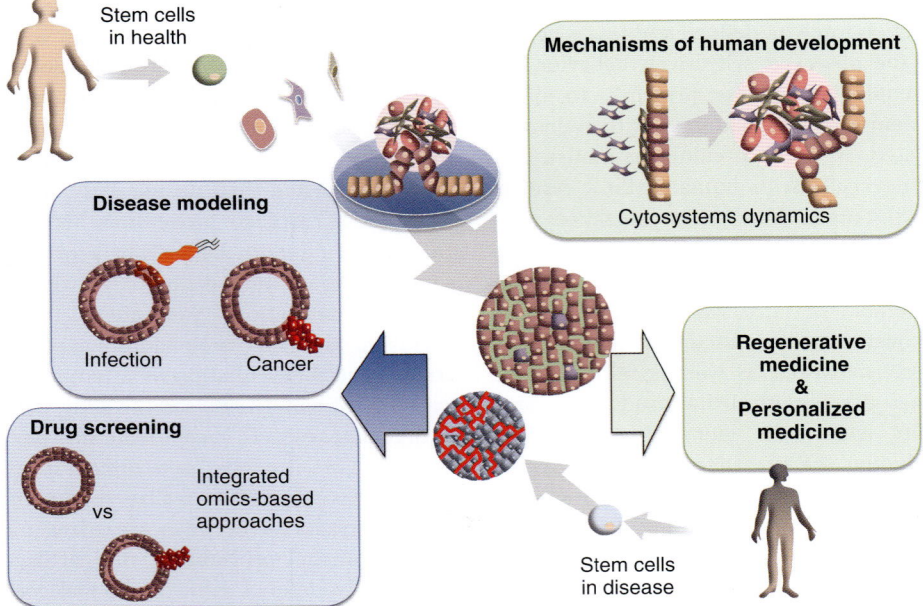

Fig. 5 Future application of stem cell-derived mini-organs. Future technological improvements in culturing and differentiating miniature organs will revolutionize the current paradigm of drug discovery and development in the pharmaceutical industry, facilitate the present understanding of human developmental biology and disease modeling, and hopefully offer novel therapies against end-stage organ failure.

organs *in vitro* from various human stem cells to recapitulate natural embryogenesis. *In vitro* human iPSC-derived mini-organ technology will undoubtedly become a novel evaluation platform for the assessment of drug efficacy and safety in preclinical phases (Fig. 5). For instance, applying these 3D organoid or organ bud technologies to human disease-specific stem cells is expected as an unprecedented research tool for clarifying pathological mechanism of diseases, as well as designing a therapy adjusted for each patient's condition. Recent pioneering research of intestinal stem cell-organoids has revealed much promise for personalized therapy design by performing a comparative analysis of cancer and normal organoids from an isogenic patients' biopsy samples [120].

As noted above, although most of the available methods have failed to recapitulate much of early developmental process *in vitro* so far, the creation of 3D mini-organs has started to evolve very quickly. Strikingly, with the hope of eliminating the present organ shortage, recently developed organ bud-based approaches seem very promising. Indeed, by using a liver model, the transplantation of an iPSC-derived rudimentary organ ("organ bud") from a mixture of multiple progenitors resulted in a therapeutic effect against lethal subacute liver failure. Given this proof-of-principle demonstration, the present authors and others will make further efforts to evaluate the precise therapeutic mechanisms and safety of human iPSC-derived organ bud transplantation.

Aiming at future clinical transplantation, enhancing post-transplant survival, and acquiring an appropriate tissue integrity of transplanted organoid or organ buds remain major challenges. To address these issues, it would be fascinating to study a possible combinatory approach by using a tissue engineering strategy such as decellularization method and a 3D bioprinting system [121–124]. For instance, the biofabrication of major blood vessels into organoids would enable tissues capable of efficiently anastomosing the recipient vessels, and thereby possibly enhancing post-transplant survival and maturation. In some cases, a capsulated device would help tissues to be immunologically privileged so as to prevent autoimmune destruction. These approaches are currently expected to improve the persistence and efficacy for type 1 diabetic patients. Thus, the present authors would like to emphasize the importance of future collaborative efforts between stem cell biologists and tissue engineers towards realistic clinical applications.

Collectively, the studies described here will provide novel regenerative approaches for the treatment of patients suffering from end-stage organ failure, insights into the human developmental biology, and new opportunities to predict the human reactions in the laboratory to promote the development of new drugs or design of personalized therapy.

Acknowledgments

The authors thank M. Park for comments on this manuscript. These studies were mainly supported by PRESTO, Japan Science and Technology Agency (JST) (T. Takebe) and partly by Grants-in-Aid from the Ministry of Education, Culture, Sports, Science and Technology of Japan (nos 24106510, 24689052, 26106721, and 15H05677 to T. Takebe), a grant of the Yokohama Foundation for Advanced Medical Science; and Research and Development Project III grant from Yokohama City University, Japan.

References

1. Odorico, J.S., D.S. Kaufman, J.A. Thomson, Multilineage differentiation from human embryonic stem cell lines. *Stem Cells*, 2001, **19** (3), 193–204.
2. Morrison, S.J., A.C. Spradling, Stem cells and niches: mechanisms that promote stem cell maintenance throughout life. *Cell*, 2008, **132** (4), 598–611.
3. Maumus, M., Guerit, D., Toupet, K. et al., Mesenchymal stem cell-based therapies in regenerative medicine: applications in rheumatology. *Stem Cell Res. Ther.*, 2011, **2** (2), 14.
4. Wei, X., Yang, X., Han, Z.P. et al., Mesenchymal stem cells: a new trend for cell therapy. *Acta Pharmacol. Sin.*, 2013, **34** (6), 747–754.
5. Thomson, J.A., Itskovitz-Eldor, J., Shapiro, S.S. et al., Embryonic stem cell lines derived from human blastocysts. *Science*, 1998, **282** (5391), 1145–1147.
6. Takahashi, K., Tanabe, K., Ohnuki, M. et al., Induction of pluripotent stem cells from adult human fibroblasts by defined factors. *Cell*, 2007, **131** (5), 861–872.
7. Si-Tayeb, K., Noto, F.K., Nagaoka, M. et al., Highly efficient generation of human hepatocyte-like cells from induced pluripotent stem cells. *Hepatology*, 2010, **51** (1), 297–305.
8. Zhang, J., Wilson, G.F., Soerens, A.G. et al., Functional cardiomyocytes derived from human induced pluripotent stem cells. *Circ. Res.*, 2009, **104** (4), e30–e41.
9. Swistowski, A., Peng, J., Liu, Q. et al., Efficient generation of functional dopaminergic neurons from human induced pluripotent

stem cells under defined conditions. *Stem Cells*, 2010, **28** (10), 1893–1904.
10. Sato, T., H. Clevers, Growing self-organizing mini-guts from a single intestinal stem cell: mechanism and applications. *Science*, 2013, **340** (6137), 1190–1194.
11. Gaspard, N., Bouschet, T., Hourez, R. et al., An intrinsic mechanism of corticogenesis from embryonic stem cells. *Nature*, 2008, **455** (7211), 351–357.
12. Eiraku, M., Watanabe, K., Matsuo-Takasaki, M. et al., Self-organized formation of polarized cortical tissues from ESCs and its active manipulation by extrinsic signals. *Cell Stem Cell*, 2008, **3** (5), 519–532.
13. Kadoshima, T., Sakaguchi, H., Nakano, T. et al., Self-organization of axial polarity, inside-out layer pattern, and species-specific progenitor dynamics in human ES cell-derived neocortex. *Proc. Natl Acad. Sci. USA*, 2013, **110** (50), 20284–20289.
14. Mariani, J., Simonini, M.V., Palejev, D. et al., Modeling human cortical development *in vitro* using induced pluripotent stem cells. *Proc. Natl Acad. Sci. USA*, 2012, **109** (31), 12770–12775.
15. Lancaster, M.A., Renner, M., Martin, C.A. et al., Cerebral organoids model human brain development and microcephaly. *Nature*, 2013, **501** (7467), 373–379.
16. Rothermel, A., Biedermann, T., Weigel, W. et al., Artificial design of three-dimensional retina-like tissue from dissociated cells of the mammalian retina by rotation-mediated cell aggregation. *Tissue Eng.*, 2005, **11** (11-12), 1749–1756.
17. Nakano, T., Ando, S., Takata, N. et al., Self-formation of optic cups and storable stratified neural retina from human ESCs. *Cell Stem Cell*, 2012, **10** (6), 771–785.
18. Nistor, G., Seiler, M.J., Yan, F. et al., Three-dimensional early retinal progenitor 3D tissue constructs derived from human embryonic stem cells. *J. Neurosci. Methods*, 2010, **190** (1), 63–70.
19. Zhu, Y., Carido, M., Meinhardt, A. et al., Three-dimensional neuroepithelial culture from human embryonic stem cells and its use for quantitative conversion to retinal pigment epithelium. *PLoS One*, 2013, **8** (1), e54552.
20. Mellough, C.B., Collin, J., Khazim, M. et al., IGF-1 signaling plays an important role in the formation of three-dimensional laminated neural retina and other ocular structures from human embryonic stem cells. *Stem Cells*, 2015, **8**, 2416–2430.
21. Kuwahara, A., Ozone, C., Nakano, T. et al., Generation of a ciliary margin-like stem cell niche from self-organizing human retinal tissue. *Nat. Commun.*, 2015, **6**, 6286.
22. Meyer, J.S., Howden, S.E., Wallace, K.A. et al., Optic vesicle-like structures derived from human pluripotent stem cells facilitate a customized approach to retinal disease treatment. *Stem Cells*, 2011, **29** (8), 1206–1218.
23. Assawachananont, J., Mandai, M., Okamoto, S. et al., Transplantation of embryonic and induced pluripotent stem cell-derived 3D retinal sheets into retinal degenerative mice. *Stem Cell Rep.*, 2014, **2** (5), 662–674.
24. Eiraku, M., Takata, N., Ishibashi, H. et al., Self-organizing optic-cup morphogenesis in three-dimensional culture. *Nature*, 2011, **472** (7341), 51–56.
25. Gonzalez-Cordero, A., West, E.L., Pearson, R.A. et al., Photoreceptor precursors derived from three-dimensional embryonic stem cell cultures integrate and mature within adult degenerate retina. *Nat. Biotechnol.*, 2013, **31** (8), 741–747.
26. Zhong, X., Gutierrez, C., Xue, T. et al., Generation of three-dimensional retinal tissue with functional photoreceptors from human iPSCs. *Nat. Commun.*, 2014, **5**, 4047.
27. Tanaka, T., Yokoi, T., Tamalu, F. et al., Generation of retinal ganglion cells with functional axons from human induced pluripotent stem cells. *Sci. Rep.*, 2015, **5**, 8344.
28. Suga, H., Kadoshima, T., Minaguchi, M. et al., Self-formation of functional adenohypophysis in three-dimensional culture. *Nature*, 2011, **480** (7375), 57–62.
29. Ogawa, M., Oshima, M., Imamura, A. et al., Functional salivary gland regeneration by transplantation of a bioengineered organ germ. *Nat. Commun.*, 2013, **4**, 2498.
30. Wei, C., Larsen, M., Hoffman, M.P. et al., Self-organization and branching morphogenesis of primary salivary epithelial cells. *Tissue Eng.*, 2007, **13** (4), 721–735.
31. Nanduri, L.S., Baanstra, M., Faber, H. et al., Purification and *ex vivo* expansion of fully functional salivary gland stem cells. *Stem Cell Rep.*, 2014, **3** (6), 957–964.

32. Campbell, J.J., Davidenko, N., Caffarel, M.M. et al., A multifunctional 3D co-culture system for studies of mammary tissue morphogenesis and stem cell biology. *PLoS One*, 2011, **6** (9), e25661.
33. Dontu, G., Abdallah, W.M., Foley, J.M. et al., In vitro propagation and transcriptional profiling of human mammary stem/progenitor cells. *Genes Dev.*, 2003, **17** (10), 1253–1270.
34. Dey, D., Saxena, M., Paranjape, A.N. et al., Phenotypic and functional characterization of human mammary stem/progenitor cells in long term culture. *PLoS One*, 2009, **4** (4), e5329.
35. Shackleton, M., Vaillant, F., Simpson, K.J. et al., Generation of a functional mammary gland from a single stem cell. *Nature*, 2006, **439** (7072), 84–88.
36. Koehler, K.R., Mikosz, A.M., Molosh, A.I. et al., Generation of inner ear sensory epithelia from pluripotent stem cells in 3D culture. *Nature*, 2013, **500** (7461), 217–221.
37. Hsu, S.H., P.S. Hsieh, Self-assembled adult adipose-derived stem cell spheroids combined with biomaterials promote wound healing in a rat skin repair model. *Wound Repair Regen.*, 2015, **23** (1), 57–64.
38. Asakawa, K., Toyoshima, K.E., Ishibashi, N. et al., Hair organ regeneration via the bioengineered hair follicular unit transplantation. *Sci. Rep.*, 2012, **2**, 424.
39. Toyoshima, K.E., Asakawa, K., Ishibashi, N. et al., Fully functional hair follicle regeneration through the rearrangement of stem cells and their niches. *Nat. Commun.*, 2012, **3**, 784.
40. Oshima, M., Mizuno, M., Imamura, A. et al., Functional tooth regeneration using a bioengineered tooth unit as a mature organ replacement regenerative therapy. *PLoS One*, 2011, **6** (7), e21531.
41. Hirayama, M., Ogawa, M., Oshima, M. et al., Functional lacrimal gland regeneration by transplantation of a bioengineered organ germ. *Nat. Commun.*, 2013, **4**, 2497.
42. Ootani, A., Li, X., Sangiorgi, E. et al., Sustained *in vitro* intestinal epithelial culture within a Wnt-dependent stem cell niche. *Nat. Med.*, 2009, **15** (6), 701–706.
43. Sato, T., Vries, R.G., Snippert, H.J. et al., Single Lgr5 stem cells build crypt-villus structures *in vitro* without a mesenchymal niche. *Nature*, 2009, **459** (7244), 262–265.
44. Koo, B.K., Stange, D.E., Sato, T. et al., Controlled gene expression in primary Lgr5 organoid cultures. *Nat. Methods*, 2012, **9** (1), 81–83.
45. Sato, T., van Es, J.H., Snippert, H.J. et al., Paneth cells constitute the niche for Lgr5 stem cells in intestinal crypts. *Nature*, 2011, **469** (7330), 415–418.
46. Schwank, G., Koo, B.K., Sasselli, V. et al., Functional repair of CFTR by CRISPR/Cas9 in intestinal stem cell organoids of cystic fibrosis patients. *Cell Stem Cell*, 2013, **13** (6), 653–658.
47. Drost, J., van Jaarsveld, R.H., Ponsioen, B. et al., Sequential cancer mutations in cultured human intestinal stem cells. *Nature*, 2015, **521** (7550), 43–47.
48. Montgomery, R.K., H.M. Zinman, B.T. Smith, Organotypic differentiation of trypsin-dissociated fetal rat intestine. *Dev. Biol.*, 1983, **100** (1), 181–189.
49. Fordham, R.P., Yui, S., Hannan, N.R. et al., Transplantation of expanded fetal intestinal progenitors contributes to colon regeneration after injury. *Cell Stem Cell*, 2013, **13** (6), 734–744.
50. Watson, C.L., Mahe, M.M., Munera, J. et al., An *in vivo* model of human small intestine using pluripotent stem cells. *Nat. Med.*, 2014, **20** (11), 1310–1314.
51. Spence, J.R., Mayhew, C.N., Rankin, S.A. et al., Directed differentiation of human pluripotent stem cells into intestinal tissue *in vitro*. *Nature*, 2011, **470** (7332), 105–109.
52. Matano, M., Date, S., Shimokawa, M. et al., Modeling colorectal cancer using CRISPR-Cas9-mediated engineering of human intestinal organoids. *Nat. Med.*, 2015, **21** (3), 256–262.
53. Sato, T., Stange, D.E., Ferrante, M. et al., Long-term expansion of epithelial organoids from human colon, adenoma, adenocarcinoma, and Barrett's epithelium. *Gastroenterology*, 2011, **141** (5), 1762–1772.
54. Yui, S., Nakamura, T., Sato, T. et al., Functional engraftment of colon epithelium expanded in vitro from a single adult Lgr5(+) stem cell. *Nat. Med.*, 2012, **18** (4), 618–623.
55. Barker, N., Huch, M., Kujala, P. et al., Lgr5(+ve) stem cells drive self-renewal in the stomach and build long-lived gastric units in vitro. *Cell Stem Cell*, 2010, **6** (1), 25–36.

56. Stange, D.E., Koo, B.K., Huch, M. et al., Differentiated Troy+ chief cells act as reserve stem cells to generate all lineages of the stomach epithelium. *Cell*, 2013, **155** (2), 357–368.
57. Bartfeld, S., Bayram, T., van de Wetering, M. et al., In vitro expansion of human gastric epithelial stem cells and their responses to bacterial infection. *Gastroenterology*, 2015, **148** (1), 126–136, e6.
58. McCracken, K.W., Cata, E.M., Crawford, C.M. et al., Modelling human development and disease in pluripotent stem-cell-derived gastric organoids. *Nature*, 2014, **516** (7531), 400–404.
59. Greggio, C., De Franceschi, F., Figueiredo-Larsen, M. et al., Artificial three-dimensional niches deconstruct pancreas development in vitro. *Development*, 2013, **140** (21), 4452–4462.
60. Huch, M., Bonfanti, P., Boj, S.F. et al., Unlimited in vitro expansion of adult bipotent pancreas progenitors through the Lgr5/R-spondin axis. *EMBO J.*, 2013, **32** (20), 2708–2721.
61. Lumelsky, N., Blondel, O., Laeng, P. et al., Differentiation of embryonic stem cells to insulin-secreting structures similar to pancreatic islets. *Science*, 2001, **292** (5520), 1389–1394.
62. Nakanishi, M., Hamazaki, T.S., Komazaki, S. et al., Pancreatic tissue formation from murine embryonic stem cells in vitro. *Differentiation*, 2007, **75** (1), 1–11.
63. Micallef, S.J., Li, X., Janes, M.E. et al., Endocrine cells develop within pancreatic bud-like structures derived from mouse ES cells differentiated in response to BMP4 and retinoic acid. *Stem Cell Res.*, 2007, **1** (1), 25–36.
64. Van Hoof, D., Mendelsohn, A.D., Seerke, R. et al., Differentiation of human embryonic stem cells into pancreatic endoderm in patterned size-controlled clusters. *Stem Cell Res.*, 2011, **6** (3), 276–285.
65. Shim, J.H., Kim, J., Han, J. et al., Pancreatic islet-like three dimensional aggregates derived from human embryonic stem cells ameliorate hyperglycemia in streptozotocin-induced diabetic mice. *Cell Transplant.*, 2014. E-pub ahead of print. doi:10.3727/096368914X685438.
66. Saito, H., Takeuchi, M., Chida, K. et al., Generation of glucose-responsive functional islets with a three-dimensional structure from mouse fetal pancreatic cells and iPS cells in vitro. *PLoS One*, 2011, **6** (12), e28209.
67. Pagliuca, F.W., Millman, J.R., Gurtler, M. et al., Generation of functional human pancreatic beta cells in vitro. *Cell*, 2014, **159** (2), 428–439.
68. Antonica, F., Kasprzyk, D.F., Opitz, R. et al., Generation of functional thyroid from embryonic stem cells. *Nature*, 2012, **491** (7422), 66–71.
69. Huch, M., Dorrell, C., Boj, S.F. et al., In vitro expansion of single Lgr5+ liver stem cells induced by Wnt-driven regeneration. *Nature*, 2013, **494** (7436), 247–250.
70. Huch, M., Dorrell, C., Boj, S.F. et al., Long-term culture of genome-stable bipotent stem cells from adult human liver. *Cells*, 2015, **160** (1-2), 299–312.
71. Takebe, T., Sekine, K., Enomura, M. et al., Vascularized and functional human liver from an iPSC-derived organ bud transplant. *Nature*, 2013, **499** (7459), 481–484.
72. Mondrinos, M.J., Koutzaki, S., Jiwanmall, E. et al., Engineering three-dimensional pulmonary tissue constructs. *Tissue Eng.*, 2006, **12** (4), 717–728.
73. Rock, J.R., Onaitis, M.W., Rawlins, E.L. et al., Basal cells as stem cells of the mouse trachea and human airway epithelium. *Proc. Natl Acad. Sci. USA*, 2009, **106** (31), 12771–12775.
74. Lee, J.H., Bhang, D.H., Beede, A. et al., Lung stem cell differentiation in mice directed by endothelial cells via a BMP4-NFATc1-thrombospondin-1 axis. *Cell*, 2014, **156** (3), 440–455.
75. Dye, B.R., Hill, D.R., Ferguson, M.A. et al., In vitro generation of human pluripotent stem cell derived lung organoids. *Elife*, 2015, **4**, e05098.
76. Karthaus, W.R., Iaquinta, P.J., Drost, J. et al., Identification of multipotent luminal progenitor cells in human prostate organoid cultures. *Cell*, 2014, **159** (1), 163–175.
77. Takebe, T., Enomura, M., Yoshizawa, E. et al., Vascularized and complex organ buds from diverse tissues via mesenchymal cell-driven condensation. *Cell Stem Cell*, 2015, **16** (5), 556–565.
78. Zhang, X., K.T. Bush, S.K. Nigam, In vitro culture of embryonic kidney rudiments and

isolated ureteric buds. *Methods Mol. Biol.*, 2012, **886**, 13–21.
79. Taub, M., Wang, Y., Szczesny, T.M. *et al.*, Epidermal growth factor or transforming growth factor alpha is required for kidney tubulogenesis in matrigel cultures in serum-free medium. *Proc. Natl Acad. Sci. USA*, 1990, **87** (10), 4002–4006.
80. Rosines, E., Johkura, K., Zhang, X. *et al.*, Constructing kidney-like tissues from cells based on programs for organ development: toward a method of in vitro tissue engineering of the kidney. *Tissue Eng. Part A*, 2010, **16** (8), 2441–2455.
81. Unbekandt, M., J.A. Davies, Dissociation of embryonic kidneys followed by reaggregation allows the formation of renal tissues. *Kidney Int.*, 2010, **77** (5), 407–416.
82. Ganeva, V., M. Unbekandt, J.A. Davies, An improved kidney dissociation and reaggregation culture system results in nephrons arranged organotypically around a single collecting duct system. *Organogenesis*, 2011, **7** (2), 83–87.
83. Chang, C.H., J.A. Davies, An improved method of renal tissue engineering, by combining renal dissociation and reaggregation with a low-volume culture technique, results in development of engineered kidneys complete with loops of Henle. *Nephron Exp. Nephrol.*, 2012, **121** (3-4), e79–e85.
84. Lawrence, M.L., C.H. Chang, J.A. Davies, Transport of organic anions and cations in murine embryonic kidney development and in serially-reaggregated engineered kidneys. *Sci. Rep.*, 2015, **5**, 9092.
85. Qiao, J., H. Sakurai, S.K. Nigam, Branching morphogenesis independent of mesenchymal-epithelial contact in the developing kidney. *Proc. Natl Acad. Sci. USA*, 1999, **96** (13), 7330–7335.
86. Kitamura, S., H. Sakurai, H. Makino, Single adult kidney stem/progenitor cells reconstitute three-dimensional nephron structures in vitro. *Stem Cells*, 2015, **33** (3), 774–784.
87. Taguchi, A., Kaku, Y., Ohmori, T. *et al.*, Redefining the *in vivo* origin of metanephric nephron progenitors enables generation of complex kidney structures from pluripotent stem cells. *Cell Stem Cell*, 2014, **14** (1), 53–67.
88. Morizane, R., Monkawa, T., Fujii, S. *et al.*, Kidney specific protein-positive cells derived from embryonic stem cells reproduce tubular structures *in vitro* and differentiate into renal tubular cells. *PLoS One*, 2014, **8** (6): e64843.
89. Xia, Y., Nivet, E., Sancho-Martinez, I. *et al.*, Directed differentiation of human pluripotent cells to ureteric bud kidney progenitor-like cells. *Nat. Cell Biol.*, 2013, **15** (12), 1507–1515.
90. Takasato, M., Er, P.X., Becroft, M. *et al.*, Directing human embryonic stem cell differentiation toward a renal lineage generates a self-organizing kidney. *Nat. Cell Biol.*, 2014, **16** (1), 118–126.
91. Halbert, S.P., R. Bruderer, T.M. Lin, In vitro organization of dissociated rat cardiac cells into beating three-dimensional structures. *J. Exp. Med.*, 1971, **133** (4), 677–695.
92. Bauer, M., Kang, L., Qiu, Y. *et al.*, Adult cardiac progenitor cell aggregates exhibit survival benefit both *in vitro* and *in vivo*. *PLoS One*, 2012, **7** (11): e50491.
93. Messina, E., De Angelis, L., Frati, G. *et al.*, Isolation and expansion of adult cardiac stem cells from human and murine heart. *Circ. Res.*, 2004, **95** (9), 911–921.
94. Bartosh, T.J., Wang, Z., Rosales, A.A. *et al.*, 3D-model of adult cardiac stem cells promotes cardiac differentiation and resistance to oxidative stress. *J. Cell. Biochem.*, 2008, **105** (2), 612–623.
95. Zhang, M., Schulte, J.S., Heinick, A. *et al.*, Universal cardiac induction of human pluripotent stem cells in two and three-dimensional formats: implications for *in vitro* maturation. *Stem Cells*, 2015, **33** (5), 1456–1469.
96. van der Schaft, D.W., van Spreeuwel, A.C., Boonen, K.J. *et al.*, Engineering skeletal muscle tissues from murine myoblast progenitor cells and application of electrical stimulation. *J. Vis. Exp.*, 2013, **73**: e4267.
97. Albini, S., Coutinho, P., Malecova, B. *et al.*, Epigenetic reprogramming of human embryonic stem cells into skeletal muscle cells and generation of contractile myospheres. *Cell Rep.*, 2013, **3** (3), 661–670.
98. Scotti, C., Piccinini, E., Takizawa, H. *et al.*, Engineering of a functional bone organ through endochondral ossification. *Proc. Natl Acad. Sci. USA*, 2013, **110** (10), 3997–4002.
99. Ferro, F., Falini, G., Spelat, R. *et al.*, Biochemical and biophysical analyses of

tissue-engineered bone obtained from three-dimensional culture of a subset of bone marrow mesenchymal stem cells. *Tissue Eng. Part A*, 2010, **16** (12), 3657–3667.
100. Marolt, D., Campos, I.M., Bhumiratana, S. *et al.*, Engineering bone tissue from human embryonic stem cells. *Proc. Natl Acad. Sci. USA*, 2012, **109** (22), 8705–8709.
101. de Peppo, G.M., Marcos-Campos, I., Kahler, D.J. *et al.*, Engineering bone tissue substitutes from human induced pluripotent stem cells. *Proc. Natl Acad. Sci. USA*, 2013, **110** (21), 8680–8685.
102. Sasai, Y., Cytosystems dynamics in self-organization of tissue architecture. *Nature*, 2013, **493** (7432), 318–326.
103. Gasic, G.J., N.L. Galanti, Proteins and disulfide groups in the aggregation of dissociated cells of sea sponges. *Science*, 1966, **151** (3707), 203–205.
104. Turner, A., Snape, A.M., Wylie, C.C. *et al.*, Regional identity is established before gastrulation in the *Xenopus* embryo. *J. Exp. Zool.*, 1989, **251** (2), 245–252.
105. Dan-Sohkawa, M., H. Yamanaka, K. Watanabe, Reconstruction of bipinnaria larvae from dissociated embryonic cells of the starfish, *Asterina pectinifera*, *J. Embryol. Exp. Morphol.*, 1986, **94**, 47–60.
106. Perez-Castro, C., Renner, U., Haedo, M.R. *et al.*, Cellular and molecular specificity of pituitary gland physiology. *Physiol. Rev.*, 2012, **92** (1), 1–38.
107. Nakata, S., E. Phillips, V. Goidts, Emerging role for leucine-rich repeat-containing G-protein-coupled receptors LGR5 and LGR4 in cancer stem cells. *Cancer Manage. Res.*, 2014, **6**, 171–180.
108. Simons, B.D., H. Clevers, Stem cell self-renewal in intestinal crypt. *Exp. Cell. Res.*, 2011, **317** (19), 2719–2724.
109. Peek, R.M., *Helicobacter pylori* infection and disease: from humans to animal models. *Dis. Model. Mech.*, 2008, **1** (1), 50–55.
110. Watson, C.J., The current challenges for pancreas transplantation for diabetes mellitus. *Pharmacol. Res.*, 2015, **98**, 45–51.
111. Suzuki, A., Sekiya, S., Onishi, M. *et al.*, Flow cytometric isolation and clonal identification of self-renewing bipotent hepatic progenitor cells in adult mouse liver. *Hepatology*, 2008, **48** (6), 1964–1978.
112. Krause, M., Rak-Raszewska, A., Pietila, I. *et al.*, Signaling during kidney development. *Cells*, 2015, **4** (2), 112–132.
113. Hirt, M.N., A. Hansen, T. Eschenhagen, Cardiac tissue engineering: state of the art. *Circ. Res.*, 2014, **114** (2), 354–367.
114. Matsumoto, K., Yoshitomi, H., Rossant, J. *et al.*, Liver organogenesis promoted by endothelial cells prior to vascular function. *Science*, 2001, **294** (5542), 559–563.
115. Lammert, E., O. Cleaver, D. Melton, Induction of pancreatic differentiation by signals from blood vessels. *Science*, 2001, **294** (5542), 564–567.
116. Ding, B.S., Nolan, D.J., Butler, J.M. *et al.*, Inductive angiocrine signals from sinusoidal endothelium are required for liver regeneration. *Nature*, 2010, **468** (7321), 310–315.
117. Hollander, A.P., S.C. Dickinson, W. Kafienah, Stem cells and cartilage development: complexities of a simple tissue. *Stem Cells*, 2010, **28** (11), 1992–1996.
118. Lach, M., Trzeciak, T., Richter, M. *et al.*, Directed differentiation of induced pluripotent stem cells into chondrogenic lineages for articular cartilage treatment. *J. Tissue Eng.*, 2014, **5**, 2041731414552701.
119. Takebe, T., Kobayashi, S., Suzuki, H. *et al.*, Transient vascularization of transplanted human adult-derived progenitors promotes self-organizing cartilage. *J. Clin. Invest.*, 2014, **124** (10), 4325–4334.
120. van de Wetering, M., Francies, H.E., Francis, J.M. *et al.*, Prospective derivation of a living organoid biobank of colorectal cancer patients. *Cell*, 2015, **161** (4), 933–945.
121. Lee, V.K., Lanzi, A.M., Haygan, N. *et al.*, Generation of multi-scale vascular network system within 3D hydrogel using 3D bio-printing technology. *Cell. Mol. Bioeng.*, 2014, **7** (3), 460–472.
122. Lee, W., Lanzi, A.M., Haygan, N. *et al.*, Three-dimensional bioprinting of rat embryonic neural cells. *NeuroReport*, 2009, **20** (8), 798–803.
123. Pati, F., Jang, J., Ha, D.H. *et al.*, Printing three-dimensional tissue analogues with decellularized extracellular matrix bioink. *Nat. Commun.*, 2014, **5**, 3935.
124. Sakaguchi, K., Shimizu, T., Horaguchi, S. *et al.*, *In vitro* engineering of vascularized tissue surrogates. *Sci. Rep.*, 2013, **3**, 1316.

5
Hemoglobin

Maurizio Brunori and Adriana Erica Miele
"Sapienza" University of Rome, Department of Biochemical Sciences, P. le Aldo Moro 5, 00185 Rome, Italy

1	**General Aspects** 162	
2	**Structural Features and Basic Terminology** 163	
3	**Human Hemoglobin** 167	
3.1	Hemoglobin Components in Erythrocytes 167	
3.2	The Globin Genes and Their Biosynthesis 168	
4	**Derivatives with Heme Ligands** 172	
4.1	Ferrous Derivatives 172	
4.2	Ferric Derivatives 172	
5	**The Reaction with Heme Ligands** 173	
5.1	Functional Properties at Equilibrium 173	
5.2	Kinetic Aspects 176	
6	**Mechanisms of Cooperative Binding and Allostery** 177	
6.1	Concerted Model of Allostery 178	
6.2	Structural Changes Associated to Oxygen Binding 180	
6.3	Ligand-Binding Pathways 183	
6.4	Molecular Dynamics (MD) Simulations 184	
7	**Assembly of Globin Monomers** 185	
8	**Reactions with Nitric Oxide** 186	
8.1	Reactions of NO with Various Hb Derivatives 186	
8.2	Physiological Implications 188	
9	**Hemoglobin Variants** 189	
9.1	Nature of Mutants 189	

Translational Medicine: Molecular Pharmacology and Drug Discovery
First Edition. Edited by Robert A. Meyers.
© 2018 Wiley-VCH Verlag GmbH & Co. KGaA. Published 2018 by Wiley-VCH Verlag GmbH & Co. KGaA.

9.2	Molecular Basis of Hemoglobin Diseases	189
9.3	Hemoglobin Solutions as Oxygen Carriers	191
10	**Evolutionary Considerations**	**194**
	References	197

Keywords

Allosteric protein
A protein whose active site reactivity can be altered by the binding of a ligand (the allosteric effector) at a nonoverlapping site.

Blood substitute
Any material (e.g., human plasma, solutions of substances such as serum albumin or dextrans) used for transfusion, as a part of treatment of hemorrhage and/or circulatory shock. Materials capable of volume replacement as well as oxygen transport (e.g., those based on chemically modified or genetically engineered hemoglobins) are more correctly termed *hemoglobin-based oxygen carriers* (HBOCs). Plasma expander is a synonym used for solutions aimed exclusively at preventing hypovolemic (low blood volume) shock.

Bohr effect
The influence of pH on the oxygen affinity of hemoglobin.

Cooperativity
Any process in which the initial event (e.g., the binding of one molecule of oxygen to the heme iron) facilitates subsequent similar events (e.g., the binding of other identical ligands) at other sites of the same oligomer (e.g., one tetrameric hemoglobin molecule). Cooperativity is phenomenologically demonstrated by the sigmoidal (S-shaped) nature of the ligand binding curve (quite different from a rectangular hyperbola), the implication being that reaction intermediates (i.e., partially saturated tetramers) are not populated to any significant extent.

Geminate recombination
The rebinding of a ligand molecule to its original site (the heme iron) *after* photolytic rupture of the bond and *before* escaping to the solvent to mingle with other ligand molecules. The geminate recombination speed, which is typically a picosecond- to nanosecond-process, is independent of bulk ligand concentration.

Globin fold
Indicates one representative of the all-α helical folds in the classification of tertiary structures of proteins. It consists of four to eight helices organized, respectively, in a 2-over-2 or a 3-over-3 topology. The globin fold was first observed in myoglobin, when the three-dimensional structure was resolved in 1958; hence, the fold was named after this protein.

Heme group
The prosthetic group comprising protoporphyrin IX and one iron atom. The protoporphyrin IX ring is an aromatic heterocycle composed by four pyrrolic groups joined by methine bridges. The four pyrrolic nitrogens coordinate the ionic iron atom located in the center of the ring. The heme group in hemoglobin and myoglobin belongs to the so-called group B; hence, it has two vinyls, four methyls, and two propionyls as substituents.

Kinetics and mechanism
The study of reaction rates (*chemical kinetics*) is concerned not only with the speed of conversion of the reactants to products but also with the sequence of events involved in this conversion (the *reaction mechanism*).

Molecular diseases
A disease whose clinical manifestations arise from alterations in the structure and function of a specific biomolecule (such as an enzyme).

Molecular evolution
The process of cumulative changes in the structure, and thereby function, of a protein or a nucleic acid, due to random mutagenesis of the corresponding gene. Any given change (such as point mutations, insertions, tandem duplication, and circular permutations) is fixed during evolution due to natural selection.

Oxygen transport
All living organisms need dioxygen for survival. Thus, multicellular organisms have evolved specialized macromolecules to transport dioxygen to the tissues, as physical diffusion would be inadequate. In the lungs of mammals, inhaled dioxygen is dissolved in the capillaries and thereby bound by deoxyhemoglobin contained in the red blood cells (erythrocytes) to be transported to the periphery.

> Hemoglobin (Hb) is the generic name for a vital protein that is basic to oxygen (O_2) metabolism in all vertebrates, almost all invertebrates, plants and many microorganisms, from bacteria to fungi, including some pathogens. The only vertebrates that survive without Hb in their blood are Antarctic fishes (such as *Chaenocephalus aceratus Lonnberg*), which carry the Hb genes but do not express the proteins. The structural and functional properties of Hb vary widely among different species, according to physiological requirements. The multifaceted behavior of Hb has been a challenge for many scientists with different backgrounds (e.g., biochemists, physiologists, geneticists, biophysicists, medical doctors), who have investigated this protein from extremely diverse points of view, ranging from the quantum chemistry of the heme iron to the unloading of O_2 into the swimbladder of fish, when needed to control buoyancy. Hb was the first demonstration of a molecular disease in humans, with the discovery by Linus

Pauling of the sickle cell hemoglobinopathy, and in this chapter attention is focused on human Hb. Hb also exhibits biochemical and biophysical behaviors in its interaction with substrates and cofactors similar to that of many multisubunit enzymes, and thus it has served as a prototype for exploring the molecular basis of protein intracommunication (i.e., the structural basis of cooperativity and allostery). Hence, it is not surprising that tetrameric Hb has been called the *hydrogen molecule of biochemistry*, because understanding its function is basic to protein chemistry at large, while myoglobin (Mb), the monomeric hemeprotein contained in red muscles, has been called the hydrogen atom.

1
General Aspects

Hemoglobin is a protein capable of undergoing a reversible reaction with O_2 because it contains heme (Fe-protoporphyrin IX) as the prosthetic group. Classically, hemoproteins present in the blood or the hemolymph of vertebrates and invertebrates are named hemoglobins (Hbs), while those present in the striated or smooth muscles of all animals are called myoglobins (Mbs) [1–4]. However, Hbs dissolved in the blood of invertebrates are often called erythrocruorins. Recently, simple intracellular hemoproteins different from Mb have been discovered in the tissues of humans, including the liver (cytoglobin) and brain (neuroglobin).

Proteins corresponding to the foregoing definition are also found in the tissues of invertebrates other than blood and muscle (e.g., in the cells of the tracheal wall in *Gastrophylus* larvae or in the nerve cells of marine mollusks), in plants (e.g., lupin and soybean), and in single-cell organisms (e.g., protozoa, fungi, and bacteria). In all these cases, the term Hb has been adopted. In addition, a number of microorganisms (such as *Mycobacterium tuberculosis* or *Candida albicans*) express Hbs, which are important for virulence by helping to counteract the host defense mechanisms.

Examination of their three-dimensional (3D) structure clearly shows that these other hemoproteins belong to the extended globin superfamily, despite the fact that their polypeptidic chain is approximately 30% shorter than that of classical Hbs; this is why they have been designated as *truncated* Hbs. In contrast, the O_2 carrier present in the blood of several polychaete worms is named chlorocruorin, as its prosthetic group is a chloroporphyrin, characterized by a formyl group substituted for the vinyl group in position 2 of the protoporphyrin IX. In contrast to protoporphyrin, which is red, the chloroporphyrin is bright green.

In the present chapter, a short overview will be presented of the distribution of globins in living organisms, with attention mainly focused on human Hb, the intraerythrocyte protein that is involved not only in the transport of O_2 from lungs to tissues but also in the reverse transport of carbon dioxide. Typically, Hb concentrations in normal human erythrocytes are extremely high ($34 \, \text{g dm}^{-3}$), which corresponds to a 0.5 mM solution, considering a molecular mass of $65\,000 \, \text{g mol}^{-1}$. One erythrocyte contains approximately 300 million Hb tetramers (which are spheroids of dimension $6.4 \times 5.5 \times 5 \, \text{nm}^3$). Although single Hb molecules in the erythrocyte are, on average, only 1 nm apart, they can rotate and flow past one another without hindrance. All general ideas that originated from studies on human Hb have been successfully transferred to other vertebrate and invertebrate Hbs and, in general, to more complex respiratory proteins, which are sometimes very large and extracellular.

2
Structural Features and Basic Terminology

Hemoglobin consists of four polypeptide chains (globins). These apoproteins carry as a prosthetic group heme, an Fe(II) complex of protoporphyrin IX (synthesized in the bone marrow from glycine and acetate). In free heme, Fe(II) is rapidly oxidized to Fe(III) by O_2 and water, whereas in Hb, Fe(II) combines reversibly with O_2 and remains in the ferrous state. Reversibility in binding with O_2 is achieved in Hb by a folding of the polypeptide chain around the heme group, which is enclosed in a hydrophobic pocket. When O_2 is bound at the Fe(II) (in a 1:1 stoichiometry), an electron rearrangement occurs, with the Fe(II) changing from high spin to low spin and the adduct acquiring the Fe(III)O_2^- character. The four hemoglobin subunits are identical in pairs: thus, the tetrameric protein of a normal adult individual is usually indicated as $\alpha_2\beta_2$. Each polypeptide chain of the two subunits designated α_1 and α_2 (where the indices denote relative locations in the tetramer) has 141 amino acid residues; each of the β_1- and β_2-subunits has 146 amino acid residues. The α- and β-subunits have different amino acid sequences (see Table 1), yet they fold spontaneously into a similar 3D structure (see Fig. 1), called the globin fold. The α- and β-subunits, respectively, are composed of seven and eight α-helical segments, interrupted by nonhelical loops. The helices are named from A to H, and the nonhelical segments lying in between are lettered AB, BC, and so on; the nonhelical portions at the subunit ends are named NA (at the amino terminus) and HC (at the carboxyl terminus). Residues within each segment (helical as well as nonhelical) are numbered from the amino end (e.g., A1–A16, EF1–EF8, etc.). Therefore, each amino acid (usually represented with a three-letter code) is uniquely identified by its topology, its position in the sequence (in parentheses), and the corresponding subunit (e.g., PheCD1(42)β).

The heme, sandwiched in between the E and F helices, is bound to the globin by a number of weak hydrophobic interactions and by a coordination bond between Fe(II) and the imidazole-Nε of histidine F8, known as *proximal*. In the β-subunits, such a residue is followed by a reactive cysteine [CysF9(93)β], which faces the surface of the molecule. Most, but not all, globins also have a histidine on the O_2-combining side of the heme (HisE7, known as *distal*). In the native protein this residue is too far (>0.4 nm) from the iron to be directly coordinated, even though under certain conditions it does bind to the metal (see Sect. 4.2); usually, the Nε atom of HisE7 makes a hydrogen bond with bound O_2 (see Fig. 2). In the field, it is the norm to divide the heme pocket into two halves along the plane of the heme itself: the side where gaseous ligand binds is called *distal*, while the other side, facing helix F, is called *proximal*. The packing of subunits into the Hb tetramer is such that there are close interlocking connections of side groups between unlike subunits (i.e., α and β), but little contact between those that are alike (see Table 2) [2, 5, 6].

Tab. 1 Primary structure of human globin subunits aligned by reference to their topological positions.

Helix	α	ζ	β	δ	γ	ε	Myoglobin
NA1	—	—	1 Val	Val	Gly	Val	1 Gly
NA2	1 Val	1 Ser	2 His	His	His	His	2 Leu
NA3	2 Leu	Leu	3 Leu	Leu	Phe	Phe	—
A1	3 Ser	Thr	4 Thr	Thr	Thr	Thr	3 Ser
A2	4 Pro	Lys	5 Pro	Pro	Glu	Ala	4 Glu
A3	5 Ala	Thr	6 Glu	Glu	Glu	Glu	5 Gly
A4	6 Asp	Glu	7 Glu	Glu	Asp	Glu	6 Glu
A5	7 Lys	Arg	8 Lys	Lys	Lys	Lys	7 Trp
A6	8 Thr	Thr	9 Ser	Thr	Ala	Ala	8 Gln
A7	9 Asn	Ile	10 Ala	Ala	Thr	Ala	9 Leu
A8	10 Val	Ile	11 Val	Val	Ile	Val	10 Val
A9	11 Lys	Val	12 Thr	Asn	Thr	Thr	11 Leu
A10	12 Ala	Ser	13 Ala	Ala	Ser	Ser	12 Asn
A11	13 Ala	Met	14 Leu	Leu	Leu	Leu	13 Val
A12	14 Trp	Trp	15 Trp	Trp	Trp	Trp	14 Trp
A13	15 Gly	Ala	16 Gly	Gly	Gly	Ser	15 Gly
A14	16 Lys	Lys	17 Lys	Lys	Lys	Lys	16 Lys
A15	17 Val	Ile	18 Val	Val	Val	Met	17 Val
A16	18 Gly	Ser	—	—	—	—	18 Glu
AB1	19 Ala	Thr	—	—	—	—	19 Ala
B1	20 His	Gln	19 Asn	Asn	Asn	Asn	20 Asp
B2	21 Ala	Ala	20 Val	Val	Val	Val	21 Ile
B3	22 Gly	Asp	21 Asp	Asp	Glu	Glu	22 Pro
B4	23 Glu	Thr	22 Glu	Ala	Asp	Glu	23 Gly
B5	24 Thr	Ile	23 Val	Val	Ala	Ala	24 His
B6	25 Gly	Gly	24 Gly	Gly	Gly	Gly	25 Gly
B7	26 Ala	Thr	25 Gly	Gly	Gly	Gly	26 Gln
B8	27 Glu	Glu	26 Glu	Glu	Glu	Glu	27 Asp
B9	28 Ala	Thr	27 Ala	Ala	Thr	Ala	28 Val
B10	29 Leu	Leu	28 Leu	Leu	Leu	Leu	29 Leu
B11	30 Glu	Glu	29 Gly	Gly	Gly	Gly	30 Ile
B12	31 Arg	Arg	30 Arg	Arg	Arg	Arg	31 Arg
B13	32 Met	Leu	31 Leu	Leu	Leu	Leu	32 Leu
B14	33 Phe	Phe	32 Leu	Leu	Leu	Leu	33 Phe
B15	34 Leu	Leu	33 Val	Val	Val	Val	34 Lys
B16	35 Ser	Ser	34 Val	Val	Val	Val	35 Gly
C1	36 Phe	His	35 Tyr	Tyr	Tyr	Tyr	36 His
C2	37 Pro	Pro	36 Pro	Pro	Pro	Pro	37 Pro
c03	38 Thr	Gln	37 Trp	Trp	Trp	Trp	38 Glu
C4	39 Thr	Thr	38 Thr	Thr	Thr	Thr	39 Thr
C5	40 Lys	Lys	39 Gln	Gln	Gln	Gln	40 Leu
C6	41 Thr	Thr	40 Arg	Arg	Arg	Arg	41 Glu
C7	42 Tyr	Tyr	41 Phe	Phe	Phe	Phe	42 Lys
CD1	43 Phe	Phe	42 Phe	Phe	Phe	Phe	43 Phe
CD2	44 Pro	Pro	43 Glu	Glu	Asp	Asp	44 Asp

Tab. 1 (*Continued*)

Helix	α	ζ	β	δ	γ	ε	Myoglobin
CD3	45 His	His	44 Ser	Ser	Ser	Ser	45 Lys
CD4	46 Phe	Phe	45 Phe	Phe	Phe	Phe	46 Phe
CD5	47 Asp	Asp	46 Gly	Gly	Gly	Gly	47 Lys
CD6	—	—	47 Asp	Asp	Asn	Asn	48 His
CD7	48 Leu	Leu	48 Leu	Leu	Leu	Leu	49 Leu
CD8	49 Ser	His	49 Ser	Ser	Ser	Ser	50 Lys
D1	—	—	50 Thr	Ser	Ser	Ser	51 Ser
D2	—	—	51 Pro	Pro	Ala	Pro	52 Glu
D3	—	—	52 Asp	Asp	Ser	Ser	53 Asp
D4	—	—	53 Ala	Ala	Ala	Ala	54 Glu
D5	—	—	54 Val	Val	Ile	Ile	55 Met
D6	50 His	Pro	55 Met	Met	Met	Leu	56 Lys
D7	51 Gly	Gly	56 Gly	Gly	Gly	Gly	57 Ala
E1	52 Ser	Ser	57 Asn	Asn	Asn	Asn	58 Ser
E2	53 Ala	Ala	58 Pro	Pro	Pro	Pro	59 Glu
E3	54 Gln	Gln	59 Lys	Lys	Lys	Lys	60 Asp
E4	55 Val	Leu	60 Val	Val	Val	Val	61 Leu
E5	56 Lys	Arg	61 Lys	Lys	Lys	Lys	62 Lys
E6	57 Gly	Ala	62 Ala	Ala	Ala	Ala	63 Lys
E7	58 His	His	63 His	His	His	His	64 His
E8	59 Gly	Gly	64 Gly	Gly	Gly	Gly	65 Gly
E9	60 Lys	Ser	65 Lys	Lys	Lys	Lys	66 Ala
E10	61 Lys	Lys	66 Lys	Lys	Lys	Lys	67 Thr
E11	62 Val	Val	67 Val	Val	Val	Val	68 Val
E12	63 Ala	Val	68 Leu	Leu	Leu	Leu	69 Leu
E13	64 Asp	Ala	69 Gly	Gly	Thr	Thr	70 Thr
E14	65 Ala	Ala	70 Ala	Ala	Ser	Ser	71 Ala
E15	66 Leu	Val	71 Phe	Phe	Leu	Phe	72 Leu
E16	67 Thr	Gly	72 Ser	Ser	Gly	Gly	73 Gly
E17	68 Asn	Asp	73 Asp	Asp	Asp	Asp	74 Gly
E18	69 Ala	Ala	74 Gly	Gly	Ala	Ala	75 Ile
E19	70 Val	Val	75 Leu	Leu	Ile, Thr	Ile	76 Leu
E20	71 Ala	Lys	76 Ala	Ala	Lys	Lys	77 Lys
EF1	72 His	Ser	77 His	His	His	Asn	78 Lys
EF2	73 Val	Ile	78 Leu	Leu	Leu	Met	79 Lys
EF3	74 Asp	Asp	79 Asp	Asp	Asp	Asp	80 Gly
EF4	75 Asp	Asp	80 Asn	Asn	Asn	Asp	81 His
EF5	76 Met	Ile	81 Leu	Leu	Leu	Leu	82 His
EF6	77 Pro	Gly	82 Lys	Lys	Lys	Lys	83 Glu
EF7	78 Asn	Gly	83 Gly	Gly	Gly	Pro	84 Ala
EF8	79 Ala	Ala	84 Thr	Thr	Thr	Ala	85 Glu
F1	80 Leu	Leu	85 Phe	Phe	Phe	Phe	86 Ile
F2	81 Ser	Ser	86 Ala	Ser	Ala	Ala	87 Lys
F3	82 Ala	Lys	87 Thr	Gln	Gln	Lys	88 Pro
F4	83 Leu	Leu	88 Leu	Leu	Leu	Leu	89 Leu

(*continued overleaf*)

Tab. 1 (Continued)

Helix	α	ζ	β	δ	γ	ε	Myoglobin
F5	84 Ser	Ser	89 Ser	Ser	Ser	Ser	90 Ala
F6	85 Asp	Glu	90 Glu	Glu	Glu	Glu	91 Gln
F7	86 Leu	Leu	91 Leu	Leu	Leu	Leu	92 Ser
F8	87 His	His	92 His	His	His	His	93 His
F9	88 Ala	Ala	93 Cys	Cys	Cys	Cys	94 Ala
F10	—	—	—	—	—	—	95 Thr
FG1	89 His	Tyr	94 Asp	Asp	Asp	Asp	96 Lys
FG2	90 Lys	Ile	95 Lys	Lys	Lys	Lys	97 His
FG3	91 Leu	Leu	96 Leu	Leu	Leu	Leu	98 Lys
FG4	92 Arg	Arg	97 His	His	His	His	99 Ile
FG5	93 Val	Val	98 Val	Val	Val	Val	—
G1	94 Asp	Asp	99 Asp	Asp	Asp	Asp	100 Pro
G2	95 Pro	Pro	100 Pro	Pro	Pro	Pro	101 Val
G3	96 Val	Val	101 Glu	Glu	Glu	Glu	102 Lys
G4	97 Asn	Asn	102 Asn	Asn	Asn	Asn	103 Tyr
G5	98 Phe	Phe	103 Phe	Phe	Phe	Phe	104 Leu
G6	99 Lys	Lys	104 Arg	Arg	Lys	Lys	105 Glu
G7	100 Leu	Leu	105 Leu	Leu	Leu	Leu	106 Phe
G8	101 Leu	Leu	106 Leu	Leu	Leu	Leu	107 Ile
G9	102 Ser	Ser	107 Gly	Gly	Gly	Gly	108 Ser
G10	103 His	His	108 Asn	Asn	Asn	Asn	109 Glu
G11	104 Cys	Cys	109 Val	Val	Val	Val	110 Cys
G12	105 Leu	Leu	110 Leu	Leu	Leu	Met	111 Ile
G13	106 Leu	Leu	111 Val	Val	Val	Val	112 Ile
G14	107 Val	Val	112 Cys	Cys	Thr	Ile	113 Val
G15	108 Thr	Thr	113 Val	Val	Val	Ile	114 Val
G16	109 Leu	Leu	114 Leu	Leu	Leu	Leu	115 Leu
G17	110 Ala	Ala	115 Ala	Ala	Ala	Ala	116 Gln
G18	111 Ala	Ala	116 His	Arg	Ile	Thr	117 Ser
G19	112 His	Arg	117 His	Asn	His	His	118 Lys
GH1	113 Leu	Phe	118 Phe	Phe	Phe	Phe	119 His
GH2	114 Pro	Pro	119 Gly	Gly	Gly	Gly	120 Pro
GH3	115 Ala	Ala	120 Lys	Lys	Lys	Lys	121 Gly
GH4	116 Glu	Asp	121 Glu	Glu	Glu	Glu	122 Asp
GH5	117 Phe	Phe	122 Phe	Phe	Phe	Phe	123 Phe
H1	118 Thr	Thr	123 Thr	Thr	Thr	Thr	124 Gly
H2	119 Pro	Ala	124 Pro	Pro	Pro	Pro	125 Ala
H3	120 Ala	Glu	125 Pro	Gln	Glu	Glu	126 Asp
H4	121 Val	Ala	126 Val	Met	Val	Val	127 Ala
H5	122 His	His	127 Gln	Gln	Gln	Gln	128 Gln
H6	123 Ala	Ala	128 Ala	Ala	Ala	Ala	129 Gly
H7	124 Ser	Ala	129 Ala	Ala	Ser	Ala	130 Ala
H8	125 Leu	Trp	130 Tyr	Tyr	Trp	Trp	131 Met
H9	126 Asp	Asp	131 Gln	Gln	Gln	Gln	132 Asn
H10	127 Lys	Lys	132 Lys	Lys	Lys	Lys	133 Lys

Tab. 1 (Continued)

Helix	α	ζ	β	δ	γ	ε	Myoglobin
H11	128 Phe	Phe	133 Val	Val	Met	Leu	134 Ala
H12	129 Leu	Leu	134 Val	Val	Val	Val	135 Leu
H13	130 Ala	Ser	135 Ala	Ala	Thr	Ser	136 Glu
H14	131 Ser	Val	136 Gly	Gly	Gly, Ala	Ala	137 Leu
H15	132 Val	Val	137 Val	Val	Val	Val	138 Phe
H16	133 Ser	Ser	138 Ala	Ala	Ala	Ala	139 Arg
H17	134 Thr	Ser	139 Asn	Asn	Ser	Ile	140 Lys
H18	135 Val	Val	140 Ala	Ala	Ala	Ala	141 Asp
H19	136 Leu	Leu	141 Leu	Leu	Leu	Leu	142 Met
H20	137 Thr	Thr	142 Ala	Ala	Ser	Ala	143 Ala
H21	138 Ser	Glu	143 His	His	Ser	His	144 Ser
H22	—	—	—	—	—	—	145 Asn
H23	—	—	—	—	—	—	146 Tyr
H24	—	—	—	—	—	—	147 Lys
H25	—	—	—	—	—	—	148 Glu
H26	—	—	—	—	—	—	149 Leu
HC1	139 Lys	Lys	144 Lys	Lys	Arg	Lys	150 Gly
HC2	140 Tyr	Tyr	145 Tyr	Tyr	Tyr	Tyr	151 Phe
HC3	141 Arg	His	146 His	His	His	His	152 Gln
HC4	—	—	—	—	—	—	153 Gly

Tab. 2 Surface area buried at the interfaces between subunits in deoxy and oxy human hemoglobin.

Intersubunit contacts	Deoxy Hb (nm^2)	Oxy Hb (nm^2)
Buried in contact		
$\alpha_1\beta_1$ or $\alpha_2\beta_2$	18.76	20.48
$\alpha_1\beta_2$ or $\alpha_2\beta_1$	14.72	11.60
$\alpha_1\alpha_2$	7.91	9.66
$\beta_1\beta_2$	0.00	7.06
Accessible to solvent	72.55	80.32

3
Human Hemoglobin

3.1
Hemoglobin Components in Erythrocytes

Besides $\alpha_2\beta_2$ (also called HbA or HbA$_0$), human erythrocytes contain other Hb components, some as products of additional globin genes and others as post-translational modifications of all these components. Among the hemoproteins belonging to the first group are HbA$_2$ and HbF (or fetal Hb), constituted by normal α-chains combined with subunits whose structure significantly differs from that of normal β-chains; these non-β subunits are designated δ-chains (HbA$_2$ or $\alpha_2\delta_2$) and γ-chains (HbF or $\alpha_2\gamma_2$). Humans have also two embryonic Hbs using ζ- (α-like) and ε- (β-like) subunits: Hb Gower-1 of composition $\zeta_2\varepsilon_2$, Hb Gower-2 or $\alpha_2\varepsilon_2$, and Hb Portland or $\zeta_2\gamma_2$. Amino acid sequences of these human subunits are reported in Table 1 [8].

Many small molecules (e.g., hexoses, cyanate ions derived from urea, and penicillin) are capable of forming covalent adducts with Hb and may reflect

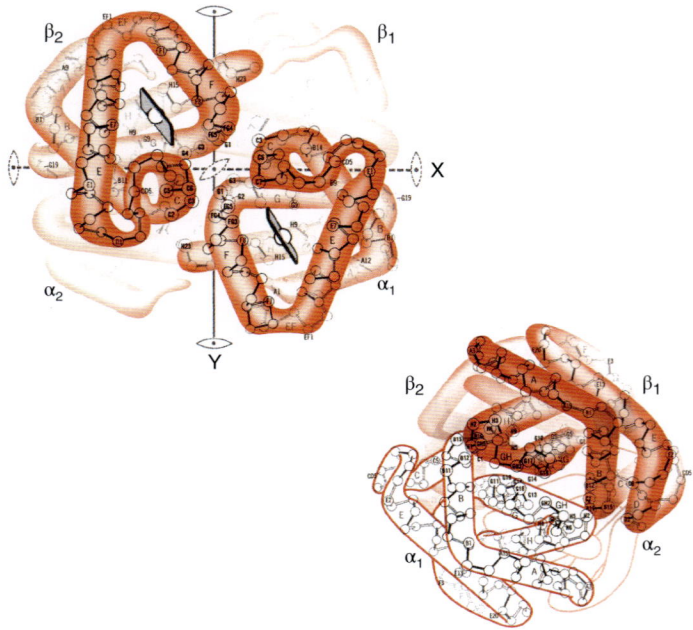

Fig. 1 Overview of ferric Hb, which is similar to the oxygenated derivative (HbO$_2$), in two orientations rotated by 90° over the Y-axis. Only the α-carbons of the polypeptide chain are shown. Residues whose side chains are involved in contacts between subunits are circled and given boldface numbers. There are two types of unlike subunit contacts: $α_1β_1$ (or $α_2β_2$) and $α_1β_2$ (or $α_2β_1$), the first interface being more extensive than the second (see also Table 2). The $α_1β_1$ contact involves the B, C, and H helices and the GH corner, whereas the $α_1β_2$ contact concerns mainly helices C and G and the FG corner. Reproduced with permission from Ref. [2]. © 1983, Benjamin/Cummings, Menlo Park, CA.

metabolic perturbation. Of significance for their clinical implications in diabetes are glycated Hbs (called Hb AIa1, AIa2, AIb, and AIc, according to their order of chromatographic elution) in which the amino terminus of each β-chain is bound to a molecule of glucose, glucose 6-phosphate, or fructose 1,6-bisphosphate. Glucose can also form a ketamine linkage with the amino-terminal group of the α-chains as well as the ε-amino group of several lysyl residues of both α- and β-subunits. HbAIc is the most abundant among the minor post-translationally modified Hb components in human erythrocytes, reaching about 4% of the total.

3.2
The Globin Genes and Their Biosynthesis

In humans, the initial ζ (α-like) and ε (β-like) embryonic subunits, synthesized in the yolk sac during three to eight weeks of gestation, are gradually replaced by α- and γ-chains so that the original Hb Gower-1 is replaced, first by Hb Gower-2 and Hb Portland and finally by HbF (see above for definitions). At about 7 weeks of gestation, when the liver and spleen become the major sites of erythropoiesis, HbF accounts for about 50% of total Hb. The synthesis of β-chains, which initially is very limited, begins to accelerate around weeks 30–40 of

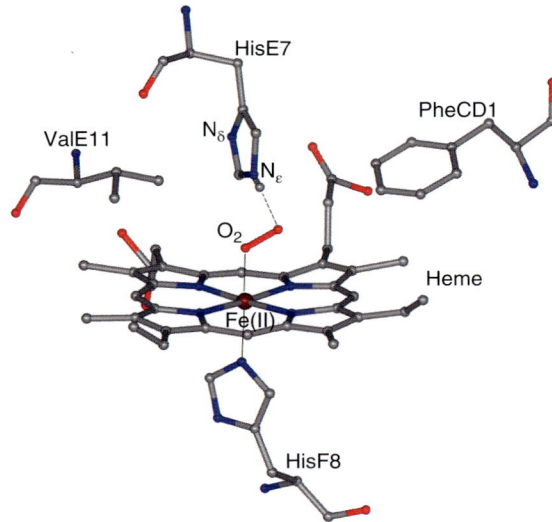

Fig. 2 Schematic representation of the heme pocket, with the *distal* side positioned above and the *proximal* side below the heme. Phenylalanine CD1 and the proximal histidine F8 are the only totally invariant residues present in all hemoglobins. In vertebrates, valine E11 and the distal histidine E7 are involved in the reversible O_2 binding to the heme Fe(II). Note the hydrogen bond between the Nε of the distal HisE7 and the bound O_2. Neutron diffraction, NMR, and computational chemistry have shown that, in MbO$_2$, HisE7 is protonated only on Nε, whereas in the CO derivative protonation also occurs on Nδ, which faces the bulk water, thereby favoring a more perpendicular geometry by releasing part of the steric hindrance [5–7].

pregnancy (see Fig. 3a). Within eight weeks after birth, when erythrocyte production has shifted from the liver, spleen and bone marrow to bone marrow only, mainly α- and β-chains are synthesized. At six months after birth, the adult $\alpha_2\beta_2$ HbA has essentially taken over, along with the minor $\alpha_2\delta_2$ component (Fig. 3b). Each of these Hbs, synthesized during the development from embryo to fetus to infant (Fig. 3b), consists of two α-type and two β-type subunits. The α-type (α, ζ) and β-type (β, γ, ε, δ) families of human genes are clustered, respectively, near the telomere of the short arm of chromosome 16 and on the short arm of chromosome 11. Each globin gene in a cluster contains three coding blocks (exons) separated by two intervening sequences (introns) of DNA (Fig. 4). It is worth noting that the gene encoding neuroglobin (located in chromosome 14 [9]), as well as that encoding cytoglobin (located on chromosome 17), have three introns (and thereby four exon), like the gene of plant Hb.

In the α-family genes, the three exons code for residues 1–31, 32–99, and 100–141 respectively; in the β-family genes, the corresponding positions are 1–30, 31–104, and 105–146. Even though the breaks occur in all chains at identical topological positions along the helical segments (i.e., between residues B12 and B13 and between G6 and G7), they do not mark the boundaries of clearly defined structural domains; they do, however, correspond to smaller compact modules. Thus, the central exon (exon 2 in Fig. 4) encodes a module

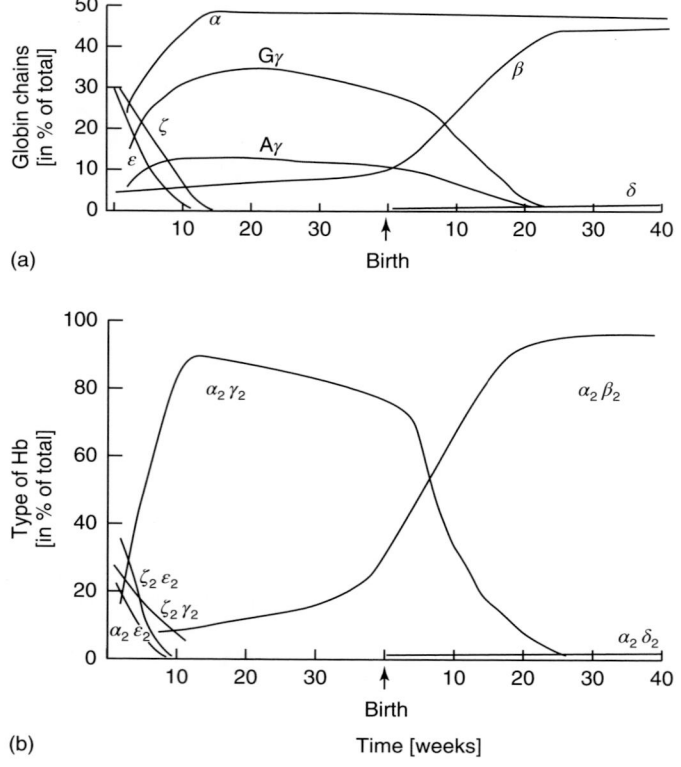

Fig. 3 Relative amounts of globin subunits (top) and Hb tetramers (bottom) at different stages in the development of the human embryo, fetus, and infant. The pattern of globin gene expression is controlled by a regulatory element, called locus control region (LCR). In the absence of an interaction with LCR, individual globin genes are either inactive or very poorly expressed [8].

that provides a competent heme-binding site (helices E and F) and the $\alpha_1\beta_2$ sliding contact (helix C and FG corner); such a module binds heme tightly and specifically, but ligand-linked structural fluctuations (as observed in the protein fragment encoded by the central exon of the present-day Mb gene) are large enough to jeopardize a stable O_2 complex. Addition of the two terminal polypeptide modules (coded by exon 1, which corresponds to helices A and B and constitutes a scaffolding for the heme pocket, and exon 3, which provides helices G and H and contributes most of the $\alpha_1\beta_1$ packing contacts; see Fig. 4) reduces the structural degrees of freedom of the central module, and stabilizes the holoprotein as an independent unit.

The α-type gene cluster contains three functional genes (ζ, α_1, and α_2) and three pseudogenes indicated by the prefix ψ ($\psi\alpha_1$, $\psi\alpha_2$, and $\psi\zeta$). Pseudogenes have sequence homology to the functional genes but are defective in some essential coding and/or regulatory regions and thus cannot be expressed in a globin chain. The β-type gene cluster has five active genes (ε, Gγ, Aγ, δ, and β) and one pseudogene ($\psi\beta$). The two fully functional and nearly identical α (α_1 and α_2) and γ (Gγ and Aγ) genes possibly

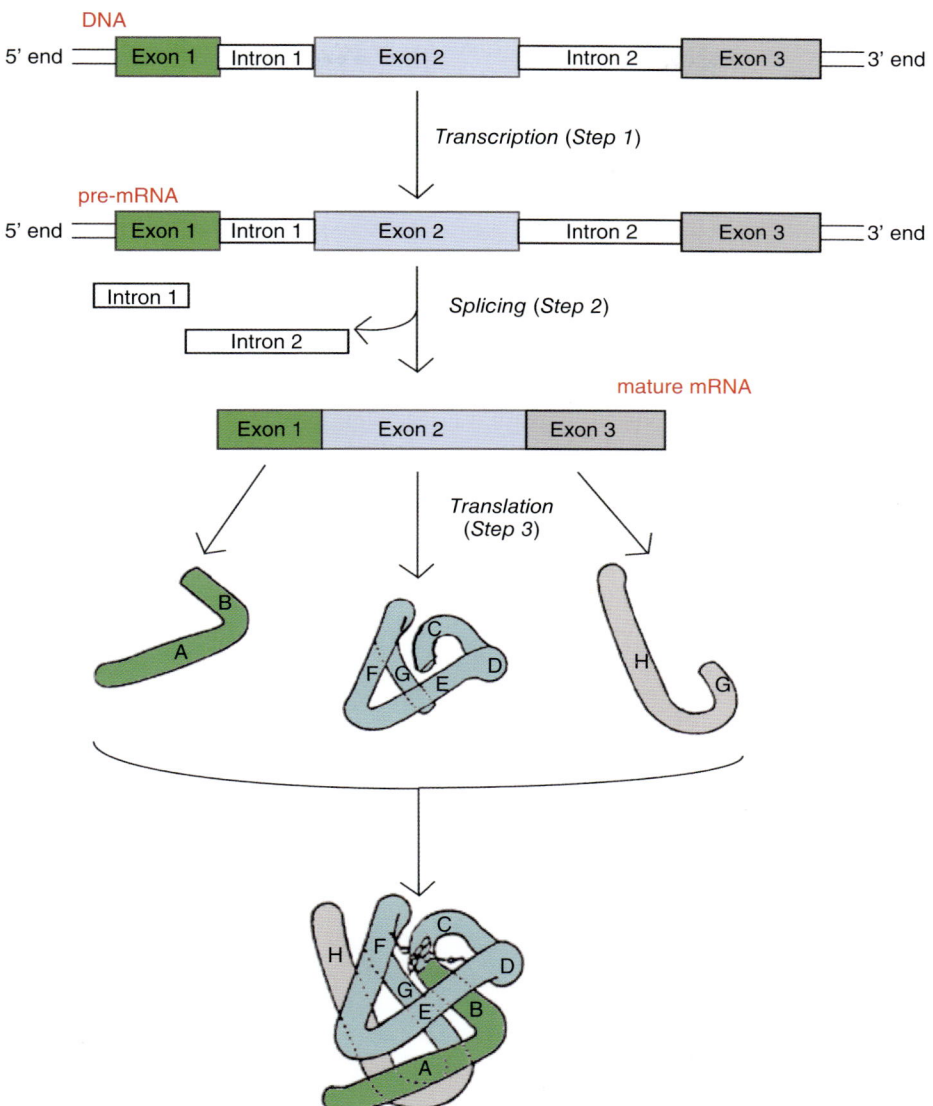

Fig. 4 The exon–intron structure of a globin gene encoding a typical polypeptide chain of mammalian Hb or Mb. From the gene, a primary transcript or pre-mRNA is synthesized (step 1); the intervening sequences or introns are then excised (step 2) by splicing to yield the mature mRNA; this is translated (step 3) into the polypeptide chain, which spontaneously adopts the typical globin fold. Three globin "modules" corresponding to the three exonic transcripts are schematically illustrated in green, light cyan, and gray, respectively. The central exon (the biggest) binds the heme in a native configuration.

reflect their duplication during evolution. In these gene families, all vertebrate globins are encoded in the same order in which they are expressed during the development: $\varepsilon \to \gamma \to \delta$, β in the β-cluster, and $\zeta \to \alpha$ in the α-cluster. Throughout development from embryo to infant, a balance in expression between α-like and β-like globin genes is maintained, yielding fully functional Hb tetramers (Fig. 3b). Alteration of the balanced synthesis of globins in the two clusters leads to a very important type of molecular disease, called thalassemia. In β-thalassemia, a partial or complete deficiency of β-chains is associated with a prolonged biosynthesis of the fetal γ-chains (yielding HbF or $\alpha_2\gamma_2$), and possibly with the accumulation of free α-chains, which associate in the insoluble aggregates and cause toxic effects on the development of erythroblasts. In α-thalassemia, a relative deficiency in α-subunit biosynthesis results in an excess of γ- or β-chains, which assemble into γ_4 (Hb Bart's) and β_4 (HbH) homotetramers, respectively.

4
Derivatives with Heme Ligands

4.1
Ferrous Derivatives

The state of ferrous Hb with no ligand attached to the heme iron has been named "unligated" Hb (or deoxy Hb). It may be prepared by removal of the heme ligand from the corresponding ferrous ligated form. The small ligands that bind reversibly to the ferrous heme iron are O_2, CO, NO, alkylisocyanides, and nitroso aromatic compounds, with NO being the strongest and bulky alkylisocyanides the weakest. When the starting derivative is oxygenated Hb (oxy Hb or HbO_2), deoxygenation is achieved by exposure to vacuum or equilibration with an inert gas (e.g., N_2, Ar) or by the addition of sodium dithionite ($Na_2S_2O_4$) which removes O_2 from solution. Deoxy Hb may also be prepared from the ferric derivatives by reduction of the Fe(III) with borohydride, dithionite, or ascorbate [1, 10].

4.2
Ferric Derivatives

In the presence of O_2, the Fe(II) is subject to a slow spontaneous oxidation, yielding ferric Hb or met Hb, which cannot combine with O_2, CO, and all of the other above-mentioned ligands, except NO. In blood, the rate of autoxidation is about 3% per day, but the oxidized iron-porphyrin is reduced to its active form by reductase systems present in the erythrocytes. *In vitro* ferrous derivatives (such as oxy and deoxy Hb) are oxidized rapidly by ferricyanide or by nitrite. The water molecule that in met Hb at neutral pH is bound to the Fe(III) can be replaced by F^-, OCN^-, SCN^-, N_3^-, imidazole, or CN^-, where F^- is the weakest and CN^- the strongest ligand.

An endogenous ligand, such as the imidazole of distal HisE7, may bind to Fe(III) under special conditions. This state is named bis-histidine complex, since the fifth and sixth axial positions of the Hb-Fe(III) are coordinated by the proximal and the distal histidines respectively. It may also result from a large alteration of the geometry within the heme pocket; in this case, the low-spin derivative is called a hemichrome, a term connecting a denatured state. A more pronounced denaturation of the protein may cause the formation of other hemichromes; in one of these, one ligand of the Fe(III) is the sulfur of CysF9(93)β. The addition of dithionite to the hemichromes results in a reduction of heme Fe(III) to Fe(II), with the production of low-spin compounds called hemochromes.

5
The Reaction with Heme Ligands

The reaction of Hb with O_2 or other ligands of the ferrous heme (see above) is associated with large modifications of the spectral (see Fig. 5) and magnetochemical properties, which change in proportion to the amount of ligand bound.

5.1
Functional Properties at Equilibrium

The fundamental representation of the reversible combination of Hb with O_2 is the ligand-binding curve wherein the fraction of O_2-bearing heme groups (Y) is measured as a function of the O_2 partial pressure (p). The sigmoid shape of the curve is in marked contrast to the rectangular hyperbola found for the isolated α- or β-subunits, and also for Mb (see Fig. 6).

The sigmoidal curve reflects a complex phenomenon called "cooperativity",

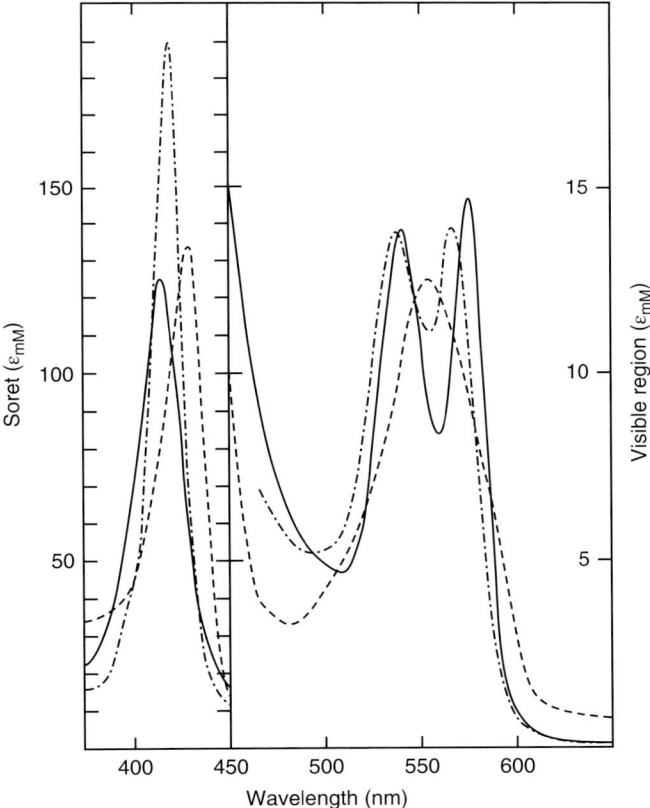

Fig. 5 Optical absorption spectra of deoxy (dashed line), oxy (solid line), and CO (dot-dashed line) hemoglobin in the visible and near-ultraviolet (the so-called Soret band) regions at pH 7.4: ε_{mM}, millimolar extinction coefficient in heme (i.e., referred to an equivalent molecular mass of $16\,250\,g\,mol^{-1}$).

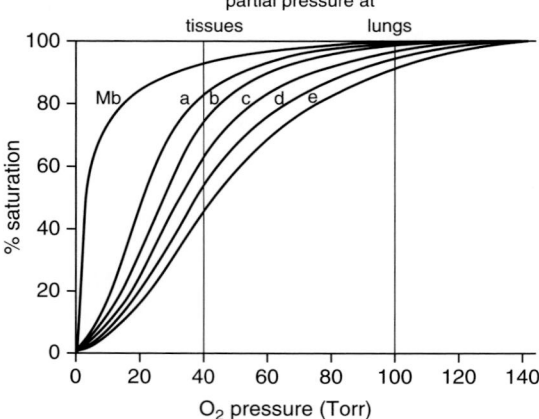

Fig. 6 O_2 equilibrium curves Mb and Hb: % saturation with O_2, reported as a function of O_2 partial pressure [1]. The saturation curve for Mb is hyperbolic (left), while the typical shape of the O_2 equilibrium curve for Hb is sigmoid (a to e). The curves from a to e represent the O_2-binding isotherms at various pH values (reflecting the Bohr effect): from left to right, pH 7.6, 7.4, 7.2, 7.0, and 6.8.

whereby the initial O_2 binding favors more binding (as occurs in the lungs), or – from the opposite perspective – some O_2 release favors more release (as occurs in the peripheral tissues) [1, 11–13]. Cooperativity is a fundamental property of all vertebrate Hbs, the only exception so far being the Hbs from *Sphenodon punctatus* (Tuatara), a reptilian relict from the Triassic period. A useful approach to analyze the experimental data is based on Eq. (1) proposed by A.V. Hill, who assumed that n was the number of Hb molecules aggregating upon O_2 binding:

$$Hb + nO_2 \rightleftharpoons Hb(O_2)_n \quad (1)$$

Assuming that the ratio of activities of Hb and $Hb(O_2)_n$ is given by their concentration ratio, the association equilibrium constant, K, can be written as follows:

$$Y = \frac{[Hb(O_2)_n]}{[Hb]P_n} \quad (2)$$

where the activity of O_2 is expressed in terms of its partial pressure p. The average fraction of hemes oxygenated is

$$Y = \frac{[Hb(O_2)n]}{[Hb] + [Hb(O_2)n]} \quad (3)$$

Therefore, by replacing $[Hb(O_2)_n]$ as expressed in Eq. (2), Eq. (3) becomes

$$\frac{Y}{1-Y} = Kp^n \quad (4)$$

or in logarithmic terms (the so-called "Hill equation"; see Fig. 7),

$$\log\left(\frac{Y}{1-Y}\right) = \log(K) + n\log(p) \quad (5)$$

With a suitable choice of K and n, Eq. (5) allows the experimental data to be fitted with only two parameters: (i) $p_{1/2}$, the value of O_2 pressure required to yield half-saturation of the hemes (p-value at $Y = 0.5$), which indicates the overall affinity of Hb for O_2 (the higher the affinity, the lower the $p_{1/2}$); and (ii) n, the slope in the Hill plot (Fig. 7), which is related to the shape of the ligand-binding curve and therefore to cooperativity (the higher its value above 1, the higher the cooperativity).

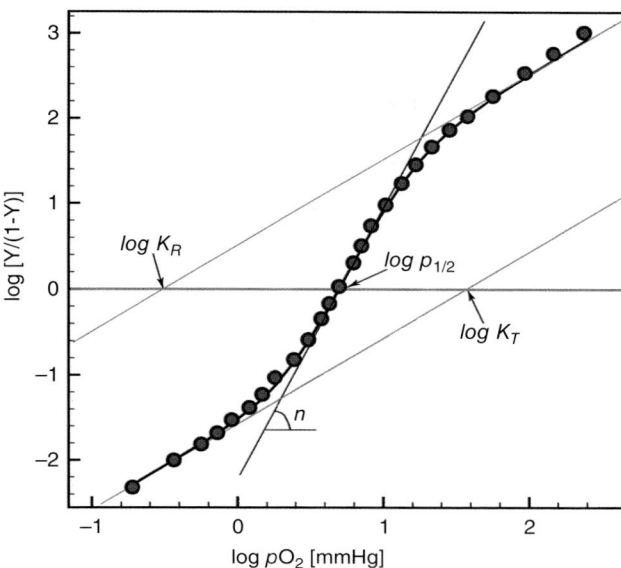

Fig. 7 Hill plot of the O_2 binding curve of Hb, as $\log[Y/(1-Y)]$ versus $\log pO_2$. The intercept of the upper asymptote with the x-axis, when the ordinate is 0, allows the calculation of K_R, the O_2 equilibrium constant of the high-affinity state; likewise, the intercept of the lower asymptote with the x-axis allows the calculation of K_T, the O_2 equilibrium constant of the low-affinity state (see allostery). The slope in the central part of the curve yields the maximum Hill coefficient, and its intercept with the x-axis at $Y = 0.5$ yields the overall O_2 affinity. The Hill equation accounts for the data between approx. $Y = 0.1$ and $Y = 0.9$, and helps to describe the behavior of vertebrate Hbs in terms of cooperativity (n) and affinity ($p_{1/2}$).

A value of $n = 1$ (as observed for the isolated α and β-subunits) corresponds to a hyperbolic binding curve, indicating equivalence of hemes and an absence of functional interactions. An observation of $n < 1$ does not necessarily imply negative cooperativity (i.e., that the binding of a ligand diminishes the affinity for subsequent binding) in that it can also arise from various types of heterogeneity: intramolecular (e.g., a large difference in affinity for O_2 between the α- and β-subunits in the tetramer) or intermolecular (e.g., a mixture of various Hb species or conformations not at equilibrium). The O_2 affinity of Hb is lowered by binding to the protein moiety of H^+, Cl^-, CO_2, and 2,3-bisphosphoglycerate (BPG), all of which are termed "heterotropic ligands" [2, 8, 10]. In particular, modulation of O_2 binding by H^+ is known as the "Bohr effect", and reflects the release of protons upon Hb oxygenation at physiological pH (the alkaline Bohr effect), whereas protons are taken up upon oxygenation at pH values below ≤6 (the acid or reverse Bohr effect). An unusually large Bohr effect, which is typical of some Hbs from many teleost fishes, is called the "Root effect". By increasing the concentration of any one of the heterotropic ligands, the O_2 equilibrium curve is shifted to the right – that is, towards a lower affinity (see Fig. 6) – thus facilitating O_2 release. The O_2 binding parameters at equilibrium of human Hb, *in vitro* under different solvent conditions, are listed in Table 3 [14].

Tab. 3 The functional parameters of human HbA, at pH 7.4, 20 °C, and Cl⁻ = 0.1 M [1, 14–16].

Conditions	Oxygen equilibria in the presence and in the absence of BPG					
	$p_{1/2}$ (Torr)	n	K_T (µM^{-1})	K_R (µM^{-1})	L_0	ΔG_0 (kJ mol^{-1})
0.05 M Tris	5.3	2.8	0.024	3.31	8.7×10^4	13.0
+2 mM BPG	14.0	3.1	0.008	3.0	3.0×10^6	16.3

Ligand	Kinetics of binding of various ligands			
	$k'_{overall}$ (M^{-1} s^{-1})	$k_{overall}$ (s^{-1})	k'_4/k'_1	k_1/k_4
O_2	4.7×10^6	35	5	150
CO	2.0×10^5	0.015	40	10
NO	5.0×10^7	0.00005	1	100

From a phenomenological viewpoint, mammalian Hbs can broadly be divided into two groups: (i) those with an intrinsically high O_2 affinity, which is lowered in the erythrocyte by BPG; and (ii) those with an intrinsically low O_2 affinity, which is lowered little, or not at all, by BPG [17, 18]. Rodents, pigs, dogs, camels, horses, marsupials and most primates belong to the first category, while cats, sheep, goats, deer, cows and one primate (the lemur) belong to the second. Typically, Hbs with high intrinsic O_2 affinity have $p_{1/2}$ values between 4 and 6 Torr, while those with low affinity have $p_{1/2}$ values between 10 and 20 Torr (measured at 20–25 °C in 0.05 M bis-Tris, with a Cl⁻ concentration of 0.1 M, pH 6.5–7.5) [14].

5.2
Kinetic Aspects

Hemoglobin kinetics with ferrous heme ligands approximates second-order behavior in the combination direction, and first-order behavior in the dissociation direction. Among the gaseous ligands that bind to the ferrous derivatives, at neutral pH and room temperature, NO binds fastest, O_2 not quite as fast, and CO significantly more slowly; the dissociation reactions follow a different order (see Table 3) [1, 19]. When the rate constants of binding (k') or dissociation (k) of the first and fourth gaseous ligands are compared, the kinetic manifestation of cooperativity is observed (Table 3). In this connection, it has been shown that cooperative binding results from a large increase in the association rate constants in the case of CO and a large decrease in the dissociation rate constants for O_2 and NO; accordingly, the transition state along the reaction pathway, is more reactant-like for O_2 and NO, and more product-like for CO.

Since light breaks the bond between the heme–Fe(II) and all heme ligands, the kinetic behavior of rebinding has been investigated by rapid photodissociation, illuminating the protein–ligand complex with a laser pulse. The relaxation kinetics over all chemically meaningful timescales (from picoseconds to seconds) demonstrates for the various heme ligands the presence of different rate-limiting steps and, therefore, the existence of intermediates in the reaction pathway [19–21].

A four-state scheme has been proposed as the most plausible approximation:

$$\begin{array}{ccc} \text{HbX} & \text{P} & \text{M} \\ \text{ligand} \leftrightarrows & \text{heme} \leftrightarrows & \text{protein} \\ \text{bound} & \text{pocket} & \text{matrix} \\ & \updownarrow & \nearrow\swarrow \\ & \text{Hb} + \text{X} & \\ & \text{bulk} & \end{array} \quad (6)$$

where: **HbX** is the ligand-bound ground state; **Hb** + **X** represents an unligated state with both species separated in the bulk; and **P** and **M** are two unligated intermediate states with **X** trapped inside the protein either in the heme pocket (**P**) or elsewhere in the protein matrix (**M**). The momentarily trapped ligand may either rebind (to yield **HbX**) in a geminate process, or escape from the matrix and diffuse into the solvent (to yield **Hb** + **X**).

The structural features of states **P** and **M** have been investigated by time resolved X-ray diffraction with picosecond resolution, in single crystals of carbonylated myoglobin (MbCO). The short-lived photolyzed state was shown to be structurally complex in so far as the photolyzed CO migrates inside the protein matrix while in state M, before escaping into the bulk or eventually recombining in what has been called a "geminate rebinding" event. Room-temperature, time-resolved Laue crystallography has highlighted that, during its pathway inside the protein matrix, CO momentarily occupies small hydrophobic internal cavities which have been shown previously to bind one atom of xenon each [21].

If the data are forced into the four-step scheme, as in Eq. (6), the tetrameric system (e.g., HbCO) starts in state HbX in which the ligand (e.g., CO) is bound to the heme Fe. When the laser flash breaks the Fe–ligand bond, the system relaxes in about 3 ps to state P where the ligand molecule is in the heme pocket. At this point, the three gaseous ligands (O_2, CO, NO) behave differently. On a picosecond to nanosecond timescale, all ligands except CO display a substantial fraction of geminate rebinding. Thus, at approximately 100 ps, HbX is the predominant state after photolysis of NO, while M predominates after photolysis of CO, and O_2 is more evenly distributed in between HbX and M. Further evolution of species M is orders of magnitude slower (nanoseconds), the overall rates being much the same for O_2, CO, and NO; however, the amplitude of this second phase is different for the three ligands, the value being smallest for NO (since in this case picosecond rebinding is dominant and, accordingly, little ligand is left to react in the nanosecond geminate process). Without discussing the chemistry responsible for the difference between NO and CO, it should be noted that the rate-limiting step in bimolecular NO binding is diffusion of the ligand into and within the protein, since reactivity with the heme is so high that almost every NO molecule that approaches will bind; on the other hand, assuming a similar rate of diffusion, only a small fraction of the photodissociated CO rebinds in a geminate mode.

6 Mechanisms of Cooperative Binding and Allostery

Cooperative binding of O_2 by Hb is the classical example of what is referred to as an allosteric phenomenon, which defines the concept that the binding of a ligand at one site on a macromolecule affects the binding of a second ligand at a distant site via a conformational change or allosteric transition [22–24]. As a result, the uptake of one ligand (e.g., one O_2) by an oligomeric protein (e.g., tetrameric Hb) influences the ligand affinities of the remaining unfilled binding sites; in human Hb, the fourth O_2 molecule binds with a 200-fold higher affinity compared to the first one. Moreover, binding of O_2 at the heme is affected by the association of other ligands (e.g., H^+ or BPG) at different nonheme sites on the protein.

The detailed structural basis of cooperativity continues to elicit contrasting views. A number of models have been proposed in the past that in various ways hybridize the two classical theories developed during the 1960s by J. Monod *et al.* (often referred to as the "Monod–Wyman–Changeux"; MWC model) [23], and by D. E. Koshland *et al.* (often referred to as the "Koshland–Nemethy–Filmer"; KNF model) [25]. Actually, the MWC and KNF models are the limiting cases of a more general scheme that can be depicted as a square array (Fig. 8), such that the number of allosteric conformers (in a row) equals the number of binding species in a given column, including the unligated or reference species.

The entire system of chemical species (macromolecule and small ligand, X) is considered to be in equilibrium. The reactions within the array are described by appropriate mass action laws, and involve the addition of iX ligands to the unligated molecule (Hb_{00}) and a change of the cth molecular conformation, as follows:

$$Hb_{00} + iX \rightarrow Hb_c X_i \qquad (7)$$

For tetrameric human Hb (and, in general, for all vertebrate Hbs), each conformation can bind a maximum number of four X ligands, and, accordingly, there are $4+1$ conformations in the square array (i.e., four ligated and one unligated). The MWC two-state model postulates that the two extreme allosteric states (the first and fifth columns of Fig. 8) are the dominant ones; the KNF model envisages the species along one of the diagonals to be preferentially populated (see Fig. 8). Since high cooperativity is always associated with a reduction of the population of the intermediate species (partially oxygenated Hb intermediates never exceed a small percentage of total Hb), these two alternative mechanisms are, to all intents and purposes, indistinguishable by analysis of O_2 binding data which, in fact, are often described by Eq. (5), the empirical Hill equation (Fig. 7).

6.1
Concerted Model of Allostery

Among the various approaches emphasizing the role of subunit–subunit interactions, the concerted (or MWC) model has a very simple algebraic description [23]. Its basic idea is that the intrinsic O_2 affinity at the heme is controlled by the quaternary structure of the protein rather than by the number of ligand molecules already bound to the tetramer. This model assumes that there are only two different quaternary structures: one with fewer and weaker interactions among the subunits (fully ligated Hb or relaxed R); and another with more and stronger bonds between the subunits (deoxy Hb or tense T). In the R or high-affinity state, Hb displays the ability to bind O_2 tightly; this functional property is damped, but not abolished, in the T or low-affinity state. On the other hand, the T state interacts more strongly than the R state with heterotropic ligands (H^+, Cl^-, BPG, and CO_2). The transition from T to R is an all-or-none process: the symmetry of the molecule is conserved (i.e., hybrid, room temperature states cannot exist) and the O_2 affinity of all subunits in a tetramer

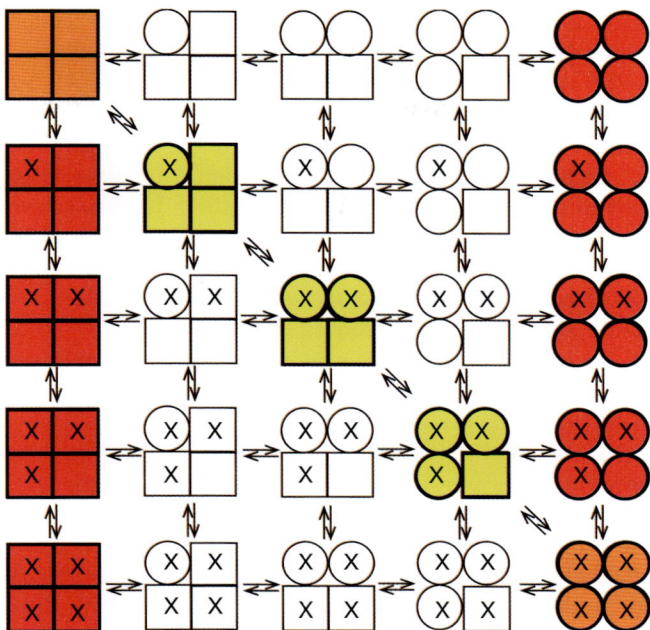

Fig. 8 General representation of ligand (X) binding and ligand-linked conformational changes, illustrating the alternative ways of generating cooperativity according to the sequential KNF model [25] (indicated by the diagonal in boldface) and the concerted MWC model [23] (represented by the equilibria between the two extreme columns, all intermediates being absent or negligible or undetectable). Each subunit within the tetramer can have only two distinct tertiary conformations, one (circle) with high affinity for heme ligands or R state; and the other (square) with low affinity or T state. In the sequential model, heme–ligand binding induces an isomerization in each ligated subunit, which assumes the high-affinity conformation; the unligated subunits retain their low-affinity structure, even though the contacts with the ligated neighbor(s) are altered, thereby increasing their tendency to switch to the high-affinity state. An important property of the sequential model is that the extent of binding and structural changes at different ligand saturations must be identical. In the concerted model, the tetramer is postulated to exist in only two quaternary states (R and T) in which all four protomers are in either one of two conformations (squares or circles). The quaternary states are in equilibrium with each other at every degree of saturation: the fully ligated tetramer is predominantly in the R state, whereas the T state is favored in the fully unligated species. The conformational transition of the subunits (between square and circle) is concerted, involving all protomers within the tetramer, with the result that the molecule conserves structural symmetry. This situation corresponds to an extreme case of cooperativity among the protomers, characterized by a virtual absence of the hybrid states (i.e., coexistence of circles and squares in one tetramer), which, on the contrary, are present in the sequential model. The sequential model can account for both positive and negative cooperativity, whereas the concerted model can accommodate only for positive cooperativity.

would be either equally low or equally high [12, 15, 23, 26–28]. Oxygenation of the subunits within each quaternary state is associated with small local structural changes such that the T state, with one O_2 molecule bound, and the R state, with three O_2 molecules bound, will have conformations that are locally different from those of the unligated T and fully ligated R, respectively.

The assumption of the concerted model in its original formulation is that these local structural differences are not transmitted to neighboring subunits and thus do not alter their O_2 affinity; rather, they affect the relative stability of the R and T quaternary states. Therefore, although binding to the four hemes in each of the two allosteric states is itself noncooperative, the shift in the T-to-R equilibrium gives rise to a sigmoid binding curve.

The MWC model allows the equilibrium ligand binding by Hb to be described with only three parameters: the two microscopic dissociation constants characteristic of the protein in low-affinity (K_T) and high-affinity (K_R) states, often reported in terms of a ratio c ($=K_T/K_R$); and the population ratio of the two quaternary states in the absence of ligand, expressed by the constant $L_0=[T]_0/[R]_0$. The total free energy of cooperativity, ΔG_0, can be calculated from the binding of the last and first ligand molecules, which is approximately equal to $\Delta G_0 = -RT \ln (K_T/K_R)$ (i.e., the difference in free energy of binding O_2 to the R state and the T state).

Values of the parameters expressing the cooperative effects in terms of the MWC model are reported in Table 3. The effect of heterotropic ligands (H^+, Cl^-, BPG, and CO_2) is likewise described in terms of the concerted allosteric model since they shift the equilibrium in favor of the T structure, with an increase of L_0: the classical example is the effect of BPG which binds to the Hb tetramer in a 1:1 ratio and stabilizes the T state.

The original MWC model assumed K_T and K_R to be intrinsic parameters independent of heterotropic ligands. However, detailed O_2-binding data, coupled with structural information obtained by X-ray crystallography and spectroscopy in solution, have indicated that in human HbA_0, K_T varies over a wide range as a function of H^-, Cl^-, CO_2, and BPG concentration, whereas K_R is little affected by heterotropic ligands (except possibly by protons, and only below pH 7.0). Thus, the original MWC model had to be slightly modified [12, 15, 27, 28].

6.2
Structural Changes Associated to Oxygen Binding

The T state is assumed to have quaternary interactions between subunits typical of deoxy Hb, while the R state is believed to have those of oxy Hb (as determined by protein crystallography). In deoxy Hb, the Fe is displaced from the plane of the porphyrin nitrogens toward the proximal histidine by 0.06 nm. There are two reasons for this: (i) the larger radius of the Fe in its high spin state; and (ii) steric repulsions between the Nε of the proximal histidine and the porphyrin nitrogens. Moreover, the heme is domed – that is, the Fe-bound nitrogens of the porphyrin are out of the plane of the carbon atoms of the heme (see Fig. 2) by 0.016 nm in the α-subunit and by 0.01 nm in the β-subunit. Upon single-subunit oxygenation, the Fe atom barely moves toward the porphyrin, which remains domed within the T state; this prevents the heme from adopting the optimal geometry with the bound O_2 (i.e., planar, with Fe in "the plane"), thus producing strain. Relief of strain at the active site requires the Fe to move toward the porphyrin plane, accompanied by a displacement of the proximal histidine together with the helix F and the FG corner (the so-called "allosteric core"). In particular, in both subunits there is a shift of helix F across the heme plane by about

0.1 nm, accompanied by a rotation; the overall effect is to allow some reorientation of the proximal histidine so that the Fe can move toward the heme plane [2, 5, 26].

To understand the connection between the foregoing tertiary modifications and quaternary structural changes, the Hb tetramer can be thought of as a pair of dimers, $\alpha_1\beta_1$ and $\alpha_2\beta_2$, whose relative orientation has to alter; with the contacts $\alpha_1\beta_1$ (and $\alpha_2\beta_2$) remaining rigid. The tertiary changes make these dimers a misfit in the quaternary T structure, favoring the T-to-R transition which consists of a 12–15° rotation of the dimer $\alpha_1\beta_1$ relative to the dimer $\alpha_2\beta_2$ and a translation of one dimer relative to the other by 0.08 nm (Fig. 9). Helix C β_2, which is in contact with the corner FG α_1, forms a ball-and-socket joint around which the dimers turn, while helix C α_1 slides relative to FG β_2. Together, the two contacts form a two-way switch that ensures that the αβ dimers click back and forth between the two stable quaternary states (T and R).

The pre-existing equilibrium between the two states is prone to a shift because the free energy landscape of proteins is dynamic, and binding events can bias the relative populations by selecting preferentially one conformer. Max F. Perutz in Cambridge was the first to attempt an explanation of the functional data based on the structures of liganded and unliganded HbA, though at low resolution. The so-called "stereochemical mechanism" postulated the role of a finite number of residues and their molecular interactions in the arising of homotropic and heterotropic effects in HbA [12, 14, 28–32]. In fact, all heterotropic ligands at physiological pH lower the O$_2$ affinity by forming additional hydrogen bonds that specifically stabilize the T structure. For example, BPG binds in a cavity on the dyad axis of the tetramer, limited by the amino termini and helices H of the β-subunits; it forms hydrogen bonds with ValNA1(1), HisNA2(2), LysEF6(82), and HisH21(143). The functionally relevant protons (i.e., those released upon O$_2$ uptake responsible for the Bohr effect) are discharged in deionized solution only from HisHC3(146)β, a residue that forms a hydrogen bond with AspFG1(94)β in the T structure. In the presence of Cl$^-$ and/or BPG, other residues (e.g., ValNA1(1)β and LysEF6(82)β) contribute additional Bohr protons.

Although Perutz's stereochemical interpretation was in line with the MWC model, the constraints on the inter-dimer interface could not explain the ligation intermediates. Subsequently, Szabo and Karplus presented a rigorous thermodynamic derivation of this elegant structural model, that proved seminal for an understanding of cooperativity [32]. These authors made use of the experimental evidence published by Gibson, Brunori, and Antonini such that:

- The affinity of the isolated subunits, of the dissociated αβ dimers and the "default" affinity of Hb is high and corresponds to that of ^{R}Hb.
- All ligation intermediates freely equilibrate with each other according to their equilibrium constants.
- Constraining inter-subunit interactions in unligated Hb forces the tetramer into the ^{T}Hb structure and lowers O$_2$ affinity.

Subsequently, Szabo and Karplus [32] derived that: (i) this stabilization of the T state links K_T to L_0 and reduces the affinity from K_R to K_T; and (ii) some of those constraining interactions might not be present under all experimental conditions. To be more specific, a crucial role in the energetics of the T-to-R transition was assigned to eight salt bridges per tetramer present in the deoxy T

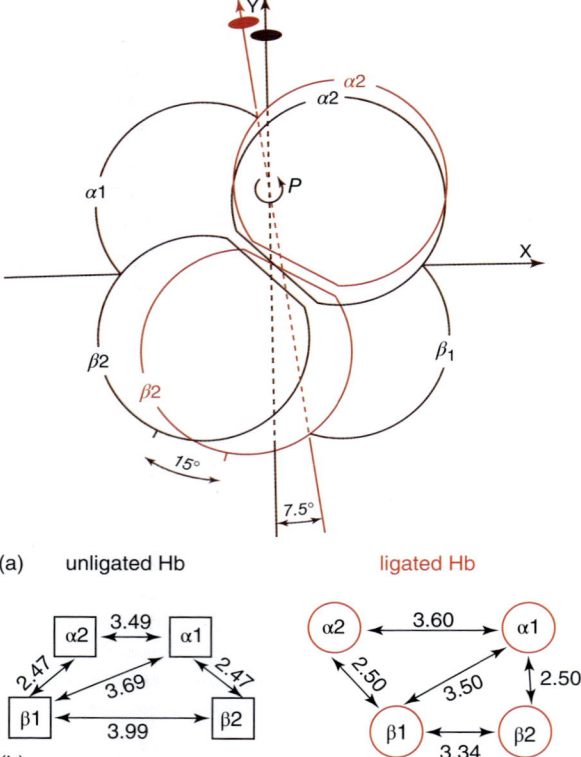

Fig. 9 Schematic diagram illustrating the ligand-linked changes in quaternary structure of Hb (top) and the distances (in nm) between heme groups (bottom). Both structures have a dyad axis (Y) relating the $\alpha_1\beta_1$ dimer to the $\alpha_2\beta_2$ dimer, and in both cases the molecular contacts between α_1 and β_1 (as well as those between α_2 and β_2) change very little; accordingly, the positions of ligated and unligated $\alpha_1\beta_1$ dimers have been superimposed. In going from deoxy to oxy Hb, the main differences in conformation are: (i) a rotation of 15° (θ) about a pivot (P) of the $\alpha_2\beta_2$ dimer relative to the $\alpha_1\beta_1$ dimer so that the two β chains are about 0.6 nm further apart in the former than in the latter; and (ii) a shift of the $\alpha_2\beta_2$ dimer along the P axis by about 7.5° into the page. Major conformational changes in the T-to-R transition involve the $\alpha_1\beta_2$ (and $\alpha_2\beta_1$) subunit interfaces. As a result of the quaternary switch, the distances between the heme groups change, bringing into closer contact the two β subunits, and further apart the two inter-dimer interfaces. This concerted movement is such that the tetramer is ready to catch the message whenever a ligand binds/unbinds from the subunits. Modified from [2, 5, 29].

structure, but not in the fully oxygenated R state. Six of those salt bridges are inter-subunits: four between the two α subunits (ArgHc3(141)α_1 facing AspH9(126)α_2 and LysH10(127)α_2); two along the inter-dimer interface [LysC5(40)α_1–HisHC3(146)β_2]; the other two being intra-β-subunit. The rupture of these interactions in going from deoxy to liganded Hb will by-and-large explain the difference in stability between the two allosteric states. The main consequence of this elegant model was to assign a precise physical meaning to the MWC parameters (K_R, K_T, and L_0), and to quantify the intuitive structural explanation of Perutz. Nevertheless, these salt-bridges alone are insufficient to fully account for the difference in stability between T_0 and R_0, under all conditions and for all tetrameric Hbs. Possibly, different arrangements in the inter-dimer interface, transmitting the iron-linked movements of HisF8 through the F-helix and the FG corner, might play a role in the so-called "allosteric core."

6.3
Ligand-Binding Pathways

All hemoglobins have a heme-binding pocket that protects the Fe(II) from the solvent. This feature has a biological importance because, in water, the ferrous heme iron is rapidly oxidized into the inactive ferric state [7, 16, 19, 33, 34]. The stability of the Fe(II)O_2 complex is controlled by the presence of a hydrogen bond with the distal histidine (located at E7; see Table 1 and Fig. 2). This crucial interaction, in fact, slows down the rate of O_2 dissociation from microseconds (as obtained in sperm whale Mb mutant with valine at E7) to many milliseconds, and thus increases the O_2 affinity to physiologically compatible values; consequently, it reduces the rate of autoxidation of the heme Fe(II) which, otherwise, would occur much too rapidly.

Examinations of the X-ray structures of many hemoproteins show that the protein environment of the heme, if rigid, would prevent the entrance and exit of even small gaseous ligands (e.g., O_2, CO, NO). Empirical energy function calculations based on the crystallographic coordinates have shown that if the protein were rigid, possible pathways in the neighborhood of the heme pocket would have barriers of the order of 400 kJ mol^{-1}; such energy barriers would lead to extremely slow escape times of ligand from the heme pocket of more than a billion seconds at room temperature. Thus, protein fluctuations are required for physiological O_2 uptake and release [35]. The dynamics of the heme pocket residues has been unveiled using nuclear magnetic resonance (NMR) spectroscopy, kinetics, and molecular dynamics (MD) simulations, showing that PheCD1(43) and PheCD4(46) – which wedge the prosthetic group into its pocket and are packed tightly between the heme and the distal helix E – flip over at rates faster than 10^4 s^{-1}; however, this is possible only if the entire heme pocket breathes very rapidly.

Dihedral rotations of key side chains explored using MD indicate that, in a fluctuating protein, energy barriers of $\approx$40 kJ mol^{-1} would be present; these are compatible with those estimated from rebinding studies after photolysis of HbCO. This dilemma has been addressed by: (i) examining X-ray structures with small (e.g., CO) and bulky (e.g., alkyl-isocyanides) ligands bound to the heme Fe; (ii) temperature-dependent X-ray diffraction data; (iii) MD simulations; (iv) time-resolved Laue crystallography on photolyzed MbCO crystals; and (v) geminate rebinding on site-directed and random mutants.

Most of these investigations were carried out initially on Mb. Nine clusters of residues have been identified as possible escape pathways for a ligand leaving the interior of the protein. The most direct pathway is defined by residues ArgCD3(45), HisE7(64), ThrE10(67), and ValE11(68), between the CD loop, the helix E, and the heme; a recent analysis of kinetic data on approximately 100 Mb mutants, later confirmed on a selection of six Hb mutants on HisE7, indicated that the ligand escape route involves, by-and-large, the so-called "histidine gate" [16, 19].

6.4 Molecular Dynamics (MD) Simulations

Both, Hb and Mb have served as models for a number of biophysical approaches, including the dynamical behavior of proteins [16, 19–21, 33, 34]. Static high-resolution studies using protein crystallography have provided a good description of thermodynamically stable states, but have only hinted at the wide spectrum of conformations that a native protein can explore. However, static pictures do not offer the possibility of identifying the pathway of a ligand to its active site. Rather, MD computer simulations highlighted movements on the picosecond timescale and above, that were subsequently accepted by the scientific community. One of the pioneers of such powerful computational chemistry exercises is M. Karplus, who was awarded the 2013 Nobel Prize in Chemistry together with M. Levitt and A. Warshel. MD simulations allowed to calculate protein relaxations between different conformational states, to couple them with ligand migration through the protein matrix, and to identify regions with enhanced flexibility, the transient opening of crevices and/or cavities, and interactions of the side chains with the solvent. This method is powerful in that it is based on two or more fixed conformations (typically, the coordinates of the protein solved at high resolution), each with its own potential energy; one conformation being considered as the starting point and the other as the end-point. During the calculation a constant term is added to the potential energy so that the protein is forced to relax to a new position by a series of displacements, each consisting of either one or several picosecond movements. At the end of the simulation a series of trajectories is produced, each consisting of a heating phase (the input energy), an equilibration phase, a sampling phase (also known as productive MD), and a displacement toward the end point. Today, Karplus' software CHARMM implemented on globins is widely used by the scientific community for MD simulations and energy minimization on a variety of macromolecules. Overall, this procedure can be compared to a morphing from one state to another.

MD studies on Hb and Mb, combined with mutational data (natural and engineered variants) and many time-resolved experimental studies, have demonstrated, among other features: (i) the role of the residues in the distal heme pocket to facilitate the entry/exit of the gaseous ligands in the protein matrix; and (ii) the role of hydrophobic cavities, previously defined as packing defects, present in the matrix in modulating ligand (re)binding dynamics.

MD simulations have proved powerful in finding the detailed relationships, at the atomic level, between structural changes induced by ligation and the thermodynamic and kinetic manifestations. Simulations have made clear that proteins are relatively soft polymers and, consequently, have significant structural fluctuations at room temperature. Moreover, the simulations provided a satisfactory description of the cooperativity in ligand binding, the influence of allosteric effectors (protons or BPG) on ligand affinity, and the influence of chemical modifications and certain mutations on these properties. A combination of functional, structural and computational data showed the presence of two quaternary structures (deoxy and oxy) for the tetramer, two tertiary structures (liganded and unliganded) for the subunits in each allosteric state, and a few flexible structural elements that couple the stabilities of the tertiary and quaternary structures which, by-and-large are composed by mostly rigid parts.

7
Assembly of Globin Monomers

The globin fold, which has been conserved throughout evolution, leads to a variety of possible quaternary structures. Although vertebrate Hbs are tetrameric, in the animal kingdom globin chains can assemble into several different quaternary structures from stable dimers to very large macromolecular aggregates, such as the extracellular invertebrate Hbs which have a molecular mass on the order of millions of daltons [1, 2, 17, 18].

Subunit interactions that stabilize the oligomeric states occur in stages and have been described as a series of coupled equilibria that are often ligand-dependent [1, 10, 16, 22, 26]. In the case of tetrameric Hb, dissociation is described in terms of the simplified scheme:

$$\text{tetramer} \rightleftharpoons \text{dimers};$$

$$\text{dimer} \rightleftharpoons \text{monomers}$$

When dissociation is limited to the dimer stage, only symmetrical splitting is observed (i.e., $\alpha\beta$ dimers and not α_2 and β_2 dimers), reflecting the nature of the molecular contacts: namely, the contacts between like chains ($\alpha\alpha$ or $\beta\beta$) are considerably less extensive than those between unlike chains (see Table 2). Cleavage occurs only along the $\alpha_1\beta_2$ interface (yielding $\alpha_1\beta_1$ and $\alpha_2\beta_2$ dimers), and the thermodynamic stability is different for the two quaternary states – that is, it is ligand-linked. In fact, the tetramer–dimer equilibrium constant of human Hb, at physiological ionic strength, is about 10^{-6} M per heme for the oxy derivative, and about 10^{-11} M per heme for deoxy Hb. A variability of the tetramer–dimer dissociation constant has been reported from comparative studies on a number of Hbs from different vertebrate species.

Further dissociation of the $\alpha\beta$ dimer to produce free α- and β-subunits is negligible under physiological conditions. The isolated subunits can exist as dimers (in the case of the α-chains) and even tetramers (in the case of the β-chains). It is generally assumed that the dissociation of these oligomeric subunits into monomers must occur before they can combine to form $\alpha\beta$ dimers, which then aggregate to produce tetrameric Hb. The overall scheme describing formation of cooperative Hb tetramer is therefore as follows:

$$\begin{array}{cc} 2\alpha_2 & \beta_4 \\ \downarrow\uparrow & \downarrow\uparrow \\ 4\alpha + 4\beta \rightarrow 4(\alpha\beta) \leftrightarrows 2(\alpha_2\beta_2) \end{array} \quad (8)$$

Data relating to the association–dissociation reaction for human oxy Hb are summarized in Table 4.

The major manifestations of the relationships between cooperativity in O_2 binding and subunit interactions in tetrameric Hbs are as follows: (i) dissociated $\alpha\beta$ dimers bind O_2 with high affinity, which is almost equal to the affinity of the R-state Hb and of the isolated α- and β-subunits, but without any detectable cooperativity; and (ii) the assembly of deoxy dimers into tetramers produces a large reduction in overall affinity for O_2. As the tetramer binds the four heme ligands with successively increasing affinity constants, the number of ligated hemes must control the interaction energy

Tab. 4 Subunit dissociation constants (K, in M per heme) for the isolated α- β-chains, and for the $\alpha\beta$ dimer and the $\alpha_2\beta_2$ tetramer, in all cases as oxygenated derivatives.

Reaction	K	k_{ass} (M^{-1} s^{-1})	k_{diss} (s^{-1})
$\alpha_2 \rightleftarrows 2\alpha$	1.1×10^{-4} M	9.1×10^5	—
$\beta_4 \rightleftarrows 4\beta$	4.8×10^{-16} M^3	—	0.05
$\alpha\beta \rightleftarrows \alpha + \beta$	4.4×10^{-13} M	5×10^5	8.9×10^{-7}
$\alpha_2\beta_2 \rightleftarrows 2(\alpha\beta)$	7.1×10^{-7} M	2.8×10^7	0.3

The table also reports the kinetic association (k_{ass}) and dissociation (k_{diss}) rate constants. All data in 0.1 M Tris-HCl + 0.1 M NaCl + 1 mM Na$_2$EDTA, pH 7.3–7.5 at 21.5–25 °C.

at the dimer–dimer interface. When a heme ligand is bound by Hb, the binding site becomes subjected to strains that are propagated to the boundary between subunits. This phenomenon of ligand-linked dissociation can be quite extreme.

8
Reactions with Nitric Oxide

Nitric oxide (NO) has been widely used as a ligand of the ferrous heme iron, similar to O$_2$ and CO,. and over the years comparisons between the three gases have provided a wealth of information on the structure and function of Hb. The chemistry of NO is much more complex than that of other gaseous ligands of the ferrous heme iron, however, and considerable interest has arisen since NO was found to be a universal second messenger regulating circulation, brain alertness and defense against invaders (see Nitric Oxide in the Encyclopedia).

8.1
Reactions of NO with Various Hb Derivatives

It is well known that NO can bind to the ferrous heme iron of all known hemoproteins, yielding a simple adduct that is characterized by a very high stability constant ($K \sim 10^{10}$ M^{-1}). The combination of NO with ferrous Hb weakens the bond between the iron and the proximal His (this is the opposite of what occurs with CO and O$_2$). Moreover, NO reacts not only with deoxy Hb or deoxy Mb but also with the oxygenated and the met derivatives, yielding a complex network of reactions that are summarized in Fig. 10. Because bound NO has a bent configuration, it is more similar to O$_2$ than to CO. Several studies with site-directed mutants have shown a direct correlation between increasing the steric hindrance of the distal pocket and decreasing the combination rate constant, which is very high and essentially diffusion-limited [1, 16, 28].

The reaction of NO with oxy Hb, forming met Hb and nitrate, is also a very rapid process ($k_{on} \sim 6-9 \times 10^7$ M^{-1} s^{-1} at 20 °C) competing with binding to deoxy Hb ($k_{on} \sim 5 \times 10^7$ M^{-1} s^{-1} at 20 °C). The latter reaction is considered the main route to scavenge NO that is produced continuously in the circulating blood through the physiological metabolic pathway (see Sect. 8.2). Moreover, the same reaction occurring with oxymyoglobin (MbO$_2$) in striated muscles and in the heart has been considered an important defense mechanism against the inhibitory effects of NO on cytochrome *c* oxidase (and thus on

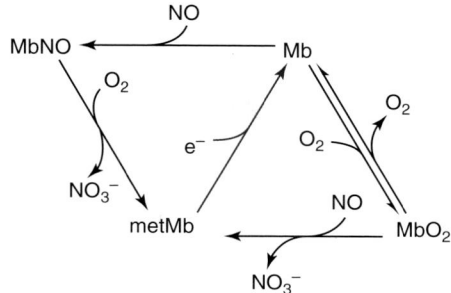

Fig. 10 The reactions of myoglobin with O_2 and NO. The primary physiological function involves the reversible binding of O_2 to deoxy-myoglobin (Mb) to yield oxy-myoglobin (MbO$_2$), which facilitates the transport of O_2 from the periphery of the cell to the mitochondria for use in respiration. MbO$_2$ reacts rapidly (and irreversibly) with nitric oxide (NO) to yield nitrate and ferric myoglobin (met Mb), thereby quenching free NO that might otherwise inhibit cytochrome c oxidase. Met Mb is reduced to Mb by met Mb reductase (e^- arrow). Moreover, Mb binds rapidly and reversibly with NO, yielding nitrosyl-myoglobin (MbNO). This pathway might be of greater significance in cellular compartments with low O_2 concentrations, for example, in the immediate environment of mitochondria. In the presence of O_2, MbNO can be converted back to met Mb at a rate limited by the thermal dissociation of NO, again yielding nitrate [36].

cellular respiration). This reaction of MbO$_2$ with NO is intrinsically very rapid and is made efficient by the high concentration of the hemeprotein in the cytoplasm of red muscle cells (0.3–0.6 mM).

The cycle reproduced in Fig. 10 also includes the reduction of the met Mb, or met Hb, catalyzed by specific reductases. However, contrary to O_2 and CO, NO also binds the oxidized form of Hb to produce an iron–nitrosyl complex HbFe(III)NO ($k_{on} \sim 4 \times 10^3\,M^{-1}\,s^{-1}$ and $k_{off} \sim 1 \times 10^3\,s^{-1}$), not indicated in Fig. 10. In any case, the kinetic rate constant of NO combination to ferric Hb is three to four orders of magnitude lower than that of either ferrous Hb or ferrous HbO$_2$, the most prevalent Hb derivatives in blood. Therefore, it is unlikely that the ferric Hb complex with NO forms *in vivo* or has any significant physiological relevance. *In vitro*, and under an excess of NO, this complex undergoes reductive nitrosylation to finally generate the reduced nitrosyl adduct and nitrite, as in Eq. (9).

It should be recalled that the reaction of free NO with free O_2 in solution ($k \sim 6\,M^{-1}\,s^{-1}$) is too slow to compete with the direct bimolecular reactions of NO with deoxy or oxy Hb.

$$\text{HbFe(III)} + \text{NO} \rightleftharpoons [\text{HbFe(III)NO} \rightleftharpoons \text{HbFe(II)}] \xrightarrow[H_2O]{NO} \text{Fe(II)NO} + NO_2^- + 2H^+ \quad (9)$$

Furthermore, alternative reactions may come into play under different conditions, such as the paraphysiological condition where the NO concentration is 10^3- to 10^6-fold lower than that of O_2 and Hb. In particular, when oxygenated hemes in the quaternary R state of Hb are $\geq 95\%$, and when NO is present at concentrations lower than the level of vacant hemes, the unoccupied binding sites will be extremely reactive towards NO. On the other hand, NO may follow an alternative chemistry, by reversibly reacting with CysF9(93)β, when Hb is in the quaternary R state (e.g., oxy Hb) but not in the quaternary deoxy state. This reaction of NO to the reactive Cys on the β-subunits (see below) should be incorporated into the MWC two-state model; hence, NO would represent a third gaseous molecule (in addition to O_2 and CO_2) carried by Hb under physiological conditions, through a different chemical mechanism [37, 38].

8.2
Physiological Implications

Nitric oxide is produced continuously in the endothelial cells through the enzymatic degradation of L-arginine by the enzyme NO-synthase. The NO gas acts as a signal transduction molecule that mediates a variety of physiological responses, including the dilation of blood vessels, achieved by the activation of soluble guanylate cyclase (a heme-containing enzyme), with an increase in the intracellular cyclic guanosine monophosphate (cGMP) concentration and thus, a relaxation of vascular smooth muscles. The supposedly rapid conversion of NO to biologically inactive metabolites in blood (e.g., nitrate) formed the rationale for inhalation NO therapy, because the short half-life ($\sim$2 ms) of NO in blood should confine its effects to the pulmonary circulation. Erythrocytes may be considered to be a major sink for NO by virtue of the quick oxidation reaction of NO with oxy Hb (see Sect. 8.1) to produce met Hb and nitrate, due to: (i) the speed of this bimolecular reaction (half-life ca. 1 µs, according to extrapolation of the *in-vitro* data); and (ii) the high concentration of oxy Hb in blood, which is in the millimolar range (2 mM in monomers) to be compared with the endothelium-derived local level of NO that reaches only nanomolar concentrations. Moreover, NO oxidation has been shown to be an important reaction in 2/2 truncated Hb of pathogens, such as Mycobacteria, to contrast the production of NO by host granulocytes and macrophages [37, 38].

Another peculiarity of NO is its ability to combine with the -SH group to form a class of compounds called *S*-nitrosothiols (SNOs). The *S*-nitrosothiol of cysteine (Cys-SNO) and *S*-nitrosothiol of glutathione (GS-SNO) can be easily produced *in vitro* and have vasodilating activity *in vivo*, although their physiological relevance to blood flow regulation is still incompletely assessed.

$$R - S^- + NO^+ \rightleftharpoons R - SNO \quad (10)$$

This reaction occurs quite easily not only with reduced glutathione (GSH), which is at high concentration in the erythrocytes ($\sim$2.5 mM), but also with oxidized glutathione (GSSG). The same reaction was shown to occur with CysF9(93)β of human HbA, which is accessible in ^{R}Hb and partially buried in ^{T}Hb. SNO-Hb has been unequivocally detected in freshly drawn Hb, to a concentration of 10–100 nM; despite accounting for a very low fraction of Hb (<1/1000), it constitutes a significant reserve of the otherwise labile NO.

According to this hypothesis, NO binds cooperatively to the minor ($\sim$1%) population of deoxygenated heme in oxygenated erythrocytes present in the arterial circulation, forming iron–nitrosyl–Hb. Although the NO dissociation rate from deoxyHb is much slower than the transit time of the red blood cells in the lungs, it is accepted that NO reacts with CysF9(93)β (through a one-electron oxidation) to produce its *S*-nitro derivative (<50 nM *in vivo*). This species dissociates NO upon deoxygenation in the peripheral tissues and elicits vasorelaxation via transnitrosation reactions to low-molecular-mass thiols (such as GSH) or high-molecular-mass sulfhydryls (specifically, the anion exchanger or band 3 in the erythrocyte membrane). Recently, experiments on mice carrying a Hb with CysF9(93)β mutated into Ala showed a higher damage of brain and heart after reperfusion following hypoxia.

9 Hemoglobin Variants

9.1 Nature of Mutants

Several hundred mutant Hbs (1594, including 463 mutations associated with thalassemia, updated in 2014) within the human population have been isolated and chemically characterized: variants of α-subunits relative to those of β-subunits are approximately in a ratio of 1:2. In most cases, the abnormality involves the replacement of a single amino acid residue per αβ dimer; all of these modifications are consistent with single-base substitution in DNA coding for the globin subunits. Some abnormal Hbs have residues deleted or inserted (271 variants), while in others the subunits are cut short or elongated (12 variants), others are constituted by the fusion of two different subunits (nine variants), and yet others show more than one point mutation in the same polypeptide chain (15 variants) [8, 31, 39, 40]. Of the large number of Hb mutants, a significant fraction has deleterious effects on health (see Table 5). Various databases of Hb variants have been developed, with increased capacity for sophisticated queries and prompt updating (see for example, *http://globin.bx.psu.edu*).

9.2 Molecular Basis of Hemoglobin Diseases

Pathological symptoms in the carriers of abnormal Hbs have been associated with various effects:

- An altered oxygen affinity (causing polycythemia if it is raised, or cyanosis if it is lowered).
- A decreased stability of the protein (causing hemolysis and clumps of denatured Hb called Heinz bodies).
- An increased tendency to form ferricHb (mainly producing cyanosis).
- Abnormal intracellular polymerization, inducing a distorted sickle-shaped erythrocyte [39–42].

In many cases, these biochemical abnormalities have been interpreted with the help of the 3D structure of the mutant compared to wild-type normal HbA.

Many of the structural abnormalities that give rise to clinical symptoms are clustered around the heme pockets, or are in the vicinity of the $\alpha_1\beta_2$ interface which is so important in allosteric transitions. Some of the typical structural changes leading to abnormal Hbs include:

1) A common structural pattern causing an increase in O_2 affinity is the loss of hydrogen bonds, salt bridges, or nonpolar interactions that stabilize the Hb tetramer and the T state in particular. Moreover, some mutations increase O_2 affinity because they impair BPG interaction by modifying its binding site (caused by 94 different point mutations, mostly on β-chains).

2) Unstable Hbs (146 variants) usually involve: (i) an opening of the heme pocket, a loss of heme, and subsequent precipitation of globin; (ii) the insertion of a proline into a helix, introducing a kink that usually disrupts the tertiary structure; or (iii) the unfavorable removal of a nonpolar side chain on the molecular surface (which excludes the interior of the protein from the surrounding water), the introduction of a charged group into the interior of

Tab. 5 Selected list of point mutations determining abnormal human Hbs.

Residue	Mutation	Common name	Effects
CE3(45)α	His → Arg	Fort de France	Increased oxygen affinity; R stabilization
G2(95)α	Pro → Ser	Rampa	Increased oxygen affinity; tetramer dissociation
HC3(141)α	Arg → His	Suresnes	Increased oxygen affinity; T destabilized
H19(136)α	Leu → Pro	Bibba	Unstable; helix H disrupted
E7(58)α	His → Tyr	M Boston	Fe(III) bound to Tyr
E18(102)β	Asn → Thr	Kansas	Decreased oxygen affinity; R destabilized
CD1(42)β	Phe → Ser	Hammersmith	Unstable; loss of heme
B14(32)β	Leu → Arg	Castilla	Unstable; positive charge in the protein core
FG4(92)α	Arg → Leu	Chesapeake	Increased oxygen affinity, low cooperativity; R stabilized
E14(70)β	Ala → Asp	Seattle	Fe(III) stabilized by the negative charge
A3(6)β	Glu → Val	HbS sickling	Fit of Val into the EF cleft of another deoxygenated tetramer; it forms fibers
A3(6)β	Glu → Lys	HbC	Decreased oxygen affinity, reduction of clinical malaria (*P. falciparum* hemoglobinase cannot efficiently digest β-chains)
GH4(121)β	Glu → Lys	HbO Arab	Enhancement of sickling anemia when in heterozygous S/O
F8(92)β	His → Gln	St. Etienne	Increased affinity; tetramer with two hemes
E7(63)β	His → Arg	Zurich	Increased affinity
FG5(98)β	Val → Met	Kholn	Increased affinity
C1(35)β	Tyr → Phe	Philly	Noncooperative
G10(108)β	Asn → Lys	Presbyterian	Decreased oxygen affinity
G1(99)β	Asp → Asn	Kempsey	Increased oxygen affinity, noncooperative
ΔHC3(146)β	des-His	McKees Rocks	Increased oxygen affinity, noncooperative
Hc03(146)β	His → Leu	Cowtown	Increased of KT
B2(20)β	Val → Met	Olimpia	Hb fibers
B9(27)β	Ala → Val	Grange Blanche	Increased oxygen affinity
B9(27)β	Ala → Ser	Knossos	Decreased oxygen affinity
EF6(82)β	Lys → Asn/Asp	Providence	Decreased oxygen affinity
EF6(82)β	Lys → Thr	Rahere	Insensitive to BPG
EF6(82)β	Lys → Met	Helsinki	Insensitive to BPG

For a more complete list, please visit the atlas of human Hb mutations at *http://globin.bx.psu.edu*.

the molecule, or the deletion of single residues or segments of polypeptide chains. All of these events can either disrupt or heavily perturb the native folding of Hb.

3) The M (for met Hb) mutants are the class of Hbs (nine in total) in which the Fe atom of one type of subunit is permanently oxidized as a result of the replacement of a heme contact residue. The most common mutations are the substitution of either the proximal histidine (HisF8) or the distal histidine (HisE7) by a tyrosine in the α-, the β-, or the γ-subunit.

4) In general, the mutations that lead to abnormal Hb behavior are not located on the exterior of the tetramer.

However, the replacement of a surface glutamate at position A3(6)β with a hydrophobic valine in HbS or sickle-cell mutant (the most prevalent Hb variant worldwide) has deleterious effects. Upon deoxygenation of a concentrated deoxygenated Hb solution, deoxy HbS tetramers aggregate into a fiber, distorting and rigidifying the erythrocytes into a variety of bizarre shapes. As a consequence, the red cells may not be able to transverse the microcirculation vessels, with transient to permanent blockage of local blood flow. When this occurs the resulting organ damage is a major cause of the morbidity and mortality of sickle-cell anemia. The key to erythrocyte sickling in HbS is the presence of a hydrophobic pocket located between helices E and F and made up of PheF1(85)β and LeuF4(88)β, which binds ValA3(6)β; upon oxygenation, the helix F motion of the β-subunits pulls the PheF1(85)β of one tetramer away from potential contact with the ValA3(6)β side chain of a second tetramer, leading to dissociation of the fibrils and formation of oxygenated HbS tetramers.

A final consideration is dedicated to the 463 forms of thalassemia, in which a deletion/insertion or nucleotide mutation leads either to a change in the frameshift or to an early stop codon, resulting in the loss of expression of the α-, β-, δ-, and γ-chains [43].

The severity of the pathology associated with these anemias (which are autosomal recessive) depends on the nature of the mutation, and symptoms can extend from iron overload to splenomegaly, from impaired growth to heart problems, to higher rates of infections.

9.3
Hemoglobin Solutions as Oxygen Carriers

Many blood substitutes should be more properly called synthetic O_2 carriers as they provide O_2 transport and volume replacement but do not perform other functions of the blood (e.g., facilitating coagulation and immune response) [44–46]. Such O_2 carriers are life-support systems able to dissolve (and thereby transport) large amounts of O_2; this mode of operation is at variance with simple plasma expanders, which contain dissolved O_2 at low concentration (≈ 0.3 mM). Among potential O_2 carriers, attention has been focused on perfluorocarbons (synthetic compounds that reversibly dissolve O_2) and a variety of Hb preparations, either free in solution or encapsulated into lipid vesicles. In order to be efficient, perfluorocarbons must be used under high O_2 pressure (see Fig. 11), otherwise the amount of O_2 transported will be much too low; moreover, they are immunotoxic.

Unprocessed solutions of Hb are unsuitable as O_2 carriers for several reasons:

- Hb quickly disappears from the plasma, partly through the formation of a complex with the plasma-protein haptoglobin (mainly cleared by the liver, with a half-life of 10–30 min), but mostly through direct filtering off by the kidney, with a half-life of 1 h. Direct elimination through the kidney is possible because, in dilute solutions, tetrameric Hb easily dissociates into free $\alpha\beta$ dimers; the kidneys are also permeable to proteins of molecular mass $<40 000$ g mol^{-1}, unless very highly charged.
- O_2 release to tissues is poor as a result of the high affinity of Hb dissolved in plasma, which is stripped from BPG, the allosteric effector that modulates the efficiency of transport. At 37 °C and

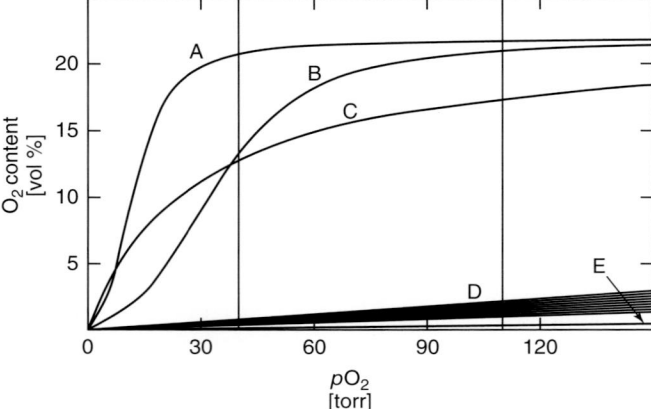

Fig. 11 The O_2 content of blood and some O_2 carriers as a function of O_2 partial pressure under physiological conditions (pH 7.4, 37 °C). **A**, native Hb stripped from BPG; **B**, whole blood; **C**, pyridoxylated, glutaraldehyde-polymerized Hb; **E**, plasma. Area **D** comprises the range of commonly employed perfluorocarbons (compounds forming fine and stable emulsions with water). The vertical lines indicate venous (40 Torr) and arterial (105 Torr) O_2 pressures. It should be noted that the release of O_2 by a solution of stripped, native Hb is very small, whereas for polymerized pyridoxylated Hb it is approximately half that of blood.

pH 7.4, $p_{1/2}$ would decrease from the physiologically desirable level of about 26 Torr to about 10 Torr, which is too low for effective delivery (see Fig. 11).
- Unprocessed solutions of Hb are not suitable as O_2 carriers because the isosmotic concentration of Hb dissolved in plasma (70 g dm^{-3}) is about half that of normal blood (150 g dm^{-3}).

To overcome these drawbacks, several covalent modifications of Hb have been prepared to produce hemoglobin-based oxygen carriers (HBOCs), with the idea of capitalizing on their potential advantages, as the Hb solutions can be used immediately without the need for blood group typing, they can be stored in a ready-for-use formulation for long periods of time, and they can be heat-treated, thus eliminating the risk of viral transmission. Consequently, a first generation of HBOCs was designed to prevent dissociation into dimers and, therefore, to prolong the half-life of Hb within the circulation. The molecular mass of chemically engineered O_2 carriers has been increased by: (i) intra- and inter-tetramer crosslinking with glutaraldehyde or glycolaldehyde; (ii) conjugation to dextran and polyethylene glycol (PEG) derivatives; and (iii) protein engineering by fusion of the two α- and β-subunits.

Moreover, to decrease the O_2 affinity of Hb – and therefore to improve unloading to the tissues – covalent attachment to the α-subunits of pyridoxal 5′-phosphate (which is known to mimic the functional effect of BPG) or specific chemical changes (e.g., Asn G10(108)$\beta \rightarrow$ Lys) obtained by site-directed mutagenesis, were realized. Other compounds, such as bis(3,5-dibromosalicyl)fumarate and *nor*-2-formylpyridoxal-5′-phosphate, reduce the O_2 affinity while simultaneously stabilizing the tetramer.

Large-scale polymerization (such as that obtained by glutaraldehyde treatment) not only prolonged the vascular retention of the Hb molecules but also lowered the osmotic activity by decreasing the number of particles per unit volume. In particular, the isosmotic concentration in plasma was increased from $70\,g\,dm^{-3}$ (of free Hb) up to $150\,g\,dm^{-3}$ (i.e., the Hb level in the blood of healthy subjects). The latter approach showed some disadvantages, however, as the solution viscosity was much less than that of blood. Under chosen conditions, glutaraldehyde-treated Hb forms polymers containing two to eight tetramers, having an isosmotic concentration of $100\,g\,dm^{-3}$ and a viscosity lower than that of blood; the vascular retention of this compound in rabbits was about 20 h, which was sufficient for several clinical applications. In fact, when the Hb derivative is used solely for the purpose of O_2 transport, as in the prevention of cardiac ischemia (e.g., during balloon angioplasty), a relatively short half-life (5–10 h) may actually be useful. The identification of alternative sources of Hb has been an additional (though not secondary) problem as supplies of the human protein are limited. Hence, the use of Hb from other mammals (e.g., ox, pig, mouse) has been considered as the Hb molecule is scarcely antigenic. Unfortunately, the possibility of anaphylactic reactions after repeated applications is not unlikely, even though the antigenic response in humans can be minimized by a modification of Hb with PEG.

Severely adverse outcomes were noted in clinical trials with first-generation HBOCs which, as a consequence, were dropped in favor of developing second-generation products. In fact, outside the erythrocyte Hb appears to be inherently toxic, being able to generate highly reactive O_2 species (mainly, superoxide and peroxide) and to interfere with the modulation of blood pressure. Thus, met Hb levels of polymerized Hb can increase in 24 h from an initial 3% to a final 40%. Harmful reactive O_2 species, which are byproducts of Hb autoxidation, are known to produce cellular damage and constitute an important concern in the use of modified Hb as reperfusion agents. Moreover, the reaction of met Hb with H_2O_2, resulting from auto-oxidation, is known to proceed via the formation of the highly reactive ferryl Hb (i.e., HbFe(IV)), a short-lived chemical species that peroxidizes lipids, degrades carbohydrates, and modifies proteins as well as nucleic acids. Moreover, extracellular oxy Hb – in contrast to that encapsulated in erythrocytes – is highly vasoconstrictive, and the resultant hypertension is a deleterious side effect of most HBOCs. Thus, some of the unique toxicological effects associated with the use of HBOCs in the clinic can be understood.

Strategies aimed at solving the above-reported toxicity were conceived to produce second-generation Hb prototypes of blood substitutes capable of safely carrying O_2. Thus, intra-erythrocyte antioxidant enzymes (e.g., superoxide dismutase and catalase) or water-soluble vitamin E analogs have been crosslinked to Hb in an attempt to control Hb oxidative side reactions, thereby obtaining an amelioration of free-radical-mediated injury to cellular systems. Moreover, the active sites of Hb subunits have been engineered (e.g., LeuB10 was replaced by large aromatic amino acids) in order to reduce the rate of NO binding (by ≥30-fold), and consequently the NO-scavenging activity, while retaining an efficient O_2 transport. At present, these second-generation HBOCs are only used to keep explanted organs that are waiting to be transplanted alive and well-oxygenated [44–46].

10 Evolutionary Considerations

The essential features of the globin fold (i.e., the overall 3D structure of the protein, with its ensemble of α-helical and nonhelical stretches; see Fig. 1) from very diverse organisms – ranging from mammals to insects to plant root nodules to bacteria – are highly conserved, even when the amino acid sequences have diverged until similarity is virtually undetectable. This finding suggested to M.F. Perutz that the spontaneous folding of the globin chain implies stereochemical similarities in sequence, which satisfy the overall structural configuration in spite of large differences in amino acid composition [31]. This intuition led to the hypothesis that the globin family has evolved by divergence from a very remote common ancestral gene (see Fig. 12), as suggested by the discovery of Hbs in virtually all kingdoms of organisms.

Evolutionary studies have led to a plausible organization of the globin's tree into three main families, based on the topology of the helices and (roughly) on their function: (i) the 3/3 globin family, including Hb, Mb, and bacterial flavoHb; (ii) the 3/3 family of oxygen sensors and protoglobins, found only in prokaryotes and Archea; and (iii) the 2/2 truncated Hb found in bacteria, green algae, plants, protozoans, and Archea [47]. Various estimates place the time of duplication of the globin genes from their single 3/3 ancestor approximately 800 million years ago (see Fig. 12). As bacteria do

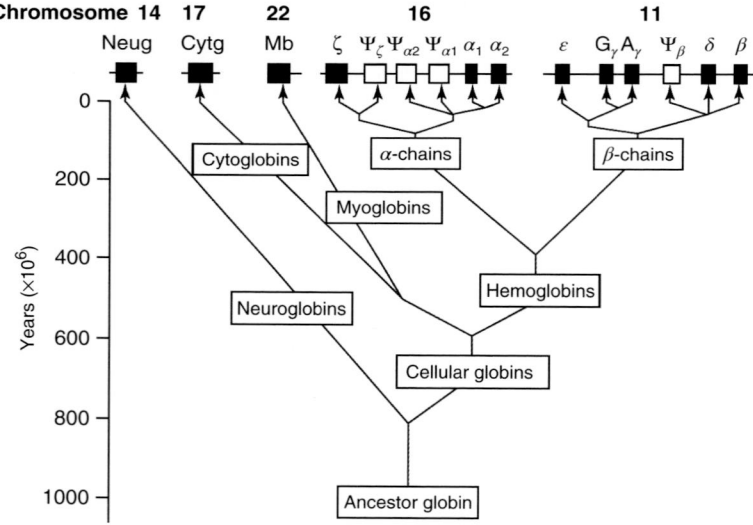

Fig. 12 Possible evolution of vertebrate globin genes, as deduced from DNA and amino acid sequence differences. The arrangement of the globin genes and pseudogenes (denoted by the prefix ψ) in humans is shown at the top, together with the indication of the chromosome where the gene is located. Estimated times of divergence in millions of years are only approximate. The high similarity between two α-genes (α_1 and α_2) and between genes Gγ and Aγ indicates that their duplication must have occurred quite recently in evolutionary time. The Gγ and Aγ genes produce chains with glycine or alanine, respectively, at position H14(136) (see also Table 1) [2, 3, 9, 18, 40, 47, 48].

not need O_2 transport, and the primordial conditions pointed towards an oxygen-free atmosphere, the initial function of these ancestors was most likely catalytic (e.g., radical scavenging). All three classes share the following structural features: the heme binding site between helices E and F; the helical topology (3/3 or 2/2); and the invariant position of the proximal His that coordinates the heme iron [2, 3, 9, 47, 48].

Although the resolution of the phylogenetic tree is poor, evolutionary analysis suggests that neuroglobins do not group with vertebrate Hbs and Mbs (which are thought to have diverged 500–600 million years ago) and must be older.

The very early ancestor to vertebrates contained a single ancestral gene. On the basis of the antiquity of neuroglobin, it has been proposed that the last common ancestor to all vertebrates most likely possessed two globin loci [48]. Thus, duplication of the ancestral locus resulted in two loci, developing into two different types of globin: neuroglobin and cellular globin (see Fig. 12). A further duplication of cellular globin locus allowed development of the Mb and cytoglobin loci on the one hand, and of the Hb locus on the other hand. The Hbs obtained their function in the circulatory system of the jawed vertebrates (fish, amphibians, and reptiles) after their divergence from the lineage leading to the Mb and cytoglobins, an event which was most likely related to an increase in body size and to the development of an efficient circulatory apparatus. Myoglobins and cytoglobins separated later. In the Hb locus, gene duplication gave rise to a cluster encoding several monomeric Hbs, as found in agnathans.

This allowed the specialization of individual genes in α-type or β-type Hbs that are found closely linked in the same locus (as in poikilothermic jawed vertebrates).

It is thought that a common ancestral globin gene gave rise to distinct α- and β-globin genes through duplication events, followed by their separation into different chromosomes, as they are in homeothermic vertebrates (birds and mammals).

The original α-gene in turn subdivided into α and ζ somewhere in the reptilian line; on the other hand, the original β-gene differentiated into adult β and fetal γ during the rise of mammals.

The original γ-gene itself later produced both γ and ε, whereas primates generated a δ-gene from the β-gene. During the long evolution of globin, only two positions have remained invariant in all Hbs: the proximal histidine F8 and phenylalanine CD1, which wedges the heme into its pocket (see Fig. 2). This means that, in order to ensure the satisfaction of similar stereochemical requirements, the vast majority of amino acid replacements have been conservative – that is, at a given position in the sequence a residue has been replaced by many others, provided these are of the same general class. As these mutations do not appear to be correlated with any fundamental change in the functional properties of Hb, they are called "neutral" and simply reflect an accumulation of random nucleotide substitutions that occur at low frequency during DNA replication. It is expected, of course, that highly deleterious mutations will have been weeded out by natural selection.

The analysis of available vertebrate sequences shows that evolutionary divergence has been constrained primarily by an almost absolute conservation of the hydrophobicity of the residues buried in the helix-to-helix and helix-to-heme contacts. The survival of mutant proteins in the globin family has been restricted to those that maintain the basic globin fold, tolerating variations not only in residues that occupy surface positions but also in those buried in the molecule. Changes in surface residues are sometimes reflected in radically different quaternary structures with novel subunit contacts, typical of some invertebrate Hbs, which maintain reversible oxygen binding and cooperativity.

As for other proteins, the more distant two globins are in the phylogenetic tree, the more dissimilar the amino acid sequences of shared proteins will be. Thus, the number of common residues among vertebrate Hbs (based on the amino acid sequence of 32 species) is relatively high: 27 amino acids in the α-subunits and 18 amino acids in the β-subunits have remained invariant. Most of these are located at sequence positions responsible for efficient oxygen delivery: 14 are heme contacts, 16 are intersubunit contacts essential for allostery, eight are involved in hydrogen bonds or salt bridges within the subunits or at the $\alpha_1\beta_1$ interface, six are involved in internal nonpolar contacts, and a single proline serves to turn the BC corner in the β-subunits.

It has been estimated that changes in the Hb genes accumulated at an average rate of one residue per α- or β-subunit every two to three million years. This notion is in line with the general ideas of evolution, and illustrated by the observation that, in the β-chains, there are eight differences between man (*Homo sapiens*) and rhesus monkey (*Macaca* spp.), 24 differences between man and cow, 45 differences between man and chicken, and 91 differences between man and shark. This could be explained by the neutralist theory, according to which most evolutionary changes at the molecular level and most of the variability within a species are caused not by selection but by a random drift of mutant genes that are selectively equivalent. In other words, evolutionary rates are essentially determined by the structure and function of the molecules and not just by environmental conditions or population size. The approximate constancy of the evolutionary rate for globin genes is supported by a comparison of the divergence between human α- and β-chains and the divergence between the α-chain of carp and the β-chain of man (see Table 6). The two α-chains and the human β-chain differ to roughly the same extent. As the sequence of the α-chain of human differs from that of carp in about half of the positions, the α-subunits of the two distinct lineages (one leading to carp and the other to human) appear to have accumulated mutations independently and at approximately the same rate over a span of 500 million years (i.e., at the time gene duplication between α- and β-chains arose). Although the biological-clock hypothesis proved to be a useful tool for the quantitative investigation of the time evolution of biological processes at the level of proteins, the clock itself does not appear to be as regular as a simple Poisson process. For example, the clock seems to have changed its rate during evolution while (or when) a tetrameric Hb was first assembled, starting from an ancestral monomeric globin. In fact, an analysis of the reconstructed genealogical tree of globin chains shows an acceleration−deceleration pattern of molecular evolution between ancestor and descendants, evidence favoring positive selection rather than neutral fixation.

According to R. E. Dickerson and I. Geis [2], in the evolution of globin there seem to have been periods when sequence changes were predominantly a consequence of natural selection; on the other hand, after satisfactory selection of a globin suited to its physiological role, the rate of accumulation of sequence changes appears to have settled down to a nearly constant value; a

Tab. 6 Number of differences between the amino acid sequences of human α- and β-chains compared with the number of differences between the sequences of α-chain from carp and human β-chain.

Type of change in the genetic code	Human α- versus human β-chain	Carp α- versus human β-chain
No change	62	61
One nucleotide	55	49
Two nucleotides	21	29
Addition or deletion of a residue	9	10
Total	147	149

behavior that agrees with random fixation of near-neutral mutations.

In conclusion, both positive selection and neutral drift seem to be the relevant driving forces directing changes at different time spans during the evolution of globin (see [48] and references therein).

References

1. Antonini, E., Brunori, M. (1971) Hemoglobin and Myoglobin in their Reactions with Ligands, North Holland Publishing Company, Amsterdam, The Netherlands.
2. Dickerson, R.E., Geis, I. (1983) Hemoglobin: Structure, Function, Evolution and Pathology, Benjamin/Cummings, Menlo Park, CA.
3. Everse, J., Vandegriff, K.D., Winslow, R.M. (Eds) (1994) Methods in Enzymology: Hemoglobins, Vols **230, 231, and 232**, Academic Press, New York.
4. Rossi-Fanelli, A., Antonini, E., Caputo, A. (1964) Hemoglobin and myoglobin. *Adv. Protein Chem.*, **19**, 73–222.
5. Fermi, G., Perutz, M.F. (1981) Atlas of Molecular Structures in Biology: Hemoglobin and Myoglobin, Clarendon Press, Oxford.
6. Park, S.Y., Yokoyama, T., Shibayama, N., Shiro, Y., Tame, J.R. (2006) 1.25 Å resolution crystal structures of human haemoglobin in the oxy, deoxy and carbonmonoxy forms. *J. Mol. Biol.*, **360**, 690–701.
7. Yuan, Y., Simplaceanu, V., Ho, N.T., Ho, C. (2010) An investigation of the distal histidyl hydrogen bonds in oxyhemoglobin: effects of temperature, pH, and inositol hexaphosphate. *Biochemistry*, **249**, 10606–10615.
8. Bunn, H.F., Forget, B.G. (1986) Hemoglobin: Molecular, Genetic and Clinical Aspects. W.B. Saunders, Philadelphia, PA.
9. Burmester, T., Weich, B., Reinhardt, S., Hankeln, T. (2000) A vertebrate globin expressed in the brain. *Nature*, **478**, 461–462.
10. Antonini, E., Rossi Bernardi, L., Chiancone, E., (Eds) (1981) Methods in Enzymology: Hemoglobins, Vol. **76**, Academic Press, New York.
11. Pauling, L. (1935) The oxygen equilibrium of hemoglobin and its structural interpretation. *Proc. Natl Acad. Sci. USA*, **21**, 186–191.
12. Eaton, W.A., Henry, E.R., Hofrichter, J., Mozzarelli, A. (1999) Is cooperative oxygen binding by hemoglobin really understood? *Nat. Struct. Biol.*, **6**, 351–358.
13. Wyman, J., Gill, S.L. (1990) Binding and Linkage: Functional Chemistry of Biological Macromolecules, University Science Books, Mill Valley, CA.
14. Imai, K. (1982) Allosteric Effects in Hemoglobin, Cambridge University Press, Cambridge.
15. Bellelli, A. (2010) Hemoglobin and cooperativity: experiments and theories. *Curr. Protein Pept. Sci.*, **11**, 2–36.
16. Birukou, I., Schweers, R.L., Olson, J.S. (2010) Distal histidine stabilizes bound O_2 and acts as a gate for ligand entry in both subunits of adult human hemoglobin. *J. Biol. Chem.*, **285**, 8840–8854.
17. Riggs, A.F. (1998) Self-association, cooperativity and supercooperativity of oxygen binding to hemoglobins. *J. Exp. Biol.*, **201**, 1073–1084.
18. Royer, W.E., Jr, Knapp, J.E., Strand, K., Heaslet, H.A. (2001) Cooperative hemoglobins: conserved fold, diverse quaternary assemblies and allosteric mechanisms. *Trends Biochem. Sci.*, **26**, 297–304.
19. Gibson, Q.H. (1989) Hemoproteins, ligands and quanta. *J. Biol. Chem.*, **264**, 20155–20158.
20. Schotte, F., Lim, M., Jackson, T.A., Smirnov, A.V., Soman, J., Olson, J.S., Phillips, G.N. Jr, Wulff, M., Anfinrud, P.A. (2003) Watching a protein as it functions with 150-ps time

resolved X-ray crystallography. *Science*, **300**, 1944–1947.
21. Brunori, M., Bourgeois, D., Vallone, B. (2008) Structural dynamics of myoglobin. *Methods Enzymol.*, **437**, 397–416.
22. Wyman, J. (1964) Linked functions and reciprocal effects in hemoglobin: a second look. *Adv. Protein Chem.*, **19**, 233–286.
23. Monod, J., Wyman, J., Changeux, J.-P. (1965) On the nature of allosteric transitions: a plausible model. *J. Mol. Biol.*, **12**, 88–118.
24. Brunori, M. (2014) Variations on the theme: allosteric control in haemoglobin. *FEBS J.*, **281**, 633–643.
25. Koshland, D.E. Jr, Nemethy, G., Filmer, D. (1966) Comparison of experimental binding data and theoretical models in proteins containing subunits. *Biochemistry*, **5**, 365–385.
26. Perutz, M.F. (1970) Stereochemistry of cooperative effects in haemoglobin. *Nature*, **228**, 726–738.
27. Perutz, M.F. (1990) Mechanisms regulating the reactions of human hemoglobin with oxygen and carbon monoxide. *Annu. Rev. Physiol.*, **52**, 1–25.
28. Shulman, R.G., Hopfield, J.J., Ogawa, S. (1975) Allosteric interpretation of haemoglobin properties. *Q. Rev. Biophys.*, **8**, 325–420.
29. J. Baldwin and C. Chothia. Haemoglobin: the structural changes related to ligand binding and its allosteric mechanism. *J. Mol. Biol.*, 1979 **129**(2), 175–220
30. Yonetani, T., Kanaori, K. (2013) How does hemoglobin generate such diverse functionality of physiological relevance? *Biochim. Biophys. Acta*, **1834**(9), 1873–1884.
31. Perutz, M.F. (1997) Science is Not a Quiet Life – Unravelling the Atomic Mechanism of Hemoglobin, World Scientific Publishing Co. Pte Ltd. Singapore.
32. Szabo, A., Karplus, M. (1972) A mathematical model for structure-function relations in haemoglobin. *J. Mol. Biol.*, **72**, 163–197.
33. Samuni, U., Friedman, J.M. (2005) Proteins in motion: resonance Raman spectroscopy as a probe of functional intermediates. *Methods Mol. Biol.*, **305**, 287–300.
34. Cui, Q., Karplus, M. (2008) Allostery and cooperativity revisited. *Protein Sci.*, **17**, 1295–1307.
35. Frauenfelder, H., Sligar, S.G., Wolynes, P.G. (1991) The energy landscapes and motions of proteins. *Science*, **254**, 1598–1603.
36. Brunori, M. (2001) Nitric oxide, cytochrome-c oxidase and myoglobin. *Trends Biochem. Sci.*, **26**, 21–23.
37. Gladwin, M.T., Lancaster, J.R., Freeman, B.A., Schechter, A.N. (2003) Nitric oxide's reactions with hemoglobin: a view through SNO-storm. *Nat. Med.*, **9**, 496–500.
38. Jia, L., Bonaventura, C., Bonaventura, J., Stamler, J.S. (1996) S-nitrosohaemoglobin: a dynamic activity of blood involved in vascular control. *Nature*, **21**, 221–226.
39. Forget, B.G., Higgs, D.R., Steinberg, M., Nagel, R.L. (2001) Disorders of Hemoglobin: Genetics, Pathophysiology and Clinical Managements, Cambridge University Press, Cambridge.
40. Forget, B.G., Bunn, H.F. (2013) Classification of the disorders of haemoglobin. *Cold Spring Harbor Perspect. Med.*, **3**, a011684.
41. Edelstein, S.J. (1986) The Sickled Cell: From Myths to Molecules. Harvard University Press.
42. Eaton, W.A., Hofrichter, J. (1990) Sickle cell hemoglobin polymerization. *Adv. Protein Chem.*, **40**, 263–279.
43. Martin, A., Thompson, A.A. (2013) Thalassemias. *Pediatr. Clin. North Am.*, **60**, 1383–1391.
44. Winslow, R.M. (2008) Cell-free oxygen carriers: scientific foundations, clinical development, and new directions. *Biochim. Biophys. Acta*, **1784**, 1382–1386.
45. Varnado, C.L., Mollan, T.L., Birukou, I., Smith, B.J., Henderson, D.P., Olson, J.S. (2013) Development of recombinant hemoglobin-based oxygen carriers. *Antioxid. Redox Signal.*, **18**, 2314–2328.
46. Alayash, A.I. (2014) Blood substitutes: why haven't we been more successful? *Trends Biotechnol.*, **32**, 177–185.
47. Wittenberg, J.B., Bolognesi, M., Wittenberg, B.A., Guertin, M. (2002) Truncated hemoglobins: a new family of hemoglobins widely distributed in bacteria, unicellular eukaryotes, and plants. *J. Biol. Chem.*, **277**, 871–874.
48. Burmester, T., Hankeln, T. (2014) Function and evolution of vertebrate globins. *Acta Physiol. (Oxf)*, **211**, 501–514.

6
Immune Checkpoint Inhibitors

Laura Mansi[1,2], Franck Pagès[3], and Olivier Adotévi[1,2]
[1] *University of Franche-Comte, INSERM Unit 1098, 30 Bd Fleming, 25030 Besançon Cedex, France*
[2] *University Hospital of Besançon, Department of Medical Oncology, 30 Bd Fleming, 25030 Besançon Cedex, France*
[3] *AP-HP, Paris-Descartes University, Laboratory of Immunology, Immunomonitoring Platform of the Georges Pompidou European Hospital, 20 rue Leblanc, 75908 Paris, France*

1	**Introduction** 201	
1.1	Regulation of T-Cell Response: Role of the Costimulatory and Inhibitory Receptor 201	
2	**CTLA-4 Blockade** 202	
2.1	Biology of Receptor Ligands 202	
2.2	Anti-CTLA-4 Antibodies from Mice to Clinics 204	
2.3	Immunological Explanation of the Kinetic and Durable Objective Clinical Responses after Anti-CTLA-4 206	
2.4	Anti-CTLA-4 Antibodies Efficacy and Fc-Dependent Depletion of Tumor-Infiltrating Regulatory T Cells 208	
3	**Targeting the PD-1 PD-L1 Pathway** 208	
3.1	Receptor Ligand Biology 208	
3.1.1	PD-1 Expression and Functions on Immune Cells 208	
3.1.2	PD-1 Ligands 210	
3.1.3	PD-1/PDL-1 Regulation in Cancers: Innate versus Adaptive Immune Resistance 210	
3.2	Clinical Implication PD-1/PD-L1 Blockade 212	
3.2.1	PD-1 Inhibitors 212	
3.2.2	PD-L1 Inhibitors 213	
4	**Others Checkpoint Inhibitors** 214	
4.1	TIM-3 214	
4.2	LAG3, BTLA 215	

Translational Medicine: Molecular Pharmacology and Drug Discovery
First Edition. Edited by Robert A. Meyers.
© 2018 Wiley-VCH Verlag GmbH & Co. KGaA. Published 2018 by Wiley-VCH Verlag GmbH & Co. KGaA.

5	Safety Aspects 216
6	Future Direction: Biomarkers and Combination 216
6.1	Biomarkers 216
6.2	Combination 218
7	Conclusion 220
	References 220

Keywords

T cell
A type of white blood cell that is of key importance in the immune system. A core component of the adaptive immune system.

Cancer
Cancer or malignant tumor defines a group of diseases involving abnormal proliferative cells with the capacity for migration, neoangiogenesis, and no sensitivity to apoptotic signals.

CTLA-4
Cytotoxic T-lymphocyte-associated protein 4 (also known as cluster of differentiation 152; CD152) is a protein receptor that downregulates the immune system.

Immunotherapy
Treatment that uses certain features of a person's immune system to fight diseases such as cancer.

PD-1/PD-L1
The programmed death 1 (PD-1 or CD279) pathway is a key immune checkpoint. Expression of PD-1 on the T-cell surface exerts an immunoregulatory role in the activation of T cells. PD-ligand 1 (PD-L1) is a transmembrane protein that binds to its receptor, PD-1.

> Emerging clinical data suggest that cancer immunotherapy is likely to become a key part of the clinical management of cancer. Knowledge of the basic mechanisms of the immune system as they relate to cancer has been increasing rapidly, and such developments have accelerated the translation of novel immunotherapy techniques into medical breakthroughs for many cancer patients. The development of checkpoint-blocking antibodies, against cytotoxic T-lymphocyte antigen 4 (CTLA-4) and the programmed death 1 receptor (PD-1), has recently demonstrated significant promise in the treatment of malignancies. Ipilimumab (to block CTLA-4) and pembrolizumab (to block PD-1) have been approved by the US Food and Drug Administration for the treatment of advanced melanoma, and additional regulatory approvals are expected for a variety

of other agents such as nivolumab (to block PD-1). To combine checkpoint inhibitors approaches with other therapies such as immunomodulators (cytokines, indoleamine 2,3-dioxygenase inhibitors), cytotoxic chemotherapy, radiation therapy or molecularly targeted therapies, may hold the key to the true potential of immunotherapy in the future management of cancer patients.

1
Introduction

Emerging clinical data suggest that cancer immunotherapy is likely to become a key part of the clinical management of cancer. Knowledge of the basic mechanisms of the immune system as they relate to cancer has been increasing rapidly, with recent developments having accelerated the translation of novel immunotherapy techniques into medical breakthroughs for many cancer patients.

Genetic and cellular alterations in tumors enable the immune system to generate T-cell responses that recognize and eradicate cancer cells [1]. The destruction of cancer by T cells is only one step of the cancer immunity cycle, which manages the delicate balance between the recognition of nonself and the prevention of autoimmunity [2]. However, tumors can escape from an equilibrium state previously held in check by the immune system to become significant, as suggested by the theory of immunoediting. Tumors may utilize a variety of mechanisms to evade the immune system, in particular against T-cell attack [1]. Thus, the identification of cancer cell/T-cell inhibitory signals such as cytotoxic T-lymphocyte antigen 4 (CTLA-4) and programmed death 1 (PD-1)/PD-L1, has prompted the development of a new class of immunotherapy.

Preexisting anticancer immune responses can be potentially reinvigorated and possibly expanded by immunotherapy, as the latter specifically hinders immune effector inhibition. A wide variety of approaches exist to elicit an antitumor immune response, involving therapeutic cancer vaccines, adoptive T-cell therapy, antitumor antibodies, and the emerging immune checkpoint blockade.

Moreover, the presence of suppressive factors in the tumor microenvironment (TME) may explain the limited activity observed with previous immune-based therapies. To combine these approaches with other therapies such as immunomodulators (cytokines, indoleamine 2,3-dioxygenase (IDO) inhibitors), cytotoxic chemotherapy, radiation therapy or molecularly targeted therapies, may hold the key to the true potential of immunotherapy in the future management of cancer patients [2, 3].

1.1
Regulation of T-Cell Response: Role of the Costimulatory and Inhibitory Receptor

The adaptive immune response is tightly regulated by multiple costimulatory and coinhibitory pathways. Naïve T cells are activated when the T-cell receptor (TCR) binds to the major histocompatibility complex (MHC)–antigen complex (signal 1) but, unfortunately, signal 1 is insufficient to generate and maintain an adaptive immune response. Rather, the full activation of a T cell also requires the simultaneous engagement of positive costimulatory molecules present on activated antigen-presenting cells (APCs); this is known

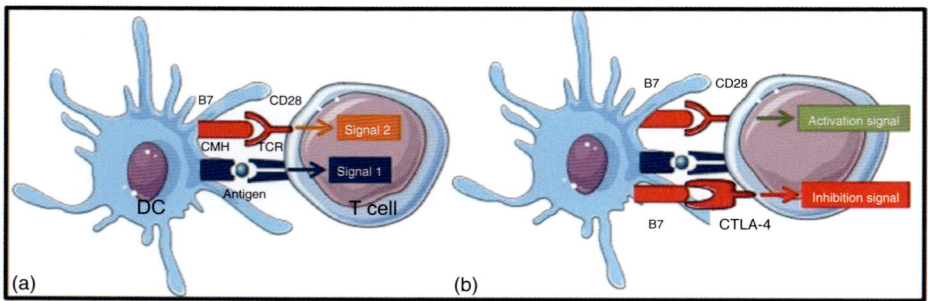

Fig. 1 Immune checkpoints regulate the evolution of immune response. (a) Stimulation of the TCR by MHC/peptide on dendritic cell (DC) complexes delivers signal 1, while interactions between costimulatory ligands on the antigen-presenting cell (APC) and CD28 on the T cell provide signal 2; (b) Activation signal is initiated by binding of B7 molecules on the APC (DC) to cluster of differentiation 28 (CD28) receptors on the T cell. Cytotoxic T-lymphocyte antigen 4 (CTLA-4), expressed on the T cell, binds B7 molecules, resulting in T-cell inactivation. Adapted from Ref. [4].

as signal 2 (Fig. 1). These costimulatory molecules are not present on quiescent APCs, tumor cells or normal host cells. A classic example of a costimulatory pathway is the interaction between B7 expressed on an APC and CD28 expressed on T cells [1] (Figs 1 and 2). When both the MHC–antigen complex and the TCR (signal 1), and the costimulatory pathway (signal 2) are activated, T cells proliferate and are activated. As both costimulatory and coinhibitory molecules may be present at the same time, signal 2 is perhaps more appropriately conceptualized as the sum of both costimulatory signals and coinhibitory signals that determine the T-cell phenotype. After activation, T cells express coinhibitory receptors such as CTLA-4, PD-1, TIM-3 and LAG-3; these compensatory coinhibitors attenuate the immune response and are often co-opted by tumors to evade the host's natural antitumor immune response. Tumor cells and macrophages within the tumor environment can also express coinhibitory receptors and promote immune tolerance (Fig. 3).

2 CTLA-4 Blockade

2.1 Biology of Receptor Ligands

CTLA-4 (CD152) was the first molecule to be identified as being coinhibitory, and is the counterpart of the costimulatory CD28 axis. It is mainly expressed on T cells, where it primarily regulates the amplitude of the early stages of T-cell activation by dendritic cells and other APCs [6–8].

CD28 and CTLA-4 share identical ligands, namely CD80 (also known as B7.1) and CD86 (also known as B7.2) [9–11]. Following activation, the T cells upregulate the surface expression of CTLA-4 that binds B7 molecules with a higher avidity, and thus counteracts CD28's positive costimulatory signal (Fig. 4). This dominance of negative signals results in a reduced T-cell proliferation and decreased interleukin 2 (IL-2) production [1]. Thus, the CTLA-4 inhibitory pathway can modulate an immune response and also avert autoimmunity [12].

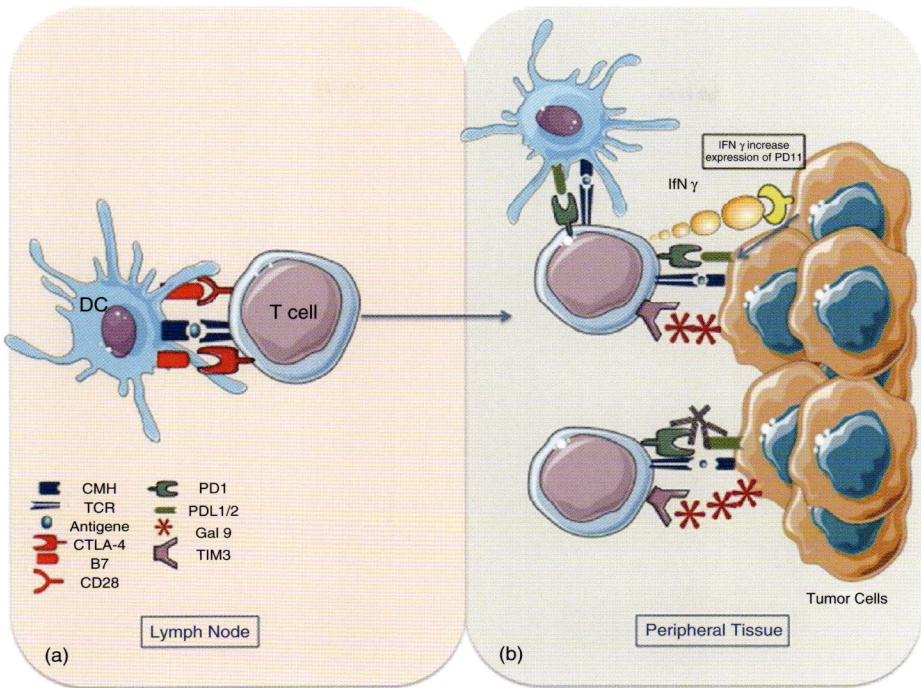

Fig. 2 Targets of antibody immune modulators. (a) The interaction of T-cell receptor (TCR) with a major histocompatibility complex (MHC) molecule expressed by antigen-presenting cells (DC). The interaction of the CD28 receptor on T cell with B7 costimulatory molecules (B7-1 and B7-2) on the DC is necessary to complete T-cell activation. This phase occurs primarily within the lymph nodes. To prevent inappropriate T-cell activation, negative regulators of T-cell immunity, including CTLA-4 and PD-1, are required. CTLA-4 competes with CD28 for the interaction with B7, and it is upregulated after T-cell activation; (b) In the peripheral tissues other negative regulators participate in T-cell inactivation. Moreover, two general mechanisms of expression of immune-checkpoint ligands on tumor cells are identified. (i) Innate immune resistance: Constitutive oncogenic signaling can upregulate PDL1 expression on all tumor cells, independently of inflammatory signals in the tumor microenvironment; (ii) Adaptive immune resistance: PDL1 is induced in response to inflammatory signals that are produced by an active antitumor immune response. For example, the secretion of interferon-γ by activated T cells increases the expression of PDL1 on tumor cells. Adapted from Ref. [4].

The specific signaling pathways by which CTLA-4 blocks T-cell activation are still under investigation, although it has been suggested in a number of studies that activation of the protein phosphatases, the SH2-domain containing tyrosine phosphatase (SHP2; also known as PTPN11) and PP2A, are important in counteracting kinase signals that are induced by TCRs and CD28 [7]. However, CTLA-4 also confers T-cell inhibition through the sequestration of CD80 and CD86 from CD28 engagement, as well as an active removal of CD80 and CD86 from the APC surface [13]. The central role of CTLA-4 for keeping T-cell activation in check is demonstrated by the lethal

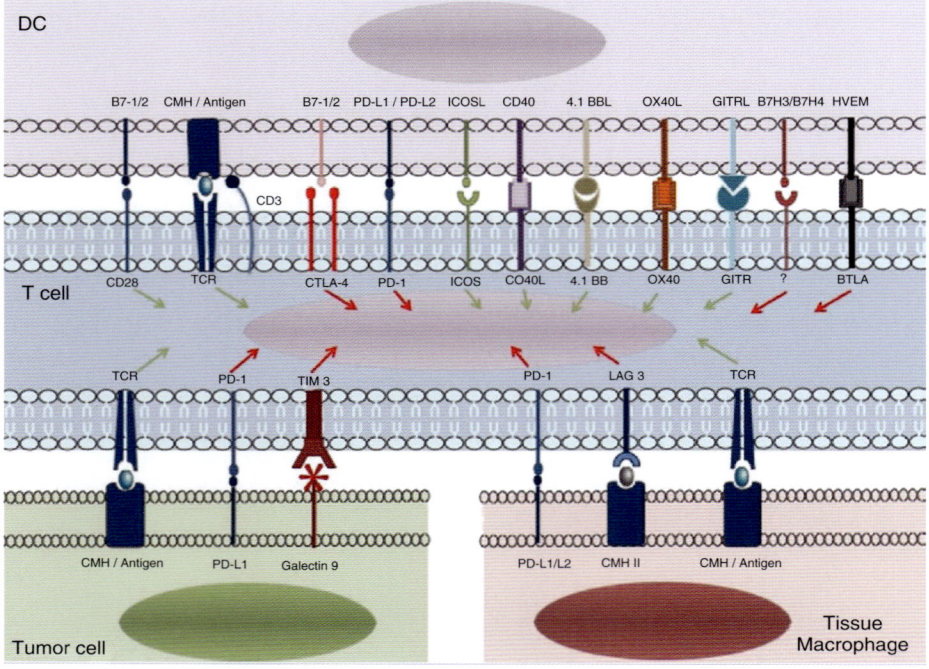

Fig. 3 Multiple costimulatory and coinhibitory ligand–receptor interaction between T cell and dendritic cell (DC), tumor cell, and tissue macrophage in the tumor microenvironment. CTLA-4: cytotoxic T-lymphocyte-associated antigen 4; GAL9, galectin 9; HVEM, herpesvirus entry mediator; ICOS, inducible T cell costimulator; IL, interleukin; LAG3, lymphocyte activation gene 3; PD1, programmed cell death protein 1; PDL, PD1 ligand; TCR, T-cell receptor; TIM3, T-cell membrane protein 3. Adapted from Refs [4, 5].

systemic immune hyperactivation phenotype of *CTLA-4*-knockout mice [14–16].

CTLA-4 is also expressed by regulatory T cells (T-reg) and memory CD4 T cells. These cells may also be targeted by CTLA-4 blockade, and both the enhancement of effector CD4 T cell activity and the inhibition of cell-dependent immunosuppression are probably important factors [17, 18].

2.2
Anti-CTLA-4 Antibodies from Mice to Clinics

The blocking CTLA-4 was initially questioned, as expression of the CTLA-4 ligands is not tumor-specific and the dramatic lethal autoimmune phenotype of CTLA-4-knockout mice predicted a high degree of immune toxicity associated with blockade of this receptor. However, the first preclinical models showed that a therapeutic window could be achieved when CTLA-4 was partially blocked with antibodies [19]. The initial studies demonstrated significant antitumor responses without overt immune toxicities when mice bearing partially immunogenic tumors were treated with CTLA-4 antibodies as single agents. Poorly immunogenic tumors did not respond to anti-CTLA-4 as a single agent, but did respond when anti-CTLA-4 was combined with a granulocyte–macrophage

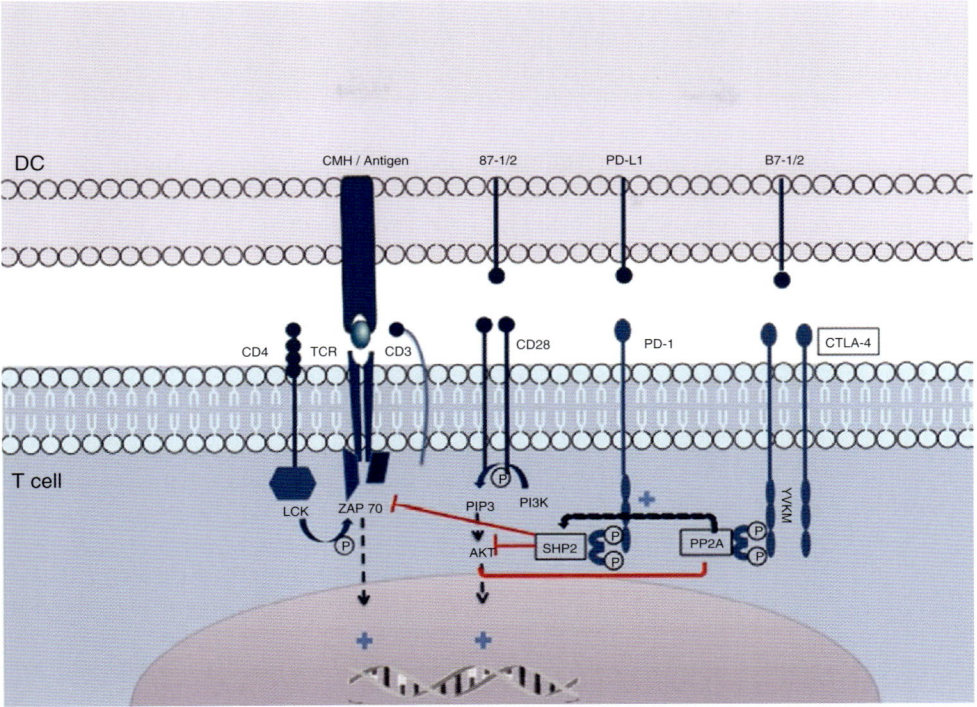

Fig. 4 Cytotoxic T-lymphocyte-associated antigen 4 (CTLA-4) pathways. CTLA-4 recruits the phosphatases SHP2 and PP2A via the YVKM motif in its cytoplasmic domain. SHP2 recruitment results in an attenuation of TCR signaling by dephosphorylating the CD3ζ chain. PP2A recruitment results in downstream dephosphorylation of AKT, further dampening the T-cell activation pathway. Adapted from Refs. [7, 11, 9].

colony-stimulating factor (GM-CSF)-transduced cellular vaccine [20]. These findings suggested that, if there is an endogenous antitumor immune response in the animals after tumor implantation, then CTLA-4 blockade could enhance that endogenous response which, ultimately, could induce tumor regression. In the case of poorly immunogenic tumors, which do not induce substantial endogenous immune responses, the combination of a vaccine and a CTLA-4 antibody could induce a strong enough immune response to slow tumor growth and, in some cases, eliminate established tumors.

On the basis of several preclinical murine models showing improved tumor control after CTLA-4 blockade [19], human monoclonal antibodies that block CTLA-4 have been developed and are currently being tested. Two such CTLA-4-targeting antibodies – ipilimumab and tremelimumab – have been tested in two Phase III clinical studies [5].

Response rates were between 11% with ipilimumab alone in previously treated advanced patients with melanoma, and 15% with ipilimumab plus dacarbazine in treatment-naïve patients [21, 22]. Nevertheless, an improved overall survival with ipilimumab was clearly demonstrated in both of these trials, with a decrease in the probability of death by 34% (compared to vaccine) and 28% (compared to dacarbazine alone). When ipilimumab was combined with dacarbazine, the median duration of best overall response was 19.3 months compared to 8.1 months with dacarbazine monotherapy. The consistent separation of the survival curves by approximately 10% until almost four years of follow-up suggests that the impact on overall survival with ipilimumab is mainly driven by the longlasting clinical benefit achieved in a small proportion of patients. Notably, tremelimumab was also shown to induce durable objective responses in patients with metastatic melanoma; in a Phase III trial, the median response duration was almost threefold longer than the response duration with chemotherapy (35.8 versus 13.7 months; P = 0.0011) [23, 24]. However, it should be stressed that tremelimumab is not yet approved for clinical application.

A long-term follow-up was recently reported on 177 patients with advanced melanoma who participated in three of the earliest ipilimumab trials conducted at the Surgery Branch of the National Cancer Institute [25–29]. In these earlier studies, the median duration of objective responses was up to 88 months (Table 1), with complete responses not being achieved on average until after 30 months of treatment; in fact, one patient received treatment for six years until a complete response was achieved. The durability of complete responses in the long-term analysis is impressive: 14 out of 15 complete responses were ongoing at data cutoff (all six complete responses in patients who were treated with concurrent ipilimumab and IL-2, and eight of nine complete responses in patients who were treated with ipilimumab and gp100 vaccine). The durability of complete responses observed even in the trials that did not include IL-2 led the authors to speculate that a cure may be achievable with this drug. The long-term clinical benefit in a subset of patients with melanoma was also confirmed in earlier Phase II ipilimumab studies, documenting flat survival curves between the four- and five-year marks [30–32].

To date, more than 100 ongoing trials with anti-CTLA-4 drugs for several cancers are recorded on the site of *www.clinicaltrials.gov*.

2.3
Immunological Explanation of the Kinetic and Durable Objective Clinical Responses after Anti-CTLA-4

An important feature of the anti-CTLA-4 clinical responses is their kinetics. Although responses to chemotherapies and tyrosine kinase inhibitors (TKIs) commonly occur within weeks of the initial administration, the response to immune-checkpoint blockers is slower and, in many patients, delayed (for up to six months after treatment initiation) [5]. Recently, at the 2013 European Cancer Congress, Hodi and colleagues presented some updated long-term results from ipilimumab on 1861 melanoma patients. An analysis of these patients showed the median overall survival to be 11.4 months, and 22% of the patients were still alive after three years. There were no deaths among patients who survived beyond seven years, at which time the overall survival rate was 17%. The longest overall survival follow-up in the database was 9.9 years. The analysis showed a plateau of 22% survival at seven years (the notion of "plateau" is a new paradigm to evaluate efficacy in immunotherapy, and its ultimate goal [33]).

Tumor-infiltrating lymphocytes (TILs), including CD8 and CD4 cells, along with

Tab. 1 Responses with CTLA-4 antibodies in clinical trials.

Drug	Disease	Trial (reference)	No. of patients	Median response duration (months)	RR(%)	CR(%)	PR(%)	SD(%)	Median OS (months)
Ipilimumab	Melanoma	Phase I							
		Ipi + gp100 [29]	56	88	13	7	6	NR	14
		Ipi + IL-2 [29]	36	79	25	17	8	NR	16
		Ipi DE ± gp100 [29]	85	42	20	6	14	NR	13
		Phase II							
		Ipi 10 mg kg^{-1} [30]	155	NR	5.8	0	5.8	21	10.2
		Ipi 10 mg kg^{-1} [31]	71	NR	11.1	2.7	8.3	18.1	11.4
		Ipi 3 mg kg^{-1} [31]	71	NR	4.2	0	4.2	22	8.7
		Ipi 10 mg kg^{-1} + budenoside [34]	58	NR	12	2	10	19	17.7
		Ipi 10 mg kg^{-1} + placebo [34]	57	NR	10	0	16	19	19.3
		Phase III							
		Ipi + dacarbazine [22]	250	19.3	15.2	1.6	13.6	18	11.2
		Ipi + gp100 [21]	403	11.5	5.7	0.2	5.5	14.4	10
		Ipi + placebo [21]	137	NR	11	1.5	9.5	17.5	10.1
	Prostate	Phase I [35]	24	NR	NR	NR	NR	12	NR
		Ipi							
		Phase I/II [36]							
		Ipi 10 mg kg^{-1} ± radiotherapy	50	NR	18	2	16	57	11.2
		Phase III [37]							
		Ipi 10 mg kg^{-1} versus placebo	799	NR	NR	NR	NR	NR	NR
	Pancreas	Phase I/II [38]	27	NR	0	0	0	1	NR
		Ipi							
Tremelimumab	Melanoma	Phase II [23]	84	NR	9.5	2.4	7.1	31	10
		Treme 10–15 mg kg^{-1}							
		Phase III							
		Treme 15 mg kg^{-1} [24]	328	35.8	11	3	8	NR	12.6
	Prostate	Phase I [39]	11	NR	NR	NR	NR	NR	NR
		Treme 15 mg kg^{-1}							
	Pancreas	Phase I [40]	34	NR	2	0	2	NR	NR
		Treme 15 mg kg^{-1}							

Ipi, ipilimumab; Treme, tremelimumab; DE, deticene; RR, response rate; CR, complete response; PR, partial response; SD, stable disease; NR, not reported; OS, overall survival.

tumor necrosis have been clearly identified in biopsies of melanoma metastases obtained after treatment with ipilimumab [41, 42]. In melanoma patients treated with ipilimumab, the NY-ESO-1 specific antibody and CD8 T responses correlated with clinical benefit [43]. However the implication of these endogenous tumor-reactive T cells remains the subject of investigation.

Immunologically, such delay likely reflects the interval during which the pre-existing, tumor-specific T cells are being activated after uncoupling from the CTLA-4-mediated inhibitory signal, and expanding and infiltrating into the tumor. The observation of an ongoing or even improving tumor response many months or years after the last dose of ipilimumab suggests that the anti-CTLA-4 blockade allows for sufficient activation and expansion of tumor-reactive T cells to control the tumor [5]. Given the multitude of immune escape mechanisms that may impede tumor responses to checkpoint blockade, it is quite remarkable that CTLA-4 blockade as monotherapy induces partial to complete tumor responses that can persist for many years. This result suggests that CTLA-4 is likely one of the predominant inhibitory drivers of pre-existing, endogenous tumor-reactive T cells in these responding patients. The observation that a subset of patients showed delays before achieving a best overall response underlies the characteristic response kinetics associated with ipilimumab. These findings require the re-evaluation of response criteria for immunotherapeutics away from the conventional time-to-progression or Response Evaluation Criteria in Solid Tumors (RECIST) objective response criteria, which were developed on the basis of experiences with chemotherapeutic agents and as the primary measure of drug efficacy [4, 44].

2.4
Anti-CTLA-4 Antibodies Efficacy and Fc-Dependent Depletion of Tumor-Infiltrating Regulatory T Cells

Murine models have demonstrated that deficient CTLA-4 T-reg have an impaired suppressive ability. Moreover, recent preclinical reports have shown that anti-CTLA-4 antibodies have a dual mechanism activity that involves blockade of the inhibitory activity of CTLA-4 on the T-cell effector and also on T-reg. The activity of the antibody on T-reg is mediated via a depletion of T-reg within tumor lesions, and this depletion is dependent on the presence of Fcγ receptor-expressing macrophages within the TME [45–47]. Although, ipilimumab is a human IgG1 antibody, which binds best to most human Fc receptors [48], T-reg depletion has not been yet demonstrated in clinical trials. However, few data are currently available on the effect of ipilimumab on intratumoral T-reg.

3
Targeting the PD-1 PD-L1 Pathway

The PD-1 (CD279) pathway is also a key immune checkpoint and is rapidly attracting a growing interest as a therapeutic target in cancer. The expression of PD-1 on the T-cell surface is a negative regulator of T-cell activity [49].

3.1
Receptor Ligand Biology

3.1.1 PD-1 Expression and Functions on Immune Cells

In contrast to CTLA-4, the major action of PD-1 is to limit autoimmunity and T-cell effector activity in the peripheral tissues

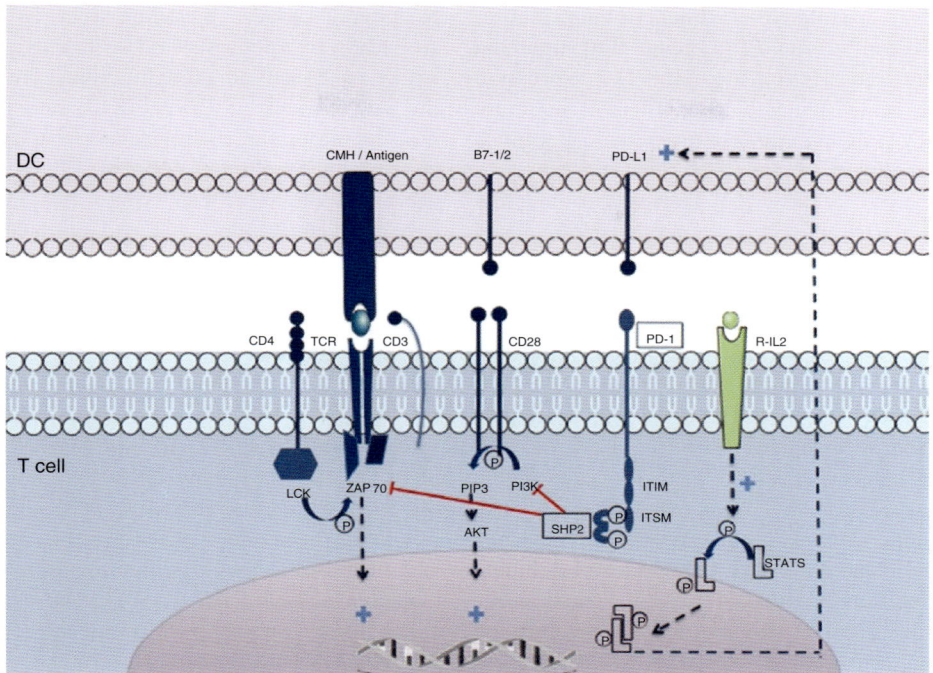

Fig. 5 Program death 1 (PD1) pathways. PD-1 contains in the cytoplasmic domain both an immune-receptor tyrosine-based inhibitory motif (ITIM) and an immune-receptor tyrosine-based switch motif (ITSM). The negative signal of PD-1 is based on the recruitment of SH2-domain containing tyrosine phosphatase (SHP2) through ITSM. SHP2 inhibits phosphatidylinositide 3-kinase (PI3K) phosphorylation, a ZAP 70 pathway that results in T-cell exhaustion. IL-2 signaling via the interleukin-2 receptor (R-IL2) induces signal transducer and activator of transcription-5 phosphorylation (pSTAT-5). Dimerization and nuclear translocation of phosphorylated STAT-5 leads to PD-1 upregulation.

at the time of inflammatory response [4, 50–55] (Fig. 2). The basis of this physiology is that the ligands for PD-1 are upregulated in response to inflammation.

PD-1 is a surface molecule of the immunoglobulin superfamily, and is expressed on activated T cells [51, 52, 56]. The interaction between PD-1 and its ligands, leads to the deactivation of signaling events under CD3-TCR and B28, resulting in a reduced T-cell activity and a decreased duration of T-cell–APC or T-cell–target contact [4, 56–58]. In the cytoplasmic domain, PD-1 contains both an immune-receptor tyrosine-based inhibitory motif (ITIM) and an immunoreceptor tyrosine-based switch motif (ITSM). The negative signal of PD-1 is based on the recruitment of SHP2 through ITSM [56, 59, 60]. T-cell exhaustion results from an inhibition of SHP2 phosphatidylinositide 3-kinase (PI3K)-mediated phosphorylation (Fig. 5) [61].

PD-1 expression is induced by the activation of T cells, and so by the interaction between MHC–antigen and TCR–CD3 [62–64]. The activation of T cells by first signal leads to a nuclear translocation of

NFAT, while NFAT2 binds to the promoter region of the *PD-1* gene, thus inducing its transcription [64, 65]. PD-1 expression is downregulated when there is no more TCR signaling. More precisely, T-bet binds upstream of the *PD-1* gene and represses its transcription [65]. During chronic infection, lymphocytes are persistently exposed to their antigens, and the continuous TCR ligation leads to a downregulation of T-bet, thus maintaining PD-1 transcription [65, 66]. In viral models, the chronicity of viral infection induces a high level of PD-1, which leads to a state of exhaustion, or anergy-specific activated T cells [4, 64].

Like CTLA-4, PD-1 is also highly expressed on T-reg cells, where it may enhance their proliferation in the presence of a ligand [67]. As many tumors are highly infiltrated with T-reg cells, which probably further suppresses effector immune responses, the blockade PD-1 pathway could also enhance antitumor response by reducing both the number and the suppressive activity of intra-tumoral T-reg cells [67].

PD-1 is also more widely expressed than CTLA-4; notably, it is expressed on other activated immune cells such as B cells, monocytes and natural killer (NK) cells which limits their lytic activity [64, 68, 69]. Hence, although PD-1 blockade is typically viewed as enhancing the activity of effector T cells in tissues and in the TME, it also probably enhances NK cell activity in tumors and tissues, and may also enhance antibody production either indirectly or through direct effects on PD-1+ B cells [70].

3.1.2 PD-1 Ligands
PD-1 interacts with its ligands, PD-L1 (B7-H1 or CD274) and PD-L2 (B7-DC or CD273), both of which are members of the B7 family [4, 71]. The interaction of PD-1/PD-L1 occurs predominantly in peripheral tissues, including the TME, and leads to apoptosis and a downregulation of T-cell effector function. Thus, the engagement of PD-1/PD-L1 interaction decreases the risk of collateral tissue damage by T cells [53, 72–74]. PD-L1 is expressed on many cell types, including hematopoietic, endothelial and epithelial, in response to proinflammatory cytokines, notably interferon (IFN) gamma, while B7-DC/PD-L2 is restricted largely to APCs on dendritic cells and macrophages in response to different proinflammatory cytokines [75, 76]. Recently, a molecular interaction between PD-L1 and CD80 was identified [71], whereby CD80 expressed on T cells (and possibly on APCs) could potentially behave as a receptor rather than as a ligand by delivering inhibitory signals when engaged by PD-L1 [77]. The relevance of this interaction in tumor immune resistance has not yet been determined.

3.1.3 PD-1/PDL-1 Regulation in Cancers: Innate versus Adaptive Immune Resistance
Much like chronic viral infections, cancers represent an example of continuous antigen exposure, and consequently it might be reasoned that tumor-specific lymphocytes would express multiple immune checkpoints.

PD-1 is expressed on a large proportion of TILs from many different tumor types. Some of the enhanced PD-1 expression among CD4+ TILs reflects a generally high level of PD-1 expression on T-reg cells which, as noted above, can represent a large proportion of intratumoral CD4+ T cells. An increased PD-1 expression on CD8+ TILs may either reflect an anergic or an exhausted state, as has been suggested by the decreased cytokine production by PD-1+ compared to PD-1– TILs from melanomas [78]. Thus, the expression of PD-1 on TILs is one of the mechanisms of tumor use to evade an active immune response.

Accordingly, the PD-1 ligands are upregulated on the tumor cell surfaces of many different human cancers [78]. PD-L1 is the major PD-1 ligand to be expressed on cells from solid tumors such as melanoma, lung, hepatocellular carcinoma, glioblastoma, kidney, breast, ovarian, pancreatic, and esophageal cancers. In addition to tumor cells, PD-L1 is commonly expressed on myeloid cells within the TME [78].

Although an initial report in renal cancer demonstrated that the expression of PD-L1 on tumor cells has been associated with a poor prognosis, other studies performed on a variety of cancers have suggested that PD-L1 status can either correlate with a worse prognosis, a better prognosis, or show no correlation with clinical outcome [4, 78]. The most probable factors contributing to the wide range of reported outcomes among different patient cohorts are variations in immunohistochemistry technique, cancer type, stage of cancer analyzed, and the treatment history [4, 79–81].

PD-L2 has also been reported to be upregulated in various tumors. PD-L2 is highly upregulated on certain B-cell lymphomas, such as primary mediastinal B-cell lymphoma, follicular cell B-cell lymphoma and Hodgkin's disease [82]. The upregulation of PD-L2 on these lymphomas is commonly associated with gene amplification or rearrangement with the class II MHC transactivator (CIITA) locus, which is highly transcriptionally active in B-cell lymphomas [83].

Two general mechanisms for the regulation of PD-L1 by tumor cells have emerged, namely innate immune resistance and adaptive immune resistance. For some tumors (e.g., glioblastomas) it has been shown that PD-L1 expression is driven by constitutive oncogenic signaling pathways in the tumor cell (this is termed constitutive or innate immune resistance). PD-L1 expression on glioblastomas is enhanced by the deletion or silencing of phosphatase and tensin homolog (PTEN), which involves the PI3K–AKT pathway [84]. Similarly, constitutive anaplastic lymphoma kinase (ALK) signaling, which has been observed in certain lymphomas and occasionally in lung cancer, has been reported to drive PD-L1 expression through signal transducer and activator of transcription 3 (STAT3) signaling [85].

A second alternative mechanism for PD-L1 regulation on tumor cells reflects their adaptation to endogenous tumor-specific immune responses, a process termed adaptive immune resistance [86]. In this model, tumor cells use the natural physiology of PD-1 ligand induction that normally occurs to protect a tissue against immune-mediated damage, in order to protect itself from an antitumor immune response. The expression of PD-L1 represents an adaptive response to endogenous antitumor immunity, and can occur because PD-L1 is induced on most tumor cells in response to IFNs, notably IFN-γ, which also occurs in epithelial and stromal cells in normal tissues [87–89]. Various preclinical and clinical studies support the adaptive immune resistance hypothesis. A strong correlation between cell-surface PD-L1 expression on tumor cells and both lymphocytic infiltration and intratumoral IFN-γ expression has been reported in melanoma. This correlation was seen not only among tumors but within individual PD-L1+ tumors at the regional level, in which regions of lymphocyte infiltration were also located exactly where PD-L1 was expressed on both tumor cells and TILs [86, 90]. These findings suggest the occurrence of a negative feedback loop whereby IFN-γ induces PD-L1 expression, which in turn suppresses the activity of PD-1+ T cells. The mechanism of adaptive immune resistance intrinsically implies that immune surveillance exists even in advanced cancers. However, the tumor ultimately resists immune elimination by upregulating the expression of ligands for inhibitory receptors on tumor-specific lymphocytes, which consequently inhibit antitumor immune responses in the TME [4].

A significant inverse correlation was recently found between PD-L1 expression and intraepithelial CD8 T lymphocyte count, indicating that PD-L1 on tumor cells inhibit the host-tumor immunity and facilitate tumor invasion [91–93]. These data suggest that PD-L1 expression might be a reliable prognostic parameter.

3.2 Clinical Implication PD-1/PD-L1 Blockade

This pathway appears to play an important function in modulating T-cell activity in peripheral tissues. The importance of this pathway in T-cell activity was confirmed in experimental models, which showed that mice deprived of the *PD-1* gene present an enhanced immunity [54]. In support of these findings, knockout models showed that this blockade would result in less immune toxicity than can be explained by the peripheral expression of PD-1 [4, 94].

Data acquired from mice and a human model of chronic viral infections suggested that this exhaustion state is partially reversible by PD-1 pathway blockade [4, 49]. The administration of anti-PD-L1 to mice chronically infected by lymphocytic choriomeningitis virus (LCMV) caused an increased T-CD8-specific response and induced viral control [49, 64]. Recently, humanized murine models with HIV demonstrated excellent results with the PD-L1 antibody, whereby the antibody suppressed HIV viremia and restored CD4 counts [95]. The PD-1 pathway has also been implicated in fungal and parasitic infections.

Many murine tumors models validate the efficacy of blockade PD-1 pathway (PD-1 or its ligands) by enhancing antitumor immunity. Currently, several PD-1 and PD-L1 inhibitors are at early or late stages of clinical development for the treatment of a wide variety of tumor models.

3.2.1 PD-1 Inhibitors

Nivolumab (BMS-936558/MDK-1106) is a fully human IgG4 monoclonal antibody targeting PD-1. In an initial Phase I clinical trial, the safety of nivolumab was evaluated in 39 patients treated for solid cancer [79, 96], while a larger Phase I study (296 patients) demonstrated not only the drug's safety but also data of its efficacy, with a good objective response rate. Some 65% of the patients also had a durable response [97, 98].

The preliminary results of a Phase II trial were presented recently at the 2014 American Society of Clinical Oncology (ASCO) annual meeting, whereby nivolumab as monotherapy in patients with advanced renal cancer achieved a 20% response rate and a median overall survival of more than two years [99]. Moreover, a recent Phase III study evaluating nivolumab versus dacarbazine in patients with previously untreated BRAF wild-type advanced melanoma was stopped early because of a superior overall survival in patients receiving nivolumab compared to the control arm. At one year, the overall survival rate was 72.9% in the nivolumab group, compared to 42.1% in the dacarbazine group (hazard ratio for death, 0.42; 99.79% confidence interval 0.25–0.73; $p < 0.001$) [100]. Nivolumab

Tab. 2 Clinical trials on going with anti-PD-1 and anti-PD-L1 antibodies.

	Clinical trial number[a]	Comparative arm
Anti-PD-1		
Nivolumab		
Melanoma	NCT01721772	DTIC
Melanoma BRAF V600	NCT02224781	Sequential ipilimumab/dabrafenib and trametinib
NSCLC	NCT01673867	Docetaxel
	NCT02066636	No
	NCT02041533	Chemotherapy investigator choice
RCC	NCT02231749	Combination with ipilimumab versus sunitinib
Head and neck cancer	NCT02105636	Treatment investigator choice
Pembrolizumab		
NSCLC	NCT02142738	Platinum-based chemotherapy
	NCT02220894	Platinum-based chemotherapy, PDL1 + tumors
Head and neck cancer	NCT02252042	Chemotherapy investigator choice
Urothelial cancer	NCT02256436	Taxane regimen, vinflumine
Anti PD-L1		
MPDL3280 A		
NSCLC	NCT02008227	Docetaxel
Urothelial	NCT02302807	Chemotherapy investigator choice
MEDI4736		
NSCLC	NCT02273375	Placebo

[a] Trials actually recruiting Phase III only (ClinicalTrials.gov.); NSCLC, non-small-cell lung cancer; RCC, renal cell carcinoma.

also demonstrated a substantial efficacy in a Phase I study of patients who had been heavily treated for relapsed Hodgkin's lymphoma, with an objective response rate of 87% being reported, including 17% of patients with complete response [101].

Pembrolizumab (MK-3475) is a highly selective, humanized IgG4 kappa monoclonal antibody targeting PD-1. This novel monoclonal antibody is designed to block PD-1 interaction with its ligands PD-L1 and PD-L2, and shows no cytotoxic [antibody-dependent cell-mediated cytotoxicity (ADCC)/complement-dependent cytotoxicity (CDC)] activity. Pembrolizumab is currently being investigated in a large Phase 1B trial for patients with several solid tumors, such as advanced non-small-cell lung cancer (NSCLC), melanoma, gastric cancer, urothelial cancer, and head and neck carcinoma. The preliminary results have shown promising clinical activity and tolerability with this novel monoclonal antibody. Pembrolizumab was approved by the US Food and Drug Administration in September 2014 for the treatment of patients with unresectable or metastatic melanoma and disease progression following ipilimumab treatment.

Pizilimumab (CT-011) is also a humanized IgG1 monoclonal antibody targeting PD-1, for which a Phase I trial in 17 patients showed clinical benefit in 33% of cases [102].

Currently, several clinical trials are ongoing with these antibodies (only Phase III trials are listed in Table 2).

3.2.2 PD-L1 Inhibitors

Clinical activity has also been observed with anti-PD-L1 inhibitor antibodies, which also inhibit the interaction between PD-L1 and CD80. To date, however, the therapeutic impact of these interactions remains unclear.

BMS-936559 (MDX-1105) is a fully human IgG4 monoclonal antibody, and was the first PD-L1 antibody to demonstrate objective tumor responses. The first results obtained from a Phase I trial of BMS-936559 in 207 patients with advanced solid cancers demonstrated a durable response, with an objective response rate of between 6% and 17%, but stable disease was also prolonged (12–41% at 24 weeks) [103]. Currently, several trials (Phases I–III) are ongoing with additional anti-PD-L1 agents (Table 2).

MPDL3280A is an engineered human monoclonal antibody for which a correlation was demonstrated between response rate and PD-L1-expressing tumors in a Phase I clinical trial; the overall response rate was 39%, and 12% of the patients had a progressive disease, whereas patients with PD-L1-negative tumors had an overall response rate of 13% and 59% had a progressive disease [104, 105]. Encouraging results were also reported at the 2014 ASCO annual meeting for this antibody in a Phase I trial in patients with urinary bladder cancer. In this case, Powles and colleagues described preliminary but promising data in patients with urinary bladder metastatic cancer, with 43% of those who had been treated with the antibody obtaining a positive response in terms of tumor size [106].

MSB0010718C is a fully human IgG1 monoclonal antibody for which Heery et al. have presented Phase I trial data in advanced solid malignancies. Notably, it was shown that MSB0010718C can be safely administered intravenously at doses of up to 20 mg kg^{-1} every two weeks until progression [107].

4
Others Checkpoint Inhibitors

Lymphocyte activation gene 3 (LAG3), T-cell immunoglobulin and mucin domain 3 (TIM-3), B- and T-lymphocyte attenuator (BTLA), B7-H3 (CD276), and the family of killer inhibitory receptors (KIRs) have also been associated with the inhibition of lymphocyte activation and an induced lymphocyte anergy.

4.1
TIM-3

(TIM-3) was discovered in 2002 and first investigated in the setting of autoimmunity. In a study conducted by Monney et al., the anti-TIM-3 antibody was shown to exacerbate experimental autoimmune encephalomyelitis [108].

Further studies are needed to confirm that TIM-3 is another inhibitory receptor that specifically causes activated T cell (Th1 and CD8) exhaustion [109–111]. For example, the upregulation of Tim-3 and PD-1 expression is associated with a tumor antigen-specific CD8+ T-cell dysfunction in melanoma patients [112]. Remarkably, the coblockade of Tim-3 and PD-1 has been shown to restore tumor-specific T-cell functions and to promote an effective antitumor immunity in a wide variety of cancers [109, 111, 112]. The C-type lectin galectin-9 (Gal9) is the best known ligand of TIM-3. An interaction between TIM-3 and Gal 9 induces phosphorylation of the Tim-3 cytoplasmic tail at Tyr256 and Tyr263, releasing Bat3 from the TIM-3 cytoplasmic tail and thus causing an accumulation of inactive LcK [113].

Furthermore, the expression of TIM-3 on T cells has been shown to promote myeloid-derived suppressor cells (MDSCs) via Gal9 interaction [114]. Recent studies

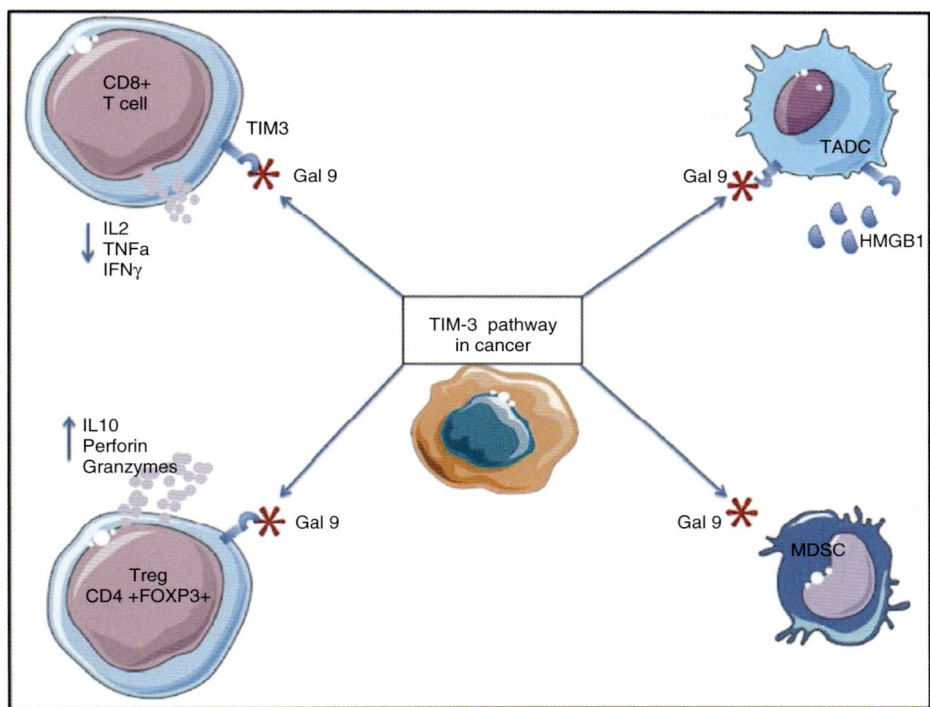

Fig. 6 TIM-3 pathways and immune cells implication. TIM-3 is expressed on CD8 T cell exhausted. TIM-3+ T regulator cells are potent suppressors of immune responses in tumor tissue, and may promote the development of dysfunctional phenotype in intratumoral CD8 T cells. Tumor-associated dendritic cells (TADCs) express TIM-3 and lead to a decrease in free HMGB1 available to bind the nucleic acids released by dying tumor cells. TIM-3 expression also promotes the recruitment of myeloid-derived suppressor cells (MDSCs). IFNγ, interferon-gamma; IL-2, interleukin-2; IL-10, interleukin-10; Gal9, Galectin 9. Adapted from Ref. [113].

have also shown that Tim-3+ T-reg is more suppressive than TIM-3-T-reg in the TME [110]. The results of other studies have also suggested that TIM-3 is upregulated on tumor-associated dendritic cells (TADCs) [115]. Collectively, Tim-3 can function as a determinant of antitumor immunity and thus may emerge as an attractive target in the immunotherapy landscape [1, 116] (Fig. 6).

A better understanding of the complexity of the TIM-3 pathway supports the use of TIM-3-targeted therapies alone, and also in combination with other checkpoint immune inhibitors. However, no clinical data are yet available with anti TIM-3 antibodies.

4.2 LAG3, BTLA

Although the LAG3 (CD223) was previously discovered as a CD4 homolog [117], its role in the immune checkpoint has only recently been defined [118]. LAG3 was shown to enhance the suppressive function of T-reg cells and to impact on CD8+ T-cell function [119, 120]. Like TIM-3,

LAG3 expression is also found on anergic or exhausted T cells, such as those infiltrating the TME [119, 121]. The ligands of LAG3 are MHC class II molecules that are expressed on tumor-infiltrating APCs and which are upregulated on some epithelial cancers (mainly in response to IFN-γ, as does PDL-1). LAG3 binds to MHC class II molecules with a higher avidity than CD4, and this is known to transduce inhibitory signals [122]. Recent preclinical models using antibody blockade LAG3 for cancer treatment have shown an enhanced activation of specific T cells at the tumor site [119, 123], suggesting that LAG3 blockade might be a potential cancer treatment.

The inhibitory function of BTLA on T cells was identified with a BTLA-knockout mouse model [124]. BTLA is constitutively expressed by naïve CD4+ and CD8+ T cells, and is upregulated when T cells are activated [124–126]. Herpes virus entry mediator (HVEM), a member of the tumor necrosis factor receptor (TNFR) superfamily, is the only known ligand of BTLA [127, 128]. BTLA–HVEM interactions lead to BTLA^{hi} T-cell inhibition, and may be involved in tumor progression and resistance to the specific immune response [112, 129].

5
Safety Aspects

The toxicity of checkpoint inhibitors is a limiting problem for the development of this new cancer immunotherapy (Table 3). The major toxicities are immune-related adverse events (IRAEs), which are related to the infiltration of activated CD4 and CD8 T cells, and the increase in inflammatory cytokine production in normal tissues [57, 130]. Both, preclinical models and clinical trials have provided evidence of the role of CTLA-4 blockade in breaking the tolerance to human cancer antigens and also self-antigens [19, 20, 57]. Various studies have been conducted to evaluate the infiltration of depigmentation lesions, revealing the presence of high levels of polymorphonuclear cells and the deposition of antibodies. Major immune toxicities can be explained by the central role of CTLA-4 for keeping T-cell activation in check. The most common IRAEs involve the skin, gastrointestinal tract, liver and endocrine system [105]. In clinical trials, up to 60% of patients treated with ipilimumab reported toxicity, with the toxic effects being severe (grade 3 or 4) in 10–15% of cases [44]. Clinical trials in Phases II and III have shown that the early administration of corticosteroids is essential to manage IRAEs, and to prevent their progression to a more serious toxicity [30, 32, 44, 105].

The most frequent adverse events with agents targeting PD-1 pathway were asthenia, anorexia, diarrhea, nausea, dyspnea, rash, and headache [32, 57, 97]. In clinical trials, 14% of patients treated with nivolumab had serious adverse events (grade 3 or 4).

A better understanding of the mechanism of action of these drugs alone, or in combination, might be useful to improve the toxicity profile. Last, but not least, recent data have suggested a possible correlation between toxicity and clinical efficacy [132, 133].

6
Future Direction: Biomarkers and Combination

6.1
Biomarkers

Due to the cost of immunotherapy, the chance of adverse events, and the

Tab. 3 Check-point inhibitors and safety.

SAE (Grade 3–4)	CTLA-4 blockade [22, 98, 130, 131]		PD-1 blockade [96, 97, 102]	PDL-1 blockade [104, 106]	
	Ipilimumab (%)	Tremelimumab (%)	Nivolumab, Pembrolizumab (%)	BMS-936559 (%)	MPDL 3280A (%)
Dermatologic					
Rash	3–4	2.5–18	1–4	<1	<1
Gastrointestinal					
Colitis	2–21	2.1–18	2	0	<1
Diarrhea	4–5.3	5–21	1–3	<1	<1
Nausea	<5	8–12.8	0	0	<1
Endocrine					
Hypothyroidism	0	1	0	<1	<1
Hypophysitis	<1	2	1	0	0
Hypopituitarism	<1	1	0	0	0
Adrenal insufficiency	1.5	1	0	0	0
Hepatic					
Elevation in ALAT	1.5–22	NA	1–7	2	0
Elevation in ASAT	0.8–19	NA	1–6	1-6	0
Hepatitis	<3	1	NA	3	2
Asthenia	6–10	2–13	2	2	2
Pneumonitis	NA	1	1–3	1–3	NA

SAE, serious adverse events; ALAT, alanine aminotransferase; ASAT, aspartate aminotransferase; NA, non-available.

heterogeneity of individual tumors it is important for clinicians and research groups to be able to predict a patient's likelihood of response and adverse events. One method of achieving this is to utilize biomarkers to predict the likelihood of a response so that patients can be stratified prior to therapy. To date, however, no such pretreatment biomarkers have been validated to a point at which they could be applied as part of standard-of-care therapeutic decision-making.

In the case of anti-CTLA-4 therapy, biomarkers such as absolute lymphocyte count, lactate-dehydrogenase (LDH) and C-reactive protein (CRP) are the most frequently identified markers correlating with improved long-term outcome upon treatment with ipilimumab [5, 134, 135]. Interestingly, most of these correlations have not (or to a lesser extent) been found to be predictive in the clinical trials, emphasizing the need for a meta-analysis of all currently available patient data. As these markers are also known to be prognostic markers, their predictive value remains unclear so far. A recent report illustrated the importance of tumor genetics in defining the basis of the clinical benefit from CTLA-4 blockade. The sequences of the whole genome have revealed mutations that lead to neoantigens, which may be immunologically relevant in responses to immune checkpoint blockade. A high mutational load is associated with a benefit from CTLA-4 blockade, but is not sufficient to impart a clinical benefit. These findings provide a rationale to consider examining the exomes of patients treated by CTLA-4 inhibitors [136].

Reports on the anti-PD-1 nivolumab have shown that patients with PD-L1-positive tumors had a greater chance to respond to the therapy than did those with PD-L1-negative tumors [97, 105, 137]. These findings suggest that PD-L1, when detected in biopsy samples by immunohistochemistry, might be capable of predicting the activity of nivolumab in advanced cancer. In addition, in a recent study it was noted that PD-L1 tumor expression and T-cell gene signature correlated with responses to the anti-PD-L1 therapy MPDL3280A [106, 138]. Recent data have also reported objective responses to nivolumab in patients with PD-L1-negative tumors [139], though this may have been due to a heterogeneous tumor expression of PD-1, and a single negative biopsy would be insufficient to determine if a tumor were truly PD-L1-negative. PD-L1 expression may also undergo changes driven by alterations in the TME, for example, infiltrations with immune cells, potentially limiting its use as a clear-cut predictive biomarker as opposed to, for example, BRAFV600 mutational status in melanoma patients [137]. Clearly, further randomized clinical trials using validated immunohistochemical assessments of PD-L1 expression are needed.

As many current immune-based therapies, including immune checkpoint inhibitors, depend on augmenting pre-existing antitumor T-cell immunity in cancer patients, there is a strong rationale to evaluate parameters associated with tumor-specific T-cell responses as potential biomarkers. NY-ESO-1 antibody and CD8 T-cell responses to the cancer/testis antigen NY-ESO-1 before treatment were associated with clinical benefit after ipilimumab treatment [43, 140]. Recently, Meyer et al. reported that the frequencies of circulating MDSCs correlated with the clinical outcome of melanoma patients treated with ipilimumab [141].

The use of biomarkers to profile immune cells that have infiltrated a tumor may also prove valuable to clinicians [2, 142]. These profiling tools could be used to understand the dynamic state of the immune system in the individual tumor, and to tailor the immunotherapy selection accordingly.

Large studies have shown that tumor-immune infiltrates in solid tumors such as colorectal cancer can serve as significant prognostic biomarkers, even after adjusting for the tumor–node–metastasis (TNM) classification and tumor molecular biomarkers [143, 144]. A potential clinical translation of these observations is the establishment of an Immunoscore, which is based on the numeration of two lymphocyte populations CD3 and CD8 in the core, and the invasive margin of the tumors [145]. Considering the important role of the host immune reaction in controlling tumor progression, it is now important to initiate the incorporation of an immune parameter such as the Immunoscore as a component of cancer classification and a prognostic tool. Whether the immune infiltrate of the primary tumor evaluated by a robust technology adapted to clinical practice such as the Immunoscore could predict therapeutic responses to immune checkpoint inhibitors is important in patient clinical management. The theranostic value of the Immunoscore is currently under investigation in research laboratories. Recently, Tumeh et al. showed that tumor regression after therapeutic PD-1 blockade requires pre-existing CD8 T cells that are negatively regulated by PD-1/PD-L1-mediated adaptive immune resistance [146].

6.2
Combination

Many preclinical models have suggested synergistic effects when checkpoint

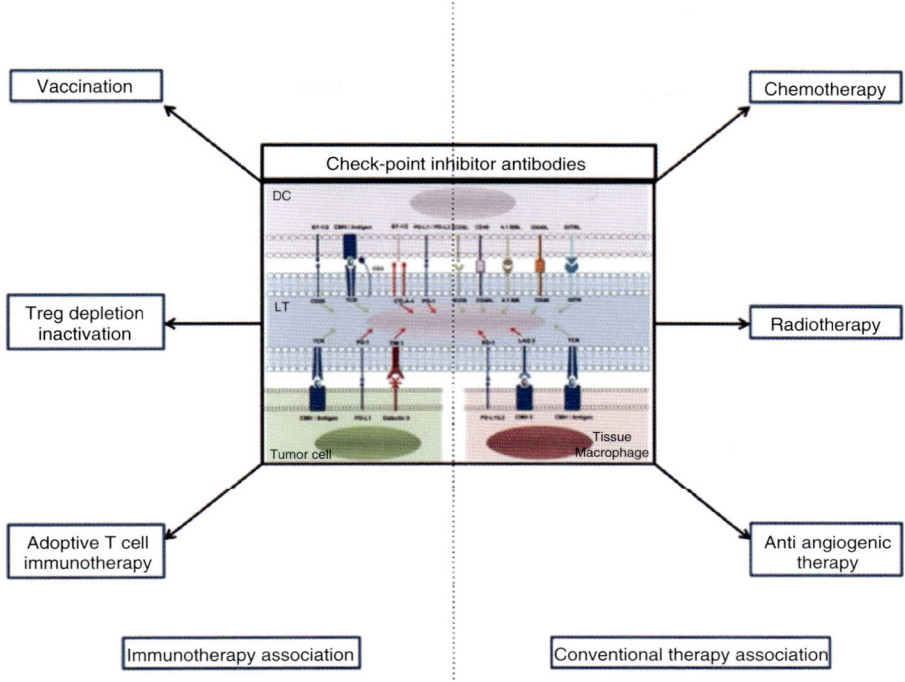

Fig. 7 Check-point inhibitor antibodies and combination currently under development. Adapted from Ref. [56].

inhibitor antibodies were used in combinations (Fig. 7), and the CTLA-4 and PD-1 antibodies association was the first such combination to be investigated. This dual blockade showed promising results of antitumor synergy [121, 147]. For example, in a Phase I clinical trial evaluating the association of ipilimumab with ascending doses of nivolumab, more than half of the patients experienced more than 80% durable tumor reductions. Notably, grade 3 or 4 adverse events related to therapy occurred in 53% of those patients in the combination regimen group [147]. This profile of toxicity suggested a probable switch in upcoming trials to using sequential rather than concomitant immunotherapies. PD-1 blockade has also shown signs of enhanced antitumor efficacy when combined with monoclonal antibody-blocking LAG-3 [123] or TIM-3 [110], which exert only modest effects as monotherapies.

Preclinical and clinical results have also demonstrated synergistic effects of combination checkpoint inhibitor antibodies and other immunotherapy, such as therapeutic vaccines, cytokines, indolamine-2,3-dioxygenase inhibitors, agonists of costimulatory receptors, or adoptive T-cell transfer [137, 145, 148–153].

Finally, synergistic effects have also been demonstrated in preclinical models with conventional anticancer therapy such chemotherapy, radiotherapy, antiangiogenic, and molecular targeted therapies [56]. On the basis of promising results with radiation association, several clinical trials are ongoing to assess the efficacy of checkpoint inhibitor antibodies with radiotherapy [148, 154]. The early clinical data obtained suggest that combined CTLA-4 blockade and vascular endothelial growth factor (VEGF) inhibition can induce durable responses in patients with melanoma [131, 155].

Furthermore, there is emerging evidence for favorable effects of BRAF/MEK pathway inhibition on the endogenous tumor immune response, as well as for potential synergies between mitogen-activated protein kinase (MAPK)-targeted therapy. In fact, immunotherapy clinical trials combining MAPK pathway inhibition with CTLA-4 and PD-L1 blockade in patients with melanoma are currently under way [156–158].

7
Conclusion

The role of checkpoint molecules in regulating immune responses has been demonstrated in preclinical models of chronic infection, self-tolerance and tumor tolerance. Trials with checkpoint inhibitor (CTLA-4, PD-1, and PD-L1) antibodies have demonstrated real evidence of clinical activity in many malignancy types, even in cancers that are not traditionally viewed as amenable to immunotherapy, such as gastric tumors. These encouraging data support the future development of checkpoint antibodies, combinatorial approaches with conventional therapies, and also other immunotherapies and predictive biomarker discovery.

References

1. Vesely, M.D., Kershaw, M.H., Schreiber, R.D., and Smyth, M.J. (2011) Natural innate and adaptive immunity to cancer. *Annu. Rev. Immunol.*, **29**, 235–271.
2. Chen, D.S. and Mellman, I. (2013) Oncology meets immunology: the cancer-immunity cycle. *Immunity*, **39**, 1–10.
3. Motz, G.T. and Coukos, G. (2013) Deciphering and reversing tumor immune suppression. *Immunity*, **39**, 61–73.
4. Pardoll, D.M. (2012) The blockade of immune checkpoints in cancer immunotherapy. *Nat. Rev. Cancer*, **12**, 252–264.
5. Ott, P.A., Hodi, F.S., and Robert, C. (2013) CTLA-4 and PD-1/PD-L1 blockade: new immunotherapeutic modalities with durable clinical benefit in melanoma patients. *Clin. Cancer Res.*, **19**, 5300–5309.
6. Lenschow, D.J., Walunas, T.L., and Bluestone, J.A. (1996) CD28/B7 system of T cell costimulation. *Annu. Rev. Immunol.*, **14**, 233–258.
7. Rudd, C.E., Taylor, A., and Schneider, H. (2009) CD28 and CTLA-4 coreceptor expression and signal transduction. *Immunol. Rev.*, **229**, 12–26.
8. Schwartz, R.H. (1992) Costimulation of T lymphocytes: the role of CD28, CTLA-4, and B7/BB1 in interleukin-2 production and immunotherapy. *Cell*, **71**, 1065–1068.
9. Azuma, M., Ito, D., Yagita, H., Okumura, K. et al. (1993) B70 antigen is a second ligand for CTLA-4 and CD28. *Nature*, **366**, 76–79.
10. Freeman, G.J., Gribben, J.G., Boussiotis, V.A., Ng, J.W. et al. (1993) Cloning of B7-2: a CTLA-4 counter-receptor that costimulates human T cell proliferation. *Science*, **262**, 909–911.
11. Hathcock, K.S., Laszlo, G., Dickler, H.B., Bradshaw, J. et al. (1993) Identification of an alternative CTLA-4 ligand costimulatory for T cell activation. *Science*, **262**, 905–907.
12. Schneider, H., Downey, J., Smith, A., Zinselmeyer, B.H. et al. (2006) Reversal of the TCR stop signal by CTLA-4. *Science*, **313**, 1972–1975.
13. Qureshi, O.S., Zheng, Y., Nakamura, K., Attridge, K. et al. (2011) Trans-endocytosis of CD80 and CD86: a molecular basis for the cell-extrinsic function of CTLA-4. *Science*, **332**, 600–603.
14. Kuehn, H.S., Ouyang, W., Lo, B., Deenick, E.K. et al. (2014) Immune dysregulation in human subjects with heterozygous

germline mutations in CTLA4. *Science*, **345**, 1623–1627.
15. Tivol, E.A., Borriello, F., Schweitzer, A.N., Lynch, W.P. et al. (1995) Loss of CTLA-4 leads to massive lymphoproliferation and fatal multiorgan tissue destruction, revealing a critical negative regulatory role of CTLA-4. *Immunity*, **3**, 541–547.
16. Waterhouse, P., Penninger, J.M., Timms, E., Wakeham, A. et al. (1995) Lymphoproliferative disorders with early lethality in mice deficient in Ctla-4. *Science*, **270**, 985–988.
17. Hill, J.A., Feuerer, M., Tash, K., Haxhinasto, S. et al. (2007) Foxp3 transcription-factor-dependent and -independent regulation of the regulatory T cell transcriptional signature. *Immunity*, **27**, 786–800.
18. Hori, S., Nomura, T., and Sakaguchi, S. (2003) Control of regulatory T cell development by the transcription factor Foxp3. *Science*, **299**, 1057–1061.
19. Leach, D.R., Krummel, M.F., and Allison, J.P. (1996) Enhancement of antitumor immunity by CTLA-4 blockade. *Science*, **271**, 1734–1736.
20. Van Elsas, A., Hurwitz, A.A., and Allison, J.P. (1999) Combination immunotherapy of B16 melanoma using anti-cytotoxic T lymphocyte-associated antigen 4 (CTLA-4) and granulocyte/macrophage colony-stimulating factor (GM-CSF)-producing vaccines induces rejection of subcutaneous and metastatic tumors accompanied by autoimmune depigmentation. *J. Exp. Med.*, **190**, 355–366.
21. Hodi, F.S., O'Day, S.J., McDermott, D.F., Weber, R.W. et al. (2010) Improved survival with ipilimumab in patients with metastatic melanoma. *N. Engl. J. Med.*, **363**, 711–723.
22. Robert, C., Thomas, L., Bondarenko, I., O'Day , S. et al. (2011) Ipilimumab plus dacarbazine for previously untreated metastatic melanoma. *N. Engl. J. Med.*, **364**, 2517–2526.
23. Camacho, L.H., Antonia, S., Sosman, J., Kirkwood, J.M. et al. (2009) Phase I/II trial of tremelimumab in patients with metastatic melanoma. *J. Clin. Oncol.*, **27**, 1075–1081.
24. Ribas, A., Kefford, R., Marshall, M.A., Punt, C.J. et al. (2013) Phase III randomized clinical trial comparing tremelimumab with standard-of-care chemotherapy in patients with advanced melanoma. *J. Clin. Oncol.*, **31**, 616–622.
25. Attia, P., Phan, G.Q., Maker, A.V., Robinson, M.R. et al. (2005) Autoimmunity correlates with tumor regression in patients with metastatic melanoma treated with anti-cytotoxic T-lymphocyte antigen-4. *J. Clin. Oncol.*, **23**, 6043–6053.
26. Maker, A.V., Phan, G.Q., Attia, P., Yang, J.C. et al. (2005) Tumor regression and autoimmunity in patients treated with cytotoxic T lymphocyte-associated antigen 4 blockade and interleukin 2: a phase I/II study. *Ann. Surg. Oncol.*, **12**, 1005–1016.
27. Maker, A.V., Yang, J.C., Sherry, R.M., Topalian, S.L. et al. (2006) Intrapatient dose escalation of anti-CTLA-4 antibody in patients with metastatic melanoma. *J. Immunother.*, **29**, 455–463.
28. Phan, G.Q., Yang, J.C., Sherry, R.M., Hwu, P. et al. (2003) Cancer regression and autoimmunity induced by cytotoxic T lymphocyte-associated antigen 4 blockade in patients with metastatic melanoma. *Proc. Natl Acad. Sci. USA*, **100**, 8372–8377.
29. Prieto, P.A., Yang, J.C., Sherry, R.M., Hughes, M.S. et al. (2012) CTLA-4 blockade with ipilimumab: long-term follow-up of 177 patients with metastatic melanoma. *Clin. Cancer Res.*, **18**, 2039–2047.
30. O'Day, S.J., Maio, M., Chiarion-Sileni, V., Gajewski, T.F. et al. (2010) Efficacy and safety of ipilimumab monotherapy in patients with pretreated advanced melanoma: a multicenter single-arm phase II study. *Ann. Oncol.*, **21**, 1712–1717.
31. Wolchok, J.D., Neyns, B., Linette, G., Negrier, S. et al. (2010) Ipilimumab monotherapy in patients with pretreated advanced melanoma: a randomised, double-blind, multicentre, phase 2, dose-ranging study. *Lancet Oncol.*, **11**, 155–164.
32. Wolchok, J.D., Weber, J.S., Maio, M., Neyns, B. et al. (2013) Four-year survival rates for patients with metastatic melanoma who received ipilimumab in phase II clinical trials. *Ann. Oncol.*, **24**, 2174–2180.
33. McDermott, D., Lebbé, C., Hodi, F.S., Maio, M. et al. (2014) Durable benefit and the potential for long-term survival with immunotherapy in advanced melanoma. *Cancer Treat. Rev.*, **40**, 1056–1064.
34. Weber, J., Thompson, J.A., Hamid, O., Minor, D. et al. (2009) A randomized, double-blind, placebo-controlled, phase II study comparing the tolerability and efficacy

of ipilimumab administered with or without prophylactic budesonide in patients with unresectable stage III or IV melanoma. *Clin. Cancer Res.*, **15**, 5591–5598.
35. Madan, R.A., Mohebtash, M., Arlen, P.M., Vergati, M. et al. (2012) Ipilimumab and a poxviral vaccine targeting prostate-specific antigen in metastatic castration-resistant prostate cancer: a phase 1 dose-escalation trial. *Lancet Oncol.*, **13**, 501–508.
36. Slovin, S.F., Higano, C.S., Hamid, O., Tejwani, S. et al. (2013) Ipilimumab alone or in combination with radiotherapy in metastatic castration-resistant prostate cancer: results from an open-label, multicenter phase I/II study. *Ann. Oncol.*, **24**, 1813–1821.
37. Kwon, E.D., Drake, C.G., Scher, H.I., Fizazi, K. et al. (2014) Ipilimumab versus placebo after radiotherapy in patients with metastatic castration-resistant prostate cancer that had progressed after docetaxel chemotherapy (CA184-043): a multicentre, randomised, double-blind, phase 3 trial. *Lancet Oncol.*, **15**, 700–712.
38. Royal, R.E., Levy, C., Turner, K., Mathur, A. et al. (2010) Phase 2 trial of single agent Ipilimumab (anti-CTLA-4) for locally advanced or metastatic pancreatic adenocarcinoma. *J. Immunother.*, **33**, 828–833.
39. McNeel, D.G., Smith, H.A., Eickhoff, J.C., Lang, J.M. et al. (2012) Phase I trial of tremelimumab in combination with short-term androgen deprivation in patients with PSA-recurrent prostate cancer. *Cancer Immunol. Immunother.*, **61**, 1137–1147.
40. Aglietta, M., Barone, C., Sawyer, M.B., Moore, M.J. et al. (2014) A phase I dose escalation trial of tremelimumab (CP-675,206) in combination with gemcitabine in chemotherapy-naive patients with metastatic pancreatic cancer. *Ann. Oncol.*, **25**, 1750–1755.
41. Hodi, F.S., Mihm, M.C., Soiffer, R.J., Haluska, F.G. et al. (2003) Biologic activity of cytotoxic T lymphocyte-associated antigen 4 antibody blockade in previously vaccinated metastatic melanoma and ovarian carcinoma patients. *Proc. Natl Acad. Sci. USA*, **100**, 4712–4717.
42. Hodi, F.S., Butler, M., Oble, D.A., Seiden, M.V. et al. (2008) Immunologic and clinical effects of antibody blockade of cytotoxic T lymphocyte-associated antigen 4 in previously vaccinated cancer patients. *Proc. Natl Acad. Sci. USA*, **105**, 3005–3010.
43. Yuan, J., Adamow, M., Ginsberg, B.A., Rasalan, T.S. et al. (2011) Integrated NY-ESO-1 antibody and CD8+ T-cell responses correlate with clinical benefit in advanced melanoma patients treated with ipilimumab. *Proc. Natl Acad. Sci. USA*, **108**, 16723–16728.
44. Hoos, A., Eggermont, A.M.M., Janetzki, S., Hodi, F.S. et al. (2010) Improved endpoints for cancer immunotherapy trials. *J. Natl Cancer Inst.*, **102**, 1388–1397.
45. Bulliard, Y., Jolicoeur, R., Windman, M., Rue, S.M. et al. (2013) Activating Fc γ receptors contribute to the antitumor activities of immunoregulatory receptor-targeting antibodies. *J. Exp. Med.*, **210**, 1685–1693.
46. Selby, M.J., Engelhardt, J.J., Quigley, M., Henning, K.A. et al. (2013) Anti-CTLA-4 antibodies of IgG2a isotype enhance antitumor activity through reduction of intratumoral regulatory T cells. *Cancer Immunol. Res.*, **1**, 32–42.
47. Simpson, T.R., Li, F., Montalvo-Ortiz, W., Sepulveda, M.A. et al. (2013) Fc-dependent depletion of tumor-infiltrating regulatory T cells co-defines the efficacy of anti-CTLA-4 therapy against melanoma. *J. Exp. Med.*, **210**, 1695–1710.
48. Bruhns, P., Iannascoli, B., England, P., Mancardi, D.A. et al. (2009) Specificity and affinity of human Fcgamma receptors and their polymorphic variants for human IgG subclasses. *Blood*, **113**, 3716–3725.
49. Barber, D.L., Wherry, E.J., Masopust, D., Zhu, B. et al. (2006) Restoring function in exhausted CD8 T cells during chronic viral infection. *Nature*, **439**, 682–687.
50. Freeman, G.J., Long, A.J., Iwai, Y., Bourque, K. et al. (2000) Engagement of the PD-1 immunoinhibitory receptor by a novel B7 family member leads to negative regulation of lymphocyte activation. *J. Exp. Med.*, **192**, 1027–1034.
51. Ishida, Y., Agata, Y., Shibahara, K., and Honjo, T. (1992) Induced expression of PD-1, a novel member of the immunoglobulin gene superfamily, upon programmed cell death. *EMBO J.*, **11**, 3887–3895.
52. Keir, M.E., Liang, S.C., Guleria, I., Latchman, Y.E. et al. (2006) Tissue expression of PD-L1 mediates peripheral T cell tolerance. *J. Exp. Med.*, **203**, 883–895.

53. Keir, M.E., Butte, M.J., Freeman, G.J., and Sharpe, A.H. (2008) PD-1 and its ligands in tolerance and immunity. *Annu. Rev. Immunol.*, **26**, 677–704.
54. Nishimura, H., Nose, M., Hiai, H., Minato, N. *et al.* (1999) Development of lupus-like autoimmune diseases by disruption of the PD-1 gene encoding an ITIM motif-carrying immunoreceptor. *Immunity*, **11**, 141–151.
55. Okazaki, T. and Honjo, T. (2007) PD-1 and PD-1 ligands: from discovery to clinical application. *Int. Immunol.*, **19**, 813–824.
56. Perez-Gracia, J.L., Labiano, S., Rodriguez-Ruiz, M.E., Sanmamed, M.F. *et al.* (2014) Orchestrating immune checkpoint blockade for cancer immunotherapy in combinations. *Curr. Opin. Immunol.*, **27**, 89–97.
57. Gelao, L., Criscitiello, C., Esposito, A., Goldhirsch, A. *et al.* (2014) Immune checkpoint blockade in cancer treatment: a double-edged sword cross-targeting the host as an "innocent bystander". *Toxins*, **6**, 914–933.
58. Zinselmeyer, B.H., Heydari, S., Sacristán, C., Nayak, D. *et al.* (2013) PD-1 promotes immune exhaustion by inducing antiviral T cell motility paralysis. *J. Exp. Med.*, **210**, 757–774.
59. Okazaki, T., Maeda, A., Nishimura, H., Kurosaki, T. *et al.* (2001) PD-1 immunoreceptor inhibits B cell receptor-mediated signaling by recruiting src homology 2-domain-containing tyrosine phosphatase 2 to phosphotyrosine. *Proc. Natl Acad. Sci. USA*, **98**, 13866–13871.
60. Saunders, P.A., Hendrycks, V.R., Lidinsky, W.A., and Woods, M.L. (2005) PD-L2:PD-1 involvement in T cell proliferation, cytokine production, and integrin-mediated adhesion. *Eur. J. Immunol.*, **35**, 3561–3569.
61. Nirschl, C.J. and Drake, C.G. (2013) Molecular pathways: coexpression of immune checkpoint molecules: signaling pathways and implications for cancer immunotherapy. *Clin. Cancer Res.*, **19**, 4917–4924.
62. Agata, Y., Kawasaki, A., Nishimura, H., Ishida, Y. *et al.* (1996) Expression of the PD-1 antigen on the surface of stimulated mouse T and B lymphocytes. *Int. Immunol.*, **8**, 765–772.
63. Chemnitz, J.M., Parry, R.V., Nichols, K.E., June, C.H. *et al.* (2004) SHP-1 and SHP-2 associate with immunoreceptor tyrosine-based switch motif of programmed death 1 upon primary human T cell stimulation, but only receptor ligation prevents T cell activation. *J. Immunol.*, **173**, 945–954.
64. Kamphorst, A.O. and Ahmed, R. (2013) Manipulating the PD-1 pathway to improve immunity. *Curr. Opin. Immunol.*, **25**, 381–388.
65. Oestreich, K.J., Yoon, H., Ahmed, R., and Boss, J.M. (2008) NFATc1 regulates PD-1 expression upon T cell activation. *J. Immunol.*, **181**, 4832–4839.
66. Kao, C., Oestreich, K.J., Paley, M.A., Crawford, A. *et al.* (2011) Transcription factor T-bet represses expression of the inhibitory receptor PD-1 and sustains virus-specific CD8+ T cell responses during chronic infection. *Nat. Immunol.*, **12**, 663–671.
67. Francisco, L.M., Salinas, V.H., Brown, K.E., Vanguri, V.K. *et al.* (2009) PD-L1 regulates the development, maintenance, and function of induced regulatory T cells. *J. Exp. Med.*, **206**, 3015–3029.
68. Fanoni, D., Tavecchio, S., Recalcati, S., Balice, Y. *et al.* (2011) New monoclonal antibodies against B-cell antigens: possible new strategies for diagnosis of primary cutaneous B-cell lymphomas. *Immunol. Lett.*, **134**, 157–160.
69. Terme, M., Ullrich, E., Aymeric, L., Meinhardt, K. *et al.* (2011) IL-18 induces PD-1-dependent immunosuppression in cancer. *Cancer Res.*, **71**, 5393–5399.
70. Velu, V., Titanji, K., Zhu, B., Husain, S. *et al.* (2009) Enhancing SIV-specific immunity in vivo by PD-1 blockade. *Nature*, **458**, 206–210.
71. Butte, M.J., Keir, M.E., Phamduy, T.B., Sharpe, A.H. *et al.* (2007) Programmed death-1 ligand 1 interacts specifically with the B7-1 costimulatory molecule to inhibit T cell responses. *Immunity*, **27**, 111–122.
72. Carreno, B.M. and Collins, M. (2002) The B7 family of ligands and its receptors: new pathways for costimulation and inhibition of immune responses. *Annu. Rev. Immunol.*, **20**, 29–53.
73. Greenwald, R.J., Freeman, G.J., and Sharpe, A.H. (2005) The B7 family revisited. *Annu. Rev. Immunol.*, **23**, 515–548.
74. Pentcheva-Hoang, T., Corse, E., and Allison, J.P. (2009) Negative regulators of T-cell

activation: potential targets for therapeutic intervention in cancer, autoimmune disease, and persistent infections. *Immunol. Rev.*, **229**, 67–87.

75. Latchman, Y., Wood, C.R., Chernova, T., Chaudhary, D. et al. (2001) PD-L2 is a second ligand for PD-1 and inhibits T cell activation. *Nat. Immunol.*, **2**, 261–268.

76. Tseng, S.Y., Otsuji, M., Gorski, K., Huang, X. et al. (2001) B7-DC, a new dendritic cell molecule with potent costimulatory properties for T cells. *J. Exp. Med.*, **193**, 839–846.

77. Paterson, A.M., Brown, K.E., Keir, M.E., Vanguri, V.K. et al. (2011) The programmed death-1 ligand 1:B7-1 pathway restrains diabetogenic effector T cells in vivo. *J. Immunol.*, **187**, 1097–1105.

78. Gros, A., Robbins, P.F., Yao, X., Li, Y.F. et al. (2014) PD-1 identifies the patient-specific CD8+ tumor-reactive repertoire infiltrating human tumors. *J. Clin. Invest.*, **124**, 2246–2259.

79. Dolan, D.E. and Gupta, S. (2014) PD-1 pathway inhibitors: changing the landscape of cancer immunotherapy. *Cancer Control J. Moffitt Cancer Cent.*, **21**, 231–237.

80. Mellman, I. (2001) Setting logical priorities. *Nature*, **410**, 1026.

81. Zitvogel, L. and Kroemer, G. (2012) Targeting PD-1/PD-L1 interactions for cancer immunotherapy. *Oncoimmunology*, **1**, 1223–1225.

82. Rosenwald, A., Wright, G., Leroy, K., Yu, X. et al. (2003) Molecular diagnosis of primary mediastinal B cell lymphoma identifies a clinically favorable subgroup of diffuse large B cell lymphoma related to Hodgkin lymphoma. *J. Exp. Med.*, **198**, 851–862.

83. Steidl, C., Shah, S.P., Woolcock, B.W., Rui, L. et al. (2011) MHC class II transactivator CIITA is a recurrent gene fusion partner in lymphoid cancers. *Nature*, **471**, 377–381.

84. Parsa, A.T., Waldron, J.S., Panner, A., Crane, C.A. et al. (2007) Loss of tumor suppressor PTEN function increases B7-H1 expression and immunoresistance in glioma. *Nat. Med.*, **13**, 84–88.

85. Marzec, M., Zhang, Q., Goradia, A., Raghunath, P.N. et al. (2008) Oncogenic kinase NPM/ALK induces through STAT3 expression of immunosuppressive protein CD274 (PD-L1, B7-H1). *Proc. Natl Acad. Sci. USA*, **105**, 20852–20857.

86. Taube, J.M., Anders, R.A., Young, G.D., Xu, H. et al. (2012) Colocalization of inflammatory response with B7-h1 expression in human melanocytic lesions supports an adaptive resistance mechanism of immune escape. *Sci. Transl. Med.*, **4**, 127ra37.

87. Kim, J., Myers, A.C., Chen, L., Pardoll, D.M. et al. (2005) Constitutive and inducible expression of b7 family of ligands by human airway epithelial cells. *Am. J. Respir. Cell Mol. Biol.*, **33**, 280–289.

88. Lee, S.-K., Seo, S.-H., Kim, B.-S., Kim, C.D. et al. (2005) IFN-gamma regulates the expression of B7-H1 in dermal fibroblast cells. *J. Dermatol. Sci.*, **40**, 95–103.

89. Wilke, C.M., Wei, S., Wang, L., Kryczek, I. et al. (2011) Dual biological effects of the cytokines interleukin-10 and interferon-γ. *Cancer Immunol. Immunother.*, **60**, 1529–1541.

90. Gajewski, T.F., Louahed, J., and Brichard, V.G. (2010) Gene signature in melanoma associated with clinical activity: a potential clue to unlock cancer immunotherapy. *Cancer J. Sudbury Mass.*, **16**, 399–403.

91. Hamanishi, J., Mandai, M., Iwasaki, M., Okazaki, T. et al. (2007) Programmed cell death 1 ligand 1 and tumor-infiltrating CD8+ T lymphocytes are prognostic factors of human ovarian cancer. *Proc. Natl Acad. Sci. USA*, **104**, 3360–3365.

92. Mahnke, Y.D., Devevre, E., Baumgaertner, P., Matter, M. et al. (2012) Human melanoma-specific CD8(+) T-cells from metastases are capable of antigen-specific degranulation and cytolysis directly ex vivo. *Oncoimmunology*, **1**, 467–530.

93. Matsuzaki, J., Gnjatic, S., Mhawech-Fauceglia, P., Beck, A. et al. (2010) Tumor-infiltrating NY-ESO-1-specific CD8+ T cells are negatively regulated by LAG-3 and PD-1 in human ovarian cancer. *Proc. Natl Acad. Sci. USA*, **107**, 7875–7880.

94. Dong, H., Strome, S.E., Salomao, D.R., Tamura, H. et al. (2002) Tumor-associated B7-H1 promotes T-cell apoptosis: a potential mechanism of immune evasion. *Nat. Med.*, **8**, 793–800.

95. Palmer, B.E., Neff, C.P., Lecureux, J., Ehler, A. et al. (2013) In vivo blockade of the PD-1 receptor suppresses HIV-1 viral loads and improves CD4+ T cell levels in humanized mice. *J. Immunol.*, **190**, 211–219.

96. Brahmer, J.R., Drake, C.G., Wollner, I., Powderly, J.D. et al. (2010) Phase I study of single-agent anti-programmed death-1 (MDX-1106) in refractory solid tumors: safety, clinical activity, pharmacodynamics, and immunologic correlates. *J. Clin. Oncol.*, **28**, 3167–3175.
97. Topalian, S.L., Hodi, F.S., Brahmer, J.R., Gettinger, S.N. et al. (2012) Safety, activity, and immune correlates of anti-PD-1 antibody in cancer. *N. Engl. J. Med.*, **366**, 2443–2454.
98. Topalian, S.L., Sznol, M., McDermott, D.F., Kluger, H.M. et al. (2014) Survival, durable tumor remission, and long-term safety in patients with advanced melanoma receiving nivolumab. *J. Clin. Oncol.*, **32**, 1020–1030.
99. Motzer, R.J., Rini, B.I., McDermott, D.F., Redman, B.G. et al. (2014) Nivolumab for metastatic renal cell carcinoma (mRCC): results of a randomized, dose-ranging phase II trial. *J. Clin. Oncol.*, **32**, 5s.
100. Robert, C., Long, G.V., Brady, B., Dutriaux, C. et al. (2015) Nivolumab in previously untreated melanoma without BRAF mutation. *N. Engl. J. Med.*, **372**, 320–330.
101. Ansell, S.M., Lesokhin, A.M., Borrello, I., Halwani , A. et al. (2015) PD-1 blockade with nivolumab in relapsed or refractory Hodgkin's lymphoma. *N. Engl. J. Med.*, **372**, 311–319.
102. Berger, R., Rotem-Yehudar, R., Slama, G., Landes, S. et al. (2008) Phase I safety and pharmacokinetic study of CT-011, a humanized antibody interacting with PD-1, in patients with advanced hematologic malignancies. *Clin. Cancer Res.*, **14**, 3044–3051.
103. Lu, J., Lee-Gabel, L., Nadeau, M.C., Ferencz, T.M. et al. (2014) Clinical evaluation of compounds targeting PD-1/PD-L1 pathway for cancer immunotherapy. *J. Oncol. Pharm. Pract.* doi: 10.1177/1078155214538087
104. Herbst, R.S., Gordon, M.S., Fine, G.D., Sosman, J.A. et al. (2013) A study of MPDL3280A, an engineered PD-L1 antibody in patients with locally advanced or metastatic tumors. *J. Clin. Oncol.*, **31** (31 suppl.), abstract 3000.
105. Weber, J.S., Kudchadkar, R.R., Yu, B., Gallenstein, D. et al. (2013) Safety, efficacy, and biomarkers of nivolumab with vaccine in ipilimumab-refractory or -naive melanoma. *J. Clin. Oncol.*, **31**, 4311–4318.
106. Powles, T., Eder, J.P., Fine, G.D., Braiteh, F.S. et al. (2014) MPDL3280A (anti-PD-L1) treatment leads to clinical activity in metastatic bladder cancer. *Nature*, **515**, 558–562.
107. Heery, C.R., Coyne, G.H.O., Madan, R.A. et al. (2014) Phase I open-label, multiple ascending dose trial of MSB0010718C, an anti-PD-L1 monoclonal antibody, in advanced solid malignancies. *J. Clin. Oncol.*, **32**, 5s.
108. Monney, L., Sabatos, C.A., Gaglia, J.L., Ryu, A. et al. (2002) Th1-specific cell surface protein Tim-3 regulates macrophage activation and severity of an autoimmune disease. *Nature*, **415**, 536–541.
109. Baitsch, L., Baumgaertner, P., Devêvre, E., Raghav, S.K. et al. (2011) Exhaustion of tumor-specific CD8$^+$ T cells in metastases from melanoma patients. *J. Clin. Invest.*, **121**, 2350–2360.
110. Sakuishi, K., Apetoh, L., Sullivan, J.M., Blazar, B.R. et al. (2010) Targeting Tim-3 and PD-1 pathways to reverse T cell exhaustion and restore anti-tumor immunity. *J. Exp. Med.*, **207**, 2187–2194.
111. Zhou, Q., Munger, M.E., Veenstra, R.G., Weigel, B.J. et al. (2011) Coexpression of Tim-3 and PD-1 identifies a CD8+ T-cell exhaustion phenotype in mice with disseminated acute myelogenous leukemia. *Blood*, **117**, 4501–4510.
112. Fourcade, J., Sun, Z., Benallaoua, M., Guillaume, P. et al. (2010) Upregulation of Tim-3 and PD-1 expression is associated with tumor antigen-specific CD8+ T cell dysfunction in melanoma patients. *J. Exp. Med.*, **207**, 2175–2186.
113. Hastings, W.D., Anderson, D.E., Kassam, N., Koguchi, K. et al. (2009) TIM-3 is expressed on activated human CD4+ T cells and regulates Th1 and Th17 cytokines. *Eur. J. Immunol.*, **39**, 2492–2501.
114. Dardalhon, V., Anderson, A.C., Karman, J., Apetoh, L. et al. (2010) Tim-3/galectin-9 pathway: regulation of Th1 immunity through promotion of CD11b + Ly-6G+ myeloid cells. *J. Immunol.*, **185**, 1383–1392.
115. Chiba, S., Baghdadi, M., Akiba, H., Yoshiyama, H. et al. (2012) Tumor-infiltrating DCs suppress nucleic acid-mediated innate immune responses through interactions between the receptor TIM-3

and the alarmin HMGB1. *Nat. Immunol.*, **13**, 832–842.
116. Ngiow, S.F., Teng, M.W.L., and Smyth, M.J. (2011) Prospects for TIM3-targeted antitumor immunotherapy. *Cancer Res.*, **71**, 6567–6571.
117. Triebel, F., Jitsukawa, S., Baixeras, E., Roman-Roman, S. et al. (1990) LAG-3, a novel lymphocyte activation gene closely related to CD4. *J. Exp. Med.*, **171**, 1393–1405.
118. Goldberg, M.V. and Drake, C.G. (2011) LAG-3 in cancer immunotherapy. *Curr. Top. Microbiol. Immunol.*, **344**, 269–278.
119. Grosso, J.F., Kelleher, C.C., Harris, T.J., Maris, C.H. et al. (2007) LAG-3 regulates CD8+ T cell accumulation and effector function in murine self- and tumor-tolerance systems. *J. Clin. Invest.*, **117**, 3383–3392.
120. Huang, C.-T., Workman, C.J., Flies, D., Pan, X. et al. (2004) Role of LAG-3 in regulatory T cells. *Immunity*, **21**, 503–513.
121. Blackburn, S.D., Shin, H., Haining, W.N., Zou, T. et al. (2009) Coregulation of CD8+ T cell exhaustion by multiple inhibitory receptors during chronic viral infection. *Nat. Immunol.*, **10**, 29–37.
122. Hannier, S., Tournier, M., Bismuth, G., and Triebel, F. (1998) CD3/TCR complex-associated lymphocyte activation gene-3 molecules inhibit CD3/TCR signaling. *J. Immunol.*, **161**, 4058–4065.
123. Woo, S.-R., Turnis, M.E., Goldberg, M.V., Bankoti, J. et al. (2012) Immune inhibitory molecules LAG-3 and PD-1 synergistically regulate T-cell function to promote tumoral immune escape. *Cancer Res.*, **72**, 917–927.
124. Watanabe, N., Gavrieli, M., Sedy, J.R., Yang, J. et al. (2003) BTLA is a lymphocyte inhibitory receptor with similarities to CTLA-4 and PD-1. *Nat. Immunol.*, **4**, 670–679.
125. Gonzalez, L.C., Loyet, K.M., Calemine-Fenaux, J., Chauhan, V. et al. (2005) A coreceptor interaction between the CD28 and TNF receptor family members B and T lymphocyte attenuator and herpesvirus entry mediator. *Proc. Natl Acad. Sci. USA*, **102**, 1116–1121.
126. Han, L., Wang, W., Lu, J., Kong, F. et al. (2014) AAV-sBTLA facilitates HSP70 vaccine-triggered prophylactic antitumor immunity against a murine melanoma pulmonary metastasis model in vivo. *Cancer Lett.*, **354**, 398–406.
127. Cheung, T.C., Oborne, L.M., Steinberg, M.W., Macauley, M.G. et al. (2009) T cell intrinsic heterodimeric complexes between HVEM and BTLA determine receptivity to the surrounding microenvironment. *J. Immunol.*, **183**, 7286–7296.
128. Steinberg, M.W., Turovskaya, O., Shaikh, R.B., Kim, G. et al. (2008) A crucial role for HVEM and BTLA in preventing intestinal inflammation. *J. Exp. Med.*, **205**, 1463–1476.
129. Gertner-Dardenne, J., Fauriat, C., Orlanducci, F., Thibult, M.L. et al. (2013) The co-receptor BTLA negatively regulates human Vγ9Vδ2 T-cell proliferation: a potential way of immune escape for lymphoma cells. *Blood*, **122**, 922–931.
130. Kaehler, K.C., Piel, S., Livingstone, E., Schilling, B. et al. (2010) Update on immunologic therapy with anti-CTLA-4 antibodies in melanoma: identification of clinical and biological response patterns, immune-related adverse events, and their management. *Semin. Oncol.*, **37**, 485–498.
131. Hodi, F.S., Friedlander, P.A., Atkins, M.B., McDermott, D.F.et al. (2011) A phase I trial of ipilimumab plus bevacizumab in patients with unresectable stage III or stage IV melanoma. *J. Clin. Oncol.*, **29** (Suppl.)(abstract 8511).
132. Downey, S.G., Klapper, J.A., Smith, F.O., Yang, J.C. et al. (2007) Prognostic factors related to clinical response in patients with metastatic melanoma treated by CTL-associated antigen-4 blockade. *Clin. Cancer Res.*, **13**, 6681–6688.
133. Hamid, O., Schmidt, H., Nissan, A., Ridolfi, L. et al. (2011) A prospective phase II trial exploring the association between tumor microenvironment biomarkers and clinical activity of ipilimumab in advanced melanoma. *J. Transl. Med.*, **9**, 204.
134. Blank, C.U. and Enk, A. (2015) Therapeutic use of anti-CTLA-4 antibodies. *Int. Immunol.*, **27**, 3–10.
135. Wang, C.J., Kenefeck, R., Wardzinski, L., Attridge, K. et al. (2012) Cutting edge: cell-extrinsic immune regulation by CTLA-4 expressed on conventional T cells. *J. Immunol.*, **189**, 1118–1122.
136. Snyder, A., Makarov, V., Merghoub, T., Yuan, J. et al. (2014) Genetic basis

137. Kohrt, H., Kowanetz, M., Gettinger, S., Powderly, J. et al. (2013) Intratumoral characteristics of tumor and immune cells at baseline and on-treatment correlated with clinical responses to MPDL3280A, an engineered antibody against PD-L1. *J. Immunother. Cancer*, **1**, O12.
138. Herbst, R.S., Soria, J.-C., Kowanetz, M., Fine, G.D. et al. (2014) Predictive correlates of response to the anti-PD-L1 antibody MPDL3280A in cancer patients. *Nature*, **515**, 563–567.
139. Sarnaik, A.A., Yu, B., Yu, D., Morelli, D. et al. (2011) Extended dose ipilimumab with a peptide vaccine: immune correlates associated with clinical benefit in patients with resected high-risk stage IIIc/IV melanoma. *Clin. Cancer Res.*, **17**, 896–906.
140. Meyer, C., Cagnon, L., Costa-Nunes, C.M., Baumgaertner, P. et al. (2014) Frequencies of circulating MDSC correlate with clinical outcome of melanoma patients treated with ipilimumab. *Cancer Immunol. Immunother.*, **63**, 247–257.
141. Couzin-Frankel, J. (2013) Breakthrough of the year 2013. Cancer immunotherapy. *Science*, **342**, 1432–1433.
142. Fridman, W.H., Pagès, F., Sautès-Fridman, C., and Galon, J. (2012) The immune contexture in human tumours: impact on clinical outcome. *Nat. Rev. Cancer*, **12**, 298–306.
143. Galon, J., Costes, A., Sanchez-Cabo, F., Kirilovsky, A. et al. (2006) Type, density, and location of immune cells within human colorectal tumors predict clinical outcome. *Science*, **313**, 1960–1964.
144. Galon, J., Mlecnik, B., Bindea, G., Angell, H.K. et al. (2014) Towards the introduction of the "Immunoscore" in the classification of malignant tumours. *J. Pathol.*, **232**, 199–209.
145. Curran, M.A., Montalvo, W., Yagita, H., and Allison, J.P. (2010) PD-1 and CTLA-4 combination blockade expands infiltrating T cells and reduces regulatory T and myeloid cells within B16 melanoma tumors. *Proc. Natl Acad. Sci. USA*, **107**, 4275–4280.
146. Tumeh, P.C., Harview, C.L., Yearley, J.H., Shintaku, I.P. et al. (2014) PD-1 blockade induces responses by inhibiting adaptive immune resistance. *Nature*, **515**, 568–571.
147. Wolchok, J.D., Kluger, H., Callahan, M.K., Postow, M.A. et al. (2013) Nivolumab plus ipilimumab in advanced melanoma. *N. Engl. J. Med.*, **369**, 122–133.
148. Deng, L., Liang, H., Burnette, B., Beckett, M. et al. (2014) Irradiation and anti-PD-L1 treatment synergistically promote antitumor immunity in mice. *J. Clin. Invest.*, **124**, 687–695.
149. Li, B., VanRoey, M., Wang, C., Chen, T.H. et al. (2009) Anti-programmed death-1 synergizes with granulocyte macrophage colony-stimulating factor--secreting tumor cell immunotherapy providing therapeutic benefit to mice with established tumors. *Clin. Cancer Res.*, **15**, 1623–1634.
150. Melero, I., Grimaldi, A.M., Perez-Gracia, J.L., and Ascierto, P.A. (2013) Clinical development of immunostimulatory monoclonal antibodies and opportunities for combination. *Clin. Cancer Res.*, **19**, 997–1008.
151. Peng, W., Lizée, G., and Hwu, P. (2013) Blockade of the PD-1 pathway enhances the efficacy of adoptive cell therapy against cancer. *Oncoimmunology*, **2**, e22691.
152. Ribas, A., Hodi, F.S., Callahan, M., Konto, C. et al. (2013) Hepatotoxicity with combination of vemurafenib and ipilimumab. *N. Engl. J. Med.*, **368**, 1365–1366.
153. Tarhini, A.A., Cherian, J., Moschos, S.J., Tawbi, H.A. et al. (2012) Safety and efficacy of combination immunotherapy with interferon alfa-2b and tremelimumab in patients with stage IV melanoma. *J. Clin. Oncol.*, **30**, 322–328.
154. Demaria, S. and Formenti, S.C. (2013) Radiotherapy effects on anti-tumor immunity: implications for cancer treatment. *Front. Oncol.*, **3**, 128.
155. Hodi, F.S., Lawrence, D., Lezcano, C., Wu, X. et al. (2014) Bevacizumab plus ipilimumab in patients with metastatic melanoma. *Cancer Immunol. Res.*, **2**, 632–642.
156. Frederick, D.T., Piris, A., Cogdill, A.P., Cooper, Z.A. et al. (2013) BRAF inhibition is associated with enhanced melanoma antigen expression and a more favorable tumor microenvironment in patients with metastatic melanoma. *Clin. Cancer Res.*, **19**, 1225–1231.

157. Jiang, X., Zhou, J., Giobbie-Hurder, A., Wargo, J. *et al.* (2013) The activation of MAPK in melanoma cells resistant to BRAF inhibition promotes PD-L1 expression that is reversible by MEK and PI3K inhibition. *Clin. Cancer Res.*, **19**, 598–609.

158. Wilmott, J.S., Long, G.V., Howle, J.R., Haydu, L.E. *et al.* (2012) Selective BRAF inhibitors induce marked T-cell infiltration into human metastatic melanoma. *Clin. Cancer Res.*, **18**, 1386–1394.

7
Molecular Mediators: Cytokines

Jean-Marc Cavaillon
Institut Pasteur, Unit Cytokines and Inflammation, Infection and Epidemiology Department, 28 Rue Dr Roux, 75015 Paris, France

1	**Cytokines: The Historical Record**	232
2	**A Universal Language of Cells**	234
2.1	Definitions	234
2.2	Families	235
2.3	Artificial Cytokines	237
3	**Receptors**	237
3.1	Families	237
3.2	Signaling	239
3.3	Soluble Receptors	240
4	**Functions**	241
4.1	Hematopoiesis	241
4.2	Immune Response	242
4.2.1	Innate Immunity	242
4.2.2	Adaptive Immunity	243
4.2.3	T-Cell Subsets and Cytokines	243
4.3	Cell Survival and Cell Death	244
4.4	Links with the Central and Peripheral Nervous Systems	245
4.5	Reproduction	246
4.6	Inflammation	247
5	**Life without Cytokines (What Knock-Out Mice Tell Us)**	249
6	**Cytokine Synthesis**	249
6.1	Homeostasis	249
6.2	Induction	252
6.3	Parameters That Affect Production	252
6.4	Measurements	252

Translational Medicine: Molecular Pharmacology and Drug Discovery
First Edition. Edited by Robert A. Meyers.
© 2018 Wiley-VCH Verlag GmbH & Co. KGaA. Published 2018 by Wiley-VCH Verlag GmbH & Co. KGaA.

7	**The Cytokine Network** 253
8	**Individual Heterogeneity** 254
9	**Cytokines and Infections** 256
9.1	Half-Angel–Half-Devil 256
9.2	The Strategies of Microbes against Cytokines 256
10	**Cytokines and Diseases** 257
10.1	Autoimmune Diseases 257
10.2	Allergy and Asthma 258
10.3	Sepsis 258
10.4	Chronic Inflammatory Diseases 258
10.5	Cancer 259
10.6	Hepatitis C 259
10.7	Transplantation 259
10.8	Multiple Sclerosis 259
10.9	Neutropenia 260
10.10	Osteoporosis 260
10.11	Autoinflammatory Disorders 260
11	**Conclusions** 260
	References 260

Keywords

Cytokines
Cytokines are soluble mediators (sometimes membrane-bound) that are constitutively released or produced upon cell activation, and which bind to receptors on target cells and induce, modulate or inhibit cellular functions. Most cytokines are produced by immune cells and act on immune cells; however, cytokines are produced by all cell types and act on a great variety of cells, thus displaying a wide spectrum of activities. Cytokines are characterized by their redundancy, their pleiotropy, their synergistic action, and their regulatory loops; those produced constitutively contribute to tissue and systemic homeostasis. Cytokines are elements of a universal language used by cells to communicate with each other.

Chemokines
Chemokines are a subfamily of cytokines with structural similarities; they have the property of recruiting cells within the extravascular space of the tissues (chemotaxis), either at homeostasis or during inflammatory processes.

Interferons (IFNs)
Interferons are a subfamily of cytokines sharing an antiviral bioactivity. Interferons are characterized by different subgroups (types I, II and III), depending on their nature. In addition to their common property, some interferons (e.g., IFNγ) are involved at different stages of the onset and further development of the immune responses, or they may play a role during gestation in mammals (e.g., IFNτ in ovines and bovines).

Interleukin (IL)
Interleukin is a convenient name for classifying cytokines. The IL-numbers have been allocated in the order of their discovery through the years. This term does not define a family, as few are biochemically or biologically related. The most recent ILs have been identified after gene banks, and are thus related to previously well-known ILs. In 1979, when the term was coined, only IL-1 and IL-2 were recognized.

Receptors
The receptors of cytokines allow the cells to respond to the signals delivered by cytokines. Receptors may be composed of one, two or three different chains. Most chains are constituted by one extracellular domain, one transmembrane domain, and one intracellular domain; exceptions to this are chemokine receptors, which are constituted by seven transmembrane domains.

> Cytokines are proteins or glycoproteins which are produced by cells and act on other cells that display on their surface specific cytokine receptors. Cytokines are used by cells to communicate. Within a determined sequence, these mediators lead the responding cells to modify their function (e.g., secretion, proliferation, induction, inhibition, enhanced or reduced function, migration, apoptosis). Despite cytokines having mainly been discovered by immunologists as a product of cells of leukocyte lineage, they are now recognized as elements of a universal language used by most cells of any other lineage. Accordingly, they are essential for many events through life, such as reproductive tissue remodeling, embryogenesis, steady-state and adaptive hematopoiesis, surveillance and maintenance of tissue structure, functions of the immune system, inflammation, cell survival, and cell death. Cytokines also allow a dialog with the central nervous system, and some may modify different behaviors (e.g., fever, anorexia, sleep). Cytokines usually act within their vicinity, but they can also act *via* an endocrine fashion. Their production is tightly controlled within a complex network of positive and negative loops. They are a prerequisite for the control of infection by invasive microorganisms, though their exacerbated production may be deleterious at local or systemic levels. Their "half-angel/half-devil" aspect has rendered cytokines difficult to use for therapeutic purposes, though some recombinant cytokines or cytokine-neutralizing strategies have been used successfully in different pathological conditions.

1
Cytokines: The Historical Record

The history of cytokines can be divided into four main stages. The first stage, dating from the late 1940s to the early 1970s, included an identification of the biological activities of factors present in cell culture supernatants or within the bloodstream. During the second stage, from the late 1970s to the early 1980s, the purification and biochemical characterization of these factors allowed the definition of molecular entities, while during the third stage from the mid-1980s to the mid-1990s the cloning of coding and noncoding sequences, the production of recombinant cytokines and the production of genetically manipulated mice extended knowledge of the complexity of the cytokine-mediated language. During the final stage, in the late 1990s, the discovery of new cytokines relied on the availability of different genome sequences and different gene banks, and the bioactivities of newly identified cytokines were discovered shortly afterwards.

The first reported bioactivity of cytokines was most likely the induction of fever, and during the late 1940s and early 1950s this led to the quest for an endogenous pyrogen. The first efforts at characterizing the pyrogen, made by Valy Menkin (1901–1960), resulted in "pyrexin" [1], the activities of which were probably due to endotoxin contamination. In 1953, Ivan L. Bennet Jr (1922–1990) and Paul Beeson (1908–2006) extracted a fever-producing substance from rabbit polymorphonuclear leucocytes [2], while in 1955 Elisha Atkins (1921–2005) and W. Barry Wood Jr (1910–1971) identified a circulating endogenous pyrogen in the blood of rabbits following the injection of typhoid vaccine [3]. The purification of a human endogenous pyrogen was reported in 1977 by Charles Dinarello [4], and this allowed the endogenous pyrogen to be identified as interleukin (IL)-1. As a result, it was established definitively that the fever was due to endogenous factor(s) and was not directly induced by microbial products. Following the cloning in 1984 of human IL-1β by Dinarello and colleagues [5], and of murine IL-1α by Mizel *et al.* [6], this effect was confirmed when an injection of recombinant IL-1 was shown to induce fever in rabbits. Later, genes encoding different IL-1 homologs were identified by means of their significant sequence similarities to IL-1.

Another earlier-described bioactivity was reported in 1957 by Isaacs and Lindenmann [7], who showed that resistance to virus development could be conferred upon a cell by another virus-infected cell. This bioactivity, which interfered with virus spreading, was termed "interferon" (IFN). However, it emerged subsequently that Isaacs and Lindenmann were not the discoverers of IFN; in fact, IFN-mediated bioactivity had been identified three years earlier by Nagano and Kojima, in Tokyo, who noted that rabbit skin or testis which had been previously inoculated with an ultra-violet light-inactivated virus exhibited an inhibition of viral growth when reinfected at the same site with live virus [8]. The fact these authors failed to create a new term and also reported their findings in a French-speaking journal did not benefit the international recognition of their discovery [9].

During the 1960s, numerous bioactivities were detected in the supernatants of lymphocytes and, accordingly, were identified as non-antibody mediators. Hence, in 1969 Dumonde *et al.* first proposed the term "lymphokine" [10], and during the ensuing years the term "monokine" was used to characterize other bioactivities found in monocyte/macrophage supernatants. In the 1970s the most famous lymphokine, the macrophage migration inhibitory factor (MIF), owed its name to its *in-vitro* properties (it was first identified in 1966 by Bloom and Bennet as a T-lymphocyte-derived product associated with delayed-type hypersensitivity [11]). In the absence of cell lines capable of producing large amounts of MIF, the biochemical characterization of MIF was delayed and the molecule was not cloned until 1989, long after its discovery [12]. In the meantime, MIF-like activity was reported in the culture supernatants of virus-loaded fibroblasts by Cohen and coworkers [13]. At this point it seemed evident that the lymphokines had perhaps been incorrectly named, because many cell types other than lymphocytes could produce similar bioactivities. Accordingly, in 1974, Cohen coined the term "cytokine" [14]: *"It is interesting that a concept I initially had trouble getting published took on such an astounding life of its own."* Indeed, MIF is a good example of a cytokine produced by many cell types, including macrophages, dendritic leukocytes, endothelial cells, and fibroblasts. Perhaps the most fascinating point was the rediscovery of MIF as a product of the pituitary gland [15], which further reinforced the concept of cytokines and illustrated that, on occasion, the borderline between cytokines and hormones may be difficult to draw.

Although, traditionally, the names of factors reflected their various biologic activities as identified in different laboratories, this approach resulted in an imprecise and redundant nomenclature. In 1979, during the Second International Lymphokine Workshop at Ermatingen, Switzerland, the decision was taken to give the name "interleukin" to two different characterized factors. Under the new terminology of IL-1 were gathered numerous names or acronyms defined according to one of the properties of this cytokine, including osteoclast-activating factor (OAF), lymphocyte-activating factor (LAF), B-cell-activating factor (BAFF), epidermal thymocyte-activating factor (ETAF), T-cell-replacing factor III (TRFIII), leukocyte endogenous mediator, endogenous pyrogen, mitogenic protein, catabolin, and hemopoietin-1.

The term interleukin (meaning "between leukocytes") was clearly not the best choice, as these factors were not restricted to leukocytes; unfortunately, the term had been mainly selected by immunologists who were too self-absorbed and considered that the immune system functioned independently of other systems. In an interview with Stanley Cohen on this topic, he specified: *"It is a common misconception that the term interleukin was accepted for adoption by the nomenclature committee. When it was proposed, both Byron Waksman and I, as Co-chairs, felt it would turn out to be too restrictive (even then) and there was general agreement by the committee that it would be premature to refer to mediators as interleukins. This view was conveyed to the proposers. However, shortly thereafter, an editorial appeared in Journal of Immunology stating that the term "interleukin" had been discussed and considered by the committee (true) and that henceforth the term would be used in scientific communications by the scientists writing the editorial. This was a true-true-unrelated kind of position to take, but it caught on in the general scientific community"* [16].

The final neologism – "chemokine" – was created in 1992 in an attempt to gather under one name a large family of small, biochemically related cytokines, all of which possessed chemoattractant properties.

The availability of the first cloned molecules allowed major progress to be made, not only in fundamental research but also for clinical applications. While growth hormone was the first recombinant molecule to be used in humans in 1980, IFN-alpha was the first recombinant cytokine to be injected into humans in 1982 [17], and IL-2 the second in 1985 [18, 19]. The development of methods employed to produce interleukins was astonishing. In 1984, a total volume of 10 000 liters of supernatants of activated Jurkat cells was required to produce 30 mg of natural IL-2, yet one year later, in 1985, only 10 liters of supernatants from recombinant *Escherichia coli* expressing the IL-2 gene were sufficient to prepare 1 g of recombinant IL-2.

2
A Universal Language of Cells

2.1
Definitions

Cytokines have been defined by immunologists as cooperative factors produced by activated lymphocytes, natural killer (NK) cells, dendritic cells, macrophages, neutrophils, mast cells that induce, modulate, favor, increase and inhibit cell proliferation, cell differentiation, or cell activities such as antibody production or mediator production. The onset of the messages delivered by these soluble (or membrane-bound) molecules relies on the binding to their specific receptors displayed on the surface of the immune cells, a process which leads to an intracellular signaling cascade. However, far from being limited to the leukocyte lineages, this is a universal language of cells of the whole body, independently of their lineages, their nature, and their localization. This is probably the main difference between cytokines and other soluble molecules such as neuromediators, which are mainly produced by the brain and the neurons, and hormones, which are mainly produced by the endocrine system (Table 1).

Although, like hormones, cytokines can act in an endocrine fashion, their actions mainly occur within the cellular microenvironment in a paracrine fashion. However, whilst some cytokines can be expressed on the cell surface of the producer cells, cytokines can act in a juxtacrine fashion within a close contact between cells in the immunological synapse. Cytokines can also act as an autocrine factor for cells that produce cytokines and also display their receptors. Finally, some rare cytokines – such as IL-1α, IL-33,

Tab. 1 Comparison between hormones and cytokines.

	Sources	Targets	Activities	Action
Hormones	Secreted by a unique cell lineage	Specificity rather limited to one single type of target cell	Single action	Endocrine
Cytokines	Produced by many cell types	Numerous target cells	Wide spectrum of activities (redundancy)	Paracrine Juxtacrine Autocrine Endocrine Intracrine

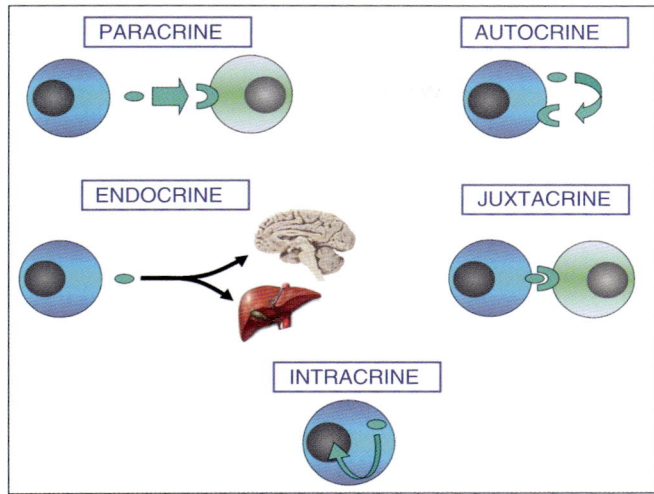

Fig. 1 Mode of action of cytokines.

IL-37 and IL-1Ra (interleukin-1 receptor antagonist – can act in an intracrine fashion, moving directly from the cytoplasm to the nucleus (Fig. 1).

Cell interactions involve more than one cytokine. Typically, each cytokine can be considered as a word, and the full sentence that will finally emerge is a reflection of the delivery of different cytokine-mediated signals. Together, these signalings lead to orientate the activities of the target cell; thus, the order of delivery of cytokines has major consequences on the nature of the message. Furthermore, the concentration of cytokines, their localization, the nature of the target cell, and the timing, are all important parameters that affect the exact nature of the cytokine-initiated processes.

In most cases, the constitutive production of cytokines is limited and most cytokines are produced upon the activation of cells. However, highly sensitive readout assays have allowed a constitutive production to be determine at homeostasis in numerous tissues and cells.

2.2 Families

The classification of cytokines can be achieved according to either:

- A common bioactivity: antiviral activity (IFNs), chemoattraction (chemokines), hematopoiesis (colony-stimulating factors; CSFs).
- A common biochemical structure, suggesting a common ancestral molecule (chemokines, tumor necrosis factor (TNF), IL-1, or IL-10 families) [20, 21] or a similar heterodimeric structure [22].
- The sharing of a common chain of a receptor (gp130 cytokine family) (Table 2).

Different subfamilies of IFNs are known. On human chromosome 9, 13 genes of IFNα have been identified among the 17 type I IFN genes [23]. All human IFNα exhibit >50% identity at the amino acid level and share with IFNβ the same receptor, interferon-alpha receptor (IFNAR), comprising two transmembrane subunits. IFNγ produced during an immune response

Tab. 2 Families of cytokines.

Interferons	Type I: IFNα, IFNβ, IFNκ, IFNω, (IFNτ); Type II: IFNγ; Type III: IFNλ
IL-1α,	IL-1β, IL-1Ra, IL-18, IL-33, IL-36α,β,γ, IL-36Ra, IL-37, IL-38
	IL-2, IL-15, IL-21 IL-3, IL-5, IL-7, IL-9, IL-34 IL-4, IL-13
	IL-6, IL-11, IL-31 IL-10, IL-19, IL-20, IL-22, IL-24, IL-26
	IL-12, IL-23, IL-27, IL-35 IL-17A-E, IL-25 IL-28, IL-29
	IL-8 (CXCL8), IL-14, IL-16, IL-30, IL-32
Hematopoietic factors	M-CSF[b], G-CSF, GM-CSF, stem cell factor
Chemokines	CCL1 … CCL28; CXCL1 … CXCL17; XCL1 and 2; CX3CL1
TNF superfamily	TNF, Ltα, Ltβ, NGF, CD27L, CD30L, CD40L, CD137L, APRIL, BAFF, EDA-A1 and A2, FasL, GITRL, LIGHT, OPGL, OX40L, TRAIL, RANKL, TWEAK, VEGI
Growth factors	TGFβ$_{1,2,3}$
Gp130 cytokine family	IL-6, IL-11, IL-31, CNTF, LIF, oncostatin-M, cardiotrophin-1

a) Biochemically related interleukins or interleukins sharing similar bioactivities are framed.
b) Abbreviations: APRIL: apoptosis-inducing ligand; BAFF: B-cell-activating factor; CCL: chemokine CC ligand; CNTF: ciliary neurotrophic factor; CSF: colony-stimulating factor; CXCL: chemokine CXC ligand; EDA: ectodysplasin-A; GITRL: glucocorticoid-induced TNF receptor family-related gene ligand; IL-1F: IL-1 family; LIF: leukemia inhibitory factor; LIGHT: lymphotoxins, inducible expression, competes with herpes simplex virus glycoprotein D for HVEM, a receptor expressed on T lymphocytes; Lt: lymphotoxin; NGF: nerve growth factor; OPGL: osteoprotegerin ligand; RANKL: receptor activator of NF-κB ligand; TGFβ: transforming growth factor-β; TRAIL: TNF-related apoptosis-inducing ligand; TWEAK: TNF-like weak inducer of apoptosis; VEGI: vascular endothelial cell growth inhibitor.

is a potent macrophage activator; IFNω is induced by virus; IFNτ is produced by trophoblasts only in ovines and bovines; and IFNκ is expressed by keratinocytes. Among chemokines, four different subfamilies can be identified based on the highly conserved presence in the N-terminal region of two cysteine residues, which are either or not separated by one or three amino acids: the CC, CXC, CX3C, and the C chemokines. The nomenclature has added the letter "L" for ligand: CXCL1 → 17, CCL1 → 28, XCL1 → 2, CX$_3$CL1. The assigned numbers of ILs only reflect the timing of their discovery. Thus, some ILs are also hematopoietic factors (e.g., IL-3, previously named multi-CSF), chemokines (IL-8, also named CXCL8), or even IFN (IL-28A, IL-28B, and IL-29 are identical to IFNλ2, -3, and -1, respectively). The first ILs have been discovered as bioactive factors, and there is no similarity in their amino acid sequences; however, there are some homologies in the tertiary and quaternary structures and similar helical structures are found. For example, the four-helix bundle is found in IL-2, IL-4, IL-6, granulocyte-macrophage colony-stimulating factor (GM-CSF), granulocyte-colony stimulating factor (G-CSF) and macrophage colony-stimulating factor (M-CSF), and is shared with the growth hormone. The more recently discovered ILs have been identified in gene banks according to sequence homology to previously identified ILs. The TNF ligand superfamily gathers 19 distinct members that include soluble molecules with biochemical homology as well as membrane-bound molecules [24]. Other cytokines that can be released and expressed at the plasma membrane of the producing cells include IL-1α, IL-10, IL-15, and IFNγ. The TNF ligand superfamily also includes a growth factor (i.e., nerve growth factor). Usually, growth factors have a rather narrow spectrum of activity, are produced constitutively, are not acting on hematopoietic cells, and thus are distinct from cytokines. An exception here is the transforming growth factor-β family (TGFβ), particularly because at least one member – namely TGFβ1 – displays both anti-inflammatory and pro-inflammatory properties.

The localization of genes on chromosomes is known, and often genes of a family are gathered on the same chromosome; for example, IL-1α, IL-1β and IL-1 receptor antagonist are on human chromosome 2; most CXCL chemokines are on human chromosome 4; and most chemokine CC ligand (CCL) chemokines are on human chromosome 17. Most cytokines are glycosylated and their molecular weights range between 7 and 30 kDa. Most cytokines are monomers (e.g., IL-1, IL-2, IL-3, IL-4), and few are homodimers (e.g., IL-8) or homotrimers (e.g., TNF). Some are heterodimers, such as IL-12, IL-23, IL-27, and IL-35. In this case, IL-12 and IL-23 share the same p40 chain; IL12 and IL-35 share the p35 chain; and IL-27 and IL-35 share the EBI3 chain [25]. Some are heterotrimers, such as lymphotoxin-β (Ltβ), which is an association of the Ltα and Ltβ chains.

2.3
Artificial Cytokines

Recently, new cytokines have been created with the aim of improving their properties. For example, an IL-2 "superkine" has been produced which was rendered CD25 (IL-2Rα) -independent and displayed a better antitumoral activity, with fewer secondary effects [26]. The other approach involved the fusing of two cytokines, with the emerging "fusokine" (such as that linking GM-CSF and IL-15) displaying unexpected immunosuppressive properties [27].

3
Receptors

3.1
Families

Although most cytokine receptors were characterized once their ligands had been identified, there were a few receptors – termed orphan receptors – that were known before their ligands; examples include: c-kit, the receptor of stem cell factor (SCF); CD40, the receptor of CD40L; and IL-1 receptor-related protein-1, part of the IL-18 receptor. Cytokine receptors are characterized by domains derived from ancestral common genes. In terms of evolution, it would be interesting to decipher which mutations in genes coding for cytokines or for cytokine receptors were first fixed. A common domain, called the "hemopoietin receptor superfamily domain" and characterized by the presence of four cysteine residues and a conserved sequence tryptophan-serine-x-tryptophan-serine, defines a family of numerous cytokine receptors (Fig. 2) [28]. Most interestingly, this family of receptors also includes receptors for hormones (e.g., growth hormone, prolactin) or hematopoietic factor (e.g., erythropoietin). This domain can be duplicated (the common β-chain of IL-3R, IL-5R, and GM-CSFR) or may be associated with other domains such as an immunoglobulin-like domain (IL-6Rα) or a fibronectin type III-like domain (gp130, G-CSF-R). In most cases, receptors are the association of more than one transmembrane chain. Usually, one chain is essential for binding (α-chain),

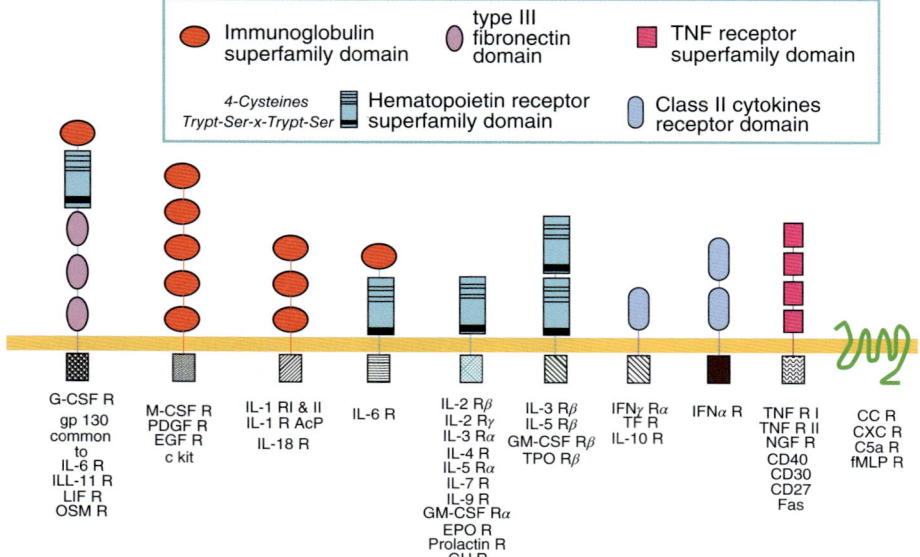

Fig. 2 Families of cytokine receptors.

while a second chain is required to initiate the intracellular signaling (β- or γ-chain). Some chains are common for receptors of a few cytokines; this is the case for the gp130 chain of the receptors of IL-6, IL-11, IL-27, ciliary neurotrophic factor (CNTF), leukemia inhibitory factor (LIF), Oncostatin-M, and Cardiotrophin-1, or the gamma chain of IL-2, IL-4, IL-7, IL-9, IL-13, IL-15 and IL-21 (Fig. 3).

Another important family is the TNF receptor family. Activation by TNF is consecutive to the bridging of three identical chains of the TNF receptor by the cytokine. Each cytokine receptor is highly specific for its ligand, except for tumor necrosis factor receptors (TNFRs) p55 and p75 which bind both TNF and Ltα, and the chemokine receptors that can bind various chemokines. Chemokine receptors, constituted by a seven-transmembrane chain, are similar to receptors for other chemoattractant molecules such as the anaphylatoxin C5a or the bacterial fMet-Leu-Phe peptide.

Cytokine receptors are often internalized after ligation to their ligands, and their expression can be induced, downregulated, or upregulated.

In some circumstances, two cells can cooperate, allowing a transpresentation of cytokines [29]. One cell expressing the α-chain can capture the cytokine and present it to another cell expressing the β- and γ-chains. This can be the case for cytokines such as IL-2 and IL-15. An interesting concept of reverse signaling has recently emerged enabling a two-way communication in cell-to-cell signaling. This property has mainly been observed within the TNF superfamily for which the membrane form of the ligand, following interaction with its receptor on another cell, transmits a signal to the cell [30].

Finally, certain receptors behave as decoy receptors as they are able to bind the cytokines, but this interaction will not initiate any signaling. This is the case for IL-1RII, some chemokine receptors (DARC,

Molecular Mediators: Cytokines | 239

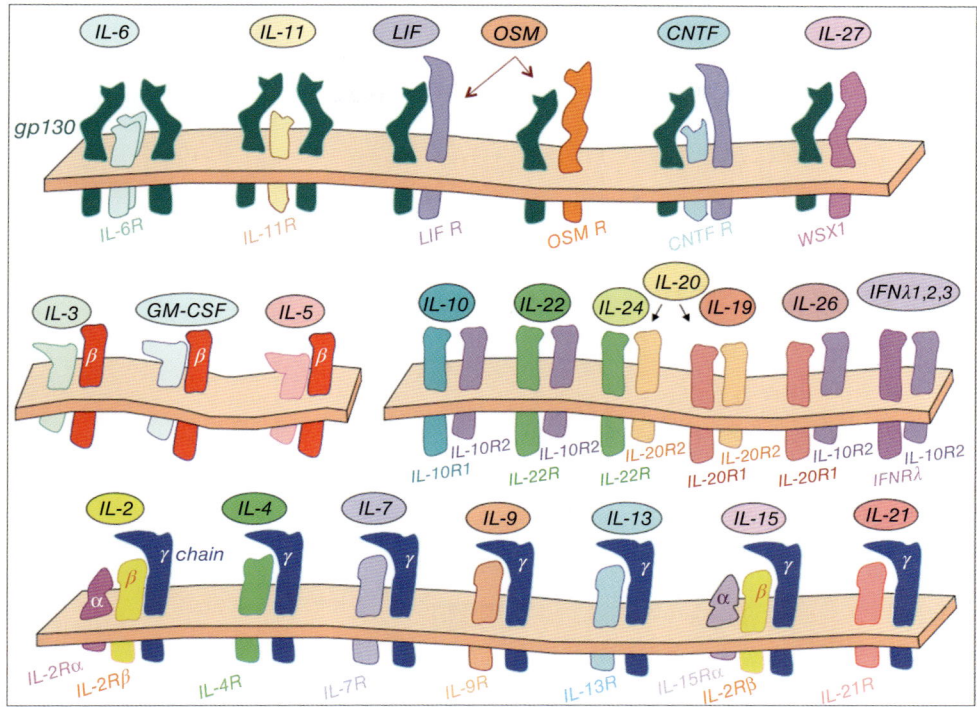

Fig. 3 Cytokine receptors share common chains.

D6, CCX-CKR) [31], and some receptors of the TNF superfamily (DcR1, 2, and 3, osteoprotegerin). Their expression on tumor cells can prevent them from being killed by the TNF-related apoptosis-inducing ligand (TRAIL) or FasL.

3.2
Signaling

Upon binding of the cytokine to the extracellular domain of receptors, the intracellular receptor domains become associated with a variety of signaling molecules. A cascade of phosphorylation of various adaptor proteins occurs following interaction of the intracellular domain of the receptors with cytoplasmic tyrosine kinases (e.g., Janus kinases; Jaks). The phosphorylation of latent cytoplasmic transcriptional activators (e.g., signal transducers and activators of transcription; STAT) allows their dimerization and translocation into the nucleus, where they bind to specific sequences present in the promoters of certain genes. Other pathways involving the mitogen-activated protein kinase (MAPK) and other kinases lead to the activation of various transcription factors (e.g., c-fos/c-jun, NF-κB). Some signaling pathways are shared with other receptors; this is true in the case of IL-1 and IL-18 receptors, the intracellular domain of which is similar to the Toll-like receptors (TLRs) (Toll-IL-1 Receptor; TIR) domain. Thus, similar adaptors (e.g., MyD88, IRAK)

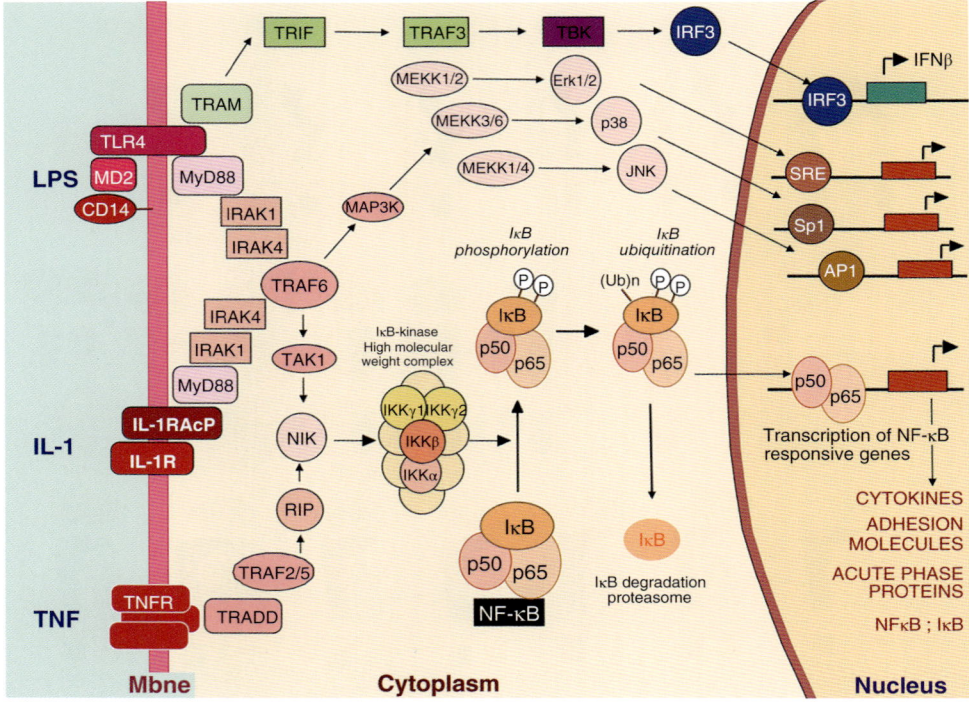

Fig. 4 Example of signaling cascade initiated by TNF and IL-1, the later sharing some common pathways with endotoxin. LPS: lipopolysaccharide.

are involved after interaction of the respective receptors with their ligands, either members of the IL-1 family or pathogen-associated molecular patterns (PAMPs) such as endotoxin lipopolysaccharide (LPS) (Fig. 4). The intracellular domain of some receptors (TNF R p55; Fas; TRAIL R) possesses a death domain that initiates a specific signaling cascade leading to apoptosis.

3.3
Soluble Receptors

Receptors are an integral part of the cytokine network (cf. Sect. 7), as not only their membrane expression can be modulated but they also exist as soluble molecules. Indeed, most cytokine receptors (except chemokine receptors) can be shed from the cell surface after enzymatic cleavage following (or not) a neosynthesis. As shown in Fig. 5, depending on the nature of the receptor its soluble form can behave as an inhibitor (e.g., soluble IL-1R, soluble TNFR) so as to prevent the cytokine from reaching the membrane form. The receptor can also be a carrier, protecting the cytokine within a complex and increasing its half-life in the plasma or other biological fluids. Dissociation of the cytokine–soluble receptor complex can then allow a release of the cytokine close to a membrane receptor (e.g., TNFR). Lastly, in the case of a receptor constituted by an α-chain that binds the cytokine and a β-chain that recognizes the bound cytokine, the soluble α-chain receptor can enhance the responsiveness to the

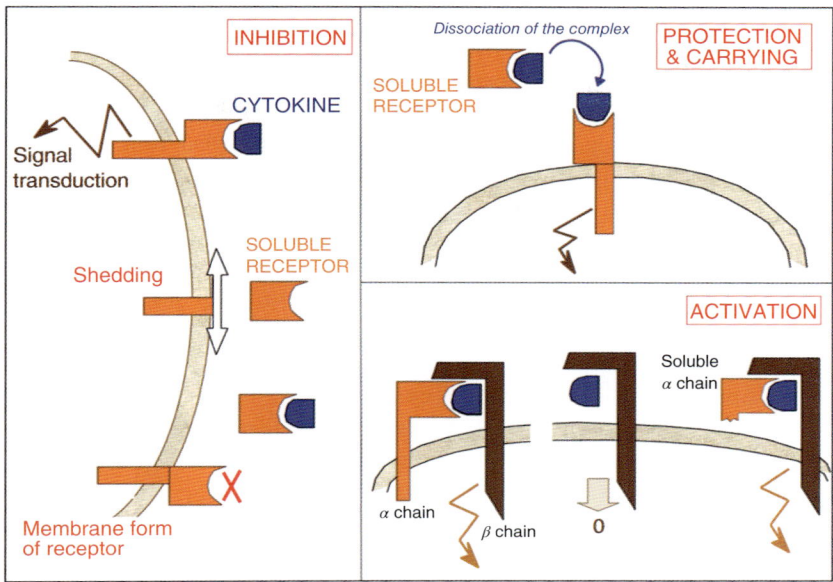

Fig. 5 Different properties of soluble receptors.

cytokine, or even allow the responsiveness of cells missing the α-chain but expressing the β-chain. An example of the latter case is provided by the soluble IL-6R that, together with IL-6, controls the switch of leukocyte recruitment during inflammation [32].

4
Functions

4.1
Hematopoiesis

The production of leukocytes in the bone marrow is a very active and efficient process (4×10^8 leukocytes produced per hour in humans). Numerous cytokines, expressed at homeostasis in human bone marrow, contribute to hematopoiesis whereas others (e.g., GM-CSF) are only active during infection in order to further increase the renewal of available leukocytes. The maturation, proliferation and differentiation of bone marrow progenitor cells are under the control of different cytokines acting concomitantly. For example, IL-1 alone is unable to allow the differentiation of bone marrow precursors cells, but it acts in synergy with other hematopoietic cytokines. IL-3 and SCF (also termed c-kit ligand) are cytokines with a wide spectrum of activity, acting on pluripotent hematopoietic stem cells (HSCs). GM-CSF favors the differentiation of myeloid progenitors cells, while M-CSF or G-CSF are specific for monocyte/macrophage and granulocyte lineages, respectively. IL-6, IL-11, and mainly thrombopoietin, favor thrombocytopoiesis, leading in turn to megakaryocyte development and platelet production. IL-5 has been identified as a major regulator of eosinophil development and function, while IL-7 is the main cytokine of lymphopoiesis, and is required for the development of B and T lymphocytes [33]. IL-7, together with

IL-33, IL-23 or IL-25, favors the differentiation of innate lymphoid cells. IL-15 is the prerequisite cytokine for the generation of NK cells and cytotoxic CD8 T cells [34]. Most of these hematopoietic cytokines act also as activators of mature cells and as anti-apoptotic ligands.

4.2 Immune Response

4.2.1 Innate Immunity

Innate immunity is characterized by a response localized within the site of infection (Fig. 6). Epithelial cells, leukocytes residing in the extravascular space of the tissue (mast cells, resident macrophages, and immature dendritic leukocytes) and endothelial cells of the micro-vessels are triggered by microorganisms and microbial-derived products, causing them to release inflammatory cytokines (e.g., IL-1, TNF, IL-12, IL-18). Of note, mast cells contain preformed cytokines and play a critical protective role during infection [35]. These cytokines further activate the endothelial cells of the postcapillary venules, leading to an increased adherence of circulating leukocytes and to the coagulation process. Other cytokines, such as IFNγ, are produced within an amplificatory loop. A local production of chemokines allows the recruitment of new leukocytes from the blood compartment to further act against the infectious process. Cytokines contribute to enhance the antimicrobial activities of newly recruited neutrophils, for example by the production of reactive oxygen species (ROS) and nitric oxide (NO), the release of proteases, the release of defensins, neutrophil extracellular trap (NET) formation and enhanced phagocytosis, and the antiviral activity of NK cells.

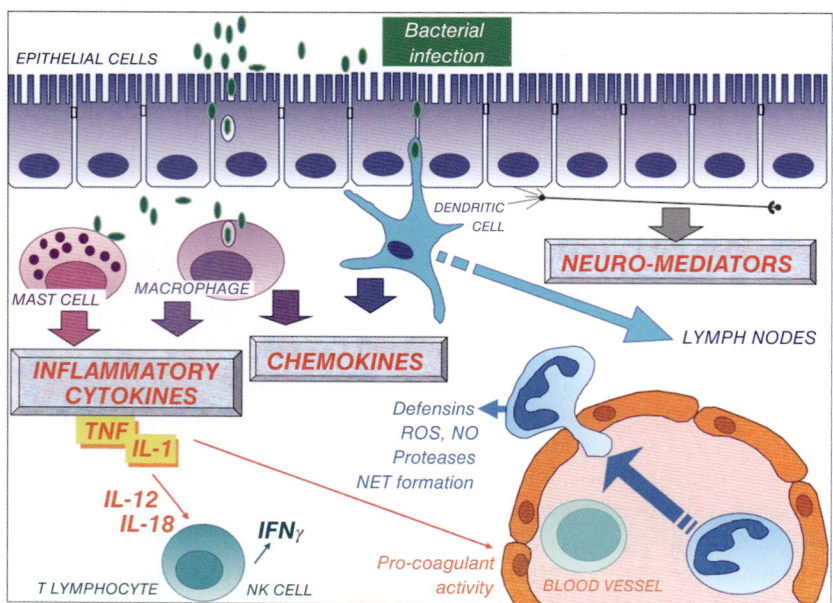

Fig. 6 Cytokines and soluble mediators as coordinators of regulated processes taking place in a bacteria-loaded site.

Neuromediators help to control the inflammatory process. In addition to local responsiveness, a systemic response occurs that includes fever which in turn reflects the presence of pyrogenic cytokines (e.g., IL-1, TNF, IL-6) in the plasma. Hyperthermia contributes to the anti-infectious process by enhancing some immune cell activities, and by reducing bacterial growth or viral replication. Finally, the production of hematopoietic cytokines (e.g., GM-CSF) further increases the number of leukocytes that otherwise are known to contribute to the clearance of microorganisms and to an initiation of the healing process.

4.2.2 Adaptive Immunity

Adaptive immunity is characterized by processes which rely on the unique properties of secondary lymphoid organs. Briefly, in secondary lymphoid organs such as the peripheral lymph nodes, professional antigen-presenting cells (APCs), including dendritic cells and mononuclear phagocytes, after processing of the native antigen, display on their membranes antigenic peptides and interact with T lymphocytes expressing a T-cell receptor that is capable of specifically recognizing these peptides. B lymphocytes recognize the native antigen, and as a result of the crosstalks between the APCs and T and B lymphocytes, the T lymphocytes proliferate and differentiate as primed effector or memory T cells, while B lymphocytes proliferate and differentiate as antibody-secreting plasma cells. All of these events are controlled by cytokines. For example, within the secondary lymphoid organs, IL-2 and IL-4 favor T-cell proliferation, IL-12 activates cytotoxic T cells, IL-2 and IL-4 allow B-cell proliferation, while IL-4, IL-5, IL-6, IL-10, IL-13, IL-21 and TGFβ influence plasma cell differentiation and favor the synthesis of different immunoglobulin classes [36, 37]. Within the microorganism-loaded tissues, the effector T lymphocytes – once reactivated – can release IFNγ and GM-CSF that activate professional phagocytes, conferring on them microbicidal functions.

4.2.3 T-Cell Subsets and Cytokines

In 1986, Mosmann et al. described two types of murine helper CD4$^+$ T cell (Th) clones that were defined according to the profile of the released cytokines [38]. Th1 cells produce IL-2, IFNγ and Ltα, while Th2 cells produce IL-4, IL-5, IL-6, IL-10 and IL-13. Both subpopulations produce IL-3 and GM-CSF. Their differentiation from a naïve T cell occurs depending upon the cytokine environment (Fig. 7). Th1 and Th2 cells control each other: IFNγ derived from Th1 neutralizes Th2 cells, while IL-4 and IL-10 both inhibit Th1 cells. Th1 cells are involved in cellular immunity and activate macrophages, but Th2 cells are involved in humoral immunity and activate B lymphocytes. The Th1/Th2 dichotomy also exists in humans, as illustrated clearly by the early detection of Th2 cytokine mRNA expressed in response to an intradermal allergen challenge (immediate hypersensitivity) and the delayed detection of Th1 cytokine mRNA after tuberculin injection (delayed-type hypersensitivity) [39]. Naïve T cells, Th1 and Th2 cells can also be distinguished by the nature of their chemokine receptors [40].

A subset of T cells termed T follicular helper (TFH) cells provides a helper function to B cells within the germinal center of lymphoid tissues by producing IL-21 that promotes isotype switching and also acts within an autocrine loop on TFH [41]. Other subsets have been described and named according to the main IL that they produce (Th9, Th17, or Th22). These T-cell

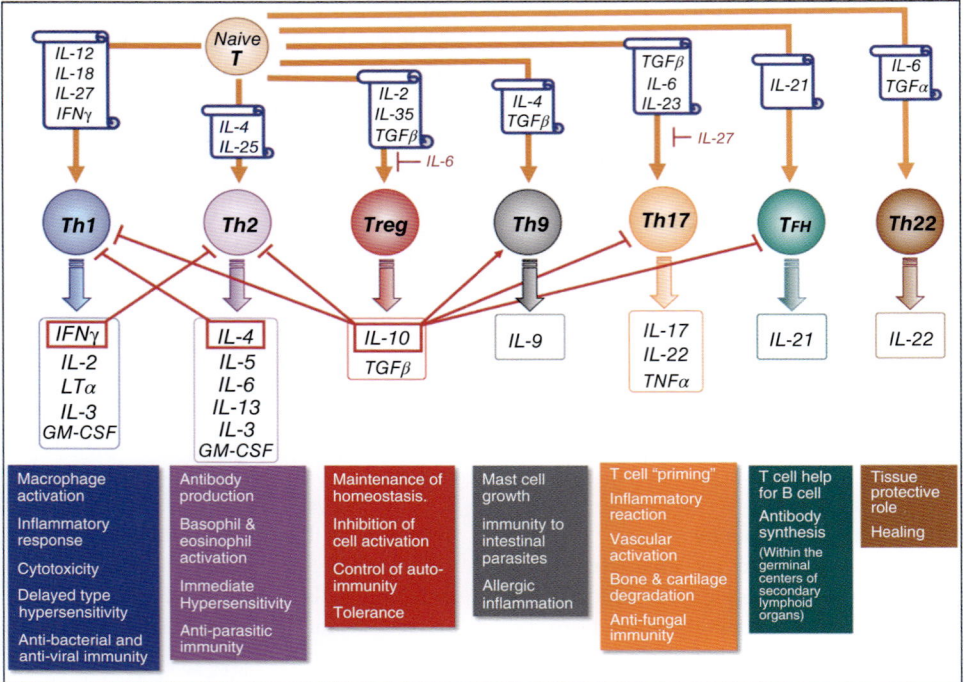

Fig. 7 T-cell subpopulations, the nature of environmental cytokines required for their differentiation, the nature of the produced cytokines and their respective role in immunity and inflammation.

subsets have been described as the key element for some specific innate immunity properties, and also for their contribution to inflammatory processes [42–44].

Other T-cell subsets display immunosuppressive properties, and this is certainly the case for Th3 cells that primarily secrete TGFβ, provide help for IgA switching, and also have suppressive properties for Th1 and other immune cells. Regulatory T cells (Treg) are also able to secrete TGFβ [45], their main product (IL-10) having a wide spectrum of activity as well as an ability to turn off most other T-cell lineages and numerous immune responses. However, the borders between these subsets are not very strict; for example, it has been reported that Th2 can also produce IL-17, while Th17 can also produce IL-22.

Once stably differentiated as long-lived clones, some chemokine receptors allow a distinction to be made among between these subpopulations: CXCR3 is mainly expressed on naïve T cells and Th1 cells, CCR5 and CCR9 are expressed on Th1, CCR3 is expressed on Th2, CCR4 is expressed on Th2, Th17 and Th22 cells, CCR6 is expressed on Th17 and Th22, and CXCR5 is expressed on TFH cells.

4.3
Cell Survival and Cell Death

Cytokines are able not only to serve as maturation and differentiation factors but can also contribute to cell survival. For example, IL-2, IL-4, IL-6, IL-7 and IL-15 have all been shown to inhibit resting T-cell death *in vitro*. B-cell survival factors include IL-4 and BAFF of the TNF family, while plasma cell longevity depends on a combination of IL-5, IL-6, stromal cell-derived factor-1α (SDF-1/CXCL12) and TNFα. Typically, IL-15 can sustain NK cell survival in the absence of serum; indeed, *in-vitro* serum starvation often leads to apoptosis that can be prevented by cytokines such as TGFβ (e.g., survival of epithelial cell). Indeed, these properties reflect an anti-apoptotic mechanism that is initiated by cytokines, such as the suppression of caspase activity or the maintenance or induction of the anti-apoptotic Bcl-2 protein. The receptor activator of NF-κB ligand (RANKL) suppresses the apoptosis of primary cultured endothelial cells; SDF-1/CXCL12 enhances the survival of myeloid progenitor cells *in vitro*; and SCF regulates the survival of cellular lineages by suppressing apoptosis. In addition to SCF, most hematopoietic factors (e.g., IL-3, GM-CSF) allow the maintenance of hematopoietic progenitors. In addition to sustaining cell survival, cytokines can allow cell growth, which is very much the case for IL-2 for T cells, or IL-8 and some other chemokines that favor angiogenesis. Cytokines can also serve as growth factors for tumor cells; an example is the case of IL-6 in malignant myeloma plasma cells. In contrast, some cytokines can induce the apoptosis of a variety of tumor cells and normal cells. Indeed, TNF was first identified in 1975 as a cytokine with antitumor effects [46], and other members of the TNF superfamily can induce proliferation and survival as well as cell death: specifically, this includes TNFα, FasL, TRAIL and VEGI (vascular endothelial cell growth inhibitor). The intracytoplasmic portion of their receptors contains a "death domain" which, upon activation by ligands, recruits intracytoplasmic adaptors expressing a death domain: TNF receptor-associated factor (TRAF) and Fas-associated death domain (FADD), which in turn recruit the pro-form of caspase-8. The auto-activation of caspase-8 leads to a subsequent activation of caspase-3, a pro-apoptotic enzyme. Other mechanisms of cell destruction associated with cytokines can reflect an indirect process, such as the key role played by RANKL in the differentiation and activation of osteoclasts that resorb bone tissue.

4.4
Links with the Central and Peripheral Nervous Systems

A crosstalk exists between immune cells and the central nervous system (CNS). Pro-inflammatory cytokines – particularly IL-1 and TNF – induce fever, anorexia and slow-wave sleep, with part of their activities being conveyed by the vagal nerve. Fever involves the local production of IL-6 and prostaglandin E2 (PGE2) [47]. Certain chemokines are also pyrogenic (e.g., IL-8, MIP-1α, MIP-1β, RANTES), but do not induce PGE2. Both, IL-1 and TNF also induce a neuroendocrine loop whereby, in response to their signal, the hypothalamus produces a corticotropin-releasing factor (CRF) that induces the release of adrenocorticotropic hormone (ACTH) from the pituitary gland. In turn, ACTH induces the release of glucocorticoids from the adrenal glands; this inhibits the production of most cytokines and also antagonizes many of their activities. The vagal nerve releases acetylcholine, which acts on adrenergic splenic neurons that in turn release noradrenaline. The latter acts via $β_2$-adrenoreceptors expressed by splenic T-lymphocytes, which in turn produce acetylcholine on a local basis. Acetylcholine then represses inflammatory cytokine production by the macrophages via α7-nicotinic receptors [48, 49]. Other neural and neuronal mediators, such as adrenaline, vasoactive intestinal peptide (VIP), pituitary adenylate cyclase-activating polypeptide (PACAP) and α-melanocyte-stimulating hormone (α-MSH) are inhibitors of IL-1 and TNF production, while noradrenaline and substance P are potentiators.

4.5
Reproduction

Cytokines are involved in spermatogenesis, ovogenesis and gestation [50]. They play an important regulatory role in the development and normal function of the testis. The cytokines are produced by Leydig cells, Sertoli cells and germinal cells. Proinflammatory cytokines, including IL-1 and IL-6, have direct effects on spermatogenic cell differentiation and testicular steroidogenesis. SCF and LIF, two cytokines which normally are involved in hematopoiesis, also play a role in spermatogenesis. Anti-inflammatory cytokines of the TGFβ family are involved in testicular development, while TNFα, a secretory product of round spermatids, increases the expression of endogenous androgen receptors in primary cultures of Sertoli cells. Given the requirement of testosterone for spermatogenesis, and the importance of androgen receptors in mediating Sertoli cell responsiveness to testosterone, the stimulation of androgen receptor expression by TNFα may represent an important regulatory mechanism required to maintain efficient spermatogenesis. M-CSF is the principal growth factor regulating macrophage populations in the testis, male accessory glands, ovary and uterus. Both, male and female M-CSF-deficient (op/op) mice have fertility defects, with males having a low spermatozoid number and libido as a consequence of dramatically reduced circulating testosterone, and females having extended estrous cycles and poor ovulation rates.

Normal ovarian tissue is rich in cytokines, including GM-CSF, M-CSF, TGFβ and TNFα, all of which are important in the physiology of ovarian function and of ovulation. The actual rupture of a follicle during ovulation may be dependent on tissue remodeling that shares some features with an acute inflammatory process. TNF and IL-1 are each implicated in ovarian follicular development and atresia, ovulation, steroidogenesis and corpus luteum function (including formation, development, and regression). Chemokines such as MCP-1 are also involved in luteolysis.

Among the large number of cytokines produced by the endometrium and placenta, IL-15 and IL-11 play key roles during decidualization, IL-5 and IL-10 play an important role in placental development, and IL-33 is a critical growth factor for the placenta [51]. Cytokines released at the fetomaternal interface play an important role in regulating embryo survival, as they control not only the maternal immune system but also angiogenesis and vascular remodeling. For example, LIF plays a role in the preimplantation, peri-implantation and post-implantation of the embryo, while IFNτ is involved in gestation in bovine and ovine species. IFNτ is produced constitutively by the embryonic trophectoderm during the period immediately prior to implantation, and acts on the uterine epithelium to suppress the transcription of genes for both estrogen receptors and oxytocin receptors. IFNτ blocks the development of the uterine luteolytic mechanism, while TGFβ is involved in regulating placental development and function. The abortive influences of TNFα and IFNγ may cause a pregnancy to be terminated

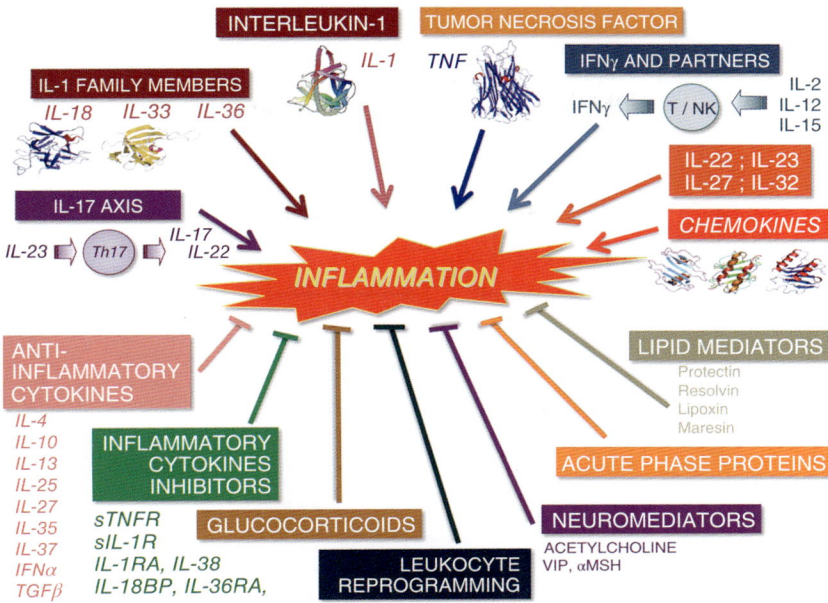

Fig. 8 Inflammation is under the control of pro-inflammatory cytokines, whereas its resolution involves anti-inflammatory cytokines and other signals that turn off the production of pro-inflammatory cytokines.

during infection of the uteroplacental unit. The slight dominance of proinflammatory cytokines in the fetal membranes and decidua suggest that inflammatory processes occur modestly with parturition and term-labor, but much more robustly in preterm delivery, particularly in the presence of intrauterine infection.

4.6
Inflammation

During the first century BC, Celsus wrote: "*Notae vero inflammationis sunt quatuor: rubor and tumor cum calore et dolore*" (Redness, swelling, heat, and pain are the four main parameters that characterize inflammation). These parameters reflect the action of numerous inflammatory mediators and proinflammatory events that are orchestrated by inflammatory cytokines, mainly IL-1 and TNF, and also by the other members of the IL-1 superfamily, namely the cytokines that favor IFNγ production (i.e., IL-2, IL-12, IL-15 and IL-18) and the Th17 axis (Fig. 8). These cytokines are produced following tissue or systemic steady-state disruption such as tissue loading by pathogens, ischemia, hypoxia or trauma. Acting on target cells, they induce a cascade of mediators, including other cytokines and chemokines, proteases, lipid mediators (e.g., prostaglandins, platelet-activating factor), free radicals (e.g., superoxide anion, nitric oxide) that contribute to the inflammatory process and tissue injury (Fig. 9). Acting on endothelial cells, and especially those of the post-capillary venules, they increase vascular permeability and plasma transudation, they increase circulating leukocyte adherence and margination, and they initiate the

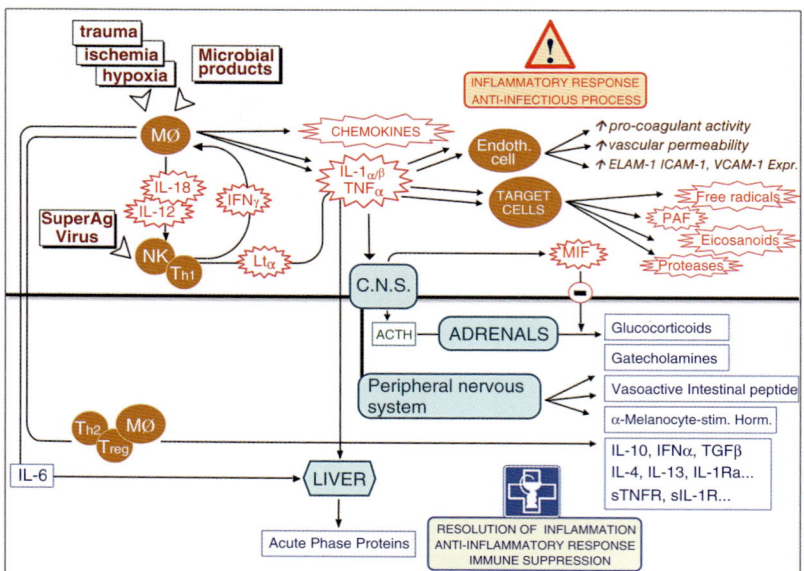

Fig. 9 Inflammation is the consequence of a cascade of events initiated by IL-1 and/or TNF. The action of these cytokines on various target cells leads to the release of numerous inflammatory mediators. Anti-inflammatory cytokines, the activation of the neuroendocrine pathway and the effects of glucocorticoids as well as the enhanced production of acute phase proteins exert a negative control on the inflammatory process. ACTH: adrenocorticotropic hormone; CNS: central nervous system; PACAP: pituitary adenylate cyclase-activating polypeptide; PAF: platelet-activating factor; VIP: vasoactive intestinal peptide.

coagulation process by inducing tissue factor expression. Chemokines contribute to the recruitment of leukocytes that further maintain the inflammatory process. Some cytokines enhance the production of IL-1 and TNF (e.g., IFNγ, GM-CSF, IL-3), while others inhibit it and behave as anti-inflammatory cytokines (e.g., IL-4, IL-10, IL-13, IL-25, IL-27, IL-35, IL-37, TGFβ, IFNα) [52–56]. The later contribute to attenuate the inflammatory process and to allow the tissue-repair process to occur. Other anti-inflammatory mediators include specific inhibitors of the major inflammatory cytokines (e.g., soluble IL-1 and TNF receptors, IL-1Ra), some neuropeptides (e.g., adrenaline, acetylcholine), some resolving lipid mediators, glucocorticoids induced after activation of the neuroendocrine loop by IL-1 and TNF, and the acute-phase proteins produced by the hepatocytes in response to IL-1, TNF, and particularly, IL-6. Acute-phase proteins contribute to eliminate debris due to cell lysis, to favor phagocytosis, to neutralize toxic mediators and to inhibit proteases. Another mechanism, termed leukocyte reprogramming (or TLR agonist tolerance), modifies the intracellular machinery to prevent cells from continuing to produce proinflammatory cytokines [57]. Furthermore, IL-6, together with its soluble receptor, favors a switching of the pattern of the recruited leukocyte within the inflammatory focus (from neutrophils to macrophages) [32].

5
Life without Cytokines (What Knock-Out Mice Tell Us)

The deletion of genes encoding for a cytokine or a cytokine receptor has led to the generation of many different knock-out (KO) mice that has further allowed the contribution of cytokines to different physiological events to be deciphered. In most cases, viable, fertile and clinically healthy mice could be obtained, but in a few cases the deletion of the genes led to death *in utero*, illustrating a major role of certain cytokines during embryogenesis. This was the case for SDF1 (CXCL12) and its receptor CXCR4, LIF, Cardiotrophin-1, and the gp130 chain of their receptors (see Table 3). The deletion of certain signaling molecules (e.g., STAT3) is also lethal during embryonic life. Other deletions have revealed the role of certain cytokines for organogenesis during embryonic life. This was the case for Lt-α, Lt-β and LtβR in the development of lymph nodes and Peyer's patches [58], and for RANKL in the development of lymph nodes and mammary glands. The organization of hematopoietic compartments is under the control of numerous cytokines, and an absence of TNFα or the γ-chain of the receptor of IL-2, -4, -7, -9, -13, -15 and -21 may have profound perturbations on tissue organization and leukocyte differentiation. This is also true for the deletion of chemokines that contribute to tissue colonization at homeostasis. While studies of KO mice have often provided supportive data, there were some surprises. One such surprise was the discovery that an absence of the hematopoietic cytokine GM-CSF did not alter hematopoiesis (the mice had normal numbers of peripheral leukocytes, bone marrow progenitors and tissue hematopoietic populations), but rather led to a profound change in lung status, thereby confirming a role of GM-CSF in pulmonary homeostasis [59]. Of major interest was a demonstration that the deletion of anti-inflammatory cytokines (IL-10, TGFβ, IL-1Ra) was deleterious at homeostasis in the absence of experimental exogenously delivered inflammatory signals, thus establishing their crucial role in preventing potential on-going inflammatory processes at homeostasis [60–62]. As expected, many deletions had major effects on the quality of the anti-infectious process. For example, in the absence of IL-6, mice produce lower levels of IgG antibodies against stomatitis vesicular virus, display a lower cytotoxic T-cell activity during vaccinia virus infection (leading to an increased number of viral particles in the lungs), and failed to control the *Listeria monocytogenes* load in the tissues that the organisms reached [63]. In contrast, the same deficiencies may appear beneficial in certain experimental models of septic shock in response to endotoxin, while being deleterious to combat infection (e.g., TNFα, TNFRp55, caspase-1) [64, 65].

In humans, natural mutations occur that may affect cytokines, their receptors, and the signaling molecules of receptors. The most famous mutation leads to a deletion of the γ-chain of the receptor of IL-2, -4, -7, -9, -13, -15 and -21, which is associated with a severe combined immunodeficiency. The first successful gene therapy was applied to this genetic deficiency in 2000 [66], but this had to be interrupted because of the occurrence of tumors.

6
Cytokine Synthesis

6.1
Homeostasis

Based on the application of highly sensitive techniques such as reverse-transcription

Tab. 3 Cytokine or cytokine receptor knockout mice reveal the relative contributions of cytokines to life processes.

Biological event	Deleted gene	Consequences
Gestation	LIF	Absence of LIF in female prevents implantation of blastocyst
	IL-11	Impaired decidualization
	GM-CSF	Reduced litter size, reduced neonate weight
Embryogenesis and organogenesis	Cardiotrophin-1 and gp130 receptor chain	Heart development blocked; death *in utero* on day 16
	SDF-1 and CXCR4	Defects in heart and brain development, intestinal vascularization, and B-cell hematopoiesis. Death *in utero*
	Ltα, Ltβ, LtβR	Absence of lymph nodes and Peyer's patches
	TNFα	Altered spleen and lymph node organization
	RANKL	Absence of lymph nodes, mammary gland defects
	IL-8R	Splenomegaly, lymphadenopathy, increased circulating neutrophils
Hematopoiesis	G-CSF	Chronic neutropenia
	Stem cell factor	Absence of mast cells
	M-CSF	Altered function of monocytes and osteoclast
	IL-7	Reduced number of lymphocytes
Leukocyte differentiation	Ltα, Ltβ	Absence of follicular dendritic cells
	IL-15	Absence of NK and NK-T cells
	γ-chain of the common receptor	Impaired B, T, and NK cell development
Osteoclastogenesis	RANK and RANKL	Defect in the differentiation of hematopoietic osteoclast progenitor
Pulmonary homeostasis	GM-CSF and β-chain of GM-CSF R	Accumulation of surfactant lipids and proteins in the alveolar space. Lymphoid hyperplasia
Gut homeostasis	IL-10	Intestinal mucosal hyperplasia, inflammatory cell infiltration, MHC class II expression on colonic epithelium
	IL-2	Colitis
	IL-15	Reduced intestinal epithelial B and T lymphocytes
Fever	IL-6	No pyrogenic response
Acute phase response	IL-6	Profound alteration of acute phase proteins production

Tab. 3 (Continued)

Biological event	Deleted gene	Consequences
Inflammation	TGFβ	Systemic lethal inflammation (day 24)
	IL-1-Ra	Chronic arthropathy, spontaneous dermatosis, Exacerbated delayed-type hypersensitivity
	IL-5	Reduced eosinophilia
Antibody production	IL-4, IL-6, IL-13, CD40L IFNγR	Altered antibody production
Immune function	IL-2, IL-4, IL-6, IL-12, IL-13, IL-15, IL-17, IL-18, IL-35, TNF, IFNγ, IFNα/β, stem cell factor, CXCL, TNF R, IFNγR, IFNAR	Altered anti-infectious response
Chemotaxis	Chemokines, and their receptors	Altered cell migration and anti-infectious response

polymerase chain reaction (RT-PCR) and *in-situ* hybridization (ELISpot), it has become possible to demonstrate the presence of cytokine mRNA in various types of cell, or to demonstrate the presence of cytokine-producing cells in the absence of activation. For example, IL-6 is produced spontaneously by 0.5% of bone marrow cells, 0.1% of spleen cells, and 0.01% of mesenteric lymph node cells. IL-6 is also produced in the absence of exogenous stimuli by enterocytes, eosinophils, neutrophils, epidermal cells, smooth muscle cells, bone marrow stromal fibroblasts, anterior pituitary cells and trophoblast cells, among others. Furthermore, at homeostasis certain cells contain cytokines, and this is particularly the case for keratinocytes that contain large amounts of IL-1α. Mast cells contain a large panel of preformed cytokines (e.g., IL-1, IL-4, IL-6, IL-13, TNF), and MIF is preformed in numerous leukocytes. Furthermore, the presence of cytokines in biological fluids has been reported in the absence of any infection or inflammatory diseases: IL-1α and IL-8 are found in sweat; IL-1α, IL-1β, IL-6, IL-8, GM-CSF and TGFβ have been reported in tears; IL-2, IL-8, TNFα, TGFβ and soluble TNFR are present in saliva; IL-2, IL-6, IL-8, IL-10, IL-12, TNF and soluble IL-2R and IL-6R have been detected in seminal fluid; and a great number of different cytokines have been identified in human colostrum. In the latter case, all of these cytokines most likely act on the oropharyngeal and gut-associated lymphoid tissue of the newborn, and favor the development and maturation of the immune system; they may also protect the newborn against invasive microorganisms delivered by the oral route. Finally, certain chemokines are present in tissues where they contribute to leukocyte recruitment at homeostasis (e.g., CCL17 and CCL25 in thymus; CCL21 and CXCL13 in lymph nodes and Peyer's patches; CXCL12 in numerous tissues).

6.2 Induction

The production of cytokines by professional antigen-presenting leukocytes and activated T cells occurs during the cellular cooperation required to initiate the adaptive immunity. This production is representative of the dialog between immune leukocytes. In contrast, innate immunity is associated with the production of cytokines in response to exogenous activators, mainly the PAMPs, and endogenous danger signals delivered by the damage-associated molecular patterns (DAMPs). Following their interaction with the "pattern recognition receptors" (PRRs), various intracellular signaling cascades lead to the synthesis and release of cytokines. The best known PRRs are the TLRs that share with IL-1 and IL-18 a homologous TIR domain. Ten different TLRs have been identified in humans that recognize bacterial, viral, parasitic or fungal conserved structures expressed on the surface or within the microbes (e.g., DNA, single- or double-strand RNA). Endotoxin produced by Gram-negative bacteria (LPS) is among the most potent PAMPs, and only picograms of this molecule are sufficient to induce the production of a whole panel of cytokines by macrophages. LPS is trapped by the CD14 molecule on the cell surface, and triggers the cell via a MD2/TLR4 complex. Other PAMPs and DAMPs can be sensed by members of the Nod-like receptors (NLRs). A few of these, such as the NLR family pyrin domain-containing (NLRP) 1, 3, 7, and the NLR family Card-containing (NLRC)-4, belong to the so-called inflammasome, a molecular platform which triggers activation of the caspase-1 required to process the precursor forms of IL-1β and IL-18.

In the case of Gram-positive bacteria, some organisms release exotoxins (also known as superantigens) which trigger the production of cytokines by both T lymphocytes and macrophages.

6.3 Parameters That Affect Production

In addition to the genetic polymorphism (see Sect. 8), there are numerous parameters that influence the level of production of cytokines. Of course, *in-vitro* experiments require careful sampling of the cells, the use of a correct medium and agents free from endotoxin are key issues, and the choice of serum added to the culture can greatly influence the results. Both, *in-vivo* and *ex-vivo* studies of cytokine production by human cells are influenced by factors such as age, nutritional status, drug and alcohol use, smoking habit, seasonal and circadian rhythms, physical exercise, altitude exposure, microgravity, social and psychological stress, physical stress, and gender. In the latter cases, neuromediators and sexual hormones can modify the cellular production of cytokines. All parameters that influence cytokine production and action contribute towards rendering the dichotomy of pro-inflammatory versus anti-inflammatory cytokines rather oversimplistic [67].

6.4 Measurements

Measurements of cytokines in humans can be performed with natural biological fluids, in induced biological fluids (e.g., bronchoalveolar or peritoneal lavages), on tissue biopsies, and/or with blood leukocytes. The detection of cytokines in biological fluids may represent the "tip of the iceberg," as they can be detected in the case of an exacerbated production; this is because when cytokines have been produced, they are efficiently trapped by their specific receptors on environmental cells [68]. Cytokine mRNA analysis can be achieved using Northern blots, *in-situ* hybridization, or RT-PCR. Cytokine-producing cells can be monitored using ELISpot or flow cytometry. Today, biological assays are no longer performed, and measurements of cytokines are mainly achieved using enzyme-linked immunosorbent assays (ELISAs). As cytokines constitute a tightly regulated network (see Sect. 7), information relating to cytokines should be acquired through the analysis and simultaneous quantification of several cytokines. Indeed, it is the cytokine milieu that influences the cellular response rather than the action of any single cytokine. Recently available techniques such as microarrays, combined with real-time PCR or multiplex immunological detection, allow the analysis of cytokines within very small sample volumes, providing additional precious information.

7
The Cytokine Network

The interactions between cytokine-producing cells and target cells lead to a cytokine network being defined (see Fig. 10). Once produced in the context of a specific immune response, or following danger signals activation, cytokines act on target cells and induce the synthesis of new cytokines. The induction of IL-2 release by lymphocytes, IL-6 by fibroblasts, G-CSF by macrophages, and IL-8 by endothelial cells in response to IL-1 is an example of cytokine cascades. Cytokines such as IL-1 can induce their own synthesis, or that of others such as IFNγ, further enhancing the productions and leading to amplificatory loops. Another key word used to define this network is synergy. Due to their great redundancy, two cytokines can share similar activities such that, when acting together on the same target cell, the effect will be far greater than the additive effect

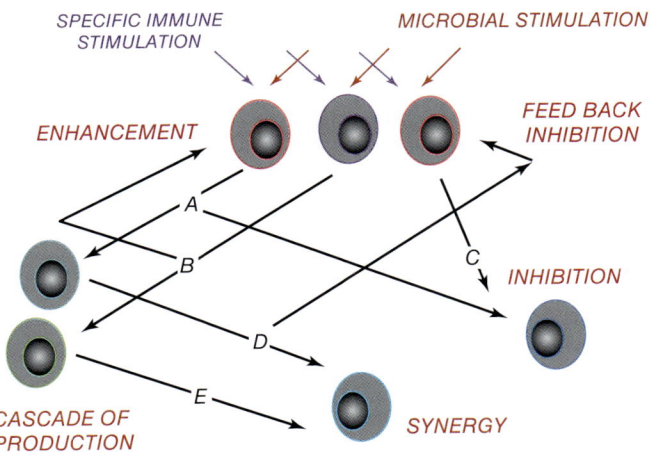

Fig. 10 The cytokine network (for details, see the text).

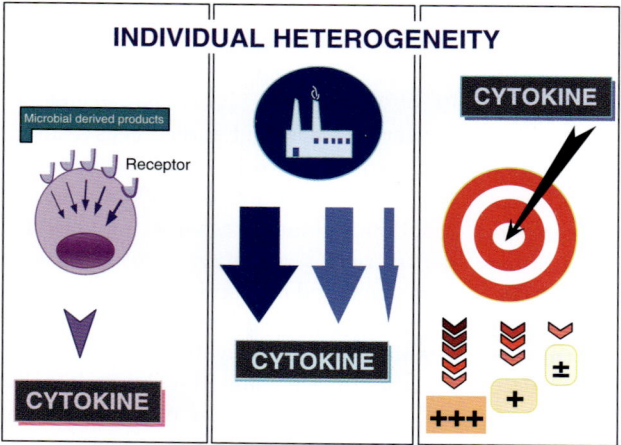

Fig. 11 Individual heterogeneity for cytokine production and responsiveness.

of each individual cytokine. For example, the production of complement factor C3 by endothelial cells in response to both TNF and IL-1 is far higher than that induced by each cytokine alone. Such synergy may be the consequence of an induction of cytokine receptors, which allows cells to respond to a second cytokine that they would not do normally because they previously lacked the specific receptor. The network also involves negative loops as a consequence of the action of cytokines, blocking the production or the effects of others. IL-10 is an example of a cytokine blocking the production and action of other cytokines. However, the nature of the target cell may lead to opposing observations; for example, IL-10 represses the LPS-induced production of IL-8 by monocytes but enhances the production by endothelial cells.

8
Individual Heterogeneity

Among genetic predisposition to diseases, premature death due to uncontrolled infection has the highest relative risk linked to heritability. The individual heterogeneity reflects in part the genetic polymorphisms of cytokines, for which three levels can be involved (Fig. 11):

- A genetic polymorphism exists for the receptors of PAMPs and DAMPs. These sensors are the first essential elements in initiating intracellular signaling cascades that lead to cytokine production. Certain single nucleotide polymorphisms (SNPs) or mutations can directly modify the intensity of the responsiveness and influence disease susceptibility, as shown by the existence of rare mutations of TLR4 among patients with meningococcal infection.
- SNPs or mutations can affect cytokine genes, particularly when expressed within gene promoters or other regulatory sequences. Accordingly, high, intermediate or low producers are reported when assessing the levels of cytokines produced in response to a given activator. A correlation between TNF genotypes and levels of released TNF in response to LPS has been demonstrated.
- Depending on the donors, target cells can react intensively, moderately or weakly

Tab. 4 Some examples of cytokine or cytokine receptor gene polymorphisms associated with diseases.

Disease	Gene polymorphism
Acute renal failure mortality	TNF, IL-10
Allergy	RANTES promoter; MIF promoter; IL-13 promoter; TNFα; IFNγ IL-16 promoter; IL-1 gene complex; IL-18; IL-12 promoter
Asthma	IL4R α-chain; IL-10; IL-15; IL-18; TGFβ1 promoter; eotaxin
Alzheimer's disease	IL-1α, TNFR2, IL-6
Breast cancer	IL-6
Cerebral malaria	TNFα
Chronic periodontitis	IL-10 promoter, IL-1B +3953, and TNF-A-308 allele 2 positive
Coronary disease	IL-1Ra
Crohn's disease	IL-10, MCP-1 (CCL2); IL-16 promoter; TNFR1
HIV resistance	CCR5 deletion
Idiopathic pulmonary fibrosis	TNF-alpha (-308 A) allele
Infectious nephropathy	Low CXCR1 expression
Infertility	Functional mutations in the LIF gene
Lupus nephritis	MCP-1 (CCL2) promoter
Multiple sclerosis	Microsatellite allele of TNF gene; IL-2 promoter
Parkinson's disease	Promoter region of IL-8
Rheumatoid arthritis	Non-coding region of IFNγ gene; TGFβ1; MIF promoter; IL-1 gene cluster; IL-4; TNFα
Schizophrenia	IL-1 gene complex
Sepsis susceptibility and/or mortality	TNF; CXCL2
Systemic lupus erythematosus	IL-1α; TNFα
Sudden infant death syndrome	IL-10
Susceptibility to cerebral malaria	TNF
Transplantation	TNF, IL-10

to a fixed amount of a cytokine. This has been well demonstrated with genetically distinct endothelial cells, which express various levels of adhesion molecules in response to similar amounts of IL-1 or TNF [69].

In all of these situations the genetic polymorphism could also reflect SNPs or mutations among the genes of the adaptors and signaling molecules involved in the signaling cascades.

As shown in Table 4, numerous cytokine gene polymorphisms have been associated with the occurrence or severity of diseases. The reported SNPs or mutations can be associated with protection against the disease or, in contrast, with a greater susceptibility. More recently, similar gene polymorphisms have been associated with the efficiency or a lack of efficiency of certain therapeutic approaches, particularly when the treatment is targeted at cytokines.

9
Cytokines and Infections

9.1
Half-Angel–Half-Devil

As noted in Sect. 4.2.1, cytokines play a major role in innate immunity during the early processes that occur in sites of microorganisms invasion. The antiviral properties of IFNs during viral infection have been established for more than 50 years. Following the interaction of IFNα or IFNβ with its receptor, the activation of 2',5'-oligoadenylate synthetase leads to the production of an endoribonuclease that degrades viral RNA, while the activation of protein kinase P1 contributes to the capacity of the cell to inhibit viral protein synthesis. During bacterial infection, the beneficial effects of pro-inflammatory cytokines have been widely demonstrated. The injection of cytokines such as IL-1, IL-6, IL-12, IL-18, TNF, G-CSF, M-CSF, GM-CSF and IFNγ before the injection of bacteria has been shown to promote a faster clearance of bacteria, and to enhance host survival. The use of cytokine-neutralizing antibodies or antagonists molecules further showed that endogenous cytokines produced during the course of infections play an essential role in controlling the microorganism invasion, and cytokine neutralization can be highly deleterious in this respect. Finally, mice rendered deficient for the expression of a cytokine or a cytokine receptor display an enhanced "susceptibility" to infection, such as an enhanced microbial load in tissues and reduced survival.

Despite their undoubted beneficial effects of clearing microorganisms, some cytokines can also contribute to death in certain clinical situations, such as septic shock in response to endotoxin injection or after a lethal bacterial infection [70]. Antibodies against TNF protect mice against lethal injections of LPS, and baboons against lethal injections of *Escherichia coli* [71]. The ambivalent role of cytokines is illustrated in TNF receptor p55 KO mice: these mice are more sensitive to *Listeria monocytogenes* infection than normal mice, but are more resistant to toxic shock induced by endotoxin or staphylococcal enterotoxin B [64]. Furthermore, synergy – which already has been described as a keyword to define the cytokine network – is also valid for lethality. For example, a nonlethal dose of TNF, when injected with a nonlethal dose of IL-1 or IFNγ, will lead to lethality in mice [72].

9.2
The Strategies of Microbes against Cytokines

The fact that pathogenic microorganisms have developed numerous subterfuges to counteract the action of cytokines further illustrates that cytokines are crucial in the fight against the infectious process. The hijacking of genes for cytokines [13] or cytokine receptors has been perpetrated by virus to elaborated soluble molecules that mimic or neutralize cytokines of the host (e.g., IFNγR and TNFR homologs of Myxoma virus and Shope fibroma virus, IL-1R homolog of Vaccinia virus, M-CSFR homolog of Epstein–Barr virus) [73, 74]. Another hijacking concerns the gene of IL-10, allowing viruses (Epstein–Barr virus, equine herpes virus, parapoxvirus orf virus) to create an immunosuppressive environment [75].

Another strategy has been developed by vaccinia virus to prevent the maturation of biologically active IL-1 and IL-18.

This virus produces an inhibitor (serpine) of caspase-1, the enzyme required for the cleavage of the pro-forms of these cytokines. Other mechanisms include the production of viral factors that downregulate cytokine synthesis (e.g., human T-lymphotropic virus (HTLV), adenovirus, West Nile virus, papilloma virus, respiratory syncytial virus), the use of cytokines to further enhance the viral replication (e.g., HIV), the use of chemokine receptors to enter the cell (e.g., HIV), the synthesis of viral chemokines that behave as antagonists (e.g., Kaposi's sarcoma-associated virus, human herpesvirus 8, stealth virus, etc.), and the capacity to block the signaling cascade induced by cytokines (hepatitis C virus, adenovirus) [76]. Certain bacteria also display strategies to block cytokine production (e.g., *Yersinia enterocolitica*, enteropathogenic *E. coli*, *Vibrio cholera*), to degrade them (*Streptococcus pyogenes*, *Treponema denticola*, *Porphyromonas gingivalis*), to favor IL-10 production (*Bordetella* sp.), to prevent intracellular signaling (*Yersinia pestis*, (genusSpecies) *Shigella flexneri*, *Salmonella* sp.), and to use them as growth factors (*Staphylococcus aureus*, *Pseudomonas aeruginosa*, *Acinetobacter*) [77].

10
Cytokines and Diseases

While the discovery of new cytokines was continuing, great hopes to cure diseases emerged, often followed by great disillusion. For example, although TNF was discovered for its antitumor effects it was soon shown to be one of the most potent pro-inflammatory cytokines, which obviously limited its therapeutic use. One cytokine is rarely the sole key effector underlying the pathogenic process, and the complexity of the cytokine network has rendered difficult their use as therapeutic tools, or as therapeutic targets. In addition, the identification of an overexpressed cytokine at one particular stage of the disease does not allow the discrimination of whether this cytokine is a marker or an actor. Nevertheless, there exist certain clinical situations where the use of recombinant cytokines or the use of neutralizing antibodies has allowed significant therapeutic progress. Some examples of diseases where the involvement of cytokines is well established and clinical applications have been developed are summarized in the following subsections.

10.1
Autoimmune Diseases

Among autoimmune diseases, systemic lupus erythematosus (SLE) has been investigated with the aim of neutralizing cytokines known to activate B lymphocytes. This is the case of IL-6 when counteracted by a humanized anti-IL6 receptor monoclonal antibody (tocilizumab) that reduces the numbers of circulating plasma cells and anti-dsDNA [78]. IL-10 is also a factor that activates B-cells, and one promising study has been conducted aimed at its neutralization. After two clinical studies involving 1684 patients with SLE had demonstrated its safety and effectiveness, elimumab, an anti-BAFF (a cytokine that belongs to the TNF superfamily) was approved in 2011 by the Food and Drug Administration (FDA) for the treatment of SLE patients [79]. This was the first new drug to treat SLE to be approved in 56 years, though another anti-BAFF (tabalumab) is also under investigation. Other promising approaches involve humanized anti-IFNα monoclonal antibodies.

10.2
Allergy and Asthma

Allergen-specific IgE antibodies, mast cell degranulation releasing inflammatory mediators such as histamine, and an influx of eosinophils into the airways mucosa and airways lumen, are hallmarks of atopy and allergic asthma. Thus, Th2 cytokines remain important candidates for a role in the pathogenesis of atopy and allergic asthma: IL-4 and IL-13 are required for IgE production [80], IL-9 is involved in mast cell growth, and IL-5 is required for eosinopoiesis. Humanized monoclonal antibodies (hMAbs) against IL-5, anti-IL-4, a recombinant soluble human IL-4 receptor, anti-IL-9, CCR3 antagonists (which block eosinophil chemotaxis) and CXCR2 antagonists (which block neutrophil and monocyte chemotaxis) have been developed as possible therapeutic interventions. Anti-IL-5 (mepolizumab) and anti-IL-13 (lebrikizumab) have demonstrated some efficacy to reducing the exacerbation of asthma [81, 82].

10.3
Sepsis

Sepsis is associated with an exacerbated production of cytokines, known as a cytokine storm [83]. It is worth noting that a cytokine storm can occur in the absence of any infection, as seen in human volunteers treated with anti-CD28 [84]. Animal models of sepsis have shown clearly that TNF, IL-1 or MIF are involved in sepsis-related lethality, and antibodies against these cytokines are highly protective. Unfortunately, anti-TNF antibodies and IL-1Ra have been used in numerous placebo double-blind controlled assays in humans, but without success. This failure may reflect the fact that antibodies in animal models and in humans are not used with similar timings, and that mice, which are often used as experimental models, are 10^5-fold less sensitive to Gram negative bacteria-derived endotoxin than humans. Clearly, mice are not the most relevant laboratory animals to model the human setting of sepsis.

10.4
Chronic Inflammatory Diseases

Fortunately for the companies that had developed these anti-TNF antibodies, successful treatments have been achieved for Crohn's disease [85, 86], psoriasis [87], and rheumatoid arthritis [88]. The significant improvements identified in patients allowed these antibodies to be marketed, although in none of these studies was successful improvement seen in all patients.

In the case of Crohn's disease, the anti-TNF treatment was approved in 1998 and a clinical response was still observed at 12 weeks after the infusion. After 12 weeks, however, remission occurred in half of the patients following the administration of a single dose.

The first successful treatment of rheumatoid arthritis was reported in 1994, when clinical improvements in responding patients were associated with biological changes that reflected an attenuation of the inflammatory process [89]. However, there was no clinical response in 30% of the cases. While the effect of a single injection of anti-TNF is longlasting (a few months), it must be repeated and, as a consequence, the occurrence of microbial pathogenic processes (tuberculosis, *Pneumocystis*(/genus Species)-triggered pneumonia, histoplasmosis, listeriosis, sepsis) has been deplored. Other approaches have been proposed to neutralize TNF with recombinant soluble

TNF receptors. IL-1Ra has also been approved by the FDA for the treatment of rheumatoid arthritis. For this disease, the targeting of IL-17 has also been suggested but appeared disappointing.

In contrast, in the treatment of psoriasis anti-IL-17 or anti-IL-17 receptors led to a significant improvement of the patients [90], while anti-IL-12/IL23 p40 led to similarly beneficial effects [91].

10.5
Cancer

IL-2 has been studied extensively in cancer patients, either alone or associated with chemotherapy, with activated NK cells, lymphokine-activated killer (LAK) cells, with tumor-infiltrating lymphocytes (TILs), with IFNα and, more recently, with tumor-pulsed dendritic cells. In some specific cases (melanoma, metastatic renal cancer) the success remained modest, with major achievements having been obtained with IFNα in hairy cell leukemia and chronic myelogenous leukemia. CXCR4, the sole receptor of CXCL12, plays a central role in cancer cell proliferation, invasion and metastasis. Accumulating evidence suggests a key role of the CXCL12/CXCR4 axis as their enhanced expression correlates with a poor outcome. Various companies have developed inhibitors of chemokines as anticancer therapies.

Angiogenesis is another parameter that favors tumor growth and, accordingly, the targeting of vascular endothelial growth factor (VEGF) has already shown some promising effects [92]. Because some tumor cells employ cytokines as growth factors, slowly progressive myeloma has been treated with IL-1Ra, while multiple myeloma has been treated with anti-IL-6 [93].

10.6
Hepatitis C

IFNα has been used successfully in a subset of hepatitis C virus-infected patients [94]. More recently, a polymerized form of IFNα, the half-life of which is prolonged and which once was associated with ribavirin, has been shown to lead to sustained virological response rates of >50% in chronic hepatitis C patients [95].

10.7
Transplantation

The α-chain of the IL-2R (CD25) is a specific peptide against which monoclonal antibodies have been raised, with the aim of blunting the immune response by means of inhibiting proliferation and inducing apoptosis in primed lymphocytes. One such antibody has proved to be effective in reducing episodes of acute rejection after kidney, liver and pancreas transplantation [95]. The use of this antibody was associated with a significant reduction in the incidence of treated rejection episodes after kidney transplantation in two major randomized European and US studies [96].

10.8
Multiple Sclerosis

In 1993, IFNβ was approved in the USA for the treatment of relapsing-remitting multiple sclerosis [97]. The use IFNβ and glatiramer acetate to treat multiple sclerosis has, to some extent, changed the course of the disease, as the annual relapse rate of patients treated with these drugs is lower than that of placebo-treated patients; moreover, a greater proportion of treated patients remain relapse-free when compared to untreated patients. The anti-CD25 antibody has been also shown to improve IFNβ treatment, and also to inhibit disease activity in patients that had failed to respond to IFNβ [98].

10.9
Neutropenia

Following chemotherapy for leukemia, neutropenia renders patients more susceptible to developing infection. The treatment of patients with G-CSF and GM-CSF accelerates the recovery of normal neutrophil counts and significantly reduces the occurrence of documented infections.

10.10
Osteoporosis

The RANK-ligand is an effector of osteoclastogenesis and bone resorption. A fully human monoclonal antibody with a high affinity and specificity for RANKL (denosumab) has been shown to be efficient in preventing fractures in postmenopausal women with osteoporosis [99].

10.11
Autoinflammatory Disorders

In some autoinflammatory diseases, such as type 2 diabetes, gout and Behçet disease, IL-1 has been shown to be the main contributor to the inflammatory process. Thus, blocking the effect of IL-1 with IL-1Ra (anakinra) has provided an efficient means of reducing the intensity of these diseases [100–102].

11
Conclusions

Despite the huge number of molecules that belong to the cytokine family, the complexity of the cytokine network, and the vast number of parameters that can affect the biological properties of cytokines, a host of investigations have allowed the "language of cells" to be deciphered. Moreover, despite these difficulties, new and promising uses of cytokines or cytokine antagonists can be expected in the future to overcome certain diseases.

References

1. Menkin, V. (1944) Chemical basis of fever. *Science*, **100**, 337–338.
2. Bennett, I.L., Jr, Beeson, P.B. (1953) Studies on the pathogenesis of fever. II. Characterization of fever-producing substances from polymorphonuclear leukocytes and from the fluid of sterile exudates. *J. Exp. Med.*, **98**, 493–508.
3. Atkins, E., Wood, W.B., Jr, (1955) Studies on the pathogenesis of fever. II. Identification of an endogenous pyrogen in the blood stream following the injection of typhoid vaccine. *J. Exp. Med.*, **102**, 499–516.
4. Dinarello, C.A., Renfer, L., Wolff, S.M. (1977) Human leukocytic pyrogen, purification and development of a radioimmunoassay. *Proc. Natl Acad. Sci. USA*, **74**, 4624–4627.
5. Auron, P.E., Webb, A.C., Rosenwasser, L.J., Mucci, S.F. et al. (1984) Nucleotide sequence of human monocyte interleukin 1 precursor cDNA. *Proc. Natl Acad. Sci. USA*, **81**, 7907–7911.
6. Lomedico, P.T., Gubler, U., Hellmann, C.P., Dukovich, M. et al. (1984) Cloning and expression of murine interleukin-1 cDNA in *Escherichia coli*. *Nature*, **312**, 458–462.
7. Isaacs, A., Lindenmann, J. (1957) Virus interference. I. The interferon. *Proc. R. Soc. Lond. B Biol. Sci.*, **147**, 258–267.
8. Nagano, Y., Kojima, Y., Sawai, Y. (1954) Immunity and interference in vaccinia; inhibition of skin infection by inactivated virus. *C. R. Seances Soc. Biol. Fil.*, **148**, 750–752.
9. Cavaillon, J.-M. (2007) Who discovered interferons? *The Scientist*, **21**, 14.
10. Dumonde, D.C., Wolstencroft, R.A., Panayi, G.S., Matthew, M. et al. (1969) "Lymphokines", non-antibody mediators of cellular immunity generated by lymphocyte activation. *Nature*, **224**, 38–42.

11. Bloom, B.R., Bennett, B. (1966) Mechanism of a reaction *in vitro* associated with delayed-type hypersensitivity. *Science*, **153**, 80–82.
12. Weiser, W.Y., Temple, P.A., Witek-Giannotti, J.S., Remold, H.G. *et al.* (1989) Molecular cloning of a cDNA encoding a human macrophage migration inhibitory factor. *Proc. Natl Acad. Sci. USA*, **86**, 7522–7526.
13. Bigazzi, P., Yoshida, T., Ward, P., Cohen, S. (1975) Production of lymphokine-like factors (cytokines) by simian virus 40-infected and simian virus 40-transformed cells. *Am. J. Pathol.*, **80**, 69–78.
14. Cohen, S., Bigazzi, P.E., Yoshida, T. (1974) Commentary. Similarities of T cell function in cell-mediated immunity and antibody production. *Cell. Immunol.*, **12**, 150–159.
15. Bernhagen, J., Calandra, T., Mitchell, R.A., Martin, S.B. *et al.* (1993) MIF is a pituitary-derived cytokine that potentiates lethal endotoxemia. *Nature*, **365**, 756–759.
16. Cohen, S. (2004) Cytokine, more than a new word, a new concept proposed by Stanley Cohen thirty years ago. *Cytokine*, **28**, 242–247.
17. Horning, S.J., Levine, J.F., Miller, R.A., Rosenberg, S.A. *et al.* (1982) Clinical and immunologic effects of recombinant leukocyte A interferon in eight patients with advanced cancer. *J. Am. Med. Assoc.*, **247**, 1718–1722.
18. Rosenberg, S.A., Lotze, M.T., Muul, L.M., Leitman, S. *et al.* (1985) Observations on the systemic administration of autologous lymphokine-activated killer cells and recombinant interleukin-2 to patients with metastatic cancer. *N. Engl. J. Med.*, **313**, 1485–1492.
19. Kern, P., Toy, J., Dietrich, M. (1985) Preliminary clinical observations with recombinant interleukin-2 in patients with AIDS or LAS. *Blut*, **50**, 1–6.
20. Garlanda, C., Dinarello, C.A., Mantovani, A. (2013) The interleukin-1 family, back to the future. *Immunity*, **39**, 1003–1018.
21. Zdanov, A. (2010) Structural analysis of cytokines comprising the IL-10 family. *Cytokine Growth Factor Rev.*, **21**, 325–330.
22. Hunter, C.A. (2005) New IL-12-family members, IL-23 and IL-27, cytokines with divergent functions. *Nat. Rev. Immunol.*, **5**, 521–531.
23. Trinchieri, G. (2010) Type I interferon, friend or foe? *J. Exp. Med.*, **207**, 2053–2063.
24. Aggarwal, B.B., Gupta, S.C., Kim, J.H. (2012) Historical perspectives on tumor necrosis factor and its superfamily, 25 years later, a golden journey. *Blood*, **119**, 651–665.
25. Collison, L.W., Delgoffe, G.M., Guy, C.S., Vignali, K.M. *et al.* (2012) The composition and signaling of the IL-35 receptor are unconventional. *Nat. Immunol.*, **13**, 290–299.
26. Levin, A.M., Bates, D.L., Ring, A.M., Krieg, C. *et al.* (2012) Exploiting a natural conformational switch to engineer an interleukin-2 "superkine". *Nature*, **484**, 529–533.
27. Rafei, M., Wu, J.H., Annabi, B., Lejeune, L. *et al.* (2007) A GMCSF and IL-15 fusokine leads to paradoxical immunosuppression in vivo via asymmetrical JAK/STAT signaling through the IL-15 receptor complex. *Blood*, **109**, 2234–2242.
28. Bazan, J.F. (1990) Structural design and molecular evolution of a cytokine receptor superfamily. *Proc. Natl Acad. Sci. USA*, **87**, 6934–6938.
29. Stonier, S., Schluns, K. (2010) Transpresentation, a novel mechanism regulating IL-15 delivery and responses. *Immunol. Lett.*, **127**, 85–92.
30. Sun, M., Fink, P.J. (2007) A new class of reverse signaling costimulators belongs to the TNF family. *J. Immunol.*, **179**, 4307–4312.
31. Mantovani, A., Bonecchi, R., Locati, M. (2006) Tuning inflammation and immunity by chemokine sequestration, decoys and more. *Nat. Rev. Immunol.*, **6**, 907–918.
32. Hurst, S.M., Wilkinson, T.S., McLoughlin, R.M., Jones, S. *et al.* (2001) IL-6 and its soluble receptor orchestrate a temporal switch in the pattern of leukocyte recruitment seen during acute inflammation. *Immunity*, **14**, 705–714.
33. Ceredig, R., Rolink, A. (2012) The key role of IL-7 in lymphopoiesis. *Semin. Immunol.*, **24**, 159–164.
34. Verbist, K.C., Klonowski, K.D. (2012) Functions of IL-15 in anti-viral immunity, multiplicity and variety. *Cytokine*, **59**, 467–478.
35. Echtenacher, B., Männel, D., Hültner, L. (1996) Critical protective role of mast

cells in a model of acute septic peritonitis. *Nature*, **381**, 75–77.

36. Howard, M., Farrar, J., Hilfiker, M., Johnson, B. *et al.* (1982) Identification of a T cell-derived B cell growth factor distinct from interleukin 2. *J. Exp. Med.*, **155**, 914–923.

37. Llorente, L., Zou, W., Levy, Y., Richaud-Patin, Y. *et al.* (1995) Role of interleukin 10 in the B lymphocyte hyperactivity and autoantibody production of human systemic lupus erythematosus. *J. Exp. Med.*, **181**, 839–844.

38. Mosmann, T.R., Cherwinski, H., Bond, M.W., Giedlin, M.A. *et al.* (1986) Two types of murine helper T cell clone. I. Definition according to profiles of lymphokine activities and secreted proteins. *J. Immunol.*, **136**, 2348–2357.

39. Tsicopoulos, A., Hamid, Q., Haczku, A., Jacobson, M.R. *et al.* (1994) Kinetics of cell infiltration and cytokine messenger RNA expression after intradermal challenge with allergen and tuberculin in the same atopic individuals. *J. Allergy Clin. Immunol.*, **94**, 764–772.

40. Sallusto, F., Lenig, D., Mackay, C.R., Lanzavecchia, A. (1998) Flexible programs of chemokine receptor expression on human polarized T helper 1 and 2 lymphocytes. *J. Exp. Med.*, **187**, 875–883.

41. Tangye, S., Ma, C., Brink, R., Deenick, E. (2013) The good, the bad and the ugly – TFH cells in human health and disease. *Nat. Rev. Immunol.*, **13**, 412–426.

42. Iwakura, Y., Ishigame, H. (2006) The IL-23/IL-17 axis in inflammation. *J. Clin. Invest.*, **116**, 1218–1222.

43. Kaplan, M. (2013) Th9 cells, differentiation and disease. *Immunol. Rev.*, **252**, 104–115.

44. Sabat, R., Ouyang, W., Wolk, K. (2014) Therapeutic opportunities of the IL-22-IL-22R1 system. *Nat. Rev. Drug Discov.*, **13**, 21–38.

45. Caridade, M., Graca, L., Ribeiro, R. (2013) Mechanisms underlying CD4+ Treg immune regulation in the adult, from experiments to models. *Front. Immunol.*, **4**, 378.

46. Carswell, E.A., Old, L.J., Kassel, R.L., Green, S. *et al.* (1975) An endotoxin-induced serum factor that causes necrosis of tumors. *Proc. Natl Acad. Sci. USA*, **72**, 3666–3670.

47. Chai, Z., Gatti, S., Toniatti, C., Poli, V. *et al.* (1996) Interleukin (IL)-6 gene expression in the central nervous system is necessary for fever response to lipopolysaccharide or IL-1 beta, a study on IL-6-deficient mice. *J. Exp. Med.*, **183**, 311–316.

48. Wang, H., Yu, M., Ochani, M., Amella, C.A. *et al.* (2003) Nicotinic acetylcholine receptor alpha7 subunit is an essential regulator of inflammation. *Nature*, **421**, 384–388.

49. Andersson, U., Tracey, K.J. (2012) Neural reflexes in inflammation and immunity. *J. Exp. Med.*, **209**, 1057–1068.

50. Ingman, W., Jones, R. (2008) Cytokine knockouts in reproduction, the use of gene ablation to dissect roles of cytokines in reproductive biology. *Hum. Reprod. Update*, **14**, 179–192.

51. Fock, V., Mairhofer, M., Otti, G.R., Hiden, U. *et al.* (2013) Macrophage-derived IL-33 is a critical factor for placental growth. *J. Immunol.*, **191**, 3734–3743.

52. de Waal Malefyt, R., Abrams, J., Bennet, B., Figdor, C.G. *et al.* (1991) Interleukin 10 (IL-10) inhibits cytokine synthesis by human monocytes, an autoregulatory role of IL-10 produced by monocytes. *J. Exp. Med.*, **174**, 1209–1220.

53. Gérard, C., Bruyns, C., Marchant, A., Abramowicz, D. *et al.* (1993) Interleukin-10 reduces the release of tumor necrosis factor and prevents lethality in experimental endotoxemia. *J. Exp. Med.*, **177**, 547–550.

54. de Waal, M.R., Figdor, C.G., Huijbens, R., Mohan-Peterson, S. *et al.* (1993) Effects of IL-13 on phenotype, cytokine production, and cytotoxic function of human monocytes. Comparison with IL-4 and modulation by IFN-gamma or IL-10. *J. Immunol.*, **151**, 6370–6381.

55. Caruso, R., Stolfi, C., Sarra, M., Rizzo, A. *et al.* (2009) Inhibition of monocyte-derived inflammatory cytokines by IL-25 occurs via p38 Map kinase-dependent induction of Socs-3. *Blood*, **113**, 3512–3519.

56. Guarda, G., Braun, M., Staehli, F., Tardivel, A. *et al.* (2011) Type I interferon inhibits interleukin-1 production and inflammasome activation. *Immunity*, **34**, 213–223.

57. Cavaillon, J.M., Adib-Conquy, M. (2006) Bench-to-bedside review, endotoxin

tolerance as a model of leukocyte reprogramming in sepsis. *Crit. Care*, **10**, 233.
58. De Togni, P., Goellner, J., Ruddle, N., Streeter, P. et al. (1994) Abnormal development of peripheral lymphoid organs in mice deficient in lymphotoxin. *Science*, **264**, 703–707.
59. Dranoff, G., Crawford, A., Sadelain, M., Ream, B, et al. (1994) Involvement of granulocyte-macrophage colony-stimulating factor in pulmonary homeostasis. *Science*, **264**, 713–716.
60. Shull, M.M., Ormsby, I., Kier, A.B., Pawlowski, S. et al. (1992) Targeted disruption of the mouse transforming growth factor-beta 1 gene results in multifocal inflammatory disease. *Nature*, **359**, 693–699.
61. Kuhn, R., Lohler, J., Rennick, D., Rajewsky, K. et al. (1993) Interleukin-10-deficient mice develop chronic enterocolitis. *Cell*, **75**, 263–274.
62. Horai, R., Saijo, S., Tanioka, H., Nakae, S. et al. (2000) Development of chronic inflammatory arthropathy resembling rheumatoid arthritis in interleukin 1 receptor antagonist-deficient mice. *J. Exp. Med.*, **191**, 313–320.
63. Kopf, M., Baumann, H., Freer, G., Freudenberg, M. et al. (1994) Impaired immune and acute-phase responses in interleukin-6-deficient mice. *Nature*, **368**, 339–342.
64. Pfeffer, K., Matsuyama, T., Kundig, T.M., Wakeham, A. et al. (1993) Mice deficient for the 55 kd tumor necrosis factor receptor are resistant to endotoxic shock, yet succumb to *L. monocytogenes* infection. *Cell*, **73**, 457–467.
65. Amiot, F., Fitting, C., Tracey, K.J., Cavaillon, J.-M. et al. (1997) LPS-induced cytokine cascade and lethality in Lta/TNFa deficient mice. *Mol. Med.*, **3**, 864–875.
66. Cavazzana-Calvo, M., Hacein-Bey, S., de Saint, B.G., Gross, F. et al. (2000) Gene therapy of human severe combined immunodeficiency (SCID)-X1 disease. *Science*, **288**, 669–672.
67. Cavaillon, J.-M. (2001) Pro- versus anti-inflammatory cytokines; myth or reality. *Cell. Mol. Biol.*, **47**, 695–702.
68. Cavaillon, J.M., Muñoz, C., Fitting, C., Misset, B. et al. (1992) Circulating cytokines, the tip of the iceberg? *Circ. Shock*, **38**, 145–152.
69. Bender, J.R., Sadeghi, M.M., Watson, C., Pfau, S. et al. (1994) Heterogeneous activation thresholds to cytokines in genetically distinct endothelial cells, evidence for diverse transcriptional responses. *Proc. Natl Acad. Sci. USA*, **91**, 3994–3998.
70. Tracey, K.J., Beutler, B., Lowry, S.F., Merryweather, J. et al. (1986) Shock and tissue injury induced by recombinant human cachectin. *Science*, **234**, 470–474.
71. Tracey, K.J., Fong, Y., Hesse, D.G., Manogue, K.R. et al. (1987) Anti-cachectin/TNF monoclonal antibodies prevent septic shock during lethal bacteraemia. *Nature*, **330**, 662–664.
72. Doherty, G.M., Lange, J.R., Langstein, H.N., Alexander, H.R. et al. (1992) Evidence for IFN-gamma as a mediator of the lethality of endotoxin and tumor necrosis factor-alpha. *J. Immunol.*, **149**, 1666–1670.
73. Smith, C.A., Davis, T., Wignall, J.M., Din, W.S. et al. (1991) T2 open reading frame from the Shope fibroma virus encodes a soluble form of the TNF receptor. *Biochem. Biophys. Res. Commun.*, **176**, 335–342.
74. Upton, C., Mossman, K., McFadden, G. (1992) Encoding of a homolog of the IFN-gamma receptor by myxoma virus. *Science*, **258**, 1369–1372.
75. Moore, K.W., Vieira, P., Fiorentino, D.F., Trounstine, M.L. et al. (1990) Homology of cytokine synthesis inhibitory factor (IL-10) to the Epstein–Barr virus gene BCRFI. *Science*, **248**, 1230–1234.
76. Bahar, M.W., Graham, S.C., Chen, R.A., Cooray, S. et al. (2011) How vaccinia virus has evolved to subvert the host immune response. *J. Struct. Biol.*, **175**, 127–134.
77. Navarro, L., Alto, N.M., Dixon, J.E. (2005) Functions of the Yersinia effector proteins in inhibiting host immune responses. *Curr. Opin. Microbiol.*, **8**, 21–27.
78. Alten, R., Maleitzke, T. (2013) Tocilizumab, a novel humanized anti-interleukin 6 (IL-6) receptor antibody for the treatment of patients with non-RA systemic, inflammatory rheumatic diseases. *Ann. Med.*, **45**, 357–363.
79. Hahn, B.H. (2013) Belimumab for systemic lupus erythematosus. *N. Engl. J. Med.*, **368**, 1528–1535.

80. Coffman, R.L., Ohara, J., Bond, M.W., Carty, J. et al. (1986) B cell stimulatory factor-1 enhances the IgE response of lipopolysaccharide-activated B cells. *J. Immunol.*, **136**, 4538–4541.
81. Nair, P., Pizzichini, M.M., Kjarsgaard, M., Inman, M.D. et al. (2009) Mepolizumab for prednisone-dependent asthma with sputum eosinophilia. *N. Engl. J. Med.*, **360**, 985–993.
82. Corren, J., Lemanske, R.F., Hanania, N.A., Korenblat, P.E. et al. (2011) Lebrikizumab treatment in adults with asthma. *N. Engl. J. Med.*, **365**, 1088–1098.
83. Cavaillon, J.M., Adib-Conquy, M., Fitting, C., Adrie, C. et al. (2003) Cytokine cascade in sepsis. *Scand. J. Infect. Dis.*, **35**, 535–544.
84. Suntharalingam, G., Perry, M.R., Ward, S., Brett, S.J. et al. (2006) Cytokine storm in a phase 1 trial of the anti-CD28 monoclonal antibody TGN1412. *N. Engl. J. Med.*, **355**, 1018–1028.
85. van Dullemen, H.M., van Deventer, S.J., Hommes, D.W., Bijl, H.A. et al. (1995) Treatment of Crohn's disease with anti-tumor necrosis factor chimeric monoclonal antibody (cA2). *Gastroenterology*, **109**, 129–135.
86. Targan, S.R., Hanauer, S.B., van Deventer, S.J., Mayer, L. et al. (1997) A short-term study of chimeric monoclonal antibody cA2 to tumor necrosis factor alpha for Crohn's disease. Crohn's Disease cA2 Study Group. *N. Engl. J. Med.*, **337**, 1029–1035.
87. Chaudhari, U., Romano, P., Mulcahy, L.D., Dooley, L.T. et al. (2001) Efficacy and safety of infliximab monotherapy for plaque-type psoriasis, a randomised trial. *Lancet*, **357**, 1842–1847.
88. Elliot, M.J., Maini, R.N., Feldmann, M., Kalden, J.R. et al. (1994) Randomised double-blind comparison of chimeric monoclonal antibody to tumor necrosis factor versus placebo in rheumatoid arthritis. *Lancet*, **344**, 1105–1110.
89. Olsen, N.J., Stein, C.M. (2004) New drugs for rheumatoid arthritis. *N. Engl. J. Med.*, **350**, 2167–2179.
90. Papp, K.A., Leonardi, C., Menter, A., Ortonne, J.P. et al. (2012) Brodalumab, an anti-interleukin-17-receptor antibody for psoriasis. *N. Engl. J. Med.*, **366**, 1181–1189.
91. Papp, K.A., Langley, R.G., Lebwohl, M., Krueger, G.G. et al. (2008) Efficacy and safety of ustekinumab, a human interleukin-12/23 monoclonal antibody, in patients with psoriasis, 52-week results from a randomised, double-blind, placebo-controlled trial (PHOENIX 2). *Lancet*, **371**, 1675–1684.
92. Coleman, R.L., Duska, L.R., Ramirez, P.T., Heymach, J.V. et al. (2011) Phase 1-2 study of docetaxel plus aflibercept in patients with recurrent ovarian, primary peritoneal, or fallopian tube cancer. *Lancet Oncol.*, **12**, 1109–1117.
93. Fulciniti, M., Hideshima, T., Vermot-Desroches, C., Pozzi, S. et al. (2009) A high-affinity fully human anti-IL-6 mAb, 1339, for the treatment of multiple myeloma. *Clin. Cancer Res.*, **15**, 7144–7152.
94. Scagnolari, C., Antonelli, G. (2013) Antiviral activity of the interferon alpha family, biological and pharmacological aspects of the treatment of chronic hepatitis C. *Expert Opin. Biol. Ther.*, **13**, 693–711.
95. Lupo, L., Panzera, P., Tandoi, F., Carbotta, G. et al. (2008) Basiliximab versus steroids in double therapy immunosuppression in liver transplantation, a prospective randomized clinical trial. *Transplantation*, **86**, 925–931.
96. Sheashaa, H.A., Bakr, M.A., Rashad, R.H., Ismail, A.M. et al. (2011) Ten-year follow-up of basiliximab induction therapy for live-donor kidney transplant, a prospective randomized controlled study. *Exp. Clin. Transplant.*, **9**, 247–251.
97. Verweij, C.L., Vosslamber, S. (2013) Relevance of the type I interferon signature in multiple sclerosis towards a personalized medicine approach for interferon-beta therapy. *Discov. Med.*, **15**, 51–60.
98. Bielekova, B., Richert, N., Howard, T., Blevins, G. et al. (2004) Humanized anti-CD25 (daclizumab) inhibits disease activity in multiple sclerosis patients failing to respond to interferon beta. *Proc. Natl Acad. Sci. USA*, **101**, 8705–8708.
99. Cummings, S.R., San Martin, J., McClung, M.R., Siris, E.S. et al. (2009) Denosumab for prevention of fractures in postmenopausal women with osteoporosis. *N. Engl. J. Med.*, **361**, 756–765.
100. Larsen, C.M., Faulenbach, M., Vaag, A., Volund, A. et al. (2007) Interleukin-1-receptor antagonist in type 2

diabetes mellitus. *N. Engl. J. Med.*, **356**, 1517–1526.
101. So, A., De Smedt, T., Revaz, S., Tschopp, J. (2007) A pilot study of IL-1 inhibition by anakinra in acute gout. *Arthritis Res. Ther.*, **9**, R28.
102. Botsios, C., Sfriso, P., Furlan, A., Punzi, L. *et al.* (2008) Resistant Behçet disease responsive to anakinra. *Ann. Intern. Med.*, **149**, 284–286.

8
Neural Transplantation: Evidence from the Rodent Cerebellum[1)]

Ketty Leto[1,2] and Ferdinando Rossi[1,2]
[1] *University of Turin, Rita Levi-Montalcini Department of Neuroscience, Corso Raffaello, 30, 10025 Turin, Italy*
[2] *University of Turin, Azienda Ospedaliero-Universitaria San Luigi Gonzaga, Neuroscience Institute Cavalieri Ottolenghi (NICO), Neuroscience Institute of Turin, Regione Gonzole 10, 10043 Orbassano, Turin, Italy*

1	Introduction	268
2	The Murine Cerebellum: A Suitable Model to Study Neural Transplantation	269
2.1	The Cerebellar Circuitry: A Point-to-Point System	270
2.2	Development of Cerebellar Phenotypes	271
2.3	Fate Potential of Cerebellar Progenitors: Evidence from Transplantation Experiments	272
2.3.1	Temporal Restriction of Progenitor Potential	273
2.3.2	Regional Specification of Neural Progenitors	274
2.3.3	Environmental Influences	274
3	Integration of Cerebellar Progenitors in the Adult Cerebellum: The Case of Purkinje Cells	276
4	Stem Cell Therapies for Cerebellar Disorders: Issues and Perspectives	279
4.1	*In-Vitro* Derivation of Cerebellar Phenotypes	280
5	Conclusions	281
	References	282

Keywords

Neural transplantation
The grafting of neural tissue or neural progenitor cells to the nervous system of individuals of the same species, or individuals of different species.

[1)] This article is in memoriam of Ferdinando Rossi.

Translational Medicine: Molecular Pharmacology and Drug Discovery
First Edition. Edited by Robert A. Meyers.
© 2018 Wiley-VCH Verlag GmbH & Co. KGaA. Published 2018 by Wiley-VCH Verlag GmbH & Co. KGaA.

Cell replacement
The appropriate substitution of damaged or lost cellular elements of a subject by the transplantation of healthy cells able to integrate in the host tissue.

Fate potential
The competence of progenitor cells to acquire defined cell identities. Fate potential is influenced by intrinsic factors and environmental cues.

Cell specification
Processes underlying the final fate choice of progenitor cells toward mature phenotypes.

Integration
Processes through which grafted cells reach their final correct positions and establish appropriate morpho-functional relationships within the recipient tissue. In the case of neural transplantation, correct integration requires the formation of anatomically correct and functionally meaningful connections with host neurons.

■ The transplantation of neural progenitors or stem cells has been attempted for several decades, in many regions of the central nervous system (CNS), to repair tissue damage or cell degeneration. In most circumstances the success of cell replacement in the brain depends on the capability to rewire precisely patterned connections. Several concurrent factors determine the outcome and efficacy of this approach, including the intrinsic properties of donor progenitors, the environmental cues provided by the recipient environment, and the complexity and precision of the connection patterns that must be restored in order to recover adaptive function. In this chapter, some major concepts concerning these issues are reviewed, with special attention being paid to studies involving neural transplantation in the rodent cerebellum and in animal models of cerebellar disorders. These approaches are considered the most suitable for the design and testing of cell-based therapeutic approaches aimed at reconstructing specific connections in point-to-point circuits.

1
Introduction

The literature on neural transplantation contains a high number of contributions that, over the past decades, have analyzed the potentialities and limits of this approach in different regions of the central nervous system (CNS), in the context of regenerative medicine or developmental neurobiology. In the latter field of research, neural transplantation has been usefully employed to elucidate the relative role of intrinsic determinants and environmental influences underlying the processes of cell specification and settlement in different brain areas [1–6]. At the same time, owing to the poor abilities for any spontaneous regeneration of the mammalian CNS, cell transplantation has been extensively tested as a way to repair brain damage following traumatic injuries or neurodegeneration.

In this context, a major challenge is to obtain full functional recovery of damaged circuitries through the insertion of healthy cells endowed with specific developmental and neurochemical properties [7–10]. While this goal can be achieved in non-neural tissues, characterized by a high cell turnover, in the adult CNS – where neurogenic capabilities are almost absent – the achievement of successful cell replacement still requires a more profound knowledge of the basic biological processes that regulate the integration of new elements in neural networks. Accordingly, important information to develop translatable cell-based therapies can also be derived from studies focusing on basic ontogenetic mechanisms.

In general, the efficacy of cell-based reparative attempts has been influenced by several concurrent factors, such as the type of injury or pathologic condition of the damaged system, the age, the receptive properties of the host milieu, and the intrinsic properties of the donor cells. Notably, the effective application of cell therapy requires deep knowledge and mastering of the mechanisms underlying cell survival, specification (i.e., cell fate choice), differentiation (i.e., acquisition of mature phenotypic traits) and integration (i.e., establishment of specific interactions with host elements) of new neurons in the recipient system [11, 12]. In spite of the significant improvements recently obtained by transplantation experiments in both animal models and selected cohorts of patients [8, 13–15], the most critical point remains the establishment of meaningful reciprocal connections between grafted cells and the host elements. In this context, encouraging results have been obtained in systems, such as the nigrostriatal pathway, where functional recovery may be achieved by restoring a diffuse connectivity that modulates the activity of target circuitries. In contrast, much less progress has been obtained in the repair of specific sensory–motor systems, in which adaptive function requires specifically patterned point-to-point connections [9]. In this chapter, a number of studies dealing with cell transplantation to the cerebellum, as a model of specific point-to-point system, will be reviewed, and the possibility of integrating new neurons in the recipient network in both physiologic and pathologic conditions will be discussed.

2
The Murine Cerebellum: A Suitable Model to Study Neural Transplantation

The cerebellum represents the most appropriate ground to study the mechanisms that underlie neuronal integration and circuit rewiring by cell replacement (for reviews, see Refs [10, 11]), for a variety of reasons:

- The cerebellar network is composed of a limited number of morphologically and molecularly distinct neuronal cell-types [16–19].
- These cells are integrated in a well-established anatomo-functional architecture [16, 19–22].
- Cerebellar circuits are organized in a highly precise topographic pattern, highlighted by the differential expression of positional markers and molecular identity codes that guide the reciprocal recognition between pre- and postsynaptic partners [22–27].
- The main processes that regulate phenotype specification and the assembly of the cerebellar intracortical and corticonuclear networks have been elucidated [16, 21, 22, 28, 29].

- Numerous strains of spontaneous or experimentally induced mutant mice are available, which are characterized by the selective loss of specific cerebellar populations [30, 31]. These mice provide suitable models to elucidate the pathogenesis of cerebellar disorders, to investigate the primary and secondary effects induced by the loss of specific neural populations, and to test approaches to promote anatomical repair and functional recovery.

2.1
The Cerebellar Circuitry: A Point-to-Point System

The cerebellum mediates a wide repertoire of complex functions through a relatively simple and stereotyped architecture. Afferent inputs from mossy fiber brainstem and spinal cord nuclei, together with climbing fibers from the inferior olive [32–35], convey fundamental information to control the coordination of movement, posture and gaze [21, 36–38]. In addition, the cerebellum plays important roles in motor learning, emotional behavior and fear memory [39–44]. Finally, recent findings have indicated that the cerebellum is also involved in a broad range of cognitive functions, such as working memory, language, executive function and neurocognitive development [45].

The cerebellum comprises three pairs of deep cerebellar nuclei (DCN), immersed in a mass of white matter, covered by a highly lobulated cortex [16, 17]. The latter is characterized by a typical three-layered organization: the deepest layer is the granular layer (GL), containing granule cells and other classes of interneurons (Golgi, Lugaro, and unipolar brush cells), while the most superficial layer is the molecular layer (ML), which contains the dendrites of Purkinje cells (PCs), the axons of granule neurons (the parallel fibers) and inhibitory basket and stellate interneurons. At the interface between these two layers, the single row of PC bodies forms the Purkinje cell layer (PCL). Inputs to the cortex converge on PCs through two principal afferent systems: (i) the climbing fibers, which originate from the inferior olive and directly innervate PCs; and (ii) the mossy fibers, which terminate on granule cells and which in turn are connected to PCs [19, 22]. The PC axons represent the only cortical output, relaying information to DCN neurons that, in turn, convey the output to extracerebellar structures (Fig. 1).

Another salient feature of the cerebellar organization is the presence of parasagittally oriented cortical modules, defined by neurochemically heterogeneous subsets of PCs [22, 24, 47] that can be highlighted by the differential expression of specific markers, such as zebrins [48–50]. Interestingly, these stripes of PCs are involved in the formation of specific microcircuits, as they project to defined DCN regions which receive direct inputs from collaterals of the climbing and mossy fibers innervating the same cortical module [27, 51, 52].

Considering this highly specific cerebellar organization, it is conceivable that a major challenge of cell replacement strategies in this system refers to the reconstruction of specific input–output relationships in the cortical network, as well as in the corticonuclear projection. To achieve this result, it is necessary to understand the processes that regulate the genesis of the different classes of cerebellar neurons and drive their assembly in the cerebellar architecture.

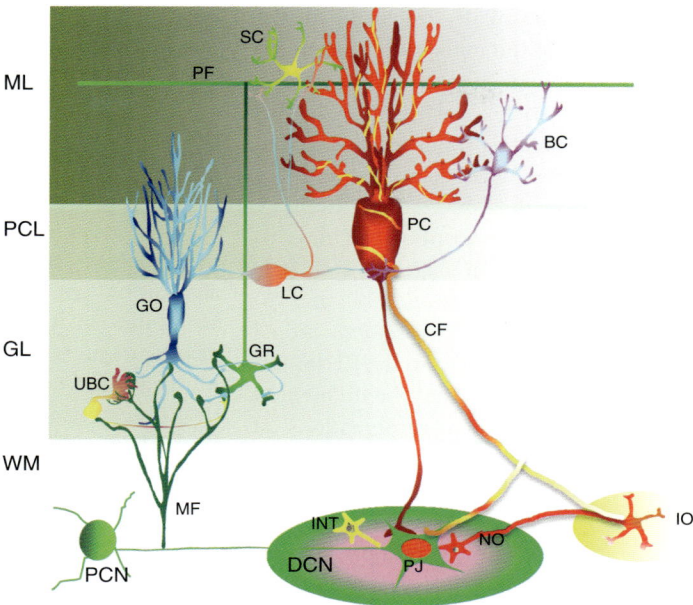

Fig. 1 Adult cerebellar cytoarchitecture. Wiring diagram for a cerebellar corticonuclear microcircuit. Abbreviations: BC, basket cell; CF, climbing fiber; DCN, deep cerebellar nuclei; GL, granule cell layer; GR, granule cell; GO, Golgi cell; IO, inferior olive; INT, interneuron; LC, Lugaro cell; MF, mossy fiber; ML, molecular layer; N-O, nucleo-olivary inhibitory projection; PC, Purkinje cell; PCL, Purkinje cell layer; PCN, precerebellar neuron; PF, parallel fiber; PJ, projection neuron; SC, stellate cell; UB, unipolar brush cell; WM, white matter. Modified from Ref. [46].

2.2 Development of Cerebellar Phenotypes

The cerebellar anlage is derived from the dorsal portion of rhombomere 1, at the junction between the caudal half of the mesencephalon and the rostral half of the metencephalon, as demonstrated by the study of cross-species transplantation of specific segments of the neural tubes between chick and quail [53–55]. Genetic analyses and transplantation studies have shown that this region at the met-mesencephalic border, termed the *isthmic organizer*, is endowed with inductive properties and controls the regionalization and differentiation of the anterior hindbrain, as well as midbrain structures [22, 56–58].

In the mouse, starting from embryonic day 9 (E9), two symmetric primordia are detectable on either side of the dorsal aspect of the neural tube. During the following days the two separated bulges grow and fuse together, giving rise to the unitary cerebellar plate comprising the vermis and the two hemispheres. These morphogenetic processes are mainly sustained by a transient structure called the *germinal trigone* or rhombic lip (RL) which, particularly through its rostral portion, leads the fusion establishing a substantial bridge between the two bilateral primordia [21]. The RL is located at the outer aspect of the cerebellar plate, adjacent to the roof-plate, while the inner side is occupied by the ventricular zone (VZ), covering the fourth ventricle. The RL and the VZ are two germinative neuroepithelia that generate all cerebellar neurons, and are distinguished by the expression of two basic helix-loop-helix transcription factors: (i) pancreas transcription factor 1-a (Ptf1a), which is expressed in the VZ [59]; and (ii) the mouse homolog of *Drosophila* atonal (Atoh-1), which is present in the RL [60]. It is now well established that the differential expression of particular combinations of transcription factors is responsible for the neurochemical compartmentation of cerebellar precursors (for reviews, see Refs [29, 61]). All GABAergic phenotypes (PCs, nucleo-olivary projection DCN neurons, and all inhibitory interneurons) originate from Ptf1-a-positive precursors of the VZ [59], whereas glutamatergic neurons (large projection neurons of DCN, unipolar brush cells, and granule neurons) derive from Atoh-1-expressing progenitors of the RL [62–66]. Typically, these two germinative epithelia disappear around birth; however, dividing VZ precursors emigrate into the cerebellar prospective white matter (PWM), whereas those of the RL move along the pial cerebellar surface, where they form the external granular layer (EGL).

The temporal sequence of the appearance of all different classes of cerebellar neurons is also finely regulated. Birthdating studies have shown that projection neurons (both PCs and DCN neurons) are the first to be born at the onset of cerebellar embryogenesis, whereas local interneurons, both GABAergic and glutamatergic, are produced during late embryonic and early postnatal life, up to the third postnatal week [18, 21, 67].

2.3
Fate Potential of Cerebellar Progenitors: Evidence from Transplantation Experiments

A successful use of neural progenitors to promote anatomical repair and functional recovery requires extensive knowledge of their developmental potentialities, and of the processes that direct their phenotype acquisition and homing in the maturing structure. In the context of cerebellar development, two main alternative (but not mutually exclusive) mechanisms regulate the specification of cerebellar neurons. Some lineages derive from fate-restricted progenitors that follow an intrinsic developmental route, while others are generated from uncommitted precursors that acquire adult identities in response to instructive environmental signals [68]. One of the best methods for evaluating the intrinsic potential of progenitor cells is to expose them to temporally or spatially unusual environmental conditions, by heterotopic/heterochronic transplantation [10, 11].

Pioneer studies using lines of immortalized neural precursors, derived from the embryonic hippocampus [69] or the postnatal cerebellar EGL [70], showed that multipotent progenitors grafted to the neonatal cerebellum maintained the capability to integrate in the host tissue and differentiate according to their site of engraftment. These findings suggested that progenitor cells are endowed with broad developmental potentialities that could be exploited for therapeutic purposes. However, another study, in which the potentialities of primary and immortalized progenitors were directly compared, yielded different conclusions [71]. Indeed, while oncogene-immortalized EGL cells acquired a variety of neuronal and glial phenotypes, their primary counterparts exclusively differentiated into granule neurons. These findings suggested that granule cells originate from fate-committed precursors, and that oncogene-induced immortalization subverts the mechanisms that direct their physiological specification [71]. Later experiments confirmed that progenitors arising from the RL are specified to a granule cell fate as early as E13 in the mouse [62, 63]. On the whole, these studies indicated that progenitor cells become committed to a mature phenotype at early stages of their development, and their ensuing differentiation is accomplished by unfolding a cell-autonomous program. Nonetheless, even strictly committed cells, such as the trigone-derived precursors, require environmental cues to progress in their maturation [62, 72]. In addition, further experiments focusing on other categories of cerebellar progenitors (as discussed in the following section) have revealed a wider variety of ontogenetic mechanisms.

2.3.1 Temporal Restriction of Progenitor Potential

It has been proposed that a precise ontogenetic rhythm is intrinsically encoded in progenitor cells so that, in addition to positional information, they are also endowed with an internal temporal program that regulates their responsiveness to external signals [73]. The choice of multipotent progenitors toward specific fates, which leads to morphologically, neurochemically and functionally distinct neuronal types, results from a progressive restriction of cell potential that can occur at different ontogenetic stages. Some neurons are already committed to specific phenotypes early in their mitotic history, as in the case of granule cell progenitors in the RL [53, 62, 63, 71], whereas other types, such as projection neurons of the cerebral cortex,

become specified during their last mitosis [1, 74, 75]. Finally, in some instances, for example in the case of cerebellar inhibitory interneurons, fate choices may occur after the cell cycle exit [76] (see Sect. 2.3.3).

Heterochronic transplantation has been used to ask when and how the potential of cerebellar progenitors becomes restricted during cerebellar development. In particular, a first question was whether postnatal progenitors, which normally are fated to generate excitatory or inhibitory interneurons, could acquire the identities of projection neuron following exposure to the environment of the embryonic cerebellar primordium, when these types are generated. To address this issue, Jankovski and colleagues [77] mixed together cerebellar fragments deriving from embryonic (E12) and postnatal (P3 to P8) mouse cerebella and cografted them to the cerebellum of adult hosts. Postnatal donor cells were derived from several transgenic mouse lines, each expressing the *lacZ* reporter gene in different sets of neuronal populations, so that the fate of postnatal progenitors could be ascertained. In such mixed grafts, embryonic donor tissue developed typical minicerebellar structures, with tri-layered cortical folia and DCN [78, 79], whereas postnatal cells appeared to be individually integrated within these structures. Importantly, the cells derived from postnatal donors differentiated exclusively into granule cells and ML interneurons, indicating that they were strictly committed toward late-generated phenotypes and had lost the sensitivity to inductive signals emerging from the embryonic milieu [77].

The progressive restriction of progenitor potential during development has been further demonstrated by comparing the phenotypic repertoires generated by embryonic (E12) or postnatal (P4) mouse cerebellar progenitors, tagged with green fluorescent protein (GFP) and transplanted *in utero* to the cerebral ventricles of E15 rat embryos [80]. In these experiments, small amounts of dissociated donor cell suspensions were used so that the individual donor cells were fully exposed to the recipient milieu [81, 82]. Grafted cells generated mature cerebellar neurons, the morphological and neurochemical profiles of which often indicated a high degree of integration in the host tissue. However, whereas E12 donor cells were able to generate the complete repertoire of cerebellar projection neurons and interneurons, their postnatal counterparts exclusively produced the late-born phenotypes (i.e., granule cells and ML interneurons). The same donor cell populations were also grafted to organotypic slices from the brainstem/cerebellum of E12 mouse embryos. Also in this case, postnatal donor cells were only capable of generating late-born phenotypes, further indicating that cerebellar progenitors undergo progressive restriction of their potentialities over time [10, 11, 80].

2.3.2 Regional Specification of Neural Progenitors

In-utero transplantation also revealed the strict regional specification of cerebellar progenitors. Indeed, cerebellar donors engrafted and differentiated in wide areas of the embryonic CNS, but exclusively acquired cerebellar identities [80]. These observations were in contrast with those reported previously, which interpreted that EGL progenitors may become granule neurons of the dentate gyrus [83], although no clear evidence for re-specification was provided. However, a strict regional identity was also observed for progenitor cells derived from different sites along the neuraxis [84, 85]. Progenitors isolated from E12 neocortex and grafted to embryonic hosts *in utero*, also engrafted in many different CNS regions [84]. Interestingly, these cells acquired host-specific phenotypes in telencephalic sites, such as the neocortex, basal ganglia, olfactory bulb and hippocampus, but failed to adopt local identities in the other engraftment sites. In addition, whilst in the forebrain these cells mostly generated neurons, in more posterior locations they more frequently acquired glial identities. In another set of experiments, progenitors derived from several neurogenic sites (telencephalic subventricular zone, medial ganglionic eminence, ventral mesencephalon, and dorsal spinal cord), were transplanted to the postnatal cerebellar PWM, at the time when local interneurons and glia are generated [85]. The results of another study, in which progenitors from the subventricular zone were grafted to the postnatal cerebellum, suggested that the donor cells might switch to local identities [86]. However, the experiments carried out by Rolando *et al.* [85] showed that that although extracerebellar cells are able to engraft and differentiate in the host cerebellum, being sensitive to local constraints that influence their mature traits, they actually failed to develop cerebellar phenotypes. Taken together, the results of these studies highlighted the strict regional specification of progenitor cells derived from different germinal sites along the neuraxis. Interestingly, however, positional identities are sharply defined along the rostrocaudal axis, whereas a certain degree of plasticity is allowed along the dorsoventral direction, as displayed by the behavior of neocortical progenitors grafted to the embryonic CNS *in utero* [84].

2.3.3 Environmental Influences

The findings described in the previous section depict a scenario in which the fate choice of progenitor cells is determined by progressive steps of potential restriction in space and time. Nonetheless, a different picture emerges when the development of GABAergic interneurons is considered [87–90]. Inhibitory interneurons of the cerebellar cortex and DCN comprise different phenotypes, characterized by distinctive morphological, neurochemical and functional properties, with specific positions and connectivity [21, 87]. In other CNS regions, such as the spinal cord [91], cerebral cortex [92, 93] and olfactory bulb [94, 95], GABAergic interneurons are produced according to precise spatiotemporal patterns, from fate-restricted progenitors residing in distinct germinal sites. In addition, interneuron progenitors are usually committed to specific phenotypes and laminar positions during their last mitosis, when they still reside in the germinal neuroepithelia [74, 95, 96]. The results of transplantation experiments have shown clearly that this common strategy does not apply to the case of cerebellar GABAergic interneurons.

The entire repertoire of cerebellar GABAergic interneurons derive from a population of Pax-2-expressing progenitors, continuously generated from E12.5 to the second postnatal week [3, 97, 98]. Pax-2-positive progenitors delaminate from the VZ of the embryonic cerebellar primordium and continue their proliferation in the postnatal PWM [99–101]. In order to assess whether interneuron progenitors located in different regions of the developing cerebellum are already committed toward specific fates, single-cell suspensions prepared from the cortex or the DCN region of P1 βactin-GFP rats were grafted to E15 embryos *in utero* or to P7 pups *in vivo* [3]. The results showed that interneuron progenitors are not spatially segregated, as both donor suspensions were able to generate a phenotypic repertoire consistent with the host age: all types of interneurons in embryonic recipients and late-born ML interneurons in postnatal cerebella. In addition, to assess whether a progressive narrowing of progenitor potential could occur through time, E14 and P7 donor progenitors were compared after grafting in hosts of different ages. Again, the interneuron repertoire generated by the two donor sources was exclusively dependent on the recipient age, showing that late-postnatal cells could also be re-specified towards earlier interneuron identities when exposed to neurogenic signals of younger hosts [3]. Altogether, these findings showed that GABAergic interneurons are produced by a common pool of progenitors, which maintain their multipotency up to late developmental phases and adopt mature identities in response to environmental instructive cues.

Later experiments provided further information regarding the dynamics of interneuron development and specification. It was shown that heterochronically grafted interneuron progenitors acquired the same phenotypes and laminar positions of host interneurons generated at the time of transplantation. The short-term evaluation of grafted material revealed that donor cells engrafted in the host PWM and acquired mature traits according to the times and modes of local interneurons [76]. Interestingly, postmitotic interneurons grafted to younger environments were also still able to switch their fate and adopt early-generated phenotypic and laminar identities, showing that young inhibitory interneurons are still sensitive to instructive signals of the host milieu [76]. The instructive cues that direct interneuron specification have still to be identified; nonetheless, they appear to be present in the microenvironment of the cerebellar

PWM. Indeed, when solid blocks of P1 or P7 PWM are grafted to the non-neurogenic milieu of the adult cerebellum, the repertoire of donor-derived interneurons is related to the age of the grafted PWM specimen [76].

3
Integration of Cerebellar Progenitors in the Adult Cerebellum: The Case of Purkinje Cells

In addition to cell survival and specification, another crucial issue to obtain a full therapeutic effect of cell replacement concerns the ability of grafted cells to incorporate in specific positions of the recipient architecture and form the correct afferent and efferent connections with the host neurons [9, 102]. The successful integration of donor cells depends not only on their ability to actively induce permissive conditions in the recipient environment, but also on the intrinsic receptive properties of the host elements [10, 11, 103]. Several lines of evidence have indicated that the successful reconstruction of highly organized circuits is strictly related to the age of the host [104–111]. In the cerebellum, this is particularly true in the case of PCs that are able to develop their mature phenotype accordingly to a cell-autonomous manner, whereas their ability to integrate in the recipient cerebellar cortex changes according to the age of the host [68, 78, 80, 112, 113]. In contrast, other cerebellar types – such as granule cells – are capable of full maturation and correct integration in the host cortex of any age [114].

The earliest experiments aimed at studying the integration of donor PCs in normal adult hosts showed that, in spite of their remarkable abilities for navigating into the adult cerebellar environment, their incorporation was defective in terms of position, orientation and establishment of long-distance connections [79, 98, 112, 113, 115, 116]. Similar conclusions were obtained after grafting PCs in Purkinje Cell Degeneration (pcd) mutant mice [78, 117]. This mouse line shows normal cerebellar development until the end of the second postnatal week, whereas selective PC degeneration occurs from P17 to P45 [118–120]. Cell transplantation experiments highlighted the ability of PC navigation and incorporation in the adult pcd environment [12, 30, 103, 112, 113, 121, 122], and suggested that this procedure may also ameliorate the motor function of the recipient animal [33, 99, 100, 123]. Concurrent observations showed that donor PCs preferentially settle in PC-depleted regions of the host cortex, and that a PC phenotype is favored over that of other cells [78, 117]. In later experiments, the proliferation of donor cells after grafting in both wild-type and pcd hosts was analyzed, and revealed that virtually all engrafted PCs derived from cells that were already postmitotic at the time of transplantation [12]. These data indicate that the adult cerebellar environment, whether wild-type or mutant, does not provide inductive cues to direct the fate choice of multipotent progenitors towards the PC phenotype (see also Ref. [124] for further evidence supporting this conclusion). Rather, the higher frequency of PCs over the other phenotypes observed in pcd hosts compared to wild-type mice actually reflected a selective mechanism, in which the survival and integration of donor PCs would be favored in the PC-depleted environment of the pcd cerebellum [12].

Another important feature that influences the outcome of transplantation to the cerebellum is an ability of the recipient tissue to undergo specific modifications that favor the integration of donor cells.

This phenomenon, termed *"adaptive rejuvenation"* [103], was originally applied to the re-expression of developmentally regulated molecules in the host Bergmann glia during the migration of grafted PCs. However, similar concepts can be applied to the ability of adult climbing fibers to bud new branches and innervate donor PCs by faithfully recapitulating all phases of the physiological development of climbing fibers, including the different stages of climbing fiber development and the transition from multi- to monoinnervation [30, 78, 79, 112, 117, 125].

In spite of the good receptive properties and adaptive capabilities of the mature cerebellum, the actual conditions are not sufficient to allow a complete integration of PCs, and in particular the corticonuclear connection cannot be rewired (Fig. 2) [12, 78, 117, 123]. Indeed, when embryonic PCs are grafted to embryonic or early postnatal hosts they entrain into the ontogenetic mechanisms of the recipient cerebellum

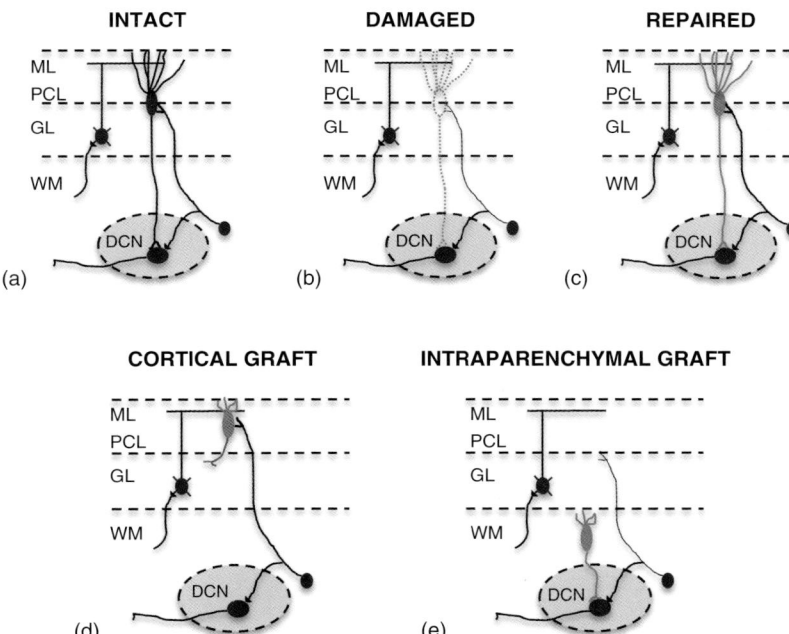

Fig. 2 Circuit rewiring after graft in the adult damaged cerebellum. (a) The cartoon shows a simplified scheme of the cerebellar cortical circuitry and corticonuclear projection. (b) In the case of selective Purkinje cell degeneration, the cortical afferents are deprived of their targets and the corticonuclear connection is interrupted. (c) Efficient cell replacement strategies should replace lost Purkinje cells through new neurons (gray) able to rewire damaged connections. (d) When cerebellar grafts are placed in the host cortex, Purkinje cells eventually settle in the recipient ML, restoring a normal relationship with their afferents, but failing to send their axons to the deep cerebellar nuclei. (e) In the case of intraparenchymally positioned transplants, many Purkinje cells remain within the graft, establishing nuclear innervation but failing to restore normal connectivity with host cortical targets. Abbreviations: ML, molecular layer; GL, granular layer, DCN, deep cerebellar nuclei; PCL, Purkinje cell layer; WM, white matter. Modified from Ref. [35].

and mature according to local schedules together with their host counterparts, leading to an efficient integration [80, 126] and lifelong survival, even in the case of allogeneic transplants [127]. This ability is lost soon after birth, although at a later stage grafted PCs settle in the host ML and become correctly innervated by host neurons, but fail to send their axons to their natural targets in the DCN (Fig. 2).

It has been shown that this progressive loss of PC integration capabilities in older recipients is due mostly to their migratory pathways and to the concomitant evolution of the maturing host cerebellar architecture. Transplanted cells that migrate to the cerebellar cortex may follow two major routes: (i) they can enter through the cortical surface into the ML (the so-called "inward path"); or (ii) they can migrate across the central white matter and reach their final destination by crossing the GL (the "outward path"; this is the path followed by PCs during physiological development). By comparing the fate of embryonic PCs grafted to recipient cerebella at the embryonic, postnatal and adult stages of maturation, it has been shown that correct integration is easily achieved when donor cells adopt the outward migratory route to reach their final destination in the cortex [126]. However, the outward migratory path is only accessible in embryonic or early postnatal hosts and, at later ages, donor cells are forced to follow the inward path. The latter, which is the path followed by granule cells during the radial phase of their developmental migration [16, 21], is still available in adulthood for both granule cells and PCs [114, 126]. During development, the PCs migrate along the scaffold provided by radial glia, which turns into Bergmann glia in mature cerebella. Hence, the outward migratory path is not viable in adult hosts due to the lack of a conductive substrate. In addition, the mature layering of the cortex represents an additional obstacle to the navigation of transplanted PCs.

Although, in adult hosts a minority of transplanted PCs take the outward migratory route, their movement is arrested at the edge between the white matter and the internal granule layer (IGL). These cells are able to send their axons to the DCN, but fail to be integrated in the host cortical network [126, 128, 129]. These findings indicate that the IGL environment hampers the migration of PCs, an idea which is further corroborated by the observation that PCs transplanted to P8 hosts, when a thick EGL borders the cortical surface, fail to penetrate into the ML along the inward route [126].

Taken together, these findings indicate that both the EGL and the IGL arrest the migration of PCs. Indeed, the transient ablation of the EGL by means of the antimitotic agent methylazoxy-methanol (MAM) allows the penetration of donor PCs into the host cerebellar cortex after grafting in P8 recipients [126]. Interestingly, in-vitro assays have demonstrated that the migration of PCs across the EGL is feasible following blockade of the signaling of reelin [126], which is expressed and released by EGL cells in order to drive the navigation and settlement of maturing PCs in vivo [130, 131]. Reelin is expressed in the IGL during postnatal development and also in the adult [130, 132], so that this layer also represents an impassable obstacle for PCs that migrate along the outward route, and also for the axons of the inwardly migrating neurons that may attempt to elongate across the IGL to reach the DCN [10, 11, 78]. Therefore, in spite of a good receptivity of the adult cerebellum for grafted neurons, the integration of PCs is strongly influenced by constraints imposed by the cortical cytoarchitecture.

Another relevant aspect that may influence the extent of functional integration of grafted PCs refers to the presence of neurochemically heterogeneous subpopulations of PCs, which are typically organized in parasagittally oriented cortical compartments, defined by the expression of subset-specific markers such as zebrins [26]. Several studies have shown that embryonic cerebellar grafts spontaneously develop distinct cortical modules [133, 134], which also direct the specific innervation from host olivary axons [134]. With regards to the integration of grafted PCs in the cerebellar cortex, observations carried out following the *in-utero* transplantation of embryonic cerebellar cells showed that 85% of donor PCs were correctly located in the correspondent zebrin-positive or zebrin-negative compartments [126]. The proportion of correctly positioned neurons might be lower when grafts are made to adult hosts ([133]; careful quantitative analysis is still missing). On the whole, the results of these studies have indicated that donor PCs settle preferentially in the appropriate compartments, but it is also clear that donor PCs retain their original neurochemical identity in the host cerebellum and may be also able to stably integrate into the "wrong" cortical modules. Whether this has any consequence on the function of these neurons remains to be elucidated.

4
Stem Cell Therapies for Cerebellar Disorders: Issues and Perspectives

The results of studies discussed above have shown that a functional integration of cerebellar progenitors in the recipient cortical network may be feasible. Nevertheless, the translation of these approaches to clinical practice has been severely hampered by a lack of adequate sources of donor cells. The recent development of techniques to produce efficient neural progenitors from different types of stem cell has disclosed new perspectives towards the development of cell-based therapies for cerebellar disorders, together with other neurodegenerative diseases [7, 135–137].

Stem cells are cells that can self-renew for the entire life of the organism, maintaining the potential to generate immature precursors and mature cells of different lineages. Among the different types of stem cell, neural stem (NS) cells were isolated as an *ad-infinitum* self-renewing population of multipotent cells, with the properties of neurogenic radial glia [135, 136]. NS cells have been efficiently derived from mouse embryonic stem (ES) cells, from fetal and adult mouse CNS, and from fetal human brain [138–142]. In addition, they have been produced from induced pluripotent stem (iPS) cells derived from mouse fibroblasts [143]. On the whole, NS cells offer a promising system for dissecting the mechanisms of neural differentiation, provide a suitable tool for genetic and chemical high-throughput screening, and represent a potential source of donor cells for cell replacement therapies.

A number of studies have addressed the possibility of applying stem cell therapies to the treatment of cerebellar neurodegeneration in both murine models and selected human patients (for details, see Ref. [143]). For instance, the transplantation of NS cells isolated from specific regions of the adult murine CNS (the subventricular zone lining the lateral ventricles and the subgranular zone of the hippocampal dentate gyrus [144, 145]) has been shown to enhance the motor skills of transgenic spinocerebellar ataxia type-1 (SCA1) mice, when compared to sham-operated controls [146]. Even if the grafted cells did not differentiate in cerebellar PCs (which represent the phenotype selectively affected in SCA1 disease), they induced behavioral amelioration, favored the survival of endogenous PCs, improved the cerebellar morphology, and restored normal PC excitability [146]. In another study, both mouse and human NS cells were injected into different groups of knockout mice that were homozygous for a target deletion of the acid sphingomyelinase (ASM) gene [147] and represented a model of human type A Niemann–Pick disease. These mice showed a massive degeneration of cerebellar PCs, accompanied by a severe motor impairment that could not be reversed by stem cell application [148]. In contrast, it has been shown that in a mouse model of type C Niemann–Pick disease, the intracerebellar transplantation of NS cells improved the neuropathological features of graft-recipient mice, promoting functional synaptic transmission in the surviving PCs [149].

Controversial results have also characterized the outcome of stem cell applications to human patients with cerebellar disorders. In a single case study, a patient affected by ataxia telangiectasia was seen to develop a tumor after grafting of NS cells derived from human fetal tissue [150]. Alternatively, in a clinical study involving 12 patients with inherited cerebellar atrophy, the effects of transplantation of *in-vitro*-cultured NS cells from human fetal cerebellum (8–10 weeks' gestation) were evaluated within the first week after graft [151]. Although there was a total absence of procedural complications due to cell rejection, the long-term efficacy and clinical outcome of the procedure remain to be elucidated [151].

4.1
In-Vitro Derivation of Cerebellar Phenotypes

Stem cells have limited capabilities to acquire spontaneously cerebellar phenotypes following transplantation. As a consequence, a convenient way to obtain efficient donor cells is to induce their differentiation *ex vivo*, prior to grafting [152]. As most cerebellar ataxias are primarily associated with PC loss, systematic investigations have been oriented towards the generation of *ex-vivo* protocols to produce PCs from mouse or human ES cells, by replicating crucial ontigenetic processes *in vitro* [135–137]. Although scant information is still available regarding the mechanisms of neural patterning and specification in the human cerebellum, the application of inductive developmental signals, followed by the administration of other mitogens and neurotrophins, directed the differentiation of human ES cells towards cerebellar phenotypes, most of which showed morphologies and marker expression typical of cerebellar granule cells [153]. In contrast, the main dynamics of cerebellar development have been elucidated in the mouse (see Sect. 2.2). Recent studies have shown that mimicking the *in-vivo* signaling pathways may allow the generation of functional cerebellar neurons from mouse ES cells [154, 155]. In a more recent study, Sasai and coworkers showed that PCs originate from Ptf1-a-positive VZ progenitors that express the cell-surface markers Neph3 and E-cadherin [156]. Subsequently, this emerging knowledge of early cerebellar development was applied to the efficient generation of PCs from mouse ES cells *in vitro* [157]. This protocol involved the application of fibroblast growth factor 2 (FGF2) and insulin to promote the development of En2-positive hindbrain tissue from murine embryonic stem (mES) cell cultures, followed by the application of cyclopamine (a Sonic Hedgehog, Shh, antagonist), to create precise dorsoventral cues for the generation of Ptf1-a$^+$ Neph3$^+$ VZ progenitors. The Neph3-positive cells were subsequently purified in a time-prospective experiment and differentiated to mature L7 and Calbindin-positive PCs. The neurons obtained were also able to differentiate and integrate in the recipient cerebellum, as shown previously for primary progenitors [156]. This protocol improved the efficiency of PC generation from mouse ES cells some 30-fold, compared to previous studies [154–156].

Along the same line, the results of another recent study disclosed some additional perspectives towards the generation of specific cerebellar phenotypes from stem cells [158]. These authors showed that human hindbrain neuroepithelial stem (hbNES) cells could be derived and expanded from early human embryos (weeks 5–7 of gestation), using adherent culture techniques, in the presence of epithelial growth factor (EGF) and fibroblast growth factor 2 (FGF2). These cells retained the stage-specific characteristics of early neuroepithelial cells, such as SOX1 expression, the formation of rosette-like structures and high neurogenic capacity, even after long-term expansion. Accordingly, hbNES cells demonstrated a higher neurogenic potential and a broader developmental capacity than previously described human NS cells [159], and provided a platform for generating region-specific neuronal populations, such as cerebellar neurons. Interestingly, the hbNES cells appeared to respond to developmental cues *in vitro* [158]. Following the introduction of dorsalizing agents, such as bone morphogenetic protein (BMP) and Wnt1, a proportion of the cells expressed dorsal hindbrain markers such as ATOH1, PAX3 and MSX1, and generated RL derivatives. When these cells were grafted into neonatal (P1–3) rat cerebellum and embryonic rat hindbrain (E13–15), granule-like cells expressing ATOH1, ZIC1, ZIC2 and NeuN were produced [158], demonstrating that cerebellar phenotypes can be generated from these cells *in vivo*. Ongoing studies are aimed at improving the efficiency of these protocols, and the ability of the cells to integrate in the host cerebellum.

5
Conclusions

As discussed in this chapter, cell transplantation approaches have provided substantial contributions to the understanding of basic mechanisms of cerebellar development, and have also provided proof-of-principle that cell replacement may be a possible approach to cure cerebellar degeneration. Most importantly, the knowledge derived from basic studies has been crucial to refining the approaches aimed at producing specific cerebellar phenotypes by donor cells and integrating new neurons into cerebellar networks. In spite of these results, however, cell-based therapies for cerebellar degeneration are still far from being translatable to clinical practice. Further preclinical analyses are required to improve the capability of donor neurons to navigate and integrate into the adult cortex and to rewire the corticonuclear connections. Likewise, the methods used to direct the *ex-vivo* differentiation of stem cells towards specific cerebellar neurons in desired quantities still need to be refined and validated. Finally, specific assessments should be carried out to demonstrate that cell replacement actually leads to functional recovery. Indeed, the positive results obtained to date (e.g., Refs [146, 160]) are likely related to the re-establishment of tonic inhibitory control on the firing of DCN neurons. In no case, however, has a complete rewiring of the corticonuclear connection been obtained, which may be mandatory to restore more sophisticated functional skills such as the fine coordination of voluntary movements or procedural learning. Further studies will be aimed at tackling these issues.

References

1. Desai, A.R., McConnell, S.K. (2000) Progressive restriction in fate potential by neural progenitors during cerebral cortical development. *Development*, **127**, 2863–2872.
2. Yang, H., Mujtaba, T., Venkatraman, G., Wu, Y.Y. et al. (2000) Region-specific differentiation of neural tube-derived neuronal restricted progenitor cells after heterotopic transplantation. *Proc. Natl Acad. Sci. USA*, **97** (24), 13366–13371.
3. Leto, K., Carletti, B., Williams, I. M., Magrassi, L., Rossi, F. (2006) Different types of cerebellar GABAergic interneurons originate from a common pool of multipotent progenitor cells. *J. Neurosci.*, **26**, 11682–11694.
4. McMahon, S.S., McDermott, K.W. (2007) Developmental potential of radial glia investigated by transplantation into the developing rat ventricular system in utero. *Exp. Neurol.*, **203** (1), 128–136.
5. Bovetti, S., Peretto, P., Fasolo, A., De Marchis, S. (2007) Spatio-temporal specification of olfactory bulb interneurons. *J. Mol. Histol.*, **38** (6), 563–639.
6. Wonders, C.P., Taylor, L., Welagen, J., Mbata, I.C. et al. (2008) A spatial bias for the origins of interneuron subgroups within the medial ganglionic eminence. *Dev. Biol.*, **314** (1), 127–136.
7. Björklund, A., Lindvall, O. (2000) Cell replacement therapies for central nervous system disorders. *Nat. Neurosci.*, **3**, 537–544.
8. Björklund, A., Dunnett, S.B., Brundin, P., Stoessl, A.J. et al. (2003) Neural transplantation for the treatment of Parkinson's disease. *Lancet Neurol.*, **2**, 437–445.
9. Rossi, F., Cattaneo, E. (2002) Neural stem cell therapy for neurological diseases: dreams and reality. *Nat. Rev. Neurosci.*, **3**, 401–409.
10. Grimaldi, P., Carletti, B., Rossi, F. (2005) Neuronal replacement and integration in the rewiring of cerebellar circuits. *Brain Res. Rev.*, **49**, 330–342.
11. Grimaldi, P., Carletti, B., Magrassi, L., Rossi, F. (2005) Fate restriction and developmental potential of cerebellar progenitors. Transplantation studies in the developing CNS. *Prog. Brain Res.*, **148**, 57–68.
12. Carletti, B., Rossi, F. (2005) Selective rather than inductive mechanisms favour specific replacement of Purkinje cells by embryonic cerebellar cells transplanted to the cerebellum of adult Purkinje cell degeneration (pcd) mutant mice. *Eur. J. Neurosci.*, **22**, 1001–1012.
13. Thompson, L.H., Grealish, S., Kirik, D., Björklund, A. (2009) Reconstruction of the nigrostriatal dopamine pathway in the adult mouse brain. *Eur. J. Neurosci.*, **30** (4), 625–638.
14. Kriks, S., Shim, J.W., Piao, J., Ganat, Y.M. et al. (2011) Dopamine neurons derived from human ES cells efficiently engraft in animal models of Parkinson's disease. *Nature*, **480** (7378), 547–551.
15. Ramsden, C.M., Powner, M.B., Carr, A.J., Smart, M.J. et al. (2013) Stem cells in retinal regeneration: past, present and future. *Development*, **140** (12), 2576–2585.
16. Ramón y Cajal, S. (1911) *Histologie du Système Nerveux de l'Homme et des Verteébreés*. Maloine, Paris.
17. Palay, S., Chan-Palay, V. (1974) *Cerebellar Cortex*. Springer-Verlag, Berlin, Heidelberg, New York.
18. Miale, I.R., Sidman, R.L. (1961) An autoradiographic analysis of histogenesis in the mouse cerebellum. *Exp. Neurol.*, **4**, 277–296.
19. Ito, M. (1984) *The Cerebellum and Neural Control*. Raven Press, New York.
20. Eccles, J.C., Ito, M., Szentagothai, J. (1967) *The Cerebellum as a Neuronal Machine*, Springer, Berlin.
21. Altman, J., Bayer, S.A. (1997) *Development of the Cerebellar System in Relation to its*

Evolution, Structures and Functions. CRC Press, Boca Raton, FL.

22. Sotelo, C. (2004) Cellular and genetic regulation of the development of the cerebellar system. *Prog. Neurobiol.*, **72**, 295–339.
23. Hawkes, R., Brochu, G., Doré, L., Gravel, C., Leclerc, N. (1992) Zebrins: molecular markers of compartmentation in the cerebellum, in: Llinàs, R., Sotelo, C. (Eds), *Cerebellum Revisited*, Oxford University Press, Oxford, pp. 22–55.
24. Wassef, M., Angaut, P., Arsenio-Nunes, L., Bourrat, F., Sotelo, C. (1992) Purkinje cell heterogeneity: its role in organising the cerebellar cortex connections, in: Llinàs, R., Sotelo, C. (Eds), *Cerebellum Revisited*. Oxford University Press, Oxford, pp. 5–21.
25. Herrup, K., Kuemerle, B. (1997) The compartmentalization of the cerebellum. *Annu. Rev. Neurosci.*, **20**, 61–90.
26. Apps, R., Hawkes, R. (2009) Cerebellar cortical organization: a one-map hypothesis. *Nat. Rev. Neurosci.*, **10**, 670–681.
27. Cerminara, N.L., Aoki, H., Loft, M., Sugihara, I., Apps, R. (2013) Structural basis of cerebellar microcircuits in the rat. *J. Neurosci.*, **33** (42), 16427–16442.
28. Hatten, M.E., Heintz, N. (1995) Mechanisms of neural patterning and specification in the developing cerebellum. *Annu. Rev. Neurosci.*, **18**, 385–408.
29. Hoshino, M. (2012) Neuronal subtype specification in the cerebellum and dorsal hindbrain. *Dev. Growth Differ.*, **54** (3), 317–326.
30. Sotelo, C. (1990) Cerebellar synaptogenesis: what can we learn from mutant mice. *J. Exp. Biol.*, **153**, 225–249.
31. Mullen, R.J., Hamre, K.M., Goldowitz, D. (1997) Cerebellar mutant mice and chimeras revisited. *Perspect. Dev. Neurobiol.*, **5** (1), 43–55.
32. Desclin, J.C. (1974) Histological evidence supporting the inferior olive as the major source of cerebellar climbing fibers in the rat. *Brain Res.*, **77** (3), 365–384.
33. Sotelo, C., Hillman, D.E., Zamora, A.J., Llinàs, R. (1975) Climbing fiber deafferentation: its action on Purkinje cell dendritic spines. *Brain Res.*, **98** (3), 574–581.
34. Matsushita, M., Hosoya, Y., Ikeda, M. (1979) Anatomical organization of the spinocerebellar system in the cat, as studied by retrograde transport of horseradish peroxidase. *J. Comp. Neurol.*, **184** (1), 81–106.
35. Gould, B.B. (1980) Organization of afferents from the brain stem nuclei to the cerebellar cortex in the cat. *Adv. Anat. Embryol. Cell Biol.*, **62** (5-7), 1–90.
36. Garwicz, M. (2002) Spinal reflexes provide motor error signals to cerebellar modules – relevance for motor coordination. *Brain Res. Brain Res. Rev.*, **40** (1-3), 152–165.
37. Apps, R., Garwicz, M. (2005) Anatomical and physiological foundations of cerebellar information processing. *Nat. Rev. Neurosci.*, **6** (4), 297–311.
38. Ito, M. (2008) Control of mental activities by internal models in the cerebellum. *Nat. Rev. Neurosci.*, **9** (4), 304–313.
39. Broussard, D.M., Kassardjian, C.D. (2004) Learning in a simple motor system. *Learn. Mem.*, **11** (2), 127–136.
40. Boyden, E.S., Katoh, A., Raymond, J.L. (2004) Cerebellum-dependent learning: the role of multiple plasticity mechanisms. *Annu. Rev. Neurosci.*, **27**, 581–609.
41. Sacchetti, B., Scelfo, B., Tempia, F., Strata, P. (2004) Long-term synaptic changes induced in the cerebellar cortex by fear conditioning. *Neuron*, **42** (6), 973–982.
42. Sacchetti, B., Scelfo, B., Strata, P. (2009) Cerebellum and emotional behaviour. *Neuroscience*, **162** (3), 756–762.
43. Bengtsson, F., Hesslow, G. (2006) Cerebellar control of the inferior olive. *Cerebellum*, **5** (1), 7–14.
44. Ruediger, S., Vittori, C., Bednarek, E., Genoud, C. et al. (2011) Learning-related feedforward inhibitory connectivity growth required for memory precision. *Nature*, **473** (7348), 514–518.
45. Koziol, L.F., Budding, D., Andreasen, N., D'Arrigo, S. et al. (2014) Consensus paper: the cerebellum's role in movement and cognition. *Cerebellum*, **13** (1), 151–157.
46. Leto, K., Bartolini, A., Rossi, F. (2008) Neurogenesis in the cerebellum of rodents, in: Bonfanti, L. (Ed.), *Postnatal and Adult Neurogenesis*. Research Signpost, Trivandrum, pp. 63–81.
47. Wassef, M., Zanetta, J.P., Brehier, A., Sotelo, C. (1985) Transient biochemical compartmentation of Purkinje cells during early cerebellar development. *Dev. Biol.*, **111**, 129–137.

48. Brochu, G., Maler, L., Hawkes, R. (1990) Zebrin II: a polypeptide antigen expressed selectively by Purkinje cells reveals compartments in rat and fish cerebellum. *J. Comp. Neurol.*, **291**, 538–552.
49. Hawkes, R., Gravel, C. (1991) The modular cerebellum. *Prog. Neurobiol.*, **36**, 309–327.
50. Rivlin, A., Herrup, K. (2003) Development of cerebellar modules: extrinsic control of late-phase zebrin II pattern and exploration of rat/mouse species differences. *Mol. Cell. Neurosci.*, **24** (4), 887–901.
51. Voogd, J., Glickstein, M. (1998) The anatomy of the cerebellum. *Trends Neurosci.*, **21** (9), 370–375.
52. Pijpers, A., Apps, R., Pardoe, J., Voogd, J., Ruigrok, T.J. (2006) Precise spatial relationships between mossy fibers and climbing fibers in rat cerebellar cortical zones. *J. Neurosci.*, **26** (46), 12067–12080.
53. Hallonet, M.E., Teillet, M.A., Le Douarin, N.M. (1990) A new approach to the development of the cerebellum provided by the quail-chick marker system. *Development*, **108**, 19–31.
54. Hallonet, M.E., Le Douarin, N.M. (1993) Tracing neuroepithelial cells of the mesencephalic and metencephalic alar plates during cerebellar ontogeny in quail-chick chimaeras. *Eur. J. Neurosci.*, **5**, 1145–1155.
55. Hallonet, M., Alvarado-Mallart, R.M. (1997) The chick/quail chimeric system: a model for early cerebellar development. *Perspect. Dev. Neurobiol.*, **5** (1), 17–31.
56. Martinez, S., Alvarado-Mallart, R.M. (1989) Rostral cerebellum originates from the caudal portion of the so-called "mesencephalic" vesicle: a study using chick/quail chimeras. *Eur. J. Neurosci.*, **1** (6), 549–560.
57. Martinez, S., Crossley, P.H., Cobos, I., Rubenstein, J.L., Martin, G.R. (1999) FGF8 induces formation of an ectopic isthmic organizer and isthmocerebellar development via a repressive effect on Otx2 expression. *Development*, **126** (6), 1189–1200.
58. Sato, N., Meijer, L., Skaltsounis, L., Greengard, P., Brivanlou, A.H. (2004) Maintenance of pluripotency in human and mouse embryonic stem cells through activation of Wnt signalling by a pharmacological GSK-3-specific inhibitor. *Nat. Med.*, **10**, 55–63.
59. Hoshino, M., Nakamura, S., Mori, K., Kawauchi, T. et al. (2005) Ptf1a, a bHLH transcriptional gene, defines GABAergic neuronal fates in cerebellum. *Neuron*, **47**, 201–213.
60. Acazawa, C., Ishibashi, M., Shimizu, C., Nakanishi, S., Kageyama, R. (1995). A mammalian helix-loop-helix factor structurally related to the product of Drosophila proneural gene atonal is a positive transcriptional regulator expressed in the developing nervous system. *J. Biol Chem.*, **270**, 8730–8738.
61. Hoshino, M. (2006) Molecular machinery governing GABAergic neuron specification in the cerebellum. *Cerebellum*, **5** (3), 193–198.
62. Alder, J., Cho, N.K., Hatten, M.E. (1996) Embryonic precursor cells from the rhombic lip are specified to a cerebellar granule neuron identity. *Neuron*, **17**, 389–399.
63. Wingate, R.J.T. (2001) The rhombic lip and early cerebellar development. *Curr. Opin. Neurobiol.*, **11**, 82–88.
64. Wang, V.Y., Rose, M.F., Zoghbi, H. (2005) Math1 expression redefines the rhombic lip derivatives and reveals novel lineages within the brainstem and cerebellum. *Neuron*, **48**, 31–43.
65. Fink, A.J., Englund, C., Daza, R.A.M., Pham, D. et al. (2006). Development of the deep cerebellar nuclei: transcription factors and cell migration from the rhombic lip. *J. Neurosci.*, **26**, 3066–3076.
66. Englund, C.M., Kowalczyk, T., Daza, R.A.M., Dagan, A. et al. (2006) Unipolar brush cells of the cerebellum are produced in the rhombic lip and migrate through developing white matter. *J. Neurosci.*, **26**, 9184–9195.
67. Sekerková, G., Ilijic, E., Mugnaini, E. (2004) Time of origin of unipolar brush cells in the rat cerebellum as observed by prenatal bromodeoxyuridine labeling. *Neuroscience*, **127**, 845–858.
68. Carletti, B., Rossi, F. (2008) Neurogenesis in the cerebellum. *Neuroscientist*, **14**, 91–100.
69. Renfranz, P.J., Cunningham, M.G., McKay, R.D.G. (1991) Region-specific differentiation of the hippocampal stem cell line HiB5 upon implantation into the developing mammalian brain. *Cell*, **66**, 713–729.
70. Snyder, E.Y., Deitcher, D.L., Walsh, C., Arnold-Aldea, S. et al. (1992) Multipotent neural cell lines can engraft and participate

in development of mouse cerebellum. *Cell*, **68**, 33–51.
71. Gao, W.Q., Hatten, M.E. (1994) Immortalizing oncogenes subvert the establishment of granule cell identity in developing cerebellum. *Development*, **120**, 1059–1070.
72. Alder, J., Lee, K.J., Jessell, T.M., Hatten, M.E. (1999) Generation of cerebellar granule neurons in vivo by transplantation of BMP-treated neural progenitor cells. *Nat. Neurosci.*, **2** (6), 535–540.
73. Pearson, B.J., Doe, C.Q. (2004) Specification of temporal identity in the developing nervous system. *Annu. Rev. Cell Dev. Biol.*, **20**, 619–647.
74. McConnell, S.K., Kaznowski, C.E. (1991) Cell cycle dependence of laminar determination in developing neocortex. *Science*, **254**, 282–285.
75. Frantz, G.D., McConnell, S.K. (1996) Restriction of late cerebral cortical progenitors to an upper-layer fate. *Neuron*, **17**, 55–61.
76. Leto, K., Bartolini, A., Yanagawa, Y., Obata, K. et al. (2009) Laminar fate and phenotype specification of cerebellar GABAergic interneurons. *J. Neurosci.*, **29** (21), 7079–7091.
77. Jankovski, A., Rossi, F., Sotelo, C. (1996) Neuronal precursors in the postnatal mouse cerebellum are fully committed cells: evidence from heterochronic transplantation. *Eur. J. Neurosci.*, **8**, 2308–2320.
78. Sotelo, C., Alvarado-Mallart, R.M. (1991) The reconstruction of cerebellar circuits. *Trends Neurosci.*, **14** (8), 350–355.
79. Rossi, F., Borsello, T., Strata, P. (1992) Embryonic Purkinje cells grafted on the surface of the cerebellar cortex integrate in the adult unlesioned cerebellum. *Eur. J. Neurosci.*, **4** (6), 589–593.
80. Carletti, B., Grimaldi, P., Magrassi, L., Rossi, F. (2002) Specification of cerebellar progenitors following heterotopic/heterochronic transplantation to the embryonic CNS in vivo and in vitro. *J. Neurosci.*, **22**, 7132–7146.
81. Cattaneo, E., Magrassi, L., Butti, G., Santi, L. et al. (1994) A short term analysis of the behaviour of conditionally immortalized neuronal progenitors and primary neuroepithelial cells implanted into the fetal rat brain. *Brain Res. Dev. Brain Res.*, **83**, 197–208.

82. Magrassi, L., Ehrlich, M.E., Butti, G., Pezzotta, S. et al. (1998) Basal ganglia precursors found in aggregates following embryonic transplantation adopt a striatal phenotype in heterotopic locations. *Development*, **125**, 2847–2855.
83. Vicario-Abejón, C., Cunningham, M.G., McKay, R.D.G. (1995) Cerebellar precursors transplanted to the neonate dentate gyrus express features characteristic of hippocampal neurons. *J. Neurosci.*, **15**, 6351–6363.
84. Carletti, B., Grimaldi, P., Magrassi, L., Rossi, F. (2004) Engraftment and differentiation of neocortical progenitor cells transplanted to the embryonic brain in utero. *J. Neurocytol.*, **33**, 309–319.
85. Rolando, C., Gribaudo, S., Yoshikawa, K., Leto, K. et al. (2010) Extracerebellar progenitors grafted to the neurogenic milieu of the postnatal rat cerebellum adapt to the host environment but fail to acquire cerebellar identities. *Eur. J. Neurosci.*, **31**, 1340–1351.
86. Milosevic, A., Noctor, S.C., Martinez-Cerdeno, V., Kriegstein, A.R., Goldman, J.E. (2008) Progenitors from the postnatal forebrain subventricular zone differentiate into cerebellar-like interneurons and cerebellar-specific astrocytes upon transplantation. *Mol. Cell. Neurosci.*, **39**, 324–334.
87. Schilling, K., Oberdick, J., Rossi, F., Baader, S.L. (2008) Besides Purkinje cells and granule neurons: an appraisal of the cell biology of the interneurons of the cerebellar cortex. *Histochem. Cell Biol.*, **130** (4), 601–615.
88. Leto, K., Bartolini, A., Rossi, F. (2008) Development of cerebellar GABAergic interneurons: origin and shaping of the "minibrain" local connections. *Cerebellum*, **7**, 523–529.
89. Leto, K., Bartolini, A., Rossi, F. (2010) The prospective white matter: an atipica neurogenic niche in the developing cerebellum. *Arch. Ital. Biol.*, **148**, 137–146.
90. Schilling, K. (2011) Specification and development of GABAergic interneurons, in: Manto, M., Koibuchi, N., Schmahmann, J., Gruol, D., Rossi, F. (Eds), *Handbook of The Cerebellum and Cerebellar Disorders*. Springer, New York, Berlin, pp. 207–235.
91. Lee, K.J., Jessell, T.M. (1999) The specification of dorsal cell fates in the vertebrate

central nervous system. *Annu. Rev. Neurosci.*, **22**, 261–294.
92. Wonders, C.P., Anderson, S.A. (2006) The origins and specification of cortical interneurons. *Nat. Rev. Neurosci.*, **7**, 687–696.
93. Fogarty, M.P., Emmenegger, B.A., Grasfeder, L.L., Oliver, T.G., Wechsler-Reya, R.J. (2007) Fibroblast growth factor blocks Sonic hedgehog signalling in neuronal precursors and tumor cells. *Proc. Natl Acad. Sci. USA*, **104** (8), 2973–2978.
94. De Marchis, S., Bovetti, S., Carletti, B., Hsieh, Y.C. et al. (2007) Generation of distinct types of periglomerular olfactory bulb interneurons during development and in adult mice: implication for intrinsic properties of the subventricular zone progenitor population. *J. Neurosci.*, **27** (3), 657–664.
95. Lledo, P.M., Merkle, F.T., Alvarez-Buylla, A. (2008) Origin and function of olfactory bulb interneuron diversity. *Trends Neurosci.*, **31** (8), 392–400.
96. Valcanis, H., Tan, S.S. (2003) Layer specification of transplanted interneurons in developing mouse neocortex. *J. Neurosci.*, **23**, 5113–5122.
97. Maricich, S.M., Herrup, K. (1999) Pax-2 expression defines a subset of GABAergic interneurons and their precursors in the developing murine cerebellum. *J. Neurobiol.*, **41**, 281–294.
98. Weisheit, G., Gliem, M., Endl, E., Pfeffer, P.L. et al. (2006) Postnatal development of the murine cerebellar cortex: formation and early dispersal of basket, stellate and Golgi neurons. *Eur. J. Neurosci.*, **24**, 466–478.
99. Zhang, L., Goldman, J.E. (1996) Generation of cerebellar interneurons from dividing progenitors in white matter. *Neuron*, **16**, 47–54.
100. Zhang, L., Goldman, J.E. (1996) Developmental fates and migratory pathways of dividing progenitors in the postnatal rat cerebellum. *J. Comp. Neurol.*, **370**, 536–550.
101. Milosevic, A., Goldman, J.E. (2002) Progenitors in the postnatal cerebellar white matter are antigenically heterogeneous. *J. Comp. Neurol.*, **452**, 192–203.
102. Patterson, M.C., Platt, F. (2004) Therapy of Niemann–Pick disease, type C. *Biochim. Biophys. Acta*, **1685** (1-3), 77–82.
103. Sotelo, C., Alvarado-Mallart, R.M., Frain, M., Vernet, M. (1994) Molecular plasticity of adult Bergmann fibers is associated with radial migration of grafted Purkinje cells. *J. Neurosci.*, **14**, 124–133.
104. Das, G.D., Altman, J. (1971) Transplanted precursors of nerve cells: their fate in the cerebellum of young rats. *Science*, **173** (3997), 637–638.
105. McConnell, S.K. (1985) Migration and differentiation of cerebral cortical neurons after transplantation into the brains of ferrets. *Science*, **229**, 1268–1271.
106. McConnell, S.K. (1988) Fates of visual cortical neurons in the ferret after isochronic and heterochronic transplantation. *J. Neurosci.*, **8**, 945–974.
107. Nikkhah, G., Cunningham, M.G., Cenci, M.A., McKay, R.D. et al. (1995) Dopaminergic microtransplants into the substantia nigra of neonatal rats with bilateral 6-OHDA lesions. I. Evidence for anatomical reconstruction of the nigrostriatal pathway. *J. Neurosci.*, **15**, 3548–3561.
108. Brüstle, O., Maskos, U., McKay, R.D.G. (1995) Hostguided migration allows targeted introduction of neurons into the embryonic brain. *Neuron*, **15**, 1275–1285.
109. Campbell, K., Olsson, M., Björklund, A. (1995) Regional incorporation and site-specific differentiation of striatal precursors transplanted to the embryonic forebrain ventricle. *Neuron*, **15**, 1259–1273.
110. Lim, D.A., Fishell, G.J., Alvarez-Buylla, A. (1997) Postnatal mouse subventricular zone neuronal precursors can migrate and differentiate within multiple levels of the developing neuraxis. *Proc. Natl Acad. Sci. USA*, **94**, 14832–14836.
111. Wichterle, H., Turnbull, D.H., Nery, S., Fishell, G. et al. (2001) In utero fate mapping reveals distinct migratory pathways and fates of neurons born in the mammalian basal forebrain. *Development*, **128**, 3759–3771.
112. Sotelo, C., Alvarado-Mallart, R.M. (1987) Embryonic and adult neurons interact to allow Purkinje cell replacement in mutant cerebellum. *Nature*, **327**, 421–423.
113. Sotelo, C., Alvarado-Mallart, R.M. (1987) Reconstruction of the defective cerebellar circuitry in adult Purkinje cell degeneration mutant mice by Purkinje cell replacement through transplantation of solid embryonic implants. *Neuroscience*, **20**, 1–22.

114. Williams, I.M, Carletti, B., Leto, K., Magrassi, L. et al. (2008) Cerebellar granule cells transplanted in vivo can follow physiological and unusual migratory routes to integrate into the recipient cortex. *Neurobiol. Dis.*, **30**, 139–149.
115. Kawamura, K., Nanami, T., Kikuchi, Y., Kitakami, A. (1988) Grafted granule and Purkinje cells can migrate into the mature cerebellum of normal and adult rats. *Exp. Brain Res.*, **70** (3), 477–484.
116. Rossi, F., Borsello, T., Strata, P. (1994) Embryonic Purkinje cells grafted on the surface of the adult uninjured rat cerebellum migrate in the host parenchyma and induce sprouting of intact climbing fibres. *Eur. J. Neurosci.*, **6**, 121–136.
117. Sotelo C., Alvarado-Mallart R.M. (1992) Cerebellar grafting as a tool to analyze new aspects of cerebellar development and plasticity, in: Llina's, R., Sotelo, C. (Eds), *Cerebellum Revisited*. Oxford University Press, Oxford, pp. 84–115.
118. Mullen, R.J., Eicher, E.M., Sidman, R.L. (1976) Purkinje cell degeneration, a new neurological mutation in the mouse. *Proc. Natl Acad. Sci. USA*, **73** (1), 208–212.
119. Landis, S.C., Mullen, R.J. (1978) The development and degeneration of Purkinje cells in pcd mutant mice. *J. Comp. Neurol.*, **177** (1), 125–143.
120. Fernandez-Gonzalez, A., La Spada, A.R., Treadaway, J., Higdon, J.C. et al. (2002) Purkinje cell degeneration (pcd) phenotypes caused by mutations in the axotomy-induced gene, Nna1. *Science*, **295**, 1904–1906.
121. Gardette, R., Alvarado-Mallart, R.M., Crepel, F., Sotelo, C. (1988) Electrophysiological demonstration of a synaptic integration of transplanted Purkinje cells into the cerebellum of the adult "Purkinje cell degeneration" mutant mouse. *Neuroscience*, **24**, 777–789.
122. Gardette, R., Crepel, F., Alvarado-Mallart, R.M., Sotelo, C. (1990) Fate of grafted Purkinje cells in the cerebellum of the adult Purkinje cell degeneration mutant mouse: II. Development of synaptic responses: an in vitro study. *J. Comp. Neurol.*, **295**, 188–196.
123. Triarhou, L.C. (1996) The cerebellar model of neural grafting: structural integration and functional recovery. *Brain Res. Bull.*, **39**, 127–138.
124. Grimaldi, P., Rossi, F. (2006) Lack of neurogenesis in the adult rat cerebellum after Purkinje cell degeneration and growth factor infusion. *Eur. J. Neurosci.*, **23**, 2657–2668.
125. Tempia, F., Bravin, M., Strata, P. (1996) Postsynaptic currents and short-term synaptic plasticity in Purkinje cells grafted onto an uninjured adult cerebellar cortex. *Eur. J. Neurosci.*, **8** (12), 2690–2701.
126. Carletti, B., Williams, I.M., Leto, K., Nakajima, K. et al. (2008) Time constraints and positional cues in the developing cerebellum regulate Purkinje cell placement in the cortical architecture. *Dev. Biol.*, **317**, 147–160.
127. Magrassi, L., Leto, K., Rossi, F. (2013) Lifespan of neurons is uncoupled from organism lifespan. *Proc. Natl Acad. Sci. USA*, **110** (11), 4374–4379.
128. Keep, M., Alvarado-Mallart, R.M., Sotelo, C. (1992) New insight on the factors orienting the axonal outgrowth of grafted Purkinje cells in the pcd cerebellum. *Dev. Neurosci.*, **14**, 153–165.
129. Triarhou, L.C., Low, W.C., Ghetti, B. (1992) Intraparenchymal grafting of cerebellar cell suspensions to the deep cerebellar nuclei of pcd mutant mice, with particular emphasis on reestablishment of a Purkinje cell corticonuclear projection. *Anat. Embryol.*, **185**, 409–420.
130. Miyata, T., Nakajima, K., Aruga, J., Takahashi, S. et al. (1996) Distribution of a Reeler gene-related antigen in the developing cerebellum: an immunohistochemical study with an allogeneic antibody CR-50 on normal and reeler mice. *J. Comp. Neurol.*, **372**, 215–228.
131. Miyata, T., Nakajima, K., Mikoshiba, K., Ogawa, M. (1997) Regulation of Purkinje cell alignment by reelin as revealed with CR-50 antibody. *J. Neurosci.*, **17**, 3599–3609.
132. Pesold, C., Impagnatiello, F., Pisu, M.G., Uzunov, D.P. et al. (1998) Reelin is preferentially expressed in neurons synthesizing gamma-aminobutyric acid in cortex and hippocampus of adult rats. *Proc. Natl Acad. Sci. USA*, **95**, 3221–3226.
133. Rouse, R.V., Sotelo, C. (1990) Grafts of dissociated cerebellar cells containing Purkinje cell precursors organize into zebrin I

defined compartments. *Exp. Brain Res.*, **82**, 401–407.
134. Rossi, F., Saggiorato, C., Strata, P. (2002) Target-specific innervation of embryonic cerebellar transplants by regenerating olivocerebellar axons in the adult rat. *Exp. Neurol.*, **173**, 205–212.
135. Conti, L., Cattaneo, E. (2010) Neural stem cell systems: physiological players or in vitro entities? *Nat. Rev. Neurosci.*, **11** (3), 176–187.
136. Koch, P., Kokaia, Z., Lindvall, O., Brüstle, O. (2009) Emerging concepts in neural stem cell research: autologous repair and cell-based disease modelling. *Lancet Neurol.*, **8** (9), 819–829.
137. Conti, L., Reitano, E., Cattaneo, E. (2006) Neural stem cell systems: diversities and properties after transplantation in animal models of diseases. *Brain Pathol.*, **16** (2), 143–154.
138. Conti, L., Pollard, S.M., Gorba, T., Reitano, E. et al. (2005) Niche-independent symmetrical self-renewal of a mammalian tissue stem cell. *PLoS Biol.*, **3** (9), e383.
139. Pollard, S.M., Conti, L., Sun, Y., Goffredo, D. et al. (2006) Adherent neural stem (NS) cells from fetal and adult forebrain. *Cereb. Cortex*, **16** (Suppl. 1), i112–i120.
140. Pollard, S.M., Conti, L. (2007) Investigating radial glia in vitro. *Progr. Neurobiol.*, **83** (1), 53–67.
141. Goffredo, D., Conti, L., Di Febo, F., Biella, G. et al. (2008) Setting the conditions for efficient, robust and reproducible generation of functionally active neurons from adult subventricular zone-derived neural stem cells. *Cell Death Differ.*, **15** (12), 1847–1856.
142. Spiliotopulos, D., Goffredo, D., Conti, L., Di Febo, F. et al. (2009) An optimized experimental strategy for efficient conversion of embryonic stem (ES)-derived mouse neural stem (NS) cells into a nearly homogeneous mature neuronal population. *Neurobiol. Dis.*, **34** (2), 320–331.
143. Onorati, M., Camnasio, S., Binetti, M., Jung, C.B. et al. (2010) Neuropotent self-renewing neural stem (NS) cells derived from mouse induced pluripotent stem (iPS) cells. *Mol. Cell. Neurosci.*, **43** (3), 287–295.
144. Gage, F.H. (2002) Neurogenesis in the adult brain. *J. Neurosci.*, **22** (3), 612–613.
145. Lie, D.C., Song, H., Colamarino, S.A., Ming, G.L. et al. (2004) Neurogenesis in the adult brain: new strategies for central nervous system diseases. *Annu. Rev. Pharmacol. Toxicol.*, **44**, 399–421.
146. Chintawar, S., Hourez, R., Ravella, A., Gall, D. et al. (2009) Grafting neural precursor cells promotes functional recovery in an SCA1 mouse model. *J. Neurosci.*, **29** (42), 13126–13135.
147. Horinouchi, K., Erlich, S., Perl, D.P., Ferlinz, K. et al. (1995) Acid sphingomyelinase deficient mice: a model of types A and B Niemann–Pick disease. *Nat. Genet.*, **10** (3), 288–293.
148. Sidman, R.L., Li, J., Stewart, G.R., Clarke, J. et al. (2007) Injection of mouse and human neural stem cells into neonatal Niemann–Pick A model mice. *Brain Res.*, **1140**, 195–204.
149. Lee, J.M., Bae, J.S., Jin, H.K. (2010) Intracerebellar transplantation of neural stem cells into mice with neurodegeneration improves neuronal networks with functional synaptic transmission. *J. Vet. Med. Sci.*, **72** (8), 999–1009.
150. Amariglio, N., Hirshberg, A., Scheithauer, B.M., Cohen, Y. et al. (2009) Donor-derived brain tumor following neural transplantation in an ataxia telangiectasia patient. *PLoS Med.*, **6** (2), e1000029.
151. Tian, Z.M., Chen, T., Zhong, N., Li, Z.C. et al. (2010) Clinical study of transplantation of neural stem cells in therapy of inherited cerebellar atrophy. *Beijing Da Xue Xue Bao*, **41** (4), 456–458.
152. Erceg, S., Moreno-Manzano, V., Garita-Hernandez, M., Stojkovic, M. et al. (2011) Concise review: stem cells for the treatment of cerebellar-related disorders. *Stem Cells*, **29** (4), 564–569.
153. Erceg, S., Ronaghi, M., Zipancic, I., Lainez, S. et al. (2010) Efficient differentiation of human embryonic stem cells into functional cerebellar-like cells. *Stem Cells Dev.*, **19** (11), 1745–1756.
154. Su, H.L., Muguruma, K., Matsuo-Takasaki, M., Kengaku, M. et al. (2006) Generation of cerebellar neuron precursors from embryonic stem cells. *Dev. Biol.*, **290**, 287–296.
155. Salero, E., Hatten, M.E. (2007) Differentiation of ES cells into cerebellar neurons. *Proc. Natl Acad. Sci. USA*, **104**, 2997–3002.
156. Mizuhara, E., Minaki, Y., Nakatani, T., Kumai, M. et al. (2010) Purkinje cells originate from cerebellar ventricular zone

progenitors positive for Neph3 and E-cadherin. *Dev. Biol.*, **338**, 202–214.
157. Muguruma, K., Nishiyama, A., Ono, Y., Miyawaki, H. et al. (2010) Ontogeny-recapitulating generation and tissue integration of ES cell-derived Purkinje cells. *Nat. Neurosci.*, **13**, 1171–1180.
158. Tailor, J., Kittappa, R., Leto, K., Gates, M. et al. (2013) Stem cells expanded from the human embryonic hindbrain stably retain regional specification and high neurogenic potency. *J. Neurosci.*, **33** (30), 12407–12422.
159. Sun, Y., Pollard, S., Conti, L., Toselli, M. et al. (2008) Long-term tripotent differentiation capacity of human neural stem (NS) cells in adherent culture. *Mol. Cell. Neurosci.*, **38**, 245–258.
160. Zhang, W., Lee, W.-H., Triarhou, L.C. (1996) Grafted cerebellar cells in a mouse model of hereditary ataxia express IGF-I system genes and partially restore behavioral function. *Nat. Med.*, **2**, 65–71.
161. Triarhou, L.C. (Ed.) (1997) *Neural Transplantation in Cerebellar Ataxia*. Springer-Verlag, Heidelberg.
162. Tao, O., Shimazaki, T., Okada, Y., Naka, H. et al. (2010) Efficient generation of mature Purkinje cells from mouse embryonic stem cells. *J. Neurosci. Res.*, **88** (2), 234–247.

9
RNA Interference in Cancer Therapy

Barbara Pasculli[1,2] and George A. Calin[2,3,4]
[1] *University of Bari "Aldo Moro", Department of Biosciences, Biotechnology and Pharmacological Sciences, Via Orabona 4, Bari 70125, Italy*
[2] *The University of Texas MD Anderson Cancer Center, Department of Experimental Therapeutics, Unit 1950, 1881 East Road, Houston, TX 77054, USA*
[3] *The University of Texas MD Anderson Cancer, Department of Leukemia, 1515 Holcombe Blvd, Houston, TX 77030, USA*
[4] *The University of Texas MD Anderson Cancer Center, Center for RNA Interference and Non-Coding RNAs, Unit 1950, 1881 East Road, Houston, TX 77030, USA*

1	**Introduction: Cancer and RNAi** 293	
2	**RNA Interference: Breakthrough of the Year 2002** 294	
2.1	RNA Interference: a Brief Historical Digression 294	
2.2	Short Regulatory RNAs: miRNA and siRNA Biogenesis 295	
2.3	The RNAi Machinery 296	
3	**MicroRNAome: The Biological Toolkit for the Regulation of Gene Expression** 298	
3.1	microRNA Deregulation is a Molecular Feature of Cancer 299	
3.2	Mechanisms of microRNA Deregulation in Cancer 299	
3.3	Role of microRNAs in Cancer 302	
3.3.1	OncomiRs 302	
3.3.2	Tumor Suppressor miRs 305	
4	**Current Prospects for RNA Interference-Based Therapy for Cancer** 307	
4.1	Harnessing the Endogenous RNAi Apparatus for Clinical Purposes 309	
4.1.1	Restoring miRNA's Endogenous Regulation of Transcription: siRNAs, shRNAs, and Derivatives 309	
4.1.2	Inhibition of oncomiRs: AntimiRs, miRNA Sponges, and miR-Masks 311	
5	**Rational Design of an RNAi-Based Therapeutic Approach in Cancer** 312	
5.1	RNAi Target Selection 312	

Translational Medicine: Molecular Pharmacology and Drug Discovery
First Edition. Edited by Robert A. Meyers.
© 2018 Wiley-VCH Verlag GmbH & Co. KGaA. Published 2018 by Wiley-VCH Verlag GmbH & Co. KGaA.

5.2	*In-Vivo* RNAi Challenges	312
5.2.1	Stability and Bioavailability Issues	314
5.2.2	Multiple Off-Target Effects	314
5.2.3	Immune Response	315
5.2.4	Delivery Issues	316
5.3	RNAi Implementation with Nanomedicine	317
5.3.1	Chemical Modifications	317
5.3.2	Nanocarriers	317
5.4	Administration	324
5.5	Targeting	325
5.5.1	Passive Targeting	325
5.5.2	Active Targeting: Antibodies, Aptamers, Small Molecules	326
5.5.3	Alternative Targeting	327
5.6	RNAi-Based Drug Combination Strategy (coRNAi)	327
5.7	RNAi in the Clinical Trials	329
6	**Summary**	**332**
	References	**333**

Keywords

Cancer
Pathologic phenotype defined by abnormal and uncontrolled cell division, arising from the accumulation of epigenetic and genetic alterations affecting the expression profile of both protein-coding genes and noncoding RNAs.

RNA interference
A natural, sequence-specific mechanism of post-transcriptional gene silencing in animal and plants, triggered and mediated by small, double-stranded RNAs.

RISC (RNA-induced silencing complex)
A multicomponent ribonucleoprotein complex consisting of a member of the Argonaute (AGO) family proteins, and a mature miRNA/siRNA incorporated into AGO, together with a number of accessory factors, acting as the molecular effectors of translational repression/degradation of miRNA/siRNA -targeted transcripts.

microRNAs (miRNAs; miRs)
Small, endogenous, single-stranded, noncoding RNAs, 19–25 nucleotides in length, which enter the RNAi apparatus to mediate the post-transcriptional regulation of gene expression by specific but imperfect binding to target mRNAs.

siRNAs
Small-interfering, double-stranded RNAs, 21–23 nucleotides in length, which can be endogenously produced or exogenously provided (synthetic oligonucleotides) and, similar

to miRNAs, are able to associate with the RISC complex and mediate post-transcriptional gene silencing.

antimiRs
Antisense oligonucleotides which sequester, by complementary binding, the mature miRNA in competition with cellular target mRNAs, leading to a functional inhibition of the miRNA and derepression of target mRNA translation.

Personalized medicine
Healthcare approach which designs a therapeutic regimen and selects the patients to treat, according to the individual clinical, molecular, and environmental information.

RNAi-based therapy
The repertoire of strategies harnessing the RNAi endogenous apparatus by using small RNA molecules to target any gene and inhibit oncogenic pathways supporting the tumor.

> Current research in cancer therapeutics is leading the way in exploiting the growing amounts of genetic and molecular information on human oncobiology to design new anticancer agents to target specific tumor-related signaling pathways, and to treat patients according to their clinical characteristics and molecular profile. In this context, RNA interference (RNAi), a ubiquitous cellular pathway of post-transcriptional gene regulation, provides an intriguing tool for an innovative rational cancer drug design. Among the endogenous mediators of RNAi, microRNAs (miRNAs) represent the most important class of small RNAs whose global dysregulation is a typical feature of human tumors. Harnessing of the RNAi machinery by using small, synthetic RNAs that target or mimic endogenous miRNAs offers the opportunity to reach virtually any gene and pathway relevant to tumor maintenance. However, along with assessments of the safety profile and effectiveness, strategies for assuring the specificity of these drugs, in terms of their effective delivery to cancer cells, must be devised before RNAi formulations can be made into ideal personalized therapeutics for the clinic.

1
Introduction: Cancer and RNAi

Cancer is a leading cause of death worldwide, second only to cardiovascular disorders. It is a genetic disease involving mutational and epigenetic alterations that cumulate over time, and cause activation of oncogenes and/or loss of function of tumor suppressor genes. More importantly, the past decade of research has revealed that the widespread deregulation of physiological mechanisms supporting the transformation of normal cells into malignant cells is due not only to alterations in protein-coding genes but also to global changes in noncodingRNAs (ncRNAs), such as the microRNAs (miRNAs) profile. miRNAs are involved in the regulation of most biological functions, including development, life span and metabolism, and their deregulation contributes to abnormal and uncontrolled cell division and survival, as well as the capability of migration and metastatization that characterizes the cancer phenotype. The mechanism through which miRNAs exert their regulatory control of gene expression in cells has been recognized as endogenous RNA interference (RNAi). Originally, this pathway was first described as a process triggered by double-stranded RNAs (dsRNAs), able to induce sequence-specific gene silencing [1]. Having arisen in the context of genetic manipulation experiments, the discovery of RNAi implied the existence of specific mechanisms within the cells, from worms to mammals, mediating the unwinding of dsRNAs. Hence, the search for complementary base-pairing partners, sequestration from the intracellular pool of nucleic acid sequences, and the activation of downstream processes culminated in translational repression. Since its discovery, RNAi and its related pathways have been exploited to identify gene functions and new putative targets for cancer therapy. Indeed, the endogenous RNAi machinery can be properly harnessed by providing exogenous triggers that enter the pathway and, similar to miRNAs, promote gene-silencing. If translated into a disease context such as cancer, RNAi is likely to provide an appealing tool that can be used to switch off the causative genes and/or restore deregulated expression patterns.

Whole-genome sequencing and integrated "omics" analyses have greatly speeded up the recent advances in dissecting the molecular basis of cancer, and have pointed out the marked heterogeneity of human tumors, as well as the need to design and develop new rational therapeutic strategies according to the patient's molecular profile. In this regard, RNA interference (RNAi) represents an emerging, powerful approach for personalized cancer medicine that can be used to tailor medical intervention to an individual's genetic and physiological background.

2
RNA Interference: Breakthrough of the Year 2002

RNA interference refers to an evolutionarily conserved cellular process that is mediated by dsRNAs. In worms, plants and flies, RNAi likely provides an innate defense to both endogenous parasitic and exogenous pathogenic nucleic acids (DNA or RNA) [1], whereas its primary role in mammals and human involves regulating the expression of protein-coding genes at the post-transcriptional level [1].

The finding that a dsRNA could activate a sequence-specific silencing, and the whole set of intracellular mediators and functional activities orchestrating the process, was such a high-impact discovery that it was hailed as the biggest "Breakthrough of the Year" in 2002 by the journal *Science* [2], and honored with the 2006 Nobel Prize in Physiology or Medicine to Andrew Z. Fire and Craig C. Mello.

Following on from their studies in the nematode *Caenorhabditis elegans* [3], several research groups have moved quickly to large mammals [4, 5], to start investigating the potential for harnessing RNAi to address biological questions and treat human diseases. Indeed, by enabling reversible gene silencing and modulating the expression of any gene, RNAi provides an appealing approach for loss-of-function gene analyses in vertebrate systems and, more importantly, for clinical purposes.

2.1
RNA Interference: a Brief Historical Digression

Retracing the history of RNAi discovery, the first hints were derived from experiments on engineering transgenic petunias, when Richard Jorgensen was attempting to alter pigmentation and obtain a more intensely purple flower strain [6]. Unexpectedly, the introduction of exogenous pigment-producing transgenes resulted in totally white or variegated pigmentation. As the endogenous loci and "foreign" transgenes appeared to be turning each other off, this phenomenon was called "co-suppression." Similar outcomes were subsequently derived from other plants, fungi (*Neurospora crassa*) [7], and metazoans (*Drosophila* and *C. elegans*) [8–10], where transmission of the silencing effect was also observed through several generations.

At that time, 1995, Guo and Kemphus were investigating the function of the *C. elegans par-1* gene by using an "antisense-mediated silencing" approach, in which a small, synthetic RNA oligonucleotide, complementary to the *par-1* sequence, once delivered to the cells, was expected to bind the messenger RNA (mRNA) and block its translation. Guo and Kemphus found that both the exogenous antisense and sense oligonucleotides (the latter being used as negative control) were able to produce the same lethal effect by inducing *par-1* mRNA silencing [11]. The turning point came when Fire, Mello and colleagues, during their experiments of manipulation of gene expression in *C. elegans*, found that the dsRNA mixture was 10-fold more potent in inducing translational repression than sense or antisense RNAs were alone [3]. This effect became known as RNA interference.

Shortly thereafter, this effect – which became variously known as post-transcriptional gene silencing (PTGS) [12], co-suppression [6], quelling [7], and RNAi – was demonstrated to be even more widespread, occurring in flies [8] as well as mammals [4, 5]. The role of RNAi pathways in the normal regulation of endogenous protein-coding genes was suggested by a flurry of many results. Small RNAs were found to be produced in plants undergoing PTGS [13, 14], while studies in *C. elegans* revealed that these small RNAs – which were now known as short interfering RNAs (siRNAs) were produced by the dsRNA-processing enzyme Dicer [15] and promoted gene silencing by triggering the assembly of a nuclease effector known as the RNA-induced silencing complex (RISC) [8]. Subsequently, a class of natural hairpin dsRNAs [16, 17], now called miRNAs, was shown to be processed by Dicer [18, 19]. Finally, the role of RNAi in chromatin regulation in yeast [20], and chromosomal rearrangement during development of the somatic macronucleus in *Tetrahymena* [21], led to a definite identification of the RNAi machinery as a natural developmental gene regulatory mechanism.

2.2
Short Regulatory RNAs: miRNA and siRNA Biogenesis

When attempting to manipulate and take advantage of RNAi for any experimental application, it is critical to have an overview of the molecular basics of this pathway.

Since its discovery, more insights have been attained about the genes and the molecules implicated in the endogenous RNAi machinery.

RNAi, also defined as a sequence-specific, PTGS process [1, 12] is mainly performed by small RNAs including endogenous miRNAs [22], short interfering RNAs (siRNAs) [4], and exogenous chemically synthetized short hairpin RNAs (shRNAs) [23].

Although similar in function, a distinction can be made between miRNAs and siRNAs based on their different biogenesis [24, 25] and mechanisms of recognition of the target RNAs [26].

miRNAs are initially transcribed in the nucleus by an RNA Polymerase II or III as long primary transcripts (pri-miRNAs), containing single or clustered double-stranded hairpins bearing single-stranded 5′- and 3′-terminal overhangs, and 10 nt distal loops [26]. The pri-miRNA is processed into a 70- to 100-nt, imperfectly base-paired hairpin precursor RNAs (pre-miRNAs) by the Microprocessor Complex, which consists of the RNase III enzyme Drosha and a protein called Pasha in *Drosophila* or DGCR8 (Di George Syndrome critical region gene 8) in mammals [27–29]. This initial cleavage is then followed by the Exportin-5/RanGTP-mediated pre-miRNA translocation to the cytoplasm for further processing into a 19- to 25-nt duplex by the RNase III endonuclease Dicer and TRBP (Human immunodeficiency virus-1 transactivating response RNA-binding protein) [30]. Final processing by Dicer is likely to culminate in incorporation of the miRNA into the RISC, which acts as effector of RNAi [31] (Fig. 1).

In contrast, siRNAs are produced from long, perfectly base-paired dsRNA precursors which can be produced endogenously or provided exogenously (Fig. 1).

Indeed, siRNAs are often referred to either genetic material deriving from virus infections or artificial inhibitory products. However, large-scale analyses in plants, fungi and animals have detected small RNAs that lack hairpin-loop precursors, and thus do not belong to miRNA species [32–34]. The larger part of these RNAs is encoded by repetitive elements of the genome, as trasponsons, especially located at heterochromatic regions such as centromeres and telomeres, and confirming that the RNAi machinery contributes to the establishment of a repressed chromatin state at these regions [20]. Other siRNAs are instead transcribed from the antisense strand of chromosome regions that often contain sequences for longer noncoding RNAs [32, 35]. Although their function is far from being completely understood, several organisms (including yeast, flies, plants, and protozoa) have been found to contain many of these siRNAs.

Nevertheless, once in the cytoplasm, the siRNA processing, as well as its assembly into the RISC complex, is mostly Dicer-dependent [36] (Fig. 1).

Thus, both for miRNA and siRNA precursors, the resulting dsRNA is a duplex of 19- to 25-nt strands bearing a 2-nt overhang at its 3′ terminus and a phosphate group at its 5′ terminus.

2.3
The RNAi Machinery

The key component of the RISC complex is an Argonaute (AGO) protein. AGO proteins have been consistently found in bacteria, archaea and eukaryotes [37]. In humans, among the eight AGO-classified proteins [38], AGO2 is the only member which exhibits endonuclease activity associated with its RNaseH-like domain [39] and accomplishes an miRNA/siRNA-induced silencing of target mRNAs. AGO2, together with Dicer and TRBP, represent the minimal RISC-loading complex (RLC) [26]. Additional proteins as Vasa intronic gene (VIG) protein, the Tudor-SN protein, Fragile X-related protein, the putative RNA helicases Dmp68, and Gemin3, as well as GW182 and TTP [40–43], have been found to associate with the RLC, although their functions are still not fully understood.

Once the diced dsRNAs interact with AGO2, one strand of the duplex – the

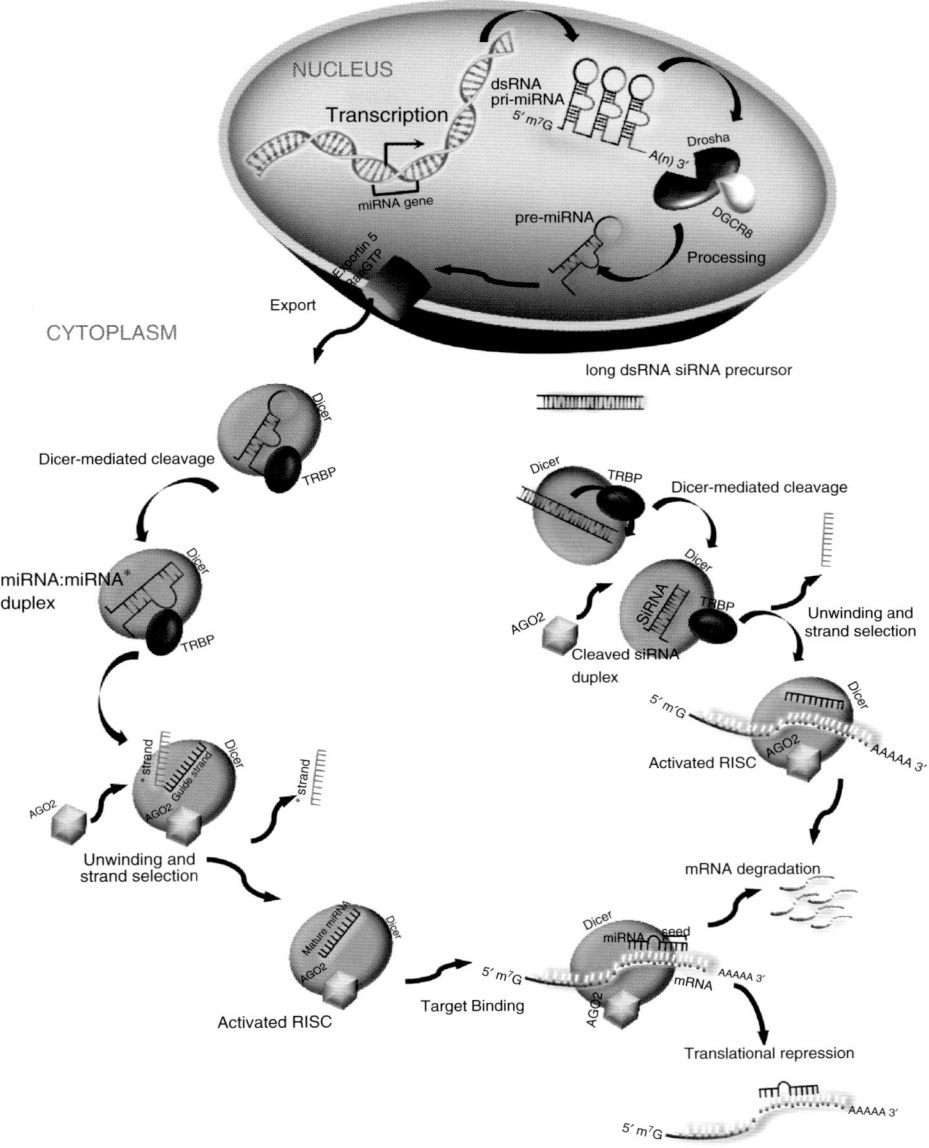

"guide" strand – is selected to be loaded into the RISC complex, while the other strand – the "passenger" strand (often indicated by *) – is discarded. Selection of the guide strand, which generally is based on thermodynamic properties, specifically addresses the silencing activity of the RISC complex to cytoplasmic single-stranded RNAs (ssRNAs), as endogenous mRNAs, through either perfect or imperfect

Fig. 1 The miRNA and siRNA pathways of RNAi. Primary miRNAs (pri-miRNAs) are transcribed in the nucleus by RNA Polymerase II as long, capped, and polyadenylated (A-(n)) precursors, and are trimmed by the microprocessor complex (including Drosha and DCGR8) into 70- to 100-nucleotide hairpin precursors, usually containing interspersed mismatches along the duplex, called pre-miRNAs. Pre-miRNAs associate with the Exportin/RanGTP complex and are translocated to the cytoplasm (left side of figure), where they are further processed into a 19- to 25-nucleotide duplex (miRNA:miRNA*) by the Dicer–TRBP complex. Finally, the duplex interacts with an Argonaute (AGO2) protein into the RISC (RNA-induced silencing complex) apparatus: one strand of the duplex (the passenger * strand) is removed, whereas the guide strand, the mature miRNA, remains stably associated within the RISC, and directs the complex to the target mRNA for the post-transcriptional gene silencing. The interactions between miRNA *seed* sequence and the 3′-UTR of the mRNA generally culminate in translation repression. However, cleavage and degradation of the mRNA have also been documented. Long, perfectly base-paired, dsRNA (right side of figure) are also processed by the Dicer–TRBP complex into small siRNAs. Once they enter the RISC complex, an Argonaute protein cleaves the passenger strand, and the guide strand is used to bind the complementary mRNA target. SiRNAs usually anneals to the mRNAs by a perfect complementarity, mainly leading to their degradation.

complementarity to the AGO2-bound guide strand. Indeed, the guide strand nucleotides 2–8 (counting from the 5′ end) constitute the "*seed* region," that is able to directly bind the target mRNAs. The degree of complementarity can be either perfect (as occurs mostly for siRNAs) or partial (as is usual for miRNAs), and determines subsequent AGO2-mediated target cleavage or, more frequently (if partial complementarity occurs) a non-endonucleolytic translational repression [26] (Fig. 1).

3
MicroRNAome: The Biological Toolkit for the Regulation of Gene Expression

A large number of miRNA genes (>1000) have been predicted to exist in the human genome, accounting for 1–5% of all predicted human genes. Altogether, miRNAs in human cells are estimated to regulate the expression of between 30% and 90% of human genes [44]. In particular, miRNAs mainly operate the PTGS by repressing translation or accelerating mRNA decay [45]. In both cases, after being loaded into the RISC complex, binding of the target mRNAs to the 3′ untranslated regions (UTRs) does not require perfect complementarity with the "*seed*" sequence. Thus, in contrast to siRNAs, a single miRNA may regulate multiple messenger RNAs and, in turn, each gene can be regulated by different miRNAs [45]. Moreover, recent studies have reported that miRNAs can also bind to the 5′ UTR or the open reading frame [46–49] and, even more surprisingly, they can upregulate translation upon growth arrest conditions [50]. It has also been noted that mature miRNAs may localize in the nucleus, where they may regulate pre-mRNA processing, act as chaperones modifying mRNA structures, or modulate mRNA–protein interactions [51]. The observations that miRNAs can be imported into the nucleus [52] or even secreted from the cells [53] suggest that the mechanistic details and the number of cellular functions of miRNAs are still far from being fully unraveled.

3.1 microRNA Deregulation is a Molecular Feature of Cancer

As data have accumulated, miRNAs have been found to be involved in a number of cellular processes such as proliferation, differentiation, survival, and apoptosis. Hence, it is not surprising that alterations affecting miRNA expression and function may promote the pathogenesis and progression of human tumors. Indeed, like classical genes, miRNAs can be either overexpressed or downregulated and act as oncogenes or tumor-suppressors, depending on their downstream targets [54]. Abnormalities in their expression or functions are associated with the typical hallmarks of cancer, including increased cell proliferation, abrogated apoptosis, enhanced cell motility and invasiveness, and neoangiogenesis [54]. For instance, miR-21 is highly upregulated in the majority of cancer tissues, and the repression of pro-apoptotic genes, such as *PTEN* (*phosphatase and tensin homolog*) or *PDCD4* (*programmed cell death 4*), stimulates proliferation and tumor initiation [55, 56]. Conversely, miR-15a and miR-16-1 were the first tumor-suppressor miRNAs to be described that were expressed from the fragile 13q14 region that is frequently deleted in patients with chronic lymphocytic leukemia (CLL) [57]. Their loss, by releasing the inhibition of tumor-promoting genes, such as *BCL2*, *BMI1*, *CCND2*, and *CCND1*, is able to promote cell growth and tumor progression [57–60].

Following the initial discovery of the miR-15a/16-1 cluster, a steadily growing number of reports have established that miRNA expression is globally deregulated in neoplastic cells compared to the corresponding normal tissue [61]. The progressive development of different high-throughput platforms for assessing miRNA expression in normal and diseased tissues (microarrays, bead-based flow cytometry, and deep sequencing) has allowed the miRNA expression profiling of several malignancies, including CLL [62], breast cancer [63], thyroid papillary carcinoma [64], lung cancer [65], glioblastoma [66], pancreatic tumors [67], hepatocellular carcinoma [68], prostate cancer [69], and gastric cancers [70]. This multi-combined approach has revealed the ability of miRNA signatures to not only differentiate between normal and cancerous tissues, but also to identify the tissue of origin. The latter point would be especially relevant when the tumor has already spread to distant metastatic sites, as it would allow the discrimination of different subtypes of a particular cancer even more successfully than previous mRNA panels [71]. Notably, the exciting discovery that miRNAs can be also detected at high levels in body fluids has spurred a rush to investigate miRNAs as potential diagnostic and prognostic biomarkers, as well as a novel class of drug targets, within the clinical setting of human malignancies [54].

3.2 Mechanisms of microRNA Deregulation in Cancer

Overall, large-scale miRNA profiling studies have revealed that the cancer phenotype is characterized by a globally reduced miRNA level relative to normal tissues [72], which suggests a role for miRNAs in the maintenance of a differentiated cell state and homeostasis.

The cause of the widespread differential expression of miRNA genes between neoplastic and normal cells can be linked to different mechanisms, including chromosomal rearrangements of miRNA genes,

DNA point mutations, epigenetic mechanisms, or alterations in the machinery responsible for miRNAs biogenesis [54, 73]. Along with the typical CLL breakpoint at the miR-15a-16-1 locus in the 13q14 chromosomal region, members of the let-7 family of tumor suppressor miRNAs map to fragile sites, as 3p2 (let-7g/let-7a-1), 9q22 (let-7f), 11q24 (let-7a-2), and 21q21 (let-7c), are frequently deleted in breast, lung, ovarian and cervical cancers [74]. Conversely, the oncogenic miR17~92 cluster in 13q31, encoding for six different miRNAs (miR-17, miR-18a, miR-19a, miR-20a, miR-19b-1, and miR-92-1), is amplified in several hematopoietic malignancies [54]. Interestingly, miR-17-5p and miR-20 are known to specifically repress the cell-cycle regulator E2F1, which mediates cell-cycle progression through G_1/S checkpoints, and whose expression is also regulated by the transcription factor c-MYC [75]. *In-vivo* studies in Eµ-Myc transgenic mice (a well-established mouse model of B-cell lymphoma) showed that the miR17~92 promoter also contains c-Myc E-box binding sites, and suggests that c-MYC is likely to fine-tune the cell cycle by regulating the expression of both mRNAs and miRNAs.

In addition to structural genetic alterations, miRNA expression in cancer can be also affected by epigenetic changes, such as altered DNA methylation [76]. Indeed, it is likely that half of the genomic sequences of miRNA genes contain CpG islands [77]. In this regard, the hypermethylation of miR-127 has been found to induce its downregulation, and to significantly increase the target proto-oncogene *BCL6* in bladder cancer cells [78]. Similarly, miR-9-1 [79] and the clustered miR-34b and miR-34c [80] were also found to be modulated by DNA methylation. Alternatively, defects in the epigenetic machinery, including DNA (cytosine-5)-methyltransferase 1 (DNMT1) and DNA (cytosine-5)-methyltransferase 3β (DNMT3β), also affect miR-124a expression, as found in colorectal cancer cells [81]. Conversely, the overexpression of putative oncogenic miRNAs in cancer can be due to DNA hypomethylation [82], and alterations affecting other chromatin-remodeling processes [83–85].

Single nucleotide polymorphisms (SNPs) in miRNAs, in either a heterozygous or homozygous configuration, represent another type of genetic variability that can affect gene and protein expression, and may be responsible for the interindividual variability of the cancer phenotype. Indeed, sequence variations in miRNA genes, including pri-miRNAs, pre-miRNAs and mature miRNAs, could also influence the processing and/or target selection of miRNAs.

A C → T germline alteration in the primary transcript of miR-15a/miR-16 reduces their expression levels, and has been found in patients with familial CLL [86]. miRNA precursor-related SNPs, such as pre-miR-196a2, pre-miR-499 and pre-miR-146a, have been associated with the risk of breast cancer [87] and papillary thyroid carcinoma [88]. Moreover, the SNP rs11614913 in pre-miR-196a2 has also shown prognostic significance correlating with survival in patients with non-small-cell lung cancer (NSCLC) [89].

SNPs in the miRNA *seed* region can impair – either by weakening/abolishing or, conversely, by strengthening – miRNA–mRNA interaction, and thus differentially affect the expression of miRNA targets [90, 91]. Despite the low (<1%) probability of SNP occurrence in a miRNA *seed* region, a polymorphism here would significantly impair a miRNA function, as compared to a polymorphism present in a 3′-mismatch-tolerant region (3′-MTR) binding region which, conversely, is very sensitive to mismatches. And last, but not least, miR-polymorphisms can alter the epigenetic regulation of a miRNA (either methylation or acetylation) and promote disease progression [79].

Several reports have also shown that changes in the miRNA biogenesis and processing machinery can be responsible for miRNA dysregulation in cancer [90]. For instance, a generalized downregulation of Dicer and/or Drosha has been significantly correlated to a poor clinical outcome in different tumors such as lung, ovarian, colon, prostate and breast cancer, nasopharyngeal carcinoma, and neuroblastoma [92–100]. In addition, the *DICER1* gene is commonly featured as a haploin-sufficient tumor suppressor because frequently it is single-copy-deleted in cancer [101]. Interestingly, a somatic hot-spot missense mutation in *DICER1* gene restricted to non-epithelial ovarian tumors has also been reported [102]. Inactivating mutations and a consequent loss of the *TARBP2* gene, encoding for TRBP, have been shown to destabilize *DICER1* and impair miRNA processing in cancers bearing microsatellite instability [103]. These tumors are also characterized by mutations of the *XPO5* gene, coding for the Expotin-5 transporter, which generate a truncated protein that is unable to associate with pre-miRNA molecules and exit the nucleus; this decreases the trafficking of pre-miRNAs in the cytoplasm and reduces the pool of diced mature miRNAs [104].

A dysregulation of AGO proteins also occurs in cancer. The loss of AGO1 (*EIF2C1*), AGO3 (*EIF2C3*) and AGO4 (*EIF2C4*) is a recurrent event in Wilms tumor of the kidney and in neuroectodermal tumors [37, 73]. However, the expression of AGO proteins, such as AGO2 (*EIF2C2*), is reported to be reduced in some cases, for example in melanoma samples (primary and metastatic) compared to normal epidermal melanocytes, but elevated in other tumors such as breast and colon cancer; this suggests that expression can be regulated in a cell context-dependent manner [105]. The RNA-binding proteins LIN28/LIN28b represent further potential oncogenic factors; they are normally highly expressed in hematopoietic stem cells (HSC), and have the unique ability to revert human somatic cells to pluripotent cells when coexpressed with the reprogramming factors OCT4, NANOG, and SOX2. The oncogenic roles of LIN28 and LIN28b have been suggested by elevated levels in cancer stem cells from various cancer types, including ovarian, breast and NSCLC [105].

Experimental findings have also documented how miRNA processing can be affected by other miRNAs, either directly, whereby the mouse miR-709 is able to bind a recognition element on pri-miR-15a/16-1 in the nucleus [106], or indirectly, whereby the miR-103-107 family targets Dicer and leads to a downstream downregulation of miRNAs as the miR-200 family, and to epithelial-to-mesenchymal transition (EMT) [100].

Finally, the deregulation of miRNA expression, as either increased or decreased transcription, can also result from an altered transcription factor activity. Indeed, miRNAs can be either positively or negatively regulated by transcription factors such as p53, activating miR-34a [107] and miR-205 [108], MYC, activating miR-17~92 cluster [75] but repressing let-7 [109] and miR-29 family members [110], and HIF-1α, regulating miR-210 [111] or ZEB1, and repressing the transcription of members of the miR-200 family, which in turn are able to directly target ZEB1 and ZEB2 [112].

3.3 Role of microRNAs in Cancer

As noted above, aberrant expressed miRNAs can act either as oncogenes (oncomiRs) or tumor suppressor genes (tumor suppressor miRNAs), and be crucially involved in the initiation and progression of the neoplastic disease in all human tumors.

Of note, their role is strictly dependent on the cellular context and tumor system: miR-221 and miR-222, for example, can function as tumor suppressor miRNAs in leukemia by targeting the *KIT* oncogene. [113]. Furthermore, they are also known for silencing tumor suppressors like *PTEN* or *TIMP3* (tissue inhibitor of metalloproteinases 3), playing the role of oncomiRs in several solid tumors including breast and lung cancer, hepatocellular carcinoma, or glioblastoma [114].

Here, a few examples of miRNAs classified as oncomiRs (Table 1) or tumor suppressor miRs (Table 2) are reported, according to the main role that they play in the majority of cancers.

3.3.1 OncomiRs

miR-155 miR-155, encoded by the B-cell integration cluster (BIC) ncRNA, exerts its role in regulating the response and adaptation of cells to the tumor microenvironment, in particular when hypoxic conditions occur. Indeed, miR-155 mediates the downregulation of FOXO3A upon HIF1α activation [115]. In this context, other targets of miR-155 are SOC1 and RhoA, which collectively act in the signaling pathways promoting EMT [116, 117].

A high expression of miR-155 has been reported in several solid tumors such as breast cancer, colon cancer, ovarian cancer, pancreatic ductal adenocarcinoma, thyroid carcinoma, and lung cancer [71], where it is also considered a marker of poor prognosis. Similarly, miR-155 levels are increased in various B-cell malignancies, including Hodgkin lymphoma, some subtypes of non-Hodgkin lymphoma, and acute myeloid leukemia (AML) [71]. Interestingly, alongside the contribution to cancer genesis and progression, *in-vitro* studies have shown that miR-155 overexpression can confer resistance to radiation therapy, and the administration of antisense oligonucleotides (ASOs) against miR-155 can overturn this effect, which suggests that miR-155 may serve as a predictive biomarker and putative therapeutic target for treatment-resistant tumors [118].

The miR-17∼92 Cluster and Paralogs The polycistronic miR-17∼92 cluster is located within approximately 1 kb of an intron of the C13orf25 locus on human chromosome 13q31.3, a region that is frequently amplified in several types of B-cell lymphoma and solid tumors [119]. The transcriptional regulation of the miR-17∼92 cluster is mostly mediated by the *MYC* oncogene [75], to inhibit the apoptotic activity of E2F1, and induce cell proliferation. Overexpression of the cluster is observed in a variety of human cancers, including small cell lung cancer (SCLC), colon and gastric cancer, retinoblastoma, neuroblastoma, medulloblastoma, and osteosarcoma [71]. The cluster, which is processed to produce mature miR-17, miR-18a, miR-19a, miR-20a, miR-19b-1, and miR-92a-1, can target several factors involved in cancer initiation (HIF1α), cell proliferation (PTEN, E2F1-3, TNFα, RAB14), survival (BIM, TGFBR2), and angiogenesis [TSP1, connective tissue growth factor (CTGF)] [54, 120, 121]. The paralog miR-106b-25 cluster, comprising highly conserved miR-106b, miR-93 and miR-25, has shown to be similarly regulated with its host gene *MCM7* by E2F1, building

Tab. 1 MicroRNAs which act mainly as oncomiRs and are upregulated in cancer.

microRNA	Locus	Cancer type(s)	microRNA deregulation effects	Key targets
miR-155	21q21.3	CLL, AML, B-lymphomas; lung, breast, colon, gastric, pancreatic, thyroid cancers	Resisting growth suppressors	FOXO3A, SOCS1, SHIP1
			Supporting inflammation	CEBPB, MEIS1, ETS1, C-MAF, CUTL1
			Mismatch repair impairment	hMSH2, hMSH6, hMLH1, WEE1
			Invasion and metastasis	RhoA
miR-17~92	13q31.3	Leukemias/lymphomas; lung, breast, colon, prostate, pancreatic, cancers; retinoblastoma, glioblastoma	Sustaining proliferation	PTEN, E2F1-3, TNF-α, RAB14, p63
			Resisting growth suppressors	BIM, TGFBR2
			Promoting angiogenesis	TSP1, CTGF, TGF-β, SMAD4
			Hypoxia response	HIF-1α
miR-221/222	Xp11.3	Lung, breast, prostate, papillary thyroid cancers; hepatocellular carcinoma; glioblastoma	Sustaining proliferation	PTEN, TIMP3, DICER
			Resisting growth suppressors	KIT, p27^{Kip1}, p57^{Kip2}, FOXO3A
			Evading apoptosis	PUMA
			EMT	TRSP1
miR-21	17q23.2	CLL, AML, myeloma; lung, breast, gastric, prostate pancreatic cancers; glioblastoma	Sustaining proliferation and ECM remodeling	PTEN, SPRY1, SPRY2
			Resisting growth suppressor	APAF1
			Evading apoptosis	PDCD4
			Metastasis	TPM-1, TPM-3, RECK, TIMP3

AML, acute myeloid leukemia; CLL, chronic lymphocytic leukemia; EMT, epithelial-to-mesenchymal transition.

a negative feedback loop where the cluster controls in turn the E2F1 intracellular level. An overexpression of miR-106b/25 has been reported in cancers of the prostate and pancreas, as well as in neuroblastoma, multiple myeloma [71], and is also likely to be responsible for the development of transforming growth factor-beta (TGF-β) resistance by suppressing p21 and BIM in gastric cancer [122].

Tab. 2 MicroRNAs which act mainly as tumor suppressor miRNAs and are downregulated in cancer.

microRNA	Locus	Cancer type(s)	microRNA deregulation effects	Key targets
miR-15a-16-1	13q14.2	CLL, multiple myeloma, lymphomas; lung, breast, colon prostate, pancreatic, ovarian cancers	Sustaining cell proliferation	CDC2, JUN, FGF-2, FGFR1 CCND-1, CHK1
			Evading apoptosis Promoting angiogenesis	BCL2, SIRT1 VEGF, VEGFR2, FGFR1
let-7 family	*LET7A1* 9q22.32	Lymphomas; lung, breast, colon, prostate, gastric, liver cancers	Sustaining cell proliferation	KRAS, NRAS, CDC25A, c-MYC
	LET7A2 11q24.1 *LET7A3,-7B* 22q13.31 *LET7C* 21q21.1 *LET7D,-7F1* 9q22.32 *LET7E* 19q13.41 *LET7F2* Xp11.22		Evading apoptosis Metastasis	BCL-XL HMGA2, TWIST1
miR-34a	1p36.22	Lymphoma; lung, breast, colon, renal, bladder pancreatic cancers; neuroblastoma, glioblastoma	Sustaining cell proliferation	CDC25A, CDK4, CDK6, c-MYC
			Evading apoptosis EMT and metastasis	BCL2, SIRT1 MET, SNAIL1
miR-200 family	*MIR200A, MIR200B, MIR429* 1p36.33	Lung, breast, endometrial, ovarian, nasopharyngeal, bladder cancers	EMT and metastasis	ZEB1, ZEB2, CTNNB1, BMI-1, PLCγ1, VEGFR1, FN1, LEPR, MALM2, MALM3
	MIR200C, MIR141 12p13.31			

FGF, fibroblast growth factor; LEPR, leptine receptor; NRAS, neuroblastoma rat sarcoma viral oncogene homolog; VEGFR1, vascular endothelial growth factor receptor 1.

The miR-222/221 Cluster Deregulation of the miR-222/221 cluster is mainly involved in promoting cell survival and the metastatic spread of cancer cells. miR-222/221 have been found highly upregulated in hepatocarcinoma, thyroid cancer, melanoma, estrogen receptor-negative breast cancer, and glioblastoma [123–127]. Cell-cycle control and anti-apoptotic activity are related to the repression of kit, p27Kip1, PTEN, and PUMA [128]. It has also been reported that miR-221 and miR-222 transcription is induced by FOSL1, leading to a repression of TRPS1 and an E-cadherin decrease, contributing to the invasive phenotype of basal-like breast cancer [129]. Recently, it was reported that miR-221 and miR-222, controlled by both MET and epidermal growth factor (EGF) receptors, may play a pivotal role in developing tyrosine kinase inhibitor resistance in NSCLC [130].

miR-21 The well-documented miR-21, transcribed from the 17q23.2 genomic locus, is typically highly expressed in virtually all human malignancies, including those derived from breast, colon, liver, brain, pancreas, and prostate.

The wide array of molecular targets, including the tumor suppressors maspin (*SERPINB5*), PDCD4, tropomyosin 1 (TPM1) and PTEN [131, 132], negative regulators of the RAS (rat sarcoma viral oncogene homolog) signaling axis, Spry1, Spry2 and Btg2 [133], or the metalloproteases RECK (reversion-inducing-cysteine-rich protein with kazal motifs) and TIMP3 (tissue inhibitor of metalloproteinase 3) [134] allocates miR-21 within the regulation of several oncogenic mechanisms supporting the aberrant growth and survival of cancer cells, as well as invasion and metastatic spread.

The central role of miR-21 is supported by a number of cell culture and animal model-based studies, which also indicate that the inhibition of miR-21 with ASOs represents a good strategy for reducing cell cancer proliferation and inducing apoptosis. For example, Medina and colleagues used Cre and Tet-off technologies to generate mice which, as a consequence of conditional expression of miR-21, developed a pre-B-cell malignant lymphoid-like phenotype, while subsequent miR-21 inactivation in the same model led to apoptosis and tumor regression [56]. This suggested that human cancers might be treated through the pharmacological inactivation of miRNAs such as miR-21.

The microarray-based profiling of tumor samples from breast cancer patients found that miR-21 was consistently upregulated in several subtypes of breast cancer, and suggested a role for this miRNA in mammary tumor initiation or progression [63]. Similarly, lung cancer profiling detected not only an elevated expression of miR-21 but also a direct correlation between its expression level and the presence of mutations in the epidermal growth factor receptor (*EGFR*) gene in patients [135]. Moreover, the blockade of miR-21 with ASO was able to enhance phosphorylation of epidermal growth factor receptor (p-EGFR) and restore apoptotic mechanisms in glioblastoma cell culture [136].

3.3.2 Tumor Suppressor miRs

The miR-15a/16-1 Cluster The link between miRNAs and cancer was first established with miR-15a and miR-16-1 [57], when it was discovered that they are often deleted and downregulated in CLL and, later, in prostate, lung cancer, and multiple myeloma [137–139], which suggested a main role in tumor suppression. Loss of their expression has been associated with the dysregulation of several pathways controlling cell proliferation, survival, migration and invasion, in which they regulate the levels of BCL-2, CDC2, ETS1, PDCD4, FGF-2, FGFR1 and other key cascade effectors. Moreover, miR-15a/16-1 was shown to regulate the response to cisplatin by targeting WEE1 and CHK1, which are commonly overexpressed in cisplatin-resistant cells, and also the restoration of normal miR-15/16 re-induced sensitivity to the drug [140], supporting the potential of a therapeutic solution to suppress aberrant growth in a variety of cancers.

let-7 Family The let-7 family includes 12 human homologs mapping to fragile sites associated with lung, breast, urothelial and ovarian cancers [71]. The let-7 family plays an important role in controlling cellular growth by regulating the Kirsten rat sarcoma viral oncogene homolog (KRAS) [141], and the specific downregulation of family members is a frequent event associated with a poor prognosis in lung cancer and the presence of lymph node metastases in breast cancer [71]. Moreover, in both types of tumor, polymorphisms in the 3′ UTR of KRAS mRNA are able to impair the let-7 regulating action and predict a worse prognosis [142, 143]. Along with KRAS, let-7 targets comprise LIN28, HMGA2 [54], MYC [144], and cell-cycle-related factors such as cyclinD2, CDK6, and CDC25A [145], allocating let-7 to every process involved in tumor growth and progression.

miR-34 Family The miR-34 family includes three miRNAs – miR-34a, miR-34b and miR-34c – under the direct transcriptional control of p53 [146]. In particular, p53 activation promotes a positive feedback loop in which an increased expression of miR-34 results in the downregulation of E2F, MET and SIRT1 and an indirect activation of p53 [141]. The promoter region of miR-34 family genes also contains CpG islands, and aberrant CpG methylation has shown to reduce miR-34 expression in multiple cancer types [80]. Other studies have progressively identified a number of targets, including BCL2, MYCN, GMNN and HDAC1 of the apoptotic cascade, cell-cycle crucial effectors such as CDC25A, CDK4 and c-MYC, or even WNT, HMGA2, and SNAIL1 that modulate migration and the invasion of cancer cells. For instance, a lower miR-34b expression in NSCLC has been correlated with a higher incidence of lymph node metastasis and a poor prognosis [54].

The miR-200 Family The miR-200 family members (miR-200a, miR-200b, miR-200c, miR-141, and miR-429) are downregulated in human tumors, where they play a critical role in the suppression of EMT, the cellular process through which epithelial cells lose their polarity and cell–cell adhesion, by targeting the key factors ZEB1 and ZEB2 [147] and promoting invasion. In lung cancer cells, an induced overexpression of the miR-200 family promotes mesenchymal-to-epithelial transition by repressing ZEB1 and increasing E-cadherin [148]. Accordingly, a loss of miR-200 family members correlates with a lack of E-cadherin expression in invasive breast cancer cell lines and breast tumor specimens, supporting an *in-vivo* role for the miR-200 family in EMT repression. Interestingly, ZEB1 and SIP1 are likely to recognize an E-box proximal minimal promoter element, generating a potential double-negative feedback loop in which the repression of a primary transcript and mature miR-200 expression allows the maintenance of high ZEB1/SIP1 levels that favor the mesenchymal phenotype of human breast cancer cells [149]. In addition, *in-vitro* studies have described a type of Akt–miR-200–E-cadherin axis, where Akt1 and Akt2 are able to control the abundance of miR-200 and E-cadherin and thus regulate breast cancer metastasis. This has been confirmed by the ratio of Akt1 to Akt2, and the abundance of miR-200 family members and E-cadherin in a set of primary and metastatic human breast cancers [150]. The deregulation of miR-220c/141 has also been linked to aberrant DNA methylation [151]. In conclusion, the miR-200 family prevalently assumes a tumor-suppressor role, and downregulation of its family members is significantly associated with an aggressive cancer cell phenotype.

When taken together, these data and the novel information that is continuously collected from the increasing number of molecular profiling studies in human cancers, highlight "miRNA addiction" as a key feature of cancer cells for developing and maintaining the malignant phenotype. Investigating miRNAs and the molecular circuitries that mediate their regulatory function in cancer can provide new clues about the nature of the cancer itself, and the molecules or mechanisms which may be targeted therapeutically to improve the clinical management of cancer patients.

4
Current Prospects for RNA Interference-Based Therapy for Cancer

To date, it is well documented that cancers can build sophisticated biological networks supporting their ability to progress and, moreover, to evade treatment. The promise of personalized medicine is to streamline clinical decision-making, according to each patient's genetic and physiological profile, distinguishing those who can benefit from a treatment from those who can experience side effects without relief. Indeed, distinct gene expression patterns may affect the body's response to medications, and an improved comprehension of the signaling cascades regulating tumor survival has determined the transition from a more promiscuously chemotherapeutic approach to highly selective targeted pharmaceutics. Initially, the antibody-based trastuzumab (Herceptin®) represented the first targeted therapy specifically for HER2-positive metastatic breast cancer (in 1997). Later, the small molecule imatinib mesylate (Gleevec™) became the first approved kinase inhibitor for targeting BCR/ABL in chronic myeloid leukemia (CML), and then for the treatment of gastrointestinal stromal tumors (GISTs), targeting c-kit.

A number of gain/loss-of-function experiments, in combination with target prediction analyses, have demonstrated that perturbation in miRNAs expression can affect not only every step of the tumorigenic process, but also the sensibility to drugs. Hence, the targeting of miRNAs, and the mechanisms orchestrating the endogenous RNAi, may undoubtedly represent a rich soil for the development of novel anti-cancer targeted therapies. RNAi-induced gene silencing mirrors the inhibitory effects of conventional drugs, such as protein-based compounds (e.g., antibodies and vaccines) and small molecules, whose inhibitory effect involves blocking the function of their targets. However, some disease-related molecules do not have enzymatic function, or they have a conformation that is not accessible to conventional drugs or small molecules, and so they are considered "undruggable." Conversely, RNAi technologies can be designed to virtually target any gene, including such undruggable molecules [152–154], with an exclusively allele-specific gene silencing [155]. Furthermore, antagonizing a target gene expression rather than its function has a more powerful downstream effect, as a single mRNA molecule is translated in multiple copies of a protein. Finally, compared to traditional drugs, the synthesis and production of RNAi modulators is more rapid, does not need the use of cellular expression systems or refolding steps, and can guarantee a higher selectivity and potency.

Nevertheless, to date, there are still some pending issues regarding the efficiency of the delivery systems, the choice of

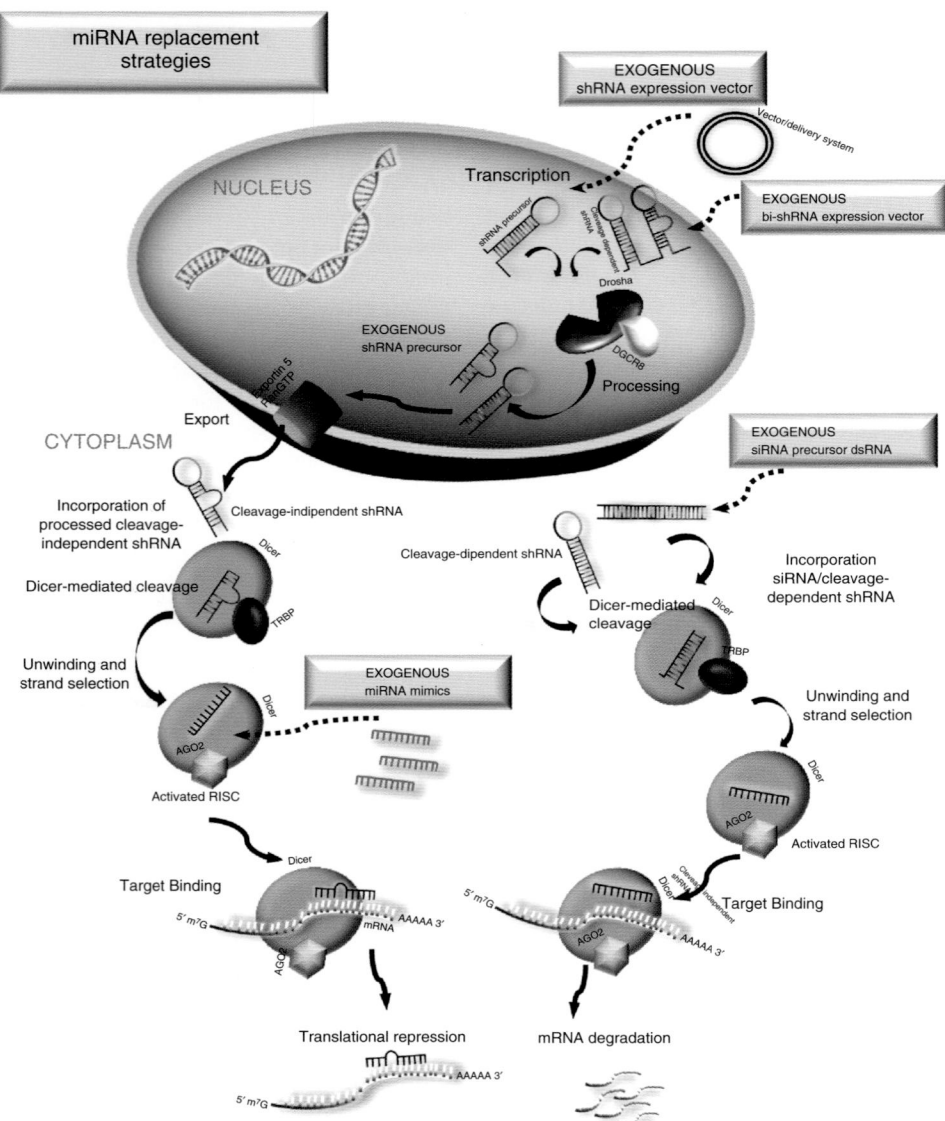

Fig. 2 MicroRNAs as targeted therapeutics. When aiming to repress oncogenic proteins, the expression levels of tumor suppressor microRNAs can be restored by using different strategies including siRNAs, shRNAs (short hairpin RNAs), bi-shRNAs (bifunctional siRNA), targeting by perfect base-complementarity oncogenic mRNAs, or miRNA mimics, synthetic analogs of endogenous miRNAs.

RNAi targets and the safety profile, that need to be addressed before allowing the widespread use of RNAi strategy *in vivo* [156].

4.1 Harnessing the Endogenous RNAi Apparatus for Clinical Purposes

The endogenous RNAi machinery can be engaged by designing exogenous triggers that enter the pathway at different points. Several synthetic modulators and strategies have been implemented to boost and optimize the effects on cancer-related pathways.

Modulating the RNAi machinery consists of either replenishing the levels of downregulated tumor-suppressor miRNAs to block tumor-promoting mRNAs, or sequestering oncogenic, overexpressed miRNAs to allow the transcription of tumor-suppressor proteins. All of the examples described below are simplified in Figures 2 and 3.

4.1.1 Restoring miRNA's Endogenous Regulation of Transcription: siRNAs, shRNAs, and Derivatives

The replacement of tumor suppressor miRNAs (let-7 family, miR-34a, miR-24, miR-26) [51, 74] finds its rationale in the global miRNA downregulation that characterizes the neoplastic phenotype. This strategy would permit the potential inhibition of oncogene transcripts that are aberrantly increased in the cancer state.

Short-interfering RNAs (siRNAs) are synthetic, short (usually 19–23 bp) dsRNAs which enter miRNA biogenesis at the cytoplasmic stage, when the pre-miRNA is chopped by Dicer, to be directly incorporated into the RISC complex. The siRNA guide-strand binds to and cleaves the complementary mRNA by a perfect-complementary match. Importantly, the guide-strand–RISC complex has the ability to be recycled, allowing the translational repression of several mRNA molecules, and propagation of the induced gene-silencing activity.

However, as the half-life of a siRNA is relatively short [157], shRNAs have been designed to enter the nucleus and be transcribed by RNA polymerase II or III from an external expression vector bearing a pre-designed, short, double-stranded DNA sequence. This means that shRNAs are constantly synthesized in the host cells, and achieve a longer gene-silencing effect. The primary transcript generated from the RNA polymerase II promoter contains a hairpin-like stem–loop structure that is regularly processed by Drosha, and translocated into the cytoplasm as pre-shRNA, to be cleaved by Dicer and incorporated into RISC, followed by the same cytoplasmic RNAi process as in siRNA [157].

A new class of bi-functional short hairpin RNAs (bi-shRNAs) has been developed for a more efficient and prolonged target-gene knockdown [158]. These consist of a vector construct for the expression of two (bi) stem–loop

Fig. 3 MicroRNAs as therapeutic targets. Aberrantly overexpressed microRNAs (oncomiRs) can conversely be inhibited by using: antimiRs (ASO, antisense oligonucleotides; LNAs, locked nucleic acids), designed to be complementary to the endogenous miRNAs; miRNA sponges, which contain multiple binding sites for microRNAs and prevent their loading onto the RISC complex; miRNA masks, small antisense oligonucleotides, which bind the target mRNA at the 3′-UTR miRNA-target sites, avoiding the interaction between mRNA and endogenous miRNAs, and transcriptional repression.

310 | RNA Interference in Cancer Therapy

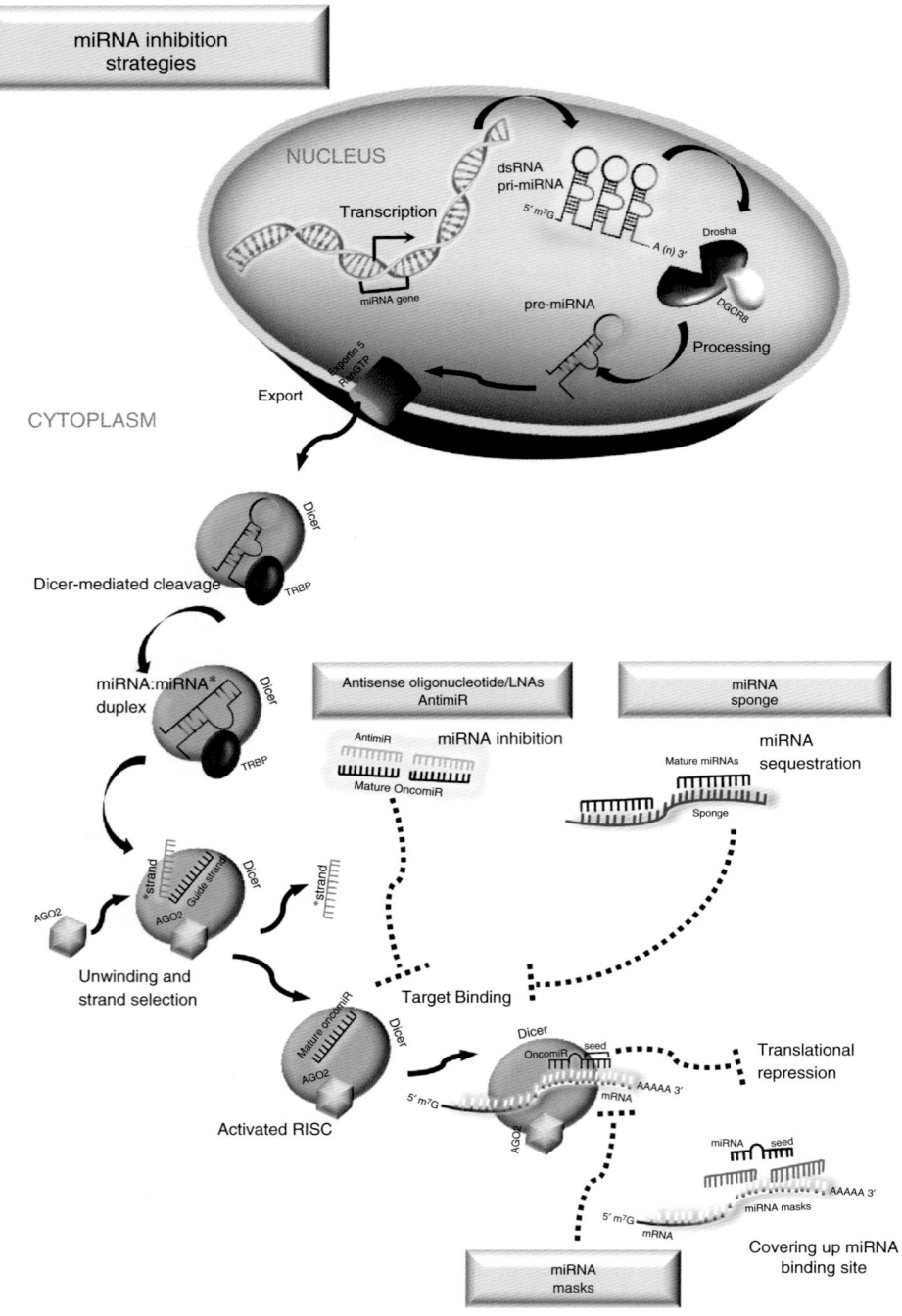

shRNAs for each targeted mRNA: one with perfect-matching stem sequences (as siRNAs), and a second with mismatches at the central location (bases 9–12) and additional locations of the stem. Compared to siRNAs, this strategy would promote the loading of mature shRNAs onto both cleavage-dependent (AGO2-mediated) and cleavage-independent (blocked-AGO2 by the mismatch) RISCs, thus achieving a higher and more effective shut down of the target mRNA by the concurrent involvement of both degradation and translation inhibition mechanisms [158].

ssRNAs may also be used as RNAi triggers but, although they have a greater structural flexibility (ssRNA versus dsRNA) and a more amphiphilic nature [159], they are not efficiently recognized by the miRNA biogenesis apparatus (~100-fold less efficient) as siRNAs and shRNAs.

Moreover, a new method to restore the expression levels of endogenous, downregulated miRNAs (single or cluster) employs double-stranded, short artificial miRNAs, termed miRNA *mimics*, which are analogs of endogenous miRNAs and repress mRNA molecules by an imperfect binding to the 3′-UTR following the same mechanism of action [51]. Indeed, the ectopic expression of synthetic miRNAs mimics with tumor suppressor function as miR-15-a and miR-29 in cancer cells has shown to induce apoptosis in prostate and AML cell lines, respectively [160]. The first *in-vivo* evidence derives from Kota and colleagues, who recently cloned tumor suppressor miR-26 into an adeno-associated virus (AAV) vector for intravenous injection in an established MYC-dependent liver cancer mouse model. As expected, the restoration of miR-26 expression, which generally was reduced in liver cancer cells compared to normal cells, resulted in a suppression of tumorigenicity through the downregulation of cyclins D2 and E2 and a consequent suppression of tumor growth, without signs of toxicity [161].

Of note, it is likely that the mimic strategy would prevent the development of resistance to an RNAi approach by cancer cells better than siRNA formulations. Indeed, by mimicking endogenous miRNAs, mimics would impair the downstream function of multiple genes at the same time; thus, it would be more difficult for cancer cells to rearrange the sequence of several genes simultaneously to escape the mimic targeting and inhibition.

4.1.2 Inhibition of oncomiRs: AntimiRs, miRNA Sponges, and miR-Masks

The inverse approach consists of the inhibition of overexpressed miRNAs with oncogenic roles as miR-17~92 cluster, miR-155, by using small, single-stranded ASO drugs or the new generation of locked nucleic acids (LNAs) [51, 74]. These compounds belong to the antimiRs category, the properties of which are complementary to a target miRNA, with an ability to bind with high affinity and specificity and to produce a functional inhibition, thereby derepressing the mRNA targets of that miRNA [162].

A novel technology for managing miRNAs involves the use of expressing vector-carrying tandem miRNA target sites, to saturate an endogenous miRNA, and prevent the binding to its natural targets. This approach, known as "decoy," "sponge," or "eraser," employs different systems of gene delivery such as plasmids and adenovirus-, lentivirus- or retrovirus-based vectors that are normally used for gene knock-down studies. Since the inhibition of one miRNA can be compensated by the expression of other miRNA genes or family members, the tandem target sites built on these vectors can be assembled to catch different miRNAs sharing the same binding site, thus amplifying the effect of miRNA inhibition. Moreover, the sponges have been shown to be stable, they do not require repetitive rounds of administration, and they assure cell-type specificity compared to ASOs. Notably, the problem of a safety profile (i.e., toxicity), the partial interference that can occur, and the unknown targets that still must be identified for each putative miRNA, has led to the use of these constructs being a major challenge that is still in progress [163].

As an alternative to miRNA sponges, MiRNA-Masking Antisense oligonucleotides (miR-Masks) are single-stranded 2′-O-methyl-modified ASOs, that are fully complementary to predicted miRNA binding sites in the 3′-UTR of the target mRNA [164]. They compete with an endogenous miRNA by covering up the miRNA binding site and de-repressing its target mRNA. As this approach has the advantage of being gene-specific, it can significantly reduce any off-target effects [164]; however, it does not allow the targeting of multiple pathways.

the predictivity of tumor biological and clinical responses.

Whole-genome RNAi screening has been applied to human research for the first time for the identification of host genes affecting influenza A virus replication [165] and West Nile virus infection [166]. Using this approach, Tiedemann and colleagues identified crucial survival genes for the non-curative myeloma cancer after transfection of siRNA constructs of myeloma human cell [167]. To date, *in-vitro* whole-genome RNAi screening, combined with mutational signatures from different types of human cancer [156], has provided valuable clues on putative molecules to conveniently hit for disturbing oncogenic pathways and inducing tumor regression. For instance, GATA2 and BRCA1 have been identified as potential therapeutic targets in tumors with *KRAS* or *CCNE-1* mutations [168, 169], respectively. Furthermore, RNAi screening in animal models can provide information about the normal physiological functions of the putative RNAi-targets, and predict toxic effects that can derive from their incorrect targeting, compromising the clinical outcome. In the near future, the implementation of RNAi screening with "omics" data from clinical samples will provide a robust framework for prioritizing targets according to their translational potentials (Fig. 4).

5
Rational Design of an RNAi-Based Therapeutic Approach in Cancer

5.1
RNAi Target Selection

A successful RNAi clinical pipeline first relies on the rational selection of targets, according to parameters such as tumor specificity, tumor survival relevance, and

5.2
In-Vivo RNAi Challenges

Overall, RNAi has important advantages over traditional pharmaceuticals such as protein-based drugs and small molecules for treating cancer, as it can be used to silence any target gene involved in tumor initiation, growth, angiogenesis, metastases formation, and chemoresistance [170]. The design of specific RNAi-effectors can

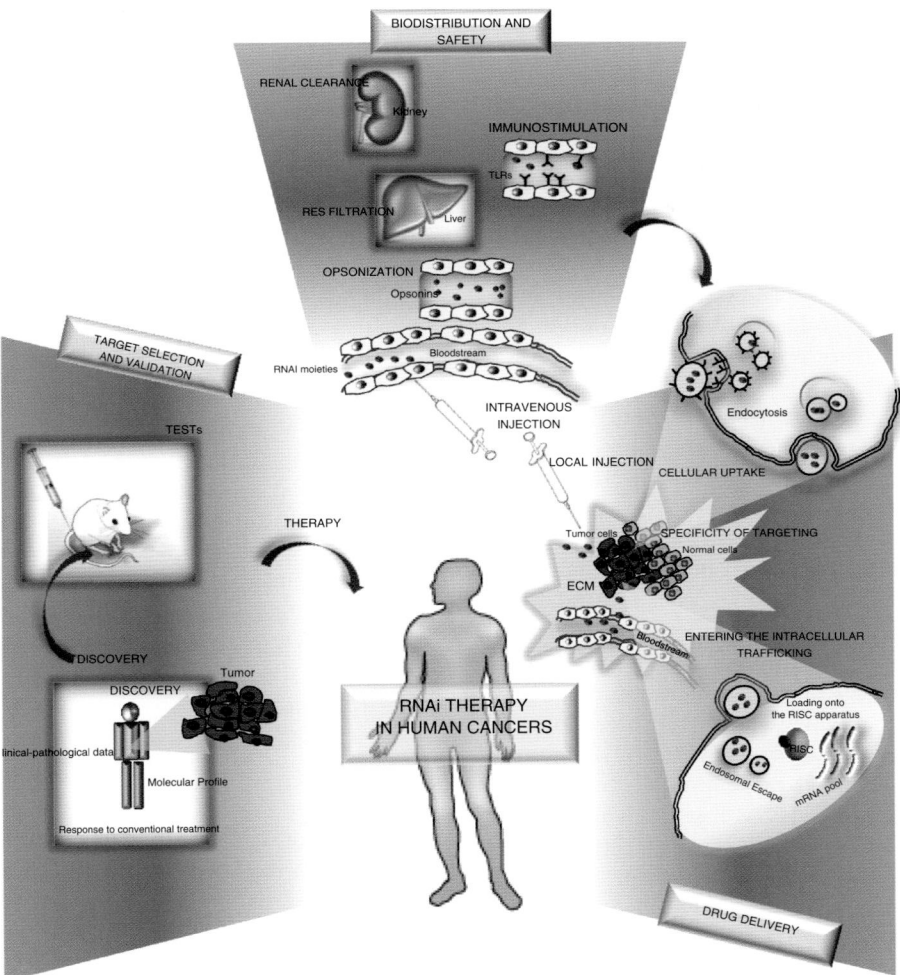

Fig. 4 Rationale design and challenges of RNAi therapeutics for human cancers. Schematic showing the critical issues to be taken into account when designing an RNAi-based strategy for cancer treatment: selection of the targets according to the tumor characteristics and patients' information from the clinic; the biological barriers affecting biostability and safety profile of RNAi therapeutics; and an efficient drug delivery culminating in the correct cellular uptake by tumor cells.

rely on the growing amount of information that is continuously being collected on potential mRNA targets, in terms of corresponding sequences and mechanisms of action. Moreover, their robustness and specificity makes them more efficient and less toxic than traditional drugs. However, despite a number of optimistic preclinical studies having been conducted and ongoing clinical trials, several barriers must first be overcome before RNAi molecules can be moved to the clinical setting. For instance, the efficient intratumoral delivery and correct cellular uptake of RNAi molecules remain critical issues; in addition, despite their sequence-specificity, the potential to knock-down non-targeted genes is an off-targeting effect that must be estimated before *in-vivo* administration. Finally, RNAi has also been shown to induce an innate immune response and to saturate the endogenous miRNA pathway, leading to cytokine-associated damage and cytotoxicity, respectively [170].

5.2.1 Stability and Bioavailability Issues

When systemic delivery methods are adopted, and before reaching their target tissue and cells, small RNA drugs need to overcome several biological barriers that derive from: (i) their intrinsic nature (superficial negative charge, and high molecular weight for crossing cellular membranes); (ii) the bloodstream, as renal clearance, nucleases digestion, opsonization, and consequent reticuloendothelial system (RES)-mediated elimination; and (iii) the target tissue itself, as the extracellular matrix (ECM), cellular uptake and endosome internalization [171] (Fig. 4).

Double-stranded RNAs are more stable compared to other oligonucleotides (ssRNA or ssDNA) [172] but once they reach the circulation, unmodified (naked) siRNAs have shown to be susceptible to nuclease degradation. Some studies have also shown that when administered, naked siRNAs accumulate preferentially in the kidney to achieve levels 40-fold higher than in other organs, and are excreted into the urine within 1 h, due to their relatively small size and effective glomerular pore size. Hence, the correct target-site accumulation of an effective dosage of RNAi agents may be one of several major concerns for their therapeutic use [173].

To this end, the insertion of chemical modifications has been explored as a strategy to increase siRNA stability without compromising the siRNA silencing potency. Most of the chemical modifications have been carried out on the 2'-OH group of the ribose moiety as it is not essential for siRNA silencing activity [174]. Such modifications are commonly known as 2'-modifications, and mainly referred to as 2'-O, 2'-Methyl-nucleoside (2'-OMe) and 2'-deoxyflouridine (2'-F) modifications [175, 176]. Another modification of the 2' position of the ribose consisting of a methylene linkage between the 2' and 4' positions generated LNAs [177]. Indeed, LNA-based oligonucleotides, as miRNA mimics or inhibitors, appear to be highly resistant to nuclease degradation and displays low toxicity in biological systems. An additional type of chemical modification is the phosphoromonothioate (PS) linkage to the 3' terminus of siRNA's phosphate backbone [178]. AntagomiRs are among the first antimiRs with proven *in vivo* efficacy and, similar to siRNAs, are 2'-O-Me oligonucleotide PS-substitution-modified constructs with the addition of a 3'-cholesteryl moiety [179].

5.2.2 Multiple Off-Target Effects

Although the mRNA cleavage induced by siRNAs and the translational repression by miRNAs are distinct pathways with different requirements [9], the two RNAi mechanisms share common cofactors and entry points, such as the RISC processing machinery. A "miRNA-like off-target gene silencing" can occur when siRNAs or shRNAs are loaded onto the RISC complex in a miRNA-like conformation, due to similarities between the core region at the 5' end of a siRNA and the *seed* region of miRNAs, both of which guide recognition of the complementary 3′ UTR of the target mRNA and stimulate the expression of often hundreds of off-target genes [180, 181]. Such adverse side effects are a safety concern, especially when RNAi therapeutics are administered over extended periods of time [180]. A substantial reduction in off-targeting may be achieved by combining and implementing different strategies of rational drug design. Bioinformatics, which considers potential interactions between the siRNA *seed* region and the genome-wide repertoire of 3′-UTRs during the *in-silico* stage of RNAi agent selection, can be considered a rational approach to bypass such off-targeting. In addition, chemical and structural modifications can reduce guide strand-mediated off-targeting, by penalizing *seed*-only interactions via more extensive base-pairing (e.g., by including 2′-O-methyl, unlocked nucleic acid-modifications in the *seed*), and eliminate passenger strand-mediated off-targeting by 5′ blocking or by shortening the length of the passenger strand, without compromising the specific gene knockdown [180, 182].

The risk of saturation of the endogenous miRNA machinery by exogenous small RNAs [183] is another type of potential RNAi toxicity. The saturation mechanism involves several components, such as the Exportin-5 [184], and the RISC complex [183], which become less accessible to native miRNAs, especially in the case of a sustained expression of shRNAs [183]. Currently, the best strategy to avoid such complications refers to the use of lower concentrations of siRNA or, even better, to siRNA pools rather than single siRNAs [185, 186].

5.2.3 Immune Response

Besides delivery, acute immune responses from activating innate immune receptors or the complement system are critical "checkpoints" when assessing the safety profile of RNAi compounds. Both, RNAi-related "flu-like symptoms" with inflammatory cytokine elevations and the activation of alternative complement pathways have been observed as adverse effects in different clinical trials [187].

Indeed, synthetic, exogenous, RNAi modulators can induce an innate immune response via interaction with RNA-binding proteins such as Toll-like receptors (TLRs) and protein kinase receptors (PKRs) located on the cell surface or within the cell (in the cytoplasm or endosome compartments) [188] (Fig. 4). The innate immunity activation leads to a release of type I interferon (IFN) [189] and pro-inflammatory cytokines, such as interleukin (IL)-6 and tumor necrosis factor-alpha (TNFα).

It is now known that siRNA immune-stimulation can occur in a sequence-independent manner, as in the induction of TLR3 [190] and IFN pathways [189], as well as a sequence-dependent manner, when shRNAs or siRNAs contain a 5′-GUCCUUCAA-3′ motif or similar GU-rich sequences, and meet TLR7 and TLR8 [191, 192]. Moreover, the length of dsRNAs seems to influence the reactivity of the immune system independently of their nucleotide sequence; that is, a longer dsRNA will markedly increase cytokine production.

Adjustments of the RNAi trigger structure by chemical modifications, or reducing the concentration of siRNA preparation to mitigate immunogenicity, have been applied with positive results [193], although occasionally activation of the immune system can be partly beneficial as an adjuvant effect and improve the efficacy of a treatment against viral infections and some cancers.

5.2.4 Delivery Issues

The appropriate delivery of RNAi therapeutics to the target cells represents the major challenge in the development and *in-vivo* application of RNAi-based approaches. Once they have evaded the main systemic barriers (renal clearance, protein binding, and RES filtration), the RNAi particles must pass through the polysaccharide and fibrous network of the ECM to reach the cell surface of the inner tissues. Although the sinusoidal vasculature supporting organs such as liver, spleen, lymph nodes and bone marrow can facilitate the passive passage of particles up to 100 nm in size from the blood, continuous capillaries are more tightly interwoven (brain, muscle) and require active vascular escape strategies. Some of these strategies take advantage of the natural transvascular transportation that occurs within the endothelial cells, by conjugating specific ligands or bifunctional antibodies for receptors and molecular shuttles to RNAi moieties [187]. Subsequently, in order to allow the localization of RNAi compounds within close proximity of the target cells, some systemic delivery systems for the treatment of solid cancers benefit from an enhanced permeability and retention (EPR) effect, but an incorrect cellular uptake can still affect the success of the intervention. For example, RNAi drugs such as naked moieties do not readily cross the cell membranes due to their negative charge and large size [171]. Compared to direct cell penetration by lytic chemistries that risk inducing cell damage, receptor-mediated endocytosis is likely to be the more convenient and efficient pathway to be exploited. ApoE/LDL (low-density lipoprotein)- or folate/folate–receptor interactions in the case of stable nucleic acid particle lipid-based nanoparticle (SNALP LNP; see below) have demonstrated very productive internalizing pathways because of the wide distribution and elevated turnover of the receptors involved. On the other hand, cationic polyplexes/lipoplexes generally enter the cells by non-receptor-mediated physical association with the outer membrane and macropinocytosis [187]. Finally, internalization through endocytosis is normally followed by trapping in the endosome trafficking, culminating in an enzyme-mediated degradation into the lysosome. The RNAi agents must escape into the cytoplasm to reach and be loaded onto the RISC. To this end, delivery strategies including ionizable liposomes (long cationic polymers) are designed to enable endosomal release before maturation into lysosomes. Most of these strategies exploit the acidification that occurs naturally through endosomal maturation because of proton influx. The neutral charge of some compounds at physiologic pH in the blood becomes positive in the endosomes, facilitating disruption of the endosomal membrane. The low pH can also unmask membrano-lytic groups conjugated to the backbone of the RNAi compounds. Moreover, enzymes in the early endosomes can be harnessed to shed any chemistries linked to the RNAi trigger and that might otherwise interfere with RISC incorporation (e.g., ligands) [187].

In conclusion, to allow the implementation of RNAi-based therapy into the clinic, delivery strategies must overcome several technical and biological challenges. Herein are reported disparate systems that have been developed so far in attempts to satisfy not only efficacy but also safety requirements to support and expand the therapeutic application of RNAi molecules.

5.3 RNAi Implementation with Nanomedicine

To summarize, RNAi regulators of downstream gene expression include miRNA mimics and (mostly employed) siRNAs/shRNAs, whose purpose is to restore downregulated tumor suppressor miRNAs. Alternatively, antimiRs can be used for specific miRNA-silencing when the target miRNAs show oncogenic activity in the tumor.

The rational design of RNAi therapeutics for *in-vivo* applications aims to provide safety and effectiveness, by reducing renal clearance and immunogenicity, and optimizing stability, biocompatibility and site-specific delivery of the synthetic agents. These criteria are imperative not only for naked RNAi compounds but also (as explained later) carrier-formulated systems. In the latter case a low tendency for aggregation, which otherwise could induce toxicity and impair delivery activity, solubility in water and endosomal escape ability, are additional features of adequate drug design.

5.3.1 Chemical Modifications

As noted above, chemical modifications to siRNA strands (ribose, nucleotide bases, phosphorous acid, strand termini or backbone of both sense and antisense strands) can significantly improve the performance of naked RNAi molecules, preventing nuclease-based degradation and prolonging the siRNA half-life in the serum from a few minutes to two to three days [175]. Accordingly, conjugation to the strand ends of an siRNA of small molecules (such as cholesterol, folic acid, and galactose), polymers such as polyethylenimine (PEI) and polyethyleneglycol (PEG), antibodies, peptides [e.g., cell-penetrating peptides (CPPs) and transcriptional activator (TAT) peptide] or aptamers is able to increase cellular uptake by promoting caveolin or clathrin (glycoreceptor-mediated) endocytosis [175], thus enhancing the potency of siRNAs. For instance, the intravenous injection of cholesterol-siRNA conjugates targeting apolipoprotein B (ApoB) in transgenic mice led to a successful downregulation of plasma ApoB protein levels, as well as a total reduction of cholesterol levels in the liver and jejunum [194, 195]. Furthermore, siRNA–PEG conjugates have shown active function, prolonged circulation times, and reduced siRNA urinary excretion following intravenous injection [196]. Collectively, these chemical alterations, along with their combination [197], have been shown to ameliorate the *in-vitro* and *in-vivo* stability, bioavailability and duration of action of RNAi molecules, while remaining associated with a discrete effect of gene silencing.

5.3.2 Nanocarriers

Carrier-mediated transport represents a more robust strategy to overcome biological barriers and optimize uptake by the target cells, correctly locating the RNAi compounds into the expected site of action.

Carrier-mediated delivery strategies can be categorized into viral and nonviral vectors. The viral category includes retroviruses, lentiviruses, baculoviruses, adenoviruses, and AAVs, which are mostly known for their high delivery and transfection efficiency, and are distinguished by size and transgene capacity, target, performance, and duration of required expression (Table 3).

Nevertheless, critical limitations that include the immunogenicity and safety profile of viral carriers have boosted the development of alternative technologies based on nonviral vectors which fulfill the safety, because of a low risk of infection, and

Tab. 3 Viral-delivery systems.

Vectors	Description	Target cells	Integration into the host genome	Expression status	
Lentiviral vectors (LVs)	Retroviral-based vector, modified to infect several types of cells. The new generation include replication defective and self-inactivating viruses	Dividing cells	Yes	Transient or stable	**Pros:** Effective in a wide variety of cell lines and tissues
		Nondividing cells			**Cons**: Immunogenicity and oncogenic potentiality following integration
Retrovirus vectors (RVs)	Enveloped viruses containing a single-stranded RNA molecule converted into a double-stranded DNA once in the infected cell	Dividing cells	Yes	Stable	**Pros:** Long-lasting gene expression
					Cons: Limited to dividing cells, receptor-mediated internalization, and oncogenic potentiality due to random integration
Adenoviral vectors (AVs)	Non-enveloped virus (d = 60–90 nm) containing a linear double-stranded DNA genome	Dividing cells	No	Transient	**Pros:** Efficient transfection in several cell types
		Nondividing cells			**Cons:** High immunogenicity, strong inflammatory stimulation, and low loading capacity (7.5 kb)
Adenovirus-associated vectors (AVVs)	Single-stranded DNA human parvoviruses	Dividing cells	No	Transient or stable	**Pros:** Nonpathogenic
		Nondividing cells			**Cons:** Small packaging size (4.5 kb)

Tab. 4 Non-viral delivery systems: NPs-based technologies.

Delivery system	Description	Common formulations	Targeting mechanisms	Pros and Cons
Lipid-based delivery				
Lipid-based nanoparticles (LNPs)/lipoplexes (LPPs)	Lipid-based neutral micelles (d = 100 nm), consisting of lipid bilayer surrounding a hydrophilic aqueous core	• DOPC liposomes • DOPE liposomes • DOTAP liposomes	EPR	*Pros:* • Protection of nucleic acids from enzymatic degradation • Low immunogenicity • Low toxicity • Biocompatibility • Ease of preparation
Cationic LNPs	"Stealth-coated" cationic lipid-based micelles (d = 100–150 nm) able to interact with the negative charge of RNAi moieties. The net surface charge, due to high risk of opsonization and immunogenicity, is generally neutralized by coating with PEG polymer	• SNALP • SLN	MPS	*Pros:* • Increased half-life • Enhanced cellular uptake • Enhanced endosomal escape *Cons:* • Opsonization (SNALP) • Potential immunogenicity (SNALP)
Lipidoids	Non-glycerol cationic lipid-based micelles derived from conjugation with amines to acrylate and acrylamide	• Epoxide-based lipidoids	MPS	*Pros:* • Suitability for siRNAs pool
Bubble liposomes (BLs)	PEG-modified liposomes containing echo-contrast gas, which can function as a gene and siRNA delivery tool with ultrasound exposure	• DOTAP liposomes • DSDAP liposomes	EPR	*Pros:* • Ultrasound-enhanced delivery • Stability and long half-life • Biocompatibility

(*continued overleaf*)

Tab. 4 (Continued)

Delivery system	Description	Common formulations	Targeting mechanisms	Pros and Cons
Polymer-based delivery				
Polyethylenimines (PEIs)	High cationic charge density polymers, presenting linear or branched structures with variable molecular weights (MWs)	• PEG-PEI • DA-PEI	Mucoadhesion	**Pros:** • High transfection efficiency (high MW and branches forms > low MW and linear forms) • Endosomal pH resistance (proton sponge effect). • Physical protection of nucleic acid from degradation. **Cons:** • Proven cytotoxicity: cell apoptosis (high MW and branches forms > low MW and linear forms) • Activation of the complement system
Chitosan	High positive-charged mucopolysaccharide. The degree of deacetylation and MW can influence the capability to interact with nucleic acid. Both low degree of acetylation and MW decrease the interaction with RNAi moieties and the stability of nanocomplexes	Chitosan	Mucoadhesion	**Pros:** • Biodegradability • Low immunogenicity

Poly($_{D,L}$-lactic-co-glycolic acid) (PLGA)	Small lactic and glycolic acids polymer	• PLGA • PLGA-PEG	Mucoadhesion

Pros:
- Biodegradability
- Rapid tissue penetration (small size)
- Physical protection of RNAi molecules from RNase degradation

Dendrimers and dendritic polymers	Highly branched, globular, nanoscaled macromolecules consisting of a core, an interior dendritic backbone (the branches), and an exterior surface with functional groups	• PAMAM • PPI • Polyglycerols • Peptide dendritic polymers	EPR

Pros:
- Solubility, and very low intrinsic viscosities
- High transfection efficiency, and drug delivery (endosomal escape)
- Ease of preparation and commercial availability

Cons:
- Low degradation rate and toxicity

Magnetic nanoparticles (MNPs)	A class of NP consisting of a magnetic core of a pure metal (Co, Mn, Ni, Fe) or their alloys and oxides usually coated by a shell, isolating the core against the environment	Coated MNPs (silane shell, gold shell, polymeric shell, dendrimeric shell): SPIONs, CPMNs	EPR

Pros:
- Tumor targeting by the aid of an external magnetic field
- Visualization opportunity by MRI
- Enhanced cellular uptake

(continued overleaf)

Tab. 4 (Continued)

Delivery system	Description	Common formulations	Targeting mechanisms	Pros and Cons
				Cons: • Tendency to aggregation • Opsonization • Liver, spleen, and brain accumulation
Others				
Carbon nanotubes (CNTs)	Nonspherical nano-cylindrical particles, 50–100 nm in length (d = 1–100 nm)	• SWCNT • MWCNT	MPS	**Pros:** • Ease of functionalization (PEGylation, Tween-20 coating, Ab-conjugation) • Good biodistribution (ease of shape manipulation to escape renal clearance) **Cons:** • Non-biodegradable (pending issue)

DOPC, 1,2-oleoyl-*sn*-glycero-3-phosphatidylcholine; DOPE, dioleoylphosphatidylethanolamine; DOTAP, *N*-[1-(2,3-dioleoyloxy)-propyl]-*N,N,N*-trimethylammoniummethylsulfate; EPR, enhanced permeability and retention; MPS, mononuclear phagocytic system; SNALP, stable lipid-nucleic acid particle; SLNs, solid lipid nanoparticles; PEG, polyethyleneglycol; DSDAP, 1,2-distearoyl-3-dimethylammonium-propane; PEI, polyethyleneimine; DA, deoxycholic acid; PAMAM, polyamidoamine; PPI, poly(propyleneimine); SPIONs, superparamagnetic iron oxide nanoparticles; CPMNs, cell-penetrating magnetic nanoparticles; MRI, magnetic resonance imaging; SWCNTs, one-dimensional single-walled carbon nanotubes; MWCNTs, multiwalled carbon nanotubes.

large-scale production requirements for their use. Promising nanocarrier systems for the clinical applications of RNAi include metallic, cationic lipid-based and polymer-based systems, collectively referred as nanoparticles (NPs). Nanoparticles are nanoscaled devices (generally in the range of 1 to 1000 nm), the size, shape and chemical surface configuration of which can be modified according to the purpose of their use [171]. To date, several types of NPs have been engineered to protect RNAi agents from serum nuclease degradation and the immune system, and to allow their uptake and cytosolic release by target cells. The most extensively studied systems are detailed in Table 4.

Additional intriguing alternatives for the targeted and intracellular delivery of RNAi compounds are represented by extracellular vesicles (EVs) and minicells. The EVs function as natural carriers of cellular components, including miRNAs, which mediate cell-to-cell communication and systemic trafficking of molecules [198]. They include exosomes, activation- or apoptosis-induced microvesicles (MVs)/microparticles, and apoptotic bodies. Although exosomes and microvesicles differ in size and biogenesis, both are produced by different cell types such as dendritic cells, intestinal epithelial cells, T cells and B cells, as well as cancer cells [199]. The detection of miRNAs in body fluids such as plasma, saliva and breast milk (despite the presence of RNase in the circulation) suggests that they can be (selectively) released by the cells into EVs in order to prevent degradation and reach distant tissues. Thus, compared to other conventional devices, EVs possess natural transporting properties that may be relevant when searching for a biocompatible and efficient system of delivery. Moreover, they are amenable to surface and content manipulation, for improving stability, reducing potential immunogenicity, and favoring the release of the RNAi moieties within the cells. An example of this was reported by Ohno and colleagues, who demonstrated that HEK293-derived EVs, expressing GE11 or EGF on their surface, were able to release let-7a to EGFR-expressing xenografted breast cancer tissue in immune-deficient mice. Similar results were obtained when a siRNA was used as alternative to miRNA [198]. Hence, despite some pending clues – such as stability in the circulation, purification protocols, mechanisms of internalization, composition, and immunogenicity in host organisms – the first experimental data seemed to fully encourage the application of EV formulations to the design of delivery systems for RNAi therapeutics.

Minicells are non-living, anucleate, bacteria-derived cells that are approximately 400 nm in diameter and are produced by mutants in which division is uncoupled from chromosomal replication [200]. They can be emptied of their endogenous RNA and protein contents, and packaged with therapeutic amounts of drugs that include chemotherapeutics, siRNAs and shRNAs. Compared to other carriers (such as liposomes), minicells have the properties of a simple drug packaging with different compounds (independent of their structure, charge, hydrophobicity and solubility) following a one-step coincubation, a higher payload capacity ($1-10 \times 10^6$ molecules versus 10 000 within each liposome), an ability to deliver the drug within a tumor cell without leakage during systemic circulation, an ease of conjugation to affinity ligands for tumor targeting, and no off-target effects [201]. Importantly, as minicells are of bacterial origin, purification procedures to eliminate free endotoxin and free bacterial components must be carried out to minimize the potential for toxic side effects that can derive from the inflammatory responses activated by TLRs.

In their earlier report, MacDiarmid et al. successfully demonstrated the *in-vivo* efficacy of minicells as targeting vehicles for an array of chemotherapeutic drugs, including 5-fluoracil, carboplatin, cisplatin, doxorubicin (DOX), irinotecan, paclitaxel (PTX), and vinblastine. Targeting was mediated by bispecific antibody (BsAb) conjugates that recognized both O-polysaccharide and a tumor antigen. As a consequence, targeted minicell-mediated drug delivery was shown to produce a highly significant inhibition and even a regression of tumor growth in different xenograft models of human breast, ovarian and lung cancer, and also leukemia [201]. More recently, the same authors demonstrated the feasibility of using minicells for the siRNA-mediated treatment of mouse xenograft models. In an attempt to treat drug-resistant tumors, they applied a dual sequential treatment that consisted of an initial administration of si/shRNA-containing minicells linked to BsAb that targeted a known drug resistance mechanism. After allowing a sufficient knockdown of the drug resistance-mediating protein, BsAb-targeted minicells packaged with cytotoxic drug were then administered intravenously. This approach, which required a markedly small amount of drug, siRNA and antibody than was needed for a conventional systemic administration of free therapeutics, led to the complete survival of mice with tumor xenografts, with no signs of toxicity [202]. Overall, similar to EVs (caution must be taken here regarding the immunogenic profile and *in-vivo* efficacy of this system), the data acquired supported the potential of a minicell-based anticancer therapeutic for transfer to the clinical setting.

5.4
Administration

Both, local and systemic modalities of RNAi administration have been assessed in animal models, and are currently under investigation in several clinical trials [202]. Local delivery through intraperitoneal, intranasal, intramuscular, intrahepatic, intraretinal, intratesticular or subcutaneous injection, and through inhalation certainly offers the advantage of site-specificity and high bioavailability with small amounts of even naked RNAi drugs in the target tissue [203]. For instance, effective gene silencing and tumor suppression have been achieved by the direct subcutaneous and intraperitoneal injection of siRNA complexed with a commercial transfection reagent to induce specific gene silencing of human papillomavirus (HPV) 16 E6 oncogene in a model of cervical cancer [172]. Nonetheless, such a methodology is manageable only for certain technically accessible sites, such as mucosal and subcutaneous tissues. Indeed, the RNAi molecules must be distributed throughout a sufficiently large interstitial space, as occurs in the skin. Moreover, a local delivery and passive diffusion of the drugs can be achieved if energy-dependent mechanisms such as pressure, local heat gradients or muscle-related pressure gradients are present. More importantly, local delivery cannot be successfully applied to hematological malignancies or solid tumors that have already metastasized.

Alternatively, a less well-documented modality of administration is represented by electropermeabilization (EP), which involves the application of a calibrated electric field pulse to target cells or tissues, which in turn allows the RNAi molecules to rapidly "catch" their targets. To date, this approach has been used for various *in-vitro* and *in-vivo* evaluations of drug and nucleic acid deliveries. Similar to local administration, it allows the cellular uptake to bypass external barriers such as clearance, and internal barriers such as endosomes/lysosome internalization. EP-mediated siRNA or the introduction of LNA/DNA oligomers has been shown not only to be effective in silencing the gene expression [204] but also to allow a homogeneous distribution of the siRNA in the cytoplasm, facilitating its entry to the RISC complex as it is directly localized in the cytoplasm of the recipient cell [205, 206]. Successful results have also been obtained by *in-vivo* direct siRNA delivery into the tissues of different animal models [207]. However, a lack of knowledge of the biophysical mechanisms that support reorganization of the cellular membranes after induced permeabilization can limit the wide application of this methodology.

In this regard, the systemic delivery of RNAi molecules (mainly by intravenous infusion) is the best-suited and multifaceted approach, and disparate formulations and targeting systems have been explored in this regard. Along with the above-mentioned chemical modifications, strategies such as the physical encapsulation of RNAi molecules, including nanocarrier functionalization with target moieties for intracellular diffusion, have been assessed as a means of breaking through the biological barriers and maintaining the targeting of diseased tissues. But even in this context, precautionary measures must be taken into account; indeed, as with other small molecules the intravenous route of administration of RNAi therapeutics can be accompanied by infusion reactions. Experience obtained from clinical trials has shown that these effects are dose-dependent and are especially related to the first infusion; consequently, a slowing rate of infusion, coupled with one-monthly optimized formulations (such as SNALP LNPs), have been shown to make adverse reactions significantly more manageable and clinically endurable [187].

5.5
Targeting

5.5.1 Passive Targeting

The unique abnormal architecture of the tumor vasculature can be exploited to enable nanodrugs to specifically accumulate in tumor tissues. Typically, tumor vessels are poorly aligned, with defective endothelial cells and wide fenestrations among them, together with a compromised lymphatic drainage. The "leaky" vascularization, which relates to the EPR effect, allows the extravasation and accumulation of macromolecules up to 400 nm greater in diameter in tumor tissue than in normal tissue [208]. The coating of cationic or hydrophobic nanocarriers with hydrophilic polymers allows them to elude opsonization and serum nuclease degradation, and also to localize in the tumor vessels. PEGylated liposomes (as SNALPs), which represent advanced liposome vehicles for passive siRNA delivery, show a much longer half-life in plasma than do traditional cationic liposomes [208].

As noted above, once the RNAi carriers access in the ECM they need to reach the target cells. The higher expression level of $\alpha_v\beta_3$ integrin in neoplastic cells compared to normal epithelial cells can be exploited by linking RGD (Arg-Gly-Asp) peptides to the NP surface to facilitate its binding to the tumor ECM, with consequent accumulation [168]. Lastly, because of the high metabolic rate of rapidly growing tumor cells, the microenvironment surrounding the tumor tissue is characterized by a lower pH; hence, some liposome preparations have been designed to be stable at physiological pH (7.4), but will degrade to release drug molecules at an acidic pH [209].

Despite these promising manipulations, the passive delivery process for RNAi therapeutics is difficult to control, and the lack of EPR in certain tumors as well as the heterogeneous permeability of vessels throughout a single tumor make passive delivery difficult to be applied extensively [209].

5.5.2 Active Targeting: Antibodies, Aptamers, Small Molecules

This strategy involves the attachment of affinity ligands (antibodies, peptides, aptamers, or small molecules) to naked RNAi or NPs–RNAi complex molecules [205–213], for molecules that are highly expressed on the cell surface of the diseased tissue. This would improve the bioavailability of the exogenous compounds, minimize nonspecific delivery, and induce cellular uptake with minimal doses of compounds.

Immuno-NP with monoclonal antibodies (mAbs), and engineered fragment antibodies (Fabs) or single chain fragment variable antibodies (scFv) has been widely used, and represents an effective method for targeting specific cells HIV-1 gp160 and HER2 represent peculiar examples. A protamine–antibody fusion protein (F105-P) against the HIV-1 envelope specifically delivered a fluorescein isothiocyanate (FITC)-silencing siRNA to HIV-1 infected and HIV-1 envelope-transfected cells, and induced silencing only in those cells enriched in gp160 (+) [205]. Similarly, a Polo-like Kinase 1 (PLK1)-targeting siRNA was injected intravenously into HER2(+) breast cancer xenografts by using an anti-Her2 ScFv-protamine fusion protein. When using this approach, the complex demonstrated a correct delivery of the siRNA to the HER-positive cells, without releasing to bystander tissues, while the siRNA reached a full PLK1 silencing, culminating in tumor growth retardation, metastasis reduction, and prolonged survival of mice [214].

Aptamers are synthetic, short, single-strand oligonucleotides with several advantages compared to full-size or single-chain antibodies that include a small size that allows for a better tissue internalization, a low immunogenicity and a high stability, as well as an amenability for modification and thus a capability to bind specific and different ligands with secondary or tertiary structures (proteins, receptors) [174, 208]. Aptamers can be easily linked covalently or noncovalently to both naked RNAi molecules (i.e., Aptamer siRNA chimera) and NPs-formulated constructs. For example, in a xenograft model of human prostate cancer, an A10 aptamer–NP complex was shown to recognize and bind the prostate membrane-specific antigen (PMSA) with high specificity, allowing its conjugated siRNAs into the cells to induce BCL2 and PLK1 silencing [215]. In another study, a dual inhibitory anti-gp120 aptamer/siRNA chimera, in which both the aptamer and the siRNA were directed against HIV genes, demonstrated effective silencing.

Cell-penetrating peptides (CPPs) are cationic oligopeptides, derived from natural sequences of viral, insect or mammalian proteins [216], that can be linked to RNAi moieties or NPs by covalent bonds or by electrostatic interactions, in order to accelerate their internalization by increasing cellular plasma membrane permeability. A typical example is the HIV–TAT family, penetratin and the chimeric peptide transportan, and oligoarginines (arginine-repeated unit) [217], whose ability to mediate siRNA silencing has been proven in the central nervous system of mice.

Small molecules can also function as effective ligands for increasing siRNA stability and cellular uptake. Cholesterol and its derivatives increase the binding of RNAi compounds to serum albumin, but reduce the interaction with LDLs and high-density lipoproteins (HDLs), with a consequent improved biodistribution in certain target tissues, such as liver [194, 197, 218]. The folate receptor is a membrane glycoprotein that is highly overexpressed in a number of human tumors, including ovarian, colorectal and breast cancer, but is almost absent in all other normal tissues, except kidney [197, 219]. Similarly, transferrin receptor-mediated delivery has been extensively studied in a variety of targets, including tumors, brain and endothelial cells [220]. Their several advantages include small size, a lack of immunogenicity, convenient availability, and easy chemical conjugation [218, 219].

5.5.3 Alternative Targeting

Alternative approaches include magnetic targeting and ultrasound (US)-enhanced delivery. Magnetic targeting utilizes the magnetic properties of metallic NPs (e.g., silica or gold NPs) to induce their specific accumulation in a target tissue by applying external, local, strong magnetic fields following systemic injection [221]. Once in the circulation, magnetic nanoparticles (MNPs) can be easily monitored using real-time magnetic resonance imaging (MRI) to follow their distribution and cellular uptake [222]. Disparate formulations of siRNA nanovectors exploiting magnetic targeting have been developed for the *in-vivo* imaging of delivered siRNA and gene silencing in tumors, and many others are currently under investigation. For instance, cell-penetrating magnetic nanoparticles (CPMNs) containing CPP-coated MNPs, conjugated with a low-molecular-weight protamine peptide, have consistently been shown to improve transfection efficiency and intracellular siRNA release *in vitro* [223]. In addition, functionalized superparamagnetic iron oxide nanoparticles (SPIONs), combined with chitosan and siRNAs, have been used to couple the SPION suitability for MRI and/or thermal therapy to the ability of chitosan to increase the stability of associated siRNA [224].

A combination of microbubbles and US has been initially proposed as a less-invasive and tissue-specific method for gene delivery. Following infusion, the application of US produces transient changes in the permeability of the cell membrane, due to pressure oscillations, and allows for the site-specific intracellular delivery of combined molecules. Due to the limits of the microbubbles, in terms of their size, stability and targeting functionality, Suzuki and colleagues have developed different formulations of so-called "bubble liposomes"; these are PEG-modified liposomes in which an US imaging gas is trapped and can function as a novel plasmid DNA (pDNA), siRNA and, more recently, as a miRNA delivery tool when used with US exposure both *in vitro* and *in vivo* [225–228].

5.6
RNAi-Based Drug Combination Strategy (coRNAi)

Initial studies in viral infection-related diseases [hepatitis B virus, hepatitis C virus (HCV), human immunodeficiency virus, and other human viruses] have prompted the evaluation of a coRNAi technology for the treatment initially of metabolic or blood disorders, and then of cancer [229].

Combination therapy has been recommended in cancer treatment because it allows the use of lower doses of each drug, compared to single-drug administration, but reaches a higher efficacy based on additive or synergistic anticancer effects. This would prevent the toxic side effects that are often related to common chemotherapeutics, as well as the development of multidrug resistance (MDR) within the tumor. In fact, as human tumors tend to accumulate mutations and acquire resistance towards single-agent-based therapies, the concurrent targeting of multiple key factors and pathways would impair the development of secondary resistance. The *MDR-1* gene is primarily responsible for activating mechanisms for the expulsion of chemotherapeutics upon expression of the components of the P-gp pump (as MRP) within the cell membrane, which is likely to be induced following the repeated administration of drugs to oppose the cytotoxic action [230]. In addition, cancer cells can activate anti-apoptotic proteins such as Bcl2 in further protective attempts at survival over chemotherapy [230].

To date, one of the more widely investigated coRNAi strategies to combat chemoresistance in cancer has employed siRNAs targeting key components of the MDR pathways, in combination with standard chemotherapeutics. As an example, an *in-vitro* sensitization of cervical cancer cells has been obtained following the administration of silica NPs containing both an siRNA against the *MRP* gene and DOX, to compromise assembly of the P-gp pump and allow DOX to cross the cell membrane and inhibit cellular growth [231]. Similarly, cationic micelles carrying anti-BCL2 siRNA and docetaxel (DTX) have been used to target the anti-apoptotic pathway in an *in-vivo* model of human breast cancer [232]. A synergistic effect of RNAi triggers, when combined with PTX, has also been documented [233]; in this case, the coadministration of PTX and an siRNA against midkine (MKsiRNA) in human prostate cancer xenografts led to a significant augmentation of the antitumor therapeutic effect.

A more promising and recently emerging coRNAi approach relies on the coadministration of siRNAs and miRNAs (siRNA/miRNA-based coRNAi). Several studies have reported the successful delivery of both molecules into the tumor tissue, with surprising synergistic anticancer effects. Nishimura *et al.* showed that the dual inhibition of EphA2 (which is overexpressed in ovarian cancer cells and patients) by using an anti-EphA2 siRNA, combined with a miR-520d-3p mimic, targeting EphA2 and with high expression levels, correlated with a better prognosis and significantly augmented the therapeutic effect in inhibiting cell proliferation, migration and invasion *in vitro*, and of tumor growth *in vivo*, compared to monotherapy with each of the drugs [234].

Similar results with lung cancer have confirmed the effectiveness of this approach, as well as the relevance of nanotechnology in supporting such experimental and therapeutic design. Indeed, a new class of polymer-based NPs, 7C1, has been used to encapsulate an equal molar ratio of anti-KRAS siRNAs and miR-34a mimic in a single NP formulation, suggesting an easier and more flexible approach for coRNAi application. Both, *in-vitro* and *in-vivo* assessments confirmed the synergistic anticancer effect derived from the RNAi compound combination, and this was even more enhanced after coadministration with cisplatin (CP) [235].

Although miRNA-based combination therapy is still in its infancy, initial experiments seem to have demonstrated great potential. Kasinski *et al.* assessed the efficacy and safety of miRNA mimics coadministration in an aggressive *KRAS*; *TP53*NSCLC mouse model for replacing the expression of two lung tumor suppressor miRNAs, let-7 and miR-34. The systemic delivery of miR-34a and let-7 by lung-targeting NPs – either neutral lipid emulsion (NLE) or NOV34 (a miRNA delivery agent already undergoing clinical trials) – resulted in tumor growth suppression and a survival advantage [236].

In conclusion, although delivery issues and the risk of reciprocal competition between codelivered RNAi drugs still needs to be fully resolved [237], the abovementioned results and ongoing clinical trials have provided great encouragement for coRNAi translation in the clinical setting of human cancer. In particular, it is a less-toxic and more direct method for targeting multiple biologically pathways that are relevant to the genesis and progression of tumors.

5.7
RNAi in the Clinical Trials

The first RNAi-based clinical trial began in 2004, with an intravitreal injection of naked siRNA targeting the vascular endothelial growth factor (VEGF) pathway in patients with age-related macular degeneration (AMD) and diabetic macular edema (DME). Most of the naked siRNAs were locally administered to treat topical diseases that included AMD, DME and non-arteritic anterior ischemic optic neuropathy (NAION) or virus infections such as respiratory syncytial virus and HCV. The clinical was trial terminated due to low efficacy, off-target effects, or company issues.

Following the application of nanotechnology to the design of siRNA delivery vehicles, new RNAi agents based on systemic delivery entered clinical trials. To date, the terminated siRNA agents in clinical trials include naked delivered bevasiranib, AGN-745, and PF-655 (NCT0030694; NCT00363714; NCT1445899; www:ClinicalTrials.gov), for the treatment of eye-related disorders, and the SNALP-based delivery agent TKM–ApoB (NCT00927459) for cardiovascular diseases. Bevasiranib administration was terminated because of low efficacy at Phase III in a clinical trial to treat for AMD/DME, while AGN-745 was terminated during the treatment of AMD due to off-target effects at Phase II of clinical testing. TKM–ApoB was terminated because only a transient reduction of cholesterol level was observed at Phase I in the clinical study. One common adverse effect among these agents was the activation of TLRs; however, although most terminated RNAi agents were naked siRNAs, those to be administered locally will still have the chance to enter further testing as long as they demonstrate appropriate safety and efficacy.

CALAA-01 (Calando Pharmaceuticals; NCT006889065) has been the first active targeted delivery agent to enter clinical trials for the treatment of solid tumors. The agent consisted of an unmodified siRNA targeting Ribonucleotide Reductase M2 (RRM2) into cyclodextrin-containing polymer (CDP) NPs, decorated with transferrin, and administered intravenously to patients with metastatic melanoma. Despite the promising data obtained, the CALAA-01 trial was recently terminated, without any clear explanations.

Tab. 5 siRNA-based RNAi cancer interventions in clinical trials.

Drug	Formulation	siRNA-target	Cancer disease	Status	Identifier at: ClinicalTrials.gov
CALAA-01	Systemic CDP-NPs	RRM2: cell proliferation	Advanced solid tumors	Phase I: terminated	NCT00689065
ALN-VSP02	Systemic SNALP liposomes (short circulating lipid). coRNAi with two siRNAs	VEGF: angiogenesis; KSP: cell proliferation	Liver cancer and advanced solid tumors with liver metastases	Phase I: completed	NCT01158079
iP siRNA	*Ex vivo* transfection of mature dendritic cells and vaccination transfected mature dendritic cells	LMP2, LMP7, MECL1: proteasome-mediated antigen-presenting processing	Metastatic melanoma	Phase I: completed	NCT00672542
TKM-PLK1	Systemic SNALP liposomes (long-circulating lipid)	PLK1: cell proliferation	Solid tumors, lymphomas, primary and secondary liver metastases	Phase I/II: recruiting	NCT01262235
Atu027	Systemic PEGylated lipoplex (Atuplex)	PKN3: angiogenesis and metastasis	Advanced solid tumors	Phase I: completed	NCT00938574
Atu 027 + Gemcitabine			Advanced pancreatic adenocarcinoma	Phase I: recruiting	NCT01808638
SIG12D LODER	Local LODER polymer (PLGA-based slow-release formulation)	Mutated KRAS	Advanced pancreatic adenocarcinoma	Phase I: ongoing	NCT01188785
				Phase II: not yet recruiting	NCT01676259
EPHARNA siRNA-EphA2-DOPC	Non-PEGylated DOPC liposome	EPHA2 cell proliferation	Advanced solid tumors	Phase I: not yet recruiting	NCT01591356
DCR-MYC	Systemic liposome	MYC: cell proliferation	Solid tumors Multiple myeloma Non-Hodgkins lymphoma	Phase I: recruiting	NCT02110563

Tab. 6 miRNA-based RNAi cancer interventions in clinical trials.

Drug	Formulation	Target	Cancer disease	Status	Identifier at: ClinicalTrials.gov
MRX34	Systemic NLE (neutral lipid emulsion) liposomal miR-34 mimic	BCL2: cell apoptosis	Primary liver cancer or metastatic cancer with liver involvement. lymphoma, leukemia, multiple myeloma	Phase I: recruiting	NCT01829971
SPC3649 (Miravirsen)	Systemic LNA-ASO (antisense oligonucleotide)	miR-122: inhibition of miR-122–HCV complex formation essential to the stability and propagation of HCV RNA	HCV and hepatocellular carcinoma	Phase I/II: completed	NCT01200420
EZN2968 anti-HIF-1α LNA AS-ODN	Systemic LNA-ASO (antisense oligonucleotide)	HIF-1α: hypoxia response	Advanced solid tumors or lymphoma	Phase I: completed	NCT00466583
SPC2996	LNA-ASO	BCL2: cell apoptosis	CLL	Phase I/II: completed	NCT00285103

Atu027 (Silence Therapeutic; NCT00938 574), another siRNA-containing PEG-nanoparticle (AtuPLEX) was the first siRNA to be applied for the inhibition of metastases. No drug-related side effects were observed at Phase I of the clinical trial, in which toxicity was evaluated in 27 of 33 patients with advanced solid tumor. A Phase Ib/IIa trial (NCT01808638), in which patients with locally advanced or metastatic pancreatic adenocarcinoma were recruited has since been undertaken, to evaluate the combination of Atu027 with gemcitabine.

ALN-VSP02 (Alnylam/Tekmira; NCT01158079) is the first dual-targeted RNAi formulation combining two siRNAs into a lipid-NP carrier system (SNALP) targeting both VEGF and kinesin spindle protein (KSP), respectively. A Phase I clinical evaluation (in 28 patients) reported that ALN-VSP02 was well tolerated, with neither hepatotoxicity nor any elevation of complement proteins. Although increases in cytokine levels were observed these were considered due to higher doses, and immediately resolved within 24 h after administration.

On-going and future trials include the TKM-PLK1 compound (NCT01262235) for the treatment of primary liver cancer or liver metastases, and two promising coRNAi trials for siG12D LODER (Silenceed Ltd; NCT01676259), a siRNA targeting KRAS combined with gemcitabine to treat advanced pancreatic cancer, and siRNA-EphA2-DOPC (1,2-oleoyl-sn-glycero-3-phosphatidylcholine) (EPHARNA, MD Anderson Cancer Center; NCT01591356). In this case, nonPEGylated liposomes will be administered to patients with advanced and recurrent solid tumors. Details of other, similar, studies are provided in Tables 5 and 6.

6
Summary

In combining the characterization of RNAi in humans with increasing knowledge of the molecular alterations that support the genesis of human malignancies, RNAi therapeutics represent the most promising approach towards innovative and personalized medical interventions on cancer, and a clear alternative to the limited effects of conventional treatments (e.g., chemotherapy, surgery, or radiotherapy).

During the past decades, gene expression profiling studies of human malignancies have revealed that the global dysregulation of small, noncoding RNAs – that is, miRNAs – which are the genuine users of the endogenous RNAi apparatus for fine-tuning gene expression, has a significant effect on every step of tumorigenesis. Thus, the ability to harness the RNAi machinery by restoring the regulatory function of downregulated miRNAs using synthetic miRNA mimics, siRNAs, viral expression constructs, or inhibiting an oncomiR function by chemically modified antimiR oligonucleotides, miRNA sponges, or miR-masks, offers the chance to target the expression of any gene linked to neoplastic phenomena, including those which have proved so far to be "undruggable."

Although a number of RNAi therapeutics have emerged as putative, effective anticancer drugs, the major drawback in their translation into the clinic remains a lack of safe and efficient delivery systems, and the ability to release the RNAi moieties at a therapeutic concentration specifically to the target cells, without affecting healthy tissues. In this context, during the past few years remarkable achievements have been made by designing NPs-based carriers (cationic lipids/polymers) with high transfection abilities, biocompatibilities, stabilities, and a lack of immunogenicity. Today, encouraging results appear to be emerging from ongoing clinical trials regarding the safety and efficacy of RNAi therapeutics, even when repeated routes of administration are used. However, seminal investigations into the real therapeutic potential of these strategies are still required before entering the clinical setting and supporting patient management.

References

1. Hannon, G.J. (2002) RNA interference. *Nature*, **418**, 244–251.
2. Couzin, J. (2002) Breakthrough of the year. Small RNAs make big splash. *Science*, **298**, 2296–2297.
3. Fire, A., Xu, S., Montgomery, M.K., Kostas, S.A. et al. (1998) Potent and specific genetic interference by double-stranded RNA in *Caenorhabditis elegans*. *Nature*, **391**, 806–811.
4. Elbashir, S.M., Harborth, J., Lendeckel, W., Yalcin, A. et al. (2001) Duplexes of 21-nucleotide RNAs mediate RNA interference in cultured mammalian cells. *Nature*, **411** (6836), 494–498.
5. Elbashir, S.M., Lendeckel, W., and Tuschl, T. (2001) RNA interference is mediated by 21-and 22-nucleotide RNAs. *Genes Dev.*, **15**, 188–200.
6. Jorgensen, R. (1990) Altered gene expression in plants due to trans interactions between homologous genes. *Trends Biotechnol.*, **8**, 340–344.
7. Romano, N. and Macino, G. (1992) Quelling: transient inactivation of gene expression in *Neurospora crassa* by transformation with homologous sequences. *Mol. Microbiol.*, **6**, 3343–3353.
8. Hammond, S.M., Bernstein, E., Beach, D., and Hannon, G.J. (2000) An RNA-directed nuclease mediates posttranscriptional gene silencing in *Drosophila* cells. *Nature*, **404**, 293–296.
9. Fire, A., Albertson, D., Harrison, S.W., and Moerman, D.G. (1991) Production of antisense RNA leads to effective and specific inhibition of gene expression in *C. elegans* muscle. *Development*, **113**, 503–514.
10. Dernburg, A.F., Zalevsky, J., Colaiacovo, M.P., and Villeneuve, A.M. (2000) Transgene-mediated cosuppression in the *C. elegans* germ line. *Genes Dev.*, **14**, 1578–1583.
11. Guo, S. and Kemphues, K.J. (1995) par-1, a gene required for establishing polarity in *C. elegans* embryos, encodes a putative Ser/Thr kinase that is asymmetrically distributed. *Cell*, **81**, 611–620.
12. de Carvalho, F., Gheysen, G., Kushnir, S., Van Montagu, M. et al. (1992) Suppression of beta-1,3-glucanase transgene expression in homozygous plants. *EMBO J.*, **11**, 2595–2602.
13. Waterhouse, P.M., Graham, M.W., and Wang, M.B. (1998) Virus resistance and gene silencing in plants can be induced by simultaneous expression of sense and antisense RNA. *Proc. Natl Acad. Sci. USA*, **A95** (23), 13959–13964.
14. Hamilton, A.J. and Baulcombe, D.C. (1999) A species of small antisense RNA in post-transcriptional gene silencing in plants. *Science*, **286**, 950–952.
15. Bernstein, E., Caudy, A.A., Hammond, S.M., and Hannon, G.J. (2001) Role for a bidentate ribonuclease in the initiation step of RNA interference. *Nature*, **409**, 363–366.
16. Reinhart, B.J., Slack, F.J., Basson, M., Pasquinelli, A.E. et al. (2000) The 21-nucleotide let-7 RNA regulates developmental timing in *Caenorhabditis elegans*. *Nature*, **403**, 901–906.
17. Lee, R.C., Feinbaum, R.L., and Ambros, V. (1993) The *C. elegans* heterochromic gene lin-4 encodes small RNAs with antisense complementarity to lin-14. *Cell*, **75**, 843–854.
18. Hutvágner, G., McLachlan, J., Pasquinelli, A.E., Bálint, E. et al (2001) A cellular function for the RNA-interference enzyme Dicer in the maturation of the let-7 small temporal RNA. *Science*, **293**, 834–838.
19. Grishok, A., Pasquinelli, A.E., Conte, D., Li, N. et al. (2001) Genes and mechanisms related to RNA interference regulate expression of the small temporal RNAs that control *C. elegans* developmental timing. *Cell*, **106**, 23–34.
20. Volpe, T.A., Kidner, C., Hall, I.M., Teng, G. et al. (2002) Regulation of heterochromatic

silencing and histone H3 lysine-9 methylation by RNAi. *Science*, **297**, 1833–1837.
21. Mochizuki, K., Fine, N.A., Fujisawa, T., and Gorovsky, M.A. (2002) Analysis of a piwi-related gene implicates small RNAs in genome rearrangement in *Tetrahymena*. *Cell*, **110**, 689–699.
22. Farazi, T.A., Spitzer, J.I., Morozov, P., and Tuschl, T. (2011) miRNAs in human cancer. *J. Pathol.*, **223** (2), 102–115.
23. Siolas, D., Lerner, C., Burchard, J., Ge, W. *et al.* (2005) Synthetic shRNAs as potent RNAi triggers. *Nat. Biotechnol.*, **23** (2), 227–231.
24. Carmell, M.A. and Hannon, G.J. (2004) RNase III enzymes and the initiation of gene silencing. *Nat. Struct. Mol. Biol.*, **11**, 214–218.
25. Kim, V.N. (2005) MicroRNA biogenesis: coordinated cropping and dicing. *Nat. Rev. Mol. Cell Biol.*, **6**, 376–385.
26. Wilson, R.C. and Doudna, J.A. (2013) Molecular mechanisms of RNA interference. *Annu. Rev. Biophys.*, **42**, 217–239.
27. Denli, A.M., Tops, B.B, Plasterk, R.H, Ketting, R.F. *et al.* (2004) Processing of primary microRNAs by the Microprocessor complex. *Nature*, **432**, 231–235.
28. Han, J., Lee, Y., Yeom, K.H., Kim, Y.K. *et al.* (2004). The Drosha–DGCR8 complex in primary microRNA processing. *Genes Dev.*, **18**, 3016–3027.
29. Landthaler, M., Yalcin, A., and Tuschl, T. (2004) The human Di George syndrome critical region gene 8 and its *D. melanogaster* homolog are required for miRNA biogenesis. *Curr. Biol.*, **14**, 2162–2167.
30. Lund, E., Guttinger, S., Calado, A., Dahlberg, J.E. *et al* (2004) Nuclear export of microRNA precursors. *Science*, **303**, 95–98.
31. Gregory, R.I., Chendrimada, T.P., Cooch, N., and Shiekhattar, R. (2005) Human RISC couples microRNA biogenesis and post-transcriptional gene silencing. *Cell*, **123**, 631–640.
32. Ambros, V., Lee, R.C., Lavanway, A., Williams, P.T. *et al.* (2003) MicroRNAs and other tiny endogenous RNAs in *C. elegans*. *Curr. Biol.*, **13**, 807–818.
33. Aravin, A.A., Lagos-Quintana, M., Yalcin, A., Zavolan, M. *et al* (2003) The small RNA profile during *Drosophila melanogaster* development. *Dev. Cell*, **5**, 337–350.
34. Xie, Z., Johansen, L.K., Gustafson, A.M., Kasschau, K.D. *et al.* (2004) Genetic and functional diversification of small RNA pathways in plants. *PLoS Biol.*, **2**, e104.
35. Vazquez, F., Vaucheret, H., Rajagopalan, R., Lepers, C. *et al.* (2004) Endogenous trans-acting siRNAs regulate the accumulation of *Arabidopsis* mRNAs. *Mol. Cell*, **16**, 69–79.
36. Tomari, Y., Matranga, C., Haley, B., Martinez, N. *et al.* (2004). A protein sensor for siRNA asymmetry. *Science*, **306**, 1377–1380.
37. Carmell, M.A., Xuan, Z., Zhang, M.Q., and Hannon, G.J. (2002) The Argonaute family: tentacles that reach into RNAi, developmental control, stem cell maintenance, and tumorigenesis. *Genes Dev.*, **16**, 2733–2742.
38. Hutvagner, G. and Simard, M.J. (2008) Argonaute proteins: key players in RNA silencing. *Nat. Rev. Mol. Cell Biol.*, **9**, 22–32.
39. Liu, J., Carmell, M.A., Rivas, F.V., Marsden, C.G. *et al.* (2004) Argonaute2 is the catalytic engine of mammalian RNAi. *Science*, **305**, 1437–1441.
40. Caudy, A.A., Ketting, R.F., Hammond, S.M., Denli, A.M. *et al.* (2003) A micrococcal nuclease homologue in RNAi effector complexes. *Nature*, **425**, 411–414.
41. Ishizuka, A., Siomi, M.C., and Siomi, H.A. (2002) *Drosophila* fragile X protein interacts with components of RNAi and ribosomal proteins. *Genes Dev.*, **16**, 2497–2508.
42. Mourelatos, Z., Dostie, J., Paushkin, S., Sharma, A. *et al.* (2002) miRNPs: a novel class of ribonucleoproteins containing numerous microRNAs. *Genes Dev.*, **16**, 720–728.
43. Valencia-Sanchez, M.A., Liu, J., Hannon, G.J., and Parker, R. (2006) Control of translation and mRNA degradation by miRNAs and siRNAs. *Genes Dev.*, **20**, 515–524.
44. Friedman, R.C., Farh, K.K., Burge, C.B., and Bartel, D.P. (2009) Most mammalian mRNAs are conserved targets of microRNAs. *Genome Res.*, **19**, 92–105.
45. Bartel, D.P. (2004) MicroRNAs: genomics, biogenesis, mechanism, and function. *Cell*, **116**, 281–297.
46. Ørom, U.A., Nielsen, F.C., and Lund, A.H. (2008) MicroRNA-10a binds the 5′UTR of ribosomal protein mRNAs and enhances their translation. *Mol. Cell*, **30**, 460–471.

47. Lytle, J.R., Yario, T.A., and Steitz, J.A. (2007) Target mRNAs are repressed as efficiently by microRNA-binding sites in the 5′ UTR as in the 3′ UTR. *Proc. Natl Acad. Sci. USA*, **104**, 9667–9672.
48. Moretti, F., Thermann, R., and Hentze, M.W. (2010) Mechanism of translational regulation by miR-2 from sites in the 5′ untranslated region or the open reading frame. *RNA*, **16**, 2493–2502.
49. Qin, W. Shi, Y., Zhao, B., Yao, C. *et al.* (2010) miR-24 regulates apoptosis by targeting the open reading frame (ORF) region of FAF1 in cancer cells. *PLoS One*, **5**, e9429.
50. Vasudevan, S., Tong, Y., and Steitz, J.A. (2007) Switching from repression to activation: microRNAs can up-regulate translation. *Science*, **318**, 1931–1934.
51. Filipowicz, W., Bhattacharyya, S.N., and Sonenberg, N. (2008) Mechanisms of post-transcriptional regulation by microRNAs: are the answers in sight? *Nature*, **9**, 102–114.
52. Hwang, H.W., Wentzel, E.A., and Mendell, J.T. (2007) A hexanucleotide element directs microRNA nuclear import. *Science*, **315**, 97–100.
53. Valadi, H., Ekström, K., Bossios, A., Sjöstrand, M. *et al.* (2007) Exosome-mediated transfer of mRNAs and microRNAs is a novel mechanism of genetic exchange between cells. *Nat. Cell Biol.*, **9**, 654–659.
54. Berindan-Neagoe, I., Moroig, P.d.C., Pasculli, B., and Calin, G.A. (2014) MicroRNAome genome: a treasure for Cancer Diagnosis and therapy. *CA Cancer J. Clin.*, **64** (5), 311–336.
55. Selcuklu, S.D., Donoghue, M.T., and Spillane, C. (2009) miR-21 as a key regulator of oncogenic processes. *Biochem. Soc. Trans.*, **37**, 918–925.
56. Medina, P.P., Nolde, M., and Slack, F.J. (2010) OncomiR addiction in an in vivo model of miRNA-21-induced pre-B-cell lymphoma. *Nature*, **467**, 86–90.
57. Calin, G.A., Dumitru, C.D., Shimizu, M., Bichi, R. *et al.* (2002) Frequent deletions and down-regulation of micro-RNA genes miR15 and miR16 at 13q14 in chronic lymphocytic leukemia. *Proc. Natl Acad. Sci. USA*, **99**, 15524–15529.
58. Cimmino, A., Calin, G.A., Fabbri, M., Iorio, M.V. *et al.* (2005) miR-15 and miR-16 induce apoptosis by targeting BCL2. *Proc. Natl Acad. Sci. USA*, **102**, 13944–13949.
59. Klein, U., Lia, M., Crespo, M., Siegel, R. *et al.* (2010) The DLEU2/miR-15a/16-1 cluster controls B cell proliferation and its deletion leads to chronic lymphocytic leukemia. *Cancer Cell*, **17**, 28–40.
60. Chen, R.W., Bemis, L.T., Amato, C.M., Myint, H. *et al.* (2008) Truncation of CCND1 mRNA alters miR-16-1 regulation in mantle cell lymphoma. *Blood*, **112**, 822–829.
61. Calin, G.A. and Croce, C.M. (2006) MicroRNA signatures in human cancers. *Nat. Rev. Cancer*, **6**, 857–866.
62. Calin, G.A., Ferracin, M., Cimmino, A., Di Leva, G. *et al.* (2005) A MiRNA signature associated with prognosis and progression in chronic lymphocytic leukemia. *N. Engl. J. Med.*, **353**, 1793–1801.
63. Iorio, M.V., Ferracin, M., Liu, C.G., Veronese, A. *et al.* (2005) MiRNA gene expression deregulation in human breast cancer. *Cancer Res.*, **65**, 7065–7070.
64. He, H., Jazdzewski, K., Li, W., Liyanarachchi, S. *et al.* (2005) The role of miRNA genes in papillary thyroid carcinoma. *Proc. Natl Acad. Sci. USA*, **102**, 19075–19080.
65. Yanaihara, N., Caplen, N., Bowman, E., Seike, M. *et al.* (2006) Unique miRNA molecular profiles in lung cancer diagnosis and prognosis. *Cancer Cell*, **9**, 189–198.
66. Ciafré, S.A., Galardi, S., Mangiola, A., Ferracin, M. *et al.* (2005) Extensive modulation of a set of miRNAs in primary glioblastoma. *Biochem. Biophys. Res. Commun.*, **334**, 1351–1358.
67. Roldo, C., Missiaglia, E., Hagan, J.P., Falconi, M. *et al.* (2006) MiRNA expression abnormalities in pancreatic endocrine and acinar tumors are associated with distinctive pathologic features and clinical behavior. *J. Clin. Oncol.*, **24**, 4677–4684.
68. Murakami, Y., Yasuda, T., Saigo, K., Urashima, T. *et al.* (2006) Comprehensive analysis of miRNA expression patterns in hepatocellular carcinoma and non-tumorous tissues. *Oncogene*, **25**, 2537–2545.
69. Porkka, K.P., Pfeiffer, M.J., Waltering, K.K., Vessella, R.L. *et al.* (2007) MiRNA expression profiling in prostate cancer. *Cancer Res.*, **67**, 6130–6135.

70. Chen, W., Tang, Z., Sun, Y., Zhang, Y. et al. (2012) miRNA expression profile in primary gastric cancers and paired lymph node metastases indicates that miR-10a plays a role in metastasis from primary gastric cancer to lymph nodes. *Exp. Ther. Med.*, **3**, 351–356.

71. Di Leva, G., Garofalo, M., and Croce, C.M. (2014) MicroRNAs in cancer. *Annu. Rev. Pathol.*, **9**, 287–314.

72. Lee, E.J., Baek, M., Gusev, Y., Brackett, D.J *et al.* (2008) Systematic evaluation of microRNA processing patterns in tissues, cell lines, and tumors. *RNA*, **14**, 35–42.

73. Esquela-Kerscher, A. and Slack, F.J. (2006) Oncomirs: miRNAs with a role in cancer. *Nat. Rev. Cancer*, **6**, 259–269.

74. Petrocca, F. and Lieberman, J. (2011) Promise and challenge of RNA interference-based therapy for cancer. *J. Clin. Oncol.*, **29** (6), 747–754.

75. O'Donnell, K.A., Wentzel, E.A., Zeller, K.I., Dang, C.V. *et al.* (2005) c-Myc-regulated microRNAs modulate E2F1 expression. *Nature*, **435**, 839–843.

76. Lopez-Serra, P. and Esteller, M. (2012) DNA methylation-associated silencing of tumor suppressor microRNAs in cancer. *Oncogene*, **31**, 1609–1622.

77. Weber, B., Stresemann, C., Brueckner, B., and Lyko, F. (2007) Methylation of human microRNA genes in normal and neoplastic cells. *Cell Cycle*, **6**, 1001–1005.

78. Saito, Y., Liang, G., Egger, G., Friedman, J.M. *et al* (2006) Specific activation of microRNA-127 with downregulation of the proto-oncogene BCL6 by chromatin-modifying drugs in human cancer cells. *Cancer Cell*, **9**, 435–443.

79. Lehmann, U., Hasemeier, B., Christgen, M., Müller, M. *et al* (2008) Epigenetic inactivation of microRNA gene has-miR- 9-1 in human breast cancer. *J. Pathol.*, **214**, 17–24.

80. Toyota, M. Suzuki, H., Sasaki, Y., Maruyama, R. *et al.* (2008) Epigenetic silencing of microRNA-34b/c and B-cell translocation gene 4 is associated with CpG island methylation in colorectal cancer. *Cancer Res.*, **68**, 4123–4132.

81. Lujambio, A., Ropero, S., Ballestar, E., Fraga, M.F. *et al.* (2007) Genetic unmasking of an epigenetically silenced microRNA in human cancer cells. *Cancer Res.*, **67**, 1424–1429.

82. Iorio, M.V., Visone, R., Di Leva, G., Donati, V. *et al.* (2007) MicroRNA signatures in human ovarian cancer. *Cancer Res.*, **67**, 8699–8707.

83. Scott, G.K., Mattie, M.D., Berger, C.E., Benz, S.C. *et al.* (2006) Rapid alteration of microRNA levels by histone deacetylase inhibition. *Cancer Res.*, **66**, 1277–1281.

84. Saito, Y., Suzuki, H., Tsugawa, H., Nakagawa, I. *et al.* (2009) Chromatin remodeling at Alu repeats by epigenetic treatment activates silenced microRNA-512-5p with downregulation of Mcl-1 in human gastric cancer cells. *Oncogene*, **28**, 2738–2744.

85. Rhodes, L.V., Nitschke, A.M., and Collins-Burow, B.M. (2012) The histone deacetylase inhibitor trichostatin A alters microRNA expression profiles in apoptosis-resistant breast cancer cells. *Oncol. Rep.*, **27**, 10–16.

86. Raveche, E.S., Salerno, E., Scaglione, B.J., Manohar, V. *et al.* (2007) Abnormal microRNA-16 locus with synteny to human 13q14 linked to CLL in NZB mice. *Blood*, **109**, 5079–5086.

87. Hu, Z., Liang, J., Wang, Z., Tian, T. *et al.* (2009) Common genetic variants in pre-microRNAs were associated with increased risk of breast cancer in Chinese women. *Hum. Mutat.*, **30**, 79–84.

88. Jazdzewski, K., Murray, E.L., Franssila, K., Jarzab, B. *et al.* (2008) Common SNP in pre-miR-146a decreases mature miR expression and predisposes to papillary thyroid carcinoma. *Proc. Natl Acad. Sci. USA*, **105**, 7269–7274.

89. Hu, Z., Chen, J., Tian, T., Zhou, X. *et al.* (2008) Genetic variants of miRNA sequences and non-small cell lung cancer survival. *J. Clin. Invest.*, **118**, 2600–2608.

90. Mishra, P.J., Humeniuk, R., Longo-Sorbello, G.S., Banerjee, D. *et al.* (2007) A miR-24 microRNA binding-site polymorphism in dihydrofolate reductase gene leads to methotrexate resistance. *Proc. Natl Acad. Sci. USA*, **104**, 13513–13518.

91. Tan, Z., Randall, G., Fan, J., Camoretti-Mercado, B. *et al.* (2007) Allele-specific targeting of microRNAs to HLA-G and risk of asthma. *Am. J. Hum. Genet.*, **81**, 829–834.

92. Karube, Y., Tanaka, H, Osada, H., Tomida, S. *et al.* (2005) Reduced expression of Dicer

associated with poor prognosis in lung cancer patients. *Cancer Sci.*, **96**, 111–115.
93. Merritt, W.M., Lin, Y.G., Han, L.Y., Kamat, A.A. et al. (2008) Dicer, Drosha, and outcomes in patients with ovarian cancer. *N. Engl. J. Med.*, **359**, 2641–2650.
94. Guo, X., Liao, Q., Chen, P., Li, X. et al. (2012) The microRNA-processing enzymes: Drosha and Dicer can predict prognosis of nasopharyngeal carcinoma. *J. Cancer Res. Clin. Oncol.*, **138**, 49–56.
95. Lin, R.J., Lin, Y.C., Chen, J., Kuo, H.H. et al. (2010) microRNA signature and expression of Dicer and Drosha can predict prognosis and delineate risk groups in neuroblastoma. *Cancer Res.*, **70**, 7841–7850.
96. Grelier, G., Voirin, N., Ay, A.S., Cox, D.G. et al. (2009) Prognostic value of Dicer expression in human breast cancers and association with the mesenchymal phenotype. *Br. J. Cancer*, **101**, 673–683.
97. Faber, C., Jelezcova, E., Chandran, U., Acquafondata, M. et al. (2011) Overexpression of Dicer predicts poor survival in colorectal cancer. *Eur. J. Cancer*, **47**, 1414–1419.
98. Chiosea, S., Jelezcova, E., Chandran, U., Acquafondata, M. et al. (2006) Upregulation of dicer, a component of the MicroRNA machinery, in prostate adenocarcinoma. *Am. J. Pathol.*, **169**, 1812–1820.
99. Muralidhar, B., Winder, D., Murray, M., Palmer, R. et al. (2011) Functional evidence that Drosha overexpression in cervical squamous cell carcinoma affects cell phenotype and microRNA profiles. *J. Pathol.*, **224**, 496–507.
100. Martello, G., Rosato, A., Ferrari, F., Manfrin, A. et al. (2010) A MicroRNA targeting dicer for metastasis control. *Cell*, **141**, 1195–1207.
101. Lambertz, I., Nittner, D., Mestdagh, P., Denecker, G. et al. (2010) Monoallelic but not biallelic loss of Dicer1 promotes tumorigenesis in vivo. *Cell Death Differ.*, **17**, 633–641.
102. Heravi-Moussavi, A., Anglesio, M.S., Cheng, S.W., Senz, J. et al. (2012) Recurrent somatic DICER1 mutations in nonepithelial ovarian cancers. *N. Engl. J. Med.*, **366**, 234–242.
103. Melo, S.A., Ropero, S., Moutinho, C., Aaltonen, L.A. et al. (2009) A TARBP2 mutation in human cancer impairs microRNA processing and DICER1 function. *Nat. Genet.*, **41**, 365–370.
104. Melo, S.A., Moutinho, C., Ropero, S., Calin, G.A. et al. (2010) A genetic defect in Exportin-5 traps precursor microRNAs in the nucleus of cancer cells. *Cancer Cell*, **18**, 303–315.
105. Adams, B.D., Kasinski, A.L., and Slack, F.J. (2014) Aberrant regulation and function of MicroRNAs in cancer. *Curr. Biol.*, **24**, R762–R776.
106. Tang, R., Li, L., Zhu, D., Hou, D. et al. (2012) Mouse miRNA-709 directly regulates miRNA- 15a/16-1 biogenesis at the posttranscriptional level in the nucleus: evidence for a microRNA hierarchy system. *Cell Res.*, **22**, 504–515.
107. Chang, T.C., Wentzel, E.A., Kent, O.A., Ramachandran, K. et al. (2007) Transactivation of miR-34a by p53 broadly influences gene expression and promotes apoptosis. *Mol. Cell*, **26**, 745–752.
108. Piovan, C., Palmieri, D., Di Leva, G., Braccioli, L. et al. (2012) Oncosuppressive role of p53-induced miR-205 in triple negative breast cancer. *Mol. Oncol.*, **6** (4), 458–472.
109. Chang, T.C., Zeitels, L.R., Hwang, H.W., Chivukula, R.R. et al. (2009) Lin-28B transactivation is necessary for Myc mediated let-7 repression and proliferation. *Proc. Natl Acad. Sci. USA*, **106**, 3384–3389.
110. Mott, J.L., Kurita, S., Cazanave, S.C., Bronk, S.F. et al (2010) Transcriptional suppression of mir-29b-1/mir-29a promoter by c-Myc, hedgehog, and NF-kappaB. *J. Cell. Biochem.*, **110**, 1155–1164.
111. Kulshretha, R., Davuluri, R.V., Calin, G.A., and Ivan, M. (2006) A microRNA component of the hypoxic response. *Cell Death Differ.*, **15**, 667–671.
112. Burk, U., Schubert, J., Wellner, U., Schmalhofer, O. et al. (2008) A reciprocal repression between ZEB1 and members of the miR-200 family promotes EMT and invasion in cancer cells. *EMBO Rep.*, **9**, 582–589.
113. Felli, N., Fontana, L., Pelosi, E., Botta, R. et al. (2005) MicroRNAs 221 and 222 inhibit normal erythropoiesis and erythroleukemic cell growth via kit receptor down-modulation. *Proc. Natl Acad. Sci. USA*, **102**, 18081–18086.

114. Garofalo, M., Quintavalle, C., Romano, G., Croce, C.M. et al. (2012) miR221/222 in cancer: their role in tumor progression and response to therapy. Curr. Mol. Med., **12**, 27–33
115. Babar, I.A., Czochor, J., Steinmetz, A., Weidhaas, J.B. et al. (2011) Inhibition of hypoxia-induced miR-155 radiosensitizes hypoxic lung cancer cells. Cancer Biol. Ther., **12**, 908–914.
116. Kong, W., He, L., Coppola, M., Guo, J. et al. (2010) MicroRNA-155 regulates cell survival, growth, and chemosensitivity by targeting FOXO3a in breast cancer. J. Biol. Chem., **285**, 17869–17879.
117. Kong, W., Yang, H., He, L., Zhao, J.J. et al. (2008) MicroRNA-155 is regulated by the transforming growth factor beta/Smad pathway and contributes to epithelial cell plasticity by targeting RhoA. Mol. Cell Biol., **28**, 6773–6784.
118. Mendell, J.T. (2008) miRiad roles for the miR-17-92 cluster in development and disease. Cell, **133**, 217–222.
119. Volinia, S., Calin, G.A., Liu, C.G., Ambs, S. et al. (2006) A microRNA expression signature of human solid tumors defines cancer gene targets. Proc. Natl Acad. Sci. USA, **103**, 2257–2261.
120. Taguchi, A., Yanagisawa, K., Tanaka, M., Cao, K. et al. (2008) Identification of hypoxia-inducible factor-1 alpha as a novel target for miR-17-92 microRNA cluster. Cancer Res., **68**, 5540–5545.
121. Dews, M., Homayouni, A., Yu, D., Murphy, D. et al. (2006) Augmentation of tumor angiogenesis by a Myc-activated microRNA cluster. Nat. Genet., **38**, 1060–1065.
122. Petrocca, F., Visone, R., Onelli, M.R., Shah, M.H. et al. (2008) E2F1-regulated microRNAs impair TGFβ-dependent cell-cycle arrest and apoptosis in gastric cancer. Cancer Cell, **13**, 272–286.
123. Fornari, F., Gramantieri, L., Ferracin, M., Veronese, A. et al. (2008) miR-221 controls CDKN1C/p57 and CDKN1B/p27 expression in human hepatocellular carcinoma. Oncogene, **27**, 5651–5661.
124. Pallante, P., Visone, R., Ferracin, M., Ferraro, A. et al. (2006) MicroRNA deregulation in human thyroid papillary carcinomas. Endocr. Relat. Cancer, **13**, 497–508.
125. Felicetti, F., Errico, M.C., Bottero, L., Segnalini, P. et al. (2008) The promyelocytic leukemia zinc finger–microRNA-221/-222 pathway controls melanoma progression through multiple oncogenic mechanisms. Cancer Res., **68**, 2745–2754.
126. Di Leva, G., Gasparini, P., Piovan, C., Ngankeu, A. et al. (2010) MicroRNA cluster 221-222 and estrogen receptor αinteractions in breast cancer. J. Natl Cancer Inst., **102**, 706–721.
127. Conti, A., Aguennouz, M., La Torre, D., Tomasello, C. et al. (2009) miR-21 and 221 upregulation and miR-181b downregulation in human grade II-IV astrocytic tumors. J. Neurooncol., **93**, 325–332.
128. Garofalo, M., Di Leva, G., Romano, G., Nuovo, G. et al. (2009) miR-221&222 regulate TRAIL resistance and enhance tumorigenicity through PTEN and TIMP3 downregulation. Cancer Cell, **16**, 498–509.
129. Stinson, S., Lackner, M.R., Adai, A.T., Yu, N. et al. (2011) miR-221/222 targeting of trichorhinophalangeal (TRPS1) promotes epithelial-to-mesenchymal transition in breast cancer. Sci. Signal., **4**, ra41.
130. Garofalo, M., Romano, G., Di Leva, G., Nuovo, G. et al. (2011) EGFR and MET receptor tyrosine kinase-altered microRNA expression induces tumorigenesis and gefitinib resistance in lung cancers. Nat. Med., **18**, 74–82.
131. Zhang, J.G., Wang, J.J., Zhao, F., Liu, Q. et al. (2010) MicroRNA-21 (miR-21) represses tumor suppressor PTEN and promotes growth and invasion in non-small cell lung cancer (NSCLC). Clin. Chim. Acta, **411**, 846–852.
132. Zhu, S., Si, M.L., Wu, H., and Mo, Y.Y. (2007) MicroRNA-21 targets the tumor suppressor gene tropomyosin 1 (TPM1). J. Biol. Chem., **282**, 14328–14336.
133. Hatley, M.E., Patrick, D.M., Garcia, M.R., Richardson, J.A. et al. (2010) Modulation of K-Ras-dependent lung tumorigenesis by MicroRNA-21. Cancer Cell, **18**, 282–293.
134. Gabriely, G., Wurdinger, T., Kesari, S., Esau, C.C. et al. (2008) MicroRNA 21 promotes glioma invasion by targeting matrix metalloproteinase regulators. Mol. Cell Biol., **28**, 5369–5380.
135. Seike, M., Goto, A., Okano, T., Bowman, E.D. et al. (2009) MiR-21 is an EGFR-regulated anti-apoptotic factor in lung

cancer in neversmokers. *Proc. Natl Acad. Sci. USA*, **106**, 12085–12090.
136. Chan, J.A., Krichevsky, A.M., and Kosik, K.S. (2005) MicroRNA-21 is an antiapoptotic factor in human glioblastoma cells. *Cancer Res.*, **65**, 6029–6033.
137. Ambs, S., Prueitt, R.L., Yi, M., Hudson, R.S. et al. (2008) Genomic profiling of microRNA and messenger RNA reveals deregulated microRNA expression in prostate cancer. *Cancer Res.*, **68**, 6162–6170.
138. Bandi, N., Zbinden, S., Gugger, M., Arnold, M. et al. (2009) miR-15a and miR-16 are implicated in cell cycle regulation in a Rb-dependent manner and are frequently deleted or down-regulated in non-small cell lung cancer. *Cancer Res.*, **69**, 5553–5559.
139. Roccaro, A.M., Sacco, A., Thompson, B., Leleu, X. et al. (2009) MicroRNAs 15a and 16 regulate tumor proliferation in multiple myeloma. *Blood*, **26**, 6669–6680.
140. Pouliot, L.M., Chen, Y.C., Bai, J., Guha, R. et al. (2012) Cisplatin sensitivity mediated by WEE1 and CHK1 is mediated by miR-155 and the miR-15 family. *Cancer Res.*, **72**, 5945–5955.
141. Johnson, S.M., Grosshans, H., Shingara, J., Byrom, M. et al. (2005) RAS is regulated by the let-7 microRNA family. *Cell*, **120**, 635–647.
142. Chin, L.J., Ratner, E., Leng, S., Zhai, R. et al. (2008) A SNP in a let-7 microRNA complementary site in the KRAS 3′ untranslated region increases non-small cell lung cancer risk. *Cancer Res.*, **68**, 8535–8540.
143. Paranjape, T., Heneghan, H., Lindner, R., Keane, F.K. et al. (2011) A 3′-untranslated region KRAS variant and triple-negative breast cancer: a case-control and genetic analysis. *Lancet Oncol.*, **12**, 377–386.
144. Sampson, V.B., Rong, N.H., Han, J., Yang, Q. et al. (2007) MicroRNA let-7a downregulates MYC and reverts MYC-induced growth in Burkitt lymphoma cells. *Cancer Res.*, **67**, 9762–9770.
145. Johnson, C.D., Esquela-Kerscher, A., Stefani, G., Byrom, M. et al. (2007) The let-7 microRNA represses cell proliferation pathways in human cells. *Cancer Res.*, **67**, 7713–7722.
146. Corney, D.C., Flesken-Nikitin, A., Godwin, A.K., Wang, W. et al (2007) MicroRNA-34b and microRNA-34c are targets of p53 and cooperate in control of cell proliferation and adhesion-independent growth. *Cancer Res.*, **67**, 8433–8438.
147. Gregory, P.A., Bert, A.G., Paterson, E.L., Barry, S.C. et al. (2008) The miR-200 family and miR-205 regulate epithelial to mesenchymal transition by targeting ZEB1 and SIP1. *Nat. Cell Biol.*, **10**, 593–601.
148. Hurteau, G.J., Carlson, J.A., Spivack, S.D., and Brock, G.J. (2007) Overexpression of the microRNA hsa-miR-200c leads to reduced expression of transcription factor 8 and increased expression of E-cadherin. *Cancer Res.*, **67**, 7972–7976.
149. Bracken, C.P., Gregory, P.A., Kolesnikoff, N., Bert et al. (2008) A double-negative feedback loop between ZEB1-SIP1 and the microRNA-200 family regulates epithelial-mesenchymal transition. *Cancer Res.*, **68**, 7846–7854.
150. Iliopoulos, D., Polytarchou, C., Hatziapostolou, M., Kottakis, F. et al. (2009) MicroRNAs differentially regulated by Akt isoforms control EMT and stem cell renewal in cancer cells. *Sci. Signal.*, **2**, ra62.
151. Vrba, L., Jensen, T.J., Garbe, J.C., Heimark, R.L. et al. (2010) Role for DNA methylation in the regulation of miR-200c and miR-141 expression in normal and cancer cells. *PLoS One*, **5**, e8697.
152. Shen, J., Samul, R., Silva, R.L., Akiyama, H. et al. (2006) Suppression of ocular neovascularization with siRNA targeting VEGF receptor 1. *Gene Ther.*, **13** (3), 225–234.
153. Filleur, S., Courtin, A., Ait-Si-Ali, S., Guglielmi, J. et al. (2003) SiRNA-mediated inhibition of vascular endothelial growth factor severely limits tumor resistance to antiangiogenic thrombospondin-1 and slows tumor vascularization and growth. *Cancer Res.*, **63** (14), 3919–3922.
154. Gaudilliere, B., Shi, Y., and Bonni, A. (2002) RNA interference reveals a requirement for myocyte enhancer factor 2A in activity-dependent neuronal survival. *J. Biol. Chem.*, **277** (48), 46442–46446.
155. Miller, V.M., Xia, H., Marrs, G.L., Gouvion, C.M. et al. (2003) Allele-specific silencing of dominant disease genes. *Proc. Natl Acad. Sci. USA*, **100** (12), 7195–7200.
156. Wu, S.Y., Lopez-Berestein, G., Calin, G.A., and Sood, A.K. (2014) RNAi therapies:

drugging the undruggable. *Sci. Transl. Med.*, **6** (240), 240ps7.
157. Rao, D.D., Vorhies, J.S., Senzer, N., and Nemunaitis, J. (2009) siRNA vs. shRNA: similarities and differences. *Adv. Drug Deliv. Rev.*, **61**, 746–759.
158. Lima, W.F., Prakash, T.P., Murray, H.M., Kinberger, G.A. et al. (2012) Single-stranded siRNAs activate RNAi in animals. *Cell*, **150**, 883–894.
159. Maples, P.B., Senzer, N., Kumar, P., Wang, Z. et al. (2010) Enhanced target gene knockdown by a bifunctional shRNA: a novel approach of RNA interference. *Cancer Gene Ther.*, **17** (11), 780–791.
160. Garzon, R., Marcucci, G., and Croce, C.M. (2010) Targeting microRNAs in cancer: rationale, strategies and challenges. *Nat. Rev. Drug Discov.*, **9** (10), 775–789.
161. Kota, J., Chivukula, R.R., O'Donnell, K.A., Wentzel, E.A. et al. (2009) Therapeutic microRNA delivery suppresses tumorigenesis in a murine liver cancer model. *Cell*, **137**, 1005–1017.
162. Stenvang, J., Petri, A., Lindow, M., Obad, S. et al. (2012) Inhibition of microRNA function by antimiR oligonucleotides. *Silence*, **3** (1), 1.
163. Brown, B.D. and Naldini, L. (2009) Exploiting and antagonizing microRNA regulation for therapeutic and experimental applications. *Nat. Rev.*, **10** (8), 578–585.
164. Choi, W.Y., Giraldez, A.J., and Schier, A.F. (2007) Target protectors reveal dampening and balancing of Nodal agonist and antagonist by miR-430. *Science*, **318**, 271–274.
165. Karlas, A., Machuy, N., Shin, Y., Pleissner, K.P. et al. (2010) Genome-wide RNAi screen identifies human host factors crucial for influenza virus replication. *Nature*, **463** (7282), 818–822.
166. Krishnan, M.N., Ng, A., Sukumaran, B., Gilfoy, F.D. et al. (2008) RNA interference screen for human genes associated with West Nile virus infection. *Nature*, **455** (7210), 242–245.
167. Tiedemann, R.E., Zhu, Y.X., Schmidt, J., Shi, C.X. et al. (2012) Identification of molecular vulnerabilities in human multiple myeloma cells by RNA interference lethality screening of the druggable genome. *Cancer Res.*, **72** (3), 757–768.
168. Steckel, M., Molina-Arcas, M., Weigelt, B., Marani, M. et al. (2012) Determination of synthetic lethal interactions in KRAS oncogene-dependent cancer cells reveals novel therapeutic targeting strategies. *Cell Res.*, **22**, 1227–1245.
169. Etemadmoghadam, D., Weir, B.A., Au-Yeung, G., Alsop, K. et al., Australian Ovarian Cancer Study Group (2013) Synthetic lethality between CCNE1 amplification and loss of BRCA1. *Proc. Natl Acad. Sci. USA*, **110**, 19489–19494.
170. Phalon, C., Rao, D.D., and Nemunaitis, J. (2010) Potential use of RNA interference in cancer therapy. *Expert Rev. Mol. Med.*, **12**, e26.
171. Daka, A. and Peer, D. (2012) RNAi-based nanomedicines for targeted personalized therapy. *Adv. Drug Deliv. Rev.*, **64**, 1508–1521.
172. Niu, X.Y., Peng, Z.L., Duan, W.Q., Wang, H. et al. (2006) Inhibition of HPV 16 E6 oncogene expression by RNA interference in vitro and in vivo. *Int. J. Gynecol. Cancer*, **16** (2), 743–751.
173. Guo, P., Coban, O., Snead, N.M., Trebley, J. et al (2010) Engineering RNA for targeted siRNA delivery and medical application. *Adv. Drug Deliv. Rev.*, **62**, 650–666.
174. Chiu, Y.L. and Rana, T.M. (2003) siRNA function in RNAi: a chemical modification analysis. *RNA*, **9** (9), 1034–1048.
175. Morrissey, D.V., Lockridge, J.A., Shaw, L., Blanchard, K. et al. (2005) Potent and persistent in vivo anti-HBV activity of chemically modified siRNAs. *Nat. Biotechnol.*, **23** (8), 1002–1007.
176. Morrissey, D.V., Blanchard, K., Shaw, L., Jensen, K. et al. (2005) Activity of stabilized short interfering RNA in a mouse model of hepatitis B virus replication. *Hepatology*, **41** (6), 1349–1356.
177. Elmen, J., Thonberg, H., Ljungberg, K., Frieden, M. et al. (2005) Locked nucleic acid (LNA) mediated improvements in siRNA stability and functionality. *Nucleic Acids Res.*, **33** (1), 439–447.
178. Bumcrot, D., Manoharan, M., Koteliansky, V., and Sah, D.W. (2006) RNAi therapeutics: a potential new class of pharmaceutical drugs. *Nat. Chem. Biol.*, **2** (12), 711–719.
179. Krutzfeldt, J., Rajewsky, N., Braich, R., Rajeev, K.G et al. (2005) Silencing of microRNAs in vivo with 'antagomiRs'. *Nature*, **438**, 685–689.

180. Jackson, A.L., Burchard, J., Schelter, J., Chau, B.N. et al. (2006) Widespread siRNA "off-target" transcript silencing mediated by seed region sequence complementarity. *RNA*, **12** (7), 1179–1187.
181. Doench, J.G. and Sharp, P.A. (2004) Specificity of microRNA target selection in translational repression. *Genes Dev.*, **18** (5), 504–511.
182. Dykxhoorn, D.M. and Lieberman, J. (2006) Knocking down disease with siRNAs. *Cell*, **126** (2), 231–235.
183. Grimm, D., Streetz, K.L., Jopling, C.L., Storm, T.A. et al (2006) Fatality in mice due to oversaturation of cellular microRNA/short hairpin RNA pathways. *Nature*, **441** (7092), 537–541.
184. Wang, Z., Rao, D.D., Senzer, N., and Nemunaitis, J. (2011) RNA interference and cancer therapy. *Pharm. Res.*, **28**, 2983–2995.
185. Persengiev, S.P., Zhu, X.C., and Green, M.R. (2004) Nonspecific, concentration-dependent stimulation and repression of mammalian gene expression by small interfering RNAs (siRNAs). *RNA*, **10** (1), 12–18.
186. Jackson, A.L., Bartz, S.R., Schelter, J., Kobayashi, S.V. et al. (2003) Expression profiling reveals off-target gene regulation by RNAi. *Nat. Biotechnol.*, **21** (6), 635–637.
187. Haussecker, D. (2014) Current issues of RNAi therapeutics delivery and development. *J. Control. Release*, **195**, 49–54.
188. Agrawal, S. and Kandimalla, E.R. (2004) Antisense and siRNA as agonists of Toll-like receptors. *Nat. Biotechnol.*, **22** (12), 1533–1537.
189. Sledz, C.A., Holko, M., de Veer, M.J., Silverman, R.H. et al (2003) Activation of the interferon system by short-interfering RNAs. *Nat. Cell Biol.*, **5** (9), 834–839.
190. Kariko, K., Bhuyan, P., Capodici, J., and Weissman, D. (2004) Small interfering RNAs mediate sequence-independent gene suppression and induce immune activation by signaling through toll-like receptor. *J. Immunol.*, **172** (11), 6545–6549.
191. Judge, A.D., Sood, V., Shaw, J.R., Fang, D. et al. (2005) Sequence-dependent stimulation of the mammalian innate immune response by synthetic siRNA. *Nat. Biotechnol.*, **23** (4), 457–462.
192. Hornung, V., Nemorin, J.G., Montino, C., Müller, C. et al. (2005) Sequence-specific potent induction of IFN-alpha by short interfering RNA in plasmacytoid dendritic cells through TLR7. *Nat. Med.*, **11** (3), 263–270.
193. Forsbach, A., Nemorin, J.G., Montino, C., Müller, C. et al. (2008) Identification of RNA sequence motifs stimulating sequence-specific TLR8-dependent immune responses. *J. Immunol.*, **180** (6), 3729–3738.
194. Bramsen, J.B. and Kjems, J. (2011) Chemical modification of small interfering RNA. *Methods Mol. Biol.*, **721**, 77–103.
195. Soutschek, J., Akinc, A., Bramlage, B., Charisse, K. et al. (2004) Therapeutic silencing of an endogenous gene by systemic administration of modified siRNAs. *Nature*, **432**, 173–178.
196. Di Figlia, M., Sena-Esteves, M., Chase, K., Sapp, E. et al. (2007) Therapeutic silencing of mutant huntingtin with siRNA attenuates striatal and cortical neuropathology and behavioral deficits. *Proc. Natl Acad. Sci. USA*, **104**, 17204–17209.
197. Iversen, F., Yang, C.X., Dagnaes-Hansen, F., Schaffert, D.H. et al. (2013) Optimized siRNA–PEG conjugates for extended blood circulation and reduced urine excretion in mice. *Theranostics*, **3**, 201–209.
198. Hagiwara, K., Ochiya, T., and Kosaka, N. (2014) A paradigm shift for extracellular vesicles as small RNA carriers: from cellular waste elimination to therapeutic applications. *Drug Deliv. Transl. Res.*, **4** (1), 31–37.
199. György, B., Szabó, T.G., Pásztói, M., Pál, Z. et al. (2011) Membrane vesicles, current state-of-the-art: emerging role of extracellular vesicles. *Cell. Mol. Life Sci.*, **68** (16), 2667–2688.
200. Adler, H.I., Fisher, W.D., Cohen, A., and Hardigree, A.A. (1967) Miniature *Escherichia coli* cells deficient in DNA. *Proc. Natl Acad. Sci. USA*, **57**, 321–326.
201. MacDiarmid, J.A., Mugridge, N.B., Weiss, J.C., Phillips, L. et al. (2007) Bacterially derived 400 nm particles for encapsulation and cancer cell targeting of chemotherapeutics. *Cancer Cell*, **11**, 431–445.
202. Deng, Y., Wang, C.C., Choy, K.W., Du, Q. et al. (2014) Therapeutic potentials of gene silencing by RNA interference: principles, challenges, and new strategies. *Gene*, **538**, 217–227.

203. Wu, S.Y., Yang, X., Gharpure, K.M., Hatakeyama, H. et al. (2014) 2′-OMe-phosphorodithioato-modified siRNAs show increased loading into the RISC complex and enhanced anti-tumour activity. *Nat. Commun.*, **5**, 3459.
204. Wilson, J.A., Jayasena, S., Khvorova, A., Sabatinos, S. et al. (2003) RNA interference blocks gene expression and RNA synthesis from hepatitis C replicons propagated in human liver cells. *Proc. Natl Acad. Sci. USA*, **100**, 2783–2788.
205. Paganin-Gioanni, A., Bellard, E., Escoffre, J.M., Rols, M.P. et al (2011) Direct visualization at the single-cell level of siRNA electrotransfer into cancer cells. *Proc. Natl Acad. Sci. USA*, **108**, 10443–10447.
206. Chabot, S., Orio, J., Castanier, R., Bellard, E. et al. (2012) LNA-based oligonucleotide electrotransfer for miRNA inhibition. *Mol. Ther.*, **20**, 1590–1598.
207. Chabot, S., Teissié, J., and Golzio, M. (2015) Targeted electro-delivery of oligonucleotides for RNA interference: siRNA and antimiR. *Adv. Drug Deliv. Rev.*, **81**, 161–168.
208. Zhou, Y., Zhang, C., and Liang, W. (2014) Development of RNAi technology for targeted therapy – A track of siRNA based agents to RNAi therapeutics. *J. Control. Release*, **10** (193), 270–281.
209. Bamrungsap, S., Zhao, Z., Chen, T., Wang, L. et al. (2012) Nanotechnology in therapeutics: a focus on nanoparticles as a drug delivery system. *Nanomedicine (Lond.)*, **7** (8), 1253–1271.
210. Song, E., Zhu, P., Lee, S.K., Chowdhury, D. et al. (2005) Antibody mediated in vivo delivery of small interfering RNAs via cell-surface receptors. *Nat. Biotechnol.*, **23** (6), 709–717.
211. Arap, W., Pasqualini, R., and Ruoslahti, E. (1998) Cancer treatment by targeted drug delivery to tumor vasculature in a mouse model. *Science*, **279** (5349), 377–380.
212. Wu, Y., Sefah, K., Liu, H., Wang, R. et al. (2010) DNA aptamer-micelle as an efficient detection/delivery vehicle toward cancer cells. *Proc. Natl Acad. Sci. USA*, **107**, 5–10.
213. Leamon, C.P. and Reddy, J.A. (2004) Folate-targeted chemotherapy. *Adv. Drug Deliv. Rev.*, **56**, 1127–1141.
214. Yao, Y.D., Sun, T.M., Huang, S.Y., Dou, S. et al. (2012) Targeted delivery of PLK1–siRNA by ScFv suppresses Her2+ breast cancer growth and metastasis. *Sci. Transl. Med.*, **4**, 130ra148.
215. McNamara, J.O. II, Andrechek, E.R., Wang, Y., Viles, K.D. et al. (2006) Cell type-specific delivery of siRNAs with aptamer-siRNA chimeras. *Nat. Biotechnol.*, **24** (8), 1005–1015.
216. Jones, A.T. and Sayers, E.J. (2012) Cell entry of cell penetrating peptides: tales of tails wagging dogs. *J. Control. Release*, **161**, 582–591.
217. Gupta, B.,. Torchilin, V.P. (2006) Transactivating transcriptional activator-mediated drug delivery. *Expert Opin. Drug Deliv.*, **3** 177–190.
218. Lorenz, C., Hadwiger, P., John, M., Vornlocher, H.P. et al. (2004) Steroid and lipid conjugates of siRNAs to enhance cellular uptake and gene silencing in liver cells. *Bioorg. Med. Chem. Lett.*, **14**, 4975–4977.
219. Yu, B., Zhao, X., Lee, L.J., and Lee, R.J. (2009) Targeted delivery systems for oligonucleotide therapeutics. *AAPS J.*, **11**, 195–203.
220. Wilner, S.E., Wengerter, B., Maier, K., Borba Magalhães, M.L. et al. (2012) An RNA alternative to human transferrin: a new tool for targeting human cells. *Mol. Ther. Nucleic Acids*, **1**, e21.
221. Scherer, F., Anton, M., Schillinger, U., Henke, J. et al. (2002) Magnetofection: enhancing and targeting gene delivery by magnetic force in vitro and in vivo. *Gene Ther.*, **9**, 102–109.
222. Medarova, Z., Pham, W., Farrar, C., Petkova, V. et al. (2007) In vivo imaging of siRNA delivery and silencing in tumors. *Nat. Med.*, **13** (3), 372–377.
223. Qi, L., Wu, L., Zheng, S., Wang, Y. et al. (2012) Cell-penetrating magnetic nanoparticles for highly efficient delivery and intracellular imaging of siRNA. *Biomacromolecules*, **13** (9), 2723–2730.
224. David, S., Marchais, H., Bedin, D., and Chourpa, I. (2014) Modelling the response surface to predict the hydrodynamic diameters of theranostic magnetic siRNA nanovectors. *Int. J. Pharm.*, **478** (1), 409–415.
225. Suzuki, R., Takizawa, T., Negishi, Y., Hagisawa, K. et al. (2007) Gene delivery by combination of novel liposomal bubbles with perfluoropropane and ultrasound. *J. Control. Release*, **117**, 130–136.

226. Suzuki, R., Takizawa, T., Negishi, Y., Utoguchi, N. *et al.* (2008) Tumor specific ultrasound enhanced gene transfer in vivo with novel liposomal bubbles. *J. Control. Release*, **125**, 137–144.
227. Suzuki, R., Takizawa, T., Negishi, Y., Utoguchi, N. *et al.* (2008) Effective gene delivery with novel liposomal bubbles and ultrasonic destruction technology. *Int. J. Pharm.*, **354**, 49–55.
228. Endo-Takahashi, Y., Negishi, Y., Nakamura, A., Ukai, S. *et al.* (2014) Systemic delivery of miR-126 by miRNA-loaded Bubble liposomes for the treatment of hindlimb ischemia. *Sci. Rep.*, **24** (4), 3883.
229. Grimm, D. and Kay, M.A. (2007) Combinatorial RNAi: a winning strategy for the race against evolving targets? *Mol. Ther.*, **15** (5), 878–888.
230. Gandhi, N.S., Tekade, R.K., and Chougule, M.B. (2014) Nanocarrier mediated delivery of siRNA/miRNA in combination with chemotherapeutic agents for cancer therapy: current progress and advances. *J. Control. Release*, **194C**, 238–256.
231. Meng, H., Liong, M., Xia, T., Li, Z. *et al.* (2010) Engineered design of mesoporous silica nanoparticles to deliver doxorubicin and P-glycoprotein siRNA to overcome drug resistance in a cancer cell line. *ACS Nano*, **4**, 4539–4550.
232. Zheng, C., Zheng, M., Gong, P., Deng, J. *et al.* (2013) Polypeptide cationic micelles mediated co-delivery of docetaxel and siRNA for synergistic tumor. *Biomaterials*, **34** (13), 3431–3438.
233. Takei, Y., Kadomatsu, K., Goto, T., and Muramatsu, T. (2006) Combinational antitumor effect of siRNA against midkine and paclitaxel on growth of human prostate cancer xenografts. *Cancer*, **107**, 864–873.
234. Nishimura, M., Jung, E.J., Shah, M.Y., Lu, C. *et al.* (2013) Therapeutic synergy between microRNA and siRNA in ovarian cancer treatment. *Cancer Discov.*, **3** (11), 1302–1315.
235. Xue, W., Dahlman, J.E., Tammela, T., Khan, O.F. *et al.* (2014) Small RNA combination therapy for lung cancer. *Proc. Natl Acad. Sci. USA*, **111** (34), E3553–E3561.
236. Kasinski, A.L., Kelnar, K., Stahlhut, C., Orellana, E. *et al.* (2014) A combinatorial microRNA therapeutics approach to suppressing non-small cell lung cancer. *Oncogene*. doi: 10.1038/onc.2014.282.
237. Castanotto, D., Sakurai, K., Lingeman, R., Li, H. *et al.* (2007) Combinatorial delivery of small interfering RNAs reduces RNAi efficacy by selective incorporation into RISC. *Nucleic Acids Res.*, **35**, 5154–5164.

10
RNA Interference to Treat Virus Infections

Karim Majzoub and Jean-Luc Imler
Institut de Biologie Moléculaire et Cellulaire, UPR9022 du CNRS, 15 Rue Descartes, 67000 Strasbourg, France

1	Introduction	347
2	RNA Interference: an Evolutionarily Conserved Genome Defense Mechanism	347
2.1	Discovery of RNA Interference	349
2.2	Mechanism of RNA Interference	349
2.2.1	AGO Proteins Mediate RNA Silencing	349
2.2.2	MicroRNAs	351
2.2.3	siRNAs	354
2.2.4	Piwi-Dependent Small RNAs	354
3	RNA Interference as a Mechanism of Antiviral Immunity	355
3.1	Viruses and RNAi in Plants	355
3.2	RNA Silencing in *Drosophila*, and Its Antiviral Function	358
3.3	Antiviral RNAi in *C. elegans*	360
3.4	Natural Regulation of Viral Infection by RNAi in Mammals	361
3.4.1	The Mammalian miRNA Pathway and Viruses	361
3.4.2	Natural Control of Viral Infection by siRNAs in Undifferentiated Mammalian Cells	362
4	Therapeutic Applications of RNAi	363
4.1	Programming the Mammalian RISC	363
4.2	Delivering the Promise: *In-Vivo* Administration of siRNAs for Gene Silencing	364
4.2.1	Delivery of Naked RNA	365
4.2.2	Chemical Modifications of siRNAs	365
4.2.3	Encapsulation into Nanoparticles	367
4.2.4	Viral Vectors for DNA-Directed RNAi	369

Translational Medicine: Molecular Pharmacology and Drug Discovery
First Edition. Edited by Robert A. Meyers.
© 2018 Wiley-VCH Verlag GmbH & Co. KGaA. Published 2018 by Wiley-VCH Verlag GmbH & Co. KGaA.

5	**Challenges for Clinical Applications 370**
5.1	Safety Issues 370
5.2	Efficiency of Virus Silencing with Small RNAs 371

6	**Conclusions 372**
	Acknowledgments 372
	References 372

Keywords

Dicer
An enzyme of the RNaseIII family processing double stranded RNA precursors to generate regulatory RNAs (siRNA, miRNA)

Argonaute
An effector enzyme of RNA-induced silencing pathways

MicroRNA (miRNA)
A small regulatory RNA generated from DNA-encoded precursors

Small hairpin RNA (shRNA)
An artificial small RNA designed to mimic a miRNA precursor

Small interfering RNA (siRNA)
A small regulatory RNA generated from long double-stranded RNAs

Antiviral
A molecule interfering with the replication cycle of one or several viruses

Delivery
A set of techniques to bring a therapeutic agent into the target organ or into target cells

Lipofection
A lipid-based technology to deliver molecules through the cell plasma membrane

Immunity
Biological defenses against biological invasion and infection

In spite of its young age, the field of RNA interference has already yielded major advances in the laboratory. This sequence-specific mechanism of gene regulation also holds strong promise for the development of a new generation of drugs, in particular to control the everlasting threat of viral infections. Here, the mechanisms and pathways of RNA interference are reviewed, with emphasis placed on how RNA silencing forms a potent antiviral immune mechanism in plants and invertebrates. The approaches developed to use RNA interference to control viral infections in mammals are then described. Finally, the problems encountered while translating this revolutionary technology into the clinic are presented, and the advances currently developed to overcome these limitations are discussed.

1
Introduction

The discovery of the RNA interference (RNAi) mechanism arguably constituted one of the major breakthroughs in biology of the past 20 years. Indeed, it brought to light an unappreciated role of small RNA molecules in different aspects of genome defense, genome maintenance, and gene regulation. Interestingly, well before RNAi was first described as a natural mechanism of gene regulation (see below), some investigators thought of using Watson–Crick complementary base-pairing to interfere with gene expression. Virologists, when confronted with infectious agents that are rapidly evolving and offering few targets for intervention, pioneered the field and first proposed the blockade of viral replication by using antisense nucleic acids. In two landmark reports made in 1978, Stephenson and Zamecnik described the successful inhibition of Rous sarcoma virus replication by the introduction into a chick embryo fibroblast of a synthetic 13-mer antisense oligonucleotide that was complementary to the viral terminal sequences [1, 2] (Fig. 1). These authors proposed many models, involving either a transcriptional or a translational block of the viral life cycle, to explain the successful control of the virus. Following this initial finding, several groups used artificially introduced antisense RNA transcripts to inhibit gene expression in a variety of model organisms, including the amoeba *Dictyostelium*, plants, the insect *Drosophila*, the amphibian *Xenopus*, and even mammalian cells [3–8] (Fig. 1).

Herein is presented a review of the current state of knowledge on the control of viral infection by RNAi. First, an overview is provided of the RNAi pathways and mechanisms, after which a description is given of how RNAi represents a natural and potent mechanism of host-defense in plants and animals. Finally, the therapeutic applications of RNAi to treat viral infections are described, and the promises – as well as the challenges – of the clinical use of this technology are discussed.

2
RNA Interference: an Evolutionarily Conserved Genome Defense Mechanism

The pioneering observations of Zamecnik and Stephenson on the antisense-mediated suppression of viral replication had potentially far-reaching implications, as indicated in the conclusion of one of their reports in 1978: "*If the primary sequence of a unique small segment of an RNA virus such as influenza, measles, or rabies were known, it would be possible to synthesize and test its oligonucleotide complement as a possible virus inhibitor.*" These words were visionary

348 RNA Interference to Treat Virus Infections

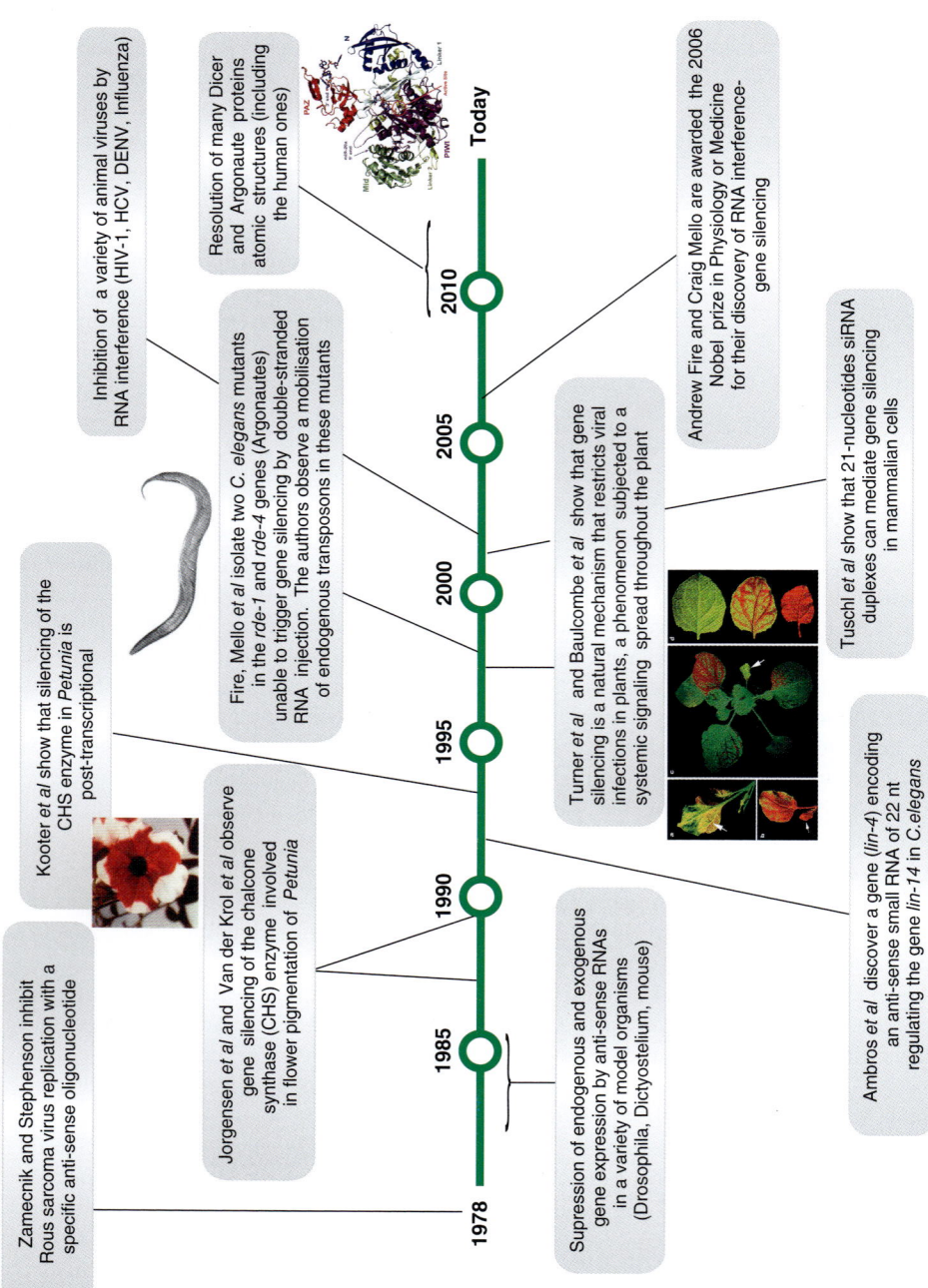

Fig. 1 Timeline of the milestones for the major advances in the RNAi field.

because, more than three decades later, it is now known that in many organisms a sequence-specific mechanism based on base complementarity is present to restrict viral infections. Importantly, the knowledge gained by deciphering these natural RNA-based systems of host-defense allowed the development of new approaches to fight and treat viral infections in human beings.

2.1 Discovery of RNA Interference

The mechanism that subsequently became known as post-transcriptional gene silencing (PTGS), RNA silencing or RNAi was first described in 1990 in plants. Initially, the groups of Jorgensen and Stuitje reported that the overexpression of chalcone synthase (a key enzyme in flavonoid synthesis) in transgenic *Petunia*, designed to manipulate the pigmentation of this ornamental plant, unexpectedly led to a silencing of expression [9, 10] (Fig. 1). Complementary studies in *Petunia* and also in *Nicotiana benthamiana* demonstrated, using nuclear run-on experiments, that the silencing phenomenon did not affect transcription of the target RNAs but rather affected a later step in gene expression [11–13]. Interestingly, towards the end of the 1990s it was realized that PTGS was a natural mechanism regulating gene expression in contexts other than artificial transgenesis in different model organisms (e.g., the plants *N. benthamiana* and *Arabidopsis thaliana*, and the worm *Caenorhabditis elegans*).

In a remarkable series of studies, David Baulcombe and colleagues at Norwich (UK) showed that viruses could initiate an RNA silencing response [14–17] (Fig. 1). Importantly, they showed that this response to infection is maintained by a cellular mechanism that is able to amplify such a response, leading to its being spread throughout the plant (Fig. 1) [18]. The same authors also observed that viruses counter this antiviral mechanism by expressing suppressors of RNA silencing. Taken together, these results provided compelling evidence that PTGS represents a natural mechanism for plant protection against viruses [17, 19]. At the same time, Andy Fire, Craig Mello, and their colleagues showed that double-stranded RNA was able to induce potent and specific genetic interference in the nematode *C. elegans* [20]. The same group also identified two *C. elegans* mutant strains that were resistant to what was referred to as RNAi. The mapping of the genes affected by these mutations (*RNAi-deficient* genes) enabled the first genes (*rde-1* and *rde-4*) regulating this new silencing mechanism to be identified [21]. Subsequently, molecular characterization of the *rde-1* gene revealed that it encodes a member of the piwi/argonaute/zwille protein family, which is conserved from plants to vertebrates. Interestingly, while characterizing the *rde-1* RNAi-deficient lines, Fire and Mello observed a mobilization of endogenous transposons, which led them to propose that transposon silencing was a natural function for RNAi [21] (Fig. 1). Hence, two independent lines of study in plants and nematodes led during the late 1990s to the realization that RNAi provided a sequence-based host-defense mechanism to control the spread of foreign nucleic acids in host cells.

2.2 Mechanism of RNA Interference

2.2.1 AGO Proteins Mediate RNA Silencing

The mechanism of RNA silencing is centered on a family of proteins named Argonautes (AGO) (Fig. 2). These proteins were first identified in plants, and their name reflects the phenotype of a

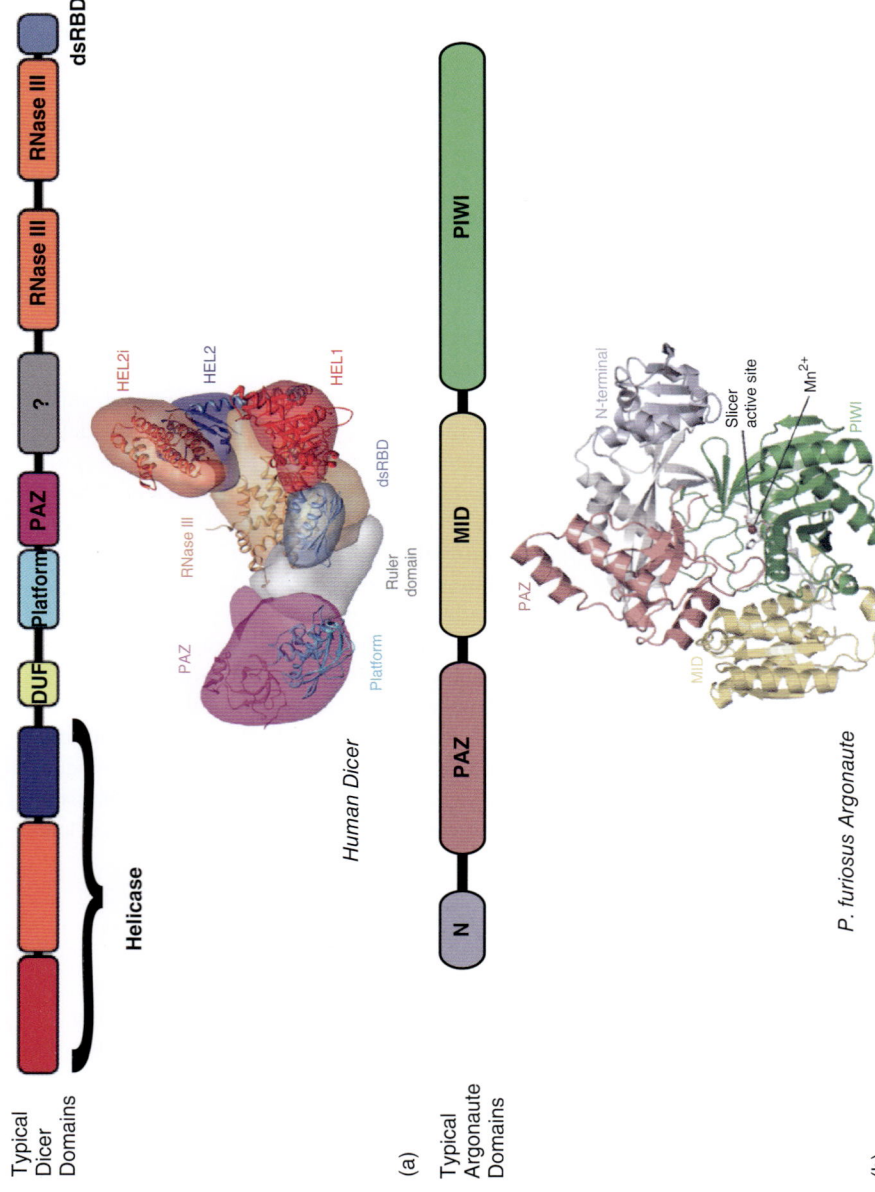

Fig. 2 (a) Domain architecture of a typical Dicer protein and the crystal/cryoelectron microscopy structures of the human Dicer protein. Adapted from Ref. [23]; (b) Domain architecture of a typical Argonaute protein and the crystal structure of *P. furiosus* Argonaute. Adapted from Ref. [24].

mutant *A. thaliana* strain deficient for AGO1, which exhibits tubular-shaped leaves that are evocative of the tentacles of an Argonaut squid [22]. Both, plants and animals encode several AGO proteins, with functional specialization (see below). The AGO proteins associate with small RNAs, which guide them towards complementary RNA targets. Small RNA-loaded Argonaute constitutes the core component of a large molecular complex known as the RNA-induced silencing complex or "RISC." The RISC is the effector complex responsible for RNA silencing, leading ultimately to either endonucleolytic cleavage of the target RNAs (slicer activity), or to the recruitment of cellular factors [GW (glycine/tryptophan-rich) proteins, exonucleases, decapping enzymes] that interfere with the translation of target RNAs (Fig. 3). The 90–110 kDa AGO proteins exhibit two main structural domains, the PAZ domain and the PIWI domain. The PAZ domain (originally identified in the proteins Piwi, AGO, Zwille, hence its name) binds the 3′ end of the guide small RNA, and is essential for RISC activation [25]. The larger Piwi domain adopts an RNaseH fold [26]. In some AGO family members, the residues forming the catalytic triad motif of this domain are present, enabling them to cleave target RNAs [27]. In other family members, these residues have been substituted and the catalytic activity has been lost. These AGOs can, nevertheless, silence gene expression, acting as adaptors or scaffold that recruit factors interfering with RNA translation. Inserted between the PAZ and the Piwi domains, the middle or MID domain is responsible for the accommodation of the 5′ extremity of the guide small RNA [28, 29]. The biological activity of AGO proteins is driven by the small RNAs to which they associate. These small RNAs have multiple origins, and define different pathways of RNA silencing.

2.2.2 MicroRNAs

A first category of small regulatory RNAs are the microRNAs (miRNAs). The latter are endogenous, host-encoded small RNAs that regulate the expression of proteins involved in a variety of basic biological mechanisms including cell cycle, apoptosis, differentiation, and development. The miRNAs are therefore essential for the general physiology of a variety of eukaryotic organisms, from unicellular to plants and animals [30–32].

MiRNAs derive from precursor transcripts called primary miRNAs (pri-miRNAs), which are typically transcribed by RNA polymerase II (RNA Pol II). The pri-miRNA, which has a hairpin structure, is then sequentially cleaved by two RNAse III endonucleases into a ~22 nt miRNA. First, while still in the nucleus, the pri-miRNA is processed into a 60- to 70-nt pre-miRNA by a complex known as the *microprocessor*. The latter comprises the RNaseIII Drosha and a double-stranded RNA (dsRNA) binding domain partner protein named DGCR8 in mammals, or Pasha in flies (see Fig. 3). Processing by Drosha/DGCR8 shortens the stem of the pri-miRNA, and generates a hairpin with a characteristic 3′OH-overhanging dinucleotide. Exportin-5 then mediates the export of this pre-miRNA to the cytoplasm through the nuclear pore, in a Ran-GTP-dependent manner [33]. Once in the cytoplasm, the pre-miRNA undergoes another cleavage to generate the mature miRNA, this time by the action of an enzyme of the Dicer family. Dicer enzymes first recognize defined RNA structures such as double-strandedness and 3′ overhanging dinucleotides; they then cleave at a fixed distance from the free extremities of the pre-miRNA, thus

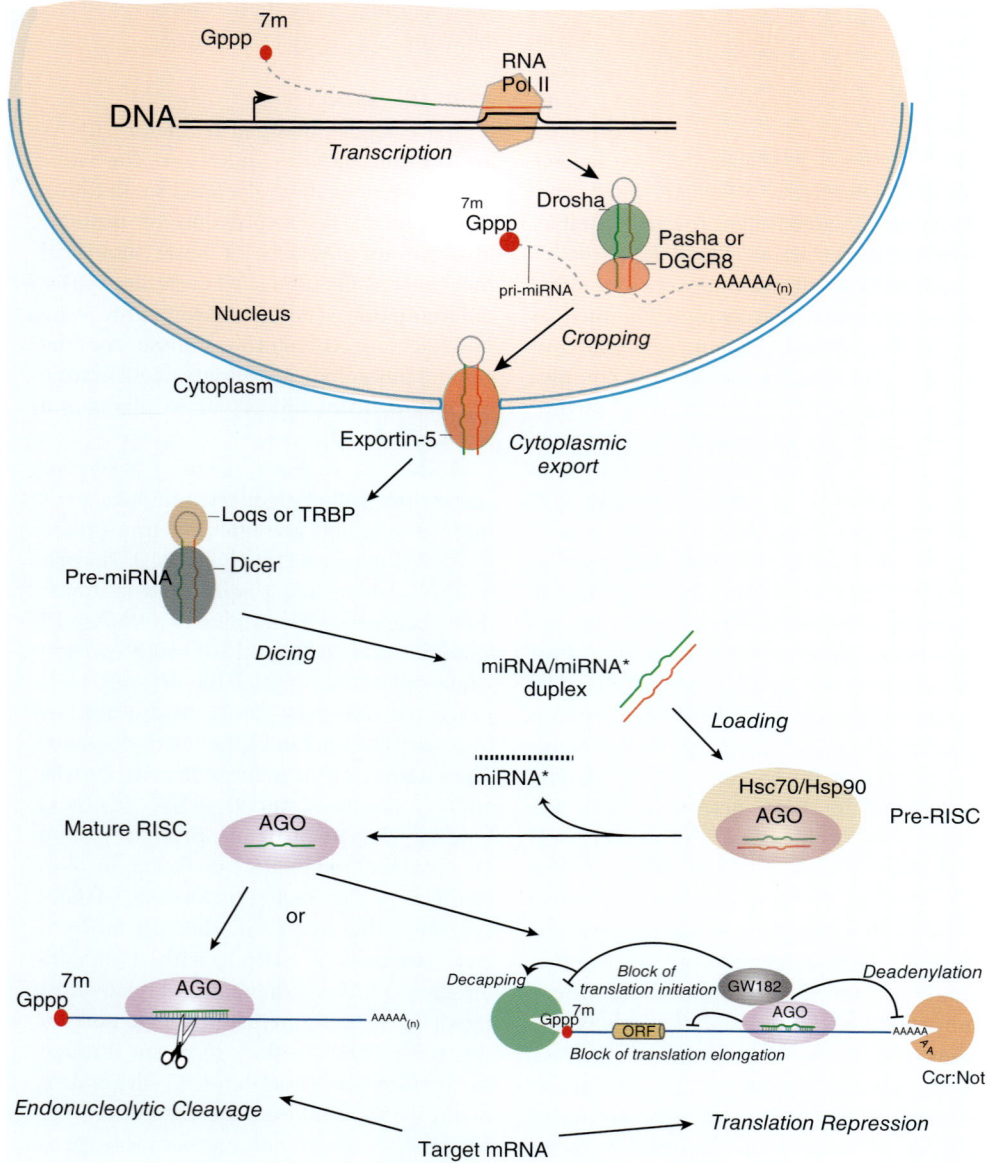

Fig. 3 The canonical microRNA pathway. The pathway is initiated by the Pol II transcription of an RNA molecule that folds back to form a hairpin-like structure. This molecule is processed by nuclear proteins (Drosha and DGCR8), and then exported to the cytoplasm by Exp-5. In the cytoplasm, the small RNA is further processed by Dicer and associated dsRNA binding proteins (e.g., TRBP) and loaded onto the RISC complex. The mature miRISC will target complementary RNA and cause either endonucleolytic cleavage (perfect complementarity) or translational arrest (presence of bulges).

releasing the single-stranded loop. Besides the two RNAse III domains, Dicer enzymes contain a PAZ domain (see Fig. 2), which allows them to accommodate the two 3′-overhanging nucleotides at the extremity of the pre-miRNA. Structural analysis of Dicer indicates that the PAZ domain is located 60–70 Å from the RNaseIII domains, which corresponds to the length of 20–25 nucleotides, thus defining the size of the dicing products [23]. Most Dicer enzymes also contain an N-terminal DExD/H box helicase domain, and two types of double-stranded RNA binding domain (dsRBD) [31].

Cleavage of pre-miRs by Dicer generates a ~22 nt duplex containing two strands, termed *miRNA* and *miRNA**; importantly, this cleavage leaves two 3′OH-overhanging nucleotides. The duplex is then loaded onto a member of the AGO family, where one of the strands (usually the miRNA*) will be discarded while the other functions as the guide strand (Fig. 3). Of note, in mammals the miRNA mir-451 does not require Dicer for its maturation; instead, the pre-miRNA is loaded onto the slicer competent AGO2 and is cleaved to generate an intermediate 3′ end, which is then further trimmed [34].

The loading of the miRNAs onto AGO proteins is assisted by dsRBD proteins associated with Dicer, namely Loquacious (Loqs) in flies and TRBP (transactivating response element RNA-binding protein) and PACT (PKR-activating protein) in mammals. This involves a sorting step, as several AGO family members are coexpressed. For example, in *Drosophila*, both AGO1 and AGO2 are coexpressed in somatic tissues. AGO1 has lost its endonucleolytic RNase-H residues, and mediates translational repression, whereas AGO2 can degrade target RNAs through its slicer activity. Most miRNAs are loaded onto AGO1, because of their thermodynamic features. Two major parameters influence sorting to AGO1: first, the complementarity status of the 9th and 10th base pair of the miRNA/miRNA* duplex, and second, the presence of a uridine (U) at the 5′ end of one strand [35–38]. As a result, a duplex with a 9th and 10th base pair mismatch or G:U wobble and with a 5′ end unpaired U, is preferentially loaded onto AGO1. By contrast, a duplex presenting more base-pair complementarity in general, and complementary sequences at the 9th and 10th position in particular, is more likely loaded onto AGO2. As the majority of miRNAs form bulgy imperfect stems, and harbor at their 5′ end an unpaired U, most of them are loaded onto AGO1. Rare miRNAs forming near-perfect complementary duplexes, like mir-277 in flies, are however loaded onto AGO2.

Mammalian genomes encode four closely related AGO subfamily proteins (AGO1, -2, -3, and -4), which share extensive sequence homology. For example, both human AGO2 and AGO3 share the same conserved catalytic cleavage motif. Curiously however, only AGO2 is capable of cleaving the target mRNA when loaded with a miRNA with perfect complementarity to the target [27, 39, 40]. The non-nucleolytic AGOs, namely AGO1, -3, and -4, presumably play a predominant role in miRNA-mediated translational silencing in mammals [27]. By contrast, mammalian AGO2 is unique and has an essential role in the small interfering RNA (siRNA) -mediated repression (see below).

The sorting decision on the appropriate AGO is an important checkpoint, dictating the subsequent silencing fate. Once loaded onto an AGO, the small RNA duplex is unwound, and one strand is kept and the other discarded. The small RNA/AGO complex can bind to either the untranslated regions (UTRs) or the open reading frame (ORF) of target mRNAs [41]. Most miRNAs pair with their targets through only a limited region at their 5′ end, which is called the *"seed region."* A minimal 7–8 bp region, followed by a bulge and a region of variable complementarity, is characteristic of most miRNA/target interactions [42]. These miRNAs are loaded onto AGO proteins (mostly AGO1 in flies, although some miRNAs can be loaded on AGO2; and AGO1, -2, -3, -4 in mammals), forming the miRISC (Fig. 3) [33]. This complex will inhibit translation of the target RNAs either at the initiation or at the elongation step, upon recruitment of the protein GW182. The results of *in-vitro* experiments performed by different groups agree on a model where GW182 inhibits the initiation of translation at a step after 5′ CAP recognition [43], or recruits deadenylases and exonucleases such as the CCR4/Not complex, resulting in RNA decay [44–46].

2.2.3 siRNAs

A second category of small regulatory RNAs, the siRNAs, participate in RNA silencing. These RNA species are characterized by extensive base-pair complementarity between the two strands of the duplex. The siRNAs are naturally produced by a Dicer protein, and therefore share with miRNAs the 2 nt 3′OH overhangs. siRNAs are processed from long dsRNA precursors that may be of exogenous origin (e.g., viral replication intermediates) or of endogenous origin, such as long hairpins or overlapping convergent transcription units. In both flies and in plants, specialized Dicer proteins are assigned to the production of siRNAs. For example, in the model plant organism *A. thaliana*, besides DCL-1 that produces miRNAs, DCL-2, DCL-3, and DCL-4

produce predominantly 22-, 24-, and 21-nt siRNAs, respectively [47]. Likewise, in insects, Dicer-1 produces 22- to 23-nt miRNAs, whereas Dicer-2 is responsible for the production of 21-nt siRNAs. In the worm *C. elegans*, which has a compact genome, a single Dicer is present which is responsible for the production of both mi- and siRNAs. The single Dicer protein encoded in mammalian genomes, and involved in miRNA production in differentiated cells [33], can also produce siRNAs *in vitro* and in stem cells [48, 49] (see also below).

One hallmark of siRNA-mediated silencing is the extensive complementarity of the siRNAs to their targets. As mentioned above, this feature directs them preferentially to an AGO family member endowed with catalytic activity, such as AGO2. As a result, siRNAs are thought to trigger gene silencing mainly through targeted RNA cleavage.

2.2.4 Piwi-Dependent Small RNAs

The Piwi-associated RNAs (piRNAs) were first discovered in flies, and were proposed to ensure germline stability by repressing transposons. The mutant *piwi* (P element insertion with wimpy testis) was initially characterized for a strong spermatogenesis and male fertility defect [50]. Piwi was found to encode a nucleoplasmic protein with a domain organization similar to AGO proteins. Immunoprecipitation of this protein revealed that it was associated with a class of ~24- to 30-nt small RNAs derived from repeat elements such as retrotransposons [51, 52]. These "repeat associated small interfering" (rasi) RNAs, or piRNAs, are distinct from mi- and siRNAs. First, they associate to a distinct clade of AGO proteins, namely the Piwi-clade, which are mostly expressed in the germline (AGO3, Piwi, and Aubergine in flies, and MIWI, MILI, and MIWI2 in mice). Second, piRNAs do not require Dicer proteins for their production, unlike mi- and siRNAs [52, 53]. Third, their size is more variable (ranging from 24 to 30 nt) and larger than the ~21-nt siRNAs and the ~22-nt miRNAs. As in flies, mutations in Piwi-family members cause defects in germline development and sterility in many organisms [54].

Germline piRNAs are thought to be generated through an amplification loop referred to as the *"ping–pong mechanism."* The model was deduced from the sequences of piRNAs bound to Piwi, Aubergine, and AGO3 in *Drosophila*. piRNAs bound to Piwi and Aubergine are typically antisense to transposon mRNAs, and have a strong preference for a U at the first nucleotide (1U), whereas AGO3 is loaded with piRNAs corresponding to the transposon mRNAs, with a strong preference for an A at position 10 (10A). Remarkably, the first 10 nucleotides of antisense piRNAs are frequently complementary to the sense piRNAs found in AGO3. Similar findings were subsequently reported for MIWI and MILI in mice, which suggested a conservation of the ping–pong mechanism in mammals [55, 56].

The ping–pong model proposes that defective anti-sense transposon transcription results in precursor RNAs from which piRNAs complementary to transposons are made. Long precursors RNAs are cleaved into piRNAs, at one extremity by Aubergine (AUB) and/or PIWI, and by other factors at the other end; these piRNAs then target PIWI and AUB to actively transcribed transposons. The transposons are then cleaved, which results in the formation of sense piRNAs, which are then loaded onto AGO3. These sense piRNAs can then target the long antisense precursors, leading to their cleavage and the formation of more antisense piRNAs that associate with AUB and PIWI. This results in a feedback loop, amplifying the initial pool of piRNAs. Many aspects of the ping–pong model, which is based mostly on bioinformatic analysis, remain nevertheless speculative. Importantly, a distinct piRNA pathway, independent of the ping–pong mechanism and involving only Piwi, operates in germline and somatic follicle cells in *Drosophila* [57, 58].

In summary, three main small RNAs associated pathways can operate to regulate gene expression in eukaryotes. One of these, the miRNA pathway, plays an important role in development and in various aspects of cell physiology by fine-tuning endogenous gene expression. The two other pathways are involved in genome defense against endogenous or exogenous selfish genetic elements in many species. piRNAs have a well-established role in controlling transposons, particularly in germ cells. In addition, they may also contribute to the control of viral infections, as virally derived piRNAs have been detected, although their functional significance has not been investigated [59–61]. By contrast, there is extensive evidence available that the siRNA pathway plays a major role in antiviral host-defense.

3
RNA Interference as a Mechanism of Antiviral Immunity

3.1
Viruses and RNAi in Plants

All four Dicer-like proteins (DCL1–DCL4) found in the model plant *A. thaliana*, contribute to antiviral silencing. Typically, they process long dsRNA molecules of various cellular origins into siRNA populations that are 18–21, 22, 24, and 21 nt in length, respectively [62, 63]. DCL4 is the primary antiviral Dicer against different (+) single-stranded RNA viruses, but it can be replaced by DCL2 when genetically removed [64]. The combined action of both DCL2 and DCL4 in antiviral defense is best illustrated by the hypersusceptibility of plants in which both enzymes have been genetically inactivated. Once a plant is infected, DCL4 and DCL2 each exhibit specific, hierarchical antiviral activities targeting as substrate viral dsRNA intermediates. In addition to virus-derived small interfering RNAs (vsiRNAs), DCL4 generates 21 nt-long siRNAs, known as *trans*-acting (ta)-siRNAs, which mediate the post-transcriptional silencing of some endogenous genes [65] and of transgenes [66]. Similarly, in addition to vsiRNAs, DCL2 synthesizes stress-related natural-antisense-transcript (nat)-siRNAs [67]. A third Dicer family member in plants, DCL3, is involved in the production of secondary 24 nt-long siRNAs important for systemic RNA silencing (see below). In addition, DCL3-dependent siRNAs recruit AGO4 to transcriptionally silence transposons and DNA repeats in *cis*, through chromatin modifications and DNA methylation [68, 69]. Finally, the miRNA producing Dicer, DCL1, also participates in the control of viral infections, either directly through processing viral RNAs with secondary structures reminiscent of miRNAs, such as the 35S leader sequence of the DNA virus Cucumber mosaic virus (CMV) [63], or indirectly through regulating the expression of DCL4 and DCL3 [70].

The genome of *A. thaliana* encodes 10 AGO proteins (AGO1–AGO10), half of which are not well characterized. *AGO1* mutants have the most severe developmental phenotype compared to other AGO mutants, which present only limited or no obvious developmental defects [71]. The AGO1 protein is crucial for the miRNA pathway, and seems to behave as a competent Slicer [72]. AGO1 hypomorphic mutant plants are hypersusceptible to CMV infection, revealing a role for this protein in antiviral defense. Moreover, two viruses encode proteins that counter AGO1 action: the cucumovirus 2b protein inhibits the AGO1 Slicer activity [72], while the polerovirus P0 protein targets AGO1 for degradation [73]. Other AGOs such as AGO7 have been involved in ta-siRNA silencing. The function of the other Argonaute proteins is under investigation. Similar to the DCLs, the AGO proteins of *A. thaliana* are expected to have many functional redundancies and specializations [71]. Of note, whether the slicing of target viral RNA is the only antiviral mechanism employed by plant AGOs remains an open question.

One hallmark of plant antiviral RNAi is the existence of a signal amplification mechanism. At least two small RNA amplification loops, which are not mutually exclusive, are present in *A. thaliana*, both of which involve RNA-dependent RNA polymerases (RdRPs). Primary siRNA originating from the dicing of viral dsRNA intermediates recruits RdRPs to complementary single-stranded (ss) RNA. The siRNAs prime the RdRP on the target ssRNA, which is subsequently copied, generating more dsRNAs, which are diced and lead to the production of secondary siRNAs (Fig. 4). This process requires the RdRP RDR6, the coiled-coil protein SGS3, and the RNA-helicases SDE3 or SDE5, together with AGO1 [47]. Another RdRP-dependent pathway that does not seem to have any antiviral role involves RDR2-dependent amplification of DCL3 products for transposon silencing by DNA

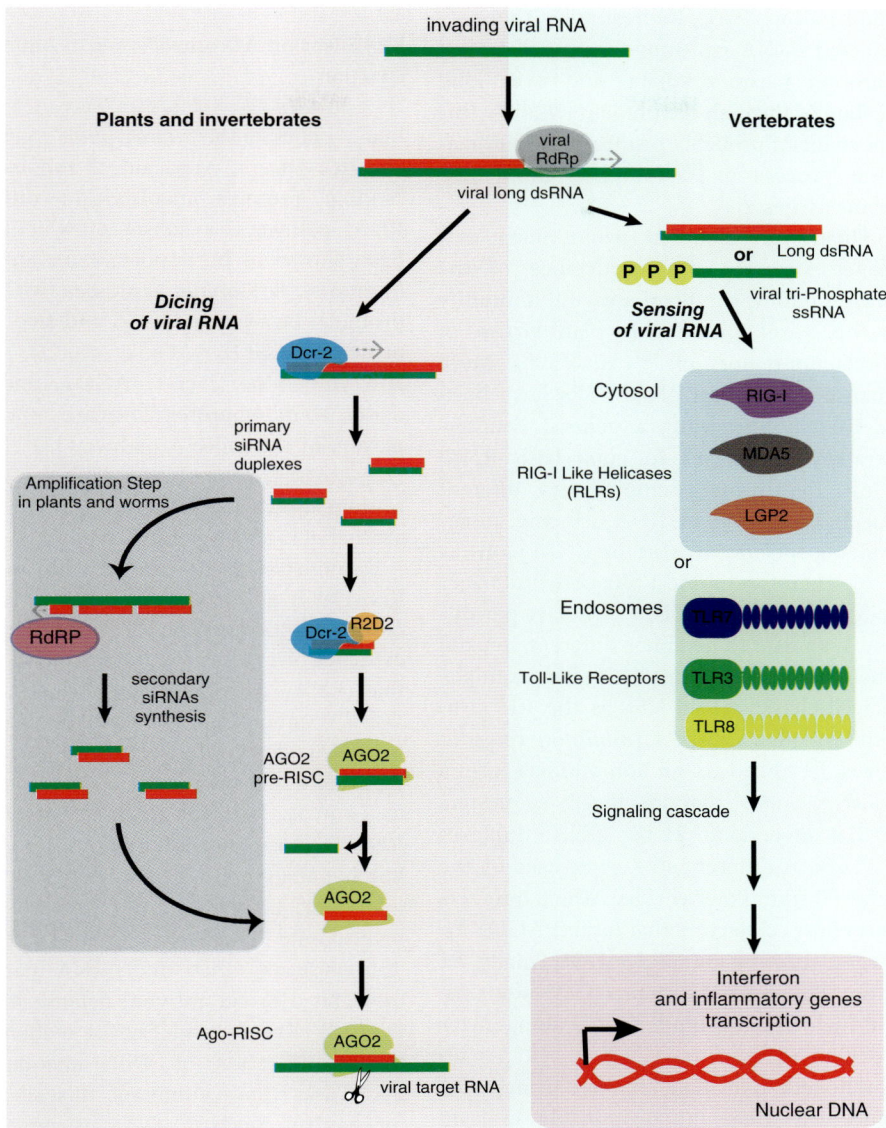

Fig. 4 Antiviral innate immunity pathways in plants, invertebrates, and vertebrates. In plants and invertebrates (left panel), long double-stranded viral replication intermediates are recognized by Dicer proteins (e.g., Dcr-2). Dicer processes these viral long dsRNA onto siRNAs. In plants and worms, host-encoded RNA-dependent RNA polymerases (RdRPs) produce secondary siRNAs, to amplify and spread the response. RISC complexes loaded with virus-derived siRNAs then target the viral RNAs. In vertebrates (right panel), viral double-stranded RNA or specific features of single-stranded RNAs (e.g., uncapped triphosphate extremities) are recognized by RIG-I-like helicases in the cytoplasm, or by TLR-3, -7, or -8 in the endosomes. Activation of these receptors triggers signaling and leads to the transcription of interferon and proinflammatory genes.

methylation [68]. Interestingly, the secondary siRNA can move from cell to cell through plasmodesmata, and over long distances through the phloem [74, 75]. This mechanism probably primes an immunization process in surrounding uninfected plant tissues [64].

Finally, an eloquent evolutionary evidence stressing the importance of the siRNA pathway in plant antiviral immunity is the expression by many plant viruses of viral suppressors of RNAi (VSRs). More than 35 individual plant VSRs have been identified, unraveling a general counter-strategy of viruses to cope with RNAi [76]. These proteins employ very different mechanisms to evade RNA silencing, antagonizing specific steps in the RNAi pathway [76]. Therefore, some VSRs constitute very valuable research tools, which have permitted scientists to decipher different steps in the RNAi mechanism [64]. For example, a well-characterized VSR is the P19 protein encoded by the *Cymbidium* ringspot virus. P19 head-to tail homodimers form a "siRNA clamp" that specifically sequesters DCL4-dependent 21-bp RNA duplexes [77, 78]. A different strategy is used by the VSR P0 from Poleroviridae, which interacts with host cell factors that targets AGO1 for degradation by autophagy [73]. Finally, P1 from Potyviridae contains GW domains, through which it binds to and antagonizes AGO1 [79].

Altogether, the data obtained from plants since the 1990s unquestionably indicate that RNAi is a major antiviral immune pathway in these organisms. This observation led research groups working on other model organisms to consider whether orthologous RNAi components (Dicers, Argonautes) are also antiviral in organisms such as flies, worms, and mammals (Fig. 4).

3.2 RNA Silencing in *Drosophila*, and Its Antiviral Function

Like plants, the *Drosophila* genome encodes Dicers, Argonautes, and dsRNA binding proteins involved in different RNAi pathways, a subset of which have been shown to be important for antiviral immunity. *Drosophila* expresses two Dicer proteins, Dcr-1 and Dcr-2, and five AGO proteins, AGO1, AGO2, AGO3, Piwi, and Aubergine. Dcr-1, AGO1, and the dsRNA-binding protein Loqs are important players in the miRNA pathway [31], while Dcr-2, AGO2, and another dsRNA-binding protein known as *R2D2* constitute the main arsenal of the siRNA pathway.

Demonstration of the critical role of RNAi as a potent antiviral mechanism in *Drosophila* is based on three main lines of evidence. First, genetic experiments show that RNAi pathway mutants ($AGO2^{-/-}$, $Dcr\text{-}2^{-/-}$, and $R2D2^{-/-}$) are hypersensitive to RNA virus infections, and succumb with increased viral loads when compared to controls [80–82]. *Dicer-2*, *AGO2*, or *R2D2* homozygous null mutants are viable, but hypersusceptible to a variety of viral infections, including the DNA virus, invertebrate iridescent virus 6 (IIV-6). This indicates that the siRNA pathway mediates a broad antiviral defense in flies [83, 84]. The second line of evidence is the identification of VSRs encoded by fly viruses that counteract this silencing mechanism. Several VSRs have now been identified in insect RNA viruses [85–87], and these are important contributors of virulence as their ectopic expression in Sindbis virus (SINV) increases the lethality [86–88] while, most importantly, their inactivation renders the virus harmless for the host [89, 90]. Two of these, B2 from Flock house virus (FHV) and 1A from

Drosophila C virus (DCV), interact with dsRNA, preventing the dsRNA recognition and cleavage by Dicer-2. Unlike DCV-1A, which only binds long dsRNAs, FHV-B2 can interact with both long dsRNA and siRNAs. Thus, FHV-B2 can potentially inhibit both the dicing and the loading steps, whereas DCV-1A only blocks the dicing of long dsRNAs. This may partly explain the stronger suppression of RNAi exerted by B2 compared to DCV-1A [91, 85]. A third example of VSR is the protein 1A from cricket paralysis virus (CrPV), which antagonizes AGO2 [86]. Likewise, Nora virus encodes a VSR that binds to AGO2 and antagonizes its slicing activity [87]. Interestingly, the existence of VSRs in insect viruses targeting key players of the siRNA pathway is indirectly supported by the rapid evolution of these host genes. Indeed, *Dicer-2*, *AGO2*, and *R2D2* are among the 3% fastest evolving genes in *Drosophila*, which contrasts with the slow evolution of their paralogs from the miRNA pathway *Dicer-1*, *AGO1*, and *Loqs* [92]. The third and final line of evidence for the antiviral function of RNAi is the presence of siRNAs of viral origin (vsiRNAs) in infected cells and flies [93, 94]. High-throughput sequencing of small RNA libraries from virus-infected cells or flies detected the presence of Dcr-2 derived 21 nt-siRNA [90, 93–97]. As in plants, these siRNAs are able to confer specific resistance to viral replication. Indeed, flies carrying a transgene directing the expression of FHV dsRNA are protected against a challenge by FHV, but not by DCV [81], while the injection of dsRNA into flies can immunize them against viral infection [98] (see below). In a natural infectious context, several lines of evidence indicate that these vsiRNAs are produced by Dcr-2, and originate from long dsRNA viral replication intermediates. First, high-throughput sequencing of small RNAs produced during the course of a viral infection in flies and mosquitoes confirms that the vsiRNAs have a size of 21 nt, as would be expected for products of the Dicer-2 enzyme [88, 93, 94]. Second, the production of vsiRNAs is strongly reduced or abolished in *Dicer-2*$^{-/-}$ mutant flies. Third, for many viruses, the vsiRNAs cover the whole length of the viral genome, and the ratio between the number of siRNAs matching the (+) strand and the (−) strand of the genome is close to 1 [93, 94, 99].

VsiRNAs form perfectly complimentary duplexes, and the Dcr-2/R2D2 heterodimer assists their preferential loading onto AGO2 [100–102]. Once loaded onto AGO2, these vsiRNAs are 2′ O-methylated at their 3′ end by the methyltransferase Hen1 [93]; this methylation is thought to protect the vsiRNAs from mechanisms such as tailing and trimming, which are responsible for recycling the miRNAs [103]. The vsiRNAs then guide the siRISC complex towards target viral RNA molecules, thus enabling AGO2 to cleave the viral RNA through its slicer activity. Interestingly, this mechanism may also be used to immunize cells that have not yet been infected. Indeed, *Drosophila* cells in culture can take up exogenous dsRNA molecules, through scavenger receptor-mediated endocytosis [104]. Viral dsRNA molecules released following the lysis of infected cells can be taken up by noninfected cells and induce RNAi, protecting them against future challenges with the virus [98]. This process does not seem to involve an amplification step, since the *Drosophila* genome does not encode RdRPs and is therefore different from the systemic RNAi described above for plants. Intriguingly, however, a recent study has shown that some parts of the RNA viruses genome can be retrotranscribed to DNA by an unknown mechanism [105], raising the possibility that a DNA-dependent amplification of vsiRNAs operates in antiviral immunity in *Drosophila* [106].

3.3
Antiviral RNAi in *C. elegans*

Although the first discovered RNAi pathway enzyme and the first microRNA (*lin-4*) were discovered in *C. elegans* [21, 30], the involvement of RNAi in antiviral defense against a natural worm virus has not been proven until very recently [107]. The results of previous studies hinted strongly at a role for RNAi in the control of viral replication in *C. elegans* RNAi. The latter studies made use of a transgenic FHV replicon system, mimicking a positive-strand RNA virus replication. The replicon comprised the coding regions of the FHV RdRP and green fluorescent protein (GFP) and, when expressed, the RdRP recognized its viral RNA template and replicated it. Components of the RNAi pathway, namely DICER and RDE-1, were shown to be required to limit the accumulation of GFP [108]. Similarly, vesicular stomatitis virus (VSV) was shown to induce antiviral silencing in adult worms and embryonic cells, respectively, and the detection of VSV-specific viRNAs was reported [109, 110].

The recent isolation of a worm virus named Orsay, which showed a ssRNA bipartite genome of (+) polarity, constituted a valuable asset to testing the susceptibility to natural infection of mutants from the RNAi pathways. Orsay virus can naturally infect wild and laboratory *C. elegans* species [107]. Two lines of evidence have suggested involvement of the worm's RNAi pathway in the defense against Orsay virus infection. First, worms mutant for the AGO protein RDE-1 have approximately 100-fold more viral copies than the wild-type (WT) strain.

(Of note, RDE-1 is essential for the production of secondary siRNAs in the worm.) Second, in infected N2 animals, both primary (~23 nt) and secondary (~22 nt) virus-derived siRNA were detected [107]. The latter were particularly abundant, and are produced by the host-encoded RdRP RRF-1. These findings, in addition to the equivalent sensitivity to Orsay virus infection of mutants for RDE-1 and RRF-1, indicated that both primary and secondary siRNAs are important for antiviral defense in *C. elegans* [111]. The only Dicer gene present in *C. elegans* is believed to mediate the production of primary vsiRNAs, in addition to its role in miRNA production [109]. This dual function of the single worm Dicer is controlled by members of the Dicer-related helicases (DRHs).

Besides DICER, the genome of *C. elegans* encodes three Dicer-related helicase proteins (DRH-1, -2, and -3). These molecules share with Dicer a highly conserved DExD/H box helicase domain, and function in the RNAi silencing pathway. Two of them have been molecularly characterized. DRH-3 is a core component of a cellular RdRP complex that produces secondary siRNAs [112], and may be also involved in the release of siRNA duplexes from Dicer [113]. By contrast, DRH-1 acts upstream of Dicer and is believed to mediate the sensing of viral RNAs [111, 114, 115]. The role of DRH-2 is molecularly less clearly defined, but this factor appears to play a negative regulatory role in antiviral defenses [115].

Interestingly, the worm DRHs share extensive sequence homology with cytoplasmic mammalian viral sensors, the RIG-I like Receptors (RLRs) [116]; this family comprises three members, RIG-I, MDA5, and the less-characterized LGP2 (Fig. 4). The homology covers not only the DExD/H box helicase domain (see Fig. 2) but also the C-terminal domain, which is important for viral RNA recognition in mammals. The biological significance of these homologies is best illustrated by recent experiments which showed that the helicase and C-terminal domains of DRH-1 can be functionally substituted by the corresponding domains of the human RIG-I protein [114]. In mammals, RLRs initiate a signaling cascade through their N-terminal CARD domains, leading to antiviral interferon (IFN) synthesis. DRHs in *C. elegans* do not contain CARD domains; however, the N-terminal domain of DRH-1 is indispensable for antiviral RNAi [114], which suggests that the C-terminal domains of DRH/RLRs carry similar evolutionarily ancient functions in viral detection in worms and mammals, whereas their divergent N-terminal domains are responsible for the activation of distinct antiviral pathways (RNAi in worms versus IFN synthesis in mammals). This intriguing similarity also raises the question as to whether RNAi may still be part of the arsenal of antiviral defenses in mammals.

3.4
Natural Regulation of Viral Infection by RNAi in Mammals

The vertebrate genome, and more particularly the mammalian genome, encode only one Dicer protein. This molecule is involved in the miRNA pathway, which carries out essential development functions and, as a result, a null homozygous mutation of Dicer in mice will be embryonic lethal. Viable mice with hypomorphic mutations in this gene exist and can be used to assess the role of Dicer in antiviral immunity [117]. The mammalian genome also encodes four Argonaute proteins (AGO1–AGO4). The involvement of AGO2 (the only mammalian AGO with a Slicer activity) in the miRNA pathway has also been proven [27]. Although the other mammalian AGOs have not been well characterized to date, they are the subject of intense current investigation. Indeed, the question of whether some of these AGOs participate in antiviral immunity in mammals has been a matter of interest and/or controversy for many biologists during the past decade.

3.4.1 The Mammalian miRNA Pathway and Viruses

Undoubtedly, the miRNA pathway is involved in the life cycle of some mammalian viruses. Examples of virus regulation by miRNAs involve: viral miRNAs fine-tuning viral gene expression during infection; viral miRNAs interfering with host gene expression to favor the virus replication cycle or escape detection by the immune system; and cellular miRNAs regulating viral replication, either indirectly through their effect on cellular genes or directly upon targeting viral RNAs [118]. For instance, direct viral targeting is exemplified by mir-32, which targets a sequence in the genome of the primate foamy virus (PFV-1), and miR24 and miR93 which target the RNA genome of VSV [117, 119]. Examples of cellular miRNAs, targeting cellular cofactors contributing to viral replication, include mir-17-5p and mir-20a, which target the gene encoding the histone acetyltransferase P300/CBP-associated factor (PCAF), which is an important cofactor of the human immunedeficiency (HIV)-1 transactivator Tat. Interestingly, HIV-1 actively suppresses the expression of these two miRNAs [120].

Similarly, many DNA viruses, and especially those from the Herpesviridae family, encode miRNAs that target cellular genes, helping the virus in its replication strategy [121, 122]. For instance, viruses such as Kaposi's sarcoma-associated herpesvirus

(KSHV) and Epstein–Barr virus (EBV), encode miRNAs that can play a subtle role in preventing apoptosis by targeting pro-apoptotic host genes, thus extending the longevity of infected cells [123]. In another case, the KSHV-encoded miRNAs miR-K12-9-5p and miR-K12-7-5p were seen to regulate the latent state by targeting the transcript of the master lytic switch protein (Replication Transactivation Activator; RTA), which controls viral reactivation [124–126]. Another function of virally encoded miRNAs is the promotion of persistent viral infections, through evasion of the immune system. For example, the deletion of two murine cytomegalovirus (MCMV) miRNAs resulted in reduced viral titers in the salivary glands of mice, in both a natural killer (NK)-cell and a CD4+ T-cell-dependent manner [127].

In contrast to DNA viruses, no RNA virus-encoded miRNAs have yet been observed, although it is possible to engineer an artificial miRNA in the 3′ end of an infectious SINV [128]. Some RNA viruses, on the other hand, use cellular miRNAs for their own purpose, an eloquent example of this scenario being the interaction between the abundant liver miR-122 and the hepatitis C virus (HCV) RNA. Indeed, HCV replication is greatly facilitated by the presence of this miRNA. miR-122 can seed-match at two distinct positions of the very beginning of the 5′-UTR of HCV. This interaction is thought to have a dual role: the first role is to stimulate the internal ribosome entry site (IRES)-dependent translation of HCV proteins, while the second role is to mask the 5′-end triphosphate of the virus so as to prevent it being recognized by the innate immunity receptor RIG-I, which in turn enables it to evade immune detection [129, 130].

In all cases, the role of miRNAs in viral life cycles appears to be modulatory, and this is consistent with the suggestion that the miRNA pathway exerts fine-tuning regulations. Furthermore, in most of these examples the miRNA pathway is pro-viral rather than antiviral.

3.4.2 Natural Control of Viral Infection by siRNAs in Undifferentiated Mammalian Cells

High-throughput sequencing technologies have provided new opportunities to characterize the small RNA profiles of virus-infected mammalian cells. Indeed, small RNA sequencing of cells infected with HIV-1, Dengue virus (DENV), VSV, poliovirus, HCV and West Nile virus (WNV), led to the detection of viral small RNAs [131–133]. However, their number was small and the virus-specific RNAs constituted between about 0.1% and about 1% of the total small RNA population in mammalian-infected cells, whereas virus-specific RNAs can accumulate up to about 20% of the total of small RNA pool in plants or in invertebrate cells [88, 134]. The functional relevance of such non-abundant virus-derived small RNA species in silencing the viral genome was, nevertheless, not addressed by these studies. In fact, until very recently there was no direct evidence that virally derived small RNAs in mammals were either antiviral or Dicer-dependent. The lack of such evidence was not surprising since: (i) it is well known that an elaborate IFN system constitutes a major antiviral defense in mammalian cells; and (ii) the single mammalian Dicer is unable to cleave long dsRNA into siRNAs in differentiated mammalian cells [48]. However, two recent reports described the antiviral function of RNAi in mammalian cells lacking a functional IFN system. For example, Maillard *et al.* reported that undifferentiated mouse embryonic stem cells (mESCs), which lack the classical innate immune response against viruses (namely

IFN), produce virus-derived siRNAs in a Dicer-dependent manner when infected with encephalomyocarditis virus (EMCV) [135]. The siRNAs derived from this picornavirus were detected when loaded onto AGO2, which suggested that they play a role in silencing the viral RNAs. Interestingly, the accumulation of EMCV siRNAs was reduced upon cell differentiation, which implied that the establishment of an IFN response as the cells differentiated would replace RNAi as the major antiviral defense pathway. In an accompanying report [136], the resistance of 7-day-old suckling mice to Nodamura virus (NoV) was investigated. Infection by the wild-type virus led to a strong viral replication and lethality, but when the animals were infected with NoVΔB2 (a mutant NoV lacking the dsRNA binding protein B2), which functions as a potent VSR in invertebrate nodaviruses (e.g., FHV), the viral RNA abundance was reduced 1000-fold. Interestingly, deep sequencing revealed an accumulation of virus-derived small RNAs with features of canonical siRNAs in suckling mice infected with the mutant virus, but not with the wild-type virus. Thus, newborn mice seem to mount a protective antiviral RNAi response to infection by a mutant virus defective for its VSR. In contrast, NoV-infected adult mice cleared the virus, supposedly on the basis of an antiviral IFN response [137]. These new findings confirmed that the machinery to produce and exploit siRNAs capable of targeting viral RNAs is present in mammals, and indicate that this can be used – at least in some contexts, such as undifferentiated cells – to counter viral infection when the IFN response is defective. The presence of this machinery in mammals might be exploited to block viral replication *in vivo* by introducing *in vitro*-produced small RNAs into mammalian cells.

4
Therapeutic Applications of RNAi

The discovery of RNAi as a natural system to silence gene expression in plants and invertebrates immediately prompted scientists to try to take advantage of this sequence-specific mechanism for therapeutic applications. For example, the silencing of pro-survival genes in cancer cells can lead to cell death and tumor regression [138], whereas silencing of cell cycle regulatory molecules in leukocytes can reverse intestinal inflammation [139]. In the field of infectious diseases, the targeting of viral genomes appeared particularly promising because, unlike other pathogens such as bacteria, viruses offer few protein targets for the development of antiviral therapeutics. Moreover, as mentioned above, RNAi is a natural antiviral defense in many organisms, which indicates that it should possible to develop RNA-based, sequence-specific therapeutic agents to inhibit viral replication in humans.

4.1
Programming the Mammalian RISC

As RNAi represents a natural antiviral mechanism in plants and invertebrates, it is easy to program the RISC complex with any sequences of choice in these organisms. This can be achieved by mimicking a viral replication, for example, by introducing *in vitro*- or *in vivo*-transcribed long dsRNAs into cells or whole organisms [98, 140]. These will be recognized by Dicer proteins as a viral pattern, and processed into siRNAs. As mentioned above, the situation is different in differentiated mammalian cells, where dsRNAs will be recognized by RLRs or Toll-like receptors, and trigger IFN production (Fig. 4). However, it is possible to bypass the dicing step by providing the mammalian cells directly with siRNAs. This was first demonstrated by Thomas Tuschl and colleagues, who reported that the transfection of siRNA duplexes in several mammalian cell lines (including human cells) could be used to specifically silence gene expression [141]. This provided the proof of principle that the mammalian RISC is programmable by artificially introduced siRNAs, paving the way for the development of RNA-based therapeutics. Soon thereafter, this strategy was adopted by virologists, who showed that the growth of poliovirus is efficiently suppressed in HeLa cells upon transfection of sequence-specific siRNAs, independently of the IFN-induced response [142]. Similar results were reported for other viruses, such as HCV [143]. One limitation of this approach is that siRNAs have to be synthesized and annealed *in vitro*, before being introduced into cells where they exert a transient effect. A solution to overcome this problem came from the design of a mammalian expression vector that directs the synthesis of siRNA-like transcripts. This vector expresses a short transcript which is designed to fold back on itself to form a 19-bp stem-loop structure evocative of miRNAs (small hairpin RNA; shRNA). Upon transfection into mammalian cells, this plasmid enables efficient and lon-glasting gene silencing, without toxicity [144].

Thus, it is possible to use RNAi to silence viral RNAs in infected cells. A direct targeting of viral RNAs with sequence-specific small RNAs is not the only option to block viral replication; indeed, cellular factors used by the virus to replicate may also be targeted. This can represent an interesting alternative if the host cell factors are not essential for the cell viability, especially because the cellular RNAs are less prone to variability than the viral RNAs (see below).

Indeed, small-scale and, later, genome-wide RNAi screens, were used successfully in tissue culture cells to define the repertoire of host cell factors required for the replication of important human viral pathogens such as HCV [145], HIV [146, 147], WNV [148], or influenza virus [149]. Si- or shRNAs targeting such host factors can then be used to block viral replication. Alternatively, these screens can prove useful to identify novel targets for the development of small molecule inhibitors [148]. Of note, cellular RNAs required for viral replication are not limited to protein-coding mRNAs; as mentioned above, some miRNAs (e.g., miR122 and HCV) are required for viral replication. These miRNAs also represent putative targets that can be successfully targeted with miRNA antagonists (antagomirs) to block viral replication [150].

In summary, there is now ample proof of principle that it is possible to bypass the crucial dicing step in human cells by feeding the RISC complex with small RNAs designed to target either viral RNAs or cellular RNAs required for efficient viral replication. Although they represent a very promising new class of drugs, siRNAs suffer drawbacks connected to their physicochemical properties; typically, their high molecular weight, hydrophilicity and net negative charge prevent these molecules from passively diffusing across the plasma membrane. In a therapeutic setting, artificial strategies must be designed not only to transport the siRNAs across the plasma membrane but also to deliver them to the infected tissues.

4.2

Delivering the Promise: *In-Vivo* Administration of siRNAs for Gene Silencing

Several strategies have been tested to trigger the uptake of siRNAs into target cells, with encouraging results (Fig. 5). These strategies include the local delivery of naked RNA, chemical modifications of the siRNA compound, encapsulation into lipid-based particles, and viral transduction. Of course, the tissue to be targeted has a strong influence on the optimal delivery option: surface epithelia are directly accessible and may be targeted by naked siRNAs, whereas systemic delivery requires protective strategies improving the siRNA's half-life in the circulation. In addition, while delivery to filtering organs such as the liver or the kidneys will be favored, the targeting of other tissues is more challenging, and therefore modifications must be built into the delivery system to allow for the targeting of the correct organ. This is especially relevant for cells such as leukocytes, which form a prime target for the delivery of siRNAs with anti-inflammatory properties, designed to treat the clinical symptoms associated with most viral infections. Indeed, leukocytes are not only dispersed throughout the organism but also are notoriously difficult to transfect.

4.2.1 Delivery of Naked RNA

The simplest strategy to silence gene expression is to use naked siRNAs in saline buffer. A successful silencing of viral RNAs has been reported in the lungs of humans or rhesus macaques by using siRNAs in saline solution. The mixing of siRNAs targeting the severe acute respiratory syndrome (SARS) coronavirus with 5% D-glucose in water was found to be an efficient formulation for intranasal instillation in rhesus macaques. Such administered siRNAs conferred significant protection when injected prior to or simultaneously with the virus infection, and also – most interestingly in a therapeutic context – when instilled up to 5 days post-infection [151]. A similar strategy was used in humans for the first randomized, double-blind, placebo-controlled trial of an siRNA-based therapy for the respiratory syncytial virus (RSV). In this case, when aerosolized siRNAs in sterile phosphate-buffered saline were administered daily for 5 days, starting on the day before the viral challenge, they conferred a reduction of infection by RSV and provided proof of principle that RNAi therapeutics can be effective in humans [152]. Naked siRNAs can also be administered systemically, and can silence a viral sequence in the mouse liver [153]. However, the efficiency of silencing with naked siRNAs is lower than with plasmid DNA encoding an shRNA and targeting the same sequences, thus highlighting the limitations of this minimalist delivery strategy.

4.2.2 Chemical Modifications of siRNAs

Sugar, base, or backbone modifications of the siRNAs can increase their potency, target affinity and specificity, improve their pharmacokinetics, and decrease their degradation by nucleases. A variety of chemical modifications have been explored, such as phosphorothioate backbone modification, or substitutions at the 2′ position of the ribose (e.g., 2′-O-methyl, 2′-fluoro, 2′-O-methoxy-ethyl) [154]. In addition, the conjugation of steroid, lipid (lauric acid, lithocholic acid, and cholesterol derivatives), or sugar (GlcNAc) moieties has been shown to increase cellular uptake [155, 156]. Some companies, such as RXi Pharmaceuticals, have also modified siRNAs for optimal uptake, not only by conjugating lipophilic moieties but also by extending one of the overhangs. The resulting single-stranded part of the small RNA improves its tissue distribution and cellular uptake [157]. Another innovative approach which relies on the extension of one strand of the small RNA duplex involves grafting an

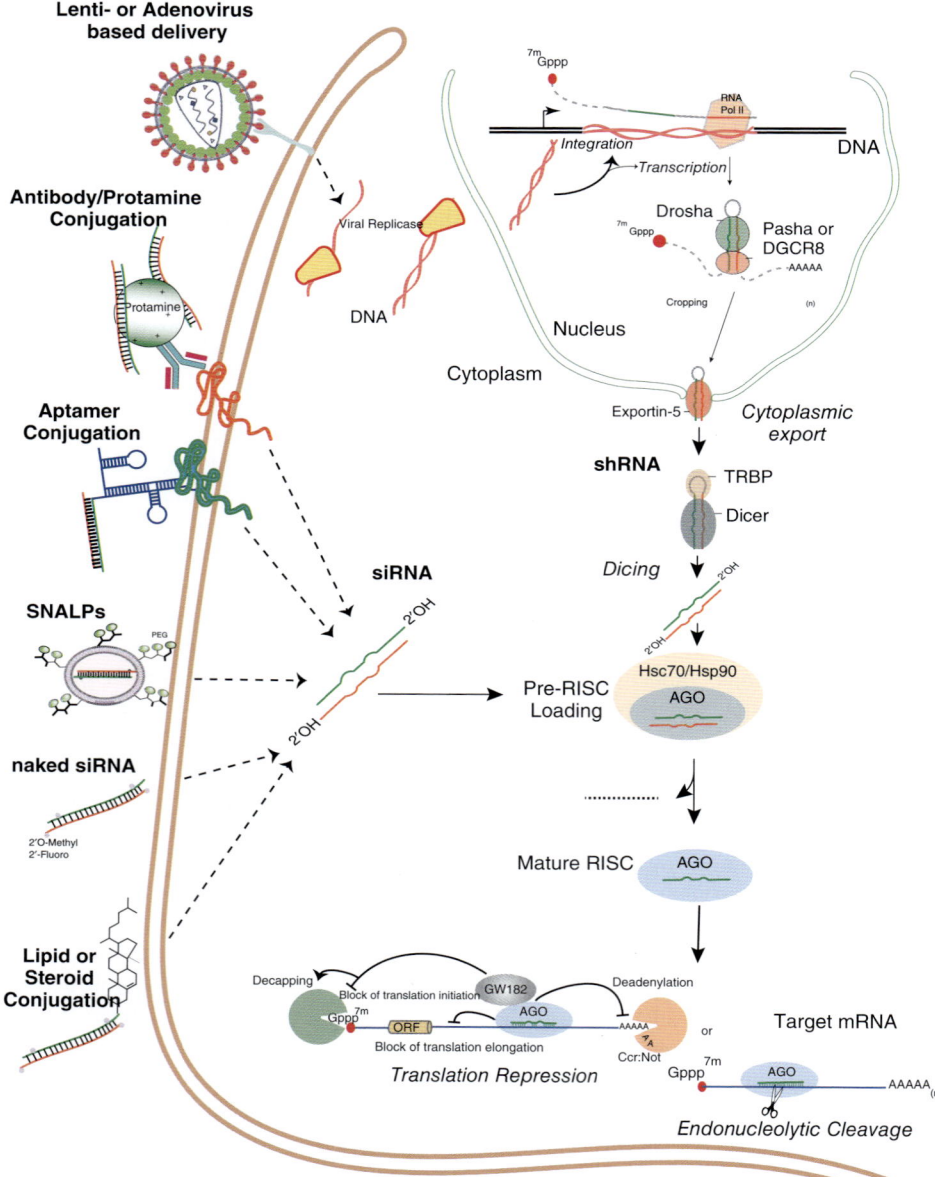

Fig. 5 Strategies for the delivery of shRNA or siRNA in mammalian cells. Adenoviral or lentiviral vectors are used to deliver DNA molecules encoding precursors of shRNAs. Upon transcription in the nucleus, these RNAs will be processed and exported to the cytoplasm, like endogenous miRNAs. In the case of siRNAs, different strategies to deliver them to the RISC complex in the cytosol are illustrated. See text for details. SNALP: stable nucleic-acid lipid particle.

aptamer to the small RNA. Aptamers are synthetic nucleic acids that have survived an *in-vitro* evolution process selecting for specific ligand-binding properties. Chimeric aptamer-siRNA molecules can be specifically targeted to tissue-specific membrane proteins (such as tumor antigens) or to virus-infected cells, therefore allowing a selective delivery of the siRNA (Fig. 5) [138, 158]. For example, an RNA aptamer selected for high-affinity binding to the gp120 envelope protein from HIV-1, and having virus neutralization properties of its own, acquires further antiviral activity when attached to an siRNA triggering a sequence-specific degradation of HIV-1 RNA [159].

4.2.3 Encapsulation into Nanoparticles

The complexing of siRNAs with carrier molecules to form delivery-competent nanoparticles represents a powerful technology to deliver small regulatory RNAs to target cells. One important issue when designing these formulations for *in vivo* use is to take into account not only the efficiency of delivery, but also the proinflammatory properties of the complexes. Indeed, positively charged lipids, which interact tightly with negatively charged siRNAs, represent efficient vectors for delivery in tissue culture cells but are associated with significant toxicity *in vivo*, which impairs their efficiency. Systemic delivery platforms for siRNAs can be based on either lipids, sugars, or proteins.

Lipid-based nanoparticles represent a widely used strategy for the *in-vivo* delivery of siRNAs, that has been used successfully in animals, including nonhuman primates [160–164]. For example, siRNAs mixed with the lipid-based transfection reagent Oligofectamine™ were efficiently taken up by epithelial cells and lamina propria cells upon vaginal instillation. Most interestingly, the intravaginal lipofection-mediated delivery of siRNAs targeting two genes from herpes simplex virus (HSV)-2 was well tolerated, did not induce inflammation, and protected mice against a lethal challenge by HSV-2 [165]. Cholesterol groups have also been used to promote the delivery of siRNAs to the liver and the gut [156].

Lipid nanoparticles vary in their lipid composition, structure and size, and two broad categories can be distinguished (Fig. 5). On the one hand, small nucleic acid lipid particles (SNALPs), which consist of monolamellar liposomes that are neutrally charged at physiological pH and stabilized with polyethylene glycol (PEG), represent the most advanced systemic delivery technology at this stage. They have, for example, been used successfully *in vivo* to control infection by hepatitis B virus [166]. On the other hand, multilamellar, positively charged, lipid-based formulations such as AtuPLEX represent a promising alternative for siRNA delivery into tissues, in particular the vascular endothelium [167]. Currently, a number of other lipid-based formulas are in the pipeline to improve siRNA delivery and decrease the inflammatory response frequently associated with such delivery [162].

The sugar polymer chitosan, which is used in many applications ranging from food to biomedical and pharmaceutical applications (including cosmetics), may also be used for siRNA delivery. Chitosan is a derivative of chitin, of which the exoskeleton of arthropods, the endoskeleton of cephalopods, and the cell wall of fungi are composed. This polyoside, which is composed of D-glucosamine and *N*-acetyl-D-glucosamine, has mucoadhesive properties that can be useful for delivering siRNAs to the respiratory tract. Indeed, chitosan–siRNA nanoparticles were shown to promote an efficient *in-vivo* gene silencing in bronchial cells following intranasal administration to mice [168]. Cyclodextrins, a family of cyclic oligosaccharides, can form stable aqueous complexes with many molecules, including siRNAs, and have also been used *in vivo* to deliver siRNAs to tumor cells and hence to silence gene expression. In this particular example, the conjugation of a protein ligand, transferrin, improved the uptake of siRNAs by cells expressing the transferrin receptor [163].

Finally, conjugation to proteins provides an effective means of improving the efficiency of siRNA delivery, through receptor-mediated endocytosis (Fig. 5). In particular, the association of siRNAs with antibodies provides an efficient way of selectively delivering the silencing RNAs to cells expressing the antigen, following binding and internalization of the antibody–siRNA complex. Protamines, which are small arginine-rich nuclear proteins involved in packaging of the genome in spermatozoa, can be used to condense siRNAs for delivery. These small proteins are largely used in the clinic, as an antidote for heparin, and therefore represent interesting molecules for *in-vivo* delivery. Furthermore, they can be fused to antibodies to improve the efficiency and specificity of targeting. Thus, it was possible to inhibit HIV-1 replication in difficult-to-transfect primary T cells by using siRNAs that targeted the *gag* gene of the virus complexed with protamine fused to a single chain Fv (scFv) fragment from an antibody raised against the Env protein of the virus [169]. An alternative to using protamine to couple siRNAs to proteins would be to employ an oligopeptide composed of nine Arg residues (9R), coupled to the antibody. For example, a scFv recognizing the T-cell-specific marker CD7 and triggering its internalization was coupled to the 9R peptide, and used to deliver siRNAs targeting the HIV-1 genes *vif* and *tat* into T cells. This treatment was able to control HIV-1 replication and to prevent the loss of $CD4^+$ T cells (a major hallmark of the symptoms associated with the pathology caused by this virus), in a hematopoiesis-deficient mouse model reconstituted with human lymphocytes or human hematopoietic stem cells (HSCs) [170]. The scFvCD7-9R moiety enabled a more efficient delivery into T cells than lipofectamine and ensured a highly specific delivery, as neither B lymphocytes nor macrophages could be transfected. The therapeutic potential of this strategy was further highlighted by a lack of apparent toxicity in this preclinical model.

One tissue associated with severe symptoms upon viral infection is the brain, in which virus-mediated encephalitis can have serious consequences. Drug delivery to the brain – including, of course, siRNA delivery – is hampered by the so-called blood–brain barrier (BBB) that results from the extremely tight junctions established between the endothelial cells forming the blood capillaries in the brain. Interestingly, the strategies set by neurotropic viruses to enter the brain parenchyma can be used to design strategies to overcome the BBB. For example, the rabies virus glycoprotein (RVG) interacts with the nicotinic acetylcholine receptor and increases cerebral microvascular permeability, thus allowing the rabies virus to enter the brain and trigger the rabies disease. A 29-amino acid peptide derived from RVG specifically binds to the acetylcholine receptor and can be used to cross the BBB, and deliver siRNAs to the brain. A fusion peptide between this 29-residue RVG peptide and the 9R oligomer was used successfully for the intravenous delivery of siRNAs that would target the Japanese encephalitis virus (JEV) to the central nervous system. This noninvasive strategy was safe, did not induce IFN production, and efficiently protected mice against a challenge with a lethal dose of JEV [171]. Moreover, the strategy could be extended to other situations. For example, the 9R oligomer was fused to DC3, a dodecamer peptide selected by phage display for its capacity to bind specifically to a ligand on dendritic cells. Association of the chimeric peptides to siRNAs mapping to a highly conserved sequence of the envelope gene of the DENV suppressed replication of the virus in these cells [172]. The pathology associated with DENV infection, which can be life-threatening in the case of dengue hemorrhagic fever and dengue shock syndrome, is associated with the production by infected dendritic cells of proinflammatory cytokines, including tumor necrosis factor α (TNFα), which plays an important role in dengue fever pathogenesis. Interestingly, the DC3-9R-mediated delivery of intravenously injected siRNAs targeting the mRNA of TNFα suppressed induction of the cytokine in dendritic cells. Thus, by using siRNAs that would target both viral RNAs and mRNAs encoding cytokines associated with deleterious side effects, it would be possible to target simultaneously the virus itself and the host genes associated with pathogenesis.

In summary, a variety of strategies has been proposed in recent years to efficiently and specifically deliver siRNAs to cells. In effect, these strategies all concur to encapsulate the siRNA to increase its circulation time in a way that will decrease the toxic effects of inflammation while promoting the siRNA's cellular uptake. While these synthetic strategies mimicking viral particles hold great promise, one alternative that is still being considered is to use *bona fide* engineered viruses to deliver shRNAs into the cells.

4.2.4 Viral Vectors for DNA-Directed RNAi

The possibility to silence gene expression with DNA-encoded shRNAs opened the way to strategies using recombinant viral vectors to introduce the nucleic acid into target cells [173]. This technology benefits from progress made over the years in the field of gene therapy, and allows for the efficient delivery – and, in some cases, long-term expression – of the therapeutic gene encoding one or several shRNAs. Classical gene therapy vectors (based on adenovirus, adenoassociated virus, lentivirus) have been used to provide proof-of-principle that virus-delivered shRNAs can be used to treat viral infections [174]. In particular, the stable expression of an shRNA targeting the *tat/rev* genes from HIV-1 in human lymphocytes was reported following the transplantation of autologous hematopoietic progenitor cells transduced *ex vivo* with a recombinant lentivirus [175, 176]. The treatment of other chronic viral diseases, such as hepatitis C [177], may also benefit from the long-term effects of virus-mediated delivery of shRNA coding sequences. Whereas, the advantages of viral delivery in terms of efficient single dose administration and prolonged expression are obvious, this technology also suffers some caveats, essentially in terms of safety. Indeed, any adverse effects caused by the silencing small RNA will be more difficult to revert. Insertional mutagenesis and immunogenicity represent other concerns for the viral vector-based delivery of silencing RNAs.

5 Challenges for Clinical Applications

The excitement elicited by the RNAi technology from the beginning was based not only on the powerful experimental tool it provided for biological research but also on the important promise it held for the silencing of genes involved in human disease. In spite of the remarkable progress made in this field during the past decade, a number of issues remain to be addressed before the technology can be used routinely in the clinic to treat viral infections. These issues are related to safety and efficacy.

5.1 Safety Issues

Several safety concerns gradually arose as the application of RNAi was increasingly adopted in *in-vivo* settings. A first concern is target specificity. Early on, genome-wide expression profile experiments revealed that the introduction of foreign siRNAs into the cell alters the expression of target genes, and also of nontarget genes. Indeed, as a seed of 7–8 nt bp between the small RNA and its target is sufficient to cause silencing, unspecific cross-reactions with targets of limited sequence similarity can occur. This collateral phenomenon, causing decreases in the RNA and protein levels of untargeted genes, was termed the "*off-target*" effect [178, 179]. The 2′-O-Me modification or DNA base substitutions in siRNA duplexes have been shown to significantly reduce such off-target effects [180, 181]. Thus, chemical modifications of small RNAs, coupled with prediction algorithms to detect partial homology to unintended targets and to exclude regions of low complexity, can overcome such a problem [182].

A second serious safety issue is the induction of inflammatory symptoms. These adverse effects of the siRNA technology stem from the double-stranded RNA nature of these molecules, which is reminiscent of viral pathogen-associated molecular patterns (PAMPs) (Fig. 4). Indeed, the mammalian immune system can detect siRNAs double-strandedness or immunostimulatory motifs (e.g., 5′-GUCCUUCAA-3′) via Toll-like receptor (TLR) 3, TLR7 or TLR8, and trigger signaling leading to a subsequent immune activation and production of IFNs and TNF [183–185]. For example, activation of TLR3 signaling upon siRNA treatment suppressed blinding choroidal neovascularization (CNV), a symptom of age-related macular degeneration, in a sequence- and target-independent manner [186]. One solution to overcome this harmful side effect is again chemical modification. For example, the simple incorporation of 2′-O-methyl (2′OMe) uridine or guanosine nucleosides into one strand of the siRNA duplex abrogates the siRNA-dependent immune stimulation [187, 188]. A locked nucleic acid (LNA) modification of the passenger strand also abrogates IFN induction [189]. Of note, DNA substitutions in siRNA duplexes, which prevent the activation of TLRs recognizing RNAs (TLR3, 7, 8), may activate the DNA sensing TLR9, especially if they contain CpG motifs [190]. It should be borne in mind, however, that in the particular case of antiviral therapy the adjuvant-like effect of small silencing RNAs may be beneficial. Indeed, in some cases the sequence-independent proinflammatory effects of RNA therapeutics were taken advantage of to improve the efficiency of the treatment. For example, the addition of an uncapped 5′ triphosphate (a characteristic feature of viral RNAs) at the extremity of an siRNA targeting the anti-apoptotic gene

Bcl-2, triggers a dual effect in melanoma tumor cells, inducing apoptosis (silencing of Bcl-2) and the production of IFN (through activation of the cytosolic receptor RIG-I) [191]. In summary, the spectacular progress made during recent years in understanding the mechanisms involved in the sensing of nucleic acids now makes it possible to envisage a modulation of the immunostimulatory properties of silencing RNAs.

A third safety issue arose when it was observed that the sustained expression of high levels of shRNAs in the livers of adult mice resulted in significant morbidity which was associated with the downregulation of abundant liver miRNAs. This observation revealed that an excessive amount of shRNAs may compete with endogenous factors of the RNAi pathway, such as the nuclear Exportin-5 [192]. Thus, besides the unwanted activation of inflammation, the issue of competition for endogenous factors stresses the importance of avoiding the use of excessive amounts of siRNAs and shRNAs. This can be achieved by designing more specific and potent RNAi triggers, permitting the downregulation of target genes with minimal expression. In-silico approaches based on algorithms taking into account the thermodynamic properties of a given sequence can be used to improve silencing potency [193]. Alternatively, the best approach so far for identifying extremely potent antisense sequences has been based on empirical in-vivo testing [194]. An elegant "Sensor Assay" that enabled the biological identification of effective shRNAs on a large scale was designed and used to evaluate 20 000 RNAi reporters covering every possible target site in nine mammalian transcripts. This provided enlightening information on the rules and sequence requirements for effective RNA silencing [194].

5.2
Efficiency of Virus Silencing with Small RNAs

Besides the delivery issues discussed above, viral escape represents a significant challenge when targeting viral genomes with RNAi. Indeed, it has been shown that poliovirus readily escapes highly effective siRNAs through unique point mutations within the targeted regions, resulting in the emergence of resistant viruses [142]. Likewise, HIV was able to escape RNAi either directly, by mutations in the targeted region, or indirectly by mutations in promoter regions that compensated for the RNAi by upregulating viral transcription [195]. It is important to note that such a mechanism of viral escape is not confined to mammalian viruses but has been also observed in plants and mosquitoes with the turnip mosaic virus (TuMV) and Dengue virus (DENV2), respectively. This phenomenon is not surprising since viruses – especially RNA viruses – possess a high ability to mutate on the basis of their low-fidelity RdRP. A solution to this problem would be to use a combination of small RNAs targeting different regions of the genome, to increase efficacy and decrease the chances of escape (e.g., the targeting of SARS with two siRNAs [151] or Ebola virus with three siRNAs targeting different structural and nonstructural proteins [196]). Another option that has been used successfully in preclinical studies to prevent HIV-1 infection was to combine one siRNA targeting the virus and one siRNA targeting a host molecule, the viral coreceptor CCR5 [170]. Of note, the issue of viral escape can be addressed by targeting only a host molecule that is essential for the viral replication cycle. Such a strategy, based on inhibition of the hepatocellular microRNA miR122 by LNAs, offers a very promising perspective to treat HCV infections [150].

6
Conclusions

The discovery of RNA silencing, a major finding of the late twentieth century, has opened new avenues of investigation in both applied and fundamental scientific research. Interestingly, the rationale design of antisense RNAs to inhibit viral replication predates the discovery of the role of RNA silencing in a natural context in the regulation of viral infection. As a result, initial attempts to inhibit viral replication by antisense oligonucleotides were mainly seen as 'groping in the dark.' On the business side, the initial striking results obtained with model organisms and tissue culture cells raised high hopes and led to a flurry of patents and new companies [155]. Unfortunately, however, unmet expectations in terms of delivery technologies, in addition to the potential of some silencing RNA formulations to trigger responses in a sequence-independent manner, through the activation of innate immunity, raised suspicion for investors and slowed down the development of the industry.

The past 15 years have yielded spectacular progresses in the understanding of the molecular basis of antiviral RNA silencing, from genetic networks to atomic resolution structures. As a result, a more comprehensive view is now available of the molecular actors at play, and of their regulation, and this in turn brought to light the toolbox for designing efficient and specific antisense RNA inhibitors. It is interesting to note that, during the same time period, the field of antiviral innate immunity in vertebrates made a big leap forward, with the identification of pattern recognition receptors sensing viral RNAs providing a set of rules to avoid the activation of inflammation by silencing RNAs.

Thus, several issues raised in preclinical studies have already been addressed, and RNAi therapies remain a highly promising strategy to counter viral infections. Besides the development of new, targeted strategies to deliver siRNAs or shRNAs systemically, the field as a whole can also expect to see some exciting new findings deriving from model organisms in the future. But, it must be borne in mind that the RNAi pathways – and in particular the piRNA pathway – have not yet given away all of their secrets. The recent development of a revolutionary genome editing technology based on a bacterial antiviral system, the CRISPR/Cas9 system [197, 198], provides another spectacular example of the reasons for looking ahead with much optimism.

Acknowledgments

The authors acknowledge financial support from NIH (PO1 AI070167), ANR (VIRVEC) and Region Alsace.

References

1. Stephenson, M.L. and Zamecnik, P.C. (1978) Inhibition of Rous sarcoma viral RNA translation by a specific oligodeoxyribonucleotide. *Proc. Natl Acad. Sci. USA*, **75**, 285–288.
2. Zamecnik, P.C. and Stephenson, M.L. (1978) Inhibition of Rous sarcoma virus replication and cell transformation by a specific oligodeoxynucleotide. *Proc. Natl Acad. Sci. USA*, **75**, 280–284.
3. Crowley, T.E., Nellen, W., Gomer, R.H., and Firtel, R.A. (1985) Phenocopy of discoidin I-minus mutants by antisense transformation in *Dictyostelium. Cell*, **43**, 633–641.
4. Izant, J.G. and Weintraub, H. (1985) Constitutive and conditional suppression of exogenous and endogenous genes by antisense RNA. *Science*, **229**, 345–352.
5. Izant, J.G. and Weintraub, H. (1984) Inhibition of thymidine kinase gene expression

by anti-sense RNA: a molecular approach to genetic analysis. *Cell*, **36**, 1007–1015.
6. Kim, S.K. and Wold, B.J. (1985) Stable reduction of thymidine kinase activity in cells expressing high levels of anti-sense RNA. *Cell*, **2**, 129–138.
7. Melton, D.A. (1985) Injected anti-sense RNAs specifically block messenger RNA translation in vivo. *Proc. Natl Acad. Sci. USA*, **82**, 144–148.
8. Rosenberg, U.B., Preiss, A., Seifert, E., Jäckle, H., and Knipple, D.C. (1985) Production of phenocopies by Krüppel antisense RNA injection into *Drosophila* embryos. *Nature*, **313**, 703–706.
9. Napoli, C., Lemieux, C., and Jorgensen, R. (1990) Introduction of a chimeric chalcone synthase gene into petunia results in reversible co-suppression of homologous genes in trans. *Plant Cell*, **2**, 279–289.
10. Van der Krol, A.R., Mur, L.A., Beld, M., Mol, J.N., and Stuitje, A.R. (1990) Flavonoid genes in petunia: addition of a limited number of gene copies may lead to a suppression of gene expression. *Plant Cell*, **2**, 291–299.
11. de Carvalho Niebel, F., Frendo, P., VanMontagu, M., Cornelissen, M. (1995) Post-transcriptional cosuppression of beta-1,3-glucanase genes does not affect accumulation of transgene nuclear mRNA. *Plant Cell*, **7**, 347–358.
12. Elmayan, T. and Vaucheret, H. (1996) Expression of single copies of a strongly expressed 35S transgene can be silenced post-transcriptionally. *Plant J.*, **9**, 787–797.
13. Stam, M., de Bruin, R., Kenter, S., van der Hoorn, R.A.L., van Blokland, R., Mol, J.N.M., and Kooter, J.M. (1997) Post-transcriptional silencing of chalcone synthase in *Petunia* by inverted transgene repeats. *Plant J.*, **12**, 63–82.
14. Covey, S.N., Al-Kaff, N.S., Lángara, A., and Turner, D.S. (1997) Plants combat infection by gene silencing. *Nature*, **385**, 781–782.
15. Ratcliff, F., Harrisson, B.D., and Baulcombe, D.C. (1997) A similarity between viral defense and gene silencing in plants. *Science*, **276**, 1558–1560.
16. Ruiz, M.T., Voinnet, O., and Baulcombe, D.C. (1998) Initiation and maintenance of virus-induced gene silencing. *Plant Cell*, **10**, 937–946.
17. Voinnet, O. (1999) Suppression of gene silencing: a general strategy used by diverse DNA and RNA viruses of plants. *Proc. Natl Acad. Sci. USA*, **96**, 14147–14152.
18. Voinnet, O. and Baulcombe, D.C. (1997) Systemic signalling in gene silencing. *Nature*, **389**, 553.
19. Voinnet, O., Li, W.X., Ji, L.H., Ding, S.W., and Baulcombe, D.C. (1998) Viral pathogenicity determinants are suppressors of transgene silencing in *Nicotiana benthamiana*. *EMBO J.*, **17**, 6739–6746.
20. Tabara, H., Grishok, A., and Mello, C.C. (1998) RNAi in *C. elegans*: soaking in the genome sequence. *Science*, **282**, 430–431.
21. Tabara, H., Sarkissian, M., Kelly, W.G., Fleenor, J., Grishok, A., Timmons, L., Fire, A., and Mello, C.C. (1999) The rde-1 gene, RNA interference, and transposon silencing in *C. elegans*. *Cell*, **99**, 123–132.
22. Bohmert, K., Camus, I., Bellini, C., Bouchez, D., Caboche, M., and Benning, C. (1998) AGO1 defines a novel locus of *Arabidopsis* controlling leaf development. *EMBO J.*, **17**, 170–180.
23. Lau, P.W., Guiley, K.Z., De, N., Potter, C.S., Carragher, B., and MacRae, I.J. (2012) The molecular architecture of human Dicer. *Nat. Struct. Mol. Biol.*, **2012** (19), 436–440.
24. Jinek, M., Doudna, J.A. (2009) A three-dimensional view of the molecular machinery of RNA interference. *Nature*, **457**, 405–412.
25. Gu, S., Jin, L., Huang, Y., Zhang, F., and Kay, M.A. (2012) Slicing-independent RISC activation requires the argonaute PAZ domain. *Curr. Biol.*, **22**, 1536–1542.
26. Rivas, F.V., Tolia, N.H., Song, J.J., Aragon, J.P., Liu, J., Hannon, G.J., and Joshua-Tor, L. (2005) Purified Argonaute2 and an siRNA form recombinant human RISC. *Nat. Struct. Mol. Biol.*, **12**, 340–349.
27. Liu, J., Carmell, M.A., Rivas, F.V., Marsden, C.G, et al. (2004) Argonaute2 is the catalytic engine of mammalian RNAi. *Science*, **305**, 1437–1441.
28. Boland, A., Huntzinger, E., and Schmidt, S. (2011) Crystal structure of the MID-PIWI lobe of a eukaryotic Argonaute protein. *Proc. Natl Acad. Sci. USA*, **108**, 10466–10471.
29. Frank, F., Sonenberg, N., and Nagar, B. (2010) Structural basis for 5′-nucleotide

base-specific recognition of guide RNA by human AGO2. *Nature*, **65**, 818–822.
30. Ambros, V. (2003) MicroRNA pathways in flies and worms: growth, death, fat, stress, and timing. *Cell*, **113**, 673–676.
31. Ghildiyal, M. and Zamore, P.D. (2009) Small silencing RNAs: an expanding universe. *Nat. Rev. Genet.*, **10**, 94–108.
32. Li, X., Cassidy, J.J., Reinke, C.A., Fischboeck, S., and Carthew, R.W. (2009) MicroRNA imparts robustness against environmental fluctuation during development. *Cell*, **137**, 273–282.
33. Ameres, S.L. and Zamore, P.D. (2013) Diversifying microRNA sequence and function. *Nat. Rev. Mol. Cell Biol.*, **14**, 475–488.
34. Cheloufi, S., Santos Dos, C.O., Chong, M.M.W., and Hannon, G.J. (2010) A dicer-independent miRNA biogenesis pathway that requires Ago catalysis. *Nature*, **465**, 584–589.
35. Czech, B., Zhou, R., Erlich, Y., Brennecke, J. *et al.* (2009) Hierarchical rules for argonaute loading in *Drosophila*. *Mol. Cell*, **36**, 445–456.
36. Förstemann, K., Horwich, M.D., Wee, L., Tomari, Y., and Zamore, P.D. (2007) *Drosophila* microRNAs are sorted into functionally distinct argonaute complexes after production by dicer-1. *Cell*, **130**, 287–297.
37. Kawamata, T., Seitz, H., and Tomari, Y. (2009) Structural determinants of miRNAs for RISC loading and slicer-independent unwinding. *Nat. Struct. Mol. Biol.*, **16**, 953–960.
38. Okamura, K., Liu, N., and Lai, E.C. (2009) Distinct mechanisms for microRNA strand selection by *Drosophila* argonautes. *Mol. Cell*, **36**, 431–444.
39. Meister, G. and Tuschl, T. (2004) Mechanisms of gene silencing by double-stranded RNA. *Nature*, **431**, 343–349.
40. Rand, T.A., Petersen, S., Du, F., and Wang, X. (2005) Argonaute2 cleaves the anti-guide strand of siRNA during RISC activation. *Cell*, **123**, 621–629.
41. Hausser, J., Syed, A.P., Bilen, B., and Zavolan, M. (2013) Analysis of CDS-located miRNA target sites suggests that they can effectively inhibit translation. *Genome Res.*, **23**, 604–615.

42. Bartel, D.P. (2009) MicroRNAs: target recognition and regulatory functions. *Cell*, **136**, 215–233.
43. Iwasaki, S., Kawamata, T., and Tomari, Y. (2009) *Drosophila* Argonaute1 and Argonaute2 employ distinct mechanisms for translational repression. *Mol. Cell*, **34**, 58–67.
44. Eulalio, A., Rehwinkel, J., Stricker, M., Huntzinger, E. *et al.* (2007) Target-specific requirements for enhancers of decapping in miRNA-mediated gene silencing. *Genes Dev.*, **21**, 2558–2570.
45. Fabian, M.R. and Sonenberg, N. (2010) Regulation of mRNA translation and stability by microRNAs. *Annu. Rev. Biochem.*, **79**, 351–379.
46. Fabian, M.R., Mathonnet, G., Sundermeier, T., Mathys, H. *et al.* (2009) Mammalian miRNA RISC recruits CAF1 and PABP to affect PABP-dependent deadenylation. *Mol. Cell*, **35**, 868–880.
47. Ding, S.-W. and Voinnet, O. (2007) Antiviral immunity directed by small RNAs. *Cell*, **130**, 413–426.
48. Babiarz, J.E., Ruby, J.G., Wang, Y., Bartel, D.P., and Blelloch, R. (2008) Mouse ES cells express endogenous shRNAs, siRNAs, and other Microprocessor-independent, Dicer-dependent small RNAs. *Genes Dev.*, **22**, 2773–2785.
49. Zhang, H., Kolb, F.A., Jaskiewicz, L., Westhof, E., and Filipowicz, W. (2004) Single processing center models for human dicer and bacterial RNase III. *Cell*, **118**, 57–68.
50. Lin, H. and Spradling, A.C. (1997) A novel group of pumilio mutations affects the asymmetric division of germline stem cells in the *Drosophila* ovary. *Development*, **24**, 2463–2476.
51. Aravin, A.A., Sachidanandam, R., Girard, A., and Fejes-Toth, K. (2007) Developmentally regulated piRNA clusters implicate MILI in transposon control. *Science*, **316**, 744–747.
52. Saito, K., Nishida, K.M., Mori, T., Kawamura, Y. *et al.* (2006) Specific association of Piwi with rasiRNAs derived from retrotransposon and heterochromatic regions in the *Drosophila* genome. *Genes Dev.*, **20**, 2214–2222
53. Vagin, V.V., Sigova, A., Li, C., Seitz, H., Gvozdev, V., and Zamore, P.D. (2006) A

distinct small RNA pathway silences selfish genetic elements in the germline. *Science*, **313**, 320–324.

54. Girard, A., Sachidanandam, R., Hannon, G.J., and Carmell, M.A. (2006) A germline-specific class of small RNAs binds mammalian Piwi proteins. *Nature*, **442** (7099), 199–202.

55. Aravin, A., Gaidatzis, D., Pfeffer, S., and Lagos-Quintana, M. (2006) A novel class of small RNAs bind to MILI protein in mouse testes. *Nature*, **442**, 203–207.

56. Brennecke, J., Aravin, A.A., Stark, A., Dus, M., Kellis, M., Sachidanandam, R., and Hannon, G.J. (2007) Discrete small RNA-generating loci as master regulators of transposon activity in *Drosophila*. *Cell*, **128**, 1089–1103.

57. Olivieri, D., Sykora, M.M., Sachidanandam, R., Mechtler, K., and Brennecke, J. (2010) An in vivo RNAi assay identifies major genetic and cellular requirements for primary piRNA biogenesis in *Drosophila*. *EMBO J.*, **29**, 3301–3317.

58. Saito, K., Inagaki, S., Mituyama, T., Kawamura, Y. *et al.* (2009) A regulatory circuit for piwi by the large Maf gene traffic jam in *Drosophila*. *Nature*, **461**, 1296–1299.

59. Wu, Q., Luo, Y., Lu, R., Lau, N., Lai, E.C., Li, W.X., and Ding, S.W. (2010) Virus discovery by deep sequencing and assembly of virus-derived small silencing RNAs. *Proc. Natl Acad. Sci. USA*, **107**, 1606–1611.

60. Morazzani, E.M., Wiley, M.R., Murreddu, M.G., Adelman, Z.N., and Myles, K.M. (2012) Production of virus-derived ping-pong-dependent piRNA-like small RNAs in the mosquito soma. *PLoS Pathog.*, **8**, e1002470.

61. Vodovar, N., Bronkhorst, A.W., van Cleef, K.W., Miesen, P., Blanc, H., van Rij, R.P., and Saleh, M.C. (2012) Arbovirus-derived piRNAs exhibit a ping-pong signature in mosquito cells. *PLoS ONE*, **7**, e30861.

62. Brodersen, P. and Voinnet, O. (2006) The diversity of RNA silencing pathways in plants. *Trends Genet.*, **22**, 268–280.

63. Moissiard, G. and Voinnet, O. (2006) RNA silencing of host transcripts by cauliflower mosaic virus requires coordinated action of the four *Arabidopsis* Dicer-like proteins. *Proc. Natl Acad. Sci. USA*, **103**, 19593–19598.

64. Deleris, A., Gallego-Bartolome, J., Bao, J., Kasschau, K.D., Carrington, J.C., and Voinnet, O. (2006) Hierarchical action and inhibition of plant Dicer-like proteins in antiviral defense. *Science*, **313**, 68–71.

65. Gasciolli, V., Mallory, A.C., Bartel, D.P., and Vaucheret, H. (2005) Partially redundant functions of *Arabidopsis* DICER-like enzymes and a role for DCL4 in producing *trans*-acting siRNAs. *Curr. Biol.*, **15**, 1494–1500.

66. Dunoyer, P., Himber, C., and Voinnet, O. (2005) DICER-LIKE 4 is required for RNA interference and produces the 21-nucleotide small interfering RNA component of the plant cell-to-cell silencing signal. *Nat. Genet.*, **37**, 1356–1360.

67. Borsani, O., Zhu, J., Verslues, P.E., Sunkar, R., and Zhu, J.K. (2005) Endogenous siRNAs derived from a pair of natural *cis*-antisense transcripts regulate salt tolerance in *Arabidopsis*. *Cell*, **123**, 1279–1291.

68. Marí-Ordóñez, A., Marchais, A., Etcheverry, M., Martin, A., Colot, V., and Voinnet, O. (2013) Reconstructing de novo silencing of an active plant retrotransposon. *Nat. Genet.*, **45**, 1029–1039.

69. Matzke, M.A. and Birchler, J.A. (2005) RNAi-mediated pathways in the nucleus. *Nat. Rev. Genet.*, **6**, 24–35.

70. Qu, F., Ye, X., and Morris, T.J. (2008) *Arabidopsis* DRB4, AGO1, AGO7, and RDR6 participate in a DCL4-initiated antiviral RNA silencing pathway negatively regulated by DCL1. *Proc. Natl Acad. Sci. USA*, **105**, 14732–14737.

71. Vaucheret, H. (2008) Plant argonautes. *Trends Plant Sci.*, **13**, 350–358.

72. Zhang, X., Yuan, Y.R., Pei, Y., Lin, S.S., and Tuschl, T. (2006) Cucumber mosaic virus-encoded 2b suppressor inhibits *Arabidopsis* Argonaute1 cleavage activity to counter plant defense. *Genes Dev.*, **20**, 3255–3268.

73. Derrien, B., Baumberger, N., Schepetilnikov, M., Viotti, C. *et al.* (2012) Degradation of the antiviral component ARGONAUTE1 by the autophagy pathway. *Proc. Natl Acad. Sci. USA*, **109**, 15942–15946.

74. Molnar, A., Melnyk, C.W., Bassett, A., Hardcastle, T.J., Dunn, R., and Baulcombe, D.C. (2010) Small silencing RNAs in plants are mobile and direct epigenetic modification in recipient cells. *Science*, **328**, 872–875.

75. Dunoyer, P., Schott, G., Himber, C., Meyer, D., Takeda, A., Carrington, J.C., and Voinnet, O. (2010) Small RNA duplexes function as mobile silencing signals between plant cells. *Science*, **328**, 912–916.
76. Li, F. and Ding, S.W. (2006) Virus counterdefense: diverse strategies for evading the RNA-silencing immunity. *Annu. Rev. Microbiol.*, **60**, 503–531.
77. Vargason, J.M., Szittya, G., Burgyan, J., and Hall, T. (2003) Size selective recognition of siRNA by an RNA silencing suppressor. *Cell*, **115**, 799–811.
78. Ye, K., Malinina, L., and Patel, D.J. (2003) Recognition of small interfering RNA by a viral suppressor of RNA silencing. *Nature*, **426**, 874–878.
79. Giner, A., Lakatos, L., García-Chapa, M., and López-Moya, J.J. (2010) Viral protein inhibits RISC activity by argonaute binding through conserved WG/GW motifs. *PLoS Pathog.*, **6**, e1000996.
80. Wang, X.H., Aliyari, R., Li, W.X., Li, H.W., Kim, K., Carthew, R., Atkinson, P., and Ding, S.W. (2006) RNA interference directs innate immunity against viruses in adult *Drosophila*. *Science*, **312**, 452–454.
81. Galiana-Arnoux, D., Dostert, C., Schneemann, A., Hoffmann, J.A., and Imler, J.-L. (2006) Essential function in vivo for Dicer-2 in host defense against RNA viruses in *Drosophila*. *Nat. Immunol.*, **7**, 590–597.
82. van Rij, R.P., Saleh, M.C., Berry, B., Foo, C., Houk, A., Antoniewski, C., and Andino, R. (2006) The RNA silencing endonuclease Argonaute 2 mediates specific antiviral immunity in *Drosophila melanogaster*. *Genes Dev.*, **20**, 2985–2995.
83. Bronkhorst, A.W., van Cleef, K.W., Vodovar, N., Ince, I.A. *et al.* (2012) The DNA virus Invertebrate iridescent virus 6 is a target of the *Drosophila* RNAi machinery. *Proc. Natl Acad. Sci. USA*, **109**, E3604–E3613.
84. Kemp, C., Mueller, S., Goto, A., Barbier, V. *et al.* (2013) Broad RNA interference-mediated antiviral immunity and virus-specific inducible responses in *Drosophila*. *J. Immunol.*, **190**, 650–658.
85. Chao, J.A., Lee, J.H., Chapados, B.R., Debler, E.W., Schneemann, A., and Williamson, J.R. (2005) Dual modes of RNA-silencing suppression by Flock House virus protein B2. *Nat. Struct. Mol. Biol.*, **12**, 952–957.
86. Nayak, A., Berry, B., Tassetto, M., Kunitomi, M. *et al.* (2010) Cricket paralysis virus antagonizes Argonaute 2 to modulate antiviral defense in *Drosophila*. *Nat. Struct. Mol. Biol.*, **17**, 547–554
87. van Mierlo, J.T., Bronkhorst, A.W., Overheul, G.J., Sadanandan, S.A. *et al.* (2012) Convergent evolution of argonaute-2 slicer antagonism in two distinct insect RNA viruses. *PLoS Pathog.*, **8**, e1002872.
88. Myles, K.M., Wiley, M.R., Morazzani, E.M., and Adelman, Z.N. (2008) Alphavirus-derived small RNAs modulate pathogenesis in disease vector mosquitoes. *Proc. Natl Acad. Sci. USA*, **105**, 19938–19943.
89. Petrillo, J.E., Venter, P.A., Short, J.R., Gopal, R. *et al.* (2013) Cytoplasmic granule formation and translational inhibition of nodaviral RNAs in the absence of the double-stranded RNA binding protein B2. *J. Virol.*, **87**, 13409–13421.
90. Wang, X.H., Aliyari, R., Han, C., Li, W.X., and Ding, S.W. (2011) RNA-based immunity terminates viral infection in adult *Drosophila* in the absence of viral suppression of RNA interference: characterization of viral small interfering RNA populations in wild-type and mutant flies. *J. Virol.*, **85**, 13153–13163.
91. Berry, B., Deddouche, S., Kirschner, D., Imler, J.-L., and Antoniewski, C. (2009) Viral suppressors of RNA silencing hinder exogenous and endogenous small RNA pathways in *Drosophila*. *PLoS ONE*, **4**, e5866.
92. Obbard, D.J., Gordon, K.H.J., Buck, A.H., and Jiggins, F.M. (2009) The evolution of RNAi as a defence against viruses and transposable elements. *Philos. Trans. R. Soc. Lond. B. Biol. Sci.*, **364**, 99–115.
93. Aliyari, R., Wu, Q., Li, H.-W., Wang, X.-H. *et al.* (2008) Mechanism of induction and suppression of antiviral immunity directed by virus-derived small RNAs in *Drosophila*. *Cell Host Microbe*, **4**, 387–397.
94. Mueller, S., Gausson, V., Vodovar, N., Deddouche, S. *et al.* (2010) RNAi-mediated immunity provides strong protection against the negative-strand RNA vesicular stomatitis virus in *Drosophila*. *Proc. Natl Acad. Sci. USA*, **107**, 19390–19395.
95. Flynt, A., Liu, N., Martin, R., and Lai, E.C. (2009) Dicing of viral replication intermediates during silencing of latent *Drosophila*

viruses. *Proc. Natl Acad. Sci. USA*, **106**, 5270–5275.

96. Marques, J.T., Wang, J.P., Wang, X., and de Oliveira, K. (2013) Functional specialization of the small interfering RNA pathway in response to virus infection. *PLoS Pathog.*, **9**, e1003579.

97. Sabin, L.R., Zheng, Q., Thekkat, P., Yang, J., Hannon, G.J., Gregory, B.D., Tudor, M., and Cherry, S. (2013) Dicer-2 processes diverse viral RNA species. *PLoS ONE*, **8**, e55458.

98. Saleh, M.-C., Tassetto, M., van Rij, R.P., Goic, B. et al. (2009) Antiviral immunity in *Drosophila* requires systemic RNA interference spread. *Nature*, **458**, 346–350.

99. Adelman, Z.N., Anderson, M.A.E., Morazzani, E.M., and Myles, K.M. (2008) A transgenic sensor strain for monitoring the RNAi pathway in the yellow fever mosquito, *Aedes aegypti*. *Insect Biochem. Mol. Biol.*, **38**, 705–713.

100. Cenik, E.S., Fukunaga, R., Lu, G., Dutcher, R., Wang, Y., Hall, T.M.T., and Zamore, P.D. (2011) Phosphate and R2D2 restrict the substrate specificity of dicer-2, an ATP-driven ribonuclease. *Mol. Cell*, **42**, 172–184.

101. Liu, Q. (2003) R2D2, a bridge between the initiation and effector steps of the *Drosophila* RNAi pathway. *Science*, **301**, 1921–1925.

102. Liu, X., Jiang, F., Kalidas, S., Smith, D., and Liu, Q. (2006) Dicer-2 and R2D2 coordinately bind siRNA to promote assembly of the siRISC complexes. *RNA*, **12**, 1514–1520.

103. Ameres, S.L., Horwich, M.D., Hung, J.H., Xu, J., Ghildiyal, M., Weng, Z., and Zamore, P.D. (2010) Target RNA-directed trimming and tailing of small silencing RNAs. *Science*, **328**, 1534–1539.

104. Saleh, M.-C., van Rij, R.P., Hekele, A., Gillis, A., Foley, E., O'Farrell, P.H., and Andino, R. (2006) The endocytic pathway mediates cell entry of dsRNA to induce RNAi silencing. *Nat. Cell Biol.*, **8**, 793–802.

105. Goic, B., Vodovar, N., Mondotte, J.A., Monot, C.E.M. et al. (2013) RNA-mediated interference and reverse transcription control the persistence of RNA viruses in the insect model *Drosophila*. *Nat. Immunol.*, **14**, 394–403.

106. Voinnet, O. (2013) How to become your own worst enemy. *Nat. Immunol.*, **14**, 315–317.

107. Félix, M.A., Ashe, A., Piffaretti, J., Wu, G., Nuez, I., and Bélicard, T. (2011) Natural and experimental infection of *Caenorhabditis* nematodes by novel viruses related to nodaviruses. *PLoS Biol.*, **9** (1), e1000586.

108. Lu, R., Maduro, M., Li, F., Li, H.W., Broitman-Maduro, G., Li, W.X., and Ding, S.W. (2005) Animal virus replication and RNAi-mediated antiviral silencing in *Caenorhabditis elegans*. *Nature*, **436**, 1040–1043.

109. Schott, D.H., Cureton, D.K., and Whelan, S.P. (2005) An antiviral role for the RNA interference machinery in *Caenorhabditis elegans*. *Proc. Natl Acad. Sci. USA*, **102**, 18420–18424.

110. Wilkins, C., Dishongh, R., Moore, S.C., Whitt, M.A., Chow, M., and Machaca, K. (2005) RNA interference is an antiviral defence mechanism in *Caenorhabditis elegans*. *Nature*, **436**, 1044–1047.

111. Sarkies, P., Frézal, L., Lehrbach, N.J., Félix, M.A., and Miska, E.A. (2013) A deletion polymorphism in the *Caenorhabditis elegans* RIG-I homolog disables viral RNA dicing and antiviral immunity. *Elife*, **2**, e00994.

112. Gu, W., Shirayama, M., Conte, D., Vasale, J. et al. (2009) Distinct argonaute-mediated 22G-RNA pathways direct genome surveillance in the *C. elegans* germline. *Mol. Cell*, **36**, 231–244.

113. Matranga, C. and Pyle, A.M. (2010) Double-stranded RNA-dependent ATPase DRH-3 insight into its role in RNA silencing in *Caenorhabditis elegans*. *J. Biol. Chem.*, **285**, 25363–25371.

114. Guo, X., Zhang, R., Wang, J., Ding, S.-W., and Lu, R. (2013) Homologous RIG-I-like helicase proteins direct RNAi-mediated antiviral immunity in *C. elegans* by distinct mechanisms. *Proc. Natl Acad. Sci. USA*, **110**, 16085–16090.

115. Lu, R., Yigit, E., Li, W.-X., and Ding, S.-W. (2009) An RIG-I-Like RNA helicase mediates antiviral RNAi downstream of viral siRNA biogenesis in *Caenorhabditis elegans*. *PLoS Pathog.*, **5**, e1000286.

116. Kato, H., Takahasi, K., and Fujita, T. (2011) RIG-I-like receptors: cytoplasmic sensors

for non-self RNA. *Immunol. Rev.*, **243**, 91–98.
117. Otsuka, M., Jing, Q., Georgel, P., New, L. *et al.* (2007) Hypersusceptibility to vesicular stomatitis virus infection in Dicer1-deficient mice is due to impaired miR24 and miR93 expression. *Immunity*, **27**, 123–134.
118. Müller, S. and Imler, J.L. (2007) Dicing with viruses: microRNAs as antiviral factors. *Immunity*, **27**, 1–3.
119. Lecellier, C.H. (2005) A cellular MicroRNA mediates antiviral defense in human cells. *Science*, **308**, 557–560.
120. Triboulet, R., Mari, B., Lin, Y.L., Chable-Bessia, C. *et al.* (2007) Suppression of microRNA-silencing pathway by HIV-1 during virus replication. *Science*, **315**, 1579–1582.
121. Pfeffer, S., Sewer, A., Lagos-Quintana, M., Sheridan, R. *et al.* (2005) Identification of microRNAs of the herpesvirus family. *Nat. Methods*, **2**, 269–276.
122. Umbach, J.L., Kramer, M.F., Jurak, I., Karnowski, H.W., Coen, D.M., and Cullen, B.R. (2008) MicroRNAs expressed by herpes simplex virus 1 during latent infection regulate viral mRNAs. *Nature*, **454**, 780–783.
123. Subramanian, S. and Steer, C.J. (2010) MicroRNAs as gatekeepers of apoptosis. *J. Cell Physiol.*, **223**, 289–298.
124. Bellare, P. and Ganem, D. (2009) Regulation of KSHV lytic switch protein expression by a virus-encoded microRNA: an evolutionary adaptation that fine-tunes lytic reactivation. *Cell Host Microbe*, **6**, 570–575.
125. Lin, X., Liang, D., He, Z., Deng, Q., Robertson, E.S., and Lan, K. (2011) miR-K12-7-5p encoded by Kaposi's sarcoma-associated herpesvirus stabilizes the latent state by targeting viral ORF50/RTA. *PLoS One*, **6**, e16224.
126. Grundhoff, A. and Sullivan, C.S. (2011) Virus-encoded microRNAs. *Virology*, **411**, 325–343.
127. Dölken, L., Krmpotic, A., Kothe, S., and Tuddenham, L. (2010) Cytomegalovirus microRNAs facilitate persistent virus infection in salivary glands. *PLoS Pathog.*, **6**, e1001150.
128. Varble, A., Benitez, A.A., Schmid, S., Sachs, D. *et al.* (2013) An in vivo RNAi screening approach to identify host determinants of virus replication. *Cell Host Microbe*, **14**, 346–356.
129. Machlin, E.S., Sarnow, P., and Sagan, S.M. (2011) Masking the 5′ terminal nucleotides of the hepatitis C virus genome by an unconventional microRNA-target RNA complex. *Proc. Natl Acad. Sci. USA*, **108**, 3193–3198.
130. Roberts, A.P.E., Lewis, A.P., and Jopling, C.L. (2011) miR-122 activates hepatitis C virus translation by a specialized mechanism requiring particular RNA components. *Nucleic Acids Res.*, **39**, 7716–7729.
131. Lefebvre, G., Desfarges, S., Uyttebroeck, F., Muñoz, M. *et al.* (2011) Analysis of HIV-1 expression level and sense of transcription by high-throughput sequencing of the infected cell. *J. Virol.*, **85**, 6205–6211.
132. Parameswaran, P., Sklan, E., Wilkins, C., Burgon, T. *et al.* (2010) Six RNA viruses and forty-one hosts: viral small RNAs and modulation of small RNA repertoires in vertebrate and invertebrate systems. *PLoS Pathog.*, **6**, e1000764.
133. Yeung, M.L., Bennasser, Y., Watashi, K., Le, S.-Y., Houzet, L., and Jeang, K.-T. (2009) Pyrosequencing of small non-coding RNAs in HIV-1 infected cells: evidence for the processing of a viral-cellular double-stranded RNA hybrid. *Nucleic Acids Res.*, **37**, 6575–6586.
134. Qi, X., Bao, F.S., and Xie, Z. (2009) Small RNA deep sequencing reveals role for *Arabidopsis thaliana* RNA-dependent RNA polymerases in viral siRNA biogenesis. *PLoS One*, **4**, e4971.
135. Maillard, P.V., Ciaudo, C., Marchais, A., Li, Y., Jay, F., Ding, S.W., and Voinnet, O. (2013) Antiviral RNA interference in mammalian cells. *Science*, **342**, 235–238.
136. Li, Y., Lu, J., Han, Y., Fan, X., and Ding, S.W. (2013) RNA interference functions as an antiviral immunity mechanism in mammals. *Science*, **342**, 231–234.
137. Sagan, S.M. and Sarnow, P. (2013) RNAi, antiviral after all. *Science*, **342**, 207–208.
138. McNamara, J.O., Andrechek, E.R., Wang, Y., Viles, K.D. *et al.* (2006) Cell type-specific delivery of siRNAs with aptamer-siRNA chimeras. *Nat. Biotechnol.*, **24**, 1005–1015.
139. Peer, D., Park, E.J., Morishita, Y., Carman, C.V., and Shimaoka, M. (2008) Systemic leukocyte-directed siRNA delivery revealing cyclin D1 as an anti-inflammatory target. *Science*, **319**, 627–630.

140. Hamilton, A.J. and Baulcombe, D.C. (1999) A species of small antisense RNA in post-transcriptional gene silencing in plants. *Science*, **286**, 950–952.
141. Elbashir, S.M., Harborth, J., Lendeckel, W., Yalcin, A., Weber, K., and Tuschl, T. (2001) Duplexes of 21-nucleotide RNAs mediate RNA interference in cultured mammalian cells. *Nature*, **411**, 494–498.
142. Gitlin, L., Karelsky, S., and Andino, R. (2002) Short interfering RNA confers intracellular antiviral immunity in human cells. *Nature*, **418**, 430–434.
143. Kapadia, S.B., Brideau-Andersen, A., and Chisari, F.V. (2003) Interference of hepatitis C virus RNA replication by short interfering RNAs. *Proc. Natl Acad. Sci. USA*, **100**, 2014–2018.
144. Brummelkamp, T.R., Bernards, R., and Agami, R. (2002) A system for stable expression of short interfering RNAs in mammalian cells. *Science*, **296**, 550–553.
145. Randall, G., Panis, M., Cooper, J.D, Tellinghuisen, T.L. et al. (2007) Cellular cofactors affecting hepatitis C virus infection and replication. *Proc. Natl Acad. Sci. USA*, **104**, 12884–12889
146. Brass, A.L., Dykxhoorn, D.M., Benita, Y., Yan, N., Engelman, A., Xavier, R.J., Lieberman, J., and Elledge, S.J. (2008) Identification of host proteins required for HIV infection through a functional genomic screen. *Science*, **319**, 921–926.
147. Zhou, H., Xu, M., Huang, Q., Gates, A.T. et al. (2008) Genome-scale RNAi screen for host factors required for HIV replication. *Cell Host Microbe*, **4**, 495–504.
148. Krishnan, M.N., Ng, A., Sukumaran, B., Gilfoy, F.D., and Uchil, P.D. (2008) RNA interference screen for human genes associated with West Nile virus infection. *Nature*, **455**, 242–245.
149. Karlas, A., Machuy, N., Shin, Y., Pleissner, K.P. et al. (2010) Genome-wide RNAi screen identifies human host factors crucial for influenza virus replication. *Nature*, **463**, 818–822.
150. Lanford, R.E., Hildebrandt-Eriksen, E.S., Petri, A., Persson, R. et al. (2010) Therapeutic silencing of microRNA-122 in primates with chronic hepatitis C virus infection. *Science*, **327**, 198–201.
151. Li, B.J., Tang, Q., Cheng, D., Qin, C. et al. (2005) Using siRNA in prophylactic and therapeutic regimens against SARS coronavirus in Rhesus macaque. *Nat. Med.*, **11**, 944–951.
152. DeVincenzo, J. and Lambkin-Williams, R. (2010) A randomized, double-blind, placebo-controlled study of an RNAi-based therapy directed against respiratory syncytial virus. *Proc. Natl Acad. Sci. USA*, **107**, 8800–8805.
153. McCaffrey, A.P., Meuse, L., Pham, T., Conklin, D.S., Hannon, G.J., and Kay, M.A. (2002) Gene expression: RNA interference in adult mice. *Nature*, **418**, 38–39.
154. Shukla, S., Sumaria, C.S., and Pradeepkumar, P.I. (2010) Exploring chemical modifications for siRNA therapeutics: a structural and functional outlook. *ChemMedChem*, **5**, 328–349.
155. Haussecker, D. (2012) The business of RNAi therapeutics in 2012. *Mol. Ther. Nucleic Acids*, **1**, e8.
156. Lorenz, C., Hadwiger, P., John, M., Vornlocher, H.-P., and Unverzagt, C. (2004) Steroid and lipid conjugates of siRNAs to enhance cellular uptake and gene silencing in liver cells. *Bioorg. Med. Chem. Lett.*, **14**, 4975–4977.
157. Khvorova, A., Salomon, W., Kamens, J. (2013) Reduced size self-delivering rnai compounds. US Patent Current U.S. Class: 514/44.0A; Nucleic Acid Expression Inhibitors 536/24.5.
158. Chu, T.C., Twu, K.Y., Ellington, A.D., and Levy, M. (2006) Aptamer mediated siRNA delivery. *Nucleic Acids Res.*, **34**, e73.
159. Neff, C.P., Zhou, J., Remling, L., Kuruvilla, J. et al. (2011) An aptamer-siRNA chimera suppresses HIV-1 viral loads and protects from helper CD4+ T cell decline in humanized mice. *Sci. Transl. Med.*, **3**, 66ra6.
160. Akinc, A., Zumbuehl, A., Goldberg, M., and Leshchiner, E.S. (2008) A combinatorial library of lipid-like materials for delivery of RNAi therapeutics. *Nat. Biotechnol.*, **26**, 561–569.
161. Frank-Kamenetsky, M. and Grefhorst, A. (2008) Therapeutic RNAi targeting PCSK9 acutely lowers plasma cholesterol in rodents and LDL cholesterol in nonhuman primates. *Proc. Natl Acad. Sci. USA*, **105** (33), 11915–11920.
162. Semple, S.C., Akinc, A., Chen, J., Sandhu, A.P. et al. (2010) Rational design of

cationic lipids for siRNA delivery. *Nat. Biotechnol.*, **28**, 172–176.

163. Bartlett, D.W., Su, H., Hildebrandt, I.J., Weber, W.A., and Davis, M.E. (2007) Impact of tumor-specific targeting on the biodistribution and efficacy of siRNA nanoparticles measured by multimodality in vivo imaging. *Proc. Natl Acad. Sci. USA*, **104**, 15549–15554.

164. Zimmermann, T.S., Lee, A.C., Akinc, A., Bramlage, B. *et al.* (2006) RNAi-mediated gene silencing in non-human primates. *Nature*, **441**, 111–114.

165. Palliser, D., Chowdhury, D., Wang, Q.Y., and Lee, S.J. (2005) An siRNA-based microbicide protects mice from lethal herpes simplex virus 2 infection. *Nature*, **439**, 89–94.

166. Morrissey, D.V., Lockridge, J.A., Shaw, L., Blanchard, K. *et al.* (2005) Potent and persistent in vivo anti-HBV activity of chemically modified siRNAs. *Nat. Biotechnol.*, **23**, 1002–1007.

167. Santel, A., Aleku, M., Keil, O., Endruschat, J., and Esche, V. (2006) RNA interference in the mouse vascular endothelium by systemic administration of siRNA-lipoplexes for cancer therapy. *Gene Ther.*, **13**, 1360–1370.

168. Howard, K.A., Rahbek, U.L., Liu, X., Damgaard, C.K. *et al.* (2006) RNA interference in vitro and in vivo using a chitosan/siRNA nanoparticle system. *Mol. Ther.*, **14**, 476–484.

169. Song, E., Zhu, P., Lee, S.K., Chowdhury, D. *et al.* (2005) Antibody-mediated in vivo delivery of small interfering RNAs via cell-surface receptors. *Nat. Biotechnol.*, **23**, 709–717.

170. Kumar, P., Ban, H.S., Kim, S.S., Wu, H. *et al.* (2008) T cell-specific siRNA delivery suppresses HIV-1 infection in humanized mice. *Cell*, **134**, 577–586.

171. Kumar, P., Wu, H., McBride, J.L., Jung, K.-E., Hee Kim, M., Davidson, B.L., Kyung Lee, S., Shankar, P., and Manjunath, N. (2007) Transvascular delivery of small interfering RNA to the central nervous system. *Nature*, **448**, 39–43.

172. Subramanya, S., Kim, S.S., Abraham, S., Yao, J. *et al.* (2010) Targeted delivery of small interfering RNA to human dendritic cells to suppress dengue virus infection

and associated proinflammatory cytokine production. *J. Virol.*, **84**, 2490–2501.

173. Boudreau, R.L. and Davidson, B.L. (2012) Generation of hairpin-based RNAi vectors for biological and therapeutic application. *Methods Enzymol.*, **507**, 275–296.

174. O'Reilly, M., Shipp, A., Rosenthal, E., Jambou, R., Shih, T., Montgomery, M., Gargiulo, L., Patterson, A., and Corrigan-Curay, J. (2012) NIH oversight of human gene transfer research involving retroviral, lentiviral, and adeno-associated virus vectors and the role of the NIH recombinant DNA advisory committee. *Methods Enzymol.*, **507**, 313–335.

175. DiGiusto, D.L., Stan, R., Krishnan, A., Li, H., and Rossi, J.J. (2013) Development of hematopoietic stem cell based gene therapy for HIV-1 infection: considerations for proof of concept studies and translation to standard medical practice. *Viruses*, **5**, 2898–2919.

176. Zhou, J. and Rossi, J.J. (2011) Current progress in the development of RNAi-based therapeutics for HIV-1. *Gene Ther.*, **18**, 1134–1138.

177. Suhy, D.A., Kao, S.C., Mao, T., Whiteley, L. *et al.* (2012) Safe, long-term hepatic expression of anti-HCV shRNA in a nonhuman primate model. *Mol. Ther.*, **20**, 1737–1749.

178. Jackson, A.L., Bartz, S.R., Schelter, J., Kobayashi, S.V., Burchard, J., Mao, M., Li, B., Cavet, G., and Linsley, P.S. (2003) Expression profiling reveals off-target gene regulation by RNAi. *Nat. Biotechnol.*, **21**, 635–637.

179. Scacheri, P.C., Rozenblatt-Rosen, O., Caplen, N.J., Wolfsberg, T.G. *et al.* (2004) Short interfering RNAs can induce unexpected and divergent changes in the levels of untargeted proteins in mammalian cells. *Proc. Natl Acad. Sci. USA*, **101**, 1892–1897.

180. Jackson, A.L., Burchard, J., Leake, D., Reynolds, A. *et al.* (2006) Position-specific chemical modification of siRNAs reduces "off-target" transcript silencing. *RNA*, **12**, 1197–1205.

181. Ui-Tei, K., Naito, Y., Zenno, S., Nishi, K., Yamato, K., Takahashi, F., Juni, A., and Saigo, K. (2008) Functional dissection of siRNA sequence by systematic DNA substitution: modified siRNA with a DNA seed arm is a powerful tool for mammalian gene silencing with significantly reduced

off-target effect. *Nucleic Acids Res.*, **36**, 2136–2151.
182. Horn, T. and Boutros, M. (2010) E-RNAi: a web application for the multi-species design of RNAi reagents – 2010 update. *Nucleic Acids Res.*, **38** (Web Server issue), W332–W339.
183. Agrawal, S., Agrawal, A., Doughty, B., Gerwitz, A., Blenis, J., Van Dyke, T., and Pulendran, B. (2003) Cutting edge: different Toll-like receptor agonists instruct dendritic cells to induce distinct Th responses via differential modulation of extracellular signal-regulated kinase-mitogen-activated protein kinase and c-Fos. *J. Immunol.*, **171**, 4984–4989.
184. Karikó, K., Bhuyan, P., Capodici, J., and Weissman, D. (2004) Small interfering RNAs mediate sequence-independent gene suppression and induce immune activation by signaling through toll-like receptor 3. *J. Immunol.*, **172**, 6545–6549.
185. Marques, J.T. and Williams, B.R. (2005) Activation of the mammalian immune system by siRNAs. *Nat. Biotechnol.*, **23**, 1399–1405.
186. Kleinman, M.E., Yamada, K., Takeda, A., Chandrasekaran, V. *et al.* (2008) Sequence- and target-independent angiogenesis suppression by siRNA via TLR3. *Nature*, **452**, 591–597.
187. Judge, A., Bola, G., Lee, A., and MacLachlan, I. (2006) Design of noninflammatory synthetic siRNA mediating potent gene silencing in vivo. *Mol. Ther.*, **13**, 494–505.
188. Sioud, M. (2007) RNA interference and innate immunity. *Adv. Drug Delivery Rev.*, **59**, 153–163.
189. Hornung, V., Guenthner-Biller, M., Bourquin, C., Ablasser, A. *et al.* (2005) Sequence-specific potent induction of IFN-α by short interfering RNA in plasmacytoid dendritic cells through TLR7. *Nat. Med.*, **11**, 263–270.
190. Sen, G., Flora, M., Chattopadhyay, G., Klinman, D.M., Lees, A., Mond, J.J., and Snapper, C.M. (2004) The critical DNA flanking sequences of a CpG oligodeoxynucleotide, but not the 6 base CpG motif, can be replaced with RNA without quantitative or qualitative changes in Toll-like receptor 9-mediated activity. *Cell. Immunol.*, **232**, 64–74.
191. Poeck, H., Besch, R., Maihoefer, C., Renn, M. *et al.* (2008) 5′-triphosphate-siRNA: turning gene silencing and Rig-I activation against melanoma. *Nat. Med.*, **14**, 1256–1263.
192. Grimm, D., Streetz, K.L., Jopling, C.L., Storm, T.A. *et al.* (2006) Fatality in mice due to oversaturation of cellular microRNA/short hairpin RNA pathways. *Nat. Cell Biol.*, **441**, 537–541.
193. Wang, L., Huang, C., and Yang, J.Y. (2010) Predicting siRNA potency with random forests and support vector machines. *BMC Genomics*, **11** (Suppl. 3), S2.
194. Fellmann, C., Zuber, J., McJunkin, K., Chang, K. *et al.* (2011) Functional identification of optimized RNAi triggers using a massively parallel sensor assay. *Mol. Cell*, **41**, 733–746.
195. Westerhout, E.M., Ooms, M., Vink, M., Das, A.T., and Berkhout, B. (2005) HIV-1 can escape from RNA interference by evolving an alternative structure in its RNA genome. *Nucleic Acids Res.*, **33**, 796–804.
196. Geisbert, T.W., Lee, A., Robbins, M., and Geisbert, J.B. (2010) Postexposure protection of non-human primates against a lethal Ebola virus challenge with RNA interference: a proof-of-concept study. *Lancet*, **375**, 1896–1905.
197. Aach, J., Guell, M., DiCarlo, J.E., Norville, J.E., and Church, G.M. (2013) RNA-guided human genome engineering via Cas9. *Science*, **339**, 823–826.
198. Jinek, M., Chylinski, K., Fonfara, I., Hauer, M., Doudna, J.A., and Charpentier, E. (2012) A programmable dual-RNA-guided DNA endonuclease in adaptive bacterial immunity. *Science*, **337**, 816–821.

11
Stem Cell Therapy for Alzheimer's Disease

Rahasson R. Ager[1] and Frank M. LaFerla[1,2]

[1] University of California, Institute for Memory Impairments and Neurological Disorders, Irvine, CA 92697-4545, USA

[2] University of California, Department of Neurobiology and Behavior, Irvine, CA 92697-4545, USA

1	**Introduction**	385
1.1	Alzheimer's Dementia	385
1.2	Amyloid Cascade Hypothesis	387
1.3	Current Therapeutic Approaches for the Treatment of Alzheimer's Disease (AD)	387
1.3.1	Cholinesterase Inhibitors and NMDA Antagonists	387
1.3.2	Aβ Vaccines and Anti-Aβ Immunotherapy	388
1.3.3	β- and γ-Secretase Inhibitors	388
2	**Stem Cell Therapy for Alzheimer's Disease**	390
2.1	Potential Therapeutic Stem Cell Populations	392
2.1.1	Embryonic Stem Cells	392
2.1.2	Induced Pluripotent Stem Cells	392
2.1.3	Adult-Derived Neural Stem Cells	393
2.1.4	Adult-Derived Non-Neuronal Stem Cells	393
2.1.5	Endogenous Neural Stem Cells	394
2.2	Therapeutic Mechanisms	394
2.2.1	Enhancing Endogenous Neurogenesis	394
2.2.2	Exogenous Stem Cell Replacement Therapy	396
2.2.3	Gene Replacement Therapy	397
2.3	Road Blocks to Clinical Application	399
2.3.1	Potential Effects of AD Pathology on Stem Cell Physiology	399
2.3.2	Translational Potential of Transgenic Models of AD	399
3	**Conclusions**	400

Translational Medicine: Molecular Pharmacology and Drug Discovery
First Edition. Edited by Robert A. Meyers.
© 2018 Wiley-VCH Verlag GmbH & Co. KGaA. Published 2018 by Wiley-VCH Verlag GmbH & Co. KGaA.

Glossary 400

References 401

Keywords

Alzheimer's disease
The most common cause of dementia the disease is characterized by memory loss, impairments in daily living activities, and behavioral disturbances. The pathological hallmarks of Alzheimer's disease are extracellular amyloid-beta-laden plaques, dystrophic neuritis, and intracellular neurofibrillary tangles (NFTs) composed of tau protein.

Stem cell therapy
An emerging therapy that seeks to introduce new cells into damaged tissue with the goals of either replacing dead cells or providing support for surviving cells

Neural stem cells
A stem cell population that is multipotent that is, it can generate the main phenotypes of the nervous system and are self-renewing

Amyloid-beta (Aβ)
Aβ is derived by endoproteolysis of the Amyloid Precursor Protein, and is the principle component of the amyloid-beta plaque

Transgenic models
Transgenic models are organisms that have DNA from a different species entered into their genome. Transgenic models can be designed to model human diseases and assist with the development of potential therapies or used as tools to understand how the disease operates

The escalating prevalence of Alzheimer's disease (AD) is on course to exact an immense medical, social, and economic impact on the United States, and most of the western world. Current therapies are palliative, providing only marginal symptomatic relief in some patients, and there is an urgent need to develop more effective interventions, particularly those that exert disease-modifying effects. Several emerging treatment strategies designed to inhibit disease progression by directly targeting the production and degradation of the amyloid-beta peptide have been tested in clinical trials. Unfortunately, many of these new strategies have failed to significantly ameliorate cognitive deficits, and some have actually exacerbated cognitive decline. The emerging field of stem cell biology offers tremendous therapeutic potential for chronic neurological conditions such as AD, particularly if combined with a multitargeted therapeutic approach. The current advances on stem cells therapies in AD animal models are summarized in this chapter, and potential roadblocks for transitioning this novel treatment into AD patients are discussed.

1
Introduction

1.1
Alzheimer's Dementia

The most common degenerative disease affecting brain function is Alzheimer's disease (AD), a progressive disorder that is the leading cause of dementia worldwide. Typically, AD patients suffer from a progressive neurodegeneration that gradually impairs their memory, and their ability to learn and to carry out daily activities. The United States Alzheimer's Association estimates that 5.4 million Americans are currently afflicted, with a new case developing every 68 s. By 2050, it is estimated that the prevalence of AD in the United States will be around 16 million cases, with the majority occurring in individuals aged over 65 years [1]. The incidence of AD further increases with age, with one in eight individuals aged over 65 years and as many as half of individuals aged over 85 years afflicted with the disease [2, 3].

Alzheimer's disease is associated with well-defined neuropathological features, including extensive synaptic and neuronal loss, neuroinflammation and gliosis, cortical atrophy, cerebrovascular dysfunction, extracellular (which can form either diffuse or neuritic plaques) and intracellular aggregates of amyloid-beta (Aβ), and intracellular aggregates of hyperphosphorylated tau protein known as neurofibrillary tangles (NFTs). These pathological alterations are mainly found within the temporal lobe, including the basal forebrain cholinergic system and limbic structures such as the hippocampus and amygdala [4].

The Aβ peptide that accumulates in amyloid plaques and in the cerebrovasculature of the AD brain is derived from the endoproteolysis of amyloid precursor protein (APP) through serial actions of the β- and γ-secretase enzymes, liberating the N and C termini of Aβ, respectively [5]. Although the physiological function of APP is still unresolved, there are reports of the protein being involved in neurite outgrowth, neuronal selection, and neuronal differentiation during development [6–9].

Tau is a component of the microtubule machinery, and functions to provide structural integrity and axonal transport in neurons. The phosphorylation of tau is critical for regulating its normal physiological function; however, hyperphosphorylation of tau can cause a breakdown of the microtubule structure, which in turn leads to the release of tau and subsequent aggregation of the protein into filamentous bundles known as NFTs [10]. It is widely hypothesized that the neuropathological features of AD contribute to the clinical manifestations, as opposed to simply being a marker of disease progression, and it is notable that tau aggregates correlate better with cognitive evaluation scores [11].

The majority of AD cases are sporadic in nature, with less than 5% of patients displaying an autosomal dominant, or familial, inheritance pattern. Currently, the most substantial genetic link to sporadic AD is apolipoprotein-E (ApoE), which affects disease onset [12, 13]. ApoE is located on chromosome 19, and is important for the transportation of cholesterol in the body, and is secreted by glia within the central nervous system (CNS) [14]. The *ApoE* gene is polymorphic, existing in three allelic variants: ε2, ε3, and ε4. The ε4 allele is overrepresented in AD, with 45–60% of patients possessing at least one copy. Inheritance of the ε2 allele may confer protection, whereas persons who inherit one ε4 allele have a fourfold higher risk of developing AD compared to ε3 homozygous individuals, and the risk is further increased 19-fold in ε4 homozygous individuals [15, 16].

In autosomal dominant familial Alzheimer's disease (FAD) cases, three genes cause disease onset when mutated. The first mutation was identified in the gene for APP, the parental protein of Aβ [17]; however, the majority of FAD cases are caused by missense mutations in two other genes encoding the catalytic unit of the gamma-secretase complex, presenilin 1 (PS1) and presenilin 2 (PS2) [18–21]. All FAD cases have a high penetrant rate and an average age of onset ranging from 28 to 60 years.

The clinical symptoms of AD can be divided into three groups: cognitive symptoms; psychiatric symptoms; and instrumental activities symptoms (involving activities of daily living) [22]. The degree of mental and emotional impairment associated with AD can be partially attributed to the widespread degeneration of multiple types of neuronal network within the brain [23]. The loss of memory – particularly episodic memory – is commonly the earliest reported symptom in AD. A diagnosis of dementia requires memory loss plus dysfunction in another cognitive domain, and patients who are not yet demented but experience memory loss may be classified as amnesic mild cognitive impairment (a-MCI), a recognized pre-AD stage. Cognitive impairment continues throughout the course of the disease, and instrumental deficits appear later as individuals progress

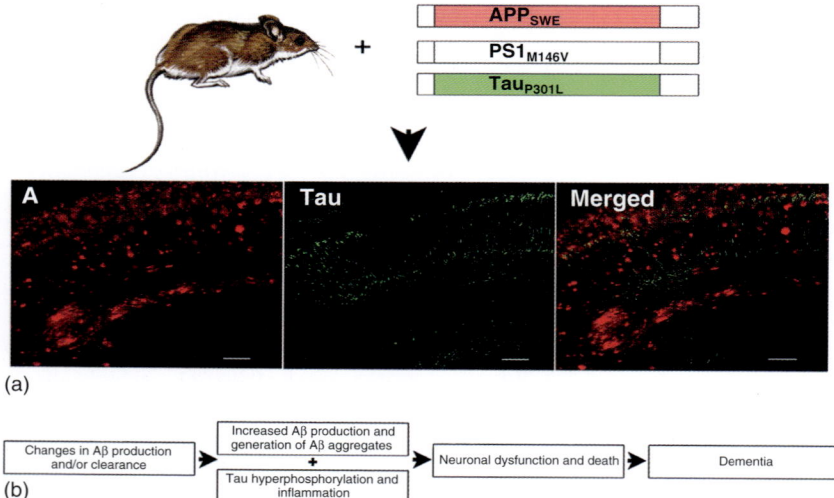

Fig. 1 Aβ and Tau in Alzheimer's disease (AD). (a) Aβ and tau comprise the hallmark neuropathological alternations associated with AD. Transgenic mouse models such as the 3xTg-AD model, which contains three distinct human mutated genes that allow for the age-dependent development of both Aβ and tau pathology, have been valuable tools in deciphering the role that Aβ and tau play in the development and progression of AD; (b) Cumulative studies using human samples and transgenic models, such as the 3xTg-AD model, have found that mutations which lead to early AD onset result in changes in Aβ levels, particularly $Aβ_{1-42}$. The elevated Aβ levels result in Aβ aggregation that can also lead to a series of downstream events that result in the initiation of tau hyperphosphorylation, neuronal injury, and the death of susceptible neurons. These findings comprise the working theory of AD progression known as the *amyloid cascade hypothesis*.

into moderate AD. The psychiatric symptoms of AD generally manifest in later disease stages and are associated with advanced cognitive decline [24]. Typical symptoms include, paranoia, hallucinations and depression, and can result in increased emotional distress in family members and caregivers [25].

1.2
Amyloid Cascade Hypothesis

Introduced almost two decades ago, the amyloid cascade hypothesis continues to be the most referenced hypothesis explaining the pathophysiology of AD (Fig. 1) [26]. The hypothesis originates from the overwhelming genetic and molecular evidence linking Aβ to autosomal AD (and also to Down syndrome cases). The derivative of APP, Aβ, was first isolated and sequenced from the blood vessels of AD patients during the 1980s [27]. The peptide was subsequently determined to be the primary component of the neuritic plaques that are found characteristically in the brain parenchyma of AD patients [28]. Once the *APP* gene had been cloned and found to exist on chromosome 21 [29–31], Aβ processing and deposition became a focal point in AD research, in part, due to prior knowledge that Down syndrome individuals contained a trisomy at chromosome 21 and displayed an AD-like pathology [32]. The first mutations in *APP* were identified from the analysis of FAD patients [17]. It was later discovered that most of the mutations found in *APP* lead to either an imbalance in secretase processing by favoring β-/γ-secretase cleavage over α-/γ-secretase cleavage, or to an increased propensity for aggregation of Aβ into more toxic species. Subsequent genetic analysis studies also demonstrated that mutations in PS1 and PS2 could further alter APP processing by increasing the Aβ42/Aβ40 ratio, causing an accelerated aggregation of the peptide.

The other major aggregated protein found in the AD brain, tau, has no known genetic mutations associated with AD, although there are several mutations in tau that can cause frontotemporal dementia (FTD), another severe neurodegenerative disease [33, 34]. Typically, FTD is characterized by NFT deposition but, unlike AD, overt Aβ pathology is absent. Additionally, transgenic mice containing mutations in both human APP and tau, demonstrate that Aβ pathology emerges prior to tau pathology [35]. Taken together, this compounding evidence adds additional support to the amyloid cascade hypothesis, which proposes that an imbalance between Aβ production, aggregation, and/or clearance initiates the neuropathology and clinical phenotype found in AD.

1.3
Current Therapeutic Approaches for the Treatment of Alzheimer's Disease (AD)

1.3.1 Cholinesterase Inhibitors and NMDA Antagonists

Currently, there are just five US Food and Drug Administration (FDA)-approved medications for the treatment of AD. Four of these drugs inhibit the enzyme acetylcholinesterase (AChE), which is found in the synaptic cleft and functions to degrade acetylcholine. It has long been documented that AD is associated with an extensive loss of cholinergic neurons that occurs in affected brain regions [36], and therefore increasing the available acetylcholine may reverse some of AD-associated clinical symptoms. AChE inhibitors are used to slow the degradation of released acetylcholine, thereby increasing the acetylcholine concentration in the brain [37]. The four AChE inhibitors approved by the FDA – donepezil, galantamine, rivastigmine, and tacrine – are only approved for mild to moderate AD and are exclusively symptomatic treatments; that is, they do not function as curative agents and are not considered disease-modifying interventions.

There is currently one FDA-approved N-methyl-D-aspartate (NMDA) antagonist for the treatment of AD. Unlike the AChE inhibitors, memantine (Namenda®) has been approved for the treatment of moderate to severe AD [38, 39]. In the brains of AD patients, dysfunctional regulation of neuronal excitatory transmission can lead to excess glutamate release, which repeatedly binds to the NMDA receptor, causing neuronal excitotoxicity [40]. In 3xTg-AD model mice, memantine was shown to restore cognition and attenuate Aβ and tau pathology [41]. Memantine interacts with the NMDA receptor and may block abnormal signaling by glutamate (an excitatory neurotransmitter), and thus allow for a normal functioning of the receptor [42]. Memantine has been shown to provide further additional benefits when administered to patients already receiving an AChE [43].

1.3.2 Aβ Vaccines and Anti-Aβ Immunotherapy

While the AChE inhibitors and memantine transiently restore some cognitive symptoms in AD patients, they do not halt or reverse the progression of the disease. Hence, an urgent need persists to develop disease-modifying strategies that target the underlining causes to better treat or cure AD. Aβ, as the hypothesized initiator of AD, has been the primary target of immunotherapy since the late 1990s, with several therapeutic strategies aimed at neutralizing or eliminating Aβ. By the start of the last decade, immunization experiments targeting Aβ, using synthetic full-length Aβ_{1-42}, had been conducted in transgenic AD model mice. The results showed that the immune response of mice immunized with full-length Aβ_{1-42} was primed to prevent further deposition of extracellular Aβ as well as the formation of dystrophic neuritis and astrogliosis [44, 45]. Subsequent studies on active immunization with Aβ_{1-42}, using a variety of Aβ immunogens and various administration routes, successfully reduced the cerebral Aβ burden, increased Aβ plasma levels, and improved cognitive performance in animal models [46]. Further experiments with the 3xTg-AD mouse model also demonstrated that the targeting of Aβ, through passive immunization, can clear interneuronal Aβ and attenuate tau pathology [47, 48].

The encouraging results obtained from Aβ immunization studies in mouse models of AD were successfully translated to a Phase I safety trial using the AN1792 vaccine developed by Elan and Wyeth. However, the trial was halted in early 2002 because several patients began to show signs of meningoencephalitis and leukoencephalopathy [49]. Despite the occurrence of these adverse effects during the initial AN1792 vaccine trails, several new passive and active immunotherapies are currently being tested in clinical trials [50], including bapineuzimab (Elan and Wyeth), solanezumab (Eli Lilly), IVIG (Baxter Healthcare), ACC-001 (Elan and Wyeth), and CAD-106 (Novartis). Elan/Wyeth and Novartis are both also pursuing active vaccine strategies. All of these new immunotherapy strategies have passed early safety studies, and Phase II and III trials are currently underway.

1.3.3 β- and γ-Secretase Inhibitors

The secretase inhibitors, which target APP processing, represent another class of emerging therapeutics aimed at decreasing

Aβ levels [51]. The β-secretase enzyme BACE1 (beta-site APP-cleaving enzyme 1B-secretase) is the rate-limiting first step in Aβ production, and has emerged as a possible prime drug target [52]. Interestingly, the results of genetic studies using BACE1-deficient mice have suggested that the blocking of β-secretase activity eliminates Aβ accumulation in the CNS, without any substantial negative biochemical or behavioral consequences [53, 54]. However, another group found recently that BACE1 knockout was associated with the development of a schizophrenia-like behavior [55]. In addition, it has been shown that BACE1 cleaves substrates other than APP, such as neuregulin−1 (Nrg−1) [56].

Unfortunately, due to BACE1 having a multitude of targets, there is a great potential for its inhibitors to cause adverse toxic effects. Ideally, a BACE1 inhibitor used to treat AD should selectively target APP processing alone. In a recent study conducted by Fukumoto and colleagues, using the nonpeptidic compound TAK-070 (a noncompetitive BACE1 inhibitor), a decrease in soluble Aβ levels and an increase in sAPPα were observed; this demonstrated the effectiveness of selective BACE1 inhibition against APP while sparing enzymatic activity towards its other targets [57]. Yang et al. also reported that the peptide compound S1 can selectively inhibit the hydrolytic activity of BACE1 by binding directly to APP [58]. The latter authors went on to show that S1 could reduce the brain Aβ burden and also improve spatial memory in APPswe/PS1dE9 double transgenic mice.

Although great strides were subsequently made in the discovery of more selective BACE1 inhibitors, most of these initial compounds were potent inhibitors *in vitro* but were unable to adequately penetrate the blood–brain barrier (BBB); moreover, they also demonstrated a variety of adverse side effects due to their nonselective nature. Consequently, other research groups began to explore the possibility of reducing BACE1 activity that would provide a cognitive benefit but limit any off-target toxicities [59, 60]. One example of this involved crossing an AD transgenic mouse model with a BACE1-deficient model; this led to a significant reduction in Aβ burden in these PDAPP/BACE1± mice [61], and suggested that a partial inhibition of BACE1 could safely reduce the Aβ burden.

γ-Secretase, a membrane-associated protease complex with a catalytic core comprised of PS1 and PS2 [62], has for many years been a highly sought-after therapeutic target for the treatment of AD [63]. In addition to APP, γ-secretase targets a diverse range of type-1 integral membrane proteins. For example, Notch proteolysis by γ-secretase results in the generation of the Notch intracellular domain fragment (NICD), which regulates many essential cellular functions [64]. A major challenge to the development of a successful γ-secretase inhibitor has been to create an inhibitor of APP proteolysis, without interfering with the cleavage of other substrates, especially Notch cleavage. Early during the last decade, the first results were reported of *in-vivo* testing in transgenic mice with a γ-secretase inhibitor, the compound DAPT, as developed by Elan and Eli Lilly [65]. During recent years, γ-secretase inhibitors have been designed and synthesized by several groups, but clinical trials have revealed serious adverse side effects. For example, in 2011 semagacestat (Eli Lilly) was halted in Phase III clinical trials as it exacerbated cognitive deficits in some participants, potentially through off-target effects.

In contrast to γ-secretase inhibition, γ-secretase modulators are compounds that can shift Aβ production away from $Aβ_{1-42}$ towards shorter and less-pathogenic species, or selectively target APP processing only and thus spare Notch cleavage. The first class of $Aβ_{1-42}$-lowering γ-secretase modulators consisted of anti-inflammatory compounds such as ibuprofen [66], though since their discovery additional compounds have been developed and tested in both preclinical and clinical settings [67]. For example, the anti-inflammatory arylacetic acid-related compound, CHF5074, has been found to lower the plaque burden and reverse the cognitive phenotype found in AD transgenic mice [68, 69]. While the pursuit of improved γ-secretase modulators continues, the successful transition of a γ-secretase targeted therapeutic that inhibits APP processing without interfering with other γ-secretase targets continues to be a much sought-after goal.

2
Stem Cell Therapy for Alzheimer's Disease

Historically, the predominant strategy targeting the restoration of cognitive function, in response to neurological insult, has involved the administration of small-molecule-based neuroprotective drugs. However, results obtained from clinical trials with AD patients have revealed several common limitations of the classical small-molecule treatment strategy:

- Current AchE inhibitors target only a single neuronal population, but the disease affects several different populations.
- Penetration of the BBB by the candidate drug is poor (this is particularly true for β-secretase inhibitors).
- The candidate drug often causes harmful side effects.

Due to the limitations of conventional therapeutics, AD treatment has remained stagnant for many years. In addition, the successes of the past decade in generating Aβ-modifying therapies using preclinical models, have encountered difficulties when translating to human clinical trials. Therefore, there remains an imperative demand for novel treatment strategies that may target the underlining causes of AD. As a result, stem cell therapy has emerged as an alternative strategy to the classical small-molecule approach to drug development and, indeed, over the past decade investigations in the field of regenerative medicine have provided a proof of principle for stem cell-based therapies in several CNS disorders, including Parkinson's disease, Huntington's disease, spinal cord injury, and AD. It is important to note, however, that the majority of these data has been acquired from only a few animal models, and additional preclinical studies must be conducted to verify the efficacy of stem cells (Table 1).

One critical step towards the development of a successful stem cell therapy to treat any disease is to identify an optimal stem cell population that can be used for clinical application. Several criteria are important when selecting a potential stem cell candidate for transplantation into AD patients; notably, a stem cell population should be selected that is negative for the ε4 allele. In addition, because AD is a disease of the CNS, stem cells that can directly generate neurons and glia might offer certain advantages over stem cell populations that do not. Moreover, with the AD population predicted to more than triple in size by the mid-2000s, an optimal stem cell source should also be capable of generating sufficient material to treat millions of potential patients. The different stem cell populations and their therapeutic potential

Tab. 1 Studies using stem cell-based therapy in animal models of Alzheimer's disease.

Cell type	Model	Treatment duration	Immunosuppression used	Transplant survival percentage	Changes in pathology	Functional recovery	References
Mouse, NS cells	Mouse, nbM ibotenic acid lesion	8, 12 weeks	None	NR	NR	Improved RAM performance	Wang et al. [127]
Mouse, NS cells	Mouse, diphtheria toxin Hp lesion	1, 3 months	None	NR	Reported increase in synaptic density	Improved NOR-P performance	Yamasaki et al. [129]
Mouse, NS cells	Rat, $A\beta_{1-40}$ injection Hp	4, 16 weeks	CsA, 10 mg kg^{-1}	6.65% (1 month) 3.01% (4 months)	NR	Improved MWM performance	Tang et al. [164]
Mouse, BMSCs	Rat, $A\beta_{1-40}$ injection into Hp	15 days	None	NR	Reported increase in neuron number	Improved MWM performance	Li et al. [146]
Mouse, NS cells	Rat, nbM ibotenic acid lesion	4 weeks	CsA, 10 mg kg^{-1}	NR	NR	Improved MWM performance	Moghadam et al. [128]
Mouse, NS cells	Mouse, 3xTg-AD	4 weeks	None	NR	Reported increase in synaptic density	Improved MWM and NOR-CD performance	Blurton-Jones et al. [141]
Rat, NS cells	Rat, $A\beta_{1-42}$ injection into Hp	7 days	None	NR	Reported reduction in gliosis and neurite loss	NR	Ryu et al. [151]
Mouse, BM-MSCs	Mouse, APP/PS1	2 months	None	NR	Reported decreased amyloid load and less phosphorylated tau	Improved MWM performance	Lee et al. [152]
Mouse, NS cells	Mouse, APP/PS1	1 month	None	NR	Reported decreased plaque load	NR	Nije et al. [153]

APP, amyloid precursor protein; BM-MSC, bone marrow mesenchymal stem cell; BMSC, bone marrow stromal cell; CsA, cyclosporine; Hp, hippocampus; MWM, Morris water maze; nbM, nucleus basalis Meynert; NOR-CD, context-dependent novel object recognition; NR, not reported; NOR-P, place-dependent novel object recognition; NS, neural stem; PS1, presenilin 1; RAM, radial arm maze.

are discussed in the following section, and the most prominent mechanisms of action for stem cell therapies (focusing on AD preclinical studies) are reviewed.

2.1 Potential Therapeutic Stem Cell Populations

Stem cells differ from adult somatic cells due to their ability to give rise to multiple cell types, and also their capacity for self-renewal. Three distinct classes of stem cells may be used for transplantation therapy, namely embryonic stem (ES) cells, adult stem cells, and induced pluripotent stem (iPS) cells. In addition, most adult tissues have regions that contain endogenous stem cells sources, which may also offer therapeutic potential without the need for surgical transplantation.

The identification of an optimal stem cell source is one of several critical questions that need to be addressed before any large-scale clinical applications can proceed in humans. Despite the need to further explore the putative mechanisms behind the reported regenerative capabilities of stem cells, the latter continue to represent a revolutionary approach towards the treatment of neurodegenerative disorders which warrants intensive exploration.

2.1.1 Embryonic Stem Cells

ES cells represent the first identified pluripotent stem cell source, and have been increasingly studied over the past few decades [70]. Typically, ES cells are derived from a collection of cells found in the inner cell mass of the blastocyst during embryogenesis, and can be expanded indefinitely *in vitro*. These cells are pluripotent, and thus can give rise to a wide variety of progeny representative of the three germ layers. Experiments utilizing ES cells in animal models of CNS insult have shown promising results [71–73]. However, undifferentiated ES cells have been shown to possess a high tumorigenic potential [74]. In order to circumvent the tumorigenicity of undifferentiated ES cells, methods have been devised to differentiate ES cells to more injury relevant cell types. ES cells can be successfully differentiated into neural stem (NS) cells with the addition of neurotrophic factors such as fibroblast growth factor (FGF) and epidermal growth factor (EGF) [75–77]. The differentiated progeny of ES cells have also been shown to possess tumorigenic potential [78], though this observation has been postulated to depend on the presence of undifferentiated ES cells within the transplanted population [79]. Clearly, further studies are required before cells derived from ES cells can be considered safe for transplantation into patients. Along with their potential tumorigenic capacity, the use of ES cells as a source for stem cell therapy has been impeded due to not only supply limitations but also ethical and political issues.

2.1.2 Induced Pluripotent Stem Cells

One way of circumventing the ethical and political issues facing the use of pluripotent ES cells is through the creation of iPS cells. Unlike ES cells, iPS cells are created *in vitro* by reprogramming somatic cells to a pluripotent state. In 2006, a research group in Japan was the first to successfully convert mouse fibroblasts into iPS cells using a retroviral approach [80], and subsequent studies led to the development of human iPS cells [81, 82]. The reprogramming of somatic cells to iPS cells provides a means of generating patient-specific cells, thereby circumventing most of the problems of immune rejection that are encountered when transplanting foreign cells into a patient. Notably, iPS cells can also be differentiated into neurons and neural precursor cells, and have been shown to provide functional recovery, after transplantation, in rat models of Parkinson's disease [83, 84]. The derivation of iPS cells has been described for FAD patients, followed by differentiation into neurons [85]; however, no iPS cell-derived neurons have yet been tested in transplantation studies in AD mouse models.

Induced pluripotent stem cells possess many biological qualities in common with ES cells, including the expression of stem cell surface markers and the ability to proliferate indefinitely [86]; however, they continue to demonstrate some genetic and epigenetic differences [87]. Although iPS cells possess unique genetic properties compared to ES cells, they are also prone to tumor formation after transplantation [88], and therefore additional studies are required before clinical transplantation studies can be commenced.

2.1.3 Adult-Derived Neural Stem Cells

Deriving stem cells from a patient's own tissue circumvents one of the major problems associated with transplantation, namely immune rejection. In contrast to the reprogramming of somatic cells is the direct isolation from the brain of adult Neural Stem Cells (NSCs), the collection of which raises very few ethical or political issues associated with the use of human embryos, if fetal sources of tissue can be avoided. When adult NS cells are isolated and grown *in vitro*, they demonstrate the capacity for expansion and differentiation into all known CNS lineages: neurons, astrocytes, and oligodendrocytes [89]. Adult NS cells also display the capacity for differentiation even after *in vivo* transplantation [90]. Thus, adult NS cells may offer a therapeutic tool for cell replacement that is free from many of the potential setbacks associated with ES cell and iPS cell usage. Unfortunately, adult NS cells tend to differentiate into interneurons, when transplanted to neurogenic regions, or astrocytes if transplanted elsewhere [91], which makes the replacement of specific neuronal subtypes potentially difficult. In addition, the accessibility of adult NS cells is arduous, and alternative pools of cells may be necessary for broad clinical use. One such alternative would be to use adult-derived non-neuronal somatic stem cells.

2.1.4 Adult-Derived Non-Neuronal Stem Cells

Recently, several non-neuronal stem cell types have emerged as potential candidates for cell transplantation therapy for CNS disorders. For example, bone marrow is more accessible than the CNS as a source of adult stem cells and contains both hematopoietic stem cells (HSCs) and mesenchymal stem cells (MSCs). In fact, HSCs have for decades been used successfully in the treatment of various autoimmune diseases [92, 93], which illustrates their resistance to immune rejection and means that they are well tolerated by the recipient. However, more attention has recently been focused on the potential use of MSCs for the treatment of CNS disorders. Once the bone marrow has been collected, dissociated and plated onto a culture dish, the MSCs will adhere directly to the plastic surfaces and can differentiate into osteoblasts and chondrocytes [94]. Once isolated and propagated *in vitro*, many more MSCs could then be harvested for clinical use.

Previously, MSCs have been reported to generate both neurons and astrocytes *in vivo* [95–97]. In addition, MSCs have been shown to express neurotransmitter-related genes; for example, in 2002, MSC-derived neuron-like cells were shown to express choline acetyltransferase (ChAT) [98], while in a subsequent study a combination of trophic factors, including sAPPα, was used to differentiate MSCs into a cholinergic-like neuronal phenotype [99]. Unfortunately, it is unknown whether MSC differentiation is a significant contributor to the beneficial effects observed after transplantation. Although most *in vitro* studies have shown the detection of neural cell surface markers, the establishment of functional neuronal properties has not been clearly demonstrated. It has been shown that MSCs are capable of secreting growth factors known to promote neuronal protection and rescue, including neurotrophin-3 (NT-3), nerve growth factor (NGF), and brain-derived neurotrophic factor (BDNF) [100, 101], which may underlie the primary mechanism of the observed beneficial effects after transplantation with MSCs. Regardless of the precise mechanism of action, MSCs represent a potential non-neuronal stem cell population that is free from most ethical concerns and should be further explored for its neuroregenerative properties.

2.1.5 Endogenous Neural Stem Cells

Part of the early dogma concerning the properties of the adult mammalian brain was the notion that the generation of new neurons could no longer occur. During its development, it was recognized that the CNS arises from stem cells that possessed the ability to differentiate and self-renew. However, during the 1960s proliferating neurons were first identified in the adult rodent brain [102] and eventually confirmed in the adult human hippocampus [103], since which time it has become widely accepted that NS cells exist within specialized areas of the adult brain. When embryonic neurogenesis occurs, uncommitted endogenous NS cells can be isolated from virtually the entirety of the CNS. This broad distribution pattern of early neurogenesis is greatly restricted in the adult brain, being concentrated predominantly in two localized areas [103]: (i) an area localized around the lateral ventricles, referred to as the subventricular zone (SVZ); and (ii) a layer of the hippocampal dentate gyrus (DG), referred to as the subgranular zone (SGZ). These two germinal niches contain a population of heterogeneous NS cells that demonstrate many of the same characteristics observed in NS cells isolated from the developing brain, such as an ability to self-renew and also multipotency. In addition to the SVZ and SGZ, reports have been made of NS cells residing in both the gray and white matter of the cerebral cortex and spinal cord and the olfactory mucosa [104–106]. The limited number of specialized germinal niches within the CNS suggests that the unique properties of the microenvironment surrounding the germinal centers must be retained after development. Although, unfortunately, the exact constitution of the microenvironment is currently unknown, determination of the germinal center microenvironment will provide vital information towards the molecular cues involved in NS cell maintenance and also assist with propagation and differentiation *in vitro*, as well as potentially upregulating neurogenesis *in vivo*. Several soluble factors have been identified which promote neuronal development, including FGF-2, insulin-like growth factor-1 (IGF-1), and BDNF [107–109].

2.2 Therapeutic Mechanisms

2.2.1 Enhancing Endogenous Neurogenesis

Neurogenesis is defined as the process by which new neurons are born and introduced into the pre-existing neuronal networks. The rate of neurogenesis has been reported to decline with age [110]. Although the full physiological function of neurogenesis in the adult CNS is unknown, emerging data suggest that new neurons may play a role in hippocampal-dependent learning and memory [111]. Therefore, the reduced neurogenesis observed in the aging brain may predispose individuals to an onset of early memory impairment.

In the case of AD, the hippocampus is affected at an early stage of the disease, and progressive and irreversible neuronal loss continues as the disease progresses. The hippocampus is one of two major brain regions in which neurogenesis continues well into adulthood; hence, neurogenesis may be significantly altered through the pathological process occurring in AD. Interestingly, there have been conflicting accounts regarding the rate of neurogenesis in AD brain. Initially, neurogenesis was reported to be enhanced in the AD brain [112], although subsequent studies reported that neurogenesis was significantly reduced in AD patients [113, 114]. Results from transgenic animal models have also been conflicting. The results of a study conducted by Lopez-Toledano and Shelanski indicated that young APP transgenic mice had increased NS cell proliferation and differentiation, which later decreased with age [115]. In contrast, the results of other studies described continuous disruptions in neurogenesis in a variety of transgenic AD models [116, 117]. Haughey et al. found that both NS cell proliferation and survival was reduced in the SGZ of the hippocampus [118], while Rodriquez et al. demonstrated an impaired neurogenesis, also in the SGZ of the hippocampus, in the presence of both tau and Aβ pathology in 3xTg-AD mice [119]. One factor that may contribute to the disparities in neurogenesis reported in transgenic models of AD is the presence of autosomal dominant FAD-linked genes that may directly affect neurogenesis [120].

For example, PS1 and the cleavage products of APP have been demonstrated to affect various pathways of neurogenesis [121, 122] (for an in-depth review on the impact of APP on neurogenesis, see Ref. [123]). In addition to the potential effects of the AD mutation within the transgenic model, previous studies examining neurogenesis in the hippocampus have shown that the rate of mature neuronal generation differs between particular mouse strains [124]. The latter authors found that, among four inbred strains commonly used to generate AD models, the mouse strains C57BL/6J, BALB/cJ, and 129/SvJ possessed varying degrees of neurogenesis, with C57BL/6J mice having the highest levels of newly proliferating cells.

Nevertheless, the natural response of endogenous NS cells to AD pathology does not appear to significantly alter disease progression, or result in any significant functional recovery. The inability of endogenous NS cells to fully compensate for the loss or dysfunction of established neural circuitry could be due to several reasons, including an insufficient mobilization response. The activation of neurogenesis requires a number of factors, including trophic factors such as BDNF [125]. For example, mice with decreased levels of BDNF (as would occur in AD) responded poorly to environmental-induced increases in neurogenesis [126]. In addition, toxins in the surrounding microenvironment of the injured tissue may affect endogenous neurogenesis. One potential approach to address this limitation would be to stimulate neurogenesis via pharmacologic intervention, utilizing also neurotrophic factors such as BDNF, or potentially, ciliary neurotrophic factor (CNTF) and vascular endothelial growth factor (VEGF) [109, 127, 128] (Fig. 2). Allopregnanolone, a metabolite of progesterone, has also been

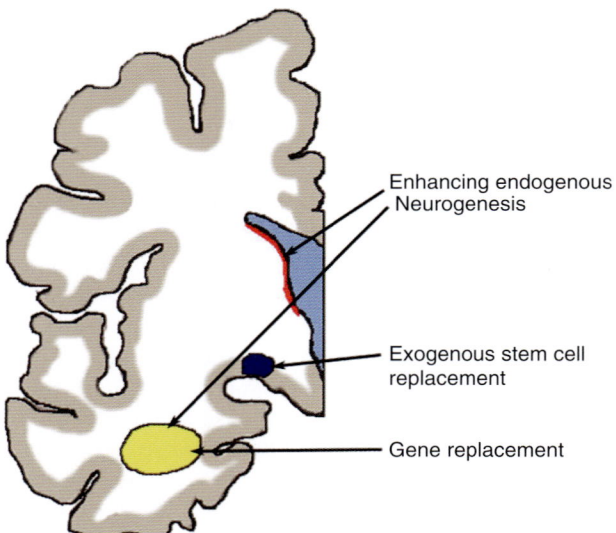

Fig. 2 Potential mechanisms for stem cell-based therapy in the Alzheimer's disease brain. Stem cell-based therapy could be used to promote endogenous neurogenesis in either the subventricular zone (red) or the subgranular zone of the hippocampus (yellow) by administering small molecules that can cross the blood–brain barrier. In addition, cell replacement strategies, utilizing exogenous stem cells sources could be targeted to regions of the brain susceptible to extensive neuronal loss such as the basal forebrain cholinergic system (blue). Alternative strategies could use exogenous stem cells to deliver growth factors or anti-inflammatory mediators to other highly damaged regions, such as the hippocampus.

shown to increase neurogenesis in the DG of 3xTg-AD mice, which suggests the potential of non-neurotrophic molecules for enhancing neurogenesis [121, 129]. Protective measures may not be met through an enhancement of endogenous neurogenesis alone, however, and it may be necessary to first expand NS cells *ex vivo* and to then transplant them into damaged regions, in the hope that exogenous NS cells could overcome the limitations predicted for endogenous populations.

2.2.2 Exogenous Stem Cell Replacement Therapy

Stem cell replacement therapy is defined as a means to replace injured or lost resident cells by introducing new cells that could integrate into the existing network and restore function (Fig. 2). Cell replacement in neurodegenerative disorders is aimed specifically at replacing damaged neurons and/or glial cells with normal cells. In the past, various research groups have attempted to either prevent the loss of neurons or glia or to replace them, using either NS cells or other somatic stem cells. Unfortunately, many technical problems (e.g., cell survival and controlled differentiation) continue to impair progress on this potential therapeutic approach, and Parkinson's disease is currently the only neurodegenerative disease in which cell replacement therapy has been extensively investigated on a clinical basis [130].

It is well documented in AD that several neurotransmitter systems are affected by the disease neuropathology [131–133]. However, the earliest symptoms of the disease correlate with a substantial degeneration of the forebrain cholinergic projection system, primarily in the nucleus basalis of Meynert [134]. In addition, although neuronal and synaptic loss occurs throughout the cortex and hippocampal formation, neurons in layer II of the entorhinal cortex and CA1 of the hippocampus are substantially affected [135, 136]. Unlike PD, there have been no previous (nor current) attempts to transplant fetal-derived NS cells, or NS cells from alternative sources, into AD patients. However, several reports have indicated that fetal-derived cholinergic neurons can improve behavior in animal models of cholinergic disruption [137, 138], as well as one report of ES cells being transplanted into the nucleus basalis of Meynert and assuming a cholinergic phenotype and promoting functional recovery [139]. In a second study performed by Moghadam *et al.*, ES cell-derived neural precursor cells that were primed with Sonic Hedgehog were shown to better differentiate into cholinergic neurons and to improve behavior in a rat model of cholinergic disruption [140]. In addition, it has been shown in the present authors' laboratory that mouse fetal-derived NS cells, when transplanted into CA1, can improve memory function in a transgenic model of hippocampal CA1 neuronal loss. Interestingly, however, in this study only a small percentage of the NS cells was differentiated into neurons once transplanted, while the remaining cells assumed a glial fate [141].

2.2.3 Gene Replacement Therapy

While recipient cell replacement remains a desired goal for stem cell therapies, the emerging narrative from most preclinical studies suggests that the benefits observed after stem cell transplantation into the CNS are not dependent on either neuronal or glial replacement. The characterization of stem cell behavior *in vivo* has resulted in the observation that stem cells possess intrinsic neuroprotective capacities due to a collection of neurotrophic and immunomodulatory factors, which they can secrete. Hence, the current rationale for transplanting exogenous stem cells as an intervention for neurodegenerative diseases is to introduce an essential protein or enzyme, the deficiency of which is a causal or contributing factor of the disease via the stem cells (Fig. 2). The CNS contains innate neurotrophic factors, which constitute a class of polypeptides involved in a multitude of functions including the support of neurite outgrowth and neuronal maturation and specification. The current hypothesis regarding many neurodegenerative disorders suggests that neurotrophic support is impacted and, indeed, evidence of neuronal protection has been observed following the administration of neurotrophic factors in animal models of neurodegenerative disorders [142–144]. Likewise, NS cells have been shown to express baseline amounts of several different neurotrophins, while additional reports have shown that NS cells can be genetically modified to produce specific neuroprotective factors, which results in significant improvements in disease outcome [145, 146].

The direct delivery of neurotrophic factors to patients with neurodegenerative disorders is limited in part by their short half-life and in part by their restricted diffusion across the BBB. Consequently, several research groups have developed different approaches to deliver neurotrophic factors directly to the site of injury or degeneration, one such example being lentiviral delivery [147]. Another avenue of research has focused on the use of stem cells as a vector for the production and delivery of specific recombinant neurotrophic factors, as transplanted NS cells possess the ability to migrate throughout the brain (specifically to areas of damage), thus highlighting their value as delivery vectors [148, 149].

The progression of AD has been correlated with the reduction of several neurotrophic factors. For example, BDNF, which normally is generated in the hippocampal formation during adulthood, becomes deficient in AD patients [150, 151]. By using a viral vector delivery, BDNF has been reported to improve learning and memory in several different animal models of AD [152]. Indeed, it has been shown in the present authors' laboratory that BDNF, when released by mouse NS cells transplanted into the hippocampus of 3xTg-AD mice, can improve memory despite there being no significant changes in Aβ or tau pathology [153]. In addition to BDNF, NGF – which can target the basal forebrain cholinergic system – has also been shown to be affected in AD [154]. However, the direct delivery of NGF has unfortunately been associated with many adverse side effects. It has been shown in several studies that NGF can increase acetylcholine (via ChAT) production and prevent basal forebrain cholinergic atrophy under normal aging or diseased processes, which suggests that NGF might be of therapeutic value in the treatment of AD [155, 156]. Similar results have been reported using acute non-transgenic rodent models of AD in which Aβ is injected directly into the brain parenchyma. Xuan et al. found that NS cells obtained from neonatal rats could increase the number of cholinergic neurons in the medial septum, increase $p75^{NTRf}$ expression, and also improve behavior in a Y-maze task, all in a rat-transected basal forebrain model of AD [157]. By using another acute Aβ injection AD rat model, Li et al. showed that MSCs which had been genetically modified to release NGF, once transplanted, could differentiate into cholinergic neurons and thereby improve learning and memory [158]. In a small human clinical study, autologous fibroblasts which had been genetically modified to secrete NGF were shown to improve cognitive scores, as assessed using the Mini-Mental Status Examination and AD Assessment Scale-Cognitive task [159]. At autopsy, one patient was found to have robust neuronal growth in response to NGF release.

Along with neurotrophic factors, stem cells can be engineered to overexpress additional factors such as anti-inflammatory molecules and extracellular matrix proteins, which may assist with the treatment of neurodegenerative diseases [160, 161]. In a study conducted by Yang et al., NS cells were engineered to secrete interleukin-10 (IL-10); once transplanted into an experimental autoimmune encephalitis (EAE) mouse model, the IL-10-secreting NS cells were able to produce an enhanced anti-inflammatory and neuronal repair response when compared to unmodified NS cells [162]. Others have shown that neural progenitors and MSCs may possess intrinsic factors that could help to decrease the inflammatory response after transplantation [163, 164]. In addition to secreting anti-inflammatory mediators, stem cells may also be modified to secrete enzymes capable of degrading Aβ, such as insulin degrading enzyme (IDE) or neprilysin. In a recent study, Njie et al. modified mouse NS cells to express matrix metalloproteinase 9 (MMP9), and transplanted these into AD model mice [165]. The modified MMP9-secreting NS cells did not significantly lower the Aβ plaque load compared to unmodified NS cells. However, this observation was most likely due to an inability of the exogenous MMP9 to reach a concentration significantly above endogenous levels, and should be further investigated using stem cells with a more robust MMP9 secretion or alternative Aβ degradation enzyme.

2.3 Road Blocks to Clinical Application

2.3.1 Potential Effects of AD Pathology on Stem Cell Physiology

There is mounting evidence that APP and its derivatives play a role in neurogenesis and NS cell proliferation and migration [166, 167]. As such, there is the potential for exogenous stem cells to function differently if transplanted into human AD patients compared to transgenic models, due to differences in APP expression and levels of the APP derivatives. For example, the APP cleavage product sAPPα has been shown to influence neurogenesis, neurite outgrowth, and neuroprotection [168, 169]. Porayette et al. reported that human ES (hES) cells express APP, and that the processing of APP to sAPPα drives the hES cells to differentiate into neural precursor cells [9]. Whilst less attention has been paid to the effects that sAPPβ might have on neurogenesis and neuroprotection, Freude et al. reported that sAPPβ is also a potent inducer of hES cell differentiation to neural precursors [170]. Taken together, these results suggest a critical role for the initial cleavage products of APP in the generation and differentiation of new neurons.

In addition to the observed differentiation effects mediated by sAPP, others have reported both neurogenic and neurotrophic effects mediated by the Aβ peptide itself [171, 172]. Heo et al. studied the effects of three varying forms of $A\beta_{1-42}$ on NS cell proliferation and differentiation, and found that low oligomeric $A\beta_{1-42}$ concentrations could enhance both the differentiation and proliferation of NS cells [173]. In contrast, both pre-fibrillar $A\beta_{1-40}$ and $A\beta_{1-42}$ have been shown to block the development of neuronal colonies and impair neurogenesis [174, 175]. One potential mechanism by which Aβ may affect NS cell physiology is through the dysregulation of cellular calcium homeostasis [118].

2.3.2 Translational Potential of Transgenic Models of AD

Another potential roadblock shares the translational relevance of using rodent models of AD – both non-transgenic and transgenic – to successfully test the efficacy of stem cell transplantation. One potential criticism here is the lack of cell loss found in rodent models of AD compared to human patients. If stem cell-based therapies are to be used to replace dying cells, this mechanism cannot be properly addressed using most rodent AD models. Thus, an attempt was made by the present authors to address this issue with a transgenic model of neuronal loss [141], although at the post-transplantation times analyzed no signs were observed of the transplanted NS cells replacing or being integrated into the endogenous circuitry. However, other groups have reported exogenous NS cells and NP cells integrating into the endogenous circuitry in the adult CNS [176, 177]. Whether or not cell replacement can occur, the majority of experimental data gathered to date have suggested that the predominant mechanism underlying the beneficial effects of stem cell transplantation has been related to the secretion of trophic factors in a paracrine-based manner [178]. Hence, the replacement of lost neurons may not need to be the underlying goal of stem cell transplantation for the treatment of AD.

3 Conclusions

Stem cell therapy undoubtedly holds great promise for the field of regenerative medicine. By using various animal models of aging and AD, substantial progress has been made in providing a proof-of-principle support that stem cells may be useful for treating neurodegenerative disorders. Nevertheless, many key issues remain to be addressed before a successful translation can be achieved from preclinical experiment to clinical application – for example, the type and source of stem cells (i.e., NS cells versus non-NS cells), or whether to genetically enhance the appropriate cell line. In the event of non-autologous stem cell sources, a variety of immune suppression compounds should be compared to obtain optimal survival. In addition, the exact mechanism by which stem cell proliferation and differentiation can be controlled *in vivo* is as yet unknown, as are the potential effects of AD-related pathology on stem cell function.

In contrast to exogenous stem cell transplantation, and the potential drawbacks associated with its use, it may be possible to enhance endogenous neurogenesis to replace dying and dead neurons and glia, particularly in the hippocampus and adjacent cortical areas. In the case of endogenous stem cell germinal centers, pharmacological means could be utilized to target the direct mobilization of NS cells as a safer alternative to invasive surgery. As neurogenesis in the hippocampus is believed to play an important role in certain aspects of learning and memory – and particularly for memory that is vulnerable to the earliest stages of AD – the targeting of neurogenesis at the onset of the disease may attenuate its further progression.

While the debate of endogenous versus exogenous stem cell therapy continues, the successful clinical translation of either therapeutic approach into human patients remains a substantial roadblock left unmet. A considerable proportion of the initial preclinical studies using exogenous stem cells has been conducted in acute models of AD – that is, in models that either completely lack Aβ and tau pathology or relevant physiological levels of these. Consequently, in order to continue the development of successful stem cell therapies for AD, additional studies with more complete AD animal models should form the primary focus.

Glossary

3xTg-AD mice
  A transgenic mouse model that displays age-dependent plaque and tangle pathology, as well as synaptic loss and cognitive dysfunction all traits similar to those found in Alzheimer's disease patients.

Amyloid precursor protein (APP)
  An integral membrane protein that is highly expressed in the synapses of neurons, and is the precursor molecule whose proteolysis generates Aβ.

Apolipoprotein E (ApoE)
  A group of proteins that function to transport lipoproteins and cholesterol through the blood. While principally synthesized in the liver, they are also extensively synthesized in the brain.

Beta-site APP-cleaving enzyme 1 or Beta-secretase 1 (BACE1)
  An aspartic acid protease important for the extracellular cleavage of APP into Aβ.

Blood–brain barrier
An all-important permeability barrier surrounding the CNS that restricts the entrance of selected molecules between the CNS and the bloodstream.

Brain-derived neurotrophic factor (BDNP)
A growth factor that is distributed heterogeneously throughout the brain, and has several functions classically associated with growth factors such as neuronal development and neurite outgrowth. BDNP belongs to a class of growth factors known as *neurotrophins*, which also includes NGF.

Immunization
The deliberate incitation of an adaptive immune response by introducing a foreign antigen into the body.

Immune Rejection
An immunological response to the introduction of non-self or altered-self tissue (or single cell suspensions).

Nerve growth factor (NGF)
A small growth factor, which is important for the growth and survival of specific neuronal cell types.

Presenilin
A family of transmembrane proteins that form part of the γ-secretase protease complex.

γ-Secretase
A multisubunit integral membrane protease complex important for the cleavage of APP into Aβ that contains presenilin as a portion of its catalytic core.

References

1. Alzheimer's Association (2008) 2008 Alzheimer's disease facts and figures. *Alzheimers Dement.*, **4**, 110–133.
2. Grossman, H., Bergmann, C., Parker, S. (2006) Dementia: a brief review. *Mt Sinai J. Med.*, **73**, 985–992.
3. Evans, D.A., Funkenstein, H.H., Albert, M.S., Scherr, P.A. et al. (1989) Prevalence of Alzheimer's disease in a community population of older persons. *Higher than previously reported. JAMA*, **262**, 2551–2556.
4. Querfurth, H.W., LaFerla, F.M. (2010) Alzheimer's disease. *N. Engl. J. Med.*, **362**, 329–344.
5. Haass, C., Selkoe, D.J. (2007) Soluble protein oligomers in neurodegeneration: lessons from the Alzheimer's amyloid beta-peptide. *Nat. Rev. Mol. Cell Biol.*, **8**, 101–112.
6. Hoe, H.S., Lee, K.J., Carney, R.S., Lee, J. et al. (2009) Interaction of reelin with amyloid precursor protein promotes neurite outgrowth. *J. Neurosci.*, **29**, 7459–7473.
7. Qiu, W.Q., Ferreira, A., Miller, C., Koo, E.H. et al. (1995) Cell-surface beta-amyloid precursor protein stimulates neurite outgrowth of hippocampal neurons in an isoform-dependent manner. *J. Neurosci.*, **15**, 2157–2167.
8. Nikolaev, A., McLaughlin, T., O'Leary, D.D., Tessier-Lavigne, M. (2009) APP binds DR6 to trigger axon pruning and neuron death via distinct caspases. *Nature*, **457**, 981–989.
9. Porayette, P., Gallego, M.J., Kaltcheva, M.M., Bowen, R.L. et al. (2009) Differential processing of amyloid-beta precursor protein directs human embryonic stem cell proliferation and differentiation into neuronal precursor cells. *J. Biol. Chem.*, **284**, 23806–23817.
10. Morris, M., Maeda, S., Vossel, K., Mucke, L. (2011) The many faces of tau. *Neuron*, **70**, 410–426.
11. Wallin, A.K., Blennow, K., Andreasen, N., Minthon, L. (2006) CSF biomarkers for Alzheimer's disease: levels of beta-amyloid, tau, phosphorylated tau relate to clinical symptoms and survival. *Dement. Geriatr. Cogn. Disord.*, **21**, 131–138.
12. Strittmatter, W.J., Saunders, A.M., Schmechel, D., Pericak-Vance, M. et al. (1993) Apolipoprotein E: high-avidity binding to beta-amyloid and increased frequency of type 4 allele in late-onset familial Alzheimer disease. *Proc. Natl Acad. Sci. USA*, **90**, 1977–1981.
13. Roses, A.D., Saunders, A.M. (1997) Apolipoprotein E genotyping as a diagnostic adjunct for Alzheimer's disease.

Int. Psychogeriatr., **9** (Suppl. 1), 277–288; discussion 317–21.
14. Han, S.H., Hulette, C., Saunders, A.M., Einstein, G. *et al*. (1994) Apolipoprotein E is present in hippocampal neurons without neurofibrillary tangles in Alzheimer's disease and in age-matched controls. *Exp. Neurol.*, **128**, 13–26.
15. Bertram, L., McQueen, M.B., Mullin, K., Blacker, D. *et al*. (2007) Systematic meta-analyses of Alzheimer disease genetic association studies: the AlzGene database. *Nat. Genet.*, **39**, 17–23.
16. Strittmatter, W.J., Roses, A.D. (1996) Apolipoprotein E and Alzheimer's disease. *Annu. Rev. Neurosci.*, **19**, 53–77.
17. Chartier-Harlin, M.C., Crawford, F., Houlden, H., Warren, A. *et al*. (1991) Early-onset Alzheimer's disease caused by mutations at codon 717 of the beta-amyloid precursor protein gene. *Nature*, **353**, 844–846.
18. Rogaev, E.I., Sherrington, R., Rogaeva, E.A., Levesque, G. *et al*. (1995) Familial Alzheimer's disease in kindreds with missense mutations in a gene on chromosome 1 related to the Alzheimer's disease type 3 gene. *Nature*, **376**, 775–778.
19. Sherrington, R., Rogaev, E.I., Liang, Y., Rogaeva, E.A. *et al*. (1995) Cloning of a gene bearing missense mutations in early-onset familial Alzheimer's disease. *Nature*, **375**, 754–760.
20. Gomez-Isla, T., Wasco, W., Pettingell, W.P., Gurubhagavatula, S. *et al*. (1997) A novel presenilin-1 mutation: increased beta-amyloid and neurofibrillary changes. *Ann. Neurol.*, **41**, 809–813.
21. Levy-Lahad, E., Wijsman, E.M., Nemens, E., Anderson, L. *et al*. (1995) A familial Alzheimer's disease locus on chromosome 1. *Science*, **269**, 970–973.
22. Burns, A., Iliffe, S. (2009) Alzheimer's disease. *Br. Med. J.*, **338**, b158.
23. Delbeuck, X., Van der Linden, M., Collette, F. (2003) Alzheimer's disease as a disconnection syndrome? *Neuropsychol. Rev.*, **13**, 79–92.
24. Steele, C., Rovner, B., Chase, G.A., Folstein, M. (1990) Psychiatric symptoms and nursing home placement of patients with Alzheimer's disease. *Am. J. Psychiatr.*, **147**, 1049–1051.
25. Doody, R.S., Massman, P., Mahurin, R., Law, S. (1995) Positive and negative neuropsychiatric features in Alzheimer's disease. *J. Neuropsychiatr. Clin. Neurosci.*, **7**, 54–60.
26. Hardy, J., Selkoe, D.J. (2002) The amyloid hypothesis of Alzheimer's disease: progress and problems on the road to therapeutics. *Science*, **297**, 353–356.
27. Glenner, G.G., Wong, C.W. (1984) Alzheimer's disease: initial report of the purification and characterization of a novel cerebrovascular amyloid protein. *Biochem. Biophys. Res. Commun.*, **120**, 885–890.
28. Masters, C.L., Simms, G., Weinman, N.A., Multhaup, G. *et al*. (1985) Amyloid plaque core protein in Alzheimer disease and Down syndrome. *Proc. Natl Acad. Sci. USA*, **82**, 4245–4249.
29. St. George-Hyslop, P.H., Tanzi, R.E., Polinsky, R.J., Haines, J.L. *et al*. (1987) The genetic defect causing familial Alzheimer's disease maps on chromosome 21. *Science*, **235**, 885–890.
30. Watkins, P.C., Tanzi, R.E., Cheng, S.V., Gusella, J.F. (1987) Molecular genetics of human chromosome 21. *J. Med. Genet.*, **24**, 257–270.
31. Kang, J., Lemaire, H.G., Unterbeck, A., Salbaum, J.M. *et al*. (1987) The precursor of Alzheimer's disease amyloid A4 protein resembles a cell-surface receptor. *Nature*, **325**, 733–736.
32. Wisniewski, K.E., Wisniewski, H.M., Wen, G.Y. (1985) Occurrence of neuropathological changes and dementia of Alzheimer's disease in Down's syndrome. *Ann. Neurol.*, **17**, 278–282.
33. Hutton, M., Lendon, C.L., Rizzu, P., Baker, M. *et al*. (1998) Association of missense and 5'-splice-site mutations in tau with the inherited dementia FTDP-17. *Nature*, **393**, 702–705.
34. D'Souza, I., Poorkaj, P., Hong, M., Nochlin, D. *et al*. (1999) Missense and silent tau gene mutations cause frontotemporal dementia with parkinsonism-chromosome 17 type, by affecting multiple alternative RNA splicing regulatory elements. *Proc. Natl Acad. Sci. USA*, **96**, 5598–5603.
35. Oddo, S., Caccamo, A., Kitazawa, M., Tseng, B.P. *et al*. (2003) Amyloid deposition precedes tangle formation in a triple

transgenic model of Alzheimer's disease. *Neurobiol. Aging*, **24**, 1063–1070.
36. Auld, D.S., Kornecook, T.J., Bastianetto, S., Quirion, R. (2002) Alzheimer's disease and the basal forebrain cholinergic system: relations to beta-amyloid peptides, cognition, and treatment strategies. *Prog. Neurobiol.*, **68**, 209–245.
37. Kosasa, T., Kuriya, Y., Matsui, K., Yamanishi, Y. (1999) Effect of donepezil hydrochloride (E2020) on basal concentration of extracellular acetylcholine in the hippocampus of rats. *Eur. J. Pharmacol.*, **380**, 101–107.
38. Reisberg, B., Doody, R., Stoffler, A., Schmitt, F. et al. (2003) Memantine in moderate-to-severe Alzheimer's disease. *N. Engl. J. Med.*, **348**, 1333–1341.
39. Robinson, D.M., Keating, G.M. (2006) Memantine: a review of its use in Alzheimer's disease. *Drugs*, **66**, 1515–1534.
40. Cacabelos, R., Takeda, M., Winblad, B. (1999) The glutamatergic system and neurodegeneration in dementia: preventive strategies in Alzheimer's disease. *Int. J. Geriatr. Psychiatr.*, **14**, 3–47.
41. Martinez-Coria, H., Green, K.N., Billings, L.M., Kitazawa, M. et al. (2010) Memantine improves cognition and reduces Alzheimer's-like neuropathology in transgenic mice. *Am. J. Pathol.*, **176**, 870–880.
42. Chohan, M.O., Iqbal, K. (2006) From tau to toxicity: emerging roles of NMDA receptor in Alzheimer's disease. *J. Alzheimers Dis.*, **10**, 81–87.
43. Weycker, D., Taneja, C., Edelsberg, J., Erder, M.H. et al. (2007) Cost-effectiveness of memantine in moderate-to-severe Alzheimer's disease patients receiving donepezil. *Curr. Med. Res. Opin.*, **23**, 1187–1197.
44. Schenk, D., Barbour, R., Dunn, W., Gordon, G. et al. (1999) Immunization with amyloid-beta attenuates Alzheimer-disease-like pathology in the PDAPP mouse. *Nature*, **400**, 173–177.
45. Games, D., Bard, F., Grajeda, H., Guido, T. et al. (2000) Prevention and reduction of AD-type pathology in PDAPP mice immunized with A beta 1–42. *Ann. N. Y. Acad. Sci.*, **920**, 274–284.
46. Lemere, C.A., Maier, M., Jiang, L., Peng, Y. et al. (2006) Amyloid-beta immunotherapy for the prevention and treatment of Alzheimer disease: lessons from mice, monkeys, and humans. *Rejuvenation Res.*, **9**, 77–84.
47. Oddo, S., Billings, L., Kesslak, J.P., Cribbs, D.H. et al. (2004) Abeta immunotherapy leads to clearance of early, but not late, hyperphosphorylated tau aggregates via the proteasome. *Neuron*, **43**, 321–332.
48. Billings, L.M., Oddo, S., Green, K.N., McGaugh, J.L. et al. (2005) Intraneuronal Abeta causes the onset of early Alzheimer's disease-related cognitive deficits in transgenic mice. *Neuron*, **45**, 675–688.
49. Orgogozo, J.M., Gilman, S., Dartigues, J.F., Laurent, B. et al. (2003) Subacute meningoencephalitis in a subset of patients with AD after Abeta42 immunization. *Neurology*, **61**, 46–54.
50. Lemere, C.A., Masliah, E. (2010) Can Alzheimer disease be prevented by amyloid-beta immunotherapy? *Nat. Rev. Neurol.*, **6**, 108–119.
51. Chow, V.W., Mattson, M.P., Wong, P.C., Gleichmann, M. (2010) An overview of APP processing enzymes and products. *Neuromol. Med.*, **12**, 1–12.
52. Cole, S.L., Vassar, R. (2008) BACE1 structure and function in health and Alzheimer's disease. *Curr. Alzheimer Res.*, **5**, 100–120.
53. Cai, H., Wang, Y., McCarthy, D., Wen, H. et al. (2001) BACE1 is the major beta-secretase for generation of Abeta peptides by neurons. *Nat. Neurosci.*, **4**, 233–234.
54. Luo, Y., Bolon, B., Kahn, S., Bennett, B.D. et al. (2001) Mice deficient in BACE1, the Alzheimer's beta-secretase, have normal phenotype and abolished beta-amyloid generation. *Nat. Neurosci.*, **4**, 231–232.
55. Savonenko, A.V., Melnikova, T., Laird, F.M., Stewart, K.A. et al. (2008) Alteration of BACE1-dependent NRG1/ErbB4 signaling and schizophrenia-like phenotypes in BACE1-null mice. *Proc. Natl Acad. Sci. USA*, **105**, 5585–5590.
56. Willem, M., Garratt, A.N., Novak, B., Citron, M. et al. (2006) Control of peripheral nerve myelination by the beta-secretase BACE1. *Science*, **314**, 664–666.
57. Fukumoto, H., Takahashi, H., Tarui, N., Matsui, J. et al. (2010) A noncompetitive BACE1 inhibitor TAK-070 ameliorates Abeta pathology and behavioral deficits in a mouse model of Alzheimer's disease. *J. Neurosci.*, **30**, 11157–11166.

58. Yang, S.G., Wang, S.W., Zhao, M., Zhang, R. et al. (2012) A peptide binding to the beta-site of APP improves spatial memory and attenuates abeta burden in Alzheimer's disease transgenic mice. *PloS ONE*, 7, e48540.
59. Hussain, I., Hawkins, J., Harrison, D., Hille, C. et al. (2007) Oral administration of a potent and selective non-peptidic BACE-1 inhibitor decreases beta-cleavage of amyloid precursor protein and amyloid-beta production in vivo. *J. Neurochem.*, 100, 802–809.
60. Sankaranarayanan, S., Holahan, M.A., Colussi, D., Crouthamel, M.C. et al. (2009) First demonstration of cerebrospinal fluid and plasma A beta lowering with oral administration of a beta-site amyloid precursor protein-cleaving enzyme 1 inhibitor in nonhuman primates. *J. Pharmacol. Exp. Ther.*, 328, 131–140.
61. McConlogue, L., Buttini, M., Anderson, J.P., Brigham, E.F. et al. (2007) Partial reduction of BACE1 has dramatic effects on Alzheimer plaque and synaptic pathology in APP transgenic mice. *J. Biol. Chem.*, 282, 26326–26334.
62. Steiner, H. (2008) The catalytic core of gamma-secretase: presenilin revisited. *Curr. Alzheimer Res.*, 5, 147–157.
63. Wolfe, M.S. (2008) Inhibition and modulation of gamma-secretase for Alzheimer's disease. *Neurotherapeutics*, 5, 391–398.
64. Selkoe, D., Kopan, R. (2003) Notch and Presenilin: regulated intramembrane proteolysis links development and degeneration. *Annu. Rev. Neurosci.*, 26, 565–597.
65. Dovey, H.F., John, V., Anderson, J.P., Chen, L.Z. et al. (2001) Functional gamma-secretase inhibitors reduce beta-amyloid peptide levels in brain. *J. Neurochem.*, 76, 173–181.
66. Weggen, S., Eriksen, J.L., Das, P., Sagi, S.A. et al. (2001) A subset of NSAIDs lower amyloidogenic Abeta42 independently of cyclooxygenase activity. *Nature*, 414, 212–216.
67. Wolfe, M.S. (2012) gamma-Secretase inhibitors and modulators for Alzheimer's disease. *J. Neurochem.*, 120 (Suppl. 1), 89–98.
68. Imbimbo, B.P., Hutter-Paier, B., Villetti, G., Facchinetti, F. et al. (2009) CHF5074, a novel gamma-secretase modulator, attenuates brain beta-amyloid pathology and learning deficit in a mouse model of Alzheimer's disease. *Br. J. Pharmacol.*, 156, 982–993.
69. Imbimbo, B.P., Giardino, L., Sivilia, S., Giuliani, A. et al. (2010) CHF5074, a novel gamma-secretase modulator, restores hippocampal neurogenesis potential and reverses contextual memory deficit in a transgenic mouse model of Alzheimer's disease. *J. Alzheimers Dis.*, 20, 159–173.
70. Evans, M.J., Kaufman, M.H. (1981) Establishment in culture of pluripotential cells from mouse embryos. *Nature*, 292, 154–156.
71. Acharya, M.M., Christie, L.A., Lan, M.L., Donovan, P.J. et al. (2009) Rescue of radiation-induced cognitive impairment through cranial transplantation of human embryonic stem cells. *Proc. Natl Acad. Sci. USA*, 106, 19150–19155.
72. Wei, L., Cui, L., Snider, B.J., Rivkin, M. et al. (2005) Transplantation of embryonic stem cells overexpressing Bcl-2 promotes functional recovery after transient cerebral ischemia. *Neurobiol. Dis.*, 19, 183–193.
73. Cui, L., Jiang, J., Wei, L., Zhou, X. et al. (2008) Transplantation of embryonic stem cells improves nerve repair and functional recovery after severe sciatic nerve axotomy in rats. *Stem Cells*, 26, 1356–1365.
74. Knoepfler, P.S. (2009) Deconstructing stem cell tumorigenicity: a roadmap to safe regenerative medicine. *Stem Cells*, 27, 1050–1056.
75. Bain, G., Kitchens, D., Yao, M., Huettner, J.E. et al. (1995) Embryonic stem cells express neuronal properties in vitro. *Dev. Biol.*, 168, 342–357.
76. Fraichard, A., Chassande, O., Bilbaut, G., Dehay, C. et al. (1995) In vitro differentiation of embryonic stem cells into glial cells and functional neurons. *J. Cell Sci.*, 108 (Pt 10), 3181–3188.
77. Carpenter, M.K., Inokuma, M.S., Denham, J., Mujtaba, T. et al. (2001) Enrichment of neurons and neural precursors from human embryonic stem cells. *Exp. Neurol.*, 172, 383–397.
78. Arnhold, S., Klein, H., Semkova, I., Addicks, K. et al. (2004) Neurally selected embryonic stem cells induce tumor formation after long-term survival following engraftment

into the subretinal space. *Invest. Ophthalmol. Vis. Sci.*, **45**, 4251–4255.
79. Bjorklund, L.M., Sanchez-Pernaute, R., Chung, S., Andersson, T. et al. (2002) Embryonic stem cells develop into functional dopaminergic neurons after transplantation in a Parkinson rat model. *Proc. Natl Acad. Sci. USA*, **99**, 2344–2349.
80. Takahashi, K., Yamanaka, S. (2006) Induction of pluripotent stem cells from mouse embryonic and adult fibroblast cultures by defined factors. *Cell*, **126**, 663–676.
81. Takahashi, K., Tanabe, K., Ohnuki, M., Narita, M. et al. (2007) Induction of pluripotent stem cells from adult human fibroblasts by defined factors. *Cell*, **131**, 861–872.
82. Park, I.H., Lerou, P.H., Zhao, R., Huo, H. et al. (2008) Generation of human-induced pluripotent stem cells. *Nat. Prot.*, **3**, 1180–1186.
83. Wernig, M., Zhao, J.P., Pruszak, J., Hedlund, E. et al. (2008) Neurons derived from reprogrammed fibroblasts functionally integrate into the fetal brain and improve symptoms of rats with Parkinson's disease. *Proc. Natl Acad. Sci. USA*, **105**, 5856–5861.
84. Cai, J., Yang, M., Poremsky, E., Kidd, S. et al. (2010) Dopaminergic neurons derived from human induced pluripotent stem cells survive and integrate into 6-OHDA-lesioned rats. *Stem Cells Dev.*, **19**, 1017–1023.
85. Yagi, T., Ito, D., Okada, Y., Akamatsu, W. et al. (2011) Modeling familial Alzheimer's disease with induced pluripotent stem cells. *Hum. Mol. Genet.*, **20**, 4530–4539.
86. De Miguel, M.P., Fuentes-Julian, S., Alcaina, Y. (2010) Pluripotent stem cells: origin, maintenance and induction. *Stem Cell Rev.*, **6**, 633–649.
87. Chin, M.H., Mason, M.J., Xie, W., Volinia, S. et al. (2009) Induced pluripotent stem cells and embryonic stem cells are distinguished by gene expression signatures. *Cell Stem Cell*, **5**, 111–123.
88. Fong, C.Y., Gauthaman, K., Bongso, A. (2010) Teratomas from pluripotent stem cells: a clinical hurdle. *J. Cell. Biochem.*, **111**, 769–781.
89. Gage, F.H., Coates, P.W., Palmer, T.D., Kuhn, H.G. et al. (1995) Survival and differentiation of adult neuronal progenitor cells transplanted to the adult brain. *Proc. Natl Acad. Sci. USA*, **92**, 11879–11883.
90. Shihabuddin, L.S., Horner, P.J., Ray, J., Gage, F.H. (2000) Adult spinal cord stem cells generate neurons after transplantation in the adult dentate gyrus. *J. Neurosci.*, **20**, 8727–8735.
91. Suhonen, J.O., Peterson, D.A., Ray, J., Gage, F.H. (1996) Differentiation of adult hippocampus-derived progenitors into olfactory neurons in vivo. *Nature*, **383**, 624–627.
92. Tyndall, A., Gratwohl, A. (2009) Adult stem cell transplantation in autoimmune disease. *Curr. Opin. Hematol.*, **16**, 285–291.
93. Thomas, E.D. (1999) A history of haemopoietic cell transplantation. *Br. J. Haematol.*, **105**, 330–339.
94. Pittenger, M.F., Mackay, A.M., Beck, S.C., Jaiswal, R.K. et al. (1999) Multilineage potential of adult human mesenchymal stem cells. *Science*, **284**, 143–147.
95. Lee, J., Kuroda, S., Shichinohe, H., Ikeda, J. et al. (2003) Migration and differentiation of nuclear fluorescence-labeled bone marrow stromal cells after transplantation into cerebral infarct and spinal cord injury in mice. *Neuropathology*, **23**, 169–180.
96. Kopen, G.C., Prockop, D.J., Phinney, D.G. (1999) Marrow stromal cells migrate throughout forebrain and cerebellum, and they differentiate into astrocytes after injection into neonatal mouse brains. *Proc. Natl Acad. Sci. USA*, **96**, 10711–10716.
97. Sanchez-Ramos, J., Song, S., Cardozo-Pelaez, F., Hazzi, C. et al. (2000) Adult bone marrow stromal cells differentiate into neural cells in vitro. *Exp. Neurol.*, **164**, 247–256.
98. Woodbury, D., Reynolds, K., Black, I.B. (2002) Adult bone marrow stromal stem cells express germline, ectodermal, endodermal, and mesodermal genes prior to neurogenesis. *J. Neurosci. Res.*, **69**, 908–917.
99. Chen, C.W., Boiteau, R.M., Lai, W.F., Barger, S.W. et al. (2006) sAPPalpha enhances the transdifferentiation of adult bone marrow progenitor cells to neuronal phenotypes. *Curr. Alzheimer Res.*, **3**, 63–70.
100. Chen, X., Katakowski, M., Li, Y., Lu, D. et al. (2002) Human bone marrow stromal cell cultures conditioned by traumatic brain tissue extracts: growth factor production. *J. Neurosci. Res.*, **69**, 687–691.
101. Lu, P., Jones, L.L., Tuszynski, M.H. (2005) BDNF-expressing marrow stromal cells

support extensive axonal growth at sites of spinal cord injury. *Exp. Neurol.*, **191**, 344–360.
102. Altman, J., Das, G.D. (1965) Autoradiographic and histological evidence of postnatal hippocampal neurogenesis in rats. *J. Comp. Neurol.*, **124**, 319–335.
103. Eriksson, P.S., Perfilieva, E., Bjork-Eriksson, T., Alborn, A.M. et al. (1998) Neurogenesis in the adult human hippocampus. *Nat. Med.*, **4**, 1313–1317.
104. Kehl, L.J., Fairbanks, C.A., Laughlin, T.M., Wilcox, G.L. (1997) Neurogenesis in postnatal rat spinal cord: a study in primary culture. *Science*, **276**, 586–589.
105. Reynolds, B.A., Tetzlaff, W., Weiss, S. (1992) A multipotent EGF-responsive striatal embryonic progenitor cell produces neurons and astrocytes. *J. Neurosci.*, **12**, 4565–4574.
106. Murrell, W., Feron, F., Wetzig, A., Cameron, N. et al. (2005) Multipotent stem cells from adult olfactory mucosa. *Dev. Dyn.*, **233**, 496–515.
107. Taupin, P., Ray, J., Fischer, W.H., Suhr, S.T. et al. (2000) FGF-2-responsive neural stem cell proliferation requires CCg, a novel autocrine/paracrine cofactor. *Neuron*, **28**, 385–397.
108. Arsenijevic, Y., Weiss, S., Schneider, B., Aebischer, P. (2001) Insulin-like growth factor-I is necessary for neural stem cell proliferation and demonstrates distinct actions of epidermal growth factor and fibroblast growth factor-2. *J. Neurosci.*, **21**, 7194–7202.
109. Pencea, V., Bingaman, K.D., Wiegand, S.J., Luskin, M.B. (2001) Infusion of brain-derived neurotrophic factor into the lateral ventricle of the adult rat leads to new neurons in the parenchyma of the striatum, septum, thalamus, and hypothalamus. *J. Neurosci.*, **21**, 6706–6717.
110. Cameron, H.A., McKay, R.D. (1999) Restoring production of hippocampal neurons in old age. *Nat. Neurosci.*, **2**, 894–897.
111. Kempermann, G., Gage, F.H. (2002) Genetic determinants of adult hippocampal neurogenesis correlate with acquisition, but not probe trial performance, in the water maze task. *Eur. J. Neurosci.*, **16**, 129–136.
112. Jin, K., Galvan, V., Xie, L., Mao, X.O. et al. (2004) Enhanced neurogenesis in Alzheimer's disease transgenic (PDGF-APPSw,Ind) mice. *Proc. Natl Acad. Sci. USA*, **101**, 13363–13367.
113. Boekhoorn, K., Joels, M., Lucassen, P.J. (2006) Increased proliferation reflects glial and vascular-associated changes, but not neurogenesis in the presenile Alzheimer hippocampus. *Neurobiol. Dis.*, **24**, 1–14.
114. Ziabreva, I., Perry, E., Perry, R., Minger, S.L. et al. (2006) Altered neurogenesis in Alzheimer's disease. *J. Psychosom. Res.*, **61**, 311–316.
115. Lopez-Toledano, M.A., Shelanski, M.L. (2007) Increased neurogenesis in young transgenic mice overexpressing human APP(Sw, Ind). *J. Alzheimers Dis.*, **12**, 229–240.
116. Chevallier, N.L., Soriano, S., Kang, D.E., Masliah, E. et al. (2005) Perturbed neurogenesis in the adult hippocampus associated with presenilin-1 A246E mutation. *Am. J. Pathol.*, **167**, 151–159.
117. Donovan, M.H., Yazdani, U., Norris, R.D., Games, D. et al. (2006) Decreased adult hippocampal neurogenesis in the PDAPP mouse model of Alzheimer's disease. *J. Comp. Neurol.*, **495**, 70–83.
118. Haughey, N.J., Nath, A., Chan, S.L., Borchard, A.C. et al. (2002) Disruption of neurogenesis by amyloid beta-peptide, and perturbed neural progenitor cell homeostasis, in models of Alzheimer's disease. *J. Neurochem.*, **83**, 1509–1524.
119. Rodriguez, J.J., Jones, V.C., Tabuchi, M., Allan, S.M. et al. (2008) Impaired adult neurogenesis in the dentate gyrus of a triple transgenic mouse model of Alzheimer's disease. *PloS ONE*, **3**, e2935.
120. Karkkainen, V., Magga, J., Koistinaho, J., Malm, T. (2012) Brain Environment and Alzheimer's disease mutations affect the survival, migration and differentiation of neural progenitor cells. *Curr. Alzheimer Res.*, **9**, 1030–1042.
121. Veeraraghavalu, K., Choi, S.H., Zhang, X., Sisodia, S.S. (2010) Presenilin 1 mutants impair the self-renewal and differentiation of adult murine subventricular zone-neuronal progenitors via cell-autonomous mechanisms involving notch signaling. *J. Neurosci.*, **30**, 6903–6915.
122. Gakhar-Koppole, N., Hundeshagen, P., Mandl, C., Weyer, S.W. et al. (2008) Activity requires soluble amyloid precursor protein alpha to promote neurite outgrowth

123. in neural stem cell-derived neurons via activation of the MAPK pathway. *Eur. J. Neurosci.*, **28**, 871–882.
123. Lazarov, O., Demars, M.P. (2012) All in the family: how the APPs regulate neurogenesis. *Front. Neurosci.*, **6**, 81.
124. Kempermann, G., Kuhn, H.G., Gage, F.H. (1997) Genetic influence on neurogenesis in the dentate gyrus of adult mice. *Proc. Natl Acad. Sci. USA*, **94**, 10409–10414.
125. Sairanen, M., Lucas, G., Ernfors, P., Castren, M. *et al.* (2005) Brain-derived neurotrophic factor and antidepressant drugs have different but coordinated effects on neuronal turnover, proliferation, and survival in the adult dentate gyrus. *J. Neurosci.*, **25**, 1089–1094.
126. Rossi, C., Angelucci, A., Costantin, L., Braschi, C. *et al.* (2006) Brain-derived neurotrophic factor (BDNF) is required for the enhancement of hippocampal neurogenesis following environmental enrichment. *Eur. J. Neurosci.*, **24**, 1850–1856.
127. Jin, K., Zhu, Y., Sun, Y., Mao, X.O. *et al.* (2002) Vascular endothelial growth factor (VEGF) stimulates neurogenesis in vitro and in vivo. *Proc. Natl Acad. Sci. USA*, **99**, 11946–11950.
128. Blanchard, J., Wanka, L., Tung, Y.C., Cardenas-Aguayo Mdel, C. *et al.* (2010) Pharmacologic reversal of neurogenic and neuroplastic abnormalities and cognitive impairments without affecting Abeta and tau pathologies in 3xTg-AD mice. *Acta Neuropathol.*, **120**, 605–621.
129. Choi, S.H., Veeraraghavalu, K., Lazarov, O., Marler, S. *et al.* (2008) Non-cell-autonomous effects of presenilin 1 variants on enrichment-mediated hippocampal progenitor cell proliferation and differentiation. *Neuron*, **59**, 568–580.
130. Ganz, J., Lev, N., Melamed, E., Offen, D. (2011) Cell replacement therapy for Parkinson's disease: how close are we to the clinic? *Expert Rev. Neurother.*, **11**, 1325–1339.
131. Garcia-Alloza, M., Gil-Bea, F.J., Diez-Ariza, M., Chen, C.P. *et al.* (2005) Cholinergic-serotonergic imbalance contributes to cognitive and behavioral symptoms in Alzheimer's disease. *Neuropsychologia*, **43**, 442–449.
132. Morrison, J.H., Hof, P.R. (1997) Life and death of neurons in the aging brain. *Science*, **278**, 412–419.
133. Garcia-Alloza, M., Tsang, S.W., Gil-Bea, F.J., Francis, P.T. *et al.* (2006) Involvement of the GABAergic system in depressive symptoms of Alzheimer's disease. *Neurobiol. Aging*, **27**, 1110–1117.
134. Bartus, R.T. (2000) On neurodegenerative diseases, models, and treatment strategies: lessons learned and lessons forgotten a generation following the cholinergic hypothesis. *Exp. Neurol.*, **163**, 495–529.
135. West, M.J., Coleman, P.D., Flood, D.G., Troncoso, J.C. (1994) Differences in the pattern of hippocampal neuronal loss in normal ageing and Alzheimer's disease. *Lancet*, **344**, 769–772.
136. Gomez-Isla, T., Price, J.L., McKeel, D.W. Jr, Morris, J.C. *et al.* (1996) Profound loss of layer II entorhinal cortex neurons occurs in very mild Alzheimer's disease. *J. Neurosci.*, **16**, 4491–4500.
137. Gage, F.H., Bjorklund, A. (1986) Cholinergic septal grafts into the hippocampal formation improve spatial learning and memory in aged rats by an atropine-sensitive mechanism. *J. Neurosci.*, **6**, 2837–2847.
138. Dickinson-Anson, H., Aubert, I., Gage, F.H., Fisher, L.J. (1998) Hippocampal grafts of acetylcholine-producing cells are sufficient to improve behavioural performance following a unilateral fimbria-fornix lesion. *Neuroscience*, **84**, 771–781.
139. Wang, Q., Matsumoto, Y., Shindo, T., Miyake, K. *et al.* (2006) Neural stem cells transplantation in cortex in a mouse model of Alzheimer's disease. *J. Med. Invest.*, **53**, 61–69.
140. Moghadam, F.H., Alaie, H., Karbalaie, K., Tanhaei, S. *et al.* (2009) Transplantation of primed or unprimed mouse embryonic stem cell-derived neural precursor cells improves cognitive function in Alzheimerian rats. *Differentiation*, **78**, 59–68.
141. Yamasaki, T.R., Blurton-Jones, M., Morrissette, D.A., Kitazawa, M. *et al.* (2007) Neural stem cells improve memory in an inducible mouse model of neuronal loss. *J. Neurosci.*, **27**, 11925–11933.
142. Georgievska, B., Kirik, D., Bjorklund, A. (2002) Aberrant sprouting and downregulation of tyrosine hydroxylase in lesioned nigrostriatal dopamine neurons induced by long-lasting overexpression of glial cell line derived neurotrophic factor in the striatum

by lentiviral gene transfer. *Exp. Neurol.*, **177**, 461–474.
143. Kaspar, B.K., Llado, J., Sherkat, N., Rothstein, J.D. et al. (2003) Retrograde viral delivery of IGF-1 prolongs survival in a mouse ALS model. *Science*, **301**, 839–842.
144. McBride, J.L., Ramaswamy, S., Gasmi, M., Bartus, R.T. et al. (2006) Viral delivery of glial cell line-derived neurotrophic factor improves behavior and protects striatal neurons in a mouse model of Huntington's disease. *Proc. Natl Acad. Sci. USA*, **103**, 9345–9350.
145. Park, S., Kim, H.T., Yun, S., Kim, I.S. et al. (2009) Growth factor-expressing human neural progenitor cell grafts protect motor neurons but do not ameliorate motor performance and survival in ALS mice. *Exp. Mol. Med.*, **41**, 487–500.
146. Cao, Q., He, Q., Wang, Y., Cheng, X. et al. (2010) Transplantation of ciliary neurotrophic factor-expressing adult oligodendrocyte precursor cells promotes remyelination and functional recovery after spinal cord injury. *J. Neurosci.*, **30**, 2989–3001.
147. Nanou, A., Azzouz, M. (2009) Gene therapy for neurodegenerative diseases based on lentiviral vectors. *Prog. Brain Res.*, **175**, 187–200.
148. Imitola, J., Raddassi, K., Park, K.I., Mueller, F.J. et al. (2004) Directed migration of neural stem cells to sites of CNS injury by the stromal cell-derived factor 1alpha/CXC chemokine receptor 4 pathway. *Proc. Natl Acad. Sci. USA*, **101**, 18117–18122.
149. Carbajal, K.S., Schaumburg, C., Strieter, R., Kane, J. et al. (2010) Migration of engrafted neural stem cells is mediated by CXCL12 signaling through CXCR4 in a viral model of multiple sclerosis. *Proc. Natl Acad. Sci. USA*, **107**, 11068–11073.
150. Narisawa-Saito, M., Wakabayashi, K., Tsuji, S., Takahashi, H. et al. (1996) Regional specificity of alterations in NGF, BDNF and NT-3 levels in Alzheimer's disease. *NeuroReport*, **7**, 2925–2928.
151. Connor, B., Young, D., Yan, Q., Faull, R.L. et al. (1997) Brain-derived neurotrophic factor is reduced in Alzheimer's disease. *Brain Res. Mol. Brain Res.*, **49**, 71–81.
152. Nagahara, A.H., Merrill, D.A., Coppola, G., Tsukada, S. et al. (2009) Neuroprotective effects of brain-derived neurotrophic factor in rodent and primate models of Alzheimer's disease. *Nat. Med.*, **15**, 331–337.
153. Blurton-Jones, M., Kitazawa, M., Martinez-Coria, H., Castello, N.A. et al. (2009) Neural stem cells improve cognition via BDNF in a transgenic model of Alzheimer disease. *Proc. Natl Acad. Sci. USA*, **106**, 13594–13599.
154. Cattaneo, A., Capsoni, S., Paoletti, F. (2008) Towards noninvasive nerve growth factor therapies for Alzheimer's disease. *J. Alzheimers Dis.*, **15**, 255–283.
155. Tuszynski, M.H., Sang, H., Yoshida, K., Gage, F.H. (1991) Recombinant human nerve growth factor infusions prevent cholinergic neuronal degeneration in the adult primate brain. *Ann. Neurol.*, **30**, 625–636.
156. Kordower, J.H., Winn, S.R., Liu, Y.T., Mufson, E.J. et al. (1994) The aged monkey basal forebrain: rescue and sprouting of axotomized basal forebrain neurons after grafts of encapsulated cells secreting human nerve growth factor. *Proc. Natl Acad. Sci. USA*, **91**, 10898–10902.
157. Xuan, A.G., Luo, M., Ji, W.D., Long, D.H. (2009) Effects of engrafted neural stem cells in Alzheimer's disease rats. *Neurosci. Lett.*, **450**, 167–171.
158. Li, L.Y., Li, J.T., Wu, Q.Y., Li, J. et al. (2008) Transplantation of NGF-gene-modified bone marrow stromal cells into a rat model of Alzheimer's disease. *J. Mol. Neurosci.*, **34**, 157–163.
159. Tuszynski, M.H., Thal, L., Pay, M., Salmon, D.P. et al. (2005) A phase 1 clinical trial of nerve growth factor gene therapy for Alzheimer disease. *Nat. Med.*, **11**, 551–555.
160. Pluchino, S., Zanotti, L., Brambilla, E., Rovere-Querini, P. et al. (2009) Immune regulatory neural stem/precursor cells protect from central nervous system autoimmunity by restraining dendritic cell function. *PloS ONE*, **4**, e5959.
161. Polacek, M., Bruun, J.A., Elvenes, J., Figenschau, Y. et al. (2011) The secretory profiles of cultured human articular chondrocytes and mesenchymal stem cells: implications for autologous cell transplantation strategies. *Cell Transplant.*, **20**, 1381–1393.
162. Yang, J., Jiang, Z., Fitzgerald, D.C., Ma, C. et al. (2009) Adult neural stem

cells expressing IL-10 confer potent immunomodulation and remyelination in experimental autoimmune encephalitis. *J. Clin. Invest.*, **119**, 3678–3691.
163. Ryu, J.K., Cho, T., Wang, Y.T., McLarnon, J.G. (2009) Neural progenitor cells attenuate inflammatory reactivity and neuronal loss in an animal model of inflamed AD brain. *J. Neuroinflammation*, **6**, 39.
164. Lee, J.K., Jin, H.K., Endo, S., Schuchman, E.H. *et al.* (2010) Intracerebral transplantation of bone marrow-derived mesenchymal stem cells reduces amyloid-beta deposition and rescues memory deficits in Alzheimer's disease mice by modulation of immune responses. *Stem Cells*, **28**, 329–343.
165. Njie e, G., Kantorovich, S., Astary, G.W., Green, C. *et al.* (2012) A preclinical assessment of neural stem cells as delivery vehicles for anti-amyloid therapeutics. *PloS ONE*, **7**, e34097.
166. Thinakaran, G., Koo, E.H. (2008) Amyloid precursor protein trafficking, processing, and function. *J. Biol. Chem.*, **283**, 29615–29619.
167. Herms, J., Anliker, B., Heber, S., Ring, S. *et al.* (2004) Cortical dysplasia resembling human type 2 lissencephaly in mice lacking all three APP family members. *EMBO J.*, **23**, 4106–4115.
168. Perez, R.G., Zheng, H., Van der Ploeg, L.H., Koo, E.H. (1997) The beta-amyloid precursor protein of Alzheimer's disease enhances neuron viability and modulates neuronal polarity. *J. Neurosci.*, **17**, 9407–9414.
169. Caille, I., Allinquant, B., Dupont, E., Bouillot, C. *et al.* (2004) Soluble form of amyloid precursor protein regulates proliferation of progenitors in the adult subventricular zone. *Development*, **131**, 2173–2181.
170. Freude, K.K., Penjwini, M., Davis, J.L., LaFerla, F.M. *et al.* (2011) Soluble amyloid precursor protein induces rapid neural differentiation of human embryonic stem cells. *J. Biol. Chem.*, **286**, 24264–24274.
171. Yankner, B.A., Duffy, L.K., Kirschner, D.A. (1990) Neurotrophic and neurotoxic effects of amyloid beta protein: reversal by tachykinin neuropeptides. *Science*, **250**, 279–282.
172. Lopez-Toledano, M.A., Shelanski, M.L. (2004) Neurogenic effect of beta-amyloid peptide in the development of neural stem cells. *J. Neurosci.*, **24**, 5439–5444.
173. Heo, C., Chang, K.A., Choi, H.S., Kim, H.S. *et al.* (2007) Effects of the monomeric, oligomeric, and fibrillar Abeta42 peptides on the proliferation and differentiation of adult neural stem cells from subventricular zone. *J. Neurochem.*, **102**, 493–500.
174. Mazur-Kolecka, B., Golabek, A., Nowicki, K., Flory, M. *et al.* (2006) Amyloid-beta impairs development of neuronal progenitor cells by oxidative mechanisms. *Neurobiol. Aging*, **27**, 1181–1192.
175. Lu, J., Esposito, G., Scuderi, C., Steardo, L. *et al.* (2011) S100B and APP promote a gliocentric shift and impaired neurogenesis in Down syndrome neural progenitors. *PloS ONE*, **6**, e22126.
176. Englund, U., Bjorklund, A., Wictorin, K., Lindvall, O. *et al.* (2002) Grafted neural stem cells develop into functional pyramidal neurons and integrate into host cortical circuitry. *Proc. Natl Acad. Sci. USA*, **99**, 17089–17094.
177. Alvarez-Dolado, M., Calcagnotto, M.E., Karkar, K.M., Southwell, D.G. *et al.* (2006) Cortical inhibition modified by embryonic neural precursors grafted into the postnatal brain. *J. Neurosci.*, **26**, 7380–7389.
178. Martino, G., Pluchino, S. (2006) The therapeutic potential of neural stem cells. *Nat. Rev. Neurosci.*, **7**, 395–406.

12
Immunotherapy with Autologous Cells*

Andrew D. Fesnak and Bruce L. Levine
University of Pennsylvania, Department of Pathology and Laboratory Medicine, The Perelman School of Medicine, M6.40 Maloney 3400 Spruce Street, Philadelphia, PA 19104, USA

1 **Introduction** 413

2 **Selecting and Isolating Effector Cell Types** 415

3 **Modifying the Effector Cell Population** 417
3.1 Gene Delivery Techniques 417
3.1.1 Viral Vectors 418
3.1.2 Transposon-Mediated Transduction 418
3.1.3 RNA Electroporation 418
3.2 Enrichment of Antigen-Specific Effector Cells 419
3.2.1 Enrichment of Virus-Specific Lymphocytes 420
3.2.2 Enrichment of Tumor-Specific Lymphocytes 420
3.3 Expression of Novel Cell-Surface Receptors 421
3.3.1 Non-Endogenous T-Cell Receptors 421
3.3.2 Chimeric Antigen Receptors 422
3.4 Generation of Resistance to Viral Infection 423

4 **Manufacturing Cells for Adoptive Transfer** 424
4.1 Preparing Cells for Culture 424
4.2 Artificial Antigen-Presenting Cells for *Ex-Vivo* Expansion 424
4.3 Polarizing Effector Cell Phenotypes 426
4.4 Expanding Immune Suppressors 426

5 **Clinical Course of Adoptive Transfer** 427
5.1 Preconditioning 427
5.2 Managing Reinfusion of Autologous Cells 427
5.3 Promoting Long-Lived Persistence 428

*This chapter has previously been published in: Meyers, R.A. (Ed.) Translational Medicine Cancer, 2016, 978-3-527-33569-5.

Translational Medicine: Molecular Pharmacology and Drug Discovery
First Edition. Edited by Robert A. Meyers.
© 2018 Wiley-VCH Verlag GmbH & Co. KGaA. Published 2018 by Wiley-VCH Verlag GmbH & Co. KGaA.

6	NextGen Cell Therapy	429
	Disclosure Statement	430
	References	430

Keywords

Immunotherapy
The use of immune cells, or the manipulation of cells of the immune system, usually through the administration of antibodies or biologic ligands, to effect and augment or suppress the immune system as a treatment for an underlying disease.

Cell therapy
Therapy through the use of cells removed from the body and readministered following minimal manipulation, or following more than minimal manipulation to impart nonhomologous function.

Gene therapy
The use of genetic modification of cells to impart nonhomologous function, or the use of DNA or RNA within a viral vector delivered to cells or directly injected, or the use of DNA or RNA directly administered for the treatment of disease.

Autologous
Customarily used in the context of cell therapies derived from a patient's own cells, in contrast to allogeneic, where cell therapies are derived from a donor.

Adoptive immunotherapy
Commonly used to refer to the transfer of immune cells for the treatment of disease, that may be either autologous or allogeneic.

Transferred autologous cells have been investigated and developed over the past few decades to reconstitute immune function and augment immunity following stem cell transplantation, for the treatment of viral infections, and more recently for the treatment of a wider array of diseases. Cellular immunotherapy begins with either the patient's own (autologous) or donor (allogeneic) cells that have been isolated, from tumor tissue, bone marrow biopsy or apheresis mononuclear cell collection. These cells may then be modified *ex vivo*, expanded and, ultimately, adoptively transferred to the patient. Cell function may be manipulated *ex vivo* through selection, differentiation, and the addition of biologic or chemical mediators. Over the past 25 years, methods of gene delivery to cells to modulate function have improved significantly in efficacy. More recently, the translation of gene editing techniques have enabled targeted gene knockout or the insertion of genetic elements to modulate cell function. With substantial clinical benefit demonstrated in a number of diseases, regulatory approval for more autologous cell therapies can be anticipated. Wide adoption of autologous cell therapies will occur with the further development of *ex-vivo* engineering, automation and the implementation of a source to manufacturing to delivery logistics systems that is an evolution of existing transfusion medicine and organ and tissue transfer networks.

1
Introduction

The term cellular immunotherapy encompasses a diverse field of therapies designed to treat a number of clinical indications. In general, cellular immunotherapy begins with either the patient's own (autologous) or donor (allogeneic) cells that have been isolated, from tumor tissue, from bone marrow biopsy, or by apheresis mononuclear cell collection (Fig. 1). These cells may then be modified *ex vivo*, expanded and, ultimately, adoptively transferred to the same patient, in the case of autologous transfer. Autologous cellular therapies can provide targeted killing of malignant or virally infected cells, and even restore broad immunologic reconstitution in immunodeficient recipients.

Mitchison first demonstrated the antitumor effects of adoptive transfer in rodent tumor models in 1955 [1]. However, it took several decades to appreciate the full potential of adoptive immunotherapy. It was not until the 1970s that the anti-leukemic effect of the reconstituting allogeneic immune system was noted in the post-transplant setting [2]. This observation prompted the transfer of mature donor lymphocytes (donor lymphocyte infusion; DLI) to treat relapsed leukemia following bone marrow transplant [3, 4]. While these allogeneic therapies demonstrate powerful graft-versus-tumor activity, graft-versus-host disease (GVHD) effects limit their use. Therefore, it was postulated that if the specificity of autologous cells could be enriched or altered, it would be possible to selectively target tumor or other antigens, without causing GVHD.

Transferred autologous cells can also be used to combat uncontrolled viral illness. In fact, the first successful clinical use of adoptively transferred antigen specific T cells targeted cytomegalovirus (CMV)-infected cells [5]. In doing so, Greenberg *et al.* demonstrated two important principles that have guided further cell therapy development: (i) that *ex-vivo* expansion was necessary to achieve sufficient clinical effect; and (II) that the specificity of a T-cell product could be manipulated *ex vivo*, in a way that endures after transfer [5]. Virus-specific T cells have also been used to treat Epstein–Barr virus (EBV)-related post-transplant lymphoproliferative disorders [6], systemic adenovirus infection [7], and even to combat human immunodeficiency virus 1 (HIV-1) infection [8, 9].

Adoptive cellular immunotherapy has also been used to restore normal immune function in some immunodeficient hosts. A subset of patients with severe combined immunodeficiency (SCID) possesses a mutation in the common gamma chain, blocking the lymphoid development of T and NK (natural killer) cells. Hematopoietic CD34+ stem cells from these patients were transduced with a retroviral vector expressing this common gamma chain, and the patients experienced a reconstitution of the previously absent immune cell

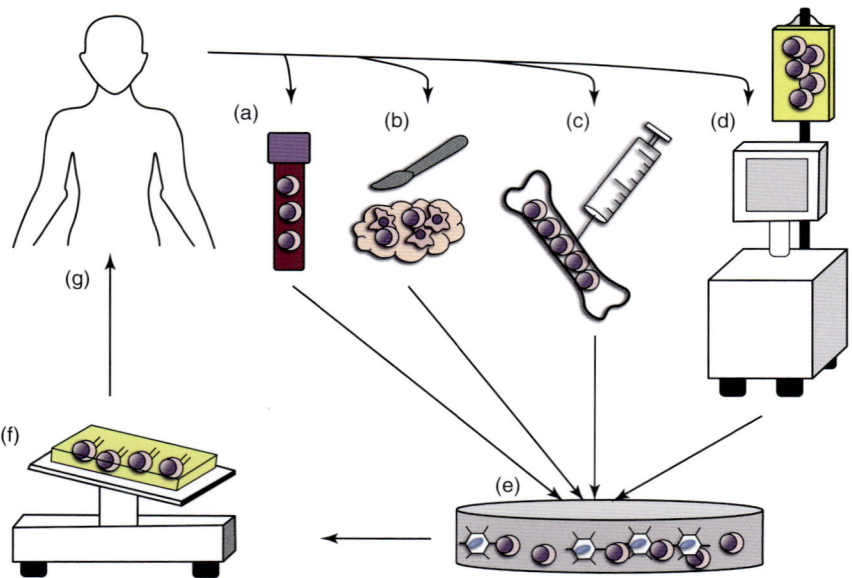

Fig. 1 Overview of approach to adoptive immunotherapy. The patient's own lymphocytes are obtained by: (a) peripheral blood draw; (b) excision of tumor bearing lymphocytes; (c) direct bone marrow biopsy; or (d) apheresis collection of mononuclear cells. Depending on the desired effect, these cells may then undergo (e) *ex-vivo* modifications including gene transfer and (f) expansion prior to (g) reinfusion.

subsets [10, 11]. Though this reconstitution was robust and durable, insertional mutagenesis was seen in some patients, leading to the development of leukemia [12, 13]. While this toxicity has prompted a reconsideration of transduction techniques as well as transduced cell populations, it is important to note that this type of adverse event has not been observed in T-cell therapy trials to date [14, 15]. This is not unexpected given that gene transfer into hematopoietic stem cells has the potential to alter not just lymphoid but rather all hematopoietic lineages. As evidence of this, Ott *et al.* have used gene transfer to correct X-linked chronic granulomatous disease, a defect in myeloid phagocytes [16]. In addition, Braun *et al.* have used gene transfer to correct Wiskott–Aldrich protein expression in lymphoid and myeloid lineages and in platelets [17]. The ability to modify multiple cell lineages is a major advantage of targeting the stem cell compartment for gene transfer.

Adoptive cellular immunotherapy has entailed many different approaches to treat many different diseases over the years. This diversity speaks to the vast potential role of the immune system in combating human disease. As the field moves forward, clinical findings will guide the quest for a deeper understanding of basic biology and laboratory advances translated to improve patient care. In this way, adoptive cellular immunotherapy represents a true bench-to-bedside science.

2
Selecting and Isolating Effector Cell Types

The effector cell type to be isolated, modified, and expanded may vary depending on the desired effect. Two cell lineages have been implicated in transferable immune-mediated cytotoxicity, namely T lymphocytes and NK cells, and both cell types may be targeted to malignant cells or virally infected cells. While it has also been reproducibly observed that the activation and proliferation of both T and NK cells can increase antitumor capacity [18–20], to date the majority of *in-vivo* tumor clearance success has been seen with T lymphocytes. The adoptive transfer of NK cells has also been shown to be safe and feasible [21, 22], and NK cells have been used to treat patients with renal cell carcinoma [23] and malignant glioma [24]. Overall, however, the resulting tumor eradication has been inconsistent [25].

One potential explanation for these findings is that, upon adoptive transfer, NK cells can demonstrate an adjustment of their activation threshold, termed "tuning" [26, 27]. Despite this limitation, active investigations to extend and enhance NK cell *in vivo* antitumor cytotoxicity are being pursued, given the unique ability of NK cells to recognize and kill major histocompatibility complex (MHC) class I-deficient targets [21]. Downregulation of MHC class I is a commonly employed mechanism by which viruses and malignancies have attempted to avoid immunosurveillance. While NK cells may be well suited to clear MHC class I-deficient targets, NK cell activation may need to be better understood and production methods enhanced before this becomes a widely applicable therapy.

Gamma delta T cells – like NK cells but unlike alpha beta T cells – are able to respond to non-MHC-restricted peptide ligands. This rare subset of T cells is able to be activated by MHC-like, stress-induced self-antigens that may be upregulated on the surface of some tumors, such as human glioblastoma multiforme. Consequently, gamma delta T cells represent one potential cellular therapeutic agent to clear these tumors. Currently, efforts are under way to optimize the *ex-vivo* expansion protocols of gamma delta T cells [28, 29], and to date gamma delta T cells have been used in at least 13 human clinical trials [30].

Dendritic cells are professional antigen-presenting cells (APCs) that, while not demonstrating direct cytotoxicity, can present target antigens and direct other target-specific native immune cells to proliferate, activate, and eradicate target cells. Autologous dendritic cells can be loaded with specific antigens or engineered *ex vivo* to present antigens. Once reinfused, these cells can direct tumor-specific immune cells within the patient's existing immune system to target the tumor. Dendritic cell therapy is, therefore, in essence a cellular-based vaccination strategy, and differs in approach to T and NK cell therapy. This cellular vaccine approach forms the basis for Sipuleucel-T (Provenge®), the first Food and Drug Administration (FDA)-approved cell therapy for cancer, and the first FDA-approved dendritic cell therapy. While groundbreaking, this therapy was associated with only a 4.1-month increase in median survival over placebo, leaving room for further progress in the field of anticancer cell therapy [31].

The isolation of T, NK or dendritic cells for autologous immunotherapy can be achieved through a variety of mechanisms. Tumor infiltrating lymphocytes (TILs) can be directly extracted from a patient's tumor tissue; in fact, the tumor may serve as an *in-vivo* enrichment tissue, selecting and preferentially expanding tumor-specific lymphocytes. Immunotherapies targeting diseases with easily accessible and abundant viable tumor tissue, such as melanoma, may derive starting effector cells for culture in this manner. Bone marrow biopsy can provide another source of lymphocytes to use as a starting cell population [32]. Finally, circulating mononuclear cells, including lymphocytes as well as dendritic cells, can be collected peripherally via a phlebotomy or the leukapheresis procedure. Once adequate starting tissue has been collected, the desired cell population can be further enriched (see Sect. 4.1; Preparation of Cells for Culture).

While traditional phlebotomy can provide a small amount of cellular material, the volume required to establish a starting culture is very often in excess of what the patient can tolerate. In those cases, the collection of mononuclear cells by apheresis can provide a large starting cell count. Apheresis collection involves establishing a high blood flow (often through a central catheter) and mixing the blood with citrate anticoagulant; centrifugation is then used to separate the blood components on the basis of their densities (see Fig. 2). The mononuclear cell layer, containing lymphocytes, can be selectively removed during

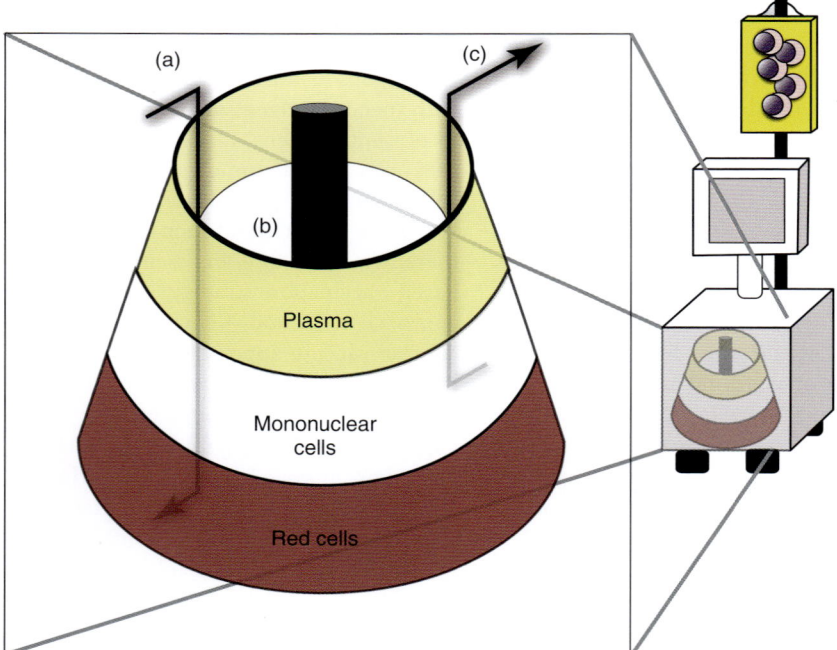

Fig. 2 Mechanism of leukapheresis for mononuclear cell collection. (a) Anticoagulated whole blood enters the machine housing traveling in sterile tubing; (b) Centrifugal force separates the various blood components by their density. In general, separation occurs between plasma (including platelet-rich and platelet-poor plasma), mononuclear cells, and red cell components; (c) By selectively aspirating the desired layer, specific blood components, may be isolated. Mononuclear cell layer removal is depicted here. Adapted from University of Utah School of Medicine WebPath Tutorial (http://library.med.utah.edu/WebPath/TUTORIAL/BLDBANK/BBAPHER.html)

leukapheresis. The latter process is advantageous because the uncollected cellular components and plasma can be returned to the patient, allowing the processing of a large blood volume. A typical collection may range from 5 to 15 liters of whole blood, or approximately one to three blood volumes for the average 70-kg patient. Leukapheresis can be performed once daily for several days if necessary, as is the case when collecting circulating hematopoietic stem cells, and the collected products may be pooled to increase the starting cell input. Typically, the procedure is well tolerated, although rare adverse effects associated with the apheresis catheter and/or citrate anticoagulant have been reported.

The efficiency of mononuclear cells removed per unit volume of processed blood ranges from approximately 30% to 60% [33, 34]. Several factors may influence the mononuclear cell yield, including the total volume of processed whole blood, the efficiency of the blood cell separators, the mobilization agents, and the blood flow rate [35–37]. Importantly, contamination with platelets has been shown to limit dendritic cell culture [38]. These data demonstrate that apheresis collection for the purpose of obtaining a cell therapy product is safe and well tolerated. However, further studies are needed to determine the optimal collection parameters for a range of cell therapy products.

At present, none of the techniques used to isolate starting cellular material is able to distinguish amongst the cell subsets, and consequently the final collected cell product will contain a mixture of the patient's mononuclear cells. These will include T, NK and dendritic cells, as well as B cells, other monocytes, neutrophils and, potentially, even malignant cells. Hence, further purification (as described in Sect. 4.1) may be necessary for the optimal culture of a desired subset of cells.

3
Modifying the Effector Cell Population

Collection methods for autologous cell therapy products provide a starting cell population consisting of a mixture of cell types and specificities. The spectrum of available modifications is quite broad. Modern molecular biologic techniques such as viral vector transduction, transposon-mediated transduction and electroporation now allow the alteration of lymphocyte specificity by engineering APCs or by directly re-engineering the lymphocyte cell surface expression. In addition, modifications can be made to a cellular product to generate resistance to viral infection. Finally, these techniques allow the introduction of inducible cell death proteins, providing an "off" switch for the therapy if desired.

3.1
Gene Delivery Techniques

Enhancing the function of primary human cells often requires the introduction of nucleic acid, either DNA or RNA. Several platforms have been shown to be effective in the transfer of genetic material into these cells:

- *Viral vectors* are the most widely used method to transfer large payloads of nucleic acid to primary human cells. This approach is particularly well suited to *ex-vivo* applications, wherein prior immunity to the viral capsid does not limit an efficient transfer.
- *Transposon-mediated transduction* is an alternative method of gene transfer that

utilizes transposase-based integration into the host genome. While transposon biology has been studied for many years, the use of transposons to introduce transgenes into cell therapy products is a relatively new approach.

- *Electroporation*, as opposed to viral vectors or transposon cassettes, employs an electrical pulse to transiently permeabilize the cell membrane, allowing hydrophilic molecules to transfer into the intracellular space [39]. Similar to viral vector transfer, once introduced, the nucleic acids are processed by the cell machinery to express protein.

3.1.1 Viral Vectors

Viral vectors are constructed by the removal of most virally encoded genes and the insertion of a gene of interest. The vector utilizes the viral machinery to introduce the genetic material and, once introduced, the gene of interest is transcribed and translated by the transduced cell machinery. Retrovirus family members are a common choice for the viral vector transduction of human cells. Gamma-retroviral vectors have been investigated in human gene therapy trials [40] and in human T-cell immunotherapy trials [41] since the early 1990s. In this case, the gamma-retrovirus integrates into the host genome, providing long-term gene expression. Gamma-retroviruses must transduce actively dividing cells, however, and insertional mutagenesis has been observed after the transduction of hematopoietic stem cells [13, 42, 43]. To address these issues, lentiviral vectors were developed for human primary cell transduction. Lentiviral vectors transduce nondividing cells [44], appear to have a safer integration site profile [45, 46], and also have the capacity to deliver transgenic material equivalent to gamma-retrovirus vectors. Other viral vectors derived from adenovirus [47, 48] and adeno-associated virus [49] are less commonly employed.

3.1.2 Transposon-Mediated Transduction

Integration-competent, replication-incompetent transposon cassettes allow for the efficient and permanent introduction of transgenes. A DNA transposon system such as Sleeping Beauty (SB) consists of flanking inverted repeats that act as transposase-binding sites [50]. Wild-type transposons encode their own transposes, allowing for replication and integration throughout the genome. For the purpose of cell therapy applications, the SB system has been engineered to replace the coding sequence for the transposase with an expression cassette. During *ex-vivo* processing this cassette, along with *in trans* provision of the transposase protein, allows cut-and-paste transposition of the transgene into the host genome. Of particular note, sites of integration appear to be nearly random. This is a clear advantage over other integrating viral vectors, which are predisposed to integrate at sites of active transcription since integration at actively transcribed sites makes insertional mutagenesis much more likely. The feasibility of transposon-mediated transduction of human peripheral blood lymphocytes has been demonstrated [51], and this technology is currently in use in clinical trials [52].

3.1.3 RNA Electroporation

The newest technology to be explored as a method of modifying human primary T-cell specificity is electroporation. The RNA electroporation of messenger RNA (mRNA) into target cells is a rapid and efficient method for generating transient expression. In humans, this approach is particularly well suited to investigational transgenes if the off-target toxicity is unknown. In addition, genotoxicity is not of concern with RNA electroporation, given that mRNA does not integrate into the genome. For some constructs, RNA electroporation is at least as efficient as retrovirus at gene transfer [53]. Potent anti-tumor activity of RNA-modified cells has been demonstrated against both liquid tumors such as advanced leukemia [54], and solid tumors such as mesothelioma [55] and neuroblastoma [56] in mice. In human trials of chimeric antigen receptor (CAR)-based therapies with potential for off-target toxicity, an RNA-based approach has been shown to be safe [57]. Clinical trials using RNA–CAR-electroporated T cells are currently under way and will provide valuable further information regarding the safety and efficacy of the procedure.

3.2 Enrichment of Antigen-Specific Effector Cells

Within a population of adaptive immune cells, typically a broad repertoire of specificities is observed. During a physiological immune response, those immune cells with sufficient specificity to respond to the presented antigen will preferentially activate and expand. In the context of cellular therapy, advantage can be taken of this situation by selecting a tissue, or recreating an environment, that is rich in the target antigen (Fig. 3). In doing so, target-specific clones will be preferentially expanded and the final product will be enriched for cells with the desired specificity. Enrichment of antigen-specific effector cells enables the production of both virus- and tumor-specific cell products for reinfusion.

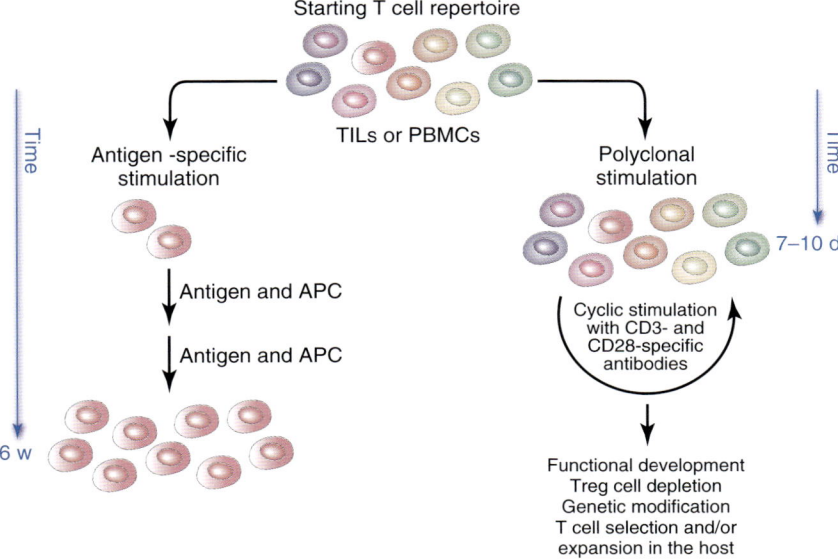

Fig. 3 Selection of antigen-specific clones and expression of engineered T-cell receptors. Reproduced with permission from Ref. [58].

3.2.1 Enrichment of Virus-Specific Lymphocytes

In the setting of bone marrow transplant, viral infection or reactivation represents an enormous cause of morbidity and mortality. Therefore, virus-specific lymphocytes – specifically cytotoxic T cells – are considered one potential avenue for restoring antiviral immunity. However, demonstration of the feasibility of production of these virus-specific T lymphocytes for large clinical trials has been a major challenge. Leen et al. have shown that, by modifying an EBV-transformed B-cell line to express viral peptides, a single culture technique could yield CD4+ and CD8+ T lymphocytes with multiple desired specificities [59]. Others have shown the feasibility of virus-specific T-cell products from cord blood products [60]. Importantly, virus-specific T cells produced in this manner are not associated with GVHD [61].

3.2.2 Enrichment of Tumor-Specific Lymphocytes

Within the tumor microenvironment, the presentation of tumor peptides generates an enriched population of tumor-specific TILs. This already enriched starting cellular material may be cultured to increase the number of tumor-specific lymphocytes. Further preferential expansion of TILs *ex vivo* and reactivation of their effector function allows for the reinfusion of a highly specific and effective product. Investigations pioneered by the Rosenberg group at the National Cancer Institute have employed this strategy in the context of melanoma, with striking clinical responses [62, 63]. While early studies demonstrated a limited persistence, the addition of pre-reinfusion recipient lymphodepletion has been shown to increase engraftment and effector function of adoptively transferred cells [64] Patients with melanoma, treated with lymphodepletion, followed by reinfusion of a TIL-based cell therapy, demonstrated clinical response rates between 49% and 72% [64–66], with some patients demonstrating complete remission beyond five years [67].

One potential limitation of this approach is that while enriched for tumor-specific lymphocytes, any tumor tissue will be a mixture of cells. This population could include suppressive T-regulatory cells (T-regs) which may limit the *ex-vivo* expansion of the desired cell effector population or worse, tolerize the tumor. Depleting the tumor tissue of T-regs prior to culture is actively being pursued.

An extended *ex-vivo* clonal expansion may result in cells with limited *in vivo* persistence following transfer and a reduced clinical efficacy [69, 70]. One hypothesis for this decreased efficacy is that the cultured T cells are "exhausted" due to persistent stimulation. In other contexts, such as chronic lymphocytic leukemia (CLL), the attenuation of T-cell function can be reversed via CD28 stimulation [71, 72]. The mechanism, however, of *ex-vivo*-stimulated exhaustion is likely to be very different than that seen in CLL-associated attenuation of function. In contrast, Heslop et al. used a co-culture technique to generate functional EBV-specific CTLs capable of persisting up to nine years *in vivo* [6]. Although this discrepancy could be explained by a number of factors, these results demonstrate that the limitations of persistence seen with culture-stimulated cells are not insurmountable.

Many tumors, however, are capable of inhibiting the antigen presentation of tumor peptides, thereby limiting the expansion and activation of anti-tumor lymphocytes. This inhibition can be overcome by forcing APCs, such as dendritic cells, to present specific tumor antigens and drive the expansion of tumor-specific lymphocytes. Upon reinfusion, these modified APCs preferentially stimulate tumor-specific lymphocytes and, in essence, vaccinate the patient against the tumor. Marked clinical and immunologic responses against HER-2/neu (c-erbB2) -positive ductal carcinoma *in situ* breast cancer have been observed after reinfusion of dendritic cells pulsed *ex vivo* with tumor peptides [68].

Several dendritic cell platforms are currently being used to treat human malignancies by augmenting the number and function of antigen-specific T cells. Sipuleucel-T (Provenge®), the first FDA-approved (April 2010) therapeutic cancer vaccine, makes use of such a platform. Sipuleucel-T was approved to treat asymptomatic or minimally symptomatic metastatic castrate-resistant (hormone-refractory) prostate cancer, and is generated from autologous leukapheresis-derived peripheral blood mononuclear cells. The patient's mononuclear cells were co-incubated with a prostate acid phosphatase (PAP)–granulocyte-macrophage colony stimulating factor (GM-CSF) fusion protein. GM-CSF activates the APCs to present PAP, and in response PAP-specific autologous T cells are preferentially expanded; PAP is then expressed on prostate cancer cells. The final product is washed to remove the fusion protein and then reinfused into the patient. In a double-blind, placebo-controlled study in humans, a 50% improvement in three-year mortality was observed over the placebo control (31.7% versus 21.7%, respectively) [31]. This success has driven many groups in both industry [73–77] and academia [78, 79] to expand on the dendritic cell vaccine approach to treat glioblastoma multiforme, prostate cancer, and ovarian cancer.

3.3
Expression of Novel Cell-Surface Receptors

The expression of novel cell-surface receptors can serve to re-engineer lymphocyte specificity. In the case of a modified T cell, this approach can utilize viral vectors or electroporation to express complete T-cell receptors (TCRs) or synthetic CARs. In either case, retargeting cell specificity can limit activation to the desired ligand and even drive the long-term persistence of stably expressed constructs [80].

3.3.1 Non-Endogenous T-Cell Receptors

Specific TCRs have the significant advantage of being able to target a T cell to an intracellular antigen, presented in the context of MHC. On the other hand, this requires that the transduced TCRs be compatible with the patient's human leukocyte antigen (HLA) type. In addition, experiments in mice have raised concerns with regard to transduced-endogenous heterodimer TCR expression. Because the TCR consists of a paired alpha and beta chain, the potential exists for a transduced cells to express a TCR composed of a transduced alpha chain pairs with an endogenous beta chain, or vice versa. The specificity of these heterodimers is unknown, and could generate an autoreactive clone as seen in animal models [81], though this has not been reported in clinical trials to date. Investigators have demonstrated several methods to preferentially express the transgenic TCR and conversely to knock down endogenous TCR expression.

Despite these concerns, the major *in vivo* limitations to T cells redirected with specific TCR has not been autoimmunity, but rather has been on-target, off-tumor activity [82, 83]. The findings of these groups and others [84, 85] demonstrate current limitations in predicting cross-reactive

TCR affinity. However, investigations conducted by Linette *et al.* have provided a strategy to circumvent this limitation by re-engineering induced pluripotent stem cells [84]. Early trials in the adoptive transfer of HLA-restricted, specific TCR-redirected T cells have shown efficacy at tumor clearance [86].

3.3.2 Chimeric Antigen Receptors

Modified T cells, expressing whole, non-endogenous TCRs demonstrate several limitations. These receptors can only recognize peptide presented in the context of MHC, while many tumors may downregulate MHC as an immune evasion mechanism. In addition, a single TCR transmits only the primary activation signal, which is insufficient to drive proliferation. Synthetic receptors termed chimeric antigen receptors are MHC-independent and do not require a second costimulatory ligand for activation. CARs are a highly specific receptors with: (i) an extracellular, single chain variable fragment (scFv); (ii) a TCR-derived transmembrane domain; and (iii) a TCR intracellular signaling domain (often CD3zeta), the primary signal transducing element of the TCR complex. A scFv consists of a fusion of one heavy and one light chain variable domain, linked by a flexible peptide linker. As noted above, unlike native TCRs scFvs can bind to antigen in an MHC-independent manner [87]. First-generation CARs were created by fusing the CD3zeta domain to the transmembrane domain (Fig. 4). Although cells expressing these CARs demonstrated on-target cytotoxicity [88], the *in-vivo* responses were short-lived. It has long been known that primary signal in the absence of costimulation causes the cell to become anergic or unable to activate in response to ligand [89]. CAR-modified T cells also require costimulation to demonstrate a durable response [90]. Second-generation

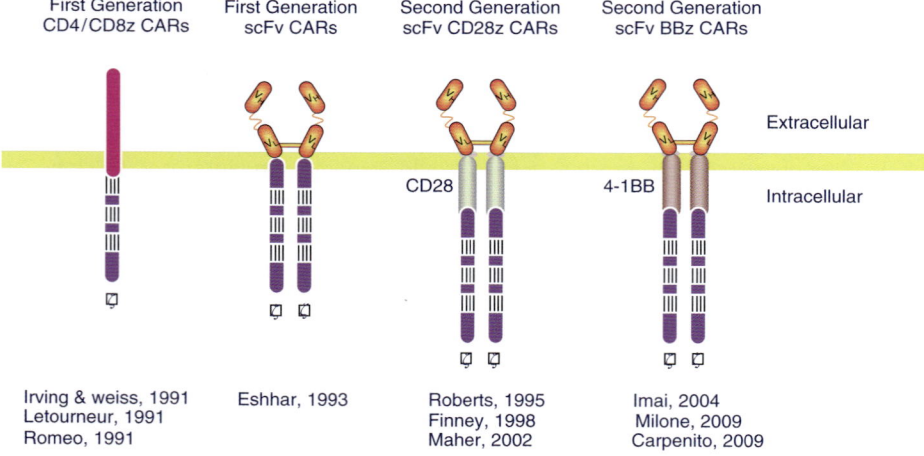

Fig. 4 Multiple generations of CARs. The first-generation CARs were composed of a single extracellular domain with a scFV [88]. Second-generation CARs incorporated intracellular domains of costimulatory molecules such as CD28 [92, 97, 98] or 4-1BB [94, 95, 99] *in cis* with the remainder of the receptor.

CARs address this problem by adding a costimulatory domain, such as the CD28 intracellular domain [91–93] or a tumor necrosis factor (TNF) receptor family domain such as the 4-1BB (CD137) intracellular domain [94] *in cis* with the other CAR domains. CARs with built-in costimulatory domains generate both primary and costimulatory signal in response to a single ligand. Further modifications to CARs, including the addition of multiple costimulatory domains, have demonstrated even greater tumor eradication in animal models [95–97].

CARs have been designed to target a variety of cell-surface proteins. CD19 has been one of the most targeted cell-surface molecules, and one that has seen the most success. CD19 is expressed on B-cell leukemias and lymphomas, and also on healthy B cells. An expansion of CD19-targeted cell therapies, including non-Hodgkin lymphoma (NHL), follicular lymphoma and large cell lymphoma, is being pursued. In addition, receptors with affinity for molecules such as CD20 (acute lymphoblastic leukemia, CLL, NHL), CD30 (NHL and Hodgkin lymphoma), CD33 (acute myeloid leukemia), CD138, and immunoglobulin-kappa light chain (multiple myeloma) are all worthy approaches for future cell therapy trials. Pule *et al.* have reported that EBV-specific T cells, expressing synthetic anti-disialoganglioside GD2 receptors, target neuroblastoma [100], while others have demonstrated the efficacy of anti-mesothelin CAR T cells [57]. CAR-based cell therapy requires an appropriate tumor-specific cell-surface target. In addition, tumor downregulation of target epitopes or the selection of target-negative clones will allow for escape mutants. Future approaches will likely return to a multi-target approach to limit tumor evasion of these engineered products.

3.4
Generation of Resistance to Viral Infection

Human viruses employ several mechanisms to gain entry to a host cell. Binding to a cell-surface receptor (in some cases to multiple receptors) is often required in order for a virus to infect a cell. The HIV-1 virion, for example, binds to CD4 and either the chemokine receptors CXCR4 or CCR5 to infect a T cell. Blocking of viral receptors is a commonly employed approach for the prevention and treatment of systemic viral illness. A permanent depletion of viral entry receptors has the potential to provide an effective therapy for viral infections.

In 1% of Caucasians, there is a naturally occurring mutation termed delta32 that results in the loss of CCR5 expression. An absence of CCR5 is associated with few pathogenic findings, including an increased risk of complications following West Nile virus infection [101]. CCR5 acts as a coreceptor for HIV-1 entry into CD4 T cells, and CCR5 blockade by the drug maraviroc has shown great promise in the treatment of HIV1 [102]. A stem cell transplant from a CCR5-null delta32 donor to an HIV-positive male patient to treat his leukemia resulted in an apparent functional cure [103]. With respect to cell therapy, CCR5-null T cells can be generated from CCR5-positive cells *ex vivo* by zinc finger nuclease gene disruption at the CCR5 locus [104]. These HIV-1 resistant cells can then be reinfused into a patient with HIV-1, providing them with cells that can restore CD4+ function and remain uninfected. To date, Phase I clinical trials have demonstrated the safety and feasibility of adoptively transfer autologous CCR5-null T cells into HIV1-positive subjects [105].

Several novel strategies to combat HIV infection are currently being developed. MazF is an *Escherichia coli*-derived endoribonuclease that recognizes and cleaves the sequence adenine–cytosine–adenine in single-stranded RNA. This sequence is particularly enriched in HIV1 mRNA, and MazF expressing cells could be resistant to infection. *Ex-vivo*-modified MazF-expressing CD4 T cells have been adoptively transferred into Rhesus macaques with success [106], and human trials are currently underway (www:Clinicaltrials.gov, NCT01787994). Recently, several other strategies have been employed to modify autologous cells to become resistant to infection or viral replication, and have shown great promise [107].

4
Manufacturing Cells for Adoptive Transfer

Once a mononuclear cell product has been obtained from a patient via apheresis collection, the product must undergo processing to expand the desired effector cell population. In addition, an appropriate specificity must either be selected or engineered into the cell population to be transferred. Targeted specificity can be obtained by enriching for antigen-specific cells or by using vector-based systems to introduce non-endogenous TCRs or synthetic CARs. The ultimate goal of the manufacturing process is to start with a limited polyclonal population and to generate a large, targeted cell population capable of potent on-target killing and persistence upon transfer.

4.1
Preparing Cells for Culture

Prior to initiating the culture of autologous cells for adoptive transfer obtained by apheresis, the collection product must be washed. If a T- or NK-cell product is to be manufactured, the cells will then undergo monocyte depletion by magnetic beads and/or counterflow centrifugal elutriation. If dendritic cells are being isolated, however, the monocytic fraction will be retained. In order to achieve an isolation of lymphocyte subsets, the cells are incubated with magnetic beads that are covalently linked to monoclonal antibodies specific for cell-surface markers. The cells and beads are passed through a magnetic field and the flow-through, which is now depleted of the magnetically bound cells, is collected and sent for culture. This type of depletion can efficiently remove monocytes as well as any additional undesired cell population prior to culture via anti-CD4, anti-CD8, anti-CD25, and other antibody–bead combinations.

Magnetic bead purification is an ideal approach in the setting of a relatively frequent target cell population, and in which yield and purity are both considerations. Generally, bead purification to provide the starting culture material is sufficient, though interferon-gamma capture can also be considered to enrich infrequent cell populations [108]. Other techniques, such as cell sorting by flow cytometry have been employed, but depending on the number of cells to be prepared this form of isolation can be quite lengthy [109, 110]. Once the desired cell subset has been isolated, this population must be expanded to obtain an effective dose.

4.2
Artificial Antigen-Presenting Cells for *Ex-Vivo* Expansion

Cell expansion is typically achieved by mimicking *in vivo* proliferation signals. In the case of T cells, this includes the provision of both a primary signal through the TCR and a simultaneous costimulatory signal through the CD28 receptor [111]. Simultaneous signaling through immobilized CD3 and CD28 ligands has been shown to induce telomerase activity in CD4+ T cells [112], which suggests that costimulation not only allows proliferation but that the progeny will also retain further proliferative capacity. With other cell types, different protocols are required. For example, the expansion of gamma delta T cells requires CD137L, interleukin (IL)-2 and IL-21 [29], and can be enhanced by the addition of bisphosphonates or CD2 stimulation [113–115]. NK cells, on the other hand, have been stimulated to expand *ex vivo* by whole irradiated tumor cells, among other methods [116].

Ex vivo, the provision of proliferation signals can be achieved by immobilizing ligands on either beads or cell scaffolds. Bead-bound monoclonal antibodies directed against CD3 and CD28 have been used to provide T-cell proliferative signals. These beads are logistically convenient in that they can easily be removed after coincubation and prior to reinfusion. This approach is particularly well suited to the expansion of human polyclonal naïve and central memory T cells, allowing for a more than 10^9-fold expansion of the input cell population [117]. In addition, anti-CD3/anti-CD28 antibody-coated beads do not cross-react with CTLA-4 (a negative costimulatory signal), and this allows for the efficient expansion of T cells. Good manufacturing practice (GMP)-compliant anti-CD3/anti-CD28 beads for use in T-cell expansion have been used in clinical trials since 1996 [118, 119].

Newer cell-based artificial antigen-presenting cell (aAPC) platforms have been developed in the chronic myeloid leukemia (CML) K562 cell line [120–122]. Although K562 cells lack MHC or T costimulatory ligands, they do express intercellular adhesion molecule (ICAM) and lymphocyte function-associated antigen 3 (LFA-3), both of which are enhancers of T cell–APC interaction. In addition, by employing lentiviral vector transduction this cell line can be engineered to express a library of T-cell signaling molecules that includes, but is not limited to, CD32, CD64, CD80, CD83, CD86, CD137L, and CD252 [120–122].

In addition to an ability to coexpress a wide variety of stimulatory molecules, aAPC platforms are more efficient than bead-based techniques at activating and expanding CD8+ T cells and antigen-specific T cells [120–122]. Cell-based aAPCs have been shown to expand the starting cell population approximately 10 000-fold and to maintain effector T-cell function and antigen specificity for more than two months [120, 122]. The efficient expansion of antigen-specific or CD8+ T cells is ideal for many cell therapy applications that aim to target and kill with specificity.

In addition, K562-based aAPCs can be used to expand NK cells for adoptive transfer. Engineered aAPCs expressing CD64/FcγRI, CD86/B7-2, CD137L/4-1BBL, truncated CD19, and membrane-bound IL-21 were shown to expand NK cells from peripheral blood mononuclear cells (PBMCs) from healthy human donors, with a more than 2000-fold expansion over a 14-day culture [123]. Other groups have developed current Good Manufacturing Practices-compliant K562-mediated NK cell expansion systems using IL-15 (mbIL15) and 4-1BB ligand [124]. Whilst the ideal expansion with maintenance of cytotoxicity upon transfer is still being elucidated, *ex-vivo* culture in the presence of membrane-bound IL-21 appears to be important [125, 126].

To enable a wide availability of autologous cell therapies, the next generation of *ex-vivo* expansion is likely to focus on scaling up and automating existing protocols. The transduction efficiency of available retroviral and lentiviral vectors is high even for large products [127–129]. Dynamic wave action bioreactors can accommodate up to 10^{10} or more cells for culture, while closed systems are also available at harvest for the large-scale removal of beads. The clinical-scale magnetic depletion of cell subsets is also available within a closed system. Clinical trials employing protocols for modifying cell products in this fashion are currently under way [130, 131]. These novel protocols may offer further advancements of antitumor cell therapy, or even allow T-cell therapy to branch out for the treatment of nonmalignant conditions such as autoimmune disease.

4.3
Polarizing Effector Cell Phenotypes

As lymphocyte subset analyses have become more refined, it has become clear that effector cell subpopulations demonstrate different capacities. In light of this, it has become desirable at times to skew cell therapy products towards specific cell subsets. Culture conditions represent one method to achieve a skewed cell population at the time of harvest. The *ex-vivo* culture of T cells with rapamycin (a mammalian target of rapamycin; mTOR inhibitor) and IL-4 has been shown to favor T-helper 2 (Th2) subsets, which upon transfer demonstrate favorable graft-versus-tumor outcomes with decreased GVHD in stem cell transplant patients [132]. Others have successfully generated tumor-specific, IL-17-producing primary human T cells by the introduction of a CAR with the inducible costimulatory (ICOS) intracellular domain. These Th17-CAR T cells may demonstrate enhanced persistence [133].

4.4
Expanding Immune Suppressors

CD4+ CD25+ forkhead box protein 3 (FoxP3)-positive T-regs are thymic-derived, *in vivo* suppressors. These cells have the ability to suppress active immune responses and to induce tolerance towards specific antigens [134], thereby playing a crucial role in preventing autoimmunity [135]. In mouse models, *ex-vivo*-expanded donor-derived T-regs have been shown to prevent acute and chronic GVHD while preserving graft-versus-tumor effects [136, 137]. However, in a different clinical setting T-regs may tolerize tumors and inhibit endogenous or adoptively transferred antitumor cytotoxic responses [138, 139]. For these reasons, a great deal of interest has been expressed in both expanding T-reg cell therapies to combat autoimmune disease, and depleting T-regs from antitumor cytotoxic T-cell products prior to reinfusion.

Generating T-regs *ex vivo* poses several obstacles. Typically, peripheral blood T-regs are scarce and there are no definitive specific cell-surface markers. Although most culture conditions favor non-T-reg growth, limiting *ex vivo* expansion [140], Golovina *et al.* showed that, in the presence of CD28 costimulation and mTOR inhibition by rapamycin, T-regs can be expanded more than 1000-fold while retaining their suppressive capacity [141]. Several trials have demonstrated the safety and feasibility of infusing *ex-vivo*-expanded T-regs in humans. However, although a clinical-scale expansion of human T-regs using rapamycin and artificial APCs has been developed, many of these protocols did not employ autologous cells.

Of note, given the capacity for T-regs to suppress, a primary concern is the potential for an increased risk of infection. In a clinical trial involving double umbilical cord blood transplants, 23 patients received enriched, *ex-vivo*-expanded T-regs. Notably, there was no reported increased risk of infection compared to historical controls, and the patients also showed a reduced incidence of grades II–IV acute GVHD [142]. Promising results such as these warrant further investigation into adoptively transferred T-regs as a potential therapy for autoimmune diseases.

5
Clinical Course of Adoptive Transfer

5.1
Preconditioning

The transfer of modified cells into a host with a moderately intact immune system is generally associated with a limited persistence of the transferred cells – an observation that seriously hampered the early clinical trials. In response, many groups began to lymphodeplete recipients immediately prior to reinfusion [65, 66, 143–145], with standard lymphodepletion regimens including cyclophosphamide with or without fludarabine or, at other sites, total body irradiation. Lymphodepletion prior to reinfusion is associated with improvements in the persistence of transferred T cells and of clinical outcome [66, 143, 146]. Several hypotheses exist as to the mechanism of improved outcome. Host lymphodepletion is likely to deplete the malignant cell population, increasing the effector:target ratio. In addition, host T-regs may be responsible for inhibiting the antitumor response by the transferred T cells. Finally, conditioning regimens are associated with an increase in IL-7 and IL-15, both of which cytokines have been associated with improved T-cell expansion post transfer [143]. Pre-transfer lymphodepletion may relieve this host-T-reg-derived inhibition, as well as contributing to long-lived persistence.

5.2
Managing Reinfusion of Autologous Cells

The reinfusion of autologous cells can have a number of on- and off-target effects requiring acute clinical management. Concurrently administered cytokines, such as high-dose IL-2, has resulted in flu-like symptoms and capillary leak with fluid retention [147]. Because modified cells may bear foreign epitopes, these cells may elicit a severe allergic response [148]. One report raised the possibility of adoptively transferred cells exacerbating an existing infection in a patient who demonstrated a sepsis-like syndrome [149]. In addition, screening for both off-target and on-target, off-tumor effects with bioinformatics or pre-clinical studies is imperfect. Unpredicted adverse effects have been observed wherein adoptively transferred, modified cells targeted to tumor antigens, demonstrated an *in vivo* affinity with cardiac muscle [84, 85] or pulmonary endothelium [150].

Once reinfused, adoptively transferred engineered CAR cells appear to be highly effective at lysing leukemic cells. In a clinical trial of two patients with acute lymphoblastic leukemia who were treated with anti-CD19 CAR-T cells, both patients demonstrated an *in vivo* expansion accounting for between 34% and 72% of peripheral CD3+ cells at peak [151]. The patients developed a cytokine release syndrome characterized by fever, with elevations of several cytokines including IL-1β, IL-6, interferon-γ, and TNF-α. The patients experienced a simultaneous reduction in tumor burden and an increase in lactate dehydrogenase consistent with tumor lysis syndrome. Both patients entered clinical remission [146]. This course, including the onset of *in-vivo*-modified cell expansion, cytokine release syndrome and tumor lysis, has been reproducibly observed at a number of centers [151–153]. Among a cohort of 30 relapsed and refractory ALL patients, a 90% complete response rate was observed with anti-CD19 CAR-T cells as a stand-alone therapy [153].

The management of patients as they endure severe tumor lysis and cytokine release syndromes continues to be a challenge. Anti-cytokine therapy has been used with some success [151], though further investigation of the role of these drugs is needed. Advances in cellular engineering may provide to key to safely tolerating the post-reinfusion period of tumor clearance. Di Stasi *et al.* used retroviral transduction to introduce an inducible caspase which, upon exposure to a synthetic dimerization agent, caused transduced cells to rapidly die [154]. One potential caveat to this approach may be that downstream inflammatory events may no longer be dependent on the transferred cells. For example, robust cytokine production initially in response to activated transferred cells may continue even after the elimination of those cells from the patient. Nonetheless, as more targeted cell therapies move into human trials, transduced "suicide genes" are likely to be employed in conjunction with already established anti-cytokine therapies.

While proof of principle has been established with regards to many adoptive cell therapies, further studies are needed to expand and refine targets. Engineered receptors confer a great deal of specificity, but *in-vitro* assays of target specificity have known limitations. At present, the true scope of off-target reactivity *in vivo* cannot be known until transfer to the patient [84, 85]. Even with high specificity, on-target/off-tumor adverse effects have also been observed [150]. Finally, even in the case of the transfer of cells with highly tumor-specific on-target activity, escape mutants have developed [151].

5.3
Promoting Long-Lived Persistence

Adoptively transferred cells have been shown to persist for more than two decades for modified hematopoietic stem cells [155], and for 10 years in the case of CAR-bearing T cells [15]. Not surprisingly, it has also been shown that a longer persistence of transferred cells correlates with a better clinical response [156]. For tumor-targeted modified cells, the *in-vivo* magnitude and expansion kinetics of recipient nonmalignant immune cells, the tumor burden and likely microenvironment, and the subtype of transferred cells influence the degree of persistence. As noted above, preconditioning has been associated with a greatly increased persistence of transferred cells [157, 158]. Several hypotheses exist as to the mechanism of this improved persistence. One possibility is that preconditioning lymphodepletion decreases the number of recipient nonmalignant immune cells, perhaps giving transferred cells a proliferative advantage over the recipient nonmalignant cells. Long-term persistence may also be influenced by the cell subset that is transferred. For instance, in animal models the long-term persistence of adoptively transferred T cells is associated with central memory T cells differentiation. Upon antigen activation, naïve T cells can generate effector memory T cells (TEM; CD8+ CD62L−) and/or long-lived central memory T cells (TCM; CD8+ CD62L−). Most clinical trials with T-cell products have transferred CD8+ TEM cells due to cell culture techniques favoring TEM expansion [159–161]. *In vivo*, however, TCM cells demonstrate superior tumor cytotoxicity [131, 159] and persistence compared to TEM cells [162]. Future studies in humans will likely optimize culture techniques to selectively expand the most effective subset cells for transfer.

6
NextGen Cell Therapy

During recent years, autologous cellular therapy has achieved enormous success, and this has generated an extensive development of clinical trials in humans. Indeed, as of October 2014, over 200 adoptive cellular therapy trials are listed (www.clinicaltrials.gov). Many trials have witnessed dramatic response rates in patients with refractory disease, and cellular therapy is now poised to make a monumental leap forward to the next generation of therapies capable of treating a myriad of diseases and large numbers of patients. Future studies will likely focus on: (i) further improving existing protocols; (ii) applying early success to new diseases; and (iii) making all cell therapies available to as many patients as possible.

An understanding of the management of cytokine release syndrome is critical in order for adoptive cellular therapies to become more widely available. It has yet to be determined which cytokines need to be blocked, and what effect this will have on the long-term persistence of the modified cells. Safety concerns may require a reconsideration of transient expression, as with RNA electroporated cells and/or the introduction of suicide genes. At present, a close monitoring of modified cell persistence and relapse is recommended, as the long term consequences of these therapies are not yet well understood.

One of the most exciting areas to observe over the next decade will be that of adoptive cellular therapies targeting solid tumors. In order to clear solid tumors, the adoptively transferred cells must first traffic and infiltrate the tissue, and then generate cytotoxicity against tumor cells within a solid mass. Each of these requirements for successful eradication present unique challenges to solid tumor-targeting cells, as opposed to their liquid tumor-targeting counterparts. It is very likely that these new demands will require the reconsideration of each step in the process of adoptive cell therapy development, from cell selection to reinfusion and the incorporation of additional immune modulators such as newly approved checkpoint blockade drugs. The inhibition of CTLA4 and/or PD1 has already been shown to be successful in combination with other therapies, and this would serve as a natural starting point to augment adoptive cell therapy [163].

Finally, with the goal of providing hope to those patients whose disease is refractory to current therapies, and regardless of the disease target, automation will emerge as a major challenge during the coming years. As noted above, various equipment already exists for large-scale production in closed systems, though all components have yet to be developed. Some autologous cell therapies may transition to allogeneic or banked products as available methods of gene and cell engineering advance. Likewise, clinical protocols will need to be developed and staff trained to perform techniques that, to date, have only been attempted at the laboratory bench. In addition, regulations may need to be revised or created as new technologies for autologous cell engineering are developed in the laboratory and translated to the clinic. Essentially, the infrastructure of producing and administering a therapeutic will need to be created to form a "next generation" of transfusion medicine. Clearly, this shifting paradigm will require a new generation of physicians and scientists that is committed to translating biology into medicine, and vice versa.

Disclosure Statement

Bruce Levine declares financial interest due to intellectual property and patents in the field of cell and gene therapy. Conflict of interest is managed in accordance with University of Pennsylvania policy and oversight. Dr Levine has sponsored research grant support from Novartis.

References

1. Mitchison, N.A. (1955) Studies on the immunological response to foreign tumor transplants in the mouse. *J. Exp. Med.*, **102** (2), 157–177.
2. Weiden, P.L., Flournoy, N., Thomas, E.D., Prentice, R. *et al.* (1979) Antileukemic effect of graft-versus-host disease in human recipients of allogeneic-marrow grafts. *N. Engl. J. Med.*, **300** (19), 1068–1073.
3. Porter, D.L., Roth, M.S., McGarigle, C., Ferrara, J. *et al.* (1994) Induction of graft-versus-host disease as immunotherapy for relapsed chronic myeloid leukemia. *N. Engl. J. Med.*, **330** (2), 100–106.
4. Kolb, H.J., Mittermuller, J., Clemm, C., Holler, E. *et al.* (1990) Donor leukocyte transfusions for treatment of recurrent chronic myelogenous leukemia in marrow transplant patients. *Blood*, **76** (12), 2462–2465.
5. Greenberg, P.D., Reusser, P., Goodrich, J., and Riddell, S. (1991) Development of a treatment regimen for human cytomegalovirus (CMV) infection in bone marrow transplantation recipients by adoptive transfer of donor-derived CMV-specific T cell clones expanded in vitro. *Ann. N. Y. Acad. Sci.*, **636**, 184–195.
6. Heslop, H.E., Slobod, K.S., Pule, M.A., Hale, G.A. *et al.* (2010) Long-term outcome of EBV-specific T-cell infusions to prevent or treat EBV-related lymphoproliferative disease in transplant recipients. *Blood*, **115** (5), 925–935.
7. Feuchtinger, T., Matthes Martin, S., Richard, C., Lion, T. *et al.* (2006) Safe adoptive transfer of virus-specific T-cell immunity for the treatment of systemic adenovirus infection after allogeneic stem cell transplantation. *Br. J. Haematol.*, **134** (1), 64–76.
8. Borrow, P., Lewicki, H., Hahn, B.H., Shaw, G.M. *et al.* (1994) Virus-specific CD8+ cytotoxic T-lymphocyte activity associated with control of viremia in primary human immunodeficiency virus type 1 infection. *J. Virol.*, **68** (9), 6103–6110.
9. Riddell, S.R., Elliott, M., Lewinsohn, D.A., and Gilbert, M.J. (1996) T-cell-mediated rejection of gene-modified HIV-specific cytotoxic T lymphocytes in HIV-infected patients. *Nat. Med.*, **2** (2), 216–223.
10. Cavazzana-Calvo, M., Hacein-Bey, S., de Saint Basile, G., Gross, F. *et al.* (2000) Gene therapy of human severe combined immunodeficiency (SCID)-X1 disease. *Science*, **288** (5466), 669–672.
11. Hacein-Bey-Abina, S., Le Deist, F., Carlier, F., Bouneaud, C. *et al.* (2002) Sustained correction of X-linked severe combined immunodeficiency by ex vivo gene therapy. *N. Engl. J. Med.*, **346** (16), 1185–1193.
12. Hacein-Bey-Abina, S., von Kalle, C., Schmidt, M., Le Deist, F. *et al.* (2003) A serious adverse event after successful gene therapy for X-linked severe combined immunodeficiency. *N. Engl. J. Med.*, **348** (3), 255–256.
13. Hacein-Bey-Abina, S., Garrigue, A., Wang, G.P., Soulier, J. *et al.* (2008) Insertional oncogenesis in 4 patients after retrovirus-mediated gene therapy of SCID-X1. *J. Clin. Invest.*, **118** (9), 3132–3142.
14. McGarrity, G.J., Hoyah, G., Winemiller, A., Andre, K. *et al.* (2013) Patient monitoring and follow-up in lentiviral clinical trials. *J. Gene Med.*, **15** (2), 78–82.
15. Scholler, J., Brady, T.L., Binder-Scholl, G., Hwang, W.T. *et al.* (2012) Decade-long safety and function of retroviral-modified chimeric antigen receptor T cells. *Sci. Transl. Med.*, **4** (132), 132ra53.
16. Ott, M.G., Schmidt, M., Schwarzwaelder, K., Stein, S. *et al.* (2006) Correction of X-linked chronic granulomatous disease by

gene therapy, augmented by insertional activation of MDS1-EVI1, PRDM16 or SETBP1. *Nat. Med.*, **12** (4), 401–409.

17. Braun, C.J., Boztug, K., Paruzynski, A., Witzel, M. *et al.* (2014) Gene therapy for Wiskott–Aldrich syndrome – long-term efficacy and genotoxicity. *Sci. Transl. Med.*, **6** (227), 227ra33.

18. Korngold, R. and Sprent, J. (1978) Lethal graft-versus-host disease after bone marrow transplantation across minor histocompatibility barriers in mice. *J. Exp. Med.*, **148** (6), 1687–1698.

19. Ruggeri, L., Mancusi, A., Capanni, M., Urbani, E. *et al.* (2007) Donor natural killer cell allorecognition of missing self in haploidentical hematopoietic transplantation for acute myeloid leukemia: challenging its predictive value. *Blood*, **110** (1), 433–440.

20. Venstrom, J.M., Pittari, G., Gooley, T.A., Chewning, J.H. *et al.* (2012) HLA-C–dependent prevention of leukemia relapse by donor activating KIR2DS1. *N. Engl. J. Med.*, **367** (9), 805–816.

21. Miller, J.S., Soignier, Y., Panoskaltsis-Mortari, A., McNearney, S.A. *et al.* (2005) Successful adoptive transfer and in vivo expansion of human haploidentical NK cells in patients with cancer. *Blood*, **105** (8), 3051–3057.

22. Ljunggren, H.-G. and Malmberg, K.-J. (2007) Prospects for the use of NK cells in immunotherapy of human cancer. *Nat. Rev. Immunol.*, **7** (5), 329–339.

23. Arai, S., Meagher, R., Swearingen, M., Myint, H. *et al.* (2008) Infusion of the allogeneic cell line NK-92 in patients with advanced renal cell cancer or melanoma: a phase I trial. *Cytotherapy*, **10** (6), 625–632.

24. Ishikawa, E., Tsuboi, K., Saijo, K., and Harada, H. (2004) Autologous natural killer cell therapy for human recurrent malignant glioma. *Anticancer Res.*, **24** (3b), 1861–1871.

25. Parkhurst, M.R., Riley, J.P., Dudley, M.E., and Rosenberg, S.A. (2011) Adoptive transfer of autologous natural killer cells leads to high levels of circulating natural killer cells but does not mediate tumor regression. *Clin. Cancer Res.*, **17** (19), 6287–6297.

26. Joncker, N.T., Fernandez, N.C., Treiner, E., Vivier, E. *et al.* (2009) NK cell responsiveness is tuned commensurate with the number of inhibitory receptors for self-MHC class I: the rheostat model. *J. Immunol.*, **182** (8), 4572–4580.

27. Vivier, E., Raulet, D.H., Moretta, A., Caligiuri, M.A. *et al.* (2011) Innate or adaptive immunity? The example of natural killer cells. *Science*, **331** (6013), 44–49.

28. Lamb, L.S. Jr., (2009) γδ T cells as immune effectors against high-grade gliomas. *Immunol. Res.*, **45** (1), 85–95.

29. Deniger, D.C., Maiti, S., Mi, T., Switzer, K. *et al.* (2014) Activating and propagating polyclonal gamma delta T cells with broad specificity for malignancies. *Clin. Cancer Res.*, **20** (22), 5708–5719.

30. Buccheri, S., Guggino, G., and Caccamo, N. (2014) Efficacy and safety of gamma-delta-T cell-based tumor immunotherapy: a meta-analysis. *J. Biol. Regul. Homeost. Agents*, **28** (1), 81–90.

31. Kantoff, P.W., Higano, C.S., Shore, N.D., Berger, E.R. *et al.* (2010) Sipuleucel-T immunotherapy for castration-resistant prostate cancer. *N. Engl. J. Med.*, **363** (5), 411–422.

32. Noonan, K., Matsui, W., Serafini, P., Carbley, R. *et al.* (2005) Activated marrow-infiltrating lymphocytes effectively target plasma cells and their clonogenic precursors. *Cancer Res.*, **65** (5), 2026–2034.

33. Strasser, E.F., Dittrich, S., Weisbach, V., Zimmermann, R. *et al.* (2004) Comparison of two mononuclear cell program settings on two apheresis devices intended to collect high yields of CD14+ and CD3+ cells. *Transfusion*, **44** (7), 1104–1111.

34. Strasser, E.F., Berger, T.G., Weisbach, V., Zimmermann, R. *et al.* (2003) Comparison of two apheresis systems for the collection of CD14+ cells intended to be used in dendritic cell culture. *Transfusion*, **43** (9), 1309–1316.

35. Brown, R.A., Adkins, D., Goodnough, L.T., Haug, J.S. *et al.* (1997) Factors that influence the collection and engraftment of allogeneic peripheral-blood stem cells in patients with hematologic malignancies. *J. Clin. Oncol.*, **15** (9), 3067–3074.

36. Moog, R. (2004) Apheresis techniques for collection of peripheral blood progenitor cells. *Transfus. Apheresis Sci.*, **31** (3), 207–220.

37. Gašová, Z., Marinov, I., Vodvářková, Š., Böhmová, M. *et al.* (2005) PBPC collection techniques: standard versus large

volume leukapheresis (LVL) in donors and in patients. *Transfus. Apheresis Sci.*, **32** (2), 167–176.
38. Glaser, A., Schuler Thurner, B., Feuerstein, B., Zingsem, J. *et al.* (2001) Collection of MNCs with two cell separators for adoptive immunotherapy in patients with stage IV melanoma. *Transfusion*, **41** (1), 117–122.
39. Rols, M.-P. (2008) Mechanism by which electroporation mediates DNA migration and entry into cells and targeted tissues, in *Electroporation Protocols*, Methods in Molecular Biology (ed. S. Li), Humana Press, Totowa, NJ, pp. 19–33.
40. Bordignon, C., Notarangelo, L.D., Nobili, N., Ferrari, G. *et al.* (1995) Gene therapy in peripheral blood lymphocytes and bone marrow for ADA-immunodeficient patients. *Science*, **270** (5235), 470–475.
41. Rosenberg, S.A., Aebersold, P., Cornetta, K., Kasid, A. *et al.* (1990) Gene transfer into humans — immunotherapy of patients with advanced melanoma, using tumor-infiltrating lymphocytes modified by retroviral gene transduction. *N. Engl. J. Med.*, **323** (9), 570–578.
42. Stein, S., Ott, M.G., Schultze-Strasser, S., Jauch, A. *et al.* (2010) Genomic instability and myelodysplasia with monosomy 7 consequent to EVI1 activation after gene therapy for chronic granulomatous disease. *Nat. Med.*, **16** (2), 198–204.
43. Howe, S.J., Mansour, M.R., Schwarzwaelder, K., Bartholomae, C. *et al.* (2008) Insertional mutagenesis combined with acquired somatic mutations causes leukemogenesis following gene therapy of SCID-X1 patients. *J. Clin. Invest.*, **118** (9), 3143–3150.
44. Naldini, L., Blömer, U., Gallay, P., Ory, D. *et al.* (1996) In vivo gene delivery and stable transduction of nondividing cells by a lentiviral vector. *Science*, **272** (5259), 263–267.
45. Schröder, A.R.W., Shinn, P., Chen, H., Berry, C. *et al.* (2002) HIV-1 integration in the human genome favors active genes and local hotspots. *Cell*, **110** (4), 521–529.
46. Biffi, A., Bartolomae, C.C., Cesana, D., Cartier, N. *et al.* (2011) Lentiviral vector common integration sites in preclinical models and a clinical trial reflect a benign integration bias and not oncogenic selection. *Blood*, **117** (20), 5332–5339.
47. Song, W., Kong, H.-L., Carpenter, H., Torii, H. *et al.* (1997) Dendritic cells genetically modified with an adenovirus vector encoding the cDNA for a model antigen induce protective and therapeutic antitumor immunity. *J. Exp. Med.*, **186** (8), 1247–1256.
48. Shayakhmetov, D.M., Papayannopoulou, T., Stamatoyannopoulos, G., and Lieber, A. (2000) Efficient gene transfer into human CD34+ cells by a retargeted adenovirus vector. *J. Virol.*, **74** (6), 2567–2583.
49. Sun, J.-Y., Krouse, R.S., Forman, S.J., Senitzer, D. *et al.* (2002) Immunogenicity of a p210BCR-ABL fusion domain candidate DNA vaccine targeted to dendritic cells by a recombinant adeno-associated virus vector in vitro. *Cancer Res.*, **62** (11), 3175–3183.
50. Ivics, Z., Hackett, P.B., Plasterk, R.H., and Izsvák, Z. (1997) Molecular reconstruction of sleeping beauty, a Tc1-like transposon from fish, and its transposition in human cells. *Cell*, **91** (4), 501–510.
51. Peng, P.D., Cohen, C.J., Yang, S., Hsu, C. *et al.* (2009) Efficient nonviral sleeping beauty transposon-based TCR gene transfer to peripheral blood lymphocytes confers antigen-specific antitumor reactivity. *Gene Ther.*, **16** (8), 1042–1049.
52. Singh, H., Moyes, J.S.E., Huls, M.H., and Cooper, L.J.N. (2015) Manufacture of T cells using the sleeping beauty system to enforce expression of a CD19-specific chimeric antigen receptor. *Cancer Gene Ther.*, **22** (2), 95–100.
53. Birkholz, K., Hombach, A., Krug, C., Reuter, S. *et al.* (2009) Transfer of mRNA encoding recombinant immunoreceptors reprograms CD4+ and CD8+ T cells for use in the adoptive immunotherapy of cancer. *Gene Ther.*, **16** (5), 596–604.
54. Barrett, D.M., Zhao, Y., Liu, X., Jiang, S. *et al.* (2011) Treatment of advanced leukemia in mice with mRNA engineered T cells. *Hum. Gene Ther.*, **22** (12), 1575–1586.
55. Zhao, Y., Moon, E., Carpenito, C., Paulos, C.M. *et al.* (2010) Multiple injections of electroporated autologous T cells expressing a chimeric antigen receptor mediate regression of human disseminated tumor. *Cancer Res.*, **70** (22), 9053–9061.
56. Singh, N., Liu, X., Hulitt, J., Jiang, S. *et al.* (2014) Nature of tumor control by permanently and transiently-modified GD2 chimeric antigen receptor T cells in

57. Beatty, G.L., Haas, A.R., Maus, M.V., Torigian, D.A. et al. (2013) Mesothelin-specific chimeric antigen receptor mRNA-engineered T cells induce antitumor activity in solid malignancies. *Cancer Immunol. Res.*, **2** (2), 112–120.
58. June, C.H. (2007) Principles of adoptive T cell cancer therapy. *J. Clin. Invest.*, **117** (5), 1204–1212.
59. Leen, A.M., Myers, G.D., Sili, U., Huls, M.H. et al. (2006) Monoculture-derived T lymphocytes specific for multiple viruses expand and produce clinically relevant effects in immunocompromised individuals. *Nat. Med.*, **12** (10), 1160–1166.
60. Hanley, P.J., Cruz, C., Savoldo, B., Leen, A.M. et al. (2009) Functionally active virus-specific T cells that target CMV, adenovirus, and EBV can be expanded from naive T-cell populations in cord blood and will target a range of viral epitopes. *Blood*, **114** (19), 4283–4292.
61. Melenhorst, J.J., Leen, A.M., Bollard, C.M., Quigley, M.F. et al. (2010) Allogeneic virus-specific T cells with HLA alloreactivity do not produce GVHD in human subjects. *Blood*, **116** (22), 4700–4702.
62. Rosenberg, S.A., Yannelli, J.R., Yang, J.C., Topalian, S.L. et al. (1994) Treatment of patients with metastatic melanoma with autologous tumor-infiltrating lymphocytes and interleukin 2. *J. Natl Cancer Inst.*, **86** (15), 1159–1166.
63. Rosenberg, S.A., Packard, B.S., Aebersold, P.M., Solomon, D. et al. (1988) Use of tumor-infiltrating lymphocytes and interleukin-2 in the immunotherapy of patients with metastatic melanoma. A preliminary report. *N. Engl. J. Med.*, **319** (25), 1676–1680.
64. Dudley, M.E., Wunderlich, J.R., Robbins, P.F., Yang, J.C. et al. (2002) Cancer regression and autoimmunity in patients after clonal repopulation with antitumor lymphocytes. *Science*, **298** (5594), 850–854.
65. Dudley, M.E., Yang, J.C., Sherry, R., Hughes, M.S. et al. (2008) Adoptive cell therapy for patients with metastatic melanoma: evaluation of intensive myeloablative chemoradiation preparative regimens. *J. Clin. Oncol.*, **26** (32), 5233–5239.
66. Wrzesinski, C., Paulos, C.M., Kaiser, A., Muranski, P. et al. (2010) Increased intensity lymphodepletion enhances tumor treatment efficacy of adoptively transferred tumor-specific T cells. *J. Immunother.*, **33** (1), 1–7.
67. Rosenberg, S.A., Yang, J.C., Sherry, R.M., Kammula, U.S. et al. (2011) Durable complete responses in heavily pretreated patients with metastatic melanoma using T-cell transfer immunotherapy. *Clin. Cancer Res.*, **17** (13), 4550–4557.
68. Czerniecki, B.J., Koski, G.K., Koldovsky, U., Xu, S. et al. (2007) Targeting HER-2/neu in early breast cancer development using dendritic cells with staged interleukin-12 burst secretion. *Cancer Res.*, **67** (4), 1842–1852.
69. Ramos, C.A. and Dotti, G. (2011) Chimeric antigen receptor (CAR)-engineered lymphocytes for cancer therapy. *Expert Opin. Biol. Ther.*, **11** (7), 855–873.
70. Berger, C., Turtle, C.J., Jensen, M.C., and Riddell, S.R. (2009) Adoptive transfer of virus-specific and tumor-specific T cell immunity. *Curr. Opin. Immunol.*, **21** (2), 224–232.
71. Riches, J.C., Davies, J.K., McClanahan, F., Fatah, R. et al. (2013) T cells from CLL patients exhibit features of T-cell exhaustion but retain capacity for cytokine production. *Blood*, **121** (9), 1612–1621.
72. Bonyhadi, M., Frohlich, M., Rasmussen, A., Ferrand, C. et al. (2005) In vitro engagement of CD3 and CD28 corrects T cell defects in chronic lymphocytic leukemia. *J. Immunol.*, **174** (4), 2366–2375.
73. Fishman, M. (2009) A changing world for DCvax: a PSMA loaded autologous dendritic cell vaccine for prostate cancer. *Expert Opin. Biol. Ther.*, **9** (12), 1565–1575.
74. Wheeler, C.J. and Black, K.L. (2009) DCVax®-Brain and DC vaccines in the treatment of GBM. *Expert Opin. Biol. Ther.*, **18** (4), 509–519.
75. Phuphanich, S. and Wheeler, C.J. (2013) Glioma-associated antigens associated with prolonged survival in a phase I study of ICT-107 for patients with newly diagnosed glioblastoma. *J. Immunother.*, **36** (2), 152–157.
76. Phuphanich, S., Wheeler, C.J., Rudnick, J.D., Mazer, M. et al. (2013) Phase I trial of a multi-epitope-pulsed dendritic cell vaccine for patients with

newly diagnosed glioblastoma. *Cancer Immunol. Immunother.*, **62** (1), 125–135.
77. Hawkins, E.S., Alken, R., Chandler, J., Fink KL, et al. (2012) A randomized, double-blind, controlled phase IIb study of the safety and efficacy of ICT-107 in newly diagnosed patients with glioblastoma multiforme following resection and chemoradiation. *J. Clin. Oncol.*, **30** (15), Suppl.; abstract TPS2107). Available at: *http://meetinglibrary.asco.org/content/100393-114*.
78. Kandalaft, L.E., Chiang, C.L., Tanyi, J., Motz, G. et al. (2013) A phase I vaccine trial using dendritic cells pulsed with autologous oxidized lysate for recurrent ovarian cancer. *J. Transl. Med.*, **11**, 149.
79. Kandalaft, L.E., Powell, D.J. Jr., Chiang, C.L., Tanyi, J. et al. (2013) Autologous lysate-pulsed dendritic cell vaccination followed by adoptive transfer of vaccine-primed ex vivo co-stimulated T cells in recurrent ovarian cancer. *Oncoimmunology*, **2** (1), e22664.
80. Cruz, C.R.Y., Micklethwaite, K.P., Savoldo, B., Ramos, C.A. et al. (2013) Infusion of donor-derived CD19-redirected virus-specific T cells for B-cell malignancies relapsed after allogeneic stem cell transplant: a phase 1 study. *Blood*, **122** (17), 2965–2973.
81. Bendle, G.M., Linnemann, C., Hooijkaas, A.I., Bies, L. et al. (2010) Lethal graft-versus-host disease in mouse models of T cell receptor gene therapy. *Nat. Med.*, **16** (5), 565–570.
82. Johnson, L.A., Morgan, R.A., Dudley, M.E., Cassard, L. et al. (2009) Gene therapy with human and mouse T-cell receptors mediates cancer regression and targets normal tissues expressing cognate antigen. *Blood*, **114** (3), 535–546.
83. Morgan, R.A., Chinnasamy, N., Abate-Daga, D.D., Gros, A. et al. (2013) Cancer regression and neurologic toxicity following anti-MAGE-A3 TCR gene therapy. *J. Immunother.*, **36** (2), 133–151.
84. Linette, G.P., Stadtmauer, E.A., Maus, M.V., Rapoport, A.P. et al. (2013) Cardiovascular toxicity and titin cross-reactivity of affinity-enhanced T cells in myeloma and melanoma. *Blood*, **122** (6), 863–871.
85. Cameron, B.J., Gerry, A.B., Dukes, J., Harper, J.V. et al. (2013) Identification of a titin-derived HLA-A1-presented peptide as a cross-reactive target for engineered MAGE A3-directed T cells. *Sci. Transl. Med.*, **5** (197), 197ra103.
86. Chapuis, A.G., Ragnarsson, G.B., Nguyen, H.N., Chaney, C.N. et al. (2013) Transferred WT1-reactive CD8+ T cells can mediate antileukemic activity and persist in post-transplant patients. *Sci. Transl. Med.*, **5** (174), 174ra27.
87. Chmielewski, M., Hombach, A.A., and Abken, H. (2013) Antigen-specific T-cell activation independently of the MHC: chimeric antigen receptor-redirected T cells. *Front. Immunol.*, **4**, 154–160.
88. Eshhar, Z., Waks, T., Gross, G., and Schindler, D.G. (1993) Specific activation and targeting of cytotoxic lymphocytes through chimeric single chains consisting of antibody-binding domains and the gamma or zeta subunits of the immunoglobulin and T-cell receptors. *Proc. Natl Acad. Sci. USA*, **90** (2), 720–724.
89. Bretscher, P. and Cohn, M. (1970) A theory of self-nonself discrimination paralysis and induction involve the recognition of one and two determinants on an antigen, respectively. *Science*, **169** (3950), 1042–1049.
90. Brocker, T. (2000) Chimeric Fv-ζ or Fv-ε receptors are not sufficient to induce activation or cytokine production in peripheral T cells. *Blood*, **96** (5), 1999–2001.
91. Finney, H.M., Akbar, A.N., and Lawson, A.D.G. (2004) Activation of resting human primary T cells with chimeric receptors: costimulation from CD28, inducible costimulator, CD134, and CD137 in series with signals from the TCRζ chain. *J. Immunol.*, **172** (1), 104–113.
92. Finney, H.M., Lawson, A.D.G., Bebbington, C.R., and Weir, A.N.C. (1998) Chimeric receptors providing both primary and costimulatory signaling in T cells from a single gene product. *J. Immunol.*, **161** (6), 2791–2797.
93. Krause, A., Guo, H.-F., Latouche, J.-B., Tan, C. et al. (1998) Antigen-dependent CD28 signaling selectively enhances survival and proliferation in genetically modified activated human primary T lymphocytes. *J. Exp. Med.*, **188** (4), 619–626.
94. Imai, C., Mihara, K., Andreansky, M., Nicholson, I.C. et al. (2004) Chimeric

receptors with 4-1BB signaling capacity provoke potent cytotoxicity against acute lymphoblastic leukemia. *Leukemia*, **18** (4), 676–684.

95. Carpenito, C., Milone, M.C., Hassan, R., Simonet, J.C. et al. (2009) Control of large, established tumor xenografts with genetically retargeted human T cells containing CD28 and CD137 domains. *Proc. Natl Acad. Sci. USA*, **106** (9), 3360–3365.

96. Zhong, X.-S., Matsushita, M., Plotkin, J., Rivière, I. et al. (2010) Chimeric antigen receptors combining 4-1BB and CD28 signaling domains augment PI3kinase/AKT/Bcl-XL activation and CD8+ T cell-mediated tumor eradication. *Mol. Ther.*, **18** (2), 413–420.

97. Roberts, M.R. (1998) Chimeric receptor molecules for delivery of co-stimulatory signals. US Patent Office 5,712,149; Cell Genesys, Inc., original assignee.

98. Maher, J., Brentjens, R.J., Gunset, G., Rivière, I. et al. (2002) Human T-lymphocyte cytotoxicity and proliferation directed by a single chimeric TCRζ/CD28 receptor. *Nat. Biotechnol.*, **20** (1), 70–75.

99. Milone, M.C., Fish, J.D., Carpenito, C., Carroll, R.G. et al. (2009) Chimeric receptors containing CD137 signal transduction domains mediate enhanced survival of T cells and increased antileukemic efficacy in vivo. *Mol. Ther.*, **17** (8), 1453–1464.

100. Pule, M.A., Savoldo, B., Myers, G.D., Rossig, C. et al. (2008) Virus-specific T cells engineered to coexpress tumor-specific receptors: persistence and antitumor activity in individuals with neuroblastoma. *Nat. Med.*, **14** (11), 1264–1270.

101. Lim, J.K., McDermott, D.H., Lisco, A., Foster, G.A. et al. (2010) CCR5 deficiency is a risk factor for early clinical manifestations of West Nile virus infection but not for viral transmission. *J. Infect. Dis.*, **201** (2), 178–185.

102. Fätkenheuer, G., Pozniak, A.L., Johnson, M.A., Plettenberg, A. et al. (2005) Efficacy of short-term monotherapy with maraviroc, a new CCR5 antagonist, in patients infected with HIV-1. *Nat. Med.*, **11** (11), 1170–1172.

103. Hütter, G., Nowak, D., Mossner, M., Ganepola, S. et al. (2009) Long-term control of HIV by CCR5Delta32/Delta32 stem-cell transplantation. *N. Engl. J. Med.*, **360** (7), 692–698.

104. Perez, E.E., Wang, J., Miller, J.C., Jouvenot, Y. et al. (2008) Establishment of HIV-1 resistance in CD4+ T cells by genome editing using zinc-finger nucleases. *Nat. Biotechnol.*, **26** (7), 808–816.

105. Tebas, P., Stein, D., Tang, W.W., Frank, I. et al. (2014) Gene editing of CCR5 in autologous CD4 T cells of persons infected with HIV. *N. Engl. J. Med.*, **370** (10), 901–910.

106. Saito, N., Chono, H., Shibata, H., Ageyama, N. et al. (2014) CD4+ T cells modified by the endoribonuclease MazF are safe and can persist in SHIV-infected Rhesus macaques. *Mol. Ther. Nucleic Acids*, **3** (6), e168.

107. DiGiusto, D.L., Krishnan, A., Li, L., Li, H. et al. (2010) RNA-based gene therapy for HIV with lentiviral vector-modified CD34+ cells in patients undergoing transplantation for AIDS-related lymphoma. *Sci. Transl. Med.*, **2** (36), 36ra43.

108. Becker, C., Pohla, H., Frankenberger, B., Schüler, T. et al. (2001) Adoptive tumor therapy with T lymphocytes enriched through an IFN-gamma capture assay. *Nat. Med.*, **7** (10), 1159–1162.

109. Edinger, M. and Hoffmann, P. (2011) Regulatory T cells in stem cell transplantation: strategies and first clinical experiences. *Curr. Opin. Immunol.*, **23** (5), 679–684.

110. Putnam, A.L., Safinia, N., Medvec, A., Laszkowska, M. et al. (2013) Clinical grade manufacturing of human alloantigen-reactive regulatory T cells for use in transplantation. *Am. J. Transplant.*, **13** (11), 3010–3020.

111. Riley, J.L. and June, C.H. (2005) The CD28 family: a T-cell rheostat for therapeutic control of T-cell activation. *Blood*, **105** (1), 13–21.

112. Weng, N.P., Levine, B.L., June, C.H., and Hodes, R.J. (1995) Human naive and memory T lymphocytes differ in telomeric length and replicative potential. *Proc. Natl Acad. Sci. USA*, **92** (24), 11091–11094.

113. Lopez, R.D., Xu, S., Guo, B., Negrin, R.S. et al. (2000) CD2-mediated IL-12-dependent signals render human γδ-T cells resistant to mitogen-induced apoptosis, permitting the large-scale ex vivo expansion of functionally distinct lymphocytes: implications for the development of adoptive immunotherapy strategies. *Blood*, **96** (12), 3827–3837.

114. Sato, K., Kimura, S., Segawa, H., Yokota, A. et al. (2005) Cytotoxic effects of γδ T cells expanded ex vivo by a third generation bisphosphonate for cancer immunotherapy. Int. J. Cancer, 116 (1), 94–99.
115. Kondo, M., Sakuta, K., Noguchi, A., Ariyoshi, N. et al. (2008) Zoledronate facilitates large-scale ex vivo expansion of functional γδ T cells from cancer patients for use in adoptive immunotherapy. Cytotherapy, 10 (8), 842–856.
116. Lim, S.A., Kim, T.-J., Lee, J.E., Sonn, C.H. et al. (2013) Ex vivo expansion of highly cytotoxic human NK cells by cocultivation with irradiated tumor cells for adoptive immunotherapy. Cancer Res., 73 (8), 2598–2607.
117. Levine, B.L., Bernstein, W.B., Connors, M., Craighead, N. et al. (1997) Effects of CD28 costimulation on long-term proliferation of CD4+ T cells in the absence of exogenous feeder cells. J. Immunol., 159 (12), 5921–5930.
118. Carroll, R.G., Riley, J.L., Levine, B.L., Blair, P.J. et al. (1998) The role of co-stimulation in regulation of chemokine receptor expression and HIV-1 infection in primary T lymphocytes. Semin. Immunol., 10 (3), 195–202.
119. Levine, B.L., Bernstein, W.B., Aronson, N.E., Schlienger, K. et al. (2002) Adoptive transfer of costimulated CD4+ T cells induces expansion of peripheral T cells and decreased CCR5 expression in HIV infection. Nat. Med., 8 (1), 47–53.
120. Maus, M.V., Thomas, A.K., Leonard, D.G.B., Allman, D. et al. (2002) Ex vivo expansion of polyclonal and antigen-specific cytotoxic T lymphocytes by artificial APCs expressing ligands for the T-cell receptor, CD28 and 4-1BB. Nat. Biotechnol., 20 (2), 143–148.
121. Thomas, A.K., Maus, M.V., Shalaby, W.S., June, C.H. et al. (2002) A cell-based artificial antigen-presenting cell coated with anti-CD3 and CD28 antibodies enables rapid expansion and long-term growth of CD4 T lymphocytes. Clin. Immunol., 105 (3), 259–272.
122. Suhoski, M.M., Golovina, T.N., Aqui, N.A., and Tai, V.C. (2007) Engineering artificial antigen-presenting cells to express a diverse array of co-stimulatory molecules. Mol. Ther., 15 (5), 981–988.
123. Liu, Y., Wu, H.-W., Sheard, M.A., Sposto, R. et al. (2013) Growth and activation of natural killer cells ex vivo from children with neuroblastoma for adoptive cell therapy. Clin. Cancer Res., 19 (8), 2132–2143.
124. Fujisaki, H., Kakuda, H., Shimasaki, N., Imai, C. et al. (2009) Expansion of highly cytotoxic human natural killer cells for cancer cell therapy. Cancer Res., 69 (9), 4010–4017.
125. Denman, C.J., Senyukov, V.V., Somanchi, S.S., Phatarpekar, P.V. et al. (2012) Membrane-bound IL-21 promotes sustained ex vivo proliferation of human natural killer cells. PLoS One, 7 (1), e30264.
126. Wang, X., Lee, D.A., Wang, Y., Wang, L. et al. (2013) Membrane-bound interleukin-21 and CD137 ligand induce functional human natural killer cells from peripheral blood mononuclear cells through STAT-3 activation. Clin. Exp. Immunol., 172 (1), 104–112.
127. Mitsuyasu, R.T., Anton, P.A., Deeks, S.G., Scadden, D.T. et al. (2000) Prolonged survival and tissue trafficking following adoptive transfer of CD4ζ gene-modified autologous CD4+ and CD8+ T cells in human immunodeficiency virus- infected subjects. Blood, 96 (3), 785–793.
128. Levine, B.L., Humeau, L.M., and Boyer, J. (2006) Gene transfer in humans using a conditionally replicating lentiviral vector. Proc. Natl Acad. Sci. USA, 103 (46), 17372–17377.
129. Walker, R.E., Bechtel, C.M., Natarajan, V., Baseler, M. et al. (2000) Long-term in vivo survival of receptor-modified syngeneic T cells in patients with human immunodeficiency virus infection. Blood, 96 (2), 467–474.
130. Wang, X., Naranjo, A., Brown, C.E., Bautista, C. et al. (2012) Phenotypic and functional attributes of lentivirus-modified CD19-specific human CD8+ central memory T cells manufactured at clinical scale. J. Immunother., 5 (9), 689–701.
131. Berger, C., Jensen, M.C., Lansdorp, P.M., Gough, M. et al. (2008) Adoptive transfer of effector CD8+ T cells derived from central memory cells establishes persistent T cell memory in primates. J. Clin. Invest., 118 (1), 294–305.
132. Fowler, D.H., Mossoba, M.E., Steinberg, S.M., Halverson, D.C. et al. (2013) Phase

2 clinical trial of rapamycin-resistant donor CD4+ Th2/Th1 (T-Rapa) cells after low-intensity allogeneic hematopoietic cell transplantation. *Blood*, **121** (15), 2864–2874.
133. Guedan, S., Chen, X., Madar, A., Carpenito, C. *et al.* (2014) ICOS-based chimeric antigen receptors program bipolar TH17/TH1 cells. *Blood*, **124** (7), 1070–1080.
134. Sakaguchi, S., Yamaguchi, T., Nomura, T., and Ono, M. (2008) Regulatory T cells and immune tolerance. *Cell*, **133** (5), 775–787.
135. Sakaguchi, S., Takahashi, T., and Nishizuka, Y. (1982) Study on cellular events in post-thymectomy autoimmune oophoritis in mice. II. Requirement of Lyt-1 cells in normal female mice for the prevention of oophoritis. *J. Exp. Med.*, **156** (6), 1577–1586.
136. Taylor, P.A., Lees, C.J., and Blazar, B.R. (2002) The infusion of ex vivo activated and expanded CD4+CD25+ immune regulatory cells inhibits graft-versus-host disease lethality. *Blood*, **99** (10), 3493–3499.
137. Edinger, M., Hoffmann, P., Ermann, J., Drago, K. *et al.* (2003) CD4+CD25+ regulatory T cells preserve graft-versus-tumor activity while inhibiting graft-versus-host disease after bone marrow transplantation. *Nat. Med.*, **9** (9), 1144–1150.
138. Lee, J.C., Hayman, E., Pegram, H.J., Santos, E. *et al.* (2011) In vivo inhibition of human CD19-targeted effector T cells by natural T regulatory cells in a xenotransplant murine model of B cell malignancy. *Cancer Res.*, **71** (8), 2871–2881.
139. Kofler, D.M., Chmielewski, M., Rappl, G., Hombach, A. *et al.* (2011) CD28 costimulation impairs the efficacy of a redirected T-cell antitumor attack in the presence of regulatory T cells which can be overcome by preventing lck activation. *Mol. Ther.*, **19** (4), 760–767.
140. Bluestone, J.A. and Tang, Q. (2005) How do CD4+CD25+ regulatory T cells control autoimmunity? *Curr. Opin. Immunol.*, **17** (6), 638–642.
141. Golovina, T.N., Mikheeva, T., Suhoski, M.M., Aqui, N.A. *et al.* (2008) CD28 costimulation is essential for human T regulatory expansion and function. *J. Immunol.*, **181** (4), 2855–2868.
142. Brunstein, C.G., Miller, J.S., Cao, Q., McKenna, D.H. *et al.* (2011) Infusion of ex vivo expanded T regulatory cells in adults transplanted with umbilical cord blood: safety profile and detection kinetics. *Blood*, **117** (3), 1061–1070.
143. Gattinoni, L., Finkelstein, S.E., Klebanoff, C.A., Antony, P.A. *et al.* (2005) Removal of homeostatic cytokine sinks by lymphodepletion enhances the efficacy of adoptively transferred tumor-specific CD8+ T cells. *J. Exp. Med.*, **202** (7), 907–912.
144. Powell, D.J., Dudley, M.E., Hogan, K.A., Wunderlich, J.R., and Rosenberg, S.A. (2006) Adoptive transfer of vaccine-induced peripheral blood mononuclear cells to patients with metastatic melanoma following lymphodepletion. *J. Immunol.*, **177** (9), 6527–6539.
145. Muranski, P., Boni, A., Wrzesinski, C., Citrin, D.E. *et al.* (2006) Increased intensity lymphodepletion and adoptive immunotherapy – how far can we go? *Nat. Rev. Clin. Oncol.*, **3** (12), 668–681.
146. Porter, D.L., Levine, B.L., Kalos, M., Bagg, A. *et al.* (2011) Chimeric antigen receptor-modified T cells in chronic lymphoid leukemia. *N. Engl. J. Med.*, **365** (8), 725–733.
147. Rosenberg, S.A., Restifo, N.P., Yang, J.C., Morgan, R.A. *et al.* (2008) Adoptive cell transfer: a clinical path to effective cancer immunotherapy. *Nat. Rev. Cancer*, **8** (4), 299–308.
148. Maus, M.V., Haas, A.R., Beatty, G.L., Albelda, S.M. *et al.* (2013) T cells expressing chimeric antigen receptors can cause anaphylaxis in humans. *Cancer Immunol. Res.*, **1** (1), 26–31.
149. Brentjens, R., Yeh, R., Bernal, Y., Rivière, I. *et al.* (2010) Treatment of chronic lymphocytic leukemia with genetically targeted autologous T cells: case report of an unforeseen adverse event in a phase I clinical trial. *Mol. Ther.*, **18** (4), 666–668.
150. Morgan, R.A., Yang, J.C., Kitano, M., Dudley, M.E. *et al.* (2010) Case report of a serious adverse event following the administration of T cells transduced with a chimeric antigen receptor recognizing ERBB2. *Mol. Ther.*, **18** (4), 843–851.
151. Grupp, S.A., Kalos, M., Barrett, D., Aplenc, R. *et al.* (2013) Chimeric antigen receptor-modified T cells for acute

lymphoid leukemia. *N. Engl. J. Med.*, **368** (16), 1509–1518.
152. Lee, D.W., Kochenderfer, J.N., Stetler-Stevenson, M., Cui, Y.K. *et al.* (2014) T cells expressing CD19 chimeric antigen receptors for acute lymphoblastic leukaemia in children and young adults: a phase 1 dose-escalation trial. *Lancet*, **385**, 517–528.
153. Maude, S.L., Frey, N., Shaw, P.A., Aplenc, R. *et al.* (2014) Chimeric antigen receptor T cells for sustained remissions in leukemia. *N. Engl. J. Med.*, **371** (16), 1507–1517.
154. Di Stasi, A., Tey, S.-K., Dotti, G., Fujita, Y. *et al.* (2011) Inducible apoptosis as a safety switch for adoptive cell therapy. *N. Engl. J. Med.*, **365** (18), 1673–1683.
155. Candotti, F., Shaw, K.L., Muul, L., Carbonaro, D. *et al.* (2012) Gene therapy for adenosine deaminase-deficient severe combined immune deficiency: clinical comparison of retroviral vectors and treatment plans. *Blood*, **120** (18), 3635–3646.
156. Robbins, P.F., Dudley, M.E., Wunderlich, J., El-Gamil, M. *et al.* (2004) Cutting edge: persistence of transferred lymphocyte clonotypes correlates with cancer regression in patients receiving cell transfer therapy. *J. Immunol.*, **173** (12), 7125–7130.
157. Rosenberg, S.A.,. and Dudley, M.E. (2009) Adoptive cell therapy for the treatment of patients with metastatic melanoma. *Curr. Opin. Immunol.*, **21** (2), 233–40.
158. Brentjens, R.J., Rivière, I., Park, J.H., Davila, M.L. *et al.* (2011) Safety and persistence of adoptively transferred autologous CD19-targeted T cells in patients with relapsed or chemotherapy refractory B-cell leukemias. *Blood*, **118** (18), 4817–4828.
159. Klebanoff, C.A., Gattinoni, L., Torabi-Parizi, P., Kerstann, K. *et al.* (2005) Central memory self/tumor-reactive CD8+ T cells confer superior antitumor immunity compared with effector memory T cells. *Proc. Natl Acad. Sci. USA*, **102** (27), 9571–9576.
160. Monteiro, J., Batliwalla, F., Ostrer, H., and Gregersen, P.K. (1996) Shortened telomeres in clonally expanded CD28-CD8+ T cells imply a replicative history that is distinct from their CD28+CD8+ counterparts. *J. Immunol.*, **156** (10), 3587–3590.
161. Van den Hove, L.E., Van Gool, S.W., Vandenberghe, P., Boogaerts, M.A. *et al.* (1998) CD57+/CD28- T cells in untreated hemato-oncological patients are expanded and display a Th1-type cytokine secretion profile, ex vivo cytolytic activity and enhanced tendency to apoptosis. *Leukemia*, **12** (10), 1573–1582.
162. Louis, C.U., Savoldo, B., Dotti, G., Pule, M. *et al.* (2011) Antitumor activity and long-term fate of chimeric antigen receptor-positive T cells in patients with neuroblastoma. *Blood*, **118** (23), 6050–6056.
163. Pardoll, D.M. (2012) The blockade of immune checkpoints in cancer immunotherapy. *Nat. Rev. Cancer*, **12** (4), 252–264.

13
Targeted Therapy: Genomic Approaches[*]

Tim N. Beck[1,2], Linara Gabitova[1,3], and Ilya G. Serebriiskii[1]
[1]*Fox Chase Cancer Center, Developmental Therapeutics Program, 333 Cottman Avenue, Philadelphia, PA 19111, USA*
[2]*Drexel University College of Medicine, Program in Molecular and Cell Biology and Genetics, 2900 Queen Lane, Philadelphia, PA 19129, USA*
[3]*Kazan Federal University, 18 Kremlevskaya Street, Kazan, Republic of Tatarstan, 420008 Russia*

1	**Principles of Genomics-Driven Targeted Therapy** 441	
1.1	The Promise of Targeted Therapy 441	
1.2	Molecular Basis of Targeted Therapy 446	
1.3	Purpose of Genomic Approaches and Available Platforms 448	
1.3.1	Identification of Genomic Alterations 449	
1.4	Epigenetics 457	
1.5	Targeting Biological Pathogens 459	
1.6	Targeting Low-Density Lipoprotein 460	
1.7	Caveats and Pitfalls 461	
2	**Therapeutic Agents Used in Targeted Therapy** 462	
2.1	Small Molecules 462	
2.2	Biologicals 464	
2.2.1	Antibody-Based Therapeutics 465	
2.3	Targeted Immunotherapy 467	
2.4	Gene Therapy 468	
2.5	Emerging Therapeutic Approaches 469	
3	**Targeted Therapy and Precision Medicine** 472	
3.1	Principles of Precision/Genomic Medicine 472	
3.2	Integrating Genomic Data for Clinical Strategies, Diagnosis, Treatment, and Prognosis 472	
3.3	Multidrug Therapies 473	

[*]This chapter has previously been published in: Meyers, R.A. (Ed.) Translational Medicine Cancer, 2016, 978-3-527-33569-5.

Translational Medicine: Molecular Pharmacology and Drug Discovery
First Edition. Edited by Robert A. Meyers.
© 2018 Wiley-VCH Verlag GmbH & Co. KGaA. Published 2018 by Wiley-VCH Verlag GmbH & Co. KGaA.

3.4 Integrative Bioinformatics 475

4 Concluding Remarks 476

Acknowledgments 476

References 476

Keywords

Targeted therapy
A biomedical approach to identify and therapeutically exploit entities specific to pathological processes (e.g., a gene, a protein, or RNA), in an effort to minimize adverse effects and maximize therapeutic impact

Precision medicine
A personalized treatment approach based on data collected on an individual basis and thus specific to the individual and a given malady or condition of concern

Genomics
The study of canonical and aberrant nucleotide sequences, gene expression, and chromosome integrity within an organism, and the roles they play in health and disease

Epigenomics
The study of how DNA modifiers and modifications influence gene expression in the absence of DNA sequence alterations

Transcriptomics
The study of RNA, coding and noncoding, within a living organism

Metabolomics
The study of chemical processes involved in the breakdown and utilization of metabolites within an organism and within cells

Proteomics
The study of all proteins within a living organism

Metagenomics
The study of genetic material from the environment, including microorganisms

Bioinformatics
A scientific discipline focused on the research, development, and application of computational tools to acquire, analyze, store, and visualize data, particularly "-omics" data

The crux of targeted therapy is the inhibition, enhancement, correction, reversal, or negation of events to overcome pathological processes, by targeting features or entities specific to the pathology. Genomics has been a critical component of target identification, and has greatly amplified the impact and potential of targeted therapy. Incredibly, genomics has only scratched the surface of this endeavor. Today, progress is driven by pan-omics, integrating genomics, epigenomics, transcriptomics, proteomics, metabolomics, and metagenomics data using bioinformatics and system biology approaches. The advent of advanced "-omics" has provided the molecular foundation for the development of new-targeted therapeutics, including small molecules, biologicals, tools for genomic editing, and more. Targeted therapy is used to treat an array of diverse diseases, ranging from cancer and HIV to atherosclerosis and malaria. In this chapter, the essential aspects of targeted therapy are discussed, including many of the remarkable successes as well as some of the shortcomings. How "-omics" approaches have contributed to the evolution of targeted therapy and what can be expected in the future is also described.

1
Principles of Genomics-Driven Targeted Therapy

1.1
The Promise of Targeted Therapy

The promise of targeted therapy has never been greater, and innovation on this front continues to be driven by the desire to treat and prevent disease with precision and efficacy. Targeted therapy is based on the idea of identifying a unique, pathology-specific, and targetable entity, be it cancer-associated aberrations (Sect. 1.3), foreign organisms (Sect. 1.5), or a normal protein that modulates a disease-associated process (Sect. 1.6). Frequently, a dominant feature necessary for the establishment and maintenance of a disease is targeted. At other times, it is neither possible nor practical to directly target the focal points of a pathological process, and instead modulating entities are targeted (Fig. 1b). A unique target, in theory, provides the opportunity to specifically inhibit – or potentially to enhance – a given process within the pathological milieu, ideally while avoiding the collateral damage of healthy tissue.

Targeted therapy is frequently described in a simplistic linear cause-and-effect schema. A pathological process arises due to a specific target activity; the therapeutic agent inhibits this activity and downstream events are returned to the *status quo ante* (Fig. 1b). The more likely scenario acknowledges the somewhat ambiguous specificity of most targeted agents and the complexity of intracellular processes. Figure 1c suggests that the target marked by the crosshair is indeed solely required for therapeutic benefit; however, it may be the molecular ripple effects of targeted intervention that are critical. In order to achieve a therapeutic effect, factors must add-up to perturb the greater signaling in the correct way. The red circle in Fig. 1d represents the point of optimal therapeutic effectiveness, where drug combination, treatment schedule and dosing are ideal to maximize the impact on the signaling network. It is important to realize that many of the drugs used as targeted therapeutics in actuality have multiple targets (Fig. 2a) that can be divided into four general categories (Fig. 2b). Interestingly, for the vast majority of drugs only a single target is known (Fig. 2c), presumably due to a lack of initial investigation of additional targets,

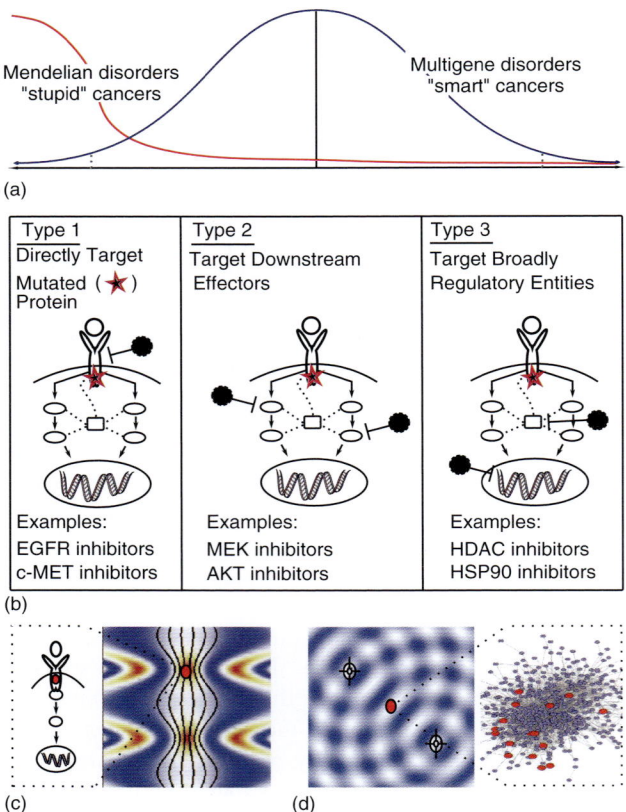

Fig. 1 Targeted therapy approaches. (a) The hypothetical distribution of disorders by the number of underlying interaction is represented by the Poisson curve (blue), with the chances of meaningfully using targeted therapy depicted as a red line. The distribution goes from asymptotically close to zero for complex disorders, and asymptotically closer to one for Mendelian disorders (see text for details); (b) Three different types of targeted therapy approaches are commonly applied; (c,d) As physicists about a century ago moved from considering light as rays to its interpretation as waves (sometimes, at their convenience, using either description), we are moving from the paradigm of thinking about molecular life in terms of linear biochemical and signaling pathways (c, which can be disrupted by a drug with a predetermined consequences), to the more complex picture of biochemical "fabric" (d), where the drugs produce waves or ripples, interacting with those from other drugs or environmental inputs; (c) Common theoretical approach to targeted therapy, wherein signaling events are considered as linear with proposed precision targeting (red circle) and narrowly focused downstream effects; (d) Ripple effects of targeted therapy and the complexity of perturbing the signaling milieu: precisely hitting targets (crosshairs) in a few cases may have the intended therapeutic effect. However, commonly, the ripples effects, perturbation of alternative targets (white spots), under the correct conditions (e.g., correct dosage and drug combination, optimal treatment schedule) induce signaling-network-wide changes for the therapeutic effect (red circles). Wave image (c) modified from the Wolfram Demonstrations Project *http://demonstrations.wolfram.com/TheQuantumHarmonicOscillatorWithTimeDependentBoundaryConditi/*; wave image (d) generated using Wolfram (*http://demonstrations.wolfram.com/WaveInterference/*); network (d) generated using Cytoscape [1].

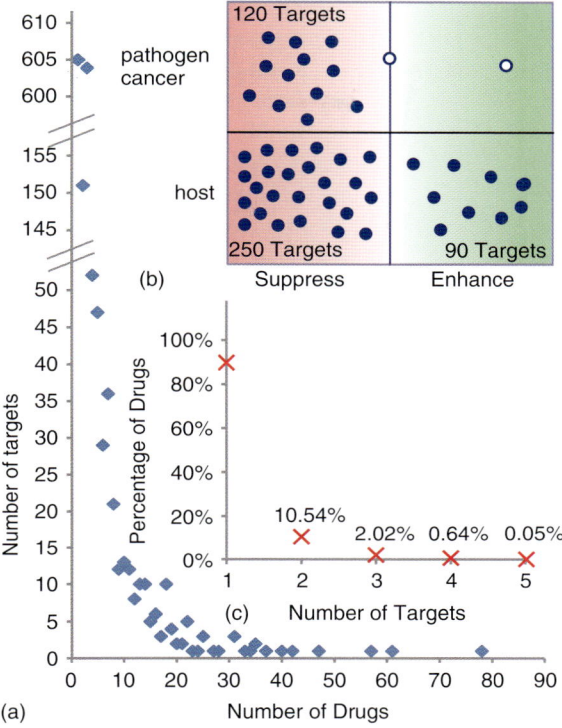

Fig. 2 Target specificity and type based on Therapeutic Targets Database (TTD) data. (a,c) Relationship between the number of drugs and targets reveals that most targets correspond to only a few drugs (a), and, reciprocally, the overwhelming majority of drugs have at most two known targets (c); (b) Targets can be divided into four categories: suppressive effect on cancer/pathogen; suppressive effect on host; enhancing effect on pathogen/cancer; enhancing effect on host. The blue circles represent 10 targets based on TTD data; open blue circles represent manually curated data from the literature, indicating cases where compounds are used either to just visualize the cancer cells to enable the subsequent surgical removal ("neutral" position on the panel; see Sect. 2.5 for details), or to restore/enhance their antigen-presenting function, enabling their subsequent destruction by the immune system (see Sect. 2.3). All information was retrieved from the TTD [2] and sorted as human versus cancer/pathogen based on the target UniProt ID/disease type, and by functionality based on "agonist/activator" versus "inhibitor/antagonist" drug classifiers.

sometimes referred to as "off targets". Commonly, knowledge about the specificity of a given drug in terms of target selectivity changes over time as the drug continues to be studied (Fig. 3), adding to the challenge of optimizing treatment conditions. The advent of technologies that make it possible to reliably and rapidly interrogate genomes, transcriptomes, proteomes, metabolomes, and microbiomes, and the necessary bioinformatics resources to analyze the gargantuan amount of data [4–11], have started to outline how pan-omics approaches guide targeted therapy and its paradigmatic impact on precision medicine (Sect. 3) [12–16].

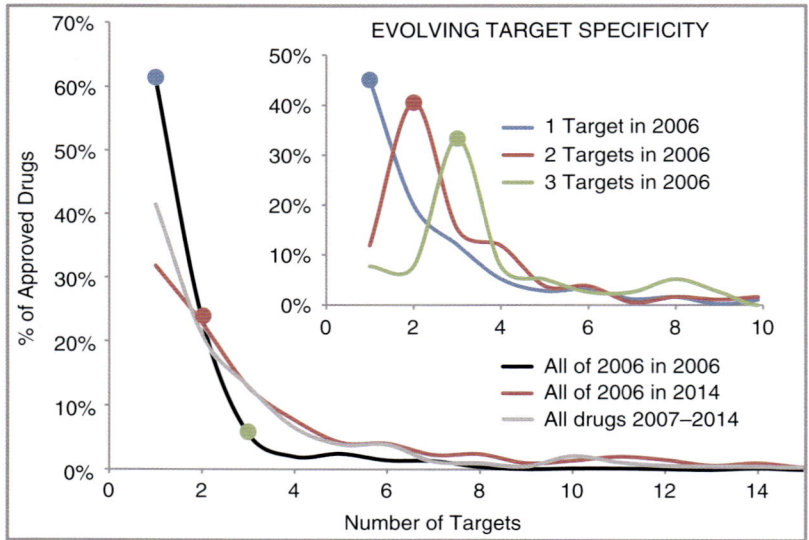

Fig. 3 Evolving target specificity. The number of targets for drugs approved by 2006 and analyzed in 2006 (black line) compared to the same set of drugs reanalyzed in 2014 (red line) highlights the changes in terms of target specificity as drugs are continue to be studied. The gray line represents the number of targets for drugs approved from 2006–2014. The inset shows the reassessment of number of drug targets for drugs originally described to have one target in 2006 (blue line), two targets in 2006 (red line), and three targets in 2006 (green line) when reanalyzed in 2014 (indicated by the lines emanating from each solid sphere). Drug target data were obtained from DrugBank [3].

Although "targeted therapy" is the buzzword, and many prominent examples of targeted drugs exist (Table 1), the specificity of approved drugs is not easy to confirm [81], nor is it easy to identify a trend suggestive of an increase in the development of drugs with greater specificity (Fig. 3). Comparing the distribution of the number of targets per drug for drugs approved by 2006 to drugs approved from 2007 to 2014 demonstrates a higher fraction of drugs with fewer targets. However, a confounding variable is the ever-increasing understanding of a drug's true specificity after it has been approved (Fig. 3). Back in 2006, the distribution of targets per drug showed a higher prevalence of drugs with only one described target compared to data from 2014. In fact, many of these drugs were later demonstrated as having multiple targets (up to 28; see Fig. 3).

Charles Heidelberger and colleagues first introduced targeted therapy to the world with the synthesis of 5-fluorouracil (5-FU), a compound targeting a specific biochemical pathway, namely increased uracil uptake by malignant cells [82, 83]. The number of targeted therapeutics developed since then is extensive, and targeted therapy has seeped into the full spectrum of medical fields and is not only relevant in oncology (Table 1; Sect. 2). Nevertheless, oncology, with the implementation of precision oncology [84, 85], remains a poster child of targeted therapy and is covered extensively in this chapter. The following sections will describe the fundamentals of

Tab. 1 Targeted therapeutics.

Target type	Target	SMDs	Biologicals
Signal transduction cascades	EGFR	Erlotinib [17]	Pertuzumab [18]
		Gefitinib [19, 20]	
		Afatinib [21]	Bevacizumab [22]*
		Sorafenib [23]	
		CO-1686 [24]	Cetuximab [20, 25]
		Lapatinib [26]	
	ERBB2		Trastuzumab [27–29]
	MET	Tivantinib	Onartuzumab
		Cabozantinib	Rilotumumab
		Foretinib [30]	Ficlatuzumab [30]
	MAP2K7	Selumetinib [31]	
		Trametinib [31]	
		PD-0325901 [31]	
		MEK162 [31]	
		PDA-66 [32]*	
	GSK3B	Tideglusib [33]	
	KDR		Bevacizumab [34]
	ESR1		Tamoxifen [35]
			Raloxifen [36]
	ERBB2, FCGR1A		MDX-210 [37]
Cluster of differentiation molecules	CD3		Muromonab-CD3 [38]
	TNFRSF8		Brentuximab vedotin [39]
	CD40		CP-870,893 [40]
			Dacetuzumab [40]
			Chi Lob 7/4 [40]
	TNFRSF8, FCGR3A		HRS-3/A9 bsAb [41]
	CTLA4		Ipilimumab [42, 43]
Fusion proteins	BCR-ABL1	Imatinib [44]	
		Bosutinib [45]	
		Dasatinib [46, 47]	
		Nilotinib [48, 49]	
	EML4-ALK	Crizotinib [50]	
		Ceritinib [51–53]	
Transcription factors	SREBF1, SREBF2	Betulin [54]	
		Fatostatin [55]*	
Metabolic pathways	26S proteasome	Bortezomib [56]	
	20S proteasome	Carfilzomib [57]	
	PCSK9		Evolocumab [58]
			Alirocumab [58]
			SPC5001 [59, 60]
Cell cycle regulators	RNA component of human telomerase RNA (hTR)		Imetelstat [61]
	CDK4, CDK6		PD-0332991 [62]

(*continued overleaf*)

Tab. 1 (Continued)

Target type	Target	SMDs	Biologicals
Chaperones	HSP90	Ganetespib [63]	
		17-DMAG [64]	
Post-translational protein modification	FNTA	Tipifarnib	
		Lonafarnib [65]	
Epigenetic therapy	DNMT1	Azacitidine [66]	
	HDAC9	Vorinostat	
		Romidepsin [67, 68]	
	BET	JQ1 [69–72]	
	DOT1L	EPZ-5676 [67, 73, 74]	
Membrane channels and transporters	CFTR	Ataluren [75, 76]	
Structural proteins	TUBA1A		Paclitaxel
			Docetaxel [77]
Extracellular molecules	*Clostridium difficile* endotoxins		Actoxumab/bezlotoxumab [78, 79]
	Bacillus anthracis anthrax toxin		Raxibacumab [80]

The listed therapeutics are all FDA-approved or are currently being evaluated in clinical trials (as of July 2014). FDA, U.S. Food and Drug Administration; EGFR, epidermal growth factor receptor; ERBB2, erb-b2 receptor tyrosine kinase 2; MET, MET proto-oncogene, receptor tyrosine kinase; MAP2K7, mitogen-activated protein kinase kinase 7; GSK3B, glycogen synthase kinase 3 beta; KDR, kinase insert domain receptor; ESR1, estrogen receptor 1; FCGR1A, Fc fragment of IgG, high affinity Ia, receptor (CD64); CD3, CD3 molecule; TNFRSF8, tumor necrosis factor receptor superfamily, member 8; CD40, CD40 molecule, TNF receptor superfamily member 5; FCGR3A, Fc fragment of IgG, low affinity IIIa, receptor (CD16a); CTLA4, cytotoxic T-lymphocyte-associated protein 4; BCR, breakpoint cluster region; ABL1, ABL proto-oncogene 1, non-receptor tyrosine kinase; EML4, echinoderm microtubule associated protein like 4; ALK, anaplastic lymphoma receptor tyrosine kinase; SREBF1 and SREBF2, sterol regulatory element binding transcription factor 1 and 2, respectively; PCSK9, proprotein convertase subtilisin/kexin type 9; CDK4 and CDK6, cyclin-dependent kinase 4 and 6, respectively; HSP90, heat shock protein 90kDa; FNTA, farnesyltransferase, CAAX box, alpha; DNMT1, DNA (cytosine-5-)-methyltransferase 1; HDAC9, histone deacetylase 9; BET, bromodomain and extra-terminal domain protein family; DOT1L, DOT1-like histone H3K79 methyltransferase; CFTR, cystic fibrosis transmembrane conductance regulator (ATP-binding cassette sub-family C, member 7); TUBA1A, tubulin, alpha 1a; SMDs, small-molecule drugs; *, promising, but not yet in clinical trials.

targeted therapy, its molecular basis, the role of genomics, and pitfalls of current approaches (Sects. 1.2–1.6). Next, the most relevant targeted agents are described: from a mechanistic standpoint as well as in terms of the therapeutic niche each class of agents occupies (Sect. 2). Lastly, the specifics of precision medicine are described (Sect. 4).

1.2
Molecular Basis of Targeted Therapy

The genomic and proteomic landscape of cancer is particularly well suited to describe the molecular basis of targeted therapy. There are approximately 140 individual mutations that are known to initiate tumorigenesis – that is, mutations

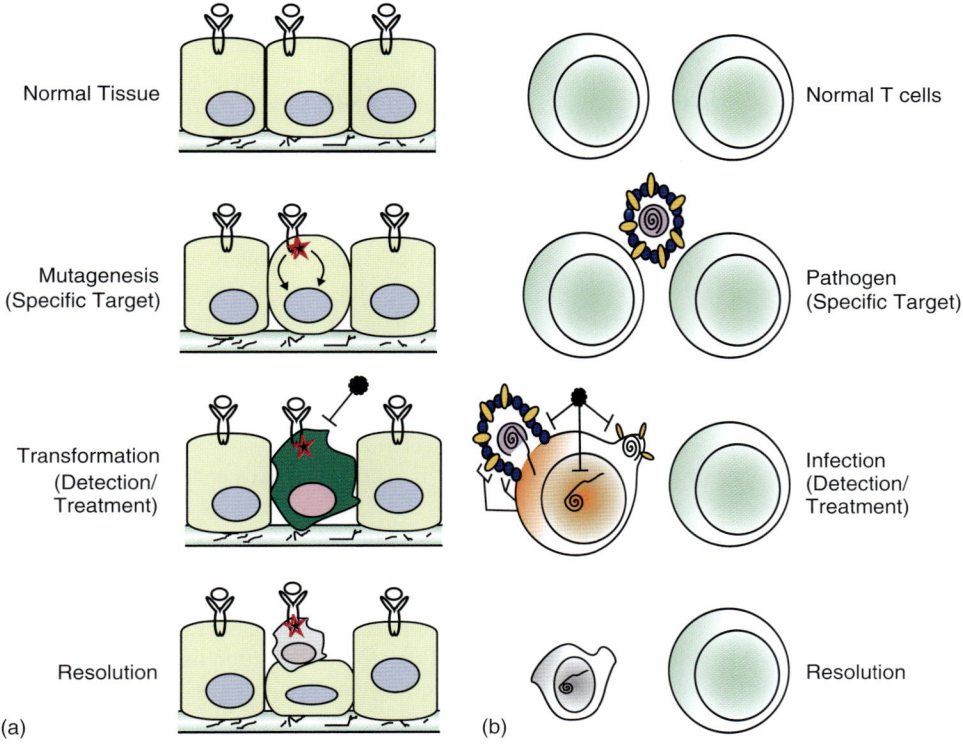

Fig. 4 General biological aspects of targeted therapy. (a) Mutation leads to malignant cell transformation, and at the same time can provide a unique target for therapeutic intervention; (b) Pathogens can present unique targets for therapeutic intervention.

that have the potential to drive normal cells towards the malignant state [86]. Of critical relevance in terms of targeted therapy is the fact that these mutations are unique to the cancer cell; thus, in some cases, providing a target not present in normal cells (Fig. 4a). As an example, the epidermal growth factor receptor (EGFR) can be considered. This receptor is commonly mutated in lung cancer [87, 88] in a way that induces the receptor to constitutively send pro-proliferative signals exclusively within the cancer cells [89]. Human immunodeficiency virus (HIV) is a second example that demonstrates the emerging molecular basis of targeted therapy. In the case of an HIV infection, therapeutics have been designed to specifically target the viral life cycle without harming normal host cells (Fig. 4b) [90]. On a molecular level, both, cancer cells and the viral life cycle, present features that have been successfully targeted (see Table 1).

With regards to targeted therapy, several conceptual components critical to understanding this category of therapeutic agents should be realized (Sect. 2). First, the area of practical applicability and sustained success of targeted therapy is more tilted towards disorders caused by alterations in one or a few genes (Fig. 1a). That being said, even knowing a unique and pharmacologically amenable target does not always make targeted therapy optimal. Cholera, the scourge of humankind for centuries, is a case in point. Its causative agent, *Vibrio cholera*, presents several potentially targetable entities, including the key virulence factors cholera toxin and the toxin coregulated pilus (TCP, required for colonization; [91]). However, simple persistent rehydration lowers the mortality rate almost to zero [92], an achievement unlikely to be improved by targeted agents.

On the other end of the spectrum shown in Fig. 1a are the cases of multiple simultaneous gene alterations, as is commonly observed in cancer [93]. However, this certainly does not preclude the use of targeted therapy. Signaling hubs, points of pro-survival signaling convergence, can potentially be targeted, providing an opportunity to achieve therapeutic goals without individually targeting each upstream mutation (Fig. 1b) [94, 95]. Today, multidrug combinations (Sect. 3.4) are being studied with the same intended goals.

Finally, even knowing a target at the submolecular level is not the same as knowing the underlying phenotypic mechanism(s). The anticoagulant ichorcumab epitomizes this notion. Ichorcumab binds to exosite 1 of thrombin and thus prevents thrombin from cleaving fibrinogen and producing clots; surprisingly, however, other functions of thrombin seem to be retained via mechanisms not understood, significantly reducing the excessive bleeding commonly seen with other anticoagulants [96].

1.3
Purpose of Genomic Approaches and Available Platforms

The advent of next-generation sequencing instruments truly revolutionized the ability to investigate the human genome [7], a fact particularly relevant for the efforts to identify novel therapeutic targets. The first human genome was partially sequenced in 2001 [97, 98], a monumental undertaking that took over a decade to complete and cost around US$ 3 billion. During the second decade of the twenty-first century, it is possible to sequence the entire human genome for close to US$ 1000 and to obtain the data in days rather than years [99], a truly remarkable achievement. Yet, in spite of these incredible achievements, several hurdles must still be overcome (Sect. 1.7).

The crux of next-generation sequencing is the construction of a genome DNA fragment library with synthetic DNA adapters attached to each fragment. The library fragments may then be polymerase-amplified, if necessary, and sequenced [100]. Multiple different next-generation sequencing approaches exist, including reliance on reversible dye terminators (using fluorescent tags) and pH change monitoring (this detects the release of hydrogen ions as nucleotides are being incorporated) [100]. Incredibly, sequencing can now even be performed on the single-cell level, allowing an analysis of the functional state of single cells and determination of cell lineage relationships [101]. From a therapeutic standpoint, the purpose of genomic approaches is to define differences between normal tissue and pathologically relevant entities (Fig. 4), and to anticipate how a pathogen or cancer cell will respond to treatment, which is particularly relevant in the case of biologicals (Sect. 2.2), such as oncolytic viruses [102, 103] that are capable of adjustments after having been administered. Not surprisingly, sequencing is emerging as a cornerstone of clinical care (see Sect. 3; [11, 16]) and drug development [13].

1.3.1 Identification of Genomic Alterations

Cancer can be defined as a disease of the genome, arguably also of the epigenome (Sect. 1.4). Thus, biological studies to elucidate the mechanisms that drive such genomic and epigenomic alterations in cancer, and also an identification of the exact set of alterations in each cancer case, are of immense importance [84]. Next-generation sequencing has emerged as the dominant technique used to identify DNA sequence alterations. The ability to identify point mutations is at an advanced level, and at this point is much more reliable than the identification of large genomic alterations (*https://www.synapse.org/#!Synapse:syn312572*; [7]). The genomic instability of many cancer cells adds to this challenge, by diversifying the genomic landscape of the disease not just from patient to patient (interpatient heterogeneity), but also within a patient (intrapatient heterogeneity), and even within a given tumor (intratumor heterogeneity). Furthermore, heterogeneity is not static, it evolves over time; especially, under the selective pressure of therapeutic treatment [104–107]. The identification of somatic mutations in protein-coding regions (exons) has been a main focus of cancer genomics, and much time and effort has been spent to reliably and accurately identify these mutations. A dominant challenge now, in terms of somatic mutations, is to distinguish between driver events that contribute to oncogenesis and stochastic passenger mutations [108]. The identification of driver genes can be particularly problematic in cases of larger chromosomal alterations, not only because of the large number of genes that are affected simultaneously but also because of technical limitations in terms of detecting large continuous changes in chromosomes [86].

Commonly, individual base-pair alterations cause somatic mutations (Sect. 1.3.1.1), some of which can lead to significant survival and proliferative advantages. These alterations are categorized based on the resulting changes: missense mutations (amino acid substitution resulting in no functional changes, functional hyperactivity, or reduced function); nonsense mutation (producing a premature stop-codon in the mRNA, resulting in a truncated protein); and splice-site mutations (splicing variants). The inferred functional impact of a given mutation type is at times somewhat complicated. Missense mutations, for example, may or may not result in protein expression level changes, depending on the specifics of a given alteration. Even synonymous mutations (the amino acid sequence is not altered, thus, no discernable change in protein sequence occurs) are no longer considered to always be "silent," as they may indeed alter transcription, splicing, mRNA transport, or translation, ultimately leading to phenotypically relevant outcomes [11, 86, 109, 110].

Frame-shift mutations – mutations that are caused by insertions or deletions (indels) of nucleotides that disrupt the normal three-nucleotide codon arrangements – are also observed in cancer [108, 109, 111]. Chromosomal aberrations (Sect. 1.3.1.2) as well as gains and losses in gene copy number (Sect. 1.3.1.3) are additional highly relevant events that contribute to malignant phenotypes [112, 113]. Genomic alterations are further considered to part take in oncogene addiction (e.g., EGFR mutations) or non-oncogene addiction (e.g., CHEK1). Oncogenic addiction describes gene alterations that are essential for the establishment and maintenance of the oncogenic state, while non-oncogenic addiction refers to the broader network of mutations that collectively support the malignant phenotype [114].

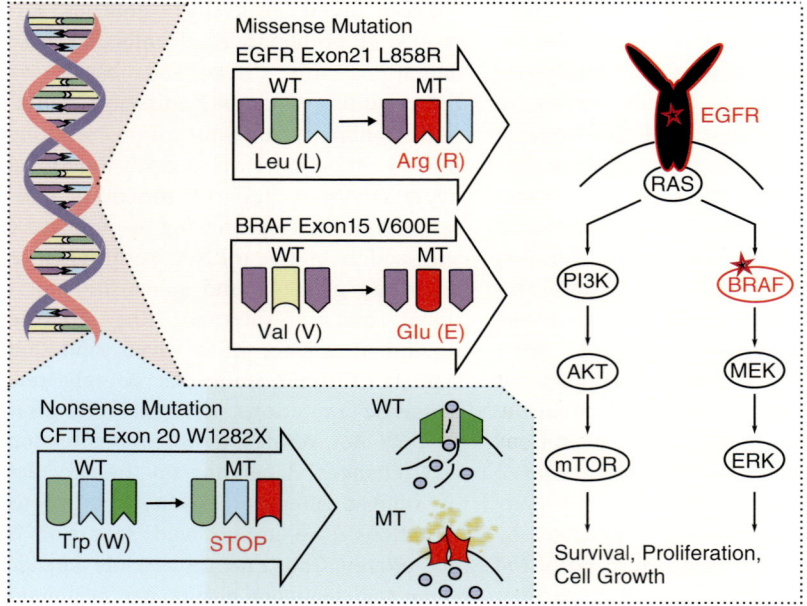

Fig. 5 Pathological point mutations. The missense mutation in EGFR exon 21 (L858R) produces constitutively active receptors; a similar effect is seen with the V600E missense mutation in exon 15 of BRAF. Both mutations result in proteins targeted therapeutically. One of the causes of cystic fibrosis is a nonsense mutation in exon 20 (W1282X) of the cystic fibrosis transmembrane conductance regulator (CFTR).

In summary, the purpose of identifying genomic alterations for targeted therapy is threefold: (i) they can be exploited directly as targets (Fig. 1b – Type 1; mutated proteins or fusion proteins, as discussed below); (ii) they can suggest signaling pathways therapeutically relevant (Fig. 1b – Type 2); or (iii) they can be regarded as base points of vulnerability to explore synthetically lethal combination therapy (Sect. 3.3).

Somatic Mutations Informing Targeted Therapy Among a given tumor type, only very few cancer genes are mutated at high proportions (>20%), with most mutations occurring at intermediate frequencies (2–20%) and only a fraction of these providing druggable targets [84, 86, 93]. Somatic mutations most suited for targeted therapy are activating mutations leading to oncogenic addiction. BRAF (v-raf murine sarcoma viral oncogene homolog B) mutations are an example of oncogenic addiction (Fig. 5). According to the catalog of somatic mutations (COSMIC; [115]), the vast majority (>99%) of BRAF mutations are missense mutations, with codon 600 being one of the most commonly affected sites. $BRAF^{V600}$ mutations are activating mutations and constitutively activate the mitogen-activated protein kinase (MAPK) survival pathway [116, 117] (Fig. 5). The BRAF mutation drives a dominant survival and proliferative signal and presents an attractive therapeutic target, with two targeted agents, vemurafenib [118] and dabrafenib [119], clinically available. Not to oversimplify the targeting of mutant drivers, melanoma with BRAF mutations is quite responsive to BRAF inhibitors, whereas colorectal cancer with BRAF mutations is not [120]. As is the case with most targeted agents, resistance is almost guaranteed, thus necessitating multi-inhibitor treatment approaches. In the case of BRAF mutations, the targeting of additional signaling proteins in the MAPK pathway has provided some additional survival benefits; although, even with dual inhibitors, eventual resistance is observed, most commonly through additional mutations that reactivate the MAPK pathway (Fig. 5) [121, 122].

The altered EGFR in some cases takes on a role similar to that described above for BRAF (Fig. 5). EGFR is a transmembrane receptor tyrosine kinase (RTK) and an example of a class of entities that are frequently targeted therapeutically (Table 1) [123]. Substitutions in exon 21 render the receptor constitutively active, establishing a high cellular dependency on its signaling and providing a sensitive target for targeted therapeutics such as erlotinib or gefitinib [19, 20]. Not surprisingly, resistance to treatment rapidly occurs. Unlike the situation with BRAF mutations, EGFR inhibitor resistance does not necessarily occur via a circumventive reactivation of the associated signaling pathway; instead, a secondary single mutation in the "gatekeeper" residue, T790, which is necessary for inhibitor binding, causes drug resistance [124, 125]. In response, a novel compound – CO-1686 – has been developed to specifically target the T790M drug-resistance mutation (Table 1) [24], epitomizing the continuous race against the Darwinian character of cancer (i.e., selective pressures favoring cancer subclones with the most advantageous alterations) [126, 127].

Nonsense mutations are not frequently targeted therapeutically, mostly because the protein product is frequently non-functional. An exception to this generality involves the cystic fibrosis transmembrane conductance regulator (CFTR; (Fig. 5)). Nonsense mutations in CFTR are the cause of 5–10% of cases (85% in Ashkenazi Jews) of cystic fibrosis (CF) [128]. Several targeted approaches have been proposed as treatment strategies, an important undertaking considering that about 10% of all genetic diseases are due to nonsense mutations [129]. One approach focuses on an induced "readthrough" of premature stopcodons, as has been described for aminoglycosides [130]. The potential "readthrough" drug ataluren has been reported to increase the amount of full-length, functional CFTR [75, 76], and is currently under clinical investigation for the treatment of CF, although its exact mechanism has not been entirely elucidated [131–134]. Gene therapy (see Sect. 2.4), an approach still in its infancy, may be an alternative [135, 136].

Chromosomal Aberrations Extensive aneuploidies are striking characteristics of the cancer genome and generally are caused by chromosomal instability. Deletions and amplifications, as well as chromosomal translocations and chromothripsis (hundreds of genomic rearrangements caused by a single event [137, 138]), are all important mechanisms contributing to chromosomal aberrations [139]. Chromosomal translocations (Fig. 6) arise through chromosome structure instability that is believed to be brought about by the incorrect repair of DNA damage [142]. Cancer-associated chromosomal alterations are of particular interest, as these translocations or inversions potentially result in fusion proteins not found in normal cells, presenting cancer-specific targets [143, 144]. The perhaps best-known chromosomal translocation is termed the Philadelphia

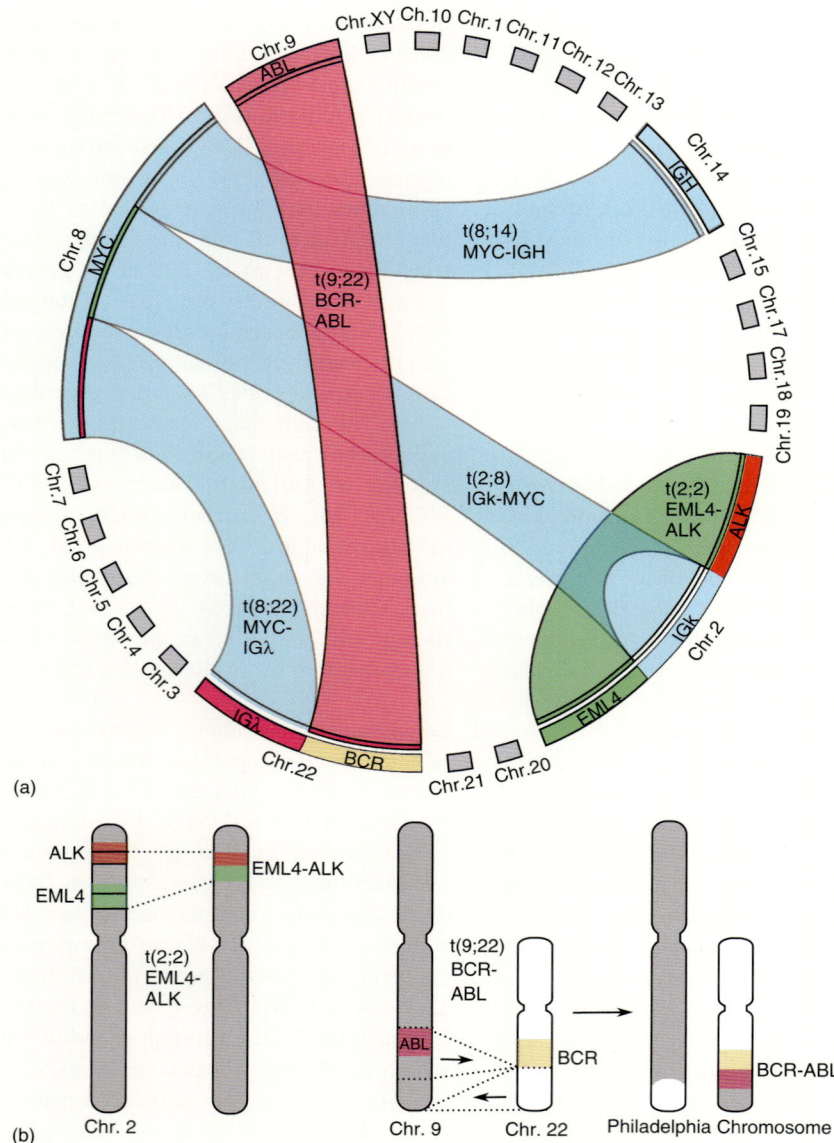

Fig. 6 Chromosomal aberration. (a) Three chromosomal translocations involving MYC are associated with Burkitt's lymphoma (BL): t(8;14), t(2;8), and t(8;22). t(8;14) is associated with 85–90% of BL cases. t(2;8) and t(8;22) are found in 10–15% of cases [140]. (a,b) EML4-ALK is caused by an inversion of Chr. 2 (t(2;2)); the Philadelphia Chromosome comes into existence when a translocation between Chr. 9 and Chr. 22 occurs. BCR-ABL is the oncogenic protein product of the Philadelphia Chromosome. The Circos plot (a) was generated using Circos (http://circos.ca/) [141].

Chromosome (Fig. 6), a translocation between the long arms of chromosomes 9 and 22 (t9;22), first described by Norwell and Hungerford [145–147]. The resulting fusion protein of breakpoint cluster region-Abelson tyrosine kinase (BCR-ABL1) is constitutively active and is not found in normal cells. Subsequently, the first selective BCR-ABL1 targeting tyrosine-kinase inhibitor, imatinib, was approved in 2001 for the treatment of cancer [44]. Imatinib has had a remarkable impact on patients with chronic myelogenous leukemia (CML), a cancer that is heavily dependent on BCR-ABL1. Unfortunately, different mutations in the kinase domain of BCR-ABL1 can render cancer cells resistant to imatinib, but this in retrospect not entirely surprising observation led to the development of two additional BCR-ABL1 targeting therapeutics: dasatinib [46, 47] and nilotinib [48, 49], both of which target most imatinib-resistant BCR-ABL1 variants. All three drugs target the ATP-binding pocket, and it seems critical to further diversify available therapeutics to address the threat of broad resistance induced by mutations of the ATP-binding pocket. The first steps in this direction have been taken by investigating the BCR-ABL1 SH2-kinase domain as a potential therapeutic target [148].

Another fusion-gene of high clinical relevance is the echinoderm microtubule-associated protein-like 4-anaplastic lymphoma kinase (EML4-ALK) protein tyrosine kinase produced by an inversion of the short arm of chromosome 2 (Fig. 6) [149, 150]. Importantly, the EML4-ALK fusion protein is indeed only found in malignant cells, thereby providing a unique pathology-specific target. The small-molecule tyrosine kinase inhibitor crizotinib (see Table 1) effectively targets EML4-ALK (as well as several other kinases, including MET and ROS1), and provides significant survival benefits to patients with EML4-ALK-positive lung cancer compared to nontargeted chemotherapy [50]. Regrettably, cancer cells respond rapidly to the change in selective pressure, and secondary mutations that render the cancer cells resistant to crizotinib frequently occur [151]. The next-generation ALK-inhibitor ceritinib has been shown to overcome crizotinib resistance to some degree, although mutations that cause cells to become resistant to ceritinib have already been identified [51–53]. An alternative approach to managing the resistance of ALK-positive lung cancer to ALK-inhibitors is to target the molecular chaperone heat shock protein 90 (HSP90), a protein that is necessary for the correct folding – and thus function – of EML4-ALK (Fig. 6) [152].

Burkitt's lymphoma (BL) [153] is a non-Hodgkin lymphoma derived from germinal center B cells [140]. Remarkably, nearly 100% of BL cases have a translocation of the MYC oncogene (Fig. 6a) [154]. BL is characterized by three translocations, with the most common (85–90% of cases) involving the distal region of the long arm of chromosome 14 and the long arm of chromosome 8 (t(8;14)) [155–157]. All three translocations result in the juxtaposition of MYC to the enhancer element of one of the immunoglobulin genes, increasing the expression of MYC. BL can often be cured with traditional (nontargeted) chemotherapeutics [140, 158], but these are extremely toxic and are all too frequently unavailable in developing countries at the correct doses or combinations. Considering that MYC is one of the most highly amplified oncogenes, targeting MYC directly could potentially have enormous impact, and not only for patients with BL [159]. As a transcription factor, MYC does not present traditionally targetable moieties, which makes specific

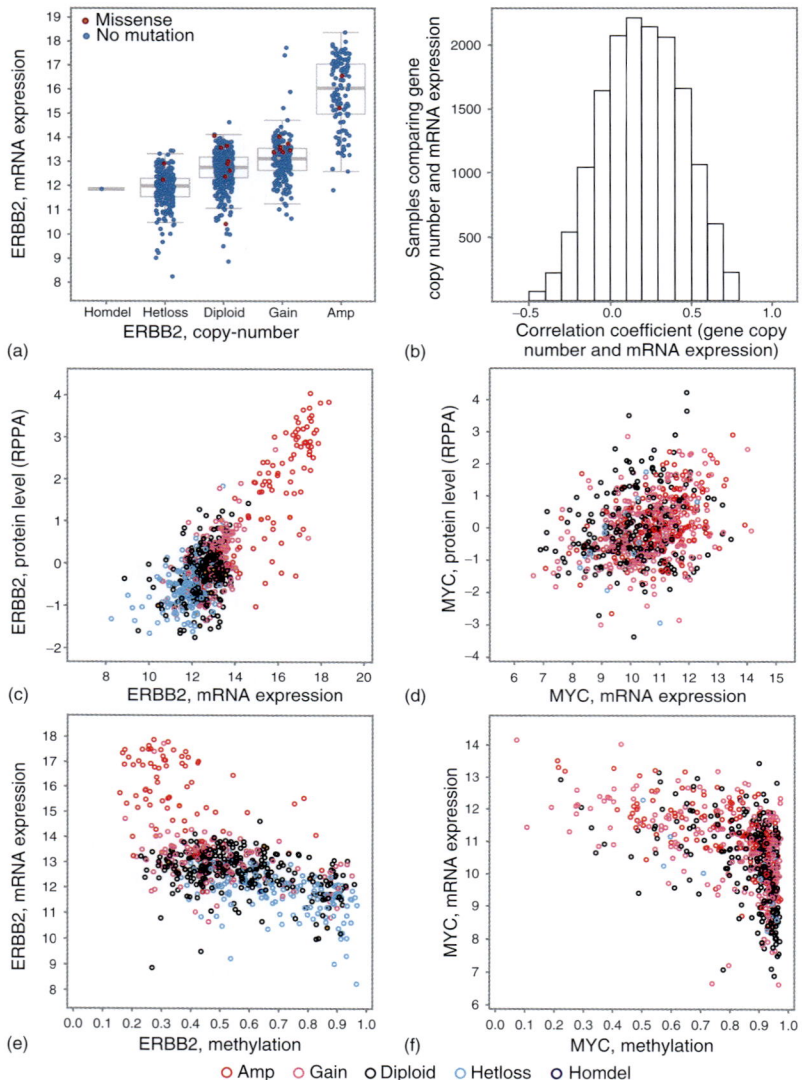

Fig. 7 Variations in gene expression. (a) In the case of ERBB2, gene amplification (Amp) leads to increased mRNA expressions; however, in general, the correlation between gene copy number and mRNA expression is low; (b) Distribution of correlation coefficients between copy number and mRNA levels in lung adenocarcinoma is shown; (c,d) Protein levels for ERBB2 correlate with mRNA expression levels: no such correlation is detectable for MYC; (e) Increased methylation correlates with decreased mRNA expression in the case of ERBB2; in the case of MYC, methylation patterns, and mRNA expression levels do not correlate (f). Data were obtained from the The Cancer Genome Atlas (TCGA; http://cancergenome.nih.gov/) to generate panels (a, c–e) using cBioPortal [162, 163] and panel (b) using the Genome Data Analysis Center (https://confluence.broadinstitute.org/display/GDAC/Home). Gain, gained; Diploid, normal number of chromosomes; Hetloss, heterozygously deleted; Homdel, homozygously deleted.

drug development particularly challenging [160]. One approach to circumvent the issue of directly targeting MYC would be to focus on associated signaling proteins for which inhibitors are available, such as phosphatidylinositol-4,5-bisphosphate 3-kinase (PI3K), Src family kinases or mTOR [155]. Another interesting approach (see Sect. 1.4) is to target MYC at the epigenetic level [69, 155].

Amplification and Overexpression Amplification on the genomic level is the selective increase in copies of DNA sometimes resulting in overexpression of a cancer gene [161]. (Definitions vary, but amplification is generally considered to be a four- or fivefold increase in copy number relative to an adjacent nonamplified marker on the same chromosome.) A number of databases (e.g., cBioPortal [162, 163], COSMIC [115], Oncomine [164]) that describe the relationship between amplification status and expression levels of a given gene are available to research groups. In the case of cancer genomic amplifications, there generally is not a pathology-specific target; rather, there is a lot more of a given gene product relative to normal cells, and the cancer cells tend to rely on the overexpressed gene product to a greater extent. It is the disproportional presence and dependence that provokes therapeutic interest and creates opportunities for targeted therapy. An important caveat regarding gene copy-number and mRNA expression is that the correlation between the two varies significantly from gene to gene (Fig. 7a and b).

A receptor tyrosine kinase ERBB2 (also known as HER2) is an important example of cancer gene amplification. ERBB2 was the first therapeutic target to be defined as such based on genomic alterations [165], and can be effectively targeted with several therapeutic agents [166, 167] including the monoclonal antibody trastuzumab, which was approved by the FDA in 1998 and initiated the age of targeted cancer therapy [27–29]. Trastuzumab is remarkably efficient when ERBB2 is amplified; although, as is the case for many – if not all – targeted therapeutics (see Sect. 2), resistance inadvertently arises [168]. Mechanisms of resistance include the upregulation of downstream signaling proteins, such as PI3K, increased signaling from related RTKs, or a resistance-associated mutation of ERBB2 itself [168].

ERBB2 is amplified in a high percentage (15–20%) of breast cancer cases [169, 170]. Importantly, there is a significant correlation between the ERBB2 gene copy number and ERBB2 mRNA expression levels (Fig. 7a); however, for the majority of other genes the correlation coefficient relating gene copy number and mRNA expression levels is less than 0.5 (Fig. 7b). The story is further complicated when translation is considered, where ERBB2 is again close to the ideal, most straightforward scenario, with mRNA expression levels correlating with amplification status and protein expression levels subsequently correlating with mRNA levels (Fig. 7c). However, this is frequently not the case (Fig. 7d), and it is important to be aware of potential discordance between gene copy number, mRNA expression levels, and protein expression levels.

Another important example of amplification involves the hepatocyte growth factor receptor (HGFR, or more commonly MET), a receptor of robust therapeutic interest. It has been observed that a significant number of cases of EGFR inhibitor resistance (Sect. 1.3.1.1) acquired MET amplification [171], which specifically increases the sensitivity of cancer cells to MET inhibitors (Table 1) [30] and provides

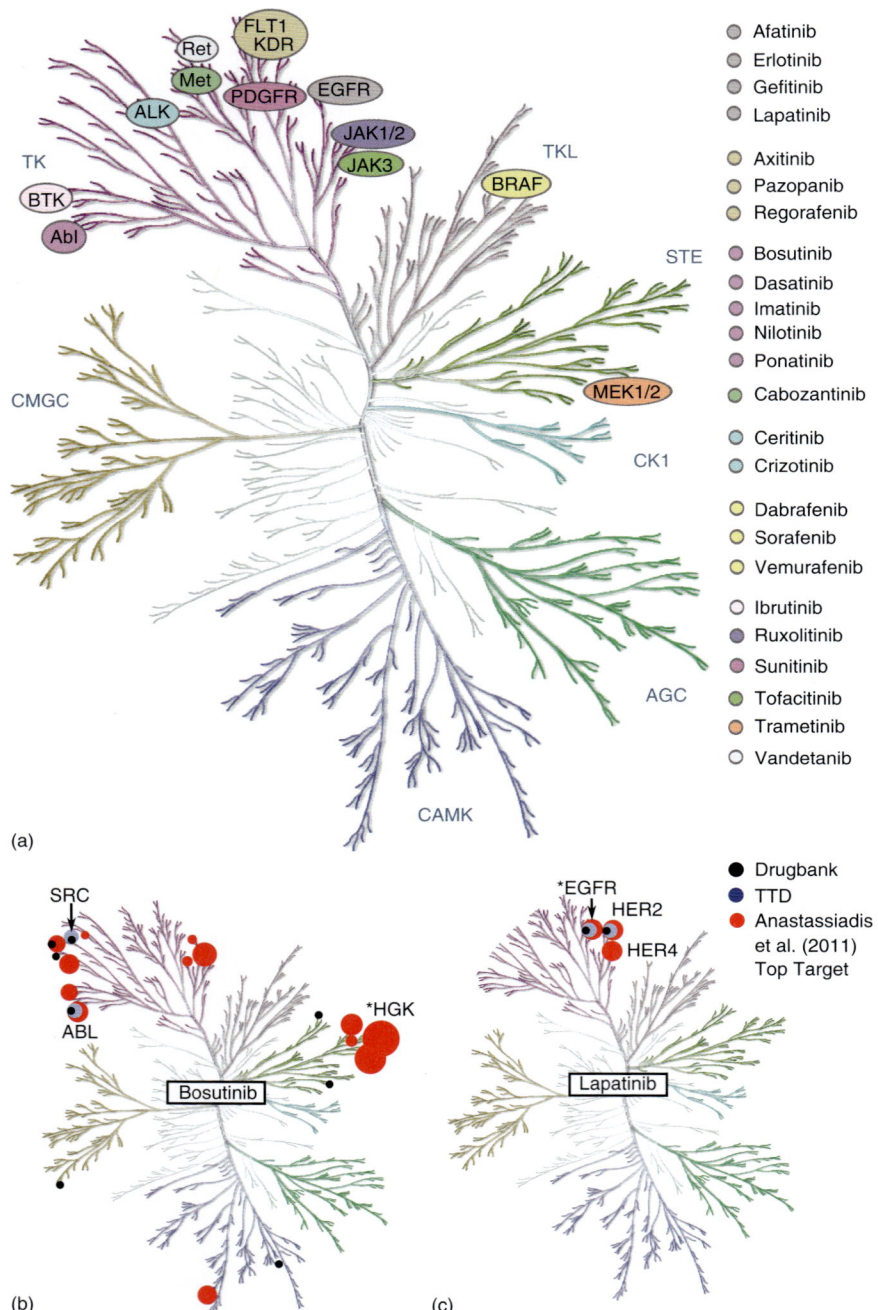

Fig. 8 Targeting the human kinome. (a) All FDA-approved small-molecule kinase inhibitors are shown, and the dominant kinase targeted by each inhibitor is indicated on the kinome tree; (b,c) Numbers of drug targets for kinase inhibitors vary widely between different inhibitors; in addition, these numbers are reported differently in different databases and studies. Targets for bosutinib and lapatinib are indicated, as retrieved from the DrugBank (black dots) [3], the Therapeutic Targets Database (TTD; blue circles [2]) and an *in-vitro* study [81] (red circles); (b) Kinases targeted by bosutinib are shown. The size of the red circles indicates the inhibitory activity of the drug against the target (kinase) [81]; (c) Kinase targets for lapatinib are shown (red circle sizes are arbitrary). The human kinome illustration is reproduced courtesy of Cell Signaling Technology, Inc. (*www.cellsignal.com*) [180].

an avenue for targeted therapy, particularly in combination with EGFR-inhibitors if EGFR is expressed [171–173].

Tumor gene expression patterns are important and can be useful in guiding initial therapy [174–176] and understanding biological changes, for example, when cells undergo epithelial-to-mesenchymal transition (a process thought to be necessary for cancer metastasis) [177–179]. Furthermore, and not surprisingly, gene expression patterns and the activation status of proteins will change in response to therapy and are frequently linked to resistance and disease relapse. This is particularly relevant for kinase inhibitors, which represent an important and well-established class of targeted therapeutics (Fig. 8a; Table 1). These compounds generally target the highly conserved ATP binding pocket of kinases [81], and as a result many kinase inhibitors lack the initially envisioned specificity and target many diverse kinases, as is highlighted by the example of bosutinib (Fig. 8b); other kinases inhibitors, such as lapatinib, are more specific (Fig. 8c). The target multiplicity of kinase inhibitors is emblematic of the difficulties of understanding the true relationship between molecular targets and therapeutic effect (see Fig. 1d), and such difficulties are amplified by the still-evolving understanding of many targeted therapeutic agents (see Fig. 3). Interestingly, kinome "reprogramming" has been observed as cells under therapeutically induced selective pressure dramatically change their signaling patterns and gene expression [181, 182]. Indeed, the interpretation and anticipation of these changes in signaling and expression pattern changes represents the biggest challenges to be overcome in order to improve therapeutic outcomes.

1.4 Epigenetics

Epigenetics describes chromatin-based events that regulate DNA-templated processes, that is, the regulation of gene expression changes rather than direct DNA codon alterations [183]. Epigenetics has emerged as an important consideration from a cancer therapeutics standpoint, and impressive advances have recently been made, as exemplified by the creation of the Encyclopedia of DNA Elements (ENCODE) project [184, 185]. Modifications on the epigenetic level modulate DNA events such as transcription, DNA repair, and replication [186]. Four specific epigenomic regulatory processes have been robustly defined, all of which are involved in regulating access to DNA:

- Adding DNA methylation and histone modifiers (performed by so-called enzymatic "writers") [187].
- Removing DNA methylation and histone modifiers (performed by so-called "erasers").

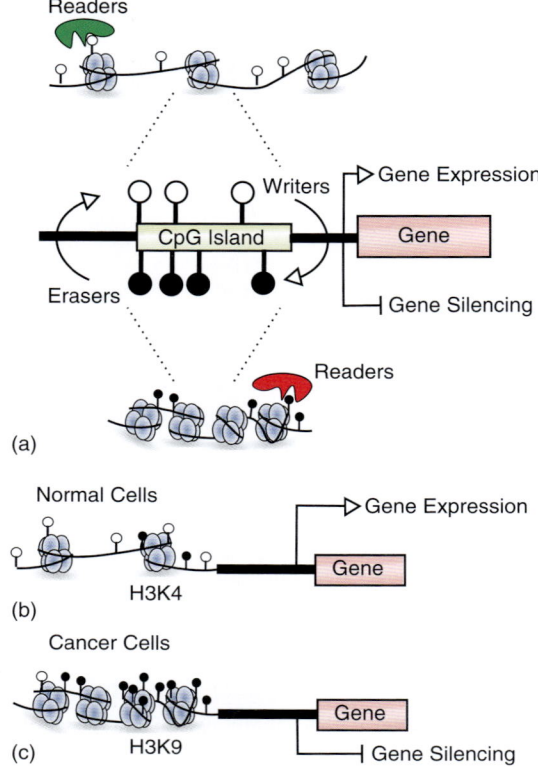

Fig. 9 Epigenetic modulation. (a) Epigenetic modulation of gene expression via methylation status changes are shown; "writers" methylate methylation sites, causing condensation of chromatin and silencing of genes; "erasers" remove methylation, opening up the chromatin and allowing gene expression; (b) Methylation of histone H3 lysine 4 (H3K4) in normal cells is associated with gene expression; (c) Methylation of histone H3 lysine 9 (H3K9) in cancer cells is associated with gene silencing.

- Binding to epigenetic modifiers (performed by so-called "readers").
- Repositioning of nucleosomes across the genome (performed by so-called "movers") (Fig. 9a) [188–190].

DNA promoter cytosine-phosphate-guanine (CpG) island (cytosine in CpG regions can be methylated [191]) hypermethylation, resulting in repressed gene transcription, is particularly frequently observed in cancer (Fig. 9b and c) [192]. Alternatively, the hypomethylation of oncogenes is an additional mechanism of realizing oncogenic potential [188]. ERBB2 expression is an example where a clear correlation exists in terms of methylation status, amplification status, and mRNA expression levels. In cases of ERBB2 amplification, hypomethylation seems to predominate, and this results in increased mRNA expression levels (see Fig. 7e) but, again, this type of robust pattern is infrequent (Fig. 7f). Malignant cells dysregulate epigenetic processes through alterations in DNA/histone modifications, somatic alterations in epigenetic proteins, or altered expression levels of epigenetic proteins [67]. By taking into consideration the fact that the malignant epigenome is indeed very much different from the epigenome of normal tissue, it has been possible to exploit these differences in order to develop targeted therapeutics (Table 1).

Targets of epigenetic therapies in cancer include DNA methyltransferases (DNMTs), histone deacetylases (HDACs), bromodomain and extra-terminal (BET) proteins, DOT1-like histone H3K79 methyltransferase (DOT1-like; DOT1L) and enhancer of zeste homolog 2 (EZH2) [184]. Azacitidine, which originally was designed as a general chemotherapeutic, has been repurposed to function as a targeted agent since the discovery that, at low concentrations, it inhibits all three known DNMTs [66].

The HDACs are another class of attractive epigenetic target, and in general, histone deacetylation is associated with gene repression. Two HDAC inhibitors, vorinostat and romidepsin, are FDA-approved and have achieved some successes in the clinic [67, 68]. EZH2 and DOT1L are both involved in the enzymatic methylation of histone H3, and both have been described as being mutated in cancer [73, 74]; efforts to develop inhibitors to EZH3 and DOT1L are ongoing (Table 1) [67]. Lastly, BET proteins bind to acetylated histones to promote transcription, and interest in this epigenetic regulator was initiated by the discovery of a fusion protein composed of bromodomain containing protein 4 (BRD4) and nuclear protein in testis (NUT) in nuclear protein in testis midline carcinoma (NMC) [184]. The potential to inhibit a novel cancer-specific target led to the development of a specific BET bromodomain inhibitor that currently is undergoing clinical trials [70, 71]. The BET inhibitor JQ1 has shown particular promise by targeting MYC (see Sect. 1.3.1.2) on a transcriptional level rather than directly. Although, from a mechanistic standpoint, the role of JQ1 is not yet completely understood, it has been shown to interfere with RNA polymerase, thus reducing MYC transcription [69, 72].

1.5
Targeting Biological Pathogens

Targeted therapy is not only relevant in the oncological setting but is also a critical tool in the effort to curb bacterial and parasitic infections. Penicillin, discovered by Sir Alexander Fleming [193], revolutionized the treatment of bacterial infections and can be considered a targeted therapeutic because it specifically inhibits transpeptidase proteins that are not found in the human host; thus, penicillin disables the crosslinking of peptidoglycan strands within the bacterial cell wall, resulting in cytolysis, while sparing host cells [194, 195]. As seen with cancer, resistance is quick to arise and multiple means of resistance to penicillin have been described, such as increased efflux of the drug out of bacterial cells, alterations in the transpeptidase structure or, most interesting and most commonly, via the expression of enzymes called β-lactamases that specifically cleave part of the molecular structure of penicillin (the β-lactam backbone) [196]. Again, similar to the efforts described for targeted therapy in cancer, next-generation β-lactam antibiotics are being developed to subvert the impact of β-lactamases [195]. To extend the list of viable targets while limiting the impact of resistance – which is seen as a global threat – genomics approaches have been applied to sequence 74 Gram-negative pathogens and integrate the information with experimental annotations and computational analysis in a platform referred to as antibacTR [197]. genomics approaches were also used to identify PfATP4, a P-Type ATPase candidate on the plasma membrane of *Plasmodium falciparum* (one of the causative agents of malaria [198]) that is needed to enable Na^+ efflux out of the parasite. Spiroindolones specifically disrupt this process and are currently being investigated as novel components of anti-malarial therapeutic regimens [199–201]. Extensive efforts are also ongoing to identify unique targets for other pathogens through comparative genomics (e.g., against *Candida albicans* [202], other fungal pathogens [203], and *Mycobacterium ulcerans* [204]).

Additional targeted approaches to substantiate the currently available antibacterial arsenal include anti-infective monoclonal antibodies (mAbs) (see Sect. 2.2.1) and phage therapy. mAbs against infectious agents are being actively developed, with particular successes achieved by targeting bacterial toxins [205]. *Clostridium difficile* is an example of an infectious agent that is immensely difficult to treat due to its frequent resistance to current treatments. The combination of two antibodies – actoxumab/bezlotoxumab (MK-3415A) – effectively neutralizes the two main pathogenic endotoxins produced by *C. difficile*, and represents a promising new treatment strategy [78, 79]. The first antibody to be approved by the FDA for an antibacterial indication was raxibacumab (ABthrax) [80], in the treatment of inhalation anthrax. Raxibacumab functions by binding to the protective antigen of the anthrax toxin produced by *Bacillus anthracis*; the bound toxin is subsequently incapable of binding to the host cell receptor and therefore is unable to trigger any adverse events [206, 207].

Today, phage therapy is re-emerging as a potential alternative to traditional (and in some cases, rapidly failing) antibiotics [208]. Bacteriophages are viruses that infect bacteria, causing bacterial lysis and the release of virion progeny, and thereby naturally amplifying the therapeutic agent in the presence of bacteria [209]. The exquisite specificity of some bacteriophages to bacteria makes them a class of targeted agent in the truest sense, since bacteriophages neither target host cells nor interfere with the normal gut flora, which is a common problem with traditional antibiotics [209, 210]. Minor immunogenic reactions to bacteriophages are possible, however, and bacteriophages are (at least in theory) capable of modifying bacteria in ways that increase the pathogenicity of the microorganism [209]. However, taking into consideration the recent rise in numbers of multidrug-resistant bacteria, the benefit of further developing bacteriophages as targeted therapeutics should not be underestimated.

1.6
Targeting Low-Density Lipoprotein

Targeted therapy can also be employed to specifically inhibit completely normal proteins in order to prevent or alter pathological processes. Cardiovascular disease caused by atherosclerosis (an imbalanced lipid metabolism and maladaptive immune response that damages the arterial walls) affects millions of people, and elevated levels of low-density lipoprotein (LDL) are a significant contributing factor [211, 212]. Genomic studies have identified an alteration in a gene among a subpopulation of individuals with extraordinarily low blood-cholesterol levels [13, 213–215]. The identified gene is proprotein convertase subtilisin/kexin type 9 (PCSK9), and the realization that loss-of-function mutations in this particular gene lead to low LDL levels has prompted great

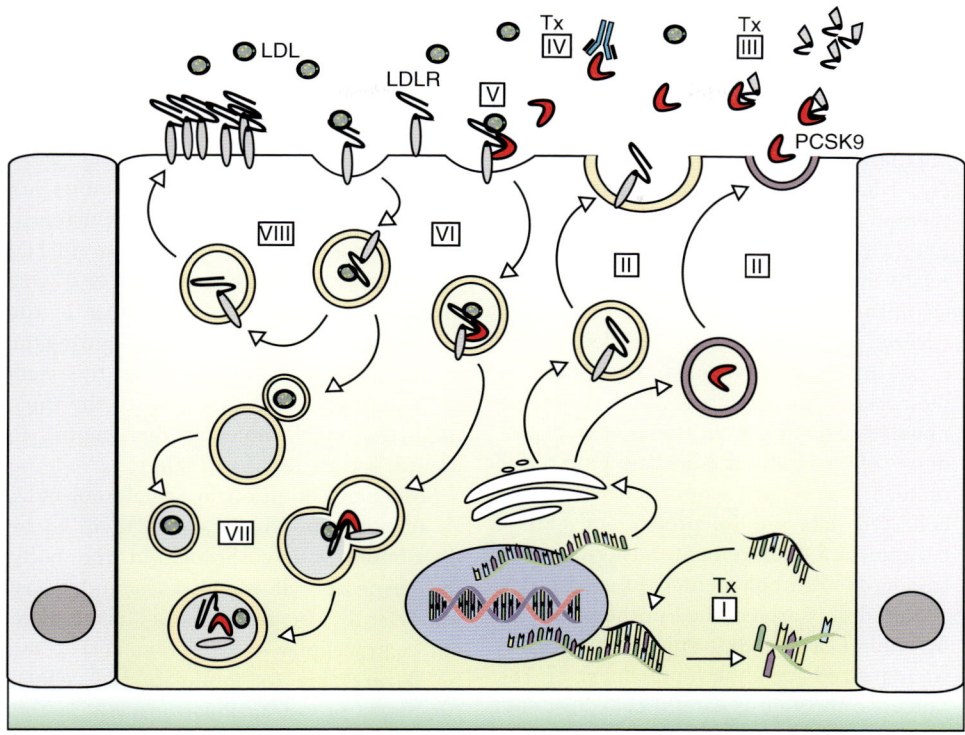

Fig. 10 Targeted therapy against PCSK9. (I) Targeting PCSK9 mRNA with oligonucleotides; (II) exocytosis of LDLR and PCSK9 (red); (III) peptide mimetic binding to PCSK9; (IV) antibody binding to PCSK9; (V) LDL bound to LDLR and PCSK9; (VI) endocytosis of LDL/LDLR/PCSK9 and LDL/LDLR; (VII) lysosomal degradation of LDL and LDL/LDLR/PCSK9; (VIII) recycling of LDLR in the absence of bound PCSK9. PCSK9, proprotein convertase subtilisin/kexin type 9; LDL, low-density lipoprotein; LDLR, low-density lipoprotein receptor; Tx, treatment.

interest in developing PCSK9 antagonists [13, 213, 214]. A reduction in PCSK9 leads to decreased lysosomal destruction of low-density lipoprotein receptors (LDLRs), with a subsequent increase in the absorption of LDL (Fig. 10) [216, 217]. The most promising approach for targeting PCSK9 has been with antibodies (Fig. 10) [218], although additional approaches have included targeting PCSK9 on the transcriptional level with antisense nucleotides or by using mimetic peptides that contain the PCSK9 binding site of LDLR [218]. Currently, SPC5001 is the most advanced of the PCSK9 mRNA targeting antisense therapeutics, and has proven efficacious in lowering LDL levels in nonhuman primates when given subcutaneously (Fig. 10) [59, 60].

1.7
Caveats and Pitfalls

The impact of the age of genomics on targeted therapy development has been rather remarkable (Table 1), though ample challenges remain. The biological effect of a therapeutic agent changes over time (partially due to biological factors, partially due to an initially incomplete understanding of the biology), and the interaction between targeted therapeutics in the context of pathology or when given in combination is often immensely complex (see Figs 1 and 3). Molecularly targeted cancer therapy remains plagued by a high failure rate, a limited therapeutic efficacy, and a relatively small number of patients that significantly benefit from it (10–20%) [29]. Clearly, more targets are needed, better drugs must be developed, and adequate biomarkers are required [219]. Unbiased massive sequencing efforts inherently produce copious amounts of data. One pitfall of such major data acquisition is that the collected data require proper and careful curation and management. It is critical that standards are implemented to ensure that acquired data are handled in the same way by different research groups (e.g., by using standardized analysis pipelines) [178]. Equally important is careful collective data interpretation to minimize the propagation of sequencing and computational errors. Moreover, as noted above, genotypic changes do not always impact phenotypic events as expected, and any conclusions drawn from sequencing studies must consistently be validated via other experimental approaches. Lastly, it is critical that all acquired data are made universally accessible and are comprehensible to the entire research and medical communities [5, 11, 178, 220–223].

2
Therapeutic Agents Used in Targeted Therapy

2.1
Small Molecules

The largest class of targeted agents are the small-molecule drugs (SMDs), also referred to as new chemical entities (NCEs). This class includes molecularly defined nonpeptide organic compounds of low molecular weight (typically <900 Da) that are used to treat a wide range of diseases and conditions (Table 1) [224]. The aforementioned kinase inhibitors are all examples of SMDs (Fig. 8). Currently, the Therapeutic Target Database (TTD) [2] contains information relating to 17 012 SMDs, compared to 14 170 SMDs in the TTD of 2012 [225], thereby highlighting the strong and sustained interest in this class of therapeutics [2]. Examples of the most important small-molecule therapeutics in clinical use are listed in Table 1.

SMDs, by definition, are small molecules – an attribute which allows them to be distributed widely throughout the body, to permeate cell membranes and, most importantly, to reach both intracellular and extracellular targets (Fig. 11) [224]. SMDs are able to: modulate the cellular functions of extracellular growth factors [227], membrane channels and transporters [228], intracellular transcriptional factors [229], proteins lacking enzymatic activities [230], and metabolic pathways [231]; disrupt protein–protein interactions in intracellular signal transduction cascades [31, 32] and cell cycle regulation reactions [232]; deregulate post-translational protein modifications [233]; activate or inactivate autophagy [234]; and cause cell types to adopt features characteristic of other cell types (a process known as transdifferentiation) [235].

Another advantage of SMDs in a general sense is the possibility for chemical modifications and synthesis of core-structure derivatives [236]. Chemical modifications are essential to: (i) improve target specificity once a promising overall molecular structure has been identified [237]; (ii) develop next-generation compounds; (iii) overcome

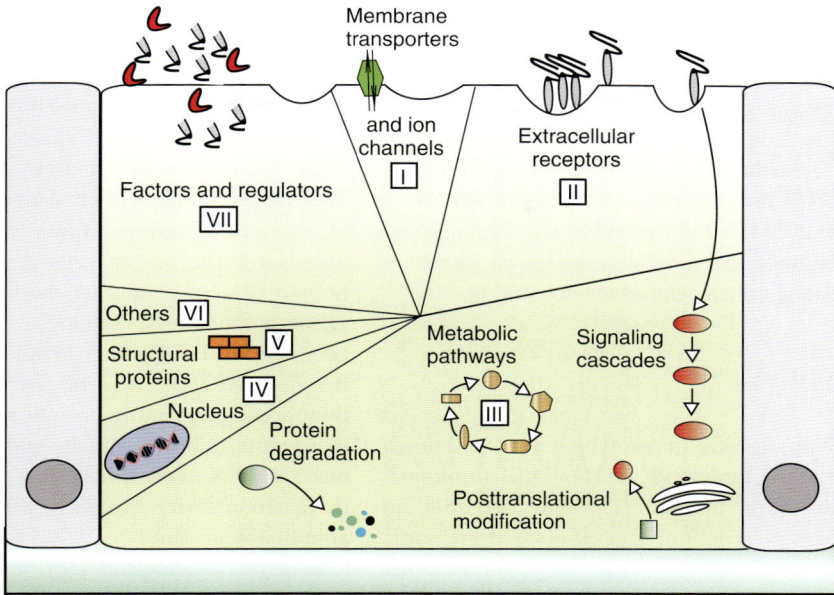

Fig. 11 Distribution of current successful targets. About 6% of targets are membrane transporters and ion channels (I); about 22% are extracellular receptors (II); enzymes, which participate in important cellular processes such as metabolism, protein modification, intracellular signaling, protein degradation, comprise the largest number of targets, approximately 44% (III); only about 4% of targets are contained within the nucleus (IV); and about 4% of targets are structural proteins (V); 19% of targets include factors and regulators (VI); and about 1% of targets are still undetermined (VII). This distribution is based on data from 2006 [226].

acquired resistance [201]; (iv) increase chemical stability [238]; (v) improve means of administration [239]; and (vi) optimize absorption, distribution, metabolism, excretion, and toxicity (ADMET) characteristics [240]. Some SMDs are able not only to act on their direct targets but can also stimulate the immune system to recognize and destroy malignant cells; this process is known as immunogenic cell death [241]. For example, the antitumor effects of the kinase inhibitor dasatinib on c-KIT mutant P815 mastocytoma tumors *in vivo* consist not only of direct tumoricidal activity. Importantly, dasatinib also induces immunomodulatory effects by boosting an antitumor T-cell response [242]. However, unlike other classes of therapeutics (particularly protein-based therapeutics), SMDs are – for the most part – rarely immunogenic [224, 243].

Whilst the advantageous characteristics of SMDs make them potent therapeutic agents, there are several disadvantages. SMDs can be less selective than protein-based therapeutics and can accumulate in specific organs, such as the kidney and liver, resulting in severely toxic side effects [244]. Furthermore, cytochrome P450 enzymes metabolize many of the SMDs, sometimes producing reactive metabolites that can cause unwanted interactions with other medications, such as macrolide antibiotics, azole antifungals, anticonvulsants, protease inhibitors, and others [245]. Another disadvantage is the relatively short half-life (though sometimes this is cited as an advantage as it allows a rapid clearance of the drug) of many SMDs, which commonly necessitates daily administration of the drug [245]. Some of the disadvantages associated with SMDs are avoided by the "biologicals," another extensively investigated group of therapeutic agents (see Sect. 2.2).

2.2
Biologicals

Biomolecular drugs, which are also known as new biological entities (NBEs), biotech drugs, or "biologicals," have emerged as an extremely important class of therapeutic agents, and advances in biotechnology continue to refine the members of this class [224, 246]. Biologicals are substances with fairly large molecular weights (5–200 kDa) that are either prepared from living organisms or derived from substances produced by living organisms [224, 246, 247]. This class of agent is used for the prevention, diagnosis, and treatment of various maladies. Biologicals, as a class, consist of vaccines, cell or gene therapies, hormones, cytokines, tissue growth factors, viruses, and mAbs [103, 224, 248].

The production of biologicals ranges from a reliance on recombinant DNA, to controlled gene expression, to advanced antibody production [249, 250]. For example, modern DNA technology allows the introduction of a functional gene (that encodes the desirable product) into a living organism (such as a bacterium, yeast, or mammalian cells), thus priming the organism to produce copious amounts of the protein [251, 252]. Biologicals are also manufactured using plants and animals. This approach has several important advantages, including correct eukaryotic protein modification (which is difficult to accomplish synthetically), low cost (depending on the specific agent), and high-speed production once a pipeline has been established [253–255].

The development of new biologicals does raise a number of challenges in terms of safety and ethics, however, because it requires the use of cells derived from human tumors or animal models, as well as some unproven technologies to facilitate production [256, 257]. A primary concern here is the limitations of currently used methodologies for the detection of microbial agents and oncogenicity and infectivity of DNA [258], which may pose a potential threat to research workers, patients, and animal models alike.

In general, biologicals are often less well-defined compared to SMDs, predominantly due to post-translational modifications (PTMs) and the presence of impurities inherent to the biological source [259]. These factors make biologicals difficult to analyze both qualitatively and quantitatively [224, 247, 260]. Similar to SMDs, biologicals are useful as either agonists or antagonists of different biological processes, and are generally highly selective and quite potent [259, 261]; however, due to the large molecular size of biologicals, they commonly cannot bypass extracellular barriers and thus can only target cell-surface proteins (Fig. 12) [224, 259]. Biologicals are frequently less stable than SMDs and prone to degradation within the gastrointestinal tract; consequently, they must often be administered by either injection or infusion [262, 263]. In contrast to SMDs, most biologicals are immunogenic. The elicitation of an immune response may not only attenuate the therapeutic effect of a given biological (though in specific cases the immune response may be beneficial [264]), but it can also lead to high toxicity [265].

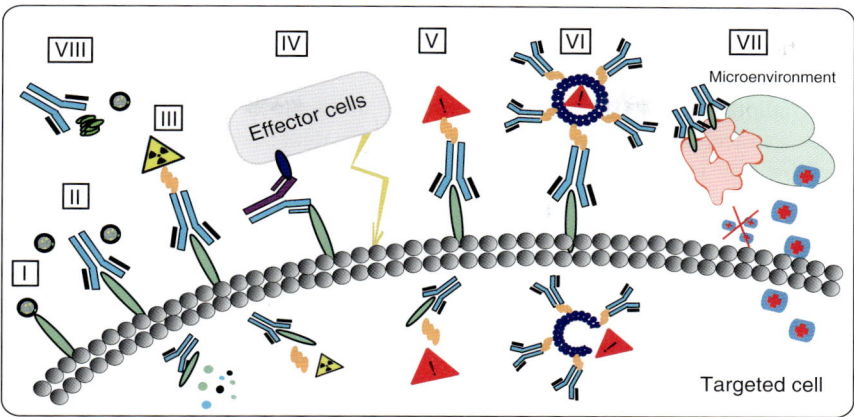

Fig. 12 Monoclonal antibody (mAb)-based therapy. (I) Successful ligand binding to the extracellular receptor; (II) mAb binds to receptor with consequent internalization and degradation of the complex; (III) Immunoconjugate of mAb, linker and radioactive agent binds to receptor, and after internalization the complex is degraded, causing intracellular release of the radioactive agent (radioactive probes can also cause cell death from the membrane surface and do not necessarily need to be internalized); (IV) bispecific monoclonal antibody (bsAb) simultaneously binds to receptors on targeted and immune effector cells; (V) cytotoxic agent is released from the immunoconjugate (similar to (III)); (VI) mAb-covered liposome with cytotoxic cargo is guided to the cell; (VII) targeting and altering the tumor microenvironment using mAbs; (VIII) targeting of extracellular cytokines, toxins and ligands can also be accomplish using mAbs.

2.2.1 Antibody-Based Therapeutics

One important subgroup of biologicals is the large group of antibody-based therapeutics (Table 1). Over the past 30 years, mAb-based therapy has been established as an important therapeutic strategy for the treatment of cancer [266, 267]. One of the first treatment successes, which was reported in 1981, involved a murine mAb against a normal T-cell differentiation antigen to treat T-cell leukemia [268]. In 1986, the first mAb, muromonab-CD3 (Orthoclone OKT3), was approved by the FDA not for the treatment of cancer, but for the prevention of acute organ rejection after transplantation, by blocking T-cell function [38]. An initial major obstacle to the increased clinical use of mAbs was the mouse-based production pipeline and the associated high immunogenicity and short half-life of the antibodies [267]. However, the development of new techniques to humanize or chimerize mAbs addressed such obstacles, and this resulted in clinically suitable mAbs that behaved similarly to natural human antibodies [269]. The high molecular weight of mAbs is another obstacle at times, and is the likely explanation for the inefficient crossing of the blood–brain barrier by mAbs; hence, therapeutic mAbs to treat brain cancer are usually delivered intratumorally [270].

mAbs target specific antigens found on the cell surface, such as transmembrane receptors or extracellular growth factors (Fig. 12) [245]. Upon binding, mAbs can exert direct effects [271], with some antibodies having multiple activities. For example, cetuximab blocks the binding of epidermal growth factor (EGF) and of transforming growth factor-α (TGF-α) to EGFR, which in turn leads to an inhibition of receptor activation and cell proliferation, induction of apoptosis, and the sensitization of tumor cells to chemotherapy and radiotherapy [20, 25]. Since mAbs target extracellular antigens, it is presumed that they are more effective against liquid tumors than solid tumors [271].

In some cases, mAbs are conjugated to radioisotopes or toxins (the latter are referred to as "cargo" or "payload"; see Fig. 12) for precision delivery to a specific target [272]. When an antibody binds to an internalizing receptor on the cell surface, receptor-mediated uptake occurs and the antibody is subsequently transferred into the cell, together with its cargo [273]. The cargo is conjugated to the antibody via a linker, which is a critically important part of many immunoconjugates as it should not allow premature release of the cargo yet ensure that, once the target is reached, efficient dissociation occurs [272]. Three mechanisms have been identified by which the cargo can be released upon endocytosis:

- The presence of a reducing environment (intracellular glutathione reduces disulfide bonds).
- A low pH in the lysosomal compartment (this disrupts the hydrazone linkage).
- Lysosomal enzyme cleavage in the cases of peptide linkers [272].

In some cases the linker is noncleavable (e.g., thioether or hindered disulfide bonds), and the whole antibody undergoes degradation within the lysosome in order to release the cytotoxic cargo. Internalization and cleavage are not required if the cargo is a radioisotope, because emitted α- or β-particles can penetrate the cell membrane unaided [39]. An alternative option for utilizing mAbs as drug-delivery vehicles is to cover nanoparticles (e.g., liposomes, which are closed spherical vesicles formed from a lipid bilayer) containing the drug cargo with antibodies or antibody fragments [274].

Antibody-based drug-delivery systems allow the delivery of general cytotoxics with greater accuracy, to maximize efficacy while minimizing side effects [39]. Brentuximab vedotin, which exemplifies the successful clinical application of this type of delivery system, is composed of a chimeric anti-CD30 antibody linked to the microtubule inhibitor monomethyl auristatin E via a protease-cleavable linker [275]. Brentuximab vedotin was approved in 2011 for the treatment of relapsed/refractory Hodgkin lymphoma and systemic anaplastic large cell lymphoma (Table 1) [39].

The direct targeting of cancer cells (see Fig. 12) is only one of the ways in which mAbs are used to treat cancer, as they have also been shown to successfully target and modulate the tumor microenvironment [276]. CD40 agonists (CD40 is expressed by immune cells in the tumor microenvironment [277]) are a prominent example of this treatment strategy. In pancreatic ductal adenocarcinoma (PDA), mAb CD40 agonists induce macrophage activation that subsequently leads to destruction of the PDA tumor stroma [40]. Another successful example is bevacizumab (see Table 1), which binds to vascular endothelial growth factor (VEGF) and prevents tumor-initiated angiogenesis. This limits the tumor's access to vital nutrients and in turn slows down or even stops tumor growth [34].

Another approach to further improve the efficiency of mAbs-based treatments is through the development of bispecific monoclonal antibodies (bsAbs). The bsAbs are mAbs which have been engineered to bind two different targets simultaneously, such that their selectivity is increased by co-targeting two tumor-associated antigens [278]. An example is co-targeting ERBB2/ERBB3 heterodimers with a bispecific single-chain antibody, A5-linker-ML3.9 bs-scFv, which showed high binding selectivity both *in vitro* and *in vivo* [279].

bsAbs can also be engineered to bind both tumor-associated antigen and an antigen on immune effector cells (Fig. 12) [280]. For example, HRS-3/A9 bsAb targets the CD30 and CD16 expressed by lymphoma cells and lymphocytes (natural killer cells and macrophages), respectively [41]. Bispecific T-cell engager (BiTE) antibodies are a second example of bsAbs. In the case of BiTE antibodies, antigens on the surfaces of cancer cells (including antigen CD19, epithelial cell adhesion molecule (EpCAM), ERBB2, EGFR, CD66e, CD33, EphA2, MCSP (melanoma chondroitin sulfate proteoglycan) [281]), and CD3 on T cells are targeted simultaneously [282]. This leads to the recruitment of cytotoxic T cells independent of T-cell receptor selectivity. Upon BiTE-initiated activation, T cells subdue the cancer cells via membrane perforation and subsequent induction of apoptosis [282, 283].

As noted above, one difference between SMDs and mAbs is the ability of mAbs to induce an immune response. Under the correct circumstances, this allows mAbs to exert not only direct inhibitory effects on tumor growth but also to activate indirect accessory antitumor activities, such as antibody-dependent cell-mediated cytotoxicity (ADCC) and complement-dependent cytotoxicity (CDC) [284]; these factors are perhaps responsible for different efficacy profiles of mAbs *in vitro* and *in vivo* [264]. The immunogenicity of mAbs is also responsible for many of the side effects, which include allergic reactions, the induction of autoimmune diseases, organ-specific adverse events such as cardiotoxicity, infections, platelet, and thrombotic disorders, while some mAbs can even contribute to tumor progression [285].

Although the development of therapeutic mAbs involves complex and expensive processes compared to the development of SMDs, interest in mAbs is growing and they remain a promising group of therapeutic agents [286].

2.3
Targeted Immunotherapy

Over the past few years, targeted immunotherapy has emerged as an innovative and successful strategy for the treatment of a multitude of cancers. Immunotherapy generally primes the immune system in different ways: (i) by stimulating a host immune response specifically against cancer cells; or (ii) by broadly boosting the immune system. Cancer cells often have cancer-specific antigens present on the surface (these molecules differ subtly from those expressed by non-cancer cells). These cancer antigens can serve as targets for the immune system, and immunotherapy often focuses on stimulating this process [287].

In addition to antibodies (see Sect. 2.2.1), immunotherapy includes cell-based and cytokine therapies. Cell-based therapies, or cancer vaccines, involve the removal of immune cells from an individual with cancer, followed by *in vitro* activation of the collected immune cells to recognize tumor-specific antigens. Next, *in vitro* expansion of the immune cell population is required, prior to re-injecting the cells into the individual, where the immune cells ideally trigger an immune response against the tumor [288]. Several cell types can be used for this technique, including natural killer cells, lymphokine-activated killer cells, cytotoxic T cells, and dendritic cells [289–291]. Vaccines are often combined with other substances or cells (termed adjuvants) which help to boost the immune response even further [292]. One successful example of a cell-based immunotherapy is the Sipuleucel-T cancer vaccine [293], the successful application of which first requires that dendritic cells are harvested from the patient and then transfected with a viral vector to increase the ability of the dendritic cells to properly present tumor antigens. When the dendritic cells are reintroduced into the patient they interact with lymphocytes to prompt the latter to specifically target cancer cells [294].

Any further advancement of immunotherapy necessitates that several critical points are addressed, specifically the intrinsic heterogeneity and genomic instability of tumors and the immune-suppressive activities of tumors and their microenvironments [295]. To combine therapeutics that affect different aspects of the immune response is a potential solution to some of these challenges [296]. One successful example is "immune checkpoint blockade," which focuses on blocking cytotoxic T-lymphocyte-associated antigen 4 (CTLA-4) and programmed cell death protein 1 (PD1) with mAbs, in combination with autologous granulocyte–macrophage colony-stimulating factor (GM-CSF)-secreting tumor cell vaccines [297]. Both, CTLA-4 and PD1 are co-inhibitory molecules that are expressed on T cells and decrease the antitumor immune response, thus blunting the efficacy of any vaccine-induced effect [297, 298]. "Immune checkpoint blockade" enables an increased activation and effector function of vaccine-induced, tumor-specific T cells [296].

A combination of immunotherapy with gene-based therapy (see Sect. 2.4) offers another means of addressing some of the shortcomings of originally contrived immunotherapy. This strategy, for example, can be directed to restore the correct activity level of transporters associated with antigen processing (TAPs), which are frequently aberrant within tumors [299]. TAPs are necessary for assembly of the major histocompatibility complex (MHC) class I, which is responsible for cell-surface antigen presentation to cytolytic T lymphocytes [300]. Successful reintroduction of the *TAP1* gene into melanoma B16F10 tumors *in vitro* and *in vivo*, using a vector system, had two significant effects: (i) reduction in the number of immunosuppressive CD^{3+}/interleukin (IL)-10-secreting lymphocytes; and (ii) restoration of MHC I on the tumor cell surface. As a result, the treatment of B16F10 tumors in mice with a vaccinia virus vector expressing *TAP1* led to a significant decrease in tumor growth [301].

Currently, plenty of scientific and clinical challenges remain to be overcome before immunotherapy can be considered a clinical mainstay. These include the heterogeneity of antigen distribution of malignant cells, antigen density variation, and the possibility of an unwanted immune response to the treatment [302]. An example of a severe side effect of immunotherapy is the so-called "cytokine storm," which is a massive and systemic release of proinflammatory cytokines that can lead to life-threatening multiorgan failure [303]. Nevertheless, immunotherapy is a rapidly expanding new class of cancer therapeutics, and in 2013 targeted immunotherapy was named "Breakthrough of the Year" in *Science* magazine – an annual and prestigious award reserved for the top scientific achievement worldwide [304].

2.4
Gene Therapy

Gene therapy was defined by the US FDA as products "… that mediate their effects by transcription and/or translation of transferred genetic material and/or by integrating into the host genome and that are administered as nucleic acids, viruses, or genetically engineered microorganisms" [305]. It would not be far-fetched to describe gene therapy as the most precise form of targeted therapy, in cases when applicable. Unlike targeting kinases with SMDs (see Fig. 8) or triggering unforeseen ripple effects with inhibitors (see Fig. 1c), gene therapy replaces the aberrant target with its normal counterpart [306]. According to data made accessible by *The Journal of Gene Medicine* (*http://www.wiley.com/legacy/wileychi/genmed/clinical/*), the majority of gene therapy approaches currently in clinical trials are related to cancer (63.8%), though other important applicable fields include monogenic diseases, infectious diseases, and cardiovascular diseases (Fig. 13). In 2013, a total of 81 clinical trials categorized as involving gene therapy were approved, although, the majority of these are still in Phase I (Fig. 13).

The potential of gene therapy has been demonstrated in a number of elegant studies [306]. For example, in 2009 Aiuti *et al.* used gene therapy to restore the immune function of nine out of ten patients with severe combined immunodeficiency (SCID) due to lack of adenosine deaminase (ADA). This was achieved by transfecting blood stem cells removed from the patients with a retroviral vector containing a functional copy of the *ADA* gene. Vector-treated cells were then returned to the patients [307].

Clustered regularly interspaced short palindromic repeats-associated system (CRISPR-Cas9) genome engineering is a new and very promising approach for targeted gene therapy that has recently been introduced to the world and has already been used to correct defective genes in model organisms and cultured cells [308]. Genome engineering of this nature relies on bacterial type II CRISPR-Cas9 to perform RNA-guided genome editing [309]. Single guided RNA (sgRNA) targets Cas9 to genomic regions that contain a protospacer-adjacent motif and that are complementary to the target region of the sgRNA [308, 310]. Cas9 generates double-stranded DNA breaks at the target loci, and the breaks are subsequently repaired by nonhomologous end-joining or homology-directed repair [308]. Yin *et al.*, in an elegant breakthrough study showed that CRISPR-Cas9 technology can be used to correct a genetic defect in the liver of adult mice [311]. In this study, Yin *et al.* treated mice with hereditary tyrosinemia type I, a disease caused by the lack of a metabolic enzyme required for tyrosine metabolism, using CRIPR-Cas9 technology [311]. Unfortunately, the CRISPR-Cas9 technology is still years away from clinical application, and measures must first be taken to reduce off-target effects and to increase safety by using a system that ensures only a temporary expression of the Cas9 nuclease. Further improving the delivery methods is also quite important, since at present only intravenous administration is possible and only genes found in the liver (where filtration of the blood occurs) can be targeted [309]; this issue is shared with RNA interference (RNAi)-based therapeutics (see Sect. 2.5).

2.5
Emerging Therapeutic Approaches

In addition to gene therapy, nucleic acids are being developed as targeted therapeutic agents in multiply innovative ways. These include the development of vaccines, antisense oligonucleotides, aptamers,

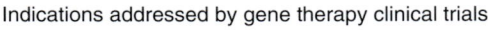

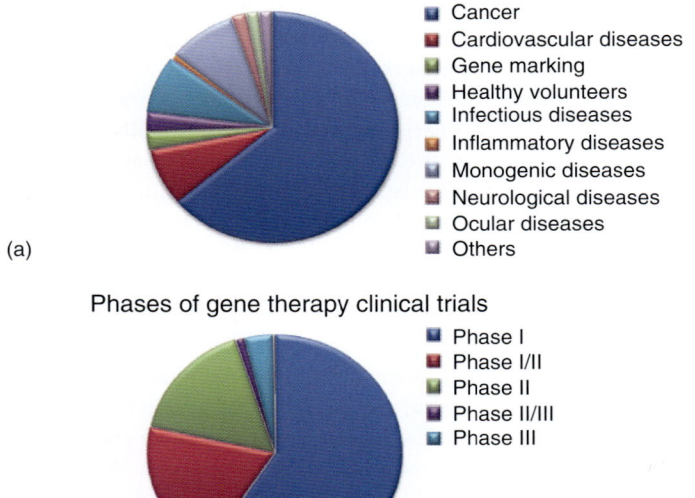

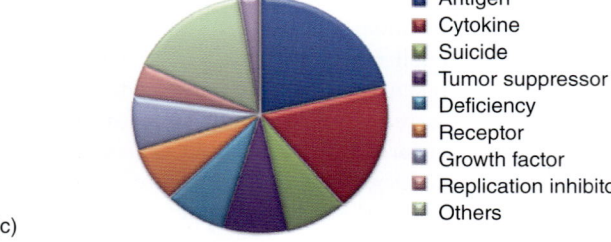

Fig. 13 Gene therapy in clinical trials. (a) Indication treated with gene therapy; (b) phase of gene therapy clinical trials; (c) gene types targeted by gene therapy. Data source: http://www.wiley.com/legacy/wileychi/genmed/clinical/.

ribozymes, and DNAzymes [312]. One example of a new therapeutic nucleic acid-based approach with advanced clinical potential is RNAi. This approach is based on the ability of small interfering RNA (siRNA) to identify its complementary mRNAs in the cell's cytoplasm. After binding to the siRNA, the target mRNA undergoes degradation by the RNA-induced silencing complex (RISC), which possesses endonuclease activity [313]. RNAi silences the target genes indirectly, via transcript degradation, with high efficiency and specificity [314]. During the past decade, major advances have been made in the field of RNAi, and its clinical application is ongoing (see Fig. 10). The introduction of new oligonucleotide chemistries, such as 2′-O-methyl ribose groups and 2′-fluoro-β-D-arabinonucleotides, significantly improved the serum stability of nucleic acids *in vivo*; new algorithms for siRNA design increased its potency and reduced off-target effects [315]. Encouraging clinical responses in cancer patients after RNAi therapies have been reported, specifically for therapies focused on silencing the translation of mRNAs that encode VEGF and kinesin spindle protein [316], the KRAS G12D mutant [317], and the M2 subunit of nucleotide reductase (RRM2) [318]. However, despite some successes in clinical trials, enthusiasm for RNAi-therapy has been somewhat hampered by the low delivery efficiency of siRNA into tumors, the difficulties in determining appropriate siRNA targets, and the significant risk of adverse side effects [314].

Oncolytic virotherapy is based on using oncolytic viruses as targeted therapy to selectively infect and damage cancer cells [102]. This selectivity is driven by the natural or enhanced cellular tropism of different types of viruses [102]. Recently, a randomized Phase III clinical trial designed to investigate the efficacy of talimogene laherparepvec (T-VEC) therapy to treat metastatic melanoma completed patient accrual. T-VEC is a herpes simplex virus that encodes GM-CSF (stimulates the immune system) and, based on preliminary results, is expected to be the first virotherapeutic agent approved for clinical use in the US [319]. The most relevant remaining obstacles are the optimization and enhancement of systemic virus delivery, intratumoral virus spread, and cross-priming of anticancer immunity [102].

The versatility of SMDs (see Sect. 2.1) has prompted interest in developing compounds akin to the immunoconjugates described in Sect. 2.2.1 (see Fig. 12). These small-molecule drug conjugates (SMDCs) are designed essentially to perform the same function as antibody–drug conjugates, but with different payloads being attached to small molecules that can then deliver the payload to a specific target. Thus far, carbonic anhydrase IX (CAIX) has been identified as a particularly promising target that is robustly expressed by many cancer types. Possible CAIX-focused tumor homing has been demonstrated using an SMDC [320]. Alternatively, an early development of delivering a payload using inhibitors of heat shock protein 90 (HSP90) in the form of an HSP90–inhibitor drug conjugate (HDC) has been reported (*http://www.syntapharma.com/*). HSP90 is selectively targetable in cancer cells [321].

The early detection and adequate visualization of cancer is a critical aspect of precision medicine (see Sect. 3), and targeted agents are increasingly being developed to accommodate this need. Different fluorescent dye probes [322], metallic nanoparticles [323] and semiconductor quantum dots [324] can be conjugated to antibodies or other ligands specific to markers expressed by cancer cells. For example, a rapidly activatable cancer-selective fluorescence imaging probe, γ-glutamyl hydroxymethyl rhodamine green (gGlu-HMRG), was synthesized and tested by Urano *et al.* [325]. gGlu-HMRG is activated by the cleavage of glutamate by the enzyme γ-glutamyltranspeptidase, which is not expressed in normal tissue but is commonly overexpressed on the cell membrane of various cancer cells. This provides targeted specificity of the probe [325] and covers a requirement for successful implementation of the topic addressed in the next section, namely precision medicine.

3 Targeted Therapy and Precision Medicine

3.1 Principles of Precision/Genomic Medicine

A consistent theme throughout the evolution of targeted therapy has been: (i) the need to identify specific targets essential to a pathogen or pathological process; and (ii) that the essentiality of the targeted entity is maintained or that therapy is adjusted to target the next entity which emerges as essential [326]. The over-riding goal of precision medicine is to establish an ability to identify disease susceptibility, to precisely map out the disease course, to personalize treatment strategies, and to optimize the timing of treatment [327]. Currently, several disease-targeted genetic tests are already available, specifically for cancer, cardiac disease, neurological diseases and others, and the number of available tests is likely to increase [16]. It is difficult to predict how precision medicine will impact medical care. Precision medicine may continue to be applied case-by-case as an auxiliary tool, or it may eventually emerge as a lifelong companion, and truly revolutionary theme [12], with the first round of sequencing initiated at birth, followed by periodic updates similar to vaccinations and physical examinations, and being a cardinal tool during times of maladies.

Molecular targeted therapy applications guided by molecular profiling of tumors is currently undergoing clinical investigation and has been determined as safe, feasible, and accomplishable in a timely manner (4–5 weeks) [328]. Current clinical trials have highlighted the upside of precision medicine, in that patients receiving precision medicine-guided targeted therapy tended to fare better than those treated without the guided approach. However, shortcomings such as limited overall responses to therapy, marginally increased survival rates and a significant number of patients who did not respond at all to precision medicine approaches, were more than evident [328–331]. Some of the most important aspects (other than political and financial) concerning precision/genomic medicine are addressed in Sects 3.2–3.4.

3.2 Integrating Genomic Data for Clinical Strategies, Diagnosis, Treatment, and Prognosis

Information integration on a community-wide level (ideally, global) as a basis for understanding information on the individual level is critical. Targeted therapeutics, in many cases, fail inadvertently (see Sect. 3.3), especially, when used as monotherapy [29, 332]. Part of the problem is an inability to appreciate the full complexity of a given disease, be it cancer or an infection caused by microorganisms. For example, as noted above, cancer is an immensely heterogeneous disease with different mutation profiles from case to case, and to some degree from cancer cell subclone population to cancer cell subclone population [93, 333–335]. Encouragingly, the tools available today may suffice to overcome many of these problems, with further refinement and increased precision. The tools currently available accomplish an impressive array of tasks, including genome/exome sequencing, cancer genome sequencing, epigenome analysis, RNA sequencing (transcriptome), proteome analysis (mass spectrometry), metabolome analysis (mass spectrometry), and integrative systems bioinformatics (see Sect. 3.4) [4].

Several solutions addressing disease heterogeneity have recorded some successes, and hold promise to change clinical strategies, diagnosis and, ultimately, treatment design. The detection of circulating nucleic acid [CNA; including cell-free DNA (cfDNA) and RNA], circulating tumor cells (CTCs), and the sequencing of single cells are beginning to emerge as clinically viable options, simplifying the multisample collection process necessary to grasp the heterogeneity of a given disease and to track treatment efficacy [336–340]. It is still being debated how representative CTCs are of a present pathology; nevertheless, the analysis of CTCs from collected blood samples is much more practical than serial tissue biopsies that are separated spatially and temporally, as is necessary to accurately track disease progression [12, 341, 342]. Today, CTCs are actively being evaluated as prognostic tools and as biomarkers for treatment response [343, 344]. Efforts are also ongoing to capture single CTCs for sequencing [345]. Single-cell genome, transcriptome, and even proteome analyses, are already at an advanced stage [338, 339, 346–348], and it has been shown that complex tumor genomics and expression patterns can be inferred from a single cell [349, 350]. The role of cfDNA in the context of precision medicine is similar to that of CTCs. cfDNA is shed by cells during apoptosis and necrosis, and its presence in the bloodstream correlates with tumor stage and can serve as a prognostic factor [337, 351]. A large amount of data can be obtained from cfDNA, including information regarding the presence of single-point mutations, amplifications, rearrangements, and aneuploidy [337, 352]. cfDNA has even been used as an aid in the diagnosis of heart transplant rejection [353]. Once adopted for clinical use, the paradigm guiding precision/genomic medicine is likely to include cfDNA, CTCs, and massive sequential data acquisition.

3.3
Multidrug Therapies

The shortcomings of targeted therapy are evident and monotherapy is seldom applicable if long-term therapeutic effects are desirable [29, 86, 332]. HIV therapy is the prototypical approach for multidrug therapy; it has rendered a disease akin to a death sentence essentially to a mere inconvenience when drugs are available (the collective moral failure of ongoing HIV-related suffering is self-evident; [90]). Based on the mathematical modeling of cancer, it has been proposed that in most cases mono-targeted therapy failing is inevitable and that simultaneous administration of, at a minimum, dual-targeted therapy is required [332]. The same model also proposes that even dual-therapy is bound to fail if a single mutation induces resistance to both agents. Thus, it is necessary to target multiple different signaling pathways simultaneously (Fig. 14), to minimize the probability of resistance; furthermore, it is necessary to track therapy response and disease progression as close to real time as possible (see Sects 3.1 and 3.2). Targeting multiple parallel pathways can be described as "horizontal" targeting; conversely, targeting members of the same pathway is "vertical" targeting (Fig. 14a,b) [354].

Targeted therapeutics are also commonly used in combination with traditional chemotherapeutics in an effort to induce a synergistic effect [355]. For example, combining the EGFR inhibitor lapatinib with the chemotherapeutic capecitabine results in significantly increased progression-free survival for patients with HER2-positive advanced breast cancer [356]. Importantly, the combination has a tolerable toxicity profile, which is one of the benefits commonly attributed to targeted therapeutics. Radiotherapy has also been proposed to

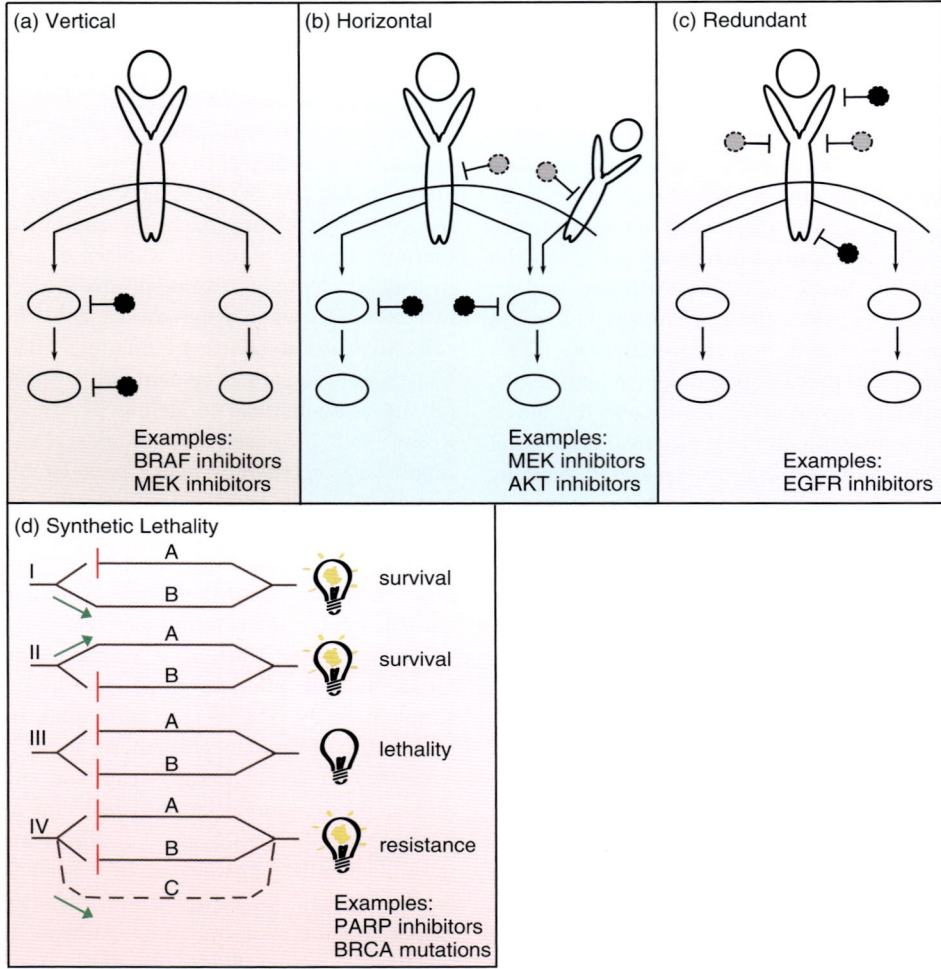

Fig. 14 Multidrug targeted therapy approaches. (a) Vertical treatment approach – inhibition of two or more proteins in the same signaling cascade; (b) Horizontal treatment approach – inhibition of two or more proteins in parallel signaling cascades; (c) Redundant treatment approach – inhibition of the same protein with two or more different therapeutic agents; (d) Synthetic lethality approach: (I and II) single gene disabled (mutation, RNAi, or inhibitor), survival possible; (III) combined inhibition of A and B, synthetic lethality; and (IV) activation of alternative pathways induces resistance (alternatively, A and B can also be reactivated).

benefit from combinatory application in conjunction with targeted therapy. In a murine model, it was suggested that radiation increases EGFR activity, given that EGFR is expressed at high levels [357].

In patients with brain metastases from non-small-cell lung cancer it was observed that, in the presence of EGFR mutations, patients robustly benefited from the combination of erlotinib (Table 1) and whole-brain radiotherapy [358]. The first steps have also been taken to explore the possibility of combining immunotherapy with radiotherapy, with initial interest sparked by observations of the abscopal effect, as first described in 1953 [359]. The abscopal effect is the phenomenon that irradiation of a primary tumor can lead to a size reduction of metastases far outside the irradiated area [360, 361]. Presumably, the immunogenic effect of radiation is the induction of damage-associated molecular patterns (DAMPs) recognizable by pattern recognition receptors expressed by cells of the immune system [362]. The clinical potential of combining radiation with immunotherapy is being explored by combining ipilimumab, a mAb that blocks CTLA-4 (an immune checkpoint molecule that is involved in the downregulation of T-cell activation [42, 43]), with radiotherapy [361].

Multidrug therapies were also anticipated to put into practice the theoretical potential of synthetic lethality. This approach, which was first described by Calvin Bridges, occurs when two genes or gene products are simultaneously perturbed to cause cellular death (Fig. 14d) [363]. The identification of synthetically lethal combinations has predominantly relied on genetic tools (e.g., siRNA, shRNA, and CRISPR [364–368]) and chemical tools [369, 370], generally in high-throughput settings. The most promising domain for synthetic lethality has been in the treatment of BRCA-mutated breast cancer. BRCA1/2 are involved in DNA repair, and cancer cells lacking this critical component of an essential function are close to three orders of magnitude more sensitive to poly (ADP-ribose) polymerase (PARP) inhibitors compared to normal cells (Fig. 14d) [371]. PARP inhibitors, in theory, deliver the second hit, presumably by also interfering with DNA damage repair, which is only lethal in the presence of BRCA1/2 loss-of-function mutations [372, 373]. Disappointingly, thus-far the successes of synthetic lethality observed in model organisms and in high-throughput screens have not been transferrable to the clinic, and even the therapeutic efficacy of PARP inhibitors has been limited [371]. An ability to perfectly align or anticipate the ripple effects of two simultaneous perturbations (see Fig. 1d) to harness the potential of synthetic lethality is likely necessary.

3.4
Integrative Bioinformatics

The apparently insurmountable volume of data produced by the explosion of high-throughput technologies requires adequate bioinformatics systems in place to analyze, normalize, quantify, and help visualize the data and to ensure adequate reproducibility [374]. A vast array of different acquisition and analytical platforms exists to organize and analyze data in different ways [178, 222, 375]. The proposed workflow for molecular profiling involves:

- Obtaining biological samples from the patient (e.g., biopsy, blood sample, CTCs, cfDNA, sputum).
- Data acquisition (proteomics, genomics, epigenomics).
- Bioinformatics (integrate patient data, cross-compare with available data, identify enrichment patterns, identify signaling nodes, and identify appropriate therapeutic approaches).
- Visualizing the data.
- A panel of experts considers the individual report, and the findings are discussed with the patient.
- A treatment regimen is then implemented.

- The whole process is then repeated if needs be.

Bioinformatics resources are plentiful, versatile, and frequently updated. Impressive and near-complete lists of resources are accessible through the Online Bioinformatics Resources Collection (*http://www.hsls.pitt.edu/obrc/*) [376] and through ExPASy (*http://expasy.org/vg/welcome*) [377].

4
Concluding Remarks

At present it is too early to say if targeted therapy will truly revolutionize medicine, or if it will become simply another treatment approach among many. The sobering fact is that the number of drugs approved per billion US dollars spent on research and development has halved approximately every nine years since 1950 [378], partially putting into question the cost–benefit ratio of developing targeted therapeutics. This is of particular concern when considering that only a minority of patients benefit from targeted therapy [29]. There is already a growing understanding that the concept of "one disease, one target, one drug" is not adequate, and the philosophy is turning to the idea of "one drug, multiple targets" – also known as polypharmacology. However, it is also undeniable that individualized treatments based on targeted therapy has already left a significant mark, and that there is no desire to turn back to the "one glove fits all" approach [379]. Genomics specifically, as well as the other "-omics," have started to elucidate the vast complexity of biological systems with remarkable resolution, potentiating the development and applicability of targeted therapy in ways previously unimaginable. The results obtained have led to the first attempts to establish protocols and procedures for true precision medicine, where physicians analyze comprehensive datasets for each patient and then utilize this information to provide personalized, targeted, and more effective care.

Acknowledgments

First and foremost, the authors would like to thank Dr. Erica Golemis for her continuous help and support, and Magda Andrews-Hoke for her assistance. The authors were supported by grant F30 CA180607 from the NIH (to T.N.B.), NIH core grant CA06927 (to the Fox Chase Cancer Center), and Kazan Federal University (to L.G.).

References

1. Saito, R., Smoot ME, Ono K, Ruscheinski J, et al., A travel guide to Cytoscape plugins. *Nat. Methods*, 2012. **9**(11), 1069–1076.
2. Qin, C., Zhang C, Zhu F, Xu F, et al., Therapeutic target database update 2014: a resource for targeted therapeutics. *Nucleic Acids Res.*, 2014, **42** (Database issue), D1118–D1123.
3. Law, V., Knox C, Djoumbou Y, Jewison T, et al., DrugBank 4.0: shedding new light on drug metabolism. *Nucleic Acids Res.*, 2014, **42** (Database issue), D1091-D1097.
4. Meyer, U.A., U.M. Zanger, and M. Schwab, Omics and drug response. *Annu. Rev. Pharmacol. Toxicol.*, **53**, 2013, 475–502.
5. Berger, B., J. Peng, and M. Singh, Computational solutions for omics data. *Nat. Rev. Genet.*, 2013, **14**(5), 333–346.
6. Patti, G.J., O. Yanes, and G. Siuzdak, Innovation: metabolomics: the apogee of the omics trilogy. *Nat. Rev. Mol. Cell Biol.*, 2012, **13**(4), 263–269.
7. Koboldt, D.C., Steinberg KM, Larson DE, Wilson RK, et al., The next-generation

sequencing revolution and its impact on genomics. *Cell*, 2013, **155**(1), 27–38.
8. Schwabe, R.F. and C. Jobin, The microbiome and cancer. *Nat. Rev. Cancer*, 2013, **13**(11), 800–812.
9. Ferte, C., Trister AD, Huang E, Bot BM, *et al.*, Impact of bioinformatic procedures in the development and translation of high-throughput molecular classifiers in oncology. *Clin. Cancer Res.*, 2013, **19**(16), 4315–4325.
10. Hu, Y., Yang X, Qin J, Lu N, *et al.*, Metagenome-wide analysis of antibiotic resistance genes in a large cohort of human gut microbiota. *Nat. Commun.*, 2013, **4**, 2151.
11. Chmielecki, J. and M. Meyerson, DNA sequencing of cancer: what have we learned? *Annu. Rev. Med.*, 2013, **65**, 63–79.
12. Garraway, L.A., Genomics-driven oncology: framework for an emerging paradigm. *J. Clin. Oncol.*, 2013, **31**(15), 1806–1814.
13. Kamb, A., S. Harper, and K. Stefansson, Human genetics as a foundation for innovative drug development. *Nat. Biotechnol.*, 2013, **31**(11), 975–978.
14. Mwenifumbo, J.C. and M.A. Marra, Cancer genome-sequencing study design. *Nat. Rev. Genet.*, 2013, **14**(5), 321–332.
15. Van Allen, E.M., Wagle, N, Stojanov, P, Perrin, DL, *et al.*, Whole-exome sequencing and clinical interpretation of formalin-fixed, paraffin-embedded tumor samples to guide precision cancer medicine. *Nat. Med.*, 2014, **20**(6), 682–688.
16. Rehm, H.L., Disease-targeted sequencing: a cornerstone in the clinic. *Nat. Rev. Genet.*, 2013, **14**(4), 295–300.
17. Wang, P., An F, Zhuang X, Liu J, *et al.*, Chronopharmacology and mechanism of antitumor effect of erlotinib in Lewis tumor-bearing mice. *PLoS One*, 2014, **9**(7), e101720.
18. Reynolds, K., Sarangi S, Bardia A, Dizon DS, Precision medicine and personalized breast cancer: combination pertuzumab therapy. *Pharmgenomics Pers. Med.*, 2014, **7**, 95–105.
19. Pao, W. and J. Chmielecki, Rational, biologically based treatment of EGFR-mutant non-small-cell lung cancer. *Nat. Rev. Cancer*, 2010, **10**(11), 760–774.
20. Liu, H., Cracchiolo, J.R., Beck, T.N., Serebriiskii, I.G, *et al.*, EGFR Inhibitors as Therapeutic Agents in Head and Neck Cancer, in: Burtness, B., Golemis, E.A. (Eds), Molecular Determinants of Head and Neck Cancer, 2014, Springer, *New York, pp.* 55–90.
21. Wislez, M., Malka D, Bennouna J, Mortier L, *et al.*, A new perspective in the treatment of non-small-cell lung cancer (NSCLC). *Role of afatinib: An oral and irreversible ErbB family blocker. Bull. Cancer*, 2014, **101**(6), 647–652.
22. Lauro, S., Onesti CE, Righini R, Marchetti P., The use of bevacizumab in non-small cell lung cancer: an update. *Anticancer Res.*, 2014, **34**(4), 1537–1545.
23. Kim, S. and G.K. Abou-Alfa, The role of tyrosine kinase inhibitors in hepatocellular carcinoma. *Clin. Adv. Hematol. Oncol.*, 2014, **12**(1), 36–41.
24. Walter, A.O., Sjin RT, Haringsma HJ, Ohashi K, *et al.*, Discovery of a mutant-selective covalent inhibitor of EGFR that overcomes T790M-mediated resistance in NSCLC. *Cancer Discovery*, 2013, **3**(12), 1404–1415.
25. Carter, P.J., Potent antibody therapeutics by design. *Nat. Rev. Immunol.*, 2006, **6**(5), 343–357.
26. de Martino, M., Zhuang D, Klatte T, Rieken M, *et al.*, Impact of ERBB2 mutations on in vitro sensitivity of bladder cancer to lapatinib. *Cancer Biol. Ther.*, 2014, **15**(9), 1239–1247.
27. Hudis, C.A., Drug therapy: trastuzumab – mechanism of action and use in clinical practice. *N. Engl. J. Med.*, 2007, **357**(1), 39–51.
28. Hudziak, R.M., Lewis GD, Winget M, Fendly BM, *et al.*, P185her2 monoclonal antibody has antiproliferative effects in vitro and sensitizes human breast tumor cells to tumor necrosis factor. *Mol. Cell. Biol.*, 1989, **9**(3), 1165–1172.
29. Huang, M., Shen A, Ding J, Geng M., Molecularly targeted cancer therapy: some lessons from the past decade. *Trends Pharmacol. Sci.*, 2014, **35**(1), 41–50.
30. Fasolo, A., Sessa C, Gianni L, Broggini M., *et al.*, Seminars in clinical pharmacology: an introduction to MET inhibitors for the medical oncologist. *Ann. Oncol.*, 2013, **24**(1), 14–20.

31. Shen, C.T., Z.L. Qiu, and Q.Y. Luo, Efficacy and safety of selumetinib compared with current therapies for advanced cancer: a meta-analysis. *Asian Pac. J. Cancer Prev.*, 2014, **15**(5), 2369–2374.
32. Kretzschmar, C., Roolf C, Langhammer TS, Sekora A, *et al.*, The novel arylindolylmaleimide PDA-66 displays pronounced antiproliferative effects in acute lymphoblastic leukemia cells. *BMC Cancer*, 2014, **14**, 71.
33. Kramer, T., B. Schmidt, and F. Lo Monte, Small-molecule inhibitors of GSK-3: structural insights and their application to Alzheimer's disease models. *Int. J. Alzheimers Dis.*, 2012, **2012**, 381029.
34. Wu, B., Wu, H., Liu, X., Li, J. *et al.* Ranibizumab versus bevacizumab for ophthalmic diseases related to neovascularisation: a meta-analysis of randomised controlled trials. *PLoS One*, 2014, **9**(7), e101253.
35. Jaitak, V., Drug target strategies in breast cancer treatment: recent developments. *Anticancer Agents Med. Chem.*, 2014, **14**(10), 1414–1427.
36. van Halteren, H., D. Mulder, and E. Ruijter, Oestrogen receptor-beta as a potential target for treatment in advanced colorectal cancer: a pilot study. *Histopathology*, 2013, **64**, 787–790.
37. Chames, P. and D. Baty, Bispecific antibodies for cancer therapy: the light at the end of the tunnel? *MAbs*, 2009, **1**(6), 539–547.
38. Sinnott, J.T., Cullison JP, Sweeney MS, Weinstein SS., Infections in patients receiving OKT3 monoclonal antibody for cardiac rejection: results of a small clinical trial. *Texas Heart Inst. J.*, 1988, **15**(2), 102–106.
39. Palanca-Wessels, M.C. and O.W. Press, Advances in the treatment of hematologic malignancies using immunoconjugates. *Blood*, 2014, **123**(15), 2293–2301.
40. Vonderheide, R.H., Bajor DL, Winograd R, Evans RA, *et al.*, CD40 immunotherapy for pancreatic cancer. *Cancer Immunol. Immunother.*, 2013, **62**(5), 949–954.
41. Hartmann, F., Renner C, Jung W, da Costa L, *et al.*, Anti-CD16/CD30 bispecific antibody treatment for Hodgkin's disease: role of infusion schedule and costimulation with cytokines. *Clin. Cancer Res.*, 2001, **7**(7), 1873–1881.
42. Hodi, F.S., O'Day SJ, McDermott DF, Weber RW, *et al.*, Improved survival with ipilimumab in patients with metastatic melanoma. *N. Engl. J. Med.*, 2010, **363**(8), 711–723.
43. Weber, J., Review: Anti-CTLA-4 antibody ipilimumab: case studies of clinical response and immune-related adverse events. *Oncologist*, 2007, **12**(7), 864–872.
44. Druker, B.J., Sawyers, CL, Kantarjian, H, Resta, DJ, *et al.*, Activity of a specific inhibitor of the BCR-ABL tyrosine kinase in the blast crisis of chronic myeloid leukemia and acute lymphoblastic leukemia with the Philadelphia chromosome. *N. Engl. J. Med.*, 2001, **344**(14), 1038–1042.
45. Hsyu, P.H., Mould DR, Abbas R, Amantea M, Population pharmacokinetic and pharmacodynamic analysis of bosutinib. *Drug Metab. Pharmacokinet.*, 2014, advance publication.
46. Talpaz, M., Shah NP, Kantarjian H, Donato N, *et al.*, Dasatinib in imatinib-resistant Philadelphia chromosome-positive leukemias. *N. Engl. J. Med.*, 2006, **354**(24), 2531–2541.
47. Lombardo, L.J., Lee FY, Chen P, Norris D, *et al.*, Discovery of N-(2-chloro-6-methylphenyl)-2-(6-(4-(2-hydroxyethyl)-piperazin-1-yl)-2-methylpyrimidin-4-ylamino)thiazole-5-carboxamide (BMS-354825), a dual Src/Abl kinase inhibitor with potent antitumor activity in preclinical assays. *J. Med. Chem.*, 2004, **47**(27), 6658–6661.
48. Kantarjian, H., Giles F, Wunderle L, Bhalla K, *et al.*, Nilotinib in imatinib-resistant CML and Philadelphia chromosome-positive ALL. *N. Engl. J. Med.*, 2006, **354**(24), 2542–2551.
49. Weisberg, E., Manley PW, Breitenstein W, Brüggen J, *et al.*, Characterization of AMN107, a selective inhibitor of native and mutant Bcr-Abl. *Cancer Cell*, 2005, **7**(2), 129–141.
50. Shaw, A.T., Kim DW, Nakagawa K, Seto T, *et al.*, Crizotinib versus chemotherapy in advanced ALK-positive lung cancer. *N. Engl. J. Med.*, 2013, **368**(25), 2385–2394.
51. Shaw, A.T., Kim DW, Mehra R, Tan DS, *et al.*, Ceritinib in ALK-rearranged non-small-cell lung cancer. *N. Engl. J. Med.*, 2014, **370**(13), 1189–1197.
52. Friboulet, L., Li N, Katayama R, Lee CC, *et al.*, The ALK inhibitor ceritinib

overcomes crizotinib resistance in non-small cell lung cancer. *Cancer Discovery*, 2014, **4**(6), 662–673.

53. Ramalingam, S.S. and F.R. Khuri, Second-generation ALK inhibitors: filling the Non "MET" Gap. *Cancer Discovery*, 2014, **4**(6), 634–636.

54. Orchel, A., Kulczycka A, Chodurek E, Bebenek E, *et al.*, Influence of betulin and 28-O-propynoylbetulin on proliferation and apoptosis of human melanoma cells (G-361). *Postepy Hig. Med. Dosw. (Online)*, 2014, **68**, 191–197.

55. Li, X., Chen YT, Hu P, Huang WC, Fatostatin displays high antitumor activity in prostate cancer by blocking SREBP-regulated metabolic pathways and androgen receptor signaling. *Mol. Cancer Ther.*, 2014, **13**(4), 855–866.

56. Weathington, N.M. and R.K. Mallampalli, Emerging therapies targeting the ubiquitin proteasome system in cancer. *J. Clin. Invest.*, 2014, **124**(1), 6–12.

57. Jakubowiak, A.J., Dytfeld D, Griffith KA, Lebovic D, *et al.*, A phase 1/2 study of carfilzomib in combination with lenalidomide and low-dose dexamethasone as a frontline treatment for multiple myeloma. *Blood*, 2012, **120**(9), 1801–1809.

58. Cicero, A.F., E. Tartagni, and S. Ertek, Safety and tolerability of injectable lipid-lowering drugs: a review of available clinical data. *Expert Opin. Drug Saf.*, 2014, **13** (8), 1023–1030.

59. Lindholm, M.W., Elmén J, Fisker N, Hansen HF, *et al.*, PCSK9 LNA antisense oligonucleotides induce sustained reduction of LDL cholesterol in nonhuman primates. *Mol. Ther.*, 2012, **20**(2), 376–381.

60. Krieg, A.M., Targeting LDL cholesterol with LNA. *Mol. Ther. Nucleic Acids*, 2012, **1**, e6.

61. Burchett, K.M., Y. Yan, and M.M. Ouellette, Telomerase inhibitor Imetelstat (GRN163L) limits the lifespan of human pancreatic cancer cells. *PLoS One*, 2014, **9**(1), e85155.

62. Li, C., Qi, L, Bellail, AC, Hao, C, *et al.*, PD-0332991 induces G1 arrest of colorectal carcinoma cells through inhibition of the cyclin-dependent kinase-6 and retinoblastoma protein axis. *Oncol. Lett.*, 2014, **7**(5), 1673–1678.

63. Nagaraju, G.P., Long TE, Park W, Landry JC, *et al.*, Heat shock protein 90 promotes epithelial to mesenchymal transition, invasion, and migration in colorectal cancer. *Mol. Carcinog.*, 2014, (in press).

64. Tukaj, S., Grüner D, Zillikens D, Kasperkiewicz M., Hsp90 blockade modulates bullous pemphigoid IgG-induced IL-8 production by keratinocytes. *Cell Stress Chaperones*, 2014, **19**(6), 887–894.

65. Appels, N.M., J.H. Beijnen, and J.H. Schellens, Development of farnesyl transferase inhibitors: a review. *Oncologist*, 2005, **10**(8), 565–578.

66. Issa, J.P.J., H.M. Kantarjian, and P. Kirkpatrick, Azacitidine. *Nat. Rev. Drug Discovery*, 2005, **4**(4), 275–276.

67. Campbell, R.M. and P.J. Tummino, Cancer epigenetics drug discovery and development: the challenge of hitting the mark. *J. Clin. Invest.*, 2014, **124**(3), 1419.

68. West, A.C. and R.W. Johnstone, New and emerging HDAC inhibitors for cancer treatment. *J. Clin. Invest.*, 2014, **124**(1), 30–39.

69. Delmore, J.E., Issa GC, Lemieux ME, Rahl PB, *et al.*, BET bromodomain inhibition as a therapeutic strategy to target c-Myc. *Cell*, 2011, **146**(6), 904–917.

70. Filippakopoulos, P., Qi J, Picaud S, Shen Y, *et al.*, Selective inhibition of BET bromodomains. *Nature*, 2010, **468**(7327), 1067–1073.

71. Parikh, S.A., French, C.A, Costello, B.A, Marks, R.S, *et al.*, NUT midline carcinoma: an aggressive intrathoracic neoplasm. *J. Thorac. Oncol.*, 2013, **8**(10), 1335–1338.

72. Fowler, T., Ghatak, P., Price, D.H., Conaway, R. *et al.*, Regulation of MYC expression and differential JQ1 sensitivity in cancer cells. *PLoS One*, 2014, **9**(1), e87003.

73. Feng, Q., Wang H, Ng HH, Erdjument-Bromage H, *et al.*, Methylation of H3-lysine 79 is mediated by a new family of HMTases without a SET domain. *Curr. Biol.*, 2002, **12**(12), 1052–1058.

74. Kuzmichev, A., Nishioka, K., Erdjument-Bromage, H., Tempst, P., *et al.*, Histone methyltransferase activity associated with a human multiprotein complex containing the Enhancer of Zeste protein. *Genes Dev.*, 2002, **16**(22), 2893–2905.

75. Du, M., Liu, X., Welch, E.M., Hirawat, S., *et al.*, PTC124 is an orally bioavailable compound that promotes suppression of the human CFTR-G542X nonsense allele in

a CF mouse model. *Proc. Natl Acad. Sci. USA*, 2008, **105**(6), 2064–2069.
76. Sermet-Gaudelus, I., Boeck KD, Casimir GJ, Vermeulen F, *et al.*, Ataluren (PTC124) induces cystic fibrosis transmembrane conductance regulator protein expression and activity in children with nonsense mutation cystic fibrosis. *Am. J. Respir. Crit. Care Med.*, 2010, **182**(10), 1262–1272.
77. Loong, H.H. and W. Yeo, Microtubule-targeting agents in oncology and therapeutic potential in hepatocellular carcinoma. *Oncol. Targets Ther.*, 2014, **7**, 575–585.
78. Reichert, J.M., Which are the antibodies to watch in 2013? *MAbs*, 2013, **5**(1), 1–4.
79. Orth, P., Xiao L, Hernandez LD, Reichert P, *et al.*, Mechanism of action and epitopes of *Clostridium difficile* toxin B-neutralizing antibody bezlotoxumab revealed by X-ray crystallography. *J. Biol. Chem.*, 2014, **289**(26), 18008–18021.
80. Fox, J.L., Anthrax drug first antibacterial mAb to win approval. *Nat. Biotechnol.*, 2013, **31**(1), 8.
81. Anastassiadis, T., Deacon SW, Devarajan K, Ma H, *et al.*, Comprehensive assay of kinase catalytic activity reveals features of kinase inhibitor selectivity. *Nat. Biotechnol.*, 2011, **29**(11), 1039–1045.
82. DeVita, V.T., Jr, and E. Chu, *A history of cancer chemotherapy*. Cancer Res., 2008, **68**(21), 8643–8653.
83. Heidelberger, C., Chaudhuri NK, Danneberg P, Mooren D, *et al.*, Fluorinated pyrimidines, a new class of tumour-inhibitory compounds. *Nature*, 1957, **179**(4561), 663–666.
84. Wang, L. and D.A. Wheeler, Genomic sequencing for cancer diagnosis and therapy. *Annu. Rev. Med.*, 2014, **65**, 33–48.
85. Shrager, J. and J.M. Tenenbaum, Rapid learning for precision oncology. *Nat. Rev. Clin. Oncol.*, 2014, **11**(2), 109–118.
86. Vogelstein, B., Papadopoulos N, Velculescu VE, Zhou S, *et al.*, Cancer genome landscapes. *Science*, 2013, **339**(6127), 1546–1558.
87. da Cunha Santos, G., F.A. Shepherd, and M.S. Tsao, EGFR mutations and lung cancer. *Annu. Rev. Pathol.*, 2011, **6**, 49–69.
88. Sharma, S.V., Bell DW, Settleman J, Haber DA, Epidermal growth factor receptor mutations in lung cancer. *Nat. Rev. Cancer*, 2007, **7**(3), 169–181.
89. Gazdar, A.F., Activating and resistance mutations of EGFR in non-small-cell lung cancer: role in clinical response to EGFR tyrosine kinase inhibitors. *Oncogene*, 2009, **28** (Suppl. 1), S24–S31.
90. Bock, C. and T. Lengauer, Managing drug resistance in cancer: lessons from HIV therapy. *Nat. Rev. Cancer*, 2012, **12**(7), 494–501.
91. Nelson, E.J., Harris JB, Morris, JG, Jr, Calderwood, SB, *et al.*, Cholera transmission: the host, pathogen and bacteriophage dynamic. *Nat. Rev. Microbiol.*, 2009, **7**(10), 693–702.
92. Harris, J.B., LaRocque, RC, Qadri, F, Ryan, ET, Calderwood, SB, *Cholera. Lancet*, 2012, **379**(9835), 2466–2476.
93. Lawrence, M.S., Stojanov, P, Mermel, CH, Robinson, JT, *et al.*, Discovery and saturation analysis of cancer genes across 21 tumour types. *Nature*, 2014, **505**(7484), 495–501
94. Behar, M., Barken D, Werner SL, Hoffmann A., The dynamics of signaling as a pharmacological target. *Cell*, 2013, **155**(2), 448–461.
95. Trepel, J., Mollapour M, Giaccone G, Neckers L., *et al.*, Targeting the dynamic HSP90 complex in cancer. *Nat. Rev. Cancer*, 2010, **10**(8), 537–549.
96. Bouchie, A., Allison M, Webb S, DeFrancesco L., *Nature Biotechnology*'s academic spinouts of 2013. *Nat. Biotechnol.*, 2014, **32**(3), 229–238.
97. Lander, E.S., Linton LM, Birren B, Nusbaum C, *et al.*, Initial sequencing and analysis of the human genome. *Nature*, 2001, **409**(6822), 860–921.
98. Venter, J.C., Adams, MD, Myers, EW, Li, PW *et al.*, The sequence of the human genome. *Science*, 2001, **291**(5507), 1304–1351.
99. Hayden, E.C., Technology: the $1,000 genome. *Nature*, 2014, **507**(7492), 294–295.
100. Mardis, E.R., Next-generation sequencing platforms. *Annu. Rev. Anal. Chem. (Palo Alto)*, 2013, **6**, 287–303.
101. Shapiro, E., T. Biezuner, and S. Linnarsson, Single-cell sequencing-based technologies will revolutionize whole-organism science. *Nat. Rev. Genet.*, 2013, **14**(9), 618–630.
102. Russell, S.J., K.W. Peng, and J.C. Bell, Oncolytic virotherapy. *Nat. Biotechnol.*, 2012, **30**(7), 658–670.

103. Bell, J. and G. McFadden, Viruses for tumor therapy. *Cell Host Microbe*, 2014, **15**(3), 260–265.
104. Almendro, V., A. Marusyk, and K. Polyak, Cellular heterogeneity and molecular evolution in cancer. *Annu. Rev. Pathol.*, 2013, **8**, 277–302.
105. Gerlinger, M., Rowan, AJ, Horswell, S, Larkin, J, *et al.*, Intratumor heterogeneity and branched evolution revealed by multiregion sequencing. *N. Engl. J. Med.*, 2012, **366**(10), 883–892.
106. Meacham, C.E. and S.J. Morrison, Tumour heterogeneity and cancer cell plasticity. *Nature*, 2013, **501**(7467), 328–337.
107. Burrell, R.A., McGranahan, N, Bartek, J, Swanton, C, The causes and consequences of genetic heterogeneity in cancer evolution. *Nature*, 2013, **501**(7467), 338–345.
108. Zack, T.I., Schumacher, SE, Carter, SL, Cherniack, AD, *et al.*, Pan-cancer patterns of somatic copy number alteration. *Nat. Genet.*, 2013, **45**(10), 1134–U257.
109. Hartl, D. and M. Ruvolo, *Genetics: Analysis of Genes and Genomes*. 8th edn 2011, Burlington, MA, Jones & Barlett Learning.
110. Supek, F., Miñana, B, Valcárcel, J, Gabaldón T, *et al.*, Synonymous mutations frequently act as driver mutations in human cancers. *Cell*, 2014, **156**(6), 1324–1335.
111. Weinstein, J.N., Collisson EA, Mills GB, Shaw KR, *et al.*, The cancer genome atlas pan-cancer analysis project. *Nat. Genet.*, 2013, **45**(10), 1113–1120.
112. Watson, I.R., Takahashi K, Futreal PA, Chin L., Emerging patterns of somatic mutations in cancer. *Nat. Rev. Genet.*, 2013, **14**(10), 703–718.
113. Garraway, L.A. and E.S. Lander, Lessons from the cancer genome. *Cell*, 2013, **153**(1), 17–37.
114. Luo, J., N.L. Solimini, and S.J. Elledge, Principles of cancer therapy: oncogene and non-oncogene addiction. *Cell*, 2009, **136**(5), 823–837.
115. Forbes, S.A., Bindal, N, Bamford, S, Cole, C, *et al.*, COSMIC: mining complete cancer genomes in the Catalogue of Somatic Mutations in Cancer. *Nucleic Acids Res.*, 2011, **39** (Database issue), D945–D950.
116. Bollag, G., Tsai, J, Zhang, J, Zhang, C, *et al.*, Vemurafenib: the first drug approved for BRAF-mutant cancer. *Nat. Rev. Drug Discov.*, 2012, **11**(11), 873–886.
117. Berger, M.F., Hodis E, Heffernan TP, Deribe YL, *et al.*, Melanoma genome sequencing reveals frequent PREX2 mutations. *Nature*, 2012, **485**(7399), 502–506.
118. Sosman, J.A., Kim KB, Schuchter L, Gonzalez R, *et al.*, Survival in BRAF V600-mutant advanced melanoma treated with vemurafenib. *N. Engl. J. Med.*, 2012, **366**(8), 707–714.
119. Hauschild, A., Grob JJ, Demidov LV, Jouary T, *et al.*, Dabrafenib in BRAF-mutated metastatic melanoma: a multicentre, open-label, phase 3 randomised controlled trial. *Lancet*, 2012, **380**(9839), 358–365.
120. Prahallad, A., Sun C, Huang S, Di Nicolantonio F, *et al.*, Unresponsiveness of colon cancer to BRAF(V600E) inhibition through feedback activation of EGFR. *Nature*, 2012, **483**(7387), 100–103.
121. Konieczkowski, D.J., Johannessen, CM, Abudayyeh, O, Kim, JW, *et al.*, A melanoma cell state distinction influences sensitivity to MAPK pathway inhibitors. *Cancer Discovery*, 2014, **4**(7), 816–827.
122. Wagle, N., Van Allen EM, Treacy DJ, Frederick DT, *et al.*, MAP kinase pathway alterations in BRAF-mutant melanoma patients with acquired resistance to combined RAF/MEK inhibition. *Cancer Discovery*, 2014, **4**(1), 61–68.
123. Lemmon, M.A. and J. Schlessinger, Cell signaling by receptor tyrosine kinases. *Cell*, 2010, **141**(7), 1117–1134.
124. Yun, C.H., Mengwasser KE, Toms AV, Woo MS, *et al.*, The T790M mutation in EGFR kinase causes drug resistance by increasing the affinity for ATP. *Proc. Natl Acad. Sci. USA*, 2008, **105**(6), 2070–2075.
125. Arteaga, C.L. and J.A. Engelman, ERBB receptors: from oncogene discovery to basic science to mechanism-based cancer therapeutics. *Cancer Cell*, 2014, **25**(3), 282–303.
126. Nowell, P.C., The clonal evolution of tumor cell populations. *Science*, 1976, **194**(4260), 23–28.
127. Greaves, M. and C.C. Maley, Clonal evolution in cancer. *Nature*, 2012, **481**(7381), 306–313.
128. Rowe, S.M., Varga K, Rab A, Bebok Z, *et al.*, Restoration of W1282X CFTR activity by enhanced expression. *Am. J. Respir. Cell Mol. Biol.*, 2007, **37**(3), 347–356.

129. Roberts, R.G., A read-through drug put through its paces. *PLoS Biol.*, 2013, **11**(6), e1001458.
130. Kellermayer, R., Translational readthrough induction of pathogenic nonsense mutations. *Eur. J. Med. Genet.*, 2006, **49**(6), 445–450.
131. Sheridan, C., Doubts raised over 'read-through' Duchenne drug mechanism. *Nat. Biotechnol.*, 2013, **31**(9), 771–773.
132. Peltz, S.W., Morsy, M., Welch, E.M., Jacobson, A., Ataluren as an agent for therapeutic nonsense suppression. *Annu. Rev. Med.*, **64**, 2013, 407–425.
133. Kerem, E., Konstan MW, De Boeck K, Accurso FJ, *et al.*, Ataluren for the treatment of nonsense-mutation cystic fibrosis: a randomised, double-blind, placebo-controlled phase 3 trial. *Lancet Respir. Med.*, 2014, **2**(7), 539–547.
134. Derichs, N., Targeting a genetic defect: cystic fibrosis transmembrane conductance regulator modulators in cystic fibrosis. *Eur. Respir. Rev.*, 2013, **22**(127), 58–65.
135. Schwank, G., Koo BK, Sasselli V, Dekkers JF, *et al.*, Functional repair of CFTR by CRISPR/Cas9 in intestinal stem cell organoids of cystic fibrosis patients. *Cell Stem Cell*, 2013, **13**(6), 653–658.
136. Alton, E.W., Boyd AC, Cheng SH, Cunningham S, *et al.*, A randomised, double-blind, placebo-controlled phase IIB clinical trial of repeated application of gene therapy in patients with cystic fibrosis. *Thorax*, 2013, **68**(11), 1075–1077.
137. Stephens, P.J., Greenman CD, Fu B, Yang F, *et al.*, Massive genomic rearrangement acquired in a single catastrophic event during cancer development. *Cell*, 2011, **144**(1), 27–40.
138. Kloosterman, W.P., Tavakoli-Yaraki M, van Roosmalen MJ, van Binsbergen E, *et al.*, Constitutional chromothripsis rearrangements involve clustered double-stranded DNA breaks and nonhomologous repair mechanisms. *Cell Rep.*, 2012, **1**(6), 648–655.
139. Davoli, T., Xu AW, Mengwasser KE, Sack LM, *et al.*, Cumulative haploinsufficiency and triplosensitivity drive aneuploidy patterns and shape the cancer genome. *Cell*, 2013, **155**(2), 948–962.
140. Molyneux, E.M., Rochford R, Griffin B, Newton R, *et al.*, Burkitt's lymphoma. *Lancet*, 2012, **379**(9822), 1234–1244.
141. Krzywinski, M., Schein J, Birol I, Connors J, *et al.*, Circos: an information aesthetic for comparative genomics. *Genome Res.*, 2009, **19**(9), 1639–1645.
142. Thompson, S.L. and D.A. Compton, Chromosomes and cancer cells. *Chromosome Res.*, 2011, **19**(3), 433–444.
143. Kirkpatrick, M., How and why chromosome inversions evolve. *PLoS Biol.*, 2010, **8**(9), e1000501.
144. Nambiar, M., V. Kari, and S.C. Raghavan, Chromosomal translocations in cancer. *Biochim. Biophys. Acta*, 2008, **1786**(2), 139–152.
145. Nowell, P.C. and D.A. Hungerford, Chromosome studies on normal and leukemic human leukocytes. *J. Natl Cancer Inst.*, 1960, **25**, 85–109.
146. Rowley, J.D., Letter: A new consistent chromosomal abnormality in chronic myelogenous leukaemia identified by quinacrine fluorescence and Giemsa staining. *Nature*, 1973, **243**(5405), 290–293.
147. Nowell, P. and D. Hungerford, A minute chromosome in human chronic granulocytic leukemia. *Science*, 1960, **132**, 1497.
148. Grebien, F., Hantschel O, Wojcik J, Kaupe I, *et al.*, Targeting the SH2-kinase interface in Bcr-Abl inhibits leukemogenesis. *Cell*, 2011, **147**(2), 306–319.
149. Soda, M., Choi YL, Enomoto M, Takada S, *et al.*, Identification of the transforming EML4-ALK fusion gene in non-small-cell lung cancer. *Nature*, 2007, **448**(7153), 561–566.
150. Choi, Y.L., Soda M, Yamashita Y, Ueno T, *et al.*, EML4-ALK mutations in lung cancer that confer resistance to ALK inhibitors. *N. Engl. J. Med.*, 2010, **363**(18), 1734–1739.
151. Katayama, R., Shaw AT, Khan TM, Mino-Kenudson M, *et al.*, Mechanisms of acquired crizotinib resistance in ALK-rearranged lung cancers. *Sci. Transl. Med.*, 2012, **4**(120), 120ra17.
152. Sang, J., Acquaviva J, Friedland JC, Smith DL, *et al.*, Targeted inhibition of the molecular chaperone Hsp90 overcomes ALK inhibitor resistance in non-small cell lung cancer. *Cancer Discovery*, 2013, **3**(4), 430–443.

153. Burkitt, D., A sarcoma involving the jaws in African children. *Br. J. Surg.*, 1958, **46**(197), 218–223.
154. Jaffe, E.S. and S. Pittaluga, Aggressive B-cell lymphomas: a review of new and old entities in the WHO classification. *Hematology Am. Soc. Hematol. Educ. Program*, 2011, **2011**, 506–514.
155. Schmitz, R., Ceribelli M, Pittaluga S, Wright G, et al., Oncogenic mechanisms in Burkitt lymphoma. *Cold Spring Harbor Perspect. Med.*, 2014, **4**(2), a014282.
156. Dalla-Favera, R., Bregni M, Erikson J, Patterson D, et al., Human c-myc oncogene is located on the region of chromosome 8 that is translocated in Burkitt lymphoma cells. *Proc. Natl Acad. Sci. USA*, 1982, **79**(24), 7824–7827.
157. Taub, R., Kirsch I, Morton C, Lenoir G, et al., Translocation of the C-myc gene into the immunoglobulin heavy-chain locus in human Burkitt lymphoma and murine plasmacytoma cells. *Proc. Natl Acad. Sci. USA*, 1982, **79**(24), 7837–7841.
158. Patte, C., Auperin A, Michon J, Behrendt H, Leverger G, et al., The Societé Francaise d'Oncologie Pediatrique LMB89 protocol: highly effective multiagent chemotherapy tailored to the tumor burden and initial response in 561 unselected children with B-cell lymphomas and L3 leukemia. *Blood*, 2001, **97**(11), 3370–3379.
159. Beroukhim, R., Mermel CH, Porter D, Wei G, et al., The landscape of somatic copy-number alteration across human cancers. *Nature*, 2010, **463**(7283), 899–905.
160. Dang, C.V., MYC on the path to cancer. *Cell*, 2012, **149**(1), 22–35.
161. Santarius, T., Shipley J, Brewer D, Stratton MR, et al., Epigenetics and genetics: a census of amplified and overexpressed human cancer genes. *Nat. Rev. Cancer*, 2010, **10**(1), 59–64.
162. Cerami, E., Gao J, Dogrusoz U, Gross BE, et al., The cBio cancer genomics portal: an open platform for exploring multidimensional cancer genomics data. *Cancer Discovery*, 2012, **2**(5), 401–404.
163. Gao, J.J., Aksoy BA, Dogrusoz U, Dresdner G, et al., Integrative analysis of complex cancer genomics and clinical profiles using the cBioPortal. *Sci. Signal.*, 2013, **6**(269), pl1.
164. Rhodes, D.R., Kalyana-Sundaram S, Mahavisno V, Varambally R, et al., Oncomine 3.0: genes, pathways, and networks in a collection of 18,000 cancer gene expression profiles. *Neoplasia*, 2007, **9**(2), 166–180.
165. King, C.R., M.H. Kraus, and S.A. Aaronson, Amplification of a novel V-erbb-related gene in a human mammary carcinoma. *Science*, 1985, **229**(4717), 974–976.
166. Dziadziuszko, R. and J. Jassem, Epidermal growth factor receptor (EGFR) inhibitors and derived treatments. *Ann. Oncol.*, 2012, **23** (Suppl. 10), x193–x196.
167. Nielsen, D.L., Kümler I, Palshof JA, Andersson M., Efficacy of HER2-targeted therapy in metastatic breast cancer. Monoclonal antibodies and tyrosine kinase inhibitors. *Breast*, 2013, **22**(1), 1–12.
168. Gajria, D. and S. Chandarlapaty, HER2-amplified breast cancer: mechanisms of trastuzumab resistance and novel targeted therapies. *Expert Rev. Anticancer Ther.*, 2011, **11**(2), 263–275.
169. Owens, M.A., B.C. Horten, and M.M. Da Silva, HER2 amplification ratios by fluorescence in situ hybridization and correlation with immunohistochemistry in a cohort of 6556 breast cancer tissues. *Clin. Breast Cancer*, 2004, **5**(1), 63–69.
170. Shiu, K.K., Wetterskog D, Mackay A, Natrajan R, et al., Integrative molecular and functional profiling of ERBB2-amplified breast cancers identifies new genetic dependencies. *Oncogene*, 2014, **33**(5), 619–631.
171. Engelman, J.A., Zejnullahu K, Mitsudomi T, Song Y, et al., MET amplification leads to gefitinib resistance in lung cancer by activating ERBB3 signaling. *Science*, 2007, **316**(5827), 1039–1043.
172. Seiwert, T.Y., Jagadeeswaran R, Faoro L, Janamanchi V, et al., The MET receptor tyrosine kinase is a potential novel therapeutic target for head and neck squamous cell carcinoma. *Cancer Res.*, 2009, **69**(7), 3021–3031.
173. Seiwert, T.Y., T.N. Beck, and R. Salgia, The Role of HGF/c-MET in Head and Neck Squamous Cell Carcinoma, in: Burtness, B., Golemis, E.A. (Eds), *Molecular Determinants of Head and Neck Cancer*, 2014, Springer, New York, pp. **91–111**.

174. O'Neill, F., Madden SF, Clynes M, Crown J, et al., A gene expression profile indicative of early stage HER2 targeted therapy response. *Mol. Cancer*, 2013, **12**, 69.
175. Yang, W., Soares J, Greninger P, Edelman EJ, et al., Genomics of Drug Sensitivity in Cancer (GDSC): a resource for therapeutic biomarker discovery in cancer cells. *Nucleic Acids Res.*, 2013, **41** (Database issue), D955–D961.
176. Bateman, A.R., El-Hachem N, Beck AH, Aerts HJ, et al., Importance of collection in gene set enrichment analysis of drug response in cancer cell lines. *Sci. Rep.*, 2014, **4**, 4092.
177. Tam, W.L. and R.A. Weinberg, The epigenetics of epithelial-mesenchymal plasticity in cancer. *Nat. Med.*, 2013, **19**(11), 1438–1449.
178. Beck, T.N., Chikwem AJ, Solanki NR, Golemis EA. Bioinformatic approaches to augment study of Epithelial-to-Mesenchymal Transition (EMT) in lung cancer. *Physiol. Genomics*, 2014, **46**(19), 699–724.
179. Creighton, C.J., D.L. Gibbons, and J.M. Kurie, The role of epithelial-mesenchymal transition programming in invasion and metastasis: a clinical perspective. *Cancer Manag. Res.*, 2013, **5**, 187–195.
180. Manning, G., Whyte DB, Martinez R, Hunter T, et al., The protein kinase complement of the human genome. *Science*, 2002, **298**(5600), 1912–1934.
181. Graves, L.M., Duncan JS, Whittle MC, Johnson GL., The dynamic nature of the kinome. *Biochem. J.*, 2013, **450**(1), 1–8.
182. Duncan, J.S., Whittle MC, Nakamura K, Abell AN, et al., Dynamic reprogramming of the kinome in response to targeted MEK inhibition in triple-negative breast cancer. *Cell*, 2012, **149**(2), 307–321.
183. Dawson, M.A. and T. Kouzarides, Cancer epigenetics: from mechanism to therapy. *Cell*, 2012, **150**(1), 12–27.
184. Jones, P.A., At the tipping point for epigenetic therapies in cancer. *J. Clin. Invest.*, 2014, **124**(1), 14–16.
185. Bernstein, B.E., Birney E, Dunham I, Green ED, et al., An integrated encyclopedia of DNA elements in the human genome. *Nature*, 2012, **489**(7414), 57–74.
186. Goldberg, A.D., C.D. Allis, and E. Bernstein, Epigenetics: a landscape takes shape. *Cell*, 2007, **128**(4), 635–638.
187. Lister, R., Pelizzola M, Dowen RH, Hawkins RD, et al., Human DNA methylomes at base resolution show widespread epigenomic differences. *Nature*, 2009, **462**(7271), 315–322.
188. Ahuja, N., H. Easwaran, and S.B. Baylin, Harnessing the potential of epigenetic therapy to target solid tumors. *J. Clin. Invest.*, 2014, **124**(1), 56–63.
189. Kelly, T.K., Miranda TB, Liang G, Berman BP, et al., H2A.Z maintenance during mitosis reveals nucleosome shifting on mitotically silenced genes. *Mol. Cell*, 2010, **39**(6), 901–911.
190. Maxmen, A., Cancer research: open ambition. *Nature*, 2012, **488**(7410), 148–150.
191. Jones, P.A., Functions of DNA methylation: islands, start sites, gene bodies and beyond. *Nat. Rev. Genet.*, 2012, **13**(7), 484–492.
192. Baylin, S.B. and P.A. Jones, A decade of exploring the cancer epigenome – biological and translational implications. *Nat. Rev. Cancer*, 2011, **11**(10), 726–734.
193. Fleming, A., On the antibacterial action of cultures of a *Penicillium*, with special reference to their use in the isolation of *B. influenzæ*. *Br. J. Exp. Pathol.*, 1929, **10**(3), 226–236.
194. Yocum, R.R., J.R. Rasmussen, and J.L. Strominger, The mechanism of action of penicillin. Penicillin acylates the active site of *Bacillus stearothermophilus* D-alanine carboxypeptidase. *J. Biol. Chem.*, 1980, **255**(9), 3977–3986.
195. Kostova, M.B., Myers CJ, Beck TN, Plotkin BJ, et al., C4-Alkylthiols with activity against *Moraxella catarrhalis* and *Mycobacterium tuberculosis*. *Bioorg. Med. Chem.*, 2011, **19**(22), 6842–6852.
196. Davies, J. and D. Davies, Origins and evolution of antibiotic resistance. *Microbiol. Mol. Biol. Rev.*, 2010, **74**(3), 417–4133.
197. Panjkovich, A., Gibert, I., and Daura, X., antibacTR: dynamic antibacterial-drug-target ranking integrating comparative genomics, structural analysis and experimental annotation. *BMC Genomics*, 2014, **15**, 36.
198. Crompton, P.D., Moebius J, Portugal S, Waisberg M, et al., Malaria immunity in

man and mosquito: insights into unsolved mysteries of a deadly infectious disease. *Annu. Rev. Immunol.*, 2014, **32**, 157–187.
199. Spillman, N.J., Allen RJ, McNamara CW, Yeung BK, et al., Na$^+$ regulation in the malaria parasite *Plasmodium falciparum* involves the cation ATPase PfATP4 and is a target of the spiroindolone antimalarials. *Cell Host Microbe*, 2013, **13**(2), 227–237.
200. Horn, D. and M.T. Duraisingh, Antiparasitic chemotherapy: from genomes to mechanisms. *Annu. Rev. Pharmacol. Toxicol.*, 2014, **54**, 71–94.
201. Rottmann, M., McNamara C, Yeung BK, Lee MC, et al., Spiroindolones, a potent compound class for the treatment of malaria. *Science*, 2010, **329**(5996), 1175–1180.
202. Tripathi, H., Luqman S, Meena A, Khan F., Genomic identification of potential targets unique to *Candida albicans* for the discovery of antifungal agents. *Curr. Drug Targets*, 2014, **15**(1), 136–149.
203. Abadio, A.K., Kioshima ES, Teixeira MM, Martins NF, et al., Comparative genomics allowed the identification of drug targets against human fungal pathogens. *BMC Genomics*, 2011, **12**, 75.
204. Butt, A.M., Nasrullah I, Tahir S, Tong Y, Comparative genomics analysis of *Mycobacterium ulcerans* for the identification of putative essential genes and therapeutic candidates. *PLoS One*, 2012, **7**(8), e43080.
205. Fox, J.L., Anti-infective monoclonals step in where antimicrobials fail. *Nat. Biotechnol.*, 2013, **31**(11), 952–954.
206. Kummerfeldt, C.E., Raxibacumab: potential role in the treatment of inhalational anthrax. *Infect. Drug Resist.*, 2014, **7**, 101–109.
207. Migone, T.S., Subramanian GM, Zhong J, Healey LM, et al., Raxibacumab for the treatment of inhalational anthrax. *N. Engl. J. Med.*, 2009, **361**(2), 135–144.
208. Reardon, S., Phage therapy gets revitalized. *Nature*, 2014, **510**(7503), 15–16.
209. Abedon, S.T., Kuhl SJ, Blasdel BG, Kutter EM., Phage treatment of human infections. *Bacteriophage*, 2011, **1**(2), 66–85.
210. Loc-Carrillo, C. and S.T. Abedon, Pros and cons of phage therapy. *Bacteriophage*, 2011, **1**(2), 111–114.
211. Brautbar, A. and C.M. Ballantyne, Pharmacological strategies for lowering LDL cholesterol: statins and beyond. *Nat. Rev. Cardiol.*, 2011, **8**(5), 253–265.
212. Weber, C. and H. Noels, Atherosclerosis: current pathogenesis and therapeutic options. *Nat. Med.*, 2011, **17**(11), 1410–1422.
213. Cohen, J., Pertsemlidis A, Kotowski IK, Graham R, et al., Low LDL cholesterol in individuals of African descent resulting from frequent nonsense mutations in PCSK9. *Nat. Genet.*, 2005, **37**(2), 161–165.
214. Zhao, Z., Tuakli-Wosornu Y, Lagace TA, Kinch L, et al., Molecular characterization of loss-of-function mutations in PCSK9 and identification of a compound heterozygote. *Am. J. Hum. Genet.*, 2006, **79**(3), 514–523.
215. Varret, M., Rabès JP, Saint-Jore B, Cenarro A, et al., A third major locus for autosomal dominant hypercholesterolemia maps to 1p34.1-p32. *Am. J. Hum. Genet.*, 1999, **64**(5), 1378–1387.
216. Grefhorst, A., McNutt MC, Lagace TA, Horton JD. et al., Plasma PCSK9 preferentially reduces liver LDL receptors in mice. *J. Lipid Res.*, 2008, **49**(6), 1303–1311.
217. Lagace, T.A., Curtis DE, Garuti R, McNutt MC, et al., Secreted PCSK9 decreases the number of LDL receptors in hepatocytes and in livers of parabiotic mice. *J. Clin. Invest.*, 2006, **116**(11), 2995–3005.
218. Do, R.Q., R.A. Vogel, and G.G. Schwartz, PCSK9 inhibitors: potential in cardiovascular therapeutics. *Curr. Cardiol. Rep.*, 2013, **15**(3), 345.
219. Mendelsohn, J., Personalizing oncology: perspectives and prospects. *J. Clin. Oncol.*, 2013, **31**(15), 1904–1911.
220. Garraway, L.A. and P.A. Janne, Circumventing cancer drug resistance in the era of personalized medicine. *Cancer Discovery*, 2012, **2**(3), 214–226.
221. Lawrence, M.S., Stojanov P, Polak P, Kryukov GV, et al., Mutational heterogeneity in cancer and the search for new cancer-associated genes. *Nature*, 2013, **499**(7457), 214–218.
222. Liu, H., Beck TN, Golemis EA, Serebriiskii IG, Integrating in silico resources to map a signaling network. *Methods Mol. Biol.*, 2014, **1101**, 197–245.

223. Mudge, J.M., A. Frankish, and J. Harrow, Functional transcriptomics in the post-ENCODE era. *Genome Res.*, 2013, **23**(12), 1961–1973.
224. Cho, M.J. and R. Juliano, Macromolecular versus small-molecule therapeutics: drug discovery, development and clinical considerations. *Trends Biotechnol.*, 1996, **14**(5), 153–158.
225. Zhu, F., Shi Z, Qin C, Tao L, et al., Therapeutic target database update 2012: a resource for facilitating target-oriented drug discovery. *Nucleic Acids Res.*, 2012, **40** (Database issue), D1128–D1136.
226. Zheng, C.J., Han LY, Yap CW, Ji ZL, et al., Therapeutic targets: progress of their exploration and investigation of their characteristics. *Pharmacol. Rev.*, 2006, **58**(2), 259–279.
227. Incorvati, J.A., Shah S, Mu Y, Lu, J Targeted therapy for HER2 positive breast cancer. *J. Hematol. Oncol.*, 2013, **6**, 38.
228. Subasinghe, N.L., Wall M.J., Winters M.P., Qin N., et al., A novel series of pyrazolylpiperidine N-type calcium channel blockers. *Bioorg. Med. Chem. Lett.*, 2012, **22**(12), 4080–4083.
229. Pencheva, N., Buss CG, Posada J, Merghoub T, et al., Broad-spectrum therapeutic suppression of metastatic melanoma through nuclear hormone receptor activation. *Cell*, 2014, **156**(5), 986–1001.
230. Peng, T., Wu, J.-R., Tong, L.-J., Li, M.-Y., et al., Identification of DW532 as a novel anti-tumor agent targeting both kinases and tubulin. *Acta Pharmacol. Sin.*, 2014, **35**(7), 916–928.
231. Lim, J.H., Luo, C, Vazquez, F, Puigserver, P., Targeting mitochondrial oxidative metabolism in melanoma causes metabolic compensation through glucose and glutamine utilization. *Cancer Res.*, 2014, **74**, 3535–3545.
232. Caldwell, J.T., Edwards H, Buck SA, Ge Y,et al., Targeting the wee1 kinase for treatment of pediatric Down syndrome acute myeloid leukemia. *Pediatr. Blood Cancer*, 2014, **61**, 1767–1773.
233. Reikvam, H., Brenner AK, Nepstad I, Sulen A, et al., Heat shock protein 70 – the next chaperone to target in the treatment of human acute myelogenous leukemia? *Expert Opin. Ther. Targets*, 2014, **18**, 1–16.
234. Sarkar, S., Perlstein EO, Imarisio S, Pineau S, et al., Small molecules enhance autophagy and reduce toxicity in Huntington's disease models. *Nat. Chem. Biol.*, 2007, **3**(6), 331–338.
235. Fomina-Yadlin, D., Kubicek S, Walpita D, Dancik V, et al., Small-molecule inducers of insulin expression in pancreatic alpha-cells. *Proc. Natl Acad. Sci. USA*, 2010, **107**(34), 15099–15104.
236. Schreiber, S.L., Organic synthesis toward small-molecule probes and drugs. *Proc. Natl Acad. Sci. USA*, 2011, **108**(17), 6699–6702.
237. Ledford, H., Complex synthesis yields breast-cancer therapy. *Nature*, 2010, **468**(7324), 608–609.
238. Lourido, S., Zhang C, Lopez MS, Tang K, et al., Optimizing small molecule inhibitors of calcium-dependent protein kinase 1 to prevent infection by *Toxoplasma gondii*. *J. Med. Chem.*, 2013, **56**(7), 3068–3077.
239. Ishikawa, M. and Y. Hashimoto, Improvement in aqueous solubility in small molecule drug discovery programs by disruption of molecular planarity and symmetry. *J. Med. Chem.*, 2011, **54**(6), 1539–1554.
240. Huttunen, K.M., H. Raunio, and J. Rautio, Prodrugs – from serendipity to rational design. *Pharmacol. Rev.*, 2011, **63**(3), 750–771.
241. Vacchelli, E., Aranda F, Eggermont A, Galon J, et al., Trial watch: chemotherapy with immunogenic cell death inducers. *Oncoimmunology*, 2014, **3**(1), e27878.
242. Yang, Y., Liu C, Peng W, Lizee, G. et al., Antitumor T-cell responses contribute to the effects of dasatinib on c-KIT mutant murine mastocytoma and are potentiated by anti-OX40. *Blood*, 2012, **120**(23), 4533–4543.
243. Bugelski, P.J., Genetic aspects of immune-mediated adverse drug effects. *Nat. Rev. Drug Discov.*, 2005, **4**(1), 59–69.
244. Atici, S., Cinel I, Cinel L, Doruk N, et al., Liver and kidney toxicity in chronic use of opioids: an experimental long term treatment model. *J. Biosci.*, 2005, **30**(2), 245–252.
245. Gerber, D.E., Targeted therapies: a new generation of cancer treatments. *Am. Fam. Physician*, 2008, **77**(3), 311–319.
246. Sathish, J.G., Sethu S, Bielsky MC, de Haan L, et al., Challenges and approaches for the

246. development of safer immunomodulatory biologics. *Nat. Rev. Drug Discov.*, 2013, **12**(4), 306–324.
247. Shi, S., Biologics: an update and challenge of their pharmacokinetics. *Curr. Drug Metab.*, 2014, **15**(3), 271–290.
248. Ferreira, I. and D. Isenberg, Vaccines and biologics. *Ann. Rheum. Dis.*, 2014, **73**(8), 1446–1454.
249. Kelley, B., Industrialization of mAb production technology: the bioprocessing industry at a crossroads. *MAbs*, 2009, **1**(5), 443–452.
250. Niederwieser, D. and S. Schmitz, Biosimilar agents in oncology/haematology: from approval to practice. *Eur. J. Haematol.*, 2011, **86**(4), 277–288.
251. Liu, D.T., F.T. Gates, III, and N.D. Goldman, Quality control of biologicals produced by r-DNA technology. *Dev. Biol. Stand.*, 1985, **59**, 161–166.
252. Kamionka, M., Engineering of therapeutic proteins production in *Escherichia coli*. *Curr. Pharm. Biotechnol.*, 2011, **12**(2), 268–274.
253. Chen, Q., L. Santi, and C. Zhang, Plant-made biologics. *Biomed. Res. Int.*, 2014, **2014**, 418064.
254. Huang, W., He T, Chai C, Yang Y, *et al.*, Triptolide inhibits the proliferation of prostate cancer cells and down-regulates SUMO-specific protease 1 expression. *PLoS One*, 2012, **7**(5), e37693.
255. Virulizin. BioDrugs, 2002, **16**(5), 374–375. Available at: http://www.ncbi.nlm.nih.gov/pubmed/12408741.
256. Grachev, V., Magrath D, Griffiths E, Petricciani JC, *et al.*, WHO requirements for the use of animal cells as in vitro substrates for the production of biologicals – (requirements for biological substances no. 50) (Reprinted from WHO Technical Report Series, No. 878, 1998). *Biologicals*, 1998, **26**(3), 175–193.
257. Cardoso, C.V. and A.E. de Almeida, Laboratory animal: biological reagent or living being? *Braz. J. Med. Biol. Res.*, 2014, **47**(1), 19–23.
258. Knezevic, I., Stacey G, Petricciani J, Sheets R, *et al.*, Evaluation of cell substrates for the production of biologicals: revision of WHO recommendations. Report of the WHO study group on cell substrates for the production of biologicals, 22-23 April 2009, Bethesda, USA. *Biologicals*, 2010, **38**(1), 162–169.
259. Morrow, T. and L.H. Felcone, Defining the difference: what makes biologics unique. *Biotechnol. Healthcare*, 2004, **1**(4), 24–29.
260. Gatto, B., Biologics targeted at TNF: design, production and challenges. *Reumatismo*, 2006, **58**(2), 94–103.
261. Yao, S., Y. Zhu, and L. Chen, Advances in targeting cell surface signalling molecules for immune modulation. *Nat. Rev. Drug Discov.*, 2013, **12**(2), 130–146.
262. Chung, S.W., T.A. Hil-lal, and Y. Byun, Strategies for non-invasive delivery of biologics. *J. Drug Target.*, 2012, **20**(6), 481–501.
263. Skalko-Basnet, N., Biologics: the role of delivery systems in improved therapy. *Biologics*, 2014, **8**, 107–114.
264. Kurai, J., Chikumi H, Hashimoto K, Takata, M., *et al.*, Therapeutic antitumor efficacy of anti-epidermal growth factor receptor antibody, cetuximab, against malignant pleural mesothelioma. *Int. J. Oncol.*, 2012, **41**(5), 1610–1618.
265. Wolbink, G.J., L.A. Aarden, and B.A. Dijkmans, Dealing with immunogenicity of biologicals: assessment and clinical relevance. *Curr. Opin. Rheumatol.*, 2009, **21**(3), 211–215.
266. Oldham, R.K., Monoclonal antibodies in cancer therapy. *J. Clin. Oncol.*, 1983, **1**(9), 582–590.
267. Oldham, R.K. and R.O. Dillman, Monoclonal antibodies in cancer therapy: 25 years of progress. *J. Clin. Oncol.*, 2008, **26**(11), 1774–1777.
268. Miller, R.A., Maloney DG, McKillop J, Levy R., In vivo effects of murine hybridoma monoclonal antibody in a patient with T-cell leukemia. *Blood*, 1981, **58**(1), 78–86.
269. Ahmadzadeh, V., Farajnia, S., Feizi, M.A.H., Nejad, R.A.K., Antibody humanization methods for development of therapeutic applications. *Monoclon. Antib. Immunodiagn. Immunother.*, 2014, **33**(2), 67–73.
270. Butowski, N. and S.M. Chang, Small molecule and monoclonal antibody therapies in neurooncology. *Cancer Control*, 2005, **12**(2), 116–124.
271. Imai, K. and A. Takaoka, Comparing antibody and small-molecule therapies for cancer. *Nat. Rev. Cancer*, 2006, **6**(9), 714–727.

272. Perez, H.L., Cardarelli, P.M., Deshpande, S., Gangwar, S., et al., Antibody–drug conjugates: current status and future directions. *Drug Discov. Today*, 2014, **19**(7), 869–881.
273. Teicher, B.A. and R.V. Chari, Antibody conjugate therapeutics: challenges and potential. *Clin. Cancer Res.*, 2011, **17**(20), 6389–6397.
274. Zhang, J. and T.H. Rabbitts, Intracellular antibody capture: a molecular biology approach to inhibitors of protein–protein interactions. *Biochim. Biophys. Acta*, 2014, **1844**(11), 1970–1976.
275. Senter, P.D. and E.L. Sievers, The discovery and development of brentuximab vedotin for use in relapsed Hodgkin lymphoma and systemic anaplastic large cell lymphoma. *Nat. Biotechnol.*, 2012, **30**(7), 631–637.
276. Scott, A.M., J.P. Allison, and J.D. Wolchok, Monoclonal antibodies in cancer therapy. *Cancer Immun.*, 2012, **12**, 14.
277. Franco, G., Guarnotta C, Frossi B, Piccaluga PP, et al., Bone marrow stroma CD40 expression correlates with inflammatory mast cell infiltration and disease progression in splenic marginal zone lymphoma. *Blood*, 2014, **123**(12), 1836–1849.
278. Vallera, D.A., Chen H, Sicheneder AR, Panoskaltsis-Mortari A, et al., Genetic alteration of a bispecific ligand-directed toxin targeting human CD19 and CD22 receptors resulting in improved efficacy against systemic B cell malignancy. *Leuk. Res.*, 2009, **33**(9), 1233–1242.
279. Robinson, M.K., Hodge KM, Horak E, Sundberg AL, et al., Targeting ErbB2 and ErbB3 with a bispecific single-chain Fv enhances targeting selectivity and induces a therapeutic effect in vitro. *Br. J. Cancer*, 2008, **99**(9), 1415–1425.
280. Kontermann, R.E., Dual targeting strategies with bispecific antibodies. *MAbs*, 2012, **4**(2), 182–197.
281. Baeuerle, P.A., P. Kufer, and R. Bargou, BiTE: teaching antibodies to engage T-cells for cancer therapy. *Curr. Opin. Mol. Ther.*, 2009, **11**(1), 22–30.
282. Baeuerle, P.A. and C. Reinhardt, Bispecific T-cell engaging antibodies for cancer therapy. *Cancer Res.*, 2009, **69**(12), 4941–4944.
283. Haas, C., Krinner E, Brischwein K, Hoffmann P, et al., Mode of cytotoxic action of T-cell-engaging BiTE antibody MT110. *Immunobiology*, 2009, **214**(6), 441–453.
284. Derer, S., Beurskens FJ, Rosner T, Peipp M, et al., Complement in antibody-based tumor therapy. *Crit. Rev. Immunol.*, 2014, **34**(3), 199–214.
285. Hansel, T.T., Kropshofer H, Singer T, Mitchell JA, et al., The safety and side effects of monoclonal antibodies. *Nat. Rev. Drug Discov.*, 2010, **9**(4), 325–338.
286. Weiner, L.M., M.V. Dhodapkar, and S. Ferrone, Monoclonal antibodies for cancer immunotherapy. *Lancet*, 2009, **373**(9668), 1033–1040.
287. Ascierto, P.A. and F.M. Marincola, What have we learned from cancer immunotherapy in the last 3 years? *J. Transl. Med.*, 2014, **12**(1), 141.
288. Janikashvili, N., N. Larmonier, and E. Katsanis, Personalized dendritic cell-based tumor immunotherapy. *Immunotherapy*, 2010, **2**(1), 57–68.
289. Badoual, C., Bastier, PL, Roussel, H, Mandavit, M, et al., An allogeneic NK cell line engineered to express chimeric antigen receptors: a novel strategy of cellular immunotherapy against cancer. *Oncoimmunology*, 2013, **2**(11), e27156.
290. Batchu, R.B., O.V Gruzdyn, S.T Kung, D.W. Weaver, et al, Dendritic cell-based immunotherapy of cancer with cell penetrating domains. Indian *J. Surg. Oncol.*, 2014, **5**(1), 3–4.
291. Yano, Y., Ueda Y, Itoh T, Fuji N, et al., A new strategy using autologous dendritic cells and lymphokine-activated killer cells for cancer immunotherapy: efficient maturation of DCs by co-culture with LAK cells *in vitro*. *Oncol. Rep.*, 2006, **16**(1), 147–152.
292. Petrovsky, N. and J.C. Aguilar, Vaccine adjuvants: current state and future trends. *Immunol. Cell Biol.*, 2004, **82**(5), 488–496.
293. Madan, R.A. and J.L. Gulley, Sipuleucel-T: harbinger of a new age of therapeutics for prostate cancer. *Expert Rev. Vaccines*, 2011, **10**(2), 141–150.
294. Di Lorenzo, G., C. Buonerba, and P.W. Kantoff, Immunotherapy for the treatment of prostate cancer. *Nat. Rev. Clin. Oncol.*, 2011, **8**(9), 551–561.
295. Zigler, M., A. Shir, and A. Levitzki, Targeted cancer immunotherapy. *Curr. Opin. Pharmacol.*, 2013, **13**(4), 504–510.

296. Vanneman, M. and G. Dranoff, Combining immunotherapy and targeted therapies in cancer treatment. *Nat. Rev. Cancer*, 2012, **12**(4), 237–251.
297. Hodi, F.S., Butler M, Oble DA, Seiden MV, et al., Immunologic and clinical effects of antibody blockade of cytotoxic T lymphocyte-associated antigen 4 in previously vaccinated cancer patients. *Proc. Natl Acad. Sci. USA*, 2008, **105**(8), 3005–3010.
298. Li, B., VanRoey M, Wang C, Chen TH, et al., Anti-programmed death-1 synergizes with granulocyte macrophage colony-stimulating factor-secreting tumor cell immunotherapy providing therapeutic benefit to mice with established tumors. *Clin. Cancer Res.*, 2009, **15**(5), 1623–1634.
299. Vermeulen, C.F., Jordanova ES, ter Haar NT, Kolkman-Uljee SM, et al., Expression and genetic analysis of transporter associated with antigen processing in cervical carcinoma. *Gynecol. Oncol.*, 2007, **105**(3), 593–599.
300. Blum, J.S., P.A. Wearsch, and P. Cresswell, Pathways of antigen processing. *Annu. Rev. Immunol.*, 2013, **31**, 443–473.
301. Zhang, Q.J., Seipp RP, Chen SS, Vitalis TZ, et al., TAP expression reduces IL-10 expressing tumor infiltrating lymphocytes and restores immunosurveillance against melanoma. *Int. J. Cancer*, 2007, **120**(9), 1935–1941.
302. Zettl, U.K. and E. Mix, Potentials and limitations of current and emerging immunotherapies. *J. Neurol.*, 2008, **255** (Suppl. 6), 1.
303. Frigault, M.J. and C.H. June, Predicting cytokine storms: it's about density. *Blood*, 2011, **118**(26), 6724–6726.
304. Couzin-Frankel, J., Breakthrough of the year 2013. Cancer immunotherapy. *Science*, 2013, **342**(6165), 1432–1433.
305. Wirth, T., N. Parker, and S. Yla-Herttuala, History of gene therapy. *Gene*, 2013, **525**(2), 162–169.
306. Kaufmann, K.B., Büning H, Galy A, Schambach A, et al., Gene therapy on the move. *EMBO Mol. Med.*, 2013, **5**(11), 1642–1661.
307. Aiuti, A., Cattaneo F, Galimberti S, Benninghoff U, et al., Gene therapy for immunodeficiency due to adenosine deaminase deficiency. *N. Engl. J. Med.*, 2009, **360**(5), 447–458.
308. Hsu, P.D., E.S. Lander, and F. Zhang, Development and applications of CRISPR-Cas9 for genome engineering. *Cell*, 2014, **157**(6), 1262–1278.
309. High, K., P.D. Gregory, and C. Gersbach, CRISPR technology for gene therapy. *Nat. Med.*, 2014, **20**(5), 476–477.
310. Jinek, M., Chylinski K, Fonfara I, Hauer M, et al., A programmable dual-RNA-guided DNA endonuclease in adaptive bacterial immunity. *Science*, 2012, **337**(6096), 816–821.
311. Yin, H., Xue W, Chen S, Bogorad RL, et al., Genome editing with Cas9 in adult mice corrects a disease mutation and phenotype. *Nat. Biotechnol.*, 2014, **32**(6), 551–553.
312. Saraswat Pushpendra, P.A., Bhandari A., Nucleic Acids as Therapeutics, in: Erdmann, V.A., Barciszewski, J. (Eds), *From Nucleic Acids Sequences to Molecular Medicine*, 2012, *Springer*, Berlin, Heidelberg.
313. Motavaf, M., S. Safari, and S.M. Alavian, Therapeutic potential of RNA interference: a new molecular approach to antiviral treatment for hepatitis C. *J. Viral Hepat.*, 2012, **19**(11), 757–765.
314. Wu, S.Y., Lopez-Berestein G, Calin GA, Sood AK., RNAi therapies: drugging the undruggable. *Sci. Transl. Med.*, 2014, **6**(240), 240ps7.
315. Is this really the RNAissance? *Nat. Biotechnol.*, 2014, **32**(3), 201. Available at: http://www.nature.com/nbt/journal/v32/n3/full/nbt.2853.html.
316. Tabernero, J., Shapiro GI, LoRusso PM, Cervantes A, et al., First-in-humans trial of an RNA interference therapeutic targeting VEGF and KSP in cancer patients with liver involvement. *Cancer Discovery*, 2013, **3**(4), 406–417.
317. Golan, T., Hubert A, Shemi A, Segal A, et al., A phase I trial of a local delivery of siRNA against k-ras in combination with chemotherapy for locally advanced pancreatic adenocarcinoma. *J. Clin. Oncol.*, 2013, **31**(15), 4037.
318. Davis, M.E., Zuckerman JE, Choi CH, Seligson D, et al., Evidence of RNAi in humans from systemically administered siRNA via targeted nanoparticles. *Nature*, 2010, **464**(7291), 1067–1070.
319. Senzer, N.N., Kaufman HL, Amatruda T, Nemunaitis M, et al., Phase II clinical trial of a granulocyte-macrophage

colony-stimulating factor-encoding, second-generation oncolytic herpesvirus in patients with unresectable metastatic melanoma. *J. Clin. Oncol.*, 2009, **27**(34), 5763–5771.
320. Krall, N., Pretto, F., Decurtins, W., Bernardes, G.J. L., *et al.*, A small-molecule drug conjugate for the treatment of carbonic anhydrase IX-expressing tumors. *Angew. Chem. Int. Ed.*, 2014, **53**(16), 4231–4235.
321. Proia, D.A. and R.C. Bates, Ganetespib and HSP90: translating preclinical hypotheses into clinical promise. *Cancer Res.*, 2014, **74**(5), 1294–1300.
322. Pu, Y., Wang WB, Achilefu S, Alfano RR, Study of rotational dynamics of receptor-targeted contrast agents in cancerous and normal prostate tissues using time-resolved picosecond emission spectroscopy. *Appl. Opt.*, 2011, **50**(10), 1312–1322.
323. Xue, J., Shan L, Chen H, Li Y, *et al.*, Visual detection of STAT5B gene expression in living cell using the hairpin DNA modified gold nanoparticle beacon. *Biosens. Bioelectron.*, 2013, **41**, 71–77.
324. Xing, Y., Smith AM, Agrawal A, Ruan G, *et al.*, Molecular profiling of single cancer cells and clinical tissue specimens with semiconductor quantum dots. *Int. J. Nanomedicine*, 2006, **1**(4), 473–481.
325. Urano, Y., Sakabe M, Kosaka N, Ogawa M, *et al.*, Rapid cancer detection by topically spraying a gamma-glutamyltranspeptidase-activated fluorescent probe. *Sci. Transl. Med.*, 2011, **3**(110), 110ra119.
326. Chabner, B.A., L.W. Ellisen, and A.J. Iafrate, Personalized medicine: hype or reality. *Oncologist*, 2013, **18**(6), 640–643.
327. Chisholm, R.L., The opportunities and challenges of implementing genomics-informed personalized medicine. *Clin. Pharmacol. Ther.*, 2013, **94**(2), 179–180.
328. Le Tourneau, C., Paoletti X, Servant N, Bièche I, *et al.*, Randomised proof-of-concept phase II trial comparing targeted therapy based on tumour molecular profiling vs conventional therapy in patients with refractory cancer: results of the feasibility part of the SHIVA trial. *Br. J. Cancer*, 2014, **111**(1), 17–24.
329. Kim, E.S., Herbst RS, Wistuba II, Lee JJ, *et al.*, The BATTLE trial: personalizing therapy for lung cancer. *Cancer Discovery*, 2011, **1**(1), 44–53.
330. Tsimberidou, A.M., Iskander NG, Hong DS, Wheler JJ, *et al.*, Personalized medicine in a phase I clinical trials program: the MD Anderson Cancer Center initiative. *Clin. Cancer Res.*, 2012, **18**(22), 6373–6383.
331. Tsimberidou, A.M., Wen, S., Hong, DS, Wheler, J.J., *et al.*, Personalized medicine for patients with advanced cancer in the Phase I program at MD Anderson: validation and landmark analyses. *Clin. Cancer Res.*, 2014, **20**, 4827–4836.
332. Bozic, I., Reiter JG, Allen B, Antal T, Evolutionary dynamics of cancer in response to targeted combination therapy. *Elife*, 2013, **2**, e00747.
333. Almendro, V., A. Marusyk, and K. Polyak, Cellular heterogeneity and molecular evolution in cancer. *Annu. Rev. Pathol.: Mech. Dis.*, **8**, 2013, 277–302.
334. Aparicio, S. and C. Caldas, The implications of clonal genome evolution for cancer medicine. *N. Engl. J. Med.*, 2013, **368**(9), 842–851.
335. Kandoth, C., McLellan MD, Vandin F, Ye K, *et al.*, Mutational landscape and significance across 12 major cancer types. *Nature*, 2013, **502**(7471), 333–339.
336. Balic, M., Williams A, Lin H, Datar R, *et al.*, Circulating tumor cells: from bench to bedside. *Annu. Rev. Med.*, 2013, **64**, 31–44.
337. Diaz, L.A., Jr, and A. Bardelli, Liquid biopsies: genotyping circulating tumor DNA. *J. Clin. Oncol.*, 2014, **32**(6), 579–586.
338. Willison, K.R. and D.R. Klug, Quantitative single cell and single molecule proteomics for clinical studies. *Curr. Opin. Biotechnol.*, 2013, **24**(4), 745–751.
339. Eberwine, J., Sul JY, Bartfai T, Kim J., The promise of single-cell sequencing. *Nat. Methods*, 2014, **11**(1), 25–27.
340. Yu, J., Zhou J, Sutherland A, Wei W, *et al.*, Microfluidics-based single-cell functional proteomics for fundamental and applied biomedical applications. *Annu. Rev. Anal. Chem. (Palo Alto)*, 2014, **7**, 275–295.
341. Krebs, M.G., Metcalf RL, Carter L, Brady G, *et al.*, Molecular analysis of circulating tumour cells-biology and biomarkers. *Nat. Rev. Clin. Oncol.*, 2014, **11**(3), 129–144.
342. Plaks, V., C.D. Koopman, and Z. Werb, Cancer. Circulating tumor cells. *Science*, 2013, **341**(6151), 1186–1188.
343. Smerage, J.B., Barlow WE, Hortobagyi GN, Winer EP, *et al.*, Circulating tumor cells

and response to chemotherapy in metastatic breast cancer: SWOG S0500. *J. Clin. Oncol.*, 2014, **32**(31), 3483–3489.
344. Muller, V., Riethdorf S., Rack B., Janni W., *et al.*, Prognostic impact of circulating tumor cells assessed with the CellSearch System™ and AdnaTest Breast™ in metastatic breast cancer patients: the DETECT study. *Breast Cancer Res.*, 2012, **14**(4), R118.
345. Adalsteinsson, V.A. and J.C. Love, Towards engineered processes for sequencing-based analysis of single circulating tumor cells. *Curr. Opin. Chem. Eng.*, 2014, **4**, 97–104.
346. Macaulay, I.C. and T. Voet, Single cell genomics: advances and future perspectives. *PLoS Genet.*, 2014, **10**(1), e1004126.
347. Streets, A.M., Zhang X, Cao C, Pang Y, *et al.*, Microfluidic single-cell whole-transcriptome sequencing. *Proc. Natl Acad. Sci. USA*, 2014, **111**(19), 7048–7053.
348. Wu, A.R., Neff NF, Kalisky T, Dalerba P, *et al.*, Quantitative assessment of single-cell RNA-sequencing methods. *Nat. Methods*, 2014, **11**(1), 41–46.
349. Heitzer, E., Auer M, Gasch C, Pichler M, *et al.*, Complex tumor genomes inferred from single circulating tumor cells by array-CGH and next-generation sequencing. *Cancer Res.*, 2013, **73**(10), 2965–2975.
350. Navin, N., Kendall J, Troge J, Andrews P, *et al.*, Tumour evolution inferred by single-cell sequencing. *Nature*, 2011, **472**(7341), 90–94.
351. Murtaza, M., Dawson SJ, Tsui DW, Gale D, *et al.*, Non-invasive analysis of acquired resistance to cancer therapy by sequencing of plasma DNA. *Nature*, 2013, **497**(7447), 108–112.
352. Dawson, S.J., Tsui DW, Murtaza M, Biggs H, *et al.*, Analysis of circulating tumor DNA to monitor metastatic breast cancer. *N. Engl. J. Med.*, 2013, **368**(13), 1199–1209.
353. De Vlaminck, I., Valantine HA, Snyder TM, Strehl C, *et al.*, Circulating cell-free DNA enables noninvasive diagnosis of heart transplant rejection. *Sci. Transl. Med.*, 2014, **6**(241), 241ra77.
354. Al-Lazikani, B., U. Banerji, and P. Workman, Combinatorial drug therapy for cancer in the post-genomic era. *Nat. Biotechnol.*, 2012, **30**(7), 679–692.
355. Bagnyukova, T., Serebriiskii IG, Zhou Y, Hopper-Borge EA, *et al.*, Chemotherapy and signaling how can targeted therapies supercharge cytotoxic agents? *Cancer Biol. Ther.*, 2010, **10**(9), 843–857.
356. Geyer, C.E., Forster J, Lindquist D, Chan S, *et al.*, Lapatinib plus capecitabine for HER2-positive advanced breast cancer. *N. Engl. J. Med.*, 2006, **355**(26), 2733–2743.
357. Akimoto, T., Hunter NR, Buchmiller L, Mason K, *et al.*, Inverse relationship between epidermal growth factor receptor expression and radiocurability of murine carcinomas. *Clin. Cancer Res.*, 1999, **5**(10), 2884–2890.
358. Welsh, J.W., Komaki R, Amini A, Munsell MF, *et al.*, Phase II trial of erlotinib plus concurrent whole-brain radiation therapy for patients with brain metastases from non-small-cell lung cancer. *J. Clin. Oncol.*, 2013, **31**(7), 895–902.
359. Mole, R.H., Whole body irradiation; radiobiology or medicine? *Br. J. Radiol.*, 1953, **26**(305), 234–241.
360. Durante, M., N. Reppingen, and K.D. Held, Immunologically augmented cancer treatment using modern radiotherapy. *Trends Mol. Med.*, 2013, **19**(9), 565–582.
361. Kalbasi, A., June CH, Haas N, Vapiwala N., Radiation and immunotherapy: a synergistic combination. *J. Clin. Invest.*, 2013, **123**(7), 2756–2763.
362. Krysko, D.V., Garg AD, Kaczmarek A, Krysko O, *et al.*, Immunogenic cell death and DAMPs in cancer therapy. *Nat. Rev. Cancer*, 2012, **12**(12), 860–875.
363. Nijman, S.M.B., Synthetic lethality: general principles, utility and detection using genetic screens in human cells. *FEBS Lett.*, 2011, **585**(1), 1–6.
364. Astsaturov, I., Ratushny V, Sukhanova A, Einarson MB, *et al.*, Synthetic lethal screen of an EGFR-centered network to improve targeted therapies. *Sci. Signal.*, 2010, **3**(140), ra67.
365. Hoffman, G.R., Rahal, R., Buxton, F., Xiang, K., *et al.*, Functional epigenetics approach identifies BRM/SMARCA2 as a critical synthetic lethal target in BRG1-deficient cancers. *Proc. Natl Acad. Sci. USA*, 2014, **111**(8), 3128–3133.
366. Bassik, M.C., Kampmann, M., Lebbink, R.J., Wang, S., *et al.*, A systematic mammalian genetic interaction map reveals pathways underlying ricin susceptibility. *Cell*, 2013, **152**(4), 909–922.

367. Koike-Yusa, H., Li, Y., Tan, E-P, Velasco-Herrera, M.D.C., *et al.*, Genome-wide recessive genetic screening in mammalian cells with a lentiviral CRISPR-guide RNA library. *Nat. Biotechnol.*, 2014, **32**(3), 267–273.
368. Luo, J., Emanuele, M.J., Li, D., Creighton, C.J., *et al.*, A genome-wide RNAi screen identifies multiple synthetic lethal interactions with the ras oncogene. *Cell*, 2009, **137**(5), 835–848.
369. Matheny, C.J., Wei, M.C., Bassik, M.C., Donnelly, A.J., *et al.*, Next-generation NAMPT inhibitors identified by sequential high-throughput phenotypic chemical and functional genomic screens. *Chem. Biol.*, 2013, **20**(11), 1352–1363.
370. Chan, D.A. and A.J. Giaccia, Harnessing synthetic lethal interactions in anticancer drug discovery. *Nat. Rev. Drug Discovery*, 2011, **10**(5), 351–364.
371. Nijman, S.M.B. and S.H. Friend, Potential of the synthetic lethality principle. *Science*, 2013, **342**(6160), 809–811.
372. Lord, C.J. and A. Ashworth, The DNA damage response and cancer therapy. *Nature*, 2012, **481**(7381), 287–294.
373. Polyak, K. and J. Garber, Targeting the missing links for cancer therapy. *Nat. Med.*, 2011, **17**(3), 283–284.
374. Servant, N., Roméjon J, Gestraud P, La Rosa P, *et al.*, Bioinformatics for precision medicine in oncology: principles and application to the SHIVA clinical trial. *Front. Genet.*, 2014, **5**, 152.
375. Gehlenborg, N., O'Donoghue SI, Baliga NS, Goesmann A, *et al.*, Visualization of omics data for systems biology. *Nat. Methods*, 2010, **7**(3), S56–S68.
376. Chen, Y.B., Chattopadhyay A, Bergen P, Gadd C, *et al.*, The online bioinformatics resources collection at the University of Pittsburgh Health Sciences Library System – a one-stop gateway to online bioinformatics databases and software tools. *Nucleic Acids Res.*, 2007, **35**, D780–D785.
377. Artimo, P., Jonnalagedda M, Arnold K, Baratin D, *et al.*, ExPASy: SIB bioinformatics resource portal. *Nucleic Acids Res.*, 2012, **40**(W1), W597–W603.
378. Scannell, J.W., Blanckley A, Boldon H, Warrington B., Diagnosing the decline in pharmaceutical R&D efficiency. *Nat. Rev. Drug Discov.*, 2012, **11**(3), 191–200.
379. Anighoro, A., J. Bajorath, and G. Rastelli, Polypharmacology: challenges and opportunities in drug discovery. *J. Med. Chem.*, 2014, **57**(19), 7874–7887.

**Translational Medicine
Molecular Pharmacology and
Drug Discovery**

*Edited by
Robert A. Meyers*

Translational Medicine

Molecular Pharmacology and Drug Discovery

Edited by
Robert A. Meyers

Volume 2

Verlag GmbH & Co. KGaA

The Editor

Dr. Robert A. Meyers
Editor-in-Chief
Ramtech Limited
122, Escalle Lane
Larkspur, CA 94939
USA

Cover

Cover picture based on an abstracted version of figure 5 from chapter 10 "RNA Interference to Treat Virus Infections" by Karim Majzoub and Jean-Luc Imler.

All books published by **Wiley-VCH** are carefully produced. Nevertheless, authors, editors, and publisher do not warrant the information contained in these books, including this book, to be free of errors. Readers are advised to keep in mind that statements, data, illustrations, procedural details or other items may inadvertently be inaccurate.

Library of Congress Card No.: applied for

British Library Cataloguing-in-Publication Data
A catalogue record for this book is available from the British Library.

Bibliographic information published by the Deutsche Nationalbibliothek
The Deutsche Nationalbibliothek lists this publication in the Deutsche Nationalbibliografie; detailed bibliographic data are available on the Internet at <http://dnb.d-nb.de>.

© 2018 Wiley-VCH Verlag GmbH & Co. KGaA, Boschstr. 12, 69469 Weinheim, Germany

All rights reserved (including those of translation into other languages). No part of this book may be reproduced in any form – by photoprinting, microfilm, or any other means – nor transmitted or translated into a machine language without written permission from the publishers. Registered names, trademarks, etc. used in this book, even when not specifically marked as such, are not to be considered unprotected by law.

Print ISBN: 978-3-527-33659-3
ePDF ISBN: 978-3-527-68719-0
ePub ISBN: 978-3-527-68721-3
Mobi ISBN: 978-3-527-68722-0

Cover Design Adam Design, Weinheim, Germany

Typesetting SPi Global, Chennai, India

Printing and Binding C.O.S. Printers Pte Ltd, Singapore

Printed on acid-free paper

10 9 8 7 6 5 4 3 2 1

Contents

Preface   IX

Volume 1

Part I   Biopharmaceuticals   1

1   **Analogs and Antagonists of Male Sex hormones**   3
    Robert W. Brueggemeier

2   **Annexins**   77
    Carl E. Creutz

3   **Genetic Engineering of Antibody Molecules**   85
    Cristian J. Payés, Tracy R. Daniels-Wells, Paulo C. Maffía, Manuel L. Penichet,
    Sherie L. Morrison, and Gustavo Helguera

4   **Growing Mini-Organs from Stem Cells**   137
    Hiroyuki Koike, Tamir Rashid, and Takanori Takebe

5   **Hemoglobin**   159
    Maurizio Brunori and Adriana Erica Miele

6   **Immune Checkpoint Inhibitors**   199
    Laura Mansi, Franck Pagès, and Olivier Adotévi

7   **Molecular Mediators: Cytokines**   229
    Jean-Marc Cavaillon

8   **Neural Transplantation: Evidence from the Rodent Cerebellum**   267
    Ketty Leto and Ferdinando Rossi

9   **RNA Interference in Cancer Therapy**   291
    Barbara Pasculli and George A. Calin

10	RNA Interference to Treat Virus Infections Karim Majzoub and Jean-Luc Imler	345
11	Stem Cell Therapy for Alzheimer's Disease Rahasson R. Ager and Frank M. LaFerla	383
12	Immunotherapy with Autologous Cells Andrew D. Fesnak and Bruce L. Levine	411
13	Targeted Therapy: Genomic Approaches Tim N. Beck, Linara Gabitova, and Ilya G. Serebriiskii	439

Volume 2

Part II	Drug Discovery Methods and Approaches	493
14	Pharmaceutical Process Chemistry Michael T. Williams	495
15	High-Performance Liquid Chromatography of Peptides and Proteins Reinhard I. Boysen	531
16	Hit-to-Lead Medicinal Chemistry Simon E. Ward and Paul Beswick	575
17	Mass Spectrometry-Based Methods of Proteome Analysis Mihir Jaiswal, Michael P. Washburn, and Boris L. Zybailov	595
18	Natural Products Based Drug Discovery Shoaib Ahmad	649
19	Neurological Biomarkers Henrik Zetterberg	683
20	Pharmacokinetics of Peptides and Proteins Chetan Rathi and Bernd Meibohm	697
21	Physical Pharmacy and Biopharmaceutics M. Sherry Ku	725
22	Prions Vincent Béringue	773
23	RNA Metabolism and Drug Design Eriks Rozners	827
24	Structure-Aided Drug Design and NMR-based Screening Lee Quill, Michael Overduin, and Mark Jeeves	871
25	Tuberculosis Drug Development Kingsley N. Ukwaja	899

Part III Nanomedicine   941

26 Microfluidics in Nanomedicine   943
YongTae Kim and Robert Langer

27 Nanoparticle Conjugates for Small Interfering RNA Delivery   969
Timothy L. Sita and Alexander H. Stegh

28 Quantum Dots for Biomedical Delivery Applications   995
Abolfazl Akbarzadeh, Sedigheh Fekri Aval, Roghayeh Sheervalilou, Leila Fekri, Nosratollah Zarghami, and Mozhdeh Mohammadian

Index   1009

Preface

The approach we pursued in this compendium is to provide the latest insights into cutting edge methodology, approaches and results in the molecular and cellular basis of human disease as well as the discovery, evaluation, formulation and production of new drugs across the widest possible range of diseases. Two important factors are increasingly recognized in the field of translational medicine: i) despite considerable progress in the field, we still don't know enough about the mechanisms of many if not all of the critical diseases and conditions (Alzheimer's is an example), and ii) the new drug pipeline is in a shrinking mode with a concomitant decrease in R&D productivity. Our compendium aimed at capturing the *state-of-the-art* in the field and is designed to offer both answers and pathways to drug discovery and testing.

The *Biopharmaceuticals* section covers the latest developments and experimental approaches in the field from biologics extracted from living systems (*e.g.,* hormones, annexins, hemoglobin, cytokines), recombinant protein and stem cell therapies (*e.g.,* neural transplantation), as well as RNA interference in cancer therapy (nanodelivery of microRNAs), to growing organs from stem cells utilizing *in vitro* approaches to model human organogenesis producing self-organizing three-dimensional (3D) tissues so-called organoids or organ buds. This section also includes immunotherapy with autologous cells and targeted genomic approaches. Biologics approaches as applied to specific diseases such as cancer (immune checkpoint inhibitors and RNA interference), viral infections, atherosclerosis and malaria and Alzheimer's are covered in detail.

The *Drug Discovery* section provides the latest on advanced methodologies for drug discovery from cutting edge analytical techniques, *e.g.,* multidimensional HPLC and also Mass Spectrometry-Based Methods of Proteome Analysis), to drug discovery methodology, *e.g.* Hit-to-Lead and Structure-Aided Drug Design and NMR spectroscopy-based Screening as well as Neurological Biomarkers. A number of articles in this section are directed to drugs for specific diseases such as the prion family of diseases, cancer, tuberculosis, Parkinson's, Huntington's, schizophrenia, frontotemporal dementia, and Alzheimer's. *Natural Products Based Discovery* covers extraction processes from plants, animals, bacteria and associated "smart screening" methods, robotic separation with structural analysis, metabolic engineering, and synthetic biology. This section is completed with chapters on preformulation, biopharmaceutics, drug absorption, nanotechnology, pharmacokinetics and drug delivery systems design and performance including targeted drug delivery—application of physical chemistry principles to the area of pharmacy in the design of drug molecules and drug products.

The emerging field of nanomedicine is the medical application of nanotechnology for the treatment and prevention of major ailments. The *Nanomedicine* section concerns the preparation of nanomaterials based drug delivery systems. These are nanoparticles *"which have been formulated using a variety of materials that includes lipids, polymers, inorganic nanocrystals, carbon nanotubes, proteins, and DNA origami. The ultimate goal of nanomedicine is to achieve a robust, targeted delivery of complex assemblies that contain sufficient amounts of multiple therapeutic and diagnostic agents for highly localized drug release, but with no adverse side effects and a reliable detection of any site-specific therapeutic response"*—Kim and Langer (authors of our article on Microfluidics in Nanomedicine). The Microfluidics article provides an overview of highly compatible platforms to create new nanomedicine development pipelines that include the required methodologies. Importantly, microfluidics presents a number of useful capabilities to manipulate very small quantities of samples, and to detect substances with a high resolution for a wide range of applications. Then there are articles on two exciting cutting edge nanomedicine approaches *Nanoparticle Conjugates for Small Interfering RNA Delivery* and *Quantum Dots for Biomedical Delivery Applications*, where these therapeutic approaches are discussed for cancer, hereditary disorders, heart disease, inflammatory conditions, and viral infections. In addition, quantum dots probes accumulate at tumors, due both to an enhanced permeability and retention at tumor sites, and also by the binding of antibodies to cancer-specific cell-surface biomarkers.

Larkspur, California, October 2017

Robert A. Meyers
Editor-in Chief
RAMTECH LIMITED

Part II
Drug Discovery Methods and Approaches

14
Pharmaceutical Process Chemistry

Michael T. Williams
Chemistry, Manufacturing, and Controls Consultant, 133, London Road, Deal, Kent CT14 9TY, UK

1	**Setting the Scene** 497	
2	**Historical Perspective** 497	
3	**What Oral Drugs Look Like** 499	
4	**Early Development Activities** 501	
5	**Late Development Activities** 502	
5.1	Route Selection 503	
5.2	Reagent and Solvent Selection 504	
5.3	Reaction Development and Optimization 506	
6	**Reaction Process Safety** 508	
6.1	Background 508	
6.2	Screening for Early Scale-Up 509	
6.3	Larger-Scale Hazard Assessment 509	
6.4	Large-Scale Production 511	
7	**The Analytical Interface** 512	
8	**Green Chemistry** 514	
8.1	Introduction 514	
8.2	Green Chemistry Metrics 515	
8.3	Green Solvent Selection 516	
8.4	Green Chemistry Case Histories 517	
9	**To Market** 519	
9.1	Technology Transfer and Process Validation 519	
9.2	Regulatory Filing 522	

Translational Medicine: Molecular Pharmacology and Drug Discovery
First Edition. Edited by Robert A. Meyers.
© 2018 Wiley-VCH Verlag GmbH & Co. KGaA. Published 2018 by Wiley-VCH Verlag GmbH & Co. KGaA.

| 10 | Future Challenges | 524 |
| | References | 526 |

Keywords

New chemical entity (NCE)
A drug candidate molecule undergoing development, with the aim of having the active moiety approved by regulatory agencies for the treatment of a disease.

Process safety
The design of safe chemical reactions, and the identification, understanding, and control of the inherent hazards involved in the scaling-up of synthetic processes.

Green chemistry
The design of chemical products and processes that reduce or eliminate the use or generation of environmentally hazardous substances.

Critical process parameters (CPPs)
The key variables which affect the manufacturing process and the quality of the product, and hence need to be controlled.

Process analytical technology (PAT)
A range of techniques used to provide real-time analytical information about reactions (often with the aid of chemometric techniques), for increased process understanding and control.

Quality by design (QbD)
Pharmaceutical QbD is focused on the concept that quality should be built into the product with an understanding of the process by which it is developed and manufactured, together with knowledge of the associated risks and their mitigating factors.

> Pharmaceutical process chemistry is a mature discipline, but one that is rarely taught in academic programs. There is therefore relatively little appreciation of the complex, multidimensional science that transforms a sequence of reactions into a viable manufacturing process. In order to design, select and develop safe, economically viable and scalable routes to small-molecule drugs, the process chemist must work closely with chemical engineers, analysts and material scientists, amongst others. In the course of this endeavor, for drugs that successfully reach the market, three disparate customers will be served: bulk drug substance will be produced to fuel the development program; a robust manufacturing process will be developed for the production organization; and evidence will be presented to the regulatory agencies that a drug substance of assured quality is reproducibly produced.

1 Setting the Scene

The discovery of a potential drug is the first step in a long development pathway, with the vast majority of drug candidates falling by the wayside (see Fig. 1) [1]. The cost of discovering and developing each new chemical entity (NCE) that does reach the market has been estimated to be in the range of US$500–2000 million, depending on the therapeutic area and developing company [2–4]. During the development process the demands for material increase, as drug substance is required not only for toxicology studies and clinical trials but also for analytical, stability, formulation, and a range of other studies. Whilst the total amount of drug substance required depends on its activity, effect, and physical properties, typical requirements are 1–5 kg at Phase I, 50–100 kg at Phase II, and up to 1000 kg during Phase III and registration activities. This chapter examines the range of tasks carried out by process chemists during the course of developing an NCE, and will illustrate how the high cost and attrition during the overall development process influences both the way and the rate in which process chemistry effort is applied during the program.

The remit of the process chemist as the program proceeds, is to discover and scale-up new synthetic processes from the laboratory, through to the pilot plant and ultimately into full-scale commercial manufacture, if the candidate succeeds. In the course of this effort, the pharmaceutical process R&D scientist serves three disparate customers: (i) providing escalating quantities of suitably pure bulk active pharmaceutical ingredient (API) to fuel the development process; (ii) developing a robust manufacturing process for the production organization; and (iii) contributing to the chemistry, manufacturing, and controls (CMC) section of the filing to the regulatory agencies.

Developing routes and processes to safely, affordably, and sustainably produce APIs is a complex, challenging and exciting endeavor that crosses the boundaries between synthetic organic chemistry, analytical chemistry, process technology, and chemical engineering. The process chemist must interact with scientists from several other disciplines, and good communications and teamwork are crucial for success. For the reader seeking more detailed coverage of process chemistry in the pharmaceutical industry, together with illustrative case histories, there are a number of excellent text books available [5–7].

2 Historical Perspective

The pharmaceutical industry has its origins in Europe in the seventeenth and eighteenth centuries, with companies such as Merck in Germany, Geigy in Switzerland, and Allen and Hanbury's in England. These companies started out making and selling herbal or plant medicines, moving later into chemically manufactured drugs, while others (such as Hoechst, Bayer, and later ICI) had their origins in the dyestuffs and chemicals industry, coming later to pharmaceuticals as an extension of these businesses. Some examples of synthetic antipyretics and analgesics discovered in Germany and introduced to the market in the last quarter of the nineteenth century are shown in Fig. 2. As pharmaceutical final products could not be patented in Germany until 1891, it was particularly important to patent the process for making APIs, and German chemists were already

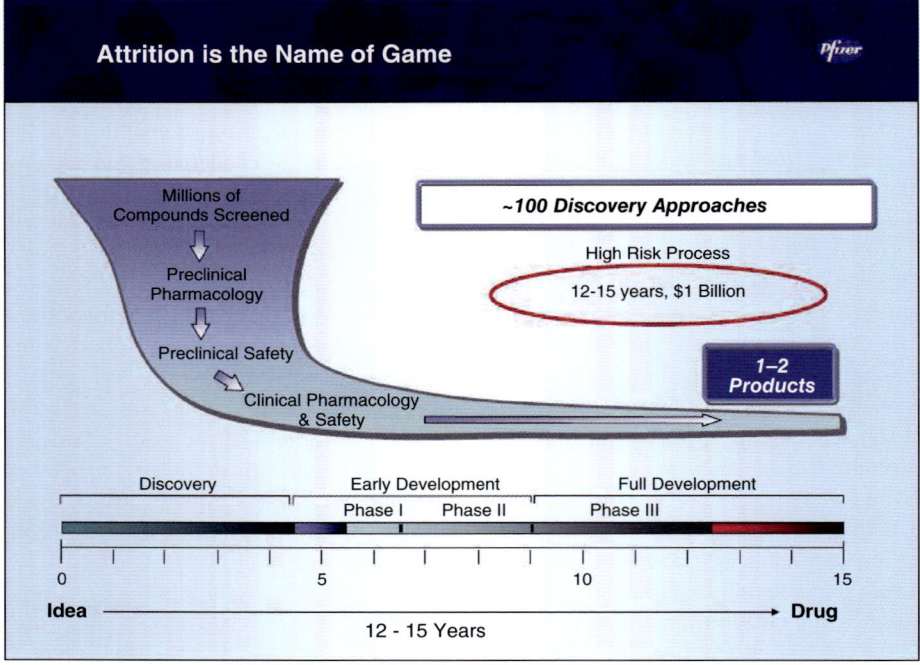

Fig. 1 Attrition and timescale of the R&D process. Reproduced with permission from Pfizer.

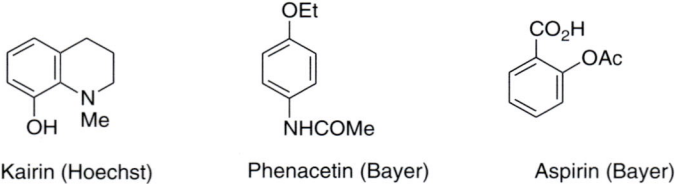

Kairin (Hoechst)   Phenacetin (Bayer)   Aspirin (Bayer)

Fig. 2 Examples of late nineteenth century pharmaceuticals.

expert by this time at process R&D and optimization.

Some early twentieth century milestones for the pharmaceutical industry include:

- The discovery and introduction of barbiturates in the first decade of the century.
- The collaboration with Toronto University that led to Eli Lilly's large-scale production and launch of insulin in the 1920s.
- The launch by Bayer in 1935 of prontosil (Fig. 3), the first general purpose antibiotic used in modern medicine; discovered by Gerhard Domagk in 1932, and leading to his 1939 Nobel Prize. The origin of this first sulfonamide from the dyestuffs industry is readily apparent.

The major growth in the industry in the USA, Western Europe and Japan occurred in the second half of the twentieth century, however, fueled by both increasing scientific knowledge and increasing healthcare requirements. As synthetic chemistry developed, more complex pharmaceutical

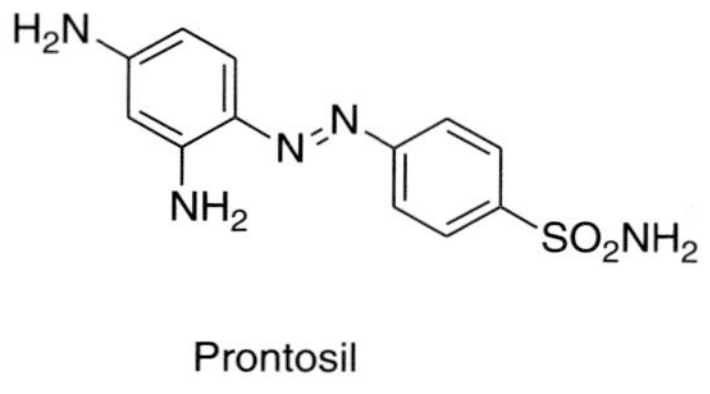

Prontosil

Fig. 3 Structure of prontosil.

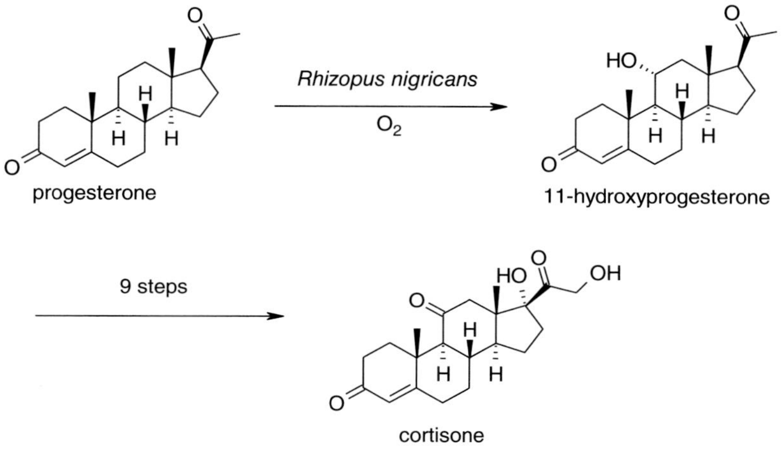

Scheme 1 The Upjohn cortisone synthesis.

products became available on a large scale, as exemplified by the 1952 Upjohn synthesis of cortisone [8] (depicted in Scheme 1). With its roots in natural products, the pharmaceutical industry has always developed complex chemical products. In the early days of the industry most of these complex materials were produced by extraction and purification, or by fermentation, rather than by the synthesis or semi-synthesis that is now used to produce approximately 80% of modern APIs. Over the past two decades the number of natural products (of plant or marine origin) launched has shrunk to less than one per year, while the number of biomolecules (vaccines, antibodies, and peptide- and nucleotide-based agents) has slowly risen and now represents almost 20% of the total. However, these macromolecules are expensive agents, with many biologics costing more than US$100 000 per patient-year of therapy. These biologic agents require parenteral delivery, and are thus often targeted for acute life-threatening conditions. Because of the molecular weight cut-off of around 550 Da for oral absorption, the majority of agents in clinical use will continue to be synthetically prepared small-molecular entities, which are therefore the focus of this chapter.

3
What Oral Drugs Look Like

As the majority of small-molecule APIs are intended for oral administration their oral

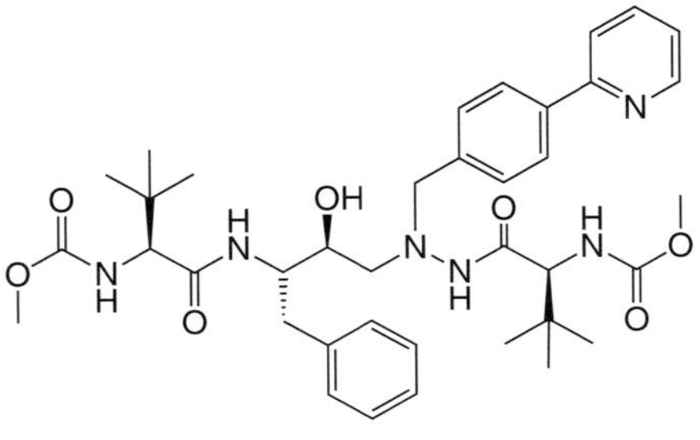

Atazanavir (HIV)

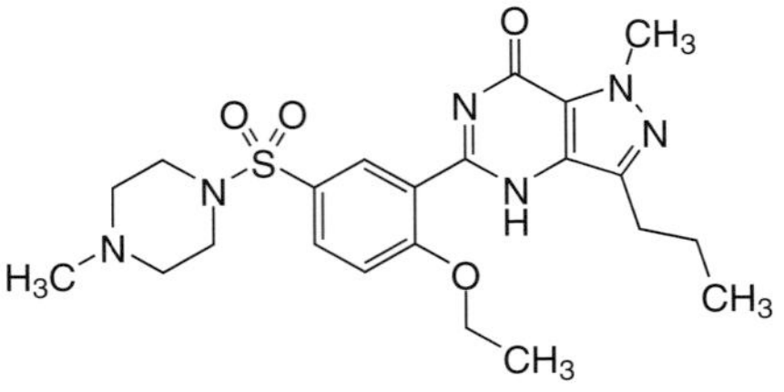

Sildenafil (erectile dysfunction)

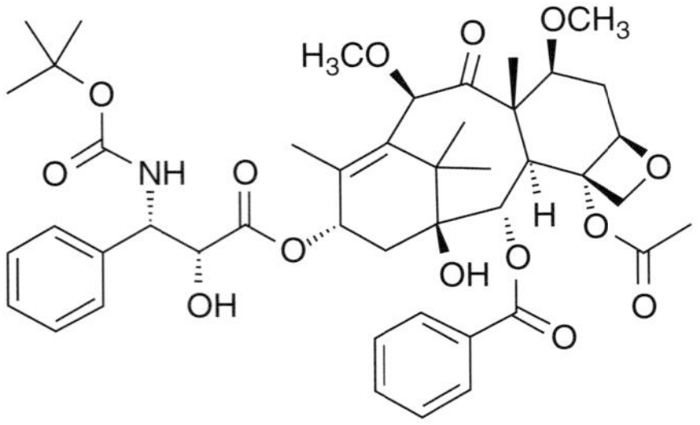

Carbazitaxel (prostate cancer)

Fig. 4 Examples of oral drugs.

bioavailability is key, and this is dependent upon the physico-chemical properties of the molecule. Lipinski articulated [9] the important structural features in his "Rule of Five," covering the agent's molecular weight, the number of H-bond donors and acceptors, and the logarithm of its partition coefficient (log P). The APIs encountered by the process chemist are therefore usually small molecules (<550 Da), generally contain both N and O atoms, and have a lipophilic region (usually an aromatic ring). The H-bond donors and acceptors, required for API substrate binding, are usually provided by groups such as alcohols, amines, amides, and sulfonamides. These restrictions nevertheless allow a tremendous variety of structural features, as is illustrated by the structures of the orally active agents atazanavir, sildenafil, and carbazitaxel (Fig. 4).

A recent survey of candidate drugs in the development pipelines of AstraZeneca, GlaxoSmithKline (GSK) and Pfizer [10] described the common features of APIs. For example, it showed that there are on average two aromatic or heteroaromatic rings per candidate drug, and that approximately 50% are chiral. The aromatic rings are frequently substituted, for example, with F, Cl or Br, in order to block metabolic pathways and or to modify physicochemical characteristics. Nearly all of the APIs are crystalline solids, and have the potential to form salts. With the wider availability of asymmetric synthetic methods [11], and the more stringent regulatory requirements, by the 1990s most chiral drugs had been developed in single-enantiomer forms [12].

4
Early Development Activities

Typical pharmaceutical industry success rates are around 60% during preclinical evaluation, 70% in Phase I, and 40% in Phase IIa clinical evaluation. Hence, any compound nominated for development has a less than one-in-five chance of achieving successful proof of concept (POC) in its clinical indication. Because of these early toxicological and clinical hurdles and high attrition rates, rapid and fit-for-purpose design and enabling of synthetic routes is a high early development priority. The imperative at this development stage is to reach key milestones as quickly as possible (to provide rapid feedback to ongoing research activities), to identify potential attrition early, and to accelerate drug candidates through to early decision points. This fit-for-purpose approach also holds down early development costs and manning, with resource-intensive activities to seek and develop the ultimate commercial route triggered only after key milestones are achieved.

For any early process, the prime concern is the safety of operating staff, plant facilities, and the environment. Many reactions that can be carried out in the laboratory with adequate safety can present a significant thermal hazard on a larger scale, and will therefore be eliminated from synthetic routes wherever possible. Likewise, early development work will attempt to remove reactions that are identified as presenting unacceptable toxicity and worker safety exposure issues, or reactions which have a particularly deleterious environmental impact.

Another important factor to consider during the early stages of a project is the process efficiency, particularly any bottlenecks that could significantly hamper the speed of delivery, or drive excessive cost. Examples of potential bottleneck processes include high-dilution or low-yielding reactions and chromatographic purifications. If feasible, linear synthetic routes will be replaced by more convergent ones. There are also many operations routinely carried out on small laboratory scale (for example, stripping to dryness, trituration, and decanting) that cannot be readily used on scale, as described in Anderson's textbook [5]. Finally, the reliability and reproducibility of reactions will be examined, to ensure that they are "robust," and will not present problems of unreliable quality and yield, or control upon scale-up.

An example of the way that synthetic routes are changed during the development process is provided by the potential antithrombotic agent, lotrafiban. The route A medicinal chemistry synthesis of lotrafiban **1**, started from an aryl Grignard reagent, and the chiral center was introduced using L-aspartic acid. This synthesis [13] involved 11 linear steps in an overall yield of 9% (Scheme 2), and the level of waste in this route was judged to have a significant environmental impact. Route B (Scheme 3) was therefore quickly developed to support early clinical requirements, and involved a one-pot procedure converting 2-nitrobenzyl alcohol **2** to intermediate **3**. Enzymatic resolution of **4** using an immobilized form of *Candida antarctica* lipase B gave the desired (*S*) stereochemistry [14]. This route was successfully scaled-up as an expedient, early bulk supply route to give kilogram quantities of **1**. However, this route involved a wasteful late stage resolution and low-yielding preparation of mono *N*-Cbz-4,4'-bipiperidine, and needed to be changed later in development (see next section).

During the early development phase, technologies such as high-throughput

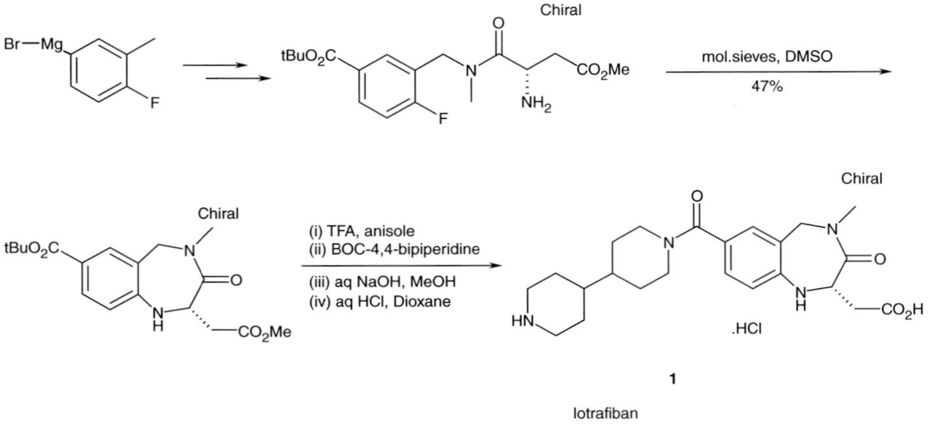

Scheme 2 Route A synthesis of lotrafiban. Reproduced with permission from Ref. [13]; © 2003, American Chemical Society.

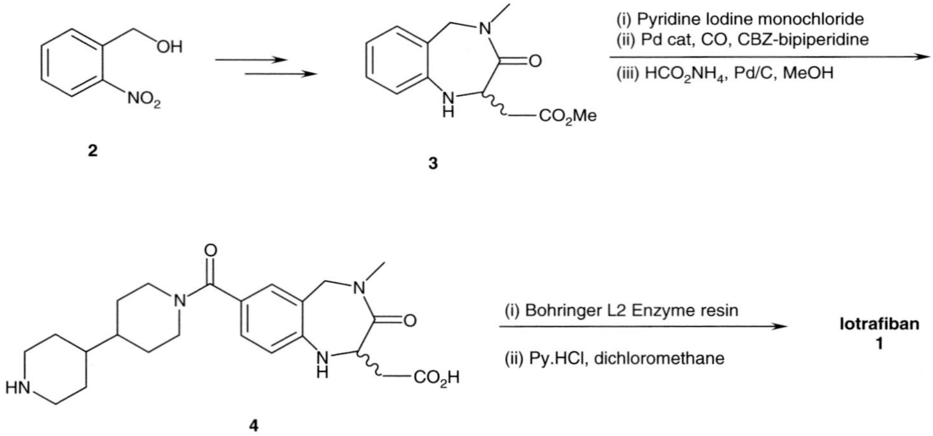

Scheme 3 Route B synthesis of lotrafiban. Reproduced with permission from Ref. [14]; © 2003, American Chemical Society.

screening are used where appropriate, and process chemists play a critical role in delivering material rapidly to support toxicology and clinical studies. Once confidence in the drug candidate is achieved through a successful POC clinical study, the role of the process chemist then changes, as described in the next section.

5
Late Development Activities

The pharmaceutical process chemist is subject to a large number of constraints, but has the opportunity to create significant economic value. Once clinical studies have confirmed that an NCE is both sufficiently safe and efficacious in its chosen indication, the process chemist will typically investigate a large number of potential synthetic routes to the API target and strive to identify the most efficient option. Because early pre-POC development programs are frequently lean, the entry into the "full development" phase is typically a period of intensive chemical effort, with an element of catch-up for deferred activities. During these studies the chemist will need to select the chemical route, select the best reagents and solvents for the chosen route, and then develop and optimize the process.

5.1
Route Selection

When designing potential commercial routes to APIs, a range of considerations come into play. The process chemist will seek convergent rather than linear routes, and will favor routes with lower step counts. In particular, there will be a focus on steps that directly construct the target NCE, and the minimization of protection/deprotection cycles, potentially redundant oxidation/reduction cycles and functional group interconversions in the route. For chiral NCEs there will be the important issue of what method is used to introduce the chirality, and at what point in the synthesis. Finally, the process chemist favors routes in which the late process stages use clean, high-yielding chemical transformations. The last process step is often a salt formation, as salts typically exhibit good crystallinity, and may help achieve good isolation, purification, and stability of the API.

Evaluation and comparison of different route options requires thought to be given of a range of criteria including safety, environmental, legal, economics, control, and throughput (SELECT) considerations.

These criteria for choosing a commercial route have been given the SELECT mnemonic [15], and are discussed below in the context of route design.

Safety is the most important of the SELECT criteria, and is covered in Section 6 (Reaction Process Safety), while the **Environmental** criterion is covered in Section 8 (Green Chemistry). In order to commercialize a pharmaceutical product, the key **Legal** considerations are that the developer is not breaking any laws (for example, using controlled substances or breaching transportation legislation for hazardous materials), or infringing valid intellectual property (IP). There is little point in a process chemist developing a route to an NCE that proceeds via an intermediate claimed in a third-party patent. In some cases a licensing option may overcome such a "freedom to operate" consideration, and advice would need to be sought from a patent attorney.

The **Economic** criterion is an important one for two main reasons. First, the economic viability of a new drug launched on the market is assessed using its cost of goods (CoG). The term CoG is used to describe the total costs involved in the manufacturing process (API production, formulation, and packaging of the drug product) as a percentage of the drug's selling price. An Office of Technology Assessment analysis [16] suggests that CoG across the industry is 25% on average, and the process chemist's route will need to meet a company CoG target. Second, the formal costing of a process highlights which steps, reagents or intermediates contribute most heavily to the NCE cost. A costing exercise can therefore not only help in selecting between rival routes but will also indicate where future development and optimization effort on the route is best focused.

The reproducible scale-up of an API process without adversely affecting API quality is a key challenge for the process chemist. **Control** of API quality is attained by careful attention to the chemical and physical parameters in the process. The process chemist's assessment of a route for control issues includes identification of:

- Side reactions that generate process related impurities; this includes any chemo-, regio-, or stereoselective steps.
- The number and efficiency of purification points in the synthetic route.
- The chemical stability (toward heat, moisture, and oxygen) and physical properties of each intermediate, with particular attention to labile stereocenters and functional groups.

The final criterion is the **Throughput** of the process, which is defined as the weight of material that can be manufactured in unit time. Process throughput can be increased in a number of ways, including:

- Avoiding specialist techniques such as chromatography or cryogenics, which increase processing time or may use bottleneck equipment.
- Increasing reaction concentrations, for example, by changing solvent or using more soluble derivatives.
- Favoring the use of lower-molecular-weight protecting groups or salts, to increase the molar size of campaigns in equipment of fixed size.
- Critically examining and minimizing time-consuming process operations such as solvent replacements, or drying operations.

Any selected commercial manufacturing process must satisfy these critical criteria, in addition to subsequent development of in-depth process understanding and control of critical parameters.

5.2
Reagent and Solvent Selection

Having selected the bond-forming chemistry, and the route intermediates in a process, the next step is to choose the most appropriate reagents and solvent for each step. The practical considerations for the suitability of a reagent for scale-up include reactivity and selectivity, safety, ready availability on scale with relatively low cost and reliable quality, ease of work-up operations, and waste minimization. High reagent reactivity has the advantage of shortened reaction times, but may come at the expense of selectivity, with the need for a balance to be sought. Safety encompasses considerations of reagent toxicity and handling issues for staff, as well as the reaction hazards (covered in the next section). The bulk availability of reagents from multiple vendors (reliance on a single source can be risky) needs to take account of any shipping restrictions and reagent stability (preferably at ambient temperature) so that materials have adequate shelf-lives. The molar cost of reagents can be a key factor in decision making, and unless consistent batch-to-batch reagent quality can be assured, it may be necessary to run use tests. Finally, preferred reagents enable the easy isolation of pure step product, avoid toxic byproducts and minimize waste disposal issues.

In the past, most pharmaceutical processes have relied heavily on classical stoichiometric chemical transformations. However, catalytic transformations are now increasingly used for their ability to not only decrease reagent and waste disposal costs but also, in many cases, to increase overall productivity. Catalytic hydrogenation has been widely used across the industry for many years [17], but many other chemocatalytic transformations that

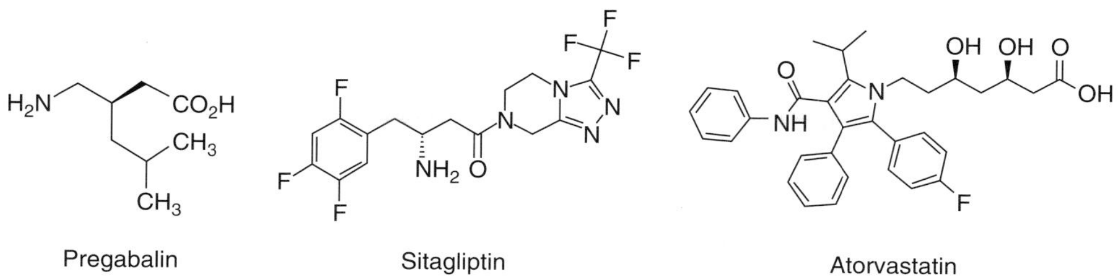

Fig. 5 Pharmaceuticals manufactured using biocatalytic reactions.

form C–C or C-heteroatom bonds [18–20] are now routinely practiced. Biocatalysts are also being increasingly used, especially to prepare optically active synthons and APIs [21–23], as exemplified by the marketed agents shown in Fig. 5. Pregabalin was developed for the treatment of nervous system disorders including epilepsy and neuropathic pain, and is now manufactured using a lipolase-catalyzed enzymatic resolution [24]. The antidiabetic agent sitagliptin is now produced using a transaminase to convert its ketone precursor to the chiral amine [25], while atorvastatin and several other statins are manufactured with the aid of alcohol dehydrogenase or deoxyribose aldolase enzymes [26].

Because the bulk supply route to lotrafiban shown earlier in Scheme 3 involved a wasteful late-stage resolution and a low-yielding preparation of a key bipiperidine, the route depicted in Scheme 4 was introduced for the later development of this agent. This manufacturing route contained 11 chemical transformations, of which eight were promoted by catalysts. The early enzymatic resolution proved advantageous, as the unwanted enantiomer could be recycled. This route also avoided the need for classical protections or deprotections, since the pyridine group acted as a masked piperidine [14].

Selection of the appropriate solvent can increase the rate, reproducibility, and ease of running a reaction, and also ensure that the requisite product yield and quality are achieved. The key physical properties of solvents have been tabulated, as part of a good discussion of how they can be used to help select solvents for reaction scale-up [5, 27]. Good dissolution of the reaction components is usually important, and sometimes a mixture of solvents may dissolve a compound better than the individual solvents. However, to ease efficient solvent recovery and reuse, and hence decrease waste (see Sect. 8), the use of single solvents is usually preferred. Many solvents commonly used in the laboratory have toxicity concerns that disfavor their use for scale-up, and these have been surveyed [28]. For example, n-hexane is neurotoxic, the ethers 1,4-dioxan and 1,2-dimethoxyethane are male reprotoxins, and many chlorinated solvents are either known or suspected carcinogens. Regulatory authorities have reviewed solvent toxicity data, and International Conference on Harmonisation (ICH) publish guidelines on the residual levels (in ppm) that are allowable in pharmaceuticals for human use [29]. Solvents that are therefore rarely used on scale include diethyl ether, 1,4-dioxan, 1,2-dimethoxyethane, hexane, benzene, chloroform, and hexamethyl phosphortriamide (HMPA). A final consideration in solvent selection is the reactivity of solvents, and their potential to react with reaction components or reagents. Thus, dimethylformamide (DMF) reacts exothermically

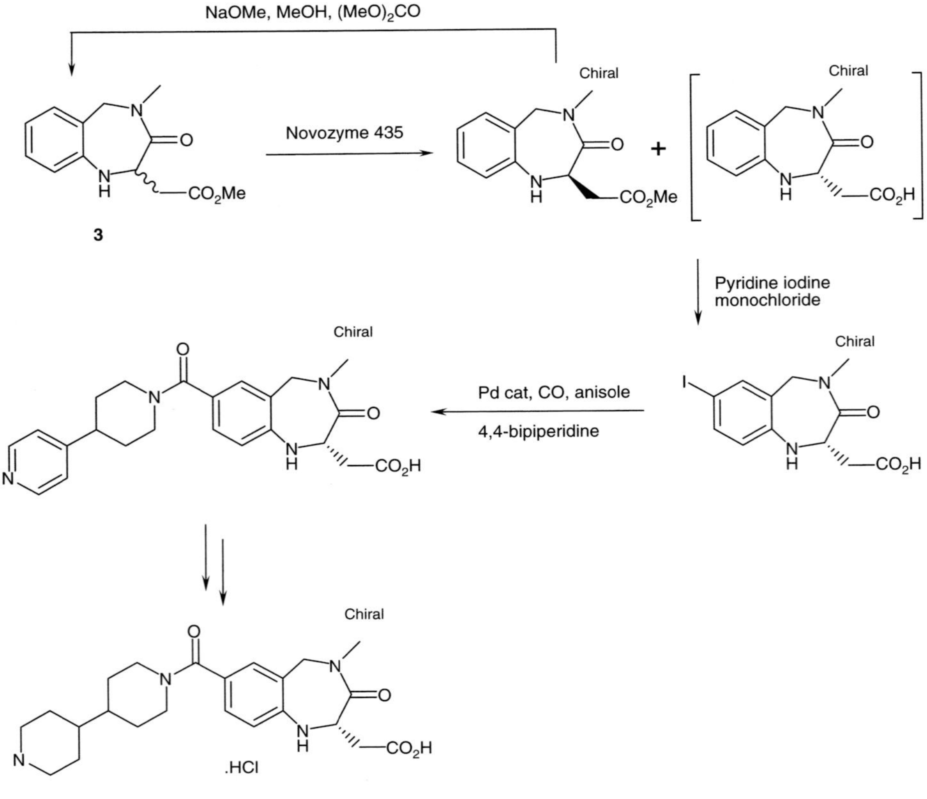

Scheme 4 Route C synthesis of lotrafiban. Reproduced with permission from Ref. [14]; © 2003, American Chemical Society.

with NaH [30] and can formylate amines at higher temperatures, while the electrophilic reactivity of methylene chloride [31] is frequently not appreciated.

5.3
Reaction Development and Optimization

Having selected the reagents and solvents for each process step, the next goal is to maximize the isolated yields of step products, with control of their quality. This involves not only the development of both the reaction conditions but also the work-up and isolation conditions for the product. Determining the best processing conditions for a reaction at scale is a multidisciplinary endeavor taking in synthetic, physical organic chemical, and chemical engineering viewpoints. The aim of this exercise is to obtain an in-depth understanding of the reaction mechanism, and those parameters that are critical for reaction performance. A variety of technologies and techniques are available to the process chemist to help achieve this.

Understanding the kinetic parameters affecting the formation of the desired

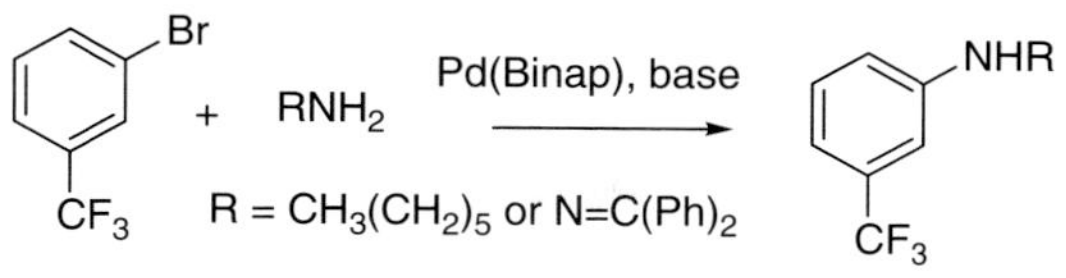

Scheme 5 Pd(binap)-catalyzed amination of 3-bromobenzotrifluoride.

product and any undesired impurities, and identifying their dependency on mass and heat transfer effects, is a key step in the successful scale-up of the process. Kinetic profiling, to examine the rate behavior of both substrate disappearance and product/impurity formations, is a powerful tool for fast and efficient process development. As the reaction rate is the rate of change of concentration, the key is to examine the shape of the concentration profiles and the steepness of the curve slopes at different periods, as these indicate the rate of formation/depletion over that reaction period. Efficient and accurate analytical techniques are key to this process development effort, and in-line analytical techniques such as Fourier transform infrared (FT-IR) spectroscopy are increasingly being used to monitor reactions and build reaction profiles.

Reaction calorimetry is widely used for hazard evaluation (see Sect. 6) but is emerging as another powerful process development tool. The thermal data obtained is a direct representation of the reaction rate at any particular moment, and calorimetric data can help to quickly identify anomalous reaction behavior, including reaction induction periods. Catalytic aminations of aryl halides are frequently used in pharmaceutical syntheses, and small-scale reaction calorimeter heat flow curves furnished rapid insights into the kinetic trends of this transformation. The Pd(binap)-catalyzed amination of 3-bromobenzotrifluoride with two different amines – hexylamine and benzophenone hydrazone (Scheme 5) – was examined and revealed a great deal about this reaction [32]. Although the catalyst and other processing parameters were similar, the kinetics of the reaction for these two amines are completely different, contributing much to both process understanding and process development. When benzophenone hydrazone was used as the amine component, the reaction showed net zero-order kinetics, whereas with hexylamine the reaction showed overall first-order reaction kinetics and the rate of reaction was considerably higher. This clearly shows that, even though the reaction mechanism and the catalytic cycle might be similar in the two cases, the rate-determining step within the catalytic cycle might be different. From a process development viewpoint, for a process with benzophenone hydrazone as the amine, changing the concentration of either of the reactants would have no influence on the reaction rate, and hence modifying these concentrations cannot be used to increase the productivity. However, in the case of the reaction with hexylamine, two further experiments varying the concentration of each substrate would determine whether the observed net first-order kinetics was due to the first-order dependence on one of the reactants and zero-order dependence on the other, or partial-order dependence on both reactants. Identifying this would provide a clearer picture of how to optimize the reactant concentrations to obtain the best operating conditions. These data-rich experiments also provided valuable heat-generation information, and indicated how the feed rate and heat-transfer parameters would need to be controlled to run the process in a safe and efficient manner in the plant.

As process knowledge and understanding is built for a reaction step, a picture emerges of which parameters are likely to be critical for the performance of the transformation. Because many processes are complex, the classical approach of modifying one variable at a time has now been superseded by statistical methods of development, referred to as "design of experiments" (DoE). Statistical DoE allows the optimization of conditions involving many parameters, and the frequent cases where the reaction variables are interdependent [33]. Computer programs are available to assist in DoE optimization [34], and software packages allow the resulting response surfaces to be plotted, allowing ready visual assessment of optimization runs. The ability to automate reaction set-up and analysis has greatly assisted the widespread application of DoE to pharmaceutical process development. A well-designed DoE package will confirm the critical process parameters (CPPs) for a reaction, and may also identify the "edges of failure," the points at which the yield and/or quality of the product rapidly fall off. The ideal optimized process should be rugged (or "forgiving"), in that it will provide product of the expected yield and quality over a relatively wide range of operating conditions. Having eliminated all unnecessary process operations, such solvent exchanges and washes, simplicity is the goal of the developed process. As most pharmaceutical intermediates are solids, crystallization is the work-horse isolation and purification technique. The minimization of process impurities is therefore important, as each impurity tends to reduce the recovery in the crystallization of the reaction product. An increasingly frequently used approach is the "direct drop" (sometimes called "bottom-up") process, in which the selected reaction solvent dissolves the starting material(s), but not the product, enabling direct filtration to be used for product isolation without the need for a work-up procedure [35]. Such an isolation has the added advantage that water quenches or washes are avoided, enabling the more efficient recovery and recycling of organic solvents.

For fast reactions, and those with mixing and heat-transfer issues upon scale-up, continuous-flow processing presents an attractive alternative to the conventional use of batch reactors. An analysis of the kinetics of reactions carried out in the fine chemical and pharmaceutical industries suggested that up to 50% of these reactions had the potential to be carried out using continuous processes [36]. Recent examples have been presented [37] of the use of continuous processing in the pharmaceutical industry, including its application to APIs such as celecoxib and naproxcinod.

6
Reaction Process Safety

6.1
Background

When scaling-up batch reactions beyond a laboratory scale of about 1 liter, process safety needs to be considered, as the ratio of surface area to volume and the ability to remove heat both decrease. Consequent auto-heating of the reaction can lead to the decomposition of any thermally unstable materials present in the reactor, runaway reactions, and potentially to an explosion. An industry-wide change was triggered during the 1970s, after several runaway reactions in the UK [38], USA, Germany and elsewhere, mostly involving nitration chemistry or the reduction of nitro compounds. A combination of legislation in some countries, the setting up of company hazard evaluation laboratories, and the commercialization of reaction calorimeters such as the RC1 then led to rapid change across the industry. It is now accepted that safety is an integral part of chemical development, with safety issues assessed at all stages of development.

6.2
Screening for Early Scale-Up

When a drug candidate enters development and decisions need to be taken about the route to be used for early bulk supply, desk screening is first carried out. Routes which contain potentially highly energetic materials or reactions are disfavored, as extra hazard testing will be required to assess the magnitude of the safety issue, and appropriate control measures will probably be needed. A list of functional groups known to exhibit high-energy decomposition (Fig. 6) is therefore consulted. The modern comprehensive literature on the subject of thermal hazards, including Bretherick's excellent compendium [39], and the special safety editions annually published in the journal *Organic Process R&D* [40] will also be consulted as good information sources. Desktop thermochemical calculations of heats of reaction are often carried out by subtracting the heats of formation of the reactants from those of the products [41]. These estimated heats of reaction give an idea of the total energy being released and the maximum potential adiabatic temperature rise, but the calculations give no indication of how fast this energy will be released.

A typical early testing strategy supplements desk screening results with basic tests using small-scale, quick-turnaround techniques. The most commonly used of these tests is differential scanning calorimetry (DSC), which typically only uses 10 mg samples to provide energy traces for endothermic (e.g., melting) or exothermic (e.g., reaction or decomposition) activity. The point where an exotherm occurs gives an indication of the onset of the event, while the area under the peak gives a value for the heat released. However, as DSC is a small-scale dynamic thermal measurement, chemical processes are not operated close to identified exothermic events, with a safety margin of about 100 °C usually applied to DSC test results. Other basic screening tests include the TS^U (thermal screening unit), which uses a larger sample but has the advantage that a pressure transducer is attached [42, 43]. The TS^U output plots therefore also measures the total pressure generated in the reaction, as well as the rate of pressure rise. As many chemical reactions generate non-condensable gas as they progress, it is important to quantify both the amount and the rate of the gas being formed, to prevent vessel over-pressurization and provide information toward the design of an appropriate pressure-relief system. The TS^U provides onsets for both temperature and pressure, and for this technique a safety margin of 60–70 °C is usually applied.

6.3
Larger-Scale Hazard Assessment

As new routes are investigated for potential longer-term use, the above safety assessment will be repeated. With further scale-up of a selected route into the pilot plant, more detailed process safety studies are integrated into the chemical development activities. Further evaluation of the thermal stability of all raw materials,

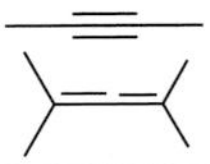 ⩚ (acetylenic structure)	acetylenic, metal acetylides, haloacetylene derivatives, allenes
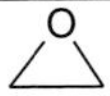 epoxide structure	epoxides
—N=N— —N⁺≡N	all substances with N-N double or triple bonds, i.e. azo, diazonium salts, azides, diazirines and other high nitrogen containing compounds like triazoles, triazenes, tetrazoles etc.
—O—O—	all substances with O-O bonds, i.e. peroxides, peroxyacids and their salts, hydroperoxides, peroxyesters
—N=O —N–O	all substances with N-O bonds, like nitro, nitroso, hydroxylamines, nitrite, nitrate, fulminates, oximes, oximates
—N–X	halogen azides, N-halogen compounds, N-haloimides
—O–X	alkyl perchlorates, aminium perchlorates, chlorite saltes, halogen oxides, hypohalites, perchloryl compounds (bromates, iodates, as well)
—N–M	metal nitrides, amides, hydrazides, imides, cyanamide. main concern is the pyrophoric nature of the pure solid material. dilute solutions of metal amides and substituted amides (i.e. LDA, LiHMDS) are generally acceptable depending on use and fate of excess quantities
Ar—M—X X—Ar—M	non-catalytic use of haloarylmetals, haloarenemetal π–complexes Note: Grignards of concern are only halo-phenyl Grignards containing trifluoromethyl moeities.

Fig. 6 High-energy functional groups.

intermediates and products is carried out, and the use of an adiabatic calorimeter is recommended. These calorimeters, such as the accelerated rate calorimeter (ARC) [44], examine the exothermic potential of individual materials and reactions. The sample is heated until thermal activity is detected, and then switches to tracking mode, when external heaters are set to match the sample temperature. As the sample self-heats due to the exothermic event, the heaters maintain an adiabatic environment, allowing the thermal event to progress under worst case conditions. More meaningful onset data and maximum temperature and pressure rate rises are therefore obtained from such instruments, for plant design and the safe operation of processes. A thermal runaway may result if the rate of heat produced by the chemical reaction exceeds the available rate of heat removal. The surplus heat raises the temperature of the reaction mass, further accelerating the reaction rate, with the rate of heat production then increasing exponentially, whilst the rate

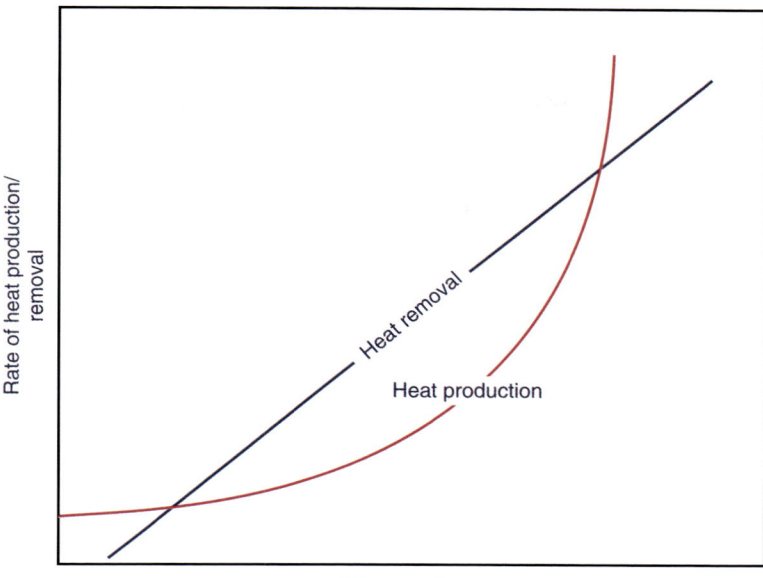

Fig. 7 Heat production versus heat removal.

of heat removal only increases linearly (Fig. 7).

Having relied earlier on calculated estimates of heats of reaction, reaction calorimetry can now be used for their direct measurement. Calorimeters such as the RC1 [45] can determine not only heats of reaction but also give a heat flow profile throughout the reaction, allowing the reaction rate to be followed, induction periods recorded, and heats of crystallization observed. Adding a gas flow meter to the calorimetry experiment enables gas flow rates to be obtained, and the calorimeter can also provide data on the amount of reaction accumulation. It is worth reiterating the point made earlier, that during the process of safety testing much data is collected which provides reaction kinetic information that improves the success of scale-up programs.

As processes are scaled-up in Pilot Plants it is important that potential hazards are carefully reviewed and documented. Should any high-energy functional groups remain in the process, further characterization testing such as impact sensitivity, friction, burning and detonation tests may be required. A safety review team (usually consisting of a project chemist, a process safety specialist, and plant supervisor or engineer) will review the process to be operated and the safety data generated, and agree to either run the chemistry, to implement other safety controls, or request more safety data for a further review cycle.

6.4
Large-Scale Production

By the time the process is moved into production equipment most of the required safety data should have been collected. However, a full-scale hazard and operability study (HAZOP) will be undertaken, revisiting the hazards as the process enters a new environment. Despite a thorough understanding of the process and its safe limits of operation, at the larger scale a further layer of control and protective measures are likely to be considered. Additional process control measures may include the use of sensors, alarms and trips that either trigger automatic action or prompt manual intervention to avoid the occurrence of conditions leading to uncontrolled reaction. Further protective measures may then be put in place to reduce the impact of a runaway scenario, should the control measures fail to prevent its occurrence. These measures include relief venting, crash cooling, and reaction dumping.

7
The Analytical Interface

The increased complexity of NCEs, and the need to identify and characterize even small amounts of new impurities, has meant that modern process groups need to have access to the best instrumentation and analytical services. The analytical function is therefore a critical support function for the pharmaceutical process chemist throughout the development program. This is well exemplified by the exercise, jointly carried out by process and analytical chemists, of identifying step impurities, and tracking them through the process to demonstrate their fate and purge. Very few transformations are totally free from side-reactions, and as the program progresses the byproducts generated in each step are identified and in-process control (IPC) assays are developed to help minimize their formation. Assays on step products then determine whether or not byproducts are totally purged during the work-up and isolation processes. The fate of any unpurged impurities in the subsequent process and isolation steps are then tracked, and analytical specifications for isolated step intermediates are developed in this exercise. Should the API reach regulatory filing, this impurity identification and the tracking of fate and purge in the synthesis is key data demonstrating reproducible control of API quality in the manufacturing process.

Chromatographic separations are the mainstay of analytical testing for IPC assays and assaying of intermediates and API. Because of the nonvolatile and often polar nature of pharmaceuticals, high-performance liquid chromatography (HPLC) is generally the separation technique of choice for monitoring impurity profiles. Many advances in HPLC column and hardware technology during the past decade have enabled reduced HPLC run times whilst maintaining the same level of efficiency [46], with reverse-phase liquid chromatography (RPLC) emerging as the most usual tool of choice for achiral separations. For the assessment of enantiomeric purity throughout the development life cycle, chiral HPLC is essential for the analysis of chiral APIs and their chiral intermediates. Polysaccharide-based chiral stationary phases (CSPs) have become a favorite of the pharmaceutical industry owing to their high success rate, with three (cellulose- and amylase-based) CSPs having demonstrated complementary separations of acidic, basic, and neutral compounds [47].

For the analysis of volatile organic impurities, gas chromatography (GC) is the favored technique. The most frequent use tends to be for the determination of residual solvents in APIs, which is the subject of ICH guidance [29]. It is unusual for the solvents selected for the manufacture of APIs to be completely removed by practical manufacturing techniques. The ICH Q3C guideline therefore recommends the use of less-toxic solvents, and describes levels considered to be toxicologically acceptable for the safety of patients. For solvents with low toxicological potential, such as acetone, acetic acid, ethanol and ethyl acetate, levels of up to 5000 ppm in API are generally considered acceptable. For many commonly used processing solvents, including toluene, DMF, acetonitrile, methylene chloride and tetrahydrofuran (THF), the recommended concentration limits are in the 100–1000 ppm range, while for chloroform or 1,2-dimethoxyethane the limits are 100 ppm, or less. Finally, certain solvents (e.g., benzene, carbon tetrachloride and 1,2-dichloroethane) are listed as to be avoided in API manufacture, and if their use cannot be overcome their levels are restricted to <10 ppm. Very sensitive GC detection methods are thus frequently needed, and as with all chromatographic methods the accuracy, linearity and precision of the assay must be demonstrated.

Even more stringent controls are needed in the case of reagents, intermediates, or byproducts that are recognized as having carcinogenic or genotoxic potential. One commonly adopted approach is to control these potential genotoxic impurities (PGIs) to the "Threshold of Toxicological Concern" (TTC) level, defined as 1.5 µg per day. This recognizes that complete elimination of PGIs is not feasible and that at the TTC level, exposure to the compound will not pose a significant carcinogenic risk [48]. Knowledge-based computer systems, which compare molecular structures to a database of structural alerts, are used across the pharmaceutical industry to identify PGIs. This desk screening picks out functional groups that may be of concern, and the process chemist must assess the reaction conditions for their potential to generate PGIs. For example, the formation of sulfonate salts in the presence of an alcohol has the potential to form sulfonate esters which are known to be genotoxic [49].

The use of off-line techniques such as HPLC for reaction monitoring and IPC testing can be resource-intensive and time-consuming, as the samples tend to require processing (e.g., filtration, quench, and dilution) before assay. There has therefore been a recent trend toward the use of on-line spectroscopic techniques to monitor reactions in real time. Near-infra red (NIR) spectroscopy and FT-IR are proving valuable for *in-situ* qualitative and quantitative analysis, without the need for sample preparation. A remote fiber optic probe, either in the vessel or via a pumped loop, is used as the measuring device to differentiate components of the reaction mixture by monitoring the appropriate characteristic wavelengths. The method has the added advantage that it can analyze compounds lacking suitable chromophores for HPLC-UV analysis. There is the limitation that spectroscopic in-line analyses generally require external calibration using a chromatographic technique, but once a calibration has been completed for a late development program the return on investment can be considerable. The example in Fig. 8 shows the formation of the required mono-hydrochloride salt of an API, which readily forms a bis-HCl salt. In-line FT-IR readily differentiated between the mono- and bis-salts, allowing a very accurate determination of the reaction end point [50].

The analytical evaluation of the suitability of salts during the screening and selection process is important, as an extensive range of physical and chemical properties need to be assessed. The relative usage of the most commonly utilized acidic and basic counterions in APIs, together with the key salt properties and the analytical

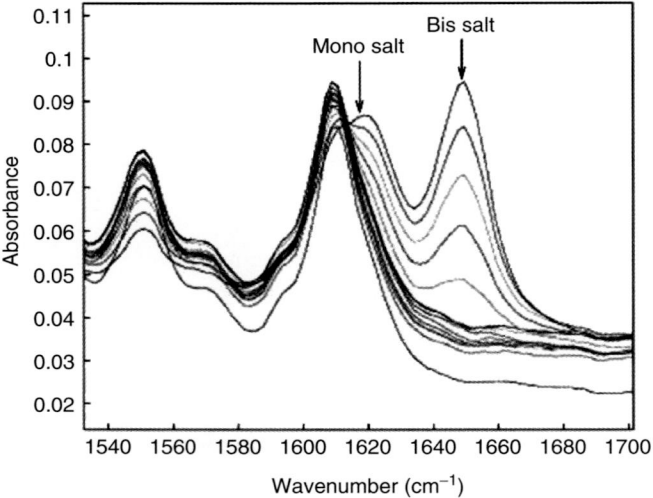

Fig. 8 FT-IR reaction monitoring of mono-HCl salt formation. The hydrochloric acid charge is optimized to minimize the formation of the bis-HCl salt. Reproduced with permission from Ref. [50]; © 2006, Elsevier.

techniques used to evaluate them, have been summarized [51]. As relatively complex molecular materials which typically have multiple H-bond donors and acceptors, pharmaceutical agents frequently display polymorphism [52, 53]. As the different solubilities and dissolution rates of polymorphs can ultimately affect the bioavailability of the API, it is important to identify the most stable polymorph early in the development program. The late emergence of a more thermodynamically stable polymorph can impact development timelines due to the potential need to repeat clinical and stability studies. In extreme cases a new polymorph has necessitated the temporary withdrawal of an approved API from the market, as was the case with the anti-human immunodeficiency virus (HIV) agent ritonavir [54]. Most pharmaceutical companies therefore incorporate polymorph screening and selection programs into their development strategies, using analytical techniques such as DSC, powder X-ray diffraction and Raman microscopy.

8
Green Chemistry

8.1
Introduction

The Environmental Protection Agency (EPA) has defined green, or sustainable, chemistry as "…the design of chemical products and processes that reduce or eliminate the use or generation of hazardous substances" [55]. In the mid-1990s, Anastas and coworkers at the EPA were considering the design of environmentally benign chemical products and processes [56], and this led to the articulation of the 12 Principles of Green Chemistry [57]. The increasing focus on green chemistry across the pharmaceutical industry has been particularly evident in the past 20 years,

Tab. 1 Green chemistry principles deliver economic and environmental benefit

	Thinking environmentally	*Thinking economically*
Atom economy	Minimal byproduct formation **Reduced environmental burden**	More from less Incorporate total value of materials **Reduced cost**
Solvent reduction	Less solvent required, less solvent waste **Reduced environmental burden**	Reduced capacity requirements, less energy required **Reduced cost**
Reagent optimization	Catalytic, low stoichiometry, recyclable **Reduced environmental burden**	Higher efficiency, higher selectivity **Reduced cost**
Convergency	**Reduced environmental burden** related to improved process efficiency	Higher efficiency, fewer operations **Reduced cost**
Energy reduction	**Reduced environmental burden** related to power generation, transport, and use	Increased efficiency, shorter processes, milder conditions, fewer heating or cooling steps **Reduced cost**
In-situ **analysis**	Reduced potential for exposure or release to the environment	Real-time data increases throughput and efficiency, fewer reworks **Reduced cost**
Safety	Nonhazardous materials and processes reduce risk of exposure, release, explosions, and fires	Worker safety and reduced downtime. Reduced special control measures **Reduced cost**

although low-waste, efficient process chemistry clearly existed before this time. Focus on many of the green chemistry principles delivers economic, as well as environmental benefits (see Table 1). The journal *Green Chemistry* was established in 1999 for publications in this area, and several good textbooks are available for readers seeking greater detail [19, 58, 59].

8.2
Green Chemistry Metrics

To track progress in the design of chemical processes that minimize both resource use and waste generation, a number of measures of "greenness" have been proposed.

The early concept of atom efficiency was first proposed by Trost in 1991 [60], and has gained widespread acceptance in both academia and industry. Atom economy is simply calculated from the reaction scheme, and defined (Eq. 1) as:

$$\text{Atom Economy} = \frac{\text{Molecular Wt. of the desired product}}{\text{Molecular Wt. of all products}} \times 100\% \quad (1)$$

Reactions such as the Diels–Alder reaction, in which all of the atoms in the two starting materials are incorporated into the product, will have 100% atom economy. However, most reactions generate

Tab. 2 E-factors in the chemical industry

Industry segment	Volume (tonnes/year)	E factor
Bulk chemicals	10^4-10^6	<1–5
Fine chemicals industry	10^2-10^4	5 to >50
Pharmaceutical industry	$10-10^3$	25 to >100

Tab. 3 Comparison of environmental metrics for lotrafiban routes

Route	Yield (%)	E-factor	RME (%)
A	9	1429	1.3
B	17	1173	2.6
C	29	262	7.6

coproducts that can significantly lower the atom efficiency. Although simple, this measure does not take account of reaction stoichiometry or yield, or make allowance for solvent usage or any materials used in the process work-up.

A second effective measure is the environmental factor (E-factor), which Sheldon introduced in 1992 [61]. The E-factor is simply the kg of waste per kg of product produced, and was shown by Sheldon to vary widely across different sectors of the chemical industry (Table 2). The relatively high E-factors for the pharmaceutical industry are partly due to the complexity of the products and their syntheses, but also reflect the lower percentage of catalytic reactions compared to the bulk chemical sector. E-factors take account of solvent usage, and can be calculated with or without consideration of process water. It is readily apparent that more concise syntheses, more selective transformations (chemo-, diastereo-, or enantio-selective), and the greater use of catalytic reactions will all minimize waste and reduce E-factors.

A third metric, the reaction mass efficiency (RME), was introduced by GSK in 2002 [62]. The RME is defined as the "mass of desired product as a percentage of the mass of all reactants" (as shown in Eq. 2). It thus takes account of reaction yield, stoichiometry, and use of reagents, keeping the simplicity of the atom economy concept whilst avoiding the E-factor's high solvent impact.

Reaction Mass Efficiency

$$= \frac{\text{Mass of desired product}}{\text{Mass of all reactants}} \times 100\% \quad (2)$$

Measures of the environmental impact of the process changes for the three lotrafiban routes (A, B, and C) depicted above in Schemes 2–4 are summarized in Table 3. Manufacture using route A would generate over 1.4 tonnes of waste per kg API, while route C reduced this waste burden by over 1.1 tonnes per kg API, a 5.4-fold improvement.

8.3
Green Solvent Selection

Solvent selection is a major issue when designing a synthetic process, since solvents typically make up over 80% of the materials used for API manufacture, consume about 60% of the overall energy, and account for 50% of greenhouse gas emissions [63, 64]. Organic solvents are thus a major contributor to the overall toxicity potential associated with API production processes. Solvents should be found for processes that can be easily recovered, separated, and purified for reuse; otherwise, spent solvents (or solvent mixtures) will have to be disposed of as waste. Solvent selection guides generally involve assessments across several criteria, including

Preferred	Usable	Undesirable
Water	Cyclohexane	Pentane
Acetone	Heptane	Hexane(s)
Ethanol	Toluene	Di-isopropyl ether
2-Propanol	Methylcyclohexane	Diethyl ether
1-Propanol	TBME	Dichloromethane
Ethyl Acetate	Isooctane	Dichloroethane
Isopropyl acetate	Acetonitrile	Chloroform
Methanol	2-MeTHF	NMP
MEK	THF	DMF
1-Butanol	Xylenes	Pyridine
t-Butanol	DMSO	DMAc
	Acetic Acid	Dioxane
	Ethylene Glycol	Dimethoxyethane
		Benzene
		Carbon tetrachloride

Fig. 9 The Pfizer solvent selection guide. Reproduced with permission from Ref. [65]; © 2008, The Royal Society of Chemistry.

health, safety, environmental impact and life cycle assessment. The GSK guide [64] reports a score for the various criteria, and presents the results in a color-coded table. The Pfizer approach [65] carries out a similar assessment, but then presents to end users the much simpler summary shown in Fig. 9.

Unfortunately, the dipolar aprotic solvents DMF, N,N-dimethylacetamide (DMAC) and N-methylpyrrolidone (NMP), which are so useful in some crosscoupling and nucleophilic substitution reactions, are all reprotoxins. Good replacements for these amide solvents have been difficult to find, so the least-volatile NMP is often chosen because of its reduced exposure risk. Water is a cheap, safe and environmentally benign solvent, and many reactions take place in water (or more correctly "on water") [66], though on closer examination many of these reactions use excesses of liquid reagents and are in fact biphasic. Furthermore, on an industrial scale there can be a considerable cost and environmental burdens associated with remediating wastewater streams [67].

8.4
Green Chemistry Case Histories

The following two examples serve to illustrate the value that the greening of pharmaceutical processes can bring, by demonstrating the industry's commitment to meet healthcare needs through sustainable practices. They also showcase the breadth of thinking and techniques that are required for successful implementation. The first example is provided by paclitaxel, which was initially isolated as the natural product taxol from the bark of the Pacific yew tree (*Taxus brevifolia*), and subsequently marketed by Bristol-Myers Squibb (BMS) for the treatment of ovarian and breast cancer. As the yew bark contains only 0.0004% paclitaxel, and stripping the bark kills the trees, a sustainable source of this API was required. Despite the molecular complexity of paclitaxel, academic total syntheses were achieved in 40 steps and ~2% overall yield [68, 69], but were not commercially viable. The commercial breakthrough came with the realization that the alternative natural product 10-deacetylbaccatin III (DAB) [70], contains most of the structural

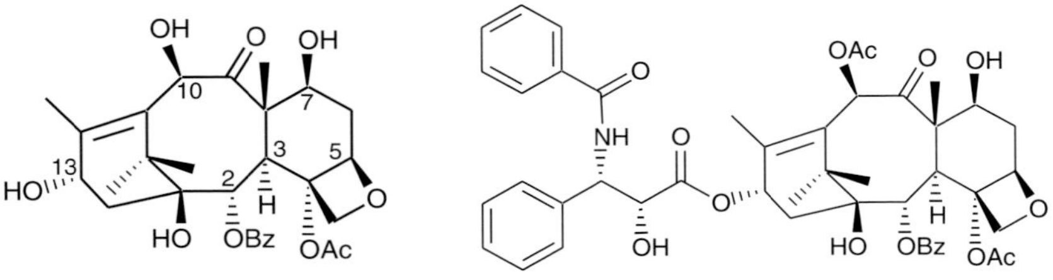

Fig. 10 Structures of deacetylbaccatin (DAB) and paclitaxel.

elements (the tetracycle and eight of the chiral centers) of paclitaxel (Fig. 10). Furthermore, DAB was present at 0.1% levels in the dried leaves and twigs of the European yew tree (*Taxus baccata*), representing a renewable source of this chiral tetracycle. The semisynthetic process [71] from DAB to paclitaxel was complex, involving 11 chemical transformations and seven isolations, but enabled the commercialization of this API.

Although the semisynthetic commercial process avoided destruction of the Pacific yew tree forests, it still presented significant environmental challenges. The process used 13 different solvents, 13 organic reagents (some with toxicity and waste stream issues), and was energy-intensive. BMS therefore licensed plant cell fermentation (PCF) technology from Phyton Biochem GmbH, and developed an aqueous-based PCF process to paclitaxel which used cells cultured from the Chinese yew tree (*Taxus chinensis*). In the cell fermentation stage of the PCF process, calluses are propagated in a wholly aqueous medium in large fermenters under controlled conditions, at ambient pressure and temperature. The feedstock for cell growth consists of renewable nutrients, sugars, vitamins, amino acids and trace elements. Commercial implementation of the PCF process overcame the need for large plantations of *Taxus baccata*, avoided the annual generation of about 240 MT of biomass waste from the isolation of DAB, and eliminated the need for chemical transformations and six isolated intermediates. BMS now manufactures paclitaxel only using the PCF process, which also eliminates the use of 10 solvents and the need for six drying steps, with a considerable energy saving. This impressive example of the application of the principles of green chemistry was recognized by a 2004 Presidential Green Chemistry Challenge Award [72].

The second case history is provided by pregabalin, an API that is marketed by Pfizer for several central nervous system disorders, including epilepsy and neuropathic pain. The first commercial synthesis (Scheme 6) prepared the key β-cyano diester **5** by condensing isovaleraldehyde with diethyl malonate, followed by the addition of cyanide using KCN. The diester **5** was then hydrolyzed and decarboxylated, and the β-cyano acid hydrogenated with Raney Ni catalyst to give the racemic agent. Final resolution with (S)-mandelic acid then furnished pregabalin [73]. This was an efficient synthesis of racemic pregabalin, and enabled the 2005 launch of this API, but the final stage classical resolution process led to the loss of well over half of the material at the end of the synthesis.

An enzymatic screen, using previously described [74] methodology and a 96-well format, subsequently identified lipases

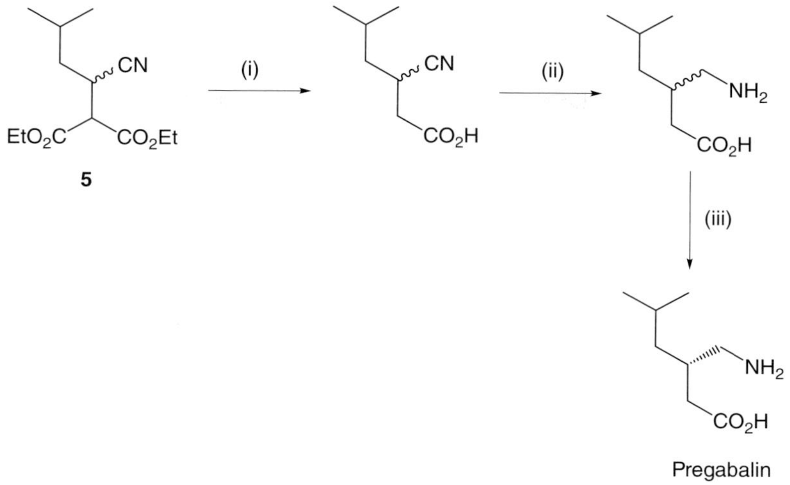

Scheme 6 Classical resolution route to pregabalin. (i) KOH, MeOH, H_2O, reflux; (ii) H_2 RaNi, EtOH, H_2O; then HOAc: then isopropyl alcohol (IPA) wash; (iii) (S)-mandelic acid, IPA, H_2O, recrystallize from IPA, H_2O, salt break, recrystallize from IPA, H_2O.

that enabled a biocatalytic resolution to be carried out earlier in the synthesis using **5** as substrate. The enzyme selected for scale-up was *Thermomyces lanuginosus* lipase (sold commercially as Lipolase), based on its high activity, enantioselectivity and diastereoselectivity, factors which translated into lower biocatalytic loadings, and hence less waste. Lipolase tolerated concentrations of **5** of up to $255\,g\,l^{-1}$ in reactions at room temperature and neutral pH, and a product inhibition issue with the biocatalytic resolution process was overcome by the addition of calcium acetate. The overall process shown in Scheme 7 was designed so that every step was performed in water, and was introduced for the manufacture of pregabalin in late 2006. The next advance was the development of an efficient base-catalyzed epimerization process to recycle the unwanted (R)-cyano diester **6** from the resolution. This required the design, building and validation of a new continuous-processing plant by Pfizer, and the filing of the new chemistry with regulatory agencies. Following regulatory approval, the commercial manufacture of pregabalin using this latest, most environmentally friendly process, was started in 2010. The E-factors calculated for each of the three commercial pregabalin manufacturing processes are shown in Table 4. The reduction in energy usage upon moving from the classical resolution to the biocatalytic route has also been calculated and reported [24].

9
To Market

9.1
Technology Transfer and Process Validation

As an API nears the end of its clinical assessment phase, and registration and commercialization is being planned, it becomes necessary to scale the synthetic process from pilot plant facilities into production equipment suitable for the

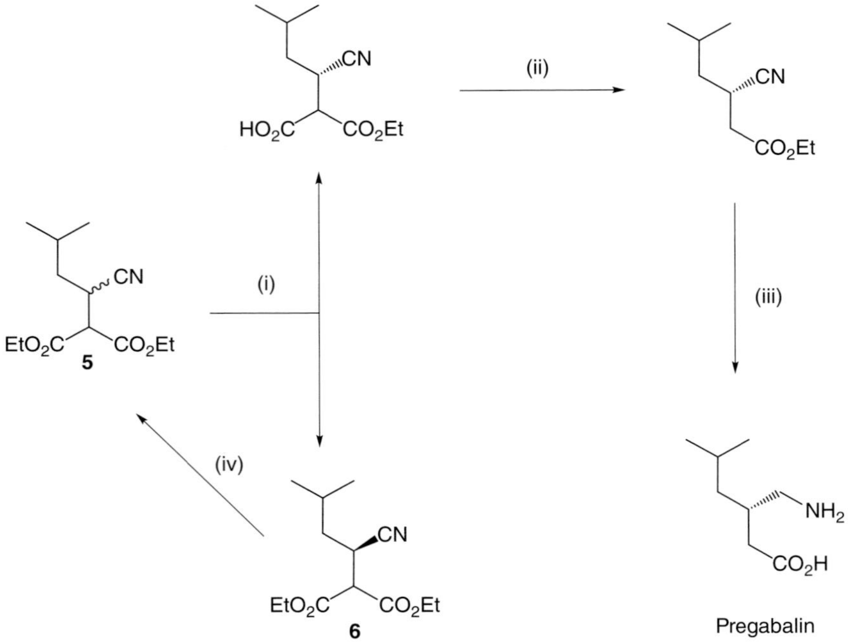

Scheme 7 Biocatalytic route to pregabalin. (i) Lipolase-catalyzed enzymatic resolution performed in water; (ii) thermal decarboxylation performed in water; (iii) hydrolysis (KOH) and hydrogenation (H_2/RaNi), both reactions performed in water; (iv) base-catalyzed epimerization.

Tab. 4 E-factors for commercial pregabalin processes

Year	Process used to supply major markets	E factor[a]
2005–2006	Classical resolution route	86
2006–2009	Biocatalysis route	17
2010 to present	Biocatalysis route with recycle of the (R)-cyanodiester	8

a) Process water not included in the calculations.

meeting of the commercial volume needs of the product. By this stage the process chemist will have a detailed knowledge of the process, including the points at which a step process can be held in processing equipment without affecting quality and yield. This "technology transfer" (TT) generally takes place during Phase 3 clinical studies to ensure a continuity of API supply and also to validate the process in facilities relevant to the commercial production scale. The transfer requires an evaluation of the understanding of the process, and also of the equipment needs of the process. This knowledge transfer is carried out using a process risk assessment, which examines the current state of process knowledge, and defines specific acceptance criteria by which success will be measured.

A TT team will be set up with donor and receiving site representation from a range of disciplines, including chemistry, process engineering, analytical, environmental health and safety (EHS), procurement and regulatory and quality. In preparation for the transfer the team's risk assessment should take account of:

- The complexity of the process being transferred.
- What could go wrong with the transfer (with mitigating actions to be put in place).
- Details of IPCs and in-process assays. Early analytical TT is key to ensuring that measurement systems are in place to enable success criteria to be demonstrated [75].
- What is known, and not known, about the process (including descriptions of what has been shown not to work).
- Any processing and safety concerns with further scale-up.
- The impact of the transfer on the receiving site, including the need for equipment modification or purchase.
- Any environmental, staff training, or work process changes that will be needed at the receiving site.

As the transfer process is a complex exercise, subteams are generally needed to assist with the drawing up of a detailed plan. This TT plan will include actions identified during the risk assessments, documentation and TT deliverables and timelines, strategies for material sourcing and dealing with any EHS issues, an analytical and quality plan, agreement on the "acceptance criteria," and the point at which responsibility is formally transferred to the receiving site. TT implementation starts with the completion of receiving site readiness actions, such as engineering work and equipment installation, analytical TT, safety risk assessments, and staff training. Manufacturing at the new site is then conducted, with careful monitoring of progress to ensure that the process is under control. Statistical or quality tools are used to evaluate and review the data, and to assess performance with respect to process-driven control limits and process parameters, as well as the product specifications and quality attributes. To close the TT process, the initially set success criteria will be reviewed, the level of process understanding should be re-examined, and the TT team will identify both what went smoothly with the handover as well as identifying key areas where the process needs to be modified as part of the future continuous improvement program. Several good textbooks or book chapters are available for readers seeking greater detail about TT [76–78].

During the TT process a plant trial will often be carried out at the new location prior to a formal validation of the API synthetic process. Process validation is then carried out to demonstrate that the plant, equipment, process and control measures provide a quality product that is fit for purpose. The FDA 2011 Guidance defines process validation as "… the collection and evaluation of data, from the process design stage through commercial production, which establishes scientific evidence that a process is capable of consistently delivering quality product" [79, 80]. This Guidance describes three stages of process validation as part of a life cycle approach:

- Stage 1: process design, based on knowledge gained through development and scale-up activities.
- Stage 2: process qualification, in which the process design is evaluated and assessed to determine if the process is capable of reproducible commercial manufacturing.

- Stage 3: process verification, in which continued routine production provides ongoing assurance that the process remains in a state of control.

9.2 Regulatory Filing

Having selected, developed and optimized the API process, and seen it scaled into production equipment, the pharmaceutical process chemist now helps compile the CMC section of the filing to the regulatory agencies. The regulatory authorities require processes to be validated, CPPs to be assessed, and proven acceptable ranges to be established within which the process is robust. The CMC dossier is therefore focused on providing the requisite evidence of control. The traditional approach had been to present an envelope of conditions (referred to as "proven acceptable ranges") within which the manufacturing process could be operated to give API of the required quality. However, regulatory agencies have shifted their focus from ensuring that quality is assured through appropriate testing ("quality by testing"), to a "quality by design" (QbD) philosophy. This QbD paradigm centers upon enhanced process and product understanding, and appropriate risk assessments [81], and favors filings in which knowledge, control and robustness are core values. The use of QbD principles in regulatory filings is still optional, and some key differences between the current minimal approach to pharmaceutical development, and the enhanced QbD approach are summarized in Fig. 11 (this is based on Appendix 1 in the ICH Q8 document) [82].

The concept of a multivariate design space lends itself well to the application of statistical DoE, frequently making use of automated laboratory reactor systems [83]. DoE enables not only the achievement of

Aspects	Traditional (minimal) Approach	Enhanced (QbD) Approach
Overall pharmaceutical development	Mainly empirical Typically uses univariate experiments	Systematic approach Multivariate experiments for process understanding Understand CQAs
Manufacturing process	Fixed Validation on 3 initial batches Focus on optimization/reproducibility	Adjustable within design space Continuous verification within design space Focus on control strategy & robustness
Process control	IPCs for go/no go decisions Off-line analysis (slow response)	PAT used with feed forward/feedback controls Tracking/trending of process operations for post-approval continuous approval effort
Product specifications	Primary means of quality control (QC) Based on registration batch data	Part of overall QC strategy Based on desired product performance (safety and efficacy)
Control strategy	Primarily by intermediate and end product testing	Quality ensured by risk-based strategy with product & process understanding QCs shifted upstream, with potential real-time release or reduced end product testing
Lifecycle management	Reactive to problems, with post-approval changes needed	Continual improvement enabled within the design space

Fig. 11 Options for pharmaceutical development.

true optimization of reactions, but also the assessment of CPPs, and the establishment of proven acceptable ranges within which the process is robust. The changing regulatory authority emphasis toward QbD, in which increased process control and understanding are required, should increase the use of DoE for fine-tuning and quantitating the effect of parameter change on chemical reactions. In addition to DoE, process analytical technology (PAT) is a critical tool with which to gather the relevant data that constitute the fundamentals of the design space.

PAT describes a range of techniques that are used to provide analytical information about a reaction or process in real time, without the need for sample preparation or work-up [84]. The most common technique is on-line or *in-situ* IR spectroscopy, which is a good tool to monitor the changes in chemical bonding in solution, such as the reactions at carbonyls and deprotonation or metallation of intermediates. As well as following the appearance and disappearance of starting materials, reactive intermediates can be monitored directly and reaction systems calibrated to allow quantitative measures of reaction completion or impurity level. On-line IR can also be used to give a much better understanding of the reaction system, helping the process chemist to direct problem-solving efforts at key issues which may not be apparent from off-line analysis by HPLC, thin-layer chromatography or nuclear magnetic resonance. The application of PAT to the control and understanding of crystallization processes is becoming increasingly important. A simple turbidity probe can provide valuable data on nucleation and the onset of crystallization. A number of techniques, including use of the focused beam reflectance method (FBRM), are then available for on-line determination of particle size and distribution in crystallization slurries [85]. In an elegant application of PAT, the Eli Lilly authors detailed how a wide range of on-line techniques (including ReactIR and FBRM) helped them develop, understand, simplify and control the process to the peroxisome proliferator-activated receptor-α agonist LY518674 [86] (Fig. 12).

The understanding of impurities formed at each stage of the regulatory synthesis is important. Evidence of how step impurities track through the synthesis, and how they are purged during the process, is therefore a key element of demonstrating the control of the process. Hence, sufficient control points need to be incorporated into the synthetic sequence to ensure that the final API meets all of the quality criteria. A specification, listing the tests, analytical procedures and appropriate acceptance criteria then establishes the set of criteria to which the API must conform to be considered acceptable for its intended use. Most purifications, either of intermediates or the API, incorporate crystallization operations. Materials science has thus emerged as a foundation of QbD, with the solid form, crystallization and particle engineering being core elements linking the final steps

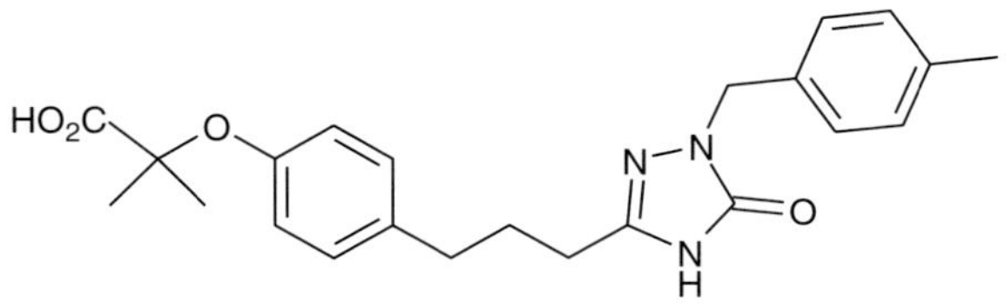

Fig. 12 Peroxisome proliferator-activated receptor-α agonist LY518674.

of the API synthetic pathway to the drug product attributes.

10
Future Challenges

Despite the huge scientific and technological advances made during the past 60 years, there has by a number of measures been a decline in productivity across the pharmaceutical industry. The number of new drugs approved by the US FDA per billion US dollars (inflation-adjusted) spent on research and development has halved roughly every nine years, resulting in an approximate 80-fold fall over this period [87]. In Fig. 13, the number of drugs launched per US$ billion of industry R&D spend is plotted against time on a logarithmic scale. There have been many proposed reasons for, and solutions to, this problem of declining R&D efficiency [88]. There is a growing realization that the traditional drug discovery and development model is not working well, because of three inter-related pressures.

First among these pressures felt within the industry is the issue of rising drug development costs. The worldwide pharmaceutical market continues to grow at a compound rate of over 5% per annum, with projected global sales predicted to rise to about US$1200 billion in 2016 [89]. However, the average cost of developing each NCE through to market as a successful drug has also risen inexorably, and is now estimated at a capitalized cost of about US$1 billion [90]. So, although the rewards of success are high, the discovery and development of NCEs is an extremely risky business.

The second major pressure stems from the decline in productivity across the pharmaceutical industry, as measured by the number of NCEs reaching their first market each year. The decline in both total launches and the number of small-molecule (synthetic and semi-synthetic) NCEs reaching the market during the period 1985–2004, with the balance of the total made up natural products isolated by extraction/fermentation and biomolecules [12], is shown in Table 5. In the four years between 2005 and 2008 the average annual number of small-molecule NCEs reaching the market was 19 [91], compared with approximately 45 during the late 1980s and 34 throughout the 1990s. This suggests that the downward trend is not reversing. The period of 2000–2009 was a particularly woeful one for new products emerging from the biopharmaceutical industry [92]. During this time the FDA approved an average of only 24 new drugs per year – a significant decrease from the 1990s, when 31 new medicines were being approved annually. However, over the past few years this trend has begun to reverse and the FDA approved 30 NCEs in 2011, 39 in 2012, 27 in 2013, and 41 in 2014 [93].

The third area of pressure comes from the regulatory environment, with an estimate that since 1980 the number of clinical trials

Tab. 5 Number of NCEs reaching the market in four-year periods

	1985–1988	1989–1992	1993–1996	1997–2000	2001–2004
Synthetic agents	187	128	144	129	90
Total NCEs	204	141	160	136	113

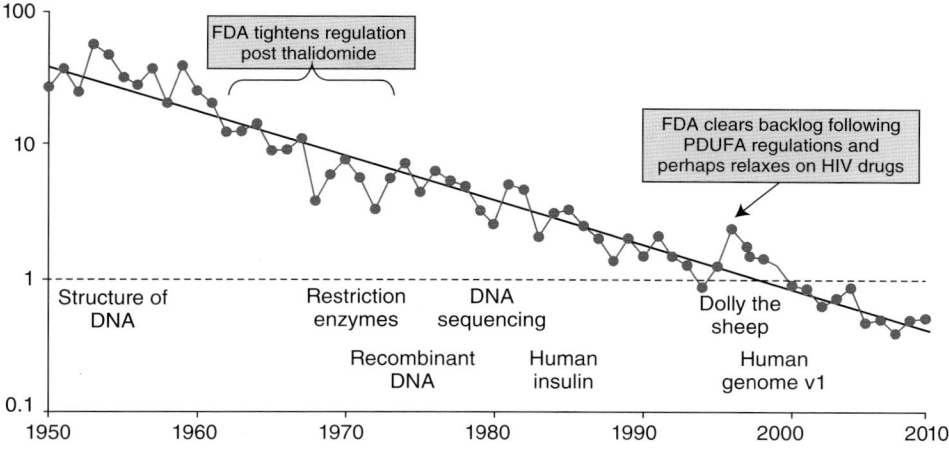

Fig. 13 Inflation-adjusted trend in R&D efficiency. Reproduced with permission from Ref. [88]; © 2012, Nature Publishing Group.

required to support a new-drug application has more than doubled, while almost three times as many patients are needed in each clinical trial. Expensive late-stage failures of agents in Phase III (and even post-launch) are becoming increasingly commonplace, and are often the result of demands made by regulatory authorities for tighter safety requirements, and for clear product differentiation from existing therapies. Analyses by the Centre for Medicines Research International in 2010 and 2011 showed that the number of Phase 3 projects killed off in 2007–2009 was double the rate seen in the previous three years, while the average for the combined success rate at Phase 3 and submission had fallen to about 50% in recent years [94]. A separate analysis showed that since 1990 the rate of attrition has been rising in all phases of drug development, so that the overall survival rates of drug candidates has fallen markedly over that period [95]. A recent attrition analysis of a cohort of 812 oral small-molecule candidates from AstraZeneca, Eli Lilly, GSK and Pfizer confirms that high drug development attrition rates remain a huge concern [96].

The steady fall in the number of new agents reaching the market, and the rise in associated costs, is shown graphically in Fig. 14. Over the past decade, process chemists have made increasing use of productivity-enhancing technologies such as automation, PAT and calorimetric investigations of reaction mechanisms. However, the timelines in Fig. 14 suggest that these technologies have merely helped cope with increased workloads and lower project staffing levels (that is, to run faster to stay where we were). In the future, more will doubtless be done across the industry to routinely use these tools to select, develop and optimize processes more efficiently. A major opportunity to use technology to effect a transformative change does appear to be provided by continuous processing (including for second-generation processes to approved products) which is still heavily under-utilized in pharmaceuticals, compared to other chemical manufacturing areas. Although recent examples of the use

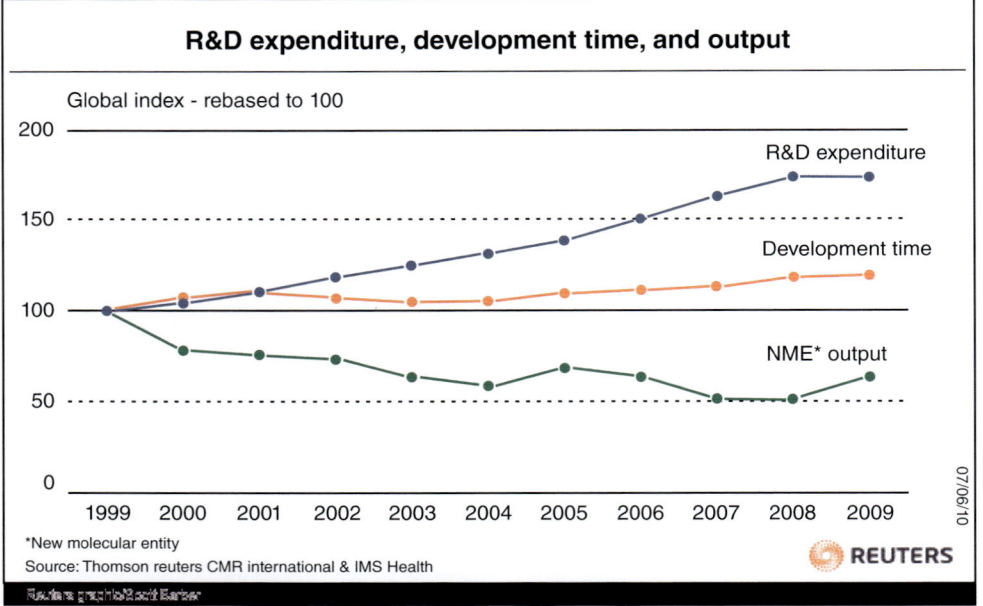

Fig. 14 Pharmaceutical R&D expenditure, timelines, and output (1999–2009). Reproduced by permission of CMR International, a Thomson Reuters business.

of continuous processing in the pharmaceutical industry have been presented [37], the pharmaceutical industry has barely scratched the surface of the opportunities that continuous processing offers. A broader implementation of continuous processing seems inevitable, but this will require close collaboration between chemists, engineers and, ultimately, a mindset change amongst the chemists devising early synthetic routes.

Yet, despite a difficult past decade the industry as a whole is not in decline, with global demand for better healthcare continuing to grow. It seems likely that in future an increasing proportion of the industry's drug candidates will be discovered by, or in collaboration with, smaller "biotech" units or academic laboratories [97]. However, because of the advantages of scale during the expensive development phases, most drug candidates will probably continue to be developed by process R&D scientists in large companies. The growth in orphan drugs and personalized medicines, together with indications that the long awaited impact of the genomics revolution is imminent, should ensure a sufficient flow of drug candidates to be developed. However, a key challenge will be to find ways of bringing the next generation of NCEs (especially the niche products) through to the market more cost effectively.

References

1. Smith C.G. and J. Y. O'Donnell (eds) (2006) *The Process of New Drug Discovery and Development*, 2nd edn, Informa Healthcare USA Inc., New York.
2. Adams C.P. and Brantner, V.V. (2010) Spending on new drug development. *Health Econ.*, **19**, 130–141.
3. DiMasi, J.A., Hansen, R.W., and Grabowski H.G. (2003) The price of innovation: new

estimates of drug development costs. *J. Health Econ.*, **22**, 151–185.
4. DiMasi, J.A., Grabowski, H.G., and Vernon, J. (2004) R&D costs and returns by therapeutic category. *Drug Inf. J.*, **38**, 211–223.
5. Anderson, N.G. (2000) *Practical Process Research & Development*, Academic Press, San Diego, CA.
6. Gadamasetti, K. and Braish, T. (eds) (2008) *Process Chemistry in the Pharmaceutical Industry*, Vol. 2, CRC Press, Boca Raton, FL.
7. Blacker, A.J. and Williams, M.T. (eds) (2011) *Pharmaceutical Process Development, Current Chemical and Engineering Challenges*, RSC Publishing, Cambridge.
8. Peterson, D.H. and Murray, H.C. (1952) Microbiological oxidation of steroids at carbon 11. *J. Am. Chem. Soc.*, **74**, 1871–1872.
9. Lipinski, C.A. (2004) Lead and drug-like compounds: the rule-of-five revolution. *Drug Discovery Today Technol.*, **1**, 337–341.
10. Carey, J.S., Laffan, D., Thomson, C., and Williams, M.T. (2006) Analysis of the reactions used for the preparation of drug candidate molecules. *Org. Biomol. Chem.*, **4**, 2337–2347.
11. Farina, V., Reeves, J.T., Senanayake, C.H., and Song, J.J. (2006) Asymmetric synthesis of active pharmaceutical ingredients. *Chem. Rev.*, **106**, 2734–2793.
12. Murakami, H. (2007) From racemates to single enantiomers – chiral synthetic drugs over the last 20 years. *Top. Curr. Chem.*, **269**, 273–299.
13. Andrews, I.P., Atkins, R.J., Breen, G.F., Carey, J.S., *et al.* (2003) The development of a manufacturing route to the GPIIb/GPIIIa receptor antagonist SB-214857. Part 1: synthesis of the key intermediate 2,3,4,5-tetrahydro-4-methyl-3-oxo-1*H*-1,4-benzodiazepine-2-acetic acid methyl ester, SB-235349. *Org. Process Res. Dev.*, **7**, 655–662.
14. Atkins, R.J., Banks, A., Bellingham, R.K., Breen, G.F., *et al.* (2003) The development of a manufacturing route to the GPIIb/GPIIIa receptor antagonist SB-214857. Part 2: conversion of the key intermediate SB-235349 to SB-214857-A. *Org. Process Res. Dev.*, **7**, 663–675.
15. Butters, M., Catterick, D., Craig, A., Curzons, A., *et al.* (2006) Critical assessment of pharmaceutical processes – a rationale for changing the synthetic route. *Chem. Rev.*, **106**, 3002–3027.
16. Office of Technology Assessment (1993) Pharmaceutical R&D, Costs, Risks and Rewards, Appendix G, Office of Technology Assessment, US Congress, Washington, DC, p. 312.
17. Rylander, P. (1979) *Catalytic Hydrogenation in Organic Synthesis*, Academic Press, San Diego, CA.
18. Blaser, H.U. and Schmidt, E. (eds) (2004) *Asymmetric Catalysis on Industrial Scale: Challenges, Approaches and Solutions*, Wiley-VCH Verlag GmbH, Weinheim.
19. Sheldon, R.A., Arends, I.W.C.E., and Hanefeld, U. (eds) (2007) *Green Chemistry and Catalysis*, Wiley-VCH Verlag GmbH, Weinheim.
20. Dunn, P.J., Hii, K.K., Krische, M.J., and Williams, M.T. (eds) (2013) *Sustainable Catalysis: Challenges and Practices for the Pharmaceutical and Fine Chemical Industries*, John Wiley & Sons, Inc., Hoboken, NJ.
21. Patel, R.N. (2008) Synthesis of chiral pharmaceutical intermediates by biocatalysis. *Coord. Chem. Rev.*, **252**, 659–701.
22. Tao, J., Zhao, L., and Ran, N. (2007) Recent advances in developing chemoenzymatic processes for active pharmaceutical ingredients. *Org. Process Res. Dev.*, **11**, 259–267.
23. Patel, R.N. (2006) Biocatalysis: synthesis of chiral intermediates for pharmaceuticals. *Curr. Org. Chem.*, **10**, 1289–1321.
24. Dunn, P.J., Hettenbach, K., Kelleher, P., and Martinez, C.A. (2010) The development of a green, energy efficient, chemoenzymatic manufacturing process for pregabalin, in *Green Chemistry in the Pharmaceutical Industry* (eds P.J. Dunn, A.S. Wells, and M.T. Williams), Wiley-VCH Verlag GmbH, Weinheim, pp. 161–177.
25. Balsells, J., Hsiao, Y., Hansen, K.B., Xu, F., *et al.* (2010) Synthesis of sitagliptin, the active ingredient in Januvia® and Janumet®, in *Green Chemistry in the Pharmaceutical Industry* (eds P.J. Dunn, A.S. Wells, and M.T. Williams), Wiley-VCH Verlag GmbH, Weinheim, pp. 101–126.
26. Kierkels, H., Panke, S., Schurmann, M., and Wolberg, M (2010) The development of short, efficient, economic, and sustainable chemoenzymatic processes for statin side chains, in *Green Chemistry in the Pharmaceutical Industry*, (eds P.J. Dunn, A.S.

Wells and M.T. Williams), Wiley-VCH Verlag GmbH, Weinheim, pp. 127–144.
27. Anderson, N.G. (2000) Solvent Selection, in *Practical Process Research & Development*, Academic Press, San Diego, CA, pp. 81–111.
28. Ashcroft, C.P., Dunn, P.J., Hayler, J.D., Wells, A.S. (2015) Survey of solvent usage in papers published in *Organic Process Research and Development* 1997–2012. *Org. Process Res. Dev.*, **19**, 740–747.
29. ICH Harmonized Tripartite Agreement (2011, step 4 version) Impurities: Guidelines for Residual Solvents, Q3C.
30. Buckley, J., Webb, R.L., Laird, T., and Ward, R.J. (1982) Report on thermal reaction. *Chem. Eng. News*, **60** (28), 5.
31. Mills, J.E., Maryanoff, C.A., Cosgrove, R.A., Scott, A., et al. (1984) The reactions of amines with methylene chloride. A brief review. *Org. Prep. Proc. Int.*, **16**, 97–114.
32. Ferretti, A.C., Mathew, J.S., Ashworth, I., Purdy, M., et al. (2008) Mechanistic inferences derived from competitive catalytic reactions Pd(binap) – catalyzed amination of aryl halides. *Adv. Synth. Catal.*, **350**, 1007–1012.
33. Carlson, R. and Carlson, J.E. (2005) *Design and Optimisation in Organic Synthesis*, 2nd edn, Elsevier, Amsterdam.
34. Rooney, T.A. (1998) Statistical design: how to optimize chemical experiments and processes. *Today's Chem. Work*, **7** (12), 21–32.
35. Chen, C.-K. and Singh, A.K. (2001), The "bottom-up" approach to process development: application of physicochemical properties of reaction products toward the development of direct drop processes. *Org. Process Res. Dev.*, **5**, 508–513.
36. Roberge, D.M. (2004), An integrated approach combining reaction engineering and design of experiments for optimizing reactions. *Org. Process Res. Dev.*, **8**, 1049–1053.
37. Proctor, L., Dunn, P.J., Hawkins, J.M., Wells, A.S. and Williams, M.T. (2010) Continuous processing in the pharmaceutical industry, in *Green Chemistry in the Pharmaceutical Industry* (eds P.J. Dunn, A.S. Wells and M.T. Williams), Wiley-VCH Verlag GmbH, Weinheim, pp. 221–242.
38. Nolan, P.F. and Barton, J.A. (1987) Some lessons from thermal-runaway incidents. *J. Hazard. Mater.*, **14**, 233–239. A survey of over 200 UK runaway reactions in the 1970s and 1980s.
39. Urben, P.G. (2005) *Bretherick's Handbook of Reactive Chemical Hazards*, 7th edn, Academic Press, Burlington, MA.
40. For the latest special safety issue see, *Org. Process Res. Dev.*, 2014, **18**, pp 1777–1849.
41. Weisenburger, G.A., Barnhart, R.W., Clark, J.D., Dale, D.J., et al (2007) Determination of reaction heat: a comparison of measurement and estimation techniques. *Org. Process Res. Dev.*, **11**, 1112–1125.
42. Dale, D.J. (2002) A comparison of reaction hazard screening techniques. *Org. Process Res. Dev.*, **6**, 933–937.
43. McIntosh, R.D. and Waldram, S.P. (2003) Obtaining more, and better, information from simple ramped temperature screening tests. *J. Therm. Anal. Calorim.*, **73**, 35–52.
44. Townsend, D. and Tou, J.C. (1980) Thermal hazard evaluation by an accelerating rate calorimeter. *Thermochim. Acta*, **37**, 1–30.
45. Mettler http://www.mettler.com (accessed 29 July 2015).
46. Hamilton, S. and Bertrand, A. (2011) The analytical interface and the impact on pharmaceutical process development, in *Pharmaceutical Process Development: Current Chemical and Engineering Challenges* (eds A.J. Blacker and M.T. Williams), RSC Publishing, Cambridge, pp. 260–285.
47. Perrin, C., Vu, V.A., Matthijs, N., Maftouh, M., et al. (2002) Screening approach for chiral separation of pharmaceuticals: Part 1. Normal phase liquid chromatography. *J. Chromatogr. A*, **947**, 69–83.
48. Pierson, D.A., Olsen, B.A., Robbins, D.K., DeVries, K.M., et al. (2009) Approaches to assessment, testing decisions, and analytical determination of genotoxic impurities in drug substances. *Org. Process Res. Dev.*, **13**, 285–291.
49. Elder, D.P. and Snodin, D.J. (2009) Drug substances presented as sulfonic acid salts: overview of utility, safety and regulation. *J. Pharm. Pharmacol.*, **61**, 269–278.
50. Lin, Z., Zhou, L., Mahajan, A., Song, S., et al. (2006) Real-time endpoint monitoring and determination for a pharmaceutical salt formation process with in-line FT-IR spectroscopy. *J. Pharm. Biomed. Anal.*, **41**, 99–104.
51. Roberts, K., Docherty, R., and Taylor, S. (2011) Materials science: solid form design

52. Hilfiker, R. (ed.) (2006) *Polymorphism in the Pharmaceutical Industry*, Wiley-VCH Verlag GmbH, Weinheim.
53. Brittain, H.G. (ed.) (2009) *Polymorphism in Pharmaceutical Solids*, Informa Healthcare, New York.
54. Chemburkar, S.R., Bauer, J., Deming, K., Spiwek, H., et al. (2000) Dealing with the impact of ritonavir polymorphs on the late stages of bulk drug process development. *Org. Process Res. Dev.*, **4**, 413–417.
55. Environmental Protection Agency www.epa.gov/greenchemistry (accessed 29 July 2015).
56. Anastas, P.T. and Farris, P.A. (eds) (1994) *Benign by Design: Alternative Synthetic Design for Pollution Prevention*, ACS Symposium Series, Vol. 577, American Chemical Society, Washington, DC.
57. Anastas, P.T. and Warner, J.C. (eds) (1998) *Green Chemistry: Theory and Practice*, Oxford University Press, Oxford.
58. Clark, J.H. and Macquarrie, D.J. (eds) (2002) *Handbook of Green Chemistry and Technology*, Blackwell, Abingdon.
59. Dunn, P.J., Wells, A.S., and Williams, M.T. (eds) (2010) *Green Chemistry in the Pharmaceutical Industry*, Wiley-VCH Verlag GmbH, Weinheim.
60. Trost, B.M. (1991) The atom economy – a search for synthetic efficiency. *Science*, **254**, 1471–1477.
61. Sheldon, R.A. (1992) Organic synthesis – past, present and future. *Chem. Ind. (London)*, **23**, 903–906.
62. Constable, D.J.C., Curzons, A.D., and Cunningham, V.L. (2002) Metrics to "green" chemistry – which are best? *Green Chem.*, **4**, 521–527.
63. Slater, C.S., Savelski, M.J., Carole, W.A., and Constable, D.J.C. (2010) Solvent use and waste issues, in *Green Chemistry in the Pharmaceutical Industry* (eds P.J. Dunn, A.S. Wells, and M.T. Williams), Wiley-VCH Verlag GmbH, Weinheim, pp. 49–82.
64. Jiminez-Gonzalez, J., Curzons, A.D., Constable, D.J.C., and Cunningham, V.L. (2005) Expanding GSK's Solvent Selection Guide – application of life cycle assessment to enhance solvent selections. *Clean Technol. Environ. Policy*, **7**, 42–50.
65. Alfonsi, K., Colberg, J., Dunn, P.J., Fevig, T., et al. (2008) Green chemistry tools to influence a medicinal chemistry and research chemistry based organisation. *Green Chem.*, **10**, 31–36.
66. Herrerias, I., Yao, X., Li, Z., and Li, C.-J. (2007) Reactions of C–H bonds in water. *Chem. Rev.*, **107**, 2546–2562.
67. Blackmond, D., Armstrong, A., Coombe, V., and Wells, A. (2007) Water in organocatalytic processes: debunking the myths. *Angew. Chem. Int. Ed.*, **46**, 3798–3800.
68. Holton, R.A., Somoza, C., Kim, H.-B., Liang, F., et al. (1994) First total synthesis of taxol. 1. Functionalization of the B ring. *J. Am. Chem. Soc.*, **116**, 1597–1598.
69. Holton, R.A., Somoza, C., Kim, H.-B., Liang, F., et al. (1994) First total synthesis of taxol. 2. Completion of the C and D rings. *J. Am. Chem. Soc.*, **116**, 1599–1600.
70. Denis, J.N., Greene, A.E., Guenard, D., Gueritte-Voegelein, F., et al. (1988) A highly efficient, practical approach to natural taxol. *J. Am. Chem. Soc.*, **110**, 5917–5919.
71. Holton, R.A. (1992) Method for preparation of taxol using an oxazinone. US Patent 5,136,060, issued Aug. 4, 1992.
72. Mountford, P.G. (2010) The taxol story-development of a green synthesis via plant cell fermentation, in *Green Chemistry in the Pharmaceutical Industry* (eds P.J. Dunn, A.S. Wells, and M.T. Williams), Wiley-VCH Verlag GmbH, Weinheim.
73. Hoekstra, M.S., Sobieray, D.M., Schwindt, M.A., Mulhern, T.A., et al. (1997) Chemical development of CI-1008, an enantiomerically pure anticonvulsant. *Org. Process Res. Dev.*, **1**, 26–38.
74. Yazbeck, D.R., Tao, J., Martinez, C.A., Kline, B.J., et al. (2003) Advanced enzyme screening methods for the preparation of enantiopure pharmaceutical intermediates. *Adv. Synth. Catal.*, **345**, 524–532.
75. Perry, S. (2010) Tech transfer: do it right first time. *Pharm. Manuf.*, **9** (1), 16–17.
76. Green, S. and Warren, P. (2002) *Technology Transfer in Practice*, Horwood, Storrington.
77. Gibson, M. (ed.) (2005) *Technology Transfer: An International Good Practice Guide for Pharmaceutical and Allied Industries*, DHI Publishing, River Grove, IL.

78. McGhie, S. and Young, S. (2011) Technology transfer of an active pharmaceutical ingredient, in *Pharmaceutical Process Development: Current Chemical and Engineering Challenges* (eds A.J. Blacker and M.T. Williams), RSC Publishing, Cambridge, pp. 317–330.
79. Katz, P. and Campbell, C. (2012) FDA 2011 process validation: process validation revisited. *J. GXP Compliance*, **16** (4), 18–29.
80. Carson, P. and Dent, N. (eds) (2007) *Good Clinical, Laboratory and Manufacturing Practices: Techniques for the QA Professional*, Royal Society of Chemistry, Cambridge.
81. FDA (May 2007) Pharmaceutical Quality for the 21st Century: A Risk Based Approach Progress Report.
82. ICH Harmonised Tripartite Guideline (2009 step 4 version) Pharmaceutical Development Q8.
83. Kirchhoff, E.W., Anderson, D.R., Zhang, S., Cassidy, C.S., *et al.* (2001) Automated process research and the optimization of the synthesis of 4(5)-(3-pyridyl)imidazole. *Org. Process Res. Dev.*, **5**, 50–53.
84. Mojica, C.A., St Pierre-Berry, L., and Sistare, F. (2008) Process analytical technology in the manufacture of bulk active pharmaceuticals-promise, practice and challenges, in *Process Chemistry in the Pharmaceutical Industry*, Vol. 2 (eds K. Gadamasetti and T. Braish), CRC Press, Boca Raton, FL.
85. Heath, A.R., Fawell, P.D., Bahri, P.A., and Swift, J.D. (2002) Estimating average particle size by focused beam reflectance measurement (FBRM). *Part. Part. Syst. Charact.*, **19**, 84–95.
86. Argentine, M.D., Braden, T.M., Czarnik, J., Conder, E.W., *et al.* (2009) The role of new technologies in defining a manufacturing process for PPARα agonist LY518674. *Org. Process Res. Dev.*, **13**, 131–143.
87. Bernstein Research Report (2010) The Long View – R&D Productivity.
88. Scannell, J.W., Blankley, A., Boldon, H., and Warrington, B. (2012) Diagnosing the decline in pharmaceutical R&D efficiency. *Nat. Rev. Drug Discovery*, **11** (3), 191–200.
89. Report by IMS Institute for Healthcare Informatics (2012) The Global Use of Medicines: Outlook through 2016. As discussed in: Scannell, J.W., Blanckley, A., Bolden, H., and Warrington, B. (2012) Diagnosing the decline in pharmaceutical R&D efficiency. *Nat. Rev. Drug Discov.*, **11**, 191–200.
90. Adams, C.P. and Brantner, V.V. (2006) Estimating the cost of new drug development: is it really $802 million? *Health Aff.*, **25**, 420–428.
91. (a) Hegde, S. and Schmidt, M. (2006) To Market, to Market – 2005. *Annu. Rep. Med. Chem.*, **41**, 439–477; (b)Hegde, S. and Schmidt, M. (2007) To Market, to Market – 2006. *Annu. Rep. Med. Chem.*, **42**, 505–554; (c) Hegde, S. and Schmidt, M. (2008) To Market, to Market – 2007. *Annu. Rep. Med. Chem.*, **43**, 455–497; (d)Hegde, S. and Schmidt, M. (2009) To Market, to Market – 2008. *Annu. Rep. Med. Chem.*, **44**, 577–632
92. Kaitin, K.I. and DiMasi, J.A. (2011) Pharmaceutical innovation in the 21st century: new drug approvals in the decade 2000–2009. *Clin. Pharmacol. Ther.*, **89** (2), 183–188.
93. Mullard, A. (2015) 2014 FDA drug approvals. *Nat. Rev. Drug Discovery*, **14**, 77–81.
94. Arrowsmith, J. (2011) Trial watch: phase III and submission failures: 2007–2010. *Nat. Rev. Drug Discovery*, **10**, 87.
95. Pammolli, F., Riccaboni, M., and Magazzini, L. (2010) The Productivity Crisis in Pharmaceutical R&D. Working Paper Series, University of Verona.
96. Waring, M.J., Arrowsmith, J., Leach, A.R., Leeson, P.D., *et al.* (2015) An analysis of the attrition of drug candidates from four major pharmaceutical companies. *Nat. Rev. Drug Discovery*, **14**, 475–486.
97. Kneller, R. (2010) The importance of new companies for drug discovery: origins of a decade of new drugs. *Nat. Rev. Drug Discovery*, **9**, 867–882.

15
High-Performance Liquid Chromatography of Peptides and Proteins

Reinhard I. Boysen
Monash University, Australian Centre for Research on Separation Science, School of Chemistry, Wellington Road, Melbourne, 3800 Victoria, Australia

1	**Introduction** 533	
2	**Instrumentation** 534	
3	**Fundamental Terms and Concepts** 535	
3.1	Liquid Chromatography 535	
3.2	Principles of Green Analytical Chemistry 539	
4	**Physico-Chemical Properties of Peptides and Proteins** 540	
5	**Chromatographic Separation Modes for Peptides and Proteins** 543	
5.1	High-Performance Size-Exclusion Chromatography (HP-SEC) 544	
5.2	High-Performance Reversed-Phase Chromatography (HP-RPC) 545	
5.3	High-Performance Normal-Phase Chromatography (HP-NPC) 546	
5.4	High-Performance Hydrophilic Interaction Chromatography (HP-HILIC) 547	
5.5	High-Performance Aqueous Normal Phase Chromatography (HP-ANPC) 548	
5.6	High-Performance Hydrophobic Interaction Chromatography (HP-HIC) 549	
5.7	High-Performance Ion-Exchange Chromatography (HP-IEX) 551	
5.8	High-Performance Affinity Chromatography (HP-AC) 552	
6	**Method Development from Analytical to Preparative Scale** 553	
6.1	Development of an Analytical Method 554	
6.1.1	Optimization of Column Efficiency 555	
6.1.2	Optimization of Selectivity 555	
6.1.3	Optimization of Retention Factors 556	
6.2	Scaling Up to Preparative Chromatography 557	
6.3	Fractionation 559	
6.4	Analysis of Fractions 559	
7	**Multidimensional HPLC** 559	

Translational Medicine: Molecular Pharmacology and Drug Discovery
First Edition. Edited by Robert A. Meyers.
© 2018 Wiley-VCH Verlag GmbH & Co. KGaA. Published 2018 by Wiley-VCH Verlag GmbH & Co. KGaA.

7.1	Purification of Peptides and Proteins using MD-HPLC Methods 560
7.2	Fractionation of Complex Peptide and Protein Mixtures using MD-HPLC 561
7.3	Strategies for MD-HPLC Methods 562
7.3.1	*Off-Line* Mode 562
7.3.2	*On-Line* Mode 562
7.4	Design of an Effective MD-HPLC Scheme 562

8 Final Remarks 564

Abbreviations 565

Symbols 566

References 567

Keywords

Peptides/Proteins
Peptides and proteins are biological molecules composed of amino acids.

Chromatography
A separation method where the soluble components of a sample are distributed between two phases, one of which is stationary, while the other – the mobile phase – moves either in either an axial or a radial direction, depending on the column design features, with the analytes migrating in a predetermined, defined direction.

Elution
The process of a mobile phase (eluent) passing through a column to transport solutes.

Gradient elution
Elution mode where the composition of the eluents is changed during the separation in order to increase the elution strength of the mobile phase over time.

High-performance liquid chromatography
Fully instrumented form of liquid chromatography that uses columns with small chromatographic particles or monoliths under high operating pressures.

Isocratic elution
Elution mode where the composition of the eluents remains constant during the separation.

Multidimensional HPLC
The use of chromatographic techniques in succession to achieve a better separation. It offers the possibility of separating the eluate into consecutive fractions, which can be treated independently in subsequent separations.

Orthogonality
Statistical independence of the retention times of the same set of analytes separated by two chromatographic modes that differ in their retention mechanisms or selectivities.

Peak capacity
The number of well-resolved peaks that can be produced between the first and the last peak in a separation of defined resolution.

Retention factor
The retention factor k is used to describe the retention of an analyte, independent of the column dimensions or the flow rate, and relates to the number of additional column volumes required (beyond the dead volume) to elute a compound.

Resolution
The resolution, R_S, of two adjacent peaks in a chromatogram is defined by the ratio of peak distance to their peak widths at baseline. Two peaks can be either unseparated ($R_S < 1$), partially overlapping ($R_S = 1$), or baseline separated ($R_S > 1.5$).

Selectivity
The selectivity, α, of a chromatographic system describes the ability of that system to separate two compounds based on their differing retention factors.

> High-performance liquid chromatography (HPLC) is a fully automated form of liquid chromatography that allows separations of high speed, high resolution, and high sensitivity. HPLC, when coupled to mass spectrometry, can be used for the separation, detection, identification and quantification of a very large number of individual components in a sample, thus enabling challenging peptide and protein analyses in complex matrices. In this chapter, the major separation modes, including size-exclusion, reversed-phase, normal-phase, hydrophilic interaction, aqueous normal phase, hydrophobic interaction, ion-exchange, and affinity chromatography, and their applications to the analysis of peptides and proteins, are described. Method development approaches for analytical HPLC procedures and their upscaling to a preparative level are discussed in the context of analytical "green chemistry" and process analytical technologies. Various options for multidimensional HPLC, involving different separation modes in sequence for the efficient analysis and purification complex peptide and protein mixtures, are presented. Recognition of the opportunities for HPLC in systems biology, medical diagnostics, personalized medicine, and in the manufacture of biotherapeutics are currently the drivers of interdisciplinary research in sorbent development, surface chemistry, mathematical modeling of chromatographic processes, protein engineering, and bioinformatics.

1
Introduction

Over the past four decades, high-performance liquid chromatography (HPLC) has become an indispensable tool for the separation, purification and characterization of chemical and biological molecules [1]. Despite this progress, the use of HPLC for the isolation of naturally occurring peptides or proteins from complex biological matrices, and for their subsequent unambiguous structural elucidation, still presents considerable challenges, for example in the fields of peptidomics, proteomics, and degradomics.

Since proteins can occur as structural isoforms (as the result of chemical or biological post-translational modifications) or protein variants (from genetic modifications), their identification requires the use of mass spectrometry (MS) and bioinformatic analysis. Such techniques may also be required for the purification of synthetic peptides and recombinantly produced proteins. Further structural elucidation can be performed using multiple-stage MS, that is, with ion-trap mass spectrometry, matrix-assisted laser desorption/ionization time of flight (MALDI TOF) mass spectrometry, orbitrap, quadrupole or Fourier transform mass spectrometry, or with proton, carbon or heteronuclear nuclear magnetic resonance (NMR)-spectroscopy.

In order to develop a purification strategy tailored to a target peptide or protein, an in-depth knowledge of the available instrumentation, of the fundamental terms and concepts in chromatography, and of the basic physico-chemical characteristics of peptides and proteins is required. These aspects are considered in Sects 2–4. The major chromatographic modes for the isolation and purification of peptides and proteins are described in Sect. 5. In order to take full advantage of a specific HPLC mode and to utilize time and resources effectively, a comprehensive method development is always recommended. Guidelines for the method development from analytical to preparative scale in high-performance reversed-phase chromatography (HP-RPC), the major chromatographic mode employed in peptide and protein analysis, are given in Sect. 6. As peptides and proteins are generally isolated from complex biological matrices, more than one chromatographic step is very often needed for their purification. The successive application of several suitable different chromatographic modes must consider their applicability to the peptide or protein of interest, as well as their compatibility with each other and with the detection procedures. The concepts and implementations of two-dimensional (2D) or higher-dimensional separation schemes for peptide and protein purification are discussed in Sect. 7.

2
Instrumentation

The configuration of a liquid chromatography (LC) system to be optimally suited for a particular application is of considerable importance in the field of bioanalysis. Modern LC instrumentation provides many possibilities to configure a separation system using individual modules or, alternatively a compact, preconfigured chromatographic system. A simple LC system consists of a binary pump, an injection system, a column, and a detector. The mobile phase can be pumped through the column at atmospheric pressure (open systems) or at elevated pressure, in medium-, high-, or ultrahigh-pressure systems. The development of HPLC resulted in significant advances in speed, resolution, and automation, and advanced to an important chromatographic method in analytical biochemistry and systems biology. According to need, additional components can be added to a chromatographic system, including an eluent degasser, a (thermostated) autosampler, a column oven with integrated valves for column switching, a T-junction for eluent splitting, a micro-spotter, a fraction collector, and so on. The individual components can be controlled manually or with a personal computer, which can also be used to record and analyze the chromatograms. A chromatographic system can be either tailored for a particular application, for example, for high-throughput (which may require a high-throughput autosampler and a robot which loads the autosampler), for method development, for trace analysis (which may require a fluorescence detector), for 2D separation (which requires an additional pump and switching valves), or for preparative purification, whereby the individual system components are integrated accordingly.

Three recent developments have had an immense impact on the field of bioanalysis, namely the coupling (hyphenation) of LC with MS, the miniaturization of separation columns, and the general availability of bioinformatics, and these represent important milestones. The LC/MS-coupling enables the high-sensitivity detection of molecular masses of peptides and proteins, the application of tandem or high-resolution MS procedures for the identification of peptides and associated post-translational modifications, and the *de-novo* sequencing of peptides as well as the relative (or absolute) quantification of peptides directly after a chromatographic separation. Following chromatographic separation, the eluate from the column is infused into an electrospray ionization mass spectrometry (ESI-MS) system, where the separated peptides and proteins are sequentially desolvated, ionized, led through the ion optics, and then detected with a mass analyzer. The optimum utilization of the electrospray technique requires a constant and low flow rate of the infused eluent stream, which can be conveniently produced by micro- or nano- LC systems. Since the mass spectrometer is a concentration-dependent detector, a decrease in the LC flow rate will result in an increase in detection sensitivity. Furthermore, the utilization of a micro-spotter allows the deposition of eluate (and matrix) on sample plates suitable for subsequent MALDI TOF MS. The development of analytic micro- and nano-LC, which can be categorized as capillary-LC, and the decreasing flow rates and column diameters, reflect a trend of increasing miniaturization. This has resulted in the development of HPLC chips with integrated enrichment and separation columns [2, 3], with the individual components of these capillary HPLC systems being designed for specific tasks. The micro- or nano-LC pumps are designed for very small flow rates, and the diameter of the connecting capillary tubing, valve volumina, and detector cell volumina are adapted accordingly. For complex samples, as occur in proteomics, the HPLC systems are necessarily multidimensional, which means that they integrate two or more different chromatographic separation modes, for example, ion-exchange chromatography (IEX) and reversed-phase chromatography (RPC). The analysis of mass spectroscopic data from such a multidimensional (multistage, multicolumn) high-performance liquid chromatography (MD-HPLC) system requires access to modern bioinformatics software and sequence data banks.

3
Fundamental Terms and Concepts

3.1
Liquid Chromatography

Chromatography can be defined as a separation method, whereby the soluble components of a sample are distributed between two phases, one of which is stationary while the other, the mobile phase, moves either in an axial or a radial direction depending on the column design features, with the analytes migrating in a predetermined, defined direction.

A chromatographic separation begins with the choice of the mobile phase conditions, followed by injection of the sample onto the column. The individual analytes in the sample, carried by the mobile phase, migrate through the chromatographic column with differing velocities, depending on the elution methods employed, and are detected, for example, by ultraviolet (UV) spectroscopy or MS, in the order in which they leave the column. The resulting chromatogram plots the response of a detector to the concentration of the compound (peak) versus the length of time it takes to elute it from the column (retention time) (see Fig. 1). The time that an unretained compound needs to migrate through the column is the column void time t_0, and this is related to the void volume V_0, which is the sum of the interstitial volume between the particles of the stationary phase and the available volume within the particle pores. The elution of sample components are delayed through their interaction with the stationary phase, which is described by the retention time t_R or by the retention volume V_R.

The retention factor k is used to describe the retention independently of the column dimensions or flow rate:

$$k = \frac{t_R - t_0}{t_0} \quad (1)$$

where t_R is the retention time and t_0 is the column void time. Alternatively, the retention factor can be expressed in terms of elution volumes, since the retention times, t_R and t_0, are related to the elution volumes and the flow rate F of the chromatographic system through the relationships:

$$V_R = t_R F \text{ and } V_0 = t_0 F \quad (2)$$

hence

$$k = \frac{V_R - V_0}{V_0}. \quad (3)$$

Thus, the retention factor relates to the number of additional column volumes beyond V_0 required to elute a compound. The retention factor can have values between $k = 0$ (no retention) and $k = \infty$ (irreversible adsorption), with values between 1 and 20 mostly preferred for practical and economic reasons. The retention factor can also be defined as the ratio n_s/n_m, where n_s is the total number of moles of the solute associated with the stationary phase,

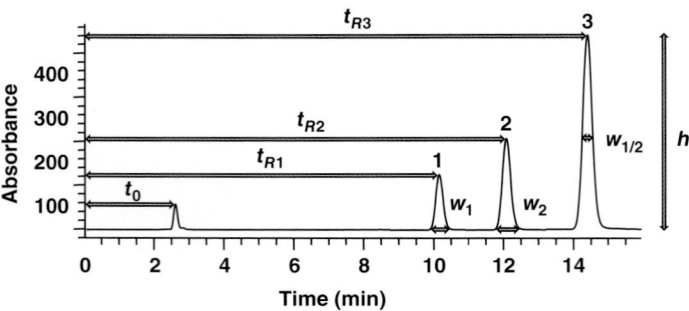

Fig. 1 Schematic depiction of a chromatogram. Abbreviations: t_0 = column void time (time from injection to detection of unretained compound), t_{R1}, t_{R2}, t_{R3} = retention times (time from injection to detection of retained compounds), h = peak height, w = peak width (measured at baseline), and $w_{1/2}$ = peak width at half height (measured to characterize column efficiency).

and n_m is the total number of moles of the solute in the mobile phase:

$$k = \frac{n_s}{n_m}. \quad (4)$$

In order to resolve two components, their retention factors must be different. The selectivity α of a chromatographic system describes the ability of a chromatographic system to separate two compounds (1 and 2) based on their different retention factors k_1 and k_2. For a selectivity of $\alpha = 1$ no separation is possible. The selectivity is defined as:

$$\alpha = \frac{k_2}{k_1}. \quad (5)$$

In order to evaluate the quality of a separation, not only the peak distance between the two components (measured at maximum height) must be considered but also their respective peak width. The resolution, R_S, of two adjacent peaks in a chromatogram is defined by the ratio of peak distance and their peak widths:

$$R_S = \frac{t_{R2} - t_{R1}}{(1/2)(w_1 + w_2)} \quad (6)$$

with retention times, t_{R1} and t_{R2}, of two adjacent peaks and the respective peak widths, w_1 and w_2, of the two peaks. Two peaks can be either unseparated ($R_S < 1$), partially overlapping ($R_S = 1$), or baseline separated ($R_S > 1.5$).

Dispersion effects of solutes in the chromatographic system are one cause of peak broadening. The extent of peak broadening depends on the column efficiency, which is usually expressed as the plate number, N, or as the plate height, H (also called the height equivalent to one theoretical plate, $HETP$),

$$N = \frac{L}{H}, \quad (7)$$

where L is the column length.

The concept of "theoretical plates" goes back to the number of distillation plates during fractional distillation. The higher the number of theoretical plates a column has at a particular column length L, the better the quality of the column and the narrower the peaks.

The value of the plate number N is dependent on a variety of chromatographic and solute parameters, including the column length, L, the chromatographic particle diameter, dp, the linear flow velocity, u, and the solutes' diffusivities (D_m and D_s) in the bulk mobile phase and within the stationary phase, respectively. The plate number can be defined as:

$$N = \left(\frac{t_R^2}{\sigma_t^2}\right) \text{ or } N = 16\left(\frac{t_R}{w}\right)^2, \quad (8)$$

where t_R is the retention time, and σ is the peak variance of the eluted zone in time units. For practical convenience, it is often replaced with the peak width w. For Gaussian peaks, w approximately corresponds to 4σ (4× peak standard variation).

In order to permit a comparison of column efficiencies between columns with identical bed dimensions packed with sorbent particles of different physical or chemical characteristics (e.g., different average diameter, ligand type), the plate height H is defined through the reduced plate height, h, while the linear flow velocity $u = L/t_0$, can be defined in terms of reduced mobile phase velocity, v:

$$h = \frac{H}{d_P} \text{ and } v = \frac{ud_P}{D_m}, \quad (9)$$

where L is the column length, d_P the particle diameter, and D_m the diffusivity of the solute in the mobile phase.

The peak zone band broadening which is caused by various mass transport effects in the column is described through the dependency of the reduced plate height, h on the linear velocity, u or the reduced velocity v through the van Deemter–Knox equation:

$$h = A + \frac{B}{u} + Cu \text{ or } h = Av^{1/3} + \frac{B}{v} + Cv, \quad (10)$$

where the A-term expresses the Eddy-diffusion and mobile phase mass transfer effects, and is a measure of the packing quality of the chromatographic bed (and is constant for a given column), the B-term entails the longitudinal molecular diffusion effects, and the C-term incorporates mass transfer resistances within the stationary phase microenvironment which describes the interaction of the solutes with the stationary phase.

In order to achieve optimal separation performance, the reduced plate height, h, needs to be as small as possible. The optimal flow velocity can be deduced from the minimum in h in the van Deemter–Knox plots (Fig. 2).

The van Deemter–Knox equation can be used to mathematically describe the chromatographic behavior of small-molecular-weight compounds such as peptides with a good approximation, whereas for large molecules such as proteins it has limitations, as the proteins vary considerably in shape and surface properties. As a consequence, the chromatographic effects can only be described with approximations and average values. It was established that:

- The behavior of small molecules is determined by their diffusion, however for large molecules, the influence of the B-term (longitudinal molecular diffusion) is negligible, particularly at higher flow velocities.
- The interaction with the stationary phase (C-term) of large molecules results in band broadening.
- The optimal plate height for large molecules can be obtained at lower flow velocities.

Besides zone band broadening within the column, band broadening can also arise from extra-column band broadening due to instrument characteristics:

$$\sigma_t^2 = \sigma_{column}^2 + \sigma_{extra}^2, \quad (11)$$

where σ_{column}^2 and σ_{extra}^2 are the peak variances arising due to column and extra-column effects, respectively. Careful attention must be paid to possible sources of extra-column effects (e.g., the choice of tubing in terms of length and diameter, the type of fittings, frits, the choice of detector cell volume) that can reduce the impact of σ_{extra}^2 on the overall h value.

The zone band broadening or peak dispersion, expressed as reduced plate height, h, arises from kinetic, time-dependent phenomena. In the absence of secondary effects (e.g., slow chemical equilibria, pH

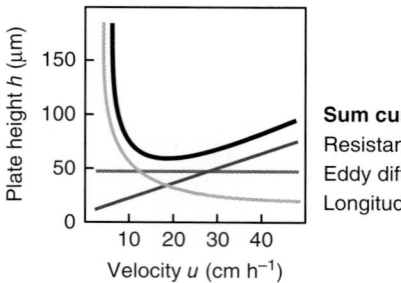

Sum curve
Resistance to mass transfer (Cu)
Eddy diffusion (A)
Longitudinal diffusion (B/u)

Fig. 2 Van Deemter–Knox plots (top curve) and the plots of individual contributions of the A-term, B/u term, and the Cu-term to the plate height.

effects, conformational changes) that could influence the chromatographic process, the resolution R_S can be expressed as:

$$R_S = (1/4)N^{1/2}(\alpha - 1)(k/(1 + k)). \quad (12)$$

This equation links three essential parameters that determine the quality of a chromatographic separation, namely the retention factor k, the selectivity α, and the plate number N, and therefore describes the extent to which the zone spreading may cause the loss of separation performance. As will be demonstrated further below, this equation can be used to guide systematic method development for resolution optimization.

3.2
Principles of Green Analytical Chemistry

Conventional analytical processes comprise sample collection, sample pretreatment, transportation (from the sampling site to the analytical laboratory), sample preparation, measurement, disposal, and reporting. These seven individual process steps can be evaluated according to criteria of green analytical chemistry which mirror the 12 principles of green chemistry originally developed to allow such evaluation for synthetic chemical processes [4, 5]. The main objectives of green analytical chemistry are [6, 7]:

- The elimination of or significant reduction in the consumption of reagents, preferably the elimination of organic solvents, from analytical procedures.
- The reduction of emission of vapors and gases, as well as liquid and solid wastes generated in an analytical laboratory.
- The elimination of hazardous reagents with high ecotoxicity.
- The reduction of labor and energy consumption in the analytical procedure (per single analyte).

Examples of how such objectives of green analytical chemistry can be met in HPLC have been recently reviewed [8].

For example, reductions in the use of organic solvents in the measurement step can be achieved by using fast chromatography with ultra-high-performance liquid chromatography (UHPLC) in conjunction with short columns of small diameter [9], the employment of capillary/nano-LC which uses less eluent, or modes of chromatography which can be employed with high-water-content eluents, such as per aqueous liquid chromatography (PALC) or aqueous normal-phase chromatography (ANPC) [10]. Hazardous waste can be reduced by using "green solvents" such as ethanol and other reagents [11]. Reductions in labor and energy consumption can be achieved by applying either automation (e.g., instruments which perform automated dilutions) or miniaturization (e.g., using HPLC chip systems). Similar considerations are valid for other steps of the analytical process [7, 12].

At the level of an individual sample, the benefits of green analytical chemistry are both economic (enhanced knowledge gain per total analysis time and reagent use) and environmental. With the integration of green analytical chemistry into real-time, in-process monitoring methods, additional benefits can be obtained. These relate to the economic streamlining of processes, and the reduction of, for example, side products and solvent and water usage. One of the drivers for this development is the acknowledged need for advancements in the scientific risk-based framework implicit to the use of process analytical technology (PAT) methods with biopharmaceuticals, based on green chemistry concepts [4]. At each process step – and particularly at the downstream/recovery and purification stages, where most challenges are currently encountered – new green analytical methods are required to validate the integrity, authenticity, and concentration of, for example, the recombinant protein product and its contaminants (i.e., host cell proteins), and to document process efficiencies. Furthermore, a new method is needed to link multidimensional analytical capabilities with the upscaling (or downscaling) of the recovery process so as to allow products to be obtained in the most economically viable manner. Such considerations are particularly relevant to the large-scale production of biopharmaceuticals (e.g., insulin), where many process steps (often in excess of 20) are required to obtain the product at the necessary level of purity.

A careful examination of these needs reveals that dedicated platform technologies will be required to provide the level of significant advancement in the future, particularly when the repetitive acquisition of key analytical data is crucial in the upscaling from the laboratory to the industrial stage. Central to such process intensification issues are robust analytical technologies, which can be integrated into on-line (or off-line) analyses with timescales matching process demands and production efficiencies. Thus, in the case of the industrial production of pharmaceuticals, chemicals or recombinant proteins, despite significant advances leading to greater overall atom-efficiency, these processes typically generate complex mixtures which must be analyzed by highly selective methods, preferably in near-real time.

Due to their multidimensional potential, HPLC, high-performance capillary electrophoresis (HP-CE) [13] and high-performance capillary electrochromatography (HP-CEC) [14], in conjunction with MS, currently offer the greatest scope for such monitoring tasks. Currently, many natural biological or semi-synthetic compounds can be analyzed using these techniques in tandem, either as the parent molecules or as derivatives, while proteins are more easily identified structurally at the peptide level, using off-line proteolysis before separation [15]. The current challenge is to progressively miniaturize these systems, to reduce the number of sample-handling steps, and to increase their limits of detection. Accordingly, for such advanced PAT technologies, microfabricated devices may play an increasing role in the future.

The concepts and potentials of green chemistry, particularly in areas associated with in-line, miniaturized analytical HPLC, HP-CE and HP-CEC, are thus powerful drivers for further significant developments in PATs in the field of recombinant protein manufacture.

4
Physico-Chemical Properties of Peptides and Proteins

A chromatographic separation of a peptide or protein is designed around its primary structure (amino acid sequence) and the folded structure (i.e., the secondary, tertiary, and quaternary structures). The 21 naturally occurring L-α-amino acids found in peptides and proteins vary with respect to the properties of their side chains. This chemical diversity is further increased when some of these side chains have been post-translationally modified with a variety of chemical and biological modifications (e.g., acetylation, deamidation, glycosylation, lipidation, and phosphorylation). The side chains are generally classified according to their polarity (e.g., nonpolar or hydrophobic versus polar or hydrophilic). The polar side chains are divided into three groups: uncharged; positively charged or basic; and negatively charged or acidic. All N- and C-terminally unblocked peptides and proteins contain several ionizable basic and acidic functionalities, and therefore typically exhibit characteristic isoelectric points with the net charges and polarities in aqueous solutions varying with pH, solvent composition, and temperature. Cyclic peptides without ionizable side chains are an exception, as they have a zero net charge. The number and distribution of charged groups influences the polarizability and ionization status of a peptide or protein, as well as the hydrophobicity. The characteristic data for the most common L-α-amino acids found in peptides and proteins, together with a summary of the N- and C-terminal groups, are listed in Table 1. This information can be used to evaluate the impact of amino acid composition on retention behavior, and thus guide the selection of optimal separation conditions for the resolution of peptide and protein mixtures; an example of this would be the choice of eluent composition or the gradient range in RP-HPLC. The information can also point to the impact of amino acid substitution or deletion in small peptides on retention; alternatively, it can be used to guide the identification of peptide fragments derived from the proteolytic (either chemical or enzymatic) digestion of proteins.

In solution, a polypeptide or protein can, in principle, explore a relatively large array of conformational space. For small peptides (e.g., fewer than 15 amino acid residues) a defined secondary structure (α-helical, β-sheet, or β-turn motif) is generally absent. However, with increasing polypeptide chain length, and depending on the nature of the amino acid sequence, specific regions/domains of a polypeptide or protein can adopt preferred secondary, tertiary, or quaternary structures. In aqueous solution this folded structure, which internalizes the hydrophobic residues and thus stabilizes the polypeptide structure, becomes a significant feature of peptides and proteins in chromatographic separations. A critical factor to be considered in the selection of a HPLC procedure is that the experimental conditions will inevitably cause perturbations of the conformational status of these biomacromolecules. In most cases, an integrated biophysical experimental strategy – including ^{1}H 2D NMR, Fourier transform infrared (FTIR) spectroscopy, circular dichroism-optical rotatory dispersion (CD-ORD) spectroscopy, and ESI-MS – is required in order to determine the secondary and higher-order structures of a polypeptide or protein in solution or in the presence of specific ligands or to detect possible self–self-aggregation effects with peptides or proteins which may have occurred during the HPLC separation.

As the peptide bond absorbs strongly in the far UV region of the spectrum ($\lambda = 205-215$ nm), UV detection is the most widely used method for detecting peptides and proteins in HPLC [21]. Besides absorbing in the far-UV range, aromatic amino acid residues also absorb light above 250 nm, due to their delocalized π-systems. Knowledge of the characteristic UV-spectra, and in particular the extinction coefficients of the non-overlapping absorption maxima of these amino acids, allows – in conjunction with UV-diode array detection (DAD) and second-derivative or difference UV-spectroscopy – the verification of peak purity and a determination of the aromatic amino acid content of peptides and proteins. Knowledge of the relative UV/visible absorbancy of a peptide or protein is thus crucial, as the choice of detection wavelength in RP-HPLC (and in the other HPLC modes) depends on the different UV wavelength cut-offs of the eluents

Tab. 1 Properties of the common L-α-amino acid residues and termini.

Three-letter code	One-letter code	Name	Mono-isotopic mass (amu)	Partial specific volume [16] (Å³)	Accessible surface area [17] (Å²)	pK_a of side-chain [18] or termini [19]	pI	Relative hydrophobicity [20]
Ala	A	Alanine	71.03711	88.6	115	—	6.02	0.06
Arg	R	Arginine	156.10111	173.4	225	12.48	10.76	−0.85
Asn	N	Asparagine	114.04293	117.7	160	—	5.41	0.25
Asp	D	Aspartic acid	115.02694	111.1	150	3.9	2.87	−0.20
Cys	C	Cysteine	103.00919	108.5	135	8.37	5.02	0.49
Gln	Q	Glutamine	128.05858	143.9	180	—	5.65	0.31
Glu	E	Glutamic acid	129.04259	138.4	190	4.07	3.22	−0.10
Gly	G	Glycine	57.02146	60.1	75	—	5.97	0.21
His	H	Histidine	137.05891	153.2	195	6.04	7.58	−2.24
Ile	I	Isoleucine	113.08406	166.7	175	—	6.02	3.48
Leu	L	Leucine	113.08406	166.7	170	—	5.98	3.50
Lys	K	Lysine	128.09496	168.6	200	10.54	9.74	−1.62
Met	M	Methionine	131.04049	162.9	185	—	5.75	0.21
Phe	F	Phenylalanine	147.06841	189.9	210	—	5.98	4.8
Pro	P	Proline	97.05276	122.7	145	—	6.10	0.71
Ser	S	Serine	87.03203	89	115	—	5.68	−0.62
Thr	T	Threonine	101.04768	116.1	140	—	6.53	0.65
Trp	W	Tryptophan	186.07931	227.8	255	—	5.88	2.29
Tyr	Y	Tyrosine	163.06333	193.6	230	10.46	5.65	1.89
Val	V	Valine	99.06841	140	155	—	5.97	1.59
Sec	U	Selenocysteine	150.95363	—	—	5.73	5.47	1.59
α-Amino						7.7–9.2		
α-Carboxyl						2.75–3		

Monoisotopic masses (amu) of N- and C-terminal groups: Hydrogen (H) 1.00782, N-formyl (HCO) 29.00274, N-acetyl (CH₃CO) 43.01839, Free acid (OH) 17.00274, amide (NH₂) 16.01872.

used. The common use of $\lambda = 215$ nm as the preferred detection wavelength for most analytical applications with peptides and proteins is a good compromise between detection sensitivity and potential detection interference due to light absorption by the eluent. However, wavelengths between 230 and 280 nm are frequently employed in preparative applications, where the use of more sensitive detection wavelengths could result in an overloading of the detector response (usually above an absorbance value of 2.0–2.5 absorbance units). It should be noted that three aromatic amino acids – phenylalanine, tryptophan, and tyrosine – demonstrate fluorescence that varies considerably according to the folded/unfolded state of the protein. Although such fluorescence can be used to monitor changes in protein folding status, the intensity of the fluorescence is dependent on the solvational environment and decreases in inverse proportion to the polarity of the solvent.

5
Chromatographic Separation Modes for Peptides and Proteins

Currently used separation modes of HPLC for peptide and protein analysis include size-exclusion chromatography (SEC), RPC, normal-phase chromatography (NPC), hydrophilic interaction chromatography (HILIC), ANPC, hydrophobic interaction chromatography (HIC), IEX and affinity chromatography (which includes immobilized metal ion-affinity chromatography and biospecific/biomimetic affinity chromatography). The principles of each method are explained in the following subsections. These and also a number of less-frequently used chromatographic modes, including hydroxyapatite-chromatography, mixed-mode chromatography, charge-transfer chromatography and ligand-exchange chromatography, can be operated under isocratic (i.e., fixed eluent composition), step-gradient or gradient elution conditions, in which the eluent conditions are changed either in variable steps or continuously. The exception here is high-performance size-exclusion chromatography (HP-SEC), which is usually only performed under isocratic conditions. All modes can be used in analytical, semi-preparative, or preparative situations [22–29]. An historical perspective of the developments in analytical and preparative LC, with milestones and advances of LC in the life sciences in particular, has been produced [1].

In order to achieve optimal selectivity – and hence resolution of peptides and proteins – in high-performance chromatographic separations, and irrespective of whether the task at hand is of analytical or preparative nature, the choice of chromatographic mode must be guided by the properties of the analytes (i.e., their molecular size/shape hydrophobicity/hydrophilicity, net charge, isoelectric point, solubility, function, antigenicity, carbohydrate content, content of free –SH groups, exposed histidine residues, exposed metal ions). The chromatographic modes and the molecular properties of the target compounds that form the basis of each separation mode are listed in Table 2.

In addition to the above-mentioned functional characteristics of these chromatographic systems, other chemical and physical parameters of the mobile and stationary phases impact on the resolution, mass recovery, and bioactivity preservation in separations of polypeptides or proteins during liquid chromatographic separations [30]. These parameters are listed in Table 3.

Tab. 2 Chromatographic separation modes for peptides and proteins.

Chromatographic mode	Acronym	Exploited molecular properties
Size exclusion/gel permeation chromatography	SEC/GPC	Molecular mass, hydrodynamic volume
Reversed-phase chromatography	RPC	Hydrophobicity
Normal-phase chromatography	NPC	Polarity
Hydrophilic interaction chromatography	HILIC	Hydrophilicity
Aqueous normal phase chromatography	ANPC	Hydrophilicity
Hydrophobic interaction chromatography	HIC	Hydrophobicity
Anion-exchange chromatography	AEX	Net negative charge
Cation-exchange chromatography	CEX	Net positive charge
Affinity chromatography	AC	Specific interaction
Immobilized metal ion affinity	IMAC	Complexation

Tab. 3 Chemical and physical factors of the mobile and stationary phase that contribute to variation in the resolution, mass recovery, and bioactivity preservation of polypeptides, proteins, and other biomacromolecules in HPLC.

Mobile phase	Stationary phase
Buffer composition including additives/ion pair reagents	Ligand composition
Ionic strength	Ligand density
pH	Surface heterogeneity
Organic solvents	Particle size
Temperature	Particle size distribution
Metal ions	Particle compressibility
Chaotropic reagents	Pore diameter
Oxidizing or reducing reagents	Pore diameter distribution
Sample loading concentration and volume	Surface area

5.1 High-Performance Size-Exclusion Chromatography (HP-SEC)

HP-SEC, which is also known as high-performance gel-permeation chromatography (HP-GPC), is performed using porous stationary phases and is able to separate analytes according to their molecular mass or, more precisely, their hydrodynamic volume [31]. The separation of analytes is based on the concept that molecules of different hydrodynamic volume (Stokes radius) will permeate to different extents into the porous HP-SEC separation media and thus exhibit different permeation coefficients according to differences in their molecular masses/hydrodynamic volumes [32]. Analytes with a molecular weight larger than the exclusion limit (such limits are normally available in the technical information provided by the column manufacturer) will be excluded from the pores and eluted in the void volume of the column. As a non-retentive separation mode, HP-SEC is usually operated with isocratic elution using aqueous low-salt mobile phases.

HP-SEC can be used for group separation or high-resolution fractionation. In group separation mode, HP-SEC removes small molecules from large molecules and is also suitable for buffer exchange or desalting. In the high-resolution fractionation mode, HP-SEC separates the various components of a sample based on their different hydrodynamic volumes, or it can be used to perform a molecular weight distribution analysis.

As HP-SEC columns (ideally) have no adsorption capacity and dilute the sample upon elution, they are not normally used in the initial capture, or for intermediate purification in multistep chromatographic processes. However, they are suitable for final polishing, an example being the final removal of unwanted aggregates or multimeric forms of proteins or other impurities of significantly different molecular weight [33]. HP-SEC can be performed directly after high-performance affinity chromatography (HP-AC), high-performance hydrophobic interaction chromatography (HP-HIC), or high-performance ion-exchange chromatography (HP-IEX), without a buffer exchange.

5.2
High-Performance Reversed-Phase Chromatography (HP-RPC)

HP-RPC is the most frequently used analytical mode for peptide and protein analysis [34]. In RPC, the polarity of the stationary and mobile phase is the reverse of that used in NPC. The peptides and proteins are loaded onto the column under aqueous conditions and eluted with a mobile phase containing an organic solvent. The column contains a porous or nonporous stationary phase with immobilized nonpolar ligands.

HP-RPC separates compounds according to their relative hydrophobicity [35]. The most commonly accepted theory of the retention mechanism in HP-RPC is based on the solvophobic theory, which describes the hydrophobic interaction between the nonpolar surface regions of the analytes and the nonpolar ligands of the stationary phase [36–38]. According to the solvophobic theory, the interaction of peptides and proteins in RPC occurs through solvophobic exclusion of the solute from the aqueous mobile phase and binding to the nonpolar surface of the immobilized ligand (solvophobic effect) [39]. The retention then depends on the size of the contact area ΔA of the hydrophobic surface of the solute and the hydrophobic area of the immobilized ligand, as well as on the surface tension (γ) of the eluents. The surface tension γ depends on the composition of the mobile phase (volume fraction of the organic solvent, e.g., acetonitrile, expressed as % B), the parameter that is varied or, more precisely, lowered during gradient elution:

$$\ln k = A + N_A/RT\Delta A\gamma, \quad (13)$$

where k is the retention factor, A is a constant, N_A is the Avogadro constant, R is the gas constant, and T is the absolute temperature.

Typically, the nonpolar ligands are immobilized onto the surface of spherical, porous or nonporous silica particles, although nonpolar polymeric sorbents (e.g., those derived from crosslinked polystyrene–divinylbenzene) can also be employed. Silica-based packing materials of 3–10 μm average particle diameter and 70–1000 Å pore size, with n-butyl, n-octyl, or n-octadecyl ligands, are widely used for the separation of peptides and proteins. Silica particles of 1 μm to more than 65 μm have been developed in various size distributions and configurations (e.g., spherical, irregular, with various pore geometries and pore connectivities; and in pellicular, fully porous or monolithic structures [40]) by a variety of routes of manufacture and with different silica types, and are grouped into type I, type II, or type III silica according to their purity and metal content [41]. For low-molecular-mass (<4000 Da) polypeptides, silica materials of 70–80 Å pore size and 3–5 μm average particle diameters are often used, which maximizes the loading capacity and retention. For proteins in the mass range of 4000–500 000 Da, a pore size of 300 Å allows the maintenance of high efficiency, whereby the loading capacity can be increased by increasing the column diameter. Macroporous HP-RPC columns of 1000 Å pore size are increasingly used for the fractionation of very complex proteins samples.

In HP-RPC, an organic solvent (e.g., acetonitrile, methanol, ethanol, n-propanol, tetrahydrofuran) is used as a surface tension modifier in the chromatographic eluent, which has a particular elution strength, viscosity, and UV cut-off. Mobile phase additives such as acetic acid (AA), formic acid (FA), trifluoroacetic acid (TFA) and hepta-fluorobutyric acid (HBFA) are used to obtain a particular pH value, typically at low pH (e.g., ~pH 2 for silica-based materials) with the exception of polymeric stationary phases, which have an extended pH range from pH 1 to 12. Some mobile phase additives may also function as ion pair reagents, which interact with the ionized analytes to form overall neutral eluting species and also suppress silanophilic interactions between free silanol groups on the silica surface and basic functional groups of the analytes. The properties of the additives determine their suitability for use with ESI-MS. Strong ion-pair interactions between analytes and mobile phase additives can suppress the ionization of analytes in ESI-MS.

HP-RPC can be operated in isocratic, step gradient or continuous gradient elution mode, and is frequently used as an intermediate or final polishing step in a multistep purification. It is ideally positioned after HP-IEX because it allows desalting and the separation of the sample in a single step.

As the majority of peptides and proteins possess some degree of hydrophobicity, HP-RPC techniques dominate the separation of peptides and proteins at the analytical and semi-preparative levels [42].

5.3 High-Performance Normal-Phase Chromatography (HP-NPC)

Chromatographic systems in which the stationary phase is more polar than the mobile phase were first developed at the start of the modern era of LC, and were known under the acronym "normal-phase" LC. High-performance, normal-phase chromatography (HP-NPC) can be performed on unmodified silica, and separates analytes according to their intrinsic polarity. The retaining mobile phase contains less-polar organic solvents, while the eluting mobile phase consists of more polar organic solvents. Water, due to its extreme polarity, is adsorbed to most NP stationary phases, and can significantly affect separation reproducibility. In contrast to HP-RPC with immobilized n-alkyl ligands, where interaction of the solute and stationary phase is based on solvophobic phenomena, the interaction in HP-NPC is based on adsorption. The retention behavior of peptides and proteins in HP-NPC is often described in terms of the classical concepts inherent to multisite displacement and site occupancy theory [43].

HP-NPC is mainly used for the separation of, for example, polyaromatic hydrocarbons (PAHs), heteroaromatic compounds, nucleotides, and nucleosides, and much less frequently for protected synthetic peptides, deprototected small peptides in "flash chromatography" mode, and protected amino acid derivatives used in peptide synthesis [44]. Originally, HP-NPC was limited to unmodified silica columns, but recent studies have utilized polar bonded phases such as amino ($-NH_2$), cyano ($-CN$), or diol ($-COHCOH-$)-coated sorbents. Such modified normal phase packing materials were suitable for polar-bonded phase chromatography (PBPC), which was used for the separation of peptides and proteins [45, 46]. Today, one of the main applications of modified normal phase silica materials is in HPLC-integrated solid-phase extraction (SPE) procedures [47]. These types of sorbent, particularly when used as pre-column packing materials in LC–LC column-switching settings, in conjunction with restricted access materials (RAMs), allow multiple injections of untreated complex biological samples, including hemolyzed blood, plasma serum, fermentation broth and cell tissue homogenates, for the isolation of bioactive peptides. Typically, with RAMs, hydrophilic, electroneutral diol groups are immobilized onto the outer surface of spherical particles. This layer prevents nonspecific interactions between the support matrix and protein(s) or other high-molecular-weight biomolecules, which are thus excluded from the interior regions of the particle and eluted as nonretained components. The inner surfaces of the porous RAM particles are, however, chemically modified with n-alkyl ligands, which are only freely accessible for low-molecular-weight analytes, such as peptides. As a consequence, a significant enrichment or partial resolution of peptide analytes can be achieved. HP-NPC can be operated in isocratic, step gradient, or gradient elution mode.

5.4
High-Performance Hydrophilic Interaction Chromatography (HP-HILIC)

HP-HILIC is performed on porous stationary phases with immobilized hydrophilic ligands [48], and separates analytes according to their hydrophilicity [49]. This variant of the HP-NPC mode was introduced by Alpert in 1990, based on polyaspartic acid immobilized onto silica and used for the separation of amino acids, small peptides, and simple maltoglycosides with mobile phases of high organic solvent content [50]. A pseudo-HILIC separation of simple saccharides was performed on a BondaPak-NH_2 column as early as 1975 [51]. The HP-HILIC mode has since been applied to the separation of various analytes, that is, simple carbohydrates and amino acids as well as peptides [52–54].

In HILIC, polar sorbents with amide-, aminopropyl-, cyanopropyl-, diol-, cyclodextrin-, poly(succinimide)-, and sulfoalkylbetaine phases are employed, and the nonaqueous mobile phases of NPC are replaced with high-organic, low-aqueous eluents [55–57]. The initial mobile phase has a high organic solvent and a low water content. The elution of compounds from HP-HILIC columns is achieved by increasing the water content in the mobile phase. The elution order was initially thought to be more or less opposite to that seen in HP-RPC separation. Although, intuitively, it would seem that retention in HILIC would simply be the "reverse" of that in HP-RPC, studies on the orthogonality of separations in 2D LC have shown that HP-HILIC (with a bare silica sorbent) and HP-RPC can be a suitable combination for proteomic analysis in 2D systems [58, 59]. Compared to the nonaqueous (organic) mobile phase in HP-NPC, the partly aqueous mobile phase used in HP-HILIC allows a greater solubility of many polar and hydrophilic compounds, and a fast separation of polar compounds can be achieved due to the low viscosity of the highly organic mobile phase. Moreover, the high content of organic solvent in the mobile phase favors ionization of polar compounds in subsequent ESI-MS, and thus provides enhanced detection sensitivity for these compounds [60].

Considerable scientific debate persists regarding the physical basis of the separation mechanism in HP-HILIC [61]. The roles of thermodynamic or kinetic effects in controlling resolution and separation efficiencies have yet to be fully explored. Although it was proposed that the retention of polar compounds in HILIC is through partitioning between the bulk of the mostly organic mobile phase and a stagnant water-enriched layer semi-immobilized on the surface of silica, processes of partitioning or adsorption (or combinations of it) have also been suggested to be responsible for generating retention in HP-HILIC [50, 57, 62, 63].

Mixed-mode hydrophilic interaction chromatography/cation-exchange chromatography (HILIC/CEX) offers another means of separating peptides. The contact region concept developed for HP-RPC [36, 64] can be used to rationalize the retention of amphipathic α-helical peptides, for example, based on experiments with a poly(2-sulfoethyl aspartamide)-silica (PolySulfoethyl A) strong cation-exchange column in HILIC/CEX mode, and with a Zorbax SB300-C8 reversed-phase column for separating amphipathic α-helical peptides [65–67]. A substitution in the hydrophilic face of the peptide resulted in a substantive effect on retention in HP-HILIC/CEX. This was not observed in HP-RPC, whereas a substitution in the hydrophobic face of the peptide resulted in a distinct effect on retention in HP-RPC which was not observed in HP-HILIC.

HP-HILIC can be operated in isocratic, step gradient or gradient elution mode, where the retaining mobile phase is organic and the eluting mobile phase is aqueous. Being more suited to the isolation of polar substances, HP-HILIC, when linked to electrospray MS, has mainly found application for the analysis of phosphopeptides [68, 69] and glycopeptides [70–73].

5.5
High-Performance Aqueous Normal Phase Chromatography (HP-ANPC)

Recently, a new chromatographic mode, high-performance aqueous normal-phase chromatography (HP-ANPC) has been developed on stationary phases based on silica hydride surfaces [74–77]. Silica hydride stationary phases, which are based on high-purity, low-metal-content type-B silica, were developed 10 years ago to complement conventional silica materials. The fabrication of silica hydride utilizes a silanization reaction that generates a surface predominantly populated with silicon-hydride groups (Si–H), which are nonpolar in nature and stable in most aqueous–organic environments. One unique feature of silica hydride stationary phases is their ability to be employed over a broad range of mobile phase compositions from 100% aqueous to pure organic solvents [78]. A unique advantage of HP-ANPC is that it allows the separation of compounds with a broad range of hydrophilicities in the same mixture, using predominantly water-rich mobile phase compositions.

The retention principle in HP-ANPC is analogous to that found HP-NPC, but the mobile phase contains water as part of the binary eluent [79]. The difference between HP-ANPC and HP-HILIC is that, in HP-HILIC, retention is determined by an adsorbed water layer on the surface, which is either not present or is substantially smaller on silica hydride surfaces employed in HP-ANPC. The precise separation mechanisms that operate in HP-ANPC are still subject to ongoing research, and the interchangeable use of the terms HP-HILIC and HP-ANPC has led to some confusion in the literature. However, due to its versatility, HP-ANPC has been applied for the separation of peptides in isocratic and gradient elution modes [80].

5.6
High-Performance Hydrophobic Interaction Chromatography (HP-HIC)

HP-HIC is used for protein separations involving hydrophobic sorbents [81–84]. In HP-HIC, the binding of proteins to the stationary phase occurs at high salt concentrations of the mobile phase, and elution is performed through a decrease in salt concentration [85–89]. In HP-HIC the two most widely used base matrices are hydrophilic carbohydrates, for example, crosslinked agarose or synthetic copolymer materials [90]. The type of immobilized ligand (alkyl- or aryl ligand) influences the selectivity of the stationary phase [81, 91]. In HP-HIC, nonpolar ligands with lower hydrophobicity and lower ligand density (approximately one-tenth of that of HP-RPC sorbents) are employed. These differences between HP-RPC and HP-HIC have fundamental effects on the recovery of proteins in the bioactive state, as well as on the selectivity of the system. The protein-binding capacities of HP-HIC sorbents increases with increasing n-alkyl chain length at a given ligand density [92] as well as with increasing ligand density, but will reach a plateau for very high ligand density

levels. HP-HIC sorbents should be selected on the basis of the critical hydrophobicity concept [34, 93].

HP-HIC separates proteins according to their hydrophobicity differences. The retention mechanism is based on the reversible interaction between a protein and a hydrophobic surface of a chromatographic support, depending on changes in the microscopic surface tension associated with the composition of the mobile phase [94]. Similar to HP-RPC, where the decrease in surface tension of the eluent is achieved through an increase in the organic solvent content of the mobile phase, in HP-HIC this is achieved with a decreasing salt concentration, that is, by increasing the water content of the eluent.

The selectivity of protein separations in HP-HIC can be influenced through stationary phase parameters, such as the type and density of the immobilized hydrophobic ligand type of base matrix, and mobile phase parameters, such as the type and concentration of salt [95], the addition of organic modifiers or surfactants [96] and the pH. The temperature of the column will also influence the selectivity of protein separations.

In HIC, the hydrophobic interactions between polypeptides or proteins and a sorbent are influenced by the use of different salts in the mobile phase (various salts have differing molal surface tension increment values; see Table 4), or by different concentrations of salt. The surface tension of the mobile phase, γ, can be related to the molal surface tension increment, σ, and the molal concentration, m, of the salt by the equation:

$$\gamma = \gamma^\circ + \sigma m \qquad (14)$$

where $\gamma^\circ = 72 \, \text{dyn} \, \text{cm}^{-1}$ for water.

Parameters of the surface tension increment, σ, the initial mobile phase

Tab. 4 Salts used in the mobile phase of HP-HIC.

Salt	Molal surface tension increment ($\sigma \times 10^3$ dyn g cm·mol^{-1})
Calcium chloride	3.66
Magnesium chloride	3.16
Potassium citrate	3.12
Sodium sulfate	2.73
Potassium sulfate	2.58
Ammonium sulfate	2.16
Magnesium sulfate	2.10
Sodium dihydrogen phosphate	2.02
Potassium tartrate	1.96
Sodium chloride	1.64
Potassium perchlorate	1.40
Ammonium chloride	1.39
Sodium bromide	1.32
Sodium nitrate	1.06
Sodium perchlorate	0.55
Potassium thiocyanate	0.45

concentration, m, and the surface tension values, γ, of common aqueous salt buffers are listed in Table 5.

The minimum surface tension reached in the HP-HIC of polypeptides or proteins with binary water–salt systems corresponds to the surface tension of pure water, that is, 72 dyn cm^{-1}.

The effect of different salts on the hydrophobic interaction follows the (lyotropic) Hofmeister series for the precipitation of proteins from aqueous solution [97, 98] (see Table 6). The Hofmeister series ranks the effect of anions and cations in promoting protein precipitation [99]; ions with a higher salting-out effect promote binding to hydrophobic interaction sorbents, whereas ions with a higher salting-in effect promote elution from hydrophobic interaction sorbents.

The salts at the start of the series promote hydrophobic interactions and protein precipitation (salting-out effect) [100]; these are termed anti-chaotropic (or kosmotropic) salts, and are considered to be water-structuring. The addition of kosmotropic salts to the equilibration buffer and sample buffer promote protein immobilized ligand interaction in HIC. The salts at the end of the series (salting-in or chaotropic ions) randomize the structure of liquid water and tend to decrease the strength of hydrophobic interactions. The chloride anion is considered to be approximately neutral with respect to the water structure.

In general, the effect of salt cations on the preferential interaction parameters is less pronounced than that of salt anions, particularly when the cation is monovalent, although divalent cations tend to bind to proteins. Typically, kosmotropic (antichaotropic) salts (i.e., ammonium sulfate, sodium sulfate, magnesium chloride) of

Tab. 5 Parameters of the surface tension of frequently used aqueous salt buffers.

Salt	σ ($\times 10^3$ dyn g cm·mol^{-1})	m ($\times 10^3$ mol g^{-1})	γ (dyn cm^{-1})
Ammonium sulfate	2.16	2	77.31
Sodium chloride	1.64	2	76.29
Magnesium sulfate	2.1	1.4	75.95
Sodium sulfate	2.73	1	75.74
Sodium perchlorate	0.55	2	74.11
Sodium phosphate	2.02	0.05	73.01

Tab. 6 Hofmeister series.

	← Salting-out effect (precipitation)
Anions	PO_4^{3-}, SO_4^{2-}, CH_3COO^-, Cl^-, Br^-, NO_3^-, ClO_4^-, I^-, SCN^-
Cations	NH_4^+, K^+, Na^+, Cs^+, Li^+, Mg^{2+}, Ca^{2+}, Ba^{2+}
	Salting-in effect →

high molal surface tension increment are to be preferred in HP-HIC applications for polypeptides and proteins.

Since protein–hydrophobic surface interactions are enhanced by high-ionic strength buffer solutions, HP-HIC is a suitable next step after ammonium sulfate precipitation or HP-IEX elution with a high-salt buffer. In combination with nondenaturing mobile phases, proteins can potentially be eluted in their native conformation from HP-HIC sorbents.

5.7
High-Performance Ion-Exchange Chromatography (HP-IEX)

HP-IEX is performed on stationary phases with immobilized charged ligands, and separation occurs according to the electrostatic interactions between the charged surface of the analyte(s) and the complementarily charged surface of the sorbent [101–109]. In high-performance anion-exchange chromatography (HP-AEX), peptides and proteins are separated according to their net negative charge, whereby the retaining mobile phase is aqueous, of high pH and low salt concentration, and the eluting mobile phase is either aqueous, of high pH and high salt concentration, or aqueous, and of low pH. In contrast, high-performance cation-exchange chromatography (HP-CEX) separates analytes according to their net positive charge, whereby the retaining mobile phase is aqueous, of low pH and low salt concentration, and the eluting mobile phase is either aqueous, of low pH and high salt concentration, or aqueous, and of high pH.

Both, weak and strong cation exchangers (e.g., based on carboxymethyl- or sulfonopropyl-ligands), as well as weak and strong anion exchangers (e.g., dimethylamino- or quaternary ammonium ligands) are commercially available and highly suitable for the HP-IEX of peptides and proteins. For peptide and protein separations, the use of a strong cation-exchange column has a considerable advantage over other ion-exchange separation modes [110], as this column can retain its negative charge character over a large pH range, from acidic to neutral. In peptides and proteins, at neutral pH, the side-chain carboxyl groups of the acidic amino acid residues (glutamic acid and aspartic acid) are completely ionized, but below pH 3 they are almost completely protonated. Thus, a change of pH allows the retention of peptides and proteins to be varied according to the modified net charge of these biosolute(s).

The "net charge" concept has been widely used as a predictive basis to anticipate the retention behavior of proteins with both, anion- and cation-exchange stationary phases [111, 112]. According to this model, a protein will be retained on a cation-exchange column if the eluent pH is lower than the pI value of the protein, since under these conditions the protein will carry positive net charges. Conversely, a protein will be retained on an anion-exchange column when the eluent pH is above the pI of the protein. Finally, with a mobile phase of a pH that equals the pI of the protein, the surface of the protein can be considered as electrostatically neutral and the protein should not be retained on either cation- or anion-exchange columns. This classical model is now considered as simplistic, however, with recent investigations having revealed that the magnitude of electrostatic interactions between a protein and the stationary phase surface in HP-IEX is dependent on the charge density of the stationary phase, the mobile phase composition, and on the number and distribution of charged sites on the protein molecule, since these define its surface topography and electrostatic contact area with the stationary phase [113–121]. As a consequence, a variation in chromatographic parameters can alter the affinity of the protein for the stationary phase through changes in the overall electrostatic surface charge, or through specific electrostatic interactions of the displacer co-ions and counter-ions with surface charge groups on the protein or with the immobilized charged ligand.

In addition, possible changes in the three-dimensional structure of the proteins may have significant effects on protein retention. Studies on the influence of experimental parameters on the number of charged interactive sites of the proteins involved in its binding to stationary phases have resulted in the development of the concept of an electrostatic interactive area (or ionotrope) through which the protein is thought to bind to the stationary phase [122]. Peptides and proteins can be separated in HP-IEX by either isocratic, step or by gradient elution at high resolution, and with high capacity.

5.8
High-Performance Affinity Chromatography (HP-AC)

HP-AC is performed on stationary phases containing immobilized biomimetic or biospecific ligands, and separates proteins according to principles of molecular recognition [123]. HP-AC is highly selective and usually has a high capacity for the protein of interest. In HP-AC, analytes are eluted by step gradient or gradient elution, where the capture (loading) mobile phase is aqueous and of low ionic strength and the eluting mobile phase is aqueous and of higher ionic strength or of different pH value; alternatively, the mobile phase might contain an additive that competes with the target compound for binding to the immobilized biospecific ligand. In terms of achieving maximal selectivity and highest affinity in the interaction between the target substance(s) and the chromatographic sorbent, the performance of HP-AC excels that of all other modes, though each affinity sorbent must be tailored to the specific target compound.

Immobilized metal-chelate affinity chromatography (IMAC) exploits the affinities of the side-chain moieties of specific surface-accessible amino acids in peptides and proteins for the coordination sites of immobilized transition metal ions [124–126]. The majority of investigations have employed tri- or tetra-dentate ligands, such as iminodiacetic acid (IDA), nitrilotriacetic acid (NTA), tris-(carboxy methyl)ethylenediamine (TED), O-phosphoserine (OPS), or carboxy-methylaspartic acid (CMA) [125]. The retaining mobile phases are aqueous with neutral pH and high ionic strength, the eluting mobile phases are of low pH, and contain competing ligands or EDTA. Novel immobilized chelate systems,

Tab. 7 Chromatographic modes used for peptide and protein separation, and their stationary and mobile phase characteristics.

Chromatographic mode	Stationary phase	Retaining mobile phase	Eluting mobile phase
SEC/GPC	Porous	(non-retentive)	Aqueous, low salt
RPC	Hydrophobic	Aqueous	Organic solvents
NPC	Polar	Nonpolar organic solvents	Polar organic solvents
HILIC	Hydrophilic	Nonpolar organic solvents	Polar organic solvents, water
ANPC	Polar	Organic (or aqueous)	Aqueous (or organic)
HIC	Mildly hydrophobic	Aqueous, high ionic strength	Aqueous, low ionic strength
AEX	Charged	Aqueous, high pH, low ionic strength	Aqueous, high pH, high ionic strength (or low pH), high selectivity counter ion
CEX	Charged	Aqueous, low pH, low ionic strength	Aqueous, low pH, high ionic strength (or high pH)
AC	Biomimetic, biospecific	Low ionic strength	High ionic strength, competing ligand
IMAC	Metal chelate	Aqueous, neutral pH, high ionic strength	Low pH, competing ligand, EDTA

such as 1,4,7-triazo-cyclononane (TACN), however, show different chromatographic properties compared to the IMAC behavior of traditional chelating ligands [126, 127]. These affinity chromatography (AC) techniques, in conjunction with soft gel matrices, have been applied in diverse analytical and preparative protein purifications. HP-AC separations can be performed with immobilized chemical or biological ligands, or with molecular imprinted polymers (MIPs).

Novel procedures to immobilize an IMAC ligand at the surface of silica supports have provided guidelines for the design of highly stable HP-IMAC systems for peptides and proteins [128].

HP-AC can be used for the initial capture of proteins, as an intermediate step in a multiple step purification procedure, or for the affinity-removal of unwanted high-abundance proteins, provided that a suitable affinity ligand for the target protein is available.

The nature of the retaining and eluting mobile phases for the above-described chromatographic modes are summarized in Table 7. Guidelines on how to select a mode, or a combination of modes, for a specific separation task are provided in Sect. 7.

6
Method Development from Analytical to Preparative Scale

Currently, HP-RPC is the most frequently used HPLC mode for the analysis and preparative purification of peptides and proteins, and in particular for applications that involve off-line or on-line ESI-MS. The development of a method for preparative HP-RPC purification for the purpose of isolation of one or more component(s) from a peptide or protein product sample (or alternatively the purification of a synthesized product) is usually performed in four steps:

1) The development, optimization, and validation of an analytical method.
2) Scaling-up of this method to a preparative chromatographic system.
3) Application of the preparative method to the fractionation of the product.
4) Analysis of the individual fractions.

6.1
Development of an Analytical Method

The development of an analytical method for the separation of a peptide or protein encompasses selection of the stationary and mobile phases, taking into consideration the analyte properties (hydrophobicity/hydrophilicity, acid–base properties, charge, temperature stability, molecular size), and is followed by a systematic optimization of the (isocratic or gradient) separations, using either aliquots of the crude extract or, if available, analytical standards. Quality by design (QbD) approaches that define the design space of a HPLC method [129, 130] and quality assurance/quality control (QA/QC) approaches that allow estimating the level of uncertainty in analytical LC results [131], are further important considerations in method development.

For selection of the stationary and mobile phases, a variety of chemical and physical factors of the chromatographic system that may contribute to variations in the resolution and recovery of peptides and proteins must be considered. The stationary phase contributions relate to the ligand composition, ligand density, surface heterogeneity, particle size, particle size distribution, particle compressibility, surface area, pore diameter, and pore diameter distribution. Typically, a particular HP-RPC material will be selected empirically as the starting point for the separation, taking into consideration its suitability for the separation task at hand, published procedures for similar types of peptides/proteins, the availability of the stationary phase material for preparative chromatography and, if information is available, on the analyte properties. The mobile phase contributions relate to the type of organic solvents, eluent composition (including additives/ion pair reagents), ionic strength, pH, temperature, sample loading concentration, and volume.

Since the quality of a separation is determined by the resolution of individual peak zones, method development always aims at an optimization of the resolution. The method development for analytical separations focuses on the least well-resolved peak pair(s) of interest, the so-called "critical" peak pair. The resolution depends on the column efficiency or plate number N, the selectivity α, and the retention factor k, all of which can be experimentally influenced through systematic changes in individual chromatographic parameters. In the isocratic mode of separation, resolution is determined by:

$$R_S = (1/4)N^{1/2}(\alpha - 1)(k/(1 + k)). \quad (15)$$

As detailed above, the plate number, N, represents the efficiency of the column and is a measure of the column performance. The selectivity, α, describes the selectivity of a chromatographic system for a defined peak pair, and is the ratio of the k-values of the second peak to the first peak. The retention factor k is a dimensionless parameter and is defined as $k = (t_R - t_0)/t_0$ where t_R is the retention time of a particular peak and t_0 is the column void time. In this manner, normalization of the relative retention can be achieved for columns of different dimensions. Whilst N and α change only slightly during the solute migration through the column, the value of k can be manipulated through changes in the eluotropicity of the mobile phase by a factor of 10 or more. The best chromatographic separations for low- or mid-molecular-weight analytes such as peptides are generally achieved with mobile phase–stationary phase combinations that result in k values of between 1 and 20.

In gradient elution, in contrast to isocratic elution, $\overline{N}$, $\overline{\alpha}$, and $\overline{k}$ are the median values for N, α, and k, since they change during the separation as the shape and duration of the gradient changes.

The "gradient" plate number $\overline{N}$ has no influence on the selectivity or the retention (except for temperature change). The selectivity $\overline{\alpha}$ and the retention factor $\overline{k}$ usually have only a minor influence on $\overline{N}$. While $\overline{N}$ and $\overline{\alpha}$ change only slightly during the solute migration through the column, the $\overline{k}$-value can change by a factor of 10 or more, depending on the gradient steepness. Again, the best chromatographic separation is generally achieved with a k-value between 1 and 20. Although resolution in isocratic and gradient elution is mainly influenced by the mobile phase variables α (or $\overline{\alpha}$) and k (or $\overline{k}$) and is nearly independent of N (or $\overline{N}$), for a given column, an optimization strategy should nevertheless start with an appropriate selection of the stationary phase. Many initial choices (e.g., column dimensions, choice of ligand, particle size, pore size, etc.) are determined by the overall strategy (i.e., separation optimization for quantification of several analytes or separation optimization for planned *scaling-up* to preparative purification of specific target compounds), and by the purification goals. Currently, a number of computer-assisted, expert systems are commercially available to guide this selection [132]. When the column has been selected, under consideration of the above equation, the separation optimization is performed in three steps: (i) optimization of the column efficiency N; (ii) optimization of the selectivity α; and (iii) optimization of the $\overline{k}$-values.

6.1.1 Optimization of Column Efficiency

The optimization of column efficiency, expressed as the theoretical plate number, N, requires an independent optimization of each of the contributing factors that influence the band-broadening of the peak zones due to column and the extra-column effects. With a particular sorbent (ligand type, particle size, and pore size) and column configuration, this can be achieved through optimization of the linear velocity (flow rate), the temperature, detector time constant, column packing characteristics, and by minimizing extra-column effects by, for example, using zero-dead volume tubing and connectors. The flow rate (or alternatively the linear flow velocity) to achieve the minimum plate height, H, for a particular column can be taken from the literature, or determined experimentally according to published procedures. The temperature of the column and the eluents should be thermostatically controlled in order to facilitate the reproducible determination of the various column parameters, and to ensure resolution reproducibility.

6.1.2 Optimization of Selectivity

Changing the selectivity α of the separation is the most effective way to influence resolution. This is mainly achieved by changing the chemical nature or concentration of the organic solvent modifier (acetonitrile, methanol, ethanol, isopropanol, etc.), in conjunction with an appropriate choice of mobile phase additive(s). However, if different organic solvents are used, different eluotropic strengths must be considered in order to allow elution of the analytes of the sample in the appropriate retention factor range [133]. The interconversion of isocratic data to gradient data, and vice versa, can be achieved through the use of algorithms [134] based on linear and nonlinear solvent strength theory [135].

6.1.3 Optimization of Retention Factors

Further optimization should focus on achieving the most appropriate retention factor for the different peptides or proteins in the mixture. In the isocratic elution mode of HP-RPC, resolution optimization can take advantage of the relationship between the retention time of an analyte (expressed as the retention factor k) and the volume fraction of the organic solvent modifier, ϕ. Although typically these dependencies are curvilinear, for practical convenience they are often treated as linear relationships. Thus, the change in retention factor as a function of ϕ can be represented by:

$$\ln k = \ln k_0 - S\phi, \quad (16)$$

where k_0 is the retention factor of the solute in the absence of the organic solvent modifier, and S is the slope of the plot of $\ln k$ versus ϕ. The values of $\ln k_0$ and S can be calculated by linear regression analysis. A greater precision in the quality of fit of the experimental data, and thus improved reliability in the prediction of the retention behavior of analytes in HP-RPC systems for mobile phases of different solvent composition, can be achieved [30] through the use of an expanded form of Eq. (16):

$$\ln k = \ln k_0 - S\phi + S'\phi^2 - S''\phi^3 + \ldots \quad (17)$$

Similarly, in gradient elution HP-RPC, resolution optimization can take advantage of the relationship between the gradient retention time of an analyte (expressed as the median retention factor $\bar{k}$) and the median volume fraction of the organic solvent modifier, $\bar{\phi}$, in regular HP-RPC systems based on the concepts of the linear-solvent-strength theory [36, 135], such that:

$$\ln \bar{k} = \ln k_0 - S\bar{\phi}. \quad (18)$$

A mapping of the dependence of analyte retention (expressed as the natural logarithm of the retention factor, k) on the mobile phase composition (expressed as the volume fraction of solvent in the mobile phase, ϕ) in isocratic elution (or as $\bar{k}$ versus $\bar{\phi}$ in gradient elution) with a minimum of two initial experiments, can be used to define the useful range of mobile phase conditions, and can indicate the mobile phase composition at which the band spacing is optimal (Fig. 3).

Irrespective of whether the data are obtained through isocratic or gradient elution techniques employing two initial experiments (differing only by their mobile phase composition or gradient run times, respectively), with tracking and assignment of the peaks, a relative resolution map (RRM) can be established, which plots resolution R_S against the separation time (or gradient run time, t_G). In the case of gradient elution, the RRM then allows determination of the optimal gradient run time (and gradient range). Such a procedure can be performed in any laboratory using Microsoft Excel spreadsheets, or through software packages (e.g., DryLab, LabExpert, etc.) [136]. Such strategies greatly reduce the time to achieve an optimal separation, as well as saving on solvent, reagent, and analyte consumption. Moreover, if fully exploited, this strategy would permit the instrumentation to be operated in a nearly fully automated, unattended fashion [132].

Optimization can also be performed via computer simulation software (e.g., Simplex methods, multivariate factor analysis

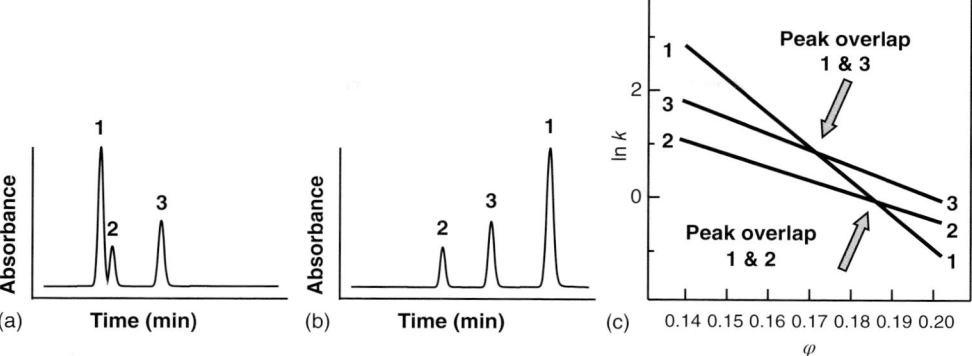

Fig. 3 Optimization of isocratic elution. Two chromatograms obtained for (a) 19% and (b) 14% (v/v) of organic solvent modifier in the mobile phase (corresponding to $\varphi = 0.19$ and 0.14, respectively) can be used to plot (c) the corresponding logarithmic retention factors ln k versus the volume fraction of the organic solvent modifier in order to identify the mobile phase composition resulting in optimal peak spacing. Trend lines can determine mobile phase compositions which result in peak overlap or excellent peak resolution.

programs, DryLab G/plus, LabExpert, etc.). In such procedures, the resolution of peak zones is optimized through a systematic adjustment of the mobile phase composition by successive changes in the φ-value (or equivalent parameters, such as the concentration of the ion-pairing reagent employed).

In gradient elution, advantage is taken of a strategy with the following eight steps:

1) Performing initial experiments.
2) Peak tracking and assignment of the peaks.
3) Calculation of ln k_0-values and S-values from the initial chromatograms.
4) Optimization of gradient run time t_G over the whole gradient range.
5) Determination of a new gradient range.
6) Calculation of new gradient retention times t_g.
7) Change of gradient shape (optional).
8) Verification of results.

Examples where such systematic method development has been used for the analytical separation of peptides and proteins can be found in the literature [137–141].

6.2
Scaling Up to Preparative Chromatography

While analytical HPLC aims at the quantification and/or identification of compounds (with the sample going to waste after the detector), preparative chromatography aims at the capture of compounds (with the sample going to the fraction collector). For separations in preparative chromatography, method development also focuses on the peaks of interest, and the two adjacent eluting peaks. Optimization of the resolution of the peaks of interest from the adjacent peaks must take into account the sample size and the relative abundances of the components in the separation.

Once an analytical method has been established, it can be scaled-up to a preparative separation by taking into consideration the operating ranges of the column (Table 8) or by using deliberate column overloading [27, 142].

Tab. 8 Operating conditions of HPLC column types.

Column type	Sample quantity range	Column diameter) (mm)	Flow rate range (mL min^{-1})	Column length (mm)	Particle size (μm)
Nano chip	<pg	<0.1	<0.0005	50	Monolithic
Nano LC	pg	0.075–0.1	0.0002–0.0005	150	3
Capillary LC	pg–ng	0.3–0.5	0.005–0.01	50–150	3–5
Microbore LC	ng–μg	1	0.05	30–250	3–5
Analytical LC	μg–mg	2–4.6	0.2–1	30–250	1.8–5
Preparative LC	mg–g	21–50	5–120	75–250	10–50

The concept of scaling up or down implies that the performance features (e.g., selectivity behavior) of the stationary phase materials used for the analytical and the preparative separation are identical, with the exception of particle size. Various studies have established the fundamentals of scaling up and have developed experimental methods for their validation. In order to obtain an equivalent elution profile, the flow rate needs to be adjusted for columns with different internal diameters, according to the following formula:

$$F_{preparative} = \left[\frac{r_{preparative}}{r_{analytical}}\right]^2 \times F_{analytical}, \quad (19)$$

where F is the flow rate and r is the column radius of the preparative or analytical column.

Estimates of the loading capacity of a particular column material can usually be obtained from the manufacturer. The mass load ability for a scaled-up separation can be calculated with following formula:

$$M_{preparative} = \left[\frac{r_{preparative}}{r_{analytical}}\right]^2 \times M_{analytical} \times C_L, \quad (20)$$

where M is the mass, r is column radius of the preparative or analytical column, and C_L is the column length ratio.

In some cases, despite some loss of resolution, column overloading is an economic and practical method for compound purification. In analytical LC, the ideal peak shape is a Gaussian curve. If, under analytical conditions, a higher amount of sample is injected, the peak height and area will change but not the peak shape or the retention factor. However, when more than the recommended amount of sample is injected onto the column, the adsorption isotherm becomes nonlinear, and as a direct consequence the resolution will be decreased and the peak retention times and peak shapes may change.

Two methods of column overloading can be employed, namely volume overloading and concentration overloading:

- In *volume overloading*, the concentration of the sample is maintained, but the sample volume is increased. The retention factor of the compound(s) increase(s), possessing a broadened peak shape.
- In *concentration overloading*, the volume of the injected sample is maintained, while the sample concentration is increased. The retention factors of the compound(s) decrease(s), and the peak may become *fronting* or *tailing* with a triangular peak shape. The applicability of this method is limited by the solubility

of the target compound(s) in the mobile phases employed.

In preparative HPLC, volume overloading is the method of choice, as it allows the separation of larger sample amounts than does concentration overloading. However, for most commercial or scientific applications, a combination of both methods is used.

6.3
Fractionation

The process of fraction collection can be: (i) manual, with a button to start and stop collection; (ii) automated, with fixed pre-programmed time intervals; (iii) peak based, based on a chosen threshold of the up- and down-slope of a detector signal; or (iv) mass-based, with fraction collection occurring only if the specific mass of a trigger ion is detected by MS [143]. Whichever type of fraction collection is employed, a fraction collection delay time measurement must be performed. For a peak with start time t_0 and end time t_E, the fraction collection needs to be started when the start of the peak arrives at the diverter valve $(t_0 + t_{D1})$, and ended when the end of the peak arrives at the needle tip $(t_E + t_{D1} + t_{D2})$, where t_{D1} is the delay time between detector and valve and t_{D2} is the delay time between valve and needle tip. In addition, a recovery collection can be performed, whereby everything that is not collected as a fraction is passed into a dedicated container.

6.4
Analysis of Fractions

When the fractions have been collected, the solvent must be removed either by using a freeze-dryer, rotary evaporator, or high-throughput parallel evaporator. Nonvolatile components can be removed using reversed-phase SPE procedures prior to solvent removal if the aqueous portion of the buffer is sufficiently large. In the absence of on-line MS, fractionation is usually accompanied by an off-line mode of quality analysis, comprising a pre-preparative analysis of the unpurified material and a post-preparative analysis of the individual fractions. The pre- and post-preparative analyses can be performed with analytical HPLC, MS, and activity testing of the biological compounds, if an assay is available.

7
Multidimensional HPLC

Although HPLC is a powerful separation technique, very often more than one chromatographic step is necessary to achieve the required degree of purity of the target compounds. In practice, this is achieved through a series of purification steps. As there are material losses associated with each purification step, the overall recovery of the product must be optimized during these procedures, and this can be achieved if the number of employed purification steps is minimized. MD-HPLC offers the possibility of cutting the elution profiles into consecutive fractions, where these fractions can be treated independently from each other. One important consequence of this strategy is the gain in peak capacity, defined as the "number of peaks which can be accommodated between the first and the last peak in a separation of defined resolution" [144]. MD-HPLC has the potential for an independent optimization of the separation conditions for each fraction, and allows a relative enrichment or depletion of components. MD-HPLC can be applied to the purification of a particular

Tab. 9 Optimization priorities at individual stages for a multidimensional, three-stage purification process after initial sample extraction exemplified for peptides and proteins and suitable chromatographic modes.

Purification stage	High priority	Lower priority	Chromatographic mode
Enrichment	High-speed, high-capacity	Resolution, recovery	AC, IMAC, IEX, HIC
Intermediate purification	High-capacity, high-resolution	Speed, recovery	IEX, HIC, SEC
Final purification	High-resolution, high-recovery	Speed, capacity	RPC, SEC

peptide, or to comprehensive fractionations of complex peptide and protein mixtures [145–147].

7.1 Purification of Peptides and Proteins using MD-HPLC Methods

Peptides and proteins are generally isolated from complex biological matrices starting with a fractionation of the crude extract. The chemical and physical properties of the target compound, and of the matrix, determine the choice of separation method. In some cases this is preceded by a protein precipitation with salt or organic solvents, after which the crude extract is clarified (e.g. by filtration, centrifugation) to free the sample from any particulate matter and make it suitable for chromatography. For the subsequent separation step, an appropriate sample buffer that is compatible with the mobile phase(s) of the particular chromatographic mode used in the next step is chosen. This is followed by an enrichment step, preferably with SPE or restricted access materials (RAMs) in a step-elution mode in order to eliminate the majority of low-molecular-weight materials and to drastically reduce the volume of the sample. The next step comprises an intermediate purification of the target compound(s) and a final chromatographic separation using one of the high-performance chromatographic modes of different selectivity (i.e., separating the analytes according to their molecular size, hydrophobicity/hydrophilicity, charge, biospecificity, etc.) according to the principles of chromatography summarized in Table 9.

When purifying a particular peptide or protein, it is often possible to select complementary chromatographic modes that allow the target compound to be obtained in high purity with only a few separation steps (preferably three or less). In such non-comprehensive MD-HPLC, only a part of the analytes (as a single fraction) eluting from the first column is transferred to a second column for further purification (this is conventionally expressed by a hyphen, i.e., IEX-RPC). Although such techniques are fast, they are not comprehensive, as the majority of analytes are not subjected to a separation in a second dimension. The main advantage of the technique is the improved resolution of compounds in the second dimension that co-elute in the first dimension. A key requirement for such a purification scheme is that subsequent stages of the separation are orthogonal, with the two separation modes not correlated to each other in relation to their retention characteristics (i.e., selectivity). For a single chromatographic dimension, the partly contradictory objectives of speed, resolution, capacity and recovery can usually not

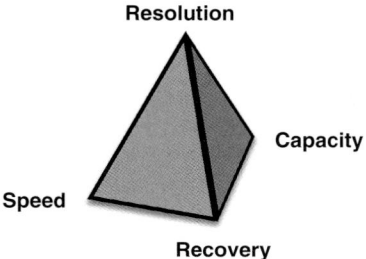

Fig. 4 Optimization goals of speed, resolution, capacity, and recovery for a chromatographic purification, and their inter-relationship.

be maximized simultaneously; that is, a high resolution can be achieved, but at the expense of speed (a high-speed separation can reduce resolution) (see Fig. 4).

A three-stage MD-HPLC protein purification process allows the overall purification objectives to be met by first placing the emphasis for each purification stage on a different pair of objectives, and then choosing a chromatographic mode which is particularly well suited to the task and in a sequence that avoids time-consuming buffer exchanges (see Table 9).

At the enrichment or capture stage, the emphasis is on speed and capacity, employing HP-AC, HP-IEX or HP-HIC (possibly in a SPE format) as a low-resolution step. This stage aims at an initial isolation of the target product from the crude sample, at its concentration, and also at the removal of any major or critical contaminants.

At the intermediate purification stage, emphasis is placed on capacity and resolution, employing chromatographic modes with intermediate resolution, such as HP-AC, HP-IEX, HP-HIC, or HP-SEC. This objectives of this stage are to remove the majority of impurities, such as other proteins, nucleic acids, viruses, and endotoxins.

At the final chromatographic polishing step, emphasis is placed on resolution and recovery, employing high-resolution modes (typically HP-RPC). The aim of this stage is to remove trace amounts of impurities or closely related compounds, so as to obtain the final pure product. HP-SEC may also be used as final polishing step, in order to remove any unwanted multimeric forms of the target protein.

In some cases, the capture and intermediate purification – or, for that matter, the intermediate purification and final polishing step – may be achievable with a single separation step, resulting in a two-stage purification process. In other cases, such as the purification of therapeutic proteins, four or more stages may be required to achieve the desired degree of protein purity.

7.2
Fractionation of Complex Peptide and Protein Mixtures using MD-HPLC

If the objective of a purification scheme is the comprehensive fractionation of a complex, multicomponent peptide mixture (as is required in proteomics), it is advantageous to use orthogonal chromatographic modes, though such extensive fraction collection will require additional (sometimes substantial) infrastructure, including a second HPLC pump, a thermostated autosampler, a robotic autosampler loader, switching valves, thermostated fraction collector, and high-throughput evaporator. In comprehensive MD-HPLC, the entire analyte pool of the first column is transferred to the second column (expressed by a cross, i.e., IEX × RPC) as sequential aliquots, either successively onto one column or alternating onto two parallel columns. The resulting data can be represented as three-dimensional contour plots, with the retention times of the second dimension plotted against the retention times of the first dimension. The information content of such comprehensive 2D chromatograms is higher than the information content of individual, one-dimensional chromatograms.

7.3 Strategies for MD-HPLC Methods

Regardless of which operational mode (off-line or on-line) is used, the compatibility of the mobile phases between successively employed chromatographic modes in a separation scheme must be considered. As a consequence, it may be necessary to process the fractions between two separation stages (e.g., through buffer exchange, concentration, or dilution) to enhance the compatibility of eluent composition of fractions from the first chromatographic dimension with the retaining mobile phase of the second chromatographic dimension. If a non-retentive chromatographic mode such as SEC is employed in conjunction with a retentive chromatographic mode, such as RPC or IEX, it is usually performed first. This allows the relatively large eluent volumes that stem from isocratic elution in the non-retentive mode to be reduced through the capture of analytes under the retaining mobile phase conditions of the subsequent retentive chromatographic mode, and a reduction in extra-column band broadening with its loss of resolution.

7.3.1 *Off-Line* Mode

In the *off-line* mode, the eluent of the first column is collected as fractions, either manually or with an automated fraction collector, and re-injected onto the second column. Typical processing steps may include volume reduction by freeze-drying or automated high-throughput parallel evaporation systems, taking into account the boiling point(s) or volatility of the target analyte(s) and organic solvent if these are contained within the eluates. The use of volatile mobile phase additives then allows a relative fast buffer exchange.

7.3.2 *On-Line* Mode

The *on-line* mode uses high-pressure, multi-position, multi-port switching valves which allows the selection of pathways for single fractions from the first chromatographic dimension to subsequent column(s) of the second chromatographic dimension. The fractions from the first dimension are either transferred directly, or through one (or more) intermediate trapping columns for the purpose of concentration and automated buffer exchange. This approach requires complex instrumentation, but results in an increased optimization time and a reduced system flexibility. However, it has numerous advantages in terms of reproducibility, recovery, speed, and automation.

7.4 Design of an Effective MD-HPLC Scheme

MD-HPLC for peptides and proteins requires a thoughtful selection of orthogonal and complementary separation modes, of the order of their utilization and independent optimization in respect to the chromatographic goals (speed, resolution, capacity, recovery). Furthermore, besides the mobile phase composition of the employed chromatographic modes, the elution mode (isocratic, step, or gradient elution) and flow rates and mobile phase temperatures must be considered.

In order to exploit the maximum peak capacity of a 2D system [148], it is advantageous when the applied chromatographic modes are orthogonal. The dimensions in a 2D separation system are generally considered orthogonal, when the separation mechanism of the two dimensions are independent – that is, the distribution coefficients of analytes in the first dimension are not correlated to those in the second dimension. An example of such orthogonality of different separation modes in HPLC is IEX (either CEX or AEX) and RPC, as these separate according to net charge and hydrophobicity, respectively. A very coarse classification of chromatographic modes commonly applied in the MD-HPLC of peptides and proteins according to their separation principles is depicted in Fig. 5.

For an ideal orthogonal, 2D separation, the overall peak capacity, PC, is defined as the product of the peak capacities in each dimension:

$$PC_{\text{2D-System}} = PC_{\text{First Dimension}} \times PC_{\text{Second Dimension}}. \quad (21)$$

However, if two non-identical chromatographic modes with some degree of similarity are used in a 2D system, the increase in peak capacity and the total number of analytes that can be separated is much lower than the product of peak capacities of the individual dimensions. The peak capacity also depends on the elution mode. Gradient elution provides a higher peak capacity than isocratic elution, and is advantageous in 2D LC. It should be noted, however, that as selectivity in chromatography depends not only on the stationary phase but also on the mobile phase, orthogonal separations can be achieved through fine-tuning of the separation conditions, even if the principal separation mechanisms of both dimensions are similar. In addition, the structure of analytes has an effect on the peak capacity. In many separation systems, the contribution of structural units (especially the repeating units) to the Gibbs free energy of association of the analytes with the immobilized chromatographic ligands are additive [149]. Such structural repeating units can be either hydrophobic or polar.

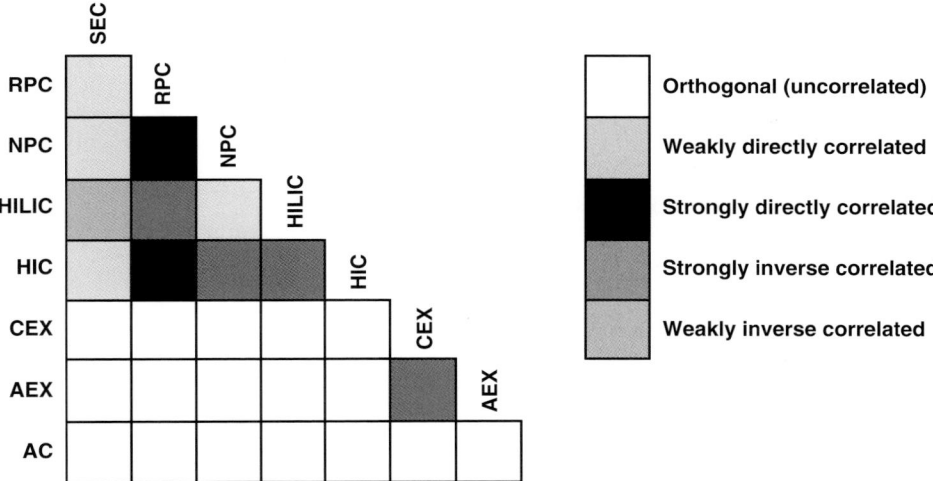

Fig. 5 Degree of orthogonality of major chromatographic modes employed in the separation of peptides and proteins.

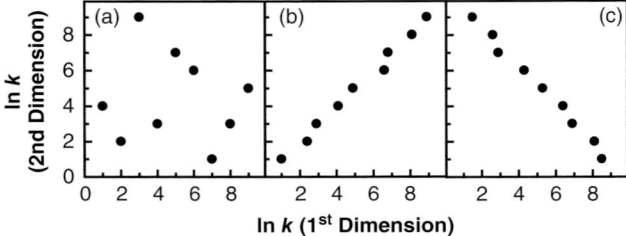

Fig. 6 Two-dimensional separation space for a set of peptides and proteins (circles) utilizing separation systems that are: (a) uncorrelated (orthogonal); (b) completely correlated; and (c) inversely correlated, where the retention factors ln$k$ obtained in the second dimension are plotted versus the retention factors obtained in the first dimension.

If one chromatographic system in a 2D LC has no selectivity for a structural element, then the first and second dimensions will be non-correlated (orthogonal) with respect to the repeating structural unit (see Fig. 6a). In a completely correlated separation system, with correlated retention factors in the two dimensions, the separations space is not utilized and thus not ideal (see Fig. 6b). In inversely correlated 2D LC×LC separation systems, the retention time increases in the first dimension, but decreases in the second dimension (see Fig. 6c). Neither correlated or inversely correlated 2D LC×LC increase the peak capacity significantly. The peak capacity in 2D LC×LC decreases with increasing correlation of the selectivity between the first and second chromatographic dimensions. In practice however, 2D LC×LC systems are rarely fully orthogonal with respect to each structure distribution type (i.e., hydrophobic, polar) [150]. Many partially orthogonal systems use only part of the theoretically available 2D separation space, but can be evaluated using analytes that differ in their numbers of hydrophobic or polar structural units, or by their quantitative structure–retention relationship (QSRR). Orthogonal systems with non-correlated selectivities provide the highest peak capacity, and therefore the highest number of resolved peaks.

Although the suitability of chromatographic modes employed in 2D LC×LC separations depends on the selectivity of the employed stationary phase, the selection of the mobile phase for each chromatographic dimension is of fundamental importance, in order to achieve maximal utilization of the 2D separation space. In contrast to off-line 2D LC procedures, where the collected fraction can be subjected to evaporation, dilution or extraction before injection onto the column of the second dimension, the compatibility of the mobile phases in on-line 2D LC×LC in terms of miscibility, solubility, viscosity and eluotropic strength is much more important. The compatibility of commonly used mobile phases of various chromatographic modes used for the separation of peptides and proteins is depicted in Fig. 7.

8
Final Remarks

The applications of HPLC modes, as presented in this chapter, may give reason to believe that these procedures have already been exhausted with respect to their possibilities for the analytical and preparative separation of peptides and proteins. Yet, numerous unresolved challenges remain in

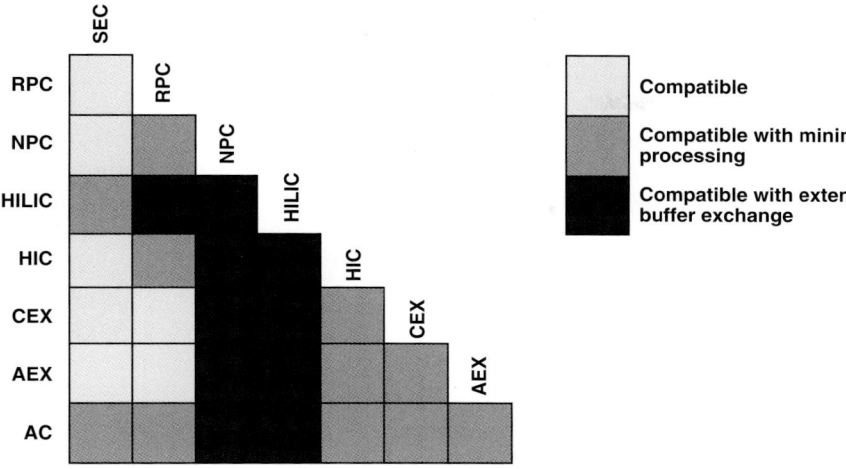

Fig. 7 Compatibility between commonly used mobile phases of chromatographic modes based on miscibility, solubility, and eluotropic strength.

the fields of proteomics, in process analytical technology, and medical diagnostics. The anticipated advances of even more sophisticated methods in the HPLC of peptides and proteins may thus lead to new discoveries in biology and medicine.

It will also be the responsibility of analytical chemists and biochemists to ensure that the development of these new separation and preparative process methods will occur according to the principles of green analytical chemistry, saving energy and materials, reducing waste, and minimizing hazards. Although the first steps in that direction have been taken, there remains a tremendous potential to improve efficiency and sustainability in method development, and this hopefully has been encouraged by the information provided in this chapter.

Abbreviations

2D	two-dimensional
AA	acetic acid
amu	atomic mass unit
CMA	carboxy-methylaspartic acid
ESI	electrospray ionization
FA	formic acid
FTIR	Fourier transform infrared
HBFA	hepta-fluorobutyric acid
HETP	height equivalent to one theoretical plate
HP-AC	high-performance affinity chromatography
HP-AEX	high-performance anion-exchange chromatography
HP-ANPC	high-performance aqueous normal-phase chromatography
HP-CE	high-performance capillary electrophoresis
HP-CEC	high-performance capillary electrochromatography
HP-CEX	high-performance cation-exchange chromatography
HP-GPC	high-performance gel-permeation chromatography

HP-HIC	high-performance hydrophobic interaction chromatography	RPC	reversed-phase chromatography
HP-HILIC	high-performance hydrophilic interaction chromatography	RRM	relative resolution map
		SPE	solid-phase extraction
		TACN	1,4,7-triazo-cyclononane
HP-IEX	high-performance ion-exchange chromatography	TED	tris-(carboxy methyl)ethylene-diamine
		TFA	trifluoroacetic acid
HPLC	high-performance liquid chromatography	UHPLC	ultra-high-performance liquid chromatography
HP-NPC	high-performance normal-phase chromatography		

Symbols

HP-RPC	high-performance reversed-phase chromatography	A	constant in the van Deemter–Knox equation (A-term)
HP-SEC	high-performance size exclusion chromatography	α	selectivity
		$\bar{\alpha}$	median selectivity
IDA	iminodiacetic acid	B	constant in the van Deemter–Knox equation (B-term)
LC	liquid chromatography		
MALDI TOF	matrix-assisted laser desorption/ionization time-of-flight	C_L	column length ratio
		C	constant in the van Deemter–Knox equation (C-term)
MD-HPLC	multidimensional high-performance liquid chromatography	D_m	solute diffusivity in the bulk mobile phase
MIP	molecularly imprinted polymer	D_s	solute diffusivity within the stationary phase
MS	mass spectrometry		
NTA	nitrilotriacetic acid	dp	particle diameter
OPS	O-phosphoserine	ΔA	contact area of the solute and the immobilized ligand
ORD	optical rotatory dispersion		
PAH	polyaromatic hydrocarbon	F	flow rate
PALC	per aqueous liquid chromatography	γ	surface tension of the mobile phase
PAT	process analytical technology	H	plate height
		h	peak height
PBPC	polar bonded phase chromatography	h	reduced plate height
		k	retention factor
PC	peak capacity	$\bar{k}$	median retention factor
QA/QC	quality assurance/quality control	k_0	retention factor of the solute in the absence of organic solvent modifier
QbD	quality by design		
RAM	restricted access material	L	column length

m	molal concentration
N	plate number
N_A	Avogadro's constant
$\overline{N}$	median plate number
n_m	number of moles of the solute in the mobile phase
n_s	number of moles of the solute associated with the stationary phase
v	reduced velocity
ϕ	volume fraction of the organic solvent modifier
$\overline{\phi}$	median volume fraction of the organic solvent modifier
pI	isoelectric point
R	gas constant
R_S	resolution
r	column radius
S	gradient parameter, slope of the plot of ln$k$ versus φ
σ	surface tension increment
σ_t^2	peak variance
σ_{column}^2	peak variances arising due to column effects
σ_{extra}^2	peak variances arising due to extra-column effects
T	absolute temperature
t_0	void time
t_R	retention time
u	linear flow velocity
V_0	void volume
V_R	retention volume
w	peak width at baseline
$w_{1/2}$	peak width at half height

References

1. Unger, K.K., Ditz, R., Machtejevas, E., and Skudas, R. (2010) Liquid chromatography. Its development and key role in life science applications. *Angew. Chem. Int. Ed.*, **49**, 2300–2312.
2. Reichmuth, D.S., Shepodd, T.J., and Kirby, B.J. (2005) Microchip HPLC of peptides and proteins. *Anal. Chem.*, **77**, 2997–3000.
3. Yin, H. and Killeen, K. (2007) The fundamental aspects and applications of Agilent HPLC-Chip. *J. Sep. Sci.*, **30**, 1427–1434.
4. Anastas, P. and Warner, J. (1998) *Green Chemistry: Theory and Practice*, Oxford University Press, Oxford, New York, Tokyo.
5. Anastas, P.T. (1999) Green chemistry and the role of analytical methodology development. *Crit. Rev. Anal. Chem.*, **29**, 167–175.
6. Namiesnik, J. (2001) Green analytical chemistry – some remarks. *J. Sep. Sci.*, **24**, 151–153.
7. Galuszka, A., Migaszewski, Z., and Namiesnik, J. (2013) The 12 principles of green analytical chemistry and the SIGNIFICANCE mnemonic of green analytical practices. *Trends Anal. Chem.*, **50**, 78–84.
8. Welch, C.J., Wu, N., Biba, M., Hartman, R. et al. (2010) Greening analytical chromatography. *Trends Anal. Chem.*, **29**, 667–680.
9. Cielecka-Piontek, J., Zalewski, P., Jelinska, A., and Garbacki, P. (2013) UHPLC: the greening face of liquid chromatography. *Chromatographia*, **76**, 1429–1437.
10. Pereira, S.A., David, F., Vanhoenacker, G., and Sandra, P. (2009) The acetonitrile shortage: is reversed HILIC with water an alternative for the analysis of highly polar ionizable solutes? *J. Sep. Sci.*, **32**, 2001–2007.
11. Safaei, Z., Bocian, S., and Buszewski, B. (2014) Green chromatography-carbon footprint of columns packed with core-shell materials. *RSC Adv.*, **4**, 53915–53920.
12. Plotka, J., Tobiszewski, M., Sulej, A.M., Kupska, M. et al. (2013) Green chromatography. *J. Chromatogr. A*, **1307**, 1–20.
13. Yang, Y., Boysen, R.I., and Hearn, M.T.W. (2006) Optimization of field-amplified sample injection for analysis of peptides by capillary electrophoresis-mass spectrometry. *Anal. Chem.*, **78**, 4752–4758.
14. Yang, Y., Boysen, R.I., and Hearn, M.T.W. (2005) Use of mixed-mode sorbents for the electrochromatographic separation of thrombin receptor antagonistic peptides. *J. Chromatogr. A*, **1079**, 328–334.
15. Hearn, M.T.W. (2001) Peptide analysis by rapid, orthogonal technologies with high separation selectivities and sensitivities. *Biologicals*, **29**, 159–178.
16. Zamyatnin, A.A. (1972) Protein volume in solution. *Prog. Biophys. Mol. Biol.*, **24**, 107–123.

17. Chothia, C. (1975) Structural invariants in protein folding. *Nature*, **254**, 304–308.
18. Dawson, R.M.C., Elliot, D.C., Elliot, W.H., and Jones, K.M. (1986) *Data for Biomedical Research*, 3rd edn, Clarendon Press, Oxford.
19. Rickard, E.C., Strohl, M.M., and Nielsen, R.G. (1991) Correlation of electrophoretic mobilities from capillary electrophoresis with physicochemical properties of proteins and peptides. *Anal. Biochem.*, **197**, 197–207.
20. Wilce, M.C.J., Aguilar, M.I., and Hearn, M.T.W. (1995) Physicochemical basis of amino acid hydrophobicity scales – evaluation of four new scales of amino acid hydrophobicity coefficients derived from RP-HPLC of peptides. *Anal. Chem.*, **67**, 1210–1219.
21. Walla, P.J. (2009) *Modern Biophysical Chemistry – Detection and Analysis of Biomolecules*, Wiley-VCH Verlag GmbH, Weinheim.
22. Bidlingmeyer, B.A. (ed) (1987) *Preparative Liquid Chromatography*, Journal of Chromatography Library, vol. 38, Elsevier, Amsterdam.
23. Grushka, E. (1989) *Preparative-Scale Chromatography*, Chromatographic Science Series, vol. 46, CRC Press.
24. Unger, K.K. (ed) (1994) *Handbook of HPLC, Part 2: Preparative Liquid Column Chromatography*, GIT-Verlag, Darmstadt.
25. Hostettmann, K., Marston, A., and Hostettmann, M. (1997) *Preparative Chromatography Techniques: Applications in Natural Product Isolation*, 2nd edn, Springer-Verlag, London.
26. Boysen, R.I., Hearn, M.T.W. (2001) HPLC of peptides and proteins, in *Current Protocols in Protein Science* (eds J.E. Coligan, B.M. Dunn, H.L. Ploegh, D.W. Speicher *et al.*), John Wiley & Sons, Inc, New York, pp. 1–40.
27. Rathore, A.S. and Velayudhan, A. (eds) (2003) *Scale-Up and Optimization in Preparative Chromatography: Principles and Biopharmaceutical Applications*, Chromatographic Science Series, vol. 88, Marcel Dekker, Inc, New York.
28. Burgess, R.R. (2006) Protein purification, in *Encyclopedia of Molecular Cell Biology and Molecular Medicine* (ed. R.A. Meyers), John Wiley & Sons, Inc, pp. 183–198.
29. Lee, T.-H. and Aguilar, M.-I. (2006) HPLC of peptides and proteins, in *Encyclopedia of Molecular Cell Biology and Molecular Medicine* (ed. R.A. Meyers), John Wiley & Sons, Inc, pp. 248–295.
30. Hearn, M.T.W. (2000) Physicochemical factors in polypeptide and protein purification and analysis by high-performance liquid chromatographic techniques. Current status and challenges for the future, in *Handbook of Bioseparations*, Academic Press, San Diego, CA, pp. 71–235.
31. Duong-Ly, K.C. and Gabelli, S.B. (2014) Gel filtration chromatography (size-exclusion chromatography) of proteins. *Methods Enzymol.*, **541**, 105–114.
32. Gooding, K.M. and Regnier, F.E. (2002) Size-exclusion chromatography. *Chromatogr. Sci. Ser.*, **87**, 49–79.
33. Hong, P., Koza, S., and Bouvier, E.S.P. (2012) A review size-exclusion chromatography for the analysis of protein biotherapeutics and their aggregates. *J. Liq. Chromatogr. Relat. Technol.*, **35**, 2923–2950.
34. Hearn, M.T.W. (2002) Reversed-phase and hydrophobic interaction chromatography of proteins and peptides, in *HPLC of Biological Macromolecules*, Marcel Dekker, Inc, New York, pp. 99–245.
35. Hancock, W.S., Bishop, C.A., Prestidge, R.L., Harding, D.R. *et al.* (1978) Reversed-phase, high-pressure liquid chromatography of peptides and proteins with ion-pairing reagents. *Science*, **200**, 1168–1170.
36. Horvath, C., Melander, W., and Molnar, I. (1976) Solvophobic interactions in liquid chromatography with nonpolar stationary phases. *J. Chromatogr.*, **125**, 129–156.
37. Horvath, C., Melander, W., and Molnar, I. (1977) Liquid chromatography of ionogenic substances with nonpolar stationary phases. *Anal. Chem.*, **49**, 142–154.
38. Horvath, C. and Melander, W. (1977) Liquid chromatography with hydrocarbonaceous bonded phases; theory and practice of reversed phase chromatography. *J. Chromatogr. Sci.*, **15**, 393–404.
39. Aguilar, M.I. and Hearn, M.T.W. (1991) Reversed-phase and hydrophobic-interaction chromatography of proteins, in *HPLC Proteins, Peptides and Polynucleotides* (ed. M.T.W. Hearn), VCH Publishers, New York, Weinheim, Cambridge, pp. 247–275.

40. Hennessy, T.P., Boysen, R.I., Huber, M.I., Unger, K.K. et al. (2003) Peptide mapping by reversed-phase high-performance liquid chromatography employing silica rod monoliths. *J. Chromatogr.*, **1009**, 15–28.
41. Unger, K.K. (1979) *Porous Silica, Its Properties and Use as Support in Column Liquid Chromatography*, Journal of Chromatography Library, vol. 16, Elsevier, Amsterdam.
42. Fekete, S., Veuthey, J.-L., and Guillarme, D. (2012) New trends in reversed-phase liquid chromatographic separations of therapeutic peptides and proteins: theory and applications. *J. Pharm. Biomed. Anal.*, **69**, 9–27.
43. Snyder, L.R. (1970) Adsorption chromatography: scope, technique and equipment. *Methods Med. Res.*, **12**, 11–36.
44. Ballschmiter, K. and Wössner, M. (1998) Recent developments in adsorption liquid chromatography (NP-HPLC) – a review. *Fresenius J. Anal. Chem.*, **361**, 743–755.
45. Buchholz, K., Gödelmann, I., and Molnar, I. (1982) High-performance liquid chromatograph of proteins: analytical applications. *J. Chromatogr.*, **238**, 193–202.
46. Yoshida, T. (1998) Calculation of peptide retention coefficients in normal-phase liquid chromatography. *J. Chromatogr.*, **808**, 105–112.
47. Papadoyannis, I.N., Zotou, A.C., and Samanidou, V.F. (1995) Solid-phase extraction study and RP-HPLC analysis of lamotrigine in human biological fluids and in antiepileptic tablet formulations. *J. Liq. Chromatogr. Relat. Technol.*, **18**, 2593–2609.
48. Jandera, P. (2011) Stationary and mobile phases in hydrophilic interaction chromatography: a review. *Anal. Chim. Acta*, **692**, 1–25.
49. Buszewski, B. and Noga, S. (2012) Hydrophilic interaction liquid chromatography (HILIC) – a powerful separation technique. *Anal. Bioanal. Chem.*, **402**, 231–247.
50. Alpert, A.J. (1990) Hydrophilic-interaction chromatography for the separation of peptides, nucleic acids and other polar compounds. *J. Chromatogr.*, **499**, 177–196.
51. Linden, J.C. and Lawhead, C.L. (1975) Liquid chromatography of saccharides. *J. Chromatogr.*, **105**, 125–133.
52. Yoshida, T. (2004) Peptide separation by hydrophilic-interaction chromatography: a review. *J. Biochem. Biophys. Methods*, **60**, 265–280.
53. Mant, C.T. and Hodges, R.S. (2008) Mixed-mode hydrophilic interaction/cation-exchange chromatography (HILIC/CEX) of peptides and proteins. *J. Sep. Sci.*, **31**, 2754–2773.
54. Yang, M., Thompson, R., and Hall, G. (2009) Some insights on retention and selectivity for hydrophilic interaction chromatography. *J. Liq. Chromatogr. Relat. Technol.*, **32**, 628–646.
55. Boutin, J.A., Ernould, A.P., Ferry, G., Genton, A. et al. (1992) Use of hydrophilic interaction chromatography for the study of tyrosine protein kinase specificity. *J. Chromatogr., Biomed. Appl.*, **583**, 137–143.
56. Yoshida, T. (1997) Peptide separation in normal phase liquid chromatography. *Anal. Chem.*, **69**, 3038–3043.
57. Hemström, P. and Irgum, K. (2006) Hydrophilic interaction chromatography. *J. Sep. Sci.*, **29**, 1784–1821.
58. Gilar, M., Olivova, P., Daly, A.E., and Gebler, J.C. (2005) Orthogonality of separation in two-dimensional liquid chromatography. *Anal. Chem.*, **77**, 6426–6434.
59. Nguyen, H.P. and Schug, K.A. (2008) The advantages of ESI-MS detection in conjunction with HILIC mode separations: fundamentals and applications. *J. Sep. Sci.*, **31**, 1465–1480.
60. Naidong, W. (2003) Bioanalytical liquid chromatography tandem mass spectrometry methods on underivatized silica columns with aqueous/organic mobile phases. *J. Chromatogr. B: Anal. Technol. Biomed. Life Sci.*, **796**, 209–224.
61. Hao, Z., Xiao, B., and Weng, N. (2008) Impact of column temperature and mobile phase components on selectivity of hydrophilic interaction chromatography (HILIC). *J. Sep. Sci.*, **31**, 1449–1464.
62. Zhu, B.Y., Mant, C.T., and Hodges, R.S. (1991) Hydrophilic-interaction chromatography of peptides on hydrophilic and strong cation-exchange columns. *J. Chromatogr.*, **548**, 13–24.
63. Wu, J., Bicker, W., and Lindner, W. (2008) Separation properties of novel and commercial polar stationary phases in hydrophilic interaction and reversed-phase liquid chromatography mode. *J. Sep. Sci.*, **31**, 1492–1503.

64. Fausnaugh-Pollitt, J., Thevenon, G., Janis, L., and Regnier, F.E. (1988) Chromatographic resolution of lysozyme variants. *J. Chromatogr.*, **443**, 221–228.
65. Mant, C.T., Litowski, J.R., and Hodges, R.S. (1998) Hydrophilic interaction/cation-exchange chromatography for separation of amphipathic alpha-helical peptides. *J. Chromatogr.*, **816**, 65–78.
66. Mant, C.T., Kondejewski, L.H., and Hodges, R.S. (1998) Hydrophilic interaction/cation-exchange chromatography for separation of cyclic peptides. *J. Chromatogr.*, **816**, 79–88.
67. Hodges, R.S., Chen, Y., Kopecky, E., and Mant, C.T. (2004) Monitoring the hydrophilicity/hydrophobicity of amino acid side-chains in the non-polar and polar faces of amphipathic alpha-helices by reversed-phase and hydrophilic interaction/cation-exchange chromatography. *J. Chromatogr.*, **1053**, 161–172.
68. McNulty, D.E. and Annan, R.S. (2008) Hydrophilic interaction chromatography reduces the complexity of the phosphoproteome and improves global phosphopeptide isolation and detection. *Mol. Cell. Proteomics*, **7**, 971–980.
69. McNulty, D.E. and Annan, R.S. (2009) Hydrophilic interaction chromatography for fractionation and enrichment of the phosphoproteome. *Methods Mol. Biol.*, **527**, 93–105.
70. Hägglund, P., Bunkenborg, J., Elortza, F., Jensen, O.N. *et al.* (2004) A new strategy for identification of N-glycosylated proteins and unambiguous assignment of their glycosylation sites using HILIC enrichment and partial deglycosylation. *J. Proteome Res.*, **3**, 556–566.
71. Picariello, G., Ferranti, P., Mamone, G., Roepstorff, P. *et al.* (2008) Identification of N-linked glycoproteins in human milk by hydrophilic interaction liquid chromatography and mass spectrometry. *Proteomics*, **8**, 3833–3847.
72. Wang, C., Jiang, C., and Armstrong, D.W. (2008) Considerations on HILIC and polar organic solvent-based separations: use of cyclodextrin and macrocyclic glycopeptide stationary phases. *J. Sep. Sci.*, **31**, 1980–1990.
73. Wuhrer, M., de Boer, A.R., and Deelder, A.M. (2009) Structural glycomics using hydrophilic interaction chromatography (HILIC) with mass spectrometry. *Mass Spectrom. Rev.*, **28**, 192–206.
74. Pesek, J.J., Matyska, M.T., Hearn, M.T.W., and Boysen, R.I. (2009) Aqueous normal-phase retention of nucleotides on silica hydride columns. *J. Chromatogr.*, **1216**, 1140–1146.
75. Pesek, J.J., Matyska, M.T., Loo, J.A., Fischer, S.M. *et al.* (2009) Analysis of hydrophilic metabolites in physiological fluids by HPLC-MS using a silica hydride-based stationary phase. *J. Sep. Sci.*, **32**, 2200–2208.
76. Pesek, J.J. and Matyska, M.T. (2010) Silica hydride – chemistry and applications. *Adv. Chromatogr. (Boca Raton)*, **48**, 255–288.
77. Pesek, J.J., Matyska, M.T., Boysen, R.I., Yang, Y. *et al.* (2013) Aqueous normal-phase chromatography using silica-hydride-based stationary phases. *Trends Anal. Chem.*, **42**, 64–73.
78. Pesek, J.J., Boysen, R.I., Hearn, M.T.W., and Matyska, M.T. (2014) Hydride-based HPLC stationary phases: a rapidly evolving technology for the development of new bio-analytical methods. *Anal. Methods*, **6**, 4496–4503.
79. Kulsing, C., Yang, Y., Munera, C., Tse, C. *et al.* (2014) Correlations between the zeta potentials of silica hydride-based stationary phases, analyte retention behaviour and their ionic interaction descriptors. *Anal. Chim. Acta*, **817**, 48–60.
80. Yang, Y., Boysen, R.I., Kulsing, C., Matyska, M.T. *et al.* (2013) Analysis of polar peptides using a silica hydride column and high aqueous content mobile phases. *J. Sep. Sci.*, **36**, 3019–3025.
81. Shaltiel, S. and Er-El, Z. (1973) Hydrophobic chromatography: use for purification of glycogen synthetase. *Proc. Natl Acad. Sci. USA*, **70**, 778–781.
82. Hjerten, S. (1973) General aspects of hydrophobic interaction chromatography. *J. Chromatogr.*, **87**, 325–331.
83. Hofstee, B.H.J. (1973) Hydrophobic affinity chromatography of proteins. *Anal. Biochem.*, **52**, 430–448.
84. Shaltiel, S. (1974) Hydrophobic chromatography. *Methods Enzymol.*, **34**, 126–140.
85. Fausnaugh, J.L., Pfannkoch, E., Gupta, S., and Regnier, F.E. (1984) High-performance hydrophobic interaction chromatography of proteins. *Anal. Biochem.*, **137**, 464–472.

86. Gooding, K.M. (1986) High-performance liquid chromatography of proteins – a current look at the state of the technique. *BioChromatography*, **1**, 34–40.
87. Wu, S.L., Benedek, K., and Karger, B.L. (1986) Thermal behavior of proteins in high-performance hydrophobic-interaction chromatography. On-line spectroscopic and chromatographic characterization. *J. Chromatogr.*, **359**, 3–17.
88. Melander, W.R., El Rassi, Z., and Horvath, C. (1989) Interplay of hydrophobic and electrostatic interactions in biopolymer chromatography. Effect of salts on the retention of proteins. *J. Chromatogr.*, **469**, 3–27.
89. Antia, F.D., Fellegvari, I., and Horvath, C. (1995) Displacement of proteins in hydrophobic interaction chromatography. *Ind. Eng. Chem. Res.*, **34**, 2796–2804.
90. McCue, J.T. (2014) Use and application of hydrophobic interaction chromatography for protein purification. *Methods Enzymol.*, **541**, 51–65.
91. Hofstee, B.H.J. and Otillio, N.F. (1973) Immobilization of enzymes through non-covalent binding to substituted agaroses. *Biochem. Biophys. Res. Commun.*, **53**, 1137–1144.
92. Hofstee, B.H.J. (1979) Non-ionic adsorption chromatography and adsorptive immobilization of proteins. *Pure Appl. Chem.*, **51**, 1537–1548.
93. Wu, S.L., Figueroa, A., and Karger, B.L. (1986) Protein conformational effects in hydrophobic interaction chromatography. Retention characterization and the role of mobile phase additives and stationary phase hydrophobicity. *J. Chromatogr.*, **371**, 3–27.
94. Melander, W.R., Corradini, D., and Horvath, C. (1984) Salt-mediated retention of proteins in hydrophobic-interaction chromatography. Application of solvophobic theory. *J. Chromatogr.*, **317**, 67–85.
95. Fausnaugh, J.L., Kennedy, L.A., and Regnier, F.E. (1984) Comparison of hydrophobic-interaction and reversed-phase chromatography of proteins. *J. Chromatogr.*, **317**, 141–155.
96. Wetlaufer, D.B. and Koenigbauer, M.R. (1986) Surfactant-mediated protein hydrophobic-interaction chromatography. *J. Chromatogr.*, **359**, 55–60.
97. Pahlman, S., Rosengren, J., and Hjerten, S. (1977) Hydrophobic interaction chromatography on uncharged Sepharose derivatives. Effects of neutral salts on the adsorption of proteins. *J. Chromatogr.*, **131**, 99–108.
98. Melander, W. and Horvath, C. (1977) Salt effects on hydrophobic interactions in precipitation and chromatography of proteins: an interpretation of the lyotropic series. *Arch. Biochem. Biophys.*, **183**, 200–215.
99. Roettger, B.F., Myers, J.A., Ladisch, M.R., and Regnier, F.E. (1989) Adsorption phenomena in hydrophobic interaction chromatography. *Biotechnol. Progr.*, **5**, 79–88.
100. Porath, J., Sundberg, L., Fornstedt, N., and Olsson, I. (1973) Salting-out in amphiphilic gels as a new approach to hydrophobic adsorption. *Nature*, **245**, 465–466.
101. Chang, S., Noel, R., and Regnier, F.E. (1976) High-speed ion-exchange chromatography of proteins. *Anal. Chem.*, **48**, 1839–1845.
102. Kopaciewicz, W. and Regnier, F.E. (1983) Mobile phase selection for the high-performance ion-exchange chromatography of proteins. *Anal. Biochem.*, **133**, 251–259.
103. Regnier, F.E. (1984) High-performance ion-exchange chromatography. *Methods Enzymol.*, **104**, 170–189.
104. Kopaciewicz, W., Rounds, M.A., and Regnier, F.E. (1985) Stationary phase contributions to retention in high-performance anion-exchange protein chromatography: ligand density and mixed mode effects. *J. Chromatogr.*, **318**, 157–172.
105. Kopaciewicz, W. and Regnier, F.E. (1986) Synthesis of cation-exchange stationary phases using an adsorbed polymeric coating. *J. Chromatogr.*, **358**, 107–117.
106. Hearn, M.T.W., Hodder, A.N., and Aguilar, M.I. (1988) High-performance liquid chromatography of amino acids, peptides and proteins. LXXXVII. Comparison of retention and bandwidth properties of proteins eluted by gradient and isocratic anion-exchange chromatography. *J. Chromatogr.*, **458**, 27–44.
107. Heinitz, M.L., Kennedy, L., Kopaciewicz, W., and Regnier, F.E. (1988) Chromatography of proteins on hydrophobic interaction and ion-exchange chromatographic matrices: mobile phase contributions to selectivity. *J. Chromatogr.*, **443**, 173–182.

108. Hodder, A.N., Aguilar, M.I., and Hearn, M.T.W. (1990) High-performance liquid chromatography of amino acids, peptides and proteins. XCVII. The influence of the gradient elution mode and displacer salt type on the retention properties of closely related protein variants separated by high-performance anion-exchange chromatography. *J. Chromatogr.*, **506**, 17–34.
109. Jungbauer, A. and Hahn, R. (2009) Ion-exchange chromatography. *Methods Enzymol.*, **463**, 349–371.
110. Mant, C.T. and Hodges, R.S. (1985) Separation of peptides by strong cation-exchange high-performance liquid chromatography. *J. Chromatogr.*, **327**, 147–155.
111. Boardman, N.K. and Partridge, S.M. (1955) Separation of neutral proteins on ion-exchange resins. *Biochem. J.*, **59**, 543–552.
112. Himmelhoch, S.R. (1971) Chromatography of proteins on ion-exchange adsorbents. *Methods Enzymol.*, **22**, 273–286.
113. Kopaciewicz, W. and Regnier, F.E. (1983) A system for coupled multiple-column separation of proteins. *Anal. Biochem.*, **129**, 472–482.
114. Kopaciewicz, W., Rounds, M.A., Fausnaugh, J., and Regnier, F.E. (1983) Retention model for high-performance ion-exchange chromatography. *J. Chromatogr.*, **266**, 3–21.
115. Rounds, M.A. and Regnier, F.E. (1984) Evaluation of a retention model for high-performance ion-exchange chromatography using two different displacing salts. *J. Chromatogr.*, **283**, 37–45.
116. Gooding, D.L., Schmuck, M.N., and Gooding, K.M. (1984) Analysis of proteins with new, mildly hydrophobic high-performance liquid chromatography packing materials. *J. Chromatogr.*, **296**, 107–114.
117. Gooding, K.M. and Schmuck, M.N. (1985) Comparison of weak and strong high-performance anion-exchange chromatography. *J. Chromatogr.*, **327**, 139–146.
118. Stout, R.W., Sivakoff, S.I., Ricker, R.D., and Snyder, L.R. (1986) Separation of proteins by gradient elution from ion-exchange columns. Optimizing experimental conditions. *J. Chromatogr.*, **353**, 439–463.
119. Hearn, M.T.W., Hodder, A.N., Stanton, P.G., and Aguilar, M.I. (1987) High-performance liquid chromatography of amino acid peptides and proteins. *Chromatographia*, **24**, 769–776.
120. Hearn, M.T.W., Hodder, A.N., and Aguilar, M.I. (1988) High-performance liquid chromatography of amino acids, peptides and proteins. LXXXVI. The influence of different displacer salts on the retention and bandwidth properties of proteins separated by isocratic anion-exchange chromatography. *J. Chromatogr.*, **443**, 97–118.
121. Hodder, A.N., Aguilar, M.I., and Hearn, M.T.W. (1989) High-performance liquid chromatography of amino acids, peptides, and proteins. LXXXIX. The influence of different displacer salts on the retention properties of proteins separated by gradient anion-exchange chromatography. *J. Chromatogr.*, **476**, 391–411.
122. Aguilar, M.I., Hodder, A.N., and Hearn, M.T.W. (1991) High-performance ion-exchange chromatography of proteins, in *HPLC Proteins, Peptides and Polynucleotides* (ed. M.T.W. Hearn), VCH Publishers, New York, Weinheim, Cambridge, pp. 199–245.
123. Rowe, L., El Khouty, G., and Lowe, C.R. (2012) Affinity chromatography: Historical and prospective overview, in *Biopharmaceutical Production Technology* (ed. G. Subramanian), Wiley-VCH Verlag GmbH, Weinheim, pp. 225–282.
124. Porath, J., Carlsson, J., Olsson, I., and Belfrage, G. (1975) Metal chelate affinity chromatography, a new approach to protein fractionation. *Nature*, **258**, 598–599.
125. Zachariou, M., Traverso, I., and Hearn, M.T. (1993) High-performance liquid chromatography of amino acids, peptides and proteins. CXXXI. O-phosphoserine as a new chelating ligand for use with hard Lewis metal ions in the immobilized-metal affinity chromatography of proteins. *J. Chromatogr.*, **646**, 107–120.
126. Jiang, W., Graham, B., Spiccia, L., and Hearn, M.T.W. (1998) Protein selectivity with immobilized metal ion tacn sorbents – chromatographic studies with human serum proteins and several other globular proteins. *Anal. Biochem.*, **255**, 47–58.
127. Petzold, M., Coghlan, C.J., and Hearn, M.T.W. (2014) Studies with an immobilized metal affinity chromatography cassette system involving binuclear triazacyclononane-derived ligands: automation of batch

adsorption measurements with tagged recombinant proteins. *J. Chromatogr. A*, **1351**, 61–69.
128. Wirth, H.J. and Hearn, M.T.W. (1993) High-performance liquid chromatography of amino acids, peptides and proteins. CXXX. Modified porous zirconia as sorbents in affinity chromatography. *J. Chromatogr.*, **646**, 143–151.
129. Molnar, I., Rieger, H.J., and Monks, K.E. (2010) Aspects of the "Design Space" in high-pressure liquid chromatography method development. *J. Chromatogr. A*, **1217**, 3193–3200.
130. Monks, K., Molnar, I., Rieger, H.J., Bogati, B. et al. (2012) Quality by design: multidimensional exploration of the design space in high performance liquid chromatography method development for better robustness before validation. *J. Chromatogr. A*, **1232**, 218–230.
131. Konieczka, P. and Namiesnik, J. (2010) Estimating uncertainty in analytical procedures based on chromatographic techniques. *J. Chromatogr. A*, **1217**, 882–891.
132. Ting-Po, I., Smith, R., Guhan, S., Taksen, K. et al. (2002) Intelligent automation of high-performance liquid chromatography method development by means of a real-time knowledge-based approach. *J. Chromatogr.*, **972**, 27–43.
133. Patel, H.B. and Jefferies, T.M. (1987) Eluotropic strength of solvents. Prediction and use in reversed-phase high-performance liquid chromatography. *J. Chromatogr.*, **389**, 21–32.
134. Schoenmakers, P.J., Billiet, H.A.H., and Galan, L.D. (1979) Influence of organic modifiers on the retention behaviour in reversed-phase liquid chromatography and its consequences for gradient elution. *J. Chromatogr.*, **185**, 179–195.
135. Snyder, L.R. (1980) Gradient elution, in *HPLC – Advances and Perspectives* (ed. C. Horvath), Academic Press, New York, pp. 208–316.
136. Snyder, L.R. (1996) Optimization of peptide mapping via computer simulation, in *New Methods in Peptide Mapping for the Characterization of Proteins* (ed. W.S. Hancock), CRC Press, Boca Raton, FL, pp. 31–55.
137. Stadalius, M.A., Gold, H.S., and Snyder, L.R. (1984) Optimization model for the gradient elution separation of peptide mixtures by reversed-phase high-performance liquid chromatography. Verification of retention relationships. *J. Chromatogr.*, **296**, 31–59.
138. Ghrist, B.F.D., Coopermann, B.S., and Snyder, L.R. (1988) Design of optimized high-performance liquid chromatographic gradients for the separation of either small or large molecules. Minimizing errors in computer simulations. *J. Chromatogr.*, **459**, 1–23.
139. Ghrist, B.F.D. and Snyder, L.R. (1988) Design of optimized high-performance liquid chromatographic gradients for the separation of either small or large molecules. II. Background and theory. *J. Chromatogr.*, **459**, 25–41.
140. Ghrist, B.F.D. and Snyder, L.R. (1988) Design of optimized high-performance liquid chromatographic gradients for the separation of either small or large molecules. III. An overall strategy and its application to several examples. *J. Chromatogr.*, **459**, 43–63.
141. Boysen, R.I., Erdmann, V.A., and Hearn, M.T.W. (1998) Systematic, computer-assisted optimisation of the isolation of *Thermus thermophilus* 50S ribosomal proteins by reversed-phase high-performance liquid chromatography. *J. Biochem. Biophys. Methods*, **37**, 69–89.
142. Mazzei, J.L. and d'Avila, L.A. (2003) Chromatographic models as tools for scale-up of isolation of natural products by semi-preparative HPLC. *J. Liq. Chromatogr. Relat. Technol.*, **26**, 177–193.
143. Rosentreter, U. and Huber, U. (2004) Optimal fraction collecting in preparative LC/MS. *J. Comb. Chem.*, **6**, 159–164.
144. Martin, M. (1995) On the potential of two- and multi-dimensional separation systems. *Fresenius J. Anal. Chem.*, **352**, 625–632.
145. Fournier, M.L., Gilmore, J.M., Martin-Brown, S.A., and Washburn, M.P. (2007) Multidimensional separations-based shotgun proteomics. *Chem. Rev.*, **107**, 3654–3686.
146. Motoyama, A. and Yates, J.R. (2008) Multidimensional LC separations in shotgun proteomics. *Anal. Chem.*, **80**, 7187–7193.
147. Zhang, X., Fang, A., Riley, C.P., Wang, M. et al. (2010) Multi-dimensional liquid chromatography in proteomics – A review. *Anal. Chim. Acta*, **664**, 101–113.

148. Giddings, J.C. (1995) Sample dimensionality: a predictor of order-disorder in component peak distribution in multidimensional separation. *J. Chromatogr.*, **703**, 3–15.

149. Martin, A.J.P. (1949) Partition chromatography. *Annu. Rep. Prog. Chem.*, **45**, 267–283.

150. Jandera, P. (2006) Column selectivity for two-dimensional liquid chromatography. *J. Sep. Sci.*, **29**, 1763–1783.

16
Hit-to-Lead Medicinal Chemistry

Simon E. Ward[1] and Paul Beswick[2]
[1] *University of Sussex, Translational Drug Discovery Group, Falmer, Brighton BN1 9QJ, UK*
[2] *University of Sussex, External Collaboration Manager and Project Leader, Brighton BN1 9QJ, UK*

1	Introduction to the Hit-to-Lead Phase	576
2	Confidence in the Hit Matter	577
2.1	Where Do the Hits Come From?	577
2.2	Hit Confirmation	578
2.2.1	Common Causes of False Hits	579
2.3	The Quality of Hit Matter is of Paramount Importance	580
3	Goal of Hit-to-Lead Medicinal Chemistry	581
3.1	Screening Cascades	583
3.2	Target Physico-Chemical Properties	583
3.2.1	Property Guidelines	584
3.3	Starting Hit Optimization	585
4	Strategies of Hit-to-Lead Medicinal Chemistry	586
4.1	Physico-Chemical Properties	586
4.2	Optimizing Target Affinity While Controlling Lipophilicity	586
4.3	Optimizing Cellular Activity	587
4.4	Selectivity	588
4.5	Consideration of Developability	588
4.5.1	ADME Properties	588
4.5.2	Safety Properties	589
4.5.3	Solubility	590
5	Fragment-Based Lead Discovery	590
6	Conclusions	591
	References	592

Translational Medicine: Molecular Pharmacology and Drug Discovery
First Edition. Edited by Robert A. Meyers.
© 2018 Wiley-VCH Verlag GmbH & Co. KGaA. Published 2018 by Wiley-VCH Verlag GmbH & Co. KGaA.

Keywords

Hit
A compound which forms the starting point for a drug discovery program. Typically, hits arise from campaigns such as high-throughput screening (HTS) or fragment screening. Hits have a reproducible but often low potency in the micromolar range.

Lead
A compound which forms the basis for the second stage of the drug discovery process, termed lead optimization. A lead is typically a representative of a family of structurally related active compounds.

Lead optimization
The third and final stage of the research phase of the drug discovery process, with the ultimate goal of selecting a candidate drug for early development.

Fragment
A very low-molecular-weight molecule (typically <250 Da) screened at high concentration. Active fragments are developed into the leads through the use of structural biology techniques, notably X-ray crystallography and nuclear magnetic resonance.

Ligand efficiency
A measure of the contribution of each (non-hydrogen) atom to the overall binding of a drug molecule to its target protein. Ligand efficiency is frequently used by medicinal chemists to aid potency optimization, without compromising developability properties.

Hit-to-Lead medicinal chemistry is the process through which "hits" are converted into "leads." Fundamental to the process is to gain confidence, first that a hit compound can be developed into a robust (lead) series, and second that the leads have the potential to be converted into drug candidates. Such confidence is gained in a stepwise fashion: initially, hit compounds must have their activity confirmed, and second it must be demonstrated that structural modification of the hit can result in changes in biological activity, thus generating a series. Third, members of the series must pass a series of so-called "developability filters," thus ensuring that there is a high chance of success in the subsequent phase of lead optimization. The hit-to-lead phase is highly important, as chemists are frequently presented with multiple hits from which to identify a small number of series to take forward. Selection at this point will have a direct impact on the success of lead optimization.

1 Introduction to the Hit-to-Lead Phase

The generation of chemical "hit" matter against a biological target of interest is an important, often challenging, step during the early stages of a drug discovery and development project. With the intent of outlining the process of hit-to-lead optimization, attention in this chapter is focused exclusively on the generation of small-molecule "hit" matter that requires optimization to generate a "lead" structure. Analogous early strategies for generation of macromolecular, protein, or other biological therapeutic approaches, are not discussed.

For a classical small-molecule approach, a "hit" would represent a molecule which has been identified as having the preliminary biological profile required. Examples would be an inhibitor of a particular enzyme in a biochemical assay; a fragment cocrystallized into a protein; or a molecule causing the required phenotypic response in a cellular assay. Following hit confirmation (as described below), the hit molecule would then undergo a process of optimization to introduce further drug-like properties to reach "lead" status, and to then initiate the full lead optimization to develop into a preclinical candidate molecule which can be progressed through the preclinical development steps required to deliver a molecule ready for clinical evaluation, as depicted in Fig. 1.

The definitions "hit," "lead," and various "candidate" terms are arbitrary labels that are used to help discussions and provide a mutual understanding of the level of data and confidence behind a molecule progressing through drug discovery and development phases. They should not be used to create silos of activities, nor to consider the discovery steps as simply a process. The transformation of a molecule into a potential new drug requires tremendous creativity, perseverance, and often also luck. It is an intellectually and practically demanding activity which should not be viewed simply as a "process" – as the classic chevron diagrams tend to imply.

2
Confidence in the Hit Matter

2.1
Where Do the Hits Come From?

The chemical starting points or "hits" within a drug discovery project can be identified by various means:

1) Screening: this can be by a range of assays (biochemical, biophysical, cellular, phenotypic, etc.), and can be on large, diverse corporate collections of up to millions of molecules or on smaller focused sets, for example, fragments, ATP-competitive kinase inhibitors, and natural products.

2) Knowledge of natural ligands: for example, the modification of endogenous ligands such as neurotransmitters or enzyme substrates to identify new molecules which bind to the same biological target

3) Knowledge of the biological target structure: *de-novo* design of ligands from a detailed understanding of the binding sites within, for example, protein crystal structures. This can also be used to generate pharmacophoric maps of the protein surface, which can in turn be used for the virtual screening of commercial molecule datasets to identify molecules for screening.

4) Using molecules already known to bind to the target, for example, from competitor patents or from the serendipitous cross-screening of other molecules.

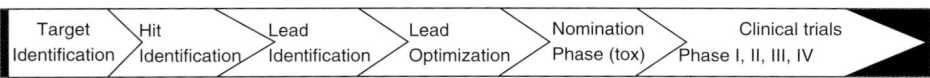

Fig. 1 Drug discovery chevron diagram.

5) Selective optimization of side activities (SOSA) of drug molecules. This technique uses clinical-stage molecules that already have proven drug-like properties, and seeks to identify potentially off-target pharmacologies that can be optimized to deliver molecules with new biological profiles [1]. Figure 2 shows an example where a weak M1 muscarinic activity was observed in the pyridazine-based antidepressant minaprine, whereby a subsequent minor structural modification led to the selective nanomolar affinity M1 partial agonist (**4**).

Clearly, only some of these options are available depending, on the nature of the biological target and its level of current investigation – that is, are there protein crystal structures/existing hit molecules? Regardless of the route used to identify the hits, it is imperative that the hits are validated or confirmed prior to the initiation of hit-to-lead medicinal chemistry. This is particularly important as it is very easy to see some initial hits emerging from a screening strategy, to be over-confident in their validity, and consequently to waste valuable resources on following up hits that are not genuine.

2.2
Hit Confirmation

Prior to the start of hit-to-lead chemistry, is important that the following steps are undertaken:

1) Retest in the biological assay that identified the hit with fresh solution.
2) Preparation of a fresh screening solution from a solid sample that has been either re-purified or at least re-analyzed to confirm its chemical structure and purity. Repeat of the biological assay (and if possible quantification of compound concentration within the assay and confirmation of chemical stability to assay conditions).
3) Confirmation of activity in an orthogonal assay. For example, a molecule found to be active in an enzyme inhibitor assay using a fluorescence read-out could be evaluated in an alternative assay with a calorimetric

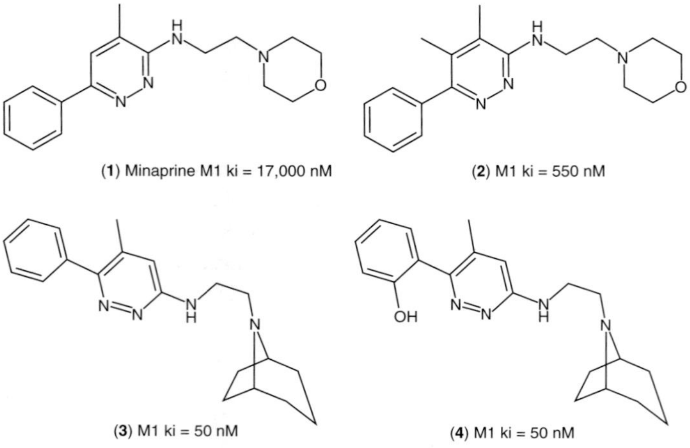

(**1**) Minaprine M1 ki = 17,000 nM

(**2**) M1 ki = 550 nM

(**3**) M1 ki = 50 nM

(**4**) M1 ki = 50 nM

Fig. 2 An example of the selective optimization of side activities (SOSA) approach.

read-out. The principle here is to ensure that there has been no interference with the assay itself (e.g., through molecule auto-fluorescence, nonspecific aggregation of protein, redox interference, etc.).

4) Ideally, molecules of closely related structure also exist which show varying levels of activity within the same assays, that is, structure–activity relationships (SARs).

With these steps complete, it would be reasonable to have confidence that the molecules represent genuine hits, although it should be expected that many of the hits identified by the various methods above will not make it through this confirmation phase. It can also be challenging to develop truly orthogonal assays, though it should be stressed that this step is vital to build confidence in the validity of the hits obtained.

2.2.1 Common Causes of False Hits

In addition to the steps above, it is important to be aware of the common causes of false positives: to be able to better curate the compound collections used for screening, and also to be able to quickly eliminate false-positive structures when identified. This is particularly important as a number of these structures can appear genuine, as well as having concentration-dependence and sometimes reproducibility in alternative assay formats. If not effectively identified, false hits can lead to a significant waste of time and resources. For example, in certain cases they are not detected until exploratory chemistry has begun, where failure to generate a meaningful SAR is indicative of issues regarding the original hit. Clearly, this is not an effective way to identify false hits. Four main classes of common false positives exist:

1) Assay interferers: these are molecules which cause direct interference with the assay read-out, perhaps through intrinsic molecule fluorescence or luminescence, or by interfering with a biological component of a coupled assay read-out.

2) Protein aggregators and solubility: this is a poorly understood phenomenon that was described for the first time in 2002 [2] and is an area of intense recent interest.

3) Non-specific chemical reactivity such as covalent binding or metal ion chelation.

4) Redox: certain compounds have the ability to oxidize protein targets, leading to their inactivation. In some cases, compounds (particularly oxidoreductase enzymes) can indirectly inhibit target function by removing electrons from the system. This activity is clearly undesirable and can lead to a lack of specificity.

All screens produce false positives, but the most efficient way to reduce their occurrence is to remove so-called "interference compounds" (commonly called PAINS or pan-assay interference compounds) from screening collections, and this has been the focus of intensive efforts by screening groups during recent years. A comprehensive list of such molecules was recently published [3] which has aided not only organizations in removing interference compounds from compound collections, but also chemists in detecting false positives in screening outputs. Inevitably, however, screens will continue to produce false hits, and it is important that measures are in place to remove them from a screening set as efficiently as possible. In some cases, such as fluorescent or colored compounds, the molecules can be eliminated simply by determining if their absorbance lies in the same region as the assay read-out. For chemically reactive compounds, detection often relies on a chemist's intuition, aided by literature reports [3].

The detection of aggregators is more problematic, as this is difficult to predict from the structure of a compound alone, but recently developed assays will help to detect these compounds in future [4–6]. Similarly, the detection of redox-based inhibitors is not straightforward, as certain structural types (e.g., quinones) are prone to this activity, while other less-reported molecules have also been shown to cause this form of interference. Simple assays are available and should be considered whenever redox activity is suspected [7, 8].

2.3
The Quality of Hit Matter is of Paramount Importance

From the classical chevron flowchart in Fig. 1, there are two early decisions which will have a profound impact on the potential of a drug discovery and development project being able to deliver a new therapeutic agent to patients: first, the choice of biological target; and second, the chemical series chosen for optimization to the clinical candidate. During recent years, the industry has recognized the need to improve target validation, selection and disease association, and many articles and analyses on this area have been published [9]. However, of equal importance is the choice of the best chemical series to maximize the potential of a molecule making it through all the development phases. Again, many analyses have been published on the attrition rates and reasons for the development of small-molecule drugs [10, 11]. As the reasons and risk factors for attrition are becoming better understood,

it is self-evident that the quality of the chemical start point is the most important factor in both reaching the clinical evaluation phase and successfully progressing through it.

Hit selection cannot be based on a single criterion, but must consider multiple factors including knowledge of the target, the intended route of administration, and physico-chemical properties. The generic lead profiles which are discussed in Sect. 3 act as profiles for the chemist to aim for within a project, but are much more useful when significant data packages have been generated. At the outset of a project a chemist is frequently faced with multiple hits, and the decision as to which hit to progress can significantly affect the ultimate outcome of a project, and is therefore critical.

There are a number of factors the chemist can use to inform a decision

- *Physico-chemical properties of the hit/hit series:* This is an important factor highlighted by many authors in recent reports [10–12]. During recent years, chemists have been increasingly choosing high-molecular-weight, lipophilic hits from screens [12]. Optimization increases these values further and has led to increased attrition. Thus, it is now recommended that hits should have relatively low molecular weight (ideally close to 300) and lipophilicity (ClogP 1–3).
- *Chemical structure – absence of alerts:* Recent publications have highlighted a growing list of templates and functional groups that should be avoided because they: (i) frequently appear as false positives in screens [3]; or (ii) cause issues regarding toxicity, either directly or through bioactivation [13]. Chemists should be aware of all of these undesirable

features and take them into account when selecting a hit for further progression.
- *Synthetic tractability:* In order to evaluate a hit series it is important that the chemist is able to prepare analogs rapidly. If a hit is amenable to rapid synthesis and fulfills all other requirements, then synthetic tractability can be used to prioritize it over similar hits. Long complex synthetic routes represent an issue at all stages of drug discovery, and can ultimately impact on the cost of a new medicine. Thus, choosing synthetically tractable hits and lead series is highly important.
- *Existing SAR:* On occasions when libraries of close analogs are screened with high throughput screening (HTS), series of analogs can be identified with a preliminary SAR directly following confirmation. This provides a high degree of confidence in the tractability of the hits series, and also allows the chemist to proceed with confidence.
- *Novelty:* Compound novelty is highly important in all drug discovery program, given that one of the key objectives is to generate intellectual property. The novelty of hit compounds is frequently used as a criterion for the prioritization of a hit series arising from a screen. Many modern screening collections are comprised of large numbers of widely available, commercial compounds, and it is therefore important to check that a hit has not been reported by another group prior to embarking on preliminary chemistry. For groups engaged in hit generation via performing chemistry on previously reported templates (the so-called "fast follower" approach), it is highly important to identify significant structural diversity. Computer-aided design using scaffold-hopping software [14] is frequently applied in these approaches.

3
Goal of Hit-to-Lead Medicinal Chemistry

During the hit-to-lead phase, the aim is to deliver a molecule which has the skeleton of a profile for onward optimization to a drug candidate. Typically, this would mean having confidence across the following parameters:

- Initial biological profile: for example, activity in a biochemical assay confirmed by effects in a cellular model. Appropriate selectivity and orthogonal assays complete.
- If possible, a biophysical measure of ligand binding, for example, cocrystallization, NMR, surface plasmon resonance (SPR), and isothermal titration calorimetry (ITC).
- Physical properties compatible with potential drug properties, including acceptable solubility, permeability, and stability.
- Calculated physico-chemical properties in line with required drug and with room for optimization, including molecular weight, ClogP, polar surface area (PSA), and numbers of hydrogen bond donors (HBD) and hydrogen bond acceptors (HBA).
- Preliminary off-target selectivity profile, sufficient to give confidence that molecules are functioning via the proposed biological target.
- Preliminary SARs against the biological target identified, giving confidence in the biological profile and ability to optimize to the required profile.
- Preliminary structure–property relationships against physical, selectivity, developability profiles to give confidence that optimization to a candidate is possible.
- Preliminary drug metabolism and pharmacokinetics (DMPK) profile (any

Tab. 1 Indicative hit, lead, and candidate profiles.

Property	Hit	Lead	Candidate
Target potency (µM)	10	≤1	≪1
Selectivity	Evidence desirable	>10-fold	>100-fold
Cell activity	Target-dependent	Target-dependent	Target-dependent
Physical properties	LogD < 3, MW <300 Da	LogD < 4, MW <400 Da	LogD < 4, MW <500 Da
	Evidence of aqueous solubility	Aqueous solubility > 50 µg ml^{-1}	Aqueous solubility > 100 µg ml^{-1}
SAR	Desirable	Essential	Essential
Structural alerts	Absent	Absent	Absent
In vitro ADME	Evidence of metabolic stability *in vitro*	Members of a series have <50% turnover in microsomes following 5 min incubation	Stable in microsomes
		Compounds passively permeable	Cyp IC$_{50}$ > 10 µM
			No evidence of Cyp induction
			No evidence of reactive metabolites
			Compounds passively permeable with no transporter interaction
In vivo ADME	N/A	Clearance < liver blood flow	Low clearance, F > 30%
Ease of synthesis	Simple – ideally amenable to parallel synthesis	Suitable for analog generation	Acceptable manufacturing cost
Novelty	Essential	Essential	Essential

ADME, absorption, distribution, metabolism, and excretion; Cyp, cytochrome P450.

of microsomal turnover, hepatocyte turnover, cytochrome P450 inhibition, initial rodent pharmacokinetic profile).
- Preliminary safety profile: key ion channel cross-screen assays, including human ether-à-go-go-related-gene (hERG), cellular toxicity.

Examples of hit, lead, and candidate profiles are provided in Table 1, for comparative purposes. Whilst this list is neither prescriptive nor exhaustive, it does at least serve to uncover the main potential pitfalls in the onward optimization process. Clearly, knowledge of the chemotype and/or the risks associated with the biological target may introduce key additional criteria to the list above (e.g., a key selectivity assay or reactive intermediate characterization). It is also worth bearing in mind the end-point of the overall optimization process, namely the development candidate, which in the ideal situation would have in addition to the profile shown in Table 1:

- *Potency*: minimum concentration required to be maintained for efficacy is low (e.g., less than 50 nM).
- *Predicted half-life in human*: dose required to maintain C_{min} does not result

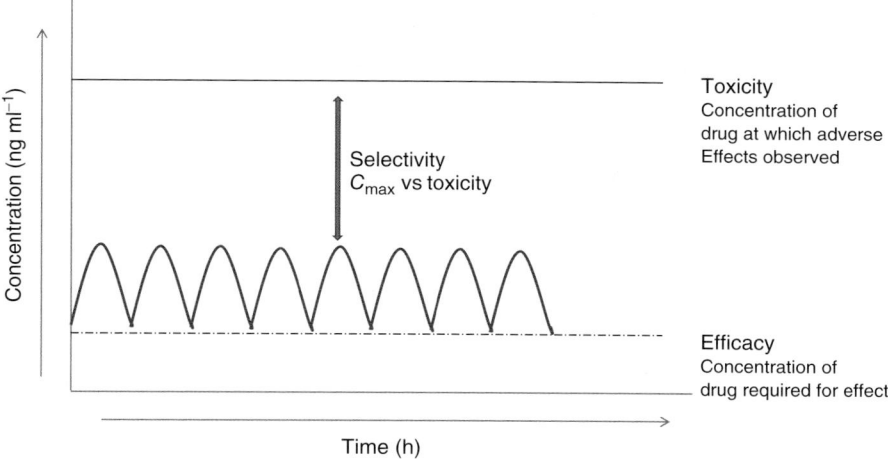

Fig. 3 Selectivity: the separation between drug concentration required for pharmacological effect and that which produces toxicity.

in a much higher C_{max} (e.g., a $t_{1/2}$ of 24 h will give a C_{max}/C_{min} ratio = 2).

- *Selectivity*: toxicity is not seen until much higher plasma levels than C_{max} (e.g., levels >10-fold above C_{max}), as depicted in Fig. 3.

3.1 Screening Cascades

Having determined the target lead molecule profile, new analogs are purchased and/or synthesized with the aim of improving on the original hit profile and addressing the issues present in the molecules. Not all of the assays can, or should, be run in parallel, and as such it is important for a screening cascade to be constructed to allow appropriate filtering of compounds for further testing. This does not mean that a prescriptive sequence of assays has to be routinely followed in order, but rather consideration is given to the throughputs and decision-making abilities of the various assays and an initial approach determined. The order of the assays, or the removal of certain assays or the inclusion of new ones, will be considered throughout the optimization phase. Particular consideration should be given to including assays which provide quantitative data, thus enabling a more informed decision making. Having determined those criteria that make up the lead profile, and the assays that will deliver those data within the screening cascade, it is important to profile exemplars from various hit series through these assays to fully understand the potential and liabilities of the various chemical series. Figure 4 shows an example of a typical screening cascade designed to efficiently filter compounds at each stage, resulting in a lead declaration.

3.2 Target Physico-Chemical Properties

It should be recognized that the physico-chemical properties of a molecule are fixed from the moment of its conception, and that many properties can be predicted and analyzed prior to synthesis. Given the cost and resources required to engage in

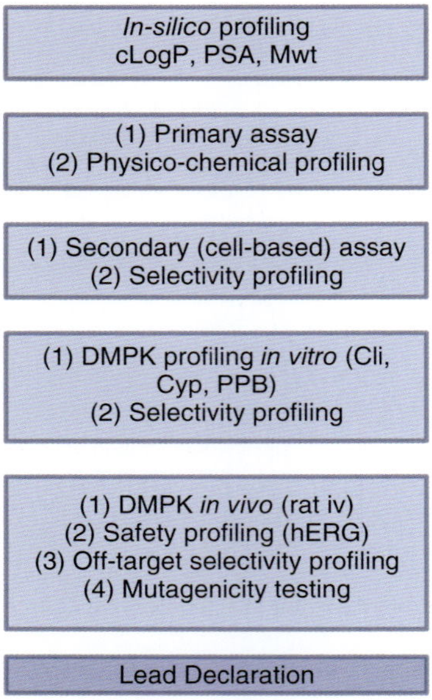

```
┌─────────────────────────────┐
│      In-silico profiling     │
│      cLogP, PSA, Mwt         │
└─────────────────────────────┘

┌─────────────────────────────┐
│   (1) Primary assay          │
│   (2) Physico-chemical profiling │
└─────────────────────────────┘

┌─────────────────────────────┐
│  (1) Secondary (cell-based) assay │
│  (2) Selectivity profiling   │
└─────────────────────────────┘

┌─────────────────────────────┐
│ (1) DMPK profiling in vitro (Cli, │
│     Cyp, PPB)                │
│ (2) Selectivity profiling    │
└─────────────────────────────┘

┌─────────────────────────────┐
│  (1) DMPK in vivo (rat iv)   │
│  (2) Safety profiling (hERG) │
│  (3) Off-target selectivity profiling │
│  (4) Mutagenicity testing    │
└─────────────────────────────┘

┌─────────────────────────────┐
│      Lead Declaration        │
└─────────────────────────────┘
```

Fig. 4 A typical screening cascade depicting several levels of assays used to filter compounds on the path to lead declaration.

the preparation of new molecules, it is very important that the key properties of the molecule are predicted and that the molecule is only prepared either if the properties lie within target criteria or the molecule can be used to refine the lead optimization hypothesis. Clearly, there are exceptions to every rule, however, the balance of probabilities means that the routine preparation of molecules which do not meet target physico-chemical criteria will most likely provide leads with a low probability of delivering successful drug candidates.

3.2.1 Property Guidelines

Many reports have described hits for their biological targets as being of high quality as they are Lipinski-compliant. This is the first of all of the many suggested guidelines for molecules to say that, for a molecule to achieve good oral bioavailability, it needs to satisfy at least three of the four properties: [15].

- HBAs ≤10
- HBDs ≤5
- ClogP ≤5
- Molecular weight ≤500 Da.

During the early stages of lead optimization, however, more stringent criteria are recommended to allow for both the inflation of molecular size, lipophilicity and polarity that result from lead optimization, and also to recognize that the lower-risk drug physico-chemical space for attrition through development is significantly more restricted than above. Alternative hit criteria can be considered from the "Rule of 3" that was originally proposed by scientists at Astex Pharmaceuticals [16, 17] for fragment criteria, but which also proved valuable at the early hit stage. Indeed, it is now common practice to consider applying the Rule of 3 for hit prioritization. Many organizations have modified their small-molecule screening collections such that they contain many compounds which are Rule of 3-compliant.

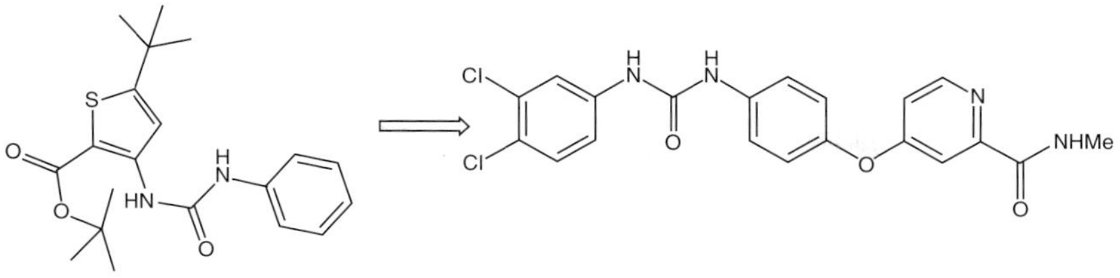

Fig. 5 Molecular weight gain involved in the development of hit (**5**) to the drug sorafenib (**6**).

- HBAs ≤3
- HBDs ≤3
- ClogP ≤3
- Molecular weight ≤300 Da
- Topological polar surface area (TPSA) ≤60 Å^2.

An example illustrating the changes in molecular size that occur on optimizing a HTS hit to a drug molecule is that of the discovery of sorafenib (**6**) (Fig. 5) [18]. In this case, the original screening hit (**5**) had a low molecular mass which was increased by approximately 50% during the optimization process.

Currently, many visualization tools are available to assist the use of physico-chemical properties, including the Golden Triangle [19] (molecular weight plotted versus ClogD) and the Pfizer multi-parameter optimization (MPO) tool [20]. The latter is of particular benefit when targeting central nervous system (CNS)-penetrant molecules, and when faced with the challenge of balancing the required physico-chemical criteria to achieve this (typically, ClogP 2–3; TPSA ≤75 Å^2) with matching the overall property profile to that displayed by currently marketed CNS drugs.

As the main focus of this chapter is orally administered drugs, it is important to highlight the point that, for different routes of administration, the compound property requirements will vary and that the guidelines described above will not necessarily apply. For example, inhaled drugs often have a higher molecular weight and lipophilicity [21], whereas intravenously administered drugs have a lower lipophilicity, given their need for a high aqueous solubility. It is also important to highlight that, for many drug discovery projects centered on non-oral routes of administration, to start with HTS is rare; frequently, the starting points are advanced molecules which have progressed for an oral program, with the focus being on physico-chemical property modulation with a retention of activity.

3.3 Starting Hit Optimization

At the start of hit optimization, two different strategies are generally employed. Either a hit series has been identified with a specific liability which is essential to address – for example, being able to remove a toxicophore such as an electron-rich aniline or an electrophile from a molecule, that is, to create a clear go/no-go decision point. Alternatively, a wider optimization is required of a given chemical series that will encompass a range of properties. In this situation, molecules might be made to expand the SAR and also refine the design hypothesis, as well as to test the hypotheses themselves.

To facilitate this, it is important that more data are generated at an early stage, either by screening hits through other available assays and/or generating structural data of the hit molecule bound to the target protein (or modeling this interaction if protein crystal data are unavailable). At this stage, the portfolio of hit matter can also be expanded by purchasing all related chemical structure analogs (so-called "SAR by catalog"). This will enable the weaknesses in the chemical series to be identified and, often through specific pair-wise comparisons, for observations of specific structural features or functional groups with identified issues (e.g., low solubility, poor selectivity, etc.) to be made.

4 Strategies of Hit-to-Lead Medicinal Chemistry

The ultimate objective of hit-to-lead medicinal chemistry is to identify one or more robust lead series with significant potential for delivering a drug candidate at the end of lead optimization. Typically, multiple hit series are investigated in parallel, with the aim of identifying two series for the next phase. It is very important that the medicinal chemistry strategy focuses on the full target profile, ensuring that issues are identified early and allowing an efficient selection of the preferred series on which to focus. Chemists at all stages should also bear in mind that the ultimate goal is to achieve as low a clinical dose as possible, both to minimize costs and side-effect burden. The key areas of focus are discussed in the following subsections.

4.1 Physico-Chemical Properties

As discussed in Sect. 3.2, the physico-chemical properties of a drug are key to its successful discovery at all stages, and the influence of these properties on a variety of parameters will now be discussed. It is paramount that the medicinal chemist considers the physico-chemical properties of every molecule. Many properties such as molecular weight and PSA can only be generated *in silico*; however, when considering lipophilicity, although LogP or LogD can be calculated this is an empirical figure and an important part of any hit to lead strategy is to determine the correlation between calculated and measured LogD values. At the early stage of a project, large differences can often exist between the measured and calculated values of LogD, and data need to be acquired in order to correct the algorithms used to generate the *in-silico* values. However, once a good correlation has been established, the chemist can apply *in-silico* lipophilicity predictions with confidence. It is worth emphasizing here that, in almost all cases, when examining data from developability assays a correlation between lipophilicity and the data in question will exist, and that by controlling lipophilicity the chemist will be able to have a strong influence on developability data.

4.2 Optimizing Target Affinity While Controlling Lipophilicity

Medicinal chemists are often guilty of placing too much emphasis on potency and focusing less on other aspects of the target profile. This common mistake frequently leads to issues and delays later in the lifetime of a project. It cannot be overemphasized that compound potency, though important, is only a constituent component of the overall profile and should never be considered alone. During the 1990s, medicinal chemists became increasingly focused on potency, and subsequent analyses [12] have shown that an increased potency was frequently achieved through increases in molecular weight and lipophilicity. This resulted in the development of compounds with suboptimal physico-chemical properties, and has been proposed as a contributing factor to the increased attrition observed during recent years. The concept of ligand efficiency (see below), which was proposed [22, 23] as a useful measure of the estimated binding energy of a compound per atom, enabled chemists to determine whether a change in potency was due to additional interactions with the target protein, or was simply a consequence of increased size:

$$\text{Ligand efficiency (LE)} = \frac{\Delta G}{n \text{ Heavy Atoms}}$$
$$\text{or } \frac{pXC_{50}}{n \text{ Heavy Atoms}}$$
$$\text{or } \frac{pXC_{50}}{\text{MW}t}$$

Although the introduction of ligand efficiency has greatly aided medicinal chemists in the analysis of SAR and subsequent design, it only takes in to account molecular weight. Subsequent studies have proposed variations which take into account lipophilicity, PSA, and combinations of physico-chemical properties [24]. One of the most commonly used indices in current use is lipophilic ligand efficiency [11]:

$$\text{Lipophilic ligand efficiency (LLE)}$$
$$= pXC_{50}$$
$$- \text{ClogP (or ClogD at pH 7.4)}$$

Many authors have cited high lipophilicity as possibly the most common cause of attrition in development [10, 12, 15], and it is therefore important in hit-to-lead chemistry to ensure that potency can be increased without significantly increasing lipophilicity. For optimized compounds, Leeson *et al.* have suggested that a LLE value in the range of 5–7 is predictive of a compound having a significant chance of success in development [12]. For hit-to-lead programs, LLE values in this range are difficult to achieve, although it is generally recommended that the value should be as high as possible.

4.3 Optimizing Cellular Activity

Although many drug targets are intracellular, most HTS methodologies do not use cell-based assays. Nonetheless, in these situations it is important to have a cell-based assay high up in the screening cascade. One frequently encountered issue in early hit-to-lead chemistry is a poor correlation between the activity of a compound in a cell-based assay and that in a cell-free assay, and several possible reasons have been identified for this:

- Poor passive permeability.
- The role of transporters.
- Differences in substrate/cofactor concentration between assays.
- Differences in activity between an isolated protein assay and a cellular assay.

One of the most common reasons for the observed difference in activity is poor passive permeability. If compounds are synthesized that the Rule of 3 (Sect. 3.2.1), then the majority of those compounds should have a good chance of being permeable. Other important factors which can affect permeability are PSA, which should ideally be <120 Å^2 [25], while the rotatable bond count should be minimized [26]. However, since not all compounds within this area of physico-chemical space will be permeable, it is important to incorporate permeability assays into a screening cascade, and a number of these are commonly used [27]. If compounds are passively permeable, an alternative explanation for a mismatch could be the involvement of an active transport mechanism, and this is an area of intense current interest [28]. Issues with respect to differences in substrate/cofactor concentrations between assays and between an isolated protein assay and a cellular assay can be avoided by using a physiologically relevant substrate and cofactor concentration in cell-free assays and, where possible, using a full-length protein for the target of interest.

4.4
Selectivity

Selectivity is an important factor at all stages of the drug discovery process. During the hit-to-lead phase it is important that chemists are able to demonstrate divergent SARs between the target of interest and related targets – that is, selectivity. Often, initial hits possess only weak activity, and given the limitations of many biological assays it is not possible to determine selectivity accurately. It is only when potency has been increased during the first phases of a hit-to-lead process that a true estimate of selectivity be obtained. Once known, however, a key objective of the chemist must be to determine the potential for a hit series and to demonstrate selectivity through the generation of a divergent SAR. In many projects, some hit series will be more tractable with respect to selectivity than others, and this point can be applied as a decision-making criterion when selecting preferred series to take forward. If protein crystal structures are available they can greatly aid the design of selective ligands. However, before progressing to lead optimization it is important to know the broader selectivity profile of selected hits, and consequently advanced representatives from each series will need to be screened against broader panels of different target classes.

4.5
Consideration of Developability

The hit-to-lead phase also offers an initial opportunity to test compounds in a number of developability screens. These data can: (i) provide valuable information to allow an informed decision regarding the selection of preferred series to progress; and (ii) identify potential issues at an early stage, giving the chemist a chance to address them. If such issues prove intractable, they can in some cases provide evidence on which to base an early "no-go" decision. Having discussed permeability in Sect. 4.2, other developability assays will be detailed in the following subsections.

4.5.1 ADME Properties

A panel of absorption, distribution, metabolism, and excretion (ADME) assays is frequently employed that includes microsomal stability, cytochrome P450 (Cyp) inhibition, and plasma protein binding together with passive permeability and transporter assays (as previously discussed). Microsomal stability assays [29] determine the potential for a drug to undergo oxidative metabolism in the liver. If compounds display poor stability in these assays then they will have little chance of demonstrating a satisfactory pharmacokinetic profile *in vivo*. Obtaining this type of data in the hit-to-lead phase provides the chemist with an early opportunity to understand the metabolic liabilities within a series and, if a liability is observed, to investigate strategies toward preparing more stable compounds. Common strategies employed in overcoming poor metabolic stability are: (i) a reduction in lipophilicity; (ii) introducing blocking groups to protect vulnerable sites from metabolism; (iii) manipulating the electronics of labile groups to reduce lability; and (iv) the removal of vulnerable groups if they are not required for potency.

Cyps are the main class of oxidative enzymes responsible for the metabolism of drugs in the liver [30]. Hence, it is important that new drugs do not inhibit these enzymes, as this could lead to drug–drug interaction in patients taking more than one medicine concomitantly. As hit compounds advance toward the desired lead profile, it is important to understand if there are any issues regarding Cyp inhibition. Simple in-vitro assays are available to determine the potential of compounds to inhibit these enzymes. As with other ADME parameters, early data on hit compounds allows the chemist to investigate strategies to resolve any issues, should they exist. Currently, a range of tools are available to aid chemistry strategies, including X-ray structures [31] and software packages [32]. A large volume of SARs has also been published, describing the requirements of inhibitors of the common Cyp isoforms [33].

In addition to investigating Cyp inhibition, it is becoming increasingly important to investigate hits for their potential to interact with drug transporters, not only with respect to effecting permeability but also to cause drug–drug interactions [34]. Although, unfortunately, assays are only available for a limited number of these proteins at present, it is expected that their availability will increase during the coming years.

The importance of plasma protein binding in drug discovery is currently a matter of debate [35]; however, it is useful during the hit-to-lead stage for chemists to be aware of the extent to which compounds of interest bind to proteins, and a number of simple assays are available to assess this situation [36].

In the case of CNS drugs, it is also important to examine the potential for a compound to penetrate the brain, and this can be achieved by employing in-silico techniques [20]. Additionally, in-vitro assays have been recently described [37] that are able to guide chemistry at the hit-to-lead stage.

For advanced leads prior to full lead optimization, it is good practice to generate some preliminary in vivo data, typically in the form of a rodent pharmacokinetic study. These data provide confidence that the lead series is robust, and highlight any weaknesses that might be the focus of the subsequent lead optimization phase. Typically, compounds are administered intravenously to determine if there is a correlation between the in-vitro clearance and that observed in vivo. Occasionally, if the in-vivo clearance is satisfactory a compound may be progressed to an oral study to determine if there are any issues with its oral absorption.

4.5.2 Safety Properties

In addition to generating preliminary ADME data to assist hit-to-lead chemistry, it is also important to generate early safety data. The assays most often considered by chemists involve cardiac ion channel inhibition and gene toxicity.

Cardiac ion channel inhibition, and in particular the hERG potassium (K^+) channel, must be avoided in drug candidates given the potential for inhibitors to produce severe cardiovascular side effects. It is currently standard practice to determine the potential of hit compounds to inhibit cardiac ion channels [38]. In addition to hERG, Nav1.5, Kv1.5 and Cav1.2 are often included in screening panels for early stage hits. If issues are encountered, *in-silico* tools are available to help the chemists design out these undesired activities [39].

Until recently, genotoxicity assays were only employed when a development candidate had been identified. Unfortunately, however, this resulted in many molecules being terminated based on positive findings in bacterial mutagenicity assays, and this in turn led to a large commitment of resources and considerable expense. Consequently, it is now common practice to test advanced leads for mutagenicity before committing them to lead optimization [40].

4.5.3 Solubility

Aqueous solubility is vital to a successful drug discovery project. A lack of solubility can cause issues right from the outset, for example, the precipitation of compounds during screening frequently leads to false negatives or, at best, to an inaccurate representation of a compound's true activity. Indeed, when working with poorly soluble compounds, confidence in data from any assay will be lower than that generated on a soluble compound. It is therefore important that chemists: (i) prioritize hits with high aqueous solubility; and (ii) strive to maintain high solubility throughout the optimization process, as failure to do so can result in problems and even attrition later in the drug discovery process [41]. There are no clear rules with respect to controlling or designing the solubility of an organic molecule; however, the careful use of calculated physical properties to ensure that compounds stay within a particular range of polarity and lipophilicity can greatly aid design. Additionally, minimizing the aromatic ring content of a compound has also been suggested to confer increased aqueous solubility [10]. During hit-to-lead chemistry it is important to establish trends that can be used to aid design at later stages of the project.

5
Fragment-Based Lead Discovery

Until now, this chapter has described traditional approaches to hit discovery, such as HTS and the modification of known ligands; however, during recent years fragment-based screening [42] has become increasingly popular and successful. Fragment-based screening involves the examination of low-molecular-weight (generally around 250 Da) compounds at high concentrations, with hit compounds typically having an affinity for the target protein in the millimolar range. The screening libraries are small, typically containing 1000–2000 compounds. The fragment-based technique is very different from HTS, in which large libraries of higher-molecular-weight compounds are screened and the hits are typically in the micromolar range. The key differences between the two methodologies are summarized in Table 2.

Fragment screening was initially successfully employed by groups investigating kinase targets [18], but has since been applied successfully to a number of other target classes including G protein-coupled receptors (GPCRs) [43], protein–protein interactions [44], and ion channels [45]. Today, the procedure is used widely by both academic and industrial drug discovery groups. Several alternative screening approaches have been adopted by different groups, including X-ray crystallography, biophysical screening, and biochemical assays for the primary screen.

One of the key differences between fragment-based lead discovery and

Tab. 2 Differences between fragment screening and high-throughput screening.

Screening strategy	High-throughput screening	Fragment screening
Requires structural information on target	No	Yes
Typical library size	100 000 – 3 000 000 compounds (MW $\geq$ 400 Da)	1000 – 2000 fragments (MW $\sim$250 Da)
Hit confirmation techniques	Alternative biochemical assay	Biophysical screening and X-ray crystallography
Typical potency of hits	µM	mM
Infrastructure requirements	Robotics for compound handling and screening	Biophysical screens and protein production and crystallization facilities

conventional approaches is the need to have structural information, ideally an X-ray crystal structure of a hit with the target protein.

A recent example (Fig. 6) demonstrates the successful application of fragment-based screening to the discovery of inhibitors of replication protein A (ROA) [46]. In this case, a NMR screen was used to identify two weakly active fragments (**7** and **8**) which bound to adjacent sites on the protein. By chemically linking the fragments together in a manner that did not disrupt the key interactions with the protein produced (**9**), a 26 µM ligand was created that was further optimized through structure-based design giving submicromolar inhibitor (**10**).

6
Conclusions

The importance of hit-to-lead medicinal chemistry to the overall drug discovery process cannot be overemphasized. This is the first stage that medicinal chemists are actively involved in the process in a significant role, and the decisions made strongly influence the overall success of the entire process. When embarking on a hit-to-lead project it is therefore important that lead generation is not considered as an isolated exercise. From the outset, scientists need to consider the overall drug discovery process, as failure to do this can results in leads being selected with little chance of progressing further, and the early termination of a project that may have succeeded if the leads had been selected more diligently.

During recent years, hit-to-lead chemistry has changed significantly in a number of ways. First, it is no longer a simple matter of identifying the most potent series of compounds. Indeed, detailed analyses from a number of groups have highlighted reasons for attrition over previous decades, and have taught chemists that multiple parameters must be considered when developing hits into leads; consequently, many of the assays that were previously employed only during the later stages of drug discovery are now routine. Second, advances in technology have opened up new methodologies for hit identification and, given the choice, the discovery scientists need to carefully select the most appropriate technique for the target of interest. The new technologies have also allowed

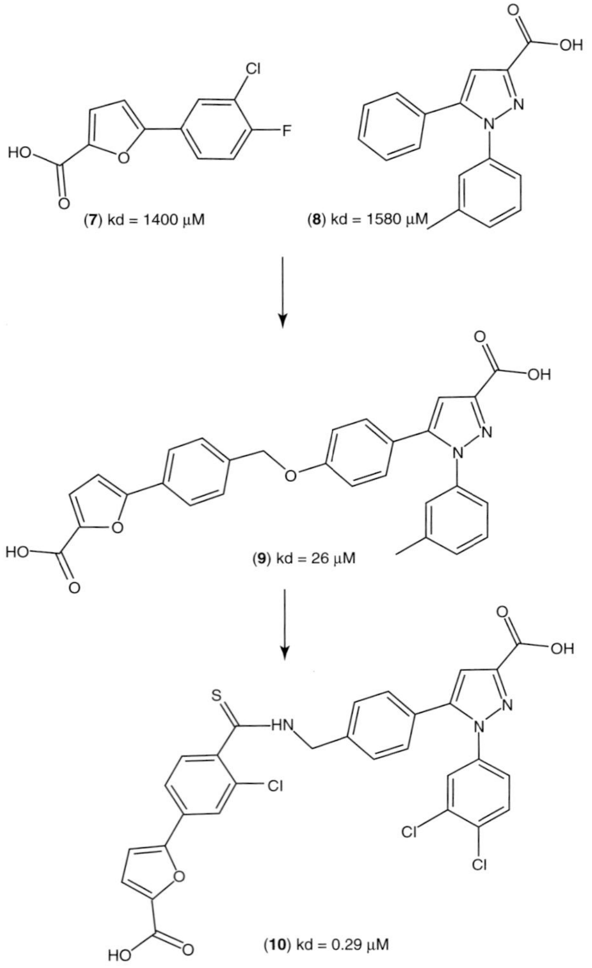

Fig. 6 An example of the evolution of fragment hits to generate a potent ROA inhibitor (**10**).

entry into target classes previously thought intractable, such as voltage-gated ion channel and protein–protein interactions, with chemists having had to learn how to prosecute these different target classes.

Recent reports have suggested that modern chemists are now applying the findings of published analyses focused on attrition and applying the recommendations made. It will be very interesting to see if the changes in strategy that are now considered common practice will lead to greater successes at later stages of the drug discovery process.

References

1. Wermuth, C.G. (2006) Selective optimization of side activities: the SOSA approach. *Drug Discov. Today*, **11**, 160–164.
2. McGovern, S.L., Caselli, E., Grigorieff, N., and Shoichet, B.K. (2002) A common mechanism underlying promiscuous

inhibitors from virtual and high-throughput screening. *J. Med. Chem.*, **45**, 1712–1722.
3. Baell, J. and Walters, M.A. (2014) Chemistry: chemical con artists foil drug discovery. *Nature*, **513** (7519), 481–483.
4. Seidler, J., McGovern, S.L., Doman, T.N., and Shoichet, B.K. (2003) Identification and prediction of promiscuous aggregating inhibitors among known drugs. *J. Med. Chem.*, **46**, 4477–4486.
5. Feng, B.Y., Shelat, A., Doman, T.N., Guy, R.K. et al. (2005) High-throughput assays for promiscuous inhibitors. *Nat. Chem. Biol.*, **1**, 146–148.
6. Ryan, A.J., Gray, N.M., Lowe, P.N., and Chung, C. (2003) Effect of detergent on "promiscuous" inhibitors. *J. Med. Chem.*, **46**, 3448–3451.
7. Lor, L.L., Schneck, J., McNulty, D.E., Diaz, E. et al. (2007) A simple assay for detection of small-molecule redox activity. *J. Biomol. Screening*, **12**, 881–890.
8. Johnston, P.A., Soares, K.M., Shinde, S.N., Foster, C.A. et al. (2008) Development of a 384-well colorimetric assay to quantify hydrogen peroxide generated by the redox cycling of compounds in the presence of reducing agents. *Assay Drug Dev. Technol.*, **6**, 505–518.
9. Trist, D.G. (2011) Scientific process, pharmacology and drug discovery. *Curr. Opin. Pharmacol.*, **11** (5), 528–533.
10. Leeson, P. (2012) Drug discovery – chemical beauty contest. *Nature*, **481** (7382), 455–456.
11. Leeson, P.D. and Springthorpe, B. (2007) The influence of drug-like concepts on decision-making in medicinal chemistry. *Nat. Rev. Drug Discov.*, **6**, 881–890.
12. Hann, M.M. and Keserü, G.M. (2012) Finding the sweet spot: the role of nature and nurture in medicinal chemistry. *Nat. Rev. Drug Discov.*, **11**, 355–365.
13. Kalgutkar, A.S., Fate, G., Didiuk, M.T., and Bauman, J. (2008) Toxicophores, reactive metabolites and drug safety: when is it a cause for concern? *Expert Rev. Clin. Pharmacol.*, **1** (4), 515–531.
14. Sun, H., Tawa, G., and Wallqvist, A. (2012) Classification of scaffold-hopping approaches. *Drug Discov. Today*, **17** (7-8), 310–324.
15. Lipinski, C.A. (2004) Lead- and drug-like compounds: the rule-of-five revolution. *Drug Discov. Today Technol.*, **1**, 337–341.
16. Congreve, M., Carr, R., Murray, C., and Jhoti, H. (2003) A "rule of three" for fragment-based lead discovery? *Drug Discov. Today*, **8**, 876–877.
17. Jhoti, H., Williams, G., Rees, D.C., and Murray, C.W. (2013) The "rule of three" for fragment-based drug discovery: where are we now? *Nat. Rev. Drug Discov.*, **12**, 644–645.
18. Erlanson, D.A. (2009) Fragment-based drug discovery of kinase inhibitors, in *Kinase Inhibitor Drugs* (eds R. Li and J.A. Stafford), John Wiley & Sons, Inc., Hoboken, pp. 461–483.
19. Johnson, T.W., Dress, K.R., and Edwards, M. (2009) Using the golden triangle to optimize clearance and oral absorption. *Bioorg. Med. Chem. Lett.*, **19**, 5560–5564.
20. Wager, T.T., Hou, X., Verhoest, P.R., and Villalobos, A. (2010) Moving beyond rules: the development of a central nervous system multiparameter optimization (CNS MPO) approach to enable alignment of drug-like properties. *ACS Chem. Neurosci.*, **1**, 435–449.
21. Ritchie, T.J., Luscombe, C.N., and MacDonld, S.J.F. (2009) Analysis of the calculated physicochemical properties of respiratory drugs: can we design for inhaled drugs yet? *J. Chem. Inf. Model.*, **49** (4), 1025–1032.
22. Kuntz, I.D., Chen, K., Sharp, K.A., and Kollman, P.A. (1999) The maximal affinity of ligands. *Proc. Natl Acad. Sci. USA*, **96**, 9997–10002.
23. Hopkins, A.L., Groom, C.R., and Alex, A. (2004) Ligand efficiency: a useful metric for lead selection. *Drug Discov. Today*, **9**, 430–431.
24. Hopkins, A.L., Keserü, G.M., Leeson, P.D., Rees, D.C. et al. (2014) The role of ligand efficiency metrics in drug discovery. *Nat. Rev. Drug Discov.*, **13**, 105–121.
25. Hou, T.J., Zhang, W., Xia, K., Qiao, X.B. et al. (2004) ADME evaluation in drug discovery. 5. Correlation of Caco-2 permeation with simple molecular properties. *J. Chem. Inf. Comput. Sci.*, **44** (5), 1585–1600.
26. Goswami, T., Kokate, A., Jasti, B.R., and Li, X. (2013) In silico model of drug permeability across sublingual mucosa. *Arch. Oral Biol.*, **58** (5), 545–551.
27. Balimane, P.V., Han, Y.-H., Chong, S. (2012) Permeability and Transporter Models in Drug Discovery and Development, in *ADME-Enabling Technologies in Drug Design and*

Development, (eds Zhang, D., Surapaneni, S.), John Wiley & Sons, Inc., Hoboken, New Jersey, pp. 161–168.
28. Ward, P. (2008) Importance of drug transporters in pharmacokinetics and drug safety. *Toxicol. Mech. Methods*, **18** (1), 1–10.
29. Di, L., Kerns, E.H., Ma, X.J., Huang, Y. et al. (2008) Applications of high throughput microsomal stability assay in drug discovery. *Comb. Chem. High Throughput Screening*, **11** (6), 469–476.
30. Lin, J.H. and Guo, J. (2012) Clinical implications of CYP induction-mediated drug–drug interactions. *Encycl. Drug Metab. Interact.*, **4**, 429–449.
31. De Montellano, P.R.O. (2012) Structure and function of cytochrome P450 enzymes. *Encycl. Drug Metab. Interact.*, **1**, 161–179.
32. Refsgaard, H.H.F., Jensen, B.F., Christensen, I.T., Hagen, N. et al. (2006) In silico prediction of cytochrome P450 inhibitors. *Drug Dev. Res.*, **67** (5), 417–429.
33. Masimirembwa, C.M., Ridderstroem, M., Zamora, I., and Andersson, T.B. (2002) Combining pharmacophore and protein modeling to predict CYP450 inhibitors and substrates. *Methods Enzymol.*, **357**, 133–144.
34. Koenig, J., Mueller, F., and Fromm, M.F. (2013) Transporters and drug-drug interactions: important determinants of drug disposition and effects. *Pharmacol. Rev.*, **65** (3), 944–966.
35. Roberts, J.A., Pea, F., and Lipman, J. (2013) The clinical relevance of plasma protein binding changes. *Clin. Pharmacokinet.*, **52** (1), 1–8.
36. Chuang, V.T.G., Maruyama, T., and Otagiri, M. (2009) Updates on contemporary protein binding techniques. *Drug Metab. Pharmacokinet.*, **24** (4), 358–364.
37. Summerfield, S.G. and Dong, K.C. (2013) In vitro, in vivo and in silico models of drug distribution into the brain. *J. Pharmacokinet. Pharmacodyn.*, **40** (3), 301–314.
38. Kireeva, N., Kuznetsov, S.L., Bykov, A.A., and Tsivadze, A.Y. (2013) Towards in silico identification of the human ether-a-go-go-related gene channel blockers: discriminative vs. generative classification models. *SAR QSAR Environ. Res.*, **24** (2), 103–117.
39. Rayan, A., Falah, M., Raiyn, J., Da'adoosh, B. et al. (2013) Indexing molecules for their hERG liability. *Eur. J. Med. Chem.*, **65**, 304–314.
40. Escobar, P.A., Kemper, R.A., Tarca, J., Nicolette, J. et al. (2013) Bacterial mutagenicity screening in the pharmaceutical industry. *Mutat. Res.*, **752** (2), 99–118.
41. Di, L., Fish, P.V., and Mano, T. (2012) Bridging solubility between drug discovery and development. *Drug Discov. Today*, **17** (9-10), 486–495.
42. Mazanetz, M., Law, R., Whittaker, M. (2014) Hit and lead identification from fragments, in *De novo Molecular Design*, Wiley-VCH Verlag GmbH & Co. KGaA, Weinheim, pp. 143–200.
43. Andrews, S.P., Brown, G.A., and Christopher, J.A. (2014) Structure-based and fragment-based GPCR drug discovery. *ChemMedChem*, **9** (2), 256–275.
44. Bower, J.F. and Pannifer, A. (2012) Using fragment-based technologies to target protein–protein interactions. *Curr. Pharm. Des.*, **18** (30), 4685–4696.
45. Thompson, A.J., Verheij, M.H.P., Leurs, R., De Esch, I.J.P. et al. (2010) An efficient and information-rich biochemical method design for fragment library screening on ion channels. *Biotechniques*, **49** (5), 822–829.
46. Frank, A.O., Feldkamp, M.D., Kennedy, J.P., Alex, G. et al. (2013) Discovery of a potent inhibitor of replication protein a protein–protein interactions using a fragment-linking approach. *J. Med. Chem.*, **56** (22), 9242–9250.

17
Mass Spectrometry-Based Methods of Proteome Analysis

Mihir Jaiswal[1,3], Michael P. Washburn[2], and Boris L. Zybailov[1,3]
[1] *University of Arkansas at Little Rock, University of Arkansas for Medical Sciences Joint Bioinformatics Program, Little Rock, AR 72204, USA*
[2] *Stowers Institute for Medical Research, Kansas City, MO 64110, USA*
[3] *University of Arkansas for Medical Sciences, Little Rock, AR 72205, USA*

1	**Principles and Instrumentation** 597	
1.1	Proteome and Proteomics 597	
1.2	Need for Large-Scale Analyses of Gene Products at the Protein Level 597	
1.3	General Problems in Proteome Analyses 598	
1.4	MS Instrumentation 599	
1.4.1	Ionization Methods 600	
1.4.2	Mass Analyzers 603	
1.4.3	Ion-Mobility Spectrometry 604	
1.4.4	Tandem MS (MS/MS) 605	
1.5	Methods of Sample Fractionation 606	
1.5.1	2D-Electrophoretic Separation of Complex Protein Mixtures 607	
1.5.2	Liquid-Phase Separation Methods 608	
1.5.3	Affinity Methods (Epitope Tagging) 610	
1.6	Protein Identification by MS 611	
1.7	Imaging Mass Spectrometry 614	
2	**Quantitative Methods of Proteome Analysis Using MS** 615	
2.1	Metabolic Labeling 615	
2.2	Isotope-Coded Affinity Tags and Isobaric Tags 617	
2.3	^{18}O Labeling 620	
2.4	Absolute Quantitation 621	
2.5	Label-Free Methods of Quantitation 621	
2.6	Quantitative Analysis of Protein–Protein Interactions 623	
3	**Specific Examples of Applications** 623	
3.1	Global Proteome Sampling 623	
3.1.1	Global Proteome Sampling Based on 2D PAGE 623	

Translational Medicine: Molecular Pharmacology and Drug Discovery
First Edition. Edited by Robert A. Meyers.
© 2018 Wiley-VCH Verlag GmbH & Co. KGaA. Published 2018 by Wiley-VCH Verlag GmbH & Co. KGaA.

3.1.2	Global Proteome Sampling Based on Multidimensional LC	624
3.1.3	Single Cell Proteomics	625
3.2	Analysis of Protein Modifications by Mass Spectrometry	626
3.2.1	Phosphorylated Proteins	627
3.2.2	Glycosylated Proteins	629
3.2.3	Ubiquitinated Proteins	630
3.2.4	Deamidation of Asparagines and Glutamines	630
3.2.5	Histone Modifications	631
3.3	Analysis of Protein–Protein Interactions by Mass Spectrometry	631
3.3.1	Computational Methods of Protein–Protein Interaction Prediction	632
3.3.2	Yeast Two-Hybrid Arrays	632
3.3.3	Direct Analysis of Large Protein Complexes; Composition of the Yeast Ribosome	632
3.3.4	Analysis of Multisubunit Protein Complexes Involved in Ubiquitin-Dependent Proteolysis by Mass Spectrometry	632
3.3.5	Proteomics of the Nuclear Pore Complex	633
3.3.6	High-Throughput Analyses of Protein–Protein Interactions	635
3.3.7	Quantitative Proteomics Methods in Studies of Protein Complexes	636
3.4	Large-Scale Chemical Crosslinking	638
3.4.1	Nomenclature of Crosslinking Products	638
3.4.2	Overcoming the Crosslinking Bottlenecks	639
4	Conclusions	640
	References	640

Keywords

2D PAGE
Two-dimensional polyacrylamide gel electrophoresis; a technology that is used to separate intact proteins according to their isoelectric point in the first dimension, and their molecular weight in the second dimension.

MS/MS
A mass spectrometry technique in which precursor ions are selectively fragmented, thus allowing detailed structural information to be obtained.

MudPIT
A multidimensional protein identification technology; a proteomic method based on liquid chromatography of peptide mixtures, with subsequent identification using tandem mass spectrometry.

Quantitative proteomics
A group of methods that allow the large-scale quantitative assessment of protein expression and identification.

Orbitrap: A uniquely shaped electrostatic ion trap, where the oscillation frequency of the trapped ions is a function of their mass and charge. Orbitrap is one of the most popular mass analyzers used in modern proteomics laboratories.

> The aim of proteomics is to identify, characterize and map gene functions at the protein level for whole cells or organisms. A typical experimental scheme for a large-scale proteomics inquiry involves the fractionation of a complex protein mixture by electrophoretic or chromatographic means, followed by subsequent identification of the components in individual fractions, using mass spectrometry (MS). Owing to continuous and rapid improvements in instrument sensitivity, throughput capacity, software versatility, and techniques of statistical validation, MS-based approaches have during recent years become mainstream methods for proteome analysis.

1
Principles and Instrumentation

1.1
Proteome and Proteomics

The term "proteome" was first introduced in mid-1990s to name the functional complement of a genome [1–3]. By analogy with genomics, the term proteomics refers to studies of a gene's function at the protein level. Both of these terms have a large-scale flavor to them; indeed, studies that fall under the category "proteomics" frequently deal with large-scale analyses of proteins on the level of the whole organism, tissue, cell, or subcellular compartments. Examples of typical biological questions addressed by modern proteomics experiments include – but are not limited to – establishing "news of difference" between healthy and pathological states, monitoring global gene expression during growth and development, establishing cellular localizations of a particular subset of an expressed genome, and identifying networks of protein–protein interactions. Proteomics studies are usually classified according to the type of the addressed biological questions into "profiling proteomics," "functional proteomics," and "structural proteomics" [4]. Profiling proteomics amounts to a large-scale identification of the proteins in a cell or tissue present at a certain physiological conditions. Functional proteomics refers to the studies of functional characteristics of proteins, whether post-translational modifications (PTMs), protein–protein interactions, and/or cellular localizations. Structural proteomics includes studies of protein tertiary structure, typically made by a combination of X-ray, nuclear magnetic resonance (NMR), and computational techniques. In adopting this classification, this chapter is focused primarily on profiling and functional proteomics.

1.2
Need for Large-Scale Analyses of Gene Products at the Protein Level

The rapid success of various genome-sequencing programs [5] is one of the major factors that led to the development of large-scale proteomics methods. As of today (March, 2014), according to the online genome database [6], 328 archael, 17 569 bacterial and 906 Eukaryal genomes have been sequenced. In fact, searching the genome-derived sequence databases is usually an intrinsic part of mass spectrometry (MS)-based proteome analysis. While of tremendous value, genomic information is, in principle, insufficient for understanding the complex processes of cellular functions. Indeed, in addition to a projection of the information from genes into proteins, cells of the living organisms are extracting useful metabolic information from the environment. Therefore, the total informational content necessarily increases in a proteome compared to the corresponding genome. An immediate consequence of this increase in the informational content is that all of the following – protein isoforms, protein PTMs, protein conformational states, and protein abundance levels, as well as dynamical changes of these properties – have their own important functional implications in living cells. In certain cases, the genomic sequence still can be used to assess some of the functional properties of the corresponding proteins. For example, the codon adaptation index (CAI) and codon bias can be used to predict the expression level of genes [7]. Another important case is a global sequence-based prediction of protein–protein interactions, an example of which is discussed in Sect. 3. Also, to some degree, the functional properties of proteins can be assessed by the analysis of mRNA transcripts. Specifically, mRNA expression profiles are frequently used to estimate the corresponding protein levels. Most common methods that are used to measure global mRNA expression are microarrays and Next Generation Sequencing (RNA-Seq) [8]. However, in many cases, the correlation between mRNA and protein is insufficient to quantitatively predict protein expression [9, 10]. This fact provides motivation to improve the old and develop new technologies for the large-scale analyses of gene products at the protein level.

1.3
General Problems in Proteome Analyses

In a given organism, there are several-fold more different proteins than there are genes. Estimates for humans give such numbers as about 20 000 different genes and more than 100 000 different proteins. Factors such as alternative splicing, PTMs, isoforms, and expression levels increase the complexity and interconnectedness of a proteome compared to a genome. Problems associated with modern proteomics methods include: (i) difficulties in the detection of low-abundance proteins; (ii) difficulties in obtaining quantitative information; (iii) biases in a proteome coverage; (iv) difficulties in the characterization of PTMs and, generally, overall heterogeneity in protein molecules; (v) difficulties in data analysis and interpretation in the large-scale experiments; and (vi) poor reproducibility.

1) Protein levels as low as several dozen copies per cell can be of functional significance. This is true for certain receptors, signaling proteins, and regulatory proteins. However, detection of the low-abundance proteins is often an issue in proteomics experiments. The ability of a proteomics method to identify low-abundance proteins in the presence of high-abundance counterparts is characterized by the dynamic range of a method. The dynamic range of a proteomics method is defined as the "ratio of concentration of the most abundant to the concentration of the least abundant proteins identified by a method." Proteomics methods based

on two-dimensional (2D) polyacrylamide gel-electrophoresis (PAGE) separations typically have a lower dynamic range than methods based on affinity separation of multidimensional liquid chromatography (LC).

2) Quantitative information on the protein abundance in different physiological contexts is the goal of comparative proteomics studies. 2D PAGE methods are good at solving these types of problem, when combined with scintillation counting or fluorescence imaging spectroscopy. In the MS-based methods, quantitation is achieved by differential isotopic labeling or by label-free approaches (see Sect. 2).

3) Biases in proteome coverage can be inherent to a particular method, or be related to the experimental design. For example, 2D PAGE methods are biased against highly hydrophobic proteins. At the same time, there is no inherent bias against hydrophobic proteins in MS identification. However, sample-enrichment or fractionation steps that necessarily precede the MS analysis are often biased against one or the other class of proteins.

4) Information on types and degrees of PTMs and protein isoforms is a necessary component of comprehensive functional description of a proteome. Unfortunately, it is difficult to design a proteomics method that would simultaneously give both good proteome coverage and identify all PTMs. Methods, such as 2D PAGE, which separate intact proteins, are good for detecting different PTMs and isoforms present at the same time. However, owing to large diversity in protein masses and shapes, biases in proteome coverage are inherent in the intact protein methods.

5) During a large-scale proteome analysis, thousands of data points are generated; hence, there are inherent problems related to the data reduction and extraction of the useful information. Additionally, the obtained information needs to be presented in an accessible format to the scientific community. These issues often require significant computational support and software development.

6) Most of the high-throughput, large-scale methods suffer from poor reproducibility. This is especially true in the case of 2D PAGE and MS-based schemes when several fractionation and separation steps are involved. Possible ways to improve reproducibility include a reduction in the number of separation and fractionation steps, adhering to standard protocols, and the use of automation whenever possible.

Whilst most of these challenges are merely technical limitations, modern technologies will continue to progress towards higher sensitivity, higher dynamic range, and higher reproducibility. Moreover, even if it is impossible to have 100% proteome coverage with a single method, the correct design of a proteomics program capable of taking advantage of several different methods can achieve impressive results.

1.4
MS Instrumentation

Mass spectrometry (MS) serves as a platform technology for all proteomics, with the instrument involved – the mass spectrometer – being used to identify proteins present in samples. A variety of MS approaches is currently available that can be used to achieve this goal, and the general

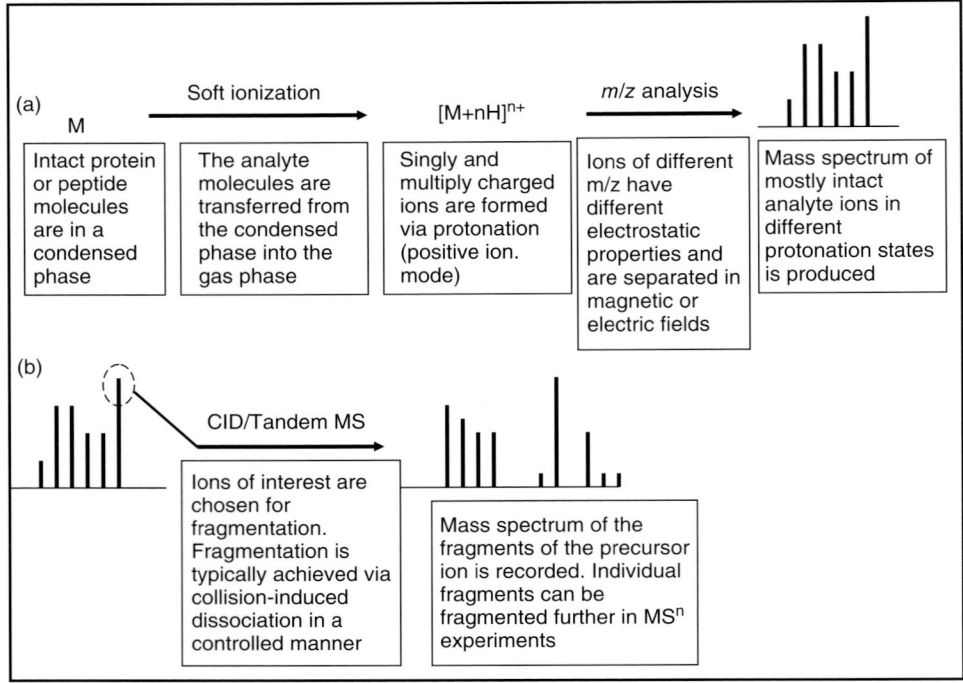

Fig. 1 Principle scheme of (a) single MS and (b) MS/MS analyses. (a) Intact proteins or peptides are transferred from the condensed phase into the gas phase and are ionized through capture (positive mode) or loss (negative mode) of protons. In high-throughput protein identification experiments, a positive mode of ionization is used. A negative mode of ionization can be used for the analysis of carbohydrates and nucleic acids. Ions of different m/z are discriminated by magnetic or electric fields; (b) Ions of which the structure needs to be determined are chosen for fragmentation via CID reactions in the MS/MS experiment. The fragmentation patterns obtained can be searched against protein databases for identification.

concept of MS analysis in proteomics is illustrated in Fig. 1.

Generally, MS analysis involves the creation of gaseous ions from an analyte, followed by separation of the produced ions according to their mass-to-charge ratio (m/z). The principal components of a mass spectrometer are: an ion source, where ionization takes place; a mass analyzer, where the m/z is measured; and a detector, where the amount of the ions corresponding to a particular m/z is recorded [11]. Large macromolecules such as proteins are nonvolatile and consequently, owing to a long-term lack of effective ionization techniques, MS was initially limited to the analysis of smaller molecules. In fact, the protein ionization methods which make MS-proteomics feasible today were not developed until the 1980s.

1.4.1 Ionization Methods

The two most commonly used methods of protein ionization are electrospray ionization (ESI) [12] and matrix-assisted laser desorption ionization (MALDI) [13]. In both techniques, ionization occurs through either the uptake (positive mode) or loss (negative mode) of one or several protons. Whilst the positive mode is typically used in large-scale proteomics studies, the negative mode finds its uses in the analyses of carbohydrates and nucleic acids. In none of the available ionization methods is there any definite relationship between the amount of ions formed and the amount of analyte – a fact which has led to the mass spectrometer being an inherently nonquantitative device without the use of an internal standards (such as stable isotope-labeled peptides).

ESI produces mostly multiply charged ions (for an average peptide) by creating a fine spray of charged droplets in a strong electric field. With the application of a dry gas, the droplets evaporate and electrostatic repulsion causes transfer of the analyte ions into the gas phase (Fig. 2a). The multiple charging allows the analysis of very large molecules with analyzers that have a relatively small m/z range. Another advantage of such multiple charging is that more accurate molecular weight values can be obtained from an analysis of the distribution of multiply charged peaks. Although ESI is used in a wide range of proteomics applications, it is limited by a susceptibility to high salt concentrations and to contaminants in the sample.

MALDI produces predominantly singly charged ions for an average peptide by aiming laser pulses at a sample which has been cocrystallized with a molecular matrix on a metal plate under high voltage (~20 kV). Typical matrices used in MALDI-assisted protein analyses include α-cyano-4-hydroxycinnamic acid and 2-(4-hydroxyphenylazo)-benzoic acid. The laser is first tuned to the absorption maximum of the matrix, after which the sample ions are preformed in the condensed phase in sufficient quantities. The matrix then absorbs some of the laser pulse energy, thus minimizing sample damage. Both, the sample ions and matrix molecules gain sufficient kinetic energy and are ejected into a gas phase (Fig. 2b). The principal characteristics of these ionization methods are outlined and compared in Table 1 [14–17].

Importantly, in both the MALDI and ESI sources, ion fragmentation occurs rarely and the analyte molecules remain largely intact. It is this nondestructiveness which causes these methods to be so attractive for the characterization of biomolecules although, for certain applications (e.g., peptide sequencing) it may be necessary to fragment the molecular ions (preferably in a predictable manner) in order to extract additional information. This is typically achieved via collision-induced dissociation (CID). Such types of MS analyses are referred to as tandem mass spectrometry (MS/MS); this is often denoted as MS^n, where n is the number of generations of fragment ions analyzed (see Fig. 1).

During recent years, hybrid ionization methods that combine some features of both ESI and MALDI have been developed. In desorption electrospray ionization (DESI), which is a typical example of such a hybrid method, an electrospray is directed at the surface of a sample and the analyte molecules are ejected into the inlet of a mass spectrometer. As only minute quantities of the solvent are sufficient for this process, DESI is the least destructive of all the ionization methods and allows the analysis of live tissues and organisms in real time [18]. To date, DESI methods have been used extensively in the surface analysis of small molecules, metabolites and lipids, and their use in proteomics research remains limited.

(a) ESI

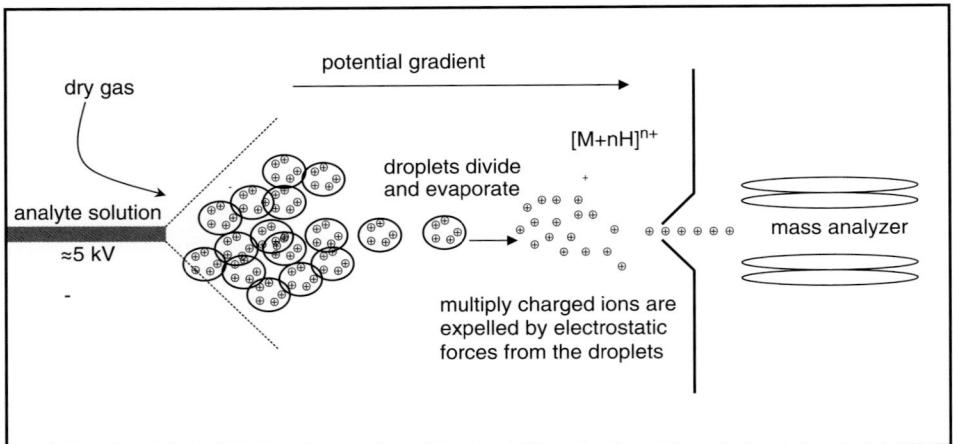

(b) MALDI

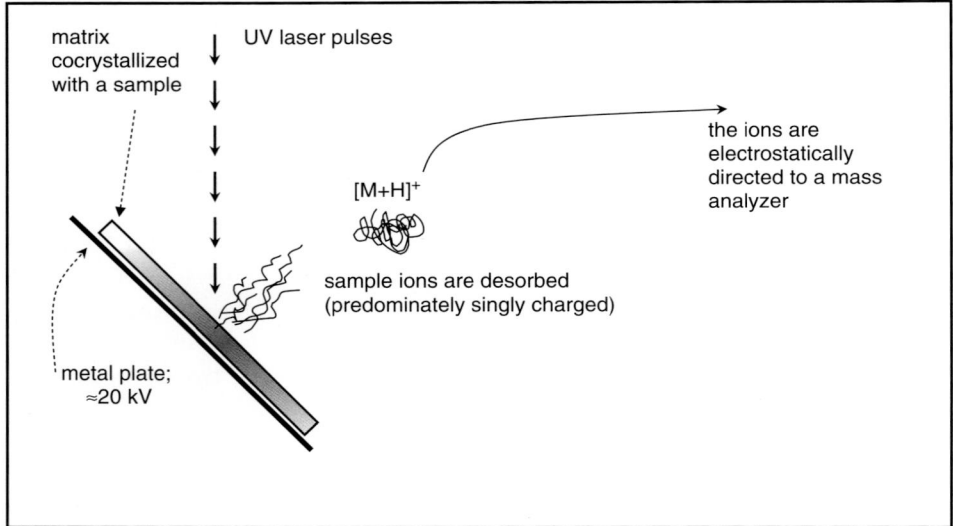

Fig. 2 Conceptual schemes of (a) electrospray ionization (ESI) and (b) matrix-assisted laser desorption ionization (MALDI). (a) During ESI, a fine spray of charged droplets is created. The droplets evaporate and the multiply-charged ions are expelled by electrostatic forces. ESI is frequently used for the analysis of complex peptide mixtures via coupling to multidimensional liquid chromatography; (b) During MALDI, the analyte molecules are ejected by laser pulses from the sample cocrystallized with a matrix. During MALDI, mostly singly charged ions are produced. Because the analyte molecules are ejected in bundles, the MALDI method is ideally suited for coupling to TOF mass analyzers. MALDI-TOF instruments are frequently used to analyze individual spots on 2D PAGE.

Tab. 1 Comparison of ESI and MALDI.

Parameter	ESI	MALDI
Principle of action	Uses electric field to produce spray of fine droplets; as the droplets evaporate ions are formed	Uses laser pulses to desorb and ionize analyte molecules cocrystallized with a matrix on a metal surface
Ions formed	Multiply charged (the larger the analyte molecule, the more likely it is to acquire multiple charges)	Singly charged
Mass range (kDa)	>100	>100
Resolution	~2500 (with IT/QD mass analyzers)	~10 000 (with TOF mass analyzers and ion reflectors)
Sensitivity (mol)	10^{-15}	10^{-15}
Typical application	LC/MS of peptide mixtures; tandem MS; protein identification by comparing experimental MS/MS spectra with theoretical MS/MS spectra produced from protein databases	Analysis of spots on 2D PAGE; determination of molecular weight; protein identification by "peptide mass fingerprinting"

ESI, electron spray ionization; MALDI, matrix-assisted laser desorption ionization; IT/QD, ion trap/quadrupole; TOF, time of flight; 2D PAGE, two-dimensional polyacrylamide gel electrophoresis; LC/MS, liquid chromatography/mass spectrometry; MS/MS, tandem mass spectrometry.

1.4.2 Mass Analyzers

Mass analyzers separate ions according to their m/z in electric or magnetic fields. The most common types of mass analyzer used in proteomics research are quadrupole (QD), ion trap (IT), time-of-flight (TOF), and Fourier transform ion cyclotron resonance (FTICR). In 2005, another Fourier transform analyzer was introduced – the Orbitrap – which is cheaper but almost equally effective as FTICR for the majority proteomics applications [19]. Often in modern instruments, several analyzers of the same or different types are combined together to achieve maximum performance; a combination of several different mass analyzers is termed a "hybrid instrument." The most important performance characteristics when comparing different analyzers and instruments are the resolving power, accuracy, and sensitivity:

- For a given mass M, the resolving power is calculated as $\frac{M}{\Delta M}$, where ΔM is the minimum separation such that masses M and $M + \Delta M$ are resolved as separate peaks. Resolving power is usually reported for the mass of 400.
- Accuracy (which is typically reported in parts per million) is a measure of difference between experimentally determined and theoretical masses, and is close to the reciprocal of the resolving power: $Accuracy = (M_{experiment} - M_{theory})/M_{theory}$.
- Sensitivity is the ability of an analyzer to generate signals for low-abundance ions.

QD mass analyzers are frequently used in conjunction with an ESI source, and offer moderate resolution (up to 2500) and moderate sensitivity. The QD analyzer consists of four parallel rods with a hyperbolic cross-section, with diagonally opposite rods being connected to radiofrequency and direct-current voltage sources, thus establishing the quadrupole field. The ions produced at an ion source are accelerated electrostatically into this quadrupole field, and mass-selection is achieved by proportional changes in the amplitudes of the radiofrequency and direct current in such a fashion that, for any given pair of these amplitudes, only ions with a specific m/z can reach the detector. QD analyzers are well known for their tolerance of high pressures (up to 10^{-4} Torr), which makes them attractive for use with ESI for LC-coupled applications.

IT mass analyzers function by confining ions to a small volume via radiofrequency oscillating electric fields, and offer a moderate sensitivity at a relatively low monetary cost. They are effective for MS/MS applications, when they are often used in conjunction with QD analyzers [20]. The mass accuracy of regular IT analyzers is rather low, as only a limited amount of ions can be accumulated in a small volume. Mass accuracy, as well as resolution, can be significantly improved by using two-dimensional (2D) IT analyzers (sometimes also called linear analyzers), which increases the ion storage volume [21].

TOF mass analyzers measure the time traveled by an ion from a source to a detector. The longer the flight path, the better the resolving power, although longer flight paths will also increase the scan time. In commercial TOF analyzers, a compromise is achieved by having the length of the flight path on the order of several meters. The resolving power of a simple TOF analyzer is poor – several to 10-fold less than that of its QD counterpart. A significant improvement in the resolving power of a TOF analyzer can be offered by the addition of one or several ion reflectors. Upon reflection, the velocity distribution of ions at a particular m/z narrows, thus increasing the resolution and sensitivity [22]. As a result, modern TOF and double-TOF spectrometers can achieve resolving powers of 10 000 and more [23]. Measurement of the ion's TOF can be achieved only if the analyte ions are presented to the TOF analyzer in discrete bundles, and because of this need TOF analyzers are particularly suited to the pulsed nature of MALDI. In fact, MALDI-TOF is one of the most common types of mass spectrometer currently used in proteomics research.

FTICR mass analyzers are based on the resonance absorption of energy by ions that process in a magnetic field [24]. The recorded array of the precession time curves is Fourier-transformed to obtain the component frequencies of the different ions. Next, the component frequencies are related to the ion's m/z. Because frequency is a parameter that is easy to measure with high precision and accuracy, FTICR has the highest resolving power amongst MS analyzers – up to 10^6 and more. With FTICR, it is also particularly easy to perform MS^n experiments.

Orbitrap is only slightly inferior to FTICR in its resolving power (as high as 2×10^5 on commercially available instruments), while being less expensive and far easier to use and maintain. Orbitrap is a special type of ion trap where the electrical field is shaped in such a way that any trapped ions rotate and oscillate inside the trap [25]. Similar to FTICR, the frequency of the oscillations is related to the mass-over-charge of the trapped ions, and is easily measured through changes in image current.

1.4.3 Ion-Mobility Spectrometry

Ion-mobility spectrometry (IMS) is a gas-phase separation method where charged ions are allowed to drift through a gas-filled tube, guided by appropriate electric potentials. The method is conceptually similar to the TOF system, but the drift times are longer and the gas pressures higher. The drift time in IMS depends not only on mass and charge but also on the cross-sectional area of the analyte; the larger the cross-section, the slower the analyte moves through the drift tube. Therefore, in IMS an unfolded protein would typically drift more slowly (i.e., a higher average cross-sectional area) compared to its compact folded form (lower cross-sectional area).

IMS is a useful tool for the analysis of protein conformations in the gas phase, and also for separating differently charged precursors. IMS is often coupled to a mass spectrometer to perform further analysis following an initial separation of the precursor molecules. While not a mainstream in proteomics analysis, IMS-MS is often used in structural proteomics and for the analysis of intact proteins [26]. A specific type of IMS – field asymmetric waveform ion mobility spectrometry (FAIMS) – is often available as an add-on for modern instruments, and is particularly useful for the analysis of modifications that alter the peptide net charge (e.g., phosphorylation; see Ref. [27]).

1.4.4 Tandem MS (MS/MS)

In some applications, it is useful to fragment molecular ions produced by ESI or MALDI further to obtain additional structural information. Figure 3a illustrates a possible way to do this with a triple-QD ESI mass spectrometer [28]. In order to obtain a tandem spectrum, the first quadrupole scans across a set m/z range and selects any ions of interest. In the second quadrupole, a CID reaction takes place whereby ions that were selected by the first quadrupole undergo collisions with argon gas, and subsequently fragment. Finally, the third quadrupole analyzes the resulted fragments. Importantly, the fragmentation via CID occurs in a predictable manner,

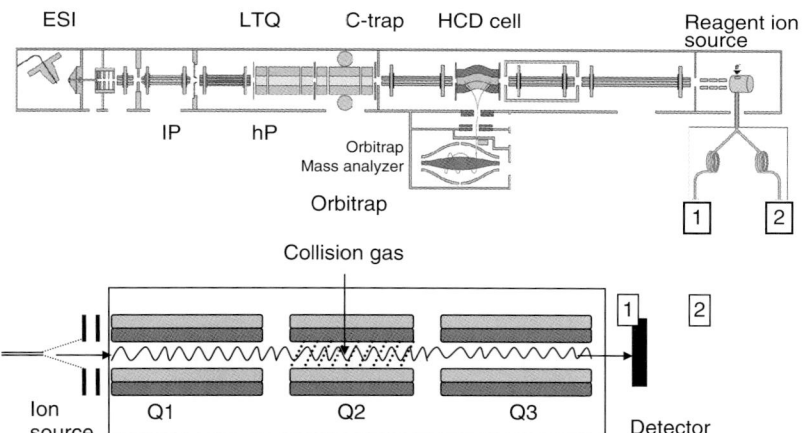

Fig. 3 MS/MS experiments. (a) A triple quadrupole mass spectrometer with MS/MS capability. Q1 selects the ion to be fragmented and allows the selected ion to pass into Q2. The ions in Q2 are fragmented via collisions with an inert gas (typically argon). Q3 analyzes the fragments generated in Q2. MS^n ($n \leq 4$) experiments can also be performed with this set-up. To achieve this, fragmentation must be increased at the level of ionization. In modern instruments, quadrupole-ion trap mass analyzers are used for MS^n ($n > 8$) experiments. Reproduced with permission from Ref. [30]; (b) The LTQ-Orbitrap instrument.

with protein and peptide ions breaking mostly at peptide bonds [29], which makes large-scale sequencing feasible.

During recent years, in addition to CID, electron transfer dissociation (ETD) and higher energy collisional dissociation (HCD) have become widely available. ETD involves a gas-phase transfer of electrons from an electron-rich molecule to the analyte cation, thereby reducing the charge by 1 and causing a predictable fragmentation. HCD produces ions similar to CID, but broadens the m/z range, albeit at the cost of analysis speed. The hybrid instrument platform that can easily combine all of these fragmentation modes is LTQ-Orbitrap, which today is becoming the instrument of choice in modern proteomics laboratories [31]. The key components of the LTQ-Orbitrap instrument with multiple fragmentation modes are shown in Fig. 3b.

In another example of a hybrid instrument, the bench-top quadrupole-Orbitrap mass spectrometer (Q-exactive) lacks the multistage MS/MS, which in turn allows for a rapid quantitative analysis of complex mixtures. Q-exactive is suitable for both bottom-up proteomics and metabolomics. An HCD cell is included in the instrument set-up, as well as an ability for the simultaneous detection of up to 10 precursors in the Orbitrap analyzer. Q-exactive might be a better alternative to the LTQ-Orbitrap when routine proteomics experiments that do not require multiple fragmentation steps are being conducted.

1.5
Methods of Sample Fractionation

The great complexity of protein mixtures from biological samples poses additional challenges for proteomics experiments. Depending on the scope of a particular proteomics task, it is desirable to introduce sample enrichment and fractionation steps prior to MS identification. The scope of a particular task can range either from the analysis of a subset of a proteome – for example, an analysis of proteins specific to a particular organelle and proteins modified in a certain manner – to a simultaneous analysis of all the proteins in the proteome. In both cases, a correct choice of separation strategy determines the overall throughput and sensitivity of the proteomics experiment. While it may be desirable to fractionate protein mixture down to individual proteins (the aim of 2D PAGE separation), a significant gain in throughput can be achieved when mixtures of proteins are introduced to the mass spectrometer. The MS analysis of intact proteins (i.e., "top-down" approach) is impractical in most high-throughput tasks with currently available instrumentation because of the large range in protein masses. To simplify the measurements, in most MS analyses the proteins are digested enzymatically (either individually purified or in mixtures) to produce peptide fragments (so-called "bottom-up" experiments). One of the most common ways to perform a bottom-up proteomics experiment is to fractionate the cell lysate on a sodium dodecyl sulfate (SDS)-PAGE, to split the gel into equal slices, and then perform in-gel enzymatic digestion. The resultant peptide mixtures are often fractionated further, normally using in-LC methods. The isolation of a particular subcellular compartment or organelle is usually achieved by a combination of centrifugation and solubilization steps [32]. Proteins modified post-translationally in a specific fashion (e.g., phosphoproteins) can be isolated using either chemical- or immuno-affinity methods (see examples in Sect. 3). The large-scale isolation of protein complexes can be achieved with epitope tagging and subsequent immunoaffinity purification (see examples in Sect. 3). Another useful affinity method is that of protein chips; this method uses large arrays of antibodies or other binding factors to isolate proteins of interest [33]. Different proteins can also be spotted onto a solid surface, followed by sequestration of their binding factors from a biological sample and subsequent analysis with MS or other methods. The technique of surface-enhanced laser desorption/ionization time-of-flight MS (SELDI-TOF) is frequently used for the analysis of protein chips in the context of biomarker discovery [34].

1.5.1 2D-Electrophoretic Separation of Complex Protein Mixtures

2D PAGE separation methods were introduced in mid-1970s and used extensively for the analysis of complex protein mixtures. In 2D PAGE (the principle of which is illustrated in Fig. 4), proteins are separated in the first dimension by their isoelectric point (pI), and in the second dimension by

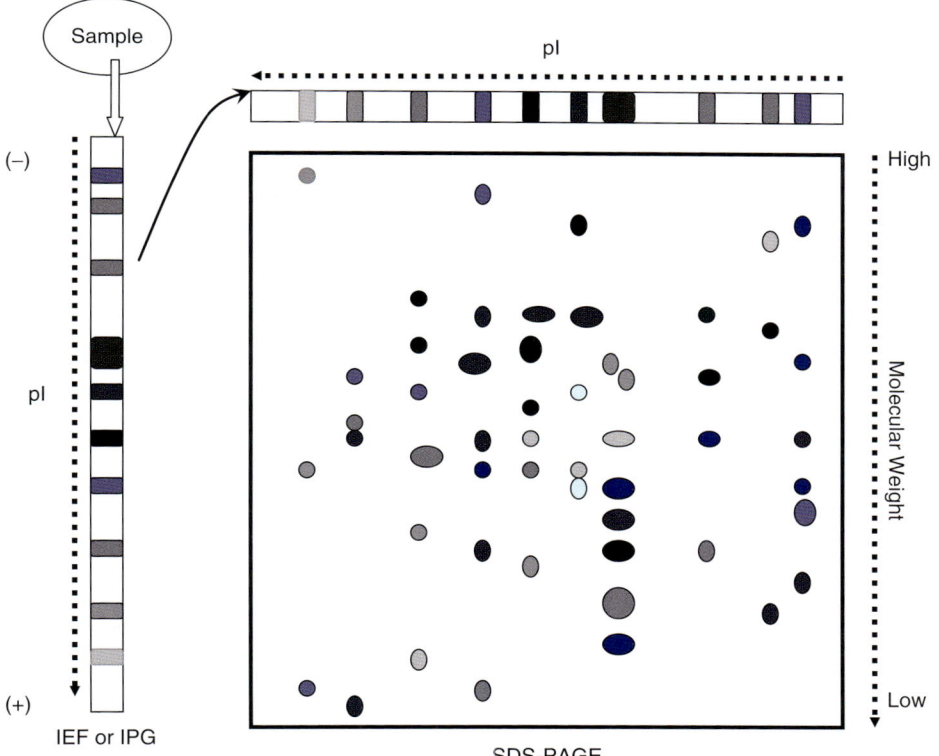

Fig. 4 Principle of two-dimensional electrophoretic separation of intact proteins. The sample is loaded onto IEF or IPG strip, where the proteins are separated according to their isoelectric points (pI). The strip is then loaded on SDS–PAGE, where the proteins are separated according to their molecular weight. Visualization of the protein spots is achieved via chemical staining methods. Individual spots can be further excised and analyzed using MS.

their molecular weight. Prior to the development of MS-identification techniques suitable for macromolecules, the identification of individual proteins in the gel spots was difficult. Typically, the identification of individual proteins in the gel spots was achieved using immunostaining methods or by N-terminal degradation sequencing. Today, however, in most laboratories the analysis of 2D PAGE spots is achieved with MS, whereby the individual spots on the gel are excised, digested, and analyzed using MALDI-MS. Protein identification is then achieved by peptide mapping, which involves comparisons of the observed peptide peak patterns with the predicted digest fragments of proteins in a database. Unfortunately, if some of the spots on the 2D PAGE are overlapping, the method of peptide mapping can produce incorrect results. As a separation technique, 2D PAGE offers high resolution and is capable of distinguishing between different protein isoforms, and also between different PTMs. However, since 2D PAGE is a denaturing technique it is not suitable for the direct analysis of protein complexes and protein–protein interactions. Other limitations of 2D PAGE include biases against proteins with pI-values outside the range of 2–10, and biases against heavy proteins (>100 kDa). In this case, instead of using the pI, the first dimension could be replaced with a non-denaturing molecular weight dimension, thereby enabling the analysis of intact protein complexes (such as in the 2D blue native SDS-PAGE method; see Ref. [35] for a detailed review). A negative point here is that 2D PAGE is a time-consuming and labor-intensive technique, and alternative methods of protein separation that employ LC are available and are often superior to the 2D gels in large-scale proteome analyses.

1.5.2 Liquid-Phase Separation Methods

An ESI source allows the MS characterization of proteins and peptides that are present in solutions, and is therefore ideally suited for online coupling with liquid-phase separations. Liquid-phase separation can be achieved by using high-performance liquid chromatography (HPLC), capillary electrophoresis (CE), or other methods. When two or more distinct liquid phases are used in a separation, with each of the phases relying on a unique independent physical property of the analyte, this liquid-phase separation is termed multidimensional [36]. Among the independent physical properties that correspond to "dimensions" can be included size, charge, hydrophobicity, or an affinity to a particular substrate. The use of several independent dimensions significantly increases the resolving power of a separation. In-liquid separation methods can be used both for intact proteins and for peptide mixtures.

One frequently used technique for the separation and analysis of intact proteins is CE, followed by MS [37–40]. In CE, the charged analytes are separated in an applied electric field according to their electrophoretic mobility; in this case, the use of capillaries increases the surface-to-volume ratio and reduces overheating. Two main variations of CE are often used for the separation of intact proteins and peptides, namely capillary zone electrophoresis (CZE) and capillary isoelectric focusing (CIEF).

CZE is a kinetic-based method of separation that is performed in solution, where the fused silica capillaries have free $-SiOH$ groups that become negatively charged to $-SiO^-$. These negatively charged groups then attract positive ions, which are dragged against this attraction by the applied electric field according to the ions' electrophoretic mobilities.

CIEF, in contrast is a steady-state technique, which imparts separation according to the pI-values of the analytes. During CIEF separation, the capillary is filled with the sample in solution and a pH gradient is applied by placing the anodic end in a low-pH solution and the cathodic end in a high-pH solution. No dragging against the walls of the capillary occurs (as in CZE), because a coated fused silica is used. A recent report by Wang *et al.* described a good example of this separation method, whereby intact proteins were separated by CIEF at the first step and then digested via an online immobilized trypsin microreactor, followed by identification using MS [40]. The final results described the identification of 101 proteins from *Escherichia coli* extracts.

Intact protein approaches are usually limited to proteins of smaller sizes, mostly because the resolving power of mass analyzers drops in line with the increase in the ions masses. Biases towards one or the other types of intact proteins are also unavoidable with any chromatographic separation method. A conceptually different approach, in which the purification of intact proteins is avoided, involves the proteolytic digestion of protein mixtures whereby protein mixtures are transformed into more uniform (in terms of mass and chromatographic properties) mixtures of peptides. Because of this increased uniformity the analysis of peptide mixtures essentially eliminates biases in proteome coverage. A drawback of this approach is that an absence of information on intact proteins makes the characterization of PTMs more difficult. The chemical or enzymatic cleavage of proteins into peptides, followed by multidimensional liquid-phase separation, is typically used in a global proteome survey-type experiments [41]. In this case, proteins from whole cells or tissue homogenates are first digested enzymatically, after which the resulting peptide mixtures are separated using electrophoresis or LC in a microcapillary format. For this, a microcapillary column is attached directly to the ionization source so that the eluting fractions are ionized and introduced to a mass analyzer. While fragmentation of the peptide ions is optional with the MALDI-MS/peptide mapping analysis of 2D PAGE spots, it is essential with the liquid-phase/MS analysis of complex peptide mixtures. The sequence information obtained by MS/MS for individual peptide ions is used to match peptides to the corresponding proteins in a database. For example, in a recent report, Zhu *et al.* [38] studied differentiating PC12 cells by preparing trypsin digests and then using off-line reversed-phase LC followed by CIEF coupled to a LTQ-Orbitrap mass spectrometer. The authors reported the identification of 2329 unique peptides and 835 proteins using this method.

Multidimensional protein identification technology (MudPIT) is another good example of a liquid-phase/MS method [42, 43]. In MudPIT, a biphasic strong cation exchange (SCX)/reverse-phase (RP) microcapillary column is used to fractionate a complex peptide mixture (Fig. 5), and the peptide fractions are eluted from the column by a series of HPLC gradients into the MS/MS. The MS/MS spectra obtained are then used to search a sequence database via the SEQUEST algorithm [44].

Previously, MudPIT was used to detect and identify approximately 500 proteins from the *Saccharomyces cerevisiae* proteome [43], about 2500 proteins from the *Oryza sativa* proteome [45], and about 2500 proteins from the *Plasmodium falciparum* proteome [46]. In a recent report, which employed a higher number of SCX steps

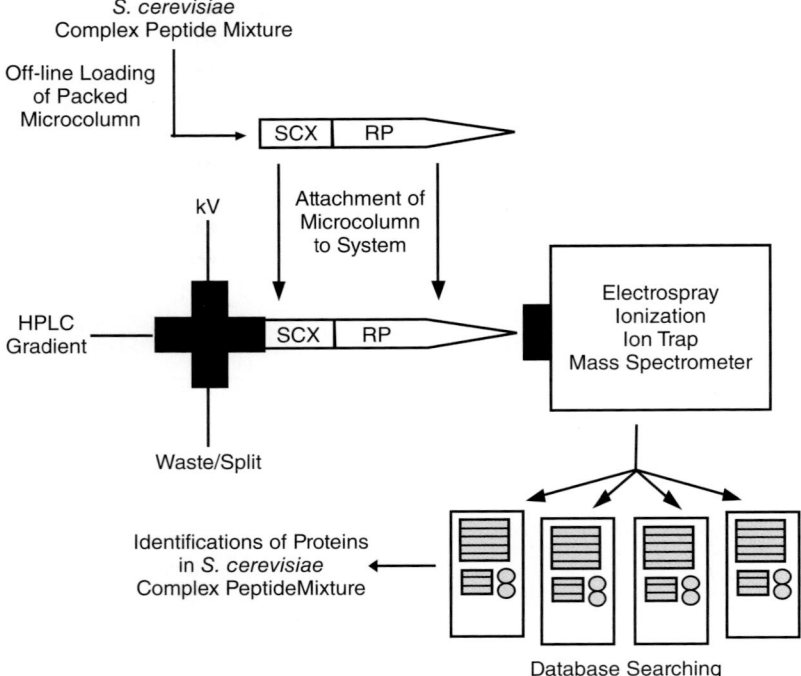

Fig. 5 Multidimensional protein identification technology (MudPIT). Complex peptide mixtures are loaded onto a biphasic microcapillary column packed with strong cation exchange (SCX) and reverse-phase (RP) materials. Peptides are eluted directly into the tandem mass spectrometer because a voltage (kV) supply is directly interfaced with the microcapillary column. Peptides are first displaced from the SCX to the RP by a salt gradient, and are eluted from the RP into the MS/MS. The tandem mass spectra generated are correlated to theoretical mass spectra generated from protein or DNA databases by the SEQUEST algorithm. Reproduced with permission from Ref. [43]; © 2001, Nature Publishing Group.

and used the LTQ-Orbitrap Velos platform, a near-complete coverage of the yeast proteome was achieved in a MudPIT experiment [47], where the authors identified 4269 proteins from 4189 distinguishable protein families in yeast.

1.5.3 Affinity Methods (Epitope Tagging)

Affinity-purification methods are frequently employed in those proteomics programs that analyze specific subsets of a proteome, such as phosphorylated or glycosylated proteins. Affinity methods are also often used in large-scale studies of protein–protein interactions and protein complexes. Affinity methods are based on targeted interactions between proteins and antibodies, or between proteins and other protein-specific molecules, immobilized in a column or in the form of an array. For example, in order to isolate phosphorylated proteins from the remainder of the proteome, an immuno-affinity column prepared with antibodies specific to the phosphorylated amino acids can be used [48].

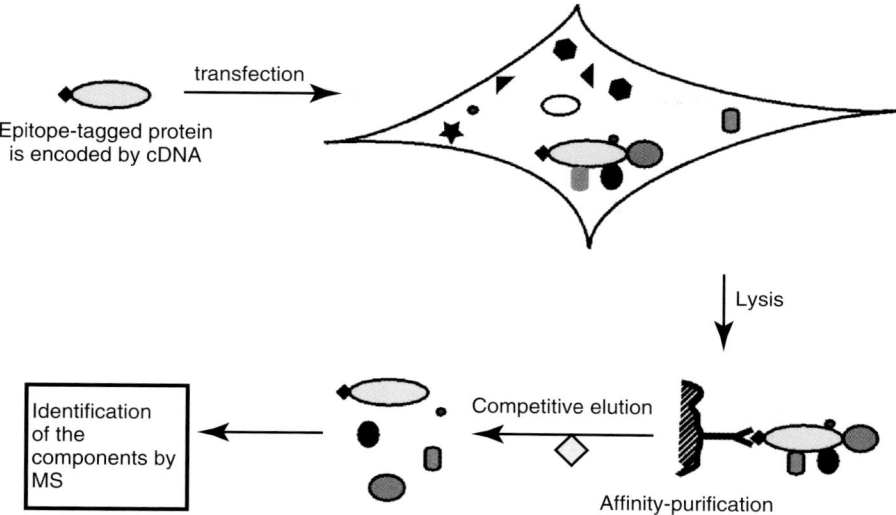

Fig. 6 Principle of affinity purification with epitope tagging. Bait proteins are expressed as a fusion with a motif recognizable by a certain antibody (epitope). The cells are lysed and proteins purified by immunoaffinity chromatography. The bound fraction is then competitively eluted from the column, and the proteins are analyzed by LC/MS or one-dimensional PAGE/MS.

Epitope tagging is used in the large-scale analyses of protein–protein interactions and protein complexes (Fig. 6). In this strategy, the proteins to be isolated are fused with a motif that is recognizable by a specific antibody. This fusion is typically achieved by incorporating the epitope sequence into the C-terminal of genes that encode proteins of interest. The fused proteins are then expressed and purified along with their interaction partners with immunoaffinity chromatography, using antibodies specific to this particular epitope. If a particular protein is of low abundance, it can be overexpressed to increase the overall recovery.

1.6
Protein Identification by MS

The accurate identification of proteins is an essential requirement of proteomics studies. The high complexity of protein mixtures derived from tissues, whole cells or subcellular compartments adds an additional requirement for a high-throughput approach in the identification of proteins. The common strategies that match MS data with proteins present in a biological sample are reviewed in the following sections.

Generally, protein identification amounts to the deduction of sequence-specific information from MS data, followed by searches of the sequence databases. The databases against which MS data can be matched are protein, expressed sequence tag (EST), and genomic databases [49]. Currently, high-throughput sequencing by MS is performed more easily in the bottom-up format, when intact proteins are first fragmented into peptides. Depending on a particular type of proteomics scheme and the MS instrumentation used, protein sequence information can be obtained by: (i) peptide mass fingerprinting; (ii) accurate mass tags (AMTs); (iii) peptide fragmentation in MS/MS; and (iv) sequence tags; or by a combinations of these four methods.

Peptide mass fingerprinting is usually adopted in those cases where individual proteins are separated during the purification steps, such as in 2D PAGE when individual spots are picked, digested, and the masses of the peptides recorded. The experimentally obtained peptide masses are then matched against theoretical peptide libraries generated from protein sequence databases [50]. For a protein mixture of higher complexity (e.g., gel bands from 1D SDS-PAGE), MS/MS fragmentation is typically required, though MS/MS data can also be obtained along with mass fingerprinting, and today this is common practice. With MALDI-TOF instrumentation – which is used with the peptide mass fingerprinting type of analysis – several peptide masses are needed for the unambiguous identification of a protein.

By using more accurate instrumentation, such as FTICR-MS or Orbitrap, it is possible to identify proteins based on the mass of a single peptide, without MS/MS data. In this case, a peptide that corresponds uniquely to a protein is termed an AMT. Conrads et al. evaluated the use of AMTs for identifying proteins from S. cerevisiae and Caenorhabditis elegans [51], and showed that up to 85% of the predicted tryptic peptides from these two organisms could be used as AMTs at mass accuracies typical of FTICR-MS instruments (<1 ppm). These authors also discussed the use of AMTs with highly accurate mass measurements in the detection of phosphorylated proteins, arguing that because the mass defect of P is larger than that of H, C, and O, then the average mass of phosphopeptides would be slightly lower than the mass of unmodified peptides of the same nominal weight. This would enable the identification of phosphorylated peptides if the mass measurement accuracy was sufficient [52].

Whenever it is possible to match a single (unique) peptide to a protein, it is no longer necessary to purify samples down to individual proteins. Instead, fractions containing mixtures of proteins can be digested with trypsin, followed by an analysis of the resultant peptide fragments. For example, the above-discussed AMT method can be used for the analysis of complex protein mixtures because it matches a single peptide to a unique protein. However, the requirement for high mass accuracy, coupled to the fact that there are still some sequences which lack unique AMTs, rather limits the use of AMT [53].

Far superior in terms of proteome coverage per unit monetary cost, are those methods that employ peptide sequence analysis using MS/MS. The key fact that enables high-throughput peptide sequencing via MS/MS is the predictability of the peptide precursor ion fragmentation in the CID, HCD, or ETD reactions. Figure 7 shows the adopted nomenclature for the fragment ions series. In CID, ions that form through the dissociation of peptide bonds are the most abundant if moderate collision energies are used (30–50 V). Those ions that retain their charge on the N-terminal part after fragmentation of the precursor are termed b ions, while those that retain their charge on the C-terminal part after fragmentation of the precursor are termed y ions. In Fig. 7, the subscript to the right of the b and y symbols represent the numbers of amino acid residues in the corresponding fragment. The difference in m/z values between consecutive ions within a given series corresponds to the difference in the sequences of the two fragments.

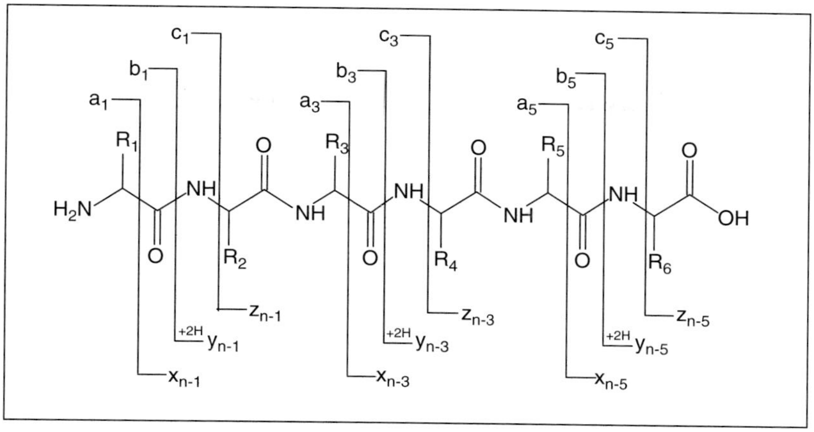

Fig. 7 Nomenclature of ions formed during peptide fragmentation. In a typical MS/MS experiment via CID reactions, y and b ions are predominant. These two types are used for computer-assisted peptide identification. Through the loss of CO fragment, a ions can be produced from b ions. While not important for identification, the a ion series can be used for independent result validation. Other types of ions are not produced in CID reactions.

Because the consecutive ions within a series represent peptide fragments that differ by exactly one amino acid, and each amino acid residue has a unique nominal weight (except I and L), the pattern of m/z values of the y and b ions corresponds to the amino acid sequence of the precursor peptide. Unfortunately, some expected peaks could be missing from the MS/MS depending on a sequence context, peptide modifications, and ion-suppression effects. Experimental spectra can also be complicated by unwanted fragmentation, and it is not always possible to unambiguously determine from which series a particular ion fragment is derived. For these reasons, the manual interpretation of MS/MS spectra can be tedious and ineffective. Instead, in the current proteomics approaches, MS/MS analyses are performed automatically by either sequence-to-database matching or by the derivation of sequence tags (e.g., in *de novo* methods). ETD and also electron-capture dissociation (ECD) fragmentation methods produce z and c series of ions, corresponding to the C-terminal and N-terminal fragments, respectively. The analysis of ETD and ECD fragmentation data employs the same principles and tools as the analysis of CID data.

A classic example of an algorithm that identifies the best matches to MS spectra is SEQUEST, as developed by Eng *et al.* [44]. The analysis of MS data by SEQUEST starts with a reduction of the tandem spectra complexity, with only a certain number of the most abundant ions being considered and the remainder discarded. The unfragmented precursor ion is also removed from the spectra in order to prevent its misidentification as a fragment. Next, SEQUEST selects sequences from a database by: (i) creating a list of peptides that have masses at or near the mass of the precursor ion; followed by (ii) the generation of a theoretical MS/MS spectra for each of the candidates and comparing these to the observed spectra. As the result of this comparison, a cross-correlation score is produced (X_{corr}), while another score, DeltaCN, reflects the difference in correlation of the second ranked match from the first one. The higher this score, the more likely it is that the first match is the correct one. SEQUEST also has the ability to detect simple PTMs, in which case it looks for increased masses of amino acids. For example, cysteines are typically carboxymethylated by iodoacetamide prior to enzymatic digestion, and in this case the masses of all cysteine residues are considered to increase by 57 Da. In order to seek amino acids that can be either modified or not, the database size is increased by allowing different weights to represent the same amino acid. Unfortunately, only known PTMs can be detected with SEQUEST, and the types of modification being sought need to be specified explicitly in the input parameters.

Other popular algorithms which are functionally similar to SEQUEST are Mascot (a product of Matrix Science), X!Tandem (which is freely available from The Global Proteome Machine), and OMSSA. The latter is a free search engine developed by NCBI, and which was particularly well suited to the analysis of multiply charged precursors; however, its development was suspended in 2013 due to NIH budgetary constraints. Recently, a number of proteomics software environments have been developed which merge the results of these and other search engines into a single output [54]. This is a useful feature, as different algorithms have different biases against certain types of peptide, and often complement each other in proteomics analysis. One notable example of such a software environment is the Search GUI/Peptide Shaker open-source platform [55].

When sequence information is derived directly from the MS/MS, without relying on sequence database, the process is termed *de-novo* sequencing. In a typical approach, a spectral graph is constructed where the vertices represent observed masses and the edges represent a possibility to explain a mass difference between vertices by using a particular amino acid or amino acid combinations. Different paths through the graphs then are analyzed and scored, and the best choice is selected. Open-source programs for this purpose include Pep-Novo/UniNovo [56, 57], and one of the high-end commercial products is the Peaks Studio software [58]. Peaks Studio is also an environment which can integrate the results of other search engines, similar to SearchGUI. At present, *de-novo* approaches are becoming as effective (if not more so) as the sequence-database-matching methods. *De-novo* methods also offer a significant advantage in that the identification of new isoforms, amino acid substitutions and novel PTMs becomes possible. Notably, as soon as the sequence tags are derived from MS/MS data, they can be used to query sequence databases, thereby mimicking sequence-database-matching methods such as SEQUEST. The unmatched tags can be further combined and searched by BLAST-type methods against all available sequences (e.g., MS-BLAST; see Refs [59, 60]). This is a common method in proteomics of organisms with unsequenced genomes.

1.7
Imaging Mass Spectrometry

The aim of imaging mass spectrometry is to visualize the spatial distribution of a given analyte in a biological sample (e.g., an organ, which is sliced into many individual cross-sections). The most appropriate ionization method for this is MALDI, where an organ slice is mounted directly onto a modified

MALDI plate and a high-resolution laser is moved across it so as to eject the analytes into a mass analyzer.

An alternative to MALDI is DESI, which may be more attractive for a surface analysis as it causes minimal damage to the sample. In order to use ESI for imaging, the initial sample would have to be cut into multiple pieces and each piece analyzed separately. An on-plate digestion could also be performed to aid the analysis, and MS/MS fragments used in addition to the MS data to construct the final images. Typically, imaging MS is limited to the most abundant proteins or metabolites. In addition, quantitative information obtained by imaging MS is limited by the high dependence of the ion signal on local environmental variables such as salt concentration, the presence of other analytes competing for charge, and morphological differences influencing ionization efficiency. Possible solutions to the problems encountered in imaging MS have been recently reviewed by Ellis *et al.* [61].

2
Quantitative Methods of Proteome Analysis Using MS

Quantitative proteome analysis aims at the large-scale identification of differences in protein expression. Two-dimensional difference gel electrophoresis (2D-DIGE) can be used to achieve such an aim, although while acknowledging the importance of gel-based methods in modern quantitative proteomics it is believed that methods based on multidimensional chromatography have greater potential, and consequently any further discussions in this chapter will be limited to quantitation within the LC/LC/MS/MS paradigm. Even though absolute quantification (i.e., a clear relationship between the peak intensity and amount of the analyte) is a challenge for the modern mass spectrometer, methods of finding relative abundances exist that are based on labeling through mass modifications of the whole proteome, or some of its subsets. Labeling through mass modification is particularly suited to quantitative analysis in differential expression profiling, where the two states under investigation are differentially labeled via "light" or "heavy" mass modifications [17, 41]. A quantitative proteomics approach is depicted schematically in Fig. 8. Mass modification can be introduced at different steps in the sample preparation, such as during growth, after growth, during digestion, or after digestion. Labeling at the early stages of sample preparation minimizes any losses of the analytes. Although the instrumentation for quantitative studies is essentially the same as that for qualitative studies, quantitative analysis is more laborious and typically achieves a lower proteome coverage.

2.1
Metabolic Labeling

Metabolic labeling methods are widely used in structural biology for the preparation of samples for NMR analysis [62, 63], and have also become very useful tools for quantitative proteomics. Metabolic labeling is a mass modification method that is carried out very early in the experiment, at the stage of cell growth. The principle behind this strategy is illustrated schematically in Fig. 8. The goal here is to compare protein abundances in cells grown under different conditions, and to do this the growth media from one of the conditions is enriched with stable, low-abundance isotopes. Such enrichment can be achieved either by labeling all amino acids with ^{15}N [41, 64] or by supplementation with

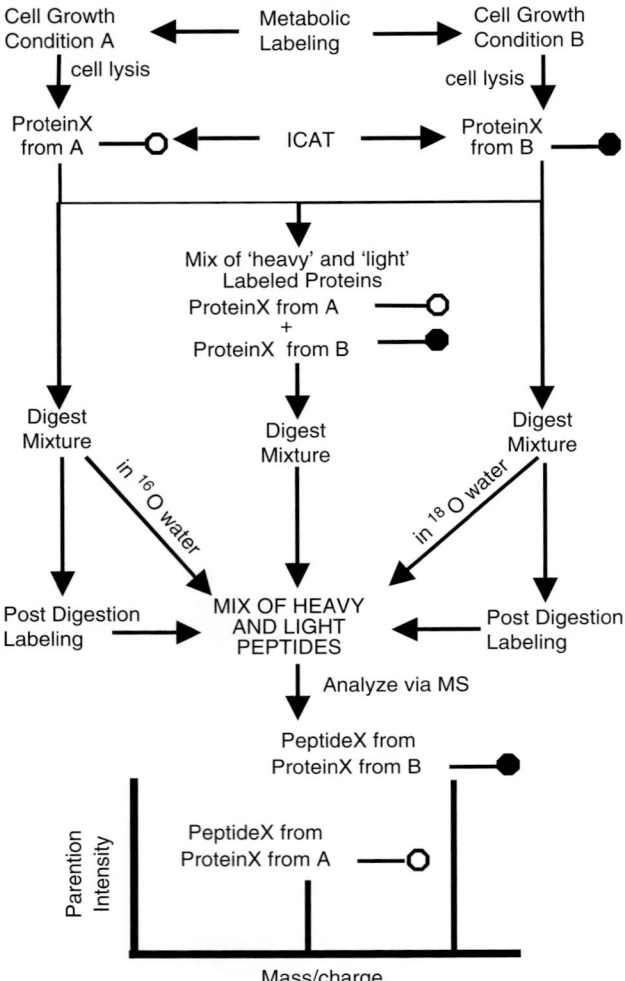

Fig. 8 Quantitative proteomics approach. When carrying out a quantitative proteomic analysis, the key is for the same peptide from two unique growth conditions to have unique masses when being analyzed by a mass spectrometer. "Heavy" and "light" peptides may be generated at many points in a sample preparation pathway. Metabolic labeling introduces a label during the growth of the organism, and is therefore the earliest point of introduction of "heavy" and "light" labels. Metabolic labeling is followed by ICAT, digestion in ^{16}O and ^{18}O water, and lastly post-digestion labeling. Only after a label has been introduced can the samples be mixed and further processed. Reproduced with permission from Ref. [41]; © 2002, American Chemical Society Publications.

a single labeled amino acid [65, 66]. The processed and digested cell extracts from the two different conditions are combined in a one-to-one ratio and analyzed using liquid-phase/MS/MS methods. In the resultant MS spectra, peaks corresponding to the same peptide from the different conditions are offset according to the degree of labeling, and the ratio between the two peaks corresponds to the difference in abundances. The first demonstrations of metabolic labeling in quantitative proteomics used gel electrophoresis and spot excision as the protein isolation method [64, 67]. For example, Oda et al. identified proteins that were altered in expression between two strains of *S. cerevisiae* grown in ^{14}N or ^{15}N media, and determined the phosphorylation levels of a specific protein in the same sample. Complete metabolic labeling has been demonstrated using chromatography-based proteomics methods with *Deinococcus radiodurans* [68, 69], mouse B16 cells [68, 70], and *S. cerevisiae* [41].

Isotopically enriched single amino acids may also be used for the selective metabolic labeling of a cell type for a quantitative proteomic analysis. This method is referred to as stable isotope labeling by amino acids in cell culture (SILAC), and is currently one of the most frequently used labeling methods in quantitative proteomics. In *S. cerevisiae*, Jiang and English described the single amino acid isotopic enrichment of *S. cerevisiae* with deuterated leucine [71], while Berger et al. described the comparative analysis of *S. cerevisiae* cultured in media containing either ^{13}C-lysine or unlabeled lysine [65].

2.2
Isotope-Coded Affinity Tags and Isobaric Tags

Metabolic labeling with stable isotopes, while analytically advantageous compared to other methods, is available only in cases when the studied cells or organisms are cultivable. For this reason, metabolic labeling is not suitable for diagnostic and clinical applications. When cultivation or controlling growth is difficult, different strategies need to be used for quantitative proteome analysis. One possible approach, isotope-coded affinity tags (ICATs), employs cysteine-specific reagents to differentially label proteomics samples. ICATs have three functional elements: (i) a cysteine-specific reactive group; (ii) an isotopically labeled linker; and (iii) an affinity group (Fig. 9). For the comparative analysis, all cysteine residues

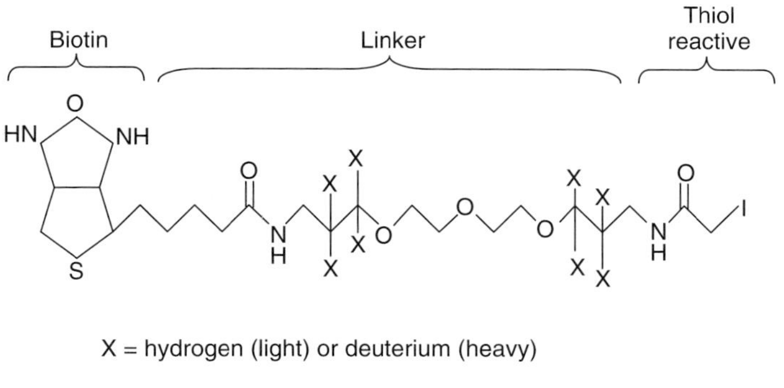

X = hydrogen (light) or deuterium (heavy)

Fig. 9 Structure of isotope-coded affinity tags (ICATs). The ICAT reagent consists of a biotin group linked to a cysteine reactive group. The linker may be deuterated eight times or protonated at each site, allowing for the generation of D_0- or D_8-ICAT. The differential masses of the linker group allow for the use of ICAT in a quantitative proteomic scheme. Reproduced with permission from Ref. [75]; © 2003, Marcel Dekker, Inc.

in the samples are separately labeled with either labeled or unlabeled ICAT. The derivatized samples are then combined in a one-to-one ratio and digested with a protease, which results in both labeled and unlabeled peptide fragments. The labeled peptide fragments are then purified by affinity chromatography, fractionated using reversed-phase chromatography, and analyzed with MS/MS. This provides both a qualitative analysis and the relative abundance of the peptide isoforms in the samples. The MS analysis is analogous to the metabolic labeling case, where the ratio of "heavy" to "light" peptide ions correlates with their relative abundance. Unfortunately, the ICAT approach has one obvious conceptual limitation, in that only peptides which contain cysteine can be detected. Consequently, since it has been shown that, in yeast, only about 10% of all tryptic peptides contain cysteine, a full sequence coverage may not be possible even for the most abundant proteins.

A significant improvement to the ICAT method is the cleavable isotope coded affinity tag (cICAT), which is presently supplied by Applied Biosystems in a kit format. The cICAT reagent has four essential structural elements:

- A protein-reactive group (iodoacetamide) that covalently links the isotope-coded affinity tag to the protein through alkylation of cysteines.
- A structural element of a biotin affinity tag, which allows enrichment of the tagged peptides.
- An isotopically labeled linker ($C_{10}H_{17}N_3O_3$). Nine carbon atoms of the linker can be either ^{12}C or ^{13}C, providing light and heavy version of the tag, respectively. The light and heavy molecules have the same chromatographic properties, but differ in mass (by 9 Da). When the sample is subjected to mass spectrometric analysis, the ratio of intensities between the heavy and light peptides provides a relative quantitation of the proteins in the original sample.
- An acid-cleavage site that allows the removal of part of the tag prior to MS analysis.

Following avidin-affinity purification of the cICAT-labeled peptides, the biotin portion of the label and part of the linker can be removed by adding trifluoroacetic acid. This reduces the overall mass of the tag on the peptides and improves the overall peptide fragmentation efficiency. The ICAT methodology has been successfully applied to a variety of biological questions, including studies of several cell types, organelles, and different classes of proteins.

Quantitative proteomic analysis via ICAT has been coupled with cDNA array analysis to investigate the galactose utilization pathway in *S. cerevisiae* [72], as well as the changes in mRNA and protein expression brought about by culturing *S. cerevisiae* in either galactose or ethanol [73]. In addition, ICAT was used to detect changes in the protein expression of peripheral and integral membrane proteins by analyzing the effect on 12-phorbol 13-myristate acetate on the microsomes of HL-60 cells [74]. Despite the ICAT approach reaching peak usage during recent years, its application is now on the decline.

An alternative approach, isobaric tags for relative and absolute quantitation (iTRAQ), which relies on quantitative information obtained from the MS/MS fragmentation rather than from precursors, has recently been developed and has rapidly become one of the most widely used quantitation methods [76]. In the iTRAQ method (as illustrated schematically in Fig. 10), the key feature is that iTRAQ-labeled peptides release intense reporter ions during the MS/MS fragmentation process. Typically,

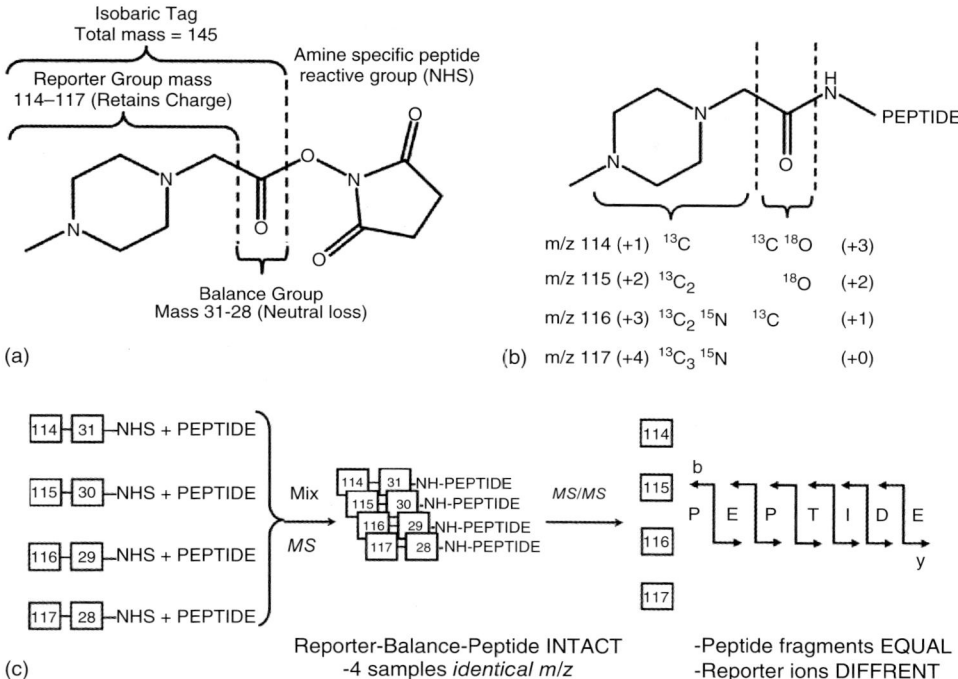

Fig. 10 iTRAQ approach for quantitative proteomics. The iTRAQ reagent uses a reporter group, a mass balance group (carbonyl), and a peptide-reactive group (NHS ester). The mass is kept constant using differential combinations of ^{13}C, ^{15}N, and ^{18}O atoms. (a) The reporter group ranges in mass from m/z 114.1 to 117.1, and the balance group ranges in mass from 28 to 31 Da, such that the combined mass remains constant (145.1 Da) for each of the four reagents; (b) When reacted with a peptide, the tag forms an amide linkage to any peptide amine (N-terminal or ε-amino group of lysine). These amide linkages fragment in a similar fashion to backbone peptide bonds when subjected to CID. Following fragmentation of the tag amide bond, however, the balance (carbonyl) moiety is lost (neutral loss), while charge is retained by the reporter group fragment. The numbers in parentheses indicate the number of enriched centers in each section of the molecule; (c) Illustration of the isotopic tagging used to arrive at four isobaric combinations with four different reporter group masses. A mixture of four identical peptides each labeled with one member of the multiplex set appears as a single, unresolved precursor ion in MS (identical m/z). Following CID, the four reporter group ions appear as distinct masses (114–117 Da). All other sequence-informative fragment ions (b-, y-, etc.) remain isobaric, and their individual ion current signals (signal intensities) are additive. This remains the case even for those tryptic peptides that are labeled at both the N terminus and lysine side chains, and those peptides containing internal lysine residues due to incomplete cleavage with trypsin. The relative concentration of the peptides is thus deduced from the relative intensities of the corresponding reporter ions. In contrast to ICAT and similar mass-difference labeling strategies, quantitation is thus performed at the MS/MS stage rather than in MS. Reproduced with permission from Ref. [76]; © 2004, American Society for Biochemistry, Molecular Biochemistry and Molecular Biology.

the MS/MS spectra are less noisy compared to the liquid chromatography–mass spectrometry (LC-MS) profile, and this has led to iTRAQ becoming the superior method. The ability of iTRAQ to multiplex is also attractive, as this minimizes variation and increases confidence in the results obtained. To date, multiplexing of eight and higher has been achieved with modified iTRAQ reagents. Because the reporter ions have a low mass (ca. 115 Da), iTRAQ initially was not available for ion trap users as the ion traps encountered problems when detecting in the lower mass range. However, with the appearance of HCD fragmentation and pulsed Q collision-induced dissociation (PQD) this problem was quickly resolved by the ion trap manufacturers, and iTRAQ is now accessible to the majority of proteomics users.

2.3
^{18}O Labeling

The global modification of all proteolytic peptides in a mixture may be carried out via the labeling of carboxyl groups that occurs through the incorporation of ^{18}O from $H_2^{18}O$ during proteolytic hydrolysis [77, 78] (Fig. 11). In order to introduce a 4 Da

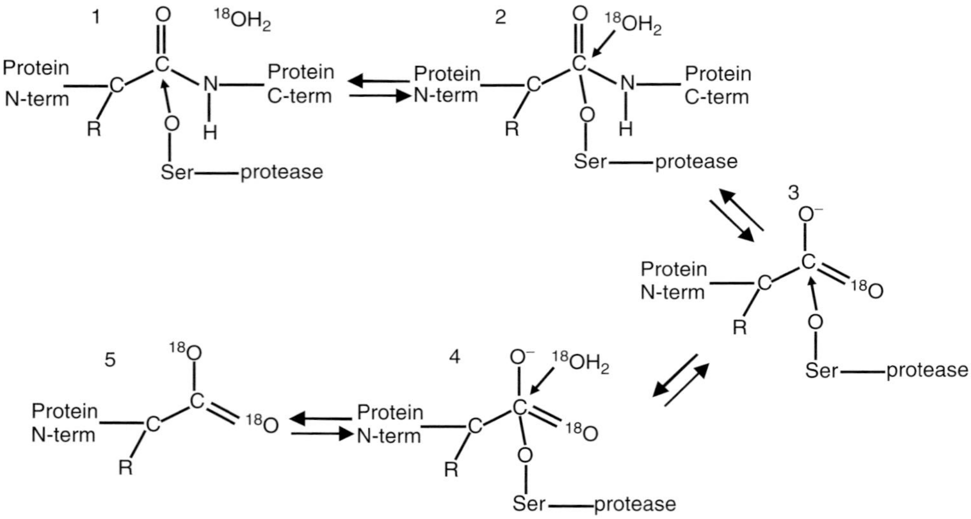

Fig. 11 C-terminal digestion modification with ^{18}O. ^{18}O may be incorporated into the C terminus of a peptide during digestion with enzymes such as trypsin, endoproteinase Lys-C, and endoproteinase Glu-C. A simplified version of this reaction scheme is shown. Step 1: Initially, the peptide must be digested in $H_2^{18}O$ in order to then incorporate ^{18}O. The serine in the active site of the proteases listed attacks the carbonyl carbon in a peptide bond. Step 2: The $^{18}OH_2$ attacks the protein–protease intermediate also at the carbonyl carbon, displacing the NH group on the peptide bond. Step 3: As a result, a peptide with a single ^{18}O has been generated. Step 4: A repeat of steps 1 and 2 is needed to drive the reaction to completion, as shown in Step 5, where two atoms of ^{18}O have now been incorporated into the peptide C terminus. Labeling of one sample with ^{18}O by digesting in $^{18}OH_2$ and mixing this with the other sample digested in ^{18}O-depleted water allows for the determination of the relative abundance of peptides from a mixture. Reproduced with permission from Ref. [75]; © 2003, Marcel Dekker, Inc.

mass shift into the C terminus of a peptide, proteins may be digested in the presence of $H_2{}^{18}O$. Proteases such as trypsin will carry out this reaction during the process of enzymatic cleavage. Hence, by mixing a sample with proteins digested in the presence and absence of ^{18}O, a pairwise comparison may be made to determine the relative abundance of peptides in a sample.

^{18}O-labeling is also useful for the analysis of crosslinked peptides, since when two peptides are crosslinked they will have two C termini, and during the ^{18}O labeling an 8 Da isotopic shift will be observed, in contrast to 4 Da shift in the case of simple linear peptides. The potential difficulties to be aware of here with regards to the ^{18}O-labeling are the possibility of incomplete label incorporation, and an introduction of the label into the side chains of asparagines and glutamines during their nonenzymatic deamidation to aspartates and glutamates.

2.4
Absolute Quantitation

In order to determine the exact concentration of a protein in a complex mixture, several methods are currently in use, all of which rely on stable isotopic labeling. The first method relies on identifying those proteotypic peptides that unambiguously identify a given protein and are the most MS friendly, that have the best ionization properties, and are easily detected in multiple proteomics experiments. When such a proteotypic peptide has been identified, its isotopically labeled version is prepared, after which the digested protein mixture is spiked with a known quantity of the isotopically labeled synthetic peptide, and the protein concentration is then inferred from the light-to-heavy ratio.

The second approach, QconCAT, involves a clever modification of the first method [79] that can be extended to several proteins. A recombinant chimeric polypeptide is first prepared in *E. coli* in such a way that, after its digestion, a mixture of proteotypic peptides is obtained, each of which has the same molar concentration as the chimeric protein before digestion. The label can be introduced either by growing *E. coli* on a ^{15}N-nitrogen source, by using labeled amino acids (e.g., SILAC), or by post-digestion labeling via trypsin-catalyzed ^{18}O incorporation. When using this method, two proteotypic peptides are typically chosen for each of the proteins to be quantified, and between six and ten proteins can be simultaneously and reliably quantified.

2.5
Label-Free Methods of Quantitation

In parallel to the stable-isotope labeling methods, label-free methods have also been developed and have become widespread among large-scale proteomics analyses. Two basic methods exist: (i) the use of ion chromatograms by integrating the LC-MS profiles of peptides of interest (extracted ion chromatography; XIC); and (ii) a spectral counting approach, using the number of observations of peptides representing a particular protein as a quantitative measure of that protein's abundance.

The first method requires a robust reproducible LC-MS separation platform, and is more difficult to use with an increasing complexity of the starting protein mixture. The second method requires appropriate normalizations and statistical testing to establish significant differences in spectral counts between the samples of interest. By definition, only those MS/MS spectra that lead to a reliable peptide identification will contribute to the spectral count for a given protein. It is considered that, in time, spectral counting will become a more useful approach for the analysis of complex protein mixtures compared to both XIC and isotope labeling methods for the following reasons: (i) no special hardware or chemical treatment is required, with spectral counts simply derived from the large-scale protein identification; (ii) when the protein has different representative peptides it is being quantified across different points in its chromatographic profile, thereby minimizing any separation biases and charge-suppression effects; and (iii) it has been proven, using standard mixtures, that the dynamic range of spectral counting and overall performance is comparable to the isotopic labeling approaches.

Pavelka *et al.* showed that a power law global error model, which is typically applied to the analysis of DNA microarrays, can also be adapted to the analysis of spectral count data [80]. The same group further showed that the spectral count-derived statistic, normalized spectral abundance factor (NSAF), conforms to the power law global error model (PLGEM), where the standard deviation of a protein's NSAF depends on the value of the NSAF itself. In order to correctly quantify homologous proteins, which between them have certain stretches of sequence identity, the shared spectral counts must be split between the different protein forms. To achieve this, the "adjusted spectral count" approach has been developed [81], whereby shared counts are split in the proportion of the unique spectral counts. For instance, if two proteins A and B have 90% peptide identity, and the spectral counts for the remaining unique peptides are 5 and 1, respectively, the spectral count for the 90% of shared peptides will be split in a 5:1 proportion between A and B. Finally, in order to estimate the confidence of differences in protein abundance, a G-test of independence can be applied to the raw or adjusted spectral counts [81], or a T-test applied to the NSAF values [82].

Arguably, spectral counting is a superior method of quantitation compared to stable isotopic labeling, especially on the large scale, although stable isotopic labeling can also potentially allow for multiplexing, which minimizes technical noise. It is also recommended that stable isotopic labeling be adopted for experiments where quantitation is targeted at a few low- to medium-abundance proteins, as well as to methods that involve the separation of intact proteins.

When an analytical task such as the quantification of a set of known analytes within a complex mixture is to be performed, another label-free quantification method, termed selected reaction monitoring (SRM), may be useful. In SRM, an analyte in question is identified and quantified by a well-defined "transition," as a pairing of the precursor and its most intense fragment. This experiment can be realized in straightforward fashion in a triple-quadrupole mass analyzer where the first quadrupole measures the precursor, the second quadrupole measures the fragments, and the third records the fragments. By default, the instrument will be tuned to fragment only one precursor of a certain, predefined mass, and consequently the intensity of the defined fragment, as specified by the SRM transition, will reflect the quantity of the analyte in the mixture. When several precursors or several transitions are monitored in the same setting, such an experiment is termed multiple reaction monitoring (MRM) [83]. Both, SRMs and MRMs, performed on triple-quadrupole instruments, are often used in biomarker research and clinical diagnostics.

2.6
Quantitative Analysis of Protein–Protein Interactions

Stable isotopic labeling and label-free approaches have each been adapted for the analysis of protein complexes. Typically, the quantitative information being sought in an analysis of protein–protein interactions is: (i) the stoichiometry of a protein complex; (ii) changes in the stoichiometry and composition under different conditions; and (iii) the probability of an identified protein being a subunit of a purified complex versus being a contaminant. The quantitative proteomics approaches discussed in the previous section, namely QconCAT, isotopic labeling, and spectral counting, can all be applied to the analysis of protein complex composition and stoichiometry.

Because the copurification of nonspecific interactors is almost unavoidable in affinity pulldowns, major problems of contamination and the establishment of true subunits frequently arise when analyzing protein complexes. A further problem arises due to a need to distinguish between contaminants and transient interactions. In an attempt to address the problem of contamination, Tackett *et al.* developed an isotopic differentiation of interactions as random or targeted (i-DIRT) [84], in which the protein affinity pulldown is compared to a nonspecific control using stable isotopic labeling. In other words, the i-DIRT method allows a distinction to be made between specific and nonspecific protein–protein interactions.

As an example of a label-free approach, Sardiu *et al.* employed statistical methods to reconstruct protein complexes from the NSAF data [85]. In this case, probabilistic methods were used to estimate the likelihood of different proteins being subunits of a given protein complex, using affinity pulldowns, where multiple proteins that were known to be subunits of the protein complex were used as baits.

The analysis of weak and transient protein–protein interactions is a difficult problem, and chemical crosslinking is the first point to come to mind as a possible solution. Chemical crosslinking is discussed in detail in Sect. 3.4.

3
Specific Examples of Applications

3.1
Global Proteome Sampling

The goal of global proteome sampling is the simultaneous identification of proteins in a cell or tissue at a given condition. Data obtained in such experiments can be further used to answer more specific biological questions, such as differences between healthy and pathological states. Typically, the proteins are identified using MS and grouped into functional categories.

3.1.1 Global Proteome Sampling Based on 2D PAGE

Global proteome analysis using 2D PAGE method is difficult because each spot needs to be picked and identified individually, which greatly increases the time and costs of the analysis. It was noted earlier that 2D PAGE separation suffers from biases, with certain classes of proteins (e.g., hydrophobic or high molecular weight) being difficult to detect.

Despite these limitations, 2D PAGE has one important advantage over other methods, in that it can easily resolve protein isoforms. In their analysis of the *Hemophilus influenzae* proteome, Langen et al. used several techniques to maximize the 2D PAGE performance [67]. To increase proteome coverage, immobilized pH gradient strips were used that covered several pH regions, while to visualize low-copy-number proteins a series of protein extractions (e.g., heparin chromatography, chromato-focusing, hydrophobic interaction chromatography) were performed. In order to detect cell envelope-bound proteins, immobilized pH gradient strips were used to provide a good illustration of the capabilities of 2D PAGE in global proteome sampling-type experiments. An analysis of the mouse brain proteome illustrated these capabilities [86] when, using 2D PAGE, a comparative analysis was performed of two distantly related mouse strains, *Mus musculus* c057Bl/6 (B6) and *Mus spretus* (SPR). In this case, about 8700 proteins from the cytosolic fraction of the brain proteome were compared between the two species, whereby 2D PAGE analyses of the B6 and SPR strains, and also of F_1 (B6 × SPR) and B_1 (F_1 × SPR) hybrids, a total of 1324 species-specific polymorphisms was detected. Among these polymorphisms, 466 proteins were identified using MALDI-TOF/MS with peptide mass fingerprinting. In order to detect the polymorphisms, the authors considered variations in electrophoretic mobility, spot intensity, and the number of different isoforms corresponding to one protein. Additionally, through analyses of the F_1 and B_1 generations, it was established which polymorphisms were genetically dominant. The key feature that enabled this comprehensive study to be completed was the high quality of the 2D PAGE analysis, with the need for a large-gel 2D PAGE, isoelectric focusing (IEF) gel incubation and a large (46 × 30 cm) format. When implemented in this way, 2D PAGE provided both high resolution and high sensitivity, with more than 10 000 protein spots from mouse tissues being visualized simultaneously.

3.1.2 Global Proteome Sampling Based on Multidimensional LC

When it is necessary to catalog proteins present in a cell or an organism in a given environmental context, the multidimensional LC separation of peptide mixtures, followed by MS/MS, is the most convenient method to use. Whilst this approach is less effective at determining protein isoforms and PTMs than 2D PAGE, the biases in proteome coverage are greatly reduced.

When Florens et al. conducted proteomics studies of the life cycle of the human pathogen *Plasmodium falciparum* (malaria) life cycle [46], they identified 2415 proteins and assigned them to functional groups at four stages of the cycle (sporozoites, merozoites, trophozoites, and gametocytes). The sporozoite is the form in which *P. falciparum* is injected by the mosquito, the merozoite is the form that invades erythrocytes, the trophozoite is the form that multiplies in the erythrocytes, and the gametocyte is the sexual stage of parasite's life cycle. The analysis was performed using MudPIT, and showed that approximately 50% of the sporozoite proteins were unique to that stage, while about 25% were shared with other stages. The trophozoites, merozoites and gametocytes each had 20–30% unique proteins, while 40–60% of their proteins were shared. Only 6% of all identified proteins were shared between all four stages, these being mainly histones, ribosomal proteins, and transcription factors. Among the 2415 proteins identified, 51% had been previously annotated as hypothetical.

When Koller et al. [45] used both 2D PAGE and MudPIT to analyze the *Oryza sativa* (rice) proteome, the analyses were performed on protein extracts from leaf, root, and seed tissue. The aim of this study was to determine the tissue-specific expression of proteins, and 2D PAGE separation followed by MS/MS yielded 556 unique protein identifications that comprised 348 proteins from the leaf, 199 from the root, and 152 from the seed. The MudPIT analysis resulted in a significantly larger coverage, with 2363 total proteins, with 867 from leaf, 1292 from root, and 822 from seed. A total of 165 proteins was uniquely detected by 2D PAGE, whereas 1972 proteins were uniquely detected by MudPIT. The authors next searched the nonredundant protein database by BLAST and grouped the identified proteins into functional categories. The largest category (32.8%) included proteins that had no homology to the predicted proteins, while proteins classified as "involved in metabolism" comprised 20.8% of all identified proteins. Of the 2528 proteins detected, 189 were shared among all three tissues; these included housekeeping proteins involved in transcription, mRNA biosynthesis, translation, and protein degradation. However, most of the proteins had a tissue-specific expression, with 622 specific to leaf, 862 specific to root, and 512 specific to seeds.

In order to characterize the proteome of *S. cerevisiae* mitochondria, Sickmann et al. combined four separation methods: IEF-incubated 2D PAGE; digestion with four different proteases, followed by multidimensional LC/MS/MS; SDS–PAGE combined with multidimensional LC/MS/MS; and the treatment of mitochondria with trypsin, followed by SDS–PAGE or HPLC and MS/MS [87]. Subsequently, a total of 750 mitochondrial proteins was identified which, when classified into functional categories, showed 24.9% to be of unknown function, 24.9% to be involved in genome maintenance and gene expression, and 14.1% to be involved in energy metabolism. The remainder of the identified proteins were involved in metabolism, transport, and cell rescue.

In another example of organelle proteomics, using the LTQ-Orbitrap platform, Zybailov et al. performed a proteomics analysis of 10 independent chloroplast preparations, identified 1325 unique proteins, and calculated abundance levels of 550 chloroplast stromal proteins [88]. In addition, N-terminal modification status and chloroplast targeting peptide cleavage data for the chloroplast proteins were obtained.

3.1.3 Single Cell Proteomics

One of the central questions of systems cell biology is the physiological impact of cell-to-cell variability, and if there is any regulation of the expression noise. It could be appreciated how such cellular heterogeneity may become important during drug development, if a drug target's accumulation levels were to vary dramatically from cell to cell. A good example of this is response to antibiotics where, at certain doses, many bacteria die but some are resistant. Similarly, in cancer drug therapy there is a well-known phenomenon when otherwise identical cancer cells respond differently to chemotherapy drugs or radiation. It is therefore desirable to develop methods that allow for the quantification of protein accumulation in an individual cell.

Although, at present, flow cytometry/microfluidics-based methods are available that can be used to quantify one or several proteins at a time in individual cells [89, 90], MS-based technologies are not yet applicable for single cells, and are certainly not suited to the whole-proteome level. So far, what "single-cell proteomics" commonly refers to is the analysis of small groups of cells (hundreds to thousands), prepared from tissues by laser-capture microdissection or other sorting methods. These methods, especially those developed for application to cancer stem cells, have been reviewed elsewhere (see Ref. [91]).

As an example of a proteomics study performed on a small group of cells, Abiko et al. investigated rice egg and sperm cells and obtained a proteomic profile on as few as 50 cells in some samples [92]. For this, a one-dimensional gel separation of miniscule protein amounts (~60 ng) was performed that yielded a reliable identification, with 102 proteins being preferentially expressed in the egg and 77 in sperm, among 2138 in the egg and 2179 in sperm of the total expressed proteins identified. In another report, MALDI-MS was used in combination with iTRAQ reagents to analyze neuropeptides in single neurons [93]. A good example of using microfluidic cell lysis and capillary electrophoretic separation, combined with MS, was described by Mellors et al., with a high-throughput detection of hemoglobin in individual erythrocytes [94].

3.2
Analysis of Protein Modifications by Mass Spectrometry

In living organisms, protein activity is regulated in part by covalent modifications, which occur either co- or post-translationally. The identification of types of modifications and their locations is often a necessary requirement to understand the regulation and function of a given protein. To date, hundreds of protein modifications have been identified [95, 96], among which phosphorylation is perhaps the most important and widespread; indeed, about one-third of all proteins from mammalian genomes are thought to be phosphorylated.

A second functionally important modification is glycosylation, with glycosylated proteins being ubiquitous components of cellular surfaces where their oligosaccharide groups participate in a wide range of cell–cell recognition events. A comprehensive analysis of glycosylated proteins is more challenging than an analysis of other modifications, mainly because the structures of oligosaccharides vary.

Other commonly occurring modifications involved in protein regulation and function are disulfide bonds, acetylations, sumoylations, and ubiquitinations (see Table 2). Changes in protein length, either as a result of alternative splicing or protein truncations, may also be considered as protein modifications. Generally, it is difficult to identify protein truncations using methods that involve protein/peptide mixtures, and often purification down to individual proteins is required in such cases (e.g., using 2D PAGE). Currently available methods for the large-scale analysis of modified proteins can be grouped into two major classes: (i) those that use sample enrichment or chemical treatment prior to MS; and (ii) those that rely on MS data alone. Enrichment methods include affinity purification, chemical tagging followed by affinity purification, and immunoprecipitation, while MS methods of detection and the identification of modified peptides include neutral loss scan, precursor

Tab. 2 Common protein modifications.

Modification	Monoisotopic/average mass change
Phosphorylation[a]	+79.9663/79.9799
Acetylation	+42.0106/42.0373
N-Acetylglucosamine (GlcNAc)[a]	+203.0794/203.1950
Disulfide bond	−2.01565/2.0159
Methylation	+14.0157/14.0269
Hydroxylation	+15.9949/15.9994
Oxidation of methionine	+15.9949/15.9994
Ubiquitination of lysines	+114.0429/114.1040[b]
Deamidation of asparagines	+0.984/0.9847[c]
Pyroglutamate from N-terminal glutamic acid	−18.011/18.015[c]

a) Occurs on Tyr, Ser, Thr. Widespread throughout the proteome. Functions include protein regulation, signal transduction.
b) Mass change is due to Gly-Gly residue, which is left on ubiquinated lysines after trypsin digestion.
c) Frequently occurs during sample preparation.

ion scan, post-source decay, and others. Sometimes, it is important to obtain information on several types of modification at once, and in that computer programs such as Inspect and PepNovo (see Sect. 2) can be used to analyze the MS/MS data. Additionally, if mass changes introduced by modifications are known, the search for modified proteins can be made using SEQUEST or Mascot, with input parameters modified in accordance with the mass changes. Another option would be to use an error-tolerant search, which consecutively searches for modifications from a predefined modification list that might contain hundreds of modifications. Such analyses are implemented in Mascot and Peaks software platforms.

In general, it is difficult to analyze modified peptides on a background of nonmodified peptides; consequently, when it is clear what type of modification needs to be analyzed the fractionation steps that enrich that particular modification need to be introduced into the experimental scheme. Several recent illustrative examples of the analyses of phosphorylated, glycosylated, ubiquitinated and deamidated proteins are discussed in the following sections. The tools and techniques used for the analysis of phosphorylation are generally applicable to the analysis of many other modifications, such as methylation and acetylation, whereas the analysis of glycosylation poses additional analytical and instrumental challenges due to the variable structures of the oligosaccharide groups. Ubiquitination, while less frequent than phospho- and glycol-modifications, is important for protein degradation in proteasomes [97].

3.2.1 Phosphorylated Proteins

The main tools employed in the large-scale identification of phosphoproteins are enrichment by immobilized metal affinity chromatography (IMAC), chemical tagging, and immunoprecipitation by phosphor-specific antibodies. IMAC technology is based on methods developed by Andersson et al. [98, 99], and relies on the interaction of phosphate group with immobilized Fe^{3+} ions. Ficarro et al. used IMAC combined with LC/MS/MS to characterize the phosphorylated portion of the yeast proteome [100], and showed that the carboxylic acid interfered with IMAC purification and needed to be protected. Such protection was achieved by esterification with methanol in the presence of HCl, the phosphorylated tryptic peptides were then identified via SEQUEST. From the whole-cell lysate, Ficarro et al. detected more than 1000 phosphopeptides, from which 383 sites of phosphorylation were determined.

A potential improvement to this analysis would be to use other proteinases in parallel with trypsin to increase the sequence coverage. An enrichment technique that could also complement IMAC is immunoprecipitation, using antibodies that could bind to any protein containing phosphorylated residues. When Pandey et al. used phosphotyrosine immunoprecipitation to study phosphorylation in HeLa cells in response to epidermal growth factor (EGF), the phosphopeptides were immunoprecipitated from untreated and EGF-treated cell lysates and resolved by electrophoresis. Individual gel bands were then excised and studied using MALDI MS and ESI-MS/MS, allowing the authors to identify Vav-2 as a substrate of the EGF-receptor [101].

In another report by Ficarro et al., the authors used anti-phosphotyrosine immunoblots to study the capacitation (a cAMP-dependent process necessary for fertilization) of human sperm [102]. For this, a comparative analysis was first performed of capacitated versus noncapacitated sperm, after which the sperm proteins were separated with 2D PAGE and Western-blotted with anti-phosphotyrosine antibodies. The next step was to excise and digest any spots that exhibited phosphorylation, followed by IMAC to enrich for phosphopeptides with subsequent MS/MS analysis. Ultimately, the authors were able to pinpoint several proteins that would undergo phosphorylation upon sperm capacitation, and also used differential isotopic labeling to quantify such phosphorylation. In this case, the labeling was achieved at the stage of protecting the carboxy groups prior to IMAC, by treatment of the peptide mixtures from capacitated and noncapacitated digests with CD_3OD/DCl and CH_3OH/HCl, respectively. These two peptide mixtures were further combined in a one-to-one ratio and analyzed with IMAC/LC/MS/MS. Such quantitation demonstrated the presence of 20 unique peptides that exhibited different phosphorylation levels between capacitated and noncapacitated sperm.

A metabolic labeling strategy was first described to quantitate changes in phosphorylation [64], whereby cells from two batches having potentially different levels of phosphorylated proteins were metabolically labeled with ^{14}N and ^{15}N. The cells were then lysed, and the target proteins purified, digested, and analyzed using MS. Changes in peak intensities that corresponded to modified and unmodified peptides from the two conditions provided a quantitation of phosphorylation. Another means of quantifying phosphorylation levels was to use a modified ICAT strategy whereby the phosphate groups in phosphor-peptides derived from two different conditions were chemically replaced with either labeled or unlabeled tags [103]. The tagging involved the following steps: (i) a beta-elimination of the phosphate groups; (ii) the addition of 1,2-ethanedithiol containing either four hydrogens (EDT-D_0) or four deuteriums (EDT-D_4); and (iii) biotinylation of the EDT group using (+)-biotinyl-iodoacetamidyl-3,6-dioxaoctanediamine. The tagged peptides were further affinity-purified using an avidin column and analyzed with LC/MS.

During recent years, titanium oxide enrichment chromatography has emerged as an alternative to IMAC and was quickly adopted for phosphoproteomics studies [104]. Wang et al. compared the effectiveness of three phosphogroup-enrichment strategies, namely IMAC, titanium oxide, and calcium phosphate precipitation, in the analysis of phosphoproteins in androgen-repressed human prostate cancer cells [105]. In this case, a hybrid LTQ/FTCIR instrument was used, with calcium phosphate precipitation followed by a titanium oxide column being the most sensitive method, leading to the identification of 385 phosphoproteins.

3.2.2 Glycosylated Proteins

The importance of protein glycosylation, especially during cell–cell communication in multicellular organisms, is often acknowledged by using the terms glycobiology and glycomics [106]. Owing to a wide range of possible polysaccharide structures, the analysis of glycosylation is less straightforward than an analysis of other PTMs. Although, currently, no satisfactory methods are available for a global, proteome-wide analysis of all glycoprotein forms, it is possible to characterize glycoproteins with glycogroups of constant structure. It is also possible to globally map glycosylation sites. An example of this would be to consider a broad research question, such as the identification and characterization of glycosylated proteins in a given biological system. In such a case, a hypothetical analysis could include the following steps: (i) proteolytic digestion; (ii) the enrichment of glycopeptides; (iii) the identification of glycopeptides by MS and MS/MS; and (iv) structure determination of the polysaccharide groups using MS^n. Alternatively, the proteins could be separated using 2D PAGE, followed by glyco-specific staining to identify any spots of interest. During the MS phase of the analyses it could also be useful to separate any constant glyco structures from any variable structures, as well as N-linked structures from their O-linked counterparts [107].

One problem in the analysis of glycopeptides is that the glycogroups are very labile, which in turn leads to a reduction in peptide fragmentation in CID reactions, making sequencing by MS/MS more difficult. An alternative approach would be to chemically (e.g., with β-elimination of O-linked-oligosaccharides) or enzymatically (e.g., with N-glycosidase F) remove the glycogroups prior to analysis and to use chemical tags. In eukaryotes, the most widespread constant type of glycosylation is O-linked N-acetylglucosamine (O-GlcNAc), which is found on many nuclear and cytoplasmic proteins [108–110]. The glycosylation of serine and threonine residues by O-GlcNAc is believed to compete with and complement phosphorylation in mediating protein–protein interactions [111]. Those proteins which are glycosylated by O-GlcNAc include RNA polymerase II, transcription factors, chromatin-associated proteins, nuclear pore proteins, proto-oncogenes, tumor suppressors, and proteins involved in translation [111]. Because O-GlcNAc and phospho groups modify essentially the same amino acids, it is of special interest to establish methods that can be used to characterize these modifications simultaneously. With this in mind, Wells *et al.* developed a method based on beta-elimination followed by Michael addition (BEMAD) of dithiothreitol (DTT) or biotinepentylamine (BAP) [110]. This method relies on the fact that O-GlcNAc groups are more prone to elimination than phosphate groups. With the correct conditions of elimination, O-GlcNAc can be tagged selectively. The DTT and BAP tags also allow enrichment by affinity chromatography and are stable in MS/MS fragmentation, thus allowing the identification of any modified sites. The authors first tested this method on synthetic peptides, and then performed an analysis of several biological samples, including Synapsin I from rat brain and nuclear pore complex (NPC). In Synapsin I, three novel O-GlcNAc sites, as well as three previously known sites, were mapped, thus validating the method. In the NPC,

BEMAD also mapped novel O-GlcNAc sites in the LaminB receptor and Nup155. By using BEMAD, along with modification-specific antibodies and enzymes, the authors were able to distinguish between O-GlcNAc- and phosphopeptides. During the CID reactions, oligosaccharide moieties fragment mainly at glycosidic bonds [112, 113], which potentially allows discernment of the primary structure of an oligosaccharide in MS^n experiments. While the technology for a large-scale structural analysis of glycoforms has not yet matured, an oligosaccharide structure determination is certainly possible for individually isolated glycopeptides. Notable examples include the characterization of lipo-oligosaccharides from *H. influenza* [114] and *Neisseria gonorrhoeae* [115].

3.2.3 Ubiquitinated Proteins

The degradation of proteins in living organisms is a complex, highly regulated process, which plays important roles in many cellular pathways. The first step in protein degradation is the attachment of ubiquitin moieties to lysines of the substrate; the second step is proteolysis of the tagged protein by the proteasome [97]. Peng et al. described a systematic approach based on LC/LC/MS/MS to analyze protein ubiquitination in yeast [116], whereby His-tagged ubiquitin was expressed in *S. cerevisiae* cells, followed by purification over a Ni-chelating resin. Denaturing conditions were used at the enrichment step in order to minimize the copurification of proteins that were not ubiquitinated but rather formed complexes with ubiquitin. The enriched ubiquitinated proteins were digested with trypsin and analyzed using SCX/RP/MS/MS, followed by identification with SEQUEST. The tryptic digestion of ubiquitinated peptides resulted in glycine–glycine fragments at the sites of modification, and the corresponding increase in mass (by 114 Da) of the modified lysine residues allowed localization of the sites of ubiquitination. As a result, the authors identified 110 ubiquitination sites present in 72 ubiquitinated proteins. Some of the known ubiquitinations were not detected, however, most likely due to the fact that the method was biased towards more abundant species. Indeed, a precise localization of a PTM via SEQUEST requires a high sequence coverage. Consequently, in their analysis the authors identified 1075 proteins in total after Ni-resin purification, but were able to confirm ubiquitination in only 10% of them.

3.2.4 Deamidation of Asparagines and Glutamines

The deamidation of asparaginyl and glutaminyl residues into respective aspartyl and glutamyl residues is a widespread nonenzymatic modification, which can occur as a part of protein aging [117, 118]. A number of examples have been identified where deamidation serves as a molecular timer, or a counter, to target over-used protein towards degradation [119–121]. Aspartate is similar to phosphoserine in both volume and electrostatic properties (both have a negative charge at physiological pH), and hence deamidation is frequently compared to phosphorylation. Interestingly, during the deamidation of asparaginyls, iso-aspartatyl is a preferred product, which is converted to aspartyl by iso-aspartyl methyltransferase, a highly conserved enzyme [122–124]. Notably, upon deamidation a negative charge is introduced into the protein because the conversion of aspartyl back to asparaginyls never occurs. A thorough analysis of deamidation, distinguishing between the different deamidation products, is a major challenge for MS-based proteomics, due mainly to the fact that the mass shift is rather small and is easily confused with ^{13}C isotopic peaks. Either high-resolution instruments that use FTCIR, or Orbitrap analyzers, are required for the analysis of deamidation. ECD or ETD fragmentation has been used to distinguish iso-aspartate from aspartate, using characteristic ions in the fragment spectra [125–127].

3.2.5 Histone Modifications

Histones, the major protein constituents of chromatin, are highly heterogeneous in their PTMs. Indeed, various histone PTMs are responsible for the positive and negative regulation of gene expression, and include phosphorylation, methylation, acetylation, and ubiquitination, among others. The MS-based approaches are highly effective in analyzing histone heterogeneity, and include top-down, bottom-up, and mid-down (limited digestion) methods. A recent review by Zee *et al.* provided a comprehensive account of the quantitative MS-based methods for histone PTM analysis [128]. For a simultaneous analysis of the different types of histone modifications, the separation of individual proteins by LC and subsequent top-down or mid-down characterization is the preferred method, as it allows for a better definition of a set of PTMs for a given histone isoform.

Recently, novel methods of targeted chromatin enrichment at specific genomics locations have been developed. Excellent examples are the two reports by Byrum *et al.* [129, 130] who, in 2012, introduced chromatin affinity purification mass spectrometry (ChAP-MS) and demonstrated its proof-of-concept by using the GAL1 promoter under conditions of positive or negative transcriptional regulation. The key idea behind ChAP-MS is to introduce the LexA-binding DNA sequence into the genomic position of interest, after which the chromatin region of interest is purified via affinity pulldown using the LexA–Protein A fusion. In order to account for nonspecific binders, the authors used a comparison with the control non-LexA strain, using stable isotopic labeling, similar to the i-DIRT method. In 2013, the same group further enhanced the utility of ChAP-MS by using a modified transcription activator-like (TAL) protein instead of LexA fused to Protein A. The TAL protein contains DNA-binding motifs which can be mutated to bind any given 18 nt stretch of DNA. The use of TAL eliminates the need to introduce a LexA-binding DNA sequence, thereby simplifying the analysis (TAL-ChAP-MS). As a method, ChAP-MS provides a significant advance as it allows a given histone PTM pattern to be related to a specific genomics context, and also provides a functional insight into various PTM combinations. In fact, Byrum *et al.* identified numerous histone methylations and acetylations in the vicinity of the GAL1 region, correlating with either positive or negative regulation.

3.3 Analysis of Protein–Protein Interactions by Mass Spectrometry

In a cell, proteins exert their functions through interactions with other proteins. Proteomics methods developed during the past few years can be applied directly to the analysis of protein–protein interactions and protein complexes. Protein–protein interactions can be studied either on the level of individual protein complexes, or on the level of the whole proteome. Examples of different types of analysis include the possibility of predicting interactions from amino acid sequences, focused mass spectrometric analyses of individual multiprotein complexes such as yeast ribosome, SAGA-like complex (SLIK), Pol II preinitiation complex (PIC), NPC, and proteome-wide analyses of protein–protein interactions. In addition, quantitative proteomic methods may be used to analyze the dynamics of protein complexes.

3.3.1 Computational Methods of Protein–Protein Interaction Prediction

To some degree, protein–protein interactions can be deduced indirectly from the amino acid sequences [131], and in pursuit of this suggestion Bock and Gough created a learning algorithm that was trained to recognize and predict protein–protein interactions. The training was based on experimentally known interactions from a variety of organisms, and as an outcome a decision function was constructed that was statistically evaluated using unseen test data. As a result, on average, about 80% of the interactions were predicted accurately from the unseen datasets. Clearly, computational methods alone do not provide exhaustive descriptions, and the results obtained need to be validated experimentally. Nevertheless, the interaction datasets obtained *in silico* provide useful reference points for experimental types of analyses.

3.3.2 Yeast Two-Hybrid Arrays

One of the most common approaches to analyze protein–protein interactions is the yeast two-hybrid (Y2H) screen, which allows the detection of protein–protein interactions in the yeast nucleus [132]. This is a well-developed technique that can be easily optimized for a high-throughput analysis [133]. When the Y2H screen was successfully applied to the mapping of a large-scale protein interaction network in the yeast *S. cerevisiae*, the Y2H was shown to have a good resolution and to detect weak and transient interactions [134, 135]. Unfortunately, the Y2H method detects only binary interactions between proteins, and cannot be used to study transcriptional activators. Another drawback of the Y2H is that interactions occur in the nucleus, so that for many proteins the interactions take place outside of their native environment.

3.3.3 Direct Analysis of Large Protein Complexes; Composition of the Yeast Ribosome

Link *et al.* employed multidimensional LC coupled with MS/MS to study the composition of the yeast ribosome [42], the method involving a direct analysis of large protein complexes, with a wide range of applications. In this case, the ribosomes were first purified using sucrose gradients and then denatured. The ribosomal RNA was then removed, the ribosomes were digested enzymatically, and the digests loaded onto a 2D separation column consisting of the SCX and RP dimensions. After separation, the peptides were eluted directly into the mass spectrometer for the MS/MS analysis. Finally, the SEQUEST algorithm was used to search nucleotide databases and to match the peptide fragmentation patterns. The authors demonstrated the high-throughput capacity of this approach, with more than 100 proteins being identified in a single experiment, and new protein components of yeast and human 40S subunits being discovered as a result.

3.3.4 Analysis of Multisubunit Protein Complexes Involved in Ubiquitin-Dependent Proteolysis by Mass Spectrometry

In a number of reports, Deshaies *et al.* used MS-based strategies to characterize the composition of various protein complexes involved in proteolysis [136]. For this, sequential epitope tagging, affinity purification, and mass spectrometry (SEAM) were used to study the regulation and function of SCF ubiquitin ligases. In applying this method, the SCF subunits Skp1 and Cul1 were C-terminally tagged with a Myc epitope, after which the cells expressing the tagged proteins were lysed and the soluble fractions affinity-purified, digested, and analyzed using MS. As a result, totals of 16 Skp1- and Cul1-interacting proteins were detected, several of which had not previously been identified, including Hrt1, Rav1, and Rav2. These new proteins were further subjected to SEAM, which in turn led to the identification of the new complex Rav1/Rav2/Skp1. Subsequently, the Rav1/Rav2/Skp1 complex was found to interact with the V1 component of V-ATPase, the vacuolar membrane ATPase.

In other studies using biochemical methods, it was further determined that Rav1/Rav2/Skp1 regulated the assembly of V-ATPase from V1 and V0 domains. The same group used MudPIT to identify proteins that would interact with the 26S proteasome [137]. As noted above, MudPIT allows the analysis of immunoprecipitated fractions without a gel-separation step; rather, the immunoprecipitated fractions are digested and the peptide mixtures separated in two dimensions (SCX and RP), followed by MS analysis. By using MudPIT, every known subunit within the affinity-purified 26S proteasome was identified, as well as one subunit that was not previously known. Additionally, a set of proteins which potentially interacted with the proteasome (PIP) was found. By using immunoblotting methods, six of these PIPs were further confirmed as being associated with the proteasome.

3.3.5 Proteomics of the Nuclear Pore Complex
MS-based protein identification is a useful tool that can efficiently determine the composition of a given multiprotein complex. However, this method by itself does not necessarily provide information on the complex's spatial architecture, and nor does it directly answer questions about the multiprotein complex function. Consequently, when such questions arise the mass spectrometric tools need to be complemented with other techniques, such as immunoblotting, immunofluorescence, and electron microscopy.

Several studies of NPCs (as discussed here) have provided good illustrations of these types of integrative strategy. During the early 1990s, several studies using 3D cryoelectron microscopy revealed the basic shape and architecture of NPCs in the *Xenopus* nucleus [138]. The NPCs were determined as proteinaceous structures situated in the double membrane of the nuclear envelope, with estimated sizes ranging from ~125 MDa (*Xenopus*) to 66 MDa (*Saccharomyces*). The NPCs have an eightfold rotational symmetry with the rotational axis normal to the nuclear envelope membrane, and a twofold mirror symmetry with the symmetry plane parallel to the nuclear envelope membrane.

Current research efforts are aimed at understanding the mechanism of the biological function of the NPCs, namely nucleotransport, and for this purpose it is useful to catalog all of the protein components of NPCs from different organisms. Ideally, this should lead to a testable hypothesis on how these components contribute to the overall structure and function of NPCs. Biological problems of this type can be addressed using proteomics methods, as was elegantly demonstrated for yeast [139] and mammalian [140] NPCs. When studying yeast, Rout *et al.* prepared a highly enriched fraction of NPC proteins, followed by separation on ceramic hydroxyapatite HPLC, which gave an efficient recovery of the loaded proteins. Reverse-phase TFA-HPLC was then used in parallel to resolve the low-molecular-mass proteins from the NPC fraction. The next step in the separation involved SDS–PAGE, with visualization of protein bands by copper staining. Subsequently, peptide mixtures were prepared from bands of interest via in-gel trypsin digestion, followed by analysis with MALDI-MS.

A peptide mass-matching method was used to search a nonredundant protein sequence database. Previously known nucleoporins were genomically tagged with a protein A (ProtA) epitope, which allowed further immunolocalization by fluorescence microscopy. Subsequently, the authors identified 29 nucleoporins and 11 transport factors and NPC-associated proteins, and also determined the stoichiometry and position of each of the nucleoporins found within the NPC by quantitative immunoblotting and immuno-electron microscopy. In the latter analysis, the ProtA-tagged proteins were labeled using gold-conjugated antibodies, which aided visualization. On the basis of the deciphered architecture of the NPC, Rout *et al.* proposed a model of nucleotransport called a Brownian affinity gating model, the core idea of which was that translocation through the NPC would occur via diffusion. Diffusive movements of the filamentous nucleoporins on the cytoplasmic face of the NPC would exclude macromolecules that do not bind to them; however, when binding does occur (as when a cargo molecule is associated with its transport factor) the residence time of the cargo at the NPC gate is increased, which in turn facilitates diffusion of the cargo into the nucleus. In a mammalian study, Cronshaw *et al.* enriched NPC fractions from rat liver nuclei by sequential solubilization, using electron microscopy, SDS–PAGE and immunoblotting at each step of the enrichment to confirm that the NPCs had remained intact. After enrichment, the NPCs were treated with a detergent, producing a mixture of monomeric nucleoporins in which the individual proteins were separated using C4 reverse-phase chromatography, followed by SDS–PAGE. Protein identification was performed using single-MS and MS/MS. In addition to the previously known 23 nucleoporins, Cronshaw *et al.* identified six novel nucleoporins and four proteins containing WD repeats. One of the latter WD-containing proteins was ALADIN, the gene of which is mutated in Allgrove syndrome.

The spatial organization of a protein complex can be assessed using the crosslinking method, whereby a protein complex is affinity-purified and then treated with a crosslinking agent that introduces new covalent bonds between the neighbor proteins. Any new bands that appear on SDS–PAGE following this treatment can be excised and identified using MS. If a pair of the proteins becomes crosslinked this usually means that the two proteins are located close to each other within the complex. A good example of this strategy is the study by Rappsilber *et al.*, in which a crosslinking/MS method was used to deduce the spatial composition of the six-member subcomplex Nup84p of the yeast NPC [141]. These authors emphasized the generic applicability of this approach, although one of the significant challenges in the application of this method was the choice of correct crosslinking reaction conditions. Usually, several different crosslinkers need to be screened before the correct degree of crosslinking is obtained; additionally, interactions between subunits hidden deep inside a complex may be inaccessible to crosslinkers. Because of these and other difficulties, the crosslinking method is limited to complexes of small sizes, and is difficult to apply on a broader scale. Nevertheless, it may prove useful in studies of conformational or compositional changes of individual protein complexes, when the specific structural states can be "frozen" through interaction with the crosslinking agents.

3.3.6 High-Throughput Analyses of Protein–Protein Interactions

In most cases, MS-based analytical schemes similar to those employed in the studies of individual complexes can be redesigned for use at the proteome-wide scale. A report by Gavin et al. was one of the first examples of MS-based proteome-wide analyses of the protein complexes [142]. For the large-scale isolation of protein complexes from yeast, Gavin et al. used a tandem affinity purification (TAP) method whereby a gene-specific fusion cassette, which contains a calmodulin-binding domain, a specific protease cleavage site, and a ProtA domain, is introduced at the C- or N-terminal of the yeast's open reading frames (ORFs) of interest [143]. Then, assuming that the expression of fusion proteins is maintained close to the natural level, the first affinity purification is performed when the fusion proteins, along with their interaction partners (so-called protein assemblies), are isolated from the cell extract by affinity selection on an immunoglobulin G (IgG) matrix. The bound proteins are then released by the addition of a protease. Finally, a second affinity purification is performed that involves incubation with calmodulin beads in the presence of calcium. The advantage of TAP over a standard epitope tagging approach is that it removes most of the nonspecific interactions. Of the 1548 yeast strains generated by Gavin et al., 1167 expressed the fusion proteins at detectable levels. After purification of the protein complexes using TAP, the complexes were subjected to an electrophoretic separation, followed by trypsin digestion and subsequent analysis with MALDI-MS. Overall, by MS analysis, Gavin et al. identified 1440 gene products (ca. 25% of the genome) from various organelles, although the identification of membrane proteins proved to be difficult, with only 40 membrane proteins purified successfully from a total of 293 detected. The authors then grouped the identified proteins into complexes by analyzing the overlaps in the composition of the pulled-down assemblies from 589 different bait proteins. Thus, a total of 245 purifications that corresponded to 98 known complexes from the yeast protein database (YPD) was reported, and another 242 purifications out of the 589 were assembled into 134 new complexes. In this way, it was possible to identify proteins with such low-abundance as 15 copies per cell, demonstrating the high sensitivity of the TAP method. Unfortunately, the reproducibility of the system was rather poor, with the probability of finding the same protein from the same bait in two purifications being estimated at about 70%. Another weakness of the TAP method derives from possibility of interference of the TAP tag with complex assembly and protein function. In fact, Gavin et al. found that when the essential genes were TAP-tagged, nonviable strains were obtained in about 20% of cases. The authors also reported a significant bias against proteins with molecular weight <15 kDa.

Another notable report of high-throughput protein complex identification was made by Ho et al. [144], where the bait proteins contained the Flag epitope tag and were overexpressed from *GAL1* or *tet* promoters. The protein assemblies were then isolated in a one-step immunoaffinity purification, followed by resolution on SDS–PAGE, digestion, and analysis using MS and/or MS/MS. The authors termed this method "high-throughput mass spectrometric protein complex identification" (HMS-PCI).

The immunoaffinity purification of complexes assembled around overexpressed baits should, in theory, generate more false-positives than would the TAP method, because of the nonphysiological concentrations of the baits. On the other hand, weak and transient interactions that would not be detected in the TAP method could be captured by HMS-PCI. In fact, Ho *et al.* were able to assess certain regulatory and signaling pathways by studying complexes pulled-down with phosphatases and kinases used as baits. As of today, none of the methods for mapping protein–protein interactions within a proteome is sufficiently comprehensive to provide full coverage, and consequently it is useful to compare datasets obtained with different approaches. In their study, von Mering *et al.* evaluated all available interaction datasets obtained in yeast [145], including the data from MS-based studies (as discussed above) as well as data from Y2H, correlated mRNA expression, synthetic lethality, and *in-silico* predictions. The evaluation was made through comparisons with the reference dataset (MIPS and YPD) and, as a result, the percentage coverage (fraction of the reference covered) and accuracy (fraction of data confirmed by the reference) were estimated for every method. According to the authors' analysis, the TAP method provides both a higher coverage and a higher accuracy than either Y2H or HMS-PCI. The analysis also showed that HMS-PCI was the least accurate method among the three.

In a series of reports, Bader and Hogue developed algorithmic approaches for identifying molecular complexes from datasets obtained in different interaction studies. By analyzing the combined data from TAP and HMS-PCI studies, a novel nucleolar complex of 148 proteins was found that included 39 proteins with unknown function. Further, a graph-theoretic clustering algorithm molecular complex detection (MCODE) was described that would allow the detection of dense regions (potential complexes) within the interaction networks. Most importantly, the authors showed that the MCODE algorithm was not affected by a high rate of false-positives in datasets from the high-throughput experiments.

To summarize, none of the current high-throughput experimental schemes provides sufficient coverage and accuracy, and therefore integrative approaches that take advantage of the different methods available are necessary. In addition, all of the discussed cases have involved cells in certain growth conditions. It is of special interest, however, to study the dynamics within protein interaction networks in response to environmental stimuli, in progression through the cell cycle, or in pathological states. Some of these questions can be addressed by applying quantitative proteomics techniques.

3.3.7 Quantitative Proteomics Methods in Studies of Protein Complexes

Methods of quantitative proteomics can be used to study dynamical changes in the abundance, composition, and activities of multiprotein complexes. A good example of this was reported by Ranish *et al.*, who assessed the composition of a large RNA polymerase preinitiation complex (PIC) using the ICAT method [146].

The ICAT approach, as employed by Ranish et al., consists of four major steps:

- The samples from two different conditions are labeled with either "light" or "heavy" tags.
- The labeled samples are combined together and digested enzymatically.
- The mixture is fractionated using SCX chromatography, after which the labeled peptides are isolated using avidin-affinity chromatography, followed by separation using reversed-phase microcapillary chromatography.
- The labeled peptides that are eluted from the reversed-phase column are analyzed using ESI-MS/MS.

In the MS analysis, the peak intensity ratios of the differentially labeled peptides on the ion chromatogram are related to the relative abundances of the corresponding proteins in the two different environments. The peptide identities are established in the MS/MS spectra, and the corresponding proteins identified by SEQUEST. With this quantitative ICAT method, the authors were able to distinguish between components of the PIC and the copurifying background proteins, some of which had higher abundances. Thus, the high analytical power of this approach could be demonstrated.

In their analysis, Ranish et al. identified a total of 326 proteins, 42% of which participated in the Pol II-mediated transcription. The authors also used the ICAT method to monitor changes in PIC composition in the presence or absence of the TATA-box binding protein (TBP). The latter is a transcription factor that binds to the TATA element and is required for functional PIC assembly. According to Ranish et al., most of the Pol II components are increased in abundance by a factor of at least 1.9 upon the addition of TBP, but several Pol II components showed no increase in abundance. In addition, a potentially new component of the PIC was discovered. A limitation of this approach, as noted by the authors, was that the use only of cysteine-specific tags omits any tryptic peptides that do not contain cysteines, and in this respect strategies that employ metabolic labeling or N-terminal labeling may be more promising. As an example, Blagoev et al. used SILAC to study EGF signaling [147], whereby control and EGF-stimulated HeLa cell populations were labeled with ^{12}C-arginine and ^{13}C-arginine, respectively, via metabolic incorporation. The combined cell lysates from these two conditions were affinity-purified, with the SH2 domain of the GSTSH2 fusion protein used as bait (the SH2 domain specifically binds phosphorylated EGF receptors). Protein complexes obtained via this purification were digested with trypsin and the peptide mixtures analyzed using MS. This led to the authors identifying 228 proteins, 28 of which were enriched upon EGF simulation.

A recent report by Olinares et al. was a good illustration of the use of Qcon-CAT to determine multisubunit complex stoichiometry [148]. These authors used QconCAT, followed by LTQ-Orbitrap MS, to determine the stoichiometry of a ClpP/R chroloplast protease complex that consisted of different catalytic (P) and noncatalytic (R) subunits organized into two heptameric rings. Defining the exact positioning of the P and R subunits within these rings is important to understand the ClpP/R function. As a result, the authors showed that one heptameric ring contained P subunits 3, 4, 5, and 6 in a $1:2:3:1$ ratio, while the other ring contained P subunit 1 and R subunits 1, 2, 3, and 4 in a $3:1:1:1:1$ ratio, which resulted in only three catalytic sites.

3.4 Large-Scale Chemical Crosslinking

With advances in MS instrumentation, and especially with the introduction of HCD fragmentation, crosslinking studies were made much more straightforward. In a review, Andrea Sinz described the power and bottlenecks of crosslinking coupled with MS [149]. In the case of a bottom-up approach, the size of protein under crosslinking analysis is theoretically unlimited, while the analysis requires femtomolar amounts of proteins and is very fast. Many bifunctional chemicals can be used as crosslinkers, and crosslinkers can also be designed for specific analysis. A large amount of data, a higher charge of the precursor ions, and an unavailability of efficient computational tools are the main bottlenecks of crosslinking analysis. However, different analytical strategies such as the enrichment of crosslinking species, cleavable crosslinkers, and isotope-labeling can reduce the complexities of the data and make the identification of crosslinking species much easier.

Crosslinking reactions are performed on proteins to incorporate crosslinkers into proteins, resulting in either intramolecular (within same protein molecule) or intermolecular (among different protein molecules) crosslinked species. In the bottom-up approach, the crosslinked proteins are digested enzymatically and then subjected to MS analysis, preferably using MALDI-TOFMS or LC/ESI-MS/MS. In the case of a top-down approach, the crosslinked proteins are not digested and are analyzed using MS, preferably ESI-FTICR. The crosslinked products are identified by analyzing the mass spectra using appropriate software.

Currently, amine-reactive crosslinkers are the most popular due to the high abundance of lysine in proteins and the ease of analysis. N-hydroxysuccinimide (NHS) esters, imidoesters, and carbodiimides are some of the available amine-reactive crosslinkers, while maleimides in the pH range of 6.5–7.5 can be used as sulfhydryl-reactive crosslinkers. Photoreactive crosslinkers are useful for crosslinking any amino acid indiscriminately (e.g., aryl azides, diarizines, and benzophenones), and are generally heterobifunctional with one amine-reactive or sulfhydryl-reactive end. Trifunctional crosslinkers have also been used in specific analyses, but their use is very limited.

As an example of large-scale *in-vivo* crosslinking, Weisbrod *et al.* used MS-labile reporter-encoded crosslinking agents and multistage MS/MS to identify protein–protein interaction networks from *E. coli* lysates. The method was referred to as a real-time crosslinked peptide identification strategy (ReACT), and used to identify 708 crosslinked peptide pairs, 40% of which were derived from known interactions [150]. A detailed review of other crosslinking approaches is available elsewhere [151].

3.4.1 Nomenclature of Crosslinking Products

Crosslinking reactions produce many different types of crosslinked peptides, and different research groups have used different terms and conventions to address them. Schilling *et al.* proposed a uniform nomenclature for crosslinked peptides [152], as shown in Fig. 12, where three distinct crosslinking scenarios involving only one crosslinker molecule are denoted as type 0, type 1, and type 2 crosslinking. Type 0 crosslinking involves only one peptide sequence, where one end of the crosslinker is modified but not attached to any amino acid. Type 0 has also been referred to as "dead-end," "decorated," "end-capped,"

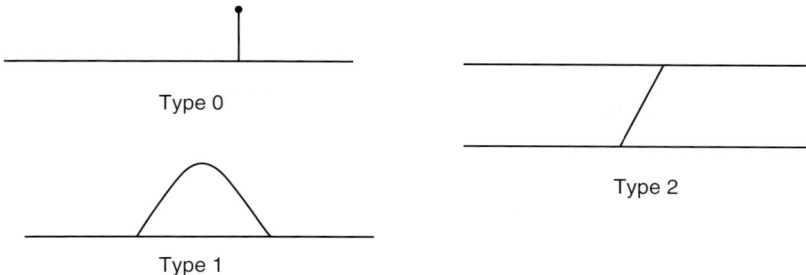

Fig. 12 Crosslinking nomenclature involving a single crosslinker.

"mono-link," and "single chain with a derivatized lysine." Internally crosslinked peptides or linking of two amino acids of the same peptide via a crosslinker is referred to as type 1 crosslinking, and is also known as loop-link. Crosslinking that involves two different peptide chains, where the crosslinker binds to amino acids of different peptide chains at either end, is termed type 2 crosslinking. Various combinations of type 0, type 1, and type 2 crosslinking are also evident, and their nomenclature is shown in Fig. 13. In type 2 crosslinking, the higher weight peptide chain is annotated as the α-chain and the lighter weight peptide chain as the β-chain.

3.4.2 Overcoming the Crosslinking Bottlenecks

The complexity of crosslinking data and the intricacies involved in its analysis have been addressed in many studies, either by developing new algorithms for data analysis or by implementing improved experimental conditions. One such way is to use ^{18}O-labeled water. For this, the crosslinked products are digested in ^{16}O-labeled water and ^{18}O-labeled water separately, and analyzed using MS. The crosslinked peptides

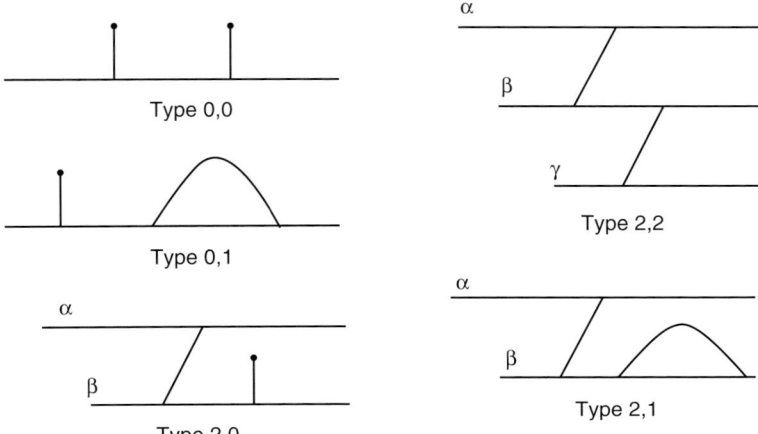

Fig. 13 Crosslinking nomenclature involving multiple crosslinkers.

can be characterized in the MS data by a shift of 8 Da, which is the result of two ^{18}O being incorporated at each C terminus, compared to one ^{18}O in the non-crosslinked peptide. El-shafey et al. demonstrated the use of ^{18}O labeling in crosslink analysis by analyzing homo-oligomeric complexes of glutathione-S-transferase (GST) [153].

Singh et al. employed a modification of the data to bypass the use of a custom analytical algorithm [154] by considering a crosslinked peptide as a single-chain peptide that had been modified with an unknown attachment at an unknown amino acid. In this case, the crosslinked products were enriched by acquiring the mass spectra at a high mass accuracy and a high-charge-state. The spectra were deconvoluted to a precursor charge state of 2 and a fragment ions charge state of 1. An open modification search was used to identify any modification ranging from 50 to 3000 Da. This pipeline was validated on a chemically crosslinked CYP2EI-b5 complex.

Nadeau et al. developed the program CrossSearch, which is capable of dealing with large data from crosslinked peptides derived from a large protein complex [155]. For this, FTICR was used as the mass spectrometric method and a database search strategy with CrossSearch to detect crosslinked peptides. xComb, developed by Panchaud et al., is another program that generates a list of all theoretical possible crosslinks from input protein sequences, which then can be used as queries for traditional database search engines [156]. StavroX, developed by Gotze et al. [157], and MeroX, developed by Muller et al. [158], are highly customizable and efficient tools for the analysis of crosslinked spectra. StavroX was developed to analyze traditional crosslinked peptides, while MeroX can be used also to analyze cleavable crosslinks. These two tools have a large array of settings available, and can handle many different types of crosslink data. Further information is available in the reviews of Singh et al. [159] and of Mayne and Patterton [160].

4
Conclusions

In this chapter, the progress made in the field of proteomics field during the past 15 years has been reviewed, and the conclusion reached that MA-based methods of proteome analysis have become an integral part of fundamental biological and biomedical research, as well as an important tool in clinical research and practice. Whilst biological proteomics is a mature technology, its development is expected to continue, both in terms of instrumentation and novel workflows and data analysis pipelines. It is possible to forecast that, during the near future, progress will be made especially in large-scale, top-down analysis, in single-cell proteomics, and in large-scale crosslinking methods.

References

1. Hochstrasser DF, Appel KD, Williams KL (1997) Proteome Research: New Frontiers in Functional Genomics (Principles and Practice). New York: Springer-Verlag.
2. Wasinger VC, Cordwell SJ, Cerpa-Poljak A, Yan JX, et al., (1995) Progress with gene-product mapping of the Mollicutes: *Mycoplasma genitalium. Electrophoresis*, **16**, 1090–1094.
3. Wilkins MR, Sanchez JC, Gooley AA, Appel RD, et al., (1996) Progress with proteome projects: why all proteins expressed by a genome should be identified and how to do it. *Biotechnol. Genet. Eng. Rev.*, **13**, 19–50.
4. Figeys D, (2003) Proteomics in 2002: a year of technical development and wide-ranging applications. *Anal. Chem.*, **75**, 2891–2905.

5. Koonin EV, (2001) Computational genomics. *Curr. Biol.*, **11**, R155-R158.
6. Pagani I, Liolios K, Jansson J, Chen IM, et al., (2012) The Genomes OnLine Database (GOLD) v.4: status of genomic and metagenomic projects and their associated metadata. *Nucleic Acids Res.*, **40**, D571-D579.
7. Jansen R, Bussemaker HJ, Gerstein M (2003) Revisiting the codon adaptation index from a whole-genome perspective: analyzing the relationship between gene expression and codon occurrence in yeast using a variety of models. *Nucleic Acids Res.*, **31**, 2242–2251.
8. Malone JH, Oliver B, (2011): Microarrays, deep sequencing and the true measure of the transcriptome. *BMC Biol.*, **9**, 34.
9. Vogel C, Marcotte EM (2012) Insights into the regulation of protein abundance from proteomic and transcriptomic analyses. *Nat. Rev. Genet.*, **13**, 227–232.
10. Maier T, Guell M, Serrano L (2009) Correlation of mRNA and protein in complex biological samples. *FEBS Lett.*, **583**, 3966–3973.
11. Watson JT (1997) Introduction to Mass Spectrometry. Philadelphia, PA: Lippincott-Raven.
12. Fenn JB, Mann M, Meng CK, Wong SF, et al., (1989) Electrospray ionization for mass spectrometry of large biomolecules. *Science*, **246**, 64–71.
13. Karas M, Hillenkamp F (1988) Laser desorption ionization of proteins with molecular masses exceeding 10000 daltons. *Anal. Chem.*, **60**, 2299–2301.
14. Siuzdak G (1994) The emergence of mass spectrometry in biochemical research. *Proc. Natl Acad. Sci. USA*, **91**, 11290–11297.
15. Bakhtiar R, Nelson RW (2000) Electrospray ionization and matrix-assisted laser desorption ionization mass spectrometry. Emerging technologies in biomedical sciences. *Biochem. Pharmacol.*, **59**, 891–905.
16. Chalmers MJ, Gaskell SJ (2000) Advances in mass spectrometry for proteome analysis. *Curr. Opin. Biotechnol.*, **11**, 384–390.
17. Aebersold R, Mann M (2003) Mass spectrometry-based proteomics. *Nature*, **422**, 198–207.
18. Laskin J, Heath BS, Roach PJ, Cazares L, et al., (2012) Tissue imaging using nanospray desorption electrospray ionization mass spectrometry. *Anal. Chem.*, **84**, 141–148.
19. Hu Q, Noll RJ, Li H, Makarov A, Hardman M, et al., (2005) The Orbitrap: a new mass spectrometer. *J. Mass Spectrom.*, **40**, 430–443.
20. Schwartz JC, Jardine I (1996) Quadrupole ion trap mass spectrometry. *Methods Enzymol.*, **270**, 552–586.
21. Schwartz JC, Senko MW, Syka JE (2002) A two-dimensional quadrupole ion trap mass spectrometer. *J. Am. Soc. Mass Spectrom.*, **13**, 659–669.
22. Moskovets E (2000) Optimization of the mass reflector parameters for direct ion extraction. *Rapid Commun. Mass Spectrom.*, **14**, 150–155.
23. Moskovets E, Karger BL (2003) Mass calibration of a matrix-assisted laser desorption/ionization time-of-flight mass spectrometer including the rise time of the delayed extraction pulse. *Rapid Commun. Mass Spectrom.*, **17**, 229–237.
24. Comisarow MB, Marshall AG (1996) The early development of Fourier transform ion cyclotron resonance (FT-ICR) spectroscopy. *J. Mass Spectrom.*, **31**, 581–585.
25. Zubarev RA, Makarov A (2013) Orbitrap mass spectrometry. *Anal. Chem.*, **85**, 5288–5296.
26. Williams DM, Pukala TL (2013) Novel insights into protein misfolding diseases revealed by ion mobility-mass spectrometry. *Mass Spectrom. Rev.*, **32**, 169–187.
27. Creese AJ, Smart J, Cooper HJ (2013) Large-scale analysis of peptide sequence variants: the case for high-field asymmetric waveform ion mobility spectrometry. *Anal. Chem.*, **85**, 4836–4843.
28. Yost RA, Boyd RK (1990) Tandem mass spectrometry: quadrupole and hybrid instruments. *Methods Enzymol.*, **193**, 154–200.
29. Smith RD, Loo JA, Edmonds CG, Baringa CJ, et al., (1990) New developments in biochemical mass spectrometry: electrospray ionization. *Anal. Chem.*, **62**, 882–899.
30. Makarov A, Denisov E, Lange O, Horning S. (2006) Dynamic range of mass accuracy in LTQ Orbitrap hybrid mass spectrometer. *J. Am. Soc. Mass Spectrom.* **17** (7), 977–982
31. Michalski A, Damoc E, Lange O, Denisov E, et al., (2012) Ultra high resolution linear

ion trap Orbitrap mass spectrometer (Orbitrap Elite) facilitates top down LC MS/MS and versatile peptide fragmentation modes. *Mol. Cell. Proteomics*, **11**, O111.013698.
32. Dreger M (2003) Subcellular proteomics. *Mass Spectrom. Rev.*, **22**, 27–56.
33. Lee YS, Mrksich M (2002) Protein chips: from concept to practice. *Trends Biotechnol.*, **20**, S14-S18.
34. Caffrey RE (2010) A review of experimental design best practices for proteomics based biomarker discovery: focus on SELDI-TOF. *Methods Mol. Biol.*, **641**, 167–183.
35. Reisenger, V., Eichacker, L.A. (2007) How to analyze protein complexes by 2D blue native SDS-PAGE. *Proteomics*, **7** (Suppl.1), 6–16.
36. Link AJ (2002) Multidimensional peptide separations in proteomics. *Trends Biotechnol.*, **20**, S8-S13.
37. Li Y, Compton PD, Tran JC, Ntai I, et al., (2014) Optimizing capillary electrophoresis for top-down proteomics of 30–80 kDa proteins. *Proteomics* **14** (10), 1158–1164.
38. Zhu G, Sun L, Keithley RB, Dovichi NJ (2013) Capillary isoelectric focusing-tandem mass spectrometry and reversed-phase liquid chromatography-tandem mass spectrometry for quantitative proteomic analysis of differentiating PC12 cells by eight-plex isobaric tags for relative and absolute quantification. *Anal. Chem.* **85**, 7221–7229.
39. Sun L, Knierman MD, Zhu G, Dovichi NJ (2013) Fast top-down intact protein characterization with capillary zone electrophoresis-electrospray ionization tandem mass spectrometry. *Anal. Chem.*, **85**, 5989–5995.
40. Wang T, Ma J, Wu S, Yuan H, et al., (2011) Integrated platform of capillary isoelectric focusing, trypsin immobilized enzyme microreactor and nanoreversed-phase liquid chromatography with mass spectrometry for online protein profiling. *Electrophoresis* **32**, 2848–2856.
41. Washburn MP, Ulaszek R, Deciu C, Schieltz DM, et al., (2002) Analysis of quantitative proteomic data generated via multidimensional protein identification technology. *Anal. Chem.*, **74**, 1650–1657.
42. Link AJ, Eng J, Schieltz DM, Carmack E, et al., (1999) Direct analysis of protein complexes using mass spectrometry. *Nat. Biotechnol.*, **17**, 676–682.
43. Washburn MP, Wolters D, Yates JR, III, (2001) Large-scale analysis of the yeast proteome by multidimensional protein identification technology. *Nat. Biotechnol.*, **19**, 242–247.
44. Eng JK, McCormack AL, Yates JR (1994) An approach to correlate tandem mass spectral data of peptides with amino acid sequences in a protein database. *J. Am. Soc. Mass Spectrom.*, **5**, 976–989.
45. Koller A, Washburn MP, Lange BM, Andon NL, et al., (2002) Proteomic survey of metabolic pathways in rice. *Proc. Natl Acad. Sci. USA*, **99**, 11969–11974.
46. Florens L, Washburn MP, Raine JD, Anthony RM, et al., (2002) A proteomic view of the *Plasmodium falciparum* life cycle. *Nature*, **419**, 520–526.
47. Webb KJ, Xu T, Park SK, Yates JR, III, (2013) Modified MuDPIT separation identified 4488 proteins in a system-wide analysis of quiescence in yeast. *J. Proteome Res.*, **12**, 2177–2184.
48. Imam-Sghiouar N, Laude-Lemaire I, Labas V, Pflieger D, et al., (2002) Subproteomics analysis of phosphorylated proteins: application to the study of B-lymphoblasts from a patient with Scott syndrome. *Proteomics*, **2**, 828–838.
49. Choudhary JS, Blackstock WP, Creasy DM, Cottrell JS (2001) Matching peptide mass spectra to EST and genomic DNA databases. *Trends Biotechnol.* **19** (10 Suppl.), S17-S22.
50. Pappin DJ (2003) Peptide mass fingerprinting using MALDI-TOF mass spectrometry. *Methods Mol. Biol.*, **211**, 211–219.
51. Conrads TP, Anderson GA, Veenstra TD, Pasa-Tolic L, et al., (2000) Utility of accurate mass tags for proteome-wide protein identification. *Anal. Chem.*, **72**, 3349–3354.
52. Conrads TP, Issaq HJ, Veenstra TD (2002) New tools for quantitative phosphoproteome analysis. *Biochem. Biophys. Res. Commun.*, **290**, 885–890.
53. Smith RD, Anderson GA, Lipton MS, Pasa-Tolic LS et al., (2002) An accurate mass tag strategy for quantitative and high-throughput proteome measurements. *Proteomics*, **2**, 513–523.
54. Shteynberg D, Nesvizhskii AI, Moritz RL, Deutsch EW (2013) Combining results of multiple search engines in proteomics. *Mol. Cell. Proteomics*, **12**, 2383–2393.

55. Vaudel M, Barsnes H, Berven FS, Sickmann A, et al., (2011) SearchGUI: an open-source graphical user interface for simultaneous OMSSA and X!Tandem searches. *Proteomics*, **11**, 996–999.
56. Frank A, Pevzner P (2005) PepNovo: de novo peptide sequencing via probabilistic network modeling. *Anal. Chem.*, **77**, 964–973.
57. Jeong K, Kim S, Pevzner PA (2013) UniNovo: a universal tool for de novo peptide sequencing. *Bioinformatics*, **29**, 1953–1962.
58. Ma B, Zhang K, Hendrie C, Liang C, et al., (2003) PEAKS: powerful software for peptide de novo sequencing by tandem mass spectrometry. *Rapid Commun. Mass Spectrom.*, **17**, 2337–2342.
59. Sunyaev S, Liska AJ, Golod A, Shevchenko A, et al., (2003) MultiTag: multiple error-tolerant sequence tag search for the sequence-similarity identification of proteins by mass spectrometry. *Anal. Chem.*, **75**, 1307–1315.
60. Liska AJ, Popov AV, Sunyaev S, Coughlin P, et al., (2004) Homology-based functional proteomics by mass spectrometry: application to the *Xenopus* microtubule-associated proteome. *Proteomics*, **4**, 2707–2721.
61. Ellis SR, Bruinen AL, Heeren RM (2013) A critical evaluation of the current state-of-the-art in quantitative imaging mass spectrometry. *Anal. Bioanal. Chem.* **406** (5), 1275–1289
62. Dotsch V, Wagner G (1998) New approaches to structure determination by NMR spectroscopy. *Curr. Opin. Struct. Biol.*, **8**, 619–623.
63. Hansen AP, Petros AM, Mazar AP, Pederson TM, et al., (1992) A practical method for uniform isotopic labeling of recombinant proteins in mammalian cells. *Biochemistry*, **31**, 12713–12718.
64. Oda Y, Huang K, Cross FR, Cowburn D, et al., (1999) Accurate quantitation of protein expression and site-specific phosphorylation. *Proc. Natl Acad. Sci. USA*, **96**, 6591–6596.
65. Berger SJ, Lee SW, Anderson GA, Pasa-Tolic L, et al., (2002) High-throughput global peptide proteomic analysis by combining stable isotope amino acid labeling and data-dependent multiplexed-MS/MS. *Anal. Chem.*, **74**, 4994–5000.
66. Ong SE, Blagoev B, Kratchmarova I, Kristensen DB, et al., (2002) Stable isotope labeling by amino acids in cell culture, SILAC, as a simple and accurate approach to expression proteomics. *Mol. Cell. Proteomics*, **1**, 376–386.
67. Lahm HW, Langen H (2000) Mass spectrometry: a tool for the identification of proteins separated by gels. *Electrophoresis*, **21**, 2105–2114.
68. Conrads TP, Alving K, Veenstra TD, Belov ME, et al., (2001) Quantitative analysis of bacterial and mammalian proteomes using a combination of cysteine affinity tags and ^{15}N-metabolic labeling. *Anal. Chem.*, **73**, 2132–2139.
69. Lipton MS, Pasa-Tolic' L, Anderson GA, Anderson DJ, et al., (2002) Global analysis of the *Deinococcus radiodurans* proteome by using accurate mass tags. *Proc. Natl Acad. Sci. USA*, **99**, 11049–11054.
70. Pasa-Tolic L, Harkewicz R, Anderson GA, Tolic N, et al., (2002) Increased proteome coverage for quantitative peptide abundance measurements based upon high performance separations and DREAMS FTICR mass spectrometry. *J. Am. Soc. Mass Spectrom.*, **13**, 954–963.
71. Jiang H, English AM (2002) Quantitative analysis of the yeast proteome by incorporation of isotopically labeled leucine. *J. Proteome Res.*, **1**, 345–350.
72. Ideker T, Thorsson V, Ranish JA, Christmas R, et al., (2001) Integrated genomic and proteomic analyses of a systematically perturbed metabolic network. *Science*, **292**, 929–934.
73. Griffin TJ, Gygi SP, Ideker T, Rist B, et al., (2002) Complementary profiling of gene expression at the transcriptome and proteome levels in *Saccharomyces cerevisiae*. *Mol. Cell. Proteomics*, **1**, 323–333.
74. Han, D.K., Eng, J., Zhou, H., Aebersold, R. (2001) Quantitative profiling of differentiation-induced microsomal proteins using isotope-coded affinity tags and mass spectrometry. *Nat. Biotechnol.*, **19** (10), 946–951.
75. Hunter, T.C., Washburn, M.P. (2003) The integration of chromatography and peptide mass modification for quantitative proteomics. *J. Liq. Chromatogr. Related Technol.*, **26**, 2285–2301

76. Ross PL, Huang YN, Marchese JN, Williamson B, et al., (2004) Multiplexed protein quantitation in *Saccharomyces cerevisiae* using amine-reactive isobaric tagging reagents. *Mol. Cell. Proteomics*, **3**, 1154–1169.
77. Mirgorodskaya OA, Kozmin YP, Titov MI, Korner R, et al., (2000) Quantitation of peptides and proteins by matrix-assisted laser desorption/ionization mass spectrometry using (18)O-labeled internal standards. *Rapid Commun. Mass Spectrom.*, **14**, 1226–1232.
78. Reynolds KJ, Yao X, Fenselau C (2002) Proteolytic ^{18}O labeling for comparative proteomics: evaluation of endoprotease Glu-C as the catalytic agent. *J. Proteome Res.*, **1**, 27–33.
79. Pratt JM, Simpson DM, Doherty MK, Rivers J, et al., (2006) Multiplexed absolute quantification for proteomics using concatenated signature peptides encoded by QconCAT genes. *Nat. Protoc.*, **1**, 1029–1043.
80. Pavelka N, Fournier ML, Swanson SK, Pelizzola M, et al., (2008) Statistical similarities between transcriptomics and quantitative shotgun proteomics data. *Mol. Cell. Proteomics*, **7**, 631–644.
81. Zybailov B, Friso G, Kim J, Rudella A, et al., (2009) Large scale comparative proteomics of a chloroplast Clp protease mutant reveals folding stress, altered protein homeostasis, and feedback regulation of metabolism. *Mol. Cell. Proteomics*, **8**, 1789–1810.
82. Zybailov B, Coleman MK, Florens L, Washburn MP (2005) Correlation of relative abundance ratios derived from peptide ion chromatograms and spectrum counting for quantitative proteomic analysis using stable isotope labeling. *Anal. Chem.*, **77**, 6218–6224.
83. Fu Q, Schoenhoff FS, Savage WJ, Zhang P, et al., (2010) Multiplex assays for biomarker research and clinical application: translational science coming of age. *Proteomics Clin. Appl.*, **4**, 271–284.
84. Tackett AJ, DeGrasse JA, Sekedat MD, Oeffinger M, et al., (2005) I-DIRT, a general method for distinguishing between specific and nonspecific protein interactions. *J. Proteome Res.*, **4**, 1752–1756.
85. Sardiu ME, Cai Y, Jin J, Swanson SK, et al., (2008) Probabilistic assembly of human protein interaction networks from label-free quantitative proteomics. *Proc. Natl Acad. Sci. USA*, **105**, 1454–1459.
86. Klose J (1999) Large-gel 2-D electrophoresis. *Methods Mol. Biol.*, **112**, 147–172.
87. Sickmann A, Reinders J, Wagner Y, Joppich C, et al., (2003) The proteome of Saccharomyces cerevisiae mitochondria. *Proc. Natl Acad. Sci. USA*, **100**, 13207–13212.
88. Zybailov B, Rutschow H, Friso G, Rudella A, et al., (2008) Sorting signals, N-terminal modifications and abundance of the chloroplast proteome. *PLoS ONE*, **3**, e1994.
89. Perfetto SP, Chattopadhyay PK, Roederer M (2004) Seventeen-colour flow cytometry: unravelling the immune system. *Nat. Rev. Immunol.*, **4**, 648–655.
90. Perez OD, Nolan GP (2002) Simultaneous measurement of multiple active kinase states using polychromatic flow cytometry. *Nat. Biotechnol.*, **20**, 155–162.
91. Skvortsov S, Debbage P, Skvortsova II (2013) Proteomics of cancer stem cells. *Int. J. Radiat. Biol.*, **12**, 1–6.
92. Abiko M, Furuta K, Yamauchi Y, Fujita C, et al., (2013) Identification of proteins enriched in rice egg or sperm cells by single-cell proteomics. *PLoS ONE*, **8**, e69578.
93. Rubakhin SS, Sweedler JV (2008) Quantitative measurements of cell-cell signaling peptides with single-cell MALDI MS. *Anal. Chem.*, **80**, 7128–7136.
94. Mellors JS, Jorabchi K, Smith LM, Ramsey JM (2010) Integrated microfluidic device for automated single cell analysis using electrophoretic separation and electrospray ionization mass spectrometry. *Anal. Chem.*, **82**, 967–973.
95. Krishna RG, Wold F (1993) Post-translational modification of proteins. *Adv. Enzymol. Relat. Areas Mol. Biol.*, **67**, 265–298.
96. Garavelli JS (2004) The RESID database of protein modifications as a resource and annotation tool. *Proteomics*, **4**, 1527–1533.
97. Glickman MH, Ciechanover A (2002) The ubiquitin-proteasome proteolytic pathway: destruction for the sake of construction. *Physiol. Rev.*, **82**, 373–428.
98. Andersson L, Porath J (1986) Isolation of phosphoproteins by immobilized metal (Fe3+) affinity chromatography. *Anal. Biochem.*, **154**, 250–254.

99. Muszynska G, Andersson L, Porath J (1986) Selective adsorption of phosphoproteins on gel-immobilized ferric chelate. *Biochemistry*, **25**, 6850–6853.
100. Ficarro SB, McCleland ML, Stukenberg PT, Burke DJ, et al., (2002) Phosphoproteome analysis by mass spectrometry and its application to *Saccharomyces cerevisiae*. *Nat. Biotechnol.*, **20**, 301–305.
101. Pandey A, Fernandez MM, Steen H, Blagoev B, et al., (2000) Identification of a novel immunoreceptor tyrosine-based activation motif-containing molecule, STAM2, by mass spectrometry and its involvement in growth factor and cytokine receptor signaling pathways. *J. Biol. Chem.*, **275**, 38633–38639.
102. Ficarro S, Chertihin O, Westbrook VA, White F, et al., (2003) Phosphoproteome analysis of capacitated human sperm. Evidence of tyrosine phosphorylation of a kinase-anchoring protein 3 and valosin-containing protein/p97 during capacitation. *J. Biol. Chem.*, **278**, 11579–11589.
103. Goshe MB, Conrads TP, Panisko EA, Angell NH, et al., (2001) Phosphoprotein isotope-coded affinity tag approach for isolating and quantitating phosphopeptides in proteome-wide analyses. *Anal. Chem.*, **73**, 2578–2586.
104. Thingholm TE, Larsen MR (2009) The use of titanium dioxide micro-columns to selectively isolate phosphopeptides from proteolytic digests. *Methods Mol. Biol.*, **527**, 57–66, xi.
105. Wang X, Stewart PA, Cao Q, Sang QX, et al., (2011) Characterization of the phosphoproteome in androgen-repressed human prostate cancer cells by Fourier transform ion cyclotron resonance mass spectrometry. *J. Proteome Res.*, **10**, 3920–3928.
106. Rudiger H, Gabius HJ (2001) Plant lectins: occurrence, biochemistry, functions and applications. *Glycoconj. J.*, **18**, 589–613.
107. Carr SA, Huddleston MJ, Bean MF (1993) Selective identification and differentiation of N- and O-linked oligosaccharides in glycoproteins by liquid chromatography-mass spectrometry. *Protein Sci.*, **2**, 183–196.
108. Hart GW, Haltiwanger RS, Holt GD, Kelly WG (1989) Nucleoplasmic and cytoplasmic glycoproteins. *Ciba Found Symp.*, **145**, 102–112, discussion 112-8.
109. Vosseller K, Wells L, Hart GW (2001) Nucleocytoplasmic O-glycosylation: O-GlcNAc and functional proteomics. *Biochimie*, **83**, 575–581.
110. Wells L, Whelan SA, Hart GW (2003) O-GlcNAc: a regulatory post-translational modification. *Biochem. Biophys. Res. Commun.*, **302**, 435–441.
111. Comer FI, Hart GW (1999) O-GlcNAc and the control of gene expression. *Biochim. Biophys. Acta*, **1473**, 161–171.
112. Bacher G, Korner R, Atrih A, Foster SJ, et al., (2001) Negative and positive ion matrix-assisted laser desorption/ionization time-of-flight mass spectrometry and positive ion nano-electrospray ionization quadrupole ion trap mass spectrometry of peptidoglycan fragments isolated from various *Bacillus* species. *J. Mass Spectrom.*, **36**, 124–139.
113. Sheeley DM, Reinhold VN (1998) Structural characterization of carbohydrate sequence, linkage, and branching in a quadrupole ion trap mass spectrometer: neutral oligosaccharides and N-linked glycans. *Anal. Chem.*, **70**, 3053–3059.
114. Gaucher SP, Cancilla MT, Phillips NJ, Gibson BW, et al., (2000) Mass spectral characterization of lipooligosaccharides from *Haemophilus influenzae* 2019. *Biochemistry*, **39**, 12406–12414.
115. Leavell MD, Leary JA, Yamasaki R (2002) Mass spectrometric strategy for the characterization of lipooligosaccharides from *Neisseria gonorrhoeae* 302 using FTICR. *J. Am. Soc. Mass Spectrom.*, **13**, 571–576.
116. Peng J, Schwartz D, Elias JE, Thoreen CC, et al., (2003) A proteomics approach to understanding protein ubiquitination. *Nat. Biotechnol.*, **21**, 921–926.
117. Zhang Z, Smith DL, Smith JB (2003) Human beta-crystallins modified by backbone cleavage, deamidation and oxidation are prone to associate. *Exp. Eye Res.*, **77**, 259–272.
118. Zabrouskov V, Han X, Welker E, Zhai H, et al., (2006) Stepwise deamidation of ribonuclease A at five sites determined by top down mass spectrometry. *Biochemistry*, **45**, 987–992.
119. Zhao R, Oxley D, Smith TS, Follows GA, et al., (2007) DNA damage-induced Bcl-xL deamidation is mediated by NHE-1 antiport regulated intracellular pH. *PLoS Biol.*, **5**, e1.
120. Solstad T, Carvalho RN, Andersen OA, Waidelich D, et al., (2003) Deamidation of

120. labile asparagine residues in the autoregulatory sequence of human phenylalanine hydroxylase. *Eur. J. Biochem.*, **270**, 929–938.
121. Robinson NE, Robinson AB (2004) Molecular Clocks: Deamidation of Asparaginyl and Glutaminyl Residues in Peptides and Proteins. Althouse Press, Cave Junction, OR.
122. Zhu Y, Qi C, Cao WQ, Yeldandi AV, *et al.*, (2001) Cloning and characterization of PIMT, a protein with a methyltransferase domain, which interacts with and enhances nuclear receptor coactivator PRIP function. *Proc. Natl Acad. Sci. USA*, **98**, 10380–10385.
123. Zhu JX, Aswad DW (2007) Selective cleavage of isoaspartyl peptide bonds by hydroxylamine after methyltransferase priming. *Anal. Biochem.*, **364**, 1–7.
124. Xu Q, Belcastro MP, Villa ST, Dinkins RD, *et al.*, (2004) A second protein L-isoaspartyl methyltransferase gene in *Arabidopsis* produces two transcripts whose products are sequestered in the nucleus. *Plant Physiol.*, **136**, 2652–2664.
125. Sargaeva NP, Lin C, O'Connor PB (2011) Unusual fragmentation of beta-linked peptides by ExD tandem mass spectrometry. *J. Am. Soc. Mass Spectrom.*, **22**, 480–491.
126. Sargaeva NP, Lin C, O'Connor PB (2009) Identification of aspartic and isoaspartic acid residues in amyloid beta peptides, including Abeta1-42, using electron-ion reactions. *Anal. Chem.*, **81**, 9778–9786.
127. O'Connor PB, Cournoyer JJ, Pitteri SJ, Chrisman PA, *et al.*, (2006) Differentiation of aspartic and isoaspartic acids using electron transfer dissociation. *J. Am. Soc. Mass Spectrom.*, **17**, 15–19.
128. Zee BM, Young NL, Garcia BA (2011) Quantitative proteomic approaches to studying histone modifications. *Curr. Chem. Genomics*, **5**, 106–114.
129. Byrum SD, Taverna SD, Tackett AJ (2013) Purification of a specific native genomic locus for proteomic analysis. *Nucleic Acids Res.*, **41**, e195.
130. Byrum SD, Raman A, Taverna SD, Tackett AJ (2012) ChAP-MS: a method for identification of proteins and histone posttranslational modifications at a single genomic locus. *Cell Rep.*, **2**, 198–205.
131. Bock JR, Gough DA (2001) Predicting protein–protein interactions from primary structure. *Bioinformatics*, **17**, 455–460.
132. Fields S, Song O (1989) A novel genetic system to detect protein–protein interactions. *Nature*, **340**, 245–246.
133. Cagney G, Uetz P, Fields S (2000) High-throughput screening for protein–protein interactions using two-hybrid assay. *Methods Enzymol.*, **328**, 3–14.
134. Ito T, Chiba T, Ozawa R, Yoshida M, *et al.*, (2001) A comprehensive two-hybrid analysis to explore the yeast protein interactome. *Proc. Natl Acad. Sci. USA*, **98**, 4569–4574.
135. Uetz P, Giot L, Cagney G, Mansfield TAJ *et al.*, (2000) A comprehensive analysis of protein–protein interactions in *Saccharomyces cerevisiae*. *Nature*, **403**, 623–627.
136. Deshaies RJ, Seol JH, McDonald WH, Cope G, *et al.*, (2002) Charting the protein complexome in yeast by mass spectrometry. *Mol. Cell. Proteomics*, **1**, 3–10.
137. Verma R, Chen S, Feldman R, Schieltz D, *et al.*, (2000) Proteasomal proteomics: identification of nucleotide-sensitive proteasome-interacting proteins by mass spectrometric analysis of affinity-purified proteasomes. *Mol. Biol. Cell*, **11**, 3425–3439.
138. Akey CW, Radermacher M (1993) Architecture of the *Xenopus* nuclear pore complex revealed by three-dimensional cryo-electron microscopy. *J. Cell Biol.*, **122**, 1–19.
139. Rout MP, Aitchison JD, Suprapto A, Hjertaas KZ *et al.* (2000) The yeast nuclear pore complex: composition, architecture, and transport mechanism. *J. Cell Biol.*, **148**, 635–651.
140. Cronshaw JM, Krutchinsky AN, Zhang W, Chait BT, *et al.* (2000) Proteomic analysis of the mammalian nuclear pore complex. *J. Cell Biol.* **158** (5), 915–927.
141. Rappsilber J, Siniossoglou S, Hurt EC, Mann M (2000) A generic strategy to analyze the spatial organization of multiprotein complexes by cross-linking and mass spectrometry. *Anal. Chem.*, **72**, 267–275.
142. Gavin AC, Bosche M, Krause R, Grandi P, *et al.*, (2002) Functional organization of the yeast proteome by systematic analysis of protein complexes. *Nature*, **415**, 141–147.
143. Puig O, Caspary F, Rigaut G, Rutz B, *et al.*, (2001) The tandem affinity purification (TAP) method: a general procedure of protein complex purification. *Methods*, **24**, 218–229.

144. Ho Y, Gruhler A, Heilbut A, Bader GD, et al., (2002) Systematic identification of protein complexes in *Saccharomyces cerevisiae* by mass spectrometry. *Nature*, **415**, 180–183.
145. von Mering C, Krause R, Snel B, Cornell M, et al., (2002) Comparative assessment of large-scale data sets of protein–protein interactions. *Nature*, **417**, 399–403.
146. Ranish JA, Yi EC, Leslie DM, Purvine SO, et al., (2003) The study of macromolecular complexes by quantitative proteomics. *Nat. Genet.*, **33**, 349–355.
147. Blagoev B, Kratchmarova I, Ong SE, Nielsen M, et al., (2003) A proteomics strategy to elucidate functional protein–protein interactions applied to EGF signaling. *Nat. Biotechnol.*, **21**, 315–318.
148. Olinares PD, Kim J, Davis JI, van Wijk KJ (2011) Subunit stoichiometry, evolution, and functional implications of an asymmetric plant plastid ClpP/R protease complex in *Arabidopsis*. *Plant Cell*, **23**, 2348–2361.
149. Sinz A (2003) Chemical cross-linking and mass spectrometry for mapping three-dimensional structures of proteins and protein complexes. *J. Mass Spectrom.*, **38**, 1225–1237.
150. Weisbrod CR, Chavez JD, Eng JK, Yang L, et al., (2013) In vivo protein interaction network identified with a novel real-time cross-linked peptide identification strategy. *J. Proteome Res.*, **12**, 1569–1579.
151. Zybailov BL, Glazko GV, Jaswail M, Raney KD (2013) Large scale chemical cross-linking mass spectrometry perspectives. *J. Proteomics Bioinf.*, **S2**, 1–11.
152. Schilling B, Row RH, Gibson BW, Guo X, et al., (2003) MS2Assign, automated assignment and nomenclature of tandem mass spectra of chemically crosslinked peptides. *J. Am. Soc. Mass Spectrom.*, **14**, 834–850.
153. El-Shafey A, Tolic N, Young MM, Sale K, et al., (2006) "Zero-length" cross-linking in solid state as an approach for analysis of protein–protein interactions. *Protein Sci.*, **15**, 429–440.
154. Singh P, Shaffer SA, Scherl A, Holman C, et al., (2008) Characterization of protein cross-links via mass spectrometry and an open-modification search strategy. *Anal. Chem.*, **80**, 8799–8806.
155. Nadeau OW, Wyckoff GJ, Paschall JE, Artigues A, et al., (2008) CrossSearch, a user-friendly search engine for detecting chemically cross-linked peptides in conjugated proteins. *Mol. Cell. Proteomics*, **7**, 739–749.
156. Panchaud A, Singh P, Shaffer SA, Goodlett DR (2010) xComb: a cross-linked peptide database approach to protein-protein interaction analysis. *J. Proteome Res.*, **9**, 2508–2515.
157. Gotze M, Pettelkau J, Schaks S, Bosse K, et al., (2012) StavroX – a software for analyzing crosslinked products in protein interaction studies. *J. Am. Soc. Mass Spectrom.*, **23**, 76–87.
158. Muller MQ, Dreiocker F, Ihling CH, Schafer M, et al., (2010) Cleavable cross-linker for protein structure analysis: reliable identification of cross-linking products by tandem MS. *Anal. Chem.*, **82**, 6958–6968.
159. Singh P, Panchaud A, Goodlett DR (2010) Chemical cross-linking and mass spectrometry as a low-resolution protein structure determination technique. *Anal. Chem.*, **82**, 2636–2642.
160. Mayne SL, Patterton HG (2011) Bioinformatics tools for the structural elucidation of multi-subunit protein complexes by mass spectrometric analysis of protein–protein cross-links. *Brief. Bioinform.*, **12**, 660–671.

18
Natural Product-Based Drug Discovery

Shoaib Ahmad
Rayat & Bahra Institute of Pharmacy, Sahauran, 140104 Punjab, India

1	**Introduction** 651	
2	**Drug Discovery Process** 651	
3	**Established Drugs from Natural Sources** 652	
3.1	Plant Sources 654	
3.2	Animal Sources 654	
3.3	Microorganisms 654	
3.4	Marine Organisms 654	
4	**Drug Discovery from Natural Sources** 656	
4.1	Anticancer Drugs 656	
4.2	Antidiabetic Drugs 657	
4.3	Drugs to treat Dementia 658	
4.4	Antidepressant Drugs 659	
4.5	Anti-Inflammatory Drugs 660	
4.5.1	Drugs to treat Conjunctivitis 660	
4.5.2	Drugs to treat Hepatitis 660	
4.5.3	Drugs to treat Nephritis 661	
4.6	Drugs to treat Respiratory Disorders 662	
4.6.1	Drugs to treat Bronchitis 662	
4.6.2	Drugs to treat Asthma 663	
4.7	Drugs to treat Infective Disorders 665	
4.7.1	Drugs to treat AIDS 666	
4.7.2	Drugs to treat Candidiasis 667	
4.7.3	Antimalarial Drugs 667	
4.7.4	Drugs to treat Tuberculosis 669	
4.8	Antiobesity Drugs 669	
5	**Drugs to Prevent Skin Aging** 672	

Translational Medicine: Molecular Pharmacology and Drug Discovery
First Edition. Edited by Robert A. Meyers.
© 2018 Wiley-VCH Verlag GmbH & Co. KGaA. Published 2018 by Wiley-VCH Verlag GmbH & Co. KGaA.

6   **Conclusions** 672

     Acknowledgments 672

     References 672

Keywords

Natural drugs
Drugs derived from natural sources such as plants, animals, microorganisms.

Drug discovery
Process of discovering the chemical or biological substances with potential for human use.

Clinical trials
Studies involving experimentation of drugs on human beings (healthy volunteers or patients).

Plants
Organisms capable of producing their own food by photosynthesis.

Animals
Organisms not capable of producing their own food and having dependence on other organisms (e.g. plants) for nutrition.

Microbes
Organisms which are not visible to naked eye and can only be seen with the help of microscope.

Marine organisms
Organisms growing in oceanic environments.

> Natural products are obtained from Nature, and can be obtained from either living sources, including plants, animals and microorganisms (bacteria, virus, fungi), and/or nonliving sources, though the number produced in the latter case is very small. While progressing in terms of evolution and civilization, humankind has shifted focus from natural therapeutics to orthodox medicines. In addition to providing new drug candidates and drug products, science has also raised concerns with regards to the safety, efficacy and toxicity of those products used in modern medicine. In many cases, modern medicine has no solutions to some health problems, while the solutions may be less efficacious in metabolic and chronic disorders. Hence, the quest for new drugs to treat diseases has brought humankind closer to Nature, and the quest for new drugs has

begun to benefit from ethnopharmacological and/or local health traditions. Clearly, interest in natural products and drug-delivery systems based on these materials has gained much importance during recent years.

1
Introduction

Ever since the evolution of life on earth, humankind has been dependent on natural resources to provide the basic needs of food, clothing, and shelter. As civilization progressed, however, and traditional wisdom was accumulated by generations of experimentation with plants, animals, microbes, and minerals, humankind began to explore the option of using natural sources of drugs to treat bodily diseases and disorders. These pathophysiological conditions range from simple fevers or pyrexia to complex syndromes such as AIDS and cancer.

Following a brief and well-documented era of orthodox medicine, humankind has now begun to return to Mother Nature to identify remedies which could be used to treat or prevent diseases and disorders, or to mitigate the symptoms associated with such pathophysiological conditions. Whilst many of these diseases (e.g., diabetes mellitus) appear to metabolic disorders, many – including malignant disorders such as leukemias and carcinomas – seem to be incurable.

2
Drug Discovery Process

The process of drug discoveryis complex and involves multiple stages. Most drugs obtained from natural resources are derived from plants, with the traditional route of drug discovery including selection of the plant material, followed by the extraction, isolation and structural elucidation of the active compound, using a combination of chromatographic and spectroscopic methods (Fig. 1). The compound is then subjected to pharmacological testing using *in-vivo* and *in-vitro* animal models. If the results are considered positive, animal toxicity studies are then conducted, after which a suitable dosage form (e.g., tablet, capsule, ointment, syrup, injection) is formulated.

The next stage is the clinical trial, where the drug dose to be used is determined in healthy human volunteers. The lead molecules are tested in a limited number of patients, and if the results are fruitful then multicentric trials in a larger patient population are conducted. When the efficacy has been proven in these multicentric trials, the drug is marketed after obtaining the necessary authorization from the appropriate national drug regulatory agency. Typically, it takes 10–15 years of Research and Development activity, and an expenditure in the region of US$ 500–800 million to complete the development process for each new compound (Fig. 2).

The modern approach to drug discovery from plants relies on information received from ethnomedicinal claims. The claimed drug is identified and the reportedly active extract is tested in *in-vitro*, and later in *in-vivo*, animal models. The extract which has shown activity in scientific studies is then subjected to fractionation, after which the bioactive constituent is isolated, identified and investigated using computational molecular modeling and quantitative structure–relationship(QSAR) programs to generate a compound library. The next

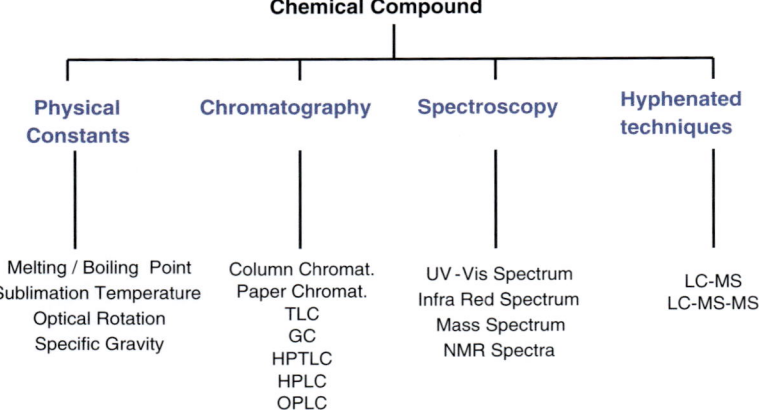

Fig. 1 Identification of a novel chemical compound.

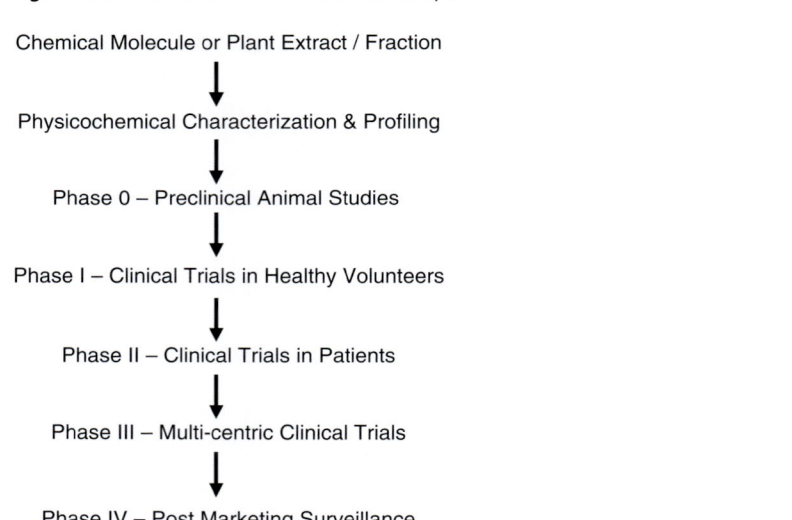

Fig. 2 Stages in drug discovery & development.

step is to synthesize the compound in the laboratory and to re-evaluate it in experimental biological screening models. A suitable dosage form is designed, after which the compound is subjected to clinical trials. This approach is understood to reduce the time for drug discovery and development process and, accordingly, the related expenditure is also reduced. The determination of a new drug's toxicity and cellular/molecular mechanism of action is a challenging area that requires special attention and great expertise.

3 Established Drugs from Natural Sources

The natural sources of drugs include plants, animals, microbes and marine organisms.

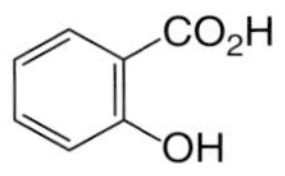

Salicylic Acid

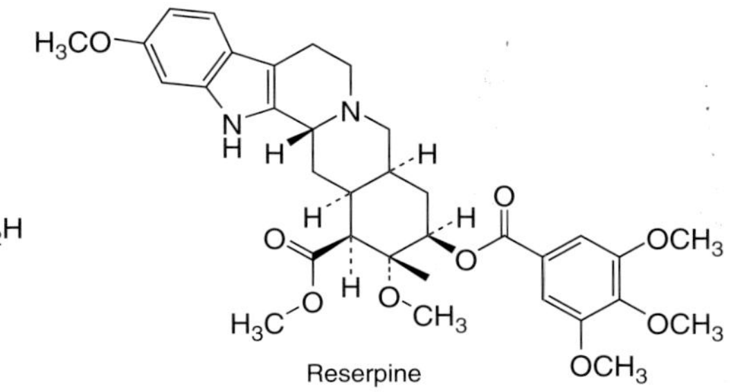

Reserpine

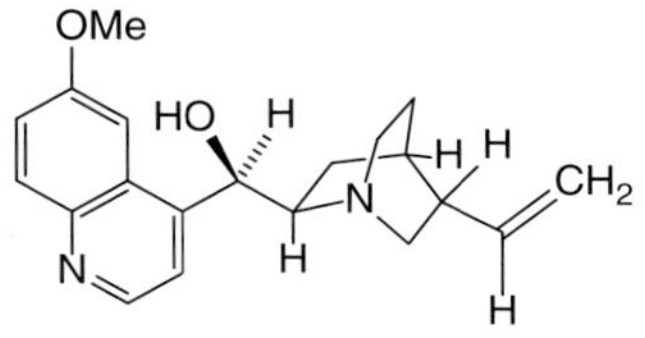

Quinine

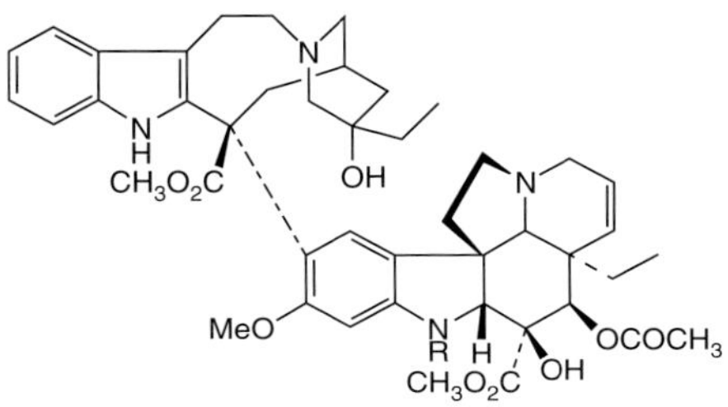

Vinblastine, R = CH₃
Vincristine, R = COH

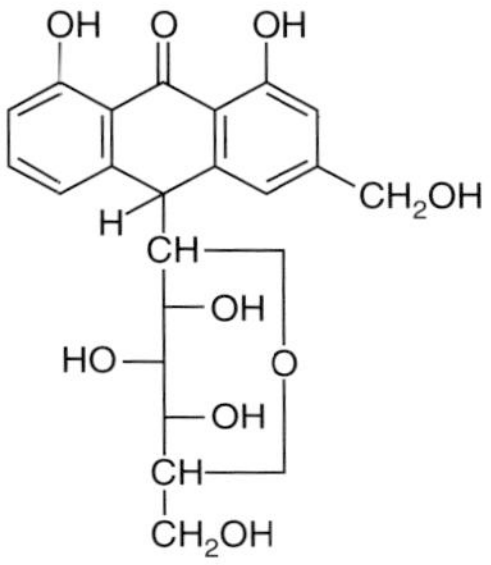

Aloin

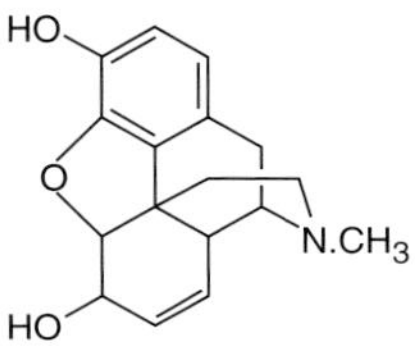

Morphine

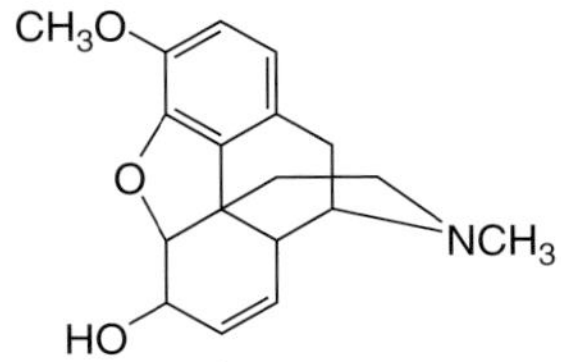

Codeine

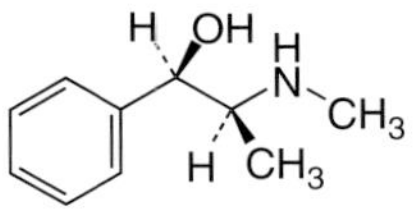

Ephedrine

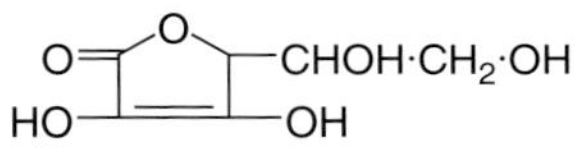

Vitamin C or ascorbic acid

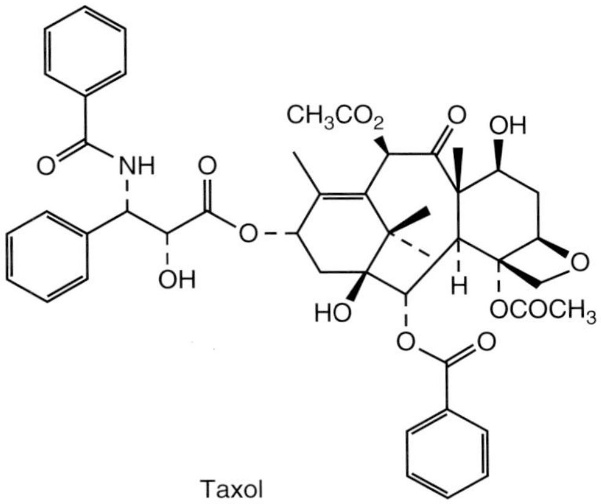

Taxol

3.1 Plant Sources

Plants are the first and foremost source of drug natural products, with compounds having been produced primarily from higher plants, that is, angiosperms. The plant families, including Apocynaceae, Asclepiadaceae, Berberidaceae, Dioscoreaceae, Ephedraceae, Euphorbiaceae, Graminae, Labiatae, Leguminosae, Liliaceae, Malvaceae, Ranunculaceae, and Umbelliferae, are well-known for their production of important phytochemicals, some of which are illustrated below and listed in Table 1.

3.2 Animal Sources

Animals sources of drug compounds include vertebrates and nonvertebrates, though vertebrates are the major producers. These include fishes, amphibians, reptiles and terrestrial mammals, with cod liver oil and anti-snake venoms among the best-known products. Some drug products derived from animals are listed in Table 2.

3.3 Microorganisms

Microorganism sources of drug materials include bacteria, fungi, and viruses. Bacteria and fungi are known to produce several drug molecules (Table 3), with antibiotics of different categories among the most important products from microbes.

3.4 Marine Organisms

Drugs from marine organisms (Table 4) are derived from plants and animals found primarily in the seas and oceans. Although

Tab. 1 Natural products from plants in medicinal use.

S. No.	Drug	Source	Use
1.	Salicylic acid	*Salix alba*	Fever
2.	Reserpine	*Rauwolfia serpentina*	Antihypertensive
3.	Ajmalicine	*Rauwolfia serpentina*	Antihypertensive
4.	Quinine	*Cinchona ledgeriana*	Antimalarial
5.	Vincristine	*Catharanthus roseus*	Anticancer
6.	Vinblastine	*Catharanthus roseus*	Anticancer
7.	Aloin	*Aloe barbadensis*	Purgative
8.	Digoxin	*Digitalis purpurea*	Antiarrythmia
9.	Digitoxin	*Digitalis purpurea*	Antiarrythmia
10.	Rutin	*Ruta graveolens*	Antifragility
11.	Oleandrin	*Nerium oleander*	Anticancer
12.	Morphine	*Papaver somniferum*	Narcotic analgesic
13.	Codeine	*Papaver somniferum*	Expectorant
14.	Ephedrine	*Ephedra sinensis*	Hay fever
15.	Ascorbic acid	*Emblica officinalis*	Scurvy
16.	Atropine	*Atropa belladonna*	Mydriatic agent
17.	Pilocarpine	*Pilocarpus jaborandi*	Myotic agent
18.	Taxol	*Taxus brevifolia*	Anticancer
19.	Hyoscine	*Hyoscyamus niger*	Antispasmodic
20.	Artemisinin	*Artemisia annua*	*Falciparum* malaria

Tab. 2 Animal origin-products in medicinal use.

S. no.	Drug	Source	Use
1.	Retinol	Fish liver oils (e.g., *Gadus callarias*)	Night blindness and xerophthalmia
2.	Dehydroretinal	Fish liver oils (e.g., *Gadus callarias*)	Night blindness and xerophthalmia
3.	Hirudin	Leech (*Hirudo medicinalis*)	Bruises
4.	Cardioactive principles	Frog and toad-skins	Treatment of dropsy (before the discovery of digitalis)
5.	Insulin	Pigs	Diabetes mellitus
6.	Snake venoms	Snakes	Snake bites

Tab. 3 Natural products from microorganisms.

S. no.	Drug	Source	Use
1.	Benzylpenicillin	*Penicillium notatum*	Antibacterial agent
2.	Cephalosporin	*Cephalosporium acremonium*	Antibacterial
3.	Chlortetracycline	*Streptomyces aureofaciens*	Bacteriostatic used in chronic bronchitis
4.	Oxytetracycline	*Streptomyces rimosus*	Chronic bronchitis
5.	Chloramphenicol	*Streptomyces venezuelae*	Typhoid, meningitis
6.	Streptomycin	*Streptomyces griseus*	Tuberculosis
7.	Gentamicin	*Micromonospora purpurea*	Ocular infections
8.	Neomycin	*Streptomyces fradiae*	Suppression of bowel flora
9.	Nystatin	*Streptomyces* species	Treatment of infections by *Candida albicans*
10.	Bacitracin	*Bacillus subtilis*	Skin infections
11.	Colistin	*Bacillus colistinus*	Bowel sterilization
12.	Polymyxin B	*Bacillus polymyxa*	Eye and ear infections
13.	Actinomycin D	*Streptomyces* species	Anticancer therapy
14.	Daunorubicin	*Streptomyces* species	Anticancer therapy
15.	Griseofulvin	*Penicillium griseofulvum*	Fungal infections

Tab. 4 Marine natural products with medicinal use potential.

S. no.	Drug	Source	Use
1.	Eudistomins	*Eudistoma olivaceum*	Antiviral
2.	Avarone and avarol	*Disidea avara*	Anti-AIDS
3.	Patellazole	*Lissoclinum patella*	Herpes simplex infections
4.	Bryostatins	*Bugula neritina*	Neoplastic bone marrow failure
5.	Fucoidan	*Fucus vesiculosus*	Anticoagulant
6.	Alpha kainic acid	*Digenia simplex*	Anthelmintic agent (round/tape/whip worms)

these organisms have attracted less attention than other sources due to their comparatively difficult accessibility, they do nevertheless produce drugs that can be used in a variety of pathophysiological conditions ranging from simple fevers to malignancies.

4
Drug Discovery from Natural Sources

The recent decades have witnessed much higher investments in the field of scientific research, and interest in the natural sources of drugs has increased considerably. Enormous advances in scientific instrumentation have taken place, such that the modern era of drug discovery involves not only genomic and metabolic technologies but also *in-silico* testing procedures. The first stage of the process involves the generation of lead molecules using molecular modeling techniques, after which compound libraries are created to identify those molecules with the highest levels of efficacy. The intention of this chapter is to provide an insight into the discovery of therapeutic modalities for some well-known diseases, with efforts focused on plant sources.

4.1
Anticancer Drugs

Cancer represents a group of malignancies that are characterized generally by uncontrolled cell division and are broadly categorized as osteoma, sarcoma, lymphoma, and leukemia. All cancers have three main symptoms of pain, depression, and fatigue. During recent years cancer therapy has undergone extensive investigation, with *in-vivo* hollow-fiber assays used for the discovery of anticancer natural products [1]. It has been predicted that the once-popular natural products used to treat cancers will make a come-back [2–5], and this has led to extensive discussions on the future of anticancer drug discovery [6]. One useful approach to cancer chemotherapy has been the development of microtubule-stabilizing natural products [7], while the use of nanoparticles of natural drugs have been discussed [8], in addition to anticancer vaccines [9].

Anticancer natural products identified since the 1960s have been reviewed [4, 5], as have those developed after 2000 [3]. *Digitalis* is known to have antineoplastic activities [10], while the cytotoxicity of some Nigerian plants has justified their use in the treatment of some cancers [11]. Medicinal plants from the island of Socotra, in Yemen, have also been subjected to pharmacological testing [12], while osthole – a coumarin used in Traditional Chinese Medicine (TCM) has been recognized as a promising lead compound for anticancer drug discovery [13].

Extracts from the bark of the tree *Ulmus laevis* have been shown to inhibit endometrial carcinoma [14], berberine and rhizoma coptidis represent novel antineoplastic agents [15], gossypol from the cotton plant (genus *Gossypium*) has an anticancer action [16], and annonaceous acetogenins from the seeds *Annona* plants have exhibited cytotoxic, antitumoral and immunosuppressive activities [17]. Both, duocarmycins and yatakemycin may lead to the development of other natural products [18]; an extract of the fruit from *Xylopia aethiopica* demonstrated an antiproliferative action on human cervical cancer cells [19], while extracts of the leaves of *Brysocarpus coccineus* have suggested their potential use in cancer chemotherapy [20].

Among microbial secondary metabolites from all types of organisms (for a review, see Ref. [21]), epothilones A and B from the myxobacterium *Sorangium cellulosum*

Tab. 5 Approved anticancer drugs based on natural products.

S. no.	Trade name	Source	Source	Reference
1.	Ixempra	Plants	Ixabepilone	[30]
2.	Yondelis	Plants	Trabectedin	[30]
3.	Torisel	Plants	Temsirolimus	[30]
4.	Prialt	Snails	Ziconotide	[28]

showed potent cytotoxicity for both paclitaxel-sensitive and paclitaxel-resistant cells [22]. Marine products, especially bacterial products, have also been evaluated for their anticancer properties [23, 24].

Antitumor activity has been demonstrated for gambogic acid and its analogs [25], while fluorinated analogs of salicylihalamide have been synthesized and evaluated for their cytotoxic effects against human tumor cell lines [26]. To date, 29 Amaryllidaceae alkaloids and their derivatives have been evaluated, providing several lead molecules [27, 28]. 2′-Hydroxycinnamaldehyde (HCA) and its analog, 2′-benzoyloxycinnamaldehyde, have emerged as potential anticancer agents [29], while the semisynthetic product psymberin acts through cell apoptosis [30]. Several anticancer drugs have been approved for clinical use during recent years (Table 5).

4.2 Antidiabetic Drugs

Diabetes mellitus is a metabolic disorder that clinically characterized by polydipsia, polyphagia, and polyurea. The condition may be either insulin-dependent (IDDM; type 1) or non-insulin-dependent (NIDDM; type 2), and in the adult population has been associated with a sedentary lifestyle. Diabetic complications include retinopathy, neuropathy, nephropathy, and impotence/infertility. The condition is very often associated with urban populations. Traditionally, *Syzygium cuminii*, *Momordica charantia*, *Gymnema sylvestre*, *Swertia chirata*, and *Azadirachta indica* have been used in the treatment of diabetes.

In addition to a review of antidiabetic drugs of natural origin [31], the hypoglycemic activity of antidiabetic plants from Mexico was also noted [32]; nopal (*Opuntia* sp.) is used traditionally in Mexico to treat diabetes mellitus, with *Opuntia streptacantha* showing hypoglycemic properties [33]. Brazilian medicinal plants (*Epidendrum monsenii*, *Marrubium vulgare*, *Rubus imperialis*, and *Wedelia paludosa*) have also exhibited beneficial effects on alloxan-induced diabetic rats [34]. A total of 30 hypoglycemia-inducing medicinal plants used in the Ayurvedic, Unani, and Siddha systems of medicines have been evaluated for antidiabetic activity [35]. Korean plants (viz. *Acer ginnala*, *Illicium religiosum*, and *Cornus macrophylla*) exhibited a strong inhibition of aldose reductase, and may be potential candidates for the treatment of diabetic retinopathy [36].

A root and leaf ginseng tincture, *Echinopanax* tincture, extracts of *Eleutherococcus*, *Rhodiola*, and *Leuzia* have each displayed antidiabetic, insulinotropic, and hypoglucagonemic effects [37]. An aqueous extract of the root from *M. tridentate* has been shown to possess significant antidiabetic activity [38], as has that of *Syzygium jambolanum* [39].

Quercetin has been proven useful in preventing alloxan diabetes [40]. Likewise, the natural product oleanolic acid has an antidiabetes action; its derivative (3-beta-(2-carboxybenzoyloxy)-oleanolic acid) upregulated the expression of GLUT4 in 3T3-L1 adipocytes [41]. Cryptolepine analogs based on *Cryptolepis sanguinolenta* have antihyperglycemic activities [42, 43], while terpenoid-type quinones from *Pycnanthus angolensis* may be beneficial in type 2 diabetes [44]. Several new plants and phytoconstituents have shown potential in the treatment of diabetes (Tables 6 and 7).

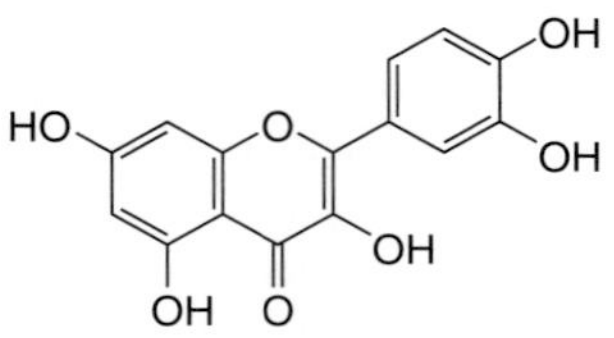

Quercetin

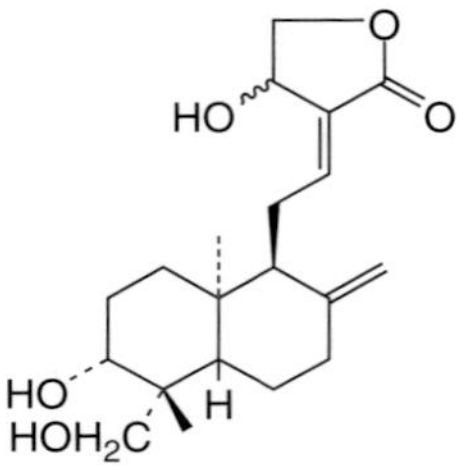

Andrographolide

4.3
Drugs to treat Dementia

Dementia is a mental illness involving memory loss. Alzheimer's disease is a form of primary dementia for which current treatments have been discussed [64], with natural products having been projected as a rich source of drugs to treat this condition [65].

Whilst the development of drugs to treat dementia poses a serious challenge [66], anti-Abeta vaccination has shown much promise as therapy for Alzheimer's disease [67]. Among geriatric populations, Alzheimer's disease manifests as the most common form of dementia, and anti-Abeta vaccine human clinical trials have been proposed in this patient group such as septic shock, circulatory disorders, migraine, gastric ulcers, nephritis, eclampsia, respiratory diseases [68]. Alzheimer's disease is a neurodegenerative disorder that causes dementia, notably in the aging population. The anti-acetylcholinesterase activities of 26 herbs used in TCM to treat

Tab. 6 Plants affecting glucose metabolism.

S. no.	Plant	Parts/extracts	Effect	Reference
1.	*Laportea ovalifolia*	Aerial parts	Antidiabetic and hypolipidemic effects	[45]
2.	*Gymnema montanum*	Leaves	Antihyperglycemic and antihyperlipidemic	[46]
3.	*Vernonia colorata*	Leaves	Hypoglycemic and antidiabetic activity	[47]
4.	*Cordyceps sinensis*	Fruits	Antihyperglycemic activity	[48]
5.	*Albizzia lebbeck*	Aqueous extract	Antioxidant	[49]
6.	*Cynodon dactylon*	Aqueous extract	Hypoglycemic and hypolipidemic antidiabetic	[50]
7.	*Brassica nigra*	Aqueous extract	Insulinotropic effect	[51]
8.	*Cichorium intybus*	Alcoholic extract	Antidiabetic effects	[52]
9.	*Artemisia dracunculus*	Alcoholic extract	Improves carbohydrate metabolism	[53]
10.	*Viscum album*	Aqueous and alcoholic extracts	Acute hypoglycemic effect	[54]

Tab. 7 Phytoconstituents affecting glucose metabolism.

S. no.	Plant	Phytoconstituent(s)	Effect	Reference
1.	Siegesbeckia glabrescens	Kaurane diterpenes: ent-16betaH, 17-isobutyryloxy-kauran-19-oic acid (1), and ent-16betaH, 17-acetoxy-18-isobutyryloxy-kauran-19-oic acid (2)	Protein tyrosine phosphatase 1B inhibition	[55]
2.	Curcuma longa	Curcumin	Antidiabetic nephropathy	[56]
3.	Cephalotaxus sinensis	Apigenin-5-O-[alpha-L-rhamnopyranosyl-(1→4)-6-O-beta-D-acetylglucopyranoside] (1), apigenin (2), and apigenin-5-O-[alpha-L-rhamnopyranosyl-(1→4)-6-O-beta-D-glucopyranoside] (3).	Antihyperglycemic activity	[57]
4.	Panax notoginseng	20(S)-25-Methoxyl-dammarane-3-beta, 12-beta, 20-triol	Antioxidant	[58]
5.	Eriobotrya japonica	Nerolidol-3-O-alpha-L-rhamnopyranosyl(1→4)-alpha-L-rhamnopyranosyl(1→2)-[alpha-L-rhamnopyranosyl(1→6)]-beta-D-glucopyranoside	Hypoglycemic	[59]
6.	Andrographis paniculata	Andrapholide	In-vitro α-glucosidase and α-amylase enzyme inhibition	[60]
7.	Sambucus nigra	Polyphenols	Antioxidant	[61]
8.	Costus speciosus	Costunolide	Normoglycemic and hypolipidemic effect	[62]
9.	Evodia rutaecarpa	Rhetsinine	Inhibition of aldose reductase	[63]

Alzheimer's disease has been reported [69], as have the effects of fuzhisan (also used in TCM) on learning & memory of aged rats [70].

4.4
Antidepressant Drugs

Depression is a major clinical problem in both developed and developing nations, and may be reactive, endogenous, or associated with bipolar mania. Traditionally, *Withania somnifera*, *Rauwolfia serpentine*, *Panax ginseng*, and *Hypericum perforatum* have been used in the treatment of depression. The protection of endogenous enkephalin catabolism has also been explored in the development of antidepressant drugs [71].

Anisodamine has been proposed as an antidepressant drug [72], and both aqueous and ethanolic extracts of *Terminalia bellerica* have been shown to possess significant antidepressant-like effects [73]. *Lepidium meyenii* is known to reduce depression in postmenopausal women, the effect being independent of estrogenic and androgenic activities [74]. Rosmarinic acid and caffeic acid, obtained from the leaves of *Perilla frutescens* Britton produce antidepressive- and/or anxiolytic-like effects [75], while polysaccharides from a TCM formulation – that is, a *Banxia-Houpu* decoction – have exhibited antidepressant-like effects [76]. *Schisandra chinensis* Bail has been proven to exert multiple effects, including an antidepressant action [77], and *Cynanchum auriculatum* has exhibited an antidepressant action in mice [78]. Marine natural products with neuropsychiatric pharmacophores have also been reviewed for their antidepressant activities [79].

4.5
Anti-Inflammatory Drugs

Inflammation is a pathophysiological condition characterized by rubor (redness), dolor (pain), calor (increased heat), and tumor (swelling) in the concerned area. Inflammation can be caused by a wide array of physical, chemical and biological factors; common examples are conjunctivitis, hepatitis, and nephritis. An Indian medicinal plant (*Lawsonia inermis*) and its phytoconstituent lawsone have demonstrated anti-inflammatory properties, as has *Boswellia serrata* [80].

4.5.1 Drugs to treat Conjunctivitis

Conjunctivitis, which involves inflammation of the conjunctiva of the eye, can also be caused by *Chlamydia* infestations [81]. Chronic allergic conjunctivitis is believed to be partly mediated by nitric oxide derived and activated by mediators other than histamine. Histamine-independent allergic conjunctivitis may be valuable for studying cases of chronic conjunctivitis [82]; cellular systems and computer models have been suggested to reduce the extent of (and even replace) whole-animal testing [83].

Caribbean herbal remedies have potential for use in conjunctivitis because of their antibacterial properties [84]. The antimicrobial activity of kalanchoe juice has been evaluated in keratoconjunctivitis, and was also active against *Shigella* [85]. Ribosomal preparations of *Shigella sonnei* were also tested against keratoconjunctivitis in guinea pigs [86]. Herpes simplex virus (HSV) types 1 and 2 also lead to keratoconjunctivitis, and a glycoprotein D adjuvant HSV vaccine is currently under development [87].

4.5.2 Drugs to treat Hepatitis

Hepatitis is inflammation of the liver, and a number of drugs are used traditionally when treating cases of hepatitis. Most often, preparations of *Emblica officinalis* and *Silybum marianum* are used, but vitamin C and silymarin-based products are all thought to be beneficial against hepatitis.

Lawsone

Ellagic acid

Tab. 8 Phytoconstituents for use in hepatitis.

S. no.	Compound	Plant	Type of hepatitis	Reference
1.	Wogonin	*Scutellaria radix*	Hepatitis B	[88]
2.	Pseudoguaianolides	*Parthenium hispitum*	Hepatitis C	[89]
3.	Curcumin	*Curcuma longa*	Hepatitis B	[90]
4.	Ellagic acid	—	Hepatitis B	[90]

Recently, wogonin, curcumin, and ellagic acid have been shown to be useful for the treatment of hepatitis (Table 8). An ethanolic extract of *Meconopsis quintuplinervia* has an antioxidant action and potential for use in the treatment of hepatitis [91].

4.5.3 Drugs to treat Nephritis

Nephritis, which involves the inflammation of nephrons in the kidney, incorporates the action of platelet-activating factor (PAF) in proteinuria and other nephropathies. The development of PAF antagonists may be useful for the treatment of nephritis [92]; indeed, when evaluated in sexually mature and immature rats, azathioprine was found to be beneficial in nephrotoxic nephritis by successfully reducing the intensity of not only proteinurias but also of hypo- and dysproteinemia and hyper-β-lipoproteinemia. Azathioprine was most effective during the active phase of developing nephritis [93].

Acetoside from *Stachys sieboldii* is known to exert an antinephritic action (Table 9) on crescentic-type anti-glomerular basement membrane (GBM) nephritis in rats, by inhibiting the elevation of proteinuria. Acetoside has also been shown useful against rapidly progressive glomerulonephritis [94].

When the antinephritic effects of berberine (Table 10) and coptisine from Coptidis rhizome were tested against original-type anti-GBM nephritis in rat models, both compounds effectively inhibited urinary protein excretion and reduced the histopathological changes in the glomerular structures. Both compounds also inhibited platelet aggregation *in vitro* and *in vivo*. Berberine was shown to

Tab. 9 Pharmacological effects of acetoside.

S. no.	Parameter	Effect
1.	Elevation of proteinuria	Inhibited
2.	Cholesterol	Reduced
3.	Creatinine	Reduced
4.	Antibody production against rabbit γ-globulin	Reduced
5.	Hypercellularity	Suppressed
6.	Crescent formation	Suppressed
7.	Capillary wall adherence to Bowman's capsule	Suppressed
8.	Fibrinoid necrosis in the glomeruli	Suppressed

Tab. 10 Pharmacological effects of berberine.

S. no.	Parameter	Observed effect
1.	Urinary protein excretion	Inhibited
2.	Elevation of serum cholesterol	Inhibited
3.	Creatinine level	Reduced
4.	Platelet aggregation	Inhibited
5.	Decline of renal blood flow	Inhibited
6.	Increase in thromboxane B_2 level	Inhibited
7.	Formation of 6-keto-prostaglandin F1α	Increased

inhibit thromboxane-β2 formation, while simultaneously enhancing the formation of 6-keto-prostaglandin F1α. From these findings it may be inferred that the antinephritic effects of these phytoconstituents may be partly attributable to antiplatelet action, although improvements in renal hemodynamics effected by an altered prostanoid synthesis may also be involved [95].

Cordyceps sinensis is a TCM remedy with a claimed reputation for immunomodulatory effects, and a pure compound (H1-A) exhibited the capability of inhibiting the mesangial proliferation in lupus nephritis in a mouse model [96].

Anisodamine, an atropine derivative developed in China, is a non-specific anticholinergic agent which is less potent than atropine and has less central nervous system (CNS) toxicity than scopolamine. It has been suggested that anisodamine may be useful for the treatment of several clinical conditions such as septic shock, circulatory disorders, migraine, gastric ulcers, nephritis, eclampsia, respiratory diseases [72]. A new substance (codenamed S6) was also evaluated for efficacy in acute experimental nephritis in an animal model (rabbits) [97].

Antigens from *Escherichia coli* purified fimbriae (H-2946 strain) have been shown to immunoprotect guinea pigs against experimental acute pyelonephritis [98].

4.6
Drugs to treat Respiratory Disorders

Respiratory disorders represent a major health problem as they can originate from allergic, infection, or a variety of other reasons; however, the most common presentations are bronchitis and asthma.

4.6.1 Drugs to treat Bronchitis

Bronchitis involves inflammation of the bronchi, as frequently occurs with the common cold. Currently, there is no effective treatment for bronchitis, and a combination of antiviral and anti-inflammatory therapies is considered the best modality. A prophylactic regimen has been considered useful for disease prevention, with such methods probably being useful for treating at-risk-populations suffering from asthma or chronic bronchitis [99]. One recognized TCM-based remedy for bronchitis and pharyngitis with a severe dry cough is Bakumondo-to [100].

Fennel (*Foeniculum vulgare* Mill.) is strongly recommended for the treatment of diabetes, bronchitis, and chronic coughs, as well as for kidney stones. The fennel shoots were found to have the highest radical-scavenging activity, and also exhibited the highest lipid peroxidation inhibition capacity. These two actions can be attributed to the high contents of phenolics and ascorbic acid in fennel, as well as a high concentration of tocopherols (Table 11) [101].

The leaves of *Elaeagnus pungens* thunb. (Family Elaeagnaceae) are traditionally used in TCM for the treatment of asthma and chronic bronchitis. The antiasthmatic, antitussive and expectorant activities of the ethanolic extract from the *E. pungens* leaves were evaluated *in vivo* in animal models. Following the induction of asthma in guinea pigs by a combination of histamine and acetylcholine chloride, a 70% ethanolic extract of *E. pungens* leaves was shown to increase the preconvulsive time of asthma.

Tab. 11 Bioactive constituents of *Foeniculum vulgare* Mill. shoots.

S. no.	Constituent	Concentration
1.	Phenolics	$65.85 \pm 0.74 \text{ mg g}^{-1}$
2.	Ascorbic acid	$570.89 \pm 0.01 \text{ μg g}^{-1}$
3.	Tocopherols	$34.54 \pm 1.28 \text{ μg g}^{-1}$

In addition, the aqueous fraction of the leaf extract produced the following effects in guinea pigs and mice models:

1) A significant prolongation of the preconvulsive time.
2) An increase in the latent period of the cough.
3) A reduction in cough frequency.
4) An enhancement of tracheal phenol red output.

The petroleum ether fraction of the *E. pungens* leaf extract was also found to cause a significant reduction in cough frequency in guinea pigs, and to enhance tracheal phenol red output in mice. It may be concluded that the petroleum ether and aqueous fractions of the *E. pungens* leaf have antiasthmatic, antitussive, and expectorant activities [102]. *Buddleja thyrsoides* Lam. leaves also have potential for use in the treatment of bronchitis [103].

4.6.2 Drugs to treat Asthma

Asthma is a respiratory disorder that is characterized clinically by chronic airways inflammation but can also be caused by air pollutants. Chronic airways inflammation is typically detected using the exhaled NO test [104].

An unwanted activation of the complement immune system is involved in asthma and allergies. It has been envisaged that a range of complement system inhibitors might exist that could prevent destructive effects on the body, and this concept has formed the basis of the main methods of drug design. Approaches to the design of drugs to treat asthma include the screening and modification of natural products, leading to structure-based ligand design [105].

Allergic diseases involve inappropriate immune responses to environmental antigens, although the reasons for the current increased prevalence of such diseases are not fully understood. The use of a DNA vaccine against house dust or mite allergen-related asthma has been proposed [106]. A low-cost pulmonary surfactant consisting of a mixture of lipids or lipid-peptides with aliphatic alcohol or soy lecithin has also been developed for asthma treatment, and considered worthy of clinical trial for the possible treatment of respiratory disorders, as exemplified by asthma [107].

During recent years, the pharmaceutical industry has focused largely on developing small-molecule inhibitors of phosphoinositide 3-kinase (PI3K) delta and gamma as therapeutic options for the treatment of inflammatory and autoimmune diseases [108].

In South Asian countries (e.g., China and Japan), ryo-kan-kyomi-sin-ge-nin-to (RKSG) is traditionally used for the treatment of asthma. This compound has been shown to suppress the degranulation of rat mast cells, as well as histamine release from these cells, in a dose-dependent manner, with the results of animal studies confirming that RKSG might indeed be valuable for the treatment of type I allergy-related diseases [109].

In Latin American countries, *Cecropia glazioui* Sneth is used as an antihypertensive, cardiotonic and antiasthmatic agent. An aqueous extract of *C. glazioui* has been shown to promote an effect in mice that was similar to that of anxiolytics, and which involves the presence of the plant's flavonoids and terpenes [110].

Traditionally, *Isodon japonicas* Hara (Labiatae) has been used in Korea as an anti-inflammatory agent. In fact, when an aqueous extract of the plant was administered to mice, it caused an inhibition of compound 48/80-induced systemic reactions, as well as plasma histamine release. Subsequent studies led to the conclusion

Tab. 12 Pharmacological studies on Lyprinol.

S. no.	Experimental model	Method	Effect
1.	In vitro	LTB_4 biosynthesis by PMNs in vitro	Inhibited
2.	In vitro	PGE_2 production by activated macrophages	Inhibited
3.	In vivo (animal)	Adjuvant-induced polyarthritis	No development of arthritis
4.	In vivo (animal)	Collagen(II)-induced autoallergic arthritis	
5.	In vivo (animal)	Disease-stressed rats	No gastrotoxicity at 300 mg kg^{-1} (oral)
6.	Human beings (patients)	Clinical trials	Anti-inflammatory action

PMN, polymorphonucleocyte.

that this extract would inhibit mast cell-derived immediate-type allergies, and that tumor necrosis factor-alpha (TNF-α) is involved in the production of these effects [111].

Menthol, a well-known compound obtained from *Mentha* species, belongs to the category of monoterpenes and is commonly used in confectionary items. L-Menthol and mint oil each affect arachidonic acid metabolism, an effect that was examined using leukotriene (LT) subset B4 and prostaglandin (PG) subset E2. L-Menthol was shown to efficiently inhibit the production of three inflammation mediators by monocytes in *in-vitro* conditions, thus highlighting the superior anti-inflammatory effects of L-menthol compared to mint oil. The therapeutic efficacy of L-menthol in chronic inflammatory conditions (e.g., bronchial asthma, colitis, and allergic rhinitis) requires further investigation in clinical trials [112].

A lipid-rich extract of freeze-dried, stabilized powder from the New Zealand green-lipped mussel *Perna canaliculus* (Lyprinol) has exhibited considerable anti-inflammatory in animal (Wistar and Dark Agouti rats) and human models. Lyprinol treatment completely inhibited adjuvant-induced polyarthritis, and the same effect was visible in collagen(II)-induced autoallergic arthritis. Lyprinol inhibited LTB_4 biosynthesis as well as PGE_2 production. Lyprinol has been established as non-gastrotoxic in disease-stressed rats, and clinically was found to be beneficial in patients with inflammatory conditions such as osteoarthritis, rheumatoid arthritis, and asthma [113] (Table 12).

Verbascum thapsus L. is a medicinal plant used for the treatment of inflammation, asthma, spasmodic coughs, and other pulmonary diseases. An aqueous extract of the plant showed significant antibacterial activity against *Klebsiella pneumonia*, *Staphylococcus aureus*, *Staphylococcus epidermidis*, and *Escherichia coli*. Extracts of this plant were also able to inhibit *Agrobacterium tumefaciens* -induced tumors in potato disc tissue systems [114] (Table 13).

An aqueous extract of *Mosla dianthera* (Maxim.) has been found to inhibit mast cell-derived immediate-type allergies; the inhibition of proinflammatory cytokines and NF-κB was seen to be involved in these effects [115]. A methanolic extract of *Vitis amurensis* Rupr. (Vitaceae) fruit was also shown to inhibit mast cell-derived, immediate-type allergic reactions, involving the proinflammatory cytokines, p38 mitogen-activated protein kinase (MAPK),

Tab. 13 Pharmacological effects of *Verbascum thapsus* L.

S. no.	Experimental model	Observation
1.	*Klebsiella pneumonia*	Suppression of growth
2.	*Staphylococcus aureus*	
3.	*Staphylococcus epidermidis*	
4.	*Escherichia coli*	
5.	*Agrobacterium tumefaciens*-induced tumors	Tumor inhibition
6.	Brine shrimp assay	Toxicity
7.	Radish seed assay	Reduced germination and growth

Tab. 14 Constituents of essential oil of *Nepeta cataria* L. identified by GC-EIMS.

S. no.	Phytoconstituent	Proportion (%)
1.	1,8-Cineol	21.00
2.	Alpha-humulene	14.44
3.	Alpha-pinene	10.43
4.	Geranyl acetate	8.21

and NF-κB [116]. Petroleum ether and aqueous fractions of *Elaeagnus pungens* (Elaeagnaceae) leaves have been found to have antiasthmatic, antitussive, and expectorant activities [102].

Essential oil of *Nepeta cataria* L. (Limiaceae) is used traditionally for treating gastrointestinal and respiratory disorders. This oil has exhibited spasmolytic and myorelaxant activities. These are activities result from dual inhibition of calcium channels and phosphodiesterase (PDE) [117] (Table 14).

Gleditsia sinensis fruits saponins also have a possibility of use in bronchitis [118]. GC-EIMS, Gas chromatography-electron impact mass spectrometry.

Chitinase Inhibitors in Asthma Family 18 chitinases are overexpressed in the asthmatic lung and promote the pathogenic process via a recruitment of inflammatory cells. Subsequently, a virtual screening-based approach led to a novel, purine-based, chitinase inhibitor which consisted of two linked caffeine moieties exhibiting pan-family 18 chitinase inhibitory actions and acted in competitive mode [119]. Immediate-type allergic reactions are involved in certain diseases (e.g., asthma and allergic rhinitis). *Artemisia iwayomogi* exhibited an inhibitory effect on compound 48/80-induced systemic reactions in a mouse model, and was also found to attenuate the secretion of phorbol 12-myristate 13-acetate-stimulated TNF-α and interleukin (IL)-6 in human mast cells [120]. Human acidic mammalian chitinase (hAMCase) serves as a good target for the development of anti-asthma drugs, notably because its binding to the natural product cyclopentapeptide (Argifin) produces an inhibitory activity against hAMCase. Consequently, the search for selective Argifin derivatives is continuing [121].

Argifin, which has demonstrated competitive inhibitory actions on family 18 chitinases, inhibited the following enzymes in particular [122]:

1) Chitinase B1 from *Aspergillus fumigatus*.
2) Human chitotriosidase.
3) Chitinase activity in lung homogenates.

4.7
Drugs to treat Infective Disorders

Infective disorders are caused by a number of organisms including protozoa, bacteria,

Tab. 15 Antibacterial agents from plants [123].

S. no.	Phytoconstituent	Plant	Remarks
1.	Allicin	*Allium sativum*	Broad-spectrum
2.	Salicylic acid	*Anacardium pulsatilla*	Active against *P. acnes*
3.	Catechin	*Camellia sinensis*	Broad-spectrum
4.	Asiatocoside	*Centella asiatica*	*Mycobacterium leprae*
5.	Cocaine	*Erythroxylum coca*	Bacteria
6.	Colchicina	*Gloriosa superba*	Broad-spectrum
7.	Berberine	*Hydrastis canadensis*	Bacteria
8.	Berberine	*Berberis vulgaris*	Bacteria
9.	Phloretin	*Malus sylvestris*	Broad-spectrum
10.	Anthemic acid	*Matricaria chamomilla*	*M. tuberculosis*
11.	Anthemic acid	*Matricaria chamomilla*	*S. typhimurium*
12.	Hexanal	*Olea europaea*	Broad-spectrum
13.	Eugenol	*Pimenta dioica*	Broad-spectrum
14.	Cathecol	*Piper betel*	Broad-spectrum
15.	Totarol	*Podocarpus nagi*	Active against *P. acnes*
16.	Fabatin	*Vicia faba*	Bacteria
17.	Reserpine	*Vinca minor*	Broad-spectrum
18.	Curcumin	*Curcuma longa*	Bacteria
19.	Menthol	*Mentha piperita*	Broad-spectrum
20.	Carvacrol	*Satureja montana*	Broad-spectrum

fungi, and viruses. To date, a phenomenal number of studies has been conducted on various aspects of infective disorders, and many antibacterial and antifungal agents have been identified that can be obtained from plants (see Tables 15 and 16).

4.7.1 Drugs to treat AIDS

Acquired Immunodeficiency Syndrome (AIDS), which is caused by the human immunodeficiency virus (HIV), is a clinically complex syndrome that is also socially taboo. The complete care of AIDS patients is a major challenge [125], as fungal infections invariably strike because the patients are inevitably immunocompromised. Nonetheless, several plant remedies have been evaluated for the treatment of these conditions [124].

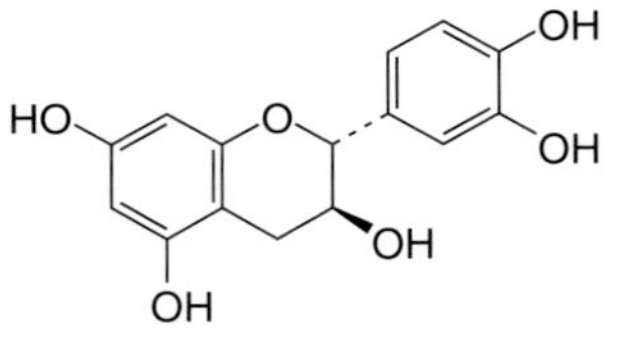

(+)-Catechin

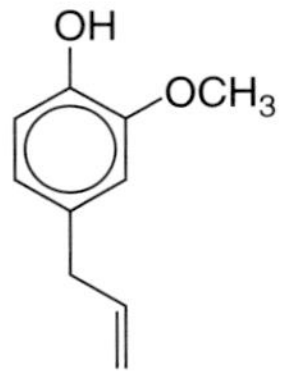

Eugenol

Elderberry, green tea, and cinnamon extracts have been shown to block HIV-1 entry and infection in GHOST cells [126], while phytocannabinoids from *Cannabis sativa* L. also have shown potential for use in AIDS treatment [127]. The use of cyclic peptides has also been debated [128]. For some time, investigations into HIV vaccines have raised optimistic concerns [129–144], and the discovery of a vaccine to treat HIV has long been anticipated

[145, 146]. Marine cyanobacteria have also been projected as a potential source of lead molecules in this respect [147].

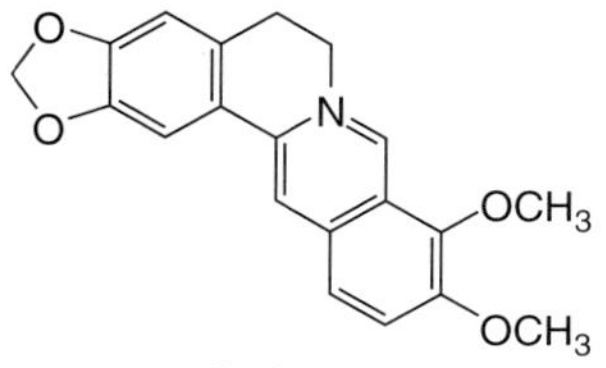

Berberine

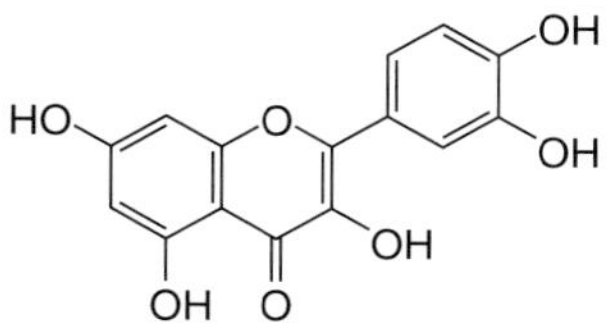

Quercetin

4.7.2 Drugs to treat Candidiasis
Candidiasis, which occurs in AIDS patients and is caused by the fungus *Candida albicans*, is commonly manifested by hairy leukoplakia. However, several plants and compounds (Tables 17 and 18) have shown promising antifungal actions against candidiasis.

4.7.3 Antimalarial Drugs
Malaria is a protozoal disease caused by *Plasmodium vivax* and *P. falciparum*. It is a quite common disease in tropical areas, and its eradication is a major challenge [130]. *Cinchona ledgeriana* and its quinoline alkaloid, quinine, have long been used for the treatment of malaria caused by *P. vivax*, while *Artemisia annua* and its sesquiterpene lactone, artemisinin, have shown promising curative effects in malaria caused by *P. falciparum*.

The production of antiparasitic compounds from higher plants via modern approaches has been reviewed [153], as has the development of antimalarial drugs [154], the naturally occurring antimalarials discovered between 1998 and 2008 [155], and plant-based antimalarial agents (non-alkaloidal natural products) [156]. Several methods have been suggested for the testing of both natural products and synthetic molecules for their antimalarial activities [157], and protocols for high-performance liquid chromatography (HPLC)-based activity profiling for antimalarial natural products have been established [158].

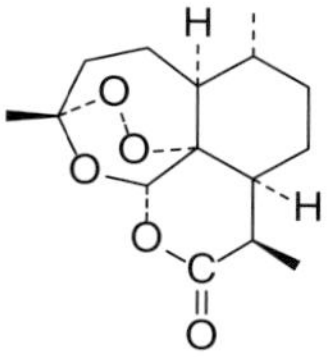

Artemisinin

The examination of natural products with carbon–phosphorus bonds led to the antimalarial clinical drug candidate 1-(4-hydroxybenzyl)-6,7-methylenedioxy-2-methylisoquinolinium trifluoroacetate [159], a benzylisoquinoline alkaloid from the tree *Doryphora sassafras* that demonstrated a clear antimalarial activity [160]. Likewise, three bis-indole alkaloids from *Flindersia* species – flinderoles A, B, and C – showed antimalarial actions [161]. The azafluorenone alkaloids, 5,8-dihydroxy-6-methoxyonychine and 5-hydroxy-6-methoxyonychine, from the Australian tree *Mitrephora diversifolia* also showed antimalarial activity [162], while febrifugine, an alkaloid from *Dichroa febrifuga* Lour demonstrated remarkable activity against *P. falciparum* [163].

Bidens pilosa, a weed commonly used in traditional Fijian local health, was shown to contain two new compounds – polyacetylenic diol and its glucoside – both of which had highly potent antimalarial and antibacterial activities [164]. Gossypol, from the cotton

Tab. 16 Antifungal agents from plants [124].

S. no.	Phytoconstituent	Botanical source
1.	Tannins and salicylic acid	*Gaultheria procumbens*, *Rhammus purshiana*, and *Anacardium pulsatilla*
2.	3-Hydroxy-4-geranyl-5-methoxybiphenyl	*Garcinia mangostana*
3.	4-Hydroxyphenyl-6-O-[(3R)-3,4-dihydroxy-2-methylenebutanoyl]-D-glucopyranoside	*Toronia toru*
4.	Amentoflavone	*Selaginella tamariscina*
5.	Eupomatenoid-3, eupomatenoid-5, conocarpan, and orientin	*Piper solmsianum*
6.	2-Hydroxy maackiain	*Hildegardia barteri*
7.	Scandenone, tiliroside, quercetin-3,7-O-α-L-dirhamnoside and kaempferol-3,7-O-α-L-dirhamnoside	
8.	Clausenidin, dentatin, nor-dentatin, and clauszoline J	*Clausena excavata*
9.	1-Tigloyloxy-8βH, 10βH-eremophil-7(11)-en-8α,12-olide	*Senecio poepigii*
10.	Kigelinone, isopinnatal, dihydro-α-lapachone, and lapachol	*Kigelia pinnata*
11.	1,3-Dihydroxy-2-methyl-5,6-imethoxyanthraquinone	*Prismatomeris fragrans*
12.	Smilagenin 3-O-β-D-glucopyranoside and disporoside A	*Smilax medica*
13.	Ypsilandroside B, ypsilandroside A, isoypsilandroside A, isoypsilandroside B, and isoypsilandrogaine	*Ypsilandra thebetica*
14.	Eight saponins	*Tribulus terrestris*
15.	Caledonixanthone	*Calophyllum caledonicum*
16.	1,3,6-Trihydroxy-2,5-dimethoxyxanthone	*Monnina obtusifolia*
17.	Seven xanthanolides	*Xanthium macrocarpum*
18.	Cudrafrutixanthone	*Cudrania fruticosa*
19.	2-(3,4-Dimethyl-2,5-dihydro-1H-pyrrol-2-yl)-1-methylethylpentanoate	*Datura metel*
20.	6,8-Didec-(1Z)-enyl-5,7-dimethyl-2,3-dihydro-1H-indolizinium	*Aniba panurensis*
21.	Cinnamodial and cinnamosmolide	*Pleodendron costaricense*
22	Cycleanine, cocsoline, and N-desmethylcycleanine	*Albertisia villosa*
23.	α- and β-Basrubrins	*Basella rubra*
24.	AFP-J	*Solanum tuberosum*
25.	Estragole	*Agastache rugosa*
26.	8-Acetylheterophyllisine, panicutine, and 3-hydroxy-2-methyl-4H-pyran-4-one	*Delphinium denudatum*
27.	16α-Hydroxy-cleroda-3,13(14)-Z-diene-15,16-olide and 16-oxo-cleroda-3,13(14)-E-diene-15-oic acid	*Polyalthia longifolia*
28.	Hyalodendrosides A and B	*Hyalodendron* species

Tab. 17 Plant extracts active against candidiasis.

S. no.	Plant	Extract	Reference
1.	*Schinus terebintifolius*	Methanolic extract	[148]
2.	*O. gratissimum*	Methanolic extract	[148]
3.	*Cajanus cajan*	Methanolic extract	[148]
4.	*Piper aduncum*	Methanolic extract	[148]
5.	*Swietenia mahagoni*	Methanolic extract	[149]

Tab. 18 Compounds active against candidiasis.

S. no.	Source	Compound	Reference
1.	*Lippia sidoides* Cham.	Essential oil (thymol and carvacrol)	[150]
2.	Tropical sponges (*Dysidea* sp.)	Sesterterpene sulfates	[151]
3.	*Sommera sabiceoides*	6-Nonadecynoic acid	[152]

Tab. 19 Natural products with potential for use in tuberculosis.

S. no.	Compound(s)	Source	Reference
1.	Essential oil (carvacrol, thymol, p-cymene, 1,8-cineole, limonene, and β-pinene)	*Achyrocline alata*	[171]
2.	Essential oil (carvacrol, thymol, p-cymene, 1,8-cineole, limonene, and β-pinene)	*Swinglea glutinosa*	[171]
3.	Myxopyronin A	—	[172]
4.	Piperidinol	—	[173]
5.	Chlorophenol	*Helichrysum aureonitens*	[174]

3-Methoxynordomesticine isolated from Colombian plants acts as an inhibitor of mure ligase of *M. tuberculosis* [175].

plant, also has antimalarial activity [16], while antiplasmodial protostane triterpenoids have been isolated from *Alisma plantago-aquatica* L. [165].

Marine compounds with antimalarial activity have also been reviewed in detail [166], and drug discoveries from marine antimalarial natural products (between 2003 and 2008) have been extensively described [167]. Cyclodepsipeptides from marine sponges, bacteria and fungi can also be used in the antimalarial drug development process [168].

Vaccine design for the human malaria parasite *P. falciparum* has also been proposed [169] although, unfortunately, lessons learned from the design of vaccine to treat HIV had ramifications for antimalarial vaccines [141]. A wheat germ cell-free protein production system was also developed for antimalarial vaccine production [170].

4.7.4 Drugs to treat Tuberculosis

Tuberculosis, caused by *Mycobacterium tuberculosis*, affects the lungs, kidneys, and bones, and tends to have a higher incidence in people of low socioeconomic status. Encouragingly, many natural products have shown promising results in the treatment of tuberculosis (Table 19).

4.8 Antiobesity Drugs

Obesity is highly prevalent in developed and developing countries, and has been associated with diabetes mellitus, eating habits, an unbalanced (lipid-rich) diet, and

Tab. 20 Phytoconstituents from natural resources against skin aging.

S. no.	Phytoconstituent	Source	Mechanism of action	Reference
1.	Aloin A and B	*Aloe vera* L.	Inhibition of *Clostridium histolyticum* collagenase	[185]
2.	Berberine	*Berberis aristata* DC	Inhibition of basal and TPA-induced expression and activity of MMP-9 Suppression of TPA-induced IL-6 expression, ERK activation, and AP-1 DNA binding activity (UV-induced skin inflammation) and degradation of extracellular matrix proteins	[186]
3.	Polyphenols (catechin, epiglactocatechin, epiglactocatechin-3-gallate, etc.)	*Camellia sinensis* L.	Reduction in the adverse effects of UV radiation-induced inflammation, oxidative stress, and DNA damage	[187]
4.	Triterpenes including asiatic acid, madecassic acid, and asiaticoside	*Centella asiatica* L. Urban	Increase in collagen synthesis	[188]
5.	Ascorbic acid	*Citrus sinensis* L.	Modulation of cellular responses exemplified by NF-κB and AP-1 translocation and procaspase-3 cleavage to UV-B in human keratinocytes	[189]
6.	Curculigoside	*Curculigo orchioides* Gaetrn.	Strong inhibition of MMP-1	[190]
7.	Curcumin	*Curcuma longa* L.	Prevention of the formation of wrinkles and melanin Increase in the dimensions of skin blood vessels Reduced expression of MMP-2	[191]
8.	Xanthorrhizol	*Curcuma xanthorrhiza* Roxb.	Decreased expression of MMP-1 protein Enhanced expression of type-1 procollagen	[192]
9.	Diosgenin	*Dioscorea composita* or *Dioscorea villosa* L.	Restoration of keratinocyte proliferation in aged skin	[193]
10.	Esculetin	*Fraxinus chinensis* Roxb.	Free radical scavenging active Decreased expression of MMP-1 mRNA and Protein in UV-inflamed dermal fibroblasts	[194]

Tab. 20 (Continued.)

S. no.	Phytoconstituent	Source	Mechanism of action	Reference
11.	Anthocyanin	*Glycine max* L. Merr.	Downregulation of UVB induced reactive oxygen species and apoptotic Cell death mediated by preventing caspase-3 pathway activation	[195]
12.	*Meso*-dihydroguaiaretic acid	*Machilus thunbergii* Sieb and Zucc	Inhibition of MMP-1 in primary human fibroblasts	[196]
13.	Magnolol	*Magnolia ovovata* Thunb.	Inhibition of MMP-1	[197]
14.	1,2,4,6-Tetra-*O*-galloyl-D-glucopyranose and gallic acid	*Melothria heterophylla* (Lour.) Cogn.	Degradation of extracellular matrix proteins	[198]
15.	Ginsenosides	*Panax ginseng* L.	Prevention of MMP-9 gene induction and elongation of the fibrillin-1 fiber length	[199]
16.	Allylpyrocatechol (18) and chavibetol	*Piper betel* L.	Prevention of photosensitization-mediated lipid peroxidation of liver mitochondria	[200]
17.	β-Carotene	*Tagetes erecta* L.	Photoprotection	[201]
18.	1,2,3,4,6-penta-*O*-galloyl-D-glucose	*Terminalia chebula* Retz.	Inhibition of antielastase and antihyaluronidase activity Induction of type II collagen expression	[202]

AP, Activator Protein; MMP, matrix metalloprotease; TPA, Tissue plasminogen activator.

reduced physical exercise, as well as having a genetic basis. 'Fast food' consumption by children has also raised suspicions regarding the higher incidence of childhood obesity. To date, numerous anti-obesity drugs have been investigated, particularly from nature, and many of these have been reviewed in detail [176]. Previously, melanotropins have been viewed as potential drugs for the treatment of obesity [177].

An ethanolic extract of *Morus alba* leaves has shown anti-obesity effects in diet-induced obese mice [178], while black and green teas have exhibited effects against the metabolic syndrome and cardiovascular disease in rat models of human obesity [179]. The selective elevation of adiponectin production by astragaloside II and isoastragaloside I has raised hopes in this respect [180]. BHUx, a patented polyherbal formulation that has been claimed to prevent hyperlipidemia [181] via an inhibition of the key enzyme, 11-β-hydroxysteroid dehydrogenase 1 by 18-alpha-glycyrrhetinic acid, has attracted much interest for antiobesity treatment [182]. Peptide-based vaccines to treat obesity have also been debated in the past [183].

5
Drugs to Prevent Skin Aging

The innate feeling of many human beings that they must have a good physical appearance relates primarily to the skin. Yet, unfortunately, in many physiological and pathophysiological conditions the skin demonstrates symptoms of aging that are characterized by a wrinkled appearance which involves degradation of the extracellular matrix in the epidermal and dermal layers. Whereas, chronological aging occurs with the passage of time, premature aging is often the result of exposures to environmental factors such as sunlight and harmful radiation. However, several plant-based chemical constituents have been shown useful in preventing the visible effects of skin aging [184]. These compounds belong to diverse chemical classes, and exert their characteristic actions via several mechanisms (Table 20).

Asiaticoside

6
Conclusions

It is a well-proven fact that natural sources provide therapeutic solutions or lead molecules for drug discovery for both known and unknown diseases. There is also a clear need to take a rational approach towards faster drug discoveries from natural resources.

Acknowledgments

The author is greatly indebted to his teachers at Hamdard University, New Delhi (particularly Dr Shibli Jameel and Dr Rasheeduz Zafar) who provided a source of constant inspiration. The author also acknowledges the selfless moral support by Dr Sultan Anjum and Mrs Zeba Ahmad.

References

1. Mi, Q., Pezzuto, J.M., Farnsworth, N.R., Wani, M.C. et al. (2009) Use of the in vivo hollow fiber assay in natural products anticancer drug discovery. *J. Nat. Prod.*, **72** (3), 573–580.
2. Bailly, C. (2009) Ready for a comeback of natural products in oncology. *Biochem. Pharmacol.*, **77** (9), 1447–1457.
3. Coseri, S. (2009) Natural products and their analogues as efficient anticancer drugs. *Mini Rev. Med. Chem.*, **9** (5), 560–571.
4. Kinghorn, A.D., Carcache De Blanco, E.J., Chai, H.B., Orjala, J. et al. (2009) Discovery of anticancer agents of diverse natural origin. *Pure Appl. Chem.*, **81** (6), 1051–1063.
5. Kinghorn, A.D., Chin, Y.W., Swanson, S.M. (2009) Discovery of natural product anticancer agents from biodiverse organisms. *Curr. Opin. Drug Discovery Dev.*, **12** (2), 189–196.
6. Ma, X., Wang, Z. (2009) Anticancer drug discovery in the future: an evolutionary perspective. *Drug Discov. Today*, **14** (23-24), 1136–1142.
7. Zhao, Y., Fang, W.S., Pors, K. (2009) Microtubule stabilising agents for cancer chemotherapy. *Expert Opin. Ther. Pat.*, **19** (5), 607–622.
8. Jin, Y., Li, H., Bai, J. (2009) Homogeneous selecting of a quadruplex-binding ligand-based gold nanoparticle fluorescence resonance energy transfer assay. *Anal. Chem.*, **81** (14), 5709–5715.
9. Zhu, J., Warren, J.D., Danishefsky, S.J. (2009) Synthetic carbohydrate-based anticancer vaccines: the Memorial Sloan-Kettering experience. *Expert Rev. Vaccines*, **8** (10), 1399–1413.
10. Khan, M.I., Chesney, J.A., Laber, D.A., Miller, D.M. (2009) Digitalis, a targeted

therapy for cancer? *Am. J. Med. Sci.*, **337** (5), 355–359.

11. Sowemimo, A., Van De Venter, M., Baatjies, L., Koekemoer, T. (2009) Cytotoxic activity of selected Nigerian plants. *Afr. J. Tradit. Complement. Altern. Med.*, **6** (4), 526–528.

12. Mothana, R.A., Lindequist, U., Gruenert, R., Bednarski, P.J. (2009) Studies of the in vitro anticancer, antimicrobial and antioxidant potentials of selected Yemeni medicinal plants from the island Soqotra. *BMC Complement. Altern. Med.*, **9**, 7.

13. You, L., Feng, S., An, R., Wang, X. (2009) Osthole: a promising lead compound for drug discovery from Traditional Chinese Medicine (TCM). *Nat. Prod. Commun.*, **4** (2), 297–302.

14. Paschke, D., Abarzua, S., Schlichting, A., Richter, D.U. *et al*. (2009) Inhibitory effects of bark extracts from *Ulmus laevis* on endometrial carcinoma: an in-vitro study. *Eur. J. Cancer Prev.*, **18** (2), 162–168.

15. Tang, J., Feng, Y., Tsao, S., Wang, N. *et al*. (2009) Berberine and coptidis rhizoma as novel antineoplastic agents: a review of traditional use and biomedical investigations. *J. Ethnopharmacol.*, **126** (1), 5–17.

16. Wang, X., Howell, C.P., Chen, F., Yin, J., Jiang, Y. (2009) Gossypol – a polyphenolic compound from cotton plant. *Adv. Food Nutr. Res.*, **58**, 215–263.

17. Yang, H., Li, X., Tang, Y., Zhang, N. *et al*. (2009) Supercritical fluid CO_2 extraction and simultaneous determination of eight annonaceous acetogenins in *Annona* genus plant seeds by HPLC-DAD method. *J. Pharm. Biomed. Anal.*, **49** (1), 140–144.

18. Ghosh, N., Sheldrake, H.M., Searcey, M., Pors, K. (2009) Chemical and biological explorations of the family of CC-1065 and the duocarmycin natural products. *Curr. Top. Med. Chem.*, **9** (16), 1494–1524.

19. Adaramoye, O.A., Sarkar, J., Singh, N., Meena, S. *et al*. (2011) Antiproliferative action of *Xylopia aethiopica* fruit extract on human cervical cancer cells. *Phytother. Res.*, **25** (10), 1558–1563.

20. Adedosu, O.T., Adejoke, T.T., Salako, O.O., Olorunsogo, O.O. (2013) Effects of extracts of the leaves of *Brysocarpus coccineus* on rat liver mitochondrial membrane permeability transition (MMPT) pore. *Afr. J. Med. Med. Sci.*, **41** (Suppl.), 125–132.

21. Newman, D.J., Cragg, G.M. (2009) Microbial antitumor drugs: natural products of microbial origin as anticancer agents. *Curr. Opin. Invest. Drugs*, **10** (12), 1280–1296.

22. Hunt, J.T. (2009) Discovery of ixabepilone. *Mol. Cancer Ther.*, **8** (2), 275–281.

23. Gulder, T.A., Moore, B.S. (2009) Chasing the treasures of the sea – bacterial marine natural products. *Curr. Opin. Microbiol.*, **12** (3), 252–260.

24. Nagle, D.G., Zhou, Y.D. (2009) Marine natural products as inhibitors of hypoxic signaling in tumors. *Phytochem. Rev.*, **8** (2), 415–429.

25. Wang, J., Zhao, L., Hu, Y., Guo, Q. *et al*. (2009) Studies on chemical structure modification and biology of a natural product, gambogic acid (i): synthesis and biological evaluation of oxidized analogues of gambogic acid. *Eur. J. Med. Chem.*, **44** (6), 2611–2620.

26. Sugimoto, Y., Konoki, K., Murata, M., Matsushita, M. *et al*. (2009) Design, synthesis, and biological evaluation of fluorinated analogues of salicylihalamide. *J. Med. Chem.*, **52** (3), 798–806.

27. Evidente, A., Kireev, A.S., Jenkins, A.R., Romero, A.E. *et al*. (2009) Biological evaluation of structurally diverse amaryllidaceae alkaloids and their synthetic derivatives: discovery of novel leads for anticancer drug design. *Planta Med.*, **75** (5), 501–507.

28. Molinski, T.F., Dalisay, D.S., Lievens, S.L., Saludes, J.P. (2009) Drug development from marine natural products. *Nat. Rev. Drug Discovery*, **8** (1), 69–85.

29. Gan, F.F., Chua, Y.S., Scarmagnani, S., Palaniappan, P. *et al*. (2009) Structure–activity analysis of 2′-modified cinnamaldehyde analogues as potential anticancer agents. *Biochem. Biophys. Res. Commun.*, **387** (4), 741–747.

30. Huang, X., Shao, N., Huryk, R., Palani, A. *et al*. (2009) The discovery of potent antitumor agent c051-deoxypsymberin/irciniastatin a: total synthesis and biology of advanced psymberin analogs. *Org. Lett.*, **11** (4), 867–870.

31. Ivorra, M.D., Paya, M., Villar, A. (1989) A review of natural products and plants as potential antidiabetic drugs. *J. Ethnopharmacol.*, **27** (3), 243–275.

32. Roman Ramos, R., Lara Lemus, A., Alarcon Aguilar, F., Flores Saenz, J. L.

(1992) Hypoglycemic activity of some antidiabetic plants. *Arch. Med. Res.*, **23** (3), 105–109.

33. Ibanez-Camacho, R., Roman-Ramos, R. (1979) Hypoglycemic effect of Opuntia cactus. *Arch. Invest. Med. (Mex.)*, **10** (4), 223–230.
34. Novaes, A.P., Rossi, C., Poffo, C., Pretti Junior, E. et al. (2001) Preliminary evaluation of the hypoglycemic effect of some Brazilian medicinal plants. *Therapie*, **56** (4), 427–430.
35. Kar, A., Choudhary, B.K., Bandyopadhyay, N.G. (2003) Comparative evaluation of hypoglycaemic activity of some Indian medicinal plants in alloxan diabetic rats. *J. Ethnopharmacol.*, **84** (1), 105–108.
36. Kim, H.Y., Oh, J.H. (1999) Screening of Korean forest plants for rat lens aldose reductase inhibition. *Biosci. Biotechnol. Biochem.*, **63** (1), 184–188.
37. Molokovskii, D.S., Davydov, V.V., Tiulenev, V.V. (1989) The action of adaptogenic plant preparations in experimental alloxan diabetes. *Prob. Endokrinol. (Mosk.)*, **35** (6), 82–87.
38. Arunachalam, K., Parimelazhagan, T. (2012) Antidiabetic activity of aqueous root extract of *Merremia tridentata* (l.) Hall. F. In streptozotocin-induced-diabetic rats. *Asian Pac. J. Trop. Med.*, **5** (3), 175–179.
39. Baliga, M.S., Fernandes, S., Thilakchand, K.R., D'souza, P., Rao, S. (2013) Scientific validation of the antidiabetic effects of *Syzygium jambolanum* DC (black plum), a traditional medicinal plant of India. *J. Altern Complement. Med.*, **19** (3), 191–197.
40. Nuraliev Iu, N., Avezov, G.A. (1992) The efficacy of quercetin in alloxan diabetes. *Eksp. Klin. Farmakol.*, **55** (1), 42–44.
41. Lin, Z., Zhang, Y., Shen, H., Hu, L. et al. (2008) Oleanolic acid derivative NPLC441 potently stimulates glucose transport in 3 t3-l1 adipocytes via a multi-target mechanism. *Biochem. Pharmacol.*, **76** (10), 1251–1262.
42. Bierer, D.E., Dubenko, L.G., Zhang, P., Lu, Q. et al. (1998) Antihyperglycemic activities of cryptolepine analogues: an ethnobotanical lead structure isolated from *Cryptolepis sanguinolenta*. *J. Med. Chem.*, **41** (15), 2754–2764.
43. Bierer, D.E., Fort, D.M., Mendez, C.D., Luo, J. et al. (1998) Ethnobotanical-directed discovery of the antihyperglycemic properties of cryptolepine: its isolation from *Cryptolepis sanguinolenta*, synthesis, and in vitro and in vivo activities. *J. Med. Chem.*, **41** (6), 894–901.
44. Luo, J., Cheung, J., Yevich, E.M., Clark, J.P. et al. (1999) Novel terpenoid-type quinones isolated from *Pycnanthus angolensis* of potential utility in the treatment of type 2 diabetes. *J. Pharmacol. Exp. Ther.*, **288** (2), 529–534.
45. Momo, C.E., Oben, J.E., Tazoo, D., Dongo, E. (2006) Antidiabetic and hypolipidaemic effects of a methanol/methylene-chloride extract of *Laportea ovalifolia* (urticaceae), measured in rats with alloxan-induced diabetes. *Ann. Trop. Med. Parasitol.*, **100** (1), 69–74.
46. Ananthan, R., Latha, M., Ramkumar, K.M., Pari, L. et al. (2003) Effect of *Gymnema montanum* leaves on serum and tissue lipids in alloxan diabetic rats. *Exp. Diabesity Res.*, **4** (3), 183–189.
47. Sy, G.Y., Cisse, A., Nongonierma, R.B., Sarr, M., Mbodj, N.A., Faye, B. (2005) Hypoglycaemic and antidiabetic activity of acetonic extract of *Vernonia colorata* leaves in normoglycaemic and alloxan-induced diabetic rats. *J. Ethnopharmacol.*, **98** (1-2), 171–175.
48. Lo, H.C., Hsu, T.H., Tu, S.T., Lin, K.C. (2006) Anti-hyperglycemic activity of natural and fermented *Cordyceps sinensis* in rats with diabetes induced by nicotinamide and streptozotocin. *Am. J. Chin. Med.*, **34** (5), 819–832.
49. Resmi, C.R., Venukumar, M.R., Latha, M.S. (2006) Antioxidant activity of *Albizzia lebbeck* (Linn.) Benth. in alloxan diabetic rats. *Indian J. Physiol. Pharmacol.*, **50** (3), 297–302.
50. Singh, S.K., Kesari, A.N., Gupta, R.K., Jaiswal, D., Watal, G. (2007) Assessment of antidiabetic potential of *Cynodon dactylon* extract in streptozotocin diabetic rats. *J. Ethnopharmacol.*, **114** (2), 174–179.
51. Anand, P., Murali, Y.K., Tandon, V., Murthy, P.S., Chandra, R. (2009) Insulinotropic effect of aqueous extract of *Brassica nigra* improves glucose homeostasis in streptozotocin induced diabetic rats. *Exp. Clin. Endocrinol. Diabetes*, **117** (6), 251–256.

52. Pushparaj, P.N., Low, H.K., Manikandan, J., Tan, B.K., Tan, C.H. (2007) Antidiabetic effects of *Cichorium intybus* in streptozotocin-induced diabetic rats. *J. Ethnopharmacol.*, **111** (2), 430–434.
53. Wang, Z.Q., Ribnicky, D., Zhang, X.H., Raskin, I. *et al.* (2008) Bioactives of *Artemisia dracunculus* L. enhance cellular insulin signaling in primary human skeletal muscle culture. *Metabolism*, **57** (7 Suppl. 1), S58–S64.
54. Orhan, D.D., Aslan, M., Sendogdu, N., Ergun, F., Yesilada, E. (2005) Evaluation of the hypoglycemic effect and antioxidant activity of three *Viscum album* subspecies (European mistletoe) in streptozotocin-diabetic rats. *J. Ethnopharmacol.*, **98** (1-2), 95–102.
55. Kim, S., Na, M., Oh, H., Jang, J. *et al.* (2006) Ptp1b inhibitory activity of kaurane diterpenes isolated from *Siegesbeckia glabrescens*. *J. Enzyme Inhib. Med. Chem.*, **21** (4), 379–383.
56. Sharma, S., Kulkarni, S.K., Chopra, K. (2006) Curcumin, the active principle of turmeric (*Curcuma longa*), ameliorates diabetic nephropathy in rats. *Clin. Exp. Pharmacol. Physiol.*, **33** (10), 940–945.
57. Li, W., Dai, R.J., Yu, Y.H., Li, L. *et al.* (2007) Antihyperglycemic effect of *Cephalotaxus sinensis* leaves and glut-4 translocation facilitating activity of its flavonoid constituents. *Biol. Pharm. Bull.*, **30** (6), 1123–1129.
58. Zhao, Y., Wang, W., Han, L., Rayburn, E.R. *et al.* (2007) Isolation, structural determination, and evaluation of the biological activity of 20(S)-25-methoxyl-dammarane-3beta, 12beta, 20-triol [20(S)-25-och3-ppd], a novel natural product from *Panax notoginseng*. *Med. Chem.*, **3** (1), 51–60.
59. Chen, J., Li, W.L., Wu, J.L., Ren, B.R., Zhang, H.Q. (2008) Hypoglycemic effects of a sesquiterpene glycoside isolated from leaves of loquat (*Eriobotrya japonica* (Thunb.) Lindl.). *Phytomedicine*, **15** (1-2), 98–102.
60. Subramanian, R., Asmawi, M.Z., Sadikun, A. (2008) In vitro alpha-glucosidase and alpha-amylase enzyme inhibitory effects of *Andrographis paniculata* extract and andrographolide. *Acta Biochim. Pol.*, **55** (2), 391–398.
61. Ciocoiu, M., Miron, A., Mares, L., Tutunaru, D. *et al.* (2009) The effects of *Sambucus nigra* polyphenols on oxidative stress and metabolic disorders in experimental diabetes mellitus. *J. Physiol. Biochem.*, **65** (3), 297–304.
62. Eliza, J., Daisy, P., Ignacimuthu, S., Duraipandiyan, V. (2009) Normo-glycemic and hypolipidemic effect of costunolide isolated from *Costus speciosus* (koen ex. Retz.)sm. in streptozotocin-induced diabetic rats. *Chem.-Biol. Interact.*, **179** (2-3), 329–334.
63. Kato, A., Yasuko, H., Goto, H., Hollinshead, J., Nash, R.J., Adachi, I. (2009) Inhibitory effect of rhetsinine isolated from *Evodia rutaecarpa* on aldose reductase activity. *Phytomedicine*, **16** (2-3), 258–261.
64. Forette, F., Hauw, J.J. (2008) Alzheimer's disease: from brain lesions to new drugs. *Bull. Acad. Natl. Med.*, **192** (2), 363–378; discussion 378–380.
65. Calcul, L., Zhang, B., Jinwal, U.K., Dickey, C.A., Baker, B.J. (2012) Natural products as a rich source of tau-targeting drugs for Alzheimer's disease. *Future Med. Chem.*, **4** (13), 1751–1761.
66. Ji, H.F., Li, X.J., Zhang, H.Y. (2009) Natural products and drug discovery. Can thousands of years of ancient medical knowledge lead us to new and powerful drug combinations in the fight against cancer and dementia? *EMBO Rep.*, **10** (3), 194–200.
67. Okura, Y., Matsumoto, Y. (2007) Development of anti-abeta vaccination as a promising therapy for Alzheimer's disease. *Drug News Perspect.*, **20** (6), 379–386.
68. Vasilevko, V., Head, E. (2009) Immunotherapy in a natural model of abeta pathogenesis: the aging beagle. *CNS Neurol. Disord. Drug Targets*, **8** (2), 98–113.
69. Lin, H.Q., Ho, M.T., Lau, L.S., Wong, K.K. *et al.* (2008) Anti-acetylcholinesterase activities of traditional Chinese medicine for treating Alzheimer's disease. *Chem.-Biol. Interact.*, **175** (1-3), 352–354.
70. Li, X.L., Wang De, S., Zhao, B.Q., Li, Q. *et al.* (2008) Effects of Chinese herbal medicine Fuzhisan on aged rats. *Exp. Gerontol.*, **43** (9), 853–858.
71. Noble, F., Roques, B.P. (2007) Protection of endogenous enkephalin catabolism as natural approach to novel analgesic and antidepressant drugs. *Expert Opin. Ther. Targets*, **11** (2), 145–159.

72. Poupko, J.M., Baskin, S.I., Moore, E. (2007) The pharmacological properties of anisodamine. *J. Appl. Toxicol.*, **27** (2), 116–121.
73. Dhingra, D., Valecha, R. (2007) Evaluation of antidepressant-like activity of aqueous and ethanolic extracts of *Terminalia bellirica* roxb. fruits in mice. *Indian J. Exp. Biol.*, **45** (7), 610–616.
74. Brooks, N.A., Wilcox, G., Walker, K.Z., Ashton, J.F. et al. (2008) Beneficial effects of *Lepidium meyenii* (Maca) on psychological symptoms and measures of sexual dysfunction in postmenopausal women are not related to estrogen or androgen content. *Menopause*, **15** (6), 1157–1162.
75. Tsuji, M., Miyagawa, K., Takeuchi, T., Takeda, H. (2008) Pharmacological characterization and mechanisms of the novel antidepressive- and/or anxiolytic-like substances identified from Perillae Herba. *Nihon Shinkei Seishin Yakurigaku Zasshi*, **28** (4), 159–167.
76. Yi, L.T., Zhang, L., Ding, A.W., Xu, Q. et al. (2009) Orthogonal array design for antidepressant compatibility of polysaccharides from Banxia-Houpu decoction, a traditional Chinese herb prescription in the mouse models of depression. *Arch. Pharm. Res.*, **32** (10), 1417–1423.
77. Panossian, A., Wikman, G. (2008) Pharmacology of *Schisandra chinensis* Bail.: an overview of Russian research and uses in medicine. *J. Ethnopharmacol.*, **118** (2), 183–212.
78. Ji, C.X., Li, X.Y., Jia, S.B., Liu, L.L. et al. (2012) The antidepressant effect of *Cynanchum auriculatum* in mice. *Pharm. Biol.*, **50** (9), 1067–1072.
79. Diers, J.A., Ivey, K.D., El-Alfy, A., Shaikh, J. et al. (2008) Identification of antidepressant drug leads through the evaluation of marine natural products with neuropsychiatric pharmacophores. *Pharmacol. Biochem. Behav.*, **89** (1), 46–53.
80. Abdel-Tawab, M., Werz, O., Schubert-Zsilavecz, M. (2011) *Boswellia serrata*: an overall assessment of in vitro, preclinical, pharmacokinetic and clinical data. *Clin. Pharmacokinet.*, **50** (6), 349–369.
81. Igietseme, J.U., Eko, F.O., Black, C.M. (2003) Contemporary approaches to designing and evaluating vaccines against *Chlamydia*. *Expert Rev. Vaccines*, **2** (1), 129–146.
82. Fukushima, Y., Nabe, T., Mizutani, N., Nakata, K., Kohno, S. (2003) Multiple Cedar pollen challenge diminishes involvement of histamine in allergic conjunctivitis of guinea pigs. *Biol. Pharm. Bull.*, **26** (12), 1696–1700.
83. Wilhelmus, K.R. (2001) The Draize eye test. *Surv. Ophthalmol.*, **45** (6), 493–515.
84. Luciano-Montalvo, C., Boulogne, I., Gavillan-Suarez, J. (2013) A screening for antimicrobial activities of Caribbean herbal remedies. *BMC Complement. Altern. Med.*, **13**, 126.
85. Suptel, E.A., Lapchik, V.F., Zagornaia, N.B., Shirobokov, V.P. (1980) Antimicrobial activity of Kalanchoe juice relative to dysentery bacteria. *Mikrobiol. Zh.*, **42** (1), 86–90.
86. Levenson, V.I., Rukhadze, E.Z., Belkin, Z.P., Essaulova, M.E. (1988) Specificity of the protective action of a ribosomal *Shigella* vaccine and the absence of activity in the ribosomes from R mutants. *Zh. Mikrobiol. Epidemiol. Immunobiol.*, **10**, 55–59.
87. Bernstein, D. (2005) Glycoprotein D adjuvant herpes simplex virus vaccine. *Expert Rev. Vaccines*, **4** (5), 615–627.
88. Guo, Q., Zhao, L., You, Q., Yang, Y. et al. (2007) Anti-hepatitis B virus activity of wogonin in vitro and in vivo. *Antiviral Res.*, **74** (1), 16–24.
89. Hu, J.F., Patel, R., Li, B., Garo, E. et al. (2007) Anti-HCV bioactivity of pseudoguaianolides from *Parthenium hispitum*. *J. Nat. Prod.*, **70**(4), 604–607.
90. Girish, C., Pradhan, S.C. (2008) Drug development for liver diseases: focus on picroliv, ellagic acid and curcumin. *Fundam. Clin. Pharmacol.*, **22** (6), 623–632.
91. He, J., Huang, B., Ban, X., Tian, J. et al. (2012) In vitro and in vivo antioxidant activity of the ethanolic extract from *Meconopsis quintuplinervia*. *J. Ethnopharmacol.*, **141** (1), 104–110.
92. Pirotzky, E., Colliez, P., Guilmard, C., Braquet, P. (1989) Renal diseases and platelet activating factor. *Pediatrie*, **44** (3), 163–167.
93. Samoilova, Z.T., Kliukina, S.S., Khaliutina, L.V., Shchitkov, K.G. (1978) Experimental study of azathioprine in nephrotoxic nephritis in rats. *Farmakol. Toksikol.*, **41** (2), 218–222.

94. Hayashi, K., Nagamatsu, T., Ito, M., Hattori, T., Suzuki, Y. (1994) Acetoside, a component of *Stachys sieboldii* Miq, may be a promising antinephritic agent: effect of acteoside on crescentic-type anti-GBM nephritis in rats. *Jpn. J. Pharmacol.*, **65** (2), 143–151.
95. Hattori, T., Furuta, K., Nagao, T., Nagamatsu, T., Ito, M., Suzuki, Y. (1992) Studies on the antinephritic effect of plant components (4): reduction of protein excretion by berberine and coptisine in rats with original-type anti-GBM nephritis. *Jpn. J. Pharmacol.*, **59** (2), 159–169.
96. Yang, L.Y., Chen, A., Kuo, Y.C., Lin, C.Y. (1999) Efficacy of a pure compound h1-a extracted from *Cordyceps sinensis* on autoimmune disease of MRL LPR/LPR mice. *J. Lab. Clin. Med.*, **134** (5), 492–500.
97. Bezruk, P.I., Larianovskaia Iu, B. (1977) Effect of the new s6 substance on the course of acute experimental nephritis in rabbits. *Farm. Zh.*, **5**, 75–78.
98. Dima, V.F., Petrovici, A., Dima, S.V., Petrovici, M., Burghelea, B. (1989) Immunoprotection of guinea pigs against experimental acute pyelonephritis after the administration of *Escherichia coli* purified fimbriae. *Arch. Roum. Pathol. Exp. Microbiol.*, **48** (3), 215–225.
99. Johnston, S.L. (1997) Problems and prospects of developing effective therapy for common cold viruses. *Trends Microbiol.*, **5** (2), 58–63.
100. Miyata, T. (2003) Novel approach to respiratory pharmacology – pharmacological basis of cough, sputum and airway clearance. *Yakugaku Zasshi*, **123** (12), 987–1006.
101. Barros, L., Heleno, S.A., Carvalho, A.M., Ferreira, I.C. (2009) Systematic evaluation of the antioxidant potential of different parts of *Foeniculum vulgare* Mill. from Portugal. *Food Chem. Toxicol.*, **47** (10), 2458–2464.
102. Ge, Y., Liu, J., Su, D. (2009) In vivo evaluation of the anti-asthmatic, antitussive and expectorant activities of extract and fractions from *Elaeagnus pungens* leaf. *J. Ethnopharmacol.*, **126** (3), 538–542.
103. Mahlke, J.D., Boligon, A.A., Machado, M.M., Athayde, M.L. (2012) In vitro toxicity, antiplatelet and acetylcholinesterase inhibition of *Buddleja thyrsoides* Lam. leaves. *Nat. Prod. Res.*, **26** (23), 2223–2226.
104. Bernard, A., Carbonnelle, S., Nickmilder, M., De Burbure, C. (2005) Non-invasive biomarkers of pulmonary damage and inflammation: application to children exposed to ozone and trichloramine. *Toxicol. Appl. Pharmacol.*, **206**(2), 185–190.
105. Bureeva, S., Andia-Pravdivy, J., Kaplun, A. (2005) Drug design using the example of the complement system inhibitors' development. *Drug Discovery Today*, **10** (22), 1535–1542.
106. Chua, K.Y., Huangfu, T., Liew, L.N. (2006) DNA vaccines and allergic diseases. *Clin. Exp. Pharmacol. Physiol.*, **33** (5-6), 546–550.
107. Yukitake, K., Nakamura, Y., Kawahara, M., Nakahara, H. *et al.* (2008) Development of low-cost pulmonary surfactants composed of a mixture of lipids or lipids-peptides using higher aliphatic alcohol or soy lecithin. *Colloids Surf., B*, **66** (2), 281–286.
108. Ameriks, M.K., Venable, J.D. (2009) Small molecule inhibitors of phosphoinositide 3-kinase (pi3k) delta and gamma. *Curr. Top. Med. Chem.*, **9** (8), 738–753.
109. Shibata, T., Kono, T., Tanii, T., Mizuno, N., Hamada, T. (1991) Effects of Ryokan-kyomi-sin-ge-nin-to extract on degranulation of and histamine release from rat mast cells. *Am. J. Chin. Med.*, **19** (3-4), 243–249.
110. Rocha, F.F., Lapa, A.J., De Lima, T.C. (2002) Evaluation of the anxiolytic-like effects of *Cecropia glazioui* Sneth in mice. *Pharmacol. Biochem. Behav.*, **71** (1-2), 183–190.
111. Shin, T.Y., Kim, S.H., Choi, C.H., Shin, H.Y., Kim, H.M. (2004) *Isodon japonicus* decreases immediate-type allergic reaction and tumor necrosis factor-alpha production. *Int. Arch. Allergy Immunol.*, **135** (1), 17–23.
112. Juergens, U.R., Stober, M., Vetter, H. (1998) The anti-inflammatory activity of l-menthol compared to mint oil in human monocytes in vitro: a novel perspective for its therapeutic use in inflammatory diseases. *Eur. J. Med. Res.*, **3** (12), 539–545.
113. Halpern, G.M. (2000) Anti-inflammatory effects of a stabilized lipid extract of *Perna canaliculus* (lyprinol). *Allerg. Immunol. (Paris)*, **32** (7), 272–278.
114. Turker, A.U., Camper, N.D. (2002) Biological activity of common mullein, a

medicinal plant. *J. Ethnopharmacol.*, **82** (2–3), 117–125.
115. Lee, D.H., Kim, S.H., Eun, J.S., Shin, T.Y. (2006) *Mosla dianthera* inhibits mast cell-mediated allergic reactions through the inhibition of histamine release and inflammatory cytokine production. *Toxicol. Appl. Pharmacol.*, **216** (3), 479–484.
116. Kim, S.H., Kwon, T.K., Shin, T.Y. (2008) Antiallergic effects of *Vitis amurensis* on mast cell-mediated allergy model. *Exp. Biol. Med. (Maywood)*, **233** (2), 192–199.
117. Gilani, A.H., Shah, A.J., Zubair, A., Khalid, S. *et al.* (2009) Chemical composition and mechanisms underlying the spasmolytic and bronchodilatory properties of the essential oil of *Nepeta cataria* l. *J. Ethnopharmacol.*, **121** (3), 405–411.
118. Lian, X.Y., Zhang, Z. (2013) Quantitative analysis of *Gleditsia* saponins in the fruits of *Gleditsia sinensis* Lam. by high performance liquid chromatography. *J. Pharm. Biomed. Anal.*, **75**, 41–46.
119. Schuttelkopf, A.W., Andersen, O.A., Rao, F.V., Allwood, M. *et al.* (2006) Screening-based discovery and structural dissection of a novel family 18 chitinase inhibitor. *J. Biol. Chem.*, **281** (37), 27278–27285.
120. Shin, T.Y., Park, J.S., Kim, S.H. (2006) *Artemisia iwayomogi* inhibits immediate-type allergic reaction and inflammatory cytokine secretion. *Immunopharmacol. Immunotoxicol.*, **28** (3), 421–430.
121. Gouda, H., Terashima, S., Iguchi, K., Sugawara, A. *et al.* (2009) Molecular modeling of human acidic mammalian chitinase in complex with the natural-product cyclopentapeptide chitinase inhibitor argifin. *Bioorg. Med. Chem.*, **17** (17), 6270–6278.
122. Andersen, O.A., Nathubhai, A., Dixon, M.J., Eggleston, I.M., Van Aalten, D.M. (2008) Structure-based dissection of the natural product cyclopentapeptide chitinase inhibitor argifin. *Chem. Biol.*, **15** (3), 295–301.
123. Mendonça-Filho R.R. (2006) Bioactive Phytocompounds: New Approaches in the Phytosciences, in: Ahmad, I. , Aqil, F. , Owai, M. (Eds), *Modern Phytomedicine. Turning Medicinal Plants into Drugs*, Wiley-VCH Verlag GmbH & Co. KGaA, Weinheim, pp. 1–24.
124. Arif, T., Bhosale, J.D., Kumar, N., Mandal, T.K. *et al.* (2009) Natural products – antifungal agents derived from plants. *J. Asian Nat. Prod. Res.*, **11** (7), 621–638.
125. Richman, D.D., Margolis, D.M., Delaney, M., Greene, W.C. *et al.* (2009) The challenge of finding a cure for HIV infection. *Science*, **323** (5919), 1304–1307.
126. Fink, R.C., Roschek, B., Jr, Alberte, R.S. (2009) HIV type-1 entry inhibitors with a new mode of action. *Antivir. Chem. Chemother.*, **19** (6), 243–255.
127. Galal, A.M., Slade, D., Gul, W., El-Alfy, A.T. *et al.* (2009) Naturally occurring and related synthetic cannabinoids and their potential therapeutic applications. *Recent Pat. CNS Drug Discov.*, **4** (2), 112–136.
128. Liu, W.T., Ng, J., Meluzzi, D., Bandeira, N. *et al.* (2009) Interpretation of tandem mass spectra obtained from cyclic nonribosomal peptides. *Anal. Chem.*, **81** (11), 4200–4209.
129. Bradac, J., Dieffenbach, C.W. (2009) HIV vaccine development: lessons from the past, informing the future. *IDrugs*, **12** (7), 435–439.
130. Breman, J.G. (2009) Eradicating malaria. *Sci. Prog.*, **92** (Pt 1), 1–38.
131. Burke, B., Gomez-Roman, V.R., Lian, Y., Sun, Y. *et al.* (2009) Neutralizing antibody responses to subtype b and c adjuvanted HIV envelope protein vaccination in rabbits. *Virology*, **387** (1), 147–156.
132. Cafaro, A., Macchia, I., Maggiorella, M.T., Titti, F., Ensoli, B. (2009) Innovative approaches to develop prophylactic and therapeutic vaccines against HIV/AIDS. *Adv. Exp. Med. Biol.*, **655**, 189–242.
133. Flores, J. (2009) Seeking new pathways for HIV vaccine discovery. *Future Microbiol.*, **4** (1), 1–7.
134. Harris, J.E. (2009) Why we don't have an HIV vaccine, and how we can develop one. *Health Aff. (Millwood)*, **28** (6), 1642–1654.
135. Kaufmann, S.H., Meinke, A.L., Von Gabain, A. (2009) Novel vaccination concepts on the basis of modern insights into immunology. *Bundesgesundheitsblatt Gesundheitsforschung Gesundheitsschutz*, **52** (11), 1069–1082.
136. Lapelosa, M., Gallicchio, E., Arnold, G.F., Arnold, E., Levy, R.M. (2009) In silico vaccine design based on molecular simulations of rhinovirus chimeras presenting

137. Lucchese, A., Serpico, R., Crincoli, V., Shoenfeld, Y., Kanduc, D. (2009) Sequence uniqueness as a molecular signature of HIV-1-derived b-cell epitopes. *Int. J. Immunopathol. Pharmacol.*, **22** (3), 639–646.
138. Mcenery, R. (2009) In with the new ... the AIDS vaccine field considers ways to encourage innovation and recruit new minds to the effort. *IAVI Rep.*, **13** (1), 9–13.
139. Mcmurray, D.N., Ly, L.H. (2009) TB vaccines: the paradigms they are shifting. *Expert Rev. Vaccines*, **8** (12), 1615–1618.
140. Ochieng, W., Sauermann, U., Schulte, R., Suh, Y.S. *et al.* (2009) Susceptibility to simian immunodeficiency virus ex vivo predicts outcome of a prime-boost vaccine after SIVMAC239 challenge. *J. Acquir. Immune Defic. Syndr.*, **52** (2), 162–169.
141. Sullivan, M. (2009) Moving candidate vaccines into development from research: lessons from HIV. *Immunol. Cell Biol.*, **87** (5), 366–370.
142. Watkins, D.I. (2009) HIV vaccine development. *Top. HIV Med.*, **17** (2), 35–36.
143. Yang, Q.E. (2009) IgM, not IgG, a key for HIV vaccine. *Vaccine*, **27** (9), 1287–1288.
144. Zhao, J., Lai, L., Amara, R.R., Montefiori, D.C. *et al.* (2009) Preclinical studies of human immunodeficiency virus/AIDS vaccines: inverse correlation between avidity of anti-env antibodies and peak postchallenge viremia. *J. Virol.*, **83** (9), 4102–4111.
145. Appay, V. (2009) 25 years of HIV research! ... and what about a vaccine? *Eur. J. Immunol.*, **39** (8), 1999–2003.
146. Barre-Sinoussi, F. (2009) A significant discovery, now a dream for a cure. *Bull. World Health Org.*, **87** (1), 10–11.
147. Uzair, B., Tabassum, S., Rasheed, M., Rehman, S.F. (2012) Exploring marine cyanobacteria for lead compounds of pharmaceutical importance. *Sci. World J.*, **7**, 179782–179792.
148. Braga, F.G., Bouzada, M.L., Fabri, R.L., De, O.M.M. *et al.* (2007) Antileishmanial and antifungal activity of plants used in traditional medicine in Brazil. *J. Ethnopharmacol.*, **111** (2), 396–402.
149. Sahgal, G., Ramanathan, S., Sasidharan, S., Mordi, M.N., Ismail, S., Mansor, S.M. (2009) Phytochemical and antimicrobial activity of *Swietenia mahagoni* crude methanolic seed extract. *Trop. Biomed.*, **26** (3), 274–279.
150. Botelho, M.A., Nogueira, N.A., Bastos, G.M., Fonseca, S.G. *et al.* (2007) Antimicrobial activity of the essential oil from *Lippia sidoides*, carvacrol and thymol against oral pathogens. *Braz. J. Med. Biol. Res.*, **40** (3), 349–356.
151. Lee, D., Shin, J., Yoon, K.M., Kim, T.I. *et al.* (2008) Inhibition of candida albicans isocitrate lyase activity by sesterterpene sulfates from the tropical sponge *Dysidea* sp. *Bioorg. Med. Chem. Lett.*, **18** (20), 5377–5380.
152. Li, X.C., Jacob, M.R., Khan, S.I., Ashfaq, M.K. *et al.* (2008) Potent in vitro antifungal activities of naturally occurring acetylenic acids. *Antimicrob. Agents Chemother.*, **52** (7), 2442–2448.
153. Dhanawat, M., Das, N., Nagarwal, R.C., Shrivastava, S.K. (2009) Antimalarial drug development: past to present scenario. *Mini Rev. Med. Chem.*, **9** (12), 1447–1469.
154. Queiroz, E.F., Wolfender, J.L., Hostettmann, K. (2009) Modern approaches in the search for new lead antiparasitic compounds from higher plants. *Curr. Drug Targets*, **10** (3), 202–211.
155. Batista, R., Silva Ade, J. Jr, De Oliveira, A.B. (2009) Plant-derived antimalarial agents: new leads and efficient phytomedicines. Part II. Non-alkaloidal natural products. *Molecules*, **14** (8), 3037–3072.
156. Kaur, K., Jain, M., Kaur, T., Jain, R. (2009) Antimalarials from nature. *Bioorg. Med. Chem.*, **17** (9), 3229–3256.
157. Krettli, A.U., Adebayo, J.O., Krettli, L.G. (2009) Testing of natural products and synthetic molecules aiming at new antimalarials. *Curr. Drug Targets*, **10** (3), 261–270.
158. Adams, M., Zimmermann, S., Kaiser, M., Brun, R., Hamburger, M. (2009) A protocol for HPLC-based activity profiling for natural products with activities against tropical parasites. *Nat. Prod. Commun.*, **4** (10), 1377–1381.
159. Metcalf, W.W., Van Der Donk, W.A. (2009) Biosynthesis of phosphonic and phosphinic acid natural products. *Annu. Rev. Biochem.*, **78**, 65–94.
160. Buchanan, M. S., Davis, R. A., Duffy, S., Avery, V. M., Quinn, R. J. (2009) Antimalarial benzylisoquinoline alkaloid from the

rainforest tree *Doryphora sassafras*. *J. Nat. Prod.*, **72**(8), 1541–1543.
161. Fernandez, L.S., Buchanan, M.S., Carroll, A.R., Feng, Y.J. *et al.* (2009) Flinderoles A–C: antimalarial bis-indole alkaloids from *Flindersia* species. *Org. Lett.*, **11** (2), 329–332.
162. Mueller, D., Davis, R.A., Duffy, S., Avery, V.M., Camp, D., Quinn, R.J. (2009) Antimalarial activity of azafluorenone alkaloids from the Australian tree *Mitrephora diversifolia*. *J. Nat. Prod.*, **72** (8), 1538–1540.
163. Zhu, S., Zhang, Q., Gudise, C., Wei, L., Smith, E., Zeng, Y. (2009) Synthesis and biological evaluation of febrifugine analogues as potential antimalarial agents. *Bioorg. Med. Chem.*, **17** (13), 4496–4502.
164. Tobinaga, S., Sharma, M.K., Aalbersberg, W.G., Watanabe, K. *et al.* (2009) Isolation and identification of a potent antimalarial and antibacterial polyacetylene from *Bidens pilosa*. *Planta Med.*, **75** (6), 624–628.
165. Adams, M., Gschwind, S., Zimmermann, S., Kaiser, M., Hamburger, M. (2011) Renaissance remedies: antiplasmodial protostane triterpenoids from *Alisma plantago-aquatica* L. (Alismataceae). *J. Ethnopharmacol.*, **135** (1), 43–49.
166. Mayer, A.M., Rodriguez, A.D., Berlinck, R.G., Hamann, M.T. (2009) Marine pharmacology in 2005–6: marine compounds with anthelmintic, antibacterial, anticoagulant, antifungal, anti-inflammatory, antimalarial, antiprotozoal, antituberculosis, and antiviral activities; affecting the cardiovascular, immune and nervous systems, and other miscellaneous mechanisms of action. *Biochim. Biophys. Acta*, **1790** (5), 283–308.
167. Peach, K.C., Linington, R.G. (2009) New innovations for an old infection: antimalarial lead discovery from marine natural products during the period 2003–2008. *Future Med. Chem.*, **1** (4), 593–617.
168. Lemmens-Gruber, R., Kamyar, M.R., Dornetshuber, R. (2009) Cyclodepsipeptides – potential drugs and lead compounds in the drug development process. *Curr. Med. Chem.*, **16** (9), 1122–1137.
169. Stern, L.J., Calvo-Calle, J.M. (2009) HLA-DR: molecular insights and vaccine design. *Curr. Pharm. Des.*, **15** (28), 3249–3261.
170. Tsuboi, T., Takeo, S., Torii, M. (2009) Breakthrough for the post-genome malaria vaccine candidate discovery: wheat germ cell-free protein production system. *Tanpakushitsu Kakusan Koso*, **54** (8 Suppl.), 1041–1046.
171. Bueno-Sanchez, J.G., Martinez-Morales, J.R., Stashenko, E.E., Ribon, W. (2009) Antitubercular activity of eleven aromatic and medicinal plants occurring in Colombia. *Biomedica*, **29** (1), 51–60.
172. Haebich, D., Von Nussbaum, F. (2009) Lost in transcription–inhibition of RNA polymerase. *Angew. Chem. Int. Ed.*, **48** (19), 3397–3400.
173. Sun, D., Scherman, M.S., Jones, V., Hurdle, J.G. *et al.* (2009) Discovery, synthesis, and biological evaluation of piperidinol analogs with anti-tuberculosis activity. *Bioorg. Med. Chem.*, **17** (10), 3588–3594.
174. Ziaratnia, S.M., Ohyama, K., Hussein, A.A., Muranaka, T. *et al.* (2009) Isolation and identification of a novel chlorophenol from a cell suspension culture of *Helichrysum aureonitens*. *Chem. Pharm. Bull. (Tokyo)*, **57** (11), 1282–1283.
175. Guzman, J.D., Gupta, A., Evangelopoulos, D., Basavannacharya, C. *et al.* (2011) Anti-tubercular screening of natural products from Colombian plants: 3-methoxynordomesticine, an inhibitor of mure ligase of *Mycobacterium tuberculosis*. *J. Antimicrob. Chemother.*, **65** (10), 2101–2107.
176. Oh, K.S., Ryu, S.Y., Lee, S., Seo, H.W. *et al.* (2009) Melanin-concentrating hormone-1 receptor antagonism and anti-obesity effects of ethanolic extract from *Morus alba* leaves in diet-induced obese mice. *J. Ethnopharmacol.*, **122** (2), 216–220.
177. Cai, M., Nyberg, J., Hruby, V. J. (2009) Melanotropins as drugs for the treatment of obesity and other feeding disorders: potential and problems. *Curr. Top. Med. Chem.*, **9** (6), 554–563.
178. Oh, S., Kim, K.S., Chung, Y.S., Shong, M., Park, S.B. (2009) Anti-obesity agents: a focused review on the structural classification of therapeutic entities. *Curr. Top. Med. Chem.*, **9** (6), 466–481.
179. Ramadan, G., El-Beih, N.M., Abd El-Ghffar, E.A. (2009) Modulatory effects of black v. Green tea aqueous extract on hyperglycaemia, hyperlipidaemia and liver dysfunction in diabetic and obese rat models. *Br. J. Nutr.*, **102** (11), 1611–1619.

180. Xu, A., Wang, H., Hoo, R.L., Sweeney, G. *et al.* (2009) Selective elevation of adiponectin production by the natural compounds derived from a medicinal herb alleviates insulin resistance and glucose intolerance in obese mice. *Endocrinology*, **150** (2), 625–633.
181. Tripathi, Y.B. (2009) Bhux: a patented polyherbal formulation to prevent hyperlipidemia and atherosclerosis. *Recent Pat. Inflamm. Allergy Drug Discov.*, **3** (1), 49–57.
182. Classen-Houben, D., Schuster, D., Da Cunha, T., Odermatt, A. *et al.* (2009) Selective inhibition of 11beta-hydroxysteroid dehydrogenase 1 by 18alpha-glycyrrhetinic acid but not 18beta-glycyrrhetinic acid. *J. Steroid Biochem. Mol. Biol.*, **113** (3-5), 248–252.
183. Kanduc, D. (2009) "Self-nonself" peptides in the design of vaccines. *Curr. Pharm. Des.*, **15** (28), 3283–3289.
184. Mukherjee, P.K., Maity, N., Nema, N.K., Saerkar, B.K. (2011) Bioactive compounds from natural resources against skin aging, *Phytomedicine*, **19**, 64–73
185. Barrantes, E., Guinea, M. (2003) Inhibition of collagenase and metalloproteinases by aloins and aloe gel. *Life Sci.*, **72**, 843–850.
186. Kim, S., Kim, Y., Kim, J.E., Cho, K.H., Chung, J.H. (2008) Berberine inhibits TPA-induced MMP-9 and IL-6 expression in normal human keratinocytes. *Phytomedicine*, **15**, 340–347.
187. Nichols, J.A., Katiyar, S.K. (2010) Skin photoprotection by natural polyphenols: antiinflammatory, antioxidant and DNA repair mechanisms. *Arch. Dermatol. Res.*, **302**, 71–83.
188. Maquart, F.X., Bellon, G., Gillery, P., Wegrowski, Y., Borel, J.P. (1990) Stimulation of collagen synthesis in fibroblast cultures by a triterpene extracted from *Centella asiatica*. *Connect. Tissue Res.*, **24**, 107–120
189. Cimino, F., Cristani, M., Saija, A., Bonina, F.P., Virgili, F. (2007) Protective effects of a red orange extract on UVB-induced damage in human keratinocytes. *Biofactors*, **30**, 129–138.
190. Lee, S.Y., Kim, M.R., Choi, H.S., Moon, H.I. *et al.* (2009). The effect of curculigoside on the expression of matrix metalloproteinase-1 in cultured human skin fibroblasts. *Arch. Pharm. Res.*, **32**, 1433–1439.
191. Sumiyoshi, M., Kimura, Y. (2009) Effects of a turmeric extract (*Curcuma longa*) on chronic ultraviolet B irradiation-induced skin damage in melanin-possessing hairless mice. *Phytomedicine*, **16**, 1137–1143
192. Oh, H.I., Shim, J.S., Gwon, S.H., Kwon, H.J., Hwang, J.K. (2009) The effect of xanthorrhizol on the expression of matrix metalloproteinase-1 and type-I procollagen in ultraviolet-irradiated human skin fibroblasts. *Phytother. Res.*, **23**, 1299–1302.
193. Tada, Y., Kanda, N., Haratake, A., Tobiishi, M., Uchiwa, H., Watanabe, S. (2009) Novel effects of diosgenin on skin aging. *Steroids*, **74**, 504–511.
194. Lee, B.C., Lee, S.Y., Lee, H.J., Sim, G.S. *et al.* (2007) Anti-oxidative and photoprotective effects of coumarins isolated from *Fraxinus chinensis*. *Arch. Pharm. Res.*, **30**, 1293–1301.
195. Tsoyi, K., Hyung, B.P., Young, M.K., Jong, I.L.C. *et al.* (2008) Protective effect of anthocyanins from black soybean seed coats on UVB-induced apoptotic cell death in vitro and in vivo. *J. Agric. Food Chem.*, **56**, 10600–10605.
196. Moon, H.I., Lee, J., Ok, P.Z., Chung, J.H. (2005) The effect of flavonol glycoside on the expressions of matrix metalloproteinase-1 in ultraviolet-irradiated cultured human skin fibroblasts. *J. Ethnopharmacol.*, **101**, 176–179.
197. Tanaka, K., Hasegawa, J., Asamitsu, K., Okamoto, T. (2007) *Magnolia ovovata* extract and its active component magnolol prevent skin photoaging via inhibition of nuclear factor-B. *Eur. J. Pharmacol.*, **565**, 212–219.
198. Cho, Y.H., Kim, J.H., Sim, G.S., Lee, B.C. *et al.* (2006) Inhibitory effects of antioxidant constituents from *Melothria heterophylla* on matrix metalloproteinase-1 expression in UVA-irradiated human dermal fibroblasts. *J. Cosmet. Sci.*, **57**, 279–289.
199. Cho, S., Won, C.H., Lee, D.H., Lee, M.J. et al. (2009) Red ginseng root extract mixed with Torilus fructus and Corni fructus improves facial wrinkles and increases type I procollagen synthesis in human skin: a randomized, double-blind, placebo-controlled study. *J. Med. Food.*, **12**, 1252–1259.
200. Mula, S., Banerjee, D., Patro, B.S., Bhattacharya, S. *et al.* (2008) Inhibitory

property of the *Piper betel* phenolics against photosensitization-induced biological damages. *Bioorg. Med. Chem.*, **16**, 2932–2938.

201. Del, V.A.A., Vanegas-Espinoza, P.E., Paredes-López, O. (2010) Marigold regeneration and molecular analysis of carotenogenic genes. *Methods Mol. Biol.*, **589**, 213–222.

202. Kim, S.J., Sancheti, S.A., Sancheti, S.S., Um, B.H. *et al.* (2010). Effect of 1,2,3,4,6-penta-O-galloyl-β-D-glucose on elastase and hyaluronidase activities and its type II collagen expression. *Acta Polon. Pharm.: Drug Res.*, **67**, 145–150.

19
Neurological Biomarkers

Henrik Zetterberg[1,2]
[1] *The Sahlgrenska Academy at the University of Gothenburg, Department of Psychiatry and Neurochemistry, Clinical Neurochemistry Laboratory, Institute of Neuroscience and Physiology, Sahlgrenska University Hospital, Mölndal, S-431 80, Sweden*
[2] *UCL Institute of Neurology, Department of Molecular Neuroscience, Queen Square, London WC1N 3BG, UK*

1	**Introduction** 685	
2	**Fluid Biomarkers for Alzheimer's Disease** 686	
2.1	CSF Aβ42 686	
2.2	CSF T-Tau 687	
2.3	CSF P-Tau 687	
2.4	Diagnostic Performance of Combined CSF T-Tau, P-Tau, and Aβ42 Tests 688	
2.5	Longitudinal Changes in CSF AD Biomarkers, and use in Clinical Trials 688	
2.6	Other Potential AD Biomarkers 688	
3	**Fluid Biomarkers for Parkinson's Disease** 689	
3.1	α-Synuclein 689	
3.2	Other Potential PD Biomarkers 690	
4	**Fluid Biomarkers for Traumatic Brain Injury (TBI)** 690	
4.1	Fluid Biomarkers for Acute TBI 690	
4.2	Fluid Biomarkers for Chronic Traumatic Encephalopathy (CTE) 691	
5	**Concluding Remarks** 691	
	Acknowledgments 691	
	References 691	

Translational Medicine: Molecular Pharmacology and Drug Discovery
First Edition. Edited by Robert A. Meyers.
© 2018 Wiley-VCH Verlag GmbH & Co. KGaA. Published 2018 by Wiley-VCH Verlag GmbH & Co. KGaA.

Keywords

Biomarker
An objective measure that reflects biological or pathophysiological processes.

Cerebrospinal fluid
A fluid located in the cerebral ventricles and around the brain and spinal cord. It communicates freely with the brain interstitial fluid, and may thereby serve as a "biochemical window" into the brain.

Alzheimer's disease
The most common of neurodegenerative diseases, characterized by extracellular senile plaques, intraneuronal neurofibrillary tangles, and neuronal loss that cause memory problems and other cognitive symptoms.

Parkinson's disease
The second most common neurodegenerative disease, in which dopaminergic brain circuits degenerate, causing movement disturbances.

Traumatic brain injury
Injury to brain cells by external mechanical forces.

Chronic traumatic brain injuryn
Chronic neurological and/or psychiatric symptoms following repetitive brain injury.

▎Neurological disorders are characterized by neuronal impairment that eventually may lead to neuroaxonal degeneration and death. In spite of the brain's known capacity for regeneration, neuronal loss is an irreversible phenomenon linked to many neurodegenerative disorders. Therefore, drugs aimed at inhibiting injurious processes to the brain are likely to be most effective if the treatment is initiated as early as possible. However, clinical manifestations in early disease stages are often variable, subtle, and difficult to diagnose. This is where biomarkers that reflect onset of pathology may have a profound impact on clinical diagnosis making. A triplet of cerebrospinal fluid biomarkers for Alzheimer's disease (AD), the most common neurodegenerative disease, total and phosphorylated tau and the 42-amino acid isoform of amyloid β, has already been established for early detection of AD. However, more biomarkers are needed for both AD and other neurological disorders, such as Parkinson's disease and traumatic brain injury. The background and recent developments in the field of neurological biomarkers in body fluids are detailed in this chapter.

1 Introduction

During recent years, there has been considerable excitement about the possibility of the early or even preclinical detection of several neurological disorders through the use of biomarkers. Although it is now well known that the human adult brain has a certain capacity of regeneration [1], the process is ineffective and dead neurons are difficult to replace. Thus, a major aim of biomarker research in neurology has been to provide clinicians with biochemical and neuroimaging tools that enable accurate diagnosis before widespread neuronal death has occurred. Such biomarkers are a prerequisite for effective secondary prevention strategies, and may also facilitate therapeutics development by: (i) providing measures of desired biochemical effects of a compound in the preclinical drug discovery phase, in animal studies and in Phase I clinical trials; (ii) allowing the specific inclusion of early cases that are very likely to develop the disease of interest; and (iii) identifying subgroups of patients according to biochemical patterns to individualize treatment. In addition, biomarker changes during disease progression and correlations between different biomarkers may provide clues on pathogenic mechanisms. Several proof-of-concept studies have already been conducted, most of them using the Alzheimer's disease (AD) cerebrospinal fluid (CSF) biomarkers total and phosphorylated tau proteins [T-tau (total tau) and P-tau (phospho-epitopes on tau), respectively] that reflect the axonal component of the disease and the 42-amino acid fragment of amyloid β (Aβ42) that reflects plaque pathology in the brain to diagnose AD before onset of dementia [2]. Whether these biomarkers can be used to predict response to treatment is still unknown, however, and there is also a lack of biomarkers that reflect disease progression [3]. Further, several studies have explored biomarkers for α-synuclein pathology associated with Parkinson's disease (PD), dementia with Lewy bodies (DLB), and multiple system atrophy (MSA), and a consensus has emerged that these so-called "synucleinopathies" may be detected by reduced levels of α-synuclein in the CSF [4]. Another neurological condition in which considerable progress has been made in the development of biomarkers in both CSF and blood is traumatic brain injury (TBI).

In this chapter, an overview is provided of recent developments in the fluid biomarker field of neurological diseases, exemplified by AD, PD, and TBI biomarkers, and their potential clinical implementation and role in drug development and prevention are discussed. AD, PD, and TBI were not selected because they share pathogenic mechanisms; indeed, most available data suggest that these conditions have distinct etiologies and affect different neuronal subpopulations, at least in their early stages. However, they all share neuronal injury and the brain's response to such injury as parts of their pathologic processes. Further, they are the neurodegenerative diseases that have been the most extensively studied from a biomarker perspective. The most well established fluid biomarkers for AD, PD and TBI are summarized in Fig. 1. Much of what is reviewed in regards to concepts is also relevant to other neurodegenerative conditions, although some of the pathology-related molecules may be different. The reader is referred to focused overviews regarding the vast field

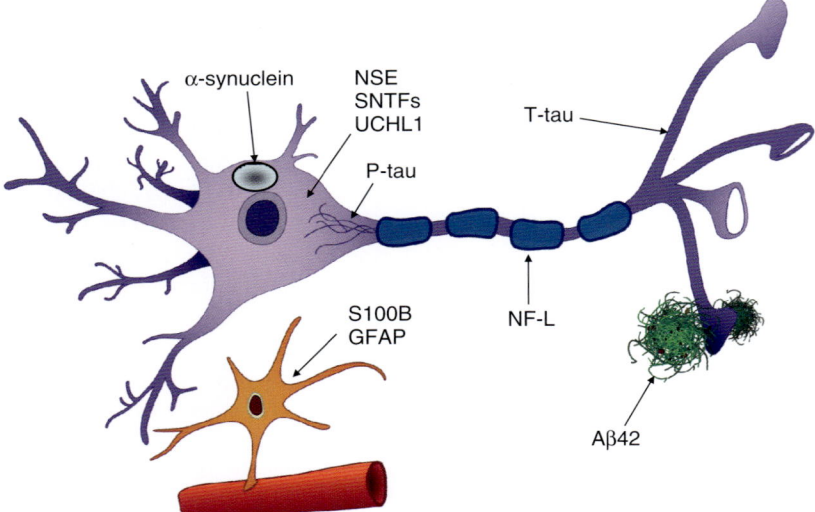

Fig. 1 Established and novel CSF and blood biomarkers for neuropathological changes in Alzheimer's and Parkinson's disease and traumatic brain injury. The 42-amino acid isoform of amyloid β (Aβ42), phospho-tau (P-tau), and total tau (T-tau) reflect plaque pathology, neurofibrillary tangle pathology, and cortical axonal degeneration, respectively, in Alzheimer's disease. α-Synuclein may reflect Lewy body pathology in Parkinson's disease, dementia with Lewy bodies and multiple system atrophy. Neurofilament light (NF-L) is a marker of large-caliber myelinated axons and has proven useful in a wide range of neurological diseases. S100B and glial fibrillary acidic protein (GFAP) are markers of astroglial activation and/or injury. Neuron-specific enolase (NSE), ubiquitin C-terminal hydrolase L1 (UCHL1), and α-spectrin N-terminal fragments (SNTFs) may indicate traumatic brain injury-related injury to the neuronal soma.

of imaging biomarkers for neurological diseases.

2
Fluid Biomarkers for Alzheimer's Disease

Alzheimer's disease, the most common neurodegenerative disease, is characterized by an accumulation of aggregation-prone Aβ42 in extracellular senile plaques, an intraneuronal accumulation of P-tau in neurofibrillary tangles, and the degeneration of brain circuits involved in memory consolidation and other higher brain functions [5]. Currently, the disease affects more than 20 million people worldwide, and its prevalence is expected to increase due to increased life expectancy [5].

2.1
CSF Aβ42

Patients with AD have reduced CSF levels of the 42-amino acid aggregation-prone variant of amyloid β (Aβ42) [6]. The reduction reflects Aβ42 retention in senile plaques in the brain, as evidenced by both autopsy and *in-vivo* brain biopsy and imaging studies [7–10]. The first report of a reduction in CSF Aβ42 levels in patients with mild cognitive impairment (MCI) who later progressed to AD dementia was made in 1999 [11]. Since then, numerous studies have verified that low levels of CSF Aβ42 are highly predictive of future AD, in both MCI patients [12–15] and cognitively normal individuals [16–18]. CSF Aβ42 is the earliest recognized non-genetic biomarker in AD, and although progress is currently being made in the development of reference methods and materials to standardize the Aβ42 biomarker [19], it is already possible to monitor Aβ42 levels in the CSF as a clinical routine. Moreover, this can be achieved in a manner that reliably reflects cerebral β-amyloidosis with longitudinal stability, by using commercially available assays, provided that certain quality control measures are taken into account [20]. Aβ42 levels can also be measured in plasma, but this has no relation to Aβ homeostasis in the brain, most likely due to its production in peripheral compartments such as blood platelets [21], and its active degradation by proteases in the blood [22, 23].

2.2
CSF T-Tau

Tau is an intra-axonal, microtubule-stabilizing protein that is highly expressed in the cerebral cortex [24]. T-tau denotes tau measured using assays employing antibodies to the mid-domain of tau, which is common to all known tau isoforms, irrespective of the phosphorylation state. Dying neurons release T-tau to the brain extracellular fluid, which communicates freely with the CSF. AD patients have elevated T-tau levels [25, 26], a finding that has been confirmed in hundreds of reports, using different assays and in several clinical contexts [6]. CSF T-tau levels have been shown to correlate with imaging measures of hippocampal atrophy [27] and gray matter degeneration [28]. In response to acute brain injury, CSF T-tau levels are dynamic; they increase during the first few days following the injury and remain elevated for a few weeks until they normalize [29, 30]. This has led to the view that elevated CSF T-tau levels reflect ongoing axonal degeneration, which in turn may indicate disease intensity. CSF T-tau predicts the malignancy of the clinical course in AD; the higher the levels, the more rapid the clinical disease progression [31]. CSF T-tau also predicts progression from normal cognition to MCI and dementia due to AD [32]. The most rapidly progressive neurodegenerative condition currently known – Creutzfeldt–Jakob disease – is characterized by very pronounced CSF T-tau elevations that are often orders of magnitude higher than what is typically seen in AD [33, 34].

2.3
CSF P-Tau

Neurofibrillary tangles are composed of hyperphosphorylated tau proteins [35], and assays using antibodies against different P-tau have been developed [36]. CSF P-tau levels correlate with neurofibrillary tangle pathology in AD [10, 37], but are normal in other tauopathies such as frontotemporal dementia or progressive supranuclear palsy (PSP) [36]. It is possible that these disorders show disease-specific tau phosphorylation, or that tau is processed or truncated in a way that is not recognized by available assays (this is an area in need of further research). There are at present only three conditions in addition to AD in which elevated CSF P-tau levels have been reported:

- In term and pre-term newborns, possibly reflecting physiological tau phosphorylation during brain development [38].
- In herpes encephalitis, where the virus undergoes lytic replication in temporal lobe neurons [39].

- In superficial siderosis, where hemosiderin deposits in the central nervous system due to prolonged or recurrent low-grade bleeding into the CSF [40, 41].

Clearly, these conditions are not important differential diagnoses to AD, but they may shed light on the mechanisms behind CSF P-tau increases. Intriguingly, tau is phosphorylated during hibernation [42, 43] and anesthesia [44], which has led to the hypothesis that tau phosphorylation may reflect a downregulation of neuronal activity.

2.4
Diagnostic Performance of Combined CSF T-Tau, P-Tau, and Aβ42 Tests

Multiple studies have been conducted to investigate the diagnostic accuracy of combined CSF tests for T-tau, P-tau, and Aβ42 [2]. These results obtained collectively point to sensitivities and specificities of 85–95% in cross-sectional AD-control studies, as well as in longitudinal studies of patients fulfilling MCI criteria [2]. The results have been validated in autopsy-confirmed AD patients and healthy controls [45]. Whilst superior diagnostic performances are typically seen in single-center studies [12, 46], multicenter studies tend to report slightly lower sensitivities and specificities [14, 47], indicating a need for standardization. Typically, AD-associated brain changes occur predominantly in subjects with positive results for both Aβ and tau biomarkers, emphasizing the need to use the two in concert [48].

2.5
Longitudinal Changes in CSF AD Biomarkers, and use in Clinical Trials

Recently acquired data have shown that it is possible to identify longitudinal changes in CSF Aβ42, T-tau, and P-tau in cognitively healthy controls followed with repeated lumbar punctures over several years [49–51], though CSF AD biomarkers have generally been found to be essentially stable in symptomatic AD [52–54]. Such biomarker stability may be useful in clinical trials to help identify the biochemical effects of interventions. Biomarker measurements may be used track drug-induced changes in specific Aβ and amyloid precursor protein (APP) metabolites [55–58], as well as downstream drug effects on secondary phenomena, such as reduced axonal degeneration as indicated by CSF tau levels [59, 60]. These data would facilitate evaluation of the overall results of the trials and also help in the design of new trials. Specifically, biomarkers would allow to address whether the results of a negative trial (which is a major problem at present in the AD research field [61]) could be explained by a lack of drug effects on the intended target, or if the intended target had changed in the expected direction but had not resulted in any clinical benefit. A clinical benefit in a positive trial would clearly be a considerably stronger finding if it were backed by expected biomarker changes.

2.6
Other Potential AD Biomarkers

In the amyloidogenic pathway, Aβ is produced through the proteolytic processing of APP by β-and γ-secretases. The major β-secretase in the brain is the β-site APP-cleaving enzyme 1 (BACE1) [62], and increased BACE1 activity has been monitored not only in the postmortem brain samples of AD patients [63] but also in their CSF [64, 65]. CSF BACE1 activity appears to be highest during the pre-dementia stages of the disease but to fall in end-stage AD dementia [66, 67]. These changes are less clear than those for CSF tau and Aβ markers and lack diagnostic utility, but CSF BACE1 may still serve as a valuable marker in clinical trials of potential BACE1 inhibitors. In addition to these Aβ-related biomarkers, there is a growing body of literature on AD-associated changes of CSF biomarkers for neuroinflammation, microglial activation, and synaptic degeneration [68, 69]. These biomarker changes are less well established than the plaque- and tangle-associated biomarkers, but collectively most studies support low-grade inflammation, microglial activation and synaptic degeneration as important elements of the AD process [68, 69]. Although, to date, no reliable blood biomarkers for AD exist [2], promising plasma lipidomics pilot data have recently been reported, but were in need of replication [70].

3
Fluid Biomarkers for Parkinson's Disease

Parkinson's disease is a progressive neurodegenerative movement disorder that is characterized by bradykinesia and at least one of three other cardinal signs: resting tremor; rigidity; and impaired postural reflexes [71]. There is usually a good and sustained response to levodopa treatment, which is contrasted by its often limited therapeutic efficacy in atypical parkinsonian syndromes – a heterogeneous group of neurodegenerative disorders that are distinct from PD but share its central characteristic of akinetic rigidity [72]. The "atypical" descriptor indicates the presence of features, such as early autonomic failure and cerebellar/pyramidal signs in MSA, supranuclear down gaze palsies and dysexecutive syndrome in PSP, dystonia, myoclonus, and cortical sensory loss in corticobasal degeneration (CBD), and a more rapid deterioration with earlier postural instability and falls in all three disorders [72]. The duration of disease to death in most atypical cases is usually only a few years. Based on their underlying pathologies, parkinsonian syndromes can be differentiated into synucleinopathies (PD, MSA) and tauopathies (PSP, corticobasal syndrome).

3.1
α-Synuclein

Predominantly expressed in the presynapse, the 140-amino acid-long α-synuclein has been found to be the major constituent of the intracellular aggregates in Lewy bodies, the pathological hallmark of PD and DLB, and in the glial cytoplasmic inclusions of MSA [73]. Aggregated α-synuclein is often acetylated and C-terminally truncated [74], and the pathology appears to transmit from cell to cell [75]. Full-length α-synuclein is secreted and has been detected in the CSF, plasma, cell media and, most recently, in saliva [76].

The quantification of extracellular α-synuclein has been proposed as a potential biomarker for synuclein-related diseases. Discrepant findings have been reported by a number of investigators using several different platforms, assays and standard operating procedures, but most studies have shown a reduction in CSF total α-synuclein in the synuclein-related disorders PD, DLB, and MSA [4], which may reflect sequestration of the protein in Lewy body inclusions. However, the overlap is too wide to provide diagnostically useful information on a case-by-case basis, and assays that are more specific for the presence of aggregated α-synuclein are warranted. Reports have been made on the detection of oligomerized α-synuclein in CSF, but these results are difficult to interpret due to high inter-individual variation [77, 78].

3.2 Other Potential PD Biomarkers

Other than CSF α-synuclein, there is a lack of promising fluid biomarkers for PD, although CSF levels of neurofilament light (NF-L) have emerged as a potential biomarker to differentiate PD from atypical parkinsonian disorders [79]. Normally, NF-L acts as an integral part of the neural cytoskeleton, providing structural support for predominantly large-caliber myelinated axons. Injury to such axons in atypical parkinsonian disorders results in an elevated CSF level of NF-L, which is not seen in PD [79, 80]. An increased risk of cognitive deterioration in PD may be detected using CSF Aβ42, reduced levels of which indicate a risk of future dementia in PD patients [81, 82]. It would be logical to assume that CSF T-tau and P-tau levels would differentiate between tauopathies and synucleinopathies in movement disorders, but no such changes have been detected [83]; however, some promising pilot data exist for the selective reduction of the four-repeat tau isoform in PSP [84]. As with PD, reliable diagnostic or prognostic biomarkers in blood are lacking; however, a well-replicated finding in PD is a reduced serum level of urate (a natural antioxidant and iron chelator) [85], which implicates a role for oxidative stress in PD pathogenesis.

4 Fluid Biomarkers for Traumatic Brain Injury (TBI)

Moderate to severe TBI, in which intracranial bleedings and/or mass lesions occur following a blow to the head, is easily diagnosed by clinical examination and using standard neuroimaging techniques. Mild TBI or concussion (defined as "a head trauma resulting in brief loss of consciousness and/or alteration of mental state") is much more difficult to diagnose, however. Typically, concussion causes no gross pathology, such as hemorrhage, and no abnormalities on a conventional computed tomography scan of the brain [86], but rather a rapid-onset neuronal dysfunction that resolves in a spontaneous manner over a few days to a few weeks. Approximately 15% of concussion patients suffer persisting cognitive dysfunction [87, 88], with diffuse axonal injury (DAI) appearing to be the most important underlying pathology in such cases [89]. It has also been shown that repetitive concussions increase the risk of chronic traumatic encephalopathy (CTE), a condition that is described in boxers and other sports athletes as well as in military veterans, and is characterized by chronic and sometimes progressive neurological and/or psychiatric symptoms following repetitive brain injury [90]. Fluid biomarkers for neuronal, axonal and astroglial damage would be valuable to diagnose mild TBI in patients with head trauma, to predict short- and long-term clinical outcome, and to determine when the brain has recovered following TBI.

4.1 Fluid Biomarkers for Acute TBI

It is possible to monitor axonal injury in concussion using CSF and plasma levels of the axonal proteins NF-L and tau [30, 91–93]. A recent case report on a knocked-out amateur boxer showed that it took eight months before his CSF NF-L levels normalized [94]. This result resonates with neuropathological analyses showing that axonopathy can continue for years after TBI

[95]. Other candidate blood biomarkers for brain injury in concussion include neuron-specific enolase (NSE), ubiquitin C-terminal hydrolase L1 (UCHL1), and α-spectrin N-terminal fragments (SNTFs), as well as the glia-enriched S100B and glial fibrillary acidic protein [96]. Unfortunately, none of these markers has a prognostic relationship with patient outcomes for mild TBI with negative head computed tomography findings [96].

4.2
Fluid Biomarkers for Chronic Traumatic Encephalopathy (CTE)

There are no established fluid biomarkers for CTE, although the gross morphological changes seen in advanced CTE may be visualized using standard neuroimaging techniques [97]. Biomarker studies of the molecular pathology of CTE are needed.

5
Concluding Remarks

CSF T-tau, P-tau, and Aβ42 are clinically useful markers of AD pathology. In the absence of definitive results from ongoing standardization efforts, it is still possible to upgrade currently available tests for routine clinical laboratory use [20]. In the case of PD, there is currently a lack of diagnostically useful biomarkers, though some markers – notably NF-L – may be helpful when differentiating PD from atypical parkinsonian disorders. CSF Aβ42 levels may pinpoint PD patients at increased risk of future cognitive decline. Several CSF biomarkers for concussion exist and, by employing ultrasensitive techniques, these are now being developed into blood tests. Reliable blood tests for major chronic neurological disorders still need to be developed, and the blood–brain barrier is a major obstacle in this quest. However, for some disorders that show not only neurological but also peripheral involvement (e.g., variant Creutzfeldt–Jakob disease, Huntington's disease), a consistent detection of pathological proteins in the blood has been achieved [98, 99].

Acknowledgments

Studies conducted at the author's laboratory were supported by the Swedish Research Council, Alzheimer's Association, Swedish State Support for Clinical Research and the Wolfson Foundation.

References

1. Falk, A. and Frisen, J. (2005) New neurons in old brains. *Ann. Med.*, **37** (7), 480–486.
2. Blennow, K., Hampel, H., Weiner, M., and Zetterberg, H. (2010) Cerebrospinal fluid and plasma biomarkers in Alzheimer disease. *Nat. Rev. Neurol*, **6** (3), 131–144.
3. Hampel, H., Frank, R., Broich, K., Teipel, S.J. et al. (2010) Biomarkers for Alzheimer's disease: academic, industry and regulatory perspectives. *Nat. Rev. Drug Discov.*, **9** (7), 560–574.
4. Zetterberg, H., Petzold, M., and Magdalinou, N. (2014) Cerebrospinal fluid alpha-synuclein levels in Parkinson's disease – changed or unchanged? *Eur. J. Neurol.*, **21** (3), 365–367.
5. Blennow, K., De Leon, M.J., and Zetterberg, H. (2006) Alzheimer's disease. *Lancet*, **368** (9533), 387–403.
6. Rosen, C., Hansson, O., Blennow, K., and Zetterberg, H. (2013) Fluid biomarkers in Alzheimer's disease – current concepts. *Mol. Neurodegener.*, **8**, 20.
7. Strozyk, D., Blennow, K., White, L.R., and Launer, L.J. (2003) CSF Abeta 42 levels correlate with amyloid-neuropathology in a population-based autopsy study. *Neurology*, **60** (4), 652–656.
8. Fagan, A.M., Mintun, M.A., Mach, R.H., Lee, S.Y. et al. (2006) Inverse relation between in vivo amyloid imaging load and cerebrospinal

fluid Abeta42 in humans. *Ann. Neurol.*, **59** (3), 512–519.
9. Forsberg, A., Engler, H., Almkvist, O., Blomquist, G. *et al.* (2008) PET imaging of amyloid deposition in patients with mild cognitive impairment. *Neurobiol. Aging*, **29** (10), 1456–1465.
10. Seppala, T.T., Nerg, O., Koivisto, A.M., Rummukainen, J. *et al.* (2012) CSF biomarkers for Alzheimer disease correlate with cortical brain biopsy findings. *Neurology*, **78** (20), 1568–1575.
11. Andreasen, N., Minthon, L., Vanmechelen, E., Vanderstichele, H. *et al.* (1999) Cerebrospinal fluid tau and Abeta42 as predictors of development of Alzheimer's disease in patients with mild cognitive impairment. *Neurosci. Lett.*, **273** (1), 5–8.
12. Hansson, O., Zetterberg, H., Buchhave, P., Londos, E. *et al.* (2006) Association between CSF biomarkers and incipient Alzheimer's disease in patients with mild cognitive impairment: a follow-up study. *Lancet Neurol.*, **5** (3), 228–234.
13. Shaw, L.M., Vanderstichele, H., Knapik-Czajka, M., Clark, C.M. *et al.* (2009) Cerebrospinal fluid biomarker signature in Alzheimer's disease neuroimaging initiative subjects. *Ann. Neurol.*, **65** (4), 403–413.
14. Visser, P.J., Verhey, F., Knol, D.L., Scheltens, P. *et al.* (2009) Prevalence and prognostic value of CSF markers of Alzheimer's disease pathology in patients with subjective cognitive impairment or mild cognitive impairment in the DESCRIPA study: a prospective cohort study. *Lancet Neurol.*, **8** (7), 619–627.
15. Buchhave, P., Minthon, L., Zetterberg, H., Wallin, A.K. *et al.* (2012) Cerebrospinal fluid levels of beta-amyloid 1-42, but not of tau, are fully changed already 5 to 10 years before the onset of Alzheimer dementia. *Arch. Gen. Psychiatry*, **69** (1), 98–106.
16. Skoog, I., Davidsson, P., Aevarsson, O., Vanderstichele, H, *et al.* (2003) Cerebrospinal fluid beta-amyloid 42 is reduced before the onset of sporadic dementia: a population-based study in 85-year-olds. *Dement. Geriatr. Cogn. Disord.*, **15** (3), 169–176.
17. Fagan, A.M., Head, D., Shah, A.R., Marcus, D. *et al.* (2009) Decreased cerebrospinal fluid Abeta(42) correlates with brain atrophy in cognitively normal elderly. *Ann. Neurol.*, **65** (2), 176–183.
18. Gustafson, D.R., Skoog, I., Rosengren, L., Zetterberg, H. *et al.* (2007) Cerebrospinal fluid beta-amyloid 1-42 concentration may predict cognitive decline in older women. *J. Neurol. Neurosurg. Psychiatry*, **78** (5), 461–464.
19. Carrillo, M.C., Blennow, K., Soares, H., Lewczuk, P. *et al.* (2013) Global standardization measurement of cerebral spinal fluid for Alzheimer's disease: an update from the Alzheimer's Association Global Biomarkers Consortium. *Alzheimers Dement.*, **9** (2), 137–140.
20. Palmqvist, S., Zetterberg, H., Blennow, K., Vestberg, S. *et al.* (2014) Accuracy of brain amyloid detection in clinical practice using cerebrospinal fluid beta-Amyloid 42: a cross-validation study against amyloid positron emission tomography. *JAMA Neurol.*, **71**, 1282–1289.
21. Zetterberg, H. and Blennow, K. (2006) Plasma Abeta in Alzheimer's disease-up or down? *Lancet Neurol.*, **5** (8), 638–639.
22. Tiribuzi, R., Crispoltoni, L., Porcellati, S., Di Lullo, M. *et al.* (2014) miR128 up-regulation correlates with impaired amyloid beta(1-42) degradation in monocytes from patients with sporadic Alzheimer's disease. *Neurobiol. Aging*, **35** (2), 345–356.
23. Saido, T. and Leissring, M.A. (2012) Proteolytic degradation of amyloid beta-protein. *Cold Spring Harbor Perspect. Med.*, **2** (6), a006379.
24. Trojanowski, J.Q., Schuck, T., Schmidt, M.L., and Lee, V.M. (1989) Distribution of tau proteins in the normal human central and peripheral nervous system. *J. Histochem. Cytochem.*, **37** (2), 209–215.
25. Vandermeeren, M., Mercken, M., Vanmechelen, E., Six, J. *et al.* (1993) Detection of tau proteins in normal and Alzheimer's disease cerebrospinal fluid with a sensitive sandwich enzyme-linked immunosorbent assay. *J. Neurochem.*, **61** (5), 1828–1834.
26. Blennow, K., Wallin, A., Agren, H., Spenger, C. *et al.* (1995) Tau protein in cerebrospinal fluid: a biochemical marker for axonal degeneration in Alzheimer disease? *Mol. Chem. Neuropathol.*, **26** (3), 231–245.
27. Wang, L., Fagan, A.M., Shah, A.R., Beg, M.F. *et al.* (2012) Cerebrospinal fluid proteins predict longitudinal hippocampal degeneration in early-stage dementia of the Alzheimer

type. *Alzheimer Dis. Assoc. Disord.*, **26** (4), 314–321.
28. Glodzik, L., Mosconi, L., Tsui, W., de Santi, S. *et al.* (2012) Alzheimer's disease markers, hypertension, and gray matter damage in normal elderly. *Neurobiol. Aging*, **33** (7), 1215–1227.
29. Hesse, C., Rosengren, L., Andreasen, N., Davidsson, P. *et al.* (2001) Transient increase in total tau but not phospho-tau in human cerebrospinal fluid after acute stroke. *Neurosci. Lett.*, **297** (3), 187–190.
30. Zetterberg, H., Hietala, M.A., Jonsson, M., Andreasen, N. *et al.* (2006) Neurochemical aftermath of amateur boxing. *Arch. Neurol.*, **63** (9), 1277–1280.
31. Wallin, A.K., Blennow, K., Zetterberg, H., Londos, E. *et al.* (2010) CSF biomarkers predict a more malignant outcome in Alzheimer disease. *Neurology*, **74** (19), 1531–1537.
32. Toledo, J.B., Weiner, M.W., Wolk, D.A., Da, X. *et al.* (2014) Neuronal injury biomarkers and prognosis in ADNI subjects with normal cognition. *Acta Neuropathol. Commun.*, **2** (1), 26.
33. Sanchez-Juan, P., Sanchez-Valle, R., Green, A., Ladogana, A. *et al.* (2007) Influence of timing on CSF tests value for Creutzfeldt–Jakob disease diagnosis. *J. Neurol.*, **254** (7), 901–906.
34. Skillback, T., Rosen, C., Asztely, F., Mattsson, N. *et al.* (2014) Diagnostic performance of cerebrospinal fluid total tau and phosphorylated tau in Creutzfeldt–Jakob disease: results from the Swedish Mortality Registry. *JAMA Neurol.*, **71** (4), 476–483.
35. Grundke-Iqbal, I., Iqbal, K., Tung, Y.C., Quinlan, M. *et al.* (1986) Abnormal phosphorylation of the microtubule-associated protein tau (tau) in Alzheimer cytoskeletal pathology. *Proc. Natl Acad. Sci. USA*, **83** (13), 4913–4917.
36. Hampel, H., Blennow, K., Shaw, L.M., Hoessler, Y.C. *et al.* (2010) Total and phosphorylated tau protein as biological markers of Alzheimer's disease. *Exp. Gerontol.*, **45** (1), 30–40.
37. Buerger, K., Ewers, M., Pirttila, T., Zinkowski, R. *et al.* (2006) CSF phosphorylated tau protein correlates with neocortical neurofibrillary pathology in Alzheimer's disease. *Brain*, **129** (Pt. 11), 3035–3041.
38. Mattsson, N., Savman, K., Osterlundh, G., Blennow, K. *et al.* (2010) Converging molecular pathways in human neural development and degeneration. *Neurosci. Res.*, **66**, 330–332.
39. Grahn, A., Hagberg, L., Nilsson, S., Blennow, K. *et al.* (2013) Cerebrospinal fluid biomarkers in patients with varicella-zoster virus CNS infections. *J. Neurol.*, **260** (7), 1813–1821.
40. Kondziella, D. and Zetterberg, H. (2008) Hyperphosphorylation of tau protein in superficial CNS siderosis. *J. Neurol. Sci.*, **273** (1-2), 130–132.
41. Ikeda, T., Noto, D., Noguchi-Shinohara, M., Ono, K. *et al.* (2010) CSF tau protein is a useful marker for effective treatment of superficial siderosis of the central nervous system: two case reports. *Clin. Neurol. Neurosurg.*, **112** (1), 62–64.
42. Williams, C.T., Barnes, B.M., Richter, M., and Buck, C.L. (2012) Hibernation and circadian rhythms of body temperature in free-living Arctic ground squirrels. *Physiol. Biochem. Zool.*, **85** (4), 397–404.
43. Hartig, W., Stieler, J., Boerema, A.S., Wolf, J. *et al.* (2007) Hibernation model of tau phosphorylation in hamsters: selective vulnerability of cholinergic basal forebrain neurons – implications for Alzheimer's disease. *Eur. J. Neurosci.*, **25** (1), 69–80.
44. Whittington, R.A., Bretteville, A., Dickler, M.F., and Planel, E. (2013) Anesthesia and tau pathology. *Prog. Neuropsychopharmacol. Biol. Psychiatry*, **47**, 147–155.
45. Struyfs, H., Molinuevo, J.L., Martin, J.J., De Deyn, P.P. *et al.* (2014) Validation of the AD-CSF-index in autopsy-confirmed Alzheimer's disease patients and healthy controls. *J. Alzheimers Dis.* **41** (3), 903–909.
46. Johansson, P., Mattsson, N., Hansson, O., Wallin, A. *et al.* (2011) Cerebrospinal fluid biomarkers for Alzheimer's disease: diagnostic performance in a homogeneous mono-center population. *J. Alzheimers Dis.*, **24** (3), 537–546.
47. Mattsson, N., Zetterberg, H., Hansson, O., Andreasen, N. *et al.* (2009) CSF biomarkers and incipient Alzheimer disease in patients with mild cognitive impairment. *JAMA*, **302** (4), 385–393.
48. Fortea, J., Vilaplana, E., Alcolea, D., Carmona-Iragui, M. *et al.* (2014) Cerebrospinal fluid beta-amyloid and phospho-tau biomarker interactions affecting brain

structure in preclinical Alzheimer disease. *Ann. Neurol.*, **76** (2), 223–230.

49. Toledo, J.B., Xie, S.X., Trojanowski, J.Q., and Shaw, L.M. (2013) Longitudinal change in CSF Tau and Abeta biomarkers for up to 48 months in ADNI. *Acta Neuropathol.*, **126** (5), 659–670.

50. Mattsson, N., Insel, P., Nosheny, R., Zetterberg, H. *et al.* (2013) CSF protein biomarkers predicting longitudinal reduction of CSF beta-amyloid42 in cognitively healthy elders. *Transl. Psychiatry*, **3**, e293.

51. Moghekar, A., Li, S., Lu, Y., Li, M. *et al.* (2013) CSF biomarker changes precede symptom onset of mild cognitive impairment. *Neurology*, **81**, 1753–1758.

52. Zetterberg, H., Pedersen, M., Lind, K., Svensson, M. *et al.* (2007) Intra-individual stability of CSF biomarkers for Alzheimer's disease over two years. *J. Alzheimers Dis.*, **12** (3), 255–260.

53. Blennow, K., Zetterberg, H., Minthon, L., Lannfelt, L. *et al.* (2007) Longitudinal stability of CSF biomarkers in Alzheimer's disease. *Neurosci. Lett.*, **419** (1), 18–22.

54. Mattsson, N., Portelius, E., Rolstad, S., Gustavsson, M. *et al.* (2012) Longitudinal cerebrospinal fluid biomarkers over four years in mild cognitive impairment. *J. Alzheimers Dis.*, **30** (4), 767–778.

55. Mattsson, N., Rajendran, L., Zetterberg, H., Gustavsson, M. *et al.* (2012) BACE1 inhibition induces a specific cerebrospinal fluid beta-amyloid pattern that identifies drug effects in the central nervous system. *PLoS One*, **7** (2), e31084.

56. Lannfelt, L., Blennow, K., Zetterberg, H., Batsman, S. *et al.* (2008) Safety, efficacy, and biomarker findings of PBT2 in targeting Abeta as a modifying therapy for Alzheimer's disease: a phase IIa, double-blind, randomised, placebo-controlled trial. *Lancet Neurol.*, **7** (9), 779–786.

57. May, P.C., Dean, R.A., Lowe, S.L., Martenyi, F. *et al.* (2011) Robust central reduction of amyloid-beta in humans with an orally available, non-peptidic beta-secretase inhibitor. *J. Neurosci.*, **31** (46), 16507–16516.

58. Portelius, E., Dean, R.A., Gustavsson, M.K., Andreasson, U. *et al.* (2010) A novel Abeta isoform pattern in CSF reflects gamma-secretase inhibition in Alzheimer's disease. *Alzheimers Res. Ther.*, **2** (2), 7.

59. Gilman, S., Koller, M., Black, R.S., Jenkins, L. *et al.* (2005) Clinical effects of Abeta immunization (AN1792) in patients with AD in an interrupted trial. *Neurology*, **64** (9), 1553–1562.

60. Blennow, K., Zetterberg, H., Rinne, J.O., Salloway, S. *et al.* (2012) Effect of immunotherapy with bapineuzumab on cerebrospinal fluid biomarker levels in patients with mild to moderate Alzheimer disease. *Arch. Neurol.*, **69** (8), 1002–1010.

61. Karran, E. and Hardy, J. (2014) A critique of the drug discovery and phase 3 clinical programs targeting the amyloid hypothesis for Alzheimer disease. *Ann. Neurol.*, **76** (2), 185–205.

62. Andreasson, U., Portelius, E., Andersson, M.E., Blennow, K. *et al.* (2007) Aspects of beta-amyloid as a biomarker for Alzheimer's disease. *Biomarkers Med.*, **1** (1), 59–78.

63. Fukumoto, H., Cheung, B.S., Hyman, B.T., and Irizarry, M.C. (2002) Beta-secretase protein and activity are increased in the neocortex in Alzheimer disease. *Arch. Neurol.*, **59** (9), 1381–1389.

64. Holsinger, R.M., Lee, J.S., Boyd, A., Masters, C.L. *et al.* (2006) CSF BACE1 activity is increased in CJD and Alzheimer disease versus other dementias. *Neurology*, **67** (4), 710–712.

65. Holsinger, R.M., Mclean, C.A., Collins, S.J., Masters, C.L. *et al.* (2004) Increased beta-secretase activity in cerebrospinal fluid of Alzheimer's disease subjects. *Ann. Neurol.*, **55** (6), 898–899.

66. Zetterberg, H., Andreasson, U., Hansson, O., Wu, G. *et al.* (2008) Elevated cerebrospinal fluid BACE1 activity in incipient Alzheimer disease. *Arch. Neurol.*, **65** (8), 1102–1107.

67. Rosen, C., Andreasson, U., Mattsson, N., Marcusson, J. *et al.* (2012) Cerebrospinal fluid profiles of amyloid beta-related biomarkers in Alzheimer's disease. *Neuromolecular Med.*, **14** (1), 65–73.

68. Galasko, D. and Montine, T.J. (2010) Biomarkers of oxidative damage and inflammation in Alzheimer's disease. *Biomarkers Med.*, **4** (1), 27–36.

69. Rosen, C. and Zetterberg, H. (2013) Cerebrospinal fluid biomarkers for pathological processes in Alzheimer's disease. *Curr. Opin. Psychiatry*, **26**, 276–282.

70. Mapstone, M., Cheema, A.K., Fiandaca, M.S., Zhong, X. *et al.* (2014) Plasma phospholipids

identify antecedent memory impairment in older adults. *Nat. Med.*, **20** (4), 415–418.
71. Samii, A., Nutt, J.G., and Ransom, B.R. (2004) Parkinson's disease. *Lancet*, **363** (9423), 1783–1793.
72. Stamelou, M. and Hoeglinger, G.U. (2013) Atypical parkinsonism: an update. *Curr. Opin. Neurol.*, **26** (4), 401–405.
73. McCann, H., Stevens, C.H., Cartwright, H., and Halliday, G.M. (2014) alpha-Synucleinopathy phenotypes. *Parkinsonism Relat. Disord.*, **20** (Suppl. 1), S62–S67.
74. Ohrfelt, A., Zetterberg, H., Andersson, K., Persson, R. et al. (2011) Identification of novel alpha-synuclein isoforms in human brain tissue by using an online nanoLC-ESI-FTICR-MS method. *Neurochem. Res.*, **36** (11), 2029–2042.
75. Moreno-Gonzalez, I. and Soto, C. (2011) Misfolded protein aggregates: mechanisms, structures and potential for disease transmission. *Semin. Cell Dev. Biol.*, **22** (5), 482–487.
76. Malek, N., Swallow, D., Grosset, K.A., Anichtchik, O. et al. (2014) Alpha-synuclein in peripheral tissues and body fluids as a biomarker for Parkinson's disease – a systematic review. *Acta Neurol. Scand.*, **130** (2), 59–72.
77. Tokuda, T., Qureshi, M.M., Ardah, M.T., Varghese, S. et al. (2010) Detection of elevated levels of alpha-synuclein oligomers in CSF from patients with Parkinson disease. *Neurology*, **75** (20), 1766–1772.
78. Hansson, O., Hall, S., Ohrfelt, A., Zetterberg, H. et al. (2014) Levels of cerebrospinal fluid alpha-synuclein oligomers are increased in Parkinson's disease with dementia and dementia with Lewy bodies compared to Alzheimer's disease. *Alzheimers Res. Ther.*, **6** (3), 25.
79. Magdalinou, N., Lees, A.J., and Zetterberg, H. (2014) Cerebrospinal fluid biomarkers in Parkinsonian conditions: an update and future directions. *J. Neurol. Neurosurg. Psychiatry*, **85** (10), 1065–1075.
80. Holmberg, B., Rosengren, L., Karlsson, J.E., and Johnels, B. (1998) Increased cerebrospinal fluid levels of neurofilament protein in progressive supranuclear palsy and multiple-system atrophy compared with Parkinson's disease. *Mov. Disord.*, **13** (1), 70–77.
81. Alves, G., Lange, J., Blennow, K., Zetterberg, H. et al. (2014) CSF Abeta42 predicts early-onset dementia in Parkinson disease. *Neurology*, **82** (20), 1784–1790.
82. Parnetti, L., Farotti, L., Eusebi, P., Chiasserini, D. et al. (2014) Differential role of CSF alpha-synuclein species, tau, and Abeta42 in Parkinson's disease. *Front. Aging Neurosci.*, **6**, 53.
83. Constantinescu, R. and Mondello, S. (2012) Cerebrospinal fluid biomarker candidates for parkinsonian disorders. *Front. Neurol.*, **3**, 187.
84. Luk, C., Compta, Y., Magdalinou, N., Martí, M.J. et al. (2012) Development and assessment of sensitive immuno-PCR assays for the quantification of cerebrospinal fluid three- and four-repeat tau isoforms in tauopathies. *J. Neurochem.*, **123** (3), 396–405.
85. Constantinescu, R. and Zetterberg, H. (2011) Urate as a marker of development and progression in Parkinson's disease. *Drugs Today (Barc.)*, **47** (5), 369–380.
86. McCrory, P., Meeuwisse, W., Johnston, K., Dvorak, J. et al. (2009) Consensus statement on concussion in sport – the Third International Conference on Concussion in Sport held in Zurich, November 2008. *Phys. Sportsmed.*, **37** (2), 141–159.
87. Roe, C., Sveen, U., Alvsaker, K., and Bautz-Holter, E. (2009) Post-concussion symptoms after mild traumatic brain injury: influence of demographic factors and injury severity in a 1-year cohort study. *Disabil. Rehabil.*, **31** (15), 1235–1243.
88. Williams, W.H., Potter, S., and Ryland, H. (2010) Mild traumatic brain injury and Post-concussion Syndrome: a neuropsychological perspective. *J. Neurol. Neurosurg. Psychiatry*, **81** (10), 1116–1122.
89. Kirov, I.I., Tal, A., Babb, J.S., Reaume, J. et al. (2013) Proton MR spectroscopy correlates diffuse axonal abnormalities with post-concussive symptoms in mild traumatic brain injury. *J. Neurotrauma*, **30** (13), 1200–1204.
90. Stein, T.D., Alvarez, V.E., and Mckee, A.C. (2014) Chronic traumatic encephalopathy: a spectrum of neuropathological changes following repetitive brain trauma in athletes and military personnel. *Alzheimers Res. Ther.*, **6** (1), 4.
91. Neselius, S., Brisby, H., Theodorsson, A., Blennow, K. et al. (2012) CSF-biomarkers in Olympic boxing: diagnosis and effects of repetitive head trauma. *PLoS One*, **7** (4), e33606.

92. Neselius, S., Zetterberg, H., Blennow, K., Randall, J. et al. (2013) Olympic boxing is associated with elevated levels of the neuronal protein tau in plasma. *Brain Inj.*, **27** (4), 425–433.
93. Shahim, P., Tegner, Y., Wilson, D.H., Randall, J. et al. (2014) Blood biomarkers for brain injury in concussed professional ice hockey players. *JAMA Neurol.*, **71** (6), 684–692.
94. Neselius, S., Brisby, H., Granholm, F., Zetterberg, H. et al. (2014) Monitoring concussion in a knocked-out boxer by CSF biomarker analysis. *Knee Surg. Sports Traumatol. Arthrosc.*, (in press).
95. Johnson, V.E., Stewart, W., and Smith, D.H. (2013) Axonal pathology in traumatic brain injury. *Exp. Neurol.*, **246**, 35–43.
96. Zetterberg, H., Smith, D.H., and Blennow, K. (2013) Biomarkers of mild traumatic brain injury in cerebrospinal fluid and blood. *Nat. Rev. Neurol.*, **9** (4), 201–210.
97. Dekosky, S.T., Blennow, K., Ikonomovic, M.D., and Gandy, S. (2013) Acute and chronic traumatic encephalopathies: pathogenesis and biomarkers. *Nat. Rev. Neurol.*, **9** (4), 192–200.
98. Edgeworth, J.A., Farmer, M., Sicilia, A., Tavares, P. et al. (2011) Detection of prion infection in variant Creutzfeldt–Jakob disease: a blood-based assay. *Lancet*, **377** (9764), 487–493.
99. Weiss, A., Trager, U., Wild, E.J., Grueninger, S. et aln (2012) Mutant huntingtin fragmentation in immune cells tracks Huntington's disease progression. *J. Clin. Invest.*, **122** (10), 3731–3736.

20
Pharmacokinetics of Peptides and Proteins

Chetan Rathi and Bernd Meibohm
University of Tennessee Health Science Center, College of Pharmacy, Department of Pharmaceutical Sciences, 881 Madison Ave, Room 444, Memphis, TN 38138, USA

1	**Introduction** 699	
2	**Administration Pathways** 699	
2.1	Administration by Injection or Infusion 703	
2.2	Inhalational Administration 704	
2.3	Intranasal Administration 705	
2.4	Transdermal Administration 705	
2.5	Peroral Administration 706	
3	**Administration Route and Immunogenicity** 707	
4	**Distribution** 708	
5	**Elimination** 710	
5.1	Proteolysis 711	
5.2	Gastrointestinal Elimination 712	
5.3	Renal Elimination 712	
5.4	Hepatic Elimination 714	
5.5	Receptor-Mediated Endocytosis 715	
6	**Interspecies Scaling** 715	
7	**Drug–Drug Interaction** 717	
7.1	Pharmacokinetic Based Drug–Drug Interactions 717	
7.2	Pharmacodynamic Based Drug–Drug Interactions 717	
8	**Conclusions** 718	
	References 718	

Translational Medicine: Molecular Pharmacology and Drug Discovery
First Edition. Edited by Robert A. Meyers.
© 2018 Wiley-VCH Verlag GmbH & Co. KGaA. Published 2018 by Wiley-VCH Verlag GmbH & Co. KGaA.

Keywords

Pharmacokinetics
The characteristic interactions of a drug and the body in terms of its absorption, distribution, metabolism, and excretion (ADME).

Peptide therapeutics
Short polymers typically with fewer than 50 amino acids and with a molecular weight less than 10 kDa.

Immunogenicity
An unwanted immune response towards a protein therapeutic agent.

Allometry
Allometry is the extrapolation of animal data to predict pharmacokinetic parameters in other species.

Drug delivery
Formulation approaches for the administration of a peptide therapeutic to achieve efficacy.

ADME
Absorption, Distribution, Metabolism and Excretion are processes which describes the disposition of a protein therapeutic.

Drug–drug interaction
This interaction occurs when one protein therapeutic affects the activity of another drug, when both are administered together.

> Protein-based therapeutics represent a rapidly growing class of new therapeutics for numerous indications. Low permeability and susceptibility to proteolytic degradation limit their oral bioavailability, and hence they are commonly administered by intravenous, subcutaneous or intramuscular routes. The distribution of protein therapeutics is usually limited to the extracellular space and takes place mainly by the processes of convective transport. Proteins are mainly eliminated by nonspecific catabolic degradation carried out by the ubiquitous presence of proteases and peptidases in the body, with intracellular uptake as a major prerequisite. Immunogenicity is a frequently encountered challenge of protein therapeutics that can affect their disposition, efficacy and toxicity. Combination therapy with protein therapeutics may potentially lead to drug–drug interactions, though such clinical events have rarely occurred. Rapid scientific advancements, unmet medical needs and high profit margins have been some of the motivating factors for the rapid emergence of next-generation protein therapeutics. An understanding of the absorption, distribution, metabolism and excretion processes of protein therapeutics is critical for their successful development.

1
Introduction

Although a small number of peptide- and protein-based therapeutics has been used in medical practice for a long time (e.g., calcitonin and glucagon), it is the advances in biotechnology and their application to drug development systems that have propelled peptides and proteins from niche products to mainstream therapeutics. Recombinant human (RH) insulin, as approved in 1982, was the first of these biotechnologically derived drug products, and many more have followed during the past 30 years such that, today, peptide- and protein-based drug products constitute a sizable fraction of all clinically used medications. There are, nevertheless, distinct variations between these drug products and small-molecule-based therapeutics that may require the use of different technologies, methodological approaches and experimental designs during their preclinical and clinical development, including assessments of their pharmacokinetics (PK) and exposure–response characteristics.

Pharmacokinetic and exposure–response concepts impact on every stage of the drug development process, starting from lead optimization to the design of Phase III pivotal trials [1]. An understanding of the concentration–effect relationship is crucial to any drug – including peptides and proteins – as it lays the foundations for dosing regimen design and rational clinical application. General PK principles are equally applicable to protein- and peptide-based drugs as they are to traditional small-molecule drugs. This includes PK-related recommendations for drug development such as exposure–response guidance documents of the U.S. Food and Drug Administration (FDA) and the ICH E4 guideline of the International Conference on Harmonization [2, 3].

The assessment and interpretation of the PK of peptides and proteins frequently poses extra challenges, however, and requires additional resources compared to small-molecule drug candidates. One such challenge arises from the fact that most peptide- and protein-based drugs are identical or similar to endogenous molecules, and need to be identified and quantified next to a myriad of structurally similar molecules. The following sections of this chapter provide a general discussion on the pharmacokinetics of peptide- and protein-based therapeutics. In addition, they outline some of the associated challenges and obstacles encountered during the drug development process, and illustrate these problems with examples of approved and experimental drugs. An overview of the basic PK parameters of selected FDA-approved protein and peptide drugs is provided in Table 1, with attention focused on non-antibody-based therapeutics.

2
Administration Pathways

Peptides and proteins, unlike conventional small-molecule drugs, are generally not therapeutically active upon oral administration [4–6]. The lack of systemic bioavailability is due mainly to two factors: (i) high gastrointestinal (GI) enzyme activity; and (ii) low permeability through the GI mucosa. In fact, the substantial peptidase and protease activities in the GI tract means that the gut is the most efficient body compartment for peptide and protein metabolism. The GI mucosa also presents a major absorption barrier for water-soluble macromolecules such as peptides and proteins [4], with absorption – at least for

Tab. 1 Pharmacokinetic parameters of select FDA-approved non-antibody-based protein/peptide drugs as reported in the prescribing information.

Generic name/trade name	Class	Manufacturer	Route	Bioavailability	Clearance[a]	Volume of distribution[a]	Half-life
Abarelix/Plenaxis	LHRH antagonist	Praecis	IM	—	$208 \pm 48 \, l \, d^{-1}$	$4040 \pm 1607 \, l$	$13.22 \pm 3.2 \, d$
Agalsidase/Fabrazyme	α-Galactosidase	Genzyme	IV	—	Nonlinear PK	—	45–102 min
Aldesleukin/Proleukin	Antineoplastic (IL-2)	Chiron	IV	—	$268 \, ml \, min^{-1}$	—	85 min
Alefacept/Amevive	Immunosuppressant	Biogen	IV, IM	63% (IM)	$0.25 \, ml \, h^{-1} \, kg^{-1}$	$94 \, ml \, kg^{-1}$ PLV[b]	270 h
Alteplase/Activase	Thrombolytic	Genentech	IV	—	380–$570 \, ml \, min^{-1}$	—	<5 min
Anakinra/Kineret	Antirheumatic (IL-1Ra)	Amgen	SC	95%	—	—	4–6 h
Asparaginase/Elspar	Antineoplastic	Merck	IV, IM	—	—	70–80% PLV	8–30 h (IV) 39–49 h (IM)
Belatacept/Nulojix	Immunosuppressant	BMS	IV	—	$0.39 \pm 0.07 \, ml \, h \, kg^{-1}$	$0.09 \pm 0.02 \, l \, kg^{-1}$	$9.8 \pm 2.8 \, d$
Carfilzomib/Kyprolis	Proteasome inhibitor	Onyx	IV	—	151–$263 \, l \, h^{-1}$	$28 \, l$	≤1 h
Cetrorelix/Cetrotide	LHRH antagonist	Serono	SC	85%	$1.28 \, ml \, min^{-1} \, kg^{-1}$	$1.16 \, l \, kg^{-1}$	62.8 h (range: 38.2–108 h)
Cyclosporine Neoral	Immunosuppressant	—	—	—	—	—	—
—	—	Novartis	PO	10–89%	5–$7 \, ml \, min^{-1} \, kg^{-1}$	3–$5 \, l/kg$	8.4 (range: 5–18 h)
Sandimmune	—	Novartis	IV, PO	30 (PO)	—	—	19 h (range: 10–27 h)
Darbepoetin-α/Aranesp	Antianemic	Amgen	IV, SC	37 (30–50)%	—	—	21 h (IV) 49 h (range: 24–72 h) (SC)
Degarelix/Firmagon	GnRH antagonist	Ferring	SC	40–60%	35–$50 \, ml \, h^{-1} \, kg^{-1}$	0.65–$0.82 \, l/kg$	43 d (range: 27–73 d)
Denileukin difftitox/Ontak	Antineoplastic	Ligand	IV	—	1.5–$2.0 \, ml \, min^{-1} \, kg^{-1}$	0.06–$0.08 \, l/kg$	70–80 min
Desmopressin/DDAVP	Antidiuretic	Aventis	PO, IN	3.2% (IN) 0.16% (PO)	—	—	75.5 min (IN) 1.5–2.5 h (PO)
Drotrecogin-α/Xigris	Activated protein C	Eli Lilly	IV	—	$40 \, l \, h^{-1}$	—	—
Ecallantide/Kalbitor	Plasma kallikrein inhibitor	Dyax	SC	—	$153 \pm 20 \, ml \, min^{-1}$	$26.4 \pm 7.8 \, l$	$2.0 \pm 0.5 \, h$

Name	Category	Company	Route	Bioavailability	Clearance	V_d	Half-life
Epoetin-α/Epogen, Procrit	Antianemic	Amgen, Ortho	IV, SC	—	—	—	4–13 h (IV) 16.3 ± 3.0 h (SC)
Eptifibatide/Integrilin	GPIIb/IIIa inhibitor	Millennium	IV	—	55–58 ml h^{-1} kg^{-1}	—	2.5 h
Etanercept/Enbrel	Antirheumatic	Amgen	SC	—	89 ml h^{-1}	—	115 h (range: 98–300 h)
Exenatide/Byetta	Incretin mimetic	Amylin	SC	—	9.1 l h^{-1}	28.3 l	2.4 h
Follitropin/Gonal-f	Follicle stimulating hormone	Serono	SC	66 ± 39%	0.7 ± 0.2 l h^{-1}	10 ± 3 l	32 h
Follitropin beta/Follistim AQ	Follicle stimulating hormone	Organon	SC	77.8%	0.01 l h^{-1} kg^{-1}	8 l	33.4 h
Ganirelix/Antagon	LHRH antagonist	Organon	SC	91%	2.4 ± 0.2 l h^{-1}	43.7 ± 11.4 l	12.8 ± 4.3 h
Glucagon/Glucagon	rh Glucagon	Eli Lilly	IV, IM, SC	—	13.5 ml min^{-1} kg^{-1}	0.25 l kg^{-1}	8–18 min (IV)
Goserelin/Zoladex	LHRH agonist	AstraZeneca	SC	—	110.5 ± 47.5 ml min^{-1} (men) 163.9 ± 71.0 ml min^{-1} (women)	44.1 ± 13.6 l (men) 20.3 ± 4.1 l (women)	4.2 ± 1.1 h (men) 2.3 ± 0.6 h (women)
Insulins							
Lispro (Humalog)	Insulin analog	Eli Lilly	SC	55–77%	—	0.26–0.36 l kg^{-1}	60 min (IV)
Aspart (Novolog)	Insulin analog	Novartis	SC	55–77%	1.22 l h^{-1} kg^{-1}	—	81 min (SC)
rh (Humulin R)	—	Eli Lilly	SC	55–77%	—	0.26–0.36 l kg^{-1}	90 min (IV)
Glargine (Lantus)	Insulin analog	Aventis	SC	—	—	—	11 h ($T_{25\%}$)
Inhaled rh (Exubera)	—	Pfizer	IH	—	Assumed to be identical to rh insulin		
Interferon β-1b/Betaseron	Biological response modifier	Berlex/Chiron	SC	50%	9.4–28.9 ml min^{-1} kg^{-1}	0.25–2.88 l kg^{-1}	8 min–4.3 h (IV)
Interferon γ-1b/Actimmune	Immunomodulator	InterMune	SC	>89%	1.4 l min^{-1}	—	38 min (IV) 2.9 h (IM) 5.9 h (SC)
Leuprolide/Eligard	LHRH agonist	Atrix	SC	—	8.36 l h^{-1}	27 l	3 h
Laronidase/Aldurazyme	Lysosomal hydrolase	Genzyme	IV	—	1.7–2.7 ml min^{-1} kg$_{-1}$	0.24–0.61 l kg^{-1}	1.5–3.6 h
Octreotide/Sandostatin	Somatostatin analog	Novartis	IV, SC	100%	7–10 l h^{-1}	13.6 l	1.7–1.9 h

(continued overleaf)

Tab. 1 (Continued)

Generic name/trade name	Class	Manufacturer	Route	Bioavailability	Clearance[a]	Volume of distribution[a]	Half-life
Oprelvekin/Neumega	Thrombopoietic stimulant (IL-11)	Wyeth	SC	>80%	—	—	6.9 ± 1.7 h
Pasireotide/Signifor	Somatostatin analog	Novartis	SC	—	$7.6 \, \text{l h}^{-1}$ (healthy)	>100 l	12 h
Pegaspargase/Oncaspar	Antineoplastic	Enzon	IM, IV	—	—	—	5.69 ± 3.25 d
Pegfilgrastim/Neulasta	Hematopoietic stimulant	Amgen	SC	—	Nonlinear PK	—	15–80 h
Peginesatide/Omontys	Erythropoiesis-stimulating agent	Takeda	SC	46%	$0.5 \pm 0.2 \, \text{ml h} \cdot \text{kg}^{-1}$	$34.9 \pm 13.8 \, \text{ml kg}^{-1}$	53.0 ± 17.7 h
Peginterferon α-2a/Pegasys	Biological response modifier	Roche	SC	84%	$94 \, \text{ml h}^{-1}$	—	80 h (range: 50–140 h)
Pramlintide/Symlin	Amylin analog	Amylin	SC	30–40%	1 l/min	15–27 l	48 min
Reteplase/Retavase	Thrombolytic	Centocor	IV	—	$250–450 \, \text{ml min}^{-1}$	—	13–16 min
Sargramostim/Leukine	Hematopoietic stimulant	Berlex	IV, SC	—	$431 \, \text{ml min}^{-1} \, \text{m}^{-2}$ (IV) $549 \, \text{ml min}^{-1} \, \text{m}^{-2}$ (SC)	—	60 min (IV) 162 min (SC)
Teduglutide/Gattex	Glucagon like peptide-2 receptor agonist	NPS	SC	88%	$123 \, \text{ml h}^{-1} \, \text{kg}^{-1}$	$103 \, \text{ml kg}^{-1}$	2 h (healthy)
Tenecteplase/TNKase	Thrombolytic	Genentech	IV	—	$99–119 \, \text{ml min}^{-1}$	PLV[b]	90–130 min
Teriparatide/Forteo	Parathyroid hormone	Eli Lilly	SC	95%	$94 \, \text{l h}^{-1}$ (men) $62 \, \text{l h}^{-1}$ (women)	$0.12 \, \text{l kg}^{-1}$ (IV)	5 min (IV) 1 h (SC)
Triptorelin/Trelstar	LHRH agonist	Pfizer	IM	83%	$212 \pm 32 \, \text{ml min}^{-1}$	30–33 l	2.81 ± 1.21 h
Urokinase/Abbokinase	Thrombolytic	Abbott	IV	—	—	11.5 l	12.6 ± 6.2 min
Glucarpidase/Voraxaze	Treat toxic methotrexate concentration	BTG International	IV	—	$7.5 \, \text{ml min}^{-1}$	3.6 l	5.6 h

a) Clearance and volume of distribution terms for drugs not administered intravenously usually reflect clearance and volume terms divided by bioavailability (i.e., CL/F or V/F).
b) PLV, plasma volume (40 ml kg^{-1}); rh, recombinant human; IV, intravenous; IM, intramuscular; IN, intranasal; SC, subcutaneous.

peptides – perhaps being further impeded through presystemic metabolism mediated by the functional system of cytochrome P450 3A and p-glycoprotein (P-gp) [7–9].

Due to a lack of activity of most peptides and proteins after their oral administration, administration by injection or infusion – that is, intravenous (IV), subcutaneous (SC) or intramuscular (IM) – is frequently the preferred route of delivery for these drug products. Other non-oral administration routes of drug delivery have also been utilized, including nasal, buccal, rectal, vaginal, transdermal, ocular and/or pulmonary. Some of these delivery routes are discussed in the following sections, in order of increasing biopharmaceutic challenge to obtain adequate systemic exposure.

2.1
Administration by Injection or Infusion

The injectable administration of peptides and proteins offers the advantage of circumventing presystemic degradation, thereby achieving the highest concentrations in the biological system. FDA-approved proteins given via the IV route include, for example, the tissue plasminogen activator (t-PA) analogs alteplase and tenecteplase (TNKase®), the recombinant human erythropoietin epoetin-α, and the macrophage colony-stimulating factor (M-CSF), filgrastim. Unfortunately, IV administration as either a bolus dose or constant rate infusion may not always provide the desired concentration–time profile, depending on the biological activity of the product, and IM or SC injections may be more appropriate alternatives. For example, luteinizing hormone-releasing hormone (LH-RH) secreted in bursts stimulates the release of follicle-stimulating hormone (FSH) and luteinizing hormone (LH), but a continuous baseline level will suppress the release of these hormones [10]. In order to avoid the high peaks resulting from an IV administration of leuprorelin (an LH-RH agonist), a long-acting monthly depot injection of the drug has been approved for the treatment of prostate cancer and endometriosis [11]. In a study comparing SC versus IV administration of epoetin-α in patients receiving hemodialysis, it was found that the SC route could maintain the hematocrit over a desired target range with a lower average weekly dose of epoetin-α compared to IV administration [12].

The drawbacks of SC and IM injections include a potentially decreased bioavailability that is secondary to variables such as local blood flow, injection trauma, protein degradation at the site of injection, and limitations of uptake into the systemic circulation related to effective capillary pore size and diffusion. The bioavailability of numerous peptides and proteins is, for example, markedly reduced after SC or IM administration compared to their IV administration. The pharmacokinetically derived apparent absorption rate constant is thus a combination of absorption into the systemic circulation and presystemic degradation at the absorption site. The true absorption rate constant k_a can then be calculated as:

$$k_a = F \cdot k_{app}$$

where F is the bioavailability compared to IV administration. A rapid apparent absorption rate constant, k_{app}, can thus be the result of a slow absorption and a fast presystemic degradation – that is, a low systemic bioavailability [13].

Several approved peptides and proteins such as insulin, enfuvirtide [14] and pramlintide [15] are administered as SC injections. Following such administration, peptide and protein therapeutics can enter the systemic circulation either via the blood capillaries or through lymphatic vessels [16]. In general, macromolecules larger than 16 kDa are predominantly absorbed into the lymphatics, whereas those under 1 kDa are mostly absorbed into the blood circulation. Studies with recombinant human interferon α-2a (rhIFN α-2a) have indicated that, following SC administration, high concentrations of the recombinant protein are found in the lymphatic system, which drains into regional lymph nodes [17]. Physiologically based PK modeling used to describe the absorption process after SC or IM administration has also supported the suggestion that the lymphatic uptake of biologics dominates other delivery pathways to the systemic circulation [18]. Furthermore, there appears to be a linear relationship between the molecular weight of the protein and the proportion of the dose absorbed by the lymphatics [19]. This is of particular importance for those agents whose therapeutic targets are lymphoid cells, such as interferons (IFNs) and interleukins (ILs) [17]. Clinical studies have shown that palliative low-to-intermediate-dose SC recombinant interleukin-2 (rIL-2), in combination with rhIFN α-2a, can be administered to patients in an ambulatory setting with efficacy and safety profiles comparable to the most aggressive IV rIL-2 protocol against metastatic renal cell cancer [20].

2.2
Inhalational Administration

The inhalational delivery of peptides and proteins offers the advantage of ease of administration, the presence of a large surface area (75 m^2) available for absorption, a high vascularity of the administration site, and an ability to bypass the hepatic first-pass metabolism. New devices have also been explored which can target drug-containing aerosols to the deep lung compartments [21].

Although the inhalational delivery of therapeutic protein provides numerous advantages, its long-term use may be associated with irritation and irreversible lung damage [22]. Other disadvantages include the presence of certain proteases in the lung, potential local side effects of the inhaled agents on the lung tissues (i.e., growth factors and cytokines), and molecular weight limitations [23, 24].

The feasibility of inhaled peptide and protein drugs can be exemplified by inhaled recombinant human insulin products, of which Exubera® was the first to be approved in 2006, though several others are currently in clinical development. Inhaled insulin offers the advantages of ease of administration and rapid onset with a shorter duration of action for a tighter postprandial glucose control as compared to subcutaneously administered regular insulin [25]. In a study that incorporated 26 patients with type II diabetes mellitus, inhalational insulin treatment for three months was shown to significantly improve glycemic control compared to baseline, as assessed by hemoglobin A1c levels [26]. A second clinical study in 249 type II diabetic subjects showed comparable efficacy between inhaled insulin and the conventional SC insulin regimen [27]. Together, these clinical trials demonstrated that the non-invasive administration of inhaled insulin offers a similar efficacy but a better patient compliance than subcutaneously administered insulin. Following the introduction of Exubera to the market, however, ex-smokers were found to exhibit a higher risk of lung cancer when treated with the drug. This uncertainty led to its low-level sales and subsequent withdrawal in October 2007 [22].

Dornase-α, which is indicated for the treatment of cystic fibrosis, is another example of a protein drug that can be successfully administered via the inhalational route. In a multicenter, two-year clinical study in pediatric cystic fibrosis patients, inhaled dornase-α was shown to cause a significant improvement in lung function and to reduce the risk of respiratory exacerbations [28].

Lucinactant, a bioengineered peptide, is indicated for the treatment of respiratory distress syndrome (RDS) in neonates caused by the deficiency of surfactants in the lungs. Lucinactant mimics the lung surfactant-associated protein B, and is administered intratracheally to premature infants with RDS [29].

2.3
Intranasal Administration

Similar to the inhalational route, the intranasal administration of peptides and proteins offers advantages of ease of dosing, delivery to a surface area that has a rich vascular and lymphatic network, and bypass of hepatic first-pass metabolism [30]. The intranasal absorption of a variety of peptide and protein drugs, including calcitonin, oxytocin, LH-RH, growth hormone, interferons and even vaccines, has been investigated extensively over the past decade. In general, polypeptides with a molecular weight of up to 2000 Da have been found to be pharmacologically active via the intranasal route. In contrast, pharmaceutical peptides with molecular weights of 2000–6000 Da (e.g., insulin, calcitonin, LH-RH) require the addition of absorption enhancers in order to achieve an adequate bioavailability [10]. Limitations that may preclude the use of intranasal administration include high variability associated with the site of deposition, the type of delivery system used, changes in mucus secretion and mucociliary clearance, as well as the presence of allergy, hay fever or the common cold among the target population [31].

Intranasal administration has recently been proposed as a means to deliver protein therapeutics directly to the central nervous system (CNS), thereby bypassing the blood–brain barrier [32]. In particular, one study reported higher concentrations of ^{125}I-labeled recombinant human insulin-like growth factor-I (rhIGF-I) in the CNS after intranasal administration than after IV administration [33]. The efficacy of intranasally delivered insulin-like growth factor-I (IGF-1) to treat stroke has also been reported in a rat model, with 150 μg of intranasal IGF-1 effectively reducing the induced infarct size and neurologic deficits compared to controls [33]. Currently available peptide drugs that employ nasal delivery include Miacalcin® (salmon calcitonin), Synarel® [nafarelin, a gonadotropin-releasing hormone (GnRH) agonist] and DDAVP® (desmopressin synthetic analog). The relatively low doses of these therapeutic proteins that are needed for efficacy helps overcome their limitation of low intranasal bioavailability (typically ≤3%), and allows them to be marketed as intranasal dosage forms [34].

2.4
Transdermal Administration

Transdermal drug delivery offers the advantages of bypassing metabolic and chemical degradation in the GI tract, as well as hepatic first-pass metabolism. Methods used to facilitate transdermal delivery include sonophoration and iontophoresis, both of which increase the permeability of the skin to ionic compounds – sonophoration by applying low-frequency ultrasound, and iontophoresis by applying a low-level electric current. Therapeutic doses of insulin, interferon-γ and epoetin-α have all been successfully delivered transdermally via sonophoresis [35]. Transdermal iontophoresis has also been applied for the delivery of a host of proteins and peptides, including leuprolide [36], insulin [37], growth hormone-releasing factor [38], calcitonin [39], and parathyroid hormone [40].

2.5
Peroral Administration

Most peptides and proteins are currently formulated as parenteral formulations because of their poor oral bioavailability. Nevertheless, the oral delivery of peptides and proteins would be the preferred route of administration if bioavailability issues could be overcome, as this would offer the advantage of a convenient, pain-free administration. Although various factors such as permeability, chemical and metabolic stability and GI transit time can affect the rate and extent of absorption of orally administered peptides and proteins, molecular size is generally considered to be the ultimate obstacle [41].

Recently, several promising strategies have emerged from intensive investigations into the oral delivery of peptides and proteins [6, 41, 42]. Absorption enhancers may be used in two ways: (i) to temporarily disrupt the intestinal barrier so that drug penetration is increased; or (ii) to serve as transport carriers for the protein via complex formation. The oral coadministration of parathyroid hormone (an 84-amino acid protein) with N-aminocaprylic acid (a transport carrier) resulted in positive bioactivity, as demonstrated in a rodent model of osteoporosis [43]. While parathyroid hormone has no oral bioavailability when administered alone, its coadministration with N-aminocaprylic acid resulted in 2.1% oral bioavailability relative to SC administration in monkeys [43]. An increasing intestinal paracellular absorption was demonstrated in a study involving insulin and immunoglobulins coadministered with zonula occludens toxin (Zot), another permeation enhancer. In animal models, Zot reversibly increased the intestinal absorption of both insulin and immunoglobulins in a time-dependent manner [5].

Encapsulation in microparticles or nanoparticles may be used to shield peptides and proteins from enzymatic degradation. Such solid particles may be taken up via endocytosis by the intestinal cells, or by passage through paracellular tight junctions. Due to their stability in the GI tract, the particles appear more favorable for oral delivery than liposomes. In particular, the gut-associated lymphoid tissue (GALT) organized in Peyer's patches has been suggested as a useful oral delivery target for encapsulated peptides and proteins. Peyer's patches cover approximately 25% of the GI mucosal surface area, and are characteristically high in phagocytic activity but have limited lysosomal activity. More importantly, protein and peptide delivery through GALT offers the advantage of bypassing the hepatic first-pass metabolism [6]. This concept has been successfully demonstrated by the oral delivery of glucagon-like peptide-1 (GLP-1) in poly(lactic-co-glycolic acid) (PLGA) microspheres to diabetic mice, when mice treated with the microsphere preparation had a lower glycemic response to oral glucose challenge than those treated with GLP-1 that had not been encapsulated into microspheres [44].

Because Peyer's patches contain a large number of IgA-committed cells that can be stimulated by antigens absorbed through membranous cells, they have also been targeted for oral vaccine delivery [45]. Whilst significant GI degradation complicates oral vaccine delivery, various studies have demonstrated successful chitosan microparticle absorption for oral vaccine delivery through Peyer's patches, using ovalbumin as a model vaccine [46].

Other strategies of oral peptide and protein delivery include amino acid backbone modification, alternate formulation design, chemical conjugation to improve their resistance to degradation, and inhibition of enzymatic degradation by the coadministration of protease inhibitors [6, 47]. Unfortunately, novel approaches to improve oral protein and peptide drug delivery may not always be ideal. For example, the use of absorption enhancers (including EDTA, bile salts and surfactants) can actually cause a disaggregation of insulin and increase its rate of degradation. Consequently, despite numerous approaches having been investigated, the development of orally administered proteins and peptides continues to pose a major challenge and remains a key area in drug delivery research [6].

Occasionally, low oral bioavailability can be advantageous for therapeutic proteins. For example, linaclotide (a guanylate cyclase-C agonist acting as a secretagog and pain modulator) is effective for the treatment of chronic constipation and irritable bowel syndrome with constipation, mainly because the drug's poor bioavailability promotes a localized delivery to the intestinal epithelium [48].

3
Administration Route and Immunogenicity

Immunogenicity refers to the unwanted immune response towards an antigenic protein drug. Numerous factors have been identified that are responsible for the immunogenicity of protein therapeutics, including: (i) factors associated with the product, such as variation from the endogenous protein and post-translational modification; (ii) production processes, including changes due to storage conditions, production, purification and formulation; and (iii) patient characteristics, such as the immune status of the patient due to existing disease or infectious agents and their genetic background [49]. The chronic or repeated administration of protein therapeutics can evoke the formation of an anti-drug antibody (ADA) that can bind to the protein therapeutic through the active site (or other portion). In this way immunecomplexes are formed that can lead to a modulation and/or even neutralization of the protein's therapeutic effects. The immunecomplex can either enhance or reduce the clearance of protein therapeutics compared to its unbound form. One of the mechanisms hypothesized in this situation relates the size and structure of the immunecomplex with its impact on clearance. A larger-sized complex will trigger the Fcγ-mediated uptake by the reticuloendothelial system (RES), followed by its lysosomal breakdown. This degradation process serves as an additional route of elimination for therapeutic proteins, and results in a decreased systemic exposure; for example, the clearance of glucuronidase was increased by 50% because of the induced ADA [50]. On the other hand, a smaller-sized complex can evade the RES, thereby sustaining circulation of the protein therapeutic. A recycling of immunecomplexes through the neonatal Fc-receptor (FcRn) has also been proposed for the development of sustained protein therapeutics [51].

Immunogenicity may be affected by the route of administration. For example, the extravascular injection of a protein has been shown to stimulate antibody formation more than IV application, most likely due to an increased immunogenicity of the protein aggregates and precipitates formed at the extravascular injection site [52]. An investigation was also made into the effect of route of administration of IFN-β preparations on inducing anti-IFN-β antibodies in multiple sclerosis patients. The results indicated that IM injections appeared less immunogenic than SC injections, and this resulted not only in a lower serum level of anti-IFN-β antibodies but also a delay in their appearance [53].

The challenges associated with immunogenicity have been successfully tackled using different approaches. One method involves the administration of immunomodulatory agents (e.g., methotrexate) that is, for example, coadministered with infliximab in rheumatoid arthritis and helps (besides other therapeutic effects) to suppress the unwanted ADA response towards the protein therapeutic [54]. The second approach involves the administration of a higher dose of the agent, taking into account the neutralizing effect of the existing ADA. For example, the IV administration of a higher dose of IFN-β overcame the loss in therapeutic activity due to neutralizing ADA [55]. A third approach would be to chemically modify the molecule, for example by conjugating it with polyethylene glycol (PEG); this process is termed PEGylation [56, 57]. Through steric hindrance, PEGylation can be used to shield antigenic determinants on the protein drug against being detected by the immune system [58]. This concept was applied successfully to overcome the high rate of allergic reactions towards L-asparaginase, resulting in pegaspargase [59].

Other major advantages of PEGylation include its ability to manipulate the PK and physico-chemical properties of the protein drug. The conjugation of protein drugs with PEG chains increases their hydrodynamic volume, which in turn can result in a reduced renal clearance, restricted biodistribution, and prolonged residence time. PEGylation can also protect against proteolytic degradation and increase drug solubility [57]. Pegfilgrastim is the PEGylated version of the M-CSF filgrastim, which is administered for the management of chemotherapy-induced neutropenia. In this case, PEGylation minimizes the renal clearance of filgrastim by glomerular filtration, such that neutrophil-mediated clearance becomes the predominant route of elimination. As a consequence, the PEGylation of filgrastim will result in so-called "self-regulating pharmacokinetics," since pegfilgrastim will have a reduced clearance (and thus a prolonged half-life) as well as a more sustained duration of action in a neutropenic setting because few mature neutrophils are available to mediate its elimination [60].

4
Distribution

Whole-body distribution studies are essential when studying classical small-molecule drugs, in order to exclude the tissue accumulation of potentially toxic metabolites. This problem does not apply to protein drugs whose catabolic degradation products are amino acids and are recycled in the endogenous amino acid pool. Hence, biodistribution studies of peptides and proteins are performed primarily to assess the targeting of a drug towards specific tissues, as well as to identify the major organ(s) of its elimination [4].

The volume of distribution of a peptide or protein drug is determined largely by its physico-chemical properties (e.g., charge, lipophilicity), protein binding, and dependency on active transport processes. Due to their large size – and therefore limited mobility through biomembranes – most therapeutic proteins have small volumes of distribution that typically are limited to the volumes of the extracellular space [30, 61].

Unlike small-molecule drugs, the transport of therapeutic proteins from the extracellular space into the interstitial space of tissues occurs by convection rather than by diffusion, with fluid flow in one direction from the extravascular space through the paracellular pores into the interstitial space. This is followed by a transfer of the therapeutic protein from tissues into the lymph, which ultimately drains back into the systemic circulation. The extended lymphatic transit time of therapeutic proteins is a consequence of the very low flow rate of lymph (15 ml per day) [62], and this in turn results in a delayed uptake of therapeutic proteins into the systemic circulation after SC or IM administration. Drug clearance during lymphatic transport may also have an influence on the bioavailability of a therapeutic protein [18]. Another, usually minor, route for the distribution of therapeutic proteins is transcellular migration via endocytosis.

After IV application, peptides and proteins usually follow a biexponential plasma concentration–time profile that can best be described by a two-compartment PK model [13]. The central compartment in this model represents primarily the vascular space and the interstitial space of well-perfused organs with permeable capillary walls (especially the liver and kidneys), while the peripheral compartment comprises the interstitial space of poorly perfused tissues such as the skin and (inactive) muscle [4].

In general, the volume of distribution of the central compartment (V_c) in which peptides and proteins initially distribute after IV administration is typically equal to or slightly larger than the plasma volume of 3–8 l (approximate body water volumes for a 70 kg person: interstitial 12 l, intracellular 27 l, intravascular 3 l). Furthermore, the steady-state volume of distribution (V_{ss}) is usually no more than twice the initial volume of distribution, or approximately 14–20 l [13, 42, 63]. This distribution pattern has been described for the somatostatin analog octreotide (V_c 5.2–10.2 l; V_{ss} 18–30 l), and t-PA analog TNKase (V_c 4.2–6.3 l; V_{ss} 6.1–9.9 l) [64]. Epoetin-α also has a volume of distribution estimated to be close to the plasma volume at 0.0558 l kg^{-1} after IV administration to healthy volunteers [65]. Similarly, V_{ss} for darbepoetin-α has been reported as 0.0621 l kg^{-1} after IV administration to patients undergoing dialysis [66], and the distribution of thrombopoietin has also been reported to be limited to the plasma volume (~3 l) [67].

Active tissue uptake and binding to intravascular and extravascular proteins, however, can lead to a substantial increase in the volume of distribution of peptide and protein drugs, as is observed with atrial natriuretic peptide (ANP) [68].

There is a tendency for V_{ss} and V_c to correlate with each other, which implies that the volume of distribution is determined predominantly by distribution in the vascular and interstitial space, as well as unspecific protein binding in these distribution spaces. The distribution rate is inversely correlated with molecular size, and is similar to that of inert polysaccharides, which suggests that passive diffusion through aqueous channels is the primary mechanism of distribution [69].

In contrast to conventional drugs, the distribution, elimination and pharmacodynamics (PDs) of peptides and proteins are frequently interrelated. The generally low volume of distribution should not necessarily be interpreted as low tissue penetration; in fact, receptor-mediated specific uptake into the target organ (as one mechanism) can result in therapeutically effective tissue concentrations, despite a relatively small volume of distribution. For example, nartograstim – a recombinant derivative of G-CSF – is characterized by a specific, dose-dependent and saturable tissue uptake into the target organ bone marrow, presumably via receptor-mediated endocytosis [70].

Another factor that can influence the distribution of therapeutic peptides and proteins is binding to endogenous protein structures. Physiologically active endogenous peptides and proteins frequently interact with specific binding proteins involved in their transport and regulation.

A wide range of protein drugs, including growth hormone [71], recombinant human DNases which are used as mucolytics in cystic fibrosis [72], and recombinant human vascular endothelial growth factor (rhVEGF) [73] have all been shown to associate with specific binding proteins. Protein binding not only affects whether a peptide or protein drug will exert any pharmacological activity, but also whether it may have an inhibitory or stimulatory effect on the biological activity of the agent [71]. When injected into the bloodstream, recombinant cytokines may encounter various cytokine-binding proteins, including soluble cytokine receptors and anti-cytokine antibodies. In either case, the binding protein may either prolong the cytokine circulation time by acting as a storage depot, or it may enhance cytokine clearance [74]. Another example, growth hormone (GH), has at least two binding proteins in plasma [75]; such protein binding causes a substantial reduction in the elimination of GH, with a 10-fold smaller clearance of total compared to free GH, as well as decreasing hormonal activity via a reduction of receptor interactions.

Apart from these specific bindings, peptides and proteins may also bind non-specifically to plasma proteins. For example, the binding of metkephamid (a metenkephalin analog) to albumin was reported as 44–49% [76], while that of octreotide (a somatostatin analog) to lipoproteins was up to 65% [31].

Aside from the physico-chemical properties and protein binding of peptides and proteins, site-specific and target-oriented receptor-mediated uptake can also influence biodistribution. In the case of rhVEGF, the administration of high doses of the protein results in nonlinear PK that has been attributed to saturable binding, internalization, and degradation of VEGF mediated by high-affinity receptors that line the vasculature (see Sect. 5.5) [73].

5
Elimination

In general, peptides and protein drugs are eliminated almost exclusively by metabolism via the same catabolic pathways as endogenous or dietary proteins, resulting in amino acids that are reutilized in the endogenous amino acid pool for the *de-novo* biosynthesis of structural or functional body proteins. This has, for example, been described for enfuvirtide, a 36-amino acid synthetic peptide used in the treatment of HIV-1 infection [77].

Nonmetabolic elimination pathways, such as renal or biliary excretion, are generally negligible for most peptides and proteins, although biliary excretion has been described for some peptides and proteins (e.g., immunoglobulin A and octreotide) [31]. Clearance through biliary excretion has also been observed for the opioid peptides DPDPE ([D-pen2, D-pen5]enkephalin, where pen = penicillamine)) [78] as well as the prodrug form of DADLE [(H(2)N-Tyr-D-Ala-Gly-Phe-D-Leu-COOH)] [79]. If the biliary excretion of peptides and proteins does occur it generally results in the subsequent metabolism of these compounds in the GI tract (see Sect. 5.2) [13].

General tendencies for the *in-vivo* elimination of proteins and peptides may often be predicted from their physiological function. Peptides, for example, frequently have hormonal activity and usually have short elimination half-lives, which are desirable for a close regulation of their endogenous levels and thus function. In contrast, transport proteins such as albumin or α-1 acid glycoprotein have elimination half-lives of several days, which enables and ensures the continuous maintenance of necessary concentrations in the bloodstream.

The elimination of peptides and proteins can occur unspecifically almost everywhere in the body, or it can be limited to a specific organ or tissue. The locations of intensive peptide and protein metabolism include not only the liver, kidneys and GI tissue but also the vascular endothelial system, especially of the skin and other body tissues. The determining factors for the rate and mechanism of protein and peptide clearance include molecular weight as well as a molecule's physico-chemical properties, including size, overall charge, lipophilicity, functional groups, glycosylation pattern, secondary and tertiary structure and propensity for particle aggregation. The metabolism rate generally increases with decreasing molecular weight, from large to small proteins to peptides. Due to the unspecific degradation of numerous peptides and proteins in blood and vascular endothelium, clearance can exceed cardiac output, that is $>5 \, l \, min^{-1}$ for blood clearance and $>3 \, l \, min^{-1}$ for plasma clearance [13]. Investigations on the detailed metabolism of peptides and proteins are relatively difficult because of the myriad of molecule fragments that may be formed [80, 81].

A model example of the dependency of clearance on the physico-chemical properties of a protein is given by regular human insulin and its rapid-acting analogs insulin lispro and insulin aspart. These insulin analogs differ structurally from regular insulin through amino acid substitutions on the B chain, leading to conformational changes that result in a shift in binding to the C-terminal portion. The structural alterations allow the rapid-acting insulin analogs to have an onset of action of 5–15 min and an effective duration lasting at most 6 h, as compared to regular human insulin, which has a much later onset of 30–60 min and a longer duration up to 8–10 h. These properties of the rapid-acting insulin analogs allow them to facilitate a much tighter control of postprandial hyperglycemia in diabetic patients compared to regular human insulin [82].

5.1
Proteolysis

Proteolytic enzymes such as proteases and peptidases are ubiquitous throughout the body. Sites capable of extensive peptide and protein metabolism are not only limited to the liver, kidneys and GI tissues but also include blood and vascular endothelium as well as other organs and tissues. As proteases and peptidases are also located within cells, intracellular uptake is *per se* more an elimination rather than a distribution process [13]. Whereas, peptidases and proteases in the GI tract and in lysosomes are relatively unspecific, soluble peptidases in the interstitial space and exopeptidases on the cell surface have a higher selectivity and determine the specific metabolism pattern of an organ. The proteolytic activity of SC tissue, for example, results in a partial loss of activity of subcutaneously administered IFN-γ compared to intravenously administered IFN-γ.

5.2
Gastrointestinal Elimination

For orally administered peptides and proteins, the GI tract is the major site of metabolism; indeed, the presystemic metabolism of peptides and proteins is the primary reason for their lack of oral bioavailability. However, parenterally administered peptides and proteins may also be metabolized in the intestinal mucosa following intestinal secretion. In fact, at least 20% of the degradation of endogenous albumin takes place in the GI tract [13].

5.3
Renal Elimination

The kidneys are the major elimination organ for parenterally administered and endogenous peptides and proteins if they are smaller than the glomerular filtration limit of ~60 kDa. However, controversy persists with regard to glomerular filtration selectivity regarding size, molecular conformation and the charge of the peptide or protein. The importance of the kidneys as elimination organs could, for example, be shown for IL-2, macrophage colony-stimulating factor (M-CSF) and IFN-α [69, 75].

Complex mathematical models have been developed in order to calculate the sieving coefficient, or the average filtrate-to-plasma concentration ratio, along the length of a representative capillary [83]. In glomerular size-selectivity studies, dextran and Ficoll (a copolymer of sucrose and epichlorohydrin) have been used as test macromolecules over a wide range of molecular sizes [84]. Despite several existing models of size selectivity having provided basic frameworks when relating structure to actual functional properties of the glomerular barrier, future studies are still warranted to elucidate the effects of shape and deformability, as well as the steric hindrance of protein and peptide molecules [83–86]. In addition to size-selectivity, glomerular charge-selectivity has also been observed where anionic polymers pass through the capillary wall less readily than neutral polymers, which in turn pass through less readily than cationic polymers. Many of the charge-selectivity studies have utilized combinations of dextran (neutral)/dextran sulfate (anionic) and neutral horseradish peroxidase (nHPR)/anionic horseradish peroxidase (aHPR) [86].

The renal metabolism of peptides and small proteins is mediated through three highly effective processes (Fig. 1), with the result that only minuscule amounts of intact protein are detectable in the urine:

- The first mechanism involves the glomerular filtration of larger, complex peptides and proteins, followed by reabsorption into endocytic vesicles in the proximal tubule and subsequent hydrolysis into small peptide fragments and amino acids [87]. This mechanism of elimination has been described for IL-2 [88], IL-11 [89], GH [90], and insulin [91].
- The second mechanism entails glomerular filtration followed by intraluminal

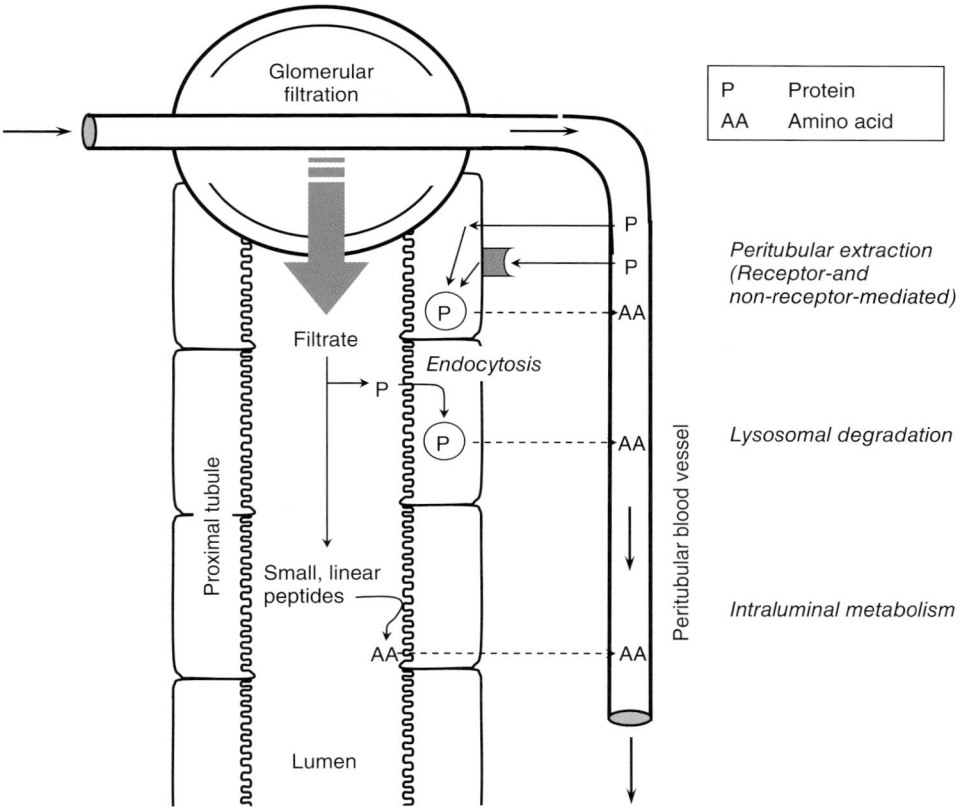

Fig. 1 Renal elimination processes of peptides and proteins. Glomerular filtration is followed by either intraluminal metabolism or tubular reabsorption with intracellular lysosomal metabolism, and peritubular extraction with intracellular lysosomal metabolism. From Ref. [21].

metabolism, predominantly by exopeptidases in the luminal brush border membrane of the proximal tubules. The resulting peptide fragments and amino acids are reabsorbed into the systemic circulation. This route of disposition applies to small linear peptides such as angiotensin I and II, bradykinin, glucagon and LH-RH [92, 93]. Recent studies have implicated the proton-driven peptide transporters PEPT1 and PEPT2 as the main route of cellular uptake of small peptides and peptide-like drugs from the glomerular filtrates [94]. These high-affinity transport proteins seem to exhibit a selective uptake of dipeptides and tripeptides, which implicates their role in renal amino acid homeostasis [95].

For both mechanisms, glomerular filtration is the dominant, rate-limiting step as subsequent degradation processes are not saturable under physiologic conditions [13, 87]. Hence, the renal contribution to the overall elimination of peptides and proteins is reduced if the metabolic activity of these proteins is high in other body regions, and becomes negligible in the presence of unspecific degradation throughout the body. In contrast, the contribution to total clearance approaches 100% if the metabolic activity is low in other tissues, or if distribution is limited. For recombinant IL-10, for instance, elimination correlates closely with the glomerular filtration rate, which means that dosage adjustments will be necessary in patients with impaired renal function [96].

- The third mechanism is the peritubular extraction of peptides and proteins from postglomerular capillaries and intracellular metabolism. Experiments using iodinated growth hormone (^{125}I-rGH) have shown that, while reabsorption into endocytic vesicles at the proximal tubule is still the dominant route of disposition, a small percentage of the hormone may be extracted from the peritubular capillaries [90, 97]. Peritubular transport of proteins and peptides from the basolateral membrane has also been shown for insulin [98] and the mycotoxin ochratoxin A [99].

Renal impairment may affect the PK of therapeutic proteins having molecular weights less than 60 kDa. Recombinant human IL-10, with a molecular weight of 18 kDa, is mainly cleared by glomerular filtration, and an increase in the area under the curve (AUC) was observed – from 548 ng·h ml^{-1} in patients with a creatinine clearance (CrCL) of >80 ml min 1.73 m^{-2} to 1500 ng·h ml^{-1} in patients with a CrCL of <15 ml min 1.73 m^{-2}. Thus, adjustments to the dose may be required in patients with chronic renal failure in order to prevent toxic effects. Recombinant human GH with a molecular weight of 22 kDa also undergoes extensive glomerular filtration, followed by proteolytic degradation. A decrease in GH clearance was observed in patients with renal insufficiency, leading to an increased systemic exposure and a higher C_{max}. In contrast, Pegfilgrastim (a recombinant methionyl human M-CSF with 20 PEG molecules attached to the N terminus) has a molecular weight of 39 kDa but is not renally filtered because of its large hydrodynamic radius; hence, no dose adjustment is required in renally impaired patients [100].

5.4
Hepatic Elimination

Aside from GI and renal metabolism, the liver may also contribute substantially to the metabolism of peptide and protein drugs. Proteolysis usually starts with endopeptidases that attack the middle part of the protein, with the resultant oligopeptides being further degraded by exopeptidases. The ultimate metabolites of proteins, amino acids and dipeptides are finally reutilized in the endogenous amino acid pool. The rate of hepatic metabolism is largely dependent on specific amino acid sequences in the protein.

Substrates for hepatic metabolism include insulin, glucagon and t-PAs [101, 102]. In the case of insulin, an acidic endopeptidase termed endosomal acidic insulinase appears to mediate an internalized insulin proteolysis at a number of sites [103]. Specifically, the endosomal activity results from cathepsin D, an aspartic acid protease [104]. Similarly, the proteolysis of glucagon has also been attributed to membrane-bound forms of cathepsins B and D [105].

An important first step in the hepatic metabolism of proteins and peptides is uptake into hepatocytes. Small peptides may cross the hepatocyte membrane via passive diffusion if they have sufficient hydrophobicity, but the uptake of larger proteins (such as t-PA; 65 kDa) is facilitated via receptor-mediated transport processes. Studies using radio-iodinated tissue plasminogen activator (^{125}I-t-PA) have implicated the role of mannose and asialoglycoprotein receptors in the liver for facilitating t-PA uptake and clearance [102]. Evidence also points to another hepatic membrane receptor – the low-density lipoprotein receptor-related protein – for contributing to overall t-PA clearance [102, 106].

5.5
Receptor-Mediated Endocytosis

For conventional small-molecule drugs, the amount of drug bound to receptors is usually negligible compared to the total amount of drug in the body, and rarely has any effect on the PK profile. In contrast, a substantial fraction of a peptide and protein dose can be bound to receptors, with such binding perhaps leading to elimination via receptor-mediated uptake and subsequent intracellular metabolism. The process of endocytosis is not limited to hepatocytes but can occur also in other cells, including the therapeutic target cells. Binding and subsequent degradation via interaction with these generally high-affinity, low-capacity binding sites is a typical example of a pharmacologic target-mediated drug disposition, where binding to the PD target structure affects drug disposition [107]. As the number of receptors is limited, drug binding and uptake can usually be saturated within therapeutic concentrations, or more specifically at relatively low molar ratios between the protein drug and the receptor. As a consequence, the PK for these drugs frequently does not follow the rule of superposition – that is, drug clearance, and potentially other PK parameters, are dose-dependent. Thus, receptor-mediated elimination constitutes a major source for the nonlinear pharmacokinetic behavior of numerous peptide and protein drugs, resulting in a lack of dose proportionality [4].

M-CSF, besides linear renal elimination, undergoes a nonlinear elimination pathway that follows Michaelis–Menten kinetics and is linked to a receptor-mediated uptake into macrophages. At low concentrations, M-CSF follows linear PK, whereas at high concentrations the nonrenal elimination pathways are saturated and this results in a nonlinear PK behavior [108, 109]. Similarly, a PK analysis of the *in-vivo* disposition of the G-CSF derivative nartograstim revealed a nonlinear clearance process by the bone marrow and spleen with increasing doses of nartograstim [70]. Further studies with nartograstim suggested that nonlinearity in the early-phase bone marrow clearance might be due to a downregulation of G-CSF receptors on the cell surface [110]. For rhVEGF, a mechanism-based target-mediated drug distribution model had to be developed in order to accurately describe the drug's nonlinear PK in patients with coronary artery disease [73]. The nonlinear elimination of rhVEGF has been shown to be caused by binding to saturable high-affinity receptors, followed by internalization and degradation.

6
Interspecies Scaling

Peptides and proteins exhibit distinct species specificity with regards to structure and activity. Peptides and proteins with identical physiological function may have different amino acid sequences in different species, and may have no activity – or may even be immunogenic – if used in a different species.

The extrapolation of animal data to predict PK parameters by allometric scaling is an often-used tool in drug development, with multiple approaches available at variable success rates [111–113]. In the most frequently used approach, PK parameters between different species are related via body weight using a power function:

$$P = a \cdot W^b$$

where P is the PK parameter scaled, W is the body weight (in kg), a is the allometric coefficient, and b is the allometric exponent. a and b are specific constants for each parameter of a compound. General tendencies for the allometric exponent are 0.75 for rate constants (i.e., clearance, elimination rate constant), 1 for volumes of distribution, and 0.25 for half-lives.

For most traditional, small-molecule drugs, allometric scaling is often imprecise, especially if hepatic metabolism is a major elimination pathway and/or if there are interspecies differences in metabolism. For peptides and proteins, however, allometric scaling has frequently proven to be much more precise and reliable, probably because of the similarity in handling peptides and proteins between different mammalian species [4, 63]. The prediction of human clearance worked reasonably well with simple allometry and a fixed exponent of 0.8: Predicted values were found to be within twofold for most of the protein therapeutics evaluated when using clearance from multiple species or by using clearance only from monkeys [114, 115]. Interspecies scaling can lay the foundation for determining dosing as it relates to both efficacy and toxicity. Clearance and volume of distribution of numerous therapeutically used proteins such as GH or t-PA follow a well-defined, weight-dependent physiologic relationship between laboratory animals and humans [96, 116]. This information often provides the basis for quantitative predictions for toxicology and dose-ranging studies based on preclinical findings.

In a study where the allometric relationships of pharmacokinetic parameters for five therapeutic proteins were investigated, the allometric equations for clearance and volumes of distribution were found to be different for each protein [116]. This variability was attributed to possible species specificity and immune-mediated clearance mechanisms. Species specificity refers to the inherent differences in structure and activity across species. Minute differences in the amino acid sequence may render an agent inactive when administered to foreign species, and may even generate an immunogenic response. Immunogenicity has been clearly demonstrated in a study with the tumor necrosis factor receptor–immunoglobulin fusion protein, lenercept. This all-human sequence protein elicits an immune response in laboratory animals which ultimately results in a rapid clearance of the protein [117].

To further complicate the matter, some studies have shown that the extent of glycosylation and/or sialylation of a protein molecule also exhibits species specificity that may affect drug efficacy, clearance and immunogenicity. This is especially important if the production of human proteins is performed using bacterial cells [75]. While epoetin-α, IFN-α and FSH are naturally glycosylated, others (such as insulin and GH) are not and this requires a careful selection of the animal model to assure an adequately intended pharmacological effect. Despite such differences, allometric scaling can still form the basis of estimating effective regimens for initial human studies. This is especially significant given that materials are often in limited supply for preclinical and early clinical development where dose optimization is crucial [116].

7
Drug–Drug Interaction

Therapeutic proteins are often given in combination with other therapeutic proteins and small-molecule drugs. This may potentially lead to therapeutic protein–small-molecule drug interactions or therapeutic protein–therapeutic protein interactions, and subsequently have an impact on their safety and efficacy. Drug interactions can be mainly classified as two broad categories: (i) pharmacokinetics-based drug–drug interaction (PK-DDI); and (ii) pharmacodynamics-based drug–drug interaction (PD-DDI). Briefly, PK-DDI results from interference with the absorption/distribution/metabolism/excretion (ADME) process of the drug, while PD-DDI involves the modulation at the receptor, signaling pathway or the effector level leading to either enhanced or diminished drug response without any change in the systemic exposure [118].

7.1
Pharmacokinetic Based Drug–Drug Interactions

The chances of PK-DDI between a therapeutic protein and small-molecule drugs are modest because of the underlying differences in the clearance mechanisms. Most small molecules are cleared through hepatic metabolism or renal elimination, whereas therapeutic proteins primarily undergo proteolytic degradation which is usually not saturated at efficacious doses. Similarly, combination therapy with therapeutic proteins does usually not result in the alteration of PK because the clearance pathway remains unsaturated [119].

7.2
Pharmacodynamic Based Drug–Drug Interactions

Some immunosuppressive drugs may inhibit the humoral immune response, decreasing the immunogenicity rate of the therapeutic protein and consequently decreasing its clearance. The target-mediated drug disposition clearance route of some protein drugs may be modulated by immunosuppressive agents, leading to PD-DDI; this is mainly observed in cases when target-mediated drug disposition is the major process of elimination at the therapeutic dose. Therapeutic proteins which have target-mediated drug disposition as their major elimination pathway may undergo a reduced clearance in combination with immunosuppressive agents because of a suppression in baseline receptor expression levels. However, these clinical situations rarely occur, as most of the therapeutic proteins at effective dose levels are mainly eliminated via the unsaturated nonspecific clearance pathway rather than by target-mediated drug disposition.

Cases have also been noted where therapeutic proteins are the perpetrators and small-molecule drugs are the victims. Some immunomodulatory agents (such as exogenous IFN-α) upregulates cytokine levels which in turns inhibits cytochrome P450 (CYP) activity and, consequently, increases the exposure of small-molecule drugs that are CYP substrates. For example, IFNα-2b increased the serum concentration of theophylline, which is a CYP1A2, substrate by 100%. In contrast, therapeutic proteins which are indicated for inflammation or infectious disease states may downregulate the elevated cytokine levels, normalizing the CYP activity and consequently decreasing the exposure of CYP substrates. Interestingly, expressions of various CYP enzymes are modulated to different extent by the cytokine levels [120].

PD-DDI may also cause adverse effects without having any impact on the exposure of drugs in combination. This can be explained by the example of combination between Anakinra, a recombinant IL-1R antagonist protein which is eliminated by renal excretion, and Etanercept, a dimeric fusion protein which is eliminated by FcRn-mediated recycling. Although there was no PK-DDI because of the different clearance pathways of these proteins, the combination still posed a serious risk of infection and neutropenia.

8
Conclusions

Peptide and protein drugs are subject to the same general principles of PK and exposure–response correlations as conventional small-drug molecules. Due to their similarity to protein nutrients and/or especially regulatory endogenous peptides and proteins, however, numerous caveats and pitfalls related to PK have to be considered and addressed during the development of these drugs. Furthermore, PK/PD correlations are frequently complicated due to the close interaction of peptide and protein drugs with endogenous substances and receptors, as well as regulatory feedback mechanisms. Additional investigations and resources are necessary to overcome some of these difficulties in order to ensure a rapid and successful drug-development process. Nevertheless, PK evaluations provide a cornerstone in the development of dosage regimens for a rational, scientifically based clinical application of peptide and protein therapeutics that, ultimately, will ensure the success of this class of compounds in applied pharmacotherapy.

References

1. Meibohm, B., Derendorf, H. (2002) Pharmacokinetic/pharmacodynamic studies in drug product development. *J. Pharm. Sci.*, **91**, 18–31.
2. CDER/FDA (2003) *Exposure Response Relationships: Guidance for Industry Rockville:* US Department of Health and Human Services, Food and Drug Administration, Center for Drug Evaluation and Research.
3. International Conference on Harmonization (1994) ICH E4: Dose–Response Information to Support Drug Registration London: European Agency for the Evaluation of Medicinal Products.
4. Tang, L., Persky, A.M., Hochhaus, G., Meibohm, B. (2004) Pharmacokinetic aspects of biotechnology products. *J. Pharm. Sci.*, **93**, 2184–2204.
5. Fasano, A. (1998) Novel approaches for oral delivery of macromolecules. *J. Pharm. Sci.*, **87**, 1351–1356.
6. Mahato, R.I., Narang, A.S., Thoma, L., Miller, D.D. (2003) Emerging trends in oral delivery of peptide and protein drugs. *Crit. Rev. Ther. Drug Carrier Syst.*, **20**, 153–214.
7. Lan, L.B., Dalton, J.T., Schuetz, E.G. (2000) Mdr1 limits CYP3A metabolism in vivo. *Mol. Pharmacol.*, **58**, 863–869.
8. Meibohm, B., Beierle, I., Derendorf, H. (2002) How important are gender differences in pharmacokinetics? *Clin. Pharmacokinet.*, **41**, 329–342.
9. Wacher, V.J., Silverman, J.A., Zhang, Y., Benet, L.Z. (1998) Role of P-glycoprotein and cytochrome P450 3A in limiting oral absorption of peptides and peptidomimetics. *J. Pharm. Sci.*, **87**, 1322–1330.
10. Handelsman, D.J., Swerdloff, R.S. (1986) Pharmacokinetics of gonadotropin-releasing hormone and its analogs. *Endocr. Rev.*, **7**, 95–105.
11. Periti, P., Mazzei, T., Mini, E. (2002) Clinical pharmacokinetics of depot leuprorelin. *Clin. Pharmacokinet.*, **41**, 485–504.

12. Kaufman, J.S., Reda, D.J., Fye, C.L., Goldfarb, D.S. et al. (1998) Subcutaneous compared with intravenous epoetin in patients receiving hemodialysis. Department of Veterans Affairs Cooperative Study Group on Erythropoietin in Hemodialysis Patients. *N. Engl. J. Med.*, **339**, 578–583.
13. Colburn, W. (1991) Peptide, Peptoid, and Protein Pharmacokinetics/Pharmacodynamics, in: Garzone, P., Colburn, W., Mokotoff, M. (Eds), *Petides, Peptoids, and Proteins*, Harvey Whitney Books, Cincinnati, OH, pp. 94–115.
14. Fuzeon (2004) *Prescribing Information*, Roche Pharmaceuticals, Nutley, NJ.
15. Symlin (2005) *Prescribing Information*, Amylin Pharmaceuticals, Inc., San Diego, CA.
16. Porter, C.J., Charman, S.A. (2000) Lymphatic transport of proteins after subcutaneous administration. *J. Pharm. Sci.*, **89**, 297–310.
17. Supersaxo, A., Hein, W., Gallati, H., Steffen, H. (1988) Recombinant human interferon alpha-2a: delivery to lymphoid tissue by selected modes of application. *Pharm. Res.*, **5**, 472–476.
18. Zhao, L., Ji, P., Li, Z., Roy, P. et al. (2013) The antibody drug absorption following subcutaneous or intramuscular administration and its mathematical description by coupling physiologically based absorption process with the conventional compartment pharmacokinetic model. *J. Clin. Pharmacol.*, **53**, 314–325.
19. Supersaxo, A., Hein, W.R., Steffen, H. (1990) Effect of molecular weight on the lymphatic absorption of water-soluble compounds following subcutaneous administration. *Pharm. Res.*, **7**, 167–169.
20. Schomburg, A., Kirchner, H., Atzpodien, J. (1993) Renal, metabolic, and hemodynamic side-effects of interleukin-2 and/or interferon alpha: evidence of a risk/benefit advantage of subcutaneous therapy. *J. Cancer Res. Clin. Oncol.*, **119**, 745–755.
21. Tang, L., Meibohm, B. (2006) Pharmacokinetics of Peptides and Proteins, in: Meibohm, B. (Ed.), *Biopharmaceutics and Pharmacodynamics of Biotech Drugs*, Wiley-VCH Verlag GmbH, Weinheim, pp. 1–30.
22. Antosova, Z., Mackova, M., Kral, V., Macek, T. (2009) Therapeutic application of peptides and proteins: parenteral forever? *Trends Biotechnol.*, **27**, 628–635.
23. Laube, B.L. (2001) Treating diabetes with aerosolized insulin. *Chest*, **120**, 99S–106S.
24. Cleland, J.L., Daugherty, A., Mrsny, R. (2001) Emerging protein delivery methods. *Curr. Opin. Biotechnol.*, **12**, 212–219.
25. Patton, J.S., Bukar, J.G., Eldon, M.A. (2004) Clinical pharmacokinetics and pharmacodynamics of inhaled insulin. *Clin. Pharmacokinet.*, **43**, 781–801.
26. Cefalu, W.T., Skyler, J.S., Kourides, I.A., Landschulz, W.H. et al. (2001) Inhaled human insulin treatment in patients with type 2 diabetes mellitus. *Ann. Intern. Med.*, **134**, 203–207.
27. Hollander, P.A., Blonde, L., Rowe, R., Mehta, A.E. et al. (2004) Efficacy and safety of inhaled insulin (Exubera) compared with subcutaneous insulin therapy in patients with type 2 diabetes: results of a 6-month, randomized, comparative trial. *Diabetes Care*, **27**, 2356–2362.
28. Quan, J.M., Tiddens, H.A., Sy, J.P., McKenzie, S.G. et al. (2001) A two-year randomized, placebo-controlled trial of dornase alpha in young patients with cystic fibrosis with mild lung function abnormalities. *J. Pediatr.*, **139**, 813–820.
29. Jordan, B.K., Donn, S.M. (2013) Lucinactant for the prevention of respiratory distress syndrome in premature infants. *Expert Rev. Clin. Pharmacol.*, **6**, 115–121.
30. Zito, S.W. (1997) *Pharmaceutical Biotechnology: A Programmed Text*, Technomic Publishing Company, Lancaster, PA.
31. Chanson, P., Timsit, J., Harris, A.G. (1993) Clinical pharmacokinetics of octreotide. Therapeutic applications in patients with pituitary tumours. *Clin. Pharmacokinet.*, **25**, 375–391.
32. Lawrence, D. (2002) Intranasal delivery could be used to administer drugs directly to the brain. *Lancet*, **359**, 1674.
33. Liu, X.F., Fawcett, J.R., Thorne, R.G., DeFor, T.A. et al. (2001) Intranasal administration of insulin-like growth factor-I bypasses the blood–brain barrier and protects against focal cerebral ischemic damage. *J. Neurol. Sci.*, **187**, 91–97.
34. Diao, L., Meibohm, B. (2013) Pharmacokinetics and pharmacokinetic-pharmacodynamic correlations of

therapeutic peptides. *Clin. Pharmacokinet.*, **52**, 855–868.

35. Mitragotri, S., Blankschtein, D., Langer, R. (1995) Ultrasound-mediated transdermal protein delivery. *Science*, **269**, 850–853.

36. Kanikkannan, N. (2002) Iontophoresis-based transdermal delivery systems. *BioDrugs*, **16**, 339–347.

37. Pillai, O., Panchagnula, R. (2003) Transdermal iontophoresis of insulin. V. Effect of terpenes. *J. Controlled Release*, **88**, 287–296.

38. Kumar, S., Char, H., Patel, S., Piemontese, D. et al. (1992) Effect of iontophoresis on in vitro skin permeation of an analogue of growth hormone releasing factor in the hairless guinea pig model. *J. Pharm. Sci.*, **81**, 635–639.

39. Chang, S.L., Hofmann, G.A., Zhang, L., Deftos, L.J. et al. (2000) Transdermal iontophoretic delivery of salmon calcitonin. *Int. J. Pharm.*, **200**, 107–113.

40. Suzuki, Y., Iga, K., Yanai, S., Matsumoto, Y. et al. (2001) Iontophoretic pulsatile transdermal delivery of human parathyroid hormone (1-34). *J. Pharm. Pharmacol.*, **53**, 1227–1234.

41. Widera, A., Kim, K.J., Crandall, E.D., Shen, W.C. (2003) Transcytosis of GCSF-transferrin across rat alveolar epithelial cell monolayers. *Pharm. Res.*, **20**, 1231–1238.

42. Kageyama, S., Yamamoto, H., Nakazawa, H., Matsushita, J. et al. (2002) Pharmacokinetics and pharmacodynamics of AJW200, a humanized monoclonal antibody to von Willebrand factor, in monkeys. *Arterioscler. Thromb. Vasc. Biol.*, **22**, 187–192.

43. Leone-Bay, A., Sato, M., Paton, D., Hunt, A.H. et al. (2001) Oral delivery of biologically active parathyroid hormone. *Pharm. Res.*, **18**, 964–970.

44. Joseph, J.W., Kalitsky, J., St-Pierre, S., Brubaker, P.L. (2000) Oral delivery of glucagon-like peptide-1 in a modified polymer preparation normalizes basal glycaemia in diabetic db/db mice. *Diabetologia*, **43**, 1319–1328.

45. O'Hagan, D.T. (1992) Oral delivery of vaccines. Formulation and clinical pharmacokinetic considerations. *Clin. Pharmacokinet.*, **22**, 1–10.

46. Van Der Lubben, I.M., Konings, F.A., Borchard, G., Verhoef, J.C. et al. (2001) In vivo uptake of chitosan microparticles by murine Peyer's patches: visualization studies using confocal laser scanning microscopy and immunohistochemistry. *J. Drug Target.*, **9**, 39–47.

47. Pauletti, G.M., Gangwar, S., Siahaan, T.J., Jeffrey, A. et al. (1997) Improvement of oral peptide bioavailability: peptidomimetics and prodrug strategies. *Adv. Drug Delivery Rev.*, **27**, 235–256.

48. Vazquez-Roque, M.I., Bouras, E.P. (2013) Linaclotide, novel therapy for the treatment of chronic idiopathic constipation and constipation-predominant irritable bowel syndrome. *Adv. Ther.*, **30**, 203–211.

49. Perez Ruixo, J.J., Ma, P., Chow, A.T. (2013) The utility of modeling and simulation approaches to evaluate immunogenicity effect on the therapeutic protein pharmacokinetics. *AAPS J.*, **15**, 172–182.

50. Genzyme (2009) *Myozyme Prescribing Information*, Genzyme, Cambridge, MA.

51. Chirmule, N., Jawa, V., Meibohm, B. (2012) Immunogenicity to therapeutic proteins: impact on PK/PD and efficacy. *AAPS J.*, **14**, 296–302.

52. Working, P., Cossum, P. (1991) Clinical and Preclinical Studies with Recombinant Human Proteins: Effect of Antibody Production, in: Garzone, P., Colburn, W., Mokotoff, M. (Eds), *Petides, Peptoids, and Proteins*, Harvey Whitney Books, Cincinnati, OH, pp. 158–168.

53. Perini, P., Facchinetti, A., Bulian, P., Massaro, A.R. et al. (2001) Interferon-beta (IFN-beta) antibodies in interferon-beta1a- and interferon-beta1b-treated multiple sclerosis patients. Prevalence, kinetics, cross-reactivity, and factors enhancing interferon-beta immunogenicity in vivo. *Eur. Cytokine Netw.*, **12**, 56–61.

54. Baert, F., Noman, M., Vermeire, S., Van Assche, G. et al. (2003) Influence of immunogenicity on the long-term efficacy of infliximab in Crohn's disease. *N. Engl. J. Med.*, **348**, 601–608.

55. Millonig, A., Rudzki, D., Holzl, M., Ehling, R. et al. (2009) High-dose intravenous interferon beta in patients with neutralizing antibodies (HINABS): a pilot study. *Mult. Scler.*, **15**, 977–983.

56. Harris, J.M., Martin, N.E., Modi, M. (2001) Pegylation: a novel process for modifying pharmacokinetics. *Clin. Pharmacokinet.*, **40**, 539–551.

57. Molineux, G. (2003) Pegylation: engineering improved biopharmaceuticals for oncology. *Pharmacotherapy*, **23**, 3S–8S.
58. Walsh, S., Shah, A., Mond, J. (2003) Improved pharmacokinetics and reduced antibody reactivity of lysostaphin conjugated to polyethylene glycol. *Antimicrob. Agents Chemother.*, **47**, 554–558.
59. Graham, M.L. (2003) Pegaspargase: a review of clinical studies. *Adv. Drug Delivery Rev.*, **55**, 1293–1302.
60. Zamboni, W.C. (2003) Pharmacokinetics of pegfilgrastim. *Pharmacotherapy*, **23**, 9S–14S.
61. Reilly, R.M., Sandhu, J., Alvarez-Diez, T.M., Gallinger, S. et al. (1995) Problems of delivery of monoclonal antibodies. Pharmaceutical and pharmacokinetic solutions. *Clin. Pharmacokinet.*, **28**, 126–142.
62. Ezan, E. (2013) Pharmacokinetic studies of protein drugs: past, present and future. *Adv. Drug Delivery Rev.*, **65**, 1065–1073.
63. Meibohm, B. (2013) Pharmacokinetics and Pharmacodynamics of Peptide and Protein Drugs, in: Crommelin, D.J.A., Sindelar, R.D., Meibohm, B. (Eds), *Pharmaceutical Biotechnology*, Springer-Verlag, New York, pp. 101–132.
64. Tanswell, P., Modi, N., Combs, D., Danays, T. (2002) Pharmacokinetics and pharmacodynamics of tenecteplase in fibrinolytic therapy of acute myocardial infarction. *Clin. Pharmacokinet.*, **41**, 1229–1245.
65. Ramakrishnan, R., Cheung, W.K., Wacholtz, M.C., Minton, N. et al. (2004) Pharmacokinetic and pharmacodynamic modeling of recombinant human erythropoietin after single and multiple doses in healthy volunteers. *J. Clin. Pharmacol.*, **44**, 991–1002.
66. Allon, M., Kleinman, K., Walczyk, M., Kaupke, C. et al. (2002) Pharmacokinetics and pharmacodynamics of darbepoetin alfa and epoetin in patients undergoing dialysis. *Clin. Pharmacol. Ther.*, **72**, 546–555.
67. Jin, F., Krzyzanski, W. (2004) Pharmacokinetic model of target-mediated disposition of thrombopoietin. *AAPS Pharm. Sci.*, **6**, E9.
68. Tan, A.C., Russel, F.G., Thien, T., Benraad, T.J. (1993) Atrial natriuretic peptide. An overview of clinical pharmacology and pharmacokinetics. *Clin. Pharmacokinet.*, **24**, 28–45.
69. McMartin, C. (1992) Pharmacokinetics of peptides and proteins: opportunities and challenges. *Adv. Drug Res.*, **22**, 39–106.
70. Kuwabara, T., Uchimura, T., Kobayashi, H., Kobayashi, S. et al. (1995) Receptor-mediated clearance of G-CSF derivative nartograstim in bone marrow of rats. *Am. J. Physiol.*, **269**, E1–E9.
71. Toon, S. (1996) The relevance of pharmacokinetics in the development of biotechnology products. *Eur. J. Drug Metab. Pharmacokinet.*, **21**, 93–103.
72. Mohler, M., Cook, J., Lewis, D., Moore, J. et al. (1993) Altered pharmacokinetics of recombinant human deoxyribonuclease in rats due to the presence of a binding protein. *Drug Metab. Dispos.*, **21**, 71–75.
73. Eppler, S.M., Combs, D.L., Henry, T.D., Lopez, J.J. et al. (2002) A target-mediated model to describe the pharmacokinetics and hemodynamic effects of recombinant human vascular endothelial growth factor in humans. *Clin. Pharmacol. Ther.*, **72**, 20–32.
74. Piscitelli, S.C., Reiss, W.G., Figg, W.D., Petros, W.P. (1997) Pharmacokinetic studies with recombinant cytokines. Scientific issues and practical considerations. *Clin. Pharmacokinet.*, **32**, 368–381.
75. Wills, R.J., Ferraiolo, B.L. (1992) The role of pharmacokinetics in the development of biotechnologically derived agents. *Clin. Pharmacokinet.*, **23**, 406–414.
76. Taki, Y., Sakane, T., Nadai, T., Sezaki, H. et al. (1998) First-pass metabolism of peptide drugs in rat perfused liver. *J. Pharm. Pharmacol.*, **50**, 1013–1018.
77. Patel, I.H., Zhang, X., Nieforth, K., Salgo, M. et al. (2005) Pharmacokinetics, pharmacodynamics and drug interaction potential of enfuvirtide. *Clin. Pharmacokinet.*, **44**, 175–186.
78. Hoffmaster, K.A., Zamek-Gliszczynski, M.J., Pollack, G.M., Brouwer, K.L. (2005) Multiple transport systems mediate the hepatic uptake and biliary excretion of the metabolically stable opioid peptide [D-penicillamine2,5]enkephalin. *Drug Metab. Dispos.*, **33**, 287–293.
79. Yang, J.Z., Chen, W., Borchardt, R.T. (2002) In vitro stability and in vivo pharmacokinetic studies of a model opioid peptide, H-Tyr-D-Ala-Gly-Phe-D-Leu-OH (DADLE), and its cyclic prodrugs. *J. Pharmacol. Exp. Ther.*, **303**, 840–848.

80. Brugos, B., Hochhaus, G. (2004) Metabolism of dynorphin A(1-13). *Pharmazie*, **59**, 339–343.
81. Muller, S., Hutson, A., Arya, V., Hochhaus, G. (1999) Assessment of complex peptide degradation pathways via structured multicompartmental modeling approaches: the metabolism of dynorphin A1–13 and related fragments in human plasma. *J. Pharm. Sci.*, **88**, 938–944.
82. Hirsch, I.B. (2005) Insulin analogues. *N. Engl. J. Med.*, **352**, 174–183.
83. Edwards, A., Daniels, B.S., Deen, W.M. (1999) Ultrastructural model for size selectivity in glomerular filtration. *Am. J. Physiol.*, **276**, F892–F902.
84. Oliver, J.D., 3rd, Anderson, S., Troy, J.L., Brenner, B.M. et al. (1992) Determination of glomerular size-selectivity in the normal rat with Ficoll. *J. Am. Soc. Nephrol.*, **3**, 214–228.
85. Venturoli, D., Rippe, B. (2005) Ficoll and dextran vs. globular proteins as probes for testing glomerular permselectivity: effects of molecular size, shape, charge, and deformability. *Am. J. Physiol. Renal Physiol.*, **288**, F605–F613.
86. Deen, W.M., Lazzara, M.J., Myers, B.D. (2001) Structural determinants of glomerular permeability. *Am. J. Physiol. Renal Physiol.*, **281**, F579–F596.
87. Maack, T., Park, C., Camargo, M. (1985) Renal Filtration, Transport and Metabolism of Proteins in: Seldin, D., Giebisch, G. (Eds), *The Kidney*, Raven Press, New York, pp. 1773–1803.
88. Anderson, P.M., Sorenson, M.A. (1994) Effects of route and formulation on clinical pharmacokinetics of interleukin-2. *Clin. Pharmacokinet.*, **27**, 19–31.
89. Takagi, A., Masuda, H., Takakura, Y., Hashida, M. (1995) Disposition characteristics of recombinant human interleukin-11 after a bolus intravenous administration in mice. *J. Pharmacol. Exp. Ther.*, **275**, 537–543.
90. Johnson, V., Maack, T. (1977) Renal extraction, filtration, absorption, and catabolism of growth hormone. *Am. J. Physiol.*, **233**, F185–F196.
91. Rabkin, R., Ryan, M.P., Duckworth, W.C. (1984) The renal metabolism of insulin. *Diabetologia*, **27**, 351–357.
92. Carone, F.A., Peterson, D.R. (1980) Hydrolysis and transport of small peptides by the proximal tubule. *Am. J. Physiol.*, **238**, F151–F158.
93. Carone, F.A., Peterson, D.R., Flouret, G. (1982) Renal tubular processing of small peptide hormones. *J. Lab. Clin. Med.*, **100**, 1–14.
94. Inui, K., Terada, T., Masuda, S., Saito, H. (2000) Physiological and pharmacological implications of peptide transporters, PEPT1 and PEPT2. *Nephrol. Dial. Transplant.*, **15** (Suppl. 6), 11–13.
95. Daniel, H., Herget, M. (1997) Cellular and molecular mechanisms of renal peptide transport. *Am. J. Physiol.*, **273**, F1–F8.
96. Andersen, S., Lambrecht, L., Swan, S., Cutler, D. et al. (1999) Disposition of recombinant human interleukin-10 in subjects with various degrees of renal function. *J. Clin. Pharmacol.*, **39**. 1015–1020.
97. Krogsgaard Thomsen, M., Friis, C., Sehested Hansen, B., Johansen, P. et al. (1994) Studies on the renal kinetics of growth hormone (GH) and on the GH receptor and related effects in animals. *J. Pediatr. Endocrinol.*, **7**, 93–105.
98. Nielsen, S., Nielsen, J.T., Christensen, E.I. (1987) Luminal and basolateral uptake of insulin in isolated, perfused, proximal tubules. *Am. J. Physiol.*, **253**, F857–F867.
99. Groves, C.E., Morales, M., Wright, S.H. (1998) Peritubular transport of ochratoxin A in rabbit renal proximal tubules. *J. Pharmacol. Exp. Ther.*, **284**, 943–948.
100. Meibohm, B., Zhou, H. (2012) Characterizing the impact of renal impairment on the clinical pharmacology of biologics. *J. Clin. Pharmacol.*, **52**, 54S–62S.
101. Authier, F., Posner, B.I., Bergeron, J.J. (1996) Endosomal proteolysis of internalized proteins. *FEBS Lett.*, **389**, 55–60.
102. Smedsrod, B., Einarsson, M. (1990) Clearance of tissue plasminogen activator by mannose and galactose receptors in the liver. *Thromb. Haemost.*, **63**, 60–66.
103. Authier, F., Danielsen, G.M., Kouach, M., Briand, G. et al. (2001) Identification of insulin domains important for binding to and degradation by endosomal acidic insulinase. *Endocrinology*, **142**, 276–289.
104. Authier, F., Metioui, M., Fabrega, S., Kouach, M. et al. (2002) Endosomal

proteolysis of internalized insulin at the C-terminal region of the B chain by cathepsin D. *J. Biol. Chem.*, **277**, 9437–9446.
105. Authier, F., Mort, J.S., Bell, A.W., Posner, B.I. et al. (1995) Proteolysis of glucagon within hepatic endosomes by membrane-associated cathepsins B and D. *J. Biol. Chem.*, **270**, 15798–15807.
106. Bu, G., Williams, S., Strickland, D.K., Schwartz, A.L. (1992) Low density lipoprotein receptor-related protein/alpha 2-macroglobulin receptor is an hepatic receptor for tissue-type plasminogen activator. *Proc. Natl Acad. Sci. USA*, **89**, 7427–7431.
107. Levy, G. (1994) Mechanism-based pharmacodynamic modeling. *Clin. Pharmacol. Ther.*, **56**, 356–358.
108. Bartocci, A., Mastrogiannis, D.S., Migliorati, G., Stockert, R.J. et al. (1987) Macrophages specifically regulate the concentration of their own growth factor in the circulation. *Proc. Natl Acad. Sci. USA*, **84**, 6179–6183.
109. Bauer, R.J., Gibbons, J.A., Bell, D.P., Luo, Z.P. et al. (1994) Nonlinear pharmacokinetics of recombinant human macrophage colony- stimulating factor (M-CSF) in rats. *J. Pharmacol. Exp. Ther.*, **268**, 152–158.
110. Kuwabara, T., Uchimura, T., Takai, K., Kobayashi, H. et al. (1995) Saturable uptake of a recombinant human granulocyte colony-stimulating factor derivative, nartograstim, by the bone marrow and spleen of rats in vivo. *J. Pharmacol. Exp. Ther.*, **273**, 1114–1122.
111. Mahmood, I., Balian, J.D. (1999) The pharmacokinetic principles behind scaling from preclinical results to phase I protocols. *Clin. Pharmacokinet.*, **36**, 1–11.
112. Mahmood, I. (2002) Interspecies scaling: predicting oral clearance in humans. *Am. J. Ther.*, **9**, 35–42.
113. Boxenbaum, H. (1982) Interspecies scaling, allometry, physiological time, and the ground plan of pharmacokinetics. *J. Pharmacokinet. Biopharm.*, **10**, 201–227.
114. Wang, W., Prueksaritanont, T. (2010) Prediction of human clearance of therapeutic proteins: simple allometric scaling method revisited. *Biopharm. Drug Dispos.*, **31**, 253–263.
115. Offman, E., Edginton, A.N. (2013) Contrasting toxicokinetic evaluations and interspecies pharmacokinetic scaling approaches for small molecules and biologics: applicability to biosimilar development. *Xenobiotica*, **43**, 561–569.
116. Mordenti, J., Chen, S.A., Moore, J.A., Ferraiolo, B.L. et al. (1991) Interspecies scaling of clearance and volume of distribution data for five therapeutic proteins. *Pharm. Res.*, **8**, 1351–1359.
117. Richter, W.F., Gallati, H., Schiller, C.D. (1999) Animal pharmacokinetics of the tumor necrosis factor receptor-immunoglobulin fusion protein lenercept and their extrapolation to humans. *Drug Metab. Dispos.*, **27**, 21–25.
118. Hinder, M. (2011) Pharmacodynamic Drug–Drug Interactions in: Vogel, H., Maas, J., Gebauer, A. (Eds), *Drug Discovery and Evaluation: Methods in Clinical Pharmacology*, Springer, Berlin, Heidelberg, pp. 367–376.
119. Lu, D., Girish, S., Theil, F.-P., Joshi, A. et al. (2010) Pharmacokinetic and Pharmacodynamic-Based Drug Interactions for Therapeutic Proteins in: Zhou, H., Meibohm, B. (Eds), *Drug–drug Interaction for Therapeutic Biologics*, John Wiley & Sons, Inc., Hoboken, NJ, pp. 5–37.
120. Huang, S.M., Zhao, H., Lee, J.I., Reynolds, K. et al. (2010) Therapeutic protein–drug interactions and implications for drug development. *Clin. Pharmacol. Ther.*, **87**, 497–503.

21
Physical Pharmacy and Biopharmaceutics

M. Sherry Ku
Saviorlifetec Corp., No.29, Kejhong Rodd, Chunan township, Miaoli county, 35053 Taiwan

1	**Definitions** 728	
1.1	Physical Pharmacy 728	
1.2	Biopharmaceutics 729	
1.3	Pharmacokinetics 729	
1.4	Pharmacodynamics 730	
2	**History of Physical Pharmacy and Biopharmaceutics** 730	
3	**Application of Thermodynamics in Pharmacy** 731	
3.1	Solubility 732	
3.1.1	Equilibrium Solubility 732	
3.1.2	Kinetic Solubility 733	
3.1.3	Heat of Solution 734	
3.2	Ionic Equilibrium 735	
3.2.1	pK_a and pH Solubility Profile 735	
3.2.2	Zwitterionic Compounds 735	
3.2.3	Solubility Product Equilibrium Constant K_{sp} 736	
3.2.4	Common Chloride Ion Effect 737	
3.3	Partition Coefficient and Distribution Coefficient 737	
3.4	Dissolution (Rate of Solution) 739	
3.4.1	Biorevelant Dissolution and Food Effect 740	
3.4.2	Drug Stability in the Gastrointestinal Tract 740	
4	**Oral Absorption Mechanism** 742	
4.1	Anatomy of the Human Intestine 742	
4.2	Passive Diffusion 743	
4.2.1	Paracellular Permeation 743	
4.2.2	pH Partition Theory 743	
4.2.3	Polar Surface Area 743	
4.3	Regional Intestine Permeability 743	

Translational Medicine: Molecular Pharmacology and Drug Discovery
First Edition. Edited by Robert A. Meyers.
© 2018 Wiley-VCH Verlag GmbH & Co. KGaA. Published 2018 by Wiley-VCH Verlag GmbH & Co. KGaA.

4.4	Active Transport 744	
4.5	Role of Enterocyte Efflux Pumps 744	
4.6	Liaison between Metabolic Enzymes and Transporters 745	
5	**Biopharmaceutics Classification System (BCS) 746**	
5.1	BCS History and FDA Regulation 746	
5.2	BCS Definition 746	
5.3	Significance of a Drug's BCS Class 747	
5.4	Methodology for Classifying Drug Solubility 747	
5.4.1	Dose Solubility Ratio 747	
5.4.2	Drug Instability in the GI Tract 748	
5.5	Methodology for Classifying Drug Permeability 748	
5.5.1	Direct Human Studies 748	
5.5.2	Indirect Methods Requiring Proof of Passive Diffusion 748	
5.5.3	Caco-2 Permeability Assay 749	
5.5.4	Rat Intestinal Perfusion 750	
5.5.5	Human GI Permeability 750	
5.5.6	Permeability Method Suitability 750	
5.5.7	Use of Partition/Distribution Coefficient in Lieu of Permeability 750	
5.6	Biopharmaceutics Drug Disposition Classification System (BDDCS) 751	
5.7	BCS Classification of Old and New Drugs 752	
6	**Biopharmaceutical Application in Drug Molecule Design 753**	
6.1	Structural Modification in Drug Design and Therapy 755	
6.2	Salt Formation: An Enabling Technology to Improve Solubility and Stability 757	
6.3	Polymorphism Screening 758	
6.4	Polymorphism Regulatory Guidance Linked with BCS Classification 759	
6.5	Thermodynamically Stable Polymorphs 760	
6.5.1	Slurry Tests 760	
6.5.2	Polymorphic Conversion 761	
6.6	BCS-Based Polymorph Salt Form Screening Strategy 761	
7	**Biopharmaceutics Application in Drug Product Design 763**	
7.1	Animal Formulation 763	
7.2	Human Formulation 765	
7.3	BCS Class and Food Effect 765	
8	**Conclusions 768**	
	References 768	

Keywords

Pharmaceutical sciences
A group of interdisciplinary areas of study concerned with the design, action, delivery, disposition, inorganic, physical, biochemical and analytical biology (anatomy and physiology,

biochemistry, cell biology, and molecular biology), epidemiology, statistics, chemometrics, mathematics, physics, and chemical engineering, and the application of their principles to the study of drugs

Solubility
The property of a solid, liquid or gaseous chemical substance (the solute) to dissolve in a solid, liquid or gaseous solvent to form a homogeneous solution of the solute in the solvent

Intestinal permeability
The phenomenon of the gut wall in the gastrointestinal tract exhibiting a semipermeable membrane that allows nutrients to pass through the gut, while maintaining a barrier function to prevent potentially harmful substances (e.g., antigens) from leaving the intestine and migrating widely to the body

Thermodynamics
A branch of physics concerned with heat and temperature, and their relation to energy and work. It defines macroscopic variables, such as internal energy, entropy and pressure, that partly describe a body of matter or radiation

Bioavailability
The fraction of an administered dose or unchanged drug that reaches the systemic circulation which is one of the principal pharmacokinetic properties of a drug

Pharmacokinetics
The characteristic interactions of a drug and the body in terms of its absorption, distribution, metabolism, and excretion

Pharmacodynamics
Studies of the biochemical and physiological effects of drugs on the body, and the mechanisms of drug action and the relationship between drug concentration and effect

Biopharmaceutics Classification System
A system used to differentiate drugs on the basis of their solubility and permeability. It is a guide for predicting the intestinal drug absorption provided by the US Food and Drug Administration

Physical pharmacy and biopharmaceutics is the study of the relationship between the physical chemical and biological properties and the fate of a drug *in vivo*, in order to design an optimized dosage form. Such studies are critical to drug and drug product design. The selection of new clinical molecules suitable for drug development is heavily dependent on the scientific discipline of physical pharmacy and biopharmaceutics. The term "physical pharmacy" stems from the application of physical chemistry principles to the area of pharmacy in the design of drug molecules and drug products. Pharmacy is an applied science, composed of principles and methods borrowed from basic sciences such as physics, quantum mechanics, thermodynamics, ionization, equilibrium, chemical stability, kinetics, diffusion, permeation, adsorption and complexation. These principles allow pharmaceutical scientists to better predict quantitatively the solubility, stability, compatibility, manufacturability of drugs, and dissolution, absorption, distribution, metabolism, and elimination of drug products. When the physical chemical and biological properties of drug molecules (i.e. preformulation) are understood, it is possible to design dosage forms for designated routes of administration in humans or animals (i.e., formulation). Collectively, the scientific principles applied in the preformulation and formulation processes is termed "physical pharmacy," and the application of this is termed "pharmaceutics." Biopharmaceutics is the study of the factors that influence the bioavailability of a drug in humans and animals, and this information is used to optimize a drug's therapeutic window by enhancing its efficacy and/or reducing side effects. Thus, biopharmaceutics involves the effect of a formulation on drug pharmacokinetics. In this chapter, attention will be focused on a subset of physical pharmacy pertaining to biopharmaceutics in areas of solubility, dissolution and gastrointestinal permeability. Pharmacokinetics and pharmacodynamics based on chemical kinetic principles will also be emphasized.

1
Definitions

1.1
Physical Pharmacy

The term "physical pharmacy" stems from the application of physical chemistry principles to the area of pharmacy in the design of drug molecules and drug products. Pharmacy is an applied science, composed of principles and methods borrowed from basic sciences such as physics, quantum mechanics, thermodynamics, ionization, equilibrium, chemical stability, kinetics, diffusion, permeation, adsorption, and complexation. These scientific principles allow pharmaceutical scientists to better predict – on a quantitative basis – the solubility, stability, compatibility and manufacturability of drugs, and the dissolution, absorption, distribution, metabolism, and elimination of drug products. When pharmaceutical scientists understand the physical, chemical and biological properties of drug molecules (i.e., preformulation), it is possible for them to design dosage forms for the designated routes of administration for humans or animals (i.e., formulation). The scientific principles applied to the preformulation and formulation processes are termed collectively "Physical Pharmacy," and the application of physical pharmacy to the pharmaceutical industry is termed "pharmaceutics." In this chapter, attention will be focused on a subset of physical pharmacy pertaining to biopharmaceutics in the areas of solubility, dissolution and gastrointestinal (GI) permeability. Pharmacokinetics and pharmacodynamics based on chemical kinetic principles will also be emphasized. Details of other areas of physical pharmacy, such as rheology, surface chemistry and flowability, are available in other text books such as *Martin's Physical Pharmacy* [1].

1.2 Biopharmaceutics

Biopharmaceutics is the study of factors that influence the bioavailability of a drug in humans and animals, and the use of this information to optimize a therapeutic window by enhancing therapeutic effects and/or reducing adverse side effects. Biopharmaceutics, together with physical pharmacy, involves studying the relationship between the physical, chemical and biological properties of a drug, as well as the fate of that drug *in vivo* in order to design an optimized dosage form. In short, biopharmaceutics relates to studies of the effect of a drug's formulation on its pharmacokinetics [2].

By now, some confusion may have arisen about the similarity and differences between biopharmaceutics and pharmaceutics. These two disciplines do overlap but have a different emphasis: pharmaceutics emphasizes the physical chemical characterization of a drug and dosage forms, whereas biopharmaceutics emphasizes the effect of that drug *in vivo* from a given dosage form. Therefore, pharmaceutics teaches the use of physical chemical principles in the design and production of acceptable/optimized dosage forms, including such disciplines as: (i) compounding, granulation, coating, encapsulation and tableting for oral dosage forms; (ii) mixing, filtration, filling, sterilization and lyophilization for parenteral dosage forms; and (iii) emulsifying, gelling, milling, sheeting and curing for topical dosage forms such as lotions, creams and transdermal devices.

1.3 Pharmacokinetics

Pharmacokinetics relates to the kinetics of the absorption, disposition, metabolism, and excretion (ADME) of drugs, clinically in humans and preclinically in animals, while pharmacodynamics involves the kinetics of drug effects in relationship to dose or concentration *in vivo*. In other words, pharmacokinetics defines what the body does to the drug, while pharmacodynamics defines what the drug does to the body [3]. The development of mathematical models is essential for the interpretation of kinetic phenomena. Both, pharmacokinetics and pharmacodynamics studies require knowledge of kinetics, and the utilization of mathematical models to a varying degree.

Pharmacokinetics can be described either by the plasma drug concentration–time profile or the time course of appearance of metabolites. This time course is governed by the extent and rate of drug input into the bloodstream, and the rate of the drug leaving the bloodstream. The drug input may be through oral ingestion, permeation through the skin, inhalation via the lungs, direct injection into the bloodstream or into a subcutaneous or muscular pocket. Bioavailability is defined as the extent and rate of drug input into the bloodstream. Thus, a direct injection into the bloodstream (most commonly via an intravenous route) is deemed 100% bioavailable, while all other routes of administration will have a certain degree of bioavailability, being a percentage of the amount of drug dosed and the rate at which it reaches the blood circulation, which is reflected in the drug's pharmacokinetic profile. Occasionally, data for direct injection may not be available, and a relative bioavailability – which compares the bioavailability of an oral solution or a reference formulation with known bioavailability – can be used instead. The percentage relative bioavailability is the ratio of the area under the plasma concentration–time curve (AUC) for the test and reference formulations.

Drugs exit the bloodstream in several different ways:

- Drugs are distributed into organs and tissues to varying degrees, depending on their ability to bind to a certain tissue. Tissue distribution is usually a reversible process, in that drugs in tissues can also be distributed back into the bloodstream.
- Metabolism is usually the primary route of drug elimination. The liver is the primary organ where drugs are metabolized, although metabolism may also occur in the blood itself or in other organs such as the intestine, lungs, and even kidneys.
- Drugs can be excreted by the kidney into the urine, or by the liver into the biliary duct, from where they are eliminated via the feces.

It is this combination of processes, controlling the appearance of metabolites and disappearance of drugs in the bloodstream, which culminates in the pharmacokinetic profile of a drug and/or its metabolite(s).

1.4
Pharmacodynamics

Pharmacodynamics is described by the "effect versus dose," the "effect versus concentration," or even "effect versus AUC." After having reached the bloodstream, it will take some time for a drug to reach the site of action where it can bind to a cell constituent (e.g., a receptor, a protein, a cytokine, or a nucleic acid) in order to create a biochemical and/or physiological effect, typically a therapeutic or toxic action. The rate and extent at which a drug reaches and subsequently leaves the target site and binds to a target receptor to exert a pharmacological action is known as the drug's pharmacodynamic profile.

It is not uncommon that the pharmacological action of a drug starts later than and lasts longer than the drug is present in the bloodstream; this is termed the latency effect of a drug molecule. For example, a drug with tight tissue binding may exhibit a slow redistribution from the tissues to the bloodstream; this in turn prolongs the pharmacological effect because the drug remains in the target tissue while drug levels in blood fall substantially below the effective concentration threshold. Therefore, for any drug it is important to know not only the pharmacokinetic profile but also the pharmacodynamics profile in order to complete the story. The summation of these two kinetic terms, which is referred to as pharmacokinetics pharmacodynamics (PK/PD) modeling, describes the total kinetics from a drug's input through to its eventual output as therapeutic, toxic, or side (exaggerated pharmacology) effects. Each of these effects is based on the kinetic principles used in many disciplines, including pharmaceutical sciences.

2
History of Physical Pharmacy and Biopharmaceutics

Pharmaceutics began early in 1906 with the emergence of the Federal Food, Drug, and Cosmetic (FD&C) Act that required the use of labels showing drug content and purity, as a guarantee of drug product quality. In 1937, an elixir of the antibacterial agent sulfanilamide that contained ethylene glycol (a solubilizer with toxic effects) had caused the poisoning of patients (including many children) and led to 107 deaths. Ethylene glycol had previously been used as a common solvent in vanilla extract for cakes, and was assumed to be safe. As a result, shortly afterwards, in 1938, the Food Drug and Cosmetic Act was amended to demand that toxicity testing be carried out prior to the introduction of any medicine into the marketplace. By 1962, FD&C Act was further amended such that, in addition to toxicological data, evidence of efficacy had to be submitted to the Food and Drug Administration (FDA; these were the Kefauver–Harris Amendments) following the appearance of major birth defects associated with the dosing of thalidomide (as an anti-emetic) to pregnant women in Europe [4]. It was during this time that it first became realized that different formulations of the same drug may not be equivalent in terms of therapeutic effectiveness. In fact, a drug in the same formulation but which had been produced using different manufacturing processes may have vastly different clinical effects. As a consequence of these findings, bioavailability became an essential requirement for the commercialization of any drug product, whether this was a new drug or a generic form of an older drug. Within a very short time, biopharmaceutics – studying the impact of formulation on bioavailability – and pharmaceutics – studying the impact of a drug's physical and chemical properties on formulation – both became mainstream in the field of pharmaceutical sciences.

3
Application of Thermodynamics in Pharmacy

Thermodynamics is a branch of physical chemistry concerned with heat and temperature, and their relationship to energy and work. Thermodynamics originated during the advent of steam engines in the late eighteenth/early nineteenth century, which defines not only microscopic but also macroscopic phenomena, such as internal energy, entropy, Gibb's free energy, and the pressure of a body of matter. The microscopic interactions between individual particles and their collective motions, in terms of classical or of quantum mechanics, are treated statistically to achieve a bulk behavior. The bulk behavior of the body – but not the microscopic behaviors – is then governed by natural laws of pressure, volume and temperature that constitute the science of thermodynamics. A pharmaceutical product will experience at least three temperatures during its life cycle: production temperature; storage temperature; and body temperature. It is important to understand the effects of temperature on equilibrium processes such as solubility, partition, and ionization, as these are critical themes in biopharmaceutics. It is also important in other branches of the pharmaceutical sciences, as the life cycle of a drug product is subject to heating or cooling during its manufacture, to environmental temperature fluctuations and, ultimately, exposure to body temperature.

However, there are exceptions to the rules of thermodynamics in Nature. Classical thermodynamics dictates volume as a function of temperature at constant pressure – that is, the volume expands as temperature rises. But this rule does not apply to water close to 4 °C, which has a density minimum or volume maximum as a result of its tetrahedron structure via hydrogen-bonding. Water is ever-present in biological systems and plays a central role in pharmaceutical sciences; however, this deviation from ideal conditions by no means undermines the importance of thermodynamics in the field of pharmacy.

In addition to the statistical nature of bulk phenomena, thermodynamics describes a state of equilibrium where no macroscopic change is occurring, and where every microscopic process is balanced by its opposite. A prime example in pharmacy is in the definition of equilibrium solubility. A solution process involves the surface molecules of a solid particle leaving the surface and diffusing into a pocket surrounded by the solvent molecules. The solution process is counteracted by a precipitation process, by which the solvated solute molecules diffuse back to the solid surface of the solute to become once more a part of the solid. When these two processes reach equilibrium, such that the rate of solution equals the rate of precipitation, the bulk phenomenon appears a constant solubility; this is termed constant solubility or equilibrium solubility. In pharmacy, equilibrium solubility is so important that all drugs are classified as per the *United States Pharmacopeia* (USP), ranging from soluble, sparingly soluble to practically insoluble and insoluble. The USP solubility terms have historical significance but are broad in nature and antiquated in light of the low solubility spectrum of drugs discovered currently. Most drugs will fall into the category of practically insoluble or insoluble in water or buffer solutions (more than 10 000 parts of solvent are required for 1 part of solute, or less than $0.1\,\text{mg ml}^{-1}$). Currently, solubility is reported in standard fashion as milligrams, micrograms or nanograms of solute per millimeter of solvent.

Most systems found in Nature or in the human body are not in thermodynamic equilibrium. Natural processes are also usually irreversible and not at equilibrium, but have a tendency towards thermodynamic equilibrium. However, if a small confined area is considered where transfers of energy and matter into or out of the system proceed so slowly that, within a reasonable time frame, then the system can be considered near thermodynamic equilibrium. This state, known as quasi-static thermodynamics, has wide application in solid-state kinetics such as polymorphic conversions, or in pharmacokinetics such as reaching a steady state with multiple dose regimens.

Thermodynamics states a set of four laws:

- *Zeroth law of thermodynamics*: If two systems are each in thermal equilibrium with a third, they are also in thermal equilibrium with each other.
- *First law of thermodynamics*: The increase in internal energy of a closed system is equal to the difference of the heat supplied to the system and the work done by it.
- *Second law of thermodynamics*: Heat cannot flow spontaneously from a colder location to a hotter location.
- *Third law of thermodynamics*: As a system approaches absolute zero, the entropy of the system approaches a minimum value.

In all, thermodynamic laws are explained by statistical mechanics of millions of molecules in a confined system, which exhibit varied behaviors but, as a whole, a general trend that can be explained by thermodynamic principles. The application of thermodynamic principles in solubility, ionic equilibrium, partition, and dissolution – the four key characteristics of drugs and drug products – are discussed in the following sections.

3.1 Solubility

3.1.1 Equilibrium Solubility

When a solid surface is exposed to a liquid, molecules on the surface of the solid leave the surface while molecules in the liquid return to the solid surface; this continues until an equilibrium is established where the rate of molecules leaving is equal to the rate of molecules returning. In other words,

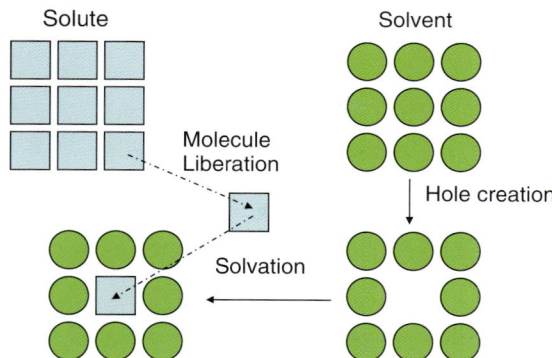

Fig. 1 Schematic representation of the solution process.

equilibrium solubility is reached when the thermodynamic activity of the molecule in the solid surface equals that of the molecule dissolved in the solvent.

To further dissect the solution process, Fig. 1 depicts the three steps that comprise the solution process as follows:

1) Breaking of the crystalline bonds holding the molecules to the solute.
2) Breaking the solvent–solvent bonds to create a cavity in the solvent.
3) Occupying the solvent cavity by the solute molecule to form solute–solvent bonds.

Step 1 involves bond breaking from the crystal surface, similar to the melting process, which is directly related to melting points and heat of fusion. Step 1 is usually the dominant component of solution energy, and is why solubility often correlates with the heat of fusion or melting point. Step 1 explains why polymorphic forms of the same compound give rise to different solubility values, because polymorphic forms have different bonding energies and thus varied melting points. Steps 2 and 3 explain why the solubility of the same compound is different in different solvents; this is because the solution process depends not only on the heat of fusion (step 1) but also on the energy of solvation, as depicted in steps 2 and 3.

Since solubility varies among different polymorphs of a given compound, it is necessary to determine an absence of crystalline form changes when determining the equilibrium solubility of a specified polymorph. If a form change begins to occur, then solubility will not be equilibrated until the solid form transformation is complete. Therefore, the initial part of the solubility–time curve approximates the solubility of the old polymorph, whereas the later part of solubility–time curve approximates the solubility of the new polymorph. Equilibrium solubility should always be reported by referencing a particular crystalline form with a specific sample/batch number.

3.1.2 Kinetic Solubility

With the advent of high-throughput screening (HTS), kinetic solubility is routinely generated for discovery compounds nowadays [5]. The compound is first dissolved in an organic solvent (typically dimethylsulfoxide; DMSO) and diluted gradually in an aqueous buffer (typically a pH 7.4 phosphate buffer) until precipitation is registered by in-line instruments such as ultraviolet detectors or laser nephelometers

[6]. The concentration where the turbidity occurs is reported as the kinetic solubility.

Kinetic solubility is generally greater than equilibrium solubility due to the lack of crystalline surfaces as the compound is first dissolved in DMSO; this often leads to supersaturation until nucleation occurs. Consequently, kinetic solubility is less reproducible and is dependent on factors affecting nucleation such as cleanliness, the contact surface texture, and any impurities present.

Kinetic solubility can serve as a useful approximation of the solubility for amorphous materials which require less energy input because of the lack of crystalline bonding energies. Kinetic solubility may also reflect the in-gut concentration when in-gut precipitation occurs. Basic compounds, which have a high solubility in stomach acid, may precipitate out extensively when traveling down to the lower GI tract, where the pH returns to neutral. This phenomenon is particularly important for controlled-release dosage forms, as a long absorption duration through the small and large intestines is usually required for once-daily dosage forms.

3.1.3 Heat of Solution

The Gibb's free energy of solution may be expressed in two parts, as the heat related enthalpy, ΔH, and the order-related entropy, ΔS. The free energy is related to equilibrium constant that is solubility (K_s) in this case. The overall equation may be expressed as:

$$\Delta G = \Delta H - T\Delta S = -2.3RT \log K_s$$

where T is absolute temperature and R is the gas constant. The equation may be rearranged into the Van't Hoff equation, which describes a linear relationship of $\log K_s$ versus $1/T$ with a slope equal to $-\Delta H/2.3RT$ and an intercept equal to ΔS as:

$$Log\ K_s = -1/2.3R\ (\Delta H\ /T - \Delta S)$$

When ΔH is positive, heat is absorbed upon dissolution and solubility increases with increasing temperature, and the plot of $\log K_s$ versus $1/T$ has a negative slope (see Fig. 2). Once solubilities are determined at several temperatures, the heat of solution is estimated from the slope.

Alternatively, solution calorimetry can be used to measure the heat of solution directly. The latter parameter, if determined from calorimetry, differs to some extent from that determined via the Van't Hoff equation, as it is related to the volume changes in the solution process ($P\Delta V$). Calorimetry is a more direct method, whereas the Van't Hoff method provides additional information on the transition temperature for

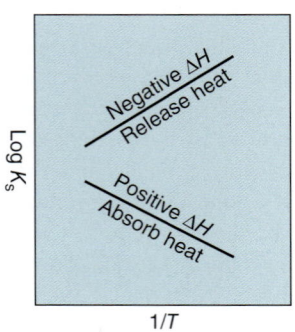

Fig. 2 Effect of temperature on equilibrium solubility.

polymorphic conversion (as described in Sect. 6.5).

3.2 Ionic Equilibrium

3.2.1 pK_a and pH Solubility Profile

The most widely studied drug property influencing oral bioavailability is probably the pH dependence of aqueous solubility of ionic compounds. The ionization of a compound alters its physical behavior and macroscopic properties, such as solubility and lipophilicity (log P). For example, the ionization of a compound will increase its solubility in water but decrease its lipophilicity. The GI tract has a pH gradient that varies from the extremes of the stomach (pH 1) to the colon (pH 8), depending on food intake, and has large intra- and inter-subject variability. Therefore, oral absorption may take place at a point of the GI tract where the compound is ionized if solubility is the limiting factor, or when the compound is not ionized if permeability is the limiting factor. This difference can be exploited in drug development to increase lead compound oral bioavailability by adjusting the pK_a of an ionizable group on the side chain distant from the pharmacophoic responsible for the pharmacologic activities.

An acid dissociation constant, K_a is a quantitative measure of the strength of an acid in solution. It is the equilibrium constant for a chemical reaction: $HA \leftrightarrow H^+ + A^-$, where HA is a generic acid that dissociates by splitting into A^-, known as the conjugate base of the acid, and the hydrogen ion or proton, H^+. A more convenient expression, pK_a, is $-\log_{10}(K_a)$. The larger the K_a value (smaller pK_a value), the more dissociation of the molecules in solution and thus the stronger the acid.

For a basic compound, the hydrolysis reaction equation is $B + H_2O \leftrightarrow BH^+ + OH^-$, where BH^+ is the conjugate acid with an acid dissociation constant K_a for the reverse reaction: $BH^+ + OH^- \leftrightarrow B + H_2O$. In this case, the larger the K_a value (smaller pK_a value), the more dissociation of BH+ and thus less hydrolysis of B, the weaker the base.

The acid dissociation equilibrium is commonly expressed in the Henderson–Hasselbalch equation: $pH = pK_a + \log [A^-]/[HA]$. At half-neutralization $[A^-]/[HA] = 1$; since $\log(1) = 0$, $pH = pK_a$ when the concentration of HA is equal to the concentration of A^-. Similarly when $[A^-]/[HA] = 10$, $pH = pK_a + 1$ or when $[A^-]/[HA] = 1/10$, $pH = pK_a - 1$.

A more useful form of the equation is:

$$S_T = S_0(1 + 10^{(pK_a - pH)})$$

where S_0 is the solubility of the unionized form and the ionized form is 10-fold of S_0 when pH is one unit less than pK_a and the total dissolved amount S_T is 11-fold that of S_0.

The determination of pK_a values is commonly achieved by: (i) calculation from chemical structures based on the Hammet and Taft equation; (ii) pH titration with acid or base and species calculations; (iii) measurement of a physical chemical property such as UV absorption that is dependent on pH and species concentration; and (iv) from the pH–solubility profile based on the above equation. Mathematical modeling and regression analysis are usually associated with these methods. More than one method may be used to verify the true pK_a values.

3.2.2 Zwitterionic Compounds

The situation is more complicated when more than one pK_a value exists, and this can be illustrated with a model zwitterionic

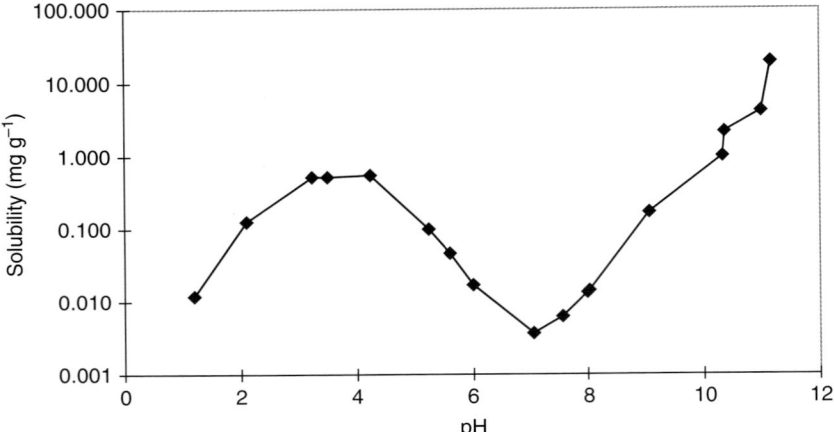

Fig. 3 pH solubility profile of a model zwitterion with pK_a of 6.6 and 7.4 for the hydroxamic acid and piperidine functional groups, respectively.

compound that has both acidic and basic functional groups. The two pK_a values are 6.6 and 7.4 for the hydroxamic acid and piperidine groups, respectively. The pH solubility profile was determined in HCl/NaOH buffers in the pH range of 1.2 to 11.2, and is presented in Fig. 3. The lowest solubility was found at the isoelectric pH of 7 (pH ≈ 7) of 0.004 mg ml^{-1}. In the basic region, the molecule has a negative charge, while solubility is increased logarithmically with increasing pH. In the acidic region, the molecule has a positive charge showing an initial logarithmical increase with decreasing pH; this reaches a plateau at pH 4.25, achieving a maximum solubility of 0.558 mg ml^{-1}. When more HCl was added to lower the pH further, the solubility decreased and indicated an interaction between the HCl salt and the chloride ions; this is known as the common ion effect. The solubility was finally reduced to 0.012 mg ml^{-1} at the stomach pH of 1.2. Because the solubility is low in most of the GI tract, two crystalline salt forms were made to increase the solubility over the free form. When simple capsule formulations of the two salts were manufactured and dosed to dogs, the acetate salt was found to be 40% more bioavailable than the HCl salt, which latter had a depressed solubility in the stomach acid due to the common ion effect.

3.2.3 Solubility Product Equilibrium Constant K_{sp}

The equilibrium between a solid salt and dissolved ionic species is more complex than that for free acid or base as described above. NaCl crystals can serve as an example here. The NaCl molecules, when leaving the solid surface, will first dissociate to form Na$^+$ and Cl$^-$ ions; the reverse process – a re-association of the two ions – must occur before NaCl molecules are deposited onto the NaCl crystalline surface. The solubility product equilibrium constant (K_{sp}) is the product of the equilibrium concentrations of the Na$^+$ and Cl$^-$ ions in a saturated solution of NaCl salt. It can imagine that if a KBr salt also present, there will be an additional K_{sp} between Na$^+$ and Br$^-$ and between K$^+$ and Cl$^-$. This is why buffers and salts have a great effect on solubility, especially on ionic compounds. The effect of buffers and salts on drug solubility can be explained by the following possible four mechanisms:

1) Reducing the amount of solvent available to dissolve the compound due to salt–solvent interaction ("salting out").
2) Association of salt with the compound, leading to increased solubility ("salting in").
3) Solubility product (K_{sp}) between the compound and the buffer species is low (K_{sp} control).
4) Buffer works as hydrogen donor or acceptor to cause ionization of the compound (pH control).

This salt/buffer phenomenon has led to a large discrepancy in solubility values reported in the literature. Wu and Benet [7] found conflicting reports of solubility data during the process to establish a Biopharmaceutics Classification System (BCS) for common drugs such as furosemide, hydrochlorothiazide, methotrexate, ciprofloxacin, and erythromycin. These are reported as having high solubility by one author, but low solubility by another author. Hence, it is very important to report the exact composition of the solvent and its salt content when reporting solubility values.

3.2.4 Common Chloride Ion Effect

The common ion effect with chloride ion which is present in stomach acid, is a demonstration of the solubility product (K_{sp})-controlled solubility phenomenon. This has physiological significance in reducing drug dissolution and solubility in the stomach, and it can also further reduce bioavailability, as noted for the zwitterionic compound described in Sect. 3.2.2. The acetate salt was found to be 40% more bioavailable than the HCl salt, which had a depressed solubility in the stomach acid. A simulated pH–solubility profile for a HCl salt of a base with $pK_a = 7$, free base solubility of 0.00001 M and a $K_{sp} = 0.000001$ M^2 with chloride counterion is shown in Fig. 4a. Above pH 5, solubility is controlled by the dissociation constant K_a exhibiting one log increase of solubility per pH unit decrease. Below pH 5, the solubility product constant is in control, where the protonated base is precipitated as HCl salt as the K_{sp} is reached. Below pH 3, the solubility is further depressed due to the increase in chloride ion concentration.

A common ion effect can easily be detected by measuring solubility in various percentages of saline solutions, as illustrated in Fig. 4b. To clarify, the common chloride ion effect historically refers to the interaction between HCl salt and chloride ion, thus with common chloride ions in both species. However, the slowdown of dissolution is seen for free base and other acid salts, and is not limiting to the HCl salt. More recent studies have demonstrated a fast interfacial counterion exchange, and the presence of chloride ions quickly inhibited dissolution of the free base and other acid salts not limited to the HCl salt. For the lack of a better term, the common chloride ion effect is used indiscriminately for all solid forms not limited to the HCl salt.

3.3
Partition Coefficient and Distribution Coefficient

Biological membranes are predominantly lipophilic, and unionized drugs can penetrate these barrier membranes through passive diffusion; this process has been referred to as the pH-partition theory since the 1950s [1]. Other absorptive processes, such as the transport of ionic drugs via sodium channels, of small molecules via

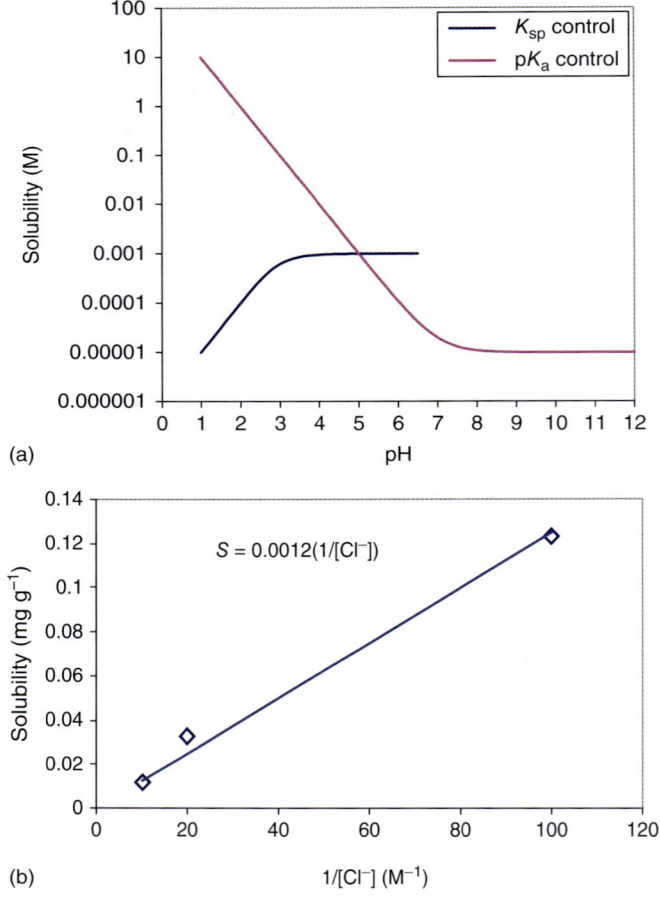

Fig. 4 (a) pH solubility profile of a model HCl salt of a base with $pK_a = 7$, So $= 0.00001$ M and $K_{sp} = 0.000001\,M^2$; (b) Common ion effect: solubility as a function of chloride ion concentration.

the aqueous pore or tight junction between enterocytes, of facilitated lipid absorption via chylomicrons, of active transport via transporters and the interplay between transporter and enterocyte microsomal enzymes, have since been discovered in quick fashion, and have rendered the pH partition theory much less important. Nevertheless, pH-partition still plays a role and correlates with intestine permeability for some drugs.

The partition coefficient (the logarithmic term is $\log P$) is the ratio of the concentrations of a compound in oil and water phases of a mixture of two immiscible liquids at equilibrium. Normally, the oil phase chosen is n-octanol, which simulates a lipid bilayer; cyclohexene has also been used to approximate brain penetration. In the case of ionic compounds, a more useful term here is the distribution coefficient (the logarithmic term is $\log D$), which is the ratio of the sum of the concentrations of all forms of the compound (ionized plus unionized) in each of the two phases. $\log D$ can be calculated from the estimated $\log P$ and the ionization constant (pK_a) using the following equations:

For acids, $\log D = \log P - \log(1 + 10^{pH - pK_a})$

For bases, $\log D = \log P - \log(1 + 10^{pK_a - pH})$

According to the pH-partition theory, the ionized form has negligible partition, and therefore $\log D$ is a fraction of $\log P$, representing partition of the unionized species in the water phase. $\log D$ at pH 7.4 is most relevant to most biological systems, except for the intestine where absorption takes place between pH 5–8. $\log D$ at pH 6.5 has been used to approximate intestinal permeability.

Methods to determine $\log P$, including calculations from chemical structures [8] and the experimental measurement of $\log D$ using the shake-flask method or reverse-phase high-performance liquid chromatography (HPLC) are available elsewhere [9]. The calculated partition coefficient (clog P) has gained popularity, and the calculation of clogP is usually embedded in most physical chemical software. clogP can be calculated based on the group contribution theory for any compound with a known chemical structure, where the group is defined as non-overlapping molecular fragments. In addition, Hammett-type corrections are included to account for electronic and steric effects, rendering a high degree of accuracy to clogP. Nowadays, $\log D$ measurements are often automated in a HTS scheme, although this may lead to erroneous results due to a lack of fine-tuning of the experimental conditions. Most often, precipitation occurs when solubility is exceeded to create a third solid phase in addition to the oil and water phases, thus upsetting the equilibrium. The pros and cons of the chemical calculation and experimental methods are detailed elsewhere [10]. Overall, the experimental procedure for the determination of partition coefficient must be tailored to the individual compound.

3.4
Dissolution (Rate of Solution)

It is important to understand the physical model of solid particle dissolution, which applies to most oral drug products. The model is depicted as a solid particle surrounded by a stagnant layer, with a concentration gradient of the solute in regard to the surrounding bulk solvent (Fig. 5). According to Fick's law, the rate of solution (dm/dt) or flux (J) is proportional to the particle surface area (A), and the concentration gradient ($C_s - C_b$) and is inversely proportional to the stagnant layer thickness (h). This is expressed by the equation:

$$dm/dt = J = D A (C_s - C_b) / h$$

where D is the diffusion coefficient, C_s is the saturation concentration at the particle surface, and C_b is the concentration in the bulk solution. It is clear that a larger number of particles with smaller particle sizes, yielding a larger total surface area with an equivalent mass, increases the dissolution rate and that an increase in the stir rate will reduce the stagnant layer thickness and thus increase the dissolution rate. It is less clear what happens to drugs in the intestinal lumen when a pH modifier such as carbonic ion or a surfactant such as bile salt is present. An acid–base reaction may occur at the interface between the stagnant and bulk layer. Similarly, micellar formation may also take place at the interface, so that the drug is carried away from the remaining solid [11]. The manipulation of dissolution

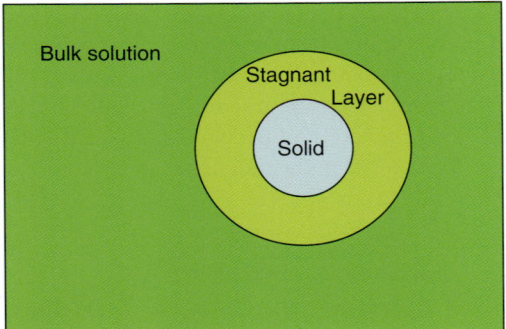

Fig. 5 Fick's law diffusion model.

rates by drug particle sizes and excipients such as buffering agents, surfactants and micelle formers, is an important technique in the formulation of solid oral dosage forms in pharmaceutical sciences.

3.4.1 Biorevelant Dissolution and Food Effect

The use of biorelevant solubility and dissolution tests has gained popularity in recent years. A clinically relevant food effect has been linked with increased drug solubility in the intestinal fluid under fed conditions (with food intake) compared to fasting conditions (without food intake). Several intestine surfactants, including bile salts and lecithin, are known to increase drug solubility. Upon food intake, more surfactants are secreted to alter the intestinal fluid composition, in addition to acid secretion which alone could have a profound impact on the solubility of ionic drugs. Furthermore, a high-fat meal alone can cause a modification of the intestinal fluid composition.

In 1997, Dessman et al. [12] proposed the addition of sodium lauryl sulfate to the USP/National Formulary simulated gastric fluid (SGF) as fast-state simulated gastric fluid (FaSGF), and sodium taurocholate and lecithin to the simulated intestinal fluid (SIF) as fast-simulated intestinal fluid (FaSIF). The feed-simulated intestinal fluid (FeSIF) is characterized by a higher amount of sodium taurocholate and lecithin. Charman et al. later added one more medium as the fed-state simulated gastric fluid (FeSGF) by increasing the pH from pH 2 to 5 [13]. In the case of once-daily formulations, the majority of the GI transit occurs in the colon, and consequently an understanding of colonic fluid composition is important. Simulated fluids for the upper and lower colon have also been devised [14].

The various simulated gastric and intestinal fluid compositions are summarized in Table 1. These biorelevant solubility tests have been used in a physiologically based absorption model (GastroPlus) to predict the clinical food effect, with limited success [15]. A high ratio between solubility in the fed- and fast-state simulated GI fluids may suggest the presence of food effect. In conclusion, comparisons of solubility in these simulated biological fluids, in addition to the pH solubility profile, will help to predict and mitigate (via formulation means) drug–food effects in the clinic.

3.4.2 Drug Stability in the Gastrointestinal Tract

Instability in the GI tract has long been a major hurdle for the oral delivery of

Tab. 1 Physiological buffer system.

Substance (mM)	SGF	SIF	FaSSGF	FeSSGF	FaSSIF	FeSSIF	FeSSIF (HF)	SUCF	SLCF
NaCl	3.4	—	50	80	100	200	102	17	17
HCl	867	—	20	—	—	—	—	—	—
KH_2PO_4, monobasic	—	3.9	—	—	—	—	50	—	—
NaOH	—	38	—	—	9	100	—	—	—
Sodium lauryl sulfate	—	—	—	—	—	—	—	—	—
Acetic acid	—	—	—	10	—	144	—	70	25
$NaH_2PO_4 \cdot H_2O$	—	—	—	—	1.7	—	—	—	—
Sodium taurocholate	—	—	—	—	8	12	—	—	—
Lecithin	—	—	—	—	0.75	2	3.7	—	—
Glycocholic acid	—	—	—	—	—	—	150	—	—
Oleic acid	—	—	—	—	—	—	43	—	—
Propionic acid	—	—	—	—	—	—	—	30	10
Final pH	1.2	6.8	1.7	5.0	6.5	5.0	6.5	6.2	7.0

SGF = simulated gastric fluid without 0.32% w/v pepsin (Sigma, P-7000); SIF = simulated intestinal fluid without 1% w/v pancreatin (Sigma, P-1625); FaSSGF = fasted-state simulated gastric fluid; FeSSGF = fed-state-simulated gastric fluid; FeSSIF(HF) = fed-state-simulated intestinal fluid (high fat); SUCF = simulated upper colon fluid; SLCF = simulated lower colon fluid.

early-generation antibiotics, and now of peptide drugs. The β-lactam ring of penicillin, cephalosporin and carbapenem is acid-labile and are known to be degraded before the molecule can permeate into the enterocyte. Today, peptide drugs face a similar issue, in that digestive enzymes such as pepsin, chymotrypsin and trypsin are Nature's way of degrading foreign peptides and proteins down to their building blocks of amino acids in readiness for absorption.

The determination of possible degradation and subsequent stabilization in the GI tract is one of the studies needed to identify clinical drug candidates. In this case, the drug is dissolved in simulated gastric and intestinal fluids[1] incubated at 37 °C for a period of time that is representative of *in vivo* drug contact with these fluids; for example, 1 h in gastric fluid and 3 h in intestinal fluid. Drug concentrations are then determined using a stability-indicating assay method. A significant degradation (>5%) of a drug in this protocol could suggest potential instability. However, drug solubility in the GI tract may limit degradation. For example, an acid-labile drug may not degrade in the stomach acid because of its limited solubility, and thus present in the stomach as a solid that is resistant to degradation. Acidic drugs such as penicillin may be acid-labile but insoluble in stomach acid; consequently, in infants or elderly patients, where the stomach pH is elevated, the drug may dissolve in the stomach and become susceptible to degradation, leading to poor oral bioavailability [71].

Drug stability in the large intestine is further complicated by the presence of colonic microorganisms. The colon, which forms the last part of the digestive system, extracts water and salts from solid wastes before they are eliminated from the body, and is the site at which the microbial flora-aided fermentation of unabsorbed material occurs. Unlike the small intestine,

---

[1] USP37/NF32 general notices: test and assay.

the colon does not play a major role in the absorption of foods and nutrients, although it does absorb water, sodium and some fat-soluble vitamins. The colon microbial flora break down dietary fiber and some of the carbohydrates that reach the colon for their own nourishment, creating acetate, propionate and butyrate as waste products; these in turn are used by the cell lining of the colon for nourishment. Although most drugs are absorbed in the upper intestine, the colon is occasionally employed for the delivery of drugs for either local action or once-daily prolonged-release dosage forms. Drug stability towards the colonic flora can be studied using liquefied feces from candidate subjects. A 24 h incubation at 37 °C would suffice to detect any sign of degradation by the microorganisms.

4
Oral Absorption Mechanism

The three obligatory steps required for a drug to be absorbed are:

1) Release of the drug from dosage forms.
2) Maintenance of a dissolved state throughout the GI tract.
3) Permeation of drug molecules through the GI membrane into the hepatic circulation.

Drugs reaching the hepatic circulation must pass through liver, such that a portion of the drug may be metabolized; only that portion that has survived hepatic metabolism can reach the systemic circulation and be distributed into the target organs. The elimination by metabolism and/or any other processes prior to reaching the systemic circulation is termed the first-pass effect. The proportion of a drug reaching the systemic circulation is termed bioavailability; this is usually less than the proportion absorbed, as the first-pass effect is rarely insignificant.

4.1
Anatomy of the Human Intestine

Physiologically, the basic function of the GI tract is to digest food and absorb the nutrients thus provided. Functionally, the stomach serves as a large sac where food is stored and acidified to effect a partial sterilization of the food content. Likewise, the colon is employed in the further processing of food waste materials, with a final removal of water from the waste, together with the degradation of fibrous materials by the gut flora. The absorptive function of the GI tract mainly occurs in the small intestine, which comprises the jejunum and ileum. The jejunum is composed of absorptive enterocytes in the form of microvilli, some mucus-producing goblet cells, endocrine cells, and Paneth cells. The ileum has fewer microvilli and a less-mucous glycocalyx, but has a follicle-associated epithelium (FAE) that contains the so-called M cells and gut-associated lymphoid tissue (GALT) with Peyer's patches. The M cells play a role of gatekeeper, continuously taking up and internalizing materials from the lumen and transporting them into the underlying lymphoid tissues [16]. Thus, it is not difficult to appreciate that the jejunum is the major site of absorption, while the ileum is important for immune responses in the GI tract.

The mucin layer, which lines the entire intestinal tract, forms a significant barrier to the penetration of drug molecules before they can reach the enterocytes. Mucin carbohydrate contains sulfate, sialic acid and sugar moieties, thereby imparting a highly negative charge that aids in the binding of positively charged compounds. Mucin fibers can also entrap particles sterically; these are eventually taken up by the M cells.

4.2 Passive Diffusion

The absorption mechanism by which drug molecules permeate through the enterocyte is multifaceted, and includes passive diffusion and active transport. Passive diffusion follows the second law of thermodynamics that drug molecules at a higher concentration in the gut have a natural tendency to move to the bloodstream, where there is a lower concentration. Passive diffusion can be further divided into transcellular and paracellular permeation. The efficiency of transcellular permeation is dependent on the ability of the drug molecule to partition from the aqueous GI fluids into the lipoidal membrane of the enterocytes. This is why a reasonable oral drug candidate should have an n-octanol partition coefficient >1.0.

4.2.1 Paracellular Permeation

There is a polar tight junction between two enterocytes about $4\,\text{Å}$ in diameter where polar molecules such as mannitol can permeate through the pores into the bloodstream. This second route of absorption applies only to drugs with small enough molecular volume to pass through the narrow pores between enterocytes. Highly hydrated molecules with many hydrogen bonds can have a large molecular volume with a slow diffusion rate, and are unable to penetrate into the tight junction or the pores between enterocytes.

4.2.2 pH Partition Theory

pH-partition theory dictates that an ionic species has a poorer partition into lipid membrane than its uncharged species [17]. Evidence of such behavior is plentiful, and the esterification of acid groups to increase lipid partition, and thus enhance oral absorption, is a common technique used in medicinal chemistry. Two prime examples of esterification are valacyclovir and enalapril. Valacyclovir, a valine ester prodrug of acyclovir, has a greater oral bioavailability (ca. 55%) than acyclovir (10–20%), while enalapril, an ethanol ester of the intravenously administered drug enalaprilat, enables oral administration. Both examples demonstrate the use of pH-partition theory in removing unfavored ionization characteristics so as to allow sufficient oral absorption [18].

4.2.3 Polar Surface Area

The polar surface area of a molecule provides an additional measurement of a drug's ability to permeate cells. It is defined as the surface sum over all polar atoms, primarily oxygen and nitrogen, also including their attached hydrogens. Molecules with a polar surface area $>140\,\text{Å}^2$ tend to be poor at permeating cell membranes. For drugs targeted at the brain, a polar surface area $<60\,\text{Å}^2$ is usually needed to penetrate the blood–brain barrier [19].

4.3 Regional Intestine Permeability

Based on intestinal anatomy, the absorption of nutrients takes place largely from the jejunum, although absorption in the ileum and colon can be important for drugs with a lower solubility and/or permeability than that of the nutrients. Insoluble drugs may not fully dissolve until they reach the lower GI tract, where absorption takes place, while drugs with a low permeability may require the entire length of the GI tract for their absorption. Moreover, a delayed-release formulation such as a colonic delivery systems is designed not to release the drug until it reaches the lower GI tract.

For a once-daily delivery that requires an absorption phase of about 12 hours, colonic drug release and absorption is essential because the typical transit time form mouth to ileocecal junction is about 4.5 hours, while colonic transit may take 24–36 hours, though small intestine transit is highly variable among individuals, ranging from 4.5 to 12 hours. Such variation occurs because GI tract motility is affected by a multitude of physiological, hormonal and neurological conditions, as well as diet and exercise.

High-permeability drugs usually exhibit a reasonable colon permeability that is suitable for colonic delivery. However, for low-permeability drugs, permeability follows a descending rank order of duodenum > jejunum > ileum > colon. Many drugs that are well absorbed from an immediate-release formulation will exhibit a poor absorption from a colonic delivery system, because the colon permeability is too low for complete absorption, rendering once-daily delivery unfeasible. Regional intestine permeability can be measured using rat or human perfusion models, as described in Sect. 5.5.

4.4
Active Transport

Whilst most drugs are passively absorbed in the intestine, some are neither lipid-soluble to allow partition into the enterocyte, nor small enough to pass through polar tight junctions. The active transport of these drugs occurs via pathways borrowed from Nature for the absorption of nutrients such as minerals, amino acids, sugars and nucleic acids. An example is the anticancer drug, 5-fluorouracil, which utilizes the pyrimidine transport system in the gut. The characteristics of active transport are multifold:

1) Active transports occur against the concentration gradient in violation of the second law of thermodynamics. Namely, drug molecules transfer from a region of low concentration to a region of high concentration.
2) The transport requires drug binding to a cell-surface receptor, followed by internalization.
3) The transport consumes energy involving ATP or ion exchange.
4) As the number of transfer receptors is limited, the mass transfer is saturable and follows zero-order rate kinetics. A passive diffusion instead follows first-order rate kinetics and is not saturable.

Active transport receptors are concentrated in the small intestine to facilitate the absorption of nutrients. Drugs absorbed via active transport are poor candidates for once-daily sustained-release dosage forms, as the drug is released further down the large intestine where there are no active transport receptors. Consequently, the late-released drug is not absorbed and a prolonged absorption is not achieved. This phenomenon is often referred to as the drug having an "absorption window."

4.5
Role of Enterocyte Efflux Pumps

The existence of efflux pumps was first reported in 1976 by Juliano and Ling, with the discovery of P-glycoprotein (P-gp). This subject has since become an area of intense research, and drug transporters are now known to exist throughout the body. Those impacting on drug pharmacokinetics are found in the epithelial cells of the intestine and kidney, in hepatocytes, and in brain capillary endothelial cells [20], while transporters expressed in the brush-border membranes of intestinal epithelial cells are involved in the efficient absorption of nutrients or the elimination of compounds foreign to the body. In particular, intestinal efflux is recognized as a major determinant for the oral absorptivity of drugs. In general, drugs identified as efflux pump substrates exhibit a high pharmacokinetic variability. For example, the oral bioavailability of digoxin (a narrow therapeutic index drug with high variability) has been inversely correlated with the intestinal P-gp content in patients [21]. Another example, tacrolimus, is used as an immunosuppressant in organ transplant patients and its effectiveness has been linked with levels of P-gp efflux pump expression in the gut [22]. Many more examples exist of the negative effect of P-gp on drug efficacy, which is why P-gp was initially referred to as multidrug resistance protein 1 (MDR1). Another multidrug resistance protein, breast cancer resistance protein (BCRP), is also expressed in the intestine brush border membrane and has been shown to play a role in not only the absorption but also the biliary secretion of topotecan [23]. A modern drug discovery strategy of active avoidance – that is, to design a drug around the efflux pumps – will hopefully help to increase oral bioavailability while at the same time reducing pharmacokinetic variability.

4.6
Liaison between Metabolic Enzymes and Transporters

The cytochrome P450 (CYP) enzymes are heavily involved in drug metabolism, accounting for about 75% of the total number of different metabolic reactions. The term P450 is derived from the spectrophotometric peak at the wavelength of the absorption maximum of the enzyme (450 nm). CYPs are membrane-associated proteins mainly located in the microsomes of the liver, lungs and intestine. The Human Genome Project has identified 57 human genes coding for the various cytochrome P450 enzymes. For instance, CYP3A4 denotes the gene family number 3, the subfamily A, and the individual gene number 4. The CYP3A4 enzyme is the major enzyme responsible for a majority of drug metabolic reactions. Recently, CYP3A4 enzymes and P-gp were found to colocate at high levels in the villus of enterocytes in the small intestine. Cummins et al. have shown that an inhibition of intestinal efflux decreases intestinal metabolism, even though the inhibitor had no direct effect on the enzyme itself [24]. This most likely occurs because efflux takes the drug back to the lumen for additional absorption cycles, providing the enterocyte enzyme with more chances to metabolize the drug. In 2001, a coordinated induction of CYP3A4 and P-gp was demonstrated independently for rifampin [25] and ritonavir [26], both of which are ligands for the human steroid xenobiotic receptor (SXR) that is responsible for regulating the expression of CYP3A4 and P-gp. Conversely, if a compound is found simultaneously to inhibit CYP3A4 and P-gp, it should cause less gut metabolism in a way that can synergistically increase systemic drug concentrations. It is, therefore, not surprising that drugs removed from the market at the FDA's recommendation due to drug–drug interactions (DDIs) are predominantly orally dosed drugs that are substrates for both CYP3A and P-gp [27]. In conclusion, intestinal drug efflux is not only a major determinant of low or variable oral absorption, but also of the extent of drug metabolism leading to DDI liability.

5
Biopharmaceutics Classification System (BCS)

5.1
BCS History and FDA Regulation

In 1995, Amidon *et al.*, under a grant from the US FDA, developed a BCS that categorized drugs into four classes according to their aqueous solubility and intestinal permeability [28]. The BCS takes into account the two intrinsic properties of a drug that govern the rate and extent of its oral absorption; dissolution is a third (but extrinsic) factor that pharmaceutical scientists can optimize to improve bioavailability or prolong absorption for a longer duration of drug action. The BCS appeared in the same year (1995) in a FDA guidance document on post-approval scale-up changes of immediate-release solid oral dosage forms [29]. The guidance stipulates that, for an immediate-release formulation, human bioequivalence data are needed for low-solubility and/or low-permeability drugs when filing level 2 formulation changes that impact on formulation quality and performance. Should both solubility and permeability be high, then only dissolution data are required to support the changes. The guidance proceeded to define solubility and permeability high/low boundaries according to the BCS definition. In 1997, a second guidance followed for extended-release dosage forms, where the FDA used BCS in the development of *In-Vitro/In-Vivo* correlations [30]. In 2000, the FDA expanded the use of the BCS concept from post-approval changes to abbreviated new drug applications (ANDAs) to allow the waiver of human bioequivalence testing of BCS class 1 drugs with high solubility and high permeability [31]. This biowaiver applied only to immediate-release solid dosage forms exhibiting a rapid dissolution – that is, 85% release within 30 min. The 2000 guidance also included an application of the BCS Class waiver concept in new drug development [investigational new drugs (INDs) and new drug applications (NDAs)]. In other words, the FDA recommended using the BCS concept in risk management when making changes in clinical formulations or manufacturing processes during drug development. Currently, BCS guidelines are provided worldwide including the World Health Organization (WHO) [32], the European Medicinal Agency (EMA) [33], and many individual countries.

5.2
BCS Definition

In the BCS, drugs are classified based on their solubility and permeability into four classes:

Class 1: High Solubility – High Permeability
Class 2: Low Solubility – High Permeability
Class 3: High Solubility – Low Permeability
Class 4: Low Solubility – Low Permeability.

A drug substance is considered highly soluble when the highest dose strength is soluble in ≤250 ml of an aqueous medium over the pH range of 1 to 7.5. The volume estimate of 250 ml is derived from the typical stomach volume. This is why clinical and bioequivalence (BE) study protocols usually prescribe the administration of a drug product to fasting human volunteers with a glass (ca. 250 ml; 8 oz.) of water.

The permeability class boundary is based on the extent of absorption in humans being ≥90% of an administered dose. The fraction of dose absorbed can be estimated based on a mass balance study, or if oral bioavailability is not less than 90%. Since neither cases is applicable to most drugs, an indirect comparison of intestinal membrane permeability between a reference and candidate drugs is a more popular method to estimate the extent of absorption. A lack of direct measurements of the extent of absorption has led to confusion in the assignment of BCS class for many common drugs, and may also have hampered the application of BCS in the pharmaceutical industry.

5.3
Significance of a Drug's BCS Class

In the case of BCS Class 1 drugs with high solubility and high permeability, oral absorption should not be a problem provided that the drug is stable in the GI tract. The majority of marketed drugs appear to be BCS Class 1 (see Sect. 5.7). To qualify a drug for BCS 1 in the US, a petition must be made to the FDA BCS committee before a biowaiver can be applied. In order to pass the committee scrutiny, data relating to solubility, permeability, GI stability and dissolution must be generated to the FDA's particular specifications. Once a drug is assigned BCS 1, a BE study is no longer required for market entry of generic copies of drugs and consequently this route of biowaiver is not popular as there is little incentive for the innovator to go for the BCS 1 petition. The same scenario applies to generic companies, where BCS 1 means an easy entry into the market for all players and thus little profit for all.

BCS Class 2 drugs have high permeability and low solubility; an example is the common nonsteroidal anti-inflammatory drug (NSAID), diclofenac. A majority of new drugs under development appears to be BCS class 2, as if the "low-hanging fruits" (easy drugs with BCS 1) have already been picked (see Sect. 5.7). The bioavailability of BCS 2 drugs is limited by their dissolution rate; where a correlation between the *in-vivo* bioavailability and the *in-vitro* dissolution can usually be found.

BCS Class 3 drugs have low permeability and high solubility; an example is the GI antacid drug, cimetidine. The absorption of BCS 3 drugs is limited by their permeation rate, as dissolution is relatively fast. Efforts are currently under way to urge the regulatory agency to accept biowaiver for BCS 3 drugs, the rationale being that if the formulation does not change the permeability or GI transit time, then a BE study is not necessary as all that can be measured is variations in individual intestine permeability, which is the rate-limiting step for the absorption of BSC Class 3 drugs [34].

BCS Class 4 drugs have low permeability and low solubility; an example is the antifungal drug amphotericin B which is, by name, amphoteric with both carboxylic and amino groups at either end of the large molecule (molecular weight 924.09 Da). Amphotericin B has low organic solvent and aqueous solubility and a low intestine permeability, and is unable to transport paracellularly, possibly with extensive P-gp efflux. The absorption of this BCS Class 4 drug is poor, with an oral bioavailability of ~0.3% in either surfactant solution or in liposomal preparations [35]. Thus, amphotericin B is administered intravenously. BCS Class 4 drugs generally have very poor bioavailability; with high variability; hence, an alternative route of administration other than oral should be considered.

5.4
Methodology for Classifying Drug Solubility

5.4.1 Dose Solubility Ratio
A compound is classified as highly soluble when the highest dose strength is soluble

in ≤250 ml of aqueous media over the pH range of 1 to 7.5; this is referred to as the dose solubility ratio. The first step is to determine the equilibrium solubility under conditions of pH 1 to 7.5. Equilibrium solubility is typically determined using a shake flask method, though alternative methods such as acid or base titration method are allowed by the FDA (but rarely used). As the solution process is sometimes accompanied by degradation, a stability-indicating assay that can distinguish the drug substance from its degradation products must be in place for drug solubility determination. The solubility should be measured with a minimum of three replicates at $37 \pm 1\,°C$ in aqueous media with a pH range of 1 to 7.5. These include water, 0.1 N HCl (pH 1), and acetate (pH 4.5) and phosphate (pH 6.8 and 7.5) buffers. As the solution pH may change after equilibration if the drug solubility is very high, it is necessary to confirm and report the final pH after equilibration; this should be the reported pH of the determination. The highest dose strength is referred to the largest quantity of drug taken orally at any one time. Therefore, if the tablet strength is 100 mg but the FDA-approved drug label states that two tablets should be taken at one time, a dose strength of 200 mg needs to be used for BCS calculations.

5.4.2 Drug Instability in the GI Tract

Drugs need not only to be soluble but also stable in the GI tract in order to be absorbed. To monitor such stability, a drug is dissolved in a simulated GI fluid to produce a concentration equivalent to the highest dose in 250 ml. This solution should then be incubated at $37\,°C$ for a period that is representative of *in-vivo* drug contact with these fluids; for example, 1 h in gastric fluid and 3 h in intestinal fluid. Drug concentrations should then be determined using a stability-indicating assay method. A significant degradation (>5%) of the drug in this protocol could suggest potential instability in the GI tract.

5.5 Methodology for Classifying Drug Permeability

5.5.1 Direct Human Studies

The amount (as a percentage) of a drug that is absorbed can be determined directly using human data of mass-balance, or an absolute bioavailability study. A 90% oral absorption can be assured when the absolute bioavailability is 90% or higher, or when at least 90% of the administered drug is recovered in the urine. Alternatively, a mass-balance study using a radiolabeled drug can demonstrate drug recovery in the bile, in addition to the urine or an absence of radioactivity in the feces as evidence of the extent of oral absorption. In Europe, a third method is permitted by using dose linearity in a human pharmacokinetic study as evidence of high permeability. Depending on the variability of these studies, a sufficient number of subjects should be enrolled to provide a reliable estimate of the extent of absorption.

5.5.2 Indirect Methods Requiring Proof of Passive Diffusion

Indirect methods measuring intestinal permeability against calibration compounds with known human absorption can also be used as a surrogate of percentage absorbed. For indirect methods the FDA requires a demonstration of passive diffusion, since *in-vitro* cell permeation systems may have varied levels of influx or efflux pump expression compared to humans. A drug with efflux may appear highly permeable if the testing system lacks the correct efflux pumps, so leading to erroneous conclusions. Passive diffusion can be demonstrated by a lack of concentration dependence of the measured permeability. The FDA recommends using initial drug concentrations at 0.01×, 0.1× and 1× the highest dose strength dissolved in 250 ml; passive diffusion is deemed to have occurred if no statistically significant difference is found among the three concentrations. A second method is to use bidirectional permeability measurements, where actively absorbed drugs have a higher apical-to-basolateral (a → b) than basolateral-to-apical (b → a) permeability. Conversely, effluxed drugs have a higher b → a than a → b permeability.

5.5.3 Caco-2 Permeability Assay

The human colon adenocarcinoma (Caco-2) cell line was developed at the Sloan-Kettering Institute for Cancer Research in 1975 [36], and applied subsequently (during the 1980s) to oral drug absorption studies by Hidalgo and Borchardt at the University of Kansas. At the time, Borchardt was a consultant for Upjohn Company and, in conjunction with Raub (a cell biologist), began the first industrial applications of Caco-2 cells [37]. The correlation between the Caco-2 apparent permeability (P_{app}) and the percentage absorbed was subsequently established [38] such that, by the 1990s, automated Caco-2 cell monolayer cultures grown onto a filter (transwell) in 24- to 96-well plates were widely used to measure intestinal permeability in the pharmaceutical profiling of new clinical lead compounds.

In culture, the Caco-2 cells differentiate and polarize into basolateral and apical sides such that their phenotype – both morphologically and functionally – resembles the enterocytes. Caco-2 cell monolayers express tight junctions, microvilli and transporters including P-glycoprotein, as well as uptake transporters for amino acids, bile acids and carboxylic acids [39]. Enzymes such as peptidases and esterases are also present in the system. Consequently, the Caco-2 system is particularly useful for detecting efflux and metabolism issues of drug candidates in their early development stage.

For compounds transported via a passive transcellular route, Caco-2 permeability is a valid method for predicting permeability in humans. However, for those compounds transported via paracellular or active transport, and also for highly insoluble compounds, Caco-2 permeability tends to underestimate human permeability for the following reasons: (i) there is an overexpression of P-gp efflux pumps; (ii) there is a reduction of the paracellular route of transport due to an absence of liquid pores; and (iii) the nonspecific binding of insoluble compounds onto the filter support and plastic components leads to a reduction in apparent permeability. In these cases, an *in-situ* rat gut perfusion method (see Sect. 5.5.4) may be used to correct for the false-negative bias seen with Caco-2 [40].

Because the Caco-2 cells take 21 days to grow into a monolayer of integrity, an alternative dog kidney (Madin Darby canine kidney; MDCK) cell line that requires only three to seven days to develop has gained popularity during recent years. Unfortunately, the MDCK cells suffer the same drawbacks as Caco-2 cells in generating false-negative results of low permeability. The MDCK cells have an additional drawback of a low expression of various efflux pumps, and are not suitable for the screening of chemical series with known efflux problems. Both, the *in-vitro* Caco-2/MDCK and the *in-situ* rat perfusion method (see below) are accepted by FDA for the purpose of BCS classification.

5.5.4 Rat Intestinal Perfusion

The *in-situ* rat perfusion model utilizes a section of intestine in an anesthetized animal [41]. For the single-pass (open loop) perfusion method, the rat intestine is cannulated proximally and distally, rinsed with buffer solution, and then perfused with a drug solution; amounts of drug in the perfusate are then measured at defined time points. The intestinal permeability (P_{eff}) is calculated using the steady-state mass loss into the intestine using the Fick's diffusion law equation. The rat perfusion model is inherently more robust as it is conducted under steady-state conditions, when parallel processes such as nonspecific drug binding is saturated. Fagerholm *et al.* correlated the *in-situ* rat intestinal perfusion data with *in-vivo* perfusion values in humans, rendering validity to the use of rat intestinal permeability in the prediction of human absorption [42].

5.5.5 Human GI Permeability

Human intestine permeability can be measured using perfusion or intubation techniques in human volunteers [43]. This involves the passage of a tube into the GI tract via the nose, mouth or rectum; the drug solution/suspension is then pumped through the tube for site-specific release, followed by absorption. An alternative approach involves the use of a capsule engineered with a gamma-emitting radionuclide and a drug reservoir. When this device has reached the specified GI region the drug is released by a radiofrequency-triggered gate opening. Such a device, the Enterion™ capsule, was developed at Nottingham University in the UK during the 1990s [44]. Both techniques are used to assess not only jejunum permeability (as per the BCS classification), but also regional intestine permeability for sustained-release dosage forms.

5.5.6 Permeability Method Suitability

In order to demonstrate permeability method suitability, model drugs representing a range of low (e.g., <50%), moderate (e.g., 50–89%) and high (≥90%) absorption must be correlated with the measured permeability in graphical form (as shown in Fig. 6). After suitability has been demonstrated, subsequent studies need to include a high-permeability drug as an internal standard which is included in the perfusion fluid or donor fluid along with the test drug. A test drug may be determined as being highly permeable when its permeability value is equal to or greater than that of the selected high-permeability internal standard.

5.5.7 Use of Partition/Distribution Coefficient in Lieu of Permeability

Amidon *et al.* used the partition coefficient ($\log P$), calculated partition coefficient ($c\log P$), or distribution coefficient ($\log D$) as a surrogate for intestinal permeability to classify the WHO essential medicine [45] and the US top 200 drugs [46]. Ku *et al.* adopted this methodology and classified 108 clinical compounds collected by the Pharmaceutical Research and Manufacturers of America (PhRMA) from 12 member companies, under an initiative to predict human pharmacokinetics from animal data [47]. This stepwise procedure employed was as follows:

1) The drug is classified as high-permeability if human permeability is not less than $1.34\,\mu m\,s^{-1}$, which is the human jejunal permeability of metoprolol, the FDA high-permeability calibration compound.
2) The drug is classified as high-permeability if human permeability is not available but Caco-2 shows an absence of efflux (the ratio of basolateral to apical fluxes is <3)

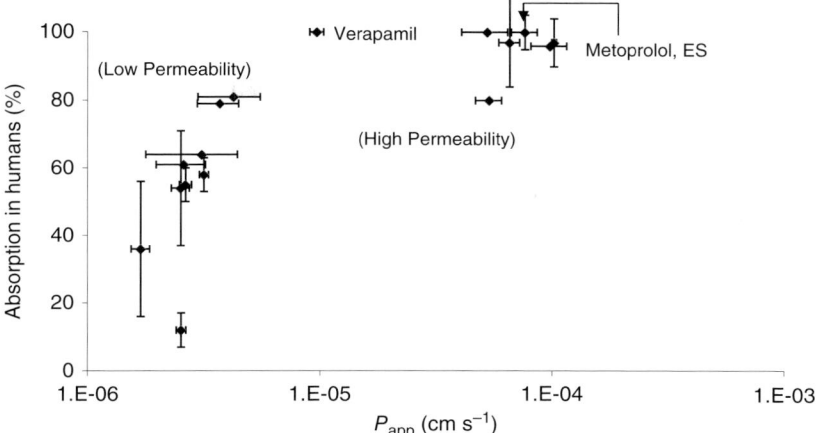

Fig. 6 Permeability method suitability in correlating permeability with the extent of human absorption. Among the 16 drug compounds with known human absorption, the external standard used in transport studies of test compounds, metoprolol, is indicated in the "high-permeability" range. Verapamil (100% FA, P_{app} = 97 nm s^{-1}) marks the lower boundary of the high permeability range.)

and the permeability is not less than 100 nm s^{-1}, the generally recognized cut-off for high permeability.

3) The drug is classified high-permeability if human and Caco-2 data are not available but the log D at pH 6.5 is not less than −1.48, which is the log D of metoprolol at pH 6.5.

Among the 108 clinical compounds tested, 36 were assigned a BCS class with certainty, and an additional 11 were assigned a question mark by the member company submitting the data. Human permeability data were available for 51 compounds. The Caco-2 permeability is available for 76 compounds, while the log P/clog$P$ and pK_a values are available for all compounds. The three-step algorithm was first tested using internal validation before application to all drugs. The internal validation is conducted using the 36 compounds for which BCS class is reported by the member company, and the reported BCS class is then compared with the system-generated BCS class. A complete (100%) agreement was found among the 36 compounds, providing validity to the algorithm. This algorithm was then used for the remainder of the compounds, which allowed BCS classification of additional 56 compounds. There were seven drugs with no solubility data, and one drug with no human dose data, which prevented the BCS classification being made for these compounds. Overall, a BCS classification is provided for 100 PhRMA compounds. In conclusion, the use of log D at pH 6.5 as a surrogate for intestinal permeability can be useful when experimental measurements of permeability are not feasible due to a lack of knowledge, experience, or resources.

5.6
Biopharmaceutics Drug Disposition Classification System (BDDCS)

In 2005, Benet *et al.* devised a second classification scheme, the Biopharmaceutics Drug Disposition Classification System

(BDDCS), in which permeability was replaced with metabolism [7]. These authors examined drugs in the four BCS classes and recognized that, in humans, the major route of elimination for high-permeability drugs was via metabolism, whereas low-permeability drugs were primarily eliminated as unchanged drug in the urine and bile. The same group also suggested – at least for drugs currently available but where the extent of metabolism was always known but the extent of absorption may not have been quantified – that metabolism may serve as a surrogate for permeability. The metabolism cut-off is ≥50% for high-metabolism drugs.

In 2006, Amidon *et al.* compared the two classification systems, using the top 200 drugs [46]. The agreement between the BDDCS and BCS for Class 2 and Class 4 drugs was good (87%), but for Class 1 and Class 3 drugs it was less (64%). These authors also found that the agreement could be improved by lowering the permeability cutoff values – that is, by using calibration compounds with ≥90% oral absorption but a lower intestinal permeability value. In particular, the agreement between the BCS and BDDCS classifications for Class 1 drugs was increased from 56% (BCS/metoprolol) to 75% (BCS/cimetidine) to 86% (BCS/atenolol). The original choice of metoprolol was purposely conservative for regulatory biowaiver purposes. In 2006, Ku classified hundreds of clinical lead compounds and found the criteria for permeability intended for waiver on human BE study to be too high to serve as a gate for clinical lead nomination in pharmaceutical companies. In fact, it was so high that only a very small percentage of the pipeline compounds passed the criteria to be BCS Class 1, whereas many Class 4 compounds with permeability and solubility somewhat lower than the criteria may have been perfectly acceptable for pharmaceutical development [48].

In conclusion, the correlation between the BDDCS and BCS was surprisingly good, since permeability and metabolism are generally considered to be different biological processes at the molecular level – that is, enzyme binding versus membrane permeation. The major organs for metabolism and absorption are also different, namely liver versus intestine. However, there is a liaison between metabolic enzymes and cell membrane transporters (as reported in Sect. 4.6), and the first step in either process is cell internalization by passive partition or carrier-mediated active transport. The observed good agreement may stem from the fact that high-permeability compounds partition better into the hepatocytes, within which ready access to the metabolizing enzymes cytochrome P450 leads to a high degree of metabolism.

5.7
BCS Classification of Old and New Drugs

With the application of BCS being hampered by a lack of freely available and accurate data summarizing the solubility and permeability data of drugs, several groups have attempted to assign BCS classes to common oral drugs. In 2004, Dressman *et al.* classified 61 out of 130 WHO drugs [49], while Amidon *et al.* expanded the number of drugs to 114 by utilizing clog$P$ in lieu of permeability [45]. Later in 2004, the WHO Expert Committee published (and updated in 2006) its Technical Report Series 937 list BCS classes for 123 oral drugs [45], and Benet *et al.* further expanded the list to 141 compounds [7]. In 2011, Ku *et al.* classified the PhRMA dataset which contains not commercial drugs but rather

108 clinical compounds from the 12 largest pharmaceutical companies worldwide [47]. Simultaneously, a BCS distribution of 178 Wyeth research compounds was published and compared with literature data [40]. A similar BCS distribution of Glaxo Smith Kline (GSK) new leads was reported by Baldoni at the 2007 AAPS Workshop, "BE, BCS, and Beyond" [50]. In addition, in 2011, the database of Benet *et al.* classified 698 reported compounds according to BDDCS criteria, using metabolism in place of permeability [51].

The WHO compound BCS assignment is shown in Table 2, where drugs with dual assignment are shown in *italic* text. Sixteen drugs listed in Class 1 had mixed reports of low permeability and may be Class 3, while 18 drugs listed in Class 2 with mixed reports of low permeability may be Class 4, and seven drugs listed in Class 3 with mixed reports of low solubility may be Class 4. Overall, the permeability assignments seem less certain due to the choice of high-permeability calibration compound and interlaboratory variability. The percentage distribution in BCS classification of compounds in the diverse datasets described above is summarized in Table 3. This confirms the continuing trend in the industry towards the development of lead compounds that are less soluble and less permeable than those in previous years. The percentage of BCS 2 compounds increased from 17% to 31% of marketed drugs (WHO), to 41–62% for new molecular entities (PhRMA, Wyeth, and GSK). The Class 4 compounds also increased from below 10% to more than 20%. The increases in BCS Class 2 and 4 compounds were at the expense of BCS Class 1 and 3 compounds. Since, for the past two decades, solubility and permeability have been the two leading parameters used for lead optimization, the undeniable increasing percentages of Class 2 and 4 compounds among PhRMA companies during the immediate past few years is disappointing, albeit the situation could be worse without pharmaceutical profiling. Consequently, new formulation technologies are in high demand that enable the delivery of these insoluble/nonpermeable compounds when lead optimization fails to produce compounds with desirable biopharmaceutical properties. This two-pronged approach – that is, pharmaceutical profiling to improve lead quality and the enabling formulation technology when choice optimization is exhausted – can hopefully improve the probability of success in new drug discovery and development.

6
Biopharmaceutical Application in Drug Molecule Design

With the advent of HTS around 1990 [5], a shift of lead compound biopharmaceutical characteristics into less drug-like has propelled the discovery departments in pharmaceutical companies to start pharmaceutical profiling to test the lead compounds for solubility, permeability, metabolism, and the potential for DDI [52]. Typically, a clinical candidate is not declared until the compound has passed a majority of gate criteria established in the company pharmaceutical profiling program. As a result, Phase 1 failure has been reduced for the past 10 years, whereas the Phase 2 proof of concept (POC) has become the major hurdle for drug development [53]. As a result, both the National Institutes of Health (NIH) and Harvard have declared Systems Pharmacology as a route to reduce POC failures [54]. Nevertheless, it is critical for the industry to continuously integrate

Tab. 2 Biopharmaceutics classification system (BCS) assignment of WHO compounds (2006).

Class 1	Class 2	Class 3	Class 4
Acetylsalicylic acid	*Acetazolamide*	Abacavir	Nelfinavir mesylate
Allopurinol	*Albendazole*	Aciclovir	Erythromycin stearate + ethyl succinate
Amiloride hydrochloride	*Azithromycin*	*Artemether*	Sulfasalazine
Amitriptyline hydrochloride	Carbamazepine	Atenolol	
Amlodipine	*Cefixime*	Benznidazole	
Amodiaquine (base)	Dapsone	*Clofazimine*	
Amoxicillin	*Diloxanide furoate*	Chloramphenicol	
Amoxicillin anhydrous	*Efavirenz*	Chloropromazine hydrochloride	
Ascorbic acid	*Furosemide*	Cloxacillin (as sodium salt)	
Biperiden hydrochloride	Glibenclamide	Codeine phosphate	
Carbidopa	Griseofulvin	Didanosine	
Chlorophenamine hydrogen maleate	Ibuprofen	Enalapril	
Chloroquine phosphate or sulfate	*Indinavir sulfate*	Ergocalciferol	
Ciprofloxacin hydrochloride	Iopanoic acid	Ethambutol	
Clavulanic acid	Ivermectin	Ethambutol hydrochloride	
Clomifene citrate	*Lopinavir*	Ferrous salt	
Clomipramine hydrochloride	Mebendazole	*Haloperidol*	
Diazepam	Mefloquine hydrochloride	Hydralazine hydrochloride	
Digoxin	Nevirapine	Hydrochlorothiazide	
DL-Methionine	Niclosamide	Levothyroxine sodium salt	
Doxycycline hydrochloride	Nifedipine	*Lumefantrine*	
Ethinylestradiol	Nitrofurantoin	Metformin hydrochloride	
Fluconazole	Phenytoin sodium salt	Methyldopa	
Folic acid	Praziquantel	Metoclopramide hydrochloride	
Glyceryl trinitrate	Pyrantel embonate	Neostigmine bromide	
Isoniazid	Rifampicin	Nifurtimox	
Isosorbide dinitrate	*Ritinol palmitate*	Penicillamine	
Lamivudine	*Ritonavir*	Pyrimethamine	
Levamisole hydrochloride	*Saquinavir*	Ranitidine hydrochloride	
Levodopa	Sulfamethoxazole	*Spironolactone*	

Tab. 2 (Continued.)

Class 1	Class 2	Class 3	Class 4
Levonorgestrel	Trimethoprim	Thiamine hydrochloride	
Lithium carbonate	Verapamil hydrochloride	*Triclabendazole*	
Metronidazole		Zinc sulfate	
Morphine sulfate			
Nicotinamide			
Norethisterone			
Paracetamol			
Phenobarbital			
Phenoxymethylpenicillin (as potassium salt)			
Potassium iodide			
Prednisolone			
Primaquine diphosphate			
Proguanil hydrochloride			
Promethazine hydrochloride			
Propranolol hydrochloride			
Propylthiouracil			
Pyrazinamide			
Pyridoxine hydrochloride			
Quinine bisulfate or sulfate			
Riboflavin			
Salbutamol sulfate			
Stavudine			
Valproic acid sodium salt			
Warfarin sodium salt			
Zidovudine			

Note: Drugs shown in italic text have dual BCS Class assignment. Sixteen drugs listed in Class 1 with mixed report of low permeability may be Class 3. Eighteen drugs listed in Class 2 with mixed report of low permeability may be Class 4. Seven drugs listed in Class 3 with mixed report of low solubility may be Class 4.

the biopharmaceutical principles into new drug discoveries.

6.1
Structural Modification in Drug Design and Therapy

In addition to pharmacological considerations, molecular modifications to yield favorable biopharmaceutical properties are critical for the successful development of commercial products. The primary objective is to improve the physical, chemical and biopharmaceutical properties without impacting on pharmacological effects such as receptor binding affinity and target selectivity. The specific goals can include the following:

- Increase water-solubility.
- Increase GI permeability.
- Reduce enterocyte efflux.
- Improve metabolic stability.
- Reduce first-pass effect.
- Prolong biological half-life.

Tab. 3 BCS classification across diverse datasets.

Database	Percentage of compound				Author(s), reference and number of compounds
	BCS 1	BCS 2	BCS 3	BCS 4	
WHO	34	17	39	10	Dressman et al. [49] ($n = 61$)
WHO[a)]	23.6	19.1	31.7	10.6	Amidon et al. [45] ($n = 114$)
WHO	44.7	12.2	28.5	14.6	WHO Expert Committee [25] ($n = 123$)
WHO	39	31	23	7	Benet et al. [7] ($n = 14$)
PhRMA	25	54	9	12	Ku et al. [47] ($n = 100$)
Wyeth	14.0	41.6	21.3	23.1	Ku and Dulin [40] ($n = 178$)
GSK	19.9	62.6	8.4	9.2	Baldoni [50] (NA)
Literature	40	33	21	6	Benet et al. [51] ($n = 698$)

a) The remaining 17.1% of the drugs could not be classified because of the inability to calculate log P values because of missing fragments.
NA = not available.

- Decrease formation of toxic metabolites.
- Reduce peak blood concentration.

All of the above are aimed at improving bioavailability and safety margins with enhanced effectiveness and a reduced toxicity of the drug. Typically, a structure–activity relationship (SAR) is established in the identification of structural elements that bind to the target receptor. Chemical modifications can then take place away from these structural elements, so that the pharmacological activities would not be impacted. A classic example of this is the chemical modification of penicillin G into ampicillin, which results in a dramatic increase in oral bioavailability due to the improved stability of ampicillin in gastric juice. Stabilization of the β-lactam ring is achieved by adding an electron-withdrawing group, a protonated amine. Ampicillin's acid degradation half-life of 660 min compares favorably with that of 3.5 min for penicillin G.

Another such example is the chemical modification of the protease inhibitor and antiretroviral drug amprenavir into fosamprenavir, a water-soluble prodrug. The oral bioavailability of amprenavir was initially enabled using a solubilized formulation with a powerful surfactant, vitamin E TPGS (D-α-tocopheryl polyethylene glycol 1000 succinate). The amprenavir-vitamin E TPGS solution is filled into a liquid gel capsule and given to patients. (TPGS, originally marketed by Eastman Kodak, is a powerful solubilizer that had never been used in pharmaceutical products prior to this event.) In conjunction with Eastman Kodak, GSK acquired approval to use this novel excipient, based on unmet medical needs, although patients needed to swallow eight very large capsules twice daily with food to achieve a daily amprenavir dose level of 1200 mg. Subsequently, fosamprenavir, a chemically modified phosphate ester of amprenavir with an enhanced water-solubility, was created five years after the commercial launch of amprenavir. The hydrolysis of fosamprenavir, to release amprenavir, took hours such that fosamprenavir would serve as a slow-release version of amprenavir, and patients needed to take only two 700 mg fosamprenavir tablets once daily, irrespective of food intake. This greatly enhanced patient convenience and led to an improved compliance for medicine intake.

Structural modification is powerful, but not without limitations. At times, one property is improved at the expense of other properties, so that a compromise must be reached among chemists, biologists and pharmaceutical scientists to nominate a clinical lead compound with the best chance of succeeding in the clinic. The present author experienced first-hand such a compromise in the development of Tigacil® (tigecycline) [55], an oral tetracycline analog that contained a central four-ring carbocyclic skeleton and was structurally similar to minocycline. Tigecycline is produced by the substitution at the 9-amino D-ring position of a *t*-butyl-glycine, which is believed to confer a broad spectrum of activity against resistant strains of bacteria such as methicillin-resistant *Staphylococcus aureus* (MRSA) [56]. The glycine substitution was originally conceived to provide active absorption by the GI dipeptide transporter. Surprisingly, however, the oral absorption was not enhanced but totally suppressed. This occurred because addition of the positive charge of the glycylamino group shifted the isoelectric point from pH 5 of minocycline to pH 8 of tigecycline [57]. Consequently, passive diffusion by pH partition was lost as only the charged species was present in the upper GI tract. Conversely, this positive charge was found essential to overcome the efflux pump of the resistant bacteria. Ultimately, tigecycline was developed as a niche injectable product instead of an oral antibiotic for broad usage.

Drug discovery has more recently been aided by pharmaceutical profiling, a systematic study of the physical, chemical and biopharmaceutical properties of lead compounds in the selection and optimization of clinical compounds in order to yield better developability. Success in pharmaceutical profiling depends on the research group's understanding of biopharmaceutical principles in order to design delivery-friendly molecules in a rational fashion.

6.2
Salt Formation: An Enabling Technology to Improve Solubility and Stability

Salt forms of many basic and acidic drugs, based on the improved solubility in the intestinal pH, have exhibited improved bioavailability over the free form and thus have optimized pharmacologic or therapeutic activity. Bighley *et al.* provided a systematic compilation of various drug salts and their effects on pharmacokinetics and bioavailability [58]. Salt formation will not only increase bioavailability but also improve solid-state stability whenever stability of the free base or free acid is in question. This is because salt formation involves proton transfer with a change in electron density distribution, so that the reactivity of the nucleophilic or electrophilic functional groups are modified. Throughout the years, the author has been able to stabilize many clinical lead compounds via salt formation. Typically, approximately 70% of pipeline compounds are ionic in nature, and about 10% of these compounds offer the opportunity for stabilization and/or solubilization. Exemplary mechanisms of stabilization are the prevention of decarboxylation of carboxylic acid, the hydrolysis of sulfonamide and hydroxamic acids, dimer and adduct formation by nucleophilic amines, and the oxidation of free amine groups. Another application of salt form involves the modification of solid-state properties such as hygroscopicity, crystal morphology, surface static force, and cohesive or adhesiveness. More detailed discussions on the impact of solid-state properties on stability are available in two excellent reviews by Yoshioka and Stella [59] and Guillory and Poust [60].

Lastly, maximizing the bioavailability or stability or solid-state properties via salt formation should be considered at an early stage in the development cycle, since regulatory agencies recognize salt as a different chemical entity that necessitates the repeat of preclinical and clinical studies to a significant extent. Determination of the opportune time for salt formation and selection is discussed in Sect. 6.6.

6.3
Polymorphism Screening

During the past decades, pharmaceutical manufacturers have experienced disrupted supply or have recalled commercial products when polymorph changes occurred in the manufacture of an active pharmaceutical ingredient (API), or during drug product storage. Polymorphic forms differ in crystal lattice energy, and this results in differences in physical properties (density, color, and refractive index), thermal properties (melting point and transition temperature), solubility, and chemical stability. In particular, when a more stable polymorph is formed during the manufacture and storage of a drug product, its lower solubility and dissolution may lead to a reduced oral bioavailability. The most infamous case is probably the late discovery of a thermodynamically stable form of ritonavir, with a lower solubility and lower oral bioavailability. This resulted in a six-month disruption in drug supply, a loss of millions of dollars in sales, and suboptimal therapy for thousands of patients [61]. As a consequence, most pharmaceutical companies have implemented polymorph screening to select a thermodynamically stable form for commercialization. Regulatory agencies have also started to demand X-ray diffraction (XRD) patterns for successive lots of drug material during and after its development, to insure against the accidental inclusion of a new or different polymorph during drug manufacture.

The key objective of polymorph screening is to identify the thermodynamically stable form in the target formulation environment during its manufacture and storage. Even for the first animal pharmacokinetic study, administering an aqueous suspension by oral gavage, hydrate formation in the suspension can result in a significant change in solubility, uniformity, and suspendability, which would lead to nonreproducible pharmacokinetic data. This is why the potential for polymorphic conversion needs to be evaluated ahead of animal studies.

The first step in polymorph screening is to obtain various solid forms. Typically, a dozen solvents and antisolvents utilizing four different crystallization methodologies (i.e., evaporative, cooling, gas diffusion, and antisolvent addition) are investigated at 100 mg of drug per solvent. The samples are examined for birefringence using polarized microscopy, while the crystalline sample is subjected to differential scanning calorimetry (DSC) and thermal gravimetric analysis (TGA) to determine whether it is a mixture or a pure polymorph. The pure polymorphs are scaled up to the multigram level to evaluate scalability, solubility and hygroscopicity. At this point, a fingerprint of powder XRD pattern, near infrared/Raman, or solid-state nuclear magnetic resonance is obtained to positively identify the polymorphic form. In order to thoroughly screen the polymorphic potential, hundreds – if not thousands – of samples could be generated, and several high-throughput methods with automated or semiautomated sample handling have recently been reported for this purpose [62, 63]. While these new methodologies provide extra capacities for solid form discovery, they cannot replace the detailed characterization studies required to understand the relationships between various polymorphic forms.

6.4
Polymorphism Regulatory Guidance Linked with BCS Classification

The FDA recognizes the importance of polymorphism. In 2000, on issuing BCS guidance, the FDA immediately issued a draft polymorphism guidance highlighting the utility of BCS in determining whether a polymorph specification is required for the drug substance and/or drug product. In the final guidance issued in July 2007 [64], the FDA stated: "For a drug whose absorption is only limited by its dissolution, large differences in the apparent solubilities of the various polymorphic forms are likely to affect bioavailability/bioequivalence (BA/BE). On the other hand, for a drug whose absorption is only limited by its intestinal permeability, differences in the apparent solubilities of the various polymorphic forms are less likely to affect BA/BE." The FDA also stated: "Furthermore, when the apparent solubilities of the polymorphic forms are sufficiently high and drug dissolution is rapid in relation to gastric emptying, differences in the solubilities of the polymorphic forms are unlikely to affect BA/BE." In other words, the FDA is saying that polymorphism is not critical for BCS Class 1 and 3 compounds, but it is for BCS Class 2 and 4 compounds.

The decision trees in FDA's guidance stipulate for the drug substance; if there are known polymorphs with different apparent solubilities and not all polymorphs are highly soluble as defined by BCS criteria, then a polymorph specification is necessary. The guidance takes one step further to stipulate "For the drug product, if there is sufficient concern that a polymorph specification in the drug product be established or if the drug product performance testing (e.g., dissolution testing) does not provide adequate controls when the polymorph ratio changes, a polymorph specification is necessary." The FDA recommends considering only those polymorphs that are likely to form during manufacture of the drug substance and the drug product, or while the drug substance or drug product is in storage. The guidance clarifies that there may not be a concern if the most thermodynamic stable polymorphic form is used, or the same form is used in a previously approved product of the same dosage form. Finally, the FDA states that drug product performance testing (e.g., dissolution) can generally provide adequate control of polymorph ratio changes for poorly soluble drugs, which may influence drug product BA/BE. It is clear that the FDA expects only in rare cases would polymorphic form characterization in the drug product be necessary. It can be concluded that the polymorph control is for the purpose of controlling oral bioavailability, and that for highly soluble and highly permeable compounds, polymorphism is of little concern. The FDA guidance has applied BCS nicely in the establishment of polymorph specification.

Similar decision trees are in the ICH Q6A Guidance [65], including the following four questions:

- Can different polymorphs be formed?
- Do the forms have different properties (solubility, stability, melting point)?

- Does drug product performance testing provide adequate control if the polymorph ratio changes (e.g., dissolution)?
- Is drug product safety, performance, or efficacy affected?

All of these questions lead to a bottom line where, for BCS Class 1 drug with high solubility and high permeability, polymorphism is unlikely to impact on drug product performance, and therefore specification is unnecessary.

In conclusion, regulators worldwide recognize the potential impact of polymorphism in drug quality, and demand drug manufacturers to control polymorphism throughout the product's life cycle. The burden of polymorphism control may be lessened if the drug substance is of BCS Class 1 with high solubility and high permeability. The burden of polymorphism control may also be lessened if a thermodynamically stable polymorph is utilized in the manufacture of drug product, since this is unlikely to change to a less-stable polymorph.

6.5
Thermodynamically Stable Polymorphs

The advantages of using a thermodynamically stable polymorph to lessen regulatory burden and to avoid the potential of disruptive product supply – both of which could cost tens, if not hundreds of millions, of dollars were discussed in the last two sections. The approach to create a thermodynamically stable polymorph will now be discussed.

A polymorph is termed thermodynamically stable if the potential for conversion to another polymorph is negligible. It is because a thermodynamically stable polymorph is at its lowest energy state among all polymorphs that polymorphic conversion will not occur during manufacture or storage. Consequently, the product quality may be maintained throughout the product life cycle. The determination of a thermodynamically stable polymorph can be achieved by comparing equilibrium solubility among different polymorphs, with the polymorph having the lowest solubility generally being the most stable form. However, pure forms are sometimes not available for solubility comparison, or the solubility is too low to be determined accurately among different forms. Moreover, polymorphic conversion sometimes occurs so rapidly that solubilities determined among different polymorphs are identical as all are converted to the thermodynamically stable form instantaneously. In these cases, a solubility comparison is of no use and a slurry test should be employed instead.

6.5.1 Slurry Tests

A slurry test uses a mixture of two polymorphs slurried in a solvent. The more-soluble form dissolves faster than the less-soluble form, and the dissolved molecules deposit back to the crystalline surfaces of two polymorphs at rates that are proportional to the remaining surface area; this leads to the disappearance of the more-soluble form and the appearance of the less-soluble form. After equilibration for a period of time, the remaining solid is isolated by filtration, followed by washing and drying. Powder XRD patterns of the isolated solid are determined to identify the thermodynamically stable polymorph. The slurry test is typically performed using an organic solvent that is employed in the chemical manufacture process. The identified stable polymorph must then be tested using aqueous vehicles employed in animal testing in order to ensure that no additional form changes have occurred, such as conversion to the hydrated forms.

6.5.2 Polymorphic Conversion

Polymorphic conversion in the solid state is different from that in slurry. The polymorph pairs that convert in the solid state are referred to as enantiotropic, while the polymorph pairs that do not convert are referred to as monotropic. Thermal analysis with a mixture of two polymorphs is the best way to discern if there is conversion below the melting points. In the case when polymorph conversion is seen around the melting point, it is difficult to discern whether the material melts first, or converts first.

The construction of Van't Hoff equations for the polymorph pairs is another good method for thermodynamically stable polymorph determination. In Fig. 7, lines A and B can be seen to intersect at the transition temperature where A converts to B. Form A is the thermodynamically stable form below the transition temperature, whereas Form B is the thermodynamically stable form above the transition temperature. Forms A and B are enantiotropic. Line C is parallel to line A. Forms A and C are monotropic with no polymorph conversion, but in a slurry Form C can convert to the more stable Form A, which is at a lower energy state.

6.6 BCS-Based Polymorph Salt Form Screening Strategy

Although polymorph/salt screening should ideally be performed to select the optimum solid form upon selection of the clinical lead compound prior to animal pharmacokinetic studies, these screening studies can be costly and time-consuming. Salt selection is typically done to improve the physical chemical and biopharmaceutical properties of the free form. The use of salt form could originate at an early stage, from Discovery or from Development upon nomination of clinical leads. Discovery chemists, when facing issues such as amorphous free form and lack of oral efficacy *in vivo*, often try to generate more-soluble or crystalline salt forms to overcome these hurdles. Development scientists, when evaluating lead compounds from Discovery, may find deficiencies in the free form in terms of hygroscopicity, stability, flowability, bioavailability, and manufacturability. Again, conversion to salt

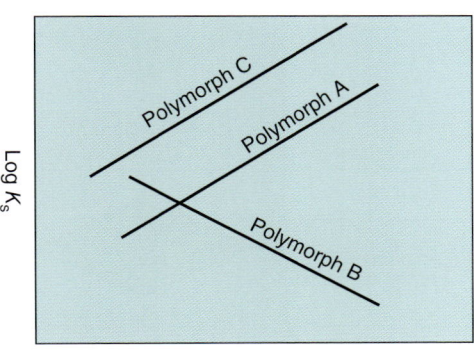

Fig. 7 Van't Hoff plot of enantiotropic (A and B) and monotropic (A and C) pairs.

forms may improve the developability of the candidate compound going forward. In all, salt screening is typically carried out early at the Discovery–Development interface, based on regulatory requirement to use the same salt form across preclinical and clinical studies.

Unlike salt screening, which is done in early development, polymorph screening is typically performed late in the development cycle. However, this could create problems, particularly for insoluble compounds such as ritonavir, for which a polymorphic conversion to a less-soluble form led to a lower bioavailability. It is not unusual that polymorphism issues have delayed either toxicological or clinical programs in the industry. With the shortened development cycle, some companies have recognized the need to move polymorph screening into early development next to salt screening. In this way, a consistent pharmacokinetic profile may be obtained from one study to another, from the outset. The overriding goal is to select a thermodynamically stable, formulatable and bioavailable polymorph/salt form right from the start, prior to significant animal pharmacokinetic and toxicology studies.

Since preliminary BCS classification with solubility and permeability data are available via HTS or pharmaceutical profiling [52], BCS can be used to rationalize the timing of polymorph/salt screening studies [66]. A decision tree is shown in Fig. 8 as an efficient way to prioritize salt and polymorph screening studies prior to conducting animal studies. For BCS 1 compounds, polymorphism or salt form is unlikely to impact on bioavailability and can be dealt with after animal studies. However, a salt form may still be of value if the physical properties or stability is improved for a BCS 1 compound. A typical scenario is to convert an oily free base to a free-flowing powdered salt. On the other hand, for non-BCS 1 compounds, the use of a salt to increase solubility/bioavailability

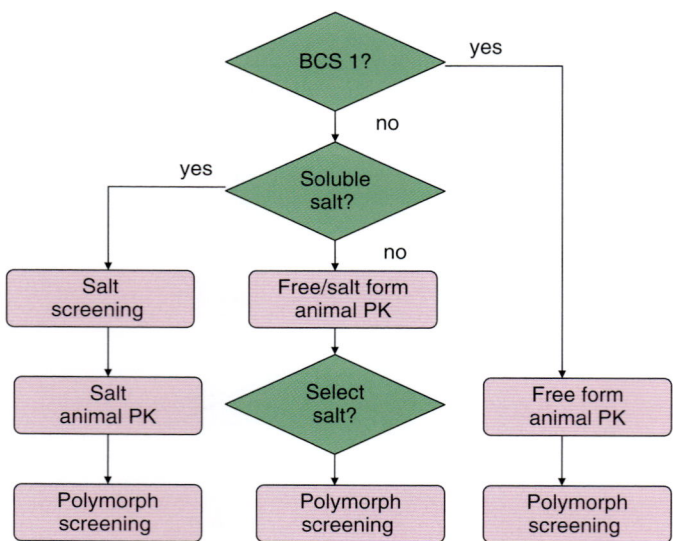

Fig. 8 Polymorph salt decision tree.

should be considered at an early stage. In Fig. 8, "Soluble Salt" is defined as a salt form that enables the human dose soluble in 250 ml water. Animal pharmacokinetic studies may be started prior to polymorph screening if a soluble salt is discovered that provides improved bioavailability. However, should the salt be still not soluble, the value of the salt should be gaged in consideration of the additional cost in the salt conversion step.

7
Biopharmaceutics Application in Drug Product Design

The BCS classification is useful not only in the rationalization of salt/polymorph selection strategy but also in the design of animal and human formulations. Since 2006, the à priori selection of a suitable formulation platform has been proposed based on the BCS class of the drug candidate [48]. A decision tree in the selection of solubilization technique for animal formulations was published in 2008 [67]. A BCS-based right-first-time formulation approach for first-in-human (FIH) studies was then described in 2010 [40]. The objective is to use the correct formulation approaches to reduce human pharmacokinetic variability in clinical Phase I studies. The right-first-time approach will then reduce the project cycle time in order to reach clinical proof-of-concept in an expedient manner.

7.1
Animal Formulation

Animal formulation development is often more challenging than human formulation development, due to the higher dose range required to explore toxicity when establishing adequate animal safety margins prior to human studies. It is not uncommon to dose animals up to 1000 mg kg^{-1} for dogs or 2000 mg kg^{-1} for rats, whereas human doses are rarely as high as 1000 mg per person, or 14 mg kg^{-1} body weight for a 70 kg person. The hurdle to solubilize the high animal doses in a small volume often requires the use of vehicles containing organic solvents that can elicit toxicity by themselves.

A BCS-based animal formulation development decision tree is shown in Fig. 9. For a Class 1 compound with high solubility and high permeability, the preferred approach is to use the API in a capsule for dogs, and in a simple solution or suspension for mice, rats, and monkeys. For a BCS Class 3 compound with high solubility but low permeability, absorption enhancers may be considered for human formulations, but perhaps not for animal formulation. This is because, for a dose of 1000 mg kg^{-1} body weight, a modest 1 : 2 drug : excipient ratio will require a dose of the excipient at 2000 mg kg^{-1} body weight, an amount which is often too high for the excipient to be safe without eliciting an excipient-related-toxicity. Placing the API in a capsule, in a simple solution, or in suspension remains the choice of toxicological formulations for BCS Class 3 compounds.

For BCS Class 2 compounds with low solubility but high permeability, a wide range of formulation choices exists (Fig. 9). First, the API should be micronized to increase its surface area in order to speed up dissolution. Second, surfactants or solubilizers should be added to improve wetting and dissolution. In cases where the compound is ionizable, a pH modifier (acid or base) may be used to enhance solubility, although the use of pH modifiers should be evaluated against in-gut precipitation due to the GI pH gradient. For instance,

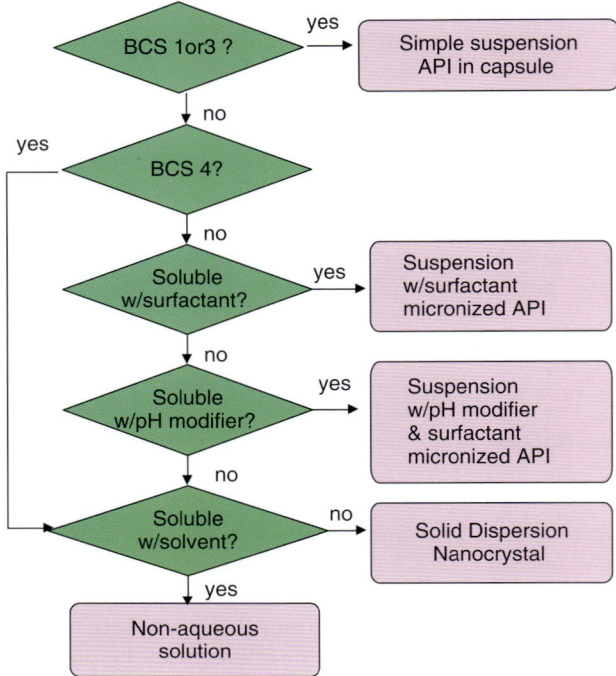

Fig. 9 Animal formulation development decision tree.

a basic compound with pK_a around 7 can be solubilized 1000-fold at pH 4 with the addition of citric acid. However, if the compound is dissolved initially in the citric acid vehicle it may precipitate out at pH 7 in the lower GI tract. This phenomenon often accompanies incomplete absorption, with a truncated absorption phase of short T_{max} (less than 1 h after dosing at fasted state). The addition of surfactants may inhibit nucleation and slow down in-gut precipitation, provided that the surfactant travels together with the drug down the GI tract. A combined use of pH modifiers and surfactants may be more effective in the enhancement of absorption than a pH modifier alone. The same scenario applies to acid salts of weak bases that precipitate as the free base at the lower GI tract pH of 7.5. A surfactant can also help to keep the salt from converting to the insoluble free base.

In some cases, animal exposure from suspensions containing a surfactant/pH modifier is still too low for toxicology studies, a high–energy solid or a non-aqueous solution may be considered. A high-energy solid can be manufactured either via dispersion in a polymer matrix as a solid dispersion, or by extensive grinding to nanosized particles. The amorphous-like state of nanoparticles has an inherent higher energy, with faster dissolution. Alternatively, if a compound is very soluble in pharmaceutical solvents such as polyols, glycerides and phospholipids, a nonaqueous solution may be considered. Again, a high drug loading in an animal vehicle that limits the dose of the excipient is critical for the vehicle not to interfere with toxicology testing. For BCS Class 3 and 4 compounds with low permeability, the high driving force of the drug in solution formulations offers an added advantage in overcoming the low permeability barrier. The iteration of formulation and animal testing may take considerable time. Collaboration with toxicologists is necessary to ensure that the placebo vehicle does not produce adverse effects by itself.

7.2
Human Formulation

The current trend in the pharmaceutical industry follows two approaches for FIH formulations: (i) active-only formulations, for example, powder-in-bottle or powder-in-capsule; or (ii) bona fide formulations that have been optimized for pharmacokinetic performance [40]. There are advantages and disadvantages of both approaches. While the active-only approach can move quickly into Phase I clinical trial, it may not perform and thus require reformulation and repeated testing in humans. Since it cannot be manufactured in scale and is not feasible beyond Phase I, the Phase II clinical study will inevitably be delayed. The development of a scaleable clinical Phase II formulation would need to start as soon as Phase I begins. A bridging pharmacokinetic study in humans may also be necessary to ensure the consistent safety and efficacy with the new formulation. For compounds with poor solubility and/or poor permeability, bioequivalence between the above two types of formulation may be disparate, rendering the adjustment of dose levels necessary. A bona fide Phase I formulation approach that provides a scalable Phase II-enabling formulation can reduce the transition time between clinical phases and shorten the overall drug development cycle.

Human formulation has different requirements than animal formulation. A solid formulation (capsule/tablet), rather than a bottle of solution/suspension, is essential for patient acceptability except in certain life-saving conditions. The amount of excipient used in a solid oral formulation is very limited (<1 g), and therefore excipient toxicity is rarely seen in humans. Thus, the choice of excipients for human formulations is wider than those for animal formulations. However, the low weight of a tablet or the small volume of a capsule requires very high drug loadings. The solubilized formulation is often not feasible for high-dose/low-potency compounds. The formulation platforms recommended for each of the four BCS classes are presented in Table 4. A set of four decision trees for the four BCS classes is presented in Fig. 10; these decision trees take into consideration bulk drug density/flowability and solution stability, in addition to solubility and permeability. In particular, criteria for the use of permeation enhancers for low-permeability compounds or a decision to pursue a non-oral route of administration is discussed in more detail in the original report [48].

7.3
BCS Class and Food Effect

Literature on the effect of food on common medicines was extensively reviewed by Leopold, who divided the drugs into four categories with reduced, delayed, unaffected, and increased drug absorption [68]. By comparing the solubility of these drugs, a trend can be discerned in that water-soluble compounds tend to have reduced bioavailability with food, whereas

Tab. 4 BCS as a guide to formulation strategy.

BCS Class	Solubility	Permeability	Formulation strategy
1	High	High	Conventional capsule or tablet
2	Low	High	Micronized API and surfactant
			Nanoparticle technology
			Solid dispersion
			Melt granulation/extrusion
			Liquid or semisolid filled capsule
			Coating technology
3	High	Low	Conventional capsule or tablet
			Absorption enhancers
4	Low	Low	Combination of BCS Class 2 and absorption enhancers

water-insoluble compounds tend to have increased bioavailability with food [10]. As Nature's way to promote the absorption of nutrients, food intake tends to slow down GI motility, increase blood flow, and promote the secretion of gastric acid and enzymes. These changes benefit compounds with low solubility and/or low permeability in the enhancement of absorption. On the other hand, food dilutes the concentration of those compounds that are well absorbed and therefore slow down absorption. Similar trends have been observed by the present author, in that food tends to have little effect on BCS Class 1 compounds, delays or reduces the bioavailability of BCS Class 3 compounds, but improves the bioavailability of BCS Class 2 and 4 compounds [69]. Benet et al. have discussed extensively the possible change of bioavailability or delay of C_{max} for compounds in each of the four BDDCS classes [7]. Benet's group reported an inhibition of efflux and influx pumps by high-fat meals, resulting in a no-food effect for the BCS Class 3 compound, acyclovir, a substrate for both pumps.

Moreover, drugs with low and pH-dependent solubility tend to have a greater food effect due to increased gastric secretion and prolonged gastric emptying. Food initially raises the stomach pH from 1–2 to 5–7 after which, with increased gastric secretion, the stomach pH is lowered and an emptying of the acidic food lowers the duodenal pH. An understanding of this complex interplay between gastric emptying and pH changes is critical to the prediction of food effect.

The US FDA and most other regulatory agencies require or recommend an evaluation of the effect of food intake on the bioavailability of controlled-release dosage forms. In the 2002 Guidance for Industry [70], the FDA states "We recommend that food-effect BA and fed BE studies be performed for all modified-release dosage forms." In the same Guidance for Industry, the FDA further states "We recommend that a food-effect BA study be conducted for all new chemical entities (NCEs) during the Investigational New Drug (IND) period." These statements highlight the importance of delineating the food effect in early drug development. It would be advantageous if the possible food effect can be predicted based on the physical chemical and biopharmaceutical properties of the drug substance so that, for molecules predicted to have a food effect, a

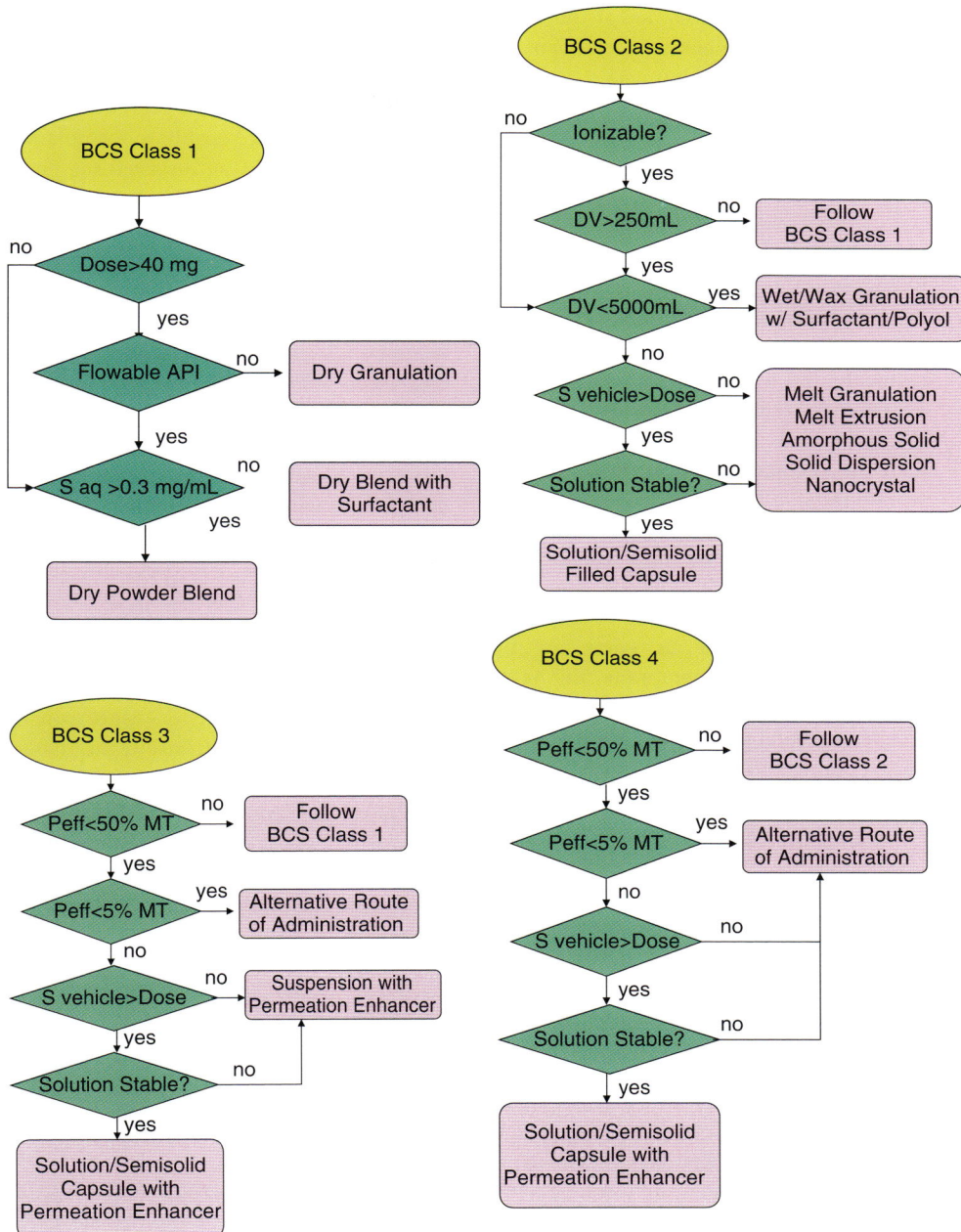

Fig. 10 Formulation decision tree for compounds from the four BCS classes.

formulation may be designed at the outset of clinical development to circumvent the potential food effect in humans.

8
Conclusions

In conclusion, physical pharmacy and biopharmaceutics are important disciplines for the pharmaceutical industry. The use of pharmaceutical profiling technology to delineate the solubility and permeability of clinical lead candidates early in the discovery and development interface will facilitate the identification of developable compounds. With the advent of HTS and the quest for nanomolar potency in IC_{50} or EC_{50}, increasing numbers of lipophilic and hydrophobic compounds are being discovered. These pipeline compounds tend to be insoluble in water, are not permeable to the GI membrane, and have a large molecular size with a high molecular weight exceeding 500 Da. Therefore, an early characterization of physical pharmacy and biopharmaceutical properties is essential for decision making in the declaration of new clinical candidates. Should the molecule exhibit less than desirable pharmaceutical properties, an early intervention in either chemical modification or enabling formulation technology can make a difference in the eventual success for commercialization.

When a simple formulation with traditional excipients fails to elicit biological potency and/or sufficient toxicology exposure for a healthy safety margin, the use of sophisticated formulations with nonconventional excipients is often considered ahead of chemical modification. These enabling formulation technologies have thus gained popularity during recent years. However, a broad access of these enabling formulation technologies may mislead discovery scientists in the selection of even less-soluble/less-permeable candidates for development. The present author believes that these enabling formulation technologies should be reserved for First-in-Class or Best-in-Class compounds with clear therapeutic advantages, and should only be used when lead optimization fails to produce compounds with desirable biopharmaceutical properties. Because of the technology challenges and the added cost often associated with these formulations, compounds with this liability often take longer and more cost to develop, and are at a disadvantage in competing with compounds without this liability.

Once entered into development, the human formulation is nothing but a continuum from the animal discovery and toxicology formulations. It should be designed based on BCS principles considering human physiology. The right-first-time formulation is essential to ensure consistence in the clinical database from Phase I to Phase III to commercial. The reduction in bioequivalence testing due to formulation changes will reduce R&D cycle time. All of these factors will help to speed up new drug development through quality by design for pharmaceutical products.

References

1. Martin, A. (2011) *Physical Pharmacy*, 64th edn. Lea & Febiger.
2. Editorial (1973) Pharmacokinetics and biopharmaceutics: a definition of terms. *J. Pharmacokinet. Biopharm.*, **1** (1), 3.
3. Derendorfand, H., Meibohm, B. (1999) Modeling of pharmacokinetic/pharmacodynamic (PK/PD) relationships: concepts and perspectives. *Pharm. Res.*, **16** (2), 176–185.
4. Skelly, J.P. (2010) A history of biopharmaceutics in the food and drug administration 1968–1993. *AAPS J.*, **12** (1), 44–50.

5. Avdeef, A. (2001) High-Throughput Measurements of Solubility Profiles, in: Testa, B., van de Waterbeemd, H., Folkers, G., Guy, R. (Eds) *Pharmacokinetic Optimization in Drug Research: Biological, Physiological, and Computational Strategies*, Verlag Helvetica Chimica Acta, Zurich, pp. 305–326.
6. Pan, L., Ho, Q., Tsutsui, K., Takahashi, L. (2001) Comparison of chromatographic and spectroscopic methods used to rank compounds for aqueous solubility. *J. Pharm. Sci.*, **90**, 521–529.
7. Wu, C.Y., Benet, L.Z. (2005) Predicting drug disposition via application of BCS: transport/absorption/elimination interplay and development of a biopharmaceutics drug disposition classification system. *Pharm. Res.*, **22**, 11–23.
8. Leo, A.J. (1993) Calculating log P_{oct} from structures. *Chem. Rev.*, **93**, 1281–1306.
9. Dunn, W.J., Block, J.H., Pearlman, R.S. (1986) *Partition Coefficient: Determination and Estimation*. Pergamon Press, Oxford.
10. Ku, M.S. (2010) Preformulation Consideration for Drugs in Oral CR Formulation, in: Wen, H., Park, K. (Eds) *Oral Controlled Release Formulation Design and Drug Delivery: Theory to Practice*, John Wiley & Sons, Ltd, Chichester.
11. Farag, S.I., Munir, B., Hussain, A. (2007) Minireview, microenvironmental pH modulation in solid dosage forms. *J. Pharm. Sci.*, **96**, 948–959.
12. Dressman, J.B., Amidon, G.L., Reppas, C., Shah, V. et al. (1998) Dissolution testing as a prognostic tool for oral drug absorption: immediate release dosage forms. *Pharm. Res.*, **15** (1), 11–22.
13. Charman, W.N., Porter, C.J.H., Mithani, A., Dressman, J.B. (1997) Physicochemical and physiological mechanisms for the effects of food on drug absorption: the role of lipids and pH. *J. Pharm. Sci.*, **86** (3), 269–282.
14. Galia, E., Nicolaides, E., Dressman, J.B. (1998) Evaluation of various dissolution media for predicting in vivo performance of class I and II drugs. *Pharm. Res.*, **15** (5), 698–705.
15. Jones, H.M., Parrott, N., Ohlenbusch, G., Lave, T. (2006) Predicting pharmacokinetic food effects using biorelevant solubility media and physiologically based modelling. *Clin. Pharmacokinet.*, **45**, 1–15.
16. Brayden, D.J., Jepson, M.A., Baird, A.W. (2005) Keynote review: intestinal Peyer's patch M cells and oral vaccine targeting. *Drug Discovery Today*, **10**, 1145–1157.
17. Hogben, C.A.M., Schanker, L.S., Tocco, D.J., Brodie, B.B. (1957) Absorption of drugs from the stomach II. The human. *J. Pharmacol. Exp. Ther.*, **120**, 540–554.
18. Stella, V.J., Charman, W.NA., Naringrekar, V.H. (1985) Prodrugs. Do they have advantages in clinical practice? *Drugs*, **2**, 455–473.
19. Rohde, P., Selzer, B., Ertl, P. (2000) Fast calculation of molecular polar surface area as a sum of fragment based contributions and its application to the prediction of drug transport properties. *J. Med. Chem.*, **43**, 3714–3717.
20. Mizuno, N., Niwa, T., Yotsumoto, Y., Sugiyama, Y. (2003) Impact of drug transporter studies on drug discovery and development. *Pharmacol. Rev.*, **55** (3) 425–461.
21. Greiner, B., Eichelbaum, M., Fritz, P., Kreichigauer, H.P., von Richter, O., Zundler, J., Kroemer, H.K. (1999) The role of intestinal P-glycoprotein in the interaction of digoxin and rifampin. *J. Clin. Invest.*, **104**, 147–153.
22. Masuda, S., Uemoto, S., Hashida, T., Inomata, Y., Tanaka, K., Inui, K. (2000) Effect of intestinal P-glycoprotein on daily tacrolimus trough level in a living-donor small bowel recipient. *Clin. Pharmacol. Ther.*, **68**, 93–103.
23. Jonker, J.W., Smit, J.W., Brinkhuis, R.F., Maliepaaed, M., Beijnen, J.H., Schellens, J.H.M., Schinkel, A.H. (2000) Role of breast cancer resistance protein in the bioavailability and fetal penetration of topotecan. *J. Natl Cancer Inst.*, **92**, 1651–1656.
24. Cummins, C.L., Salphati, L., Reid, M.J., Benet, L.Z. (2003) In vivo modulation of intestinal CYP3A metabolism by P-glycoprotein: studies using the rat single-pass intestinal perfusion model. *J. Pharmacol. Exp. Ther.*, **305**, 306–314.
25. Geick, A., Eichelbaum, M., Burk, O. (2001) Nuclear receptor response elements mediate induction of intestinal MDR1 by rifampin. *J. Biol. Chem.*, **276**, 14581–14587.
26. Dussault, I., Lin, M., Hollister, K., Wang, E.H., Synold, T.W., Forman, B.M, (2001) Peptide mimetic HIV protease inhibitors are

ligands for the orphan receptor SXR. *J. Biol. Chem.*, **276**, 33309–33312.
27. Huang, S.M., Lesko, L.J. (2004) Drug–drug, drug–dietary supplement and drug–citrus fruit and other food interactions: what have we learned? *J. Clin. Pharmacol.*, **44**, 559–569.
28. Amidon, G.L., Lennernas, H., Shah, V.P., Crison, J.R. (1995) A theoretical basis for a biopharmaceutics drug classification: the correlation of in vitro drug product dissolution and *in vivo* bioavailability. *Pharm. Res.*, **12**, 413–420.
29. US FDA (1995) Guidance for Industry: Immediate Release Solid Oral Dosage Forms Scale-Up and Post Approval Changes: Chemistry Manufacturing and Controls, In Vitro Dissolution Testing and *In Vivo* Bioequivalence Documentation.
30. US FDA (1997) Guidance for Industry: Extended Release Oral Dosage Forms: Development Evaluation and Application of *In Vitro/In Vivo* Correlations.
31. US FDA (2000) Guidance for industry. Waiver of In Vivo Bioavailability and Bioequivalence Studies for Immediate-Release Solid Oral dosage Forms Based on a Biopharmaceutics Classification System.
32. World Health Organization (2006) WHO Expert Committee on Specifications For Pharmaceutical Preparations. Technical Report Series 937.
33. CPMP Committee for Proprietary Medicinal Products (2001) Note for Guidance on the Investigation of Bioavailability and Bioequivalence.
34. Blume, H.H., Schug, B.S. (1999) The biopharmaceutics classification system (BCS): class III drugs-better candidates for BA/BE waiver? *Eur. J. Pharm. Sci.*, **9**, 117–121.
35. Torrado, J.J., Espada, R., Ballesteros, M.P., Torrado-Santiago, S. (2008) Amphotericin B formulations and drug targeting. *J. Pharm. Sci.*, **97**, 2405–2425.
36. aFogh, A.J., Trempe, G. (1975) New human tumor cell lines, in: Fogh, J. (Ed.) *Human Tumor Cells In Vitro*, Plenum Publishing, New York, pp. 115–141.
37. bHidalgo, I.J., Raub, T.J., Borchardt, R.T. (1989) Characterization of the human colon carcinoma cell line (Caco-2) as a model system for intestinal epithelial permeability. *Gastroenterology*, **96** (3), 736419.
38. Artursson, P., Karlsson, J. (1991) Correlation between oral drug absorption in humans and apparent drug permeability coefficients in human intestinal epithelial (Caco-2) cells. *Biochem. Biophys. Res. Commun.*, **175** (3), 880–885.
39. Artursson, P., Palm, K., Luthman, K. (2001) Caco-2 monolayers in experimental and theoretical predictions of drug transport. *Adv. Drug Delivery Rev.*, **46** (1-3), 27–43.
40. Ku, M.S., Dulin, W. (2010) A biopharmaceutical classification-based right-first-time formulation approach to reduce human pharmacokinetic variability and project cycle time from first-in-human to clinical proof-of-concept. *Pharm. Dev. Technol.*, **11**, 1–18.
41. Doluisio, J.T., Billups, N.F., Dittert, L.W., Sugita, E.T., Swintosky, J.V. (1969) Drug absorption. I. An in situ rat gut technique yielding realistic absorption rates. *J. Pharm. Sci.*, **58**, 1196–1200.
42. Fagerholm, U., Johansson, M., Lennernäs, H. (1996) Comparison between permeability coefficients in rat and human jejunum. *Pharm. Res.*, **13**, 1336–1342.
43. Lennernäs, H. (1998) Human intestinal permeability. *J. Pharm. Sci.*, **87** (4), 403–410.
44. Wilding, I., Hirst, P., Connor, A. (2000) Development of a new engineering-based capsule for human drug absorption studies. *Pharm. Sci. Technol. Today*, **3** (11), 385–392.
45. Kasim, N.A., Whitehouse, M., Ramachandran, C., Bermejo, M., Lennernäs, H., Hussain, A.S., Junginger, H.E., Stavchansky, S.A., Midha, K.K., Shah, V.P., Amidon, G.L. (2004) Molecular properties of WHO essential drugs and provisional biopharmaceutical classification. *Mol. Pharm.*, **1**, 85–96.
46. Takagi, T., Ramachandran, C., Bermejo, M., Yamashita, S., Yu, L.X., Amidon, G.L. (2006) A provisional biopharmaceutical classification of the top 200 oral drug products in the United States, Great Britain, Spain, and Japan. *Mol. Pharm.*, **3**, 631–643.
47. Poulin, P., Jones, H.M., Jones, R.D.O., Yates, J.W.T., Gibson, R., Chien, J.Y., Ring, B.J., Adkison, K.K., He, H., Vuppugalla, R., Marathe, P., Fischer, V., Dutta, S., Sinha, V.K., Björnsson, T., Lavé, T., Ku, M.S. (2011) PhRMA CPCDC initiative on predictive models of human pharmacokinetics. 1. Goals, properties of the PhRMA dataset and

comparison with literature datasets. *J. Pharm. Sci.*, **100** (10), 4050–4073.

48. Ku, M.S. (2006) An Oral Formulation Decision Tree Based on the Biopharmaceutical Classification System for First in Human Clinical Trials in: L.Z. Benet (Ed.) *Bulletin Technique Gattefosse 99*, Gattefosse, Saint-Priest Cedex France, pp. 89–102.
49. Lindenberg, M., Kopp, S., Dressman, J.B. (2004) Classification of orally administered drugs on the world health organization model of essential medicines according to the biopharmaceutics classification system. *Eur. J. Pharm. Biopharm.*, **58**, 265–278.
50. Baldoni, J. (2007) BCS Experience in GSK. AAPS/BE BCS & Beyond. Conference Presentation.
51. Benet, L.Z., Broccatelli, F., Oprea, T.I. (2011) BDDCS applied to over 900 drugs. *AAPS J.*, **13**, 519–547.
52. Kerns, E.H., Di, L. (2002) Multivariate pharmaceutical profiling for drug discovery. *Curr. Top. Med. Chem.*, **2**, 87–98.
53. Kola, I., Landis, J. (2004) Can the pharmaceutical industry reduce attrition rates? *Nat. Rev. Drug Discovery*, **3** (8), 711–715.
54. Sorger, P.K., Allerheiligen, S.R.B. (2011) NIH White Paper: New Approaches to Discovering Drugs and Understanding Therapeutic Mechanisms, in: *QSP Workshop Group Quantitative and Systems Pharmacology in the Post-genomic Era*, NIH, Bethesda, Maryland.
55. Projan, S.J. (2010) Francis Tally and the discovery and development of tigecycline: a personal reminiscence. *Clin. Infect. Dis. (US)*, **50** (Suppl. 1), S24–S25.
56. Mitscher, L.A. (1978) *The Chemistry of The Tetracycline Antibiotics*. Marcel Dekker, New York.
57. Chow, D., Morton, G., Ku, M.S. (1994) pK_a values of tetracyclines from H-NMR. *Pharm. Res.*, **11** (10), S270.
58. Bighley, L.D., Berge, S.M., Monkhouse, D.C. (1995) Salt Forms of Drugs and Absorption, in: Swarbrick, J., Boylan, J.C. (Eds) *Encyclopedia of Pharmaceutical Technology*, Vol. 13, Marcel Dekker, New York, pp. 453–499.
59. Yoshioka, S., Stella, V.J. (2000) *Stability of Drugs and Dosage Forms*, Kluwer Academic/Plenum Publishers, New York.
60. Guillory, J.K., Poust, R.I. (2002) Chemical Kinetics and Drug Stability, in: Banker, G.S., Rhodes, C.T. (Eds), *Modern Pharmaceutics*, 4th edn, Drugs and the Pharmaceutical Sciences, Vol. 121, Marcel Dekker, New York, pp. 139–166.
61. Chemburkar, S.R., Bauer, J., Deming, K., Spiwek, H., Patel, K., Morris, J., Henry, R., Spanton, S., Dziki, W., Porter, W., Quick, J., Bauer, P., Donaubauer, J., Narayanan, B.A., Soldani, M., McFarland, D., McFarland, K. (2000) Dealing with the impact of ritonavir polymorphs on the late stages of bulk drug process development, *Org. Process Res. Dev.*, **4**, 413–417.
62. Morissette, S.L., Almarsson, Ö., Peterson, M.L., Remenar, J.F., Read, M.J., Lemmo, A.V., Ellis, S., Cima, M.J., Gardner, C.R. (2004) High-throughput crystallization: polymorphs, salts, co-crystals and solvates of pharmaceutical solids. *Adv. Drug Delivery Rev.*, **56**, 275–300.
63. Almarsson, O., Gardner, C.R. (2003) Novel approaches to issues of developability. *Curr. Drug Discov.*, 3, 21–26
64. US FDA (2007) *Guidance for Industry, ANDAs: Pharmaceutical Solid Polymorphism*, Food and Drug Administration, Rockville, MD.
65. ICH International Conference on Harmonization (1999) Q6A Guideline: Specifications for New Drug substances and Products: Chemical Substances.
66. Ku, M.S. (2010) Salt and polymorph selection strategy based on the biopharmaceutical classification system for early pharmaceutical development. *Am. Pharm. Rev.*, **13**, 22–30.
67. Ku, M.S. (2008) Use of the biopharmaceutical classification system in early drug development. *AAPS J.*, **10**, 208–212.
68. Leopold, G. (1984) Experimental Factors Influencing the Results of Drug Product Bioavailability/Bioequivalency Studies in Humans, in: Smolen, V.F., Ball, L.A. (Eds) *Controlled Drug Bioavailability: Bioavailability Methodology and Regulation*, Vol. 2, John Wiley & Sons, Ltd, Chichester, pp. 7–9.
69. Ku, M.S. (2009) Presentation "Food Effect and Poorly Soluble Compounds: Can We Predict It? Can We Overcome It?" Eastern Pharmaceutical Technology Meeting.
70. US FDA (2002) Guidance for Industry Food-Effect Bioavailability and Fed Bioequivalence Studies.
71. Schwartz, M.A., Buckwalter, F.H. (1962) Pharmaceutics of penicillin. *J. Pharm. Sci.*, **51**, 1119–1127.

22
Prions

Vincent Béringue
INRA (Institut National de la Recherche Agronomique), UR892 Virologie Immunologie Moléculaires, Domaine de Vilvert, F-78350 Jouy-en-Josas, France

1	**Prions** 775	
2	**Prions Are Distinct from Viruses** 776	
3	**Disease Paradigms** 778	
4	**Nomenclature** 778	
5	**Discovery of the Prion Protein** 779	
6	**PrP Gene Structure and Organization** 780	
6.1	The Prion Protein Family 780	
6.2	The PRNP Gene 781	
6.3	Expression of the PrP Gene 781	
7	**Cellular form of PrP: Primary Structure, Expression, and Putative Function(s)** 781	
7.1	PrP Primary Structure and Species Variations 781	
7.2	Metabolism, Expression, and Distribution of PrP^C 783	
7.3	PrP^C Putative Function(s) 784	
7.3.1	PrP-Deficient Mice 784	
7.3.2	PrP Partners 784	
7.3.3	PrP and Copper 784	
7.3.4	PrP and Neuroprotection 785	
7.3.5	PrP and Stemness 785	
7.3.6	PrP Involvement in Other Pathologies 785	
8	**Structures of PrP Isoforms** 786	
8.1	PrP Primary to Quaternary Structure 786	
8.2	Topological Models of PrP Isoforms 788	

Translational Medicine: Molecular Pharmacology and Drug Discovery
First Edition. Edited by Robert A. Meyers.
© 2018 Wiley-VCH Verlag GmbH & Co. KGaA. Published 2018 by Wiley-VCH Verlag GmbH & Co. KGaA.

8.2.1	3-D Structure of Recombinant Monomeric PrP 788
8.2.2	PrP^{Sc} Structural Models 789
8.2.3	PrP Domain(s) Involved in the Conversion 790

9	**Prion Replication 791**

10	**Cell Biology of PrP^{Sc} Formation 792**

11	**Prion Peripheral Pathogenesis 793**

12	**Prion Toxicity 794**

13	**Strains of Prions 794**

14	**Interplay between the Species and Strains of Prions 796**

15	**Human Prion Diseases 796**
15.1	Sporadic Human Prion Diseases 796
15.2	Heritable Human Prion Diseases 798
15.3	Infectious Human Prion Diseases 799
15.3.1	Human Growth Hormone 799
15.3.2	Variant Creutzfeldt–Jakob Disease (vCJD) 799

16	**Prion Diseases of Animals 800**
16.1	Scrapie in Sheep and Goats 800
16.2	Chronic Wasting Disease 801
16.3	Bovine Spongiform Encephalopathy 802
16.4	Transmission of BSE Prions to Humans, and the Zoonotic Potential of Non-BSE Animal Prions 802

17	**Fungal Prions 804**

18	**Prevention and Therapeutics for Prion Diseases 804**

19	**Implications for Common Neurodegenerative Diseases 805**

References 805

Keywords

Prion
An infectious agent primarily composed of a protein in a misfolded form

Infectious pathogen
A microorganism that causes diseases in the infected host

Neurodegeneration
A process leading to a loss of structure or function of neurons

Protein
A biological molecule consisting of long chains of amino acid residues

Protein aggregation
A biological process by which abnormally folded proteins aggregate

Amyloid
Fibrous, abnormally folded protein aggregates that are associated with pathologies referred to as amyloidopathies

Transgenic mouse
A genetically engineered mouse commonly used for research purposes

Prions are infectious proteins that have been described in both mammals and fungi. That prions are composed solely of proteins makes them unprecedented infectious pathogens. Prions are composed of proteins that can misfold and are prone to aggregation, and multiply by forcing the precursor protein to acquire the corrupted conformation. Different conformations of proteins in the prion state encipher distinct strains with specific patterns of pathology. In mammals, prions accumulate to high levels in the nervous system, where they cause dysfunction and fatal degeneration. Both, mammalian and fungal prions have been produced in cell-free systems. Synthetic prion protein (PrP) peptides and recombinant PrP fragments have been used to form mammalian prions, while N-terminal regions – called prion domains – that are rich in glutamine and asparagine have been used to form fungal prions. Mammalian prions cause a group of invariably fatal, neurodegenerative diseases. Prion diseases may present as genetic, infectious, or sporadic disorders, all of which involve the modification of PrPs. The tertiary and quaternary structures of PrPs are profoundly altered as prions are formed and, as such, prion diseases represent disorders of protein conformation. Prion diseases are present in humans and farmed animals. Cross-species transmission of prions can occur; indeed, dietary exposure of the human population to the prions responsible for bovine spongiform encephalopathy (BSE) have led to the emergence of variant Creutzfeldt–Jakob disease (vCJD). Prions can persist in the environment for years and are extremely difficult to inactivate, thus posing a threat to animal and human health.

1
Prions

In mammals, prions are primarily composed of prion proteins (PrP^{Sc}) (Table 1), an aggregated, misfolded conformer of the ubiquitously expressed, host-encoded prion protein (PrP^{C}). During pathogenesis, prions replicate by recruiting PrP^{C} and stimulating its conversion into PrP^{Sc}. This conversion, presumably through a nucleated polymerization mechanism, involves the refolding of soluble, α-helix-rich PrP^{C} into β-sheet-enriched PrP^{Sc} polymers that

Tab. 1 Glossary of mammalian prion terminology.

Term	Description
Prion	A proteinaceous infectious particle that lacks nucleic acid. Prions are composed largely, if not entirely, of PrP^{Sc} molecules
PrP^{Sc}	Abnormal, pathogenic isoform of the prion protein that causes illness. This protein is the only identifiable macromolecule in purified preparations of prions
PrP^{C}	Cellular isoform of the prion protein
PrP^{res} or PrP 27–30 or rPrP^{Sc}	Truncated PrP^{Sc}, generated by digestion with proteinase K
sPrP^{Sc}	Forms of PrP^{Sc} digested by proteinase K, but not by thermolysin or pronase
PRNP	Human PrP gene located on chromosome 20
Prnp	Mouse PrP gene located on syntenic chromosome 2. Prnp controls the length of the prion incubation time and is congruent with the incubation-time genes Sinc and Prn-i PrP-deficient (Prnp$^{0/0}$) mice are resistant to prion infection
PrP amyloid	Fibril of PrP fragments derived from PrP^{Sc} by proteolysis. Plaques containing PrP amyloid are found in the brains of some mammals with prion disease
Prion rods or scrapie-associated fibrils	An amyloid polymer composed of PrP^{res} molecules. Created by detergent extraction and limited proteolysis of PrP^{Sc} and collected by high-speed-centrifugation

form deposits in prion-infected brains and are responsible for the observed neurodegenerative disorders.

Solution structures of recombinant mammalian PrPs produced in bacteria showed conserved three-dimensional (3D) structures. The N-terminal region of the protein is unstructured, while the C-terminal moiety is a globular domain, with three α-helices and two short β-strands [1–3]. The results of low-resolution structural studies have suggested that the minimal size of the infectious prion is a trimer of PrP^{Sc} molecules (for a review, see Ref. [4]). Usually, these PrP^{C} to PrP^{Sc} conformational changes make PrP^{Sc} resistant to limited proteolysis [5] (Fig. 1). Generally, limited proteolysis truncates the N terminus of PrP^{Sc} to produce PrP 27–30, which consists of the C-terminal ~142 amino acids. PrP 27–30 can polymerize into amyloid fibrils that are indistinguishable from fibrils found in amyloid plaques of the brains of mammals with prion disease [6].

2
Prions Are Distinct from Viruses

The major features that distinguish prions from viruses are: (i) the ability to create prion infectivity by modifying the conformation of a polypeptide devoid of nucleic acid; and (ii) PrP is encoded by a chromosomal gene, designated PRNP in humans and Prnp in mice. The PrP gene is located on the short arm of chromosome 20 in humans and in the syntenic region of chromosome 2 in mice. Prions differ from viruses and viroids in that they lack a nucleic acid genome that directs the

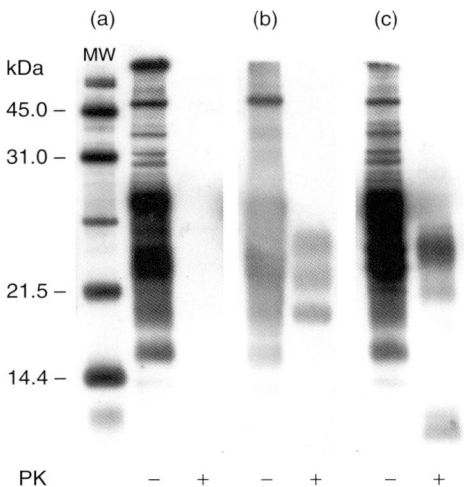

Fig. 1 Prion protein isoforms. Western immunoblot of brain homogenates from (a) uninfected and (b,c) prion-infected transgenic mice overexpressing ovine PrP (tg338 line). These mice were infected with either classical (b) or atypical (c) scrapie. Samples were digested or not with proteinase K (PK) before analysis as indicated. After PK digestion, PrP^C was completely hydrolyzed, whereas PrP^{Sc} was partially digested to generate protease-resistant PrP^{Sc} or PrP^{res}. Note the PrP^{res} different migration pattern between classical and atypical scrapie which is indicative of two different prion strains. After polyacrylamide gel electrophoresis and electrotransfer, the blot was developed with anti-PrP monoclonal antibody Sha31. Molecular weight markers (MW) are depicted in kilodaltons.

synthesis of their progeny. No nucleic acid has been found within the infectious prion particle, despite intensive searches using a wide variety of techniques and approaches [7]. On the basis of a wealth of evidence, it is reasonable to assert that such a nucleic acid has not been found because it does not exist. Prions are composed of an alternative isoform of a cellular protein, whereas most viral proteins are encoded by a viral genome, and viroids are devoid of protein.

In contrast to viruses, prions are poorly immunogenic. They do not elicit an immune response because the host is tolerant to PrP^C, which prevents an immune response to PrP^{Sc} being mounted. In contrast, foreign proteins of viruses that are encoded by the viral genome often elicit a profound immune response.

When prions and viruses are passaged from one host species to another, the consequences can be very different. The crossing of prions from one species to another is restricted by what has been called the "species barrier." The closer the conformational compatibility is between the PrP^{Sc} from the host in which the prions last replicated and the PrP^C of the newly infected animal, the more likely replication will occur in the new host. Strains of prions play a pivotal role in these interactions. For example, variant Creutzfeldt–Jakob disease (vCJD) prions from humans replicate more readily in wild-type mouse than do sporadic Creutzfeldt–Jakob disease (CJD) prions. The issue of prion strains posed a profound conundrum for many years. How could an infectious pathogen composed only of protein encipher biological information? This riddle was solved when prion strains with different physical properties were isolated, and strains were noted to be formed of different self-templating PrP^{Sc} conformers, at the level of the tertiary and quaternary structures.

3
Disease Paradigms

Despite some similarities between prion and viral illnesses, these disorders are very different. Prion diseases are uniformly fatal, with no human or animal ever recovering from a prion disease once neurologic dysfunction is manifest. No host defenses are mounted in response to prion infection: neither are humoral nor cellular immunity elicited to the replicating prion.

While prions can be spread from one host to another, the most common form of prion disease is spontaneous or sporadic. In these illnesses, prions arise endogenously and replicate, and the generated prions are truly infectious, either experimentally or accidentally. The generation of synthetic prions demonstrated that only PrP is required to generate prion infectivity [8]. These findings contrast with those in viruses, where exogenous infection is required, except in the case of latent retroviral genomes. For example, after infection with exogenous HIV the virus may disappear but its RNA genome may be reverse transcribed into DNA and the DNA copies may remain dormant for years.

The dramatically different principles that govern prion biology have often been poorly understood, and this lack of understanding has led to some regrettable decisions of great economic, political – and possibly public health – importance. For example, scrapie and bovine spongiform encephalopathy (BSE) have different names, yet they are the same disease in two different species. Scrapie and BSE differ in only two respects: (i) the PrP sequence in sheep differs from that of cattle at seven or eight positions of 270 amino acids, which results in different PrP^{Sc} molecules; and (ii) some aspects of each disease are determined by the particular prion strain that infects the respective host. Scrapie and BSE are different conformational variants of PrP^{Sc}. Understanding prion strains and the "species barrier" is of paramount importance with respect to the circulation and zoonotic potential of animal prions.

4
Nomenclature

The generic term PrP^{Sc} is usually employed to designate the pathogenic PrP isoform, whatever the affected species and the considered prion strain (Table 1). The "Sc" superscript of PrP^{Sc} was initially derived from the term "scrapie" because scrapie was the prototypic prion disease and because all of the known prion diseases (Table 2) of mammals involve aberrant metabolism of PrP similar to that observed in scrapie. The terms "PrP^{res}" and "PrP-res" were derived to describe the protease resistance of PrP^{Sc}, and have sometimes been used interchangeably with PrP^{Sc}. The use of these terms became particularly problematic with the discovery that a variable, strain-dependent proportion of PrP^{Sc} (designated sPrP^{Sc}) can be protease-sensitive [9–12]. These forms may [12, 13] or may not [14, 15] be infectious, but whether sPrP^{Sc} oligomers are smaller than PrP^{res} with respect to size remains to be established [16–19]. Examples also exist of the presence of infectivity in the absence of PrP^{res} [20] and presence of PrP^{res} [21] in the absence of apparent transmissibility. Protease resistance, insolubility, and a high β-sheet content should be considered only as surrogate markers of PrP^{Sc} infectivity, because not all characteristics may be present.

In mice, *Prnp* is now known to be identical to two genes, *Sinc* and *Prn-i*, that are known to control the length of the incubation time in mice inoculated with prions [22].

For most inherited prion diseases (Table 2), the disease is specified by the respective mutation, such as familial CJD (fCJD)(E200K) and GSS(P101L). In the case of fatal familial insomnia (FFI), a description of the D178N mutation and M129 polymorphism seems unnecessary because this is the only known mutation–polymorphism combination that results in the FFI phenotype. The sporadic form of fatal insomnia is denoted sFI.

5
Discovery of the Prion Protein

The discovery of PrP 27–30 in Syrian hamster brain fractions progressively enriched for scrapie infectivity transformed research on scrapie and related diseases [5, 23, 24]. PrP 27–30 was so named because it has an apparent molecular weight (Mr) of 27–30 kDa (Fig. 1). PrP 27–30 is derived from the larger PrP^{Sc} by N-terminal

Tab. 2 The prion diseases.

Disease	Host	Mechanism of pathogenesis
Kuru	Fore tribe of Papua New Guinea	Infection through ritualistic cannibalism
Iatrogenic CJD	Human	Infection from prion-contaminated HGH, dura mater grafts, etc.
Variant CJD	Human	Food-borne infection from bovine prions
Familial CJD	Human	Germline mutations in *PRNP* gene
GSS	Human	Germline mutations in *PRNP* gene
FFI	Human	Germline mutations in *PRNP* gene (D178N, M129)
Sporadic CJD	Human	Somatic mutation or spontaneous conversion of PrP^C into PrP^{Sc}?
sFI	—	Somatic mutation or spontaneous conversion of PrP^C into PrP^{Sc}?
Scrapie	Sheep, goat	Infection or spontaneous disease (atypical scrapie)
Bovine spongiform encephalopathy (BSE)	Cattle	Infection with prion-contaminated MBM ("classical"), spontaneous disease (atypical L-type and H-type), germline mutation in PrP gene (E211K)
Transmissible mink encephalopathy	Mink	Infection with prions from cattle (L-Type)?
Chronic wasting disease	Mule deer, elk	Unknown infection
Feline spongiform encephalopathy	Cat	Infection with prion-contaminated bovine tissues or MBM
Exotic ungulate encephalopathy	Greater kudu, nyala, oryx	Infection with prion-contaminated MBM

Notes: CJD: Creutzfeldt–Jakob disease, FFI: fatal familial insomnia, sFI: sporadic fatal insomnia, GSS: Gerstmann–Sträussler–Scheinker disease, HGH: human growth hormone, MBM: meat and bone meal.

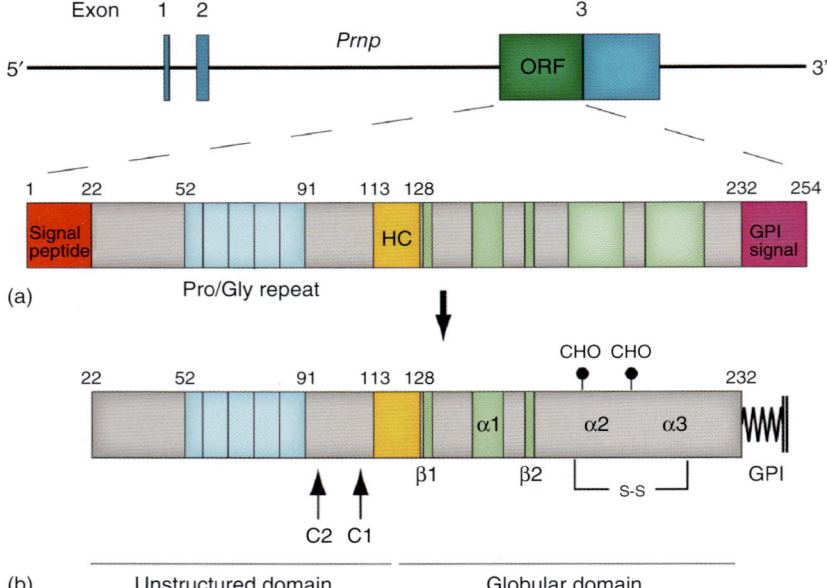

Fig. 2 Organization of the cellular prion protein. (a) Bar diagram of the mouse *Prnp* gene that encodes the prion protein PrP of 254 amino acids; (b) Domains architecture outline of full-length forms of the cellular prion protein, including post-translational modifications (octapeptide repeat, hydrophobic core (HC), disulfide bride (S-S), glycosylation sites (CHO), and GPI anchor (GPI)) and elements of secondary structure (alpha-helices (α), beta-sheet (β)). The position of prion protein cleavage into C1 and c04 fragments are indicated. The position of the respective amino acids (mouse no.) is also indicated.

truncation; both PrP 27–30 and PrP^{Sc} are infectious.

The molecular biology and genetics of prions began with the purification of PrP 27–30 that allowed the determination of its N-terminal, amino acid sequence. Multiple signals in each cycle of the Edman degradation suggested that either multiple proteins were present in these "purified fractions," or a single protein with a ragged N terminus was present [25, 26]. When the signals in each cycle were grouped according to their strong, intermediate and weak intensities, it became clear that a single protein with a ragged N terminus was being sequenced. The determination of a single, unique sequence for the N terminus of PrP 27–30 permitted the synthesis of isocoding mixtures of oligonucleotides that were used subsequently to identify incomplete PrP cDNA clones from hamster and mouse [27, 28]. cDNA clones encoding the entire open reading frames (ORFs) of Syrian hamster and mouse PrP were eventually recovered.

6
PrP Gene Structure and Organization

6.1
The Prion Protein Family

Prnp is a member of the so-called PrP gene family. During the last decade, two mammalian PrP paralogs – Doppel (Dpl) and Shadoo (Sho) – have been identified [29]. These three loci are likely to derive from an ancestral Zrt- and Irt-related protein family (ZIP) metal ion transporter gene [30]. Dpl is encoded by the *Prnd* gene that lies approximately 19 kb downstream from the PrP locus on chromosome 2 in mouse (chromosome 20 in human). In contrast to PrP, which is expressed in many different tissues, Dpl expression is essentially confined to the testis of adult mammals; its absence can cause male sterility [31], while its ectopic expression can cause neurodegeneration in the central nervous system (CNS) [32]. Sho is encoded by *Sprn*, is found on chromosomes 7 and 10 in mouse and human, respectively, and is expressed in the CNS. Although their expression patterns are not superimposable, both Shadoo and PrP share neuroprotective properties, notably against Doppel and N-terminally truncated PrP neurotoxicities [33]. Sho knockout mice are viable [34], but whether double PrP/Sho invalidation is lethal (*vide infra*) is debated [34, 35].

6.2
The *PRNP* Gene

The *PRNP* gene is highly conserved in mammalian species (over 80%), and genes with similarities are present in birds, reptiles, amphibians, and fish [36]. The PrP gene has either three (rat, mouse, bovine, sheep) or two exons (hamster, human). The entire ORF of all known mammalian and avian PrP genes resides within the last exon, which eliminates the possibility that PrP^{Sc} arises from alternative RNA splicing (Fig. 2) [37, 38]. The *Prnp* gene promoter contains multiple copies of G–C-rich repeats and is devoid of TATA boxes [39]. These G–C nonamers represent a motif that may function as a canonical binding site for several transcription factors such as Sp1 [40]. Based on these observations, *Prnp* is considered as a housekeeping gene and, like *Prnp*, the entire ORFs of *Prnd* and *Sprn* are encoded within a single exon.

6.3
Expression of the PrP Gene

Both, PrP^C mRNA and protein are developmentally and regionally regulated, although postnatal expression is higher than embryonic expression. During mouse embryonic development, *Prnp* gene expression has been detected from embryonic days 6.5 and 8.5 onwards [41–43]. *Prnd* mRNAs are expressed in embryos, in peripheral tissues, and at low levels in the adult CNS [32]. The development of *Sprn* reporter mice revealed that Sho is expressed in mouse embryos, most notably in extraembryonic annexes, and also in various adult tissues including the brain [44].

7
Cellular form of PrP: Primary Structure, Expression, and Putative Function(s)

7.1
PrP Primary Structure and Species Variations

Native PrP^C is composed of 235–264 amino acids depending on the considered species (Fig. 2). PrP is post-translationally processed to remove a 22-amino acid, N-terminal signal peptide, which allows translocation in the endoplasmic reticulum. The extreme N-terminal part of the protein contains a polybasic region that may contribute to PrP endocytosis [45, 46]. Mice expressing PrP deleted for this short region displayed a dramatically reduced susceptibility to prion infection, suggesting a role for these residues in PrP conversion [47].

The N-terminal domain of mammalian PrP contains five copies of a P(H/Q)GGG (G)WGQ octarepeat sequence and occasionally more, as in the case of one sequenced bovine allele, which has six copies (Fig. 3). These repeats are remarkably conserved between species, which implies a functionally important role. The chicken sequence contains a different repeat, PGYP(H/Q)N. Although insertions of extra repeats have been found in patients with familial prion disease (Fig. 3), naturally occurring deletions of single octarepeats do not appear to cause disease, and the deletion of all these repeats does not prevent PrP^C from undergoing a

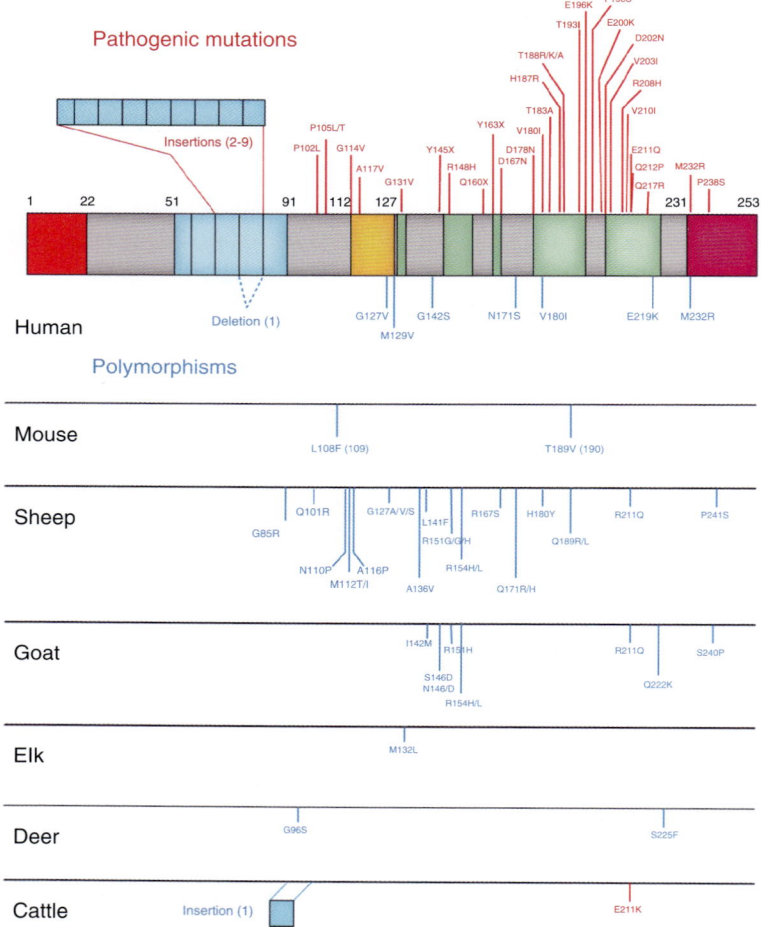

Fig. 3 Mutations and polymorphisms of the prion protein gene. PrP mutations causing inherited prion diseases (red) and PrP polymorphisms (blue) are found in humans, mice, sheep, goats, elk, deer, and cattle. Most of the polymorphisms shown have been reported to influence the onset and/or phenotype of disease.

conformational transition into PrP^{Sc} [48]. The histidine residues in the octarepeats might bind metal ions with high affinity [49, 50].

The C-terminal 120 amino acids contain two conserved disulfide-bonded cysteines and a sequence that beckons the addition of a glycosylphosphatidylinositol (GPI) anchor (see Fig. 2). Twenty-three residues are removed during the addition of this GPI moiety, which anchors the protein to the cell membrane [51, 52]. GPI anchoring allows the subcellular localization of PrP in membrane domains enriched in lipids classes (i.e., cholesterols, sphingolipids, and saturated fatty acids) called lipid rafts or detergent-resistant domains, which are important for cell signaling and trafficking [53]. Some minor cytosolic or transmembrane forms of PrP have also been reported [54, 55].

Contributing to the mass of the protein are two Asn side chains linked to large oligosaccharides with multiple structures that have been shown to be complex and diverse [56]. Glycosylated isoforms of PrP^C may be differentially distributed in the CNS [57]. The physiological significance of PrP glycosylation is unknown, but in cell cultures it favors addressing of the protein at the cell surface [58].

Although many species variants of PrP have been sequenced, only the chicken sequence has been found to differ greatly from the human sequence. Alignment of the translated sequences from more than 40 PrP genes shows a striking degree of conservation between the mammalian sequences, and is suggestive of the retention of some important function through evolution [59]. The cross-species conservation of PrP sequences makes it difficult to draw conclusions about the functional importance of many of the individual residues in the protein.

7.2
Metabolism, Expression, and Distribution of PrP^C

In cell models, PrP^C is synthesized in just a few minutes and reaches the cell surface within an hour [60]. The results of pulse–chase experiments have indicated a half-life of only a few hours, which is relatively short compared to other GPI-anchored proteins [61–64]. Depending on the cell type, PrP^C can then be endocytosed through either caveolae-coated [65, 66] or clathrin-coated pit [67]-dependent pathways.

Before or after reaching the cell surface, a substantial part of PrP^C can be cleaved by endogenous proteases at positions 111/112, or near histidine at position 96, to generate fragments that are referred to as C1 and c04 (see Fig. 2), respectively [68–70]. C1 generation may be dependent on ADAM (A Disintegrin And Metalloproteinase) activity [71, 72], and the proportion of full-length versus C1 fragments is both cell- and tissue-dependent [57, 73].

A minor proportion of PrP can also be anchorless [62, 74, 75] and excreted into the extracellular space. The respective roles or contribution to the cell biology of PrP of these PrP variants remain to be established. At variance with C1, which is not convertible but may act as a dominant-negative inhibitor of PrP^{Sc} conversion [76], anchorless PrP is convertible [77] and can even misfold spontaneously into prions when overexpressed in the brain of transgenic mouse lines [78].

In mammalian tissues, PrP^C expression is nearly ubiquitous [79, 80]. The protein is highly expressed within the nervous system, albeit at variable levels among distinct brain regions, cell types, and neurons. In the CNS of adult rodents and primates, there is an anteroposterior gradient of expression

[57, 81, 82]. PrP^C is also detected in the peripheral nervous system. Many cellular components of the lymphoid tissue, bone marrow, blood, skeletal muscles and mammary glands also express detectable levels of PrP^C. Follicular dendritic cells (FDCs), which accumulate prions in the lymphoid tissues (see Sect. 11) and dendritic cells, express the highest levels of PrP^C [83, 84]. Levels of PrP^C and mRNA expression may vary among maturation and cellular activation in various cell types such as astrocytes, granulocytes, T cells, monocytes, and dendritic cells [85].

7.3
PrP^C Putative Function(s)

PrP expression in a broad range of vertebrate tissue and the conservation of PrP primary and tertiary structures among mammals would lend support for a strong biological role for PrP. However, some 30 years after its discovery the precise biological function(s) of PrP (in the uninfected host) remains an enigma. The invalidation or overexpression of the protein *in vivo* or *in vitro* and the search for PrP interactants has allowed a number of possible functions to be proposed.

7.3.1 PrP-Deficient Mice
Ablation of the *Prnp* ORF in mice [86, 87], cattle [88], and goat [89] has been obtained, with no drastic developmental phenotype, except for some subtle abnormalities [90, 91], among which the involvement of PrP in synaptic hippocampal activity has been debated [92–95]. Whilst postnatal knockout in mouse adulthood showed no discernable deficits [96, 97], a knockdown of PrP proteins in zebrafish proved to be lethal and but could be rescued by mouse PrP [98]. In these animals, PrP would regulate embryonic cell adhesion, a function consistent with the cell mobility and angiogenesis pathways identified by comparative transcriptomic analysis of PrP and PrP knockout mouse embryos [43]. It has been hypothesized that a host-encoded gene (possibly *Sprn*?) could compensate for the lack of PrP [99].

In two *Prnp* knockout lines, Purkinje cell loss was accompanied by ataxia beginning at approximately 70 weeks of age [100, 101]. Crossing one of these lines with transgenic mice overexpressing mouse PrP rescued the ataxic phenotype. With the discovery of Dpl, it emerged that Dpl ectopic expression in the brains of these *Prnp* knockout mice provoked cerebellar degeneration [32]. The only clear phenotype associated with *Prnp* knockout mice is their absolute resistance to transmissible spongiform encephalopathy (TSE) and to the neurotoxicity associated with PrP^{Sc} [102, 103].

7.3.2 PrP Partners
Systematic searches for PrP partners have been made following different strategies, such as yeast two-hybrid screen, coimmunoprecipitation, and crosslinking (for reviews, see Refs [36, 85, 104]). The functionality of the interaction between PrP and the candidates has not always been validated. Among the physiological ligands are the neural cell adhesion molecule (NCAM) [105], which is recruited into lipid rafts [106] and activates the Fyn kinase, an important partner of PrP^C, with caveolin, in neuronal cell signaling [107]. The interaction of PrP^C with several extracellular matrix components (e.g., laminin, laminin receptor, heparin, heparan sulfate proteoglycans, ST1) would lend support for a role in cell adhesion and neuritogenesis.

7.3.3 PrP and Copper
The fixation of copper to PrP^C octarepeats has been clearly established both *in vivo*

and *in vitro*, with binding affinities within the micromolar range [108]. PrP overexpression increases cellular copper binding [109, 110], and other copper-binding sites have been identified near the C1 and c04 cleavage sites (see Fig. 2) in human PrP [50]. The binding of Cu^{2+} promotes PrP conformational changes [50, 111] and endocytosis [112, 113], yet the role of PrP^C in brain copper homeostasis, established using PrP knockout mice, remains controversial [49, 114].

7.3.4 PrP and Neuroprotection
A wealth of data suggests that PrP^C may (neuro)protect against oxidative damage [49, 115–117] or apoptosis-inducing insults. Thus, cultured hippocampal neurons from *Prnp* knockout mice were more susceptible to ischemic brain injury [118, 119] or to serum starvation [120, 121] than their wild-type counterparts. The latter effect could be rescued by the expression of anti-apoptotic Bcl-2 protein [120]. Mechanistically, PrP^C could protect cells against tumor necrosis factor alpha (TNF-α) [122] or Bax-mediated cell death [123–125]. PrP octarepeats, but not anchoring at the cell surface, would be necessary to elicit PrP^C neuroprotective functions [123, 126–129].

Independently of Dpl ectopic expression, transgenic mouse lines expressing a partially N-terminally deleted mutant PrP (Δ32–121 or 32–134) on a PrP null background spontaneously developed a rapidly fatal neurodegenerative disease [99, 130, 131]. Perinatal lethality was even observed with certain deleted residues within the PrP hydrophobic stretch (Δ105–125) [132], but the reintroduction of a normal copy of the *Prnp* gene generally abrogated the observed neurodegenerative phenotype. Thus, certain PrP domains may have neuroprotective functions, while others may serve as binding domains for the transduction of toxic signals.

7.3.5 PrP and Stemness
Converging evidence has indicated that PrP^C might regulate the proliferation/differentiation balance of a number of stem and progenitor cells, such as neural precursors [133], embryonic stem cells (ESCs) [134], dental mesenchymal cells [135], and muscle precursors [136]. The protein has also been reported to be involved in the self-renewal of hematopoietic stem cells (HSCs) [137], and to regulate the self-renewal and differentiation towards the three germ layers of ESCs in early embryogenesis [138, 139]. However, only limited *in vivo* evidence is available supporting a link between PrP^C and stemness.

7.3.6 PrP Involvement in Other Pathologies
PrP^C has been linked with other pathological situations, most likely because of its involvement in the aforementioned ubiquitous cascades. The protein is overexpressed in a variety of human cancers, including gastric carcinoma, osteosarcoma, pancreatic, melanoma, and breast cancers [140]. In pancreatic, colorectal and breast cancer, the levels of PrP^C expression are associated with an unfavorable evolution [140, 141].

Beyond the obvious pathogenic commonalities between Alzheimer's disease and prion disease (see Sect. 19), PrP^C has emerged as a potential key player in this pathology. Aβ peptides oligomers and fibrils, which accumulate in the brain of the affected patients, were shown to interact with PrP^C [142–144]. PrP^C partners, such as the co-chaperone STI1, may also participate or be incorporated in these PrP^C/Aβ oligomers interactions [145]. PrP^C also mediated (at least partly) soluble

Aβ oligomer-induced synaptotoxicity [146–148]. However, the role of PrP in the toxic effect of olimeric Aβ remains a subject of controversy [149–152], possibly because of the use of different animal models and the preparation of Aβ oligomers. In contrast, PrP^C may exert a protective role in Alzheimer's disease, as PrP has been shown to inhibit the cleavage of Alzheimer's amyloid precursor protein (APP) into Aβ peptides by beta-secretase [153–155]. As secretion of the N1 fragment after C1 cleavage (see Fig. 2) may protect against Aβ toxicity [156], it becomes increasingly apparent that PrP^C and its potential deregulation have implications far beyond the prion field.

8
Structures of PrP Isoforms

8.1
PrP Primary to Quaternary Structure

Following their discovery, mass spectrometry and gas-phase sequencing failed to identify post-translational chemical modifications that might explain the differences in the biochemical properties of PrP^C and PrP^{Sc} [157, 158]. These observations forced the consideration of the possibility that a conformational change distinguishes the two PrP isoforms. Prior to comparative studies on the structures of PrP^C and PrP^{Sc}, metabolic labeling studies showed that the acquisition of protease resistance in PrP^{Sc} is a post-translational process. Both PrP isoforms were found to carry GPI anchors, and only the proportion of glycoforms varied [159]. Currently, a large body of experimental evidence indicates that the secondary, tertiary and quaternary structures of the two PrP isoforms differ profoundly.

When the secondary structures of PrP^C and PrP^{Sc} were compared using optical spectroscopy, they were found to be markedly different. Both, Fourier transform infrared (FTIR) and circular dichroism (CD) spectroscopy studies showed that brain PrP^C contains approximately 40% α-helix and minimal β-sheet, while brain PrP^{Sc} (or PrP^{res}) is composed of approximately 30% α-helix and 45% β-sheet [6, 160–162]. Similar ratios were obtained with unglycosylated, anchorless PrP, suggesting a poor participation of these post-translational modifications to the overall PrP^{Sc} secondary structure [163].

Indirect experimental evidence has also indicated important structural differences between PrP^C and PrP^{Sc}. For example, PrP^{Sc} exhibits a pronounced resistance to proteolysis [25], denaturation by chaotropic agents [161, 164], or heat [161], and is inaccessible to many anti-PrP antibodies with various epitopes [10, 26].

Under nondenaturing solubilizing conditions, fractionation techniques such as gradient ultracentrifugation [16–19, 165–167], gel filtration [16], asymmetric field flow fractionation [168, 169] and or ultrafiltration [170] have indicated that PrP^{Sc} and/or the infectious particles are aggregated, while PrP^C is mostly monomeric. It emerges that PrP^{Sc} is not a collection of multimers with a regular continuum of size, but rather that the size distribution of (solubilized) PrP^{Sc} oligomers is finite. Depending on the techniques employed, the nature of the molecular mass markers used as reference, the solubilization conditions and the type of prion strain studied, the mean size of PrP^{res} would vary between 150 and 20 000 kDa. It must be noted that the specific infectivity and templating activities of PrP^{Sc} can be largely heterogeneous. For certain

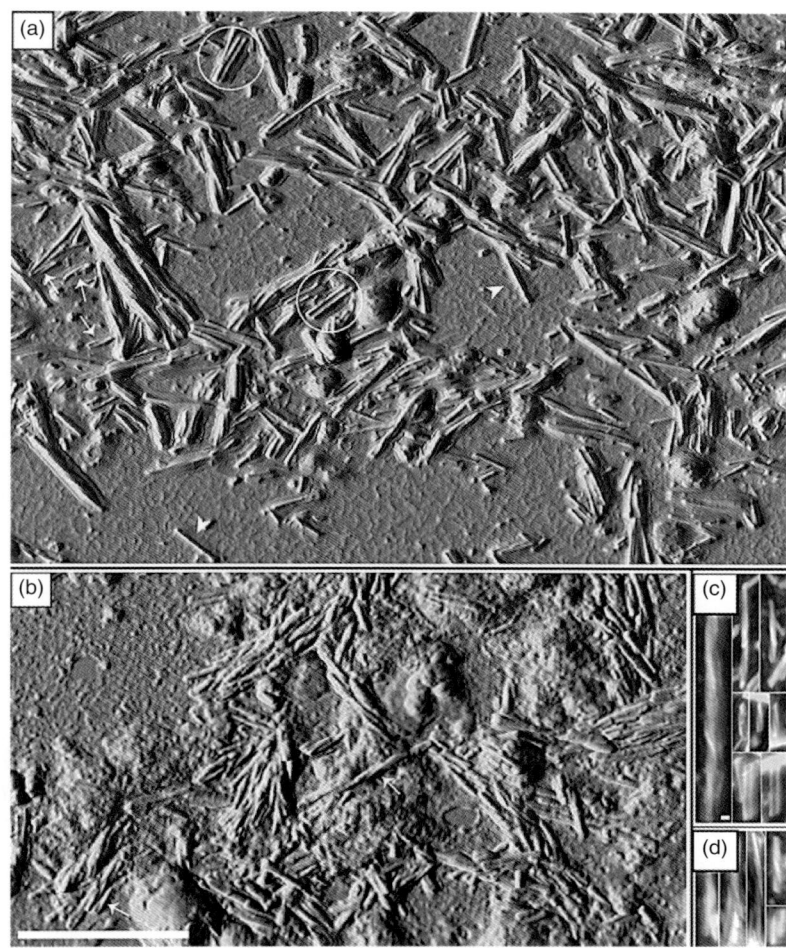

Fig. 4 Atomic force microscopy of abnormal prion protein fibrils. (a,b) Images of anchorless PrP^{Sc} fibrils from (a) 22L and (b) RML mouse scrapie strains by atomic force microscopy. Circles indicate examples of straight fibrils, arrowheads designate individual protofilaments, and arrows highlight twisting fibrils. Scale bar = 500 nm; (c,d) Height images of individual (c) anchorless 22L and (d) anchorless RML fibrils. Scale bar = 20 nm. Reproduced with permission from Ref. [172]; © 2009, Elsevier.

strains, a subset of PrP assemblies may be the best template for prion replication [18, 19, 169].

The ultrastructural imaging by electron microscopy [24, 171] or atomic force microscopy [172] of semi-purified PrP^{Sc} fractions from prion-infected brain (expressing wild-type or anchorless PrP) allowed the visualization of fibrillar structures of heterogeneous size and shape (Fig. 4). Non-PrP elements such as lipids [173] or sugars [174, 175] may participate in the PrP^{Sc} assemblies.

8.2 Topological Models of PrP Isoforms

8.2.1 3-D Structure of Recombinant Monomeric PrP

Most high-resolution studies of "normal" PrP have been performed with recombinant PrP by using nuclear magnetic resonance (NMR) or X-ray crystallography (Fig. 5), and more than 10 PrPs from different species have been determined as being quite similar [2, 3, 176–180]. It is noteworthy that recombinant PrP produced in *Escherichia coli* have neither N-linked sugar chains nor a GPI anchor, both of which are attached to PrP^C. However, a comparison of spectroscopic values of recombinant and extractive PrP has suggested that these modifications had no effect on the final refolding of the protein [181]. Conversely, adding glycans or anchoring recombinant PrP to lipid membranes did not alter its topology [182].

Full-length PrP is a three-helix-bundle protein with two short antiparallel β-strands. While the three helices form a globular C-terminal domain, the N-terminal domain is highly flexible and lacks an identifiable secondary structure under the experimental conditions employed. There are, however, transient interactions

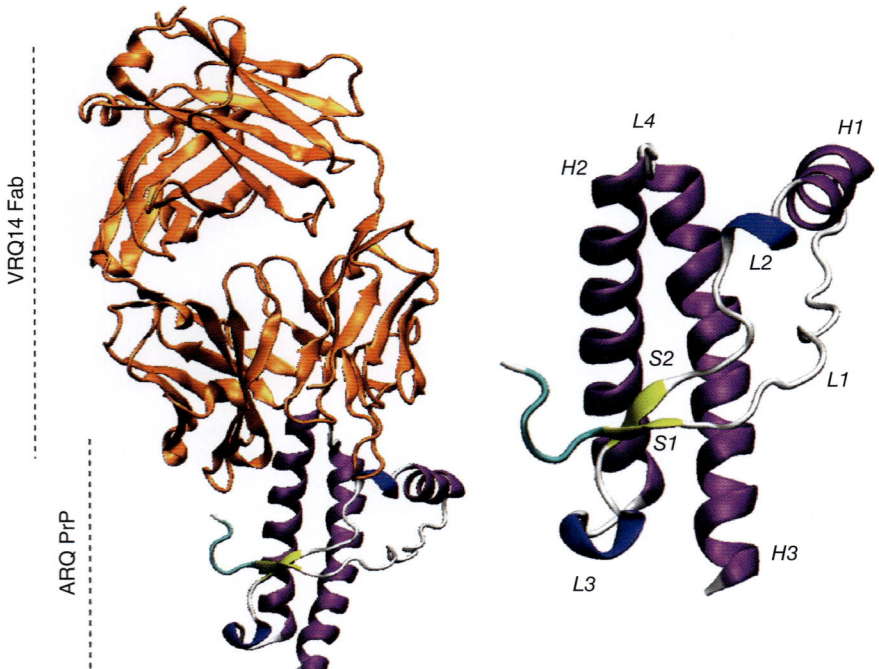

Fig. 5 Structure of PrP^C. X-ray crystallography 3D structure of PrP obtained by cocrystallization of VRQ14 Fab fragment with sheep recombinant PrP (ARQ variant, residues 114–234) [3]. The analysis revealed two short β-strands (S1, S2) and three α-helices. The loops (L) are indicated. Protein database accession number: 1TXP. Figure prepared by Dr Human Rezaei.

between these two domains (see Figs 2 and 5).

8.2.2 PrP^{Sc} Structural Models

To date, PrP^{Sc} insolubility has frustrated high-resolution structural analysis using X-ray crystallography or NMR spectroscopy. Although structural models have been developed from data obtained with low-resolution techniques (Fig. 6), none has been shown consensual.

In a first model, structural information was gained from electron crystallography analyses of isomorphous, two-dimensional crystals of PrP 27-30 [183, 184]. Thus, it was determined that if PrP^{Sc} follows a known protein fold, it would adopt a parallel left-handed β-helical architecture. Left-handed β-helices readily form trimers, providing a natural template for a trimeric model of PrP^{Sc} (Fig. 6). This trimeric model accommodates the PrP sequence from residues 89–175 in a β-helical conformation, with the C terminus (residues 176–227) retaining the disulfide-linked α-helical conformation observed in PrP^C.

A second model was based on molecular dynamics simulating, under amyloidogenic conditions, the conversion of PrP [185]. The resultant conformation was first enriched in beta-structure and then used to model a protofibril by docking hydrophobic patches of the template structure to form hydrogen-bonded sheets spanning adjacent subunits. The resulting trimeric oligomer would stack in a spiral-like manner to form higher-order protofibrillar aggregates

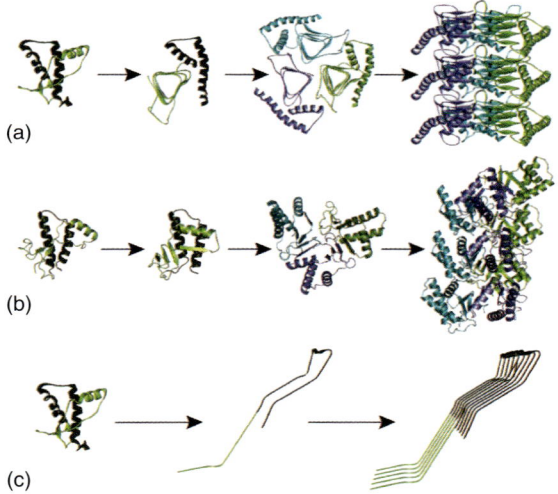

Fig. 6 Topological models proposed for PrP^{Sc} structure. (a) In the β-helical model, a major refolding of the N-terminal region of PrP 27–30 into a β-helix motif from residues 90–177 (light green) is proposed. The C-terminal region (residues 178–230, dark green) maintains the α-helical secondary structure organization, as in PrP^C; (b) The β-spiral model developed by molecular dynamics simulation consists of a spiraling core of extended sheets comprising short β-strands, spanning residues 116–119, 129–132, 135–140, and 160–164. In this model, the three α-helices in PrP^C maintain this conformational motif; (c) The parallel in-register extended β-sheet model of PrP^{Sc} proposes a thorough refolding of PrP^C into a β-sheet-enriched structure. Reproduced with permission from Ref. [4]; © 2012, Macmillan Publishers Ltd.

(Fig. 6). At variance with the first model, the β-amyloid core would consist of a three-β-sheet (residues 116–119, 129–132, and 160–164) and an isolated strand at residues 135–140, with three α-helices in PrP^C retaining their native conformation (Fig. 6).

The last model is derived from structural studies by hydrogen–deuterium exchange and site-directed spin labeling on recombinant and extractive fibrillar PrP [186, 187]. The whole protein would be restructured, such that residues within the 160–200 region would form single-molecule layers that stacked on top of one another with a parallel in-register alignment of β-strands (Fig. 6).

None of these divergent structural models was fully compatible with experimental data. For example, in the first two models, the conversion would rather involve the N-terminal part of the protein, while the C-terminal rearrangements would rather be minor, which is in apparent contradiction with the proteolysis resistance of this segment [6]. The whole remodeling of PrP^{Sc} in the third model was difficult to accommodate, with a significant proportion of α-helices found in PrP^{Sc} by spectroscopic studies [6, 160, 161]. In contrast, a number of studies suggested a whole remodeling in the strands and loop [163, 187, 188]. These models may also reflect the conformational landscape of PrP oligomers associated with prion strain diversity (see Sect. 13).

Some local variations in PrP structure might be difficult to assign structurally. For example, large peptides or tags can be inserted into the loop between the second and third α-helices, without altering prion conversion in cell culture [189]. This suggests that this part remains unstructured while remaining inaccessible to proteolysis.

8.2.3 PrP Domain(s) Involved in the Conversion

The definition of a canonical prion domain in PrP (i.e., a critical region involved in the conformational change leading to PrP), as with yeast prions (see Sect. 17), has mostly relied on mutating or deleting PrP amino acid(s) and studying the consequence of susceptibility to prion infection in cell culture or transgenic mouse models. Unfortunately, this strategy has shown only limited relevance, as a single substitution in many parts of the PrP sequence may dramatically alter prion replication [190–196].

Proteolytic access to the N terminus of PrP^{Sc}, without loss of infectivity, would suggest that this part of the protein is not essential for the conversion process [158, 197, 198]. However, the disease phenotype was significantly altered in transgenic mice deleted of the 32–92 [199] or of the 29–32 polybasic region of PrP [47], which suggests that it may somehow participate in the replication dynamics.

Besides developing spontaneous degeneration, transgenic mouse lines expressing PrP deleted in the globular core of PrP proved resistant to prion infection [129, 130, 199], except for mice expressing a so-called "miniprion" deleted from residues 23–88 and residues 141–176 [200]. The absence of a requirement for the first α-helix and the second β-sheet for prion conversion was an apparent contradiction with the findings of other *in-vivo* or biophysical studies [201–204].

PrP^C conversion necessitates presence of the disulfide bond [205, 206], but not of the GPI anchor [77, 207, 208], and whether prion conversion necessitates PrP glycosylation remains the subject of debate. Aglycosylated PrP is convertible in cell-free systems [207, 209, 210] but not (or less efficiently) in cell culture or transgenic mouse models [58, 210–212]. Worthy of note, an absence of glycans may adversely affect PrP trafficking to the cell membrane and thus indirectly impede its conversion [58], as PrP^{Sc} formation occurs after PrP^C reaches the cell surface (see Sect. 10).

Cells or mice expressing C-terminally tagged-PrP are fully susceptible to prion infection, which suggests that this region is not pivotal to prion conversion [61, 213, 214].

To reconcile these discrepant data, the entire protease-resistant domain of PrP may be considered as a prion domain, in accordance with the in-register PrP structural model.

9
Prion Replication

Most available data indicate that the formation of nascent prions is associated with PrP^{Sc} aggregation processes. Prion replication may proceed through a nucleation–polymerization (NP) reaction [215], according to which the key element would be a PrP^{Sc} oligomer acting as a stable nucleus (Fig. 7); monomeric PrP^C would be incorporated into this nucleus and adopt the structure of PrP^{Sc}. Nucleation is the rate-limiting step, and would be responsible for the lag phase observed during the spontaneous formation of prions (Fig. 7b), but could be bypassed by seeding with a small amount of exogenous PrP^{Sc} aggregates, as in the case of an acquired infection (Fig. 7a). The elongation or polymerization would then be straightforward. Mathematical modeling has indicated that the fragmentation of aggregated PrP^{Sc} is key to the sustainability/dynamics of the process [216].

Cell-free conversion assays supported the view that PrP^{Sc} formation most likely

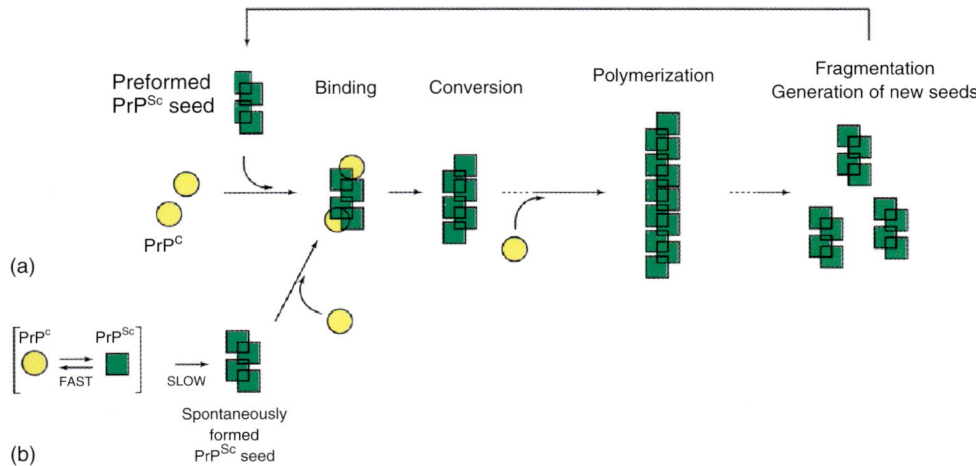

Fig. 7 PrP^{Sc} polymerization mechanism. According to the nucleation polymerization process, prion replication may proceed through the binding of PrP^C (or a partially unfolded species) onto aggregated PrP^{Sc} seed, (a) acquired by infection or (b) formed spontaneously, at a very slow rate. PrP^C would then be converted into PrP^{Sc} conformation. Further incorporation would occur due to constant PrP^C renewal, and would concur to the growth of the aggregate. Degradation may also occur (not shown). At a point, due for example to physical constraint, the polymer may break and generate new seeds, contributing to the autocatalytic nature of the reaction.

occurs through a NP polymerization process rather than via a heterodimer mechanism between PrP^C and a putative PrP^{Sc} monomer [217]. In its simpler principle, pioneered by Caughey and colleagues, these assays consisted of incubating a purified, metabolically labeled PrP^C with semi-purified PrP^{Sc} and then monitoring the conversion efficacy over time [207, 218]. Kinetically, the cell-free conversion was described as a two-step process [219, 220]. In the first step, PrP^C binds to PrP^{Sc} and becomes sedimentable but remains protease-sensitive. Most importantly, this binding step seems highly sequence-specific, that is, it is not intrinsically related to the sticky properties of PrP^{Sc} [219, 221]. Heterologous binding might be as efficient as homologous binding, whether conversion occurs, or not [221, 222]. The second step is the conversion *sensu stricto*, and this follows at a slower pace, whereby PrP^C undergoes a conformational change and acquires a protease-resistant state [219–221, 223]. This second step necessitates further intermolecular interactions between PrP^C and PrP^{Sc}, and the newly formed PrP^{Sc} will remain tightly bound to the PrP^{Sc} template.

The low-converting yield and an absence of generated infectivity in these early conversion assays impeded any further mechanistic studies of prion conversion. However, the protein misfolding cyclic amplification (PMCA) technique has subsequently emerged as a highly efficient procedure to amplify prions *in vitro* [224]. PMCA exploits the ability of PrP^{Sc} to template the conversion of PrP^C by repetitive cycles of incubation and sonication, leading to the amplification of minute amounts of PrP^{Sc} [225, 226]. Mechanistically, analogous with the NP reaction, the incubation of PrP^{Sc} seeds with PrP^C-containing substrate is thought to favor PrP^C conversion and the growth of PrP^{Sc} aggregates. The sonication stage is thought to fragment the polymers, providing new seeds for the conversion. The PMCA-generated products are infectious and share (in general) similar biochemical and structural properties and biological strain properties with the prion strain seed that serves for amplification [227, 228].

The formation of infectious prions from recombinant PrP can occur in a minimal substrate composed of lipids and RNA, which suggests that no auxiliary protein would be required to produce prion infectivity [229]. However, a non-PrP-component might act as a scaffold during the NP process [8, 230].

10
Cell Biology of PrP^{Sc} Formation

The low immunoreactivity of PrP^{Sc} aggregates and an absence of methods to distinguish between PrP^C and PrP^{Sc} without denaturation, makes identification of the conversion site(s) difficult by live cell imaging [231]. The plasma membrane and the lipid rafts, the endolysosomial compartment and the endoplasmic reticulum have all been proposed as being involved in the conversion process [232, 233]. Whatever the site in prion-infected cells, PrP^C molecules destined to become PrP^{Sc} must exit to the cell surface prior to their conversion to PrP^{Sc} [214, 234]. Subsequently, PrP^{Sc} is trimmed at the N terminus in an acidic compartment in prion-infected cultured cells to form PrP 27-30. In contrast, the N-terminal truncation of PrP^{Sc} is usually minimal in the brain, where little PrP 27-30 is found [73]. The results of initial pulse–chase experiments have suggested that the conversion of PrP^C into PrP 27–30 was a slow process, taking between 3 and 15 h to complete [63].

However, the *de-novo* formation of PrP^{Sc} in cells expressing tagged PrP^C seems to occur within minutes after prion exposure [214].

11
Prion Peripheral Pathogenesis

Prion contamination most frequently occurs via a peripheral route, and food-borne oral contamination has been observed in both humans (Kuru, vCJD) and animals (BSE, TME). These diseases have also been transmitted via intravenous, intraperitoneal, or intramuscular infection, through the use of prion-contaminated vaccines or hormones (Table 2). The question then arises of how prions reach the CNS in order to elicit their toxic effects. Following an (experimental) oral infection and passage through the gastrointestinal tract, the direct transport of prions to the brain via the vagus nerve was observed [235], though strain-dependent prion replication was also shown to occur in lymphoid tissues very soon after peripheral infection. Following transepithelial passage from the gut lumen, presumably through the M cells [236], lymphocompetent prions were seen to replicate primarily in the gut-associated lymphatic tissue, notably in Peyer's patches and mesenteric lymph nodes, before they reached the secondary lymphoid organs such as the spleen and lymph nodes [237]. Lymphoinvasion was previously recognized in sheep [238, 239] and mice [240, 241] that had been infected experimentally with scrapie, long before PrP was identified.

Within the lymphoid tissue, prion accumulation occurs primarily in the FDCs [242, 243], which are post-mitotic cells that play a pivotal role in antigen trapping and the capture of immune complexes. The complement cascade most likely also plays a role in the early steps of prion/FDC interactions. Prions have been identified in lymph nodes [244, 245] and granulomas [246] lacking mature FDC, and can also replicate in ectopic, tertiary lymphoid follicles in excretory tissues such as the kidneys, liver, and pancreas in case of chronic lymphocytic inflammation [247]; thus, they may contribute to a higher infection prevalence due to their being shed in the excreta. The results of studies of prion pathogenesis in immunodeficient or splenectomized mouse models have indicated that, although the peripheral phase of prion replication can be bypassed, this dramatically decreases the incidence of clinical disease, notably at low-dose infection [248, 249].

In mouse models, prion infectivity rises rapidly in lymphoid tissues soon after peripheral inoculation and reaches a plateau until the terminal stage of the disease; however, such accumulation is mostly innocuous. In both natural and experimental models, the lymphoid tissues are approximately 100-fold less infectious than the brain, and consequently these tissues are systematically removed from ruminant carcasses before their consumption by humans.

The way in which prions pass the neuroimmune interface at the level of the sympathetic nervous system [250] is not well understood, though the distance between the germinal centers and terminal nerves appears to determine the onset of neuroinvasion [251]. Membrane-anchored PrP^C expression is required for prion transport through the peripheral nervous system to the CNS [252], but exactly how prions spread along the peripheral nerves remains unclear. The axon that ensheathes myelinating Schwann cells does not appear to play a role [253, 254], despite the ability of these cells to replicate prions [255]. Notably, the fact that the neuroinvasive phase is greatly shortened after peripheral inoculation compared to intracerebral inoculation suggests an adequate topological addressing of prions to the upmost sensitive areas of the brain with respect to toxicity [256].

12
Prion Toxicity

While PrP^C and its conversion is key to prion replication [102, 103], its role in prion-induced neurotoxicity remains highly debated, as well as the nature of the potentially toxic species themselves [257]. The exposure of primary cultured neurons expressing, or not, PrP^C to purified preparations of PrP^{Sc} aggregates, recombinant PrP oligomers or PrP-derived synthetic fragments (most often at supraphysiological doses) provided contradictory elements with regards to the involvement of PrP^C [258–261]. In vivo, while certain transmission studies in transgenic mice expressing PrP variants indicated that PrP expression and anchoring at the cell surface was key to neurodegeneration [77, 103, 262], others found that extraneural PrP^{Sc}, either provided by astrocytes [263] or "floating" in the extracellular space [208], can be toxic per se.

Whether the PrP^C conversion step itself is involved in neurodegeneration remains uncertain. PrP^C has been proposed to mediate toxic signaling of β-enriched protein conformers, including amyloid β responsible for Alzheimer's disease [146, 264, 265]; crosslinking of the protein by anti-PrP antibody was also shown to be detrimental [266] (although these observations are discussed [267]), suggesting that a direct disruption of PrP^C functions could be sufficient to induce neurodegeneration (see above). In contrast, prion neurotoxicity dramatically relied on the expression and efficient conversion of neuronal PrP^C into PrP^{Sc} in cocultures of neurons and astrocytes [268].

Following the experimental transmission of prions to mice expressing variable PrP^C levels in brain, it has been shown that the generation of prion infectivity in brain proceeds first exponentially and then switches to a plateau until the terminal stage of disease. While the first phase was not rate-limited by PrP^C concentration, the duration of the second phase was inversely correlated to PrP^C concentration [269]. It was concluded that neurotoxic species are produced once prion propagation saturates, due to a switch from autocatalytic production of infectivity to a toxic pathway. Alternatively, PrP^C may mediate toxic signaling, at least in case of intraspecies transmissions of prions [270, 271]. The downregulation of PrP^C in response to infection with slowly evolving prions has been observed, and would further support the inverse relationship between prion disease tempo and PrP^C levels [272].

13
Strains of Prions

The finding that goats with scrapie prions can manifest two different syndromes – one syndrome in which the goats became hyperactive and the other in which they became drowsy – raised the possibility that strains of prions might exist. Subsequent seminal studies with mice documented the existence of multiple strains in the same host species through careful measurements of incubation times and the distribution of vacuoles in the CNS [273]. A large set of (physio)pathological or biochemical criteria can now be used to distinguish between prion strains (Table 3).

Tab. 3 An overview of the criteria used to distinguish between prion strains.

Phenotype	Technique/protocol used	Reference(s)
Incubation period in recipient animals	Time period between inoculation and disease onset or terminal stage	[274, 275]
Disease presentation	Behavioral tests	[276]
	Clinical signs in animals	[277, 278]
Biochemical properties of PrP^{Sc}	Protease-induced cleavage (proteinase K/thermolysin/pronase)	[5, 279, 280]
	Ratio of glycosylated species	[281, 282]
	Denaturation with chaotropes	[283]
	Resistance to heat	[284]
	Ratio of native versus denatured PrP	[10]
	Infrared spectroscopy, circular dichroism, Fourier-transform infrared spectroscopy	[6, 160, 163, 285]
PrP^{Sc} distribution in brain	Immunohistochemistry/histoblotting	[286, 287]
Nature of PrP^{Sc} deposits	Histology, amyloid dye (Congo Red, thioflavin S/T) binding	[288, 289]
	PrP deposits epitope mapping	[290]
	Other binding probes	[291]
Distribution and intensity of spongiosis	Histology (so-called "lesion profile")	[292]
Tropism	For lymphoreticular system	[256, 293, 294]
	For cultures	[14, 295]

The search for an explanation of how biological information encrypting the disease phenotype could be enciphered within the prion posed a conundrum. Many investigators argued for a small nucleic acid, but none could be found [7]. Today, a large body of evidence indicates that this information must be enciphered within the structure of PrP^{Sc}. The first evidence supporting the hypothesis that strain-specific information is enciphered in PrP^{Sc} came from studies with prions causing TME, which were passaged into Syrian hamsters. On serial passaging, two strains emerged: one strain (HY) produced hyperactivity in Syrian hamsters, and the other (DY) was manifest as a drowsy syndrome. PrP^{Sc} present in the HY strain and DY strain strongly differed with respect to protease resistance, electrophoretic, and sedimentation properties [277, 296, 297].

A large set of criteria allows PrP^{Sc} to be distinguished biochemically among strains by using spectroscopic methods (Table 3, Fig. 4). The demonstration that, in cell-free assays, the PrP^{C} conversion product usually retains a PrP^{res} molecular signature similar to that of the seeding material [207, 298] further suggested that prion strain specificity might be encoded at the level of protein conformation, particularly the PrP^{Sc} tertiary structure. However, because PrP^{Sc} has not been amenable to high-resolution structural studies (see Sect. 8), the conformational underpinnings of the prion strain phenomenon remain largely elusive. Conversely, how specific

PrP^{Sc} conformations dictate specific and tightly determined tempo, disease presentation, and tropism for tissue or cells (Table 3) remains unknown. The PrP^{Sc} quaternary structure was suggested to participate in strain-dependent prion replication dynamics in the infected host [18, 19].

14
Interplay between the Species and Strains of Prions

Prions can transmit between species and present a zoonotic risk, as exemplified by the emergence of vCJD in the BSE-exposed human population (see Sect. 16). However, such events are restricted by a species or a transmission barrier, the strength of which depends on interactions between host PrP^C and the infecting prion strain type(s). While the transmission barrier is routinely gauged by the appearance of disease-specific, clinical signs, and/or PrP^{Sc} in the brain of the new host, it is becoming increasingly apparent that it can be broken insidiously in the CNS [271, 299] or in the lymphoid tissue [293] of the infected host.

Prion interspecies transmission can be associated with the acquisition of new biological strain properties [293, 294, 300–303], a phenomenon referred to as a "mutation." Monitoring variations in PrP^{Sc} relative conformational stability [283] during such events allowed the conclusion to be drawn that a change in PrP^{Sc} conformation was intimately associated with the emergence of a new prion strain type [304]. The question arises as to whether the new agent is created *de novo* by passage onto a new host, or whether distinct strain subvariants preexist in the transmitted prion and emerge upon selective pressure.

It has been proposed that prions could exist as quasispecies [305, 306], and may thus be endowed with subvariants adapted to the new host, as observed with viral pathogens. The newly emerging prion agent can exhibit phenotypic properties identical to other prions adapted to the host-species, raising the issue of the limits of prion conformational diversity on a given PrP sequence.

To accommodate all these observations, the conformational selection model [305] proposes that prion interspecies transmission is essentially constrained by the degree of steric compatibility between the incoming PrP^{Sc} and the spectrum of conformations that the exposed host PrP^C primary structure can adopt in the PrP^{Sc} state (Fig. 8).

15
Human Prion Diseases

15.1
Sporadic Human Prion Diseases

In most patients with CJD, there is neither an infectious nor an heritable etiology. How prions arise in patients with sporadic Creutzfeldt–Jakob disease (sCJD) is unknown, but the main hypotheses include: (i) horizontal transmission from humans or animals; (ii) somatic mutation of the *PRNP* ORF; and (iii) spontaneous conversion of PrP^C into PrP^{Sc}. Limited attempts to establish an infectious link between sCJD and a preexisting prion disease in animals or humans have been unrewarding (*vide infra*). Studies demonstrating the spontaneous formation of prions in transgenic mice [78, 307] or the formation of synthetic prions from recombinant PrP [8] show that only PrP^C is needed for an animal to produce infectious prions. Thus,

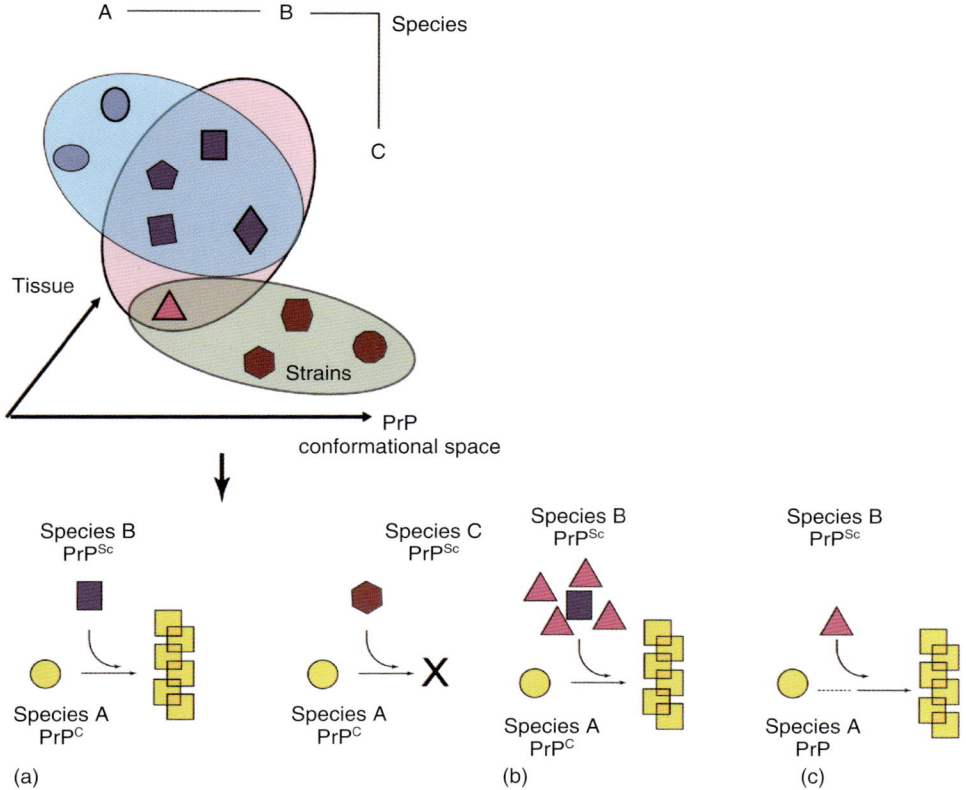

Fig. 8 Conformational model and strain evolution on interspecies transmission. (a) Venn diagram illustrating the spectrum of conformations that PrP^C of species A, B, or C can adopt in the PrP^{Sc} state, and the potential overlap. Transmission of most species B PrP^{Sc} to species A may occur because species A PrP^C can adopt a conformation compatible with the infecting strain. The transmission barrier would therefore be relatively low, except for the conformation represented by a triangle. In contrast, conformations of PrP^{Sc} molecules of species C are not compatible with those allowed by species A PrP^C. The transmission barrier between these species will be high. The tissue might play a role due to specific environmental factors or distinct PrP^C conformational landscape; (b) Compatible conformations between species A and B may be selected because the species B PrP^{Sc} is an ensemble of heterogeneous PrP^{Sc} conformations (or a "quasi-species"). In the example shown, a compatible PrP^{Sc} subvariant is preferentially selected, resulting phenotypically in a strain shift; (c) During molecular interactions between species A PrP^C and conformationally incompatible PrP^{Sc} of species B, a compatible PrP^{Sc} conformation may emerge *ex abrupto* or by iterative interactions, a phenomenon referred to as "mutation."

the spontaneous conversion of PrP^C to PrP^{Sc} seems reasonable, and any mammal expressing PrP^C may be considered capable of forming prions [308]. In countries with specific human prion disease monitoring and reporting systems, the incidence of sCJD ranges between 1 and 1.5 cases per million and per year, and affects mainly elderly persons [309]. Studies of Caucasian patients with sCJD have shown that most are homozygous for Met or Val at codon 129 (see Fig. 3), but this contrasts with the general Caucasian population, in which Met/Val heterozygosity is the most frequent [310].

In 2010, a novel PrP-associated, neurodegenerative disease affecting elderly people was reported in several countries [311–313], in which the major difference from sCJD was the sensitivity of the disease-associated PrP isoforms to protease digestion. The disease, which was referred to as variably protease-sensitive prionopathy, seems to affect the three genotypes at codon 129, and whether the disease is experimentally transmissible is currently being investigated.

15.2
Heritable Human Prion Diseases

About 15% of human prion diseases are associated with an autosomal dominant pathogenic mutation in the *PRNP* gene. As with scrapie, the relative contributions of genetic and infectious etiologies in the human prion diseases remain problematic, with over 30 different mutations in the *PRNP* gene having been described to date [314] (see Fig. 3). The brains of humans dying of inherited prion disease contain infectious prions that have been transmitted to experimental animals [315–317]. The discovery that Gerstmann–Sträussler–Scheinker disease (GSS), which is a familial disease, could be transmitted to apes and monkeys was first reported when many still thought that scrapie, CJD, and related disorders were caused by viruses. With the discovery that the P102L mutation of the human PrP gene was genetically linked to GSS, prion diseases were concluded to have both genetic and infectious etiologies. This mutation has been found in many families in numerous countries, including the first family to be identified as having GSS. Subsequent molecular genetic investigations revealed that the unusually high incidence of CJD among Israeli Jews of Libyan origin was due to a PrP gene point mutation at codon 200, resulting in a Glu-to-Lys substitution (see Fig. 3). The D178N mutation has been linked to the development of FFI, and more than 30 families worldwide with FFI have been recorded. Studies of inherited human prion diseases have shown that the amino acid at polymorphic residue 129 with the D178N pathogenic mutation alters the clinical and neuropathologic phenotype, and that the D178N mutation combined with M129 results in FFI. In this disease, adults generally aged over 50 years present with a progressive sleep disorder and usually die within one year. Within the patients' brains, the deposition of PrP^{Sc} is confined largely to the anteroventral and dorsal medial nuclei of the thalamus. In contrast, the same D178N mutation with V129 produces fCJD, in which patients present with dementia and a widespread deposition of PrP^{Sc} is found postmortem (Table 2).

The transgenic modeling of inherited prion disease by generations of mice expressing only pathogenic mutations allowed the point to be established that the mutations favor PrP deposition and/or the development of spontaneous neurodegeneration [318, 319], although the contribution of PrP overexpression in this process was much debated [191]. Knock-in mice engineered to express mouse PrP mutations associated with FFI developed a spontaneous diseases that resembled FFI and produced *de novo* infectious prions [193], suggesting that PrP mutations favor PrP^{C} to form PrP^{Sc}.

15.3 Infectious Human Prion Diseases

Prions from different sources have infected humans, and human prions have in turn been transmitted to others, both by ritualistic cannibalism and iatrogenic means. Kuru, which occurs in the highlands of New Guinea, was transmitted by ritualistic cannibalism, as people in the region attempted to immortalize their dead relatives by eating their brains [320]. Iatrogenic transmissions include prion-tainted human growth hormone (hGH) and gonadotropin, dura mater grafts, and corneal transplants from people who had died from CJD. In addition, CJD cases have been recorded after neurosurgical procedures in which ineffectively sterilized depth electrodes or instruments were used. As with the inherited and sporadic forms, the Met/Val polymorphism at PrP codon 129 affects the development and expression of acquired prion diseases (see Fig. 3) [310].

15.3.1 Human Growth Hormone

Since the mid-1980s, more than 165 young adults have been diagnosed with iatrogenic CJD (iCJD) between four and 30 years after receiving hGH or gonadotropin from cadaveric pituitaries. The longest incubation periods (20–30 years) are similar to those associated with more recent cases of Kuru. Since 1985, recombinant HGH produced in *E.coli* has been used in place of cadaveric hGH, and with recombinant hGH no cases of iCJD have been identified.

15.3.2 Variant Creutzfeldt–Jakob Disease (vCJD)

The first cases of CJD in teenagers and young adults that were eventually labeled variant CJD (vCJD) occurred in Great Britain in 1994 [321]. To date, about 230 clinical cases of vCJD have been reported, mostly among teenagers and young adults from the UK, France, Spain, and Ireland. When compared to sCJD, the average age of vCJD patients is 28 years (as opposed to 68 years), and the disease duration is slightly longer (median 14 months versus 4.5 months). Additionally, vCJD is characterized by numerous PrP amyloid plaques surrounded by a halo of intense spongiform degeneration in the brain, and by a distinctive PrP^{Sc} electrophoretic pattern (Fig. 9). These unusual neuropathological changes have not been seen in CJD cases in the United States, Australia, or Japan. The majority of vCJD patients presented with psychiatric symptoms, including dysphoria, withdrawal, anxiety, insomnia, and loss of interest. The peripheral pathogenesis of vCJD was also strikingly distinct from sporadic or acquired forms of the disease, with a pronounced and uniform contamination of lymphoreticular tissues [322]. All clinical cases involved patients homozygous for PrP with Met at position 129 [321]. From both epidemiologic and experimental studies, evidence is now quite compelling that vCJD is the result of prions being transmitted from cattle with BSE to humans through the consumption of contaminated beef products (*vide infra*). Today, studies of the prion diseases have taken on a new significance with the identification of vCJD, and these tragic cases have generated a continuing discourse concerning "mad cows," prions, and the safety of the human and animal food supplies worldwide.

The number of cases of vCJD caused by bovine prions that will occur during the years ahead is unknown. Whereas, the number of definite cases has remained low and is in steady decline, the estimated number of asymptomatic carriers may be considerably higher, approaching – in the UK – 1 in 2000 individuals born between

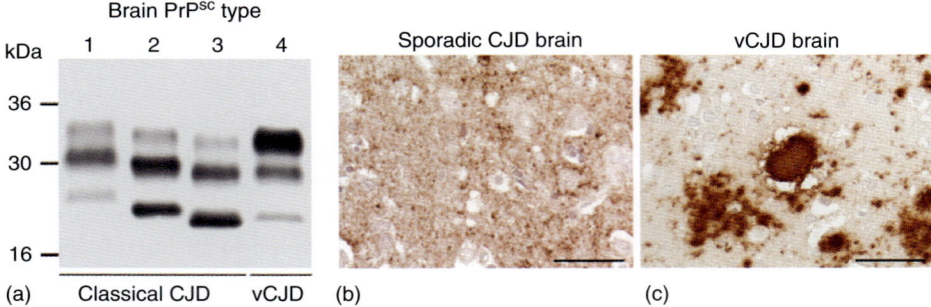

Fig. 9 Distinctive strain features of variant Creutzfeldt–Jakob disease (vCJD). (a) Western blot of proteinase K-digested brain material showing PrP^{Sc} electrophoretic profiles associated with CJD (either sporadic or iatrogenic) and vCJD cases. Type 4, which is only seen in vCJD cases, is characterized by a predominance of diglycosylated PrP^{Sc} (upper band). The anti-PrP monoclonal antibody 3F4 was used; (b,c) Immunohistochemical staining of brain sections from sporadic (b) CJD and (c) vCJD, showing PrP^{Sc} deposition in the brain. While PrP^{Sc} deposition in sporadic CJD is mostly diffuse, vCJD is distinguished by the presence of florid PrP plaques consisting of a round amyloid core surrounded by a ring of spongiform vacuoles. Scale bars = 50 mm. The anti-PrP ICSM35 was used to reveal the sections. Reproduced with permission from Ref. [323]; © 2010, Neuropathology and Applied Neurobiology, British Neuropathological Society, Wiley.

1941 and 1985. This estimate is based on surveys of tissues collected during appendicectomies and tonsillectomies and screened for the presence of PrP^{Sc} [324, 325]. The interindividual transmission of vCJD can occur through blood transfusions or via the use of blood products from asymptomatic donors that later died from vCJD [326–328]. That one such patient was heterozygous at position 129 indicated that substantial levels of endogenous vCJD infectivity are present in the blood at the preclinical stage of disease. A bioassay titration of vCJD blood fractions from a vCJD-confirmed patient suggested a low infectious titer [329], as in natural and experimental models of prion diseases [330–332]. Currently, there is significant concern that a secondary transmission of vCJD could also occur through contaminated surgical and medical instruments. Prions are known to resist conventional inactivation methods, and the accidental transmission of CJD has been reported to occur during diagnostic or surgical procedures [333].

16
Prion Diseases of Animals

The prion diseases of animals include scrapie of sheep and goats, BSE, TME, chronic wasting disease (CWD) of mule deer and elk, feline spongiform encephalopathy, and exotic ungulate encephalopathy (Table 2).

16.1
Scrapie in Sheep and Goats

Discovered more than two centuries ago in the UK, scrapie is the generic term that designates prion diseases of sheep and goats. The disease is not uniform, but the number of prion strain(s) actually involved has so far remained elusive [334] and cases have been reported worldwide.

As with other prion diseases, the clinico-pathological presentation and peripheral tissue (lymphoid tissue, muscle, and placenta) contamination of scrapie vary greatly with the contaminating prion strain type(s) and the animals' genetic background. The disease is thought to propagate horizontally and vertically, and its prevalence may attain 10 in 10 000 animals, with high variations among different countries [335]. The PrP genetics of naturally acquired or experimental scrapie in sheep has been extensively investigated, and polymorphism in the PrP gene is known to greatly influence susceptibility to the disease and its incubation period (see Fig. 3). Most notably, polymorphisms at codons 136, 154, and 171 of the PrP gene that produce amino acid substitutions have been studied with respect to the incidence of scrapie [336]. The pronounced resistance of animals homozygous for the ARR allele has been the basis of genetic selection of breeds resistant to disease in some European countries such as France, the UK and the Netherlands. An "atypical" form of scrapie was discovered in the early 2000s in European sheep and goat flocks through active surveillance programs based on the rapid biochemical detection of PrP^{res} in animal brains. Also known as Nor98 [337], this atypical scrapie is characterized by a PrP^{res} pattern that shows a lower degree of resistance to protease digestion compared to classical scrapie (see Fig. 1).

Atypical scrapie has been detected worldwide, and its prevalence can exceed that of classical scrapie in some European countries [335, 338]. Most worryingly, animals homozygous for the ARR are fully susceptible to atypical scrapie [339]. Although the etiology of atypical scrapie is still speculative, a sporadic, spontaneous origin is currently favored (Table 2) [340, 341].

16.2
Chronic Wasting Disease

Mule deer, white-tailed deer and elk have all been reported to develop CWD, which is unique among the prion diseases because it is the only such disease identified in free-ranging animals [342]. First discovered in Colorado in 1967, and reported in 1978 as a spongiform encephalopathy, based on histopathology in the brain, the disease has expanded geographically to a large part of North America, including Canada. Imported cases have also been notified in South Korea.

The origin of CWD is unknown, but several aspects of its physiopathology are reminiscent of sheep scrapie. A methionine/leucine polymorphism at codon 132 (the site corresponding to codon 129 in humans) of elk PrP gene influences the animal's susceptibility to CWD (see Fig. 3). Sick animals homozygous for Met at codon 132 are overrepresented [343], an effect which has been reproduced in transgenic mice [344]. Some endemic pockets have been noted. In captive cervid (deer) herds, up to 90% of mule deer have been reported to be positive for prions, while the incidence of CWD in cervids living in the wild has been estimated to up to 15%.

The mode of transmission of CWD prions among mule deer, white-tailed deer and elk is unresolved, but horizontal transmission through the contamination of grass with prions excreted in fecal matter [345] or bodily fluids such as saliva [346] seems to be a likely source. At variance with scrapie, the placenta is poorly infectious, which suggests that transmission from mother-to-lamb might be rather low [347]. CWD prions exhibit a pronounced tropism for lymphoid tissues, and the muscles are infectious [348, 349]. Of particular importance is an ability to assess CWD for potential cross-species transmission to other farm animals that might come into contact with the cervids (e.g., cattle, sheep), wild-life scavengers (racoons, etc.), and humans (in particular, hunters and venison meat eaters). It is essential in this respect to document any strain variation of CWD agents, though this task is complicated by the multiplicity of the affected species.

16.3
Bovine Spongiform Encephalopathy

Almost 30 years after the first cases were diagnosed in the UK, the BSE epidemic is under control in most European countries. Since 1985, more than 180 000 cattle (primarily dairy cows) have died of BSE, and it is estimated that between one and two million cattle were infected with prions.

BSE is a massive common-source epidemic caused by meat and bone meal (MBM) being fed primarily to dairy cows. The MBM was prepared from the offal of sheep, cattle, pigs and chickens as a high-protein nutritional supplement. During the late 1970s, the hydrocarbon-solvent extraction method used to render offal began to be abandoned, and this resulted in MBM with a much higher fat content. It is now thought that this change allowed the initial prions – regardless of whether they originated from sheep or spontaneously from cattle – to survive the rendering process and to be passed into cattle. Additional species, including domestic cats and zoological species, have contracted a spongiform encephalopathy due to the consumption of BSE-contaminated MBM. The practice of feeding MBM to ruminants was banned in July 1988, and although the BSE epidemic is fading it is unclear whether cases will arise "spontaneously" in the future. There is good evidence to indicate that a unique major prion strain is responsible for the BSE epidemic [334].

Although no reliable, specific test for prion disease in live animals is currently available, immunoassays for PrP^{Sc} in the brainstems of cattle currently provide the best available approach for preventing BSE prions from entering the human food supply. This active and systematic surveillance brought to light the existence of new, "atypical" BSE prion types that were distinct from "classical" (i.e., epidemic) BSE (see Table 2). These were termed H-BSE and L-BSE, which referred to their PrP^{res} signature as compared to BSE [334]. Both, H-BSE and L-BSE have been detected in rare cases of aged cattle, worldwide, at a low prevalence that was consistent with the possibility of sporadic forms of TSE in cattle [350]. As with classical BSE, these cases can be transmitted experimentally to a number of foreign species and/or to transgenic mice expressing non-bovine PrP [334], which suggests that their potential to propagate in other species is high.

16.4
Transmission of BSE Prions to Humans, and the Zoonotic Potential of Non-BSE Animal Prions

The restricted geographic occurrence and chronology of vCJD rapidly raised the possibility that BSE prions have been transmitted to humans. There is no set of dietary habits that can distinguish vCJD patients from apparently healthy people; moreover, there is no explanation for the predilection of vCJD for teenagers and young adults. In animal tests, cynomolgus macaques infected experimentally with BSE developed brain lesions similar to those observed in human vCJD cases, including notably the presence of florid plaques [351]

(Fig. 9). Further evidence for a common link between prions from cattle with BSE and those from humans with vCJD was provided by the experimental transmission to a panel of inbred mouse lines and the observation of similar incubation times, neuropathology and PrP^{Sc} molecular signature, at variance with either scrapie or sCJD [352–354]. However, these studies may have suffered from the transmission of both BSE and vCJD prions to a heterologous host, that is, with a species barrier (see Sect. 14). By using transgenic mice expressing bovine PrP or the Met allele of human PrP at codon 129, compelling evidence was provided for the transmission of bovine prions to humans. Both, BSE and human vCJD prions were transmitted readily to a number of different bovine PrP mice and exhibited similar phenotypes [355, 356]. In some human PrP (Met$_{129}$) mice, both BSE and vCJD prions exhibited a similar phenotype but serial passaging was necessary to observe this [300, 356, 357]. During these transmissions, a divergent evolution of BSE/vCJD prions occasionally occurred with the emergence of a prion having a distinct phenotype, which suggested that some substrain components may be present in the BSE prions (see Sect. 14). Worthy of note, BSE prions may gain virulence in "humanized" and "bovinized" mice after their intermediate passage in small ruminants, or on the ovine/caprine PrP sequence [356, 358]. Whilst BSE prions are usually transmitted inefficiently to human PrP (Met$_{129}$) mice on primary passage (as gauged by the absence of any detectable replication in the brain of most inoculated animals), it appears that the species barrier was more readily crossed in the lymphoid tissues, asymptomatically [293]. There might, in the lymphoid tissue, be a better conformational compatibility between the invading prion and the cellular PrP, or a favored environment for prion conversion. These findings have implications for public health risk assessments, with regard to the real prevalence of subclinically infected vCJD individuals [324].

The recent identification of new "atypical" forms of BSE and scrapie pose many questions about their zoonotic potential. In the case of classical scrapie, there is a widely shared consensus among the risk management community that this disease is not zoonotic, based on the apparent lack of any spatiotemporal correlation between the frequency of human prion diseases and the recognized presence/absence of prion diseases in animals. However, the low incidence of human and animal prion diseases renders the epidemiological analysis of the data collected over short periods delicate. The recent discovery of atypical scrapie in New Zealand and Australia – two countries that were considered free of animal prions – also clearly illustrates the limits of this argument [359, 360]. A very limited number of experimental results have suggested that non-BSE animal prions could potentially be zoonotic. One sheep scrapie isolate [361] and one atypical L-BSE case [362] were transmitted to primates models with similar efficacy as cattle BSE, and L-BSE was also transmitted successfully to lemurs via an oral challenge [363]; however, CWD was transmissible to squirrel monkeys but not to cynomolgus macaques [364]. L-type BSE propagated without apparent species barrier, and was more virulent than classical BSE in different human PrP (Met$_{129}$) mice lines [357, 365]. While CWD and scrapie do not seem to transmit to certain humanized mouse models [366–369], it must be noted that the possibility of asymptomatic peripheral replication has not been addressed [293]. Furthermore, the restricted number and diversity of the inoculated cases and the absence of experimental transmission to all lines expressing the human PrP polymorphism at codon 129 does not allow the hypothesis that these prions might be zoonotic to be ruled out.

17
Fungal Prions

Although prions were originally defined in the context of an infectious pathogen, it is now becoming widely accepted that they are elements that impart and propagate variability through multiple conformers of a normal, cellular protein. Such a mechanism is not restricted to a single class of transmissible pathogens. Important cellular functions such as long-term memory or immune response to viral infection have been found to be governed by a similar mechanism of information transfer with cytoplasmic polyadenylation element binding (CPEB) and mitochondrial antiviral signaling (MAVS) proteins, respectively [370–373].

Two notable prion-like genetic determinants, [URE3] and [PSI], have been initially described in the yeast *Saccharomyces cerevisiae* [374], and a growing number of proteins are now recognized with prion properties in yeast [375, 376]; worthy of note, none of these shares any sequence homology to PrP^C. As these proteins under the prion form transmit biological information (usually a loss of function of the protein in the non-prion conformation) both vertically (to the progeny) and horizontally (to neighbor cells), they can be considered as epigenetic factors [377]. Another prion-like determinant [Het-s] has been described in the filamentous fungus *Podospora anserine*. Under its prion form, HET-s controls a cell death program termed heterokaryon incompatibility [378].

Both, Ure2p (a regulator of nitrogen catabolism) and Sup35p (a translation-termination factor) have a C terminally located functional domain and an N-terminally defined, canonical prion domain enriched in Q and N, as do most yeast prions. Structurally, the prions are amyloid forms of the respective proteins, and the prion domain becomes structured, constituting a cross-β spine, a structure that may be common to many amyloid fibrils [379]. At variance with PrP prions, chaperones such as the disaggregase Hsp104 are required for the propagation of yeast prions [375].

The observation of prion conversion and aggregation by infection with amyloids of recombinant Sup35p [380] or HET-s proteins [381] was key to demonstrating that a protein can act as an "infectious" agent by causing self-propagating, conformational changes. As in mammals, yeast prions can exist as strains, exhibiting different biochemical phenotypes or an ability to cross transmission barriers due to sequence differences [382–384]. Overall, studies of prions in fungi have been extremely helpful in establishing the prion concept.

18
Prevention and Therapeutics for Prion Diseases

Because people at risk of inherited prion diseases can now be identified decades before any neurologic dysfunction becomes evident, the development of an effective therapy for these fully penetrant disorders is imperative. With systems for predicting the number of individuals who may develop variant CJD being currently uncertain, the identification of an effective therapy now seems most prudent, and interfering with the conversion of PrP^C into PrP^{Sc} would appear to be the most attractive therapeutic target.

To define the pathogenesis of prion disease is an important issue with respect to developing an effective therapy. The issue of whether oligomers of misprocessed PrP rather than large or amyloid aggregates cause CNS degeneration, as in other neurodegenerative diseases [385, 386], has been addressed in several studies of prion diseases in humans, as well as in transgenic mice.

Various compounds have been proposed as potential therapeutics for the treatment of prion diseases [387], including polyanion, dextrans, antibiotics (e.g., amphotericin B and derivatives) and cyclic tetrapyrroles, all of which have been shown in rodents to increase survival times when administered at about the time of prion infection. Unfortunately, however, very few of these agents were efficient when administered after neuroinvasion has been established [388]. In addition to studies in rodents, cells which have been chronically infected with prions were used to identify candidate antiprion drugs, but none has proved effective in halting prion diseases in either animals or humans. Nonetheless, the successful use of yeast prions to screen for anti-mammalian prion drugs would suggest that targeting pathways involved in prion formation or maintenance might represent a promising approach [389, 390].

It is notable that anti-PrP antibodies delivered by transgenetics [391] or by injection [392] were successful in blocking prion replication in mice inoculated intraperitoneally, yet the same anti-PrP antibodies were ineffective in treating prion infections initiated by intracerebral inoculation. Whether antibodies can be effectively administered to humans to prevent and treat prion diseases is unclear, as neither antibodies nor Fab fragments can cross the blood–brain barrier in sufficiently high concentrations. Consequently, the delivery of these proteins to the CNS remains a critical issue.

19
Implications for Common Neurodegenerative Diseases

An understanding of how PrP^C misfolds into PrP^{Sc} will undoubtedly open new approaches to deciphering the causes of – and developing effective therapies for – some common neurodegenerative diseases, including Alzheimer's disease, Parkinson's disease, Huntington's disease and amyotrophic lateral sclerosis. Indeed, there is mounting evidence that a prion-like scenario takes place during the pathogenesis of each of these diseases, with APP, α-synuclein, huntingtin, and SOD-1 amyloid-prone proteins, respectively [393–397]. Many of the basic principles of prion biology, which are becoming increasingly well understood, will undoubtedly have much wider implications and benefit in the near future.

References

1. Wuthrich, K., and Riek, R. (2001) Three-dimensional structures of prion proteins. *Adv. Protein Chem.* **57**, 55–82.
2. Riek, R., Hornemann, S., Wider, G., Billeter, M., Glockshuber, R., and Wuthrich, K. (1996) NMR structure of the mouse prion protein domain PrP(121–231). *Nature*, **382**, 180–182.
3. Eghiaian, F., Grosclaude, J., Lesceu, S., Debey, P., Doublet, B., Treguer, E., Rezaei, H., and Knossow, M. (2004) Insight into the PrPC-->PrPSc conversion from the structures of antibody-bound ovine prion scrapie-susceptibility variants. *Proc. Natl Acad. Sci. USA*, **101**, 10254–10259.

4. Diaz-Espinoza, R., and Soto, C. (2012) High-resolution structure of infectious prion protein: the final frontier. *Nat. Struct. Mol. Biol.*, **19**, 370–377.
5. Prusiner, S.B. (1982) Novel proteinaceous infectious particles cause scrapie. *Science*, **216**, 136–144.
6. Pan, K.M., Baldwin, M., Nguyen, J., Gasset, M., Serban, A., Groth, D., Mehlhorn, I., Huang, Z., Fletterick, R.J., Cohen, F.E., et al. (1993) Conversion of alpha-helices into beta-sheets features in the formation of the scrapie prion proteins. *Proc. Natl Acad. Sci. USA*, **90**, 10962–10966.
7. Safar, J.G., Kellings, K., Serban, A., Groth, D., Cleaver, J.E., Prusiner, S.B., and Riesner, D. (2005) Search for a prion-specific nucleic acid. *J. Virol.*, **79**, 10796–10806.
8. Colby, D.W., and Prusiner, S.B. (2011) De novo generation of prion strains. *Nat. Rev. Microbiol.* **9**, 771–777.
9. Cronier, S., Gros, N., Tattum, M.H., Jackson, G.S., Clarke, A.R., Collinge, J., and Wadsworth, J.D. (2008) Detection and characterization of proteinase K-sensitive disease-related prion protein with thermolysin. *Biochem. J.* **416**, 297–305.
10. Safar, J., Wille, H., Itri, V., Groth, D., Serban, H., Torchia, M., Cohen, F.E., and Prusiner, S.B. (1998) Eight prion strains have PrP(Sc) molecules with different conformations. *Nat. Med.*, **4**, 1157–1165.
11. Safar, J.G., DeArmond, S.J., Kociuba, K., Deering, C., Didorenko, S., Bouzamondo-Bernstein, E., Prusiner, S.B., and Tremblay, P. (2005) Prion clearance in bigenic mice. *J. Gen. Virol.*, **86**, 2913–2923.
12. Sajnani, G., Silva, C.J., Ramos, A., Pastrana, M.A., Onisko, B.C., Erickson, M.L., Antaki, E.M., Dynin, I., Vazquez-Fernandez, E., Sigurdson, C.J., et al. (2012) PK-sensitive PrP is infectious and shares basic structural features with PK-resistant PrP. *PLoS Pathog.*, **8**, e1002547.
13. Berardi, V.A., Cardone, F., Valanzano, A., Lu, M., and Pocchiari, M. (2006) Preparation of soluble infectious samples from scrapie-infected brain: a new tool to study the clearance of transmissible spongiform encephalopathy agents during plasma fractionation. *Transfusion*, **46**, 652–658.
14. Cronier, S., Béringue, V., Bellon, A., Peyrin, J.M., and Laude, H. (2007) Prion strain- and species-dependent effects of antiprion molecules in primary neuronal cultures. *J. Virol.*, **81**, 13794–13800.
15. Deleault, A.M., Deleault, N.R., Harris, B.T., Rees, J.R., and Supattapone, S. (2008) The effects of prion protein proteolysis and disaggregation on the strain properties of hamster scrapie. *J. Gen. Virol.*, **89**, 2642–2650.
16. Tzaban, S., Friedlander, G., Schonberger, O., Horonchik, L., Yedidia, Y., Shaked, G., Gabizon, R., and Taraboulos, A. (2002) Protease-sensitive scrapie prion protein in aggregates of heterogeneous sizes. *Biochemistry*, **41**, 12868–12875.
17. Pastrana, M.A., Sajnani, G., Onisko, B., Castilla, J., Morales, R., Soto, C., and Requena, J.R. (2006) Isolation and characterization of a proteinase K-sensitive PrPSc fraction. *Biochemistry*, **45**, 15710–15717.
18. Laferriere, F., Tixador, P., Moudjou, M., Chapuis, J., Sibille, P., Herzog, L., Reine, F., Jaumain, E., Laude, H., Rezaei, H., et al. (2013) Quaternary structure of pathological prion protein as a determining factor of strain-specific prion replication dynamics. *PLoS Pathog.*, **9**, e1003702.
19. Tixador, P., Herzog, L., Reine, F., Jaumain, E., Chapuis, J., Le Dur, A., Laude, H., and Béringue, V. (2010) The physical relationship between infectivity and prion protein aggregates is strain-dependent. *PLoS Pathog.*, **6**, e1000859.
20. Lasmezas, C.I., Deslys, J.P., Robain, O., Jaegly, A., Béringue, V., Peyrin, J.M., Fournier, J.G., Hauw, J.J., Rossier, J., and Dormont, D. (1997) Transmission of the BSE agent to mice in the absence of detectable abnormal prion protein. *Science*, **275**, 402–405.
21. Piccardo, P., Manson, J.C., King, D., Ghetti, B., and Barron, R.M. (2007) Accumulation of prion protein in the brain that is not associated with transmissible disease. *Proc. Natl Acad. Sci. USA*, **104**, 4712–4717.
22. Moore, R.C., Hope, J., McBride, P.A., McConnell, I., Selfridge, J., Melton, D.W., and Manson, J.C. (1998) Mice with gene targetted prion protein alterations show that Prnp, Sinc and Prni are congruent. *Nat. Genet.*, **18**, 118–125.
23. Bolton, D.C., McKinley, M.P., and Prusiner, S.B. (1982) Identification of a protein that purifies with the scrapie prion. *Science*, **218**, 1309–1311.

24. DeArmond, S.J., McKinley, M.P., Barry, R.A., Braunfeld, M.B., McColloch, J.R., and Prusiner, S.B. (1985) Identification of prion amyloid filaments in scrapie-infected brain. *Cell*, **41**, 221–235.
25. McKinley, M.P., Bolton, D.C., and Prusiner, S.B. (1983) A protease-resistant protein is a structural component of the scrapie prion. *Cell*, **35**, 57–62.
26. Bendheim, P.E., Barry, R.A., DeArmond, S.J., Stites, D.P., and Prusiner, S.B. (1984) Antibodies to a scrapie prion protein. *Nature*, **310**, 418–421.
27. Oesch, B., Westaway, D., Walchli, M., McKinley, M.P., Kent, S.B., Aebersold, R., Barry, R.A., Tempst, P., Teplow, D.B., Hood, L.E., et al. (1985) A cellular gene encodes scrapie PrP 27–30 protein. *Cell*, **40**, 735–746.
28. Locht, C., Chesebro, B., Race, R., and Keith, J.M. (1986) Molecular cloning and complete sequence of prion protein cDNA from mouse brain infected with the scrapie agent. *Proc. Natl Acad. Sci. USA*, **83**, 6372–6376.
29. Westaway, D., Daude, N., Wohlgemuth, S., and Harrison, P. (2011) The PrP-like proteins Shadoo and Doppel. *Top. Curr. Chem.*, **305**, 225–256.
30. Ehsani, S., Tao, R., Pocanschi, C.L., Ren, H., Harrison, P.M., and Schmitt-Ulms, G. (2011) Evidence for retrogene origins of the prion gene family. *PLoS ONE*, **6**, e26800.
31. Behrens, A., Genoud, N., Naumann, H., Rulicke, T., Janett, F., Heppner, F.L., Ledermann, B., and Aguzzi, A. (2002) Absence of the prion protein homologue Doppel causes male sterility. *EMBO J.*, **21**, 3652–3658.
32. Moore, R.C., Lee, I.Y., Silverman, G.L., Harrison, P.M., Strome, R., Heinrich, C., Karunaratne, A., Pasternak, S.H., Chishti, M.A., Liang, Y., et al. (1999) Ataxia in prion protein (PrP)-deficient mice is associated with upregulation of the novel PrP-like protein doppel. *J. Mol. Biol.*, **292**, 797–817.
33. Watts, J.C., Drisaldi, B., Ng, V., Yang, J., Strome, B., Horne, P., Sy, M.S., Yoong, L., Young, R., Mastrangelo, P., et al. (2007) The CNS glycoprotein Shadoo has PrP(C)-like protective properties and displays reduced levels in prion infections. *EMBO J.*, **26**, 4038–4050.
34. Daude, N., Wohlgemuth, S., Brown, R., Pitstick, R., Gapeshina, H., Yang, J., Carlson, G.A., and Westaway, D. (2012) Knockout of the prion protein (PrP)-like Sprn gene does not produce embryonic lethality in combination with PrP(C)-deficiency. *Proc. Natl Acad. Sci. USA*, **109**, 9035–9040.
35. Young, R., Passet, B., Vilotte, M., Cribiu, E.P., Béringue, V., Le Provost, F., Laude, H., and Vilotte, J.L. (2009) The prion or the related Shadoo protein is required for early mouse embryogenesis. *FEBS Lett.*, **583**, 3296–3300.
36. Aguzzi, A., Baumann, F., and Bremer, J. (2008) The prion's elusive reason for being. *Annu. Rev. Neurosci.*, **31**, 439–477.
37. Kretzschmar, H.A., Stowring, L.E., Westaway, D., Stubblebine, W.H., Prusiner, S.B., and Dearmond, S.J. (1986) Molecular cloning of a human prion protein cDNA. *DNA*, **5**, 315–324.
38. Liao, Y.C., Lebo, R.V., Clawson, G.A., and Smuckler, E.A. (1986) Human prion protein cDNA: molecular cloning, chromosomal mapping, and biological implications. *Science*, **233**, 364–367.
39. Basler, K., Oesch, B., Scott, M., Westaway, D., Walchli, M., Groth, D.F., McKinley, M.P., Prusiner, S.B., and Weissmann, C. (1986) Scrapie and cellular PrP isoforms are encoded by the same chromosomal gene. *Cell*, **46**, 417–428.
40. Mahal, S.P., Asante, E.A., Antoniou, M., and Collinge, J. (2001) Isolation and functional characterisation of the promoter region of the human prion protein gene. *Gene*, **268**, 105–114.
41. Manson, J., West, J.D., Thomson, V., McBride, P., Kaufman, M.H., and Hope, J. (1992) The prion protein gene: a role in mouse embryogenesis? *Development*, **115**, 117–122.
42. Tremblay, P., Bouzamondo-Bernstein, E., Heinrich, C., Prusiner, S.B., and DeArmond, S.J. (2007) Developmental expression of PrP in the post-implantation embryo. *Brain Res.*, **1139**, 60–67.
43. Khalife, M., Young, R., Passet, B., Halliez, S., Vilotte, M., Jaffrezic, F., Marthey, S., Béringue, V., Vaiman, D., Le Provost, F., et al. (2011) Transcriptomic analysis brings new insight into the biological role of the prion protein during mouse embryogenesis. *PLoS ONE*, **6**, e23253.
44. Young, R., Bouet, S., Polyte, J., Le Guillou, S., Passet, B., Vilotte, M., Castille, J.,

Béringue, V., Le Provost, F., Laude, H., et al. (2011) Expression of the prion-like protein Shadoo in the developing mouse embryo. *Biochem. Biophys. Res. Commun.*, **416**, 184–187.

45. Nunziante, M., Gilch, S., and Schatzl, H.M. (2003) Essential role of the prion protein N terminus in subcellular trafficking and half-life of cellular prion protein. *J. Biol. Chem.*, **278**, 3726–3734.

46. Sunyach, C., Jen, A., Deng, J., Fitzgerald, K.T., Frobert, Y., Grassi, J., McCaffrey, M.W., and Morris, R. (2003) The mechanism of internalization of glycosylphosphatidylinositol-anchored prion protein. *EMBO J.*, **22**, 3591–3601.

47. Turnbaugh, J.A., Unterberger, U., Saa, P., Massignan, T., Fluharty, B.R., Bowman, F.P., Miller, M.B., Supattapone, S., Biasini, E., and Harris, D.A. (2012) The N-terminal, polybasic region of PrP(C) dictates the efficiency of prion propagation by binding to PrP(Sc). *J. Neurosci.*, **32**, 8817–8830.

48. Aguzzi, A., and Calella, A.M. (2009) Prions: protein aggregation and infectious diseases. *Physiol. Rev.*, **89**, 1105–1152.

49. Brown, D.R., Qin, K., Herms, J.W., Madlung, A., Manson, J., Strome, R., Fraser, P.E., Kruck, T., von Bohlen, A., Schulz-Schaeffer, W., et al. (1997) The cellular prion protein binds copper in vivo. *Nature*, **390**, 684–687.

50. Jackson, G.S., Murray, I., Hosszu, L.L., Gibbs, N., Waltho, J.P., Clarke, A.R., and Collinge, J. (2001) Location and properties of metal-binding sites on the human prion protein. *Proc. Natl Acad. Sci. USA*, **98**, 8531–8535.

51. Stahl, N., Borchelt, D.R., Hsiao, K., and Prusiner, S.B. (1987) Scrapie prion protein contains a phosphatidylinositol glycolipid. *Cell*, **51**, 229–240.

52. Stahl, N., Borchelt, D.R., and Prusiner, S.B. (1990) Differential release of cellular and scrapie prion proteins from cellular membranes by phosphatidylinositol-specific phospholipase C. *Biochemistry*, **29**, 5405–5412.

53. Lingwood, D., and Simons, K. (2010) Lipid rafts as a membrane-organizing principle. *Science*, **327**, 46–50.

54. Hegde, R.S., Mastrianni, J.A., Scott, M.R., DeFea, K.A., Tremblay, P., Torchia, M., DeArmond, S.J., Prusiner, S.B., and Lingappa, V.R. (1998) A transmembrane form of the prion protein in neurodegenerative disease. *Science*, **279**, 827–834.

55. Ma, J., Wollmann, R., and Lindquist, S. (2002) Neurotoxicity and neurodegeneration when PrP accumulates in the cytosol. *Science*, **298**, 1781–1785.

56. Rudd, P.M., Merry, A.H., Wormald, M.R., and Dwek, R.A. (2002) Glycosylation and prion protein. *Curr. Opin. Struct. Biol.*, **12**, 578–586.

57. Béringue, V., Mallinson, G., Kaisar, M., Tayebi, M., Sattar, Z., Jackson, G., Anstee, D., Collinge, J., and Hawke, S. (2003) Regional heterogeneity of cellular prion protein isoforms in the mouse brain. *Brain*, **126**, 2065–2073.

58. Salamat, M.K., Dron, M., Chapuis, J., Langevin, C., and Laude, H. (2011) Prion propagation in cells expressing PrP glycosylation mutants. *J. Virol.*, **85**, 3077–3085.

59. Prusiner, S.B., Scott, M.R., DeArmond, S.J., and Cohen, F.E. (1998) Prion protein biology. *Cell*, **93**, 337–348.

60. Caughey, B., Race, R.E., Ernst, D., Buchmeier, M.J., and Chesebro, B. (1989) Prion protein biosynthesis in scrapie-infected and uninfected neuroblastoma cells. *J. Virol.*, **63**, 175–181.

61. Taguchi, Y., Shi, Z.D., Ruddy, B., Dorward, D.W., Greene, L., and Baron, G.S. (2009) Specific biarsenical labeling of cell surface proteins allows fluorescent- and biotin-tagging of amyloid precursor protein and prion proteins. *Mol. Biol. Cell*, **20**, 233–244.

62. Parizek, P., Roeckl, C., Weber, J., Flechsig, E., Aguzzi, A., and Raeber, A.J. (2001) Similar turnover and shedding of the cellular prion protein in primary lymphoid and neuronal cells. *J. Biol. Chem.*, **276**, 44627–44632.

63. Borchelt, D.R., Scott, M., Taraboulos, A., Stahl, N., and Prusiner, S.B. (1990) Scrapie and cellular prion proteins differ in their kinetics of synthesis and topology in cultured cells. *J. Cell Biol.*, **110**, 743–752.

64. Lehmann, S., Milhavet, O., and Mange, A. (1999) Trafficking of the cellular isoform of the prion protein. *Biomed. Pharmacother.*, **53**, 39–46.

65. Vey, M., Pilkuhn, S., Wille, H., Nixon, R., DeArmond, S.J., Smart, E.J., Anderson, R.G., Taraboulos, A., and Prusiner, S.B. (1996) Subcellular colocalization of the cellular and scrapie prion proteins in caveolae-like

membranous domains. *Proc. Natl Acad. Sci. USA*, **93**, 14945–14949.
66. Peters, P.J., Mironov, A., Jr, Peretz, D., van Donselaar, E., Leclerc, E., Erpel, S., DeArmond, S.J., Burton, D.R., Williamson, R.A., Vey, M., et al. (2003) Trafficking of prion proteins through a caveolae-mediated endosomal pathway. *J. Cell Biol.*, **162**, 703–717.
67. Shyng, S.L., Heuser, J.E., and Harris, D.A. (1994) A glycolipid-anchored prion protein is endocytosed via clathrin-coated pits. *J. Cell Biol.*, **125**, 1239–1250.
68. Shyng, S.L., Huber, M.T., and Harris, D.A. (1993) A prion protein cycles between the cell surface and an endocytic compartment in cultured neuroblastoma cells. *J. Biol. Chem.*, **268**, 15922–15928.
69. Harris, D.A., Huber, M.T., van Dijken, P., Shyng, S.L., Chait, B.T., and Wang, R. (1993) Processing of a cellular prion protein: identification of N- and C-terminal cleavage sites. *Biochemistry*, **32**, 1009–1016.
70. Walmsley, A.R., Watt, N.T., Taylor, D.R., Perera, W.S., and Hooper, N.M. (2009) alpha-cleavage of the prion protein occurs in a late compartment of the secretory pathway and is independent of lipid rafts. *Mol. Cell. Neurosci.*, **40**, 242–248.
71. Mange, A., Beranger, F., Peoc'h, K., Onodera, T., Frobert, Y., and Lehmann, S. (2004) Alpha- and beta- cleavages of the amino-terminus of the cellular prion protein. *Biol. Cell*, **96**, 125–132.
72. Cisse, M.A., Sunyach, C., Lefranc-Jullien, S., Postina, R., Vincent, B., and Checler, F. (2005) The disintegrin ADAM9 indirectly contributes to the physiological processing of cellular prion by modulating ADAM10 activity. *J. Biol. Chem.*, **280**, 40624–40631.
73. Dron, M., Moudjou, M., Chapuis, J., Salamat, M.K., Bernard, J., Cronier, S., Langevin, C., and Laude, H. (2010) Endogenous proteolytic cleavage of disease-associated prion protein to produce c04 fragments is strongly cell- and tissue-dependent. *J. Biol. Chem.*, **285**, 10252–10264.
74. Toni, M., Massimino, M.L., Griffoni, C., Salvato, B., Tomasi, V., and Spisni, E. (2005) Extracellular copper ions regulate cellular prion protein (PrPC) expression and metabolism in neuronal cells. *FEBS Lett.*, **579**, 741–744.
75. Parkin, E.T., Watt, N.T., Turner, A.J., and Hooper, N.M. (2004) Dual mechanisms for shedding of the cellular prion protein. *J. Biol. Chem.*, **279**, 11170–11178.
76. Westergard, L., Turnbaugh, J.A., and Harris, D.A. (2011) A naturally occurring C-terminal fragment of the prion protein (PrP) delays disease and acts as a dominant-negative inhibitor of PrPSc formation. *J. Biol. Chem.*, **286**, 44234–44242.
77. Chesebro, B., Trifilo, M., Race, R., Meade-White, K., Teng, C., LaCasse, R., Raymond, L., Favara, C., Baron, G., Priola, S., et al. (2005) Anchorless prion protein results in infectious amyloid disease without clinical scrapie. *Science*, **308**, 1435–1439.
78. Stohr, J., Watts, J.C., Legname, G., Oehler, A., Lemus, A., Nguyen, H.O., Sussman, J., Wille, H., DeArmond, S.J., Prusiner, S.B., et al. (2011) Spontaneous generation of anchorless prions in transgenic mice. *Proc. Natl Acad. Sci. USA*, **108**, 21223–21228.
79. Bendheim, P.E., Brown, H.R., Rudelli, R.D., Scala, L.J., Goller, N.L., Wen, G.Y., Kascsak, R.J., Cashman, N.R., and Bolton, D.C. (1992) Nearly ubiquitous tissue distribution of the scrapie agent precursor protein. *Neurology*, **42**, 149–156.
80. Moudjou, M., Frobert, Y., Grassi, J., and La Bonnardiere, C. (2001) Cellular prion protein status in sheep: tissue-specific biochemical signatures. *J. Gen. Virol.*, **82**, 2017–2024.
81. Sales, N., Rodolfo, K., Hassig, R., Faucheux, B., Di Giamberardino, L., and Moya, K.L. (1998) Cellular prion protein localization in rodent and primate brain. *Eur. J. Neurosci.*, **10**, 2464–2471.
82. Moya, K.L., Sales, N., Hassig, R., Creminon, C., Grassi, J., and Di Giamberardino, L. (2000) Immunolocalization of the cellular prion protein in normal brain. *Microsc. Res. Tech.*, **50**, 58–65.
83. McBride, P.A., Eikelenboom, P., Kraal, G., Fraser, H., and Bruce, M.E. (1992) PrP protein is associated with follicular dendritic cells of spleens and lymph nodes in uninfected and scrapie-infected mice. *J. Pathol.*, **168**, 413–418.
84. Li, R., Liu, D., Zanusso, G., Liu, T., Fayen, J.D., Huang, J.H., Petersen, R.B., Gambetti, P., and Sy, M.S. (2001) The expression and potential function of cellular prion protein

in human lymphocytes. *Cell. Immunol.*, **207**, 49–58.
85. Linden, R., Martins, V.R., Prado, M.A., Cammarota, M., Izquierdo, I., and Brentani, R.R. (2008) Physiology of the prion protein. *Physiol. Rev.*, **88**, 673–728.
86. Bueler, H., Fischer, M., Lang, Y., Bluethmann, H., Lipp, H.P., DeArmond, S.J., Prusiner, S.B., Aguet, M., and Weissmann, C. (1992) Normal development and behaviour of mice lacking the neuronal cell-surface PrP protein. *Nature*, **356**, 577–582.
87. Manson, J.C., Clarke, A.R., Hooper, M.L., Aitchison, L., McConnell, I., and Hope, J. (1994) 129/Ola mice carrying a null mutation in PrP that abolishes mRNA production are developmentally normal. *Mol. Neurobiol.*, **8**, 121–127.
88. Richt, J.A., Kasinathan, P., Hamir, A.N., Castilla, J., Sathiyaseelan, T., Vargas, F., Sathiyaseelan, J., Wu, H., Matsushita, H., Koster, J., et al. (2007) Production of cattle lacking prion protein. *Nat. Biotechnol.*, **25**, 132–138.
89. Yu, G., Chen, J., Xu, Y., Zhu, C., Yu, H., Liu, S., Sha, H., Xu, X., Wu, Y., Zhang, A., et al. (2009) Generation of goats lacking prion protein. *Mol. Reprod. Dev.*, **76**, 3.
90. Tobler, I., Gaus, S.E., Deboer, T., Achermann, P., Fischer, M., Rulicke, T., Moser, M., Oesch, B., McBride, P.A., and Manson, J.C. (1996) Altered circadian activity rhythms and sleep in mice devoid of prion protein. *Nature*, **380**, 639–642.
91. Bremer, J., Baumann, F., Tiberi, C., Wessig, C., Fischer, H., Schwarz, P., Steele, A.D., Toyka, K.V., Nave, K.A., Weis, J., et al. (2010) Axonal prion protein is required for peripheral myelin maintenance. *Nat. Neurosci.*, **13**, 310–318.
92. Collinge, J., Whittington, M.A., Sidle, K.C., Smith, C.J., Palmer, M.S., Clarke, A.R., and Jefferys, J.G. (1994) Prion protein is necessary for normal synaptic function. *Nature*, **370**, 295–297.
93. Criado, J.R., Sanchez-Alavez, M., Conti, B., Giacchino, J.L., Wills, D.N., Henriksen, S.J., Race, R., Manson, J.C., Chesebro, B., and Oldstone, M.B. (2005) Mice devoid of prion protein have cognitive deficits that are rescued by reconstitution of PrP in neurons. *Neurobiol. Dis.*, **19**, 255–265.
94. Lledo, P.M., Tremblay, P., DeArmond, S.J., Prusiner, S.B., and Nicoll, R.A. (1996) Mice deficient for prion protein exhibit normal neuronal excitability and synaptic transmission in the hippocampus. *Proc. Natl Acad. Sci. USA*, **93**, 2403–2407.
95. Lipp, H.P., Stagliar-Bozicevic, M., Fischer, M., and Wolfer, D.P. (1998) A 2-year longitudinal study of swimming navigation in mice devoid of the prion protein: no evidence for neurological anomalies or spatial learning impairments. *Behav. Brain Res.*, **95**, 47–54.
96. Mallucci, G.R., Ratte, S., Asante, E.A., Linehan, J., Gowland, I., Jefferys, J.G., and Collinge, J. (2002) Post-natal knockout of prion protein alters hippocampal CA1 properties, but does not result in neurodegeneration. *EMBO J.*, **21**, 202–210.
97. White, M.D., Farmer, M., Mirabile, I., Brandner, S., Collinge, J., and Mallucci, G.R. (2008) Single treatment with RNAi against prion protein rescues early neuronal dysfunction and prolongs survival in mice with prion disease. *Proc. Natl Acad. Sci. USA*, **105**, 10238–10243.
98. Malaga-Trillo, E., Solis, G.P., Schrock, Y., Geiss, C., Luncz, L., Thomanetz, V., and Stuermer, C.A. (2009) Regulation of embryonic cell adhesion by the prion protein. *PLoS Biol.*, **7**, e55.
99. Shmerling, D., Hegyi, I., Fischer, M., Blattler, T., Brandner, S., Gotz, J., Rulicke, T., Flechsig, E., Cozzio, A., von Mering, C., et al. (1998) Expression of amino-terminally truncated PrP in the mouse leading to ataxia and specific cerebellar lesions. *Cell*, **93**, 203–214.
100. Rossi, D., Cozzio, A., Flechsig, E., Klein, M.A., Rulicke, T., Aguzzi, A., and Weissmann, C. (2001) Onset of ataxia and Purkinje cell loss in PrP null mice inversely correlated with Dpl level in brain. *EMBO J.*, **20**, 694–702.
101. Sakaguchi, S., Katamine, S., Nishida, N., Moriuchi, R., Shigematsu, K., Sugimoto, T., Nakatani, A., Kataoka, Y., Houtani, T., Shirabe, S., et al. (1996) Loss of cerebellar Purkinje cells in aged mice homozygous for a disrupted PrP gene. *Nature*, **380**, 528–531.
102. Bueler, H., Aguzzi, A., Sailer, A., Greiner, R.A., Autenried, P., Aguet, M., and Weissmann, C. (1993) Mice devoid of

PrP are resistant to scrapie. *Cell*, **73**, 1339–1347.
103. Brandner, S., Isenmann, S., Raeber, A., Fischer, M., Sailer, A., Kobayashi, Y., Marino, S., Weissmann, C., and Aguzzi, A. (1996) Normal host prion protein necessary for scrapie-induced neurotoxicity. *Nature*, **379**, 339–343.
104. Schneider, B., Pietri, M., Pradines, E., Loubet, D., Launay, J.M., Kellermann, O., and Mouillet-Richard, S. (2011) Understanding the neurospecificity of prion protein signaling. *Front. Biosci. (Landmark Ed.)*, **16**, 169–186.
105. Schmitt-Ulms, G., Legname, G., Baldwin, M.A., Ball, H.L., Bradon, N., Bosque, P.J., Crossin, K.L., Edelman, G.M., DeArmond, S.J., Cohen, F.E., et al. (2001) Binding of neural cell adhesion molecules (N-CAMs) to the cellular prion protein. *J. Mol. Biol.*, **314**, 1209–1225.
106. Santuccione, A., Sytnyk, V., Leshchyns'ka, I., and Schachner, M. (2005) Prion protein recruits its neuronal receptor NCAM to lipid rafts to activate p59fyn and to enhance neurite outgrowth. *J. Cell Biol.*, **169**, 341–354.
107. Mouillet-Richard, S., Ermonval, M., Chebassier, C., Laplanche, J.L., Lehmann, S., Launay, J.M., and Kellermann, O. (2000) Signal transduction through prion protein. *Science*, **289**, 1925–1928.
108. Millhauser, G.L. (2007) Copper and the prion protein: methods, structures, function, and disease. *Annu. Rev. Phys. Chem.*, **58**, 299–320.
109. Urso, E., Manno, D., Serra, A., Buccolieri, A., Rizzello, A., Danieli, A., Acierno, R., Salvato, B., and Maffia, M. (2012) Role of the cellular prion protein in the neuron adaptation strategy to copper deficiency. *Cell. Mol. Neurobiol.*, **32**, 989–1001.
110. Rachidi, W., Vilette, D., Guiraud, P., Arlotto, M., Riondel, J., Laude, H., Lehmann, S., and Favier, A. (2003) Expression of prion protein increases cellular copper binding and antioxidant enzyme activities but not copper delivery. *J. Biol. Chem.*, **278**, 9064–9072.
111. Thakur, A.K., Srivastava, A.K., Srinivas, V., Chary, K.V., and Rao, C.M. (2011) Copper alters aggregation behavior of prion protein and induces novel interactions between its N- and C-terminal regions. *J. Biol. Chem.*, **286**, 38533–38545.
112. Pauly, P.C., and Harris, D.A. (1998) Copper stimulates endocytosis of the prion protein. *J. Biol. Chem.*, **273**, 33107–33110.
113. Jones, C.E., Abdelraheim, S.R., Brown, D.R., and Viles, J.H. (2004) Preferential Cu^{2+} coordination by His96 and His111 induces beta-sheet formation in the unstructured amyloidogenic region of the prion protein. *J. Biol. Chem.*, **279**, 32018–32027.
114. Waggoner, D.J., Drisaldi, B., Bartnikas, T.B., Casareno, R.L., Prohaska, J.R., Gitlin, J.D., and Harris, D.A. (2000) Brain copper content and cuproenzyme activity do not vary with prion protein expression level. *J. Biol. Chem.*, **275**, 7455–7458.
115. Herms, J., Tings, T., Gall, S., Madlung, A., Giese, A., Siebert, H., Schurmann, P., Windl, O., Brose, N., and Kretzschmar, H. (1999) Evidence of presynaptic location and function of the prion protein. *J. Neurosci.*, **19**, 8866–8875.
116. Klamt, F., Dal-Pizzol, F., Conte da Frota, M.L., Jr, Walz, R., Andrades, M.E., da Silva, E.G., Brentani, R.R., Izquierdo, I., and Fonseca Moreira, J.C. (2001) Imbalance of antioxidant defense in mice lacking cellular prion protein. *Free Radical Biol. Med.*, **30**, 1137–1144.
117. White, A.R., Collins, S.J., Maher, F., Jobling, M.F., Stewart, L.R., Thyer, J.M., Beyreuther, K., Masters, C.L., and Cappai, R. (1999) Prion protein-deficient neurons reveal lower glutathione reductase activity and increased susceptibility to hydrogen peroxide toxicity. *Am. J. Pathol.*, **155**, 1723–1730.
118. McLennan, N.F., Brennan, P.M., McNeill, A., Davies, I., Fotheringham, A., Rennison, K.A., Ritchie, D., Brannan, F., Head, M.W., Ironside, J.W., et al. (2004) Prion protein accumulation and neuroprotection in hypoxic brain damage. *Am. J. Pathol.*, **165**, 227–235.
119. Spudich, A., Frigg, R., Kilic, E., Kilic, U., Oesch, B., Raeber, A., Bassetti, C.L., and Hermann, D.M. (2005) Aggravation of ischemic brain injury by prion protein deficiency: role of ERK-1/-2 and STAT-1. *Neurobiol. Dis.*, **20**, 442–449.
120. Kuwahara, C., Takeuchi, A.M., Nishimura, T., Haraguchi, K., Kubosaki, A., Matsumoto, Y., Saeki, K., Yokoyama, T., Itohara, S., and Onodera, T. (1999) Prions prevent

neuronal cell-line death. *Nature*, **400**, 225–226.
121. Kim, B.H., Lee, H.G., Choi, J.K., Kim, J.I., Choi, E.K., Carp, R.I., and Kim, Y.S. (2004) The cellular prion protein (PrPC) prevents apoptotic neuronal cell death and mitochondrial dysfunction induced by serum deprivation. *Brain Res. Mol. Brain Res.*, **124**, 40–50.
122. Diarra-Mehrpour, M., Arrabal, S., Jalil, A., Pinson, X., Gaudin, C., Pietu, G., Pitaval, A., Ripoche, H., Eloit, M., Dormont, D., *et al.* (2004) Prion protein prevents human breast carcinoma cell line from tumor necrosis factor alpha-induced cell death. *Cancer Res.*, **64**, 719–727.
123. Bounhar, Y., Zhang, Y., Goodyer, C.G., and LeBlanc, A. (2001) Prion protein protects human neurons against Bax-mediated apoptosis. *J. Biol. Chem.*, **276**, 39145–39149.
124. Roucou, X., Guo, Q., Zhang, Y., Goodyer, C.G., and LeBlanc, A.C. (2003) Cytosolic prion protein is not toxic and protects against Bax-mediated cell death in human primary neurons. *J. Biol. Chem.*, **278**, 40877–40881.
125. Roucou, X., Gains, M., and LeBlanc, A.C. (2004) Neuroprotective functions of prion protein. *J. Neurosci. Res.*, **75**, 153–161.
126. Sakudo, A., Lee, D.C., Saeki, K., Nakamura, Y., Inoue, K., Matsumoto, Y., Itohara, S., and Onodera, T. (2003) Impairment of superoxide dismutase activation by N-terminally truncated prion protein (PrP) in PrP-deficient neuronal cell line. *Biochem. Biophys. Res. Commun.*, **308**, 660–667.
127. Sakudo, A., Hamaishi, M., Hosokawa-Kanai, T., Tuchiya, K., Nishimura, T., Saeki, K., Matsumoto, Y., Ueda, S., and Onodera, T. (2003) Absence of superoxide dismutase activity in a soluble cellular isoform of prion protein produced by baculovirus expression system. *Biochem. Biophys. Res. Commun.*, **307**, 678–683.
128. Drisaldi, B., Coomaraswamy, J., Mastrangelo, P., Strome, B., Yang, J., Watts, J.C., Chishti, M.A., Marvi, M., Windl, O., Ahrens, R., *et al.* (2004) Genetic mapping of activity determinants within cellular prion proteins: N-terminal modules in PrPC offset pro-apoptotic activity of the Doppel helix B/B′ region. *J. Biol. Chem.*, **279**, 55443–55454.
129. Li, A., Barmada, S.J., Roth, K.A., and Harris, D.A. (2007) N-terminally deleted forms of the prion protein activate both Bax-dependent and Bax-independent neurotoxic pathways. *J. Neurosci.*, **27**, 852–859.
130. Baumann, F., Tolnay, M., Brabeck, C., Pahnke, J., Kloz, U., Niemann, H.H., Heikenwalder, M., Rulicke, T., Burkle, A., and Aguzzi, A. (2007) Lethal recessive myelin toxicity of prion protein lacking its central domain. *EMBO J.*, **26**, 538–547.
131. Li, A., Piccardo, P., Barmada, S.J., Ghetti, B., and Harris, D.A. (2007) Prion protein with an octapeptide insertion has impaired neuroprotective activity in transgenic mice. *EMBO J.*, **26**, 2777–2785.
132. Li, A., Christensen, H.M., Stewart, L.R., Roth, K.A., Chiesa, R., and Harris, D.A. (2007) Neonatal lethality in transgenic mice expressing prion protein with a deletion of residues 105–125. *EMBO J.*, **26**, 548–558.
133. Steele, A.D., Emsley, J.G., Ozdinler, P.H., Lindquist, S., and Macklis, J.D. (2006) Prion protein (PrPc) positively regulates neural precursor proliferation during developmental and adult mammalian neurogenesis. *Proc. Natl Acad. Sci. USA*, **103**, 3416–3421.
134. Peralta, O.A., Huckle, W.R., and Eyestone, W.H. (2011) Expression and knockdown of cellular prion protein (PrPC) in differentiating mouse embryonic stem cells. *Differentiation*, **81**, 68–77.
135. Zhang, Y., Kim, S.O., Opsahl-Vital, S., Ho, S.P., Souron, J.B., Kim, C., Giles, K., and Den Besten, P.K. (2011) Multiple effects of the cellular prion protein on tooth development. *Int. J. Dev. Biol.*, **55**, 953–960.
136. Stella, R., Massimino, M.L., Sandri, M., Sorgato, M.C., and Bertoli, A. (2010) Cellular prion protein promotes regeneration of adult muscle tissue. *Mol. Cell. Biol.*, **30**, 4864–4876.
137. Zhang, C.C., Steele, A.D., Lindquist, S., and Lodish, H.F. (2006) Prion protein is expressed on long-term repopulating hematopoietic stem cells and is important for their self-renewal. *Proc. Natl Acad. Sci. USA*, **103**, 2184–2189.
138. Lee, Y.J., and Baskakov, I.V. (2013) The cellular form of the prion protein is involved in controlling cell cycle dynamics, self-renewal, and the fate of human embryonic stem cell differentiation. *J. Neurochem.*, **124**, 310–322.

139. Miranda, A., Ramos-Ibeas, P., Pericuesta, E., Ramirez, M.A., and Gutierrez-Adan, A. (2013) The role of prion protein in stem cell regulation. *Reproduction*, **146**, R91-R99.
140. Mehrpour, M., and Codogno, P. (2010) Prion protein: from physiology to cancer biology. *Cancer Lett.*, **290**, 1–23.
141. Li, Q.Q., Sun, Y.P., Ruan, C.P., Xu, X.Y., Ge, J.H., He, J., Xu, Z.D., Wang, Q., and Gao, W.C. (2011) Cellular prion protein promotes glucose uptake through the Fyn-HIF-2alpha-Glut1 pathway to support colorectal cancer cell survival. *Cancer Sci.*, **102**, 400–406.
142. Schmitt-Ulms, G., Hansen, K., Liu, J., Cowdrey, C., Yang, J., DeArmond, S.J., Cohen, F.E., Prusiner, S.B., and Baldwin, M.A. (2004) Time-controlled transcardiac perfusion cross-linking for the study of protein interactions in complex tissues. *Nat. Biotechnol.*, **22**, 724–731.
143. Morales, R., Estrada, L.D., Diaz-Espinoza, R., Morales-Scheihing, D., Jara, M.C., Castilla, J., and Soto, C. (2010) Molecular cross talk between misfolded proteins in animal models of Alzheimer's and prion diseases. *J. Neurosci.*, **30**, 4528–4535.
144. Nieznanski, K., Surewicz, K., Chen, S., Nieznanska, H., and Surewicz, W.K. (2014) Interaction between prion protein and Abeta amyloid fibrils revisited. *ACS Chem. Neurosci.*, **5**(5), 340–345.
145. Ostapchenko, V.G., Beraldo, F.H., Mohammad, A.H., Xie, Y.F., Hirata, P.H., Magalhaes, A.C., Lamour, G., Li, H., Maciejewski, A., Belrose, J.C., et al. (2013) The prion protein ligand, stress-inducible phosphoprotein 1, regulates amyloid-beta oligomer toxicity. *J. Neurosci.*, **33**, 16552–16564.
146. Lauren, J., Gimbel, D.A., Nygaard, H.B., Gilbert, J.W., and Strittmatter, S.M. (2009) Cellular prion protein mediates impairment of synaptic plasticity by amyloid-beta oligomers. *Nature*, **457**, 1128–1132.
147. Barry, A.E., Klyubin, I., Mc Donald, J.M., Mably, A.J., Farrell, M.A., Scott, M., Walsh, D.M., and Rowan, M.J. (2011) Alzheimer's disease brain-derived amyloid-beta-mediated inhibition of LTP in vivo is prevented by immunotargeting cellular prion protein. *J. Neurosci.*, **31**, 7259–7263.
148. Freir, D.B., Nicoll, A.J., Klyubin, I., Panico, S., Mc Donald, J.M., Risse, E., Asante, E.A., Farrow, M.A., Sessions, R.B., Saibil, H.R., et al. (2011) Interaction between prion protein and toxic amyloid beta assemblies can be therapeutically targeted at multiple sites. *Nat. Commun.*, **2**, 336.
149. Balducci, C., Beeg, M., Stravalaci, M., Bastone, A., Sclip, A., Biasini, E., Tapella, L., Colombo, L., Manzoni, C., Borsello, T., et al. (2010) Synthetic amyloid-beta oligomers impair long-term memory independently of cellular prion protein. *Proc. Natl Acad. Sci. USA*, **107**, 2295–2300.
150. Calella, A.M., Farinelli, M., Nuvolone, M., Mirante, O., Moos, R., Falsig, J., Mansuy, I.M., and Aguzzi, A. (2010) Prion protein and Abeta-related synaptic toxicity impairment. *EMBO Mol. Med.*, **2**, 306–314.
151. Gimbel, D.A., Nygaard, H.B., Coffey, E.E., Gunther, E.C., Lauren, J., Gimbel, Z.A., and Strittmatter, S.M. (2010) Memory impairment in transgenic Alzheimer mice requires cellular prion protein. *J. Neurosci.*, **30**, 6367–6374.
152. Bate, C., and Williams, A. (2011) Amyloid-beta-induced synapse damage is mediated via cross-linkage of cellular prion proteins. *J. Biol. Chem.*, **286**, 37955–37963.
153. Parkin, E.T., Watt, N.T., Hussain, I., Eckman, E.A., Eckman, C.B., Manson, J.C., Baybutt, H.N., Turner, A.J., and Hooper, N.M. (2007) Cellular prion protein regulates beta-secretase cleavage of the Alzheimer's amyloid precursor protein. *Proc. Natl Acad. Sci. USA*, **104**, 11062–11067.
154. Griffiths, H.H., Whitehouse, I.J., Baybutt, H., Brown, D., Kellett, K.A., Jackson, C.D., Turner, A.J., Piccardo, P., Manson, J.C., and Hooper, N.M. (2011) Prion protein interacts with BACE1 protein and differentially regulates its activity toward wild type and Swedish mutant amyloid precursor protein. *J. Biol. Chem.*, **286**, 33489–33500.
155. McHugh, P.C., Wright, J.A., Williams, R.J., and Brown, D.R. (2012) Prion protein expression alters APP cleavage without interaction with BACE-1. *Neurochem. Int.*, **61**, 672–680.
156. Guillot-Sestier, M.V., Sunyach, C., Ferreira, S.T., Marzolo, M.P., Bauer, C., Thevenet, A., and Checler, F. (2012) Alpha-secretase-derived fragment of cellular prion, N1, protects against monomeric and oligomeric amyloid beta (Abeta)-associated cell death. *J. Biol. Chem.*, **287**, 5021–5032.

157. Stahl, N., Baldwin, M.A., and Prusiner, S.B. (1991) Electrospray mass spectrometry of the glycosylinositol phospholipid of the scrapie prion protein. *Cell Biol. Int. Rep.*, **15**, 853–862.
158. Stahl, N., Baldwin, M.A., Teplow, D.B., Hood, L., Gibson, B.W., Burlingame, A.L., and Prusiner, S.B. (1993) Structural studies of the scrapie prion protein using mass spectrometry and amino acid sequencing. *Biochemistry*, **32**, 1991–2002.
159. Turk, E., Teplow, D.B., Hood, L.E., and Prusiner, S.B. (1988) Purification and properties of the cellular and scrapie hamster prion proteins. *Eur. J. Biochem.*, **176**, 21–30.
160. Caughey, B.W., Dong, A., Bhat, K.S., Ernst, D., Hayes, S.F., and Caughey, W.S. (1991) Secondary structure analysis of the scrapie-associated protein PrP 27–30 in water by infrared spectroscopy. *Biochemistry*, **30**, 7672–7680.
161. Safar, J., Roller, P.P., Gajdusek, D.C., and Gibbs, C.J., Jr (1993) Conformational transitions, dissociation, and unfolding of scrapie amyloid (prion) protein. *J. Biol. Chem.*, **268**, 20276–20284.
162. Thomzig, A., Spassov, S., Friedrich, M., Naumann, D., and Beekes, M. (2004) Discriminating scrapie and bovine spongiform encephalopathy isolates by infrared spectroscopy of pathological prion protein. *J. Biol. Chem.*, **279**, 33847–33854.
163. Baron, G.S., Hughson, A.G., Raymond, G.J., Offerdahl, D.K., Barton, K.A., Raymond, L.D., Dorward, D.W., and Caughey, B. (2011) Effect of glycans and the glycophosphatidylinositol anchor on strain dependent conformations of scrapie prion protein: improved purifications and infrared spectra. *Biochemistry*, **50**, 4479–4490.
164. Prusiner, S.B., Groth, D.F., Cochran, S.P., Masiarz, F.R., McKinley, M.P., and Martinez, H.M. (1980) Molecular properties, partial purification, and assay by incubation period measurements of the hamster scrapie agent. *Biochemistry*, **19**, 4883–4891.
165. Safar, J., Wang, W., Padgett, M.P., Ceroni, M., Piccardo, P., Zopf, D., Gajdusek, D.C., and Gibbs, C.J., Jr (1990) Molecular mass, biochemical composition, and physicochemical behavior of the infectious form of the scrapie precursor protein monomer. *Proc. Natl Acad. Sci. USA*, **87**, 6373–6377.
166. Riesner, D., Kellings, K., Post, K., Wille, H., Serban, H., Groth, D., Baldwin, M.A., and Prusiner, S.B. (1996) Disruption of prion rods generates 10-nm spherical particles having high alpha-helical content and lacking scrapie infectivity. *J. Virol.*, **70**, 1714–1722.
167. Sklaviadis, T.K., Manuelidis, L., and Manuelidis, E.E. (1989) Physical properties of the Creutzfeldt-Jakob disease agent. *J. Virol.*, **63**, 1212–1222.
168. Sklaviadis, T., Dreyer, R., and Manuelidis, L. (1992) Analysis of Creutzfeldt-Jakob disease infectious fractions by gel permeation chromatography and sedimentation field flow fractionation. *Virus Res.*, **26**, 241–254.
169. Silveira, J.R., Raymond, G.J., Hughson, A.G., Race, R.E., Sim, V.L., Hayes, S.F., and Caughey, B. (2005) The most infectious prion protein particles. *Nature*, **437**, 257–261.
170. Tateishi, J., Kitamoto, T., Mohri, S., Satoh, S., Sato, T., Shepherd, A., and Macnaughton, M.R. (2001) Scrapie removal using Planova virus removal filters. *Biologicals*, **29**, 17–25.
171. Merz, P.A., Kascsak, R.J., Rubenstein, R., Carp, R.I., and Wisniewski, H.M. (1987) Antisera to scrapie-associated fibril protein and prion protein decorate scrapie-associated fibrils. *J. Virol.*, **61**, 42–49.
172. Sim, V.L., and Caughey, B. (2008) Ultrastructures and strain comparison of under-glycosylated scrapie prion fibrils. *Neurobiol. Aging*, **30** (12), 2031–2042.
173. Klein, T.R., Kirsch, D., Kaufmann, R., and Riesner, D. (1998) Prion rods contain small amounts of two host sphingolipids as revealed by thin-layer chromatography and mass spectrometry. *Biol. Chem.*, **379**, 655–666.
174. Appel, T.R., Dumpitak, C., Matthiesen, U., and Riesner, D. (1999) Prion rods contain an inert polysaccharide scaffold. *Biol. Chem.*, **380**, 1295–1306.
175. Dumpitak, C., Beekes, M., Weinmann, N., Metzger, S., Winklhofer, K.F., Tatzelt, J., and Riesner, D. (2005) The polysaccharide scaffold of PrP 27–30 is a common compound of natural prions and consists of alpha-linked polyglucose. *Biol. Chem.*, **386**, 1149–1155.

176. Zahn, R., Liu, A., Luhrs, T., Riek, R., von Schroetter, C., Lopez Garcia, F., Billeter, M., Calzolai, L., Wider, G., and Wuthrich, K. (2000) NMR solution structure of the human prion protein. *Proc. Natl Acad. Sci. USA*, **97**, 145–150.
177. Lopez Garcia, F., Zahn, R., Riek, R., and Wuthrich, K. (2000) NMR structure of the bovine prion protein. *Proc. Natl Acad. Sci. USA*, **97**, 8334–8339.
178. Lysek, D.A., Schorn, C., Nivon, L.G., Esteve-Moya, V., Christen, B., Calzolai, L., von Schroetter, C., Fiorito, F., Herrmann, T., Guntert, P., et al. (2005) Prion protein NMR structures of cats, dogs, pigs, and sheep. *Proc. Natl Acad. Sci. USA*, **102**, 640–645.
179. Calzolai, L., Lysek, D.A., Perez, D.R., Guntert, P., and Wuthrich, K. (2005) Prion protein NMR structures of chickens, turtles, and frogs. *Proc. Natl Acad. Sci. USA*, **102**, 651–655.
180. Gossert, A.D., Bonjour, S., Lysek, D.A., Fiorito, F., and Wuthrich, K. (2005) Prion protein NMR structures of elk and of mouse/elk hybrids. *Proc. Natl Acad. Sci. USA*, **102**, 646–650.
181. Hornemann, S., Schorn, C., and Wuthrich, K. (2004) NMR structure of the bovine prion protein isolated from healthy calf brains. *EMBO Rep.*, **5**, 1159–1164.
182. Eberl, H., Tittmann, P., and Glockshuber, R. (2004) Characterization of recombinant, membrane-attached full-length prion protein. *J. Biol. Chem.*, **279**, 25058–25065.
183. Govaerts, C., Wille, H., Prusiner, S.B., and Cohen, F.E. (2004) Evidence for assembly of prions with left-handed beta-helices into trimers. *Proc. Natl Acad. Sci. USA*, **101**, 8342–8347.
184. Wille, H., Bian, W., McDonald, M., Kendall, A., Colby, D.W., Bloch, L., Ollesch, J., Borovinskiy, A.L., Cohen, F.E., Prusiner, S.B., et al. (2009) Natural and synthetic prion structure from X-ray fiber diffraction. *Proc. Natl Acad. Sci. USA*, **106**, 16990–16995.
185. DeMarco, M.L., and Daggett, V. (2004) From conversion to aggregation: protofibril formation of the prion protein. *Proc. Natl Acad. Sci. USA*, **101**, 2293–2298.
186. Cobb, N.J., Sonnichsen, F.D., McHaourab, H., and Surewicz, W.K. (2007) Molecular architecture of human prion protein amyloid: a parallel, in-register beta-structure. *Proc. Natl Acad. Sci. USA*, **104**, 18946–18951.
187. Smirnovas, V., Baron, G.S., Offerdahl, D.K., Raymond, G.J., Caughey, B., and Surewicz, W.K. (2011) Structural organization of brain-derived mammalian prions examined by hydrogen-deuterium exchange. *Nat. Struct. Mol. Biol.*, **18**, 504–506.
188. Kunes, K.C., Clark, S.C., Cox, D.L., and Singh, R.R. (2008) Left handed beta helix models for mammalian prion fibrils. *Prion*, **2**, 81–90.
189. Salamat, K., Moudjou, M., Chapuis, J., Herzog, L., Jaumain, E., Béringue, V., Rezaei, H., Pastore, A., Laude, H., and Dron, M. (2012) Integrity of helix 2-helix 3 domain of the PrP protein is not mandatory for prion replication. *J. Biol. Chem.*, **287**, 18953–18964.
190. Priola, S.A., Caughey, B., Race, R.E., and Chesebro, B. (1994) Heterologous PrP molecules interfere with accumulation of protease-resistant PrP in scrapie-infected murine neuroblastoma cells. *J. Virol.*, **68**, 4873–4878.
191. Manson, J.C., Jamieson, E., Baybutt, H., Tuzi, N.L., Barron, R., McConnell, I., Somerville, R., Ironside, J., Will, R., Sy, M.S., et al. (1999) A single amino acid alteration (101L) introduced into murine PrP dramatically alters incubation time of transmissible spongiform encephalopathy. *EMBO J.*, **18**, 6855–6864.
192. Ott, D., Taraborrelli, C., and Aguzzi, A. (2008) Novel dominant-negative prion protein mutants identified from a randomized library. *Protein Eng. Des. Sel.*, **21**, 623–629.
193. Jackson, W.S., Borkowski, A.W., Faas, H., Steele, A.D., King, O.D., Watson, N., Jasanoff, A., and Lindquist, S. (2009) Spontaneous generation of prion infectivity in fatal familial insomnia knockin mice. *Neuron*, **63**, 438–450.
194. Sigurdson, C.J., Nilsson, K.P., Hornemann, S., Heikenwalder, M., Manco, G., Schwarz, P., Ott, D., Rulicke, T., Liberski, P.P., Julius, C., et al. (2009) De novo generation of a transmissible spongiform encephalopathy by mouse transgenesis. *Proc. Natl Acad. Sci. USA*, **106**, 304–309.
195. Sigurdson, C.J., Joshi-Barr, S., Bett, C., Winson, O., Manco, G., Schwarz, P., Rulicke, T., Nilsson, K.P., Margalith,

I., Raeber, A., et al. (2011) Spongiform encephalopathy in transgenic mice expressing a point mutation in the beta2-alpha2 loop of the prion protein. *J. Neurosci.*, **31**, 13840–13847.
196. Atarashi, R., Sim, V.L., Nishida, N., Caughey, B., and Katamine, S. (2006) Prion strain-dependent differences in conversion of mutant prion proteins in cell culture. *J. Virol.*, **80**, 7854–7862.
197. Howells, L.C., Anderson, S., Coldham, N.G., and Sauer, M.J. (2008) Transmissible spongiform encephalopathy strain-associated diversity of N-terminal proteinase K cleavage sites of PrP(Sc) from scrapie-infected and bovine spongiform encephalopathy-infected mice. *Biomarkers*, **13**, 393–412.
198. Notari, S., Strammiello, R., Capellari, S., Giese, A., Cescatti, M., Grassi, J., Ghetti, B., Langeveld, J.P., Zou, W.Q., Gambetti, P., et al. (2008) Characterization of truncated forms of abnormal prion protein in Creutzfeldt-Jakob disease. *J. Biol. Chem.*, **283**, 30557–30565.
199. Flechsig, E., Shmerling, D., Hegyi, I., Raeber, A.J., Fischer, M., Cozzio, A., von Mering, C., Aguzzi, A., and Weissmann, C. (2000) Prion protein devoid of the octapeptide repeat region restores susceptibility to scrapie in PrP knockout mice. *Neuron*, **27**, 399–408.
200. Supattapone, S., Bosque, P., Muramoto, T., Wille, H., Aagaard, C., Peretz, D., Nguyen, H.O., Heinrich, C., Torchia, M., Safar, J., et al. (1999) Prion protein of 106 residues creates an artificial transmission barrier for prion replication in transgenic mice. *Cell*, **96**, 869–878.
201. Sigurdson, C.J., Nilsson, K.P., Hornemann, S., Manco, G., Fernandez-Borges, N., Schwarz, P., Castilla, J., Wuthrich, K., and Aguzzi, A. (2010) A molecular switch controls interspecies prion disease transmission in mice. *J. Clin. Invest.*, **120**, 2590–2599.
202. Avbelj, M., Hafner-Bratkovic, I., and Jerala, R. (2011) Introduction of glutamines into the B2-H2 loop promotes prion protein conversion. *Biochem. Biophys. Res. Commun.*, **413**, 521–526.
203. Adrover, M., Pauwels, K., Prigent, S., de Chiara, C., Xu, Z., Chapuis, C., Pastore, A., and Rezaei, H. (2010) Prion fibrillization is mediated by a native structural element that comprises helices H2 and H3. *J. Biol. Chem.*, **285**, 21004–21012.
204. Chakroun, N., Prigent, S., Dreiss, C.A., Noinville, S., Chapuis, C., Fraternali, F., and Rezaei, H. (2010) The oligomerization properties of prion protein are restricted to the H2H3 domain. *FASEB J.*, **24**, 3222–3231.
205. Herrmann, L.M., and Caughey, B. (1998) The importance of the disulfide bond in prion protein conversion. *NeuroReport*, **9**, 2457–2461.
206. Welker, E., Raymond, L.D., Scheraga, H.A., and Caughey, B. (2002) Intramolecular versus intermolecular disulfide bonds in prion proteins. *J. Biol. Chem.*, **277**, 33477–33481.
207. Kocisko, D.A., Come, J.H., Priola, S.A., Chesebro, B., Raymond, G.J., Lansbury, P.T., and Caughey, B. (1994) Cell-free formation of protease-resistant prion protein. *Nature*, **370**, 471–474.
208. Bett, C., Kurt, T.D., Lucero, M., Trejo, M., Rozemuller, A.J., Kong, Q., Nilsson, K.P., Masliah, E., Oldstone, M.B., and Sigurdson, C.J. (2013) Defining the conformational features of anchorless, poorly neuroinvasive prions. *PLoS Pathog.*, **9**, e1003280.
209. Bossers, A., Belt, P., Raymond, G.J., Caughey, B., de Vries, R., and Smits, M.A. (1997) Scrapie susceptibility-linked polymorphisms modulate the in vitro conversion of sheep prion protein to protease-resistant forms. *Proc. Natl Acad. Sci. USA*, **94**, 4931–4936.
210. Piro, J.R., Harris, B.T., Nishina, K., Soto, C., Morales, R., Rees, J.R., and Supattapone, S. (2009) Prion protein glycosylation is not required for strain-specific neurotropism. *J. Virol.*, **83**, 5321–5328.
211. Browning, S., Baker, C.A., Smith, E., Mahal, S.P., Herva, M.E., Demczyk, C.A., Li, J., and Weissmann, C. (2011) Abrogation of complex glycosylation by swainsonine results in strain- and cell-specific inhibition of prion replication. *J. Biol. Chem.*, **286**, 40962–40973.
212. Tuzi, N.L., Cancellotti, E., Baybutt, H., Blackford, L., Bradford, B., Plinston, C., Coghill, A., Hart, P., Piccardo, P., Barron, R.M., et al. (2008) Host PrP glycosylation: a major factor determining the outcome of prion infection. *PLoS Biol.*, **6**, e100.
213. Rutishauser, D., Mertz, K.D., Moos, R., Brunner, E., Rulicke, T., Calella, A.M., and Aguzzi, A. (2009) The comprehensive native

interactome of a fully functional tagged prion protein. *PLoS ONE*, **4**, e4446.
214. Goold, R., Rabbanian, S., Sutton, L., Andre, R., Arora, P., Moonga, J., Clarke, A.R., Schiavo, G., Jat, P., Collinge, J., et al. (2011) Rapid cell-surface prion protein conversion revealed using a novel cell system. *Nat. Commun.*, **2**, 281.
215. Lansbury, P.T. (1994) Mechanism of scrapie replication. *Science*, **265**, 1510.
216. Masel, J., Jansen, V.A., and Nowak, M.A. (1999) Quantifying the kinetic parameters of prion replication. *Biophys. Chem.*, **77**, 139–152.
217. Prusiner, S.B. (1991) Molecular biology of prion diseases. *Science*, **252**, 1515–1522.
218. Caughey, B., Kocisko, D.A., Raymond, G.J., and Lansbury, P.T., Jr, (1995) Aggregates of scrapie-associated prion protein induce the cell-free conversion of protease-sensitive prion protein to the protease-resistant state. *Chem. Biol.*, **2**, 807–817.
219. DebBurman, S.K., Raymond, G.J., Caughey, B., and Lindquist, S. (1997) Chaperone-supervised conversion of prion protein to its protease-resistant form. *Proc. Natl Acad. Sci. USA*, **94**, 13938–13943.
220. Horiuchi, M., and Caughey, B. (1999) Specific binding of normal prion protein to the scrapie form via a localized domain initiates its conversion to the protease-resistant state. *EMBO J.*, **18**, 3193–3203.
221. Rigter, A., and Bossers, A. (2005) Sheep scrapie susceptibility-linked polymorphisms do not modulate the initial binding of cellular to disease-associated prion protein prior to conversion. *J. Gen. Virol.*, **86**, 2627–2634.
222. Horiuchi, M., Priola, S.A., Chabry, J., and Caughey, B. (2000) Interactions between heterologous forms of prion protein: binding, inhibition of conversion, and species barriers. *Proc. Natl Acad. Sci. USA*, **97**, 5836–5841.
223. Callahan, M.A., Xiong, L., and Caughey, B. (2001) Reversibility of scrapie-associated prion protein aggregation. *J. Biol. Chem.*, **276**, 28022–28028.
224. Saborio, G.P., Permanne, B., and Soto, C. (2001) Sensitive detection of pathological prion protein by cyclic amplification of protein misfolding. *Nature*, **411**, 810–813.
225. Castilla, J., Saa, P., Hetz, C., and Soto, C. (2005) In vitro generation of infectious scrapie prions. *Cell*, **121**, 195–206.
226. Castilla, J., Saa, P., and Soto, C. (2005) Detection of prions in blood. *Nat. Med.*, **11**, 982–985.
227. Castilla, J., Gonzalez-Romero, D., Saa, P., Morales, R., De Castro, J., and Soto, C. (2008) Crossing the species barrier by PrP(Sc) replication in vitro generates unique infectious prions. *Cell*, **134**, 757–768.
228. Moudjou, M., Sibille, P., Fichet, G., Reine, F., Chapuis, J., Herzog, L., Jaumain, E., Laferriere, F., Richard, C.A., Laude, H., et al. (2013) Highly infectious prions generated by a single round of microplate-based protein misfolding cyclic amplification. *MBio*, **5**, e00829-13.
229. Deleault, N.R., Harris, B.T., Rees, J.R., and Supattapone, S. (2007) Formation of native prions from minimal components in vitro. *Proc. Natl Acad. Sci. USA*, **104**, 9741–9746.
230. Supattapone, S. (2010) Biochemistry. What makes a prion infectious? *Science*, **327**, 1091–1092.
231. Rouvinski, A., Karniely, S., Kounin, M., Moussa, S., Goldberg, M.D., Warburg, G., Lyakhovetsky, R., Papy-Garcia, D., Kutzsche, J., Korth, C., et al. (2014) Live imaging of prions reveals nascent PrPSc in cell-surface, raft-associated amyloid strings and webs. *J. Cell Biol.*, **204**, 423–441.
232. Lewis, V., and Hooper, N.M. (2011) The role of lipid rafts in prion protein biology. *Front. Biosci. (Landmark Ed.)*, **16**, 151–168.
233. Grassmann, A., Wolf, H., Hofmann, J., Graham, J., and Vorberg, I. (2013) Cellular aspects of prion replication in vitro. *Viruses*, **5**, 374–405.
234. Paquet, S., Sabuncu, E., Delaunay, J.L., Laude, H., and Vilette, D. (2004) Prion infection of epithelial Rov cells is a polarized event. *J. Virol.*, **78**, 7148–7152.
235. Beekes, M., and McBride, P.A. (2007) The spread of prions through the body in naturally acquired transmissible spongiform encephalopathies. *FEBS J.*, **274**, 588–605.
236. Heppner, F.L., Christ, A.D., Klein, M.A., Prinz, M., Fried, M., Kraehenbuhl, J.P., and Aguzzi, A. (2001) Transepithelial prion transport by M cells. *Nat. Med.*, **7**, 976–977.
237. Maignien, T., Lasmezas, C.I., Béringue, V., Dormont, D., and Deslys, J.P. (1999) Pathogenesis of the oral route of infection of mice with scrapie and bovine spongiform

encephalopathy agents. *J. Gen. Virol.*, **80** (Pt 11), 3035–3042.

238. Pattison, I.H., and Millson, G.C. (1960) Further observations on the experimental production of scrapie in goats and sheep. *J. Comp. Pathol.*, **70**, 182–193.

239. Hadlow, W.J., Kennedy, R.C., and Race, R.E. (1982) Natural infection of Suffolk sheep with scrapie virus. *J. Infect. Dis.*, **146**, 657–664.

240. Eklund, C.M., Kennedy, R.C., and Hadlow, W.J. (1967) Pathogenesis of scrapie virus infection in the mouse. *J. Infect. Dis.*, **117**, 15–22.

241. Fraser, H., and Dickinson, A.G. (1970) Pathogenesis of scrapie in the mouse: the role of the spleen. *Nature*, **226**, 462–463.

242. Mabbott, N.A., and MacPherson, G.G. (2006) Prions and their lethal journey to the brain. *Nat. Rev. Microbiol.*, **4**, 201–211.

243. Aguzzi, A., Kranich, J., and Krautler, N.J. (2014) Follicular dendritic cells: origin, phenotype, and function in health and disease. *Trends Immunol.*, **35**, 105–113.

244. Prinz, M., Montrasio, F., Klein, M.A., Schwarz, P., Priller, J., Odermatt, B., Pfeffer, K., and Aguzzi, A. (2002) Lymph nodal prion replication and neuroinvasion in mice devoid of follicular dendritic cells. *Proc. Natl Acad. Sci. USA*, **99**, 919–924.

245. Race, R., Oldstone, M., and Chesebro, B. (2000) Entry versus blockade of brain infection following oral or intraperitoneal scrapie administration: role of prion protein expression in peripheral nerves and spleen. *J. Virol.*, **74**, 828–833.

246. Heikenwalder, M., Kurrer, M.O., Margalith, I., Kranich, J., Zeller, N., Haybaeck, J., Polymenidou, M., Matter, M., Bremer, J., Jackson, W.S., *et al.* (2008) Lymphotoxin-dependent prion replication in inflammatory stromal cells of granulomas. *Immunity*, **29**, 998–1008.

247. Heikenwalder, M., Zeller, N., Seeger, H., Prinz, M., Klohn, P.C., Schwarz, P., Ruddle, N.H., Weissmann, C., and Aguzzi, A. (2005) Chronic lymphocytic inflammation specifies the organ tropism of prions. *Science*, **307**, 1107–1110.

248. Lasmezas, C.I., Cesbron, J.Y., Deslys, J.P., Demaimay, R., Adjou, K.T., Rioux, R., Lemaire, C., Locht, C., and Dormont, D. (1996) Immune system-dependent and -independent replication of the scrapie agent. *J. Virol.*, **70**, 1292–1295.

249. Béringue, V., Lasmezas, C.I., Adjou, K.T., Demaimay, R., Lamoury, F., Deslys, J.P., Seman, M., and Dormont, D. (1999) Inhibiting scrapie neuroinvasion by polyene antibiotic treatment of SCID mice. *J. Gen. Virol.*, **80** (Pt 7), 1873–1877.

250. Glatzel, M., Heppner, F.L., Albers, K.M., and Aguzzi, A. (2001) Sympathetic innervation of lymphoreticular organs is rate limiting for prion neuroinvasion. *Neuron*, **31**, 25–34.

251. Prinz, M., Heikenwalder, M., Junt, T., Schwarz, P., Glatzel, M., Heppner, F.L., Fu, Y.X., Lipp, M., and Aguzzi, A. (2003) Positioning of follicular dendritic cells within the spleen controls prion neuroinvasion. *Nature*, **425**, 957–962.

252. Heikenwalder, M., Julius, C., and Aguzzi, A. (2007) Prions and peripheral nerves: a deadly rendezvous. *J. Neurosci. Res.*, **85**, 2714–2725.

253. Bradford, B.M., Tuzi, N.L., Feltri, M.L., McCorquodale, C., Cancellotti, E., and Manson, J.C. (2009) Dramatic reduction of PrP C level and glycosylation in peripheral nerves following PrP knock-out from Schwann cells does not prevent transmissible spongiform encephalopathy neuroinvasion. *J. Neurosci.*, **29**, 15445–15454.

254. Halliez, S., Chesnais, N., Mallucci, G., Vilotte, M., Langevin, C., Jaumain, E., Laude, H., Vilotte, J.L., and Béringue, V. (2013) Targeted knock-down of cellular prion protein expression in myelinating Schwann cells does not alter mouse prion pathogenesis. *J. Gen. Virol.*, **94**, 1435–1440.

255. Archer, F., Bachelin, C., Andreoletti, O., Besnard, N., Perrot, G., Langevin, C., Le Dur, A., Vilette, D., Baron-Van Evercooren, A., Vilotte, J.L., *et al.* (2004) Cultured peripheral neuroglial cells are highly permissive to sheep prion infection. *J. Virol.*, **78**, 482–490.

256. Kimberlin, R.H., and Walker, C.A. (1988) Pathogenesis of experimental scrapie. *Ciba Found. Symp.*, **135**, 37–62.

257. Caughey, B., and Lansbury, P.T. (2003) Protofibrils, pores, fibrils, and neurodegeneration: separating the responsible protein aggregates from the innocent bystanders. *Annu. Rev. Neurosci.*, **26**, 267–298.

258. Gavin, R., Braun, N., Nicolas, O., Parra, B., Urena, J.M., Mingorance, A., Soriano, E., Torres, J.M., Aguzzi, A., and del Rio, J.A. (2005) PrP(106–126) activates neuronal intracellular kinases and Egr1 synthesis through activation of NADPH-oxidase independently of PrPc. *FEBS Lett.*, **579**, 4099–4106.
259. Brown, D.R., Herms, J., and Kretzschmar, H.A. (1994) Mouse cortical cells lacking cellular PrP survive in culture with a neurotoxic PrP fragment. *NeuroReport*, **5**, 2057–2060.
260. Kunz, B., Sandmeier, E., and Christen, P. (1999) Neurotoxicity of prion peptide 106–126 not confirmed. *FEBS Lett.*, **458**, 65–68.
261. Simoneau, S., Rezaei, H., Sales, N., Kaiser-Schulz, G., Lefebvre-Roque, M., Vidal, C., Fournier, J.G., Comte, J., Wopfner, F., Grosclaude, J., et al. (2007) In vitro and in vivo neurotoxicity of prion protein oligomers. *PLoS Pathog.*, **3**, e125.
262. Mallucci, G., Dickinson, A., Linehan, J., Klohn, P.C., Brandner, S., and Collinge, J. (2003) Depleting neuronal PrP in prion infection prevents disease and reverses spongiosis. *Science*, **302**, 871–874.
263. Raeber, A.J., Race, R.E., Brandner, S., Priola, S.A., Sailer, A., Bessen, R.A., Mucke, L., Manson, J., Aguzzi, A., Oldstone, M.B., et al. (1997) Astrocyte-specific expression of hamster prion protein (PrP) renders PrP knockout mice susceptible to hamster scrapie. *EMBO J.*, **16**, 6057–6065.
264. Resenberger, U.K., Winklhofer, K.F., and Tatzelt, J. (2012) Cellular prion protein mediates toxic signaling of amyloid beta. *Neurodegener. Dis.*, **10**, 298–300.
265. Resenberger, U.K., Harmeier, A., Woerner, A.C., Goodman, J.L., Muller, V., Krishnan, R., Vabulas, R.M., Kretzschmar, H.A., Lindquist, S., Hartl, F.U., et al. (2011) The cellular prion protein mediates neurotoxic signalling of beta-sheet-rich conformers independent of prion replication. *EMBO J.*, **30**, 2057–2070.
266. Solforosi, L., Criado, J.R., McGavern, D.B., Wirz, S., Sanchez-Alavez, M., Sugama, S., DeGiorgio, L.A., Volpe, B.T., Wiseman, E., Abalos, G., et al. (2004) Cross-linking cellular prion protein triggers neuronal apoptosis in vivo. *Science*, **303**, 1514–1516.
267. Klohn, P.C., Farmer, M., Linehan, J.M., O'Malley, C., Fernandez de Marco, M., Taylor, W., Farrow, M., Khalili-Shirazi, A., Brandner, S., and Collinge, J. (2012) PrP antibodies do not trigger mouse hippocampal neuron apoptosis. *Science*, **335**, 52.
268. Cronier, S., Carimalo, J., Schaeffer, B., Jaumain, E., Béringue, V., Miquel, M.C., Laude, H., and Peyrin, J.M. (2012) Endogenous prion protein conversion is required for prion-induced neuritic alterations and neuronal death. *FASEB J.*, **26**, 3854–3861.
269. Sandberg, M.K., Al-Doujaily, H., Sharps, B., Clarke, A.R., and Collinge, J. (2011) Prion propagation and toxicity in vivo occur in two distinct mechanistic phases. *Nature*, **470**, 540–542.
270. Hill, A.F., Joiner, S., Linehan, J., Desbruslais, M., Lantos, P.L., and Collinge, J. (2000) Species-barrier-independent prion replication in apparently resistant species. *Proc. Natl Acad. Sci. USA*, **97**, 10248–10253.
271. Race, R., Raines, A., Raymond, G.J., Caughey, B., and Chesebro, B. (2001) Long-term subclinical carrier state precedes scrapie replication and adaptation in a resistant species: analogies to bovine spongiform encephalopathy and variant Creutzfeldt-Jakob disease in humans. *J. Virol.*, **75**, 10106–10112.
272. Mays, C.E., Kim, C., Haldiman, T., van der Merwe, J., Lau, A., Yang, J., Grams, J., Di Bari, M.A., Nonno, R., Telling, G.C., et al. (2014) Prion disease tempo determined by host-dependent substrate reduction. *J. Clin. Invest.*, **124**, 847–858.
273. Bruce, M.E. (1993) Scrapie strain variation and mutation. *Br. Med. Bull.*, **49**, 822–838.
274. Bruce, M.E. (2003) TSE strain variation. *Br. Med. Bull.*, **66**, 99–108.
275. Prusiner, S.B., Cochran, S.P., Groth, D.F., Downey, D.E., Bowman, K.A., and Martinez, H.M. (1982) Measurement of the scrapie agent using an incubation time interval assay. *Ann. Neurol.*, **11**, 353–358.
276. Dell'Omo, G., Vannoni, E., Vyssotski, A.L., Di Bari, M.A., Nonno, R., Agrimi, U., and Lipp, H.P. (2002) Early behavioural changes in mice infected with BSE and scrapie: automated home cage monitoring reveals prion strain differences. *Eur. J. Neurosci.*, **16**, 735–742.

277. Bessen, R.A., and Marsh, R.F. (1992) Identification of two biologically distinct strains of transmissible mink encephalopathy in hamsters. *J. Gen. Virol.*, **73** (Pt 2), 329–334.
278. Kim, Y.S., Carp, R.I., Callahan, S.M., and Wisniewski, H.M. (1987) Scrapie-induced obesity in mice. *J. Infect. Dis.*, **156**, 402–405.
279. Owen, J.P., Rees, H.C., Maddison, B.C., Terry, L.A., Thorne, L., Jackman, R., Whitelam, G.C., and Gough, K.C. (2007) Molecular profiling of ovine prion diseases by using thermolysin-resistant PrPSc and endogenous c04 PrP fragments. *J. Virol.*, **81**, 10532–10539.
280. D'Castro, L., Wenborn, A., Gros, N., Joiner, S., Cronier, S., Collinge, J., and Wadsworth, J.D. (2010) Isolation of proteinase K-sensitive prions using pronase E and phosphotungstic acid. *PLoS ONE*, **5**, e15679.
281. Somerville, R.A., and Ritchie, L.A. (1990) Differential glycosylation of the protein (PrP) forming scrapie-associated fibrils. *J. Gen. Virol.*, **71** (Pt 4), 833–839.
282. Collinge, J., Sidle, K.C., Meads, J., Ironside, J., and Hill, A.F. (1996) Molecular analysis of prion strain variation and the aetiology of 'new variant' CJD. *Nature*, **383**, 685–690.
283. Peretz, D., Scott, M.R., Groth, D., Williamson, R.A., Burton, D.R., Cohen, F.E., and Prusiner, S.B. (2001) Strain-specified relative conformational stability of the scrapie prion protein. *Protein Sci.*, **10**, 854–863.
284. Somerville, R.A., Oberthur, R.C., Havekost, U., MacDonald, F., Taylor, D.M., and Dickinson, A.G. (2002) Characterization of thermodynamic diversity between transmissible spongiform encephalopathy agent strains and its theoretical implications. *J. Biol. Chem.*, **277**, 11084–11089.
285. Spassov, S., Beekes, M., and Naumann, D. (2006) Structural differences between TSEs strains investigated by FT-IR spectroscopy. *Biochim. Biophys. Acta*, **1760**, 1138–1149.
286. Hecker, R., Taraboulos, A., Scott, M., Pan, K.M., Yang, S.L., Torchia, M., Jendroska, K., DeArmond, S.J., and Prusiner, S.B. (1992) Replication of distinct scrapie prion isolates is region specific in brains of transgenic mice and hamsters. *Genes Dev.*, **6**, 1213–1228.
287. Bruce, M.E., McBride, P.A., Jeffrey, M., and Scott, J.R. (1994) PrP in pathology and pathogenesis in scrapie-infected mice. *Mol. Neurobiol.*, **8**, 105–112.
288. Prusiner, S.B., McKinley, M.P., Bowman, K.A., Bolton, D.C., Bendheim, P.E., Groth, D.F., and Glenner, G.G. (1983) Scrapie prions aggregate to form amyloid-like birefringent rods. *Cell*, **35**, 349–358.
289. Liberski, P.P., Bratosiewicz, J., Walis, A., Kordek, R., Jeffrey, M., and Brown, P. (2001) A special report I. Prion protein (PrP)–amyloid plaques in the transmissible spongiform encephalopathies, or prion diseases revisited. *Folia Neuropathol.*, **39**, 217–235.
290. Jeffrey, M., and Gonzalez, L. (2007) Classical sheep transmissible spongiform encephalopathies: pathogenesis, pathological phenotypes and clinical disease. *Neuropathol. Appl. Neurobiol.*, **33**, 373–394.
291. Sigurdson, C.J., Nilsson, K.P., Hornemann, S., Manco, G., Polymenidou, M., Schwarz, P., Leclerc, M., Hammarstrom, P., Wuthrich, K., and Aguzzi, A. (2007) Prion strain discrimination using luminescent conjugated polymers. *Nat. Methods*, **4**, 1023–1030.
292. Fraser, H., and Dickinson, A.G. (1973) Scrapie in mice. Agent-strain differences in the distribution and intensity of grey matter vacuolation. *J. Comp. Pathol.*, **83**, 29–40.
293. Béringue, V., Herzog, L., Jaumain, E., Reine, F., Sibille, P., Le Dur, A., Vilotte, J.L., and Laude, H. (2012) Facilitated cross-species transmission of prions in extraneural tissue. *Science*, **335**, 472–475.
294. Béringue, V., Le Dur, A., Tixador, P., Reine, F., Lepourry, L., Perret-Liaudet, A., Haik, S., Vilotte, J.L., Fontes, M., and Laude, H. (2008) Prominent and persistent extraneural infection in human PrP transgenic mice infected with variant CJD. *PLoS ONE*, **3**, e1419.
295. Mahal, S.P., Baker, C.A., Demczyk, C.A., Smith, E.W., Julius, C., and Weissmann, C. (2007) Prion strain discrimination in cell culture: the cell panel assay. *Proc. Natl Acad. Sci. USA*, **104**, 20908–20913.
296. Bessen, R.A., and Marsh, R.F. (1992) Biochemical and physical properties of the prion protein from two strains of the transmissible mink encephalopathy agent. *J. Virol.*, **66**, 2096–2101.

297. Bessen, R.A., and Marsh, R.F. (1994) Distinct PrP properties suggest the molecular basis of strain variation in transmissible mink encephalopathy. *J. Virol.*, **68**, 7859–7868.
298. Bessen, R.A., Kocisko, D.A., Raymond, G.J., Nandan, S., Lansbury, P.T., and Caughey, B. (1995) Non-genetic propagation of strain-specific properties of scrapie prion protein. *Nature*, **375**, 698–700.
299. Hill, A.F., Desbruslais, M., Joiner, S., Sidle, K.C., Gowland, I., Collinge, J., Doey, L.J., and Lantos, P. (1997) The same prion strain causes vCJD and BSE. *Nature*, **389** (448-450), 526.
300. Asante, E.A., Linehan, J.M., Desbruslais, M., Joiner, S., Gowland, I., Wood, A.L., Welch, J., Hill, A.F., Lloyd, S.E., Wadsworth, J.D., *et al.* (2002) BSE prions propagate as either variant CJD-like or sporadic CJD-like prion strains in transgenic mice expressing human prion protein. *EMBO J.*, **21**, 6358–6366.
301. Béringue, V., Andreoletti, O., Le Dur, A., Essalmani, R., Vilotte, J.L., Lacroux, C., Reine, F., Herzog, L., Biacabe, A.G., Baron, T., *et al.* (2007) A bovine prion acquires an epidemic bovine spongiform encephalopathy prion-like phenotype on interspecies transmission. *J. Neurosci.*, **27**, 6965–6971.
302. Capobianco, R., Casalone, C., Suardi, S., Mangieri, M., Miccolo, C., Limido, L., Catania, M., Rossi, G., Di Fede, G., Giaccone, G., *et al.* (2007) Conversion of the BASE prion strain into the BSE strain: the origin of BSE? *PLoS Pathog.*, **3**, e31.
303. Scott, M.R., Groth, D., Tatzelt, J., Torchia, M., Tremblay, P., DeArmond, S.J., and Prusiner, S.B. (1997) Propagation of prion strains through specific conformers of the prion protein. *J. Virol.*, **71**, 9032–9044.
304. Peretz, D., Williamson, R.A., Legname, G., Matsunaga, Y., Vergara, J., Burton, D.R., DeArmond, S.J., Prusiner, S.B., and Scott, M.R. (2002) A change in the conformation of prions accompanies the emergence of a new prion strain. *Neuron*, **34**, 921–932.
305. Collinge, J., and Clarke, A.R. (2007) A general model of prion strains and their pathogenicity. *Science*, **318**, 930–936.
306. Weissmann, C., Li, J., Mahal, S.P., and Browning, S. (2011) Prions on the move. *EMBO Rep.*, **12**, 1109–1117.
307. Watts, J.C., Giles, K., Stohr, J., Oehler, A., Bhardwaj, S., Grillo, S.K., Patel, S., DeArmond, S.J., and Prusiner, S.B. (2012) Spontaneous generation of rapidly transmissible prions in transgenic mice expressing wild-type bank vole prion protein. *Proc. Natl Acad. Sci. USA*, **109**, 3498–3503.
308. Chianini, F., Fernandez-Borges, N., Vidal, E., Gibbard, L., Pintado, B., de Castro, J., Priola, S.A., Hamilton, S., Eaton, S.L., Finlayson, J., *et al.* (2012) Rabbits are not resistant to prion infection. *Proc. Natl Acad. Sci. USA*, **109**, 5080–5085.
309. Ladogana, A., Puopolo, M., Croes, E.A., Budka, H., Jarius, C., Collins, S., Klug, G.M., Sutcliffe, T., Giulivi, A., Alperovitch, A., *et al.* (2005) Mortality from Creutzfeldt-Jakob disease and related disorders in Europe, Australia, and Canada. *Neurology*, **64**, 1586–1591.
310. Collinge, J. (2001) Prion diseases of humans and animals: their causes and molecular basis. *Annu. Rev. Neurosci.*, **24**, 519–550.
311. Zou, W.Q., Puoti, G., Xiao, X., Yuan, J., Qing, L., Cali, I., Shimoji, M., Langeveld, J.P., Castellani, R., Notari, S., *et al.* (2010) Variably protease-sensitive prionopathy: a new sporadic disease of the prion protein. *Ann. Neurol.*, **68**, 162–172.
312. Head, M.W., Yull, H.M., Ritchie, D.L., Langeveld, J.P., Fletcher, N.A., Knight, R.S., and Ironside, J.W. (2013) Variably protease-sensitive prionopathy in the UK: a retrospective review 1991–2008. *Brain*, **136**, 1102–1115.
313. Rodriguez-Martinez, A.B., Garrido, J.M., Zarranz, J.J., Arteagoitia, J.M., de Pancorbo, M.M., Atares, B., Bilbao, M.J., Ferrer, I., and Juste, R.A. (2010) A novel form of human disease with a protease-sensitive prion protein and heterozygosity methionine/valine at codon 129: case report. *BMC Neurol.*, **10**, 99.
314. Wadsworth, J.D., and Collinge, J. (2011) Molecular pathology of human prion disease. *Acta Neuropathol.*, **121**, 69–77.
315. Brown, P., Gibbs, C.J., Jr, Rodgers-Johnson, P., Asher, D.M., Sulima, M.P., Bacote, A., Goldfarb, L.G., and Gajdusek, D.C. (1994) Human spongiform encephalopathy: the National Institutes of Health series of 300 cases of experimentally transmitted disease. *Ann. Neurol.*, **35**, 513–529.
316. Collinge, J., Palmer, M.S., Sidle, K.C., Gowland, I., Medori, R., Ironside, J., and Lantos, P. (1995) Transmission of fatal

familial insomnia to laboratory animals. *Lancet*, **346**, 569–570.

317. Mastrianni, J.A., Capellari, S., Telling, G.C., Han, D., Bosque, P., Prusiner, S.B., and DeArmond, S.J. (2001) Inherited prion disease caused by the V210I mutation: transmission to transgenic mice. *Neurology*, **57**, 2198–2205.

318. Nazor, K.E., Kuhn, F., Seward, T., Green, M., Zwald, D., Purro, M., Schmid, J., Biffiger, K., Power, A.M., Oesch, B., *et al.* (2005) Immunodetection of disease-associated mutant PrP, which accelerates disease in GSS transgenic mice. *EMBO J.*, **24**, 2472–2480.

319. Dossena, S., Imeri, L., Mangieri, M., Garofoli, A., Ferrari, L., Senatore, A., Restelli, E., Balducci, C., Fiordaliso, F., Salio, M., *et al.* (2008) Mutant prion protein expression causes motor and memory deficits and abnormal sleep patterns in a transgenic mouse model. *Neuron*, **60**, 598–609.

320. Gajdusek, D.C. (2008) Review. Kuru and its contribution to medicine. *Philos. Trans. R. Soc. London, Ser. B Biol. Sci.*, **363**, 3697–3700.

321. Will, R.G., Ironside, J.W., Zeidler, M., Cousens, S.N., Estibeiro, K., Alperovitch, A., Poser, S., Pocchiari, M., Hofman, A., and Smith, P.G. (1996) A new variant of Creutzfeldt-Jakob disease in the UK. *Lancet*, **347**, 921–925.

322. Wadsworth, J.D., Joiner, S., Hill, A.F., Campbell, T.A., Desbruslais, M., Luthert, P.J., and Collinge, J. (2001) Tissue distribution of protease resistant prion protein in variant Creutzfeldt-Jakob disease using a highly sensitive immunoblotting assay. *Lancet*, **358**, 171–180.

323. Wadsworth J.D.F., Asante E.A., and Collinge J. (2010) Review: contribution of transgenic models to understanding human prion disease. *Neuropathol. Appl. Neurobiol.*, **36**, 579.

324. Gill, O.N., Spencer, Y., Richard-Loendt, A., Kelly, C., Dabaghian, R., Boyes, L., Linehan, J., Simmons, M., Webb, P., Bellerby, P., *et al.* (2013) Prevalent abnormal prion protein in human appendixes after bovine spongiform encephalopathy epizootic: large scale survey. *Br. Med. J.*, **347**, 5675.

325. Hilton, D.A., Ghani, A.C., Conyers, L., Edwards, P., McCardle, L., Ritchie, D., Penney, M., Hegazy, D., and Ironside, J.W. (2004) Prevalence of lymphoreticular prion protein accumulation in UK tissue samples. *J. Pathol.*, **203**, 733–739.

326. Wroe, S.J., Pal, S., Siddique, D., Hyare, H., Macfarlane, R., Joiner, S., Linehan, J.M., Brandner, S., Wadsworth, J.D., Hewitt, P., *et al.* (2006) Clinical presentation and pre-mortem diagnosis of variant Creutzfeldt-Jakob disease associated with blood transfusion: a case report. *Lancet*, **368**, 2061–2067.

327. Llewelyn, C.A., Hewitt, P.E., Knight, R.S., Amar, K., Cousens, S., Mackenzie, J., and Will, R.G. (2004) Possible transmission of variant Creutzfeldt-Jakob disease by blood transfusion. *Lancet*, **363**, 417–421.

328. Peden, A.H., Head, M.W., Ritchie, D.L., Bell, J.E., and Ironside, J.W. (2004) Preclinical vCJD after blood transfusion in a PRNP codon 129 heterozygous patient. *Lancet*, **364**, 527–529.

329. Douet, J.Y., Zafar, S., Perret-Liaudet, A., Lacroux, C., Lugan, S., Aron, N., Cassard, H., Ponto, C., Corbiere, F., Torres, J.M., *et al.* (2014) Detection of infectivity in blood of persons with variant and sporadic Creutzfeldt-Jakob disease. *Emerg. Infect. Dis.*, **20**, 114–117.

330. Andreoletti, O., Litaise, C., Simmons, H., Corbiere, F., Lugan, S., Costes, P., Schelcher, F., Vilette, D., Grassi, J., and Lacroux, C. (2012) Highly efficient prion transmission by blood transfusion. *PLoS Pathog.*, **8**, e1002782.

331. Cervenakova, L., Yakovleva, O., McKenzie, C., Kolchinsky, S., McShane, L., Drohan, W.N., and Brown, P. (2003) Similar levels of infectivity in the blood of mice infected with human-derived vCJD and GSS strains of transmissible spongiform encephalopathy. *Transfusion*, **43**, 1687–1694.

332. Lacroux, C., Bougard, D., Litaise, C., Simmons, H., Corbiere, F., Dernis, D., Tardivel, R., Morel, N., Simon, S., Lugan, S., *et al.* (2012) Impact of leucocyte depletion and prion reduction filters on TSE blood borne transmission. *PLoS ONE*, **7**, e42019.

333. Taylor, D.M. (1999) Inactivation of prions by physical and chemical means. *J. Hosp. Infect.*, **43** (Suppl.), S69–S76.

334. Béringue, V., Vilotte, J.L., and Laude, H. (2008) Prion agent diversity and species barrier. *Vet. Res.*, **39**, 47.

335. Fediaevsky, A., Tongue, S.C., Noremark, M., Calavas, D., Ru, G., and Hopp, P. (2008) A descriptive study of the prevalence of atypical and classical scrapie in sheep in 20 European countries. *BMC Vet. Res.*, **4**, 19.
336. Goldmann, W. (2008) PrP genetics in ruminant transmissible spongiform encephalopathies. *Vet. Res.*, **39**, 30.
337. Benestad, S.L., Arsac, J.N., Goldmann, W., and Noremark, M. (2008) Atypical/Nor98 scrapie: properties of the agent, genetics, and epidemiology. *Vet. Res.*, **39**, 19.
338. Fediaevsky, A., Maurella, C., Noremark, M., Ingravalle, F., Thorgeirsdottir, S., Orge, L., Poizat, R., Hautaniemi, M., Liam, B., Calavas, D., et al. (2010) The prevalence of atypical scrapie in sheep from positive flocks is not higher than in the general sheep population in 11 European countries. *BMC Vet. Res.*, **6**, 9.
339. Le Dur, A., Béringue, V., Andreoletti, O., Reine, F., Lai, T.L., Baron, T., Bratberg, B., Vilotte, J.L., Sarradin, P., Benestad, S.L., et al. (2005) A newly identified type of scrapie agent can naturally infect sheep with resistant PrP genotypes. *Proc. Natl Acad. Sci. USA*, **102**, 16031–16036.
340. Benestad, S.L., Sarradin, P., Thu, B., Schonheit, J., Tranulis, M.A., and Bratberg, B. (2003) Cases of scrapie with unusual features in Norway and designation of a new type, Nor98. *Vet. Rec.*, **153**, 202–208.
341. Nentwig, A., Oevermann, A., Heim, D., Botteron, C., Zellweger, K., Drogemuller, C., Zurbriggen, A., and Seuberlich, T. (2007) Diversity in neuroanatomical distribution of abnormal prion protein in atypical scrapie. *PLoS Pathog.*, **3**, e82.
342. Sigurdson, C.J. (2008) A prion disease of cervids: chronic wasting disease. *Vet. Res.*, **39**, 41.
343. O'Rourke, K.I., Spraker, T.R., Zhuang, D., Greenlee, J.J., Gidlewski, T.E., and Hamir, A.N. (2007) Elk with a long incubation prion disease phenotype have a unique PrPd profile. *NeuroReport*, **18**, 1935–1938.
344. Green, K.M., Browning, S.R., Seward, T.S., Jewell, J.E., Ross, D.L., Green, M.A., Williams, E.S., Hoover, E.A., and Telling, G.C. (2008) The elk PRNP codon 132 polymorphism controls cervid and scrapie prion propagation. *J. Gen. Virol.*, **89**, 598–608.
345. Tamguney, G., Miller, M.W., Wolfe, L.L., Sirochman, T.M., Glidden, D.V., Palmer, C., Lemus, A., DeArmond, S.J., and Prusiner, S.B. (2009) Asymptomatic deer excrete infectious prions in faeces. *Nature*, **461**, 529–532.
346. Mathiason, C.K., Powers, J.G., Dahmes, S.J., Osborn, D.A., Miller, K.V., Warren, R.J., Mason, G.L., Hays, S.A., Hayes-Klug, J., Seelig, D.M., et al. (2006) Infectious prions in the saliva and blood of deer with chronic wasting disease. *Science*, **314**, 133–136.
347. Miller, M.W., and Williams, E.S. (2003) Prion disease: horizontal prion transmission in mule deer. *Nature*, **425**, 35–36.
348. Sigurdson, C.J., Williams, E.S., Miller, M.W., Spraker, T.R., O'Rourke, K.I., and Hoover, E.A. (1999) Oral transmission and early lymphoid tropism of chronic wasting disease PrPres in mule deer fawns (*Odocoileus hemionus*). *J. Gen. Virol.*, **80** (Pt 10), 2757–2764.
349. Angers, R.C., Browning, S.R., Seward, T.S., Sigurdson, C.J., Miller, M.W., Hoover, E.A., and Telling, G.C. (2006) Prions in skeletal muscles of deer with chronic wasting disease. *Science*, **311**, 1117.
350. Biacabe, A.G., Morignat, E., Vulin, J., and Calavas, D. (2008) Atypical bovine spongiform encephalopathies, France, 2001–2007. *Emerg. Infect. Dis.*, **14**, 298–300.
351. Lasmezas, C.I., Deslys, J.P., Demaimay, R., Adjou, K.T., Lamoury, F., Dormont, D., Robain, O., Ironside, J., and Hauw, J.J. (1996) BSE transmission to macaques. *Nature*, **381**, 743–744.
352. Bruce, M., Chree, A., McConnell, I., Foster, J., Pearson, G., and Fraser, H. (1994) Transmission of bovine spongiform encephalopathy and scrapie to mice: strain variation and the species barrier. *Philos. Trans. R. Soc. London, Ser. B Biol. Sci.*, **343**, 405–411.
353. Bruce, M.E., Will, R.G., Ironside, J.W., McConnell, I., Drummond, D., Suttie, A., McCardle, L., Chree, A., Hope, J., Birkett, C., et al. (1997) Transmissions to mice indicate that 'new variant' CJD is caused by the BSE agent. *Nature*, **389**, 498–501.
354. Green, R., Horrocks, C., Wilkinson, A., Hawkins, S.A., and Ryder, S.J. (2005) Primary isolation of the bovine spongiform encephalopathy agent in mice: agent definition based on a review of 150 transmissions. *J. Comp. Pathol.*, **132**, 117–131.

355. Scott, M.R., Will, R., Ironside, J., Nguyen, H.O., Tremblay, P., DeArmond, S.J., and Prusiner, S.B. (1999) Compelling transgenetic evidence for transmission of bovine spongiform encephalopathy prions to humans. *Proc. Natl Acad. Sci. USA*, **96**, 15137–15142.

356. Padilla, D., Béringue, V., Espinosa, J.C., Andreoletti, O., Jaumain, E., Reine, F., Herzog, L., Gutierrez-Adan, A., Pintado, B., Laude, H., *et al.* (2011) Sheep and goat BSE propagate more efficiently than cattle BSE in human PrP transgenic mice. *PLoS Pathog.*, **7**, e1001319.

357. Béringue, V., Herzog, L., Reine, F., Le Dur, A., Casalone, C., Vilotte, J.L., and Laude, H. (2008) Transmission of atypical bovine prions to mice transgenic for human prion protein. *Emerg. Infect. Dis.*, **14**, 1898–1901.

358. Plinston, C., Hart, P., Chong, A., Hunter, N., Foster, J., Piccardo, P., Manson, J.C., and Barron, R.M. (2011) Increased susceptibility of human-PrP transgenic mice to bovine spongiform encephalopathy infection following passage in sheep. *J. Virol.*, **85**, 1174–1181.

359. Kittelberger, R., Chaplin, M.J., Simmons, M.M., Ramirez-Villaescusa, A., McIntyre, L., MacDiarmid, S.C., Hannah, M.J., Jenner, J., Bueno, R., Bayliss, D., *et al.* (2010) Atypical scrapie/Nor98 in a sheep from New Zealand. *J. Vet. Diagn. Invest.*, **22**, 863–875.

360. Simmons, M.M., Moore, S.J., Konold, T., Thurston, L., Terry, L.A., Thorne, L., Lockey, R., Vickery, C., Hawkins, S.A., Chaplin, M.J., *et al.* (2011) Experimental oral transmission of atypical scrapie to sheep. *Emerg. Infect. Dis.*, **17**, 848–854.

361. Baker, H.F., Ridley, R.M., and Wells, G.A. (1993) Experimental transmission of BSE and scrapie to the common marmoset. *Vet. Rec.*, **132**, 403–406.

362. Comoy, E.E., Casalone, C., Lescoutra-Etchegaray, N., Zanusso, G., Freire, S., Marce, D., Auvre, F., Ruchoux, M.M., Ferrari, S., Monaco, S., *et al.* (2008) Atypical BSE (BASE) transmitted from asymptomatic aging cattle to a primate. *PLoS ONE*, **3**, e3017.

363. Mestre-Frances, N., Nicot, S., Rouland, S., Biacabe, A.G., Quadrio, I., Perret-Liaudet, A., Baron, T., and Verdier, J.M. (2012) Oral transmission of L-type bovine spongiform encephalopathy in primate model. *Emerg. Infect. Dis.*, **18**, 142–145.

364. Race, B., Meade-White, K.D., Miller, M.W., Barbian, K.D., Rubenstein, R., LaFauci, G., Cervenakova, L., Favara, C., Gardner, D., Long, D., *et al.* (2009) Susceptibilities of nonhuman primates to chronic wasting disease. *Emerg. Infect. Dis.*, **15**, 1366–1376.

365. Kong, Q., Zheng, M., Casalone, C., Qing, L., Huang, S., Chakraborty, B., Wang, P., Chen, F., Cali, I., Corona, C., *et al.* (2008) Evaluation of the human transmission risk of an atypical bovine spongiform encephalopathy prion strain. *J. Virol.*, **82** (7), 3697–3701

366. Sandberg, M.K., Al-Doujaily, H., Sigurdson, C.J., Glatzel, M., O'Malley, C., Powell, C., Asante, E.A., Linehan, J.M., Brandner, S., Wadsworth, J.D., *et al.* (2010) Chronic wasting disease prions are not transmissible to transgenic mice overexpressing human prion protein. *J. Gen. Virol.*, **91**, 2651–2657.

367. Kong, Q., Huang, S., Zou, W., Vanegas, D., Wang, M., Wu, D., Yuan, J., Zheng, M., Bai, H., Deng, H., *et al.* (2005) Chronic wasting disease of elk: transmissibility to humans examined by transgenic mouse models. *J. Neurosci.*, **25**, 7944–7949.

368. Tamguney, G., Giles, K., Bouzamondo-Bernstein, E., Bosque, P.J., Miller, M.W., Safar, J., DeArmond, S.J., and Prusiner, S.B. (2006) Transmission of elk and deer prions to transgenic mice. *J. Virol.*, **80**, 9104–9114.

369. Wadsworth, J.D., Joiner, S., Linehan, J.M., Balkema-Buschmann, A., Spiropoulos, J., Simmons, M.M., Griffiths, P.C., Groschup, M.H., Hope, J., Brandner, S., *et al.* (2013) Atypical scrapie prions from sheep and lack of disease in transgenic mice overexpressing human prion protein. *Emerg. Infect. Dis.*, **19**, 1731–1739.

370. Hou, F., Sun, L., Zheng, H., Skaug, B., Jiang, Q.X., and Chen, Z.J. (2011) MAVS forms functional prion-like aggregates to activate and propagate antiviral innate immune response. *Cell*, **146**, 448–461.

371. Si, K., Choi, Y.B., White-Grindley, E., Majumdar, A., and Kandel, E.R. (2010) Aplysia CPEB can form prion-like multimers in sensory neurons that contribute to long-term facilitation. *Cell*, **140**, 421–435.

372. Si, K., Lindquist, S., and Kandel, E.R. (2003) A neuronal isoform of the aplysia CPEB has prion-like properties. *Cell*, **115**, 879–891.

373. Cai, X., Chen, J., Xu, H., Liu, S., Jiang, Q.X., Halfmann, R., and Chen, Z.J. (2014) Prion-like polymerization underlies signal transduction in antiviral immune defense and inflammasome activation. *Cell*, **156**, 1207–1222.
374. Wickner, R.B. (1994) [URE3] as an altered URE2 protein: evidence for a prion analog in *Saccharomyces cerevisiae*. *Science*, **264**, 566–569.
375. Wickner, R.B., Edskes, H.K., Bateman, D.A., Kelly, A.C., Gorkovskiy, A., Dayani, Y., and Zhou, A. (2013) Amyloids and yeast prion biology. *Biochemistry*, **52**, 1514–1527.
376. Liebman, S.W., and Chernoff, Y.O. (2012) Prions in yeast. *Genetics*, **191**, 1041–1072.
377. Halfmann, R., and Lindquist, S. (2010) Epigenetics in the extreme: prions and the inheritance of environmentally acquired traits. *Science*, **330**, 629–632.
378. Saupe, S.J. (2011) The [Het-s] prion of *Podospora anserina* and its role in heterokaryon incompatibility. *Semin. Cell Dev. Biol.*, **22**, 460–468.
379. Sawaya, M.R., Sambashivan, S., Nelson, R., Ivanova, M.I., Sievers, S.A., Apostol, M.I., Thompson, M.J., Balbirnie, M., Wiltzius, J.J., McFarlane, H.T., et al. (2007) Atomic structures of amyloid cross-beta spines reveal varied steric zippers. *Nature*, **447**, 453–457.
380. Sparrer, H.E., Santoso, A., Szoka, F.C., Jr, and Weissman, J.S. (2000) Evidence for the prion hypothesis: induction of the yeast [PSI+] factor by in vitro- converted Sup35 protein. *Science*, **289**, 595–599.
381. Maddelein, M.L., Dos Reis, S., Duvezin-Caubet, S., Coulary-Salin, B., and Saupe, S.J. (2002) Amyloid aggregates of the HET-s prion protein are infectious. *Proc. Natl Acad. Sci. USA*, **99**, 7402–7407.
382. Tanaka, M., Chien, P., Naber, N., Cooke, R., and Weissman, J.S. (2004) Conformational variations in an infectious protein determine prion strain differences. *Nature*, **428**, 323–328.
383. Tanaka, M., Chien, P., Yonekura, K., and Weissman, J.S. (2005) Mechanism of cross-species prion transmission: an infectious conformation compatible with two highly divergent yeast prion proteins. *Cell*, **121**, 49–62.
384. Tanaka, M., Collins, S.R., Toyama, B.H., and Weissman, J.S. (2006) The physical basis of how prion conformations determine strain phenotypes. *Nature*, **442**, 585–589.
385. Bemporad, F., and Chiti, F. (2012) Protein misfolded oligomers: experimental approaches, mechanism of formation, and structure-toxicity relationships. *Chem. Biol.*, **19**, 315–327.
386. Jucker, M., and Walker, L.C. (2011) Pathogenic protein seeding in Alzheimer disease and other neurodegenerative disorders. *Ann. Neurol.*, **70**, 532–540.
387. Trevitt, C.R., and Collinge, J. (2006) A systematic review of prion therapeutics in experimental models. *Brain*, **129**, 2241–2265.
388. Demaimay, R., Adjou, K.T., Béringue, V., Demart, S., Lasmezas, C.I., Deslys, J.P., Seman, M., and Dormont, D. (1997) Late treatment with polyene antibiotics can prolong the survival time of scrapie-infected animals. *J. Virol.*, **71**, 9685–9689.
389. Tribouillard-Tanvier, D., Béringue, V., Desban, N., Gug, F., Bach, S., Voisset, C., Galons, H., Laude, H., Vilette, D., and Blondel, M. (2008) Antihypertensive drug guanabenz is active in vivo against both yeast and mammalian prions. *PLoS ONE*, **3**, e1981.
390. Bach, S., Talarek, N., Andrieu, T., Vierfond, J.M., Mettey, Y., Galons, H., Dormont, D., Meijer, L., Cullin, C., and Blondel, M. (2003) Isolation of drugs active against mammalian prions using a yeast-based screening assay. *Nat. Biotechnol.*, **21**, 1075–1081.
391. Heppner, F.L., Musahl, C., Arrighi, I., Klein, M.A., Rulicke, T., Oesch, B., Zinkernagel, R.M., Kalinke, U., and Aguzzi, A. (2001) Prevention of scrapie pathogenesis by transgenic expression of anti-prion protein antibodies. *Science*, **294**, 178–182.
392. White, A.R., Enever, P., Tayebi, M., Mushens, R., Linehan, J., Brandner, S., Anstee, D., Collinge, J., and Hawke, S. (2003) Monoclonal antibodies inhibit prion replication and delay the development of prion disease. *Nature*, **422**, 80–83.
393. Aguzzi, A., and Rajendran, L. (2009) The transcellular spread of cytosolic amyloids, prions, and prionoids. *Neuron*, **64**, 783–790.

394. Jucker, M., and Walker, L.C. (2013) Self-propagation of pathogenic protein aggregates in neurodegenerative diseases. *Nature*, **501**, 45–51.
395. Polymenidou, M., and Cleveland, D.W. (2011) The seeds of neurodegeneration: prion-like spreading in ALS. *Cell*, **147**, 498–508.
396. Polymenidou, M., and Cleveland, D.W. (2012) Prion-like spread of protein aggregates in neurodegeneration. *J. Exp. Med.*, **209**, 889–893.
397. Angot, E., Steiner, J.A., Hansen, C., Li, J.Y., and Brundin, P. (2010) Are synucleinopathies prion-like disorders? *Lancet Neurol.*, **9**, 1128–1138.

23
RNA Metabolism and Drug Design

Eriks Rozners
Binghamton University, The State University of New York, Department of Chemistry, 4400 Vestal Parkway East, Binghamton, NY 13902, USA

1	**Introduction: RNA as a Drug Target** 829	
2	**Antisense Oligonucleotides** 830	
2.1	Chemical Modifications of Antisense Oligonucleotides 830	
2.1.1	Backbone Modifications 831	
2.1.2	Sugar Modifications 833	
2.1.3	Nucleobase Modifications 836	
2.2	Antisense Oligonucleotides Using the RNase H Mechanism 837	
2.3	Antisense Oligonucleotides Using Steric Block Mechanisms 838	
2.3.1	Antisense-Mediated Splicing Regulation 838	
2.3.2	Antibacterial Antisense Oligonucleotides 840	
2.3.3	Antisense Oligonucleotides Targeting MicroRNA 842	
3	**RNA Interference** 843	
3.1	Chemical Modifications for Short Interfering RNAs 844	
3.1.1	Backbone Modifications 844	
3.1.2	Sugar Modifications 845	
3.1.3	Nucleobase Modifications 847	
3.2	Improving the Delivery of siRNAs 848	
3.2.1	Direct Conjugation of siRNAs with Delivery Enhancers 848	
3.2.2	siRNA Delivery by Noncovalent Encapsulation 849	
3.3	Single-Stranded siRNAs 850	
4	**Small Molecules as RNA-Binding Drug Leads** 851	
4.1	Targeting Disease-Related Toxic RNA Motifs 852	
4.2	Targeting microRNA 854	
4.3	Targeting Viral and Bacterial RNAs 854	
5	**Conclusions and Outlook** 857	
	References 858	

Translational Medicine: Molecular Pharmacology and Drug Discovery
First Edition. Edited by Robert A. Meyers.
© 2018 Wiley-VCH Verlag GmbH & Co. KGaA. Published 2018 by Wiley-VCH Verlag GmbH & Co. KGaA.

Keywords

Antisense oligonucleotides
Modified single-stranded DNA fragments of 8–20 nucleotides in length that modulate mRNA metabolism and protein synthesis by targeting mRNA through sequence-specific Watson–Crick hydrogen bonding.

Guide strand
The siRNA strand that guides RNAi proteins to cleave the complementary target mRNA at a unique position opposite nucleosides 10–11 of the guide strand. The guide strand is the 5′-strand of the thermally less-stable end of an siRNA duplex.

MicroRNAs (miRNAs)
Endogenous, small, noncoding RNAs that regulate many target genes by inhibiting the translation of their mRNA. In contrast to siRNAs, miRNAs are only partially complementary to their targets and require only 6–8 bp in the so-called seed region at the 5′-ends.

Passenger strand
The inactive strand of siRNA that is cleaved and discarded during the loading of siRNA in the RNA-induced silencing complex.

RNA-induced silencing complex (RISC)
A large multiprotein complex that silences the gene complementary to the guide strand of siRNA. The most common mechanism involves the endonuclease argonaute 2-catalyzed cleavage of mRNA.

RNA interference (RNAi)
Post-transcriptional silencing of gene expression mediated by 21–22 nucleotide-long double-stranded RNA molecules with sequence complementary to the silenced gene.

RNase H
An endonuclease that degrades the RNA strand of an RNA-DNA heteroduplex.

Short interfering RNA (siRNA)
Short, double-stranded RNA molecules (21–22 nucleotides) with two nucleotide overhangs at each end. siRNAs silence the gene complementary to one of the strands, the so-called guide strand.

The central role that ribonucleic acid (RNA) plays in the regulation of gene expression makes it a promising drug target. Compared to that of proteins, the therapeutic regulation of RNA metabolism has been relatively unexplored. In this chapter, the recent progress made using chemically modified oligonucleotides and small molecules to interfere with the function of therapeutically relevant RNA species is reviewed. The inhibition of disease-related protein production by the antisense oligonucleotide-mediated cleavage of mRNA is one of the most mature technologies. Recent discoveries of noncoding RNAs have initiated the development of therapeutic approaches that use RNA interference or target microRNA. The development of therapeutic oligonucleotides requires chemical modifications to improve their drug-like properties. In contrast, the key challenge in the development of small-molecule drugs is the identification of strong and selective binders for specific RNA motifs. Each approach provides unique advantages and shortcomings that will challenge chemists and biologists involved in the exciting, but challenging, field of RNA.

1
Introduction: RNA as a Drug Target

Until the early 1990s, RNA was viewed as a passive messenger in the transfer of genetic information from DNA to proteins. However, this view has changed dramatically over the past two decades, which have brought about revolutionary discoveries of the catalytic role that RNA plays in splicing, RNA maturation, and protein synthesis; the regulation of gene expression by small RNAs; and, finally, a large group of long noncoding RNAs that play important but, as of today, poorly understood biological functions [1, 2]. While less than 2% of DNA encodes for functional proteins, as much as 90% of the genome can be transcribed into RNA. Today, the functional importance of most RNA transcripts is still unknown, and it is fairly safe to predict that many more regulatory RNAs will be discovered in the near future.

Therapeutic interventions employing RNA's functions can be achieved either by small molecules or oligonucleotides. Small molecules typically require the RNA being targeted to have noncanonical (i.e., not Watson–Crick base-paired) features, such as a hairpin loop, bulge, or internal loop to bind the target RNA with some selectivity for the shape of the feature. In contrast, most of the oligonucleotide-derived compounds target single-stranded RNA and achieve remarkable sequence selectivity by forming Watson–Crick base-paired duplexes. From the perspective of drug design, small molecules have major advantages, such as easily adjustable pharmacokinetics and pharmacodynamics, simple synthesis and formulation, and also have extensive precedents for medicinal chemistry optimization. However, the discovery of small molecules that bind RNA sequences selectively and may serve as lead candidates for drug development remains a challenging goal [3, 4]. The main problem is the flexibility of noncanonical RNA features, which complicates the rational design of small molecules with both strong and specific binding. A common problem with high-affinity compounds is their low sequence selectivity, which causes indiscriminate binding to various RNAs.

In contrast, it is relatively easy to identify suitable messenger ribonucleic acid mRNA sequences and to design oligonucleotide leads that target these sequences. The shortcomings of oligonucleotides as drug candidates include poor bioavailability and pharmacokinetics, difficult delivery to specific tissues, poor cellular uptake, and low stability in biological media [5]. These problems are mostly due to the high molecular weight and negatively charged and polar phosphodiester backbone, which prevents oligonucleotides

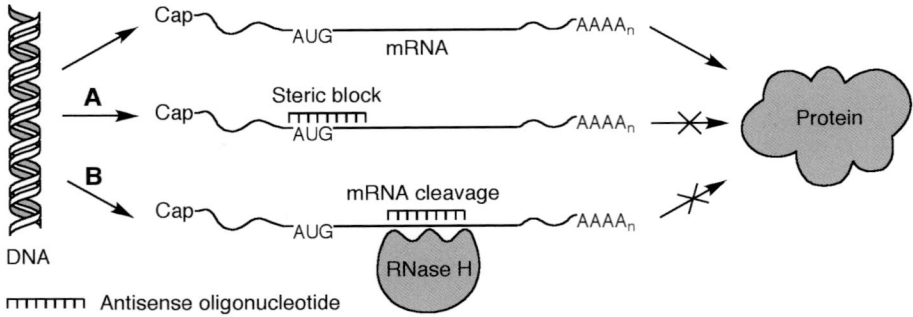

Fig. 1 Mechanism of antisense oligonucleotides. Route A: Direct steric block of translation. Route B: RNase H-mediated mRNA cleavage.

from reaching their therapeutic targets *in vivo* and results in poor efficiency in the clinic. Nevertheless, today there are more than 50 ongoing clinical trials that employ oligonucleotide-mediated gene silencing. Although the progress of oligonucleotide drugs to the clinic remains relatively slow, the recent (January 2013) approval of mipomersen (ISIS 301012; Kynamro™), a cholesterol-reducing drug developed by ISIS Pharmaceuticals [6], has increased the enthusiasm about therapeutic strategies that operate through the direct silencing of disease-related RNA by chemically modified oligonucleotides [7–12].

2
Antisense Oligonucleotides

Antisense oligonucleotides are the first – and perhaps the most direct – therapeutic approach to using oligonucleotide drugs that modulate RNA metabolism. The principle was first demonstrated in the pioneering studies of Zamecnik and Stephenson in 1978 [13]. Antisense oligonucleotides are typically 8–20 nucleotides in length and form sequence-specific Watson–Crick hydrogen-bonded double helices with their target mRNA. This allows them to modulate the function of their targets by either promoting RNA degradation or providing direct steric interference such that RNA translation is hindered [14]. The antisense principle is simple and elegant: the oligonucleotide binds the disease-related mRNA and prevents its translation either by sterically blocking the translation process or by RNase H-assisted degradation of the mRNA (Fig. 1). In contrast to traditional drugs that target disease-related proteins, antisense aims at destroying the source of the protein – the disease-related mRNA. Since binding to the mRNA occurs via the well-understood Watson–Crick base-pairing, antisense oligonucleotides – at least in principle – can be designed to be highly selective for their target.

2.1
Chemical Modifications of Antisense Oligonucleotides

For drug design purposes, antisense oligonucleotides must be chemically modified to overcome the following problems: (i) low stability in biological systems caused by DNA- and RNA-cleaving enzymes; (ii) the need for a high RNA binding affinity to compete with RNA's self-structure of mRNA and protein binding, especially when the antisense functions by the steric block mechanism; and (iii) poor bioavailability and pharmacokinetics due to weak binding to plasma proteins that leads to rapid excretion via the urine [5]. Backbone, sugar, or nucleobase residues may be used as modification points for antisense oligonucleotides [8, 14, 15].

2.1.1 Backbone Modifications

The phosphodiester internucleoside linkage is the most popular moiety for chemical modification, because even small changes of the natural linkage enhance the stability of antisense oligonucleotides in biological systems. Phosphorothioate (PS; Fig. 2) is one of the most widely used backbone modifications. Substitution of the nonbridging oxygen with sulfur enhances nuclease stability and improves the pharmacokinetic properties of oligonucleotides [16]. The latter benefit is due to increased binding to plasma proteins such as albumin and α_2-macroglobulin, which slows down the excretion. However, the introduction of PS modifications slightly reduces the thermal stability of DNA and RNA duplexes by about 0.5 °C per modification.

PS DNAs are often called first-generation antisense drugs. Although the progress of first-generation antisense drugs in entering clinical applications has been relatively slow, it should be noted that PS is still one of the most common modifications in RNA-targeting therapeutics in current clinical trials [9, 12]. In fact, both of the antisense drugs that have gained regulatory approval, fomivirsen (1998) and mipomersen (2013), contain PS modifications. Despite being the

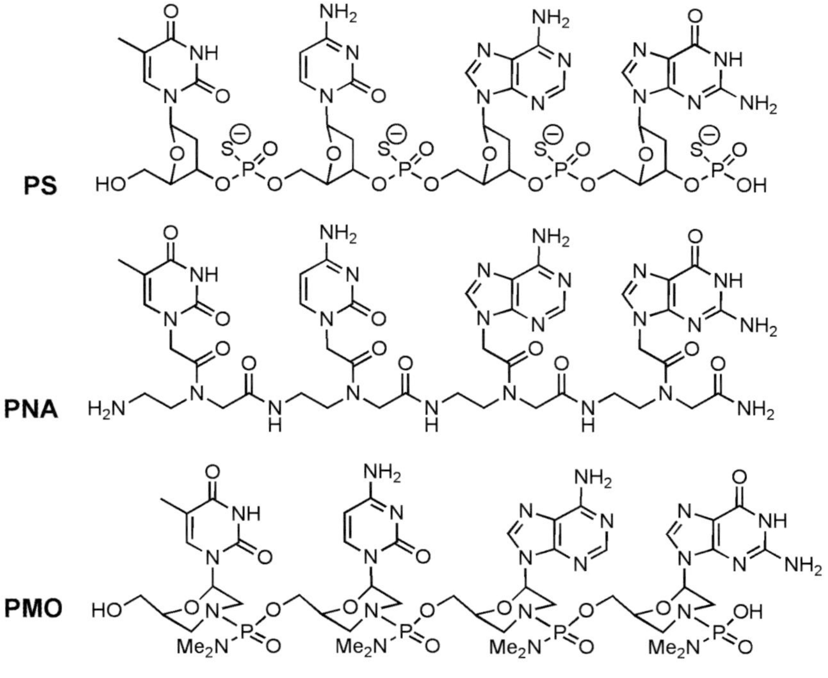

Fig. 2 Chemical structures of backbone modifications. PS, phosphorothioate; PNA, peptide nucleic acid; PMO, phosphorodiamidate morpholino oligomer.

modification of choice for improving the pharmacokinetic properties of antisense and short interfering ribonucleic acid siRNA therapeutics, PS modification is not without its drawbacks. PS-modified oligonucleotides bind to a wide variety of proteins, which causes proinflammatory and apoptosis-inducing off-target effects, as indicated by elevated levels of liver transaminases [5]. Although no serious hepatotoxicity has been observed, the potential off-target effects raise concerns about the safety of long-term applications of PS-modified oligonucleotide therapeutics [5]. While many other modifications of the phosphate linkage have been described [15], no clear advantages over the PS modification have been demonstrated.

Two chemical modifications that replace the entire sugar-phosphate backbone with nonionic linkages have been explored for antisense oligonucleotides, namely peptide nucleic acids (PNAs; Fig. 2) and phosphorodiamidate morpholino oligomers (PMOs; Fig. 2). In PNA, the entire sugar–phosphate backbone is replaced by a neutral N-(2-aminoethyl)glycine moiety that makes it highly resistant to nucleases and proteases [17, 18]. PNA forms exceptionally stable Watson–Crick base-paired helices with both complementary DNA and RNA. PNA binds to double-stranded DNA via two competing binding modes, namely Hoogsteen triple helix (PNA : DNA, 1 : 1) and strand invasion (where PNA displaces one of the DNA strands), typically followed by a triple helix formation (PNA : DNA, 2 : 1) [19–23]. The binding mode depends on the sequence of PNA; thymine-rich PNAs prefer invasion complexes whereas cytosine-rich PNAs prefer triple helix formation. The binding mode also depends on salt concentration, DNA duplex stability, and PNA concentration [22]. Recently, Rozners and coworkers showed that PNA forms exceptionally stable and sequence-selective triple helices with double-stranded RNA [24–26]. The affinity of PNA for double-stranded RNA was around two orders of magnitude higher than that for the same sequence of double-stranded DNA [26]. PNA does not support an RNase H mechanism; however, the strong binding to single-stranded RNA makes PNA a promising steric block antisense agent. In terms of antisense therapeutic development, PNA has achieved limited success to date, mostly because of poor cellular uptake and pharmacokinetics. However, promising results have been obtained with antibacterial antisense PNA oligonucleotides (as will be discussed below).

Despite its neutral backbone, PNA does not cross cellular membranes efficiently, and usually requires special delivery techniques. One of the most popular and successful solutions to this problem uses cell-penetrating peptides (CPPs) that deliver a PNA–CCP conjugate through endocytosis [27–29]. The conjugation of PNA with Transportan [30], R_6-Penetratin [30, 31], or Tat [32, 33] peptides has been shown to increase cellular delivery. An alternative approach has been to introduce cationic modifications directly into the PNA backbone, as in guanidine-modified PNAs that are derived from arginine [34–36]. Because of the low ability of PNA–CPP to escape from endosomes, relatively high concentrations of PNA–CCP are required for efficient delivery, which may cause adverse side effects *in vivo*. Various endosomolytic compounds have been explored, but unfortunately most are too toxic for *in-vivo* applications [29]. Recently, the groups of Corey [37, 38] and Gait [30, 31, 39] showed that the conjugation of PNAs with short oligolysine peptides enabled the efficient delivery of PNAs in various cancer cell lines. Four lysine residues in a Lys–PNA–Lys$_3$ conjugate were as efficient as R6-Penetratin, a CPP optimized for the cellular delivery of PNA [31]. The use of short oligolysines instead of longer CPPs led to a significant reduction in synthetic complexity and enabled efficient PNA delivery in cell cultures and in mice [40, 41]. The addition of a terminal cysteine residue further increased the cellular uptake of Cys–Lys–PNA–Lys$_3$ conjugates [27]. Although these results were encouraging for potential antisense applications, relatively high concentrations were still required to produce a pharmacological effect. Further optimization is needed to improve the potency and pharmacokinetics of antisense PNAs. As will be discussed below, PNAs have been explored as steric block antisense agents against genetic disorders (using a splicing modulation mechanism) [42] and as antibacterial agents [43].

In PMOs or Morpholinos, the riboses are replaced with morpholine rings that are linked by phosphorodiamidate linkages [44] (Fig. 2). Similar to PNAs, PMOs are nonionic DNA analogs that are not degraded by natural nucleases and do not support RNase H activity. PMOs are used either to modify splicing or to interfere with the assembly of ribosomes, so as to block translation. While unmodified PMOs have been reported to diffuse into some cells, most of the applications require further modification to achieve a useful uptake [44]. As with PNAs, the most popular solution to improve uptake is conjugation with CPPs, among which Tat and arginine-rich peptides have been found to be the most efficient [45]. Similar to PNAs, PMOs are taken up by an endocytosis mechanism, where the rate-limiting step is endosomal escape. Among various arginine-rich CPPs, the conjugation of PMOs with (R-Ahx-R)$_4$ (Ahx is 6-aminohexanoic acid) achieved the best antisense activity due to efficient endosomal escape [46, 47]. Compared to PS oligonucleotides, PMOs do not bind to plasma proteins and so far have not shown any toxicity in either animals or clinical trials [5]. Currently, morpholino oligomers are being explored as steric block antisense agents against genetic disorders (using a splicing modulation mechanism), and also as antiviral and antibacterial agents [43, 44, 48].

2.1.2 Sugar Modifications

Modification of the 2′-position has been one of the most popular approaches to increase enzymatic stability and the RNA-binding affinity of antisense oligonucleotides and siRNAs [49, 50]. In early studies, it was recognized that the introduction of electronegative substituents at the 2′-position in DNA (Fig. 3; 2′-O-methyl and 2′-fluoro) increased the affinity of modified oligonucleotides for target mRNA. Initially, this was believed to be due largely to the preference of such 2′-modified riboses toward the C3′-endo conformation. However, recent studies have shown that the stereoelectronic effects that correlate 2′-fluoro modification and duplex stability are more complex, and involve enhanced stacking as well as hydrogen bonding [51–54]. Enzymatic stability is related to steric hindrance and correlates with the size of the 2′-substituent. Steric interactions and a preference for the C3′-endo conformation make most of the 2′-modifications incompatible with the RNase H mechanism. However, this problem can be overcome by using the gapmer strategy, as will be discussed below.

2′-O-Methyl (2′-O-Me; Fig. 3) is a naturally occurring RNA modification that, along with other 2′-O-alkyl groups, is one of the most popular sugar modifications.

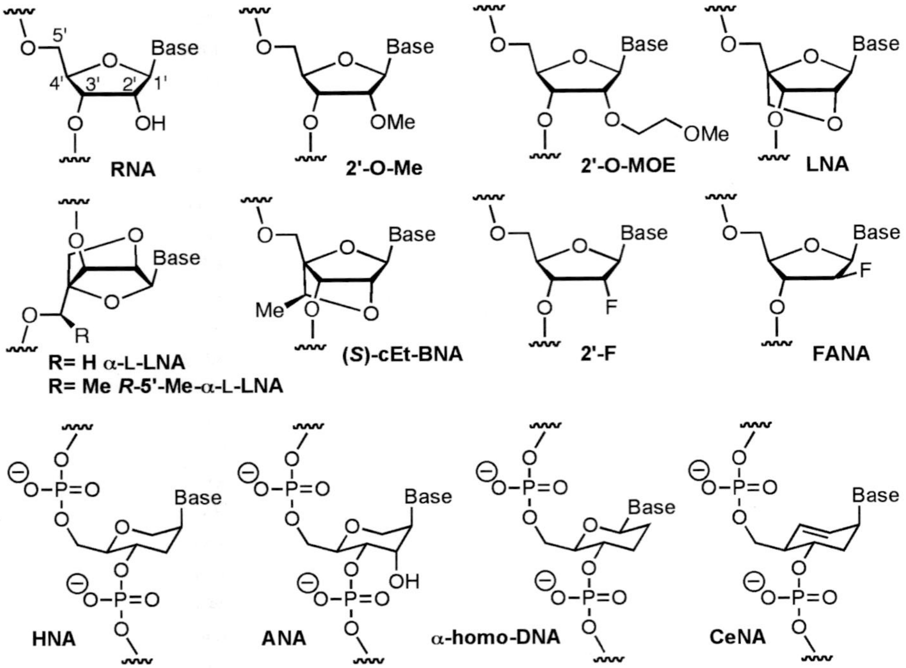

Fig. 3 Chemical structures of sugar modifications.

2′-O-Alkyl groups significantly enhance the enzymatic stability and RNA-binding affinity of antisense oligonucleotides. Among the various larger 2′-O-alkyl substituents studied [50], 2′-O-methoxyethyl (MOE; Fig. 3) has emerged as one of the most successful sugar modifications. MOE-modified oligonucleotides have an unusually high RNA affinity (an increase in t_m of about +2 °C per modification), especially when compared to oligonucleotides having the sterically similar 2′-O-butyl group [55], and show nuclease resistance similar to that of the PS-modified oligonucleotides. Crystal structures suggest that the higher enzymatic and thermal stability of 2′-O-MOE-modified oligonucleotides is due to a conformational preorganization towards C3′-endo and favorable hydration of the 2′-O-MOE-substituted ribose [56, 57]. 2′-O-MOE modification is part of the so-called second-generation antisense oligonucleotides that are currently undergoing clinical trials. As will be discussed below, mipomersen – the first of the second-generation antisense oligonucleotides to obtain regulatory approval – contains 2′-O-MOE-modified nucleosides.

An alternative approach that is related to 2′-O-alkyl modifications is to introduce a covalent methylene linkage between the 2′-O and 4′-C of ribose. These compounds, developed by Imanishi [58] and Wengel [59, 60], are termed locked nucleic acids (LNAs; Fig. 3) because the modification locks the sugar in the C3′-endo conformation. LNA modifications enhance the enzymatic stability and dramatically improve the affinity of oligonucleotides for their RNA targets. Depending on the sequence context, this increase may be as much as +5 to +9 °C per modification. This increased affinity is believed to be mostly due to the preorganization of DNA nucleotides adjacent to the LNA modification into the C3′-endo conformation, which enhances base stacking and enforces an RNA-like structure favorable for duplex formation. α-L-LNA (Fig. 3; R = H) is a stereoisomer of LNA that also forms high-affinity complexes with RNA [61, 62]. As with other 2′-O modifications, LNA and α-L-LNA do not support the RNase H mechanism and must be used in a gapmer strategy. The "constrained ethyl," (S)-cEt-BNA (Fig. 3) is a further development of LNA-like modifications that has shown promising properties in antisense gapmers [63]. (S)-cEt-BNA shows an RNA affinity similar to that of LNA, along with an improved nuclease resistance. In contrast to LNA, (S)-cEt-BNA reduces the hepatotoxicity caused by PS modifications [63].

The 2′-F modification (Fig. 3) has been studied in antisense oligonucleotides [64, 65] and siRNAs [52, 66, 67]. Substitution of the 2′-H in DNA with fluorine imparts an RNA-like conformation because the high electronegativity of fluorine strongly favors the C3′-endo conformation of ribose. The conformational preorganization towards an RNA-like structure was initially believed to be the chief reason for the increased RNA binding affinity (difference in melting temperatures, $\Delta t_m \sim +2$ °C per modification) of the 2′-F-modified oligonucleotides [64]. Recent thermodynamic and structural studies have shown that the thermal stabilization of 2′-F-modified oligonucleotides is not entropically favored, as would be expected if the conformational preorganization were the dominant factor [51]. Instead, the higher stability of 2′-F oligonucleotides is almost entirely based on favorable enthalpy gains. More recent nuclear magnetic resonance (NMR) spectroscopic and thermodynamic studies have shown that the axial 2′-fluorine substituent increases the RNA affinity of the modified oligonucleotide by favorably affecting both Watson–Crick H-bonding and base stacking [68]. In other words, the higher enthalpy-based stability of 2′-F oligonucleotides as compared to RNA is due to a combination of conformational preorganization of the modified single strand and strengthening of both the H-bonding and stacking interactions in the modified duplex.

2′-Fluoroarabinonucleic acid (FANA) is a stereoisomer of 2′-F DNA with a high binding affinity for complementary DNA and RNA [69–71]. The conformation of a FANA–RNA duplex closely resembles that of a DNA–RNA duplexes, and FANA–RNA duplexes are cleaved by RNase H with an efficiency similar to that observed for DNA–RNA duplexes [69]. Extensive studies conducted by Damha and coworkers have shown that, while in various chimeric structures FANA behaves primarily as a DNA mimic, it is also compatible with RNA [71]. NMR structural studies have revealed that FANA–RNA duplexes are stabilized by a favorable 2′-F⋯H8 pseudohydrogen bond, where optimal bonding was achieved at the pyrimidine–purine steps, where the base-stacking geometry adjusted to optimize the interaction between 2′-F of a 3′-sugar and H8 of a 5′ adjacent purine nucleoside [54, 72]. Interestingly, a more recent study identified a different type of nontraditional hydrogen bond that stabilized 2′-F-modified A-type duplexes, such as FANA–2′-F RNA [53]. In this study, C–H⋯O hydrogen bonds were found between the H2′ of a fluorine-modified sugar and its 3′-neighbor's O4′-sugar and O5′-backbone atoms. Taken together with the findings of studies on 2′-F RNA, these results illustrate how the strongly polarizing fluorine atom stabilizes nucleic acid secondary structure through a complex interplay of sugar conformation, nucleobase polarization, and nonconventional H-bonding interactions. 2′-F and 2′-FANA nucleosides are known substrates of DNA and RNA polymerases, which raises concern that the metabolites of such antisense oligonucleotides may be incorporated into native nucleic acids and cause undesired side effects.

Herdewijn and coworkers have studied nucleic acid analogs comprised of six-membered sugars [73]. Among many possible pyranose backbones, oligonucleotides built of 2-(1,5-anhydro-2,3-dideoxy-D-arabino-hexitol)nucleosides (hexitol nucleic acids, HNAs; Fig. 3) [74, 75], 2-(1,5-anhydro-2-deoxy-D-alitriol) nucleosides (alitritol nucleic acids, ANAs) [76], and 1-(2,3-dideoxy-α-D-erythro-hexopyranosyl)nucleosides (α-homo-DNA) [77] formed stable double helices with natural nucleic acids. In general, all pyranose-based nucleic acids formed more stable complexes with RNA than with DNA.

HNA is an analog of DNA with an extra methylene group inserted between O4′ and C1′. Although the incorporation of a single HNA monomer into a DNA strand resulted in a decrease in melting temperature by 2–4 °C, fully modified HNA strands formed duplexes with DNA and RNA which were more stable than their endogenous counterparts [74]. HNA formed an overall A-type duplex with complementary RNA that was not a good substrate for RNase H [78]. Although HNA was able to act as a steric block antisense, its efficiency was lower than that of a PS oligonucleotide, presumably due to an inability to activate RNase H [73]. Interestingly, HNA was completely resistant to degradation by snake venom phosphodiesterase.

ANA formed A-type duplexes with RNA and DNA, with an even higher affinity than those containing HNA [76]. The higher stability of ANA–RNA duplexes compared to HNA–RNA duplexes was suggested to be due to a better hydration involving 3′-OH and a conformational preorganization of ANA [73]. Promising results have been reported in ribonucleic acid interference (RNAi) experiments, where ANA modifications at the 3′-ends of the guide and passenger strands improved the efficacy of siRNA [79].

Cyclohexenyl nucleic acid (CeNA; Fig. 3) behaves as an RNA mimic and forms stable double helices with complementary DNA and RNA [73, 80]. CeNA is stable against enzymatic degradation, and CeNA–RNA duplexes are substrates of RNase H, although the cleavage efficiency is significantly lower than that for DNA–RNA duplexes [80]. The incorporation of one or two CeNA nucleosides into the guide or passenger strand resulted in a similar, or somewhat enhanced, RNAi activity when compared to unmodified siRNAs [81].

2.1.3 Nucleobase Modifications

Compared to modifications of the backbone and sugar moieties, nucleobase modifications have not been used extensively for developing antisense therapeutics. One concern with using nucleobase-modified oligonucleotides has been that their metabolism may create modified nucleosides that may be incorporated into native nucleic acids and interfere with the correct expression and maintenance of genetic material. Notable exceptions include C5-alkyl-modified pyrimidines. 5-Methylcytosine is a naturally occurring DNA modification present at eukaryotic cytosine–guanosine (CpG) motifs; accordingly, DNA lacking CpG methylation is identified by the immune system as foreign

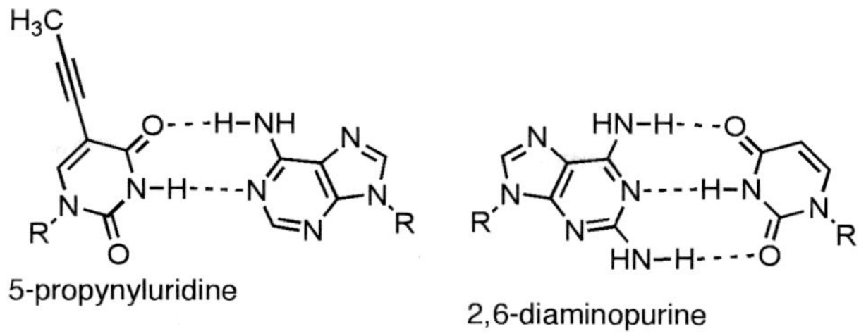

Fig. 4 Chemical structures of Watson–Crick base-pairs formed by modified nucleobases in antisense oligonucleotides.

[82]. The substitution of cytosine with 5-methylcytosine in antisense oligonucleotides eliminates immune stimulation caused by unmodified CpG motifs.

Early studies on nucleobase-modified oligonucleotides recognized the importance of RNA binding affinity for the successful development of antisense-based therapies and, therefore focused on improving the thermal stability of Watson–Crick base pairs. 5-Alkyl groups stack with heterocyclic nucleobases in the major groove, thereby stabilizing the double helix. For example, the substitution of pyrimidines with 5-propynyl pyrimidines and adenine with 2,6-diaminopurine (Fig. 4) in antisense oligonucleotides increased their affinity for target mRNA. However, the 5-propynyl modifications increased the hepatotoxicity of antisense oligonucleotides while having only a moderately beneficial effect on potency and only on the first-generation PS oligonucleotides, but not on the second generation 2′-*O*-MOE/PS compounds [83]. Thus, antisense oligonucleotides currently undergoing clinical trials do not typically have nucleobase modifications, apart from 5-methylcytosine.

2.2
Antisense Oligonucleotides Using the RNase H Mechanism

The most common and best-understood mechanism of antisense oligonucleotide action involves RNA degradation by RNase H, an endogenous enzyme that is hypothesized to participate in DNA replication and repair. RNase H recognizes an RNA–DNA heteroduplex and cleaves the RNA strand. In chemically modified antisense oligonucleotides, RNase H requires at least five consecutive DNA nucleotides to recognize and process the heteroduplex [65, 84]. In general, RNase H-activating antisense oligonucleotides can target any region of mRNA that does not have a significant secondary structure.

The RNase H mechanism imposes restrictions on what chemical modifications can be made in the antisense oligonucleotides. The PS backbone does not interfere with RNase H activity. Oligonucleotides modified at the 2′-position do not support an RNase H mechanism except, as discussed above, for FANA, which is a notable exception [69]. A popular approach in antisense drug design has been the so-called gapmers. These antisense oligonucleotides incorporate gaps of PS deoxyribonucleotides to support RNase H cleavage, which are flanked by sugar-modified nucleotides to enhance enzymatic stability and RNA-binding affinity [65]. An example of gapmer technology is mipomersen, a fully PS-modified oligonucleotide that has 10 2′-*O*-MOE-modified nucleosides (out of 20), five at the 5′-end and five at the 3′-end (Fig. 5); such a design is called a 5-10-5

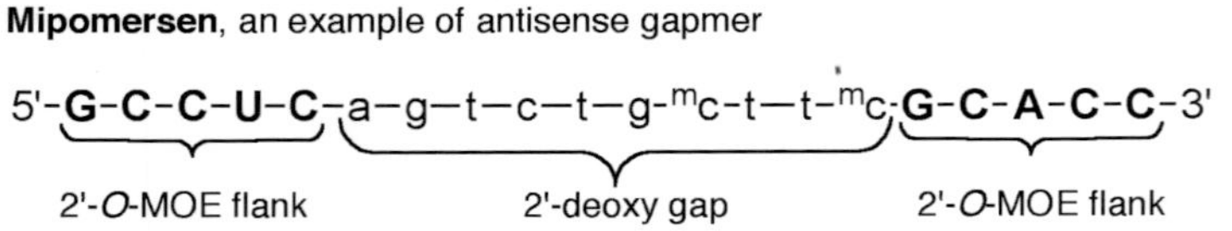

Mipomersen, an example of antisense gapmer

5'-**G**-**C**-**C**-**U**-**C**-a-g-t-c-t-g-^{m}c-t-t-^{m}c-**G**-**C**-**A**-**C**-**C**-3'

2'-*O*-MOE flank   2'-deoxy gap   2'-*O*-MOE flank

Fig. 5 Mipomersen, an example of a typically designed antisense gapmer. Bold capital letters designate the 2'-*O*-MOE nucleotides; lower-case letters indicate deoxyribonucleotides; ^{m}c designates a 5-methylcytosine base.

gapmer. The 2'-*O*-MOE modifications led to a 20-fold improvement in the potency of mipomersen compared to its PS-only counterpart.

In the gapmer context, LNA enhanced the potency of antisense oligonucleotides, but hepatotoxic off-target effects were also observed [85]. α-L-LNA-modified 2-10-2 and 2-12-2 PS antisense gapmers effectively downregulated phosphatase and tensin homolog mRNA in a mouse model [86]. In contrast to LNA, α-L-LNA did not cause hepatotoxicity but did induce some immunostimulation, as judged by a modest increase in liver and spleen weights. The introduction of a methyl group at the C5' of α-L-LNA (*R*-5'-Me-α-L-LNA; Fig. 3, R = Me) eliminated the immunostimulatory side effect [86].

A remarkable recent improvement in gapmer technology was achieved by shortening the antisense oligonucleotides (20- to 14-mer) and replacing the 2'-*O*-MOE with (*S*)-cEt-BNA modifications (see Fig. 3 for the structures of modified sugars) [63]. These changes increased the potency of second-generation 2-10-2 gapmer antisense oligonucleotides in animals by three- to fivefold, without inducing hepatotoxicity [63]. The rationale behind the use of a (*S*)-cEt-BNA modification was to combine the structural elements of LNA and 2'-*O*-MOE modifications that are known to increase potency (LNA) and decrease hepatotoxicity (2'-*O*-MOE). This technology is currently being developed by ISIS Pharmaceuticals as Generation 2.5 antisense oligonucleotides.

2.3 Antisense Oligonucleotides Using Steric Block Mechanisms

The achievement of a direct steric block of mRNA translation has proven challenging. Because the translating ribosomes easily unfold RNA structures and displace antisense oligonucleotides, such an approach is limited to targeting the translation initiation region where antisense binding may interfere with the assembly or initial movement of the ribosome. Although morpholino oligonucleotides (for structures, see Fig. 2) that act by direct steric block mechanisms are successfully used as tools for *in-vivo* studies in zebrafish [87], this approach has not been popular in the development of therapeutic agents [14].

2.3.1 Antisense-Mediated Splicing Regulation

The antisense oligonucleotide steric block mechanism has been remarkably successful in interfering with the early stages of mRNA metabolism, especially in modulating the splicing of pre-mRNA [42, 88]. Since approximately 90% of mRNA transcripts may undergo alternative splicing, this approach has tremendous therapeutic potential [88]. The idea is that the oligonucleotide binding to mRNA sites essential for splicing (splicing enhancer or repressor

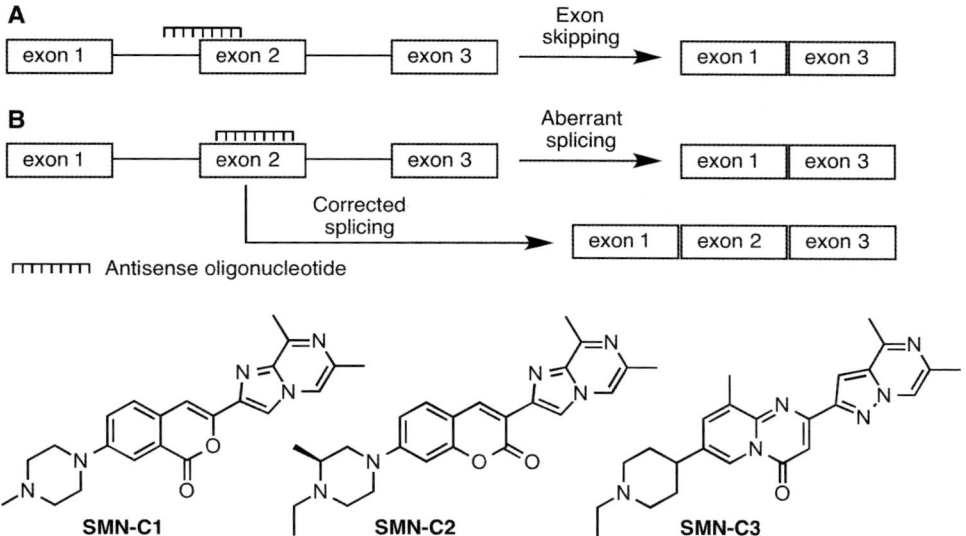

Fig. 6 Mechanisms of antisense oligonucleotide-mediated exon skipping (route A) and exon inclusion (route B) for the correction of aberrant splicing. SMN-C1, SMN-C2 and SMN-C3 are structures of recently discovered small molecules that modulate splicing.

sequences) blocks the interactions between mRNA and spliceosome components (proteins and nuclear RNAs) and directs the splicing machinery to alternative splicing sites (Fig. 6). This may result in either the skipping of an exon or the inclusion of an alternative exon, which may restore the production of a missing protein, repair a defective protein, or produce an alternative protein [88]. Oligonucleotide-mediated skipping of mutated exons and exon inclusion have been studied extensively as therapeutic approaches for neuromuscular diseases, such as Duchenne muscular dystrophy (DMD) and spinal muscular atrophy (SMA).

Steric block oligonucleotides rely entirely on strong binding to their targets, pre-mRNAs, and hence they must be chemically modified to increase their RNA binding affinity and ensure nuclease resistance. The most common sugar modifications for splicing modulations are $2'$-O-Me, $2'$-O-MOE and LNA, in combination with the PS backbone to insure the desired pharmacokinetic properties [42]. Neutral oligonucleotide analogs, such as PNA and PMO, have also shown very promising results in splicing modulation [42]. Specifically, PMOs have become the leading compounds in the development of therapeutic exon skipping and inclusion for DMD and other diseases [44, 89].

SMA is a genetic disease caused by deletion or loss-of-function mutations in the *SMN1* gene that causes a loss of the survival of motor neuron (SMN) protein. Collaborative studies conducted at ISIS Pharmaceuticals and by Krainer and colleagues at the Cold Spring Harbor Laboratory have shown that production of the SMN protein can be restored from the homologous *SMN2* gene that normally skips exon 7, resulting in a shorter and rapidly degrading protein product [90]. The approach of these authors involved the use of a uniformly 2′-O-MOE and PS-modified oligonucleotide containing 5-methylcytosines to target the intronic splicing silencer region located near the 5′-splice site of intron 7 in *SMN2* pre-mRNA, which resulted in an approximate 90% inclusion of intron 7 [91]. Interestingly, the same oligonucleotide having 2′-F modifications caused complete skipping of intron 7 due to recruiting of the interleukin enhancer-binder factor 2 and 3 complex by the modified duplex [92]. The latter study is an intriguing example of antisense-mediated recruiting of specific proteins to RNA. The systemic or direct administration of 2′-O-MOE antisense oligonucleotide in the lateral ventricle of mice resulted in the synthesis of normal SMN from the *SMN2* gene in the central nervous system (CNS) and various peripheral tissues, thereby rescuing the SMA phenotype [91, 93]. The *SMN2* targeting antisense oligonucleotide (ISIS−SMN$_{Rx}$) is currently entering Phase III clinical trials [90]. In a recent collaborative study, a structurally related series of small molecules (SMN-C1−SMN-C3; Fig. 6) was discovered that shifted the SMN2 splicing towards the production of full-length SMN protein [94]. While the exact mechanism of action remains unknown, the administration of these orally available small molecules to mice increased SMN protein levels, improved motor function, and extended the life span of the animals.

2.3.2 Antibacterial Antisense Oligonucleotides

Today, the development and spread of antibiotic resistance is becoming a global healthcare problem [95]. The introduction of penicillin in 1942 was followed by two decades of intensive antibiotic discovery such that, by late the 1960s, many concerns associated with controlling bacterial infections were considered less worrisome and antibiotic development research had lost its momentum. But, at the same time, the resistance of bacteria to existing antibiotics was steadily growing and by 2002 the first *Staphylococcus aureus* infections that were completely resistant to vancomycin – the 'last line of defense' antibiotic – were reported in the United States. Since then, antimicrobial resistance surveillance data has shown an uphill battle, with bacteria acquiring resistance faster than new antibacterial drugs can be developed using traditional methods of medicinal chemistry. In fact, there is already an alarming dearth of new antibiotics, as only a few new compounds have been introduced during the past 40 years: oxazolidinones in 2000, lipopetides in 2003, and a new diarylquinoline antituberculosis drug (Sirturo) in 2012 [96]. Unfortunately, most currently used antibiotics are close derivatives of old compounds to which bacteria have already developed resistance and, thus, it is only a matter of time before these compounds lose their activity. Consequently, there is an urgent need to develop new compounds that function via new mechanisms of antibacterial action, and especially those that can counteract and retard the development of resistance.

Bacterial RNA is a well-established target of current antibiotics, with aminoglycosides that target the ribosomal A-site being one of the first classes of antibiotics to be introduced into clinical practice [97]. Although aminoglycosides suffer from high toxicity and widespread resistance in bacteria, they are still used as one of the few remaining treatment options for multidrug-resistant bacteria [97]. The development of new RNA-binding antibiotics has been relatively slow; the challenge here is in achieving selectivity for bacterial RNA in the presence of closely related RNA (e.g., human RNA).

Antisense oligonucleotides constitute a promising alternative approach for the sequence-specific targeting of bacterial RNA. In fact, bacteria use natural antisense RNA to regulate their own gene expression. In contrast to small-molecule antibiotics, the antisense approach has the potential to eradicate pathogens with species or even strain specificity, while leaving benign bacteria unaffected. The concept of targeting bacteria with antisense oligonucleotides was first proposed more than three decades ago [98], shortly after the seminal report that antisense oligonucleotides could target viruses [13]. While PS antisense oligonucleotides are still being actively investigated as potential antibacterial agents against *Mycobacterium tuberculosis* [99, 100], two types of nucleic acid analog have emerged as more effective antisense compounds against bacteria, namely PNAs (Fig. 2), as pioneered by Good and Nielsen [101, 102], and PMOs (Fig. 2), as pioneered by Geller and coworkers [103]. The development of these compounds has been described in an excellent review [43].

In initial studies, antisense PNAs that targeted β-lactamase mRNA [101] and ribosomal RNA [102] were active in *Escherichia coli* strain AS19 that had an outer membrane permeable to macromolecules such as PNA. The inhibition of β-lactamase reduced β-lactam (e.g., penicillin) antibiotic resistance, while targeting of the ribosomal RNA inhibited the growth of *E. coli*. Later studies using PNA [104] and PMO [103] showed that the bacterial cell membrane, especially in Gram-negative pathogens, was the main barrier against the uptake of these compounds, and the bottleneck for the development of antibacterial antisense therapeutics. Currently, the best solution to the uptake problem is to reduce the length of antibacterial antisense oligonucleotides and to attach cell-membrane-penetrating peptides to the PNA [105] or PMO [103]. Among the various peptides tested, KFFKFFKFFK emerged as the most effective [103, 105, 106] as it delivered an antibacterial PNA that accumulated in *E. coli* and was not a substrate for drug efflux pumps [107]. The results of these pioneering studies suggested that nonionic nucleic acid analogs such as PNA and PMO represent promising candidates for the development of novel antibiotics.

Follow-up studies have demonstrated that PNAs and PMOs targeting essential genes unique to bacteria, such as acyl carrier protein (*acpP*), inhibit the growth of *E. coli* [103, 105]. Interestingly, shorter antisense oligomers (8–12 bases) were more efficient at inhibiting bacterial growth than the originally tried 15- and 20-mers [105, 108]. Since antisense antibacterials act by a direct steric block mechanism, the AUG start codon of prokaryotic mRNA is the best target [109]. More recent studies have targeted PNAs to 16S [110] and 23S [111] rRNA and bacterial RNase P [112]. Other bacteria, such as *S. aureus* [106], *Klebsiella pneumonia* [113], and *Mycobacterium smegmatis* [114] have also been successfully targeted by PNAs. *In vivo*, the severity of *E. coli* infection was reduced in mouse peritonitis using PNA [115] and PMO [116] antisense compounds targeting the *acpP* gene. More recently, antisense PMOs have been found to reduce the severity of bacterial infection and to increase survival in mouse models infected with *E. coli* [117, 118], *Burkholderia cepacia* [119], *Bacillus anthracis* [120], *Acinetobacter lwoffii*, and *Acinetobacter baumannii* [121]. Taken together, the results of these studies confirm that antisense PNAs and PMOs could become an alternative therapeutic approach against emerging multidrug-resistant pathogens.

The antisense approach has the advantage in its ability to target genes that are less likely to promote the development of antibacterial resistance. Moreover, if resistance develops as a result of mutation, then antisense compounds can easily be adjusted by changing their sequences. However, antisense antibacterial compounds face the same challenges as other oligonucleotide drugs, most importantly, the low permeability of bacterial membranes. Compared to other antisense approaches, however, fewer research investigations have been conducted, and progress in the field of antisense antibacterials has consequently been much slower.

2.3.3 Antisense Oligonucleotides Targeting MicroRNA

Mammalian microribonucleic acids (miRNAs) are transcribed as long hairpin structures, pri-miRNAs, which are processed into mature miRNAs in a two-step process [122–124]. First, in the cell nucleus, pri-miRNAs are cleaved by an RNase III enzyme, Drosha, into approximately 70 nt pre-miRNA hairpins that are transported into the cytoplasm. In the second step, the pre-miRNA hairpin is cleaved into ~22 nt miRNA duplexes by another RNase III enzyme, Dicer. Mature miRNAs control gene expression by inhibiting the translation of target mRNA (Fig. 7a). The deregulation of miRNA activity is implicated in various human diseases and, most profoundly, in cancer [125–128]. Thus, miRNAs represent very attractive therapeutic targets [129, 130].

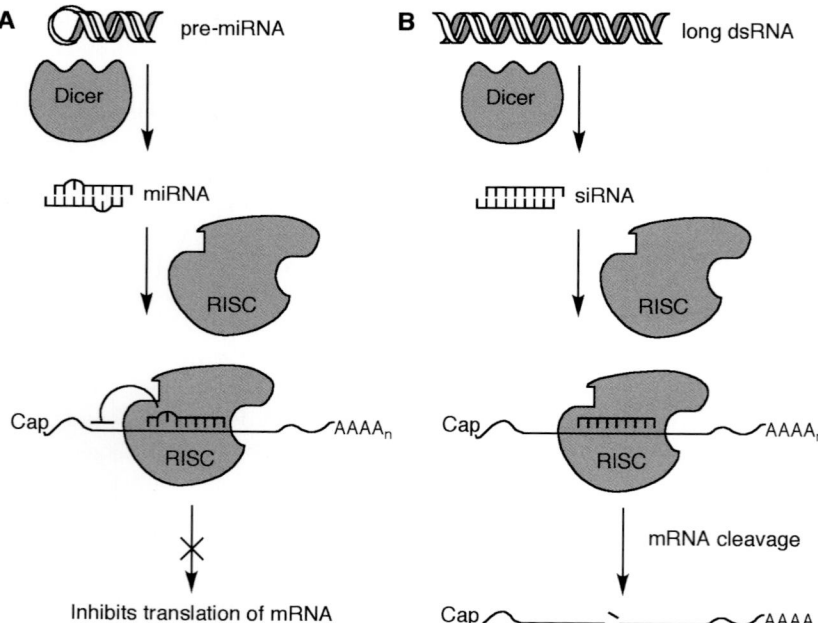

Fig. 7 (a) Mechanism of miRNA-mediated translation inhibition; (b) Mechanism of RNA interference.

The first examples of miRNA inhibition using 2′-O-Me-modified antisense oligonucleotides were reported in 2004 [131, 132]. More recent studies on the antisense inhibition of miRNA-122 [27, 133–135], miRNA-10b [136], and miRNA-155 [41] have demonstrated the therapeutic potential of miRNA silencing. The antisense oligonucleotides used to inhibit mature miRNAs were short LNAs [133, 134, 137–140], 2′-O-Me antagomirs [135, 136] 2′-O-MOE [141, 142], 2′-F [143], and PNAs [27, 32, 33, 39, 41, 144, 145]. PS modification was used to further increase the nuclease stability of sugar-modified anti-miRNA oligonucleotides [135, 143]. Antisense morpholinos targeting the Drosha and Dicer cleavage sites in pri- and pre-miRNA hairpins inhibited the maturation of miRNA-205 in zebrafish embryos [146]. Although the mechanism remains unclear, it is likely that morpholinos invaded the pri- and pre-miRNA hairpins and sterically blocked their processing. This notion was supported by the observation that morpholinos inhibited pri-miRNA-205, but were inactive against the more stable hairpin of pri-miRNA-375-1 [146].

A common problem with anti-miRNA oligonucleotides is that relatively high doses are required for *in-vivo* inhibition. Among the various current designs, a combination of LNA with other modifications has provided promising results for therapeutic applications. Miravirsen, an LNA/PS-modified antisense oligonucleotide that targets miRNA-122, is currently undergoing clinical trials for the treatment of hepatitis C virus (HCV) infections [11, 147]. Therapeutic interference with miRNA function is a relatively new field for the antisense approach, but one which holds great promise, as novel important functions for these and other regulatory RNAs will continue to be discovered. The recent progress, ongoing challenges and future prospects of miRNA targeting for therapeutic purposes have been discussed in an excellent review [129].

3
RNA Interference

RNAi is an evolutionarily conserved response of cells to double-stranded RNA that leads to the cleavage of RNA having sequence homology to the double-stranded trigger (Fig. 7b). The trigger may be either endogenous or foreign (e.g., RNA virus) in origin. Exogenous RNAi is driven by 21–22 nucleotides-long RNA duplexes, so-called siRNAs [148, 149]. The siRNA duplex is unwound and the strand that has the thermally less-stable 5′-end becomes the guide strand (Fig. 8; antisense to the target mRNA) that is incorporated into a protein complex called the RNA-induced silencing complex (RISC); the passenger strand is then cleaved and discarded [150]. The guide strand directs the RISC to cleave the complementary mRNA sequence at a unique position opposite nucleosides 10–11 of

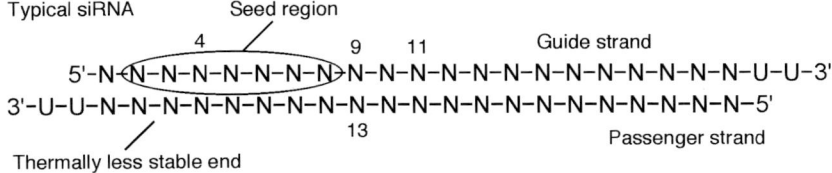

Fig. 8 Typical design of a short interfering RNA (siRNA).

the guide strand (Fig. 7b). Such cleavage prevents mRNA from being translated into protein, thus silencing the target gene. The RISC can also regulate gene expression via chromatin remodeling or translational inhibition if the guide and mRNA sequences are only partially matched. The latter mechanism is typical for miRNAs that originate from endogenous long hairpins and act as natural regulators of gene expression. In contrast to siRNAs that are perfectly complementary to the target mRNA, miRNAs require formation of only 6–8 bp in the so-called 'seed region' of the 5'-end of the guide strand (Fig. 7a). Because of this limited sequence requirement, one miRNA may regulate hundreds of genes and, a given gene may be under the control of several miRNAs.

3.1
Chemical Modifications for Short Interfering RNAs

Both, siRNAs and miRNAs may be considered as naturally occurring antisense oligonucleotides. Compared to antisense oligonucleotides, siRNAs and miRNAs use endogenous gene control pathways, which increases the effectiveness of target inhibition. Chemically synthesized short RNA duplexes can be introduced directly into cells where they bypass the Dicer cleavage step and act as external gene silencers [151]. While unmodified siRNAs are relatively stable and can be used in cell culture RNAi experiments, the development of therapeutic siRNAs requires the use of chemical modifications. Besides ensuring adequate enzymatic stability, for *in-vivo* applications, therapeutic siRNAs need chemical modifications to: (i) improve pharmacokinetic properties; (ii) optimize intracellular delivery; (iii) avoid innate immune responses; and (iv) minimize off-target effects, such as off-target silencing by the passenger strand and miRNA-like silencing at partially complementary sites. Many of the modifications developed for antisense oligonucleotides have also been used successfully in siRNAs [8, 152–158]. A general consensus is that the entire passenger strand and the 3'-end of the guide strand are more tolerant to modification than the seed region of the guide strand [66, 159].

3.1.1 Backbone Modifications

As with antisense oligonucleotides, PS modifications (see Fig. 2) of the phosphate backbone are commonly used in siRNAs to increase enzymatic stability and improve pharmacokinetic properties [67, 160]. Moderate PS modification is well tolerated in siRNAs and leads to a more persistent gene silencing [67, 161, 162]. While Corey and coworkers [67] showed that an siRNA duplex with all phosphates modified (40 PS linkages) had no toxicity, other studies have reported toxicity of siRNAs with 12 and 20 PS modifications [161, 162]. Boranophosphates (PBs; Fig. 9) [163] and phosphorodithioates (PS2) [164, 165] are also well tolerated in siRNAs, increase their enzymatic stability, and may increase silencing activity when placed at certain positions. Similar to PS, PB and PS2 slightly reduce the thermal stability of siRNA duplexes ($\Delta t_m \sim 0.5\,°C$ per modification).

Apart from PS, PB and PS2 modifications, the backbone modification of siRNAs has been explored only minimally. Iwase and coworkers reported that the introduction of two neutral amide linkages (AM1; Fig. 9) at the 3'-overhangs of siRNAs increased enzymatic stability but did not decrease silencing activity [166, 167]. AM1 modification was originally developed in DNA by Just and coworkers [168] and De Mesmaeker and coworkers [169, 170], among other backbone modifications explored for

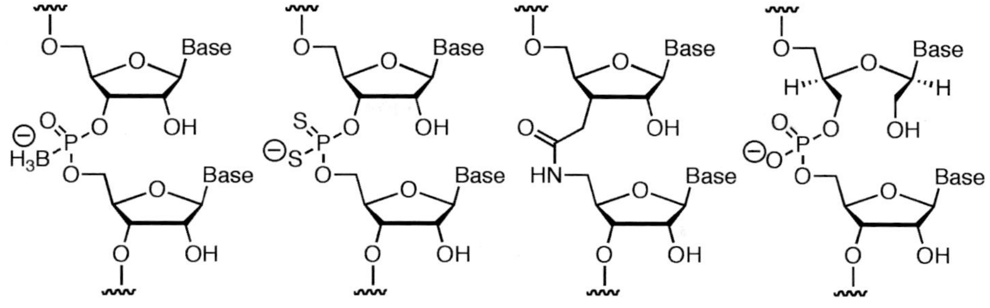

Fig. 9 Structures of backbone modifications in siRNAs.

antisense technology [15]. Later, detailed thermodynamic, NMR structural and X-ray crystallography studies performed by Rozners and co-workers showed that AM1 amides had surprisingly little effect on the structure and biological activity of RNA duplexes [171–173]. Up to three consecutive amide linkages were easily accommodated by small adjustments in the structure of the RNA duplex, and had little effect on siRNA activity when placed at the 5′-end of the passenger strand [172]. Internal amide modifications in the guide strand somewhat reduced the silencing activity; however, this could be more than compensated for by an unexpected increase in activity due to amide modification in the passenger strand [173]. Taken together, these structural and biological results suggest that amides are excellent replacements for phosphate linkages in RNA, and offer promising modifications for optimizing siRNA properties.

3.1.2 Sugar Modifications

Modifications in the ribose moiety (see Fig. 3) are very popular in the design of therapeutic siRNAs. While the 2′-O-Me modification is well tolerated at some positions in siRNAs, the use of large alkyl groups (such as 2′-O-MOE) decrease the activity of siRNA [174]. For example, whereas 2′-O-Me is well tolerated at the 3′-end of the guide strand, 2′-O-MOE is not; moreover, while 2′-O-Me is well tolerated anywhere in the passenger strand, 2′-O-MOE is tolerated only at the 3′-end [174]. The substitution of 2′-OH with 2′-O-Me slightly increases the thermal stability of the siRNA duplex by about 0.5 °C per modification. In addition to increasing enzymatic stability, the strategic use of 2′-O-Me modifications can enhance RNAi activity and decrease off-target effects. For example, introducing two 2′-O-Me groups at the 5′-end of the passenger strand increases RNAi activity and selectivity by enforcing the loading of the guide strand into RISC [175]. 2′- O-alkyl modifications, in general, inhibit the immunostimulatory off-target effects of double-stranded siRNAs [176] and, when used in conjunction with other modifications, have yielded successful *in-vivo* experiments, thereby advancing the therapeutic use of siRNA [160, 177].

2′-F modifications (see Fig. 3) have been extensively studied in siRNAs [52, 66, 67, 178, 179]. The substitution of 2′-OH with 2′-F increases the thermal stability of an siRNA duplex by about 1 °C per modification [52]. Early studies showed that 2′-F groups were well tolerated in the guide strand of siRNAs, regardless of position [174]. An siRNA duplex having all pyrimidine nucleotides of both passenger and guide strands replaced with 2′-F nucleotides and all purine nucleotides of the guide strand having 2′-O-Me modifications retained most of the RNAi activity [180]. In another study, an siRNA with alternating 2′-O-Me and 2′-F modifications was about 500-fold more potent *in vitro* than the unmodified counterpart [178]. This surprisingly high potency correlated with an increase in the thermal stability of the 2′-F modified siRNA (t_m > 20 °C higher than unmodified siRNA). More recent studies have shown that, compared to unmodified siRNAs, compounds in which all pyrimidine ribonucleotides have been replaced by 2′-F nucleotides have improved efficacy both *in vitro* and *in vivo*, and do not produce immunostimulatory effects [52]. The alternating of 2′-O-Me and 2′-F modifications has become a popular design for therapeutic siRNAs.

The combination of PS2 with 2′-O-Me (but not 2′-F) modifications at the 3′-end of the passenger strand led to an increased RISC loading that, in combination with increases in nuclease resistance and a favorable modulation of thermal stability, enhanced the silencing potency of the doubly modified siRNAs [165]. The therapeutic potential of PS2/2′-O-Me-modified siRNA was demonstrated by targeting *GRAMD1B*, a gene responsible for paclitaxel resistance in a mouse model of ovarian cancer [165].

2′-FANA -modified siRNAs (see Fig. 3) have also shown promising results. The use of 2′-FANA modifications in the passenger strand led to a significant increase in serum half-life and silencing potency [181]. This was somewhat unexpected, because the 2′-FANA modification prefers a DNA-like conformation and would be expected to distort the A-type structure of an siRNA duplex [182]. 2′-FANA, which is the arabinose analog of 2′-F ribose, slightly enhances the thermal stability of siRNA duplexes by about 0.5 °C per modification. Interestingly, the combination of 2′-FANA with 2′-F and LNA modifications led to highly potent and nuclease-stable siRNAs [183].

LNA modifications (see Fig. 3) strongly increase the thermal stability of RNA duplexes (Δt_m 2–10 °C per modification), which limits the location and number of LNA modifications that can be introduced into siRNAs [52]. For example, LNA modifications in the seed region of the guide strand decrease silencing activity, presumably by altering the strand selection process [159]. Nevertheless, LNA modifications at carefully selected sites could be used to improve specificity, enhance nuclease resistance, and reduce immunostimulatory effects [159]. In a complementary approach, destabilizing modifications – such as unlocked nucleic acid (UNA; Fig. 9) at the 3′-end of the passenger strand – favor incorporation of the guide strand into RISC, which in turn reduces the potential off-target silencing by the passenger strand and enhances on-target siRNA potency [184, 185]. A single UNA modification at one of the 5′-nucleotides (positions 1–4) enforces the loading of the other unmodified strand into RISC [185, 186]. Thus, 5′-UNA modifications may be used to improve the properties of siRNAs that have poor activity due to an incorrect selection of the guide strand. A single UNA modification in the seed region, especially at position 7 of the guide strand, strongly reduced miRNA-like off-target silencing [185, 187]. siRNAs heavily modified with stabilizing modifications (LNA, 2′-O-Me, and 2′-F) also benefited from the destabilizing effect of UNA modifications [184]. Taken together, the results of these studies have shown that

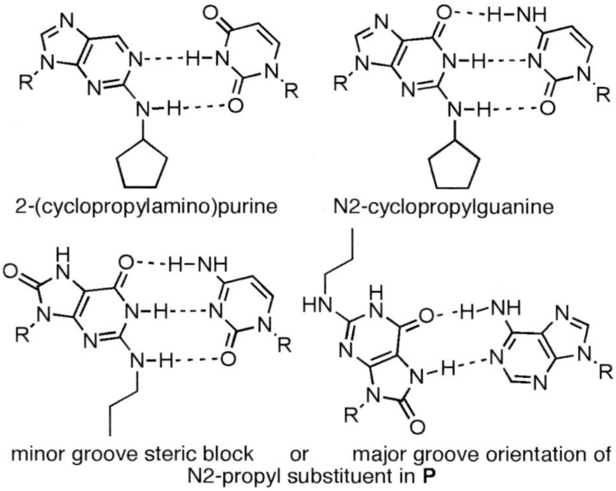

Fig. 10 Chemical structures of Watson–Crick and Hoogsteen base-pairs formed by modified nucleobases in siRNA duplexes.

UNA modifications can be used to improve siRNA specificity and mitigate miRNA-like off-target effects.

Pyranose sugars (see Fig. 3) have also been tested in siRNA design. The incorporation of HNA, ANA, and CeNA modifications at the 3′-ends of both guide and passenger strands enhances nuclease resistance and, in some cases, improves silencing potency and duration [79, 81, 188].

3.1.3 Nucleobase Modifications

In contrast to antisense, the duplex stability of siRNAs does not directly correlate with their biological activity. The effect of nucleobase modifications on siRNA activity is dependent on the position of modification along the siRNA strand [189]. For example, stabilizing 2,6-diaminopurine (see Fig. 4) modifications at the 5′-end of the guide strand actually decrease the siRNA activity [66], which is similar to the effect of LNA. In general, thermally stabilizing modifications tend to have a favorable effect when placed at the 3′-end of the guide strand, whereas thermally destabilizing modifications may have a favorable effect when placed at the 3′-end of the passenger strand.

Beal and coworkers [176] designed modified nucleobases projecting N-alkyl substituents into the minor groove of the RNA duplex (Fig. 10). The expectation was that an increased steric hindrance in the minor groove would block the interaction of siRNA with double-stranded RNA binding proteins, such as RNA-dependent protein kinase (PKR) and toll-like receptors (TLRs) that cause immune stimulation, but might be tolerated by proteins involved in RNAi. The incorporation of two such modifications, namely 2-(cyclopentylamino)purine at position 13 of the passenger strand and N2-cyclopentylguanine at position 9 of the guide strand (Fig. 8), did not decrease the RNAi activity and effectively prevented the immune stimulation of the double-stranded RNA (a mimic of miRNA-122) [176]. The beneficial effect was comparable to that of 2′-O-Me and 2′-F modifications, which are also located in the minor groove.

Burrows, Beal, and coworkers [190] introduced a switchable base, $N2$-propyl-8-oxo-7,8-dihydroguanine (P; Fig. 10) that shifted the sterically demanding $N2$-alkyl group from the minor groove into the major groove, depending on the base with which it was pairing. During siRNA delivery, P was Watson–Crick base-paired with C and projected the propyl group into the minor grove, which inhibited the binding of PKR, thus reducing potential off-target effects. When bound to target mRNA in the RISC, a Hoogsteen P–A base pair places the propyl group in the major groove. The P modification was well tolerated by RNAi at positions 4 and 11 of the guide strand (see Fig. 8). Later, the same authors introduced 8-alkoxyadenine modifications that expanded the range of switchable base-pairs [191]. While these recent studies have clearly demonstrated the potential of nucleobase modifications to improve the therapeutically relevant properties of siRNAs, so far the siRNA compounds in current clinical trials do not typically have nucleobase modifications. As noted above, modified nucleobases run the risk of being incorporated into native nucleic acids, which would limit their use in nucleic acid therapeutics.

3.2
Improving the Delivery of siRNAs

Unmodified siRNAs are rapidly eliminated from the bloodstream and do not cross cellular membranes efficiently. Indeed, the efficient delivery of siRNAs *in vivo* is a formidable problem as unmodified siRNAs are susceptible to serum nucleases, renal excretion, and inefficient biodistribution. Thus, major efforts in the chemical modification of siRNAs have focused on improving not only bioavailability but also targeted organ delivery and cellular uptake.

siRNA delivery approaches can be broadly classified as those that covalently modify siRNAs by direct conjugation, and those that rely on the noncovalent assembly of siRNA-containing delivery vehicles.

3.2.1 Direct Conjugation of siRNAs with Delivery Enhancers

The direct conjugation of siRNA with compounds that enhance organ targeting and cellular uptake remains the most straightforward and desirable approach to improve the delivery of these compounds. Direct conjugation gives well-defined individual compounds, in contrast to multicomponent encapsulation that may yield complex mixtures. In early studies, cholesterol conjugation was used to deliver 2′-O-Me/PS-modified siRNAs to the mouse liver [160]. While this approach allowed for the first demonstration of a therapeutically significant inhibition of human ApoB protein in a transgenic mouse model, the dose required ($50 \, \text{mg kg}^{-1}$) was fairly large. Attempts to use CPPs for siRNA delivery have produced mixed and controversial results [192], since it appears that the negatively charged siRNA backbone neutralizes the cationic CPPs, which in turn leads to cytotoxic effects of the poorly characterized, insoluble and large aggregates [192]. Nevertheless, new approaches using CPPs for siRNA delivery are being actively investigated and may yield promising technologies in the near future [192–194].

Most recently, Alnylam developed novel siRNA conjugates with N-acetylgalactosamine, which enabled the efficient delivery of siRNAs to hepatocytes via the asialoglycoprotein receptor (Fig. 11). In nonhuman primates, this conjugate achieved an 80% reduction of transthyretin, a protein implicated in amyloidosis, at a dose level of only $2.5 \, \text{mg kg}^{-1}$ [195, 196].

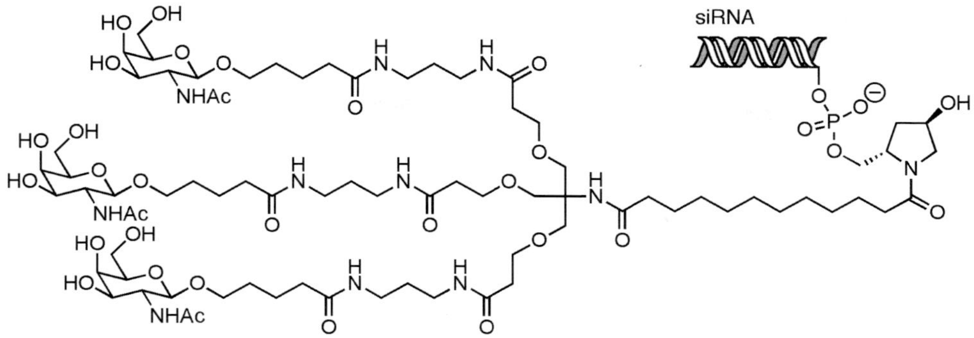

Fig. 11 Structure of N-acetylgalactosamine conjugates.

3.2.2 siRNA Delivery by Noncovalent Encapsulation

The most popular approach for siRNA delivery involves the encapsulation of siRNAs in liposomes and lipid nanoparticles [196, 197]. One of the most successful formulations has been the so-called stable nucleic-acid-lipid particle (SNALP) [179]. The siRNAs used in this study had 2′-O-Me purine (A and G) combined with 2′-F pyrimidine (C and U) modifications, along with deoxyribonucleosides and ribonucleosides at selected positions. The enhanced efficacy of SNALP-formulated siRNAs targeting hepatitis B virus (HBV) RNA *in vivo* (3 mg kg^{-1} per day in a mouse model having a replicating virus) correlated with a longer half-life of siRNAs in plasma and liver. SNALP nanoparticles are formed by combining ionizable lipids (the presence of charge depends on the pH), such as 1,2-dilinoleyloxy-*N,N*-dimethyl-3-aminopropane (DLinDMA) and 1,2-distearoyl-*sn*-glycero-3-phosphocholine (DSPC) (Fig. 12), cholesterol, and PEGylated lipids (PEG-C-DMA) in various molar ratios [196, 197]. Shielding the surface of lipid nanoparticles with polar additives, such as polyethylene glycol (PEG), reduces the particle size, prevents undesirable aggregation, increases circulation time, and reduces toxicity by minimizing any nonspecific interactions with serum proteins and components of the innate immune system. Current therapeutic implementations of SNALP nanoparticles are limited by the relatively strong stimulation of the innate immune system.

Cationic lipid-based nanoparticles deliver siRNAs primarily to the liver. Accordingly, SNALPs have been used in several recent clinical trials [196, 197] against targets causing liver disorders, such as ApoB and PCSK9, the proteins implicated in hypercholesterolemia [177, 198, 199]. A recent study reported a further optimization of SNALP by using DLin-KC2-DMA (Fig. 12) instead of DLinDMA [200].

Targeting cancer has been another area of therapeutic applications for SNALP [196]. Due to the permeable endothelia and poor lymphatic drainage of cancer cells, nanoparticles with a size range of between 50 and 200 nm have the advantages of easier access to, and accumulation in, solid tumors due to the enhanced permeation and retention effects [201]. The first example of siRNA delivery to solid cancers in humans, using cyclodextrin-based nanoparticles, was recently reported [202].

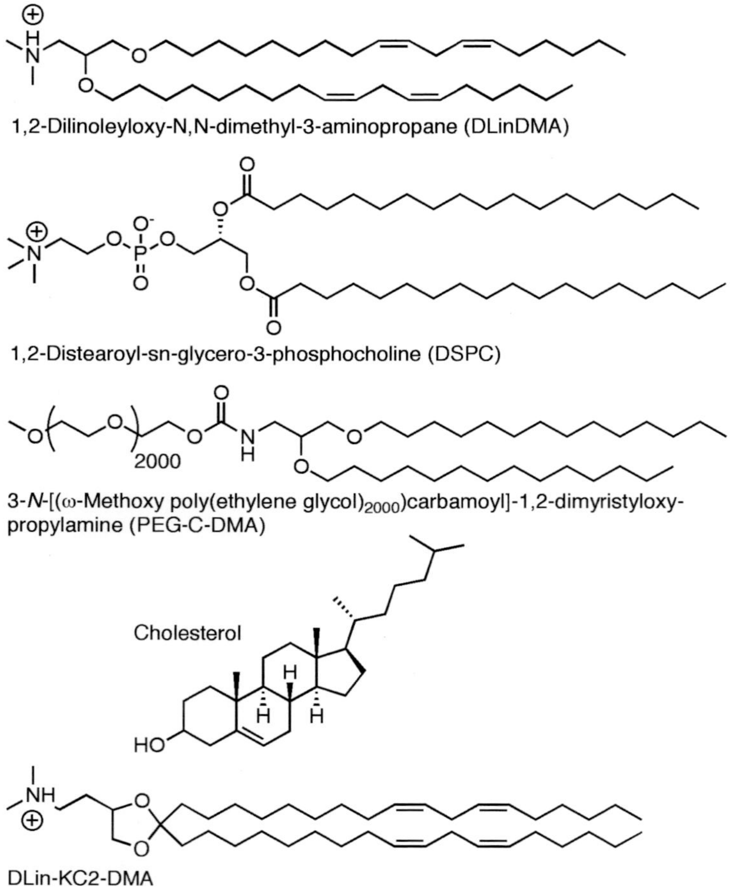

1,2-Dilinoleyloxy-N,N-dimethyl-3-aminopropane (DLinDMA)

1,2-Distearoyl-sn-glycero-3-phosphocholine (DSPC)

3-N-[(ω-Methoxy poly(ethylene glycol)$_{2000}$)carbamoyl]-1,2-dimyristyloxy-propylamine (PEG-C-DMA)

Cholesterol

DLin-KC2-DMA

Fig. 12 Structures of SNALP components.

Despite the above-described progress, the problems associated with siRNA delivery are far from being solved. One of the most common mechanisms for the uptake of siRNA conjugates and nanoparticles is endocytosis, and consequently endosomal entrapment becomes a major limiting factor for efficient delivery [203]. Another challenge is the targeting of areas other than liver, kidney and solid tumors. Clearly, the addressing of these challenges represents the future focus of chemical approaches that are likely to provide major advances in bringing siRNA technology to the clinic.

3.3
Single-Stranded siRNAs

Recent advances have defined the chemical, structural and sequence-specificity principles for designing highly potent siRNAs [204]. Despite the impressive success in using siRNAs as tools in molecular biology, double-stranded RNA is not without drawbacks, especially for *in-vivo* applications.

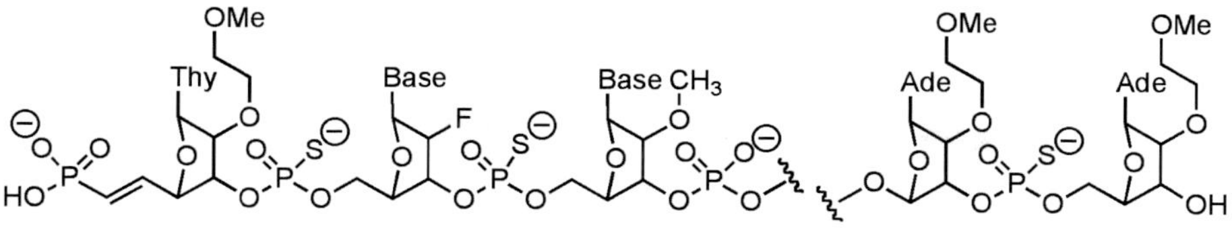

Fig. 13 Chemical modifications of single-stranded siRNAs.

As discussed above, compared to single-stranded oligonucleotides, double-stranded RNA delivery requires either lipid formulations or specialized modifications that will increase not only the complexity but also manufacturing costs. Even when siRNAs are carefully designed, a certain risk remains that the passenger strand may be incorporated into the RISC and cause undesired off-target effects. However, two recent breakthrough studies from ISIS Pharmaceuticals and David Corey's group at UT Southwestern Medical Center have demonstrated that appropriately modified single-stranded RNA molecules can inhibit gene expression through the RNAi pathway *in vitro* and also in animals [205, 206].

The success of single-stranded short interfering ribonucleic acids (ss-siRNAs) was due to a careful optimization of the chemical modifications to insure metabolic stability, RNAi activity and *in-vivo* pharmacokinetics. The ss-siRNAs feature alternating PS and a phosphate backbone linking alternating 2′-F and 2′-O-Me sugars (Fig. 13) [205]. The ends of the ss-siRNAs have MOE sugars (one on the 5′-end and two on the 3′-end), 5′-(E)-vinylphosphonate, and two adenosine nucleotides at the 3′-end. The latter were hypothesized to interact with Dicer and enhance the loading of ss-siRNA in Ago2. In contrast to siRNAs, ss-siRNAs are not phosphorylated in biological systems; on the contrary, chemically phosphorylated ss-siRNAs were found to lose their 5′-phosphates [205]. Therefore, ss-siRNAs require a metabolically stable analog of 5′-phosphate, such as 5′-(E)-vinylphosphonate, for RNAi activity *in vivo* [205].

The results of these studies showed that the passenger strand is not required for the guide strand to enter the RNAi pathway. Rather, appropriately modified lone guide strands (Fig. 13) were able to activate the RNAi mechanism in cell and animal models [205], achieving allele-selective inhibition of huntingtin expression in patient cells and an animal model of Huntington disease [206]. Interestingly, the latter study used mismatch-containing ss-RNAs that achieved a higher allele selectivity by functioning through an miRNA-like mechanism [206]. A follow-up study showed that ss-siRNAs achieved an allele-selective inhibition of mutant ataxin-3, the genetic cause of Machado–Joseph disease [207]. In this case, the ss-siRNAs were found to act through both RNAi and non-RNAi mechanisms, the latter involving splicing modulation. The future development of ss-siRNAs can be expected to yield advantages over other strategies by combining the simplicity of single-stranded antisense oligonucleotides with the efficiency of siRNAs.

4
Small Molecules as RNA-Binding Drug Leads

The RNA helix has a relatively uniform and polar surface that presents little opportunity for traditional hydrophobic shape-selective recognition. The sequence selectivity of RNA-binding intercalators, which rely on hydrophobic stacking between the nucleobases of the helix, is inherently low. The selectivity of small molecules for RNA bulges and internal loops, which are the most common targets of RNA-binding drugs, is complicated by the conformational flexibility of nonhelical RNA. Moreover, most RNA-binding ligands rely on electrostatic interactions with the negatively charged phosphate backbone to increase affinity at the expense of selectivity. This leads to an indiscriminate binding to a variety of RNA species, and is the chief cause of clinical toxicity. Clearly, to design a small molecule that binds a specific RNA sequence selectively is a highly desirable, but formidable, challenge [3, 4, 208, 209].

While many natural products, such as aminoglycosides [97], bind bacterial RNA and act as antibiotics, there are no *de-novo* human-designed RNA binding antibiotics except linezolide, a single example of a fully synthetic antibiotic. All currently used "new" RNA-binding antibiotics are derivatives of previously discovered antibiotics [3, 4]. The traditional screening and structure-aided design and docking approaches, which are commonly used to identify small molecules that bind protein targets, have not provided novel RNA-binding drug leads. The modest progress made in this area may have been due in part to the conformational flexibility of the negatively charged RNA, which presents challenges for computational techniques, and due in part to the limited chemical space of protein-oriented small-molecule libraries. Recently, several new approaches have been developed for the rational design of small molecules that bind medically relevant RNA motifs [210]. For example, Al-Hashimi and coworkers [211] have developed a method that allows small molecules to dock onto an RNA dynamic ensemble that is constructed by combining NMR spectroscopy and computational molecular dynamics. By using this approach, small molecules could be virtually screened against an ensemble of HIV transactivation response (TAR) element RNA conformations, such that the affinity of small molecules to bind different RNA conformations could be accurately predicted. One of the compounds screened in this way, netilmicin (Fig. 14), demonstrated good activity and excellent selectivity for TAR RNA, and inhibited the replication of HIV with an IC_{50} of ~23 µM in an HIV-1 indicator cell line [211]. Further developments of approaches that account for the conformational diversity of RNA are likely to have a major impact on the discovery of RNA-binding drug leads.

4.1
Targeting Disease-Related Toxic RNA Motifs

The expansion of two- to six-nucleotide-long repetitive sequences to abnormal lengths causes diseases related to the toxicity of such RNA motifs [212]. Examples of such diseases include myotonic dystrophy types 1 and 2, Huntington disease, and fragile X-associated tremor ataxia syndrome. Type 1 myotonic dystrophy is caused by an expansion of trinucleotide repeats of CUG in the mRNA of dystrophia myotonica protein kinase. RNAs containing a large number of CUG repeats (between 80 and 3000) fold into hairpin-like structures that sequester splicing factors (Fig. 14), especially muscleblind, thereby depleting these important proteins and causing toxicity as a result of aberrant splicing. In this case, a potential therapeutic approach would be to design small molecules that could bind the hairpins formed by the trinucleotide repeat RNA and thus competitively displace

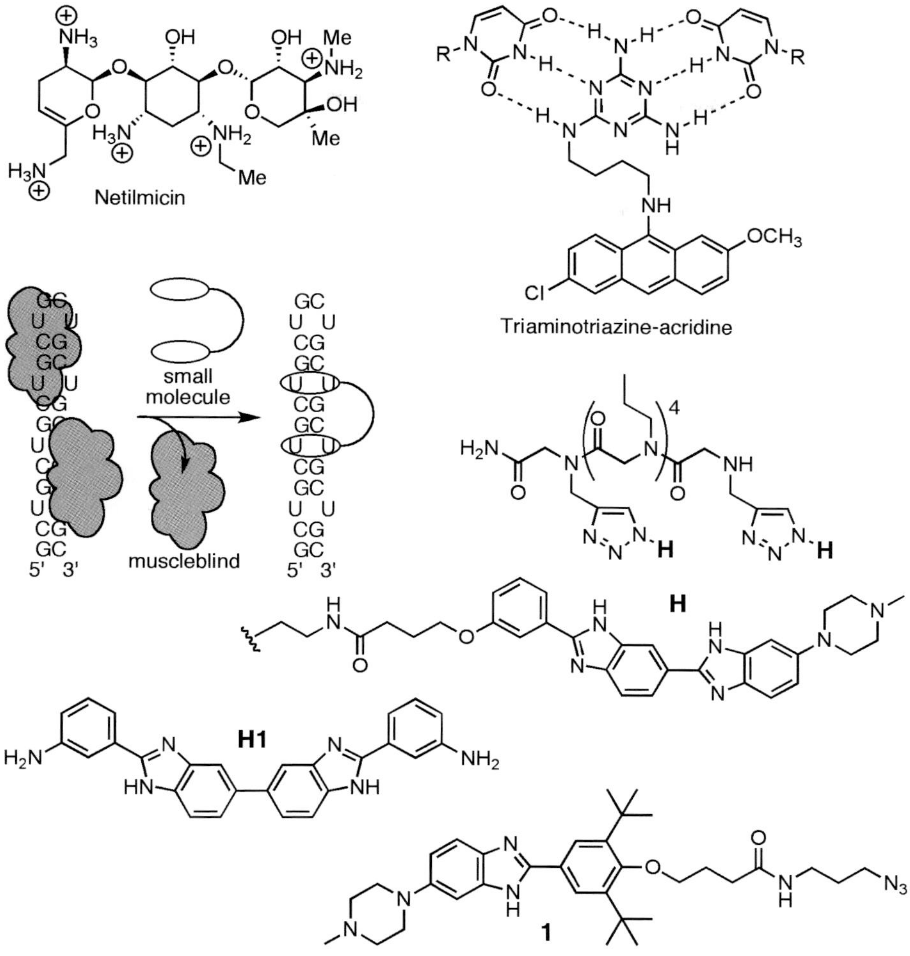

Fig. 14 Small molecules that bind therapeutically relevant RNA motifs.

muscleblind (Fig. 14). Subsequently, Baranger, Zimmerman and coworkers [213] designed a triaminotriazine–acridine conjugate that was hypothesized to bind CUG secondary structural motifs by Janus-type hydrogen bonding, with the central U–U mismatch enforced by acridine intercalation (Fig. 14). The conjugate had nanomolar affinity and good selectivity for the CUG internal loops, and destabilized the complexes between $(CUG)_{12}$ and muscleblind-like protein with an IC_{50} of ~50 μM [213]. This approach was later extended to myotonic dystrophy type 2, which is caused by expanded repeats of CCUG [214].

Disney and coworkers developed a two-dimensional (2-D) combinatorial screening approach that successfully identified small molecules capable of binding to the internal loops of toxic repetitive RNA sequences [210, 215]. This approach took advantage of the multivalent display of small molecules that bind CUG loops on a peptoid scaffold [4, 215]. Modularly assembled benzimidazole (**H**; Fig. 14) recognized the CUG motifs with a higher affinity than muscleblind protein, and reduced the myotonic dystrophy-associated defects in a cell model system at low micromolar concentrations [216]. A similar approach was used to design multivalent displays of kanamycin for targeting CCUG repeats in myotonic dystrophy type 2 [217]. Chemical similarity searching identified the ligand **H1**, which bound repeated CUG motifs cooperatively and did not need a multivalent scaffold [218]. Ligand **H1** reduced myotonic dystrophy-associated defects in cells and mouse model systems at 1 µM and 100 mg kg^{-1}, respectively [218].

4.2
Targeting microRNA

Although the targeting of miRNAs with small molecules has emerged as a potentially promising therapeutic approach [219–222] it faces the common challenge of RNA recognition, namely that it is difficult to design a specific inhibitor that will not interfere with the global processing of miRNAs [219, 220]. However, the results of recent studies have suggested that it is possible to achieve a selective binding to pre-miRNA hairpin structures and to inhibit their biogenesis to mature miRNAs. In pioneering studies, Beal and coworkers showed that helix-threading peptides could recognize the secondary structures of pre-miRNA-39 and pre-miRNA-29b [223, 224], while Maiti and colleagues recently reported that streptomycin (a well-known aminoglycoside drug used to treat tuberculosis) inhibited Dicer cleavage and the maturation of miRNA-21 [225]. Later, Luebke and coworkers identified an N-substituted oligoglycine peptide that had a low micromolar affinity for the apical loop of pri-miRNA-21 and inhibited Drosha cleavage and the processing of pri-miRNA-21, but had little effect on pri-miRNA-16 [226].

Disney and coworkers used a new lead identification strategy, Inforna, that combines 2-D combinatorial screening, statistical analysis and RNA structural information to search for small molecules that bind pre-miRNA hairpins and inhibit Drosha or Dicer cleavage. Of the many RNA–small-molecule combinations screened, benzimidazole (**1**; Fig. 14) emerged as an efficient and selective inhibitor of Drosha processing of pre-miRNA-96 [227]. A similar approach was used to identify guanidinylated neomycin B as an inhibitor of miRNA-10b [228]. The results of these studies suggest that, by using appropriate computational and screening tools, it is possible to identify small molecules that bind to biologically and therapeutically relevant RNA with high affinity and selectivity.

4.3
Targeting Viral and Bacterial RNAs

Viral RNAs feature unique functionally critical motifs that present excellent targets for small-molecule drug development [229, 230]. Seth and coworkers [231] at ISIS Pharmaceuticals performed a mass spectroscopy-assisted screening of a 180 000-compound library to identify a benzimidazole scaffold that was further optimized to structures Isis-11 and Isis-13 (Fig. 15) with low micromolar affinities and good selectivities for the 5′-untranslated region (5′-UTR) of the HCV. The 5′-UTR contains an internal ribosome entry site (IRES) that is critical for the translation of HCV RNA. The binding of Isis-13 caused a large conformational change in the IRES [232] and inhibited HCV RNA synthesis [231, 232]. The conformational change

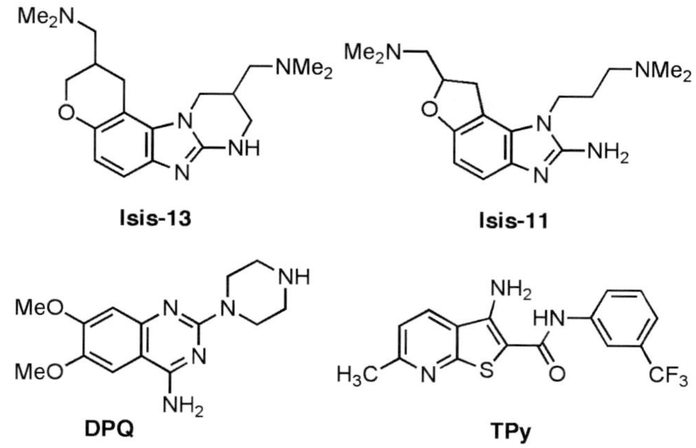

Fig. 15 Structures of small molecules that bind therapeutically relevant viral RNAs.

upon binding of benzimidazole derivatives to HCV IRES was confirmed by NMR structural studies [233]. The binding of Isis-11 to a bulge structure in the HCV IRES was shown to displace two adenosines and alter the RNA helical axis from a 90° bent to a nearly straight conformation [233]. Because the bent conformation is required for correct positioning of the ribosome, the binding of benzimidazole compounds inhibits the translation of HCV RNA. In a similar mode, quinazolinamine DPQ (Fig. 15) binds to the RNA promoter of the influenza A virus and inhibits viral replication at mid-micromolar concentrations [234].

The binding of transactivation protein (Tat) to TAR RNA, a 59-nucleotide hairpin formed at the 5′-end of the HIV transcript, enhances the translation of viral mRNA and is critical for effective viral replication. Tat–TAR interaction is unique to HIV and constitutes an attractive therapeutic target, and studies of various inhibitors of Tat–TAR interaction were recently reviewed [230]. Despite significant activity, none of these compounds has yet entered clinical trials [230], although some of the most recent and promising continuations of these studies can now be mentioned.

For example, Varani and coworkers used a structure-guided design to develop cyclic, conformationally constrained peptides that mimic the antiparallel β-sheet of Tat that is responsible for binding the TAR hairpin [235, 236]. The cyclic peptide mimetics had a nanomolar affinity and good selectivity for TAR RNA [235]. The peptide mimetics also blocked the Tat-dependent replication of diverse HIV-1 strains in cell-based reporter assays, with low micromolar IC_{50} values [237]. Surprisingly, the bulk of the cyclic peptide's antiviral activity was not due to inhibition of translation but to an unexpected blocking of reverse transcription by the TAR–peptide complex [237].

Other groups have targeted various regions of HIV RNA using aminoglycoside derivatives [238–240] and conjugates [241–243], multivalent oligoamines [244], and small molecules [245]. While many of these compounds have displayed high RNA binding affinities, the polycationic structures of typical strong RNA binders often suffer from poor selectivity and physico-chemical properties that hinder progress towards clinical implementation. Schneekloth and coworkers [246]

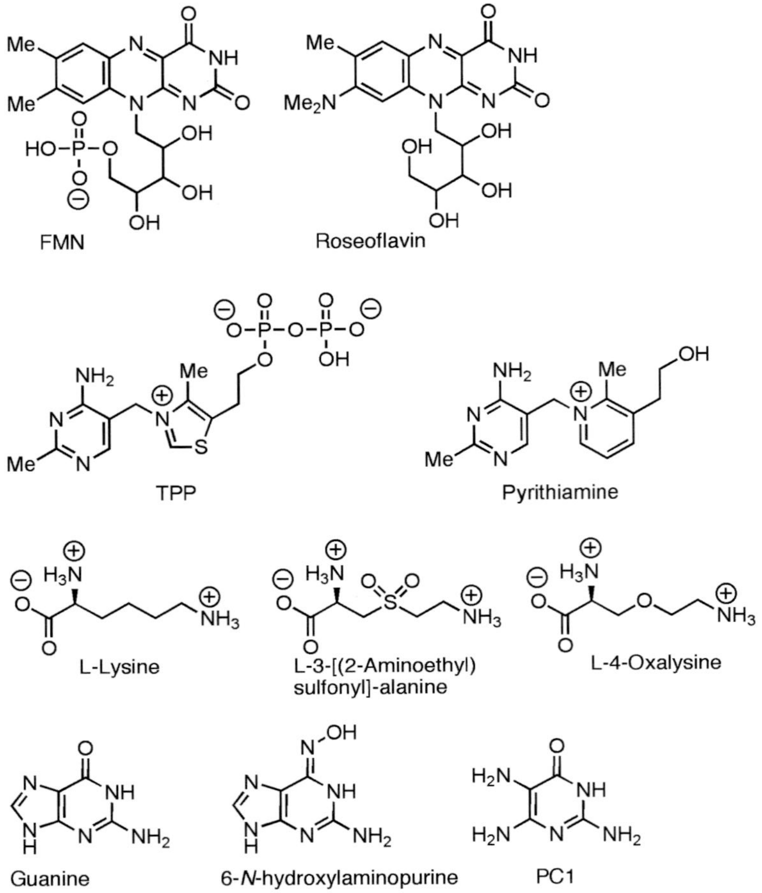

Fig. 16 Structures of riboswitch binding compounds: native ligands and analogs with antibacterial properties.

recently used a small-molecule microarray screening of a 20 000 diverse library of drug-like, commercially available molecules to identify a thienopyridine scaffold (TPy; Fig. 15) that had a K_d of 2.4 µM for TAR RNA and inhibited HIV-induced cytopathicity in T lymphocytes with an EC_{50} of 28 µM.

Riboswitches are RNA motifs that are typically found in the 5′-UTR of bacterial mRNA molecules [247, 248] and which recognize and bind small molecules (usually bacterial metabolites) with high affinity and selectivity. The binding event causes a switch in mRNA conformation, which in turn leads to the regulation of gene expression from that mRNA. Because most riboswitches are unique to bacteria, they represent promising novel drug targets [249–253]. Roseoflavin (Fig. 16) produced by *Streptomyces davawensis*, is a naturally occurring antibiotic that acts on the flavin mononucleotide riboswitch [254, 255]. Pyrithiamine, an analog of thiamine historically designed as a probe to study vitamin metabolism, was shown to have antibacterial properties due to binding to the thiamine pyrophosphate (TPP) riboswitch [256]. Breaker and coworkers [257] showed that lysine analogs with side-chain modifications (Fig. 16) inhibit the growth of *Bacillus subtilis*, despite having an order of magnitude weaker affinity for the target riboswitch than the cognate ligand, unmodified lysine. The same group later reported that 6-N-hydroxylaminopurine represses the gene expression controlled by the purine riboswitch and inhibited the growth of *B. subtilis*, though high micromolar concentrations were required to achieve an antibacterial effect [258]. Mulhbacher *et al.* [259] reported that the pyrimidine analog PC1 (Fig. 16) bound the guanosine riboswitch and showed bactericidal activity in several pathogenic species.

Despite the promising initial findings discussed above, the development of riboswitch-targeting antibiotics is a formidable task. The main challenge is to find compounds that are structurally close enough to the natural ligand to have a high binding affinity, but distinct enough to be competitive and selective inhibitors while avoiding drug resistance and off-target effects. One common problem with riboswitch-targeting compounds is that it is difficult to compete with the high affinity of the native ligands. However, riboswitch-targeting drugs may have an advantage in that they are less susceptible to the development of resistance. Mutations of the drug-binding site are also likely to decrease the affinity for the native ligand, which would lower the bacterial fitness. Efforts in developing riboswitch-targeting therapeutics have been summarized in several excellent recent reviews [249–253].

5
Conclusions and Outlook

Historically, the development of compounds that target RNA has received less attention than those that target DNA and proteins. Perhaps this is not surprising, because only two decades ago RNA was viewed mostly as a passive messenger in the flow of genetic information from DNA to proteins. However, since the discovery that RNA can catalyze chemical reactions, appreciation of the many important roles that RNAs play in cell biology and the development of disease has grown steadily. Along with the recently discovered long noncoding RNAs [2], other notable discoveries of gene expression regulators include siRNAs, miRNAs, riboswitches, and the RNA motifs involved in splicing machinery [1, 248, 260]. The significance of these discoveries continues to drive the development of oligonucleotide analogs and small molecules that recognize and interfere with the metabolism of various RNAs. It is expected that the modulation of RNA metabolism may open doors to new therapeutic approaches.

Antisense oligonucleotides represent the first examples of applying therapeutic interference to the metabolism of disease-related mRNAs. Not surprisingly, this technology is the most mature, and the recent approval of mipomersen by the FDA has increased confidence in therapeutic strategies based on the silencing of disease-related mRNAs by chemically modified oligonucleotides. Other technologies, such as RNAi, splicing modulation, and the inhibition of miRNAs using both oligonucleotide analogs and small molecules, are at relatively early stages of their development. Nevertheless, the preliminary results obtained to date have shown great promise and interest. Research activity in these areas will surely continue. Clearly, RNA is a promising drug target, and the therapeutic potential of various coding and noncoding RNA species has only just begun to be tapped.

References

1. Meister, G., Tuschl, T. (2004) Mechanisms of gene silencing by double-stranded RNA. *Nature* **431**, 343–349.
2. Lee, J.T. (2012) Epigenetic regulation by long noncoding RNAs. *Science* **338**, 1435–1439.
3. Thomas, J.R., Hergenrother, P.J. (2008) Targeting RNA with small molecules. *Chem. Rev.* **108**, 1171–1224.
4. Guan, L., Disney, M.D. (2012) Recent advances in developing small molecules targeting RNA. *ACS Chem. Biol.* **7**, 73–86.
5. Dirin, M., Winkler, J. (2013) Influence of diverse chemical modifications on the ADME characteristics and toxicology of antisense oligonucleotides. *Expert Opin. Biol. Ther.* **13**, 875–888.
6. Hair, P., Cameron, F., McKeage, K. (2013) Mipomersen sodium: first global approval. *Drugs* **73**, 487–493.
7. Watts, J.K., Corey, D.R. (2012) Silencing disease genes in the laboratory and the clinic. *J. Pathol.* **226**, 365–379.
8. Deleavey, G.F., Damha, M.J. (2012) Designing chemically modified oligonucleotides for targeted gene silencing. *Chem. Biol.* **19**, 937–954.
9. Burnett, J.C., Rossi, J.J. (2012) RNA-based therapeutics: current progress and future prospects. *Chem. Biol.* **19**, 60–71.
10. Kurreck, J. (2009) RNA interference: from basic research to therapeutic applications. *Angew. Chem. Int. Ed.* **48**, 1378–1398.
11. Lightfoot, H.L., Hall, J. (2012) Target mRNA inhibition by oligonucleotide drugs in man. *Nucleic Acids Res.* **40**, 10585–10595.
12. Sharma, V.K., Kumar, R., Rungta, P., Parmar, V.S., Prasad, A.K. (2013) Modified oligonucleotides: strides towards antisense drugs. *Trends Carbohydr. Res.* **5**, 1–7.
13. Zamecnik, P.C., Stephenson, M.L. (1978) Inhibition of Rous sarcoma virus replication and cell transformation by a specific oligodeoxynucleotide. *Proc. Natl Acad. Sci. USA* **75**, 280–284.
14. Bennett, C.F., Swayze, E.E. (2010) RNA targeting therapeutics: molecular mechanisms of antisense oligonucleotides as a therapeutic platform. *Annu. Rev. Pharmacol. Toxicol.* **50**, 259–293.
15. Freier, S.M., Altmann, K.H. (1997) The ups and downs of nucleic acid duplex stability: structure–stability studies on chemically-modified DNA:RNA duplexes. *Nucleic Acids Res.* **25**, 4429–4443.
16. Levin, A.A. (1999) A review of issues in the pharmacokinetics and toxicology of phosphorothioate antisense oligonucleotides. *Biochim. Biophys. Acta* **1489**, 69–84.
17. Egholm, M., Buchardt, O., Christensen, L., Behrens, C. et al. (1993) PNA hybridizes to complementary oligonucleotides obeying the Watson–Crick hydrogen-bonding rules. *Nature* **365**, 566–568.
18. Nielsen, P.E., Egholm, M., Berg, R.H., Buchardt, O. (1991) Sequence-selective recognition of DNA by strand displacement with a thymine-substituted polyamide. *Science* **254**, 1497–1500.
19. Nielsen, P.E. (1999) Peptide nucleic acid. A molecule with two identities. *Acc. Chem. Res.* **32**, 624–630.
20. Hyrup, B., Nielsen, P.E. (1996) Peptide nucleic acids (PNA): synthesis, properties and potential applications. *Bioorg. Med. Chem.* **4**, 5–23.
21. Uhlmann, E., Peyman, A., Breipohl, G., Will, D.W. (1998) PNA: synthetic polyamide nucleic acids with unusual binding properties. *Angew. Chem. Int. Ed. Engl.* **37**, 2796–2823.
22. Wittung, P., Nielsen, P., Norden, B. (1997) Extended DNA-recognition repertoire of peptide nucleic acid (PNA): PNA-dsDNA triplex formed with cytosine-rich homopyrimidine PNA. *Biochemistry* **36**, 7973–7979.
23. Hansen, M.E., Bentin, T., Nielsen, P.E. (2009) High-affinity triplex targeting of double stranded DNA using chemically modified peptide nucleic acid oligomers. *Nucleic Acids Res.* **37**, 4498–4507.
24. Li, M., Zengeya, T., Rozners, E. (2010) Short peptide nucleic acids bind strongly to homopurine tract of double helical RNA at pH 5.5. *J. Am. Chem. Soc.* **132**, 8676–8681.
25. Muse, O., Zengeya, T., Mwaura, J., Hnedzko, D. et al. (2013) Sequence selective recognition of double-stranded RNA at physiologically relevant conditions using

PNA–peptide conjugates. *ACS Chem. Biol.* **8**, 1683–1686.

26. Zengeya, T., Gupta, P., Rozners, E. (2012) Triple helical recognition of RNA using 2-aminopyridine-modified PNA at physiologically relevant conditions. *Angew. Chem. Int. Ed.* **51**, 12593–12596.

27. Torres, A.G., Fabani, M.M., Vigorito, E., Williams, D. et al. (2012) Chemical structure requirements and cellular targeting of microRNA-122 by peptide nucleic acids anti-miRs. *Nucleic Acids Res.* **40**, 2152–2167.

28. Said Hassane, F., Saleh, A.F., Abes, R., Gait, M.J. et al. (2010) Cell penetrating peptides: overview and applications to the delivery of oligonucleotides. *Cell. Mol. Life Sci.* **67**, 715–726.

29. Shiraishi, T., Nielsen, P.E. (2006) Enhanced delivery of cell-penetrating peptide-peptide nucleic acid conjugates by endosomal disruption. *Nat. Protoc.* **1**, 633–636.

30. Turner, J.J., Ivanova, G.D., Verbeure, B., Williams, D. et al. (2005) Cell-penetrating peptide conjugates of peptide nucleic acids (PNA) as inhibitors of HIV-1 Tat-dependent trans-activation in cells. *Nucleic Acids Res.* **33**, 6837–6849.

31. Abes, S., Turner, J.J., Ivanova, G.D., Owen, D. et al. (2007) Efficient splicing correction by PNA conjugation to an R6-Penetratin delivery peptide. *Nucleic Acids Res.* **35**, 4495–4502.

32. Oh, S.Y., Ju, Y., Kim, S., Park, H. (2010) PNA-based antisense oligonucleotides for microRNAs inhibition in the absence of a transfection reagent. *Oligonucleotides* **20**, 225–230.

33. Oh, S.Y., Ju, Y., Park, H. (2009) A highly effective and long-lasting inhibition of miRNAs with PNA-based antisense oligonucleotides. *Mol. Cells* **28**, 341–345.

34. Zhou, P., Wang, M., Du, L., Fisher, G.W. et al. (2003) Novel binding and efficient cellular uptake of guanidine-based peptide nucleic acids (GPNA). *J. Am. Chem. Soc.* **125**, 6878–6879.

35. Dragulescu-Andrasi, A., Rapireddy, S., He, G., Bhattacharya, B. et al. (2006) Cell-permeable peptide nucleic acid designed to bind to the 5′-untranslated region of E-cadherin transcript induces potent and sequence-specific antisense effects. *J. Am. Chem. Soc.* **128**, 16104–16112.

36. Sahu, B., Chenna, V., Lathrop, K.L., Thomas, S.M. et al. (2009) Synthesis of conformationally preorganized and cell-permeable guanidine-based γ-peptide nucleic acids (γGPNAs). *J. Org. Chem.* **74**, 1509–1516.

37. Hu, J., Matsui, M., Gagnon, K.T., Schwartz, J.C. et al. (2009) Allele-specific silencing of mutant huntingtin and ataxin-3 genes by targeting expanded CAG repeats in mRNAs. *Nat. Biotechnol.* **27**, 478–484.

38. Hu, J., Corey, D.R. (2007) Inhibiting gene expression with peptide nucleic acid (PNA)–peptide conjugates that target chromosomal DNA. *Biochemistry* **46**, 7581–7589.

39. Fabani, M.M., Gait, M.J. (2008) miR-122 targeting with LNA/2′-O-methyl oligonucleotide mixmers, peptide nucleic acids (PNA), and PNA-peptide conjugates. *RNA* **14**, 336–346.

40. Wancewicz, E.V., Maier, M.A., Siwkowski, A.M., Albertshofer, K. et al. (2011) Peptide nucleic acids conjugated to short basic peptides show improved pharmacokinetics and antisense activity in adipose tissue. *J. Med. Chem.* **53**, 3919–3926.

41. Fabani, M.M., Abreu-Goodger, C., Williams, D., Lyons, P.A. et al. (2010) Efficient inhibition of miR-155 function in vivo by peptide nucleic acids. *Nucleic Acids Res.* **38**, 4466–4475.

42. Järver, P., O'Donovan, L., Gait, M.J. (2014) A chemical view of oligonucleotides for exon skipping and related drug applications. *Nucleic Acid Ther.* **24**, 37–47.

43. Geller, B.L. (2005) Antibacterial antisense. *Curr. Opin. Mol. Ther.* **7**, 109–113.

44. Moulton, H.M., Moulton, J.D. (2008) Antisense Morpholino Oligomers and Their Peptide Conjugates, in: *Therapeutic Oligonucleotides* (ed. Kurreek J.) Royal Society of Chemistry, Biomolecular Sciences, Cambridge, UK, pp. 43–79.

45. Moulton, H.M., Nelson, M.H., Hatlevig, S.A., Reddy, M.T. et al. (2004) Cellular uptake of antisense morpholino oligomers conjugated to arginine-rich peptides. *Bioconjugate Chem.* **15**, 290–299.

46. Abes, S., Moulton, H.M., Clair, P., Prevot, P. et al. (2006) Vectorization of morpholino oligomers by the (R-Ahx-R)4 peptide allows efficient splicing correction in the absence

of endosomolytic agents. *J. Controlled Release* **116**, 304–313.
47. Wu, R.P., Youngblood, D.S., Hassinger, J.N., Lovejoy, C.E. et al. (2007) Cell-penetrating peptides as transporters for morpholino oligomers: effects of amino acid composition on intracellular delivery and cytotoxicity. *Nucleic Acids Res.* **35**, 5182–5191.
48. Warren, T.K., Shurtleff, A.C., Bavari, S. (2012) Advanced morpholino oligomers: a novel approach to antiviral therapy. *Antiviral Res.* **94**, 80–88.
49. Manoharan, M. (1999) 2′-carbohydrate modifications in antisense oligonucleotide therapy: importance of conformation, configuration and conjugation. *Biochim. Biophys. Acta* **1489**, 117–130.
50. Rozners, E. (2006) Carbohydrate chemistry for RNA interference: synthesis and properties of RNA analogues modified in sugar-phosphate backbone. *Curr. Org. Chem.* **10**, 675–692.
51. Pallan, P.S., Greene, E.M., Jicman, P.A., Pandey, R.K. et al. (2011) Unexpected origins of the enhanced pairing affinity of 2′-fluoro-modified RNA. *Nucleic Acids Res.* **39**, 3482–3495.
52. Manoharan, M., Akinc, A., Pandey, R.K., Qin, J. et al. (2011) Unique gene-silencing and structural properties of 2′-fluoro-modified siRNAs. *Angew. Chem. Int. Ed.* **50**, 2284–2288.
53. Martin-Pintado, N., Deleavey, G.F., Portella, G., Campos-Olivas, R. et al. (2013) Backbone FC-H...O hydrogen bonds in 2′F-substituted nucleic acids. *Angew. Chem. Int. Ed.* **52**, 12065–12068.
54. Anzahaee, M.Y., Watts, J.K., Alla, N.R., Nicholson, A.W. et al. (2011) Energetically important C-H...F-C pseudo-hydrogen bonding in water: evidence and application to rational design of oligonucleotides with high binding affinity. *J. Am. Chem. Soc.* **133**, 728–731.
55. Egli, M., Minasov, G., Tereshko, V., Pallan, P.S. et al. (2005) Probing the influence of stereoelectronic effects on the biophysical properties of oligonucleotides: comprehensive analysis of the RNA affinity, nuclease resistance, and crystal structure of ten 2′-O-ribonucleic acid modifications. *Biochemistry* **44**, 9045–9057.
56. Tereshko, V., Portmann, S., Tay, E.C., Martin, P. et al. (1998) Correlating structure and stability of DNA duplexes with incorporated 2′-O-modified RNA analogs. *Biochemistry* **37**, 10626–10634.
57. Teplova, M., Minasov, G., Tereshko, V., Inamati, G.B. et al. (1999) Crystal structure and improved antisense properties of 2′-O-(2-methoxyethyl)-RNA. *Nat. Struct. Biol.* **6**, 535–539.
58. Obika, S., Nanbu, D., Hari, Y., Morio, K.I. et al. (1997) Synthesis of 2′-O,4′-C-methyleneuridine and -cytidine. Novel bicyclic nucleosides having a fixed C3′-endo sugar puckering. *Tetrahedron Lett.* **38**, 8735–8738.
59. Koshkin, A.A., Nielsen, P., Meldgaard, M., Rajwanshi, V.K. et al. (1998) LNA (locked nucleic acid): an RNA mimic forming exceedingly stable LNA:LNA duplexes. *J. Am. Chem. Soc.* **120**, 13252–13253.
60. Wengel, J. (1999) Synthesis of 3′-C- and 4′-C-branched oligodeoxynucleotides and the development of locked nucleic acid (LNA). *Acc. Chem. Res.* **32**, 301–310.
61. Rajwanshi, V.K., Hakansson, A.E., Sorensen, M.D., Pitsch, S. et al. (2000) The eight stereoisomers of LNA (locked nucleic acid): a remarkable family of strong RNA binding molecules. *Angew. Chem. Int. Ed.* **39**, 1656–1659.
62. Sorensen, M.D., Kvrno, L., Bryld, T., Hakansson, A.E. et al. (2002) α-L-ribo-configured locked nucleic acid (α-L-LNA): synthesis and properties. *J. Am. Chem. Soc.* **124**, 2164–2176.
63. Seth, P.P., Siwkowski, A., Allerson, C.R., Vasquez, G. et al. (2009) Short antisense oligonucleotides with novel 2′-4′ conformationally restricted nucleoside analogues show improved potency without increased toxicity in animals. *J. Med. Chem.* **52**, 10–13.
64. Kawasaki, A.M., Casper, M.D., Freier, S.M., Lesnik, E.A. et al. (1993) Uniformly modified 2′-deoxy-2′-fluoro-phosphorothioate oligonucleotides as nuclease-resistant antisense compounds with high affinity and specificity for RNA targets. *J. Med. Chem.* **36**, 831–841.
65. Monia, B.P., Lesnik, E.A., Gonzalez, C., Lima, W.F. et al. (1993) Evaluation of 2′-modified oligonucleotides containing 2′-deoxy gaps as antisense inhibitors

of gene expression. *J. Biol. Chem.* **268**, 14514–14522.

66. Chiu, Y.L., Rana, T.M. (2003) siRNA function in RNAi: a chemical modification analysis. *RNA* **9**, 1034–1048.
67. Braasch, D.A., Jensen, S., Liu, Y., Kaur, K. et al. (2003) RNA interference in mammalian cells by chemically-modified RNA. *Biochemistry* **42**, 7967–7975.
68. Patra, A., Paolillo, M., Charisse, K., Manoharan, M. et al. (2012) 2′-fluoro RNA shows increased Watson–Crick H-bonding strength and stacking relative to RNA: evidence from NMR and thermodynamic data. *Angew. Chem. Int. Ed.* **51**, 11863–11866.
69. Damha, M.J., Wilds, C.J., Noronha, A., Brukner, I. et al. (1998) Hybrids of RNA and arabinonucleic acids (ANA and 2′F-ANA) are substrates of ribonuclease H. *J. Am. Chem. Soc.* **120**, 12976–12977.
70. Wilds, C.J., Damha, M.J. (2000) 2′-deoxy-2′-fluoro-2′-β-D-arabinonucleosides and oligonucleotides (2′F-ANA): synthesis and physicochemical studies. *Nucleic Acids Res.* **28**, 3625–3635.
71. Damha, M.J., Watts, J.K. (2008) 2′F-arabinonucleic acids (2′F-ANA) – History, properties, and new frontiers. *Can. J. Chem.* **86**, 641–656.
72. Watts, J.K., Martin-Pintado, N., Gomez-Pinto, I., Schwartzentruber, J. et al. (2010) Differential stability of 2′F-ANA-RNA and ANA-RNA hybrid duplexes: roles of structure, pseudohydrogen bonding, hydration, ion uptake and flexibility. *Nucleic Acids Res.* **38**, 2498–2511.
73. Herdewijn, P. (2010) Nucleic acids with a six-membered carbohydrate mimic in the backbone. *Chem. Biodivers.* **7**, 1–59.
74. Van Aerschot, A., Verheggen, I., Hendrix, C., Herdewijn, P. (1995) 1,5-anhydrohexitol nucleic acids, a new promising antisense construct. *Angew. Chem., Int. Ed. Engl.* **34**, 2165.
75. Hendrix, C., Rosemeyer, H., Verheggen, I., Seela, F. et al. (1997) 1′,5′-anhydrohexitol oligonucleotides: synthesis, base pairing and recognition by regular oligodeoxyribonucleotides and oligoribonucleotides. *Chem. Eur. J.* **3**, 110–120.
76. Allart, B., Khan, K., Rosemeyer, H., Schepers, G. et al. (1999) D-altritol nucleic acids (ANA): hybridisation properties, stability, and initial structural analysis. *Chem. Eur. J.* **5**, 2424–2431.
77. Froeyen, M., Lescrinier, E., Kerremans, L., Rosemeyer, H. et al. (2001) α-homo-DNA and RNA form a parallel oriented non-A, non-B-type double helical structure. *Chem. Eur. J.* **7**, 5183–5194.
78. Wu, T., Froeyen, M., Kempeneers, V., Pannecouque, C. et al. (2005) Deoxythreosyl phosphonate nucleosides as selective anti-HIV agents. *J. Am. Chem. Soc.* **127**, 5056–5065.
79. Fisher, M., Abramov, M., Van Aerschot, A., Xu, D. et al. (2007) Inhibition of MDR1 expression with altritol-modified siRNAs. *Nucleic Acids Res.* **35**, 1064–1074.
80. Wang, J., Verbeure, B., Luyten, I., Lescrinier, E. et al. (2000) Cyclohexene nucleic acids (CeNA): serum stable oligonucleotides that activate RNase H and increase duplex stability with complementary RNA. *J. Am. Chem. Soc.* **122**, 8595–8602.
81. Nauwelaerts, K., Fisher, M., Froeyen, M., Lescrinier, E. et al. (2007) Structural characterization and biological evaluation of small interfering RNAs containing cyclohexenyl nucleosides. *J. Am. Chem. Soc.* **129**, 9340–9348.
82. Krieg, A.M. (2002) CpG motifs in bacterial DNA and their immune effects. *Annu. Rev. Immunol.* **20**, 709–760.
83. Shen, L., Siwkowski, A., Wancewicz, E.V., Lesnik, E. et al. (2003) Evaluation of C-5 propynyl pyrimidine-containing oligonucleotides in vitro and in vivo. *Antisense Nucleic Acid Drug Dev.* **13**, 129–142.
84. Wu, H., Lima, W.F., Zhang, H., Fan, A. et al. (2004) Determination of the role of the human RNase H1 in the pharmacology of DNA-like antisense drugs. *J. Biol. Chem.* **279**, 17181–17189.
85. Swayze, E.E., Siwkowski, A.M., Wancewicz, E.V., Migawa, M.T. et al. (2007) Antisense oligonucleotides containing locked nucleic acid improve potency but cause significant hepatotoxicity in animals. *Nucleic Acids Res.* **35**, 687–700.
86. Seth, P.P., Jazayeri, A., Yu, J., Allerson, C.R. et al. (2012) Structure activity relationships of α-L-LNA modified phosphorothioate gapmer antisense oligonucleotides in animals. *Mol. Ther. Nucleic Acids* **1**, E47/1–E47/8.

87. Bill, B.R., Petzold, A.M., Clark, K.J., Schimmenti, L.A. et al. (2009) A primer for morpholino use in zebrafish. *Zebrafish* **6**, 69–77.
88. Kole, R., Krainer, A.R., Altman, S. (2012) RNA therapeutics: beyond RNA interference and antisense oligonucleotides. *Nat. Rev. Drug Discovery* **11**, 125–140.
89. Moulton, H.M., Moulton, J.D. (2010) Morpholinos and their peptide conjugates: therapeutic promise and challenge for Duchenne muscular dystrophy. *Biochim. Biophys. Acta, Biomembr.* **1798**, 2296–2303.
90. Rigo, F., Hua, Y., Krainer, A.R., Bennett, C.F. (2012) Antisense-based therapy for the treatment of spinal muscular atrophy. *J. Cell Biol.* **199**, 21–25.
91. Hua, Y., Sahashi, K., Hung, G., Rigo, F. et al. (2010) Antisense correction of SMN2 splicing in the CNS rescues necrosis in a type III SMA mouse model. *Genes Dev.* **24**, 1634–1644.
92. Rigo, F., Hua, Y., Chun, S.J., Prakash, T.P. et al. (2012) Synthetic oligonucleotides recruit ILF2/3 to RNA transcripts to modulate splicing. *Nat. Chem. Biol.* **8**, 555–561.
93. Hua, Y., Sahashi, K., Rigo, F., Hung, G. et al. (2011) Peripheral SMN restoration is essential for long-term rescue of a severe spinal muscular atrophy mouse model. *Nature* **478**, 123–126.
94. Naryshkin, N.A., Weetall, M., Dakka, A., Narasimhan, J. et al. (2014) SMN2 splicing modifiers improve motor function and longevity in mice with spinal muscular atrophy. *Science* **345**, 688–693.
95. Bush, K., Courvalin, P., Dantas, G., Davies, J. et al. (2011) Tackling antibiotic resistance. *Nat. Rev. Microbiol.* **9**, 894–896.
96. Osborne, R. (2013) First novel anti-tuberculosis drug in 40 years. *Nat. Biotechnol.* **31**, 89–91.
97. Becker, B., Cooper, M.A. (2013) Aminoglycoside antibiotics in the 21st century. *ACS Chem. Biol.* **8**, 105–115.
98. Jayaraman, K., McParland, K., Miller, P., Ts'o, P.O.P. (1981) Nonionic oligonucleoside methylphosphonates. 4. Selective inhibition of *Escherichia coli* protein synthesis and growth by nonionic oligonucleotides complementary to the 3' end of 16S rRNA. *Proc. Natl Acad. Sci. USA* **78**, 1537–1541.
99. Harth, G., Zamecnik, P.C., Tabatadze, D., Pierson, K. et al. (2007) Hairpin extensions enhance the efficacy of mycolyl transferase-specific antisense oligonucleotides targeting *Mycobacterium tuberculosis*. *Proc. Natl Acad. Sci. USA* **104**, 7199–7204.
100. Harth, G., Horwitz, M.A., Tabatadze, D., Zamecnik, P.C. (2002) Targeting the *Mycobacterium tuberculosis* 30/32-kDa mycolyl transferase complex as a therapeutic strategy against tuberculosis: proof of principle by using antisense technology. *Proc. Natl Acad. Sci. USA* **99**, 15614–15619.
101. Good, L., Nielsen, P.E. (1998) Antisense inhibition of gene expression in bacteria by PNA targeted to mRNA. *Nat. Biotechnol.* **16**, 355–358.
102. Good, L., Nielsen, P.E. (1998) Inhibition of translation and bacterial growth by peptide nucleic acid targeted to ribosomal RNA. *Proc. Natl Acad. Sci. USA* **95**, 2073–2076.
103. Geller, B.L., Deere, J.D., Stein, D.A., Kroeker, A.D. et al. (2003) Inhibition of gene expression in *Escherichia coli* by antisense phosphorodiamidate morpholino oligomers. *Antimicrob. Agents Chemother.* **47**, 3233–3239.
104. Good, L., Sandberg, R., Larsson, O., Nielsen, P.E. et al. (2000) Antisense PNA effects in *Escherichia coli* are limited by the outer-membrane LPS layer. *Microbiology* **146**, 2665–2670.
105. Good, L., Awasthi, S.K., Dryselius, R., Larsson, O. et al. (2001) Bactericidal antisense effects of peptide-peptide nucleic acid conjugates. *Nat. Biotechnol.* **19**, 360–364.
106. Nekhotiaeva, N., Awasthi, S.K., Nielsen, P.E., Good, L. (2004) Inhibition of *Staphylococcus aureus* gene expression and growth using antisense peptide nucleic acids. *Mol. Ther.* **10**, 652–659.
107. Nikravesh, A., Dryselius, R., Faridani, O.R., Goh, S. et al. (2007) Antisense PNA accumulates in *Escherichia coli* and mediates a long post-antibiotic effect. *Mol. Ther.* **15**, 1537–1542.
108. Deere, J., Iversen, P., Geller, B.L. (2005) Antisense phosphorodiamidate morpholino oligomer length and target position effects on gene-specific inhibition in *Escherichia coli*. *Antimicrob. Agents Chemother.* **49**, 249–255.
109. Dryselius, R., Aswasti, S.K., Rajarao, G.K., Nielsen, P.E. et al. (2003) The translation

start codon region is sensitive to antisense PNA inhibition in *Escherichia coli*. *Oligonucleotides* **13**, 427–433.
110. Hatamoto, M., Nakai, K., Ohashi, A., Imachi, H. (2009) Sequence-specific bacterial growth inhibition by peptide nucleic acid targeted to the mRNA binding site of 16S rRNA. *Appl. Microbiol. Biotechnol.* **84**, 1161–1168.
111. Huang, X.-W., Pan, J., An, X.-Y., Zhuge, H.-X. (2007) Inhibition of bacterial translation and growth by peptide nucleic acids targeted to domain II of 23S rRNA. *J. Pept. Sci.* **13**, 220–226.
112. Gruegelsiepe, H., Brandt, O., Hartmann, R.K. (2006) Antisense inhibition of RNase P: mechanistic aspects and application to live bacteria. *J. Biol. Chem.* **281**, 30613–30620.
113. Kurupati, P., Tan, K.S.W., Kumarasinghe, G., Poh, C.L. (2007) Inhibition of gene expression and growth by antisense peptide nucleic acids in a multiresistant β-lactamase-producing *Klebsiella pneumoniae* strain. *Antimicrob. Agents Chemother.* **51**, 805–811.
114. Kulyte, A., Nekhotiaeva, N., Awasthi, S.K., Good, L. (2005) Inhibition of *Mycobacterium smegmatis* gene expression and growth using antisense peptide nucleic acids. *J. Mol. Microbiol. Biotechnol.* **9**, 101–109.
115. Tan, X.-X., Actor, J.K., Chen, Y. (2005) Peptide nucleic acid antisense oligomer as a therapeutic strategy against bacterial infection: proof of principle using mouse intraperitoneal infection. *Antimicrob. Agents Chemother.* **49**, 3203–3207.
116. Geller, B.L., Deere, J., Tilley, L., Iversen, P.L. (2005) Antisense phosphorodiamidate morpholino oligomer inhibits viability of *Escherichia coli* in pure culture and in mouse peritonitis. *J. Antimicrob. Chemother.* **55**, 983–988.
117. Tilley, L.D., Mellbye, B.L., Puckett, S.E., Iversen, P.L. et al. (2007) Antisense peptide-phosphorodiamidate morpholino oligomer conjugate: dose-response in mice infected with *Escherichia coli*. *J. Antimicrob. Chemother.* **59**, 66–73.
118. Mellbye, B.L., Puckett, S.E., Tilley, L.D., Iversen, P.L. et al. (2009) Variations in amino acid composition of antisense peptide-phosphorodiamidate morpholino oligomer affect potency against *Escherichia coli* in vitro and in vivo. *Antimicrob. Agents Chemother.* **53**, 525–530.
119. Greenberg, D.E., Marshall-Batty, K.R., Brinster, L.R., Zarember, K.A. et al. (2010) Antisense phosphorodiamidate morpholino oligomers targeted to an essential gene inhibit *Burkholderia cepacia* complex. *J. Infect. Dis.* **201**, 1822–1830.
120. Panchal, R.G., Geller, B.L., Mellbye, B., Lane, D. et al. (2012) Peptide conjugated phosphorodiamidate morpholino oligomers increase survival of mice challenged with Ames *Bacillus anthracis*. *Nucleic Acid Ther.* **22**, 316–322.
121. Geller, B.L., Marshall-Batty, K., Schnell, F.J., McKnight, M.M. et al. (2013) Gene-silencing antisense oligomers inhibit *Acinetobacter* growth in vitro and in vivo. *J. Infect. Dis.* **208**, 1553–1560.
122. Krol, J., Loedige, I., Filipowicz, W. (2010) The widespread regulation of microRNA biogenesis, function and decay. *Nat. Rev. Genet.* **11**, 597–610.
123. Siomi, H., Siomi, M.C. (2010) Posttranscriptional regulation of microRNA biogenesis in animals. *Mol. Cell* **38**, 323–332.
124. Breving, K., Esquela-Kerscher, A. (2010) The complexities of microRNA regulation: mirandering around the rules. *Int. J. Biochem. Cell Biol.* **42**, 1316–1329.
125. Calin, G.A., Croce, C.M. (2006) MicroRNA signatures in human cancers. *Nat. Rev. Cancer* **6**, 857–866.
126. Redis, R.S., Berindan-Neagoe, I., Pop, V.I., Calin, G.A. (2012) Non-coding RNAs as theranostics in human cancers. *J. Cell. Biochem.* **113**, 1451–1459.
127. Ferracin, M., Querzoli, P., Calin, G.A., Negrini, M. (2011) MicroRNAs: toward the clinic for breast cancer patients. *Semin. Oncol.* **38**, 764–775.
128. Volinia, S., Galasso, M., Costinean, S., Tagliavini, L. et al. (2010) Reprogramming of miRNA networks in cancer and leukemia. *Genome Res.* **20**, 589–599.
129. Li, Z., Rana, T.M. (2014) Therapeutic targeting of microRNAs: current status and future challenges. *Nat. Rev. Drug Discovery* **13**, 622–638.
130. Lennox, K.A., Behlke, M.A. (2011) Chemical modification and design of anti-miRNA oligonucleotides. *Gene Ther.* **18**, 1111–1120.

131. Meister, G., Landthaler, M., Dorsett, Y., Tuschl, T. (2004) Sequence-specific inhibition of microRNA- and siRNA-induced RNA silencing. *RNA* **10**, 544–550.
132. Hutvagner, G., Simard, M.J., Mello, C.C., Zamore, P.D. (2004) Sequence-specific inhibition of small RNA function. *PLoS Biol.* **2**, 465–475.
133. Elmen, J., Lindow, M., Schuetz, S., Lawrence, M. *et al*. (2008) LNA-mediated microRNA silencing in non-human primates. *Nature* **452**, 896–899.
134. Lanford, R.E., Hildebrandt-Eriksen, E.S., Petri, A., Persson, R. *et al*. (2010) Therapeutic silencing of MicroRNA-122 in primates with chronic hepatitis C virus infection. *Science* **327**, 198–201.
135. Kruetzfeldt, J., Rajewsky, N., Braich, R., Rajeev, K.G. *et al*. (2005) Silencing of microRNAs in vivo with 'antagomirs'. *Nature* **438**, 685–689.
136. Ma, L., Reinhardt, F., Pan, E., Soutschek, J. *et al*. (2010) Therapeutic silencing of miR-10b inhibits metastasis in a mouse mammary tumor model. *Nat. Biotechnol.* **28**, 341–347.
137. Chan, J.A., Krichevsky, A.M., Kosik, K.S. (2005) MicroRNA-21 is an antiapoptotic factor in human glioblastoma cells. *Cancer Res.* **65**, 6029–6033.
138. Elmen, J., Lindow, M., Silahtaroglu, A., Bak, M. *et al*. (2008) Antagonism of microRNA-122 in mice by systemically administered LNA-antimiR leads to up-regulation of a large set of predicted target mRNAs in the liver. *Nucleic Acids Res.* **36**, 1153–1162.
139. Lennox, K.A., Behlke, M.A. (2010) A direct comparison of anti-microRNA oligonucleotide potency. *Pharm. Res.* **27**, 1788–1799.
140. Obad, S., dos Santos, C.O., Petri, A., Heidenblad, M. *et al*. (2011) Silencing of microRNA families by seed-targeting tiny LNAs. *Nat. Genet.* **43**, 371–378.
141. Esau, C.C. (2008) Inhibition of microRNA with antisense oligonucleotides. *Methods* **44**, 55–60.
142. Esau, C., Davis, S., Murray, S.F., Yu, X.X. *et al*. (2006) miR-122 regulation of lipid metabolism revealed by in vivo antisense targeting. *Cell Metab.* **3**, 87–98.
143. Davis, S., Lollo, B., Freier, S., Esau, C. (2006) Improved targeting of miRNA with antisense oligonucleotides. *Nucleic Acids Res.* **34**, 2294–2304.
144. Fabbri, E., Manicardi, A., Tedeschi, T., Sforza, S. *et al*. (2011) Modulation of the biological activity of microRNA-210 with peptide nucleic acids (PNAs). *ChemMedChem* **6**, 2192–2202.
145. Manicardi, A., Fabbri, E., Tedeschi, T., Sforza, S. *et al*. (2012) Cellular uptakes, biostabilities and anti-miR-210 activities of chiral arginine-PNAs in leukaemic K562 cells. *ChemBioChem* **13**, 1327–1337.
146. Kloosterman, W.P., Lagendijk, A.K., Ketting, R.F., Moulton, J.D. *et al*. (2007) Targeted inhibition of miRNA maturation with morpholinos reveals a role for miR-375 in pancreatic islet development. *PLoS Biol.* **5**, 1738–1749.
147. Lindow, M., Kauppinen, S. (2012) Discovering the first microRNA-targeted drug. *J. Cell Biol.* **199**, 407–412.
148. Hannon, G.J. (2002) RNA interference. *Nature* **418**, 244–251.
149. Liu, Q., Paroo, Z. (2011) Biochemical principles of small RNA pathways. *Annu. Rev. Biochem.* **79**, 295–319.
150. Rand, T.A., Petersen, S., Du, F., Wang, X. (2005) Argonaute2 cleaves the anti-guide strand of siRNA during RISC activation. *Cell* **123**, 621–629.
151. Elbashir, S.M., Harborth, J., Lendeckel, W., Yalcin, A. *et al*. (2001) Duplexes of 21-nucleotide RNAs mediate RNA interference in cultured mammalian cells. *Nature* **411**, 494–498.
152. Corey, D.R. (2007) Chemical modification: the key to clinical application of RNA interference? *J. Clin. Invest.* **117**, 3615–3622.
153. Watts, J.K., Deleavey, G.F., Damha, M.J. (2008) Chemically modified siRNA: tools and applications. *Drug Discovery Today* **13**, 842–855.
154. Bramsen, J.B., Kjems, J. (2013) Engineering Small Interfering RNAs by Strategic Chemical Modification, in *Methods in Molecular Biology*, Springer, New York, pp. 87-109.
155. Bramsen, J.B., Kjems, J. (2012) Development of therapeutic-grade small interfering RNAs by chemical engineering. *Front. Genet.: Non-Coding RNA* **3**, Article 154.
156. Watts, J.K., Corey, D.R. (2010) Clinical status of duplex RNA. *Bioorg. Med. Chem. Lett.* **20**, 3203–3207.

157. Bramsen, J.B., Grunweller, A., Hartmann, R.K., Kjems, J. (2014) Using Chemical Modification to Enhance siRNA Performance, in *Handbook of RNA Biochemistry*, 2nd edn (eds R.K. Hartmann, A.B.A. Schon, E. Westhof) Wiley-VCH Verlag GmbH, Co. KGaA, Weinheim, pp. 1243–1277.
158. Shukla, S., Sumaria, C.S., Pradeepkumar, P.I. (2011) Exploring chemical modifications for siRNA therapeutics: a structural and functional outlook. *ChemMedChem* **5**, 328–349.
159. Bramsen, J.B., Laursen, M.B., Nielsen, A.F., Hansen, T.B. et al. (2009) A large-scale chemical modification screen identifies design rules to generate siRNAs with high activity, high stability and low toxicity. *Nucleic Acids Res.* **37**, 2867–2881.
160. Soutschek, J., Akinc, A., Bramlage, B., Charisse, K. et al. (2004) Therapeutic silencing of an endogenous gene by systemic administration of modified siRNAs. *Nature* **432**, 173–178.
161. Harborth, J., Elbashir, S.M., Vandenburgh, K., Manninga, H. et al. (2003) Sequence, chemical, and structural variation of small interfering RNAs and short hairpin RNAs and the effect on mammalian gene silencing. *Antisense Nucleic Acid Drug Dev.* **13**, 83–105.
162. Amarzguioui, M., Holen, T., Babaie, E., Prydz, H. (2003) Tolerance for mutations and chemical modifications in a siRNA. *Nucleic Acids Res.* **31**, 589–595.
163. Hall, A.H.S., Wan, J., Shaughnessy, E.E., Ramsay Shaw, B. et al. (2004) RNA interference using boranophosphate siRNAs: structure–activity relationships. *Nucleic Acids Res.* **32**, 5991–6000.
164. Yang, X., Sierant, M., Janicka, M., Peczek, L. et al. (2012) Gene silencing activity of siRNA molecules containing phosphorodithioate substitutions. *ACS Chem. Biol.* **7**, 1214–1220.
165. Wu, S.Y., Yang, X., Gharpure, K.M., Hatakeyama, H. et al. (2014) 2′-OMe-phosphorodithioate-modified siRNAs show increased loading into the RISC complex and enhanced anti-tumor activity. *Nat. Commun.* **5**, 4459.
166. Iwase, R., Toyama, T., Nishimori, K. (2007) Solid-phase synthesis of modified RNAs containing amide-linked oligoribonucleosides at their 3′-end and their application to siRNA. *Nucleosides Nucleotides Nucleic Acids* **26**, 1451–1454.
167. Iwase, R., Kurokawa, R., Ueno, J. (2009) Synthesis of modified double stranded RNAs containing duplex regions between amide-linked RNA and RNA at both ends and enhanced nuclease resistance. *Nucleic Acids Symp. Ser.* **53**, 119–120.
168. Idziak, I., Just, G., Damha, M.J., Giannaris, P.A. (1993) Synthesis and hybridization properties of amide-linked thymidine dimers incorporated into oligodeoxynucleotides. *Tetrahedron Lett.* **34**, 5417–5420.
169. De Mesmaeker, A., Waldner, A., Lebreton, J., Hoffmann, P. et al. (1994) Amides as a new type of backbone modifications in oligonucleotides. *Angew. Chem., Int. Ed. Engl.* **33**, 226–229.
170. Lebreton, J., Waldner, A., Lesueur, C., De Mesmaeker, A. (1994) Antisense oligonucleotides with alternating phosphodiester-"amide-3" linkages. *Synlett*, 137–140.
171. Selvam, C., Thomas, S., Abbott, J., Kennedy, S.D., Rozners, E. (2011) Amides Are Excellent Mimics of Phosphate Linkages in RNA. *Angew. Chem. Int. Ed.* **50**, 2068–2070.
172. Tanui, P., Kennedy, S.D., Lunstad, B.D., Haas, A. et al. (2014) Synthesis, biophysical studies and RNA interference activity of RNA having three consecutive amide linkages. *Org. Biomol. Chem.* **12**, 1207–1210.
173. Mutisya, D., Selvam, C., Lunstad, B.D., Pallan, P.S. et al. (2014) Amides are excellent mimics of phosphate internucleoside linkages and are well tolerated in short interfering RNAs. *Nucleic Acids Res.* **42**, 6542–6551.
174. Prakash, T.P., Allerson, C.R., Dande, P., Vickers, T.A. et al. (2005) Positional effect of chemical modifications on short interference RNA activity in mammalian cells. *J. Med. Chem.* **48**, 4247–4253.
175. Jackson, A.L., Burchard, J., Leake, D., Reynolds, A. et al. (2006) Position-specific chemical modification of siRNAs reduces 'off-target' transcript silencing. *RNA* **12**, 1197–1205.
176. Peacock, H., Fucini, R.V., Jayalath, P., Ibarra-Soza, J.M. et al. (2011) Nucleobase and ribose modifications control immunostimulation by a MicroRNA-122-mimetic RNA. *J. Am. Chem. Soc.* **133**, 9200–9203.
177. Zimmermann, T.S., Lee, A.C.H., Akinc, A., Bramlage, B. et al. (2006) RNAi-mediated

gene silencing in non-human primates. *Nature* **441**, 111–114.
178. Allerson, C.R., Sioufi, N., Jarres, R., Prakash, T.P. *et al.* (2005) Fully 2′-modified oligonucleotide duplexes with improved in vitro potency and stability compared to unmodified small interfering RNA. *J. Med. Chem.* **48**, 901–904.
179. Morrissey, D.V., Lockridge, J.A., Shaw, L., Blanchard, K. *et al.* (2005) Potent and persistent in vivo anti-HBV activity of chemically modified siRNAs. *Nat. Biotechnol.* **23**, 1002–1007.
180. Morrissey, D.V., Zinnen, S.P., Dickinson, B.A., Jensen, K. *et al.* (2005) Chemical modification of synthetic siRNA. *Pharm. Discovery Dev.* Suppl., 16–20.
181. Dowler, T., Bergeron, D., Tedeschi, A.L., Paquet, L. *et al.* (2006) Improvements in siRNA properties mediated by 2′-deoxy-2′-fluoro-β-D-arabinonucleic acid (FANA). *Nucleic Acids Res.* **34**, 1669–1675.
182. Martin-Pintado, N., Yahyaee-Anzahaee, M., Campos-Olivas, R., Noronha, A.M. *et al.* (2012) The solution structure of double helical arabino nucleic acids (ANA and 2′F-ANA): effect of arabinoses in duplex-hairpin interconversion. *Nucleic Acids Res.* **40**, 9329–9339.
183. Deleavey, G.F., Watts, J.K., Alain, T., Robert, F. *et al.* (2010) Synergistic effects between analogs of DNA and RNA improve the potency of siRNA-mediated gene silencing. *Nucleic Acids Res.* **38**, 4547–4557.
184. Laursen, M.B., Pakula, M.M., Gao, S., Fluiter, K. *et al.* (2010) Utilization of unlocked nucleic acid (UNA) to enhance siRNA performance in vitro and in vivo. *Mol. Biosyst.* **6**, 862–870.
185. Vaish, N., Chen, F., Seth, S., Fosnaugh, K. *et al.* (2011) Improved specificity of gene silencing by siRNAs containing unlocked nucleobase analogs. *Nucleic Acids Res.* **39**, 1823–1832.
186. Snead, N.M., Escamilla-Powers, J.R., Rossi, J.J., McCaffrey, A.P. (2013) 5′ unlocked nucleic acid modification improves siRNA targeting. *Mol. Ther. Nucleic Acids* **2**, E103/1–E103/6.
187. Bramsen, J.B., Pakula, M.M., Hansen, T.B., Bus, C. *et al.* (2010) A screen of chemical modifications identifies position-specific modification by UNA to most potently reduce siRNA off-target effects. *Nucleic Acids Res.* **38**, 5761–5773.
188. Fisher, M., Abramov, M., Van Aerschot, A., Rozenski, J. *et al.* (2009) Biological effects of hexitol and altritol-modified siRNAs targeting B-Raf. *Eur. J. Pharmacol.* **606**, 38–44.
189. Peacock, H., Kannan, A., Beal, P.A., Burrows, C.J. (2011) Chemical modification of siRNA bases to probe and enhance RNA interference. *J. Org. Chem.* **76**, 7295–7300.
190. Kannan, A., Fostvedt, E., Beal, P.A., Burrows, C.J. (2011) 8-oxoguanosine switches modulate the activity of alkylated siRNAs by controlling steric effects in the major versus minor grooves. *J. Am. Chem. Soc.* **133**, 6343–6351.
191. Ghanty, U., Fostvedt, E., Valenzuela, R., Beal, P.A. *et al.* (2012) Promiscuous 8-alkoxyadenosines in the guide strand of an SiRNA: modulation of silencing efficacy and off-pathway protein binding. *J. Am. Chem. Soc.* **134**, 17643–17652.
192. Presente, A., Dowdy, S.F. (2013) PTD/CPP peptide-mediated delivery of siRNAs. *Curr. Pharm. Des.* **19**, 2943–2947.
193. Palm-Apergi, C., Dowdy, S.F. (2013) Delivery of Single siRNA Molecules, in *Pan Stanford Series on Biomedical Nanotechnology* (ed. D. Peer), Pan Stanford Publishing Pte. Ltd, pp. 93-106.
194. van den Berg, A., Dowdy, S.F. (2011) Protein transduction domain delivery of therapeutic macromolecules. *Curr. Opin. Biotechnol.* **22**, 888–893.
195. Akinc, A., Querbes, W., De, S., Qin, J. *et al.* (2010) Targeted delivery of RNAi therapeutics with endogenous and exogenous ligand-based mechanisms. *Mol. Ther.* **18**, 1357–1364.
196. Falsini, S., Ciani, L., Ristori, S., Fortunato, A. *et al.* (2014) Advances in lipid-based platforms for RNAi therapeutics. *J. Med. Chem.* **57**, 1138–1146.
197. Kanasty, R., Dorkin, J.R., Vegas, A., Anderson, D. (2013) Delivery materials for siRNA therapeutics. *Nat. Mater.* **12**, 967–977.
198. Akinc, A., Zumbuehl, A., Goldberg, M., Leshchiner, E.S. *et al.* (2008) A combinatorial library of lipid-like materials for delivery of RNAi therapeutics. *Nat. Biotechnol.* **26**, 561–569.

199. Frank-Kamenetsky, M., Grefhorst, A., Anderson, N.N., Racie, T.S. et al. (2008) Therapeutic RNAi targeting PCSK9 acutely lowers plasma cholesterol in rodents and LDL cholesterol in nonhuman primates. *Proc. Natl Acad. Sci. USA*, **105**, 11915–11920.

200. Semple, S.C., Akinc, A., Chen, J., Sandhu, A.P. et al. (2010) Rational design of cationic lipids for siRNA delivery. *Nat. Biotechnol.* **28**, 172–176.

201. Tabernero, J., Shapiro, G.I., LoRusso, P.M., Cervantes, A. et al. (2013) First-in-humans trial of an RNA interference therapeutic targeting VEGF and KSP in cancer patients with liver involvement. *Cancer Discovery* **3**, 406–417.

202. Davis, M.E., Zuckerman, J.E., Choi, C.-H.J., Seligson, D. et al. (2010) Evidence of RNAi in humans from systemically administered siRNA via targeted nanoparticles. *Nature* **464**, 1067–1070.

203. Juliano, R., Alam, M.R., Dixit, V., Kang, H. (2008) Mechanisms and strategies for effective delivery of antisense and siRNA oligonucleotides. *Nucleic Acids Res.* **36**, 4158–4171.

204. Birmingham, A., Anderson, E., Sullivan, K., Reynolds, A. et al. (2007) A protocol for designing siRNAs with high functionality and specificity. *Nat. Protoc.* **2**, 2068–2078.

205. Lima, W.F., Prakash, T.P., Murray, H.M., Kinberger, G.A. et al. (2012) Single-stranded siRNAs activate RNAi in animals. *Cell* **150**, 883–894.

206. Yu, D., Pendergraff, H., Liu, J., Kordasiewicz, H.B. et al. (2012) Single-stranded RNAs use RNAi to potently and allele-selectively inhibit mutant huntingtin expression. *Cell* **150**, 895–908.

207. Liu, J., Yu, D., Aiba, Y., Pendergraff, H. et al. (2013) ss-siRNAs allele selectively inhibit ataxin-3 expression: multiple mechanisms for an alternative gene silencing strategy. *Nucleic Acids Res.* **41**, 9570–9583.

208. Chow, C.S., Bogdan, F.M. (1997) A structural basis for RNA-ligand interactions. *Chem. Rev.* **97**, 1489–1513.

209. Sucheck, S.J., Wong, C.H. (2000) RNA as a target for small molecules. *Curr. Opin. Chem. Biol.* **4**, 678–686.

210. Disney, M.D., Yildirim, I., Childs-Disney, J.L. (2014) Methods to enable the design of bioactive small molecules targeting RNA. *Org. Biomol. Chem.* **12**, 1029–1039.

211. Stelzer, A.C., Frank, A.T., Kratz, J.D., Swanson, M.D. et al. (2011) Discovery of selective bioactive small molecules by targeting an RNA dynamic ensemble. *Nat. Chem. Biol.* **7**, 553–559.

212. Cooper, T.A., Wan, L., Dreyfuss, G. (2009) RNA and disease. *Cell* **136**, 777–793.

213. Arambula, J.F., Ramisetty, S.R., Baranger, A.M., Zimmerman, S.C. (2009) A simple ligand that selectively targets CUG trinucleotide repeats and inhibits MBNL protein binding. *Proc. Natl Acad. Sci. USA*, **106**, 16068–16073.

214. Wong, C.-H., Fu, Y., Ramisetty, S.R., Baranger, A.M. et al. (2011) Selective inhibition of MBNL1-CCUG interaction by small molecules toward potential therapeutic agents for myotonic dystrophy type 2 (DM2). *Nucleic Acids Res.* **39**, 8881–8890.

215. Disney, M.D. (2013) Rational design of chemical genetic probes of RNA function and lead therapeutics targeting repeating transcripts. *Drug Discovery Today* **18**, 1228–1236.

216. Childs-Disney, J.L., Hoskins, J., Rzuczek, S.G., Thornton, C.A. et al. (2012) Rationally designed small molecules targeting the RNA that causes myotonic dystrophy type 1 are potently bioactive. *ACS Chem. Biol.* **7**, 856–862.

217. Lee, M.M., Pushechnikov, A., Disney, M.D. (2009) Rational and modular design of potent ligands targeting the RNA that causes myotonic dystrophy 2. *ACS Chem. Biol.* **4**, 345–355.

218. Parkesh, R., Childs-Disney, J.L., Nakamori, M., Kumar, A. et al. (2012) Design of a bioactive small molecule that targets the myotonic dystrophy type 1 RNA via an RNA motif-ligand database and chemical similarity searching. *J. Am. Chem. Soc.* **134**, 4731–4742.

219. Deiters, A. (2010) Small molecule modifiers of the microRNA and RNA interference pathway. *AAPS J.* **12**, 51–60.

220. Georgianna, W.E., Young, D.D. (2011) Development and utilization of non-coding RNA-small molecule interactions. *Org. Biomol. Chem.* **9**, 7969–7978.

221. Gumireddy, K., Young, D.D., Xiong, X., Hogenesch, J.B. et al. (2008) Small-molecule

inhibitors of microRNA miR-21 function. *Angew. Chem. Int. Ed.* **47**, 7482–7484.

222. Young, D.D., Connelly, C.M., Grohmann, C., Deiters, A. (2010) Small molecule modifiers of MicroRNA miR-122 function for the treatment of hepatitis C virus infection and hepatocellular carcinoma. *J. Am. Chem. Soc.* **132**, 7976–7981.

223. Krishnamurthy, M., Simon, K., Orendt, A.M., Beal, P.A. (2007) Macrocyclic helix-threading peptides for targeting RNA. *Angew. Chem. Int. Ed.* **46**, 7044–7047.

224. Gooch, B.D., Beal, P.A. (2004) Recognition of duplex RNA by helix-threading peptides. *J. Am. Chem. Soc.* **126**, 10603–10610.

225. Bose, D., Jayaraj, G., Suryawanshi, H., Agarwala, P. *et al.* (2012) The tuberculosis drug streptomycin as a potential cancer therapeutic: inhibition of miR-21 function by directly targeting its precursor. *Angew. Chem. Int. Ed.* **51**, 1019–1023.

226. Diaz, J.P., Chirayil, R., Chirayil, S., Tom, M. *et al.* (2014) Association of a peptoid ligand with the apical loop of pri-miR-21 inhibits cleavage by Drosha. *RNA* **20**, 528–539.

227. Velagapudi, S.P., Gallo, S.M., Disney, M.D. (2014) Sequence-based design of bioactive small molecules that target precursor microRNAs. *Nat. Chem. Biol.* **10**, 291–297.

228. Velagapudi, S.P., Disney, M.D. (2014) Two-dimensional combinatorial screening enables the bottom-up design of a microRNA-10b inhibitor. *Chem. Commun.* **50**, 3027–3029.

229. Davis, D.R., Seth, P.P. (2011) Therapeutic targeting of HCV internal ribosomal entry site RNA. *Antiviral Chem. Chemother.* **21**, 117–128.

230. Mousseau, G., Valente, S. (2012) Strategies to block HIV transcription: focus on small molecule Tat inhibitors. *Biology* **1**, 668–697.

231. Seth, P.P., Miyaji, A., Jefferson, E.A., Sannes-Lowery, K.A. *et al.* (2005) SAR by MS: discovery of a new class of RNA-binding small molecules for the hepatitis C virus: internal ribosome entry site IIA subdomain. *J. Med. Chem.* **48**, 7099–7102.

232. Parsons, J., Castaldi, M.P., Dutta, S., Dibrov, S.M. *et al.* (2009) Conformational inhibition of the hepatitis C virus internal ribosome entry site RNA. *Nat. Chem. Biol.* **5**, 823–825.

233. Paulsen, R.B., Seth, P.P., Swayze, E.E., Griffey, R.H. *et al.* (2010) Inhibitor-induced structural change in the HCV IRES domain IIa RNA. *Proc. Natl Acad. Sci. USA*, **107**, 7263–7268.

234. Lee, M.-K., Bottini, A., Kim, M., Bardaro, M.F. *et al.* (2014) A novel small-molecule binds to the influenza A virus RNA promoter and inhibits viral replication. *Chem. Commun.* **50**, 368–370.

235. Davidson, A., Leeper, T.C., Athanassiou, Z., Patora-Komisarska, K. *et al.* (2009) Simultaneous recognition of HIV-1 TAR RNA bulge and loop sequences by cyclic peptide mimics of Tat protein. *Proc. Natl Acad. Sci. USA* **106**, 11931–11936.

236. Davidson, A., Patora-Komisarska, K., Robinson, J.A., Varani, G. (2011) Essential structural requirements for specific recognition of HIV TAR RNA by peptide mimetics of Tat protein. *Nucleic Acids Res.* **39**, 248–256.

237. Lalonde, M.S., Lobritz, M.A., Ratcliff, A., Chamanian, M. *et al.* (2011) Inhibition of both HIV-1 reverse transcription and gene expression by a cyclic peptide that binds the Tat-transactivating response element (TAR) RNA. *PLoS Pathog.* **7**, e1002038.

238. Luedtke, N.W., Liu, Q., Tor, Y. (2003) RNA-ligand interactions: affinity and specificity of aminoglycoside dimers and acridine conjugates to the HIV-1 Rev response element. *Biochemistry* **42**, 11391–11403.

239. Blount, K.F., Zhao, F., Hermann, T., Tor, Y. (2005) Conformational constraint as a means for understanding RNA-aminoglycoside specificity. *J. Am. Chem. Soc.* **127**, 9818–9829.

240. Kumar, S., Kellish, P., Robinson, W.E., Wang, D. *et al.* (2012) Click dimers to target HIV TAR RNA conformation. *Biochemistry* **51**, 2331–2347.

241. Ennifar, E., Aslam, M.W., Strasser, P., Hoffmann, G. *et al.* (2013) Structure-guided discovery of a novel aminoglycoside conjugate targeting HIV-1 RNA viral genome. *ACS Chem. Biol.* **8**, 2509–2517.

242. Ranjan, N., Kumar, S., Watkins, D., Wang, D. *et al.* (2013) Recognition of HIV-TAR RNA using neomycin-benzimidazole conjugates. *Bioorg. Med. Chem. Lett.* **23**, 5689–5693.

243. Das, I., Desire, J., Manvar, D., Baussanne, I. *et al.* (2012) A peptide nucleic acid-aminosugar conjugate targeting transactivation response element of HIV-1 RNA genome

shows a high bioavailability in human cells and strongly inhibits Tat-mediated transactivation of HIV-1 transcription. *J. Med. Chem.* **55**, 6021–6032.

244. Wang, D., Iera, J., Baker, H., Hogan, P. et al. (2009) Multivalent binding oligomers inhibit HIV Tat-TAR interaction critical for viral replication. *Bioorg. Med. Chem. Lett.* **19**, 6893–6897.

245. Hermann, T., Tor, Y. (2005) RNA as a target for small-molecule therapeutics. *Expert Opin. Therapeutic Patents* **15**, 49–62.

246. Sztuba-Solinska, J., Shenoy, S.R., Gareiss, P., Krumpe, L.R.H. et al. (2014) Identification of biologically active, HIV TAR RNA-binding small molecules using small molecule microarrays. *J. Am. Chem. Soc.* **136**, 8402–8410.

247. Breaker, R.R. (2011) Prospects for riboswitch discovery and analysis. *Mol. Cell* **43**, 867–879.

248. Mandal, M., Breaker, R.R. (2004) Gene regulation by riboswitches. *Nat. Rev. Mol. Cell Biol.* **5**, 451–463.

249. Blount, K.F., Breaker, R.R. (2006) Riboswitches as antibacterial drug targets. *Nat. Biotechnol.* **24**, 1558–1564.

250. Deigan, K.E., Ferre-D'Amare, A.R. (2011) Riboswitches: discovery of drugs that target bacterial gene-regulatory RNAs. *Acc. Chem. Res.* **44**, 1329–1338.

251. Penchovsky, R., Stoilova, C.C. (2013) Riboswitch-based antibacterial drug discovery using high-throughput screening methods. *Expert Opin. Drug Discovery* **8**, 65–82.

252. Lee, E.R., Blount, K.F., Breaker, R.R. (2011) Metabolite-Sensing Riboswitches as Antibacterial Drug Targets, in *Emerging Trends in Antibacterial Discovery* (eds A.A. Miller, P.F. Miller), Caister Academic Press, Norwich, pp. 107–130.

253. Luense, C.E., Schueller, A., Mayer, G. (2014) The promise of riboswitches as potential antibacterial drug targets. *Int. J. Med. Microbiol.* **304**, 79–92.

254. Lee, E.R., Blount, K.F., Breaker, R.R. (2009) Roseoflavin is a natural antibacterial compound that binds to FMN riboswitches and regulates gene expression. *RNA Biol.* **6**, 187–194.

255. Ott, E., Stolz, J., Lehmann, M., Mack, M. (2009) The RFN riboswitch of *Bacillus subtilis* is a target for the antibiotic roseoflavin produced by *Streptomyces davawensis*. *RNA Biol.* **6**, 276–280.

256. Sudarsan, N., Cohen-Chalamish, S., Nakamura, S., Emilsson, G.M. et al. (2005) Thiamine pyrophosphate riboswitches are targets for the antimicrobial compound pyrithiamine. *Chem. Biol.* **12**, 1325–1335.

257. Blount, K.F., Wang, J.X., Lim, J., Sudarsan, N. et al. (2007) Antibacterial lysine analogs that target lysine riboswitches. *Nat. Chem. Biol.* **3**, 44–49.

258. Kim, J.N., Blount, K.F., Puskarz, I., Lim, J. et al. (2009) Design and antimicrobial action of purine analogues that bind guanine riboswitches. *ACS Chem. Biol.* **4**, 915–927.

259. Mulhbacher, J., Brouillette, E., Allard, M., Fortier, L.-C. et al. (2010) Novel riboswitch ligand analogs as selective inhibitors of guanine-related metabolic pathways. *PLoS Pathog.* **6**, e1000865.

260. Plasterk, R.H.A. (2006) Micro RNAs in animal development. *Cell* **124**, 877–881.

24
Structure-Aided Drug Discovery and NMR-Based Screening

Lee Quill[1], Michael Overduin[2], and Mark Jeeves[1]
[1]*University of Birmingham, School of Cancer Sciences, Vincent Drive, Birmingham B15 2TT, UK*
[2]*Present address: University of Alberta, Faculty of Medicine & Dentistry, Department of Biochemistry, 474 Medical Sciences Building, Edmonton, Alberta T6G 2H7, Canada*

1	**Introduction** 873	
1.1	Structure-Aided Drug Design 873	
1.2	Principles of NMR 873	
1.3	NMR Screening Methods 875	
2	**Protein Tyrosine Kinases** 878	
2.1	Protein Tyrosine Kinase Superfamily 878	
2.2	NMR Studies of Protein Tyrosine Kinases 879	
2.3	Targeting BCR-Abl 879	
2.4	Use of NMR in Abl Kinase Drug Design 881	
3	**Protein Tyrosine Phosphatases** 882	
3.1	PTP Superfamily 882	
3.2	PTP1B Signaling Mechanism 883	
3.3	Therapeutic Significance of PTP1B 884	
3.4	NMR Studies of PTP1B and Rational Drug Design 884	
4	**Ras GTPases** 888	
4.1	Ras Superfamily 888	
4.2	Structural Studies of Ras 888	
4.3	Signaling Activities of Ras 889	
4.4	Targeting Ras 890	
5	**Conclusions** 893	
	Acknowledgments 893	
	References 893	

Translational Medicine: Molecular Pharmacology and Drug Discovery
First Edition. Edited by Robert A. Meyers.
© 2018 Wiley-VCH Verlag GmbH & Co. KGaA. Published 2018 by Wiley-VCH Verlag GmbH & Co. KGaA.

Keywords

Nuclear magnetic resonance (NMR)
A discipline that exploits the magnetic properties of certain atomic nuclei to molecular structures.

Drug discovery
The process through which potential new medicines are identified.

Fragment
A portion of a larger molecule, such as a drug.

Screening
A method for identifying the presence of one or more drug-like molecules that engages a target.

Kinase
An enzyme that catalyzes the transfer of phosphate groups onto specific substrates.

Phosphatase
An enzyme that removes a phosphate group from its substrate.

GTPase
A hydrolase enzymes that can bind and hydrolyze guanosine triphosphate (GTP).

Structure
The arrangement of parts within a larger complex, such as the atoms within a molecule.

■ Structure-aided drug design is a rapidly growing discipline due to the significant advances in technology, techniques, and target identification. At the forefront of this progress, nuclear magnetic resonance (NMR) is now an essential tool for modern drug discovery, providing insights at multiple stages of the design process. As a data-rich method, NMR is used to screen compound libraries and identify and validate ligands that bind to any part of proteins, with a wide range of affinities. Rapid experiments can be used to elucidate the binding modes and interaction sites of targets, and can prioritize hits for cocrystallization studies and medicinal chemistry optimization. In this chapter, three distinct cases are presented, with kinase, phosphatase and GTPase targets represented by Abl, PTP1B, and Ras, respectively. These demonstrate the range and potential of NMR in drug discovery, and how it complements other biophysical techniques. Design strategies for ligands of specific conformational states are discussed, highlighting the key role that NMR plays in the expanding field of structure-based drug design.

1
Introduction

1.1
Structure-Aided Drug Design

Atomic level information provides fine details of molecular interactions that can be used to improve drug design and aid the validation of promising drug candidates from initial screens. High-resolution structural data of the protein target provides vital information pertaining to interactions between the target and potential ligands, and is a prerequisite for the development of drug molecules with sufficient potency and selectivity. Nuclear magnetic resonance (NMR) is routinely employed as a sensitive and versatile tool which provides useful information throughout the drug discovery process. The utility of NMR-based methods arises from the ability to monitor the chemical environments of each atom in a molecule. When placed in an external magnetic field, the nucleus of each atom has a specific resonance frequency which is standardized as the chemical shift. This value is influenced by the environment of the nucleus in question, and as such provides a highly sensitive measure of small-molecule binding. Once the chemical shifts of both the target and ligand atoms have been assigned, the binding moieties of both molecules can be determined.

Structure-aided drug design provides a mechanism for the rational improvement of initial ligand hits (Fig. 1). The use of high-resolution structure determination methods such as X-ray crystallography provides an ideal starting point for the transformation and optimization of initial hit molecules into high-affinity lead molecules with excellent potency and specificity for a designated target.

1.2
Principles of NMR

NMR spectroscopy is the measurement of the behavior of certain nuclei in the presence of an external magnetic field. In biological systems the nuclei commonly studied are ^{1}H, ^{15}N, and ^{13}C, though in certain applications other nuclei such as ^{2}H, ^{31}P and ^{19}F are used. The chemical environment of individual nuclei affects the influence that the magnetic field has upon them, a phenomenon visualized in the chemical shift. The chemical shift can therefore be used to detect changes in the conformation of a macromolecule, or to detect the binding of a ligand. Only those nuclei whose chemical environment is altered show a perturbation in their chemical shift.

Nuclei that are linked through chemical bonds to other NMR-active nuclei are said to be "coupled." This coupling means that the behavior of a nucleus is affected by the quantum state of the coupled nuclei. The phenomenon is visualized by splitting of the NMR spectrum, and can be used to obtain information of the conformational state of the molecule. The extent of coupling is dependent on the nuclei involved, their chemical environment, and on the number of bonds separating the two active nuclei. Coupling can be used to transfer magnetization from one nucleus to another, as in multidimensional spectra such as the heteronuclear single quantum coherence (HSQC) spectrum.

Nuclei are not only affected by through-bond interactions but can also be influenced by the close proximity (<6 Å) of other active nuclei. This through-space interaction between nuclei is termed the nuclear Overhauser effect (NOE), and can provide the distance information required for NMR structural determination and also

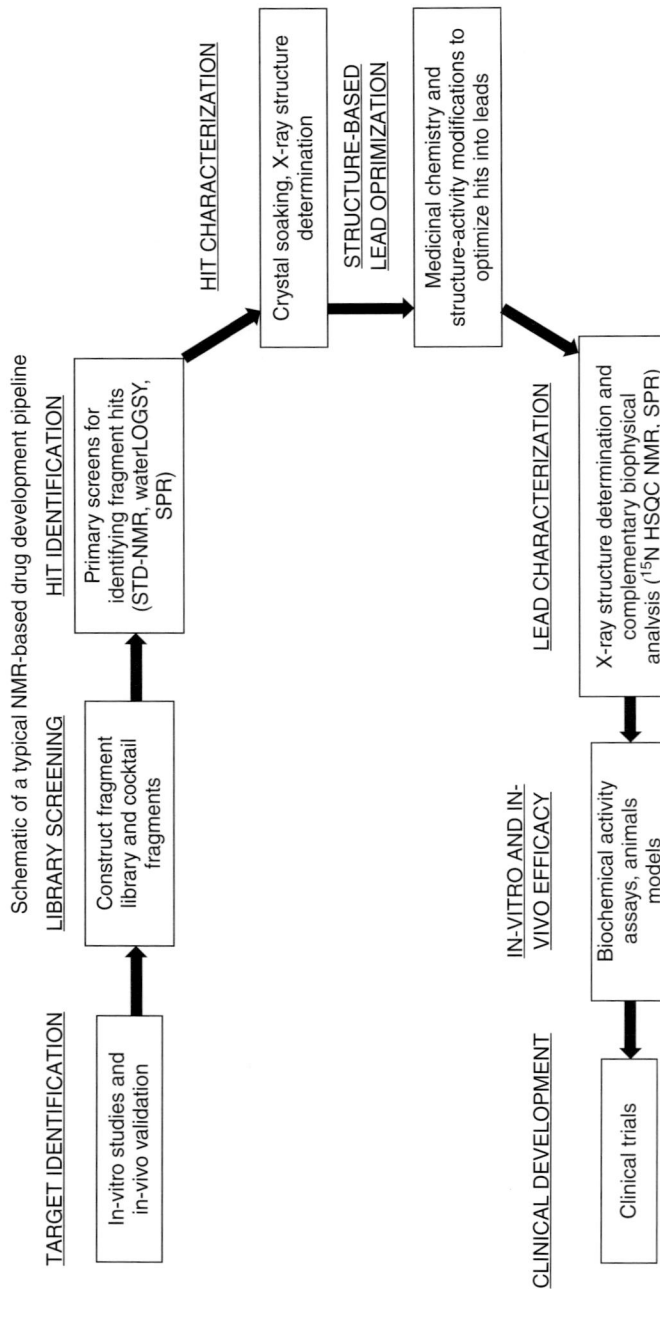

Fig. 1 The sensitivity and versatility of NMR is ideally positioned to inform multiple facets of the drug discovery process. A key precursor to the development of novel drug molecules is access to crucial information regarding the *in-vivo* significance of the target. This often includes detailed knowledge of biochemical function, its mechanistic importance in physiologically relevant processes, and the ensuing phenotype that emerges from genetic deletion studies in preclinical animal models. After target validation, the target of interest is typically screened against a diverse library of compounds or fragments where initials hits are then validated. Hit validation is often conducted using two orthogonal methods: NMR (STD-NMR and waterLOGSY) and SPR. Positive hits independently validated through both methodologies are then cocrystallized with the target of interest to characterize the binding mode and probe novel inhibitable pockets or sites of interaction. On the basis of the determined cocrystal structures the initial hits are expanded, usually through iterative rounds of medicinal chemistry, and subsequently transformed into a progressable lead candidate. The resulting lead molecule is often recrystallized with the target of interest and subjected to a combination of biophysical analysis [by ^{15}N HSQC NMR, SPR, or isothermal titration calorimetry (ITC)] and biochemical analysis (in a suitable enzymatic assay), to determine the dissociation constant. Once *in-vivo* efficacy has been established, it is then tested to observe whether addition of the lead molecule is sufficient to ameliorate symptoms in a disease-relevant animal model.

relaxation parameters that give crucial information pertaining to the molecular dynamics of the system being studied. The relaxation of a system is also dependent on the tumbling rate of the molecules involved. A large system will tumble slower, while the NMR signal will relax at a faster rate.

1.3
NMR Screening Methods

Efficient drug discovery initiatives often begin with the identification of novel starting points, for which NMR provides several screening assays. Common NMR experiments include the $T_{1\rho}$ [1], Saturation Transfer Difference-NMR (STD-NMR) [2], and the water-Ligand Observed via Gradient Spectroscopy (waterLOGSY) [3]. The $T_{1\rho}$ experiment measures the increase in spin lattice relaxation of the ligand upon macromolecule binding, leading to increase in the linewidth of the NMR signals. This is due to the decrease in average tumbling rate of the ligand as a result of transient complex formation between the ligand and the macromolecule during the lifetime of the experiment. Conversely, ligands that do not bind to the target show no increase in relaxation, and correspondingly their NMR signals remain unaffected. In the STD experiment (Fig. 2a) the methyl groups of a protein are selectively irradiated, leading to saturation. This saturation energy is rapidly transferred throughout the protein due to the efficient cross-relaxation and spin diffusion present in large macromolecules. The proton methyl region for proteins lies between −0.5 and 2 ppm, an area typically sparsely populated by small-molecule signals, which means that this saturation is only experienced by the protein. Upon binding, the saturation magnetization is transferred to the bound ligand via the NOE (non-binding molecules do not experience this transfer). When the spectrum is acquired the transferred saturation magnetization leads to a decrease in signal intensity of the compound. A reference spectrum is acquired using saturation of a region of the NMR spectrum which contains no signals from any species (typically 30 ppm), resulting in no decrease in the signals of the acquired spectrum of the compound. The transfer spectrum is then subtracted from the reference spectrum, which results in peaks remaining only for those molecules which bind to the protein. This method can be used efficiently to identify binding molecules in mixtures of compounds with non-overlapping signals. However, as the STD-NMR experiment measures the spectrum of the free small molecule, ligands which bind tightly to the protein and are not released in the timescale of the experiments will not result in a STD signal.

The WaterLOGSY experiment (Fig. 2b) relies on a selective excitation of the water molecules that are bound to the protein. A mixing time is then used to allow transfer of magnetization from the bound water to the macromolecule, and also to any bound ligands. Transfer also occurs between bulk water and molecules free in solution, without having to go via the macromolecule; these transfers are achieved via the NOE. For direct interactions between the bulk water and the small molecule, the NOE is small and positive, due to the fast tumbling rate of the species involved. NOEs resulting from the magnetization pathway via the protein will be negative due to the slow tumbling rate of the macromolecule species. As the interaction between the ligand and the macromolecule is much tighter than that between the ligand and the bulk water, the

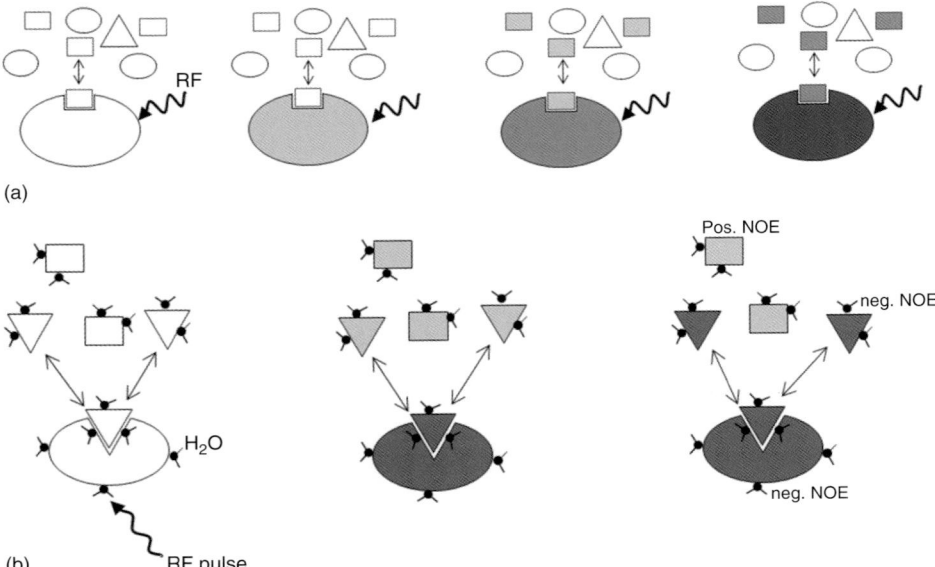

Fig. 2 Schematic representation of the STD-NMR and WaterLOGSY experiments. (a) In the STD-NMR experiment. protein resonances are selectively irradiated by a radiofrequency (RF) pulse (left) leading to saturation, as shown by darker shading (second left). This saturation is transferred to those ligands that bind to the protein (second right). Ligands that are released within the timescale of the experiment have peaks intensities as a consequence of the saturation transfer and thus can be detected in the STD-NMR experiment. Nonbinding compounds are unaffected and therefore their signals are not detected; (b) In the WaterLOGSY experiments, the bulk water molecules are selectively excited by an RF pulse (left). This magnetization is allowed to transfer to both the protein and ligands in solution via the NOE (center). The sign of the NOE signal is dependent on the size of the molecules involved. Small molecules such as free ligand give a positive NOE (as shown by the light-gray shading) and large molecules such as the protein give a negative NOE (as shown by the dark-gray shading). With longer mixing times (right) the release of the bound ligand leads to free ligands displaying a negative NOE for those compounds which bind to the protein.

transfer of magnetization is more efficient, which means that the signals from the binding ligands will dominate. The result is that binding and nonbinding ligands will have opposite signs in the WaterLOGSY spectrum. As with the STD experiment, a large excess of ligand is used and the signal for the free ligand is observed, so that those molecules which bind tightly to the macromolecule may be difficult to detect without resorting to long mixing times.

These ligand-observed methods have been widely used throughout multiple drug discovery initiatives, and are ideally suited to larger protein targets whilst still being applicable to smaller targets (>15 kDa). The use of experiments which observe the NMR spectra of the ligand are particularly advantageous in that they reduce the necessary protein concentrations to between 1 and 10 µM, and allow the screening of mixtures of compounds, thus reducing experimental time. These methods can also be used to identify ligands which possess novel binding modes and interaction sites. They can also enable epitope mapping of the ligand and detection of even very weakly binding compounds with affinities in the 100 μM to 5 mM range. Other ligand-observed methods commonly used include target-immobilized NMR screening [4], experiments using paramagnetic probes [5] and heteronuclear detection using ^{31}P and ^{19}F nuclei [6–8]. For the detection of high-affinity ligands (>1 μM), reporter screening can be used where a stronger ligand is used to compete with the test compound and displace it in a concentration-dependent manner [9]. Ligand-observed techniques are ideally suited to fragment-based drug design, which in turn provides an alternative strategy to conventional high-throughput screening. Fragment screening, a method for discovering initial lead compounds at the start of the drug design process, uses small chemical fragments (typically <250 Da) which bind (albeit weakly) to the target of interest to provide efficient starting points for subsequent rounds of medicinal chemistry elaboration to optimize affinity and specificity. NMR also performs well as a complementary method when run in parallel with orthogonal screens such as surface plasmon resonance (SPR) to filter out false positives and also to validate hits from activity screens.

The optimization of initial hit molecules from primary fragment screens, and the subsequent progression of these hits into high-affinity lead molecules, is aided by atomic resolution structures of the binding site. Such high-resolution information provides a robust platform for rational elaboration of the hit molecule into a potent lead with enhanced affinity through iterative rounds of medicinal chemistry.

Although cocrystallizing the target of interest with initial hit molecules is a common starting point for elucidating the atomic details of binding site interactions, complementary binding site information can be obtained by observing changes in the protein's NMR signals upon ligand interaction. In general, protein-observed NMR spectroscopy requires significantly higher concentrations of a soluble protein (<50 μM) which is typically isotopically labeled with either ^{15}N, ^{13}C, or both ^{15}N and ^{13}C. The maximum protein molecular weight limit is ~35 kDa due to the increased relaxation rates and spectral complexity. However, the availability of technical advances, including specific labeling of the protein, perdeuteration, and the use of transverse relaxation optimized spectroscopy (TROSY) [10], mean that the upper size limit can be increased to 100 kDa in favorable cases. One of the major techniques involving protein-observed NMR spectroscopy is the monitoring of progressive chemical shift perturbations (CSPs) in the protein resonances induced by the binding of a ligand. This can then be used to generate binding affinities by measuring the extent of perturbation versus the ligand concentration. Previously, protein-observed NMR screening was an expensive and time-consuming process, but the use of high-throughput autosamplers, fast methods such as the SOFAST-HMQC experiments [11], and low-volume cryogenic probes means that small-molecule libraries can now be cost-effectively screened within several days. In addition, these protein-detected methods become especially useful when the protein resonance assignments are available, as the perturbed signals can then be used to identify protein residues that are in close proximity to the site of ligand interaction [12]. This epitope data can then be used to model the interaction between target and ligand, using programs such as HADDOCK [13]. This is used in conjunction with X-ray structure determinations of protein–ligand complexes or as an alternative, especially when protein complexes appear refractory to crystallization. A rapid solution structure determination of protein–ligand complexes is also feasible, particularly for relatively small proteins (<15 kDa), or if an existing three-dimensional structure of the protein is available.

The full structural determination of protein–ligand complexes yields intimate details of the molecular interactions occurring in the binding site, allows an efficient analysis of the differential binding properties of ligands, and also provides important data for informing hit optimization through subsequent rounds of medicinal chemistry. An additional advantage of NMR relies on its unique ability to characterize the dynamics and flexible regions of biological molecules. In this way, NMR remains the most widely used technique for studying the molecular structure of biological molecules in the solution state.

One of the key factors critical for rational drug design is finding the most relevant protein state to target for therapeutic intervention. Indeed, NMR is ideally placed to interrogate the druggable potential of different protein states and yields crucial insights into the multiple conformations and domain orientations adopted during activation and signaling, in addition to its role in compound screening. Three examples of proteins where NMR-based screening and structure-based drug design has been used to generate novel drug-like molecules with demonstrated efficacy are discussed in the following sections. Some key insights into exciting new avenues for further drug development are also provided.

2
Protein Tyrosine Kinases

2.1
Protein Tyrosine Kinase Superfamily

Phosphorylation of proteins on tyrosine residues is a key regulatory process in the control of cellular signaling, and plays important roles in governing the activation of cellular events such as proliferation, differentiation, migration, and apoptosis [14]. The process of protein phosphorylation operates in a reversible manner, and is tightly controlled by the reciprocal activities of protein tyrosine kinase (PTK) and protein tyrosine phosphatase (PTP) enzymes. Aberrant regulation of these phosphorylation patterns through the dysregulation of kinase and phosphatase enzyme activities is known to be a critical factor in many diseases, including cancers, diabetes and inflammatory conditions. As such, phosphorylation-dependent signaling pathways harbor a promising reservoir of putative kinase and phosphatase drug targets, and provide opportunities for targeted therapeutic intervention, for diseases with defective phosphorylation signaling. Protein kinases have long been targets for the development of drugs, primarily due to their key roles in signaling networks. Indeed, about 30% of all proteins targeted by pharmaceutical companies for drug development have been either protein or lipid kinases, and several inhibitors are already available in the clinic [15]. Within the human genome a total of 518 protein kinases is encoded [16], each sharing a catalytic domain with a conserved sequence and structure. However, the regulation of kinase catalytic activity is often profoundly different. The conserved ATP-binding pocket, which is found between the two lobes of the kinase fold, was the first site to be targeted for drug development in this family of enzymes. Unfortunately, a lack of selectivity and the limited number of available chemotypes that bind the ATP-binding site hindered the drug discovery potential of some kinases. As an alternative strategy, the targeting of less-conserved surrounding pockets affords selective interactions whilst retaining the affinity generated by occupying the ATP site. The kinase catalytic mechanism requires the activation loop to adopt a conformation in which the Asp-Phe-Gly (DFG) motif is orientated into the active site [17]. The alternative DFG-out conformation results in an inactive state, and has been identified in a number of kinase structures [18]. Hence, the loop dynamics of kinases is a critical determinant of ligand binding, and resolving the structural and dynamic details of different loop conformations may yield novel insights into the design of ligands that preferentially discriminate between kinase states.

2.2
NMR Studies of Protein Tyrosine Kinases

Despite technological advances in data collection, NMR structural studies of PTKs remain challenging due to several factors. These include the relatively large size of the catalytic domain (>32 kDa), difficulties in the expression of sufficient quantities of isotopically labeled protein, the toxic effects of kinase overexpression on the host organism, and the inherent dynamics of signaling enzymes, which often leads to intermediate exchange on the NMR timescale. To reduce the dynamics, the binding of an inhibitor to the ATP site is often employed, and this serves to lock the kinase into one conformation. Despite the challenges associated with the use of NMR-based methods to probe the structure of kinases in solution,

NMR resonance assignments have been reported on several kinases, including cAMP-dependent protein kinase, both in complex with AMP–PNP and the free form [19], the p38 protein [20], Src in complex with imatinib [21], the FGFR1 protein [22], Abl with imatinib [23], the Eph receptor [24], extracellular signal-regulated kinase 2 (ERK2) [25], and VRK1 [26, 27]. The use of NMR in conjunction with X-ray crystallography integrates the complementary strengths of both techniques and builds a robust framework for studying the structural and dynamic aspects of ligands binding modes in solution. Deriving such information provides a clear rationale for transforming the ligand chemistry in order to generate lead molecules with high affinities for kinase targets.

2.3
Targeting BCR-Abl

The Abl kinase is best known for its role in chronic myelogenous leukemia (CML), and represents a paradigm for target-based therapeutic intervention. CML is caused by the t(9,22) chromosomal translocation which gives rise to the Philadelphia chromosome [28] and the subsequent fusion of parts of the breakpoint cluster region protein and Abl protein, including its catalytic domain. This leads to the high-level expression of BCR-Abl with a constitutively activated tyrosine kinase that causes uncontrolled proliferation and the replication of progenitor cells. The BCR-Abl fusion protein also constitutes a unique molecular target for the design of inhibitors selective for leukemic cells, as this form of the protein is not expressed in other tissues.

The first inhibitor designed to treat Philadelphia chromosome-positive CML by inhibiting BCR-Abl was imatinib, which is

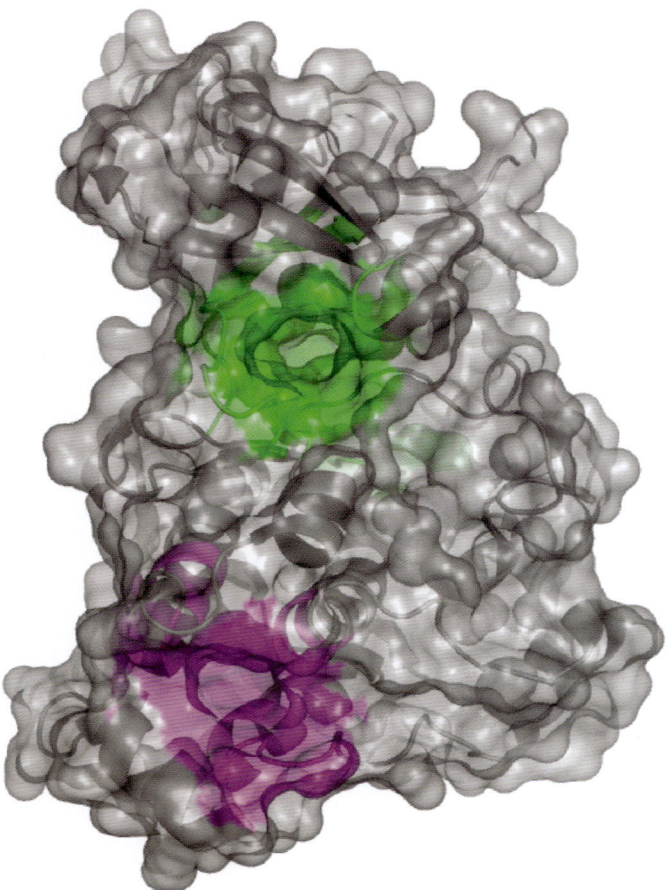

Fig. 3 The X-ray crystal structure of the catalytic domain of Abl, showing the classical kinase structure with two distinct lobes. The ATP-binding site (green), which lies at the hinge region between the two lobes, and the allosteric myristate site (magenta) are highlighted. These two sites are proven targets for independent inhibitors of BCR-Abl that have the potential to be used in conjunction in order to overcome drug-resistant mutations [29].

marketed by Novartis as Gleevec/Glivec®. The X-ray crystal structure of the kinase domain of c-Abl in complex with imatinib showed a classical two-lobe kinase fold, and subsequently identified two putative druggable pockets that could be exploited for the modulation of kinase activity (Fig. 3). The structure unequivocally revealed that imatinib was bound to the ATP site, making contacts with both the activation loop and the αC-helix (Fig. 4a) [31]. Together, this information provided a clear rationale underpinning the hyperproliferative state of cancer cells, and how the acquisition of mutations within the Abl kinase domain might confer an imatinib-resistant phenotype. Since this observation was made, further drug design has led to the development of second-generation inhibitors dasatinib and ponatinib, both of which are capable of targeting many of the common mutations in BCR-Abl, with the exception of the gatekeeper mutation T315I which results in very poor prognosis for the patient. Threonine 315 is known as the gatekeeper residue as it lies at the edge of the ATP-binding site and as such can regulate ligand entry into the deep hydrophobic pocket of the active site [32]. Titrations using both fully and specifically labeled Abl kinase domain against dasatinib and ponatinib, and monitored by HSQC NMR experiments, were performed to probe inhibitor binding. The binding modes of these second-generation drugs for CML were revealed and shown to form two distinct classes. This yielded key insights into possible routes to design improved drugs which would bypass the gatekeeper mutation at the core of the hinge region [33].

2.4
Use of NMR in Abl Kinase Drug Design

Residue-specific assignment provides essential information for NMR-based screening strategies and allows structure–activity relationship (SAR) screening using ^{15}N-HSQC experiments to be performed. Indeed, this strategy was adopted in order to resolve the critical residues responsible for governing imatinib binding to c-Abl. The assignment of 96% of the backbone resonances of the imatinib-bound form of Abl kinase was achieved using a combination of uniform and specific labeling with ^{13}C/^{15}N in baculovirus Sf9 insect cells [23]. It was shown that both imatinib and nilotinib would bind preferentially to the form of the kinase domain with the activation loop in the "out" conformation (DFG-out), which corresponds to an inactive state. In contrast, dasatinib interacts with the DFG-in structure (i.e., it binds to the active form)

as does AFN941, a staurosporine-derived generic kinase inhibitor [33]. Identification of the different ways in which these inhibitors bind to the kinase domain is a key step in the generation of novel drug compounds that are capable of overcoming CML resistance. By observing CSPs in ^{15}N-HSQC spectra, together with X-ray crystallography, mutagenesis and mass spectrometry, it has been shown that the inhibitor GNF-2 (and also its derivative GNF-5) do not bind to the ATP site but instead bind to an allosteric site of Abl kinase domain, which usually binds a myristate moiety (Fig. 4b). This opens up possibilities for the rational design of new drugs with high specificity which could be used in conjunction with ATP site inhibitors to repress c-Abl activity [30, 34]. To this end, a fragment-based NMR screen run on Abl kinase in complex with imatinib (to occlude the active site) using water-LOGSY and $T_{1\rho}$ experiments, identified a series of compounds with specificity for the myristate site that demonstrated allosteric inhibitor profiles. The conformational changes in the myristate site caused by these allosteric compounds were identified using CSPs of ^{15}N-Val-selectively labeled Abl kinase. By monitoring the length and bending of helix I, these compounds could be classified either as inhibitory antagonists or as activating agonists [29]. The state of helix I was determined by observation of the amide resonances of Val525, which in the uninhibited state is highly flexible, thus giving a sharp NMR signal; however, the formation of an α-helix in the inhibited form yielded a considerably broader signal due to a faster relaxation. The agonists identified in this screen, whilst not useful in treating CML patients, could prove beneficial for patients undergoing radiotherapy due to the key roles of c-Abl in DNA damage repair pathways. As the

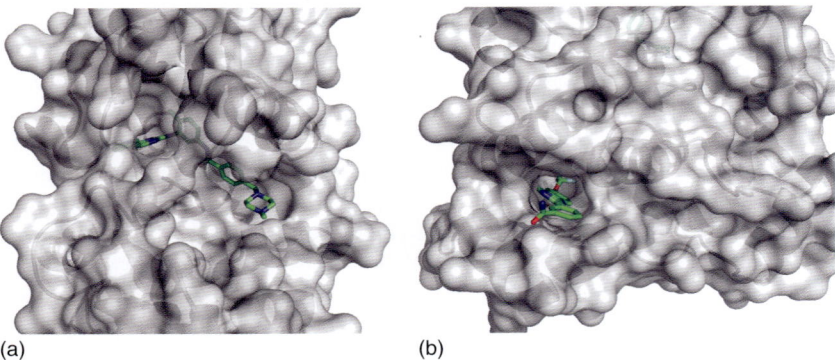

Fig. 4 (a) The front line BCR-Abl inhibitor imatinib is inserted deep into the ATP pocket. The ATP site has the potential for high-affinity binding, but specificity between kinases is an issue. In contrast, the allosteric myristate site, here shown bound by GNF-2 (b), is much shallower but offers a greater degree of selectivity [30].

potential benefits of kinase activators have been subjected to limited study, this currently remains an underexploited route for therapeutic intervention.

The binding mode of myristate to c-Abl has greatly benefited from NMR-based insights, including relaxation measurements and residual dipolar couplings, as well as small-angle X-ray scattering studies. The binding of a second drug molecule to this allosteric site for myristate, in conjunction with imatinib, has been shown to effectively restore the closed conformation of Abl [35]. Furthermore, imatinib induces an open inhibited state of the protein, but the binding of compounds to the myristate site can recreate the closed conformation seen in the apo protein. This has the effect of occluding a vital phosphorylation site key to full activation of BCR-Abl, which in turn suggests a way to overcome the T334I gatekeeper mutant that is immune to imatinib, dasatinib, and nilotinib.

An NMR study of the intermolecular dynamics within the N-terminal regulatory domains revealed that c-Abl may be targeted to the cellular membrane, leading to possible new target sites in Abl [36] that would control its intracellular location and thereby its activity. In addition, studies of interactions of Abl's regulatory SH2 domain with phosphoinosides, phosphopeptides and pyridone-based inhibitors, using HSQC experiments, have each revealed partially overlapping binding sites [37]. Thus, NMR is one of several key tools which is providing an increased understanding into how regulatory and drug molecules bind to and affect the structure and activity of Abl kinase. This, in turn, is opening up new avenues for the design of next-generation agents to combat forms of CML which are resistant to existing drug therapies.

3
Protein Tyrosine Phosphatases

3.1
PTP Superfamily

The maintenance of tightly regulated tyrosine phosphorylation is a principal mechanism for the modulation of intracellular signaling. PTPs provide the balance required for homeostatic signaling by dephosphorylating proteins that transduce signals to control cell proliferation and differentiation. Thus far, drug-discovery initiatives focusing on signal transduction pathways have largely targeted PTKs and have met with considerable success, generating inhibitors with demonstrable clinical efficacy [38]. However, the emergence of resistance against PTK inhibitors has been reported in a variety of tumors, prompting the search for additional novel therapeutic targets to circumvent acquired chemoresistance. Phosphatases represent an attractive, yet relatively unexploited, family of potential targets that govern the levels of protein phosphorylation by reversing PTK activities.

The PTP family comprises 107 enzymes, all of which are highly conserved in both sequence and structure [39]. The highly conserved, charged nature of the PTP active site, along with the catalytically crucial cysteine, present an array of technical obstacles for targeting the PTP active site with small-molecule inhibitors. As such, the development of small-molecule inhibitors with attractive efficacy has been fraught with complications, borne largely as a result of the complementary charges and highly reactive chemical groups required for efficient active site potency, rendering these ligands poorly cell-permeable, with undesirable selectivity profiles and compromised cellular bioavailability. Although they represent a potentially valuable set of drug targets, such physico-chemical caveats have led to suggestions that phosphatases are an undruggable class of enzymes that are refractory to conventional active site-targeted inhibition [40, 41]. This, in turn, has motivated a search for small molecules with alternative modes of inhibition with specificity for novel pockets and allosteric-modulatory sites that could enhance the druggable potential of these enzymes, as exemplified by efforts to develop drug leads for PTP1B.

3.2 PTP1B Signaling Mechanism

PTP1B is an abundant, ubiquitously expressed non-receptor phosphatase that plays a central role in the downregulation of insulin- and leptin-dependent signaling [42]. The full-length form of PTP1B was first purified to homogeneity from the human placenta during the late 1980s, and subsequent characterization revealed its complete sequence of 435 amino acids [43]. The structure of PTP1B displays a typical modular architecture that is consistent with topologies for classical phosphatases and other signaling proteins: it contains an N-terminal catalytic domain, two proline-rich sequences, and a C-terminal hydrophobic region. The C-terminal region is of particular functional importance to PTP1B when it acts as a retention signal sequence responsible for the sequestration of PTP1B in the inner endoplasmic reticulum (ER) membrane. Such subcellular localization to the cytoplasmic face of the ER membrane imposes conformational restraints for accessing native substrates and, as has been suggested, may pose important considerations for drug-discovery initiatives targeting PTP1B in cancers and metabolic disease [44–46].

Biochemical and genetic studies have firmly established PTP1B as a key player in the attenuation and subsequent downregulation of insulin- and leptin-triggered signaling pathways, both of which are critical for the necessary regulation of body weight, glucose homeostasis, and energy expenditure [42]. In both pathways, the activation of insulin and leptin receptors initiates an intracellular phosphorylation cascade which serves as the key signaling mechanism for modulating insulin and leptin receptor activation in the presence of fluctuating physiological stimuli. PTP1B activation is known to direct the dephosphorylation of the insulin receptor (IR) and insulin receptor substrates (IRSs) [47, 48] while simultaneously instigating the dephosphorylation of activated JAK2 and STAT3 [49, 50]. Harnessing the use of PTP1B antibodies and small-molecule inhibitors has demonstrated an increase in insulin-triggered IR, IRS and STAT3 phosphorylation, suggesting that chemotherapeutic strategies targeting PTP1B inhibition might favorably modulate the re-sensitization of insulin and leptin signaling pathways [51, 52]. This approach has been supported by studies with PTP1B-deficient mice which exhibited enhanced insulin sensitivity and leptin hypersensitivity, and a resultant phenotype of lower blood glucose and basal insulin levels [53, 54].

3.3
Therapeutic Significance of PTP1B

Diabetes and obesity pose a significant burden to the current human population. Although significant progress has been made in identifying the molecular origins driving these pathologies, the potential translational benefits have only been modest to patients. The discoveries that PTP1B-null mice were resistant to diet-induced obesity and were insulin-hypersensitive led to the idea that specific PTP1B inhibitors could be useful in the treatment of Type II diabetes and obesity [53]. Indeed, vanadate – a nonspecific PTP inhibitor – has been used for over a century to treat diabetes [55]. This idea was further enhanced by the discovery that antisense PTP1B oligonucleotide normalized glucose levels and restored insulin sensitivity in diabetic mice [56]. This triggered a surge in efforts to develop PTP1B inhibitors but so far this has been unsuccessful, with no drug proceeding beyond Phase II clinical trials [57].

In addition to well-characterized roles in regulating glucose metabolism and insulin sensitivity, PTP1B has been shown to be consistently upregulated in breast tumors, along with the HER2 receptor [58]. Mice harboring oncogenic activating mutations in the HER2 receptor in mammary glands developed multiple tumors and metastases to the lung. More significantly, when these mice were crossed with wild-type PTP1B mice, the tumor development was reduced and the induction of incipient lung metastases decreased. In the same study, a stimulated overexpression of PTP1B was sufficient for driving the oncogenic transformation of mammary cells into a malignant phenotype [59]. These observations correlated with the notion that PTP1B may play a positive role in the signaling events contributing to mammary tumorigenesis, and highlighted the potential for the pharmacological inhibition of PTP1B as a target for breast cancer. However, the development of an appropriate chemotherapeutic strategy has been confounded by the difficulties associated with targeting the phosphatase active site with small-molecule inhibitors. Consequently, novel approaches are required which will create new avenues for developing inhibitors with unique binding modes, occupancies, and specificities for phosphatase targets.

3.4
NMR Studies of PTP1B and Rational Drug Design

The development of potent, novel and highly selective PTP1B-directed inhibitors have greatly benefited from the utilization of NMR-based approaches to yield detailed insights into conformational dynamics and novel druggable sites. Indeed, multiple drug-discovery initiatives have employed strategies to exploit sensitive NMR methods to elucidate the binding modes of hits and lead molecules. These strategies have benefited from the use of combinatorial approaches where NMR-based screening methods have been used in order to identify ligands with high affinity and selectivity for PTP1B [60, 61]. These approaches have highlighted the advantages offered by systematic approaches to small-molecule inhibitor design, and demonstrated the progressive development of a lead inhibitor through a novel synthetic chemistry-based approach. NMR offers versatility due to wide range of ligand affinities that can be detected, thus proving invaluable in the screening and validation of hits. A compound library consisting of 10 000 fragments has been screened using HSQC experiments in order to find novel hit molecules [61]. In this instance, PTP1B was either uniformly ^{15}N-labeled or selectively ^{13}C-methyl labeled on Ile residues, with ligands being validated by monitoring CSPs in ^{1}H,^{15}N, or ^{1}H,^{13}C-resolved HSQC spectra. Significant perturbations were subsequently assigned and mapped to residues Val49, Gly228, Gly218, and Ile219 surrounding the PTP1B catalytic site. The results of the initial screens revealed a phosphotyrosine mimetic ($K_D = 100\,\mu$M) which was subsequently chemically modified in an attempt to yield a more potent active site ligand ($K_D = 26\,\mu$M). An X-ray crystal structure of PTP1B bound to the modified compound revealed intimate contacts between the compound and residues with the active site pocket. Contacts were also observed between the bound naphthyloxomic acid and residues Arg221 and Gln262, which precludes closing of the WPD (Trp-Pro-Asp) loop over the active site cysteine. This has the effect of locking the phosphatase in the "open" conformation and making the binding site available much larger. The highly conserved molecular architecture and chemistry of PTP active sites represents a significant obstacle for the design of specific, cell-permeable inhibitors. Circumventing these challenges requires the exploitation of additional pockets and contacts with non-homologous regions outside the active site cavity. A second, non-catalytic, aryl phosphate-binding site lies adjacent to the catalytic site and is much less-conserved throughout the protein tyrosine phosphate family [62]. The synthesis of bivalent inhibitors which occupy both the active site and the second binding site positioned nearby provides a viable strategy for optimizing inhibitor selectivity, affinity, and efficacy.

Ligands for the second pocket were discovered using selectively ^{13}C-methionine labeling and observing the induced resonance shifts in ^{1}H,^{13}C-HSQC spectra. This screening approach pinpointed a ligand for the secondary binding site which, following iterative cycles of synthetic chemistry, was joined to the initial active site binder using a chemical linker synthesized according to the predetermined co-crystal structure. This structure-guided approach yielded a potent bivalent inhibitor (Fig. 5) that exhibited improved binding parameters, with an experimentally determined K_i of 22 nM, as based on 4-nitrophenol phosphate hydrolysis activity assays. Screening against the same panel of phosphatases, which included leukocyte antigen-related (LAR), SHP-2, CD-45 and calcineurin, revealed an impressive specificity profile ranging from 36-fold to 10 000-fold selectivity with a moderate twofold selectivity observed over T-cell protein tyrosine phosphatase

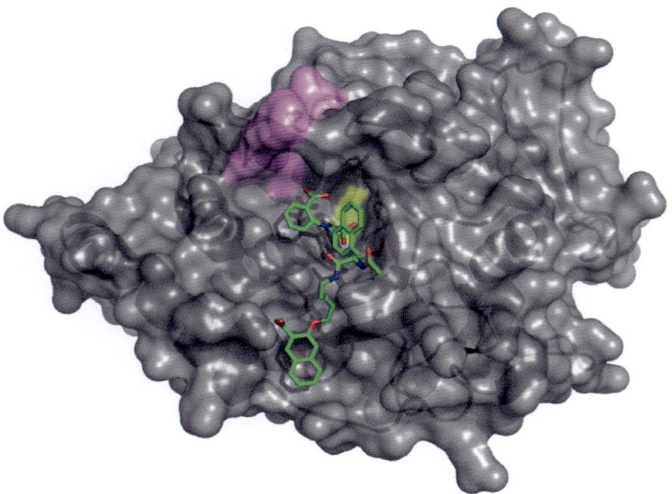

Fig. 5 The structure of PTP1B catalytic domain complexed with a bivalent inhibitor [61]. The non-selective oxamic acid part is bound in the active site (catalytic cysteine shown in yellow), and is linked to a second site ligand identified by NMR-based fragment screening, resulting in a potent inhibitor of PTP1B with increased specificity.

(TCPTP), a phosphatase that also has a second phosphotyrosine-binding site. The derived selectivity profiles for the two compounds indicated that the synthetic bivalent inhibitor, in comparison to the initial active site binder, possessed markedly improved specificity and selectivity, and confirmed that the additional contacts provided by the synthetically derived second-site ligand could be a possible explanation for the differences in the selectivity of the two compounds. As such, these observations illustrated a classic application of the SAR-by-NMR approach in collaboration with high-resolution X-ray crystallography, and helped to establish the novel paradigm that the development of a ligand into a second site can yield benefits to both affinity and specificity, thereby demonstrating the validity of this approach for the design of therapeutic agents for clinically significant phosphatases [61].

Despite being highly potent with promising selectivity profiles, bivalent oxamic acid-based PTP1B inhibitors possess undesirable physico-chemical properties, and thus exhibit poor cellular bioavailability [61]. Additional screens utilizing selective ^{13}C labeling at the methyl groups of Ile219 and monitoring subsequent chemical shifts have been conducted in an attempt stimulate the development of PTP1B inhibitors with greater cell permeability. Such screening strategies have focused principally on the utilization of monocarboxylic acid- and non-carboxylic acid-based fragments as the starting chemotypes for potential catalytic site ligands [60]. The initial selection of starting chemicals with a low charge density, such as mono- and non-carboxylic acid-based fragments not only maximizes the possibility of cell permeability but also increases the likelihood of obtaining a more tractable inhibitor. The structure-directed modification of the initial ligands of both the catalytic site and a second phosphotyrosine site yielded a potent PTP1B inhibitor with improved selectivity and a

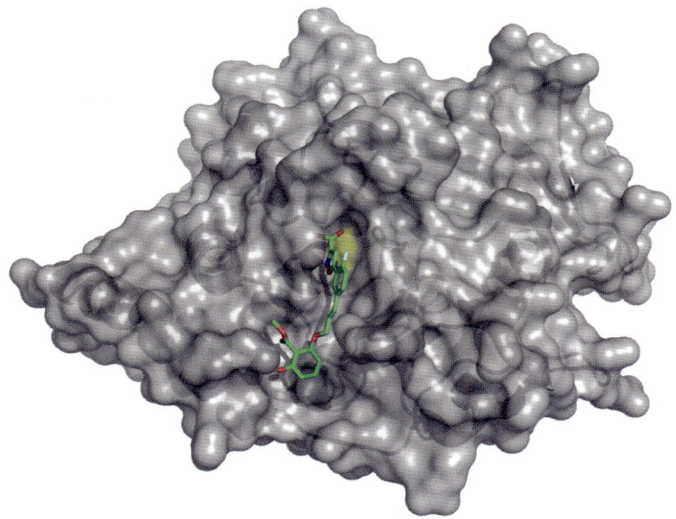

Fig. 6 PTP1B catalytic domain complexed with a cell-permeable oxalylarylaminobenzoic acid-based inhibitor [60]. The catalytic cysteine is highlighted in yellow.

cell permeability that superseded the original cell-impermeable oxamic acid-based inhibitors (Fig. 6). In addition, the design yielded compounds with micromolar potency and 30-fold selectivity over similar phosphatases, including TCPTP [60].

Another strategy for developing PTP1B inhibitors rests on exploiting the conformational mobility of the regulatory C-terminal region [63]. In this instance, two-dimensional ^{1}H,^{15}N-HSQC and three-dimensional experiments were recorded on ^{2}H/^{13}C/^{15}N-labeled protein using TROSY experiments. NMR spectroscopy was used in order to probe the flexibility of amino acid residues 300–393 comprising the C-terminal region of PTP1B, thus illuminating the key residues forming the binding site for the inhibitor MSI-1436. Resonances corresponding to residues 300–393 of PTP1B were observed in the ^{15}N-HSQC-TROSY spectrum at approximately fivefold greater intensity compared with those from PTP1B(1–301), and were found clustered in the random coil portion of the spectrum, indicating a conformationally mobile region. Superimposition of the two-dimensional TROSY spectra of the PTP1B constructs PTP1B(1–393) and PTP1B(1–301) enabled the successful assignment of the intrinsically disordered region, and provided evidence for the mode of inhibition of MSI-1436. NMR analysis combined with biochemical DiFMUP (6,8-difluoro-4-methylumbelliferyl phosphate) enzymatic assays, along with further biophysical characterization, established MSI-1436 (which is also known as trodusquemine [64]) as a bivalent, allosteric inhibitor of PTP1B that binds reversibly and selectively to both the disordered C-terminal segment of PTP1B and a unique pocket situated in close proximity to the active site. The therapeutic potential of such an approach to PTP1B inhibition was subsequently extended to more rigorous investigation by translation of these findings into *in vivo* models. The inhibition of PTP1B with MSI-1436 was shown to result in a downregulation of HER2 signaling, a significant reduction in tumor growth, and an attenuation of lung metastasis in HER2-positive mouse models of breast cancer [63]. This illustrates the unique value of NMR for developing lead candidates that target novel allosteric sites, or act through intrinsically disordered regions.

Currently, NMR is being used in conjunction with other complementary biophysical approaches to expand the present knowledge of the structure, function, and drug development potential of a range of other clinically and therapeutically significant phosphatases. These include the tyrosine-protein phosphatase non-receptor type 11 (PTPN11) that has been implicated in a wide array of developmental disorders and in tumorigenesis [65–67], and the striatal-enriched protein tyrosine phosphatase (STEP) that has been implicated in neurological disorders. With indications of these targets in Alzheimer's disease [68], schizophrenia [69], and Huntington's disease [70], NMR-based phosphatase inhibitor design is positioned to have broad clinical benefits.

4
Ras GTPases

4.1
Ras Superfamily

The Ras oncogene is a paradigm for the superfamily of small guanosine triphosphatases (GTPases) that drive cell proliferation and migration within eukaryotic cells. Over 150 similar proteins are encoded by the human genome, and share a highly conserved 19 kDa catalytic domain [71]. The members are classified into five subfamilies based on differences in their sequences, most of which are found in the termini that interact with membranes. Drug discovery efforts have concentrated on Ras due to its deregulation in various forms of cancer. The similar Arf, Rab, Ran and Rho proteins are also now emerging as targets for therapeutic intervention. Despite their potential, the Ras superfamily has presented challenges for drug design due to intrinsic dynamics and the lack of obvious druggable pockets. Several NMR spectroscopy groups have responded by developing new tools and mechanistic insights to progress GTPases as feasible targets.

4.2
Structural Studies of Ras

As a comparatively small enzyme, the Ras protein is amenable to analysis by NMR structural studies. Rational drug design efforts have been enabled by the assignment of the NMR signals of several forms of Ras, as well as of its RalB, Rheb, RhoA, Rac1, Cdc42 and Arf1 relatives. Nonetheless, resonance assignment of the dynamic loops of the active sites and membrane-binding pockets remains challenging due to line broadening. The H-Ras protein has been assigned in the presence of a guanosine triphosphate (GTP) analog and guanosine diphosphate (GDP), and revealed significant chemical shifts changes relative to the forms bearing oncogenic mutations [72]. Moreover, all of the backbone and many side-chain assignments of the catalytic domain of K-Ras complexed with GDP are available at physiological pH levels [73]. In addition, the assignments of the unprocessed and farnesylated full-length forms of cysteine-substituted H-Ras are also available. Comparison of the resonances within these states reveals that changes are induced in the GTPase domain by the extension of the dynamic terminus [74]. Despite this progress, no structure has been solved of a Ras family member bound to a membrane mimic, despite the central importance of this key oncogenic target state. Nonetheless, the impressive amount of chemical shift, structural and dynamic information which has accumulated provides a solid foundation for NMR-based drug discovery for this class of oncogenic proteins.

The three-dimensional structure of Ras was determined a quarter of a century ago [75], but since this milestone hundreds of structures of GTPases – including Arf, Cdc42, Rab, RAD, Ran, Rap, Ras, Rheb and Rho – have been solved, mostly using X-ray crystallography. These structures reveal a highly conserved fold composed of a compact domain with five α-helices flanking a central, six-stranded β-sheet. Two crucial elements display conformational changes when GTP binds: the switch I element binds with downstream effectors, while the switch II element associates with a guanine nucleotide exchange factor (GEF). Key signaling states of Ras have been stabilized by substituting Thr35 with a serine and Pro40 with an Asp, based on ^{31}P NMR studies [76], thus yielding structural insights into how GTP is bound and suggesting novel druggable pockets [77, 78]. The recognition of the farnesyl group by part of Raf, its downstream effector, is evident by CSPs, which also suggest interactions near switch I, indicating that these regions may be functionally coupled [74]. Thus, the analysis of wild-type and mutant forms of Ras using NMR spectroscopy and X-ray crystallography and substrate analogs indicates a set of inducible sites that could be exploited for drug discovery.

4.3
Signaling Activities of Ras

Isolated Ras family proteins display weak catalytic rates, but are activated by interaction with a GTP-activating protein (GAP). As an example, Ras activity is enhanced by RasGAP which orients Gln61 for nucleophilic attack, thus stabilizing the enzyme's transition state. The next step involves the exchange of a GDP molecule in the active site with GTP, and is typically slow. Recruitment of a GEF protein accelerates the latter step. The structural interaction of Ras with the GEF known as Son-of-Sevenless (SOS) shows how it contacts residues between positions 32 and 67, thus impeding the interaction with magnesium and releasing the GDP molecule [79]. This event opens this site such that GTP can bind which, due to its picomolar affinity, dislodges the GEF. Finally, the activated GTPase engages downstream effectors such as the Raf kinase, thereby signaling through the mitogen-activated protein kinase kinase (MEK)/ERK pathway. The association with the Raf protein boosts the rate by which Ras hydrolyzes GTP hydrolysis, thus releasing Ras from activation by GAP proteins [80]. The interaction with Raf is the tightest of the various competing Ras effectors based on a comparison of how the different effector complexes alter the NMR signals of activated ^{15}N-labeled Ras protein [81]. The effectors directly compete with GAP and SOS proteins for Ras binding, with the entire system of controlled access and feedback cycles being observable in real-time by NMR. Oncogenic mutations of Ras either accelerate or decelerate the hydrolysis and exchange of nucleotides [72], and alter the affinity profile of the different effector molecules, thus providing mechanistic insights for their various effects on signaling within cellular systems.

Ras family members mediate signaling events on the ER, endosomal, Golgi, mitochondrial and plasma membranes. The C terminus of the activated forms may be farnesylated, proteolytically cleaved, and methylated at a conserved cysteine residue, and may also be palmitoylated at nearby cysteines. The array of possible subcellular states and localizations presents challenges for designing agents that act on the specific form that drives cancer progression. The post-translational modifications and basic motifs, as well as interactive surfaces on the catalytic domain, presumably influence the protein's association with membrane rafts and non-raft regions, thus controlling downstream interactions with regulatory and effector proteins. When Ras is bound to GDP it associates with membrane rafts through its lipid anchors, whereas when bound to GTP the protein is found in disordered membrane regions via the adjacent linker sequence [82]. The conformational states of Ras proteins bound to membrane have not been experimentally determined, being confounded by the complex and dynamic nature of the lipid interactions and the need for full-length, lipid-modified forms. Thus, the structural mechanisms of the most relevant target state as well as the dynamic coupling between the exchange of GDP and GTP remain obscure, as do the specific protein–protein and membrane–protein interactions involved. Nonetheless, simulations of full-length H-Ras infer that the catalytic domain is positioned on the bilayer by the binding of phospholipid headgroups by the α4 helix residues Arg128 and Arg135, thus altering accessibility with downstream partners and controlling signaling [83]. This lipid-binding site could potentially represent a novel druggable site, offering new potential opportunities for drug design.

4.4
Targeting Ras

The progression of cancer is influenced by hyperactive forms of Ras proteins, with 20–30% of tumors found to be expressing mutated forms. For pancreatic cancer the rate is even higher, with 60% of cases expressing mutant K-ras. The residues that are most often altered are at positions 12, 13 and 61, and help to hydrolyze GTP. The reduced GTPase activity of Ras contributes to constitutive signaling and thus to tumorigenesis. The overexpression of mutation of receptors that are upstream of Ras such as the epidermal growth factor receptor (EGFR) also plays a role in the development of cancer, as do mutations of Ras effectors, including the Raf kinase. A common undesirable outcome of treating patients with EGFR or Raf inhibitors is Ras upregulation or activation, thus emphasizing the need to develop Ras inhibitors which could counter the drug-resistant mutant forms that often arise during treatment. In addition to Ras, other related targets are being identified. For example, Rho family members are overexpressed along with their regulators, with their structures providing new opportunities to design agents to combat the cancer progression and cardiac hypertrophy that can result [84, 85].

Several approaches have been employed to design inhibitors of Ras family members given their demonstrated importance to cancer. NMR spectroscopy has proven useful for identifying compounds which bind Ras directly, including agents that selectively interact with one of its two rapidly interconverting GTP-bound states. Another strategy has been to block aberrant signaling by Ras by identifying ligands of the GEF interface that alter the regeneration of its active state. The interfaces mediating assembly of Ras complexes with effector proteins or membrane surfaces can be exploited. In total, there are eight allosteric states of Ras which have been identified, with high-pressure and ^{31}P-NMR studies identifying four Ras conformers [86].

Together, this is exposing a range of potential druggable pockets and conformational changes, and is also yielding a variety of probe molecules for biological experiments to help with the validation of new lead molecules.

Due to the availability of protein and membrane-binding surface features [87] rather than deep hydrophobic pockets (Fig. 7), Ras proteins are best suited for NMR-based screening of drug-like fragments as efficient starting points for subsequent medicinal chemistry elaboration. This strategy has been used by several groups interested in discovering Ras inhibitors for use as therapeutic agents. The outcomes of drug discovery campaigns targeting K-Ras have been increasingly promising. A Genentech team screened 3285 fragments using one-dimensional NMR experiments that detect ligands, and this resulted in 266 fragments that bind Ras [88]. Their sites of interaction were mapped by tracking ^{1}H, ^{15}N resonances from the GDP-bound protein, confirming 25 molecules that bound to an inducible pocket where the side chain of Tyr71 normally sites, compromising the hydrogen bonds it forms with the SOS protein. The affinities of the hit molecules were in the upper micromolar range, based on progressive chemical shift changes observed in titration experiments. This assay was used to test series of derivative compounds which were then designed, allowing the affinity and activity of the ligands to be improved. The structures of several of these ligands bound to the protein identified a common binding pocket formed by Lys5, Leu6, Val7, Ile55, Leu56, and Thr74 residues. The adaptability of the site was apparent from the expansion of its size by up to 50% when occupied by some ligand molecules, as was their proximity to the Ras–SOS interface [89]. The most lead candidates included the DCAI (4,6-dichloro-2-methyl-3-aminoethyl-indole) molecule, which block SOS association thus inhibiting nucleotide release from K-Ras. This series thus provided a potential route for further lead optimization by competing with docking of GEF proteins rather than active site occupancy.

Another approach was taken by Fesik and colleagues, who used two-dimensional HSQC experiments to identify ligands of the GDP-bound K-Ras protein. These authors screened a library of 11 000 fragment molecules [90], which resulted in 140 ligands that mainly bound around the SOS site with millimolar affinities, while some interacted with another site near the switch I region. Two of these hits were chemically linked together to produce a bivalent molecule with an affinity of 190 μM that inhibits SOS association and nucleotide exchange. This strategy was expanded by the use of a S39C mutant of K-Ras to which a thiol-reactive compound was attached in order to block access to a site proximal to the first ligand [91]. This engineered complex was then used to screen for compounds that bound to distinct, second sites. This approach sidestepped the confounding need to use a saturating concentration of one ligand in order to identify a second ligand to which it can be linked to form bivalent inhibitors.

A group at Kobe University has utilized a mutant form of Ras that assumes a critical GTP-bound conformation. They obtained useful hits by virtual ligand screening of a library of over two million compounds, and validated these in a Raf-binding assay [92]. The structures of the complexes were solved using intermolecular NOEs and CSPs from H-Ras in its GTP analog-loaded form, as cocrystals proved elusive. The hits were bound to a pocket formed by the conserved Lys5, Leu56, Met67, Gln70, and Tyr 74

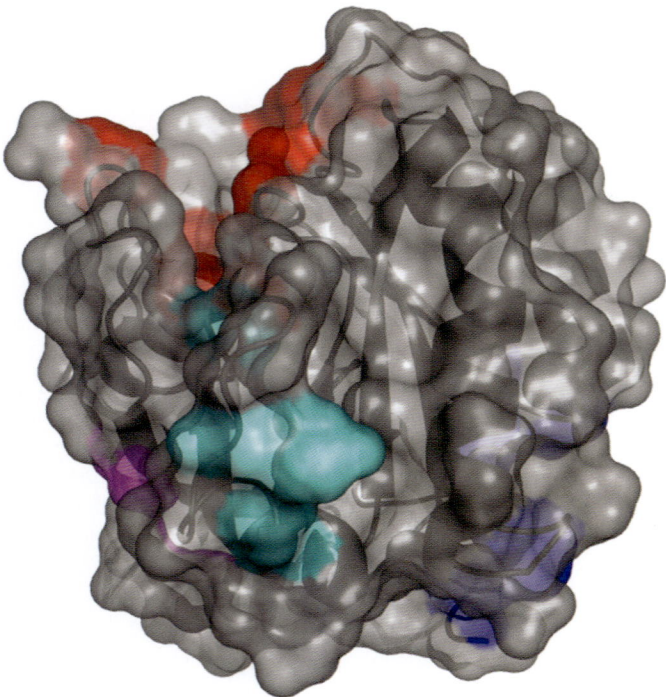

Fig. 7 The structure of activated Ras with the highest candidate binding sites highlighted [87]. The widely studied nucleotide-binding site is shown (red), together with the underexploited novel binding sites p1 (magenta), p2 (cyan), and p3 (blue). These addition sites offer the possibility of greater selectivity for novel inhibitors for Ras.

residues near switch I. Relaxation-edited proton NMR experiments indicated that the ligands also bound to the GDP-loaded state of Ras and its relatives, highlighting the need for iterative rounds of design to enhance target specificity and potency.

Kalbitzer's group identified a metal–cyclen compound that binds weakly to Ras near its GTP-binding pocket [93]. This ligand was found by monitoring the ^{31}P NMR signals of a GTP analog bound to full-length and mutant forms of H-Ras. The results indicated that when the Zn^{2+}–cyclen binds the Ras-GTP complex shifts toward an inactive conformation which has lower affinity for downstream effectors, thus explaining its signaling effects.

One of the first Ras inhibitors to be discovered resulted from studies conducted by the Schering-Plough Research Institute team. The compound, SCH-54292, was designed from a GDP analog, has a half-maximal inhibitory concentration (IC_{50}) of 700 nM, and binds to a flexible pocket that is induced under the switch II element upon nucleotide exchange. Docking of the compound to the H-Ras structure using chemical shifts and intermolecular NOE distances from NMR experiments [94] explains its specificity for the GDP-bound form of the enzyme. A similar compound displays inhibitory activity in PC-12 cells at concentrations of 10–20 μM [95], thus representing a possible candidate for further medicinal chemistry optimization. Subsequently, a range of new starting points and interfaces has emerged that illustrate the potential for designing next-generation inhibitors of this valuable cancer target.

5
Conclusions

NMR, an increasingly useful technique for structure-based drug design, occupies a unique position in that it provides atomic-level details of structure, dynamics, and ligand interactions. When combined with complementary techniques such as X-ray crystallography and SPR, NMR enables the expansion and refinement of initial hits into high-affinity and specific lead molecules with the potential for treating previously incurable conditions. NMR is also an ideal tool for searching for new binding sites, such as allosteric and inducible pockets in target proteins previously described as undruggable. There are many other examples of the use of NMR in drug discovery, including the design of Bcl-2 [96], BACE1 [97], and Hsp90 [98] inhibitors that have progressed toward clinical trials. Here, three established targets – PTP1B, Abl and Ras – have been presented, all of which have been exploited to identify new pockets for drug design and to discover and elaborate drug-like lead molecules. This has led to a deeper understanding of their mechanisms of action, and has also reinforced the promise of fragment-based NMR approaches, providing further opportunities for drug discovery for emerging targets within these major superfamilies of proteins.

Acknowledgments

The authors would like to thank the Biotechnology and Biological Sciences Research Council, Cancer Research UK and Leukaemia & Lymphoma Research for funding, and AstraZeneca for support of a BBSRC Industrial CASE PhD Studentship to L.Q.

References

1. Hajduk, P.J., Olejniczak, E.T., Fesik, S.W. (1997) One-dimensional relaxation- and diffusion-edited NMR methods for screening compounds that bind macromolecules. *J. Am. Chem. Soc.*, **119**, 12257–12261.
2. Meyer, B., Klein, J., Mayer, M., Meinecke, R. et al. (2004) Saturation Transfer Difference NMR Spectroscopy for Identifying Ligand Epitopes and Binding Specificities, in Ernst Schering Research Foundation Workshop, Vol. **44**, pp. 149–167.
3. Dalvit, C., Fogliatto, G., Stewart, A., Veronesi, M., Stockman, B. (2001) Water-LOGSY as a method for primary NMR screening: practical aspects and range of applicability. *J. Biomol. NMR*, **21**, 349–359.
4. Vanwetswinkel, S., Heetebrij, R.J., van Duynhoven, J., Hollander, J.G. et al. (2005) TINS, target immobilized NMR screening: an efficient and sensitive method for ligand discovery. *Chem. Biol.*, **12**, 207–216.
5. Jahnke, W. (2002) Spin labels as a tool to identify and characterize protein-ligand interactions by NMR spectroscopy. *ChemBioChem*, **3**, 167–173.
6. Dalvit, C., Flocco, M., Veronesi, M., Stockman, B.J. (2002) Fluorine-NMR competition binding experiments for high-throughput screening of large compound mixtures. *Comb. Chem. High Throughput Screen.*, **5**, 605–611.
7. Tengel, T., Fex, T., Emtenas, H., Almqvist, F. et al. (2004) Use of 19F NMR spectroscopy to screen chemical libraries for ligands that bind to proteins. *Org. Biomol. Chem.*, **2**, 725–731.
8. Manzenrieder, F., Frank, A.O., Kessler, H. (2008) Phosphorus NMR spectroscopy as a versatile tool for compound library screening. *Angew. Chem. Int. Ed. Engl.*, **47**, 2608–2611.
9. Zhang, X., Sanger, A., Hemmig, R., Jahnke, W. (2009) Ranking of high-affinity ligands by

NMR spectroscopy. *Angew. Chem. Int. Ed. Engl.*, **48**, 6691–6694.
10. Pervushin, K., Riek, R., Wider, G., Wuthrich, K. (1997) Attenuated T2 relaxation by mutual cancellation of dipole–dipole coupling and chemical shift anisotropy indicates an avenue to NMR structures of very large biological macromolecules in solution. *Proc. Natl Acad. Sci. USA*, **94**, 12366–12371.
11. Schanda, P., Kupce, E., Brutscher, B. (2005) SOFAST-HMQC experiments for recording two-dimensional heteronuclear correlation spectra of proteins within a few seconds. *J. Biomol. NMR*, **33**, 199–211.
12. Shuker, S.B., Hajduk, P.J., Meadows, R.P., Fesik, S.W. (1996) Discovering high-affinity ligands for proteins: SAR by NMR. *Science*, **274**, 1531–1534.
13. Dominguez, C., Boelens, R., Bonvin, A.M. (2003) HADDOCK: a protein–protein docking approach based on biochemical or biophysical information. *J. Am. Chem. Soc.*, **125**, 1731–1737.
14. Hunter, T. (1995) Protein kinases and phosphatases: the yin and yang of protein phosphorylation and signaling. *Cell*, **80**, 225–236.
15. Vieth, M., Sutherland, J.J., Robertson, D.H., Campbell, R.M. (2005) Kinomics: characterizing the therapeutically validated kinase space. *Drug Discov. Today*, **10**, 839–846.
16. Manning, G., Whyte, D.B., Martinez, R., Hunter, T., Sudarsanam, S. (2002) The protein kinase complement of the human genome. *Science*, **298**, 1912–1934.
17. Badrinarayan, P., Sastry, G.N. (2013) Rational approaches towards lead optimization of kinase inhibitors: the issue of specificity. *Curr. Pharm. Des.*, **19**, 4714–4738.
18. Garuti, L., Roberti, M., Bottegoni, G. (2010) Non-ATP competitive protein kinase inhibitors. *Curr. Med. Chem.*, **17**, 2804–2821.
19. Langer, T., Vogtherr, M., Elshorst, B., Betz, M. et al. (2004) NMR backbone assignment of a protein kinase catalytic domain by a combination of several approaches: application to the catalytic subunit of cAMP-dependent protein kinase. *ChemBioChem*, **5**, 1508–1516.
20. Vogtherr, M., Saxena, K., Grimme, S., Betz, M. et al. (2005) NMR backbone assignment of the mitogen-activated protein (MAP) kinase p38. *J. Biomol. NMR*, **32**, 175.
21. Campos-Olivas, R., Marenchino, M., Scapozza, L., Gervasio, F.L. (2011) Backbone assignment of the tyrosine kinase Src catalytic domain in complex with imatinib. *Biomol. NMR Assign.*, **5**, 221–224.
22. Vajpai, N., Schott, A.K., Vogtherr, M., Breeze, A.L. (2014) NMR backbone assignments of the tyrosine kinase domain of human fibroblast growth factor receptor 1. *Biomol. NMR Assign.*, **8**, 85–88.
23. Vajpai, N., Strauss, A., Fendrich, G., Cowan-Jacob, S.W. et al. (2008) Backbone NMR resonance assignment of the Abelson kinase domain in complex with imatinib. *Biomol. NMR Assign.*, **2**, 41–42.
24. Wiesner, S., Wybenga-Groot, L.E., Warner, N., Lin, H. et al. (2006) A change in conformational dynamics underlies the activation of Eph receptor tyrosine kinases. *EMBO J.*, **25**, 4686–4696.
25. Piserchio, A., Dalby, K.N., Ghose, R. (2012) Assignment of backbone resonances in a eukaryotic protein kinase-ERK2 as a representative example. *Methods Mol. Biol.*, **831**, 359–368.
26. Shin, J., Chakraborty, G., Bharatham, N., Kang, C. et al. (2011) NMR solution structure of human vaccinia-related kinase 1 (VRK1) reveals the C-terminal tail essential for its structural stability and autocatalytic activity. *J. Biol. Chem.*, **286**, 22131–22138.
27. Shin, J., Chakraborty, G., Yoon, H.S. (2014) Backbone (1)H, (1)(3)C and (1)(5)N resonance assignments of human vaccinia-related kinase 1 (VRK1). *Biomol. NMR Assign.*, **8**, 29–31.
28. de Klein, A., van Kessel, A.G., Grosveld, G., Bartram, C.R. et al. (1982) A cellular oncogene is translocated to the Philadelphia chromosome in chronic myelocytic leukaemia. *Nature*, **300**, 765–767.
29. Jahnke, W., Grotzfeld, R.M., Pelle, X., Strauss, A. et al. (2010) Binding or bending: distinction of allosteric Abl kinase agonists from antagonists by an NMR-based conformational assay. *J. Am. Chem. Soc.*, **132**, 7043–7048.
30. Zhang, J., Adrian, F.J., Jahnke, W., Cowan-Jacob, S.W. et al. (2010) Targeting Bcr-Abl by combining allosteric with ATP-binding-site inhibitors. *Nature*, **463**, 501–506.
31. Nagar, B., Bornmann, W.G., Pellicena, P., Schindler, T. et al. (2002) Crystal structures

of the kinase domain of c-Abl in complex with the small molecule inhibitors PD173955 and imatinib (STI-571). *Cancer Res.*, **62**, 4236–4243.
32. Zhou, T., Parillon, L., Li, F., Wang, Y. et al. (2007) Crystal structure of the T315I mutant of Abl kinase. *Chem. Biol. Drug Des.*, **70**, 171–181.
33. Vajpai, N., Strauss, A., Fendrich, G., Cowan-Jacob, S.W. et al. (2008) Solution conformations and dynamics of ABL kinase-inhibitor complexes determined by NMR substantiate the different binding modes of imatinib/nilotinib and dasatinib. *J. Biol. Chem.*, **283**, 18292–18302.
34. Fabbro, D., Manley, P.W., Jahnke, W., Liebetanz, J. et al. (2010) Inhibitors of the Abl kinase directed at either the ATP- or myristate-binding site. *Biochim. Biophys. Acta*, **1804**, 454–462.
35. Skora, L., Mestan, J., Fabbro, D., Jahnke, W., Grzesiek, S. (2013) NMR reveals the allosteric opening and closing of Abelson tyrosine kinase by ATP-site and myristoyl pocket inhibitors. *Proc. Natl Acad. Sci. USA*, **110**, E4437–E4445.
36. de Oliveira, G.A., Pereira, E.G., Ferretti, G.D., Valente, A.P. et al. (2013) Intramolecular dynamics within the N-Cap-SH3-SH2 regulatory unit of the c-Abl tyrosine kinase reveal targeting to the cellular membrane. *J. Biol. Chem.*, **288**, 28331–28345.
37. Tokonzaba, E., Capelluto, D.G., Kutateladze, T.G., Overduin, M. (2006) Phosphoinositide, phosphopeptide and pyridone interactions of the Abl SH2 domain. *Chem. Biol. Drug Des.*, **67**, 230–237.
38. Bialy, L., Waldmann, H. (2005) Inhibitors of protein tyrosine phosphatases: next-generation drugs? *Angew. Chem. Int. Ed. Engl.*, **44**, 3814–3839.
39. Tonks, N.K., Neel, B.G. (2001) Combinatorial control of the specificity of protein tyrosine phosphatases. *Curr. Opin. Cell Biol.*, **13**, 182–195.
40. Andersen, J.N., Tonks, N.K. (2004) Protein tyrosine phosphatase-based therapeutics: lessons from PTP1B. *Top. Curr. Genet.*, **5**, 201–230.
41. Li, X., Oghi, K.A., Zhang, J., Krones, A. et al. (2003) Eya protein phosphatase activity regulates Six1-Dach-Eya transcriptional effects in mammalian organogenesis. *Nature*, **426**, 247–254.
42. Zhang, Z.Y., Lee, S.Y. (2003) PTP1B inhibitors as potential therapeutics in the treatment of type 2 diabetes and obesity. *Expert Opin. Invest. Drugs*, **12**, 223–233.
43. Tonks, N.K., Diltz, C.D., Fischer, E.H. (1988) Purification of the major protein-tyrosine-phosphatases of human placenta. *J. Biol. Chem.*, **263**, 6722–6730.
44. Frangioni, J.V., Beahm, P.H., Shifrin, V., Jost, C.A., Neel, B.G. (1992) The nontransmembrane tyrosine phosphatase PTP-1B localizes to the endoplasmic reticulum via its 35 amino acid C-terminal sequence. *Cell*, **68**, 545–560.
45. Mauro, L.J., Youngren, O.M., Proudman, J.A., Phillips, R.E., el Halawani, M.E. (1992) Effects of reproductive status, ovariectomy, and photoperiod on vasoactive intestinal peptide in the female turkey hypothalamus. *Gen. Comp. Endocrinol.*, **87**, 481–493.
46. Woodford-Thomas, T.A., Rhodes, J.D., Dixon, J.E. (1992) Expression of a protein tyrosine phosphatase in normal and v-src-transformed mouse 3T3 fibroblasts. *J. Cell Biol.*, **117**, 401–414.
47. Bandyopadhyay, D., Kusari, A., Kenner, K.A., Liu, F. et al. (1997) Protein-tyrosine phosphatase 1B complexes with the insulin receptor *in vivo* and is tyrosine-phosphorylated in the presence of insulin. *J. Biol. Chem.*, **272**, 1639–1645.
48. Goldstein, B.J., Bittner-Kowalczyk, A., White, M.F., Harbeck, M. (2000) Tyrosine dephosphorylation and deactivation of insulin receptor substrate-1 by protein-tyrosine phosphatase 1B. Possible facilitation by the formation of a ternary complex with the Grb2 adaptor protein. *J. Biol. Chem.*, **275**, 4283–4289.
49. Kaszubska, W., Falls, H.D., Schaefer, V.G., Haasch, D. et al. (2002). Protein tyrosine phosphatase 1B negatively regulates leptin signaling in a hypothalamic cell line. *Mol. Cell. Endocrinol.*, **195**, 109–118.
50. Lund, I.K., Hansen, J.A., Andersen, H.S., Moller, N.P., Billestrup, N. (2005) Mechanism of protein tyrosine phosphatase 1B-mediated inhibition of leptin signalling. *J. Mol. Endocrinol.*, **34**, 339–351.
51. Ahmad, F., Li, P.M., Meyerovitch, J., Goldstein, B.J. (1995) Osmotic loading of neutralizing antibodies demonstrates a role for protein-tyrosine phosphatase 1B

in negative regulation of the insulin action pathway. *J. Biol. Chem.*, **270**, 20503–20508.
52. Zhang, S., Zhang, Z.Y. (2007) PTP1B as a drug target: recent developments in PTP1B inhibitor discovery. *Drug Discov. Today*, **12**, 373–381.
53. Elchebly, M., Payette, P., Michaliszyn, E., Cromlish, W. et al. (1999) Increased insulin sensitivity and obesity resistance in mice lacking the protein tyrosine phosphatase-1B gene. *Science*, **283**, 1544–1548.
54. Haj, F.G., Zabolotny, J.M., Kim, Y.B., Kahn, B.B., Neel, B.G. (2005) Liver-specific protein-tyrosine phosphatase 1B (PTP1B) re-expression alters glucose homeostasis of PTP1B-/-mice. *J. Biol. Chem.*, **280**, 15038–15046.
55. Thompson, K.H., Orvig, C. (2006) Vanadium in diabetes: 100 years from phase 0 to phase I. *J. Inorg. Biochem.*, **100**, 1925–1935.
56. Zinker, B.A., Rondinone, C.M., Trevillyan, J.M., Gum, R.J. et al. (2002) PTP1B antisense oligonucleotide lowers PTP1B protein, normalizes blood glucose, and improves insulin sensitivity in diabetic mice. *Proc. Natl Acad. Sci. USA*, **99**, 11357–11362.
57. Barr, A.J. (2010) Protein tyrosine phosphatases as drug targets: strategies and challenges of inhibitor development. *Future Med. Chem.*, **2**, 1563–1576.
58. Bentires-Alj, M., Neel, B.G. (2007) Protein-tyrosine phosphatase 1B is required for HER2/Neu-induced breast cancer. *Cancer Res.*, **67**, 2420–2424.
59. Julien, S.G., Dube, N., Read, M., Penney, J. et al. (2007) Protein tyrosine phosphatase 1B deficiency or inhibition delays ErbB2-induced mammary tumorigenesis and protects from lung metastasis. *Nat. Genet.*, **39**, 338–346.
60. Liu, G., Xin, Z., Pei, Z., Hajduk, P.J. et al. (2003) Fragment screening and assembly: a highly efficient approach to a selective and cell active protein tyrosine phosphatase 1B inhibitor. *J. Med. Chem.*, **46**, 4232–4235.
61. Szczepankiewicz, B.G., Liu, G., Hajduk, P.J., Abad-Zapatero, C. et al. (2003) Discovery of a potent, selective protein tyrosine phosphatase 1B inhibitor using a linked-fragment strategy. *J. Am. Chem. Soc.*, **125**, 4087–4096.
62. Puius, Y.A., Zhao, Y., Sullivan, M., Lawrence, D.S. et al. (1997) Identification of a second aryl phosphate-binding site in protein-tyrosine phosphatase 1B: a paradigm for inhibitor design. *Proc. Natl Acad. Sci. USA*, **94**, 13420–13425.
63. Krishnan, N., Koveal, D., Miller, D.H., Xue, B. et al. (2014) Targeting the disordered C terminus of PTP1B with an allosteric inhibitor. *Nat. Chem. Biol.*, **10**, 558–566.
64. Lantz, K.A., Hart, S.G., Planey, S.L., Roitman, M.F. et al. (2010) Inhibition of PTP1B by trodusquemine (MSI-1436) causes fat-specific weight loss in diet-induced obese mice. *Obesity (Silver Spring)*, **18**, 1516–1523.
65. Chan, G., Kalaitzidis, D., Neel, B.G. (2008) The tyrosine phosphatase Shp2 (PTPN11) in cancer. *Cancer Metastasis Rev.*, **27**, 179–192.
66. Chan, R.J., Feng, G.S. (2007) PTPN11 is the first identified proto-oncogene that encodes a tyrosine phosphatase. *Blood*, **109**, 862–867.
67. Tartaglia, M., Niemeyer, C.M., Fragale, A., Song, X. et al. (2003) Somatic mutations in PTPN11 in juvenile myelomonocytic leukemia, myelodysplastic syndromes and acute myeloid leukemia. *Nat. Genet.*, **34**, 148–150.
68. Zhang, Y., Kurup, P., Xu, J., Carty, N. et al. (2010) Genetic reduction of striatal-enriched tyrosine phosphatase (STEP) reverses cognitive and cellular deficits in an Alzheimer's disease mouse model. *Proc. Natl Acad. Sci. USA*, **107**, 19014–19019.
69. Carty, N.C., Xu, J., Kurup, P., Brouillette, J. et al. (2012) The tyrosine phosphatase STEP: implications in schizophrenia and the molecular mechanism underlying antipsychotic medications. *Transl. Psychiatry*, **2**, e137.
70. Saavedra, A., Giralt, A., Rue, L., Xifro, X. et al. (2011) Striatal-enriched protein tyrosine phosphatase expression and activity in Huntington's disease: a STEP in the resistance to excitotoxicity. *J. Neurosci.*, **31**, 8150–8162.
71. Wennerberg, K., Rossman, K.L., Der, C.J. (2005) The Ras superfamily at a glance. *J. Cell Sci.*, **118**, 843–846.
72. Smith, M.J., Neel, B.G., Ikura, M. (2013) NMR-based functional profiling of RASopathies and oncogenic RAS mutations. *Proc. Natl Acad. Sci. USA*, **110**, 4574–4579.
73. Vo, U., Embrey, K.J., Breeze, A.L., Golovanov, A.P. (2013) (1)H, (1)(3)C and (1)(5)N resonance assignment for the human K-Ras at physiological pH. *Biomol. NMR Assign.*, **7**, 215–219.
74. Thapar, R., Williams, J.G., Campbell, S.L. (2004) NMR characterization of full-length

farnesylated and non-farnesylated H-Ras and its implications for Raf activation. *J. Mol. Biol.*, **343**, 1391–1408.

75. Pai, E.F., Krengel, U., Petsko, G.A., Goody, R.S. *et al.* (1990) Refined crystal structure of the triphosphate conformation of H-ras p21 at 1.35 Å resolution: implications for the mechanism of GTP hydrolysis. *EMBO J.*, **9**, 2351–2359.

76. Spoerner, M., Herrmann, C., Vetter, I.R., Kalbitzer, H.R., Wittinghofer, A. (2001). Dynamic properties of the Ras switch I region and its importance for binding to effectors. *Proc. Natl Acad. Sci. USA*, **98**, 4944–4949.

77. Shima, F., Ijiri, Y., Muraoka, S., Liao, J. *et al.* (2010) Structural basis for conformational dynamics of GTP-bound Ras protein. *J. Biol. Chem.*, **285**, 22696–22705.

78. Araki, M., Shima, F., Yoshikawa, Y., Muraoka, S. *et al.* (2011) Solution structure of the state 1 conformer of GTP-bound H-Ras protein and distinct dynamic properties between the state 1 and state 2 conformers. *J. Biol. Chem.*, **286**, 39644–39653.

79. Sondermann, H., Soisson, S.M., Boykevisch, S., Yang, S.S. *et al.* (2004) Structural analysis of autoinhibition in the Ras activator Son of sevenless. *Cell*, **119**, 393–405.

80. Buhrman, G., Holzapfel, G., Fetics, S., Mattos, C. (2010) Allosteric modulation of Ras positions Q61 for a direct role in catalysis. *Proc. Natl Acad. Sci. USA*, **107**, 4931–4936.

81. Smith, M.J., Ikura, M. (2014) Integrated RAS signaling defined by parallel NMR detection of effectors and regulators. *Nat. Chem. Biol.*, **10**, 223–230.

82. Rotblat, B., Prior, I.A., Muncke, C., Parton, R.G. *et al.* (2004) Three separable domains regulate GTP-dependent association of H-ras with the plasma membrane. *Mol. Cell. Biol.*, **24**, 6799–6810.

83. Gorfe, A.A., Hanzal-Bayer, M., Abankwa, D., Hancock, J.F., McCammon, J.A. (2007) Structure and dynamics of the full-length lipid-modified H-Ras protein in a 1,2-dimyristoylglycero-3-phosphocholine bilayer. *J. Med. Chem.*, **50**, 674–684.

84. Lenoir, M., Sugawara, M., Kaur, J., Ball, L.J., Overduin, M. (2014) Structural insights into the activation of the RhoA GTPase by the lymphoid blast crisis (Lbc) oncoprotein. *J. Biol. Chem.*, **289**, 23992–24004.

85. Abdul Azeez, K.R., Knapp, S., Fernandes, J.M., Klussmann, E., Elkins, J.M. (2014) The crystal structure of the RhoA-AKAP-Lbc DH-PH domain complex. *Biochem. J.*, **464**, 231–239.

86. Kalbitzer, H.R., Rosnizeck, I.C., Munte, C.E., Narayanan, S.P. *et al.* (2013) Intrinsic allosteric inhibition of signaling proteins by targeting rare interaction states detected by high-pressure NMR spectroscopy. *Angew. Chem. Int. Ed. Engl.*, **52**, 14242–14246.

87. Grant, B.J., Lukman, S., Hocker, H.J., Sayyah, J. *et al.* (2011) Novel allosteric sites on Ras for lead generation. *PLoS One*, **6**, e25711.

88. Maurer, T., Wang, W. (2013) NMR study to identify a ligand-binding pocket in Ras. *Enzymes*, **33** (Pt. A), 15–39.

89. Maurer, T., Garrenton, L.S., Oh, A., Pitts, K. *et al.* (2012) Small-molecule ligands bind to a distinct pocket in Ras and inhibit SOS-mediated nucleotide exchange activity. *Proc. Natl Acad. Sci. USA*, **109**, 5299–5304.

90. Sun, Q., Burke, J.P., Phan, J., Burns, M.C. *et al.* (2012) Discovery of small molecules that bind to K-Ras and inhibit Sos-mediated activation. *Angew. Chem. Int. Ed. Engl.*, **51**, 6140–6143.

91. Sun, Q., Phan, J., Friberg, A.R., Camper, D.V. *et al.* (2014) A method for the second-site screening of K-Ras in the presence of a covalently attached first-site ligand. *J. Biomol. NMR*, **60**, 11–14.

92. Shima, F., Yoshikawa, Y., Matsumoto, S., Kataoka, T. (2013) Discovery of small-molecule Ras inhibitors that display antitumor activity by interfering with Ras·GTP-effector interaction. *Enzymes*, **34** (Pt. B), 1–23.

93. Rosnizeck, I.C., Graf, T., Spoerner, M., Trankle, J. *et al.* (2010) Stabilizing a weak binding state for effectors in the human ras protein by cyclen complexes. *Angew. Chem. Int. Ed. Engl.*, **49**, 3830–3833.

94. Ganguly, A.K., Wang, Y.S., Pramanik, B.N., Doll, R.J. *et al.* (1998) Interaction of a novel GDP exchange inhibitor with the Ras protein. *Biochemistry*, **37**, 15631–15637.

95. Taveras, A.G., Remiszewski, S.W., Doll, R.J., Cesarz, D. *et al.* (1997) Ras oncoprotein inhibitors: the discovery of potent, ras nucleotide exchange inhibitors and the structural determination of a drug-protein complex. *Bioorg. Med. Chem.*, **5**, 125–133.

96. Petros, A.M., Huth, J.R., Oost, T., Park, C.M. et al. (2010) Discovery of a potent and selective Bcl-2 inhibitor using SAR by NMR. *Bioorg. Med. Chem. Lett.*, **20**, 6587–6591.
97. Stamford, A., Strickland, C. (2013) Inhibitors of BACE for treating Alzheimer's disease: a fragment-based drug discovery story. *Curr. Opin. Chem. Biol.*, **17**, 320–328.
98. Roughley, S., Wright, L., Brough, P., Massey, A., Hubbard, R.E. (2012) Hsp90 inhibitors and drugs from fragment and virtual screening. *Top. Curr. Chem.*, **317**, 61–82.

25
Tuberculosis Drug Development

Kingsley N. Ukwaja
Federal Teaching Hospital, Department of Internal Medicine, FMC Road, 480241 Abakaliki, Ebonyi State, Nigeria

1	**Introduction** 901
2	**The Present** 903
2.1	Modes of Action and Activities of Currently Used Anti-TB Drugs 903
2.2	Existing Challenges Facing Tuberculosis Treatment 905
2.2.1	Unmet Needs of First-Line Anti-TB Drugs 905
2.2.2	Treatment of Drug-Resistant TB 906
2.2.3	Treatment of TB with Comorbidities 908
2.2.4	Treatment of Latent TB 909
2.2.5	Treatment of Childhood TB 909
3	**Anti-TB Drug Discovery** 910
3.1	Attributes of an Ideal Anti-TB Drug 910
3.2	Drug Discovery and Development Process 910
3.2.1	Target Selection, Identification, and Optimization of Lead Compounds 910
3.2.2	Early and Late Discovery, and Preclinical Development 912
3.2.3	Early Clinical Development and Clinical Evaluation 913
4	**Drugs in Development** 914
4.1	Novel Drug Candidates 914
4.1.1	Bedaquiline (TMC207) 914
4.1.2	SQ109 917
4.1.3	Benzothiazinones 917
4.1.4	Sudoterb (LL3858) 918
4.2	Re-dosing Current First-Line Anti-TB Drugs 918
4.2.1	High-Dose Rifamycins 918
4.2.2	High-Dose Isoniazid 919

Translational Medicine: Molecular Pharmacology and Drug Discovery
First Edition. Edited by Robert A. Meyers.
© 2018 Wiley-VCH Verlag GmbH & Co. KGaA. Published 2018 by Wiley-VCH Verlag GmbH & Co. KGaA.

4.3		Re-engineering of Antibacterials or Re-purposing Next-Generation' Chemical Relatives 920
4.3.1		Fluroquinolones (Moxifloxacin, Gatifloxacin, and DC-159a) 920
4.3.2		Nitroimidazoles (Delamanid (OPC-67683), Pretomanid (PA-824), and TBA-354) 920
4.3.3		Oxazolidinones (Linezolid, Sutezolid (PNU-100480), and AZD5847) 922
4.3.4		β-Lactams (Meropenem and Faropenem) 923
4.3.5		Riminophenazine (Clofazimine and TBI-166) 924
4.3.6		Cotrimoxazole (Trimethoprim–Sulfamethoxazole) 924
4.4		Drug Combinations/Regimens 927
5		Conclusions 927
		References 927

Keywords

Multidrug-resistant-TB (MDR-TB)
Tuberculosis (TB) caused by *Mycobacterium* sp. that is resistant to at least isoniazid and rifampicin treatment.

Extensively drug-resistant TB (XDR-TB)
Tuberculosis (TB) caused by *Mycobacterium* sp. that is resistant to rifampicin, isoniazid, plus any fluoroquinolone, and at least one of three injectable second-line drugs: amikacin, kanamycin, or capreomycin.

Drug-susceptible TB (DS-TB)
Tuberculosis (TB) that is caused by *Mycobacterium* sp. that is sensitive (responds) to first-line anti-TB drugs (rifampicin, isoniazid, pyrazinamide, and ethambutol) when administered for a period of 6–9 months.

Minimum inhibitory concentration (MIC)
The concentration of a drug or agent which will inhibit the growth of a microorganism.

Minimum bactericidal concentration (MBC)
The concentration of a drug or agent which kills 99% of a microorganism in culture.

Repurposed drugs
Drugs which have been developed and marketed to treat one clinical condition but are used to treat another condition.

Phenotypic screen
A drug screen that results in an altered phenotype of an organism or cell, for example, growth inhibition or death.

Whole-cell high-throughput screening
Trial-and-error screening of large libraries of small molecules on cultured bacteria or infected cells, using robotics.

Granuloma
A focal, compact collection of inflammatory cells in which various types of immune cells surround an undegradable product (pathogen, foreign bodies, or hypersensitivity reaction).

Cidal power
The ability or power of the drug/agent to kill a microorganism.

Anti-tuberculosis drugs currently in use have limited efficacy against drug-resistant tuberculosis (TB), have major adverse effects and have to be given as a combination of four drugs for at least 6 months or six drugs for up to 24 months for drug-susceptible and drug-resistant TB, respectively. These factors has led to increased patient default from treatment; which in turn has resulted in the emergence of additional resistance in a downward spiral that have generated forms of TB presently untreatable with existing drugs. There is an urgent need for efficacious and less toxic drugs that allow combinations with antiretroviral therapy and anti-diabetic agents, and can be used effectively for children with TB. This review covers the current TB treatment strategies, existing challenges of treatment, current concepts, and recent advances in TB drug discovery and development.

1
Introduction

Tuberculosis (TB) is caused by one of several mycobacterial species collectively called the *Mycobacterium tuberculosis* complex. Human TB is caused by *Mycobacterium tuberculosis* itself, *Mycobacterium africanum*, *Mycobacterium bovis*, *Mycobacterium caprae*, *Mycobacterium microti*, *Mycobacterium pinnipedii*, and *Mycobacterium canettii* [1, 2]. Pulmonary tuberculosis is transmitted through aerosol droplets from persons with active TB [3]. An estimated one-third of the world's population has latent TB infection – meaning that viable but dormant mycobacteria are contained in the granulomas of such persons [1]. About 10% of individuals with latent TB infection develop active TB disease at some point during their lifetime; the greatest risk of this is during the first two years after initial exposure [4]. However, immunocompromised individuals (malnourished, diabetes mellitus, steroid-treated, etc.) have a higher risk of TB reactivation, including up to 10% increased annual risk and approximately 50% over a lifetime for people with the human immunodeficiency virus (HIV) coinfection [5, 6].

Since the declaration of TB as a global public health emergency over two decades ago, several concerted international efforts have been deployed in improving TB control [7]. This has resulted in the global TB mortality rate falling by an estimated 45% between 1990 and 2013, and the

TB prevalence rate falling by 41% during the same period [8]. According to the 2014 Global Tuberculosis report by the World Health Organization (WHO) [8], the estimated number of incident TB cases globally in 2013 was nine million, including three million, 550 000, and 1.1 million new cases in women, children, and people living with HIV, respectively [8]. Most of these cases occurred in Asia (56%) and the African region (29%) [8]. The 22 high-burden countries accounted for 82% of all estimated cases globally. Furthermore, multidrug-resistant TB (MDR-TB) has emerged as another global challenge. An estimated 3.5% of new and 20.5% of previously treated TB cases had MDR-TB in 2013. In the same year, an estimated 480 000 new cases of MDR-TB occurred worldwide [8]. Extensively drug-resistant TB (XDR-TB) has been reported by 100 countries. On average, an estimated 9.0% of people with MDR-TB have XDR-TB.

The diagnosis of drug-resistant TB (DR-TB) requires TB patients to be tested for susceptibility to anti-TB drugs, either by conventional drug-susceptibility testing (DST) or rapid molecular diagnostic methods [9]. The Global Plan to Stop TB 2011–2015 calls for 20% of all new bacteriologically confirmed TB cases (i.e., considered be at high risk for MDR-TB), as well as all previously treated cases, to undergo DST to first-line TB drugs [10]. According to the WHO recommendations, all patients with MDR-TB should undergo DST to fluoroquinolones and second-line injectable agents to determine if they have XDR-TB. The current challenge is that many countries, especially in low-income settings, lack the capacity to diagnose DR-TB. Nevertheless, 136 412 cases of MDR-TB or rifampicin (RMP)-resistant TB were notified to WHO in 2013 globally [8].

Of this number, only 96 617 were enrolled for MDR-TB treatment because second-line anti-TB drugs were not available. Furthermore, the outcomes of treated DR-TB patients are disappointing. Overall, the proportion of MDR-TB patients in the 2011 cohort who successfully completed treatment (i.e., cured or treatment completed) was 48% (median: 59.5%). More so, among 1269 XDR-TB patients in 40 countries for whom outcomes were reported in the 2014 WHO TB report, 284 (22%) completed treatment successfully and 438 (35%) died; 126 (10%) failed treatment and 421 (33%) were lost to follow-up or their treatment outcome was not evaluated. Considering the apparent failure of the current anti-TB drugs in treating DR-TB, and the recent emergence and spread of TB cases that are resistant to all first- and second-line anti-TB drugs – inappropriately called totally drug-resistant TB (TDR-TB) – new anti-TB drugs and more effective regimens are urgently needed to improve the outcomes for patients with DR-TB.

During the past decade, substantial investments by research groups, international development agencies and the WHO Stop TB department has led to a new surge of activity in anti-TB drug discovery and the evaluation of alternative TB treatment regimens [11–13]. The result of these efforts culminated in the emergence and registration of new anti-TB drugs about 40 years since the last anti-TB drugs were approved. Indeed, recently, two new drugs – delamanid (previously known as OPC67683) and bedaquiline (also known as TMC207 or R207910) – were approved. In addition, several new drug candidates with novel modes of action are at the late stages of clinical evaluation (e.g., SQ109 and sutezolid). In this chapter, current TB treatment strategies, existing challenges of treatment, current concepts, and recent advances in TB drug discovery, and development are reviewed.

2
The Present

2.1
Modes of Action and Activities of Currently Used Anti-TB Drugs

Current treatment for TB relies largely on mycobacteria-specific drugs that are mainly inhibitors of the cell wall superpolymer. Based on their evidence of efficacy, potency, drug class, and clinical experience with use, existing anti-TB drugs are now divided into five groups (Table 1) [9]. The first-line anti-TB drugs (Group 1) are the most effective, consisting of RMP, isoniazid (INH), ethambutol (EMB), and pyrazinamide (PZA), and are currently administered as a combination of four drugs to treat drug-susceptible tuberculosis (DS-TB). Second-line drugs which are less effective and more toxic (Groups 2, 3, and 4) are reserved for DR-TB [14]. Group 2 agents include injectable aminoglycosides (streptomycin, kanamycin, and amikacin) and cyclic polypeptides (capreomycin, viomycin) [13–15]. Group 3 agents include mainly flouroquinolones such as ofloxacin, moxifloxacin, and gatifloxacin. Higher doses of older flouroquinolones (namely, ofloxacin and laevofloxacin) have also been safely used [16]. Group 4 agents includes mainly bacteriostatic, less efficacious, expensive, and more toxic oral agents, such as ethionamide, prothionamide, D-cycloserine, terizidone, and *para*-amino salicylic acid, that are used mainly for the treatment of MDR-TB and XDR-TB based on the susceptibility of *M. tuberculosis* strains (Table 1) [11, 14]. Third-line anti-TB drugs (Group 5) have unclear efficacy or undefined roles – they include linezolid, amoxicillin–clavulanate, imipenem plus cilastatin, meropenem–clavulanate, clofazimine, clarithromycin, and thiacetazone. Recently, efficacy has been demonstrated for some of the agents such as linezolid and clofazimine [17, 18]. The third-line agents have been used occasionally for the treatment of MDR-TB and XDR-TB, but are not recommended for use under TB control program conditions due to variable efficacy and serious side effects [14].

The modes of action and activities of currently used anti-TB agents are detailed in Table 1. A brief explanation is given for the first-line agents. INH enters the cell as a prodrug, and undergoes oxidative activation by the mycobacterial catalase-peroxidase *kat*G. This converts the drug into to an isonicotinyl-NAD adduct which inhibits *inh*A, a key bacterial enzyme in the FAS II (fatty acid biosynthesis) production of the cell-wall mycolic acid [15]. INH resistance occurs mainly by mutation in *kat*G, but less often in *inh*A, *ahp*C, and *ndh* [19, 20]. Although INH is highly bactericidal against replicating bacteria, with a minimum inhibitory concentration (MIC) of $0.05\,\mu g\,ml^{-1}$ and a high therapeutic margin, it has much slower action against Mycobacteria persisters [15]. Rifamycins, such as RMP, rifabutin (RBT) and rifapentine (RPT), are synthesized either naturally by the bacterium *Amycolatopsis rifamycinica* or artificially [15]. Rifamycins inhibit DNA-dependent RNA polymerase, blocking the transcription of proteins. It is highly bactericidal against *M. tuberculosis* throughout treatment, but it has a low therapeutic margin at very high doses [15]. Rifamycins show excellent sterilizing activity *in vivo* against semi-dormant *M. tuberculosis*, probably due to its rapid onset of action. Patient

Tab. 1 Description and activity of drugs currently in use for the treatment of tuberculosis.

Group	Chemical description	Target	Action
Group 1: First-line oral drugs			
Isoniazid	Nicotinic acid hydrazide	Enoyl-(acyl-carrier-protein) reductase	Inhibits mycolic acid synthesis
Rifampicin	Rifamycin derivatives	RNA polymerase, β-subunit	Inhibits transcription
Pyrazinamide	Pranzoic acid	30S ribosomal subunit, cytoplasm	Inhibits trans-translation, acidifies cytoplasm
Ethambutol	Ethylene diimino di-1-butanol	Arabinosyl transferases	Inhibits arabinogalactan synthesis
Group 2: Second-line, injectables			
Streptomycin	Aminoglycoside	30S ribosomal subunit	Inhibits protein synthesis
Kanamycin	Aminoglycoside	30S ribosomal subunit	Inhibits protein synthesis
Amikacin	Aminoglycoside	30S ribosomal subunit	Inhibits protein synthesis
Capreomycin	Cyclic polypeptide	50S ribosomal subunit	Inhibits protein synthesis
Group 3: Second-line, fluoroquinolones			
Ofloxacin	Fluoroquinolone	DNA gyrase and DNA topoisomerase	Inhibits DNA supercoiling
Moxifloxacin	8-Methoxy-fluroquinolone	DNA gyrase and DNA topoisomerase	Inhibits DNA supercoiling
Gatifloxacin	8-Methoxy-fluroquinolone	DNA gyrase and DNA topoisomerase	Inhibits DNA supercoiling
Group 4: Second-line, oral agents			
Ethionamide	Isoconitic acid derivative	Enoyl-(acyl-carrier-protein) reductase	Inhibits mycolic acid synthesis
Para-amino salicylic acid	*Para*-amino salicylic acid	Dihydropteroate synthase	Inhibits folate biosynthesis
D-Cycloserine	Serine derivative	D-Alanine racemase and ligase	Inhibits peptidoglycan synthesis
Group 5: Third-line, oral agents			
Linezolid	Oxazolidinone	50S ribosomal subunit	Inhibits protein synthesis
Meropenem-clavulanate	Carbapenem with β-lactamase inhibitor	β-lactam	Inhibits cell wall biosynthesis
Clofazimine	Iminophenazine derivative	Mycobacterial DNA	Inhibits replication
Clarithromycin	Macrolide	Translocation and ribosomal translation	Inhibits protein synthesis

isolates that are resistant to rifamycins almost invariably have mutations within the beta-subunit of the RNA polymerase gene (*rpoB*) [15]. RMP resistance develops from single amino acid substitutions in the beta subunit of RNA polymerase [19]. This modifies the binding of RMP, with the degree of resistance dependent on the site and nature of the amino acid substitution. The mutations leading to this resistance occur at a rate of about one in 10^8 [19, 20]. Using higher doses of RMP monotherapy do not prevent the emergence of resistance [19, 20]. PZA is also a prodrug that requires activation by a hydrolytic pyrazinamidase – a product of the *pnc*A gene – that converts it to pyrazinoic acid, its active form. Mutations in the *pnc*A gene are associated with PZA resistance. Its mechanism of action is not clear. Some report suggests that PZA blocks fatty acid synthesis, by inhibiting fatty acid synthase I (FAS-I) [21], while others suggest that the pyrazinoic acid produced reaches the exterior of the bacilli, where it is reabsorbed by passive diffusion in a highly pH-dependent process. Once inside the bacillus, the pyrazinoic acid can only be excreted by an inefficient efflux pump that requires energy. Failure of this pump results in pyrazinoic acid accumulation within the bacillus, acidifying the interior, and probably causing cell death by membrane damage or by inhibiting trans-translation in persisting cells [15, 22]. Due to this mode of action, PZA is very effective in killing dormant or near-dormant *Mycobacterium* populations which may be tolerant to other anti-TB agents, such as INH [23]. EMB interferes with the construction of the arabinogalactan layer of the mycobacterial cell wall by inhibiting arabinosyl transferases [13, 15, 24]. Arabinotransferase is involved in the polymerization of arabinofuranose, needed for the production of arabinogalactan, a structural component of the mycobacterial cell wall [25]. EMB also plays a role in inhibiting the polymerization of arabinofuranose into lipoarabinomannan (LAM), but this occurs more slowly, requiring a longer exposure to the drug [26]. Under clinical conditions, EMB is probably bacteriostatic [26]. Mutations in the *embB* region, specifically codon 306, appear to be the most common source of EMB resistance [27].

2.2 Existing Challenges Facing Tuberculosis Treatment

There are several problems with anti-TB drugs currently in use which require elucidation to highlight the urgent need for new anti-TB agents.

2.2.1 Unmet Needs of First-Line Anti-TB Drugs

The WHO treatment recommendations for newly diagnosed DS-TB require the use of a multidrug regimen for at least six months. The goal of this long duration of treatment is to cure the TB patient, reduce relapses, improve quality of life, decrease risk of death, break the chain of transmission, and prevent the emergence of DR-TB [9, 10]. The treatment regimen consists of RMP, INH, PZA, EMB for a period of two months (intensive phase), followed by only RMP and INH for an additional four months (continuation phase). If this regimen is administered based on the directly observed treatment short-course (DOTS) strategy, it has the potential of achieving a cure rate of 95% or more [28, 29]. The *M. tuberculosis* bacillus exists in varying spectrum of replication states; at one end of the spectrum are biologically active rapidly replicating bacilli, and at the other end are the nonreplicating/dormant bacilli, aptly called persisters. INH has a very high early bactericidal activity (EBA), eradicating most of the rapidly replicating bacilli within the first two weeks of therapy. Most of the benefit of the current treatment regimen for DS-TB comes during the first two months, during which the four anti-TB drugs (INH, RMP, PZA, and EMB) are given together in the intensive/bactericidal phase. During this period, active replicators are rapidly killed and eliminated, the mycobacterial bacillary load is greatly reduced, and patients become noninfectious [11–13]. Thereafter, the four-month continuation phase is required to eliminate the relatively dormant persister *M. tuberculosis* bacilli – decreasing the risk of relapse from ≥30% to <5% [30].

Furthermore, although the current treatment regimens for DS-TB have been shown to be very effective when the patients' adherence is optimal under closely supervised clinical trial conditions [12, 31, 32], treatment outcome is variable under tuberculosis program conditions, especially in resource-limited settings [33]. One of the major challenges faced by the current first-line treatment regimen is the long treatment duration of at least six months. When administered under routine TB control program conditions, this regimen is associated with a high default and death rate (about 10% each) [34–36], treatment failure, and the emergence of drug-resistant strains of *M. tuberculosis* [37, 38]. Therefore, there is a priority need for the development of new anti-TB drugs that will shorten the duration and frequency of therapy [12, 34]. In addition, because *M. tuberculosis* persisters remains until late in treatment, incomplete treatment is common, and this has been proposed to be the key to the rise in DR-TB cases [12, 37, 38]. A theory has suggested that the persister *M. tuberculosis* bacilli reside in granulomas. The granuloma microenvironment imposes several stressful conditions on the resident bacilli, including acidity, nitric oxide, hypoxia, and nutrient deprivation [12, 39–41]. Moreover, these varied conditions form the basis for the heterogeneity in the persister *M. tuberculosis* bacilli population, including their different sensitivities to anti-TB drugs. Indeed, some of the persister bacilli in patients may or may not have acid-fast staining, may adopt different morphologies, and may or may not grow in routine culture media – requiring extended incubation for even 6–9 months – or will grow rapidly only with resuscitation-promoting factors (Rpfs) [42, 43]. Populations of persister *M. tuberculosis* bacilli have been observed to survive for up to 100 days from the commencement of anti-TB treatment – during which time they are tolerant to many anti-TB drugs – resulting in the need for a long treatment duration [12, 44]. Thus, new drugs and regimens that target and eliminate these persisters early in treatment – irrespective of their developmental stage or location – are urgently needed [12, 42].

2.2.2 Treatment of Drug-Resistant TB

New anti-TB drugs are needed to control the rising burden of DR-TB. Several challenges are encountered in the current approaches used for the management of DR-TB. In ideal conditions, the treatment of DR-TB requires "individualized" regimens based on results of *in-vitro* DST for each patient's specimen isolates [37]. In settings where culture facilities for *M. tuberculosis* are available, culture-based systems for first-line DST do not provide results for several weeks, while for second-line DST the results are frequently not available for several months [11]. Therefore, the decision on which empirical TB regimen is to be used for DR-TB is determined by the local pattern of TB drug resistance, clinical experience with anti-TB drugs used previously for patients, comorbid medical conditions, and the adverse effects associated with the drugs [11]. As DST-based treatment is not available for over 95% of patients with MDR-TB, most cases are either missed or treated empirically on suspicion of active TB. Patient groups for which empirical treatment for MDR-TB is considered, and should be given, include newly diagnosed TB patients treated with first-line anti-TB drugs who showed treatment failure (i.e., who remained smear/culture-positive after at least four months of treatment with first-line agents), previously treated TB cases, symptomatic contacts of known DR-TB cases, and persons with TB who were born in countries, or reside in settings, with a high burden of DR-TB [11].

Presently, empirical MDR-TB treatment requires a combination of four to seven drugs taken for a period of 18–24 months or longer: only four of these drugs were originally developed for the treatment of TB [45]. These four drugs are the Group 4 anti-TB agents. In addition, cohort studies have shown that the addition of fluoroquinolone and injectable agents to these drugs gave better outcomes for MDR-TB patients. However, the evidence for this combination is low because there are no randomized clinical trials for this combination and duration as opposed to the four-drug combination used for DS-TB [46]. A combination of factors related to this suboptimal therapy, the use of both oral and intravenous routes of drug administration and long durations of therapy (6–8 months for injectable agents and 18–24 months for oral combinations), led to very poor treatment outcomes for DR-TB patients. In the 2014 WHO report, the treatment outcomes of the 2011 cohort of MDR-TB patients showed that 52% of them had unsuccessful outcomes. Treatment options for XDR-TB are far more limited and their outcomes worse-off, with 78% having unsuccessful outcomes [8]. Moreover, there are serious adverse effects with most MDR-TB and XDR-TB drugs, including nephrotoxicity and ototoxicity with aminoglycosides, hepatotoxicity with ethionamide, dysglycemia with gatifloxacin, and psychiatric disorders associated mainly with cycloserine and terizidone [46, 47]. Thus, the current situation suggests that containing the spread of MDR-TB and XDR-TB will be very difficult without treatment regimens that are shorter, safer, more effective, and less expensive than those that are available. In addition, there is a need to conduct clinical trials of treatment regimens used in treating MDR-TB or XDR-TB patient groups, as historically only very few clinical studies have been performed to evaluate the efficacy of drugs in MDR-TB or XDR-TB patient groups [48, 49]. A "9-month Bangladeshi regimen" consisting of a four-month intensive phase of treatment with seven drugs (gatifloxacin, kanamycin, prothionamide, clofazimine, EMB, PZA, and INH), with a five-month continuation phase regimen comprising gatifloxacin, clozimine, EMB, PZA, is currently the most up-to-date regimen accepted for the treatment of MDR-TB [50, 51]. However, the basis of this nine-month regimen was a cohort study conducted in Bangladesh and pilot-tested in Niger, which gave impressive treatment outcomes when used for MDR-TB patients [50, 51]. The Standardized Treatment Regimen of Anti-tuberculosis drugs for patients with MDR-TB known as the "STREAM" Trial will most likely provide quality evidence in support of this regimen in the next few years [52]. The improvement of treatment systems should also go hand-in-hand with improvements in diagnostics with wider coverage of DST; this will lead to early diagnosis, address the high mortality of MDR-TB/XDR-TB, and curb the emergence and spread of complex resistance [46]. New diagnostics, especially the MTB-RIF GeneXpert and its rapid roll-out to high-burden countries, may help in the detection – and hence treatment – of more cases of DR-TB.

2.2.3 Treatment of TB with Comorbidities

TB is the commonest opportunistic infection and the most important cause of morbidity and mortality in HIV-infected individuals [53]. Also, HIV is the greatest predictor for the progression of latent TB infection to TB disease, thus fuelling the TB epidemic in many countries, especially those in sub-Saharan Africa [8, 53, 54]. TB accounts for about one-quarter of deaths occurring among all HIV-infected persons, and approximately 80% of these coinfections occur in the African region [8]. The treatment of TB/HIV coinfection poses a major challenge, especially in health systems in developing countries that are not equipped to provide the specialized care required by individuals with coinfection due to their taking large quantities of medications for TB and HIV which may in turn affect compliance, immune reconstitution inflammatory syndrome, and drug–drug interactions leading to subtherapeutic concentrations of antiretrovirals, while overlapping toxic side effects increase safety concerns [11, 55]. Another major concern is the patients' readiness to maintain anti-retroviral therapy (ART) and anti-TB treatment once started [54]. On the other hand, delays in initiating treatment (anti-TB or ART) due to these concerns may result in high mortality rates for TB/HIV coinfected patients. The main drug–drug interactions between anti-TB and ART agents occur between rifamycins (RMP, RBT, and RPT) and protease inhibitors (ritonavir, indinavir, saquinavir, etc.), as well as non-nucleoside reverse transcriptase inhibitors (nevirapine) [11, 55]. RMP, which is a very potent anti-TB drug, is also a potent inducer of the hepatic cytochrome P450 3A (CYP3A) system, and reduces serum concentrations of protease inhibitors and non-nucleoside reverse transcriptase inhibitor drugs – in some cases to subtherapeutic doses [55]. Although the Starting Antiretroviral Therapy at Three Points in Tuberculosis (SAPiT) trial has shown that a once-daily ART regimen consisting of didanosine, lamivudine and efavirenz could be optimally integrated with first-line anti-TB agents and started based on their CD4+ count either early (within four weeks) or late (after four weeks) in the TB treatment, there is need for other ART and anti-TB agents that could be optimally combined for DS-TB/HIV treatment [56, 57]. Another major challenge lies with ART and anti-TB agents that could be optimally combined for the treatment of MDR-TB and HIV where mortality is high. Thus, there is a clear need for newer anti-TB agents that could optimally be combined with ART agents for TB/HIV patients.

Diabetes is another chronic disease that has been found to cause a threefold increase in the risk of active TB [55, 58]. This has been suggested to be due to a reduction in cell-mediated immunity, the dysfunction of alveolar macrophages, low levels of interferon gamma, pulmonary microangiopathy, and micronutrient deficiency in diabetic persons, and explains the higher rates of TB among diabetics [58, 59]. The biological rationale for the slower response of diabetics to anti-TB drugs and for their increased risk of developing MDR-TB is poorly understood [55]. Furthermore, the time to bacterial culture negativity, reduced anti-TB drug plasma concentrations, relapse rates, and mortality are all substantially higher in diabetic TB patients than non-diabetics [50–61]. Thus, there is a need to identify new anti-TB agents that are strongly bactericidal and have minimal drug–drug interactions with oral antidiabetic drugs [62].

2.2.4 Treatment of Latent TB

New drugs are needed for the treatment of latent TB infection. Of about two billion persons with latent tuberculosis infection (LTBI) globally, it is estimated that between 100 and 200 million will develop active disease during their lifetime [8]. Thus, individuals with LTBI are the reservoirs of TB in the community, and the hope of eradicating TB could only be achieved if LTBI were to be treated and eradicated. The present rationale of treatment of LTBI is to prevent the progression to active disease in persons at risk (contacts of TB patients and persons living with HIV) [46]. A meta-analysis showed that INH therapy for 6–9 months in HIV-infected persons decreased the risk of active TB by 60% [63]. Therefore, INH preventive therapy has now been recommended for the treatment of LTBI [64]. Also, INH alone or combined with RMP and taken for three or four months has been shown as an effective therapy for LTBI [65, 66]. However, these successes in the treatment of LTBI were recorded mainly in regions with low-burden TB, whereas in high TB/HIV settings the outcomes of treatment of patients with LTBI have largely been inconclusive and the benefits doubtful [67–69]. As the improved diagnosis and treatment of LTBI have been suggested as the way forward for the elimination of TB, there is clearly an urgent need to identify new therapeutic agents for LTBI, especially as there are limited methods of diagnosis and treatment for people who have latent infection with drug-resistant strains of *M. tuberculosis* [69].

2.2.5 Treatment of Childhood TB

Children constitute about 5% of incident cases of TB in low-burden settings, compared with 20–40% in high-burden countries [70, 71]. No clear data are currently available on the burden of MDR-TB in children, though Jenkins *et al.* estimate it to be at double the number of children with TB per year (33 000 incident DR-TB cases per year) [72]. Pediatric TB differs in that there is a rapid progression from infection with *M. tuberculosis* to disease, and the children are more likely to present with extrapulmonary manifestations (lymphadenopathy, central nervous system involvement, etc.) [71]. Treatment doses for pediatric TB are usually based on data from adult pharmacokinetic studies, and recent studies have pointed to the inadequacy of currently recommended doses of first-line anti-TB drugs. Indeed, it has been determined that children require a higher body weight-related dose ($mg\,kg^{-1}$) of these drugs to achieve serum concentrations comparable to those in adults [71, 73, 74], and this has resulted in a recent update on treatment recommendations for pediatric TB [75]. However, uncertainties remain as to how this new treatment guideline will affect the outcomes of pediatric TB, and less information is available to guide the use of second-line-drugs for pediatric cases. Data are also scant on the adverse effects of current first-line or second-line anti-TB agents in children. This underscores the need to develop new pediatric drug formulations to suit high-burden settings, and specific studies for investigating their appropriate dosing and safety [46]. An ongoing trial with delamanid is aimed at evaluating the dose in children, using a pediatric drug formulation.

3 Anti-TB Drug Discovery

3.1 Attributes of an Ideal Anti-TB Drug

Given the unmet needs of present TB treatment, what should be the important attributes of an ideal anti-TB drug? An ideal anti-TB drug should inhibit new targets and be more potent than existing drugs, have good bioavailability, have good safety and tolerability profile, be given orally, show no antagonism to other TB drugs or drug candidates so that the pill burden can be reduced by constituting a regimen comprising at least three active drugs, have a lower dosing frequency (e.g., once-weekly regimen), have a low incidence of adverse effects, have minimal drug–comorbid disease interactions, be active against drug-resistant strains of TB, show minimal interactions and be compatible with concomitant ART, be active against the various replication and physiological states of *M. tuberculosis*, including those with drug-resistant latent TB infection, and should have low costs [11, 12, 46, 55, 76, 77]. Given that only very few drugs from the discovery stage successfully enter the TB clinical pipeline, a better understanding of the challenges in the drug discovery process should facilitate the development of novel intervention strategies.

3.2 Drug Discovery and Development Process

The development of new drugs involves several phases. According to the Scientific Blueprint for Tuberculosis Drug Development from the Global Alliance for TB Drug Development, these phases broadly include: target selection and identification; the identification and optimization of lead compounds; the development and application of drug screens; a late discovery phase; preclinical development; clinical trials; regulatory approval; and technology transfer [13]. Brief details on some of these phases are provided in the following subsections.

3.2.1 Target Selection, Identification, and Optimization of Lead Compounds

Ideally, antibacterial agents display activity by targeting essential functions of bacteria. Two main strategies have been employed in pinpointing such functions (Fig. 1) [78, 79]. The first strategy, referred to as the classic, whole-cell or phenotypic approach, aims to identify compounds that kill the mycobacteria following whole-cell, high-throughput screening of several compounds. The second strategy has been exploited substantially during the past decade when whole-genome sequences of *Mycobacterium* organisms became available [80–82], and targets gene products directly with the use of expressed proteins as templates to design new inhibitors. Known as the target-based approach, the rationale here is to develop inhibitors of essential and validated enzymes so that they will kill the mycobacteria through their specific action on the targeted enzyme. Therefore, TB drug discovery can follow two main routes: (i) a drug-to-target strategy; or (ii) a target-to-drug strategy. The drug-to-target approach using whole cells for the screening of candidate compounds has largely been used to derive all the current anti-TB drugs in use, and candidate compounds currently undergoing clinical trials. However, the target-to-drug strategies for anti-TB drug discovery have so far been unsuccessful. This is because, although a compound may show very high effectiveness in inhibiting a specific enzyme or biochemical function (as indicated by

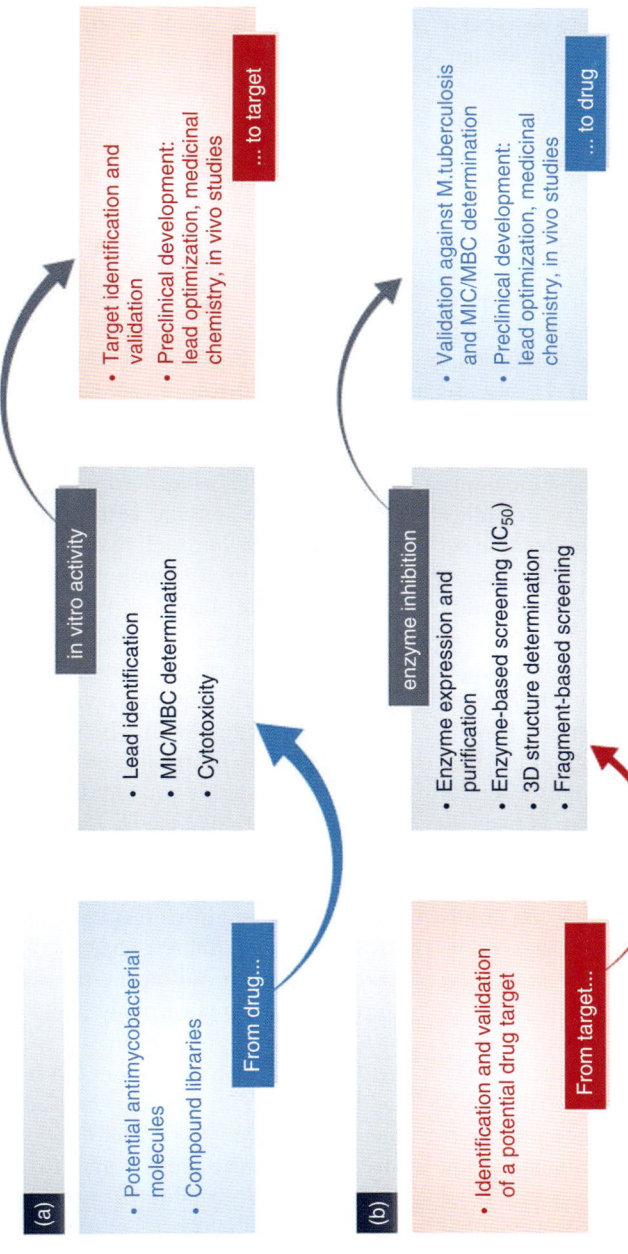

Fig. 1 Tuberculosis (TB) drug discovery: from drug-to-target (a) and from target-to-drug (b). The figure displays the two main distinct methodologies in the search for new anti-TB drugs. It is noteworthy that a target identified after a chemical screen (a) can enter the target-to-drug pipeline in enzyme-based screening (b). MIC, minimal inhibitory concentration; MBC, minimal bactericidal concentration; IC_{50}, half-maximal inhibitory concentration. Reproduced with permission from Ref. [78], under the terms of the Creative Commons Attribution License (CCBY).

the half-maximal inhibitory concentration, IC$_{50}$, the concentration of a drug/agent that is required for 50% inhibition *in vitro*), translating this ability into potency against whole-cell organisms such as mycobacteria (based on its MIC) has proved to be a major challenge against rational drug design. Thus, in the target-to-drug strategy, a lack of correlation between IC$_{50}$ (effectiveness of enzyme inhibition within the mycobacteria and MIC (effectiveness of inhibition of the whole mycobacterial organisms) of candidate compounds has so far limited its use for anti-TB drug discovery [55, 78, 83].

The ideal properties of future anti-TB drug targets within *Mycobacterium* sp. should include essentiality for growth (i.e., the target should be essential for growth and/or survival of mycobacteria under the various conditions encountered during *M. tuberculosis* infection in humans, including the latent/dormant phase); vulnerability (i.e., the target should be vulnerable to inhibition by a drug/candidate compound under these conditions); druggability (i.e., the target should have a family of drugs/compounds that could inhibit or be modified to inhibit its function); low mutability (an ideal target should have a reduced propensity to evolve and be drug resistant); and target location (a target should be located in such a way as to allow access/binding of the drugs or compounds). For example, most of the available anti-TB drugs and promising candidate compounds currently available have targets that are outside of the cytoplasm of the *Mycobacterium* sp., that is, extracellular targets [84, 85].

The process of lead compound identification has been greatly enhanced by the advent of combinatorial chemical approaches to generating compound diversity [78, 83, 84, 86]. Lead compounds are optimized through the synthesis of related substances, while maintaining the essential features of the original compound that conferred the inhibitory property [13].

3.2.2 Early and Late Discovery, and Preclinical Development

The optimized lead compounds are screened further by assessing their antimicrobial activity and cytotoxicity. The discovery of new antibiotic candidates is achieved by employing several types of screening strategy, including growth inhibition assays, intracellular growth inhibition assays, surrogate assays of growth inhibition, inhibition of target biochemical function binding to known structural targets, and the computational design of inhibitors to known targets [13]. Lead development involves the synthesis of between 200 and 300 compounds, many of which will be excluded due to poor efficacy during *in-vitro* testing. Further evaluation through *in-vivo* animal testing decreases the numbers further such that, ultimately, a shortlist of about 10 compounds is made. During the late discovery phase some of those compounds that made the shortlist are eliminated further, usually leaving only a single candidate and one compound as a back-up for further evaluation [13]. A compound with very promising anti-TB activity but with undesirable product features may undergo alterations in such a way that the desired product characteristics are present [13]. The robustness of the lead compound is also evaluated at this phase, by assessing formulation feasibility, pharmacokinetics (absorption, distribution, metabolism and excretion) and pharmacodynamics, toxicology (including mutagenicity), effects on fertility, safety (pharmacology) in humans, and patenting [13]. Once a compound has been selected for preclinical development, correctly planned animal and toxicology studies must be conducted in compliance with good laboratory practice (this is an absolute requirement for regulatory approval purposes) [13]. Animal studies for anti-TB drugs are used to evaluate *in vivo* the antimicrobial activity of a lead compound in comparison with that of existing drugs. Animal models are also used to test the compound's antagonistic, additive or synergistic effects when given in combination with other drugs, and its ability to sterilize *Mycobacterium*-mediated lesions in experimentally infected animals [13]. Due to its ease of handling in terms of size, supply, maintenance, robustness and reproducibility, the mouse is the animal model of choice for monitoring anti-TB drugs [13, 87, 88].

3.2.3 Early Clinical Development and Clinical Evaluation

All human clinical trials must conform to globally accepted standards of good clinical practice. The trials should be performed in countries with institution review boards and/or independent ethics committee, and are conducted in up to four phases (Phases I–IV). A Phase I trial is designed to assess the safety, tolerability and pharmacokinetics of a candidate drug in a small (20–100) group of healthy volunteers. The Early Bactericidal Activity (EBA) study is the next and very important early-phase clinical development screen for investigational antimycobacterial agents, where they are evaluated for microbiologic outcomes at early time points. Given that the current regimen for DS-TB could achieve up to 95% cure under clinical trial conditions, the evaluation of new drugs or combination regimens represents a challenge to clinical trials design due to the sample size required to demonstrate statistical significance between the current treatment and the new regimen group. This, combined with the two years of follow-up of such patients and the financial costs of these studies may not be attractive to the pharmaceutical industry. The EBA trial represents a faster and less-expensive method for the early evaluation of the activity of new anti-TB drugs. EBA trials evaluate the quantitative counts of viable tubercle bacilli from daily collections of sputum, and provide information on the bactericidal activity of single drugs or a regimen in clearing *M. tuberculosis* from the sputum of patients with newly diagnosed pulmonary TB [89]. Currently, the EBA trial is defined as the rate of fall of colony-forming units (cfus) during the first 2–14 days of treatment, and should be expressed as $\log_{10}$ cfu per day. The ability of a drug to reduce the viable count quickly is an important consideration, because this rapidly renders the patient noninfectious [90]. Thus, EBA trials are not intended to provide definitive treatment for patients, but rather to evaluate the antimycobacterial activity in a brief setting (7–14 days) [89]. Patients appropriate for inclusion in EBA trials would be immunocompetent, treatment-naïve adults at low risk of drug resistance or extrapulmonary disease, who can begin standard-of-care treatment for pulmonary TB on completion of the EBA trial [89]. It has been hypothesized that EBA trials may increase the risk of acquired resistance to the drugs studied by means of selecting naturally resistant bacilli found in the microbial load. However, in the results of EBA studies published over 30 years since the first such trials were performed, no cases have been reported of acquired resistance after use of the drug for a period from two days to two weeks [91, 92].

Following the EBA studies, Phase II trials are designed to assess the safety and efficacy of a candidate drug on small groups of TB patients (20–300) over a four- to eight-week period. Phase II trials are sometimes divided into Phase IIa and IIb (Phase IIa is specifically designed to assess dosing requirements; Phase IIb is specifically designed to study efficacy at the prescribed doses).

Phase III studies are randomized controlled, multicenter trials on large patient groups (300–3000 or more) that are aimed at providing the definitive assessment of a candidate drug's efficacy, in comparison with the "gold standard" treatment. The assessment of adverse events always must be included. Documenting, monitoring, and process evaluation during the study is essential to ensure good quality of the data.

Phase IV studies involves post-marketing surveillance of the drug (further details are beyond the scope of this review) [13].

A single adequate and well-controlled trial in patients with pulmonary TB, supported by other independent data (e.g., evidence of antimycobacterial activity from an EBA trial), can provide evidence of effectiveness when the single trial has demonstrated a clinically meaningful treatment effect [93]. In summary, TB drug discovery and development is a lengthy process, and with the failure of genetic target-based identification of lead compounds, the current strategy which employs whole-cell phenotypic screens have been aptly called the 3M-approach – that is, MIC/MBC (minimum bactericidal concentration) | Mouse | Man | [94]. The current 3M-approach remains the key strategy for the identification and development of lead compounds.

4
Drugs in Development

About five decades after the introduction of RMP for the treatment of TB, considerable efforts have been made in identifying and introducing new drugs that would improve TB treatment and control. In fact, a number of anti-TB drugs have entered the pipeline within the past five years, and at present there are at least 22 drugs at various stages of development for the treatment of TB (Table 2). These increases in the TB drugs portfolio are being achieved through: (i) the discovery of new novel drug candidates (TMC207, SQ109, LL3858); (ii) the re-optimization or re-dosing of known first-line anti-TB drugs, such as rifamycins (RMP, RPT); and (iii) the re-engineering of existing antibacterial compounds – that is, currently licensed drugs for other indications and "next-generation" compounds of the same chemical class being "repurposed" for TB (e.g., gatifloxacin and moxifloxacin; linezolid, PNU100480, and AZD5847; metronidazole, OPC-67683, PA-824). The mechanism of action of most drugs currently under development for TB treatments is shown in Fig. 2. Given that several excellent recent reviews of these drugs/agents are available (see Sect. 6) [11, 12, 14, 46, 55, 76, 95–102], only the salient features of these advanced development compounds are summarized below.

4.1
Novel Drug Candidates

4.1.1 Bedaquiline (TMC207)

Bedaquiline is a diarylquinoline discovered by Janssen, and is currently undergoing Phase II/III studies for both MDR-TB and DS-TB indications [78, 84, 85]. Its mechanism of action is unique, in that it inhibits the proton transfer chain by binding to subunit c (*atpE*) of ATP synthase, thus leading to intracellular ATP depletion [103]. Bedaquiline is highly potent *in vitro* against *M. tuberculosis*, having a MIC of 0.03–0.12 mg ml^{-1} [103–105]. The killing of actively dividing bacilli is initially slow and time-dependent, but is

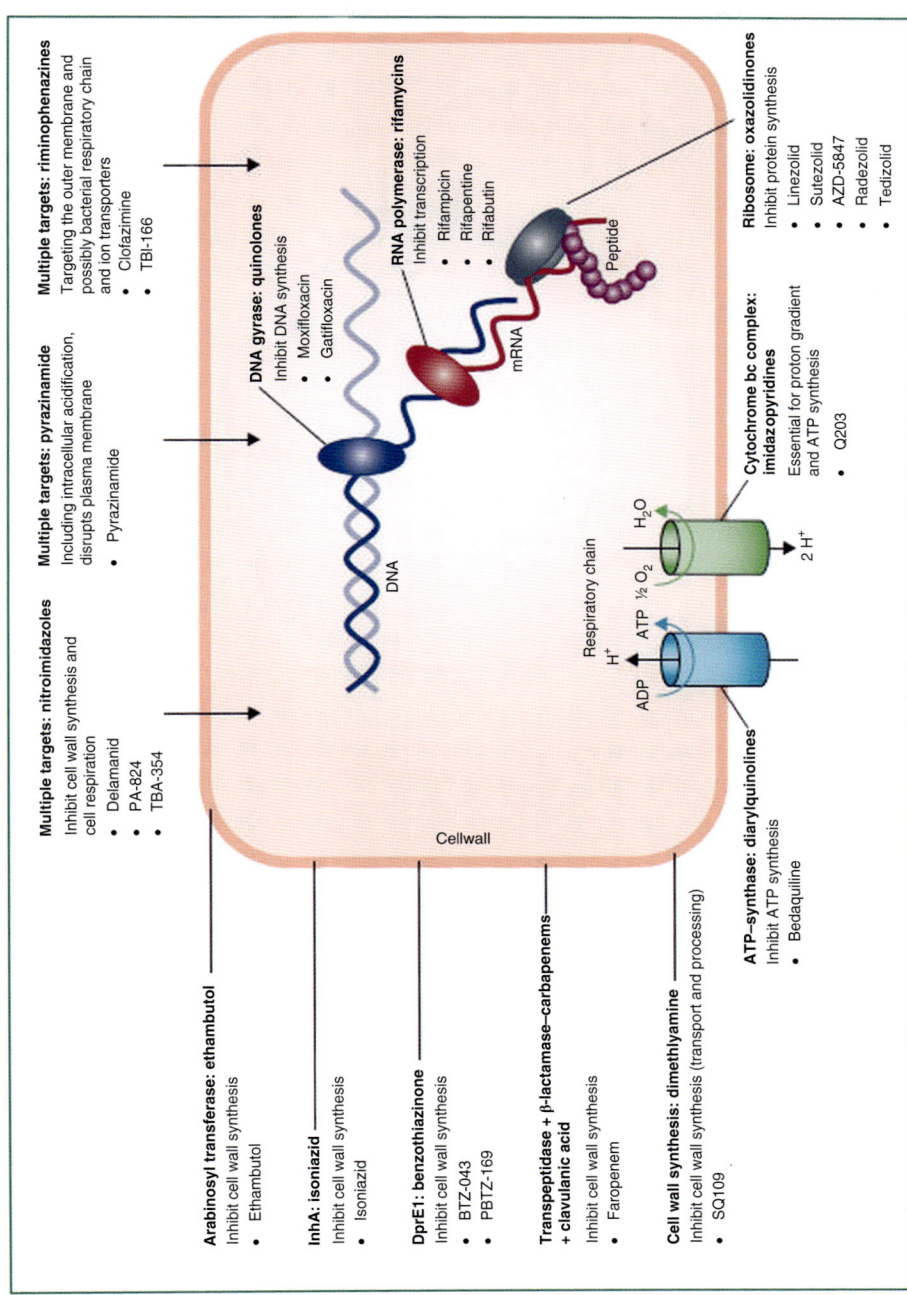

Fig. 2 Mechanisms of action of anti-tuberculosis drugs in preclinical development. Reproduced from Ref. [12]. with permission of Elsevier (License no. 3553430893684).

Tab. 2 Drugs at different clinical stages of development for the treatment of tuberculosis.

Drug	Category	Target	Stage of development
Bedaquiline (TMC207)	Novel	ATP synthase	Phase II/III
SQ109	Novel	Cell wall synthesis	Phase II
BTZ-043	Novel	Cell wall synthesis	Phase I
PBTZ-169	Novel	Cell wall synthesis	Preclinical
Sudoterb (LL3858)	Novel	Unknown	Phase IIa
High-dose rifampicin	Current anti-TB drug	RNA polymerase	Phase III
High-dose rifapentine	Current anti-TB drug	RNA polymerase	Phase III
High-dose isoniazid	Current anti-TB drug	Cell wall synthesis	Phase II
Moxifloxacin	Re-purposed	DNA gyrase	Phase III
Gatifloxacin	Re-purposed	DNA gyrase	Phase III
DC-159a	Re-purposed	DNA gyrase	Preclinical
Delamanid (OPC-67683)	Re-engineering	Multiple, including cell wall synthesis	Phase II/III
Pretomanid (PA-824)	Re-engineering	Multiple, including cell wall synthesis	Phase II/III
TBA-354	Re-engineering	Multiple, including cell wall synthesis	Preclinical
Linezolid	Repurposed	Protein synthesis initiation complex	Phase II
Sutezolid (PNU-100480)	Repurposed	Protein synthesis initiation complex	Phase II
AZD5847	Repurposed	Protein synthesis initiation complex	Phase II
Meroponem-clavulanate	Repurposed	Cell wall synthesis	Phase II
Faropenem	Repurposed	Cell wall synthesis	Phase I
Clofazimine	Repurposed	Protein synthesis	Phase I/II
TBI-166	Repurposed	Protein synthesis	Preclinical
Cotrimoxazole	Repurposed	Folate synthesis	Phase I

ATP, adenosine triphosphate.

more rapid and concentration-dependent for nonreplicating intracellular bacilli (persisters) because these dormant bacilli maintain a residual ATP synthase activity [105]. Thus, bedaquiline may shorten the treatment duration of drug-susceptible pulmonary TB and MDR-TB [106, 107]. Resistance develops through point mutations in the gene *atpE*, which encodes part of the ATP synthase, at predictable rates that are similar to those observed for RMP [102]. Mutations conferring bacillary resistance to bedaquiline do not cause cross-resistance to other anti-TB drugs, except for clofazimine, possibly due to the upregulation of a multi-substrate efflux pump [108], which has also been found in clinical isolates [109]. In the mouse model, bedaquiline has demonstrated synergy with PZA [110], and this has given birth to drug combination regimens containing bedaquiline and PZA that currently are in Phase IIa trials. In MDR-TB patients, the addition of bedaquiline to the treatment regimen reduced the time required to convert to sputum culture negativity, and also increased the proportion of patients with conversion of sputum culture at eight weeks (48% versus 9%) [111]. This, combined with other findings, led to the drug's conditional approval for treating adult MDR-TB by the US Food and Drug Administration (FDA) in December 2012, and by the European Medicines Agency (EMA) in March 2014 [112, 113]. The WHO has issued interim guidance on the use of bedaquiline for treating MDR-TB [114]. A study (ACTG 5343 trial) on the drug–drug interaction of bedaquiline and delamanid has been planned to start in the fourth quarter of 2015. However, this trial has yet to be started due to bureaucratic delays. Whether bedaquiline may shorten TB treatment, as has been suggested from animal studies, should be interpreted with caution and requires further investigation.

4.1.2 SQ109

SQ109 is a 1,2-ethylenediamine, identified by high-throughput screening of 63 238 synthetic analogs of EMB [101]. It is active *in vitro* against drug-sensitive and drug-resistant strains of *M. tuberculosis*, including EMB-resistant strains, which suggests a novel mechanism of action [101, 102]. Although its precise mechanism of action has not been elucidated, recent studies have shown that the primary target of this compound is the mycobacterial membrane protein large 3 (*MmpL3*) gene [115, 116]. Initially, investigators were unable to identify spontaneous SQ109-resistant mutants; however, by using ethylenediamine analogs resistant mutants were generated that showed cross-resistance to SQ109 and whole-genome sequencing revealed mutations in the *MmpL3* gene [78, 115]. This gene encodes a transporter belonging to the resistance, nodulation, and division (RND) family that is required for the export of trehalose monomycolate (TMM), an essential component of the mycobacteria cell wall [116], and this is compatible with the observation that TMM accumulates in cells treated with SQ109 [115]. Indeed, SQ109 inhibits mycolic acid biogenesis instead of the arabinogalactan synthesis inhibited by EMB [11, 116]. Interestingly, many other *MmpL3* inhibitors have been found which have varying activities against *Mycobacterium* sp. [11, 78, 115]. SQ109 has an MIC range of 0.11 to 0.64 μg ml^{-1} against *M. tuberculosis*, including MDR-TB strains resistant to EMB [117], and appears to be safe and well-tolerated in human studies [102]. SQ109 has been reported to have synergistic activity with INH and RMP *in vitro* and in animal models [118–121], and more recently with TMC207 *in vitro*, both in the presence and absence of RMP [109, 122]. Replacing EMB with SQ-109 improved the effectiveness of the standard first-line regimen containing INH and RMP, with or without PZA, in a murine model of TB [121]. There is a need to demonstrate similar additive and synergistic effects of SQ109 with other anti-TB agents in the human model, as this will aid in optimizing multidrug regimens for DS-TB and DR-TB [98]. Although SQ109 has completed Phase II studies, no Phase III clinical trials have been registered for this drug.

4.1.3 Benzothiazinones

Benzothiazinones (PBTZ-169 and BTZ-043) are new clinical entities that are in the late preclinical development stage [12]. Benzothiazinone [BTZ043; (2-[(2S)-2-methyl-1,4-dioxa-8-azaspiro[4.5]dec-8-yl]-8-nitro-6-(trifluoromethyl)-4H-1,3-benzothia-zin-4-one] is nearing Phase I clinical trials [123]. BTZ043 is one of the most potent inhibitors of *M. tuberculosis* yet described, displaying nanomolar bactericidal activity both *in vitro* and in *ex vivo* models of drug-susceptible and drug-resistant clinical isolates [124]. Benzothiazinones are nitroaromatic compounds, and target the essential flavoprotein subunit, DprE1, of decaprenylphosphoryl-beta-D-ribose 2-epimerase in *M. tuberculosis* [123, 124]. This enzyme produces the sole source of the D-arabinose required for biosynthesis of the key cell wall components arabinogalactan and lipoarabinomannan. Benzothiazinone serves as a "suicide" substrate, since blockade of this enzyme inhibits the synthesis of decaprenyl phosphoryl arabinose (a key precursor in the biosynthesis of the cell wall arabinans), which in turn results in cell lysis and bacterial death [12, 123–125]. The MIC of benzothiazinone against *M. tuberculosis* is $0.004\,\mu g\,ml^{-1}$ [124]. In murine models of acute and chronic TB, BTZ043 showed efficacy comparable to that of INH and RMP (though the latter compounds are far less potent with respect to their *in vitro* MIC [126, 127]) and does not appear to be mutagenic [124]. Due to its mode of action, BTZ043 is mainly active against dividing organisms, and in short-term mouse studies (28 days of monotherapy) it has activity similar to RMP [124]. In addition, BTZ043 showed additive effects, and in some cases synergistic activity against *M. tuberculosis* when used with other TB drugs or drug candidates (e.g., RMP, INH, EMB, bedaquiline, PA-824, moxifloxacin, meropenem with or without clavulanate, and SQ-109); and no antagonism was found [126, 127]. Three other drug discovery initiatives have also identified other DprE1 inhibitors with differing structures and activity against *Mycobacterium* sp., suggesting that this may be an important target for the development of a future viable anti-TB drug [12].

4.1.4 Sudoterb (LL3858)

Sudoterb (LL3858) is a pyrrole derivative that has been evaluated by Lupin Limited, India. It has activity against *M. tuberculosis* but its metabolic target is unknown [98, 101, 128–131]. It has an MIC range of $0.06-0.25\,\mu g\,ml^{-1}$ against DS-TB and DR-TB strains [128, 129]. In the mouse model, 12-week treatment with Sudoterb at $12.5\,mg\,kg^{-1}$ provided good efficacy and complete clearance from the lungs and spleen, and no relapses were observed for up to two months after the final dose [128]. Sudoterb demonstrates *in vitro* synergy with RMP, and was active against *M. tuberculosis* by itself; it also enhanced the effects of standard therapy with INH, RMP, and PZA [130]. Although, recently, Lupin Limited listed this project on the Stop TB Working Group on New Drugs webpage as no longer being aimed towards drug registration, oral sudoterb (400 mg per day for five days) has been evaluated in a single-arm, open-label EBA trial of newly diagnosed, sputum smear-positive, pulmonary TB patients [132].

4.2 Re-dosing Current First-Line Anti-TB Drugs

4.2.1 High-Dose Rifamycins

Rifamycin (e.g., RMP) is a key first-line anti-TB agent due to its sterilizing ability. As described in Sect. 2.1, these agents inhibit the mycobacterial RNA polymerase. Recent reviews of previous trials have suggested that the standard doses of RMP presently in use were introduced for safety and cost reasons, and were not at the limit of the dose–response curve [133, 134]. Higher than standard RMP dosing also led to improved bactericidal activity in both animal and human models, resulting in improved culture conversion rates and a reduction in treatment duration [134–137]. A number of Phase II trials have been conducted to confirm these findings, but with rather disappointing results [138–140]. The Rapid Evaluation of High Dose Rifampicin and other Rifamycins in Tuberculosis (HIGHRIF) 1 study, which compared increasing RMP doses against the standard dose (10 mg kg^{-1}) for safety and bactericidal activity over two weeks, showed that up to 35 mg kg^{-1} of RMP is safe, well-tolerated, and showed greater bactericidal activity with higher doses [12, 138]. In addition, although the preliminary findings of the HIGHRIF 2 study found no serious adverse events for two months' treatment with RMP at 15 and 20 mg kg^{-1}, it showed only a modest, non-significant difference in efficacy compared to standard doses [139]. Similarly, the Evaluation of High-Dose Rifampicin Toxicity in Pulmonary Tuberculosis (RIFATOX) trial showed that doses of RMP at 15 and 20 mg kg^{-1} for the first four months of the standard six-month regimen was safe, but these higher doses did not lead to any significant increase in negative sputum smear conversion after eight weeks of treatment [140]. In the recently concluded MAMS-TB-01 study, different doses of RMP (up to 35 mg kg^{-1}) were evaluated in a multi-arm, multi-stage trial involving standard first-line and combinations of newer anti-TB drugs [12, 139]. The study results showed: (i) that increasing the dose of rifampicin increases the average and minimum exposures effectively; (ii) there were no apparent effects of high-dose rifampicin on exposure to isoniazid, pyrazinamide and ethambutol; (iii) high-dose (35 mg kg^{-1}) rifampicin + concomitant drugs was reasonably well tolerated for three months in African TB patients; and (iv) high-dose rifampicin (35 mg kg^{-1}) plus 20 mg kg^{-1} moxifloxacin resulted in an increased likelihood of, and shorter time to, culture conversion in liquid media [139].

RPT, with its longer half-life and greater potency than rifamycin, is being investigated in trials aimed at reducing treatment duration [141]. Data from animal models have suggested that a RPT–moxifloxacin combination, given daily or intermittently, may shorten therapy for DS-TB from six to four months [142, 143]. However, this effect was not observed in the Phase III Rifapentine and a Quinolone in the Treatment of Pulmonary Tuberculosis (RIFAQUIN) trial [144]. Similarly, in a head-to-head comparison of RPT and RMP at 10 mg kg^{-1} in humans, no difference was noted in their negative sputum culture conversion at two months [145], in contrast to what had been observed in the mouse model [146]. Subsequently, two trials (ACTG A5311; RioMAR) to evaluate higher doses/exposures of RPT were stopped early due to higher toxicity, or withdrawal/loss to follow-up in the experimental (RPT) arm [147, 148]. Dose-escalation studies for rifamycins may be limited by drug–drug interactions that occur due to the induction of cytochrome P450, which may in turn affect the doses of other medications, especially antiretrovirals and newer anti-TB agents under development (e.g., bedaquiline) [141].

4.2.2 High-Dose Isoniazid

During *in vitro* assessments and preclinical evaluation in an animal model, higher than standard (5 mg kg^{-1} day^{-1}) doses of INH (i.e., 10 mg kg^{-1} day^{-1}) exhibited bactericidal activity against an *inhA* promoter mutant, while a dose level of 25 mg kg^{-1} day^{-1} demonstrated bactericidal activity against a *katG* mutant [149–151]. In humans, the pharmacokinetics of high-dose INH varied between individuals who were slow- versus rapid-acetylators [152]. Furthermore, in a recent trial, the addition of high-dose INH (16 mg kg^{-1}) but not standard-dose INH (5 mg kg^{-1}) to MDR-TB treatment increased the sputum culture conversion rates [153]. Thus, depending on the INH dose, patient acetylator status and degree of INH resistance, high-dose INH may be useful in the management of DR-TB [149, 154, 155] – especially in cases of fluoroquinolone-resistant MDR-TB and XDR-TB [154]. A Phase II trial (ACTG–A5132) is currently underway to determine the correct dose of high-dose INH among adult patients with different genetic variants of INH-resistant TB [155]. With these early successes of high-dose INH, seven new related compounds, now in early stages of development, have recently been discovered through quantitative structure–activity relationships (QSARs) models. Early evaluation has suggested that they may be more effective than INH against the mutated S315T-resistant strain of *M. tuberculosis* [156–158].

4.3 Re-engineering of Antibacterials or Re-purposing Next-Generation' Chemical Relatives

4.3.1 Fluroquinolones (Moxifloxacin, Gatifloxacin, and DC-159a)

Fluoroquinolones inhibit bacteria DNA gyrase and DNA topoisomerase and are used as second-line agents for treating MDR-TB [49, 141]. Findings from animal models and Phase II trials suggest that fluoroquinolones have the potential to shorten treatment duration for DS-TB from six to four months [159–162]. Moxifloxacin and gatifloxacin are 8-methoxyquinolones in advanced stages of development for TB treatment. They are both broad-spectrum antimicrobials that are now being re-purposed for the treatment of TB. Four Phase III trials with the aim of shortening DS-TB treatment from six to four months using fluoroquinolone-containing regimens has been shown to be inferior to the present standard regimen [144, 163–165]. At this stage, it is acceptable to conclude that substituting a quinolone with other first-line agents is ineffective in reducing DS-TB treatment durations. It is, however, premature to conclude that fluroquinolones cannot contribute in shortening treatment duration; however, their optimum positioning in a treatment regimen will depend on the progress of ongoing trials of its combinations with newer compounds [141]. Another methoxy-fluoroquinolone in the early stages of development is DC-159a [101].

4.3.2 Nitroimidazoles (Delamanid (OPC-67683), Pretomanid (PA-824), and TBA-354)

Nitroimidazoles – for example, the earliest, metronidazole – was found to be effective against a wide variety of anaerobes and anaerobic bacteria [141]. Since *M. tuberculosis* adapts to survive in anaerobic environments such as in hypoxic granulomas, as well as existing in its dormant form (LTBI), it appeared reasonable to evaluate the effect of nitroimidazoles on *M. tuberculosis* [166]. Metronidazole is bactericidal against *M. tuberculosis* under *in vitro* hypoxic conditions, but shows contrasting effects in animal models [141, 166]. Modification of the side chain of the nitroimidazole skeleton has resulted in two compounds – delamanid (OPC-67683) and pretomanid (PA-824) – which are in advanced stages of clinical development, and one compound which is still in early clinical (Phase I) development (TBA-354) [12]. These compounds have been found to have good bactericidal activity both under aerobic and anaerobic conditions *in vitro*, and also in the mouse model of TB when given either alone or in combination with other anti-TB agents [141].

Pretomanid (PA-824) is a nitroimidazoxazine, pro-drug that requires intracellular activation by a F420-deazaflavin-dependent nitroreductase (Ddn) present in *M. tuberculosis* [167, 168]. The activation of PA-824 produces des-nitroimidazole, that generates reactive nitrogen species, including nitric oxide (NO), leading to a decrease in intracellular ATP and its anaerobic-mediated killing [169, 170]. Besides being very bactericidal under anaerobic conditions, pretomanid kills aerobically by blocking cell-wall mycolic acid biosynthesis [170, 171]. Thus, pretomanid shows activity both against actively replicating and dormant *M. tuberculosis* as well as against DS-TB and DR-TB [172–175]. No cross-resistance with standard anti-TB drugs has been observed [172]. Mutations in the mycobacterial genes *fbiA*, *fbiB*, and *fbiC* led to impaired coenzyme F420 synthesis, and therefore resistance to PA-824 [176, 177]. Mutations have been identified in *fgd1* and *ddn* [168, 171, 178], with cross-resistance to delamanid [179]. Pretomanid has a MIC of 0.15 to 0.3 mg ml^{-1} [102, 168, 171]. A Phase IIa study has shown that pretomanid is safe at daily doses of between 100 and 200 mg [180], although it might be associated with a mild QT-interval prolongation [69]. A number of Phase III trials (STAND NC-006, ACTGPR682) and Phase II trials (MARVEL, Nix-TB, NC-005) involving pretomanid in combination with other anti-TB agents are currently in progress [181].

Delamanid is a nitro-dihydro-imidazooxazole pro-drug which requires nitroreductive activation [11, 141]. It is an inhibitor of mycolic acid biosynthesis, blocking the synthesis of methoxy and ketomycolic acids only [179, 182]. Delamanid is highly potent *in vitro* with a MIC of 0.006–0.024 µg ml^{-1} against both DS-TB and DR-TB [141, 179, 182]. Delamanid also demonstrates a concentration-dependent activity against intracellular *M. tuberculosis* [179, 183]. In the mouse model, a combination of delamanid with RMP and PZA achieved a faster eradication of the *Mycobacterium* bacilli compared to a standard first-line anti-TB regimen [179]. The optimal dose of delamanid in humans is still under investigation. In a Phase II trial, MDR-TB patients in whom delamanid was added to an optimal background regimen of second-line drugs showed a significantly higher proportion of sputum culture conversion at two months, from 29.6% (placebo) to 45.4% (delamanid 100 mg twice daily), and 41.9% (delamanid 200 mg twice daily) [102, 184]. An extension of this trial showed that delamanid use for more than six months in comparison with its use for less than two months significantly increased the proportion with successful outcomes (cure or treatment completion) from 55% to 74.5%, and significantly reduced mortality from 8.3% to 1.0% [102, 185]. Delamanid also significantly prolongs the QT-interval, and this should be considered when designing other clinical trials involving its use in combination with other anti-TB agents [184]. Delamanid was given conditional approval by the EMA in April 2014, while the WHO has released a guide for its use in MDR-TB [186, 187]. These early successes with delamanid strengthened the justification for a current Phase III study that has completed follow-up for the primary endpoint, with data analysis in progress, as well as an ongoing pediatric clinical trial. Other Phase III clinical trials involving delamanid in combination with other anti-TB agents are being planned [181].

4.3.3 Oxazolidinones (Linezolid, Sutezolid (PNU-100480), and AZD5847)

Oxazolidinones (with linezolid as "lead" agent) are a new class of antibiotics recently approved, and have been used to successfully treat Gram-positive infections [141]. Linezolid is active against *M. tuberculosis in vitro* and in animal models [188, 189]. This property has also been demonstrated by two other newer oxazolidinone agents – sutezolid (PNU-100480) and AZD5847 – that have completed Phase II and Phase I trials, respectively [141]. The mechanism of action of oxazolidinones is unique, in that they inhibit protein synthesis by binding to the 23S rRNA in the 50S ribosomal subunit [141].

Linezolid has low MIC-values for non-MDR-TB and MDR-TB strains which range from $0.12\,\mu g\,ml^{-1}$ to $0.5\,\mu g\,ml^{-1}$ [190]. It also has a modest activity in the mouse model of TB, appearing to be bacteriostatic and weakly bactericidal towards *M. tuberculosis* [191]. Linezolid appeared to be very effective in the human TB model. A systematic review and meta-analysis of the efficacy, safety and tolerability of linezolid-containing regimens for DR-TB, based on individual patient data from 12 studies performed in 11 countries worldwide, showed interesting findings [192]. In individuals who received linezolid-containing regimens, sputum smear- and culture-conversions were achieved in 92.5% and 93.5%, respectively, with a treatment success rate of 81.8%. However, its safety and tolerability profile was poor as 58.9% of the patients experienced adverse events that included anemia (38.1%), peripheral neuropathy (47.1%), gastrointestinal disorders (16.7%), optic neuritis (13.2%), and thrombocytopenia (11.8%) [192]. In a similar study among XDR-TB patients in China, individuals who received a linezolid-containing regimen had a significantly higher treatment success rate and a higher proportion of adverse events [193]. The rate of occurrence of adverse events was significantly higher in individuals who received a daily dosage of linezolid that exceeded 600 mg [192, 193]. There is currently an ongoing debate as to what the standard dose of linezolid should be, with some investigators arguing in favor of a 300 mg daily dose in order to limit adverse events [190], and others suggesting a dose of 600 mg in order to limit the emergence of resistance to linezolid due to suboptimal exposures [194]. Clearly, quality clinical research is needed to better understand the appropriate dosage of linezolid to be prescribed.

Sutezolid is a linezolid analog with increased antimycobacterial activity and less toxicity compared to linezolid [141, 189]. It has a better activity than linezolid both in *in vitro* and in the mouse model of TB [195]. It also has a better safety profile and shows better time-dependent killing in an *ex-vivo* whole-blood culture test [196]. In addition, sutezolid has activity against both DS-TB and DR-TB [197]. A sutezolid and SQ109 combination regimen had additive effects *in vitro*, and SQ109 synergistically improved the time to kill intracellular *M. tuberculosis* in an infected macrophage assay [198]. A combination regimen containing sutezolid, SQ109 and bedaquiline appeared to have cumulative activity comparable to standard TB therapy for all forms of DS-TB and DR-TB [199]. In the mouse model of TB, *in-vivo* studies showed that the addition of sutezolid to standard first-line anti-TB agents led to a significantly improved efficacy and a potential to shorten treatment duration, because it led to a relapse-free cure with a shorter duration of treatment [200]. Sutezolid is currently undergoing Phase II trials. It was safe and well-tolerated at doses of up to

1200 mg daily for 14 days, or 600 mg twice daily for 28 days, and significantly reduced the number of TB bacteria in sputum (daily log change of −0.088 cfu; 90% CI: −0.112 to −0.065, $P < 0.0001$ for the 600 mg twice-daily dose; and daily log change of −0.068 cfu; 90% CI: −0.090 to −0.045, $P < 0.0001$ for the 1200 mg once-daily dose), demonstrating that the drug is active in humans [201]. Sutezolid does not appear to cause QT-interval prolongation or bone marrow suppression [202], although the potential for neurotoxicity is still being considered [102, 202]. However, a Phase II trial showed a transient and asymptomatic elevation of alanine transaminase to occur in 14% of patients in the sutezolid arm [201].

AZD5847 is another linezolid analog which shares a similar mechanism of action [202]. It has a better bactericidal activity than linezolid *in vitro*, and appears to be safe and well tolerated in Phase I studies, with nausea being the most common adverse effect at higher doses [102, 203]. A Phase IIa EBA trial of AZD5847 has been concluded, the interim results of which suggest that it has activity against slow-growing and intracellular *M. tuberculosis*, is likely to have excellent combinability with other anti-TB and ART agents, and is potentially the safest drug in its class [204–206]; the final results of these investigations are still pending.

4.3.4 β-Lactams (Meropenem and Faropenem)

β-Lactam antibiotics, such as cephalosporins, penicillins, monobactams and carbapenems, form a group of bactericidal agents that have not been used widely to treat TB due to a lack of efficacy. The mechanism of action of β-lactam antibiotics is to block the transpeptidases required for formation of the cell-wall peptidoglycan layer [101, 141]. β-Lactam antibiotics are not effective against *M. tuberculosis* due to the presence of an extended spectrum β-lactamase, BlaC, which hydrolyzes β-lactams, leading to resistance [141, 207].The β-lactamase inhibitor, clavulanate, has been shown to irreversibly inhibit *M. tuberculosis* BlaC *in vitro* [208]. With recent interest in re-purposing known antibacterials, a combination of β-lactamase inhibitor (e.g., clavulanic acid) with β-lactam antibiotics (e.g., meroponem) showed potent activity *in vitro*, killing XDR-TB strains under both aerobic and anaerobic conditions, as well as sterilizing cultures in 14 days [209]. A meropenem–clavulanate combination also showed a MIC (meropenem) of $<1\,\mu g\,ml^{-1}$ against laboratory strains of *M. tuberculosis* [209]. In the mouse model of TB, findings have been equivocal with the meropenem–clavulanic acid combination [210, 211]. In humans, the addition of meropenem–clavulanate to one or two active second-line agents as a salvage regimen for six patients with severe XDR-TB led to a cure in one patient and sputum culture conversion in four of five additional cases [212]. In a case-control study in which meropenem–clavulanate was added to a linezolid-containing regimen for patients with DR-TB, those patients with meropenem–clavulanate added to their treatment showed significantly higher proportions of sputum-smear- and culture-conversion [213]. The study also demonstrated preliminary evidence of the effectiveness, safety, and tolerability of meropenem–clavulanate. Unfortunately, despite these early evidences of clinical efficacy, meropenem–clavulanate faces a major obstacle against its use as an anti-TB agent under program conditions in developing countries. This is because it has a poor oral availability and short half-life; thus, meropenem must be administered either via frequent intravenous injections or continuously by intravenous infusion [141], which may be operationally challenging.

Faropenem is an orally bioavailable carbapenem antibiotic in Phase IIa studies that is more resistant to hydrolysis by lactamases compared to other β-lactam and carbapenems [12, 214, 215]. The MIC of faropenem against *M. tuberculosis* is 1.3 μg ml^{-1}, with or without clavulanate [216]. However, it has a short half-life and requires the administration of several doses daily, which makes it impractical for managing DS-TB [12]. Additional studies are needed to optimize the combinations of β-lactams and β-lactamase inhibitors for DS-TB and DR-TB.

4.3.5 Riminophenazine (Clofazimine and TBI-166)

Clofazimine is a riminophenazine that have been used mainly for treating leprosy [141]. Its mode of action is through a redox cycling pathway in which it undergoes cycles of enzymatic reduction (by NDH-2, part of the primary respiratory chain NADH:quinone oxidoreductase) and subsequent spontaneous oxidation producing reactive oxygen species that are toxic to the cell [141, 217]. The MIC of clofazimine ranges from 0.5 to 2 mg l^{-1} [218]. It is active against dormant/persistent *M. tuberculosis in vitro*, and its combination with bedaquiline and PZA was effectively used in a mouse model of TB as a relapse-free, short-course regimen [219, 220]. In humans, clofazimine was included in several regimens being considered to treat MDR-TB [221]. The most effective of these was the "Bangladeshi" regimen, which resulted in a relapse-free cure rate in 84–89% of MDR-TB cases in Niger and Bangladesh, a substantial improvement from the WHO-recommended treatment, which is given for a minimum of 20 months and has only 48% treatment success rate [50, 51]. In another study, XDR-TB patients with HIV coinfection who received a clofazimine-containing regimen had a twofold increase in TB culture conversion at six months compared to those who received a non-clofazimine-containing XDR-TB treatment regimen [222]. Recently, in an experimental model of DS-TB in mice, clofazimine was added to standard first-line TB treatment and evaluated. The clofazimine-treated mice were all culture-negative by three months, whereas control mice (treated with standard regimen only) required five months of treatment to become culture-negative. Moreover, there were no relapses among mice treated with the clofazimine-containing regimen for three or four months [223]. This suggests that clofazimine has the potential to shorten the duration of treatment for DS-TB. A major adverse effect of clofazimine is skin discoloration due to its accumulation in fatty tissue [12]. In an effort to reduce this side effect, an analog TBI-166 was selected from 69 riminophenazine derivatives; early results of preclinical studies with this have suggested that it has similar antimycobacterial properties, but fewer adverse side effects [12, 224].

4.3.6 Cotrimoxazole (Trimethoprim–Sulfamethoxazole)

Cotrimoxazole contains a 1 : 5 combination of trimethoprim (TMP) and sulfamethoxazole (SMX), respectively. It is a potent antibacterial antibiotic against a variety of pathogens causing infections in humans [225]. Recent interests in re-purposing drugs for TB have led to its evaluation for activity against *M. tuberculosis*. In *in-vitro* studies, trimethoprim–sulfamethoxazole has both intracellular and extracellular activity against susceptible and multidrug-resistant *M. tuberculosis* [226–228]. Among MDR-TB patients, preliminary data showed

Tab. 3 Summary of ongoing or planned experimental regimens for DS-TB or DR-TB in clinical trials.

Trial name (funding)	Phase	Duration of experimental regimen (months)	Comparator	Experimental arm	Status
STREAM I (MRC)	III	9	WHO standard	Modified Bangladeshi regimen (4(MFX + CFZ + EMB + PZA + KM + INH + PTH)/5(MFX + CFZ + EMB + PZA))	85% enrolled
STREAM II (MRC)	III	6 versus 9	WHO standard/9 month regimen	BDQ + LFX + CFZ + PZA + 2(INH + KM)	Expected start Q1 2015
MARVEL (ACTG)	IIb	2	WHO standard regimen	BDQ + CFZ + EMB + LFX + PZA + 4(INH + PTH)	In development
(STAND-NC-006) (GATB)	III	4–6	None for MDR arm	BDQ, PA824, PZA, and sutezolid (1200 mg, four times a day) vs. BDQ, PA824, PZA, and sutezolid (600 mg twice a day) vs. BDQ, PA824, PZA, and LFX (600 mg)	Enrollment expected in Q1 2015
TB-PRACTECAL (MSF)	II and III	6	WHO-standard of care	MFX, PA-824, and PZA for 4 or 6 months compared to HRZE for 6 months	Expected start Q3 2015
NC-005 (GATB)	II	2 (followed by OBT)	None for MDR arm	BDQ + PA-824 + LZD + MXF BDQ + PA-824 + LZD + CFZ BDQ + PA-824 + LZD BDQ + PA-824 + PZA	Expected start Q4 2014
NiX-TB (GATB)	II	6–9	None	BDQ + PA-824 + MFX + PZA	Expected start Q1 2015
C213 (Otsuka)	III	24	WHO standard regimen	DLM + OBT	Completed follow-up for primary endpoint, data analysis ongoing

(*continued overleaf*)

Tab. 3 (Continued)

Trial name (funding)	Phase	Duration of experimental regimen (months)	Comparator	Experimental arm	Status
NeXT (MRC-SA)	III	6–9	SA standard	BDQ + LZD + LFX + ETA/high-dose INH + PZA	In development (waiting for MCC approval.)
End-TB (MSF)	III	9	None	BDQ + LZD + MXF + PZA BDQ + CFZ + LZD + LFX + PZA BDQ + CFZ + LFX + PZA DLM + LZD + MFX + PZA DLM + CFZ + LZD + LFX + PZA DLM + CFZ + LFX + PZA	In development (protocol completed)
TRUNCATE-TB (UCL)	II and III	12	WHO standard 6 month treatment/6-month re-treatment	Novel combination regimens for 2 months/re-treat relapse for 6 months with standard therapy	Expected start Q1 2015

OBT, optimized background therapy; BDQ, bedaquiline; LFX, levofloxacin; PA-824, Pretomanid; INH, isoniazid; PZA, pyrazinamide; EMB, ethambutol; KM, kanamycin; PTH, prothionamide; MFX, moxifloxacin; CFZ, clofazimine; LZD, linezolid; DLM, delamanid; and ETA, ethionamide.
Source: treat-TB/Resist-TB/GATB/MSF.

that trimethoprim–sulfamethoxazole was safe, well tolerated, and has a favorable pharmacokinetic profile in TB patients [225]. Future studies are needed to confirm its efficacy in the treatment of DS-TB and DR-TB.

4.4
Drug Combinations/Regimens

TB treatment must be delivered in multidrug regimens to prevent the development of drug resistance. Traditionally, as new drugs are being developed they are used to replace one or more existing drug within the regimen, or the new candidate drug is added to the existing standard treatment to evaluate its impact. As each of these trials could last for six years or longer, this means that a novel TB regimen could take decades to develop when using this model. With the recent resurgence in TB drug discovery, there is growing interest in developing new TB treatment regimens. The first novel new DS-TB drug regimen to be developed is PaMZ, which consists of Pretomanid (PA-824), Moxifloxacin, and PZA. In a two-week Phase II EBA trial, this regimen killed more bacteria than the current standard of care for DS-TB [229, 230]. Furthermore, based on positive results from an eight-week Phase IIb trial in both DS-TB and MDR-TB patients [231], PaMZ will be advanced to a global Phase III trial, known as a Shortening Treatment by Advancing Novel Drugs (STAND) trial. A number of planned and ongoing trials have been designed to develop new treatment regimens for TB, and these (for a summary, see Table 3) are generally aimed at either shortening the duration of the current treatment regimen for DS-TB/DR-TB, improving the ease of administration (via the oral route), or improving their effectiveness (cure rate).

5
Conclusions

The containment of TB remains a global health challenge, and there is an urgent need for new anti-TB drugs and treatment regimens that are shorter and safer, as well as quicker ways of evaluating new TB drugs and drug regimens. Over the past decade, increasing financial and research investments have resulted in the development of new diagnostic tools and drugs for TB. Nonetheless, further research is required on new clinical trial designs to test the efficacy of these newer agents, efficacy studies in pediatric populations, identifying predictive biomarkers for long-term TB cure, assessing the role of immunotherapy for TB treatment, and the provision of incentives to improve the availability of these newer agents in under-resourced, high-TB populations. With the resurgence in activity in TB drug discovery and development, there are good reasons for an optimism that newer effective regimens will be available within the next decade to tackle the persistent scourge of TB.

References

1. Zumla, A., Raviglione, M., Hafner, R. von Reyn, C.F. (2013) An important update of current concepts on the clinical, epidemiological and management aspects of tuberculosis. *N. Engl. J. Med.*, **368** (8), 745–755.
2. Grange, J.M. (2009) The Genus *Mycobacterium* and the *Mycobacterium tuberculosis* Complex, in *Tuberculosis: A Comprehensive Clinical Reference*, 1st edn (eds S. Schaaf, A.I. Zumla), Saunders-Elsevier, Pennsylvania, pp. 44–59.
3. Frieden, T.R., Sterling, T.R., Munsiff, S.S., Watt, C.J., Dye C. (2003) Tuberculosis. *Lancet*, **362** (9387), 887–899.
4. Millet, J.P., Moreno, A., Fina, L., del Baño, L. *et al.* (2013) Factors that influence

current tuberculosis epidemiology. *Eur. Spine J.*, **22** (Suppl. 4), 539–548.
5. Gordin, F.M., Masur, H. (2012) Current approaches to tuberculosis in the United States. *J. Am. Med. Assoc.*, **308** (3), 283–289.
6. Pawlowski, A., Jansson, M., Sköld, M., Rottenberg, M.E., Källenius, G. (2012) Tuberculosis and HIV co-infection. *PLoS Pathogen.*, **8** (2), e1002464.
7. World Health Organization (1993) WHO Declares Tuberculosis a Global Emergency. Press Release, WHO/31, April 23 1993.
8. World Health Organization (2014) *Global Tuberculosis Control: WHO Report 2014*, World Health Organization, Geneva.
9. World Health Organization (2010) *Treatment of Tuberculosis Guidelines*. 4th edn, World Health Organization, Geneva.
10. World Health Organization (2010) *The Global Plan to Stop TB, 2011–2015: Transforming the Fight Towards Elimination of Tuberculosis*. World Health Organization, Geneva, pp. 1–100.
11. Zumla, A., Nahid, P., Cole, S.T. (2013) Advances in the development of new tuberculosis drugs and treatment regimens. *Nat. Rev. Drug Discov.*, **12** (5), 388–404.
12. Zumla, A., Gillespie, S.H., Hoelscher, M., Philips, P.P. et al. (2014) New antituberculosis drugs, regimens, and adjunct therapies: needs, advances, and future prospects. *Lancet Infect. Dis.*, **14** (4), 327–340.
13. Global Alliance for TB Drug Development (2001) Tuberculosis: scientific blueprint for tuberculosis drug development. *Tuberculosis (Edinb.)*, **81** (Suppl. 1), 1–52.
14. Ahmad, S., Mokaddas, E. (2014) Current status and future trends in the diagnosis and treatment of drug-susceptible and multidrug-resistant tuberculosis. *J. Infect. Public Health*, **7** (2), 75–91.
15. Mitchison, D., Davies, G. (2012) The chemotherapy of tuberculosis: past, present and the future. *Int. J. Tuberc. Lung Dis.*, **16** (6), 724–732.
16. Yew, W.W., Chan, C.K., Leung, C.C., Chau, C.H. et al. (2003) Comparative roles of levofloxacin and ofloxacin in the treatment of multidrug-resistant tuberculosis: preliminary results of a retrospective study from Hong Kong. *Chest*, **124** (4), 1476–1481.
17. Villar, M., Sotgiu, G., D'Ambrosio, L., Raymundo, E. et al (2011) Linezolid safety, tolerability and efficacy to treat multidrug- and extensively drug-resistant tuberculosis. *Eur. Respir. J.*, **38** (3), 730–733.
18. Dey, T., Brigden, G., Cox, H., Shubber, Z. et al. (2013) Outcomes of clofazimine for the treatment of drug-resistant tuberculosis: a systematic review and meta-analysis. *J. Antimicrob. Chemother.*, **68** (2), 284–293.
19. Somoskovi, A., Parsons, L.M., Salfinger, M. (2001) The molecular basis of resistance to isoniazid, rifampin, and pyrazinamide in *Mycobacterium tuberculosis. Respir. Res.*, **2** (3) 164–168.
20. Da Silva, P.E.A., Palomino, J.C. (2011) Molecular basis and mechanisms of drug resistance in *Mycobacterium tuberculosis* classical and new drugs. *J. Antimicrob. Chemother.*, **66** (7), 1417–1430.
21. Zimhony, O., Cox, J.S., Welch, J.T., Vilchèze, C., Jacobs, W.R., Jr, (2000) Pyrazinamide inhibits the eukaryotic-like fatty acid synthetase I (FASI) of *Mycobacterium tuberculosis. Nat. Med.*, **6** (9), 1043–1047.
22. Shi, W., Zhang, X., Jiang, X., Yuan, H. et al. (2011) Pyrazinamide inhibits trans-translation in *Mycobacterium tuberculosis. Science*, **333** (6049), 1630–1632.
23. Mitchison, D.A., Zhang, Y. (2011) Recent Developments in the Study of Pyrazinamide: An Update, in: *Antituberculosis Chemotherapy Progress in Respiratory Research* (eds Donald, P., van Helden, P.D.), vol. **40**, Karger, Cape Town, pp. 32–43.
24. McNeil, M.R., Brennan, P.J. (1991) Structure, function and biogenesis of the cell envelope of mycobacteria in relation to bacterial physiology, pathogenesis and drug resistance: some thoughts and possibilities arising from recent structural information. *Res. Microbiol.*, **142** (4), 451–463.
25. Mikusova, K., Slayden, R.A., Besra, G.S., Brennan, P.J. (1995) Biogenesis of the mycobacterial cell wall and the site of action of ethambutol. *Antimicrob. Agents Chemother.*, **39** (11), 2484–2489.
26. Peloquin, C.A., Alsutan, A. (2008) Clinical Pharmacology of the Anti-Tuberculosis Drugs, in *Clinical Tuberculosis*, 4th edn (eds P.D.O. Davies, P.F. Barnes, S.B. Gordon), Hodder & Stoughton Ltd, London, pp. 205–224.
27. Srivastava. S., Garg, A., Ayyagari, A., Nyati, K.K. et al. (2006) Nucleotide polymorphism associated with ethambutol resistance in

clinical isolates of *Mycobacterium tuberculosis. Curr. Microbiol.*, **53** (5), 401–405.
28. Hong Kong Chest Service/British Medical Research Council (1979) Controlled trial of 6-month and 8-month regimens in the treatment of pulmonary tuberculosis: the results up to 24 months. *Tubercle*, **60** (4), 201–210.
29. Hong Kong Chest Service/British Medical Research Council (1991) Controlled trial of 2, 4, and 6 months of pyrazinamide in 6-month, three-times weekly regimens for smear-positive pulmonary tuberculosis, including an assessment of a combined preparation of isoniazid, rifampin, and pyrazinamide: results at 30 months. *Am. Rev. Respir. Dis.*, **143** (4, Pt. 1), 700–706.
30. Fox, W., Mitchison, D.A. (1975) Short-course chemotherapy for pulmonary tuberculosis. *Am. Rev. Respir. Dis.*, **111** (6), 845–848.
31. Nunn, A.J., Jindani, A., Enarson, D.A and the Study A Investigators (2011) Results at 30 months of a randomised trial of two 8-month regimens for the treatment of tuberculosis. *Int. J. Tuberc. Lung Dis.*, **15** (6), 741–745.
32. East African/British Medical Research Councils (1974) Controlled clinical trial of four short-course (6-month) regimens of chemotherapy for treatment of pulmonary tuberculosis. Third report. *Lancet*, **2** (7875), 237–240.
33. Ukwaja, K.N., Oshi, S.N., Alobu, I., Oshi, D.C. (2015) Six- versus eight-month antituberculosis regimen for pulmonary tuberculosis under programme conditions. *Int. J. Tuberc. Lung Dis.*, **19** (3), 295–301.
34. Kruk, M.E., Schwalbe, N.R., Aguiar, C.A. (2008) Timing of default from tuberculosis treatment: a systematic review. *Trop. Med. Int. Health*, **13** (5), 1–10.
35. Alobu, I, Oshi, S.N, Oshi, D.C., Ukwaja, K.N. (2014) Risk factors of treatment default and death among tuberculosis patients in a resource-limited setting. *Asian Pac. J. Trop. Med.*, **7** (12), 977–984.
36. Alobu, I., Oshi, S.N., Oshi, D.C., Ukwaja, K.N. (2014) Profile and determinants of treatment failure among smear-positive pulmonary tuberculosis patients in Ebonyi, Southeastern Nigeria. *Int. J. Mycobacteriol.*, **3** (2), 127–131.

37. Falzon, D., Jaramillo, E., Schunemann, H.J., Arentz, M. et al. (2011) WHO guidelines for the programmatic management of drug-resistant tuberculosis: 2011 update. *Eur. Respir. J.*, **38** (3), 516–528.
38. Davies P.D. (2001) Drug-resistant tuberculosis. *J. R. Soc. Med.*, **94** (6), 261–263.
39. Cho, S.H., Warit, S., Wan, B., Hwang, C.H. et al. (2007) Low-oxygen-recovery assay for high-throughput screening of compounds against nonreplicating *Mycobacterium tuberculosis. Antimicrob. Agents Chemother.*, **51** (14), 1380–1385.
40. Garton, N.J., Christensen, H., Minnikin, D.E., Adegbola, R.A., Barer, M.R. (2002) Intracellular lipophilic inclusions of mycobacteria in vitro and in sputum. *Microbiology*, **148** (Pt. 10), 2951–2958.
41. Shleeva, M.O., Kudykina, Y.K., Vostroknutova, G.N., Suzina, N.E. et al. (2011) Dormant ovoid cells of *Mycobacterium tuberculosis* are formed in response to gradual external acidification. *Tuberculosis (Edinb.)*, **91** (2), 146–154.
42. Zhang, Y., Yew, W.W., Barer, M.R. (2012) Targeting persisters for tuberculosis control. *Antimicrob. Agents Chemother.*, **56** (5), 2223–2230.
43. Mukamolova. G.V., Turapov, O., Malkin, J., Woltmann, G., Barer, M.R. (2010) Resuscitation-promoting factors reveal an occult population of tubercle bacilli in sputum. *Am. J. Respir. Crit. Care. Med.*, **181** (1), 174–180.
44. Wayne, L.G., Hayes, L.G. (1996) An in vitro model for sequential study of shiftdown of *Mycobacterium tuberculosis* through two stages of nonreplicating persistence. *Infect. Immun.*, **64** (6), 2062–2069.
45. Gandhi, N.R., Nunn, P., Dheda, K., Schaaf, H.S. et al. (2010) Multidrug-resistant and extensively drug-resistant tuberculosis: a threat to global control of tuberculosis. *Lancet*, **375** (9728), 1830–1843.
46. Ma, Z., Lienhardt, C., McIlleron, H., Nunn, A.J., Wang, X. (2010) Global tuberculosis drug development pipeline: the need and the reality. *Lancet*, **375** (9731), 2100–2109.
47. Wu, S., Zhang, Y., Sun, F., Chen, M. et al. (2013) Adverse events associated with the treatment of multidrug-resistant tuberculosis: a systematic review and meta-analysis. *Am. J. Ther.* (Epub ahead of print).

48. Kwon, Y.S., Jeong, B.H., Koh, W.J. (2014) Tuberculosis: clinical trials and new drug regimens. *Curr. Opin. Pulmon. Med.*, **20** (3), 280–286.
49. Ahuja, S.D., Ashkin, D., Avendano, M., Banerjee, R. *et al.* (2012) Multidrug resistant pulmonary tuberculosis treatment regimens and patient outcomes: an individual patient data meta-analysis of 9,153 patients. *PLoS Med.*, **9** (8), e1001300.
50. Piubello, A., Harouna, S.H., Souleymane, M.B., Boukary, I. *et al.* (2014) High cure rate with standardised short-course multidrug-resistant tuberculosis treatment in Niger: no relapses. *Int. J. Tuberc. Lung Dis.*, **18** (10), 1188–1194.
51. Aung, K.J., Van Deun, A., Declercq, E., Sarker, M.R. *et al.* (2014) Successful '9-month Bangladesh regimen' for multidrug-resistant tuberculosis among over 500 consecutive patients. *Int. J. Tuberc. Lung Dis.*, **18** (10), 1180–1187.
52. Nunn, A.J., Rusen, I.D., Van Deun. A., Torrea, G. *et al.* (2014) Evaluation of a standardized treatment regimen of antituberculosis drugs for patients with multidrug-resistant tuberculosis (STREAM): study protocol for a randomized controlled trial. *Trials*, **15**, 353.
53. Kwan, C.K., Ernst, J.D. (2011) HIV and tuberculosis: a deadly human syndemic. *Clin. Microbiol. Rev.*, **24** (2), 351–376.
54. Oshi, D.C., Oshi, S.N., Alobu, I., Ukwaja, K.N. (2014) Profile, outcomes, and determinants of unsuccessful tuberculosis treatment outcomes among HIV-infected tuberculosis patients in a Nigerian state. *Tuberc. Res. Treat.*, **2014**, 202983.
55. Koul, A., Arnoult, E., Lounis, N., Guillemont, J., Andries, K. (2011) The challenge of new drug discovery for tuberculosis. *Nature*, **469** (7331), 483–490.
56. Abdool Karim, S.S., Naidoo, K., Grobler, A., Padayatchi, N. *et al.* (2010) Timing of initiation of antiretroviral drugs during tuberculosis therapy. *N. Engl. J. Med.*, **362** (8), 697–706.
57. Abdool Karim, S.S., Naidoo, K., Grobler, A., Padayatchi, N. *et al.* (2011) Integration of antiretroviral therapy with tuberculosis treatment. *N. Engl. J. Med.*, **365** (16), 1492–1501.
58. Webb, E.A., Hesseling, A.C., Schaaf, H.S., Gie, R.P. *et al.* (2009) High prevalence of *Mycobacterium tuberculosis* infection and disease in children and adolescents with type 1 diabetes mellitus. *Int. J. Tuberc. Lung Dis.*, **13** (7), 868–874.
59. Ottmani, S.E., Murray, M.B., Jeon, C.Y., Baker, M.A. *et al.* (2010) Consultation meeting on tuberculosis and diabetes mellitus: meeting summary and recommendations. *Int. J. Tuberc. Lung Dis.*, **14** (12), 1513–1517.
60. Jiménez-Corona, M.E., Cruz-Hervert, L.P., García-García. L., Ferreyra-Reyes, L. *et al.* (2013) Association of diabetes and tuberculosis: impact on treatment and post-treatment outcomes. *Thorax*, **68** (3), 214–220.
61. Baghaei. P., Marjani, M., Javanmard, P., Tabarsi, P., Masjedi, M.R. (2013) Diabetes mellitus and tuberculosis facts and controversies. *J. Diabetes Metab. Disord.*, **12** (1), 58.
62. Dooley, K.E., Chaisson, R.E. (2009) Tuberculosis and diabetes mellitus: convergence of two epidemics. *Lancet Infect. Dis.*, **9** (12), 737–746.
63. Woldehanna, S., Volmink, J. (2004) Treatment of latent tuberculosis infection in HIV infected persons. *Cochrane Database Syst. Rev.*, **1**, CD000171.
64. World Health Organization (2010) *Guidelines for Intensified Tuberculosis Case-Finding and Isoniazid Preventive Therapy for People Living with HIV in Resource-Constrained Settings.* World Health Organization, Geneva.
65. Ena, J., Valls, V. (2005) Short-course therapy with rifampin plus isoniazid, compared with standard therapy with isoniazid, for latent tuberculosis infection: a meta-analysis. *Clin. Infect. Dis.*, **40** (5), 670–676.
66. Spyridis, N.P., Spyridis, P.G., Gelesme, A., Sypsa, V. *et al.* (2007) The effectiveness of a 9-month regimen of isoniazid alone versus 3- and 4-month regimens of isoniazid plus rifampin for treatment of latent tuberculosis infection in children: results of an 11-year randomized study. *Clin. Infect. Dis.*, **45** (6), 715–722.
67. Churchyard, G.J., Fielding, K.L., Lewis, J.J., Coetzee, L. *et al.* (2014) A trial of mass isoniazid preventive therapy for tuberculosis control. *N. Engl. J. Med.*, **370** (4), 301–310.
68. Mills, H.L., Cohen, T., Colijn, C. (2013) Community-wide isoniazid preventive

therapy drives drug-resistant tuberculosis: a model-based analysis. *Sci. Transl. Med.*, **5** (180), 180ra49.
69. Ukwaja, K.N., Abimbola, S.A. (2014) A trial of mass isoniazid preventive therapy for tuberculosis control. *N. Engl. J. Med.*, **370** (17), 1661.
70. Marais, B.J., Hesseling, A.C., Gie, R.P., Schaaf, H.S., Beyers, N. (2006) The burden of childhood tuberculosis and the accuracy of routine surveillance data in a high burden setting. *Int. J. Tuberc. Lung Dis.*, **10** (3), 259–263.
71. Swaminathan, S., Rekha, B. (2010) Pediatric tuberculosis: global overview and challenges. *Clin. Infect. Dis.*, **50** (Suppl. 3), S184–S194.
72. Jenkins, H.E., Tolman, A.W., Yuen, C.M., Parr, J.B. et al. (2014) Incidence of multidrug-resistant tuberculosis disease in children: systematic review and global estimates. *Lancet*, **383** (9928), 1572–1579
73. Thee, S., Seddon, J.A., Donald, P.R., Seifart, H.I. et al. (2011) Pharmacokinetics of isoniazid, rifampin, and pyrazinamide in children younger than two years of age with tuberculosis: evidence for implementation of revised World Health Organization recommendations. *Antimicrob. Agents Chemother.*, **55** (12), 5560–5567.
74. Hiruy, H., Rogers, Z., Mbowane, C., Adamson, J. et al. (2014) Subtherapeutic concentrations of first-line anti-TB drugs in South African children treated according to current guidelines: the PHATISA study. *J. Antimicrob. Chemother.*, **70** (4), 1115–1123.
75. World Health Organization (2014) *Guidance for National Tuberculosis Programmes on the Management of Tuberculosis in Children*, 2nd edn. World Health Organization, Geneva.
76. Field, S.K., Fisher, D., Jarand, J.M., Cowie, R.L. (2012) New treatment options for multidrug-resistant tuberculosis. *Ther. Adv. Respir. Dis.*, **6** (5), 255–268.
77. Friedland, J.S. (2011) Tuberculosis in the 21st century. *Clin. Med.*, **11** (4), 353–357.
78. Lechartier, B., Rybniker, J., Zumla, A., Cole, S.T. (2014) Tuberculosis drug discovery in the post-post-genomic era. *EMBO Mol. Med.*, **6** (2), 158–168.
79. Coxon, G.D., Cooper, C.B., Gillespie, S.H., McHugh, T.D. (2012) Strategies and challenges involved in the discovery of new chemical entities during early-stage tuberculosis drug discovery. *J. Infect. Dis.*, **205** (Suppl. 2), S258–S264.
80. Cole, S.T., Brosch, R., Parkhill, J., Garnier, T. et al. (1998) Deciphering the biology of *Mycobacterium tuberculosis* from the complete genome sequence. *Nature*, **393** (6685), 537–544.
81. Cole, S.T. (1998) The *Mycobacterium leprae* genome project. *Int. J. Leprol. Other Mycobact. Dis.*, **66** (4), 589–591.
82. Garnier, T., Eiglmeier, K., Camus, J., Medina, N. et al. (2003) The complete genome sequence of *Mycobacterium bovis*. *Proc. Natl Acad. Sci. USA*, **100** (13), 7877–7882.
83. Wei, J.R., Krishnamoorthy, V., Murphy, K., Kim, J.H. et al. (2011) Depletion of antibiotic targets has widely varying effects on growth. *Proc. Natl Acad. Sci. USA*, **108** (10), 4176–4181.
84. Pavan, F.R., Sato, D.N., Leite, C.Q.F. (2012) An Approach to the Search for New Drugs Against Tuberculosis, in *Understanding Tuberculosis – New Approaches to Fighting Against Drug Resistance*, 1st edn (ed. P. Cardona), IntecOpen, pp. 137–146.
85. Kana, B.D., Karakousis, P.C., Parish, T., Dick, T. (2014) Future target-based drug discovery for tuberculosis? *Tuberculosis (Edinb.)*, **94** (6), 551–556.
86. Barry, C.E. III, Slayden, R.A., Sampson, A.E., Lee, R.E. (2000) Use of genomics and combinatorial chemistry in the development of new antimycobacterial drugs. *Biochem. Pharmacol.*, **59** (3), 221–231.
87. Pierce, S.H., Dubos, R.J., Schaefer, W.B. (1953) Multiplication and survival of tubercle bacilli in the organs of mice. *J. Exp. Med.*, **97** (2), 189–205.
88. Lefford, M.J. (1984) Diseases in Mice and Rats, in *The Mycobacteria: A Source Book* (eds G.P. Kubica, L.G. Wayne), Marcel Dekker, New York, pp. 947–977.
89. U.S. Department of Health and Human Services: Food and Drug Administration (2013) Guidance for Industry Concept Paper Pulmonary Tuberculosis: Developing Drugs for Treatment. https://www.federalregister.gov/articles/2013/11/06/2013-26549/draft-guidance-for-industry-on-pulmonary-tuberculosis-developing-drugs-for-treatment-availability (accessed 3 September 2015).

90. Donald, P.R., Diacon, A.H. (2008) The early bactericidal activity of anti-tuberculosis drugs: a literature review. *Tuberculosis (Edinb.)*, **88** (Suppl. 1), S75–S83.
91. Donald, P.R., Sirgel, F.A., Venter, A., Smit, E. *et al.* (2002) The early bactericidal activity of streptomycin. *Int. J. Tuberc. Lung Dis.*, **6** (8), 696–698.
92. Donald, P.R., Sirgel, F.A., Venter, A., Smit, E. *et al.* (2001) The early bactericidal activity of amikacin in pulmonary tuberculosis. *Int. J. Tuberc. Lung Dis.*, **5** (6), 533–538.
93. U.S. Department of Health and Human Services: Food and Drug Administration (1998) Guidance for Industry Providing Clinical Evidence of Effectiveness for Human Drug and Biological Products. http://www.fda.gov/downloads/Drugs/…/Guidances/ucm078749.pdf (accessed 3 September 2015).
94. Dartois, V., Leong, F.J., Dick, T. (2009) Tuberculosis Drug Discovery: Issues, Gaps and the Way Forward, in *Antiparasitic and Antibacterial Drug Discovery: from Molecular Targets to Drug Candidates* (ed. P.M. Selzer), Wiley-VCH Verlag, Weinheim, pp. 415–440.
95. Barry, C.E., III, Blanchard, J.S. (2010) The chemical biology of new drugs in the development for tuberculosis. *Curr. Opin. Chem. Biol.*, **14** (4), 456–466.
96. Lalloo, U.G., Ambaram, A. (2010) New anti-tuberculosis drugs in development. *Curr. HIV/AIDS Rep.*, 7 (3), 143–151.
97. Palomino, J.C., Ramos, D.F., de Silva, P.A. (2009) New anti-tuberculosis drugs: strategies, sources and new molecules. *Curr. Med. Chem.*, **16** (15), 1898–1904.
98. Ginsberg, A. (2010) Drugs in development for tuberculosis. *Drugs*, **70** (17), 2201–2214.
99. Rivers, E.C., Mancera, R.L. (2008) New anti-tuberculosis drugs in clinical trials with novel mechanism of action. *Drug Discov. Today*, **13** (23-24), 1090–1098.
100. Van den Boogaard, J., Kibiki, G.S., Kisanga, E.R., Boeree, M.J., Aarnoutse, R.E. (2009) New drugs against tuberculosis: problems, progress, and evaluation of agents in clinical development. *Antimicrob. Agents Chemother.*, **53** (3), 849–862.
101. Kaneko, T., Cooper, C., Mdluli, K. (2011) Challenges and opportunities in developing novel drugs for TB. *Future Med. Chem.*, **3** (11), 1373–1400.
102. Olaru, I.D., von Groote-Bidlingmaier, F., Heyckendorf, J., Yew, W.W. *et al.* (2014) Novel drugs against tuberculosis: clinician's perspective. *Eur. Respir. J.*, **45** (4), 1119–1131.
103. Andries, K., Verhasselt, P., Guillemont, J., Göhlmann, H.W. *et al.* (2005) A diarylquinoline drug active on the ATP synthase of Mycobacterium tuberculosis. *Science*, **307** (5707), 223–227.
104. Huitric, E., Verhasselt, P., Andries, K., Hoffner, S.E. (2007) In vitro antimycobacterial spectrum of a diarylquinoline ATP synthase inhibitor. *Antimicrob. Agents Chemother.*, **51** (11), 4202–4204.
105. Koul, A., Vranckx, L., Dendouga, N., Balemans, W. *et al.* (2008) Diarylquinolines are bactericidal for dormant mycobacteria as a result of disturbed ATP homeostasis. *J. Biol. Chem.*, **283** (37), 25273–25280.
106. Grosset, J.H., Singer, T.G., Bishai, W.R. (2012) New drugs for the treatment of tuberculosis: hope and reality. *Int. J. Tuberc. Lung Dis.*, **16** (8), 1005–10014.
107. Dhillon, J., Andries, K., Phillips, P.P.J., Mitchison, D.A. (2010) Bactericidal activity of the diarylquinoline TMC207 against *Mycobacterium tuberculosis* outside and within cells. *Tuberculosis (Edinb.)*, **90** (5), 301–305.
108. Hartkoorn, R.C., Uplekar, S., Cole, S.T. (2014) Cross-resistance between clofazimine and bedaquiline through upregulation of MmpL5 in *Mycobacterium tuberculosis*. *Antimicrob. Agents Chemother.*, **58** (5), 2979–2981.
109. Andries, K., Villellas, C., Coeck, N., Thys, K. *et al.* (2014) Acquired resistance of *Mycobacterium tuberculosis* to bedaquiline. *PLoS One*, **9** (7), e102135.
110. Reddy, V.M., Einck, L., Andries, K., Nacy, C.A. (2010) In vitro interactions between new antitubercular drug candidates SQ109 and TMC207. *Antimicrob. Agents Chemother.*, **54** (7), 2840–2846.
111. Diacon, A.H., Pym, A., Grobusch, M., Patientia, R. *et al.* (2009) The diarylquinoline TMC207 for multidrug-resistant tuberculosis. *N Eng. J. Med.*, **360** (23), 2397–2405.
112. European Medicines Agency (2013) European Medicines Agency Recommends Approval of a New Medicine for Multidrug-resistant Tuberculosis – 20

December 2013. Press Release. http://www.ema.europa.eu/ema/index.jsp?curl5pages/news_and_events/news/2013/12/news_detail_001999.jsp&mid5WC0b01ac058004d5c1 (accessed 9 January 2015).

113. US Food and Drug Administration (2012) FDA News Release on Bedaquiline Approval- December 31, 2012. www.fda.gov/newsevents/newsroom/pressannouncements/ucm333695.htm (accessed 9 January 2015).

114. World Health Organization (2013) *The Use of Bedaquiline in the Treatment of Multidrug-Resistant Tuberculosis Interim Policy Guidance*. World Health Organization, Geneva.

115. Tahlan, K., Wilson, R., Kastrinsky, D.B., Arora, K. et al. (2012) SQ109 targets MmpL3, a membrane transporter of trehalose monomycolate involved in mycolic acid donation to the cell wall core of *Mycobacterium tuberculosis. Antimicrob. Agents Chemother.*, **56** (4), 1797–1809.

116. Grzegorzewicz, A.E., Pham, H., Gundi, V.A., Scherman, M.S. et al. (2012) Inhibition of mycolic acid transport across the *Mycobacterium tuberculosis* plasma membrane. *Nat. Chem. Biol.*, **8** (4), 334–341.

117. Protopopova, M., Hanrahan, C., Nikonenko, B., Samala, R. et al. (2005) Identification of a new antitubercular drug candidate, SQ109, from a combinatorial library of 1,2-ethylenediamines. *J. Antimicrob. Chemother.*, **56** (5), 968–974.

118. Sacksteder, K.A., Protopopova, M., Barry, C.E. III, Andries, K., Nacy, C.A. (2012) Discovery and development of SQ109: a new antitubercular drug with a novel mechanism of action. *Future Microbiol.*, **7** (7), 823–837.

119. La Rosa, V., Poce, G., Canseco, J.O., Buroni, S. et al. (2012) MmpL3 is the cellular target of the antitubercular pyrrole derivative BM212. *Antimicrob. Agents Chemother.*, **56** (1), 324–331.

120. Chen, P., Gearhart, J., Protopopova, M., Einck, L., Nacy, C.A. (2006) Synergistic interactions of SQ109, a new ethylenediamine, with front-line antitubercular drugs in vitro. *J. Antimicrob. Chemother.*, **58** (2), 332–337.

121. Nikonenko, B.V., Protopopova, M., Samala, R., Einck, L., Nacy, C.A. (2007) Drug therapy of experimental tuberculosis (TB): improved outcome by combining SQ109, a new diamine antibiotic, with existing TB drugs. *Antimicrob. Agents Chemother.*, **51** (4),1563–1565.

122. Engohang-Ndong, J. (2012) Antimycobacterial drugs currently in phase II clinical trials and preclinical phase for tuberculosis treatment. *Expert. Opin. Invest. Drugs*, **21** (12), 1789–1800.

123. Makarov, V., Lechartier, B., Zhang, M., Neres, J. et al. (2014) Towards a new combination therapy for tuberculosis with next generation benzothiazinones. *EMBO Mol. Med.*, **6** (3), 372–383.

124. Makarov, V., Manina, G., Mikusova, K., Möllmann, U. et al. (2009) Benzothiazinones kill *Mycobacterium tuberculosis* by blocking arabinan synthesis. *Science*, **324** (5928), 801–804.

125. Trefzer, C., Rengifo-Gonzalez, M., Hinner, M.J., Schneider, P. et al. (2010) Benzothiazinones: prodrugs that covalently modify the decaprenylphosphoryl-b-D-ribose-2′-epimerase DprE1 of *Mycobacterium tuberculosis. J. Am. Chem. Soc.*, **132** (39), 13663–13665.

126. Pasca, M.R., Degiacomi, G., Ribeiro, A.L., Zara, F. et al. (2010) Clinical isolates of *Mycobacterium tuberculosis* in four European hospitals are uniformly susceptible to benzothiazinones. *Antimicrob. Agents Chemother.*, **54** (4), 1616–1618.

127. Lechartier, B., Hartkoorn, R.C., Cole, S.T. (2012) *In vitro* combination studies of benzothiazinone lead compound BTZ043 against *Mycobacterium tuberculosis. Antimicrob. Agents Chemother.*, **56** (11), 5790–5793.

128. Arora, S.K., Sinha, N., Sinha, R.K. (2004) Synthesis and In Vitro Anti-Mycobacterial Activity of a Novel Anti-TB Composition LL4858 [Abstract No. F-1115]. *American Society for Microbiology: Program and Abstracts of the 44th Interscience Conference on Antimicrobial Agents and Chemotherapy*. American Society for Microbiology, Washington, p. 212.

129. Sinha, R.K., Arora, S.K., Sinha, N. (2004) In Vivo Activity of LL4858 Against Mycobacterium Tuberculosis [Abstract No. F-1116]. *American Society for Microbiology: program and abstracts of the 44th Interscience Conference on Antimicrobial Agents and Chemotherapy*. American Society for Microbiology, Washington, p. 212.

130. Shi, R., Sugawara I. (2010) Development of new anti-tuberculosis drug candidates. *Tohoku J. Exp. Med.*, **221** (2), 97–106.
131. Yew, W., Cynamon, M., Zhang, Y. (2011) Emerging drugs for the treatment of tuberculosis. *Expert Opin. Emerging Drugs*, **16** (1), 1–21.
132. Lupin Limited (2011) A Phase IIa, Open Label Study to Determine the EBA, Extended EBA and Pharmacokinetic Study of LL3858 for the Treatment of Newly Diagnosed Sputum Smear-Positive Pulmonary TB. http://www.newtbdrugs.org/project.php?id=155 (accessed 15 January 2015).
133. Van Ingen, J., Aarnoutse, RE., Donald, P.R., Diacon, A.H. *et al.* (2011) Why do we use 600 mg of rifampicin in tuberculosis treatment? *Clin. Infect. Dis.*, **52** (9), e194–e199.
134. Steingart, K.R., Jotblad, S., Robsky, K., Deck, D. *et al.* (2011) Higher-dose rifampin for the treatment of pulmonary tuberculosis: a systematic review. *Int. J. Tuberc. Lung Dis.*, **15** (3), 305–316.
135. Diacon, A.H., Patientia, R.F., Venter, A., van Helden, P.D. *et al.* (2007) Early bactericidal activity of high-dose rifampin in patients with pulmonary tuberculosis evidenced by positive spectrum smears. *Antimicrob. Agents Chemother.*, **51** (8), 2944–2996.
136. Jayaram, R., Gaonkar, S., Kaur, P., Suresh, B.L. *et al.* (2003) Pharmacokinetics–pharmacodynamics of rifampin in an aerosol infection model of tuberculosis. *Antimicrob. Agents Chemother.*, **47** (7), 2118–2124.
137. Kreis, B., Pretet, S., Birenbaum, J., Guibout, P. *et al.* (1976) Two three-month treatment regimens for pulmonary tuberculosis. *Bull. Int. Union Tuberc.*, **51** (1), 71–75.
138. Boeree, M., Diacon, A., Dawson, R., Venter, A. *et al.* (2013) What is the "right" dose of rifampin? 20th Conference on Retroviruses and Opportunistic Infections, Atlanta, GA, 3–6 March, p. 148LB.
139. Aarnoutse, R., Colbers, A., and the PanACEA Consortium (2015) Pharmacokinetics of high-dose rifampicin, moxifloxacin and firstline TB drugs in the PanACEA-MAMS-TB-01 trial. Presentation at the 8th International Workshop on Clinical Pharmacology of TB drugs, 17 September, San Diego, USA.
140. Jindani, A., Shrestha, B., Westernmann de Patino, I., Alvarez de Fernandes, R. *et al.* (2013) A multicentre randomised controlled clinical trial to evaluate the toxicity of high dose rifampicin in the treatment of pulmonary tuberculosis (RIFATOX). Presentation at the 44th Union World Conference on Lung Health, Paris, France, 30 October–3 November.
141. Pym, A.S. (2014) New Developments in Drug Treatment, in *Clinical Tuberculosis*, 5th edn (eds P.D.O. Davies, S.B. Gordon, G. Davies), Taylor & Francis Group, London, pp. 241–252.
142. Rosenthal, I.M., Zhang, M., Williams, K.N., Peloquin, C.A. *et al.* (2007) Daily dosing of rifapentine cures tuberculosis in three months or less in the murine model. *PLoS Med.*, **4** (12), e344.
143. Rosenthal, I.M., Zhang, M., Almeida, D., Grosset, J.H., Nuermberger, E.L. (2008) Isoniazid or moxifloxacin in rifapentine-based regimens for experimental tuberculosis? *Am. J. Respir. Crit. Care. Med.*, **178** (9), 989–993.
144. Jindani, A, Harrison, T.S., Nunn, A.J., Phillips, P.P. *et al.* (2014) High-dose rifapentine with moxifloxacin for pulmonary tuberculosis. *N. Engl. J. Med.*, **371** (17), 1599–1608.
145. Dorman, S.E., Goldberg, S., Stout. J.E., Muzanyi, G. *et al.* (2012) Substitution of rifapentine for rifampin during intensive phase treatment of pulmonary tuberculosis: study 29 of the tuberculosis trials consortium. *J. Infect. Dis.*, **206** (7), 1030–1040.
146. Rosenthal, I.M., Tasneen, R., Peloquin, C.A., Zhang, M. *et al.* (2012) Dose-ranging comparison of rifampin and rifapentine in two pathologically distinct murine models of tuberculosis. *Antimicrob. Agents Chemother.*, **56** (8), 4331–4340.
147. Dooley, K.E., Savic, R., Park, J.G. (2014) Rifapentine safety and PK with novel dosing strategies to increase drug exposures for TB: ACTG A5311 (Abstract 816). Presentation at the Conference on Retroviruses and Opportunistic Infections, Boston, MA, 5 March.
148. Conde, M.B., Cavalcante, S.C., Dalcolmo, M. (2014) A phase 2 trial of a rifapentine plus moxifloxacin-based regimen for pulmonary TB treatment (Abstract 93).

Oral Abstract Presented at: Conference on Retroviruses and Opportunistic Infections, Boston, MA, 5 March.

149. Dooley, K.E., Mitnick, C.D., Ann DeGroote, M., Obuku, E. et al. (2012) Old drugs, new purpose: retooling existing drugs for optimized treatment of resistant tuberculosis. *Clin. Infect. Dis.*, **55** (4), 572–581.

150. Schaaf, H.S., Victor, T.C., Venter, A., Brittle, W. et al. (2009) Ethionamide cross- and co-resistance in children with isoniazid-resistant tuberculosis. *Int. J. Tuberc. Lung Dis.*, **13** (11), 1355–1359.

151. Cynamon, M.H., Zhang, Y., Harpster, T., Cheng, S., DeStefano, M.S. (1999) High-dose isoniazid therapy for isoniazid-resistant murine *Mycobacterium tuberculosis* infection. *Antimicrob. Agents Chemother.*, **43** (12), 2922–2924.

152. Donald, P.R., Sirgel, F.A., Botha, F.J., Seifart, H.I. et al. (1997) The early bactericidal activity of isoniazid related to its dose size in pulmonary tuberculosis. *Am. J. Respir. Crit. Care. Med.*, **156** 3 (Pt. 1), 895–900.

153. Katiyar, S.K., Bihari, S., Prakash, S., Mamtani, M., Kulkarni, H. (2008) A randomised controlled trial of high-dose isoniazid adjuvant therapy for multidrug-resistant tuberculosis. *Int. J. Tuberc. Lung Dis.*, **12** (2), 139–145.

154. Lange, C., Abubakar, I., Alffenaar, J.W., Bothamley, G. et al. (2014) Management of patients with multidrug-resistant/extensively drug-resistant tuberculosis in Europe: a TBNET consensus statement. *Eur. Respir. J.*, **44** (1), 23–63.

155. AIDS Clinical Trials Group Network (2015) A5312: High-Dose Isoniazid. https://actgnetwork.org/study/a5312-high-dose-isoniazid (accessed 20 October 2015).

156. Yee, L.C., Wei, Y.C. (2012) Current Modeling Methods Used in QSAR/QSPR, in *Statistical Modeling of Molecular Descriptors in QSAR/QSPR* (eds M. Dehmer, K. Varmuza, D. Bonchev), Wiley-VCH Verlag GmbH & Co., Weinheim, pp. 3–31.

157. Chohan, K.K., Paine, S.W., Waters, N.J. (2008) Advancements in predictive in silico models for ADME. *Curr. Chem. Biol.*, **2** (3), 215–228.

158. Martins, F., Santos, S., Ventura, C., Elvas-Leitão, R. et al. (2014) Design, synthesis and biological evaluation of novel isoniazid derivatives with potent antitubercular activity. *Eur. J. Med. Chem.*, **81**, 119–138.

159. Nuermberger, E.L., Yoshimatsu, T., Tyagi, S., O'Brien, R.J. et al. (2004) Moxifloxacin-containing regimen greatly reduces time to culture conversion in murine tuberculosis. *Am. J. Respir. Crit. Care. Med.*, **169** (3), 421–426.

160. Nuermberger, E.L, Yoshimatsu, T., Tyagi, S., Williams, K. et al. (2004) Moxifloxacin-containing regimens of reduced duration produce a stable cure in murine tuberculosis. *Am. J. Respir. Crit. Care. Med.*, **170** (10), 1131–1134.

161. Conde, M.B., Efron, A., Loredo, C., De Souza, G.R. et al. (2009) Moxifloxacin versus ethambutol in the initial treatment of tuberculosis: a double-blind, randomised, controlled phase II trial. *Lancet*, **373** (9670), 1183–1189.

162. Rustomjee, R., Lienhardt C., Kanyok, T., Davies, G.R. et al. (2008) A phase II study of the sterilising activities of ofloxacin, gatifloxacin and moxifloxacin in pulmonary tuberculosis. *Int. J. Tuberc. Lung Dis.*, **12** (2), 128–138.

163. Jawahar, M.S., Banurekha, V.V., Paramasivan, C.N, Rahman, F. et al. (2013) Randomized clinical trial of thrice-weekly 4-month moxifloxacin or gatifloxacin containing regimens in the treatment of new sputum positive pulmonary tuberculosis patients. *PLoS One*, **8** (7), e67030.

164. Merle, C.S., Fielding, K., Sow, O.B., Gninafon, M. et al. (2014) A four-month gatifloxacin-containing regimen for treating tuberculosis. *N. Engl. J. Med.*, **371** (17), 1588–1598.

165. Gillespie, S.H., Crook, A.M., McHugh, T.D., Mendel, C.M. et al. (2014) Four-month moxifloxacin-based regimens for drug-sensitive tuberculosis. *N. Engl. J. Med.*, **371** (17), 1577–1587.

166. Wayne, L.G., Sramek, H.A. (1994) Metronidazole is bactericidal to dormant cells of *Mycobacterium tuberculosis. Antimicrob. Agents Chemother.*, **38** (9), 2054–2058.

167. Cellitti, S.E., Shaffer, J., Jones, D.H., Mukherjee, T. et al. (2012) Structure of Ddn, the deazaflavin-dependent nitroreductase from *Mycobacterium tuberculosis* involved in bioreductive activation of PA-824. *Structure*, **20** (1), 101–112.

168. Manjunatha, U.H., Boshoff, H., Dowd, C.S., Zhang, L. et al. (2006) Identification of a nitroimidazo-oxazine-specific protein involved in PA-824 resistance in *Mycobacterium tuberculosis*. *Proc. Natl Acad. Sci. USA*, **103** (2), 431–436.
169. Singh, R., Manjunatha, U., Boshoff, H.I.M. (2008) PA-824 kills nonreplicating *Mycobacterium tuberculosis* by intracellular NO release. *Science*, **322** (5906), 1392–1395.
170. Manjunatha, U., Boshoff, H.I, Barry, C.E. (2009) The mechanism of action of PA-824 novel insights from transcriptional profiling. *Commun. Integr. Biol.*, **2** (3), 215–218.
171. Somasundaram, S., Anand, R.S., Venkatesan, P., Paramasivan, C.N. (2013) Bactericidal activity of PA-824 against *Mycobacterium tuberculosis* under anaerobic conditions and computational analysis of its novel analogues against mutant Ddn receptor. *BMC Microbiol.*, **13**, 218.
172. Stover, C.K., Warrener, P., VanDevanter, D.R., Sherman, D.R. et al. (2000) A small-molecule nitroimidazopyran drug candidate for the treatment of tuberculosis. *Nature*, **405** (6789), 962–966.
173. Ginsberg, A.M., Laurenzi, M.W., Rouse, D.J., Whitney, K.D., Spigelman, M.K. (2009) Safety, tolerability, and pharmacokinetics of PA-824 in healthy subjects. *Antimicrob. Agents Chemother.*, **53** (9), 3720–3725.
174. Lenaerts, A.J., Gruppo, V., Marietta KS., Johnson, C,M. et al. (2005) Preclinical testing of the nitroimidazopyran PA-824 for activity against *Mycobacterium tuberculosis* in a series of in vitro and in vivo models. *Antimicrob. Agents Chemother.*, **49** (6), 2294–2301.
175. Tyagi, S., Nuermberger, E., Yoshimatsu, T., Williams, K. et al. (2005) Bactericidal activity of the nitroimidazopyran PA-824 in a murine model of tuberculosis. *Antimicrob. Agents Chemother.*, **49** (6), 2289–2293.
176. Choi, K.P., Bair, T.B., Bae, Y.M., Daniels, L. (2001) Use of transposon Tn*5367* mutagenesis and a nitroimidazopyran-based selection system to demonstrate a requirement for *fbiA* and *fbiB* in coenzyme F420 biosynthesis by *Mycobacterium bovis* BCG. *J. Bacteriol.*, **183** (24), 7058–7066.
177. Choi, K.P., Kendrick, N., Daniels, L. (2002) Demonstration that *fbiC* is required by *Mycobacterium bovis* BCG for coenzyme F420 and FO biosynthesis. *J. Bacteriol.*, **184** (9), 2420–2428.
178. Feuerriegel, S., Köser, C.U., Bau, D., Rushes-Gerdes, S. et al. (2011) Impact of Fgd1 and ddn diversity in *Mycobacterium tuberculosis* complex on in vitro susceptibility to PA-824. *Antimicrob. Agents Chemother.*, **55** (12), 5718–5722.
179. Matsumoto, M., Hashizume, H., Tomishige, T., Kawasaki, M. et al. (2006) OPC-67683, a nitro-dihydro-imidazooxazole derivative with promising action against tuberculosis in vitro and in mice. *PLoS Med.*, **3** (11), e466.
180. Diacon, A.H, Dawson, R., du Bois, J., Narunsky, K. et al. (2012) Phase II dose-ranging trial of the early bactericidal activity of PA-824. *Antimicrob. Agents Chemother.*, **56** (6), 3027–3031.
181. Lessem, E. (2014) Tuberculosis Drug Development Hobbles Forward, in *2014 Pipeline Report – Drugs, Diagnostics, Preventive Technologies Reach toward a Cure, and Immune-Based and Gene Therapies in Development* (ed. A, Benzacar). HIV i-Base & Treatment Action Group, London, New York, pp. 197–216.
182. Sasaki, H., Haraguchi, Y., Itotani, M., Kuroda, H. et al. (2006) Synthesis and antituberculosis activity of a novel series of optically active 6-nitro-2,3-dihydroimidazo[2,1-b]oxazoles. *J. Med. Chem.*, **49** (26), 7854–7860.
183. Saliu, O.Y., Crismale, C., Schwander, S.K., Wallis, R.S. (2007) Bactericidal activity of OPC 67683 against drug-tolerant *Mycobacterium tuberculosis*. *J. Antimicrob. Chemother.*, **60** (5), 994–998.
184. Gler, M.T., Skripconoka, V., Sanchez-Garavito, E., Xiao, H. et al. (2012) Delamanid for multidrug-resistant pulmonary tuberculosis. *N. Engl. J. Med.*, **366** (23), 2151–2160.
185. Skripconoka, V., Danilovits, M., Pehme, L., Tomson, T. et al. (2013) Delamanid improves outcomes and reduces mortality in multidrug-resistant tuberculosis. *Eur. Respir. J.*, **41** (6), 1393–1400.
186. European Medicines Agency (2014) Deltyba (Delamanid): Authorisation Details. http://www.ema.europa.eu/ema/index.jsp?curl=pages/medicines/human/medicines/002552/human_med_001699.jsp&mid=WC0b01 (accessed 20 March 2015).

187. World Health Organization (2013) *The Use of Delamanid in the Treatment of Multidrug-Resistant Tuberculosis: Interim Policy Guidance*. World Health Organization, Geneva.
188. Alcala, L., Ruiz-Serrano, M.J., Perez-Fernandez Turegano, C., Garcia de Viedma, D. et al. (2003) In vitro activities of linezolid against clinical isolates of *Mycobacterium tuberculosis* that are susceptible or resistant to first-line antituberculosis drugs. *Antimicrob. Agents Chemother.*, **47** (1), 416–417.
189. Cynamon, M.H., Klemens, S.P., Sharpe, C.A., Chase, S. (1999) Activities of several novel oxazolidinones against *Mycobacterium tuberculosis* in a murine model. *Antimicrob. Agents Chemother.*, **43** (5), 1189–1191.
190. Weiss, T., Schönfeld, N., Otto-Knapp, R., Bös, L. et al. (2015) Low minimal inhibitory concentrations of linezolid against multidrug-resistant tuberculosis strains. *Eur. Respir. J.*, **45** (1), 285–287.
191. Fattorini, L., Tan, D., Iona, E., Mattei, M. et al. (2003) Activities of moxifloxacin alone and in combination with other antimicrobial agents against multidrug-resistant *Mycobacterium tuberculosis* infection in BALB/c mice. *Antimicrob. Agents Chemother.*, **47**, 360–362.
192. Sotgiu, G., Centis, R., D'Ambrosio, L., Alffenaar, J.W. et al. (2012) Efficacy, safety and tolerability of linezolid containing regimens in treating MDR-TB and XDR-TB: systematic review and meta-analysis. *Eur. Respir. J.*, **40** (6), 1430–1442.
193. Tang, S., Yao, L., Hao, X., Zhang, X. et al. (2015) Efficacy, safety and tolerability of linezolid for the treatment of XDR-TB: a study in China. *Eur. Respir. J.*, **45** (1), 161–170.
194. Sotgiu, G., Centis, R., D'Ambrosio, L., Castiglia, P., Migliori, G.B. (2015) Low minimal inhibitory concentrations of linezolid against multidrug-resistant tuberculosis strains. *Eur. Respir. J.*, **45** (1), 287–289.
195. Barbachyn, M.R., Hutchinson, D.K., Brickner, S.J., Cynamon, M.H. et al. (1996) Identification of a novel oxazolidinone (U-100480) with potent antimycobacterial activity. *J. Med. Chem.*, **39** (3), 680–685.
196. Wallis, R.S., Jakubiec, W.M., Kumar, V., Silvia, A.M. et al. (2010) Pharmacokinetics and whole-blood bactericidal activity against Mycobacterium tuberculosis of single doses of PNU-100480 in healthy volunteers. *J. Infect. Dis.*, **202** (5), 745–751.
197. Alffenaar, J.W., van der Laan. T., Simons, S., van der Werf, T.S. et al. (2011) Susceptibility of clinical *Mycobacterium tuberculosis* isolates to a potentially less toxic derivate of linezolid, PNU-100480. *Antimicrob. Agents Chemother.*, **55** (3), 1287–1289.
198. Reddy, V.M., Dubuisson, T., Einck, L., Wallis, R.S. et al. (2012) SQ109 and PNU-100480 interact to kill *Mycobacterium tuberculosis* in vitro. *J. Antimicrob. Chemother.*, **67** (5), 1163–1166.
199. Wallis, R.S., Jakubiec, W., Mitton-Fry, M., Ladutko, L. et al. (2012) Rapid evaluation in whole blood culture of regimens for XDR-TB containing PNU-100480 (sutezolid), TMC207, PA-824, SQ109, and pyrazinamide. *PLoS One*, **7** (1), e30479.
200. Williams, K.N., Brickner, S.J., Stover, C.K., Zhu, T. et al. (2009) Addition of PNU-100480 to first-line drugs shortens the time needed to cure murine tuberculosis. *Am. J. Respir. Crit. Care. Med.*, **180** (4), 371–376.
201. Wallis, R., Dawson, R., Friedrich, S., Venter, A. et al. (2014) Mycobactericidal activity of sutezolid (PNU-100480) in sputum (EBA) and blood (WBA) of patients with pulmonary tuberculosis. *PLoS One*, **9** (4), e94462.
202. Wallis, R.S., Jakubiec, W., Kumar, V., Bedarida, G. et al. (2011) Biomarker-assisted dose selection for safety and efficacy in early development of PNU-100480 for tuberculosis. *Antimicrob. Agents Chemother.*, **55** (2), 567–574.
203. Balasubramanian, V., Solapure, S., Iyer, H., Ghosh, A. et al. (2014) Bactericidal activity and mechanism of action of AZD5847, a novel oxazolidinone for treatment of tuberculosis. *Antimicrob. Agents Chemother.*, **58** (1), 495–502.
204. Reele, S., Xiao, A.J., Das, S., Balasubramanian, V. et al. (2011) 14 day multiple ascending dose study with AZD5847 Is well tolerated at predicted exposure for treatment of tuberculosis (TB). 51st Interscience Conference on Antimicrobial Agents and Chemotherapy: Abstract A11735. http://www.abstractsonline.com/plan/View Abstract.aspx?mID=2789&sKey=0b498641-f10a-4936-b80c-d274ffe4b143&cKey

=2271e63a-2cdc-4956-b7b0-6f7dc642b346&mKey=0c918954-d607-46a7-8073-44f4b537a439 (accessed 24 January 2015).

205. ClinicalTrials.gov (2014) Phase 2a EBA Trial of AZD5847. Date last updated: April 16, 2014. http://clinicaltrials.gov/show/NCT01516203 (accessed 18 January 2015).

206. AstraZeneca (2014) AZD5847 Oxazolidinone for the Treatment of Tuberculosis. http://www.newtbdrugs.org/meetings/annual2013/downloads/presentations/11_S-Butler_AZD5847_WGND_2013.pdf (accessed 25 March 2015)

207. Flores, A.R., Parsons, L.M., Pavelka, M.S. Jr, (2005) Genetic analysis of the beta-lactamases of *Mycobacterium tuberculosis* and *Mycobacterium smegmatis* and susceptibility to beta-lactam antibiotics. *Microbiology*, **151** (Pt. 2), 521–532.

208. Hugonnet, J.E., Blanchard, J.S. (2007) Irreversible inhibition of the *Mycobacterium tuberculosis* beta-lactamase by clavulanate. *Biochemistry*, **46** (43), 11198–12004.

209. Hugonnet, J.E., Tremblay, L.W., Boshoff, H.I, Barry, C.E. III, Blanchard, J.S. (2009) Meropenem-clavulanate is effective against extensively drug-resistant *Mycobacterium tuberculosis*. *Science*, **323** (5918), 1215–1218.

210. Veziris, N., Truffot, C., Mainardi, J.L., Jarlier, V. (2011) Activity of carbapenems combined with clavulanate against murine tuberculosis. *Antimicrob. Agents Chemother.*, **55** (6), 2597–2600.

211. England, K., Boshoff, H.I., Arora, K., Weiner, D. et al. (2012) Meropenem-clavulanic acid shows activity against *Mycobacterium tuberculosis* in vivo. *Antimicrob. Agents Chemother.*, **56** (6), 3384–3387.

212. Payen, M.C., De Wit, S., Martin, C., Sergysels, R. et al. (2012) Clinical use of the meropenem-clavulanate combination for extensively drug-resistant tuberculosis. *Int. J. Tuberc. Lung Dis.*, **16** (4), 558–560.

213. De Lorenzo, S., Alffenaar, J.W., Sotgiu, G., Centis, R. et al. (2013) Efficacy and safety of meropenem-clavulanate added to linezolid-containing regimens in the treatment of MDR-/XDR-TB. *Eur. Respir. J.*, **41** (6), 1386–1392.

214. Mushtaq, S., Hope, R., Warner, M., Livermore, D.M. (2007) Activity of faropenem against cephalosporin-resistant Enterobacteriaceae. *J. Antimicrob. Chemother.*, **59** (5), 1025–1030.

215. Dalhoff, A., Nasu, T., Okamoto, K. (2003) Beta-lactamase stability of faropenem. *Chemotherapy*, **49** (5), 229–236.

216. Dhar, N., Dubée, V., Ballell, L., Cuinet, G. et al. (2014) Rapid cytolysis of *Mycobacterium tuberculosis* by faropenem, an orally bioavailable β-lactam antibiotic. *Antimicrob. Agents Chemother.*, **59** (2), 1308–1319.

217. Yano, T., Kassovska-Bratinova, S., Teh, J.S., Winkler, J. et al. (2011) Reduction of clofazimine by mycobacterial type 2 NADH:quinone oxidoreductase: a pathway for the generation of bactericidal levels of reactive oxygen species. *J. Biol. Chem.*, **286** (12), 10276–10287.

218. Cholo, M.C., Steel, H.C., Fourie, P.B., Germishuizen, W.A., Anderson, R. (2012) Clofazimine: current status and future prospects. *J. Antimicrob. Chemother.*, **67** (2), 290–298.

219. Grant, S.S., Kaufmann, B.B., Chand, N.S., Haseley, N., Hung, D.T. (2012) Eradication of bacterial persisters with antibiotic-generated hydroxyl radicals. *Proc. Natl Acad. Sci. USA*, **109** (30), 12147–12152.

220. Williams, K., Minkowski, A., Amoabeng, O., Peloquin, C.A. et al. (2012) Sterilizing activities of novel combinations lacking first- and second-line drugs in a murine model of tuberculosis. *Antimicrob. Agents Chemother.*, **56** (6), 3114–3120.

221. Van Deun, A., Maug, A.K., Salim, M.A., Das, P.K. et al. (2010) Short, highly effective, and inexpensive standardized treatment of multidrug-resistant tuberculosis. *Am. J. Respir. Crit. Care Med.*, **182** (5), 684–692.

222. Padayatchi, N., Gopal, M., Naidoo, R., Werner, L. et al. (2014) Clofazimine in the treatment of extensively drug-resistant tuberculosis with HIV coinfection in South Africa: a retrospective cohort study. *J. Antimicrob. Chemother.*, **69** (11), 3103–3107.

223. Tyagi, S., Ammerman, N.C., Li, S.Y., Adamson, J. et al. (2015) Clofazimine shortens the duration of the first-line treatment regimen for experimental chemotherapy of tuberculosis. *Proc. Natl Acad. Sci. USA*, **112** (3), 869–874.

224. Zhang, D., Lu, Y., Liu, K., Liu, B. et al. (2012) Identification of less lipophilic riminophenazine derivatives for the treatment of drug-resistant tuberculosis. *J. Med. Chem.*, **55** (19), 8409–8417.
225. Alsaad, N, van Altena, R., Pranger, A.D., van Soolingen, D. et al. (2013) Evaluation of co-trimoxazole in the treatment of multidrug-resistant tuberculosis. *Eur. Respir. J.*, **42** (2), 504–512.
226. Alsaad, N., van der Laan, T., van Altena, R., Wilting, K.R. et al. (2013) Trimethoprim/sulfamethoxazole susceptibility of *Mycobacterium tuberculosis*. *Int. J. Antimicrob. Agents*, **42** (5), 472–474.
227. Davies Forsman, L., Schön, T., Simonsson, U.S., Bruchfeld, J. et al. (2014) Intra- and extracellular activities of trimethoprim-sulfamethoxazole against susceptible and multidrug-resistant *Mycobacterium tuberculosis*. *Antimicrob. Agents Chemother.*, **58** (12), 7557–7559.
228. Huang, T.S., Kunin, C.M., Yan, B.S., Chen, Y.S. et al. (2012) Susceptibility of *Mycobacterium tuberculosis* to sulfamethoxazole, trimethoprim and their combination over a 12-year period in Taiwan. *J. Antimicrob. Chemother.*, **67** (3), 633–637.
229. Diacon, A H., Dawson, R., von Groote-Bidlingmaier, F., Symons, G. et al. (2012) 14-day bactericidal activity of PA-824, bedaquiline, pyrazinamide, and moxifloxacin combinations: a randomised trial. *Lancet*, **380** (9846), 986–993.
230. Diacon, A.H., Dawson, R., von Groote-Bidlingmaier, F., Symons, G. et al. (2015) Bactericidal activity of pyrazinamide and clofazimine alone and in combinations with pretomanid and bedaquiline. *Am. J. Respir. Crit. Care Med.*, **191** (8), 943–953.
231. Dawson, R., Diacon, A.H., Everitt, D., van Niekerk, C. et al. (2015) Efficiency and safety of the combination of moxifloxacin, pretomanid (PA-824), and pyrazinamide during the first 8 weeks of antituberculosis treatment: a phase 2b, open-label, partly randomised trial in patients with drug-susceptible or drug-resistant pulmonary tuberculosis. *Lancet*, **385** (9979), 1738–1747.

Part III
Nanomedicine

26
Microfluidics in Nanomedicine

YongTae Kim[1] and Robert Langer[2]
[1]*Georgia Institute of Technology, George W. Woodruff School of Mechanical Engineering, Wallace H. Coulter Department of Biomedical Engineering, Institute for Electronics and Nanotechnology, Parker H. Petit Institute for Bioengineering and Bioscience, 345 Ferst Drive, Atlanta, GA 30318, USA*
[2]*Massachusetts Institute of Technology, Department of Chemical Engineering, Harvard-MIT Division of Health Sciences and Technology, David H. Koch Institute for Integrative Cancer Research, 500 Main Street, Cambridge, MA 02139, USA*

1	**Introduction**	**944**
1.1	Nanomedicine Development	945
1.2	Microfluidics Technology	945
2	**Microfluidic Assembly of Nanomedicines**	**946**
3	**Microfluidic Characterization of Nanomedicines**	**949**
4	**Microfluidic Evaluation of Nanomedicines**	**951**
4.1	Mimicking Physiological Environments	952
4.2	Endothelial Cell Systems	952
4.3	"Organ-On-A-Chip"	954
4.4	Renal Toxicity and Hepatotoxicity	955
4.5	Live Tissue Explants	955
4.6	Intact Organisms	957
5	**Challenges and Opportunities**	**957**
6	**Concluding Remarks**	**959**
	Acknowledgments	959
	References	959

Translational Medicine: Molecular Pharmacology and Drug Discovery
First Edition. Edited by Robert A. Meyers.
© 2018 Wiley-VCH Verlag GmbH & Co. KGaA. Published 2018 by Wiley-VCH Verlag GmbH & Co. KGaA.

Keywords

Microfluidics
The science and technology that involves the manipulation of nanoscale amounts of fluids in microscale fluidic channels for applications that include chemical synthesis, and biological analysis and engineering.

Nanotechnology
The manipulation of matter on atomic and molecular scales.

Nanomedicine
The medical application of nanotechnology for the advanced diagnosis, treatment and prevention of a number of diseases.

Biomimetic microsystem
A microscale device that mimics biological systems and is used to probe complex human problems.

Clinical translation
Clinical translation involves the application of discoveries made in the laboratory to diagnostic tools, medicines, procedures, policies and education, in order to improve the health of individuals and the community.

Nanomedicine is the medical application of nanotechnology for the treatment and prevention of major ailments, including cancer and cardiovascular diseases. Despite the progress and potential of nanomedicines, many such materials fail to reach clinical trials due to critical challenges that include poor reproducibility in high-volume production that have led to failure in animal studies and clinical trials. Recent approaches using microfluidic technology have provided emerging platforms with great potential to accelerate the clinical translation of nanomedicine. Microfluidic technologies for nanomedicine development are reviewed in this chapter, together with a detailed discussion of microfluidic assembly, characterization and evaluation of nanomedicine, and a description of current challenges and future prospects.

1 Introduction

Nanomedicine is the medical application of nanotechnology that uses engineered nanomaterials for the robust delivery of therapeutic and diagnostic agents in the advanced treatment of many diseases, including cancer [1–3], atherosclerosis [4–6], diabetes [7–9], pulmonary diseases [10, 11] and disorders of the central nervous system [12, 13]. One key advantage of nanomedicine is the ability to deliver poorly water-soluble drugs [14–16] or plasma-sensitive nucleic acids (e.g., small interfering (si)RNA [17, 18]) into the circulation with enhanced stability. Nanomedicine is also capable of providing contrast agents for different imaging modalities and the targeting of specific sites for the delivery of drugs and/or genes [19–23]. Engineered nanomaterials, developed as particulates that are widely referred to as nanoparticles (NPs), have been formulated using a variety of materials that includes lipids, polymers, inorganic nanocrystals, carbon nanotubes, proteins, and DNA origami [24–36]. The ultimate goal of nanomedicine is to achieve a robust, targeted delivery of complex assemblies that contain sufficient amounts of multiple therapeutic and diagnostic agents for highly localized drug release, but with no adverse side effects [37, 38], and a reliable detection of any site-specific therapeutic response [39, 40].

1.1 Nanomedicine Development

Typical nanomedicine development processes for the clinical translation include benchtop syntheses, characterizations, *in-vitro* evaluations, *in-vivo* evaluations with animal models, and scaled-up production in readiness for clinical trials. Although, previously, several NPs have been reported as superior platforms, many are still far from their first stages of patient clinical trials due to several critical challenges [41, 42]. Such challenges mainly result from batch-to-batch variations of NPs produced in the benchtop synthesis process, and from insignificant outcomes in the *in-vitro* evaluation process under physiologically irrelevant conditions. These limitations ultimately lead to highly variable results in the *in-vivo* evaluation, or to failure in clinical trials. In order to address these challenges, the following methodologies need to be established in the nanomedicine development process:

- Nanomedicine needs to be continuously produced in a high-throughput fashion. The large-scale, continuous production of nanomedicines will allow a robust supply of highly reproducible materials for the *in-vitro* and *in-vivo* evaluation stages and clinical trials, ultimately increasing the success rate in clinical trials.
- Nanomedicines synthesized using large-scale, continuous production methods also need to be characterized in a high-throughput manner. Rapid characterization will create an efficient production cycle for an optimized nanomedicine via feedback loops between the synthesis and characterization stages.
- The *in-vitro* evaluation of nanomedicine must be conducted in more physiologically relevant environments. Highly repeatable results obtained from these biomimetic conditions will allow the obviation of a number of simple screening experiments in animal studies, not only saving costly animal models but also accelerating the clinical translation.

1.2 Microfluidics Technology

Microfluidics technology provides highly compatible platforms to create a new nanomedicine development pipelines that include the required methodologies introduced above. Basically, microfluidics presents a number of useful capabilities to manipulate very small quantities of samples, and to detect substances with a high resolution for a wide range of applications, including chemical syntheses [43, 44] and biological analysis [45, 46]. More importantly, the adaptability of microfluidics allows its integration with many other technologies, such as micro/nanofabrication, electronics, and feedback control systems [47–52]. Recently, microfluidic platforms integrated with control systems and advanced microfabrication technologies have been used to address the critical challenges in nanomedicine [53–57]. For example, the continuous synthesis of NPs in microfluidics has demonstrated a versatility to produce a variety of NPs with different sizes, shapes, and surface compositions [58, 59]. Several advances have recently been made in the label-free detection, characterization and identification of single NPs [60]. The confluence of microfluidics and biomimetic design has enabled the creation of physiologically relevant microenvironments for the evaluation of drug candidates [61–63]. The key microfluidic technologies in nanomedicine, including microfluidic assembly, and the characterization and evaluation of nanomedicines, are discussed in the following sections (see Fig. 1), and their current challenges and future research directions are highlighted.

2
Microfluidic Assembly of Nanomedicines

The bulk synthesis of NPs typically has strong dependencies on nonstandard multistep processes which are time-consuming, difficult to scale up, and depend heavily on specific synthetic

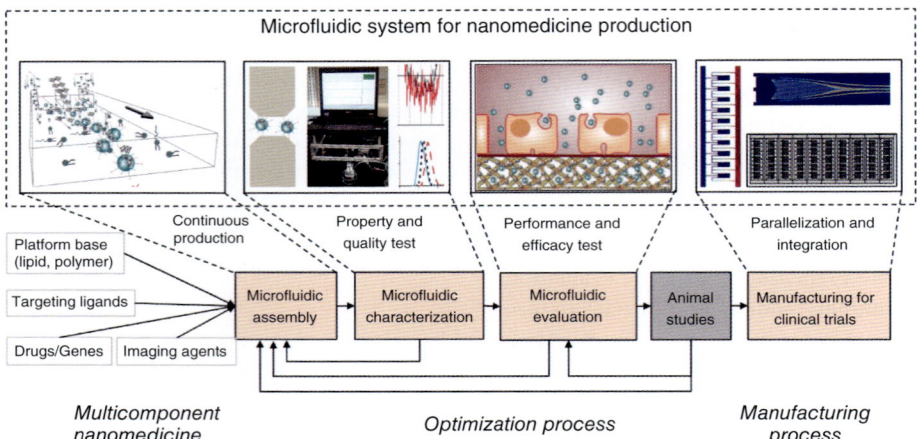

Fig. 1 A new nanomedicine development pipeline using microfluidic systems. First, a designed nanomedicine with multiple precursor components is continuously assembled through controlled strong mixing patterns in the Microfluidic Assembly stage, and the properties of the nanomedicine produced are identified at the Microfluidic Characterization stage. Only if those properties meet the nanomedicine design criteria will the performance and efficacy of a selected nanomedicine be evaluated in *in-vitro* biomimetic microsystems that recapitulate the structure and function of human organs in the Microfluidic Evaluation stage. If the targeting, therapeutic, and imaging efficacies are satisfactory in the *in-vitro* model system, the nanomedicine will then be validated with animal models. All nanomedicine candidates that are unsuccessful at the above stage will be reformulated in the Microfluidic Assembly stage and go through the iterative processes. If successful in animal models, the selected nanomedicine will then be manufactured through parallelized microfluidic platforms. The pressure and flow patterns in the integrated microfluidic system are regulated by high-precision control systems.

conditions in the laboratory. This reliance of NPs on such nonstandard multistep processes inevitably causes high batch-to-batch variations in their physico-chemical properties [64–69]. Batch size is also subject to custom protocols that vary among laboratories, leading to difficulties in screening and identifying optimal NP physico-chemical characteristics for enhanced drug delivery. Furthermore, the introduction and combination of multiple materials for creating multicomponent NPs compromises the expected functionality of the individual elements. This is largely because of an inability to precisely control the continuous assembly process in various conventional bulk syntheses that involve the macroscopic mixing of precursor solutions [58, 70]. As the micrometer- and nanometer-scale interactions of precursors will direct the characteristics of NPs, it is essential that their composition is fine-tuned in order to attain the anticipated functionalities of multicomponent NP assemblies. In general, the central challenge for the synthesis of multicomponent NPs is to establish large-scale and continuous manufacturing methodologies with high reproducibility.

Amphiphilic blocks self-assemble spontaneously into NPs through size-dependent formation mechanisms and on timescales governed by diffusion [71]. The physico-chemical properties of NPs are, at least in part, determined by the timescales at which the multiple solutions mix in the system [72], as well as the thermodynamic characteristics of the block polymers [73]. Thus, a mixing timescale that is longer than the characteristic time for chemical chain formation will result in an uncontrolled aggregation due to incomplete solvent change. Conversely, a complete solvent change through shorter mixing times in rapid precipitations can result in stable assembly kinetics that lead to the production of homogeneous NPs [74]. One critical difference between conventional bulk synthesis and microfluidic assembly is the mixing time, which occurs on the order of seconds in bulk synthesis and contrasts with those in the millisecond and microsecond range in microfluidic assembly [75]. This shorter mixing time results in more homogeneous NPs by reducing the aggregation of precursors, leading to high reproducibility, which in turn prevents the subsequent thermal and mechanical agitation needed in conventional bulk synthesis for NP homogenization. Therefore, a precise control of microfluidic flow patterns with tunable characteristic mixing times will offer a better understanding of the effect of the mixing time on NP reproducibility and homogeneity.

Microfluidic technologies have demonstrated a better control over effective mixing of the precursor solutions for assembling a range of NP types (Fig. 2a) when compared to conventional bulk methods, due to the larger contact surface areas given per unit volume of fluid in microfluidics [58, 70]. For example, typical laminar flows in microfluidics enabled the controlled syntheses of several NPs (Fig. 2b), including liposomes [76–78] and polymeric NPs [75, 79, 80], with a narrower size distribution compared to those of conventional bulk synthesis. Under laminar flow conditions at a low Reynolds number (~1),[1] mixing occurs only through diffusion across the interface between two miscible fluids moving next to each other in viscous flows. Unfortunately, NP synthesis by diffusive mixing does not allow for the development

---

[1] The Reynold's number is a dimensionless number that provides a ratio of inertial to viscous forces to quantify the relative importance of these two types of force for given flow conditions.

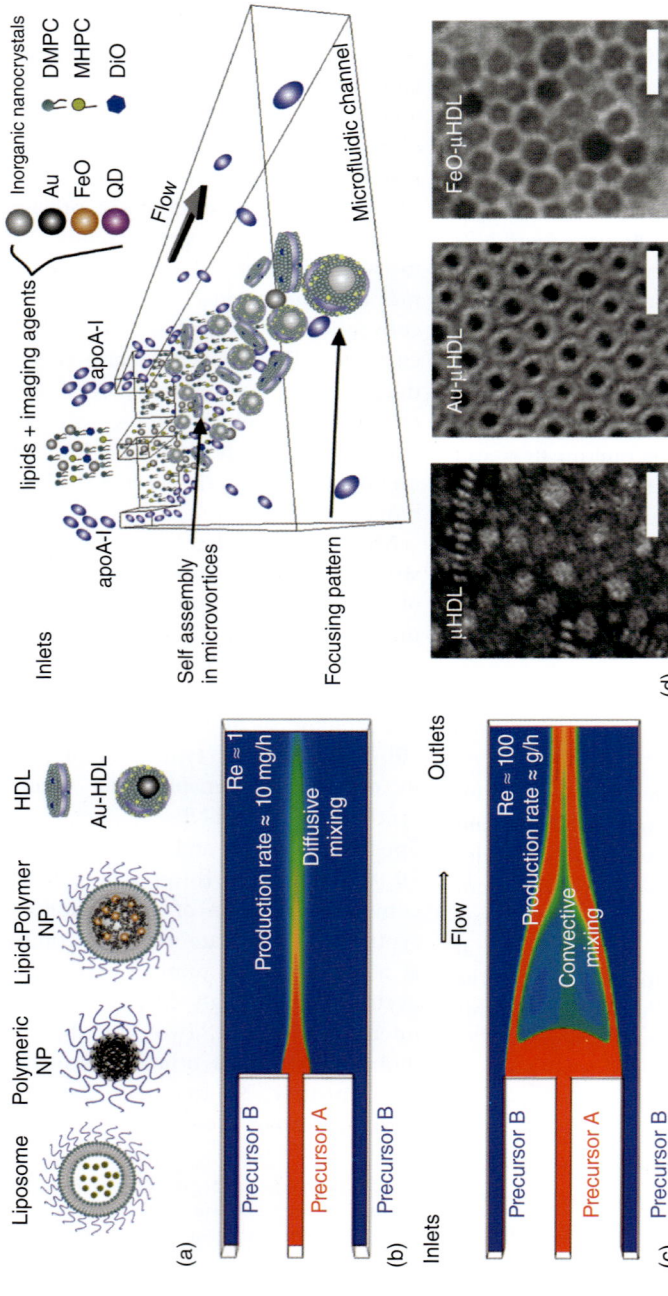

Fig. 2 (a) Representative nanomedicine types (liposome, polymeric NP, lipid-polymer NP, and high-density lipoprotein (HDL)) that have been assembled by microfluidic technology; (b) A microfluidic channel that creates diffusive mixing across laminar flow interfaces at Re ~ 1; (c) A microfluidic channel that creates convective mixing between two precursors at Re ~ 100; (d) Microfluidic reconstitution of HDL (referred to as ■ μHDL) with inorganic nanocrystals, such as gold (Au) and iron oxide (FeO), using controlled microvortices. The transmission electron microscopy images show μHDL, Au-μHDL, and FeO-μHDL (scale bars = 20 nm). Reproduced with permission from Refs [100].

of materials such as lipid–polymer hybrid NPs [81], which require a strong mixing of solutions in the aqueous and organic phases. Lipid–polymer hybrid NPs have shown a higher drug loading within the polymer core and a slower drug release due to the lipid shell when compared to pure polymeric NPs [82, 83]. Furthermore, the diffusive mixing required is difficult to scale up and ultimately leads to a limited controllability of the precursor mixing time, thereby restricting NP homogeneity and leading to a low-throughput production of NPs.

One approach to facilitate precursor mixing (i.e., shortening the mixing times) is to use convective mixing, thereby increasing the interfacial surface area between fluids and reducing the diffusion length scales. Whereas, conventional microfluidic systems exploit easy-to-control flow patterns, which are strictly laminar at low Re values (~1), an increase in Re (10 < Re < 30) generates complex flow patterns under a variety of geometric conditions of the microfluidic channel, such as local microvortices and flow separation due to an increase in inertial forces [84–86]. In order to implement convective mixing in microfluidic devices, microfluidic platforms have been designed for the rapid mixing of fluids using relatively higher inertial forces in localized regions with moderate Re values (10 < Re < 100) [87–93]. Furthermore, microvortices have demonstrated the ability to rapidly manipulate, sort, and excite particles in microfluidics [94–97]. Recently, a new generation of three-dimensional (3D) focusing patterns in a simple, single-layer microfluidic channel has allowed the development of a pattern-controllable microvortex platform (Fig. 2c). This device has been used for the highly reproducible synthesis of lipid–polymer hybrid NPs with multiple drugs and imaging agents [98, 99], and multifunctional high-density lipoprotein-derived NPs [100] with high productivity (up to $1\,g\,h^{-1}$) (Fig. 2d).

3
Microfluidic Characterization of Nanomedicines

The most important properties of NPs to be characterized before probing their interaction with biological systems are size, shape, surface chemistry/charge, and stability. The development of novel NP characterization tools will impact heavily on nanomedicine, as the lack of characterization standards and quality control tools for NPs has inhibited their clinical adoption to date. One practical obstacle to the clinical-scale commercialization of NPs is an inability to certify the stability of formulations, as even small property variations will have significant effects on *in-vivo* distribution, causing unpredictable therapy outcomes. Recently, several studies have been conducted on NP quality evaluation in microfluidics. For example, a rapid liposome quality assessment in microfluidics allows for quantitative results on liposome formulation composition and stability using dielectric spectroscopy and multivariate data analysis methods [101]. Instantaneous immobilization by ultra-rapid cooling in microfluidics reveals the formation of nonequilibrium liposomes in detail [102]. Yet, in spite of recent advances in the label-free characterization of single NPs [60], it remains difficult to effectively integrate a high-throughput microfluidic technology capable of detecting NPs (or their motion) with currently available characterization equipment that includes dynamic light scattering (DLS) [103], transmission electron microscopy (TEM) [104], atomic force microscopy (AFM) [105], Auger electron spectroscopy (AES) [106], nuclear magnetic resonance (NMR) [107], and flow cytometry [108].

Meanwhile, recent advances in nanofabrication and microfluidics have allowed for the development of high-throughput devices capable of characterizing NP properties, including size and surface charge. The most common electrical technique is to probe impedance changes in nanowire-embedded microfluidic and nanofluidic

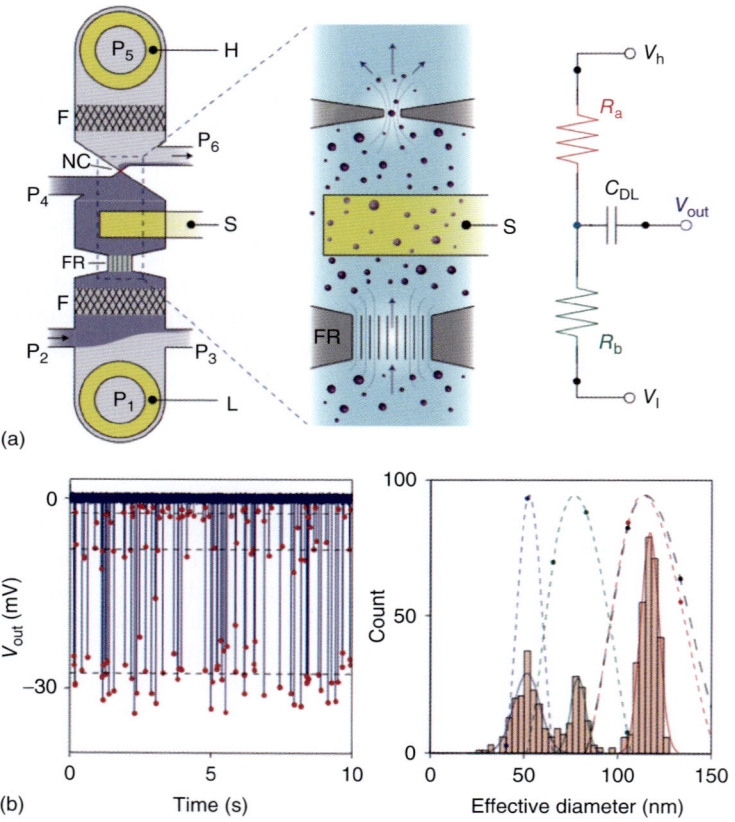

Fig. 3 (a) Overall device layout (left) depicting the electrical and fluidic components: external voltage bias electrodes (H, L) and sensing electrode (S); embedded filters (F); fluid resistor (FR); nanoconstriction (NC); pressure-regulated fluidic ports (P_1~P_6). The nanoparticle (NP) suspension enters at P2 and exits at P6. A detailed image of the dashed box area in (a) (middle) shows the key sensing parts. While NPs flow in the direction of the arrows, changes in the electrical potential in the NC are detected by the electrode S. Electrical circuit expression of the device (right): a constant bias voltage V_h (V_l); Resistors R_a and R_b represent the resistance of the nanoconstriction and the fluidic resistor, respectively; (b) An example analysis of a NP mixture with polydispersity. Left: Output voltage over time for a mixture of NPs of different diameters. Events marked with red circles cluster around three values of V (horizontal dashed black lines). Right: Histogram of effective diameters (40 s measurements). Reproduced with permission from Ref. [117].

channels, and to detect any perturbation in the local electrical properties of the nanowires in response to disturbance by NP solutions. For example, nanowire field-effect transistors allowed for the real-time sensitive detection of label-free molecules [109–111]. A combination of these technologies with nanoscale mechanical systems offers real-time, high-precision, single-molecule/NP/cell detectors, such as advanced mass spectrometry [112] and microfluidic and nanofluidic channel resonators [113–115].

With recent advances in the fundamental physical chemistry of nanoscale pore sensors, several pore-based sensors have been developed in the nanoscale range, offering a rapid and specific, yet simple, biosensing strategy with an improved measurement sensitivity over a wide particle size range [116]. An example of this is a high-throughput microfluidic analyzer that has been developed to detect and characterize unlabeled NPs in a multicomponent mixture at a rate of 500 000 particles per second (Fig. 3) [117]. In this case, a real-time single-nucleotide detection of a model G487A mutation (which is responsible for glucose-6-phosphate dehydrogenase deficiency) was achieved by leveraging the *in-situ* reaction-monitoring capability of the nanopore platform [118]. Tunable pore sensors, which can elastically adjust the size of their pore, have been used to count and detect the size and concentration of smaller NPs compared to other techniques such as flow cytometry [119]. This platform also allows for a simultaneous extraction of the size and zeta-potential of NPs from their charge density, under electrophoretic forces [120]. NP translocation was also detected using a pressure-reversal technique through a cone-shaped nanopore membrane [121]. Today, these approaches, all of which employ size-tunable pore sensors, are starting to provide a better understanding of the fundamental behavior of NPs, as well as a high-throughput characterization of their properties.

4
Microfluidic Evaluation of Nanomedicines

Nanomedicines needs to be nontoxic, biodegradable, sufficiently stable to be delivered to targeted sites, and to have a superior therapeutic advantage over the free drug [122, 123]. Conventionally, nanomedicine evaluation has been made in static cell culture plates, but unfortunately this neglects the important effects of flowing conditions and subsequent transport phenomena on the microenvironment. In contrast to static conditions, flowing conditions assist in the homogeneous distribution of NPs with no gravitational sedimentation, which is similar to the physiological conditions encountered *in vivo*. For example, microfluidic approaches have been used to measure the cytotoxicity of quantum dots (QDs) in a flowing condition [124, 125] and to examine the stability of multicomponent NPs across a laminar flow interface [126]. Compared to tests conducted in conventional static plates, these approaches have provided a more accurate approximation of nanomedicine performance *in vivo*. Microfluidic approaches were also used to evaluate the selective binding of NPs to cells while varying the fluid shear stress, the targeting ligand concentration, receptor expression on target cells, and NP size [127, 128]. The targeted delivery of a nanomedicine represents a powerful technology for the development of safer and more effective therapeutics compared to systemic delivery by nontargeted formulations [83]. Indeed, such approaches show that the targeting performance of an NP can be examined under more physiologically relevant conditions, which is preferential to examining a wide array of cell–particle interactions prior to *in-vivo* experiments. The accurate detection of biomarkers also holds significant promise for "personalized" cancer diagnostics, with more physiologically relevant 3D platforms having been developed for identifying and validating ubiquitous biomarkers [129].

4.1
Mimicking Physiological Environments

Today, an increasing number of engineered NPs requires a reliable high-throughput screening methodology with more physiologically accurate conditions. Whereas, microfluidic approaches have demonstrated the potential to closely approximate physiological environments, current preclinical studies on drug candidates mostly rely on costly and highly variable animal models, mainly because existing cell culture models fail to recapitulate the organ-level pathophysiology of humans. This lack of accurate predictive models highlights the need for better approaches to mimic the structure and function of cells, tissues and organs, as well as the dynamically changing environments *in vivo*. Recently, the evolution of microfluidics has witnessed the integration of *in-vitro* cellular approaches onto chips, which allow real-time, *in-vitro* microscopic observations to be made as well as an evaluation of cell function [130]. But, in order to probe the targeting, therapeutic and diagnostic efficacy of NPs in spatially and temporally regulated environments, it is important first to examine how the NPs interact with cells, tissues and organisms under more physiologically realistic conditions [131, 132]. Consequently, the microfluidic approaches for replicating organ-level structure and function will be discussed at this point (Fig. 4a) [133], and current applications and potentials for the *in-vitro* evaluation of nanomedicines highlighted.

4.2
Endothelial Cell Systems

The vascular endothelium is a crucial target for therapeutic intervention in pathological processes that include inflammation, atherosclerosis, and thrombosis. Endothelial cells exist under dynamically changing mechanical stresses that are generated by blood flow patterns.

Fig. 4 (a) Schematic of microengineered biomimetic systems with spatiotemporal control over physiological effectors, including mechanical cues, chemical factors, electrical signals, multi-layered platform with 3D scaffold; (b) Schematic depiction of a biomimetic model that mimics the function of the blood–brain barrier. b.END3 brain endothelial cells and C8-D1A astrocyte cells are cultured on either side of a porous membrane between two microfluidic flow chambers; (c) Schematic of a microfluidic model that mimics the permeable endothelium in artery-surrounding microvessels. Permeability is detected using microelectrodes embedded in the chip. The fluorescent image shows the disrupted endothelial connections. Adherens junctions are shown in green, and nuclei in blue (scale bar = 20 µm). Schematic and TEM image of nanoparticles used for NP translocation studies in the chip (scale bar = 100 nm); (d) Schematic of a lung-on-a-chip device showing IL-2-induced pulmonary edema (scale bar in contrast image = 200 µm). The graph shows barrier permeability in response to IL-2, with and without cyclic strain. Error bars indicate SEM. The fluorescent images show that immunostaining of epithelial occulidin (green) and vascular endothelial cadherin (VE-cadherin; red) with 10% strain with and without IL-2 (scale bars = 30 µm). Reproduced with permission from Refs [133, 143, 149, 152].

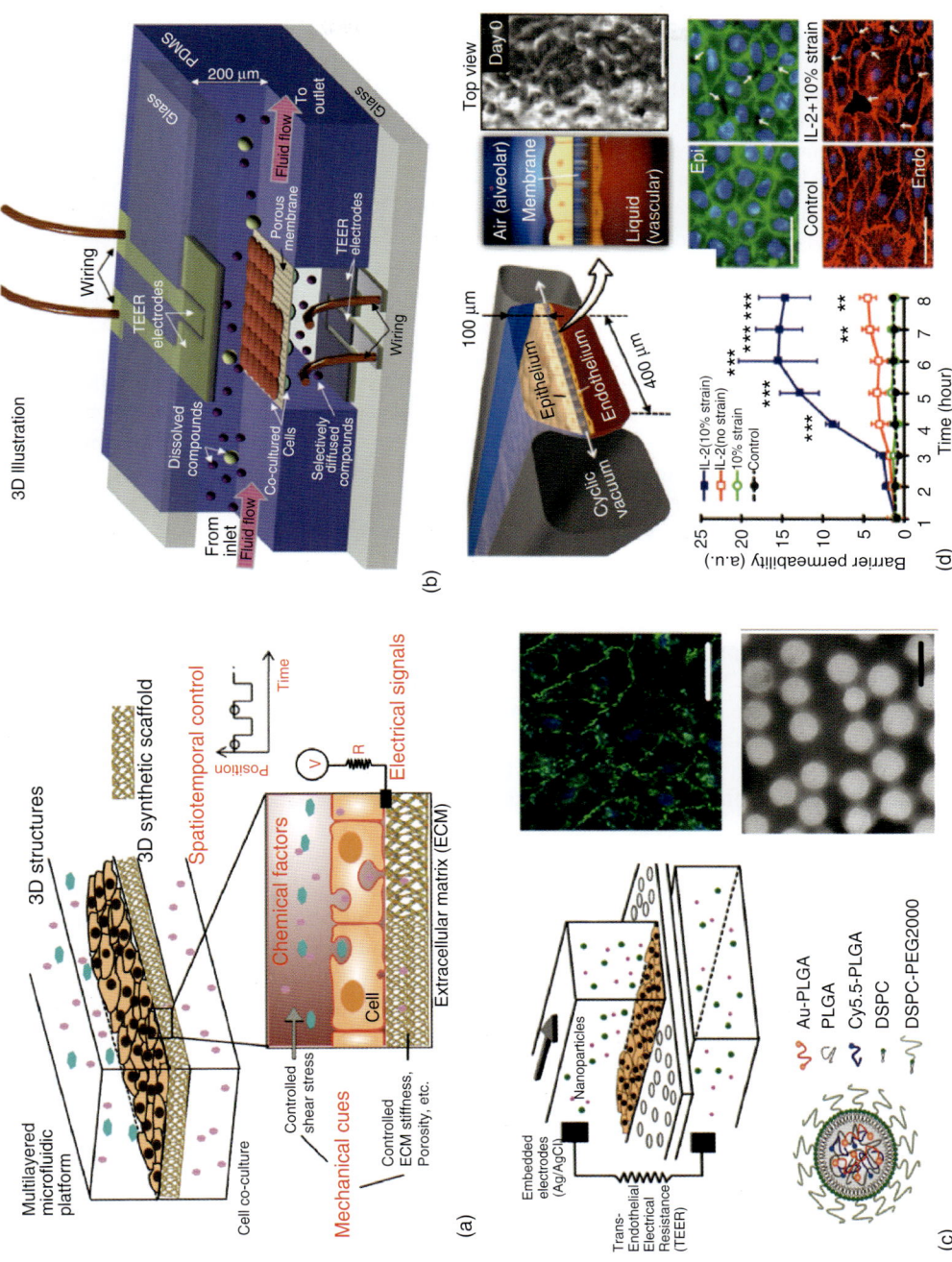

Yet, endothelial cell monolayers cultured in conventional multiwell plates fail to reproduce the complex architecture of a vascular network *in vivo* and thus fail to capture the relationship between shear stress experienced by the cells and the local concentration of the drug used. Recent developments in microengineered vascular systems have shown the potential for evaluating nanomedicines under physiologically realistic conditions. For example, replicating the structure and function of blood vessels *in vitro* can be helpful for investigating NP behavior and interaction in and around the targeted sites [134–138]. An accurate reconstitution of the geometric configuration of natural blood vessels is also important, as the interactive effects between blood flow and drug concentration were not captured by a rectangular channel coated with endothelial cells. Rather, a branching network with tubular channels was constructed in order to reproduce these effects [139]. In addition, microvessels supported by the extracellular matrix (ECM), when patterned in a tubular structure, establish the endothelial monolayer, maintain permeability, and are not prone to delamination (which was relatively common in the rectangular channels). The rectangular channels were also very susceptible to delamination that disrupted local permeability, due mainly to the poor connections between the endothelial cells and ECM at the sharp corners [140, 141].

Several additional methods to reproduce microvessels accommodating multiple cells (endothelial cells, pericytes, and astrocytes) have allowed the development of microfluidic models of the blood–brain barrier (BBB) (Fig. 4b) [142–148], as well as for endothelial dysfunction and permeability control in atherosclerosis (Fig. 4c) [149–151].

4.3 "Organ-On-A-Chip"

In combining microfabrication techniques with tissue engineering, the "lung-on-a-chip" device offered a novel *in-vitro* approach to drug screening by mimicking the mechanical and biochemical activities of the human lung (Fig. 4d) [152]. For example, a recent study using this device revealed that mechanical strains associated with physiological breathing movements play an essential role in the development of the increased vascular leakage which leads to pulmonary edema, and that circulating immune cells are not necessary for this disease to develop. The same studies also led to the identification of potential new therapeutics, including angiopoietin-1 (Ang-1) and a new transient receptor potential vanilloid 4 (TRPV4) ion channel inhibitor (GSK2193874), which might prevent the severe toxicity associated with interleukin (IL)-2 therapy. An *in-vitro* model of the intestine has also been developed, together with its crucial microbial symbionts [153, 154]. Whereas, previous *in-vitro* models of intestinal function depend on the use of epithelial cell lines (e.g., Caco-2 cells) which create polarized epithelial monolayers but fail to mimic human intestinal functions for drug development. The recent development of a "gut-on-a-chip" device recreated the gut microenvironment with low shear stress ($0.02\,\text{dyne}\,\text{cm}^{-2}$) with cyclic strain (10%; 0.15 Hz) that mimicked physiological peristaltic motions. This precise regulation allowed for an increased exposure of the intestinal surface area and a robust 3D intestinal villi morphogenesis, which mimicked the enhanced cytochrome isoform-based drug-metabolizing activity and the absorptive efficiency of the human intestine.

4.4
Renal Toxicity and Hepatotoxicity

A further use of biomimetic microfluidic platforms for nanomedicine is to evaluate renal toxicity and hepatotoxicity in pre-clinical studies. Renal excretion represents a clearance pathway for the removal of molecules from vascular compartments, during which time the circulating NPs enter the glomerular capillary and undergo a size-dependent filtration. Those NPs smaller than the pore size of glomerular filtration (~5 nm) can be filtered to enter the proximal tubule, where the brush border of the epithelial cells is negatively charged. As a consequence, positively charged NPs are readily resorbed from the luminal space compared to the negatively charged NPs. The recent development of a microfluidic device lined by human kidney epithelial cells that can be exposed to a shear rate demonstrated a significant increase in albumin transport, glucose reabsorption and brush border alkaline phosphatase activity, all of which are crucial functions of the human kidney proximal tubule [155]. This approach also confirmed that cisplatin toxicity and Pgp efflux transporter activity detected on-chip more closely mimicked the *in-vivo* responses than those obtained with cells maintained on conventional culture plates.

It should be noted that any NPs which are not cleared via the kidney are excreted via the hepatobiliary system. The hepatocytes, which are referred to as potential sites for toxicity, play an important role in liver clearance through endocytosis and the enzymatic breakdown of NPs. As NPs between 10 and 20 nm in size are efficiently eliminated via the liver, any NPs designed in this size range must be modified in order to avoid their prolonged retention in the liver as they undergo excretion. Recently, microfluidic devices have been developed that reconstitute the function of the hepatocytes; for example, a 3D hepatocyte chip has been fabricated for *in-vitro* drug toxicity examinations aimed at predicting drug hepatotoxicity *in-vivo* [156]. This device allowed for the controlled delivery of multiple drug doses to functional primary hepatocytes, while an incorporated concentration gradient generator created *in vitro* dose-dependent drug responses in order to predict *in-vivo* hepatotoxicity.

4.5
Live Tissue Explants

The introduction of *ex-vivo* live tissue explants into microfluidics may provide additional physiological conditions to be investigated. For example, embryonic tissues excised from live frog embryos were used to examine dynamic responses to time-varying chemical stimuli (Fig. 5a) [157]. Carcinoma tumor biopsies were also introduced into a reproducible glass microfluidics system to study the tumor environment, thus offering a preclinical model for the creation of "personalized" treatment regimens [158]. The culturing of brain tissue slices on transistor arrays fabricated on silicon chips may also become a novel platform for neurophysiological and pharmacological studies. Typical microfluidics and semiconductor technologies can be integrated to produce high-resolution, planar transistor arrays for mimicking neuronal structures in long-term studies of topographic mapping [159], and also for mapping evoked extracellular field potentials in organotypic brain slices of rat hippocampus [160].

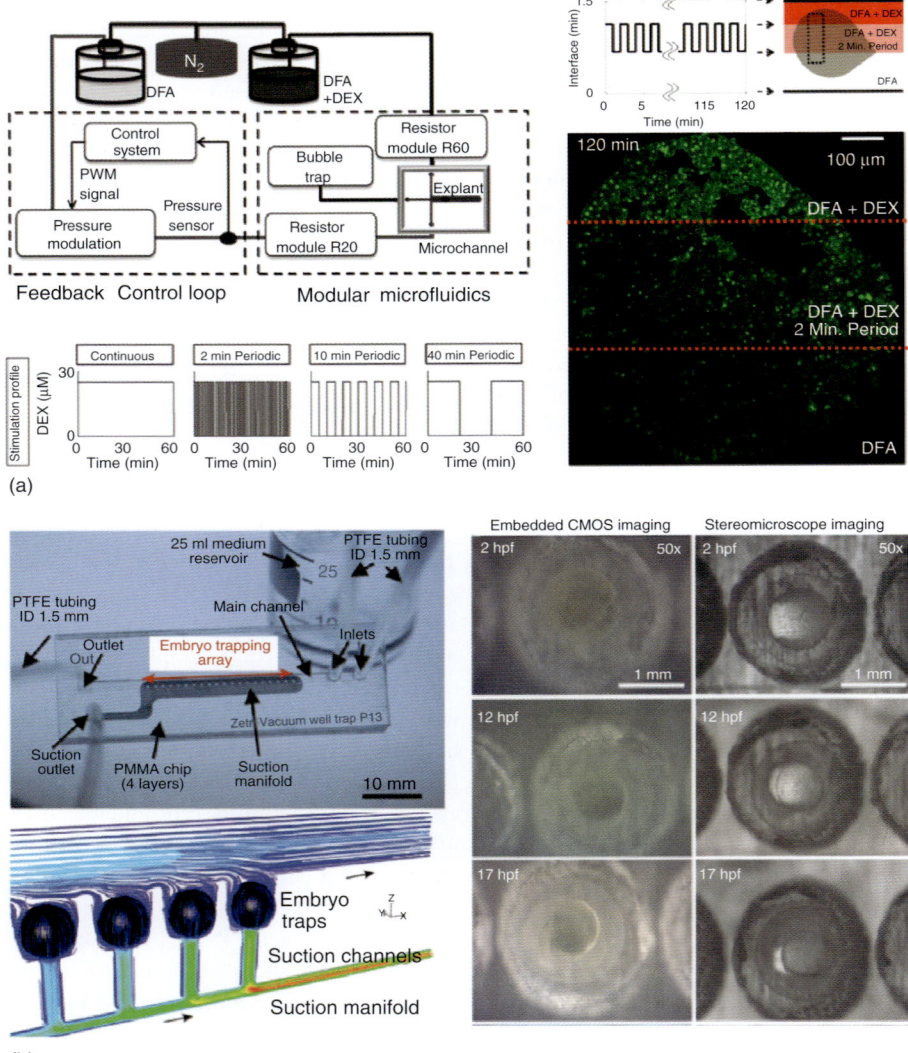

Fig. 5 (a) Schematic of feedback control system that allows for long-term culture of an embryonic tissue excised from a live frog embryo within a microfluidic channel to examine dynamic responses to spatially and temporally varying chemical stimuli. The fluorescent image show the distinct localized responses of an embryonic tissue to dynamic stimuli in a single tissue explant for 2 h, showing that stimulation with localized bursts versus continuous stimulation can result in highly distinct responses. This platform can be used to investigate the cell intercommunication in response to localized drug stimulation for toxicity tests; (b) Schematic of microfluidics-based chip integrating embedded electronic interfaces, and CFD simulation for the flow stream line prediction. Embryos cultured in the device that allowed for immobilization, culture, and treatment of developing zebrafish embryos for toxicity tests. Reproduced with permission from Refs [157] and [167].

4.6
Intact Organisms

Small multicellular organisms such as nematodes, fruit flies, clawed frogs and zebrafish, allow for toxicological screening in the normal physiological environments of intact organisms, providing substantial advantages over cell lines and extracted tissues [161–163]. While the fully automated analysis of these model systems in a high-throughput manner remains challenging, the application of microfluidics to these model organisms has demonstrated the ability to handle multicellular organisms in an efficient manner and to precisely manipulate the local conditions to allow for the assessment and imaging of these small organisms [163–166]. For example, manipulating small organisms, such as the worm *Caenorhabditis elegans*, allowed the observation of neuronal responses in order to correlate the activity of sensory neurons with the worm's behavior *in vivo* [161]. Moreover, the integration of embedded electronic interfaces with microfluidic chip-based technologies allowed for the automatic immobilization, culture, and treatment of developing zebrafish embryos during fish embryo toxicity (FET) biotests (Fig. 5b) [167].

To capture the interactions between multiple organs on microfluidic chips would potentially enable a more accurate model of how organs function and interact with one another for potential drug development applications [168, 169]. For example, the combination of a mathematical model and a multiorgan approach provided a novel platform with improved predictability for testing the toxicity of an anticancer drug, 5-fluorouracil, in a pharmacokinetics-based manner [170]. These approaches can help to achieve a better insight into the mechanisms of action of drug candidates, perhaps leading to patient-specific therapies in the future [171]. While multiorgan chips have the potential to simulate human body functions for patient-specific point-of-care devices and therapies, the inherent complexity of each organ itself hinders the development of reliable "human-on-a-chip" model systems. For example, the practical challenges include an optimization of organ size, the control of fluid volumes, the maintenance of coupled organ systems, and the development of a universal blood substitute [172]. The key question for building multiorgan systems is how to simplify the organ complexity without losing physiological accuracy.

5
Challenges and Opportunities

A new nanomedicine development pipeline using microfluidics technologies includes microfluidic assembly, characterization, evaluation, and the manufacture of nanomedicines. These technologies will allow the robust supply of highly reproducible nanomedicines to the entire development process and thereby increase the success rates in clinical trials. In addition to the stages discussed above, microfluidics technology for nanomedicine manufacture is key to the successful translation of a nanomedicine from the laboratory to the clinic. A long-term vision for the manufacture of nanomedicines is to create reliable, continuous and scalable assembly methodologies for a variety of multifunctional NPs with high reproducibility, yield, and homogeneity [173]. The development of these assembly methods requires microfluidic approaches to allow for an efficient and strong mixing of precursors, modular methods for incorporating multicomponents (e.g., therapeutic compounds, imaging agents, targeting ligands, etc.) into multifunctional NPs, and automated control systems for the large-scale integration and parallelization of microfluidic modules [117, 174]. The ability to integrate microfluidics with dynamics, control, and more complex microfabrication techniques opens the door for high-throughput, automated manufacturing.

A current challenge of nanomedicine manufacture using microfluidics is to optimize and maximize microfluidic devices with tunable mixing flow patterns that are applicable to the synthesis of a wide range of nanomedicine types, without losing the physico-chemical properties of the designed nanomedicine [175]. The key technologies required for the development of these nanomedicine manufacturing techniques are computational fluid dynamics to allow simulation of mixing flow patterns in microfluidic devices, highly reliable microfabrication techniques capable of integrating microscale pumps, valves and detecting sensors [176], and high-precision control systems that regulate parallelization and automation capability [177–180].

While the quantities of NPs synthesized by microfluidic devices are often in the microgram to milligram range, the parallelization of microfluidic channels has the potential to scale-up the synthesis by several orders of magnitude to a clinical scale of grams to kilograms. With parallel and stackable microfluidic systems, gram to kilogram scales of NPs could be prepared with the same properties as those prepared at the bench scale, as long as a precise control of either flow rates or pressure in the microfluidic platform is achieved. Pressure control is far better than flow control for controlling the flow rate into a microfluidic network because the flow rate is proportional to the inlet pressure, which can be easily measured for high bandwidth feedback control [181]. To maintain a precise control over fluid pressure with the potential to scale-up production, it is necessary to isolate the pressure-regulating mechanism from the fluid reservoir and the microfluidic device, so that larger reservoir volumes and diverse microfluidic devices can be used independently and integrated as needed. Three important features should be considered for robust and reliable parallelization:

- Fouled modules must be easily replaced or disconnected from other systems.
- Unexpected disturbances due to air bubble formation in microfluidic devices need to be compensated, as this increases the hydraulic resistance between neighboring devices in the parallelized network, leading to an imprecise regulation of the entire system.
- Bridging or networking channels that connect microfluidic modules need to be well designed with minimal secondary flows, which may affect the main bulk flow streams, leading to chip-to-chip variations.

While the microfluidic assembly of nanomedicine has demonstrated much progress using several platforms and various mixing patterns, practical development has been significantly constrained due to a lack of tools capable of detecting, characterizing, and analyzing NPs in a high-throughput manner. Although single NP detectors and characterization tools using microfluidics and nanofabrication technologies have been demonstrated, these are still limited to specific solutions and NP types and need to be generalized in order to function with multicomponent NPs. In addition, there is a need for a technique allowing the easy preparation of highly concentrated NP samples to be developed, including the purification of toxic solvents via either separation or filtration. Furthermore, while many approaches have shown the promise of combining optical systems with microfluidics [60], the precise and reliable control of light delivery to targeted areas in microfluidic platforms remains an active topic of research and engineering. A combination of microfluidics and optics – termed optofluidics – has demonstrated a synergistic effect for new capabilities in several applications including lens, colloidal suspensions, and flow cytometry [182–185]. In addition, optofluidics technology could be employed to incorporate the microfluidic characterization capabilities into NP production platforms in a high-throughput fashion.

Reliability of biomimetic microsystems that mimic the structure and function of human organs for nanomedicine evaluation is crucial. Current challenges include the reliability of long-term cultivations of multiple cell types [186], as well as real-time monitoring of nanomedicines, cellular response to nanomedicines, and critical chemical cues (e.g., reactive oxygen stress) in 3D microenvironments [187–189]. In addition, the development of synthetic biomaterials remains a critical topic of research for physiological accuracy and niches for specific cells, tissues and organs. Furthermore, the development of *in-vitro* model systems that can accurately replicate the structure and function of *in-vivo* systems necessitates a precise 3D control of dynamically changing properties, such as mechanical properties of the ECM, at a scale comparable to human cells, tissues and organs [132, 172, 190].

6
Concluding Remarks

Microfluidics in nanomedicine has demonstrated the ability to overcome critical issues with conventional approaches used for nanomedicine development. When combined with advanced nanofabrication, synthetic biomaterials and high-precision control systems, microfluidic technologies constitute a novel platform capable of replacing the entire nanomedicine production process in a scalable manner. Although microfluidics as applied to nanomedicine is still in its infancy, it will surely continue to expand to provide innovative systems at industrially relevant scales in the near future.

Acknowledgments

The authors thank the members of their laboratories for participating in stimulating discussions of these investigations.

References

1. Peer D., Karp J.M., Hong S., Farokhzad O.C., *et al.* (2007) Nanocarriers as an emerging platform for cancer therapy. *Nat. Nanotechnol.* **2**(12), 751–760.
2. Namiki Y., Fuchigami T., Tada N., Kawamura R., *et al.* (2011) Nanomedicine for cancer: lipid-based nanostructures for drug delivery and monitoring. *Acc. Chem. Res.* **44**(10), 1080–1093.
3. Seigneuric R., Markey L., Nuyten D.S., Dubernet C., *et al.* (2010) From nanotechnology to nanomedicine: applications to cancer research. *Curr. Mol. Med.* **10**(7), 640–652.
4. Lobatto M.E., Fuster V., Fayad Z.A., Mulder W.J. (2011) Perspectives and opportunities for nanomedicine in the management of atherosclerosis. *Nat. Rev. Drug Discovery* **10**(11), 835–852.

5. Psarros C., Lee R., Margaritis M., Antoniades C. (2012) Nanomedicine for the prevention, treatment and imaging of atherosclerosis. *Nanomedicine* **8** Suppl. 1, S59–S68.
6. Schiener M., Hossann M., Viola J.R., Ortega-Gomez A., et al. (2014) Nanomedicine-based strategies for treatment of atherosclerosis. *Trends Mol. Med.* **20**(5), 271–281.
7. Sung H.W., Sonaje K., Feng S.S. (2011) Nanomedicine for diabetes treatment. *Nanomedicine* **6**(8), 1297–1300.
8. Pickup J.C., Zhi Z.L., Khan F., Saxl T., et al. (2008) Nanomedicine and its potential in diabetes research and practice. *Diabetes Metab. Res. Rev.* **24**(8), 604–610.
9. Sonaje K., Lin K.J., Wey S.P., Lin C.K., et al. (2010) Biodistribution, pharmacodynamics and pharmacokinetics of insulin analogues in a rat model: oral delivery using pH-responsive nanoparticles vs. subcutaneous injection. *Biomaterials* **31**(26), 6849–6858.
10. Taratula O., Kuzmov A., Shah M., Garbuzenko O.B., et al. (2013) Nanostructured lipid carriers as multifunctional nanomedicine platform for pulmonary co-delivery of anticancer drugs and siRNA. *J. Controlled Release* **171**(3), 349–357.
11. Mansour H.M., Rhee Y.S., Wu X. (2009) Nanomedicine in pulmonary delivery. *Int. J. Nanomed.* **4**, 299–319.
12. Muldoon L.L., Tratnyek P.G., Jacobs P.M., Doolittle N.D., et al. (2006) Imaging and nanomedicine for diagnosis and therapy in the central nervous system: report of the 11th Annual Blood-Brain Barrier Disruption Consortium meeting. *Am. J. Neuroradiol.* **27**(3), 715–721.
13. Sharma H.S., Sharma A. (2011) New strategies for CNS injury and repair using stem cells, nanomedicine, neurotrophic factors and novel neuroprotective agents. *Expert Rev. Neurother.* **11**(8), 1121–1124.
14. Morgen M., Lu G.W., Du D., Stehle R., et al. Targeted delivery of a poorly water-soluble compound to hair follicles using polymeric nanoparticle suspensions. *Int. J. Pharm.* 2011, **416**(1), 314–322.
15. Sigfridsson K., Bjorkman J.A., Skantze P., Zachrisson H. (2011) Usefulness of a nanoparticle formulation to investigate some hemodynamic parameters of a poorly soluble compound. *J. Pharm. Sci.* **100**(6), 2194–2202.
16. Jia L. (2005) Nanoparticle formulation increases oral bioavailability of poorly soluble drugs: approaches experimental evidences and theory. *Curr. Nanosci.* **1**(3), 237–243.
17. Gao W., Xiao Z., Radovic-Moreno A., Shi J., et al. (2010) Progress in siRNA delivery using multifunctional nanoparticles. *Methods Mol. Biol.* **629**, 53–67.
18. Leuschner F., Dutta P., Gorbatov R., Novobrantseva T.I., et al. (2011) Therapeutic siRNA silencing in inflammatory monocytes in mice. *Nat. Biotechnol.* **29**(11), 1005–1010.
19. Soppimath K.S., Aminabhavi T.M., Kulkarni A.R., Rudzinski W.E. (2001) Biodegradable polymeric nanoparticles as drug delivery devices. *J. Controlled Release* **70**(1-2), 1–20.
20. Chen X., Schluesener H.J. (2008) Nanosilver: a nanoproduct in medical application. *Toxicol. Lett.* **176**(1), 1–12.
21. Huang X., Jain P.K., El-Sayed I.H., El-Sayed M.A. (2007) Gold nanoparticles: interesting optical properties and recent applications in cancer diagnostics and therapy. *Nanomedicine* **2**(5), 681–693.
22. Sanvicens N., Marco M.P. (2008) Multifunctional nanoparticles–properties and prospects for their use in human medicine. *Trends Biotechnol.* **26**(8), 425–433.
23. Jain K.K. (2003) Nanodiagnostics: application of nanotechnology in molecular diagnostics. *Expert Rev. Mol. Diagn.* **3**(2), 153–161.
24. Torchilin V.P. (2005) Recent advances with liposomes as pharmaceutical carriers. *Nat. Rev. Drug Discovery* **4**(2), 145–160.
25. Gu F., Zhang L., Teply B.A., Mann N., et al. (2008) Precise engineering of targeted nanoparticles by using self-assembled biointegrated block copolymers. *Proc. Natl Acad. Sci. USA* **105**(7), 2586–2591.
26. Salvador-Morales C., Zhang L., Langer R., Farokhzad O.C. (2010) Immunocompatibility properties of lipid-polymer hybrid nanoparticles with heterogeneous surface functional groups. *Biomaterials* **30**(12), 2231–2240.
27. Shi J., Xiao Z., Votruba A.R., Vilos C., et al. (2011) Differentially charged hollow core/shell lipid-polymer-lipid hybrid nanoparticles for small interfering RNA

28. Bianco A., Kostarelos K., Prato M. (2005) Applications of carbon nanotubes in drug delivery. *Curr. Opin. Chem. Biol.* **9**(6), 674–679.
29. Huang X., El-Sayed I.H., Qian W., El-Sayed M.A. (2006) Cancer cell imaging and photothermal therapy in the near-infrared region by using gold nanorods. *J. Am. Chem. Soc.* **128**(6), 2115–2120.
30. Gupta A.K., Gupta M. (2005) Synthesis and surface engineering of iron oxide nanoparticles for biomedical applications. *Biomaterials* **26**(18), 3995–4021.
31. Mulder W.J., Koole R., Brandwijk R.J., Storm G., et al. (2006) Quantum dots with a paramagnetic coating as a bimodal molecular imaging probe. *Nano Lett.* **6**(1), 1–6.
32. Kratz F. (2008) Albumin as a drug carrier: design of prodrugs, drug conjugates and nanoparticles. *J. Controlled Release* **132**(3), 171–183.
33. Skajaa T., Cormode D.P., Jarzyna P.A., Delshad A., et al. (2011) The biological properties of iron oxide core high-density lipoprotein in experimental atherosclerosis. *Biomaterials* **32**(1), 206–213.
34. Basta T., Wu H.J., Morphew M.K., Lee J., et al. (2014) Self-assembled lipid and membrane protein polyhedral nanoparticles. *Proc. Natl Acad. Sci. USA* **111**(2), 670–674.
35. Maune H.T., Han S.P., Barish R.D., Bockrath M., et al. (2010) Self-assembly of carbon nanotubes into two-dimensional geometries using DNA origami templates. *Nat. Nanotechnol.* **5**(1), 61–66.
36. Pal S., Deng Z., Ding B., Yan H., et al. (2010) DNA-origami-directed self-assembly of discrete silver-nanoparticle architectures. *Angew. Chem. Int. Ed.* **49**(15), 2700–2704.
37. Farokhzad O.C., Cheng J., Teply B.A., Sherifi I., et al. (2006) Targeted nanoparticle-aptamer bioconjugates for cancer chemotherapy in vivo. *Proc. Natl Acad. Sci. USA* **103**(16), 6315–6320.
38. Xiao Z., Levy-Nissenbaum E., Alexis F., Luptak A., et al. (2012) Engineering of targeted nanoparticles for cancer therapy using internalizing aptamers isolated by cell-uptake selection. *ACS Nano* **6**(1), 696–704.
39. Gianella A., Jarzyna P.A., Mani V., Ramachandran S., et al. (2011) Multifunctional nanoemulsion platform for imaging guided therapy evaluated in experimental cancer. *ACS Nano* **5**(6), 4422–4433.
40. Lobatto M.E., Fayad Z.A., Silvera S., Vucic E., et al. (2010) Multimodal clinical imaging to longitudinally assess a nanomedical anti-inflammatory treatment in experimental atherosclerosis. *Mol. Pharm.* **7**(6), 2020–2029.
41. Venditto V.J., Szoka F.C., Jr, (2013) Cancer nanomedicines: so many papers and so few drugs! *Adv. Drug Delivery Rev.* **65**(1), 80–88.
42. Duncan R., Gaspar R. (2011) Nanomedicine(s) under the microscope. *Mol. Pharm.* **8**(6), 2101–2141.
43. Olofsson J., Bridle H., Sinclair J., Granfeldt D., et al. (2005) A chemical waveform synthesizer. *Proc. Natl Acad. Sci. USA* **102**(23), 8097–8102.
44. Duraiswamy S., Khan S.A. (2010) Plasmonic nanoshell synthesis in microfluidic composite foams. *Nano Lett.* **10**(9), 3757–3763.
45. El-Ali J., Sorger P.K., Jensen K.F. (2006) Cells on chips. *Nature* **442**(7101), 403–411.
46. Schimek K., Busek M., Brincker S., Groth B., et al. (2013) Integrating biological vasculature into a multi-organ-chip microsystem. *Lab Chip* **13**(18), 3588–3598.
47. Andersson H., van den Berg A. (2004) Microfabrication and microfluidics for tissue engineering: state of the art and future opportunities. *Lab Chip* **4**(2), 98–103.
48. Minteer S.D., Moore C.M. (2006) Overview of advances in microfluidics and microfabrication, in: *Methods in Molecular Biology*, vol. **321**, Humana Press, pp. 1–2.
49. Ozaydin-Ince G., Coclite A.M., Gleason K.K. (2012) CVD of polymeric thin films: applications in sensors, biotechnology, microelectronics/organic electronics, microfluidics, MEMS, composites and membranes. *Rep. Progr. Phys. Phys. Soc.* **75**(1), 016501.
50. Shih S.C., Fobel R., Kumar P., Wheeler A.R. (2011) A feedback control system for high-fidelity digital microfluidics. *Lab Chip* **11**(3), 535–540.
51. Welch D., Christen J.B. (2014) Real-time feedback control of pH within microfluidics using integrated sensing and actuation. *Lab Chip* **14**(6), 1191–1197.

52. Prohm C., Stark H. (2014) Feedback control of inertial microfluidics using axial control forces. *Lab Chip* **14**(12), 2115–2123.
53. Valencia P.M., Farokhzad O.C., Karnik R., Langer R. (2012) Microfluidic technologies for accelerating the clinical translation of nanoparticles. *Nat. Nanotechnol.* **7**(10), 623–629.
54. Capretto L., Carugo D., Mazzitelli S., Nastruzzi C. (2013) Microfluidic and lab-on-a-chip preparation routes for organic nanoparticles and vesicular systems for nanomedicine applications. *Adv. Drug Delivery Rev.* **65**(11-12), 1496–1532.
55. Hashimoto M., Tong R., Kohane D.S. (2013) Microdevices for nanomedicine. *Mol. Pharm.* **10**(6), 2127–2144.
56. Bhise N.S., Ribas J., Manoharan V., Zhang Y.S., et al. (2014) Organ-on-a-chip platforms for studying drug delivery systems. *J. Controlled Release* **190**, 82–93
57. Lee J.B., Sung J.H. (2013) Organ-on-a-chip technology and microfluidic whole-body models for pharmacokinetic drug toxicity screening. *Biotechnol. J.* **8**(11), 1258–1266.
58. Song Y., Hormes J., Kumar C.S. Microfluidic synthesis of nanomaterials. *Small*, 2008 **4**(6), 698–711.
59. Marre S., Jensen K.F. (2010) Synthesis of micro and nanostructures in microfluidic systems. *Chem. Soc. Rev.* **39**(3), 1183–1202.
60. Yurt A., Daaboul G.G., Connor J.H., Goldberg B.B., Ünlü, M.S. (2012) Single nanoparticle detectors for biological applications. *Nanoscale* **4**(3), 715–726.
61. Neeves K.B., Onasoga A.A., Wufsus A.R. (2013) The use of microfluidics in hemostasis: clinical diagnostics and biomimetic models of vascular injury. *Curr. Opin. Hematol.* **20**(5), 417–423.
62. Kuo C.T., Chiang C.L., Chang C.H., Liu H.K., et al. (2014) Modeling of cancer metastasis and drug resistance via biomimetic nano-cilia and microfluidics. *Biomaterials* **35**(5), 1562–1571.
63. Domachuk P., Tsioris K., Omenetto F.G., Kaplan D.L. (2010) Bio-microfluidics: biomaterials and biomimetic designs. *Adv. Mater.* **22**(2), 249–260.
64. Jonas A. (1986) Synthetic substrates of lecithin: cholesterol acyltransferase. *J. Lipid Res.* **27**(7), 689–698.
65. Mieszawska A.J., Mulder W.J., Fayad Z.A., Cormode D.P. (2013) Multifunctional gold nanoparticles for diagnosis and therapy of disease. *Mol. Pharm.* **10**(3), 831–847.
66. Cormode D.P., Skajaa T., van Schooneveld M.M., Koole R., et al. (2008) Nanocrystal core high-density lipoproteins: a multi-modality contrast agent platform. *Nano Lett.* **8**(11), 3715–3723.
67. Chorny M., Fishbein I., Danenberg H.D., Golomb G. (2002) Lipophilic drug loaded nanospheres prepared by nanoprecipitation: effect of formulation variables on size, drug recovery and release kinetics. *J. Controlled Release* **83**(3), 389–400.
68. Boehm A.L., Martinon I., Zerrouk R., Rump E., et al. (2003) Nanoprecipitation technique for the encapsulation of agrochemical active ingredients. *J. Microencapsul.* **20**(4), 433–441.
69. Betancourt T., Brown B., Brannon-Peppas L. (2007) Doxorubicin-loaded PLGA nanoparticles by nanoprecipitation: preparation, characterization and in vitro evaluation. *Nanomedicine* **2**(2), 219–232.
70. Medina-Sanchez M., Miserere S., Merkoci A. (2012) Nanomaterials and lab-on-a-chip technologies. *Lab Chip* **12**(11), 1932–1943.
71. Johnson B.K., Prud'homme R.K. (2003) Mechanism for rapid self-assembly of block copolymer nanoparticles. *Phys. Rev. Lett.* **91**(11), 118302.
72. Capretto L., Cheng W., Carugo D., Katsamenis O.L., et al. (2012) Mechanism of co-nanoprecipitation of organic actives and block copolymers in a microfluidic environment. *Nanotechnology* **23**(37), 375602.
73. Zhu Z. (2013) Effects of amphiphilic diblock copolymer on drug nanoparticle formation and stability. *Biomaterials* **34**(38), 10238–10248.
74. Chen T., Hynninen A.P., Prud'homme R.K., Kevrekidis I.G., et al. (2008) Coarse-grained simulations of rapid assembly kinetics for polystyrene-b-poly(ethylene oxide) copolymers in aqueous solutions. *J. Phys. Chem. B* **112**(51), 16357–16366.
75. Karnik R., Gu F., Basto P., Cannizzaro C., et al. (2008) Microfluidic platform for controlled synthesis of polymeric nanoparticles. *Nano Lett.* **8**(9), 2906–2912.
76. van Swaay D., deMello A. (2013) Microfluidic methods for forming liposomes. *Lab Chip* **13**(5), 752–767.

77. Jahn A., Stavis S.M., Hong J.S., Vreeland W.N., DeVoe D.L., Gaitan M. Microfluidic mixing and the formation of nanoscale lipid vesicles. *ACS Nano* 2010 **4**(4), 2077–2087.
78. Hong J.S., Stavis S.M., DePaoli Lacerda S.H., Locascio L.E. et al. (2010) Microfluidic directed self-assembly of liposome-hydrogel hybrid nanoparticles. *Langmuir* **26**(13), 11581–11588.
79. Kolishetti N., Dhar S., Valencia P.M., Lin L.Q., et al. (2010) Engineering of self-assembled nanoparticle platform for precisely controlled combination drug therapy. *Proc. Natl Acad. Sci. USA* **107**(42), 17939–17944.
80. Rhee M., Valencia P.M., Rodriguez M.I., Langer R., et al. (2011) Synthesis of size-tunable polymeric nanoparticles enabled by 3D hydrodynamic flow focusing in single-layer microchannels. *Adv. Mater.* **23**(12), H79–H83.
81. Tan S., Li X., Guo Y., Zhang Z. (2013) Lipid-enveloped hybrid nanoparticles for drug delivery. *Nanoscale* **5**(3), 860–872.
82. Zhang L., Chan J.M., Gu F.X., Rhee J.-W., et al. (2008) Self-assembled lipid-polymer hybrid nanoparticles: a robust drug delivery platform. *ACS Nano* **2**(8), 1696–1702.
83. Shi J., Xiao Z., Kamaly N., Farokhzad O.C. (2011) Self-assembled targeted nanoparticles: evolution of technologies and bench to bedside translation. *Acc. Chem. Res.* **44**(10), 1123–1134.
84. Cheng C.M., Kim Y., Yang J.M., Leuba S.H., et al., (2009) Dynamics of individual polymers using microfluidic based microcurvilinear flow. *Lab Chip* **9**(16), 2339–2347.
85. Kim Y., Joshi S.D., Davidson L.A., LeDuc P.R., et al., (2011) Dynamic control of 3D chemical profiles with a single 2D microfluidic platform. *Lab Chip* **11**(13), 2182–2188.
86. Kim Y., Pekkan K., Messner W.C., Leduc P.R. (2010) Three-dimensional chemical profile manipulation using two-dimensional autonomous microfluidic control. *J. Am. Chem. Soc.* **132**(4), 1339–1347.
87. deMello A.J. (2006) Control and detection of chemical reactions in microfluidic systems. *Nature* **442**(7101), 394–402.
88. Lee M.G., Choi S., Park J.K. (2009) Three-dimensional hydrodynamic focusing with a single sheath flow in a single-layer microfluidic device. *Lab Chip* **9**(21), 3155–3160.
89. Mao X.L., Lin S.C.S., Dong C., Huang T.J. (2009) Single-layer planar on-chip flow cytometer using microfluidic drifting based three-dimensional (3D) hydrodynamic focusing. *Lab Chip* **9**(11), 1583–1589.
90. Chang C.C., Huang Z.X., Yang R.J. (2007) Three-dimensional hydrodynamic focusing in two-layer polydimethylsiloxane (PDMS) microchannels. *J. Micromech. Microeng.* **17**(8), 1479–1486.
91. Nguyen N.T., Wu Z.G. (2005) Micromixers – a review. *J. Micromech. Microeng.* **15**(2), R1–R16.
92. Valencia P.M., Basto P.A., Zhang L., Rhee M., et al. (2010) Single-step assembly of homogenous lipid-polymeric and lipid-quantum dot nanoparticles enabled by microfluidic rapid mixing. *ACS Nano* **4**(3), 1671–1679.
93. Fang R.H., Chen K.N., Aryal S., Hu C.M., et al. (2012) Large-scale synthesis of lipid-polymer hybrid nanoparticles using a multi-inlet vortex reactor. *Langmuir* **28**(39), 13824–13829.
94. Liu S.J., Wei H.H., Hwang S.H., Chang H.C. (2010) Dynamic particle trapping, release, and sorting by microvortices on a substrate. *Phys. Rev. E Stat. Nonlin. Soft Matter Phys.* **82**(2 Pt 2), 026308.
95. Stott S.L., Hsu C.H., Tsukrov D.I., Yu M., et al. (2010) Isolation of circulating tumor cells using a microvortex-generating herringbone-chip. *Proc. Natl Acad. Sci. USA* **107**(43), 18392–18397.
96. Hsu C.H., Di Carlo D., Chen C., Irimia D., et al. (2008) Microvortex for focusing, guiding and sorting of particles. *Lab Chip* **8**(12), 2128–2134.
97. Shelby J.P., Lim D.S.W., Kuo J.S., Chiu D.T. (2003) High radial acceleration in microvortices. *Nature* **425**(6953), 38.
98. Kim Y., Lee C.B., Ma M., Mulder W.J., et al. (2012) Mass production and size control of lipid-polymer hybrid nanoparticles through controlled microvortices. *Nano Lett.* **12**(7), 3587–3591.
99. Mieszawska A.J., Kim Y., Gianella A., van Rooy I., et al. (2013) Synthesis of polymer-lipid nanoparticles for image-guided delivery of dual modality therapy. *Bioconjugate Chem.* **24**(9), 1429–1434.
100. Kim Y., Fay F., Cormode D.P., Sanchez-Gaytan B.L., et al. (2013) Single step reconstitution of multifunctional

high-density lipoprotein-derived nanomaterials using microfluidics. *ACS Nano* **7**(11), 9975–9983.
101. Birnbaumer G., Kupcu S., Jungreuthmayer C., Richter L., et al. (2011) Rapid liposome quality assessment using a lab-on-a-chip. *Lab Chip* **11**(16), 2753–2762.
102. Jahn A., Lucas F., Wepf R.A., Dittrich P.S. (2013) Freezing continuous-flow self-assembly in a microfluidic device: toward imaging of liposome formation. *Langmuir* **29**(5), 1717–1723.
103. Hinterwirth H., Wiedmer S.K., Moilanen M., Lehner A., et al. (2013) Comparative method evaluation for size and size-distribution analysis of gold nanoparticles. *J. Sep. Sci.* **36**(17), 2952–2961.
104. Ungureanu C., Kroes R., Petersen W., Groothuis T.A., et al. (2011) Light interactions with gold nanorods and cells: implications for photothermal nanotherapeutics. *Nano Lett.* **11**(5), 1887–1894.
105. Ramachandran S., Lal R. (2010) Scope of atomic force microscopy in the advancement of nanomedicine. *Indian J. Exp. Biol.* **48**(10), 1020–1036.
106. Wang Y., Wu Q., Sui K., Chen X.X., et al. (2013) A quantitative study of exocytosis of titanium dioxide nanoparticles from neural stem cells. *Nanoscale* **5**(11), 4737–4743.
107. Vasanthakumar S., Ahamed H.N., Saha R.N. (2014) Nanomedicine I: in vitro and in vivo evaluation of paclitaxel loaded poly-(epsilon-caprolactone), poly (DL-lactide-*co*-glycolide) and poly (DL-lactic acid) matrix nanoparticles in Wistar rats. *Eur. J. Drug Metab. Pharmacokinet.*, in press.
108. Zhu G., Zheng J., Song E., Donovan M., et al. (2013) Self-assembled, aptamer-tethered DNA nanotrains for targeted transport of molecular drugs in cancer theranostics. *Proc. Natl Acad. Sci. USA* **110**(20), 7998–8003.
109. Stern E., Wagner R., Sigworth F.J., Breaker R., et al. (2007) Importance of the Debye screening length on nanowire field effect transistor sensors. *Nano Lett.* **7**(11), 3405–3409.
110. Patolsky F., Zheng G., Hayden O., Lakadamyali M., et al. (2004) Electrical detection of single viruses. *Proc. Natl Acad. Sci. USA* **101**(39), 14017–14022.
111. Sridhar M., Xu D., Kang Y., Hmelo A.B., et al. (2008) Experimental characterization of a metal-oxide-semiconductor field-effect transistor-based Coulter counter. *J. Appl. Phys.* **103**(10), 104701–10470110.
112. Naik A.K., Hanay M.S., Hiebert W.K., Feng X.L., et al. (2009) Towards single-molecule nanomechanical mass spectrometry. *Nat. Nanotechnol.* **4**(7), 445–450.
113. Burg T.P., Godin M., Knudsen S.M., Shen W., et al. (2007) Weighing of biomolecules, single cells and single nanoparticles in fluid. *Nature* **446**(7139), 1066–1069.
114. Lee J., Shen W., Payer K., Burg T.P., et al. (2010) Toward attogram mass measurements in solution with suspended nanochannel resonators. *Nano Lett.* **10**(7), 2537–2542.
115. Lee J., Chunara R., Shen W., Payer K., et al. (2011) Suspended microchannel resonators with piezoresistive sensors. *Lab Chip* **11**(4), 645–651.
116. Kozak D., Anderson W., Vogel R., Trau M. (2011) Advances in resistive pulse sensors: devices bridging the void between molecular and microscopic detection. *Nano Today* **6**(5), 531–545.
117. Fraikin J.L., Teesalu T., McKenney C.M., Ruoslahti E., et al. (2011) A high-throughput label-free nanoparticle analyser. *Nat. Nanotechnol.* **6**(5), 308–313.
118. Ang Y.S., Yung L.Y. (2012) Rapid and label-free single-nucleotide discrimination via an integrative nanoparticle-nanopore approach. *ACS Nano* **6**(10), 8815–8823.
119. Roberts G.S., Yu S., Zeng Q., Chan L.C., et al. (2012) Tunable pores for measuring concentrations of synthetic and biological nanoparticle dispersions. *Biosens. Bioelectron.* **31**(1), 17–25.
120. Kozak D., Anderson W., Vogel R., Chen S., et al. (2012) Simultaneous size and zeta-potential measurements of individual nanoparticles in dispersion using size-tunable pore sensors. *ACS Nano* **6**(8), 6990–6997.
121. Lan W.J., White H.S. (2012) Diffusional motion of a particle translocating through a nanopore. *ACS Nano* **6**(2), 1757–1765.
122. Love S.A., Maurer-Jones M.A., Thompson J.W., Lin Y.S., et al. (2012) Assessing nanoparticle toxicity. *Annu. Rev. Anal. Chem.* **5**, 181–205.

123. Kim S.T., Saha K., Kim C., Rotello V.M. (2013) The role of surface functionality in determining nanoparticle cytotoxicity. *Acc. Chem. Res.* **46**(3), 681–691.
124. Wu J., Chen Q., Liu W., Zhang Y. (2012) Cytotoxicity of quantum dots assay on a microfluidic 3D-culture device based on modeling diffusion process between blood vessels and tissues. *Lab Chip* **12**(18), 3474–3480.
125. Mahto S.K., Yoon T.H., Rhee S.W. (2010) A new perspective on in vitro assessment method for evaluating quantum dot toxicity by using microfluidics technology. *Biomicrofluidics* **4**(3), 034111.
126. Ozturk S., Hassan Y.A., Ugaz V.M. (2012) A simple microfluidic probe of nanoparticle suspension stability. *Lab Chip* **12**(18), 3467–3473.
127. Farokhzad O.C., Khademhosseini A., Jon S., Hermmann A., et al. (2005) Microfluidic system for studying the interaction of nanoparticles and microparticles with cells. *Anal. Chem.* **77**(17), 5453–5459.
128. Kusunose J., Zhang H., Gagnon M.K., Pan T., et al. (2013) Microfluidic system for facilitated quantification of nanoparticle accumulation to cells under laminar flow. *Ann. Biomed. Eng.* **41**(1), 89–99.
129. Lai Y., Asthana A., Kisaalita W.S. (2011) Biomarkers for simplifying HTS 3D cell culture platforms for drug discovery: the case for cytokines. *Drug Discovery Today* **16**(7-8), 293–297.
130. Hsieh C.C., Huang S.B., Wu P.C., Shieh D.B., et al. (2009) A microfluidic cell culture platform for real-time cellular imaging. *Biomed. Microdevices* **11**(4), 903–913.
131. Zhang X.Q., Xu X., Bertrand N., Pridgen E., et al. (2012) Interactions of nanomaterials and biological systems: implications to personalized nanomedicine. *Adv. Drug Delivery Rev.* **64**(13), 1363–1384.
132. Huh D., Hamilton G.A., Ingber D.E. (2011) From 3D cell culture to organs-on-chips. *Trends Cell Biol.* **21**(12), 745–754.
133. Sei Y., Justus K., LeDuc P., Kim Y. (2014) Engineering living systems on chips: from cells to human on chips. *Microfluid. Nanofluid.* **16**(5), 907–920.
134. Zheng Y., Chen J., Craven M., Choi N.W., et al. (2012) In vitro microvessels for the study of angiogenesis and thrombosis. *Proc. Natl Acad. Sci. USA* **109**(24), 9342–9347.
135. Borenstein J.T., Tupper M.M., Mack P.J., Weinberg E.J., et al. (2010) Functional endothelialized microvascular networks with circular cross-sections in a tissue culture substrate. *Biomed. Microdevices* **12**(1), 71–79.
136. Srigunapalan S., Lam C., Wheeler A.R., Simmons C.A. (2011) A microfluidic membrane device to mimic critical components of the vascular microenvironment. *Biomicrofluidics* **5**(1), 13409.
137. Shin Y., Jeon J.S., Han S., Jung G.-S., et al. (2011) In vitro 3D collective sprouting angiogenesis under orchestrated ANG-1 and VEGF gradients. *Lab Chip* **11**(13), 2175–2181.
138. Gunther A., Yasotharan S., Vagaon A., Lochovsky C., et al. (2010) A microfluidic platform for probing small artery structure and function. *Lab Chip* **10**(18), 2341–2349.
139. Zhang B., Peticone C., Murthy S.K., Radisic M. (2013) A standalone perfusion platform for drug testing and target validation in micro-vessel networks. *Biomicrofluidics* **7**(4), 44125.
140. Esch M.B., Post D.J., Shuler M.L., Stokol T. (2011) Characterization of in vitro endothelial linings grown within microfluidic channels. *Tissue Eng. Part A* **17**(23-24), 2965–2971.
141. Wong K.H., Truslow J.G., Khankhel A.H., Chan K.L., et al. (2013) Artificial lymphatic drainage systems for vascularized microfluidic scaffolds. *J. Biomed. Mater. Res. Part A* **101**(8), 2181–2190.
142. Cucullo L., Marchi N., Hossain M., Janigro D. (2011) A dynamic in vitro BBB model for the study of immune cell trafficking into the central nervous system. *J. Cereb. Blood Flow Metab.* **31**(2), 767–777.
143. Booth R., Kim H. (2012) Characterization of a microfluidic in vitro model of the blood-brain barrier (muBBB). *Lab Chip* **12**(10), 1784–1792.
144. Culot M., Lundquist S., Vanuxeem D., Nion S., et al. (2008) An in vitro blood-brain barrier model for high throughput (HTS) toxicological screening. *Toxicol. in Vitro* **22**(3), 799–811.
145. Cucullo L., McAllister M.S., Kight K., Krizanac-Bengez L., et al. (2002) A new dynamic in vitro model for the multidimensional study of astrocyte-endothelial

cell interactions at the blood-brain barrier. *Brain Res.* **951**(2), 243–254.
146. Cucullo L., Hossain M., Tierney W., Janigro D. (2013) A new dynamic in vitro modular capillaries-venules modular system: cerebrovascular physiology in a box. *BMC Neurosci.* **14**, 18.
147. Prabhakarpandian B., Shen M.C., Nichols J.B., Mills I.R., et al. (2013) SyM-BBB: a microfluidic blood brain barrier model. *Lab Chip* **13**(6), 1093–1101.
148. Parkinson F.E., Friesen J., Krizanac-Bengez L., Janigro D. (2003) Use of a three-dimensional in vitro model of the rat blood-brain barrier to assay nucleoside efflux from brain. *Brain Res.* **980**(2), 233–241.
149. Kim Y., Lobatto M.E., Kawahara T., Lee Chung B., et al. (2014) Probing nanoparticle translocation across the permeable endothelium in experimental atherosclerosis. *Proc. Natl Acad. Sci. USA* **111**(3), 1078–1083.
150. Estrada R., Giridharan G.A., Nguyen M.D., Prabhu S.D., et al. (2011) Microfluidic endothelial cell culture model to replicate disturbed flow conditions seen in atherosclerosis susceptible regions. *Biomicrofluidics* **5**(3), 32006–3200611.
151. Polk B.J. Stelzenmuller A., Mijares G., MacCrehan W., Gaitan M. (2006) Ag/AgCl microelectrodes with improved stability for microfluidics. *Sens. Actuators, B* **114**, 239–247.
152. Huh D., Leslie D.C., Matthews B.D., Fraser J.P., et al. (2012) A human disease model of drug toxicity-induced pulmonary edema in a lung-on-a-chip microdevice. *Sci. Transl. Med.* **4**(159), 159ra47.
153. Kim H.J., Ingber D.E. (2013) Gut-on-a-chip microenvironment induces human intestinal cells to undergo villus differentiation. *Integr. Biol.* **5**(9), 1130–1140.
154. Kim H.J., Huh D., Hamilton G., Ingber D.E. (2012) Human gut-on-a-chip inhabited by microbial flora that experiences intestinal peristalsis-like motions and flow. *Lab Chip* **12**(12), 2165–2174.
155. Jang K.J., Mehr A.P., Hamilton G.A., McPartlin L.A., et al. (2013) Human kidney proximal tubule-on-a-chip for drug transport and nephrotoxicity assessment. *Integr. Biol.* **5**(9), 1119–1129.
156. Toh Y.C., Lim T.C., Tai D., Xiao G., et al. (2009) A microfluidic 3D hepatocyte chip for drug toxicity testing. *Lab Chip* **9**(14), 2026–2035.
157. Kim Y., Joshi S.D., Messner W.C., LeDuc P.R. (2011) Detection of dynamic spatiotemporal response to periodic chemical stimulation in a *Xenopus* embryonic tissue. *PLoS ONE* **6**(1), e14624.
158. Hattersley S.M., Sylvester D.C., Dyer C.E., Stafford N.D., et al. (2012) A microfluidic system for testing the responses of head and neck squamous cell carcinoma tissue biopsies to treatment with chemotherapy drugs. *Ann. Biomed. Eng.* **40**(6), 1277–1288.
159. Hutzler M., Fromherz P. (2004) Silicon chip with capacitors and transistors for interfacing organotypic brain slice of rat hippocampus. *Eur. J. Neurosci.* **19**(8), 2231–2238.
160. Besl B., Fromherz P. (2002) Transistor array with an organotypic brain slice: field potential records and synaptic currents. *Eur. J. Neurosci.* **15**(6), 999–1005.
161. Chronis N., Zimmer M., Bargmann C.I. (2007) Microfluidics for in vivo imaging of neuronal and behavioral activity in *Caenorhabditis elegans*. *Nat. Methods* **4**(9), 727–731.
162. George S., Xia T., Rallo R., Zhao Y., et al. (2011) Use of a high-throughput screening approach coupled with in vivo zebrafish embryo screening to develop hazard ranking for engineered nanomaterials. *ACS Nano* **5**(3), 1805–1817.
163. Jimenez A.M., Roche M., Pinot M., Panizza P. (2011) Towards high throughput production of artificial egg oocytes using microfluidics. *Lab Chip* **11**(3), 429–434.
164. Crane M.M., Chung K., Stirman J., Lu H. (2010) Microfluidics-enabled phenotyping, imaging, and screening of multicellular organisms. *Lab Chip* **10**(12), 1509–1517.
165. Wlodkowic D., Khoshmanesh K., Akagi J., Williams D.E., et al. (2011) Wormometry-on-a-chip: innovative technologies for in situ analysis of small multicellular organisms. *Cytometry, Part A* **79**(10), 799–813.
166. Shi W., Wen H., Lin B., Qin J. (2011) Microfluidic platform for the study of *Caenorhabditis elegans*. *Top. Curr. Chem.* **304**, 323–338.

167. Wang K.I., Salcic Z., Yeh J., Akagi J., et al. (2013) Toward embedded laboratory automation for smart lab-on-a-chip embryo arrays. *Biosens. Bioelectron.* **48**, 188–196.
168. Moraes C., Mehta G., Lesher-Perez S.C., Takayama S. (2012) Organs-on-a-chip: a focus on compartmentalized microdevices. *Ann. Biomed. Eng.* **40**(6), 1211–1227.
169. Sung J.H., Esch M.B., Prot J.M., Long C.J., et al. (2013) Microfabricated mammalian organ systems and their integration into models of whole animals and humans. *Lab Chip* **13**(7), 1201–1212.
170. Sung J.H., Kam C., Shuler M.L. (2011) A microfluidic device for a pharmacokinetic-pharmacodynamic (PK-PD) model on a chip. *Lab Chip* **10**(4), 446–455.
171. Williamson A., Singh S., Fernekorn U., Schober A. (2013) The future of the patient-specific body-on-a-chip. *Lab Chip* **13**(18), 3471–3480.
172. Wikswo J.P., Block III, F.E., Cliffel D.E., Goodwin C.R., et al. (2013) Engineering challenges for instrumenting and controlling integrated organ-on-chip systems. *IEEE Trans. Biomed. Eng.* **60**(3), 682–690.
173. Olcum S., Cermak N., Wasserman S.C., Christine K.S., et al. (2014) Weighing nanoparticles in solution at the attogram scale. *Proc. Natl Acad. Sci. USA* **111**(4), 1310–1315.
174. Yang M., Sun S., Kostov Y., Rasooly A. (2011) A simple 96 well microfluidic chip combined with visual and densitometry detection for resource-poor point of care testing. *Sens. Actuators, B* **153**(1), 176–181.
175. Biswas S., Miller J.T., Li Y., Nandakumar K. (2012) Developing a millifluidic platform for the synthesis of ultrasmall nanoclusters: ultrasmall copper nanoclusters as a case study. *Small* **8**(5), 687–698.
176. Hong J.W., Quake S.R. (2005) Integrated nanoliter systems. *Nat. Biotechnol.* **21**(10), 1179–1183.
177. Kobayashi I., Mukataka S., Nakajima M. (2005) Novel asymmetric through-hole array microfabricated on a silicon plate for formulating monodisperse emulsions. *Langmuir* **21**(17), 7629–7632.
178. Kobayashi I., Mukataka S., Nakajima M. (2005) Effects of type and physical properties of oil phase on oil-in-water emulsion droplet formation in straight-through microchannel emulsification, experimental and CFD studies. *Langmuir* **21**(13), 5722–5730.
179. Nisisako T., Torii T. (2008) Microfluidic large-scale integration on a chip for mass production of monodisperse droplets and particles. *Lab Chip* **8**(2), 287–293.
180. Li W., Greener J., Voicu D., Kumacheva E. (2009) Multiple modular microfluidic (M3) reactors for the synthesis of polymer particles. *Lab Chip* **9**(18), 2715–2721.
181. Kim Y., LeDuc P., Messner W. (2013) Modeling and control of a nonlinear mechanism for high performance microfluidic systems. *IEEE Trans. Control Syst. Technol.* **21**(1), 203–211.
182. Psaltis D., Quake S.R., Yang C.H. (2006) Developing optofluidic technology through the fusion of microfluidics and optics. *Nature* **442**(7101), 381–386.
183. Mao X., Waldeisen J.R., Juluri B.K., Huang T.J. (2007) Hydrodynamically tunable optofluidic cylindrical microlens. *Lab Chip* **7**(10), 1303–1308.
184. Yang A.H., Erickson D. (2010) Optofluidic ring resonator switch for optical particle transport. *Lab Chip* **10**(6), 769–774.
185. Song C.L., Luong T.D., Kong T.F., Nguyen N.T. (2011) Disposable flow cytometer with high efficiency in particle counting and sizing using an optofluidic lens. *Opt. Lett.* **36**(5), 657–659.
186. Ziolkowska K., Stelmachowska A., Kwapiszewski R., Chudy M. (2013) Long-term three-dimensional cell culture and anticancer drug activity evaluation in a microfluidic chip. *Biosens. Bioelectron.* **40**(1), 68–74.
187. Lii J., Hsu W.J., Parsa H., Das A. (2008) Real-time microfluidic system for studying mammalian cells in 3D microenvironments. *Anal. Chem.* **80**(10), 3640–3647.
188. Cheah L.T., Dou Y.H., Seymour A.M., Dyer C.E., et al. (2010) Microfluidic perfusion system for maintaining viable heart tissue with real-time electrochemical monitoring of reactive oxygen species. *Lab Chip* **10**(20), 2720–2726.

189. Richter L., Charwat V., Jungreuthmayer C., Bellutti F. (2011) Monitoring cellular stress responses to nanoparticles using a lab-on-a-chip. *Lab Chip* **11**(15), 2551–2560.

190. Ahmad A.A., Wang Y., Gracz A.D., Sims C.E. (2014) Optimization of 3-D organotypic primary colonic cultures for organ-on-chip applications. *J. Biol. Eng.* **8**, 9.

27
Nanoparticle Conjugates for Small Interfering RNA Delivery

Timothy L. Sita and Alexander H. Stegh
Northwestern University Ken and Ruth Davee Department of Neurology, The Northwestern Brain Tumor Institute, The Robert H. Lurie Comprehensive Cancer Center, The International Institute for Nanotechnology, 303 East Superior, Chicago, IL 60611, USA

1	The RNA Interference Pathway	971
2	Limitations of Unmodified siRNA Delivery and Carrier Design Considerations	972
2.1	Serum Stability	972
2.2	Immunogenicity	973
2.3	Renal Clearance	974
2.4	Biodistribution	974
2.5	Intracellular Uptake	975
2.6	Endosomal Escape	975
3	siRNA Nanocarriers in Clinical Development	976
3.1	Conjugate Delivery Systems	980
3.2	Gold Nanoparticles	980
3.3	Spherical Nucleic Acids	981
3.4	Iron Oxide Nanoparticles	984
3.5	Cyclodextrin Polymer Nanoparticles	985
3.6	Liposomes and Lipid-Based Materials	986
3.6.1	Cationic Liposomes	986
3.6.2	Neutral Liposomes	987
3.6.3	Stable Nucleic Acid–Lipid Particles	987
3.6.4	Exosome-Based siRNA Carriers	987
4	The Future of Nanoparticle-Based siRNA Delivery	988
	References	989

Translational Medicine: Molecular Pharmacology and Drug Discovery
First Edition. Edited by Robert A. Meyers.
© 2018 Wiley-VCH Verlag GmbH & Co. KGaA. Published 2018 by Wiley-VCH Verlag GmbH & Co. KGaA.

Keywords

Enhanced permeation and retention (EPR) effect
The phenomenon by which nanoparticles accumulate in tumor tissue due to both increased porosity of tumor vasculature and decreased lymphatic drainage

Mononuclear phagocyte system (MPS)
Previously known as the reticuloendothelial system (RES), which consists of monocytes and macrophages in the liver, spleen, bone marrow, blood, and lymph nodes, and is required for the body's defense against microorganisms and foreign materials

Proton sponge effect (PSE)
The phenomenon by which siRNA carriers become protonated in the acidic endosomal environment, leading to an influx of additional protons and chloride ions, and resulting in an osmotic imbalance that draws water into the endosome, causing it to burst and release its contents into the cytosol

RNA-induced silencing complex (RISC)
A protein complex mediating the binding and unwinding of double-stranded siRNA within the cytoplasm and the subsequent sequence-specific recognition and degradation of target mRNA

RNA interference (RNAi)
A fundamental pathway in eukaryotic cells, by which RNA oligonucleotide sequences induce the degradation of mRNA containing a complementary sequence

Small interfering RNA (siRNA)
RNA fragments approximately 21–23 nucleotides long that are capable of inducing sequence-specific degradation of complementary mRNA

Spherical nucleic acids (SNAs)
Nanoparticles for siRNA delivery that consist of a 13 nm gold nanoparticle core functionalized with a corona of thiolated siRNA duplexes and polyethylene glycol

Stable nucleic acid–lipid particles (SNALPs)
Nanoparticles for siRNA delivery, which have a lipid bilayer composed of cationic lipids and neutral fusogenic lipids, and which are coated with polyethylene glycol to provide a neutral surface

As a therapeutic strategy, small interfering RNA (siRNA) has a remarkable potential to treat genetic diseases driven by mutated or aberrantly expressed genes, including cancer, inflammatory conditions, neurodegenerative disorders, and viral infections. Since the discovery of the RNA interference (RNAi) pathway in 1998, a multitude of siRNA delivery methods have been designed and tested. However, siRNA-based therapeutics are currently limited by the inability to safely and robustly deliver nucleic acids to target cells and tissues. With the development of nanocarriers composed of organic (lipids, liposomes, conjugated polymers) and inorganic (iron oxide, gold) materials, nanotechnology has emerged as a discipline that offers one of the most powerful solutions to enable the stable and safe delivery of siRNA oligonucleotides. Many of these nanocarriers demonstrate acceptable safety profiles, enhance the *in-vivo* stability of siRNA, promote robust tissue penetration, and can be modified with a targeting ligand to allow for tissue-specific uptake. In this chapter, the challenges of delivering siRNA are reviewed, and the most important advances in the development of nanoparticle-based siRNA delivery systems currently in preclinical and clinical development will be highlighted.

1
The RNA Interference Pathway

In 1998, Fire and Mello demonstrated siRNA-based gene silencing in *Caenorhabditis elegans*, a discovery that earned them the Nobel Prize in Physiology or Medicine [1]. Tuschl and colleagues then confirmed the RNA interference (RNAi) pathway in mammalian cells [2], and the idea of harnessing RNAi-mediated gene silencing for therapeutic purposes has garnered enormous attention from scientists ever since [3–5]. The ability to knockdown any gene of interest opened up an entirely new way to combat diseases that were minimally responsive to traditional therapeutics, including (but not limited to) cancer, hereditary disorders, heart disease, inflammatory conditions, and viral infections [6–9]. These diseases are often driven by genetic aberrations, most of which are considered "undruggable" by conventional pharmacological strategies, foremost small molecules and biotherapeutic antibodies. RNAi, however, holds promise as a novel tool for precision medicine, as it is able to efficiently and persistently silence deregulated or mutated genes through the use of customizable siRNA oligonucleotides [10, 11].

The RNAi pathway inhibits gene expression through the selective cleavage of mRNA (Fig. 1) [12–14]. The pathway is activated by the presence of double-stranded RNA (dsRNA) in the cytoplasm. The cytoplasmic enzyme Dicer cleaves longer dsRNA into smaller segments, typically 21–23 nucleotides in length. These siRNA segments are then recognized and loaded into the RNA-induced silencing complex (RISC), activating the complex. Once activated, antisense strands bind complementary mRNA. Argonaute 2, an RNA endonuclease resident in the activated RISC complex, subsequently cleaves mRNA, thereby inhibiting its translation. Because the RISC complex is recycled and cleaves mRNA continuously, knockdown is persistent, lasting between three and seven days in dividing cells, and up to three to four weeks in nondividing cells [15].

However, a number of barriers to effective delivery are encountered when naked siRNA is systemically injected, limiting its therapeutic potential *in-vivo*. Unmodified

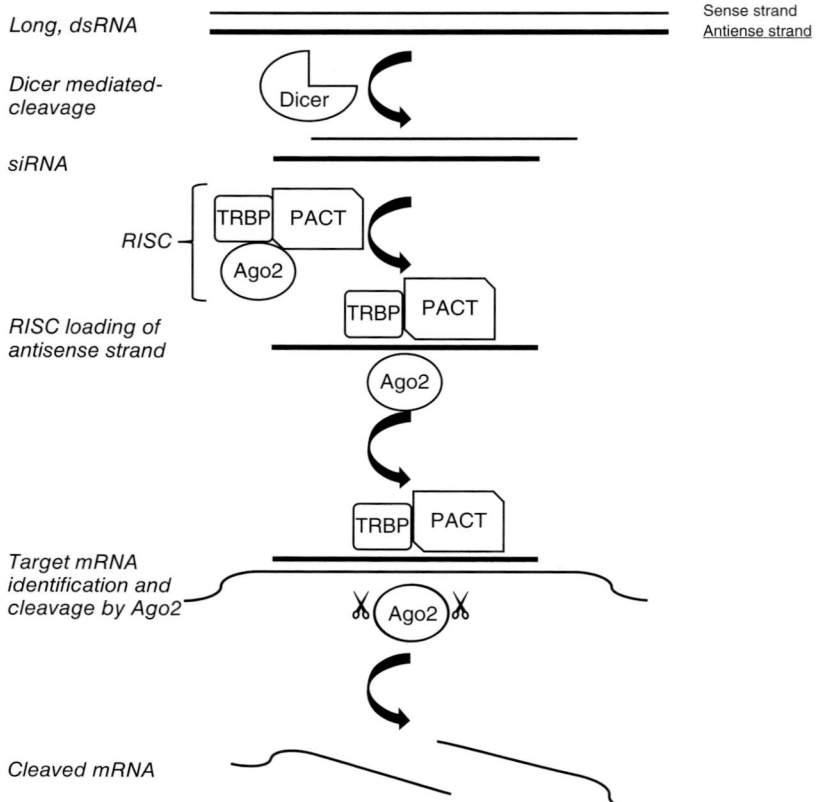

Fig. 1 The RNA interference pathway. Long, dsRNA is cleaved by the enzyme Dicer into 21- to 23-nucleotide siRNA segments. These segments are loaded into the RISC complex, which includes Protein ACTivator of the interferon-induced protein kinase (PACT), transactivation response RNA binding protein (TRBP) and Argonaute 2 (Ago2). The sense siRNA strand is degraded and the activated RISC complex recognizes mRNA complementary to the antisense siRNA strand. Target mRNA is then cleaved by Ago2, prohibiting its translation.

siRNA is subject to rapid serum cleavage, renal clearance, and distribution to nontarget organ systems. Furthermore, unmodified siRNA shows poor intracellular uptake and has been shown to trigger a cellular immune response [16–18]. The *in-vivo* obstacles to siRNA delivery, as well as the potential solutions afforded by siRNA nanocarriers currently in development, are discussed in the following sections.

2
Limitations of Unmodified siRNA Delivery and Carrier Design Considerations

2.1
Serum Stability

The phosphodiester backbone of unmodified siRNA makes it sensitive to rapid hydrolysis by serum RNases. On average, intravenously administered siRNA has an approximate serum half-life of 15 min, with some sequences hydrolyzed and rendered nonfunctional in as little as 1 min or less [19, 20]. Such rapid degradation renders therapeutic benefit of systemically delivered unmodified siRNA unlikely in humans. However, chemical modifications have demonstrated efficacy in prolonging the *in-vivo* stability of intravenously administered siRNA without affecting RISC complexation. Examples include modifications with 2′-O-methyl, 2′-fluoro, and 2′-O-methoxyethyl groups at the 2′ hydroxyl in the sugar ring, as well as at the phosphate backbone with phosphorothioate and boranophosphate (Fig. 2) [21–23]. These modifications are capable of disguising binding motifs recognized by RNases and significantly slowing the rate of siRNA degradation.

2.2 Immunogenicity

Numerous studies have demonstrated activation of the innate immune system following the systemic administration of siRNA [24]. Although this immunostimulatory potential may be advantageous in circumstances where a proinflammatory environment is desired (i.e., recent viral inoculation) [25], it is usually an unwanted outcome. Some of these immunostimulatory effects are sequence-specific, as siRNA sequences with uridine-, guanosine-, and uridine-rich regions have been shown to more robustly activate the immune system [26]. Other immunogenic responses are independent of the siRNA sequence. In particular, interferon-α induction via activation of toll-like receptor (TLR) 3 occurs independently of siRNA sequence [27].

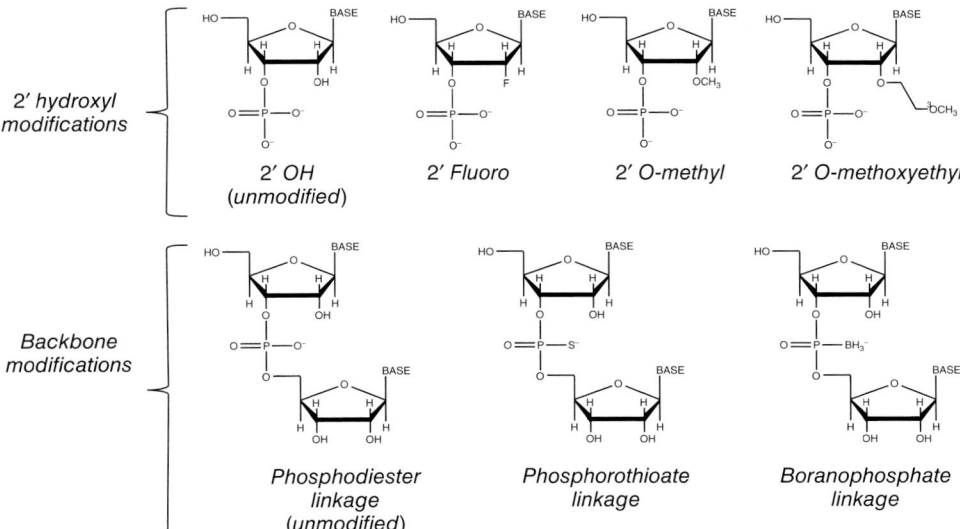

Fig. 2 Chemical modifications of siRNA. Common sites for modification include the 2′ hydroxyl in the sugar ring and the phosphate backbone. These modifications disguise binding motifs recognized by RNases.

Some of the chemical modifications intended to prolong *in-vivo* stability also reduce immune responses triggered by the intravenous administration of siRNA. The incorporation of as few as two 2′-O-methyl groups into a siRNA duplex can be sufficient to prevent TLR7 activation, as these groups serve as competitive TLR7 inhibitors [28, 29]. Other sugar ring modifications, such 2′-fluoro groups, have also been shown to confer protection against immune activation by abrogating the interaction between the siRNA and TLR7 [22]. Hence, most siRNA therapeutics that have reached clinical development utilize chemically modified siRNA to increase stability and reduce immunogenicity.

2.3
Renal Clearance

While chemically modified siRNA increases serum stability and reduces immunogenicity, it does not effectively prevent renal clearance, which is largely a function of molecular size. Physical filtration of the blood occurs at the basement membrane of the renal glomerulus through pores roughly 8 nm wide, allowing the passage of water and other small molecules into the urine for subsequent excretion [30]. This includes naked siRNA, which has been observed to pass freely through the glomerular basement membrane [31]. With the exception of urinary tract targets, renally cleared siRNA is not delivered to its target tissue and fails to engage the RNAi pathway at the desired organ site. When designing delivery systems for siRNA, a minimum size of 10–20 nm is typically required to avoid renal clearance [32, 33]. Molecular weight is an additional consideration, with polymers of >40 kDa more likely to be retained in circulation rather than be passed through the glomerular filtration barrier into urine [20].

2.4
Biodistribution

In addition to evading renal clearance, systemically delivered siRNA has to selectively reach its target site while avoiding off-target effects in nontarget tissues. Both passive and active targeting mechanisms influence the biodistribution of systemically administered siRNA. To reach organ sites, siRNA must passively leave the bloodstream by traversing fenestrations in the endothelium. Typical endothelial fenestrations are 10–50 nm in diameter, which prohibits the extravasation of larger molecules [34]. However, extravasation occurs more readily in organs such as the liver and spleen, as well as many solid tumors, which are characterized by more discontinuous endothelia [35]. During circulation and following extravasation from the bloodstream, molecules – particularly those more than 100 nm in size – are subject to engulfment by macrophages of the mononuclear phagocytic system (MPS) in the blood, liver, and spleen [36]. Thus, a suggested maximal size of 100 nm for siRNA carrier systems has been proposed to minimize engulfment by the MPS [37].

In solid tumor tissue, passive targeting can be increased with nanocarriers via the enhanced permeation and retention (EPR) effect – a phenomenon by which circulating nanoparticles preferentially accumulate in tumor tissue [38]. Uncontrolled tumor angiogenesis results in poorly formed and leaky tumor vasculature. Over time, nanoparticles accumulate due to this leaky vasculature and are retained, as the tumor microenvironment is characterized by decreased lymphatic drainage.

To achieve optimal permeation with concomitant retention in tumor tissue, molecules less than 200 nm in diameter are ideal, further supporting the notion that a particle size of 100 nm is ideal for tissue delivery [36].

In addition to size-based considerations, both carrier charge and shape influence the passive biodistribution of siRNA. For example, increasing the carrier charge through the introduction of 1,2-dioleoyl-3-tri-methylammonium-propane (DOTAP) has been shown to decrease siRNA accumulation in the spleen and normal vasculature, while increasing accumulation in the liver and tumor vasculature [39]. Furthermore, multiple studies have suggested that nonspherical particles – including filomicelles, discoidal, and rod-shaped particles – promote a greater tumor accumulation and reduced uptake via the MPS compared to spherical particles, which suggests that nonspherical carriers have a physiologic advantage for siRNA delivery [40–42].

For active targeting, siRNA and siRNA carriers may be conjugated with a variety of ligands to enable tissue-specific uptake. For example, the galactose derivative N-acetylgalactosamine (GalNAc) has been studied extensively in hepatocyte targeting. GalNAc has a strong affinity for the asialoglycoprotein receptor (ASPGR), which is highly expressed in hepatocytes [43]. Additionally, ligands with the capacity to bind the transferrin receptor (TfR) are frequently conjugated to nanoparticles for targeting purposes. While TfR is ubiquitously expressed at low levels in most human tissues, it is overexpressed on most cancer cell surfaces, and its expression on malignant cells has been shown to correlate with tumor grade and rate of progression [44]. Furthermore, TfR is overexpressed on brain capillary endothelial cells and can facilitate increased blood–brain barrier (BBB) penetration via receptor-mediated transcytosis [45]. Moreover, an abundance of other tissue-specific receptors, small molecules, antibodies, proteins, peptides and aptamers have been employed to increase target tissue uptake. These include small molecules such as folic acid, clinically relevant monoclonal antibodies (e.g., rituximab), and aptamers that are capable of specifically binding prostate-specific membrane antigen (PSMA). These targeting moieties are discussed elsewhere in detail by Sanna *et al.* [46].

2.5
Intracellular Uptake

siRNA must first cross the plasma membrane in order to engage the RNAi machinery in the cytosol. Naked, linear siRNA cannot freely diffuse across this membrane due to its negative charge and relatively large size; hence, it must be either coupled with a delivery agent able to cross or fuse with the membrane, or be presented in a spherical configuration that is recognized by scavenger receptors and internalized via receptor-mediated endocytosis (see Sect. 3.3) [47]. Common classes of delivery agents conjugated to siRNA include lipid moieties, cell-penetrating peptides, and small molecules [17]. The vast majority of these materials mediate intracellular uptake via clathrin- or lipid raft/caveolin-mediated endocytosis [48].

2.6
Endosomal Escape

As receptor-mediated endocytosis is the most common route by which siRNA can enter the cytoplasm, it must first escape the endosomal pathway to gain access to

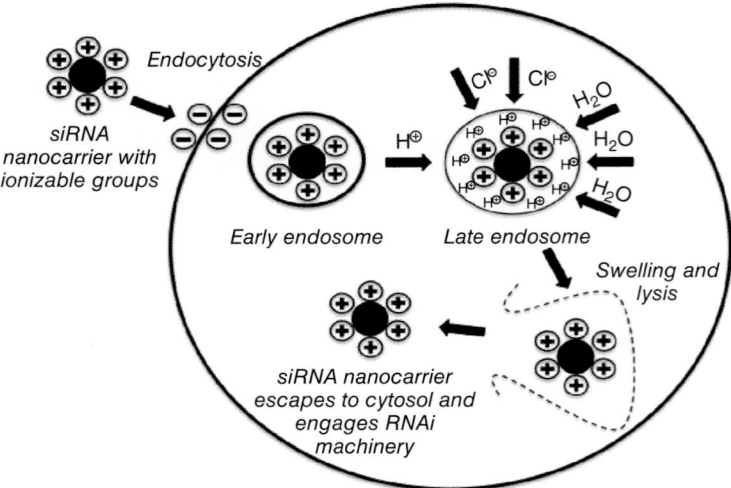

Fig. 3 The proton sponge effect. siRNA nanocarriers with ionizable groups are protonated in the acidic environment of the endosome, leading to an influx of additional protons and chloride ions. This results in an osmotic imbalance that draws water into the endosome, causing swelling and lysis of the endosome and allowing siRNA to escape to the cytosol, where it can engage the RNAi pathway.

the RNAi machinery. In the endosomal pathway, extracellular material is entrapped in membrane-bound vesicles, which then fuse with early endosomes, mature into increasingly acidic late endosomes, and finally fuse with lysosomes. If siRNA is unable to escape the endosomal pathway, it is degraded by lysosomal RNases. Certain types of siRNA carrier can promote endosomal escape, although the precise mechanism of escape is often elusive. One theory, based on the "proton sponge effect" (PSE), suggests that siRNA carriers harboring amine groups with pK_a values ranging from 5 to 7, such as those found on polyethylenimine (PEI) and β-amino esters, are protonated as the endosome becomes increasingly acidic [49, 50]. This protonation is followed by an influx of additional protons and chloride ions, resulting in an osmotic imbalance that draws water into the endosome, causing it to burst and release its contents into the cytosol (Fig. 3) [51]. Additionally, pH-sensitive groups on lipid carriers, such as the citraconic anhydride-modified phospholipid 1,2-dioleoyl-3-phosphatidylethanolamine (C-DOPE), are designed to degrade under acidic conditions and induce structural transformations that promote the endosomal escape of siRNA [17, 52].

3 siRNA Nanocarriers in Clinical Development

Currently, there are 24 different siRNA-based therapeutics in 43 different clinical trials (Table 1). The first siRNA therapeutics to enter clinical trials were based on the local delivery of naked siRNA or chemically modified siRNA [10]. These included intravitreal delivery for age-related macular degeneration and intranasal delivery to

Tab. 1 siRNA-based compounds in clinical trials.

Compound	Target	Delivery system	Condition	Phase	Status	Sponsor	ClinicalTrials.gov identifier
AGN211745 (previously known as Sirna-027)	VEGFR1	Naked siRNA	Age-related macular degeneration, choroidal neovascularization	II	Terminated	Allergan	NCT00395057
ALN-RSV01	RSV nucleocapsid	Naked siRNA	Respiratory syncytial virus infections	II	Completed	Alnylam Pharmaceuticals	NCT00496821
ALN-RSV01	RSV nucleocapsid	Naked siRNA	Respiratory syncytial virus infections	II	Completed	Alnylam Pharmaceuticals	NCT00658086
ALN-RSV01	RSV nucleocapsid	Naked siRNA	Respiratory syncytial virus infections	II	Completed	Alnylam Pharmaceuticals	NCT01065935
Bevasiranib	VEGF	Naked siRNA	Wet age related-macular degeneration	II	Completed	Opko Health, Inc.	NCT00259753
Bevasiranib	VEGF	Naked siRNA	Diabetic macular edema	II	Completed	Opko Health, Inc.	NCT00306904
I5NP	p53	Naked siRNA	Kidney injury, acute renal failure	I	Completed	Quark Pharmaceuticals	NCT00554359
I5NP	p53	Naked siRNA	Delayed graft function, kidney transplant complications	I	Active, not recruiting	Quark Pharmaceuticals	NCT00802347
PF-04523655	RTP801 (proprietary target)	Naked siRNA	Choroidal neovascularization, diabetic retinopathy, diabeter macular edema	II	Active, not recruiting	Quark Pharmaceuticals	NCT01445899
QPI-1007	CASP2	Naked siRNA	Optic atrophy, non-arteritic anterior ischemic optic neuropathy	I	Completed	Quark Pharmaceuticals	NCT01064505
SYL040012	ADRB2	Naked siRNA	Ocular hypertension, glaucoma	I, II	Completed	Sylentis, S.A.	NCT01227291
SYL040012	ADRB2	Naked siRNA	Ocular hypertension, open-angle glaucoma	II	Completed	Sylentis, S.A.	NCT01739244
SYL1001	TRPV1	Naked siRNA	Ocular pain, dry-eye syndrome (in healthy volunteers)	I	Completed	Sylentis, S.A.	NCT01438281
SYL1001	TRPV1	Naked siRNA	Ocular pain, dry-eye syndrome	I, II	Recruiting	Sylentis, S.A.	NCT01776658
TD101	K6a	Naked siRNA	Pachyonychia congenita	I	Completed	Pachyonychia Congenita Project	NCT00716014

(continued overleaf)

Tab. 1 (Continued.)

Compound	Target	Delivery system	Condition	Phase	Status	Sponsor	ClinicalTrials.gov identifier
ALN-AT3SC	Antithrombin	siRNA-GalNAc conjugate	Hemophilia A, Hemophilia B	I	Recruiting	Alnylam Pharmaceuticals	NCT02035605
ALN-TTRsc	Transthyretin	siRNA-GalNAc conjugate	Transthyretin-mediated amyloidosis	I	Recruiting	Alnylam Pharmaceuticals	NCT01814839
ALN-TTRsc	Transthyretin	siRNA-GalNAc conjugate	Transthyretin-mediated amyloidosis	II	Recruiting	Alnylam Pharmaceuticals	NCT01981837
ARC-520	Conserved regions of HBV	Dynamic polyconjugates	Chronic hepatitis B	I	Recruiting	Arrowhead Research Corporationq	NCT01872065
ARC-520	Conserved regions of HBV	Dynamic polyconjugates	Chronic hepatitis B	II	Recruiting	Arrowhead Research Corporationq	NCT02065336
RXi-109	CTGF	Self-delivering RNAi compound	Cicatrix, scar prevention	I	Active, not recruiting	RXi Pharmaceuticals, Corp.	NCT01640912
RXi-109	CTGF	Self-delivering RNAi compound	Cicatrix, scar prevention	I	Active, not recruiting	RXi Pharmaceuticals, Corp.	NCT01780077
RXi-109	CTGF	Self-delivering RNAi compound	Hypertrophic scar	II	Recruiting	RXi Pharmaceuticals, Corp.	NCT02030275
RXi-109	CTGF	Self-delivering RNAi compound	Keloid	II	Recruiting	RXi Pharmaceuticals, Corp.	NCT02079168
siG12D LODER	KRAS	LODER polymer	Pancreatic cancer	I	Active, not recruiting	Silenseed Ltd.	NCT01188785
siG12D LODER	KRAS	LODER polymer	Unresectable locally advanced pancreatic cancer	II	Not yet recruiting	Silenseed Ltd.	NCT01676259
CALAA-01	RRM2	CDP NP	Solid tumors	I	Terminated	Calando Pharmaceuticals	NCT00689065
ALN-PCS02	PCSK9	Lipid nanoparticle	Hypercholesterolemia	I	Completed	Alnylam Pharmaceuticals	NCT01437059
ALN-VSP02	KSP/VEGF	Lipid nanoparticle	Solid tumors	I	Completed	Alnylam Pharmaceuticals	NCT00882180

ALN-VSP02	KSP/VEGF	Lipid nanoparticle	Solid tumors	I	Completed	Alnylam Pharmaceuticals	NCT01158079
ND-L02-s0201	HSP47	Lipid nanoparticle	Healthy subjects	I	Completed	Nitto Denko Corporation	NCT01858935
Patisiran (ALN-TTR02)	Transthyretin	Lipid nanoparticle	Transthyretin-mediated amyloidosis	I	Completed	Alnylam Pharmaceuticals	NCT01559077
Patisiran (ALN-TTR02)	Transthyretin	Lipid nanoparticle	Transthyretin-mediated amyloidosis	II	Active, not recruiting	Alnylam Pharmaceuticals	NCT01617967
Patisiran (ALN-TTR02)	Transthyretin	Lipid nanoparticle	Transthyretin-mediated amyloidosis	III	Recruiting	Alnylam Pharmaceuticals	NCT01960348
Patisiran (ALN-TTR02)	Transthyretin	Lipid nanoparticle	Transthyretin-mediated amyloidosis	II	Recruiting	Alnylam Pharmaceuticals	NCT01961921
Patisiran (ALN-TTR02)	Transthyretin	Lipid nanoparticle	Transthyretin-mediated amyloidosis	I	Recruiting	Alnylam Pharmaceuticals	NCT02053454
Atu027	PKN3	Cationic liposome	Advanced solid tumors	I	Completed	Silence Therapeutics AG	NCT00938574
Atu027	PKN3	Cationic liposome	Advanced or metastatic pancreatic cancer	I, II	Recruiting	Silence Therapeutics AG	NCT01808638
siRNA-EphA2-DOPC	EphA2	Neutral liposome	Advanced cancers	I	Not yet recruiting	M.D. Anderson Cancer Center	NCT01591356
PRO-040201	ApoB	SNALP	Hypercholesterolemia	I	Terminated	Tekmira Pharmaceuticals Corporation	NCT00927459
TKM-080301	PLK1	SNALP	Neuroendocrine tumors, adrenocortical carcinoma	I, II	Recruiting	Tekmira Pharmaceuticals Corporation	NCT01262235
TKM-080301	PLK1	SNALP	Primary or secondary liver cancer	I	Completed	National Cancer Institute	NCT01437007
TKM-100802	VP24, VP35, Zaire Ebola L-polymerase	SNALP	Ebola virus infection	I	Recruiting	Tekmira Pharmaceuticals Corporation	NCT02041715

LODER, LOcal Drug EluterR.

treat respiratory syncytial virus (RSV) [53–59]. Subsequent generations of siRNA therapeutics were designed for systemic delivery, and are based on a variety of nanoparticle carriers capable of overcoming the delivery challenges described above.

3.1
Conjugate Delivery Systems

Conjugate delivery systems (CDSs) are the simplest of all siRNA delivery materials in design, consisting of siRNA directly conjugated to either polymers, peptides, antibodies, aptamers, and small molecules [60]. CDSs that are composed of siRNA attached to cholesterol and other lipid moieties have demonstrated efficacy *in-vivo*. One particular CDS system, termed Dynamic PolyConjugates (DPCs) technology, consists of siRNA conjugated to an amphipathic, membrane-active polymer, a shielding polyethylene glycol (PEG) that can mask the polymer until it has reached the endosome, and a GalNAc hepatocyte targeting ligand [61]. DPCs were shown to achieve a functional knockdown of two different siRNA targets in the liver – apolipoprotein B (ApoB) and peroxisome proliferator-activated receptor alpha (PPARα) [61]. The quantification of target mRNA levels, using quantitative real-time reverse transcriptase polymerase chain reaction (qRT-PCR), in murine liver tissue indicated that DPC silenced *ApoB* expression up to 74% and *PPARα* expression up to 61%, as compared to mice treated with control siRNA. Furthermore, serum liver enzymes and cytokine levels were not statistically different from saline-treated mice, indicating that DPCs were well tolerated. These constructs are currently undergoing Phase I and II trials for hepatitis B infection (Table 1).

3.2
Gold Nanoparticles

Gold nanoparticles (AuNPs) serve as excellent siRNA vectors due to their biocompatibility, tunable size, and ease of functionalization. Although AuNPs have not yet been approved by the United States Food and Drug Administration (FDA) for use in humans, gold salts have been used for many years in the clinic as treatments for arthritis; indeed, recent preclinical studies of AuNPs conjugated with siRNA have indicated little to no toxicity in mice [62–67]. Additionally, monodisperse AuNPs ranging from 1 to 150 nm in diameter can be reproducibly generated by reducing gold salts in the presence of stabilizing agents that prevent the AuNPs from aggregating. For particles ranging from ~1.5 to 6 nm in diameter, the one-pot protocol developed by Schiffrin and colleagues can be utilized, in which tetrachloroaurate ($AuCl_4^-$) is reduced with sodium borohydride ($NaBH_4$) in the presence of an alkanethiol [68]. For particles with core sizes ranging from ~10 to 150 nm, chloroauric acid ($HAuCl_4$) can be reduced with sodium citrate ($Na_3C_6H_5O_7$) [69–73].

AuNPs can be easily modified to carry siRNA through both covalent and noncovalent interactions. The strong metal–ligand interaction between gold and sulfur (S-Au bond) allows for the covalent attachment of almost any thiolated biomolecule (including siRNA) to the surface of AuNPs [74–76]. In the past, investigators have taken full advantage of this gold–thiol chemistry to develop multimodal drug delivery constructs, using the S-Au bond as a scaffold for the attachment of targeting ligands (e.g., folic acid, anti-HER2 antibody), PEG moieties for increased colloidal stability and blood circulation times, and gadolinium for magnetic resonance imaging (MRI) [77–80]. The addition of thiolated hydrophobic drugs (e.g., paclitaxel) to AuNPs has even been used as a method to increase drug solubility in blood and thus to enhance drug efficacy [81, 82]. Noncovalent AuNP–siRNA nanoconjugates are generated by electrostatically complexing negatively charged siRNA with cationic AuNPs and/or cationic polymers. The layer-by-layer assembly of positively charged PEI with siRNA on AuNPs is another technique that has been used extensively, resulting in particles that exhibit both efficient siRNA transfection and high rates of endosomal escape [83, 84]. Other groups have demonstrated the successful delivery of siRNA using positively charged biomolecules, complexing protamine, dendrons, and cysteamine to siRNA on AuNPs [62, 64, 66].

AuNPs also benefit from a distinct localized surface plasmon resonance (LSPR), which is the collective oscillation of electrons in a solid or liquid stimulated by incident light. Resonance occurs when the frequency of incident light photons matches the natural frequency of surface electrons oscillating against the restoring force of positive nuclei. Depending on the size and shape of the AuNPs, LSPR occurs in the visible and near-infrared (NIR) range of the spectrum [85]. This property opens up a multitude of applications in imaging, diagnosis, and photothermal therapy [73, 79, 86–89]. For example, Eghtedari et al. used NIR light to detect gold nanorods (50 nm × 15 nm) in mouse tissue (4 cm depth) using an optoacoustic method [90]. Likewise, Shim et al. took advantage of the LSPR of AuNPs in conjunction with siRNA delivery, by combining gene therapy and stimuli-responsive optical imaging with 15 nm AuNPs conjugated to acid-degradable ketalized linear polyethylenimine (KL-PEI) electrostatically complexed with siRNA [91]. In this construct, N-succinimidyl 3-(2-pyridyldithio)-propionate (SPDP) was directly conjugated to the AuNP surface via pyridyldithio groups, while the amine-reactive N-succinimidyl (NHS) groups on SPDP reacted with KL-PEI/siRNA polyplexes. In this way, Shim and colleagues showed that in a mildly acidic environment (e.g., tumor microenvironment), the acid-sensitive PEI/siRNA complexes would disassociate from the AuNPs, resulting in both gene knockdown and changes in optical signals, including diminished scattering intensity, increased variance of Doppler frequency, and blue-shifted ultraviolet absorbance.

Given their straightforward synthesis and conjugation protocols, biologically inert nature, and multimodal imaging and gene delivery capabilities, the preclinical outlook on AuNP–siRNA constructs is particularly encouraging [92]. Although clinical trials with AuNPs for siRNA delivery have not yet commenced, AuNPs conjugated to PEG and tumor necrosis factor alpha (TNFα) have already been investigated in Phase I trials for the treatment of cancer. It is very likely, therefore, that AuNP–siRNA constructs will shortly be enrolled in clinical trials [93, 94].

3.3
Spherical Nucleic Acids

In 1996, Mirkin and colleagues pioneered the spherical nucleic acid (SNA) platform, which originally consisted of a 13-nm citrate-capped AuNP core functionalized with thiolated DNA [95]. Over time, many variations of the original structure have been explored (e.g., adding fluorophores, modifying oligonucleotide sequence, and

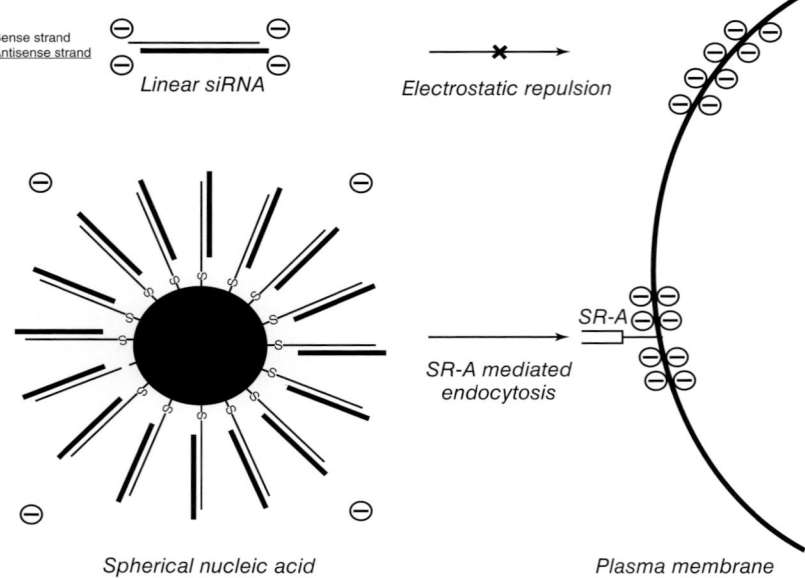

Fig. 4 Mechanism for intracellular entry of spherical nucleic acids. Negatively charged, linear siRNA duplexes are unable to independently cross the negatively charged plasma membrane. However, the spherical presentation of siRNA on SNAs allows for rapid intracellular uptake via scavenger receptor-A (SR-A) -mediated endocytosis.

length), and these constructs have proven to be incredibly useful in a vast amount of applications, including (but not limited to) the *in-vitro* biodetection of mRNAs (more than 1500 different RNA detection assays are currently marketed by Millipore as SmartFlare™ technology), DNA-based materials synthesis and engineering [96, 97], and RNAi-mediated gene regulation [67, 98, 99]. Subsequent generations of SNAs intended for *in-vivo* gene regulation exchanged thiolated DNA for thiolated siRNA duplexes, and backfilled the nanocarriers with PEG molecules for increased colloidal stability and prolonged *in-vivo* circulation times [67, 98, 100].

Unique properties arise that address siRNA delivery challenges when siRNA duplexes are arranged in this spherical fashion, including intracellular delivery into almost all cell types, increased nuclease resistance, little to no immune activation *in-vitro* and *in-vivo*, and the persistent knockdown of target genes [95, 101–105]. Importantly, these properties are retained even when the gold core is dissolved or changed to a lipid-based core, crediting the spherical three-dimensional (3D) presentation of nucleic acid for the majority of these distinctive properties [106]. Despite the highly negative surface charge (zeta potential = -34 mV) associated with presenting oligonucleotides in a spherical architecture, SNAs manage to rapidly cross almost all negatively charged cell membranes [107]. This seemingly atypical intracellular uptake of SNAs was found to be facilitated by scavenger receptor A-mediated endocytosis, and allows for the accumulation of hundreds of thousands of AuNPs per cell (Fig. 4) [47].

The SNA structure has been leveraged for the delivery of siRNA, microRNA and antisense DNA target sequences, and had its applicability expanded by cofunctionalization with targeting antibodies and chemotherapeutics for translational applications, including skin diseases, breast cancer, prostate cancer, and high-grade gliomas [67, 78, 82, 98, 108, 109]. To assess the potential of topically applied SNAs to treat skin diseases, Zhang et al. utilized SNAs with a 13 nm AuNP core conjugated to epidermal growth factor receptor (EGFR)-targeting thiolated siRNA duplexes [98]. The results showed that even without the assistance of auxiliary transfection agents, the EGFR-targeting SNAs could enter 100% of human keratinocytes (hKCs) within 2 h *in-vitro* (this was an especially impressive feature given that hKCs are traditionally a difficult cell line to transfect) [110]. The EGFR-targeting SNAs were then used to treat, topically, the skin of hairless mice three times weekly for three weeks. Relative to a control SNA treatment, the EGFR-targeting SNAs reduced Ki-67 (a proliferation marker) staining of keratinocytes in the basal layer by 40%, and reduced epidermal thickness by 40%. Following EGFR-targeting SNA treatment, EGFR expression was almost abolished, while the downstream phosphorylation of extracellular-signal-regulated kinase (ERK) 1/2 was decreased by 74% relative to control SNA treatment, as assessed by Western blotting. Importantly, the treated skin did not show any clinical or histological evidence of toxicity, including a lack of cytokine activation in mouse blood and tissue samples. In addition, after the three-week treatment, SNAs were almost undetectable in the internal organs, with 0.0003% and 0.00015% of the injected AuNP dose detected in the liver and spleen, respectively, as assessed by inductively coupled plasma mass spectrometry (ICP-MS). These data suggest that SNA conjugates may be safe and efficacious for the treatment of cutaneous tumors, skin inflammatory conditions, and dominant-negative genetic skin disorders.

In a glioblastoma multiforme (GBM) model, Jensen et al. demonstrated the ability of SNAs to cross the BBB and blood–tumor barrier (BTB) and to pervasively penetrate glioma tissue, both *in-vitro* and *in-vivo* [67]. Using an *in-vitro* coculture model of the human BBB, which consisted of human primary brain microvascular endothelial cells (huBMECs) and human astrocytes, Jensen and colleagues showed that SNAs were able to undergo transcytosis through the huBMEC layer and enter human astrocytes [111–113]. This BBB-penetrating capacity was abolished when polyinosinic acid (Poly I) was added prior to SNA treatment, which blocks SR-A (scavenger receptor-A) uptake and likely mediates transcytosis. The penetration of both BBB and glioma tissue was then investigated *in-vivo* in both healthy and glioma-bearing mice. In addition to Cy-5-labeled SNAs, gadolinium (Gd(III)) was conjugated to SNAs to visualize and quantify tissue penetration. SNA distribution was evaluated via ICP-MS, MRI, and confocal fluorescence microscopy. Following local administration, as assessed by the 3D reconstruction of MRI images and confocal fluorescence, SNAs exhibited extensive intratumoral dissemination. ICP-MS further substantiated these results with a 10-fold higher accumulation of SNAs in tumor versus nontumor brain regions, possibly due to the EPR effect. Analogous results were obtained when Cy5-labeled SNAs were injected into the tail vein of mice and fluorescence monitored with an *in-vivo* imaging system (IVIS). Both, IVIS imaging and the quantification of radiant intensities showed a 1.8-fold higher accumulation of SNAs in GBM xenograft-bearing mice compared to sham GBM-inoculated mice. Notably, SNA accumulation in the brain tissue of healthy mice was also extensive, with approximately 10^{10} SNA particles per gram of tissue, as determined with ICP-MS.

Jensen and colleagues then developed SNAs which consisted of a 13-nm AuNP core conjugated to thiolated siRNA duplexes targeting the GBM oncogene *Bcl2L12*, an effector caspase and p53 inhibitor overexpressed in the vast majority (>90%) of GBM patients. Systemically delivered Bcl2L12-targeting SNAs neutralized *Bcl2L12* expression, increased intratumoral apoptosis, reduced tumor burden, and augmented the survival of GBM-xenografted mice. Furthermore, rodent toxicity studies did not reveal any adverse side effects or signs of toxicity, as the systemically administered SNAs did not induce inflammatory cytokines and did not cause any changes in blood chemistry, complete blood counts, or histopathology compared to saline or control SNAs. With no evidence to date of toxicity, and encouraging *in-vivo* results, SNAs represent a promising construct for siRNA delivery that will shortly enter clinical testing.

3.4
Iron Oxide Nanoparticles

Iron oxide nanoparticles (IONs) possess superparamagnetic properties that enable them to exhibit magnetic interaction only in the presence of an external magnetic field. Thus, IONs can serve as contrast agents for MRI applications and also as therapeutic constructs for magnetic hyperthermia treatment [114, 115]. Dextran-coated IONs are biocompatible, biodegradable, and have been approved by the FDA for certain imaging and treatment applications, such as the imaging of liver lesions (Feridex I.V.®) and the treatment of iron-deficiency anemia in patients with chronic kidney disease (Feraheme®) [116]. For siRNA delivery purposes, IONs are commonly functionalized with oligonucleotides by first fabricating them with a polycationic layer such as PEI, polyarginine, polylysine, or cationic lipids, after which siRNA can be electrostatically adsorbed to the cationic nanoparticle surface [117]. Specifically, monodisperse IONs can be synthesized via a high-temperature organic phase decomposition of an iron precursor such as iron(III) acetylacetonate with 1,2-hexadecanediol, oleic acid, and oleylamine [118]. The oleic acid layer on these hydrophobic IONs can then be directly exchanged with a cationic molecule that has an affinity for the ION surface (e.g., PEI, which has a strong affinity for IONs due to the amine coordination of iron) through a ligand-exchange reaction to form cationic layer-coated, water-soluble IONs [119]. siRNA or other oligonucleotides may then be electrostatically tethered to the cationic nanoparticle surface for delivery purposes.

Several ION systems have successfully delivered siRNA to cells and tissues [89, 117, 119–123]. Notably, Liu *et al.* showed that PEI-coated IONs complexed to luciferase-targeting siRNA (lucsiRNA) were capable of increasing the serum stability of lucsiRNA, could efficiently release lucsiRNA in the presence of heparin, and could knockdown luciferase expression *in-vitro* and, via local intratumoral injection *in-vivo*, in a 4T1 breast tumor model [117]. Other ION systems, upon the application of an external magnetic field, can increase target site accumulation and siRNA transfection efficiency [119, 121–123]. For example, Anderson and colleagues developed an epoxide-derived lipidoid-coated ION that was capable of complexing siRNA and DNA for transfection [121]. These authors showed that green fluorescent protein (GFP)-targeting siRNA could be transfected into HeLa cells using a lipidoid–ION construct, and could silence GFP expression by ~80% at siRNA concentrations as low as 1.5 nM with external magnetic field application, compared to ~50% reduction in GFP expression at the same dose without an external magnetic field. Furthermore, when the lipoid–ION construct was complexed to plasmid DNA encoding GFP, an approximate 70% transfection efficiency of DNA was achieved at a DNA concentration as low as 0.05 nM with the application of an external magnetic field, compared to ~20% transfection efficiency without the external magnetic field. Although preliminary *in-vitro* results are encouraging, it is important to stress that there is currently no experimental evidence to support the claim that magnetic-based delivery *in-vivo* is superior to EPR-based passive delivery [89]. Despite the enticing combination of *in-vivo* imaging and siRNA delivery potential afforded by ION systems, the biodegradability and biocompatibility profiles of siRNA–ION constructs are currently undefined and more preclinical studies are required to establish their safety and efficacy *in-vivo* prior to human trials.

3.5
Cyclodextrin Polymer Nanoparticles

Cyclodextrin is a natural polymer produced during the bacterial digestion of cellulose [124]. Cyclodextrins have been incorporated into many pharmaceutical formulations as they do not activate the immune system, display low toxicity, and are not enzymatically degraded in humans [124]. Cyclodextrin polymer (CDP) nanoparticles were the first siRNA nanoparticle-based delivery system to enter clinical trials for cancer [125]. CDPs are composed of polycationic oligomers with amidine functional groups; the positively charged amidine groups serve to complex siRNA. CDP NPs can also be end-capped with imidazole groups, which can become protonated to enable endosomal escape [126–128]. For *in-vivo* targeting and stability purposes, CDP NPs can be functionalized with PEG and transferrin molecules through conjugation with adamantane, a hydrophobic molecule that stably interacts with the cyclic core of CDP NPs [15, 126, 129, 130]. CDP–PEG–transferrin nanoparticles loaded with siRNA targeting the EWS-FL11 fusion gene successfully inhibited tumor growth in a mouse model of Ewing's sarcoma, without inducing any toxicity or immune activation [126]. Another type of CDP–PEG–transferrin nanoparticle, CALAA-01, was designed with siRNA specific for the M2 subunit of ribonucleotide reductase (RRM2) and demonstrated efficacy in mouse models [131]. In cynomolgus monkeys, CALAA-01 was found to be active at doses as low as $0.6\,\mathrm{mg\,kg^{-1}}$ while being tolerated at doses as high as $27\,\mathrm{mg\,kg^{-1}}$ [132]. Phase I clinical trials with CALAA-01 are currently under way for the treatment of solid tumors shown to be refractory to standard-care therapies, and thus far nanoparticles have been detected in tumor biopsies taken from melanoma patients treated with CALAA-01. Immunohistochemical staining and qRT-PCR of tumor tissue have demonstrated the knockdown of RRM2, proving that RNAi can be achieved in humans by systemically administering a nanoparticle-based siRNA delivery system [133].

3.6
Liposomes and Lipid-Based Materials

Liposomes are the most prevalent nanocarrier for siRNA, with five different siRNA–liposomal formulations currently being investigated in clinical trials. In addition to their ability to effectively deliver siRNA into cells, there is significant past clinical experience with liposomal drug formulations. In fact, seven liposomal drug formulations have approved by the FDA, dating back to 1995 with the introduction of liposomal doxorubicin (Doxil®) [20, 33].

Analogous to the cell membrane, liposomes consist of a phospholipid bilayer. The polar, hydrophilic head groups of the phospholipids face the exterior of the liposome and the internal core of the liposome, while the nonpolar, hydrophobic phospholipid tail groups interact with each other to form the bilayer. This structure allows the entrapment of a variety of molecules in the core of the liposome, including siRNA, DNA, proteins, chemotherapeutics, and other therapeutic payloads. Additionally, liposome size, lipid composition and charge can be tailored to optimize target site accumulation. Similar to metal-based nanoconjugates, PEG molecules can also be conjugated to liposomes to prevent aggregation during the fabrication process and reduce clearance by the MPS [134, 135].

Liposomes are typically categorized by their charge, and thus can be subdivided into cationic, neutral, and anionic liposomes. As anionic liposomal formulations have only shown a limited success in delivering RNAi-active compounds due to impaired fusion with the negatively charged cell membrane, only cationic and neutral liposomes will be discussed here. In addition to cationic and neutral liposomes, stable nucleic acid–lipid particles (SNALPs; a hybrid lipid nanoparticle containing a cationic liposomal core but exhibiting a neutral exterior due to PEG conjugation) and synthetic exosomes (a mimic of naturally occurring liposomes that can be either cationic or neutral in surface charge) will be described.

3.6.1 Cationic Liposomes

Cationic liposomes composed of lipids such as DOTAP and DOTMA (N-[1-(2,3-dioleoyloxy)propyl]-N,N,N-trimethyl-ammonium methyl sulfate) can form complexes with negatively charged siRNA via electrostatic interactions, thus facilitating delivery and transfection [20, 136]. However, due to the high-level intracellular stability of cationic liposomes, siRNA often remains entrapped in the core and does not reach the cytoplasm [137, 138], resulting in only a moderate target gene knockdown. In addition, unmodified cationic liposomes are toxic on systemic administration. Possibly due to TLR4 activation, dose-dependent hepatotoxicity, pulmonary inflammation, the induction of reactive oxygen species (ROS) production and a systemic interferon type I response have been observed in mice treated with cationic liposomes [137–139]. However, liposomes containing a mixture of cationic lipids, neutral fusogenic lipids, and PEG-modified lipids have been developed that are less toxic and release their siRNA contents more effectively. One such siRNA-based liposomal carrier that targets protein kinase N3, Atu027, inhibited lymph node metastasis in prostate and pancreatic cancer mouse models, as well as pulmonary metastasis in a variety of mouse models

[140, 141], and has been enrolled into clinical trials. In Phase I trials, doses of up to 0.180 mg kg^{-1} Atu027 were well tolerated by patients, though dose escalation studies are expected to be performed in the near future [142].

3.6.2 Neutral Liposomes

Given the significant toxicity associated with cationic lipids, liposomes composed of neutral lipids have been investigated. While neutral liposomes show a greatly increased biocompatibility compared to cationic liposomes, the efficiency with which neutral liposomes can load siRNA tends to be much lower due to a poor interaction between neutral lipids and siRNA [143, 144]. One neutral liposomal formulation is composed of 1,2-dioleoyl-sn-glycero-3-phosphatidylcholine (DOPC) and involves a synthesis method that increases the loading efficiency of siRNA via conjugation of the RNA with DOPC in the presence of excess *tert*-butanol and Tween 20 [145]. These DOPC-based liposomes have been loaded with a variety of different siRNA sequences, including those targeting EphA2 [145], focal adhesion kinase (FAK) [146], neuropilin-2 [147], interleukin (IL)-8 [148], transmembrane protease, serine 2 (TMPRRS2)/ETS-related gene (ERG) [149], and Bcl-2 [150], and have been used successfully in preclinical models of ovarian, breast, pancreatic, prostate cancer, and melanoma [20, 151–153]. The constructs did not induce any adverse side effects after four weeks of repeated intravenous injection in mice and nonhuman primates [145–148, 150]. Consequently, DOPC-based liposomes packed with EphA2-targeting siRNA entered Phase I clinical trials in 2012 for patients with advanced, recurrent solid tumors.

3.6.3 Stable Nucleic Acid–Lipid Particles

SNALPs contain a lipid bilayer composed of cationic lipids and neutral fusogenic lipids that are coated with PEG-lipids to provide a neutral surface (Fig. 5).

SNALPs have been shown to facilitate the cellular uptake and endosomal escape of siRNA, and to effectively regulate gene expression *in-vivo*. SNALPs made with siRNA targeting hepatitis B virus (HBV) RNA were able to reduce serum HBV levels more than 10-fold, with effects persisting for up to one week [19]. Additionally, Zimmerman and coworkers were able to achieve up to 90% knockdown of *ApoB* mRNA in nonhuman primate livers; SNALPs containing ApoB-targeting siRNA were injected once intravenously and demonstrated RNAi-mediated gene silencing for up to 11 days [154]. Based on the results of preclinical trials, multiple SNALP-based siRNA constructs are currently undergoing Phase I clinical trials for the treatment of hypercholesterolemia, lymphomas, advanced tumors with liver involvement, and ebola-virus infection [155].

3.6.4 Exosome-Based siRNA Carriers

Exosomes are biological vesicles that have a diameter of 50–90 nm and contain a lipid bilayer composed of glycerophospholipids, sphingomyelins, and cholesterol [156, 157]. Exosomes are naturally occurring liposomes, and are secreted from cells with payloads of protein and/or nucleic acids (including mRNA and miRNA); they are important for intercellular communication [158]. As these vesicles are nonimmunogenic and serve as the body's natural carrier for RNAi, the aim to design synthetic exosomes that mimic naturally occurring exosomes is an exciting new opportunity for RNAi-based therapy development [159]. Alvarez-Erviti and colleagues

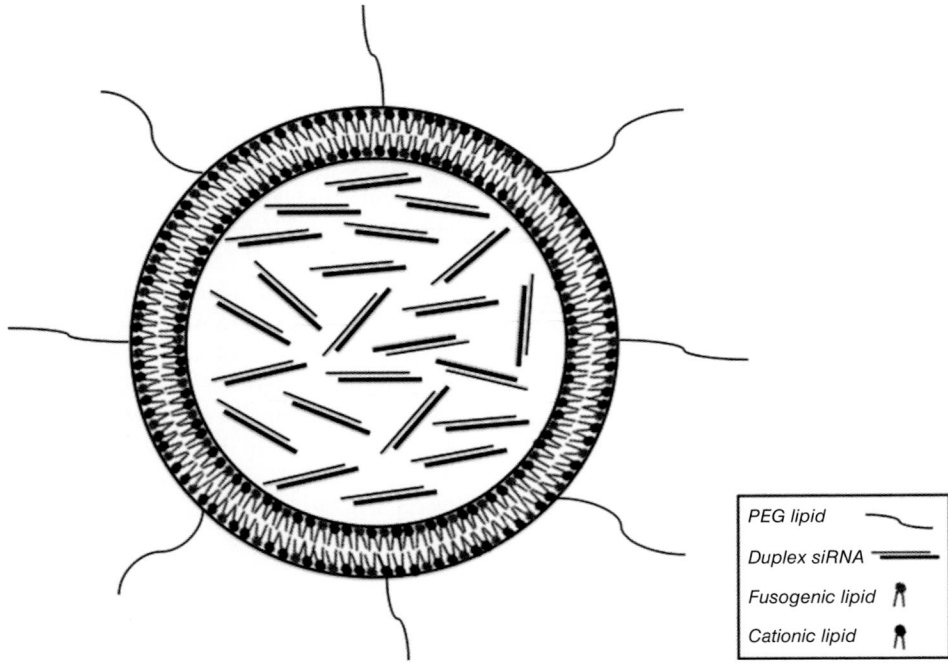

Fig. 5 Stable nucleic acid–lipid particles (SNALPs) encapsulate siRNA duplexes within a lipid bilayer composed of cationic lipids and neutral fusogenic lipids. The surface is coated with polyethylene glycol (PEG) lipids to provide a neutral exterior.

genetically engineered dendritic cells to produce targeted exosomes containing neuron-specific rabies virus glycoprotein (RVG) peptides; the exosomes were then loaded with glyceraldehyde 3-phosphate dehydrogenase (GAPDH)-targeting siRNA via electroporation [160]. These vesicles downregulated GAPDH expression in the central nervous system (CNS), but not in the liver or other organs. Additionally, Wahlgren *et al.* purified human exosomes, used them to encapsulate siRNA targeted to mitogen-activated protein kinase (MAPK)-1 via electroporation, and then demonstrated gene silencing in human monocytes and lymphocytes [161]. These promising initial studies have laid the groundwork for the future development of exosome-based siRNA carriers. Additional *in-vitro* and *in-vivo* validation studies are required, however, prior to the advancement of these conjugates into human trials.

4 The Future of Nanoparticle-Based siRNA Delivery

Nanoparticle delivery systems for siRNAs have emerged as a powerful gene regulation platform, with over 24 conjugates undergoing clinical trials just 13 years after the first demonstration of the RNAi pathway in mammalian cells. The major challenges in siRNA delivery can be addressed with nanovectors, though these systems carry their own problems that are related to scaling-up syntheses, achieving organ site-selective or site-specific delivery, an incomplete understanding of immune stimulation by these constructs, and a lack of long-term safety/toxicity profiles. However, as these different classes of constructs continue to be tested in preclinical and clinical trials, more is being learned regarding their synthesis, physicochemical characterization, biodistribution, toxicity, immunogenicity, and *in-vivo* efficacy. Design heuristics, such as the recommended size, charge, and shape of nanoparticles for optimal *in-vivo* delivery, are also beginning to emerge as investigators study the pharmacokinetics and pharmacodynamics of nanoparticle-based constructs. These design considerations will accelerate the pace of siRNA therapeutic development and decrease the time for bench-to-bedside translation.

With multiple classes of nanoparticle-based delivery systems undergoing Phase II clinical trials, the implementation of these conjugates into clinical practice is becoming close, and will bring humankind one giant step closer to realizing the concept of precision medicine to target virtually any disease-causing genetic element. If the genetic driver of a specific patient's disease can be identified, then siRNA against this sequence can be rapidly designed and loaded into one of these carriers. This adaptability becomes even more important when considering that the genetic base of many diseases is constantly changing, such as multidrug-resistant bacteria, continuously evolving viral genomes, and the acquisition of therapy-resistant phenotypes in cancer. Yet, given the speed by which advancements have been made in RNAi therapy thus far, perhaps this technology will be available sooner than expected.

Acknowledgments

This research was supported by the Center for Cancer Nanotechnology Excellence (CCNE) initiative of the National Institutes of Health (NIH) (U54 CA151880), the Dixon Translational Research Grants Initiative of the Northwestern Memorial Foundation, the James S. MacDonnell 21st Century Initiative, the Coffman Charitable Trust, the John McNicholas foundation, the American Cancer Society, and the Association for Cancer Gene Therapy. T.L.S. is supported by the NIH Ruth L. Kirschstein National Research Service Award for Predoctoral MD/PhD Fellows (F30CA174058) awarded by the National Cancer Institute (NCI) and the Northwestern Ryan Fellowship.

References

1. Fire, A., Xu, S., Montgomery, M.K., Kostas, S.A. et al. (1998) *Nature*, **391**, 806–811.
2. Elbashir, S.M., Harborth, J., Lendeckel, W., Yalcin, A. et al. (2001) *Nature*, **411**, 494–498.
3. Novina, C.D., Sharp, P.A. (2004) *Nature*, **430**, 161–164.
4. Hannon, G.J., Rossi, J.J. (2004) *Nature*, **431**, 371–378.
5. Castanotto, D., Rossi, J.J. (2009) *Nature*, **457**, 426–433.
6. Kanasty, R., Dorkin, J.R., Vegas, A., Anderson, D. (2013) *Nat. Mater.*, **12**, 967–977.
7. Burnett, J.C., Rossi, J.J. (2012) *Chem. Biol.*, **19** (1), 60–71.
8. Davidson, B.L., McCray, P.B. (2011) *Nat. Rev. Genet.*, **12**, 329–340.
9. Kim, D.H., Rossi, J.J. (2007) *Nat. Rev. Genet.*, **8**, 173–184.

10. Burnett, J.C., Rossi, J.J., Tiemann, K. (2011) *Biotechnol. J.*, **6**, 1130–1146.
11. Stegh, A.H. (2013) *Integr. Biol. (Camb.)*, **5**, 48–65.
12. Zamore, P.D., Tuschl, T., Sharp, P.A., Bartel, D.P. (2000) *Cell*, **101**, 25–33.
13. Hannon, G.J. (2002) *Nature*, **418**, 244–251.
14. McManus, M.T., Sharp, P.A. (2002) *Nat. Rev. Genet.*, **3**, 737–747.
15. Bartlett, D.W., Davis, M.E. (2006) *Nucleic Acids Res.*, **34**, 322–333.
16. Whitehead, K.A., Langer, R., Anderson, D.G. (2010) *Nat. Rev. Drug Discovery*, **9**, 412.
17. Schroeder, A., Levins, C.G., Cortez, C., Langer, R., Anderson, D.G. (2010) *J. Intern. Med.*, **267**, 9–21.
18. Kanasty, R.L., Whitehead, K.A., Vegas, A.J., Anderson, D.G. (2012) *Mol. Ther.*, **20**, 513–524.
19. Morrissey, D.V., Lockridge, J.A., Shaw, L., Blanchard, K. et al. (2005) *Nat. Biotechnol.*, **23**, 1002–1007.
20. Ozpolat, B., Sood, A.K., Lopez-Berestein, G. (2014) *Adv. Drug Delivery Rev.*, **66**, 110–116.
21. Layzer, J.M., McCaffrey, A.P., Tanner, A.K., Huang, Z. et al. (2004) *RNA*, **10** (5), 766–771.
22. Behlke, M.A. (2008) *Oligonucleotides*, **18**, 305–320.
23. Jackson, A.L., Burchard, J., Leake, D., Reynolds, A. et al. (2006) *RNA*, **12**, 1197–1205.
24. Robbins, M., Judge, A., MacLachlan, I. (2009) *Oligonucleotides*, **19**, 89–102.
25. Gantier, M.P., Tong, S., Behlke, M.A., Irving, A.T. et al. (2010) *Mol. Ther.*, **18**, 785–795.
26. Judge, A.D., Sood, V., Shaw, J.R., Fang, D. et al. (2005) *Nat. Biotechnol.*, **23**, 457–462.
27. Kleinman, M.E., Yamada, K., Takeda, A., Chandrasekaran, V. et al. (2008) *Nature*, **452**, 591–597.
28. Judge, A., Bola, G., Lee, A., Maclachlan, I. (2006) *Mol. Ther.*, **13**, 494–505.
29. Robbins, M., Judge, A., Liang, L., McClintock, K. et al. (2007) *Mol. Ther.*, **15**, 1663–1669.
30. Wartiovaara, J., Öfverstedt, L.-G., Khoshnoodi, J., Zhang, J. et al. (2004) *J. Clin. Invest.*, **114**, 1475–1483.
31. Huang, Y., Hong, J., Zheng, S., Ding, Y. et al. (2011) *Mol. Ther.*, **19** (2), 381–385.
32. Choi, H.S., Liu, W., Liu, F., Nasr, K. et al. (2009) *Nat. Nanotechnol.*, **5**, 42–47.
33. Petros, R.A., DeSimone, J.M. (2010) *Nat. Rev. Drug Discovery*, **9**, 615–627.
34. Aird, W.C. (2007) *Circ. Res.*, **100** (2), 174–190.
35. Braet, F., Wisse, E., Bomans, P., Frederik, P. et al. (2007) *Microsc. Res. Tech.*, **70**, 230–242.
36. Alexis, F., Pridgen, E., Molnar, L.K., Farokhzad, O.C. (2008) *Mol. Pharm.*, **5**, 505–515.
37. Li, W., Szoka, F.C., Jr (2007) *Pharm. Res.*, **24**, 438–449.
38. Maeda, H. (2010) *Bioconjugate Chem.*, **21**, 797–802.
39. Campbell, R.B., Fukumura, D., Brown, E.B., Mazzola, L.M. et al. (2002) *Cancer Res.*, **62** (23), 6831–6836.
40. Champion, J.A., Mitragotri, S. (2006) *Proc. Natl Acad. Sci. USA*, **103** (13), 4930–4934.
41. Liu, Z., Cai, W., He, L., Nakayama, N. et al. (2006) *Nat. Nanotechnol.*, **2**, 47–52.
42. Park, J.-H., von Maltzahn, G., Zhang, L., Schwartz, M. P. et al. (2008) *Adv. Mater.*, **20**, 1630–1635.
43. Wu, J. (2002) *Front. Biosci.*, **7**, d717–d725.
44. Daniels, T.R., Bernabeu, E., Rodríguez, J.A., Patel, S. (2012) Biochim. *Biophys. Acta*, **1820** (3), 291–317.
45. Abbott, N.J., Patabendige, A.A.K., Dolman, D.E.M., Yusof, S.R., Begley, D.J. (2010) *Neurobiol. Dis.*, **37**, 13–25.
46. Sanna, V., Pala, N., Sechi, M. (2014) *Int. J. Nanomed.*, **9**, 467–483.
47. Choi, C.H.J., Hao, L., Narayan, S.P., Auyeung, E., Mirkin, C.A. (2013) *Proc. Natl Acad. Sci. USA*, **110**, 7625–7630.
48. Lu, J.J., Langer, R., Chen, J. (2009) *Mol. Pharm.*, **6**, 763–771.
49. Reddy, J.A., Low, P.S. (2000) *J. Controlled Release*, **64** (1-3), 27–37.
50. Kichler, A., Leborgne, C., Coeytaux, E., Danos, O. (2001) *J. Gene Med.*, **3**, 135–144.
51. Sonawane, N.D., Szoka, F.C., Verkman, A.S. (2003) *J. Biol. Chem.*, **278**, 44826–44831.
52. Drummond, D.C., Daleke, D.L. (1995) Chem. Phys. *Lipids*, **75** (1), 27–41.
53. Shen, J., Samul, R., Silva, R.L., Akiyama, H. et al. (2005) *Gene Ther.*, **13**, 225–234.
54. Dejneka, N.S., Wan, S., Bond, O.S., Kornbrust, D.J., Reich, S.J. (2008) *Mol. Vision*, **14**, 997–1005.

55. Garba, A.O., Mousa, S.A. (2010) Ophthalmol. Eye Dis., **2**, 75–83.
56. Martínez, T., Wright, N., López-Fraga, M., Jimenez, A.I., Pañeda, C. (2013) Hum. Genet., **132** (5), 481–493.
57. Alvarez, R., Elbashir, S., Borland, T., Toudjarska, I. et al. (2009) Antimicrob. Agents Chemother., **53**, 3952–3962.
58. DeVincenzo, J., Lambkin-Williams, R., Wilkinson, T., Cehelsky, J. et al. (2010) Proc. Natl Acad. Sci. USA, **107**, 8800–8805.
59. Zamora, M.R., Budev, M., Rolfe, M., Gottlieb, J. et al. (2011) Am. J. Respir. Crit. Care Med., **183**, 531–538.
60. Jeong, J.H., Mok, H., Oh, Y.K., Park, T.G. (2008) Bioconjugate Chem., **20** (1), 5–14.
61. Rozema, D.B., Lewis, D.L., Wakefield, D.H., Wong, S.C. et al. (2007) Proc. Natl Acad. Sci. USA, **104**, 12982–12987.
62. DeLong, R.K., Akhtar, U., Sallee, M., Parker, B. et al. (2009) Biomaterials, **30**, 6451–6459.
63. Sashin, D., Spanbock, J., Kling, D.H. (1939) J. Bone Joint Surg. Am., **21**, 723–734.
64. Kim, S.T., Chompoosor, A., Yeh, Y.-C., Agasti, S.S. et al. (2012) Small, **8**, 3253–3256.
65. Fraser, T.N. (1945) Ann. Rheum. Dis., **4**, 71–75.
66. Lee, S.H., Bae, K.H., Kim, S.H., Lee, K.R., Park, T.G. (2008) Int. J. Pharm., **364**, 94–101.
67. Jensen, S.A., Day, E.S., Ko, C.H., Hurley, L.A. et al. (2013) Sci. Transl. Med., **5**, 209ra152.
68. Brust, M., Walker, M., Bethell, D., Schiffrin, D.J., Whyman, R. (1994) J. Chem. Soc. Chem. Commun., >801–802.
69. Frens, G. (1973) Nat. Phys. Sci., **241**, 20–22.
70. Sutherland, W.S., Winefordner, J.D. (1992) J. Colloid Interface Sci., **148**, 129–141.
71. Grabar, K.C., Freeman, R.G., Hommer, M.B., Natan, M.J. (1995) Anal. Chem., **67**, 735–743.
72. Templeton, A.C., Wuelfing, W.P., Murray, R.W. (2000) Acc. Chem. Res., **33**, 27–36.
73. Jiang, X.-M., Wang, L.-M., Wang, J., Chen, C.-Y. (2012) Appl. Biochem. Biotechnol., **166**, 1533–1551.
74. Daniel, M.C., Astruc, D. (2004) Chem. Rev., **104** (1), 293–346.
75. Pakiari, A.H., Jamshidi, Z. (2010) J. Phys. Chem. A, **114**, 9212–9221.
76. Ding, Y., Jiang, Z., Saha, K., Kim, C.S. et al. (2014) Mol. Ther., **22** (6), 1075–1083.
77. Dixit, V., Van den Bossche, J., Sherman, D.M., Thompson, D.H., Andres, R.P. (2006) Bioconjugate Chem., **17** (1), 603–609.
78. Zhang, K., Hao, L., Hurst, S.J., Mirkin, C.A. (2012) J. Am. Chem. Soc., **134**, 16488–16491.
79. Ghosh, P., Han, G., De, M., Kim, C.K., Rotello, V.M. (2008) Adv. Drug Delivery Rev., **60** (11), 1307–1315.
80. Debouttière, P.J., Roux, S., Vocanson, F., Billotey, C. et al. (2006) Adv. Funct. Mater., **16**, 2330–2339.
81. Gibson, J.D., Khanal, B.P., Zubarev, E.R. (2007) J. Am. Chem. Soc., **129**, 11653–11661.
82. Zhang, X.-Q., Xu, X., Lam, R., Giljohann, D. et al. (2011) ACS Nano, **5**, 6962–6970.
83. Elbakry, A., Zaky, A., Liebl, R., Rachel, R., Goepferich, A. (2009) Nano Lett., **9** (5), 2059–2206.
84. Guo, S., Huang, Y., Jiang, Q., Sun, Y. et al. (2010) ACS Nano, **4** (9), 5505–5511.
85. Kelly, K.L., Coronado, E., Zhao, L.L., Schatz, G.C. (2003) J. Phys. Chem. B, **107**, 668–677.
86. Alkilany, A.M., Murphy, C.J. (2010) J. Nanopart. Res., **12**, 2313–2333.
87. Zhang, Z., Wang, J., Chen, C. (2013) Theranostics, **3**, 223–238.
88. Shim, M.S., Kwon, Y.J. (2012) Adv. Drug Delivery Rev., **64**, 1046–1059.
89. Wang, Z., Liu, G., Zheng, H., Chen, X. (2014) Biotechnol. Adv., **32** (4), 831–843.
90. Eghtedari, M., Oraevsky, A., Copland, J.A., Kotov, N.A. et al. (2007) Nano Lett., **7**, 1914–1918.
91. Shim, M.S., Kim, C.S., Ahn, Y.-C., Chen, Z., Kwon, Y.J. (2010) J. Am. Chem. Soc., **132**, 8316–8324.
92. Weintraub, K. (2013) Nature, **495**, S14–S16.
93. Libutti, S.K., Paciotti, G.F., Byrnes, A.A., Alexander, H.R. et al. (2010) Clin. Cancer Res., **16**, 6139–6149.
94. Haynes, R., Gannon, W., Walker, M. (2009) J. Clin. Oncol., **27**, 3586.
95. Mirkin, C.A., Letsinger, R.L., Mucic, R.C., Storhoff, J.J. (1996) Nature, **382**, 607–609.
96. Park, S.Y., Lytton-Jean, A.K.R., Lee, B., Weigand, S. et al. (2008) Nature, **451**, 553–556.

97. Nykypanchuk, D., Maye, M.M., van der Lelie, D., Gang, O. (2008) *Nature*, **451**, 549–552.
98. Zheng, D., Giljohann, D.A., Chen, D.L., Massich, M.D. et al. (2012) *Proc. Natl Acad. Sci. USA*, **109**, 11975–11980.
99. Cutler, J.I., Auyeung, E., Mirkin, C.A. (2012) *J. Am. Chem. Soc.*, **134**, 1376–1391.
100. Giljohann, D.A., Seferos, D.S., Prigodich, A.E., Patel, P.C., Mirkin, C.A. (2009) *J. Am. Chem. Soc.*, **131**, 2072–2073.
101. Seferos, D.S., Prigodich, A.E., Giljohann, D.A., Patel, P.C., Mirkin, C.A. (2009) *Nano Lett.*, **9**, 308–311.
102. Massich, M.D., Giljohann, D.A., Seferos, D.S., Ludlow, L.E. et al. (2009) *Mol. Pharm.*, **6**, 1934–1940.
103. Massich, M.D., Giljohann, D.A., Schmucker, A.L., Patel, P.C., Mirkin, C.A. (2010) *ACS Nano*, **4**, 5641–5646.
104. Williams, S.C.P. (2013) *Proc. Natl Acad. Sci. USA*, **110**, 13231–13233.
105. Mirkin, C.A., Stegh, A.H. (2013) *Oncotarget*, **5** (1), 9–10.
106. Cutler, J.I., Zhang, K., Zheng, D., Auyeung, E. et al. (2011) *J. Am. Chem. Soc.*, **133**, 9254–9257.
107. Rosi, N.L. (2006) *Science*, **312**, 1027–1030.
108. Alhasan, A.H., Patel, P.C., Choi, C.H.J., Mirkin, C.A. (2014) *Small*, **10**, 186–192.
109. Alhasan, A.H., Kim, D.Y., Daniel, W.L., Watson, E. et al. (2012) *Anal. Chem.*, **84**, 4153–4160.
110. Dickens, S., Van den Berge, S., Hendrickx, B., Verdonck, K. et al. (2010) *Tissue Eng. Part C Methods*, **16**, 1601–1608.
111. Boveri, M., Berezowski, V., Price, A., Slupek, S. et al. (2005) *Glia*, **51**, 187–198.
112. Cecchelli, R., Dehouck, B., Descamps, L., Fenart, L. et al. (1999) *Adv. Drug Delivery Rev.*, **36**, 165–178.
113. Culot, M., Lundquist, S., Vanuxeem, D., Nion, S. et al. (2008) *Toxicol. in Vitro*, **22**, 799–811.
114. Hergt, R., Dutz, S., Müller, R., Zeisberger, M. (2006) *J. Phys.: Condens. Matter*, **18**, S2919–S2934.
115. Mahmoudi, M., Sant, S., Wang, B., Laurent, S., Sen, T. (2011) *Adv. Drug Delivery Rev.*, **63**, 24–46.
116. Tassa, C., Shaw, S.Y., Weissleder, R. (2011) *Acc. Chem. Res.*, **44**, 842–852.
117. Liu, G., Xie, J., Zhang, F., Wang, Z. et al. (2011) *Small*, **7**, 2742–2749.
118. Sun, S., Zeng, H., Robinson, D.B., Raoux, S. et al. (2004) *J. Am. Chem. Soc.*, **126**, 273–279.
119. Zhang, H., Lee, M.Y., Hogg, M.G., Dordick, J.S. (2010) *ACS Nano*, **4** (8), 4733–4743.
120. Liu, G., Wang, Z., Lu, J., Xia, C. et al. (2011) *Biomaterials*, **32**, 528–537.
121. Jiang, S., Eltoukhy, A.A., Love, K.T., Langer, R., Anderson, D.G. (2013) *Nano Lett.*, **13**, 1059–1064.
122. Shubayev, V.I., Pisanic, T.R. II, Jin, S. (2009) *Adv. Drug Delivery Rev.*, **61**, 467–477.
123. Del Pino, P., Munoz-Javier, A., Vlaskou, D., Gil, P.R. (2010) *Nano Lett.*, **10** (10), 3914–3921.
124. Davis, M.E., Brewster, M.E. (2004) *Nat. Rev. Drug Discovery*, **3**, 1023–1035.
125. Davis, M.E. (2009) *Mol. Pharm.*, **6**, 659–668.
126. Hu-Lieskovan, S. (2005) *Cancer Res.*, **65**, 8984–8992.
127. Mishra, S., Heidel, J.D., Webster, P., Davis, M.E. (2006) *J. Controlled Release*, **116** (2), 179–191.
128. Bartlett, D.W., Davis, M.E. (2007) *Bioconjugate Chem.*, **18**, 456–468.
129. Bellocq, N.C., Pun, S.H., Jensen, G.S., Davis, M.E. (2003) *Bioconjugate Chem.*, **14**, 1122–1132.
130. Bartlett, D.W., Su, H., Hildebrandt, I.J., Weber, W.A., Davis, M.E. (2007) *Proc. Natl Acad. Sci. USA*, **104**, 15549–15554.
131. Bartlett, D.W., Davis, M.E. (2008) *Biotechnol. Bioeng.*, **99**, 975–985.
132. Heidel, J.D., Yu, Z., Liu, J.Y.C., Rele, S.M. et al. (2007) *Proc. Natl Acad. Sci. USA*, **104**, 5715–5721.
133. Davis, M.E., Zuckerman, J.E., Choi, C., Seligson, D. (2010) *Nature*, **464** (7291), 1067–1070.
134. Klibanov, A.L., Maruyama, K., Torchilin, V.P., Huang, L. (1990) *FEBS Lett.*, **268** (1), 235–237.
135. Ishida, T., Harashima, H., Kiwada, H. (2002) *Biosci. Rep.*, **22**, 197–224.
136. Miller, C.R., Bondurant, B., McLean, S.D., McGovern, K.A., O'Brien, D.F. (1998) *Biochemistry*, **37**, 12875–12883.
137. Dokka, S., Toledo, D., Shi, X., Castranova, V., Rojanasakul, Y. (2000) *Pharm. Res.*, **17**, 521–525.
138. Spagnou, S., Miller, A.D., Keller, M. (2004) *Biochemistry*, **43**, 13348–13356.

139. Lv, H., Zhang, S., Wang, B., Cui, S., Yan, J. (2006) *J. Controlled Release*, **114**, 100–109.
140. Aleku, M., Schulz, P., Keil, O., Santel, A. *et al.* (2008) *Cancer Res.*, **68**, 9788–9798.
141. Santel, A., Aleku, M., Röder, N., Möpert, K. *et al.* (2010) *Cancer Res.*, **16**, 5469–5480.
142. Strumberg, D., Schultheis, B., Traugott, U., Vank, C. (2012) Int. J. Clin. Pharmacol. *Ther.*, **50** (1), 76–78.
143. Lee, J.-M., Yoon, T.-J., Cho, Y.-S. (2013) Biomed. *Res. Int.*, **2013**, 782041.
144. Wu, S.Y., McMillan, N.A.J. (2009) *AAPS J.*, **11**, 639–652.
145. Landen, C.N., Chavez-Reyes, A., Bucana, C., Schmandt, R. *et al.* (2005) *Cancer Res.*, **65** (15), 6910–6918.
146. Halder, J., Kamat, A.A., Landen, C.N., Han, L.Y. *et al.* (2006) *Clin. Cancer Res.*, **12**, 4916–4924.
147. Gray, M.J., Van Buren, G., Dallas, N.A., Xia, L. *et al.* (2008) *J. Natl Cancer Inst.*, **100** (2), 109–120.
148. Merritt, W.M., Lin, Y.G., Spannuth, W.A., Fletcher, M.S. *et al.* (2008) *J. Natl Cancer Inst.*, **100**, 359–372.
149. Shao, L., Tekedereli, I., Wang, J., Yuca, E. *et al.* (2012) *Clin. Cancer Res.*, **18**, 6648–6657.
150. Tekedereli, I., Alpay, S.N., Akar, U., Yuca, E. *et al.* (2013) *Mol. Ther. Nucleic Acids*, **2**, e121.
151. Nick, A.M., Stone, R.L., Armaiz-Pena, G., Ozpolat, B. *et al.* (2011) *J. Natl Cancer Inst.*, **103**, 1596–1612.
152. Pan, X., Arumugam, T., Yamamoto, T., Levin, P.A. *et al.* (2008) *Clin. Cancer Res.*, **14** (24), 8143–8151.
153. Villares, G.J., Zigler, M., Wang, H., Melnikova, V.O. *et al.* (2008) *Cancer Res.*, **68**, 9078–9086.
154. Zimmermann, T.S., Lee, A.C.H., Akinc, A., Bramlage, B. *et al.* (2006) *Nature*, **441**, 111–114.
155. Barros, S.A., Gollob, J.A. (2012) *Adv. Drug Delivery Rev.*, **64** (15), 1730–1737.
156. Vlassov, A.V., Magdaleno, S., Setterquist, R. (2012) Biochim. *Biophys. Acta*, **1820** (7), 940–948.
157. Subra, C., Laulagnier, K., Perret, B., Record, M. (2007) *Biochimie*, **89**, 205–212.
158. Valadi, H., Ekström, K., Bossios, A., Sjöstrand, M., Lee, J.J. (2007) *Nat. Cell Biol.*, **9** (6), 654–659.
159. Kooijmans, S.A., Vader, P., van Dommelen, S.M., van Solinge, W.W., Schiffelers, R.M. (2012) *Int. J. Nanomed.*, **7**, 1525–1541.
160. Alvarez-Erviti, L., Seow, Y., Yin, H.F., Betts, C. (2011) *Nat. Biotechnol.*, **29** (4), 341–345.
161. Wahlgren, J., Karlson, T.D.L., Brisslert, M., Vaziri Sani, F. *et al.* (2012) *Nucleic Acids Res.*, **40**, e130.

28
Quantum Dots for Biomedical Delivery Applications

Abolfazl Akbarzadeh[*,1,3], *Sedigheh Fekri Aval*[1,2,4], *Roghayeh Sheervalilou*[4,5], *Leila Fekri*[6], *Nosratollah Zarghami*[1,2], *and Mozhdeh Mohammadian*[7]

[1] *Tabriz University of Medical Sciences, Drug Applied Research Center, Daneshgah Street, Tabriz 51656-65811, Iran*
[2] *Tabriz University of Medical Sciences, Department of Medical Biotechnology, School of Advanced Medical Sciences, Golgasht Ave, Azadi Street, Tabriz 51666-14766, Iran*
[3] *Tabriz University of Medical Sciences, Department of Medical Nanotechnology, Faculty of Advanced Medical Sciences, Golgasht Ave, Azadi Street, Tabriz 51666-14766, Iran*
[4] *Tabriz University of Medical Sciences, Advanced Medical Sciences Research Center, Student's Research Committee, Golgasht Ave, Azadi Street, Tabriz 51666-14766, Iran*
[5] *Tabriz University of Medical Sciences, Department of Molecular Medicine, Faculty of Advanced Medical Sciences, Golgasht Ave, Azadi Street, Tabriz 51666-14766, Iran*
[6] *Islamic Azad University, Plasma Physics Research Center, Science and Research Branch, Shahid Sattari Highway - University Square, Shohadaye Hesarak Street, Tehran 1477893855, Iran*
[7] *Mazandaran University of Medical Sciences, Amol Faculty of Paramedical Sciences, Valiasr Street, Sari 48178-44718, Iran*

1	Introduction	996
2	Properties of QDs	997
3	Lithographically Defined QDs	997
4	Colloidal QDs	998
5	QDs Application in Optics	999
6	Transport (Electrical) Properties of QDs	1001
7	Application of QDs in Diagnostics	1001
8	Nucleic Acid Detection	1002
9	Application of QDs in Drug Treatments	1003

Translational Medicine: Molecular Pharmacology and Drug Discovery
First Edition. Edited by Robert A. Meyers.
© 2018 Wiley-VCH Verlag GmbH & Co. KGaA. Published 2018 by Wiley-VCH Verlag GmbH & Co. KGaA.

Acknowledgments 1005

References 1005

Keywords

Lithography
The most common method of quantum dot fabrication.

Quantum dot (QD)
A nanocrystal made from semiconductor materials, exhibiting unique optical properties.

Detection
Identification process of a sign, using tools such as QDs, to provide additional information and superior results.

Nanocrystals
A crystalline material with size in the nanometer range.

SiRNA
Small interfering RNA is a powerful tool for knocking down or gene silencing in most cells.

Quantum dots (QDs) are novel, photostable fluorescent semiconductor nanocrystals. The understanding of quantum-confined electrons in very small particles forms the basis for an understanding of the exceptional features described in this chapter. Lithographically defined QDs, epitaxially self-assembled QDs, and colloidal QDs are methods of QD synthesis. In this chapter, an outline is provided of the unique characteristics of QDs, such as their wide excitation spectra, narrow symmetrical emission spectra, tunable emission, superior brightness, and photostability. These properties have led to QDs becoming an ideal platform for many beneficial applications in a variety of scientific and medical fields.

1 Introduction

During the past decade, new approaches have been devised that involve the ability to fabricate, characterize and manipulate artificial structures, the features of which may be controlled at the nanometer level. These embrace areas of research as diverse as physics, engineering, chemistry, materials sciences, pharmaceuticals, and biomedicine. This rapid progress has led to an expansion of possible extensive research at the nanoscale, which has in turn led to the fabrication of materials with exceptional properties. In the case of very small crystals of nanometer dimensions – termed nanocrystals – assumptions of translational symmetry and infinite sizes of crystals are no longer valid, and consequently these systems cannot be described using the same model as can be applied to a bulk solid [1, 2]. The electronic structure of a nanocrystal should be intermediate between the discrete levels of an atomic system and the band structure of a bulk solid. Nanocrystals have demonstrated discrete energy level structures and narrow transition line widths, and it is because of these discrete energy levels that such structures are referred to as quantum dots (QDs). The density of these structures is much greater, and their spacing is smaller than for the corresponding levels of one atom or a small atomic cluster [3, 4]. The highest occupied atomic levels and lowest unoccupied levels of the atomic species interact to form the valence band and conduction band of the QD, respectively. One of the most exciting interfaces of QDs is their high efficiency in biomedical applications, and this makes them ideal donor fluorophores – a property which has led to a revolution in detection processes [5, 6]. QDs also provide a versatile platform for the design and engineering of nanoparticle-based drug delivery (NDD) vehicles [7–9].

2
Properties of QDs

QDs exhibit unique optical properties due to a combination of their material band gap energy and quantum phenomena. Colloidal semiconductor QDs can have a very narrow symmetrical intense distribution, at specific wavelengths, ranging from the ultraviolet to the infra-red [10].

Although the emission peak from a single colloidal QD can be less than 0.1 meV wide, this emission energy shifts randomly over time, and such behavior is referred to as either "spectral jumps" or "spectral diffusion" [11]. The fluorescence emission from a single QD exhibits a dramatic on/off behavior, which is referred to as "blinking"; this is another spectroscopic property that distinguishes colloidal from self-assembled QDs [12, 13].

3
Lithographically Defined QDs

Lithographically defined QDs are formed by isolating a small region of a two-dimensional (2D) system. Such two-dimensional electron systems (2DESs) can be found in metal oxide-semiconductor field-effect transistors, or in so-called semiconductor heterostructures [14, 15]. Heterostructures are composed of several thin layers of semiconductor materials grown on top of each other, using the technique of molecular beam epitaxy (MBE). The layer sequence can be chosen in such a way that all free charge carriers are confined to a thin slice of the crystal, forming essentially a 2DES. A superstructure derived from the periodic repetition of this sequence of layers is also termed a "multiple quantum well." Electrons will be repelled by the electric field of the electrodes, such that the region of the 2DES below the electrodes will be depleted of electrons; this charge-depleted region will then behave like an insulator. By applying an electric field with metal electrodes of an appropriate shape, it is possible to create an island of charges insulated from the remainder of the 2DES, and if the island within the 2DES is small enough it will behave as a QD. In the vertical geometry, a small pillar of the 2DES can be isolated by etching away the heterostructure around it, and in such an arrangement the charge carriers will again be confined in all three dimensions. To date, most of the electron transport measurements performed on QDs have employed the two types of QD in the form just described. The lateral arrangement offers a relatively high degree of freedom for the design of the structure, as this is determined by the choice of electrode geometry. In addition, it is possible to fabricate and study "artificial molecules" [16–19] composed of several QDs linked together. When using the vertical arrangement, structures with very few electrons can be fabricated [20].

At present, a variety of ongoing studies are being undertaken to investigate many-body phenomena in these QD systems. Relevant examples include studies of the Kondo effect [14, 15, 21] and the design and control of coherent quantum states, with the ultimate goal being to perform quantum information processing. One remarkable advantage of lithographically defined QDs is that their electrical connection to the "macro-world" is straightforward. Moreover, the manufacturing processes are similar to those used in chip fabrication and, at least in principle, such structures could be embedded within conventional electronic circuits. The geometry of these QDs is determined lithographically, and is limited to the usual size and resolution limits of lithographic techniques. However, even when using electron beam lithography to fabricate the QDs it is not possible to tailor their size with nanometer precision. Typically, QDs fabricated lithographically are >9 nm in size, and consequently only relatively low lateral confining energies can be achieved.

4
Colloidal QDs

Colloidal QDs can be chemically synthesized using wet chemistry, and are freestanding nanoparticles (NPs) or nanocrystals grown in solution [22]. Colloidal QDs represent only a subgroup of a broader class of materials that can be synthesized at the nanoscale level using wet chemical methods, where the reaction chamber contains a liquid mixture of compounds that control both nucleation and growth. In a general synthesis of QDs in solution, each of the atomic species that will form part of the nanocrystals is introduced into the reactor in the form of a precursor. (A precursor is a molecule or complex containing one or more of the atomic species required to grow the nanocrystals.) When the precursors are introduced into the reaction flask they decompose, forming new reactive species (the monomers) that will in turn cause nucleation and growth of the nanocrystals. The energy required to decompose the precursors is provided by the liquid in the reactor, either by thermal collisions, by a chemical reaction between the liquid medium and the precursors, or by a combination of these two mechanisms [23].

The controlling growth of colloidal nanocrystals is the presence of mobile molecular species to provide access for the addition of monomer units in the reactor, broadly termed "surfactants" [24]. Suitable surfactants include alkyl thiols, phosphines, phosphine oxides, phosphates, phosphonates, amides or amines, carboxylic acids, and QDs of nitrogen-containing aromatics. If the growth of nanocrystals is carried out at high temperatures (e.g., at 200–400 °C), the surfactant molecules must be stable under such conditions if they are to serve as suitable candidates for controlling the growth. At low temperatures or, more generally, when growth ceases, the surfactants will be more strongly bound to the surface of the nanocrystals such that the latter are soluble in a wide range of solvents. The coating process allows for a greater synthetic flexibility, in that it can be exchanged for another coating of organic molecules with different functional groups or polarity. The surfactants can also be temporarily removed such that an epitaxial layer of another material with different electronic, optical or magnetic properties can be grown on the initial nanocrystal [25, 26].

An excellent control of the size and shape of QDs is possible by controlling the mixture of surfactant molecules present during their generation and growth [23, 27, 28]. Colloidal nanocrystals, when dispersed in solution, are not bound to any solid support, as is the case for the other two QD systems described above. Rather, colloidal nanocrystals can be produced in large quantities in a reaction flask, and later transferred to any desired substrate or object. It is, for example, possible to coat the nanocrystal surfaces with biological molecules such as proteins or oligonucleotides. The fact that many biological molecules perform tasks of molecular recognition with extremely high efficiency means that the ligand molecules can bind with very high specificity to certain receptor molecules, similar to a "key-and-lock" system. Notably, if a colloidal QD is tagged with ligand molecules it can bind specifically to all of the positions where a receptor molecule is present, and in this way it has been possible to create small groupings of colloidal QDs mediated by molecular recognition [29–31], and also to label the specific compartments of a cell with different types of QD [32–34]. Although colloidal QDs are rather difficult to connect electrically, a few electron transport experiments have been reported in which nanocrystals were used as the active material in devices that behave as single-electron transistors (Fig. 1) [35, 36].

5
QDs Application in Optics

Many applications of QDs can be found in optics. As with the more general case of atoms or molecules, QDs can be excited optically, and regardless of the nature of the excitation they may emit photons as they relax from an excited state to ground state. Based on these properties, QDs may be used as lasing media, as single-photon sources, as optically addressable charge storage devices, or as fluorescent labels. Self-assembled QDs incorporated into the active layer of a quantum well laser can significantly improve the operation characteristics of the laser, due to the zero-dimensional density of states of QDs. In QD lasers, the threshold current density is reduced, its temperature stability is improved, and the differential gain is increased. The absorption operation of a QD laser structure was first demonstrated as early as 1994 [37, 38], since when the lasing characteristics have been improved by a better control of the growth of the self-organized QD layers, such that QD lasers are now approaching commercialization [39]. Optical gain and stimulated emission have also been observed from CdSe and CdS colloidal nanocrystal QDs [40]. On the basis of these results, it is conceivable that optical devices could also be built using the self-assembly of colloidal QDs.

In the past, QDs have been used not only as conventional laser sources but also as "non-classical" light sources. Photons emitted from thermal light sources have

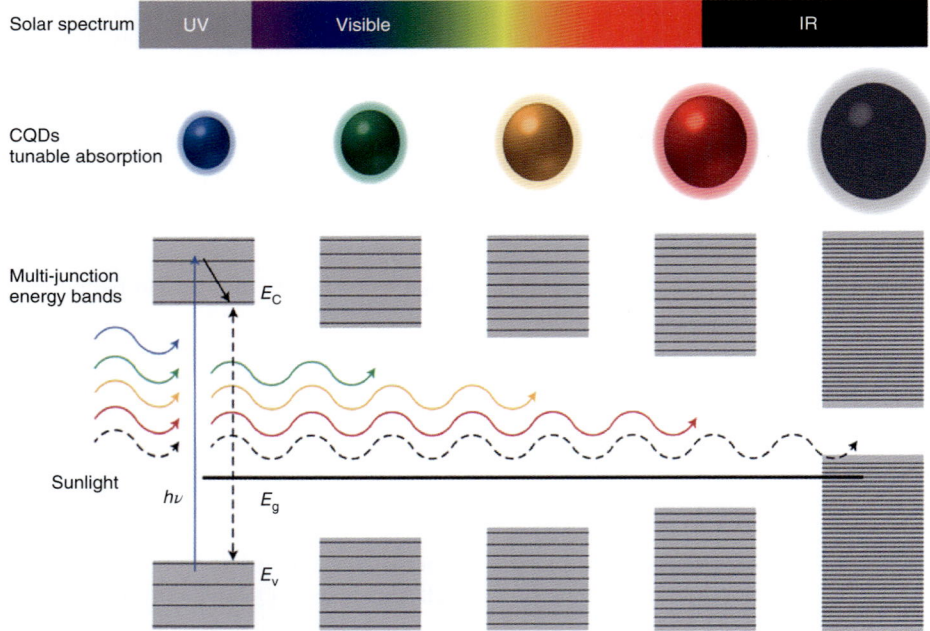

Fig. 1 Colloidal quantum dot (CQD) photovoltaics, size-dependent absorption enables CQDs to be tuned to absorb, sequentially, the constituent bands making up the Sun's broad spectrum, paving the way for the construction of multijunction solar cells that overcome the energy loss.

characteristic statistical correlations. In arrival time measurements, it is found that they tend to "bunch" together (super-poissonian counting statistics), but for applications in quantum information processing it would be desirable to emit single photons one at a time (sub-poissonian counting statistics, or anti-bunching). During the past few years it has been possible to demonstrate the first prototypes of such single-photon sources based on single QDs [41–43]. Self-assembled QDs have also been discussed as the basis of an all-optical storage device, whereby excitons would be generated optically and the electrons and holes stored separately in coupled QD pairs [44]. By applying an electric field, the electron and hole could then be forced to recombine and to generate a photon that would provide an optical read-out. Colloidal QDs have also been used in the development of light-emitting diodes (LEDs) [45, 46], in which colloidal QDs are incorporated into a thin film of a conducting polymer. Colloidal QDs have also been used in the fabrication of photovoltaic devices [47, 48].

Chemically synthesized QDs fluoresce in the visible range with a wavelength that is tunable by the size of the colloids. The possibility of controlling the onset of absorption and the color of fluorescence by tailoring the size of colloidal QDs makes them interesting objects for the labeling of biological structures [32, 33], specifically as a new class of fluorescent biomarkers. Such tenability, combined with extremely reduced photobleaching, would make colloidal QDs an interesting alternative to conventional fluorescent molecules [42]. The possible biological applications of fluorescent colloidal QDs are discussed in detail elsewhere in this encyclopedia.

6
Transport (Electrical) Properties of QDs

Electron transport through ultrasmall structures such as QDs is governed by charge and energy quantization effects. Charge quantization comes into play for structures with an extremely small capacitance. The capacitance of a nanostructure which, roughly speaking, is proportional to its typical linear dimension, may become so small that the energy required to charge the structure with one additional charge carrier (electron, hole, cooper pair) would exceed the thermal energy available. In this case, charge transport through the structure would be blocked – an effect of QDs which has appropriately been termed "Coulomb blockade" (CB).

The CB effect can be exploited to manipulate single electrons within nanostructures. In contrast to bulk structures, charge carriers within a QD are only allowed to occupy discrete energy levels, as in the case of electrons within an atom. In metallic nanostructures, CB can also occur without "quantum aspects," as the energy level spacing in these structures is usually too small to be observable. In fact, energy quantization is the main reason for using the name "quantum dot," and differentiates this from most of the metallic nanostructures. A brief history and fundamentals of the CB effect are provided in the following section [49]. Spectral diffusion is likely related to the local environment of the QDs, which creates rapidly fluctuating electric fields that may perturb the energy levels of the system. Similarly, spectral diffusion can also be observed in single organic fluorophores [50]. Conversely, self-assembled QDs embedded in a matrix do not exhibit spectral jumps, because their local environment does not change with time. In self-assembled QDs, multiexciton states can be observed and studied at high pumping power, but these have never been observed in colloidal QDs. The absence of multiexcitons in single colloidal QDs is believed to be correlated with the fluorescence intermittence observed in these systems [51]. The fluorescence emission from a single QD exhibits a dramatic on/off behavior that is referred to as "blinking," and is another spectroscopic feature that distinguishes colloidal from self-assembled QDs [12, 13].

7
Application of QDs in Diagnostics

The accurate identification of diseases requires a precise detection. The high extinction coefficient of QDs, in addition to their long lifetime and broad absorption spectra that may lead to Stokes shifts, makes them more appropriate for diagnostic purposes than organic fluorophores. The size-dependent properties of QDs such as CdSe provide the possibility for multiplex diagnosis (Fig. 2), with application as either donors or acceptors. The fluorescent properties of QDs make them ideal as donor fluorophores in hybridization-mediated Förster resonance energy transfer (FRET) [53–56], and they are also exceptionally well suited to *in-situ* hybridization, immunohistochemistry, and immunoassay. As QDs are more resistant to degradation than are other optical imaging probes, this allows them to be used for cell tracking over longer periods of time.

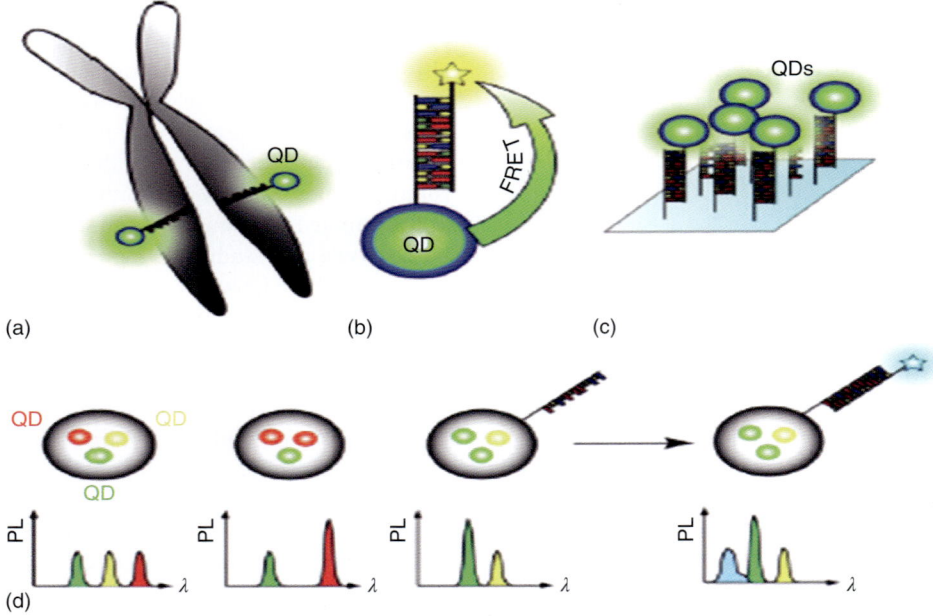

Fig. 2 Applications of quantum dots (QDs) in nucleic acid diagnostics. (a) QDs as labels for FISH; (b) QDs as donors in hybridization-mediated fluorescence resonance energy transfer (FRET); (c) QDs as labels for microarray/solid-phase hybridization; (d) Encapsulated combinations of QDs for spectral bar-coding [52].

8
Nucleic Acid Detection

Fluorescence *in-situ* hybridization (FISH) is an optical nucleic acid diagnostic method that employs QDs, with FISH-based QDs having been used to detect and localize target DNA sequences on chromosomes by hybridization with a specific QD-labeled probe DNA fragment. The QD-labeled probes were used to identify mutations in a Y-chromosome-specific sequence in human sperm, using *in-situ* hybridization. Subsequently, streptavidin-conjugated QDs were used successfully to quantify Fourier transform infra-red (FTIR) signals in human, mouse, and plants [53, 54, 57, 58]. In another study, QDs were shown to be photostable and to have twofold brightness compared to Texas Red and fluorescein dyes in the HER2 locus of breast cancer cells [57].

QD-based molecular beacons (MBs) were fabricated specifically to detect β-lactamase genes that were located in pUC18 and which were responsible for antibiotic resistance in *Escherichia coli* DH5 [59]. The comparison between QD-based and Cy3-based FISH showed, despite brightness and less photobleaching, the background signal to be higher for the QD-based system than for the Cy3-based counterpart [60]. An additional application of QDs to probe DNA is that of FRET, the process of which depends on the extent of spectral overlap between two particles and involves the transfer of fluorescence energy from a donor to an acceptor whenever the distance is typically 1–10 nm [61–63]. When QDs are used as donor particles, there is a limitation in the acceptor particles. The FRET efficiency can be improved with increasing particle as acceptors around a central QD. The limitation of QDs as an acceptor particle is inefficient FRET, as a result of the broad spectrum, that results in a small ratio of excited-state donors to ground-state QD acceptors. Donors with long excited-state lifetimes and pulsed excitation can be used to resolve this problem [60]. QD-based FRET can be applied to photochromic switching, photodynamic medical therapy, pH, and ion sensing and the sensing of enzymatic activity [64–70]. A new class of QD bioconjugates used in bioluminescence resonance energy transfer (BRET). When donors are not excited optically, self-illuminating QDs are suitable acceptors in BRET experiments for deep tissue imaging [71]. This application is indebted to the exceptional optical features of QDs (e.g., superior brightness and photostability, tunable emission, multiplexing) as well as the high sensitivity of bioluminescence imaging [72]. Recently, a BRET-based sensing system was described that was capable of detecting a nucleic acid target within 5 min, with high sensitivity and selectivity. This system as a result of adjacent binding of oligonucleotide probes labeled with Renilla luciferase (Rluc) and QD on the nucleic acid target [73]. The electrochemiluminescent resonance energy transfer (ECRET) process takes place between luminol and QDs. Resonance energy transfer (RET) with electrochemiluminescence (ECL) between CdS QDs and tris (2,2′-bipyridine)ruthenium$^{(2+)}$ ion (Ru(bpy)$_3^{2+}$) and between the ECL of 4-aminobutyl-N-ethylisoluminol (ABEI-luminol)/H$_2$O$_2$ and QDs. Compared to FRET and BRET, ECRET demonstrated an even better controllability, a higher sensitivity, and a wider dynamic range [74–76].

9
Application of QDs in Drug Treatments

Semiconductor particles act as a model platform in shaping the intricate design criteria for the engineering of NDD vehicles, and can also be considered as potential drug-delivery vehicles [77]. The small size, versatile surface chemistry, high photostability and brightness, large Stokes shift, sensitivity to microenvironment, electron-dense inorganic core (and many other properties, as noted above) enable QDs to be used as a platform for nanocarrier design (Fig. 3).

One major limitation of using QDs is the presence of toxic components such as cadmium or lead, although layers of organic and/or inorganic polymers (e.g., poly(ethylene glycol) or SiO$_2$) can be applied in solution form to minimize any possible cytotoxic effects [79–81]. The coated particles sizes typically range from 5 to 20 nm; particles less than 5 nm in size are quickly cleared via renal filtration, while larger particles may be taken up by the reticuloendothelial system before reaching targeted sites, while achieving also a limited penetration into solid tissues [78, 82]. Studies with QDs have provided a systematic assessment of appropriate measurements for efficient delivery; notably, the high surface-to-volume ratio of QDs makes it possible to link multiple functionalities on the particles while maintaining the overall size within an optimal range [78].

Variations in transfection efficiency, delivery-induced cytotoxicity and "off-target" effects at high concentrations of small interfering RNAs (siRNAs) can confound the interpretation of functional

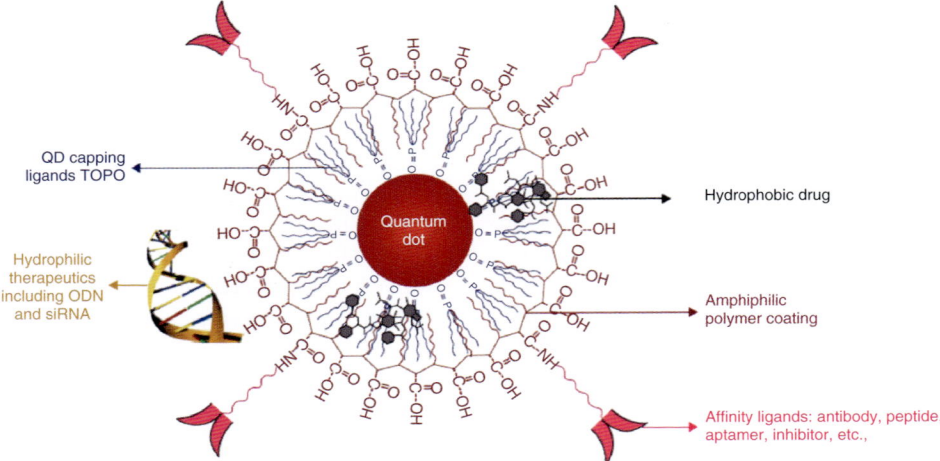

Fig. 3 The high surface-to-volume ratio of QDs makes it possible to link multiple functionalities with optimal size [78]. ODN: oligodeoxynucleotide; TOPO: tri-n-octylphosphine oxide.

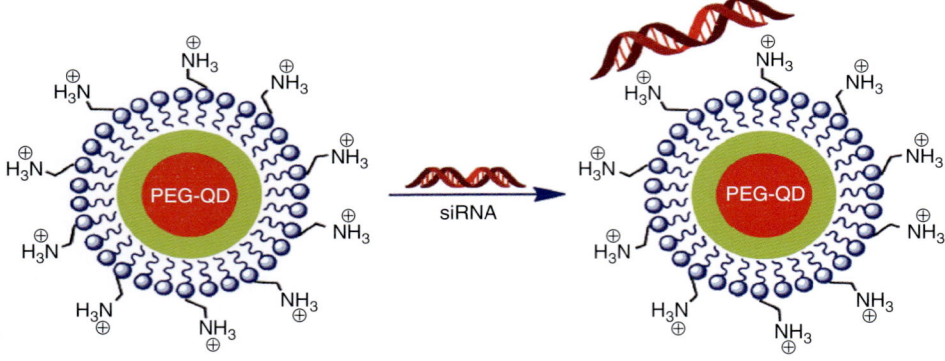

Fig. 4 Formation of QD–poly(ethylene glycol) (PEG)/siRNA complexes by electrostatic interaction.

studies. An unmodified siRNA combination with semiconductor QDs as multicolor biological probes was used to address this problem [83]. Recently, CdSe/ZnS fluorescent QDs, conjugated with amino-polyethylene glycol, were used to deliver siRNAs targeting β-secretase (BACE1) in order to achieve a high transfection efficiency of siRNAs and a reduction in β-amyloid (Aβ) levels in nerve cells (Fig. 4) [84].

Additionally, when mucin1 aptamer-doxorubicin by pH-responsive quantum dot (QD-MUC1-DOX) was applied for the chemotherapy of ovarian cancer, the system showed a preferential targeting of ovarian cancer cells while efficiently releasing doxorubicin at acidic pH; as a consequence, the anticancer efficacy of doxorubicin in ovarian cancer cells was enhanced compared to that of the free drug [85].

The results of *in-vivo* studies of human prostate cancer growing in nude mice have indicated that the QD probes accumulate at tumors, due both to an enhanced permeability and retention at tumor sites, and also by the binding of antibodies to cancer-specific cell-surface biomarkers [86]. Cancer photodynamic therapy is another application of QDs, which also have potential as photosensitizers since irradiation with ultraviolet light can induce the generation of reactive oxygen species (ROS) such as hydroxyl, superoxide radical, cytotoxic single oxygen (O_2), and toxic heavy-metal ions [87, 88]. In one of these studies, fluorescent CdTe and CdSe QDs, with red/brown, brown, or deep brown (close to black) colors, were seen to effectively convert light energy into heat both *in vitro* and *in vivo* when activated by 671 nm laser irradiation. In order to monitor QD retention at the tumor sites, the levels of Cd atoms in the tumors were monitored using inductively coupled plasma atomic emission spectroscopy following the injection of QDs [89]. In the field of immunotherapy, dual QD-magnetic imaging probes were used to trace dendritic cell migration to the lymph nodes in mice, using two-photon optical imaging and magnetic resonance imaging [90].

Acknowledgments

The authors thank the Department of Medical Nanotechnology, Faculty of Advanced Medical Science of Tabriz University for the support provided. These studies were funded by a 2014 Drug Applied Research Center Tabriz University of Medical Sciences Grant.

References

1. Atkins, P.W. (1986) *Physical Chemistry*, 3rd edn. Oxford University Press: Oxford.
2. Karplus, M., Porter, R.N. (1970) *Atoms and Molecules*, 1st edn. W.A. Benjamin, Inc., New York.
3. Kittel, C. (1989) *Einführung in die Festkörperphysik*, 8th edn. R. Oldenbourg Verlag, München, Wien.
4. Ashcroft, N.W., Mermin, N.D. (1976) *Solid State Physics*, Saunders College, Philadelphia.
5. B.L. Abrams, J.P. Wilcoxon (2005) *Crit. Rev. Solid State Mater. Sci.*, **30**, 153–182.
6. C. Burda, X. Chen, R. Narayanan, M.A. El-Sayed (2005) *Chem. Rev.*, **105**, 1025–1102.
7. J.B. Delehanty, K. Boeneman, C.E Bradburne, K. Robertson, I.L. Medintz (2009) *Expert Opin. Drug Deliv.*, **6** (10), 1091–1112.
8. D.J. Bharali, S.A. Mousa (2011) *Pharmacol. Ther.*, **128** (2), 324–335.
9. J.A. Barreto, W. O'Malley, M. Kubeil, B. Graham, H. Stephan, L. Spiccia (2011) *Adv. Mater.*, **23** (12), H18–H40.
10. T.M Samir, M.M. Mansour, S.C. Kazmierczak, H.M. Azzazy (2012) *Nanomedicine*, 7(11), 1755–1769.
11. S. Empedocles, M. Bawendi (1999) *Acc. Chem. Res.*, **32**, 389–396.
12. A.L. Efros, M. Rosen (1997) *Phys. Rev. Lett.*, **78**, 1110–1113.
13. M. Kuno, D.P. Fromm, H.F. Hamann, A. Gallagher, D.J. Nesbitt (2000) *J. Chem. Phys.*, **112**, 3117–3120.
14. J.H. Davies (1998) *The Physics of Low-Dimensional Semiconductors*, Cambridge University Press, Cambridge.
15. T. Ando, A. B. Fowler, F. Stern (1982) *Rev. Mod. Phys.*, **54**, 437–672.
16. A. Führer, S. Lüscher, T. Ihn, T. Heinzel, K. Ensslin, W. Wegscheider, M. Bichler (2001) *Nature*, **413**, 822–825.
17. M. Kemerink, L.W. Molenkamp (1994) *Appl. Phys. Lett.*, **65**, 1012.
18. F.R. Waugh, M.J. Berry, D.J. Mar, R.M. Westervelt, K.L. Campman, A.C. Gossard (1995) *Phys. Rev. Lett.*, **75**, 705.
19. M. Bayer, T. Gutbrod, J.P. Reithmaier, A. Forchel, T.L. Reinecke, P.A. Knipp, A.A. Dremin, V.D. Kulakovskii (1998) *Phys. Rev. Lett.*, **81**, 2582–2585.
20. S. Tarucha, D.G. Austing, T. Honda, R.J. Hage, L.P. Kouwenhoven (1996) *Phys. Rev. Lett.*, **77**, 3613.
21. J.E.F. Frost, D.G. Hasko, D.C. Peacock, D.A. Ritchie, G.A. Jones (1988) *J. Phys. C*, **21**, L209.
22. A.Y. Cho (1999) *J. Cryst. Growth*, **202**, 1–7.

23. R. Elbaum, S. Vega, G. Hodes (2001) *Chem. Mater.*, **13**, 2272.
24. H. Zhao, E.P. Douglas, B.S. Harrison, K.S. Schanze (2001) *Langmuir*, **17**, 8428.
25. H. Zhao, E.P. Douglas (2002) *Chem. Mater.*, **14**, 1418.
26. N. Pinna, K. Weiss, H. Sackongehl, W. Vogel, J. Urban, M.P. Pileni (2001) *Langmuir*, **17**, 7982.
27. S.C. Farmer, T.E. Patten (2001) *Chem. Mater.*, **13**, 3920.
28. Y.A. Wang, J.J. Li, H. Chen, X. Peng (2002) *J. Am. Chem. Soc.*, **124**, 2293.
29. P. Nandakumar, C. Vijayan, Y.V.G.S. Murti (2002) *J. Appl. Phys.*, **91**, 1509.
30. R.G. Ispasoiu, Y. Jin, J. Lee, F. Papadimitrakopoulos, T. Goodson, III, (2002) *Nano Lett.*, **2**, 127.
31. S.L. Cumberland, K.M. Hanif, A. Javier, G.A. Khitrov, G.F. Strouse, S.M. Woessner, C.S. Yun (2002) *Chem. Mater.*, **14**, 1576.
32. A. Henglein (1995) *Ber. Bunsen Ges. Phys. Chem.*, **99**, 903.
33. H. Weller, A. Eychmüller (1995) *Adv. Photochem.*, **20**, 165.
34. B. Dubertret, P. Skourides, D.J. Norris, V. Noireaux, A.H. Brivanlou, A. Libchaber (2002) *Science*, **298**, 1759–1762.
35. Y. Shirasaki, G.J. Supran, M.G. Bawendi, V. Bulović (2013) *Nat. Photonics*, **7**, 13–23.
36. D.L. Klein, R. Roth, A.K.L. Lim, A.P. Alivisatos, P.L. McEuen (1997) *Nature*, **389**, 699–701.
37. N.N. Ledentsov, V.M. Ustinov, A.Y. Egorov, A.E. Zhukov, M.V. Maksimov, I.G. Tabatadze, P.S. Kop'ev (1994) *Semiconductors*, **28**, 832.
38. N. Kirstaedter, N.N. Ledentsov, M. Grundmann, D. Bimberg, V.M. Ustinov, S.S. Ruvimov, V.M. Maximov, P.S. Kop'ev, Z.I. Alferov, U. Richter, P. Werner, U. Gösele, J. Heydenreich, (1994) *Electron. Lett*, **30**, 1416.
39. D. Bimberg, M. Grundmann, F. Heinrichsdorff, N.N. Ledentsov, V.M. Ustinov, A.E. Zhukov, A.R. Kovsh, M.V. Maximov, Y.M. Shernyakov, B.V. Volovik, A.F. Tsatsul'nikov, P.S. Kop'ev, Z.I. Alferov (2000) *Thin Solid Films*, **367**, 235.
40. V.I. Klimov, A.A. Mikhailovsky, S. Xu, A. Malko, J.A. Hollingsworth, C.A. Leatherdale, H.J. Eisler, M.G. Bawendi (2000) *Science*, **290**, 314–317.
41. P. Michler, A. Imamoglu, A. Kiraz, C. Becher, M.D. Mason, P.J. Carson, G.F. Strouse, S.K. Buratto, W.V. Schoenfeld, P.M. Petroff (2002) *Phys. Status Solidi B-Basic Res.*, **229**, 399–405.
42. V. Zwiller, H. Blom, P. Jonsson, N. Panev, S. Jeppesen, T. Tsegaye, E. Goobar, M.E. Pistol, L. Samuelson, G. Björk (2001) *Appl. Phys. Lett.*, **78**, 2476.
43. C. Santori, D. Fattal, J. Vuckovic, G.S. Solomon, Y. Yamamoto (2002) *Nature*, **419**, 594.
44. T. Lundstrom, W. Schoenfeld, H. Lee, P.M. Petroff (1999) *Science*, **286**, 2312.
45. V.L. Colvin, M.C. Schlamp, A.P. Alivisatos (1994) *Nature*, **370**, 354–357.
46. B.O. Dabbousi, M.G. Bawendi, O. Onotsuka, M.F. Rubner (1995) *Appl. Phys. Lett.*, **66**, 1316.
47. W.U. Huynh, J.J. Dittmer, A.P. Alivisatos (2002) *Science*, **295**, 2425–2427.
48. W.U. Huynh, X. Peng, A.P. Alivisatos (1999) *Adv. Mater.*, **11**, 923–927.
49. I. Giaever, H.R. Zeller (1968) *Phys. Rev. Lett.*, **20**, 1504.
50. Basche, T.J. (1998) *Luminescence*, **76–77**, 263–269.
51. M. Nirmal, B.O. Dabbousi, M.G. Bawendi, J.J. Macklin, J.K. Trautman, T.D. Harris, L.E. Brus (1996) *Nature*, **383**, 802–804.
52. W.R. Algar, M. Massey, U.J. Krull (2009) *Trends Anal. Chem.*, **28** (3), 292–306.
53. L. Ma, S.M. Wu, J. Huang, Y. Ding, D.W. Pang, L. Li (2008) *Chromosoma*, **117**, 181–187.
54. L.A. Bentolila, S. Weiss (2006) *Cell Biochem. Biophys.*, **45**, 59–70.
55. P. Chan, T. Yuen, F. Ruf, J. Gonzalez-Maeso, S.C. Sealfon (2005) *Nucleic Acids Res.*, **33**, e161.
56. S. Pathak, S.K. Choi, N. Arnheim, M.E. Thompson (2001) *J. Am. Chem. Soc.*, **123**, 4103.
57. Y. Xiao, P.E. Barker (2004) *Nucleic Acids Res.*, **32**, e28.
58. L.A Bentolila, Y. Ebenstein, S. Weiss (2009) *J. Nucl. Med.*, **50**, 493–496.
59. S.M. Wua, Z.Q. Tiana, Z.L. Zhang, B.H. Huanga, P. Jianga, Z.X. Xieb, D.W. Panga (2010) *Biosens. Bioelectron.*, **26**, 491–496.
60. Z. Jin, N. Hildebrandt (2012) *Trends Biotechnol.*, **30**, 7.
61. P.R Selvin (2000) *Nat. Struct. Biol.*, **7**, 730–734.
62. I.L. Medintz, H. Mattoussi (2009) *Phys. Chem.*, **11**, 17–45.

63. A. Iqbal, S. Arslan, B. Okumus, T.J. Wilson, D.G. Giraud, T. Norman, D.M. Ha, J. Lilley (2008) *Proc. Natl Acad. Sci. USA*, **105** (32), 11176.
64. I.L. Medintz, S.A. Trammell, H. Mattoussi, J.M. Mauro (2004) *J. Am. Chem. Soc.*, **126**, 30–31.
65. R. Bakalova, H. Ohba, Z. Zhelev, M. Ishikawa, Y. Baba (2004) *Nat. Biotechnol.*, **22**, 1360–1361.
66. A.C.S. Samia, X. Chen, C. Burda (2003) *J. Am. Chem. Soc.*, **125**, 15736–15737.
67. Y. Chen, Z. Rosenzweig (2002) *Anal. Chem.*, **74**, 5132–5138.
68. P.T. Snee, R.C. Somers, G. Nair, J.P. Zimmer, M.G. Bawendi, D.G. Nocera (2006) *J. Am. Chem. Soc.*, **128**, 13320–13321.
69. I.L. Medintz, A.R. Clapp, F.M. Brunel, T. Tiefenbrunn, H.T. Uyeda, E.L. Chang, J.R. Deschamps, P.E. Dawson, H. Mattoussi (2006) *Nat. Mater.*, **5**, 581–589.
70. V.M. Rotello (2008) *ACS Nano*, **2**, 4–6.
71. M.K. So, C. Xu, A.M. Loening, S.S. Gambhir, J. Rao (2006) *Nat. Biotechnol.*, **24**, 339–343.
72. W.W. Yu, E. Chang, R. Drezek, V.L. Colvin (2006) *Biochem. Biophys. Res. Commun.*, **348**, 781–786.
73. M. Kumar, D. Zhang, D. Broyles, S.K. Deo (2011) *Biosens. Bioelectron.*, **30**, 133–139.
74. L. Sun, H. Chu, J. Yan, Y. Tu (2012) *Electrochem. Commun.*, **17**, 88–91.
75. M. Wu, H. Shi, J. Xu, H. Chen (2011) *J. Chem. Soc., Chem. Commun.*, **47**, 7752.
76. L. Li, M. Li, Y. Sun, W. Jin (2011) *J. Chem. Soc., Chem. Commun.*, **47**, 8292.
77. P. Zrazhevskiy, M. Sena, X. Gao (2011) *Chem. Soc. Rev.*, **39** (11), 4326–4354.
78. L. Qi, X. Gao (2008) *Drug Deliv.*, **5**(3), 263–267.
79. I.L. Medintz, H.T. Uyeda, E.R. Goldman, H. Mattoussi (2005) *Nat. Mater.*, **4**, 435–446.
80. S.J. Clarke, C.A. Hollmann, Z. Zhang, D. Suffern, S.E. Bradforth, N.M. Dimitrijevic (2006) *Nat. Mater.*, **5**, 409–417.
81. S.C. Hsieh, F.F. Wang, S.C. Hung, Y. Chen, Y.J. Wang (2006) *J. Biomed. Mater. Res.*, **79B**, 95–101.
82. H. Soo Choi, W. Liu, P. Misra (2007) *Nat. Biotechnol.*, **25** (10), 1165–1170.
83. A.A. Chen, A.M. Derfus, S.R. Khetani, S.N. Bhatia (2005) *Nucleic Acids Res.*, **33** (22), e190.
84. S. Li, Z. Liu, F. Ji, Z. Xiao, M. Wang, Y. Peng, Y. Zhang, L. Liu, Z. Liang, F. Li (2012) *Mol. Ther. Nucleic Acids*, **1**, e20.
85. R. Savla, O. Taratula, O. Garbuzenko, T. Minko (2011) *J. Controlled Release*, **153**, 16–22.
86. X. Gao, Y. Cui, R.M. Levenson, L.W. Chung, S. Nie (2004) *Nat. Biotechnol.*, **22**(8), 969–976.
87. R. Bakalova, H. Ohba, Z. Zhelev, M. Ishikawa, Y. Baba (2004) *Nat. Biotechnol.*, **22**(11), 1360e1.
88. H.M. Azzazy, M.M. Mansour, S.C. Kazmierczak (2007) *Clin. Biochem.*, **40**(13–14), 917e27.
89. M. Chu, X. Pan, D. Zhang, Q. Wu, J. Peng, W. Hai (2012) *Biomaterials*, **33**, 7071–7083.
90. P.S. Mackay, G.J. Kremers, S. Kobukai, J.G. Cobb, A. Kuley, S.J. Rosenthal, D.S. Koktysh, J.C. Gore, W. Pham (2011) *Nanomedicine*, **7** (4), 489–496.

Index

a
abatacept 116
abbreviated new drug application 746
abiraterone 56
abrogated apoptosis 299
abscopal effect 475
absolute lymphocyte count 217
absorption 728
absorption enhancers 705, 763
absorption rate constant 703
absorption, distribution, metabolism, excretion, and toxicity 463
absorptive efficiency 954
Aβ burden 389
acceptance criteria 521
accuracy 603
accurate mass tag 612
acetoside 661
acetylation 540, 627
acetylators 920
acetylcholine 245
acetylcholinesterase 387
AChE inhibitors 387
α-acid glycoprotein 12
acid dissociation constant 735
acidic lipids 81
acne 45
acquired immunodeficiency syndrome (AIDS) treating drugs 666–667
activating agonists 881
activation loop 879
active pharmaceutical ingredient 497, 758
active site pocket 885
active site-targeted inhibition 883
active transport 709, 743
active-only formulations 765
actoxumab/bezlotoxumab 460
acute immune responses 315
acute lymphoblastic leukemia 423, 427

AD model mice 388
adalimumab 102
ADAM 783
adamantane 985
adaptive immune resistance 211
adaptive immunity 243
adaptive rejuvenation 277
adeno-associated virus 311
adenocarcinomas 45
adenovirus 311
adiabatic calorimeter 510
adiabatic environment 510
adiabatic temperature rise 509
ADME 588, 717
adoptive immunotherapy 413
adoptive transfer 413, 425
adrenal atrophy 42
adrenocorticotrophic hormone 19
adsorption isotherm 558
advanced melanoma 212
adverse events 914
affinity ceiling 113
affinity maturation 90
affinity-purification 610
age-related macular degeneration 329
aggregation 78
aggregators 580
agnathans 195
AGO proteins 351, 354
albumin 12, 32
algorithms 614
alitritol nucleic acid 836
allele 385
allele-specific gene silencing 307
allergic asthma 258
allergic diseases 663
allergic reactions 111
allergic response 427
AllergoOncology 118

Translational Medicine: Molecular Pharmacology and Drug Discovery
First Edition. Edited by Robert A. Meyers.
© 2018 Wiley-VCH Verlag GmbH & Co. KGaA. Published 2018 by Wiley-VCH Verlag GmbH & Co. KGaA.

allergy 705
allgrove syndrome 634
allogeneic transplantation 139
allometric exponent 716
allometric scaling 716
allosteric compounds 881
allosteric core 180
allosteric inhibitor 887
allosteric phenomenon 177
allosteric sites 888
allosteric states 178
Alnylam/Tekmira 332
aloin 653
alteplase 703
alternative splicing 626
alveolar macrophages 908
alzheimer's disease 385, 658, 685, 785, 888
amaryllidaceae alkaloids 657
amine-reactive crosslinker 638
6-aminohexanoic acid 833
amino acid sequences 167
amino-terminal laminin G-like domain 13
aminoglutethimide 53
aminoglycosides 840, 903
amphibians 781
amphiphilic blocks 947
amphotericin B 747
ampicillin 756
amplification 455, 473
amplification loop 355
amplification status 455, 458
amprenavir 756
amyloid β 385, 685
amyloid cascade hypothesis 387
amyloid plaques 385, 776
amyloid precursor protein 385, 688, 786
amyloidogenic pathway 688
amyotrophic lateral sclerosis 805
anabolic effects 23
anabolic stacking 45
anabolic steroids 33
anabolic-to-androgenic ratio 29
anabolic-androgenic ratios 33
anabolics 6
anakinra 260, 718
analgesics 497
analyte retention 556
analytical biochemistry 534
analytical chemistry 497
analytical method 554
anaplastic large cell lymphoma 466
anaplastic lymphoma kinase 211
anastrozole 58
ancestral gene 194

androgen antagonist 6, 46
androgen binding protein 13
androgen insensitivity syndrome 22
androgen receptor 21, 246
androgen-dependent prostate
    carcinoma 51
androgen-induced growth factor 23
androstenedione 8
androsterone 6
anemia 34, 191, 922, 984
anesthesia 688
aneuploidies 451
aneuploidy 473
angiocrine support 149
angiogenesis 118, 245, 784
animal formulation 763, 765
animal sources 651, 654, 655, 662, 664
anion-exchange column 552
anionic horseradish peroxidase 712
anisodamine 659, 662
ankylosing spondylitis 102
anlage 271
annexin 78
annexin–annexin self-association 83
anorexia 216
anorexia nervosa 33
antagomiRs 314
anterior pituitary 251
anthrax toxin 460
anti-drug antibody 707
anti-inflammatory compounds 390
anti-phosphotyrosine antibodies 628
anti-phosphotyrosine immunoblots 628
anti-abeta vaccination 658
anti-apoptotic gene 370
anti-apoptotic proteins 328
anti-chaotropic (or kosmotropic) salts 550
anti-doping detection 45
anti-inflammatory drugs
– conjunctivitis 660
– hepatitis 660–661
– nephritis 661–662
anti-obesity drugs 669, 671
anti-retroviral therapy 908
antibacterial agents 833, 910
antibiotic resistance 1002
antibiotics 805
antibodies 88, 256, 307, 499, 687, 805
antibody-based drug-delivery
    systems 466
antibody-based therapeutics 465
antibody-dependent cell-mediated cytotoxicity
    90, 213, 467

antibody-dependent cell-mediated phagocytosis   90
antibody-fusion proteins   121
antibody–cytokine fusion proteins   126
anticancer drugs   656–657
anticancer immunity   471
anticoagulants   83
antidepressant drugs   659–660
antidiabetic drugs   657–658
antigen specific T cells   413, 425
antigen-presenting cell   243, 415
antigen-specific effector cells   419
antigen-binding receptors   113
antigen-binding site   88, 119
antigen-presenting cell   114, 201
antigenic   193
antigenicity   543
antigens   88
antimalarial drugs   667, 669
antimicrobial resistance   840
antimycobacterial activity   913
antipyretics   497
antisense antibacterials   841
antisense oligonucleotide   302, 347, 372, 469, 830
antisense RNA transcript   347
antisense therapeutics   461
antisense-mediated silencing   295
antitumor antibodies   201
antitumor cell therapy   426
antitumor cytotoxicity   415
antitumor effects   245, 257, 463
antitumor immunity   120
antitumor lymphocytes   420
antitumor synergy   219
antiviral silencing   355
apheresis   413
apo protein   882
apolipoprotein B   317, 980
apoptosis   23, 83, 240, 299, 351, 362, 466, 785, 878
3M-approach   914
APP proteolysis   389
aptamers   469
aqueous normal-phase chromatography   539
arabinogalactan   905
arabinotransferase   905
arachidonic acid   83
argifin   665
argonaute   296
argonaute proteins   361
aromatase   17
aromatase inhibitors   57
aromatic rings   500
aromatization   17
artemisinin   667

artificial antigen-presenting cell   425
artificial molecules   998
ascorbic acid   653
asiaticoside   672
assay interferers   579
association equilibrium constant   174
asthenia   216
asthma treating drugs   663–665
astrocytes   784, 954
asymmetric field flow fractionation   786
ataxin-3   851
atherosclerosis   460, 944, 952, 954
atomic force microscopy   950
atomic resolution   877
atopy   258
ATP-binding pocket   453
ATP-binding site   879
atrial natriuretic peptide   709
attrition analysis   525
atypical scrapie   801
auger electron spectroscopy   950
autoimmune destruction   153
autoimmune disease   96, 257, 426, 467
autoimmunity   202
autologous cells   413
autologous fibroblasts   398
autologous transplantation   122
autoxidation   172
avidin   121
avidin-affinity purification   618
axon   793
axonal injury   690
azacitidine   459
azafluorenone alkaloids   667
azathioprine   661
AZD5847   923

b

B cells   90, 210, 323, 417
B lymphocytes   92
B-cell-activating factor   233
B-cell leukemia   423
B-cell proliferation   243
B-cell integration cluster   302
B-cell lymphoma   211, 302
BACE1 inhibitors   389
backbone modification   844
bacteria   194
bacterial cell membrane   841
bacterial cell wall   459
bacterial death   918
bacterial infections   459
bacteriophages   101, 460
baculovirus   105

baldness 47
"Bangladeshi" regimen 924
barbiturates 498
basal ganglia 274
base stacking 835
base-catalyzed epimerization 519
base-pair alterations 449
base-paired duplexes 829
base-pairing 315, 347
BCS classification 751
BCS criteria 759
bedaquiline 902
Behçet disease 260
benign prostatic hypertrophy 14
benzimidazole 854
benzothiazinones 917
2′-benzoyloxycinnamaldehyde 657
berberine 661, 667
BET proteins 459
beta-elimination 629
bevasiranib 329
BHUx 671
bi-functional short hairpin RNA 309
bicalutamide 49
bicyclic hydantoins 31
bidirectional permeability 749
bilateral primordia 271
biliary excretion 710
binding affinity 113
binding ligands 876
bioactivation 580
bioassay systems 26
bioavailability 324, 703, 729, 910
biocatalysts 505
biocatalytic resolution 519
biochemical assays 590
biodistribution 115, 708, 848, 989
bioequivalence 746
biogenesis 295
bioinformatics 315, 535
bioinformatics resources 443
biologic agents 499
biological barriers 314
biological matrices 534
biological networks 307
biological profile 578, 581
bioluminescence imaging 1003
biomarker 218, 685
biomass waste 518
biomimetic 946
biomolecular drugs 464
biomolecules 499
biopharmaceuticals 540
biopharmaceutics 729

Biopharmaceutics Classification System 737
Biopharmaceutics Drug Disposition Classification System 751
biophysical screening 590
biospecific/biomimetic affinity chromatography 543
"biotech" units 526
biotin affinity tag 618
biotinylated therapeutics 122
bis-histidine complex 172
bispecific antibody 120, 324
bispecific monoclonal antibodies 466
bivalent inhibitors 891
bladder cancer 214
BLAST-type methods 614
block polymers 947
blocking groups 589
blood 784
blood circulation 704
blood coagulation 78
blood flow 952
blood glucose 884
blood platelets 687
blood pressure 45, 193
blood substitute 191, 957
blood vessel networks 151
blood vessels 149, 954
blood–brain barrier 368, 389, 465, 691, 743, 805, 954, 975
blood–tumor barrier 983
blood-cholesterol levels 460
bloodstream 12, 729, 974
Bohr effect 175
bona fide formulations 765
bond-forming chemistry 504
bonding energies 733
bone formation 78
bone marrow 241, 251, 316, 710, 784
bone marrow biopsy 413
bone marrow suppression 923
bone marrow transplant 413, 420
bone resorption 260
bosutinib 457
bottom-up 508
"bottom-up" experiment 606
bovine spongiform encephalopathy 778
bovinization 113
bradykinesia 689
BRAF mutations 451
brain 18, 368, 685, 743, 776, 792
brain damage 268
brain proteome 624
brain-derived neurotrophic factor 394
brainstem 274, 802

Brambell receptor   93, 117
breast cancer   26, 101, 299, 475, 785, 888, 983
breast milk   323
breast tumor   984
brentuximab vedotin   466
broad-spectrum antimicrobials   920
bronchitis treating drugs   662–663
Brownian affinity gating model   634
Buddleja thyrsoides leaves   663
bubble liposomes   327
bulge   829
bulk synthesis   947
Burkitt's lymphoma   453

c

c-myc   23
C-reactive protein   217
Caco-2 permeability   751
Caenorhabditis elegans   294, 349, 957
Cajal–Retzius cell zones   143
calbindin   281
calcimedins   78
calcitonin   705
calcium   79
calcium excretion   33
calcium homeostasis   399
calcium ion   81
calcium-binding proteins   79
calcium-release channels   79
calmodulin   79
calmodulin-binding domain   635
calorimeters   511
calpactins   78
calphobindins   78
cancer   152, 293, 446, 785, 883, 849, 944, 971
cancer antigens   467
cancer cell lines   833
cancer cells   363, 455
cancer gene amplification   455
cancer genome sequencing   472
cancer genomics   449
cancer immunity cycle   201
cancer immunotherapy   201
cancer metastasis   119, 457
cancer phenotype   293
cancer photodynamic therapy   1005
cancer progression   890
cancer therapy   294
cancer treatment   93
cancer vaccine   421
candidate   577
candidiasis treating drugs   667, 669
cannibalism   799
capacitance   1001

capillary electrophoresis   608
capillary isoelectric focusing   608
capillary tubing   535
capillary walls   709
capillary zone electrophoresis   608
capon comb growth   26
capture stage   561
carbapenem   741
carbohydrate metabolism   8
carbonic anhydrase IX   471
carbonylated myoglobin   177
carboplatin   324
carcinogenesis   145
carcinogenic risk   513
carcinogens   505
carcinoma   213
CARD domains   361
cardiac differentiation   148
cardiac disease modeling   148
cardiac hypertrophy   890
cardiac ion channel inhibition   589
cardiac ischemia   193
cardiac muscle   427
cardiac output   711
cardiotrophin-1   238
cardiovascular disease   329, 460
cargo   466
carrier-mediated transport   317
cartilage tissue   151
caspase-8   245
catabolic pathways   710
catabolin   233
catalytic activity   351
catalytic aminations   507
catalytic cycle   507
catalytic domain   879
catalytic hydrogenation   504
catalytic mechanism   879
catalytic reactions   516
catalytic site   885
(+)-catechin   666
cation exchangers   551
cationic liposomes   325
cationic micelles   328
catumaxomab   120
caveolae   783
caveolin   317
cDNA array analysis   618
CDR shuffling   114
CDR-grafting   101
Cecropia glazioui Sneth   663
cell cancer proliferation   305
cell culture   951
cell cycle   273, 351

cell cycle regulation   462
cell differentiation   139, 234
cell division   293
cell envelope-bound proteins   624
cell expansion   424
cell expansion systems   425
cell fermentation   518
cell internalization   752
cell lineage relationships   448
cell lysis   918
cell membrane   78, 316, 418
cell monolayer cultures   749
cell penetration   316
cell permeability   886
cell polarity   142
cell proliferation   234, 299, 305, 399, 883, 888
cell rejection   280
cell scaffolds   425
cell signaling   783
cell surface   783
cell survival   244
cell therapy   269
cell tissue homogenates   547
cell transplantation   281
cell transplantation therapy   393
cell turnover   269
cell wall superpolymer   903
cell-surface proteins   423
cell-to-cell signaling   238
cell–cell communication   629
cell–particle interactions   952
cell-cycle progression   300
cell-free DNA   473
cell-membrane-penetrating peptides   841
cell-penetrating magnetic nanoparticle   327
cell-penetrating peptide   317, 832
cell-surface proteins   115
cell-surface receptors   93
cell-to-cell transmission   142
cell-wall mycolic acid   921
cell-wall peptidoglycan layer   923
cell–cell adhesion   306
cellular bioavailability   886
cellular globin   195
cellular heterogeneity   625
cellular immunity   243
cellular immunotherapy   413
cellular microenvironment   234
cellular protein   777
cellular response   959
cellular toxicity   582
cellular uptake   365, 832, 848
cellular-based vaccination   415
CeNA–RNA duplex   836

central memory T cells   428
central nervous system   245, 268, 327, 385, 688, 705, 781, 840, 909, 944
centromeres   296
cephalosporin   741
cerebellar architecture   270
cerebellar disorders   270
cerebellar inhibitory interneurons   273
cerebellar network   269
cerebellar neurons   270
cerebellar phenotypes   280
cerebellar plate   271
cerebellar primordium   273
cerebellum   269
cerebral β-amyloidosis   687
cerebral organoid   144
cerebrospinal fluid   685
ceritinib   453
cervical cancer   328
cetuximab   465
channel resonators   951
charge quantization   1001
charge storage devices   999
charge-transfer chromatography   543
checkpoint blockade drugs   429
chelating ligands   553
chemical crosslinking   623
chemical cues   959
chemical engineering   497
chemical reactivity   579
chemical shift perturbation   877
chemical structure   580
chemistry, manufacturing, and controls   497
chemokine   93, 233
chemotherapeutic agents   208
chemotherapeutics   323
chemotherapy   201, 328, 466
chevron diagram   577
chimeric antibodies   96
chimeric antigen receptor   419
chimeric aptamer-siRNA molecules   367
chimeric peptides   369
Chinese hamster ovary   96
chips   952
chiral drugs   500
chiral stationary phase   512
chitinase inhibitors   665
chitosan   367
chitosan–siRNA nanoparticles   368
chlorocruorin   162
chloroplast   625
chloroplast stromal proteins   625
chlorotestosterone   27
cholera   448

cholera toxin   448
cholestasis   45
cholesterol conjugation   848
cholesterol esters   9
cholesterol levels   317
choline acetyltransferase   393
choroidal neovascularization   370
chromaffin granules   82
chromatin   631
chromatin affinity purification mass spectrometry   631
chromatin modification   356
chromatin-associated proteins   629
chromatograms   535
chromatographic bed   538
chromatographic profile   622
chromatographic purifications   501
chromatographic separations   512
chromatographic support   549
chromobindins   78
chromophores   513
chromosomal aberrations   449
chromosomal alterations   449
chromosomal gene   776
chromosomal instability   451
chromosomal replication   323
chromosomal translocation   451, 879
chromosome   20, 169, 237, 776, 1002
chromothripsis   451
chronic constipation   707
chronic infection   210
chronic lymphocytic leukemia   299, 420
chronic myelogenous leukemia   259, 879
chronic myeloid leukemia   307, 425
chronic traumatic encephalopathy   690
chronic wasting disease   800
chymotrypsin   741
ciliary neurotrophic factor   238, 395
cimetidine   747
ciprofloxacin   737
circadian rhythms   252
circular dichroism-optical rotatory dispersion (CD-ORD) spectroscopy   541
circulating nucleic acid   473
circulating tumor cell   473
cisplatin   305, 324, 955
cis-unsaturated fatty acids   78
clathrin   317, 783
cleavable isotope coded affinity tag   618
climbing fibers   270
clinical assessment phase   519
clinical efficacy   216
clinical evaluation   580
clinical symptoms   386
clinical trials   651, 658, 663, 664
clinical trials   469, 685, 884
clofazimine   924
clones   18
cluster region protein   879
CMC dossier   522
CNS degeneration   805
co-suppression   295
coactivators   22
coagulation   242
coating   729
codeine   653
codon adaptation index   598
codon bias   598
cognitive decline   387
coiled-coil protein   356
coinhibitory signals   202
collision-induced dissociation   601
colloidal QDs   998
colloidal semiconductor QD   997
colon   741
colon adenocarcinoma   749
colony-forming unit   913
colostrum   251
column configuration   555
column length   537
column overloading   558
column packing characteristics   555
column void time   536
combination therapy   328, 329
combinatorial approaches   885
combinatorial screening   853
common ion effect   736
complement   254
complement cascade   111, 793
complement proteins   332
complement system   315
complement-dependent cytotoxicity   90, 213
complement-mediated lysis   121
complementarity-determining region   90
complementary binding site   877
complementary chromatographic modes   560
complementary DNA   12
complimentary duplexes   359
compound library   885
computed tomography   691
computer-aided design   581
concentration gradient   739
concentration overloading   558
concentration–effect relationship   699
concentration–time profile   703, 709
concussion   690
conduction band   997
conformational mobility   887

conjugate delivery system   980
conjugates   989
conjunctivitis treating drugs   660
connective tissue growth factor   302
continuous gradient elution   546
controlled-release dosage   734
cooper pair   1001
cooperative binding   176
cooperativity   173
copper   784
copper binding   785
copper homeostasis   785
coptisine   661
cordyceps sinensis   662
corepressors   22
corneal transplant   799
cortical folia   273
corticobasal degeneration   689
corticonuclear networks   269
corticotropin-releasing factor   245
cortisone   499
cost of goods   503
costimulatory domain   423
costimulatory signals   202
cotrimoxazole   924
Coulomb blockade   1001
coupled nuclei   873
covalent bond   163
CpG islands   300, 306
crash cooling   512
creams   729
Creutzfeldt–Jakob disease   687, 777
cricket paralysis virus   359
critical process parameter   508
"critical" peak pair   554
Crohn's disease   102, 258
cross-β spine   804
cross-correlation score   614
cross-species transmission   802
"cross-paired" dimers   119
crosscoupling   517
crosslinked peptide   638
crosslinkers   634
crosslinking analysis   638
Crosslinking reactions   638
cryoelectron microscopy   633
cryogenics   504
cryptolepine   657
crystalline bonds   733
crystallography   183
cucumber mosaic virus   356
curing   729
cutaneous tumors   983
cyclic AMP   9

cyclic guanosine-monophosphate   188
cyclic peptides   541
cyclic tetrapyrroles   805
cyclodepsipeptides   669
cyclodextrin   985
cyclodextrin polymer   985
cyclodextrin-based nanoparticles   849
cyclodextrin-containing polymer   329
cyclohexenyl nucleic acid   836
cyclophosphamide   427
cynomolgus macaques   802
Cyp inhibition   589
Cyp isoforms   589
cyproterone acetate   13
cysteine-specific tags   637
cystic fibrosis   451, 705, 710
cystic fibrosis transmembrane conductance regulator   451
cytoarchitecture   278
cytochrome c oxidase   186
cytochrome P450   9, 717, 745, 908
cytochrome P450 enzymes   463
cytochrome P450 inhibition   582
cytoglobin   162
cytokine   233
cytokine cascades   253
cytokine release syndrome   427
cytokine storm   258, 468
cytokine-mediated signals   235
cytokine-associated damage   314
cytokines   93, 122, 232, 427, 464, 710
cytolysis   459
cytomegalovirus   413
cytoplasm   235, 316, 351, 986
cytoplasmic polyadenylation element binding   804
cytoskeleton   78
cytosol receptor protein   28
cytotoxic activity   122
cytotoxic cargo   466
cytotoxic effects   1003
cytotoxic T cells   420, 467
cytotoxic T-lymphocyte antigen   4, 117, 201
cytotoxicity   117, 314, 912

d

damage-associated molecular pattern   252
danazol   41
darbepoetin-α   709
dasatinib   882
deacetylbaccatin   517
deadenylases   354
deamidation   540, 630
deazaflavin-dependent nitroreductase   921
decidualization   246

decision tree   763, 759
decoy   311
deep cerebellar nuclei   270
deep sequencing   299
deep tissue imaging   1003
degradomics   534
degranulation   258
dehydroepiandrosterone   8
delamanid   902, 921
delayed-release formulation   743
delta32   423
dementia   386, 685, 798
dementia treating drugs   658–659
de-novo sequencing   614
denaturation   172
dendrites   270
dendritic cell platforms   421
dendritic cells   93, 202, 323, 415, 424, 467
dendritic leukocytes   233
dengue hemorrhagic fever   369
dengue shock syndrome   369
Dengue virus   362
denosumab   260
dental mesenchymal cells   785
dentate gyrus   394
deoxy Hb   172
deoxygenation   172
depot effect   27
depot injection   31
design of experiments   508
"designer" anabolic steroids   45
desorption electrospray ionization   601
desoxymethyltestosterone   40, 45
detector time constant   555
detergent-resistant domains   783
development process   497
dextran   192, 805
DGCR8   351
Di George Syndrome   296
diabetes   884, 908, 944
diabetic macular edema   329
diarrhea   51, 216
dicer   296, 301, 353, 842, 971
dicer-related helicase   360
diclofenac   747
dielectric spectroscopy   949
Diels–Alder reaction   515
differential scanning calorimetry   509, 758
differentiated cells   354
differentiation   21, 281, 299, 351
diffuse axonal injury   690
diffusion coefficient   739
diffusivity   537
digestive system   741

digoxin   745
5α-dihydrotestosterone   6
DIMP   49
αβ dimers   181
7α,17α-dimethyltestosterone   37
diode array detection   541
dipolar aprotic solvents   517
direct conjugation   848
direct tumoricidal effect   122
directly observed treatment short-course   905
discovery phase   912
discovery–development interface   762
disease association   580
disease heterogeneity   472
disease modeling   139
disease progression   301, 473
disease-associated process   441
dissolution   728, 732
distribution   728
distribution coefficient   738
disulfide bonds   88
DNA base substitutions   370
DNA codon alterations   457
DNA damage   475
DNA fragment library   448
DNA hypomethylation   300
DNA methylation   300, 356, 457
DNA methyltransferase   459
DNA sequences   1002
DNA shuffling   114
DNA target sequences   983
DNA transposon system   418
DNA/histone modifications   459
DNA–RNA duplex   835
DNAzymes   470
DNMT1   300
DNMT3β   300
docetaxel   328
documentation   521
domain orientations   878
domain structure   118
donor fluid   750
donor lymphocyte infusion   413
donor-derived T-regs   426
doppel   780
dormant bacilli   916
dornase-α   705
dorsal spinal cord   274
dosage form   729
dose solubility ratio   748
dose-ranging studies   716
dosing frequency   910
double-strand RNA   252, 351, 847, 850

double-stranded DNA 309, 469
double-stranded RNA binding domain 353
Down syndrome 387
doxorubicin 324, 986, 1004
Drosha 301, 842
Drosophila 353
drug : excipient ratio 763
drug candidates 584
drug concentration–time profile 729
drug delivery 703
drug design 317
drug discovery 875
drug efficacy and safety 152
drug instability 748
drug metabolism capacity 150
drug resistance mechanism 324
drug screening 139
drug toxicity testing 148
drug-metabolizing activity 954
drug–drug interaction 745
drug-binding site 857
drug-delivery vehicles 1003
drug-resistant TB 902
drug-susceptibility testing 902
drug-susceptible tuberculosis 903
drug–comorbid disease interactions 910
drug–drug interaction 589, 908
druggability 912
druggable potential 878
druggable sites 885
dual-therapy 473
Duchenne muscular dystrophy 839
duocarmycins 656
dura mater graft 799
dynamic light scattering 949
dynamic polyConjugates 980
dynamic range 598
dynamic terminus 888
dynamic wave action bioreactors 426
dysglycemia 907
dyspnea 216
dystonia 689

e
E-factor 516
E-cadherin 305
early bactericidal activity 906, 913
Ebola virus 371
economic value 502
ecotoxicity 539
eculizumab 115
Eddy-diffusion 538
edema 32

effect versus AUC 730
effect versus concentration 730
effect versus dose 730
effector cell 415
effector functions 115
effector memory T cells 428
effector target ratio 427
efflux pump 744, 748
Elaeagnus pungens leave extract 662–663
electrochemiluminescence 1003
electron beam lithography 998
electron microscopy 633, 787
electron transfer dissociation 606
electron transport 999
electron-capture dissociation 613
electron-dense inorganic core 1003
electropermeabilization 324
electrophilic reactivity 506
electrophoretic mobilities 608
electroporation 82, 417
electrospray ionization 600
electrospray ionization mass spectrometry 535
electrostatic contact area 552
electrostatic interactions 551
electrostatic repulsion 601
elimination 728
elimumab 257
ellagic acid 660
eluotropic strength 564
elution volume 536
embryo 169, 246
embryogenesis 140, 249
embryonic expression 781
embryonic milieu 273
embryonic neurogenesis 394
embryonic stem 105, 279, 392
embryonic stem cell 139, 785
emotional behavior 270
empirical MDR-TB treatment 907
emulsifying 729
enalapril 743
enantiotropic 761
encapsulation 706, 729
encephalitis 687
encephalomyelitis 214
encephalomyocarditis virus 363
encyclopedia of DNA elements 457
end-stage organ failure 153
endocrine system 234
endocytosis 316, 359, 709, 781, 833, 850, 955, 975
endogenous long hairpins 844
endolysosomial compartment 792

endometriosis   26
endoplasmic reticulum   10, 102, 792, 883
endoribonuclease   256, 424
endosomal acidic insulinase   714
endosomal escape   833
endosomal membrane   316
endosome internalization   314
endosome trafficking   316
endosomes   976
endosomolytic compounds   832
endothelial cells   188, 233
endothelial networks   150
endothelium   974
endotoxin   249, 256
"4-ene" pathway   10
"5-ene" pathway   10
energy of solvation   733
enhanced permeability and retention   316
enhanced permeation and retention   974
enrichment   561
enterotoxin   256
entropy   731
environment   501
environmental health and safety   521
environmental impact   501
Environmental Protection Agency   514
environmental signals   272
enzalutamide   50
enzymatic activity   1003
enzymatic degradation   706
enzymatic digestion   606
enzymatic resolution   505
eosinopoiesis   258
Eph receptor   879
ephedrine   653
epidermal growth factor   23, 145, 305, 392, 465, 628
epidermal growth factor receptor   105, 305, 447, 890, 983
epidermal thymocyte-activating factor   233
epididymis   7
epigenetic changes   300
epigenetic factors   804
epigenetic modifiers   458
epigenetics   457
epigenome analysis   472
episodic memory   386
epitaxial layer   999
epithelial cancer   216
epithelial cell adhesion molecule   467
epithelial cells   323, 744
epithelial-to-mesenchymal transition   301, 457
epitope   115

epitope mapping   877
epitope tag   635
epitope tagging   607, 611
Epothilones A and B   656
epristeride   53
Epstein–Barr virus   256, 362, 413
equilibrium processes   731
equilibrium solubility   732
eraser   311
error-prone PCR   113
erythroblastic leukemia viral oncogene homolog 2   455
erythrocytes   167
erythromycin   737
erythropoiesis   8
erythropoietin   33, 237
escape mutants   428
estradiol   14
estrogen receptor-negative breast cancer   304
etanercept   718
ethambutol   903
etherification   42
etiocholanolone   15
ETS-related gene   987
eugenol   666
eukaryotic cells   106, 888
eukaryotic organisms   351
eukaryotic protein modification   464
evolution   196
evolutionary studies   194
excipients   765
exocytosis   78
exon skipping   839
exons   169, 449
exonucleases   354
exosomes   988
exportin-5   315, 351, 371
exposure–response characteristics   699
expressed sequence tag   611
expression levels   455
extensively drug-resistant TB   902
external granular layer   272
extra-column effects   555
extracellular barriers   464
extracellular growth factors   462, 465
extracellular matrix   142, 314, 784, 954
extracellular matrix proteins   398
extracellular signal-regulated kinase 2   879
extracellular space   709
extracellular vesicle   323
extracted ion chromatography   621
extrapulmonary disease   913
extravascular injection   708

f

facial hair 32
false positives 579, 877
familial Alzheimer's disease 386
Family 18 chitinases 665
β-family genes 169
FANA–RNA duplex 835
faropenem 924
"fast follower" approach 581
fast-simulated intestinal fluid 740
fatal insomnia 779
fatty acid synthase 905
Fc receptors 208
fear memory 270
febrifugine 667
fed-state simulated gastric fluid 740
feed-simulated intestinal fluid 740
feedback control systems 946
feedback loop 211
Fennel 662
fermentation 499
fermentation broth 547
fertility 912
fetal kidney cells 151
fetal kidney progenitor cells 148
fetus 169
FGFR1 protein 879
fibroblast growth factor 392
fibroblast growth factor 2 280
fibroblasts 233
field asymmetric waveform ion mobility spectrometry 605
filamentous phages 101
filgrastim 703
fimbriae (H-2946 strain) 662
finasteride 52
first-in-human 763
first-order kinetics 507
first-pass effect 742
first-line anti-TB agents 908
first-line DST 906
fish 781
flash chromatography 547
flexible peptide linker 422
flinderoles 667
floating culture system 143
Flock house virus 358
florid plaques 802
flouroquinolones 903
flow control 958
flow cytometry 299, 424, 950
fludarabine 427
5-fluoracil 324
5-fluorouracil 444, 957
fluorescence 543
fluorescence *in-situ* hybridization 1002
fluorescence emission 1001
fluorescence-activated cell sorting 114
fluorescent dye probes 471
fluorescent labels 999
2′-fluoroarabinonucleic acid 835
fluorophores 997
fluoxymesterone 27
fluroquinolones 920
flutamide 22
2′-F modification 845
focal adhesion kinase 987
focused beam reflectance method 523
folate receptor 327
folate–receptor interactions 316
follicle-associated epithelium 742
follicle-stimulating hormone 9, 703
follicular dendritic cell 784
Food and Drug Administration 699, 731
forkhead box protein 3 426
formulation changes 746
formulation feasibility 912
fosamprenavir 756
Fourier transform infrared (FTIR) spectroscopy 541
Fourier transform ion cyclotron resonance 603
fraction collection 559
fractional distillation 537
fractionation 560
fragile X-associated tremor ataxia syndrome 852
fragment antibodies 326
fragment antigen binding 88
fragment crystallizable 88
fragment-based screening 590
fragmentation efficiency 618
frame-shift mutation 449
free-radical 193, 247
frontotemporal dementia 387
full-length antibody 119
full-length protein 588
functional gene 464
functional group interconversions 503
functional groups 504
fungi 295
fungus 78
fused silica 609
Förster resonance energy transfer 1001

g

G protein-coupled receptor 590
G–C nonamers 781
GABAergic 272
gain in peak capacity 559

galectin-9 214
gambogic acid 657
gamma delta T cells 415
gamma-emitting radionuclide 750
gamma-retroviral vectors 418
gamma-secretase 386
GAP proteins 889
gapmer strategy 833
gapmers 837
gas chromatography 512
gas chromatography/mass spectrometry 45
gastric acid 766
gastric cancer 213, 299
gastric carcinoma 785
gastric emptying 766
gastric secretion 766
gastric tissue 146
gastrointestinal (GI) permeability 728
gastrointestinal disorders 922
gastrointestinal stromal tumor 307
gastrointestinal tract 464, 793
gatekeeper mutation 881
gatekeeper residue 881
"gatekeeper" residue 451
gel filtration 786
gelling 729
gemcitabine 332
geminate rebinding 177
gene alterations 448
gene cluster 170
gene control pathways 844
gene copy number 455
gene delivery 311, 981
gene expression 20, 347, 631, 829, 971
gene mutations 145
gene regulation 347, 982
gene regulatory mechanism 295
gene repression 459
gene silencing 294, 317, 830
gene therapy 369, 468
gene therapy trials 418
gene toxicity 589
gene transfer 414, 417
gene-based therapy 468
genetic aberrations 971
genetic alterations 300
genetic analysis 387
genetic disorders 833
genetic engineering 96
genetic polymorphism 252
genetics 780
genome 232, 597, 829
genome defense 347, 355
genome engineering 469

genome maintenance 347
genome-wide RNAi screen 364
genome/exome sequencing 472
genomes 443
genotoxic potential 513
genotoxicity 419, 590
geriatric patients 34
germ layers 785
germinal center 394
germinal neuroepithelia 275
germinal trigone 271
germline alteration 300
germline variable regions 114
Gerstmann–Straussler–Scheinker disease 798
gestation 168, 246
GI mucosa 699
GI tract motility 744
Gibb's free energy 731
glioblastoma 211, 299
glioblastoma multiforme 421
gliomas 983
global proteome sampling 623
global sales 524
globin fold 163
glomerular basement membrane 974
glomerular charge-selectivity 712
glomerular filtration 712, 955, 974
glucagon 714
glucocorticoids 13, 248
glucose challenge 706
glucose metabolism
– phytoconstituents affecting 659
– plants affecting 658
β-glucuronidase 23
glutamatergic 272
glutamatergic neurons 272
glutaraldehyde 192
glutathione 466
glutathione-S-transferase 23, 640
glycemic response 706
glycocalyx 742
glycoengineering 111
glycolaldehyde 192
glycopeptides 548
glycoprotein forms 629
glycoproteins 90
glycosilation pattern 711
glycosylation 96, 540, 626
glycosylation pattern 105
glycosylphosphatidylinositol 783
gold nanoparticles 980
golden triangle 585
gonadotropic hormone 9
gonadotropic receptor 46

gonadotropin   799
good clinical practice   913
gossypol   656, 667
gout   260
gradient elution   555
gradient ultracentrifugation   786
"gradient" plate number   555
graft-versus-host disease   413
graft-versus-tumor activity   413
gram-negative bacteria   105
gram-positive bacteria   105
granule aggregation   82
granule cell layer   270
granule cell progenitors   273
granule cells   276, 280
granulocyte–macrophage colony-stimulating factor   122, 205, 236
granulocytes   784
granulomas   906
green analytical chemistry   539
green chemistry   514
green fluorescent protein   360
greenness   515
growth factors   151
growth hormone   710
growth hormone-releasing factor   706
growth-promoting agent   39
GSK guide   517
GSTSH2 fusion protein   637
GTP-activating protein   889
guanine nucleotide exchange factor   889
guanosine triphosphatase   888
guide strand   297, 844
gut epithelial organoids   145
gut flora   742
gut-associated lymphoid tissue   706, 742
gut-on-a-chip   954
gynecomastia   51

h

H-bonding interactions   836
HADDOCK   878
hairpin dsRNA   295
hairpin loop   829
hairpin structure   351
hairy cell leukemia   259
half-life   118, 464
half-life in human   582
half-maximal inhibitory concentration   912
halogenated testosterone   27
haptoglobin   191
hay fever   705
hazard and operability study   511
hazard evaluation   509

Hb molecules   162
Hb tetramer   162, 181
headache   45, 216
healthcare   517
heart disease   148, 971
heart transplant rejection   473
heat gradients   324
heat shock protein   20
heat shock protein 90   453
heat-generation   507
heat-transfer   508
heats of reaction   509
heavy chain   82
heavy chain antibodies   119
heavy chain polypeptides   88
Heinz bodies   189
HeLa cells   364
α-helical sandwich   22
α-helical segments   163
helicase   361
α-helices   776
β-helices   789
helper T cells   114
hemagglutinating agents   121
hematopoietic factors   236
hematopoietic malignancies   300
hematopoietic stem cell   241, 301, 368, 393, 414, 428, 785
heme   163
heme ligand   172
heme pocket   170
heme-binding pocket   183
heme-binding site   170
hemichrome   172
hemochromes   173
hemodialysis   703
hemoglobin   162, 626
hemoglobin-based oxygen carrier   192
hemolyzed blood   547
hemopoietin receptor superfamily domain   237
hemopoietin-1   233
hemoprotein   162
hemosiderin   688
hepatic biliary ductal system   147
hepatic circulation   742
hepatic dysfunction   36
hepatic first-pass metabolism   704
hepatic function   32
hepatic metabolism   714
hepatitis B   980
hepatitis B virus   849
hepatitis C   259
hepatitis C virus   327, 362, 843
hepatitis treating drugs   660–661

hepatobiliary system   955
hepatocellular carcinoma   211, 299
hepatocyte growth factor receptor   455
hepatocyte targeting   975
hepatocyte transplantation   140
hepatocytes   714
hepatogenesis   150
hepatotoxic side effects   42
hepatotoxicity   837, 907, 955
HER2 receptor   884
hereditary disorders   971
hereditary tyrosinemia type I   469
herpes simplex virus   367
herpes virus   256
herpes virus entry mediator   216
heteroaromatic rings   500
heterochronic transplantation   273
heterocyclic nucleobases   837
heterocyclic rings   43
heteronuclear single quantum coherence   873
heterotropic ligands   175
hexitol nucleic acid   836
hibernation   688
"High Five" insect cells   111
high throughput screening   581
high-affinity receptors   710
high-density lipoprotein-derived NP   949
high-dose/low-potency compounds   765
high-energy decomposition   509
high-performance liquid chromatography   512, 608, 739
high-permeability drugs   744
high-throughput   945
high-throughput analysis   632
high-throughput mass spectrometric protein complex identification   636
high-throughput microfluidic technology   949
high-throughput screening   279, 501, 733, 952
high-affinity ligands   877
high-density lipoprotein   327
high-performance affinity chromatography   545
high-performance anion-exchange chromatography   551
high-performance aqueous normal-phase chromatography   548
high-performance capillary electrochromatography   540
high-performance capillary electrophoresis   540
high-performance cation-exchange chromatography   551
high-performance gel-permeation chromatography   544
high-performance hydrophilic interaction chromatography   547
high-performance hydrophobic interaction chromatography   545
high-performance ion-exchange chromatography   545
high-performance liquid chromatography   533
high-performance reversed-phase chromatography   534
high-performance size-exclusion chromatography   543
high-performance, normal-phase chromatography   546
high-throughput parallel evaporation   562
high-throughput screening   53, 475, 910
High-throughput sequencing   359
high-throughput technologies   475
higher energy collisional dissociation   606
HIGHRIF 1 study   919
HIGHRIF 2 study   919
Hill equation   174
hindbrain   281
hindbrain neuroepithelial stem   281
hinge region   90
hippocampal atrophy   687
hippocampal dentate gyrus   279
hippocampus   272, 274, 394, 955
hirsutism   47
histamine   258, 660
histone deacetylase   459
histone methylation   631
histone modification   631
histone modifiers   457
histones   624, 631
histoplamosis   258
hit   577
hit identification   591
hit-to-lead medicinal chemistry   578
hit-to-lead optimization   576
HIV   257
HIV viremia   212
HNA–RNA duplex   836
Hodgkin lymphoma   302, 423, 466
Hodgkin's disease   211
homeostasis   235, 299, 687
homeostatic signaling   883
homologous domain   80
homozygous null mutant   358
Hoogsteen P–A base pair   848
"horizontal" targeting   473
hormone response element   21
hormone-binding domain   21
hormone–receptor complex   19
hormone–receptor interactions   28
hormones   793
host defense mechanisms   162

host immune response 467
host-encoded prion protein 775
host-defense mechanism 349
human acidic mammalian chitinase (hAMCase) 665
human anti-chimeric antibody 101
human anti-mouse antibody 94
human anti-rat antibody 120
human artificial chromosome 113
human bioequivalence testing 746
human developmental biology 153
human embryonic retinal cell line 112
Human formulation 765
human framework 114
human genome 878
Human Genome Project 745
human growth hormone 799
human hyperimmune polyclonal serum 113
human immunodeficiency virus 327, 447, 514, 901
human immunodeficiency virus 1 413
human immunoglobulins 93
human kidney model 147
human leukocyte antigen 421
human malignancies 299
"human-on-a-chip" 957
human papillomavirus 324
humanized glycosylation patterns 106
humanized mouse models 803
humoral response 94
huntingtin 805, 851
Huntington disease 691, 805, 851, 888
hybrid hybridoma 120
hybrid instrument 603
hybridoma mRNA 96
hybridoma technology 102
hydrodynamic volume 544
hydrogen bond 163
hydrogen bond donors 581
hydrophilic interaction chromatography 543
hydrophilic interaction chromatography/cation-exchange chromatography 548
hydrophilic polymers 325
hydrophilic pore 82
hydrophobic cavities 184
hydrophobic interaction 163, 550
hydrophobic interaction chromatography 543
hydrophobic ligand 549
hydrophobic pockets 891
hydrophobic proteins 599
hydrophobic sorbents 549
hydrophobicity 608

1-(4-hydroxybenzyl)-6,7-methylenedioxy-2-methylisoquinolinium trifluoroacetate 667
11β-hydroxytestosterone 38
17β-hydroxysteroid dehydrogenase 12
4-hydroxytestosterone 38
2′-hydroxycinnamaldehyde (HCA) 657
2-hydroxymethylene-17α-methyl-5α-androstan-17β-ol-3-one 38
hydroxyapatite-chromatography 543
hypercholesterolemia 849
hyperplasia 47
hypersensitivity 233
hypertension 193
hyperthermia 243
hypogonadism 25
hypolipidemic effect 34
hypomethylation 458
hypomorphic mutations 361
hypothalamus 19, 23
hypoxia 247
hypoxic granulomas 920

i
iatrogenic CJD 799
ileum 742
in-silico lipophilicity 586
in-situ IR spectroscopy 523
in-situ hybridization 1001
in-vitro dissolution 747
in-vivo bioavailability 747
in-vivo distribution 949
in-vivo imaging system 984
imatinib 879
imatinib mesylate 307
imatinib-resistant phenotype 880
immediate-release formulation 744
immobilized metal affinity chromatography 627
immobilized metal ion-affinity chromatography 543
immortality 94
immune cells 234
immune checkpoint blockade 201, 468
immune effector cell 467
immune effector functions 93
immune evasion 422
immune reconstitution inflammatory syndrome 908
immune rejection 139
immune response 90, 426, 742, 777
immune stimulation 837
immune system 88, 201, 708, 973
immune tolerance 202
immune-mediated cytotoxicity 415
immune-checkpoint blockers 206

immune-receptor tyrosine-based inhibitory motif 209
immune-related adverse event 216
immunecomplex 707
immunedeficient mice 151
immunoadhesins 126
immunoaffinity chromatography 611
immunoaffinity purification 607
immunoassay 1001
immunoblotting 633
immunoediting 201
immunogenic tumors 204
immunogenicity 315, 317, 369, 707, 974, 989
immunogens 388
immunoglobulin 88, 209
immunoglobulin genes 453
immunohistochemistry 211, 1001
immunomodulators 201
immunomodulatory agents 708
immunomodulatory effects 463
immunoprecipitation 628
Immunoscore 218
immunostaining 608
immunostimulation 838
immunosuppression 204
immunosuppressive drugs 717
immunotherapies 388, 416
immunotherapy 122, 388, 467
immunotoxins 101
impotence 51
in-process control 512
inclusion bodies 105
"individualized" regimens 906
indoleamine 2,3-dioxygenase 201
induced pluripotent stem 279, 392
induced pluripotent stem cell 139, 422
inductively coupled plasma atomic emission spectroscopy 1005
inertial forces 949
infant 169
infected macrophage assay 922
infective disorders 665
– AIDS 666–667
– antibacterial agents 666
– antifungal agents 668
– candidiasis 667, 669
– malaria 667, 669
– tuberculosis 669
infectivity 790
inflammation 93, 209, 241, 689, 952
inflammatory cytokines 247
inflammatory response 209
influenza A virus 312, 855
influenza virus 364x

information integration 472
inhalation 729
inhalational delivery 704
inhaled insulin 704
inheritance pattern 385
inhibin 9
inhibitory antagonists 881
inhibitory signal 208
injection site reactions 117
innate immune receptors 315
innate immune resistance 211
innate immune response 314
innate immune system 849
insect cells 106
insertional mutagenesis 369, 414, 418
inspect 627
insulin 699, 883
insulin aspart 711
insulin degrading enzyme 398
insulin receptor 884
insulin receptor substrate 884
insulin secretion 147
insulin sensitivity 884
integrative systems bioinformatics 472
integrin 325
intercellular adhesion molecule 425
intercellular communication 987
interdomain linker peptide 119
interference compounds 579
interferon 23, 210, 232, 315, 704
interferon-gamma 424, 908
interferon-α 122
interleukin 23, 122, 232, 315, 425, 468, 704, 954
interleukin 2 202
internal energy 731
internal loop 829
internal ribosome entry site 362, 854
International Conference on Harmonisation 505
interneuron development 275
internucleoside linkage 831
interstitial space 709
intestinal absorption 31
intestinal barrier 706
intestinal paracellular absorption 706
intestinal permeability 739
intestinal stem cells 145
intestinal surface area 954
intestine brush border membrane 745
intestine permeability 750
intracellular metabolism 715
intracellular signaling 238, 882
intracellular signaling domain 422
intracellular signaling pathways 24
intrachain disulfide bond 90

intranasal administration 705
intrapatient heterogeneity 449
intratumor heterogeneity 449
intravitreal injection 329
intronic splicing silencer region 840
invertebrate 362
investigational new drug 746
"involved in metabolism" 625
iodoacetamide 618
ion fragmentation 601
ion pair reagents 546
ion sensing 1003
ion signal 615
ion-mobility spectrometry 604
ionic equilibrium 732
ionization 600, 735
iontophoresis 705
ipilimumab 205
iron oxide nanoparticles 984
irritable bowel syndrome 707
ischemia 247
isobaric tags for relative and absolute quantitation 618
isocratic elution 555
Isodon japonicas 663
isoelectric focusing 624
isoelectric point 543, 607
isoniazid 903
isonicotinyl-NAD adduct 903
isosmotic concentration 193
isosteres 40
isothermal titration calorimetry 581
isotope-coded affinity tag 617
isotopic shift 621
isotopically labeled linker 618
isotype 90
isozymes 16
isthmic organizer 271

j
Janus kinase 239
Japanese encephalitis virus 369
jaundice 45
jejunum 742

k
Kaposi's sarcoma 257
Kaposi's sarcoma-associated herpesvirus 361
keratinocyte growth factor 23
keratinocytes 236
keratoconjunctivitis 660
ketoconazole 54
kidney development 148
kidney epithelial cells 955

kidney proximal tubule 955
kidneys 147, 365
kinase inhibitor 457, 577, 881
kinase overexpression 879
kinesin spindle protein 332
kinetic parameters 506
kinetic profiling 507
kinetic solubility 733, 734
kinome "reprogramming" 457
Kirsten rat sarcoma viral oncogene homolog 306
klinefelter's syndrome 25
knockout mice 785
knockout models 212
"knobs-into-holes" approach 120
knowledge transfer 520
Kondo effect 998
Koshland–Nemethy–Filmer 178
Kuru 793

l
label-free methods 621
labile stereocenters 504
β-lactam antibiotics 923
β-lactamase inhibitor 923
β-lactamase mRNA 841
β-lactamases 459
lactate-dehydrogenase 217
lamina propria 146
laminar flows 947
language 270
lapatinib 457
larynx 8
laser nephelometer 733
laser-capture microdissection 626
latency effect 730
latent TB infection 909
latent tuberculosis infection 909
lawsone 660
layer-by-layer assembly 981
lead 577
lead compounds 910
lead development 912
lead molecules 873
lead optimization hypothesis 584
lead profiles 580
lentiviral vectors 418
lentivirus 311
leptin 883
leptin hypersensitivity 884
letrozole 58
leukapheresis 416
leukemia 413
leukemia inhibitory factor 238
leukocyte antigen-related 885

leukocyte reprogramming   248
leukocytes   242, 363
levator ani muscle   33
levodopa   689
lewy bodies   685
lewy body   689
LexA–Protein A fusion   631
leydig cells   8
liarozole   54
libido   51
ligand binding   581
ligand efficiency   587
ligand-binding domain   21
ligand-exchange chromatography   543
ligands   237
light chain polypeptides   88
light chain shuffling   114
light-to-heavy ratio   621
light-emitting diode   1000
lipid metabolism   8
lipid nanoparticles   367, 849
lipid rafts   783
lipid solubility   12
lipid–polymer hybrid   949
lipidation   540
lipo-oligosaccharides   630
lipoarabinomannan   905
lipocortins   78
lipofectamine   368
lipolase   519
lipophilic ligand efficiency   587
lipophilicity   585, 711, 735
lipopolysaccharide   240
liposome   323, 706, 849, 947, 986
liquid chromatography   534, 599
liquid chromatography–mass spectrometry   620
liquid-phase separation   608
lispro   711
live cell imaging   792
liver   13, 168, 316, 365, 983
liver bud formation   149
liver failure   147
liver stem cells   147
L-menthol   664
LNA/DNA oligomers   325
local androgenicity   28
localized surface plasmon resonance   981
locked nucleic acid   311, 370, 834
loop dynamics   879
loop-link   639
loss-of-function gene analyses   294
loss-of-function mutation   475, 839
lotions   729
lotrafiban   501

low-copy-number protein   624
low-density lipoprotein receptor   461
lower-molecular-weight protecting groups   504
LTQ-Orbitrap   606
luciferase-targeting siRNA   984
lucinactant   705
lung   14, 365, 954
lung cancer   299, 453
luteinizing hormone   9, 703
luteinizing hormone-releasing hormone   9, 703
lymph   709
lymph node metastases   306, 986
lymph nodes   243, 249, 316, 793, 1005
lymphadenopathy   909
lymphatic drainage   974
lymphatic system   704
lymphatic transport   709
lymphatic vessels   704
lymphocyte activation   214
lymphocyte activation gene 3   214
lymphocyte anergy   214
lymphocyte-activating factor   233
lymphocytes   210, 232
lymphocytic choriomeningitis virus   212
lymphocytic infiltration   211
lymphodepletion   420
lymphoid tissue   784
lymphoinvasion   793
lymphokine   232
lymphokine-activated killer   259
lymphokine-activated killer cells   467
lymphoproliferative disorders   413
lymphoreticular tissues   799
lymphotoxin-β   237
lyprinol   664
lysosome   316, 712

m

Machado–Joseph disease   851
macro-world   998
macromolecule binding   875
macrophage colony-stimulating factor   703
macrophage migration inhibitory factor   233
macrophages   93, 233, 715
macropinocytosis   316
macular degeneration   119, 976
"mad cows"   799
magic bullets   94
magnetic bead purification   424
magnetic field application   985
magnetic nanoparticle   327
magnetic resonance imaging   981, 1005
magnetic targeting   327
magnetization pathway   875

major groove   848
major histocompatibility class   114
major histocompatibility complex   201, 415
malaria   460
MALDI TOF mass spectrometry   534, 604
male castration   6
male contraception   47
male fertility   354
male genitalia   6
male pseudohermaphroditism   25
male reprotoxins   505
male sterility   781
malignancy   145
malignant cells   293, 444
malignant epigenome   459
malignant glioma   415
malignant myeloma   94, 245
malignant phenotype   449, 884
malnutrition   33
mammalian PrP   776
mammary carcinoma   28
mammary gland   249, 784
mammary tumors   118
MAMS-TB-01 study   919
mannose   111
maraviroc   423
marine organisms   654–657, 660, 667, 669
mascot   627
masculinizing side effects   32
mass modification   615
mass spectrometer   599
mass spectrometry   534, 599
mass-balance   748
mass-to-charge ratio   600
matrix metalloproteinase 9   398
matrix-assisted laser desorption ionization   600
maturation   278
MD simulations   184
meat and bone meal   802
mechanism-based inhibitors   57
medial ganglionic eminence   274
medical diagnostics   565
melanoma   205, 301
melting temperatures   835
membrane organization   79
membrane proteins   367, 618, 635
membrane rafts   890
membrane receptor   240
membrane trafficking   78
membrane-binding pockets   888
membrane–protein interactions   890
menopausal symptoms   26
MeroX   640
mesencephalon   271

mesenchymal stem cell   139, 393
mesenchymal-to-epithelial transition   306
mesothelioma   419
messenger ribonucleic acid   829
messenger RNA   18, 295, 418
metabolic disease   883
metabolic labeling   615
metabolism   728, 730
metabolome analysis   472
metabolomes   443
metal–cyclen compound   892
metal–ligand interaction   980
metal-chelate affinity chromatography   552
metallic nanoparticles   471
metanephric mesenchyme   148
metastases   208
metastasis   888
metastasis reduction   326
metastatic melanoma   471
metastatic sites   299
metastatic spread   304
metazoans   295
methicillin-resistant *Staphylococcus aureus*   757
methotrexate   708, 737
8-methoxyquinolones   920
methylation   627
methylation status   458
methylazoxy-methanol   278
methyltestosterone   36
methyltransferase   359
metkephamid   710
metribolone   22
metronidazole   920
MHC–antigen complex   202
Michael acceptor   57
micro-spotter   535
microarray   299, 598
microarray-based profiling   305
microbiomes   443
microbubbles   327
microcircuits   270
microengineered vascular systems   954
microfluidic cell lysis   626
microfluidics   945
microglial activation   689
micronutrient deficiency   908
microorganisms   654–656
microparticles   706
microprocessor   351
microprocessor complex   296
microribonucleic acid   842
microRNA   293, 351
microsomal stability   588
microspheres   706

Index

microtubule   687
microvesicle   323
microvilli   742
microvortex platform   949
microvortices   949
mild cognitive impairment   686
milling   729
mineralization   8
mini-organ technology   152
mini-organs   139
miniaturization   535
minibody   120
minicells   323
minimum bactericidal concentration   914
minimum inhibitory concentration   903
miniprion   790
minocycline   757
minor groove   848
miR-Mask   312, 332
miRNA mimics   311
miRNA genes   298
miRNA hairpins   843
miRNA silencing   843
miRNA sponges   312
miRNA/miRNA* duplex   353
3'-mismatch-tolerant region   300
missense mutations   449
mitochondrial antiviral signaling   804
mitochondrial proteins   625
mitogen-activated protein kinase   220, 239, 450, 988
mitogen-activated protein kinase kinase   889
mitogenic protein   233
mixed-mode chromatography   543
mobile phase   535
mobile phase temperature   562
modifications   314
molecular architecture   885
molecular beacon   1002
molecular beam epitaxy   997
molecular biology   780
molecular complex detection   636
molecular dynamics   183
molecular evolution   196
molecular imprinted polymer   553
molecular interactions   873
molecular layer   270
molecular matrix   601
molecular modifications   755
molecular profiling   307, 472
molecular recognition   552, 999
molecular targeted therapy   472
molecular weight distribution analysis   545

monoclonal antibody   93, 205, 258, 326, 424, 455, 460, 975
monocytes   210, 417, 784
Monod–Wyman–Changeux   178
mononuclear cell   413, 417
mononuclear phagocytic system   974
monooxygenase   18
monotropic   761
morphine   653
morpholino oligonucleotides   838
Morus alba leaves   671
mossy fiber brainstem   270
Mosla dianthera extract   664
motavizumab   117
motor function   840
motor learning   270
mouse embryonic stem cell   362
mouse myeloma cell line   111
mouse peritonitis   841
mouse xenograft models   324
mRNA analysis   253
mRNA expression   455, 598, 636
mucin fibers   743
mucin layer   742
mucociliary clearance   705
mucosal surface area   706
mucus secretion   705
multi-parameter optimization   585
multi-substrate efflux pump   916
multicellular organisms   957
multicentric clinical trials   651
multidimensional (multistage, multicolumn) high-performance liquid chromatography   535
multidimensional protein identification technology   609
multidrug resistance   328
multidrug resistance protein 1   745
multidrug therapy   473
multidrug-resistant bacteria   460
multidrug-resistant pathogens   842
multidrug-resistant TB   902
multiexciton states   1001
multiorgan systems   957
multiple conformations   878
multiple myeloma   122, 303, 423
multiple quantum well   997
multiple reaction monitoring   622
multiple sclerosis   259
multiple system atrophy   685
multiplex diagnosis   1001
multipotency   139
multipotent progenitors   276
multiprotein complex   633, 636
multivalent antibodies   120

multivariate design space   522
murine cytomegalovirus   362
murine models   205
muscle mass   7
muscularis mucosa   146
mutagenesis   881
mutagenicity   590, 912
mutant Hbs   189
mutation   779, 796
mutation profiles   472
mutation–polymorphism combination   779
mutational load   217
mutational status   218
mutations   90, 446, 1002
Myc epitope   633
mycobacterial catalase-peroxidase   903
mycobacterial cell wall   905, 917
mycobacterial membrane protein large 3   917
mycobacterium tuberculosis   901
myelinated axons   690
myeloid cells   211
myeloid-derived suppressor cell   214
myeloma   259
myeloma cell   94
myoclonus   689
myoglobin   162
myotonic dystrophy   852
myotrophic activity   34
myotrophic effect   7
myxoma virus   256

n

N-acetylgalactosamine   848
N-aminocaprylic acid   706
N-arylpropionamides   31
N-glycosylation   112
N-hydroxysuccinimide   638
N-terminal degradation sequencing   608
N-terminal domain   82, 361
naked siRNAs   365
nandrolone   35
nanocarrier functionalization   325
nanocarrier systems   323
nanocrystals   996
nanomedicine   944, 946
nanomedicine manufacture   957
nanoparticle   323, 367, 466, 706, 849, 945, 985
nanoparticle delivery system   988
nanoparticle-based drug delivery   997
nanotechnology   328
nartograstim   710, 715
natural killer   234
natural killer (NK)-cell   362
natural killer cells   413, 467

natural ligands   577
natural product-based drug discovery
– animals   654, 655
– cancer   656–657
– chemical compound identification   651, 652
– dementia   658–659
– depression   659–660
– diabetes   657–658
– infective disorders   665–669
– inflammation   660–662
– marine organisms   654–656
– microorganisms   654, 655
– obesity   669, 671
– plants   654
– respiratory disorders   662–665
– skin aging   670, 672
– stages   651, 652, 656
nausea   51, 216
near-infra red (NIR) spectroscopy   513
necrosis   473
neoangiogenesis   299
neocortex   274
neoplasia   47
neoplastic phenotype   309
Nepeta cataria   665
nephritis treating drugs   661–662
nephron progenitors   148
nephrotoxicity   907
nerve cells   1004
nerve growth factor   394
nerve-growth factor   23
"net charge" concept   551
neural cell adhesion molecule   784
neural cytoskeleton   690
neural stem   279
neural transplantation   268
neural tubes   271
neuroblastoma   301, 419
neurocognitive development   270
neurodegeneration   268, 385, 798
neurodegenerative disorders   776
neurofibrillary tangle   385, 686
neurofilament light   690
neurogenesis   394, 399
neuroglobin   162, 195
neuroimaging   685
neuroimmune interface   793
neuroinvasion   793, 805
neurologic dysfunction   778
neurological disorders   685, 888
neuromuscular disease   839
neuron-specific enolase   691
neuronal differentiation   385
neurons   245, 269, 394, 687, 783

neuropathological features   385
neuroprotective drugs   390
neuroprotective properties   781
neurotransmitter   397
neurotrophic factors   392
neurotrophin-3   394
neurotrophins   280, 397
neutral horseradish peroxidase   712
neutral lipid emulsion   329
neutralist theory   196
neutropenia   260, 718
neutrophils   93, 242, 417
new chemical entity   497
next generation sequencing   598
"next generation" of transfusion medicine   430
"next-generation" compounds   914
next-generation inhibitors   893
next-generation sequencing   448
nicotiana benthamiana   349
Niemann–Pick disease   280
nigrostriatal pathway   269
nilotinib   882
nilutamide   49
nitric oxide   186, 247
nitrogen balance   8
nitrogen retention   34
nitroimidazoles   920
nitroimidazoxazine   921
nivolumab   212
NMDA   388
NMDA receptor   388
NMR screening   877
NMR spectroscopy   873
NMR spectrum   875
no-food effect   766
NO-scavenging   193
NO-synthase   188
nod-like receptor   252
Nodamura virus   363
non-antibody-based therapeutics   699
non-Hodgkin lymphoma   423
non-arteritic anterior ischemic optic neuropathy   329
non-genetic biomarker   687
non-Hodgkin lymphoma   96
non-small-cell lung cancer   213, 300, 474
nonbinding ligands   876
noncodingRNA   293
nondenaturing mobile phase   551
nonenzymatic modification   630
nonhelical loops   163
nonmalignant immune cells   428
nonsense mutation   449
nonsteroidal anti-inflammatory drug   747

Nora virus   359
noradrenaline   245, 246
norbolethone   45
normal phase silica materials   547
normal-phase chromatography   543
normalized spectral abundance factor   622
NS cells   393
nuclear envelope membrane   633
nuclear magnetic resonance   183, 597, 788, 835, 873, 950
nuclear magnetic resonance (NMR)-spectroscopy   534
nuclear membrane   16
nuclear Overhauser effect   873
nuclear pore proteins   629
nuclear transcriptional factors   21
nuclear translocation   209
nuclease degradation   314
nuclease stability   831
nucleation   734, 998
nucleation–polymerization   791
nucleic acid   776
nucleic acid genome   776
nucleic acid secondary structure   836
nucleic acid sequences   294
nucleophilic substitution   517
nucleoporins   634
nucleosomes   458
nucleotide exchange   891
nucleotide substitutions   195
nucleus   235, 351

o

$2'$-O-alkyl modifications   834
^{16}O-labeled water   639
^{18}O-labeled water   639
^{18}O-labeling   621
O-linked N-acetylglucosamine   629
$2'$-O-MOE modification   834
obesity   884
octarepeats   782
octreotide   709
"off-target" effect   370
off-line analysis   523
off-target genes   315
off-target silencing   846
Office of Technology Assessment analysis   503
oleanolic acid   657
olfactory bulb   274
oligoarginines   326
oligofectamin   367
oligonucleotide probes   1003
oligonucleotides   314, 780, 829, 999
oligosaccharide groups   626

"omics" analyses   294
2′-O-Me modification   370
oncogene   293, 299, 302, 453, 984
oncogenic addiction   449, 450
oncogenic proteins   889
oncogenic transformation   884
oncogenicity   464
oncology   444
oncolytic virotherapy   471
oncostatin-M   238
online genome database   598
ontogenetic mechanism   269
ontogenetic rhythm   273
oocytes   24
open reading frame   18, 298, 635, 780
open-source programs   614
opsonization   325
optic cup   144
optic neuritis   922
optic vesicle   144
optical gain   999
optical imaging   1001
optics   999
optimal gradient run time   556
optimization   498, 503, 580
optimization phase   583
optimization process   582
optimization runs   508
optimized formulations   325
optofluidics   959
oral absorption   735
oral bioavailability   500
oral vaccine delivery   707
orbitrap   604
orchiectomy   47
organ bud transplantation therapy   150
organ buds   139
organ shortage   152
organ targeting   848
organic solvents   516
organogenesis   140, 249
organogenetic processes   143
organoids   139
orphan drugs   526
orsay   360
Orsay virus   360
orthogonal assay   578
orthogonal separations   563
orthogonality of different separation modes   563
osteoarthritis   119
osteoclast-activating factor   233
osteoclastogenesis   260
osteoporosis   34
osteosarcoma   785

osthole   656
otelixizumab   117
ototoxicity   907
ovarian cancer   112, 421, 846, 1004
ovarian function   246
ovarian tumors   301
ovary   246
overhanging nucleotides   353
ovogenesis   246
ovulation   246
oxazolidinones   922
oxidation/reduction cycles   503
oxidative enzymes   589
oxirene   57
oxy Hb   186
oxygen delivery   196
oxymyoglobin   186

p
P-glycoprotein   703, 744, 749
P-gp efflux   747
P-gp pump   328
P300/CBP-associated factor   361
p38 protein   879
p53   306
paclitaxel   324, 517
palivizumab   117
pan-assay interference compounds   579
pancreas transcription factor 1-a   272
pancreatic adenocarcinoma   332
pancreatic cancer   890
pancreatic ductal adenocarcinoma   302, 466
pancreatic tissues   146
pancreatic tumors   299
pancreatic, melanoma   785
papilloma virus   257
paracellular permeation   743
paracellular route   749
paramagnetic probes   877
parasitic infections   459
parathyroid hormone   706
parenteral formulations   706
Parkinson's disease   685, 805
paroxysmal nocturnal hemoglobinuria   116
partial pressure   173
particle compressibility   554
particle diameter   537
particle size   951
particle size distribution   554
particle surface area   739
partition coefficient   738
partition   732
Pasha   296, 351
passenger strand   297, 844

passive delivery   326
passive diffusion   13, 709, 714, 743, 748
passive immunization   388
passive immunotherapy   112
passive permeability   587
passive transcellular route   749
pathogen-associated molecular pattern   370
pathogenesis   775
pathogenic mechanisms   685
pathogenic nucleic acids   294
pathological milieu   441
pathophysiology   952
patient compliance   704
PAZ domain   351
peak broadening   537
peak capacity   563
peak dispersion   538
peak intensity ratio   637
pediatric TB   909
pedigree analysis   22
PEG   986
pegaspargase   708
pegfilgrastim   708
PEGylated lipids   849
PEGylated liposomes   325
pembrolizumab   213
penetratin   326
penicillin   741, 840
PepNovo   627
pepsin   741
peptide bonds   606
peptide drug delivery   707
peptide fragmentation   629
peptide fragments   541
peptide mapping   608
peptide mass fingerprinting   612
peptide mass-matching   634
peptide nucleic acid   832
peptide sequence analysis   612
peptides   538
peptidomics   534
per aqueous liquid chromatography   539
perfluorocarbons   191
perfusion fluid   750
pericytes   954
perinatal lethality   785
peripheral blood lymphocytes   418
peripheral blood mononuclear cell   101, 425
peripheral neuropathy   922
permeability class boundary   746
permeability   581
permeation enhancers   765
persister bacilli   906
persisters   906

personalized medicine   307
personalized treatment   955
personalized therapy   152
petunia   349
Peyer's patches   251, 706, 793
pH gradient strips   624
pH-dependent binding   117
pH-partition theory   737
phage display   101
phage libraries   101
phage therapy   460
pharmaceutical profiling program   753
pharmaceutics   728
pharmacodynamic profile   115
pharmacodynamics-based drug–drug interaction   717
pharmacodynamics   728, 729, 829, 912
pharmacokinetic profile   588, 729, 927
pharmacokinetics-based drug–drug interaction   717
pharmacokinetics   106, 115, 699, 728, 829, 912
pharmacological action   730
pharmacological testing   59
pharmacophoric map   577
Phase I clinical trials   423, 497, 763
Phase II clinical trials   497, 765, 913
Phase III trials   497, 914
phenotype   272, 349, 784
phenotypic mechanism   448
phenotypic response   577
phenylalanine   543
Philadelphia chromosome   453
Philadelphia chromosome-positive CML   879
phlebotomy   416
phloem   358
phosphatidylinositide 3-kinase   209
phosphodiester backbone   972
phosphoinositide 3-kinase (PI3K) delta and gamma   663
phosphopeptides   548, 628
phosphorodiamidate morpholino oligomer   832
phosphoromonothioate   314
phosphorothioate   831
phosphorylation   540, 627, 878
photobleaching   1001
photochromic switching   1003
photodynamic medical therapy   1003
photolysis   177
photosensitizers   1005
photostability   1003
photothermal therapy   981
phylogenetic tree   195
physical pharmacy   728
physico-chemical properties   500, 580, 711

pilot plants   511
ping–pong mechanism   355
pituitary gland   19, 144, 233
pituitary gonadotrophic inhibition   34
PIWI domain   351
Piwi-associated RNA   354
placenta   246, 801
placental transmission   90
plant cell fermentation   518
plant sources   651, 654, 656–658, 664, 666, 668, 669, 672
plant trial   521
plantibodies   112
plants   112
plasma cells   92
plasma expanders   191
plasma membrane   79, 364, 889, 975
plasma protein binding   589
plasmid DNA   327
plasmids   311
plasmodesmata   358
plate height   537
plate number   537
platelet   241
pluripotent stem cell   139, 146, 392
pneumonia   258
point mutations   117, 449, 916
Poisson process   196
polar surface area   581, 743
polar-bonded phase chromatography   547
poliovirus   362, 371
polo-like kinase 1   326
polyacrylamide gel-electrophoresis   599
polyaromatic hydrocarbon   546
polyaspartic acid   547
polybasic region   781
polyclonal antibodies   112
polyclonal population   424
polycystic kidney disease   148
polycythemia   189
polyethylene glycol   192, 317, 708
polyethylenimine   317, 976
polymerization   92
polymers   184
polymorph ratio   759
polymorph screening   514, 758
polymorph specification   759
polymorph/salt screening   762
polymorph   733
polymorphic conversion   735, 758, 760
polymorphic forms   758
polymorphic regions   22
polymorphism   21, 514, 624, 801
polymorphonuclear leucocytes   232

polypeptide chain   163
polypharmacology   476
polysaccharide structures   629
pore connectivities   546
pore diameter distribution   554
pore diameter   554
pore geometries   545
pore sensors   951
porphyrin   180
post-transcriptional gene silencing   295, 349
post-translational modification
post-translational modification   106, 464, 535, 597, 786
post-translational protein modifications   462
postmitotic interneurons   275
postprandial hyperglycemia   711
postsynaptic partners   269
potency   582
potential genotoxic impurities   513
powder X-ray diffraction   514
power law global error model   622
pre-miRNA hairpin   854
precision medicine   472
preclinical studies   313
precursor RNA   355
precursor solutions   947
pregabalin   505, 518
pregnancy   112, 169
preinitiation complex   632
preparative chromatography   557
pressure control   958
pressure gradients   324
pressure-relief system   509
presystemic metabolism   703
pretomanid   921
primary hypogonadism   25
primary miRNA   351
primate foamy virus   361
primates   195
prion infection   778
prion infectivity   778
prion replication   790, 791
prions   775
pro-inflammatory cytokine   119
pro-survival genes   363
probasin   23
process analytical technology   523, 539
process chemist   497
process design   521
process qualification   521
process R&D   498
process technology   497
process validation   521

process verification   522
process-driven control limits   521
processing conditions   506
productivity-enhancing technologies   525
progenitor cells   272
progestational activity   39
prognostic markers   217
programmed cell death   201, 299
progressive supranuclear palsy   687
prokaryotes   106
proliferation signals   425
proliferation   299
prolonged-release dosage   742
promegestrone   56
promoter mutant   919
proof of concept   501
property profile   585
prospective white matter   272
prostaglandin   245
prostate cancer   14, 22, 299, 421, 628, 983
prostate membrane-specific antigen   326
prostate-specific antigen   22, 118
prostate-specific membrane antigen   975
prostate   13
prostatic carcinoma   47
prostatic tissues   14
prosthetic group   183
protamines   368
protease D   23
protease inhibitors   707
proteasome   630
protection/deprotection cycles   503
protein A   634
protein abundance   598
protein accumulation   625
protein affinity pulldown   623
protein aggregators   579
protein assemblies   635
protein binding   710
protein catabolism   8
protein complex composition   623
protein crystal structure   577, 588
protein crystallography   180, 184
protein expression   455
protein isoforms   598
protein kinase R   847
protein kinase receptor   315
protein misfolding cyclic amplification   792
protein phosphorylation   878
protein purification process   561
protein surface   577
protein synthesis   8, 829, 922
protein target   873

protein truncation   626
protein tyrosine kinase   878
protein tyrosine phosphatase   878
protein–ligand complex   176
protein–protein interaction   598
protein-based drugs   312
protein-based therapeutics   699
protein-binding capacities   549
protein-coding genes   294, 295
protein-coding regions   449
protein–hydrophobic surface interactions   551
protein–ligand complexes   878
protein–protein interactions   462, 592
proteins   999
proteome analysis   472, 599
proteome   443, 597
proteomics   534
proteotypic peptide   621
proto-oncogenes   629
protofibrillar aggregates   789
proton sponge effect   976
proton transfer chain   914
protoporphyrin IX   163
proven acceptable ranges   522
pseudohermaphroditism   16
psoriasis vulgaris   119
psoriasis   258, 259
psychiatric disorders   907
psychiatric symptoms   799
psymberin   657
PTP1B   883
pulmonary diseases   944
pulmonary endothelium   427
pulmonary microangiopathy   908
pulmonary tuberculosis   901
pulse–chase experiments   783
pulsed Q collision-induced dissociation   620
purification points   504
purification steps   559
Purkinje cell degeneration   276
Purkinje cell   270, 784
pyranose sugars   847
pyrazinamidase   905
pyrazinamide   903
pyrazoles   41
pyrithiamine   856

q
QconCAT   621
QD acceptors   1003
QT-interval   921
quadroma   120
quadrupole field   604

quality assurance/quality control 554
quality by design 522, 554
quality by testing 522
quantitative proteomics 636
quantitative structure–activity relationship 920
quantitative structure–retention relationship 564
quantum aspects 1001
quantum dot 471, 951, 997
quaternary ammonium ligands 551
quaternary structure 178, 540, 786
quercetin 657, 667
quinine 653, 667
quinolinones 31

r
R&D efficiency 524
rabies virus glycoprotein 368
radial glia 278
radiation therapy 201
radioimmunoassays 39
radiotherapy 466
Raman microscopy 514
random mutagenesis 113
ranibizumab 118
rapamycin 426
Ras complexes 890
Ras conformers 890
Ras effectors 889
Ras family proteins 889
Ras inhibitors 891
Ras oncogene 888
Ras protein 888
Ras upregulation 890
Ras–SOS interface 891
rash 216
rat ventral prostate 19
raxibacumab 460
reaction calorimetry 507
reaction dumping 512
reaction end point 513
reaction mass efficiency 516
reaction monitoring 513
reaction profiles 507
reaction stoichiometry 516
reaction times 504
reactive oxygen species 242, 924, 986, 1005
reactivity of solvents 505
reagent reactivity 504
reagent toxicity 504
real-time monitoring 959
rearrangements 473
receptor binding affinity 755
receptor binding analyses 30

receptor molecules 999
receptor tyrosine kinase 451
receptor-mediated elimination 715
receptor-mediated specific uptake 710
receptor-mediated transport 715
receptor-mediated endocytosis 316, 368
receptors 237, 598
reciprocal recognition 269
recombinant DNA 464
recombinant human insulin 704
recombinant IgG 93
recombinant molecule 234
recombinant protein 106
recombinant viral vectors 369
redox 579
redox-based inhibitors 580
5α-reductase 7
reelin 278
regeneration 685
regenerative medicine 139, 390, 400
regulatory agencies 497
regulatory environment 524
regulatory filings 522
regulatory proteins 598
regulatory RNAs 351
regulatory T cells 244
Reifenstein syndrome 25
relative abundance 615
relative resolution map 556
relief venting 512
renal amino acid homeostasis 713
renal cancer 212
renal cell cancer 704
renal cell carcinoma 415
renal clearance 317
renal elimination 715
renal failure 714
renal filtration 1003
renal insufficiency 714
renal metabolism 712
renal organoids 147
renal toxicity 955
renilla luciferase 1003
replication protein A 591
replication states 905
replication transactivation activator 362
replicon 360
reproducibility 579
reprogramming 392
reptiles 781
reserpine 653
resistance, nodulation, and division 917
resolving power 603
resonance energy transfer 1003

respiratory disorders
- asthma   663–665
- bronchitis   662–663
respiratory distress syndrome   705
respiratory syncytial virus   117, 257, 365
restricted access material   547
resuscitation-promoting factor   906
retention behavior   556
retention factor   536
retention volume   536
reticuloendothelial system   111, 314, 707
retinitis pigmentosa   144
retinoids   21
retrotransposons   354
retroviral genomes   778
retroviral transduction   428
retrovirus   311, 418
reverse transcriptase inhibitors   908
reverse-phase liquid chromatography   512
reversed-phase chromatography   535
reversibility   163
Reynolds number   947
rheumatoid arthritis   102, 126, 258, 708
Rho-associated coiled-coil forming kinase   143
ribavirin   259
ribonucleic acid interference   836
ribonucleotide reductase M2   329
ribosomal proteins   625
ribosomal RNA   841
ribosomes   632
riboswitch   857
riboswitches   856
ribozymes   470
rifabutin   903
rifampicin   902
rifampin   745
rifapentine   903
RIFAQUIN trial   919
RIFATOX trial   919
RISC complex   309
RISC-loading complex   296
risk management   803
ritonavir   745, 758
rituximab   96
RNA binding affinity   830
RNA degradation   830
RNA interference   294, 347, 469
RNA polymerase   905
RNA sequencing   472
RNA silencing   349
RNA translation   351, 830
RNA viruses   362
RNA-induced silencing complex   971
RNA-binding affinity   837

RNA-binding antibiotics   840
RNA-binding intercalators   852
RNA-binding ligands   852
RNA-binding proteins   315
RNA-dependent RNA polymerase   356
RNA-helicase   356
RNA-induced silencing complex   351, 470, 843
RNAi modulators   307
RNAi pathway   971
RNAi pathway mutant   358
RNAi therapeutics   316
RNAi therapies   471
RNAi toxicity   315
RNAi-based therapy   987
RNAi-targets   312
RNase H mechanism   837
rodent pharmacokinetic study   589
root effect   175
roseoflavin   856
rostrocaudal axis   274
Rous sarcoma virus   347
Rule of 3   584
Rule of Five   500
ryo-kan-kyomi-sin-ge-nin-to (RKSG)   663

s
safety data   589
safety margins   756
safety profile   115, 582
safety review team   511
SAGA-like complex   632
salicylic acid   653
salicylihalamide   657
saliva   323
salivary gland   144
salt bridges   182
salt formation   503, 757
salt selection   761
salting in   737
salting out   737
salting-out effect   550
salvage receptor   93
saponins   665
saturation energy   875
saturation magnetization   875
saturation mutagenesis   113
saturation transfer difference-NMR   875
scaling up   558
schizophrenia   888
Schwann cells   793
scrapie   778
screening   577
screening cascade   583, 587
screening solution   578

seborrhea 47
second messenger 186
second pocket 885
second-generation processes 525
second-line anti-TB agents 909
second-line DST 906
secondary hypogonadism 25
secondary structure 541, 786
secosteroid 53
β-secretase 389, 688
γ-secretase 389
secretase inhibitors 388
γ-secretase modulators 390
secretory component 93
secretory vesicle membrane 78
select 503
selected reaction monitoring 622
selective androgen receptor modulator 22, 31
selectivity 537, 543, 583
self-morphogenesis 142
self-regulating pharmacokinetics 708
semen 8
semiconductor heterostructures 997
seminal vesicles 7
seminiferous fluid 13
semisynthetic process 518
senile plaques 686
sensitivity 603
separation columns 535
septic shock 256
sequence homology 360
sequence selectivity 829
sequencing instruments 448
sequential epitope tagging, affinity purification, and mass spectrometry 632
sequest 627
severe acute respiratory syndrome 365
severe combined immunodeficiency 249, 413
sex characteristics 7
sex glands 6
sex hormone binding β-globulin 12
SH2-domain containing tyrosine phosphatase 203
shadoo 781
shape-selective recognition 852
sheeting 729
shope fibroma virus 256
short hairpin RNA 295
short interfering ribonucleic acid 832
short interfering RNA 295, 309
shRNA coding sequences 369
sialic acid 111
sickle-cell 191
sickle-cell anemia 191
side chains 183
side effects 7, 122, 704
side reactions 504
side-effect burden 586
siderosis 688
signal amplification 356
signaling cascades 307
signaling enzymes 879
signaling pathways 878, 884
signaling proteins 455, 598
silencing mechanism 358
silica hydride stationary phase 548
silica-based packing materials 545
simulated gastric fluid 740
simulated intestinal fluid 740
sindbis virus 358
single chain variable fragment 422
single guided RNA 469
single nucleotide polymorphism 254, 300
single-chain peptide 640
single-electron transistors 999
single-enantiomer forms 500
single-pass (open loop) perfusion 750
single-photon sources 1000
single-point mutations 473
single-stranded RNA 297
single-stranded short interfering ribonucleic acid 851
single-subunit oxygenation 180
sinusoidal vasculature 316
siRNA clamp 358
siRNA duplexes 360
siRNA oligonucleotides 971
siRNA silencing 314
siRNA–ION constructs 985
siRNA 971
β-site APP-cleaving enzyme 1 688
site occupancy theory 546
site-directed mutagenesis 113, 192
size-exclusion chromatography 543
Sjogren's syndrome 145
skeletal calcium 8
skeletal growth 7
skeletal muscle 784
skin aging preventing drugs 670, 672
skin discoloration 924
skin diseases 983
skipping 839
slime mold 78
slurry test 760
small cell lung cancer 302
small hairpin RNA 364
small interfering RNA 353, 470, 1003
small intestine 741
small molecules 840

small nucleic acid lipid particle   367
small RNAs   351
small, noncoding RNA   332
small-molecular entities   499
small-molecule antibiotics   841
small-molecule drug conjugates   471
small-molecule drug   462, 699
small-molecule inhibitors   883
small-molecule libraries   852
small-molecule NCEs   524
snake venom phosphodiesterase   836
SNALPs   987
SNARE proteins   78
solid cancers   214
solid particle dissolution   739
solid tissues   1003
solid tumor   302, 466, 850
solid-phase extraction   547
solid-state properties   758
solubility   579
soluble salt   763
solute–solvent bonds   733
solution calorimetry   734
solvent cavity   733
solvent modifier   555
solvent selection   516
solvent–solvent bonds   733
solvents   505
solvophobic effect   545
solvophobic phenomena   546
solvophobic theory   545
somatic cells   392
somatic hypermutation   90
somatic macronucleus   295
somatic mutations   449
somatic stem cell   139
Son-of-Sevenless   889
sonophoration   705
SOS protein   891
species barrier   777
spectral counting   621, 622
spectral diffusion   997
spectral jumps   997
spectral overlap   1002
α-spectrin N-terminal fragment   691
spectroscopy   507
spermatogenesis   23, 51, 246, 354
spermatozoa   8
spermine-binding protein   23
spherical nucleic acid   981
spinal muscular atrophy   839
spinocerebellar ataxia type-1   280
spiroindolones   460
spleen   168, 251, 316, 983

splenomegaly   191
splice-site mutations   449
spliceosome   839
splicing   829
sponge   311
sporadic Creutzfeldt–Jakob disease   796
SQ109   917
stability   581
stable isotope labeling   617
stable isotopes   617
stable nucleic acid particle lipid-based nanoparticle   316
stable nucleic-acid-lipid particle   849, 986
staff training   521
stanolone   7
stanozolol   43
stat3   211
stationary phase   545
StavroX   640
stellate interneurons   270
stem cell culture   140
stem cell replacement therapy   396
stem cell surface markers   393
stem cell therapy   390
stem cell transplant   423
stem cell   279, 469
stem–loop structure   309, 364
step impurities   523
stereochemical mechanism   181
stereoelectronic effects   833
steric block   838
steric hindrance   186
steric interactions   833
steroid hormones   19
steroid nucleus   6
steroid receptor proteins   20
steroid xenobiotic receptor   745
steroid–receptor complexes   20
steroidal aromatase inhibitors   56
steroidogenesis   9, 246
stimulated emission   999
stoichiometric chemical transformations   504
stoichiometry   623
Stokes shifts   1001
stomatitis vesicular virus   249
stopcodons   451
β-strands   776, 790
strand selection   846
STREAM   907
streptavidin-conjugated QD   1002
streptavidin   121
streptomycin   854
striatal-enriched protein tyrosine phosphatase   888

stromal cell-derived factor-1   245
stromal cells   211
structural flexibility   311
3D structure   163
structure analogs   586
structure–activity relationship   26, 579, 756, 881
structure-aided drug design   873
structure–function   22
structure–property relationship   581
subcloning   102
subgranular zone   394
submucosa   146
β-subunit   163
40S-subunit   632
subunit–subunit interactions   178
subventricular zone   394
sudoterb   918
sugar modifications   834, 845
sugar-phosphate backbone   832
suicide genes   428
sulfamethoxazole   924
sulfanilamide   730
sulfate conjugates   44
sulfonamide   498
superkine   237
superparamagnetic iron oxide nanoparticle   327
surface analysis   615
surface area   554
surface epithelia   365
surface heterogeneity   554
surface plasmon resonance   581, 877
surface tension   549
surface tension modifier   546
surface topography   552
surface-enhanced laser desorption/ionization time-of-flight   607
surface-to-volume ratio   608
surfactants   998
survival   299
sutezolid   922
switchable base-pairs   848
synapsin I   629
synaptic degeneration   689
synergistic activity   917, 918
synergistic effect   218, 473
synexins   78
synonymous mutations   449
synthetic androgens   26
synthetic antibiotic   852
synthetic bivalent inhibitor   886
synthetic chemistry   885
synthetic copolymer materials   549
synthetic lethality   475
synthetic organic chemistry   497

synthetic prions   778
synthetic tractability   581
α-synuclein   805
synucleinopathies   689
systemic bioavailability   699
systemic circulation   704
systemic delivery   365
systemic exposure   703
systemic lupus erythematosus   257
systems biology   534
systems cell biology   625

t
T-cell immunotherapy trials   418
T-cell receptor   243, 421
T-cell-replacing factor III   233
T-cell activation   203
T-cell effector function   210
T-cell epitopes   114
T-cell function   465
T-cell immunoglobulin and mucin domain 3   214
T-cell leukemia   465
T-cell receptor   201, 467
T-cell therapy   201
T cells   323, 784
T lymphocytes   243
T-lymphocyte attenuator   214
T-regulatory cell   420
tabalumab   257
tableting   729
tacrolimus   745
TAL protein   631
tamoxifen   58
tandem affinity purification   635
tandem mass spectrometry   601
tandem spectra   613
tandem target sites   312
TAR–peptide complex   855
target cells   365
target RNA   295
target selectivity   755
target specificity   370, 428
target validation   580
target-mediated drug disposition   717
target-based approach   910
target-to-drug strategy   912
targeted mutagenesis   113
targeted specificity   424
targeted therapeutics   473
targeted therapy   441, 444
TATA-box binding protein   637
tau   385
tau proteins   685
tauopathies   689

Index | 1041

taxol   517, 653
TB prevalence   902
TB reactivation   901
TB treatment regimens   927
TB/HIV coinfection   908
technology transfer   520
telencephalic subventricular zone   274
telomerase   425
telomeres   296
template structure   789
tenecteplase   703
tertiary structure   795
testicular atrophy   51
testicular dysfunction   25
testicular feminization   25
testis   246
testosterone   6
testosterone glucuronide   15
testosterone pool   9
tetrahydrogestrinone   45
tetramer–dimer dissociation constant   185
thalamus   798
thalassemia   172
thalidomide   731
theophylline   717
theoretical plates   537
therapeutic agents   838
therapeutic antibodies   106, 112
therapeutic applications   363
therapeutic effect   441
therapeutic intervention   882
therapeutic margin   903
therapeutic potential   887
therapeutic protein–small-molecule drug
    interactions   717
therapeutic protein–therapeutic protein
    interactions   717
therapeutic proteins   705
therapeutic target database   462
therapeutic window   729
therapy   805
thermal analysis   761
thermal gravimetric analysis   758
thermal runaway   510
thermal screening unit   509
thermal stability   831, 845
thermal stabilization   835
thermally unstable materials   508
thermochemical calculations   509
thermodynamic equilibrium   732
thermodynamic or kinetic effects   548
thermodynamic stability   185
thermodynamic studies   835
thermodynamically stable polymorph   760

thermodynamics   731
thiamine pyrophosphate   857
thienopyridine scaffold   856
thiomesterone   38
third-line agents   903
three-dimensional contour plots   562
three-dimensional structure   552
Threshold of Toxicological Concern   513
thrombocytopenia   922
thrombopoietin   709
thromboses   83
thrombosis   952
thyroid follicles   147
thyroid hormones   13, 21
thyroid papillary carcinoma   299
tigecycline   757
tissue distribution   365, 730
tissue engineering strategy   153
tissue explants   955
tissue growth factors   464
tissue inhibitor of metalloproteinases 3   302
tissue plasminogen activator   703
titanium oxide enrichment chromatography   628
TNF-related apoptosis-inducing ligand   239
tobacco mosaic virus   112
tocilizumab   257
tolerability   923
toll-like receptor   315, 363, 847, 973
"top-down" approach   606
topology   163
topotecan   745
total body irradiation   427
totally drug-resistant TB   902
toxic components   1003
toxic effects   216, 389
toxic metabolites   708
toxic side effects   463
toxicity   312
toxicity profile   473
toxicological potential   513
toxicology   716, 912
toxicology studies   912
toxin coregulated pilus   448
trans stereochemistry   26
transactivation response   852
transaminases   832
transcellular migration   709
transcellular permeation   743
TransChromo mouse   105
transchromosomic calves   113
transcortin   32
transcription factors   625, 781
transcriptional activator   317
transcriptomes   443

transdermal drug delivery 705
transdifferentiation 462
transfection 985
transfection efficiency 985, 1003
transferrin 368
transferrin receptor 327
transferrin receptor 1 122
transformation efficiency 114
transforming growth factor-β 237
transforming growth factor-α 465
transgene 418
transgenes 356, 418
transgenic mice 794, 803
transgenic mouse 105
transgenic organisms 105
transgenic plants 112
transient expression 429
transit time 188, 744
transition temperature 761
translational repression 295
translational silencing 353
transmembrane channel 82
transmembrane domain 422
transmembrane receptors 465
transmembrane subunit 235
transmissibility 778
transmissible spongiform encephalopathy 112, 784
transmission electron microscopy 949
transplantation 259
transplantation therapy 392
transportan 326
transporters associated with antigen processing 468
transposase protein 418
transposon cassettes 418
transposon-mediated transduction 417
transposons 349
transvascular transportation 316
transverse relaxation optimized spectroscopy 877
trasponsons 296
trastuzumab 101, 118, 307, 455
trauma 247
traumatic brain injury 685
TRBP 353
treatment duration 919
treatment regimens 906
trehalose monomycolate 917
tremelimumab 205
trenbolone 45
triene analogs 36
trifunctional crosslinker 638
trimethoprim 924
triple helix 832

trophic factors 395
tropomyosin 1 305
troponin C 79
trypsin 741
trypsin microreactor 609
tryptic peptides 627
tryptophan 543
tubercle bacilli 913
tuberculosis 258, 901
tuberculosis treating drugs 669
tumor accumulation 975
tumor antigens 367
tumor burden 427
tumor cells 120, 245
tumor downregulation 423
tumor eradication 415
tumor formation 393
tumor gene expression patterns 457
tumor genomics 473
tumor growth 888
tumor growth retardation 326
tumor infiltrating lymphocyte 415
tumor initiation 312
tumor invasion 212
tumor lesions 208
tumor lysis 428
tumor microenvironment 201, 420, 466
tumor molecular biomarkers 218
tumor necrosis factor 235, 423
tumor necrosis factor receptor 216
tumor necrosis factor-alpha 315, 664, 785, 981
tumor peptides 421
tumor progression 467
tumor regression 312, 363
tumor response 208
tumor sites 1005
tumor suppressor gene 145
tumor suppressors 305, 629
tumor targeting 119, 126
tumor tissue 413
tumor vasculature 325
tumor-specific lymphocytes 420
tumor-associated antigens 466
tumor-associated dendritic cell 215
tumor-infiltrating lymphocyte 206
tumor-initiated angiogenesis 466
tumor-reactive T cells 208
tumor-specific antigens 467
tumor-specific T cells 468
tumor-suppressors 299
tumorigenesis 446, 884, 888
tumors 883, 974
turbidity probe 523

two-dimensional difference gel electrophoresis   615
two-dimensional electron system   997
two-photon optical imaging   1005
type 2 diabetes   260
type II diabetes mellitus   704
tyrosine   543
tyrosine kinase   879
tyrosine-kinase inhibitor   206, 453
tyrosine-protein phosphatase non-receptor type 11   888

u

ubiquitin   630
ubiquitin C-terminal hydrolase L1   691
ubiquitination   627
ultra-high-performance liquid chromatography   539
ultrafiltration   786
ultrasound   327
umbilical cord blood transplant   427
unlocked nucleic acid   846
5′-untranslated region   854
untranslated region   298, 354
ureteric epithelium   148
urogenital tract   16
uterine bleeding   26
uterus   246
UV detection   541

v

vaccine   205, 307, 467, 499, 705, 793
vaccinia virus   249
valacyclovir   743
valence band   997
van Deemter–Knox equation   538
Van't Hoff equations   761
vanadate   884
variably protease-sensitive prionopathy   798
variant Creutzfeldt–Jakob disease   777
vas deferens   7
vascular endothelial cell growth inhibitor   245
vascular endothelial growth factor   106, 220, 329
vascular endothelial system   711
vascular endothelium   367, 952
vasoactive intestinal peptide   246
velocity distribution   604
ventral mesencephalon   274
ventricular zone   272
Verbascum thapsus   664, 665
vertebrates   195
"vertical" targeting   473
vesicular stomatitis virus   360
vinblastine   324, 653

vincristine   653
viral entry receptors   423
viral escape   371
viral genome   777
viral infection   210, 315, 971
viral replication   257, 347, 359, 424, 855
viral RNAs   854
viral sensors   360
viral suppressors of RNAi   358
viral transcription   371
viral vector transduction   417
virilization   8
virilizing actions   45
virion   423
viroids   777
virotherapeutic agent   471
virulence factors   448
virus-derived small interfering RNA   356
viruses   776
viscosity   193
vitamin C   653
vitamin D   21
vitamin E analogs   193
vitamins   742
volatile organic impurities   512
voltage-dependent ion channels   78
voltage-gated ion channel   592
volume of distribution   709, 716
volume overloading   558
vulnerability   912

w

waste minimization   504
wastewater streams   517
water-insoluble compounds   766
water-soluble compounds   765
water-Ligand Observed via Gradient Spectroscopy   875
Watson–Crick base-pairing   830
West Nile virus   312, 362
West Nile virus infection   423
wheat germ cell-free protein production system   669
whole-cell phenotypic screens   914
whole-genome RNAi screening   312
whole-genome sequencing   294
wild-type antibody   115
Wilms tumor   301
wingless-int   143
working memory   270

x

X-linked chronic granulomatous disease   414
X-ray crystallography   788

X-ray diffraction 183
X-ray structures 183
X-ray crystallography 590, 845, 879
X-ray structure 878
XenoMouse™ 102

y

yatakemycin 656
yeast 78, 105, 804
yeast cell display technology 114
yeast prions 804
yeast protein database 635
yeast proteome 627
yeast ribosome 632
yeast two-hybrid 632
"YTE" mutant 117

z

zanoterone 48
zebrafish 838, 957
zebrins 279
zero-dead volume tubing 555
zeta-potential 951
zinc finger nuclease 423
zone band broadening 538
zoonotic risk 796
zwitterionic compound 736